# 2017 IEEE 44th Photovoltaic Specialist Conference (PVSC 2017)

Washington, DC, USA
25-30 June 2017

Pages 2833-3533

IEEE Catalog Number: CFP17PSC-POD
ISBN: 978-1-5090-5606-4

**Copyright © 2017 by the Institute of Electrical and Electronics Engineers, Inc.
All Rights Reserved**

*Copyright and Reprint Permissions*: Abstracting is permitted with credit to the source. Libraries are permitted to photocopy beyond the limit of U.S. copyright law for private use of patrons those articles in this volume that carry a code at the bottom of the first page, provided the per-copy fee indicated in the code is paid through Copyright Clearance Center, 222 Rosewood Drive, Danvers, MA 01923.

For other copying, reprint or republication permission, write to IEEE Copyrights Manager, IEEE Service Center, 445 Hoes Lane, Piscataway, NJ 08854. All rights reserved.

***\*\*\* This is a print representation of what appears in the IEEE Digital Library. Some format issues inherent in the e-media version may also appear in this print version.***

IEEE Catalog Number:      CFP17PSC-POD
ISBN (Print-On-Demand):   978-1-5090-5606-4
ISBN (Online):            978-1-5090-5605-7
ISSN:                     0160-8371


**Additional Copies of This Publication Are Available From:**

Curran Associates, Inc
57 Morehouse Lane
Red Hook, NY  12571 USA
Phone:     (845) 758-0400
Fax:       (845) 758-2633
E-mail:    curran@proceedings.com
Web:       www.proceedings.com

# 2017 IEEE 44th Photovoltaic Specialist Conference (PVSC 2017)

Washington, DC, USA
25-30 June 2017

## Pages 2833-3533

IEEE Catalog Number: CFP17PSC-POD
ISBN: 978-1-5090-5606-4

# TABLE OF CONTENTS

**OPEN CIRCUIT VOLTAGE CALCULATION USING TEMPERATURE AND IRRADIANCE** ........................... 1
Andrew Melvin

**EFFECT OF CL-DOPING IN ZNTEO ON PHOTOLUMINESCENCE AND PHOTOVOLTAIC PROPERTIES OF ZNTEO-BASED INTERMEDIATE BAND SOLAR CELLS** .................................. 3
T. Tanaka ; S. Tsutsumi ; Y. Okano ; K. Matsuo ; K. Saito ; Q. Guo ; M. Nishio ; T. Tayagaki ; K. M. Yu ; W. Walukiewicz

**TOWARD LEAD HALIDE PEROVSKITE-BASED INTERMEDIATE BAND ABSORBERS** .................. 6
Matthew D. Sampson ; Ji-Sang Park ; Richard D. Schaller ; Maria K. Y. Chan ; Alex B. F. Martinson

**TYPE-II QUANTUM DOTS FOR APPLICATION TO PHOTON RATCHET INTERMEDIATE BAND SOLAR CELLS** .................................................................................................. 10
Ryo Tamaki ; Yasushi Shoji ; Yoshitaka Okada

**AN INVESTIGATION OF THE ROLE OF RECOMBINATION PROCESSES IN THE OPERATION OF INAS/GAASL-XSBX QUANTUM DOT SOLAR CELLS** ................................ 14
Y. Cheng ; A. J. Meleco ; A. J. Roeth ; V. R. Whiteside ; M. C. Debnath ; M. B. Santos ; T. D. Mishima ; S. Hatch ; H.Y. Liu ; I. R. Sellers

**TEMPERATURE AND VOLTAGE-BIAS DEPENDENT TWO-STEP PHOTON ABSORPTION IN INAS/GAASL AL0.3GAAS QUANTUM DOT IN A WELL SOLAR CELLS** ............................... 18
Yushuai Dai ; Brittany L. Smith ; Michael A. Slocum ; Zachary S. Bittner ; Hyun Kum ; Julia D'Rozario ; Seth M. Hubbard

**INCREASING CURRENT GENERATION BY PHOTON UP-CONVERSION IN A SINGLE-JUNCTION SOLAR CELL WITH A HETERO-INTERFACE** ................................................. 23
Shigeo Asahi ; Kazuki Kusaki ; Toshiyuki Kaizu ; Takashi Kita

**RTP-ASSISTED EX-SITU ANALYSIS OF (AG,CU)(IN,GA)SE2FORMATION USING SELENIZATION** .................................................................................................. 26
Sina Soltanmohammad ; William N. Shafarman

**ROLE OF EV+0.98 EV TRAP IN LIGHT SOAKING-INDUCED SHORT CIRCUIT CURRENT INSTABILITY IN CIGS SOLAR CELLS** ....................................................................... 30
P. K. Paul ; T. Jarmar ; L. Stolt ; A. Rockett ; A. R. Arehart

**STUDY OF DEFECT PROPERTIES IN CUGASE2THIN-FILM SOLAR-CELLS USING ADMITTANCE SPECTROSCOPY** ............................................................................... 33
Muhammad Monirul Islam ; Shogo Ishizuka ; Hajime Shibata ; Shigeru Niki ; Katsuhiro Akimoto ; Takeaki Sakurai

**TRANSMISSIVE SPECTRUM-SPLITTING CONCENTRATOR PHOTOVOLTAIC CELLS AND MODULES** ...................................................................................................... 37
Yaping Ji ; Qi Xu ; Brian Riggs ; John Robertson ; Kazi Islam ; Vince Romanin ; Dimitri D. Krut ; Jim H. Ermer ; Matthew D. Escarra

**ALGAINP/GAAS TANDEM SOLAR CELLS FOR POWER CONVERSION AT 400 C AND 1000X CONCENTRATION** ............................................................................................... 42
Myles A. Steiner ; Emmett E. Perl ; John Simon ; Daniel J. Friedman ; Nikhil Jain ; Paul Sharps ; Claiborne Mcpheeters ; Minjoo L. Lee

**GALNASP SOLAR CELLS GROWN BY HYDRIDE VAPOR PHASE EPITAXY FOR ONE-SUN & LOW-CONCENTRATION III-V/SI PHOTOVOLTAICS** ................................................ 46
Nikhil Jain ; John Simon ; Kevin L. Schulte ; David R. Diercks ; Corinne E. Packard ; David Young ; Aaron J. Ptak

**PHOTO-ELECTROCHEMICAL HYDROGEN GENERATION FROM INVERTED METAMORPHIC MULTIJUNCTION III-VS** ...................................................................... 47
Todd G. Deutsch ; James L. Young ; Myles A. Steiner ; Henning Döscher ; Ryan M. France ; John A. Turner

**ADVANCED SILICON THIN FILMS FOR HIGH-EFFICIENCY SILICON HETEROJUNCTION-BASED SOLAR CELLS** ........................................................................................... 50
A. Descoeudres ; C. Allebe ; N. Badel ; L. Barraud ; J. Champliaud ; G. Christmann ; L. Curvat ; F. Debrot ; A. Faes ; J. Geissbühler ; J. Horzel ; A. Lachowicz ; J. Levrat ; S. Martin De Nicolas ; S. Nicolay ; B. Paviet-Salomon ; L.-L. Senaud ; A. Tomasi ; C. Ballif ; M. Despeisse

**MOOXAND WOXBASED HOLE-SELECTIVE CONTACTS FOR WAFER-BASED SI SOLAR CELLS** .............................................................................................................. 55
Stephanie Essig ; Julie Dréon ; Jérémie Werner ; Philipp Löper ; Stefaan De Wolf ; Mathieu Boccard ; Christophe Ballif

**METAL NANOPARTICLE HOLE CONTACTS FOR SILICON SOLAR CELLS** ................................ 59
James Bullock ; Zhaoran Xu ; Mark Hettick ; Yimao Wan ; Ali Javey

**NEAR-FIELD TRANSPORT IMAGING APPLICATION OF PHOTOVOLTAIC MATERIALS**.............62
Chuanxiao Xiao ; Chun-Sheng Jiang ; John Moseley ; John Simon ; Kevin Schulte ; Aaron J. Ptak ; Steve Johnston ; Brian Gorman ; Mowafak Al-Jassim ; Nancy M. Haegel ; Helio Moutinho

**APPLICATIONS OF DMD-BASED INHOMOGENEOUS ILLUMINATION PHOTOLUMINESCENCE IMAGING FOR SILICON WAFERS AND SOLAR CELLS**.........66
Yan Zhu ; Mattias Klaus Juhl ; Ziv Hameiri ; Thorsten Trupke

**NUMERICAL MODEL TO EXTRACT MATERIALS PROPERTIES MAP FROM SPECTRALLY RESOLVED LUMINESCENCE IMAGES**.........70
Nicolas Paul ; Vincent Le Guen ; Daniel Ory ; Laurent Lombez

**NON-DESTRUCTIVE CONTACT RESISTIVITY MEASUREMENTS ON SOLAR CELLS USING THE CIRCULAR TRANSMISSION LINE METHOD**.........74
Geoffrey Gregory ; Andrew M. Gabor ; Andrew Anselmo ; Rob Janoch ; Zhihao Yang ; Kristopher O. Davis

**RADIATION RESISTANCE OF LOW COST HIGH EFFICIENCY TRIPLE JUNCTION SOLAR CELLS**.........76
Roberta Campesato ; Erminio Greco ; Giuseppe Gabetta ; Mariacristina Casale ; Gabriele Gori ; M. Sankaran ; Suresh E. Puthanveettil ; B. R. Uma ; M. Ravindra ; Sheeja Krishnan

**AMORPHOUS SILICON CARBIDE REAR-SIDE PASSIVATION AND REFLECTOR LAYER STACKS FOR MULTI-JUNCTION SPACE SOLAR CELLS BASED ON GERMANIUM SUBSTRATES**.........83
Stefan Janz ; Charlotte Weiss ; Christian Mohr ; Rufi Kurstjens ; Bruno Boizot ; Bianca Fuhrmann ; Victor Khorenko

**HOT CARRIER TRANSPORTATION DYNAMICS IN INAS/GAAS QUANTUM DOT SOLAR CELL**.........85
Tomah Sogabe ; Kohdai Nii ; Katsuyoshi Sakamoto ; Koichi Yarnaquchi ; Yoshitaka Okada

**INTEGRATION OF CRACK-TOLERANT COMPOSITE GRIDLINES ON TRIPLE JUNCTION PHOTOVOLTAIC CELLS**.........88
Omar K. Abudayyeh ; Geoffrey K. Bradshaw ; Steven Whipple ; David M. Wilt ; Sang M. Han

**SUBCELL LIGHT CURRENT- VOLTAGE CHARACTERIZATION OF IRRADIATED MULTIJUNCTION SOLAR CELL**.........93
Don Walker ; John Nocerino ; Yao Yue ; Colin J. Mann ; Simon H. Liu

**ANALYTICAL METHOD FOR PREDICTING SPACECRAFT POWER GENERATION ON PARTIALLY SHADED SOLAR PANELS**.........96
Gordon Wu ; Bao Hoang

**EVALUATING THE EMISSIVITY OF PSEUDOMORPHIC GLASS (PMG)**.........102
Ryan D. Beauchemin ; David M. Wilt ; Paul E. Hausgen

**CHARACTERIZING THE IMPACT OF SOLAR SPECTRAL IRRADIANCE ON PV MODULE OUTPUT**.........107
M. Schweiger ; W. Herrmann

**USE OF MEASURED AEROSOL OPTICAL DEPTH AND PRECIPITABLE WATER TO MODEL CLEAR SKY IRRADIANCE**.........110
Mark M. Mikofski ; Clifford W. Hansen ; William F. Holmgren ; Gregory M. Kimbal

**RECENT ADVANCEMENTS IN THE NUMERICAL SIMULATION OF SURFACE IRRADIANCE FOR SOLAR ENERGY APPLICATIONS**.........116
Yu Xie ; Manajit Sengupta ; Chris Deline

**OPTIMAL IRRADIANCE SENSOR PLACEMENT FOR PHOTOVOLTAIC SYSTEMS USING MUTUAL INFORMATION BASED GREEDY ALGORITHM IN GAUSSIAN PROCESS**.........120
Lian Lian Jiang ; R. Srivatsan ; Douglas L. Maskell

**EVALUATING DIFFERENT UPSCALING APPROACHES TO DERIVE THE ACTUAL POWER OF DISTRIBUTED PV SYSTEMS**.........126
Sven Killinger ; Björn Müller ; Bernhard Wille-Haussmann ; Russell Mckenna

**ADVANCES IN LONG-TERM SOLAR ENERGY PREDICTION AND PROJECT RISK ASSESSMENT METHODOLOGY**.........132
Alemu Tadesse ; Adam Kankiewicz ; Alex Kubiniec ; Richard Perez ; John Dise ; Thomas Hoff

**DECOUPLING THIN FILM CDTE GROWTH FROM PACKAGING: TOWARD RECORD SPECIFIC POWER IN LOW COST POLYCRYSTALLINE PV**.........138
D. Clayton-Warwick ; M.D. Kempe ; M. S. Dabney ; T. M. Barnes ; C. A. Wolden ; M. O. Reese

**JUNCTION ACTIVATION OF CDTE/CDS SOLAR CELL USING MGCL2**.........142
G. Angeles-Ordóñez ; E. Regalado-Pérez ; M.G. Reyes-Banda ; N. R. Mathews ; X. Mathew

**VARIATION OF CU CONTENT OF SPRAYED CU(IN, GA)(S,SE)2SOLAR CELLS BASED ON A THIOL-AMINE SOLVENT MIXTURE**.........146
Panagiota Arnou ; Sona Ulicná ; Alexander Eeles ; Mustafa Togay ; Lewis D. Wright ; Andrei V. Malkov ; John M. Walls ; Jake W. Bowers

**CUINSE2 ABSORBER LAYER GROWN UNDER COPPER EXCESS WITH A COPPER POOR SURFACE FORMED BY A KF POST DEPOSITION TREATMENT** ...... 151

*Finn Babbe ; Hossam Elanzeery ; Michele Melchiorre ; Susanne Siebentritt*

**CU2ZNSNSE4SOLAR CELLS ONTO POLYIMIDE SUBSTRATES FABRICATED AT LOW TEMPERATURE** ...... 155

*Ignacio Becerril-Romero ; Simón Lopez-Marino ; Moisés Espíndola-Rodríguez ; Laura Acebo ; Markus Neuschitzer ; Yudania Sánchez ; Edgardo Saucedo ; Paul Pistor*

**AN OPTIMIZED PHOTOLITHOGRAPHY RECIPE FOR CU(IN1-X,GAX)(SY,SE1-Y)2(CIGSSE) SOLAR CELLS** ...... 160

*Xia Hao ; Shenghao Wang ; Katsuhiro Akimoto ; Takuya Kato ; Hiroki Sugimoto ; Takeaki Sakurai*

**EFFECTS OF CDCL2PASSIVATION ON THIN CDTE ABSORBERS FABRICATED BY CLOSE-SPACE SUBLIMATION** ...... 164

*Anna Wojtowicz ; Alexandra M. Huss ; Jennifer A. Drayton ; James R. Sites*

**CDS1-XSEXWINDOW LAYER FOR CDTE PREPARED BY THE EXCHANGE OF S WITH SE IN CDS FILMS** ...... 170

*Geethika K. Liyanage ; Adam B. Phillips ; Zhaoning Song ; Suneth C. Watthage ; Ramez H. Ahanzhamejhad ; Michael J. Heben*

**EFFECT OF ILLUMINATION ON THERMAL CDCL2TREATMENT OF CDTE** ...... 175

*Sudhajit Misra ; Carina E. Hahn ; Vasilios Palekis ; Christos Ferekides ; Michael A. Scarpulla*

**CHALLENGES IN THE INDUSTRIAL PRODUCTION OF CZTS MONOGRAIN SOLAR CELLS** ...... 178

*Gerhard Peharz ; Valentin Satzinger ; Sandra Pötz ; Gemot Oreski ; Theodoros Dimopoulos ; Stefan Edinger ; Wolfeanz Hackl ; Hannes Starkl ; Parichehr Esfandiari ; Peter Krabb ; Stefan Gahr ; Lukas Plessing ; Dieter Meissner*

**UNDERSTANDING INSTABILITIES AND DEGRADATION DUE TO MOISTURE INGRESS IN CU(IN, GA)SE2SOLAR CELLS** ...... 182

*Grace Rajan ; Shankar Karki ; Isaac Butt ; Krishna Aryal ; Tyler J. Grassman ; Angus Rockett ; Sylvain Marsillac*

**CONTROL OF MOSE2 FORMATION IN HYDRAZINE-FREE SOLUTION-PROCESSED CIS/CIGS THIN FILM SOLAR CELLS** ...... 186

*Sona Ulicná ; Panagiota Arnou ; Alexander Eeles ; Mustafa Togay ; Lewis D. Wright ; Ali Abbas ; Andrei V. Malkov ; John M. Walls ; Jake W. Bowers*

**GROWTH AND PROPERTIES OF EPITAXIAL CU(IN, GA)SE2THIN FILMS DEPOSITED BY THE THREE-STAGE PROCESS FOR SOLAR CELLS** ...... 192

*Takeru Yamagami ; Yuta Ando ; Ishwor Khatri ; Mutsumi Sugiyama ; Tokio Nakada*

**IMPROVEMENT OF CIS SOLAR CELLS WITH KF POSTDEPOSITION FOLLOWING A SIMPLE TWO-STEP SELENIZATION PROCESS** ...... 195

*Yang Zhang ; Robert E. Bartolo ; Sang Jik Kwon ; Mario Dagenais*

**THE TWINS STRUCTURE, ELECTRICAL PROPERTIES AND CELL PERFORMANCE OF MAGNETRON SPUTTERING DEPOSITED CHLORINE DOPED CDTE** ...... 198

*Ziyao Zhu ; Fu-Kuo Chiang ; Zhongming Du ; Yufeng Zhang ; Xiangxin Liu*

**INVESTIGATION AND MITIGATION OF SHUNTS FOR HIGHER EFFICIENCY EPITAXIAL GASB/GASB AND GASB/GAAS SOLAR CELLS** ...... 202

*George T. Nelson ; Bor-Chau Juang ; Steve Johnston ; Michael A. Slocum ; Zachary S. Bittner ; Ramesh B. Lagumavarapu ; Diana Huffaker ; Seth M. Hubbard*

**DEVELOPMENT OF GASB SOLAR CELLS ON GAAS BY MOVPE VIA INTERFACE MISFIT TECHNIQUE** ...... 206

*Michael A. Slocum ; Alessandro Giussani ; Emily Kessler ; Phil Ahrenkiel ; George T. Nelson ; Seth M. Hubbard*

**FABRICATION OF INGAASP SOLAR CELLS FOR CONCENTRATOR APPLICATIONS** ...... 210

*Mitchell F. Bennett ; Matthew P. Lumb ; Kenneth J. Schmieder ; Brent Fisher ; Eric A. Armour ; Robert J. Walters*

**DETAILED CHARACTERIZATION FOR TCAD SIMULATIONS OF GAAS0.76P0.24/SI1-YGEY/SI SINGLE JUNCTION SOLAR CELLS** ...... 213

*Sabina Abdul Hadi ; Timothy Milakovich ; Eugene A. Fitzgerald ; Ammar Nayfeh*

**COMPARATIVE STUDY OF >2 EV LATTICE-MATCHED AND METAMORPHIC (AL)GAINP MATERIALS AND SOLAR CELLS GROWN BY MOCVD** ...... 215

*Daniel J. Chmielewski ; Christine Jackson ; Jacob Boyer ; Daniel Lepkowski ; John A. Carlin ; Aaron R. Arehart ; Tyler J. Grassman ; Steven A. Ringel*

**PERFORMANCE OF GASB PHOTOVOLTAICS WITH GRAPHENE COATING** ...... 219

*Benjamin P. Conlon ; Daniel J. Herrera ; Shaimaa A. Abdallah ; Jonathan O. Okafor ; Luke F. Lester*

**HIGH EFFICIENCY SINGLE-JUNCTION INGAP PHOTOVOLTAIC DEVICES UNDER LOW INTENSITY LIGHT ILLUMINATION** ...... 222

*Yushuai Dai ; Hyun Kum ; Michael A. Slocum ; George T. Nelson ; Seth M. Hubbard*

**RADIATION RESISTANT OF UPRIGHT METAMORPHIC GAINP/GAINAS/GE TRIPLE JUNCTION SOLAR CELLS FOR SPACE USE** ... 226

Liang Fang ; Abuduwayiti Aierken ; Zhen Pan ; Qiming Zhang ; Zhanhang Li ; Heini Maliya ; Wei Gao ; Hui Gao ; Ronghua Wan ; Bao Zhang ; He Wang ; Qi Guo

**HIGH EFFICIENCY GLASS WAVEGUIDING SOLAR CONCENTRATOR** ... 229

Chehao Hu ; Yusuf Dogan ; Matthew Morrison ; A. Nanda ; D. Ma ; R. Atkins ; C. K. Madsen

**GAINASP/GAINAS TANDEM SOLAR CELL WITH 32.6% ONE-SUN EFFICIENCY** ... 232

Nikhil Jain ; Kevin L. Schulte ; John F. Geisz ; Ryan M. France ; Myles A. Steiner

**EVALUATION OF TANDEM EFFICIENCIES: DILUTE NITRIDE P-I-N (BULK OR MQWS) IN CONJUNCTION WITH PRACTICAL SI SOLAR CELLS** ... 236

Khim Kharel ; Alexandre Freundlich

**GALLIUM PHOSPHIDE NANOSTRUCTURE ON SILICON BY SILICA NANOSPHERES LITHOGRAPHY AND METAL ASSISTED CHEMICAL ETCHING** ... 240

Sangpyeong Kim ; Chaomin Zhang ; Som Dahal ; Stuart Bowden ; Christiana B. Honsberg

**EFFICIENCY ENHANCEMENT OF INGAP/INGAAS/GE SOLAR CELLS WITH GRADUALLY DOPED P-N JUNCTION ACTIVE LAYERS** ... 244

Youngjo Kim ; Sang Hyun Jung ; Chang Zoo Kim ; Kangho Kim ; Hyun-Beom Shin ; Kyung Ho Park ; Won-Kyu Park ; Jaejin Lee ; Ho Kwan Kang

**ANALYSIS OF INGAP OXIDE GROWTH RATE AT HIGH TEMPERATURES AND AMBIENT CONDITIONS FOR TERRESTRIAL PHOTOVOLTAIC APPLICATIONS** ... 247

Nicole A. Kotulak ; Matthew P. Lumb ; Raymond Hoheisel ; Erin Cleveland ; Mitchell Bennett ; Phillip P. Jenkins ; Robert J. Walters

**GRAIN BOUNDARIES IN THIN-FILM POLYCRYSTALLINE GAAS SOLAR CELLS: A SIMULATION STUDY** ... 251

Khushboo Kumari ; Sushobhan Avasthi

**TIME-RESOLVED PL MEASUREMENTS IN THE GROWTH OF HIGH VOLTAGE (AL)GAINP/GAAS SOLAR CELLS** ... 255

Xinyi Li ; Wei Zhang ; Hongbo Lu

**LOW-RESISTANCE AND HIGHLY-TRANSPARENT GASB-BASED TUNNEL JUNCTIONS** ... 259

Matthew P. Lumb ; Shawn Mack ; Maria Gonzalez ; Kenneth J. Schmieder ; Mitchell F. Bennett ; Chaffra A. Affouda ; James E. Moore ; Robert J. Walters

**MODULATED PHOTOCURRENT MEASUREMENTS IN DOUBLE JUNCTION SOLAR CELLS** ... 263

Nicolás Márquez Peraca ; Behrang H. Hamadani

**EFFECT OF ATMOSPHERIC ABSORPTION BANDS ON THE OPTIMAL DESIGN OF MULTIJUNCTION SOLAR CELLS** ... 268

William E. Mcmahon ; Daniel J. Friedman ; John F. Geisz

**EFFECTS OF CONTACT CONFIGURATION AND PERIMETER RECOMBINATION ON OPTIMAL CELL SIZE FOR HIGH CONCENTRATION PHOTOVOLTAICS** ... 272

James E. Moore ; Matthew P. Lumb ; Kenneth J. Schmieder ; Robert J. Walters ; Brent Fisher ; Matt Meitl ; Scott Burroughs

**NUMERICAL SIMULATION OF DEFECTS IN III-V PV CELLS: THE EFFECT OF VOLTAGE BIAS AND DOPING CONCENTRATION** ... 276

Vasiliki Paraskeva ; Constantinos Lazarou ; Andreas Livera ; Venizelos Venizelou ; Maria Hadjipanayi ; George E. Georghiou

**IMPROVEMENT OF OPEN-CIRCUIT VOLTAGE IN METAMORPHIC GASB CELLS GROWN ON GAAS SUBSTRATES BY USING AN INTERFACIAL MISFIT ARRAY AND AN ALSB BLOCKING LAYER** ... 281

E. J. Renteria ; S. J. Addamane ; D. M. Shima ; A. Mansoori ; A. L. Soudachanh ; G. Balakrishnan

**ENERGY YIELD EVALUATION FOR FIELD OPERATION OF SOLAR CELLS IN SINGAPORE: GAAS/GAAS TANDEM VS. GAAS SINGLE-JUNCTION SOLAR CELLS** ... 284

Maung Thway ; Zekun Ren ; Kevin Nay Yaung ; Haohui Liu ; Zhe Liu ; Samuel Raj ; Soo Jin Chua ; Armin G. Aberle ; Tonio Buonassisi ; Ian Marius Peters ; Fen Lin

**SIMULATION OF THE PERFORMANCES OF MULTIJUNCTION SOLAR CELLS WITH IMPROVED VOLTAGE BY TRANSFER AND SCATTERING MATRIX METHODS** ... 290

Gianluca Timò ; Lucio Andreani

**OPTIMIZED DESIGN OF BACK-CONTACT THIN-FILM GAAS SOLAR CELLS** ... 294

Jia-Ling Tsai ; Chung-Yu Hong ; Tien-Chien Zhan ; Yuh-Renn Wu ; Albert Lin ; Peichen Yu

**DESIGN CONSIDERATIONS ON GAINNAS SOLAR CELLS WITH BACK SURFACE REFLECTORS** ... 297

Antti Tukiainen ; Arto Aho ; Timo Aho ; Ville Polojärvi ; Mircea Guina

**QUANTITATIVE ELECTROLUMINESCENCE ANALYSIS OF TRIPLE JUNCTION SOLAR CELLS TO DETERMINE SUBCELL VOLTAGE-TEMPERATURE COEFFICIENTS** ... 301

Kevin Tyler ; Geoffrey K. Bradshaw ; Sam Wilt ; David M. Wilt ; Richard R. King

**PROGRESS TOWARDS DOUBLE-JUNCTION INGAN SOLAR CELL** ........ 305

Ehsan Vadiee ; Evan A. Clinton ; Heather Mcfavilen ; Alec M. Fischer ; Yi Fang ; Joshua J. Williams ; Christiana B. Honsberg ; William A. Doolittle ; Stephen M. Goodnick

**A PHYSICS-BASED SIMULATION TOOL FOR LEAKAGE CURRENTS IN C-SI PV MODULES** .......... 309

John M. Waddle ; Saroj Dahal ; Marco Nardone

**BROADBAND TA2O5 MOTH-EYE ANTIREFLECTION COATINGS FOR TANDEM SOLAR CELLS ON SI** ........ 315

Bo Yuan ; Brian Thibeault ; David Payne ; James Mutitu ; Ivan Perez-Wurfl ; Kevin Dobson ; Brianna Conrad ; Allen Barnett ; Robert L. Opila

**CARRIER TRANSPORT IN POLYCRYSTALLINE SILICON AT HIGH OPTICAL INJECTION: TRANSIENT PHOTOCONDUCTANCE VS. NUMERICAL MODELING** ........ 319

Uchechi Anyanwu ; Christian Harris ; Andrey Semichaevsky

**IMPROVING SILICON SURFACE PASSIVATION WITH A SILICON OXIDE LAYER GROWN VIA OZONATED DEIONIZED WATER** ........ 322

Sara Bakhshi ; Ngwe Zin ; Kristopher O. Davis ; Marshall Wilson ; Ismail Kashkoush ; Winston V. Schoenfeld

**DEPOSITION OF SIOC BY PLASMA-FREE ULTRA-LOW-TEMPERATURE ALD (ULT-ALD) AND ITS PASSIVATION ON P-TYPE SILICON** ........ 326

Meixi Chen ; Naoto Noda ; Raphael Rochat ; Abhishek Iyer ; James H. Hack ; Changhee Ko ; Christian Dussarrat ; Robert L. Opila

**A METHOD FOR QUANTITATIVELY INVESTIGATING THE REAR-SIDE PASSIVATION PERFORMANCE OF PERC CELLS** ........ 329

Tsung-Cheng Chen ; Yung-Sheng Lin ; Chen-Hao Ku ; Ting-Wei Kuo ; Cheng-Shun Hu ; Ching-Chang Wen

**FIELD-EFFECT PASSIVATION BY NEGATIVE CHARGE ON BORON EMITTER AND BORON-DOPED SURFACES BY A NOVEL LOW-COST PLASMA CHARGE INJECTION** ........ 333

Eunhwan Cho ; Young-Woo Ok ; James Hwang ; Aditi Jain ; Vijay D. Upadhyaya ; John Keith Tate ; Ajeet Rohatgi

**INDUSTRY RELEVANT RIE TEXTURING FOR MC-SI DIAMOND WIRE OR DIRECT WAFER® PRODUCT: OPTIMIZED REFLECTIVITY, UNIFORMITY, AND THROUGHPUT** ........ 337

Jose Luis Cruz-Campa ; Ray Fraser ; Rob Steeman ; John Linton

**SHORT-CIRCUIT CURRENT-DENSITY ENHANCEMENT OF SILICON SOLAR CELLS USING PLASMONICS ANTIREFLECTIVE COATING AND LUMINESCENT DOWNSHIFTING** ........ 343

Sheng-Kai Feng ; Wen-Jeng Ho ; Guan-Yi Li ; Jheng-Jie Liu ; Hao-Yu Yang ; Ta-Wei Chuang

**EXTREMELY LOW REFLECTIVITY NANOPOROUS BLACK SILICON SURFACE BY COPPER CATALYZED ETCHING FOR EFFICIENT SOLAR CELLS** ........ 346

K A S M Ehteshamul Haque ; Wenqi Duan ; Fatima Toor

**IMPACT OF FRONT SIDE PYRAMID SIZE ON THE LIGHT TRAPPING PERFORMANCE OF WAFER BASED SILICON SOLAR CELLS AND MODULES** ........ 352

Oliver Höhn ; Nico Tucher ; Benedikt Bläsi

**A STUDY OF BLISTER CONTROL OF AL2O3 THIN FILM DEPOSITED BY PLASMA-ASSISTED ATOMIC LAYER DEPOSITION AFTER FIRING PROCESS** ........ 356

Min Gu Kang ; Jeong In Lee ; Hee-Eun Song ; Myeong Sangjeong ; Kyung Taekjeong ; Hyo Sikchang

**PYPVCELL: AN OPEN-SOURCE SOLAR CELL MODELING LIBRARY IN PYTHON** ........ 359

Kan-Hua Lee ; Kenji Araki ; Omar Elleuch ; Nobuaki Kojima ; Masafumi Yamaguchi

**IMPROVEMENT IN SURFACE PASSIVATION OF C-SI USING GRADIENT-LAYERED A-SI:H FILM FOR HIGH EFFICIENCY SILICON HETEROJUNCTION SOLAR CELLS** ........ 363

Soonil Lee ; Leo Mathew ; Rajesh Rao ; Jae Hyun Kim ; Sanjay K. Banerjee ; Edward T. Yu

**PHOTOVOLTAIC PERFORMANCE ENHANCEMENT OF TEXTURED SILICON SOLAR CELLS USING LUMINESCENT DOWN-SHIFTING METHYLAMMONIUM LEAD TRIBROMIDE PEROVSKITE NANOPHOSPHORS** ........ 367

Guan-Yi Li ; Wen-Jeng Ho ; Sheng-Kai Feng ; Hao-Yu Yang ; Ta-Wei Chuang ; Bang-Jin You ; Zong-Xian Lin ; Zong-Liang Tseng ; Lung-Chien Chen

**SINX THIN FILMS WITH APPROPRIATE ANTIREFLECTION AND SHIFT-CONVERSION PROPERTIES FOR SILICON SOLAR CELLS** ........ 370

E. Men-Pérez ; J. Salazar ; A. Dutt ; J. Santoyo-Salazar ; G. Santana

**NUMERICAL SIMULATION OF CRYSTALLINE SILICON SOLAR CELLS WITH FULL AREA METAL OXIDE REAR CONTACTS** ........ 373

James E. Moore ; Woojun Yoon ; Phillip P. Jenkins ; Robert J. Walters

**INTERDIGITATED BACK CONTACT SILICON SOLAR CELL WITH PEROVSKITE LAYER FOR FRONT SURFACE PASSIVATION AND ULTRAVIOLET RADIATION STABILITY** ........ 377

Rahul Pandey ; Shivam Gupta ; Trijul Khatri ; Rishu Chaujar

**POTENTIAL OF A-SI:H/C-SI HETEROJUNCTION SOLAR CELLS WITH VERY THIN WAFERS** ........................................................................................................................................381

*Hitoshi Sai ; Hiroshi Umishio ; Takuya Matsui ; Shota Nunomura ; Tomoyuki Kawatsu ; Hidetaka Takato ; Koji Matsubara*

**MANIPULATING FIXED CHARGES IN ZRO2 BY DOPING FOR PASSIVATION AND ANTIREFLECTION ON WAFER-SI SOLAR CELLS** ..........................................................................385

*Woo Jung Shin ; Laidong Wang ; Wen-Hsi Huang ; Meng Tao*

**LOW TEMPERATURE ANTIREFLECTION COATING FOR SILICON SOLAR CELLS** ...................389

*O. S. Shinde ; Ej Schneller ; N. Dhere ; S. V. Ghaisas*

**RELATIONSHIP BETWEEN POWER LOSS AND VOLTAGE APPLIED TO SOLAR CELLS IN PID-AFFECTED SOLAR MODULES** ..........................................................................................................392

*Fumei Wang ; Baosong Duan ; Wenshuang He ; He Wang ; Hong Yang ; Chengfeng Su ; Bojie Su ; Xue Zhang ; Yunxue Cao ; Hui Zhao*

**A NEW LOW-COST AND LOW-TEMPERATURE CHEMICAL PASSIVATION PROCESS FOR LARGE AREA INDUSTRIAL SINGLE CRYSTALLINE SILICON WAFERS** ...................................396

*Tarun S. Yadav ; K. Sandeep ; Ashok K. Sharma ; B. Spandana ; K.L. Narasimhan ; B.M. Arora ; Anil Kottantharayil ; Prabir K. Basu*

**EVALUATION OF ALD PASSIVATION LAYERS FOR INDUSTRIAL PERC PROCESS** .................399

*Chang Youn Yoo ; Keunkee Hong ; Jisun Kim ; Eunjoo Lee ; Dong Seop Kim*

**QUANTITATIVE ANALYSIS OF ELECTROLUMINESCENCE AND INFRARED THERMAL IMAGES FOR AGED MONOCRYSTALLINE SILICON PHOTOVOLTAIC MODULES** ..................402

*Irene Berardone ; Juan Lopez Garcia ; Marco Paggi*

**GAP PASSIVATION STRUCTURE FOR SCALABLE N-TYPE INTERDIGITATED ALL BACK CONTACT SILICON HETERO-JUNCTION SOLAR CELL** ..............................................................408

*Lei Zhang ; Ujjwal Das ; Steven Hegedus*

**PROPOSAL OF THE BANDGAP DESIGN USING THE SUN HEIGHT OF THE CULMINATION ON THE WINTER SOLSTICE** ...........................................................................................................................412

*Kenji Araki ; Kan-Hua Lee ; Masafumi Yamaguchi*

**PHOTOEXCITED CARRIERS, PHONONS, AND THEIR SCATTERING MEASURED IN SEMICONDUCTOR JUNCTIONS BY TRANSIENT EXTREME ULTRAVIOLET SPECTROSCOPY** .......................................................................................................................................417

*Scott K. Cushing ; Brett M. Marsh ; Mihai E. Vaida ; Lucas M. Carneiro ; Ilana J. Porter ; Angela Lee ; Stephen R. Leone*

**ON THE USE OF VOLTAGE MEASUREMENTS FOR DETERMINING CARRIER LIFETIME AT HIGH ILLUMINATION INTENSITY** ...............................................................................................420

*Robert Dumbrell ; Mattias K. Juhl ; Thorsten Trupke ; Ziv Hameiri*

**HIGH RESOLUTION 3D CHEMICAL CHARACTERISATION OF A CADMIUM TELLURIDE SOLAR CELL BY DYNAMIC SIMS** ...............................................................................................................424

*Thomas Fiducia ; Kexue Li ; Chris Grovenor ; Kurt Barth ; Walajabad Sampath ; Michael Walls*

**HARSH OUTDOOR EVALUATION SETUP AND FIRST POWER PRODUCTION RESULTS FOR SI MINI-MODULES COVERED BY EU3+-BASED DOWN CONVERTERS** .....................................429

*Benjamín González-Díaz ; Carlos Montes ; Joaquín Sanchiz ; Luis Ocaña ; Carlos Quinto ; Cecilio Hernández-Rodríguez ; Mari Paz Friend ; Manuel Cendagorta-Galarza ; David Cañadillas ; Ricardo Guerrero-Lemus*

**STUDY OF MICRO-STRUCTURAL PROPERTIES OF ZNO AND TIO2THIN FILM GROWN BY SPRAY PYROLYSIS** ...........................................................................................................................................433

*G. Gordillo ; J.M. Correa ; A.A. Ramirez ; E. A. Ramírez*

**NONLINEAR RESPONSE OF SILICON SOLAR CELLS** ......................................................................437

*Behrang H. Hamadani ; Andrew Shore ; Howard W. Yoon ; Mark Campanelli*

**EXTENDED LINEAR INTERPOLATION/EXTRAPOLATION PROCEDURE FOR ACCURATE AND VERSATILE TRANSLATION OF THE I-V CURVES OF PV CELLS AND MODULES** ..........441

*Y. Hishikawa ; H. Ohshima ; M. Higa ; K. Yamagoe ; T. Takenouchi ; T. Doi*

**SEVERITY TEST WITH UNEVEN LOAD DUE TO WIND ACTION ON PHOTOVOLTAIC MODULE** .......................................................................................................................................................445

*Shu-Tsung Hsu*

**STANDARDIZED DURABILITY TEST FOR ORGANIC PHOTOVOLTAIC AND DYE SENSITIZED SOLAR CELL** ..................................................................................................................................448

*Shu-Tsung Hsu ; Yean-San Long ; Teng-Chun Wu*

**SPATIAL THICKNESS UNIFORMITY AND STRUCTURAL EVALUATION OF RF SPUTTERED ZNO THIN FILMS FOR SOLAR CELL** ...............................................................................................451

*Babar Hussain ; Taj M. Khan*

**LOCAL MEASUREMENTS OF SURFACE CAPACITANCE BY ELECTROSTATIC FORCE MICROSCOPY ON CU(IN, GA)SE2MATERIALS** ........................................................................................455

*Tomoaki Ishii ; Takashi Minemoto ; Takuji Takahashi*

**A COMPARISON OF SI-BASED CAMERAS FOR IMAGING LUMINESCENCE FROM PHOTOVOLTAIC MATERIALS AND DEVICES** .................................................................................. 459

Steve Johnston

**BLISTERING OF AL2O3/A-SINX:H STACKS: ANALYSIS OF THE SUBMERGED PART OF THE ICEBERG BY COLORED PICOSECOND ACOUSTIC MICROSCOPY** ................................... 464

Fabien Lebreton ; Arnaud Devos ; Etienne Drahi ; Patricia De Coux ; François Silva ; Sergej Filonovich ; Pere Roca I Cabarrocas

**SELF-REFERENCE PROCEDURE TO REDUCE UNCERTAINTY IN MODULE CALIBRATION** ................... 467

D.H. Levi ; C.R. Osterwald ; S. Rummel ; L. Ottoson ; A. Anderberg

**UNCERTAINTY EVALUATION OF PRIMARY REFERENCE PHOTOVOLTAIC CELL CALIBRATION UNDER OUTDOOR CONDITION IN TIBET** ........................................................ 472

Haitao Liu ; Shiyu Sang ; Guomin Zhou ; Yonghui Zhai

**REQUIREMENT OF ARTIFICIAL LIGHTING SIMULATOR FOR EVALUATION EMERGING PV PERFORMANCE RATING UNDER INDOOR ENVIRONMENT** ................................................. 476

Yean-San Long ; Shu-Tsung Hsu ; Teng-Chun Wu

**NON-CONTACT VOLTAGE MEASUREMENT OF SOLAR CELL WITH ELECTROSTATIC VOLTMETER** ...................................................................................................................... 480

Sakutaro Miyajima ; Kensuke Nishioka ; Yoshihiro Hishikawa

**NREL'S CELL AND MODULE PERFORMANCE GROUP'S ASYMPTOTIC PMAX PROTOCOL FOR PEROVSKITE DEVICES** ............................................................................................. 483

Tom Moriarty ; Dean Levi

**OUTDOOR OPERATING TEMPERATURE MODELING OF PHOTOVOLTAIC MODULES INCLUDING TRANSIENT EFFECT** ................................................................................... 487

Soo-Young Oh ; Min-Soo Kim ; Won-Shup So ; Woo Kyoung Kim ; Jae Hak Jung ; Chinho Park ; Benazzouz Aboubakr ; Ikken Badr ; Naimi Zakaria ; Benlarabi Ahmed

**PRIMARY REFERENCE CELL CALIBRATIONS WITH REDUCED MEASUREMENT UNCERTAINTY** ............................................................................................................. 490

C.R. Osterwald ; L. Ottoson ; R. Williams ; C. Mack ; T. Moriarty ; K.A. Emery ; D.H. Levi

**IMPLEMENTATION OF NOVEL PIN CONNECTION AND TEST ROUTINE FOR IMPROVED ACCURACY IN I-V MEASUREMENTS** .................................................................................. 496

Samuel Raj ; Johnson Kai Chi Wong ; Mohan Krishan Bhan ; Evan Palmer ; Jian Wei Ho ; Sumukh Ramprasad ; Wang Junci ; Thomas Mueller ; Armin G. Aberle

**A NEW METHOD TO QUANTIFY CONTACT RESISTANCE USING LOCALIZED-ILLUMINATION PHOTOLUMINESCENCE TECHNIQUE IN A SOLAR CELL** ..................................... 499

Amit Singh Rajput ; Samuel Raj ; Johnson Wong ; Armin G. Aberle

**IMPROVEMENT OF THE PROPERTIES OF CZTS THIN FILMS PREPARED BY SPRAY PYROLYSIS USING DMSO IN ACETONE AS SOLVENT** ...................................................... 503

E. A. Ramírez ; A. Ramírez ; G. Gordillo

**ASSESSMENT OF CARRIER LIFETIMES AND SURFACE RECOMBINATION VELOCITY THROUGH SPECTRAL MEASUREMENTS** .......................................................................... 508

John Roller ; Behrang H. Hamadani

**IMPACT OF SPACE RADIATION ENVIRONMENT ON CONCENTRATOR PHOTOVOLTAIC SYSTEMS** ................................................................................................................. 512

Pilar Espinet-Gonzalez ; Tatiana Vinogradova ; Michael D. Kelzenberg ; Alexander Messer ; Emily C. Warmann ; Chris Peterson ; Nina Vaidya ; Ali Naqavi ; Jing-Shun Huang ; Samuel P. Loke ; Don Walker ; Colin J. Mann ; Sergio Pellegrino ; Harry A. Atwater

**EXTRACTING THE FIXED CHARGE DENSITY IN HFOX FILMS GROWN ON HIGHLY-DOPED P-SI SAMPLES** .................................................................................................. 517

Alexander To ; Jie Cur ; Bram Hoex

**NEAR-UNITY ULTRA-WIDEBAND THERMAL INFRARED EMISSION FOR SPACE SOLAR POWER RADIATIVE COOLING** ...................................................................................... 521

Ali Naqavi ; Samuel P. Loke ; Michael D. Kelzenberg ; Emily C. Warmann ; Pilar Espinet-González ; Nina Vaidya ; Jing-Shun Huang ; Tatiana A. Roy ; Alexander J. Messer ; Tatiana G. Vinogradova ; Ali Hajimiri ; Sergio Pellegrino ; Harry A. Atwater

**LINE-FOCUS AND POINT-FOCUS SPACE PHOTOVOLTAIC CONCENTRATORS USING ROBUST FRESNEL LENSES, 4-JUNCTION CELLS, & GRAPHENE RADIATORS** ......................... 525

Mark O'Neill ; A.J. Mcdanal ; Michael Piszczor ; Matt Myers ; Paul Sharps ; Claiborne Mcpheeters ; Jeff Steinfedt

**SIMULATION OF LIGHT TRAPPING STRUCTURES FOR ENHANCING RADIATION HARDNESS IN SPACE SOLAR CELLS** .......................................................................... 531

Nizami Z. Vagidov ; Kyle H. Montgomery ; Geoffrey K. Bradshaw ; David M. Wilt

**AN ALTERNATIVE METHOD FOR SOLAR CELL INTEGRATION** .............................................. 537

Jessica Buckner ; Tracy Davis ; Eric Muskovin ; Bernard Carpenter

**NIEL DOSE ANALYSIS ON TRIPLE JUNCTION CELLS 30% EFFICIENT AND RELATED SINGLE JUNCTIONS** .......... 541

Roberta Campesato ; Erminio Greco ; Mariacristina Casale ; Massimo Gervasi ; P.G. Rancoita ; Davide Rozza ; Mauro Tacconi ; Enos Gombia ; Aldo Kingma ; Carsten Baur

**THIN AND FLEXIBLE TRIPLE JUNCTION CELLS 30% EFFICIENT: QUALIFICATION RESULTS AND FUTURE SPACE APPLICATIONS** .......... 545

Roberta Campesato ; Mariacristina Casale ; Giuseppe Gabetta ; Emilio Fernandez Lisbona ; Laurent D'Abrigeon

**PRINTED ASSEMBLIES OF MICROSCALE TRIPLE-JUNCTION (3J) INVERTED METAMORPHIC (IMM) GAINP/GAAS/INGAAS SOLAR CELLS** .......... 549

Boju Gai ; John Geisz ; Daniel Friedman ; Jongseung Yoon

**COMPARATIVE STUDY ON NONRADIATIVE RECOMBINATION CENTERS IN PROTON IRRADIATED INAS/GAAS QUANTUM DOT STRUCTURE BY TWO WAVELENGTH EXCITED PHOTOLUMINESCENCE** .......... 552

M. D. Haque ; N. Kamata ; S-I. Sato ; S. M. Hubbard

**DESIGN AND PROTOTYPING EFFORTS FOR THE SPACE SOLAR POWER INITIATIVE** .......... 558

Michael D. Kelzenberg ; Pilar Espinct-Gonzalez ; Nina Vaidya ; Tatiana A. Roy ; Emily C. Warmann ; Ali Naqvi ; Samuel P. Loke ; Jing-Shun Huang ; Tatiana G. Vinogradova ; Alexander J. Messer ; Christophe Leclerc ; Eleftherios E. Gdoutos ; Fabien Royer ; Ali Hajimiri ; Sergio Pellegrino ; Harry A. Atwater

**DEFECT CHARACTERIZATION OF III-V QUANTUM STRUCTURE SOLAR CELLS USING PHOTO-INDUCED CURRENT TRANSIENT SPECTROSCOPY** .......... 562

Shin-Ichiro Sato ; Takeyoshi Sugaya ; Tetsuya Nakamura ; Takeshi Ohshima

**EFFECT OF LUMINESCENCE COUPLING BETWEEN INGAP AND GAAS SUBCELLS TO EXTERNAL QUANTUM EFFICIENCY IN TRIPLE-JUNCTION SOLAR CELLS** .......... 567

Mitsunobu Suga ; Mitsuru Imaizumi ; Tetsuya Nakamur ; Takeshi Ohshima

**LIGHTWEIGHT CARBON FIBER MIRRORS FOR SOLAR CONCENTRATOR APPLICATIONS** .......... 572

Nina Vaidya ; Michael D. Kelzenberg ; Pilar Espinet-Gonzalez ; Tatiana G. Vinogradova ; Jing-Shun Huang ; Christophe Leclerc ; Ali Naqvi ; Emily C. Warmann ; Sergio Pellegrino ; Harry A. Atwater

**GAAS SOLAR CELLS ON V-GROOVED SILICON VIA SELECTIVE AREA GROWTH** .......... 578

Michelle Vaisman ; Nikhil Jain ; Qiang Li ; Kei May Lau ; Adele C. Tamboli ; Emily L. Warren

**HIGH TEMPERATURE ANNEALING OF IN1-XGAXN MQW SOLAR CELLS** .......... 582

Joshua J. Williams ; Heather Mcfavilen ; Steven Young ; Christiana B. Honsberg ; Stephen M. Goodnick

**SOLAR PROBE PLUS ARRAY RELIABILITY** .......... 585

Anton Yanchilin ; Edward Gaddy

**PHOTOVOLTAIC TEMPERATURE ESTIMATION MODEL FOR RAPID IRRADIANCE CHANGE CONDITIONS IN TROPICAL REGIONS USING HEURISTIC ALGORITHMS** .......... 589

R. Srivatsan ; Lian L. Jiang ; Douglas L. Maskell

**ACCURACY OF CDTE PV ENERGY PREDICTIONS USING SPECTRAL CORRECTIONS** .......... 595

Mitchell Lee ; Kendra Passow ; Paul Wolffersdorff

**PLANTPREDICT: SOLAR PERFORMANCE MODELING MADE SIMPLE** .......... 600

Kendra Passow ; Lauren Ngan ; Geoffrey Rich ; Mitch Lee ; Stephen Kaplan

**INTEGRABILITY COMPARISON BETWEEN BIPV AND BAPV IN TROPICAL CONDITIONS: A BANGALORE CASE-STUDY** .......... 604

Gayathri Aaditya ; Roshan R Rao ; Monto Mani

**A NEW PHOTOVOLTAIC SYSTEM TOPOLOGY THROUGH LOAD MANAGEMENT** .......... 608

Joseph A. Azzolini ; Meng Tao

**FIRST STEP FOR POWER GENERATION AMOUNT ESTIMATION OF SOLAR MATCHING SYSTEM** .......... 613

Kazuya Hosokawa ; Toshiaki Yachi ; Yoichi Hirata ; Yasuyuki Watanabe

**IRRADIANCE AND TEMPERATURE DISTRIBUTIONS AT HIGH LATITUDES: DESIGN IMPLICATIONS FOR PHOTOVOLTAIC SYSTEMS** .......... 619

Anne Gerdimenes ; Josefine Sclj

**STEP-BY-STEP EVALUATION OF PHOTOVOLTAIC MODULE PERFORMANCE RELATED TO OUTDOOR PARAMETERS: EVALUATION OF THE UNCERTAINTY** .......... 626

Anne Migan Dubois ; Jordi Badosa ; Fausto Calderón-Obaldía ; Olivier Atlan ; Vincent Bourdin ; Marko Pavlov ; Dae Young Kim ; Yvan Bonnassieux

**PERFORMANCE COMPARISONS OF A PV SYSTEM BY MONITORING SOLAR IRRADIANCE WITH DIFFERENT PYRANOMETERS** .......... 632

Yasuhiro Matsumoto ; J. Antonio Urbano ; Ramón Peña ; María De La Luz Olvera ; Nun Pitalúa ; Miguel A. Luna ; René Asomoza

**FINANCIAL ANALYSIS OF A GRID-CONNECTED PHOTOVOLTAIC SYSTEM IN SOUTH FLORIDA** .......... 638

Hadis Moradi ; Amir Abtahi ; Ali Zilouchian

STUDY OF PHOTOVOLTAIC SYSTEMS MONITORING METHODS..................643
    *E. Ortega ; G. Aranguren ; M.J. Sáenz ; R. Gutiérrez ; J.C. Jimeno*

GLOBAL DESIGN ASPECTS OF PERSISTENT AND AUTONOMOUS PV POWERED SYSTEMS ..................648
    *I. M. Peters ; S. Watson ; N. Sahraei ; T. Buonassisi*

HOW TO CHOOSE THE BEST EMPIRICAL MODEL FOR OPTIMUM ENERGY YIELD
PREDICTIONS ..................652
    *Steve Ransome ; Juergen Sutterlueti*

MODELING AND ANALYSIS OF PHOTOVOLTAIC ELECTROCHEMICAL SYSTEM USING
MODULE-LEVEL POWER ELECTRONICS..................658
    *Gowri M. Sriramagiri ; Nuha Ahmed ; Kevin D. Dobson ; Steven S. Hegedus*

BETAVOLTAIC GENERATION FUNCTION IN SILICON..................663
    *A.V. Sachenko ; I.O. Sokolovskyi ; M. Evstigneev*

MULTI-OBJECTIVE OPTIMIZATION FOR COLOR-TUNABILITY AND TRANSPARENCY IN
COLLOIDAL QUANTUM DOT SOLAR CELLS..................667
    *Ebuka S. Arinze ; Botong Qiu ; Nathan Palmquist ; Yan Cheng ; Yida Lin ; Gabrielle Nyirjesy ; Gary Qian ;*
    *Susanna M. Thon*

CUBIC PHASE INXGA1-XN/GAN QUANTUM WELLS FOR THEIR APPLICATION TO
TANDEM SOLAR CELLS..................670
    *C. A. Hernández-Gutiérrez ; Y. L. Casallas-Moreno ; Dagoberto Cardona ; Yu. Kudriavtsev ; A. Morales-Acevedo*
    *; G. Santana-Rodríguez ; M. López-López*

MODELING OF P-I-N GAASPN/GAP MQWS SOLAR CELL: TOWARDS LATTICE MATCHED
III-V/SI TANDEM ..................673
    *Khim Kharel ; Alexandre Freundlich*

INP QUANTUM DOT INTERMEDIATE BAND SOLAR CELL GROWN VIA MOCVD ..................677
    *Hyun Kum ; Yushuai Dai ; Michael Slocum ; Zachary Bittner ; Seth Hubbard*

MODIFIED LIMITING EFFICIENCY FOR MULTIPLE EXCITON GENERATION SOLAR
CELLS..................681
    *Jongwon Lee ; Christiana B. Honsberg*

A SIMPLE MONTE CARLO MODEL OF A HOT CARRIER CELL..................685
    *Tor Oskar Saetre*

OPTIMIZATION OF SEMICONDUCTOR QUANTUM DOTS FOR LUMINESCENT SOLAR
CONCENTRATORS: MINIMIZING REABSORPTION LOSSES..................690
    *Anatoli I. Shkrebtii ; Anatoliy V. Sachenko ; Igor O. Sokolovskyi ; Vitaliy P. Kostylyov ; Mykola R. Kulish ; Denis*
    *V. Khomcnko ; Mykhaylo A. Evstigneev*

DEVELOPMENT OF ABSORBER AND ENERGY SELECTIVE CONTACTS FOR HOT
CARRIER SOLAR CELLS..................696
    *Santosh Shrestha ; Simon Chung ; Yuanxun Liao ; Wenkai Cao ; Neeti Gupta ; Yi Zhang ; Xiaoming Wen ; Gavin*
    *Conibeer*

GAASBI DEVICES FOR THERMAL ENERGY CONVERSION ..................701
    *Margaret Stevens ; Abigail Licht ; Nicole Pfiester ; Emily Carlson ; Kevin Grossklaus ; Thomas E. Vandervelde*

ANALYTIC JV-CHARACTERISTICS OF IDEAL IMPURITY PV-CELLS ..................706
    *Rune Strandberg*

PHOTOLUMINESCENCE PROPERTIES OF IN-PLANE ULTRAHIGH-DENSITY INAS
QUANTUM DOTS ON GAASSB/GAAS(001) FOR SOLAR CELL APPLICATIONS..................712
    *Ryo Sugiyama ; Naoki Akimoto ; Tomah Sogabe ; Koichi Yamaguchi*

CARRIER SELECTIVE BACK CONTACT (CSBC) SOLAR CELL USING TRANSITION METAL
OXIDES..................716
    *Astha Tyagi ; Kunal Ghosh ; Anil Kottantharayil ; Saurabh Lodha*

ANALYSIS OF OPEN-CIRCUIT VOLTAGE AND CONVERSION EFFICIENCY IN QUANTUM-
DOT SOLAR CELLS VIA DETAILED-BALANCE-LIMIT THEORY..................721
    *Lin Zhu ; Hidefumi Akiyama ; Yoshihiko Kanemitsu*

ZINC SELENIDE SURFACE PASSIVATION LAYER FOR SINGLE-CRYSTALLINE CZTSE
SOLAR CELLS ..................726
    *Michael A. Lloyd ; Douglas Bishop ; Brian E. Mccandless ; Robert Birkmirc*

USE OF SINGLE WALL CARBON NANOTUBE FILMS DOPED WITH TRIETHYLOXONIUM
HEXACHLORANTIMONATE AS A TRANSPARENT BACK CONTACT FOR CDTE SOLAR
CELLS..................730
    *Fadhil K. Alfadhili ; Jacob M. Gibbs ; Geethika K. Liyanage ; Patrick W. Krantz ; Suneth C. Watthage ; Zhaoning*
    *Song ; Adam B. Phillips ; Michael J. Heben*

GRAIN AND GRAIN BOUNDARY GEOMETRICAL SHAPE CONSIDERATIONS ON SODIUM
AND POTASSIUM DIFFUSION THROUGH MOLYBDENUM FILMS ..................735
    *Orlando Ayala ; Chinedum Akwari ; Tasnuva Ashrafee ; Shankar Karki ; Grace Rajan ; Sylvain Marsillac*

**SOLUTION-PROCESSED NICKEL-ALLOYED IRON PYRITE THIN FILM AS HOLE TRANSPORT LAYER IN CADMIUM TELLURIDE SOLAR CELLS** ............ 738

*Ebin Bastola ; Khagendra P. Bhandari ; Randy J. Ellingson*

**USE OF CDS:O AND CDSE AS WINDOW LAYERS FOR CDTE PHOTOVOLTAICS** ............ 742

*Tom Baines ; Guillaume. Zoppi ; Ken Durose ; Jonathan D. Major*

**APPLICATIONS OF HYBRID ORGANIC-INORGANIC METAL HALIDE PEROVSKITE THIN FILM AS A HOLE TRANSPORT LAYER IN CDTE THIN FILM SOLAR CELLS** ............ 748

*Khagendra P. Bhandari ; Suneth C. Watthage ; Zhaoning Song ; Adam Phillips ; Michael J. Heben ; Randy J. Ellingson*

**MAGNESIUM-DOPED ZINC OXIDE AS A HIGH RESISTANCE TRANSPARENT LAYER FOR THIN FILM CDS/CDTE SOLAR CELLS** ............ 752

*Francesco Bittau ; Elisa Artegiani ; Ali Abbas ; Daniele Menossi ; Alessandro Romeo ; Jake W. Bowers ; John M. Walls*

**INVESTIGATION OF ZNL-XMGXO:A1 FILM BY RATIO FREQUENCY MAGNETRON CO-SPUTTERING AS TRANSPARENT CONDUCTIVE OXIDE LAYER** ............ 757

*Jakapan Chantana ; Yuya Ishino ; Takashi Minemoto*

**A NEW TCO/WINDOW-BUFFER FRONT STACK FOR CDTE SOLAR CELLS AND ITS IMPLEMENTATION** ............ 761

*Alan E. Delahoy ; Xuehai Tan ; Akash Saraf ; Payal Patra ; Surya Manda ; Yunfei Chen ; Krishnakumar Velappan ; Bastian Siepchen ; Shou Peng ; Ken K. Chin*

**SYNTHESIS OF HIGH-QUALITY AZO POLYCRYSTALLINE FILMS VIA TARGET BIAS RADIO FREQUENCY MAGNETRON SPUTTERING** ............ 767

*Zhongming Du ; Yufeng Zhang ; Xiangxin Liu*

**CLOSE-SPACE SUBLIMATED CDTE SOLAR CELLS WITH CO-SPUTTERED CDSXSE1-XALLOY WINDOW LAYERS** ............ 771

*Corey R. Grice ; Maxwell Junda ; Alexander Archer ; Jian Li ; Yanfa Yan*

**EFFECTS OF GRAPHENE OXIDE BARRIER ON CU2ZNSNSXSE4-XTHIN FILM SOLAR CELLS** ............ 777

*Woo-Lim Jeong ; Jung-Hong Min ; In-Young Kim ; Hae-Sun Kim ; Jin-Hyeok Kim ; Dong-Seon Lee*

**13% CDS/CDTE SOLAR CELL USING A NANOCOMPOSITE (CUS)X(ZNS)1-X THIN FILM HOLE TRANSPORT LAYER** ............ 781

*Kamala Khanal Subedi ; Khagendra P. Bhandari ; Ebin Bastola ; Randy J. Ellingson*

**MOLYBDENUM OXIDE AND MOLYBDENUM NITRIDE BACK CONTACTS FOR THIN-FILM CDTE SOLAR CELLS** ............ 785

*Anna Kindvall ; Jason Kephart ; Walajabad Sampath*

**INVESTIGATION AND OPTIMIZATION OF CD-FREE BUFFER LAYERS IN2S3 AND ZN(O, S) FOR CU2ZNSN(S, SE)4-BASED SOLAR CELLS** ............ 791

*Willi Kogler ; Thomas Schnabel ; Andreas Bauer ; Stefanie Spiering ; Erik Ahlswede ; Michael Powalla*

**REAR CONTACT PASSIVATION FOR HIGH BANDGAP CU(IN, GA)SE2 SOLAR CELLS WITH VARYING ABSORBER THICKNESS AND FLAT GA PROFILE** ............ 796

*Dorothea Ledinek ; Pedro Salome ; Carl Hägglund ; Marika Edoff*

**LASER ANNEALED BACK CONTACTS FOR CDTE SOLAR CELLS** ............ 802

*Vasilios Palekis ; Shamara Collins ; Imran Khan ; Vamsi Evani ; Sudhajit Misra ; Michael A. Scarpulla ; Mark Lonergan ; Don Morel ; Chris Ferekides*

**ENHANCED ANTI-REFLECTIVE COATING FOR THIN FILM SOLAR CELLS** ............ 807

*Grace Rajan ; Shankar Karki ; Robert W. Collins ; Sylvain Marsillac*

**INFLUENCE OF AGS LAYER INSERTION AT ABSORBER/ITO INTERFACE ON STRUCTURAL AND PHOTOVOLTAIC PROPERTIES OF ULTRATHIN CU(IN,GA)SE2 SOLAR CELLS** ............ 810

*Muhammad Saifullah ; Jihye Gwak ; Kihwan Kim ; Joo Hyung Park ; Junsik Cho ; Jae Ho Yun*

**NOVEL, FACILE BACK SURFACE TREATMENT FOR CDTE SOLAR CELLS** ............ 815

*Suneth C. Watthage ; Geethika K. Liyanage ; Zhaoning Song ; Fadhil K. Alfadhili ; Rabee B. Alkhayat ; Khagendra P. Bhandari ; Randy J. Ellingson ; Adam B. Phillips ; Michael J. Heben*

**OPTIMIZING CDS BUFFER LAYER FOR CIGS BASED THIN FILM SOLAR CELL** ............ 820

*Weijie Zhang ; Korhan Demirkan ; Geordie Zapalac ; David Spaulding ; Jochen Titus ; Neil Mackie*

**INVESTIGATION OF INP DEFECT CHARACTERISTICS GROWN USING NOVEL TF-VLS TECHNIQUE** ............ 823

*Abhinav Chikhalkar ; Alec Fischer ; Mark Hettick ; Ali Javey ; Richard R. King*

**INVESTIGATION OF FAST GROWTH GAAS-BASED SOLAR CELL ON REUSABLE SUBSTRATE BY METALORGANIC CHEMICAL VAPOR DEPOSITION** ............ 827

*Chaomin Zhang ; Abhinav Chikhalkar ; Ehsan Vadiee ; Richard King ; Christiana Honsberg ; Eric Armour ; Yeongho Kim*

DEVELOPMENT OF ALUMINUM EPILAYERS AS BUFFERS FOR GAINAS ............... 831

*Phil Ahrenkiel ; Nathan Smaglik ; Nikhil Pokharel ; Alessandro Giussani ; Michael A. Slocum ; Seth M. Hubbard*

LASER CRYSTALLIZATION OF AMORPHOUS GERMANIUM ON TITANIUM NITRIDE-COATED STEEL FOR LOW-COST GAAS SOLAR-CELLS ............... 837

*Saloni Chaurasia ; Srinivasan Raghavan ; Sushobhan Avasthi*

HIGH QUALITY EPITAXIAL GERMANIUM ON SI (100) FOR LOW -COST III–V SOLAR-CELLS ............... 841

*Saloni Chaurasia ; Srinivasan Raghavan ; Sushobhan Avasthi*

CRYSTALLINITY CONTROL IN LOW-TEMPERATURE GROWTH OF POLY-CRYSTALLINE GE BY ION BEAM DEPOSITION ............... 845

*S. I. Maximenko ; N. A. Mahadik ; P. P. Jenkins ; R. J. Walters ; A. Giussani ; E. L. Mcclure ; S. M. Hubbard ; C. Bailey*

HIGH EFFICIENCY GAINP/GAAS DOUBLE JUNCTION SOLAR CELL ON SI SUBSTRATE ASSISTED BY THE ELECTRON BEAM TREATMENT ............... 849

*Hyo Jin Kim ; Yong Whan Kim*

ANALYSIS OF DEPOSITED RESIDUES AND ITS CLEANING PROCESS ON GAAS SUBSTRATE AFTER EPITAXIAL LIFT-OFF ............... 854

*Tatsuya Nakata ; Kentaroh Watanabe ; Hassanet Sodabanlu ; Daiki Kimura ; Naoya Miyashita ; Yoshitaka Okada ; Yoshiaki Nakano ; Masakazu Sugiyama*

ULTRATHIN SILICON-AN-INSULATOR (SOI) WAFER FOR COMPLIANT SUBSTRATE ............... 858

*Shinyoung Noh ; Anita Ho-Baillie ; Stephen Bremner ; Martin A. Green ; Xiaojing Hao*

CHARACTERIZATION OF GAAS SOLAR CELLS GROWN BY HYDRIDE VAPOR PHASE EPITAXY IN HORIZONTAL REACTOR ............... 861

*Ryuji Oshima ; Kikuo Makita ; Takeyoshi Sugaya ; Akinori Ubukata*

FLEXIBLE GAAS SINGLE-JUNCTION SOLAR CELLS BASED ON SINGLE-CRYSTAL-LIKE THIN-FILM MATERIALS DIRECTLY GROWN ON METAL TAPES ............... 866

*Sara Pouladi ; Monika Rathi ; Mojtaba Asadirad ; Pavel Dutta ; Seung Kyu Oh ; Devendra Khatiwada ; Shahab Shervin ; Yao Yao ; Venkat Selvamanickam ; Jae-Hyun Ryou*

REDUCED DEFECT DENSITY IN SINGLE-CRYSTALLINE-LIKE GAAS THIN FILM ON FLEXIBLE METAL SUBSTRATES BY USING SUPERLATTICE STRUCTURES ............... 869

*M. Rathi ; P. Dutta ; D. Khatiwada ; Y. Yao ; Y. Gao ; Y. Li ; S. Sun ; S. Pouladi ; S. Reed ; A. Khadimallah ; J. Ryou ; V. Selvamanickam ; N. Zheng ; P. Ahrenkiel*

ECONOMIC ANALYSIS OF TRANSFER PRINTED III–V VIRTUAL SUBSTRATES ............... 873

*Kenneth J. Schmieder ; Matthew P. Lumb ; Michael K. Yakes ; Shawn Mack ; Mitchell F. Bennett ; Sergey I. Maximenko ; Laura B. Ruppalt ; Michael A. Meeker ; Chase T. Ellis ; Matthew Meitl ; Joseph G. Tischler ; Robert J. Walters*

THIN FILMS OF ZINC-DOPED GAAS BY RF MAGNETRON SPUTTERING FOR USE IN PHOTOVOLTAIC CELLS ............... 876

*Kirby Simon ; Kyle Cepeda ; Nishit Shetty ; Elijah Thimsen*

SELF ALIGNED ALUMINUM SELECTIVE EMITTER FOR N-TYPE SI CELLS ............... 881

*San Theigi ; Robert C. Reedy ; Vincenzo Lasalvia ; Paul Stradins ; Benjamin G. Lee*

HOW TO REALIZE SOLAR CELLS WITH LASER STRUCTURED PLATED NI-CU-CONTACTS WITH EXCELLENT ADHESION AND HIGH FILL-FACTORS WITHOUT PARASITIC PLATING ............... 884

*A. Büchler ; S. Kluska ; J. Bartsch ; B. Grübel ; A.A. Brand ; S. Gutscher ; M. Glatthaar*

EXPLOITING THE POTENTIALS OF THE FRONT SURFACE FIELD (FSF) INDUSTRIAL SILICON SOLAR CELL ............... 888

*Ahrar Ahmed Chowdhury ; Yu -Chen Hsu ; Veysel Unsur ; Abasifreke Ebong*

PHOTOVOLTAIC PERFORMANCE OF SILICON SOLAR CELLS ENHANCED BY PLASMONIC SILVER NANOPARTICLES OF VARIOUS DIMENSIONS DEPOSITING THROUGH ANODIC ALUMINUM OXIDE TEMPLATE ............... 893

*Ta-Wei Chuang ; Wen-Jeng Ho ; Sheng-Kai Feng ; Jheng-Jie Liu ; Guan-Yi Li ; Hao-Yu Yang ; Yun-Chie Yang ; Cho-Chun Chiang ; Yao-Hui Chen*

MITIGATION OF POTENTIAL-INDUCED DEGRADATION ............... 896

*Orry Faur ; Maria Faur*

ELECTRODEPOSITION OF SI-LAYER THROUGH REDUCTION OF DIATOMACEOUS EARTH FOR THE APPLICATION OF SOLAR-CELLS ............... 900

*Muhammad Monirul Islam ; Imane Abdellaoui ; Takeaki Sakurai ; Saad Hamzaoui ; Katsuhiro Akimoto*

EFFECT OF SI CONTENT IN A1 PASTE ON LOCAL A1 REAR CONTACTS IN PERC CELL ............... 904

*Supawan Joonwichien ; Katsuhiko Shirasawa ; Satoshi Utsunomiya ; Hidetaka Takato*

NEW SILVER PASTE METALLIZATION APPROACH ON P+ DIFFUSION ZONES OF SILICON SOLAR CELLS ............... 907

*Yunjun Li ; Mohshi Yang ; Igor Pavlovsky ; Guoping Zeng*

**INFLUENCES OF ANNEALING AND DEFECT LIMITATION ON P-TYPE SILICON SOLAR CELL** ............................................................................................................................... 911

*Yu-Hsuan Lin ; Sung-Yu Chen ; Kuen-Yi Wu ; Chien-Hsun Chen ; Chen-Hsun Du ; Chun-Ming Yeh*

**REDUCED TEMPERATURE SILVER PASTE WITH LOW CONTACT RESISTANCE FOR ADVANCED SOLAR CELL APPLICATIONS** ............................................................................ 914

*Ryan Mayberry ; Daniel Holzmann ; Gerd Schulz ; Lindsey Karpowich ; Mark Naylor ; Matthias Hoerteis*

**BSF ISLANDS FOR REDUCED RECOMBINATION IN IBC CELLS** ............................................ 917

*Agnes A. Mewe ; Nicolas Guillevin ; Ilkay Cesar ; Antonius R. Burgers*

**THERMAL STABILITY OF HYDROGENATED BORON EMITTERS** .......................................... 921

*Khaja H. Mohammed ; Larry C. Cousar ; Philip A. Mcmeans ; Garrett Z. Evans ; Douglas A. Hutchings ; Hameed A. Naseem ; Sergiu C. Pop*

**LIGHT INDUCED PLATING OF SILICON SOLAR CELLS USING BORIC ACID-FREE NICKEL CHEMISTRY** ............................................................................................................................. 925

*Krystal Munoz ; Lynne Michaelson ; Joseph Karas ; Tom Tyson ; James Rand ; Stuart Bowden*

**BAKING TEMPERATURE DEPENDENCE OF CU PASTE ON A1-BSF CELL PROPERTIES** .......................... 931

*Tomohiro Saito ; Tetsuya Fukuda ; Hoang Tri Hai ; Yuji Kurimoto ; Daisuke Ando ; Yuji Sutou ; Katsuhiko Shirasawa ; Junichi Koike*

**THE SILVER CONTACT AND FORMATION MECHANISM OF THE BORON EMITTER AND THE CURRENT FLOW MECHANISM OF THE SOLAR CELL ELECTRODE** ................................ 935

*Seunghyun Shin ; Soohyun Bae ; Sungeun Park ; Yoonmook Kang ; Hae-Seok Lee ; Donghwan Kim*

**LASER ANNEALING TO ENHANCE PERFORMANCE OF ALL-LASER-BASED SILICON BACK CONTACT SOLAR CELLS** ............................................................................................... 937

*Zeming Sun ; Mool C. Gupta*

**LARGE AREA N-TYPE SELECTIVE EMITTER CELLS USING LASER DOPING THROUGH BORON DOPED SCREEN PRINTED PASTE** ....................................................................... 940

*Ajay D Upadhyaya ; Vijaykumar D Upadhyaya ; Brian Rounsaville ; Keeya Madani ; Ajeet Rohatgi ; Toru Hanada*

**METALLIZED BORON-DOPED BLACK SILICON EMITTERS FOR FRONT CONTACT SOLAR CELLS** ............................................................................................................................... 944

*Guillaume Von Gastrow ; Hele Savin ; Eric Calle ; Pablo Ortega ; Ramón Alcubilla ; Andreana Daniil ; Elias Z. Stutz ; Anna Fontcuberta I Morral ; Sebastian Husein ; Tara Nietzold ; Mariana Bertoni*

**CONTACT RESISTANCE MEASUREMENT FOR THERMALLY DIFFUSED POINT CONTACT BY LOCALIZED DIELECTRIC BREAKDOWN SOLAR CELLS** ...................................... 948

*Qilin Ye ; Ned J. Western ; Anqi Liao ; Stephen P. Bremner*

**LOW TEMPERATURE REAR SURFACE METALLIZATION OF MULTI-CRYSTALLINE SILICON SOLAR CELLS FOR IMPROVED BULK LIFETIME** .......................................... 953

*N. J. Western ; S. P. Bremner*

**INVESTIGATION OF HIGH PERFORMANCE PEROVSKITE-BASED SOLAR CELLS GROWN BY HYBRID CHEMICAL VAPOR DEPOSITION TECHNIQUE** ............................................. 958

*Huseyin Cem Gokkaya ; Shen Qian ; Zhiwei Ren ; Annie Ng ; Charles Surya*

**ENHANCED PEROVSKITE SOLAR CELL PERFORMANCE USING FULL SPACE DEVICE OPTIMIZATION** ....................................................................................................................... 963

*Ahmer A.B. Baloch ; Shahzada P. Aly ; Mohammad I. Hossain ; Raka Jovanovic ; Nouar Tabet ; Fahhad H. Alharbi*

**MEASURING OPTICAL ABSORPTION IN ORGANIC PHOTOVOLTAICS USING MONOCHROMATED ELECTRON ENERGY-LOSS SPECTROSCOPY** ...................................... 966

*Jessica A. Alexander ; Frank J. Scheltens ; David W. Mccomb ; Lawrence F. Drummy ; Michael F. Durstock ; James B. Gilchrist ; Sandrine Hentz*

**ADVANCED DEPOSITION OF PHOTO-CATALYTIC TIO2 FILM BY ATMOSPHERIC SPPS FOR DYE SENSITIZED SOLAR CELLS** ............................................................................... 970

*Ifeanacho Anyadiegwu ; Dickson Kindole ; Geoffrey Kibiegon Ronoh ; Yoshimasa Noda ; Yasutaka Ando*

**CH3NH3PBI3-XBRXPEROVSKITE SOLAR CELLS VIA SPRAY ASSISTED TWO-STEP DEPOSITION: INFLUENCE OF BROMIDE ON THE DEVICE PERFORMANCE** ...................... 976

*Gaoda Chai ; Shiqiang Luo ; Shizhen Wang ; Hang Zhou*

**MODULATED STRUCTURE TO MAXIMIZE THE OPEN-CIRCUIT VOLTAGE WITH MODERATE BAND-GAP OF SMALL MOLECULE ORGANIC SOLAR CELLS-DFT APPROACH** ............. 980

*Saravanan Chinnusamy ; Amita Munshi ; Sukanya Santhosh Kumar ; W. S. Sampath ; Milind S. Dangate*

**PEROVSKITE GRAIN SIZE MODULATION BY ANNEALING IN METHYL-AMINE ENVIRONMENT** ....................................................................................................................... 986

*Arun Singh Chouhan ; Naga Prathibha Jasti ; Srinivasan Raghavan ; Sushobhan Avasthi ; Shreyash Hadke*

**FE2O3AS AN ELECTRON TRANSPORT MATERIAL FOR ORGANO-METAL HALIDE PEROVSKITE SOLAR CELLS** ....................................................................................... 989

*Dallas Fisher ; Pravakar P. Rajbhandari ; Tara P. Dhakal*

**OPTICAL EVALUATION OF PEROVSKITE FILMS IN AND FOR SOLAR CELL DEVICE STRUCTURES** .......... 993

Kiran Ghimire ; Dewei Zhao ; Changlei Wang ; Yanfa Yan ; Nikolas J. Printraza

**HYBRID ORGANIC-INORGANIC SOLAR CELLS WITH A BENZOQUINONE PASSIVATING LAYER** .......... 999

James Hack ; Abhishek Iyer ; Meixi Chen ; Nicole Kotulak ; Akirt Sridharan ; Robert Opila

**PRECISE 1-V CURVE MEASUREMENT PROCEDURE FOR PEROVSKITE SOLAR CELLS: APPLICATION TO VARIOUS TYPES OF DEVICES** .......... 1003

Y. Hishikawa ; M. Yoshita ; H. Shimura ; A. Sasaki ; T. Ueda

**ENHANCING THE CRYSTALLINE OF PLANAR-STRUCTURE CH3NH3PBI3PEROVSKITE SOLAR CELLS VIA SANDWICH EVAPORATION TECHNIQUE** .......... 1006

Po-Tsun Kuo ; Shang-Pang Lin ; Cheng-Shian Lin ; Ching-Fuh Lin

**TOWARD HIGH PERFORMANCE ORGANIC-SILICON HYBRID SOLAR CELLS** .......... 1009

Yi Lai ; Hong-Jhang Syu ; Ching-Fuh Lin

**NICKEL OXIDE THIN FILMS BY RADIO FREQUENCY SPUTTER FOR INVERTED PEROVSKITE SOLAR CELLS** .......... 1012

Hyeonseok Lee ; Yu-Ting Huang ; Shien-Ping Feng

**ANOMALOUS EFFICIENCY SCALING WITH DARK CURRENT IN PEROVSKITE SOLAR CELLS** .......... 1015

Vikas Nandal ; Pradeep R. Nair

**NUMERICAL SIMULATION AND PERFORMANCE OPTIMIZATION OF PEROVSKITE SOLAR CELL** .......... 1018

Sai Naga Raghuram Nanduri ; Mahbube K. Siddiki ; Ghulam M. Chaudhry ; Yahya Z. Alharthi

**PERFORMANCE PREDICTION FOR LARGE AREA PEROVSKITE SOLAR CELLS** .......... 1022

Yojak Raote ; Hitarth Choubisa ; Pradeep R. Nair

**PHOTOCONVERSION EFFICIENCY MODELING IN PEROVSKITE SOLAR CELLS** .......... 1025

A.V. Sachenko ; V.P. Kostylyov ; A.V. Bobyl ; V.M. Vlasiuk ; I.O. Sokolovskyi ; E.I. Terukov ; M. Evstigneev

**INFLUENCE OF MONO- AND DI-VALENT METAL ADDITIVES ON MORPHOLOGY AND CHARGE CARRIER DYNAMICS OF CH3NH3PBI3PEROVSKITE** .......... 1030

Niraj Shrestha ; Suneth C. Watthage ; Zhaoning Song ; Paul J. Roland ; Adam B. Phillips ; Michael J. Heben ; Randall J. Ellingson

**EFFECT OF DUAL CATHODE BUFFER LAYER ON TERNARY ORGANIC SOLAR CELL** .......... 1034

Ashish Singh ; T. Bhim Raju ; Anamika Dey ; Ritesh Kant Gupta ; Parameswar K. Iyer

**COPPER PLATED TOP ELECTRODE FOR AN INVERTED ORGANIC PHOTOVOLTAIC** .......... 1037

Malia Steward ; Zhan Shi ; Kyoung- Tae Kim ; Seungkeun Choi

**INTERFACE BAND GAP AND CHARGE TRAPPING IN BULK HETEROJUNCTION SOLAR CELLS** .......... 1040

Marian Tzolov ; Maxwell Mcintyre

**FABRICATION OF EFFICIENT CH3NH3PBI3 SOLAR CELLS IN AMBIENT AIR** .......... 1044

Feng Wang ; Ye Zhongbiao ; Hojjatollah Sarvari ; Somin Park ; Kenneth Graham ; Yuetao Zhao ; Zhi David Chen

**HIGH EFFICIENCY PEROVSKITE SOLAR CELLS BY A MODIFIED LOW-TEMPERATURE SOLUTION PROCESS INTER-DIFFUSION METHOD** .......... 1048

Yangyi Yao ; Wei-Lun Hsu ; Mario Dagenais

**INTERFACIAL MODIFICATION OF SOL-GEL ZNO/AZO BILAYER AS HIGHLY EFFICIENT ELECTRON TRANSPORT LAYER FOR PEROVSKITE SOLAR CELLS** .......... 1051

Shang-Hsuan Wu ; Ming-Yi Lin ; Sheng-Hao Chang ; Wei-Chen Tu ; Chi-Wei Chu ; Via-Chung Chang

**THE POTENTIAL OF BIFACIAL PHOTOVOLTAICS: A GLOBAL PERSPECTIVE** .......... 1055

Xingshu Sun ; Mohammad R. Khan ; Amir Hanna ; Muhammad M. Hussain ; Muhammad A. Alam

**PERFORMANCE ASSESSMENT OF STAND ALONE BIFACIAL SOLAR PANEL UNDER REAL TIME CONDITIONS** .......... 1058

Ahmer A.B. Baloch ; Maher Armoush ; Basel Hindi ; Abdelkader Bousselham ; Nouar Tabet

**OPERATION AND PERFORMANCE ASSESSMENT OF GRID-CONNECTED PV SYSTEMS IN OPERATION IN MAUI, HAWAII** .......... 1061

Severine Busquet ; Jonathan Kobayashi ; Richard E. Rocheleau

**A NOVEL MULTILEVEL SOLAR PANEL SYSTEM: IMPLEMENTATION AND VERIFICATION** .......... 1067

Tanmoy Debnath ; Syed N. Imtiaz ; Syed F. Nawaz ; Abdullah Al Mahmud ; Mosaddequr Rahman

**PREDICTING POWER LOSS DUE TO MODULE MISMATCH IN UTILITY-SCALE PHOTOVOLTAIC SYSTEMS** .......... 1071

Stephen Kaplan ; Kendra Passow

**APPLICATION OF SHAPED REFLECTORS TO INCREASE THE ENERGY HARVEST OF BIFACIAL PV SYSTEMS - ANALYZED WITH A MINIATURIZED TEST ARRAY** ......................................... 1077
*Hartmut Nussbaumer ; Markus Klenk ; Nico Keller ; Dominic Heller ; Remo Kaslin ; Thomas Baumann ; Franz Baumgartner*

**TOWARDS NEW MODULE AND SYSTEM CONCEPTS FOR LINEAR SHADING RESPONSE** ................... 1081
*Kostas Sinapis ; Tom T.H. Rooijakkers ; Lenneke H. Slooff ; Lars A.G. Okel ; Mark J. Jansen ; Anna J. Carr*

**PARTIAL SHADING ABATEMENT THROUGH CASCADED H-BRIDGE TOPOLOGY** ................................ 1086
*Steven Tidwell ; Joseph Latham ; Michael Mcintyre*

**DATA ANALYSIS FOR EFFECTIVE MONITORING OF PARTIALLY SHADED PHOTOVOLTAIC SYSTEMS** .......................................................................................................... 1090
*Odysseas Tsafarakis ; Kostas Sinapis ; Wilfried G.J.H.M. Van Sark*

**BIFACIAL PHOTOVOLTAIC MODULE ENERGY YIELD CALCULATION AND ANALYSIS** ....................... 1094
*Christopher E. Valdivia ; Chu Tu Li ; Annie Russell ; Joan E. Haysom ; Rui Li ; David Lekx ; Mohsen M. Sepeher ; Dan Henes ; Karin Hinzer ; Henry P. Schriemer*

**DESIGN AND DEVELOPMENT OF A SOLAR PHOTOVOLTAIC MODULE DETECTION CONTROL SYSTEM BASED ON PLC** ................................................................................................. 1100
*Yiwang Wang ; Jili Zhang ; Kanglin Liu ; Houjun Tang ; Hui Pan ; Yan Lin ; Peter Yang ; Rui Wang*

**DETECTING CALIBRATION DRIFT AT GROUND TRUTH STATIONS A DEMONSTRATION OF SATELLITE IRRADIANCE MODELS' ACCURACY** .................................................................... 1104
*Richard Perez ; James Schlemmer ; Adam Kankiewicz ; John Dise ; Alemu Tadese ; Thomas Hoff*

**PERFORMANCE OF SOLAR RESOURCE MONITORING STATIONS IN HOT CLIMATE REGIONS** ....................................................................................................................................... 1110
*Yahya Z. Alharthi ; Mahbube K. Siddiki ; Ghulam M. Chaudhry ; Saad Muaddi ; Ahmed Alahmed*

**FIRST RESULTS OF A LOW COST ALL-SKY IMAGER FOR CLOUD TRACKING AND INTRA-HOUR IRRADIANCE FORECASTING SERVING A PV-BASED SMART GRID IN LA GRACIOSA ISLAND** ...................................................................................................................... 1116
*David Cañadillas ; Walter Richardson ; Benjamín Gonzalez-Díaz ; Les E. Shephard ; Ricardo Guerrero Lemus*

**STATISTICAL ANALYSIS OF PV INSOLATION DATA** ............................................................................. 1122
*Abdulmunim Guwaeder ; Rama Ramakumar*

**A COMPARISON OF PV POWER FORECASTS USING PVLIB-PYTHON** .................................................. 1127
*William F. Holmgren ; Antonio T. Lorenzo ; Clifford Hansen*

**COMPARING THE TYPICAL GHI YEAR VS TYPICAL POWER YEAR** .................................................... 1132
*Alex Kubiniec ; Adam Kankiewicz ; Alemu Tadesse*

**THE HOLY GRAIL OF RESOURCE ASSESSMENT: LOW COST GROUND-BASED MEASUREMENTS WITH GOOD ACCURACY** ........................................................................................ 1134
*Bill Marion ; Benjamin Smith*

**GLOBAL COMPARISON OF THE IMPACT OF TEMPERATURE AND PRECIPITABLE WATER ON CDTE AND SILICON SOLAR CELLS** .............................................................................. 1140
*I. M. Peters ; L. Haohui ; T. Reindl ; T. Buonassisi*

**ESTIMATION OF MEAN MONTHLY GLOBAL SOLAR RADIATION USING MODEL BASED ON SUNSHINE HOURS FOR COLOMBIA** ...................................................................................... 1143
*Diego J. Rodríguez ; Johan Hernández ; Adolfo Jaramillo*

**IMPLEMENTATION OF SOLAR DIFFUSE CIE MODEL IN RAY TRACING PROGRAM FOR IRRADIANCE CALCULATIONS** ............................................................................................................ 1147
*Liliana Ruiz Diaz ; Pierre-Alexandre Blanche ; Robert A. Norwood*

**INVESTIGATION OF CITY-LEVEL SITE-PAIR CORRELATIONS OF SOLAR VARIABILITY USING EMPIRICAL SATELLITE DATA** ........................................................................................... 1151
*Rhythm Singh ; Rangan Banerje*

**ULTRA-SHORT-TERM PHOTOVOLTAIC GENERATION FORECASTING MODEL BASED ON WEATHER CLUSTERING AND MARKOV CHAIN** .................................................................... 1158
*Jin Tan ; Changhong Deng*

**DAILY SOLAR IRRADIANCE PROFILE CHARACTERIZATION AND RAMP RATE ANALYSIS AT DIFFERENT TIME RESOLUTIONS** ............................................................................ 1163
*Spyros Theocharides ; Venizelos Venizelou ; George Makrides ; George E. Georghiou*

**COMPARISON AND ANALYSIS OF INSTRUMENTS MEASURING PLANE OF ARRAY IRRADIANCE FOR ONE-AXIS TRACKING PV SYSTEMS** ........................................................... 1169
*Frank Vignola ; Chun-Yu Chiu ; Josh Peterson ; Michael Dooraghi ; Manajit Sengupta*

**A SKY IMAGE ANALYSIS SYSTEM FOR SUB-MINUTE PV PREDICTION** ......................................... 1175
*Rodrigo Verschae ; Li Li ; Shohei Nobuhara ; Takekazu Kato*

**LARGE AREA NANOSTRUCTURE INTEGRATION FOR BROAD-SPECTRUM, OMNIDIRECTIONAL ANTIREFLECTION IMPROVEMENTS ON POLYMER PACKAGED, MECHANICALLY FLEXIBLE, EPITAXIAL LIFT-OFF III-V SOLAR CELLS** ..................... 1181

*Gabriel Cossio ; Jihwan Lee ; Gautham Ragunathan ; Andre Wibowo ; Sudersena Rao Tatavarti ; Kimberly Sablon ; Edward T. Yu*

**DEVELOPMENT OF BACK SURFACE TEXTURE FOR LIGHT MANAGEMENT IN EPITAXIAL LIFT OFF (ELO) QUANTUM DOT SOLAR CELLS** ..................... 1184

*Brittany L. Smith ; George T. Nelson ; Yushuai Dai ; Michael A. Slocum ; Andre Wibowo ; Rao Tatavarti ; Seth M. Hubbard*

**ENABLING HIGH-EFFICIENCY INAS/GAAS QUANTUM DOT SOLAR CELLS BY EPITAXIAL LIFT-OFF AND LIGHT MANAGEMENT** ..................... 1189

*F. Cappelluti ; A. P. Cédola ; A. Khalili ; Farid Elsehrawy ; G. Bauhuis ; P. Mulder ; J. Schermer ; G. Bissels ; T. Aho ; T. Niemi ; M. Guina ; D. Kim ; J. Wu ; H. Liu*

**CHARACTERIZATION OF ARSENIC DOPED CDTE LAYERS AND SOLAR CELLS** ..................... 1193

*Sachit Grover ; Xiaoping Li ; Wei Zhang ; Ming Yu ; Gang Xiong ; Markus Gloeckler ; Roger Malik*

**ENHANCING P-TYPE DOPING IN POLYCRYSTALLINE CDTE FILMS** ..................... 1196

*Brian Mccandless ; Wayne Buchanan ; Gowri Sriramagiri ; Christopher Thompson ; Joel Duenow ; David Albin ; Soren Jensen ; John Moseley ; M. Al-Jassim ; Wyatt K. Metzger*

**SPECTRAL AND CONCENTRATION SENSITIVITY OF MULTIJUNCTION SOLAR CELLS AT HIGH TEMPERATURE** ..................... 1201

*Daniel J. Friedman ; Myles A. Steiner ; Emmett E. Perl ; John Simon*

**ON THE USE OF TRANSPARENT CONDUCTIVE OXIDES IN HIGH CONCENTRATOR III-V MULTIJUNCTION SOLAR CELLS** ..................... 1204

*Ignacio Rey-Stolle ; Yeonbae Lee ; Iván Garcia ; Luis Cifuentes ; Kin Man Yu ; Carlos Algora ; Wladek Walukiewicz*

**COMPONENT INTEGRATION EFFECTS IN 4-JUNCTION SOLAR CELLS WITH DILUTE NITRIDE 1EV SUBCELL** ..................... 1210

*I. García ; M. Ochoa ; I. Lombardero ; L. Cifuentes ; P. Caño ; M. Hinojosa ; I. Rey-Stolle ; C. Algora ; A. D. Johnson ; J. I. Davies ; K.H. Tan ; W.K. Loke ; S. Wicaksono ; S. F. Yoon*

**BISMUTH SURFACTANT-MEDIATED GROWTH OF GANASSB(BI) SOLAR CELLS** ..................... 1215

*Aymeric Maros ; Chaomin Zhang ; Jongwon Lee ; Hongfeng Wang ; Stephen Bremner ; Nikolai Faleev ; Christiana B. Honsberg ; Richard. R. King*

**AMORPHOUS SILICON CARBIDE FOR SILICON SURFACE PASSIVATION IN CARRIER-SELECTIVE CONTACT DEVICES** ..................... 1220

*Mathieu Boccard ; Christophe Ballif ; Zachary C. Holman*

**SURFACE PASSIVATION OF BORON DIFFUSED JUNCTIONS BY BOROSILICATE GLASS AND IN SITU GROWN SILICON DIOXIDE INTERFACE LAYER** ..................... 1222

*Valentin D. Mihailetchi ; Haifeng Chu ; Jan Lossen ; Radovan Kopecek*

**IMPROVED LIGHT INCOUPLING IN PLANAR SOLAR CELLS VIA IMPROVED TEXTURE MORPHOLOGY OF PDMS SCATTERING LAYER** ..................... 1228

*Salman Manzoor ; Zhengshan J. Yu ; Asad Ali ; Waqar Ali ; Zachary C. Holman*

**DAMAGE-FREE LASER ABLATION FOR EMITTER PATTERNING OF SILICON HETEROJUNCTION INTERDIGITATED BACK-CONTACT SOLAR CELLS** ..................... 1233

*Menglei Xu ; Twan Bearda ; Miha Filipic ; Hariharsudan Sivaramakrishnan Radhakrishnan ; Maarten Debucquoy ; Ivan Gordon ; Jozef Szlufcik ; Jef Poortmans*

**BENEFITS OF A THERMAL DRIFT DURING ATOMIC LAYER DEPOSITION OF AL2O3FOR C-SI PASSIVATION** ..................... 1237

*Fabien Lebreton ; Andy Zauner ; Pavel Bulkin ; Francois Silva ; Sergej Filonovich ; Pere Roca I Cabarrocas*

**GROWTH DIFFERENCE OF AMORPHOUS SILICON BETWEEN PLASMA ENHANCED AND CATALYTIC CVD BASED ON SILICON HETEROJUNCTION SOLAR CELLS** ..................... 1241

*Liping Zhang ; Renfang Chen ; Zhuopeng Wu ; Chenguang Sun ; Fanying Meng ; Zhengxin Liu*

**DEVELOPING AN UNDERSTANDING-BASED SELECTION OF HYBRID-PEROVSKITE COMPOUNDS AND THE CU-IN HYBRID-PEROVSKITE (CIHP) FAMILY** ..................... 1245

*Alex Zunger ; G. Dalpian ; Qihang Liu ; L.B Abdalla ; L.L. Kazmerski*

**EFFECTS OF ELECTRON AND PROTON RADIATION ON PEROVSKITE SOLAR CELLS FOR SPACE SOLAR POWER APPLICATION** ..................... 1248

*Jing-Shun Huang ; Michael D. Kelzenberg ; Pilar Espinet-González ; Colin Mann ; Don Walker ; Ali Naqavi ; Nina Vaidya ; Emily Warmann ; Harry A. Atwater*

**TOWARDS PEROVSKITE SILICON TANDEM SOLAR CELLS WITH OPTIMIZED OPTICAL PROPERTIES** ..................... 1253

*Jan Christoph Goldschmidt ; Alexander J. Bett ; Patricia S.C. Schulze ; Nico Tucher ; Martin Bivour ; Markus Kohlstädt ; Seunghun Lee ; Simone Mastroianni ; Laura Mundt ; Markus Mundus ; Paul Ndione ; Karl Wienands ; Kristina Winkler ; Uli Würfel ; Martin Hermle ; Stefan W. Glunz*

**FIRST-PRINCIPLES DENSITY FUNCTIONAL THEORY CALCULATION OF METAL-SUBSTITUTED LEAD HALIDE PEROVSKITE** ........................................................................ 1256
*Ji-Sang Park ; Matthew D. Sampson ; Alex B.F. Martinson ; Maria K.Y. Chan*

**ESTIMATING THE EFFECTS OF MODULE AREA ON THIN-FILM PHOTOVOLTAIC SYSTEM COSTS** ................................................................................................................................ 1259
*Kelsey A. W. Horowitz ; Ran Fu ; Xingshu Sun ; Tim Silverman ; Michael Woodhouse ; Muhammad A. Alam*

**COST ANALYSIS OF TANDEM MODULES** ...................................................................... 1264
*Sarah E. Sofia ; Jonathan Mailoal ; Dirk Weiss ; Tonio Buonassisi ; Ian Marius Peters*

**CAUSE OF CURRENT-COLLECTION FAILURE OBSERVED INISC-REDUCTION PHASE OF PV CELLS AND MODULES EXPOSED TO ACETIC ACID** ............................................ 1268
*Tadanori Tanahashi ; Norihiko Sakamoto ; Hajime Shibata ; Atsushi Masuda*

**COMPARISON OF PV MODULE PERFORMANCE BEFORE AND AFTER 11, 20, AND 25.5 YEARS OF FIELD EXPOSURE** ...................................................................................... 1271
*Jacob Rada ; Charles Chamberlin ; Peter Lehman ; Arne Jacobson*

**MARRYING QUALITY ASSURANCE WITH DESIGN ENGINEERING – A WINNING PARTNERSHIP! BUT, A CULTURAL DIVIDE?** ...................................................................... 1275
*Sarah Kurtz ; Govind Ramu ; Robert Cornell ; Sumanth Lokanath ; Edward Hsi ; Tony Sample ; Masaaki Yamamichi ; George Kelly ; Ted Spooner ; Jonathan Previtali ; John Wohlgemuth*

**UPDATED EVALUATION OF SHOCK HAZARDS TO FIREFIGHTERS WORKING IN PROXIMITY OF PV SYSTEMS** .................................................................................... 1280
*Olga Lavrova ; Jimmy E. Quiroz ; Jack Flicker ; Renee Gooding*

**GROWTH AND OPTIMIZATION OF GAINP/INP NANOWIRE TUNNEL DIODE** ............... 1286
*Xulu Zeng ; Gaute Otnes ; Magnus Heurlin ; Magnus T Borgström*

**CATHODOLUMINESCENCE MAPPING FOR THE DETERMINATION OF N-TYPE DOPING IN SINGLE GAAS NANOWIRES** ....................................................................................... 1289
*Hung-Ling Chen ; Chalermchai Himwas ; Andrea Scaccabarozzi ; Pierre Rale ; Fabrice Oehler ; Aristide Lemaître ; Laurent Lombez ; Jean-François Guillemoles ; Maria Tchemycheva ; Jean-Christophe Harmand ; Andrea Cattoni ; Stéphane Collin*

**OPTICAL OPTIMIZATION OF PASSIVATED GAAS NANOWIRE SOLAR CELLS** .............. 1294
*Kyle W. Robertson ; Ray R. Lapierre ; Jacob J. Krich*

**HIGH EFFICIENCY GAN NANOWIRE/SI PHOTOCATHODE FOR PHOTOELECTROCHEMICAL WATER SPLITTING** .............................................................. 1299
*Srinivas Vanka ; Sheng Chu ; Yichen Wang ; Ishiang Shih ; Hong Guo ; Zetian Mi*

**ANALYTIC DESCRIPTION OF THE IMPACT OF GRAIN BOUNDARIES ON VOC** ............. 1303
*Paul Haney ; Benoit Gaury*

**ROLE OF TELLURIUM BUFFER LAYER ON CDTE SOLAR CELLS' ABSORBER/BACK-CONTACT INTERFACE** .................................................................................................. 1308
*Tao Song ; James R. Sites*

**SIMULTANEOUS EXAMINATION OF GRAIN-BOUNDARY POTENTIAL, RECOMBINATION, AND PHOTOCURRENT IN CDTE SOLAR CELLS USING DIVERSE NANOMETER-SCALE IMAGING** ................................................................................................................... 1312
*C.S. Jiang ; H.R. Moutinho ; J. Moseley ; A. Kanevce ; J.N. Duenow ; E. Colegrove ; C. Xiao ; W.K. Metzger ; M.M. Al-Jassim*

**NANOPARTICLE/METAL REAR REFLECTORS FOR LOW- AND HIGH-TEMPERATURE SILICON SOLAR CELLS** ............................................................................................... 1317
*Syeda Qudsia ; Farah Qazi ; Mehwish Azher Javed ; Mathieu Boccard ; Zhengshan J. Yu ; Peter Firth ; Jonathan Bryan ; Zachary C. Holman*

**ABSORPTION IN EACH LAYER OF A SILICON HETEROJUNCTION SOLAR CELL** ........... 1322
*Keith R. Mcintosh ; Malcolm D. Abbott ; Benjamin A. Sudbury ; Salman Manzoor ; Zhengshan J. Yu ; Mehdi Leilaeioun ; Jiatiwei Shi ; Zachary C. Holman*

**INVESTIGATIONS ON PLASMONIC COLOR TUNING COATING ON C-SI SOLAR CELLS** ........ 1329
*Gerhard Peharz ; Wolfgang Waldhauser ; Christine Prietl ; Bettina Großschädl ; Martin C. Schubert ; Bernhard Michl*

**INVESTIGATION OF INTERFACE AND BULK LOCALIZED STATES IN A-SI:H SOLAR CELLS** ....................................................................................................................... 1333
*Adrien Bidiville ; Takuya Matsui ; Hitoshi Sai ; Koji Matsubara*

**EXPERIMENTAL AND THEORETICAL STUDY OF THE INFRARED EMISSIVITY OF CRYSTALLINE SILICON SOLAR CELLS** ........................................................................ 1339
*Alberto Riverola ; Alexander Mellor ; Diego Alonso Alvarez ; Lourdes Ferre Llin ; Ilaria Guarracino ; Christos N. Markides ; Douglas Paul ; Daniel Chemisana ; Ned Ekins-Daukes*

**HIGH PERFORMANCE MOLECULAR DONORS FOR ORGANIC SOLAR CELLS, MATERIALS DESIGN AND DEVICE OPTIMIZATION** .............................................................................. 1342
*Paul Geraghty ; Haotian Wang ; Calvin Lee ; Jegadesan Subbiah ; David Jones*

**ADVANCED OPTICAL MODELLING OF MICRO-TEXTURED SOLUTION-PROCESSED SOLAR CELLS WITH CONSIDERATION OF SMALL-AREA EFFECTS**........1346

*Benjamin Lipovšek ; Marko Jošt ; Andrej Campa ; Fei Gu ; Christoph J. Brabec ; Karen Forberich ; Janez Krc ; Marko Tonic*

**IDENTIFICATION OF DEGRADATION PATHWAYS OF ORGANIC SOLAR CELLS USING INFRARED SPECTROSCOPY**........1350

*S. Shah ; R Biswas ; T. Koschny ; V L Dalal*

**A DEVICE-INDEPENDENT SCREENING TECHNIQUE FOR RAPIDLY IDENTIFYING NEXT GENERATION OPV MATERIALS**........1354

*Bryon W. Larson ; Andrew J. Ferguson ; Bertrand J. Tremolet De Villers ; Ross E. Larsen*

**NOVEL ANTHANTHRONE AND ANTHANTHRENE CO-POLYMERS AS P-TYPE CONJUGATED SEMICONDUCTORS FOR ORGANIC PHOTOVOLTAICS**........1360

*Suru Vivian John ; Patrick Denk ; Christoph Ulbricht ; Herwig Heilbrunner ; Jean-Benoit Giguère ; Antoine Lafleur-Lambert ; Jean-Francois Morin ; Emmanuel Iwuoha ; Daniel Ayuk Mbi Egbe*

**REDUCING UV INDUCED DEGRADATION LOSSES OF SOLAR MODULES WITH C-SI SOLAR CELLS FEATURING DIELECTRIC PASSIVATION LAYERS**........1366

*Robert Witteck ; Henning Schulte-Huxel ; Boris Veith-Wolf ; Malte Ruben Vogt ; Fabian Kiefer ; Marc Kontges ; Robby Peibst ; Rolf Brendel*

**LARGE-AREA JUNCTION DAMAGE IN POTENTIAL-INDUCED DEGRADATION OF C-SI SOLAR MODULES**........1371

*Chuanxiao Xiao ; Chun-Sheng Jiang ; Steve Johnston ; Steve P. Harvey ; Peter Hacke ; Brian Gorman ; Mowafak Al-Jassim*

**SEARCH FOR MICROSTRUCTURAL DEFECTS AS NUCLEI FOR PID-SHUNTS IN SILICON SOLAR CELLS**........1376

*Volker Naumann ; Otwin Breitenstein ; Jan Bauer ; Christian Hagendorf*

**INVESTIGATING PID SHUNTING IN POLYCRYSTALLINE SILICON MODULES VIA MULTI-SCALE, MULTI-TECHNIQUE CHARACTERIZATION**........1381

*Steven P. Harvey ; John Moseley ; Adam Stokes ; Andrew Norman ; Brian Gorman ; Peter Hacke ; Steve Johnston ; Mowafak Al-Jassim*

**POTENTIAL-INDUCED DEGRADATION OF A SI NITRIDE/CRYSTALLINE SI INTERFACE OBSERVED THROUGH MINORITY CARRIER LIFETIME MEASUREMENT**........1385

*Naoyuki Nishikawa ; Seira Yamaguchi ; Keisuke Ohdaira*

**FIELD INSPECTION OF PV MODULES: QUANTIFICATION OF EVA BROWNING LEVEL USING AN IMAGE PROCESSING TOOL**........1389

*Sushanth Gudla ; Govindasamy Tamizhmani*

**PREVENTING POTENTIAL-INDUCED DEGRADATION IN CRYSTALLINE SILICON PV MODULES: RELATIONSHIP BETWEEN DEGRADATION AND BILL OF MATERIAL**........1395

*Alessandro Virtuani ; Eleonora Annigoni ; Christophe Ballif*

**IDENTIFYING REVERSE-BIAS BREAKDOWN SITES IN CUINXGA(1-X)SE2**........1400

*Steve Johnston ; Elizabeth Palmiotti ; Andreas Gerber ; Harvey Guthrey ; Lorelle Mansfield ; Timothy J. Silverman ; Mowafak Al-Jassim ; Angus Rockett*

**HIMAWARI-8 ENABLED REAL-TIME DISTRIBUTED PV SIMULATIONS FOR DISTRIBUTION NETWORKS**........1405

*Nicholas A. Engerer ; Jamie M. Bright ; Sven Killinger*

**REDUCED MEASUREMENT UNCERTAINTY IN PV MODULE BATCH TESTING**........1411

*Blagovest Mihaylov ; Bengt Jaeckel ; Juergen Arp ; Ralph Gottschalg*

**CLOUD MOTION IDENTIFICATION ALGORITHMS BASED ON ALL-SKY IMAGES TO SUPPORT SOLAR IRRADIANCE FORECAST**........1415

*Lydie Magnone ; Fabrizio Sossan ; Enrica Scolari ; Mario Paolone*

**AUTOMATIC DETECTION OF INACTIVE SOLAR CELL CRACKS IN ELECTROLUMINESCENCE IMAGES**........1421

*Sergiu Spataru ; Peter Hacke ; Dezso Sera*

**APPLYING SPATIAL DOWNSCALING AND SMART PERSISTENCE TO PROVIDE AN IMPROVED SOLAR FORECAST TO REDUCE COMMERCIAL DEMAND CHARGES**........1427

*Alex Kubiniec ; Ted Belanger ; Adam Kankiewicz ; Skip Dise ; Nate Glasgow ; Alemu Tadesse*

**THERMAL CHARACTERISTICS OF PID-AFFECTED MONOCRYSTALLINE SILICON SOLAR MODULES UNDER ILLUMINATED AND DARK CONDITIONS**........1430

*Pan Zhao ; Shuwen Guo ; He Wang ; Hong Yang ; Dengyuan Song ; Shiyu Sang ; Bojie Su ; Xue Zhang ; Yunxue Cao ; Hui Zhao*

**TARGETED EVALUATION OF UTILITY-SCALE AND DISTRIBUTED SOLAR FORECASTING**........1435

*Matthew Lave ; Robert J. Broderick ; Laurie Burnham*

**RECORD EFFICIENCIES FOR SELENIUM PHOTOVOLTAICS AND APPLICATION TO INDOOR SOLAR CELLS** ............ 1441
Douglas M. Bishop ; Teodor Todorov ; Yun Seog Lee ; Oki Gunawan ; Richard Haight

**CLOSE-SPACED SUBLIMATION FOR SB2SE3SOLAR CELLS** ............ 1445
Laurie J. Phillips ; Peter Yates ; Oliver S. Hutter ; Tom Baines ; Leon Bowen ; Ken Durose ; Jonathan D. Major

**FABRICATION OF COPPER ARSENIC SULFIDE THIN FILMS FROM NANOPARTICLES FOR APPLICATION IN SOLAR CELLS** ............ 1449
Scott A. Mcclary ; Joseph Andler ; Carol A. Handwerker ; Rakesh Agrawal

**ORIENTATION CONTROLLED GE THIN FILMS ON GLASS BY AL-INDUCED CRYSTALLIZATION** ............ 1452
Kaveh Shervin ; Khim Kharel ; Alexandre Freundlich

**IN-LINE POTASSIUM FLUORIDE TREATMENT OF CIGS ABSORBERS DEPOSITED ON FLEXIBLE SUBSTRATES IN A PRODUCTION-SCALE PROCESS TOOL** ............ 1455
Ryan Kaczynski ; Jinwoo Lee ; Jane Van Alsburg ; Baosheng Sang ; Urs Schoop ; Jeffrey Britt

**LIGHT-SOAK AND DARK-HEAT INDUCED CHANGES IN CU(IN, GA)SE2 SOLAR CELLS: A MACROSCOPIC TO MICROSCOPIC STUDY** ............ 1459
Rouin Farshchi ; Benjamin Hickey ; Dmitry Poplavskyy

**A NEW MODEL TO DETERMINE INSTALLED SYSTEM COST AND LCOE FOR ARPA-E'S MOSAIC MICRO-CONCENTRATOR PV PROGRAM** ............ 1463
Ran Fu ; Kelsey A.W. Horowitz ; Daniel W. Cunningham ; James Zahler

**FIXED-TILT 660 × CONCENTRATING PHOTOVOLTAIC SYSTEM WITH 30% EFFICIENCY** ............ 1469
Alex J. Grede ; Jared S. Price ; Baomin Wang ; Michael V. Lipski ; Brent Fisher ; Kyu-Tae Lee ; Junwen He ; Gregory S. Brulo ; Xiaokun Ma ; Scott Burroughs ; Christopher D. Rahn ; Ralph G. Nuzzo ; John A. Rogers ; Noel C. Giebink

**WAFER INTEGRATED MICRO-SCALE CONCENTRATING PHOTOVOLTAICS** ............ 1473
Tian Gu ; Duanhui Li ; Lan Li ; Bradley Jared ; Gordon Keeler ; Bill Miller ; William Sweatt ; Scott Paap ; Michael Saavedra ; Ujjwal Das ; Steve Hegedus ; Anna Tanke-Pedretti ; Juejun Hu

**TOWARD STATIONARY CONCENTRATOR PHOTOVOLTAIC PANELS** ............ 1476
Peter Kozodoy ; Christopher Gladden ; Michael Pavilonis ; Tobias Wheeler ; Christopher Rhodes ; Chadwick Casper ; Kevin Schneider

**CPV TECHNOLOGIES NOT RELYING ON PERFECTION OF TRACKERS** ............ 1479
Kenji Araki ; Yasuyuki Ota ; Kan-Hua Lee ; Kensuke Nishioka ; Masafumi Yamaguchi

**THE GETTERING EFFECT OF DIELECTRIC FILMS FOR SILICON SOLAR CELLS** ............ 1485
A. Y. Liu ; C. Sun ; V. P. Markevich ; A. R. Peaker ; J. D. Murphy ; D. Macdonald

**TABULA RASA: OXYGEN PRECIPITATE DISSOLUTION THOUGH RAPID HIGH TEMPERATURE PROCESSING IN SILICON** ............ 1491
Erin E. Looney ; Hannu S. Laine ; Mallory A. Jensen ; Amanda Youssef ; Vincenzo Lasalvia ; Paul Stradins ; Tonio Buonassisi

**TOWARD EFFECTIVE GETTERING IN BORON-IMPLANTED SILICON SOLAR CELLS** ............ 1494
Hannu S. Laine ; Ville Vähänissi ; Zhengjun Liu ; Ernesto Magaña ; Ashley E. Morishige ; Jan Krügener ; Kristian Salo ; Hele Savin ; Barry Lai ; David P. Fenning

**IMPACT OF THE INITIAL GROWTH INTERFACE ON THE GRAIN STRUCTURE IN HPMC-SI INGOT** ............ 1498
Giri Wahyu Alam ; Etienne Pihan ; Benoit Marie ; Nathalie Mangelinck-Noël

**EFFECT OF CARBON CONCENTRATION AND GROWTH CONDITIONS ON OXYGEN PRECIPITATION BEHAVIOR IN N-TYPE CZ-SI** ............ 1504
Takuto Kojima ; Ryota Suzuki ; Kosuke Kinoshita ; Kyotaro Nakamura ; Atsushi Ogura ; Yoshio Ohshita ; Isao Masada ; Shoji Tachibana

**NANO-IMAGING OF PERFORMANCE IN PHOTOVOLTAICS** ............ 1508
Elizabeth M. Tennyson ; Marina S. Leite

**IMPLICATIONS OF CONDUCTIVE GRAIN BOUNDARIES IN CHLORINE-TREATED CDTE SOLAR CELLS** ............ 1511
Mohit Tuteja ; Vasilios Palekis ; Allen Hall ; Scott Maclaren ; Chris S. Ferekides ; Angus A. Rockett

**IMAGING THE MULTI-TEMPORAL PHOTO-CARRIER DYNAMICS AT THE NANOMETER SCALE IN ORGANIC AND INORGANIC SOLAR CELLS** ............ 1516
Pablo A. Fernández Garrillo ; Lukasz Borowik ; Florent Caffy ; Renaud Demadrille ; Benjamin Grévin

**NANOSCALE TOMOGRAPHIC CHARGE TRANSPORT IN POLYCRYSTALLINE CHALCOGENIDE ABSORBERS: CDTE VERSUS CIGS** ............ 1522
Justin L. Luria ; Andrew Moore ; Sun Yu ; Mark Aindow ; Bryan D. Huey

**IMPROVING THE PV MODULE SINGLE-DIODE MODEL ACCURACY WITH TEMPERATURE DEPENDENCE OF THE SERIES RESISTANCE** ............ 1526
Kyumin Lee

CELL-TO-MODULE (CTM) ANALYSIS FOR PHOTOVOLTAIC MODULES WITH SHINGLED SOLAR CELLS ........................................................................................................... 1531

Max Mittag ; Tobias Zech ; Martin Wiese ; David Blasi ; Matthieu Ebert ; Harry Wirth

A PRACTICAL IRRADIANCE MODEL FOR BIFACIAL PV MODULES ..................................... 1537

Bill Marion ; Sara Macalpine ; Chris Deline ; Amir Asgharzadeh ; Fatima Toor ; Daniel Riley ; Joshua Stein ; Clifford Hansen

A DETAILED MODEL OF REAR-SIDE IRRADIANCE FOR BIFACIAL PV MODULES ................ 1543

Clifford W. Hansen ; Renee Gooding ; Nathan Guay ; Daniel M. Riley ; Johnson Kallickal ; Donald Ellibee ; Amir Asgharzadeh ; Bill Marion ; Fatima Toor ; Joshua S. Stein

VIEW FACTOR MODEL AND VALIDATION FOR BIFACIAL PV AND DIFFUSE SHADE ON SINGLE-AXIS TRACKERS .................................................................................................... 1549

Marc Abou Anoma ; David Jacob ; Ben C. Bourne ; Jonathan A. Scholl ; Daniel M. Riley ; Clifford W. Hansen

A FAST QUASI-STATIC TIME SERIES (QSTS) SIMULATION METHOD FOR PV IMPACT STUDIES USING VOLTAGE SENSITIVITIES OF CONTROLLABLE ELEMENTS .......................... 1555

Xiaochen Zhangl ; Santiago Grijalva ; Matthew J. Reno ; Jeremiah Deboever ; Robert J. Broderick

FAST DETERMINATION OF DISTRIBUTION-CONNECTED PV IMPACTS USING A VARIABLE-TIME-STEP QUASI-STATIC TIME-SERIES APPROACH .......................................... 1561

Barry Mather

SCALABILITY OF THE VECTOR QUANTIZATION APPROACH FOR FAST QSTS SIMULATION ................................................................................................................................. 1567

Jeremiah Deboever ; Santiago Grijalva ; Matthew J. Reno ; Xiaochen Zhang ; Robert J. Broderick

MACHINE LEARNING FOR RAPID QSTS SIMULATIONS USING NEURAL NETWORKS ............ 1573

Matthew J. Reno ; Robert J. Broderick ; Logan Blakely

ALGORITHMIC ASPECTS OF A COMMERCIAL-GRADE DISTRIBUTION SYSTEM LOAD FLOW ENGINE ...................................................................................................................... 1579

Francis Therrien ; Marc Belletête ; Jean-Sébastien Lacroix ; Matthew J. Reno

RESONANT AND NON-RESONANT DIELECTRIC COATINGS FOR HIGH EFFICIENCY SOLAR CELLS ....................................................................................................................... 1585

Dongheon Ha ; Chen Gong ; Marina S. Leite ; Jeremy N. Munday

ENHANCED LIGHT TRAPPING IN THIN SILICON SOLAR CELLS USING EFFECTIVELY TRANSPARENT CONTACTS (ETCS) ..................................................................................... 1589

Rebecca Saive ; André Augusto ; Stuart G. Bowden ; Harry A. Atwater

ENHANCED POWER CONVERSION EFFICIENCY IN SINGLE NANOWIRE DEVICES THROUGH SYMMETRY BREAKING DESIGN ............................................................................ 1594

Jian Zhou ; Yonggang Wu ; Zihuan Xia ; Xuefei Qin ; Zongyi Zhang

CDSE(TE)/CDS/CDSE RODS VS. CDTE/CDS/CDSE SPHERES: MORPHOLOGY-DEPENDENT CARRIER DYNAMICS FOR PHOTON UPCONVERSION ........................................................... 1598

Eric Y. Chen ; Zhuohui Li ; Christopher C. Milleville ; Kyle R. Lennon ; Matthew F. Doty

DRIFT-DIFFUSION INGAN/GAN SOLAR CELL SIMULATOR WITH OPTICAL MANAGEMENT ...................................................................................................................... 1603

Y. Fang ; D. Guo ; A. Fischer ; E. Vadiee ; C. Zhang ; J. Williams ; S. M. Goodnick ; D. Vasileska

PERFORMANCE ENHANCEMENT OF A GAAS SOLAR CELL WITH COLLOIDAL QUANTUM DOTS EMBEDDED IN TRENCHES ......................................................................................... 1606

Chia-Jhe Shu ; Yu-Ming Huang ; Shun-Chieh Hsu ; Jinn-Kong Shu ; Jia-Lin Tsai ; Pei-Chen Yu ; Yung-Jr Hung ; Chien-Chung Lin

ENHANCED PHOTORESPONSE OF INN DEVICES USING INDIUM-TIN OXIDE NANORODS ......... 1610

Lung-Hsing Hsu ; Yuh-Jen Cheng ; Peichen Yu ; Hao-Chung Kuo ; Chien-Chung Lin

PLASMONIC SILVER STRUCTURES FOR IMPROVED PEROVSKITE PHOTOVOLTAIC PERFORMANCE ..................................................................................................................... 1614

Arul Varman Kesavan ; Arun D Rao ; Praveen C Ramamurthy

QUANTUM CUTTING LUMINESCENT PMMA FILMS CONTAINING CE3+ - YB3+ CODOPED YAG PHOSPHOR FOR SI CONCENTRATOR SOLAR CELLS ..................................................... 1619

Lu Li ; Chaogang Lou ; Huihui Cao

NUMERICAL EVALUATION ON THE NANO-ROD ARRAY ON A N-SIDE-UP THIN-FILM GAAS SOLAR CELLS .............................................................................................................. 1623

Po-Ching Wu ; Yan-Zhang Lin ; Shun-Chieh Hsu ; Chia-Jhe Hsu ; Chien-Chung Lin

DOWN SHIFTED CONVERSION FOR ENHANCED HIT SOLAR CELL EFFICIENCY ................ 1627

Albert S. Lin ; Parag Parashar ; Wei-Ming Huang ; Yi-Wen Huang ; Ding-Rung Jian ; Ming-Hsuan Kao ; Shi-Wei Chen ; Chang-Hong Shen ; Jia-Min Shieh ; Tzu-Yu Chen ; Chien-Chung Lin ; Hao-Chung Kuo

THE PLANAR THERMOPHOTOVOLTAIC SELECTIVE NEARLY-PERFECT ABSORBERS/EMITTERS ............................................................................................................ 1631

Parag Parashar ; Ding-Rung Jian ; Weiming Huang ; Vi-Wen Huang ; Albert Lin

**HYBRID PEDOT:PSS SILICON SOLAR CELLS WITH PENCIL ROD STRUCTURES** ............................ 1635
Ruei-Ying Wu ; Liang-Chian You ; Hsin-Fei Meng ; Chun-Chi Chen ; Peichen Yu

**PL STUDY OF PHOSPHORUS-DOPED CDTE EVT FILMS** ............................................................ 1638
Shamara Collins ; Imran Khan ; Vamsi Evani ; Chih An Hsu ; Vasilios Palekis ; Don Morel ; Chris Ferekides

**CHARACTERIZATION OF SINGLE-SOURCE DEPOSITED CLOSE-SPACE SUBLIMATION CDTEXSE1-XTHIN FILM SOLAR CELLS** .................................................................................. 1643
Corey R. Grice ; Jian Li ; Yanfa Yan

**THE INFLUENCE OF THE CU-RICH/CU-POOR SEQUENCE ON THE PROPERTIES OF CU(IN, GA)SE2 FILMS DEPOSITED BY IN-LINE CO-EVAPORATION PROCESS** .................................. 1648
He Wang ; Fang Fang Liu ; Yi Tong Yang ; Li You Yao ; Peng Gao ; Zhi Bin Xiao ; Qiang Sun

**DETERMINATION AND MODELING OF INJECTION DEPENDENT SERIES RESISTANCE IN CIGS SOLAR CELLS** ...................................................................................................................... 1651
Vito Huhn ; Bart E. Pieters ; Andreas Gerber ; Yael Augarten ; Uwe Rau

**LARGE GRAIN GROWTH IN CU2ZNSNS4 THIN FILMS IN THE ABSENCE OF NA USING RAPID THERMAL ANNEALING** ................................................................................................ 1656
J. L. Johnson ; A. Bhatia ; J. G. Bolke ; M. A. Scarpulla

**CU2ZNSNS4THIN FILMS SYNTHESIZED BY COSPUTTERING AND RAPID THERMAL ANNEALING: EFFECTS OF COMPOSITION AND TEMPERATURE** ............................................ 1661
J.L. Johnson ; W.M. Hlaing Oo ; M. Karmarkar ; M.A. Scarpulla

**EARTH-ABUNDANT CZTSSE THIN FILM SOLAR CELLS ON FLEXIBLE STAINLESS STEEL FOIL SUBSTRATES** ..................................................................................................................... 1665
Hae-Sun Kim ; Woo-Lim Jeong ; Dong-Seon Lee

**COMPARISON OF MGCL2AND CDCL2ACTIVATION TREATMENT FOR CDTE SOLAR CELLS: RECRYSTALLIZATION AND DEFECTS** ............................................................................ 1669
Daniele Menossi ; Elisa Artegiani ; Ivan Rimmaudo ; Alessia Le Donne ; Simona Binetti ; Juan Luis Pena ; Fabio Piccinelli ; Alessandro Romeo

**CHARACTERIZATION OF CDTE PHOTOVOLTAIC DEVICES PASSIVATED USING HYDROGEN PLASMA** ................................................................................................................ 1674
Amit Munshi ; Piotr Kaminski ; Ali Abbas ; Shiva Tarun Chenna ; Sreeram Chandralal ; John Walls ; Walajabad Sampath

**GROUP-V DOPING IMPACT ON CD-RICH CDTE SINGLE CRYSTALS GROWN BY TRAVELING-HEATER METHOD** ................................................................................................ 1679
Akira Nagaoka ; Kenji Yoshino ; Yoshitaro Nose ; Darius Kuciauskas ; Michael A. Scarpulla

**BAND-GAP ENGINEERING IN CU2ZNSN(S,SE)4SOLAR CELLS BY POST-SULPHURIZATION OF SELENIZED ABSORBER LAYERS** ............................................................................................ 1682
Markus Neuwirth ; Elisabeth Seydel ; Heinz Kalt ; Michael Hetterich

**IMPACT OF GA/III PROFILE ON VOLTAGE-DEPENDENT COLLECTION LOSSES IN CIGS SOLAR CELLS** ................................................................................................................................ 1686
Dmitry Poplavskyy ; Jeff Bailey ; Rouin Farshchi ; David Spaulding

**CL DIFFUSION IN CDTE SOLAR CELLS ACTIVATED BY GASEOUS CHCLF2ATMOSPHERE** ................. 1691
I. Rimmaudo ; R. Mis Fernandez ; V. Rejon ; A. Abbas ; F. Lisco ; J.M. Walls ; J.L. Peña

**STABILITY OF CD1-XZNXTE ALLOYS UNDER CDTE PROCESSING CONDITIONS** ..................... 1697
Yegor Samoilenko ; Colin A. Wolden

**CIGSE ABSORBER PREPARATION: AN ALTERNATIVE TO H2SE** ............................................... 1701
O.S. Shinde ; E.J. Schenller ; S.R. Jadkar ; S.V Ghaisas ; N. Dhere

**CHARGE CONTROLLED SEQUENTIAL ELECTRODEPOSITION FOR SYNTHESIS OF CU2ZNSNS4ON MO-COATED GLASS SUBSTRATE** ................................................................... 1704
Ashish K. Singh ; Rajiv Dubey ; Manoj Neergat ; Kavaipatti R. Balasubramaniam

**EFFECT OF DEPOSITED PRESSURE ON THE CDTE THIN FILMS BY CLOSED SPACE SUBLIMATION METHOD** ................................................................................................................ 1707
Yufeng Zhang ; Zhongming Du ; Xiangxin Liu

**ANALYZING THE COST REDUCTION POTENTIAL OF III-V/SI HYBRID CONCENTRATOR PHOTOVOLTAIC SYSTEMS** ....................................................................................................... 1711
Kan-Hua Lee ; Kenji Araki ; Masafumi Yamaguchi

**GENERALIZED NUMERICAL DESIGN OF AXIALLY-ASYMMETRICAL AND GRID-ARRANGED STATIC CPV ARRAY FOR MAXIMIZING ANNUAL ENERGY GENERATION** ................... 1714
Kenji Araki ; Kan-Hua Lee ; Masafumi Yamaguchi

**SPECTRAL TRANSMITTANCE ANALYSIS OF LIQUIDS FOR HIGH CONCENTRATION III-V PHOTOVOLTAIC IMMERSION COOLING APPLICATIONS** ............................................................ 1719
Xinyue Han ; Yongjie Guo

**OPTICAL DESIGN FOR 2-TERMINAL III-V/SI SMAC MODULE** ................................................ 1724
Masaaki Baba ; Kikuo Makita ; Hidenori Mizuno ; Hidetaka Takato ; Takeyoshi Sugaya ; Noboru Yamada

**DESIGN OF OPTICAL ELEMENTS FOR LOW PROFILE CPV PANEL WITH SUN TRACKING FOR ROOFTOP INSTALLATION**.................................................................................................1728
*Xinbing Liu ; Zhou Lu ; Riccardo Leto ; Carlton Brule ; Nanu Brates*

**MICRO CHIPLET PRINTER DEVELOPMENT FOR MOSAIC PROGRAM** ...............................1733
*P.Y. Maeda ; Y. D. Wang ; S. Raychaudhuri ; J. Kalb ; D. K. Biegelsen ; R. Lujan ; Q. Wang ; Y. Wang ; J. Bert ; B. Rupp ; I. Matei ; L. Crawford ; A. Plochowietz ; E.M. Chow ; J.P. Lu ; V. Gupta*

**MICRO-OPTICAL TANDEM LUMINESCENT SOLAR CONCENTRATOR** ..............................1737
*David R. Needell ; Zach Nett ; Ognjen Ilic ; Colton R. Bukowsky ; Junwen He ; Lu Xu ; Ralph G. Nuzzo ; Benjamin G. Lee ; John F. Geisz ; A. Paul Alivisatos ; Harry A. Atwater*

**INCREASE IN MAXIMUM POWER OF A-SI, C-SI AND GAAS.76P.24 SOLAR CELLS UNDER LOW CONCENTRATION**.......................................................................................................1741
*Hiba Riaz ; Sabina Abdul Hadi ; Ammar Nayfeh*

**DESIGN AND EVALUATION OF PARTIAL CONCENTRATION III-V/SI MODULE WITH ENHANCED DIFFUSE SUNLIGHT TRANSMISSION** ..............................................................1743
*Daisuke Sato ; Noboru Yamada ; Kan-Hua Lee ; Kenji Araki ; Masafumi Yamaguchi*

**CONTAMINATION CONTROL CHALLENGES ON SHJ SOLAR CELL PROCESSING** ........1747
*G. Condorelli ; P. Rotoli ; A. Canino ; A. Battaglia ; W. Favre ; A. -S. Ozanne ; A. Moustafa ; A. Danel ; D. Muñoz ; P. -J. Ribeyron ; C. Gerardi*

**>23% SILICON HETEROJUNCTION SOLAR CELLS IN MEYER BURGER'S DEMO LINE: RESULTS OF PILOT PRODUCTION ON MASS PRODUCTION TOOLS**.................................1752
*J. Zhao ; M. König ; A. Wissen ; V. Breus ; D. Deckerl ; M. Fritzsche ; M. Schorch ; H. J. Nonnenmacher ; M. Leonhardt ; T. Große ; J. Hausmann ; A. Waltmger ; D. Landgraf ; S. Burkhardt ; H. Mehlich ; E. Vetter ; F. Schitthelm ; Y. Yao ; T. Söderström ; A. Richter ; D. Habermann ; S. Leu*

**EXPERIMENTAL AND SIMULATION STUDIES ON TIO2/SILICON HETEROJUNCTION DIODES**...................................................................................................................................1755
*Swasti Bhatia ; Neha Raorane ; Nimisha Sreekumar ; Pradeep R. Nair ; Aldrin Antony*

**A STUDY ON BLISTER FORMATION AND ELECTRICAL PROPERTIES UNDER VARIOUS ANNEALING CONDITION FOR TUNNELING OXIDE PASSIVATION LAYER**......................1758
*Sungjin Choi ; Ka-Hyun Kim ; Min Gu Kang ; Jeong In Lee ; Donghwan Kim ; Hee-Eun Song*

**PROCESSING APPROACHES AND CHALLENGES OF INTERDIGITATED BACK CONTACT SI SOLAR CELLS**.................................................................................................................1761
*Ujjwal Das ; Lei Zhang ; Steven Hegedus*

**FABRICATION OF CUI/A-SI:H/C-SI STRUCTURE FOR APPLICATION TO HOLE-SELECTIVE CONTACTS OF HETEROJUNCTION SI SOLAR CELLS**...............................................................1765
*Kazuhiro Gotoh ; Min Cui ; Nguyen Cong Thanh ; Koichi Koyama ; Isao Takahashi ; Yasuyoshi Kurokawa ; Hideki Matsumura ; Noritaka Usami*

**CHARACTERISTICS OF THIN CRYSTALLINE SILICON SOLAR CELLS WITH RIB STRUCTURE** .......................................................................................................................1769
*Yukimi Ichikawa ; Shuhei Yoshiba ; Masakazu Hirai ; Makoto Konagai*

**MEASUREMENT OF TIO2/P-SI SELECTIVE CONTACT PERFORMANCE USING A HETEROJUNCTION BIPOLAR TRANSISTOR WITH A SELECTIVE CONTACT EMITTER**......1773
*Janam Jhaveri ; Alexander Berg ; Sigurd Wagner ; James C. Sturm*

**EFFECT OF GROWTH AND POST-OXIDATION ANNEALING TEMPERATURE OF THERMALLY GROWN TUNNELING SIOX, ON THE IIMPLIED VOCOF PASSIVATED CONTACTS FOR C-SI BASED SOLAR CELLS**.........................................................................1777
*Abhijit S. Kale ; William Nemeth ; Matthew Page ; Sumit Agarwal ; Paul Stradins*

**PARTIALLY CONTACTED SURFACES WITH CONTACT SIZE IN THE 1 µM RANGE FOR C-SI PERC SOLAR CELLS**.........................................................................................................1781
*R. Khoury ; I. Martín ; G. López ; C. Jin ; J.M. López-González ; L. Zeyu ; P. Bulkin ; E.V. Johnson ; R. Alcubilla*

**ENTRANCE OF LOW COST FABRICATION OF BACK-CONTACT HETEROJUNCTION SOLAR CELLS BY USING PLASMA ION IMPLANTATION**..............................................................1787
*Koichi Koyama ; Keisuke Ohdaira ; Hideki Matsumura*

**TLM MEASUREMENTS VARYING THE INTRINSIC A-SI:H LAYER THICKNESS IN SILICON HETEROJUNCTION SOLAR CELLS**.........................................................................................1790
*Mehdi Leilaeioun ; William Weigand ; Pradyumna Muralidharan ; Mathieu Boccard ; Dragica Vasileska ; Stephen Goodnick ; Zachary Holman*

**SOLAR CELLS APPLICATION OF P-TYPE POLY-SI THIN FILM BY ALUMINUM INDUCED CRYSTALLIZATION** .........................................................................................................1794
*Shota Masuda ; Kazuhiro Gotoh ; Isao Takahashi ; Kyotaro Nakamura ; Yoshio Ohshita ; Noritaka Usami*

**A SELF - CONSISTENTLY COUPLED DRIFT DIFFUSION AND MONTE CARLO SIMULATOR TO MODEL SILICON HETEROJUNCTION SOLAR CELLS**.................................................................1797
*Pradyumna Muralidharan ; Stuart Bowden ; Stephen M. Goodnick ; Dragica Vasileska*

**DOPANT PATTERNING BY PECVD AND MECHANICAL MASKING FOR PASSIVATED TUNNELING CONTACT IBC CELL ARCHITECTURES** ........................................................ 1801
William Nemeth ; Vincenzo Lasalvia ; Benjamin G. Lee ; Abhijit Kale ; Paul Stradins

**ALD ALUMINUM OXIDE AS A HOLE SELECTIVE TUNNELING CONTACT FOR CRYSTALLINE SILICON SOLAR CELLS** ........................................................ 1804
Kortan Ögütman ; Kristopher O. Davis ; Winston V. Schoenfeld ; Michael Haslinger ; Sofie Robert ; Emanuele Cornagliotti ; Joachim John

**SCREEN PRINTED, LARGE AREA BIFACIAL N-PERT CELLS WITH TUNNEL OXIDE PASSIVATED BACK CONTACT** ........................................................ 1807
Young-Woo Ok ; Ajay D Upadhyaya ; Brian Rounsaville ; Ying-Yuan Huang ; Vijaykumar D Upadhyaya ; Ajeet Rohatgi

**CORRELATION BETWEEN ELECTROLUMINESCENCE AND PHOTOCONVERSION EFFICIENCY IN A-SI:H/C-SI HETEROJUNCTION SOLAR CELLS** ........................................................ 1811
A.V. Sachenko ; A.V. Bobyl ; V.N. Verbitskiy ; V.M. Vlasyuk ; D.M. Zhigunov ; V.P. Kostylyov ; I.O. Sokolovskyi ; E.I. Terukov ; P.A. Forsh ; M. Evstigneev

**AN ISOTOPE STUDY OF HYDROGEN PASSIVATION OF POLY-SI/SIOXPASSIVATED CONTACTS FOR SI SOLAR CELLS** ........................................................ 1817
Manuel Schnabel ; William Nemeth ; Bas W.H. Van De Loo ; Bart Macco ; Wilhelmus M.M. Kessels ; Paul Stradins ; David L. Young

**ALLEVIATING HYDROGEN PLASMA DAMAGE TO AMORPHOUS/CRYSTALLINE SILICON INTERFACE PASSIVATION** ........................................................ 1820
Jianwei Shi ; Zachary C. Holman

**LARGE-AREA N-TYPE TOPCON CELLS WITH SCREEN-PRINTED CONTACT ON SELECTIVE BORON EMITTER FORMED BY WET CHEMICAL ETCH-BACK** ........................................................ 1824
Yuguo Tao ; Felix Book ; Barbara Terheiden ; Viiaykumar Upadhvaya ; Keeya Madani ; Brian Rounsaville ; Eunhwan Cho ; Ajeet Rohatgi

**HYDROGEN PLASMA POST-DEPOSITION TREATMENT FOR PASSIVATION OF A-SI/C-SI INTERFACE FOR HETEROJUNCTION SOLAR CELL BY CORRELATING OPTICAL EMISSION SPECTROSCOPY AND MINORITY CARRIER LIFETIME** ........................................................ 1828
Anishkumar Soman ; Ugochukwu Nsofor ; Lei Zhang ; Ujjwal Das ; Tingyi Gu ; Steve Hegedus

**MEASURING DIODE RESISTIVITY OF PASSIVATED CONTACTS** ........................................................ 1832
San Theingi ; William Nemeth ; David L. Young ; Paul Stradins ; Benjamin G. Lee

**ULTRA-THIN CRYSTALLINE SILICON SOLAR CELLS WITH NICKEL OXIDE INTERLAYER AS HOLE-SELECTIVE CONTACT** ........................................................ 1835
Muyu Xue ; Raisul Islam ; Junyan Chen ; Zheng Lyu ; Yusi Chen ; Daniel Dewitt ; Albert Pleus ; Christian Tae ; Ching-Ying Lu ; Kai Zang ; Jieyang Jia ; Yijie Huo ; Ted Kamins ; Krishna Saraswat ; James Harris

**CRYSTALLINE SI SOLAR CELLS WITH PASSIVATING, CARRIER-SELECTIVE NICKEL OXIDE CONTACTS** ........................................................ 1838
Woojun Yoon ; James Moore ; David Scheiman ; Eunhwan Cho ; Young-Woo Ok ; Nicole Kotulak ; Phillip P. Jenkins ; Ajeet Rohatgi ; Robert J. Walters

**GAP/SI HETEROJUNCTION SOLAR CELLS GROWN BY MOLECULAR BEAM EPITAXY** ........................................................ 1841
Chaomin Zhang ; Ehsan Vadiee ; Richard R. King ; Christiana B. Honsberg

**SPIN COATED NICKEL OXIDE AND VANADIUM OXIDE LAYERS ON SILICON FOR A CARRIER SELECTIVE CONTACT SOLAR CELL** ........................................................ 1845
Jing Zhao ; Fa-Jun Ma, Jae-Yun ; Anita Ho-Baillie ; Stephen Bremner

**QUANTIFICATION OF PV MODULE DISCOLORATION USING VISUAL IMAGE ANALYSIS** ........................................................ 1850
Shashwata Chattopadhyay ; Chetan Singh Solanki ; Anil Kottantharayil ; K.L. Narasimhan ; Juzer Vasi ; Sai Tatapudi ; Govindasamy Tamizhmani

**TEMPERATURE AND POWER STUDY OF ADHERED AND RACKED DOUBLE GLASS PHOTOVOLTAIC MODULES** ........................................................ 1855
Volker Beutner ; Rubina Singh ; Cameron Stark

**FIELD INSPECTION OF PV MODULES: QUANTITATIVE DETERMINATION OF PERFORMANCE LOSS DUE TO CELL CRACKS USING EL IMAGES** ........................................................ 1858
Carlos A. Rodríguez Castañeda ; Shashwata Chattopadhyay ; Jaewon Oh ; Sai Tatapudi ; Govindasamy Tamizhmani ; Hailin Hu

**SCALE UP DESIGNS FOR HAND-HELD LIGHT-WEIGHT TPV DC POWER SUPPLY** ........................................................ 1863
L. M. Fraas ; J. E. Avery ; L. Minkin ; Hui She ; L. Ferguson

**HIGH EFFICIENCY ANTI-REFLECTIVE COATING FOR PV MODULE GLASS** ........................................................ 1869
Brennen M. Freiburger ; Corey S. Thompson ; Robert A. Fleming ; Douglas Hutchings ; Sergiu C. Pop

**INVESTIGATION OF EFFICIENCY FOR PID-AFFECTED SOLAR MODULE AT NONSTANDARD TEST CONDITIONS** ........................................................ 1873
Shuwen Guo ; Pan Zhao ; Weijing Huang ; Jipeng Chang ; He Wang ; Hong Yang ; Chengfeng Su ; Bojie Su ; Xue Zhang ; Yunxue Cao ; Hui Zhao

**THERMAL UNIFORMITY MAPPING OF PV MODULES AND PLANTS**............................................................1877
Ashwini Pavgi ; Jaewon Oh ; Joseph Kuitche ; Sai Tatapudi ; Govindasamy Tamizhmani

**CLIMATE-SPECIFIC THERMAL MODEL COEFFICIENTS FOR C-SI AND THIN-FILM PV MODULES**............................................................1883
Ashwini Pavgi ; Joseph Kuitche ; Jaewon Oh ; Govindasamy Tamizhmani

**EFFECT OF THE THERMOPHYSICAL PROPERTIES OF A PHASE CHANGE MATERIAL ON THE ELECTRICAL OUTPUT OF A CONCENTRATED PHOTOVOLTAIC SYSTEM**............................................................1888
Jawad Sarwar ; Ahmed E. Abbas ; Konstantinos E. Kakosimos

**PASSIVE COOLING OF PHOTOVOLTAICS WITH DESICCANTS**............................................................1893
Lin J. Simpson ; Jason Woods ; Nicolas Valderrama ; Alex Hill ; Nina Vincent ; Timothy Silverman

**MODIFIED MAXIMUM POWER EXTRACTION TECHNIQUE FOR RAPIDLY CHANGING NUI AND DYNAMIC LOADS**............................................................1898
U Aswani ; S.P. Duttagupta ; T.I. Eldho ; B.V. Rao

**REAL-TIME MONITORING OF PHOTO VOLTAIC RELIABILITY ONLY USING MAXIMUM POWER POINT - THE SUNS-VMP METHOD**............................................................1904
Xingshu Sun ; Haejun Chung ; Raghu Vamsi Krishna Chavali ; Peter Bermel ; Muhammad Ashraful Alam

**PHOTOVOLTAIC MODULE DURABILITY AND RELIABILITY: ANALYSIS OF A 23-YEAR-OLD ARRAY OPERATING IN QUEBEC, CANADA**............................................................1908
Christopher Baldus-Jeursen ; Alexandre Côté ; Naveen Goswamy ; Tanya Deer ; Yves Poissant

**ARE E-W TRACKERS A BETTER OPTION FOR FUTURE INVESTMENTS IN PV SECTOR-A DETAILED TECHNO-COMMERCIAL STUDY**............................................................1912
Rakesh Bohra ; Ramesh Rame Gowda ; Mani R. Krishnan

**EXPERIMENTAL EVALUATION OF THE PERFORMANCE OF CRYSTALLINE SI PV MODULE DEGRADATION AFTER 15-YEARS OF FIELD EXPOSURE**............................................................1917
Denio A. Cassini ; Antonia Sônia A. C. Diniz ; Marcelo Machado Viana ; Michele C. C. De Oliveira ; F. C. Lins Vanessa De ; Roberto Zilles ; Lawrence L. Kazmerski

**FIELD INVESTIGATIONS OF POTENTIAL-INDUCED DEGRADATION (PID) FOR CRYSTALLINE SILICON PV PANELS IN DIFFERENT CLIMATES**............................................................1922
Yifeng Chen ; Peter Hacke ; Yong Sheng Khoo ; Kaitlyn Vansant ; Zigang Wang ; Wei Luo ; Jing Chai ; Chris Deline ; Yan Wang ; Armin G. Aberle ; Pietro P. Altermatt ; Zhiqiang Feng ; Sarah Kurtz ; Pierre J. Verlinden

**DETERMINING THE POWER RATE OF CHANGE OF 353 PLANT INVERTERS TIME-SERIES DATA ACROSS MULTIPLE CLIMATE ZONES, USING A MONTH-BY-MONTH DATA SCIENCE ANALYSIS**............................................................1927
Alan J. Curran ; Yang Hu ; Rojiar Haddadian ; Jennifer L. Braid ; David Meakin ; Timothy J. Peshek ; Roger H. French

**PHOTOVOLTAIC ARRAY DIFFERENTIAL BACKSIDE EXPOSURE CONDITIONS: BACKSHEET DEGRADATION AND SITE DESIGN**............................................................1933
Andrew Fairbrother ; Julien Avenet ; Yadong Lyu ; Matthew Boyd ; Scott Julien ; Kai-Tak Wan ; Liang Ji ; Kenneth Boyce ; Sebastien Merzlic ; Amy Lefebvre ; Greg O'Brien ; Yu Wang ; Laura Bruckman ; Roger French ; Michael Kempe ; Brian Dougherty ; Xiaohong Gu

**STUDY ON RANDOM FAILURE OF CRYSTALLINE SILICON SOLAR MODULES IN THE FIELD**............................................................1937
Xuefang Jiang ; Fumei Wang ; Ao Wang ; Hong Yang ; He Wang ; Jie Ding ; Junjun Zhang ; Jingsheng Huang

**POTENTIAL INDUCED DEGRADATION (PID) POWER LOSS CORRELATION TO LEAKAGE AND REVERSE BIAS CURRENTS**............................................................1941
Michalis Florides ; Georgios Konstantinou ; Venizelos Venizelou ; George Makrides ; George E. Georghiou

**PERFORMANCE STUDY OF VARIOUS PV MODULE TECHNOLOGIES IN DESERT CONDITIONS**............................................................1946
Jim J John ; Ammar Elnosh ; Anwar Almheiri ; Wadhah Alzahmi ; Marco Stefancich ; Pedro Banda

**HIGH-SPEED MEASUREMENTS OF GENERATED POWER AND ITS RELATIONSHIP TO WEATHER OBSERVATIONS AT YOSHINOGARI MEGA SOLAR POWER PLANT**............................................................1950
Makoto Kasu ; Shigeomi Hara ; Takumi Uematsu

**IMPACT OF MISSING DATA ON THE ESTIMATION OF PHOTOVOLTAIC SYSTEM DEGRADATION RATE**............................................................1954
Andreas Livera ; Alexander Phinikarides ; George Makrides ; George E. Georghiou

**FIELD DEGRADATION AND FAILURES OF AGED CRYSTALLINE SILICON PV MODULES IN MEXICO**............................................................1959
D. Martínez Escobar ; P. A. Sánchez-Pérez ; Rocío De La Luz Santos Magdaleno ; José Ortega Cruz ; Sai Tatapudi ; Aarón Sánchez Juárez ; Govindasamy Tamizhmani

**RAPID SHUTDOWN WITH PANEL LEVEL ELECTRONICS-A SUITABLE SAFETY MEASURE?**............................................................1965
Adam Cordova ; Christopher Merz ; Gerd Bettenwort ; Markus Hopf ; Hannes Knopf ; Joachim Laschinski

INVESTIGATING A NEW OPERATING POINT FOR PV PANELS SEEKING MAXIMUM LIFE SPAN...........................................................................................................................................................1968

Bechara Nehme ; Nacer K. M'sirdi ; Tilda Akiki

POWER GENERATION EVALUATION OF LARGE-SCALE PHOTOVOLTAIC SYSTEMS LOCATED ON INCLINED PLANE ................................................................................................................1973

Naotaka Oka ; Yasuhito Takahashi ; Koji Fujiwara ; Kazuyuki Hidaka ; Hiroshi Morita

INVESTIGATING THE IMPACT OF SOLAR CELLS PARTIAL SHADING ON PHOTOVOLTAIC MODULES BY THERMOGRAPHY.........................................................................................1979

David Pera ; José A. Silva ; Sara Costa ; João M. Serra

ANNUAL DEGRADATION RATE AND ITS LINEARITY ANALYSIS USING METERED KWH DATA.................................................................................................................................................................1984

Christopher Raupp ; Govindasamy Tamizhmani

ELECTRICAL PERFORMANCE ANALYSIS OF A 27 KW GRID-CONNECTED PV SYSTEM WITH SOILING AND SHADING IN MORELOS MEXICO.....................................................................................1990

P. A. Sánchez-Pérez ; D. Martínez Escobar ; E. O. Ángel Ruiz ; R. Santos Magdaleno ; José Ortega Cruz ; A. Sánchez Juárez

MODIFIED STC CORRECTION PROCEDURE FOR ASSESSING PV MODULE DEGRADATION IN FIELD SURVEYS.................................................................................................................................................1995

Hemant K. Singh ; R. Dubey ; S. Zachariah ; K. L. Narasimhan ; B. M. Arora ; A. Kottantharayil ; J. Vasi

DEGRADATION MODELS OF PHOTOVOLTAIC MODULE BACKSHEETS EXPOSED TO DIVERSE REAL WORLD CONDITION..................................................................................................................2000

Yu Wang ; Sebastien Merzlic ; Andrew Fairbrother ; Scott Julien ; Lucas Fridman ; Camille Loyer ; Amy L. Lefebvre ; Gregory O'Brien ; Xiaohong Gu ; Liang Ji ; Ken Boyce ; Michael Kempe ; Kai-Tak Wan ; Roger H. French ; Laura S. Bruckman

ADDRESSING HOTSPOTS IN THE PRODUCT ENVIRONMENTAL FOOTPRINT OF CDTE PHOTOVOLTAICS.....................................................................................................................................................2005

Parikhit Sinha ; Andreas Wade

PHOTOVOLTAIC SMART HOME SYSTEM - DUBAI CASE STUDY ........................................................2011

Ammar Natsheh ; Marwa Aljaziri ; Maitha Moosa ; Gharibah Essa ; Hassa Moosa

DIRECT DRIVE PHOTOVOLTAIC MILK CHILLING EXPERIENCE IN KENYA..................................2014

Robert Foster ; Brian Jensen ; Brian Dugdill ; Wendy Hadley ; Bruce Knight ; Abudul Faraj ; Johnson Kyalo Mwove

COST OPTIMIZATION OF DECOMMISSIONING AND RECYCLING CDTE PV POWER PLANTS ...........................................................................................................................................................................2019

V. Fthenakis ; Z. Zhang ; J. -K Choi

CHALLENGES FOR DECISION MAKERS WHEN FEED-IN TARIFFS OR NET METERING SCHEMES CHANGE TO INCENTIVES DEPENDENT ON A HIGH SHARE OF SELF-CONSUMED ELECTRICITY .........................................................................................................................................2025

Mattias Gustafsson

PROCEDURES TO MAKE PROJECTS ABOUT RENEWABLE ENERGY GENERATION CONNECTED TO THE GRID IN COLOMBIA........................................................................................................2031

J. A. Hernandez ; C. A. Arredondo ; D. J. Rodriguez

A CRITICAL ANALYSIS ON THE THIN CRYSTALLINE SILICON PV MODULE OF THE LIGHTWEIGHT PV SYSTEM.......................................................................................................................................2035

Meixi Chen ; Abhishek Iyer ; Cheng-Hao Shih ; Lado Kurdgelashvili ; Robert Opila

PHOTOVOLTAIC MODULE MANUFACTURING COSTS, AVERAGE PRICES AND INDUSTRY BALANCE 2006–2016.....................................................................................................................................................2039

Paula Mints ; Zhengshan J Yu

SOLAR CELL AND WIND ENERGY REPLACEMENT OF POWER PLANTS GLOBALLY ...........................2042

Larry Partain ; Shirley Hansen ; Dirk Bennett ; Richard Hansen ; Allan Newlands ; Lewis Fraas

ANALYSIS OF LIGHT ENVIRONMENT UNDER SOLAR PANELS AND CROP LAYOUT ...........................2048

Deng Wang ; Yaojie Sun ; Yandan Lin ; Yuan Gao

INTERFACE EFFECTS OF ALKALI TREATMENT ON CU-RICH THIN FILM SOLAR CELLS .................2054

Hossam Elanzeery ; Finn Babbe ; Anastasiya Zelenina ; Michele Melchiorre ; Susanne Siebentritt

INCREASEDVOCAND FF IN ZNO1-XSX-BUFFERED CUIN1-XGAXSE2SOLAR CELLS BY CADMIUM PARTIAL ELECTROLYTE TREATMENT ...........................................................................................2058

Andreas Bauer ; Dimitrios Hariskos ; Wiltraud Wischmann

PASSIVATING AND CARRIER-SELECTIVE CONTACTS - BASIC REQUIREMENTS AND IMPLEMENTATION ...........................................................................................................................................................2064

S.W. Glunz ; M. Bivour ; C. Messmer ; F. Feldmann ; R. Müller ; C. Reichel ; A. Richter ; F. Schindler ; J. Benick ; M. Hermle

**FIRST-PRINCIPLES MODELING OF ALKALI METAL POST DEPOSITION TREATMENT EFFECTS IN CIGS SOLAR CELLS**.................................................................................................2070
*Maria Fedina ; Hannu-Pekka Komsa ; Ville Havu ; Martti J. Puska*

**EXPLORING SILICON CARBIDE- AND SILICON OXIDE-BASED LAYER STACKS FOR PASSIVATING CONTACTS TO SILICON SOLAR CELLS**.......................................................2073
*P. Löper ; G. Nogay ; P. Wyss ; M. Hyvl ; P. Procel ; J. Stuckelberger ; A. Ingenito ; I. Mack ; Q. Jeangros ; M. Ledinsky ; A. Fejfar ; C. Allebé ; J. Horzel ; M. Despeisse ; F. Crupi ; F.-J. Haug ; C. Ballif*

**EFFICIENT ELECTRON CONTACTS FORN-TYPE SILICON SOLAR CELLS USING MAGNESIUM METAL, OXIDE, AND FLUORIDE**.....................................................................2076
*Yimao Wan ; Chris Samundsett ; James Bullock ; Di Yan ; Thomas Allen ; Jun Peng ; Jie Cui ; Mark Hettick ; Ali Javey ; Andres Cuevas*

**GRADED (ALZGA1-Z)XIN1-XP WINDOW-EMITTER STRUCTURES FOR IMPROVED SHORT-WAVELENGTH RESPONSE**..................................................................................................2079
*Jacob T. Boyer ; Daniel L. Lepkowski ; Daniel J. Chmielewski ; Steven A. Ringel ; Tyler J. Grassman*

**INTEGRATION OF QUANTUM DOTS AND QUANTUM WELLS INTO INGAAS METAMORPHIC SUBCELL FOR RADIATION HARD 3-J ELO IMM PHOTOVOLTAICS**....................2084
*Zachary S. Bittner ; Hyun Kum ; Michael A. Slocum ; George T. Nelson ; Rao Tatavarti ; Andre Wibowo ; Seth M. Hubbard*

**PROTON IRRADIATION OF 3J SOLAR CELLS AT LOW TEMPERATURE**.............................2087
*Seonyong Park ; Jacques C. Bourgoin ; Olivier Cavani ; Sandrine Picard ; Jérôme Bourcois ; Victor Khorenko ; Carsten Baur ; Bruno Boizot*

**ULTRA-THIN GAAS SOLAR CELLS: RADIATION TOLERANCE AND SPACE APPLICATIONS**...............2091
*Louise C. Hirstl ; Michael K. Yakes ; Jeffery. H. Warner ; Mitchell F. Bennett ; Kenneth J. Schmieder ; Stephanie Tomasulo ; Erin Cleveland ; Sergey Maximenko ; James Moore ; Robert J. Walters ; Phillip P. Jenkins*

**LARGE AREA MULTIJUNCTION III-V SPACE SOLAR CELLS OVER 31% EFFICIENCY**...........................2094
*X.Q. Liu ; C. Fetzer ; P. Chiu ; M. Haddad ; X. Zhang ; R. Cravens ; D. Law ; J. Ermer ; J. Krogen ; S. Sharma ; J. Hanley*

**ADVANCED-ARCHITECTURE HIGH-EFFICIENCY SOLAR CELLS FOR LOW IRRADIANCE LOW TEMPERATURE (LILT) APPLICATIONS**.........................................................................2099
*Andreea Boca ; Jonathan Grandidier ; Claiborne Mcpheeters ; Paul Sharps ; Philip Chiu ; Xing-Quan Liu ; James Ermer*

**ULTRA-LIGHTWEIGHT PV MODULE DESIGN FOR BUILDING INTEGRATED PHOTOVOLTAICS**................................................................................................................2104
*Ana C. Martins ; Valentin Chapuis ; Alessandro Virtuani ; Christophe Ballif*

**DESIGN IT WITH LSCS; AN EXPLORATION OF APPLICATIONS FOR LUMINESCENT SOLAR CONCENTRATOR PV TECHNOLOGIES**.........................................................................2109
*Wouter Eggink ; Angèle Reinders*

**INVESTIGATING PV-BATTERY 3-TERMINAL INTEGRATION CONCEPT AS A SELF-SUSTAINING POWER SOLUTION**................................................................................................2114
*Solomon N. Agbo ; Oleksandr Astakhov ; Uwe Rau ; Tsvetelina Merdzhanova*

**PERFORMANCE ASSESSMENT OF A BIPV ROOFING TILE IN OUTDOOR TESTING**...............2118
*Cristina S. Polo Lopez ; Pierluigi Bonomo ; Francesco Frontini ; Vasco Medici ; Lorenzo Nespoli*

**LIFE CYCLE ASSESSMENT OF TRANSPARENT ORGANIC PHOTOVOLTAIC FOR WINDOW APPLICATIONS**........................................................................................................................2124
*Annick Anctil ; Eunsang Lee ; Jack Stephan ; Anjali Munasinghe ; Christopher Traverse ; Richard R. Lunt*

**A REDUCED ORDER MODEL FOR A TOV STUDY IN A SOLAR PV PROJECT**.........................2128
*Ahmad Abdullah ; Billy Yancey*

**CYBER SECURITY ASSESSMENT OF DISTRIBUTED ENERGY RESOURCES**.............................2135
*Cedric Carter ; Ifeoma Onunkwo ; Patricia Cordeiro ; Jay Johnson*

**EVALUATION OF FAST-FREQUENCY SUPPORT FUNCTIONS IN HIGH PENETRATION ISOLATED POWER SYSTEMS**................................................................................................2141
*Mohamed Elkhatib ; Jason Neely ; Jay Johnson*

**LOSS OF UTILITY DETECTION CAPABILITIES FOR TODAY'S UTILITY INTERCONNECTED PHOTOVOLTAIC INVERTERS**............................................................2147
*Sigifredo Gonzalez ; Gregory Kern ; Michael Ropp*

**PARAMETRIC PV GRID-SUPPORT FUNCTION CHARACTERIZATION FOR SIMULATION ENVIRONMENTS**....................................................................................................................2153
*Javier Hernandez-Alvidrez ; Jay Johnson*

**COST ANALYSIS AND COST REDUCTION OPPORTUNITIES OF RESIDENTIAL PV SYSTEM IN THE JAPAN**........................................................................................................................2159
*Izumi Kaizuka ; Haruki Yamaya ; Takashi Ohigashi ; Risa Kurihara ; Osamu Ikki*

**SUPPLY AND DEMAND CONSTRAINTS ON FUTURE PV POWER IN THE USA**....................2163
*Paul A. Basore ; Wesley J. Cole*

**RESIDENTIAL PHOTOVOLTAIC ELECTRICITY GENERATION IN THE EUROPEAN UNION 2017-OPPORTUNITIES AND CHALLENGES** ........................................................................ 2167
*Arnulf Jäger-Waldau ; Thomas Huld ; Sandor Szabo*

**INVESTIGATING NANOSCALE DETERMINANTS OF CHARGE COLLECTION IN QUASI-2D PEROVSKITE SOLAR CELLS** ........................................................................ 2170
*Yanqi Luo ; Xueying Li ; Bat-El Cohen ; Barry Lai ; Lioz Etgar ; David P Penning*

**RECENT DEVELOPMENTS OF SOLAR PHOTOVOLTAIC SYSTEMS IN INDIA** ..................... 2172
*Saravanan Vasudevan ; Arumugam Murugesan*

**OPERANDO X-RAY DIFFRACTION FOR CHARACTERIZATION OF PHOTOVOLTAIC MATERIALS** ........................................................................ 2176
*Laura T Schelhasl ; Jeffrey A. Christians ; Joseph J. Berry ; Michael F. T Oney ; Christopher J. Tassone ; Joseph M. Luther ; Kevin H. Stone*

**X-RAY BEAM INDUCED VOLTAGE: A NOVEL TECHNIQUE FOR ELECTRICAL NANOCHARACTERIZATION OF SOLAR CELLS** ........................................................................ 2179
*Michael E. Stuckelberger ; Tara Nietzold ; Bradley M. West ; Barry Lai ; Jörg M. Maser ; Volker Rose ; Mariana I. Bertoni*

**ELECTRO-LUMINESCENT REFRIGERATION ENABLED BY HIGHLY EFFICIENT PHOTOVOLTAICS** ........................................................................ 2185
*T. Patrick Xiao ; Kaifeng Chen ; Parthiban Santhanam ; Shanhui Fan ; Eli Yablonovitch*

**MULTIPLE QUANTUM WELLS AS SLOWED HOT CARRIER COOLING ABSORBERS IN HOT CARRIER CELLS** ........................................................................ 2186
*Gavin Conibeer ; Yi Zhang ; Simon Chung ; Yuaxun Liao ; Stephen Bremner ; Santosh Shrestha*

**QUANTITATIVE OPTOELECTRONIC MEASUREMENTS OF CARRIER THERMODYNAMICS PROPERTIES IN QUANTUM WELL HOT CARRIER SOLAR CELL** ........................................................................ 2192
*Dac-Trung Nguven ; Laurent Lombez ; François Gibelli ; Soline Boyer-Richard ; Alain Le Corre ; Olivier Durand ; Jean-François Guillemoles*

**ABSORPTION ENHANCEMENT IN INGAASP/INGAP QUANTUM WELL SOLAR CELLS** ......... 2195
*Islam E.H. Sayed ; Nikhil Jain ; Myles A. Steiner ; John F. Geisz ; Salah M. Bedair*

**CARRIER COLLECTION MODEL AND DESIGN RULE FOR QUANTUM WELL SOLAR CELLS** ........................................................................ 2201
*Kasidit Toprasertpong ; Boram Kim ; Yoshiaki Nakano ; Masakazu Sugiyama*

**INFLUENCE OF CONDUCTION BAND OFFSETS AT WINDOW/BUFFER AND BUFFER/ABSORBER INTERFACES ON THE ROLL-OVER OF J-V CURVES OF CIGS SOLAR CELLS** ........................................................................ 2205
*Giovanna Sozzi ; Simone Di Napoli ; Roberto Menozzi ; Florian Werner ; Susanne Siebentritt ; Philip Jackson ; Wolfram Witte*

**OVERVIEW OF SURFACE PASSIVATION SCHEMES FOR THIN FILM SOLAR CELLS** .............. 2209
*Ratan Kotipalli ; Bart Vermang*

**TOWARDS 10% STATE-OF-THE-ART PURE SULFIDE CU2ZNSNS4 SOLAR CELL BY MODIFYING THE INTERFACE CHEMISTRY** ........................................................................ 2213
*Kaiwen Sun ; Jialiang Huang ; Steve Johnston ; Chang Yan ; Fangyang Liu ; Xiaojing Hao ; Martin Green*

**BAND GAP CHANGES OF THE CDS BUFFER INDUCED BY POST-ANNEALING OF CU2ZNSN(S,SE)4SOLAR CELLS** ........................................................................ 2216
*Mario Lang ; Nicolas Schäfer ; Christian Huber ; Thomas Schnabe ; Heinz Kalt ; Michael Hetterich*

**22.61 % EFFICIENT FULLY SCREEN PRINTED PERC SOLAR CELL** ........................................ 2220
*Weiwei Deng ; Feng Ye ; Ruimin Liu ; Yunpeng Li ; Haiyan Chen ; Zhen Xiong ; Yang Yang ; Yifeng Chen ; Yongqian Wang ; Pietro P. Altermatt ; Zhiqiang Feng ; Pierre J. Verlinden*

**HOW TO ACHIEVE 23% EFFICIENT LARGE-AREA CU PLATED N-PERT CELLS?** ..................... 2227
*Monica Aleman ; Angel Uruena ; Emanuele Cornagliotti ; Patrick Choulat ; Joachim John ; Richard Russell ; Sukvhinder Singh ; Loic Tous ; Wen-Cheng Sun ; Filip Duerinckx ; Jozef Szlufcik*

**MICROSTRUCTURE AND RECOMBINATION ACTIVITY OF GRAIN BOUNDARIES FROM FRONT AND REAR SIDE DURING A LID-CYCLE OF MC-PERC SOLAR CELLS** ...................... 2232
*Tabea Luka ; Marko Turek ; Stephan Großer ; Christian Hagendorf*

**THERMODYNAMIC EFFICIENCY LIMIT OF BIFACIAL SOLAR CELLS FOR VARIOUS SPECTRAL ALBEDOS** ........................................................................ 2236
*Thomas C.R. Russell ; Rebecca Saive ; Harry A. Atwater*

**PROCESS-INDUCED DEGRADATION RESISTANT N-CZ WAFERS THROUGH TABULA RASA DEFECT ENGINEERING** ........................................................................ 2242
*Vincenzo Lasalvia ; William Nemeth ; Matthew Page ; Wooseok Nam ; Youngsik Han ; Sungsun Baik ; Amanda Youssef ; Tonio Buonassisi ; Paul Stradins*

**DETECTION OF A SHIFTING BROMINE CONCENTRATION IN HYBRID PEROVSKITES BY X-RAY FLUORESCENCE MICROSCOPY** ........................................................................ 2245
*Yanqi Luo ; Parisa Khoram ; Sarah Brittman ; Barry Lai ; Erik C. Garnett ; David P. Fenning*

**INFLUENCE OF GRAIN SIZE AND INTERFACES ON PHOTO-STABILITY OF PEROVSKITE SOLAR CELLS** .......... 2247
Istiaque Hossain ; Liang Zhang ; Ranjith Kottokkaran ; Mohamed El-Henawey ; Pranav Joshi ; Max Noack ; Vikram Dalal

**COLD THOUGHTS ON PEROVSKITE FEVER** .......... 2251
Tao Xu ; Jue Gong

**LBIC ANALYSIS OF PEROVSKITE BASED SOLAR CELLS STABILITY** .......... 2255
Carmen M. Ruiz ; Javier Ramos ; Richard Garuz ; Damien Barakel ; Jean Reusser ; Judikaël Le Rouzo

**ASSESSING JOB GROWTH AND SUSTAINABILITY IN THE US PV INDUSTRY** .......... 2258
Brion Bob

**ENSURING THE RELIABILITY OF PHOTOVOLTAIC POWER SYSTEMS USING INTERNATIONAL STANDARDS AND THE IECRE CONFORMITY ASSESSMENT SYSTEM** .......... 2263
George Kelly ; Adrian Häring ; Ted Spooner ; Greg Ball ; Sarah Kurtz ; Matthias Heinze ; Masaaki Yamamichi ; Govind Ramu

**A FRAMEWORK TO CALCULATE UNCERTAINTIES FOR LIFETIME ENERGY YIELD PREDICTIONS OF PV SYSTEMS** .......... 2267
Bjorn Muller ; Peter Bostock ; Boris Farnung ; Christian Reise

**INTEGRATED PV-RECYCLING-MORE EFFICIENT, MORE EFFECTIVE** .......... 2272
Wolfram Palitzsch ; Ulrich Loser

**ANALYSIS OF GAINP SOLAR CELLS GROWN BY HYDRIDE VAPOR PHASE EPITAXY** .......... 2275
Kevin L. Schulte ; John Simon ; David L. Young ; Aaron J. Ptak

**INVESTIGATION OF ADHESION FORCES BETWEEN DUST PARTICLES AND SOLAR GLASS** .......... 2280
H.R. Moutinho ; C.-S. Jiang ; B. To ; C. Perkins ; M. Muller ; M.M. Al-Jassim ; L. Simpson

**ANTI-REFLECTIVE AND ANTI-SOILING PROPERTIES OF A KLEANBOOST™, A SUPERHYDROPHOBIC NANO-TEXTURED COATING FOR SOLAR GLASS** .......... 2285
Illya Nayshevsky ; Qianfeng Xu ; Gil Barahman ; Alan Lyons

**MULTILAYER-GROWN ULTRATHIN NANOSTRUCTURED GAAS SOLAR CELLS** .......... 2291
Boju Gai ; Yukun Sun ; Minjoo Lee ; Jongseung Yoon

**LABORATORY STUDIES OF PARTICLE CEMENTATION AND PV MODULE SOILING** .......... 2294
Craig L. Perkins ; Matthew Muller ; Lin Simpson

**VIRTUAL SUBSTRATES FOR LOW-COST HIGH EFFICIENCY III-V PHOTOVOLTAICS** .......... 2298
Sean J. Babcock ; Marlene L. Lichty ; Shankar Karki ; Grace Rajan ; Sylvain Marsillac ; Elisabeth L. Mcclure ; Seth M. Hubbard ; Christopher G. Bailey

**SEASONAL TRENDS OF SOILING ON PHOTOVOLTAIC SYSTEMS** .......... 2301
Leonardo Micheli ; Daniel Ruth ; Matthew Muller

**INTERRELATIONSHIPS AMONG NON-UNIFORM SOILING DISTRIBUTIONS AND PV MODULE PERFORMANCE PARAMETERS, CLIMATE CONDITIONS, AND SOILING PARTICLE AND MODULE SURFACE PROPERTIES** .......... 2307
Lawrence L. Kazmerski ; Antonia Sonia A.C. Diniz ; Daniel Sena Braga ; Cristiana Brasil Maia ; Marcelo Machado Viana ; Suellen C. Costa ; Pedro P. Brito ; Cláudio Dias Campos ; Sergio De Morais Hanriot ; Leila R. De Oliveira Cruz

**PV MODULE DURABILITY -CONNECTING FIELD RESULTS, ACCELERATED TESTING, AND MATERIALS** .......... 2312
T. John Trout ; W. Gambogi ; T. Felder ; K. R. Choudhury ; L. Garreau-Iles ; Y. Heta ; K. Stika

**FEMTOSECOND VS NANOSECOND: AN ANALYSIS ON THE LASER ABLATION PROPERTIES OF DIELECTRIC LAYERS FOR SOLAR CELLS** .......... 2318
Jaffar Moideen Yacob Ali ; Vinodh Shanmugam ; Carlos D. Rodríguez-Gallegos ; Bianca Lim ; Armin Aberle ; Thomas Mueller

**GROWTH OF MOS2 THIN FILMS WITH MICRODOME TEXTURE AS OMNIDIRECTIONAL LIGHT TRAP FOR SOLAR CELL APPLICATIONS** .......... 2324
Hussain M. Abouelkhair ; Nina A. Orlovskaya ; Robert E. Peale

**STUDY OF SPATIAL DISTRIBUTION OF ELECTRICAL, OPTICAL AND STRUCTURAL PROPERTIES OF MAGNETRON SPUTTERED AZO THIN FILMS** .......... 2330
Mohit Agarwal ; Rajiv O Dusane

**MULTIBAND FORMATION IN CR DOPED CUGAS2 THIN FILMS SYNTHESIZED BY CHEMICAL SPRAY PYROLYSIS** .......... 2334
Nazmul Ahsan ; Sivaperuman Kalainatharr ; Naoya Miyashita ; Takuya Hoshii ; Yoshitaka Okada

**EFFECTS OF ANNEALING AND SUBSTRATE TEMPERATURE FOR SN-S THIN FILMS** .......... 2338
Yoji Akaki ; Kazuya Iwasaki ; Shigeyuki Nakamura ; Hideaki Araki

**MOLYBDENUM OXIDE THIN FILMS FOR HETEROJUNCTION SOLAR CELLS** .......... 2342
A. Dominguez ; Ateet Dutt ; O. De Melo ; G. Santana

**DUAL ION BEAM SPUTTERED TCO THIN FILMS: SPUTTER-INSTIGATED PLASMONIC FEATURES FOR ULTRATHIN PHOTOVOLTAICS**............2345
*Vivek Garg ; Brajendra S. Sengar ; Vishnu Awasthi ; Shailendra Kumar ; Shaibal Mukherjee*

**COMBINATORIAL STUDY OF SN-TI-W-O TRANSPARENT CONDUCTING OXIDE THIN FILMS FOR PHOTOVOLTAIC APPLICATIONS**............2349
*Michael N. Gona ; Patrick J. M. Isherwood ; Jake W. Bowers ; John M. Walls*

**BANDGAP AND ELECTRON AFFINITY OPTIMIZATION OF ZINC OXIDE FOR N-ZNO/P-SI SINGLE HETEROJUNCTION SOLAR CELL**............2355
*Babar Hussain ; Aasma Aslam*

**MODELING AND OPTIMIZING THE EFFICIENCY OF A ZNO/ZNTE SOLAR CELL USING SCAPS SOFTWARE**............2358
*Amal Kabalan ; Sam Roy ; Benjamin Chen*

**TERNARY PHOSPHIDE SEMICONDUCTOR INMG/ZN3P2SOLAR CELLS**............2361
*Ryoji Katsube ; Kenji Kazumi ; Yoshitaro Nose*

**NUMERICAL MODELING OF WSE2SOLAR CELLS**............2364
*H. Kyureghian ; M. Hilfiker ; E. Ediger ; V. Medic ; N.J. Ianno*

**BIAXIAL-TEXTURED TITANIUM NITRIDE THIN FILMS ON LOW-COST, FLEXIBLE METAL SUBSTRATE AS A CONDUCTIVE BUFFER LAYER FOR THIN FILM SOLAR CELLS**............2368
*Yongkuan Li ; Yao Yao ; Ying Gao ; Sicong Sun ; Pavel Dutta ; Monika Rathi ; Jae-Hyun Ryou ; Venkat Selvamanickam*

**SNS BY IONIZED JET DEPOSITION FOR PHOTOVOLTAIC APPLICATIONS**............2372
*Daniele Menossi ; Simone Di Mare ; Ivan Rimmaudo ; Elisa Artegiani ; Giampiero Tedeschi ; Juan Luis Pena ; Fabio Piccinelli ; Andrei Salavei ; Alessandro Romeo*

**EFFECT OF VALENCE BAND SPLITTING ON THE ABSORPTION SPECTRA OF MONOLAYER MOS2 IN PRESENCE OF SULPHUR VACANCIES**............2376
*Himani Mishra ; Sitangshu Bhattacharya*

**THE STUDY OF SOME MATERIALS AS BUFFER LAYER IN COPPER ANTIMONY SULPHIDE (CUSBS2) SOLAR CELL USING SCAPS 1-D**............2381
*Muteeu Olopade ; Adeyinka Adewoyin ; Michael Chendo ; Adewumi Bolaji*

**INFLUENCE OF HETERO-INTERFACES ON PHOTOVOLTAIC PERFORMANCE IN SOLAR CELLS BASED ON ZNSNP2BULK CRYSTAL**............2385
*Shigeru Nakatsuka ; Shunsuke Akari ; Jakapan Chantana ; Takashi Minemoto ; Yoshitaro Nose*

**JUNCTION BY DIFFUSION OF ELEMENTAL SODIUM ALONE INTO BRIDGMAN CU(IN, GA) SE2**............2388
*S. Park ; C. H. Champness ; S. Vanka ; Z. Mi ; I. Shih*

**OXYGEN SUBSTITUTION AND SULFUR VACANCIES IN NABIS2: A PB-FREE CANDIDATE FOR SOLUTION PROCESSABLE SOLAR CELLS**............2392
*Robert J Patterson ; Hongze Xia ; Long Hu ; Zhilong Zhang ; Lin Yuan ; Jianfeng Yang ; Weijian Chen ; Zihan Chen ; Yijun Gao ; Yicong Hu ; Binesh Puthen Veettil ; John A. Stride ; Gavin Conibeer ; Shujuan Huang*

**EFFECT OF ANNEALING ON PERFORMANCE OF SOLAR CELLS WITH NEW OXIDE ABSORBER MN2V2O7**............2395
*Pramod Ravindra ; Eashwer Athresh ; Rajeev Ranjan ; Srinivasan Raghavan ; Sushobhan Avasthi*

**ELECTRO-OPTICAL PROPERTIES OF ZN2MO3O8THIN-FILMS: A NOVEL LOW-BANDGAP SOLAR ABSORBER**............2399
*Pramod Ravindra ; Eashwer Athresh ; Rajeev Ranjan ; Srinivasan Raghavan ; Sushobhan Avasthi*

**LOW TEMPERATURE SOLUTION PROCESS FOR RANDOM HIGH ASPECT RATIO SILVER NANOWIRE AS PROMISING TRANSPARENT CONDUCTIVE LAYER**............2403
*Arastoo Teymouri ; Supriya Pillai ; Zi Ouyang ; Xiaojing Hao ; Martin Green*

**OXYGEN INCORPORATION INTO SI NANOCRYSTAL/SIC MULTILAYERS**............2407
*Charlotte Weiss ; Andreas Reichert ; Johannes Hofmann ; Stefan Janz*

**DESIGN OF CASCADED HETEROSTRUCTURED P-I-I-N CDS/CDSE LOW COST SOLAR CELL**............2411
*M. Zinaddinov ; S. Mil'shtein*

**FAST C-V METHOD TO MITIGATE EFFECTS OF DEEP LEVELS IN CIGS DOPING PROFILES**............2414
*P. K. Paull ; J. Bailey ; G. Zapalac ; A. R. Arehart*

**CRYSTAL GROWTH PHENOMENA IN POLYCRYSTALLINE (CU)ZNTE/CDTE/CDS VIA MOLECULAR DYNAMICS**............2419
*Rodolfo Aguirre ; Jose J. Chavez ; Xiao W. Zhou ; David Zubia*

**USING HIGH-RESOLUTION ANOMALOUS-SCATTERING X-RAY DIFFRACTION TO OBSERVE OFF-STOICHIOMETRIC CU2ZNSNS4CRYSTAL STRUCTURES**............2423
*Christopher J. Bosson ; Max T. Birch ; Douglas P. Halliday ; Chiu C. Tang ; Peter D. Hatton*

SIMULATION OF ZNMGO AS THE WINDOW LAYER FORCDTESOLAR CELLS ............................ 2427
Yunfei Chen ; Shou Peng ; Xin Cao ; Alan E. Delahoy ; Ken K. Chin

MODELING EFFECT OF DEFECTS ON EFFICIENCY OF NANOWIRE CDS-CDTE SOLAR
CELLS ....................................................................................................................................................... 2432
Hongmei Dang ; Esther Ososanya ; Nian Zhang ; Xiaohui Wang ; Hojjatollah Sarvari ; Vijay P. Singlr

ANALYTICAL DESCRIPTION OF CHARGED GRAIN BOUNDARY RECOMBINATION IN
POLYCRYSTALLINE THIN FILM SOLAR CELLS ........................................................................ 2438
Benoit Gaury ; Paul M. Haney

IMAGING THE EFFECT OF CDSE WINDOW LAYERS IN CDTE PHOTOVOLTAICS .................. 2443
John M. Howard ; Elizabeth M. Tennyson ; William B. Gunnarsson ; Naba R. Paudel ; Yanfa Yan ; Marina S.
Leite

INVESTIGATION OF TRAPS DENSITY AND POSITION IN ALKALI TREATED CU(IN,GA)SE2
THIN FILMS AND SOLAR CELLS .................................................................................................. 2446
Shankar Karki ; Pran K. Paul ; Grace Rajan ; Chinedum Akwari ; Angus Rockett ; Steven A Ringel ; Aaron R.
Arehart ; Sylvain Marsillac

THE EFFECT OF DEPOSITION STOICHIOMETRY AND POST-DEPOSITION TREATMENTS
ON DEEP DEFECTS IN CDTE ........................................................................................................ 2449
Imran S. Khan ; Vamsi Evani ; Shamara Collins ; Chih An Hsu ; Vasilis Palekis ; Chris Ferekides

TESTING THE LIMITS OF MECHANICALLY-SCRIBED CIGS MICROCELLS ......................... 2453
Ombline Lafont ; Nicolas Vandamme ; Leia Ruffini ; Jia Yu ; Philip Jackson ; Jose Alvarez ; Daniel Lincot

PHOTOLUMINESCENCE IMAGING ANALYSIS OF DOPING IN THIN FILM CDS AND
CDS/CDTE DEVICES ....................................................................................................................... 2457
C. Potamialis ; F. Lisco ; B. Maniscalco ; M. Togay ; A. Abbas ; M. Biiss ; J.W. Bowers ; J.M. Waiis ; I.
Rimmaudo ; R. Mis Fernandez ; V. Rejon ; J.L. Peña

APPLICATION OF MAPPING SPECTROSCOPIC ELLIPSOMETRY FOR CDSE/CDTE SOLAR
CELLS: OPTIMIZATION OF LOW-TEMPERATURE PROCESSED DEVICES WITH ALL-
SPUTTERED SEMICONDUCTORS ................................................................................................ 2462
Mohammed A. Razooqi ; Adam B. Phillips ; Geethika K. Liyanage ; Fadhil K. Al-Fadhili ; Maxwell M. Junda ;
Nikolas J. Podraza ; Michael J. Heben ; Robert W. Collins ; Prakash Koirala

ASSESSING THE VALIDITY AND ACCURACY OF EFFECTIVE ELECTRONIC MATERIALS:
CAN 1D SIMULATIONS PREDICT POLYCRYSTALLINE DEVICE PERFORMANCE? ............... 2467
Yubo Sun ; Allison Perna ; Sudhajit Misra ; Vasilios Palekis ; Chris Ferekides ; Jeffrey Aguiar ; Peter Bermel ;
Michael A. Scarpulla

CHARACTERIZING RECOMBINATION IN CDTE-BASED SOLAR CELLS BY THE
TEMPERATURE AND EXCITATION DEPENDENCE OF OPEN-CIRCUIT VOLTAGE AND
PHOTOLUMINESCENCE ................................................................................................................ 2473
Craig H. Swartz ; Sanjoy Paul ; Corey R. Grice ; Yanfa Yan ; Lorelle Mansfield ; Sachit Grover ; Gang Xiong ;
Jian V. Li

EXPERIMENTAL EVIDENCE FOR CDS-RELATED TRANSPORT BARRIER IN THIN FILM
SOLAR CELLS AND ITS IMPACT ON ADMITTANCE SPECTROSCOPY ................................... 2478
Florian Werner ; Anastasiya Zelenina ; Susanne Siebentritt

TRANSPARENT CONDUCTIVE ADHESIVES FOR TANDEM SOLAR CELLS ........................... 2482
Talysa R. Klein ; Benjamin G. Lee ; Manuel Schnabel ; Emily L. Warren ; Pauls Stradins ; Adele C. Tamboli ;
Maikel F.A.M. Van Hest

MODELING THREE-TERMINAL III- V LSI TANDEM SOLAR CELLS ....................................... 2488
Emily L. Warren ; Michael G. Deceglie ; Paul Stradins ; Adele C. Tamboli

WAFER BONDING APPROACHES FOR III-V ON SI MULTI-JUNCTION SOLAR CELLS........... 2492
Laura Vauche ; Elias Veinberg-Vidal ; Clément Weick ; Christophe Morales ; Vincent Larrey ; Christophe
Lecouvey ; Mickaël Martin ; Jérémy Da Fonseca ; Christophe Jany ; Thibaut Desrues ; Céline Brughera ;
Philippe Voarino ; Thierry Salvetat ; Frank Fournel ; Mathieu Baudrit ; Cécilia Dupré

DESIGN ARITHMETIC OF THE LATERAL III-V / SI HYBRID MODULE ................................... 2498
Kenji Araki ; Kyotaro Nakamura ; Kan-Hua Lee ; Takefumi Kamioka ; Yu-Cian Wang ; Nobuaki Kojima ; Yoshio
Ohshita ; Masafumi Yamaguchi

GAASP NANOWIRE SOLAR CELL DEVELOPMENT TOWARDS NANOWIRE/SI TANDEM
APPLICATIONS ................................................................................................................................ 2502
Enrique Barrigon ; Yang Chen ; Gaute Otnes ; Vilgaile Dagyte ; Nicklas Anttu ; Lars Samuelson ; Magnus
Borgström

DEMONSTRATION OF GAINP2/SI VOLTAGE MATCHED TANDEM SOLAR CELLS ................ 2506
David C. Bobela ; Kenneth J. Schmieder ; Matthew P. Lumb ; James E. Moore ; Robert J Walters ; Eric A. Armour
; Leo Matthew ; Rajesh Rao ; Angelo Mascarenhas ; Kirstin Alberi

WAFER BONDED III–V ON SILICON MULTI -JUNCTION CELL WITH EFFICIENCY
BEYOND 31% .................................................................................................................................... 2511
Romain Cariou ; Jan Benick ; Paul Beutel ; Nico Tucher ; Martin Graf ; David Lackner ; Martin Hermle ; Stefan
W. Glunz ; Andreas W. Bett ; Frank Dimroth

**INTEGRATION OF THIN AL FILMS ON IN0.18GA0.82AS METAMORPHIC GRADE STRUCTURES FOR LOW-COST III- V PHOTOVOLTAICS**................................................................2514
*Alessandro Giussani ; Michael A. Slocum ; Seth M. Hubbard ; Nathan Smaglik ; Nikhil Pokharel ; S. Phillip Ahrenkiel*

**TEMPERATURE DEPENDENT CHARACTERISTICS OF GAINP/GAAS/GAINNASSB SOLAR CELL UNDER SIMULATED AM0 SPECTRA**.....................................................................2520
*Riku Isoaho ; Arto Aho ; Antti Tukiainen ; Mircea Guina*

**EFFICIENCY OF GAAS P/SI TWO-JUNCTION SOLAR CELLS WITH MULTI-QUANTUM WELLS: A REALISTIC MODELING WITH CARRIER COLLECTION EFFICIENCY**................................2524
*Boram Kim ; Kasidit Toprasertpong ; Oliver Supplie ; Agnieszka Paszuk ; Thomas Hannappel ; Yoshiaki Nakano ; Masakazu Sugiyama*

**INVERSE METAMORPHIC III-V/EPI-SIGE TANDEM SOLAR CELL PERFORMANCE ASSESSED BY OPTICAL AND ELECTRICAL MODELING**.........................................................2528
*Raphaël Lachaurne ; Martin Foldyna ; Gwénaëlle Hamon ; Nicolas Vaissiére ; Jean Decobert ; Romain Cariou ; Pere Roca I Cabarrocas ; José Alvarez ; Jean-Paul Kleider*

**TOWARDS MONOLITHICALLY INTEGRATED GAAS ON SI TANDEM SOLAR CELL**...............2532
*Zhen Liu ; Zekun Ren ; Haohui Liu ; Tonio Buonassisi ; Ian Marius Peters*

**ZNSIP2 THIN FILM GROWTH FOR SI-BASED TANDEM PHOTOVOLTAICS**............................2536
*Aaron D. Martinez ; Elisa M. Miller ; Andrew G. Norman ; Paul Stradins ; Eric S. Toberer ; Adele C. Tamboli*

**IN SITU CONTROL OVER THE SUBLATTICE ORIENTATION OF GAP/SI(100): AS VIRTUAL SUBSTRATES FOR TANDEM ABSORBERS**.............................................................2538
*Aznieszka Paszuk ; Oliver Supplie ; Sebastian Brückner ; Matthias M. May ; Anja Dobrich ; Andreas Nägelein ; Boram Kim ; Yoshiaki Nakano ; Masakazu Sugiyama ; Peter Kleinschmidt ; Thomas Hannappel ; Thomas Hannappel*

**III-V/SI TANDEM CELL TO MODULE INTERCONNECTION - COMPARISON BETWEEN DIFFERENT OPERATION MODES**..................................................................................2543
*Henning Schulte-Huxel ; Emily L. Warren ; Manuel Schnabel ; Paul Stradins ; Daniel Friedman ; Adele C. Tamboli*

**INGAP/GAAS/ITO/SI HYBRID TRIPLE-JUNCTION CELLS WITH GAAS/ITO BONDING INTERFACES**.............................................................................................................2548
*Naoteru Shigekawa ; Tomoya Hara ; Tomoki Ogawa ; Jianbo Liang ; Takefumi Kamioka ; Kenji Araki ; Masafumi Yamaguchi*

**MEASUREMENTS OF POTENTIALS AT TAP CONTACTS AND ESTIMATION OF RESISTANCE ACROSS BONDING INTERFACES IN INGAP/GAAS/SI HYBRID TRIPLE-JUNCTION CELLS**.....................................................................................................2551
*Naoteru Shigekawa ; Jianbo Liang*

**OPTIMIZATION OF A GAASP TOP CELL FOR IMPLEMENTATION IN A III-V/SI TANDEM STRUCTURE**...............................................................................................................2554
*Amber C. Silvaggio ; Daniel L. Lepkowski ; Daniel J. Chmielewski ; Jacob T. Boyer ; Steven A. Ringel ; Tyler J. Grassman*

**THEORETICAL DESIGN OF PEROVSKITE/CDTE FOUR-TERMINAL TANDEM SOLAR CELLS**.......................................................................................................................2558
*Tao Tang ; Huan Zhang ; Xingzhi Du ; Yiming Lnr ; Hang Zhou*

**WAFER-BONDED ALGAAS///SI DUAL-JUNCTION SOLAR CELLS**.......................................2562
*Elias Veinberg-Vidal ; Laura Vauche ; Clément Weick ; Jérémy Da Fonseca ; Christophe Jany ; Christophe Morales ; Christophe Lecouvey ; Thibaut Desrues ; Philippe Voarino ; Frank Fournel ; Anne Kaminski-Cachopo ; Alejandro Datas ; Pablo Garcia-Linares ; Mathieu Baudrit ; Pierre Mur ; Cécilia Dupré*

**ENHANCEMENT OF SI PHOTOVOLTAIC MODULE BY INTRODUCING III-V/SI HYBRID CONFIGURATIONS AND COST EVALUATIONS UNDER VARIOUS COST RATIOS OF III-V/SI PHOTOVOLTAICS**...................................................................................................2566
*Yu-Cian Wang ; Kenii Araki ; Kyotaro Nakamura ; Kan-Hua Lee ; Takefumi Kamioka ; Nobuaki Kojima ; Yoshio Ohshita ; Masafumi Yamaguchi*

**NUMERICAL SIMULATION OF P-TYPE FRONT JUNCTION PERL SILICON CELL FOR III-V LSI TANDEM DEVICES**..........................................................................................2569
*Chuqi Yi ; Fa-Jun Ma ; Anita Ho-Baillie ; Stephen Bremner*

**EPITAXIAL GAP LAYERS GROWN ON SI SUBSTRATES USING MIGRATION ENHANCED AND MOLECULAR BEAM EPITAXY**..........................................................................2573
*Chaomin Zhang ; Allison Boley ; Nikolai Faleev ; David J. Smith ; Christiana B. Honsberg*

**INVESTIGATION OF CARRIER-INDUCED DEFECT BEHAVIOR IN P-TYPE MULTICRYSTALLINE SILICON**.......................................................................................2576
*Catherine E. Chan ; Tsun H. Fung ; David N.R. Payne ; Daniel Chen ; Malcolm D. Abbott ; Alison M. Ciesla ; Ran Chen ; Brett J. Hallam ; Stuart R. Wenham*

**MAGNETRON SPUTTERED HYDROGENATED SILICON THIN FILMS: ASSESSMENT FOR APPLICATION IN PHOTOVOLTAICS** ............................................................................................2582

Dipendra Adhikari ; Maxwell M. Junda ; Sylvain X. Marsillac ; Robert W. Collins ; Nikolas J. Podraza

**HIGH QUALITY AND THIN SILICON WAFER FOR NEXT GENERATION SOLAR CELLS** ........................2588

Yoshio Ohshita ; Takuto Kojima ; Ryota Suzuki ; Kosuke Kinoshita ; Tomoyuki Kawatsu ; Kyotaro Nakamura ; Atsushi Ogura

**FIRST DEMONSTRATION OF RADIAL JUNCTION SILICON NANOWIRE SOLAR MINI-MODULES PREPARED BY PECVD AND LASER SCRIBING** .........................................................................2593

Mutaz Al-Ghzaiwat ; Martin Foldyna ; Takashi Fuyuki ; Wanghua Chen ; Erik V. Johnson ; Jacques Meot ; Pere Roca I Cabarrocas

**IMPACT OF INDUCED DEFECTS ON DEVICE PERFORMANCE IN SILICON HETEROJUNCTION SOLAR CELLS** ................................................................................................................2596

Pradeep Balaji ; André Augusto ; Stuart G. Bowden

**LASER HYDROGENATION ON HEAVILY DISLOCATED CAST-MONO SILICON CELLS** ....................2600

Alison M. Ciesla ; Catherine E. Chan ; Sisi Wang ; Malcolm D. Abbott ; Cheemun Chong ; Stuart R. Wenham

**PERFORMANCE OPTIMIZATION OF SEMI-TRANSPARENT THIN-FILM AMORPHOUS SILICON CELLS** ..............................................................................................................................................2605

Yuan Gao ; Fai Tong Si ; Olindo Isabella ; Rudi Santbergen ; Guangtao Yang ; Jianfei Dong ; Guoqi Zhang ; Miro Zeman

**LOW TEMPERATURE SPALLING OF SILICON: A CRACK PROPAGATION STUDY** ...............................2610

Pablo Guimera Coll ; Tine Uberg Nærland ; Nathan Stoddard ; Michael Stuckelberger ; Mariana Bertoni

**NEW FINDINGS OF THERMAL EFFECT ON PM-SI:H SOLAR CELLS OPTOELECTRONIC PROPERTIES** ...................................................................................................................................................2614

L. Hamui ; L. A. Górnez-González ; G. Santana

**STUDY OF PV MODULE DEGRADATION RATE PREDICTION THROUGH CORRELATION OF FIELD-AGED AND ACCELERATED-AGED MODULE DEGRADATION DATA** .................................2618

Babak T. Hamzavy ; William J. Grieco ; Brian J. Fields ; Cara S. Libby ; William B. Hobbs ; Olga Lavrova ; C. Birk Jones

**ADVANCED ANALYSIS OF MULTI WIRE WAFERING PROCESSES** .................................................................2622

Ringo Koepgel ; Samuel Brinnig ; Felix Kaule ; Hartmut Schwabe ; Stephan Schoenfelder

**CONSIDERATION ON OPEN-CIRCUIT VOLTAGE OF SI HETEROJUNCTION SOLAR CELLS UNDER LOW CONCENTRATION CONDITION** .............................................................................................2627

Makoto Konagai

**CHARACTERIZATION OF MICROCRYSTALLINE SILICON THIN FILM SOLAR CELLS PREPARED BY HIGH WORKING PRESSURE PLASMA-ENHANCED CHEMICAL VAPOR DEPOSITION** ..................................................................................................................................................2631

Jung-Dae Kwon ; Dong-Ho Kim ; Ji-Hoon Lee ; Myungkwan Song ; Myunghun Shin

**ATOMIC-LAYER-DEPOSITEDV2O5-XFILMS AS A HIGHLY-EFFICIENT P-TYPE LAYER FOR THIN FILM A-SI SOLAR CELLS** ...............................................................................................................2634

Ji-Hoon Lee ; Myungkwan Song ; Dong-Ho Kim ; Jung-Dae Kwon

**A NOVEL DEFECT PASSIVATION METHOD FOR MULTICRYSTALLINE SI WAFER BY H2S REACTION** .................................................................................................................................................2637

Hsiang-Yu Liu ; Ujjwal K. Das ; Robert W. Birkmire

**CARRIER TRANSPORTATION AT NOVEL SILVER PASTE CONTACT** ............................................................2642

Takefumi Kamioka ; Satoshi Kamevama ; Kazuo Muramatsu ; Aki Tanaka ; Naotaka Iwata ; Kyotaro Nakamura ; Atsushi Ogura ; Yoshio Ohshita

**INFLUENCE OF DEPOSITION PARAMETERS ON SILICON THIN FILMS DEPOSITED BY MAGNETRON SPUTTERING** .......................................................................................................................2646

Grace Rajan ; Tejaswini Miryala ; Shankar Karki ; Robert W. Collins ; Nikolas Podraza ; Sylvain Marsillac

**MINORITY CARRIER LIFETIME VARIATIONS IN MULTICRYSTALLINE SILICON WAFERS WITH TEMPERATURE AND INGOT POSITION** ................................................................................2651

Sissel Tind Søndergaard ; Jan Ove Odden ; Rune Strandberg

**CUO NANOWIRES-BASED RADIAL HETERO-JUNCTION THIN FILM SILICON SOLAR CELLS WITH A HIGH OPEN-CIRCUIT VOLTAGE** .............................................................................2656

Xiaolin Sun ; Jiawen Lu ; Fan Yang ; Linwei Yu ; Jun Xu ; Ling Xu ; Kunji Chen

**THE EFFECT OF CHEMICAL COMPOSITION ON POROUS ETCHING FOR EPI AND LIFT-OFF WAFER PROCESS** ..............................................................................................................................2660

Teng-Yu Wang ; Peng-Wei Chen ; Han-Wen Liu

**ELECTRICAL AND OPTICAL PERFORMANCE OF SILICON SOLAR CELLS USING PLASMONICS INDIUM NANOPARTICLES LAYER EMBEDDED IN SIO2ANTIREFLECTIVE COATING** ...................................................................................................................................................2664

Hao-Yu Yang ; Wen-Jeng Ho ; Sheng-Kai Feng ; Jheng-Jie Liu ; Ta-Wei Chuang ; Guan-Yi Li ; Yun-Chie Yang ; Cho-Chun Chiang ; Yao- Hui Chen

**ELECTROLUMINESCENCE ANALYSIS FOR SEPARATION OF SERIES RESISTANCE FROM RECOMBINATION EFFECTS IN SILICON SOLAR CELLS WITH INTERDIGITATED BACK CONTACT DESIGN**.................................................................................................................2667

*Nuha Ahmed ; Lei Zhang ; Ujjwal Das ; Steven Hegedus*

**INDOOR MEASUREMENT OF ANGLE RESOLVED LIGHT ABSORPTION BY BLACK SILICON**............................................................................................................................2672

*Mekbib W. Amdemeskel ; Beniamino Iandolo ; Rasmus S. Davidsen ; Ole Hansen ; Gisele A. Dos Reis Benatto ; Nicholas Riedel ; Peter B. Poulsen ; Sune Thorsteinsson ; Anders Thorseth ; Carsten Dam-Hansen*

**IMPACT OF NON- FLAT PHOTOGENERATION AND CARRIER PROFILES ON THE LUMINESCENT EMISSION AND DETECTION OF SILICON SOLAR CELLS**.................2677

*Nekane Azkona ; Federico Recart ; Pedro Rodríguez ; Vanesa Fano ; Aloña Otaegi ; Juan Carlos Jimeno*

**DEVELOPMENT OF OUTDOOR LUMINESCENCE IMAGING FOR DRONE-BASED PV ARRAY INSPECTION**....................................................................................................2682

*Gisele A. Dos Reis Benatto ; Nicholas Riedel ; Sune Thorsteinsson ; Peter B. Poulsen ; Anders Thorseth ; Carsten Dam-Hansen ; Claire Mantel ; Soren Forchhammer ; Kenn H. B. Frederiksen ; Jan Vedde ; Michael Petersen ; Henrik Voss ; Michael Messerschmidt ; Harsh Parikh ; Sergiu Spataru ; Dezso Sera*

**CLIMBING DRUM PEEL (CDP) TEST METHOD FOR CHARACTERIZING ADHESION IN FLEXIBLE PV MODULES**.......................................................................................2688

*Venkata Bheemreddy ; Kedar Hardikar*

**ACCURACY OF SOLAR SIMULATOR SPECTRAL DETERMINATION USING BAND-PASS FILTERING METHOD**..........................................................................................2692

*Weston Dobson ; Harrison Wilterdink ; Cassidy Sainsbury ; Adrienne Blum ; Justin Dinger ; Ronald A. Sinton ; Karsten Bothe ; David Hinken ; Martin Wolf*

**CORRELATION OF I-V CURVE PARAMETERS WITH MODULE-LEVEL ELECTROLUMINESCENT IMAGE DATA OVER 3000 HOURS DAMP-HEAT EXPOSURE**...........................2697

*Justin S. Fada ; Andrew J. Loach ; Alan J. Curran ; Jennifer L. Braid ; Shuying Yang ; Timothy J. Peshek ; Roger H. French*

**A NOVEL METHOD TO INVESTIGATE STOICHIOMETRY AND PERFORMANCE OF BURIED PASSIVATED CONTACTS UTILIZING TIME-OF-FLIGHT SIMS**.................2702

*Steven P. Harvey ; William Nemeth ; Jeff Aguiar ; Craig Perkins ; Pauls Stradins*

**A COMPARISON BETWEEN QUASI-STEADY STATE AND TRANSIENT PHOTOCONDUCTANCE LIFETIMES IN SILICON INGOTS: SIMULATIONS AND MEASUREMENTS**.....................................................................................................................2707

*Mohsen Goodarzi ; Ronald Sinton ; Daniel Chung ; Bernhard Mitchell ; Thorsten Trupke ; Daniel Macdonald*

**NEW DEVELOPMENT IN GLOW DISCHARGE OPTICAL EMISSION SPECTROMETRY FOR THE CHARACTERIZATION AND THE THICKNESS MEASUREMENT OF LAYERS FOR PHOTOVOLTAIC APPLICATIONS**..................................................................................2711

*Philippe Hunault ; Matthieu Chausseau ; Patrick Chaporr ; Sofia Gaiaschi ; Anais Loubar ; Muriel Bouttcmy ; Arnaud Etcheberry*

**DEEP LEVEL TRANSIENT SPECTROSCOPY MEASUREMENTS OF SILICON HETEROJUNCTION CELLS**..................................................................................................2716

*Sanchit Khatavkar ; C. V. Kannan ; Vijay Kumar ; P. R. Nair ; B. M. Arora*

**CHARACTERIZATION OF MODULES AND ARRAYS WITH SUNS VOC**.......................2719

*Alex Killam ; Stuart Bowden*

**A STUDY OF PERFORMANCE CHARACTERIZATION WITH REAR LIGHT SOURCE IN CONVENTIONAL BIFACIAL SOLAR CELLS**.............................................................2723

*Soo Min Kim ; Sang Hoon Jung ; Rae-Won Choi ; Yong Bae Kim ; Min Gu Kang ; Hee-Eun Sonp ; Gyu-Seok Choi*

**ELECTRICAL CHARACTERIZATION OF THE CARRIER TRANSPORT PROPERTIES IN ACU(IN,GA)SE2SOLAR CELL**......................................................................................2728

*Roberto Lopez ; Sanjoy Paull ; Ingrid Repins ; Jian V. Li*

**SYSTEMATIC THERMALPHOTOVOLTAIC SOLAR CELL OPTIMIZATION**.................2732

*Zheng Lyu ; Muyu Xue ; Junyan Chen ; Jieyang Jia ; Shanhui Fan ; James Harris*

**CHARACTERIZATION OF TELLURIUM AS A BACK CONTACT FOR CDTE SOLAR CELLS**...................2736

*C.E. Moffett ; W.S. Sampath*

**ON THE DIFFERENT EXPLANATIONS OF THE RECOMBINATION CURRENTS WITH HIGH IDEALITY FACTOR IN SILICON SOLAR CELLS**.................................................2740

*A. Otaegi ; V. Fano ; N. Azkona ; J. R. Gutiérrez ; J. C. Jimeno*

**IDENTIFICATION OF SHUNTS IN A MONOLITHIC MULTIJUNCTION GAAS/GAAS DEVICE BY SPECTROMETRIC CHARACTERIZATION**....................................................2744

*Felipe Oviedo ; Liu Zhe ; Zekun Ren ; Kevin Nay Yaung ; Maung Thway ; Liu Haohui ; Tonio Buonassisi ; Ian Marius Peters*

**A SIMULATION STUDY ON RADIATIVE RECOMBINATION ANALYSIS IN CIGS SOLAR CELL** .......... 2749

Sanjoy Paul ; Roberto Lopez ; Md Dalim Mia ; Craig H. Swartz ; Jian V. Li

**SIMULATION AND SPECTROSCOPY OF CARRIER RELAXATION IN GASB AND GAAS** .......... 2755

A.C. Scofield ; A.I. Hudson ; B.L. Liang ; B.C. Juang ; D.L. Huffaker ; S.M. Hubbard ; W.T. Lotshaw

**COMPUTATIONAL DESIGN OF DOPANTS IN CDTE GRAIN BOUNDARIES FOR EFFICIENT PHOTOVOLTAICS** .......... 2759

Fatih G. Sen ; Tadas Paulauskas ; Ce Sun ; Moon Kim ; Robert F. Klie ; Maria K.Y. Chan

**ANALYSES OF PHOTOVOLTAIC POWER PLANT PERFORMANCE ESTIMATES BASED ON DETAILED LABORATORY MODULE CHARACTERIZATIONS AND TYPICAL REAL-WORLD INPUT DATA SOURCES** .......... 2762

Rajeev Singh ; John L.R. Watts ; Kellen Gillispie

**CRITICAL EVALUATION OF THE FOUNDATIONS OF SOLAR SIMULATOR STANDARDS** .......... 2765

Ronald A. Sinton ; Harrison Wilterdink ; Justin Dinger ; Adrienne L. Blum ; Weston Dobson ; Cassidy Sainsbury

**IMPACT OF INFRARED OPTICAL PROPERTIES ON CRYSTALLINE SI AND THIN FILM CDTE SOLAR CELLS** .......... 2771

Indra Subedi ; Timothy J Silverman ; Michael Deceglie ; Nikolas J. Podraza

**THE IMPACT OF IMPURITIES ON THE RELATIVE EFFICIENCIES OF SOLAR CELLS FROM DIFFERENT SILICON FEEDSTOCKS** .......... 2776

Muhammad Tayyib ; Aleksandr Dobroliubov ; Zekija Ramic ; Muhammad Nadeem Akarm ; Jan Ove Odden

**ACCURACY EVALUATION OF ABSOLUTE ELECTROLUMINESCENCE-EFFICIENCY MEASUREMENTS OF SOLAR CELLS USING A SENSITIVITY-CALIBRATED-PHOTODETECTOR CONTACT METHOD** .......... 2781

Masahiro Yoshita ; Yoshihiro Hishikawa ; Yoshihiko Kanemitsu ; Hidefumi Akiyama

**NANOMETER-SCALE CARRIER IMAGING OF POTENTIAL-INDUCED DEGRADATION IN C-SI SOLAR CELLS** .......... 2785

C.-S. Jiang ; C. Xiao ; H.R. Moutinho ; S. Johnston ; M.M. Al-Jassim ; X. Yang ; Y. Chen ; J. Ye

**NREL EFFORTS TO ADDRESS SOILING ON PV MODULES** .......... 2789

Lin J. Simpson ; Matthew Muller ; Michael Deceglie ; Helio Moutinho ; Craig Perkins ; C. S. Jiang ; David C. Miller ; Leonardo Micheli ; Govindasamy Tamizhmani ; Sai Ravi Vasista Tatapudi ; Mowafak Al-Jassim

**MODELING POTENTIAL-INDUCED DEGRADATION (PID) OF FIELD-EXPOSED CRYSTALLINE SILICON SOLAR PV MODULES: FOCUS ON A REGENERATION TERM** .......... 2794

Eleonora Annigoni ; Alessandro Virtuani ; Fanny Sculati-Meillaud ; Christophe Ballif

**SOILING LOSS ON PV MODULES AT TWO LOCATIONS IN INDIA STUDIED USING A WATER BASED ARTIFICIAL SOILING METHOD** .......... 2799

Sonali Bhaduri ; Sachin Zachariah ; Lawrence L. Kazmcrski ; Balasubramaniam Kavaipatti ; Anil Kottantharayil

**QUANTIFYING YEAR-TO-YEAR VARIATIONS IN SOLAR PANEL SOILING FROM PV ENERGY-PRODUCTION DATA** .......... 2804

Michael G. Deceglie ; Leonardo Micheli ; Matthew Muller

**ACCURATELY MEASURING PV SOILING LOSSES WITH SOILING STATION EMPLOYING PV MODULE POWER MEASUREMENTS** .......... 2808

Michael Gostein ; Bill Stueve ; Mandy Chan

**PERFORMANCE OF MONOCRYSTALLINE SILICON SOLAR CELL- INFLUENCE OF DUST ON ULTRA-VIOLET AND VISIBLE REGION DURING EARLY STAGE OF DEPOSITION** .......... 2811

Hemaprabha Elangovan ; Upasna Ranjan ; A K Jagdish ; Praveen C. Ramamurthy ; Kamanio Chattopadhyay

**A COMPREHENSIVE STUDY OF LIGHT SOAKING EFFECT IN CDTE SOLAR CELLS** .......... 2816

D. Guo ; A. Moore ; D. Krasikov ; I. Sankin ; D. Vasileska

**CORRECTION FOR METASTABILITY IN THE QUANTIFICATION OF PID IN THIN-FILM MODULE TESTING** .......... 2819

Peter Hacke ; Sergiu Spataru ; Steve Johnston

**A FINE MODEL OF POWER DEGRADATION FOR CRYSTALLINE SILICON SOLAR MODULES** .......... 2823

Wenshuang Hea ; Baosong Duan ; Fumei Wang ; Ao Wang ; Jipeng Chang ; He Wang ; Hong Yang ; Jie Ding ; Junjun Zhang ; Jingsheng Huang

**TEST METHODS FOR HYDROPHOBIC COATINGS ON SOLAR COVER GLASS** .......... 2827

Kenan Isbilir ; Biancamaria Maniscalco ; Ralph Gottschalg ; John Michael Walls

**IMPACT OF DEGRADATION RATES ON SOLAR PV FINANCING FOR PROJECTS LOCATED IN THE UNITED STATES** .......... 2833

Rounak A. Kharait ; Phil Stiles ; Jarrett Carriere ; Larry Mcclung

**ANALYSIS OF WIND DIRECTION AND SPEED MEASUREMENTS IN ARID REGION - A SITE EVALUATION USING DATA WITH LOW TEMPORAL RESOLUTION** .......... 2836

Elisabeth Klimm ; Felix Guischard ; Karl-Anders Weiss

**FORECASTING ENVIRONMENTAL DEGRADATION POWER LOSS IN SOLAR PANELS WITH A PREDICTIVE CRACK OPENING TEST** ...... 2839

Jason L. Lincoln ; Andrew M. Gabor ; Eric J. Schneller ; Hubert Seigneur ; Joseph Walters ; Rob Janoch ; Andrew Anselmo ; Victor Huayamave ; Winston Schoenfeld

**FLUORESCENCE IMAGING ON THE CROSS-SECTION OF PHOTOVOLTAIC LAMINATES AGED UNDER DIFFERENT UV INTENSITIES** ...... 2844

Yadong Lyu ; Jae Hyun Kim ; Xiaohong Gu

**STATISTICAL ANALYSIS OF DEGRADATION DATA FOR C-SI MODULES OBSERVED IN INDIA IN 2016** ...... 2849

Chiranjibi Mahapatra ; Rajiv Dubey ; Shashwata Chattopadhyay ; Sachin Zachariah ; Sanjeev Sabnis

**PROCESS INDUCED DEFLECTION AND STRESS ON ENCAPSULATED SOLAR CELLS** ...... 2854

Xiaodong Meng ; Michael Stuckelberger ; Peter Hacke ; Mariana Bertoni

**A UNIFIED GLOBAL INVESTIGATION ON THE SPECTRAL EFFECTS OF SOILING LOSSES OF PV GLASS SUBSTRATES: PRELIMINARY RESULTS** ...... 2858

Leonardo Micheli ; Eduardo F. Fernández ; Greg P. Smestad ; Hameed Alrashidi ; Nabin Sarmah ; Nazmi Sellami ; Ibrahim A. I. Hassan ; Amal Kasry ; Gustavo Nofuentes ; Neeru Sood ; Bala Pesala ; S. Senthilarasu ; Florencia Almonacid ; K.S. Reddy ; Matthew Muller ; Tapas K. Mallick

**REFERENCE: PROCEEDINGS OF THE IEEE PVSC CONF., 2017 THE DEVELOPMENT OF A DC BREAKDOWN VOLTAGE TEST FOR PHOTOVOLTAIC INSULATING MATERIALS** ...... 2864

David C. Miller ; Bernt Ake-Sultan ; Axel Borne ; Rene Eugen ; Bradley L. Givot ; Jürgen Jung ; Steven W. Macmaster ; Byron K. Mcdanold ; Ulf H. Nilsson ; Nancy H. Phillips ; Ian A. Tappan ; Nick S. Bosco

**FIELD-EVALUATION OF ELECTRODYNAMIC SCREENS FOR MAINTAINING HIGH OPTICAL EFFICIENCY OPERATION OF SOLAR COLLECTORS** ...... 2870

Cristian Morales ; Annie Bernard ; Ryan Eriksen ; Julius Yellowhair ; Sean Garner ; Ricci La Centra ; Alecia Griffin ; Alexis Lloyd ; Yujie Gao ; Ramakrishnan Lakshmanan ; Mark Horenstein ; Malay Mazumder

**EFFECT OF REVERSE BIAS VOLTAGES ON SMALL SCALE GRIDDED CIGS SOLAR CELLS** ...... 2875

Soheyl Mortazavi ; Klaas Bakker ; Jome Carolus ; Michael Daenen ; Gabriela De Amorim Soares ; Henk Steijvers ; Arthur Weeber ; Mirjam Theelen

**A METHOD TO EXTRACT SOILING LOSS DATA FROM SOILING STATIONS WITH IMPERFECT CLEANING SCHEDULES** ...... 2881

Matthew Muller ; Leonardo Micheli ; Alfredo A. Martinez-Morales

**ANALYTICAL (S)TEM STUDIES OF DEFECTS ASSOCIATED WITH PID IN STRESSED SI PV MODULES** ...... 2887

Andrew Norman ; Adam Stokes ; John Moseley ; Steven Harvey ; Steve Johnston ; Harvey Guthrey ; Mowafak Al-Jassim

**DESIGN, DEVELOPMENT, AND EVALUATION OF ELECTRODYNAMIC SCREENS FOR SELF-CLEANING SOLAR PANELS AND CONCENTRATING MIRRORS** ...... 2891

Annie Bernard ; Cristian Morales ; Ryan S. Eriksen ; Alecia C. Griffin ; Yujie Gao ; Ramakrishnan Lakshmanan ; Ricci La Centra ; Arash Sayyah ; Julius E. Yellowhair ; Sean M. Garner ; N Mark Horenstein ; Malay K. Mazumder

**EVALUATING SOLAR CELL FRACTURE AS A FUNCTION OF MODULE MECHANICAL LOADING CONDITIONS** ...... 2897

Eric J. Schneller ; Andrew M. Gabor ; Jason Lincoln ; Rob Janoch ; Andrew Anselmo ; Joseph Walters ; Hubert Seigneur

**COMPUTATIONAL STUDY OF THE EFFECT OF PHOTOVOLTAIC (PV) MODULE PARAMETERS ON STRESS DEVELOPMENT IN SILICON UNDER STATIC LOADING** ...... 2902

Saurabh Sethia ; Karan Shishir Yadav ; Sudharm Rathore ; Abhishek Shubhrant ; Aparna Singh

**A SIMPLE METHOD FOR MEASURING SOLAR RADIATION INTENSITY BY IMAGE ANALYSES** ...... 2906

Akiko Takahashi ; Akinori Moriki ; Nobuyuki Yamada ; Jun Imai ; Shigeyuki Funabiki

**DEGRADATION OF SOLDER BONDS IN FIELD AGED PV MODULES: CORRELATION WITH SERIES RESISTANCE INCREASE** ...... 2912

Abhishiktha Tummala ; Jaewon Oh ; Sai Tatapudi ; Govindasamy Tamizhmani

**PERFORMANCE OF LIGHT AND DARK CURRENT-VOLTAGE CHARACTERISTICS FOR PID-AFFECTED MONOCRYSTALLINE SILICON SOLAR MODULES** ...... 2918

He Wang ; Pan Zhao ; Shuwen Guo ; Hong Yang ; Weijing Huang ; Shiyu Sang ; Bojie Su ; Xue Zhang ; Yunxue Cao ; Hui Zhao

**SOILING RATES OF PV MODULES VS. THERMOPILE PYRANOMETERS** ...... 2923

Martin Waters ; Tejas Tirumalai ; Michael Gostein ; Bill Stueve

**GRID INTEGRATION OF BUILDING SYSTEMS AND 1 MW PHOTOVOLTAIC ARRAY USING VOLTTRON** ...... 2926

David Raker ; Andrew Sellers ; Roshan Kini ; Michael Green ; Thomas Stuart ; Randall Ellingson ; Raghav Khanna ; Michael Heben

**INTERCONNECTION STUDY OF DISTRIBUTED PV SYSTEMS BY INTERFACING MATLAB WITH OPENDSS AND GIS** ........................................................................................... 2931

Joseph A. Ahamioje ; Hariharan Krishnaswami

**NOVEL MPPT ALGORITHM FOR ACTIVE POWER CONTROL OF MULTI-LEVEL DUAL-ACTIVE BRIDGE PV CONVERTER IMPLEMENTED IN NI MYRIO** ............................... 2936

Shilpa Marti ; Hariharan Krishnaswami

**MODELING A GRID-CONNECTED PV/BATTERY MICROGRID SYSTEM WITH MPPT CONTROLLER** ......................................................................................................... 2941

Genesis Alvarez ; Hadis Moradi ; Mathew Smith ; Ali Zilouchian

**>94.5%REDUCTION IN GRID-BUY ELECTRICITY AND ELIMINATION OF AM & PM ENERGY PEAKS/SPIKES BY OPTIMIZING ENERGY USAGE AND INTEGRATION OF CUSTOMER SELF-SUPPLY ROOFTOP SOLAR PV WITH ELECTRICAL & THERMAL (HOT & COLD) STORAGE BATTERIES: A CASE STUDY FOR RESIDENTIAL HAWAII** .............. 2947

John Borland ; Jay Moore ; Corpuz Poncho ; Takahiro Tanaka ; Harumi Mcclure

**A SINGLE-STAGEC"UK-BASED TRANSFORMERLESS INVERTER FOR 1-Φ GRID-CONNECTED PV SYSTEMS** ................................................................................. 2952

Phani Kumar Chamarthi ; Amit Kumar Gupta ; Madhuwanti S. Joshi ; Vivek Agarwal

**A STATE SPACE AVERAGE MODEL FOR DYNAMIC MICROGRID BASED SPACE STATION SIMULATIONS** ............................................................................................... 2957

Rachid Darbali-Zamora ; Eduardo I. Ortiz-Rivera

**BUCK CONVERTER AND SEPIC BASED ELECTRONIC POWER SUPPLY DESIGN WITH MPPT AND VOLTAGE REGULATION FOR SMALL SATELLITE APPLICATIONS** .............. 2963

Rachid Darbali-Zamora ; Nicolás Cobo-Yepes ; John E. Salazar-Duque ; Eduardo I. Ortiz-Rivera ; Amilcar A. Rincon-Charris

**VIRTUAL POWER PLANT FEEDBACK CONTROL DESIGN FOR FAST AND RELIABLE ENERGY MARKET AND CONTINGENCY RESERVE DISPATCH** ................................... 2969

Mohamed Elkhatib ; Jay Johnson ; David Schoenwald

**INTELLIGENT SAMPLING OF PERIODS FOR REDUCED COMPUTATIONAL TIME OF TIME SERIES ANALYSIS OF PV IMPACTS ON THE DISTRIBUTION SYSTEM** ..................... 2975

Jason Galtieri ; Matthew J. Reno

**A PWM SCHEME TO REALISE TWO TIMES EFFECTIVE SWITCHING FREQUENCY WITH CONSTANT COMMON MODE VOLTAGE AND REACTIVE POWER CAPABILITY IN 1- Φ GRID-TIED TRANSFORMERLESS H6 PV INVERTER** ............................................... 2981

Amit Kumar Gupta ; Madhuwanti S. Joshi ; Vivek Agarwal

**A SOLAR PV RETROFIT SOLUTION FOR RESIDENTIAL BATTERY INVERTERS** ................ 2986

Amit Kumar Gupta ; Vaibhav Pawar ; Madhuwanti S. Joshi ; Vivek Agarwal ; Deepak Chandran

**COST BENEFIT AND ALTERNATIVES ANALYSIS OF DISTRIBUTION SYSTEMS WITH ENERGY STORAGE SYSTEMS** ............................................................................... 2991

Tom Harris ; Adarsh Nagarajan ; Murali Baggu ; Tom Bialek

**EVALUATION OF PV HOSTING CAPACITIES OF DISTRIBUTION GRIDS WITH UTILIZATION OF SOLAR-ROOF-POTENTIAL-ANALYSES** ......................................... 2996

Gerd Heilscher ; Falko Ebe ; Basem Idlbi ; Jeromie Morris ; Florian Meier

**EXPERIMENTAL DISTRIBUTION CIRCUIT VOLTAGE REGULATION USING DER POWER FACTOR, VOLT-VAR, AND EXTREMUM SEEKING CONTROL METHODS** ..................... 3002

Jay Johnson ; Sigifredo Gonzalez ; Daniel B. Arnold

**DYNAMIC SETPOINT CONTROL OF ELECTRIC HOT WATER HEATER TANKS FOR INCREASED INTEGRATION OF SOLAR PHOTOVOLTAIC SYSTEMS** ......................... 3008

C. Birk Jones ; Monte Lunacek ; Matthew Lave ; Jay Johnson ; Robert Broderick

**SPATIAL ANALYSIS OF RESIDENTIAL COMBINED PHOTOVOLTAIC AND BATTERY POTENTIAL: CASE STUDY UTRECHT, THE NETHERLANDS** ..................................... 3014

Geert Litjens ; Bala Bhavya Kausika ; Ernst Worrell ; Wilfried Van Sark

**POWER BALANCE REQUIREMENTS FOR SUSTAINED ISLANDING OF INVERTER BASED DISTRIBUTED GENERATION** ........................................................................... 3020

Gregory A. Kern ; Michael Ropp ; Sigifredo Gonzalez

**FULL-SCALE DEMONSTRATION OF DISTRIBUTION SYSTEM PARAMETER ESTIMATION TO IMPROVE LOW-VOLTAGE CIRCUIT MODELS** ............................................... 3025

Matthew Lave ; Matthew J. Reno ; Robert J. Broderick ; Jouni Peppanen

**CREATION AND VALUE OF SYNTHETIC HIGH-FREQUENCY SOLAR INPUTS FOR DISTRIBUTION SYSTEM QSTS SIMULATIONS** ..................................................... 3031

Matthew Lave ; Matthew J. Reno ; Robert J. Broderick

**A DIRECT MAXIMUM POWER POINT SEARCH USING CURRENT-VOLTAGE BASED POWER-LAW RELATION FOR PHOTOVOLTAIC SYSTEM UNDER UNIFORM IRRADIANCE** ............... 3038
*Hitesh K. Mehta ; Ashish K. Panchal*

**PASSIVITY BASED CONTROLLER FOR PHOTOVOLTAIC MODULES USING ZETA CONVERTER** ............................................................................................................................ 3044
*Daniel A. Merced Cirino ; Rachid Darbali Zamora ; Eduardo I. Ortiz Rivera*

**SIC SWITCH BASED SINGLE-STAGE BUCK-BOOST TRANSFORMERLESS MINI INVERTER WITH LOW LEAKAGE CURRENT AND NEGLIGIBLE DC INJECTION** ......................... 3050
*Soumya Ranjan Mohapatra ; Amit Kumar Gupta ; Madhuwanti S. Joshi ; Vivek Agarwal*

**OPEN SOURCE TOOLS FOR HIGH PERFORMANCE QUASI-STATIC-TIME-SERIES SIMULATION USING PARALLEL PROCESSING** ................................................................. 3055
*Davis Montenegro ; Roger C. Dugan ; Matthew J. Reno*

**MAXIMUM POWER POINT TRACKING OF PV MODULE BASED ON NEW EXPLICIT I-V RELATION** ................................................................................................................................ 3061
*Tejeswar Nukala ; A. K. Panchal*

**AN AUTOCORRELATION-BASED COPULA MODEL FOR PRODUCING REALISTIC CLEAR-SKY INDEX AND PHOTOVOLTAIC POWER GENERATION TIME-SERIES** ..................... 3067
*Joakim Munkhammar ; Joakim Widén*

**DYNAMIC RESPONSE OF MAXIMUM POWER POINT TRACKING USING PERTURB AND OBSERVE ALGORITHM WITH MOMENTUM TERM** ............................................................. 3073
*Gautam A. Raiker*

**A FRAMEWORK FOR COMPARING THE ECONOMIC PERFORMANCE AND ASSOCIATED EMISSIONS OF GRID-CONNECTED BATTERY STORAGE SYSTEMS IN EXISTING BUILDING STOCK: A NYISO CASE STUDY** ........................................................................... 3077
*Julian Do Nascimento Ricardo ; Vasilis Fthenakis*

**IMPROVING ANY ARBITRARY MPPT HILL CLIMBER WITH ANN ESTIMATIONS** ............... 3083
*Jesse Roberts ; Indranil Bhattacharya*

**INCREASING SOLAR PHOTOVOLTAIC PENETRATION USING THERMAL ENERGY STORAGE** ................................................................................................................................... 3088
*Alexander F. Routhier ; Christiana Honsberg*

**MODEL PREDICTIVE CONTROL OF GRID CONNECTED MODULAR MULTILEVEL CONVERTER FOR INTEGRATION OF PHOTOVOLTAIC POWER SYSTEMS** ...................... 3092
*Amir Shahirinia ; Amin Hajizadeh*

**MAXIMIZATION OF SELF-SUFFICIENCY WITH GRID CONSTRAINTS: PV GENERATORS, WIND TURBINES AND STORAGE TO FEED TERTIARY SECTOR USERS** ............................ 3096
*Filippo Spertino ; Jawad Ahmad ; Alessandro Ciocia ; Paolo Di Leo ; Francesco Giordano*

**SWITCHES CONTROLLING TO IMPLEMENT ADAPTIVE MULTILEVEL INVERTER ON PV SYSTEM** ................................................................................................................................. 3102
*Hadi Suhana ; Ngapuli I Sinisuka ; Muhammad Nurdin ; Yvon Besanger ; Vincent Debusschere*

**DEMAND RESPONSE FOR THE PROMOTION OF PHOTOVOLTAIC PENETRATION** ............ 3107
*Venizelos Venizelou ; Spyros Theocharides ; George Makrides ; Venizelos Efthymiou ; George E. Georghiou*

**GRIDDLER AI: NEW PARADIGM IN LUMINESCENCE IMAGE ANALYSIS USING AUTOMATED FINITE ELEMENT METHODS** ........................................................................... 3113
*Johnson Wong ; Percis Teena ; Daniel Inns*

**INTERACTION OF O2IDIMERS WITH GA IN SI AND IMPLICATIONS FOR A COMPREHENSIVE MODEL OF LIGHT- INDUCED DEGRADATION** ................................... 3119
*Yu Jin ; Scott T. Dunham*

**NUMERICAL SIMULATION OF EBIC FOR ANALYSIS OF EXTENDED DEFECTS** ............ 3123
*Marco Nardone ; John Moseley ; Saroj Dahal ; Anuja V. Parikh ; John M. Waddle*

**COLLOIDAL QUANTUM DOT SOLAR CELL ELECTRICAL PARAMETER IMAGING USING CAMERA-BASED HIGH-FREQUENCY HETERODYNE LOCK-IN CARRIEROGRAPHY** ........................... 3129
*Lilei Hu ; Mengxia Liu ; Andreas Mandelis ; Qiming Sun ; Alexander Melnikov ; Edward H. Sargent*

**A NEW PERSPECTIVE ON POTENTIAL-INDUCED DEGRADATION OF THE SHUNTING TYPE BY MICRO RAMAN-SPECTROSCOPY AND MICRO LIGHT-BEAM-INDUCED CURRENT** ................................................................................................................................ 3135
*A. Büchler ; H. Nagel ; M. Breitwieser ; S. Kluska ; F. D. Heinz ; M. C. Schubert ; M. Glatthaar ; S. Glunz*

**NANOSCALE DETECTION OF DEEP LEVELS IN CIGS USING ELECTRON ENERGY LOSS SPECTROSCOPY** .................................................................................................................... 3139
*Julia I. Deitz ; Pran K. Paul ; Shankar Karki ; Sylvain Marsillac ; Aaron R. Arehart ; Tyler J. Grassman ; David W. Mccomb*

**MEASUREMENT OF CARRIER DYNAMICS IN PHOTOVOLTAIC CZTSE BY TIME-RESOLVED TERAHERTZ SPECTROSCOPY** .................................................................................................. 3143
Siming Li ; Michael A. Lloyd ; Andrew A. Golembeski ; Brian E. Mccndless ; Jason B. Baxter

**DECOUPLING GRAIN-BOUNDARY, GRAIN-INTERIOR, AND SURFACE RECOMBINATION WITH CATHODOLUMINESCENCE** ................................................................................................... 3147
John Moseley ; Pierre Rale ; Stéphane Collin ; Ana Kanevce ; Eric Colegrove ; Joel Duenow ; Soren Jensen ; Wyatt K. Metzger ; Mowafak M. Al-Jassim

**HIGH RESOLUTION THZ SCANNING FOR OPTIMIZATION OF DIELECTRIC LAYER OPENING PROCESS ON DOPED SI SURFACES** ................................................................................. 3150
P. Spinelli ; F.J.K. Danzl ; D. Deligiannls ; N. Guillevin ; A.R. Burgers ; S. Sawallich ; M. Nage ; I. Cesar

**DEGRADATION ASSESSMENT OF FIELDED CIGS PHOTOVOLTAIC ARRAYS** ........................... 3155
Bruce H. King ; Joshua S. Stein ; Daniel Riley ; C. Birk Jones ; Charles D. Robinson

**APPLICATION OF IEC 61724 STANDARDS TO ANALYZE PV SYSTEM PERFORMANCE IN DIFFERENT CLIMATES** ................................................................................................................. 3161
Katherine A. Klise ; Joshua S. Stein ; Joseph Cunningham

**EFFECTS OF URBAN ENVIRONMENT ON SOLAR PV PERFORMANCE** .................................... 3167
Panagiotis Moraitis ; Bala Bhavya Kausika ; Wilfried G.J.H.M. Van Sark

**IRRADIANCE MEASUREMENT CONSIDERATIONS FOR SYSTEM PERFORMANCE ASSESSMENT WHEN MANAGING FLEETS OF PHOTOVOLTAIC ASSETS ACROSS ASIA** ................ 3172
André M. Nobre ; Shravan Karthik ; Chenxi Liu ; Rohit Jaswal ; Rupesh Baker ; Raghav Malhotra ; Alan Khor

**MACHINE LEARNING IN PV FAULT DETECTION, DIAGNOSTICS AND PROGNOSTICS: A REVIEW** ........................................................................................................................................ 3178
Sandy Rodrigues ; Helena Geirinhas Ramos ; F. Morgado-Dias

**OUTDOOR FIELD PERFORMANCE FROM BIFACIAL PHOTOVOLTAIC MODULES AND SYSTEMS** ...................................................................................................................................... 3184
Joshua S. Stein ; Daniel Riley ; Matthew Lave ; Clifford Hansen ; Chris Deline ; Fatima Toor

**DEFINING THRESHOLD VALUES OF ENCAPSULANT AND BACKSHEET ADHESION FOR PV MODULE RELIABILITY** ......................................................................................................... 3190
Nick Bosco ; Joshua Eafanti ; Sarah Kurtz ; Jared Tracy ; Reinhold Dauskardt

**CHARACTERIZATIONS OF AGED GLASS/ETHYLENE VINYL ACETATE/GLASS USING FLUORESCENCE SPECTROSCOPY AND INSTRUMENTED INDENTATION** ............................... 3195
Jae Hyun Kim ; Yadong Lyu ; David C. Miller ; Xiaohong Gu

**ENCAPSULANT ADHESION TO SURFACE METALLIZATION ON PHOTOVOLTAIC CELLS** ................ 3200
Jared Tracy ; Nick Bosco ; Reinhold Dauskardt

**IMPACT OF UV LIGHT INTENSITY ON PHOTODEGRADATION OF PV BACKSHEETS** ............ 3204
Xiaohong Gu ; Li-Chieh Yu ; Yadong Lyu ; Jae Hyun Kim ; Andrew Fairbrother ; Tinh Nguyen

**SURVEY OF MECHANICAL DURABILITY OF PV BACKSHEETS** ............................................... 3208
Michael D. Kempe ; David C. Miller ; Allen Zielnik ; Daniel Montiel-Chicharro ; Jiang Zhu ; Ralph Gottschalg

**SOLAR VARIABILITY REDUCTION USING OFF-MAXIMUM POWER POINT TRACKING AND BATTERY STORAGE** ........................................................................................................... 3214
Jason Galtieri ; Philip T. Krein

**INTEGRATION OF ELECTROCHEMICAL CAPACITORS ON SILICON PHOTOVOLTAIC MODULES FOR RAPID-RESPONSE POWER BUFFERING** ..................................................... 3220
Yu Jiang ; Xuanyi Shi ; Derwin Lau ; Da-Wei Wang ; Zi Ouyang ; Alison Lennon

**DESIGN & EVALUATION OF A HYBRID SWITCHED CAPACITOR CIRCUIT WITH WIDE-BANDGAP DEVICES FOR COMPACT MVDC PV POWER CONVERSION** .............................. 3224
J. Stewart ; J. Delhotal ; J. Richards ; J. Neely ; L. Rashkin ; J. D. Flicker ; R. Kaplar ; S. Gonzalez ; J. Lehr

**SOLAR ENERGY FOR CLEAN AND AFFORDABLE WATER DESALINATION** ......................... 3230
V. M. Fthenakis ; Adam A. Atia

**GLOBAL RESIDENTIAL AIR-CONDITIONING SECTOR AS A DRIVER FOR PHOTOVOLTAIC INDUSTRY GROWTH DURING THE 21ST CENTURY** ................................. 3236
Hannu S. Laine ; Jyri Salpakari ; Marius Peters ; Erin E. Looney ; Ashley E. Morishige ; Hele Savin ; Gregory Wilson ; Tonio Buonassisi

**MEASURES TO REMOVE ECONOMIC NON-MARKET FAILURE AND INSTITUTIONAL BARRIERS THAT RESTRICT PHOTOVOLTAICS SELF-CONSUMPTION AND NET-METERING IN SPAIN** ......................................................................................................... 3240
Enrique Rosalcs-Ascnsio ; Juan A. Méndez ; Benjamín Gonzálcz-Díaz ; Ricardo Guerrero Lemus

**COST COMPETITIVE CONCENTRATOR PHOTOVOLTAICS FOR SOLAR THERMAL APPLICATIONS** ......................................................................................................................... 3245
Brian C. Riggs ; Richard E. Biedenham ; Chris Dougher ; Yaping Vera Ji ; Qi Xu ; Vince Romanin ; Daniel S. Codd ; James M. Zahler ; Matthew D. Escarra

**PREDICTING THE EFFICIENCY OF THE SILICON BOTTOM CELL IN A TWO-TERMINAL TANDEM SOLAR CELL** ............ 3250
Zhengshan J. Yu ; Zachary C. Holman

**MECHANICALLY STACKED 4-TERMINAL III-V/SI TANDEM SOLAR CELLS** ............ 3254
Stephanie Essig ; Christophe Allebe ; John F. Geisz ; Myles A. Steiner ; Loris Barraud ; Antoine Descoeudres ; J. Scott Ward ; Manuel Schnabel ; David L. Young ; Matthieu Despeisse ; Christophe Ballif ; Adele Tamboli

**PEROVSKITE/SILICON TANDEM SOLAR CELLS: CHALLENGES TOWARDS HIGH-EFFICIENCY IN 4-TERMINAL AND MONOLITHIC DEVICES** ............ 3256
Jérémie Werner ; Florent Sahli ; Brett Kamino ; Davide Sacchetto ; Matthias Bräuninger ; Arnaud Walter ; Christophe Ballif ; Matthieu Despeisse ; Sylvain Nicolay ; Bjoern Niesen ; Raphäel Monnard ; Stefaan De Wolf ; Soo-Jin Moon ; Loris Barraud ; Bertrand Paviet-Salomon ; Jonas Geissbuehler ; Christophe Allebé

**THE OUTCOME OF REPLACING SN COMPLETELY BY GE IN KESTERITE CU2ZNSNSE4SOLAR CELLS** ............ 3260
S. Sahayaraj ; G. Brammertz ; B. Vermang ; T. Schnabel ; E. Ahlswede ; Z. Huang ; S. Ranjbar ; M. Meuris ; J. Vleugels ; J. Poortmans

**TRANSITION METAL OXIDES NANO-LAYERS AS EFFICIENT BACK ELECTRON REFLECTORS FOR CU2ZNSNSE4SOLAR CELLS** ............ 3265
Sergio Giraldo ; Moisés Espíndola-Rodríguez ; Florian Oliva ; Víctor Izquierdo-Roca ; Alejandro Pérez-Rodríguez ; Edgardo Saucedo

**MIXED SULFUR AND SELENIUM ANNEALING STUDY OF COMPOUND-SPUTTERED BILAYER CU2ZNSNS4/ CU2ZNSNSE4PRECURSORS** ............ 3269
N. Ross ; S. Grini ; L. Vines ; C. Platzer-Björkman

**REVEALING THE ROLE OF MN INCORPORATION IN CU2ZNSN(S, SE)4PHOTOVOLTAIC ABSORBER LAYER** ............ 3275
Stener Lie ; Joel M. R. Tan ; Wenjie Li ; Shin Woei Leow ; Oki Gunawan ; Doug Bishop ; Lydia H. Wong

**NON-VACUUM SINGLE STEP SYNTHESIS OF LARGE-GRAIN SIZE CZTS PHOTO ABSORBER FOR THIN FILM SOLAR CELLS BY FLUX ASSISTED CHEMICAL SPRAY** ............ 3279
Ratheesh R. Thankalekshmi ; Navjot Kaur Sidhu ; A.C. Rastogi

**RAMAN SCATTERING ASSESSMENT OF POINT DEFECTS IN KESTERITE SEMICONDUCTORS: UV RESONANT RAMAN CHARACTERIZATION FOR ADVANCED PHOTOVOLTAICS** ............ 3285
Florian Oliva ; Laia Arqués Farré ; Sergio Giraldo ; Mirjana Dimitrievska ; Paul Pistor ; Alejandro Martínez-Pérez ; Lorenzo Calvo-Barrio ; Edgardo Saucedo ; Alejandro Pérez-Rodríguez ; Victor Izquierdo-Roca

**ASSESSING THE DEFECT RESPONSIBLE FOR LETID: TEMPERATURE- AND INJECTION-DEPENDENT LIFETIME SPECTROSCOPY** ............ 3290
Mallory A. Jensen ; Yan Zhu ; Erin E. Looney ; Ashley E. Morishige ; Carlos Vargas ; Ziv Hameiri ; Tonio Buonassisi

**MICROSCOPIC DISTRIBUTION OF LUMINESCENCE FROM DISLOCATION CLUSTERS IN MULTICRYSTALLINE SILICON WAFERS** ............ 3295
H. T. Nguyen ; M. A. Jensen ; L. Li ; C. Samundsett ; H. C. Sio ; B. Lai ; T. Buonassisi ; D. Macdonald

**DO GRAIN BOUNDARIES MATTER? ELECTRICAL AND ELEMENTAL IDENTIFICATION AT GRAIN BOUNDARIES IN LETID-AFFECTED P-TYPE MULTICRYSTALLINE SILICON** ............ 3300
Mallory A. Jensen ; Ashley E. Morishige ; Sagnik Chakraborty ; Romika Sharma ; Hang Cheong Sio ; Chang Sun ; Barry Lai ; Volker Rose ; Amanda Youssef ; Erin E. Looney ; Sarah Wieghold ; Jeremy Poindexter ; Juan-Pablo Correa-Baena ; Daniel Macdonald ; Joel B. Li ; Tonio Buonassisi

**PERC SOLAR CELL PERFORMANCE PREDICTIONS FROM MULTICRYSTALLINE SILICON INGOT METROLOGY DATA** ............ 3304
Bernhard Mitchell ; Daniel Chung ; Qiuxiang He ; Hua Zhang ; Zhen Xiong ; Pietro P. Altermatt ; Peter Geelan-Small ; Thorsten Trupke

**PHOTOLUMINESCENCE-IMAGING-BASED EVALUATION OF NON-UNIFORM CDTE DEGRADATION** ............ 3305
Steve Johnston ; David Albin ; Peter Hacke ; Steven P. Harvey ; Helio Moutinho ; Mowafak Al-Jassim ; Wyatt K. Metzger

**MACHINE LEARNING AND CORRELATIVE MICROSCOPY: HOW 'BIG DATA' TECHNIQUES CAN BENEFIT THIN FILM SOLAR CELL CHARACTERIZATION** ............ 3309
Bradley M. West ; Michael Stuckelberger ; Tara Nietzold ; Barry Lai ; Jörg Maser ; Mariana I. Bertoni

**METAL INDUCED CONTACT RECOMBINATION MEASURED BY QUASI-STEADY-STATE PHOTOLUMINESCENCE** ............ 3315
Robert Dumbrell ; Mattias K. Juhl ; Mengjie Li ; Thorsten Trupke ; Ziv Hameiri

**USING TIME-OF-FLIGHT SIMS TO INVESTIGATE GROUP V DOPANT DISTRIBUTION IN CDTE** ............ 3319
Steven P. Harvey ; Eric Colegrove ; Brian Mccandless ; David Albin ; Mowafak Al-Jassim ; Wyatt K. Metzger

QUANTITATIVE ANALYSIS OF ACTIVE DOPANT DISTRIBUTION AND ESTIMATION OF
EFFECTIVE DIFFUSIVITY IN PHOSPHORUS- IMPLANTED EMITTER OF SI SOLAR CELL
USING SCANNING NONLINEAR DIELECTRIC MICROSCOPY ............................................. 3323
    Kotaro Hirose ; Katsuto Tanahashi ; Hidetaka Takato ; Yasuo Cho
SIMULATION OF DRIVE-LEVEL CAPACITANCE PROFILING TO INTERPRET
MEASUREMENTS ON CU(IN, GA)SE2SCHOTTKY DEVICES .............................................. 3327
    Geordie Zapalac ; Jeff Bailey
ANALYSIS OF THE IMPACT OF INSTALLATION PARAMETERS AND SYSTEM SIZE ON
BIFACIAL GAIN AND ENERGY YIELD OF PV SYSTEMS ..................................................... 3333
    Amir Asgharzadeh ; Tomas Lubenow ; Joseph Sink ; Bill Marion ; Chris Deline ; Clifford Hansen ; Joshua Stein ;
    Fatima Toor
DEPENDENCE OF STRING POWER ON ITS HEIGHT IN THE ARRAY IN YOSHINOGARI
MEGA SOLAR POWER PLANT ............................................................................................... 3339
    Shigeomi Hara ; Makoto Kasu ; Yasuki Masutomi
A BOTTOM-UP ENERGY SIMULATION FRAMEWORK TO ACCURATELY COMPARE PV
MODULE TOPOLOGIES UNDER NON-UNIFORM AND DYNAMIC OPERATING
CONDITIONS ............................................................................................................................. 3343
    Patrizio Manganiello ; Maro Baka ; Hans Goverde ; Tom Borgers ; Jonathan Govaerts ; Arvid Van Der Heide ;
    Eszter Voroshazi ; Francky Catthoor
A PERFORMANCE MODEL FOR BIFACIAL PV MODULES ................................................. 3348
    Daniel Riley ; Clifford Hansen ; Joshua Stein ; Matthew Lave ; Johnson Kallickal ; Bill Marion ; Fatima Toor
ACCURATE MODELING OF PARTIALLY SHADED PV ARRAYS ........................................ 3354
    Bennet Meyers ; Mark Mikofski
EVALUATION OF UNCERTAINTY IN PV PROJECT DESIGN: DEFINITION OF SCENARIOS
AND IMPACT ON ENERGY YIELD PREDICTIONS .............................................................. 3360
    Giorgio Belluardo ; Magnus Herz ; Ulrike Jahn ; Mauricio Richter ; David Moser
MONOCRYSTALLINE 1.7 EV MGCDTE DOUBLE-HETEROSTRUCTURE SOLAR CELL WITH
11.2% EFFICIENCY .................................................................................................................. 3366
    Calli M. Campbell ; Xin-Hao Zhao ; Yuan Zhao ; Mathieu Boccard ; Cheng- Ying Tsai ; Jacob J. Becker ; Zachary
    Holman ; Yong-Hang Zhang
MBE GROWTH OF 1.7EV AL0.2GA0.8AS AND 1.42EV GAAS SOLAR CELLS ON SI USING
DISLOCATIONS FILTERS: AN ALTERNATIVE PATHWAY TOWARD III-V/ SI SOLAR CELLS
ARCHITECTURES ..................................................................................................................... 3370
    Arthur Onno ; Mingchu Tang ; Mu Wang ; Yurii Maidaniuk ; Mourad Benamara ; Yuriy I. Mazur ; Gregory J.
    Salamo ; Lars Oberbeck ; Jiang Wu ; Huiyun Liu
III- V/SI TANDEM CELLS UTILIZING INTERDIGITATED BACK CONTACT SI CELLS AND
VARYING TERMINAL CONFIGURATIONS ......................................................................... 3371
    Manuel Schnabel ; Michael Rienacker ; Agnes Merkle ; Talysa R. Klein ; Nikhil Jain ; Stephanie Essig ; Henning
    Schulte-Huxel ; Emily Warren ; Maikel F.A.M. Van Hest ; John Geisz ; Jan Schmidt ; Rolf Brendel ; Robby Peibst
    ; Paul Stradins ; Adele Tamboli
TOWARDS HIGH-EFFICIENCY GAASP/SI TANDEM CELLS ............................................. 3376
    S. Fan ; M. Vaisman ; K. Nay Yaung ; E. Perl ; D. Martín-Martín ; M. Leilaeioun ; Z. C. Holman ; M. L. Lee
CHARACTERIZATION OF HETEROEPITAXIAL GAAS FILMS GROWN ON SI USING
SELECTIVE AREA NUCLEATION ......................................................................................... 3381
    Emily L. Warren ; Emily A. Makoutz ; Michelle Vaisman ; Benjamin F. Bachman ; William E. Mcmahon ; Jeramy
    D. Zimmerman ; Adele C. Tamboli
EFFICIENT PHOTON UPCONVERSION IN SEMICONDUCTOR NANOSTRUCTURES:
CONSTRAINTS AND OPPORTUNITIES ................................................................................ 3384
    Matthew F. Doty ; Eric Y. Chen ; Jing Zhang ; Diane G. Sellers ; Zhuohui Li ; Christopher C. Milleville ; Kyle
    Lennon ; Joshua M. O. Zide
ENHANCED ULTRA-THIN A-GE:H SOLAR CELLS BY PLASMONIC NANOPARTICLES
EMBEDDED IN THE OPTICAL RESONANT CAVITY .......................................................... 3388
    Brendan Brady ; Volker Steenhoff ; Benedikt Nickel ; Martin Vehse ; Alexander G. Brolo
NATIVE-METAL-OXIDE-COATED PLASMONIC ELECTRODE METASURFACES FOR
NANOPHOTONIC LIGHT TRAPPING AND EFFICIENT CHARGE COLLECTION ............. 3393
    Deirdre M. O'Carroll ; Christopher E. Petoukhoff ; Zhongkai Cheng ; Zeqing Shen ; Catrice M. Carter
IN-GA PRECURSOR ISLANDS FOR CU(IN, GA)SE2MICRO-CONCENTRATOR SOLAR CELLS ............. 3396
    Katharina Eylers ; Franziska Ringleb ; Berit Heidmann ; Sergiu Levcenco ; Thomas Unold ; Hagen W. Klemm ;
    Gina Peschel ; Alexander Fuhrich ; Thomas Teubner ; Thomas Schmidt ; Martina Schmid ; Torta Boeck
ADVANCES IN SILICON SURFACE TEXTURIZATION BY METAL ASSISTED CHEMICAL
ETCHING FOR PHOTOVOLTAIC APPLICATIONS ............................................................. 3402
    Sylvain Le Gall ; Raphaël Lachaume ; Encarnacion Torralba ; Mathieu Halbwax ; Vincent Magnin ; Taha El
    Assimi ; Marin Fouchier ; Joseph Harari ; Jean-Pierre Vilcot ; Christine Cachet-Vivier ; Stéphane Bastide

**SINGLE CRYSTALLINE SUBSTRATES FOR III- V GROWTH VIA EXFOLIATION OF BULK SINGLE CRYSTALS** ................................................................................................................................ 3406

Celeste L. Melamed ; Brenden R. Ortiz ; Aaron D. Martinez ; William E. Mcmahon ; Adele C. Tamboli ; Andrew G. Norman ; Eric S. Toberer

**CUZNS HOLE CONTACTS ON MONOCRYSTALLINE CDTE SOLAR CELLS** .................................. 3410

Jacob J. Becker ; Xiaojie Xu ; Rachel Woods-Robinson ; Calli M. Campbell ; Maxwell Lassise ; Joel Ager ; Yong-Hang Zhang

**THE EFFECT OF THE CDCL2 HEAT TREATMENT ON CDSEXTE1-X SOLAR CELLS** ................ 3413

Chih An Hsu ; Vasilios Palekis ; Imran Khan ; Shamara Collins ; Don Morel ; Chris Ferekides

**EFFECTS OF CDCL2TREATMENT ON THE LOCAL ELECTRONIC PROPERTIES OF POLYCRYSTALLINE CDTE MEASURED WITH PHOTOEMISSION ELECTRON MICROSCOPY** ....................................................................................................................................................... 3417

Morgann Berg ; Jason M. Kephart ; Walajabad S. Sampath ; Taisuke Ohta ; Calvin Chan

**POINT DEFECTS IN CDTE BULK SINGLE CRYSTALS GROWN IN CD-RICH CONDITIONS** ........ 3422

Tursun Ablekim ; Santosh K. Swain ; Teresa M. Barnes ; Kelvin G. Lynn

**OPTICAL PROPERTIES OFCDSE1-XSXANDCDSE1-YTEYALLOYS AND THEIR APPLICATION FOR CDTE PHOTOVOLTAICS** ................................................................................... 3426

Maxwell M. Junda ; Corey R. Grice ; Prakash Koirala ; Robert W. Collins ; Yanfa Yan ; Nikolas J. Podraza

**BLISTERING OF MAGNETRON SPUTTERED THIN FILM CDTE DEVICES** ................................. 3430

P.M. Kaminski ; S. Yilmaz ; A. Abbas ; F. Bittau ; J.W. Bowers ; R.C. Greenhalgh ; J.M. Walls

**ENERGY YIELD IN HOT & SUNNY CLIMATES: IMPACT OF SILICON SOLAR CELL ARCHITECTURE AND CELL INTERCONNECTION** ................................................................................... 3435

Jan Haschke ; Johannes P. Seif ; Yannick Riesen ; Andrea Tomasi ; Jean Cattin ; Loïc Tous ; Patrick Choulat ; Monica Aleman ; Emanuele Comagliotti ; Angel Uruena ; Richard Russell ; Filip Duerinckx ; Jonathan Champliaud ; Jacques Levrat ; Amir A. Abdallah ; Brahim Aïssa ; Nouar Tabet ; Nicolas Wyrsch ; Matthieu Despeisse ; Jozef Szlufcik ; Stefaan De Wolf ; Christophe Ballif

**NOVEL REAR SIDE METALLIZATION ROUTE FOR SI SOLAR CELLS USING A TRANSPARENT CONDUCTING ADHESIVE** ..................................................................................................... 3439

Manuel Schnabel ; Talysa R. Klein ; Benjamin G. Lee ; William Nemeth ; Vincenzo Lasalvia ; Maikel F.A.M. Van Hest ; Paul Stradins

**MULTILAYER FOIL METALLIZATION FOR ALL BACK CONTACT CELLS** ................................. 3442

David H. Levy ; David E. Carlson

**ELECTROLUMINESCENCE EXCITATION SPECTROSCOPY: A NOVEL APPROACH TO NON-CONTACT QUANTUM EFFICIENCY MEASUREMENTS** ......................................................... 3448

Kristopher O. Davis ; Greg S. Horner ; Joshua B. Gallon ; Leonid A. Vasilyev ; Kyle B. Lu ; Antonius B. Dirriwachter ; Terry B. Rigdon ; Eric J. Schneller ; Kortan Ogutman ; Richard K. Ahrenkiel

**ILLUMINATED OUTDOOR LUMINESCENCE IMAGING OF PHOTOVOLTAIC MODULES** ............. 3452

Timothy J Silverman ; Michael G. Deceglie ; Kaitlyn Vansant ; Steve Johnston ; Ingrid Repins

**ELECTROLUMINESCENT IMAGE PROCESSING AND CELL DEGRADATION TYPE CLASSIFICATION VIA COMPUTER VISION AND STATISTICAL LEARNING METHODOLOGIES** ......................................................................................................................................... 3456

Justin S. Fada ; Mohammad A. Hossain ; Jennifer L. Braid ; Shuying Yang ; Timothy J Peshek ; Roger H. French

**TOWARDS DEVELOPING A STANDARD FOR TESTING BIFACIAL PV MODULES: SINGLE-SIDE VERSUS DOUBLE-SIDE ILLUMINATION METHOD I-V MEASUREMENTS UNDER DIFFERENT IRRADIANCE AND TEMPERATURE** .............................................................................. 3462

Stefan Roest ; Witek Nawara ; Bas B. Van Aken ; Elias Garcia Goma

**ELECTRICAL TRANSPORT PROPERTIES FROM LONG WAVELENGTH ELLIPSOMETRY** ............. 3468

Prakash Uprety ; Maxwell M. Junda ; Indra Subedi ; Michael A. Slocum ; David V. Forbes ; Seth M. Rubbard ; Nikolas J. Podraza

**IN SITU RAMAN MONITORING OF KESTERITE CU2ZNSNS4 PHASE FORMATION FROM SULFURIZATION OF SOL-GEL OXIDE PRECURSORS** ........................................................................ 3473

Osama Awadallah ; Joseph Hernandez ; Andriy Durygin ; Zhe Cheng

**PERFORMANCE OF FIELD-AGED PV MODULES IN INDIA: RESULTS FROM 2016 ALL INDIA SURVEY OF PV MODULE RELIABILITY** ................................................................................... 3478

Rajiv Dubey ; Sachin Zachariah ; Shashwata Chattopadhyay ; Vivek Kuthanazhi ; Sugguna Rambabu ; Sonali Bhaduri ; Hemant K. Singh ; Archana Sinha ; Birinchi Bora ; Rajesh Kumar ; O. S. Sastry ; Chetan S. Solanki ; Anil Kottantharayil ; Brij M. Arora ; K. L. Narasimhan ; Juzer Vasi

**INFERRING THE PERFORMANCE RATIO OF PV SYSTEMS DISTRIBUTED IN AN REGION: A REAL-CASE STUDY IN SOUTH TYROL** .................................................................................................. 3482

Marco Pierro ; Giorgio Belluardo ; Philip Ingenhoven ; Cristina Cornaro ; David Moser

**QUANTIFY PHOTOVOLTAIC MODULE DEGRADATION USING THE LOSS FACTOR MODEL PARAMETERS** ............................................................................................................ 3488

C. Birk Jones ; Bruce H. King ; Joshua S. Stein ; Justin S. Fada ; Alan J. Curran ; Roger H. French ; Erdmut Schnabel ; Michael Koehl ; Olga Lavrova

**SIMULATING PV SYSTEM PERFORMANCE WITH COMPONENT RELIABILITY DISTRIBUTIONS** ........................................................................................................................... 3494

Geoffrey T. Klisel ; Janine M. Freeman ; Olga Lavrova

**LIFETIME AND DEGRADATION OF PRE-DAMAGED PV-MODULES – FIELD STUDY AND LAB TESTING** ................................................................................................................................ 3500

Claudia Buerhop ; Sven Wirsching ; Simon Gehre ; Tobias Pickel ; Thilo Winkler ; Andreas Bemrrr ; Julia Merghcim ; Christian Camus ; Jens Hauch ; Christoph J. Brabec

**IMM TRIPLE-JUNCTION SOLAR CELLS AND MODULES OPTIMIZED FOR SPACE AND TERRESTRIAL CONDITIONS** ..................................................................................................... 3506

Tatsuya Takamoto ; Hiroyuki Juso ; Kohsuke Ueda ; Hidetoshi Washio ; Hiroshi Yamaguchi ; Mitsuru Imaizumi ; Taishi Sumita ; Tetsuya Nakamura

**VERY HIGH SPECIFIC POWER ELO SOLAR CELLS (>3 KW/KG) FOR UAV, SPACE, AND PORTABLE POWER APPLICATIONS** ................................................................................................. 3511

D. Cardwell ; A. Kirk ; C. Stender ; A. Wibowo ; F. Tuminello ; M. Drees ; R. Chan ; M. Osowski ; N. Pan

**ENHANCED ENDURANCE OF A UNMANNED AERIAL VEHICLES USING HIGH EFFICIENCY SI AND III-V SOLAR CELLS** ...................................................................................... 3514

David Scheiman ; Raymond Hoheisel ; Daniel J Edwards ; Andrew Paulsen ; Justin Lorentzen ; Steve Carruthers ; Sam Carter ; Matthew Kelly ; Phillip Jenkins ; Robert Walters

**HIGH PERFORMANCE, LIGHTWEIGHT GAAS SOLAR CELLS FOR AEROSPACE AND MOBILE APPLICATIONS** ...................................................................................................................... 3520

Aarohi Vijh ; Lori Washington ; Robert C. Parenti

**THROUGH-EPITAXIAL-VIA BACK-CONTACT MULTI-JUNCTION SOLAR CELLS FABRICATED USING EPITAXIAL LIFT-OFF** .......................................................................... 3524

Rekha Reddy ; Marilyn L. Nowakowski ; David Rowell ; Christopher L. Stender ; Christopher Youtsey

**DESIGN OF INGAP/GAAS/LNGAAS MULTI-JUNCTION CELLS WITH REDUCED LAYER THICKNESSES USING LIGHT-TRAPPING REAR TEXTURE** ....................................... 3528

Lin Zhu ; Anurag Reddy ; Kentaroh Watanabe ; Masakazu Sugiyama ; Yoshiaki Nakano ; Hidefumi Akiyama

**Author Index**

# Impact of Degradation Rates on Solar PV Financing for Projects Located in the United States

Rounak A. Kharait[1], Phil Stiles[1], Jarrett Carriere[2] and Larry McClung[2]

[1]Leidos Engineering LLC, Denver, CO, 80202, USA
[2]Leidos Canada Inc., Ottawa, ON, K1P 5Y7, Canada

*Abstract* — A project's financial viability is, in part, determined using its projected net present value (NPV). The accuracy of the NPV corresponds to the accuracy of several factors, among which is the assumption of photovoltaic (PV) module degradation. This paper utilizes the PV module degradation from two peer-reviewed studies and investigates its impact on project financial viability. In order to quantify the effect, the paper will consider the NPV of a reference project. We calculated annual energy estimates (P50) for a hypothetical 100 MW-AC solar PV project at three geographically distinct locations in the United States. Further, this paper investigated the impact of degradation rates for different geographies on project finance. For all three locations, the NPV with 0.5 percent module degradation was found to be on average 5.4 percent higher than the NPV with 1 percent module degradation. The selection of degradation rates also impacts the operation and maintenance of the solar PV projects. A project with high degradation rates will have lower gross revenue over time and may experience shrinking O&M budgets in the later years of the project.

*Index Terms* — net present value, solar, PV, degradation.

## I. INTRODUCTION & BACKGROUND

Degradation in solar PV plants is a well debated topic in the PV industry. Degradation for fossil-fuel and other thermal power plants which have been operational for more than a century has been well studied. Even the earliest solar plants have been operational for no more than a couple of decades and hence the consequences of degradation are not well understood. Moreover, solar technology has rapidly evolved in the past decade. State-of-the-art PV plants employing newer technology have not been operational long enough to study actual degradation patterns. With solar PV module technology advancing every day, it is difficult to measure actual degradation for solar PV projects.

Jordan et al. [1] [2] have compiled degradation data in two published papers. The authors present the variation of degradation by technology and season, as well as its linear variability per year. Their findings confirm module degradation depends on the quality of manufacturing, type of technology, and climate in which the modules are installed, among other parameters. When a solar PV project is financed, cash flow predictions are made over 25 years, representative of the term of a project's power purchase agreement (PPA). This information is included in the pro forma financial statement that supports the financing transaction. Project developers refer to the module manufacturer's warranty statement as the basis of degradation assumption supporting the pro forma. Lenders' independent engineers have challenged this assumption, as the warranty degradation rates provided by module manufacturers are usually a commercially negotiated value rather than actual measured data than can be substantiated. This has led to developers and independent engineers recommending different degradation values for project finance. As a small difference in the degradation assumption can lead to a significant difference in the project cash flows over the debt term, lenders financing a PV power plant are concerned about this discrepancy in the developer's and independent engineer's points of view.

Fig. 1 shows the several metrics that can be used to define project financials for a PV project, such as Internal Rate of Return (IRR), Levelized Cost of Electricity (LCOE) and Net Present Value (NPV). Metrics such as IRR and LCOE are highly variable in terms of project cost, tax, power purchase price, etc. Thus, in order to isolate our analysis from the variability of these factors, we will present the impact of degradation rates on solar PV project financials using the NPV of a PV project.

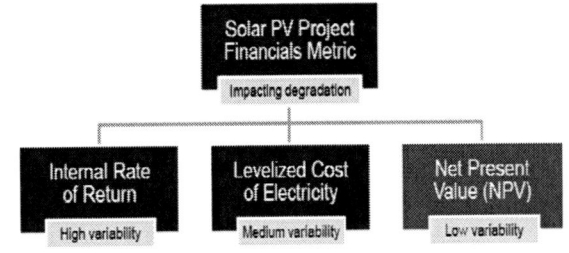

Fig. 1. Metrics defining solar PV project financial value

## II. METHODOLOGY

In order to evaluate the NPV, we calculated annual energy estimates (P50) for a hypothetical 100 megawatt alternating current (MW-AC) project. We used the assumptions provided in Table 1 for crystalline and thin film (cadmium telluride) based projects which were common to all iterations.

Utility scale solar PV projects are usually overbuilt to explore economic benefits associated with the project [3]. The DC to AC ratio is defined as the ratio of array capacity in DC-watts to inverter capacity in AC-watts. We have calculated the

978-1-5090-5606-4/17 $31.00 © 2017 IEEE

P50 values for DC to AC ratios of 1.25, 1.30, 1.35 and 1.40 by varying the DC capacity of the project.

TABLE 1
SUMMARY OF MODELING ASSUMPTIONS

| Assumptions | Crystalline | Thin Film |
|---|---|---|
| Module manufacturer | Trina Solar TSM-320PD14 | First Solar FS-4117A-3 |
| Module capacity | 320 W-DC | 117.5 W-DC |
| Fixed tilt angle | 30° | |
| GCR[1] for fixed tilt system | 44.4% | 57.7% |
| Tracking range | ± 60° | |
| Azimuth | 0° | |
| GCR[1] for tracker based system | 35.7% | 45% |
| Inverter | SMA SC 2200-US | |
| Inverter capacity | 2,000 kW-AC | |
| MV[2] transformer output | 35 kV | |
| HV[3] transformer output | 115 kV | |
| 1. Ground coverage ratio 2. Medium voltage 3. High voltage | | |

Secondly, we selected three locations in the U.S. to model each iteration of varying DC to AC ratio and degradation: a) Mojave, California, b) Colorado Springs, Colorado and c) Raleigh, North Carolina. The three locations are representative of hot and dry, cold and dry and hot and humid conditions, respectively. The reason for choosing these variants was to understand the effect of climate on selection of degradation rates and, in turn, its effect on NPV of the project. The potential P50 energy estimates of each project were modeled using version 6.5.2 of PVsyst, which is widely used by developers, contractors, and consultants to model the annual electricity production of utility scale PV systems. PVsyst was used to calculate hourly AC inverter output based on the clean power resource (CPR) solar resource dataset for each project site, equipment configurations, and loss factors related to module behavior, DC wiring, and array soiling as described in Table 1.

Thirdly, in order to calculate the NPV based on degradation, it was assumed that the degradation per year for both crystalline and thin film modules was linear [1] [2]. This assumption is based on industry-wide standard. Finally, the NPV for each iteration was calculated using (1).

$$NPV \equiv \sum \frac{PCF}{(1+r)^t} \qquad (1)$$

Where, PCF – Project cash flow at time t, and r is the discount rate. We assumed that the project will have a 25-year PPA signed at $55 per megawatt hour of electricity injected into the grid. Overall for the three locations, a total of 668 different permutations of NPV were calculated.

## III. RESULTS AND DISCUSSIONS

In order to evaluate the effect of location on the NPV for the project, we plotted the NPV for both crystalline and thin film technologies with respect to location. Fig. 2 gives the variation of NPV with respect to location for a constant DC to AC ratio of 1.3. An annual degradation rate of 0.75 was assumed for crystalline modules and 0.5 percent was assumed for thin film modules. The degradation values for both technologies were the mean value observed by Jordan et al [1]. It is observed that the highest NPV was obtained for a thin film based tracking system in California. The NPV of a tracking system is approximately 18 percent higher than a fixed tilt system in California, while about 12 percent higher in Colorado and North Carolina. This reassures the fact that the tracking systems are employed at a faster rate in the U.S. than the fixed tilt systems.

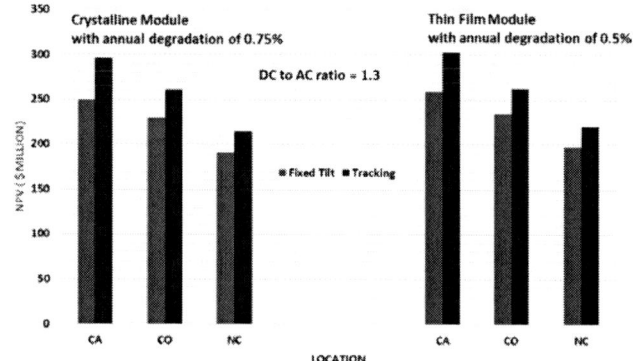

Fig. 2.   Variation of NPV with respect to location

The effect of degradation on the NPV of the project using crystalline modules for varying DC to AC ratios is demonstrated in Fig. 3. Jordan et al [1], observed a 1 percent median degradation for desert climate like California, while lower median degradation of 0.5 percent was observed for snowy and moderate locations like Colorado and North Carolina. Even though the annual degradation is assumed higher (1 percent) for California, the NPV is approximately 8 percent higher than Colorado and 30 percent higher than North Carolina for the crystalline based tracking system. We observed similar results for thin film based tracking systems.

In order to understand the effect of degradation rates on NPV, we conducted sensitivity analysis by varying the DC to AC ratio, discount rates and degradation rates for each location. For California, with fixed tilt systems, the NPV for a thin film based system was 1.1 percent higher than the NPV for a crystalline based system. However, with the tracking system, the NPV for thin film was 0.8 percent lower than the NPV for the crystalline system. This is mainly due to the lower ground coverage ratios for designs with thin film vs crystalline modules resulting in different plane of array gains for the system. For Colorado and North Carolina, for fixed tilt

978-1-5090-5606-4/17 $31.00 © 2017 IEEE          2834

and tracking systems, the crystalline based systems had a higher NPV than thin film based system. Since the ambient temperature in these locations is lower, the power coefficient of temperature for the modules does not significantly affect the energy output from the system. Thus, we are seeing more crystalline modules being implemented in temperate locations as compared to high ambient temperature climates.

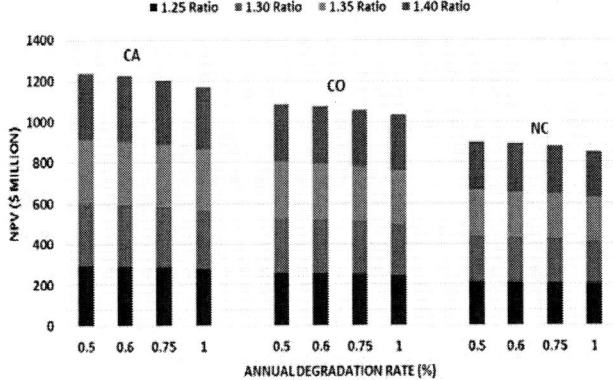

Fig. 3. Variation of NPV with degradation with varying DC to AC ratios

For all three locations, we observed that the NPV is inversely proportional to degradation for varying discount rates. There is significant variation in the NPV of a project with varying degradation rates. For California, the NPV of a project assuming 0.5 percent annual degradation is 5.8 percent higher than assuming 1 percent annual degradation. This number is 4.9 percent and 5.5 percent in Colorado and North Carolina, respectively.

When evaluating the O&M of the different permutations, we assumed the O&M costs of modules with different degradation rates to be identical. Based on this assumption, module degradation evaluation focuses on the electrical production and, therefore, revenue expectations of the project. A project that utilizes modules with high degradation rates will have lower gross revenue over time relative to a project that utilizes modules with low degradation rates and may experience shrinking O&M budgets as the project ages. Shrinking O&M budgets in later years may become increasingly problematic as maintenance costs increase with equipment age.

Higher discount rates value early-year revenues more than later-year revenues. Therefore, projects with higher discount rates are less sensitive to modules with lower degradation.

## III. CONCLUSIONS

The selection of actual (single) value degradation rate is not straightforward. Jordan et al. provide a non-normal distribution of degradation rates for PV modules, with the median degradation rate for crystalline silicon technologies within 0.5-0.6 percent per year range while the mean was in the 0.8-0.9 percent per year range. For thin film based systems, the mean was observed close to 0.5-0.6 percent range [1], [2].

The failure modes and rates affect the variation and absolute value of degradation rates for the PV modules. Geographic location also affects the degradation modes and should be considered while narrowing down the degradation rates. We propose using a Monte Carlo simulation to arrive at a probability distribution of degradation rates to allow stakeholders to make well-informed decisions for a PV project. We propose combining this non-normal probability distribution of degradation with the non-normal probability distribution of energy estimates to provide the stakeholders a better tool for decision making.

Module degradation is particularly important to O&M budgets in later years. As the equipment in a project ages, it becomes increasingly expensive to maintain. A project with high degradation may find the funding of O&M increasingly difficult. A project that initially assumed a low degradation rate in the project financial pro forma may experience increasing budget pressure in later years if the actual degradation is higher than originally modeled.

## ACKNOWLEDGEMENT

The authors wish to acknowledge Clean Power Research (CPR) for providing the typical meteorological year files for the three locations across the U.S. The authors also wish to thank Leidos for providing the necessary support for the work.

## REFERENCES

[1] D. C. Jordan, S. R. Kurtz, K. VanSant, and J. Newmiller, "Compendium of photovoltaic degradation rates," *Progress in Photovoltaics*: Re. Appl. 2016: 24: pp. 978-989.

[2] D. C. Jordan, T. J. Silverman, B. Sekulic ,and S. Kurtz, "PV degradation curves: non-linearity and failure modes." *Progress in Photovoltaics*: Re. Appl. 2016:

[3] V. Sheldon, "Optimizing array-to-inverter power ratio," *SolarPro Magazine*, Issue 7.6, 2014.

# Analysis of wind direction and speed measurements in arid region – a site evaluation using data with low temporal resolution

Elisabeth Klimm, Felix Guischard, Karl-Anders Weiss

Fraunhofer Institute for Solar Energy Systems (ISE), 79110 Freiburg, Germany

*Abstract* — In arid regions, rich in solar irradiance and with dry climate, soiling and abrasion effects on solar material surfaces are most likely. Furthermore they are induced and strongly affected by the present climatic factors, especially wind direction and speed.

The paper analyzes high resolution measured wind data, direction and speed, and possibilities how meaningful low temporal resolution data can be generated to evaluate effects of soiling and abrasion for PV modules.

## I. INTRODUCTION

To measure and analyze ambient conditions of a broadly climatic classified location can help to gain a better understanding how to ensure a more reliable photovoltaic (PV) module design. Extensive climatic monitoring can further give the opportunity to select adequate materials for high quality and efficient solar products used in arid regions. Climatic adapted PV modules and components can prevent expensive over-engineering and still conserve a service life of 25 years and more.

Amongst the climatic monitoring data the wind is, particularly in arid areas, one of the most important factors to be understood in terms of its linking to soiling and abrasion effects. Abrasion can be triggered by strong winds, carrying sand or dust, and therefore occurs most likely in arid regions. Since both effects, soiling and abrasion, can reduce the performance significantly and alter the materials optical properties irreversibly a reduction in efficiency and of the expected life-time is to be expected. Still there are differences in the probability of the occurrence of such performance decreasing effects in dependency of the weather conditions and the actual location. To determine the differences between an arid climate with and one without such strong wind events, it is necessary to look into the temporal distribution and resolution of the two parameters wind speed and wind direction. This study analyzes measured wind parameters of the Negev Desert. It investigates if a decrease of the temporal resolution of climatic data is usable and to which extent these parameters can be down-sized and are then still well suited to enable a good overview on the local conditions, enabling an appropriate comparison amongst arid locations. Basis of this study are data with a high one minute resolution. The resulting data may also allow the development of adapted tests and testing sequences for meaningful test results [1].

## II. STATISTICAL ANALYSIS OF WHAT HAS BEEN FOUND WHILE LOOKING AT THE CLIMATIC – WIND – DATA?

The 44640 values per month for each parameter, wind direction and wind speed, due to the one minute temporal resolution are analyzed for pattern and occurrence. Then we take the approach to down-size these data in order to compare a set of parameters of each month. The maximal, minimal and main values of the parameter wind speed and main wind direction are investigated.

### A. High temporal resolution

Of the high temporal measured climatic data from the year 2015 at the arid region Negev Desert are following two climatic factors wind speeds and wind direction chosen for investigation. By the example of the month December are patterns and occurrence of the parameter examined. So are approximately 90 % of the whole wind speed between 0.3 m/s and 3 m/s. In Fig. 1 are the wind direction in [°] on the z-scale and the wind speed in [m/s] on the y-scale both in high temporal resolution of 1-minute values. It shows a pattern of parameter repetition in the diurnal cycle. The lower wind speeds in blue color occur generally during night, see sections mid at the x-axis. The maximum wind speeds show their peaks at middle of the day. The given periods data shows a maximum of 6 m/s in red color, however also in December there are scarce occasion from wind speeds up to 10 m/s. The average wind direction is north-west.

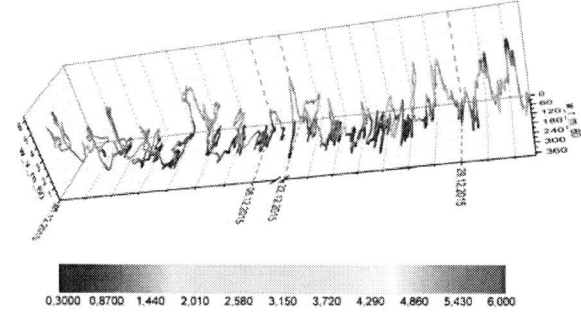

Fig. 1. Excerpt of the high temporal resolution wind data of December 2015. X-axis shows the time with 1 minute resolution, y-axis the wind speed and Z-axis wind direction at angeles 0° to 360°.

Fig. 2 shows the distribution of all 1-minute data in December 2015. The wind speed is plotted in correlation of the direction in [°] for one month. 360 ° means wind direction north. As mentioned before occur the highest speeds (> 6 m/s) at the north-west wind direction. Table 1 and Figure 4 further proof, that north-west is the predominant wind direction.

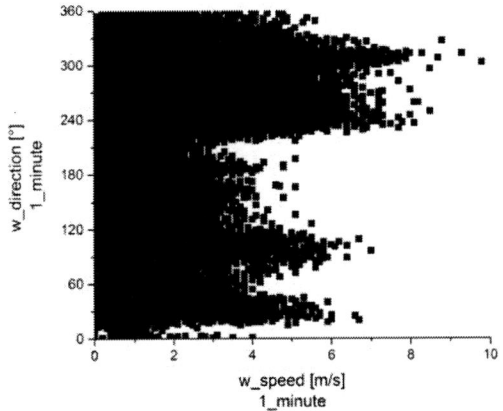

Fig. 2. Wind direction against wind speed in Dec 2015 measured in the Negev Desert. The temporal resolution is one minute.

*B. Low temporal resolution*

Recorded data are reduced to one daily set in order determine the predominant wind speed v [m/s] and wind direction wd [°]. In Table I are the reduced values for each month between June 2015 and June 2016 of all measured wind parameters shown as main wind direction and with 3 values for wind speed: minimal, maximal and main wind speed.

The values named 'main' are calculated by the modal functions and show the wind direction and speed which occurs with the highest number of single 1 minute values in the respective month. The maximum wind speeds do evenly occur with a probability of only around 0.002 % among all data. The probability of the minimum wind speed (*approximated to 0.0 and 0.1 m/s) are, with values between 0.002 and 0.3 % more common and far more inhomogeneous. Since 0.1 % means 40 1-minute values it is obvious that these extreme events are rather rare.

The question is about the significance of such minimal and maximal extrema and how important are they for the soiling and abrasion effects on PV module technology.

But before starting to answer these questions, we look at the occurrence of the main wind speed. Here we find a probability between 2 and 5 % for exact values given in Table 1. These values are also plotted in Figure 3.

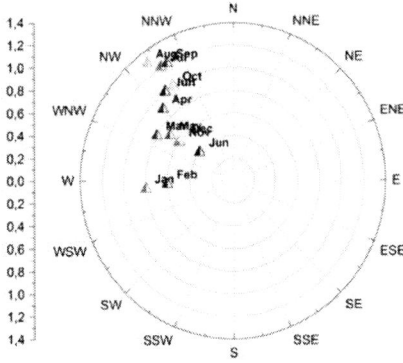

Fig. 3. Monthly main wind speed correlated to main wind direction in 2015.

TABLE I

WIND DATA – 2015 AND 2016 – NEGEV DESERT

| Month | Wind Direction [°] MAIN | Wind Speed [m/s] / | | |
| --- | --- | --- | --- | --- |
| | | MIN* | MAX | MAIN |
| Jun | 322 | 0.0 | 10.6 | 1,0 |
| Jul | 327 | 0.1 | 10.4 | 1,2 |
| Aug | 323 | 0.0 | 9.8 | 1,3 |
| Sep | 330 | 0.0 | 10.3 | 1,2 |
| Oct | 327 | 0.1 | 12.4 | 1,0 |
| Nov | 305 | 0.0 | 9.3 | 0,6 |
| Dec | 309 | 0.0 | 10.6 | 0,6 |
| Jan | 265 | 0.1 | 12.6 | 0,8 |
| Feb | 267 | 0.0 | 12.6 | 0,6 |
| Mar | 300 | 0.0 | 13.2 | 0,8 |
| Apr | 315 | 0.0 | 11.6 | 0,9 |
| May | 305 | 0.1 | 12.6 | 0,7 |
| Jun | 309 | 0.1 | 12.4 | 0,4 |

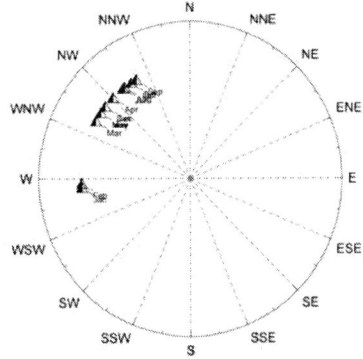

Fig. 4. Normalized monthly main wind direction in 2015.

Then only the wind direction is plotted in Fig. 4 in normalized way. It indicates that the main wind direction is north-west. There are two values, for January and February, indicating that during the winter season the wind changes the direction slightly to west.

## III. RESULTS

### A. Results of the Israelian weather data

Looking at the wind data it has been found that there are daily cycles for the wind speed. The main wind speed values are not explicitly high, usually in the range of 1 m/s. Monthly wind speed maxima are in the range of 10 m/s. There is also a predominant wind direction of north-west. Both indicate the possibility to reduce the necessary amount of data for a sufficient qualification of the site.

### B. What significance can the weather parameter Wind, as wind speed and direction, have on abrasion (and soiling)?

Now it is obvious, that during day time, at the hottest time of the day the wind speed picks up the pace, whereas when the ambient temperature drops the wind also slows down, so to see after midnight. It is also obvious that the times with low and very low wind speeds have a minor effect on abrasive effects since the wind will not pick up and transport particles under these conditions. When during night time the wind is very low the sand and dust has time to settle onto the surface. Soiling can occur. Is it known, that against the night sky the surfaces cools down more likely below the dew point, leading to condensation. The slowly settling dust, having time to build dust-water-mixtures under calm wind conditions, may now even worsen the clogging of the surface.

### C. Advantages for the use of aggregated data

The reported statistical approach shows the meaningfulness and usefulness of the aggregated data. Since high resolution climatic data sets are very elaborative and also costly and therefor rare, it is especially an issue for remote sites, e.g. in the desert [2]. Desert regions are of major interest for the installation of PV systems.
This approach builds a connection between the high resolution data sets from scientific research sites with low resolution data from application-oriented PV installation sites. Still it will remain necessary to work with the best available resolution to generate validated aggregated data sets.

## IV. CONCLUSION

A procedure was developed to reduce big data sets to predominant and meaningful data, also available from locations without high end monitoring. This procedure enables the comparison of different locations.

Now, it is extremely important for the estimation of soiling and abrasion risk for these specific locations on the bases of low resolution climatic data, like e.g. monthly maximum and main values. Therefore we found a significant importance of maximum, whereas the min values are currently of less interest for the approach of minimizing temporal data.

It is also expected that high maximum wind speeds are accompanied by more or longer events of high wind speeds. Therefore the approach is to see the monthly maximum wind speed as indicator for the abrasive load.

## V. OUTLOOK

The presented approach will also be used to analyze wind data of other exposure sites, e.g. Canarias Islands with dry maritime climate, which are equipped with the same measurement technology and also with similar material samples and PV modules. The results will be used to compare the soiling and abrasion effects of the sites and analyze correlations with the developed low resolution wind data.

This data will further be used to relate experimental abrasion and soiling results from laboratory tests to outdoor exposure sites so help select appropriate materials for locations where only low resolution climatic data is available.

It is to determine the location and the general climatic classification of the Negev desert compared to a "normal, standard" dust storm location.

## REFERENCES

[1] M. Koehl, M. Heck and S. Wiesmeier, "Evaluation of the accelerated life testing conditions for PV-modules based on measured and simulated weathering stress" *37th IEEE Photovoltaic Specialists Conference*, June 2011 pp.003166-003169

[2] S. Wiesmeier, C. Schill, K.-A. Weiss and M. Koehl, "Web-basierte Analyse und Verwaltung von Klima-, Mess- und Metadaten aus Outdoor Monitoring und Laborprüfungen von solaren Energiesystemen" *Umwelteinflüsse erfassen, simulieren, bewerten*, Jahrestagung der GUS, 42, 2013, pp.33-40

# Forecasting Environmental Degradation Power Loss in Solar Panels with a Predictive Crack Opening Test

Jason L. Lincoln[1], Andrew M. Gabor[2], Eric J. Schneller[1], Hubert Seigneur[1], Joseph Walters[1], Rob Janoch[2], Andrew Anselmo[2], Victor Huayamave[3] and Winston Schoenfeld[1].

[1]Florida Solar Energy Center, University of Central Florida, Cocoa FL, USA
[2]BrightSpot Automation LLC, Westford, MA, USA
[3]Department of Mechanical Engineering, Embry-Riddle Aeronautical University, Daytona Beach, FL, USA

*Abstract* — Manufacturing, shipping & handling, installation, and in-field loading of photovoltaic solar panels are common contributors to the creation of cracks within the cells of a panel. Many cracks initially cause little or no power loss in the panel, but such tightly closed cracks may open over time due to environmental forces, and cause significant power loss and even failure of the module. We developed a method, using the *LoadSpot* tool, to apply a mechanical load to a panel to temporarily open pre-existing cracks while also allowing for electroluminescence (EL) imaging and flash IV testing. The change in the IV and EL measurements upon loading provides a quantifiable metric that can be used to evaluate reliability and durability. Such Predictive Crack Opening (PCO) tests have value in assessing preexisting damage as well as in the correlation with the degradation due to cracked cells opening upon environmental chamber and cyclic loading. We performed finite element modeling and simulation to illustrate the stresses applied at different load and mounting conditions. We demonstrate a wide range of mechanical loading and stress testing with accompanying EL and IV measurements which not only show the narrative of damage and power loss through static mechanical load, environmental chamber testing, and cyclic loading, but also suggests potential improvements which can be made to the order of chamber and cyclic load testing within the IEC 61215 standard. next.

*Index Terms* — accelerated aging, cell fracture, electroluminescence, finite element analysis, photovoltaic modules, silicon.

## I. INTRODUCTION

Cracked solar cells are commonplace in crystalline silicon based solar panels, and NREL has assigned "degradation related to fractured cells" as the 3rd most important degradation mechanism in newer generations of panels which have thin, easily fractured cells [1]. The 2nd most important mechanism is due to hot spots which may be linked to broken cells forced into reverse bias. Anecdotal evidence abounds concerning panels in the field that have been heavily damaged due to cracked cells, although the origin of the cracking is often not clear. The problem of assessing the risk posed by cracked cells is challenging since most cracks are tightly closed prior to field exposure, with continuity of the metallization across the cracks, and little or no power degradation. It can be quite challenging to even detect tightly closed cracks using the standard electroluminescence (EL) images of the panels taken with low-resolution (~1 Megapixels) scientific EL cameras.

Extended exposure of installed solar panels to environmental conditions, including wind and snow load and thermal cycling, is known to cause cracks in silicon solar cells, and such cracks have been shown to cause significant power loss [2].

To reduce the risk posed by cracked cells, the industry could 1) shift to cell/panel designs that are less sensitive to cracking [3], 2) develop quick, non-destructive tests that quantify how the cracks may affect power degradation in field, and 3) prove that such tests correlate well to actual degradation seen during accepted environmental chamber tests and/or actual field exposure. We previously suggested a predictive crack opening (PCO) test using the *LoadSpot* mechanical load testing tool [4], and discuss various consideration in other recent works [5]. In this work, we add a finite element modeling component to help inform the PCO test definition, delve deeper into the possibilities of predictive crack opening, present data on the power loss response of modules during and after stress testing, and investigate the impact of environmental chamber degradation before and after mechanical loading.

The *LoadSpot* tool is capable of using negative pressure (*e.g.* front side load) to induce a mechanical load on a solar module while allowing the module to undergo electroluminescence (EL) and I-V characterization. A unique image can then be taken showing the previously closed cracks, now more open to varying degrees, without stressing the module to the point of creating new cracks. Additionally, a snapshot of the degradation in the I-V curve can be extracted while loaded, to show the predicted power loss due to the eventual opening of these cracks. Furthermore, we can use these snapshots to capture the narrative of power loss due to cracking through various load conditions and stress testing.

By applying a static mechanical load of 2400 Pa, as is done in the IEC 61215 standard, we may generate new cracks which might be caused in-field via snow/wind load on the front surface of the panel. Then, we subject these modules to a series of thermal cycling and humidity-freeze tests, followed by cyclic loading tests, which can permanently open those cracks which were created via mechanical load. Additionally, we can show what impact cyclic loading can have on solar panels which were weakened by environmental chamber testing. The loaded and unloaded EL and IV measurements at each major step are then compared to describe the predictive qualities of the *LoadSpot* tool, as well as its ability to assist in the optimization of module technologies by amplifying the changes in the IV and EL signals used for the characterization of mechanical reliability.

## II. MODELING OF LOADSPOT TEST

The *LoadSpot* tool is capable of loading panels while clamping at the usual two points along the long edges (four points total) as is required for IEC static and cyclic load testing. Although four-point support is more representative of field conditions, the industry may be better served by a PCO test that is applicable to a simpler tool design. Multiple groups [6] have explored full perimeter support for loading under vacuum from the rear side, and we model this approach here.

We consider four loading conditions for stress modeling in the Abaqus software. The simulations are split into two categories, perimeter support and four-point support. Each category shows a load of 800Pa, the load at which cracks are opened, but not formed if not weakened by environmental chamber testing [6], and 2400Pa, the IEC standard, known to cause cracks.

The simulations all use a uniform distributed load on the front surface of the panel. Panel dimensions are 1.5 m x 1 m. A sheet of silicon is used to simulate the layer of solar cells in the panel in order to simplify the simulation as well as to allow for analysis of the stresses at this layer. The encapsulation structure includes a 3.2 mm thick soda-lime float glass sheet followed by 0.2 mm of EVA, the 0.2 mm silicon sheet, another 0.2 mm layer of EVA and finally enclosed by a 0.325 mm backsheet. The encapsulation is surrounded by a standard aluminum frame. An encastre constraint, allowing for zero degrees of freedom, is used at four points, following typical clamp locations, for group 1, and surrounding the perimeter of the frame for group 2. These setup configurations can be seen in Fig. 1.

Because of the brittle nature of silicon, it is most relevant to consider the first principal stress as an indicator for crack creation and propagation. We generate the first principal stress profile at the silicon layer of the panel where the solar cells lie for each configuration. The results can be seen in Fig. 2.

Comparing each configuration shows an interesting narrative as load conditions are changed. At the lower load level in Fig. 2 a) and b) the differences in first principal stress area, magnitude, and shape are small. At 2400 Pa, the differences are quite large, as can be seen in Fig. 2 c) and d). This is caused by deformation of the aluminum frame at higher load levels, which is restricted in a full perimeter support.

Therefore, in performing the predictive crack test at 800 Pa, perimeter support will be used in order to more closely conform with other pressure induced loading tests on simpler equipment, such that results in this test might contribute to a formal testing standard. Note however, when performing static loading at higher pressures with the aim of creating new cracks, it is important to use the four-point support configuration so that the stresses are representative of what may be experienced by the module in the field.

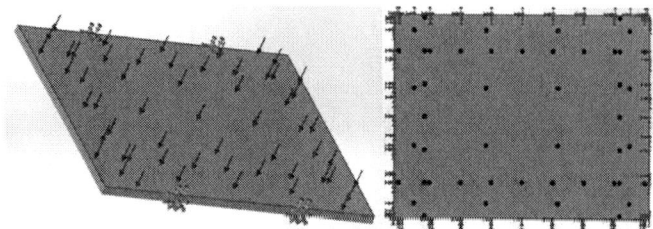

Fig. 1. Simulation setup display for both groups. Red arrows represent the uniform load, whereas green arrows represent encastre constraints. The left image shows the four-point constraint at clamp locations, while the right shows full perimeter support.

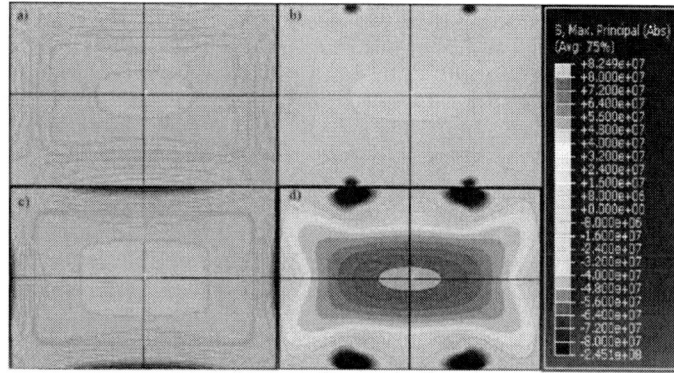

Fig. 2. First principal stress profile at the silicon layer for a) Perimeter supported panel loaded at 800 Pa, b) Four-point supported panel loaded at 800 Pa, c) Perimeter supported panel loaded at 2400 Pa, and d) Four-point supported panel loaded at 2400 Pa. Black and grey regions indicate areas of very high compressive and tensile stress, respectively.

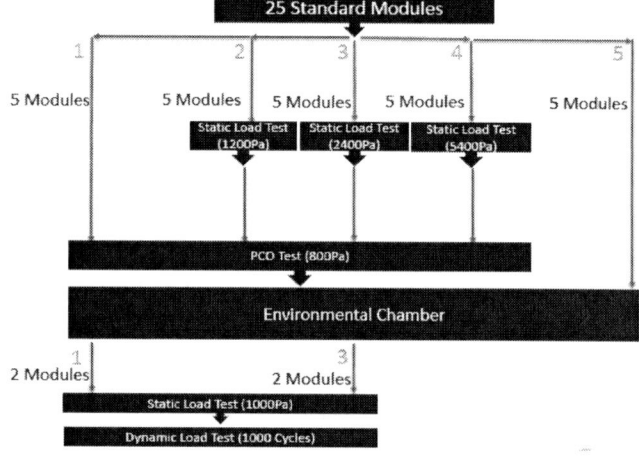

Fig. 3.. Experimental Plan – Five groups of five modules, which underwent various loading conditions prior to an environmental chamber test. Additional modules from groups 1 and 3 were subjected to additional static and cyclic loading after environmental chamber testing.

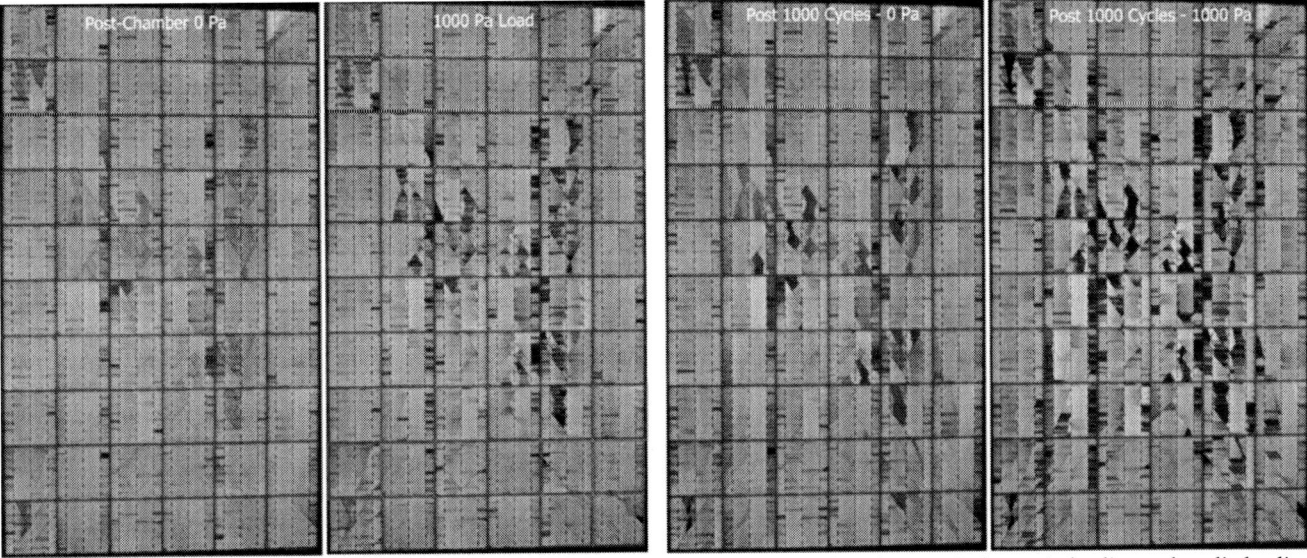

Fig. 4. Electroluminescence images of a module from group 3 (2400 Pa Static Load), which underwent further static loading and cyclic loading after environmental chamber testing. From left to right: 1) Unloaded module post environmental chamber. 2) Module after chamber testing, with a 1000 Pa load applied. Cells which had new cracks form during the 1000 Pa static load test have been highlighted. 3) Module image after 1000 cyclic load cycles, taken at 0 Pa. 4) Module after 1000 cyclic load cycles, taken at 1000 Pa.

## III. EXPERIMENT

The full experiment includes the testing of twenty-five standard mono-silicon 60-cell panels, shown in Fig. 3. The panels are split into five groups, which each include five panels. Initial IV and EL measurements are taken of each panel before any further testing. Group 1 undergoes an 800 Pa predictive crack opening test, with EL and IV measurements taken both during and after loading. Groups 2, 3, and 4 are subject to static mechanical loading on the *LoadSpot* to create varying levels of cracks using standard 4-point support at 1200 Pa, 2400 Pa, and 5400 Pa, respectively, before receiving the PCO test, which shows how newly formed cracks from mechanical loading can be detected at lower load levels. Group 5 is kept as the control and does not undergo any mechanical loading. Each group is then subject to fifty thermal cycles and ten humidity-freeze cycles (TC/HF) as has been recommended by PVQAT [7] in an environmental chamber to further open the cracks. Subsequently, four modules, two from Groups 1 and 3, were subjected to further mechanical loading in the form of a 1000 Pa load, followed by a cyclic load sequence of 1000 cycles, from 1000 to -1000 Pa. EL and IV measurements were taken at 0 and 1000 Pa for each step.

## IV. RESULTS AND DISCUSSION

### A. Crack generation and damage to modules

The snapshots of electroluminescence and IV characteristics throughout the experiment tell a story which is uniquely enhanced by the *LoadSpot* tool. Imaging the modules as we increase the static mechanical load shows that cracks can generate or propagate as early as 800 Pa. This cracking gradually gets worse as you increase mechanical load, up to the complete shattering of cells, seen at 5400 Pa. Additionally, EL images provide clear imagery which show how previously created cracks can be re-opened at lower load levels than was used to create those same cracks [6]. However, before chamber testing, the opening of cracks at lighter loads, as in the PCO test, does not provide a large signal of power loss, such that alternative failure modes may overshadow those results and obscure the role of cracking in module degradation. For instance, in this experiment, most modules tested were influenced by a disconnection of fingers, which created large sums of dark areas over the surface of the cells. This proved to show the PCO test before environmental degradation as a situational tool for module characterization when other modes of degradation are lesser in their role of power loss, and it is inconclusive how well it may perform in other circumstances as a predictive tool in this stage.

Fig. 5. shows the change in max power as four modules traversed the various stages of the experiment. Compared to just static loading, with the measurement taken at 2400 Pa, environmental chamber testing proved to cause significantly more power loss. However, those modules which saw higher mechanical load did not see much greater power loss after chamber testing than those which only underwent the PCO test.

Furthermore, it is the mechanical loading after chamber testing which has the greatest implications to our understanding of module reliability testing and power loss due to cracking. Fig. 4. shows the EL images taken at each major step after chamber testing. We can see many new cracks forming on the cells after only a load of 1000 Pa, far below the load applied in the original static mechanical load test, prior to chamber testing.

978-1-5090-5606-4/17 $31.00 © 2017 IEEE

This implies that TC/HF sequence causes a mode of weakening in the solar panel, which degrades the mechanical reliability of the encapsulated cells. As the common testing procedure stands, mechanical loading is frequently performed only before chamber testing, with the expectation of the environmental degradation to open the cracks created via mechanical load. The implication follows that because of the nature of in-field conditions, which would apply mechanical loading before, during, and after thermal cycling, it may be wise to implement tests that address this new mode of failure.

Finally, in Fig. 5, the power loss seen at 1000 Pa is near identical to that seen at 0 Pa after 1000 cycles. Although this investigation was rather limited in its scope, this data suggests that a PCO test taken to 1000 Pa may serve as a quick replacement to the slower standard cyclic load test. This result can be further reinforced when reviewing the second and third module images in Fig. 4, which have remarkably similar crack opening patterns, as illustrated by the dark regions.

### B. The LoadSpot as a tool for module optimization

By taking a measurement of the power at each step in the cyclic load test, both at 0 and 1000 Pa, we can not only get an idea of how power degrades over the course of the test, but we also get a glimpse into how else the *LoadSpot* might be used for module design and optimization. Seen in Fig. 6, the measurement of power at 1000 Pa is much less than the measurement taken at the same number of cycles at 0 Pa. In addition, the difference between the measurements taken at 0 and 1000 Pa gets larger as the number of load cycles increases. As shown previously, the signal from cracking due to mechanical load can be smaller and difficult to differentiate from other degradation mechanisms. Therefore, the *LoadSpot* test offers an opportunity to greatly increase this signal, which can lead to finer tuning of module design to improve the reliability of modules with respect to cracking in solar cells.

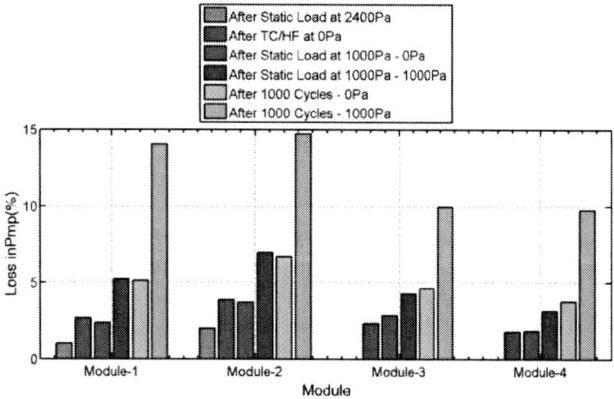

Fig. 5. Comparison of the loss in $P_{mp}$, in terms of loss percentage from initial performance, for four modules which underwent additional mechanical loading after environmental chamber testing. Module 1-2 are from Group 3 (2400 Pa static load test group), and Module 3-4 are from Group 1 (PCO only group). Each bar shows the change in $P_{mp}$ from each major IV snapshot, compared to the original $P_{mp}$.

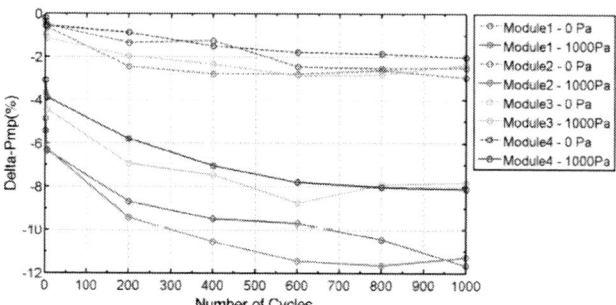

Fig. 6. Plot showing the change in $P_{mp}$, in terms of loss percentage from the initial unloaded state, over the course of the cyclic load test for four modules. Data is presented from measurements taken in the unloaded state (0Pa) and unloaded state(1000Pa) for each interval. The first data point shows the change in $P_{mp}$ after one cycle.

### V. CONCLUSION

The momentum within the PV durability testing community entails a shift to a testing sequence involving cyclic loading to generate cracks, followed by environmental chamber testing to "open up" these cracks. The data we presented here on a single module type questions this approach as the environmental chamber exposure appears to have weakened these modules such that subsequent cyclic loading caused extensive damage and power loss. Such a sequence represents real world conditions where wind and snow loading occur after years of climate exposure, and is thus of concern. Future work will attempt to understand the root causes of the climate-chamber induced sensitivity to load testing, and explore whether this sensitivity extends to other module designs from different manufacturers. If this sensitivity is seen to be commonplace and result in more damage than the reverse sequence, then modification of the testing standards may be prudent.

We have demonstrated both cyclic and static loading of solar panels with the *LoadSpot* tool where IV and EL data can be collected in both the unloaded and loaded states to clearly show crack formation and the evolution of cracks from closed to open states. In general, the solar panels tested are remarkably resilient to power loss even after extensive cell cracking. With such small changes in module power, it is difficult to optimize module designs, materials, and processing based on the changes in IV data after accelerated testing, and instead must usually lean more heavily on the EL data. The measurement of module power during application of small loads to prop open otherwise closed cracks amplifies the change in IV data to allow better optimization using the IV data. We found such a predictive crack opening test to correlate well to the degradation seen after subsequent cyclic loading of the modules weakened by chamber testing.

### REFERENCES

[1] D. C. Jordan, T. J. Silverman, B. Sekulic, and S. R. Kurtz, "PV degradation curves: non-linearities and failure modes," *Progress in Photovoltaics: Research and Applications*, 2016.

[2] M Köntges, I Kunze, S Kajari-Schröder, X Breitenmoser, B Bjørneklett. "The risk of power loss in crystalline silicon based photovoltaic modules due to micro-cracks" *Solar Energy Materials and Solar Cells* 95 (4), pp. 1131-1137, 2011

[3] A. M. Gabor, R. Janoch, A. Anselmo, H. Field. "Solar panel design factors to reduce the impact of cracked cells and the tendency for crack propagation" *NREL PV Module Reliability Workshop*, Golden, CO, USA 2015

[4] A. M. Gabor, R. Janoch, A. Anselmo, J. L. Lincoln, H. Seigneur, and C. Honeker, "Mechanical load testing of solar panels - Beyond certification testing," in *2016 IEEE 43rd PVSC*, 2016, pp. 3574-3579.

[5] E. J. Schneller, A. M. Gabor, J. L. Lincoln, R. Janoch, A. Anselmo, J. Walters, and H. Seigneur, "Evaluating Solar Cell Fracture as a Function of Module Mechanical Loading Conditions" *IEEE 43rd PVSC*, 2017

[6] C. Buerhop, S. Wirsching, S. Gehre, T. Pickel, T. Winkler, A. Bemm, J. Mergheim, C. Camus, J. Hauch, and C. J. Brabec, "Lifetime and Degradation of Pre-damaged PV-Modules – Field study and lab testing" *IEEE 44th PVSC*, 2017

[7] Kurtz, S., Sample, T., Wohlgemuth, J., Zhou, W., Bosco, N., Althaus, J., Phillips, N., Deceglie, M., Flueckiger, C., Hacke, P. and Miller, D., "Moving toward quantifying reliability-the next step in a rapidly maturing PV industry" In *42nd IEEE Photovoltaic Specialist Conference (PVSC), (2015)*, p.5

# Fluorescence imaging on the cross-section of photovoltaic laminates aged under different UV intensities

Yadong Lyu, Jae Hyun Kim, Xiaohong Gu

Engineering Laboratory, National Institute of Standards and Technology, Gaithersburg, MD 20899, USA

*Abstract* — Fluorescence imaging is used to investigate the degradation depth-profile of multilayer glass/EVA/backsheet laminates for photovoltaic (PV) applications. Samples were aged on the NIST SPHERE (Simulated Photodegradation via High Energy Radiant Exposure) at 85 °C ± 0.5 °C and 0 % RH ± 5 % RH under different UV light intensities. Based on fluorescence imaging results, non-uniform distribution of degradation species along the thickness direction is observed for EVA encapsulant and EVA middle layer in the backsheet for all exposed specimens. Two adhesive layers in backsheet also show fluorescence emission, implying the degradation of these adhesives and the propensity of the delamination within backsheet layers. It is also found that the fluorescence results are consistent with the changes in the yellowing index by micro-UV-Vis spectroscopy and the modulus by atomic force microscopy-based quantitative nanomechanical mapping (QNM-AFM). This study shows that fluorescence imaging is effective in the characterization of degradation gradient of packaging materials for PV as a function of depth.

*Index Terms* — fluorescence imaging, photovoltaic laminates, UV aging, UV light intensity.

## I. INTRODUCTION

Packaging materials for photovoltaics (PV), including encapsulant and backsheet (BS), play a critical role to guarantee that PV modules are able to survive in harsh operating environments, and provide optical coupling, mechanical support, and electrical insulation. However, these polymeric materials are susceptible to degrade due to UV exposure and invasion of water and oxygen. Common failure modes observed in the field include discoloration, delamination and cracking [1]; these could potentially impair the performance and safety of PV modules. It is important to understand the durability and degradation behaviors of these materials for a better prediction of service life and improved design for PV modules.

In view of the complicated multilayer and multicomponent construction of PV modules, it is challenging to quantify the structure and property degradation of individual components or layers during their service lifetime. Several spectroscopic imaging techniques, such as attenuated total reflection Fourier transformed infrared (ATR-FTIR) imaging and Raman imaging have been used to examine the depth profile of PV multilayered systems after degradation [2, 3]. Compared with ATR-FTIR spectroscopy, Raman spectroscopy has superior lateral resolution and is more suitable to analyze thin layers [3]. However, significantly enhanced fluorescence background emitted from aged samples could dominate Raman spectra and mask the spectroscopic vibrational information. In fact, fluorescence emission can be also used to evaluate polymer degradation [4, 5], as it is sensitive to the formed conjugated products. Nevertheless, fluorescence imaging on the cross-section of degraded PV multilayer system has been rarely reported so far. Investigation on the correlation between fluorescence intensity and properties such as yellowing and modulus of packaging material in PV is also limited [6].

In this work, depth-profile analysis on PV laminates aged under different UV intensities were conducted. A glass/EVA/backsheet laminate was prepared and aged on the NIST SPHERE [7] at 85 °C/0 % RH with UV for 3840 h. The maximum UV dose is approximately equivalent to that of 6-year exposure in Miami, Florida [8]. Laser scanning confocal microscopy (LSCM) was used to evaluate fluorescence profile on the cross-sections of glass/EVA/BS. In addition, the micro-UV-Vis spectroscopy and the atomic force microscopy-based quantitative nanomechanical mapping (QNM-AFM) were carried out on the same sample. The correlation between fluorescence profile and the changes of yellowing index and modulus as function of depth from the exposed surface was investigated.

## II. MATERIALS AND METHODS

### A. Materials and Sample Preparation

The commercial glass/EVA/EVA/polymer backsheet (labeled as PPE) laminate with a size of 180 mm × 180 mm obtained from NIST-industry PV Consortium was used as received. The glass was a two side-polished fused silica with transmittance of around 95 %. EVA encapsulant layer contained classical commercial stabilization system (Tinuvin 770 and Cyasorb 531) and curing agent lupersol® TBEC (tertbutyl peroxy 2-ethyl-hexyl carbonate). The vinyl acetate content of EVA encapsulant is about 33 mass. %. More details for the materials cannot be accessed due to proprietary information. The laminates were prepared using vacuum lamination by a commercial processing procedure with the platen temperature of 150 °C, evacuation time of 4 min, then pressing time of 1 min and crosslinking time of 13 min.

### B. Laboratory Accelerating Aging

Accelerated aging was performed with the glass side of the laminates faced to the UV light on the NIST SPHERE at 85 °C/0 % RH under different UV light intensities for about 3840 h. Different types of neutral density filters with nominal transmittance of 40 %, 60% and 100% were used to control the UV light irradiance and were placed in front of sample during exposure. UV irradiance (300 nm – 400 nm) of 60.5 W/m$^2$ (40%), 93.5 W/m$^2$ (60%) and 142.5 W/m$^2$ (100%) were obtained. To conduct cross-sectional analysis, 1 mm x 5 mm strips under each exposure condition were cut first from the large laminate after exposure using a diamond saw. Then EVA/PPE backsheet part was carefully peeled off from the glass, with minimum polymers remained on the glass. To achieve a smooth surface, cryo-microtoming was performed along the thickness direction of the EVA/PPE backsheet sample using diamond blade at -80°C. Thin films of 2 μm thick of EVA/PPE backsheet sample were cut and collected along the cross-section direction for micro-FTIR and micro-UV-Vis microscopic study.

### C. Characterization methods

*Micro-FTIR measurement* Infrared analyses of each layer in EVA/PPE backsheet sample were performed by infrared imaging microscope purged by dry air and equipped with a liquid nitrogen-cooled mercury-cadmium-telluride (MCT) detector. Spectra were collected by operating in transmission mode with an aperture size of 10 μm x 50 μm using a Micro Compression Cell with Diamond Windows. Spectra were recorded in the spectral range from 4000 cm$^{-1}$ to 650 cm$^{-1}$ at 4 cm$^{-1}$ resolution with 64 scans. To ensure the reproducibility, three different locations were inspected for each layer.

*Fluorescence measurements* Morphology and fluorescence imaging were conducted on the Zeiss LSM510 Meta (Carl Zeiss, Inc., Oberkochen, Germany) laser scanning confocal microscope. For cross-section morphology observation, an objective of the Epiplan-Neofluar 10×/0.3 was used with field view of about 840 μm × 840 μm and pixel size of 512 pixels by 512 pixels using laser wavelength of 543 nm in reflection mode. The lateral resolution of this objective with 543 nm laser is about 0.9 μm for pinhole size at 1 Airy unit. For fluorescence measurements, an objective of 5x/0.15 was used with diode laser (405 nm) to excite the sample. Scanning area is about 1700 μm × 1700 μm and the lateral resolution of this objective at 405 nm is about 1.4 μm. Both channel mode and lambda mode were used to detect emission signals with long pass (LP) 420 filter, and monitor emission spectra from 417 nm to 748 nm with 10.7 nm increment and perform spectra imaging with the META detector, respectively. During measurement, the sample was adjusted to ensure the cross-section is perpendicular to the microscopic optical axis.

*UV-Vis-NIR microspectrophometry (MSP)* UV-Vis spectra were recorded using a microspectrophotometer (CRAIC Technologies, Model MSP 121) in the UV-Vis range between 200 nm and 800 nm across the sample in transmission mode.

The microscope objective is a Davin reflecting objective 36x, NA 0.5 with aperture size of 7.9 μm × 7.9 μm. Sectioned films with thickness of 2 μm was mounted between quartz microscope slides and cover slips with glycerin as a medium.

*Quantitative nanomechanical mapping atomic force microscopy (QNM-AFM) measurements* Local modulus of EVA encapsulant was measured by the Bruker Dimension Icon AFM on the Peak Force QNM mode. Bruker AFM probe of Tap150 was used as received. This probe was selected based on the Young's moduli of EVA encapsulant. Modulus was obtained following the relative method given in the Peakforce QNM user guide [9] based on the Derjaguin-Muller-Toporov (DMT) model. This model considers the adhesion force and is applicable when the deformation of sample is lower than the radius of tip. The PDMS-SOFT02012M with nominal modulus of 3.5 MPa was used as reference sample. A (512 × 512) pixels resolution with scan rate of 0.5 Hz was performed at ambient condition.

*\*\* Certain commercial products or equipment are described in this paper to specify adequately the experimental procedure. In no case does such identification imply recommendation or endorsement by the National Institute of Standards and Technology, nor does it imply that it is necessarily the best available for the purpose.*

## III. RESULTS AND DISCUSSION

### A. Characterization of the EVA/PPE sample

Laser scanning confocal microscopy (LSCM) image of the cross-sectional morphology of fresh EVA/PPE backsheet sample is shown in Fig. 1. The thickness of the EVA encapsulant layer is (920 ± 6) μm, and multiple layers of PPE backsheet are detected with a total thickness of (300 ± 4) μm. The error represents the standard deviation obtained based on measurements at three locations on the cross-section of samples. The micro-FTIR spectra of these various layers are exhibited in Fig. 2. The characteristic peaks of EVA [10], including the asymmetrical and symmetrical stretching vibration of -CH$_2$- at 2922 cm$^{-1}$ and 2852 cm$^{-1}$, the C=O stretching at 1737 cm$^{-1}$, and C-O-C stretching vibration at 1242 cm$^{-1}$, are observed in the spectra of the EVA encapsulant and the three EVA layers (outer, middle and inner) in the PPE backsheet. The EVA encapsulant has a higher vinyl acetate amount than the backsheet EVA layers based on the relative intensity of the C=O stretching at 1737 cm$^{-1}$ with regard of -CH$_2$- stretching at 2922 cm$^{-1}$. Other typical absorption bands such as 1465 cm$^{-1}$ related to deformation vibration of -CH$_2$- and 1371 cm$^{-1}$ to deformation vibration of –CH$_3$ also appear in the spectra of these EVA layers. Meanwhile, the characteristic vibration absorption for PET, such as the aromatic skeletal stretching bands at 1410 cm$^{-1}$, 1450 cm$^{-1}$, 1505 cm$^{-1}$ and 1580 cm$^{-1}$, and the ester group absorption bands at 1723 cm$^{-1}$,

1268 cm$^{-1}$, 1100 cm$^{-1}$ and 1020 cm$^{-1}$ are observed in the micro-FTIR spectra of PET outer layer and PET core layer. Although our previous Raman results [11] have indicated that the PET outer layer is pigmented with BaSO4, but not for the PET core layer, the polymers essentially have similar chemical structures in these two layers.

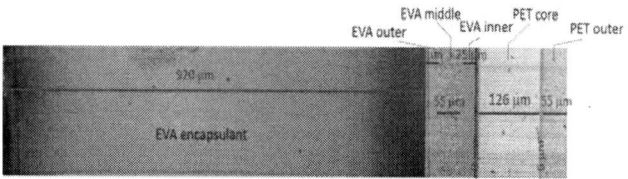

Fig. 1. Cross-sectional morphology of fresh EVA/PPE backsheet sample and the corresponding thickness.

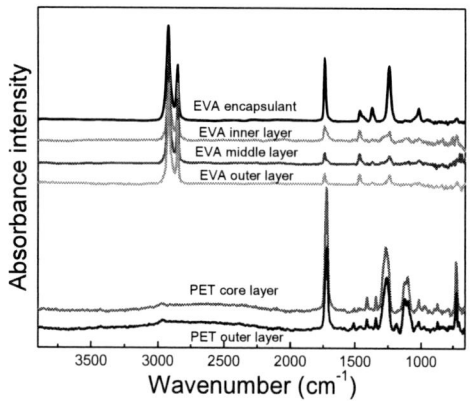

Fig. 2. FTIR spectra of different layers in EVA/PPE backsheet sample.

It is indicated that the PPE backsheet part of the fresh EVA/PPE sample is composed of five layers: EVA outer layer (about 25 μm thick), EVA middle layer (55 μm), EVA inner layer (25 μm), PET core layer (126 μm) and PET outer layer (55 μm). In addition, based on our previous Raman data [11], there are two aromatic polyester-based polyurethane adhesive layers with thickness around 8 μm and 6 μm between EVA inner and PET core layer, and PET core and PET outer layer, respectively.

### B. Influence of UV light intensities

Accelerated aging under intensified exposure conditions, especially enhanced UV light intensity, is commonly used to stimulate the degradation of photovoltaic laminates in a shorter term. The degradation produced by the UV light could be heterogeneous across the thickness, which demands that spatial analysis should be carried out at the microscopic level to understand the influence of UV light on the degradation as a function of depth. Fig. 3 shows the integral fluorescence emission intensity images of the cross-section of EVA encapsulant part in glass/EVA/PPE backsheet sample before and after exposure to UV/85 °C /0 % RH under different UV light intensities ((a)-(d)) and their corresponding line profiles

(e). Green or red regions observed near the exposed side of specimens have a higher fluorescence emission in these areas than the interior blue regions. The gradual drop in the fluorescence intensity across the thickness is suggested to be due to the attenuation of the UV light along the EVA encapsulant depth. Among three UV light intensities, the specimen exposed under 60 % UV light has the highest fluorescence intensity near the surface. The overall degradation depth under different UV light intensities is essentially similar, as shown in the integral emission line profiles (Fig. 3 (e)). The substantial changes in the fluorescence intensity appear to be within 200 μm from the exposed side of the encapsulant with over 80 % drop in the intensity.

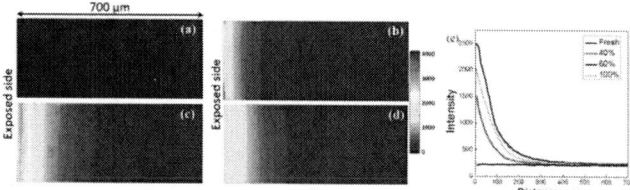

Fig. 3. Integral fluorescence intensity images of the cross-section of EVA encapsulant part in (a) fresh glass/EVA/PPE and samples exposed to UV/85°C/0 % RH under UV light intensities of (b) 40 %, (c) 60 % and (d) 100 %. (e) shows their corresponding line profiles.

Fig. 4 shows depth-dependent fluorescence spectra of same parts as shown in Fig. 3. A broad peak is observed and the peak progressively shifts to lower wavelength and a sequential drop of peak intensity from the exposed side to the interior part. Specifically, for 40 % UV light, the peak position shifts from 600 nm to 520 nm along the depth direction, and no obvious change is observed thereafter. For 60 % UV light, the emission peak position shifts consistently from 617 nm to 520 nm, while for 100 % UV light the peak position (around 635 nm) changes slowly at first and then a sharp shift to 520 nm is observed. These results indicate the existence of different degradation species distributions along the depth direction. These fluorescence emissions could be ascribed to the formation of α,β-unsaturated carbonyl species. The short chain α,β-unsaturated carbonyl species play important role for emission at shorter wavelength while the contribution of relatively long polyconjugated structures is dominated for emission at longer wavelength [5]. Moreover, the peak intensity near the exposed side increases as the UV light intensity increases from 40 % to 60 %, while it decreases again under 100 %. A similar peak intensity drop for severely discolored EVA encapsulant is also reported by Pern et al. [4].

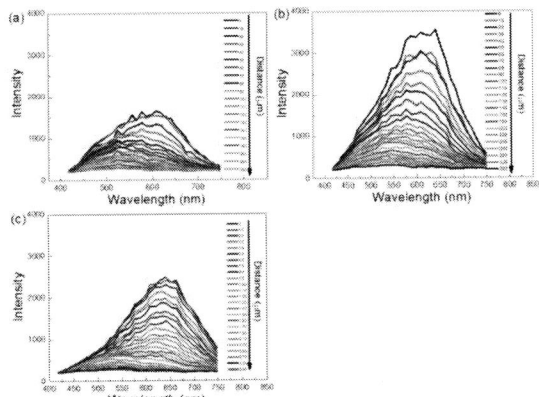

Fig. 4. Evolution of the fluorescence spectra along the depth direction of EVA encapsulant part in glass/EVA/PPE backsheet samples exposed to UV/85 °C/0 % RH under different UV light intensities of (a) 40 %, (b) 60 % and (c) 100 %.

Fig. 5 exhibits the fluorescence images and corresponding line profiles on the cross-section of the fresh PPE backsheet part and that after exposure to UV/85 °C /0 % RH. Fluorescence emission is observed on the PET outer layer of the unexposed sample, which could result from the fluorescent whitening agent added in this layer designed to brighten colors [12]. After degradation, three main changes are observed: (1) nonuniform fluorescence emission from the EVA middle layer; (2) fluorescence emission from two adhesive layers; (3) fluorescence suppression of PET outer layer. These results imply that the degradation has occurred in the EVA middle layer and two adhesive layers.

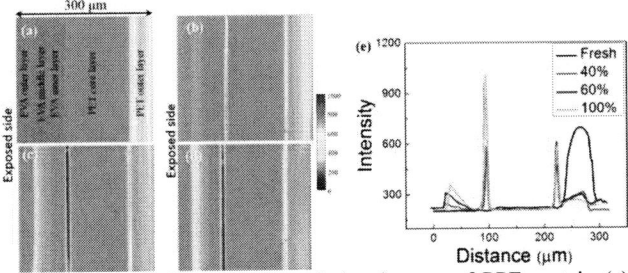

Fig. 5. Integral fluorescence emission images of PPE part in (a) fresh glass/EVA/PPE and samples exposed to UV/85 °C/0 % RH under UV light intensities of (b) 40 %, (c) 60 % and (d) 100 %. (e) shows the corresponding line profiles.

## C. Correlation with optical and mechanical properties

The aforementioned results have demonstrated that fluorescence imaging is an effective and spatially sensitive tool to monitor evolution in depth-dependent degradation of multiple-component and multilayered glass/EVA/PPE laminates under various exposure conditions. As indicated by fluorescence images, most significant changes have occurred in the EVA encapsulant and two adhesive layers of the PPE backsheet. We further examine the relationship between fluorescence profiles and the optical as well as mechanical

properties for EVA encapsulant on the same cross-sectional samples aged under 100 % UV light intensity. Depth-dependent UV-Vis absorbance spectra of EVA encapsulant are shown in Fig. 6 (a). Three typical UV absorbance peaks at (215-257) nm, (260-310) nm and (310-370) nm were observed for the fresh sample (see insert in Fig. 6 (a)). After degradation, the absorbance peak becomes broad and extends to longer wavelength in the visible region. Corresponding yellowing index is calculated based on these spectra according to ASTM E308 (Fig. 6 (b)). The yellowing index of the EVA encapsulant is reduced from the exposed side to inner part, then it reaches to a plateau. The depth profile is consistent with that of the fluorescence (Fig. 6 (d)). The DMT modulus of EVA encapsulant before and after aging along the thickness up to about 600 µm is also measured by QNM-AFM (Fig. 6 (c)). For the fresh sample, the DMT modulus is constant with small variations. Compared with the results of the fresh sample, a substantial drop in the modulus of aged EVA encapsulant is observed near the glass/EVA interface. This result is consistent with the investigation of Röder *et al.* [6], in which they found that the storage modulus of EVA encapsulant at 40 °C in a UV aged module decreases compared with the unaged sample.

Such changes in the yellowing index and the modulus across the thickness of the EVA encapsulant can be explained by the fluorescence results obtained from the same sample (Fig. 3 (d) and Fig. 4 (c)). The degradation products are distributed near the exposed side of the sample and their concentration decreases from the exposed surface to the internal part of the sample. These results indicate that a non-uniform degradation has occurred as a function of depth.

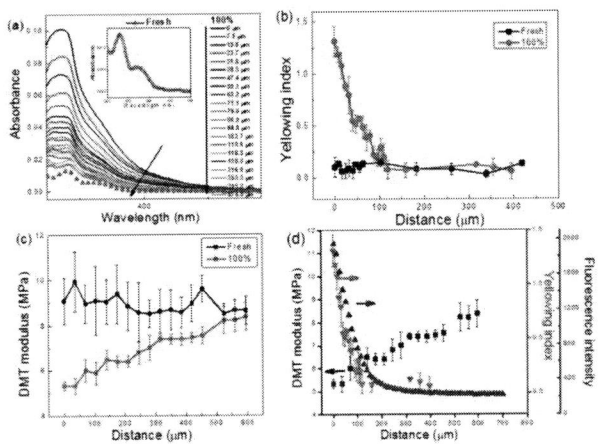

Fig. 6. Depth dependent (a) UV-Vis spectra, (b) corresponding yellowing index changes and (c) DMT modulus of fresh EVA encapsulant and sample aged under 100 %. (d) Correlation between changes of integral fluorescence intensity profile, yellowing index profile and modulus profile for EVA encapsulant aged under 100 % as a function of depth. Insert in (a) shows the enlarged view of fresh sample. (Error bars represent standard deviations from the results obtained at 3 locations on the cross-section of sample).

## IV. CONCLUSIONS

In this work, the degradation depth-profiles of the glass/EVA/PPE PV laminates have been investigated by fluorescence imaging, in combination with micro-UV-Vis spectroscopy and QNM-AFM after exposing samples under different UV light intensities with 85 °C/0 % RH. A gradual decrease in the intensity of fluorescence was observed for EVA encapsulant from the exposed surface to the internal part for all UV aged samples, indicating a non-uniform degradation has occurred after UV exposure as a function of the depth. Moreover, peak position of the fluorescence spectra shifts from longer wavelength to lower wavelength with the increase of the depth, suggesting changes in the distribution of different degradation species along the thickness direction. The spatial profile of fluorescence intensity was consistent with the change of yellowing index and modulus along the depth. Our fluorescence results also show that degradation occurs in the EVA middle layer and two adhesive layers in PPE backsheet part. This study has demonstrated that the fluorescence imaging is an effective and spatially sensitive technique to monitor the degradation depth-profile, which is important for the failure mechanism analysis and the reliability study of PV modules.

## REFERENCES

1. A. Ndiaye, A. Charki, A. Kobi, C.M. Kébé, P.A. Ndiaye, V. Sambou, Degradations of silicon photovoltaic modules: A literature review, Solar Energy 96 (2013) 140-151.

2. E. Planes, B. Yrieix, C. Bas, L. Flandin, Chemical degradation of the encapsulation system in flexible PV panel as revealed by infrared and Raman microscopies, Solar Energy Materials and Solar Cells 122 (2014) 15-23.

3. Y. Voronko, B.S. Chernev, G.C. Eder, Spectroscopic investigations on thin adhesive layers in multi-material laminates, Applied spectroscopy 68(5) (2014) 584-592.

4. F. Pern, Luminescence and absorption characterization of ethylene-vinyl acetate encapsulant for PV modules before and after weathering degradation, Polymer degradation and stability 41(2) (1993) 125-139.

5. N.S. Allen, M. Edge, M. Rodriguez, C.M. Liauw, E. Fontan, Aspects of the thermal oxidation of ethylene vinyl acetate copolymer, Polymer Degradation and Stability 68(3) (2000) 363-371.

6. J.C. Schlothauer, K. Grabmayer, G.M. Wallner, B. Röder, Correlation of spatially resolved photoluminescence and viscoelastic mechanical properties of encapsulating EVA in differently aged PV modules, Progress in Photovoltaics: Research and Applications 24(6) (2016) 855-870.

7. J. Chin, E. Byrd, N. Embree, J. Garver, B. Dickens, T. Finn, J. Martin, Accelerated UV weathering device based on integrating sphere technology, Review of scientific instruments 75(11) (2004) 4951-4959.

8. Data supplied by Atlas Weathering Services Group as published on http://www.atlaswsg.com/weath/2007.pdf.

9. PeakForce QNM User Guide, Rev. E, Burker Corporation, 2011.

10. S. Ayutthaya, J. Wootthikanokkhan, Investigation of the photodegradation behaviors of an ethylene/vinyl acetate copolymer solar cell encapsulant and effects of antioxidants on the photostability of the material, Journal of applied polymer science 107(6) (2008) 3853-3863.

11. C.-C. Lin, P.J. Krommenhoek, S.S. Watson, X. Gu, Depth profiling of degradation of multilayer photovoltaic backsheets after accelerated laboratory weathering: Cross-sectional Raman imaging, Solar Energy Materials and Solar Cells 144 (2016) 289-299.

12. H. Van Aert, I. Srivastava, F. Vangaever, A backsheet for photovoltaic modules, in: U.P. Office (Ed.) 2014.

# Statistical Analysis of Degradation Data for c-Si modules observed in India in 2016

Chiranjibi Mahapatra[1], Rajiv Dubey[2], Shashwata Chattopadhyay[3], Sachin Zachariah[3], Sanjeev Sabnis[1]

[1]Department of Mathematics, Indian Institute of Technology Bombay, Mumbai, 400076, India

[2]Department of Electrical Engineering, Indian Institute of Technology Bombay, Mumbai, 400076, India

[3]Department of Energy Science & Engineering, Indian Institute of Technology Bombay, Mumbai, 400076, India

*Abstract* — **This paper presents statistical analysis of power degradation rates of the PV modules inspected during the All India Survey of Photovoltaic Module Reliability 2016. Our analysis shows that there is a significantly higher probability of degradation rate being above 1%/year in the Hot zone as compared to the Non-hot zone, and similarly in the rooftop systems as compared to the ground mounted systems. Statistical analysis confirms that young modules are degrading at a faster rate than the old modules. However, the system size does not seem to affect the degradation rate for crystalline silicon modules.**

*Index Terms* — **Degradation, photovoltaic modules, defects, silicon, reliability.**

## I. INTRODUCTION

The modules in photovoltaic systems are often found to degrade differently even if they are made of the same material construction. As a result, almost all studies concerning degradation rates [1][2][3] have data sets that exhibit high variability. This calls for a rigorous statistical analysis of the data to find out whether perceived differences in central tendencies of the data sets are statistically significant or just an artifact of the data distribution. A joint team from the National Centre for Photovoltaic Research And Education (NCPRE) and National Institute of Solar Energy (NISE), India, undertook a survey of field-aged solar panels installed in different parts of India in 2016 [4]. Before doing any analysis on the data, the outliers in the degradation rate data are identified based on the inter-quartile range. The data points are arranged in ascending order, and then split into two halves, with the median included in both halves if the total number of sample points is odd. Then the lower fourth (lf) is the median of the lower half, while the upper fourth (uf) is the median of the upper half. A measure of the spread which is resistant to outliers is given by the difference between the upper fourth and lower fourth, which is referred to as the inter-quartile range. Any observation beyond 1.5 times of inter-quartile range from its nearest fourth is considered as an outlier. In our case, all data points above 5.12%/year degradation rate are identified as outliers, as shown in Fig. 1. Further analysis of the data is carried out after eliminating the outliers from the original data set. A high variability in the $P_{max}$ degradation rates has been observed for crystalline silicon (c-Si) modules, ranging from –2%/year to as high as +5%/year, as shown in Fig. 2. (The negative numbers most probably arise from uncertainty / under-rating on the name plate value and/or error in the translation to

STC.) Since crystalline silicon modules undergo a rapid initial degradation (caused due to exposure to light, so referred to as Light Induced Degradation) within the first few weeks, the degradation rates have been calculated after discounting for 2% LID, and this is referred to as *Linear $P_{max}$* degradation rate [4]. In this paper, we shall present a statistical analysis of the linear degradation rate data of c-Si modules (a large subset of the 2016 survey), and point out the significance of statistical evaluation of the data. This paper thus lends rigorous statistical support to the many of the conclusions presented in [4].

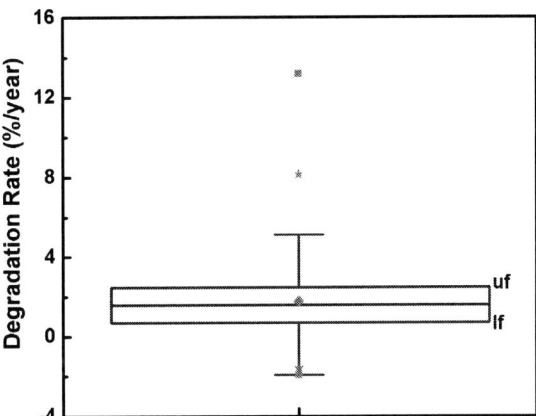

Fig. 1. Box plot for $P_{max}$ degradation rates for c-Si modules in Group A ('All') sites.

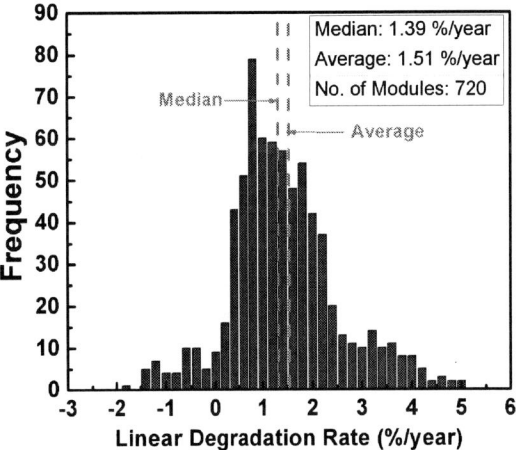

Fig. 2. Histogram for *Linear $P_{max}$* degradation rates for c-Si modules in Group A ('All') sites, after discarding outliers.

## II. METHODOLOGY

The modules inspected during the field survey can be grouped on the basis of the categories of the following explanatory variables:

   a) Climatic zone (Hot & Dry, Warm & Humid, Composite, Moderate, Cold & Sunny, Cold & Cloudy)
   b) Age (a continuous variable)
   c) Mounting configuration (ground (rack) mounted, rooftop (rack) mounted)
   d) System size (small/medium, large)

These categories have been considered as the explanatory variables for the degradation rates in the statistical analysis. Here the degradation rate is treated as a response variable. It is evident from Fig. 3 and also from the formal analysis involving the use of the Shapiro-Wilk test that the distribution of this response variable is non-normal.

Let Y = 1 if the $P_{max}$ degradation rates of the module is greater than 1%/year

     = 0 if the $P_{max}$ degradation rates of the module is less than or equal to 1%/year

This figure of "1%/year" is taken based on the internationally accepted power warranty provided by many module manufacturers.

In order to assess the impact of each of the afore-mentioned non-numeric categorized explanatory variables on this binary response random variable, a member of the generalized linear models, namely, logistic regression model is considered below. This model is given by:

$$g(EY) = \text{logit}[\pi(\underline{x})] = \log \frac{\pi(\underline{x})}{1-\pi(\underline{x})} = \alpha + \beta_1 x_1 + \ldots + \beta_p x_p$$
$$\ldots (1)$$

where,
EY denotes the expected value of response variable Y
g(EY) is a well-defined function of EY.
$\underline{x} = (x_1, x_2, \ldots, x_p)$ is a vector of p explanatory variables
     $x_1, x_2, \ldots, x_p$
$\alpha$ is the intercept,
$\beta_i$ is the coefficient associated with $i^{th}$ explanatory variable $x_i$, (i = 1,2,…,p)

$\pi(\underline{x})$ = P[Y=1] = P [the $P_{max}$ degradation rate is greater than 1%/year under $\underline{x}$]

This model in (1) yields two important expressions and they are:

(i) $\pi(\underline{x})$ = P[Y=1] = P [the $P_{max}$ degradation rate is greater than 1%/year under $\underline{x}$]

$$= \frac{\exp[\alpha + \beta_1 x_1 + \cdots + \beta_p x_p]}{1 + \exp[\alpha + \beta_1 x_1 + \cdots + \beta_p x_p]} \quad \ldots (2)$$

and,

(ii) $\frac{\pi(\underline{x})}{1-\pi(\underline{x})}$ = the odds of the $P_{max}$ degradation rate are greater than 1%/year under $\underline{x}$

$$= \exp[\alpha + \beta_1 x_1 + \cdots + \beta_p x_p] \quad \ldots (3)$$

## III. RESULTS AND DISCUSSION

### A. Nature of the distribution

The $P_{max}$ degradation rates for c-Si survey modules (histogram shown in Fig.2) do not follow any known distribution. However, by virtue of Central Limit Theorem, the asymptotic distribution of the average degradation rate per site turns out to be a normal distribution. Also, by virtue of extreme value theory, the limiting distribution of the maximum degradation rate per site is found the Gumbel distribution (shown in Fig. 3), with estimated value of the location parameter = 2.043 and scale parameter = 1.218. The advantage of having this information is that it enables one to compute probability of various events of interest and use properties of underlying probability distribution, albeit approximate, to draw inferences about physical characteristics of the phenomenon under consideration. For example, in the present context, one can determine the probability of maximum module degradation rate at a particular site exceeding 1%/year is 0.905, while the probability exceeding 2%/year is 0.645. It should be noted that this is the maximum degradation rate at the site, and most of the modules will degrade at a lower rate.

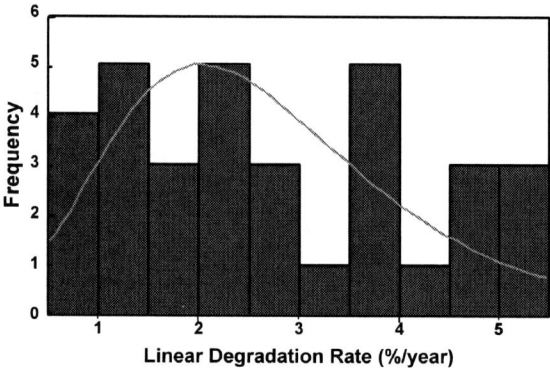

Fig. 3. Histogram for maximum *Linear $P_{max}$* degradation rates for a module at any site with Gumbel curve over histogram

### B. Determination of Confidence Intervals)

Non-normality of $P_{max}$ degradation rates (shown in Fig. 2) does not permit construction of 100(1-α) % confidence intervals using percentile points of normal distribution. Here one is required to take recourse to constructing confidence intervals based on bootstrap methodology. This method involves drawing a large number of samples of size n with replacement from the original sample of size n. For each of the bootstrap samples, test statistics of interest, in the present context the average degradation rate, is computed and this, in turn, can be used to determine the percentile points which are then used to obtain the requisite 100(1-α)% confidence interval for the unknown parameters. Fig.4 shows the histogram of the bootstrap samples and the 95% confidence interval for the population mean of $P_{max}$ degradation rates.

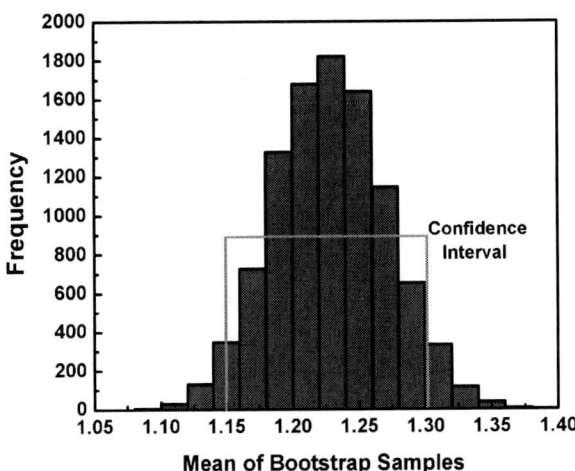

Fig. 4. Means of the bootstrap samples (No. of samples = 10000).

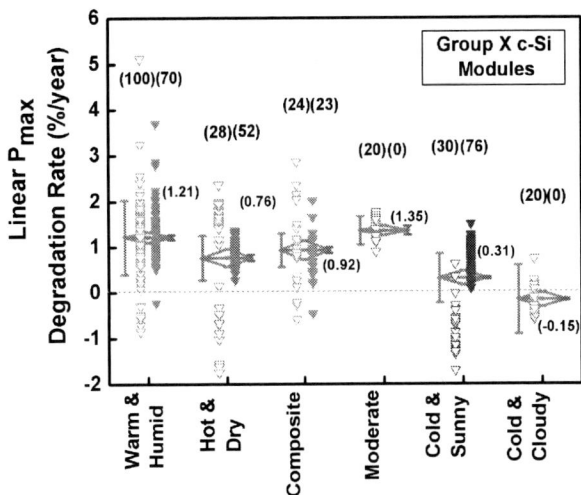

Fig. 5. Influence of climatic zone on $P_{max}$ degradation rate for modules in Group X sites[4].

## C. Model I

One of the major objectives regarding the performance of photovoltaic modules is the impact of explanatory variables such as climatic zones (categorical variable with six levels), mounting type (categorical variable with two levels), system size (categorical variable with two levels) and age (continuous variable) on the power degradation rates. The field survey conducted in 2016 showed differences in the average degradation rates in the different climatic zones (refer to Fig. 5 which is reproduced from [4]), but there was also large variability in each zone. This calls for a thorough statistical analysis of this survey data. In view of the definition of Y given in section II, the logistic regression model given in equation (1) is fitted using the afore-mentioned explanatory variables. It may be noted that for each of the categorical variables, the number of dummy variables that need to be introduced in the proposed model is equal to (number of levels of the variable minus 1) and this essentially ensures no redundancy in the levels of the given categorical variables. The equation for this model is:

$$\log \frac{\pi(x)}{1-\pi(x)} =$$
$$\alpha + \beta_1(\text{climate}_2) +$$
$$\beta_2(\text{climate}_3) + \beta_3(\text{climate}_4) + \beta_4(\text{climate}_5) + \beta_5(\text{climate}_6) +$$
$$\beta_6(\text{age}) + \beta_7(\text{mounting}) + \beta_8(\text{system size}) \qquad \dots (4)$$

where,
climate$_2$ = Hot & Dry
climate$_3$ = Moderate
climate$_4$ = Composite
climate$_5$ = Cold & Sunny
climate$_6$ = Cold & Cloudy

The parameter estimates obtained after fitting the above model to the degradation rate data are given in Table I.

### TABLE I
PARAMETER ESTIMATES AND P-VALUES FOR SIX CLIMATIC ZONE MODEL

| Covariates | Estimates | p-value |
|---|---|---|
| Intercept | 2.71410 | 3.22e-15 |
| Climate Category 2 (Hot & Dry) | 0.43965 | 0.07953 |
| Climate Category 3 (Moderate) | 1.12257 | 0.15015 |
| Climate Category 4 (Composite) | 0.76486 | 0.00706 |
| Climate Category 5 (Cold & Cloudy) | -16.52064 | 0.97544 |
| Climate Category 6 (Cold & Sunny) | -2.98704 | 1.05e-12 |
| Age | -0.03544 | 0.11454 |
| Mounting (=1 for ground mounted) | -2.61777 | < 2e-16 |
| System size (=1 for large system size) | -0.07216 | 0.77676 |

The climate categories for which p-values are greater than 0.05 (like Hot & Dry, Moderate and Cold & Cloudy) do not have any significant effect on the degradation rates, while the Cold & Sunny climate seem to have significant affect. Also, the age of the system and system size appears to have insignificant effect on the degradation rates. As these findings do not agree with the expert knowledge, the climatic zones have been regrouped into two categories, namely, Hot and 'Non-Hot'. Hot zones are the Hot & Dry, Warm and Humid, and Composite, and the 'Non-Hot' are the remaining three.

## D. Model II

The modified version of the logistic regression model of the previous section considers:

a) Climatic (a categorical variable with two categories – Hot and 'Non-Hot')
b) Age (a continuous variable)
c) Mounting type (a categorical variable with two-rooftop and ground mounting- categories)
d) System size (a categorical variable with two-large and small- categories)

as explanatory variables and attempts to fit

$$\log \frac{\pi(x)}{1-\pi(x)} = \alpha + \beta_1(\text{age}) + \beta_2(\text{mounting}) + \beta_3(\text{system size}) + \beta_4(\text{zone}) \qquad \dots (5)$$

where,
climate$_1$ (climate$_2$) = Hot (Non-Hot)
mounting$_1$ (mounting $_2$) = rooftop (ground)
system size$_1$ (system size $_2$)= large (small)

and $\underline{x}$ is a four-dimensional column vector with its elements being one level of each of the categorical explanatory variables climate, mounting, system size and a continuous variable age.

The parameter estimates obtained after fitting the above model to the degradation rate data are given in Table II. As the system size explanatory variable turns out to be statistically insignificant (p-value=o.665>0.05), Model II is refitted by dropping this explanatory variable and the resulting model is denoted by Model III and the information about its parameters estimates is summarized in Table III. It is evident from the information in this table that each of the explanatory variables

a) Climatic zone (a categorical variable with two categories – Hot and 'Non-Hot')
b) Age (a continuous variable)
c) Mounting type (a categorical variable with two-rooftop and ground mounting- categories)

is statistically significant (p-value <0.05) and affects P[Y=1], that is, each of the variables affects P[$P_{max}$ degradation rates for the module is greater than 1%/year under $\underline{x}$]

TABLE II
PARAMETER ESTIMATES AND P-VALUES FOR TWO CLIMATIC ZONE MODEL

| Covariates | Estimates of Parameters | p-value |
|---|---|---|
| Intercept | 1.88387 | 2.39e-12 |
| Age | -0.10685 | 3.32e-09 |
| Mounting (ground mounted) | -3.42402 | < 2e-16 |
| System size =1 (large system size) | -0.10387 | 0.665 |
| Zone | 2.41870 | 1.04e-15 |

TABLE III
PARAMETER ESTIMATES AND P-VALUES FOR REVISED TWO CLIMATIC ZONE MODEL

| Covariates | Estimates of Parameters | p-value |
|---|---|---|
| Intercept | 1.86730 | 1.86e-12 |
| Age | -0.10440 | 1.10e-09 |
| Mounting (ground mounted) | -3.44135 | < 2e-16 |
| Zone | 2.38193 | < 2e-16 |

The parameter estimate is negative for age and this means that the older modules have a lower degradation rate. Similarly, in view of the negative parameter estimate for mounting type, modules in the ground mounted systems have lower degradation rate than modules in rooftop systems. The sign of the parameters estimate corresponding to zone variables is positive and this means that the modules in Hot zone area have higher degradation rates than those in Non-Hot zones.

To understand the impact of type of mounting (ground versus rooftop), the above model is simplified separately for the roof mounted and ground mounted systems. Table IV shows the odds and probabilities for power degradation greater than 1%/year for modules installed in different mounting configurations and climatic zones. The odds are lowest for ground mounted systems in Non-Hot zone, and highest for roof mounted systems in Hot zone. In the Hot zone, the chances of power degradation greater than 1%/year are significantly higher in roof mounted modules than in ground mounted modules.

TABLE IV
EFFECT OF MOUNTING TYPE AND CLIMATIC ZONE

| Covariates | Odds for age=1 [π(x)/(1-π(x))] | Probability for age=1 [π(x)] | Odds for age=5 [π(x)/(1-π(x))] | Probability for age=5 [π(x)] |
|---|---|---|---|---|
| GM* systems in Hot Zone | 2.02080 | 0.66896 | 1.33097 | 0.57099 |
| GM* systems in Non-hot Zone | 0.18666 | 0.15730 | 0.12294 | 0.10948 |
| RM* systems in Hot Zone | 63.10767 | 0.98440 | 41.56514 | 0.97650 |
| RM* systems in Non-hot Zone | 5.82936 | 0.85357 | 3.83944 | 0.79336 |
| GM=Ground Mounted, RM=Roof Mounted | | | | |

## V. Conclusion

Field data obtained to assess PV module degradation are extremely valuable, but necessarily contain some in-built uncertainty and variability due to a variety of reasons. It is therefore important to perform a thorough statistical analysis on the data to ensure that correct conclusions are being drawn. With this in mind, a statistical modeling scheme has been set up, and applied to the 2016 field survey performed in India.

The power degradation rate for the surveyed c-Si modules in 2016 does not follow any known distribution, but the maximum degradation rates at the sites follow the Gumbel distribution. Statistical analysis of the data confirms that the module degradation rates in the different climates of Hot zone (Hot & Dry, Warm & Humid and Composite) are similar, and hence it is recommended that for analysis of photovoltaic degradation, only two climatic zones be considered – Hot and Non-Hot. The degradation rate in the Hot zone is higher than in the Non-Hot zone, and also the modules in the roof mounted systems are degrading faster than the modules in ground mounted systems. These results, presented in [4], are rigorously supported by the statistical analysis presented in this paper. Young modules are degrading at a faster rate than the old modules (even after discounting the initial LID). System size does not seem to have any statistically significant impact on the module degradation rates as per the present analysis.

## Acknowledgement

This research is based upon work supported in part by (a) the National Centre for Photovoltaic Research and Education funded by Ministry of New and Renewable Energy of the Government of India through the Project No. 31/09/2015-16/PVSE-R&D dated 15th June 2016 and (b) the Solar Energy Research Institute for India and the U.S. (SERIIUS) funded jointly by the U.S. Department of Energy subcontract DE AC36-08G028308 (Office of Science, Office of Basic Energy Sciences, and Energy Efficiency and Renewable Energy, Solar Energy Technology Program, with support from the Office of International Affairs) and the Government of India subcontract IUSSTF/JCERDC-SERIIUS/2012 dated 22nd Nov. 2012. The authors thank A. Kottantharayil, J. Vasi and C. S. Solanki of IIT Bombay for useful inputs

## References

[1] R. Dubey et al., "All India Survey of Photovoltaic Module Reliability 2014", National Centre for Photovoltaic Research and Education, Mumbai, India, 2016, available online at http://www.ncpre.iitb.ac.in/uploads/All_India_Survey_of_Photo voltaic Module Reliability 2014.pdf

[2] R. Dubey et al." Comprehensive study of performance degradation of field-mounted photovoltaic modules in India", Energy Science & Engineering, 5(1), 51-64 (2017)

[3] D. C. Jordan et al., "Compendium of Photovoltaic Degradation Rates," Progress in Photovoltaics: Research and Applications, online version, February 2016.

[4] R. Dubey et al., "Performance of Field-Aged PV Modules in India: Results from 2016 All India Survey of PV Module Reliability", 44th IEEE PVSC, 2017, Washington, DC.

[5] Alan Agresti, An Introduction to Categorical Data Analysis, second edition, Wiley series, Published by John Wiley & Sons, Inc., Hoboken, New Jersey, 2007

# Process Induced Deflection and Stress on Encapsulated Solar Cells

Xiaodong Meng[1], Michael Stuckelberger[1], Peter Hacke[2], Mariana Bertoni[1]

[1] Ira A. Fulton Schools of Engineering, Arizona State University, Tempe, AZ, 85287

[2] National Renewable Energy Laboratory, Golden, CO 80401

*Abstract* — Cell cracking is one of the most common factors that limit the lifetime of PV modules. Until now electroluminescence (EL) has been the tool of choice to inspect cracks in finished modules. However, there are intrinsic limitations to the size of the cracks that this technique can resolve making it complicated to study the origins of crack formation. We also argue that the process of module assembly today is optimized from the point of view of the inactive materials (e.g. encapsulant cross-linking) offering no insights into the solar cell status. To this regard, even the correlation of module degradation to cell cracking in flat modules is only evident when cracks are of detectable size and detrimental to the electrical performance as measured by EL or PL. We have shown that in-house X-ray topography (XRT) is a unique technology that provides a non-destructive way of assessing the mechanical state of encapsulated solar cells, not only the evaluation of cracks and microdefects developed during handling and outdoor operation, but also the analysis of intrinsic deflection and stress induced by materials and processes (e.g. soldering, lamination). In this contribution, we present results on the deflection of the solar cells caused by different processing temperatures, various lamination materials, accelerated testing as well as metal ribbon.

*Index Terms* — Solar cell, Deflection, Stress, X-ray topography, Ribbon mismatch, Reliability, Silicon.

## I. INTRODUCTION

The bankability of Photovoltaics relies on assuring modules' lifetime of more than 25 years. In order to achieve that target for every module as well as reach the SunShot Levelized Cost of Electricity (LCOE) goal of 3 cents/kWh by 2030, it is imperative to understand the origin of the simplest of failure modes: cracking. It has been shown that cell cracking and delamination account for a high percentage of the power loses in the midlife-failure and wear-out-failure regimes [1]-[2]. Severe cracks can separate parts of solar cell from electrical connection, resulting in mismatched cells and potential safety hazard. It has been reported that cell cracking can cause up to 16% of power loss [3]. Solar cells with visible cell processing related cracks, as detected by PL, are rejected before stringing. It is important to note two things here: (1) microcrack formation can go easily unnoticed if the crack does not propagate above the PL/EL camera detection limit (150 μm resolution), and (2) the majority of cell cracking in modules is actually developed by the thermal and mechanical stresses induced during stringing, module production, handling and outdoor operation [4]-[5].

Unlike a bare solar cell, which can be easily inspected by multiple methods, it is difficult to characterize cells under encapsulation, especially with high throughput and resolution.

Electroluminescence is the common method of inspecting finished PV modules, especially before and after accelerated stress testing [6]. We argue that while helpful EL is a "phenomenological" tool that delivers a 2D luminescence map that provides no information regarding the origin of the cracks it can image – e.g. the surrounding stress and strain fields. As the PV market moves to thinner cells [7], which are known to be more susceptible to cracking, it is more imperative that we understand the origin of microcrack formation from the point of view of the cell's stress distribution as an approach to appropriately chose materials and optimize processes.

X-ray topography (XRT) is a non-destructive method that has been utilized to inspect defects and striations on films. Our previous work has demonstrated that this technique can not only detect cell cracks, but also deliver a deflection map on encapsulated solar cell, offering another level of detailed information when compared to EL [8]. Though synchrotron based XRT has also showed the potential ability to inspect cell cracks [9], here we argue that laboratory XRT setup is sufficient for the measurement and inline systems could be easily adapted to contemplate the high throughput of a module fabrication line. The laboratory XRT method can have the resolution of < 10 μm and can easily evaluate the silicon under glass, encapsulation and through metal ribbons.

We present herein examples of the variations in deflection for different lamination processes such as lamination temperatures and different material stacks. We also present the deflection maps after accelerated testing showcasing the how the forces acting on a cell change as a function of time and environment affecting the deflection and thus the stress on the cells. Finally, we present our preliminary studies of a solar cell deflection under the metal ribbon, which will illustrate why most of the microcracks originate near the ribbon area.

## II. EXPERIMENTAL METHOD

Flat mini modules of encapsulated cells were made using heat-resistant borosilicate flat glass. The encapsulated solar cells were laser-cut-outs (32 mm x 32 mm) of heterojunction with thin intrinsic layer (HIT) silicon cells (180 μm) fabricated at the Solar Power Lab at Arizona State University. In the lamination temperature experiment, three samples A, B, C were laminated with three temperatures (100°C, 145°C and 160°C) following the PV stack: glass / EVA / Cell / EVA / Backsheet. In the lamination stack comparison, a sample D was fabricated at 145°C with PVMirror stack containing an optical filter, as shown in Fig. 2(b) [10]. Both sample B and D went through a standard damp-heat DH1000 test for accelerate module testing. The environmental test was performed at National Renewable Energy Laboratory (NREL). Another sample E was prepared with a soldered solar cell (50 mm x 50 mm) under 145°C with PV stack.

The XRT characterization was done by a HITACHI Rigaku XRT-100 instrument in transmission mode using Mo X-ray source. The acquired X-ray patterns depend on the physical status of the solar cell (curvature, rotation angles). Since all our heterojunction cell wafers are silicon <100> wafers the diffraction patterns presented hereafter correspond to the [004] diffractions of the silicon crystal. The diffraction patterns are translated into diffraction lines through watershed treatment and then transformed into a deflection map of the solar cell following the procedure described in Refs [8], [11].

## III. RESULTS AND DISCUSSION

### A. Lamination Temperature

Lamination temperature is one of the most important parameters in determining the lamination quality (gel content) of PV modules. Several failure modules such as delamination are associated with this characteristic temperature. Higher temperature favors higher gel contents, but it may cause potential higher stresses on the solar cell because of larger thermal expansion and shrinkage of the backsheet and

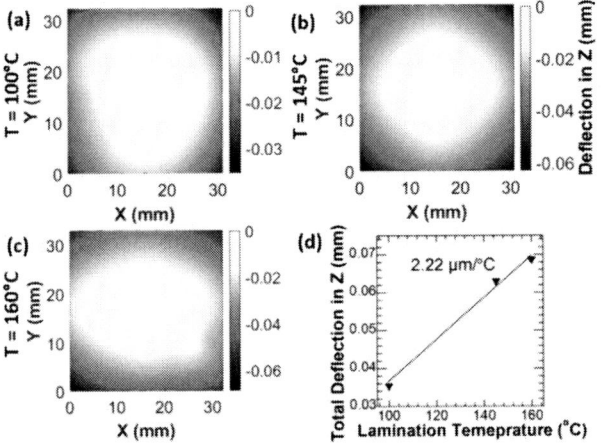

Fig. 1. XRT analyzed deflection maps of three mini modules laminated with EVA at three temperatures: (a) sample A, 100°C; (b) sample B, 145°C and (c) sample C, 160°C. The center is bent toward to the glass. Bright means high and dark means low in the Z direction (out of the plane of this paper). (d) Figure showing the correlation of total deflection and lamination temperature using the maximum cell deflection from (a), (b) and (c).

encapsulant. In this case, sample A, B and C were laminated under three different temperatures (100°C, 145°C, 160°C) for XRT analysis and the data are shown in Fig. 1. The EVA encapsulant starts curing at 100°C. As the temperature goes up, the deflection on the cell from center to corner increases; the center of the cell is bent towards the front glass while the

corners are bent toward the back. There are two possible explanations for this bending: (i) the melted EVA was pressed into the center area because the applied pressure was higher at the edges (edge pressing); (ii) the shrinkage of EVA and backsheet pulls the cell corners backward during solidification. The maximum deflection of sample A is around 0.03 mm while the deflection of sample B and C are both above 0.06 mm. The total deflection shows a linear trend of 2.22 μm/°C correlation between the maximum deflection and lamination temperature (Fig. 1(d)). This suggests that the lamination induced deflection on the cell is associated with the degree of crosslinking of the EVA as well as the thermal mismatch of the backsheet and EVA.

### B. Lamination Stack

Different lamination stacks can also induce different

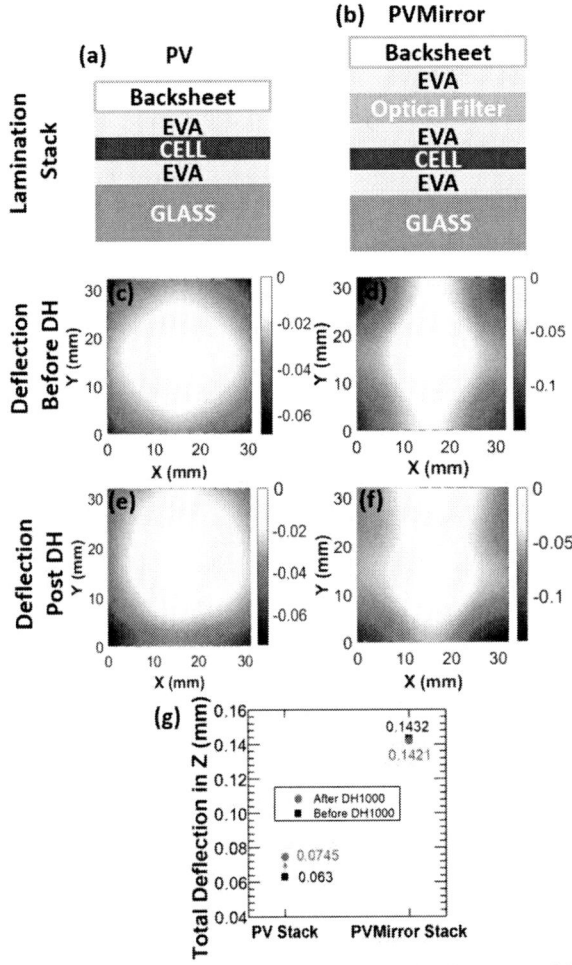

Fig. 2. Diagram illustrating the lamination stack of (a) PV and (b) PVMirror; XRT analyzed deflection maps of laminated mini modules: (c) sample B using PV stack, (d) sample D using PVMirror stack, (e) sample B post DH1000 and (f) sample D post DH1000. (g) Analyzed figure showing the total deflection changes between PV stack and PVMirror and stack, both before and after DH1000 test.

amount of stresses on solar cell. A PVMirror stack has been selected to compare the deflection generated by different stacks (Fig. 2(a) and (c)). The PVMirror stack has a plastic optical filter and an extra EVA layer on the backsheet side of the stack [10], which makes the whole stack thicker. Sample B and D were fabricated following the above mentioned stacks respectively and the deflection of two cells were characterized by XRT as shown in Fig 2 (c) and (d). As we have discussed in previous section, the deflection of the laminated solar cell is related to the EVA and backsheet. As sample D has more layers of polymer, the cell shows increased total deflection.

Figure 2 (e) and (f) show the deflection map of the sample B and D after damp heat (1000 h of 85°C/85% RH) test. The DH1000 test is designed to stress the PV module to reveal potential lamination problems such as delamination. The total cell deflection comparison before and after DH1000 (Fig. 2(g)) indicates that there is no significant change in terms of maximum cell deflection in both sample B and D. However, the deflection map of the solar cell varies a lot, suggesting that the distribution of stresses acting on the cell change through time. Both sample B and D's top half area on the deflection map show increased Z values. As we know, the curing process is a thermal crosslinking process. Even the optimized lamination recipe cannot deliver 100% gel content in EVA; the left-over uncured EVA can still melt even at a temperature of 85°C. What's more, the polymer morphology changes under elevated temperature & humidity environment. EVA has been reported to show an increase in elastic modulus values due to damp heat exposure [12]. All of those factors reduce the adhesion between polymer layers and cause delamination eventually. In our case, the stresses that the EVA puts on the cell changes causing the variations in the deflection map mentioned above. It has been reported that cell cracks can be developed during the damp heat test [12]. Though no cracks are observed in those samples, it is important to understand correlation between the cell deflection changes and cell crack formation so that we can have a better solution of minimizing the cell cracks.

*C. Metal Ribbon Induced Cell Deflection*

As mentioned before, one of the advantages of XRT is that it can see through metal ribbons, which means the deflection and stress on the silicon crystal can be evaluated under the metallization and tabbing. During soldering process, the ribbon and cell are heated up to 400°C. Due to different coefficients of thermal expansion (CET) of the silicon and the metal ribbon, huge amount of stress is developed near the ribbon areas. Moreover, the laminator puts pressure on ribbons during process (Fig. 3 (a)), which explain why cracks seem to originate near the busbars and tabbing points. To better illustrate this point, sample E with a tabbed silicon cell (2'' by 2'') was laminated with EVA in PV stack and measured by XRT as shown in Fig. 3(b) and (c). The deflection distribution indicates that the front ribbon pushed the cell downwards

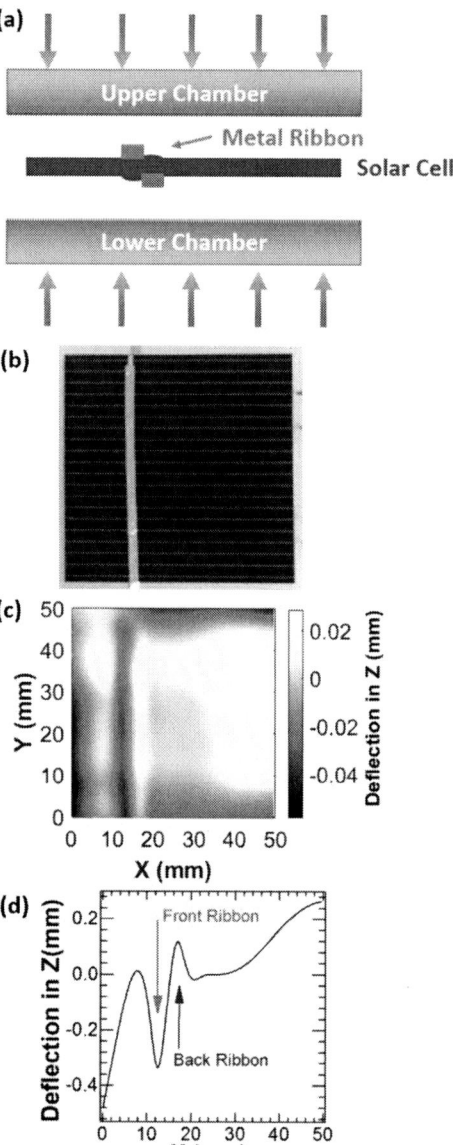

Fig. 3. (a) Diagram illustrating how cell deflection is induced by the lamination through front and back ribbons near ribbon areas. Noted that the front ribbon is slightly shifted to the left compared to the back ribbon. (b) Photo of a laminated mini module with silicon solar cell tabbed on both front and back sides. (c) Analyzed XRT data showing the deflection of the solar cell in (b). (d) Figure of cross section (passing through the cell center) of sample E along X direction.

(dark line in the middle) while the back ribbon pushed the cell upwards, causing a deflection in the silicon of about 70 μm from top to bottom. This deflection above and below the neutral axis can only mean substantial amounts of shear stress at those points, which are well known to facilitate crack formation and propagation. A deflection cross section along the cell's X-axis is presented in Fig. 3(d). The front ribbon and the back ribbon can be readily identified on the 2D deflection curve because of the obvious distortion. The reason why the left part of cell is bent downward is because of the non-zero net torque from front ribbon and back ribbon corresponding to the wafer center. To the best of our knowledge, this is the first time that this kind of analysis has been performed in encapsulated cells.

IV. SUMMARY AND OUTLOOK

In multiple experiments, we demonstrated that XRT is able to assess the status of encapsulated solar cell under the full encapsulation stack and show the changes the solar cell sees as a function of lamination temperature and lamination stack. We have also demonstrated that XRT can be utilize to inspect silicon solar cells through metal ribbons, revealing the local stress induced by soldering and subsequent lamination. Further analysis will be conducted to evaluate stress distribution and deflection induced by wafer thickness, solder joints, backsheet and encapsulants on finished PV modules.

Laboratory XRT is a promising technique capable of paving the way to longer PV module lifetimes by understanding the underlying science of crack formation and propagation.

ACKNOWLEDGEMENTS

The information, data, or work presented herein was funded in part by the Advanced Research Projects Agency-Energy (ARPA-E), U.S. Department of Energy, under Award Number DE-AR0000474

REFERENCES

[1] E. L. Meyer and E. E. van Dyk, "Assessing the reliability and degradation of photovoltaic module performance parameters," *IEEE Trans. Reliab.*, vol. 53, no. 1, pp. 83–92, Mar. 2004.

[2] M. Köntges, S. Kurtz, C. Packard, U. Jahn, K. A. Berger, and K. Kato, *Performance and reliability of photovoltaic systems: subtask 3.2: Review of failures of photovoltaic modules: IEA PVPS task 13: external final report IEA-PVPS*. Sankt Ursen: International Energy Agency, Photovoltaic Power Systems Programme, 2014.

[3] M. Köntges, I. Kunze, S. Kajari-Schröder, X. Breitenmoser, and B. Bjørneklett, "The risk of power loss in crystalline silicon based photovoltaic modules due to micro-cracks," *Sol. Energy Mater. Sol. Cells*, vol. 95, no. 4, pp. 1131–1137, Apr. 2011.

[4] X. F. Brun and S. N. Melkote, "Analysis of stresses and breakage of crystalline silicon wafers during handling and transport," *Sol. Energy Mater. Sol. Cells*, vol. 93, no. 8, pp. 1238–1247, Aug. 2009.

[5] A. M. Gabor *et al.*, "Soldering induced damage to thin Si solar cells and detection of cracked cells in modules," in *21st European Photovoltaic Solar Energy Conference, Dresden, Germany, September*, 2006, pp. 4–8.

[6] M. Sander, S. Dietrich, M. Pander, M. Ebert, and J. Bagdahn, "Systematic investigation of cracks in encapsulated solar cells after mechanical loading," *Sol. Energy Mater. Sol. Cells*, vol. 111, pp. 82–89, Apr. 2013.

[7] J. Jean, P. R. Brown, R. L. Jaffe, T. Buonassisi, and V. Bulović, "Pathways for solar photovoltaics," *Energy Env. Sci*, vol. 8, no. 4, pp. 1200–1219, 2015.

[8] X. Meng, M. Stuckelberger, L. Ding, B. West, A. Jeffries, and M. Bertoni, "Quantitative Mapping of Deflection and Stress on Encapsulated Silicon Solar Cells: A Production-line Capable Technique," *IEEE J Photovolt*, vol. Under Review.

[9] A. Colli, K. Attenkofer, B. Raghothamachar, and M. Dudley, "Synchrotron X-Ray Topography for Encapsulation Stress/Strain and Crack Detection in Crystalline Silicon Modules," *IEEE J. Photovolt.*, vol. 6, no. 5, pp. 1387–1389, Sep. 2016.

[10] Z. J. Yu, K. C. Fisher, B. M. Wheelwright, R. P. Angel, and Z. C. Holman, "PVMirror: A New Concept for Tandem Solar Cells and Hybrid Solar Converters," *IEEE J. Photovolt.*, vol. PP, no. 99, pp. 1–9, 2015.

[11] X. Meng, M. Stuckelberger, L. Ding, B. West, A. Jeffries, and M. Bertoni, "Characterization of encapsulated solar cells by x-ray topography," in *2016 IEEE 43rd Photovoltaic Specialists Conference (PVSC)*, 2016, pp. 0111–0114.

[12] G. Oreski and G. M. Wallner, "Damp heat induced physical aging of PV encapsulation materials," in *2010 12th IEEE Intersociety Conference on Thermal and Thermomechanical Phenomena in Electronic Systems*, 2010, pp. 1–6.

# A unified global investigation on the spectral effects of soiling losses of PV glass substrates: preliminary results

Leonardo Micheli[1,2*], Eduardo F. Fernández[3], Greg P. Smestad[4], Hameed Alrashidi[5], Nabin Sarmah[6], Nazmi Sellami[7], Ibrahim A. I. Hassan[8], Amal Kasry[9], Gustavo Nofuentes[3], Neeru Sood[10], Bala Pesala[11,12], S. Scnthilarasu[5], Florencia Almonacid[3], K.S. Reddy[13], Matthew Muller[1], Tapas K. Mallick[5]

[1] National Renewable Energy Laboratory, Golden (CO), USA
[2] Colorado School of Mines, Golden (CO), USA
[3] University of Jaén, Jaén, Spain
[4] Sol Ideas Technology Development, San José (CA), USA
[5] University of Exeter, Penryn, UK
[6] Tezpur University, Tezpur, India
[7] Robert Gordon University, Aberdeen, UK
[8] South Valley University, Qena, Egypt
[9] British University in Egypt, El Sherouk City, Egypt
[10] BITS Pilani, Dubai Campus, Dubai, UAE
[11] Academy of Scientific and Innovative Research, Chennai, India
[12] CSIR-Central Electronics Engineering Research Institute, Chennai, India
[13] Indian Institute of Technology Madras, Chennai, India

*Abstract* — **The present work reports on the initial results of an international collaboration aiming to investigate the spectral effects of soiling losses. Identical glass coupons have been exposed outdoors for eight weeks in different locations worldwide, and weekly direct and hemispherical transmittance (T%) measurements are compared. Maximum losses as high as 7% and 50% in hemispherical and direct transmittance, respectively, have been found during the 8-week outdoor exposure. At the end of the data collection, a preliminary analysis of the spectral impact of soiling has been performed. The results show that the blue end of the spectrum is more affected and that lower hemispherical T% correlate to larger area covered by particles.**

*Index Terms* — **photovoltaic systems, reliability, optical losses, soiling.**

## I. INTRODUCTION

The accumulation of dust is one of the major concerns for photovoltaic (PV) systems since it reduces the sunlight transmitted by the glass surface and, thus, the energy converted by the modules. The dust composition and the particle size play a fundamental role in the impact of soiling losses [1], [2]. Several works have analyzed the relation between the dust and the soiling losses in different locations [3]–[5]. All these analysis are generally site-specific or highly regionalized. Other works have instead analyzed the effect of artificially deposited dust [6], [7]. This work presents the preliminary results of an international collaboration among academic and research institutes and private partners that aims to investigate the spectral effects of soiling naturally deposited on PV glasses installed at various locations worldwide.

## II. MATERIALS AND METHODS

Tests have been conducted at eight locations worldwide, listed in Table I, chosen to represent a wide variety of climates and environmental conditions. Each partner used a spectrophotometer to weekly measure the change in transmission due to the accumulation of soiling. Seven identical 4 cm × 4 cm sized and 3 mm-thick Diamant® low-iron glass from Saint-Gobain Glass were shipped to each location. Coupons were numbered from 0 to 6: six of them (coupons 1 to 6) were installed outdoors, at zero tilt angle, using the supporting structure shown in Fig. 1. Coupon 0 was instead kept in a safe, dust-free container and used to calibrate and compare the different spectrophotometers.

Fig. 1. Supporting structure holding six glass coupons. These are held with binder clips on a horizontally-mounted aluminum plate. Weekly transmittance measurements are taken for coupons 1, 2 and 3, whereas coupons 4, 5 and 6 have been used for dust characterization.

Weekly transmission measurements were taken on coupons 1, 2 and 3. Coupon 1 was cleaned weekly, coupon 2 was cleaned every 4 weeks (twice during the data collection) and coupon 3 was never cleaned. A dry cleaning is performed by using a microfiber cleaning cloth. Coupons 4, 5 and 6 were not cleaned nor moved until the end of the data collection, since they will be used for the dust characterization analyses. Daily weather data and, where available, mean daily concentrations of particulate matter (PM) have been recorded.

## III. RESULTS

### A. Impact of soiling on broadband hemispherical transmittance

The data collection commenced in January 2017. The average and maximum weekly loss in hemispherical transmittance is reported in Table II. The results obtained by the measurements performed in Golden, Colorado, USA, are shown in Fig. 2. Coupon 1, cleaned every week, shows a weekly average reduction of 0.7% in hemispherical transmittance. In contrast, this value increases to 1.3% in Chennai. This result agrees with the expectations: mean daily concentrations of $PM_{2.5}$ of 11 $\mu g/m^3$ and 41 $\mu g/m^3$ were recorded in January 2017 from monitoring stations nearby Golden and Chennai, respectively. At the end of the data collection, Coupon 3 had lost 1.7% in Golden and 6% in Chennai respectively. During the same time period, maximum losses of 7% have been registered in Jaén, because of Saharan dust transported onto the town during Week 3 and week 5.

Fig. 2 shows that the effect of the deposits on soiled coupon 3 does not necessarily increase at the same rate as coupon 1, which is cleaned weekly. Indeed, differently soiled coupons can be differently impacted by rainfalls: if soiling has been accumulating for some time, it might be more difficult to wash. This is found to happen in all the locations investigated.

### B. Impact of soiling on broadband direct transmittance

As expected, higher losses have been found when direct transmittance is considered instead of hemispherical. Figure 3 shows the results of the weekly measurements conducted on Coupon 3 in Golden, CO. When compared in the same wavelength range (500 nm to 1100 nm), the direct transmittance is found to drop by 6% in 6 weeks, while the loss in hemispherical transmittance is limited to 1.5%.

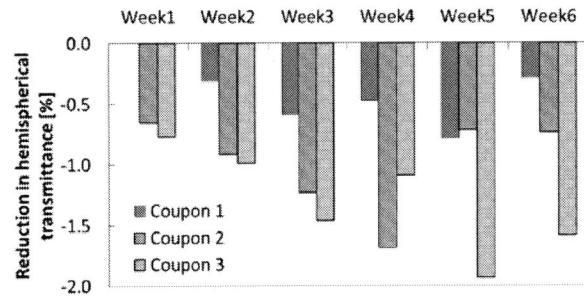

Fig. 2. Progressive absolute drop in hemispherical transmittance, compared to the initial conditions, registered in Golden, CO. Transmittance is obtained by averaging the data recorded between 200 and 1100 nm, with a 1 nm step. Coupon 1 was manually cleaned every week and coupon 2 was cleaned on week 4.

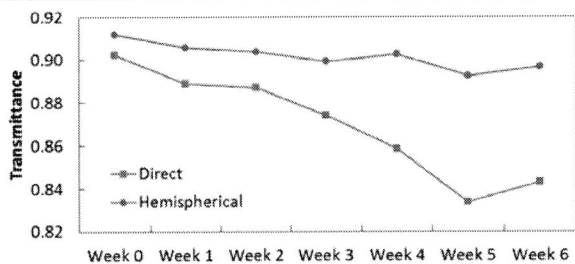

Fig. 3. Direct and hemispherical transmittance of coupon 3 in Golden. Wavelengths between 500 and 1100 nm have been averaged.

Average weekly and maximum losses recorded at three sites are shown in Table III. In El Shorouk City, weekly losses as high as 20% in direct transmittance have been registered. In Tezpur, a drop of 23% in direct transmittance was measured on coupon 3 after the longest dry period (2 weeks). The measurements taken at the end of the dry period, week 5, are shown in Fig. 4. Coupons 1 and 2 (both cleaned in week 4) have similar transmittance, although, they both show a drop in transmittance between 8 and 10%, compared to coupon 0, after only one week.

TABLE I
LIST OF MONITORED LOCATIONS AND CLIMATE CLASSIFICATIONS, SOURCED FROM [9].

| City, Country | Coordinates | Measured transmittance | Climate classification |
|---|---|---|---|
| Chennai, India | 13.08, 80.27 | Hemispherical | Equatorial savannah with dry winter (Aw) |
| Dubai, UAE | 28.36, 75.59 | Hemispherical | Desert climate (Bwh) |
| El Shorouk City, Egypt | 30.12, 31.61 | Direct | Desert climate (Bwh) |
| Golden (CO), USA | 39.74, -105.18 | Hemispherical and Direct | Snow climate, fully humid (Dfb) |
| Jaén, Spain | 37.79, -3.78 | Hemispherical | Warm temperate climate with dry summer (Csa) |
| Penryn, UK | 50.17, -5.13 | Hemispherical | Warm temperate climate, fully humid (Cfb) |
| San José (CA), USA | 37.29, -121.91 | Hemispherical | Warm temperate climate with dry summer (Csb) |
| Tezpur, India | 26.70, 92.83 | Direct | Warm temperate climate with dry winter (Cwa) |

## TABLE II
ABSOLUTE AVERAGE AND MAXIMUM WEEKLY LOSS IN BROADBAND HEMISPHERICAL TRANSMITTANCE. WEEKLY LOSSES CALCULATED FOR COUPON 1, MAXIMUM LOSSES CALCULATED FOR COUPON 3. TRANSMITTANCE WAS MEASURED WITH DIFFERENT SPECTROPHOTOMETERS AT EACH SITE.

| City, Country | Absolute average weekly loss (%) | Maximum loss (%) |
|---|---|---|
| Chennai, India | -1.3 | -7.8 |
| Golden (CO), USA | -0.7 | -2.0 |
| Jaén, Spain | -3.2 | -7.0 |
| Penryn, UK | -0.8 | -2.3 |
| San José (CA), USA | N.A. | -3.7 |

### C. Impact of soiling on spectral transmittance

At the end of the data collection, coupons 4, 5 and 6 have been shipped to NREL from each location, in order to perform the analysis using the same instrument. The variety of soiling conditions in this project allows for a general analysis of spectral losses. The normalized hemispherical transmittance spectra in the visible and NIR regions of the spectrum for coupon 5 are shown in Fig. 5 for the end of the outdoor exposure period. The transmittance is divided by that recorded for coupon 0 at each location; this way, the effect of soiling on the transmission at each wavelength can be analyzed independently of the optical nature of the glass. The initial results show that soiling has a higher impact on the blue than on the red end of the spectrum, independently of the location and of the amount of losses. This is in agreement with the conclusions of Ref. [6] where coupons artificially soiled with dust collected in Kuwait were studied.

## TABLE III
ABSOLUTE AVERAGE AND MAXIMUM WEEKLY LOSS IN BROADBAND DIRECT TRANSMITTANCE. WEEKLY LOSSES CALCULATED FOR COUPON 1, MAXIMUM LOSSES CALCULATED FOR COUPON 3. TRANSMITTANCE WAS MEASURED WITH DIFFERENT SPECTROPHOTOMETERS AT EACH SITE

| City, Country | Absolute average weekly loss (%) | Maximum loss (%) |
|---|---|---|
| Golden (CO), USA | -1.4 | -6.9 |
| El Shorouk City, Egypt | -6.0 | -46.9 |
| Tezpur, India | -11.9 | -23.6 |

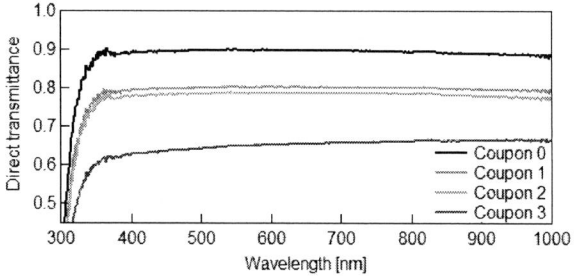

Fig. 4. Direct transmittance of coupons 1, 2, 3 in Tezpur, India, measured at week 5, after two dry weeks. Coupons 1 and 2 have similar transmittance since both have been cleaned at week 4.

For a better understanding of the spectral behavior for mild soiling conditions, the hemispherical transmittance of five sites have also been reported in Fig. 6, using a different y-axis scale. In this case, the data have been processed using a local regression technique to remove noise. $R^2$ between original and smoothed data is kept above 0.91. All the sites show similar trends at large wavelengths, whereas different behaviors have been found at wavelengths shorter than 500 nm. Indeed, Golden (CO) and San Jose (CA) have similar broadband hemispherical transmittance and the %T overlaps between 500 nm and 1000 nm. The same happens for Jaen and Tezpur. Below that value, the curves diverge, having Golden (CO) and Tezpur a steeper drop occurring at shorter wavelength than San José (CA) and Jaén. This result suggests that other factors might have an impact on the spectral losses of soiling.

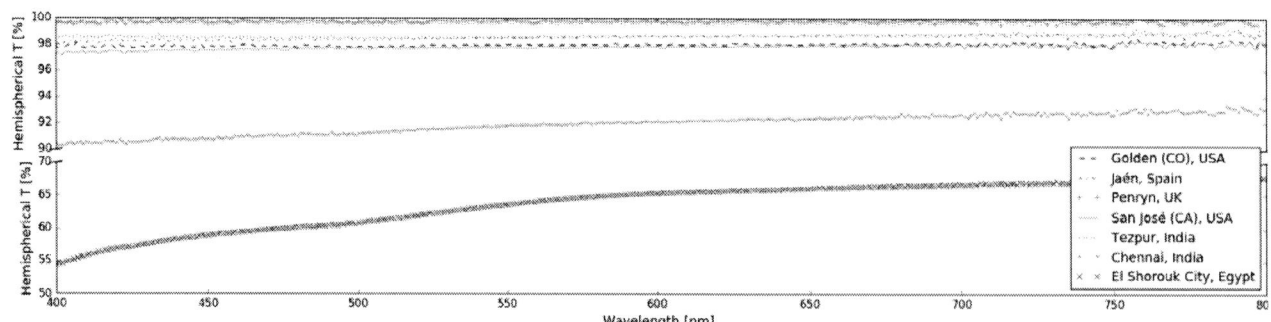

Fig. 5. Hemispherical transmittance in the visible range of coupon 5, referenced to the transmittance of coupon 0. The spectra were measured using a PerkinElmer Lambda 1050 UV/Vis spectrophotometer with a 150 mm integrating sphere.

## TABLE IV
BROADBAND HEMISPHERICAL TRANSMITTANCE (300-2500 MICROMETERS), AVERAGE PARTICLE AREA, AND PERCENTAGE OF THE SURFACE COVERED BY PARTICLES, MEASURED AT THE END OF THE DATA COLLECTION. UNSOILED GLASS TRANSMITTANCE IS 90.4%

| City, Country | Hemispherical transmittance [%] | Average particle area [$\mu m^2$] | Area coverage [%] |
|---|---|---|---|
| Chennai, India | 84.2 | 132-168 | 5.1-8.3 |
| El Shorouk City, Egypt | 63.1 | 110-194 | 21.3-22.8 |
| Golden (CO), USA | 88.8 | 55-100 | 1.7-2.4 |
| Jaén, Spain | 89.3 | 33-92 | 1.3-1.4 |
| Penryn, UK | 90.1 | N.A. | N.A. |
| San José (CA), USA | 88.5 | 206-220 | 1.9 |
| Tezpur, India | 89.6 | 47-60 | 0.3-0.4 |

From a visual inspection of the photographs in Fig. 7, it can be seen that a larger area density of particles was collected in Egypt than for any other location. This is in agreement with the results of the analysis conducted using the image processing program ImageJ [8] summarized in Table IV. The larger density may be responsible for its higher transmission losses. Indeed, a linear correlation, with $R^2$ higher than 0.99, is found by comparing the percentage area covered by particles to the hemispherical transmission. This means that, despite the spectral losses, the broadband hemispherical transmission could be directly obtained from the covered area, independently of dust type and composition. Due to the low amount of soiling that occurred at Penryn, UK site, no particle characterization was performed using optical microscopy.

## IV. CONCLUSIONS

Soiling is an issue affecting PV systems worldwide and depends on a number of site-specific factors. The main aim of this work is the comparison of naturally-accumulated soiling on PV glass at diverse regions around the world. Identical glass coupons have been exposed and cleaned at fixed time intervals. Transmissivity has been measured weekly and the preliminary results highlight similarities and differences. Weekly broadband losses are as high as 3% and 12% in

hemispherical and direct transmission, respectively, and are greater in the blue and UV portion of the spectrum. From a preliminary analysis of the data, it has been found that there is a linear correlation between the area covered by particles and the broadband hemispherical transmittance, independently of the location of soiling. A more detailed investigation of the spectral losses and their relation to the soiling type, particle size and composition should be pursued in the future.

## ACKNOWLEDGEMENTS

This work is part of the "Global investigation on the spectral effects of soiling losses" project, conceived and financed under the EPSRC SUPERGEN SuperSolar Hub's "International and industrial engagement fund". This work was partly supported by the U.S. Department of Energy under Contract No. DE-AC36-08GO28308 with Alliance for Sustainable Energy, LLC, the Manager and Operator of the National Renewable Energy Laboratory.

The U.S. Government retains and the publisher, by accepting the article for publication, acknowledges that the U.S. Government retains a nonexclusive, paid-up, irrevocable, worldwide license to publish or reproduce the published form of this work, or allow others to do so, for U.S. Government purposes.

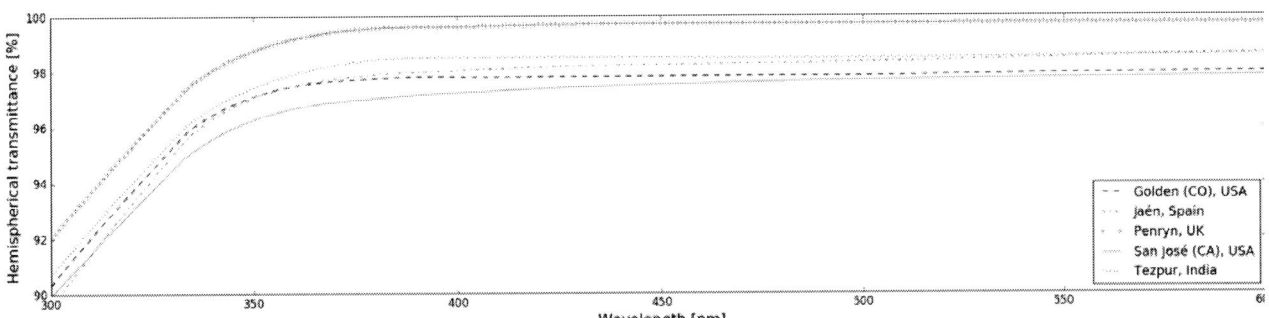

Fig. 6. Hemispherical transmittance in the visible and NIR range of coupon 5 for five low soiling sites, referenced to the transmittance of coupon 0. The spectra were measured using a PerkinElmer Lambda 1050 UV/Vis spectrophotometer with a 150 mm integrating sphere and processed using a local regression technique to remove noise.

## REFERENCES

[1] T. Sarver, A. Al-Qaraghuli, and L. L. Kazmerski, "A comprehensive review of the impact of dust on the use of solar energy: History, investigations, results, literature, and mitigation approaches," *Renew. Sustain. Energy Rev.*, vol. 22, pp. 698–733, 2013.

[2] S. C. S. Costa, A. Sonia, A. C. Diniz, and L. L. Kazmerski, "Dust and soiling issues and impacts relating to solar energy systems : Literature review update for 2012 – 2015," *Renew. Sustain. Energy Rev.*, vol. 63, pp. 33–61, 2016.

[3] W. Javed, Y. Wubu likasimu, B. Figgis, and B. Guo, "Characterization of dust accumulated on photovoltaic panels in Doha, Qatar," *Sol. Energy*, vol. 142, pp. 123–135, 2017.

[4] T. Khatib, H. Kazem, K. Sopian, F. Buttinger, W. Elmenreich, and A. Said Albusaidi, "Effect of Dust Deposition on the Performance of Multi-Crystalline Photovoltaic Modules Based on Experimental Measurements," *Int. J. Renew. Energy Res.*, vol. 3, no. 4, pp. 850–853, 2013.

[5] P. D . Burton, L. Boyle, J. J. M. Griego, and B. H. King, "Quantification of a Minimum Detectable Soiling Level to Affect Photovoltaic Devices by Natural and Simulated Soils," *IEEE J. Photovoltaics*, vol. 5, no. 4, pp. 1143–1149, 2015.

[6] H. Qasem, T. R. Betts, H. Müllejans, H. AlBusairi, and R. Gottschalg, "Dust-induced shading on photovoltaic modules," *Prog. Photovoltaics Res. Appl.*, vol. 22, no. 2, pp. 218–226, 2014.

[7] P. D. Burton and B. H. King, "Artificial soiling of photovoltaic module surfaces using traceable soil components," *Conf. Rec. IEEE Photovolt. Spec. Conf.*, pp. 1542–1545, 2013.

[8] M. D. Abramoff, P. J. Magalhaes, and S. J. Ram, "Image Processing with ImageJ," *Biophotonics Int.*, vol. 11, no. 7, pp. 36–42, 2004.

[9] M. Kottek, J. Grieser, C. Beck, B. Rudolf, and F. Rubel, "World map of the Köppen-Geiger climate classification updated," *Meteorol. Zeitschrift*, vol. 15, no. 3, pp. 259–263, 2006.

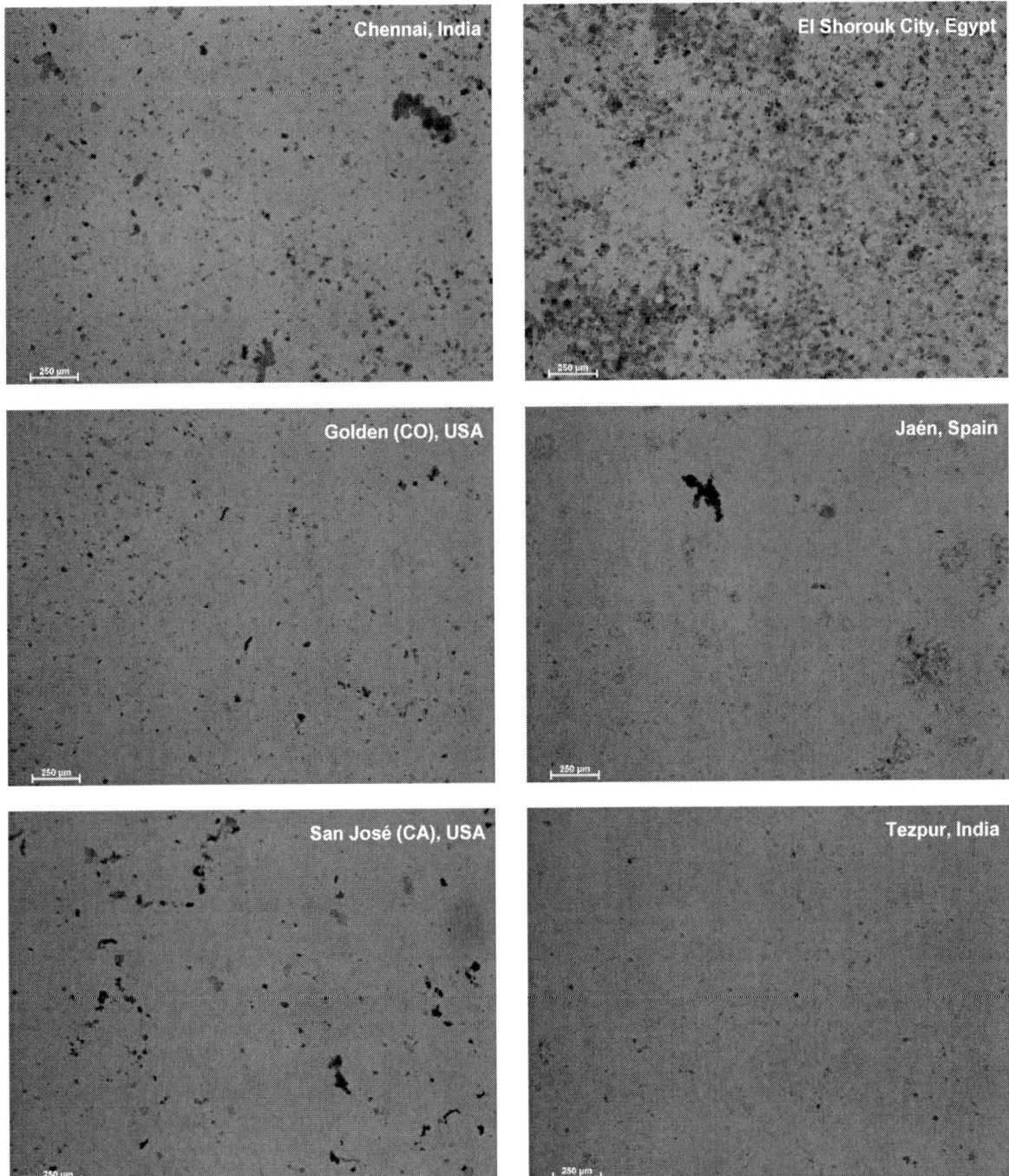

Fig. 7.    Microscope pictures of six coupons at the end of the data collection. Pictures have been taken using a Nikon SMZ 1500 stereomicroscope at a magnitude of 5×: the scale bar on the bottom left represents a length of 250 μm.

Reference: Proceedings of the IEEE PVSC Conf., 2017

# The Development of a DC Breakdown Voltage Test for Photovoltaic Insulating Materials

David C. Miller[1], Bernt Åke-Sultan[2], Axel Borne[3], Rene Eugen[4], Bradley L. Givot[5], Jürgen Jung[6], Steven W. MacMaster[7], Byron K. McDanold[1], Ulf H. Nilsson[2], Nancy H. Phillips[7], Ian A. Tappan[1], and Nick S. Bosco[1]

[1]National Renewable Energy Laboratory (NREL), 15013 Denver West Parkway, Golden, CO 80401, USA; [2]Borealis AB, 44486 Stenungsund, Sweden; [3]DuPont Photovoltaic and Advanced Materials, Meyrin, Switzerland ; [4]Isovoltaic AG, Isovoltaicstraße 1, Lebring, Austria, 8403; [5]The 3M Company, 3M Center, 201-BW-03, St. Paul MN 55144 USA; [6]Agfa-Gevaert NV, Septestraat 27, Mortsel, Belgium 2640; [7]DuPont Photovoltaic Solutions, Wilmington, DE, USA

*Abstract* — The ability of electrical insulating materials within a module to act as insulators is a key safety requirement for PV technology. Direct current breakdown voltage is therefore now specified for relied-upon insulator materials in the IEC 61730-1 safety standard. To fulfill that requirement, a new test method has been developed within the IEC TS 62788-2 backsheet standard for the measurement of breakdown voltage. The development of the test will be described, including the verification of the most critical parameters relative to factors such as defect population(s), dielectric medium, electrode size, electrode surface roughness, maximum current limit, moisture conditioning, number of replicate specimens, rate of voltage rise, specimen thickness, test polarity, and test temperature. Many of these parameters were specifically explored in discovery experiments as the test method was developed. An interlaboratory round-robin (R-R) experiment was conducted to further validate the test method by quantifying its repeatability and reproducibility. The materials examined in the R-R include the backsheet materials: polyethylene terephthalate (PET, two thicknesses), polyvinyl fluoride (PVF), and laminated PVF/PET/PVF ("TPT") as well as the encapsulants poly(ethylene-co-vinyl acetate) (EVA), and polyvinyl butyral (PVB). The precision of the test method as well as key factors contributing to the measurement will be described.

*Index Terms* — DC, dielectric strength, durability, reliability

## I. INTRODUCTION

The ability of electrical insulating materials within a module to act as insulators is a key safety requirement for photovoltaic (PV) technology. The module backsheet must protect personnel against inadvertent shock by preventing electrical-shunts or -arcs in PV systems operating up to 1.5 kV. Direct current (DC) breakdown voltage ($V_{BD}$) is therefore now specified as a required characteristic for relied-upon insulators (RUI) in the IEC 61730-1 safety standard [1]. In a pending amendment, IEC 61730-1 identifies that insulating materials (like backsheets) must withstand at least 2 kV plus the four times system voltage, *i.e.*, 8 kV for a Class II 1.5 kV system. In order to verify insulation, a test method has been developed within the IEC 62788-2 backsheet standard [2] for measurement of the $V_{BD}$.

An interlaboratory study was conducted to support the development of the $V_{BD}$ test method. Our study includes a round-robin (R-R) experiment, shared between seven institutions, as represented by the authors of this paper. Additional discovery experiments were conducted to improve understanding of the $V_{BD}$ test method. The goals of the breakdown voltage study described here include:

•Quantify the repeatability and reproducibility of the test method using an interlaboratory study.

•Quantify the significance of relevant test parameters, including: defect population(s), dielectric medium, electrode size, electrode surface roughness, maximum current limit, moisture conditioning, number of replicate specimens, rate of voltage rise, specimen thickness, test polarity, and test temperature.

## II. BACKGROUND

Electrical breakdown occurs when the dielectric insulating barrier properties are exceeded, causing an insulator to become electrically conducting [3]. The voltage level at which this occurs, the breakdown voltage, depends on a number of factors. Defects that might be present are important, including: contaminants, voids, protrusions, clusters of additives, interfaces (commonly present in PV backsheet products), and near soldering joints. In addition to this, the breakdown voltage will be affected by test parameters, such as the voltage type (AC or DC), dielectric medium, ramping rate, and temperature [4],[5]. The relative importance of these factors might be different for AC and DC breakdown.

Because breakdown is often initiated at defects, it is frequently modeled using a Weibull distribution [6]. A two-parameter model may be used to represent the cumulative distribution function (*cdf*) as shown in Equation 1, where: $V_j$ represents the breakdown voltage of the *j*th specimen {V}; α represents the Weibull scale parameter {V}; and β represents the Weibull shape parameter, characterizing the variability of the results {unitless}. A change in the slope of the Weibull fit may indicate multiple enabling defect types. IEC 62539 [7] specifies the analysis of breakdown voltage data, including the determination of α and β, the censoring of outliers, and the identification of upper- and lower-confidence limits.

$$cdf = 1 - \exp\left[-1 \cdot \frac{V_j}{\alpha}\right]^{\beta} \qquad (1)$$

Edition 1 of IEC 61730-2 contained a controversial Partial Discharge (PD) test that could be performed according to IEC 60664-1 [8] using a ramped AC signal in air or in transformer oil. The method described here, which follows IEC 60243-2 [9], is tailored towards the PV application and is intended to control relevant experimental variables to enable consistent measurements between laboratories. First, the test is performed using a controlled DC ramp because PV modules are operated with DC bias. Second, the test is performed in a dielectric medium (e.g., transformer oil) to prevent flashover (breakdown of the surrounding medium) and limit corona discharge. (Performing the $V_{BD}$ test in air can require the use of specimen sizes >1m² to avoid flashover for test voltages above 50 kV.) It is anticipated this test will be adopted and replace the PD test in Edition 1 of IEC 61730-2 in an amendment to Edition 2 of IEC 61730-1.

### III. MATERIALS AND METHODS

#### A. Round-Robin Experiment

Six materials were examined in the R-R experiment including the backsheet materials: polyethylene terephthalate (PET, 2 thicknesses), polyvinyl fluoride (PVF), and a laminated PVF/PET/PVF ("TPT"), as well as the encapsulants poly(ethylene-co-vinyl acetate) (EVA), and polyvinyl butyral (PVB). The EVA was unformulated so that the labs could examine a more similar (albeit thermoplastic) material and to avoid the influence of other additives. The encapsulants were both untextured at both surfaces and unlaminated (to prevent microvoids), which may introduce additional variability between samples, reducing the ability to determine the precision of the measurement technique itself. All specimens were at least 50 mm x 50 mm in size. Except where noted, specimens were conditioned at 23 ± 2°C and 50 ± 5% relative humidity (RH) for at least 24 h before testing. The thickness of the backsheet specimens was verified at laboratory 0, 1, and 4 using a digital micrometer. For example, the thickness was taken from the average of 12 measurements, made from the four sides of the periphery of the specimen sheet before they was diced into 50 mm² squares. Encapsulation thickness was also specifically verified during testing at laboratory 5.

The $V_{BD}$ test was performed in a dielectric medium consistent with IEC 60296 [10], e.g., transformer oil or mineral oil as identified in Table III. The test duration was at least 10 seconds, as controlled by the rate of the DC voltage ramp (specified rates include 0.1, 0.2, 0.5, 1, 2, or 5 kV·s⁻¹). Positive polarity for the test was defined as the high potential being connected to the air (outward facing) side of the backsheet with the negative terminal connected to the sun (inward facing) side. The maximum current during testing and breakdown was controlled by the test circuit and/or an external resistor.

For the R-R experiment, the participating laboratories examined five separate replicate specimens of each material. In the case of any of the results varying by more than 15% of the average, five additional replicates were tested. In these instances, the dielectric strength was determined from the median of the 10 replicates. The variance was calculated from the standard deviation (S.D.) of all measured replicates [9]. Repeatability (r) and reproducibility (R) were then determined according to ISO 5725-2 [11]. At laboratory 4, fifty replicate specimens were tested for each backsheet material and twenty replicates were tested for each encapsulant material to enable a Weibull statistical analysis of the results. To quantify the variance as a function of the number of replicate specimens, thirty separate subsets of the replicate measurements were chosen at random (with no replacement for n = 5, 10, 15, 20, 25, 30, 35, 40, 45) from the original set of fifty or twenty measurements.

In addition to the repeated material measurements conducted by the laboratories participating in the R-R using the proposed method, the independent effect of several test parameters were investigated, as described in the following sections.

#### B. Electrode Surface Roughness and Diameter

Different test fixtures were used at laboratory 4 to examine the effect of electrode surface roughness, as shown in Fig. 1 including (a) a set of brass electrodes with a rough surface, and (b) the standard smooth brass surface for the vendor test fixture (TF2, Phenix Technologies, Inc.). The rough 360 brass (62% Cu/36% Zn) electrodes were made to the same nominal dimensions as the TF2 components, but were machined with a high rate of feed to leave a coarse textured surface.

Fig. 1. Test fixtures used in the surface roughness experiment: (a) rough machined, and (b) standard smooth brass surfaces.

Equal diameter electrodes (diameter 25 mm, curvature 3 mm) were used in the R-R. A subset of additional characterizations was performed on the R-R materials with unequal diameter (25 mm and 75 mm) electrodes.

#### C. Dielectric Medium

Direct comparison of two dielectric medium (Shell Diala S2 ZX-A and Mobil Univolt N61B) was performed at laboratory

4. Both oil products fulfill the requirements of ASTM D3487 [12], but do not meet the specifications of IEC 60296 [10]. The original version of the $V_{BD}$ test method only required the dielectric medium fulfilling IEC 60296. Measurements were first performed using Shell S2 ZX-A [laboratory 4(a)], then performed again for Mobil Univolt N61B [4(b)]. Ten replicates of the backsheet materials were tested for each oil.

### D. Specimen Conditioning

Specimen conditioning was examined, including an isothermal series (65°C and 10%, 20%, 40%, 60%, 80%, or 95% *RH*), the ISO 291 test condition (23°C/50% *RH*) [13], and desiccated specimens (25°C/<1% *RH*). Separate unaged sets of ten replicate TPT specimens were maintained at the aforementioned conditions for at least 24 hours before testing.

### IV. RESULTS AND ANALYSIS

#### A. Round-Robin Experiment

Table III summarizes the results of the R-R experiment, including details of the experiment as well as the measured $V_{BD}$ (median and 2 S.D. variation) for the six materials examined. Instances where 10 replicates (rather than just 5) were required to fulfill the original ±15% variability criteria are indicated in bold. The average $r$ (variation within the laboratories) of ~10 kV is observed for all of the materials, whereas $R$ (including the variation between the laboratories) was ~15 kV. Laboratory 0 was found to approach or fall below the $h$ lower limit for an interlaboratory study, therefore it was censored from the R-R. (The gray color in Table III for laboratory 0 reflects that it was not considered for subsequent analysis, including $r$ and $R$). The low results for laboratory 0 were attributed to the use of fluorinert fluid as the dielectric medium. The dielectric constant of fluorinert fluid is 1.9, whereas that of insulating mineral oils typically ranges from 2.2-2.3. The dielectric medium was identified to reduce the breakdown voltage and increase its variability based on the concentration of electric field in Ref. [4], when the dielectric constant of the specimen was less than that of the medium. Because of the results of the round-robin, only mineral oils were allowed to be used in the final test method.

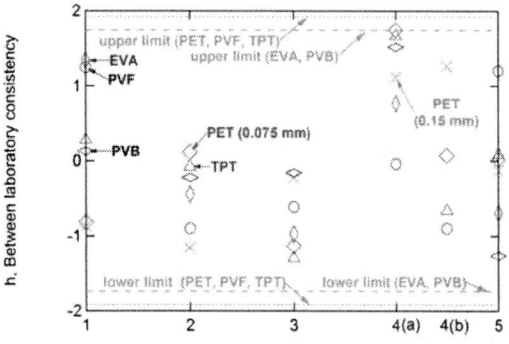

Fig. 2. Mandel analysis for between-laboratory variability. The upper and lower 95% confidence interval limits are indicated.

Fig. 2 shows the analysis of the Mandel statistics for the R-R [11]. While all measurements fall within the 95%

confidence interval limit for between laboratory variability ($h$), three of the materials measured at laboratory 4 (PET[0.075], PVB and TPT) do approach the upper limit. As shown in Fig. 3, $k$ (the within-laboratory variability) falls within the range of typical variation at four out of five laboratories. While several of the largest $k$ values were observed for laboratory 5, even exceeding the 95% confidence interval limit for PET (0.075 mm), the variability for EVA and PVB at laboratory 5 was less than at the other labs.

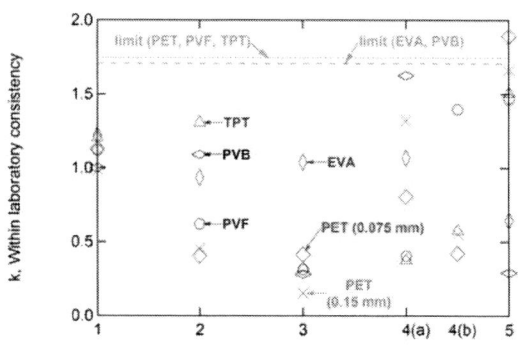

Fig. 3. Mandel analysis for within laboratory variability. The 95% confidence interval limit is indicated.

The $V_{BD}$ measured for PET (0.15 mm) at laboratory 4(a) was >100 kV for Shell S2 ZX-A oil (including several measurements at or beyond the 100 kV limit of the instrument), nearly twice the mean value measured by all other participants. This high measurement was also observed at laboratory 4(a) for a replicate set of ten PET (0.15 mm) specimens specifically conditioned to ISO 291 [13]. The result at laboratory 4(a) is consistent with a material of twice the thickness, PET (0.15mm), exhibiting double the $V_{BD}$ value of the thinner PET (0.075mm), 45±8 kV. $V_{BD}$ was found to range between 64 kV and 96 kV at laboratory 4(b). While laboratory 5 also measured the $V_{BD}$ for PET (0.15 mm) to be over 100 kV for several samples, additional low value measurements resulted in a lower mean value with a correspondingly high variation, 68±45 kV. The remaining three laboratories reported lower $V_{BD}$ values for PET (0.15 mm), however, with relatively large variabilities when compared to the other materials. These results, together with the small change between the thin and thick PET measurements, suggest that the large measurement variability for the PET (0.15 mm) may be the result of an intrinsic variability in the material and not an inconsistency in the test method between laboratories.

The measured specimen thickness average and its variation (2 S.D.) are identified in Table III for the test materials. The measured values are comparable to the nominal thickness of 0.037, 0.075, 0.15, 0.17, 0.20, and 0.45 mm for PVF, PET, PET, TPT, EVA, and PVB, respectively. Process control data from one of the backsheet manufacturers identified the standard deviation of the material thickness to be in the range of ±1.25%, with 99.7% of the thickness values being within ±4%. The spring-loaded test fixture at laboratory 5, however, reduced the thickness to 0.16 and 0.3 mm for the PVB and EVA encapsulants, respectively. The compressive test fixture (and corresponding reduced thickness) may contribute to the reduced variability for the encapsulants at laboratory 5.

A Weibull analysis of the replicate specimens is shown in Fig. 4. The Weibull analysis compares the survival function (S) to the breakdown voltage. A fit is provided to the datasets, analyzed according to the maximum likelihood estimation method recommended in Ref. [7] for $n \geq 20$. Unlike the other materials, the PET materials are labeled with an arrow at perceived inflections in their data distributions.

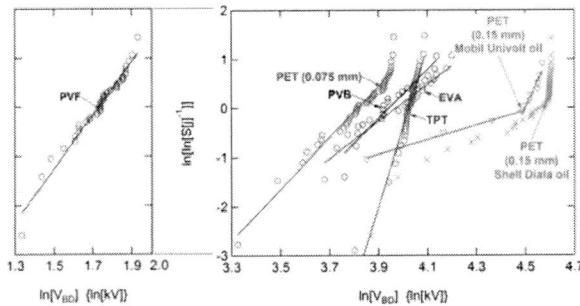

Fig. 4. Weibull analysis, for 50 replicate backsheet specimens or 20 replicate encapsulant specimens, at laboratory 4(a).

The monotonic distribution of data relative to the fit for PVF, PVB, TPT, and EVA suggests that a single population of defects causes the dielectric breakdown of those materials. The upward inflection for PET (0.075 mm) suggests a second population of defects may contribute to its breakdown at the highest test voltages. When measured in the Mobil Univolt oil at laboratory 4(b) for 10 data points only, two populations are observed for PET (0.15 mm); when measured in Shell Diala oil for laboratory 4 (a) for 50 data points, the measurements are shifted to higher voltage, exceeding the instrument capability of 100 kV. The inflection observed at laboratory 4(b) for PET (0.15 mm), suggests multiple defect populations. Except for PET, multiple defect populations cannot explain the high $h$ (between laboratory variability) for the materials examined in the R-R.

The effect of the analysis of the number of replicate specimens is examined in Fig. 5 for PET (0.075 mm) and TPT. The results are shown for the average and variation (2 S.D.), normalized relative to the result for $n = 50$. The variation in $V_{BD}$ for the materials is on the order of $\pm 3\%$ for $n = 5$ and on the order of $\pm 1.5\%$ for $n = 10$.

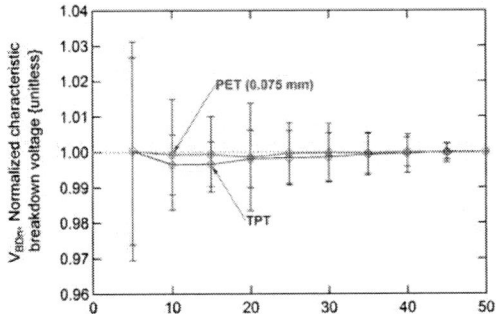

Fig. 5. Variation in $V_{BDc}$ with number of replicate specimens.

The representative results suggest that the variability in $V_{BD}$ is noticeably reduced between 5 and 10 replicate specimens. The

use of at least 10 replicates is recommended in IEC 62539 [7] to reduce the variation of measurement. Upon review of this analysis as well as the precision of the R-R experiment, the use of 10 replicate specimens was chosen for the final test method to compromise between the accuracy and cost of experiments (including for weathered specimens). Five replicates were typically examined in Table III. Improved precision is therefore expected in future experiments where 10 replicates would be examined.

### B. Electrode Surface Roughness and Diameter

The examination of electrode surface finish is summarized in Table I. Following the method in IEC 62539 [7], the table gives the lower 90% confidence limit, nominal value (for the $63^{rd}$ percentile, $\alpha = 1 - e^{-1}$, in boldface), and the upper 90% confidence limit for both $\alpha$ and $\beta$. In most cases, the results for smooth and rough electrodes were identical within statistical variation. The $V_{BD}$ is slightly reduced when the Mobil oil was used with a smooth electrode; the variance was increased when the Mobil oil was used with a rough electrode.

TABLE I: COMPARISON OF ELECTRODE SURFACE ROUGHNESS.

| Specimen | Surrounding Medium | Electrode | $\alpha$, Weibull scale parameter {kV} | $\beta$, Weibull shape parameter {unitless} |
|---|---|---|---|---|
| TPT | Shell Diala S2 ZX-A | smooth | 54-**56**-57 | 17-**25**-43 |
| | | rough | 54-**56**-57 | 15-**22**-37 |
| TPT | Mobil Univolt N61B | smooth | 49-**50**-51 | 16-**23**-39 |
| | | rough | 50-**54**-57 | 6-**9**-15 |

While a rough surface was speculated to reduce $V_{BD}$ based on localized concentration of the electric field, no systematic difference was observed. IEC 60243 does, however, caution to maintain good electrode surface quality.

The comparison of electrode diameter is summarized in Table II. Unlike Table III, the nominal thickness of the materials is identified for reference in Table II. The examination was performed on PVF at laboratory 2 and on a wider set of materials at laboratory 3. As in Table III, the results are given for the median and 2 S.D. variation. $V_{BD}$ is lesser for unequal electrodes in 5/6 instances, whereas the variation is greater for the unequal electrodes in 5/6 instances.

TABLE II: COMPARISON OF ELECTRODE DIAMETER.

| | Laboratory | 2 | | 3 | |
|---|---|---|---|---|---|
| | Electrodes, mm/mm | 25/25 | 25/75 | 25/25 | 25/75 |
| $V_{BD}$, Breakdown voltage {kV} | PVF (0.037 mm) | 5.5±0.7 | 6.0±2.8 | 5.6±0.4 | 3.8±1.6 |
| | PET (0.075 mm) | 41±4 | N/A | 38±4 | 24±7 |
| | PET (0.15 mm) | 45±12 | N/A | 65±4 | 57±46 |
| | TPT (0.17 mm) | 51±7 | N/A | 47±2 | 49±3 |
| | PVB (0.2 mm) | 42±9 | N/A | 41±2 | N/A |
| | EVA (0.45 mm) | 56±7 | N/A | 50±7 | 46±4 |

The use of unequal electrodes (e.g., 25 mm/75 mm) is popular in Europe, particularly for sheet specimens. The use of unequal electrodes, however, was found to decrease the breakdown voltage and increase its variation for backsheet materials. Based on the results, the use of unequal electrodes was not allowed in the final test method.

## C. Dielectric Medium

The effect of the dielectric medium is presented in Table III from measurements at laboratory 4(a) (for Shell Diala S2 ZX-A oil) and laboratory 4(b) (same lab and equipment, for Mobil Univolt N61B oil) as well as in Table I for TPT (also measured at laboratory 4). In both tables, $V_{BD}$ is reduced on the order of 5 kV for PET (0.075 mm) and TPT tested with the Univolt oil. $V_{BD}$ and its variability is similar for PVF with Shell Diala or Mobil Univolt oil.

This study identifies that the dielectric medium can affect the $V_{BD}$ measurement. The magnitude of the effect, however, seems to be proportional to $V_{BD}$. The effect of the dielectric may be reduced for the final test method because only transformer oil or mineral oil may be used.

## D. Specimen Conditioning

The effect of specimen conditioning for ten replicate TPT specimens is shown in Fig. 6. Following the method in IEC 62539 [7], error bars are given in the figure based on the 90% upper- and lower-confidence intervals for a Weibull distribution. The range of absolute humidity for terrestrial locations (spanning from winter in the Arctic to tropical Asia-Pacific locations) is indicated at the bottom left of the figure. A variation on the order of 25 kV for $V_{BD}$ is observed for the isothermal test series. $V_{BD}$ is, however, less for the ISO 291 and desiccator conditioned specimens.

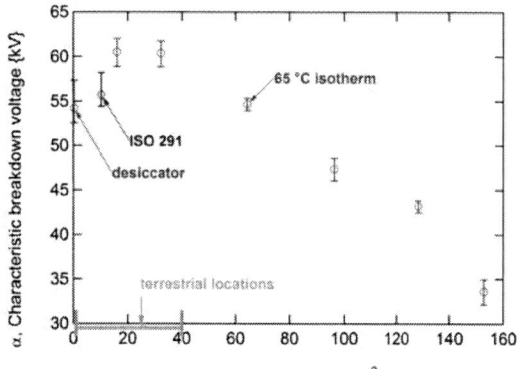

Fig. 6. Variation in α with specimen conditioning for TPT, shown relative to the range of humidity found in terrestrial locations.

These results demonstrate the importance of specimen conditioning after weathering, e.g., the Damp Heat test, where substantial moisture can be present. To explain, the dielectric constant of water at room temperature is on the order of 80, far exceeding that of the dielectric medium or specimen. Local concentration of field would be expected to affect the uniformity of the electric field, reducing $V_{BD}$ and increasing its variability. Regarding the specimens examined at conditions similar to laboratory ambient, they were not nearly as affected as specimens conditioned to elevated humidity.

## V. DISCUSSION

Regarding $h$ (the between-laboratory variability), factors including: the number of replicates; specimen moisture

conditioning; and the dielectric medium could all contribute to observed inconsistency in the R-R. Multiple defect populations could only contribute to $h$ for PET, where multiple populations were suggested in a Weibull analysis or subsequent examination of the R-R data. Based on the results of separate examinations, the variation in specimen thickness and electrode surface roughness here are not believed to contribute to $h$ in this R-R experiment. Maximum current limit and polarity were not as rigorously examined in the R-R, but do not appear to have an overt effect on the measured results.

Regarding the high variability in the measured $V_{BD}$ of PET (0.15 mm), the result at Laboratory 4 is consistent with a material of twice the thickness exhibiting twice the breakdown voltage as PET (0.075 mm). The result was also found to be repeatable with conditioning, i.e., laboratory ambient or 23 °C/50% RH (as in ISO 291 [13]). Further, factors examined in the additional experiments, including: specimen thickness; number of replicate specimens; specimen conditioning; the choice of dielectric medium; and electrode surface roughness were not identified to result in a large change in $V_{BD}$. The $V_{BD}$ of PET (0.15 mm) may have been influenced at some laboratories by multiple, and inconsistent, defect populations.

Regarding the application of the $V_{BD}$ test, the $V_{BD}$ for many of the materials in Table III well exceeds the material requirement, up to 8 kV, in IEC 61730-1. The precision of the test method may be expected to improve, e.g., $r$<10 kV or 20% and $R$<15 kV or 25%, when the number of replicates is increased from 5 to 10. This suggests that the method might be used to successfully evaluate $V_{BD}$, whether a material requirement is specified for an absolute minimum- or relative-value. Regarding a relative value, the maximum measurement limit of 100 kV is common for commercial breakdown voltage test equipment. If a 50% change from an initial value is considered, similar to a Relative Thermal Index test, a 50 kV $V_{BD}$ value is the maximum that could be evaluated using standard test equipment. A 50 kV $V_{BD}$ well exceeds the material requirements in IEC 61730-1, which may imply that the backsheet is overdesigned (e.g., unnecessarily thick) relative to the PV application. It should be noted that there are some uncontrolled variables (e.g., specimen temperature) that have not been examined in the experiments in this study. Future studies will examine field-aged and artificially weathered specimens to validate the appropriate $V_{BD}$ requirement and provide final improvement of the test method.

## VI. CONCLUSION

An interlaboratory study was conducted to quantify the precision of a recently developed test method for the DC breakdown voltage of PV materials. Additional experiments were conducted to further develop the test method and quantify the factors of greatest influence.

The round-robin experiment identifies intralaboratory repeatability on the order of 10 kV (or 25%) and interlaboratory reproducibility on the order of 15 kV (or 30%) for the DC breakdown voltage test. The precision is expected to improve for the final test method because at least 10 replicates will be required, rather the five replicates as examined in the round-robin.

Factors including the number of replicate specimens (affecting on the order of ±2%-±3%), specimen conditioning (~25 kV out of 55 kV for TPT), and dielectric medium (~5 kV out of 55 kV for TPT) were found to have a quantitative effect on the test results. Multiple defect populations were only found to be present in the PET material. The variation of specimen thickness and electrode surface roughness for the specimens and fixture in this study was not found to significantly affect the $V_{BD}$ results.

Some refinement of the original test method resulted from this study, including: 10 replicate specimens shall be used; only transformer oil or mineral oil is allowed as a dielectric medium; the use of oil qualified to ASTM D3487 may be used in addition to oil fulfilling IEC 60296; and the use of unequal diameter electrodes is not allowed. Since the study described here was completed, these refinements have been applied to the final $V_{BD}$ test method published in IEC TS 62788-2.

TABLE III: DETAILS AND RESULTS OF THE INTERLABORATORY EXPERIMENT

| | Laboratory | 0 | 1 | 2 | 3 | 4(a) | 4(b) | 5 | | |
|---|---|---|---|---|---|---|---|---|---|---|
| **Test parameters** | Instrument (Make: Model) | Phenix Technologies: 4TCE100-10/D149 | Custom | Phenix Technologies: 4TCE100-10/D149 | Haefely Hipotronics: DC Insulation Test Set | Phenix Technologies: 4TCE100-10/D149 | | Custom | | |
| | Maximum instrument voltage {kV} | 100 | 100 | 100 | 100 | 100 | | 130 | | |
| | Surrounding medium | Fluorinert-72 | Shell Diala S3 ZX-1 | Mobil Univolt N61B | Nynas Nytro 3000 | Shell Diala S2 ZX-A | Mobil Univolt N61B | Shell Diala S2 ZU-I Dried | | |
| | Ramping speed, Tedlar {kV/s} | 0.2 | 0.5 | 2 | 0.5 | 2 | 0.5 | 0.5 | | |
| | Ramping speed, other materials {kV/s} | 0.5 | 2 | 2 | 2 | 2 | | 2 | | |
| | Polarity | Negative | Positive | Negative | Negative | Negative | | Positive | r, repeatability & R, reproducibility | |
| | Current limit, mA | 11 | 0.05 | 11 | 10 | 11 | | 50 | r {kV/%} | R {kV/%} |
| **$V_{BD}$, Breakdown voltage {kV}** | PVF (0.05±0.01 mm) | 4.5±1.2 | 6.2±1.3 | 5.5±0.7 | 5.6±0.4 | 5.7±0.5 | 5.8±0.9 | 6.8±1.7 | 1.6/27 | 1.9/31 |
| | PET (0.08±0.01 mm) | 19±4 | 41±11 | 41±4 | 38±4 | 45±8 | 41±4 | 48±18 | 13/32 | 15/36 |
| | PET (0.15±0.01 mm) | 36±14 | 56±23 | 45±12 | 65±4 | >100 | 95±15 | 68±45 | N/A | N/A |
| | TPT (0.17±0.01 mm) | 47±7 | 51±6 | 51±7 | 47±2 | 56±2 | 48±3 | 50±8 | 7/14 | 11/22 |
| | PVB (0.18±0.02 mm) | 32±9 | 43±8 | 42±9 | 41±2 | 54±11 | N/A | 33±2 | 11/26 | 25/59 |
| | EVA (0.41±0.01 mm) | 40±7 | 62±9 | 56±7 | 50±7 | 59±8 | N/A | 53±5 | 16/18 | 16/29 |

## ACKNOWLEDGMENT

This work was executed by the U.S. Department of Energy under Contract No. DE-AC36-08-GO28308 with the National Renewable Energy Laboratory. Instruments and materials are identified in this paper to describe the experiments. In no case does such identification imply recommendation or endorsement by NREL. Funding provided by the U.S. Department of Energy Office of Energy Efficiency and Renewable Energy Solar Energy Technologies Office.

## REFERENCES

[1] "IEC 61730-1 Photovoltaic (PV) module safety qualification - Part 1: Requirements for constructiong," Edition 2, International Electrotechnical Commission: Geneva, 1–51 (2016).

[2] "IEC TS 62788-2 Measurement procedures for materials used in photovoltaic modules - Part 2: Polymeric materials used for frontsheets and backsheets", International Electrotechnical Commission: Geneva, in preparation.

[3] J.K. Nelson, "Breakdown Strength of Solids," in R. Bartnikas and R.M. Eichorn Engineering Dielectrics Volume IIA Electrical Properties of Solid Insulating Materials: Molecular Structure and Electrical Behavior. ASTM STP 783: Philadelphia, 1979, 445-520.

[4] I.L. Hosier, A.S. Vaughan, R.D. Chippendale, "Permittivity Mismatch and Its Influence on Ramp Breakdown Performance," Proc. IEEE Intl. Conf. Solid Dielectrics, 2013, 664-647.

[5] R. Bartnikas, "High Voltage Measurements," in R. Bartnikas and R.M. Eichorn, Engineering Dielectrics Volume IIB Electrical Properties of Solid Insulating Materials: Molecular Structure and Electrical Behavior. ASTM STP 926: Philadelphia, 1987, 157-220.

[6] C. Chauver and C. Laurent, "Weibull Statistics in Short-Term Dielectric Breakdown of Thin Film Polyethelyene Films," IEEE Trans. Elect. Insulation, 28 (1), 1993, 18-29.

[7] "IEC 62539 Guide for the Statistical Analysis of Electrical Insulation Breakdown Data," International Electrotechnical Commission: Geneva, 1–53 (2007).

[8] "IEC 60664-1 Insulation coordination for equipment within low-voltage systems - Part 1: Principles, requirements and tests," International Electrotechnical Commission: Geneva, 1–148 (2007).

[9] "IEC 60243-2 Electric strength of insulating materials - Test methods - Part 2: Additional requirements for tests using direct voltage", International Electrotechnical Commission: Geneva, 1–20 (2013).

[10] "IEC 60296 Fluids for electrotechnical applications - Unused mineral insulating oils for transformers and switchgear", International Electrotechnical Commission: Geneva, 1–17 (2003).

[11] "ASTM E691 Standard Practice for Conducting an Interlaboratory Study to Determine the Precision of a Test Method," ASTM International, West Conshohocken, PA, 1–24 (2011).

[12] "ASTM D3487 Standard Specification for Mineral Insulating Oil Used in Electrical Apparatus," ASTM International, West Conshohocken, PA, 1–6 (2016).

[13] "ISO 291 Plastics -- Standard atmospheres for conditioning and testing," International Standards Organization: Geneva, 1–12 (2008).

# Field-Evaluation of Electrodynamic Screens for Maintaining High Optical Efficiency Operation of Solar Collectors

Cristian Morales[1], Annie Bernard[1], Ryan Eriksen[1], Julius Yellowhair[2], Sean Garner[3], Ricci La Centra[1], Alecia Griffin[1], Alexis Lloyd[1]; Yujie Gao[1], Ramakrishnan Lakshmanan[1], Mark Horenstein[1], and Malay Mazumder[1]

[1]Boston University, MA, USA, [2]Sandia National Lab. NM., [3]Corning Research & Development Center, Corning. NY, USA.

*Abstract* - **Performance of the electrodynamic screen (EDS) films, laminated on solar panels and solar mirrors, is reported based on their evaluation for dust removal efficiency in simulated desert climates within an environmental chamber and in two solar fields. Design of field-test units incorporating EDS film laminated solar panels and mirrors is described. EDS samples demonstrated continued performance after months of exposure to outdoor conditions with slight decreases in specular reflectance restoration. We report here the stability of the EDS film activated by three-phase high voltage pulses against moisture and air ingress, environmental degradation, windstorm abrasion, the effects of humidity cycling, and UV radiation.**

*Keywords* - **Dust, efficiency, electrostatic, screen, self-cleaning, solar.**

## I. Introduction

Utility-scale solar energy generation facilities, both those using photovoltaics (PV) and concentrated solar power (CSP), are often located in arid, desert environments to take advantage of the greater availability of effective sun hours. Unfortunately, these desert locations also have high levels of dust in their environments, and wind can lead to these dust particles being deposited and accumulating on the surface of PV panels and CSP mirrors. This dust accumulation can decrease power generation efficiency by 5 to 40% annually. [1]

One solution being investigated to address the problem of dust deposition on PV panels and CSP mirrors is the electrodynamic screen (EDS). [2]-[3] The EDS is a series of interdigitated, parallel electrodes embedded within two transparent dielectric films, which generates an electrostatic force when a phased voltage wave is applied across the electrodes, illustrated in Fig. 1. This electrostatic force is then used to charge dust particles on the EDS surface and direct their movement off the PV panel or CSP mirror the EDS film is laminated on.

EDS films laminated on solar collectors in field operation would be exposed to environmental factors such as UV light, high wind speeds, high temperatures, and moisture. Each of these factors has the potential to damage an EDS film and adversely affect its performance: exposure to UV radiation could degrade the EDS electrode material or dielectric layers, high wind speeds can lead to high velocity dust particles impacting the EDS surface causing abrasion, high temperatures coupled with humidity and rain could create moisture ingress reaching the electrodes and creating short circuits between them.

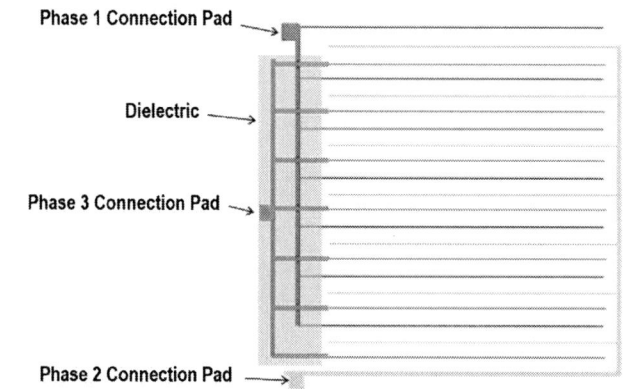

Fig. 1.    The electrode geometry of a three-phase EDS.

## II. Experimental

To determine the effects environmental factors have on EDS performance and durability, experiments were conducted both at outdoor facilities and in-lab. These tests were conducted on EDS samples produced in-lab via screen printing and on samples produced via gravure offset printing (GOP) at professional manufacturing facilities.

The outdoor tests consisted of: A.i) a four month experiment at Sandia National Labs, A.ii) a six month experiment at Sandia National Labs, A.iii) an ongoing six month experiment in Chile's Atacama Desert, and A.iv) a two month experiment in Boston.

The in-lab experiments were conducted to simulate specific, isolated environmental conditions and test for their effects, a task difficult to perform with the field experiments. These in-lab tests consisted of: B) a UV radiation test simulating accelerated UV radiation, C) a dust impact test simulating sandstorm-accelerated dust particle impacts, D) a humidity cycling test simulating the daily rise and fall of humidity in deserts, E) a water submersion test simulating rainfall ingress into EDS samples, and F) a three-month EDS operation test to simulate extended, regular use of the EDS over time.

### A. Field evaluation of EDS-laminated solar collectors

Evaluating the durability and operation of EDS samples in outdoor conditions requires testing setups durable enough to withstand environmental conditions while housing the EDS-laminated solar collector samples and a pulsed high-voltage power supply unit. These testing setups constructed for field evaluation consisted of wooden boxes coated with varnish to increase water and UV radiation resistance and sealed to prevent moisture ingress. Each box contained a 1.2 kV three-phase power supply for EDS operation. Because an EDS is a low-power device, the 1.2 kV power supply was powered by batteries which were also housed in the testing setup.

i) Sandia National Labs, Test 1 - Initial outdoor testing of EDS samples was conducted at the Sandia National Laboratories' National Solar Thermal Test Facility (NSTTF) in Albuquerque, New Mexico. This experiment consisted of three screen-printed EDS films laminated on CSP mirrors and one EDS produced on a printed circuit board, all driven by a 1.2 kV power supply, as well as two control mirrors without EDS films on their surfaces, displayed in Fig. 2. In this experiment, the EDS performance metric being evaluated was the specular reflectance restoration (SRR). The SRR is a function of the EDS film's efficiency in removing dust from an optical surface and is calculated using Equation 1.

$$SRR = \frac{SR_{Post\,EDS\,Operation}}{SR_{Clean\,Mirror\,Surface}} * 100\% \tag{1}$$

In this experiment, the EDS panels were operated and measurements were recorded at least once a week for three months. Using a contact specular reflectometer (D&S Model 15R-USB), the specular reflectance (SR) of the mirrors were measured with a 10 mm.-diameter collimated LED beam at 650 nm wavelength with a 15 milliradian acceptance angle. The following procedure was followed: 1) measure the SR values of the control mirrors, 2) measure the SR values of the 3 EDS mirrors, 3) operate the EDS films for 2 minutes, and 4) post-EDS operation, measure the SR values of the EDS mirrors were again. These measured SR values were then used to calculate the EDS films' SRR values.

This field experiment was performed during a period of abnormally frequent summer rains at the NSTTF site, which cleared away any natural dust deposition on the EDS samples. The frequent rainfall skewed the results of the experiment in a way that prevented measuring the full extent of the specular reflectance restoration derived from the operation of an EDS. Qualitative results show the three EDS-laminated mirrors were still operating normally after four months of outdoor exposure before being removed from the site.

Fig. 2. The outdoor setup at Sandia National Labs testing EDS endurance and lifetime.

ii) Sandia National Labs, Test 2 - After this initial outdoor experiment at the NSTTF, a second experiment was conducted at the site using the same testing setup and power supply with three new GOP-produced EDS mirrors. The focus of this experiment was not on SRR but instead on the endurance of the GOP-produced EDS films when exposed to outdoor environments. For six months, these GOP-produced EDS films were periodically evaluated and demonstrated successful operation. After six months, the electrodes on the three EDS samples began shorting, leading to all samples ceasing functionality.

iii) Atacama Desert, Test 1 - An experiment testing the outdoor endurance and operation of EDS-laminated PV panels has begun at a Geodrill solar facility in Chile's Atacama Desert. In this experiment, the EDS performance metric being evaluated is the optical power restoration (OPR), calculated using Equation 2 using the short circuit current ($I_{sc}$) measured before and after activation of the EDS film.

$$OPR = \frac{I_{SC,\,Post\,EDS\,Operation}}{I_{SC,\,Clean\,Panel\,Surface}} * 100\% \tag{2}$$

Using testing setups like those used in the NSTTF experiments, 4 EDS-laminated PV panels were installed at the Geodrill facility, as displayed in Fig. 3, and are to be tested three times a week for 6 months using the following procedure: 1) measure the $I_{sc}$ of each EDS panel using a multimeter, 2) operate the EDS panels for 2 minutes, and 3) re-measure the $I_{sc}$ of each sample after EDS operation. These $I_{sc}$ measurements are used to calculate the OPR values of each EDS panel. This experiment is ongoing.

Fig. 3. The outdoor setup in Chile's Atacama Desert testing EDS SRR values over time.

iv) Boston, Test 1 - Finally, because the completed Sandia National Labs tests focused primarily on the endurance of EDS samples in outdoor conditions and not on environmental exposure's effects on SRR, an outdoor test focusing on SRR was performed. In this experiment, five screen-printed EDS-laminated mirrors were left on a rooftop exposure facility in Boston for two months, exposed to natural summer weather conditions. These conditions included temperatures up to 35 °C and rain. In comparing the SRR values of the five samples before and after the two months of environmental exposure, one sample showed no statistically significant change in SRR, three samples showed SRR decreases of 1.4, 2.2, and 2.2%, and the final sample showed an anomalous increase in SRR of 3.5%. These numbers show that although the NSTTF experiments demonstrated continued EDS functionality over the course of many months, there is an associated decrease in efficiency and SRR which needs to be accounted for and further researched.

*B. Effect of UV radiation on resistance*

The effects of long term UV radiation on EDS films must be investigated if they are to be operated in outside environments for multiple years. In-door lab tests were conducted to evaluate the effect of UV exposure on the sheet resistance of screen-printed EDS films, with a goal of remaining below 200 $\Omega/\square$. UV light could cause a chemical breakdown of the electrodes, leading to an increase in the resistance, and in extreme cases, potentially creating a break in the electrodes.

Eighteen EDS films were exposed to over 2000 hours of continuous UV radiation with their sheet resistance being measured periodically. The results of these measurements were analyzed via a paired t test. The results suggest that EDS films suffered no significant adverse effects due to UV exposure; the sheet resistance of the electrodes remained far below 200 $\Omega/\square$ (Mean increase = [0.05 $\Omega/\square$], SD = [0.017]; t (16)=[-46142.6], p=[0.0001]). Overall, changes in sheet resistance were minimal for all samples tested.

*C. Dust impact testing*

The dust impact test evaluates the effect of sandstorms on the optical transmissivity of EDS films. Airborne dust particles and debris have the potential to abrade solar plant equipment if they are travelling at high speeds, i.e. in excess of 10 m/s, such as would be encountered in a sandstorm. [4] To ensure that EDS films can withstand long-term abrasion without becoming scratched or degraded is of great importance to their continued effectiveness over time.

Three types of films were subjected to abrasion testing: an EDS film (silver paste screen-printed on Corning® Willow® Glass [5]), a Willow Glass sheet without an EDS on its surface, and a polymer film (PET). Each film was laminated onto a mirror so as to measure the changes in specular reflectance (SR) caused by abrasion.

Using the gravity-based set-up described in ASTM D968-15 [6], 10.1 grams of test dust simulating desert sand were accelerated to a terminal settling velocity of over 10 m/s. This sand was then allowed to strike a test film at a 45 degree angle. The SR for each film was measured before and after the impact of the test dust and the change in SR was calculated from these measurements.

After impact, PET films showed clear signs of abrasion and SR losses between 3.8 and 6.2%. The sheet of Willow Glass and EDS film showed no visible signs of abrasion, and changes in SR were within experimental error, ±1% or smaller. Therefore, it is reasonable to assume that EDS films printed on Willow Glass will be able to withstand sandstorms with their transmissivity intact.

*D. Humidity cycling*

Experiments were conducted to simulate the humidity cycling from a low relative humidity (RH) to a high RH and back to low RH which occurs daily at solar fields in desert environments. Because humidity cycling has the potential to increase the adhesion of the dust to the EDS glass film, it is necessary to determine if an EDS can effectively remove dust having undergone RH cycling.

To examine this process, a screen-printed EDS film was tested using the following procedure: 1) clean the sample with isopropanol and weigh it, 2) deposit dust on the EDS surface to achieve a dust density of approximately 5 g/m², 3) weigh the dust-deposited sample to determine the amount of dust deposited in grams, 3) place the dust-deposited EDS sample in a chamber with 80% RH for 30 minutes, then remove the sample and place it in a chamber at RH ≈ 50% for 30 minutes, 4) activate the EDS for a minute, and 5) reweigh the EDS sample. Using these measurements, the DRE was calculated using Equation 3.

$$DRE = \frac{Dust\ on\ EDS_{Pre\ Operation} - Dust\ on\ EDS_{Post\ Operation}}{Dust\ on\ EDS_{Pre\ Operation}} * 100\% \quad (3)$$

The initial DRE of the EDS film before humidity cycling

was recorded as 90.48% under standard test conditions (STC): EDS at a 30 degree tilt angle, 28 °C and operated on a 1.5kV power supply. After 5 trials of humidity cycling, the DRE was recorded as 93.7%.

The humidity cycling increased the DRE slightly for the samples. This may indicate that fine dust particles are adhering to other dust particles forming larger particles that are more easily removed by the EDS, increasing the DRE.

### E. Water submersion test

EDS films deployed in solar fields will be exposed to rain and dew. Because moisture can create short circuits between EDS electrodes, a submersion test was performed to evaluate how well the EDS film is encapsulated against moisture ingress. The submersion test protocol was derived from ASTM F1895 − 14, "Standard Test Method for Submersion of a Membrane Switch". [7] The metric used to determine if these submersion tests affected EDS performance was sheet resistance and phase-to-phase resistance. These were chosen because if water ingressed occurred, it would create short circuits between electrodes, noticeably decreasing sheet and phase-to-phase resistance.

The experiment was conducted on 23 EDS samples, both screen-printed and GOP-produced, divided into one set of 5 samples and a second set of 18 EDS samples, both sets containing samples of both production methods. This experiment was conducted with the following procedure: 1) edge-seal EDS samples with weatherproof 3M tapes, 2) measure the set of 5 samples for sheet resistance, and the set of 18 EDS samples for phase-to-phase resistance, $R_{p-p}$, 3) submerge both sets in a bucket of tap water for two minutes, and allow them to dry overnight, and 4) remeasure the two sets for sheet resistance or phase-to-phase resistance, respectively.

A single sample t-test was conducted to determine if a statistically significant difference existed between the sheet resistance measurements before and after submersion of the first set of EDS samples. Of the samples tested, none showed any significant difference in their sheet resistance values. The sheet resistances had a mean value of 72.85 $\Omega / \square$, well within the requirements of $R_s < 200$ $\Omega / \square$. In addition to sheet resistance measurements, the $R_{p-p}$ of the second set of EDS samples were also subjected to the t test where again there was no statistically significant difference in $R_{p-p}$ after submersion. The mean $R_{p-p}$ value was 300.7 M$\Omega$, significantly higher than the minimum requirements of $R_{p-p} > 10$ M$\Omega$.

### F. Long-term EDS film testing

EDS films in the field will be expected to operate for years, so the effects of extended operation on EDS dust removal efficiency must be investigated. With this aim, a single screen-printed EDS film was operated and had its DRE measured under STC three times a week for three months. At the beginning of the three-month period, the sample had a DRE of 91.81%. In the three months that followed, the

sample's DRE decreased and then fluctuated between 91.81% and 85.1%, as seen in Fig. 4. Upon final testing, the sample had a DRE of 87.5%.

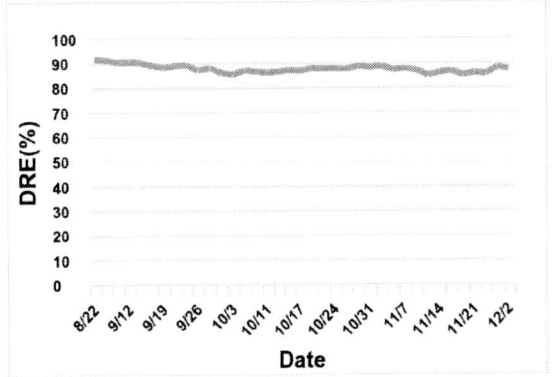

Fig. 4. The variation in DRE of a screen-printed EDS film over three months of regular operation.

### III. Summary

The two field tests at Sandia National Labs demonstrated that EDS samples can be sustain dust removal functionality throughout months of desert environmental exposure. The field test conducted in Boston demonstrated continued functionality after three months of outdoor exposure, but coupled with a small decrease in specular reflectance restoration.

The various tests conducted in-lab to simulate specific environmental conditions showed promising results for EDS environmental endurance with some environmental challenges, such as sunlight, sandstorm, or rain exposure, showing no effect on EDS functionality. One environmental challenge, humidity cycling, seemed to improve EDS dust removal efficiency. The test monitoring the effects of long-term operation on EDS dust removal efficiency did show a small decrease in EDS dust removal efficiency. This suggests that the decrease in efficiency experienced by EDS samples deployed in field situations is not due to the effects of environmental conditions, but instead due simply to the degradation of the EDS over time, independent of environmental conditions.

### Acknowledgements

We acknowledge the grant from the Department of Energy (DOE CSP APOLLO EE-0007119) for supporting this research project on "Enhancement of Optical Efficiency of CSP Mirrors for Reducing O&M Cost via Near-Continuous Operation of Self-Cleaning Electrodynamic Screens (EDS)". Support from MassCEC for a cost-sharing grant on the project is acknowledged. We also acknowledge the support from our partners including Corning Research & Development Corporation, Industrial Technology Research Institute (ITRI),

Geodrill® (EDS Chile SpA), and Sandia National Laboratories.

## References

[1] T. Sarver, A. Al-Qaraghuli, and L. L. Kazmerski, "A comprehensive review of the impact of dust on the use of solar energy: History, investigations, results, literature, and mitigation approaches". *Renewable and Sustainable Energy Reviews*, vol. 22, pp. 698–733, 2013.

[2] A. Sayyah, M. N. Horenstein, M. K. Mazumder, "Energy yield loss caused by dust deposition on photovoltaic panels". *Solar Energy*, vol. 107, pp. 576–604, 2014.

[3] M. N. Horenstein, M. K. Mazumder, R. C. Sumner, "Predicting particle trajectories on an electrodynamic screen – theory and experiment". *Journal of Electrostatics*, vol. 71(3), pp. 185–188, 2013.

[4] M. Mazumder, M. Horenstein, J. Stark, J. Hudelson, A. Sayyah, C. Heiling, J. Yellowhair, "Electrodynamic removal of dust from solar mirrors and its applications in concentrated solar power (CSP) plants". *Industry Applications Society Annual Meeting*, pp. 1–7, 2014.

[5] C. Sansom, et. al. "Predicting the effects of sand erosion on collector surfaces in CSP plants." *Energy Procedia*, vol. 69, pp. 198–207, 2015.

[6] ASTM D968-15, "Standard Test Methods for Abrasion Resistance of Organic Coatings by Falling Abrasive". ASTM International, 2016.

[7] ASTM F1895-14, "Standard Test Method for Submersion of a Membrane Switch". ASTM International, 2014

# Effect of Reverse Bias Voltages on small scale gridded CIGS Solar Cells

Soheyl Mortazavi[1], Klaas Bakker[2,3], Jorne Carolus[4], Michael Daenen[4], Gabriela de Amorim Soares[1], Henk Steijvers[1], Arthur Weeber[2,3], Mirjam Theelen[1]

[1] TNO Solliance, Thin Film Technology, High Tech Campus 21, 5656 AE Eindhoven, The Netherlands
[2] ECN Solliance, High Tech Campus 21, 5656 AE Eindhoven, The Netherlands
[3] Delft University of Technology, Photovoltaic Materials and Devices, 2628 CD Delft. The Netherlands
[4] Hasselt University, Instituut voor Materiaalonderzoek, Wetenschapspark 1, 3590, Diepenbeek, Belgium

*Abstract* Partial shading on photovoltaic modules can cause cells to operate at reverse bias conditions. An innovative experimental setup with two test protocols has been designed to simulate the effect of reverse bias voltage on small scale CIGS solar cells with a metal grid. In these tests, the effect on the performance of the solar cells exposed to either a stepwise increasing or prolonged constant negative bias voltage was studied. It was demonstrated that performance loss occurred mainly due to a reduction in shunt resistance. Moreover, some solar cells have partially recuperated after light soaking.

*Index Terms* — CIGS, reliability, partial shading, reverse bias voltage,

## I. INTRODUCTION

Partial shading on photovoltaic (PV) modules causes reverse bias on cells which can result in temporary and permanent damages. Thin Film (TFPV) technologies often differ from conventional wafer based crystalline Silicon (c-Si) PV due to the use of monolithically interconnections resulting in long narrow cells. Especially for these modules the magnitude of reverse bias is dependent on the shape of the shade, geometrical structure of the TFPV module and the intensity of the shadow [1]. For wafer based modules bypass diodes can be tailored for optimal protection against reverse bias damage [2]. Major disadvantage for the use of bypass diodes in a string of cells is that the bypass diode bypasses the contribution of the whole string to the power output of the module. This results in loss of output power of the whole string when only a fraction of the string is shaded. The monolithically interconnection scheme used in most thin film modules does not easily allow the integration of bypass diodes therefore other means to protect from or mitigate against reverse bias damage are needed.

Off all the thin film technologies Cu(In,Ga)Se$_2$ (CIGS) is one of the most promising with high efficiencies and growing sales volumes. However worst case scenario shading tests like in the IEC 61215 standard on CIGS modules showed that even very short events (~20 s) can result in a permanent relative decrease in efficiency from 4% to 14% [3].

In literature, most partial shading tests have been performed on CIGS modules rather than individual cells, which resulted in the formation of hot spots and wormlike defects [3]-[7] Additionally, from literature it has been observed that light soaking can have a recuperating effect on CIGS modules after reduction in performance due to partial shading [4],[8].

Several different studies of the reverse current voltage (IV) characteristics of CIGS solar cells and modules have been executed, experimental conditions that have been varied include the influence of light intensity [8], light spectrum [8],[9], temperature [9], sample design and composition (i.e. absorber and buffer layers thicknesses) [9] on reverse bias characteristics of solar cells. As stated by Silverman et al. [3], more information is required to be able to predict and prevent degradation due to reverse bias and to optimally profit from the light soaking effect. Currently, it is expected that the geometrical structure of solar cells, magnitude of the reverse bias, the exposure time, the relaxation time, as well as the sequence of exposures all also influence the effect of reverse bias on CIGS solar cells. However, information about the exact influence of these parameters is not yet sufficiently available. It is important to understand the effects of all these parameters on individual cells in order to effectively mitigate the effects of reverse bias in modules.

Fig. 1. (a) Iviumstat programmable potentiostat/galvanostat as central control device (b) IviSUN LED light source (c) Sample and Kelvin contact probes

The aim of this study was to define the effect of reverse bias voltages and subsequent light soaking on gridded CIGS solar cells. A setup was designed and built to investigate the parameters that influence the reverse bias behaviour, and later to define the occurring degradation mechanisms. Initial results including the influence of magnitude and exposure time of reverse bias voltage as well as the effect of light soaking obtained with this setup are presented here.

## II. Materials And Methods

### A. Experimental units

In order to execute the experiments, a commercial programmable potentiostat/galvanostat Ivium - Iviumstat has been used as the central control device of the setup. The Iviumstat is a 4-quadrant voltage/current controlled power supply that enables both forward and reverse biasing and precise four wire measurements up to ±10A and ±10V with 0.2% accuracy. (fig. 1a)

To perform measurements under illumination, a light source Ivium - IviSUN controlled by the Iviumstat has been added to the setup. It features, white LEDs (5000K) with an illuminated area of 150 x 150 mm² (fig. 1b). Due to the precise control of voltage, current, and light intensity during defined time intervals, the Iviumstat combined with the IviSUN gives the freedom to design different test protocols for different experiments.

A cooling fan has been used to continuously blow air on the surface of the cells under test to keep them at room temperature. Kelvin probes have been used to connect the cells to the Iviumstat. (fig. 1c)

For better readability all references with respect to reverse bias voltages are based on the absolute value e.g. -2V is a lower reversed bias voltage than -3V.

Due to the low intensity and non-ideal spectrum for CIGS of the LEDs utilised by the IviSUN, the intensity was estimated by comparing current densities obtained from IV and EQE measurements at standard test conditions of cells on the same substrate. This intensity of the IviSUN was roughly estimated to be 0.3 sun equivalent for the cells used and is assumed to be constant. All reported currents are calculated to the 1 sun equivalent. Series and shunt resistances are estimated from the slope at respectively the highest and lowest voltages of the IV scan that ranged from -0.3 till +0.7 Volts.

Illuminated Lock In Thermography (ILIT) pictures were made with a system from Infratec using IR LEDs for excitation. Real time infrared movies have been taken with a high speed IR camera (FLIR X6580sc) combined with a Keithley 2400 source measure unit.

### B. CIGS solar cells

Non-encapsulated CIGS solar cells were deposited on a 1 mm thick 100 mm x 100 mm soda lime glass substrate (SLG). The stack contained 500 nm DC-sputtered molybdenum (Mo)

/ 2 μm coevaporated CIGS / 50 nm chemical bath deposited cadmium sulfide (CdS) / 65 nm DC sputtered intrinsic zinc oxide (i-ZnO) / 260 nm DC sputtered aluminium doped zinc oxide (ZnO:Al) with an ebeam evaporated metal (20 nm Ni+600 nm Ag+20 nm Ni) grid on top. The metal grid was made of two tapered fingers connected to a contact island which in total covered 2% of the active area of a cell. 162 individual solar cells of 10 mm x 5 mm were obtained by mechanical division via an automated scriber utilising a diamond tipped needle to cut through the CIGS and AZO. Formation of the back contact was done by scratching away the CIGS with a scalpel and application of Ag ink as contact material on the exposed molybdenum.

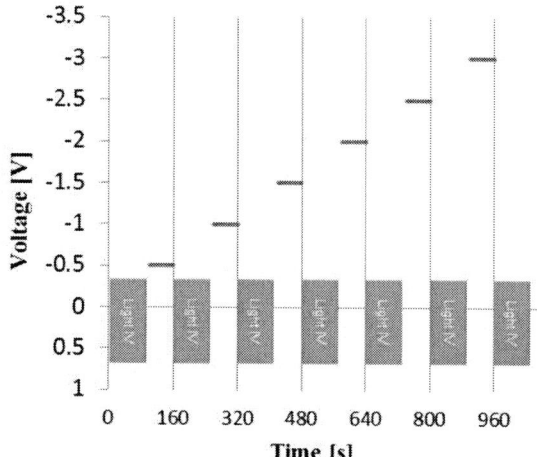

Fig. 2. Applied voltages on solar cells for the *elevating reverse bias* test protocol. Orange blocks indicate IV measurements under illumination (100 s), where the width of the block represents the duration of the measurement and the height of the block is the voltage range the IV measurements were performed (-0.3 V until 0.7 V). Blue lines are exposures to negative voltage levels (60 s).

### C. Treatments and methods

To measure the susceptibility of CIGS solar cells to negative voltages, two test protocols have been defined. The first protocol was intended to evaluate the evolution of solar cell performance as a function of the magnitude of the applied reverse bias voltages and also to find if there is a fatal breakdown voltage. For this *elevating reverse bias* protocol, the reverse bias voltage was increased in linear steps in which the upper limit was based on preliminary dark IV measurements that showed that there is a voltage at which the cell suddenly became fully shunted. This fatal breakdown voltage occurred between -3 and -4 Volt. During the *elevating reverse bias* protocol, for each step the cells were exposed to a predetermined voltage level for 60 seconds in darkness. The voltage levels varied from -0.5 to -3 V with 0.5 V increments. Before each step, an IV measurement under illumination was

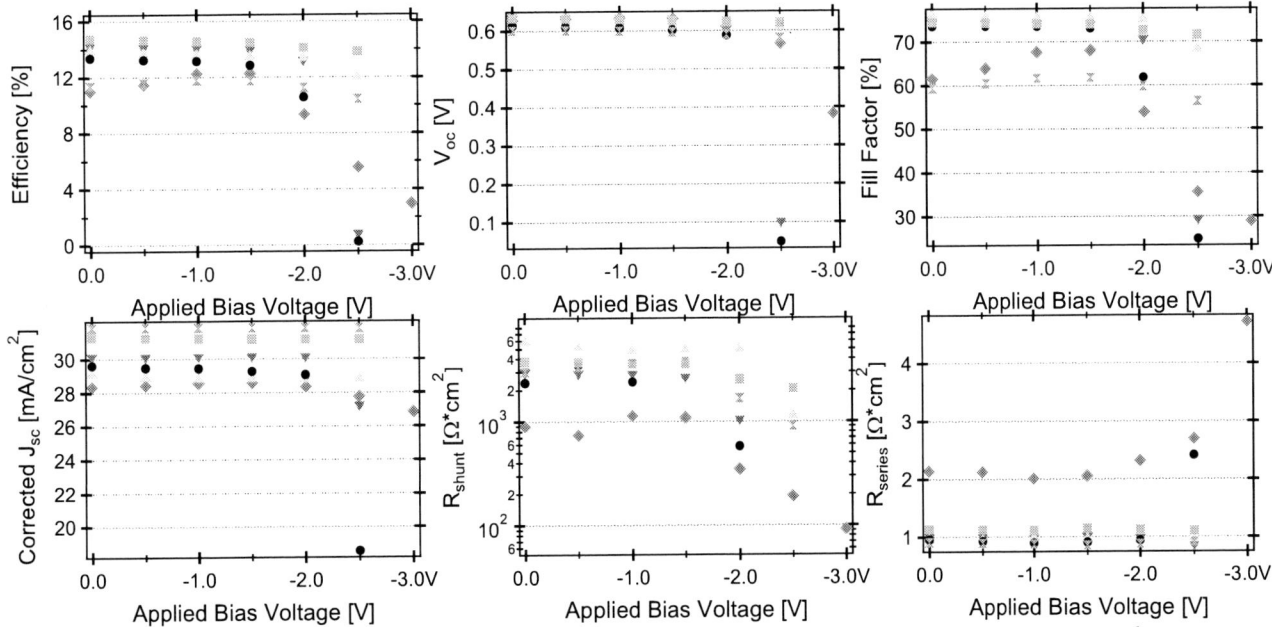

Fig. 3. IV parameters as function of applied bias voltage for 6 similar CIGS solar cells under the *elevating reverse bias* test protocol. Measurements were performed after cells have been exposed to the defined negative voltages for 60 s. The $J_{sc}$ and the efficiency were normalized to 1 sun.

performed to quantify the evolution in electrical performance. The IV measurement was executed in both forward and reverse scan direction ranging from -0.3V to 0.7V and back. The duration of the scan was 100 seconds. An illustration of the conditions applied during this test protocol is presented in fig. 2.

A second protocol has been defined to observe solar cell behaviour under elongated exposure to "mild" negative bias voltages. Two voltages were selected for this *constant negative voltage* protocol, the selection was made based on the results obtained from the *elevating reverse bias* experiment that will be described in the Results and Discussion section. For each of the 30 cycles, first an illuminated IV measurement was executed, followed by the exposure to a *constant negative voltage* for 120s in darkness.

To study the effect of light soaking after the application of reverse bias voltage, the samples have been exposed to the light for 6 hours at $V_{OC}$ conditions. During this illumination IV measurements have been performed every 10 minutes to monitor the evolution of electrical parameters.

### III RESULTS AND DISCUSSION

*Elevating reverse bias tests*

The *elevating reverse bias* protocol was executed on 6 CIGS solar cells. The evolution of electrical parameters of these 6 solar cells as a function of applied voltage is shown in fig. 3. All cells were stable to exposures to voltages smaller than -1.5V and started to display a small decrease in efficiency when exposed to -2V for 60 seconds, while larger decreases in

electrical parameters became visible after exposure to -2.5 V. Further increasing the voltage to -3V was fatal for five out of six cells. The main reason for the observed reduction in efficiency was the decrease of shunt resistance. The dramatic reduction in shunt resistance was likely also influencing other electrical parameters especially $V_{oc}$. The cell that survived the exposure to -3 V was later exposed to light soaking (not depicted). This cell displayed an increase of efficiency in absolute value from 3.0% to 7.2%. This recuperation was mainly the result of increase in shunt resistance (91 $\Omega$.cm$^2$ to 264 $\Omega$.cm$^2$) as well as in $V_{oc}$ (0.38 V to 0.6 V). None of the other cells showed any improvement after light soaking.

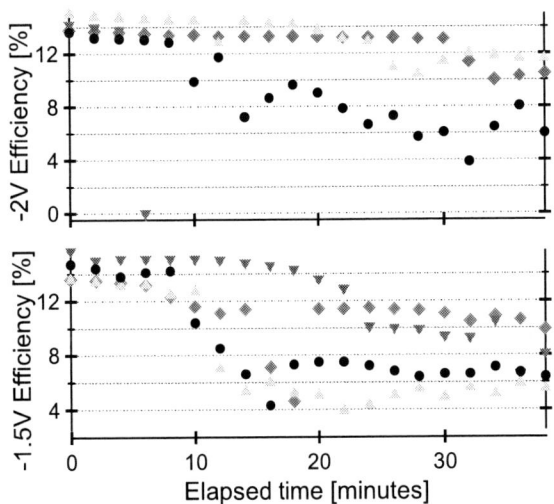

Fig. 4. Efficiency evolution for first 20 cycles for both *constant negative voltage* exposures to -2V (top graph) and -1.5V (bottom graph).

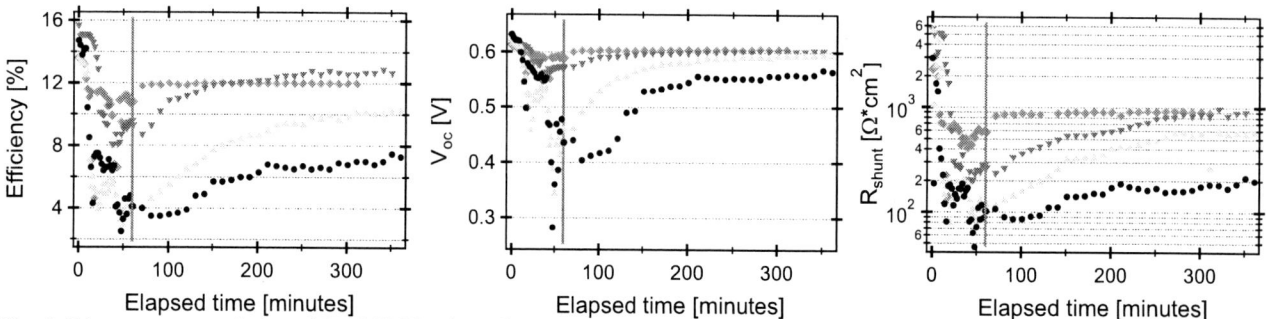

Fig. 5. IV parameters evolution of the 4 CIGS solar cells under *constant negative voltage* test protocol, exposed to -1.5V for 120 seconds after each IV measurement. The red line indicates when light soaking started.

## Constant reverse bias tests

From the results of the previous experiment two voltages were selected:

- -1.5V which was the highest voltage that cells were stable for 60s and
- -2V this was the first voltage to exhibit major changes in IV parameters.

For both voltages 4 similar samples were tested. The evolution of efficiency during the first 20 cycles for exposure to both -1.5V and -2V are graphed in fig. 4. In general a strong decrease in efficiency over time was observed for all cells exposed to constant negative bias voltages. However, every cell seems to have a point of abrupt change that occurs at different elapsed times. As with the elevating reverse bias test the changes were mostly due to a decrease in shunt resistance. Most cells exposed to constant negative voltage showed a decrease in shunt resistance already after the first one to five exposures. However, for two cells exposed to -2V it took between 10 and 18 cycles to exhibit changes in FF and shunt resistance (not depicted). Interestingly, for these cells the efficiency dropped over time due to a decrease in $V_{oc}$. The fact that no change in shunt resistance and a change in $V_{oc}$ was observed could be an indication that something is happening in the diode parameters of the solar cell. Similar observations of a separate influence of shunt resistance and diode parameters on amorphous-Silicon cells have been made by Donaonkar et al. [10]. After the first changes the behaviour of electrical parameters is not fluent and displays hills and valleys. More information is displayed in fig. 5 that shows the evolution of efficiency, $V_{oc}$ and shunt resistance for the cells exposed to -1.5V, followed by light soaking.

## Influence of light soaking

The results for all IV parameters of the cells exposed to -1.5V and consecutive light soaking over time are shown in fig. 5. The red line in fig. 5 indicates the transition from the exposure to negative bias of -1.5V to light soaking at $V_{oc}$. Not shown is that the trend for both -1.5 and -2V are roughly the same.

The reduction in efficiency due to the exposure to negative bias is partly restored by the light soaking. Cells that have been severely damaged did not recuperate or recuperated slowly. Recuperation is visible in all electrical parameters however none of the parameters came back to the initial value.

It should be mentioned that the $I_{sc}$ drops relatively about 3% after the first ten minutes and remains stable. This initial decrease is most likely due to heating of the LED light source, this drop in $I_{sc}$ however does not explain why the efficiency remains much lower than initial.

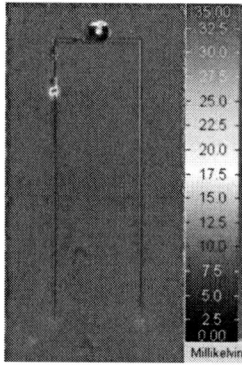

Fig. 6- Lock in Thermograph of a solar cell from the *elevating reverse bias* test protocol after final exposure to negative bias. Shunts are visible as red spots of increased temperature on the metal grid and contact island

## Identification of hotspots

Lock-in thermography has been performed on a number of cells after exposure to both measurement protocols. It was noted that especially for the cells that stopped working during the *elevating reverse bias* test hotspots were formed underneath the metal grid as well as on top and surrounding the contact islands. Fig. 6 shows a typical LIT image of a cell after completion of the *elevating reverse bias* protocol, revealing hotspots underneath the metal grid and at the point of electrical contact. An explanation of the shunts at the contact islands was revealed when recording a live thermal

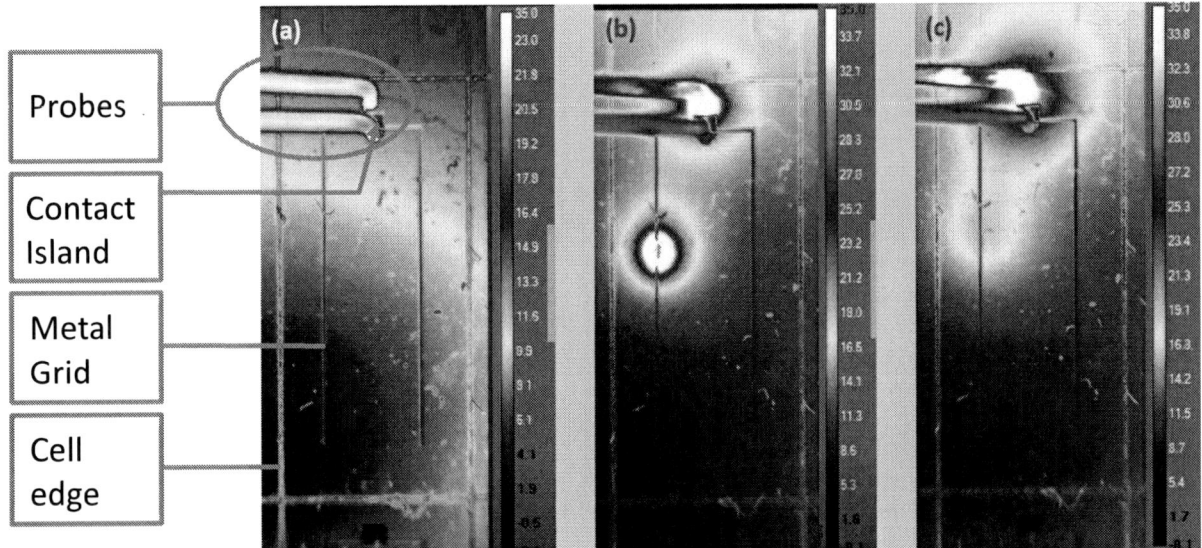

Fig. 7. IR camera images of a solar cell during dark IV characterization: (a) the IR image at low currents (b) created hot spot after reaching breakdown voltage (c) change in position of hot spot from underneath metal finger to contact probe while high current is passing through the solar cell – the size of the cell is 10 mm × 5 mm

image during dark IV measurements on some comparable cells from a different batch.

Three different stages of a real time movie made with an IR camera for an individual cell is shown in fig. 7. Note that the temperatures are not to scale and only function as an indication of temperature differences. The corresponding IV graph of the cell that acted in the movie is shown in fig. 8. Fig. 7a, shows the cell at forward bias and low currents as indicated with [A] in the IV graph of the corresponding measurement (fig. 8). The contact probes, metal grid and cell edge are indicated in the picture of fig. 7a. The spacing of the probes is slightly bigger than the contact island. It can therefore be seen that one of the two probes is located just next to the contact island.

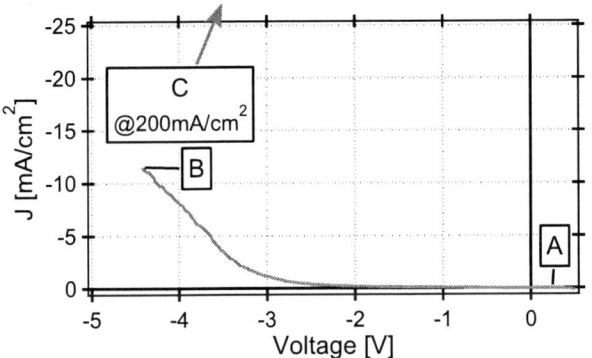

Fig. 8. JV graph of cell used for live IR imaging. Points indicated in the graph: A low current forward bias operation, B breakdown voltage point and C compliance current of 200mA/cm2 after breakdown was reached.

With increasing reverse bias voltage, the breakdown voltage was reached. The moment the breakdown appeared is shown by the creation of a hotspot underneath the metal grid in fig. 7b. This moment is indicated in the IV graph of fig. 8 with

[B]. Both real time IR images and IV measurements displayed that the creation of this hotspot was instantaneous. The moment the hotspot appeared, the current jumped to the current limit (compliance current) set on Keithley (200 mA/cm$^2$) indicated with [C] in fig. 8. After this incident, the set compliance current of 200 mA/cm$^2$ was still being injected. During the remaining time of the programmed IV sweep high current (200 mA/cm$^2$) was injected while in the thermal image, the initial hotspot underneath the metal grid disappeared and a new hotspot emerged. This new spot was formed in the area where the probe was injecting the compliance current into the TCO as seen in fig. 7c.

The movie with IR camera revealed two distinctly different mechanisms for shunt formation. First, the formation of a large shunt underneath the metal grid due to the instantaneous formation of a very intense but short living event. Second, the big contact probe injecting high current in the TCO, led to a large rise in temperature, causing the formation of a second shunt. Heat at the second hotspot can be generated both as a result of high current density in TCO and due to contact resistance between probe and TCO. One possible explanation could be that the shunt created underneath the metal grid has such a low resistance that it is not dissipating as much heat, compared to the higher resistive current path at the contact island. However it seems that both shunts are unique for cells equipped with a metal grid even if the heat generated from the contact probes could likely have been prevented with narrower probe spacing or bigger contact islands.

## IV. CONCLUSIONS

The susceptibility of non-encapsulated gridded CIGS solar cells to reverse bias voltages was monitored with an innovative experimental setup. Two test protocols have been defined to study the effect of exposure to reverse bias voltages and the recuperation due to consecutive light soaking. During the first test protocol, the effect of *elevating reverse bias* on the performance of gridded solar cells has been studied by stepwise increase of the absolute value of negative voltages in the dark, until -1.5 V no significant change has happened, while a slow decrease of electrical parameters was observed after exposure to -2 V. This decline of electrical parameters continued for increasing negative voltages after exposure to -3V when five out of six cells broke down permanently. For the second test protocol, the effect of exposure to a *constant negative voltage* on the performance of gridded CIGS solar cells has been studied. It was demonstrated that the duration that cells are exposed to negative bias has a considerable impact on loss of performance. This is due to the fact that none of the cells exposed for 60 seconds to -1.5 V degraded during the *elevating reverse bias* test. On the other hand, cells longer exposed to -1.5 V bias during the *constant negative voltage* suffered from severe degradation. Results from light soaking after exposure to reverse bias showed a recuperation of performance for all non-fatally shunted cells. This recuperation could be an indication that besides the formation of ohmic shunts another mechanism might play a role when non-encapsulated gridded CIGS cells are exposed to a reverse bias.

Real time videos made with an IR camera revealed that shunts are created rapidly. However, for the metallisation design used in combination with the contact probes, additional damage was made by the heat generated by the injection of current by the contact probe. It is noteworthy that the hotspots found with LIT mainly appeared underneath the metal grid and did not form the typical wormlike defects that are sometimes observed in thin film modules after exposure to (partial) shading.

It should be noted that large differences were observed between the solar cells, which were deposited on one substrate. Therefore, more experiments and statistics are needed for further conclusions.

In the future, the influence of variations of other parameters (i.e. applied current instead of voltage, as well as applying different light intensities) can be tested via the experimental setup described in this study. More insight in the failure mechanisms can be obtained by studying the current over time (I(t)) and IV data obtained during this and future experiments. Future studies with the setup described in this paper will help to better understand the failure mechanism and hopefully mitigate damages due to partial shading in CIGS cells and monolithically interconnected modules.

## IV. REFERENCES

[1] S. Dongaonkar, C. Deline and M. A. Alam, "Performance and Reliability Implications of Two Dimensional Shading in Monolithic Thin Film Photovoltaic Modules," *IEEE Journal Of Photovoltaics*, vol. 3 no. 4, pp. 1367-1375, 2013.

[2] S. Silvestre, A. Boronat, A. Chouder, "Study of bypass diodes configuration on PV modules," *Applied Energy*, vol. 86, pp. 1632–1640, 2009.

[3] T. J. Silverman, L. Mansfield, I. Repins and S. Kurtz, "Damage in Monolithic Thin-Film Photovoltaic," *IEEE Journal Of Photovoltaics*, vol. 6 no. 5, pp. 1333-1338, 2016.

[4] T. J. Silverman, M. G. Deceglie, C. Deline and S. Kurtz, "Partial Shade Stress Test for Thin-Film Photovoltaic Modules," in *Proc. SPIE 9563, Reliability of Photovoltaic Cells, Modules, Components, and Systems VIII, 95630F (September 23, 2015)*, 2015.

[5] N. G. Dhere, E. Schneller and A. Kaul, "Effect of shading on CIGS thin film photovoltaic modules," in 2015 *IEEE 42nd Photovoltaic Specialist Conference, PVSC 7355707*, 2015.

[6] T. J. Silverman, M. G. Deceglie, X, Sun, R. L. Garris, M. A. Alam, C. Deline and S. Kurtz, "Thermal and Electrical Effects of Partial Shade in Monolithic Thin-Film Photovoltaic Modules," *IEEE Journal Of Photovoltaics*, vol. 5 no. 6, pp. 1742-1747, 2015.

[7] P. O. Westin, U. Zimmermann, L. Stolt and M. Edoff, "Reverse Bias Damage in CIGS modules," in *24th European Photovoltaic Solar Energy Conference*, pp. 2967-2970, 2009.

[8] P. Mack, T. Walter, R. Kniese, D. Hariskos, and R. Schäffler, "Reverse Bias and Reverse Currents in CIGS Thin Film Solar Cells and Modules," *23rd European Photovoltaic Solar Energy Conference and Exhibition*, pp. 2156-2159, 2008.

[9] P. Szaniawski, J. Lindahl, T. Törndahl, U. Zimmermann and M. Edoff, "Light-enhanced reverse breakdown in $Cu(In,Ga)Se_2$ solar cells," *Thin Solid Films*, vol. 535, pp. 326-330, 2013.

[10] S. Dongaonkar, M. A. Alam, Y. Karthik, S. Mahapatra, D. Wang and M. Fei, "Identification, characterization, and implications of shadow degradation in thin film solar cells," in *2011 IEEE 49th International Reliability Physics Symposium, IRPS 5784535*, pp. 5E.4.1-5E.4.5, 2011.

# A Method to Extract Soiling Loss Data from Soiling Stations with Imperfect Cleaning Schedules

Matthew Muller[a], Leonardo Micheli[a,b], and Alfredo A. Martinez-Morales[c]

[a] National Renewable Energy Laboratory, Golden, Colorado, 80401, United States
[b] Colorado School of Mines, Golden, Colorado, 80401, United States
[c] University of California, Riverside, California, 92507, United States

*Abstract* — Typical PV soiling stations determine the natural soiling losses by comparing the output of a naturally soiled PV cell to that of a PV cell maintained in the clean state. Inadequate cleaning frequency, equipment failure, or human error provides opportunity for the cleaned cell to soil, directly resulting in error in the reported soiling ratio. This work investigates an algorithm to automatically detect and correct the data stream for errors associated with soiling of the clean cell. The methodology is tested on several soiling stations with irregular cleaning schedules as well as two soiling stations where both ideal and imperfect cleaning schedules are in place. The initial results show that the algorithm can reduce error associated with imperfect cleaning but also confirms the benefits of maintaining an optimal cleaning schedule.

*Index Terms*—dust, performance analysis, photovoltaic systems, soiling

## I. INTRODUCTION

PV soiling losses (energy generation losses due to dust on a PV surface) can be in the range of 0-6% in the United States and can be 20-80% in dusty regions of the world like the Middle East [1]. Lost energy generation directly equates to lost dollars and therefore as the PV market has expanded into the dusty regions of the world, soiling research has increased dramatically. Soiling research spans a range of topics including: soiling loss measurements and associated equipment, dust characterization, soiling adhesion mechanisms, soiling loss modeling, automated cleaning systems, and module surface coatings to reject soil adhesion. This work seeks to add to the PV community's ability to measure and report soiling loss data so the data are both reliable and well understood. Soiling stations are being used around the world to report soiling losses which are then used by various parties such as financiers, performance modelers, and planners of system operations and maintenance (O&M). While soiling stations can vary in design (i.e. cell versus module measurements, short-circuit current versus peak power tracking, tilt angle, and choices on supplementary meteorological equipment), in most stations the soiling loss measurement is the ratio of the

output of a naturally soiled PV device to the output of a clean PV device. This type of device is well described in the literature; including measurement uncertainty associated with errors in mounting, calibration drift, angle of incidence effects, and temperature effects [2-5]. The fundamental measured ratio between a dirty and clean PV device, however, is not always systematically treated. While the dirty device is allowed to soil based on natural events, the cleaned cell is maintained in the "clean" state by manual or automated cleaning practices. The authors of this work have experienced a range of soiling stations where the "clean" state is assumed because daily, weekly, biweekly, or other irregular cleaning intervals are applied to the clean cell. Although "clean" in the purest sense would imply no soil particles on the PV device, this is not realistic for any outdoor device. Rather, the amount of soiling allowed on the clean cell impacts the uncertainty of the reported soiling ratio. The level of tolerable uncertainty will generally depend on the end needs of the data user and therefore are not defined in this work. On the other hand, if nothing can be established in regards to the uncertainty of an imperfect soiling ratio (it could be in the range of 0-100% losses), the data are of no use. Therefore it is valuable to be able to somehow quantify the uncertainty of an imperfect soiling ratio and develop methods to glean valuable information from this data. Hereafter, a methodology is presented to quantify errors in an imperfect soiling ratio, correct for these errors, and quantify the uncertainty based on these errors.

## II. METHODS FOR SOILING LOSS EXTRACTION

### A. Soiling Stations and soiling metrics

This study is based on the data recorded by ten soiling stations installed in the southwestern United States. Each soiling station has two identical PV cells, where one is intended to be the clean device and the other is allowed to naturally soil. The ten sites were chosen because the cleaning was done manually and the intervals of cleaning were not consistently achieved. Using the same method employed in [3], a daily soiling ratio (SRatio) is

978-1-5090-5606-4/17 $31.00 © 2017 IEEE

calculated, obtained as the ratio between the daily mean short circuit currents of the two cells. A soiling ratio of one represents clean conditions, and its value decreases with the soiling. Short circuit currents, irradiances and weather data are recorded each minute and converted into hourly data. Only data occurring between 11AM and 1PM have been considered to minimize variation due to angle of incidence effects. Performance plots of each soiling station have been visually checked in order to remove any data due to equipment malfunction, shading, or performance outside of reasonable bounds.

*B. Algorithm for correcting imperfect soiling ratio data*

An algorithm has been developed to correct for errors in the measured soiling ratio due to infrequent cleaning and therefore soiling of the clean cell. Fig. 2 provides a flow chart of the algorithm where key points are as follows: 1) Noise is defined as two times the mean of all upward shifts in the SRatio not associated with precipitation, 2) Downward shifts in the SRatio that are twice or more than the noise are interpreted as a cleaning of the clean cell and the SRatio following the shift is an accepted data point. 3) In cases where precipitation coincides with an upward shift in the SRatio (twice or more than the noise) both cells are considered clean and the SRatio is set to one, 4) SRatios between acceptable data points are determined by connecting lines between the acceptable data points (note the majority of measured data points are rejected). Fig. 1 is a cartoon of soiling station data that helps to elucidate these key points. The black SRatio points demonstrate measurements of the SRatio over 50 days. The SRatio starts at one because both cells have been cleaned and then it has a noisy trend until day 14 where a downward shift is detected. The red triangle is considered a true measurement of the SRatio and a line is drawn between day 1 and day 14. On day 30 an upward shift and rain are detected so the SRatio is set to one. The SRatio measurement on day 29 was not acceptable because of unknown soil level on the clean cell (i.e. no downward shifts were detected beyond day 14). The SRatio slope between day 1 and day 14 is projected to day 29 because it is the most recent estimate of what is occurring at the site. No shifts are detected between the rains on day 30 and day 50 and so no new information is obtained about the rate of change in the SRatio. An estimate is then assumed by using the slope that was used in the previous soiling period. This assumes that both cells soil together between days 30 and 50 and that the 14 day cleaning schedule was missed. It is possible that soiling could have been minimal in this time period but the authors considered it more conservative to assume soiling is occurring.

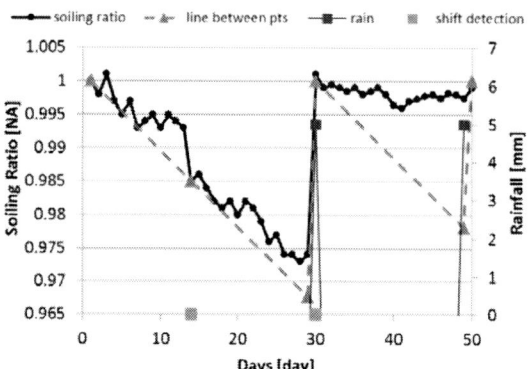

Fig. 1. A cartoon of key points in the algorithm for correcting an imperfect soiling ratio.

---

**Calculate all daily changes in the SRatio**
$\Delta SRatio_i = SRatio_i - SRatio_{i-1}$

---

**Calculate Noise using all $\Delta SRatio_i > 0$**
$Noise = \overline{\Delta SRatio_i}$

---

**Detect significant shifts in $\Delta SRatio_i$**
$\Delta SRatio_i \geq 2 \cdot Noise$
→ If also precipitation $\Delta SRatio_i$ reset to 1
→ No precip then $\Delta SRatio_i$ = constant until next significant shift
$\Delta SRatio_i \leq -2 \cdot Noise$
→ $\Delta SRatio_i$ is accepted (clean cell cleaning)

---

**Establish corrected profile for $\Delta SRatio_i$**
- For each significant shift of $\Delta SRatio_i$ an algorithm datapoint is stored; $aSRatio_j$
- Starting at $aSRatio_{j=0} = 1$ extract a slope between the j and j+1 points for $aSRatio_j$
→ If slope < 0 accept and project between the times associated with j and j+1
→ If slope > 0 and no precipitation occurs on j+1 then the slope is reset to 0 and projected forward from the $aSRatio_j$
→ If slope > 0 and precipitation occurs on j+1 then the slope from $aSRatio_{j-1}$ to $aSRatio_j$ is projected to the date immediately before the date associated with j+1.
→ The accepted slope for each period is started at the end of the previous slope so the corrected profile has continuity. For this reason $aSRatio_j$ is not necessarily on the corrected profile.

---

Fig. 2. A flow chart that explains the steps taken in the corrective algorithm.

Fig. 3 presents an example of how the SRatio correction algorithm is applied to an actual soiling station in the southwest United States. The black data points are the measured SRatio data and red lines connect each set of data points that the algorithm accepts. Days with precipitation are marked in blue and the downward shift detection is marked in green. The period from July through October shows a series of downward steps that are believed to be cell cleanings about every two weeks. The fact that the steps are easily visible indicates that significant soiling is possible at this site on the clean cell over two weeks. This site is an ideal case for the correction algorithm because the dry period is multiple months and data points every two weeks make it possible to still see the longer downward trend.

provide information about the worst case uncertainty for the average SRatio. For example, if after two weeks the clean cell is cleaned and the SRatio shifts by 10%, the worst case scenario is that a large soiling event occurred immediately after the previous cleaning and the SRatio was in 10% error for the entire 2 weeks. It is unrealistic to assume this would always happen and therefore the magnitudes of all the detected downward shifts are averaged and this value is reported as the site uncertainty. A long term data set with side-by-side PV cells cleaned at different frequencies would provide a more thorough uncertainty analysis but the above value is useful because it is conservative (the most extreme soiling loss case is estimated) and it will increase both for dirtier sites and for longer periods between cleanings. The reported

TABLE I

AVERAGE SOILING RATIOS DETERMINED FOR 10 SITES IN THE SOUTHWESTERN UNITED STATES. THE RATIOS ARE GIVEN FOR THE MEASURED VALUES ($SR_M$) AND BASED ON THE CORRECTIVE ALGORITHM ($SR_C$). ALL VALUES ARE GIVEN AS PERCENTAGES.

| SITE | 1 | 2 | 3 | 4 | 5 | 6 | 7 | 8 | 9 | 10 |
|---|---|---|---|---|---|---|---|---|---|---|
| $SR_M$ | 99.9 | 99.6 | 99.8 | 95.9 | 99.2 | 99.1 | 93.6 | 99.4 | 99.9 | 99.9 |
| $SR_C$ | 99.5 | 99.1 | 98.8 | 95.4 | 98.8 | 97.5 | 91.9 | 99.5 | 99.4 | 99.4 |
| U | 0.2 | 0.6 | 0.9 | 1.9 | 1.7 | 2.1 | 3.3 | 0.4 | 0.6 | 0.6 |

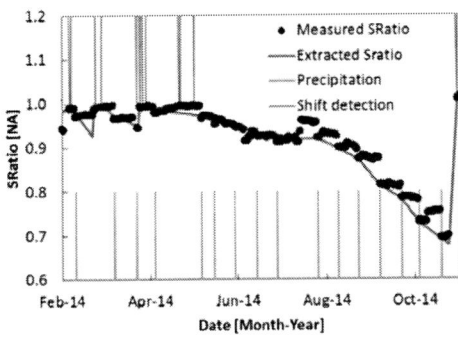

Fig. 3. Applying the correction algorithm to the measured SRatio for one site in the southwest United States.

*B. Uncertainty determination*

A major assumption of the SRatio correction algorithm is that the soiling rate (slope of the SRatio) is linear between accepted data points. It is possible that over an interval like two weeks that soiling rate could vary for many reasons (i.e. pollution from a worksite, and agricultural activity or changes in site conditions). Without actually measuring such deviations one cannot quantify the specific effects but errors in the SRatio are likely to occur that are both positive and negative (therefore canceling each other in an average result). A data set is being collected to analyze such variation but the magnitude of shifts in the SRatio at each cleaning also

uncertainty therefore provides a measure of the impact of less frequent cleanings.

## III. RESULTS AND DISCUSSION

*A. Results for average soiling ratios*

Table I provides the average SRatios for the 10 sites in this study. $SR_M$ is the average of all the daily measured SRatios while $SR_C$ is average of the daily values per the corrective algorithm. The site uncertainty, $U$, as described in the previous section, is also given in Table I.

*B. Discussion*

The results in Table I show that the 10 sites have measured soiling ratios from 93.6% to 99.9% and the corrected values range from 91.9% to 99.5%. Some sites that have measured losses less than 1% have corrections on the order of tenths of a percent, while others have corrections on the order of a few percent. Site 6 has the greatest correction, reducing from 99.1% to 97.5%. While the model has yet to go through a true validation, the fit in Fig. 3 suggests that it is possible to automatically detect soiling station cleaning events and then use these events to construct a correction for a long term soiling trend. It is interesting to note that the uncertainty values as defined in the previous section match very closely to the magnitude of the correction for each site. Although the algorithm and

978-1-5090-5606-4/17 $31.00 © 2017 IEEE

uncertainty are related to each other, the algorithm applies a series of complex steps and assumptions whereas the uncertainty represents only the average of all the errors in the SRatio for just the days where the SRatio shifts downward (believed to be cleanings of the clean cell).

### C. Validation efforts

In order to draw further conclusions on the value and uncertainty of the proposed SRatio corrective algorithm, two soiling stations were built and recently deployed to gather validation data. Fig. 4 shows a picture of one of the validation stations which each have four PV cells. One of the cells is left to naturally soil while the others are intended to be cleaned daily, weekly and every other week. One of the stations was deployed at the University of California, Riverside where past years have shown an average annual soiling losses of ~ 4-6% [6]. The second station was employed in Golden, Colorado where the average annual soiling loss is typically less than 1%. Due to this low soiling loss in Golden, artificial soiling was employed to increase the soiling rate. A leaf blower was used on an approximate daily basis to agitate soil and other matter on the ground adjacent to the soiling station, effectively creating a short term dust cloud around the soiling station. The Golden station was intended for a short term proof of concept while the Riverside station was intended to be an actual field validation.

Fig. 4. Deployed PV cells cleaned daily, weekly, every 2 weeks, and not at all to validate the results of the correction algorithm.

### Golden Station

The intent of the Golden station was to provide a first pass validation of the correction algorithm while waiting for longer term data from a field site with substantial soiling. Although some lessons were learned from the Golden station the data did not provide a simple first pass validation. The artificial soiling station presented the following challenges and lessons:

1) Leaf blower soiling resulted in losses on the order of several percent per day (greater than any typical U.S. location).

2) Leaf blower soiling rates were not as linear as what is often seen in the field and the SRatio seemed to bottom out near 0.5.

3) Precipitation frequency was on the order of every 2 weeks and therefore prevented obtaining a useful comparison between weekly and biweekly cleaning.

4) The majority of precipitation was in the form of snow. The typical experience with PV systems at NREL is that, on the first sunny day following a snow, the solar panels will be warmed and the snow will slide off all of the panels in a matter of minutes, resulting in a complete cleaning. Although the soiling station reference cells were mounted at latitude tilt, the cells did not behave the same in response to snow as full size panels. On some days following snowstorms, the snow was seen to slowly melt and partially slide on the small reference cell. The data showed that, in several cases, this resulted in partial cleaning over several days (the SRatio increases over multiple days following precipitation, see the yellow circle on Fig. 5). The corrective algorithm was not designed to correct for partial cleanings or for cleanings that occur over multiple days due to melting snow and therefore it performed poorly in these time periods.

Fig. 5 provides the time series SRatio data for the daily and weekly cleanings as well as the result of applying the algorithm to the weekly cleaning data. The actual average measured SRatio per the daily cleaning was 0.82, the uncorrected average from weekly cleaning was 0.86, and the result from applying the corrective algorithm was 0.78. The soiling loss was over estimated during the snow melt off periods and therefore the corrective algorithm over predicted the soiling losses and therefore the corrective algorithm provides no clear benefit.

### Riverside Station

At the time of preparation of this paper data was only available from the Riverside station from 3/6/2017 through 5/9/2017. Students provided the cleaning support and therefore cleaning did not occur on the weekends or during spring break (3/25/2017-4/3/2017). Further issues resulted in no cleaning occurring from 4/8/2017 – 4/25/2017. Although this results in only approximately 7 weeks of data where the true soiling ratio was measured (daily cleaning), the corrective algorithm was still run on various data streams and the results for the biweekly cleaning are shown in Fig. 6. Although the data collection period is quite short, the graph clearly demonstrates that cleaning every two weeks in Riverside does not result in an accurate SRatio. The first three periods between cleanings result in a SRatio that trends upward. This suggests the clean cell actually soils at a slightly faster rate than the cell that is only cleaned by rain or natural events. The average SRatios (for the entire data collection period) for the different cleaning frequencies are as follows: estimate of the "true" SRatio = 0.98 (daily data with corrections applied for the noted

978-1-5090-5606-4/17 $31.00 © 2017 IEEE

time period where no cleanings were performed), SRatio measured using weekly cleaning = 0.99, SRatio after applying a correction to the weekly cleaning data = 0.98, SRatio measured using biweekly cleaning = 0.99,

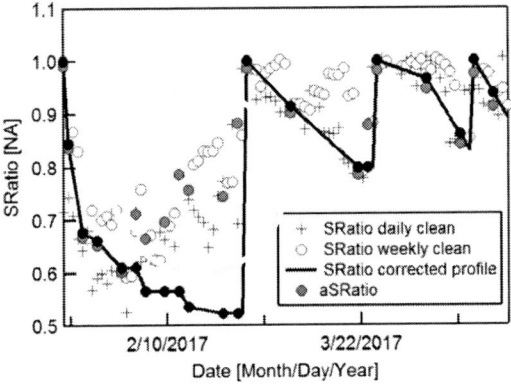

Fig. 5.    Golden, Colorado soiling station with artificial soiling. The soiling ratio based on daily and weekly cleaning is shown along with the soiling ratio correction algorithm as applied to the weekly cleaning data. The corrective algorithm is shown as a series of black lines that connect all the black dots. The black dots align with the dates for which there are also red dots. The red dots are values where shifts are detected which are used to construct the corrected SRatio. In February (yellow circle) there were several snow events (snow dates not shown on the graph) that caused partially cleaning and therefore the true SRatio (daily cleaning) improves.    Although upward shifts were recorded by the algorithm (red dots) the snow did not occur on the days of the upward shifts and therefore the algorithm fits with a zero slope. The cleaning by snow usually occurs in the days following the snow storm. No changes were made to account for the effects of snow because typical high soiling sites are desert environments where snow is not expected to be a factor.

SRatio after applying a correction to the biweekly cleaning data = 0.98.    The corrective algorithm is successful in predicting the estimate of the "true" SRatio based on both weekly and biweekly cleaning cells but caution must be taken in using these results for the following reasons: 1) The "true" value is only an estimate due to the weeks where cleaning support was not available, 2) The data only span 9 weeks and therefore should be considered preliminary, and 3) No extended soiling period has been experienced and therefore the soiling losses are generally low regardless of the cleaning frequency.

## IV. PRELIMINARY CONCLUSIONS

Soiling stations are being deployed all over the world to determine information about the soiling losses on PV systems. These soiling stations provide a SRatio between a dirty and cleaned PV device but often the SRatio has

significant error because the cleaned cell is also allowed to soil, depending on the cleaning interval. If no information is known about the level of soiling on the

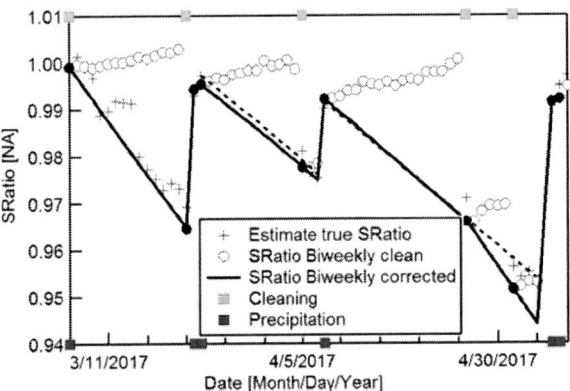

Fig. 6.    The soiling ratio measurements based on the biweekly cleaned cell are shown in comparison to the estimate of the true soiling ratio and the correction as applied to the data from biweekly cleaning.    A dashed line is shown where the true soiling ratio was estimated as daily cleaning did not occur in this time period. The data from biweekly cleaning illustrates the problem with infrequent cleanings. The trend of the soiling ratio appears to be positive between cleanings. There has been anecdotal evidence that a regularly manually cleaned cell can soil at a slightly different rate than a cell that is only cleaned by rain but the trend in this graph could also be a product of noise.

clean cell, then the true SRatio is also unknown. An algorithm has been presented that automatically detects cleaning events for the clean cell and therefore establishes trustworthy data points in the time series data set for the SRatio. These data points have been used to both determine a worst case uncertainty for the annual average measured SRatio and to estimate a time series correction for the SRatio. As should be expected, the uncertainty for a site increases when the soiling rate is greater and the time between cleanings is greater. The uncertainty metric ultimately provides a way to interpret and use data that otherwise is difficult to trust.

An initial effort was presented to validate the proposed SRatio corrective algorithm based on an outdoor artificial soiling station in Golden, Colorado and a representative field soiling station in Riverside, California. The Golden soiling station presented several limitations but it demonstrated that the corrective algorithm failed to appropriately handle partially cleanings that occur when snow melts on the surface of PV reference cells. This is not surprising as the algorithm was designed around complete cleanings; either manually performed or due to significant rainfall. Validation based on the station in Riverside, California is very limited due to the short data collection period. The initial results from the Riverside station suggest that the corrective algorithm has potential for sites when cleaning events result in full, rather than partial, cleaning. Both weekly and biweekly cleaning of the clean cell underestimated the losses and the algorithm was able to correct for this. Further validation efforts are necessary to determine if the corrective algorithm should be more broadly used.

### ACKNOWLEDGEMENTS

The authors would like to thank Ryo Huntamer and Aaron Chan for their support installing and maintaining the soiling station at the University of California, Riverside.

This work was supported by the U.S. Department of Energy under Contract No. DE-AC36-08GO28308 with the National Renewable Energy Laboratory. The U.S. Government retains and the publisher, by accepting the article for publication, acknowledges that the U.S. Government retains a nonexclusive, paid up, irrevocable, worldwide license to publish or reproduce the published form of this work, or allow others to do so, for U.S. Government purposes.

### REFERENCES

[1] T. Sarver, A. Al-Qaraghuli, L. L. Kazmerski, "A comprehensive review of the impact of dust on the use of solar energy: History, investigations, results, literature, and mitigation approaches," *Renew. Sustain. Energy Rev.*, vol. 22, pp. 698–733, 2013.

[2] L. Micheli, M. Muller, S. Kurtz, "Determining the effects of environment and atmospheric parameters on PV field performance," *IEEE 43rd Photovoltaic Specialist Conference (PVSC)*, 2016, pp. 1724 – 1729.

[3] L. Micheli, M. Muller, "An Investigation of the Key Parameters for Predicting PV Soiling Losses," *Prog. Photovoltaics Res. Appl. - Accept.*

[4] M. Gostein, J. Riley Caron, B. Littmann, "Measuring Soiling Losses at Utility-scale PV Power Plants," *IEEE Photovoltaic Specialist Conference (PVSC)*, 2014, pp. 885-890

[5] L. Dunn, B. Littmann, J. Riley Caron, M. Gostein, "PV Module Soiling Measurement Uncertainty Analysis," *IEEE Photovoltaic Specialist Conference (PVSC)*, 2013, pp. 658-663

[6] Gostein, M., Stueve, B., Brophy, B., Jung, K., Martinez-Morales, A.A., Zhang, S., Jin, Y., & Xu, J. "Soiling measurement station to evaluate anti-soiling properties of PV module coatings," *IEEE Photovoltaic Specialist Conference (PVSC)*, 2016.

# Analytical (S)TEM Studies of Defects Associated with PID in Stressed Si PV Modules

Andrew Norman,[1] Adam Stokes,[2] John Moseley,[1] Steven Harvey,[1] Steve Johnston,[1] Harvey Guthrey,[1] and Mowafak Al-Jassim[1]

[1]National Renewable Energy Laboratory, Golden, CO, 80401, USA
[2]Colorado School of Mines, Golden CO, 80401, USA

*Abstract* — We report the results of analytical transmission electron microscopy studies of defects associated with potential-induced degradation (PID) of a laboratory stressed multicrystalline silicon mini-module. Shunt defects were first identified using dark lock in thermography (DLIT) and scanning electron microscopy (SEM) electron beam induced current (EBIC) imaging. Transmission electron microscopy (TEM) cross-section samples were then prepared of selected shunt defects using conventional focused ion beam (FIB) lift out techniques and examined by TEM and scanning transmission electron microscopy (STEM). The defects were identified as {111} planar defects and compositional mapping of the defects revealed elevated levels of both Na, expected from previous work, and also surprisingly O at some of the defect cores.

*Index Terms* —photovoltaic cells, potential induced degradation, silicon, transmission electron microscopy.

## I. INTRODUCTION

Potential-induced degradation (PID) is an important issue for the long-term performance and reliability of crystalline Si photovoltaic (PV) modules deployed in the field and can even lead to catastrophic failure [1]. Migration of $Na^+$ ions and accumulation at defects such as stacking faults that cross the p/n junction causing electrical shunts is thought to play a key role in the mechanism of PID degradation, [2]-[8]. We report a detailed study of PID shunt defects in a laboratory stressed multicrystalline Si mini-module and reveal that Na and, in some cases, O segregate to the core of the {111} planar defects that are observed to be present.

## II. EXPERIMENTAL

A front junction (n+/p), multicrystalline, texture-etched Si cell with a $SiN_x$ antireflective coating was fabricated into a mini-module as described previously in [8]. The mini-module was then stressed under a constant voltage bias of -1000 V applied to the shorted module leads under the conditions detailed in [8]. This resulted in significant PID as assessed by the development of shunting in dark IV curves obtained periodically during the stress test. The mini-module was then delaminated to recover the cell. Photoluminescence (PL) and dark lock in thermography (DLIT) imaging were then performed to identify the areas of the cell that had suffered significant PID. Representative areas exhibiting PID related shunts/defects were then laser marked so that they could subsequently be identified in-situ in the scanning electron microscope (SEM) enabling the same defects to be studied using SEM electron beam induced current (EBIC) mode. Cross-section transmission electron microscopy (TEM) samples of selected defects were then extracted by conventional focused $Ga^+$ ion beam (FIB) lift-out techniques using an FEI Nova NanoLab 200 dual beam FIB workstation. The thin TEM samples were then examined by diffraction and phase contrast TEM in a FEI $G^2$30 SuperTwin operated at 300 kV. Elemental mapping was performed using an EDAX Octane T Optima 60 $mm^2$ windowless Si drift detector energy dispersive x-ray spectroscopy (EDS) system installed on an FEI F20 UltraTwin field emitting gun scanning transmission electron microscope (FEG STEM) operated at 200 kV.

## III. RESULTS AND DISCUSSION

Using SEM EBIC, performed in-situ in a FIB, we selected a shunt related defect, Fig. 1, previously identified using DLIT, that contained two linear features for subsequent preparation of a TEM cross-section sample using FIB conventional lift out techniques.

Low magnification bright-field TEM images and magnified images of defects 1 and 2 identified in the SEM EBIC image of Fig. 1 are shown in Fig. 2. Both defects show contrast suggestive of {111} planar defects in the Si. After tilting the sample to a <112> zone axis to view the {111} defect, Defect 2, end on, high-resolution TEM images were taken and revealed it to be a planar defect lying parallel to the Si {111} planes, Fig. 3. The core of the defect showed contrast typical of an amorphous material and was measured to be approximately $4d_{111}$ Si wide (1.3 nm) near the Si cell surface.

The sample was then transferred to a FEI F20 UltraTwin FEG STEM for STEM EDS mapping. The results of this analysis are shown in Fig. 4. Segregation of both Na and O is observed to occur to the core of this {111} defect near the Si cell surface. Some care has to be taken when mapping STEM EDS of Na on a FIB prepared (using $Ga^+$ ions) sample as the Ga $L\alpha$ peak is very close to the Na $K\alpha$ peak used for the mapping. However, mapping of the identical area using the Ga $K\alpha$ peak revealed no increase in Ga concentration at the defect core strongly suggesting that the observed increase in Na concentration at the defect is real and not an artifact of the $Ga^+$ ion FIB sample preparation used.

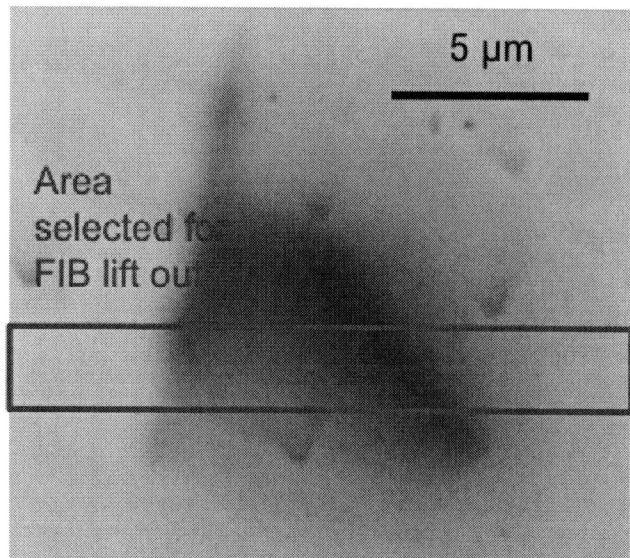

Fig. 1. SEM EBIC image of 2 line-type shunt causing defects in PID stressed multicrystalline Si solar cell. Area selected for FIB lift out of TEM sample marked by blue box.

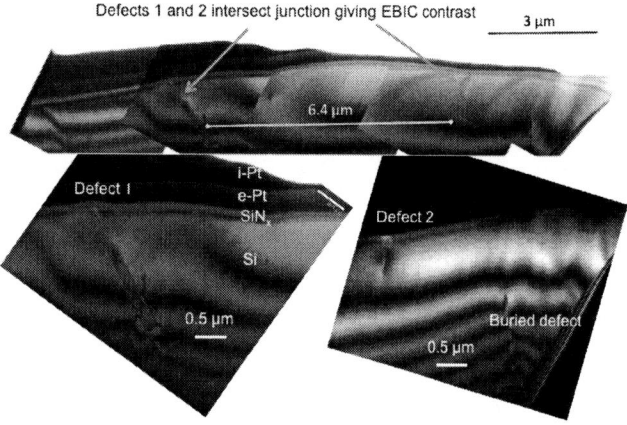

Fig. 2. Bright field diffraction contrast TEM images showing PID shunt related defects identified by SEM EBIC in Fig. 1.

Atomic % line scans across the defect extracted from the EDS maps of Fig. 4, averaged over 100 points in the Y direction and 5 points in the x direction, are shown in Fig. 5. The results show segregation of 1-2 atomic % of both Na and O to the {111} defect core.

Subsequent to (S)TEM analysis, the thin sample was returned to the FIB and a sample was prepared for local electrode atom probe (LEAP) analysis of the same defect performed at the Colorado School of Mines. These results are described in detail in another submission to this conference [9] and confirmed the segregation of both Na and O to the defect core.

SEM EBIC performed in the FIB was used to select another shunt defect from the stressed mini-module for lift out as

shown in Fig. 6. TEM (not shown) revealed this defect to also be associated with a {111} planar defect.

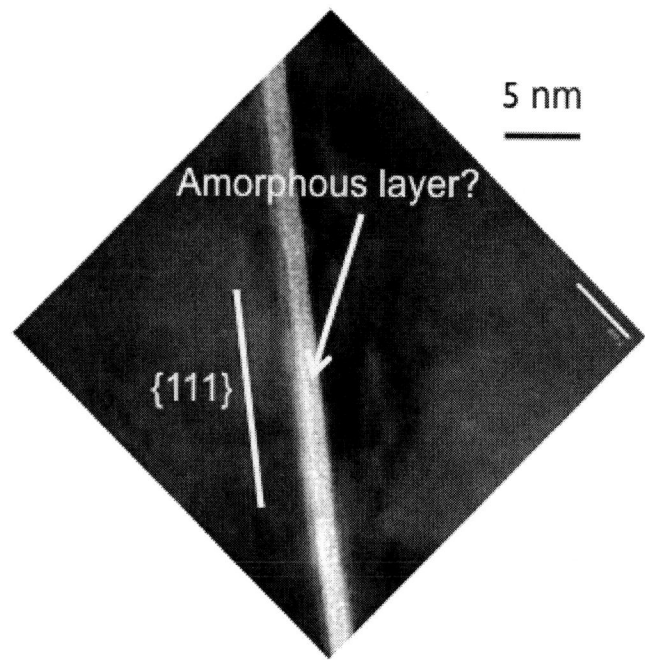

Fig. 3. HRTEM image of Defect 2 in Fig. 2 showing it lies parallel to {111} Si planes and appears to have an amorphous core.

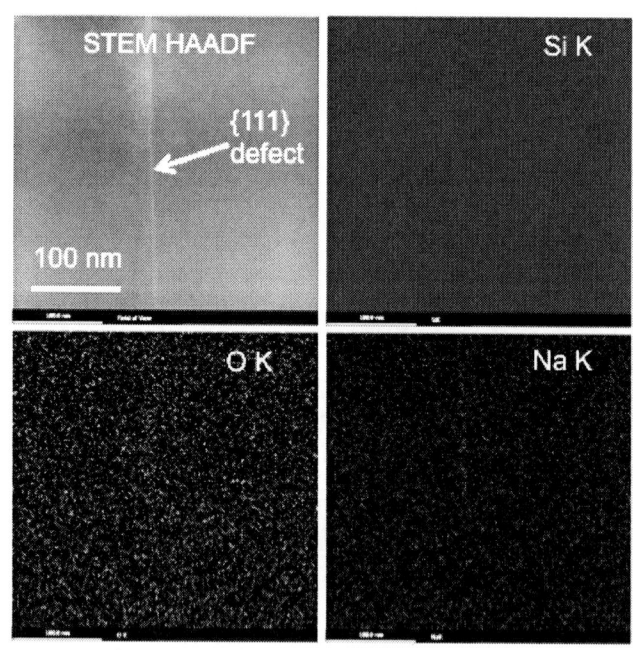

Fig. 4. STEM EDS mapping of Defect 2 showing elevated concentrations of both Na and O at the defect core.

The sample was inserted into the FEG STEM and tilted such that the defect was viewed end on at a <110> zone axis.

STEM EDS mapping was then performed and the results obtained are shown in Fig. 7 taken from an area close to the Si cell surface. Na was again observed to have segregated to the defect core but in this case no O segregation was found. The {111} planar defect extended several microns beneath the Si surface. Na segregation at the planar defect was still found to occur at a depth of 2 μm beneath the surface.

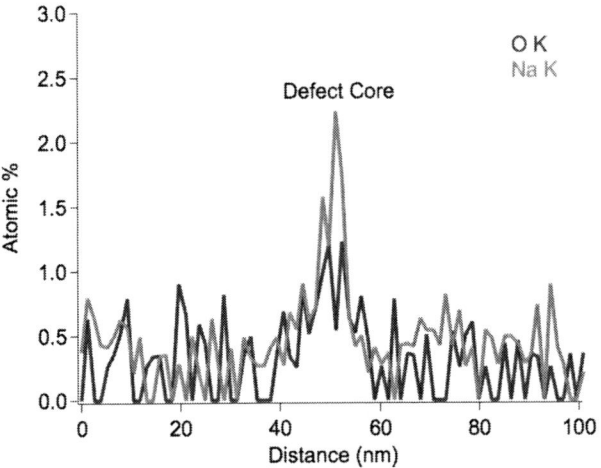

Fig. 5. Atomic % line scan across Defect 2 extracted from STEM EDS maps showing segregation of both Na and O to defect core.

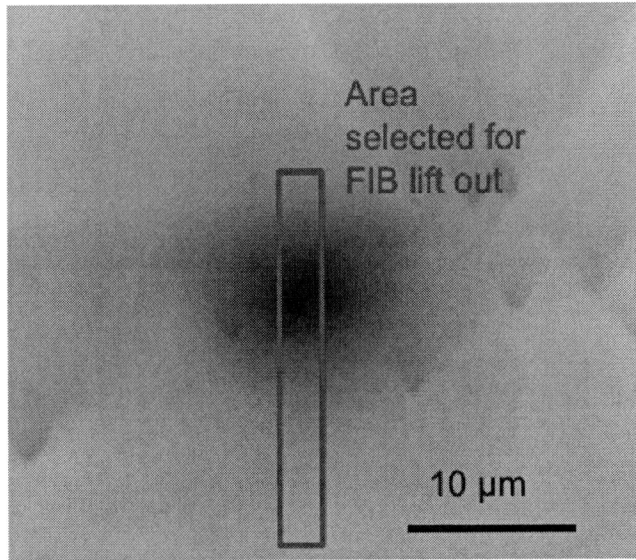

Fig. 6. SEM EBIC image of shunt defect in PID stressed multicrystalline Si solar cell. Area selected for FIB lift out of TEM sample marked by blue box.

These results are consistent with the results obtained on other defects in this stressed mini-module by time-of-flight secondary ion mass spectrometry (TOF SIMS) at NREL. This technique also reveals segregation of both Na and O to the

cores of defects in our samples [9]. Previous work, [1]-[8], has indicated that migration of Na+ ions to {111} stacking faults is the primary mechanism of shunt related PID in crystalline Si solar cells. Our experiments also indicate that Na decoration of {111} planar defects may play an important role in PID of multicrystalline Si PV modules. However, the observation in this work of the additional segregation of O to some of the defects suggests further study of the mechanism for PID is warranted.

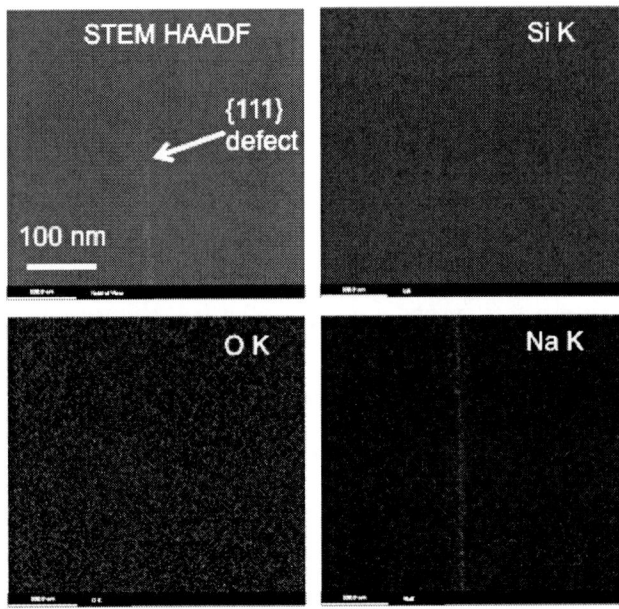

Fig. 7. STEM EDS mapping of shunt defect shown in Fig. 6 showing segregation of Na to the defect core but not O.

CONLUSIONS

A combination of techniques including SEM EBIC, FIB, TEM, and STEM have been used to study defects associated with shunting in PID degraded multicrystalline Si solar cells. The results reveal the defects to be planar and lie parallel to {111} Si planes. The core of some of the defects appears to be amorphous. Chemical analysis performed using STEM EDS mapping revealed segregation of Na and in one example O at the defect cores. The potential role of O in the PID degradation mechanism remains to be determined.

ACKNOWLEDGEMENTS

This work was supported by the U.S. Department of Energy under Contract No. DE-AC36-08GO28308 with the National Renewable Energy Laboratory. The U.S. Government retains and the publisher, by accepting the article for publication, acknowledges that the U.S. Government retains a nonexclusive, paid up, irrevocable, worldwide license to publish or reproduce the published form of this work, or allow others to do so, for U.S. Government purposes.

## REFERENCES

[1]  W. Luo, Y. S. Khoo, Khan, P. Hacke, V. Naumann, D. Lausch, S. P. Harvey, J. P. Singh, J. Chai, Y. Wang, A. C. Aberle, and S. Ramakrishna, "Potential-induced degradation in photovoltaic modules: a critical review," *Energy and Environmental Science*, vol. 10, pp. 43-68, 2017.

[2]  V. Naumann, C. Hagendorf, S. Grosser, M. Werner, and J. Bagdahn, "Micro structural root cause analysis of potential induced degradation in c-Si solar cells," *Energy Procedia*, vol. 27, pp. 1-6, 2012.

[3]  V. Naumann, D. Lausch, S. Großer, M. Werner, S. Swatek, C. Hagendorf, and J. Bagdahn, "Microstructural analysis of crystal defects leading to potential-induced degradation (PID) of Si solar cells," *Energy Procedia*, vol. 33, pp. 76-83, 2012.

[4]  V. Naumann, D. Lausch, A. Graff, M. Werner, S. Swatek, J. Bauer, A. Hähnel, O. Breitenstein, S. Großer, J. Bagdahn, and C. Hagendorf, "The role of stacking faults for the formation of shunts during potential-induced degradation of crystalline Si solar cells," *Phys. Status Solidi*, vol. 7, pp. 315-318, 2013.

[5]  D. Lausch, V. Naumann, A. Graff, A. Hähnel, O. Breitenstein, C. Hagendorf, and J. Bagdahn, "Sodium outdiffusion from stacking faults as root cause for the recovery process of potential-induced degradation (PID)," *Energy Procedia*, vol. 55, pp. 486-493, 2014.

[6]  V. Naumann, D. Lausch, A. Hähnel, J. Bauer, O. Breitenstein, A. Graff, M. Werner, S. Swatek, S. Großer, J. Bagdahn, and C. Hagendorf, "Explanation of potential-induced degradation of the shunting type by Na decoration of stacking faults in Si solar cells," *Solar Energy Materials and Solar Cells*, vol. 120, pp. 383-389, 2014.

[7]  V. Naumann, D. Lausch, A. Hähnel, O. Breitenstein, and C. Hagendorf, "Nanoscopic studies of 2D-extended defects in silicon that cause shunting of Si-solar cells," *Phys. Status Solidi*, vol. 12, pp. 1103-1107, 2015.

[8]  S. P. Harvey, J. A. Aguiar, P. Hacke, H. Guthrey, S. Johnston, and M. M. Al-Jassim, "Sodium accumulation at potential-induced degradation shunted areas in polycrystalline silicon modules," *IEEE Journal of Photovoltaics*, vol. 6, pp. 1440-1445, 2016.

[9]  S. P. Harvey, J. Moseley, A. Stokes, A. Norman, B. Gorman, P. Hacke, S. Johnston, and M. Al-Jassim, "Investigating PID shunting in polycrystalline silicon modules via multi-scale, multi-technique characterization," accepted for presentation at the 2017 IEEE PVSC-44.

# Design, Development, and Evaluation of Electrodynamic Screens for Self-Cleaning Solar Panels and Concentrating Mirrors

Annie Bernard*, Cristian Morales*, Ryan S. Eriksen*, Alecia C. Griffin*, Yujie Gao*, Ramakrishnan Lakshmanan*, Ricci La Centra*, Arash Sayyah[1], Julius E. Yellowhair[2], Sean M. Garner[3], Mark N. Horenstein*, Malay K. Mazumder*

*Department of Electrical and Computer Engineering, Boston University, Boston, MA 02215

[1]Department of Chemical Engineering, Massachusetts Institute of Technology, Cambridge, MA 02139

[2]Sandia National Laboratories, Concentrating Solar Technologies Department, Albuquerque, NM 87185

[3]Corning Incorporated, One River Front Plaza, Corning, NY 14831

*Abstract* — In this study, we report on design, construction, and evaluation of self-cleaning electrodynamic screen (EDS) films and their production via by gravure offset printing as a transition from laboratory-scale to roll-to-roll printing process. Both transparent and reflecting EDS films are fabricated for their applications to self-cleaning solar panels and concentrating mirrors, respectively. Experimental data are presented on the evaluation of the EDS films under simulated solar field environments demonstrating output power restoration of solar panels and specular reflectivity restoration for concentrating mirrors with low energy consumption without water. Electrode materials, geometry, optical modeling, and pulsed-voltage operation of the EDS film and scale-up studies are discussed.

*Index Terms* — electrodynamic screen, self-cleaning solar panels, photovoltaic cells.

## I. Introduction

Semi-arid areas and deserts receive the highest direct normal irradiance for photovoltaic (PV) and concentrated solar power (CSP) systems' operation and have the least interruptions from cloud and rain. Utility scale PV/CSP solar plants installed in these vast lands provide reliable and predictable high annual energy yield. However, dust and other particulate accumulation on solar collectors causes transmission losses in photovoltaic (PV) and concentrated photovoltaic (CPV) systems, and cause reflection losses in concentrated solar power (CSP) systems. Since dust layer build up on solar collectors in these areas causes major energy-yield loss [1], [2], ranging from 5% to 40% annually, there have been many efforts to maintain the optical surface of the solar collectors clean. In order to restore the efficiency of the solar energy harvesting systems, the use of high-pressure water jets, sometimes mixed with detergents, is the most commonly-practiced manual cleaning method in large-scale solar plants. This method, however, requires significant amounts of distilled or demineralized water which is scarce in arid zones and also is labor-intensive, as it requires a team of trained personnel to perform the cleaning task. Furthermore, water cleaning use is not scalable or sustainable. Recently, there have been some advancements in utilizing robotic devices in cleaning flat-plate PV panels yet they are still in the developmental stage, require water resources for cleaning, and their scalability in large-scale solar plants has not been well established. Robotic devices would also require frequent maintenance due to the multitude of moving parts, and dusty environment.

The method of cleaning used must be scalable and suitable for hot and dry weather conditions. Electrodynamic screen (EDS) technology represents a viable, promising solution as it does not require water resources or mechanical movement in removing dust particles, and it is an extremely low-power technology that can be fed from the harvesting device itself and does not need an external power source. We reported earlier our laboratory studies on the EDS films for electrostatic removal of dust [3]-[5], prototype developments by screen-printing electrodes on glass substrates and their performance for restoring output power of solar panels. We are studying the production of the prototype EDS film printed on ultra-thin flexible glass [6] using gravure offset printing (GOP) in an industrial production environment. The objective is to establish possible manufacturing process for EDS films.

For EDS films, transparent conducting electrodes are used for solar panel applications and reflective conducting electrodes are used for concentrating mirrors. We discuss here the effects of: (1) geometrical configurations of the electrodes,

(2) optical and electrical properties, (3) environmental stability of the electrodes and (4) the three-phase pulsed voltage operation with minimal energy requirements on dust removal efficiency. Since electrostatic charging of the dust particles and the Coulomb repulsion force effectuate the removal of dust, an analysis of the dust removal force against the associated adhesion force as a function of particle size is discussed based on our electrostatic modeling EDS operation.

## II. Electrode Geometry, Materials, and Printing Methods

The EDS consists of interdigitated electrodes deposited on a glass substrate, encapsulated by one or more transparent dielectric coating(s) that protect the electrodes from direct exposure to the outdoor environment. An EDS film consists of three components: (1) rows of conducting transparent or reflective, parallel electrodes, (2) transparent ultra-thin glass used as superstrates for printing the electrodes, and (3) optically clear adhesive (OCA) films used for lamination on the front surface of the solar collectors. The basic design of the electrode geometry for three-phase excitation is shown in Fig. 1.

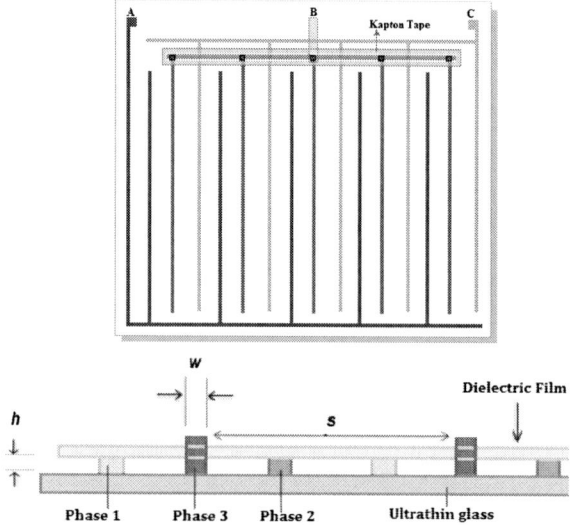

Fig. 1. Schematic diagram of the parallel electrode geometry of an EDS film (top). Three-phase connections of electrodes printed on ultrathin glass as superstrate. The configuration shown here is the upside down position where the optical path is air-dielectric film-electrode-OCA film-solar collector.

### A. Transparent Conductive Electrode Materials

For EDS film applications to solar panels the electrode materials needs to be mechanically flexible, environmentally durable, conducting with sheet resistance $R_s < 200$ $\Omega$/sq, transparent with transmission denoted as T, to be $> 80\%$ within the useful range of solar spectrum used for the solar cells and be available for roll-to-roll production in a large scale. The choice of electrode materials is limited to silver

nanowires (AgNW), Carbon nanotubes (CNT), printable metal grids, and graphene [7].

### B. Reflective Conducting Electrode Materials

Fig. 2 illustrates the optical geometry of the EDS film with conducting reflective electrodes for their applications to solar mirrors. To accomplish this objective, we use silver paste ink Loctite ECI 1011 purchased from Henkel Electronics as the electrode material. For screen printing purposes, we diluted the ink with DBE-9 obtained from Sigma-Aldrich. Silver paste electrodes, widely used in solar panels, can be made further reflective by adding a reflective coating atop the existing electrode surface. For a high specular reflectance (SR) of the electrode, the surface of the electrodes needs to have minimum roughness before coating with silver.

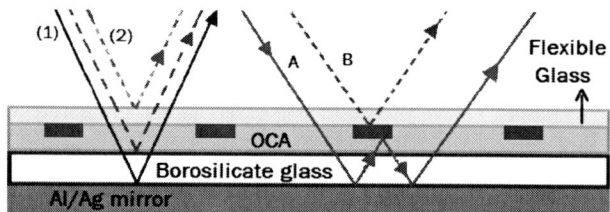

Fig. 2. Schematic diagram of the parallel electrode geometry for an EDS film where A, B and C denote phase connection pads for phases 2, 3 and 1 respectively (top) reflection pattern of incident light on different EDS layers(1) Primary reflections (2) Secondary reflections A) Reflection from bottom electrode surface and B) Reflection from top surface of electrode.

The screen-printed electrodes have s rough surface profile, hence a low SR but high diffused reflectivity. Experiments were done in order to investigate the possibility of achieving a smoother electrode surface. Table I shows an analysis of the variations tried with respect to curing temperature and period of curing. The curing technique most suitable to achieve higher SR was curing at room temperature.

For improving SR of the EDS, it is necessary to add a reflective coating on both sides of the electrodes. Possible methods of incorporating a reflective coating include the application of Ag-based polymer ink, reactive silver ink [8], and electroplating the electrodes with silver [9]. GOP based EDS film production provides convenient means for the inclusion of reflective coatings.

Table I
Summary of Variations of Curing Techniques

| Curing environment | Curing temperature ($^{\circ}C$) | Curing period | Surface roughness difference (nm) |
|---|---|---|---|
| Lab Convection Oven | 250 | 10 to 12 hours | 40 |
| Room/Lab Environment | About 28 | 10 to 12 hours | 10 |
| Freezer | > 4 | 10 to 12 hours | > 10 |

### C. Effects of Electrode Geometry

The electrode geometry, involving the electrode width (w), inter-electrode separation (s) and the electrode height or thickness (h) determines the optical transmittance or specular reflectance of the EDS films for the case of solar panels and solar mirrors respectively. It also plays a pivotal role in the calculation and the spatial distribution of the electrostatic repulsion forces on the surface of the EDS films [9].

Different values were experimented with, to determine the ideal ratio of electrode width and inter electrode spacing (w:g) that would ensure maximum dust removal efficiency (DRE) of the EDS panels. Samples of PCB-EDSs with a fixed value for w = 80μm and varied values for s ranging between 400μm to 800μm were tested with the same amount of dust deposition and were operated at different input voltages. Current designs produced in lab, based on performance of tested PCBs are of w:g ratio 80:700 as it proved to be an ideal ratio for EDS performance. The ratio is subjected to change with demands for higher SR and Specular Reflection Restoration (SRR) for the case of reflecting electrodes. For transparent electrodes, the value of w is expected to be greater than 100μm in order to ensure continuity of the silver nanowire mesh network in each printed electrode.

### D. Optical and Electrostatic Models for the EDS film Configurations

FRED optical model was used to estimate the transmittance and reflectance of different EDS films based on the (1) electrode geometry, (2) electrode materials (transmission or reflection efficiency), (3) dielectric film used for encapsulation and (4) OCA film for lamination on the front surface of solar collectors. In our optical modeling, we assumed that the ultrathin glass and the OCA film used for lamination have negligible absorption losses and calculated the optical efficiency based on the geometrical configuration and the optical properties of the electrodes. For solar panel applications, if w = 30 μm, s = 500 μm, and thickness h = 3.5 μm, then T (electrode transmission efficiency) = 80%, the overall optical transmission efficiency of the EDS film will be 94.5%; i.e., the transmission loss of the EDS film will be as

low as 1% correcting for the reflection loss of the Willow Glass. The optical model does not include diffused transmission of solar radiation to the panel.

For solar mirrors, if the same electrode geometry is used for an EDS film with 90% SR (reflectance of the electrode surfaces as shown in Fig. 2), the corresponding SR of the EDS film laminated on a mirror would be 98.3% assuming the mirror surface has 100% SR efficiency. The dust removal efficiency of the EDS film depends upon the EDS film configurations along with the thickness (t) and the dielectric constant (ε) of the Willow Glass and the electrical field distribution on the top surface of the EDS for a given amplitude and frequency of the three-phase voltage pulses applied to the electrodes [10].

An electrostatic force model of the electric field proves to have a decisive role in the understanding of the EDS and in how to increase its dust removal efficiency by optimizing the design parameters as well as material selection. The current EDS' working is interpreted based on prior COMSOL models that deal with the analysis of electric field distributions in two EDS configurations. The analytical solutions for the electric potential and electric field components and simulation results for the electric field distribution are based on the models. Fig. 3 depicts the structure of the first EDS structure modeled using COMSOL. The electrodes are encapsulated by two layers of transparent dielectric materials 1 and 2, with relative permittivities of $\varepsilon_{d1}$ and $\varepsilon_{d2}$, respectively. In manufactured devices, dielectric 1 is optically clear adhesive (OCA) and dielectric 2 is a very thin borosilicate glass.

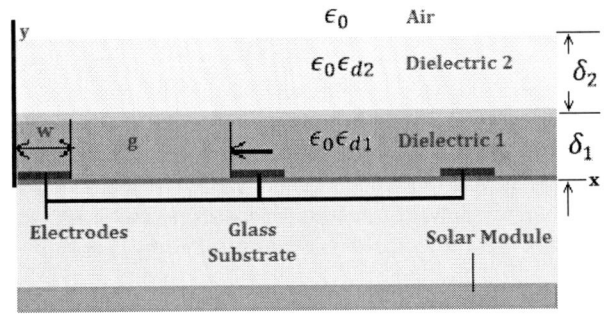

Fig. 3. Schematic representation of the EDS with two layers of transparent dielectric coatings. In a concentrated solar power application, the solar module is either a photovoltaic cell or a reflecting mirror film. The dust particles deposit on the dielectric layer 2. The electrodes are connected to a three-phase power supply.

Fig.4 represents a configuration of the EDS that is a viable candidate for dust removal from solar mirrors in CSP application. Here, the EDS consists of three layers of dielectric coatings. The electrodes are printed on the lower side of the third dielectric layer. The electrode width and inter-electrode spacing are denoted as w and g, respectively and thicknesses of transparent dielectric layers 1 to 3 are denoted as $\delta 1$, $\delta 2$ and $\delta 3$ respectively. The relative permittivities of the three dielectric layers are $\varepsilon d1$, $\varepsilon d2$, and $\varepsilon d3$. The dielectric layers used as 1, 2, and 3 are acrylic, optically clear adhesive (OCA) [11], and Corning® Willow® Glass [12], respectively. The metal layer is a thin layer, in the order of 1 to 2 mm of aluminum or silver mirror film which is adhered to the back of the borosilicate glass.

Ongoing modelling work includes utilization of various material properties for electrode parameters based on the new inks we use to print the EDS. We also incorporate new values for dielectric properties, particularly for the height depending on the OCA layer used. The different geometry parameters are used to understand and simulate the electric field distribution on the surface of the EDS.

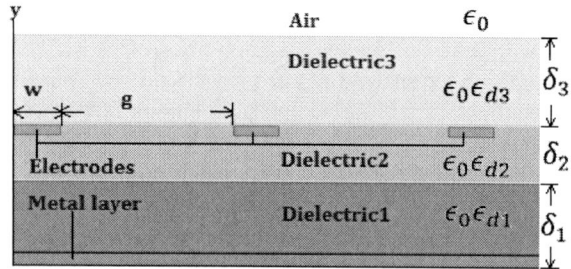

Fig. 4. Second EDS configuration wherein the electrodes are printed on the lower side of dielectric layer 3.

In the current model, the mesh used is the Extremely Fine preset with a maximum element size of 86.6_m and a minimum element size of 0.173_m. This range has been explored and we are trying to find the optimum size. In the design that is currently being developed, a layer of OCA or Willow Glass is not included. Hence there is no dielectric constants value to report. Also the electrodes are modeled as copper to study the electric field and electric lines between them. Fig.5 shows the simulations obtained by doing so.

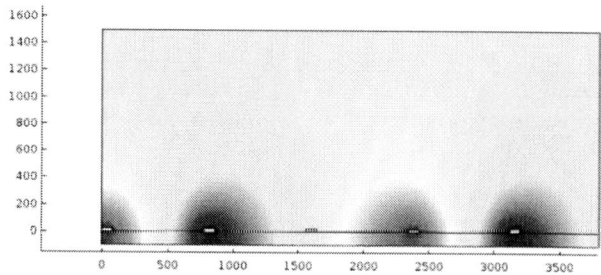

Fig. 5. (Top) Electric field norm |E| (unit: V/m) and electric vector field in one fundamental spatial period for the EDS model. (Bottom) electric potential (unit: V) in one fundamental spatial period for the EDS model.

*E. Methods of Electrode Printing for Fabrication of EDS Films*

Solution or suspension based electrodes can be printed on polymer or glass using methods that include screen-printing, gravure offset printing (GOP), and flexographic printing. We produced lab-scale EDS panels and EDS films by screen-printing electrodes on borosilicate glass plates and on the Willow Glass film and are implementing the process to GOP based production for better line resolution, improved surface morphology, higher SR, and has printing higher volume potential. Lab scale EDS panels are also produced using photolithography with highly reflective chrome, chrome and aluminum as electrode materials.

### III. EDS Film Production by Gravure Offset Printing (GOP) and Evaluation

Prototype EDS films were produced by gravure offset printing (GOP) with (1) silver paste electrodes (w = 30 μm, h = 3.5 μm, and s = 500 μm) for solar panel applications and (2) reflective electrodes (w = 30 μm, h = 3.5 μm, and s = 500 μm).

Reflectivity on both sides was accomplished by first printing a seed layer of silver followed by deposition of copper for high conductivity, followed by chemical deposition of nickel as a reflective coating on the silver/copper electrodes and for improving corrosion resistance. Fig. 6 shows a section of the EDS film produced by the GOP process.

Experimental data show top surface reflectivity is 30% and bottom surface reflectivity 40%, and the resultant SR of the EDS varied from 81 to 85%. These results agree with the optical model calculations and show clearly that if both top and bottom surface SR values are increased by silver-plating, the EDS film be able to achieve an SR higher than 97%.

Table II shows that GOP based electrodes with a nickel coating provided improved SR compared to that of screen-printed electrodes. The best SR was obtained with Al electrodes with flat surface printed by photolithography.

978-1-5090-5606-4/17 $31.00 © 2017 IEEE

Table II. SR values measured from EDS using different manufacturing techniques and electrode materials

| EDS Film with reflective electrodes | SR (STD) |
|---|---|
| GOP based nickel coated electrodes | 85.8% (1.7%) |
| Screen-printed silver ink electrodes | 72.9% (2.1%) |
| Photolithography based Al electrodes | 92% (1.2%) |

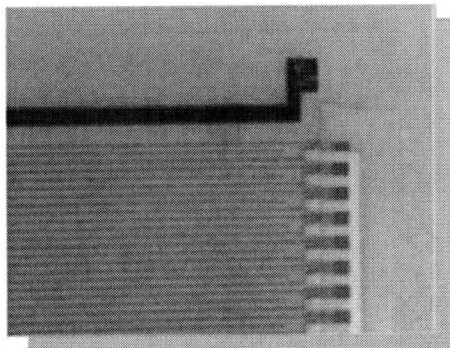

Fig. 6. An enlarged view of the GOP produced EDS film showing electrodes, crossover tape and the electrode connections.

The EDS films were laminated on solar panels and solar mirrors and the laminated solar collectors were evaluated for restoring output power of solar panels and SRR of the solar mirrors. Specular reflectivity restoration (SRR) defined as (SR after dust removal by EDS/ SR before dust deposition) was measured for EDS films produced by (1) photolithography, (2) screen-printing and (3) GOP. Results are shown in Table 2. SRR depends upon the dust removal efficiency of the EDS film; our goals are to reach SRR > 95% for solar mirrors and OPR (output power restoration) > 95%. The OPR defined by ($I_{sc}$ after dust removal by EDS / $I_{sc}$ before dust deposition) of the EDS film laminated solar panels was 95%, for both photolithographic and screen-printed EDS films. Evaluation of GOP based EDS films for OPR is under progress.

Table III. Measured values of SRR for GOP produced EDS film samples. The films were produced in different batches.

| EDS-printing process | Electrode Material | SRR values |
|---|---|---|
| Screen Printing | Silver ink | 88% |
| Photo-lithography | Al, Cr | 89% |
| Gravure-Offset Printing | Ag/Cu/Ni | 81% |

The basic electrode geometry was nearly the same (w = 80 μm and s = 800 μm); the electrode thickness (h) was 0.95 μm for photolithography and 3.5 μm for screen-printing and GOP based printing. As we are transitioning to GOP for possible scale up and advancing prototype production, we are developing methods for applying reflective coating and making the surface profile uniform.

## IV. Conclusions

To reduce O&M cost of large-scale solar plants, it is necessary to utilize a low- cost durable automated self-cleaning system with minimal water usage. Electrodynamic screen (EDS) film is an emerging technology for self-cleaning solar collectors being developed to maintain high optical efficiency of concentrating mirrors, receivers and PV modules. Electrodynamic dust removal method has a strong potential for maintaining high efficiency of solar collectors with minimum usage of water and at a low O&M cost [13]. Dust removal from solar collectors with minimum use of water will prove beneficial, if not necessary, for large-scale applications of solar energy meeting global demand. Our studies show that transparent electrodynamic screen can be integrated on the surface of solar mirrors to maintain specular reflection efficiency greater than 90%. The method is applicable for both parabolic troughs and heliostat mirrors. Production of the EDS film by gravure offset printing based production method shows the method can be used for scale up and high volume production. Evaluations performed under environmental conditions simulating desert and semi-arid climates show that the EDS based dust removal process would be able to maintain high optical efficiency with low energy requirements.

## V Acknowledgements

We acknowledge the grant from the Department of Energy (DOE CSP APOLLO EE-0007119) for supporting this research project on "Enhancement of Optical Efficiency of CSP Mirrors for Reducing O&M Cost via Near-Continuous Operation of Self-Cleaning Electrodynamic Screens (EDS)". Support from MassCEC for a cost-sharing grant on the project is acknowledged. We also acknowledge the support from our partners including Corning Research & Development Corporation, Industrial Technology Research Institute (ITRI), GeodrillR (EDS Chile SpA), and Sandia National Laboratories.

## References

1. T. Sarver, A. Al-Qaraghuli, and L. L. Kazmerski, "A comprehensive review of the impact of dust on the use of solar energy: History, investigations, results, literature, and mitigation approaches," *Renewable and Sustainable Energy Reviews*, vol. 22, pp. 698 – 733, 2013.
2 M.K.Mazumder, M.N.Horenstein, J.W.Stark, P.Girouard, R.Sumner, B. Henderson, O. Sadder, I. Hidetaka, A. S. Biris, and R. Sharma, "Characterization of electrodynamic screen performance for dust removal from solar panels

and solar hydrogen generators," *IEEE Transactions on Industry Applications*, vol. 49, no. 4, pp. 1793–1800, 2013.

3. M. N. Horenstein, M. K. Mazumder, and R. C. Sumner, "Predicting particle trajectories on an electrodynamic screen - theory and experiment," *Journal of Electrostatics*, vol. 71, no. 3, pp. 185–188, 2013.

4. M. K. Mazumder, J. W. Stark, C. Heiling, M. Liu, A. Bernard, M. N. Horenstein, S. Garner, and H. Y. Lin, "Development of transparent electrodynamic screens on ultrathin flexible glass substrates for retrofitting solar panels and mirrors for self-cleaning function," *MRS Advances*, 2016, Published on line http://dx.doi.org/10.1016/j.elstat.2016.02.0045.S.

5. Garner, S. Glaesemann, X. Li, Ultra-slim flexible glass for roll-to-roll electronic device fabrication, Applied Physics A 116 (2) (2014) 403–407. 6. Pudas, Marko, Gravure-offset printing in the manufacture of ultra-fine-line thickfilms for electronics Department of Electrical and Information Engineering, Microelectronics Laboratory, http://jultika.oulu.fi/files/isbn9514273036.pdf, University of Oulu, P.O.Box 4500, FIN-90014 University of Oulu, Finland 2004 Oulu, Finland

7. Hecht, David S., Liangbing Hu, and Glen Irvin. "Emerging transparent electrodes based on thin films of carbon nanotubes, graphene, and metallic nanostructures." Advanced Materials 23, no. 13 (2011): 1482-1513.

8. Walker, SB, and JA Lewis. 2012. "Reactive Silver Inks for Patterning High-Conductivity Features at Mild Temperatures." Journal of the American Chemical Society 134: 1419-1421.Dec 12, 2014

9. Norman L. Thomas ; Jesse D. Wolfe ; Joseph C. Farmer; Protected silver coating for astronomical mirrors. Proc. SPIE 3352, Advanced Technology Optical/IR Telescopes VI, 580 (August 25, 1998); doi:10.1117/12.319243

10. A. Sayyah, M. N. Horenstein, and M. K. Mazumder, "A comprehensive analysis of the electric field distribution in an electrodynamic screen," *Journal of Electrostatics*, vol. 76, pp. 115 − 126, 2015

11. 3M8146e2 Optically Clear Adhesive, 2014 [Online]. available:http://multimedia.3m.com/mws/media/944360O/3mtm-optically-clear-adhesive-8146-x-series-tds.pdf

12. Corning® Willow® Glass Fact Sheet, 2014 [Online]. Available:http://www.corning.com/WorkArea/downloadas set.aspx?id¼51335.

13. N. Joglekar, E. Guzelsu, M. Mazumder, A. Botts, and C. Ho, "A Levelized Cost Metric for EDS-Based Cleaning of Mirrors in CSP Power Plants," p. V001T02A026, Jun. 2014.

# Evaluating Solar Cell Fracture as a Function of Module Mechanical Loading Conditions

Eric J. Schneller[1], Andrew M. Gabor[2], Jason Lincoln[1], Rob Janoch[2], Andrew Anselmo[2], Joseph Walters[1] and Hubert Seigneur[1].

[1]Florida Solar Energy Center, University of Central Florida, Cocoa FL, USA
[2]BrightSpot Automation LLC, Westford, MA, USA

*Abstract* — Cell cracking presents a serious risk for the long term reliability of c-Si photovoltaic modules. Cracks may not initially result in performance loss, but over time performance may degrade as the module experiences stresses in the field such as temperature cycling and snow/wind loading. This performance loss is due to the formation of new cracks with front side loading, propagation of existing cracks, and the opening of existing cracks in which regions of the cell become more electrically isolated. This work utilizes a new tool, the *LoadSpot*, that allows for I-V performance characterization and electroluminescence imaging of PV modules while under mechanical load. We explore a variety of cell technologies to understand the magnitude of mechanical stress required to induce cell fracture, and assess the impact these cracks have on performance. In addition, we study the use of cyclic loading to open existing cracks. The tests used in this work have potential applications in product development, factory quality control, product evaluations, and optimization of mounting hardware and methods.

*Index Terms* — cell fracture, cyclic load testing, electroluminescence, mechanical load testing, photovoltaic modules, reliability, silicon.

## I. INTRODUCTION

Photovoltaic (PV) modules have an excellent track record of reliability that has been established from numerous studies of the degradation rates of field deployed modules. Degradation rates in the range of 0.5-0.6%/year have been demonstrated for crystalline silicon PV module, with deviations occurring based on variations in the specific module technology, the operational climate, and module mounting configuration [1]. As cell and module technology advance and manufacturing cost decline, it is essential to ensure adequate durability of PV modules in order to reduce the levelized costs of energy for PV systems.

A trend of reducing cell thickness has been established over the years in an effort to reduce manufacturing costs. This has increased the occurrence and susceptibility of cells to mechanical failure through cracking. Basic panel design is vulnerable to cell cracking. Copper wires contract more than silicon during soldering resulting in the formation of microcracks in the silicon underneath the busbars [2]. This may not initially cause performance issues, however forces applied to the cell later can cause these microcracks to propagate into full cracks. The asymmetric design of the standard panel with a thick and stiff glass sheet on the front side and thin polymer backsheet means that front side loads put the cells into tensile stress. If such stresses are high enough, the benign microcracks

under the busbars can propagate into full cracks down the length of the cell.

Initially most of these cracks are tightly closed with current transport across the metallization on both sides. Cracks in the closed state do not significantly reduce module performance and are difficult to even detect in electroluminescence (EL) or photoluminescence (PL) imaging. This presents a significant degradation risk as these crack may open up over time, inhibiting current transport across the crack and ultimately reducing overall power generation.

To study cell cracking dynamics, this work utilizes the *LoadSpot* tool developed by BrightSpot Automation. The *LoadSpot* was developed to perform mechanical load testing while leaving the front surface of the module unobstructed to allow for *in-situ* performance characterization [3]. The tool can perform the standard static and cyclic load tests for module design qualification as per IEC-61215 and IEC- DTS-62782, as well as more specialized stress sequences that may be used for product development. Loads are applied to the rear side of the module with vacuum/air-pressure, while a unique seal design allows the module to freely deflect as is depicted in Fig.1. While under load, the module can be characterized using a solar simulator or by EL imaging. In this work, a Sinton FMT-350 is used for I-V measurements and a BrightSpot EL camera system is used for EL imaging.

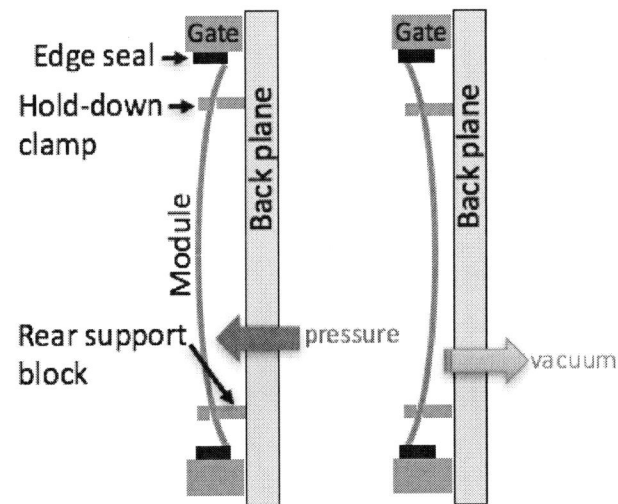

Fig. 1. Diagram showing the operation of the *LoadSpot*.

Previous studies by other groups, including Evergreen Solar, Fraunhofer CSP, SunCycle and ZAE Bayern, have investigated crack opening during module deformation using EL imaging [2, 4, 5]. In this work, we explore how characterization of a module in both the loaded and unloaded state can be used to evaluate a module design for susceptibility to cell fracture.

## II. CELL FRACTURE AS A FUNCTION OF MECHANICAL LOAD

Four module types were evaluated for this work that are representative of standard 60-cell module designs with front glass/EVA/backsheet packaging. One representative module of each type is explored in this section. Modules 1-3 were multi-crystalline technology and Module 4 was mono-crystalline technology. Fig. 2 shows the number of fractured cells as the module was exposed to increasing front side load up to a maximum of 5400Pa. A cell was classified as fractured with the presence of any crack, and therefore this data does not attempt to quantify the severity of any specific crack. In many cases cracks became more severe as the loading condition increased. Modules were supported at the manufacturer specified mounting locations so that the loading conditions would be representative of field conditions.

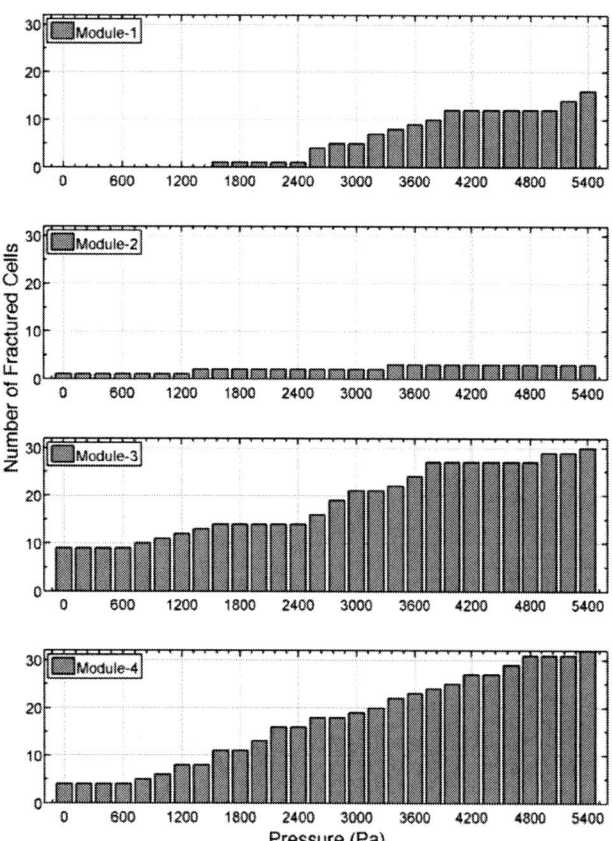

Fig. 2. Histogram showing the number of fractured cells as a function of applied pressure for 4 module types. All modules were of standard size with 60-cells, and 4-point clamping was used.

It is obvious from Fig. 2 that there is a wide range of susceptibility to cell fracture across the various module types. Module 2, for example, has only 2 new cracks formed during the entire testing sequence. In contrast, Module 3 and 4 each have more than 30 fractured cells at 5400Pa. The superior performance of Module 2 is due to the frame design in which there are two cross members, or back rails, that provide extra mechanical support. This data highlights the value that simple mechanical support structures can provide with respect to crack formation. This data also emphasizes that modules with similar electrical performance and upfront cost, may respond quite different when exposed to mechanical stress.

Above 2400Pa, the mono-crystalline cells used in this work appear to fail catastrophically, referring to the dendritic nature of any new cracks formed above this pressure. The shattering of monocrystalline cells also leads to irreversible performance loss once the load is removed. This does not appear to be the case for the multi-crystalline cells. In mono-crystalline cells, cracks can easily propagate along certain crystal planes, whereas the randomized grain structure in multi appears to limit crack propagation. In general, we observed that fracture in multi-crystalline cells was less severe as compared to the fracture observed in mono-crystalline cells. An example of this is in Fig. 3, where the difference in crack patterns that occurred during the test is shown. The dendritic crack in the mono-crystalline cell continues to severely impact performance even after the removal of applied load, whereas the multi-crystalline cell appears to fully recover. If we consider the total crack length for the cells in Fig. 3, the mono-crystalline cell would have an order of magnitude higher total length.

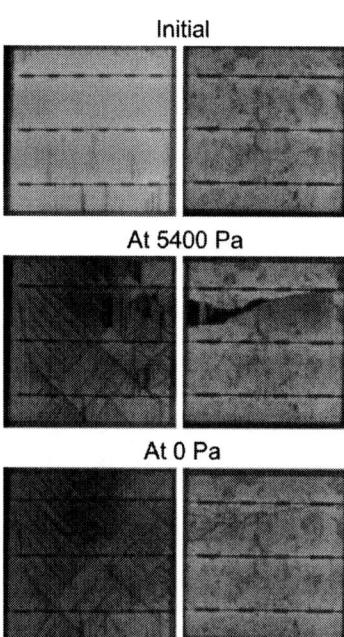

Fig. 3. Example of two cracks formed at 5400Pa, highlighting the differences between fracture in mono-crystalline (left) and multi-crystalline (right) cells.

Increasing Pressure

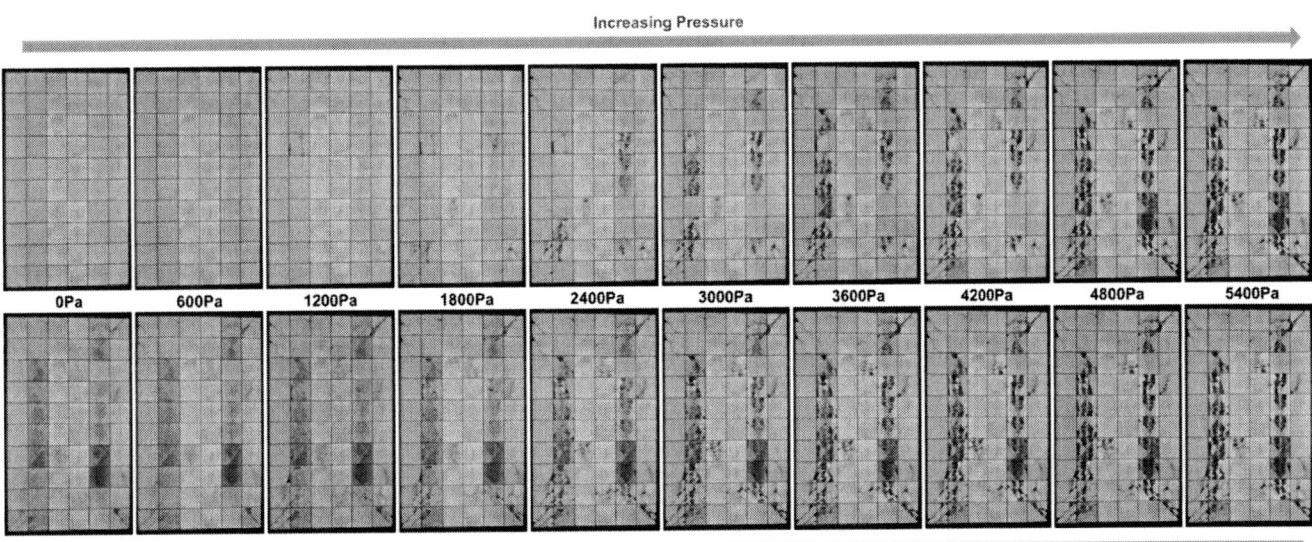

Fig. 4. Electroluminescence images of a mono-crystalline module (Module 4) as a function of applied front side mechanical load.

Increasing Pressure

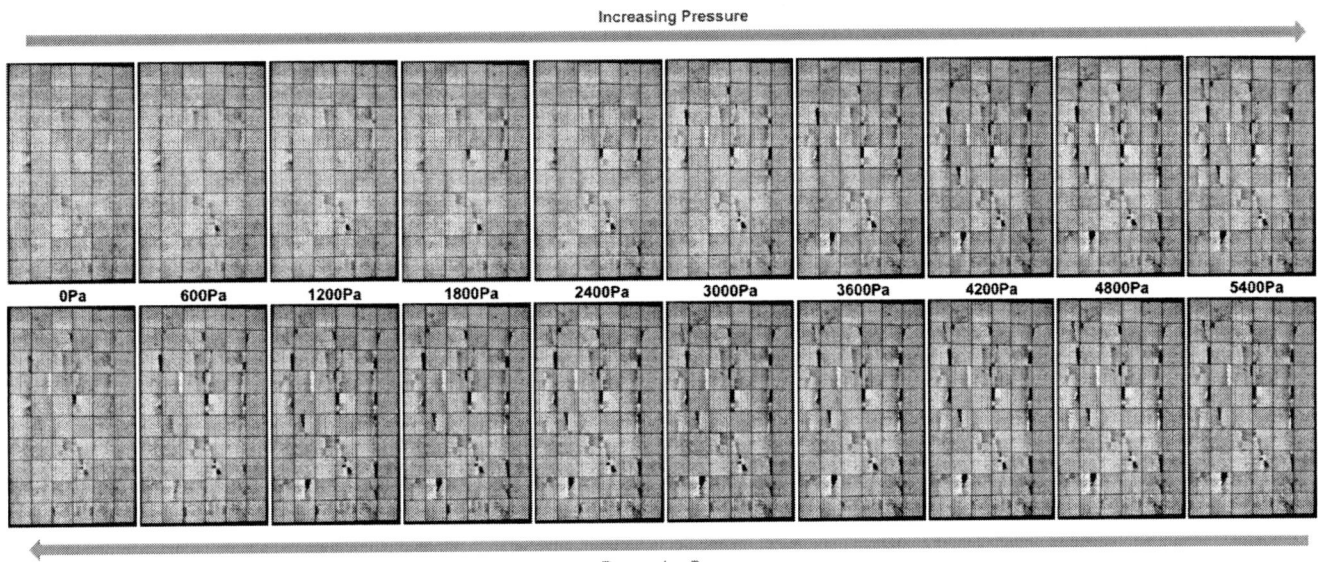

Fig. 5. Electroluminescence images of a multi-crystalline module (Module 3) as a function of applied front side mechanical load.

Fig. 4 and Fig. 5 show the complete progression of the testing sequence used in this work for both a mono-crystalline and multi-crystalline module. One can visually see that many of the cracks are formed at or above 2400Pa for both modules. There is a clear difference in the type of cracks that occur and how these cracks behave upon removal of the load between the two modules. For the mono-crystalline module we see a pattern that matches the first principle stress that is predicted from simulations. The multi-crystalline module shows a different trend where many of the cracks form parallel to the busbars.

While the cells are under tensile stress (i.e. applied front side load), cracks have a tendency to open, which can affect the electrical conductivity between the contacts on either side of the crack. This can lead to certain regions of the cell being electrically isolated. These regions appear dark in the EL images. As the tensile stress is removed, the cracks close and conduction across the cracks may increase. We see this effect in the EL images as the area fraction of these dark regions reduces dramatically from 5400Pa to 0Pa. This so called crack "healing" has been reported in other work [6]. Although the trend is clear that as one reduces front side pressure more cracks close, some portion of this effect is random.

## III. IMPACT OF MECHANICAL STRESS ON PERFORMANCE

As cracks form within a module, the power of the module will degrade. The magnitude of that power loss, however, is dependent on several factors including the directionality of the crack, the electrical resistance across the crack, and the total number of cracks [7, 8]. With the ability to measure performance while under load, we can start to quantify these different influences. In this work we focus specifically on two metrics. The first metric is the maximum power loss which is measured at 5400 Pa. A large fraction of this loss is recoverable upon removal of the load. The second metric is the irreversible power loss measured as the difference in power between the initial state and the final unloaded state. Fig. 6 and Fig. 7 show the change in maximum power ($P_{mp}$) as a function of applied load for the modules shown in Fig. 4 and Fig. 5 respectively. Although both modules had a similar number of fractured cells, the mono-crystalline module had a maximum power loss above 20% whereas the multi-crystalline module only had a maximum power loss of 5.3%.

Another significant difference between the two modules is how the power loss recovers as the pressure is reduced. The mono-crystalline module recovers almost completely by 1200Pa, whereas the multi-crystalline module only starts to recover at that pressure. This is likely due to the differences in the type of cracks that are present in the two modules.

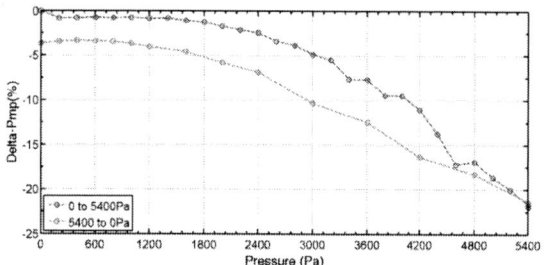

Fig. 6. Change in Maximum Power as a function of front side load for the mono-crystalline module depicted in Fig. 4.

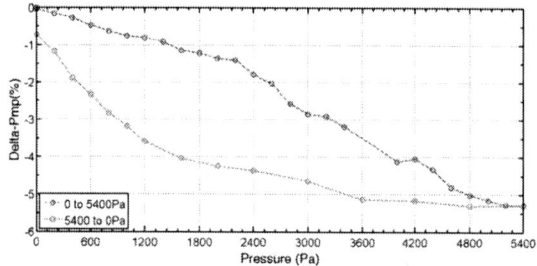

Fig. 7. Change in Maximum Power as a function of front side load for the multi-crystalline module depicted in Fig. 5.

To explore the reproducibility of these results, we performed the same test sequence on five modules of the same make and model as the mono-crystalline module. The results are shown in Fig. 8. All five modules behave in a similar manner. The irreversible damage is just under 5% for each of the modules. There is greater variability in the maximum damage measured for each module, highlighting the somewhat random nature of crack formation.

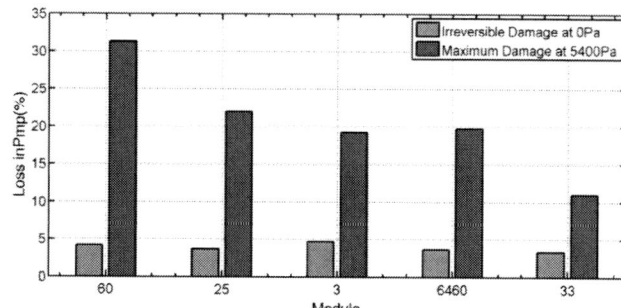

Fig. 8. Power loss for 5 mono-crystalline modules, from the same manufacturer, at an applied pressure of 5400Pa (maximum damage) and with pressure removed (irreversible damage).

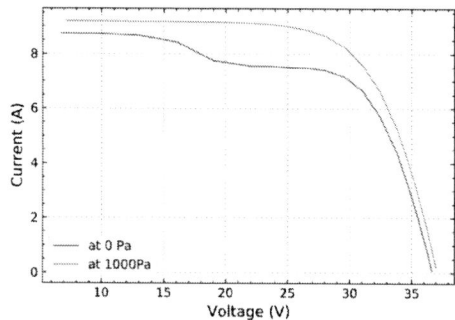

Fig. 9. Variation observed in I-V characteristics for the same module in the both the loaded (at 1000Pa) and the unloaded state (at 0Pa).

There is a significant difference in the I-V curves measured in the loaded and the unloaded state. This difference is shown in Fig. 9 for a module that underwent static and cyclic loading. The curve under load shows a step, that is the result of cell mismatch. Cell mismatch is known to cause hot-spots when the module is deployed in the field [9, 10]. These hot-spots can cause secondary degradation modes to occur, such as back sheet delamination, further impacting module performance.

It is unclear if the module will transition from the performance in the unload condition to the performance in the loaded condition while deployed in the field. This transition may occur slowly as the number of thermo-mechanical cycles continues to increase and the cracks begin to permanently open up. There is also the possibility that during particularly windy conditions the module could rapidly change from one state to the other. This rapid switching, particularly when there is severe cell mismatch, could stress the bypass diodes in the module. This cycling of the diode may lead to failure over time, presenting a serious safety hazard. If a bypass diode quickly switches from forward to reverse bias, there is also a concern that a thermal runaway type of failure may occur [11].

The use of the performance of a module during application of a front side load may be a useful indicator of long term module performance. Further testing is underway to validate the predictive nature of this type of test [12]. The strong power loss signal produced from the module while underload can be used to evaluate module designs in terms of their susceptibility to cell fracture and maximum power loss.

## IV. Impact of Cyclic Loading on Cracks

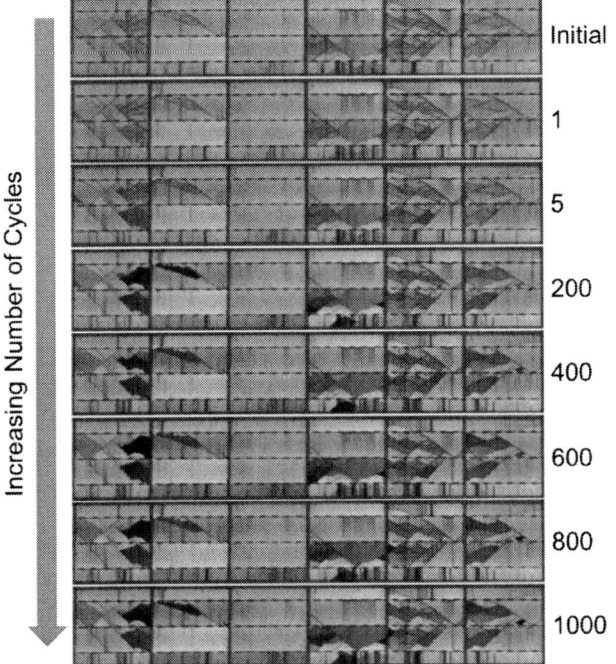

Fig. 10. Progression of cell opening, witnessed through electroluminescence imaging, as a function of loading cycles (+1000Pa to -1000Pa).

Cyclic loading is a widely used testing protocol for the purpose of qualification testing and accelerated aging. After the static loading performed in the initial section of this work, select modules where then taken through the standard IEC protocol for cyclic loading: 1000 cycles of +1000Pa to -1000Pa at a rate of 7 cycles per minute. The motivation was to explore how cracks respond to the high number of opening and closing cycles. The results for a region of one module are shown in Fig. 10. The initial state exhibited very few open cracks. There is very little change after only 5 cycles, however, significant change is observed after the first 200 cycles. The extent of permanent crack opening continues to progress as more and more cycles accumulate. This result indicates that cracks have a tendency to remain open after being exposed to a high number of loading cycles.

## IV. Conclusions

To ensure adequate reliability of PV modules, the influence that cell fracture has on the long term performance of PV modules must be understood and quantified. Current IEC test specifications, including static and cyclic loading, are effective in creating cracks but may not capture the latent power loss induced from these defects once the module is deployed. We investigate the use of module performance characterization while under load as a critical indicator for module reliability and durability. This indicator could provide value throughout the PV supply chain for module R&D, within production, and also for module sellers and module buyers. The ability to quantify performance of the module with and without crack opening provides an avenue to predict the maximum potential power loss that may occur over time in the field. This type of test could be used in module design qualification, quality control, and product evaluations

We use the proposed testing sequence to evaluate several module technologies. Modules that had additional mechanical support in the form of back-rails, performed superior to those without such support. We also identified a substantial difference between the performance of mono-crystalline and multi-crystalline modules. In this somewhat limited investigation, we identified that damage was more severe in mono-crystalline modules at similar load levels. These test cases were used to highlight the breadth of knowledge that could be obtained through performance characterization of modules while under load.

## References

[1] D. C. Jordan, S. R. Kurtz, K. VanSant, and J. Newmiller, "Compendium of photovoltaic degradation rates," *Progress in Photovoltaics: Research and Applications,* vol. 24, pp. 978-989, 2016.

[2] A. M. Gabor, M. Ralli, S. Montminy, L. Alegria, C. Bordonaro, J. Woods, *et al.,* "Soldering induced damage to thin Si solar cells and detection of cracked cells in modules," in *21st EUPVSEC,* 2006, pp. 2042-2047.

[3] A. M. Gabor, R. Janoch, A. Anselmo, J. L. Lincoln, H. Seigneur, and C. Honeker, "Mechanical load testing of solar panels - Beyond certification testing," in *IEEE 43rd Photovoltaic Specialists Conference (PVSC),* 2016, pp. 3574-3579.

[4] C. Buerhop, S. Wirsching, S. Gehre, T. Pickel, T. Winkler, A. Bemm, *et al.,* "Lifetime and Degradation of Pre-damaged PV-Modules – Field study and lab testing," in *44th IEEE Photovoltaic Specialists Conference (PVSC),* 2017.

[5] M. Sander, S. Dietrich, M. Pander, M. Ebert, and J. Bagdahn, "Systematic investigation of cracks in encapsulated solar cells after mechanical loading," *Solar Energy Materials and Solar Cells,* vol. 111, pp. 82-89, 2013.

[6] M. Paggi, I. Berardone, A. Infuso, and M. Corrado, "Fatigue degradation and electric recovery in Silicon solar cells embedded in photovoltaic modules," *Sci. Rep.,* vol. 4, 2014.

[7] S. Kajari-Schröder, I. Kunze, and M. Köntges, "Criticality of Cracks in PV Modules," *Energy Procedia,* vol. 27, pp. 658-663, 2012.

[8] S. Kajari-Schröder, I. Kunze, U. Eitner, and M. Köntges, "Spatial and orientational distribution of cracks in crystalline photovoltaic modules generated by mechanical load tests," *Solar Energy Materials and Solar Cells,* vol. 95, pp. 3054-3059, 2011.

[9] Y. Hu, W. Cao, J. Ma, S. Finney, and D. Li, "Identifying PV Module Mismatch Faults by a Thermography-Based Temperature Distribution Analysis," *Device and Materials Reliability, IEEE Transactions on,* vol. PP, pp. 1-1, 2014.

[10] E. J. Schneller, R. P. Brooker, N. S. Shiradkar, M. P. Rodgers, N. G. Dhere, K. O. Davis, *et al.,* "Manufacturing metrology for c-Si module reliability and durability Part III: Module manufacturing," *Renewable and Sustainable Energy Reviews,* vol. 59, pp. 992-1016, 2016.

[11] N. S. Shiradkar, E. Schneller, N. G. Dhere, and V. Gade, "Predicting thermal runaway in bypass diodes in photovoltaic modules," in *Photovoltaic Specialist Conference (PVSC), 2014 IEEE 40th,* 2014, pp. 3585-3588.

[12] J. Lincoln, A. M. Gabor, *et al.*" Forecasting Post-Environmental Degradation Power Loss in Solar Panels with a Predictive Crack Opening Test," *in 2017 IEEE 44th PVSC,* 2017

# Computational study of the effect of photovoltaic (PV) module parameters on stress development in silicon under static loading

Saurabh Sethia[1], Karan Shishir Yadav[1], Sudharm Rathore, Abhishek Shubhrant, Aparna Singh

Department of Metallurgical Engineering and Materials Science, Indian Institute of Technology Bombay,
Mumbai-400076, India

*Abstract* — A photovoltaic (PV) module is subjected to numerous cycles of mechanical and thermal loads during its lifetime. These loads can induce normal and shear stresses in different layers of the module. Normal stresses may lead to crack development and propagation in silicon resulting in power loss. Excessive shear stresses may culminate in delamination of different layers. Choosing appropriate materials, module-layout design and layers' dimensions may help in mitigating the stresses that induce premature failure. However, it is both expensive and time consuming to evaluate the most optimum conditions for module reliability experimentally. We have used finite element (FE) computations to determine stress development in the silicon layer while exerting a static pressure load of 5400 Pa on the PV module. This has been done for a range of glass thicknesses. Stress profiles for multiple module-layouts as well as stack designs have been evaluated as well and significant differences in the magnitude and distribution of normal stresses have been found. This has serious implications for crack development in silicon layer and the subsequent power loss. Informed decisions about the overall design of the module to improve reliability can be taken using the results in this study.

*Index Terms* — crack, finite element computation, photovoltaic module, principal stresses, silicon, static loading.

## 1. INTRODUCTION

Photovoltaic (PV) modules are expected to have a lifetime of at least 20 years in the field. Damage tolerance of the modules under mechanical and thermal loads becomes a critical criterion to ensure reliability of the PV module. During production, the soldering and lamination processes can lead to small cracks in the solar cells [1]. The most consequential mechanical loads for PV modules after manufacturing result from transportation [2]. In alpine climates, snow loading poses a significant problem to the integrity of PV modules as well [3],[4]. The weight of accumulated snow exerted on the PV modules can break the glass cover and cells leading to significant power degradation. Dolara et al. [5] have shown strong correlation between snail trail discolorations in PV modules and the presence of cracked cells. Kontges et al. [6] have systematically investigated the number, orientation and frequency of micro-cracks in PV modules in field using fluorescence. These cracks can isolate cell sections leading to a module strings power loss between 6-22 % [7].

The thickness of different layers in the modules can significantly affect not only the module's photo conversion efficiency but also the bending stresses. E.g. replacing the bottom backsheet by glass not only enables transmission of extra light through the bottom, but also brings the neutral axis of the module close to the silicon layer thus alleviating stresses in silicon. Replacing the top glass with PET and the backsheet with glass is an alternative way to capture more light while keeping the module lightweight. The implications of these designs on the stress profile are still unknown. Mechanical load test in IEC 61215 can be used to systematically study effect of stack design, material selection as well as thickness on the crack development. Multiple sand bags or air pressure have been tried to apply a uniform pressure of 5400 Pa. Not only are these experiments difficult to set-up but also load application design needs to be changed based on the modules' dimensions. Moreover, such experiments necessitate making numerous modules based on the design of experiments (DOE) study planned. The observations from such tests in terms of crack orientation and distribution is not easy to quantify and is at best statistical. Since the origins of crack nucleation and growth are linked to stress development, techniques that can provide the stress profiles as a function of material selection and design can help us in deciding the optimum module parameters for enhanced PV reliability.

We have used Finite Element (FE) Analysis to model multiple design configurations and have expressed the results in terms of the spatial distribution of maximum principal stress generated in the silicon layer since it is the most vulnerable layer to crack development resulting in power degradation. The robustness of the modules can be enhanced if lower stresses develop under the same pressure loading and that can be tailored by materials selection and design.

## 2. TEST CONDITIONS

*2.1 Geometry and material selection*

Three different stacks (Fig. 1) are chosen to represent some of the commonly used framed PV module stacks in the market:

Stack 1: Glass-EVA-silicon-EVA-back sheet: Fig. 1(a)
Stack 2: Glass-EVA-silicon-EVA-glass: Fig. 1(b)
Stack 3: PET-EVA-silicon-EVA-glass: Fig. 1(c)

The modules consist of 60 solar cells of 156 mm X 156 mm each arranged in a 10 by 6 matrix. The thickness, elastic modulus and Poisson's ratio of the different layers are shown

---

[1]These authors contributed equally to this work.

in Table 1. A maximum pressure load of 5400 Pa is applied in the FE model. The maximum principal stress developed in silicon was studied for the following cases:

a. Aspect ratio variation for Stack 1
b. Glass thickness variation for Stack 1, 2 and 3.

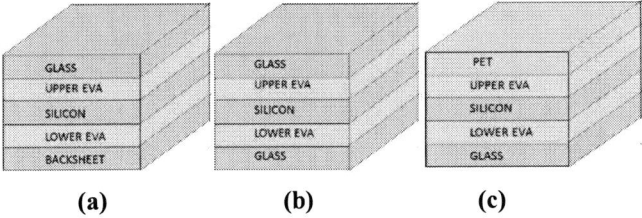

Fig. 1.   PV modules with different layers (a) Stack 1 (b) Stack 2 and (c) Stack 3

| Material | Thickness (mm) | Young's Modulus (MPa) | Poisson's Ratio |
|---|---|---|---|
| Glass | 4 | 73000 | 0.22 |
| Silicon | 0.166 | 130000 | 0.22 |
| EVA | 0.5 | 10 | 0.49 |
| Backsheet | 0.1 | 2800 | 0.29 |
| PET | 0.02 | 2500 | 0.4 |

Table 1.   Material properties and thicknesses of different layers used in the PV modules.

### 2.2 FEM model

The general purpose FEM package (ABAQUS SIMULIA, Providence, RI, USA) has been used to construct the model for the PV modules and simulate the bending response under 5400 Pa pressure load. Boundary conditions have been specified for clamping at all edges. The mesh was tested for convergence for all the cases.

## 3. RESULTS AND DISCUSSION

### 3.1 Effect of aspect ratio of modules

The most commonly used configuration of a solar module is 60 cells placed in a 10 X 6 layout. Fig. 2 shows the maximum principal stress distribution for a pressure load of 5400 Pa for alternative design layouts in terms of placing the silicon cells i.e. 12 X 5, 15 x 4 and 20 X 3. The maximum principal stress in silicon as a function of the cell layout design is shown in Fig. 3. It can be seen that silicon experiences less normal stress for design layouts with more skewed aspect ratio. The maximum principal stress in 20 X 3 layout is 75% smaller than 10 X 6 layout. This is because a more skewed configuration implies more cells being placed close to the clamped edges resulting in more constraint and hence less displacement and smaller stresses under 5400 Pa load. This will however imply larger material usage during clamping due to the higher circumference of the more skewed modules.

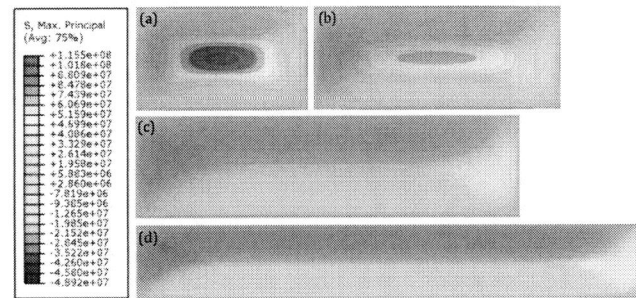

Fig. 2.   Principal Stress (Pa) in Si shown for different layouts (top to bottom): (a) 10 X 6 (b) 12 X 5 (c) 15 X 4 and (d) 20 X 3

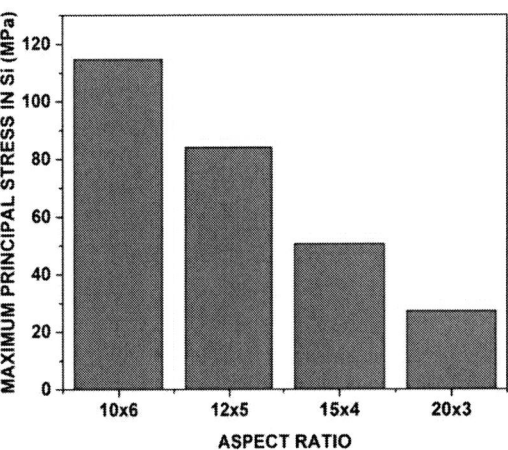

Fig. 3.   Maximum principal stress in silicon in MPa as a function of the aspect ratio of the modules.

### 3.2. Effect of glass thickness in Stack 1, 2 and 3

#### 3.2.1 Stack 1

It can be seen from Fig. 4 and Fig. 7 that the maximum principal stress increases with a decrease in the glass thickness. The region with the highest tensile stress is located close to the center of the module and turns to compressive

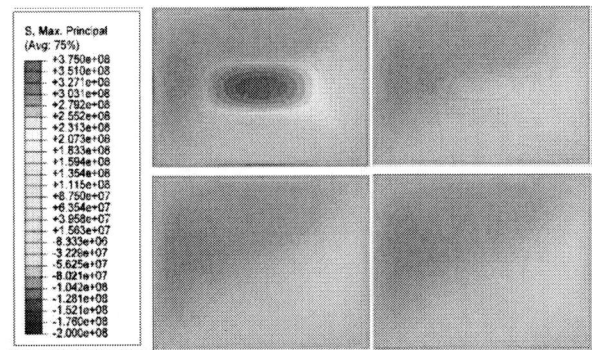

Fig. 4.   Principal Stress (Pa) in Si shown in clockwise sense for different glass thickness (Stack 1): 2 mm, 3 mm, 4 mm and 5 mm.

978-1-5090-5606-4/17 $31.00 © 2017 IEEE          2903

close to the edges. The maximum stress is close to 375 MPa when the glass thickness is 2 mm and decreases to less than 100 MPa with an increase in glass thickness to 5 mm. Thus, increasing the glass thickness will decrease the probability of crack propagation while increasing the weight of the module.

### 3.2.2 Stack 2

Fig. 5 clearly shows that the position of the maximum principal stress is at the edges of the module unlike Stack 1. Therefore, while highest probability of crack detection is at the center for Stack 1, it is expected be close to the edges for Stack 2.

A thickness of 2 mm glass on either side implies total 4 mm of glass usage and imposes a maximum principal stress of 15.7 MPa in Si as can be seen in Fig. 7 whereas for Stack 1, the corresponding stress is considerably higher ($\approx$120 MPa). This is due to the silicon layer housing the neutral axis of the module under bending thus minimizing the flexural stresses. Moreover, the principal stresses in Stack 2 are relatively insensitive to glass thickness. Thus, the glass thickness can be reduced significantly resulting in a lighter module without increasing the stresses much.

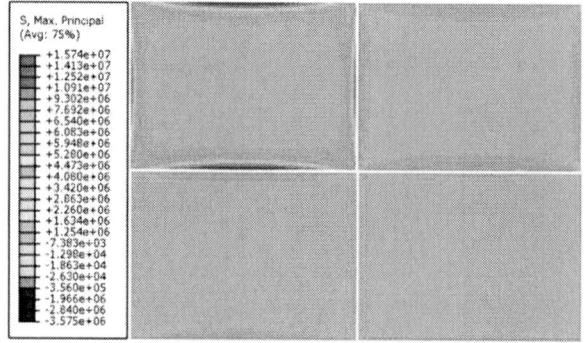

Fig. 5. Principal Stress (Pa) in Si shown in clockwise sense for different glass thickness (Stack 2): 2mm, 3mm, 4mm and 5mm.

### 3.2.3 Stack 3

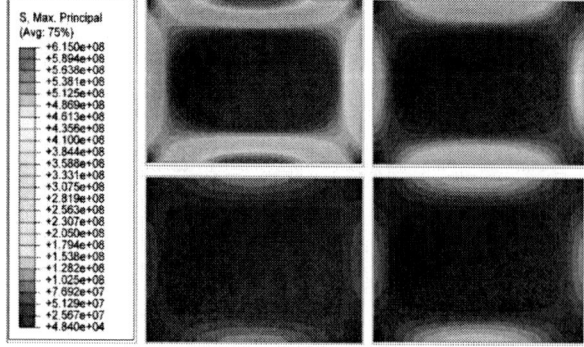

Fig. 6. Principal Stress (Pa) in Si shown in clockwise sense for different glass thickness (Stack 3): 2 mm, 3 mm, 4 mm and 5 mm.

The maximum principal stress in Si for Stack 3 is higher than Stack 1 and 2 for all glass thicknesses and the stresses rise sharply with a decrease in glass thickness below 3 mm (Fig. 6 and Fig. 7). However the regions of high stresses are localized at the edges. Therefore, cracks are expected to nucleate at the edges instead of the center for this design.

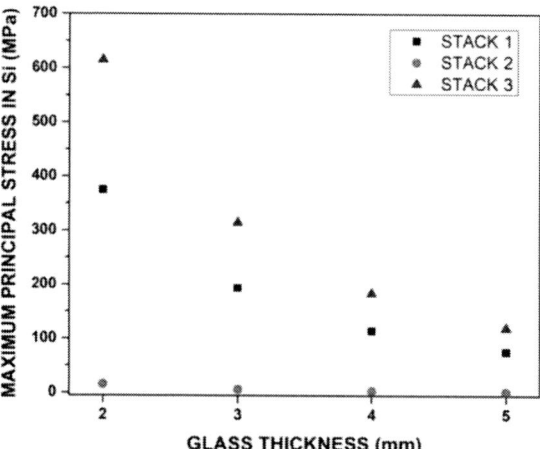

Fig. 7. Maximum principal stress (MPa) in silicon as a function of glass thickness for Stack 1, 2 and 3.

## 4. CONCLUSIONS

This study shows that the lateral and transverse design of the module significantly affects the stresses developed in Si under mechanical loading. Considerable reduction in maximum principal stress is observed when the aspect ratio of the cell layout is increased. However, more clamping supports per unit of power will be required for a more skewed module. An increase in glass thickness makes the module heavier but alleviates the principal stresses developed. However, the sensitivity of principal stress with thickness decreases with higher glass thickness. Thus, making the glass thicker cannot be used indefinitely for decreasing the principal stresses. Si in Stack 2 experiences the lowest stress whereas for Stack 3 the stress is maximum and in both cases the maximum stress is confined to the edges whereas for Stack 1 the maximum stress is towards the center. Thus, glass thickness and its placement in the stack can significantly affect the magnitude of principal stresses and hence the module reliability.

## ACKNOWLEDGEMENT

The authors will like to acknowledge the support of Department of Science and Technology- Clean Energy Research Initiative (DST-CERI-C17(G)), Solar Energy Research Institute for India and the United States (SERIIUS) and National Center for Photovoltaic Research and Education (NCPRE) towards this work.

# REFERENCES

[1] A. Gabor *et al.*, "Soldering induced damage to thin Si solar cells and detection of cracked cells in modules," *21st Eur. Photovolt. Sol. Energy Conf.*, 2006.

[2] S. Pingel, Y. Zemen, O. Frank, T. Geipel, and J. Berghold, "Mechanical stability of solar cells within solar panels," *Proc. 24th EUPVSEC*, pp. 3459–3464, 2009.

[3] M. Köntges, I. Kunze, and S. Kajari-Schröder, "Quantifying the risk of power loss in pv modules due to micro ceacks," *Photovoltaic, Eur. Energy, Sol. Conf. World Conversion, Photovolt. Energy*, no. September, pp. 6–10, 2010.

[4] K.-A. W. and M. K. Marcus Assmus, Steffen Jack, "Measurement and simulation of vibrations of PV-modules induced by dynamic mechanical loads," *Prog. photovoltaics*, no. 19, pp. 688–694, 2011.

[5] A. Dolara, G. C. Lazaroiu, S. Leva, G. Manzolini, and L. Votta, "Snail Trails and Cell Microcrack Impact on PV Module Maximum Power and Energy Production," *IEEE J. Photovoltaics*, vol. 6, no. 5, pp. 1269–1277, 2016.

[6] M. Köntges, S. Kajari-Schröder, I. Kunze, and U. Jahn, "Crack Statistic of Crystalline Silicon Photovoltaic Modules," *Eupvsec*, vol. 26, pp. 3290–3294, 2011.

[7] A. Morlier, F. Haase, and M. Kontges, "Impact of Cracks in Multicrystalline Silicon Solar Cells on PV Module Power;A Simulation Study Based on Field Data," *Photovoltaics, IEEE J.*, vol. 5, no. 6, pp. 1–7, 2015.

# A Simple Method for Measuring Solar Radiation Intensity by Image Analyses

Akiko Takahashi, Akinori Moriki, Nobuyuki Yamada, Jun Imai and Shigeyuki Funabiki

Okayama University, Okayama, 700-8530, Japan

*Abstract* — This paper proposes a simple method for measuring solar radiation intensity using pictures taken by a camera. The proposed method creates an estimation model based on the correlation between the theoretical solar radiation intensity and the color properties of pictures as the pre-processing. When measuring solar radiation intensity, a color property of the picture is substituted to the estimation model. The solar radiation intensity in a wide area can be measured from one picture using the proposed method because the proposed method uses only image analysis. The validity of the proposed method is examined using the root mean square error.

*Index Terms* — HSV color pace, image analysis, multiple regression analysis, solar radiation intensity, photovoltaic generation.

## I. INTRODUCTION

Introduction of photovoltaic generation (PV) systems have been increasing in recent years because PV systems are effective as a countermeasure of energy issues [1]. However, the output power of PV systems fluctuates with changes in temperature and solar radiation intensity, impacted by weather and installation conditions. In a large-scale PV system, the efficiency of power generation of the whole system is decreased by the partial shadow on the PV modules. The methods to change the series-parallel connections of PV modules based on the solar radiation intensity on each PV module have been proposed [2]. It is infeasible to employ pyranometers from a view of installation costs to measure the solar radiation intensity on each PV module. Therefore, the method which can measure the solar radiation intensity in a wide area is desired.

Many researchers have been investigating methods for estimating solar radiation intensity [3]–[5]. For measuring the solar radiation intensity in a wide area, the authors have proposed methods using pictures taken by a camera [6], [7]. These methods use correlations between the solar radiation intensity and color properties of pictures, and create two kinds of model for estimating solar radiation intensity as the pre-processing. Two kinds of model are created according to the weather; one of models is the clear model and another is the cloudy model. When measuring solar radiation intensity, a color property of the picture is substituted to the estimation model. The method indicated in [6] switches the estimation models to either a clear model or a cloudy model according to the weather information announced by Japan Meteorological Agency (JMA). The estimation error increases because the

weather information is updated only three times a day in this method. To improve this problem, the method for estimating solar radiation intensity by switching a clear model or a cloudy model sequentially has been proposed [7]. In these previous methods, a pyranometer is required to create the estimation models.

This paper proposes a simple method for measuring solar radiation intensity without a pyranometer. The proposed method creates an estimation model based on a correlation between the theoretical solar radiation intensity and the color property of pictures on clear days. The proposed method revises the estimation model sequentially by the brightness value of the optional area in the picture. The validity of the proposed method is evaluated using the root mean square error between the measured and the estimated solar radiation intensity.

## II. MEASURING METHODS

### A. Conventional Methods

Previous studies [6] and [7] have indicated that the solar radiation intensity and the brightness value of an analysis area in pictures show a positive correlation. The estimation models have been shown as the third-order polynomial of the brightness value as follows:

$$E = a_1 V + a_2 V^2 + a_3 V^3 \qquad (1)$$

where $E$ is the estimated solar radiation intensity, $V$ is a brightness value of an analysis area, and $a_i$ ($i$=1, 2, 3) are regression coefficients. In the conventional methods, $a_i$ are decided based on the correlation between the solar radiation intensity measured by a pyranometer and the brightness value as pre-processing. In other words, a pyranometer is required when creating the estimation models.

The conventional methods create two kinds of model for estimating solar radiation intensity according to the weather. One of models is the clear model and another is the cloudy model.

### 1) Method 1

The method indicated in [6] switches the estimation models to either a clear model or a cloudy model according to the

978-1-5090-5606-4/17 $31.00 © 2017 IEEE

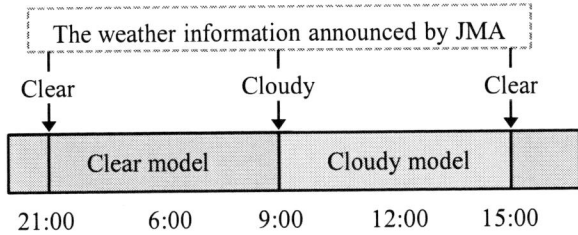

Fig. 1. Schematic diagram of Method 1.

TABLE I
RELATION BETWEEN USED ESTIMATION MODEL AND
ANNOUNCED WEATHER INFORMATION

| Weather Information | Used Model |
|---|---|
| Clear | Clear Model |
| Fine | |
| Slight Cloudy | Cloudy Model |
| Cloudy | |
| Rainy | |

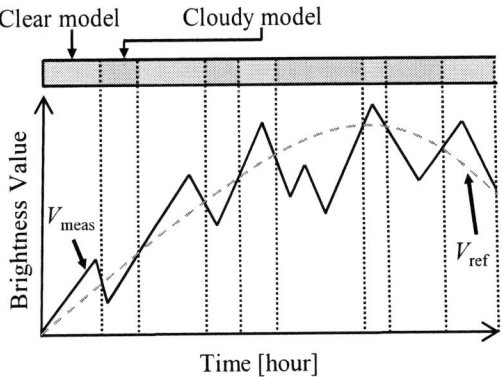

Fig. 2. Schematic diagram of Method 2.

weather information announced by JMA. This method is named as "Method 1" in this paper.

Fig. 1 shows the schematic diagram of Method 1. The weather information is announced at 9:00, 15:00, and 21:00 (JST: Japan Standard Time). Table 1 shows the relation between the used estimation model and the announced weather information. When the weather information is announced as "clear" or "fine," the clear model is used for estimating solar radiation intensity. When the weather information is announced as "slight cloudy," "cloudy" or "rainy," the cloudy model is used.

When estimating solar radiation intensity, the brightness value measured at $t$, $V_{\mathrm{meas}}(t)$, is substituted to the estimation model. The estimation error of Method 1 increases when the actual weather is not coincident with the announced weather because the weather information is updated only three times a day.

*2) Method 2*

In reference [7], the brightness value has been used as an index of switching the estimation models instead of the weather information in Method 1. This method is named as "Method 2" in this paper. The estimation error of Method 2 decreases by switching the clear model or the cloudy model sequentially.

First, the standard brightness value on clear day, $V_{\mathrm{s}}$, is defined as the next equation:

$$V_{\mathrm{s}} = b_1 E_{\mathrm{clear}} + b_2 E_{\mathrm{clear}}^2 + b_3 E_{\mathrm{clear}}^3. \qquad (2)$$

Where, $E_{\mathrm{clear}}$ is the theoretical solar radiation intensity and $b_i$ ($i$=1, 2, 3) are regression coefficients. The theoretical solar

radiation intensity, $E_{\mathrm{clear}}$, is astronomically determined. The regression coefficients of (2), $b_i$, are decided based on the correlation between the theoretical solar radiation intensity and the brightness value measured on clear days. The brightness value measured on clear days is selected from the maximum one measured at same time for clear days.

$V_{\mathrm{s}}(t)$ is calculated by substituting $E_{\mathrm{clear}}(t)$ to (2). $E_{\mathrm{clear}}$ is a value varying with time from sunrise to sunset. Next, the reference value of switching the estimation models is defined as the next equation:

$$V_{\mathrm{ref}}(t) = \alpha \cdot V_{\mathrm{s}}(t). \qquad (3)$$

Where, $\alpha$ is an impact parameter of Method 2.

Fig. 2 shows the schematic diagram of Method 2. Here, $V_{\mathrm{meas}}(t)$ is the brightness value measured at $t$. When $V_{\mathrm{meas}}(t) > V_{\mathrm{ref}}(t)$, the clear model is used. On the other hand, when $V_{\mathrm{meas}}(t) \leq V_{\mathrm{ref}}(t)$, the cloudy model is used. The clear and cloudy models are same in Method 1. Then, the estimated solar radiation intensity is given by substituting $V_{\mathrm{meas}}(t)$ to the estimation model.

*B. Proposed Method (Method 3)*

In the conventional methods, a pyranometer is required to create the estimation models. The proposed method does not require a pyranometer. This proposed method is named as "Method 3" in this paper.

Method 3 creates an estimation model shown in (1). Regression coefficients, $a_i$, are decided based on the correlation between the theoretical solar radiation intensity and the brightness value measured on clear days. When estimating solar radiation intensity, $V_{\mathrm{meas}}(t)$ is substituted to the estimation model. Furthermore, the coefficient of the next equation is multiplied sequentially to the estimated model.

$$\beta(t) = \left( \frac{V_{\mathrm{meas}}(t)}{V_{\mathrm{s}}(t)} \right)^{\{1+(V_{\mathrm{meas}}(t)-V_{\mathrm{ave}}(t))\}^m}. \qquad (4)$$

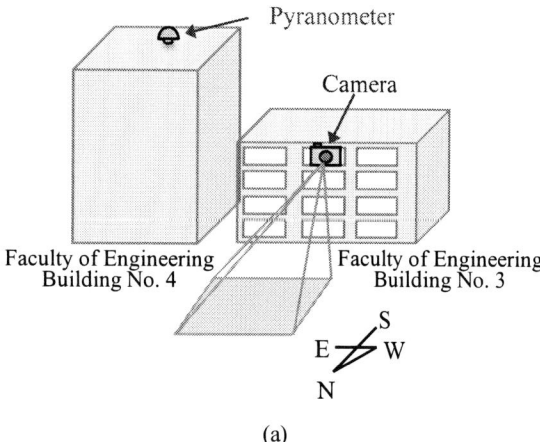

Pyranometer

Camera

Faculty of Engineering Building No. 4

Faculty of Engineering Building No. 3

S
E⇋W
N

(a)

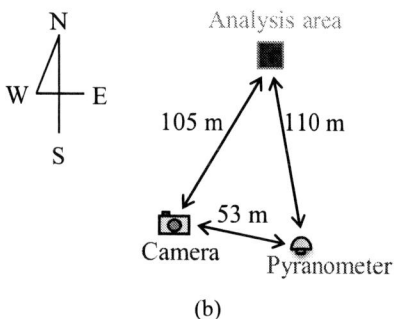

N

W⊢E

S

Analysis area

105 m    110 m

53 m

Camera    Pyranometer

(b)

Fig. 3.   Experimental condition: (a) schematic diagram and (b) physical relationship.

Where, $V_{ave}$ is the average of brightness values in the optional area and $m$ is an impact parameter of Method 3. The optional area is needed to set bigger than the analysis area in the picture.

## III. EXPERIMENT

### A. Experimental Conditions

The experimental condition is shown in Fig. 3. Fig. 3 (a) is a schematic diagram and Fig. 3 (b) is a physical relationship of a camera, pyranometer and an analysis area. A camera (Logicool HD Webcam C525) is installed on the fourth floor of Faculty of Engineering Building No. 3 at Okayama University in Japan. It faces to north, and a picture is taken every 60 seconds from 5:00 to 19:00 (JST). The size of a picture is 240x320 pixels and the format is JPEG. The parameters of the camera such as a white balance and a gain are set no automatic control.

Fig. 4 is an example of picture taken by the camera. The red quadrangle in the picture indicates the analysis area and consists of 49 pixels. The center position of the analysis area is (252, 62). RGB (R: Red, G: green and B: blue) values of the color properties are extracted from the picture. The average filter removes the noise using 48 pixels surrounding the center

Fig. 4.   Example of picture taken by the camera.

position of the analysis area and decides the RGB values of the center position. The RGB values of center position are converted into HSV values. Here, H is hue, S is saturation and V is brightness value. The blue quadrangle in Fig. 4 indicates the optional area in Method 3 and its size is 101x101 pixels focusing on the analysis area.

A pyranometer (PREDE PCM-01N) is installed on the roof of Faculty of Engineering Building No. 4 at Okayama University. It faces to south at an inclination of 20 degrees. The solar radiation intensity is measured every 60 seconds.

The weather information used for Method 1 are collected at the central point of Okayama city, 34°39.6' N, 133° 55.0' E at a 2.8 m elevation by JMA Okayama Local Meteorological Office. The weather information is announced at 9:00, 15:00 and 21:00.

### B. Determination of Estimation Models

The period for the pre-processing to decide the regression coefficients of the estimation models, the standard brightness value on clear day and the impact parameters is from Oct. 1st, 2014 to Oct. 31st, 2014. Oct. 8th, 18th, 19th and 24th are clear days.

Regression coefficients of the clear model and the cloudy model are decided based on correlation diagram on Oct. 8th and Oct. 2nd, respectively. Fig. 5 shows the correlation diagrams between the solar radiation intensity and the brightness value in the analysis area measured on Oct. 8th and Oct. 2nd. Table 1 shows the regression coefficients of the estimation models for Method 1 and Method 2.

Table 2 shows the regression coefficients of the standard brightness value on clear day. Then, Oct. 8th, 18th, 19th and 24th are used to decide the regression coefficients. Fig. 6(a) shows $V_s(t)$ and $E_{clear}(t)$ on Oct. 15th, 2014. $V_s(t)$ varies with time like $E_{clear}(t)$ and draws an arc.

Table 3 shows the regression coefficients of the estimation model for Method 3. These regression coefficients are decided based on the correlation between the theoretical solar radiation intensity and the brightness value measured on four clear days as shown Fig. 6 (b).

978-1-5090-5606-4/17 $31.00 © 2017 IEEE

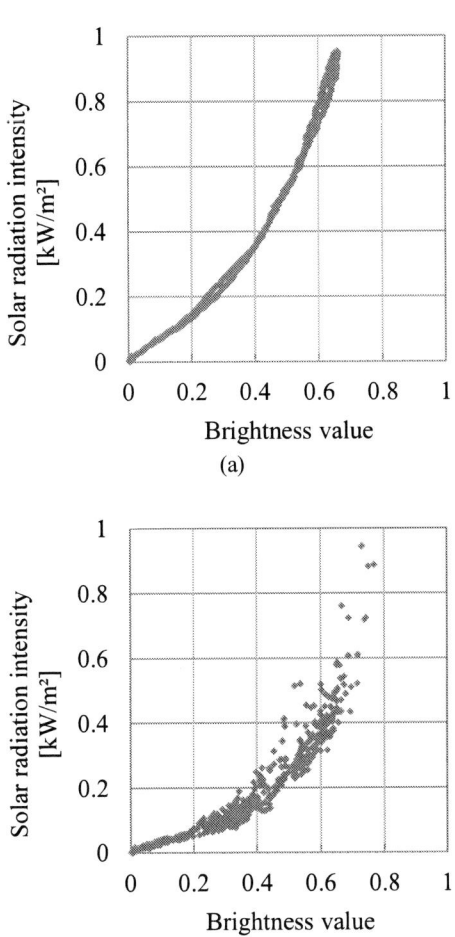

Fig. 5. Correlation diagrams between the solar radiation intensity and the brightness value: (a) a clear day on Oct. 8th, 2014 and (b) a cloudy day on Oct. 2nd, 2014.

TABLE I
REGRESSION COEFFICIENTS OF ESTIMATION MODELS
FOR METHOD 1 AND METHOD 2

| Model | Coefficients | | |
|---|---|---|---|
| | $a_1$ | $a_2$ | $a_3$ |
| Clear | 0.7584 | -0.7505 | 2.5927 |
| Cloudy | 0.4543 | -1.0951 | 2.3681 |

*C. Evaluation Method*

Estimating methods are evaluated using the root mean square error (RMSE) between the measured and the estimated solar radiation intensity. RMSE is defined as follows:

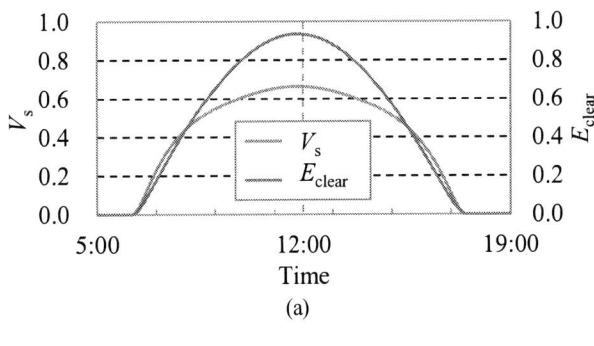

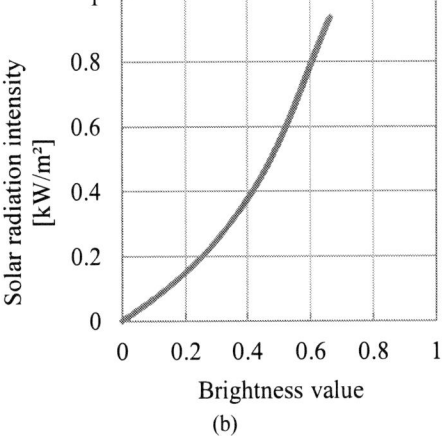

Fig. 6. $V_s$ and $E_{clear}$ on Oct. 15th, 2014: (a) $V_s(t)$ and $E_{clear}(t)$ and (b) correlation diagrams between $E_{clear}$ and $V_s$.

TABLE II
REGRESSION COEFFICIENTS OF
THE STANDARD BRIGHTNESS VALUE

| $b_1$ | $b_2$ | $b_3$ |
|---|---|---|
| 1.5495 | -1.5338 | 0.6798 |

TABLE III
REGRESSION COEFFICIENTS OF ESTIMATION MODEL
FOR METHOD 3

| $a_1$ | $a_2$ | $a_3$ |
|---|---|---|
| 0.6176 | 0.2358 | 1.4781 |

$$RMSE = \sqrt{\frac{1}{N}\sum_{n=1}^{N}\left(E_{est}(n) - E_{meas}(n)\right)^2}, \quad (5)$$

where $N$ is the number of sampling points, $E_{est}$ [kW/m$^2$] is the solar radiation intensity estimated by the conventional or proposed methods and $E_{meas}$ [kW/m$^2$] is the solar radiation intensity measured by the pyranometer. $N$ is 720 for a day

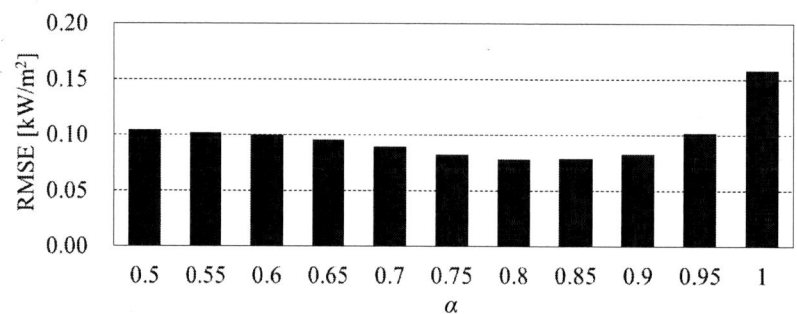

Fig. 7. Average of RMSEs from Oct. 1st to Oct. 30th, 2014 by changing value of $\alpha$.

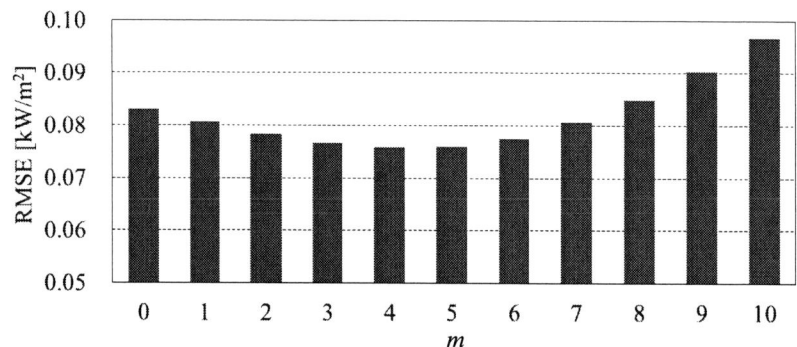

Fig. 8. Average of RMSEs from Oct. 1st to Oct. 30th, 2014 by changing value of $m$.

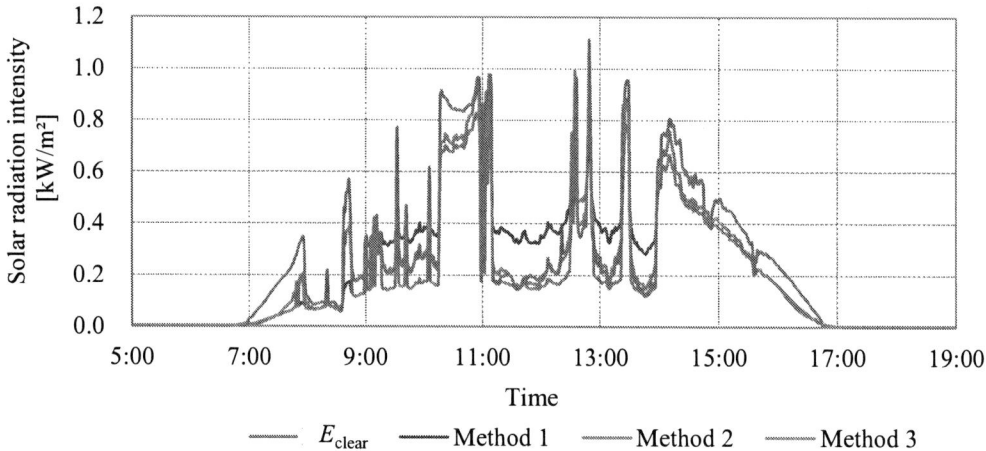

Fig. 9. Measurement results of solar radiation intensity on Nov. 27th, 2014.

because sampling time is 60 seconds and the term of measuring solar radiation intensity is from 5:00 to 19:00.

### D. Determination of Impact parameters

The impact parameter of Method 2, $\alpha$, is decided using RMSEs during the period for the pre-processing. It is impossible to switch the estimation model either to the clear model or the cloudy model if $\alpha$ is out of an appropriate range. When $\alpha$ is small ($\alpha \leq 0.45$), only the clear model is used. On the other hand, when $\alpha$ is large ($\alpha > 1$), only the cloudy model is used. Then, the optimal value of $\alpha$ is decided by changing value of $\alpha$ in steps of 0.05 from 0.5 to 1.0. Fig. 7 shows the average of RMSEs from Oct. 1st to Oct. 30th, 2014 for Method 2 by changing value of $\alpha$. Consequently, $\alpha$ is decided to 0.8 because the average of RMSE is smallest then.

The impact parameter of Method 3, $m$, is decided using RMSEs during the period for the pre-processing. The optimal value of $m$ is decided by changing value of $m$ from 1 to 10. Fig. 8 shows the average of RMSEs from Oct. 1st to Oct. 30th, 2014 for Method 3 by changing value of $m$. Consequently, $m$ is decided to 4 because the average of RMSE is smallest then.

## IV. MEASUREMENT RESULTS

Measurement results of solar radiation intensity using the pyranometer, Method 1, Method 2 and Method 3 on Nov. 27, 2014 are shown in Fig. 9. The measured solar radiation intensity of Method 1 becomes bigger than that of the pyranometer at around 12:00. This is the reason that the solar radiation intensity is estimated using the clear model from 9:00 to 15:00 in Method 1. The solar radiation intensity measured by Method 2 becomes smaller than that of the pyranometer from 9:00 to 10:00. This is affected by the value of $\alpha$. RMSEs for Method 1, Method 2 and Method 3 are 0.116, 0.090 and 0.069, respectively. RMSE for Method 3, the proposed method, is smaller than that for the conventional methods.

Fig. 10 shows the average of RMSEs for each method from Nov. 1st, 2014 to Feb. 4th, 2015. The average of RMSEs for the proposed method is 0.079. The average of RMSEs for the proposed method is improved by approximately 32% compared with the conventional Method 1 and is improved by approximately 13% compared with the conventional Method 2.

## V. CONCLUSIONS

This paper proposed the simple method for measuring solar radiation intensity without a pyranometer. The proposed method created the estimation model using the correlation between the theoretical solar radiation intensity and the brightness value measured on four clear days. The estimation model was revised sequentially based on the brightness value of the optional area in the picture.

The average of RMSEs for the proposed method was 0.079. The RMSEs for the proposed method was smaller than that for the conventional methods. The solar radiation intensity in a wide area can be measured from one picture using the proposed method because the solar radiation intensity can be measured by the image analyses.

## ACKNOWLEDGEMENT

This work was supported by JSPS KANENHI Grant Number JP16K21188.

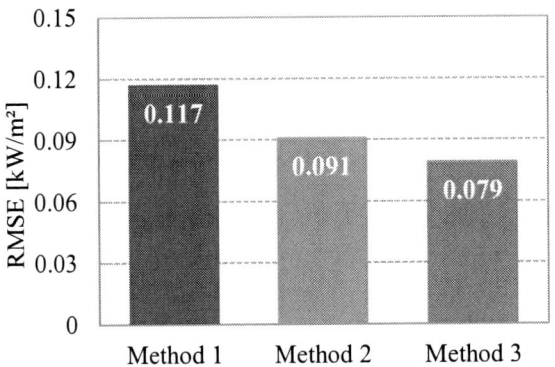

Fig. 10. Average of RMSEs for each method from Nov. 1st, 2014 to Feb. 4th, 2015.

## REFERENCES

[1] T. Aziz, S. Dahal, N. Mithulananthan and T. K. Saha, "Impact of Widespread Penetrations of Renewable Generation on Distribution System Stability," *6th International Conference on Electrical and Computer Engineering ICECE 2010*, pp. 338-341, 2010.

[2] Investigating R&D Committee on System Optimization and Benchmark Problems for Industrial Application, "The System Optimization and the Benchmark Problem for Industrial Application," *Technical Report of the Institute of Electrical Engineering of Japan*, Vol. 1635, pp. 27-31, 2016.

[3] Q. Zhang, J. Huang, Y. Hongxing and L. Chengzhi, "Development of Models to Estimation Solar Radiation for Chinese Locations," *Journal of Asian Architecture and Building Engineering*, vol. 2, no. 2, pp. 35-41, 2003.

[4] M. Rivington, G. Bellocchi, K. B. Matthews and K. Buchan, "Evaluation of three model estimations of solar radiation at 24UK stations," *Agricultural and Forest Meteorology*, vol. 132, no. 3-4, pp. 228–243, 2005.

[5] M. Boulifa, A. Adane, A.Rezagui, and et Z.Ameur, "Estimate of the global solar radiation by cloudy sky using HRV images," *Energy Procedia*, vol. 74, pp. 1079-1089, 2015.

[6] A. Takahashi, H. Morio, J. Imai and S. Funabiki, "An Estimation Method for Solar Radiation Intensity Using a Web Camera," *Journal of the Japan Institute of Energy*, vol. 94, pp. 1330-1336, 2015.

[7] A. Moriki, A. Takahashi, J. Imai and S. Funabiki, "A Novel Method of Estimating Solar Radiation Intensity Based on Camera Image Analysis by Sequentially Switching Method," *Proceedings of The International Conference on Electrical Engineering 2016*, 90031, 2016.

# Degradation of Solder Bonds in Field Aged PV Modules: Correlation with Series Resistance Increase

Abhishiktha Tummala, Jaewon Oh, Sai Tatapudi, GovindaSamy TamizhMani

Arizona State University Photovoltaic Reliability Laboratory (ASU-PRL), Mesa, AZ 85212, USA

*Abstract* - **One of the major defects that can cause significant power loss is the degradation of interconnect metallization system (IMS) comprised of cell metallization and solder bonds of cell and string interconnects. Weak cell interconnect solder bonds between copper ribbon and busbar of cells result in series resistance increase which affects the fill factor causing a power drop. In this paper, the results obtained from series resistance and peel test experiments performed on the cells extracted from modules exposed in three different climates (Arizona - Hot and Dry, Mexico - Warm and Humid, and California - Temperate) for more than 18 years are presented. Finally, climate specific thermal modelling was performed for those sites over 20 years to calculate the accumulated thermal fatigue and to evaluate its correlation, if any, with series resistance.**

*Index Terms* — **series resistance, solder bonds, degradation, reliability, photovoltaic cells, silicon.**

## I. INTRODUCTION

One of the major reasons for power loss in the field exposed photovoltaic (PV) modules over time is the increase of series resistance ($R_S$). One of the major sources for the series resistance increase is the degradation of solder bonds between copper ribbon and cell metallization in the cell strings. As a module is exposed in the field, depending on the climatic conditions, the thermomechanical fatigue or IMC (intermetallic compound) formation caused cracks develop in the solder bonds leading to increased series resistance.

Previous studies show that the field exposed modules undergo thermomechanical fatigue which results in changes in the solder-joint geometry thus causing reduction in the number of redundant solder joints and increase in series resistance [1]. A streamlined approach is needed to understand the relationship between the solder bond fatigue, series resistance between various components of a solar cell and the thermomechanical stress. The main objective of this work is to calculate the series resistance for various circuit components and interfaces of the individual cells extracted from field aged modules in three diverse climatic conditions. The cells were cut from the modules, encapsulant was chemically dissolved and the resistance measurements were performed on the IMS (interconnect metallization system). The IMS is comprised of cell metallization and solder bonds of cell strings. After the measurements, peel test experiments were performed to determine the solder bond strength between aged and fresh samples. Finally, climate specific thermal modelling was performed to observe the correlation, if any, between thermal fatigue, series resistance and peel strength.

## II. EXPERIMENTAL METHODS

To separate the cells from the field aged module, two methods were developed: chemical and mechanical. In the chemical method, trichloroethylene (TCE) was used using the concept developed by Doi *et al* [2]. In the mechanical method, a hollow metal bar was used using the concept developed by Nick Bosco at NREL (private communication). For the chemical method, after the cell containing laminate was cut from the module using a dremel tool, TCE was used to dissolve out EVA from the cut laminate piece containing broken glass, EVA, cell and back sheet. In this method, two thick stainless steel metal plates which are little larger than the size of the cell were used as shown in Fig. 1. In one of the plates, about 1 mm diameter holes were made with uniform distance between them. The cell laminate was sandwiched between these two stainless plates and tightly bolted using four corner bolts and nuts. To ensure a good contact between TCE and cell-front EVA, the plate containing holes made in contact with the broken glass superstrate. The whole sandwich containing metal plates and cell laminate was immersed in a 100% TCE solution contained

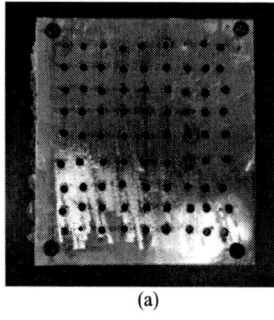

| (a) | (b) |

Fig. 1. (a) Front view of the setup (b) Back view of the setup

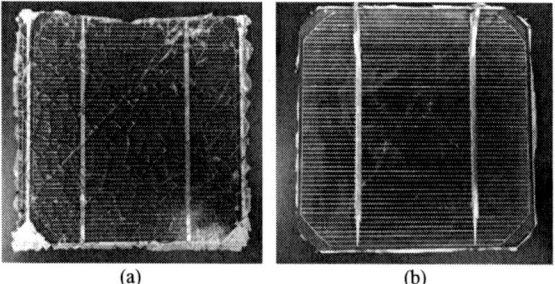

| (a) | (b) |

Fig. 1. (a) Sample before extraction using TCE method (b) sample after extraction using TCE method

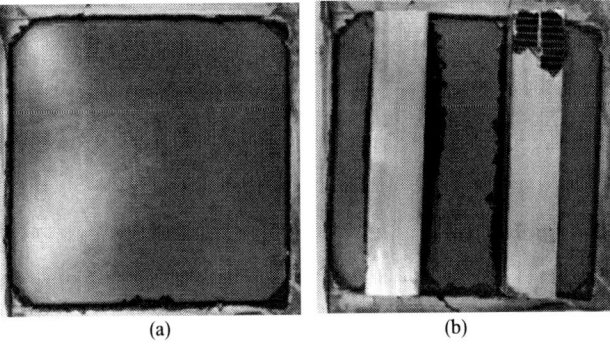

(a)                          (b)

Fig. 3. (a) Sample of cell polished on the backside using IPA and sandpaper after extraction using TCE method, (b) Metal beam glued to the cell using 3M epoxy glue

Fig. 4.   Four-point probe test setup

glass beaker. This beaker was then placed in another larger glass beaker containing water. This setup was placed on a hot plate kept inside a fume hood. The solution temperature was raised to 60-80°C and kept at this temperature for 60-90 minutes depending on the need. As shown in Fig. 2, after dissolving out EVA from the cell-front, the entire cell without EVA on the cell-front became available for the planned experiments. Cell extraction using TCE is not expected to have any damaging effect on the solder bonds as it is an organic solvent. To further confirm if TCE causes any damaging effect on the solder bond, the mechanical method was implemented to extract a cell strip along the location and direction of cell busbar. In this method, the back sheet of the desired cell was first cut using a heavy-duty razor blade and a heat gun, and then the back metallization of the cell was removed by polishing with a use of sandpaper and IPA (isopropyl alcohol) as shown in Fig. 3 (a). Once the backside of the cell was polished, a square metal tube was placed on the cell beneath the busbar/ribbon and the cell was cut along the periphery of the metal tube using a heavy-duty razor blade. As shown in Fig. 3 (b), 3M epoxy glue DP 460 was used to glue the metal beam to the backside of the cell and it was allowed to cure overnight. Once the glue was hardened and the beam was stuck to the cell firmly, a heat gun was used and heat was provided from the front side through the glass superstrate of the module over the area of interest (along the metal tube area). By providing heat for about 5 minutes, the EVA on the front side loosened up and the cell strip along metal tube was extracted from the module. It is to be noted that by providing excessive amount of heat, one may melt/affect the solder bond strength which need to be avoided by limiting heating time and temperature. This method is a very cost and time effective method when compared to the TCE method discussed above but a greater caution should be exercised during the heating steps.

One of the most common ways of measuring the resistivity of some thin, flat materials, such as semiconductors or conductive coatings, is to use a four-point collinear probe. The four-point probe technique involves bringing four equally spaced probes in contact with a material of unknown resistance. The instrumentation used for this test includes a DC current source, a highly sensitive voltmeter, and a four-point collinear probe. The four-point probe resistance measurements are done using the SMU 2450 source measurement unit, SP4 - four-point probe head and S-302 test stand as shown in Fig. 4. For the series resistance measurements, all the various combinations possible for causing the resistance were considered. The various combinations used for the resistance measurements are shown in the Fig. 5. The setup was connected to the multimeter and by using the resistance value shown on the multimeter the series resistance was calculated. For a combination between two surfaces, it was made sure that two probes were placed on one surface and the other two probes on the other surface as shown in Fig. 6.

The thermal fatigue is mainly developed due to two factors. The first factor is the daily temperature change that is the day and night temperatures which effects the solder bond gradually by the expansion and contraction of the solder bond. The second factor is the cloud cycles which occur almost every day which cause the sudden expansion and contraction in the solder ribbon

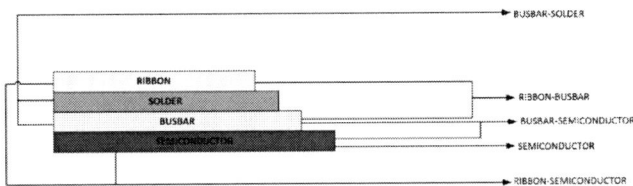

Fig. 5.   Combinations used in four-point probe measurements

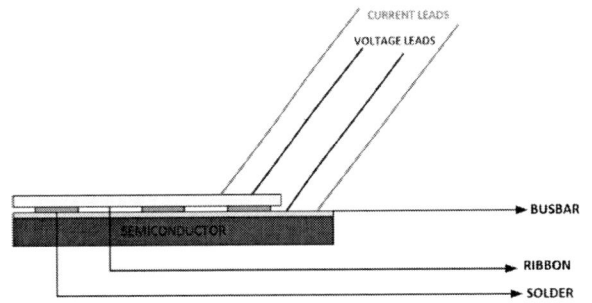

Fig. 6.   Placement of probes on the semiconductor

which might induce cracks in it as the time goes on. In this work, the thermal fatigue for 20 years was calculated from 1991 to 2010 to have a better understanding of how much fatigue a module can develop over 20 years in different climates. In order to estimate the fatigue developed, first the total irradiance was calculated using the Liu-Jordan model using Matlab software and also by using PVsyst software by converting the meteorological data into PVsyst format. Once the total plane of array (POA) irradiation is calculated, the cell temperature is calculated by using the following equation [3],

$$T_{cell} = T_{amb} + E \cdot e^{a+(b \cdot WS)} + E \cdot (\Delta T/E_0) \qquad (1)$$

where a and b were empirically determined for a glass/polymer backsheet module construction deployed in an open-rack configuration to be −3.56 and −0.075, respectively [3]. $E_0$ is the reference solar irradiance of 1000 W/m², WS stands for wind speed, $T_{amb}$ stands for the ambient temperature of the module and $\Delta T$ represents the temperature difference between the cell and module at this reference irradiance. For an open-rack configuration $\Delta T$ was determined to be 3°C; however, this offset temperature will be sensitive to racking method and module construction. The thermomechanical fatigue is calculated by using the formula [3],

$$D = C \cdot (\Delta T)^n \cdot (r(T))^b \cdot e^{\frac{Q}{k_B \cdot T_{max}}} \qquad (2)$$

where $\Delta T$ is the mean daily maximum cell temperature change, $T_{max}$ is the mean daily maximum cell temperature (calculated using the $T_{cell}$ formula above), C a scaling constant and Q and $k_B$ are activation energy and Boltzmann's constant. The temperature reversal term, r(T), is the number of times the temperature history increases or decreases across the reversal temperature, T, over the course of a year. The scaling constant C and the reversal temperature T were used to fit this model to our simulated data, while the values of the exponents n and b and the activation energy Q are shared with the Coffin-Manson and Norris-Lanzberg equations for Pb-Sn eutectic solder (C= 240, T= 56°C, n= 1.9, b= 0.33, Q= 0.12 eV).

## III. RESULTS AND DISCUSSIONS

Dark I-V measurements were performed for the three modules from 3 different climates for all the possible cells. As shown in Fig. 7, the highest drop in the fill factor was observed for the cells from Mexico module (23 years) due to their high series resistance followed by Arizona (18 years) and California module (20 years). The highest drop in the fill factor is observed for the cells from Mexico module due to their high series resistance. The high series resistance is observed due to their high field exposure (23 years) and also due to the climate in which they were exposed (warm and humid). Due to the humid conditions, the moisture ingresses through the backsheet of the

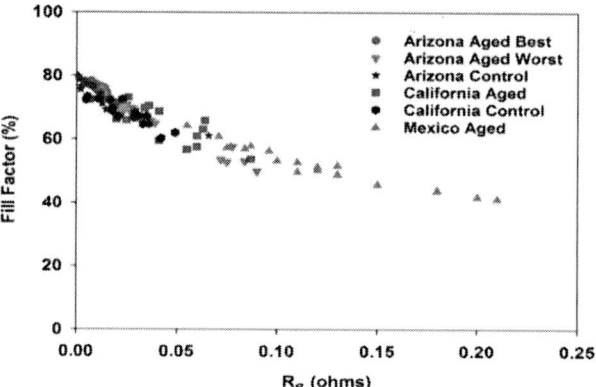

Fig. 7.    Effect of series resistance on fill factor

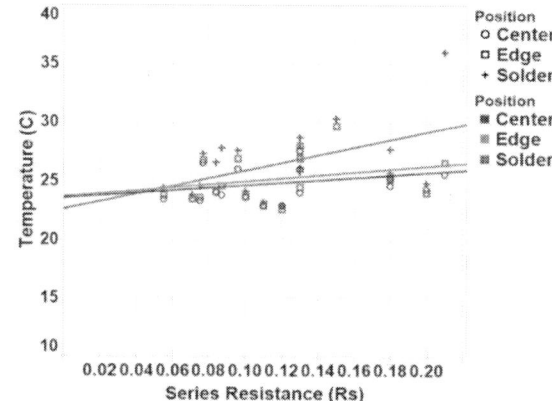

Fig. 8.    Effect of series resistance on temperature of the module

modules and creeps into the solder joints causing corrosion which decreases the ribbon contact with the busbar. When this happens, the electrons generated in the cell have to find an alternate but a narrow and long route in order to get transferred from cell to ribbon thus increasing the series resistance. After Mexico module, the worst series resistance is shown by the Arizona field exposed module (514210) which is exposed for 18 years in hot and dry climate. The other exposed module from Arizona (464185) shows series resistance values very close to that of the control module which indicates that the solder bonds in the modules are practically intact and show very less degradation. The California aged module shows higher series resistance when compared to the California control module as expected. California module are expected to have lower series resistance than Arizona modules due to their temperate climate and also due to lesser cloud cycles.

Dark current was passed through the module and the temperature of each cell was measured using infrared (IR) camera. When the temperatures at three different position of a cell are considered, the temperature at the ribbon of the cell is higher when compared to the temperatures at the center and edge of the cell respectively as shown in Fig. 8.

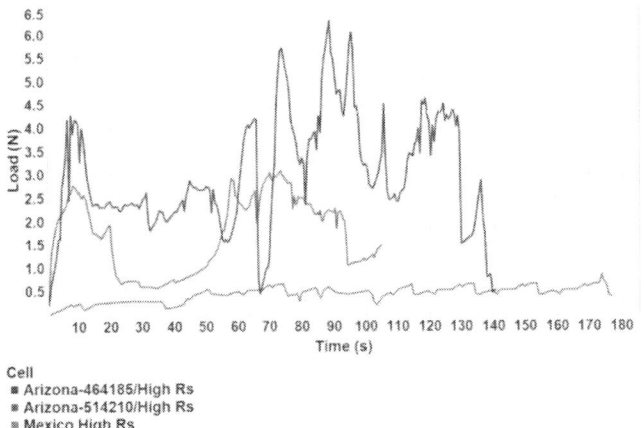

Fig. 9.    Variation of peel strength with time for Arizona and Mexico

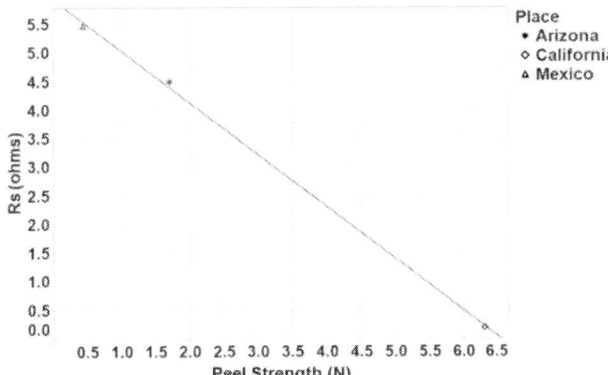

Fig. 10.    Effect of peel strength on series resistance

The peel strength of the cell with highest series resistance from each module was compared. From the Fig. 9, it can be observed that the peel strength decreases with series resistance. Cell from Mexico aged module has the highest series resistance thus experiences the lowest peel strength followed by Arizona aged modules 514210 and 464185.

When the results were processed, it was observed that that the peel strength decreases with the increase of series resistance. Cells from Mexico aged module has the highest series resistance thus experiences the lowest peel strength followed by Arizona aged modules. Fig. 10 shows the relationship between the module level $R_S$ and the average peel strength of the same module obtained from different cells. The series resistance was calculated by taking the slope of the last few points close to the $V_{OC}$ side of the light I-V curve. From Fig. 10, it can be observed that the peel strength of the module decreases with increase in the series resistance of the module which is like the trend observed when peel strength was compared with cell level series resistance taken from dark I-V curves.

Fig. 11 shows the plot between the peel strength and thermal fatigue accumulated in the module over a period of 20 years from 1991-2010. It can be observed that peel strength and fatigue have no correlation as such. It can be also concluded that lower fatigue does not necessarily imply higher bond

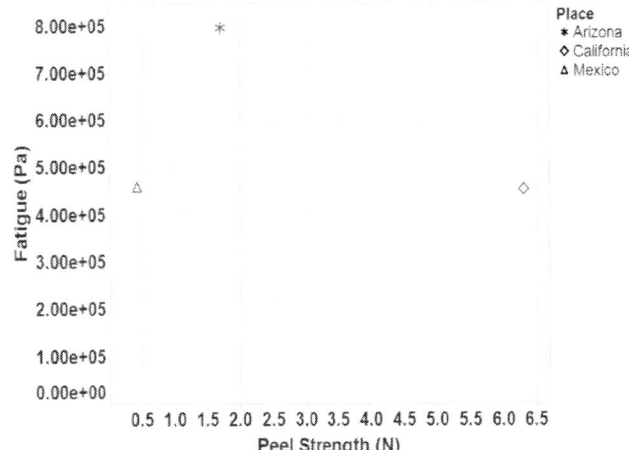

Fig. 11.    Effect of peel strength on thermal fatigue

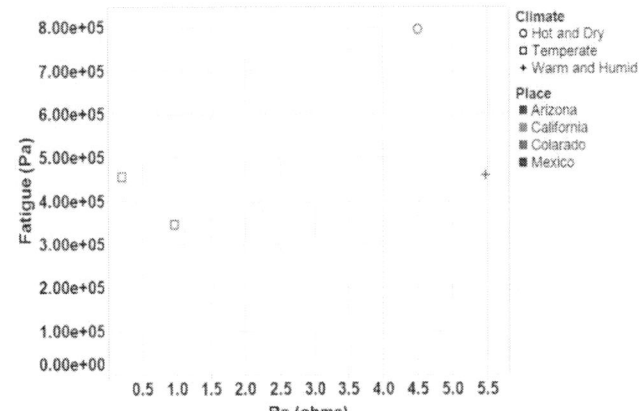

Fig. 12.    Effect of series resistance on thermal fatigue

strength. In order to fully demonstrate the absence of fatigue vs peel strength, it is recommended to pull one module every year from a plant from a single manufacturer in Arizona as there is no corrosion but only thermal fatigue and generate this plot again.

Peel strength is influenced by both material/design properties and process control as well. Since process control from one manufacturer to another manufacturer varies, no correlation between fatigue and peel strength could be expected.

Fig. 12 shows the relationship between the module level $R_S$ and the thermal fatigue accumulated by the module over a span of 20 years from 1991-2010. From the plot, it can be observed that typically higher thermal fatigue should lead to weakened bond strength due to temperature and cloud cycles which results in expansion and contraction of solder bonds and ribbons. This weakens the interface between ribbon-solder and/or solder-busbar resulting in higher series resistance. However, as shown in Fig. 7 Mexico module (warm and humid climate) has the highest series resistance but not the highest fatigue which implies that not only fatigue, but other factors like IMC formation and corrosion can also aide the increase in series resistance in the presence of humidity/moisture. Therefore,

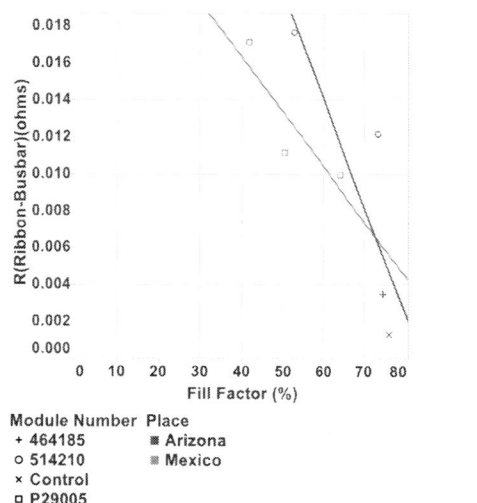

Module Number    Place
+ 464185     ■ Arizona
o 514210     ■ Mexico
× Control
□ P29005

Fig. 13. R (Ribbon-Busbar) Vs fill factor of cells from Arizona and Mexico Aged

Fig. 14. Placement of four-point probe for ribbon-busbar measurement

fatigue alone may not be considered for the series resistance increase/correlation. The four-point probe resistance measurements were performed for modules from 3 different climates. It is to be noted that the control module for both Arizona and Mexico was assumed to be the same. Fig. 13 shows the variation of $R_{(Ribbon-Busbar)}$ with Fill Factor for cells from Arizona field aged/control and Mexico field aged modules. As shown in Fig. 13, it can be observed that a decreasing trend is seen in fill factor with increasing resistance. This combination of resistance has the highest values of series resistance when compared to other busbar combinations. In this graph, a higher rate of decrease in fill factor can be seen in Arizona modules than the Mexico module. The placement of the probes for the ribbon-busbar combination is shown in Fig. 14.

It can be observed from Table I that the highest rise in resistance is observed for R $_{(Ribbon-Busbar)}$ for both Arizona and Mexico modules and it implies that the interface between the ribbon and busbar is the most affected interface resulting in fill factor drop which in turn leads to power loss. This interface

## TABLE I
FOUR-POINT PROBE RESISTANCE DATA FOR THE CELLS EXTRACTED FROM FIELD AGED ARIZONA AND MEXICO MODULES

| Combination | Control module (Ω) | Arizona modules (Ω) | % change | Mexico module (Ω) | % change |
|---|---|---|---|---|---|
| $R_{(Semiconductor)}$ | 4.60 | 8.73 | 89.78(↑) | 11.10 | 141.30(↑) |
| $R_{(Ribbon-Semic.)}$ | 6.05 | 6.88 | 13.81(↑) | 9.77 | 61.62(↑) |
| $R_{(Busbar-Solder)}$ | 0.01 | 0.01 | 144.44(↑) | 0.01 | 217.77(↑) |
| $R_{(Busbar-Semic.)}$ | 4.35 | 5.57 | 28.04(↑) | 7.02 | 61.26(↑) |
| $R_{(Ribbon-Busbar)}$ | 0.01 | 0.02 | 238.40(↑) | 0.02 | 228.8(↑) |
| $R_{(Busbar-Fingers)}$ | 0.02 | 0.07 | 209.43(↑) | 0.02 | 10.84(↑) |

degradation is attributed to the degradation of solder bonds. The resistance of the fingers remained nearly constant irrespective of the change in the fill factor.

## IV. CONCLUSIONS

The fill factor and short-circuit current of the test samples are the most affected performance parameters. The fill factor is determined to be affected by the increase of series resistance and the short-circuit current is determined to be affected by the encapsulant browning and series resistance. Temperature of the cell increases with the increase in series resistance. Also, the temperature along the solder in a cell was observed to be higher than the temperatures at the edge and center of the cells. In a module from Mexico where series resistance effect is higher, a 0.05 Ω increase in series resistance causes a 2.7°C increase in temperature near the solder region when compared to 1.07°C and 0.94°C increase in edge and center regions, respectively.

The peel strength of the ribbon-busbar interface decreases with increase in series resistance. The peel strength of the ribbon-busbar interface decreases with increase in series resistance. The major factors that might influence the degradation of the interface are the cloud cycles, IMC formation and also corrosion when the module is fielded in humid conditions. For the module from Mexico, the peel strength decreases by 47% between the lowest series resistance cell and the highest series resistance cell. For Arizona, one module (464185) which has a series resistance of 1.4 Ω had an average peel strength of 3.01N compared to another module (514210) which has a series resistance of 4.49 Ω had an average peel strength of 0.9N. The major factors that might influence the degradation of the IMS are the fatigue due to cloud cycles, IMC formation (between copper ribbon and solder bond and/or solder bond and metallization) due to higher operating temperature and corrosion when the module is fielded in humid conditions. In the four-point probe resistance measurements, it was observed that the ribbon-busbar configuration (solder bonds) was the largest part effecting the series resistance and fill factor. Thermal fatigue developed by the modules over the years due to cloud cycles was investigated to observe if there is any correlation between thermal fatigue and peel strength. Since peel strength is influenced by a combination of thermal fatigue, IMC formation and corrosion, no specific correlation between only thermal fatigue and peel strength could be established. Mexico module, despite having a lower calculated fatigue, has a high series resistance which is possibly due to the moisture ingress through the backsheet or laminate edges leading to corrosion of metallic components of the cells.

## ACKNOWLEDGEMENT

This research is based upon work supported by the Solar Energy Research Institute for India and the U.S. (SERIIUS) funded jointly by the U.S. Department of Energy subcontract DE AC36-08G028308 (Office of Science, Office of Basic Energy Sciences, and Energy Efficiency and Renewable Energy, Solar Energy Technology Program, with support from the Office of International Affairs) and the Government of India subcontract IUSSTF/JCERDC-SERIIUS/2012 dated 22nd Nov. 2012. The seed funding to initiate this project was provided by Salt River Project (SRP), Arizona. The great technical support provided by Dr. Nick Bosco of NREL is sincerely appreciated.

## REFERENCES

[1] D. King *et al.*, "Photovoltaic module performance and durability following long-term field exposure" *Progress in Photovoltaics Research and Applications*, vol. 8, pp. 241-256, 2000.

[2] T. Doi *et al.*, "Experimental study on PV module recycling with organic solvent method" *Solar Energy Materials and Solar Cells*. vol. 67, pp. 397-403, 2001.

[3] N. Bosco *et al.*, "Climate specific thermomechanical fatigue of flat plate photovoltaic module solder joints" *Microelectronics Reliability*, vol. 62, pp. 124-129, 2016.

# Performance of Light and Dark Current-Voltage Characteristics for PID-Affected Monocrystalline Silicon Solar Modules

He Wang[a,*], Pan Zhao[a], Shuwen Guo[a], Hong Yang[a], WeiJing Huang[b], Shiyu Sang[c], Bojie Su[d], Xue Zhang[d], Yunxue Cao[e], Hui Zhao[e]

[a]MOE Key laboratory for Nonequilibrim Synthese and Modulation of Condensed Matter, School of Science, Xi'an Jiaotong University, Xi'an 710049, People's Republic of China

[b]Xi'an Huanghe Photovoltaic Technology Co., Ltd, Xi'an, No.21 North Xingfu Road, Xi'an 710049, People's Republic of China

[c]Institute of Electrical Engineering of the Chinese Academy of Sciences, Beijing 100190, People's Republic of China

[d]China Quality Certification Center, Beijing 100070, People's Republic of China

[e]SPIC Power Plant Operation Technology Co., Ltd, Beijing 100190, People's Republic of China

*Corresponding author: He Wang, hw69cn@126.com

*Abstract* — **In this paper, the I-V characteristics of PID-affected solar modules were investigated under illuminated and dark conditions. Based on the measured data, it is found that the I-V characteristic of PID-affected modules shows a regular change under different irradiations. Under dark conditions, the change of I-V characteristic of PID-affected modules disassembled from a power plant is also revealed. The dark I-V characteristic of PID-affected modules becomes soft. These results indicate that the efficiency of PID-affected modules has a significant degradation under low irradiance.**

*Index Terms* — **solar module, current-voltage characteristics, potential-induced degradation (PID), performance degradation.**

## I. INTRODUCTION

The I-V characteristics measured under dark and light conditions are performed to provide an essential tool for the assessment of performance of solar cells Both light I-V and dark I-V measurements have been commonly used to analyze the electrical parameters of solar cells [1]-[3].

It is proved that potential-induced degradation (PID) that caused by high electrical potential shows more detrimental power loss to solar power plant [4]-[5]. The mechanism on PID is basically divided into two types. One is that the polarization effect based on a field effect resulting in PID because of the increased surface recombination [6]-[7]. For another, it is assumed that sodium ions diffuse into stacking faults crossing the p–n junction causing PID [8]-[14].

PID-affected solar modules reproduced in the laboratory have been investigated by many researchers. Under laboratory conditions, M. Schütze et al. studied that using different encapsulation materials and cell process can identify PID-affected cells, and obtained detailed PID characteristics [15]. Spataru et al. reported on the change rule of various PID sensitive modules degraded by means of damp-heat stress testing in the laboratory, and discovered that electrical parameters depending on the light I-V and dark I-V characteristics can identify the PID [1]. However, few studies focus on light and dark I-V characteristics of PID-affected solar modules emerging from solar plant. In this paper, irradiance and module temperature is acquired by Solar Survey 100/200R. Then the illuminated I-V characteristic of PID-affected modules is acquired outdoors by PV 200 tester. We find that, the variation of electrical parameters of three modules is reflected. And a significant degradation of illuminated I-V characteristic of PID-affected modules under low irradiance is shown. What's more, the regular change of dark I-V characteristic of PID-affected modules is revealed. Dark I-V characteristics for three substrings of PID-affected modules in the field show a significant difference for the first time

## II. EXPERIMENTAL RESULTS AND DISCUSSION

The area of the performed tests is a distributed PV power plant located in Guang Zhou City. This area is classified as high-temperature and high-humidity region. Fig. 1. is the scene of a distributed PV power plant. A temperature sensor is glued on the backsheet of the solar module to allow the temperature reading. Irradiance and module temperature is acquired by Solar Survey 100/200R. A PV 200 tester is shown to be a fast tool for the recording I-V characteristics.

Fig. 1. The scene of a distributed PV power plant.

*A. The I-V characteristics of PID-free module and PID-affected modules under illuminated conditions outdoors.*

After measuring a number of modules and taking thermographic images, some typical modules were disassembled to be characterized on a solar simulator at standard test conditions. As seen from Table I and Table II, the result illustrates the variation of electrical parameters of PID-free module 1 with 3% power loss, slight-PID module 2 with 21.9% power loss and serious-PID module 3 with 48.7% power loss. The I-V characteristics of these modules measured under different test conditions are shown in Fig. 2. For the same irradiance, it is found that the I-V characteristic of PID-affected modules shows a regular change under different irradiations, as seen from Fig. 3. More seriously, module with serious PID degrades more severely under low light irradiance.

TABLE I

THE MAIN ELECTRICAL PARAMETERS OF MODULE 1, MODULE 2 AND MODULE 3 BEFORE GRID-CONNECTED OPERATION.

| Module | $V_{oc}$(V) | $I_{sc}$(A) | $V_m$(V) | $I_m$(A) | FF(%) | $P_{max}$(W) |
|---|---|---|---|---|---|---|
| Module 1 | 44.52 | 5.72 | 36.12 | 5.39 | 75.58 | 195.39 |
| Module 2 | 44.56 | 5.73 | 36.15 | 5.43 | 75.63 | 196.33 |
| Module 3 | 44.50 | 5.72 | 36.10 | 5.40 | 75.55 | 195.22 |

TABLE II

THE MAIN ELECTRICAL PARAMETERS OF MODULE 1, MODULE 2 AND MODULE 3 AFTER GRID-CONNECTED OPERATION FOR 2 YEARS.

| Module | $V_{oc}$(V) | $I_{sc}$(A) | $V_m$(V) | $I_m$(A) | FF(%) | $P_{max}$(W) |
|---|---|---|---|---|---|---|
| Module 1 | 44.39 | 5.64 | 35.96 | 5.22 | 75.22 | 189.51 |
| Module 2 | 40.10 | 5.63 | 30.48 | 4.98 | 61.65 | 153.33 |
| Module 3 | 31.01 | 5.62 | 20.30 | 4.93 | 56.03 | 100.21 |

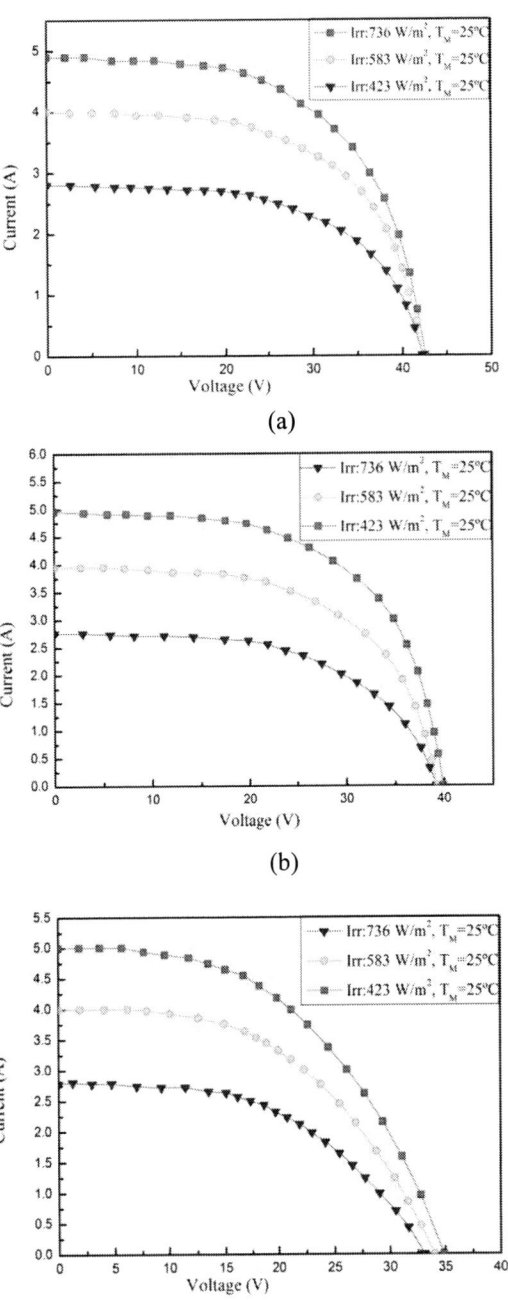

Fig. 2. I-V characteristics of (a) PID-free module 1, (b) slight-PID module 2 and (c) serious-PID module 3 measured under different test conditions.

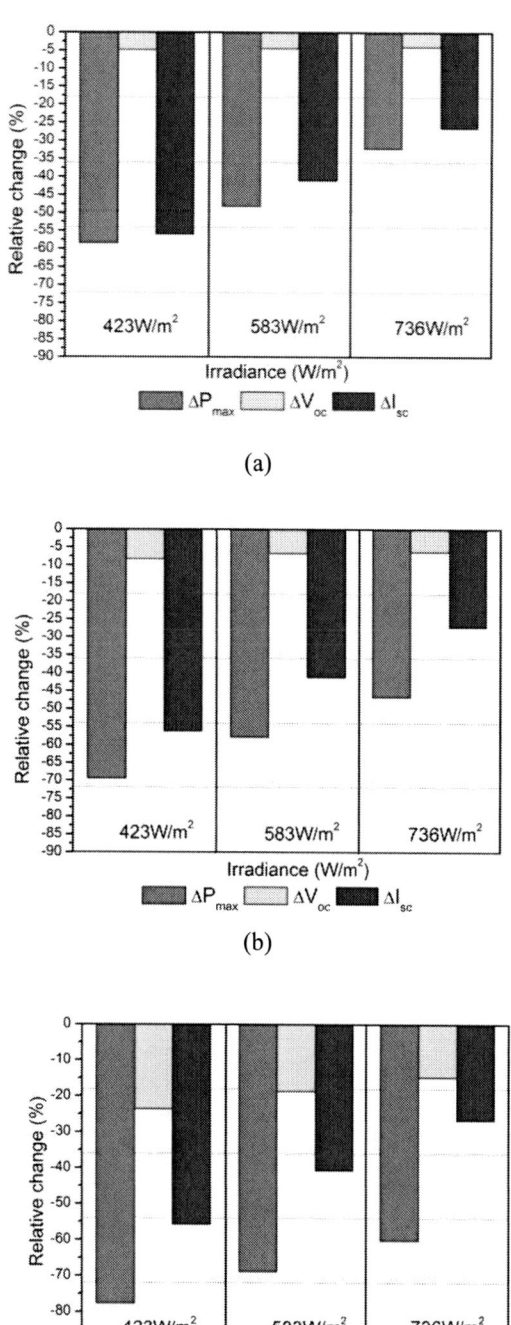

Fig. 3. Relative change in the light I–V parameters (measured at 423, 583, and 736 W/m² and 25 °C) of PID-free module 1 (a), module 2 (b) and module 3 (c). These parameters have been corrected by the temperature coefficients.

*B. Forward and reverse dark current-voltage characteristic of fielded modules*

As seen from Fig. 4, under forward and reverse bias, the forward current of module 2 and module 3 ascends faster than that of the PID-free module 1 as the growth of forward bias voltage. It is observed that the dark I-V characteristics of module 2 and module 3 become soft. What's more, the forward current of module 3 with serious PID grows more rapidly than that of module 2 with slight PID. It is concluded that the more severely the module degrades, the more quickly the forward current of the module mounts.

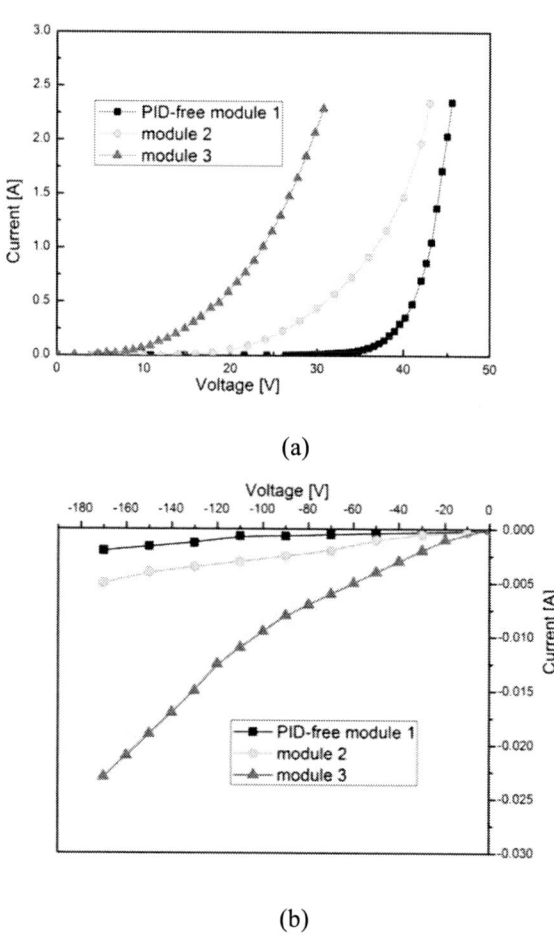

Fig. 4. Forward (a) and reverse (b) dark I-V curves for module 2, module 3, and PID-free module 1.

*C. Forward and reverse dark current-voltage characteristic for the three substrings of each fielded module*

After measuring forward and reverse dark I-V curves, we removed three bypass diodes to apply power supply connected with three substrings from the positive electrode to the negative respectively shown in Fig. 6. Fig. 5(a) gives out that the forward current of three substrings of modules 1 with PID grows faster than that of the PID-free substring which is one of three identical substrings of PID-free module 1. Fig. 5(b) exhibits the reverse current of substring 1 and substring 3 mounts quickly than that of substring 2. It is obvious that the forward and reverse current of substring 1 and substring 3

978-1-5090-5606-4/17 $31.00 © 2017 IEEE

become more rapidly than that of substring 2, which indicates that the substring 1 and substring 3 are more susceptible to PID than substring 2.

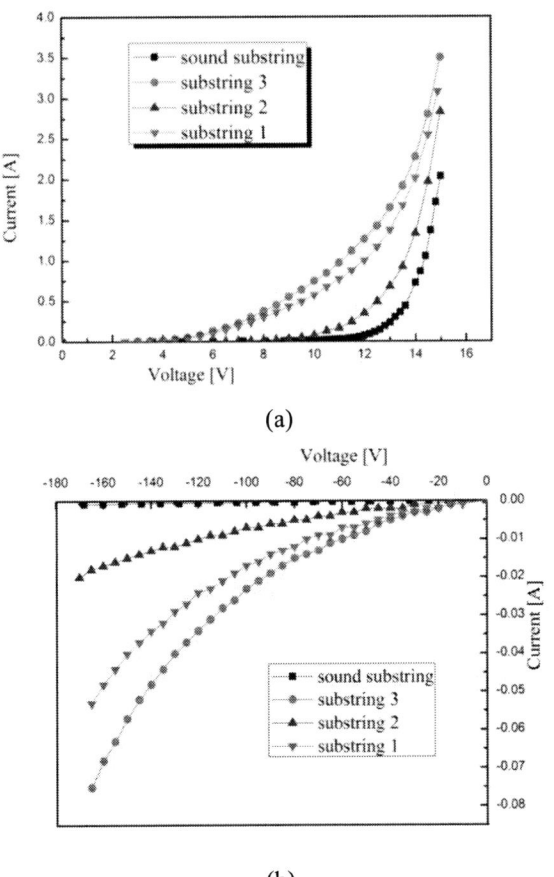

(a)

(b)

Fig. 5. Forward (a) and reverse (b) dark I-V curves for the three substrings of module 1.

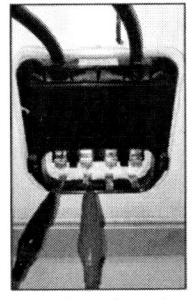 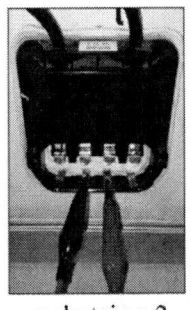 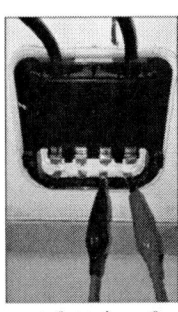

substring 1          substring 2          substring 3

Fig. 6. Each connection between positive and negative pole of three substrings after removing bypass diodes.

## III. CONCLUSION

The illuminated and dark I-V characteristics of solar modules degraded by PID in the field have been demonstrated in this study. By investigating and analyzing the illuminated and dark characteristics of solar modules at various degraded level, a significant degradation of illuminated I-V characteristic of PID-affected modules under low irradiance is shown. And the change of dark I-V characteristic of PID-affected modules is revealed. These provide a fast and reliable method for studying PID-affected solar modules.

### ACKNOWLEDGEMENTS

The authors would like to thank the support of Natural Science Foundation of China (Grant No. 61376067 and 61274050). This study was also supported by the National High Technology Research and Development Program of China (Grant No.2015AA050301).

### REFERENCES

[1] Sergiu Viorel Spataru, Dezso Sera, Peter Hacke, Tamas Kerekes, and Remus Teodorescu, "Fault identification in crystalline silicon PV modules by complementary analysis of the light and dark current-voltage characteristics", *Progress in Photovoltaics Research & Applications*, vol. 24, pp. 517-532, 2016.

[2] L. De Bernardez and R. H. Buitrago, "Dark I–V Curve Measurement of Single Cells in a Photovoltaic Module," *Progress in Photovoltaics Research & Applications*, vol. 14, pp. 321-327, 2006.

[3] D. L. King, B. R. Hansen, J. A. Kratochvil, and M. A. Quintana, "Dark current-voltage measurements on photovoltaic modules as a diagnostic or manufacturing tool," *in: Proceedings 26th IEEE Photovoltaic Specialists Conference*, Anaheim, USA, September 1997, pp. 1125-1128.

[4] Thomas Kaden, Katrin Lammers and Hans Joachim Möller, "Power loss prognosis from thermographic images of PID affected silicon solar modules," *Solar Energy Materials & Solar Cells*, vol. 142, pp. 24-28, 2015.

[5] Sergiu Spataru, Peter Hacke, Dezso Sera, Corinne Packard, Tamas Kerekes and Remus Teodorescu, "Temperature - dependency analysis and correction methods of in situ power - loss estimation for crystalline silicon modules undergoing potential - induced degradation stress testing," *Progress in Photovoltaics Research & Applications*, vol.23, pp. 1536-1549, 2015.

[6] R. Swanson, M. Cudzinovic, D. DeCeuster, V. Desai, J. Juergens, N. Kaminar, W. Mulligan, L. Rodrigues-Barbarosa, D. Rose, D. Smith, A. Terao, and K. Wilson, "The surface polarization effect in high-efficiency silicon solar cells", *in: Proceedings 15th International Photovoltaic Science and Engineering Conference*, 2005, pp. 410.

[7] Hara K, Jonai S and Masuda A. "Potential-induced degradation in photovoltaic modules based on n-type single crystalline Si solar cells," *Solar Energy Materials & Solar Cells*, vol 140, pp. 361-365, 2015.

[8] Pan Zhao, He Wang, Hong Yang, Fumei Wang, Ao Wang and Dengyuan Song, "Mechanism Analysis of Potential-induced Degradation of P-type Crystalline Si Solar Cells," *in: Proceedings 43rd IEEE Photovoltaic Specialists Conference*, 2016, pp. 2756-2760.

[9] Hong Yang, Fumei Wang, He Wang, Jipeng Chang, Dengyuan Song and Chengfeng Su, "Performance deterioration of p-type single crystalline silicon solar modules affected by potential

induced degradation in photovoltaic power plant," *Microelectronics Reliability,* vol 72, pp. 18–23, 2017.

[10] Fumei Wang, He Wang, Hong Yang, Jipeng Chang, Pan Zhao, Ao Wang and Dengyuan Song, "Effect of Potential Induced Degradation on Crystalline Silicon Solar Modules in Photovoltaic Power Plant," *in : Proceedings 43rd IEEE Photovoltaic Specialists Conferenc,* 2016, pp. 1752-1756.

[11] S. Pingel, O. Frank, M. Winkler, S. Daryan, T. Geipel, H. Hoehne and J. Berghold, "Potential Induced Degradation of solar cells and panels," *in: Proceedings of the 35th IEEE Photovoltaic Specialists Conference,* 2010, pp. 2817–2822.

[12] V. Naumann, D. Lausch, A. Hahnel, J. Bauer, O. Breitenstein, A. Graff, M. Werner, S. Swatek, S. Großer, J. Bagdahn, and C. Hagendorf, "Explanation of potential-induced degradation of the shunting type by Na decoration of stacking faults in Si solar cells," *Solar Energy Materials & Solar Cells,* vol. 120, pp. 383-389, 2013.

[13] Dominik Lausch, Volker Naumann, Otwin Breitenstein, Jan Bauer, Andreas Graff, Joerg Bagdahn, and Christian Hagendorf, "Potential-Induced Degradation (PID): Introduction of a Novel Test Approach and Explanation of Increased Depletion Region Recombination," *IEEE Journal of Photovoltaics,* vol. 5, pp. 834-840, 2014.

[14] Volker Naumann, Carlo Brzuska, Martina Werner, Stephan Großer, Christian Hagendorf, "Investigations on the formation of stacking fault-like PID-shunts," *in: Proceedings 6th International Conference on Silicon Photovoltaics,* 2016, pp. 569-575.

[15] M. Schütze, M. Junghänel, M.B. Koentopp, S. Cwikla, S. Friedrich, J.W. Müller, and P. Wawer, "Laboratory study of potential induced degradation of silicon photovoltaic modules," *in: Proceedings 37th IEEE Photovoltaic Specialists Conference,* 2011, pp. 821-826.

# Soiling Rates of PV Modules vs. Thermopile Pyranometers

Martin Waters[1], Tejas Tirumalai[1], Michael Gostein[2], Bill Stueve[2]

[1]Recurrent Energy, San Francisco, CA, 94104. [2]Atonometrics, Austin, TX, 78757.

*Abstract* — **Performance monitoring of solar facilities depends on quantifying both irradiance and soiling losses, among other factors. Frequently thermopile pyranometers are used for irradiance measurement in conjunction with PV module-based soiling monitoring stations. However, thermopile pyranometers are also subject to soiling and must be cleaned regularly to maintain accuracy. In this paper we compare the rate of soiling of a thermopile pyranometer to a PV module at a site in the western US. The results for this site indicate that the thermopile pyranometer soils at less than half the rate of the PV module. The data can be used to establish cleaning guidelines for thermopile pyranometers. In addition, the results show that soiling of thermopile pyranometers should not be used as a proxy for the soiling of a PV array.**

## I. INTRODUCTION

Accurate performance monitoring of solar facilities is highly dependent on accurately quantifying both the irradiance at the facility and any losses due to the accumulation of soiling. In order to quantify both of these at large scale solar installations it is common practice to install both pyranometers and soiling monitoring stations. However, as the number of pyranometers installed increases, maintenance on these devices becomes a non-trivial operational cost. While it is common to quantify the soiling rate of modules at a PV facility, it has not been common to quantify the rate of soiling on the pyranometers.

## II. METHODOLOGY

### A. Monitoring station description

For the purpose of monitoring both module soiling as well as pyranometer soiling, a system consisting of the following equipment was deployed in January of 2016 at a site in the western US: an Atonometrics RDE300 soiling station with one soiled PV module and one self-cleaning reference cell, a tipping bucket rain gauge, and three pyranometers. The pyranometers consist of two second class Hukseflux LP02 pyranometers and a secondary standard Hukseflux SR20 pyranometer serving as a reference measurement for the station. The LP02 pyranometers were field-calibrated to a reference device of the same model. For the purposes of this paper the two LP02 pyranometers are designated LP02-1 and LP02-2, with LP02-2 being the device that was allowed to soil. Figure 1 shows a close-up photograph of the instruments in the monitoring station.

The station received weekly maintenance consisting of leveling of all of the pyranometers, cleaning of LP02-1 and

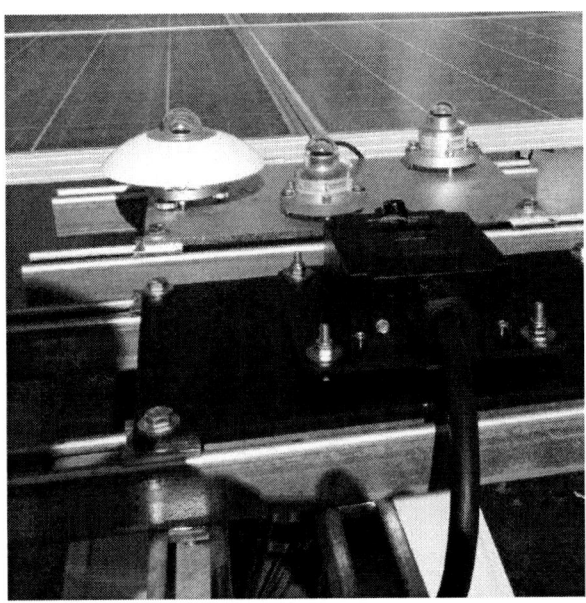

Fig. 1. Monitoring Station with SR20 (Left), LP02-1 (center), LP02-2 (right), self-cleaning reference cell (foreground), and reference module (back right)

SR20, maintenance of the soiling station and reporting of site conditions. In addition to the routine maintenance, there were three events over the test period where the station was recalibrated to a clean condition. These occurred due to an initial calibration error, bird droppings on the reference module, and a module cleaning of the entire site; each was accounted for in the analysis.

### B. Soiling ratio of module

As previously described [1][1][2][2][3][3], the RDE300 series soiling measurement station measures PV module soiling by performing an I-V sweep on the soiled PV reference module, determining the module power from the sweep, and comparing the measured power to the expected power for clean conditions. In the present work the measurement is performed once per minute, from which the system calculates the module soiling ratio

$$SR = \frac{P}{P_0 \cdot \left(1 + \gamma \cdot (T - T_0)\right) \cdot (G/G_0)} \qquad (1)$$

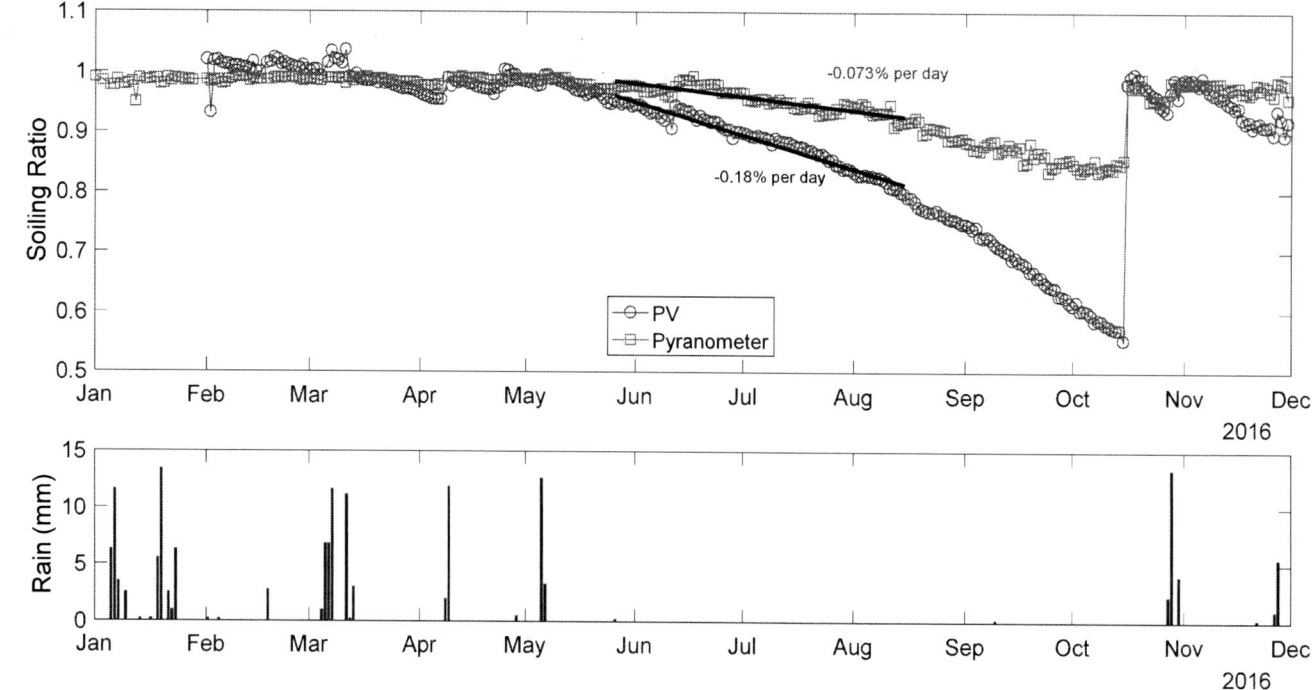

Fig. 2.    Measured daily average values of the soiling ratios for the POA PV module and thermopile pyranometer and daily rainfall.

where $P$ is the measured maximum power of the soiled module, $P_0$ is the module maximum power at STC, $\gamma$ is the temperature coefficient of maximum power, $T_0$ is 25° C, $G$ is the measured plane-of-array irradiance determined from the automatically cleaned PV reference cell, and $G_0$ is 1000 W/m². The numerator and denominator in (1) are the measured and expected power, respectively; under clean conditions the ratio should be 1.

In order to quantify a daily value for the soiling ratio, we first filter the once-per-minute reported $SR$ values, excluding periods of irradiance <500 W/m², periods of shading due to tracker misalignment, and early morning and late afternoon data that are subject to angle-of-incidence artifacts [4]. We then perform an irradiance-weighted average of the remaining points for each day, to yield the daily average values of the soiling ratio.

*C. Soiling ratio of pyranometer*

We determine the pyranometer soiling ratio by dividing the reading of LP02-2 by the reading of LP02-1. We filter the once-per-minute values of this soiling ratio using the same filter employed for the PV module analysis, and then perform an irradiance-weighted average of the remaining points for each day, yielding the daily average values of the pyranometer soiling ratio.

Due to the manual nature of the cleaning regime for the pyranometers, the data often appears to have a leveling off of the soiling ratio for the periods of time in between cleanings.

This is likely due to the accumulation of dust on LP02-1 following the cleanings at a rate similar to that of the accumulation on LP02-2. Deriving the soiling rate over a number of cleaning intervals helps to account for this effect.

III. RESULTS

*A. Module vs. pyranometer soiling*

Figure 2 shows the measured daily average values of the soiling ratios for the PV module and thermopile pyranometer as well as the daily rainfall. The summer and early fall were dry leading to an extended period of soiling where the soiling ratios steadily decrease until a module cleaning took place on Oct. 14th.

We performed a linear fit to each soiling ratio curve to extract an initial soiling rate for each device, with fit slopes shown in the plot. It is notable that the PV module soiling ratio and to a lesser extent the pyranometer soiling ratio appear to decrease nonlinearly, with the apparent soiling rate increasing in late August and September. This could be due either to higher airborne particulate concentrations or higher particulate sticking probabilities during this period.

It was found that the pyranometer in this test soiled at approximately 35% the rate of the reference PV module. The greater soiling of the PV modules relative to the LP02-2 pyranometer is also apparent in Figure 3.

The lower soiling rate of the thermopile pyranometer in comparison to the PV module is in agreement with manufacturer claims that thermopile pyranometers resist soiling due to their hemispherical domes. [5][5]

Fig. 3. Soiled pyranometer (bottom) compared to soiled modules (background).

## B. O&M indications

It can be assumed that the pyranometer will lose accuracy at a rate concurrent with the rate of soiling accumulation inducing a downward bias in the readings. Based on this assumption, it is a simple calculation to determine the frequency of maintenance based on the desired accuracy. Assuming a well calibrated secondary standard pyranometer with a calibration uncertainty of 1.2% [6], it can be seen that the downward bias caused by uniform soiling will be within the uncertainty of the device for approximately 18 days at this site. Thus at this site it may be possible to reduce the required maintenance frequency from weekly to once every two weeks, thereby reducing maintenance expense.

In addition, the results demonstrate that thermopile pyranometers should not be used as a proxy for determining soiling losses of PV arrays. In this study, the soiling rate of the thermopile pyranometer remained less than half that of the PV modules over an extended period.

## REFERENCES

[1] M. Gostein, et al, "Soiling Station to Evaluate Anti-Soiling Properties of PV Module Coatings," in *43rd IEEE Photovoltaic Specialists Conference*, Portland, OR, 2016.

[2] M. Gostein, T. Düster, and C. Thuman, "Accurately Measuring PV Soiling Losses with Soiling Station Employing Module Power Measurements," in *42nd IEEE Photovoltaic Specialists Conference*, New Orleans, LA, 2015.

[3] M. Gostein, B. Littmann, J. R. Caron, L. Dunn, "Comparing PV Power Plant Soiling Measurements Extracted from PV Module Irradiance and Power Measurements," in *39th IEEE Photovoltaic Specialists Conference*, Tampa, FL, 2013.

[4] M. Gostein, J. R. Caron, B. Littmann, "Measuring soiling losses at utility-scale PV power plants," in *40th IEEE Photovoltaic Specialists Conference*, Denver, CO, 2014.

[5] Web: https://eko-eu.com/faqs/solar-radiation-and-photonic-sensors/sun-trackers/soiling-effect-versus-measured-irradiance

[6] Web: http://www.hukseflux.com/product/sr20-pyranometer?referrer=/product_group/pyranometer

# Grid Integration of Building Systems and 1 MW Photovoltaic Array using VOLTTRON

David Raker*‡, Andrew Sellers†, Roshan Kini†, Michael Green, Thomas Stuart†, Randall Ellingson*‡,
Raghav Khanna†, and Michael Heben*‡

* School for Solar and Advanced Renewable Energy, Department of Physics and Astronomy,
The University of Toledo, Toledo, Ohio 43606
† Department of Electrical Engineering and Computer Science, The University of Toledo, Toledo, Ohio 43606
‡ Wright Center for Photovoltaics Innovation and Commercialization, The University of Toledo, Toledo, Ohio 43606

*Abstract*—We describe the goals, facets, and progress to-date of a grid integration project that seeks to control and stabilize the electric power demand/supply to mitigate intermittency and provide ancillary services to the grid. The University of Toledo features an all-electric Campus of Energy and Innovation with a 1 MW PV array. The VOLTTRON platform developed at Pacific Northwest National Laboratory will be used to integrate this array with a Building Automation System and 130 kWh capacity Li-ion storage. Irradiance forecasting will be used to predict PV generation, and algorithmic control of the building and storage systems will enable management of power output and VAR to present as flat a response as possible to the grid.

## I. Introduction

Multiple renewable energy technologies are reaching price parity with fossil fuel-based electricity. [1] As such, the cost of renewable technologies and photovoltaic (PV) systems no longer limits wide-spread deployment. Unlike other forms of electrical generation, however, terrestrial solar power is fundamentally intermittent. As these systems become widespread and occupy an increasingly larger position in our overall energy mix, it becomes urgent to address concerns, held by the power distribution industry, that transients and shifts in aggregate generation associated with intermittent sources of power pose a threat to the stability of the electrical grid. [1]

Commercial scale building HVAC systems, with characteristically large and intermittent inductive loads also present a challenge for the electrical grid. As such, utilities charge institutional customers on the basis not only of aggregate consumption, but also the bases of peak demand and reactive power. In situations where these loads exist in close proximity with large PV systems, which provide only real power, the presence of abundant real power may actually increase the reactive component of the customer draw from the grid. Coordination of loads and generation can help to mitigate these issues, in addition to diminishing the negative effects of intermittent generation from the perspective of the grid.

Algorithms exist which will allow management of peak demand from building systems, assuming that the system is sufficiently controllable and observable. In addition to algorithmic control, there is also research into the development of distributed control by a system of transactions. We are investigating application of algorithmic and, eventually,

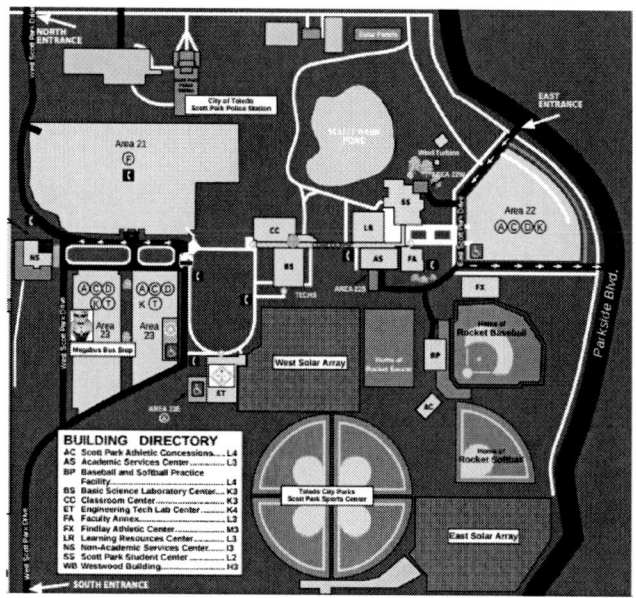

Fig. 1. University of Toledo Scott Park Campus

transactive control to integrate large building systems, PV generation, and battery energy storage with the grid.

These experiments are being carried out on the Scott Park Campus of Energy and Innovation at the University of Toledo. The campus consists of eight buildings, six of which are connected via a common corridor, which may be considered a ninth building itself. Most of these were designed and built in the late 1960s, though two have been added more recently, the newest being added in 1992. The layout of the campus is shown as Figure 1.

The campus features all-electric building systems, the vast majority of which are integrated into a network-connected Siemens Building Automation System (BAS). The campus also houses two PV arrays totaling 1 MW of capacity, based on thin film CdTe modules, which were installed and integrated into the campus electrical grid in 2010. The campus building systems include other elements which may be controllable, but which are not a part of the Siemens system, including a water chiller and LED lighting. The

978-1-5090-5606-4/17 $31.00 © 2017 IEEE

all-electric nature of the campus systems and the presence of significant PV capacity with respect to the size of the campus loads make the site well-suited to be a test-bed for advanced control strategies which can improve the integration of solar electricity installations with the wider power grid. In June of 2017, a Building Energy Storage System (BESS) is being installed on the campus, providing an energy storage component as well. The BESS is made by Johnson Controls, and features 130 kWh of Li-ion battery capacity, a four quadrant inverter, and a 125 kW energy delivery rate.

## II. Experimental Scope

In order to begin, it is first necessary to be capable of collecting data from the various systems and to gain the ability to dynamically control elements within them. To achieve this, the campus is being interconnected with an open source sensing and control system called VOLTTRON, which is developed by Pacific Northwest National Laboratory (PNNL). [2] VOLTTRON provides a common message bus on which agents may publish and subscribe to data points reporting the status of elements within the system. This also provides a common framework for controlling those elements from new or existing VOLTTRON agents by writing to control points on the message bus, which will be forwarded along to the appropriate points in the systems themselves. VOLTTRON comes with drivers to facilitate transparent communication with Modbus and BACnet devices, and can be extended to communicate with other protocols such that it allows the abstraction of divergent systems into a common, protocol agnostic, format. This allows control and monitoring agents to communicate with a wide range of systems without the need for modification, nor indeed any knowledge of the underlying protocols used by the systems with which they are communicating. As the Siemens systems on the Scott Park Campus were able to be upgraded to use BACnet and the BESS, meters, and PV inverters are all capable of Modbus communication, VOLTTRON allows these systems to interact with the same VOLTTRON agents without further modification.

Our research has four main foci, each of which alone can assist in integrating PV generation with the grid, and which will be used together to produce the eventual result of developing a transactive signal for facilitating interaction with energy markets. Each of these will be discussed in the following sections. Our first focus, of a very practical nature, is to serve as a test environment for the VOLTTRON platform and to ferret out those issues which may arise in the installation and operation of this platform outside the locations/sites around which it was originally developed. Initial integration of the building systems with VOLTTRON has been largely completed, and integration with the PV arrays and BESS system are in progress, as will be reported on here. The second focus is on the achievement of reliable day-ahead and finer resolution forecasts for the output of the PV arrays. The third is to experiment with new and existing algorithms for control of the BAS and BESS in order

to smooth the output of the array from the perspective of the grid. The fourth focus will involve transactive control of campus loads and generation, including interaction with a simulated energy market. This last will not be completed until later in the year, and so is not greatly discussed, though we are actively engaged in the first area of focus and taking preliminary steps with regard to the second and third.

### A. Campus Integration with VOLTTRON Environment

While this is not the first time VOLTTRON has been deployed at a new site, the platform is still relatively new and our experiences will be used to assist in the refinement of user experience and the development of improved documentation. The activities performed at the University of Toledo are also, of course, necessary for the implementation of the remaining foci, as this provides the infrastructure on which the several experiments will be run.

In order to identify the electrical loads within the Scott Park Campus and garner a complete understanding of the campus load distribution and makeup, two approaches were implemented simultaneously and the output of each was reconciled against the other. In one approach, a catalog of physical loads on the campus was generated from mechanical and electrical equipment schedules and blueprints dating back to the initial design and construction of the campus buildings and following through with each known update. In the other approach, the equipment being controlled by the Siemens building automation system (BAS) was analyzed to determine the campus HVAC configuration from its perspective. The reconciliation of these approaches allowed us to determine with a large degree of certainty exactly which loads exist, in which sizes and configurations they occur, and the degree to which they can be controlled.

In the document-based approach, relevant documents, including mechanical and electrical equipment schedules and building floor plans were gathered. Separate schedules and drawings exist to reflect the initial design and construction of the campus buildings as well as any renovations that occurred. Thus, there was need to develop a catalog, in a tiered manner, beginning with the initial data and building upon that with renovation data. The load catalog therefore matured in much the same way the campus did in its construction. To begin, the equipment schedules from the initial campus design were analyzed. A campus load catalog detailing the loads present in each building was generated. Relevant load information included power ratings, physical location within the building, and the load type. Once all loads present in the initial campus plans were identified and recorded, the equipment schedules for campus renovations were analyzed. For each building, the relevant renovation documents were analyzed chronologically and pertinent additions and modifications to the existing catalog data were made. In this way, the load information for each building was brought up to date. Upon the completion of an initial load catalog, additional documents were reviewed in order to fill in values missing from the equipment schedules. A total of 518 individual building loads were identified in this manner with a total power rating amounting to roughly 5.9 MW for the buildings that were assessed. This value can be compared to 1.7 MW and 0.53 MW, which are the maximum and average power draws for 2016, respectively. Additional verification of data was achieved via campus walkthroughs to visit individual loads and confirm existence, functionality, and physical location. An accounting of the types and sizes of loads in the system is shown in Table I.

The BAS is a Siemens APOGEE system which consists of three tiers of devices. At the highest level, a system called Insight forms a centralized control point for all the HVAC systems on all of the University of Toledo campuses. Hierarchically below Insight are a set of eight field panels, each controlling roughly one of the buildings on the Scott Park Campus. Larger building-wide devices such as air handlers and the chiller plant are controlled directly from field panels. They also coordinate bottom-tier, semi-autonomous Terminal Equipment Controllers (TEC) which are connected to the field panels by a serial link. Each TEC controls one end-point HVAC device such as a VAV, unit ventilator, or fan box using PID loops as a part of an one of several available applications.

The BAS system is not natively BACnet, but rather employs proprietary Siemens protocols for communication between its devices. In order for VOLTTRON to be able to interact, with these devices, a mapping to BACnet is provided by the Insight server and exposed on a network interface. As TEC devices are intended to run semi-autonomously, most of the control and data points in them are normally not reported up to Insight. In order to be exposed over BACnet therefore, it is also necessary to unbundle these points with configuration changes to Insight in order to make them usable. Once these steps been performed, it was possible for VOLTTRON to automatically detect the points in the BAS system using BACnet auto-discovery mechanisms. Due to the manner in which the mapping to BACnet is performed, however, each field panel appears as a single device. For usability, it was therefore also necessary to configure VOLTTRON to further separate these into virtual devices, each representing one actual device in the system. In total 8,536 data and control points were mapped. Through these, VOLTTRON is able to control and monitor the devices in the BAS, overriding Insight by using a higher BACnet priority. VOLTTRON is additionally configured to provide a heartbeat signal to a point visible to Insight, which will allow Insight to retake control should VOLTTRON become unexpectedly inactive.

Integration of building electric meters on the campus and the BESS system are somewhat more straightforward. These systems are already exposed via Modbus TCP on university controlled networks. Integration with VOLTTRON involves only writing a configuration file to specify the Modbus points to be queried. Control of the BESS, however, requires the development of a control application which will run as an agent in VOLTTRON. This work is ongoing. Additional work is also ongoing to integrate real time data from the meters and inverters associated with the PV arrays. These are operated by an outside company as part of a power purchase agreement, and data from these is collected outside the university network. Further, as these make use of the serial version of Modbus, which is single master, equipment is required to convert this first to a Modbus TCP signal which can be queried by more than one party as well as to connect this to the university network. This work is also ongoing.

### B. PV Performance Modeling and Forecasting

In order to achieve reliable predictions of the power generated by the photovoltaic plant, it is first necessary to have a reliable model of its performance. Currently the data being collected from the PV array has a fifteen minute resolution, which is too coarse for a full understanding of its behavior. The first task, with regard to the PV array, therefore, will be to use the VOLTTRON platform to capture performance metrics at a finer 1 minute or better resolution. With the ongoing changes discussed in the previous section, we expect to be capable of receiving data at a resolution as high as 15 seconds. Using collected power and weather data, it will then be possible to fully model the performance of the PV system from modules to inverter.

One central goal of this phase of the project is to develop a day-ahead predictive model of the power output of the array; to do so, it is also necessary to have good models of the local irradiance. Work is in progress to develop an accurate model to predict the effect of forecast weather conditions on the local irradiance at the site of the campus, which can be directly translated into PV array output power. Combined with an accurate performance model, this should allow reasonable day-ahead predictions of power output from the arrays. The degree of accuracy and precision which can be obtained from these forecasts is an important area of interest, and effort will be made to improve on existing algorithms where possible to improve their predictive power. Initial results of look-ahead modeling and comparison with actual performance is expected to be completed by early summer.

### C. Mitigation of PV Power Output Variability

Intermittent sources of power, such as PV, inherently produce transients which are of concern – when present on a large scale – to the stability of the electrical grid. The largest focus of this project is to study strategies for controlling and minimizing the variability of power produced by the PV arrays. In addition to a significant energy storage system, the building systems of the campus provide a further opportunity for control. By abstracting the disparate building systems into a single addressable environment, the VOLTTRON platform gives a great deal of flexibility for controlling the various building loads on a real-time basis.

The U.S. Department of Energys PNNL has already demonstrated in other projects the ability of an intelligent load control (ILC) agent running in a VOLTTRON environment to shave demand peaks by time shifting loads of building systems without significantly affecting occupant comfort. [3] We seek to extend this to a more complex environment involving intermittency in both demand and supply. Control

978-1-5090-5606-4/17 $31.00 © 2017 IEEE

| | | LR | AS | CC | BS/AH | ET | NS | FA | SS | Total |
|---|---|---|---|---|---|---|---|---|---|---|
| Air Handler | Fan | 78.5 | 3.7 | 29.8 | 26.1 | 18.6 | 7.5 | 14.9 | 48.5 | 227.6 |
| | Preheat Coil | 204.0 | 36.0 | 240.0 | 216.0 | 42.0 | 30.0 | | 68.6 | 836.6 |
| Roof Top Unit | Fan | | | 1.7 | | | | | | 1.7 |
| | Heating Coil | | | 46.0 | | | | | | 46.0 |
| Fan Coil | Fan | 6.3 | | | | | | 4.2 | 0.7 | 11.2 |
| | Heating Coil | 210.0 | | | 20.0 | | | 149.0 | 85.2 | 464.2 |
| Chiller | | 0.0 | | | | | 634.6 | | | 634.6 |
| Compressor | | | 1.1 | | 0.8 | | 0.6 | 37.7 | | 40.1 |
| Cooling Tower Fan | | | | | | | 29.8 | | | 29.8 |
| Condenser Unit | | | | 24.1 | | 0.8 | | | | 24.8 |
| Cold Diffuser | | | | | | 0.4 | 0.4 | | | 0.7 |
| VAV Reheat Coils | | | 901.5 | 319.5 | 409.5 | 294.0 | 386.0 | 49.5 | | 2360.0 |
| Exhaust Fan | | 2.6 | 0.4 | 0.6 | 8.6 | 6.8 | 2.0 | 0.1 | 5.7 | 26.9 |
| Return Air Fan | | 41.0 | | 14.9 | 11.2 | 7.5 | 5.6 | | 9.7 | 89.9 |
| Pump | | 7.5 | 0.6 | 3.4 | 3.5 | 1.6 | 231.7 | 3.0 | | 251.3 |
| Cabinet Heater | | 84.0 | 12.0 | 60.0 | 18.0 | 30.0 | 30.0 | | | 234.0 |
| Unit Ventilator | | | | | | 70.5 | | | | 70.5 |
| Unit Heater | | 12.0 | | 3.0 | 22.0 | 56.0 | 127.0 | 3.0 | 56.5 | 279.5 |
| Baseboard | | 12.1 | 5.8 | 74.9 | 35.3 | 5.0 | 4.2 | | 106.8 | 244.1 |
| All | | 1560.5 | 378.0 | 908.6 | 725.7 | 555.2 | 1189.6 | 174.3 | 381.7 | 5873.5 |
| | | | | | | | | BAS Controlled Loads | | 4628.99 |

TABLE I
LOAD BY BUILDING AT THE SCOTT PARK CAMPUS

is accomplished by altering set-points or other writable points within the building systems algorithmically to accomplish the goal of mitigating variations due to both demand from buildings and supply from the PV arrays. By coordinating large loads with excess generation, it should be possible to diminish the impact of both on the larger grid. This also has benefits for energy efficiency and has the potential for the building owner to avoid high-demand pricing from the utility. Eliminating demand peaks also benefits the utility, and the grid at large. The PV arrays provide the campus with real power, but do not affect the amount of reactive power involved in highly inductive loads associated with large building systems. This creates an environment wherein reactive power becomes a larger portion of the demand being made from the grid. Another goal of this project is to investigate the ability of the BESS and algorithmic control to mitigate excessive KVAR (kilo volt-ampere reactive), and thus to provide frequency regulation services to the grid. Collection of baseline data from the BAS and BESS will begin as soon as the integration of these systems with VOLTTRON is complete. Control experiments will begin by late spring and will be ongoing through the summer.

## III. CONCLUSION

It is necessary, as the world transitions to new, cleaner forms of power, to be prepared to integrate these systems with the electrical grid in a way which maximizes their efficacy and value to existing infrastructure. While it is necessary to make improvements to the grid to manage and optimize performance while accommodating the intermittency of these new sources, PV installations can serve to stabilize the grid if they are properly integrated with existing building control and energy storage systems. The project will generate data characterizing the extent to which we are able to mitigate the negative effect of these generation components, on the grid, at the point of origin. In addition, the project will enable the integration of PV energy forecasting with energy storage, forecast demand profiles, and actual energy demand intermittency. Lastly, we will describe the plans currently in place to document and quantify the implementation of these complex system components on the University of Toledos Scott Park Campus of Energy and Innovation.

## ACKNOWLEDGMENT

The authors gratefully acknowledge financial support provided by the U.S. Department of Energy, subcontracted through Pacific Northwest National Laboratory, DE-AC05-76RL01830/323688.

## REFERENCES

[1] M. Obi and R. Bass, "Trends and challenges of grid-connected photovoltaic systems a review," *Renewable and Sustainable Energy Reviews*, vol. 58, pp. 1082 – 1094, 2016. [Online]. Available: //www.sciencedirect.com/science/article/pii/S136403211501672X

[2] J. Haack, B. Akyol, B. Carpenter, C. Tews, and L. Foglesong, "Volttron: An agent platform for the smart grid," in *Proceedings of the 2013 International Conference on Autonomous Agents and Multi-agent Systems*, ser. AAMAS '13. Richland, SC: International Foundation for Autonomous Agents and Multiagent Systems, 2013, pp. 1367–1368. [Online]. Available: http://dl.acm.org/citation.cfm?id=2484920.2485228

[3] D. Stiles, "Pnnl research for a transactive energy future clean energy and transactive campus project," 2016. [Online]. Available: http://bgintegration.pnnl.gov/pdf/Clean_Energy_Transactive_Campus_Project.pdf

# Novel MPPT Algorithm for Active Power Control of Multi-Level Dual-Active Bridge PV Converter Implemented in NI myRIO

Shilpa Marti and Hariharan Krishnaswami

The University of Texas at San Antonio, San Antonio, Texas, 78249, USA

*Abstract* — To enable high penetration of PV systems into grid, it is imperative for the existing PV converters to perform load-following functions to be more dispatchable. Contrary to implementing energy storage systems or load shedding options for active power control, this paper focuses on implementing simple, inexpensive yet effective active power control strategy for two-stage multi-level dual-active bridge converter (ML-DAB), interfacing each PV string. Modified perturb and observe (P&O) MPPT algorithm is proposed to perform active power control. NI myRIO generates the desired phase shift in ML-DAB to extract either maximum power or reduced power.

*Index Terms* — MPPT, ML-DAB, PV systems, Active Power Control

## I. INTRODUCTION

There has been an increasing demand for clean and reliable power from PV systems. Variability and unpredictability of PV power output cause challenges for the grid operators. This paper addresses the challenges from frequency regulation capability. Within the framework of frequency regulation, there are two main strategies to reduce the fluctuation of PV output power:

1) Implement energy storage system method [1]
2) Operate PV systems by load shedding [2]

However, all the approaches mentioned above are complicated, labor-intensive and costly. Frequency regulation through active power control by operating below the MPPT point is a topic of current research [3]-[5]. While these works have made remarkable progress towards active power control for both single stage and two stage PV systems, significant research towards active power control for a modular multi-level PV converter remains to be explored. This paper proposes active power control [3]-[5] for a two-stage ML-DAB implemented in a NI myRIO device. Maximum Power Point Tracking (MPPT) control is a technique to vary the terminal voltage of PV panels to extract the maximum power out of them. Among various MPPT algorithms, P&O algorithm is simple and cost effective to implement. [6] gives a detailed discussion on the analysis and control of ML-DAB. Hardware test set up of ML-DAB is mentioned in [7] This ML-DAB acts as a building block that can be cascaded to meet the voltage and power requirements of large-scale PV systems. Fig. 1 shows the block diagram of the grid connected PV

system with the novel MPPT algorithm implemented in NI myRIO.

This paper demonstrates MPPT control of ML-DAB is by controlling the phase shift ($\phi$) between the two active bridges [8]. Active power control will be triggered when a signal is received to limit the PV output generation. Voltage and current sensors are used to sense the slope of PV power curve. The control voltage from the PI controller will vary the phase shift of $\phi$. Hill climbing method of the P&O algorithm is chosen to achieve the MPPT control, the same methodology is used for the active power control and implemented in the LabVIEW real-time. The phase shifted PWM pulses of $\phi$ are executed in LabVIEW FPGA.

Overview of the active power integrated control algorithm, and the flow chart is discussed in section II. Implementation of the control algorithm in NI myRIO is discussed in section III. Simulation and experimental results are given in section IV.

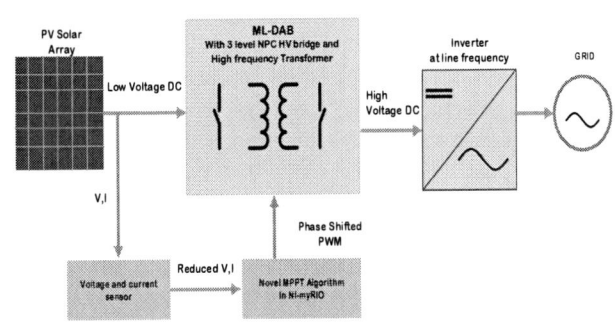

Fig.1. ML-DAB converter with MPPT control for large-scale PV systems

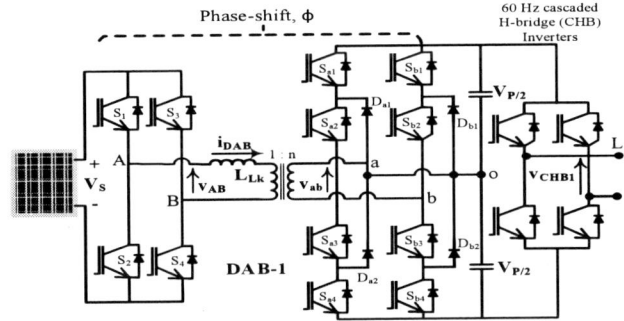

Fig.2. Circuit Schematic Diagram for ML-DAB Converter.

## II. ACTIVE POWER CONTROL INTEGRATED MPPT ALGORITHM

Various advanced control strategies are proposed in the literature [8], [9] that enables the high penetration of PV system into the grid. One of these methods is control of active power for frequency regulation. Energy storage is typically used to perform this control [10], [11]. However, cost and lifetime of the energy storage elements do not make this option effective. For grid stability during grid frequency oscillations, the PV plant can run at and below the MPP point to control its power output. This idea is implemented in many ways, but modifying the MPPT tracking control algorithm will result in fast and efficient results.

For large-scale grid-connected PV systems, there are usually two working modes:
- MPPT mode, where the PV plant operates at MPPT to extract the maximum power available
- Active power control mode, when the grid operator sends a command, to operate below the MPP point

When the grid meets its maximum demand, it can send a signal to the PV plants to enable them to operate below the MPP points, instead of shutting down the plants completely. Once the PV control system receives a signal, it alters its MPPT algorithm to operate below the maximum point.

The flowchart of the active power control integrated MPPT algorithm is shown in Fig. 3. As per the flowchart, the system initializes PV power $P_K$, phase shift increment $\delta_K$ and initial phase angle $D_K$. Then, it senses instantaneous voltage and current values. The algorithm iteratively checks for the signal from the grid operator to perform load curtailment. Once the algorithm gets the signal, it obtains the $P_{limit}$ from the grid operator and compares it with the instantaneous PV output power. If $P_{limit} < P_K$, then the algorithm performs active power control, otherwise it will operate in the typical MPPT method.

The MPPT method follows a P&O method, in which the PV operating point is perturbed with an MPPT sampling time (200 μs). After each perturbation, the control algorithm compares the values of power fed by the PV source before and after the perturbation. If after the perturbation the PV power has increased, it implies that the operating point is closer to MPP point. Consequently, the subsequent perturbation of the phase angle ϕ will have the same sign as the previous one. If after the perturbation the PV power has decreased, it implies that the operating point is away from MPP point. Thus, the sign of subsequent phase angle perturbation is reversed.

If the converter should operate in the active power control mode then, the $P_{limit}$ generated by the grid operator will be fixed as $P_k$. Then the algorithm follows the same operation as in the MPPT mode. After every perturbation PV power is compared with $P_k$ value, if the error is small, then the operating point is close to $P_k$. Consequently, the sign of the phase angle will be evaluated for reaching the pre-determined $P_K$ value.

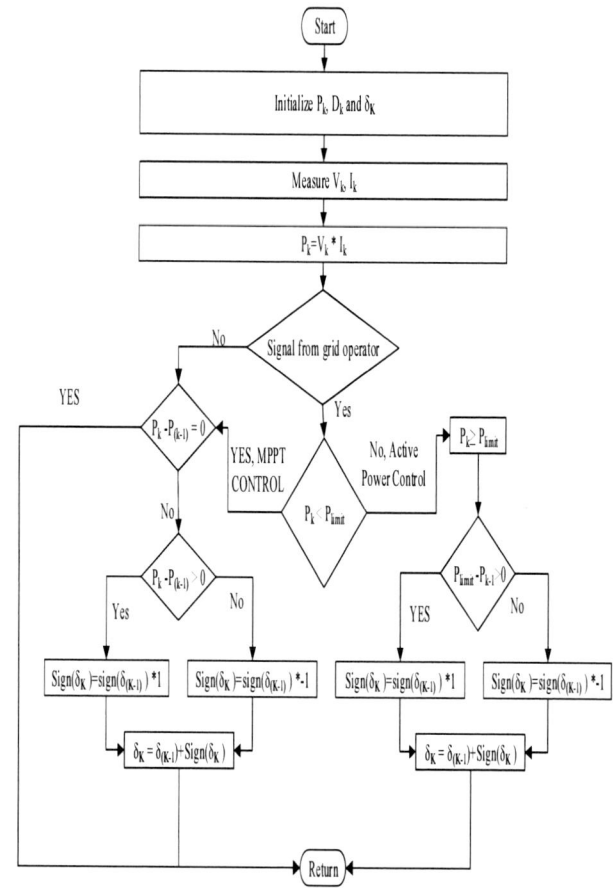

Fig. 3    Flowchart for RPM integrated P&O algorithm in NI myRIO

## III. IMPLEMENTATION OF ACTIVE POWER INTEGRATED MPPT CONTROL IN NI MYRIO

The output power of ML-DAB from [8] is given as

$$p_o = \frac{v_p v_s}{n \omega L} \cdot \left( \phi - \frac{\phi^2}{\pi} - \frac{\alpha^2}{2\pi} - \frac{\beta^2}{2\pi} \right) \qquad (1)$$

Where, $P_o$ is the average power measured at the output of the converter, with the variables defined as, $\omega = 2\pi f_s$ where $f_s$ is the switching frequency, n is the transformer turns ratio and L is the primary referred inductance ($L_{LK}$) used at the high-frequency link.

The circuit schematic is shown in Fig.2. From the figure, it can be observed that the power flow is controlled by ϕ. Thus, by varying ϕ, the desired maximum power from the solar module can be obtained.

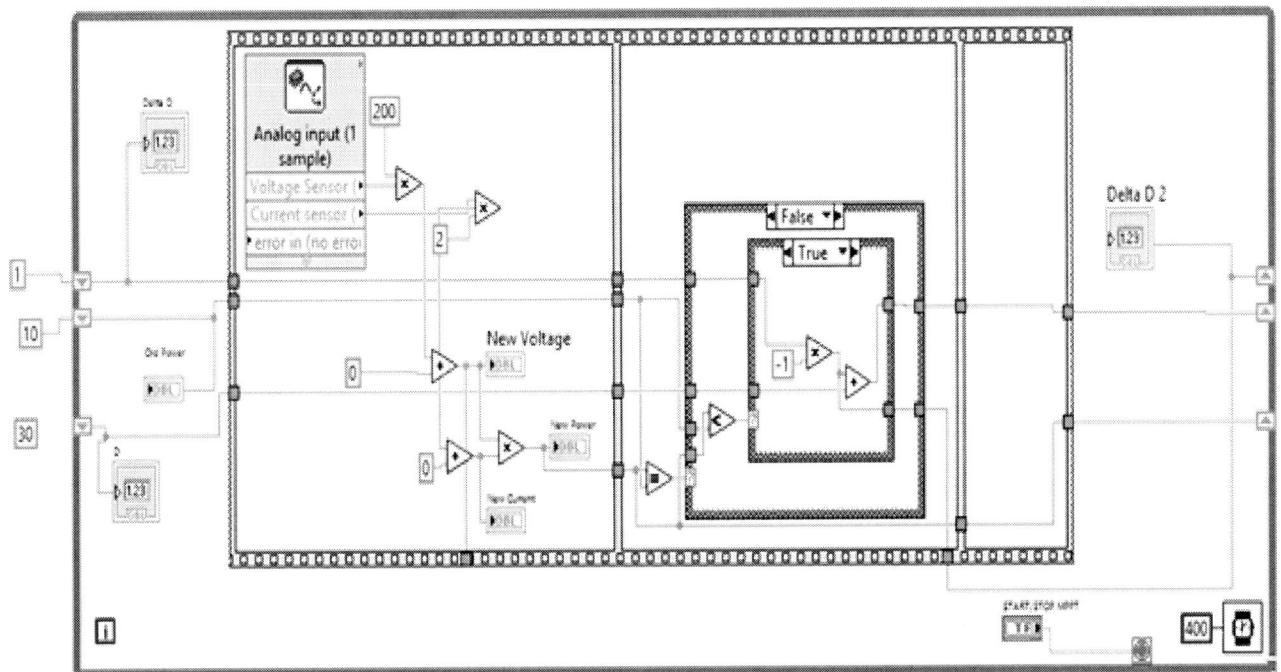

Fig.4. LabVIEW real-time VI block diagram for P&O algorithm

### A. LabVIEW Real-time Simulation

First, the initial values of the duty cycle $D_k$, Power $P_k$ and step in the duty cycle sign $\delta_k$ are defined in the while loop. There are three steps inside the while loop.

- In the first step, the values of the voltage and current sensors are acquired, and new power, new voltage, and new current are recorded.
- The second step has two concurrent for loops for implementing the P&O algorithm shown in the flowchart Fig. 3.
- The third step generates the control voltage from the PI control.

The value obtained through the PI control is translated to the corresponding phase shift angle by using a XOR gate and comparing the switching frequency (5kHz) pulses with double frequency (10kHz) triangular voltage source. These variable PWM pulses are used to phase shift the PWM pulses of $\phi$. Fig. 5. illustrates the phase shifted pulses from NI my-RIO in hardware.

### B. PWM generation in LabVIEW FPGA

Express PWM signals generated from the FPGA of NI myRIO generates the PWM signals required for controlling the phase shift of $\phi$. This Phase shifted PWM pulses of $\phi$ are given to

the switches on the primary side of ML-DAB. The power flow from PV array to DC bus is based upon the phase difference between the primary and the secondary bridges of ML-DAB.

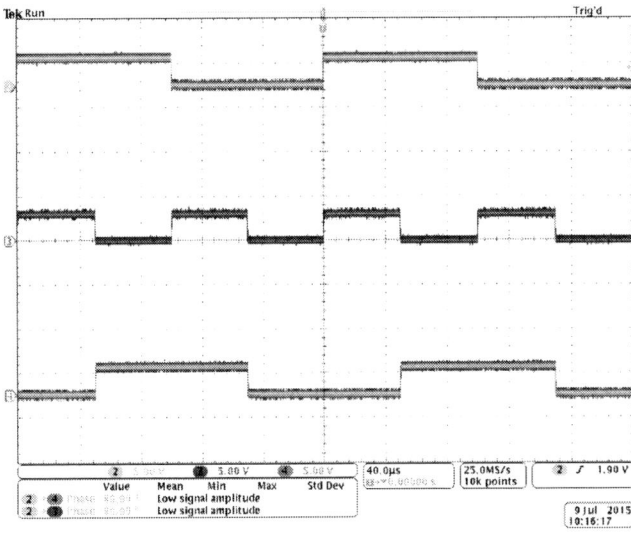

Fig. 5. Hardware results of MPPT control for ML-DAB a) represents Initial PWM pulses to the primary bridge of ML-DAB b) represents the pulses obtained from the MPPT PI controller loop c) represents the phase-shifted PWM pulses to the primary bridge of ML-DAB.

Active power control is included as shown in Fig 3. The algorithm will choose active power control mode per the reference signal from the supervisory controller. A non-zero reference signal will send the system into active power control. The initial values of the duty cycle $D_k$, step in the duty cycle sign $\delta_k$ are same as before. However, $P_k$ reflects the power mentioned by the grid operator

Fig. 6. and Fig. 7. shows the images of the front panel of the LABVIEW, that is used to implement the MPPT and active power control mode algorithm through Multisim CO – Simulation. During the active power control mode once it receives a signal from the grid operator the algorithm is modified as shown in Fig 6 to track the operating point below the MPP point as shown in the flowchart. The PWM pulses of $\phi$ with a fixed duty cycle of 0.5 is phase shifted by the rising edge of the pulses obtained from the control voltage. By varying this phase shift, desired maximum power is achieved. These varying gate pulses are given to the H-bridge of the ML-DAB to control the power flow. The MPPT integrated with the active power control mode will generate the desired control output voltage and the gate pulses are obtained from the control voltage.

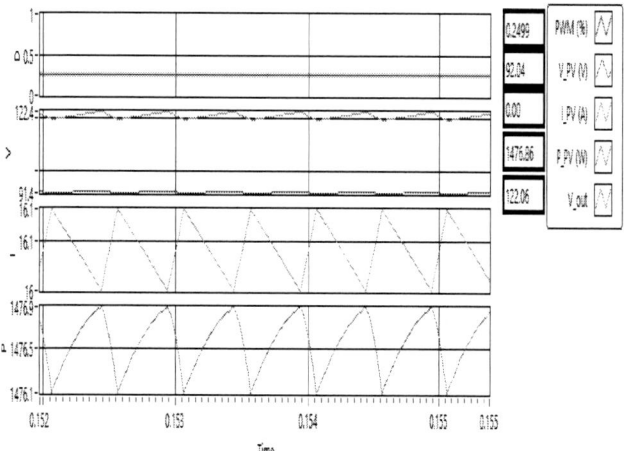

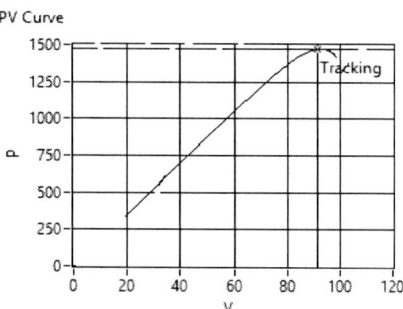

Fig. 6.    LabVIEW VI for ML-DAB under regular MPPT mode.

In Fig. 6. The converter operates in MPPT mode. Upper block depicts the simulation results, and the lower block shows the PV graph for different irradiance levels.

In simulation results channel a) represent the duty cycle percentage of pulses obtained from the MPPT controller of ML-DAB. Channel b) represents the Vdc link and Vpv values respectively channel c) represents the Ipv value and channel d) represents the Ppv value. The above simulation results are for 1000W/m$^2$ insolation and at 25°C temperature. The step size increment for the phase angle is set at 0.001

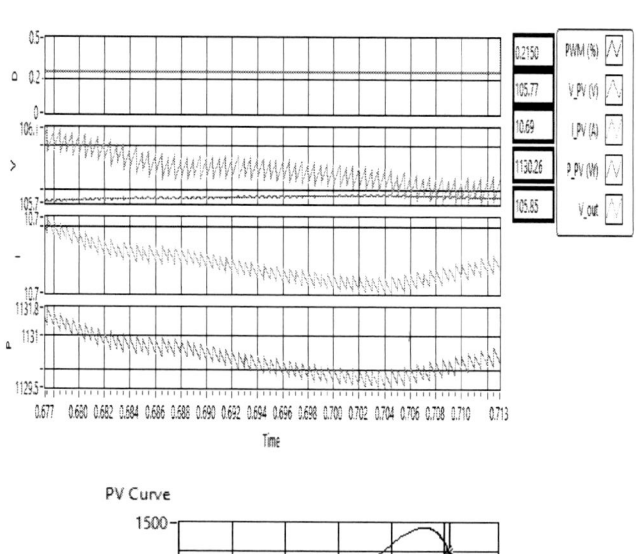

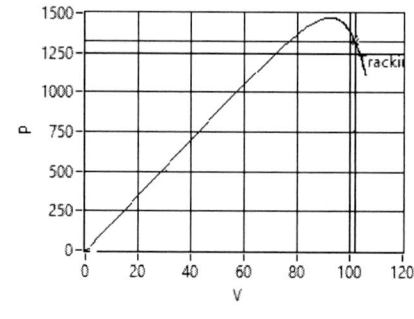

Fig. 7.    LabVIEW VI under active power control mode.

In Fig.7. The converter operates in the active power control mode; upper block shows the simulation waveforms and the lower block shows the PV graphs for different insolation. The value of Pk is assumed as 0.75 MPP. The simulation is carried out under the same conditions as of MPPT mode - 1000W/m$^2$ and 25°C.

Active power control algorithm will force the PV system to operate at 0.75MPP point rather than the maximum power point, thus following the frequency regulation instead of load shedding. From the figure, it can be observed that the duty cycle of the pulses from the PI controller was 0.6 under MPPT and the duty cycle of the pulses was reduced to 0.4 under active power control algorithm,

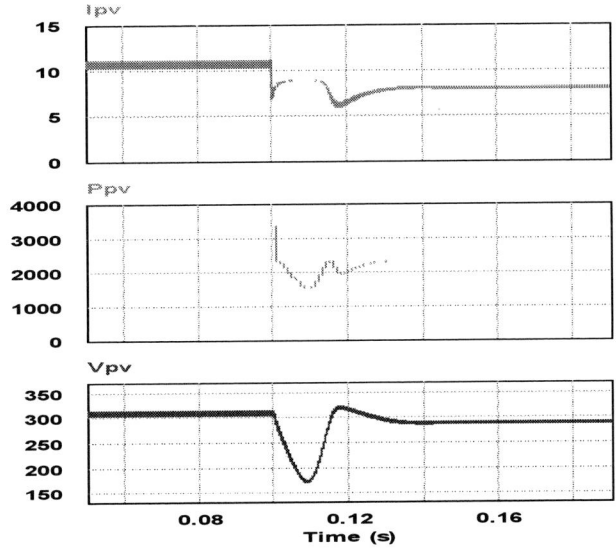

Fig. 8. Proposed MPPT control of the PV system in PSIM simulation. With the insolation change from 1000W/m² to 800W/m² $I_{PV}$ changes from 11.9A to 8.6A and $V_{PV}$ changes from 292V to 289V and $P_{PV}$ changes from 3.5KW to 2.65KW.

## V. CONCLUSION

This paper proposed an integrated active power control MPPT algorithm for frequency regulation for a two-stage ML-DAB PV converter. With the proposed control algorithm, active power control can be achieved by modifying the existing MPPT controls. Thus, it proves to be a simple and cost-effective alternative compared to the other solutions. Simulation and hardware results demonstrate the effectiveness of the proposed algorithm.

REFERENCES

[1] W. Omran, M. Kazerani, and M. Salama, "Investigation of methods for reduction of power fluctuations generated from large grid-connected photovoltaic systems," IEEE Trans. Energy Convers., vol. 26, no. 1, pp. 318–327, Mar. 2011.

[2] R. Tonkoski, L. Lopes, and D. Turcotte, "Active power curtailment of PV inverters in diesel hybrid mini-grids," in Proc. IEEE Electr. Power Energy Conf., Oct. 22–23, 2009, pp. 1–6.

[3] Hoke, Anderson, and Dragan Maksimović. "Active power control of photovoltaic power systems." *Technologies for Sustainability (SusTech), 2013 1st IEEE Conference on.* IEEE, 2013.

[4] Sangwongwanich, Ariya, Yongheng Yang, and Frede Blaabjerg. "Sensorless reserved power control strategy for two-stage grid-connected photovoltaic systems." *Power Electronics for Distributed Generation Systems (PEDG), 2016 IEEE 7th International Symposium on.* IEEE, 2016.

[5] Sangwongwanich, Ariya, Yongheng Yang, and Frede Blaabjerg. "High-performance constant power generation in grid-connected PV systems." *IEEE Transactions on Power Electronics* 31.3 (2016): 1822-1825.

[6] M. A. Moonem *, C. L. Pechacek, R. Hernandez and H. Krishnaswami, "Analysis of a Multilevel Dual Active Bridge (ML-DAB) DC-DC Converter Using Symmetric Modulation" Electronics 2015, 4, 239-26

[7] Duman, T., Marti, S., Moonem, M. A., Abdul Kader, A. A. R., & Krishnaswami, H. (2017). A Modular Multilevel Converter with Power Mismatch Control for Grid-Connected Photovoltaic Systems. Energies, 10(5), 698.

[8] National Renewable Energy Laboratory, "Industry perspectives on advanced inverters for U.S. solar photovoltaic systems: Grid benefits, deployment challenges, and emerging solutions," Tech. Rep., 2015

[9] A. Hoke, E. Muljadi, and D. Maksimovic, "Real-time photovoltaic plant maximum power point estimation for use in grid frequency stabilization," in Proc. COMPEL, pp. 1–7, July 2015

[10] H. Beltran, E. Bilbao, E. Belenguer, I. Etxeberria-Otadui, and P. Rodriguez, "Evaluation of storage energy requirements for constant production in PV power plants," IEEE Trans. on Ind. Electron., vol. 60, no. 3, pp. 1225–1234, Mar. 2013.

[11] E. Romero-Cadaval, B. Francois, M. Malinowski, and Q. C. Zhong, "Grid-connected photovoltaic plants: An alternative energy source, replacing conventional sources," IEEE Ind. Electron. Mag., vol. 9, no. 1, pp. 18–32, Mar. 2015

# Interconnection Study of Distributed PV Systems by Interfacing Matlab with OpenDSS and GIS

Joseph A. Ahamioje and Hariharan Krishnaswami

University of Texas at San Antonio, Department of Electrical and Computer Engineering,

San Antonio, Texas 78249, USA

*Abstract* — High penetration of PV on electrical grid generates voltage fluctuations and frequency variations. Innovative simulation approaches are needed to model the likely real impacts of PV to the operations and to the equipment in distribution systems. This paper proposes to use OpenDSS along with Matlab to implement advanced inverter functionalities. GridPV toolbox is used for specific circuit which was mapped using GIS features to an imaginary feeder circuit with a 170kW of solar in the campus of the University of Texas at San Antonio (UTSA). The GIS interface allows the user to draw and visualize the location of PV plant directly on the map. Moreover, it was considered in this paper that a circuit with several distributed PV can now be represented using GridPV with GIS and Quasi-Static Time Series (QSTS) simulation performed on it and further use active power control to minimize impact. IEEE 13 bus feeder is used to validate the effectiveness of the proposed procedure interfacing OpenDSS, Matlab and GIS.

*Index Terms* — *Active Power Control, GIS, GridPV, OpenDSS, Photovoltaic (PV) System.*

## I. INTRODUCTION

Solar Photovoltaic (PV) has grown progressively over the previous ten years. In 2014, the U.S. solar installed capacity increased by 54% and PV capacity then grew by 51% [1]. Consistent high growth rates over the last decade have resulted in a total of 18,305 MW of PV capacity and the end of 2014[1]. There has been a rapid increase in solar power in the last seven years. The cost per KWh has been reducing and the average price of a PV system logged a decrease of 33% since 2011[1]. There are numerous adverse effects of high PV penetration that include voltage fluctuation, frequency variation and system power losses [2].

Quasi Static Time Series Simulation (QSTS) is one of the most useful ways of analyzing the control of distributed PV on the grid i.e. voltage regulation device operations [2-3]. In time series, PV power output is modelled and a power flow solution is performed for every time instant, given into consideration that the system reaches a quasi-steady state. The results can then be plotted as a function of time.

There has been a lot of changes using Voltage regulation devices over the years and this has made available complex and customized control modes for many diverse applications. It could be difficult to find a software that can perform QSTS in all existing modes. This can only be done, if the software offers a COM interface capability to develop and implement control algorithms through an external program [4]. What is essential from a PV distributed system analytical tool with respect to PV inverters has the ability to perform Load flow analysis, solving balanced and unbalanced networks, able to create feeder configuration and equipment model directly from the GIS data and perform time series solution with the desired time interval.

Control methodologies that are currently used to address such high PV penetration include active power control to achieve frequency regulation and dynamic Volt/VAR control to reduce voltage fluctuations. The Quasi-static time series simulation approach using OpenDSS is used to model the complete system. Matlab and OpenDSS can be interfaced using COM feature thereby simulating the functionalities of active power control and Volt/VAR and detect the effect of these controls in the grid. Recently GridPV software uses the COM feature to give a detailed predictable graphic analysis when simulating the performance of system. However, a Geographic Information System (GIS) has not been fully integrated into these simulation approaches. In this paper, a novel method is proposed to integrate geographical information. GIS is applied in this paper using Google Maps to display street images. Google Maps allows Matlab to interact and download maps with specific GPS co-ordinates. The interface allows the user to draw and visualize the location of the PV plant directly on the map. IEEE 13 bus feeder is used to validate the effectiveness of the proposed procedure interfacing OpenDSS, Matlab and GIS.

Similarly, high PV penetration on the grid results in grid voltage fluctuations and frequency variations which in some cases causes voltage rise and reverse power flow in the distribution system. Adopting VAR and Active Power Control are two methods used in practice [5].

The focus on this paper will be a simulation approach that exploits the COM (Common Object Model) proficiencies of OpenDSS, an open source platform that will perform Quasi Static Time Series Analysis with dynamic Volt/VAR and active power control with integration of geographical information. Matlab is introduced as an external program to invoke OpenDSS through COM [5]. The paper advances further by introducing GridPV toolbox with GIS integration. The GridPV toolbox gives a well detailed and predictable analysis when simulating the performance of any PV system and the GIS allows Matlab to interrelate and download google maps of a specific location based on geographical coordinates

[6-7]. The interface enables the user to draw and visualize the location of the designed circuit directly on the map. The results provided after solving the circuits were used to achieve the Volt/VAR control and in regulating the active power.

## II. QUASI STATIC TIME SERIES WITH IEEE 13 NODE FEEDER

### A. Network System Description

Various load flow simulation software's are obtainable and can provide us with the simplicity of resolving complex radial networks both balanced and unbalanced. However, it is essential to offer a standard that can be used in all existing platforms and compare results. A set of defined test feeders provided by IEEE can be used in any other platform and the results compared. In this paper, IEEE 13 Bus Node Feeder will be used for analysis. The purpose of the IEEE 13 Bus Node Feeder is to validate the designed circuit feeder by checking for errors before running it.

### B. Software's Used for Simulation

It becomes essential to examine the effects of distributed energy integrated on the grid. Since this paper discusses PV systems, a simulator that is capable performing Quasi Static Time Series Simulation at a high resolution is essential. OpenDSS is a simulation software tool established by EPRI and it gives a procedure to model the feeder incorporated with distributed source of energy and performing a QSTS and finally evaluates the same. One of the important features of OpenDSS is that it can perform power or load flow analysis using the Newton Raphson method and hence solve unbalanced radial feeders.

OpenDSS runs a user friendly GUI that helps demonstrating the feeder in a vibrant way with each element defined independently. It also provides the user with different solving modes that helps in examining the same circuit in a broader range. Since it is an open source, program writers are allowed to make changes in its source code and exploit the properties of OpenDSS as desired. On the other hand OpenDSS does its analysis in frequency domain.

GridPV is a Matlab toolbox that offers a set of well predictable functions for simulating the performance of photovoltaic distribution system. It normalizes interface between Matlab and OpenDSS for easy queries. It has the capacity to plot and visualize results which on the output gives clean and clear interactive plots with many options.

On the other hand, GridPV models a solar power without difficulty and precise GUI for setting up PV plants, models solar variability for size and dispersion of PV and provides power factor and reactive power control for PV plants.

GIS is defined as a system that is designed with the intent to gather, store, control, examine and be able to and present spatial or geographical data. Matlab/GridPV tool boxes can be modified to integrate an OpenDSS circuit plot with GIS

functions through a Google Map. This will also include functions to convert between coordinate systems. GIS gives the user a visualized location of the distribution PV system power lines on a landscape map.

### C. Methodology

This paper comprises of OpenDSS interface with Matlab using which data can be transmitted in both ways from OpenDSS and Matlab through the COM interface. QSTS is performed and the feeder is analyzed with PV integrated in it. Matlab utilizes the data received from the OpenDSS to vary the parameters to perform Volt/VAR and Active Power control.

### D. Simulation of Volt/VAR and Active Power Control as proposed in the methodology.

#### 1) Volt/VAR Control

High penetration of PV causes a high degree of randomness in the operation of distribution feeder every hour and sometimes day due to inconsistent behavior of PV. The PV output depends upon the solar radiation getting to the earth's surface and being tapped by the solar panels. This has a great impact over the flow of power and the voltage profile in the feeder. If we take an example of a cloudy day, the PV output goes low and large amount of power is drawn and consumed from the feeder. This in turn causes a low voltage profile being created in the feeder, which is not suitable for the end user as it may damage appliances. The old way of controlling this is to use voltage regulators and tap changes that would maintain the voltage levels during fluctuations or switching of capacitor banks that would reduce the reactive power in the feeder when ON or boost the reactive power in the feeder when OFF.

In utilities, it is becoming an essential practice to supervise VAR in the power grid and implement the Volt/VAR control. This technique will help increase efficiency, diminish demand and encourage energy conservation. This can be made possible, if end users that have individual PV inverters and PV plants can contribute to the cause. Therefore, the utility demands the PV plants and the customers to support the cause. However this gives the end users a benefit of creating power more than needed and marketing it to the utilities which in turn will lessen their total utility bill.

This paper emphasizes on showing a VAR control simulation by using the OpenDSS-Matlab/GridPV toolboxes. However, it turns out that injecting VAR into the grid as the voltage reduces alone cannot provide geographic information on the voltage profiles.

#### 2) Active Power Control

Frequency is usually maintained at 60Hz in the grid which is supplied to the customers. However when the utilities experience high PV penetration, the frequency starts

978-1-5090-5606-4/17 $31.00 © 2017 IEEE      2937

fluctuating. This occurs as the active power of the PV is not constant and it depends on the current irradiance value, weather conditions, temperature which in general are not constant during the entire day. Trying to ensure a constant frequency becomes a big concern for the utility companies involved in running a distributed PV system. The other issue is when the PV is producing at its peak and the entire amount of power to be distributed needs to be reduced. Nevertheless taking needed actions to control the PV output becomes important.

In this paper, it is demonstrated using simulations to reduce the active power output of PV generators to minimize the total output power of PV being supplied to the grid and ensure the frequency is maintained within a certain threshold. Nevertheless energy operatives in such circumstances can mandate the PV plants to maintain the active power at a continuous level through regulating of smart inverters. OpenDSS-Matlab interface with GridPV is used to simulate the situation and detect the effects of load flow in a 2D distribution grid output graphic display.

The feeder is solved initially using OpenDSS through the interface. When the power fluctuates, it then becomes essential to regulate it and retain it within a required limit as required by the utilities for curbing. However, similar to the Volt/VAR control method, there is no information on geographic profile of frequency variations. In the following sections, results are shown with GIS integration.

## III. RESULTS

### A. Volt/VAR Control

Fig. 1. shows an analysis of maximum and minimum voltages with PV while Fig 2 shows the maximum and minimum voltages after Volt/VAR control is introduced. The simulation (for both graphs) stops at each time step for Matlab to process the state of the OpenDSS simulation. The output is further displayed using the GridPV

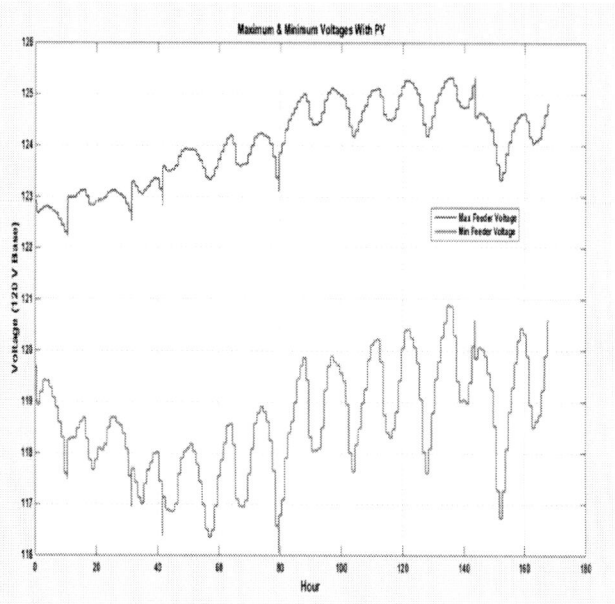

Fig 1. Maximum and minimum voltage with PV

It could be noticed that there is a drop in voltage when the PV plant is connected at the Bus. Consequently it becomes necessary to provide a VAR compensation at this stage in order to increase the voltage in the circuit to stabilize the voltage within a certain limit. Fig. 2 also shows both maximum and minimum voltage after VAR is induced into the feeder.

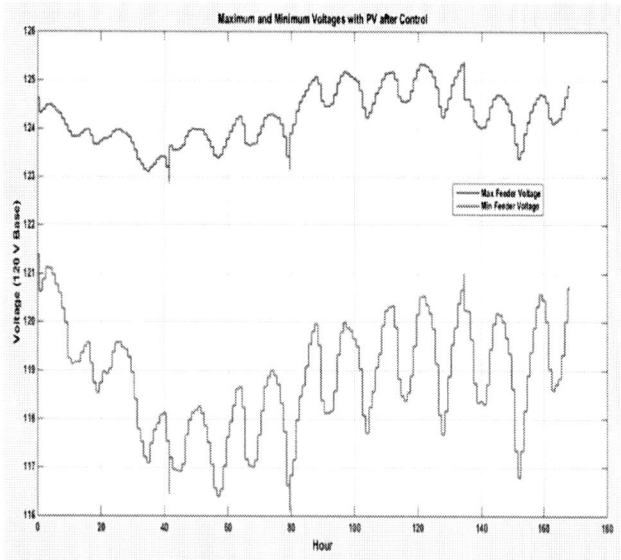

Fig 2. Maximum and minimum voltage after VAR

Similarly, Fig 2 shows that the voltage levels rises and is brought within the limits. This happened because the PV generator that is connected to the bus was instructed to inject VAR to offset the voltage levels and bring them within the ANSI limits. Matlab/GridPV comes into play as it senses this drop in voltage using COM interface and taking voltage on the

Bus as a reference, changes the VAR output of the PV generator in order to increase the voltage.

### B. Active Power Control

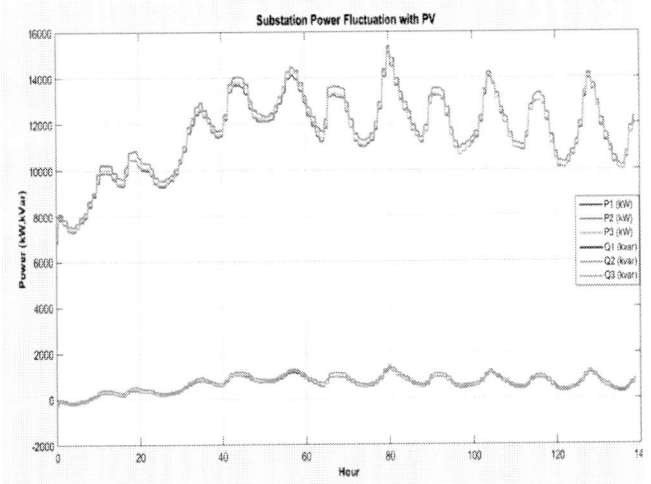

Fig 3 Active power flunctuation with PV

After the circuit is solved using OpenDSS via COM interface, the active power is then measured and plotted above as shown in Fig. 3. It can be deduced from Fig. 3 that based on the varying irradiance, the output of PV is not steady and causes the power to fluctuate. It becomes important to control and keep it within limits demanded by utilities for curbing.

However Matlab senses the change in a voltage fluctuation at the Bus and adjusts the KW value of the generator in order to respond to curtailment. Fig. 4 shows the active power after curtailment has been done and shows that the PV output has been reduced after control. The plot is executed using Matlab/GridPV and shows a detailed control mechanism.

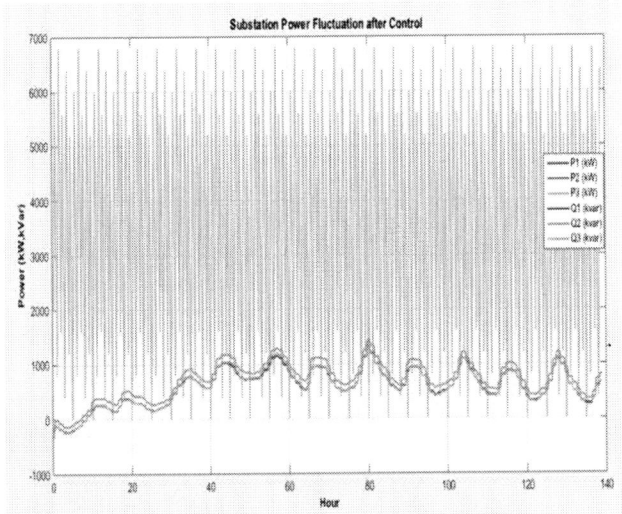

Fig 4. Active power fluctuation after control

### C. GIS

The purpose of using GIS in this paper is to give the user a visualized location of the distribution system power lines on a landscape map. This guides the users in interpreting the circuit better in real time. This in future can help the facility engineers plan adequately in terms of making adequate injected load in substation, equipment cost, minimizing the challenges faced on proposed site such as land barriers, unauthorized zones, and public facilities. Another thing the GIS can also do is to assist engineers in locating and tracing line faults.

Similarly, in this paper geographical information was created for the OpenDSS feeder that can be used to map the feeder to the real world. Nevertheless, with a known coordinate system, definite GIS features are obtainable in the toolbox to visualize the location of the distribution system power lines. The introduction of Google Maps is used to show streets, location names, and satellite images. Google Maps allows MATLAB to interrelate and then download maps with a specific location with elevation inclusive. This interface between the two software's allows the user to draw the location of the PV circuit plant directly on the map. This user-drawn PV plant gives a Clear visualized location of the PV distribution system power lines. Fig. 5 shows an imaginary feeder circuit traced on University of Texas at San Antonio (UTSA) main campus with PV and substation locations. It is possible to integrate the previous control methods with the GIS information to clearly observe the voltage and frequency profiles for a general N-bus system.

Fig 5. Sample Feeder circuit traced on UTSA Google Map

## IV. CONCLUSION

OpenDSS functions as a good diagnostic tool in demonstrating the inconsistency of PV distribution feeder. It is however imperative to interface OpenDSS with external programs for additional detailed results. GridPV being a Matlab tool box served as a good simulation tool in the sense that it gave a 2D pictorial simulation of results plotted. In a way it gives the programmer a visual concept of the circuit. However GIS can further assist the user to have a visual concept of where the plant circuit is to be placed and consider all physical challenges before setting up the plant and control in real time.

## ACKNOWLEDGEMENT

The authors gratefully acknowledge the contributions of the University of Texas at San Antonio department of Electrical Engineering.

## REFERENCES

[1] NREL, "National Renewable Energy Laboratory", www.nrel.gov

[2] Jimmy E. Quiroz, Matther J. Rieno, Robert J. Broderick, "Time Series Simulation of Voltage Regulation Device Control Modes", Sandia National Laboratories, Georgia Institute of Technology, 2013.

[3] B. A. Mather, "Quasi-static time-series test feeder for PV integration analysis on distribution systems," IEEE Power Energy Soc. Gen. Meet., pp. 1–8, 2012

[4] Distribution System Analysis Sub Committee Report, "Radial Distribution Test Feeders"

[5] Wes Sunderman, Roger C. Dugan, Jeff Smith, "Open Source Modelling of Advanced Inverter Functions for Solar Photovoltaics Installation", EPRI, Knoxville, USA, 2014.

[6] Matthew J. Reno, Kyle Coogan, "Grid Integrated Distributed PV (Grid PV)", Sandia Report, August 2013.

[7] Matthew J. Reno, "Streamlined Interconnection Analysis of Distributed PV using Advanced Simulation Methods", Georgia Institute of Technology, May 2015.

[8] Huanhai Xin, Yun Liu, Zhen Wang, Dequiang Gan, Taicheng Yang, " A New Frequency Regulation Strategy for Photovoltaics Systems Without Energy Storage", Oct 2013.

[9] Vaidyanath Ramachandran, Dr. Sarika Khushalani Solanki, "Modelling of Utility Distribution Feeder in OpenDSS and Steady State Impact Analysis of Distributed Generation", West Virginia University.

[10] Xiaoxiang Gao, "Study of FSU Campus Distribution System", Florida State University, 2016.

[11] Masoud Farivar, Russel Neal, Christopher Clarke, Steven Low, Southern California Edison, CA, USA, Department of Electrical Engineering, Caltech, CA, USA, " Optimal Inverter VAR Control in Distribution Systems with High PV Penetration", 2012.

[12] M.J.E. Alam, K.M .Muttaqi, D.Sutano, "A Multi-Mode Control Strategy for VAR support by Solar PV Inverters in Distribution Networks", 2014.

[13] J.W. Smith, W. Sunderman, R. Dugan, Brian Seal, " Smart Inverter Volt/VAR control functions for High penetration of PV on Distribution Systems", 2011.

[14] D. Montenegro, M. Hernandez, G.A. Ramos, "Real Time OpenDSS framework for Distribution Systems Simulation and Analysis", 2012

[15] Meghasai Seethamraju, "Simulation of Smart Functionalities of Photovoltaic Inverters by Interfacing Both OpenDSS and Matlab", The University of Texas at San Antonio, May 2015.

[16] Nikhar Jung Abbas, "Small Disturbance, Long Term Voltage Stabilization on a Distribution Feeder in Kathmandu, Nepal ", University of California San Diego, 2016.

# Modeling a Grid-Connected PV/Battery Microgrid System with MPPT Controller

Genesis Alvarez[1], Hadis Moradi[1], Mathew Smith[2], and Ali Zilouchian[1]

[1]Florida Atlantic University, Boca Raton, FL, 33431, USA

{genesisalvar2013, hmoradi, zilouchi} @fau.edu

[2]IEEE Smart Village Volunteer, Piscataway, NJ, 08854, USA chemicalbull03@gmail.com

*Abstract* — **This paper focuses on performance analyzing and dynamic modeling of the current grid-tied fixed array 6.84kW solar photovoltaic system located at Florida Atlantic University (FAU). A battery energy storage system is designed and applied to improve the systems' stability and reliability. An overview of the entire system and its PV module are presented. In sequel, the corresponding I-V and P-V curves are obtained using MATLAB-Simulink package. Actual data was collected and utilized for the modeling and simulation of the system. In addition, a grid-connected PV/Battery system with Maximum Power Point Tracking (MPPT) controller is modeled to analyze the system performance that has been evaluated under two different test conditions: (1) PV power production is higher than the load demand; (2) PV generated power is less than required load. A battery system has also been sized to provide smoothing services to this array. The simulation results show the effective of the proposed method. This system can be implemented in developing countries with similar weather conditions to Florida.**

## I. INTRODUCTION

The massive increase in the global demand for energy is due to the industrial development, population growth and economic development. Many people in the world are currently experiencing dramatic shifts in lifestyle as their economies make the transition from subsistence to an industrial or service base. The largest raise in energy demand will take place in developing countries where the proportion of global energy consumption is expected to increase from 46 to 58 percent between 2004 and 2030 [1-6]. As environmental issues continue to progress due to conventional energy systems. Renewable energy sources such as photovoltaic (PV) is becoming a more promising source of energy for power generation. The amount of photovoltaic installations has exponentially grown, primarily due to the decrease in cost of solar panel systems [7-9]. The electric grid is an unlimited amount of energy supply. A photovoltaic system can either be a stand-alone system or a grid-tied system. In both systems a battery storage unit is often essential to the entire system. Therefore, battery management is significant to the functional life of the battery bank. The battery bank can also be implemented in smoothing the PV output power fluctuations.

The battery bank stores energy produced by PV panels. If the ramp rate of the PV is greater than the load can tolerate, then the battery bank serves to produce or consume power in order to smooth the output. One of two methods of charging the battery bank includes using the PV panels to charge the batteries during peak sun hours. In the occasion of cloudy weather, the lack of solar radiation energy doesn't permit the battery bank to recharge [10]. In a grid-tied PV system, distribution lines is a backup power source. Specific battery requirements in designing a PV array include battery charging losses, load demand, discharge rate, battery size, and storage temperature. Weather conditions, lower solar irradiation and higher temperatures are known causes for lower energy efficiency production by the PV array. However, there is an on-going research pertaining to voltage control of a PV array with maximum power point tracking (MPPT) and battery storage [11]. The MPPT algorithm supports sustainable efficiency by dynamically adjusting the voltage to ensure power optimization [12].

A Battery Energy Storage System (BESS) will be beneficial not only on a daily saving but reducing the PV output power fluctuations. The aim of this paper is to present the performance evaluation of the FAU fixed grid-connected PV array with a MPPT algorithm and a BESS with a time moving average algorithm to reduce voltage sag.

## II. SYSTEM OVERVIEW

In this paper. The performance of a solar unit located at FAU in Boca Raton, Florida as shown in Fig.1 is analyzed. The fixed solar array is installed on an East-West axis with a 23° inclination angle. The 12x2 PV array is connected to a Sunny Boy 7000US-12 inverter and then ties to the Florida Power and Light (FPL) utility grid. Features within this inverter are arc-fault circuit interrupter, measuring channel and MPPT algorithm. It is proposed to install a sealed lithium-ion battery storage system to the current grid-connected array.

## III. PV CELL MODELING

A p-n junction fabricated in a layer of a semiconductor forms a photovoltaic cell structure. The ideal solar cell is a semiconductor diode connected in parallel to a current source with series resistance, and parallel resistance as shown in Fig. 2 [13].

978-1-5090-5606-4/17 $31.00 © 2017 IEEE

Fig. 1. Photovoltaic unit installed at FAU Boca campus

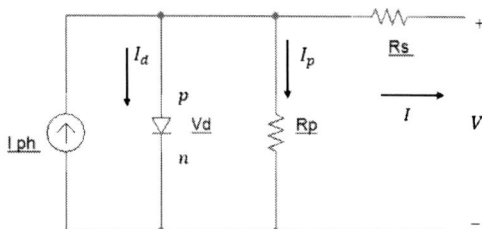

Fig. 2. Equivalent circuit of a PV cell with single-diode model

$$I = I_{ph} - I_0[e^{\frac{q(V+RsI)}{nkT}} - 1] - \frac{V+R_S I}{R_{ph}} \qquad (1)$$

$$V_d = V + IR_s \qquad (2)$$

A simple equivalent photovoltaic cell circuit model includes a diode in parallel with a current source and a resistor as shown in Fig. 2. In the characteristic equation, $I_{ph}$ represents a current source created by sunlight known as a photocurrent (A), $I_o$ is the saturated current, $V$ denotes an output voltage (V), $V_d$ is the diode voltage, q is an electron charge of $1.6 \times 10^{-19}$, K is Boltzann constant of $1.38 \times 10^{-23}$, $R_s$ is the resistance connected in series, $R_s$ denotes a series resistance ($\Omega$), $R_p$ is the shunt resistance across the diode ($\Omega$) which deals with shading problems, $T$ is consider the cell temperature, $n$ is a deviation factor from the ideal p–n junction diode. Two important parameters within photovoltaics are the short-circuit current $I_{sc}$ and the open-circuit voltage $V_{oc}$ [6].

The fixed solar array located at FAU is composed of Multicrystalline Silicon cells. Parameters provided by Trina Solar indicate that each solar cell has a nominal operating cell temperature (NOCT) of 45 C°. After substituting and rearranging (1), the simplified equation is as follows:

$$I = I_{ph} - I_0(e^{36.44Vd} - 1) \qquad (3)$$

## IV. Solar PV Array Model

Although, there are varies types of solar cell materials, silicon is currently a dominant material utilize in solar cells due to its scalability, momentum and efficiency in light absorption [5]. The fixed PV array has a minimum array size of 6.84kW. This PV array consists of 12 modules per string connected in series and 2 strings in parallel resulting to 24 modules per array. Each 285W solar panel contains 72 PV cells connected in series as shown in Fig. 3. The parameters of the solar module under study are shown in Table I.

TABLE I
MODULE MANUFACTURE SPECIFICATION

| Electrical specs | Value |
|---|---|
| Module efficiency-$\eta_m$ | 14.7 % |
| Power output tolerance | 0/+3 % |
| Maximum power voltage | 35.6 V |
| Maximum power current | 8.02 A |
| Open circuit voltage-$V_{oc}$ | 44.7 V |
| Short circuit current | 8.5 A |
| Peak power watts-$P_{max}$ | 285 W |

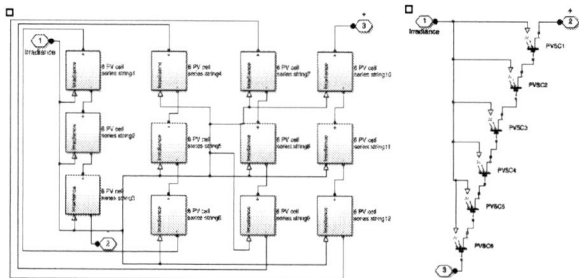

Fig. 3. (Right) Solar panel block diagram, (left) Six solar cells in series located in each subsystem

A dynamic model of the studied multicrystalline silicon cell Trina TSM 285 PA14 is presented in Fig. 4. In the I-V curve and P-V curves of the panel under 400, 600, 800 and 1000 W/m² solar radiation are presented is in Fig. 5.

The MPPT controller tracks the output power of the PV array in real time through the adjustments of the DC-DC converter. The MPP is dependent on ambient temperature and insolation shifts [14]. An algorithm is used to calculate the voltage and current of the PV output. There are varies types of algorithms; the two widely applied methods in MPPT are Perturbation and observation (P&O) [15] and incremental conductance (INC). Both methods regulate the PV voltage base on power delivered. In the P&O method since the voltage is increased the power delivered also increases, the voltage will continue to increase until the maximum power point (MPP) as shown in Fig. 6.

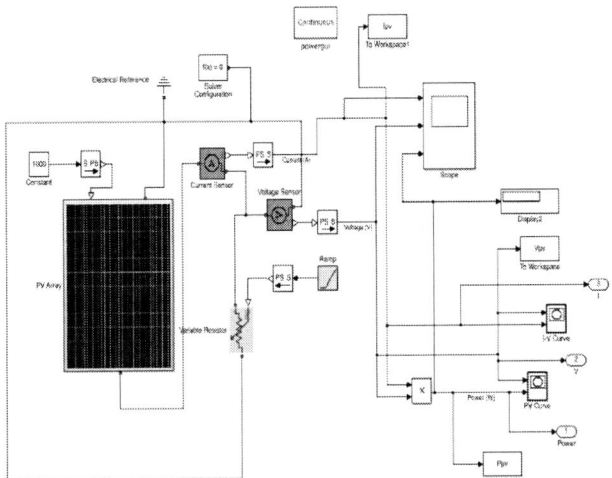

Fig. 4. The modeling of a 285W Trina solar panel

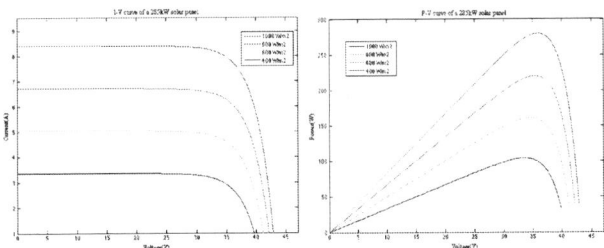

Fig. 5. I-V and P-V curves of a 285W solar model

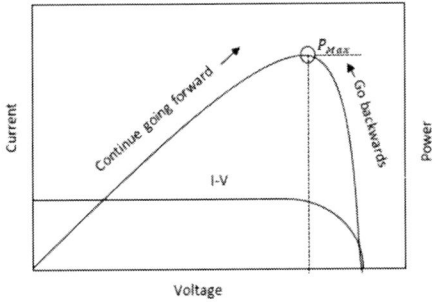

Fig. 6. Perturb - and - observe approach [7]

The incremental conductance algorithm (INC) is founded upon the fact that the MPP can be evaluated when the power versus voltage curve is equal to 0. Therefore it is determined that $dP/dV$ is equal to zero. Since $P=VI$, the MPP can be manipulated as shown in (4)-(7). The operating current and voltage is resembled in equation (1-4) as I and V. The ratio $\Delta I/\Delta V$ is consider as the incremental conductance.

$$\frac{dP}{dV} = I\frac{dV}{dV} + V\frac{dI}{dV} = I + V\frac{dI}{dV} \approx I + V\frac{\Delta I}{\Delta V} \quad (4)$$

At the MPP: $\dfrac{\Delta I}{\Delta V} = -\dfrac{1}{V}$ \quad (5)

The left of MPP: $\dfrac{\Delta I}{\Delta V} = -\dfrac{1}{V}$ \quad (6)

The right of MPP: $\dfrac{\Delta I}{\Delta V} = -\dfrac{1}{V}$ \quad (7)

The INC algorithm measures the fixed incremental change of the solar panel conductance and compares it against the instantaneous conductance, which is known to be the ratio of $I/V$. Once the to are compared, the position of the operating current and voltage changes in reference to the MPP.

## V. BATTERY STORAGE DESIGNING

The capacity and life of the battery is dependent on the rate of discharge, depth of discharge and temperature. In this particular project, lithium-ion phosphate batteries will be used due to its robust life cycle characteristics and safety record .BESS are utilized for peak shaving, peak shifting and smoothing. Smoothing is used to smooth the generated solar power fluctuation which occur during periods with transient cloud shadows on the PV array. Smoothing also regulates the battery state of charge (SOC) under proposed conditions. Moving average configuration is applied for this small-scale BEES based smoothing [16]. In this case a ±700W ramp rate will trigger the BESS to output power for smoothing.

Many industries use the statistical software Minitab which some of the features includes basic statistics, measurement analysis, regression and graphics. This program was used to create the histogram graph in Fig.7 and Fig.8. A 4kW/2kWhr lithium-ion phosphate BESS has been designed according to Fig.7 and Fig.8. The historical worst case ramp rate was used for this design criteria. The service duty of the battery in this case was found to be 3.65 kW and 0.91 kWhr. The battery has to be ready to accept charge or discharge power so a resting state of charge should be somewhere near 50%. The battery should accept up to 1 kWhr or discharge 1 kWhr thus a 2 kWhr battery was selected.

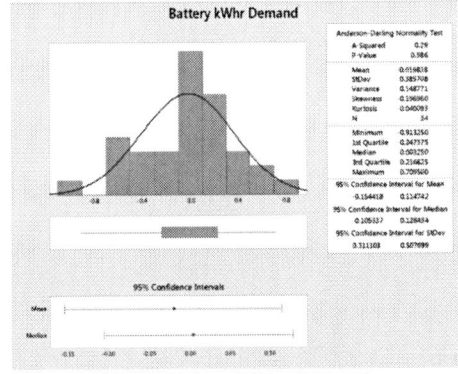

Fig. 7. Battery kWhr demand

The daily cycling on such a battery would be 5-10 partial cycles. Using vendor cycle life degradation curves it was found that cycling had less than 1% energy capacity degradation annually. A standard miner's rule fatigue calculation was used [17]. In this case, calendar degradation will dominate. The total energy capacity degradation rate will be approximately 2% annually. The battery size of 2kWhrs will support 8-10 years of the smoothing application under this scenario. Battery augmentations should be planned for year 10 to continue to serve this application. The worst-case scenario for voltage sag is during the summer in Florida. Voltage sag occurs in the occasion when the root-mean-square (rms) voltage drops below 90% nominal voltage, which causes a power disturbance. Depending on the duration of the voltage drop the voltage sag is distinguish with different titles. If the duration is 0.5-30 cycles this is consider an instantaneous sag, up to 3 seconds is consider a momentary and temporary sags the duration could go up to a minute [18].

Voltage sag is caused by abrupt increases in loads, short circuits and over-current condition, normally due to faults. In Fig. 9-11 the voltage sags are distinguished and the battery storage system is utilize to improve the effects voltage sags may have on the equipment.

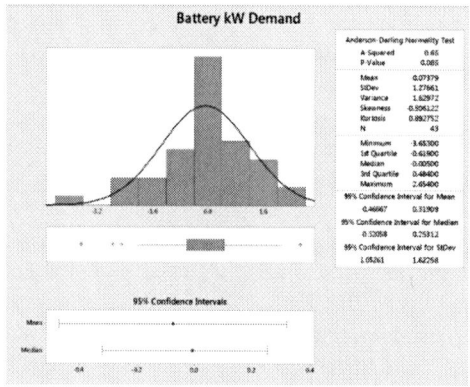

Fig. 8. Battery kW demand

Regarding the results, the battery storage smooths and shifts the system power output graph, which shows the effectiveness of BESS on system performance.

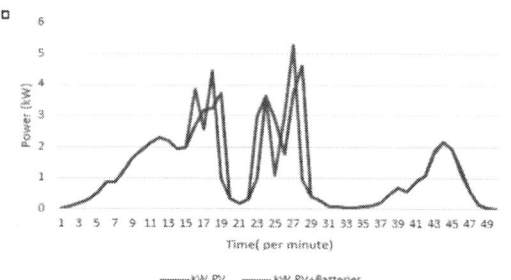

Fig. 9. PV output on June 07

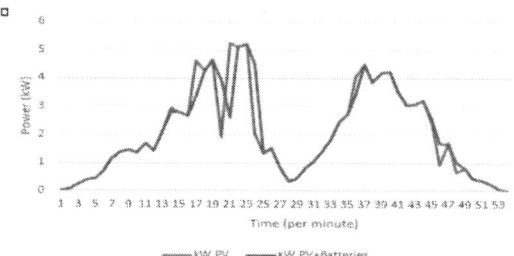

Fig. 10. PV output on June 26

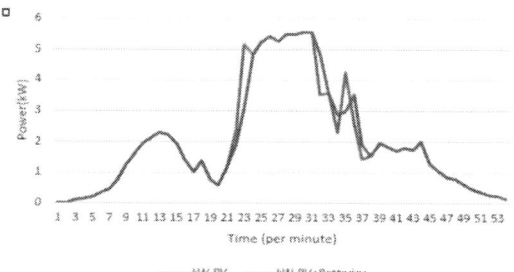

Fig. 11. PV output on August 26

## VI. SIMULATION RESULTS

The integration of the PV and battery connected to the utility grid is simulated in MATLAB/Simulink using Simscape power systems toolbox as shown in Fig. 12.

The model composed of an array of PV panels connected to the DC/DC boost convertor, MPPT controller, a battery set connected to the bidirectional DC/DC converter, a 3-phase DC/AC inverter associated with 3-phase local load and the main grid. The output from the array system is provided as input to the boost converter to produce a regulated output of 24V. In sequel, the system can be used for charging the battery or to feed a 3-phase inverter. The initial battery state of charge (SOC) is assumed to be at 100% charge capacity.

Two cases are considered for analyzing the system performance under different test conditions. First, it is assumed that the 6840W PV solar system operates with 600 W/m² radiations [19]. The system is also tested under Standard Test Condition (STC) with radiation of 1000W/m² as the second case. The model parameters are adjustable to work with various solar radiations. The simulation results of the first and second case are shown in Fig.13 and Fig.14 respectively. The 3-phase load current is fixed because of the constant load at that time. Fig.13 shows the simulation results when $P_{pv} < P_{load}$. In this case, the battery is discharged to provide the additional needed power to load, and the DC current output of the BESS is indeed positive. The main grid also works normally to feed the load deficit. In second case, $P_{pv} > P_{load}$, therefore, the PV DC current output is higher, and the battery is in charging mode. Thus, the BESS current is negative with excess power to the utility grid. As the consequence of power feeding to the grid network, the current

978-1-5090-5606-4/17 $31.00 © 2017 IEEE          2944

amplitude of both inverter and external grid are higher in comparison to the first case.

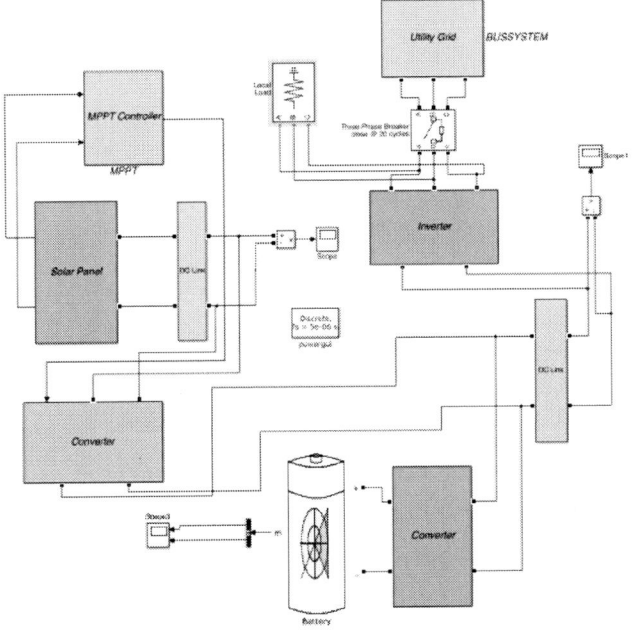

Fig. 12. Simulation model of a grid-tied PV solar system connected to battery storage

## VII. CONCLUSION

In this paper, different methods applied in MPPT were evaluated such as INC and P&O. A BESS was designed to smooth the PV array output. Based on the calculations, a lithium-ion battery with the capacity of 4kW was selected and integrated with the PV system. The results show that the PV/Battery dynamic model works properly and the system has reasonable reactions to the environmental and technical changes. In future work, other applications of battery storage will be studied such as peak shifting and voltage support. Also PV/BESS systems have the potential to be applied on residential rural areas in developing countries, where there is not much power consumption due to the absence of AC units, pool pumps and other appliances. In developing regions such as Africa, South East Asia, Latin America and the Middle East, 1,186 million people have no access to electricity. Applying renewable power allows energy service to be reliable, affordable and environmental friendly. Currently, rural committees are obligated to use polluting energy sources such as heating oil, wood and coal. Also, by improving the energy necessity, developing countries can achieve a long-term economic prosperity. This will reduce workload, increase jobs and improve education.

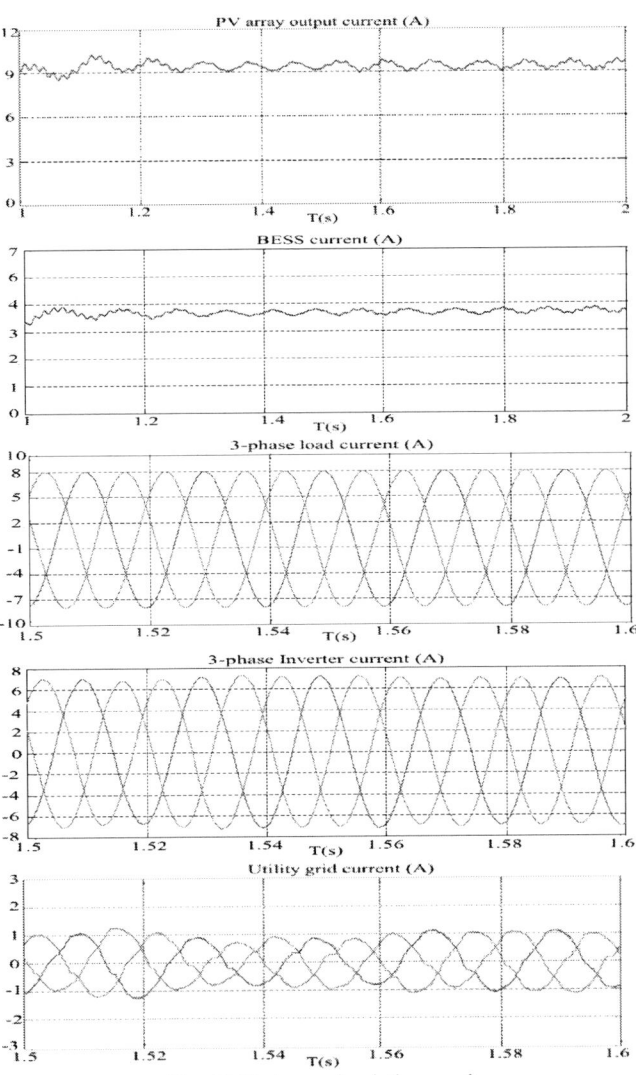

Fig. 13 First case simulation results

### ACKNOWLEDGMENTS

The authors would like to thank Dr. Vichate Ungvichian for his valuable comments and also Florida Power and Light (FPL) Company for financial and technical supports through this project.

### REFERENCES

[1] Y. Zhao and H. Khazaei, "An incentive compatible profit allocation mechanism for renewable energy aggregation," *2016 IEEE Power and Energy Society General Meeting (PESGM)*, pp. 1-5, Boston, MA, 2016.

[2] A. S. Mobarakeh, A. Rajabi-Ghahnavieh and A. Zahedian, "A game theoretic framework for DG optimal contract pricing," *IEEE PES ISGT Europe 2013*, pp. 1-5, Lyngby, 2013.

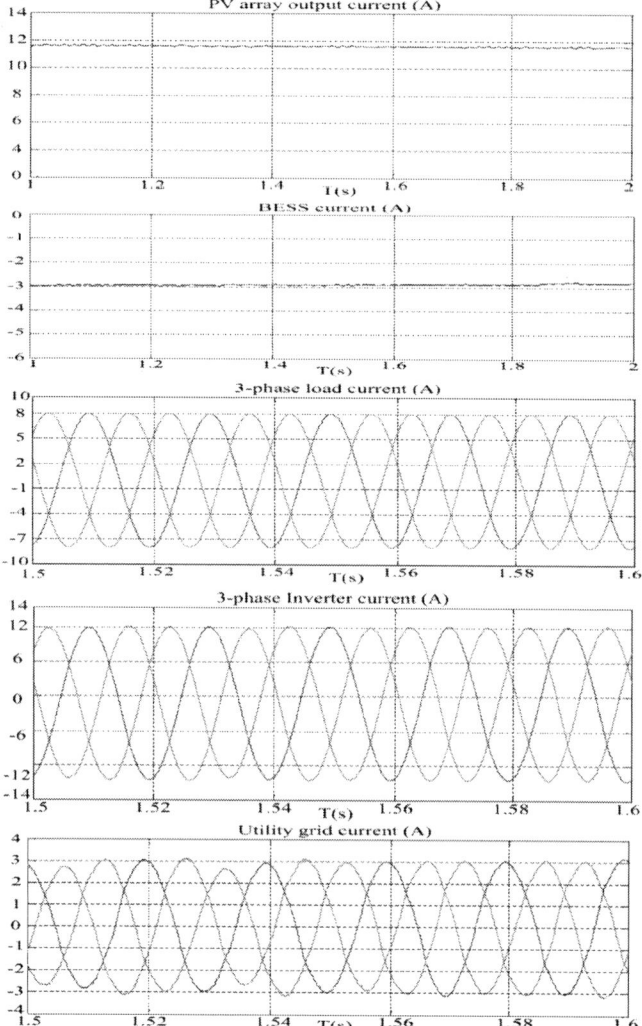

Fig. 14. Second case simulation results

[3] M. Farajollahi, M. Fotuhi-Firuzabad, A. Safdarian, "Deployment of Fault Indicator in Distribution Networks: A MIP-Based Approach," *IEEE Transactions on Smart Grid*, vol.PP, no.99, pp.1-9.

[4] M. E. Raoufat, A. Khayatian, A. Mojallal, "Performance Recovery of Voltage Source Converters With Application to Grid-Connected Fuel Cell DGs," *IEEE Transactions on Smart Grid*, vol.PP, no.99, pp.1-8.

[5] A. Shahsavari, A. Sadeghi-Mobarakeh, E. Stewart and H. Mohsenian-Rad, "Distribution grid reliability analysis considering regulation down load resources via micro-PMU data," *2016 IEEE International Conference on Smart Grid Communications (SmartGridComm)*, pp. 472-477, Sydney, NSW, 2016.

[6] H. R. Sadeghian and M. M. Ardehali, "A novel approach for optimal economic dispatch scheduling of integrated combined heat and power systems for maximum economic profit and minimum environmental emissions

based on Benders decomposition, " *Energy*, vol. 102, pp. 10– 23, 2016.

[7] M. H. Athari, Z. Wang, and S.H. Elyas, "Time-Series Analysis of Photovoltaic Distributed Generation Impacts on a Local Distributed Network," *2017 IEEE PowerTech Conference*, pp. 1–6, 2017.

[8] H. Sadeghian, M. H. Athari and Z.Wang, "Optimized Solar Photovoltaic Generation in a Real Local Distribution Network," *2017 IEEE Innovative Smart Grid Technologies Conference (ISGT)*, 2017.

[9] L. Cadavid, M. Jimenez and C. J. Franco, "Financial analysis of photovoltaic configurations for Colombian households," *IEEE Latin America Transactions*, vol. 13, no. 12, pp. 3832-3837, Dec. 2015.

[10] S. Duryea, S. Islam and W. Lawrance, "A battery management system for stand-alone photovoltaic energy systems," *IEEE Industry Applications Magazine*, vol.7, no. 3, pp. 67-72, Jun 2001.

[11] E. B. Ssekulima and A. A. Hinai, "Coordinated voltage control of solar PV with MPPT and battery storage in grid-connected and microgrid modes," *18th Mediterranean Electrotechnical Conference (MELECON)*, pp. 1-6, Lemesos 2016.

[12] L. Tang, W. Xu, C. Zeng, J. Lv and J. He, "One novel variable step-size MPPT algorithm for photovoltaic power generation," *IECON 2012 - 38th Annual Conference on IEEE Industrial Electronics Society*, pp. 5750-5755, Montreal, QC 2012.

[13] J. L. Gray, *The Physics of the Solar Cell*, in Handbook of Photovoltaic Science and Engineering, eds A. Luque, S. Hegedus, Manchester, UK: John Wiley & Sons, 2010.

[14] Renewable and Efficient Electric Power Systems, Gilbert M. Masters, Second Edition, ISBN: 978- 1-1181-4062-8, 2013, Wiley-IEEE Press

[15] M. A. G. de Brito, L. Galotto, L. P. Sampaio, G. d. A. e Melo and C. A. Canesin, "Evaluation of the Main MPPT Techniques for Photovoltaic Applications," in *IEEE Transactions on Industrial Electronics*, vol. 60, no. 3, pp. 1156-1167, March 2013.

[16] D. Shwetha and S. Ramya, "Comparison of smoothing techniques and recognition methods for online Kannada character recognition system," *2014 International Conference on Advances in Engineering & Technology Research (ICAETR - 2014)*, pp. 1-5, Unnao, 2014.

[17] M. Safari, M. Morcrette, A. Teyssot, and C. Delacourt, "Life-Prediction Methods for Lithium-Ion Batteries Derived from a Fatigue Approach: I. Introduction: Capacity-Loss Prediction Based on Damage Accumulation," *J. Electrochem. Soc.*, vol. 157, no. 6, pp. A713–A720, Jun. 2010.

[18] IEEE Guide for Voltage Sag Indices, in IEEE Std 1564-2014, pp.1-59, June 20, 2014.

[19] H. Moradi, A. Abtahi and R. Messenger, "Annual performance comparison between tracking and fixed photovoltaic arrays," *2016 IEEE 43rd Photovoltaic Specialists Conference (PVSC)*, pp. 3179-3183, Portland, OR, 2016.

# >94.5% Reduction in Grid-Buy Electricity and Elimination of AM & PM Energy Peaks/Spikes by Optimizing Energy Usage and Integration of Customer Self-Supply Rooftop Solar PV with Electrical & Thermal (Hot & Cold) Storage Batteries: A Case Study for Residential Hawaii

John Borland[1], Jay Moore[2], Corpuz Poncho[2], Takahiro Tanaka[3] and Harumi McClure[3]

[1]J.O.B. Technologies, 98-1204 Kuawa St, Aiea, Hawaii, 96701, USA
[2]Poncho's Solar, 650 Kakoi St., Honolulu, Hawaii, 96819, USA
[3]Tabuchi Electric America, 5225 Hellyer Ave, Suite 150, San Jose, California, 95138, USA

*Abstract*--- We investigated the integration of customer self-supply rooftop solar PV system with electrical battery storage and hot & cold thermal storage for residential Hawaii. Optimizing time of use for key appliances and improvements to hardware and software control system, we reduced the average daily Grid-Buy from 48.7kWh/day in April 2016 to 2.7kWh/day in April 2017 and eliminated all AM and PM energy peaks. For 12 days in April we achieved 0.0kWh/day Grid-Buy.

*Index Terms*--- grid-buy, battery storage, thermal storage, hot thermal storage, cold thermal storage, solar PV, self-supply.

## I. INTRODUCTION

At the 43rd IEEE PVSC meeting in June 2016, Reindl gave the talk "LCOE reduction of PV electricity—does technology still matter?" based on end-users perspective the cost of PV generated electricity is not determined by the PV cell technology efficiency but rather dominated by climate and environmental energy yield effects [1]. This is especially true with the end of Net Energy Metering (NEM) and export/selling of excess rooftop solar PV generated energy back to the utility Grid as is the case for residential Hawaii when NEM ended in Oct 2015. This impact can be seen in the past year as the US solar PV market has grown 95% from 7.5GW in 2015 to 14.6GW in 2016 while Hawaii has fallen 31% from its peak of 130MW in 2013 to 99MW in 2016. Future growth in the residential solar market will no longer drive the solar industry to higher solar cell efficiency because this results in the loss of PV energy generation due to dumping/curtailment between 27-48% per month based on NREL's System Advisor Model for Honolulu [2]. For customer self-supply in a post-NEM world, lower price packaging of smaller solar-PV systems integrated with battery storage (electrical & thermal) and optimized

energy usage of key household appliances will drive the next wave of residential solar deployment by maximizing the energy generation of smaller solar PV system and reducing utility Grid-Buy electricity to the best case of 0.0kWh/day. Therefore, we will report field data starting from the June 1, 2016 installation of the rooftop solar system + battery storage and steps we took to achieve the reduction in the daily utility Grid-Buy electricity going from a daily average of 48.7kWh/day for April 2016 to 2.7kWh/day for April 2017 as shown in Fig.1. Note that initially after the first month of solar PV + battery storage we only achieved a 53% reduction in Grid-Buy to 22kWh/day which was much less than expected. The Hawaii Energy website reports that the average Hawaii home with rooftop solar PV and NEM (the utility acts as a storage battery) sees a 56% reduction in utility bill from 537kWh/month to 234kWh/month [3].

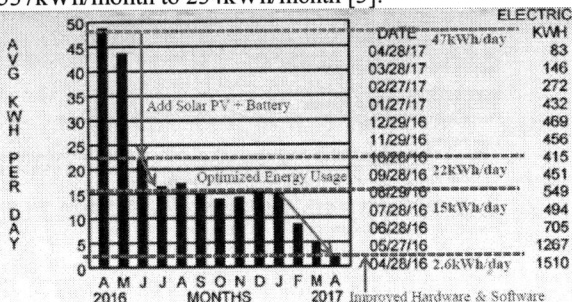

Fig.1. Hawaiian Electric April 2017 home utility bill.

## II. EXPERIMENTATION

This residential Hawaii case study is for a Poncho's Solar installed 7kWh (27 cSi modules) rooftop solar PV + Panasonic 10kWh Li-ion battery for storage and a solar thermal hot water system as shown in Fig.2. The solar PV system is controlled by a Tabuchi Electric 5.5kWh inverter that is tied to the Grid in a customer self-supply mode with no export back to the Grid due to Hawaii's post-NEM. The solar thermal hot water heater has two solar panels, the 2nd panel was installed on Nov 3, 2016 to boost water heating temperature of the 80 gallon solar hot water storage tank to >170°F (super-charged hot thermal storage battery) and on Jan 18, 2017 a 2nd hot water storage tank (40 gallons) was added in series to increase the hot thermal storage battery capacity from 80 to 120 gallons to provide a full day supply of hot water when fully thermally charged for the 4 baths/day each requiring 40 gallons of hot water. For cold thermal storage we used daytime solar PV generation to power the 3 to 7 portable air conditioning (AC) units around the house. This would chill selected rooms around the house from daytime high of >85°F to <69°F avoiding the need for running the AC after sunset or returning home after work.

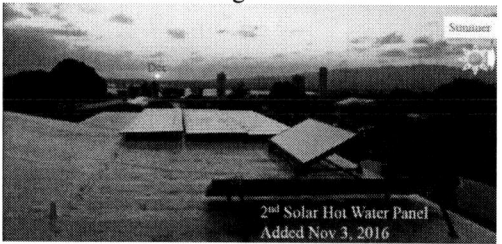

Fig.2. Poncho's Solar installed 27 cSi solar PV modules and two solar thermal hot water panels.

## III. RESULTS

### A. Max Solar PV Energy Generation:

For post-NEM, the only way to monitor the solar PV system performance is by creating a total system load/demand >7.0kW/h with 1.5kW/h for maximum battery charging and 5.5kW/h for maximum inverter DC to AC energy conversion. This was achieved by running the clothes dryer around 12:30PM and the results for the past 11 months are shown in Fig.3. The peak solar PV energy generation varied from a high of ~6.6kW/h mid-Jan to mid-Oct to a lower high of 5.6kW/h for Nov and Dec, a 16.5% output difference. The reported yearly sun radiance variation for Honolulu Hawaii at 1PM is 0.6-1.1kW/m² or 45.5% difference as shown in the Fig.3 insert. The 16.5% shift we observed is less than sun radiance variation and since the panels were never cleaned this result also includes soiling effects.

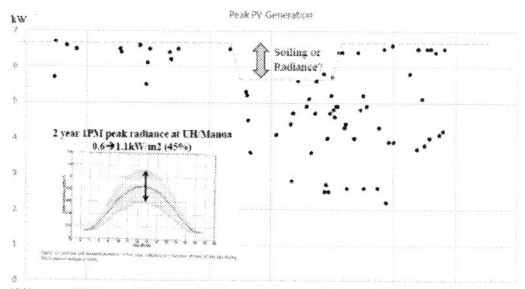

Fig.3. Past 11 months peak solar PV energy generation.

To determine the amount of solar PV generation dumping/loss and curtailment each month, we used the PVWATTS.NREL.GOV System Advisor Model simulation results based on data for Honolulu as shown in Fig.4 [2]. This shows a monthly average of ~30% solar PV energy dumping throughout the past year and the months with ~45% correspond to the 1+ week vacation mode out of town months (July, Sep, Oct & Dec).

| Month | Solar Radiation (kWh / m² / day) | | AC Energy (kWh) | Ene |
|---|---|---|---|---|
| | | | | Actual PV Results |
| January | 4.52 | =23.9kWH/Day | 584 | →383kWH (65.5%) |
| | | | | =12.4kWH/Day |
| February | 5.20 | =27.7kWh/day | 609 | →374kWh (61.4%) |
| | | | | =13.34kWh/day |
| March | 5.74 | =30.3kWH/day | 738 | →489kWh (66.2%) |
| | | | | =15.8kWh/day |
| April | 5.89 | =31.2kWh/day | 735 | →512kWh (69.7%) |
| | | | | =17.1kWh/day |
| May | 6.32 | =33.3kWH/day | 811 | →499kWh (61.5%) |
| | | | | =16.1kWh/day |
| June | 6.36 | =33.4kWH/Day | 786 | →560kWH (71.2%) |
| | | | | =18.6kWH/Day |
| July | 6.41 | =33.6kWH/Day | 818 | →461kWH (56.4%) |
| | | | | =14.8kWH/Day |
| August | 6.46 | =33.7kWH/Day | 820 | →599kWH (73.0%) |
| | | | | =19.3kWH/Day |
| September | 6.32 | =33.0kWH/Day | 777 | →452kWH (58.2%) |
| | | | | =15.1kWH/Day |
| October | 5.41 | =28.6kWH/Day | 697 | →360kWH (51.6%) |
| | | | | =11.6kWH/Day |
| November | 4.71 | =25.0kWH/Day | 589 | →368kWH (62.5%) |
| | | | | =12.5kWH/Day |
| December | 4.40 | =23.3kWH/Day | 568 | →309kWH (54.4%) |
| | | | | =10.0kWH/Day |
| Annual | 5.65 | | 8,532 | |

Fig.4. NREL solar PV generation simulation results for Honolulu.

### B. Energy Usage Monitoring for Grid-Buy Reduction

The typical home energy usage for residential Honolulu as reported by Hawaiian Electric is #1 electrical water heater at 40%, #2 refrigerator/freezer at 15%, #3 air conditioner (AC) at 12%, tied for #4 are clothes dryer, cooking and lighting at 8% and #7 dishwasher at 3%. In order to reduce Grid-Buy energy usage, daily energy usage monitoring with 1 minute and not 1 hour data interval resolution is needed. Fig.5 shows the Tabuchi Electric remote wall solar system monitor in real-time with data refresh every 5 seconds for June 12, 2016. Energy demand usage at 1:16PM was 12.0kW/h with solar PV generation of 6.37kW/h, Grid-Buy electricity of 7.22kW/h and battery charge of 1.50kW/h. Fig.6 shows the web-link data using the Laplace software analysis in real-time with hourly updates and 1 hour data averaging.

978-1-5090-5606-4/17 $31.00 © 2017 IEEE 2948

The 1PM data shows demand was only 9kW/h with PV generation of <5kW/h which is much lower than the data in Fig.5. Also, the state of battery charge plot shows the 2nd battery discharge that occurs between 9AM and 11AM. Due to the poor resolution of hourly data monitoring no information for each critical household appliance energy usage can be determined. Therefore we used off-line Excel data analysis of the Laplace solar PV monitoring system that provides daily data with 1 minute interval resolution as shown in Fig.7. Now we can identify each energy usage spike including #1 clothes dryer at 6kW/h, #2 the 7 portable AC units at 5.5kW/h, #3 the electrical hot water heater at 4.5kW/h, #4 refrigerator/freezer at 300-800W/h, #5 pool pump at 200-2000W/h and cooking lunch/dinner at 4.5kW/h.

Fig.5. Tabuchi Electric remote wall solar system monitor in real-time with data refresh every 5 seconds.

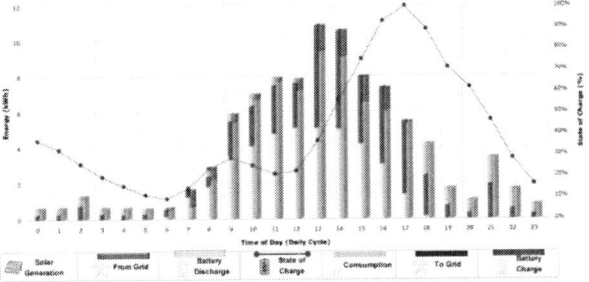

Fig.6. Laplace solar PV monitoring with 1 hour data resolution.

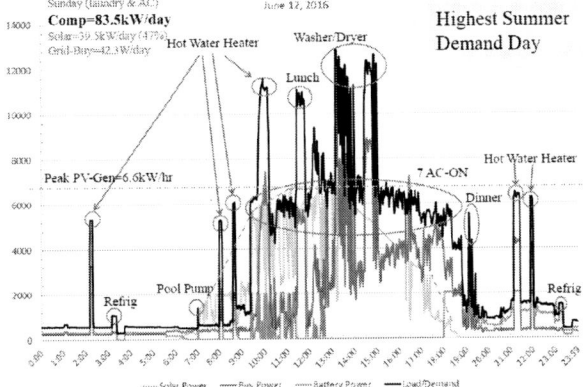

Fig.7. Off-line excel data analysis of the Laplace collected data with 1 minute resolution.

## C. Hot Thermal Storage to Eliminate AM & PM Grid-Buy Energy Peaks/Spikes

The #1 dryer and #2 AC energy usage appliances can be scheduled so their time-of-use (TOU) coincides with the daytime peak solar PV energy generation as shown in Fig.7 leaving the #3 electrical hot water heater as the key contributor to the evening and morning Grid-Buy energy spikes and peaks. Results for rainy day Nov. 7, 2016 is shown in Fig.8 with a Grid-Buy of 21.0kWh/day and 5 hot water heater energy spikes throughout the day for heating the water from 110°F to 135°F at 5:30AM for the 1st morning shower/bath, again at 2:30PM because solar thermal heating was insufficient for heating water after the 2nd morning shower/bath above 105°F requiring both Grid-Buy and battery discharge to reach 135°F, again at 5:30PM, 7PM after the 1st evening bath/shower and 9:30PM after the 2nd evening bath/shower. Nov. 5 & 6, 2016 were sunny days and the solar thermal panels heated the water to >157°F for full hot thermal charging (super charging) of the 80-gallon tank thermal storage battery so no Grid-Buy electricity was needed for morning and evening hot water heating for these 2 days as shown in Figs. 9 & 10 for Laplace and Bidgely monitoring systems. The Bidgely energy monitor is from Hawaii Blue Planet Foundation with real-time 5 minute interval energy usage collection and it also records the outside air temperature. Fig.11 shows the daily peak water temperature for the 80 gallon (up to 178°F) and 40 gallon (up to 172°F) hot thermal storage batteries and after the morning AM shower/bath (3rd hot water discharge). The 80 gallon tank drops by -35°F and the 40 gallon tank by -45°F.

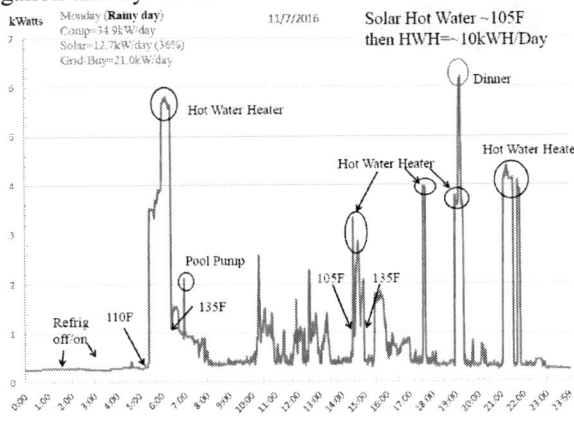

Fig.8. Laplace energy usage monitor for rainy day.

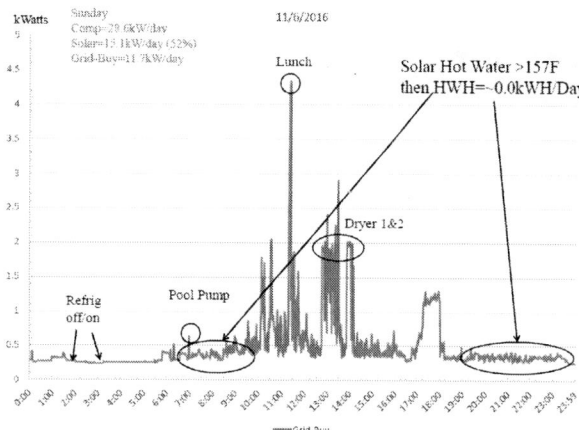

Fig.9. Laplace energy usage monitor for sunny day.

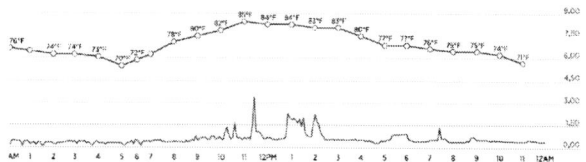

Fig.10. Bidgely energy usage monitor for sunny day.

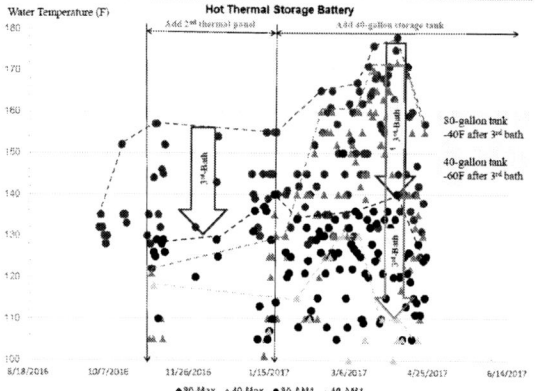

Fig. 11. Water temperature of Hot thermal storage battery.

### D. Cold Thermal Storage to Eliminate PM Grid-Buy

Grid-Buy energy usage for room/house AC cooling for late afternoon or early evening especially after returning home from work can be significantly reduced or eliminated by using the daytime solar PV energy generation to power the room/house AC units during the day since this energy is free and not being exported back to the Grid it would reduce PV energy dumping/curtailment. This reduces the room temperature from a daytime high of >85°F to <69°F. Operating the solar control system in the Max Solar Energy Mode with all 7 portable AC units on for 6 hours resulted in a Grid-Buy of 5.1kWh/day as shown in Fig.12 but if the number of AC units on is reduced to 3 as shown in Fig.13 then Grid-Buy can be reduced to 0.0kWh/day. Fig.14 shows 12 days in April with 0.0kWh/day Grid-Buy.

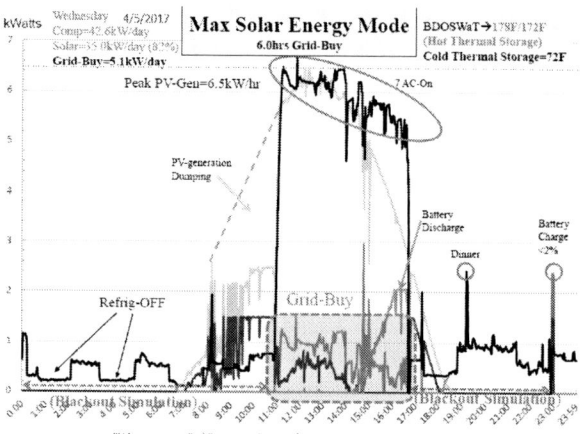

Fig.12. Max Solar Energy mode with 7 AC units on.

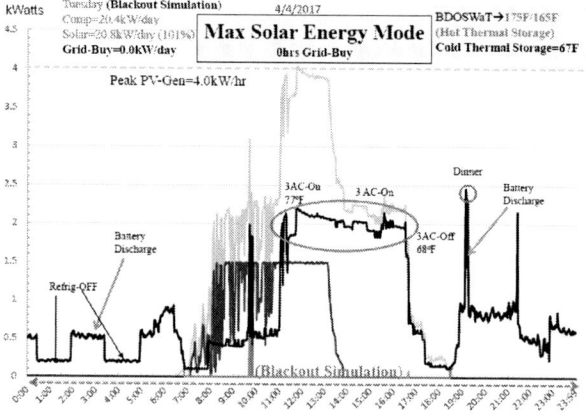

Fig.13. Max Solar Energy mode with 3 AC units on.

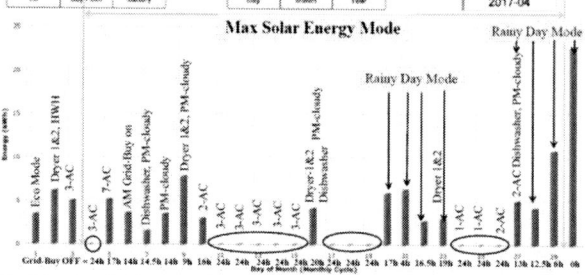

Fig.14. April 2017 Grid-Buy =0.0kWh/day for 12 days.

## IV. SUMMARY

We investigated the integration of customer self-supply rooftop solar PV system with electrical battery storage and hot & cold thermal storage for post-NEM residential Hawaii. Optimizing time of use for key appliances and improvements to hardware and software control system we reduced the average daily Grid-Buy by >94.5% from 48.7kWh/day to 2.7kWh/day eliminating the AM and PM Grid-Buy energy peaks/spikes and also achieving 12 days in April with 0.0kWh/day Grid-Buy.

## REFERENCES

[1] T. Reindl, "LCOE reduction of PV electricity—does technology still matter?", 43rd IEEE-PVSC 2016, area 8 plenary talk not published.

[2] www.pvwatts.nrel.gov website.

[3] www.Hawaiienergy.com website.

# A Single-Stage Ćuk-based Transformerless Inverter for 1-φ Grid-Connected PV Systems

Phani Kumar Chamarthi[1], Amit Kumar Gupta[1], Madhuwanti S. Joshi[2], and Vivek Agarwal[1]

[1]Indian Institute of Technology Bombay, Mumbai, India, [2]Integra Power LLC, USA

*Abstract* — In this paper, a Ćuk based single-stage, single-phase transformerless inverter topology is proposed for grid-connected PV applications with inherent buck-boost capability. The proposed topology is presented along with a dedicated proposed modulation scheme. Since the negative terminal of the PV source is directly connected to the neutral of the grid, the proposed transformerless topology incurs zero leakage current, making it highly suitable for grid-tied solar PV applications. Only one input inductor is used for both the positive and negative half cycles, which along with symmetrical operation helps in eliminating DC current injection into the grid. The topology has five switches, out of which, two switches operate at line frequency. Therefore, higher efficiency is achieved due to lower switching loss. The PWM control and operating modes of the inverter are presented along with the simulation results. The proposed topology is validated experimentally and those results also included in the paper.

*Index Terms* — Ćuk converter, grid-connected inverter, leakage current, single-stage inverter, sinusoidal PWM, solar PV inverter, transformerless inverter.

## I. INTRODUCTION

Eliminating a transformer from the inverter gives several advantages like high efficiency, low cost, lower weight and size. But the transformerless inverters have a few challenges like an injection of DC current, leakage current issue, etc. These issues have drawn the attention of many researchers, as a result of which, several transformerless topologies have been proposed in recent year for grid-connected PV applications, which ensure that the leakage current and DC current injection are kept under the limit imposed by the related standards. Limit on the DC current injection is 0.5% to 1% of the rated current, which is imposed by IEEE 929-2000 and IEC 61727 [1]-[3]. German standard DIN VDE0126-1-1 specifies the value of the leakage current beyond which the system must be disconnected within the specified time limit [4], [5]. String or central inverters have a string of the PV panels at the input, but low voltage PV source necessitates a boosting stage for grid-connected applications, which reduces the overall efficiency of the system [6], [7].

Several buck based transformerless topologies have been proposed earlier, which comply with the standards. Some of the most popular are HERIC, FB-DCBP and H5. Due to buck operation, they cannot work with low PV voltage without a boost stage [8], [9]. There are some boost topologies available, but despite several benefits, they cannot track the full MPPT curve especially at higher PV voltage [10]. Researchers are also focusing on buck-boost based topologies, which can track

MPP even in wide variation in PV voltage. Topologies proposed in [11] and [12] need two separate PV sources for each cycle of the grid voltage. The number of switches used in buck-boost topologies proposed in [13] is six and above. The asymmetrical operation during the two half cycles may have DC current injection problem as in doubly grounded topology [1]. The buck-boost inverter topologies described in [14]-[16] have an input inductor in series with the PV source to improve the MPPT performance and to reduce the input DC capacitor value. But the topologies in [14], [15] have an asymmetrical operation during the positive and negative half cycles. The topology in [16] has an symmetrical operation for the two half cycles, but it requires six power switches.

This paper proposes a novel Ćuk converter based PV inverter topology. The topology is capable of working with low as well as high input voltage, which is very helpful in PV based applications due to their intermittent nature. Two out of a total of switches operate at low frequency (50/60 Hz) and the pulses for high frequency switches are generated based on logic operations after sine-triangular comparison as shown in a later section. PV source's negative terminal and grid neutral share common ground, which ensures zero leakage current.

## II. THE PROPOSED ĆUK BASED PV INVERTER TOPOLOGY

The grid connected proposed Ćuk based PV inverter (CPVI) topology is shown in Fig. 1. In this CPVI, the negative terminal of PV source is directly connected to the neutral point of the grid, which completely eliminates the variation in common mode voltage, resulting in zero leakage current.

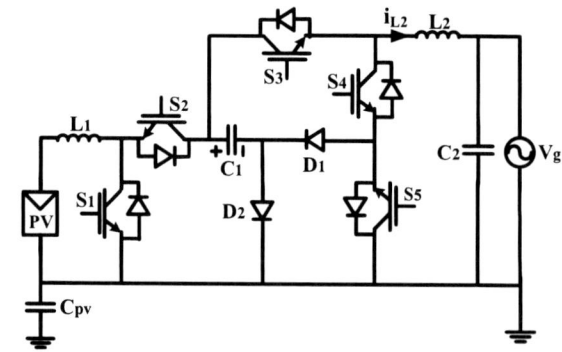

Fig. 1. The proposed grid connected Ćuk based PV Inverter.

The proposed CPVI topology consists of five controlled switches ($S_1$ to $S_5$) and two diodes ($D_1$ and $D_2$). Out of which,

978-1-5090-5606-4/17 $31.00 © 2017 IEEE

$S_1$, $S_2$, and $S_3$ operate at high frequency whereas $S_4$ and $S_5$ operate at low frequency (50/60 Hz). The input inductor ($L_1$) reduces the input current ripple, which in turn reduces the size of the input electrolytic capacitor connected across the PV source. Capacitor $C_1$ is used as an intermediate energy storage element, and $L_2$, $C_2$ act as output low pass filter.

### A. Modes of operation

The proposed CPVI topology has four modes of operation as shown in Figs. 2(a)-(d). The modes corresponding to positive half cycle are shown in Figs. 2(a) and (b), where $S_4$ is OFF and $S_5$ is ON continuously. The modes corresponding to negative half cycle are shown in Figs. 2(c) and (d), where $S_4$ is ON, and $S_5$ is OFF continuously.

***Mode-1***: This mode begins with the turning ON of the switches $S_1$, $S_3$ and $S_5$. During this mode, $L_1$ is charged through the switch $S_1$, and an auxiliary capacitor $C_1$ is discharged through $S_3$, $L_2$, $C_2$, grid, $S_5$ and $D_1$ respectively. Thick lines in Fig. 2(a) indicate the conducting paths of the circuit.

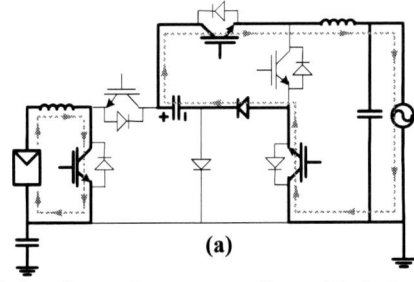

Fig. 2(a). Modes of operation corresponding to Mode-1.

***Mode-2***: This mode begins with the turning off the switches $S_1$, $S_3$ and by keeping the switch $S_5$ as it is. During this mode of operation, $L_1$, is discharged through $C_1$ and $D_2$, which in turn charges the auxiliary capacitor $C_1$. The output inductor $L_2$ is discharged (freewheeling mode) through $S_5$, $C_2$ and grid. Thick lines in Fig. 2(b) indicate the conducting paths of the circuit.

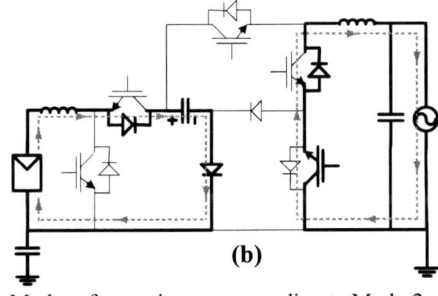

Fig. 2(b). Modes of operation corresponding to Mode-2.

***Mode-3***: This mode begins with the turning ON of the switches $S_1$, $S_2$ and $S_4$. During this mode, $L_1$ is charged through the switch $S_1$, and auxiliary capacitor $C_1$ is discharged through $S_2$, $S_1$, $C_2$, grid, $L_2$, $S_4$ and $D_1$ respectively. Thick lines in Fig. 2(c) indicate the conducting paths of the circuit.

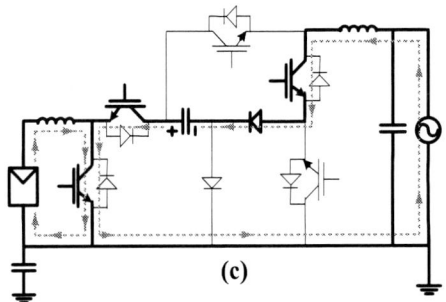

Fig. 2(c). Modes of operation corresponding to Mode-3.

***Mode-4***: This mode begins with the turning OFF of the switches $S_1$, $S_2$ and by keeping the switch $S_4$ as it is. During this mode of operation, $L_1$ is discharged through $C_1$ and $D_2$, which in turn charges the auxiliary capacitor $C_1$. The output Inductor $L_2$ is discharged (freewheeling mode) through $S_4$, $C_2$ and grid. Thick lines in Fig. 2(d) indicate the conducting paths of the circuit.

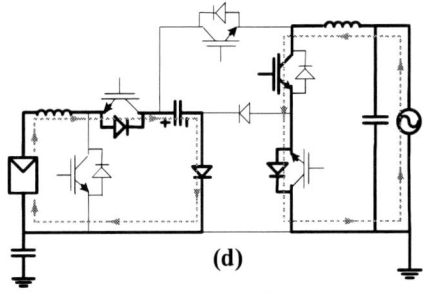

Fig. 2(d). Modes of operation corresponding to Mode-4.

The switching states corresponding to the various modes of operation are given in Table-I. The proposed CPVI has only one energy storage inductor at the input and symmetrical operation in both the half cycles of the grid voltage, which eliminates the DC current injection into the grid.

TABLE I
SWITCHES STATES CORRESPONDING TO THE VARIOUS MODES

|  | Switches states (1=ON, 0=OFF) | | | | | Mode |
|---|---|---|---|---|---|---|
|  | $S_1$ | $S_2$ | $S_3$ | $S_4$ | $S_5$ | |
| +ve half cycle | 1 | 0 | 1 | 0 | 1 | 1 |
| | 0 | 0 | 0 | 0 | 1 | 2 |
| -ve half cycle | 1 | 1 | 0 | 1 | 0 | 3 |
| | 0 | 0 | 0 | 1 | 0 | 4 |

### B. Design of the power components

The design of the power components are given in the following sub-section.

#### 1) The design of the input inductor ($L_1$)

An input inductor ($L_1$) is designed for a power rating of 1 kW at a switching frequency $f_s$ =10 kHz (i.e. $f_s = 1/T_s$). It has been assumed that $L_1$ operates in just discontinuous conduction mode and modulation index $m_I$ =1. Inductor $L_1$

stores energy from the input source when switch $S_1$ is turned on. Therefore, voltage across $L_1$ is given as follows [17]:

$$V_{in} = L_1 (di_{L_1}/dt) \qquad (1)$$

$$L_1 = V_{in}(\Delta T/\Delta i_{L_1}) = V_{in}(m_I T_S/I_{L_{1\max}}) \qquad (2)$$

Where $V_{in}$ is the source voltage and $I_{L_1\max}$ is the peak inductor current. The calculated value of $L_1$ is obtained as 100 µH.

*2) The design of the output capacitor ($C_2$)*

The value of the output capacitor is calculated by equating the energies delivered by the input source ($E_{in}$) and the input inductor ($E_{L_1}$) with the energy transferred to the output capacitor ($E_{C_2}$). The energy balance equation is given as follows:

$$E_{in} + E_{L_1} = E_{C_2} \qquad (3)$$

$$V_{in}I_{in}T_S + 0.5L_1I_{L_1\max}{}^2 = 0.5C_2V_{C_2}{}^2 \qquad (4)$$

From (4), the expression to calculate the output capacitor is obtained as given below:

$$C_2 = \frac{2}{V_{C2}^2}(V_{in}I_{in}T_S + 0.5L_1I_{L_{1\max}}^2) \qquad (5)$$

The calculated value of $C_2$ is 10 µF.

*3) The design of the output inductor ($L_2$)*

The equation for designing the output inductor ($L_2$) is given as follows:

$$L_2 = \frac{1}{C_2(2\pi f_{cut})^2} \qquad (6)$$

The cutoff frequency ($f_{cut}$) is suitably chosen (i.e. $1/10^{th}$ of the switching frequency). The calculated value of $L_2$ is 1 mH.

*4) The design of the auxiliary capacitor ($C_1$)*

The auxiliary capacitor ($C_1$) is calculated by using the following equation:

$$C_1 = \frac{1}{\omega_r^2(L_1 + L_2)} \qquad (7)$$

The resonance frequency ($\omega_r$) of $L_1$, $L_2$ and $C_1$ is chosen in such a way that $\omega_r$ is greater than the grid frequency and lesser than switching frequency (i.e. 50 Hz $<<\omega_r<f_s$). In accordance with the above-given requirements, $C_1$ is chosen to be 47 µF for a resonance frequency $\omega_r=2\times\pi\times2000$.

### III. The Modulation and Control Strategies

A modulation strategy is proposed to control the proposed CPVI. The current control strategy is used to control the power fed into the grid [1].

*Proposed modulation strategy*

In this modulation strategy, a modulating wave $V_m \sin\omega t$ (reference wave) is compared with the high-frequency carrier wave $V_{tri}$ as shown in Fig. 3. After getting the pulses from the sine-triangular comparison, some logic operations need to be performed based on the proposed modulation strategy, which

eventually generates the gate pulses for the switches ($G_{S_1}$ to $G_{S_2}$) as shown in Fig. 4. The modulation index of the proposed CPVI is defined as

$$\frac{V_0}{V_{in}} = \frac{m_I}{1-m_I} \qquad (8)$$

where $m_I$ is the amplitude modulation index, $V_{in}$ is the supply voltage and $V_0$ is the output voltage.

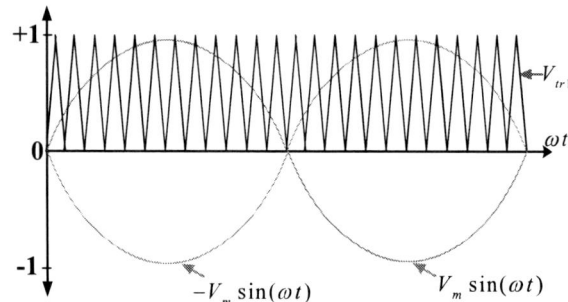

Fig. 3. Modulation strategy of the CPVI.

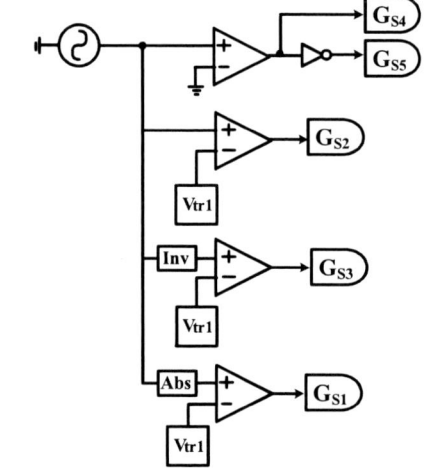

Fig. 4. Proposed modulation logic.

The proposed CPVI topology is compared with the existing buck-boost transformerless inverters [18]. Many existing transformerless topologies don't have an input inductor which increases the input current ripple. This, in turn, increases the size of the electrolytic capacitor at the input. The

TABLE II

COMPARISON OF THE PROPOSED CPVI TOPOLOGY WITH EXISTING BUCK-BOOST TRANSFORMERLESS TOPOLOGIES

| System Parameters | Proposed | Ref [1] | Ref [7] | Ref [10] |
|---|---|---|---|---|
| Input Capacitance | Less | High | Less | High |
| Number of switches | 5 | 5 | 4 | 4 |
| Number of diodes | 2 | 3 | 0 | 2 |
| Number of inductors | 2 | 3 | 3 | 2 |
| Number of capacitors | 2 | 2 | 2 | 1 |
| DC offset | No | No | Yes | Yes |
| % THD | 3.43 | 4.9 | <5 | 6 |

978-1-5090-5606-4/17 $31.00 © 2017 IEEE

proposed CPVI topology consists of an input inductor which reduces the input capacitor size and increases the reliability as well. The filtering requirement is further reduced because of the presence of output inductor and capacitor at the grid side. The detailed comparison of proposed CPVI with the existing buck-boost topologies is given in Table II.

## IV. SIMULATION RESULTS

The simulations are carried out in MATLAB-Simulink for the proposed grid connected CPVI topology at 1 kW power rating. The parameters considered for the simulation studies are $V_{in}$ =300 V, $L_1$=100 µH, $C_1$=47 µF, $L_2$=1 mH, and $C_2$=10 µF. Figs. 5(a) and 5(b) show the waveforms of the current through the input inductor ($L_1$) and voltage across the auxiliary capacitor ($C_1$) respectively. It can be understood from Fig. 5(a) that the current through $L_1$ is discontinuous. Fig. 5(c) shows the waveform of the grid voltage. The current injected into the grid by CPVI is in phase with the grid voltage. The corresponding waveform of the grid current is shown in Fig. 5(d). The THD of the grid current is observed to be 3.43% which is well within the IEEE limits.

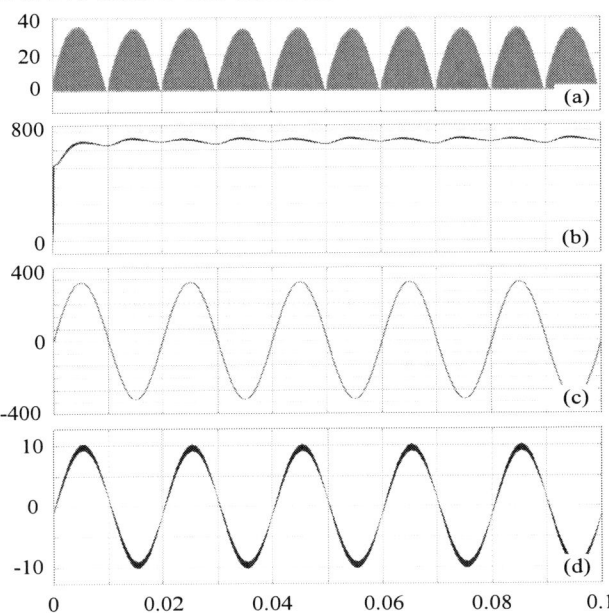

Fig. 5 (a) Current in input inductor $L_1$; (b) Voltage across $C_1$; (c) The grid voltage waveform; (d) Grid current waveform.

## V. EXPERIMENTAL RESULTS

The laboratory prototype of proposed CPVI is built for the 300 W system as shown in Fig. 6. The CPVI is tested in open loop with an input DC voltage, $V_{in}$ =75 V drawing an input current, $I_{in}$ =4 A. The waveforms of the current through input inductor ($I_{L_1}$), output voltage ($V_{C_2}$), the output current and the voltage across the auxiliary capacitor ($V_{C_1}$) are shown in Figs. 7(a)-(d). The parameters used for experimental study are

given in Table-III.

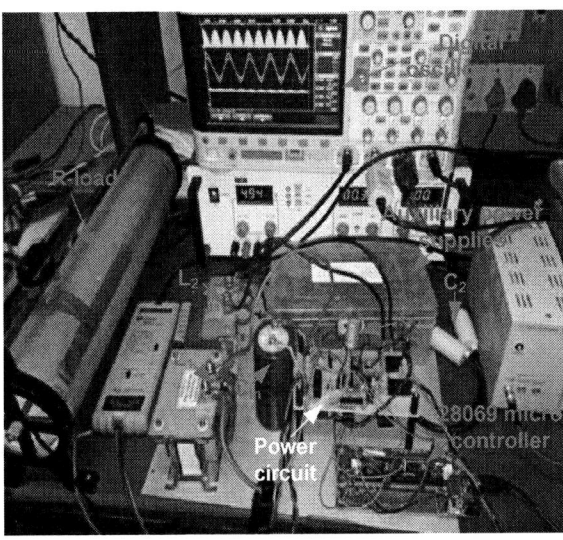

Fig. 6. Experimental setup of the proposed CPVI.

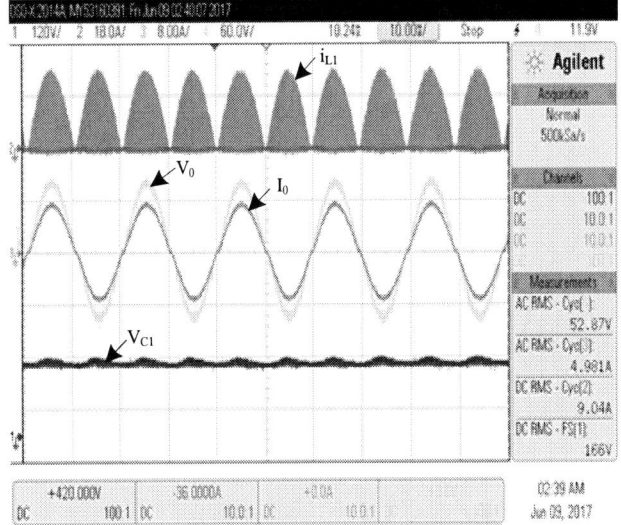

Fig. 7(a). Current in input inductor ($L_1$); (b) Voltage across output capacitor ($C_2$); (c) Output current; (d) Voltage across the auxiliary capacitor ($C_1$).

### TABLE III
PARAMETERS USED IN THE EXPERIMENTAL STUDY

| System Parameters | Value |
|---|---|
| Input Voltage ($V_{in}$) | 75 V |
| Input current ($I_{in}$) | 4 A |
| Input inductor ($L_1$) | 110 µH |
| Auxiliary capacitor ($C_1$) | 46 µF |
| Output inductor ($L_2$) | 1 mH |
| Output capacitor ($C_2$) | 10 µF |
| Switching frequency ($f_s$) | 10 kHz |

## VI. CONCLUSION

A novel transformerless Ćuk based topology has been proposed, analyzed and validated. It is verified that the proposed topology eliminates the leakage current and the DC current injection into the grid for grid-tied PV application. Since CPVI is a buck-boost derived topology, therefore it can track MPP under a wide PV voltage variation. Another advantage of the proposed inverter is the low input current ripple because of the presence of input inductor, which obviates the need for a large input capacitor across the PV source. The CPVI was tested at 10 kHz switching frequency and showed a low THD (3.43%) in the grid current. The performance of the proposed CPVI is found to comply with all the stipulated standards. The detailed literature survey for the proposed CPVI is done. The design of the energy storage elements is also presented. The experimental work has been carried out at 300 W power level, which validates the working of the proposed CPVI.

## REFERENCES

[1] H. Patel and V. Agarwal, "A single-stage, single-phase transformerless doubly grounded grid-connected PV interface," *IEEE Trans. Energy Conversion*, vol. 24, pp. 93–101, Mar. 2009.

[2] A. K. Gupta, M. S. Joshi, and V. Agarwal, "On the control and design issues of single-phase transformerless inverters for photovoltaic applications," in *IEEE 6th India International Conf. Power Electronics*, Kurukshetra, India, Dec. 2014.

[3] E. Gubia, P. Sanchis, A. Ursua, J. Lopez, and L. Marroyo, "Ground Currents in Single-phase Transformer-less Photovoltaic systems," *in Progress in Photovoltaics: Research and Applications*. New York: Wiley, pp. 629-650, Nov. 2007.

[4] G. Vazquez, T. Kerekes, A. Rolan, D. Aguilar, A. Luna and G. Azevedo, "Losses and CMV evaluation in transformer-less grid connected PV topologies", *Proc. IEEE Int. Symp. Ind. Electron.*, pp. 544-548, July 2009.

[5] Automatic Disconnection Device Between a Generator and the Public Low-Voltage Grid, VDE V 0126-1-1: 2006-02, 2006.

[6] W. Huang, K. Yen, G. Roig, and E. Lee, "Voltage divided non-inverting ĆUK converter with large conversion ratios," in *IEEE Proc., Southeast con*, pp.1005-1007, Apr. 1991.

[7] F. Schimpf and L. E. Norum, "Grid connected converters for PV, state of the art, ideas for improvement of transformerless inverters," *Nordic Workshop on Power and Industrial Electronics*, Jun. 2008.

[8] D. Barater, E. Lorenzani, C. Concari, G. Franceschini, and G. Buticchi, "Recent advances in single-phase transformerless photovoltaic inverters," *IET Renew. Power Gener.*, vol. 10, no. 2, pp. 260–273, 2015.

[9] W. Yu, 1. S. Lai, H. Qian, C. Hutchens, J. Zhang, G. Lisi, A. Diabbari, G. Smith and T. Hegarty, "High-efficiency inverter with H6-type configuration for photovoltaic non-isolated AC module applications," *Proc. IEEE Appl. Power Electron. Conf. Expo.*, pp.1056 -1061, Feb. 2010.

[10] W. Wu, J. Ji, and F. Blaabjerg, "Aalborg inverter-A new type of "buck in buck, boost in boost" grid-tied inverter, " *in IEEE Trans. Power Electron.*, vol. 30, no. 9, pp. 4784–4793, Sep. 2015.

[11] N. Kasa, H. Ogawa, T. Iida, and H. Iwamoto, "A transformerless inverter using buck-boost type chopper circuit for the photovoltaic power system," *in Proc. IEEE Int. Conf. Power Electron. Drive Syst.*, 1999, pp. 653–658.

[12] A. Kumar and P. Sensharma, "A four switch single stage single phase buck–boost inverter," *in IEEE Trans. Power Electron.*, vol. 32, no. 7, pp. 5282-5292, Jul. 2017.

[13] Y. Tang, X. Dong, and Y. He, "Active buck-boost inverter," *IEEE Trans. Ind. Electron.*, vol. 61, no. 9, pp. 4691–4697, Sep. 2014.

[14] M. Rajeev and V. Agarwal, "Novel transformer-less inverter topology for the single-phase grid-connected photovoltaic system," in *IEEE 42nd Photovolt. Spec. Conf.*, pp. 1-5, Jun. 2015.

[15] V. Gautam, A. Kumar, and P. Sensharma, "A Novel Single Stage, Transformerless PV inverter," *IEEE Int. Conf. on Ind. Tech.*, pp. 907-912, 2014.

[16] P. K. Chamarthi, M. Rajeev, and V. Agarwal, "A novel single stage zero leakage current transformer-less inverter for grid connected PV systems," in *IEEE 42nd Photovolt. Spec. Conf.*, pp. 1–5, Jun. 2015.

[17] N. Mohan, T. M. Undeland, and W. P. Robbins, *"Power Electronics*, Converters, Applications and Design". Wiley India, 2006.

[18] T. K. S. Freddy, N. A. Rahim, W. P Hew, and H. S. Che, "Comparison and Analysis of Single-phase transformer-less Grid connected PV inverters", *in IEEE Trans. Power Electron.*, vol. 29, pp. 5358-5369, Oct. 2014.

# A State Space Average Model for Dynamic Microgrid Based Space Station Simulations

Rachid Darbali-Zamora and Eduardo I. Ortiz-Rivera

Department of Electrical and Computer Engineering,
University of Puerto Rico-Mayagüez,
Mayagüez, Puerto Rico 00682, USA

*Abstract* – **Space stations are spacecrafts designed for crew members to perform space science research and exploration for prolonged periods of time. To power a space station, solar panels are employed to absorb available sunlight, storing them in batteries for later use. Their electrical distribution system functions like that of a microgrid, incorporating components such as power generation, storage, electric vehicles and household payloads. This paper presents a dynamic equation based microgrid simulation that uses an exponential analytical solar panel model as the main power generation for the microgrid and state space average models of several DC/DC converters for the conditioning of the electrical distribution system of a space station. Simulation results of the of the microgrid are also presented.**

*Index Terms – Aerospace, Space Station, Microgrids, Simulation*

## I. Introduction

Space stations are spacecrafts with the capability of allowing crew members to remain for prolonged periods of time as the spacecraft orbits the earth [1]. They are used as research platforms to perform space science and exploration. Space stations are mainly powered by solar panel arrays that transform available solar power into energy which is then stored in batteries for later use. This energy is distributed along the space stations payloads. Fig. 1 illustrates the main components that compose a space station.

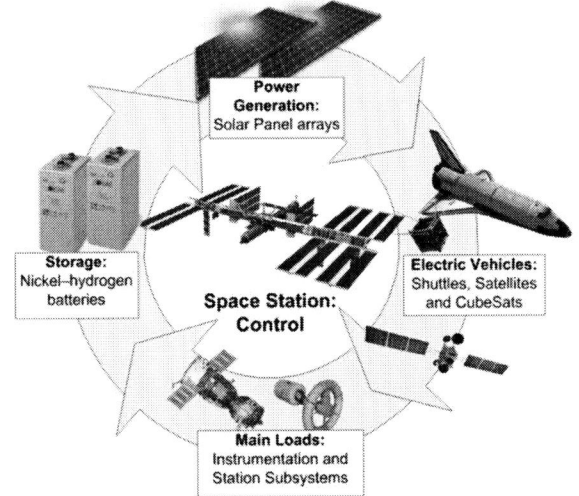

Fig. 1. The space stations components work like that of a microgrid, employing power generation, storage, primary loads and electrical vehicle integration.

The electrical distribution system of a space station functions like that of a microgrid, incorporating several components such as solar panels for their main energy generation, batteries as storage devices, electric vehicles such as shuttles and satellites as well as the space stations loads [2-3]. In addition, shuttles that dock unto the space station can use the power generated by the solar panels as their main power source. The objective of this project is to design a simulation platform that can be subjected to real case scenarios that involve energy generation and storage.

This article is organized in the following manner: section II illustrates the space stations electrical power distributions characteristics and design. Section III discusses the microgrid simulation results obtained. Finally, the conclusion is discussed in section IV.

## II. Electrical Power Distribution Details

The International Space Station (ISS) is the largest functional space station to date. To power the space station, solar panels are used as its primary energy source. These solar panel arrays are also known as solar array wings. To offer a power solution during eclipsed periods, when sunlight is not readily available, battery storage is used as an energy alternative. The entire electrical system is designed to generate from 80kW to 100kW. Fig. 2 illustrates a simplified diagram of the electrical power distribution system of the ISS.

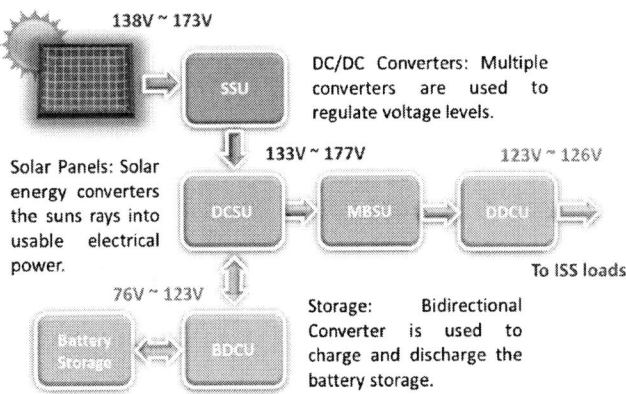

Fig. 2. Simplified diagram of the electrical power distribution system of the ISS. The electric system is separated into generation (blue), storage (green) and the DC link system (cyan).

978-1-5090-5606-4/17 $31.00 © 2017 IEEE

The Russian Segments of the ISS provides a small amount of power to the space station. With a maximum solar array output power of 16.8kW it can supply 2kW to the US segment of the space station. Including both the U.S. and Russian segments, the system can generate up to 110kW of power.

The electrical distribution system is separated into two subsystems. The primary subsystem operates at a voltage range of 137V to 173V and consists of power generation, and storage. DDCU (DC/DC Converter Units) are used to convert primary power to secondary power. The secondary subsystem operates at a voltage range of 123V to 126V and is used to supply power to the ISS payloads.

The ISS uses nickel-hydrogen batteries to store energy to provide power during eclipsed periods in orbit. For every 92 minutes that the ISS is in orbit, the battery storage provides 35 minutes of power. The batteries are rated at 81Ah and 4kWh.

Most of the power generated by the ISS is used to supply research facilities, modules and nodes. Currently, there are three laboratories in the ISS; Destiny, Columbus and Kibo. Close to 24kW of power is used for science and research purposes [4]. For non-science, related power allocation, the total housekeeping and operations payloads at the ISS are approximately 50kW. At assembly, 26kW minimum continuous and 30kW average power is required to supply thermal conditioning payloads. An additional 2.5kW of power is provided to operate ISS systems that support payload operations. During various other operating modes, payloads receive a minimum of 6.5kW of continuous power. The total consumption of the ISS can reach up to 75kW to 90kW.

*A. Photovoltaic Analytical Model*

To verify the performance of the Photovoltaic Module (PVM), an exponential model is used. This model uses the information provided by the PVM manufacturers data sheet [5]. The equation for the PVM is shown in equation (1).

$$I(V) = \frac{I_X}{1-exp(-1/b)} \cdot \left[1 - exp\left(\frac{V}{b \cdot V_X} - \frac{1}{b}\right)\right] \quad (1)$$

In this equation, $V$ is the output voltage of the PVM. $b$ is the characteristic constant of the PVM model. The variable $V_x$ and $I_x$ are obtained using equation (2) and equation (3) respectively.

$$V_x = \frac{E_{iN}}{E_i} TCv(T - T_N) + V_{max}$$
$$- (V_{max} - V_{min})exp\left[\frac{E_i}{E_{iN}}\ln\left(\frac{V_{max}-V_{oc}}{V_{max}-V_{min}}\right)\right] \quad (2)$$

$$I_x = \frac{E_i}{E_{iN}}[I_{sc} + TCi(T - T_N)] \quad (3)$$

In this equation, the variable $V_{max}$ is the open-circuit voltage at 25°C and a solar irradiance level of more than 1,200W/m². The parameter $V_{max}$ is approximately $1.03 \cdot V_{oc}$. The variable $V_{min}$ is the open-circuit voltage at 25°C with a solar irradiance level of less than 200W/m². The parameter $V_{min}$ is approximately

$0.85 \cdot V_{oc}$. The input variable $E_i$ is the effective solar irradiance in $W/m^2$. The input variable $T$ is the PVM temperature in °C. The variables $TCv$ and $TCi$ are the temperature coefficient of the $V_{oc}$ and the $I_{sc}$ measured in V/°C and in A/°C respectively. The constant $T_N$ is 25°C and the nominal effective solar irradiance $E_{iN}$ is 1,000W/m². These constants represent the standard test conditions of the PVM. The variables $V_{oc}$ and $I_{sc}$ are the open-circuit voltage and short circuit-current at 25°C and 1,000W/m². The multiplication of the PVM voltage with equation (1) yields the power output of the PVM. The mathematical relationship for the PVM power output is shown in equation (4)

$$P(V) = \frac{V \cdot p \cdot I_X}{1-exp(-1/b)} \cdot \left[1 - exp\left(\frac{V}{s \cdot b \cdot V_X} - \frac{1}{b}\right)\right] \quad (4)$$

In addition, the angle of the PVM can also be added into the model, to consider any inclinations that may occur in space [6]. Maximum Power Point Tracking (MPPT) algorithms allow the PVM to generate the most available power. This is achieved by adjusting the PVM voltage and current, so they may operate at their optimal values, yielding the maximum power. Fig. 3 illustrates the P-V curve obtained from the analytical PVM at solar irradiance levels of 400W/m², 800W/m² and of 1,200W/m² at a constant temperature of 25°C. Fig. 4 illustrates the I-V curve obtained from the analytical PVM model at varying solar irradiance levels with a constant temperature of 25°C. These parameters are used for the microgrid simulation.

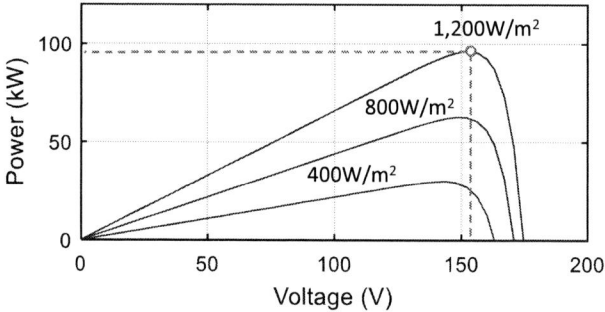

Fig. 3. The PVM P-V curve at varying solar irradiance. Tracing the maximum power output in the P-V curve it is possible to identify the optimal voltage at a solar irradiance level of 1,200W/m².

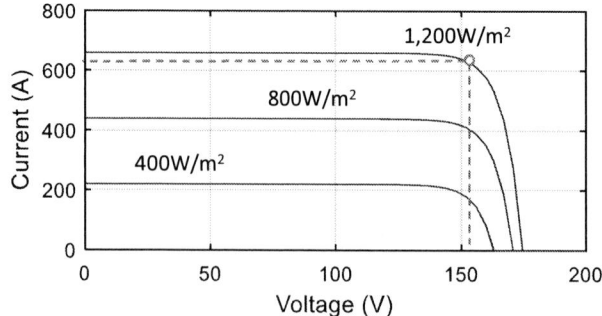

Fig. 4. The PVM I-V curve at varying solar irradiance. Tracing the optimal voltage from the P-V curve helps identify the optimal current from the I-V curve at a solar irradiance level of 1,200W/m².

## B. Unidirectional DC/DC SEPIC

The SEPIC is a DC/DC converter that can increase or decrease the voltage applied to its input. Contrary to other step-up/step-down converters, the SEPIC does not have an inverted output which in many cases is considered an advantage. Another benefit of the SEPIC is that it avoids the issue of a charged floating inductor during switching transitions, which can cause unwanted discharges. Some applications of the SEPIC range from large electric machines to smaller switching power supplies.

The SEPIC topology is composed of two inductors, two capacitors, a diode and a switching transistor. The input current of the SEPIC is provided by the exponential analytical mathematical PVM model. The input capacitor of the SEPIC represents the PVM arrays output capacitor, which defines the PVM arrays voltage. In this application, the SEPIC is used to extract the utmost available power from the PVM array by using the Optimal Duty Ratio MPPT technique [7]. Fig. 5 illustrates the SEPIC topology circuit schematic. Fig. 6 illustrates the equivalent circuits diagrams obtained from the different ON and OFF switching states of the SEPIC.

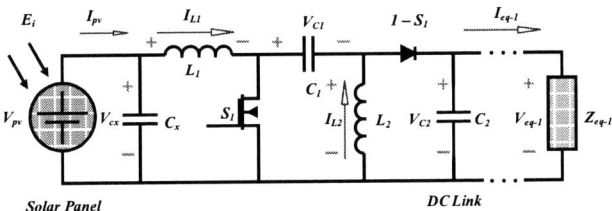

**Solar Panel**  **DC Link**

Fig. 5. The SEPIC circuit schematic. The SEPIC allows for lower, equal or greater voltage levels at its output, in a similar manner as the Buck-Boost converter without the inverted output polarity.

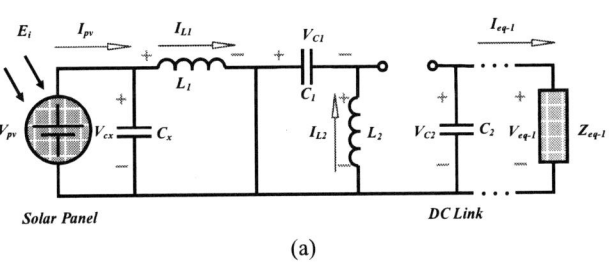

(a)

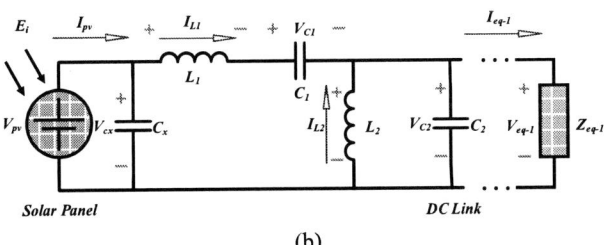

(b)

Fig. 6. Different ON and OFF switching states of the SEPIC. (a) With $S_1$ closed, current increases through $L_1$ and $C_1$ discharges increasing current in $L_2$. (b) With $S_1$ open, current through $L_1$ and current through $L_2$ produces a current that supplies the DC link.

Performing an analysis on the ON and OFF switching states of the SEPIC, the dynamic equations can be obtained. The state equations for the SEPIC are shown in equations (5) through (8).

$$\frac{dI_{L1}}{dt} = \frac{V_{Cx}}{L_1} + \frac{V_{C1}+V_{C2}}{L_1} \cdot (S_1 - 1) \tag{5}$$

$$\frac{dI_{L2}}{dt} = \frac{V_{C1}}{L_2} \cdot S_1 + \frac{V_{C2}}{L_2} \cdot (S_1 - 1) \tag{6}$$

$$\frac{dV_{C1}}{dt} = \frac{I_{L1}}{C_1} \cdot (1 - S_1) - \frac{I_{L2}}{C_1} \cdot S_1 \tag{7}$$

$$\frac{dV_{C2}}{dt} = \frac{I_{L1}+I_{L2}}{C_2} \cdot (1 - S_1) - \frac{V_{C2}}{C_2 \cdot Z_{eq-1}} \tag{8}$$

In these equations, $S_1$ is the status of the switch (1 for ON, 0 for OFF), $I_{L1}$ is the current through the inductor $L_1$ and $I_{L2}$ is the current through the inductor $L_2$. $V_{C1}$ is the voltage across the capacitor $C_1$ and $V_{C2}$ is the voltage across the capacitor $C_2$. The current $I_{eq-1}$ is the equivalent current supplied to the DC link. These dynamic state space equations can be summarized in matrix form, as shown in equation (9).

$$\begin{bmatrix} \frac{dI_{L1}}{dt} \\ \frac{dI_{L2}}{dt} \\ \frac{dV_{C1}}{dt} \\ \frac{dV_{C2}}{dt} \end{bmatrix} = \begin{bmatrix} 0 & 0 & \frac{(S_1-1)}{L_1} & \frac{(S_1-1)}{L_1} \\ 0 & 0 & \frac{S}{L_2} & \frac{(S_1-1)}{L_2} \\ \frac{(1-S_1)}{C_1} & \frac{-S_1}{C_1} & 0 & 0 \\ \frac{(1-S_1)}{C_2} & \frac{(1-S_1)}{C_2} & 0 & \frac{-1}{C_2 \cdot Z_{eq-1}} \end{bmatrix} \cdot \begin{bmatrix} I_{L1} \\ I_{L2} \\ V_{C1} \\ V_{C2} \end{bmatrix} + \begin{bmatrix} \frac{1}{L_1} \\ 0 \\ 0 \\ 0 \end{bmatrix} \cdot (V_{Cx}) \tag{9}$$

In addition, the equation to the output capacitor of the PVM is shown in equation (10).

$$\left[\frac{dV_{Cx}}{dt}\right] = \begin{bmatrix} \frac{-1}{C_x} & 0 & 0 & 0 \end{bmatrix} \cdot \begin{bmatrix} I_{L1} \\ I_{L2} \\ V_{C1} \\ V_{C2} \end{bmatrix} + \begin{bmatrix} \frac{-1}{C_x} \end{bmatrix} \cdot I_{pv} \tag{10}$$

In this equation $V_{Cx}$ is the PVM output capacitor and $C_x$ is its capacitance. The current $I_{pv}$ is equal to the current $I(V)$ from the exponential PVM mathematical model. The mathematical representation between the output and input voltages of the SEPIC is obtained by using equations (5) and (6), if the duty cycle $\bar{d}_1$ is an averaged $S_1$. The relationship between the input and output voltages of the SEPIC is shown in equation (11).

$$V_{C2} = V_{cx} \cdot \left(\frac{\bar{d}_1}{1-\bar{d}_1}\right) \tag{11}$$

In this expression, $V_{C2}$ is the voltage at the DC link, while $V_{cx}$ is the voltage in the input and $\bar{d}_1$ represents the duty cycle. Notice that the polarity at the output voltage of the converter is not inverted.

978-1-5090-5606-4/17 $31.00 © 2017 IEEE

## C. Bidirectional DC/DC Buck-Boost Converter

For the charging and discharging of the batteries, the Buck-Boost Bidirectional converter is used [8]. In theory, the conditions of the Bidirectional DC/DC converter operates depending on the battery voltage. If the DC link voltage is greater than that of the battery voltage, then the converter will operate in Boost mode, supplying power to the DC link. On the contrary, if the battery voltage is lower than that of the DC link, then the converter will operate in Buck mode and the battery will be charged by any excess power available from the DC link side of the system.

The Bidirectional converters topology is composed of an inductor, a capacitor, and two switching transistors. The diode of the switching transistor plays an important part in the circuit design, allowing the current to flow bidirectionally. Fig. 7 illustrates the Buck-Boost Bidirectional converter circuit schematic. Fig. 8 illustrates the equivalent circuits obtained for the different ON and OFF switching states of the Buck-Boost Bidirectional converter.

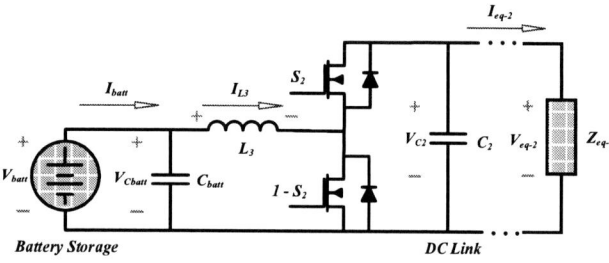

Fig. 7. The Buck-Boost Bidirectional converter circuit schematic. The Buck-Boost converter allows for charging and discharging states of the systems storage.

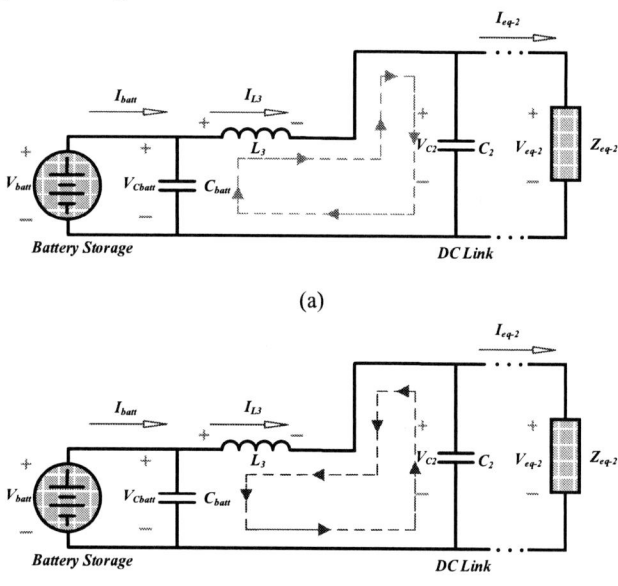

(a)

(b)

Fig. 8. Different ON and OFF switching states of the Bidirectional converter during charging and discharging operations. (a) With $S_2$ open the converter operates in Boost mode and the battery supplies current that increases through $L_3$, supplying the DC link. (b) With $S_2$ closed the converter operates in Buck mode and the battery is charged.

In Fig. 8, the output voltage of the Bidirectional Buck-Boost converter represents the DC link of the system. Capacitor $C_2$ of the circuit is also capacitor $C_2$ of the SEPIC and serves as the DC high voltage bus connection between all other DC converters in the microgird. In addition, the voltage provided by this capacitor is also the input voltage of the systems Buck converters. When excess power is generated by the PVM, the Bidirectional Buck-Boost converter begins its charging mode. When the PVM is not able to supply sufficient power to the systems loads, the Bidirectional Buck-Boost converter operates in discharging mode, providing the remaining power to the loads.

Depending on the input and output voltage, the Bidirectional converter can either operate in Buck mode or in Boost mode. Performing an analysis on the ON and OFF switching states of the Buck-Boost Bidirectional converter, the dynamic equations can be obtained. The dynamic equations of the Bidirectional Buck-Boost converter are shown in equations (12) and (14).

$$\frac{dI_{L3}}{dt} = \frac{V_{Cbatt}}{L_3} - \frac{V_{C2}}{L_3} \cdot (1 - S_2) \tag{12}$$

$$\frac{dV_{Cbatt}}{dt} = \frac{I_{batt}}{C_{batt}} - \frac{I_{L3}}{C_{batt}} \tag{13}$$

$$\frac{dV_{C2}}{dt} = \frac{I_{L3}}{C_2}(1 - S_2) - \frac{V_{C2}}{C_2 \cdot Z_{eq-2}} \tag{14}$$

In these equations, $S_2$ is the status of the switch (1 for ON, 0 for OFF), $I_{L3}$ is the current through the inductor $L_3$. $V_{Cbatt}$ is the voltage across the battery capacitor $C_{batt}$ and $V_{C2}$ is the voltage across the capacitor $C_2$. $V_{batt}$ is the battery voltage and $I_{batt}$ is the battery current. $I_{eq-2}$ is the equivalent current supplied to the DC link. The dynamic state equations for the Bidirectional Buck-Boost converter can be summarized in matrix form, shown in equation (15).

$$\begin{bmatrix} \frac{dI_{L3}}{dt} \\ \frac{dV_{Cbatt}}{dt} \\ \frac{dV_{C2}}{dt} \end{bmatrix} = \begin{bmatrix} 0 & \frac{1}{L_3} & \frac{-(1-S_2)}{L_3} \\ \frac{-1}{C_{batt}} & 0 & 0 \\ \frac{1-S_2}{C_2} & 0 & \frac{-1}{C_2 \cdot Z_{eq-2}} \end{bmatrix} \cdot \begin{bmatrix} I_{L3} \\ V_{Cbatt} \\ V_{C2} \end{bmatrix} + \begin{bmatrix} 0 \\ \frac{1}{C_{batt}} \\ 0 \end{bmatrix} \cdot I_{batt} \tag{15}$$

Solving equation (12) as well as assuming $\bar{d}_2$ is an averaged $S_2$, the mathematical relationship between the input voltage and output voltage of the Bidirectional converter can be derived. This expression is shown in equation (16).

$$V_{C2} = V_{Cbatt} \cdot \left( \frac{1}{1 - \bar{d}_2} \right) \tag{16}$$

In this equation, $V_{C2}$ is the voltage at the output terminal which represents the voltage at the DC link. $V_{batt}$ is the voltage at the input of the converter, representing the battery voltage. The variable $\bar{d}_2$ is the Bidirectional Buck-Boost converters duty cycle. Note that the output voltage of the converter depends only on the input voltage and the duty cycle.

## D. Unidirectional DC/DC Buck Converter

The Buck converter is a step-down converter with the ability to reduce voltage at its output. Although widely used, the Buck converter has several limitations to take into consideration. One of these is that to regulate its output voltage, its output voltage must be less than its input voltage. The Buck converters main application is mostly seen in voltage regulated DC power supplies, motor speed control as well as in small sized embedded systems. Some advantages of the Buck converter include its high efficiency, and low ripple. Another advantage is that its simplicity allows for easy, low cost implementation in comparison with other converters that might have additional components. The Buck converter can operate in either continues or discontinuous mode, depending on the current drawn by the load. If the current drawn causes the inductor current in the converter to drop to zero, then the converter is operating in discontinuous mode. Note that operating in discontinuous mode can lead to higher switching losses. The Buck converter topology is composed of a capacitor, an inductor, a switching transistor and a diode. In this application, the input voltage is provided by the DC link. Fig. 9 illustrates the Buck converter circuit schematic. Fig. 10 illustrates the equivalent circuits obtained for the different ON and OFF switching states of the Buck converter.

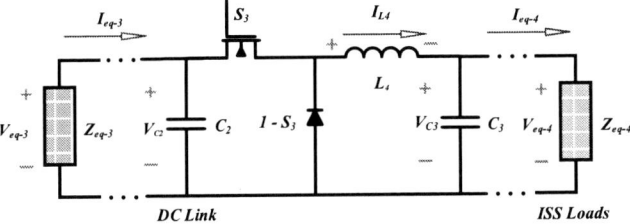

Fig. 9. The Buck converter circuit schematic. The Buck converter allows for lower voltage levels at its output.

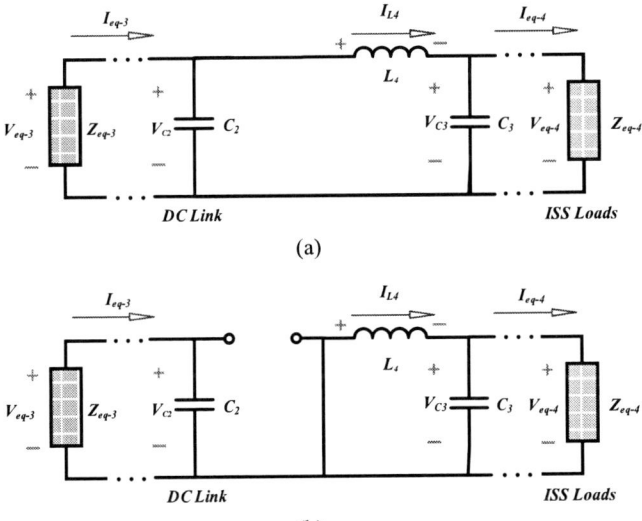

Fig. 10. Different ON and OFF switching states of the Buck converter. (a) With $S_4$ closed current increases through $L_4$ and $C_3$ discharges. (b) With $S_4$ open, current through $L_4$ producing a current through the equivalent loads.

To obtain the dynamic equations that model the Buck converter's behavior, a simple analysis of the ON and OFF states are performed on the circuit. The dynamic equations of the converter are shown in equations (17) through (19).

$$\frac{dI_{L4}}{dt} = \frac{V_{C2}}{L_4} \cdot S_3 - \frac{V_{C3}}{L_4} \tag{17}$$

$$\frac{dV_{C2}}{dt} = \frac{V_{C2}}{C_2 \cdot Z_{eq-3}} - \frac{I_{L4}}{C_2} \cdot S_3 \tag{18}$$

$$\frac{dV_{C3}}{dt} = \frac{I_{L4}}{C_3} - \frac{V_{C3}}{C_3 \cdot Z_{eq-4}} \tag{19}$$

In these equations, $S_3$ is the status of the switch (1 for ON, 0 for OFF), $I_{L4}$ is the current through the inductor $L_4$, while $V_{C3}$ is the voltage across the capacitor $C_3$. The variable $V_{C2}$ is the input voltage of the converter. The variable $Z_{eq-4}$ represents the equivalent ISS loads that can be connected at the output of the converter. The dynamic equations of the converter are shown in matrix form, shown in equation (20).

$$\begin{bmatrix} \frac{dI_{L4}}{dt} \\ \frac{dV_{C2}}{dt} \\ \frac{dV_{C3}}{dt} \end{bmatrix} = \begin{bmatrix} 0 & \frac{S_3}{L_4} & \frac{-1}{L_4} \\ \frac{-S_3}{C_2} & \frac{1}{C_2 \cdot Z_{eq-3}} & 0 \\ \frac{1}{C_3} & 0 & \frac{-1}{C_3 \cdot Z_{eq-4}} \end{bmatrix} \cdot \begin{bmatrix} I_{L4} \\ V_{C2} \\ V_{C3} \end{bmatrix} \tag{20}$$

Solving equation (17) as well as assuming the duty cycle $\bar{d}_3$ is an averaged $S_3$, the mathematical relationship between the input voltage and output voltage of the Buck converter can be derived. This expression is shown in equation (21).

$$V_{C3} = V_{C2} \cdot (\bar{d}_3) \tag{21}$$

In this equation, $V_{C3}$ is the voltage at the output terminal, while $V_{C2}$ is the voltage in the input and $\bar{d}_3$ represents the duty cycle. The output voltage varies linearly and depends only on the input voltage and the duty ratio.

### III. MICROGRID SIMULATION RESULTS

This simulation uses an exponential analytical mathematical model to represent the PVM arrays behavior. In addition, it uses the state space average model to represent each DC/DC converter. The SEPIC is used to extract the utmost available power from the PVM. The Buck-Boost Bidirectional converter charges and discharges the battery storage unto the DC link. Finally, the Buck converter is used to provide regulated voltage to the systems payloads. Two Buck converters are used to provide constant and varying power consumption. Fig. 11 through Fig. 14 illustrate the power for the PVM, battery storage, as well as the constant power and variable power consumption. Notice from the simulation results that during instances when the PVM array is not able to supply power to the loads, the battery storage provides the required power. Also, during instances when excess power generated by the PVM array, it is stored in the battery storage.

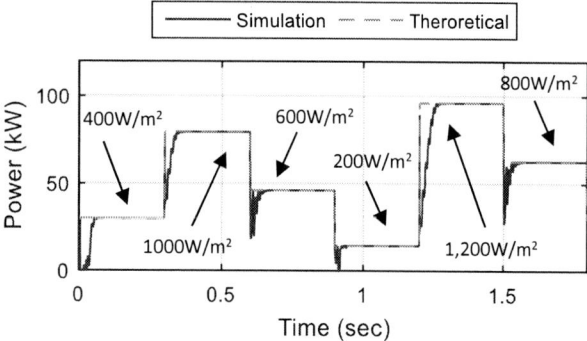

Fig. 11. Simulation results obtained from the microgrids PVM.

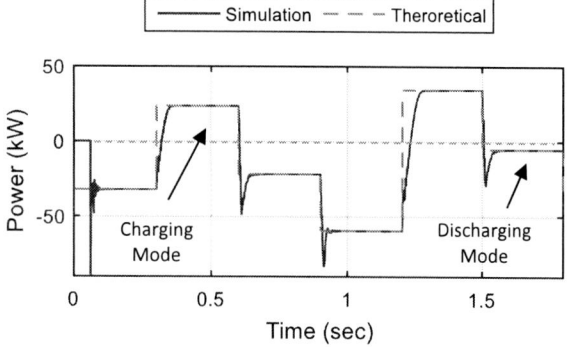

Fig. 12. Simulation results obtained from the microgrids Battery.

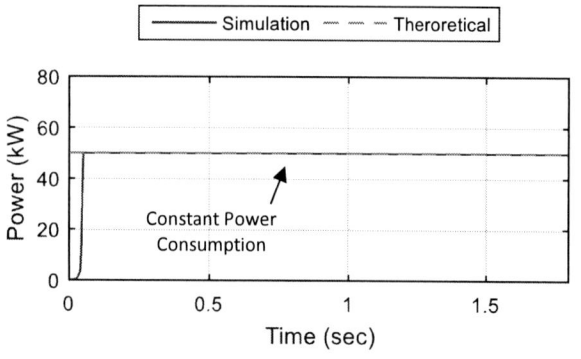

Fig. 13. Simulation results obtained from the microgrids constant housekeeping and operations payloads.

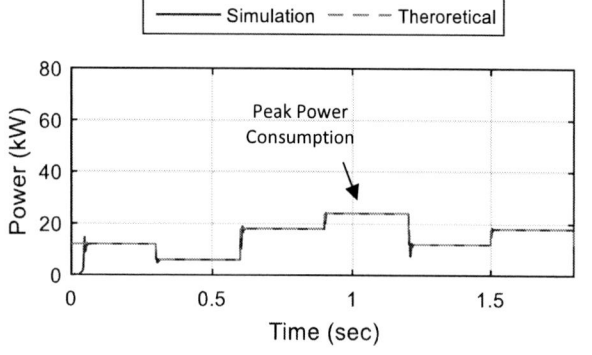

Fig. 14. Simulation results obtained from the microgrids varying research payloads.

## IV. CONCLUSION

A dynamic equation based mathematical microgrid model for space station applications is demonstrated. The model uses an exponential analytical mathematical expression to represent a PVM arrays behavior. The state space averaging equations of several DC/DC converters are used to represent their behavior. A SEPIC is used to perform the optimal duty ratio MPPT to extract the utmost available power from the PVM array. A Bidirectional Buck-Boost converter controls the charging and discharging states of the battery storage. Finally, Buck converters are used to model constant housekeeping and operations payloads as well as the varying power consumed by research processes. From the simulation results it can be noted that during instances when the PVM array is not able to supply power to the loads, the battery storage provides the required power. During instances when excess power is generated by the PVM array, it is stored in the battery storage.

## ACKNOWLEDGEMENT

The authors acknowledge the contributions made by Sandia National Laboratories (SNL), the National Nuclear Security Administration (NNSA), the US Department of Energy (DOE) and the Consortium for Integrating Energy Systems in Engineering and Science Education (CIESESE), for the sponsorship of this project, as well as the financial support of the Transformational Initiative for Graduate Education and Research (TIGER) program under grant P031M140035.

## REFERENCES

[1]. M. Savoy, T. Miller; "International space station electrical power system performance and operational lessons learned", *Intersociety Engineering Conference in Energy Conversion (IECEC'02)*, 29-31, July 2004, pp.93.

[2]. R. Darbali-Zamora, C.J. Gómez-Méndez, E.I. Ortiz-Rivera, H. Li, J. Wang, "Solar Irradiance Prediction Model based on a Statistical Approach for Microgrid Application", *Photovoltaic Specialists Conference (PVSC'15)*, June 14-19, 2015.

[3]. S.M. Sharkh, M.A. Abu-Sara; G.I. Orfanoudakis and B. Hussain, "Power Electronic Converters for Microgrids", Wiley-IEEE Press, 2014, pp.352.

[4]. M.J. Hart, R.J. Kinsey, A.S. Lee, J.S Yoshida, "International Space Station life extension", *IEEE Conference in Aerospace*, pp.1-15, 6-13 March 2010.

[5]. E.I Ortiz-Rivera, "Modeling and Analysis of Solar Distributed Generation", Ph.D. dissertation, Department of Electrical and Computer Engineering, Michigan State University, 2006.

[6]. R. Darbali-Zamora, E.I. Ortiz-Rivera, A.A. Rincon-Charris, "Analytical Photovoltaic Mathematical Model with Varying Inclination Angle for Satellite Applications", *IEEE Andean Council International Conference (ANDESCON'16)*, Oct. 19- 21, 2016.

[7]. R. Darbali-Zamora, E.I. Ortiz-Rivera, "Optimal Duty Ratio Maximum Power Point Tracking Technique Using the SEPIC Topology for Photovoltaic Systems Applications", *IEEE Andean Council International Conference (ANDESCON'16)*, Oct. 19-21, 2016.

[8]. T.H. Wu, C.S. Moo, Y.C. Hsieh, C.Y. Juan, "Operation of battery power modules with bidirectional DC/DC converters", *International Conference on System Science and Engineering (ICSSE'13)*, 2013, pp. 443-448.

# Buck Converter and SEPIC Based Electronic Power Supply Design with MPPT and Voltage Regulation for Small Satellite Applications

Rachid Darbali-Zamora[1], Nicolás Cobo-Yepes[1], John E. Salazar-Duque[1], Eduardo I. Ortiz-Rivera[1], Amilcar A. Rincon-Charris[2]

[1]Department of Electrical and Computer Engineering, University of Puerto Rico-Mayagüez,
Mayagüez, Puerto Rico 00682, USA

[2]Department of Mechanical Engineering, Inter American University of Puerto Rico-Bayamon
Bayamon, Puerto Rico 00957, USA

*Abstract* - **This paper presents the proposed power distribution system designed for the power management of the Puerto Rico CubeSat. Power is provided to the CubeSat using Gallium Arsenide (GaAs) solar panels mounted on each side of the satellite. It is vital to maximize the available electrical energy obtained from the minimal solar cell area available. SEPIC topology performs the Maximum Power Point Tracking (MPPT) to extract the utmost available power from the solar panels. The Buck converter is used to provide voltage regulation for the CubeSat payloads. The power distribution for the CubeSat is designed for higher efficiency, utilizing high-performance GaAs solar panels and DC/DC converters in combination with a MPPT technique to improve the power extraction and further increase efficiency.**

*Index Terms – CubeSat, Electrical Power Supply, Power Distribution, MPPT, Buck Converter, SEPIC.*

## I. INTRODUCTION

The interest in space exploration has increased to such lengths that there exists a need for developing advanced, low-cost tools that aid in the information collection process. CubeSat's are cube shaped micro-scaled satellites of low mass and size targeted at perform a wide range of tasks, such as collecting and transmitting data [1]. Although originally conceived as an educational tool, CubeSat's have begun to challenge the standards of traditional satellites and have been recognized for their potential utility by space and research agencies around the world [2-3]. Fig. 1 shows a standard 1U CubeSat design.

Traditionally, CubeSat's are composed of an onboard computer, electronic power supply (EPS), control and communication systems, among other structural subsystems [4-5]. The EPS is a critical subsystem in the CubeSat design. This subsystem is tasked with supplying electricity to the CubeSat's payloads. A CubeSat is powered by solar energy; solar panels located on each one of its sides. Solar cell area limitations require maximizing the available electrical energy generated. To achieve this, Maximum Power Point Tracking (MPPT) techniques can be employed to extract the utmost available power from Photovoltaic Module (PVM). This solar power can later be stored in rechargeable batteries that provide power during moments when sunlight is not available.

This article presents a Buck converter and SEPIC based EPS with voltage regulation and MPPT for CubeSat applications. A 9cm x 9cm EPS prototype is constructed and tested to guarantee it can perform the MPPT as well as provide regulated voltages of 3.3V and 5.0V

## II. CUBESAT ELECTRICAL POWER SUPPLY DESIGN

The design consists of Gallium Arsenide (GaAs) solar panels, a SEPIC tasked with performing the MPPT, a battery that provides storage and two Buck converters that function as voltage regulators that will supply the necessary 3.3V and 5.0V for the CubeSat payloads. Fig. 2 illustrates the block diagram for the proposed Puerto Rico CubeSat EPS. The EPS parameters are based on previous Puerto Rico CubeSat designs [6-7]. Table I summarizes the electrical requirements of typical payloads found in CubeSat applications.

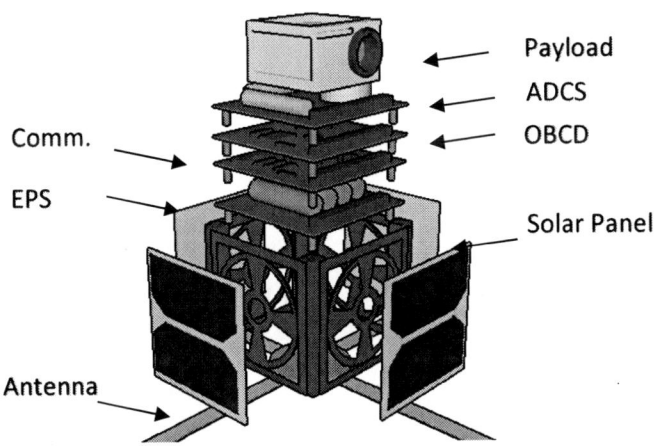

Fig. 1. A standard 1U CubeSat. CubeSat's are small scale satellites most commonly designed for space science and exploration.

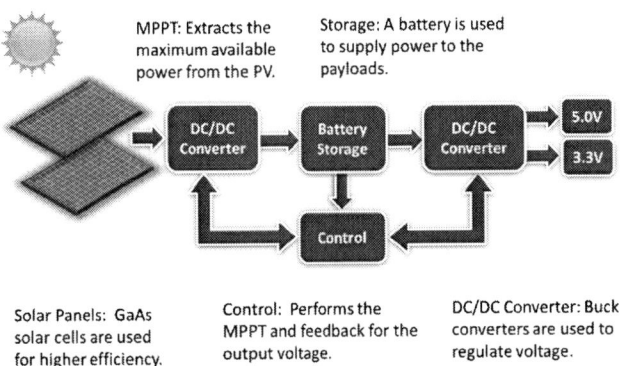

Fig. 2. Proposed CubeSat's EPS design. The blue blocks show the power flow while the red blocks illustrate the control design.

978-1-5090-5606-4/17 $31.00 © 2017 IEEE

TABLE I: Puerto Rico CubeSat Payload Ratings

| Load | Voltage (V) | Current (A) | Power (W) | Load (Ω) |
|---|---|---|---|---|
| IMU | 3.3 | 0.265 | 0.874 | 12.453 |
| GPS Receiver | 3.3 | 0.400 | 1.320 | 8.250 |
| Magnetorquers | 3.3 | 0.360 | 1.188 | 9.167 |
| Flight Computer | 3.3 | 0.071 | 0.234 | 46.479 |
| GPS Antenna | 5.0 | 0.050 | 0.250 | 100.000 |
| Transceiver | 5.0 | 0.850 | 4.250 | 5.882 |
| MADS | 5.0 | 0.200 | 1.000 | 25.000 |
| Reaction Wheel | 6.0 | 0.027 | 0.162 | 185.185 |

## A. Analytical Photovoltaic Model

To provide power to a CubeSat, solar panels are located on each one of its sides. The designed PVM array consists of improved triple-junction GaAs PVM. Compared to Si cells, these solar cells are over twice as efficient and will deliver more than twice the power for the same area (27% efficient). A mathematical model is used to verify the performance of the CubeSat's PVM array [8-9]. This model is also used to obtain the characteristic curves of the CubeSat solar panel array at different solar irradiance levels. Once obtained, these results are used to emulate the PVM arrays behavior using an *Agilent E4351B Solar Array Simulator*. The selected PVM array has a current and voltage rating of 1.3A and 5.0V at an irradiance level of 1,000W/m². Fig. 3 and Fig. 4 illustrate the I-V and P-V curves of a single GaAs solar cell obtained using the analytical model.

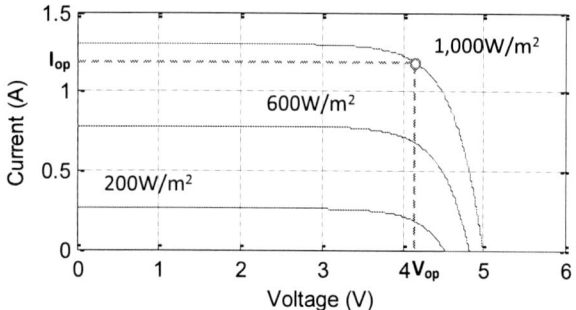

Fig. 3. The solar panels I-V curve at varying solar irradiance. Using the optimal voltage obtained from the P-V curve it is possible to obtain the optimal current from the I-V curve.

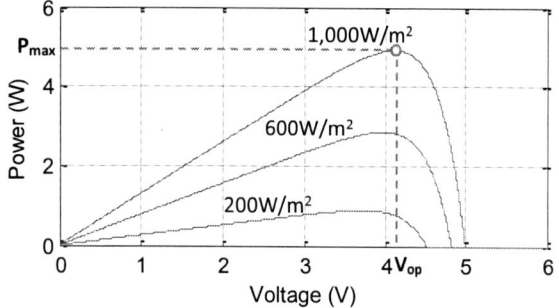

Fig. 4. The solar panels P-V curve at varying solar irradiance. By tracing the optimal power output in the P-V curve it is possible to identify the optimal voltage.

## B. DC/DC Converter Designs

DC/DC converters are circuits that can convert a DC voltage to a different regulated DC voltage level [10]. The converter topology refers to the arrangement of components that compose the DC/DC converter. In most cases, non-isolated DC/DC converters are composed of capacitors, inductors, transistors, and diodes. Two DC/DC converter topologies are used for the CubeSat EPS design. The DC/DC converter of choice for EPS CubeSat applications are the Single-Ended Primary-Inductor Converter (SEPIC) and the Buck converter [11]. The SEPIC is used to perform the MPPT to extract the maximum available power from the PVM. The Buck converter is used to provide voltage regulation for the payloads.

The SEPIC is a DC/DC converter similar in behavior to the Buck-Boost converter, it can either increase or decrease the voltage applied to its input [12]. Contrary to the Buck-Boost, the SEPIC does not have an inverted output which in many cases is a favorable trait. Another benefit is that it avoids the issue of a floating inductor during switching transitions, which can be a dangerous risk. The SEPIC topology is composed of two inductors, two capacitors, a diode and a power transistor. Fig. 5 illustrates a diagram of the SEPIC circuit schematic. Fig. 6 illustrates the equivalent circuits obtained for the different ON and OFF switching states. The red and blue dashed lines indicate the inductors charging and discharging.

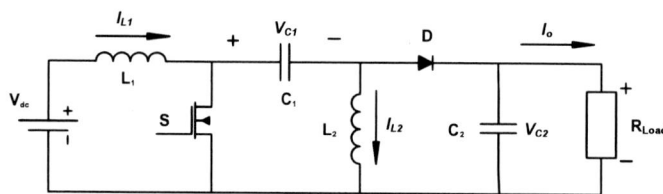

Fig. 5. SEPIC circuit schematic. The SEPIC allows for lower, equal or greater voltages at its output. Like the Buck-Boost without the inverted output and with the capability of true shut off.

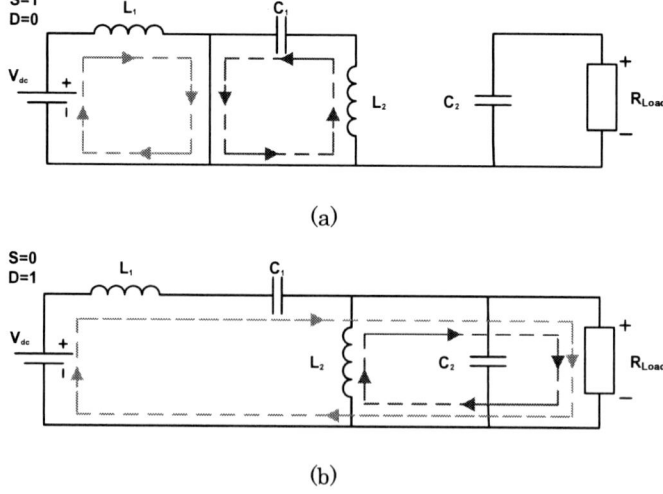

(a)

(b)

Fig. 6. Different ON and OFF switching states of the SEPIC. (a) With S closed, current increases through $L_1$ (red) and $C_1$ discharges increasing current in $L_2$ (blue). (b) With S open, current through $L_1$ (red) and current through $L_2$ (blue) produces a current through the load.

978-1-5090-5606-4/17 $31.00 © 2017 IEEE

The dynamic state equations for the SEPIC can be summarized in matrix form, shown in equation (1).

$$
\begin{bmatrix} \frac{dI_{L1}}{dt} \\ \frac{dI_{L2}}{dt} \\ \frac{dV_{C1}}{dt} \\ \frac{dV_{C2}}{dt} \end{bmatrix} = \begin{bmatrix} 0 & 0 & \frac{S-1}{L_1} & \frac{S-1}{L_1} \\ 0 & 0 & \frac{S}{L_2} & \frac{(S-1)}{L_2} \\ \frac{(S-1)}{C_1} & -\frac{S}{C_1} & 0 & 0 \\ \frac{(S-1)}{C_2} & \frac{(S-1)}{C_2} & 0 & \frac{-1}{RC_2} \end{bmatrix} \cdot \begin{bmatrix} I_{L1} \\ I_{L2} \\ V_{C1} \\ V_{C2} \end{bmatrix} + \begin{bmatrix} \frac{1}{L_1} \\ 0 \\ 0 \\ 0 \end{bmatrix} \cdot (V_{dc}) \quad (1)
$$

In these equations, $S$ is the status of the switch (1 for ON, 0 for OFF), $I_{L1}$ is the current through the inductor $L_1$ while $I_{L2}$ is the current through the inductor $L_2$. $V_{C1}$ is the voltage across the capacitor $C_1$ and $V_{C2}$ is the voltage across the capacitor $C_2$.

The Buck converter is one of the most basic DC/DC converters and the stepping stone to other, more complicated DC/DC converter topologies. It is a non-isolated converter and possesses the ability to reduce the voltage at its output [13]. The Buck converter is composed of a capacitor, an inductor, a power transistor and a diode. Fig. 7 illustrates the Buck converter circuit schematic. Fig. 8 illustrates the equivalent circuits obtained for the different on and off switching states of the Buck converter.

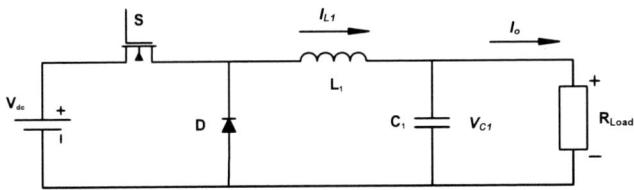

Fig. 7. Buck converter circuit schematic. The Buck converter allows for lower voltages at its output. Its simple topology allows for a compact design.

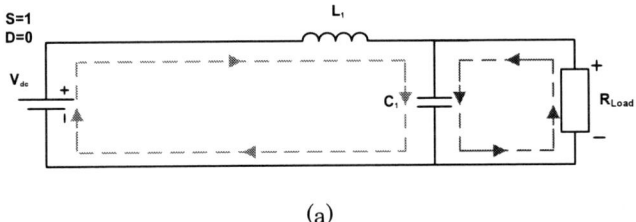

(a)

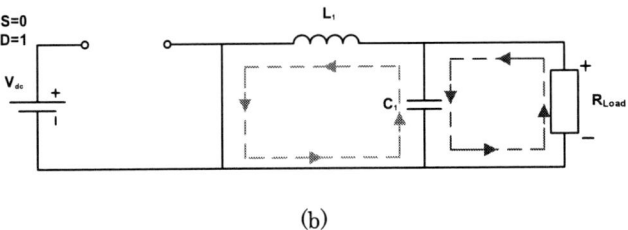

(b)

Fig. 8. Different ON and OFF switching states of the Buck converter. (a) With S closed current increases through $L_1$ (red) and $C_1$ discharges (blue). (b) With S open, current through $L_1$ (red) producing a current through the load.

The dynamic state equations of the converter are summarized in matrix form, shown in equation (2).

$$
\begin{bmatrix} \frac{dI_{L1}}{dt} \\ \frac{dV_{C1}}{dt} \end{bmatrix} = \begin{bmatrix} 0 & -\frac{1}{L_1} \\ \frac{1}{C_1} & -\frac{1}{R \cdot C_1} \end{bmatrix} \cdot \begin{bmatrix} I_{L1} \\ V_{C1} \end{bmatrix} + \begin{bmatrix} \frac{V_{dc}}{L_1} \\ 0 \end{bmatrix} \cdot S \quad (2)
$$

In these equations, $S$ is the status of the switch (1 for ON, 0 for OFF), $I_{L1}$ is the current through the inductor $L_1$, while $V_{C1}$ is the voltage across the capacitor $C_1$. The variable $V_{dc}$ is the input voltage of the converter.

### C. Optimal Duty Ratio Maximum Power Point Tracking

MPPT is defined as a technique that obtains the maximum possible power from one or more solar panels. The optimal duty ratio MPPT method is used to obtain the optimal duty ratio of a DC/DC converter that will yield the optimal current and voltage values of the solar panel [14-15]. The SEPIC is chosen for the MPPT due to its non-inverter output and ability to increase or decrease its output voltage. The relationship between the input and the output resistance is shown in equation (3).

$$
R_{out} = \frac{V_{out}}{I_{out}} \quad (3)
$$

$I_{out}$ and $V_{out}$ are the output current and voltage These can be defined as a mathematical relation between the duty ratio $D$, and the input current $I_{in}$ and output current $in_c$. This is shown in equation (4) and (5).

$$
I_{out} = I_{in} \cdot \left( \frac{1-D}{D} \right) \quad (4)
$$

$$
V_{out} = V_{in} \cdot \left( \frac{D}{1-D} \right) \quad (5)
$$

Substituting equations (4) and (5) in equation (3) yields equation (6).

$$
R_{out} = \frac{D^2 \cdot V_{in}}{(1-D)^2 \cdot I_{in}} \quad (6)
$$

Further expanding equation (6), by using ohms' law, yields equation (7).

$$
R_{out} = \frac{D^2 \cdot R_{in}}{(1-D)^2} \quad (7)
$$

Solving this equation for the duty cycle $D$, will yield equation (8).

$$
D_{op} = \frac{\sqrt{R_{out}}}{\sqrt{R_{out}} + \sqrt{R_{op}}} \quad (8)
$$

The optimal duty ratio $D$, as a relationship of the optimal voltage $V_{op}$, and output voltage $V_{out}$, is shown in equation (9).

$$
D_{op} = \frac{V_{out}}{V_{out} + V_{op}} \quad (9)
$$

## III. EXPERIMENTAL RESULTS

The CubeSat EPS is the stepping stone to develop more advanced Puerto Rico CubeSat EPS designs. It is the third-generation Puerto Rico CubeSat EPS of its kind. The CubeSat's EPS prototype is constructed on a two layered 9cm x 9cm PCB, weighing 0.077kg, to comply with the CubeSat size and weight limitations. The top layer is the main power loop, composed of a SEPIC designed to perform the MPPT using the optimal duty ratio algorithm, and two Buck converters that provide voltage regulation for the 3.3V and 5.0V payloads using a proportional control. The bottom layer is composed of gate driver circuits for the switching transistor of the DC/DC converters. An *Atmega328* microcontroller is tasked with generating the required PWM signals for the DC/DC converters as well as measuring the output voltages of each DC/DC converter to perform the optimal duty ratio MPPT technique and the closed loop voltage feedback. Fig. 9 illustrates the Puerto Rico CubeSat EPS prototype.

The CubeSat EPS testbed is composed of a *Tektronix 4034 MOS Mixed Signal oscilloscope* to measure current and voltage signals, and an *Agilent E4351B Solar Array Simulator* that emulates the CubeSat's GaAs PVM array. The evaluation process for the CubeSat EPS consists of testing the performance of the SEPIC as well as the Buck converters. The SEPIC is tested to guarantee that it executes the optimal duty ratio MPPT technique while at the same time charging the CubeSat's battery. The battery used to obtain the experimental results is rated at 8.4V. To test the optimal duty ratio MPPT technique, the *Agilent E4351B Solar Array Simulator* is programmed to operate at varying solar irradiance conditions for a period of 10s. The Buck converters are tested to guarantee voltage regulation at different loads. For each Buck converter, a payload sequence is used, incrementing the current each step for a period of 10s. These payloads are emulated using resistive loads calculated based on voltage and current parameters of the Puerto Rico CubeSat's existing payloads.

Using the optimal duty ratio MPPT technique it is possible to extract the maximum available power from the PVM by varying the duty cycle of the SEPIC. The MPPT technique samples the open circuit of the PVM every 5ms to determine the optimal duty ratio for each specific open circuit voltage present at different solar irradiance conditions. To ensure that the MPPT is extracting the optimal power, measurements are made to the PVM voltage and current at different solar irradiance conditions. For the experimental results, tests are performed at a solar irradiance level of 200W/m² and incremented by a step of 200W/m² until reaching 1,200W/m² (200W/m², 600W/m², 400W/m², 800W/m², 1,000W/m² and 1,200W/m²). Fig. 10 through 12 illustrates the experimental results obtained from the optimal current, voltage and power drawn from the PVM array using the optimal duty ratio MPPT technique.

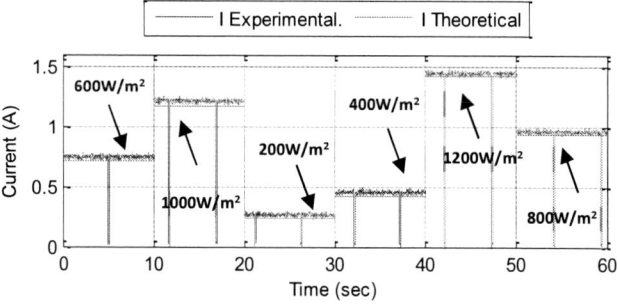

Fig. 10. SEPIC optimal PV module current at varying irradiance levels.

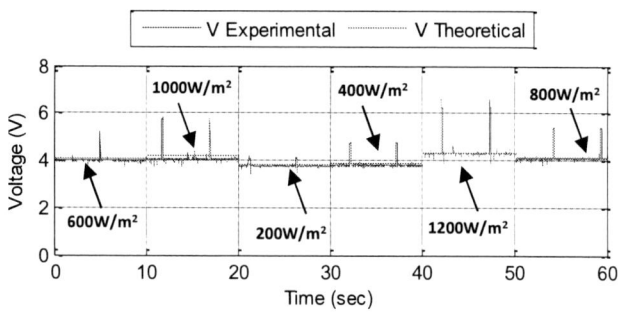

Fig. 11. SEPIC optimal PV module voltage at varying irradiance levels.

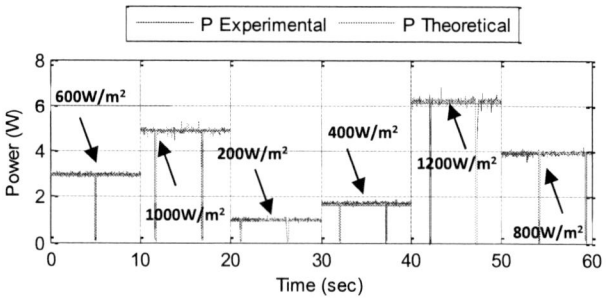

Fig. 12. SEPIC optimal PV module power at varying irradiance levels.

Fig. 9. The Puerto Rico CubeSat EPS prototype. The EPS prototype is a double-sided PCB; the top layer composed of DC/DC converters while the bottom layer is composed for gate driver circuits.

Notice from the experimental results obtained in Fig. 10 and Fig. 11, the SEPIC can execute the optimal duty ratio MPPT technique to extract the optimal current and voltage values from the PVM array. Voltage dips present every 5ms are a result of measuring the open circuit voltage of the PVM array. Fig. 12 illustrates a comparison between the theoretical and experimental values, demonstrating that the SEPIC can extract the maximum power at varying solar irradiance conditions.

For the CubeSat EPS, to test the performance of the 3.3V Buck converter, different loads are connected to the output of the converter to emulate case scenarios where the system is operating under different payloads. The output voltage and current of the Buck converter are also measured to calculate the output power. For the payload sequence of the 3.3V Buck converter, the current is increased by connecting the Flight Computer, IMU, Magnetorquers and GPS Receiver. Fig. 13 through 14 illustrates the experimental results obtained by measuring the output current and voltage of the 3.3V Buck converter with feedback. Fig. 15 illustrates the power output of the 3.3V Buck converter.

The experimental results of Fig. 13, illustrate how as the load is varied at the output of the Buck converter, the output current is kept at the desired values for all payload variations. Fig. 14 illustrates how the output voltage of the Buck converter remains at a constant 3.3V, despite these load variations being present. Fig. 15 illustrates the supplied power for each payload variation.

The performance of the 5.0V Buck converter is also tested to guarantee voltage regulation. For the payload sequence of the 5.0V Buck converter, the current is increased by connecting the reaction wheel, GPS antenna, MADS, Transceiver. Fig. 16 through 17 illustrate the experimental results obtained from the output current and voltage of the 5.0V Buck converter. Fig. 18 illustrates the power output of the 5.0V Buck converter. The experimental results of Fig. 16, illustrate how as the load is varied at the output of the converter, the output current is kept at the desired values for all payload variations. Fig. 17 illustrates how the output voltage of the Buck converter remains at a constant 5.0V, despite load variations at the output of the converter. Fig. 18 illustrates the supplied power for each payload variation.

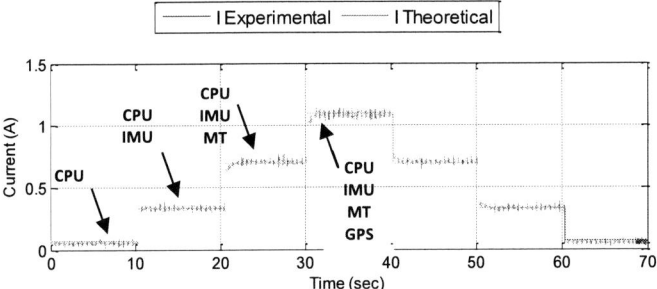

Fig. 13. Output current of the 3.3 V Buck converter.

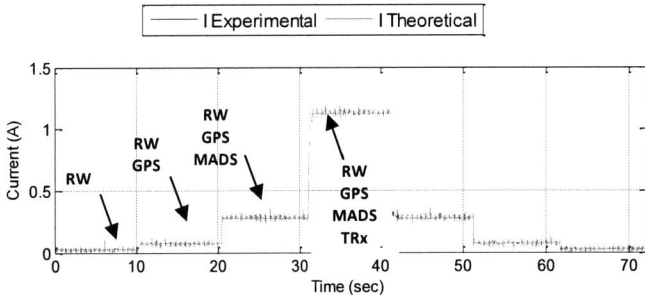

Fig. 16. Output current of the 5.0 V Buck converter.

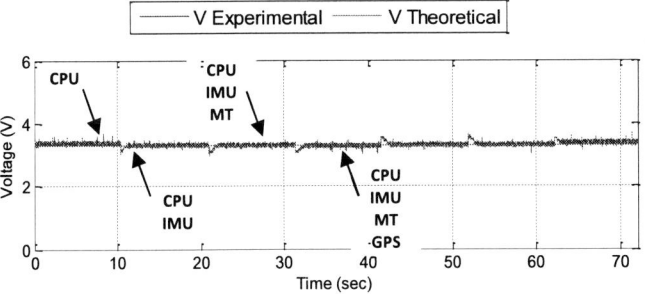

Fig. 14. Output voltage of the 3.3 V Buck converter.

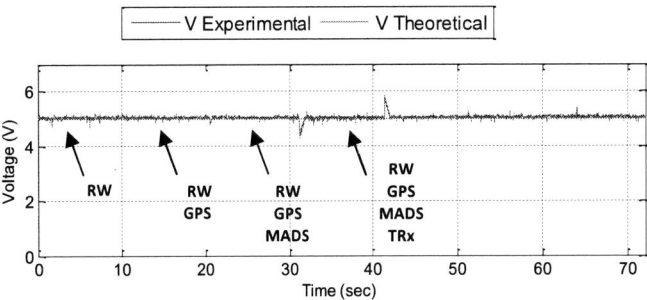

Fig. 17. Output voltage of the 5.0 V Buck converter.

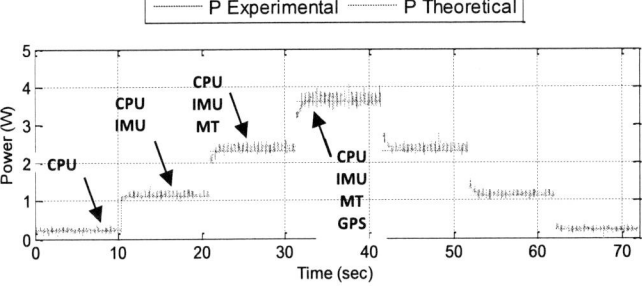

Fig. 15. Output power of the 3.3 V Buck converter.

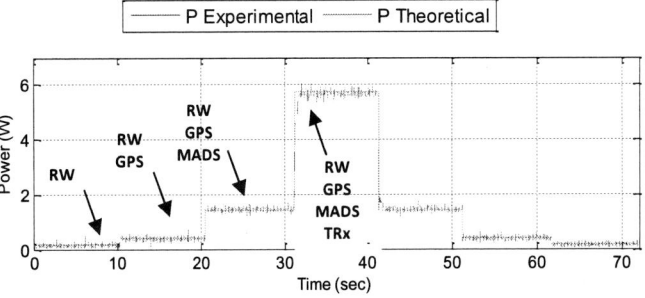

Fig. 18. Output power of the 5.0 V Buck converter.

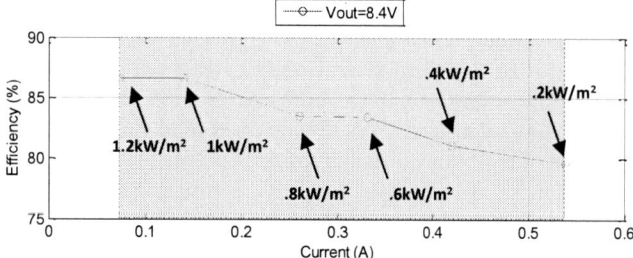

Fig. 19. SEPIC efficiency at varying output current.

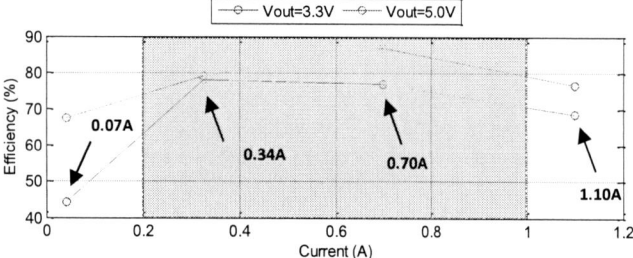

Fig. 20. Buck converter efficiency at varying output current.

Fig. 19 and Fig. 20 illustrate the efficiency results obtained for the SEPIC and Buck converters respectively. The efficiency of the SEPIC is 87% when exposed to its lowest irradiance condition, as the solar irradiance increases, the efficiency of the system drops to 80%. From these results, it can be concluded that the SEPIC operates between 87% and 80% when exposed to low and high solar irradiance levels. Results indicate that the 3.3V Buck converter operates at 43% at low currents and 69% at high currents. The 5.0V Buck converter operates at 68% at low currents and 78% at high currents. As the input voltage of the Buck converter is increased, so does the efficiency.

## IV. CONCLUSION

The design and the testing of an EPS for the Puerto Rico CubeSat is demonstrated. The design for the CubeSat EPS consists of: a SEPIC that performs the MPPT, and two Buck converters that function as load voltage regulators that will supply the 3.3V and 5.0V payloads. To extract the utmost available power from the PVM array, a MPPT method that uses the optimal duty ratio of the SEPIC is implemented. From the results, the optimal duty ratio MPPT technique can track and extract the optimal current and voltage values of the PVM. The obtained results also illustrate that the EPS can supply 3.3V and 5.0V while operating under different payloads.

## ACKNOWLEDGEMENT

The authors acknowledge the contributions made by Sandia National Laboratories (SNL), the National Nuclear Security Administration (NNSA), the US Department of Energy (DOE) and the Consortium for Integrating Energy Systems in Engineering and Science Education (CIESESE), for the sponsorship of this project, as well as the financial support of the NASA Puerto Rico Space Grant Consortium (PRSGC) under grant NX15AI11H.h

## REFERENCES

[1]. C. Edwards, "The cubed route", *Engineering & Technology*, vol.8, no.2, pp.68-72, Mar. 2013.

[2]. R. Darbali-Zamora, E.I. Ortiz-Rivera, A.A. Rincon-Charris, "The Puerto Rico CubeSat Project to attract STEM Students into the Area of Aerospace Engineering", *IEEE Frontiers in Education (FIE'15)*, Oct. 21-24, 2015, pp.1-7.

[3]. R. Rose, J. Dickinson, A. Ridley, "CubeSats to NanoSats: bridging the gap between educational tools and science workhorses", *IEEE Aerospace Conference*, Mar. 3-10, 2012, pp.1-11.

[4]. A.R. Aslan, H.B. Yağcı, M.E. Umit, A. Sofyalı, M.E. Bas, M.S. Uludag, O.E. Ozen, M.D. Aksulu, E. Yakut, C. Oran, M. Suer, İ.A. Akyol, A.B. Ecevit, M.S. Ersöz, İ. Öz, Ş. Gülgönül, B. Dinç, T. Dengiz, "Development of a LEO communication CubeSat", *6th International Conference on Recent Advances in Space Technologies (RAST'13)*, 2013, pp.637-641.

[5]. S. Notani, S. Bhattacharya, "Flexible electrical power system controller design and battery integration for 1U to 12U CubeSats", *IEEE Energy Conversion Congress and Exposition (ECCE'11)*, 2011, pp.3633-3640.

[6]. R. Darbali-Zamora, D.A. Merced-Cirino, C.S. Gonzalez-Ortiz, E.I. Ortiz-Rivera, "An Electric Power Supply design for the space plasma ionic charge analyzer (SPICA) CubeSat", *IEEE 40th Photovoltaic Specialist Conference (PVSC'14)*, June 8-13, 2014, pp.1790-1795.

[7]. R. Darbali-Zamora, D.A. Merced-Cirino J. Rivera-Alamo, E.I. Ortiz-Rivera, A.A. Rincon-Charris, "Design and Thermal Testing of a Power Supply Prototype for the Space Plasma Ionic Charge Analyzer (SPICA) CubeSat", *42th IEEE Photovoltaic Specialists Conference (PVSC'15)*, June 14-19, 2015, pp.1-6.

[8]. E.I Ortiz-Rivera, "Modeling and Analysis of Solar Distributed Generation," Ph.D. dissertation, Department of Electrical and Computer Engineering, Michigan State University, 2006.

[9]. E.I. Ortiz-Rivera, F.Z. Peng, "Algorithms to Estimate the Temperature and Effective Irradiance Level over a Photovoltaic Module using the Fixed-Point Theorem", *37th IEEE Specialists Conference in Power Electronics (PESC'06)*, June 18-22, 2006, pp.1-4.

[10]. D.W. Hart, "DC-DC Converters", in *Power Electronics*, 1st ed. New York, USA, The McGraw-Hill Companies, inc., 2011, ch. 6, sec. 2, pp.197.

[11]. S.P. Priya, A. Radhika, T.D. Vinothini, "MPPT and SEPIC based controller development for energy utilization in CubeSats", *IEEE India International Conference (INDICON'12)*, Dec. 7-9, 2012, pp.143-148.

[12]. J.E. Salazar-Duque, E.I. Ortiz-Rivera, J. Gonzalez-Llorente, "Analysis and non-linear control of SEPIC dc-dc converter in photovoltaic systems", *IEEE Workshop on Power Electronics and Power Quality Applications (PEPQA'15)*, June 2-4, 2015, pp.1-6.

[13]. N. Mohan, W.P. Robbin, T. Undeland, in *Power Electronics: Converters, Applications, and Design*, 3rd ed. New Jersey, USA, John Wiley and Sons, inc., 2003, ch. 7, sec. 3, pp.164-172.

[14]. E.I. Ortiz-Rivera, "Maximum power point tracking using the optimal duty ratio for DC-DC converters and load matching in photovoltaic applications", *23rd Annual IEEE Applied Power Electronics Conference and Exposition (APEC'08)*, Feb. 24-28, 2008, pp.987-991.

[15]. R. Darbali-Zamora, E. I. Ortiz-Rivera; *IEEE Andean Council International Conference (ANDESCON'16)*, "Optimal Duty Ratio Maximum Power Point Tracking Technique Using the SEPIC Topology for Photovoltaic Systems Applications", Oct. 19-21, 2016.

# Virtual Power Plant Feedback Control Design for Fast and Reliable Energy Market and Contingency Reserve Dispatch

Mohamed Elkhatib, Jay Johnson, and David Schoenwald

Sandia National Laboratories, Albuquerque, New Mexico, 87185, USA

*Abstract* — **An increasing number of state and national interconnection standards are requiring Distributed Energy Resources (DER) to include grid-support functionality. These capabilities along with the growing number of communications-enabled DER make it possible for 3$^{rd}$ party aggregators to provide a range of high-level grid services such as voltage regulation, frequency regulation, and contingency reserves. For the last three years, Sandia National Laboratories has been designing and testing a real-time Virtual Power Plant (VPP) optimization and control platform to provide ancillary services with interoperable DER. In this paper we address the design of feedback controllers for VPPs to meet energy market and tertiary reserve targets. The VPP controller is designed to issue set points to the fleet of DERs to maintain the VPP output within the error margin. This is accomplished by compensating for individual DER losses and output fluctuations with the remainder of the aggregation. The impact of the communication network on the controller design is discussed and simulation results are presented to validate the proposed controller design.**

*Keywords—Virtual Power Plants, Frequency regulation, Distributed Energy Resources, Ancillary services*

## I. INTRODUCTION

The national trend of increasing renewable energy (RE) penetrations is a worst-case scenario for bulk system reliability as grid inertia and governor control are displaced and frequency deviations from RE variability are increasingly common [1]. Therefore, instituting frequency response reserves with DER in accordance with utility, Independent System Operator (ISO)/Regional Transmission Organization (RTO), and NERC requirements are critical for future grid resiliency. Due to the sheer number of DERs and their small sizes, it is not practical for bulk system operators to optimize and control individual DERs. In that regard, a VPP represents a framework for cohesive optimization and control of large numbers of small DERs which are then seen as a single entity by grid operators. VPPs provide grid support services using robust communications, robust control, and efficient optimization of large and diverse sets of DER; and ultimately, this functionality may eliminate the need for dedicated ancillary service thermal plants entirely.

In this work, we choose to focus on the challenge of providing energy market power and tertiary contingency reserves with a VPP. One distinct feature of a VPP serving contingency reserves is that it does not need to have a single point of connection to the grid, but instead composed of an aggregation of different DER sources that connect to the grid at geographically diverse points of common coupling. Therefore, the VPP could be used to aggregate distributed generators, energy storage systems, entire microgrids, demand response units, electric vehicles and even entire distribution stations across an interconnection.

In general, VPPs may be composed of grid operator-owned assets or privately-owned DER that are controlled under a legal agreement. In the case of operating in regions with vertically-integrated utilities, the VPP would be economically dispatched as part of unit commitment planning [2]. In market-based jurisdictions, the VPP would submit offer into the energy or reserve markets [3].

VPP design and optimization has been the subject of number of recent studies. The European Union (EU) has sponsored projects to create a VPP composed of fuel cell DER [4] and the EU FENIX project investigated (a) technical VPPs consisting of DER in one geographical region that accounted for the local power network (e.g., voltage regulation) and (b) commercial VPPs designed to bid into wholesale and other markets [5]. Many researchers have studied VPP bidding mechanisms and market interactions. Centralized bidding strategies for VPPs have been investigated extensively [6]-[8] and a detailed optimization formulation to optimize the day-ahead thermal and electrical scheduling of large scale VPPs has been proposed [9]. Once the VPP is contracted for power delivery, a control system must issue commands to DER to produce the desired aggregate power. A few dispatch architectures have been proposed, including direct, hierarchical and distributed management architectures for VPPs [10] and decentralized multi-agent based techniques for VPP operations [11]-[12]. However, there is little emphasis in the literature on the design and implementation of real-time feedback control for VPP operations.

The main goal of the VPP control system presented herein is to ensure that the real-time total output of the VPP is maintained within an acceptable error margin. This control task is challenging for several reasons. First, the presence of renewable energy DERs in the VPP cause the VPP output to fluctuate. Second, the VPP controller should compensate in real-time for the loss of any particular DER (communication failures, DER disconnection, etc.) or the inability of any DER to attain its reference power output. Third, due to the geographical diversity of DERs in the VPP, a communication

978-1-5090-5606-4/17 $31.00 © 2017 IEEE

network must connect the VPP controller to the DERs through public internet channels. Unlike many previous VPP implementations, DERs included in this work extend down to the residential level (e.g., rooftop microinverters on homes). The presence of the internet-based communication network introduces additional difficulties to the design of the control system due to the effects of communication latencies and data loss. In this work, we present a VPP controller design utilizing PID and proportional controllers to provide fast, reliable aggregate power production.

## II. VIRTUAL POWER PLANT DESIGN

The Sandia VPP is designed with modular components which run as multi-processing servers in a Python environment. The components of the VPP interact to exchange pertinent information through a backend process. The components in the VPP are:

- A forecasting component provides long-term (24-60 hours) forecast of RE anticipated power to the *commitment engine* which provides offers to the ISO/RTO markets, and short term (0-12 hours) forecasts to the *optimization engine*.
- A stochastic *commitment engine* determines the VPP energy and reserve bids based on the maximum expected profit for the required market time period (e.g., day-ahead). A heavy penalty is applied in cases where the VPP cannot meet the power commitments so bids are conservative.
- Once the energy and reserve commitments are established, the stochastic *optimizer* minimizes the operating cost of the VPP over the next 24 hours by determining the setpoints for the DER devices. The optimizer monitors the status of the DERs and—based on short-term DER forecasts and DER availability—the optimizer will charge energy storage systems or start gensets to maintain enough headroom to always meet the commitment.
- To quickly and consistently reach the desired VPP power output, a *centralized controller* is employed to quickly adapt to changes in DER availability, RE power changes, and other DER interoperability or equipment failures.

More detail of the VPP and associated components will be forthcoming in [13].

## III. VPP CONTROLLER DESIGN

The VPP controller receives optimal dispatch setpoints for each DER from the optimization routine at a specified interval (e.g. every 15 minutes). From this starting operating condition, the VPP controller is responsible for keeping the total output of the VPP within an acceptable error margin from the VPP reference power defined by:

$$VPP_{ref} = E(t) + \alpha(t)R(t)$$

$$(1)$$

where $E$ is the energy market commitment, $R$ is the reserve commitment, and $\alpha$ is a binary variable indicating if the reserve is required at time, $t$.

The design of VPP controllers is challenging for different reasons. First, VPPs aggregate heterogeneous DERs with wide ranges of ramp rates which makes it hard to tune the controller and ensure stable response. Second, the controller has to compensate for small variations in the VPP output due to the variability of RE DER resources and respond to changes in VPP output due to unexpected DER tripping or communication failure. Third, reliance on communication network introduces significant latencies and a probability of data loss which could destabilize the controller.

Fig.1 shows a schematic overview of the proposed VPP controller structure. The optimizer (Optimization Block) resolves for the optimal DER dispatch settings every 15 minutes to account for changes in short-term forecast and other DER status changes — e.g., loss of DER communications. These new setpoints are issued to the VPP controller to re-adjust DER reference powers. The proposed controller consists of the Feedback Controller and the Re-dispatch Processor as detailed below.

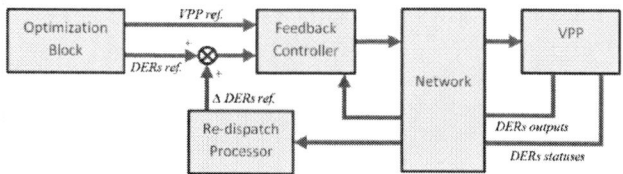

Fig. 1. VPP controller consisting of feedback control and re-dispatch processor.

### A. Feedback Controller

The feedback controller is responsible for maintaining the VPP output at the target level by compensating for changes in DERs outputs. The proposed controller structure is shown in Fig. 2 for a VPP with three DERs—though this architecture can be expanded to any number of devices. The controller uses overall VPP error to derive the output of different DERs. Due to the wide range of DER ramp rates, only one DER is equipped with PID controller and the rest of the DERs are equipped with proportional gain control to avoid output ringing.

The DER equipped with PID controller is designated the swing DER of the VPP and is responsible for smoothing the output of the VPP and eliminating any steady state errors. Typically a large storage-based DER should be used as a swing DER to ensure adequate controller response because of its fast ramp ability.

978-1-5090-5606-4/17 $31.00 © 2017 IEEE          2970

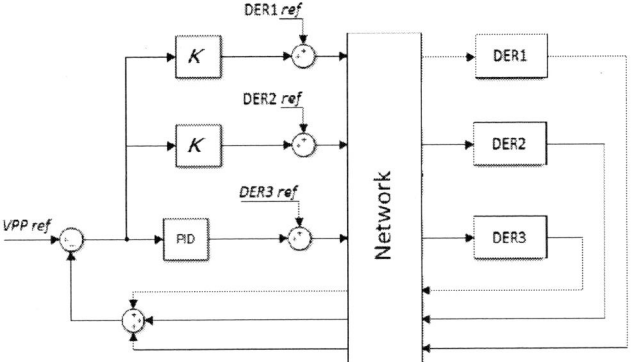

Fig. 2.    VPP feedback controller structure.

### B.  Re-dispatch Processor

As shown in Fig. 2, during real-time operation, for large VPP errors, DERs may drift significantly from their reference powers determined by the VPP optimizer. As a result, the VPP operates in a suboptimal economical state. One possible solution to this problem is to actively re-adjust DERs reference powers in real-time to ensure that the VPP output is restored using the most economical DERs. Due to time constraints of real-time operation, it is hard to formulate and solve a complete optimization problem in the re-dispatch processor. However, the initial DER reference powers from the optimizer represents the most economical solution using DER cost curves to meet the VPP bids. Therefore, we propose to dispatch DERs proportional to their initial reference powers. In other words, if $P_{error}$ is the difference between the VPP reference and actual powers—due to communication failures, renewable energy reductions, or tripping of DER $k$—then for each available DER $i$ in the VPP, the reference output power will be updated as follows.

$$\Delta P_i = P_{error} \frac{P_{i,initial}}{\sum_{m=1,m\neq k}^{N} P_{m,initial}} \tag{2}$$

$$\Delta P_{i,new} = P_{i,initial} + \Delta P_i \tag{3}$$

where, $P_{i,\,initial}$ is the initial output power of DER $i$ before the contingency $P_{m,\,initial}$ denote the output power of DER $m$.

Note that, once updated reference powers are received from the optimization engine at the beginning of the subsequent optimization period, DERs will follow the new reference powers and the re-dispatch processor will be reset.

### IV. CONTROL SIMULATIONS

In order to study the impact of different factors on the performance of the VPP controller, a simulated collection of DERs was created based on the equipment located at Mesa del Sol (MdS), Public Service Company of New Mexico (PNM) Prosperity Site, and Sandia's Distributed Energy

Technologies Laboratory (DETL) in Albuquerque. The equipment at MdS and Prosperity sites was controlled previously for PV smoothing [14]-[15] so this collection of devices could form a VPP with the correct control structures. The DER included in the simulations is listed in Table 1 with their size, dispatchable power levels, and swing settings.

In order to create a stable VPP controller first the swing controller settings were determined and then the gain was selected for the non-swing DER.

Table I: DER VPP PARAMETERS

| DER | Size (kW) | Dispatch-able Power (kW) | Swing? |
|---|---|---|---|
| Miller Cycle Genset at MdS | 240 | 200 | No |
| Diesel Genset at DETL | 250 | 90 | No |
| Battery at Prosperity Site | 500 | 300 | Yes |
| PV at Prosperity Site | 500 | 500 | No |
| Battery at MdS | 163 | 140 | No |
| Fuel cell at MdS | 80 | 40 | No |
| Rooftop PV at MdS | 100 | 100 | No |
| Eight Inverters at DETL | 8 x 3 | 24 | No |

### A.  Controller Tuning

The swing PID controller for the above VPP was tuned using the Ziegler–Nichols method [16]. The VPP scenario in Table II was simulated, shown in Fig. 3, to illustrate the basic operation of the VPP controller with different controller parameters. For each setpoint command issued to the DERs, a delay and probabatility of packet loss were simulated. The simulation time step was set to 0.01 s but the control setpoints were only recalculated and re-issued every 0.2 seconds to represent the communication delay in sending and receiving power data from the equipment. Fig. 3 shows the performance of the VPP controller which quickly reaches the VPP power reference, but with different overshoot levels and settling times. The final swing control parameters were chosen to be $K_p = 0.7$, $K_i = 1.0$, and $K_d = 0$.

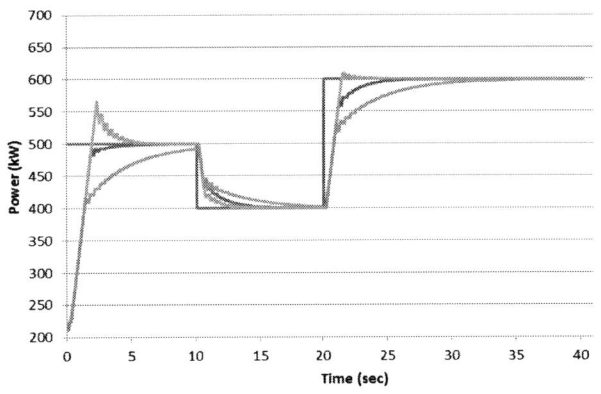

Fig. 3.    Swing PID parameters influence on the response of the VPP.

Table II: VPP OPERATION SCENARIO

| Time (s) | Energy Market Power Commitment (kW) | Reserve Market Power Commitment (kW) | Reserve Request, α |
|---|---|---|---|
| 0 | 500 | 200 | 0 |
| 10 | 400 | 200 | 0 |
| 20 | 400 | 200 | 1 |

Once the swing controller PID settings were selected, the gain for the non-swing DER was determined. All the PV systems included in Table I were simulated by replaying one of seven 24-hour AC power 1-second datasets recorded and scaled from the 500 kW Prosperity Site PV plant. The effect of $K_p$ gain on the VPP response is shown in Fig. 4. The final non-swing DER gain was selected to be 0.1.

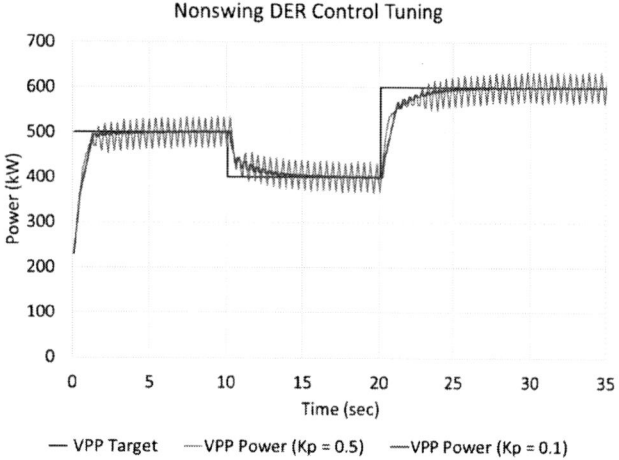

Fig 4: VPP response for two non-swing gains.

### B. Impact of Communication Rate and Delay

The scenario from Table II was repeated with different communication rates. The control rate is the speed at which new setpoints are issued to the DER and represents the aggregate time to measure the DER outputs and issue new setpoints. Communications to physical DER devices at DETL takes approximately 0.2 sec, which does not significantly influence the stability or effectiveness of the VPP controller, as shown in Fig. 5. It is clear from Fig. 6 that the slower the controller rate is, the more oscillations will appear in the swing DER response as well at the VPP power.

The impact of different communication delays on the VPP response was studied as well. After control information is issued to the DER, the device does not respond for a period of time while the data packet is routed through the communication network. Depending on the transport media, communication protocol, and network topology this time could be quite short (< 10 ms) or relatively long (seconds). In the past, this was a challenge in the MdS and Prosperity PV smoothing project [14] and was ultimately a challenge for the

VPP, as described below. Simulations of 100 and 150 ms delays showed the VPP controller was robust to some network latency. The delay in the DER output from network delay is seen when the VPP target changes in Fig. 7.

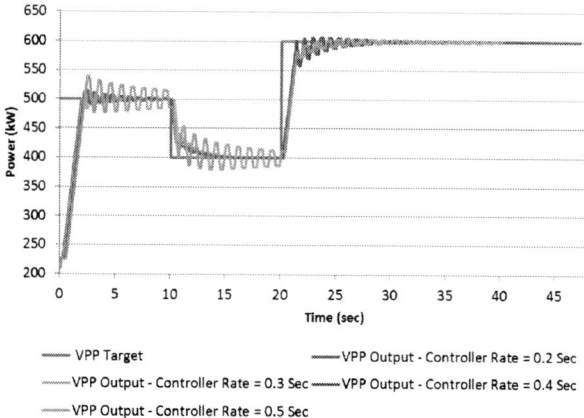

Fig. 5.    VPP Output under different controller rates.

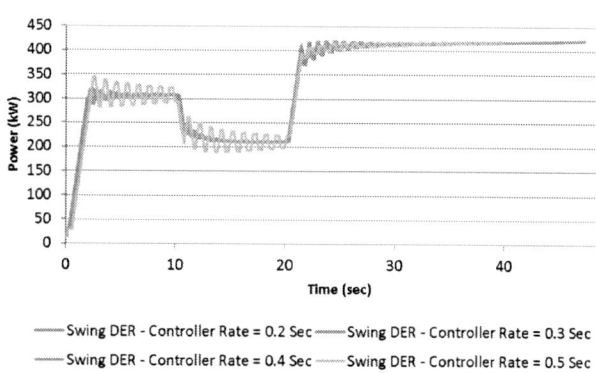

Fig. 6.    Swing DER output under different controller rates.

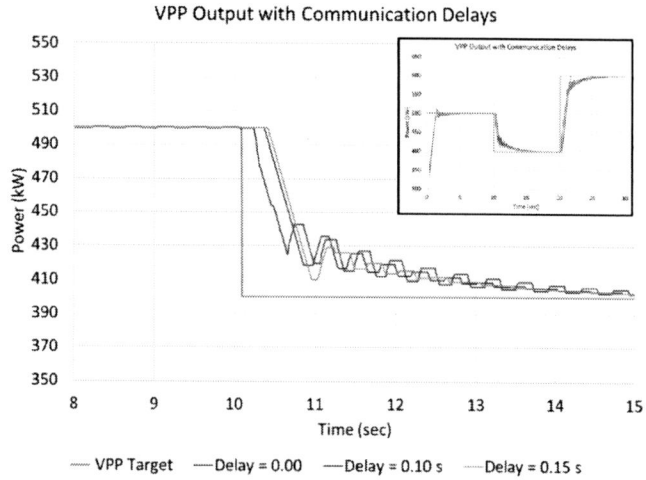

Fig. 7.    Influence of DER Delay on VPP output.

## IV. EXPERIMENTAL RESULTS

### A. VPP Control with Simulated DERs

The VPP was run with the tuned control settings, 0.2 second communication rate and no network delay for the scenario shown in Table II. The output of the VPP and the DER devices is shown in Fig. 8.

Next, the commitment and optimization engines were run to determine the energy and reserve bids and DER setpoints for a day in June 2017 based on live forecasts of the DER assets. Data from the controller was captured for 40 seconds with the reserve being called at t = 20 s. The response of the VPP is shown in Fig. 9.

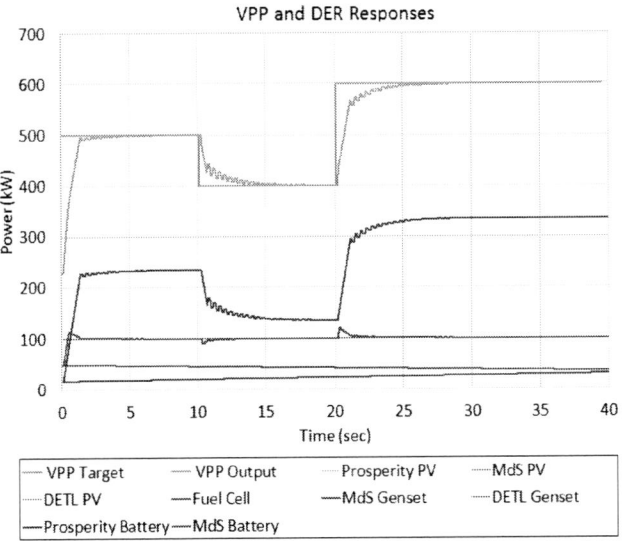

Fig. 8.   VPP and DER outputs for a commitment scenario.

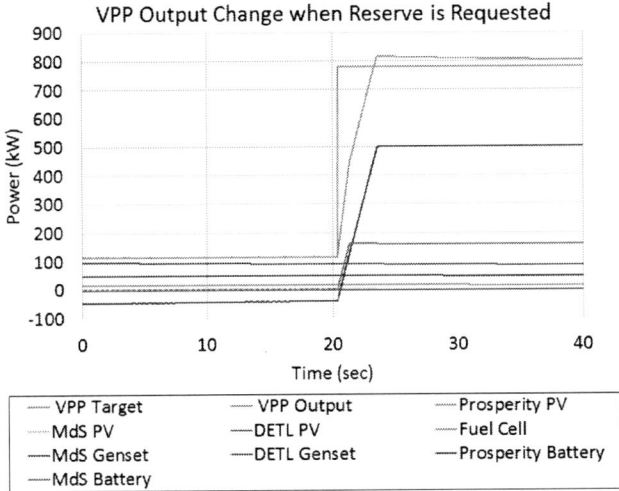

Fig. 9.   VPP and DER outputs based on commitment and optimization targets at a time when the reserve is requested.

### B. VPP Control with Real DERs

To validate the VPP control with a real communication network, three PV inverters in DETL were issued curtailment commands from the VPP dispatch controller via SunSpec Alliance Modbus TCP commands. The DER output power was sequentially read and then the level of active power curtailment of the DER equipment was adjusted. An example of the PV controls reaching a specified power level is shown in Fig. 10. In cases where there was insufficient PV power available, the active power level was not meet, as shown in Fig. 11.

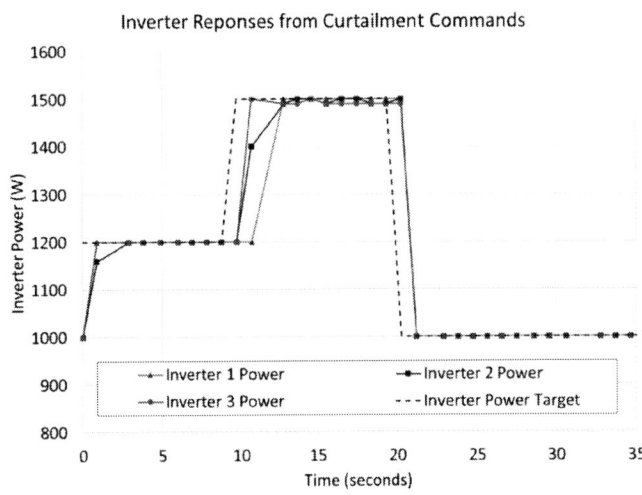

Fig. 10.   Response of three inverters to a target power signal.

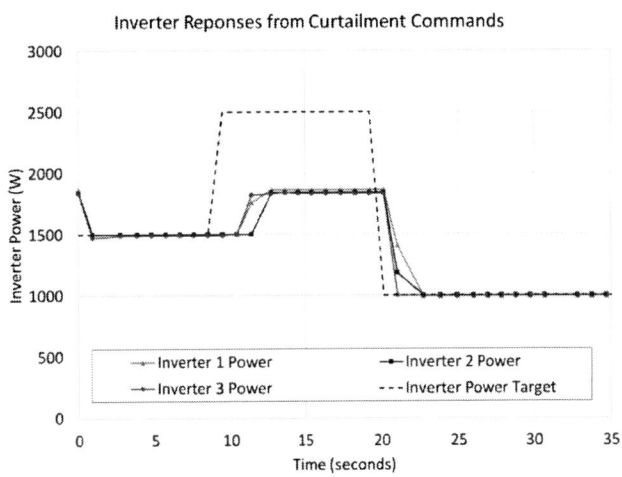

Fig. 11.   Response of three inverters to a target power signal, where the power level is above the available power of the renewable source.

Using simulated DER, the control loop was configured to execute in 0.01 seconds, but when adding the physical devices the loop time increased and the duration became variable. As shown in Fig. 12, the read times for the DER was consistently ~200 ms for the inverters, but the write times varied between

~50 and ~1200 ms and the tuned VPP controls were no longer effective. In order to have stable control, the loop time must be consistent, so the variability forced VPP operator to execute the control loop at the largest duration, i.e., 2 seconds. This control speed produced poor VPP system performance. One option to improve the VPP response would be to issue set points via parallelized communications, as opposed to sequentially. This would also allow the VPP to scale as more DER resources are added to the pool.

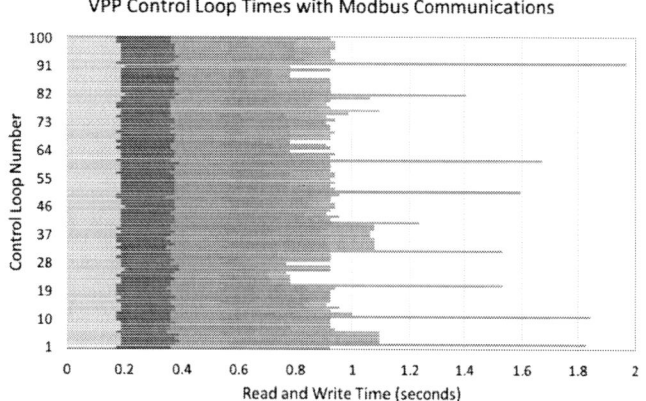

Fig. 12.    Inverter read and write rates for three physical DERs.

## V. CONCLUSIONS

In this paper, a centralized feedback control architecture for virtual power plants was proposed to maintain a reference power output in the presence of individual DER output fluctuations and losses. Simulation results demonstrate the effectiveness of the proposed method for simulated DER devices. The response time and overshoot are appropriate for providing energy and reserve power to ISO/RTO market. In the case of using physical devices to provide this service, significant communication times prevented real time operations. It is recommended to use communication dispatchers, multi-threading or multicast communications to control VPPs to avoid communications-related scaling issues.

## ACKNOWLEDGEMENTS

The authors would like to thank Anya Castillo, Jack Flicker, Cliff Hansen and the rest of the VPP team for their support completing this work.

Sandia National Laboratories is a multi-mission laboratory managed and operated by National Technology and Engineering Solutions of Sandia, LLC., a wholly owned subsidiary of Honeywell International, Inc., for the U.S. Department of Energy's National Nuclear Security Administration under contract DE-NA-0003525.

## REFERENCES

[1]  C. Martinez, S. Xue, and M. Martinez, "Review of the recent frequency performance of the Eastern, Western and ERCOT Interconnections," Lawrence Berkeley National Laboratory, Tech. Rep., December 2010.

[2]  N.P. Padhy. Unit commitment – a bibliographical survey, IEEE Transaction on Power Systems 19(2):1196–1205, 2004.

[3]  M. Shahidehpour, H. Yamin, and Z. Li. Market Operations in Electric Power Systems: Forecasting, Scheduling, and Risk Management, Wiley-IEEE Press, 2002.

[4]  A. Dauensteiner, "European virtual fuel cell power plant," Management Summary Report, Feb 2007.

[5]  D. Pudjianto, C. Ramsay, G. Strbac, and M. Durstewitz, "The virtual power plant: Enabling integration of distributed generation and demand," FENIX Bulletin 2, Feb 2008.

[6]  E. Mashhour and S. Moghaddas-Tafreshi, "Bidding strategy of virtual power plant for participating in energy and spinning reserve markets - Part I: Problem formulation," Power Systems, IEEE Transactions on, vol. 26, no. 2, pp. 949–956, May 2011.

[7]  E. Mashhour and S. Moghaddas-Tafreshi, "Bidding strategy of virtual power plant for participating in energy and spinning reserve markets - Part II: Numerical analysis," Power Systems, IEEE Transactions on, vol. 26, no. 2, pp. 957–964, May 2011.

[8]  D. Pudjianto, C. Ramsay, and G. Strbac, "Virtual power plant and system integration of distributed energy resources," Renewable Power Generation, IET, vol. 1, no. 1, pp. 10–16, March 2007.

[9]  M. Giuntoli and D. Poli, "Optimized thermal and electrical scheduling of a large scale virtual power plant in the presence of energy storages," Smart Grid, IEEE Transactions on, vol. 4, no. 2, pp. 942–955, June 2013.

[10] A. Raab, et al., "Virtual power plant control concepts with electric vehicles," in Intelligent System Application to Power Systems (ISAP), 2011 16th International Conference on, Sept 2011, pp. 1–6.

[11] M. Vasirani, R. Kota, R. Cavalcante, S. Ossowski, and N. Jennings, "An agent-based approach to virtual power plants of wind power generators and electric vehicles," Smart Grid, IEEE Transactions on, vol. 4, no. 3, pp. 1314–1322, Sept 2013.

[12] H. Yang, D. Yi, J. Zhao, and Z. Dong, "Distributed optimal dispatch of virtual power plant via limited communication," Power Systems, IEEE Transactions on, vol. 28, no. 3, pp. 3511–3512, Aug 2013.

[13] J. Johnson, et al. Design and Implementation of a Secure Virtual Power Plant, Sandia Technical Report, 2017 (forthcoming).

[14] J. Johnson, A. Ellis, A. Denda, K. Morino, T. Shinji, T. Ogata, M. Tadokoro, "PV Output Smoothing using a Battery and Natural Gas Engine-Generator," 39th IEEE Photovoltaic Specialists Conference, Tampa Bay, Florida, 16-21 Jun, 2013.

[15] J. Johnson, K. Morino, A. Denda, J. Hawkins, B. Arellano, T. Ogata, T. Shinji, M. Tadokoro, A. Ellis, "Experimental Comparison of PV-Smoothing Controllers using Distributed Generators," Sandia Technical Report SAND2014-1546, Feb 2014.

[16] K. Ogata, Modern Control Engineering, Prentice Hall, 2010.

# Intelligent Sampling of Periods for Reduced Computational Time of Time Series Analysis of PV Impacts on the Distribution System

Jason Galtieri, Matthew J. Reno

Sandia National Laboratories, Albuquerque, NM, 87185, USA

*Abstract* — In this work, a sampling method, known as intelligent sampling (IS), is presented to reduce simulation time in Quasi Static Time Series (QSTS) analysis on electrics grids with distributed PV. The sampling method decomposes a year's worth of input solar and load data into six hour intervals and bins the intervals according to irradiance and load metrics. Representative samples are chosen from the bins and simulated using standard power flow solvers. We show that when using the IS method, only a fraction of the total entries in the year need to be simulated. An example test circuit is used and the IS method achieves a 57% reduction in simulation time while meeting acceptable error margins.

*Index Terms* — distributed power generation, photovoltaic systems, power distribution, power system interconnection.

## I. INTRODUCTION

The addition of renewable and distributed generation on the electric power system has altered traditional control techniques and grid analysis methods. Historically, in distribution feeders, power flows from the substation to the various loads along the feeder length with the voltage regulators reacting to changes in load and self-correcting to maintain normal operating voltages. Increasing penetrations of distributed PV can create significant power output fluctuations and reverse power flows along specific segments of the feeder, causing voltage limit violations and additional wear and tear on voltage regulation equipment [1].

In order to model the variability of distributed PV, high-resolution quasi-static time-series (QSTS) simulations are required to simulate the grid impact at different times of year and to determine any interactions between PV and existing voltage regulation equipment [2]. To capture the interactions and seasonal variations, accurate QSTS simulation should be performed at high-resolution (<5 second time-step) and for the duration of a year [3]. Certain QSTS metrics such as extreme voltages and line losses can be approximated using relatively large time steps, but voltage regulators and capacitor switching require time-steps on the order of one to a few seconds. These types of QSTS simulations are computationally intensive and can take days to perform for large distribution system models. The computational burden limits the practicality of QSTS for parametric analysis [4] or for the hundreds of PV interconnection requests that a utility receives.

This research was supported by the DOE SunShot Initiative, under agreement 30691. Sandia National Laboratories is a multimission laboratory managed and operated by National Technology and Engineering Solutions of Sandia, LLC., a wholly owned subsidiary of Honeywell International, Inc., for the U.S. Department of Energy's National Nuclear Security Administration under contract DE-NA0003525.

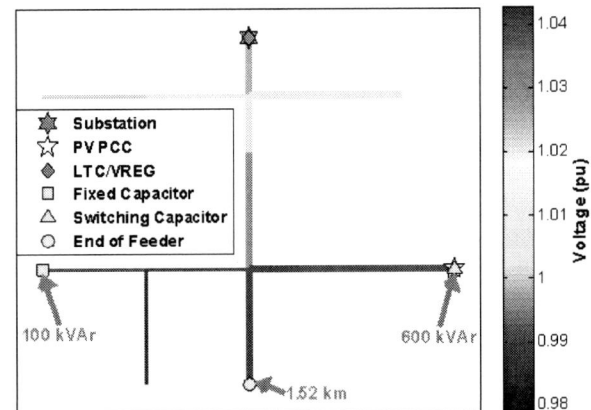

Figure 1. Diagram of the modified IEEE 13-node feeder colored by voltage.

There has been limited research into improving the speed of QSTS simulations [5]. Due to the unbalanced nonlinear nature of the distribution system power flow equations, it can be quite challenging to decrease the computational time [6]. Additionally, the voltage regulation devices have time delays, deadbands, and hysteresis that require each power flow to be solved sequentially in order [6].

This paper proposes a novel method to intelligently perform QSTS simulation for part of the year while accurately modelling the PV impacts over the entire year. The intelligent sampling (IS) algorithm works by analyzing the simulation input data and selecting a representative sample of inputs throughout the year. This reduced input list is then simulated using a powerflow solver and results are scaled to estimate the yearly simulation results. In general, [3] demonstrated that solving the QSTS for half the days randomly sampled results in situations with potentially very high error compared to running the entire year. This is explained by the logic that randomly sampling could sometimes miss certain common situations (e.g. sampling only clear days, or sampling entirely from winter months). The purpose of proposed intelligent sampling method is to ensure the full range of days throughout the year are analyzed in detail with the QSTS simulation, and then the results for the non-simulated days will be inferred by the simulation results for similar type days. The intelligent sampling method is explained in Section III, and Section IV provides a demonstration of the errors introduced by the method and the reduction in computational time.

## II. TEST SETUP

The modified IEEE 13-bus circuit shown in Figure 1 is used in this work. A 2 MW PV plant is located at the end of the feeder and accounts for up to 40% of the peak load. Measured 1-second resolution solar irradiance data and 5-minute measured load data are applied as the inputs to the simulation. The effectiveness of intelligent sampling will be determined by comparing the algorithm's outputs with the actual results attained from the yearlong QSTS simulation. The brute-force simulation is performed at 1-second resolution using OpenDSS. Due to the significant impact variable PV generation can have on distribution system voltage regulators, the number of tap changes predicted for the year is used as the simulation accuracy evaluation metric. Based on feedback from distribution system engineers, the expected accuracy of number of regulator tap changes in a year should be within 10% of the detailed brute-force QSTS simulation and will be the focus of this work [3].

## III. INTELLIGENT SAMPLING METHODOLOGY

The brute-force QSTS simulation results are saved into 6-hour periods (e.g. the number of tap changes per 6-hours), resulting in 1460 6-hour periods in a year. The objective of the intelligent sample selection is to select which of those 1460 6-hour periods are the most effective to simulate with QSTS to estimate the yearly impacts.

### A. Categorizing Input Data

The input data time-series profiles are analyzed using irradiance variability metrics found in the literature and simple load metrics such as the maximum, minimum, mean, and median over a given time period. The irradiance variability index (VI) [7] and irradiance variability score [8] were used for this work. Variability metrics were chosen, as opposed to static metrics, from the intuition that voltage regulator operations are caused by grid dynamics. Based on the simulated data, a stepwise linear regression was performed to identify the two statistics that are the mostly highly correlated with the number of regulator tap changes. For this work, it was determined that the VI and the median of the load for each period were the most highly correlated statistics from the input data timeseries.

Intelligent sampling (IS) decomposes the year into the smaller 6-hour time periods. Each of these time periods is then categorized and binned according to the calculated statistics from the input time-series during this period. The intuition is that "similar" time periods (i.e inside the same bin with the same input time-series statistics) will have closely correlated QSTS results. Due to the focus on variability, the length of the time period interval plays an important role. Too large a time interval will average out periods of high variability with idle periods, while very short intervals lack enough data to be able to differentiate points of interest. A 6-hour interval is chosen to split the day into quarters. The middle two quarters encompass

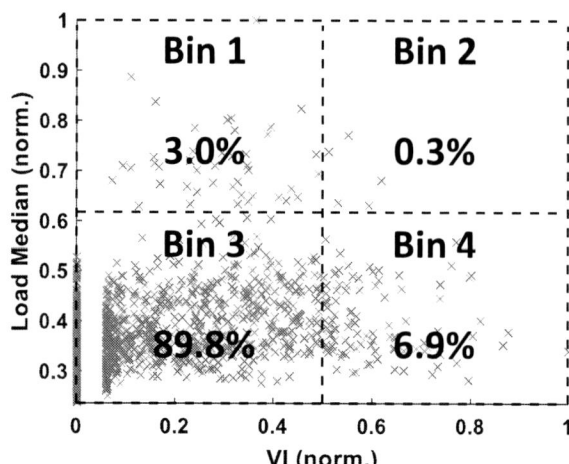

Figure 2. Four bin grid example using VI and load median as input metrics

most of the daylight hours while the first and fourth quarters cover morning and nighttime, respectively.

Several binning techniques are available to group data with two input metrics for sampling. The simplest binning technique is to lay a grid across the 2D-plane and treat the resulting, equally sized, rectangles as individual bins, where samples are pulled per bin. This method is called stratified sampling and has been previously demonstrated for sampling representative days for simulation [9]. Empty bins, with zero entries, are possible with this approach but only "filled" bins are sampled.

Another binning strategy, known as K-means clustering, employs an iterative approach to create clusters where the intra-cluster Euclidean distances between samples are minimized. Resulting clusters will have unique shapes and sizes. Many other binning techniques are possible by considering different geometric bin shapes and sizes.

A grid-based stratified sampling approach is used for this work. Being the simplest approach, it is easily adaptable to different circuit topologies, whereas K-means approaches suffer replicability and tuning issues, as the initial clusters locations are typically randomly placed. Clusters are also heavily influenced by outliers, which can be mitigated by removing outliers, but we do not want to make any assumptions beforehand on the designation or significance of outliers. Ultimately the grid-based stratified sampling approach requires the least circuit tuning, with robust and repeatable results.

After the bins are compiled, a determined number of sample periods are randomly chosen from each bin and simulated with the QSTS power flow solver. These simulation results are used to establish an estimated average bin output ($\hat{B}_i$) for each bin ($i$). The estimated total yearly tap changes ($\hat{y}$) is calculated according to (1), where $n_i$ is the total number of entries in bin $i$, $s$ are individual bin samples, and $v_i$ is the number of samples drawn from $B_i$.

$$\hat{y} = \sum n_i \hat{B}_i \qquad (1)$$

$$\hat{B}_i = \frac{s_1 + \cdots + s_v}{v_i}, v_i \leq n_i \qquad (2)$$

The accuracy of the yearly estimations will be determined by how close the estimated bin mean, $\hat{B}_i$, is to the true bin mean, $\bar{B}_i$. Simulating more bin samples will always increase $\hat{B}_i$'s accuracy, but at the expense of simulation time. From Figure 2, it is apparent samples are not spread evenly among the bins with most falling in bin 3. For an accurate representation, sampling also may not be even among the bins. One sample from bin 2 may suffice to estimate $\bar{B}_2$, but estimating $\bar{B}_3$ may require more samples. The challenge then becomes determining the number of samples to draw from each bin to establish a good estimate of the bin means, especially as the number of bins increases.

Figure 3 shows a binning example using a 21x21 grid. Figure 3a shows the average of the number of tap changes that occurred during the 6-hour periods for that bin. Figure 3b displays the number of 6-hour samples with the statistics that correlate to that bin. Although there is some correlation between input values and bin means, it is quite apparent that the relation is not linear due to the nonlinear power flow equations and piecewise discontinuities from the discrete operating states of the voltage regulators. We cannot apply a least squares fit to the input/output data, and any fit would be circuit dependent, requiring a large amount of simulation data to create the fit model. Figure 3a shows that the input metrics by themselves cannot accurately predict the tap change results, but the metrics can be used for intelligent stratified sampling.

### B. Sampling from Bins

Several sampling methods were tested to try to reduce the simulation time as much as possible while keeping errors within the 10% tolerance. A key focus was to limit the number of assumptions that may or not hold across multiple circuits. This was also the reason we chose a simple binning technique as opposed to a complex one. The complex binning processes could be tuned extensively for our given circuit and yield good results, however, there was no certainty the tuning parameters were universal.

On the same note, we did not make too many assumptions about the impact of outliers on sampling. Visually, outliers are easily identified in Figure 2 at high load and VI metrics. When grid size increases, outliers are typically the only sample in their bin and their $\bar{B}_i$ is attained by simulating the one sample. However, (1) small bin sizes (low $n_i$) may or not may contribute much significance to the overall year metric.

Even with the relatively small grid size in Figure 3, bin counts (number of samples in the bin) vary unpredictably across the range of inputs metrics. The high concentration when VI is zero is due to the dark time intervals, when the sun is not up and the solar variability is zero. Comparing Figure 3a and 3b, there is little correlation between bin mean and bin

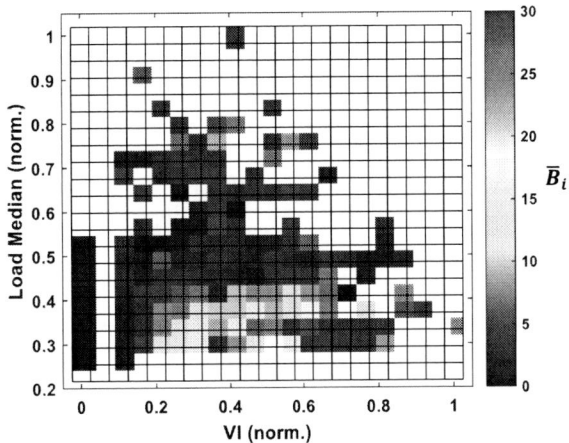

(a) Bin Mean Regulator Tap Changes

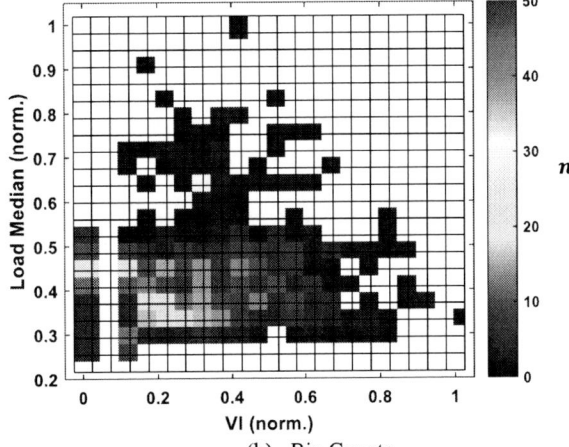

(b) Bin Counts

Figure 3. 21x21 grid bin example: (a) Bin Mean Regulator Tap Changes (b) Bin Sample Counts. White squares are empty bins

count. Outlier bins with few samples on the edge display the largest range in the average number of tap changes. Techniques such as outlier removal, undersampling or oversampling outlier bins, or combining outlier bins would have to make assumptions on the importance of outliers, which will not be consistent for different distribution systems being analyzed with QSTS.

Another idea considered was to undersample the bins with low intra-bin variability between samples in the bin. Of course, this is a very limited approach since it assumes the true means are known beforehand, but some intuition narrows the focus. Bins where the VI index is zero, corresponding to dark time periods in early morning or late at night should have low PV induced variability in their output metrics. These time intervals are also numerous and take up about half the year. Therefore, simulation time can be greatly reduced by selecting only a couple samples from these bins to estimate the bin mean. However, as given by (1), bins with high sample counts get a larger weighting to estimate the yearly averages. Even small variations in the bin estimate are amplified by the high bin count.

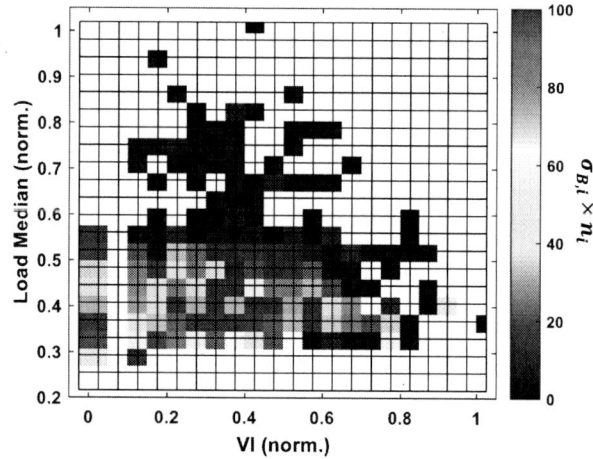

Figure 4: 21x21 grid bin example for number of regulator tap changes. Color code shows each bin's $\sigma_{B,i} \times n_i$ as an approximate error margin on the bin. White squares are empty bins

Using the same 21x21 grid as before, Figure 4 shows the individual bin's standard deviation ($\sigma_B$) of the number of tap changes recorded for all samples in the bin multiplied by the bin count. The product term gives an approximate error margin on sampling a single or low number of entries from each bin. Large error margins are seen across the spectrum of bins, independent of bin size. For the above reasons, we concluded that an accurate representative sample needs to span the entire range of samples in the year.

A considerable effort was spent trying to develop convergence methods to test whether the individual $\hat{B}_i$ were within the error tolerance of their true values. Rather than running simulations for a fixed sample size, the sample number would increase until convergence was detected. Some iterations would converge quickly while others needed additional samples. The intuition was that fixed simulation times had to be sized to bound the worse-case scenarios, while a variable method would display a shorter mean time to convergence. Some of the convergence methods considered were: tracking changes in $\hat{B}_i$ when adding additional samples, tracking changes in $\hat{y}$ yearly estimates when adding additional samples, and performing post-processing on samples to decrease sample variability. However, we were unable to find a 100% reliable convergence method and leave this to future work.

In the chosen IS method, the number of samples chosen from the bins is determined by the bin count and the total number of samples desired. For example, if 50% simulation time reduction is desired, then the number sampled from the bin is simply the bin count divided by two. We determine the number of samples ($\tau_i$) taken from the $i$th bin by

$$\tau_i = \left\lceil \frac{n_i}{x} \right\rceil \quad (3)$$

where the ceiling function guarantees at least one entry from each nonempty bin. The variable $x$ is a tuning parameter which is used to get the sum of $\tau's$ as close to the target sample size

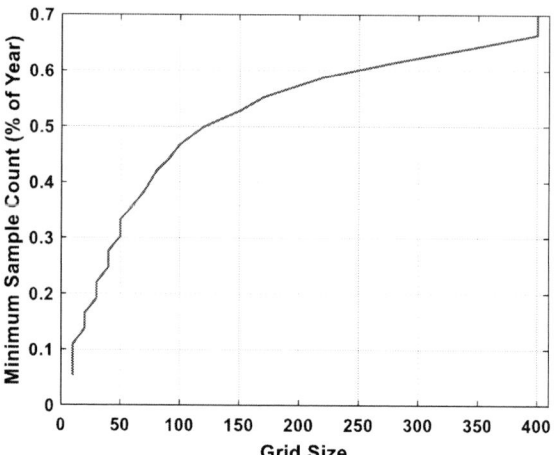

Figure 5: Minimum attainable sample count as a function of grid size

as possible. The optimal value of $x$ is dependent on the number of non-empty bins, as well as the target sample size, and is found iteratively.

For simplicity, the bins are always kept as a square grid of equal proportions. As a result, the number of bins is always equal to the square of the grid size. Grid size plays an important role in intelligent sampling. If the grid includes a large number of bins, then there are too many single entry bins, each of which must be included based on the previous outlier discussion, and too many total samples will be included. However, with few bins, there will be too much variability for the entries inside the bin and our "similar" samples hypothesis fails. The algorithm tries to find the largest grid size that can reach the target sample size.

With the sampling constraint in (3), the grid size becomes closely tied to the minimum samples drawn, as illustrated in Figure 5. Due to the relatively small 1460 sample count (number of 6-hour periods in 365 days), even the smallest grid size at 10x10 has at least 60 filled bins, corresponding to about 4% of the year. Smaller grid sizes introduce more intra-bin variability and weaken our initial binning assumption that intra-bin variability is minimal. With more intra-bin variability, additional samples needs to be drawn from the bin for $\hat{B}_i$ to accurately estimate $\bar{B}_i$. As a result, sampling smaller grid sizes becomes difficult to accurately estimate bin sizes while keeping the total sample size low. On a note, we tried a variety of sampling methods without the "one sample per bin" requirement. However, in these methods the unsampled filled bins had to be accounted for with a multiplier variable, which introduced too much error.

## IV. RESULTS

The effectiveness of the intelligent sampling method is analyzed using a Monte Carlo (MC) simulation. For each MC simulation, the mean absolute error (MAE) is calculated between the actual yearly regulator tap changes ($\hat{y}$) and the

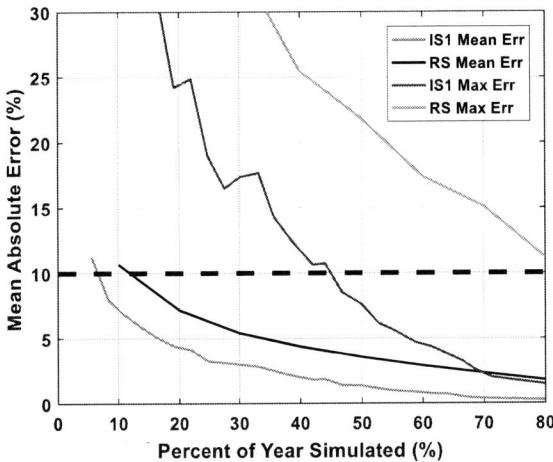

Figure 6. Mean absolute error for voltage regulators's yearly tap changes. Intelligent sampling method (IS) is compared with random sampling (RS)

estimated yearly regulator tap changes ($\bar{y}$) averaged between the three regulators.

$$\text{MAE} = \frac{1}{3}\left(abs(\hat{y}_1 - \bar{y}_1) + abs(\hat{y}_2 - \bar{y}_2) + abs(\hat{y}_3 - \bar{y}_3)\right) \quad (4)$$

The goal is to find the smallest number of days necessary where the MAE is less than 10%. Additionally, we want to always perform within the error threshold, so the maximum of the MAE from all MC simulations is calculated as well. If the worst-case MAE from the MC is not bounded, then we cannot have confidence that results from a single intelligent sampling selection are not outliers with extreme error.

The intelligent sampling results are compared to randomly sampling (RS) time periods out of the entries to ensure there is merit to binning inputs. Random sampling was tried in [2] and found to have too of large error margins to be effective.

The MC simulation is run for 100,000 iterations at different target sample sizes and the results for the sampling method is displayed in Figure 6. The MAE are calculated for each MC iteration and the means of the iterations are shown for the IS and RS algorithms. The maximum MAE among the MC simulations are also displayed for two methods. The goal is for the mean and max MAE to be below 10% which is based on feedback from distribution system engineers [2].

The mean MAE are relatively low and similar for the IS and RS algorithms. The means for IS and RS cross the 10% error tolerance at 7% and 12 %, respectively, of the year simulated. Above 30% of the year simulating, further increases in sample size has only marginal effects on the mean MAE. On the other hand, the maximum error is closely tied to sample size. The maximum error for the IS algorithm crosses the 10% threshold when approximately 43% of the time periods are selected to be simulated with QSTS. The RS algorithm's maximum error crosses the 10% threshold around 85% of the year sampled and is more heavily influenced by outliers. The 57% versus 15% reduction shows the IS algorithm has significant benefit compared to simple random sampling.

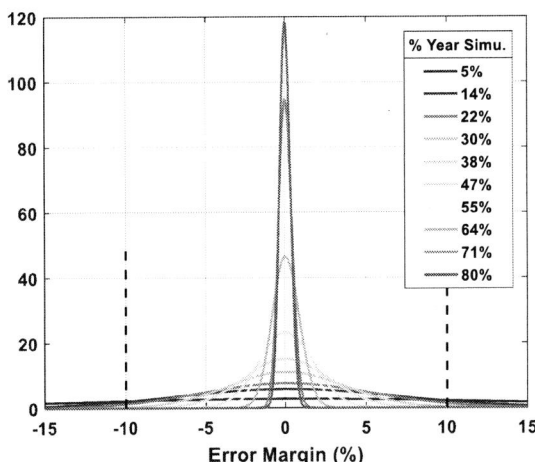

Figure 7. Error margins for regulator one tap changes using IS algorithm

Although Figure 6 shows the maximum error, it does not indicate just how many iterations are falling outside the 10% error margin. For this we plot the distribution of errors for the 100,000 MC simulations using IS in Figure 7. The low samples sizes (<20% of year) have relatively flat error distributions with large parts of the tail falling outside the 10% tolerance. However, the 20-30% samples sizes fall mainly inside the margins with a small fraction actually violating the tolerance. Although previous constraints called for error to be entirely bounded, the improvement in simulation time may be worth the slight relaxation in error constraints. For example, a large parametric QSTS, with long simulation times, may trade some accuracy for increased speed.

## V. ESTIMATING SAMPLE ERRORS

The very small percentage of MC simulations that are outside the error tolerances in Figure 7 suggests that lower samples sizes (20-30%) are possible if that small fraction of erroneous sampling scenarios is somehow detectable. Several convergence methods were experimented with, but ultimately it is difficult to accurately determine convergence for such a limited sample size. Sampling 30% of the year is approximately 400 samples, but due to the intelligent stratified sampling, each sample represents a unique set of sampling conditions based on that bin. Most bins end up with only two or three samples in them, so it is not possible to detect the difference in accuracy or convergence between two or three samples. This may be a limitation of the binning where, at low sample size and grid size, there is always a particular subset of samples that give high error. For example, all the bins have some intra-bin variability and if the chosen samples are all skewed the same way (high or low), the overall output metric will be skewed as well.

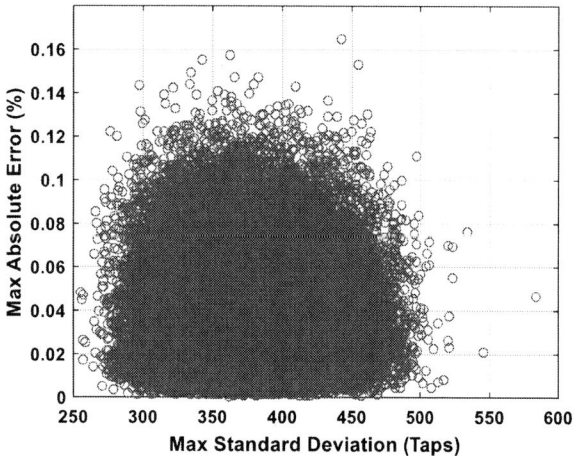

Figure 8: Bootstrapping convergence method showing sample standard deviation versus max error.

One convergence strategy tried was to post-process the samples drawn using bootstrapping to determine the variability of the sample population. The idea was that the width of the distribution of yearly predictions from bootstrapping the samples would give an indication on the convergence to the true population mean. A fixed number of samples were initially drawn with IS, with at least one sample per bin. Next, using that sample population, smaller subsets were randomly drawn from the initial sample and used to calculate bin estimates and corresponding $\hat{y}$. Bootstrapping to repeat this secondary sampling multiple times gave us a distribution for each $\hat{y}$ for which we could find the mean and standard deviation. The goal was to use the standard deviation of the bootstrapping as the convergence metric. Results for a MC simulation of the above method, where the initial sample is 28% of the year and the secondary samples are drawn from within this 28%, are shown in Figure 8. Interestingly, the MC simulations where the true error fell outside the 10% margin do not typically exhibit very large standard deviations using bootstrapping. This seems to suggest that particular sample subsets are precise but not accurate and cluster tightly around an incorrect sample mean. For this reason, the convergence test is not reliable. Future work may involve trying to detect when a skewed sample is drawn from the bins.

## VI. CONCLUSION

A new intelligent sampling algorithm has been presented to reduce simulation time for yearlong QSTS distribution system analysis with high penetrations of distributed PV. The algorithm analyzes the input irradiance and load data and selects representative time periods to simulate using QSTS. By simulating these representative samples, which accounts for a fraction of the total days of the year, the simulation time is decreased. The goal for this work was to estimate the number of regulator tap changes in a year within 10% error margins,

and to achieve a 50% reduction in simulation time. With intelligent sampling, we demonstrate a 57% reduction in simulation time that meets error tolerances with the test circuit. The algorithm is also compared with a random sampling algorithm to demonstrate a significant improvement in the maximum possible sampling errors.

Future work will continue to develop ideas to calculate sampling stopping conditions during the simulation by determining statistical convergence in the confidence interval around the true answer. While intelligent sampling only reduces the computational time of QSTS simulations to around 50% of the brute-force yearlong simulation, IS methods can easily be incorporated into other algorithms. For example, IS can select the days in the year to simulate, and then QSTS simulation of those days can use a variable time-step method [5] for additional speed. We will also investigate using the intelligently selected sample periods to train machine learning algorithms to model the correlation to the number of tap changes [10].

## REFERENCES

[1]. Palmintier, R. Broderick, B. Mather, et al., "On the Path to SunShot: Emerging Issues and Challenges in Integrating Solar with the Distribution System," National Renewable Energy Laboratory, NREL/TP-5D00-65331, 2016.

[2] R. J. Broderick, J. E. Quiroz, M. J. Reno, A. Ellis, J. Smith, and R. Dugan, "Time Series Power Flow Analysis for Distributed Connected PV Generation," Sandia National Laboratories, SAND2013-0537, 2013.

[3] M. J. Reno, J. Deboever, and B. Mather, "Motivation and Requirements for Quasi-Static Time Series (QSTS) for Distribution System Analysis." *IEEE PES General Meeting,* 2017.

[4] J. Seuss, M. J. Reno, R. J. Broderick, and S. Grijalva, "Analysis of PV Advanced Inverter Functions and Setpoints under Time Series Simulation," Sandia National Laboratories, SAND2016-4856, 2016.

[5] M. J. Reno and R. J. Broderick, "Predetermined Time-Step Solver for Rapid Quasi-Static Time Series (QSTS) of Distribution Systems," IEEE Innovative Smart Grid Technologies (ISGT), 2017.

[6] J. Deboever, X. Zhang, M. J. Reno, R. J. Broderick, S. Grijalva, and F. Therrien "Challenges in reducing the computational time of QSTS simulations for distribution system analysis," Sandia National Laboratories, SAND2017-5743, 2017.

[7] J. S. Stein, C. W. Hansen, and M. J. Reno, "The variability index: A new and novel metric for quantifying irradiance and PV output variability." World Renewable Energy Forum. 2012.

[8] M. Lave, M. J. Reno, and R. J. Broderick, "Characterizing local high-frequency solar variability and its impact to distribution studies." *Solar Energy* 118 (2015): 327-337.

[9] B. Palmintier, J. Giraldez, K. Gruchalla, et al., "Feeder Voltage Regulation with High-Penetration PV Using Advanced Inverters and a Distribution Management System: A Duke Energy Case Study," National Renewable Energy Laboratory, NREL/ TP-5D00-65551, 2016

[10] M. J. Reno, R. J. Broderick, and L. Blakely, "Machine Learning for Rapid QSTS Simulations using Neural Networks," IEEE Photovoltaic Specialists Conference (PVSC), 2017.

# A PWM Scheme to Realise Two Times Effective Switching Frequency with Constant Common Mode Voltage and Reactive Power Capability in 1-φ Grid-Tied Transformerless H6 PV Inverter

Amit Kumar Gupta[1], Madhuwanti S. Joshi[2], and Vivek Agarwal[1]

[1]Indian Institute of Technology Bombay, Mumbai, India, [2]Integra Power LLC, USA

*Abstract* — A novel scheme consisting of the unipolar PWM scheme in conjunction with an H6 type transformerless topology is proposed for the grid-connected solar photovoltaic application. The high-frequency operation of the inverter is desirable, but it leads to higher switching loss. The proposed scheme requires the switches to be operated just at half the actual switching frequency thereby reducing the switching loss. The proposed single-phase transformerless inverter concept exhibits very low leakage current due to no variation in common mode voltage. This is done with controlled reactive power feeding capability, which makes the system capable of satisfying the futuristic demand expected from the grid connected PV sources. Voltage stress on the additional two switches is half of the DC link voltage. Therefore, their voltage rating goes down to half. Also, all the switches operate at half the effective switching frequency, which results in high efficiency. The working principle of the novel PWM scheme is discussed and analyzed in detail, and the simulation results are presented. The experimental results are also included to validate the concept.

*Index Terms* — common mode voltage, double frequency PWM, effective switching frequency, grid-connected inverter, H6 topology, leakage current, PV inverter, sinusoidal PWM, transformerless inverter, unipolar PWM.

## I. INTRODUCTION

In grid-connected PV systems, transformerless inverters are being used increasingly. They have many advantages such as lower losses, low cost, lower weight and size, high efficiency and higher power density [1, 2]. These benefits, however, come with certain trade-offs like high ground leakage current, DC current injection into the grid and personal safety [1, 3]. A stray capacitance of the order of nearly 100 nF/kW exists between the ground and the PV cells [4]. Since the neutral of the grid is also grounded, if the common mode voltage is not constant, a current flows in the ground loop, which is known as leakage current. The leakage current has high-frequency components, which coupled with the grid current result in increased harmonics injection. It also gives rise to losses and electromagnetic interference (EMI) in the circuit. Ground leakage current is of particular importance since not only that it affects the personal safety but it also results in inadvertent tripping of ground fault breaker and the operation of the inverter becomes impractical [5]. The German standard DIN VDE 0126-1-1 has a mandate on the disconnection of the system within a specified time limit if leakage current goes beyond a certain limit [6] as specified in Table I.

TABLE I
LEAKAGE CURRENT ALLOWANCE

| Leakage current (mA) | Time limit (sec) |
|---|---|
| 30 (avg) | 0.3 |
| 60 (avg) | 0.15 |
| 100 (avg) | 0.04 |
| 300 (peak) | 0.03 |

Grid stability is one of the major issues in case of large PV distributed generation systems due to solar PV's intermittent nature. Controlled reactive power feeding and Low Voltage Ride Through (LVRT) operation of the inverter are some of the futuristic demands, expected even from the 1-φ grid systems to contribute to the grid stabilization [7]. Combining the reactive power ability and low ground leakage current in a transformerless inverter is a challenging task. Researchers have been working on reactive power control extensively to address this issue [8].

Traditionally, transformerless inverters using the basic full bridge (FB) inverter topology mainly work with three types of pulse width modulation (PWM) scheme, which also supports reactive power feeding viz. Bipolar, Unipolar and Hybrid PWM [9]. Bipolar PWM exhibits low leakage current, but it has low efficiency due to bipolar nature of the output voltage. Unipolar and Hybrid PWM schemes have unipolar output, thus lower filtering requirement. Unipolar PWM has a significant advantage of dual effective switching frequency, resulting in the reduction of the switching losses by half. Whereas in Hybrid PWM scheme, only two out of four switches operate at the high switching frequency, the other two switches operate at the line frequency (50/60 Hz), thereby reducing the switching losses [10]. Both Unipolar and Hybrid PWM schemes render high efficiency, but they exhibit high leakage current. Therefore, none of the three basic PWM schemes is suitable for grid-connected PV systems [2, 9].

Therefore, modifications are done in the transformerless topologies and PWM strategies to meet the requirements of the grid-connected PV systems. The well-known H5 topology has five switches out of which three switches operate at the high switching frequency, and the remaining two switches operate at line frequency. The oH5 is a modified version of H5 topology, which has six switches [11]. One additional switch than the H5, which operates at high frequency ensures constant common mode voltage. However, both the topologies have uneven loss distribution as well as no reactive power feeding capability [2]. HERIC is another popular topology having six switches where leakage current is reduced by bypassing the AC side during freewheeling periods [12]. However, HERIC too is not capable of providing reactive power. H5 and HERIC can have reactive power capabilities with the modifications in switching strategies, which leads to higher losses [2], [13]. Another HERIC derived topology, Full Bridge with Zero Vector Rectifier (FB-ZVR), has more components [9] [14], and gives bipolar output during dead time clamping, resulting in higher filter requirement and low efficiency. Araujo et al. have proposed an interesting high-efficiency topology in [2] in which one buck converter each is used for the positive and negative half cycles respectively. However, this topology may have significant DC injection problem due to unsymmetrical operation [15] of the two buck converters. Another FB topology, proposed by REFU Solar, has a high component count (six switches and five diodes), and it needs double DC-link voltage. In addition, this topology does not have reactive power capability [16].

There are several FB type H6 topologies available with DC bypass (FB-DCBP). The one proposed in [17] has six power devices and two clamping diodes with two DC-link capacitors. The same topology has been used in [15] named UniTL with modified switching strategy, which is capable of producing double the effective switching frequency with low leakage current. Another H6 type transformerless configuration is proposed in [18] with six switches and two diodes, which exhibits high efficiency. However, in spite of high efficiency and low leakage current, none of these H6 topologies is capable of feeding reactive power.

A modified version of UniTL is proposed in [19], which can provide reactive power capability, but it uses total eight number of switches, resulting in higher losses. Two PWM schemes – the Unipolar SPWM strategy and the Double frequency SPWM strategy are proposed and applied to a modified H6 topology, with the clamping diodes removed [20]. However, due to the configuration of the switches, the common mode voltage may vary from $V_{DC}/3$ to $2V_{DC}/3$ during freewheeling, giving rise to leakage current. For low power operation and due to filter inductors, leakage current may fall below the stipulated limit, but as the power levels are increased, the leakage current may increase and cross the permissible limits. All the above-mentioned topologies along with their features are also summarized in Table IV in section V.

In this paper, a novel scheme is proposed consisting of Unipolar PWM scheme and an H6 transformerless topology with diode clamping for grid-tied PV application, which produces unipolar output with an effective switching frequency that is two times the actual switching frequency of the devices. Therefore, filtering requirement is less, and core

losses and switching losses are reduced. The unipolar PWM also enables controlled reactive power feeding into the grid. It exhibits very low leakage current due to the fact that the clamping diodes are able to hold the common mode voltage constant. The common mode voltage is expressed as follows:

$$V_{CM} = \left(V_{AN} + V_{BN}\right)/2 = V_{DC}/2 \qquad (1)$$

In addition, the blocking voltage of the two additional switches is half the DC-link voltage. Thus, low rating switches can be used.

## II. PROPOSED CONCEPT

The topology used here has six switches with a diode clamp circuit at the input side as shown in Fig. 1. The figure also shows the PV-ground capacitance, which is responsible for high-frequency leakage current. Two inductors are used at the output for a symmetrical operation, which helps in reducing the differential mode voltage, eventually helping in reducing leakage current.

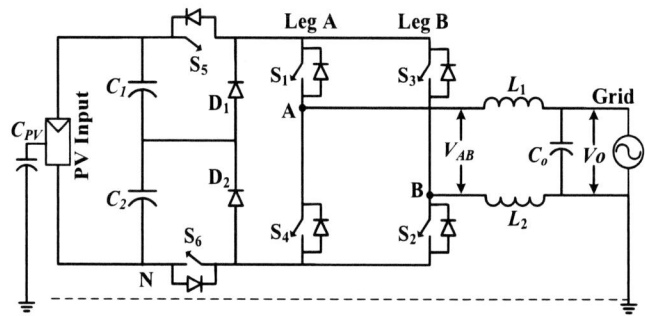

Fig. 1. H6 topology with leakage current path.

The operation of the circuit is based on the unipolar SPWM scheme, which enables it to exchange reactive power between the grid and input capacitor with double effective switching frequency and low leakage current.

### PWM Scheme

The operation of the four switches of the bridge ($S_1$, $S_2$, $S_3$, and $S_4$) is based on the conventional Unipolar PWM scheme, and the two additional switches ($S_5$ and $S_6$) modulate in coordination with the original four switches as shown in Fig. 2.

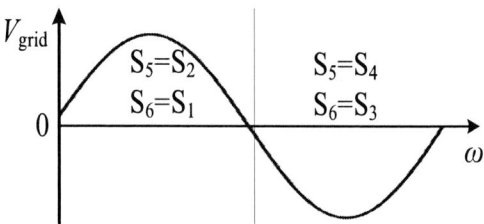

Fig. 2. Operation of the two additional switches.

978-1-5090-5606-4/17 $31.00 © 2017 IEEE

For the positive half cycle, the switch $S_5$ operates along with switch $S_2$, and the switch $S_6$ operates along with $S_1$. In other words, same gate pulses of $S_2$ and $S_1$ are applied to $S_5$ and $S_6$ respectively for the positive half cycle. Similarly, for the negative half cycle, the switch $S_5$ operates along with switch $S_4$, and the switch $S_6$ operate along with $S_3$. By doing this, additional switches share the DC-link voltage with the bridge switches. Therefore the voltage stress across them is half of the DC-link voltage. Thus, the voltage rating and the switching losses of the additional switches are reduced to half.

## III. SIMULATION RESULTS

Simulations are performed on MATLAB-Simulink platform. A simulation model of the system has been created along with the PWM generation; the specifications of the inverter model are shown in Table II.

TABLE II
SPECIFICATIONS OF THE SIMULATION MODEL

| Parameters | Value | Unit |
|---|---|---|
| Output Apparent Power | 1200 | VA |
| Output Active Power | 1100 | W |
| Output Reactive Power | 500 | VAR |
| Input DC Voltage | 400 | V |
| Maximum input current | 3 | A |
| Output Voltage (rms) | 230 | V |
| Output Current (rms) | 5.2 | A |
| Switching frequency | 20 | kHz |
| Effective switching frequency | 40 | kHz |
| Output inductor value ($L_1$ and $L_2$) | 500 | μH |
| Output Capacitor value $C_O$ | 4 | μF |

The simulation model is run in the grid-tied mode with 230 V grid voltage. Output Voltage and current waveforms obtained by simulations are shown in Figs. 3 (a) and (b) respectively. The THD measured in the grid current is 3.9%. It can be observed that the grid current is lagging behind the grid voltage by 25°, implying that the system is supplying 500 VAR. The results validate the reactive power capability.

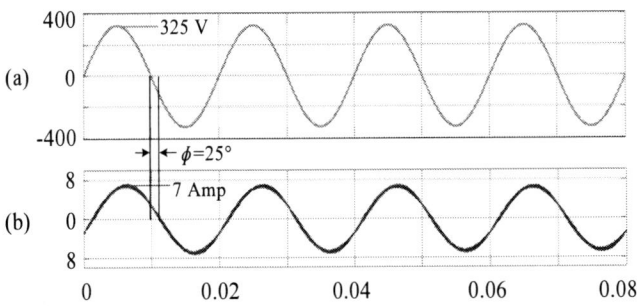

Fig. 3. Inverter Output waveforms: (a) Grid voltage and (b) Grid current.

Fig. 4 (a) shows the 12 V gate pulses of the switch $S_1$. A unipolar voltage ($V_{AB}$) across the terminal A and B is produced across the filters as shown in Fig. 4 (b). It is observed from Fig. 4 (c) that the time-period of the output voltage is half the time-period of the gate pulses ($T_s$ = 20 kHz). Hence, the effective switching frequency of the output voltage is twice the actual switching frequency of the switches. Therefore, the filters need to be designed according to 40 kHz frequency instead of 20 kHz. The cut-off frequency of the LC filter is expressed as

$$f_{cut-off} = \frac{1}{2\pi\sqrt{LC}} = 3.55 \text{ kHz} \qquad (2)$$

Even after keeping cut-off frequency so high, THD of the inverter output waveform is quite low, by keeping the reduced filters size.

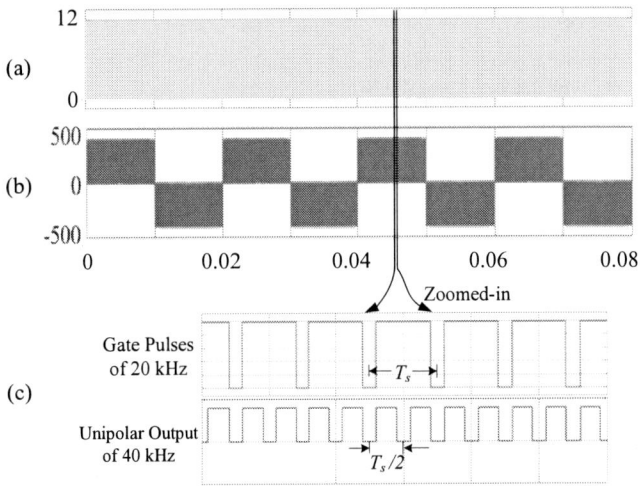

Fig. 4. (a) Gate pulses of $S_1$; (b) Unipolar output voltage before filter and (c) Zoomed in view of (a) and (b).

The leakage current is also measured in the simulation with the same specifications. The spectrum of the leakage current is shown in Fig. 5 along with its average value (= 9 mA), which is well under the limits imposed by the standards.

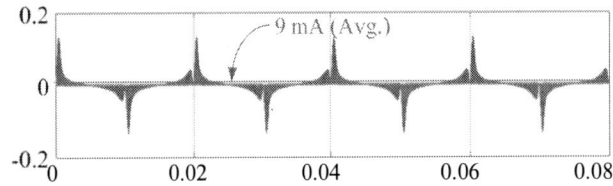

Fig. 5. Leakage current for the proposed scheme.

## IV. EXPERIMENTAL WORK

A laboratory prototype has been developed to validate the scheme, and the results are shown. The hardware

## TABLE III
### HARDWARE SPECIFICATION

| Parameters | Value | Unit |
|---|---|---|
| Output Apparent Power | 480 | VA |
| Output Active Power | 460 | W |
| Output Reactive Power | 130 | VAR |
| Input DC Voltage | 200 | V |
| Output Voltage (rms) | 110 | V |
| Output Current (rms) | 4.3 | A |
| Switching frequency | 20 | kHz |
| Effective switching frequency | 40 | kHz |
| Output inductor value ($L_1$ and $L_2$) | 500 | µH |
| Output Capacitor value $C_O$ | 4 | µF |

specifications are listed in Table III. The experiments are carried out in open loop for nearly 500 W power level. A photograph of the hardware set-up is shown in Fig. 6, which comprises of the power circuit, filter circuits, control board, auxiliary power supply and reactive load along with the DSO showing experimental results.

The experiments are carried out for a 200 V DC input, which is provided by a DC source. The circuit works on 20 kHz PWM pulses provided by a TMS320F28069 processor, which eventually produces 40 kHz output. Infineon's IKW25N120H3 IGBTs are used as power devices with HCPL3120 driver ICs. The experimental inverter setup produces 110 V rms output and 4.3 An rms output current so

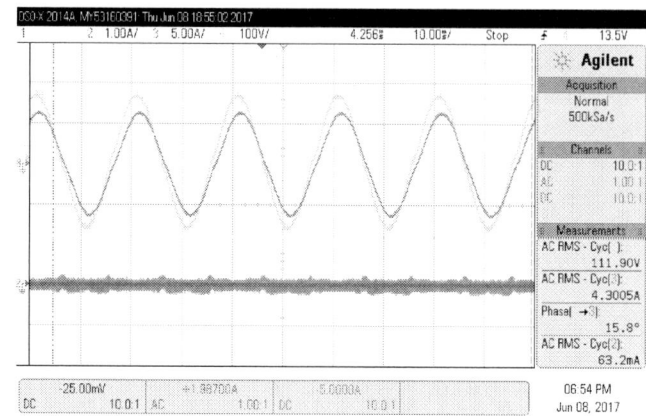

Fig. 7. Experimental results: (a) Grid voltage; (b) Grid current and (c) Leakage current.

that the total power output is more than 480 W as shown in Fig. 7. Leakage current is also measured by connecting ground resistance as 15 Ω and ground capacitance as 47 nF for 500 W. It can be observed that the rms value of the leakage current is only 63.2 mA by the transformerless topology, which is within limits imposed by the standards. A reactive load is connected at the output, which gives 16° phase shifted current. Thus, the system is supplying around 130 VAR. Also, no zero crossing distortion is observed, which validates that the topology is able to handle reactive power as well.

## V. COMPARISON AND CONCLUSION

A comparison of the transformerless H6 based PWM scheme is made with the existing schemes as shown in Table-IV. It is observed that the proposed scheme has low component count (only six switches). It is one of the very few schemes, which can produce double effective switching frequency. The switching pattern is capable of holding the common mode voltage constant, which results in negligible leakage current even at higher power levels. As shown in Table IV, there are only a few topologies, which can provide reactive power.

It is demonstrated through the simulation, and experimental results that the proposed H6 based PWM scheme is capable of increasing the effective switching frequency two times along with the reactive power feeding capability, which is the requirement of next-generation PV inverters. The proposed transformerless grid-connected PV system exhibits very low leakage current by keeping common mode voltage constant. Due to lower switching losses, the proposed concept gives the benefits in terms of overall efficiency. Cost effective devices can be used due to the low 'actual' switching frequency of the devices and less voltage stress across them. Thus, the size and cost of the system are reduced. Proposed scheme incorporated with wide bandgap power devices can give a tremendous improvement in the power density.

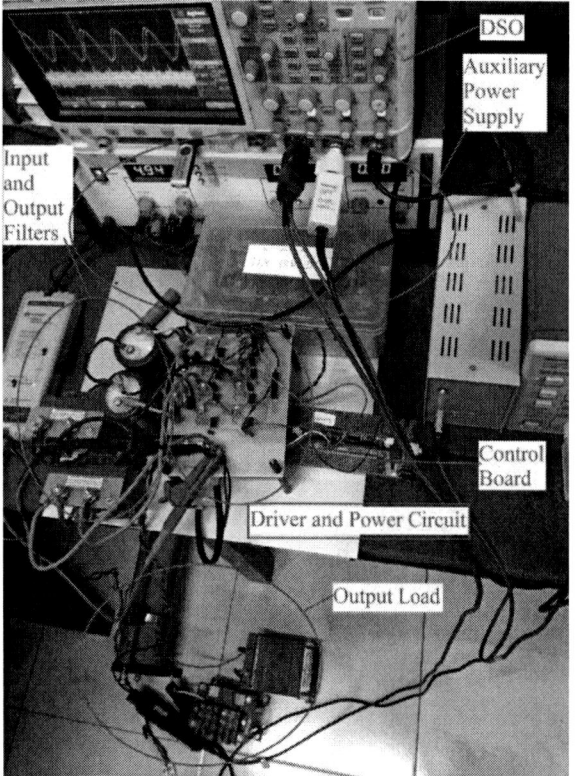

Fig. 6. Experimental setup developed for this study.

978-1-5090-5606-4/17 $31.00 © 2017 IEEE

TABLE IV

SUMMARY OF FB DERIVED 1-φ GRID CONNECTED PV INVERTER TOPOLOGIES

| Topologies | Number of Switch+Diode | Double effective switching frequency | Constant CMV | Leakage Current Elimination | Reactive Power Capability | Overall Efficiency |
|---|---|---|---|---|---|---|
| FB with Bipolar | 4+0 | No | Yes | Yes | Yes | Low |
| FB with Unipolar | 4+0 | Yes | No | No | Yes | High |
| FB with Hybrid | 4+0 | No | No | No | Yes | High |
| H5 | 5+0 | No | No | Yes | No | High |
| H5 with Modified Switching | 5+0 | No | No | Yes | Yes | Medium |
| HERIC | 6+0 | No | Yes | Yes | No | High |
| HERIC with Modified Switching | 6+0 | No | Yes | Yes | Yes | Medium |
| FB-ZVR | 5+5 | No | Yes | Yes | No | High |
| oH5 | 6+0 | No | Yes | Yes | No | High |
| Topology by Araujo et al. | 4+4 | No | No | Yes | No | High |
| Topology by REFU Solar | 6+5 | No | No | Yes | No | High |
| FB-DCBP (H6) | 6+2 | No | Yes | Yes | No | High |
| UniTL (H6) | 6+2 | Yes | Yes | Yes | No | High |
| H6 proposed in [18] | 6+2 | No | No | Yes | No | High |
| UniTL Modified | 8+0 | Yes | Yes | Yes | Yes | Medium |
| H6 with the two PWMs in [20] | 6+0 | Yes | No | Yes | Yes | High |
| **Proposed Scheme** | **6+2** | **Yes** | **Yes** | **Yes** | **Yes** | **High** |

REFERENCES

[1] A. K. Gupta, M. S. Joshi, and V. Agarwal, "On the control and design issues of single phase transformerless inverters for photovoltaic applications," in *IEEE 6th India International Conf. Power Electronics*, Kurukshetra, India, Dec. 2014.

[2] S. V. Araujo, P. Zacharias, and R. Mallwitz, "Highly efficient single-phase transformerless inverters for grid-connected photovoltaic systems," *IEEE Trans. Ind. Electron.*, vol. 57, no. 9, pp. 3118–3128, Sep. 2010.

[3] H. Patel and V. Agarwal, "A single-stage single-phase transformer-less doubly grounded grid-connected PV interface," IEEE Trans. Energy Convers., vol. 24, no. 1, pp. 93–101, Mar. 2009.

[4] H. Agrawal, A. K. Gupta, and V. Agarwal, "A novel 3-phase, transformerless H-8 topology with low variation in CMV to reduce leakage current," in *IEEE Int. Conf. on Power Electronics, Drives and Energy Systems*, Kerala, Dec. 2016.

[5] O. Lopez, F. D. Freijedo, et. al., "Eliminating ground current in a transformerless photovoltaic application," *IEEE Trans. Energy Convers.*, vol. 25, no. 1, pp. 140–147, Mar. 2010.

[6] Automatic Disconnection Device Between a Generator and the Public Low-Voltage Grid, VDE V 0126-1-1:2006-02, 2006.

[7] Y. Yang, F. Blaabjerg, and H. Wang, "Low-voltage ride-through of single-phase transformerless photovoltaic inverters," *IEEE Trans. Ind. App.*, vol. 50, no. 3, pp. 1942–1952, May/Jun. 2014.

[8] F. Blaabjerg, T. Teodorescu, M. Liserre, and A. V. Timbus, "Overview of control and grid synchronization for distributed power generation systems," *IEEE Trans. Ind. Electron.*, vol. 53, no. 5, pp. 1398–1409, Oct. 2006.

[9] R. Teodorescu, M. Liserre, and P. Rodriguez, "Grid synchronization in single-phase power converters," in *Grid Converters for Photovoltaic and Wind Power System*, London: John Wiley & Sons, Ltd., 2011, pp. 43–89.

[10] R. S. Lai, and K. D. T. Ngo, "A PWM method for reduction of switching loss in a full-bridge inverter," *IEEE Transactions on Power Electronics*, vol. 10, no. 3, pp. 326–332, May 1995.

[11] H. Xiao, S. Xie, Y. Chen, and R. Huang, "An optimized transformerless photovoltaic grid-connected inverter," *IEEE Trans. Ind. Electron.*, vol. 58, no. 5, pp. 1887–1895, May 2011.

[12] D. Schmidt, D. Siedle, and J. Ketterer, "Inverter for transforming a DC voltage into an AC current or an AC voltage," EP Patent 1, 369, 985, 2009.

[13] D. Barater, E. Lorenzani, C. Concari, G. Franceschini, and G. Buticchi, "Recent advances in single-phase transformerless photovoltaic inverters," *IET Renew. Power Gener.*, vol. 10, no. 2, pp. 260–273, 2015.

[14] T. Kerekes, R. Teodorescu, P. Rodríguez, G. Vázquez, and E. Aldabas, "A new high-efficiency single-phase transformerless PV inverter topology," *IEEE Trans. Ind. Electron.*, vol. 58, no. 1, pp. 184–191, Jan. 2011.

[15] D. Barater, G. Buticchi, A. S. Crinto, G. Franceschini, and E. Lorenzani, "Unipolar PWM strategy for transformerless PV grid-connected converters," *IEEE Trans. Energy Convers.*, vol. 27, no. 4, pp. 835–843, 2012.

[16] Hantschel, J., German Patent Application, Publication Number DE102006010694 A11, 20 September 2007.

[17] R. Gonzalez, J. Lopez, P. Sanchis, and L. Marroyo, "Transformerless inverter for single-phase photovoltaic systems," *IEEE Trans. Power Electron.*, vol. 22, no. 2, pp. 693–697, Mar. 2007.

[18] W. Yu, J. S. Lai, H. Qian et al. "High-efficiency MOSFET inverter with H6-type configuration for photovoltaic nonisolated AC-module applications," *IEEE Trans. Power Electron.*, 2011, 26, (4), pp. 1253–1260.

[19] D. Barater, G. Buticchi, E. Lorenzani, and V. Malori, "Transformerless grid-connected for PV plants with constant common mode voltage and arbitrary power factor," in *Proc. 38th Annu. Conf. IEEE Ind. Electron. Soc.*, Oct. 2012, pp. 5756–5761.

[20] B. Yang, W. Li, Y. Gu, W. Cui, and X. He, "Improved transformerless inverter with common-mode leakage current elimination for a photovoltaic grid-connected power system," *IEEE Trans. Power Electronics*, vol. 27, no. 2, pp. 752–762, Feb. 2012.

# A Solar PV Retrofit Solution for Residential Battery Inverters

Amit Kumar Gupta[1], Vaibhav Pawar[1], Madhuwanti S. Joshi[2], Vivek Agarwal[1], and Deepak Chandran[3]

[1]Indian Institute of Technology Bombay, India, [2]Integra Power LLC, USA, [3]IRIS Energy LLC, USA

*Abstract* — Frequent power cuts necessitate the use of battery fed inverters in many homes and small industries in countries like India. Due to several charge and discharge cycles, the batteries have a typical life of 3 − 4 years. In this paper, a solar PV based retrofit solution is proposed which can be used along with these inverters. This solution uses a bi-directional DC to DC converter which charges and discharges the battery in the presence and absence of solar radiation respectively. The converter has been implemented using a simple add-on circuit with the main battery inverter. The proposed solution has the potential of providing high-quality power, grid connectivity, energy saving and enhancing the battery life. An analysis showing the effective energy saving using the proposed scheme has been presented. Simulations and experimental results have been obtained, and some representative results are included.

*Index Terms* — battery fed inverter, bi-directional converter, single-phase inverter, solar PV solution, uninterrupted power supply.

## I. INTRODUCTION

Residential and commercial roof top PV market has been growing at a rapid pace. Primarily, there are two types of systems in this space. The first one is based on the grid connected PV inverters and the second one is off the grid standalone systems. Grid-connected PV inverters convert the DC voltage obtained from the PV panels to the grid compatible AC voltage. Generally, solar panels are connected in series, and parallel combinations and a high DC voltage is generated [1]. A DC to AC converter is used to convert this high voltage into AC voltage. The term grid-connected is used when the generated AC voltage is in synchronization with the grid voltage and can be physically connected to the utility grid as a parallel generator. Fig. 1 shows one example of grid-connected installation [2].

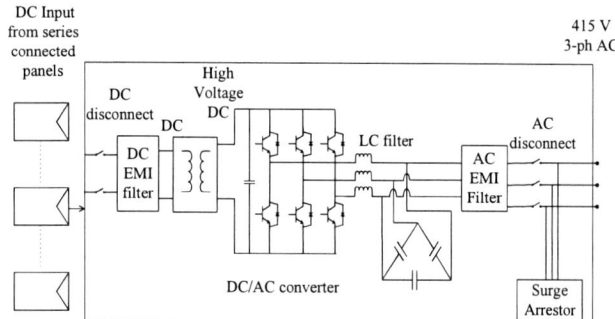

Fig. 1. Generic grid connected PV inverter configuration

One of the main problems with the rooftop grid-tied installations is that they do not work when the grid is not available. In countries like India, there is a significant deficit of electricity. Frequent intentional power cuts are widespread in many parts of the country due to the significant difference in the generation and requirement capacity, especially during hot summer days. It is not uncommon to have up to 12 hours of power outages in the rural areas during summer. To add to the problem of the load shedding, the load management at the distribution level is also not that good. There are no switches at different points in the distribution network. Most of the time the circuit breakers provided for protection at the main 11 kV feeder, which are used to shut down the power for the network in the case of any faults. This disconnects the power from a substantial number of the consumer.

In such situations, off the grid rooftop installations are a good and preferred choice. These types of installations rely on the battery backup with solar power. It has a charge controller, which operates the solar panel at its maximum power and a battery charge/discharge controller. Together, they give the DC input voltage to the inverter. The inverter converts this DC voltage into regulated AC voltage. A changeover switch is provided for the load to switch between AC mains and inverter output. Fig. 2 shows this type of installation.

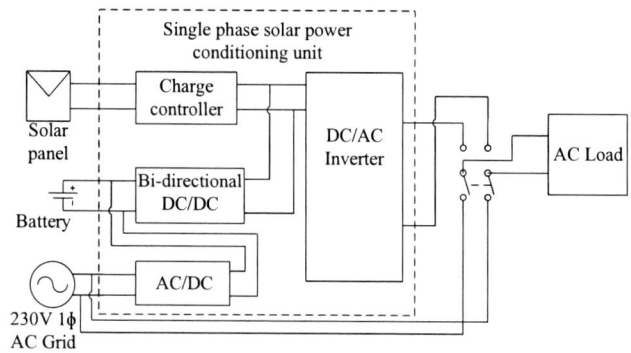

Fig. 2. Off the grid system.

Whether grid-connected or off-the-grid, there have been multiple attempts by various researchers to integrate battery storage and PV to ensure a continuous electric supply to the critical loads. For example, in [3] a bidirectional buck boost converter has been used along with a grid-connected PV inverter to interface with the battery. In [4], a three-port PV inverter for off grid applications has been designed, which has the ability to interface with the battery. A multiple input

978-1-5090-5606-4/17 $31.00 © 2017 IEEE      2986

converter has been proposed in [5] for a continuous AC supply to the critical load. Although all these efforts lead to adding storage to PV, the solutions they offer necessitate investing into an entirely new PV based system. For a common household user, this means a significant cost and paperwork. The higher upfront investment can lead to a reduced interest in the users for the PV installation. To reduce this investment and encourage common people to use PV, it is necessary to come up with a PV system, which can be easily retrofitted into the existing electric power systems [7].

In the present paper, a PV retrofit solution has been provided for the regular household battery inverters. As mentioned in the earlier paragraphs, due to the frequent power cuts, many houses in India have a battery inverter for critical load backups. If additional battery charging facility is provided using the PV panels along with a small add-on circuit, the existing battery inverter can be effectively used for supplying the backup power for a longer time and also, it can be used to reduce the consumption from the main grid. This not only results in energy saving but also reduces the surge in electricity demand when the batteries are being charged from the grid.

The main contributions of this paper are as follows. A PV based retrofit solution for the residential battery inverter power system has been proposed. This solution reduces the electricity consumption from mains and also ensures the steady electric supply to critical loads. Mathematical analysis along with the Simulink model of the system has been developed and simulated. A 250 Watt prototype circuit has been developed and experimentally tested. It has been shown that the battery inverter works well with the addition of solar PV and improves the overall battery life.

## II. PROPOSED SYSTEM

Fig. 3 shows the proposed system. The battery inverter is connected to the critical loads with an AC disconnect switch. The PV component in the system is interfaced through a bi-directional (buck and boost) type DC-to-DC converter. When the PV power is more than that required by the load, the converter directs the PV energy into the battery and battery is charged. When the PV power is not available, then the battery supplies power to the critical load and holds the DC voltage across the inverter input constant.

### A. Operating modes

The system has four distinct operating modes.
**Mode 1. PV is available, grid is off, and the PV is supplying energy to the critical load through inverter:**
This mode may occur in the daytime. In this mode, the DC-to-DC converter extracts power from the PV panels and feeds it to the inverter. If the load requirement is less than the PV generation, the excess energy is used to charge the battery. Grid power is not available in this mode of operation.

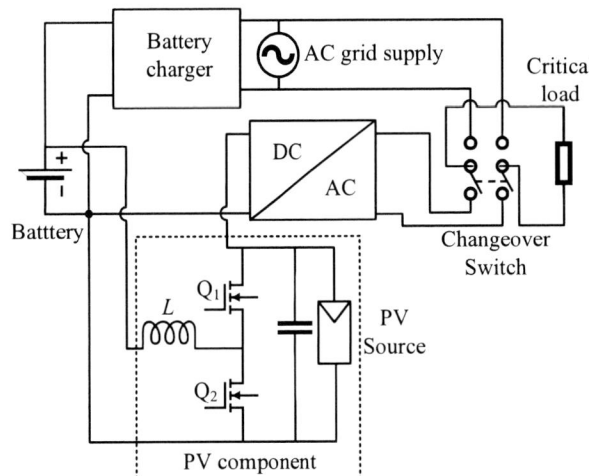

Fig. 3. Proposed system.

**Mode 2. PV power is available, and the grid is supplying the load:**
In this mode, the DC-to-DC converter extracts power from the PV and uses it to charge the battery. If the battery is fully charged, the DC to DC converter is switched off. In this mode, the grid is available and supplying power to the load.

**Mode 3. PV is intermittent, grid is off, and the battery is supplying the load through inverter:**
This mode may occur in the early morning and evening. In this mode, the DC-to-DC converter takes help from the battery to maintain a constant DC voltage at the input of the inverter. The inverter operation remains unaffected. Grid power is not available in this mode of operation.

**Mode 4. PV is intermittent, and the grid is supplying the load:**
In this mode, whenever PV power is available, the PV charges the battery. When the PV is not available, the grid supplies the load while the battery maintains its charge.

Fig. 4 shows a general operating flowchart of the system.

### B. Mathematical modeling of the PV component

The DC to DC converter used in the system is bi-directional in nature. It has two MOSFETs $Q_1$ and $Q_2$ connected in half bridge configuration. The inductor L and capacitor C are energy storage elements, which facilitate the buck-boost operation. When the PV power generation is higher than the load requirement, the converter operates as a buck or step down converter. During the night, when PV is not available, the battery supplies power to the inverter, and the converter acts a boost or step-up converter. Voltage control mode can be used to control the overall operation of the system. Two controllers are required to control the converter operation. The inner current controller controls the current flowing in to and out of the battery. Outer voltage controller controls the DC voltage input to the battery inverter. Fig. 5 shows the control block diagram of the proposed system.

978-1-5090-5606-4/17 $31.00 © 2017 IEEE

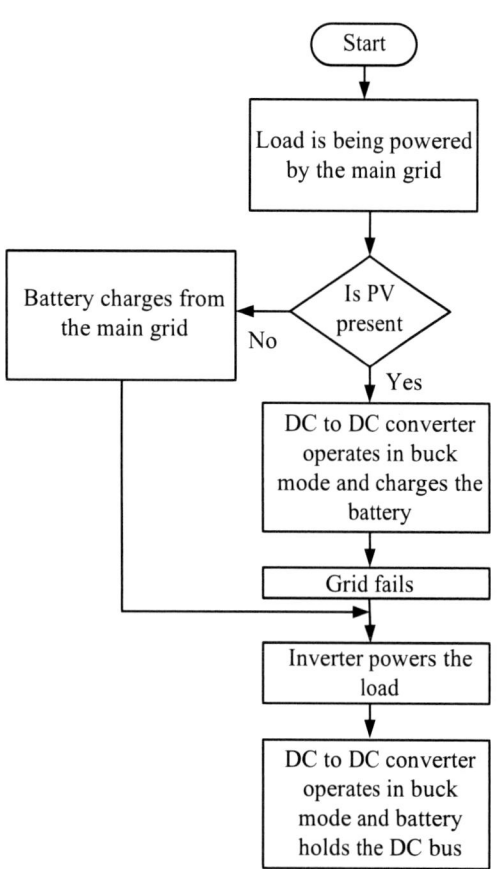

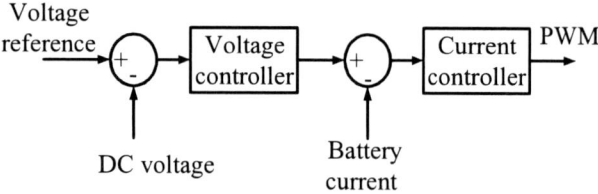

Fig. 4. Flow chart showing the operation of the proposed system.

Fig. 5. Control block diagram for the DC to DC converter.

The overall transfer function of the converter can be expressed by equations 1 and 2 [5].

$$\frac{\hat{V}_{dc}}{\hat{i}} = \frac{(1-D)V_{dc} - LIs}{CV_{dc}s + 2(1-D)I} \tag{1}$$

$$\frac{\hat{i}}{\hat{d}} = \frac{CV_{dc}s + 2(1-D)I}{LCs^2 + \dfrac{L}{R}s + (1-D)^2} \tag{2}$$

where $L$ is the inductor value in the DC to DC converter, $C$ is the capacitor value across the DC input of the battery inverter, $R$ is the effective load resistance across the DC bus, $I$ is the

current flowing through the battery and $\hat{i}$, $\hat{d}$ and $\hat{V}_{dc}$ are the perturbations in the current, duty cycle, and the dc bus voltage respectively.

## III. VALIDATION

Validation of the proposed concept is carried out using a detailed system design followed by simulation results. A proof of concept was developed to obtain the experimental results. A 250 Watt system was designed to validate the concept presented in the earlier section. Table I shows the system specifications.

TABLE I
SYSTEM SPECIFICATIONS

| Parameters | Value | Unit |
|---|---|---|
| Output Power | 250 | W |
| PV panel voltage | 30 to 44 | V |
| Battery voltage | 24 | V |
| Switching frequency | 100 | kHz |

### A. Simulations

The PV components of the system were simulated using MATLAB and Simulink software. Standard component libraries were used for the simulation.

The circuit has been simulated for a case when the PV (solar radiation) is intermittent. The simulation model has been developed to represent all the blocks of the proposed system shown in Fig. 3. Figs. 6 and 7 show various simulation results under different operating modes.

In Fig. 6, the first section represents the early morning situation, when PV power is not available. This is mode-3, when battery current is positive, which shows that the load is supplied by the battery when PV is not available. In the second section of the Fig. 6, PV power is available in day time. This represents the mode-1. The load is drawing less power than the available in PV, and hence extra available PV power is being absorbed by the battery. Thus, the measured battery current is negative. There would be a case when PV is available, and there would not be any load across inverter then the total available power will be used for battery charging. Again in mode 3, when PV power is not available during night times (or during cloud cover situation), the load is being supplied by the battery. The battery holds the input voltage of the inverter constant.

The battery current changes the direction when solar PV is not available, as shown in Fig. 6, and holds the DC voltage across the inverter input constant. Fig. 7 shows the output voltage and output current waveforms of the inverter. The inverter load is being supplied at 230 V. The load power observed is nearly 100 W.

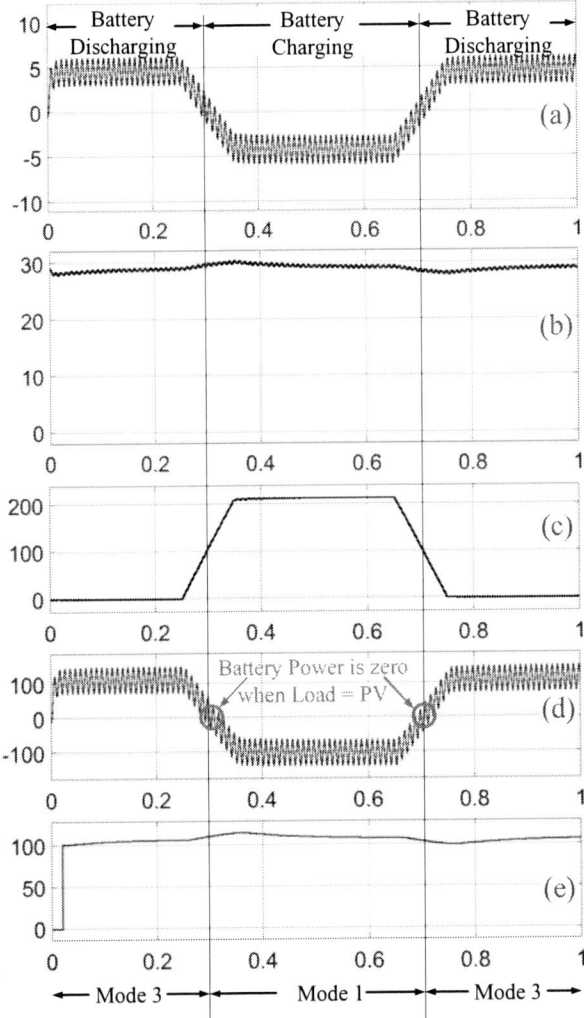

Fig. 6. Simulation results: (a) Battery current; (b) DC-Link (PV) voltage; (c) PV Power; (d) Battery Power; and (e) AC load at the inverter output.

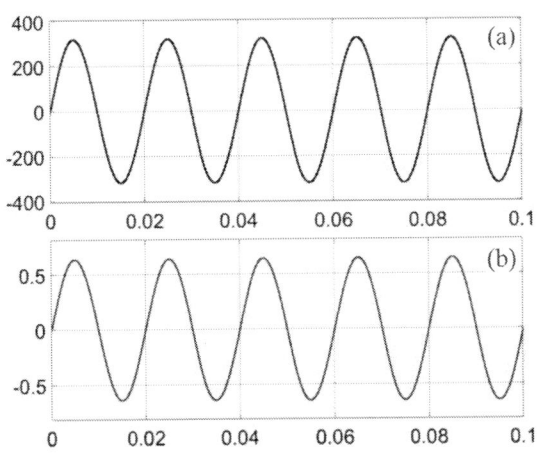

Fig. 7. Simulation results: (a) Inverter output Voltage and (b) Inverter Output Current.

## B. Experimental results

A multilayer PCB was designed and populated as per the specifications discussed in the earlier section. Custom magnetic parts were designed specifically for this converter. Fig. 8 shows a photograph of the proof of concept board. The board was tested and operated as per the modes described in the earlier sections. Agilent InfiniiVision series scope was used to capture the voltage waveforms of the converter.

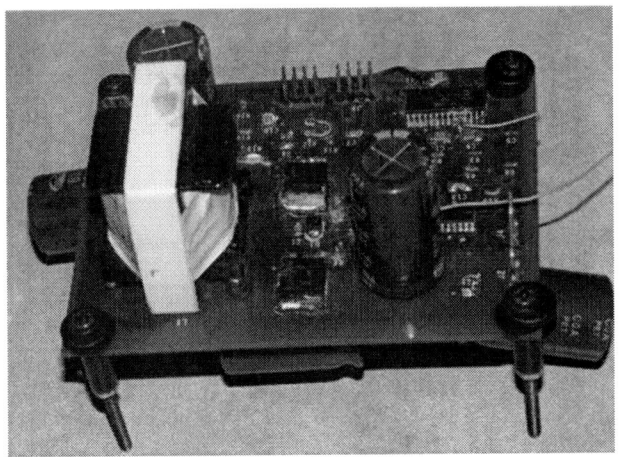

Fig. 8. Photograph of the DC to DC converter board developed in the lab.

Fig. 9 shows a screen captures of the scope. The top trace of the Fig. 9 illustrates the voltage of the PV bus, which is also the DC voltage input to the battery inverter, and the bottom trace shows the battery. Fig. 10 shows the inverter output voltage and current.

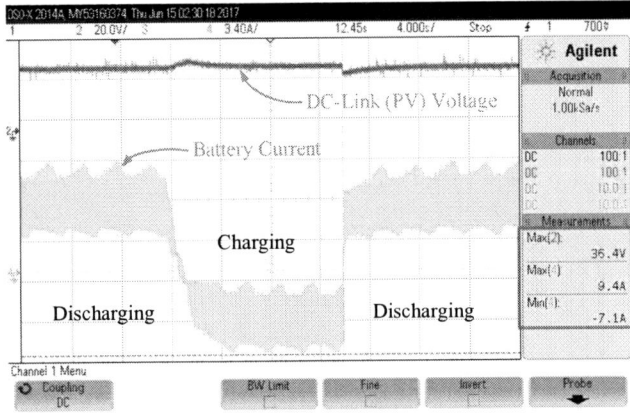

Fig. 9. Experimental results acquired from the DC to DC converter board. The bottom trace is showing the DC voltage being regulated.

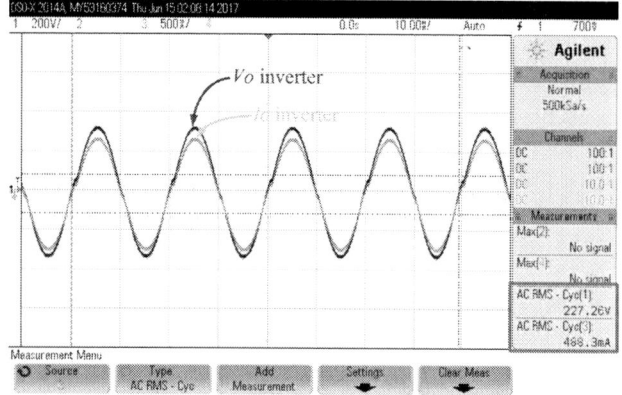

Fig. 10. Experimental results showing inverter output voltage and current waveforms.

## IV. DISCUSSION OF RESULTS

Simulation results presented in the paper indicate that the bi-directional converter works in the buck and boost modes. The DC voltage across the battery-inverter is being maintained constant. The experimental results also confirm the results obtained in simulations. The DC bus voltage is maintained constant despite fluctuations in the PV energy. The set point to the DC voltage can be derived as per the inverter requirement and/or MPPT calculations [8]. Since the circuit supports bi-directional power flow, the battery can be charged using PV energy.

This is of particular importance in the developing countries. In these countries, even a small house has battery inverters for backup power and emergency lighting for the frequent power cuts. When the grid power comes back again, all the batteries draw a lot of power from the grid for charging them again. Due to excessive load on the grid, circuit breakers may again trip and cause un-necessary shutdown situation. This stress on the grid is removed with this simple retrofit solution.

TABLE II
PROPOSED SYSTEM

| | |
|---|---|
| Number of sunny hours in a day | 8 |
| Total sunny days in a year | 300 |
| Converter rating (Watt) | 300 |
| Electricity Units generated by PV in one year | 720 |
| Electricity cost per unit (INR) | 5 |
| Savings over one year | 3600 |
| PV panel cost (INR) | 9000 |
| Electronics cost (INR) | 1500 |
| Total cost (INR) | 10500 |
| Payback period (years) | 3 |

Table II shows a sample calculation for the energy saving using the proposed system. The calculations show that there is energy saving using PV and payback period is close to three years without considering any subsidy or government rebates.

Overall battery utilization is better, and its life is improved by reducing the depths of charge and discharge cycles in daytime.

## V. CONCLUSION

A new scheme for the residential battery inverters to integrate PV has been proposed in this paper. The proposed system provides a very simple PV retrofit solution to the existing battery inverters employed in many households in the developing countries. It is based on the bi-directional buck boost converter topology. This converter forms an add-on circuit to the battery inverter. The circuit has minimum electronic components and hence the very small size. Simulation and experimental results show that the concept works well under fluctuating PV condition. Since the battery is charged using PV and the grid, the battery does not draw surge currents from the grid when the grid is on. Work on reducing the PCB size and improving the functionality of the proposed scheme is underway and will be presented in a future paper.

## ACKNOWLEDGEMENT

The simulations and experimental work presented in this paper were carried out at IIT-Bombay, India with funding from Ministry of New and Renewable Energy, Govt. of India under the National Solar Science Fellowship program. Authors thank and acknowledge this support.

## REFERENCES

[1] A. K. Gupta, M. S. Joshi, and V. Agarwal, "On the control and design issues of single phase transformerless inverters for photovoltaic applications," in *IEEE 6th India International Conf. Power Electronics*, Kurukshetra, India, Dec. 2014.

[2] Madhuwanti Joshi, "Study and development of grid interactive PV system," Report under the National Solar science Fellowship during 2012-2015, MNRE, Govt. of India

[3] Zulhani Rasin , M.F. Rahman ,"Control of bidirectional DC-DC converter for battery storage system in grid-connected quasi-Z-source pv inverter," *IEEE Conf. on Energy Conversion, 2015.*

[4] Zhijun Qian ; Osama Abdel-Rahman ; Haibing Hu ; Issa Batarseh ," An integrated three-port inverter for stand-alone PV applications," *IEEE Energy Conversion Congress and Exposition* (ECCE), Atlanta GA, USA , 2010.

[5] S. Kabeer, L. Padma Suresh, and F. Shamila, "Continuous AC supply using semi-isolated multi-input converter for hybrid PV/WIND power charger system," *International Conference on Circuit, Power and Computing Technologies* (ICCPCT), Nagarcoil, India 2016.

[6] A. K. Rathore, "Small Signal Modeling of Boost Converter" [Online]. Available:
https://www.ece.nus.edu.sg/stfpage/akr/ssmboost.pdf.

[7] D. Chandran, M. S. Joshi, and V. Agarwal, "Solar PV based retrofit solution for cell phone towers powered by diesel generators," in *IEEE Int. Tel. Energy Conf.*, Austin, TX, USA, Oct. 2016.

[8] S. Jain and V. Agarwal, "A Single-Stage Grid Connected Inverter Topology for Solar PV Systems with Maximum Power Point Tracking," *IEEE Tran. Power Electro.* 22, 1928-1940.

# Cost Benefit and Alternatives Analysis of Distribution Systems with Energy Storage Systems

Tom Harris, Adarsh Nagarajan, Murali Baggu
National Renewable Energy Laboratory

Tom Bialek
San Diego Gas & Electric

*Abstract*—This paper explores monetized and non-monetized benefits from storage interconnected to a distribution system through use cases illustrating potential applications for energy storage in California's electric utility system. This work supports SDG&E in its efforts to quantify, summarize, and compare the cost and benefit streams related to implementation and operation of energy storage on its distribution feeders. This effort develops a prototype cost benefit and alternatives analysis platform, integrates with QSTS feeder simulation capability, and analyzes use cases to explore the cost-benefit of the implementation and operation of energy storage for feeder support and market participation.

*Index Terms*—Cost benefit analysis, energy storage benefits, net present value analysis, markets participation, energy storage dispatch

## I. INTRODUCTION

California's energy storage mandate, legislated by AB 2514 and implemented through CPUC D.13-10-040, sets procurement targets for utilities for *viable and cost effective* energy storage systems. This paper focuses on end uses of storage to specific scenarios to help reduce the risk of undervaluing storage as a resource and allow for the identification of utilization opportunities. The tool Cost Benefit and Alternatives Analysis Tool (CBAAT) facilitates the inclusion of energy storage as needs are identified, such as resource adequacy, renewable portfolio standard, and long term resource planning.

### A. Contribution from this paper:

Integrating newer devices into legacy utility operations is a challenge and given that these energy storage systems are expensive assets, understanding the cost benefits plays an important role. An energy storage system is a sophisticated, expensive technology yet sensitive to charge/discharge cycle. Investing in an energy storage system is one aspect where as making an asset out of that is a challenging problem. An energy storage system can play a role of a flexible bi-directional source to accommodate issues from constantly varying loads and renewable resources. Utilizing energy storage systems to support distribution feeder operations is helpful, but every charge/discharge cycle reduces the life-cycle of the system. Additionally, the most important aspect is to ensure that the revenue from the energy storage operation should be higher

This work was supported by the U.S. Department of Energy under Cooperative Research and Development Agreement contract No. #CRD-14-562 with the National Renewable Energy Laboratory. Funding was provided by San Diego Gas & Electric and the U.S. DOE Office of Energy Efficiency and Renewable Energy's SunShot Program

than the cost ascribed to life-cycle cost.

There are commercial tools such as E3 for estimating distributed renewable and energy storage benefits. However, the CBAAT breaks the basic assumptions in cost-benefit framework by directly using the physics-of-feeder by developing a direct link with an open-source distribution system simulator from EPRI, OpenDSS. Figure 1 presents the overall data-flow diagram starting from OpenDSS to CBAAT. A real world feeder will be simulated for multiple years with an increased load each year. This set-up enables CBAAT to capture physics-of-feeder (feeder losses for each time-step, line-currents for assessing upgrades, energy-storage dispatch for evaluating costs etc.) for accurate cost calculations unlike other works in this area.

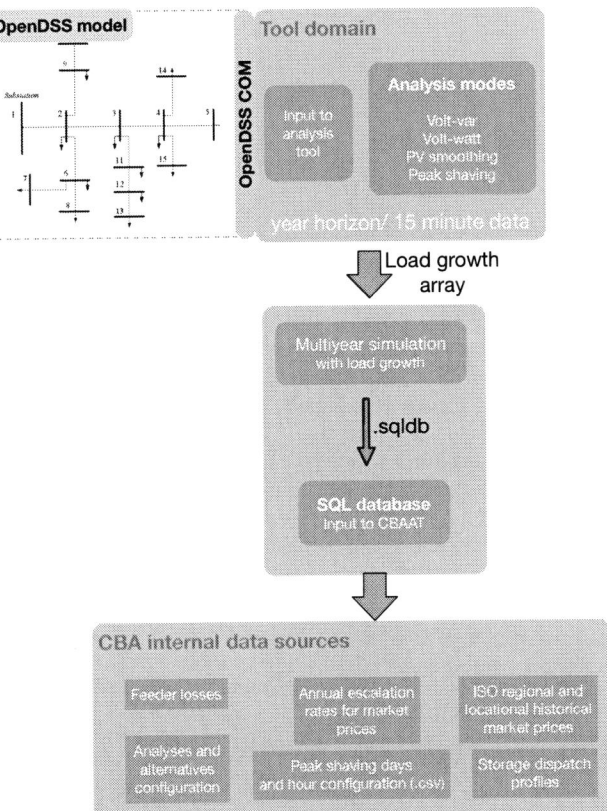

Figure 1 – Diagram representing the overall data flow and operation of CBAAT

The evolution of the process-chain proposed in this paper can be sparsely related to the authors work described in papers presented in the references [1-6]. This paper summarizes the comprehensive process-chain and focuses on the further design, functional capabilities, and exercises the CBAAT . This paper is organized as follows: Section II will describe the activity that led to the development of CBAAT, Section III will describe the process-chain for evaluating ESS operation profiles, Section IV will describe CBAAT design philosophy and architecture, and finally Section V will describe the results of the exercise of the tool-chain with application of the CBAAT to different Energy Storage Systems (ESS) dispatch profiles corresponding to different operational objectives.

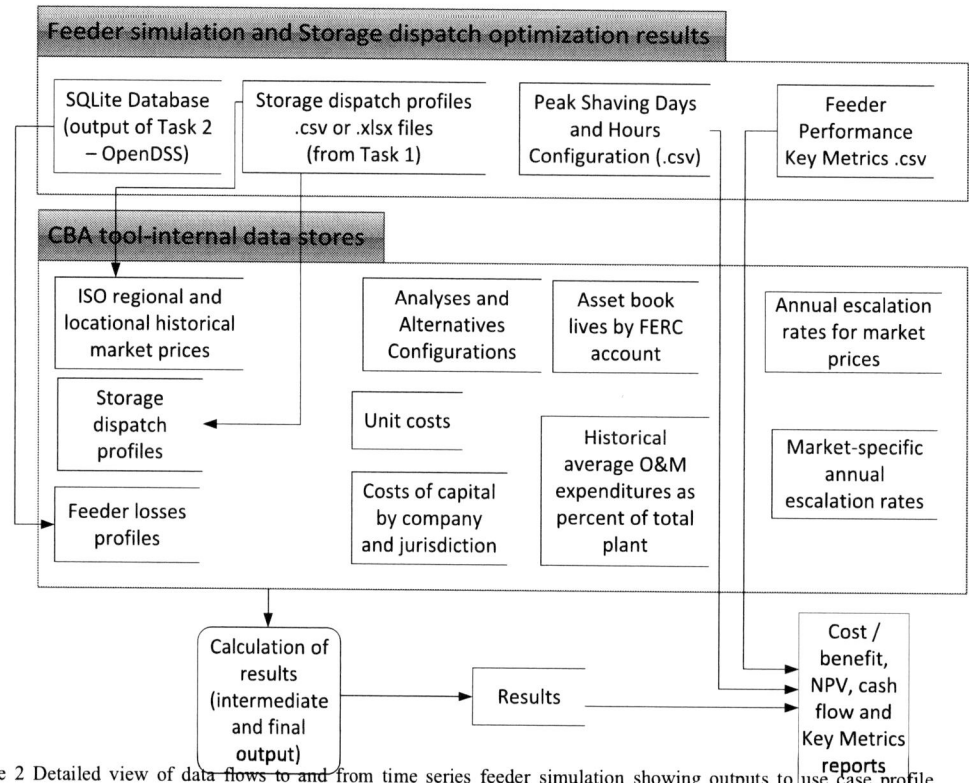

Figure 2 Detailed view of data flows to and from time series feeder simulation showing outputs to use case profile generation and cost benefit analysis

### B. Description of activity

In an effort to assess the potential costs and benefits of ESS, we developed a prototype process-chain for San Diego Gas and Electric for feeder simulation, cost benefit alternative analysis of capital investments, and operational profiles of feeders with energy storage. The specific objectives of the CBAAT work were to identify cost and benefit streams associated with ESS, formulate calculations for them, and implement this functionality in a prototype software tool to SDG&E for use in capital and operational planning.

### II. PROCESS-CHAIN DESCRIPTION

This section describes the process-chain for evaluating ESS operation profiles and their corresponding operational cost and benefit impacts. Figure 2 shows the data stores and data flows through the tool-chain.

The diagram gathers internal CBAAT data stores and processes in one group labeled "CBA tool-internal data stores," and groups data stores for result outputs of other tools into a separate categorical group labeled "Feeder simulation and Storage dispatch optimization results." The process and data flows through the tool-chain as follows (some steps are omitted to maintain relevance to the CBAAT focus of this paper):

1. CBAAT configurations are performed: Ancillary Services (AS) and energy market prices, unit costs, costs of capital, asset book lives, historical O&M figures, annual escalation/inflation rates, and framework alternatives for each use case are configured;

2. the CBAAT-external feeder simulation and ESS dispatch optimizations are run and the results are stored in the appropriate data stores;

3. results of the simulation and ESS dispatch are loaded into the framework alternatives that have been configured into the CBAAT;

4. CBAAT queries are run, generating results that are stored in a Results table;

5. cost-benefit cash flow reporting is performed based on contents of the Results table and may be integrated with key feeder performance metrics.

### III. CBAAT FUNCTIONAL CAPABILITIES, ACCOUNTED-FOR VALUE STREAMS AND DESIGN PHILOSHOPHY

Given the context of the overall tool-chain, this section describes the capabilities and design philosophy of the CBAAT. The objective for the CBAAT component is to calculate detailed cost-benefit financial metrics on multiple scenarios (alternatives).

The CBAAT calculates financial metrics based on various capital and operating cost configurations, outputs of the feeder simulation, and optimal ESS dispatch modules. It does this for plurality of simulated operating policies, each corresponding to an alternative analysis, as well as comparison and inspection.

### A. Added CBAAT Functional Capabilities

This project enhanced the CBAAT to be capable of using price data from energy and ancillary services markets, multiple market-specific annual escalation rates, historical operating and maintenance expenses, approved rates of return, and overhead rates and corresponding rules of applicability

### B. Accounted-for Value-streams

NREL and SDG&E selected value streams and agreed upon approaches relevant to the alternative analysis of distributed energy storage implementation and operation. The following cost/value streams were analyzed and approaches decided upon for modeling:

- Equipment, labor, materials
- Property tax
- Tax benefit of depreciation
- Tax benefit of business expense / tax cost of revenue
- Energy generation
- Transmission losses
- Distribution losses
- Load
- Regulation up / down
- Transmission capacity
- Distribution capacity
- Spinning reserve requirements, Non-spinning/replacement reserve requirements, wide-area black-start capability requirements, regulation reserves, contingency reserves, flexibility reserves
- Voltage control
- Increased/decreased O&M, capital expenditures, time-shifting of these
- Battery degradation

#### 1. Fixed Charge Rate (FCR)

Another evolution in tool functionality was to incorporate a proof-of-concept cost normalization capability to stand in for an SDG&E proprietary approach to calculation of the revenue requirement. The revenue requirement is important because regulations require the utility to select projects with the lowest revenue requirement possible within the constraints imposed by its obligation to meet the requirements of its provision of services obligations.

SDG&E decided that in lieu of the calculation of the full revenue requirement, the CBA methodology should utilize a fixed charge rate (FCR) calculation. The FCR is a percentage which, when applied to the total initial capital cost (book value) of an asset, gives a leveled annual revenue amount attributable to the cost of the asset. When this leveled annual revenue amount is applied as a cash flow in each year of the book life of the asset, it represents the annual leveled payments associated with the asset.

Input parameters required for the revenue requirement and/or FCR differed by FERC Uniform System of Accounts account and Sempra company, such as asset book life and federal and state tax treatment. Tracking, organization, and use of these parameters were incorporated into the functionality of the tool.

#### 2. Non-capital expenditures

The CBA tool provides the capability for the manual configuration of non-capital (O&M) expenditures within the unit cost estimating set of functions. It was decided that, in addition to the FCR approach (described above), the CBA tool be able to automatically estimate O&M using the historical figures by FERC account in the way that SDG&E currently estimates O&M for capital projects. This facilitates the use of SDG&E's current approach for capital projects of the type SDG&E had historical data on, and allows for a more detailed, custom estimation of O&M for projects for which there is not historical O&M data (such as storage projects) but for which O&M expenditures needs to be anticipated.

#### 3. Overhead expenses

Overhead calculations were also identified as a key capability for the CBA tool. NREL worked to automate the calculation and application to cash flows of overhead (loaders) utilizing user-pre-configurable loader, loading base, company, fuel type, and "activity type" tables and relationships in the CBA tool. User selections per line item of loading base, company, and fuel type were utilized in actual cost estimation.

### C. CBAAT Design Philosophy and Architecture

The design philosophy behind the prototype CBAAT and the tool architecture is addressed in this section. The design involved the management and structuring of data including built data management, user interface, and code for the generation of results.

The relational database model is a collection of stored data organized into multiple tables related to one another using key fields. The rows in a table can reference rows in other tables and can be cascaded in ways that allow the representation of complex data relationships, providing systematic way of managing data.

Though there are performance trade-offs with the use of relational databases, well-designed relational database models can eliminate duplicate data promoting the "single source of truth" best practice (data need only be loaded or updated in a single place and records are always in agreement) and reducing the likelihood of mistakes. Such a model facilitates easy and highly targeted access to information of tightly controllable types, and allows for access to this information through easily generated, potentially complex queries that can systematically join information from many tables at once.

CBAAT was enhanced with user interface elements and code to automate the import of data from other elements of the tool chain. Procedures were implemented to sequence the execution of queries to perform successive operations for value stream calculation, aggregation, accumulation, discounting, and meta-data tagging. Finally, facilities were created for reporting allowing user inspection of results at various levels of aggregation and at any time horizon.

## IV. COST BENEFIT ALTERNATIVES ANALYSIS RESULTS

This section describes the use-cases that were explored to illustrate the value streams for ESS and the results generated by the tool-chain for each use-case. SDG&E and NREL identified three use-cases for demonstrating the cost benefits as follows.

### A. Use-case 1: Baseline markets simulation

The markets algorithm generated base case battery dispatch using all the available capacity of the ESS. The results were fed in to the QSTS simulation tool to run power flows and identify technical metrics such as losses and voltage violations. The output from the QSTS simulation tool were fed into CBAAT which approximated the net present value of the revenue requirement and different planning horizons for the use case based on energy-related value streams, capitalization of case-specific asset costs, and O&M expenses.

### B. Use-case 2: Market participation and PV smoothing

PV smoothing was performed using the QSTS simulation tool and the results were passed to the markets algorithm. The markets algorithm derived a dispatch profile, constrained inverter allocation for PV smoothing, and the output was fed back to the QSTS simulation tool. The QSTS simulation tool then re-ran power flows accounting for dispatch profiles from the markets algorithm. The outputs from the QSTS simulation tool and markets algorithm were feed into CBAAT, which approximate the net present values of the revenue requirement at different planning horizons.

### C. Use-case 3: Markets participation and peak shaving:

Peak shaving was performed using the peak shaving algorithm in the QSTS tool and the results passed to the markets participation algorithm. Markets algorithm derived a dispatch profiles constrained by peak shaving requirements and results were fed to the QSTS simulation tool which was used to re-run the power flows based on the outputs from the markets and the peak shaving algorithms. The outputs from the QSTS simulation tool and markets algorithm were fed into the CBAAT which approximated cumulative net present values of the revenue requirement at various planning horizons.

TABLE 1 – CBA OUTPUT FOR THREE USE-CASES

| Category | Sub-Category | 2016 (0) | | | 2017 (1) | | |
|---|---|---|---|---|---|---|---|
| | | Use-case 1: Baseline Markets Simulation | Use-case 2: Market Participation and PV Smoothing | Use-case 3: Markets Participation and Peak Shaving | Use-case 1: Baseline Markets Simulation | Use-case 2: Market Participation and PV Smoothing | Use-case 3: Markets Participation and Peak Shaving |
| Levelized Total Capital Cost | WP: None, U: None, Co: None, Cu: All-in Capex of the Battery | ($412,651) | ($412,651) | ($412,651) | ($412,651) | ($412,651) | ($412,651) |
| Losses: Losses | Losses (NORTHCTY_6_N004 DAM LMP) | ($5,990) | ($6,009) | ($5,977) | ($7,461) | ($7,482) | ($7,445) |
| Use-case specific utilization | Total | $10,769 | $9,701 | $10,737 | $11,092 | $9,992 | $11,059 |
| | Energy Cost (Battery Charge) (NORTHCTY_6_N004 DAM LMP) | ($8,448) | ($11,522) | ($8,431) | ($8,701) | ($11,868) | ($8,684) |
| | Energy Revenue (Battery Discharge) (NORTHCTY_6_N004 DAM LMP) | $19,217 | $21,223 | $19,168 | $19,793 | $21,860 | $19,743 |
| Frequency Regulation: Frequency Regulation | Total | $5,090 | $3,807 | $5,090 | $5,243 | $3,921 | $5,243 |
| | Frequency Regulation (AS_SP26 DAM RD) | $873 | $630 | $873 | $899 | $649 | $899 |
| | Frequency Regulation (AS_SP26 DAM RU) | $4,217 | $3,177 | $4,217 | $4,344 | $3,272 | $4,344 |
| Cumulative DCF | | ($402,782) | ($405,152) | ($402,801) | ($778,781) | ($783,277) | ($778,669) |

## D. Cost benefit and alternative analysis

Table 1 shows the cost benefit analysis results of all three use-cases. CBAAT outputs a detailed breakdown of the financial results by providing category and subcategory labels for the output dollar values. In the use cases modeled in this work, there was only one capital cost modeled, the "All-in Capex of the Battery," configured as a custom cost in all scenarios. The calculated leveled cost of the $1,000,000 outlay in 2016 (analysis year zero) based on the FCR calculation was $412,651 each year over the assumed asset life of 15 years. This leveled cost did not change among the scenarios.

Dollar values of losses were aggregated over single years and reported under a single category and sub-category. The sub-category was labeled with the market name whose prices were used to calculate the dollar value of the losses.

The category label "Use-case specific utilization" was used to indicate costs and savings associated with charging and discharging the battery for all of energy arbitrage, AS markets participation, and providing PV smoothing and peak shaving. When the CBAAT output report displays more than one sub-category, it displays a total row above the sub-category rows summing the sub-category figures.

The frequency regulation category had associated regulation up and regulation down subcategories which the market participation algorithm often identified as the most lucrative market for which the battery could be used. The figures reported in these (and the total) rows report only the revenues associated with battery provision of these services to the grid. The energy costs and revenues associated with charge and discharge associated with provision of these services are rolled into the use-case specific utilization figures in the energy cost and energy revenue (and total) rows.

Finally, the Cumulative DCF (discounted cash flow) row shows the sums of the discounted cash flows across all categories; and, in years other than simulation year zero, the sums in each year of these discounted cash flows with those of all prior years. CBAAT can also display cumulative DCFs for each category and sub-category for detailed inspection of all components of cost benefit.

First, looking only at figures associated with the AS markets participation and battery charging and dispatch which are largely determined by the market participation algorithm, we see that the largest total dollar value ($15,859 vs $13,508, and $15,827, respectively for the other use cases) is associated with the market-participation-only use case because the battery is not being prioritized for any grid service and can be fully utilized for market participation.

Although the market participation algorithm, which is designed to optimize charge and dispatch of the ESS to maximize cost-benefit of energy arbitrage and AS market participation, does not account for the costs of feeder losses or operational or capital costs other than those of operation of the ESS,

the order of the overall net present values (shown in the cumulative DCF row) of the scenarios was the same as the ordering of the DCFs of the market participation subtotals. This is because, even though losses were less costly in the peak shaving use case than in the market-participation-only use case, the difference in cost of losses was not as great as the amount by which revenues from market participation and energy arbitrage were greater in the market-participation-only scenario than in the market participation and peak shaving scenario.

In this study, CBAAT did not take account of costs or savings (if they existed) of effects on other distribution system components that might result from the operating mode of the ESS. It is not known whether, had such costs and benefits existed and been accounted for, the cumulative DCFs of the market-participation-only use case would have been higher than those of the other use cases.

## V. CONCLUSION

This effort demonstrated the suitability of an application developed in a relational database to structure and manage data and the configuration information needed for cost benefit alternatives analysis of utility systems. Microsoft Access was used for relational database and "front end" tools that are well-suited for prototyping this type of application. As a summary the CBAAT prototype was: 1) easily integrated with other components of the tool chain; 2) user-friendly for data input, configuration, and usability of output; 3) capable of generating its results in a reasonable amount of time (less than five min. depending on number and complexity of alternatives); and 4) based on an architecture that is maintainable and susceptible to modification and evolution.

## REFERENCES

[1] D. Narang and J. Hambrick, "High Penetration PV deployment in the Arizona Public Service System," 37th IEEE PVSC, 2011.

[2] J. Hambrick and D. Narang, "High Penetration PV deployment in the Arizona Public Service System, Phase 1 Update," 38th IEEE PVSC, 2012.

[3] J. Cale and D. Narang, "High penetration PV deployment in the Arizona Public Service System Phase II results and update to phase III," 39th IEEE PVSC, 2013.

[4] M. Baggu, R. Ayyanar, and D. Narang, "Feeder model validation and simulation for high-penetration photovoltaic deployment in the Arizona Public Service System," 40th IEEE PVSC 2014.

[5] M. Baggu, J. Giraldez, T. Harris, N. Brunhart-Lupo, L. Lisell, and D. Narang, "Interconnection Assessment Methodology and Cost Benefit Analysis for High-Penetration PV Deployment in the Arizona Public Service System," 41st IEEE PVSC, 2015.

[6] P. Gotseff, J. Cale, M. Baggu, D. Narang, and K. Carroll, "Accurate Power Prediction of Spatially Distributed PV Systems using Localized Irradiance Measurements," IEEE Power and Energy Society General Meeting 2014.

[7] "High Penetration of Photovoltaic Generation Study – Flagstaff Community Power: Final Technical Report: Result of Phase 1," http://www.osti.gov/servlets/purl/1036532/

# Evaluation of PV Hosting Capacities of Distribution Grids with Utilization of Solar-Roof-Potential-Analyses

Gerd Heilscher[1], Falko Ebe[1], Basem Idlbi[1], Jeromie Morris[1], Florian Meier[2]

[1]Smart Grids Research Group – Ulm University of Applied Sciences, 89075 Ulm, Germany

[2]Stadtwerke Ulm/Neu-Ulm Netze GmbH, 89075 Ulm Germany

*Abstract* — Increasing distributed photovoltaic (PV) systems can lead to voltage violations and overloading of grid assets in distribution grids. This raises the necessity to consider the future growth of installed PV capacity for the planning process of distribution system operators (DSOs). This paper proposes the combination of a solar roof potential analysis and grid integration studies at the medium voltage (MV) level based on detailed input at low voltage level. Three different methods were developed to define the distribution of PV along the feeders based on the solar roof potential, and subsequently to estimate the PV hosting capacity of these feeders. The results show that the approach with an even distribution of PV systems along the feeders leads to higher hosting capacity of PV for the analyzed grids. In addition, for most analyzed feeders in this study, an overloading of MV/LV (low voltage) transformers is expected to be the limitation of hosting capacity for potential PV systems.

*Index Terms* — Solar roof potential, feeder analysis, hosting capacity, grid planning, distributed photovoltaic energy systems.

## I. INTRODUCTION

A strategic target, set by the German government, is the transition from the centralized energy system based on fossil fuels to a decentralized energy supply based on renewable resources [1], [2]. According to the high installation rate of distributed generators (DGs) in the distribution system, voltage violations and overloading of grid components has been observed already. In other words, if the installation of DGs in a distribution grid exceeds its hosting capacity, a renewal of network assets (e.g., reinforcement of lines or exchange of transformers) is required which can lead to considerable costs for the distribution system operators [3], [4]. For example, the distribution grid extension in Germany until the year 2030 is estimated to be higher than 27 billion € [4]. Considering these high costs, several studies investigate many approaches to increase the hosting capacity of distribution grids, such as active on-load tap changing of transformers, reactive power provision and power curtailment of DGs [5], [6]. The estimation of the hosting capacity of distribution grids has become of crucial importance for grid planning, cost estimations as well as decision making. The hosting capacity can be understood as the maximum installed power of DGs that a network can host without the need for grid reinforcement [5]. In order to estimate grid extension costs and to evaluate possible alternatives, DSOs have to consider future DGs installations in planning scenarios. However, the

allocation of upcoming installed capacity of DGs per medium voltage and low voltage connection points is hard to be accurately predicted. In addition to statistical or expert knowledge approaches like in [4], [7], methods of the domain of geoscience and remote sensing can be applied. By using light detection and ranging (LIDAR) methods, it is possible to create a detailed 3-dimensional map of the investigated area. Subsequently, it is possible to extract the most relevant parameters for each single potential PV-system (e.g., roof area, orientation and inclination) [8], [9]. In addition, also the hourly power output of these expected future PV-systems is available including also local shading losses. Based on the solar-roof-potential-analysis (Fig.1) the DSO becomes capable to investigate the impact of the transformation towards a decentralized renewable energy system on his distribution network.

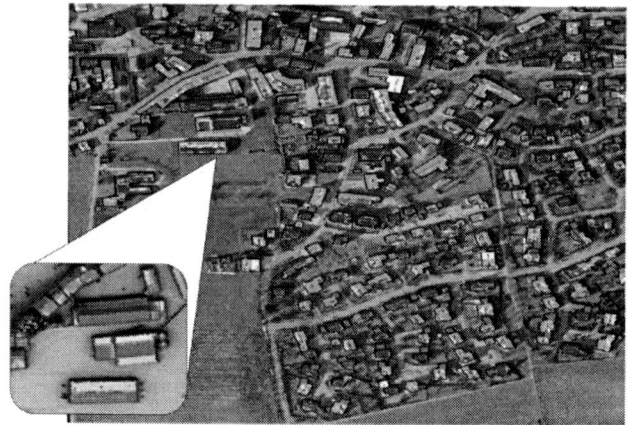

Fig. 1: Solar-Roof-Potential-Analysis for each single roof in a LV test area

This paper introduces the solar roof potential analyses and applies it to the evaluation of the hosting capacity of the distribution grid of the DSO of Ulm, Germany with a deterministic approach. When modelling a PV installation scenario within a distribution network, it is necessary to define the allocation of installed PV power to grid connection points. Some studies used the Monte Carlo analysis to represent the random process of PV installations in distribution networks. For example, [5] introduced several analyses that show the increase of the grid hosting capacity through some PV control strategies. Thus, the hosting capacity is estimated here through Monte Carlo analyses, considering random distributions of PV systems with discrete nominal powers.

The first objective of this contribution is the utilization of solar roof potential in distribution system studies in order to increase the accuracy of predicted PV scenarios. The second objective is the neutralization of the effect of random distribution of PV systems on the estimation of grid hosting capacity. Therefore, a deterministic method has been developed and evaluated based on a real MV grid.

This paper is structured in four sections. This section presents the topic and motivation of the study. The second section illustrates the methodology followed in the paper. The last two sections introduce the results and findings of the study.

## II. METHODOLOGY

The steps and assumptions defined to perform the hosting capacity analysis are introduced hereinafter.

### A. Solar Roof Potential Analysis

Up to now, four different Solar-Roof-Potentials-Analyses were evaluated. In the first analysis, different input data (stereoscopic pictures, LIDAR-data in different resolution) have been used. Best results have been obtained from LIDAR with a resolution of 5 points/m². Three geo informatics service providers have used these data. The forth analysis was based on LIDAR data with just 1 point/m². The comparison of the four analyses showed differences in the detection of the suitable roof areas and the respective roof potentials. The potential power and the annual energy potential is available for all four analyses. The analysis of inclination and orientation was published in [8]. Only two providers were able to deliver also hourly power input including local shading at the point of common coupling of the individual houses.

### B. Defined Methods

For the analyses of grid hosting capacity, a realistic set of scenarios should be assumed, considering the derived Solar Roof Potential Analysis. A typical method is the probabilistic distribution of PV systems at random connection points in the grid, then the repetition of the distribution several times in a Monte Carlo analysis. This leads to a huge number of different scenarios, and thus a simpler method is targeted in this study. The purpose here is the evaluation of the maximal and minimal hosting capacity using a deterministic method, focusing on the medium voltage feeders. The hosting capacities are obtained by increasing the utilization factor of the solar roof potential analysis per secondary MV/LV substation until a local congestion occurs (e.g., transformer overload or voltage band violation). By using a utilization factor (UF), i.e. rate of installed PV power proportional to the analysed roof potential capacity, three "Upscaling Methods" have been/will be defined as follows:

Forward PV increase: the installation of PV potential power starts from the secondary bus bar of the high voltage (HV)/MV transformer. Then the UF is increased each iteration until the maximum potential is reached.

Backward PV increase: the installation of PV potential starts from the furthest connection point of the secondary bus bar of the HV/MV transformer.

Even PV increase (All Together): an equal utilization factor for all the PV systems will be defined. The UF will be increased by a certain step at each iteration.

The 'Backward' method is expected to lead to a minimal hosting capacity since higher voltage rise can be expected. Moreover, the reverse feed-in powers go through more lines in the feeder so that overloading is more likely. Fig. 2 shows an illustration of these methods.

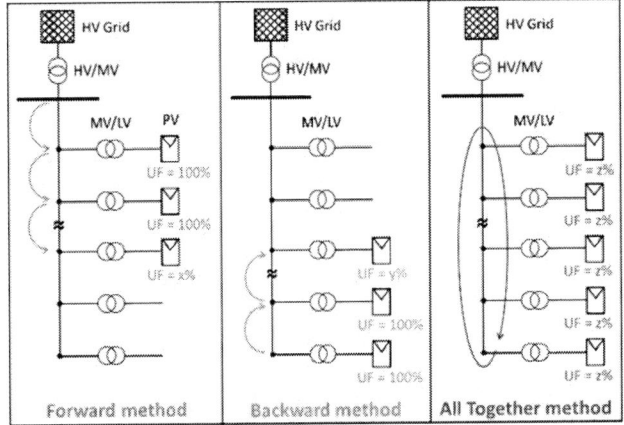

Fig. 2. Schematic overview of the implemented methods. x and y indicates the utilization factor (UF) for the directed methods. z indicates the UF for the equal utilization.

### C. Topology Analysis

One key element of this study is the topology analysis, which assigns the individual PV potential power to a level structure as a function of the nodal distance from the primary substation (HV/MV). As a part of the developed methods, the PV systems within a single level are scaled up with the same scaling factor for the three defined methods. Fig. 3 illustrates the used level concept.

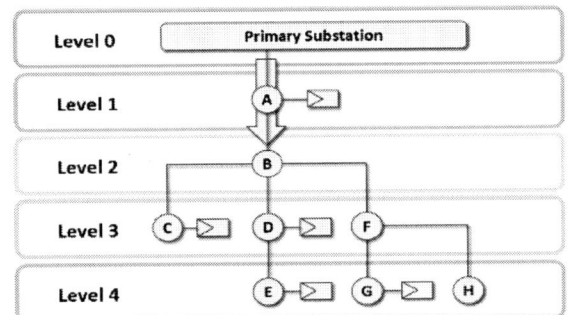

Fig. 3. Exemplary illustration of the level concept used in the topology analysis. Each circle represents an MV/LV substation.

The topology analysis as well as the overall algorithms is implemented as a Python script, which utilizes the functions and the grid model implemented in power systems analysis software (PSAS) DIgSILENT PowerFactory 2016 and uses its solver for load flow calculations.

Fig. 4 illustrates the script as a flowchart. For every single defined feeder, the script runs a series of functions in order to determine the maximum tolerated PV hosting capacity.

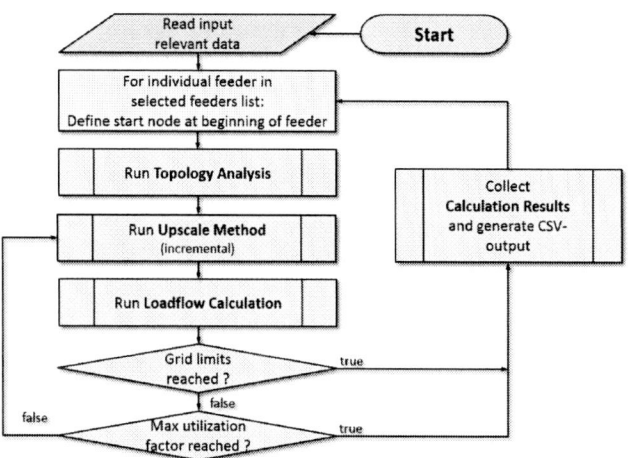

Fig. 4. Schematic overview of the key functions implemented.

The main functions blocks in Fig. 4 are:

**Input:** Defined grid limits, modification methods and feeders within the script's input provide the possibility of creating individual scenarios and thereby form the basis for every hosting capacity calculation.

**Topology Analysis:** This function sequentially runs through a determined feeder and gathers all calculation relevant data needed to define the 'PV-potential-systems'. The result is a sorted table of the 'PV-potential-systems' based on their location along the branch, beginning with the system closest to the feeder's start (see also Fig. 5 in more detail)

The basic principle of the topology analysis is to read in nodes connected to an element and evaluating the unknown node. The unknown node is applied as a new input value when calling the topology analysis function again, accordingly topology analysis runs recursive. Therefore, the analysis always moves in the direction of the undiscovered elements. In case the topology analysis detects a load switch (i.e. coupler), it assumes to have discovered a substation, which then is examined for 'PV-potential-systems'. The detected 'PV-potential-system' is stored in a dictionary along with a corresponding level, which represents the location along the branch. In case of a branching, the respective terminal is marked together with the corresponding level and processed whenever the function reaches the end of a branch.

**Loadflow Calculation:** PowerFactory's load flow calculation is utilized throughout the '**Upscaling Method**'. Whenever a PV-potential-system's power is upscaled, a load

flow calculation is executed. During the active load flow calculation, the initially determined grid limits are evaluated. These can be: line loading [%], transformer loading [%] and voltage band [p.u.]. As long as the defined grid limits are not reached, the grid is considered to be in a stable state and the upscaling of the PV-potential-systems is continued.

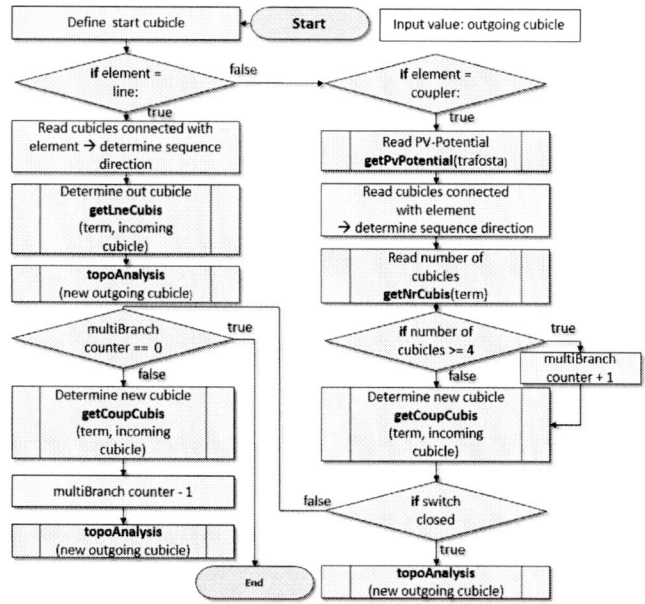

Fig. 5. Schematic overview of the topology analysis function

*D. Utilized Grid Modell and Assumptions*

Based on the grid data provided by the local grid operator in Ulm (i.e. Stadtwerke Ulm/Neu-Ulm Netze GmbH), a detailed real MV grid is modelled and analyzed. The examined part of the MV grid area consists mainly of urban and industrial areas. Table 1 provides some key figures of the examined grid.

The boundaries of the modelled grid are the MV bus bar of the main HV/MV substation and the LV bus bars of the MV/LV substations. The model assumes the connected PV plants in the LV grids as aggregated PV systems connected at the LV side of the MV/LV substations. The overlaying HV grid is modelled as a slack element connected to the medium voltage bus bar of the primary substations. Therefore, each feeder is analyzed independently from other feeders because the voltage rise across a HV/MV transformer is assumed to be neglected for simplification purposes. According to this simplification, the iterations for the load flow calculations, which are caused by the regulations of the tap-changer of the HV/MV transformer, can be avoided.

Besides the aforementioned assumptions, the technical guidelines for decentralized power plants connected to the medium voltage grid are considered [10]. The guidelines describe a simple method for the assessment of grid connection request for a DG. This study uses the limitation of the voltage rise in the MV grid with a maximum of 2 % for all connected stations as stated in the mentioned guideline. These limits imply that all potential PV-systems are connected to the MV grid. Therefore, there is no consideration of the additional 3 % voltage rise in the LV grid introduced by the relevant German technical guideline VDE AR4105 [11]. In addition, the wind power feed-in is not considered in the analyzed grid, since the wind potential in the grid area is negligible compared to the PV potential [12]. Furthermore, the voltage set point for the slack element is assumed to be a constant value of 1.0. p.u.. The derating factor for transformers, cables and overhead lines is assumed to be 1.0 p.u.

Table 1. Key Parameters of analyzed MV/LV grids

| Parameter | Value |
|---|---|
| MV Feeders | 243 pcs |
| Primary Substations | 7 pcs |
| Public Substations supplied | 543 pcs |
| Private Substations supplied | 332 pc |
| MV line length (cable / overhead line) | 892 / 79 km |
| Total MV/LV Transformer capacity | 1340 MVA |

## III. RESULTS

The calculated results as well as the main derivations are presented in the section hereinafter.

### A. Scenario and Set Point Selection

Numerous hosting capacity calculations have been carried out considering the number of feeders in Table 1, the three methods as well as the three assumed scenarios. The scenarios differ in the used set point value of the maximum loading of MV/LV transformers. The first two are used to represent the present situation and the possibility to replace a transformer with a higher rated transformer. The third scenario neglects the transformer overloading as a termination reason for the extension process. The assumed scenarios regarding the set points are illustrated in Table 2. The aim of the second and third set point is to show the hosting capacity only for the MV feeder grid components, regardless of the nominal capacity of the MV/LV transformers.

Table 2. Set point values for the assumed scenarios

| Scenario | Voltage Band | Line Overload | Transformer Overload |
|---|---|---|---|
| 1 | 1.02 p.u. | 1.0 p.u. | 1.0 p.u. |
| 2 | 1.02 p.u. | 1.0 p.u. | 2.5 p.u. |
| 3 | 1.02 p.u. | 1.0 p.u. | Unlimited |

### B. Limitation of Hosting Capacity

In order to give an overview on the limitations of the analyzed MV grid, a distribution of the MV feeders numbers based on the termination reason for the method 'All Together' is depicted in Fig. 6. The distribution is classified according to the defined scenarios in Table 2 as well as to the connected primary HV/MV substations (SS). Considering a transformer-

loading limit of 1.0 p.u., the main termination reason is the LV/MV transformer overloading except for few feeders having a line overloading or voltage violation as a termination reason. As a result, for the analyzed grids, the costs of transformer replacement should be taken into account for the future scenarios. For Example, the SS7 is directly located in the city center, hence lines are relatively short, and thus the main congestion reason is transformer overloading. When setting the transformer loading limit to 2.5 p.u., more line loading congestions and voltage band violations occur. Therefore, line replacement should also be taken into consideration for the long-term scenarios.

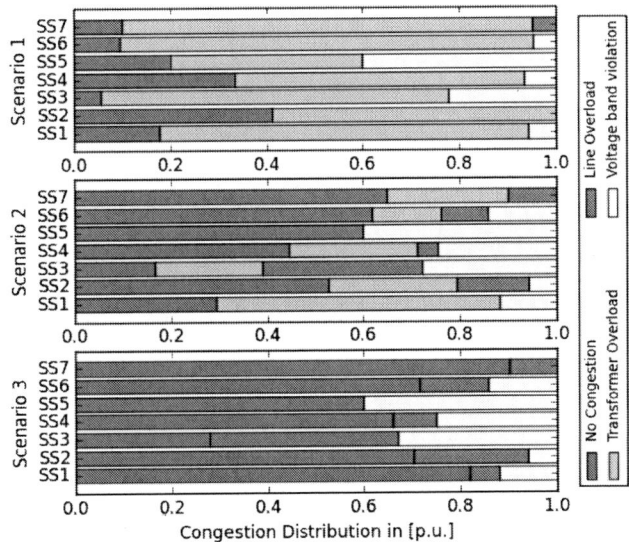

Fig. 6. Distribution of the feeder numbers according to the congestion reason, for the scenarios in Table 2, cosidering the 'All Together' method. Specific SS group the feeders.

### C. Evaluation of Calculated Hosting Capacities

The hosting capacities for the analyzed feeders are illustrated statistically in Fig. 7. Theoretically, the 'Backward' method is likely to lead to minimal hosting capacity, since voltage violation and line overloading is more probable. In contrast to what is anticipated, the 'Forward' method results in the lowest hosting capacity for Scenario 1 (orange rectangle in Fig. 7). This can be justified by the fact that the Transformer overloading is independent from the connection point of PV systems. In addition, the PV potential for the analyzed feeders can be higher close to the primary substation, so that the transformer can be overloaded. Due to the even distribution of PV used in method 'All Together', a single transformer is less burdened which leads to higher hosting capacity. A more detailed examination of this circumstance is illustrated in Fig.8.

For each feeder, the difference in hosting capacity resulted by the methods 'All Together' and 'Forward' is calculated. With the increase of the permissible transformer loading, the difference between the methods decreases. For Scenario 3, when neglecting the transformer overloading congestion, the

'Backward' method results in a lower hosting capacity mainly due to early voltage violations (see Fig. 7). For this scenario, the 'All Together' method results in higher hosting capacity due to the even distribution of PV power along the feeders.

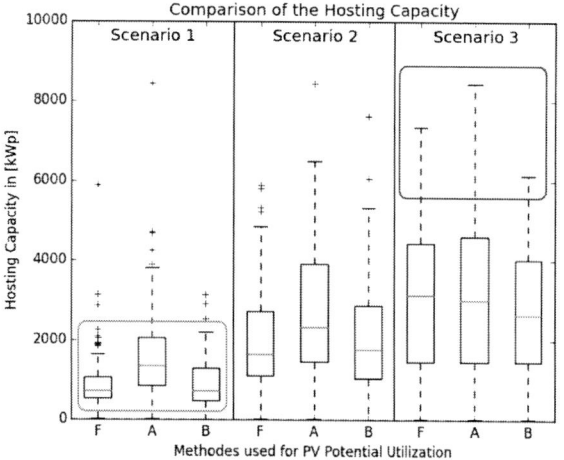

Fig. 7. Boxplots of calculated hosting capacities for the scenarios of Table 2 and utilizing the developed methods. 'A' for All Together, 'B' for Backward and 'F' for Forward. One Box represents the interquartile range (IQR) from 25 % to 75 % and whiskers are at 1.5 * IQR

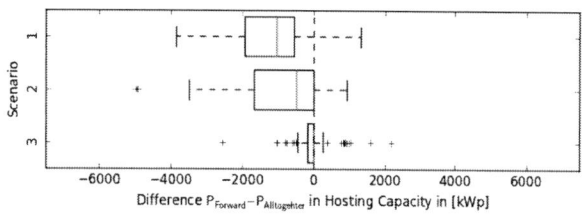

Fig. 8. Boxplots comparing the differences of the methods 'All Together' and 'Forward'. Box represents the IQR from 25 % to 75 % and whiskers are at 1.5 * IQR

*D. Illustrative Example for a Single Feeder*

Using the built-in plotting function of PowerFactory (i.e. the voltage profile plot), the profile of a single feeder is depicted in Fig. 7 in order to illustrate the effect of the three developed methods. The selected feeder consists of a main branch supplying a suburban area via cables and a side branch with overhead lines and 2 small MV/LV transformers, which supply farmsteads with a decent amount of PV already installed. The plotted data is derived from three load flow calculations at the hosting capacity power for Scenario 2 which considers a transformer loading of 2.5 p.u. as a termination reason. Fig. 9 shows the voltage at the MV lines, bus bars and the LV bus bars. The vertical lines represent the voltage rise at the transformers, while the red-colored lines represent overloaded components which are loaded more than 1.0. p.u. of their nominal capacity. For the 'All Together' method, the termination is based on the overloading of the MV/LV transformers of the side branch. A crucial voltage rise over

these tow transformers can also be observed. The 'Forward' method terminates due to the same reason but a smaller amount of PV power is integrated as the extension process directed. On the other hand, the voltage violation at the main branch is the termination reason for the 'Backward' method in this example.

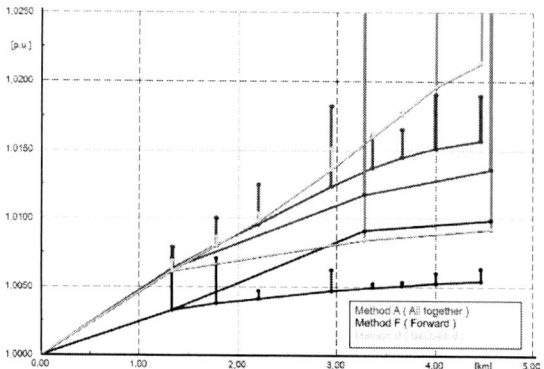

Fig. 9. Voltage profile of a single feeder according to Scenario 2 (Transformer overloading: 2.5 p.u.).

## IV. CONCLUSION

The work demonstrated that Solar-Roof-Potential-Analyses deliver detailed and realistic basic values about future local PV-potential for each individual feeder. The presented results explained a deterministic and automated approach for DSOs to investigate the impact of the future utilization of this PV-potential on his distribution network. This paper proposed three methods for the utilization of the results of a solar roof potential analysis for the calculation of PV hosting capacity on the medium voltage feeder level. For the analyzed grid area and assumptions made in this study, it is not possible to predict which method can provide the minimal hosting capacity value. However, the method with an even installation of PV systems along the feeder results in higher hosting capacity of PV for the analyzed grids.

In addition, for most analyzed feeders in this study, an overloading of MV/LV transformers was the main limitation reason for the exploitation of potential PV power. Therefore, the cost of transformer replacement should be taken into account for future scenarios.

## V. OUTLOOK

Further investigations using the proposed method are able to evaluate the impact of increasing solar module efficiency or might use additional information from social networks about the realistic utilization of the PV-potential.

Additional work will compare this method with statistical approaches and synthetic reference type grids. The developed methods will be further examined, considering the targeted increase of spatial resolution for the roof potential analysis. In a further step, the analysis of the hosting capacity can consider the effect of charging stations of electric vehicles.

## ACKNOWLEDGMENTS

This work is part of the research project NATHAN-PV funded by the German Federal Ministry of Education and Research (Grant No.: 03FH03013).

## REFERENCES

[1]  M. Kaltschmitt and A. Wiese, *Erneuerbare Energien: Systemtechnik, Wirtschaftlichkeit, Umweltaspekte*. Springer-Verlag, 2013.

[2]  Bundesministerium für Wirtschaft und Energie (BMWi), "Energiekonzept - für eine umweltschonende, zuverlässige und bezahlbare Energieversorgung", 2010.

[3]  E. J. Coster, J. M. A. Myrzik, B. Kruimer, and W. L. Kling, "Integration Issues of Distributed Generation in Distribution Grids," *Proc. IEEE*, vol. 99, no. 1, pp. 28–39, Jan. 2011.

[4]  Deutsche Energie Agentur (dena Verteilnetzstudie), "Ausbau und Innovationsbedarf der Stromverteilnetze in Deutschland bis 2030," Berlin, 2012.

[5]  T. Stetz, "Autonomous voltage control strategies in distribution grids with photovoltaic systems - technical and economical assessment," Ph.D. dissertation, Univ. Kassel, 2013.

[6]  B. Idlbi, K. Diwold, T. Stetz, H. Wang, and M. Braun, "Cost-benefit analysis of central and local voltage control provided by distributed generators in MV networks," in *PowerTech, 2013 IEEE Grenoble*, 2013, pp. 1–6.

[7]  G. Kerber, "Aufnahmefähigkeit von Niederspannungs-verteilnetzen für die Einspeisung aus Photovoltaik-kleinanlagen," Ph.D. dissertation, Univ. Munich, 2011.

[8]  H. Ruf, D. Funk, F. Meier, and G. Heilscher, "Bestimmung von Ausrichtungs-und Neigungswinkeln von Bestandsanlagen mittels LIDAR Daten," in *Proceedings of the 30. Symposium Photovoltaische Solarenergie, Bad Staffelstein, Regensburg*, 2015.

[9]  Y. Zheng and Q. Weng, "Assessing solar potential of commercial and residential buildings in Indianapolis using LiDAR and GIS modeling," in *Earth Observation and Remote Sensing Applications (EORSA), 2014 3rd International Workshop on*, 2014, pp. 398–402.

[10]  Bundesverband der Energie- und Wasserwirtschaft BDEW "Technical Guideline: Generating Plants Connected to the Medium-Voltage Network," 2008.

[11]  VDE-AR-N 4105 "Anwendungsregel Erzeugungsanlagen am Niederspannungsnetz: Technische Mindest-anforderungen für Anschluss und Parallelbetrieb von Erzeugungsanlagen am Niederspannungsnetz," 2011.

[12]  LUBW Landesanstalt für Umwelt, Messungen und Naturschutz Baden-Württemberg [Online]. Available: https://www.lubw.baden-wuerttemberg.de [Accessed: 16.01.2017]

# Experimental Distribution Circuit Voltage Regulation using DER Power Factor, Volt-Var, and Extremum Seeking Control Methods

Jay Johnson[1], Sigifredo Gonzalez[1], and Daniel B. Arnold[2]

[1]Sandia National Laboratories, Albuquerque, New Mexico, 87185, USA

[2]Lawrence Berkeley National Laboratory, Berkeley, California, 94720, USA

*Abstract* — With the rapidly increasing number of distributed energy resources (DER) on electric grids worldwide, there is a growing need to have these devices provide grid services and contribute to voltage and frequency regulation. PV inverters and other reactive-power capable DER have the capability to minimize distribution losses and provide voltage regulation with grid-support functions such as volt-var, volt-watt, and fixed power factor. The optimal settings for these functions for specific distribution topologies has been widely studied using centralized (reactive power settings) and decentralized control modes (volt-var, volt-watt). However, optimal selection of the DER operating mode and settings depends on *a priori* knowledge of system topology, DER sizes and locations, and renewable energy power generation or forecasts. In this paper, we experimentally evaluate an approach for optimal reactive power compensation that does not rely on the aforementioned information. Specifically, we study the ability of Extremum Seeking (ES) control—a decentralized, model-free control strategy—to minimize distribution losses and maintain voltage limits in a laboratory environment. The speed and performance of the ES control algorithm is compared to a typical smart inverter volt-var function for reactive power compensation utilizing local system measurements.

## I. INTRODUCTION

As the quantity of renewable energy DER interconnected to the distribution system increases, grid operators are confronted with new distribution management and voltage regulation challenges. Traditionally, utilities and other grid operators could assume unidirectional power flow and grid voltage within ANSI Standard C84.1 [1] limits using larger centralized voltage regulation equipment such as on-load tap changing transformers and STATCOMs. With increasing solar penetrations, DER generation is causing larger voltage swings on distribution systems and thereby increasing systems reactive power losses [2]-[3].

Fortunately, the United States is updating the interconnection and interoperability requirements at the state [4]-[5] and national-level [6] to require additional distributed energy resource (DER) grid-support functions, including volt-var, volt-watt, and fixed power factor. When these functions are configured correctly, the distribution hosting capacity can be drastically increased by mitigating thermal and voltage excursions, minimizing losses, and maintaining ANSI limits [7]-[8].

Many techniques for controlling fleets of DER assets to provide voltage regulation have been described in the literature. Centralized control algorithms have been designed to solve for the optimal reactive power injection for DER devices [8]-[9]. Other researchers have investigated distributed control (autonomous functions) to provide voltage regulation. The most widely studied decentralized method is the volt-var function which changes DER reactive power injection of the inverter based on local voltage measurements[1]. Shen and Baran created a gradient-based optimization for the volt-var (VV) function for unbalanced systems [10]. The National Renewable Energy Laboratory (NREL) has studied the influence of VV setpoints on power quality and energy savings HECO and PG&E distribution systems [11] and EPRI, Georgia Tech, and Sandia have extensively studied the VV function for distribution voltage regulation and increasing feeder hosting capacity [12]-[15].

In this work, we experimentally test an Extremum Seeking (ES) controller for reactive power compensation. ES control of DER, a topic of recent study in literature, has been shown to achieve optimal results when coordinating reactive power assets for voltage regulation and loss minimization [16]. Such results are achievable without having to rely on external knowledge, such as the distribution system topology and DER characteristics. In this work, we experimentally evaluate the ES approach via its implementation in an integrated control interface that is connected to a simulated distribution circuit created in the Sandia's Distributed Energy Technologies Laboratory (DETL). The testbed environment allowed the ES control performance to be compared directly to centralized (fixed power factor) and distributed (VV) voltage control techniques.

## II. ES CONTROLLER DESIGN AND IMPLEMENTATION

Extremum Seeking Control is a real-time optimization technique for multi-agent, nonlinear, and infinite-dimensional systems [17]. The algorithm operates by adjusting system

---

[1] Some voltage regulation benefit can be obtained with volt-watt function, but active power reduction is seen as less favorable from an economic accounting and regulatory standpoint.

inputs in an effort to optimize measured outputs. While there are many forms of ES control, the method discussed herein adjusts system inputs via use of a sinusoidal perturbation, demodulates system outputs to extract approximate gradients, and finally performs a gradient descent. A block diagram of the approach is shown in Fig. 1. The parameters $k$, $l$, $h$, $a$, and $\omega$ are chosen by the designer. In addition to choosing unique probing frequencies for each controllable DER, one must ensure that $l$, $h \ll \omega$ thereby ensuring proper operation of the high and low pass filters in each ES loop. The reader is invited to consult [16] for more detail regarding the configuration of ES control for this activity.

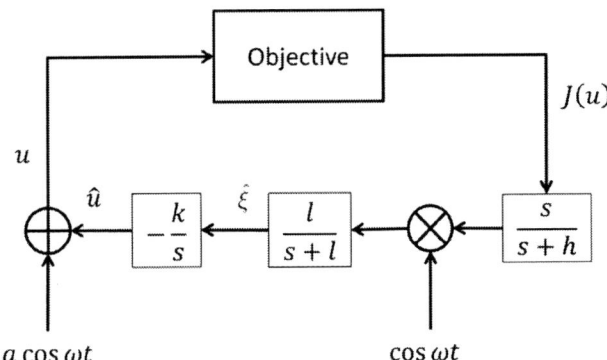

Fig. 1. Block diagram of ES controller for managing DER reactive power output.

### III. DISTRIBUTION CIRCUIT EXPERIMENTS

Experiments were conducted at DETL to evaluate voltage regulation using fixed power factor, VV, and ES control using the testbed shown in Fig. 2. The testbed consisted of nine photovoltaic inverters:

- Two 3.0 kW inverters with adjustable reactive power modes and ±0.80 power factor limits on Bus 2.
- Two 3.0 kW inverters at unity power factor on Bus 2.
- Four 3.0 kW inverters with adjustable reactive power modes and ±0.85 power factor limits on Bus 3.
- One 3.8 kW inverter at unity power factor on Bus 3.

The circuit was connected to a 180 kVA grid simulator to maintain the voltage near nominal. Additional tests were conducted connecting the test circuit to the utility, but the

variation in Bus 1 voltage was smaller using the grid simulator. The 200-foot cable between Buses 2 and 3 simulated approximately 10.4 miles of 2/0 cable on a 12.47 kV radial distribution circuit with a calculated X/R less than 0.5. While the X/R ratio may be lower than many overhead distribution runs, it represented a realistic rural installation and demonstrated undercompensated, long feeders with low X/R ratios are prone for voltage deviations with high penetrations of DER. The PV inverters were powered with nine 10 kW Ametek PV simulators with simulated PV systems configured such that their maximum power point was the DC nameplate rating at 1000 W/m² irradiance. The AC power production of the inverters depended on the reactive power output and inverter efficiency.

Three voltage regulation control strategies were conducted with the testbed. The first adjusted the power factors settings of the inverters to evaluate the influence of inverter reactive power on the circuit voltage profile. The second experiment investigated the volt-var function against a no-control baseline for a range of irradiance values. And lastly, a voltage regulation objective function was employed with the ES controller to determine if this approach was practical for voltage regulation.

### A. Fixed Power Factor

In order to assess the influence of power factor on the distribution circuit, the fixed power factor (PF) function (see the INV3 function in IEC 61850-90-7 [18]) was independently adjusted on the inverters on Bus 2 and 3 (referenced as $PF_2$ and $PF_3$) at five values from $PF_{min}$ to $PF_{max}$ for the inverters. The irradiance of all the PV systems on each bus—$G_2$ for Bus 2 and $G_3$ for Bus 3—were set to 200, 500, and 1000 W/m². Note that that power of the maximum power point of the PV simulator—and therefore the inverters—scales nearly linearly with irradiance. The 225 parametric tests were repeated 10 times to gather averages and measure the repeatability of the experiments. The maximum standard deviations on Bus 1, 2, and 3 were 0.22%, 0.47%, and 0.35% p.u. voltage for the 225 experiments, though the majority were below 0.10%. The variance in measurements was generated from a combination of the grid simulator voltage tolerance and data acquisition measurement errors. The voltage at the three buses shown in Fig. 2 were recorded to determine the voltage profile in each

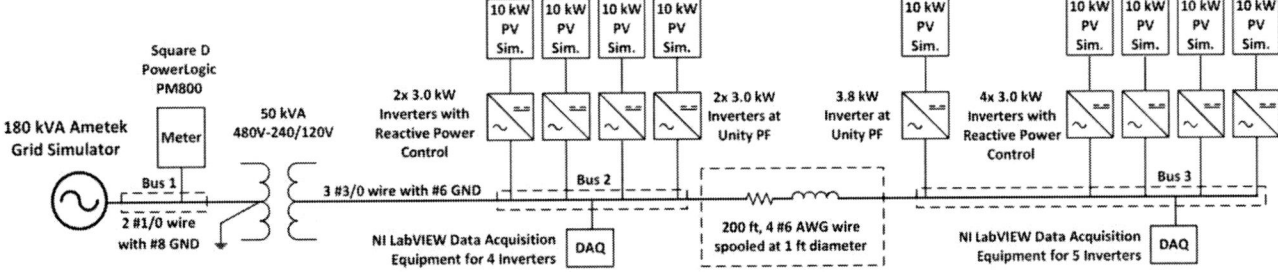

Fig. 2. Simulated distribution circuit test bed at the Distributed Energy Technologies Laboratory.

978-1-5090-5606-4/17 $31.00 © 2017 IEEE

scenario. As an example of these results, the 250 voltage profiles for one of the irradiance combination is shown in Fig. 3. The voltage is significantly larger at the end of the load-less feeder than at the grid simulator and the fixed power factor function is effective at reducing the voltage (or increasing the voltage). Notably, with larger quantities of absorptive reactive power (positive PF—according to the IEEE sign convention) the grid simulator voltage decreased as well. In order to minimize this bias, the voltage changes from Bus 1 were used as the parameters of interest for the remainder of this document.

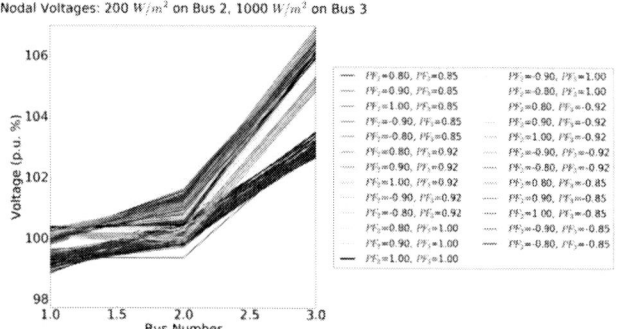

Nodal Voltages: 200 $W/m^2$ on Bus 2, 1000 $W/m^2$ on Bus 3

Fig. 3.    Voltage profile of the circuit with different power factor values for Bus 2, $PF_2$, and Bus 3, $PF_3$.

The average bus voltages for each of the 225 configurations is shown in Fig. 4 with the color indicating the power factor and the shape of the marker representing the irradiance level of the simulated PV system. The key trends in the results are:

1.  The maximum voltages on Bus 3 are much larger than

Bus 2 due to the impedance in the 200-ft cable between Bus 2 and 3.

2.  Bus 3 power factor and irradiance had the largest influence on the Bus 3 voltage. With increasing irradiance, the inverters produced greater active power which increased the voltage from resistive losses. The power factor changed the reactive power on the circuit which shifted the bus voltages.

3.  In the case of rated power for all the inverters (as indicated by the ★ symbol), a significant amount of absorptive reactive power was needed to pull the voltage to within ANSI Range A limits. In scenarios with lower irradiances, the voltage deviations were less severe.

To better visualize the influence of the power factor on the bus voltages, trajectories were created from the data in Fig. 4 in which the $PF_i$ values were plotted from {$PF_2$ = -0.80, $PF_3$ = -0.85} (trajectory start) to {$PF_2$ = 1.0, $PF_3$ = 1.0} (trajectory elbow) to {$PF_2$ = 0.80, $PF_3$ = 0.85} (trajectory end) for each of the irradiance sets. As seen in Fig. 5, these PF changes reduce the Bus 3 voltage and in all cases the final PF settings are within ANSI Range A. Interestingly, for most of the irradiance cases, the Bus 2 voltage decreases in the first step, but increases in the $2^{nd}$ step due to the reactive power flows through the testbed circuit.

### B. Volt-Var Autonomous Function

ES control and VV control were evaluated with the system in Fig. 2 with an aggressive VV curve: V points = {95%, 99%, 101%, 105%} and Var points = {100%, 0%, 0%, -100%}. All inverters were configured with reactive power priority, so active power would be reduced to provide the reactive power

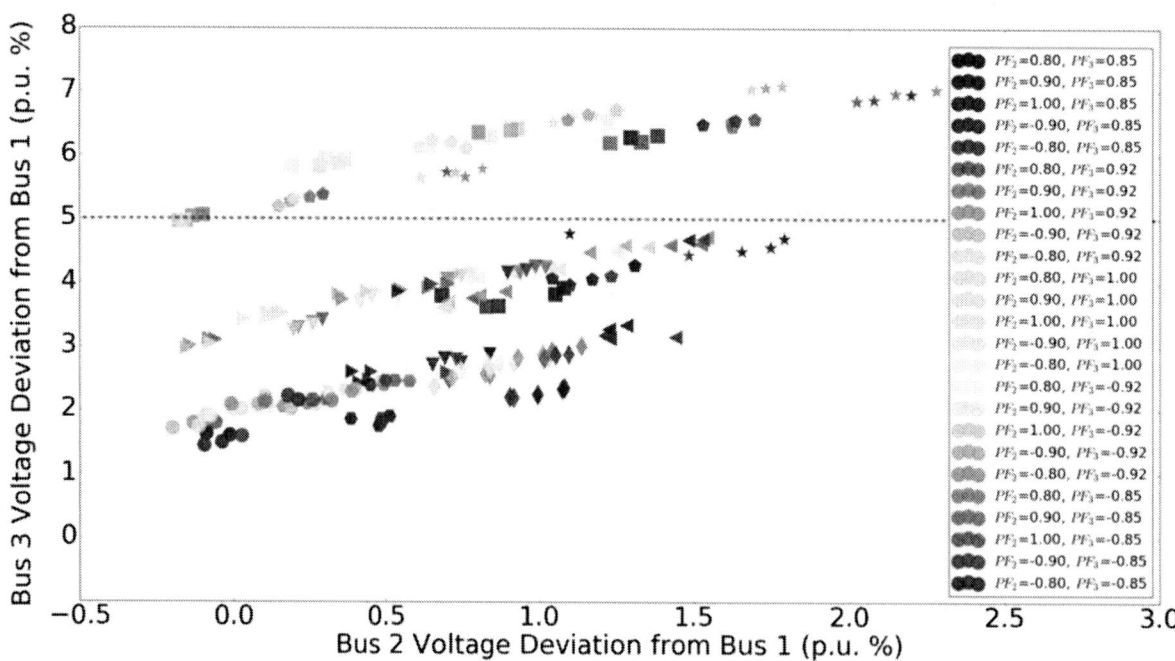

Fig. 4.    Bus 2 and 3 voltages for different inverter power factors and irradiances. ANSI Range A upper voltage limit is the dashed red line.

978-1-5090-5606-4/17 $31.00 © 2017 IEEE

requested by the VV function, if necessary. As shown in the results in Fig. 6, Bus 2 and 3 voltages were reduced using this function. In the case of Bus 3, the voltage reduction was nearly as significant as the most absorptive reactive power level supported by the devices, shown in Fig. 5. No instabilities or reactive power oscillations were observed with the volt-var function.

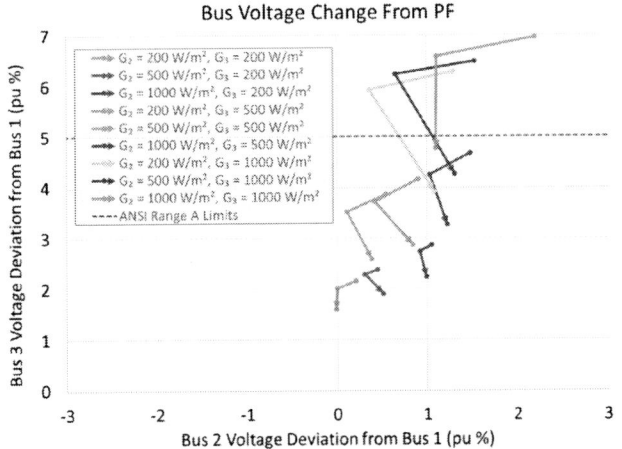

Fig. 5. Change in bus voltages with increasing reactive power absorption.

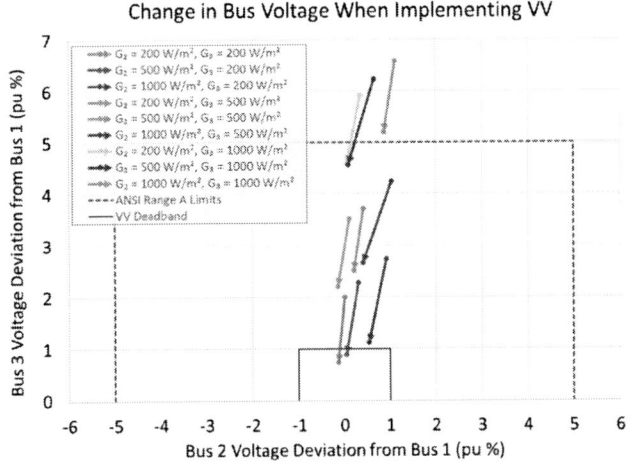

Fig. 6. Change in Bus 2 and Bus 3 voltages when implementing volt-var control with 6 inverters, starting from $PF_i = 1$ setpoints.

*C. Extremum Seeking Control*

ES control can be used to optimize the reactive power contributions of DER to minimize system resistive losses and perform voltage regulation. In order to generate the ES control probe signal a script was created with the SunSpec System Validation Platform (SVP) [19] to produce a reactive power waveforms by continuously issuing fixed power factor commands. The inverter PF settings could be updated at a maximum rate of ~5 Hz using Modbus RTU and TCP commands to the DER, however the actual implementation

was much slower (~1.5 seconds/update) because the centralized ES controller needed to communicate to the data acquisition equipment and each of the inverters sequentially. An example of different reactive power waveforms created using this method with a single inverter is shown in Fig. 1.

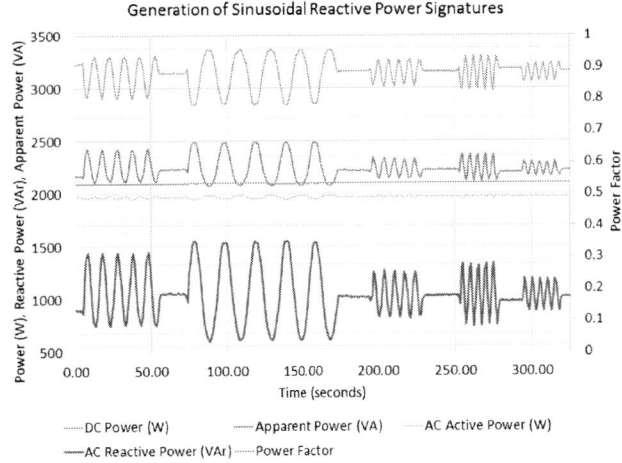

Fig. 7. Example reactive power waveforms generated by the SVP for ES control.

The ES control optimization function was based on the work in [16], but tailored to the voltage regulation problem[2]. The optimization objective function for these simulations is given by

$$ J = C_2\left(V_2 - V_n\right)^2 + C_3\left(V_3 - V_n\right)^2 \qquad (1) $$

where, $V_2$ and $V_3$ are Bus 2 and 3 voltages, $V_n$ is the nominal voltage (240 Vac), and $C_2$ and $C_3$ are voltage deviation weightings. The reactive power perturbation was evaluated at different magnitudes to determine a waveform amplitude that affected a change in $J$ and was not lost in the voltage measurement noise. One of the benefits of the ES control is that all DER can independently optimize their behavior by the selection of different probing frequencies. However, in these experiments all DER on Bus 2 probed at 0.02 Hz and all DER on Bus 3 probed at 0.03 Hz to amplify their effect on the objective function (i.e., bus voltages). Only 1000 W/m² irradiance levels were investigated with the ES controller.

In order to tune the ES control parameters, the inverters on Bus 3 were programmed with the ES control functionality. The four Bus 3 DER with fixed PF capabilities were issued power factor settings to generate sinusoidal reactive power waveforms with 200 Var amplitude and initial condition of -1000 Var (absorptive). As shown in Fig. 6, the ES control probe determined the objective function gradient and migrated

---

[2] Bus 1 active and reactive power objective functions were also investigated with encouraging results, indicating other grid operation objectives could be met with ES control.

toward the optimal solution at the power factor limit of the inverters (approximately -1800 Var).

The next experiment controlled the six programmable DER in the test bed. Bus 2 inverters could not influence the objective function as significantly as the Bus 3 inverters (as described in Section III-B), so the amplitude of the probing signal was doubled to 400 Var. As shown in Fig. 9, inverters at Bus 2 do not meaningfully contribute reactive power until the Bus 3 inverters reach their steady state solution. At this point the Bus 2 probe determines the appropriate gradient and begins to increase absorptive reactive power. This discrepancy between the actions of the controllers at Buses 2 and 3 is due to the faster probing frequency associated with Bus 3 controllers. It is worth noting that in both of these tests, conservative gains were used to ensure stability and demonstrate the approach. Proper tuning of the control parameters would decrease the response time. In the current implementation, this controller would likely only assist with PV-induced voltage deviations on low variability (clear sky) days, but would be ineffective during high variability (partly cloudy) days.

One of the disadvantages of the scripted implementation deployed here is poor scalability. Each of the PF settings were issued to the DER sequentially with a ~0.14 sec communication delay and the voltage at each of the Buses were read each time through the loop. This limitation was evidenced by the speed of the PF commands: there were roughly 15 pts/cycle for Bus 3 inverters in the first ES experiment, but only ~11 pts/cycle for Bus 3 inverters in the second when there were two additional inverters in the control loop. In this paper, the voltage was measured with a LabVIEW-based data acquisition system and transferred to the Python control script with 0.03 sec delay. This delay will also scale with the number of measured buses. When optimizing the substation or feeder active or reactive power, there will also be delays associated with these communications but they will not scale with the number of devices. Therefore, in order to effectively use this technique at scale, aggregators, multi-threading or multicast communications should be considered.

The ES control technique was compared to the results from the VV and PF control at full irradiance. As shown in Fig. 10, the ES control produced similar results to the VV technique, but did not reduce the Bus 3 voltage to below the ANSI Range A limit like the fixed power factor control. The PF outperformed the VV because the Bus 2 devices did not absorb the maximum reactive power since they did not measure high voltages that existed on Bus 3. The ES control did have the ability to reach (and was tracking towards) this global optimum, but the experiment was concluded after 2,000 seconds as shown in Fig. 9. The results indicate all the methods are effective in reducing voltage deviations on distribution circuits.

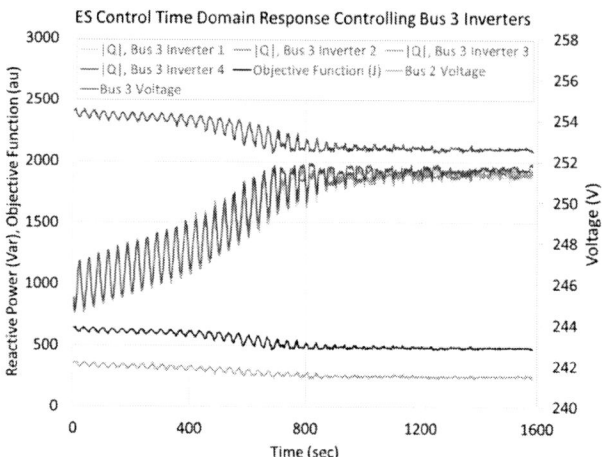

Fig. 8. Time domain response of the ES control governing DER on Bus 3.

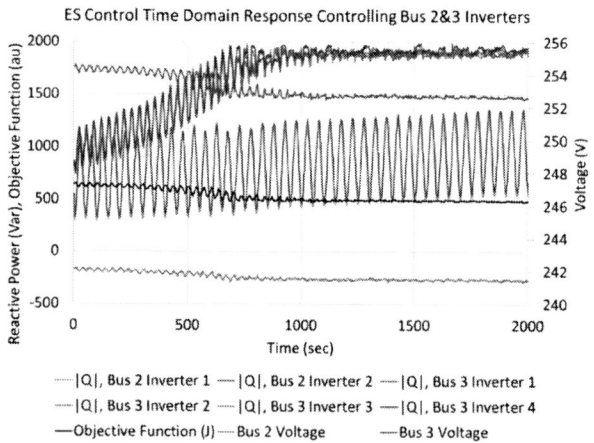

Fig. 9. Time domain response of the ES control governing DER on Buses 2 and 3.

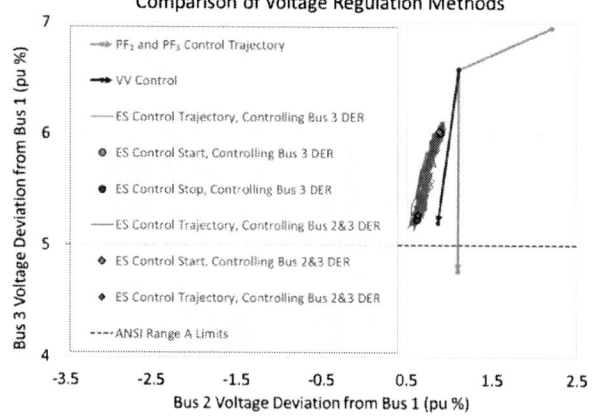

Fig. 10. Comparison of feeder voltage regulation methods with irradiance values of $G_2 = G_3 = 1000$ W/m$^2$. PF control and VV trajectories are the same as those in Figs. 5 and 6. The two ES control trajectories in Figs. 8 and 9 are shown with their start and stop positions marked.

## IV. CONCLUSIONS

Three different techniques for performing voltage regulation were implemented on a testbed representing a rural feeder with nine PV inverters. All three methods were effective in reducing voltages on the distribution circuit, but each included pros and cons. The PF control relied on a centralized control algorithm to issue setpoints to each of the DER so it was computationally and communications intensive, but it produced the greatest voltage improvement. The volt-var function was convenient because it could be preprogrammed or updated infrequently. VV did not produce as significant a reduction in voltage as PF control on Bus 3, but it was highly effective and fast. ES control required high communication speed and proper selection of ES control gains, probing frequencies, and probing magnitudes—but the method showed promise to reach optimal voltage regulation DER setpoints. Additionally, other objectives (e.g., reduced substation active power) could be included in the objective function to provide multi-objective optimization of distribution circuits. It is recommended this technique be investigated in further detail with different physical systems to determine feasibility of field deployment and to improve response time.

## ACKNOWLEDGEMENTS

Sandia National Laboratories is a multi-mission laboratory managed and operated by National Technology and Engineering Solutions of Sandia, LLC., a wholly owned subsidiary of Honeywell International, Inc., for the U.S. Department of Energy's National Nuclear Security Administration under contract DE-NA-0003525.

This material is based upon work supported by the U.S. Department of Energy Grid Modernization Laboratory Consortium (GMLC) Community Control of Distributed Resources for Wide Area Reserve Provision project. The authors would like to thank C. Birk Jones for his assistance establishing reliable meter interoperability and Matthew Reno for his analysis of the distribution circuit.

## REFERENCES

[1] ANSI C84.1-2011, American National Standard for Electric Power Systems and Equipment—Voltage Ratings (60 Hertz), National Electrical Manufacturers Association, 2011.

[2] B. Palmintier, R. Broderick, B. Mather, M. Coddington, K. Baker, F. Ding, M. Reno, M. Lave, and A. Bharatkumar, "On the Path to SunShot: Emerging Issues and Challenges in Integrating Solar with the Distribution System," National Renewable Energy Laboratory, NREL/TP-5D00-65331, 2016.

[3] R. J. Broderick, J. E. Quiroz, M. J. Reno, A. Ellis, J. Smith, and R. Dugan, "Time Series Power Flow Analysis for Distributed Connected PV Generation," Sandia National Laboratories, SAND2013-0537, 2013.

[4] Pacific Gas and Electric Co., Electric Rule No. 21, Generating Facility Interconnections, Filed with the CPUC, Jan. 20, 2015.

[5] Hawaiian Electric Company, Inc., Rule No. 14, Service Connections and Facilities on Customer's Premises, Section H: Interconnection of distributed generating facilities with the company's distribution system, effective October 21, 2015.

[6] IEEE 1547 Std. 1547-2008, "IEEE Standard for Interconnecting Distributed Resources with Electric Power Systems," Institute of Electrical and Electronics Engineers, Inc., New York, NY.

[7] J. Seuss, M. J. Reno, R. J. Broderick, and S. Grijalva, "Improving Distribution Network PV Hosting Capacity via Smart Inverter Reactive Power Support," in IEEE PES General Meeting, Denver, CO, 2015.

[8] M. Farivar, R. Neal, C. Clarke, S. Low, "Optimal inverter VAR control in distribution systems with high PV penetration," 2012 IEEE Power and Energy Society General Meeting, 1-7.

[9] M. Farivar, C.R. Clarke, S.H. Low, K.M. Chandy, "Inverter VAR control for distribution systems with renewables," IEEE International Conference on Smart Grid Communications (SmartGridComm), pp. 457-462.

[10] Z. Shen and M. E. Baran, "Gradient based centralized optimal volt/var control strategy for smart distribution system," 2013 IEEE PES Innovative Smart Grid Technologies (ISGT), pp. 1-6.

[11] F. Ding, A. Nagarajan, S. Chakraborty, M. Baggu, a. Nguyen, S. Walinga, M. McCarty, F. Bell, NREL Technical Report, NREL/TP-5D00-67296, December 2016.

[12] J. Smith, B. Seal, W. Sunderman, R. Dugan, "Simulation of Solar Generation with Advanced Volt-Var Control," 21st International Conference on Electricity Distribution, Frankfurt, Germany, 6-9 June, 2011.

[13] J. Smith, "Stochastic Analysis to Determine Feeder Hosting Capacity for Distributed Solar PV," Electric Power Research Institute, Tech. Rep., 2012

[14] J. Seuss, M. J. Reno, R. J. Broderick, and S. Grijalva, "Analysis of PV Advanced Inverter Functions and Setpoints under Time Series Simulation," Sandia National Laboratories, SAND2016-4856, 2016.

[15] J. Seuss, M. J. Reno, R. J. Broderick and S. Grijalva, "Improving distribution network PV hosting capacity via smart inverter reactive power support," 2015 IEEE Power & Energy Society General Meeting, Denver, CO, 2015, pp. 1-5.

[16] D.B. Arnold, M. Negrete-Pincetic, M. D. Sankur, D. M. Auslander, D. S. Callaway, Model-Free Optimal Control of VAR Resources in Distribution Systems: An Extremum Seeking Approach, IEEE Trans. Power Systems, Vol. 31. No. 5, Sept. 2016.

[17] S.-J. Liu, M. Krstic, Stochastic Averaging and Stochastic Extremum Seeking, Communications and Control Engineering, Springer-Verlag, London, 2012.

[18] IEC Technical Report 61850-90-7, "Communication networks and systems for power utility automation–Part 90-7: Object models for power converters in distributed energy resources (DER) systems," Edition 1.0, Feb 2013.

[19] J. Johnson, B. Fox, "Automating the Sandia Advanced Interoperability Test Protocols," 40th IEEE PVSC, Denver, CO, 8-13 June, 2014.

# Dynamic Setpoint Control of Electric Hot Water Heater Tanks for Increased Integration of Solar Photovoltaic Systems

C. Birk Jones *, Monte Lunacek †,
Matthew Lave *, Jay Johnson *, and Robert Broderick *

*Sandia National Laboratories, Albuquerque, NM, U.S.A
†National Renewable Energy Laboratory, Golden, CO, U.S.A

*Abstract*—The integration of solar photovoltaic (PV) systems onto the existing grid provides a clean source of electrical power. However, PV systems can only produce power during the day and it is often intermittent. Utility companies must implement mitigation strategies that account for large ramp rates and high variability. The strategies include energy storage and the control of dispatchable resources that can react quickly to abrupt changes in demand and generation. The charging of electric water heater tanks can be controlled dynamically to help the grid match PV generation more closely. This paper reports on a simulation effort that evaluated this potential. The experiment implemented a dynamic setpoint controller that synchronized the charging of the electric water heater (EWH) tanks with the sun and used the storage to bridge periods when no solar PV was available. The simulation results were coupled with data from an actual feeder that supported 2,900 residential homes and had a 6 megawatt PV system. The approach successfully synchronized the EWH with the PV and on average maintained a comfortable temperature for the occupants.

*Index Terms*—electric water heaters, demand response, photovoltaic

## I. INTRODUCTION

The balance between electrical consumption and production offers significant challenges for today's grid. The integration of renewable energy sources, electro-mobility, and increased demand requires sophisticated control methods to balance the overall system in real-time. Demand side management controls offer a cost effective means to optimize and temporarily reduce electrical power. For example, energy efficiency can permanently reduce demand, time-of-use rates can optimize schedules to shift demand, and demand response (DR) can shed loads quickly [1]. This paper examines the potential for advanced control of thousands of residential electric water heater (EWH) tanks to synchronize demand with solar photovoltaic (PV) generation.

The residential sector has the potential to provide about half of the total peak demand reduction in the United States [2]. Advanced control of residential systems, such as EWHs, can provide significant benefits for the electric grid. Existing and past incentive programs offered by utility companies have enticed customers to minimize their overall energy consumption and reduce power draw when needed to improve grid stability. Existing programs, described by Ericson [3], have successfully incentivized customers to allow the utility to control their EWHs. For example, Xcel Energy paid customers $2 per month for an entire year if they allowed their EWH to be

disconnected for a 6 hour period during hot summer and cold winter days. This program included 280,000 EWHs and was able to reduce demand by 330 megawatts (MW) in 2001. An Australian program was also implemented successfully and reduced demand during peak operations by 389 MW using 355,000 EWHs. Current and past incentives programs were typically designed to shed peak load at the expense of occupant comfort. This paper investigates the potential to implement a program that controls EWHs based on solar PV production while avoiding occupant discomfort.

Synchronizing the electric power demand of EWHs with the sun can improve the integration of PV on the grid by leveling the net power profile. The net power is the difference between the electrical load and the PV system generation. The net load profile can have a large valley in the middle of the day caused by the generation of electricity from PV systems [4]. As a result, there is a large increase in demand as the sun sets. The utilities must account for this significant increase in demand and often rely on expensive generation plants. Instead, utilities can implement a program that dynamically controls EWH setpoints. The proposed dynamic setpoint control algorithm uses the measured irradiance as an independent variable. The implementation of the algorithm has the potential to fill the net load profile valley and smooth intermittent solar PV generation.

This paper describes an experiment that simulated the control of EWHs using a dynamic setpoint algorithm. The paper is structured into four main sections: background, methodology, results, and conclusions. The methodology provides an overview of the EWH model and water draw profile. It also discusses the impact that solar generation has on the grid and defines the dynamic setpoint control algorithm. The results section provides a discussion and supporting graphs from the simulation effort. The conclusions section describes the key findings and potential next steps.

## II. BACKGROUND

Considerable work has been conducted to evaluate the control of EWHs to support grid services. Research studies have shown that EWH can be used for demand side management. For example, Sepulveda et al. successfully implemented a particle swarm optimization algorithm to control 200 simulated EWHs [5]. The simulation results showed that the EWH control was able to shift the residential load and reduce

978-1-5090-5606-4/17 $31.00 © 2017 IEEE

morning and evening peaks by 100 kW and 150 kW respectively. Another study evaluated the charging and discharging impacts on exergy to improve peak shaving control [6]. This study determined what customers would be most likely to draw power during the utility's peak. This helped the utility company determine which customers were eligible for the load shedding program. Another study performed by Pourmousavi et al. simulated the control of 1,000 residential EWHs for demand response (DR) by modulating the temperature setpoint [7]. The research effort defined multiple control cases that included setpoint control based on time of use pricing. The intent was to control the EWHs for peak shaving. In addition to peak shaving, studies have shown that EWH can provide balancing services for the grid. For example, Diao et al. [8] demonstrated that EWHs can be used for frequency support. The experiment modeled 147 residential hot water heaters and tested centralized and decentralized controller algorithms.

## III. METHODOLOGY

The present work used a model to simulate EWH performance. The simulated EWH results were combined with actual demand and PV data from an electric feeder. The feeder was comprised of 4 substations that support about 2,900 residential buildings. The feeder was observed to have a maximum electrical load of about 11 MW. It also had 6 MW PV array connected to one of the substations. The PV array provided about 20% of the energy on an annual basis. The experiment assumed that each of the homes had a EWH that could be controlled. The EWHs were simulated inside a control aggregation model written in Python programming language.

### A. Electric Water Heater Model

EWH models have been discussed frequently in past literature. The tanks have commonly been modeled using a single node approach where the temperature in the tank was assumed to be constant. This type of model applied a first order differential equation that solved for the tank temperature [8], [9]. The present work modeled the dynamic EWH system using a state space model. Unlike the single node approach, the state space model considered the thermal stratification. EWH storage tanks experience stratification, shown in Figure 1, where the cold water is on the bottom and the hot water is on the top.

The state space model used in this experiment was created to simulate model predictive controls by Jin et al. [10]. The approach divided the tank into vertically stacked isothermal nodes. The model then calculated the energy balance for the nodes using:

$$C_i \frac{dT_i}{dt} = Q_{e_i} - Q_{loss_i} - Q_{draw_i} \tag{1}$$

where $Q_e$ is the heating element thermal power,
$Q_{loss}$ is the thermal loss, and
$Q_{draw}$ is the thermal power lost to the house.

The state space model was used to solve the non-iterative part of Equation 1. State space models represent physical systems using a first-order differential equation and a set of inputs, outputs, and state variables. In this case, the equations were developed using:

$$\dot{x}(t) = A(t)x(t) + B(t)u(t) \tag{2}$$

$$\dot{y}(t) = Cx(t) \tag{3}$$

where the state variable x is a vector of the thermal tank node temperatures $T_i$ (x = [$T_0,T_1,...,T_{11}$]) and the control variable u is a vector that contains the ambient temperature ($T_a$), inlet temperature ($T_{in}$), and the heating element control signal represented by $s_{element}$ (u = [($T_a,T_{in},s_{element}$]). The state space model, described in Equations 2 and 3, was converted to a discrete time state space model in order to incorporate temperatures and flow rate. The simulation effort used a 50 gallon tank with two 4,500 Watt heating elements for each of the residential buildings. The model also depended on a

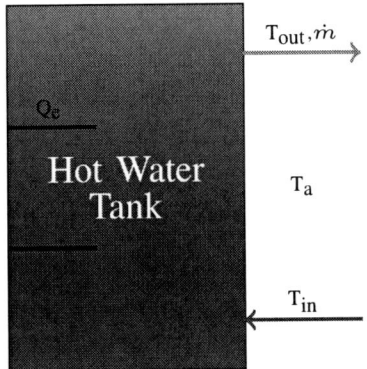

Fig. 1. The electric hot water heater was modeled using a state space model. The variables included tank mass and temperature, inlet temperature, heating element power, and mass flow rate.

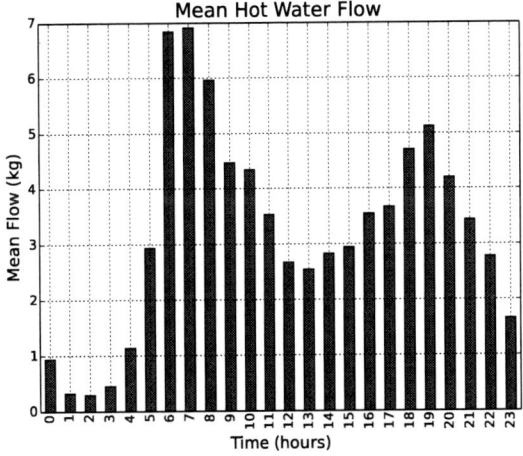

Fig. 2. The average hot water flow at each hour of the day peaked in the morning around 07:00. It then decreased during the middle of the day and then peaked again around hour 19:00.

realistic draw profile to simulate the thermal and electrical performance.

### B. Electric Hot Water Draw Profiles

The simulation of the 2,900 EWH used draw profiles that were based on past statistical analysis of residential use [11]. A generator was used to create unique profiles that represented actual use at 20 minute intervals. The mean of all the profiles for each hour of the day is shown in Figure 2. The average flow profile was very low in the early morning and quickly increased to a peak around hour 07:00. The profile then decreased during the middle of the day and peaked again around hour 19:00.

### C. Electric Grid

The integration of solar PV systems presents many complications. Locations where there is a high penetration of PV

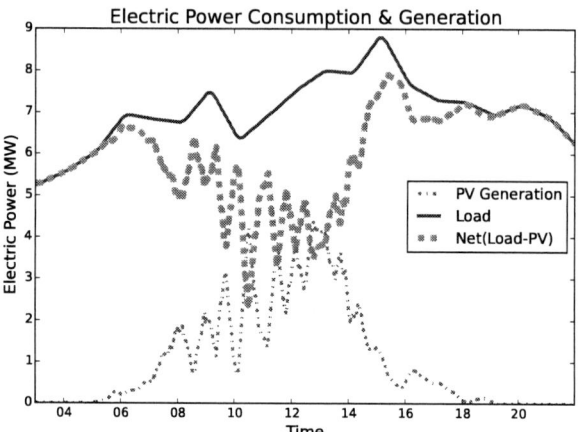

Fig. 4. The net load, which is the difference between the load and the PV generation, can have rapid fluctuations caused by solar intermittence due to clouds. The variability can cause instabilities on the grid.

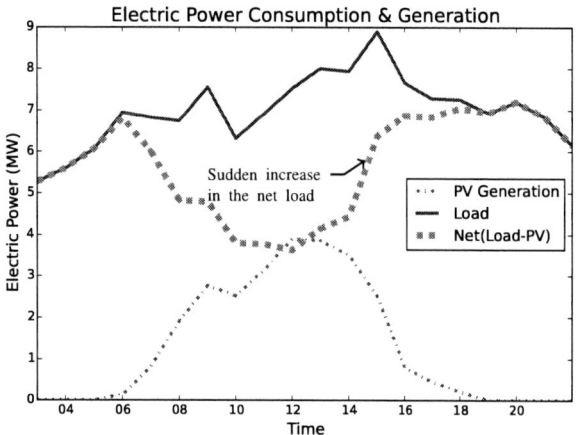

Fig. 3. The net load is the difference between the load and the PV generation. The net load has a large ramp rate at the end of the day when the sun sets and demand increases. The change may require utilities to turn on expensive generation stations.

require the utility to react quickly to large changes in the net load. The net load is the difference between the load and the PV generation as shown in Figure 3. The increase in demand coupled with the rapid reduction in solar power generation as the sun goes down at the end of the day can cause instability on the electric grid. This situation may require the deployment of an expensive generation station to rapidly come online and accommodate the load. EWHs have the potential to mitigate this issue by synchronizing their charging with the solar production.

The synchronization of the EWHs with the sun can also help with the variable generation of power caused by clouds. The variability in solar generation, shown in Figure 4, can cause large fluctuations in the net load. The variability has been addressed in past literature with batteries [12], hybrid storage [13], heating/cooling equipment [14], and others. The different approaches have attempted successfully to smooth the PV output using various storage devices. The present work used EWH to match the load with the solar generation. The

approach used a control signal for the hot water tank setpoint temperature that was dependent on the solar irradiance.

### D. Dynamic Setpoint Control

The dynamic setpoint control of EWH can help synchronize the solar power with the load and simultaneously maintain occupant comfort. The control of the simulated EWHs that typically have a static setpoint temperature were altered to have a dynamic setpoint. The setpoint temperature is the reference temperature that is compared with the actual water temperature to determine the control of the heating element inside the EWH tanks. If the water temperature ($T_{water}$) is less than the setpoint temperature ($T_{sp}$) by more than 2.7°C then the heating element is turned on as shown in Equation 4:

$$\text{Heating Element} = \begin{cases} \text{On,} & \text{if } T_{water} < T_{sp} - 2.7°C \\ \text{Off,} & \text{otherwise} \end{cases} \quad (4)$$

The dynamic setpoint, used in this experiment, was a nonlinear function:

$$T_{sp} = 12(E/1300)^3 + 45 \quad (5)$$

where E is the measured irradiance (W/m$^2$), 1300 was used to normalize the irradiance value, 12 is a multiplier, and 45 is the minimum setpoint temperature. In this experiment the static setpoint temperature was set to be 49°C. The nonlinear function, plotted in Figure 5, is a third-order equation where the setpoint temperature is set to be 45°C at low solar irradiance and slowly climbs to 49°C at 700 W/m$^2$. Between 700 and 1200 W/m$^2$ the function increases from 49°C to about 54.5°C. The nonlinear setpoint control is necessary so that the heating element in the EWH does not turn on at low irradiance unless it is necessary for occupant comfort. The setpoint temperature can exceed 52°C at high irradiance so that the tank can be overcharged in order to maintain a temperature throughout the night and early morning.

978-1-5090-5606-4/17 $31.00 © 2017 IEEE

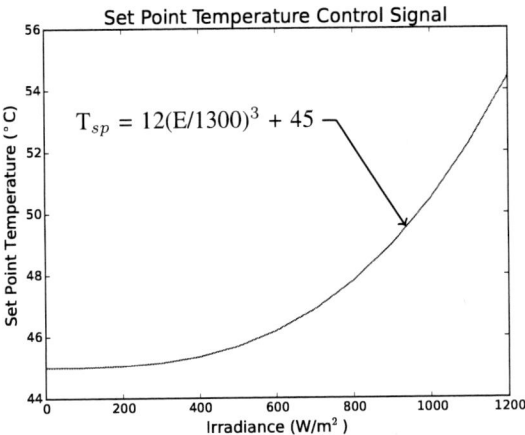

Fig. 5. The control algorithm increases the temperature setpoint based on a nonlinear function that increases from 45°C to 54.5°C exponentially.

### E. Experiment

The simulation of 2,900 EWHs was conducted in a coordinated manner to compare typical operations with two different dynamic setpoint control scenarios as described in Table I. The EWH systems were simulated at one minute intervals for each of the control scenarios. The first test was the baseline simulation that modeled all of the EWHs with a static setpoint of 49°C. The second test modeled 1,933 EWHs with a static

TABLE I
CONTROL SCENARIO TESTS

| Test | Name | Number of Electric Water Heaters | |
|---|---|---|---|
| | | Static setpoint (49°C) | Dynamic setpoint |
| 0 | Baseline | 2,900 | 0 |
| 1 | Solar Control A | 1,933 | 967 |
| 2 | Solar Control B | 967 | 1,933 |

setpoint of 49°C and 967 with the dynamic nonlinear setpoint defined by Equation 5. The third test decreased the number of EWHs controlled using the static setpoint of 49°C to 967 and applied the dynamic setpoint to 1,933 EWHs.

### IV. RESULTS

The experiment simulated three different scenarios as described in Section III-E. The EWH results from each scenario were combined with the actual PV and load data from a feeder to estimate the impact on the grid. The results include an overview of the total water draw from the EWHs and electric demand profiles from the simulated EWHs and the actual feeder. The typical operations simulation was performed first to develop a baseline profile that could be compared with the results from the proposed dynamic setpoint control algorithm. The dynamic setpoint control algorithm was simulated in two different scenarios where it was applied to 33% and 66% of the EWHs on the feeder.

### A. Typical Control of Electric Water Heater

The electric feeder had a load profile that peaked to about 10 MW in the afternoon and dropped below 4 MW in the early

morning as shown in Figure 6. Over the three day period,

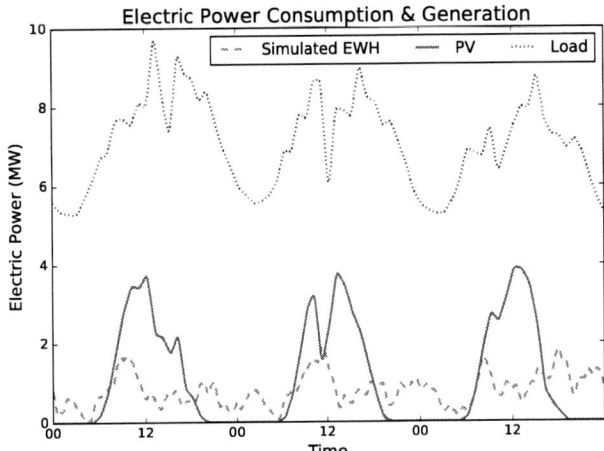

Fig. 6. Under normal operating conditions the load peaked to about 10MW in the late afternoon and dropped to around 4MW in the early morning. The PV generation reached a maximum value of 6MW during the day and dropped to zero at night. The electric water heater simulation had a peak load in the morning around 08:00 and then a smaller peak around hour 19:00.

plotted in Figure 6, the PV generation fell to zero during the night and reached a maximum of 3.8 MW at the middle of the day. The simulation results for the typical EWH operating conditions had a profile that peaked to about 2MW around hour 08:00.

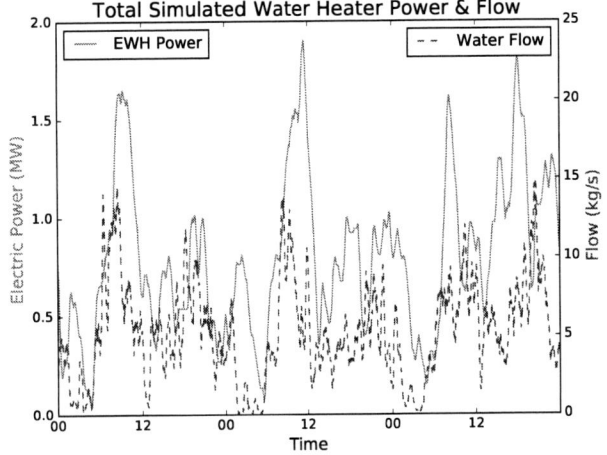

Fig. 7. The three day period in August had a peak power draw of about 1.9MW. The water flow reached a high of about 15kg/second for the 2,900 simulate EWHs.

The electric power demand for the 2,900 EWHs followed closely with the hot water draw. The total power for all the EWHs increased and decreased with the water flow, but was slightly offset as shown in Figure 7. The offset indicated that time to charge the thermal tank expanded beyond the usage of the hot water. The tanks were charged right after the water temperature dropped below the setpoint value.

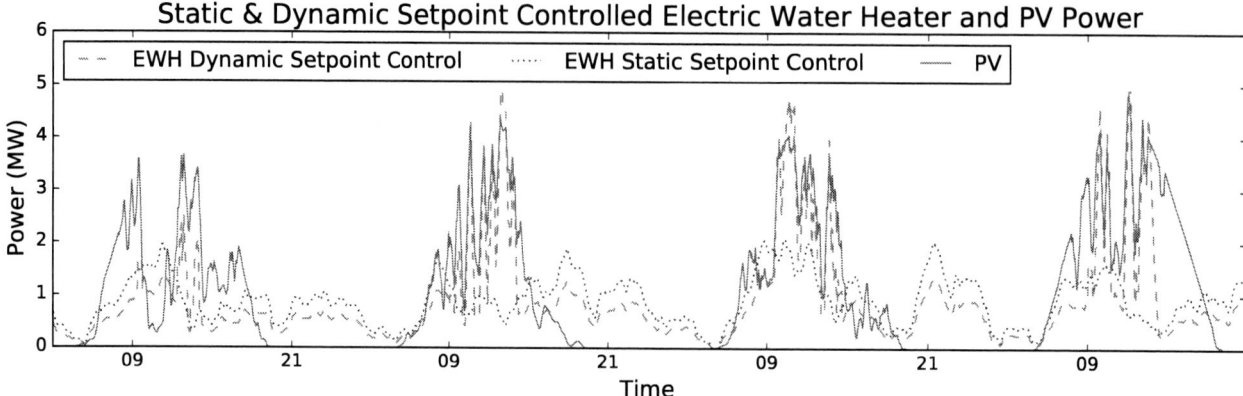

Fig. 8. The uncontrolled electric water heater charging power does not match well with the PV generation profile under normal conditions. The nonlinear setpoint control algorithm applied to 33% of the residential homes synchronizes consumption with PV generation over a four day period.

## B. Dynamic Setpoint Control of Electric Water Heater

The control algorithm was able to synchronize the EWH electrical demand with the solar generation as shown in Figure 8. In this case, the simulation controlled about 960 of the 2,900 EWHs using the dynamic setpoint control algorithm. The power consumed by the 967 EWHs increased and

with the dynamic setpoint. The new load profile, shown in Figure 9, filled in the valley and decreased the magnitude of the spikes. The approach eliminated the large ramp rate that had occurred between 14:00 and 15:00 in the baseline simulation.

The second test, which controlled 1,933 EWHs with a dynamic setpoint and 967 with a static setpoint, was able to follow the PV generation profile well (Figure 10). The net

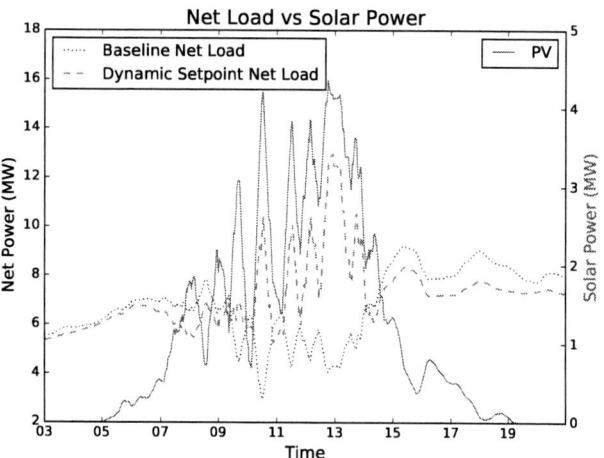

Fig. 9. The solar irradiance dependent setpoint algorithm was able to fill in the valley created by the PV production and smooth the variability.

decreased with the generated solar power. During the night the overall EWH power draw was less than the baseline control scenario that used a static setpoint. The power demand during the night was reduced because the tanks had been charged to a slightly higher setpoint than normal during the day. The EWH power demand was then combined with the net load to evaluate the impact on the variability and end of day ramp rates.

The control of the 967 EWHs affected the net load by increasing demand during the day and decreasing it during the night. This result was computed by first subtracting the baseline EWH demand from the measured net load and then adding the results from test 1 that controlled 33% of the EWHs

Fig. 10. The dynamically controlled setpoints for 67% of the EWHs allowed the net load to follow the solar PV generation profile well. It also decreased the overall load at night by about 1.1 MW between 16:00 and midnight.

load was much higher than the baseline during the day and it followed the variable output of the PV system starting at hour 11:00. The dynamic setpoint controlled net load followed the PV generation profile until about hour 14:30. At that point the net power fell below the baseline. The dynamic control increased the demand during the day and decreased it during the night by relying on the storage tank capacity.

The simulation calculated the average temperature at the top and bottom of the tank. During the same four day period, between August 4th and 7th, plotted in Figure 8 the average temperature at the top of the tank did not drop below 50°C as

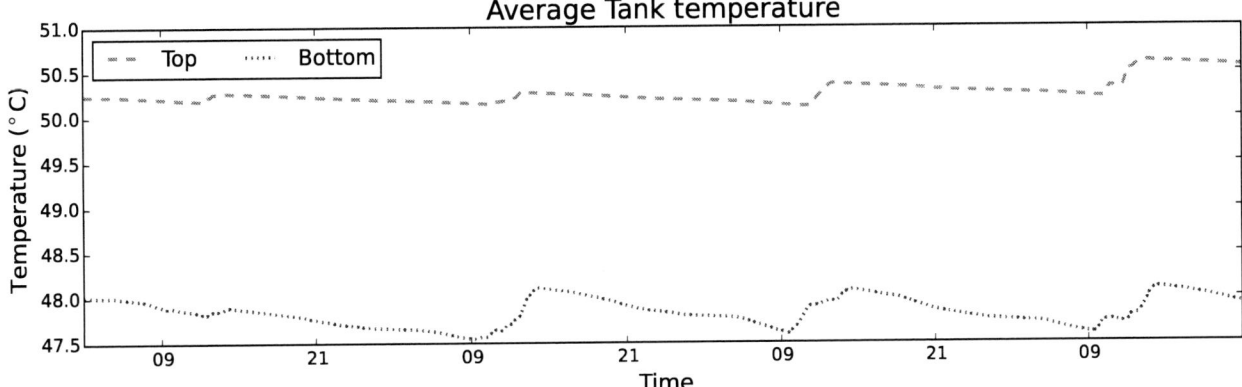

Fig. 11. The simulation effort calculated the top and bottom temperatures in the hot water tanks. The temperatures stayed within a range that maintained occupant comfort through the four day period. The top temperature did not drop below 50°C and the bottom temperature stayed above 47.5°C.

shown in Figure 11. Additionally, the bottom temperature did not drop below 47.5°C. The average temperature for the top and bottom of the tank showed that the EWHs can continue to provide desired hot water to occupants while simultaneously synchronizing with the PV generation.

## V. CONCLUSION

The present work successfully simulated the control of 2,900 EWH tanks for increased integration of PV systems onto the electric grid. The simulation effort included a baseline test where the temperature setpoint was a constant value and two new control scenarios where the setpoint was dynamic. The dynamic setpont control was based on a third order equation that used the solar irradiance as an independent variable.

The simulation results were combined with the demand and PV data from an actual feeder to evaluate the impact on the net load. The synchronization of the EWHs with the electrical generation from the PV system eliminated the large increase in demand at the end of the day. The approach also smoothed the variable generation of power caused by clouds. This investigation provided a general review of the dynamic setpoint control approach. The next step will be to perform further simulations and also apply it to an actual system.

## ACKNOWLEDGMENT

This work was supported by the U.S. Department of Energy Grid Modernization Laboratory Consortium (GMLC).

Sandia National Laboratories is a multimission laboratory managed and operated by National Technology and Engineering Solutions of Sandia, LLC., a wholly owned subsidiary of Honeywell International, Inc., for the U.S. Department of Energy's National Nuclear Security Administration under contract DE-NA0003525. SAND NO. 2017-6665C

## REFERENCES

[1] P. Palensky and D. Dietrich, "Demand Side Management: Demand Response, Intelligent Energy Systems, and Smart Loads," *IEEE Transactions on Industrial Informatics*, vol. 7, no. 3, pp. 381–388, Aug. 2011.

[2] M. Pipattanasomporn, M. Kuzlu, S. Rahman, and Y. Teklu, "Load Profiles of Selected Major Household Appliances and Their Demand Response Opportunities," *IEEE Transactions on Smart Grid*, vol. 5, no. 2, pp. 742–750, Mar. 2014.

[3] T. Ericson, "Direct load control of residential water heaters," *Energy Policy*, vol. 37, no. 9, pp. 3502–3512, Sep. 2009. [Online]. Available: http://www.sciencedirect.com/science/article/pii/S0301421509002201

[4] M. Obi and R. Bass, "Trends and challenges of grid-connected photovoltaic systems A review," *Renewable and Sustainable Energy Reviews*, vol. 58, pp. 1082–1094, May 2016. [Online]. Available: //www.sciencedirect.com/science/article/pii/S136403211501672X

[5] A. Sepulveda, L. Paull, W. G. Morsi, H. Li, C. P. Diduch, and L. Chang, "A novel demand side management program using water heaters and particle swarm optimization," in *2010 IEEE Electrical Power Energy Conference*, Aug. 2010, pp. 1–5.

[6] U. Atikol, "A simple peak shifting DSM (demand-side management) strategy for residential water heaters," *Energy*, vol. 62, pp. 435–440, Dec. 2013. [Online]. Available: http://www.sciencedirect.com/science/article/pii/S0360544213008189

[7] S. A. Pourmousavi, S. N. Patrick, and M. H. Nehrir, "Real-Time Demand Response Through Aggregate Electric Water Heaters for Load Shifting and Balancing Wind Generation," *IEEE Transactions on Smart Grid*, vol. 5, no. 2, pp. 769–778, Mar. 2014.

[8] R. Diao, S. Lu, M. Elizondo, E. Mayhorn, Y. Zhang, and N. Samaan, "Electric water heater modeling and control strategies for demand response," in *2012 IEEE Power and Energy Society General Meeting*, Jul. 2012, pp. 1–8.

[9] Z. Xu, R. Diao, S. Lu, J. Lian, and Y. Zhang, "Modeling of Electric Water Heaters for Demand Response: A Baseline PDE Model," *IEEE Transactions on Smart Grid*, vol. 5, no. 5, pp. 2203–2210, Sep. 2014.

[10] X. Jin, J. Maguire, and D. Christensen, "Model Predictive Control of Heat Pump Water Heaters for Energy Efficiency," in *18th ACEEE Summer Study on Energy Efficiency in Buildings*. Pacific Grove, CA: National Renewable Energy Laboratory (NREL), Golden, CO., 2014, pp. 133–145. [Online]. Available: https://www.osti.gov/scitech/biblio/1160190

[11] B. Hendron, J. Burch, and G. Barker, "Tool for Generating Realistic Residential Hot Water Event Schedules: Preprint," *ResearchGate*, 2010. [Online]. Available: https://www.researchgate.net/publication/239883840_Tool_for_Generating_Realistic_Residential_Hot_Water_Event_Schedules_Preprint

[12] A. Ellis, D. Schoenwald, J. Hawkins, S. Willard, and B. Arellano, "PV output smoothing with energy storage," in *2012 38th IEEE Photovoltaic Specialists Conference*, Jun. 2012, pp. 001 523–001 528.

[13] G. Wang, M. Ciobotaru, and V. G. Agelidis, "Power Smoothing of Large Solar PV Plant Using Hybrid Energy Storage," *IEEE Transactions on Sustainable Energy*, vol. 5, no. 3, pp. 834–842, Jul. 2014.

[14] A. Mammoli, H. Barsun, R. Burnett, J. Hawkins, and J. Simmins, "Using high-speed demand response of building HVAC systems to smooth cloud-driven intermittency of distributed solar photovoltaic generation," in *PES T D 2012*, May 2012, pp. 1–10.

# Spatial Analysis of Residential Combined Photovoltaic and Battery Potential: Case Study Utrecht, the Netherlands

Geert Litjens, Bala Bhavya Kausika, Ernst Worrell, Wilfried van Sark

Copernicus Institute of Sustainable Development, Utrecht University
PO Box 80.115, 3508TC Utrecht, the Netherlands

The first two authors contributed equally to this work

*Abstract*—An analysis of current share of potential, level of autarky, self-consumption rate and self-sufficiency rate have been performed for residential photovoltaics (PV) battery combined systems on neighbourhood level for the city of Utrecht in the Netherlands. PV yield potential has been assessed for each roof and scaled to neighbourhood level. Residential demand patterns were created using measured demand patterns and demographic information. All the information has been scaled to the neighbourhood level. It was found that battery systems are currently not required to improve the PV self-sufficiency on neighbourhood scale. Also, the calculated PV potential is not sufficient to meet the residential demand in the neighbourhoods. The historical city centre shows a low PV-battery potential, due to the low PV potential and high demand, whereas the suburban areas show large PV-battery potential. Batteries are useful to improve the self-sufficiency rates up to 14% in the neighbourhoods. The obtained knowledge is valuable for local governments to implement effective policies for a transition towards a sustainable city.

*Index Terms*—PV-battery potential, Spatial analyses, Self-sufficiency, Self-consumption, Neighbourhood

## I. INTRODUCTION

Solar photovoltaic systems (PV) are a promising solution for local generation of electricity, especially in urban areas. However, PV production may be limited due to suitable spaces for installation in these areas. PV energy can be directly consumed, or transported to other regions, which causes additional power flows and voltage fluctuations on the low voltage distribution grid. Also, high penetration of PV may be limited depending on the specific local grid characteristics.

Combined PV and battery systems reduce the impact on the low voltage grid, and increase the direct consumption of PV energy. PV self-consumption is seen as a major driver for new market and PV systems designs [1]. Consequently, lower investments are required in grid expansion and central back up power capacity. The optimal battery capacity depends on PV production potential and the local electricity consumption.

A study of the PV potential in eastern Slovakia showed that the potential could cover about 2/3 of the current electricity demand [2]. A review showed that the change of self-consumption rate with batteries and a 13% to 24 % -points

increase for battery capacity of 0.5-1 kWh per kWp of installed PV capacity was found [3]. PV-battery systems contribute to a reduction of PV peak power injection in the grid [4].

A spatio-temporal model, that analyzes charging of plug-in electric vehicles, provided more realistic results on the impact of charging on the local grid [5]. Another study investigated spatial clustering of electrical vehicle charging based on the expected PV surplus. It was found that 27% of the electrical vehicles not in use would be required to store the maximum PV surplus power [6]. However, no study was found that used spatio-temporal data for modelling and analyzing the PV-battery potential at neighborhood level.

Currently, the spatial relation between the impact of PV potential and PV battery capacity at neighbourhood or even postal code level has not been explored to its full potential. Therefore, this study aims to analyse and identify the spatial potential of combined PV and battery systems at neighbourhood level. We use the city of Utrecht in the Netherlands as case study to demonstrate the methodology.

PV potential, demand data and socio-economic factors like household composition and house values were used to explore the self-sufficiency rates and power flows between different neighbourhoods. Battery potential was determined for each neighbourhood using historical data on annual demand and load patterns. The final outputs show neighbourhoods which could act as potential battery sites for providing electricity to adjacent neighbourhoods.

The obtained knowledge is valuable for local governments to assisting them to design effective policies for a transition towards a sustainable city. The developed method can be easily adapted to other areas. In addition, visualization of PV potential and power flows assists distribution system operators (DSO's) for future grid planning. In addition, we compared current (2016) and potential PV yield production and identify neighbourhoods that are front runners in becoming more energy independent.

978-1-5090-5606-4/17 $31.00 © 2017 IEEE

## II. METHODOLOGY

The city Utrecht (latitude 52°05'38" North, longitude 5° 05'12" East) is divided into 101 neighbourhoods, of which 88 neighbourhoods have more than 100 households. These neighbourhoods were selected to analyse a set of indicators to give information on the relation between the current and potential PV yield, the level of autarky and the self-sufficiency rates. The level of autarky shows the maximum potential of energy autonomy, whereas the self-sufficiency rate shows the actual energy autonomy. Therefore, a potential PV yield pattern and a current demand pattern were created.

### A. Potential PV yield pattern

The maximum PV capacity potential for a total of 59,554 residential roofs was calculated using high resolution laser altimetry data [7]. The digital elevation model used for calculations has a spatial resolution of 50 cm [8]. ArcGIS Solar Radiation Tool was used to calculate irradiation (Wh/m$^2$/yr) incorporating slopes, azimuths of rooftops and shading from nearby objects from the DEM (cite solar radiation tool).

Rooftop suitability for PV siting is then estimated by categorizing the irradiation received by each rooftop. Suitable areas were computed for these categories at 40% of total available roof space. By taking the roofs for which at least 40% of the total roof area is suitable it is ensured that each rooftop investment will have a positive payback period therefore a more probable investment. In addition, this ensures that the obstacles on the roofs, likes chimneys, gable style roofs and shadows from trees are not taken into account for potential calculations. Power densities of 100 Wp/m$^2$ and 150 Wp/m$^2$ were used for estimating the potential capacities for flat and sloping roofs respectively. These values were then aggregated to neighbourhood level using simple statistics based on location.

PV yield pattern for each building was modelled using the open source package PVLIB [9]. Weather data was obtained on a 10-minute interval, measured by the Royal Netherlands Meteorological Institute (KNMI) in De Bilt, The Netherlands, (latitude: 52.11°, longitude: 5.18°). Irradiation, temperature, wind-speed and pressure data of 2013 were used as input. The PV module azimuth and tilt angles were calculated using the DEM.

The Sanyo HIP-225HDE1 module and the Enphase Energy M210 inverter were used as PV module and inverter input. The PV yield patterns were scaled to the potential roof capacity of each building, and aggregated to neighbourhood level. The total annual potential PV energy ($E_{\mathrm{PV\,pot}}$) was calculated by the summation of the PV power ($P_{\mathrm{PV\,pot}}$) of each 5 minutes ($\Delta t$) interval between timestep t=1 and $t_{\mathrm{end}}$, see Eq. (1).

$$E_{\mathrm{PV\,pot}} = \sum_{t=1}^{t_{\mathrm{end}}} P_{\mathrm{PV\,pot,t}} \cdot \Delta t \qquad (1)$$

### B. Current demand patterns

Residential electricity demand patterns were measured by a Dutch distribution system operator between 2012 and 2014 [10]. Unique demand patterns of 60 households with different dwelling types and residents were derived from these measurements. Each demand pattern contains information on the type of houschold. The demand patterns are available with a 15 minute interval for the calendar year 2013.

Neighbourhood demographic statistics were obtained from Statistics Netherlands (CBS) [11]. This dataset contains information on the percentage of people living alone, living together, of living in a family, and the average monetary value of the dwelling, for each neighbourhood. These demographics were combined with the 60 demand patterns to create a residential demand pattern for the neighbourhood.

Each neighbourhood consist of smaller areas, which are indicated by a postal code 6 level. Average annual household energy demand is available on the postal code 6 level as open data for the Netherlands for 2015 [12]. The annual energy demand per postal code 6 was multiplied with the amount of grid connections and scaled to neighbourhood level. The demand patterns were scaled to match the annual energy demand. The total energy consumed ($E_{\mathrm{TC}}$) was calculated by the summation of the demand ($P_{\mathrm{demand}}$), see Eq. ( 2).

$$E_{\mathrm{TC}} = \sum_{t=1}^{t_{\mathrm{end}}} P_{\mathrm{demand,t}} \cdot \Delta t \qquad (2)$$

### C. Spatial indicators

The annual PV produced energy ($E_{\mathrm{PV\,cur}}$) for 2016 was obtained online from Energie in Beeld [13]. Consequently, the share of potential yield that has currently been realized is defined by Eq. (3).

$$\mathrm{Share\ of\ potential} = \frac{E_{\mathrm{PV\,cur}}}{E_{\mathrm{PV\,pot}}} \qquad (3)$$

The maximum share of energy demand that can be covered with PV energy on an annual basis is defined as the level of autarky (LOA). This is the ratio between annual consumed energy and annual produced energy. It gives an indication of the maximum self-sufficiency that a neighbourhood can achieve, see Eq. (4). The LOA is calculated using the current ($E_{\mathrm{PV\,cur}}$) and potential ($E_{\mathrm{PV\,pot}}$) annual produced PV energy.

$$\mathrm{LOA} = \frac{E_{\mathrm{PV}}}{E_{\mathrm{TC}}} \qquad (4)$$

An algorithm was developed to calculate the energy demand that is directly met with energy from locally installed PV systems, or indirectly provided with battery energy storage systems. Both neighbourhood PV yield and demand patterns were linearly interpolated to a 5 minute interval, matching demand patterns with PV yield production patterns.

Self-consumption rate is defined as the ratio between the PV energy direct consumed and total produced ($E_{\mathrm{PV}}$) energy. The directly consumed energy is the PV energy consumed in

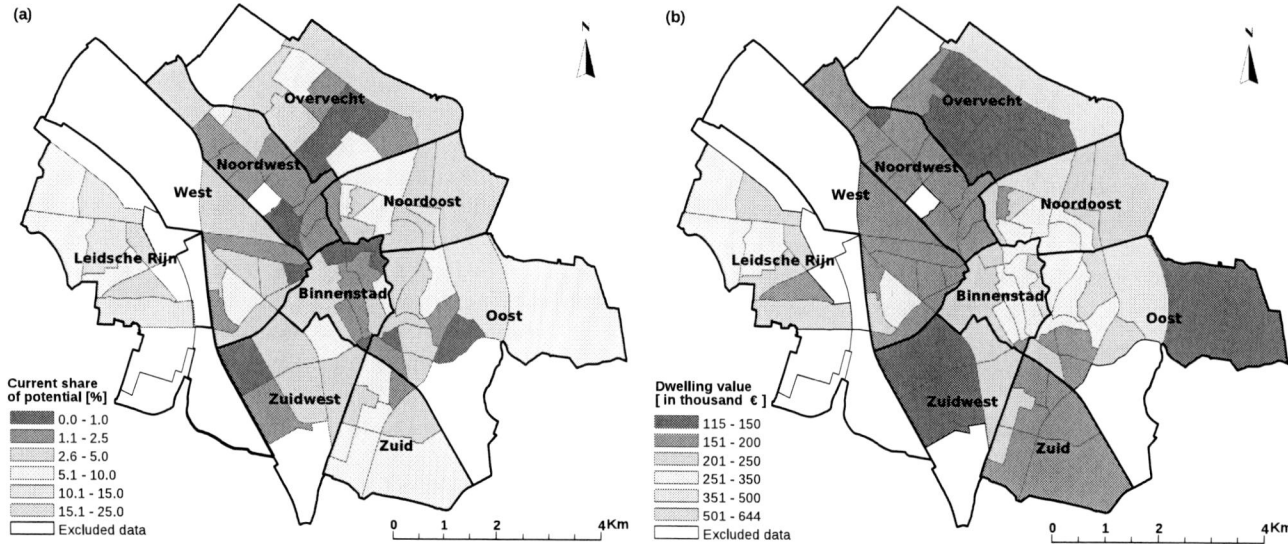

Fig. 1. Color-coded share of potential (a) and average dwelling value (b) for each neighbourhood of the city of Utrecht. The white areas are neighborhoods containing less than 100 households, and are excluded.

the building and the energy used for charging the battery, see Eq. 5.

$$
P_{\text{direct-consumed}} = \begin{cases} P_{\text{PV}} & \text{if } P_{\text{PV}} < P_{\text{demand}} \\ P_{\text{demand}} & \text{if } P_{\text{PV}} \geq P_{\text{demand}} \end{cases} \quad (5a)
$$

$$
\text{SCR} = \frac{\sum\limits_{t=1}^{t_{\text{end}}} (P_{\text{direct-consumed,t}} + P_{\text{bat charge,t}}) \cdot \Delta t}{E_{\text{PV}}} \quad (5b)
$$

Self-sufficiency rate is defined as the ratio between the PV energy directly consumed energy from the PV system and the total consumed ($E_{\text{TC}}$) energy. Consumed energy is the discharged energy coming from the battery with the energy used directly from the PV system, see Eq. 6.

$$
\text{SSR} = \frac{\sum\limits_{t=1}^{t_{\text{end}}} (P_{\text{direct-consumed,t}} + P_{\text{bat discharge,t}}) \cdot \Delta t}{E_{\text{TC}}} \quad (6)
$$

Charging and discharging of the battery was modelled using a simple control strategy, presented in previous research [14]. The battery is charged when produced PV power is larger than the demand, and discharged when demand is larger than the PV production. The battery state of charge range was set from 10% to 90%. Battery roundtrip efficiency of 92% and a C-rate of 0.5 was used. Battery sizes were normalized by the amount of connections within the neighbourhood.

## III. RESULTS

### A. Share of potential

The share of the current (2016) annual PV production from the potential, and the average monetary values of the dwellings analysed per neighbourhood are shown in Fig. 1. The map shows high current share of the potential yield in the suburb Leidsche Rijn, and in the Noordoost (North-East) and Oost

(East) areas. Low shares are shown in Noordwest (Nord-West), Binnenstad (Inner-city) and Zuidwest (South-West). The current share of the potential has an average of 4.54%, a median of 3.51% and a maximum of 24.22%. These results show a wide range between the neighbourhoods.

We find that most neighbourhoods with relatively higher dwelling values have a higher share of PV potential yield, compared to neighbourhoods with a lower dwelling value. This implies that residents with high value dwellings invest more in residential PV systems. However, Overvecht area shows high share of potential, even with relatively low dwelling values. In these areas, social housing cooperation invested in most PV systems.

### B. Level of autarky

The current (2016) and potential level of autarky for each neighbourhood is shown in Fig. 2. Current levels are between 0% and 4.4%. Potential levels of autarky can go up to 61%. This indicates that neighbourhoods in Utrecht cannot fulfil their annual energy demand with the potential energy production.

The historical centre (Binnenstad) shows a very low level of autarky, mainly due to the large demand and the relative small PV potential. Tall residential apartment buildings are located in Overvecht, resulting in a relatively low PV potential for each resident. On the other hand, dwellings in sub-urban areas like Leidsche Rijn or Zuid are one or two family terraced houses, which the roof potential could provide a substantial share of the annual energy demand. These areas also have relatively high amount of PV potential and a lower demand.

### C. Self-consumption and self-sufficiency

The potential self-consumption rate and self-sufficiency rate, without battery, for each neighbourhood is shown in Fig. 3. Self-consumption rates are between 54% and 100%, indicating

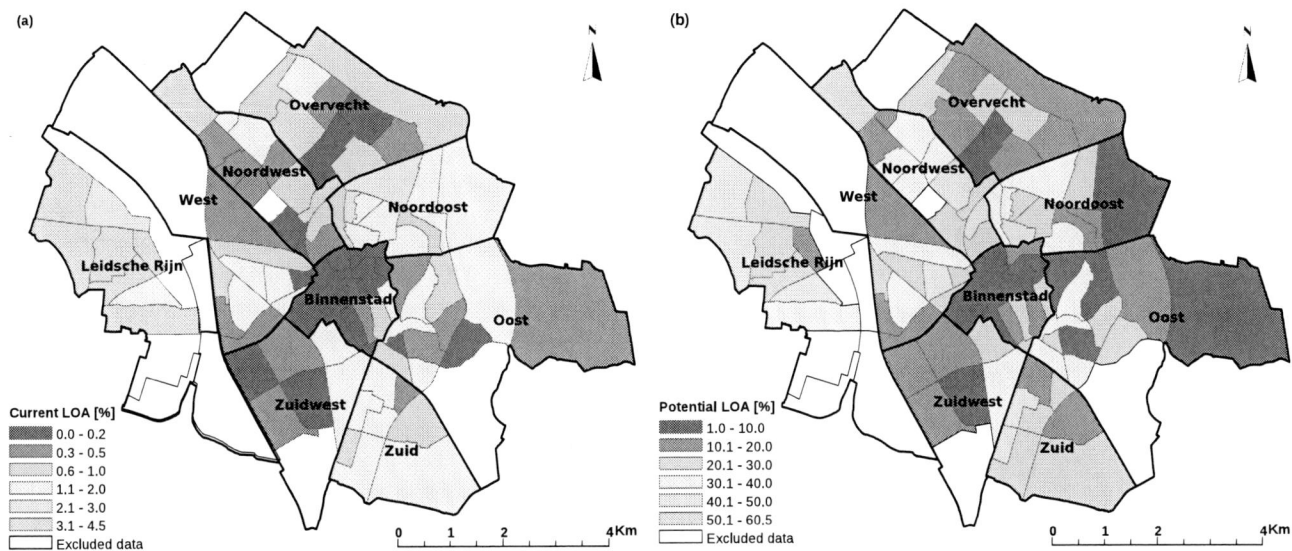

Fig. 2. Current level of autarky (a) and potential level of autarky (b) for each neighbourhood of the city of Utrecht.The white areas are neighborhoods containing less than 100 households, and are excluded.

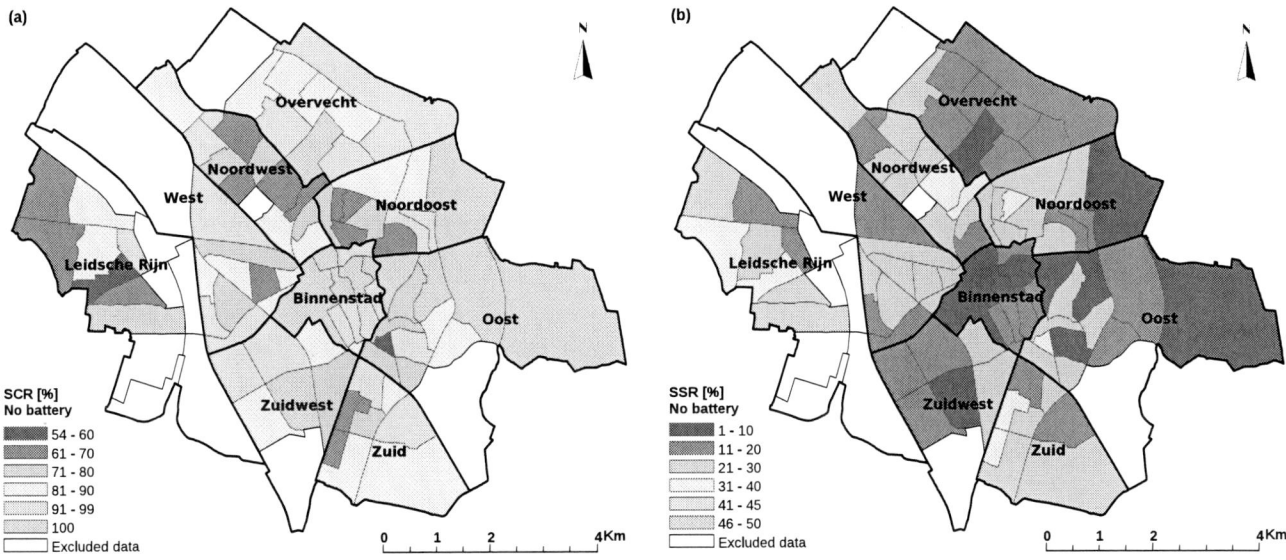

Fig. 3. Color-coded potential self-consumption rate (a) and potential self-sufficiency rate (b) for each neighborhoods of the city of Utrecht without batteries. The white areas are neighborhoods containing less than 100 households, and are excluded.

a wide difference between the neighbourhoods. Noordwest and Leidsche Rijn and other areas indicate low self-consumption ratios, whereas Binnenstad, Oost and Overvecht show large SCR.

Self-sufficiency rates are between 1% and 35%. Areas with high self-consumption rates and low self-sufficiency rates have limited PV yield potential. The PV produced energy can be used at most times within the neighbourhood. Areas with relative low self-consumption rates that have access to surplus PV energy production for some moments in time are interesting for battery development.

The relation between the potential level of autarky and the potential self-sufficiency rate for PV systems with different

batteries is shown in Fig. 4. The battery sizes were normalized using the number of grid connections within a neighbourhood. The potential level of autarky for each neighbourhood can be compared with the maximized self-sufficiency that can be achieved. This can be realized when the surplus energy is shifted towards moments of high energy demand, and charge and discharge losses of batteries are low.

Potential levels of autarky between 0% and 20% show similar self-sufficiency rates, indicating that all PV energy is directly consumed within the neighbourhood. For LOA between 20% and 30%, a battery with a capacity of 1 kWh increases the SSR with a few percent, and a 5 kWh battery reaches the maximum.

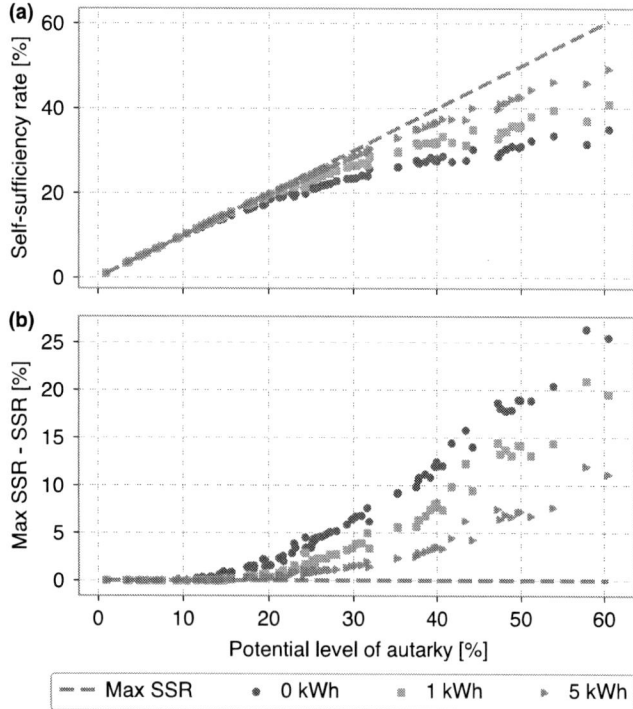

Fig. 4. Relation between the potential level of autarky and the potential self-sufficiency rate (a) and the differences between maximum SSR and the SSR (b) for PV systems with no battery, 1 and 5 kWh batteries for each neighbourhoods of the city of Utrecht. The maximum possible SSR is indicated by the dotted line. Battery sizes were normalized with the amount of connections within a neighbourhood.

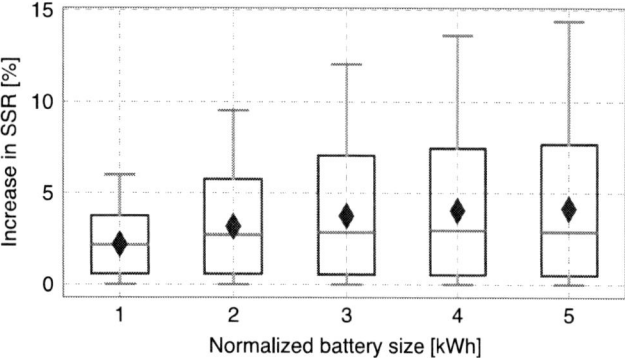

Fig. 5. Distribution of the increase in self-sufficiency for battery sizes from 1 to 5 kWh for each neighbourhood, shown using box-plots.Battery sizes were normalized with the amount of connections within a neighbourhood. Medians are indicated by the red line and means by the blue diamond.

The influence of the battery size on the increase in self-sufficiency is shown with a boxplot in Fig. 5. Around 25% of the neighbourhood self-sufficiency rates is are not affected by a PV-battery system. A 1 kWh battery has a small effect on the other 75% of neighbourhoods with an average increase of 2.2% in SSR. Furthermore, only a small amount of systems show an increase in SSR due to a battery system >3kWh. The

escalation in SSR due to the 5kWh battery is roughly doubled when compared to the 1 kWh battery.

The current level of autarky is <4.5% for all neighbourhoods, see Fig. 2. The relation of the level of autarky with battery size shows that batteries only start affecting the self-sufficiency rates with potential LOA of >10 %, see Fig. 4. Consequently, adding batteries with the current PV yield will not increase the self-sufficiency rate on neighbourhood level.

Overall 80-90% SCR can be achieved in Utrecht with a 5 kWh battery for each household, if estimated PV potential is reached. However, neighbourhoods with a low initial self-consumption can supply PV power towards areas with a high initial self-consumption. Therefore the overall PV battery potential can be reduced, leading to a more efficient electricity grid. Especially, the historical Binnenstad is surrounded with neighbourhoods that have surplus of PV power.

## IV. DISCUSSION

The results show large difference in potential levels of autarky, PV self-consumption and PV self-sufficiencies between neighbourhoods in Utrecht. The differences are mainly caused by differences in PV yield potential and local demand. Several limitations have been made in the research, which could lead to different results.

Potential PV capacities have been analysed only if 40% of the roof area was suitable. Also, the building facades were not included in the study. Including all roofs and facades would greatly increase the potential for solar PV. Furthermore, power density of 100 Wp/m$^2$ and 150 Wp/m$^2$ are assumed, but are expected to increase in the future. Both would lead to higher PV yield potential and therefore increase the level of autarky, but lower self-consumption.

Energy demand was obtained for 2013, however, it could significant increase in residential areas as the share of electrical vehicles are observed to increase. This will decrease the level of autarky, but only will lead to a small increase in direct self-consumption. Currently, charging of electrical vehicles is typically starting when residents arrive at home in the evening. Therefore, batteries are required to use surplus PV energy for charging during daytime, and they will be used in the evening and night-time to charge electrical vehicles.

Also, power and voltage constrains in the low voltage grid was excluded. Demand and PV yield patterns were aggregated on neighbourhood level, however the PV yield production source could be on a different low voltage grid than electricity demand. A detailed research on grid constrains which includes local electricity grid is recommended for further research.

Due to data availability on different spatial levels, they were aggregated to the neighbourhood level. Some of the spatial boundaries of the data parameters for e.g., postal code 6 data do exactly fit in the neighbourhood boundaries. Spatial aggregation methods were used to deal with this kind of data, which could lead to underestimation or over estimation of variables in certain neighbourhoods.

## V. CONCLUSION

The results show a large difference of current and potential level of autarky between neighbourhoods in the city Utrecht. PV yield is currently larger in areas with relatively higher dwellings values. None of the neighbourhoods achieved a 100% level of autarky. High self-consumption values are indicated in most neighbourhoods, representing limited PV potential.

The influence of the battery size shows interesting areas where self-sufficiency can be increased with relatively small battery sizes. Currently, battery systems are not required to improve the PV self-sufficiency on neighbourhood scale. These results also give an insight into areas which could act as potential storage areas which could be intermediate areas of electricity transmission. The obtained knowledge is valuable for local governments and developed method can be easily adapted to other areas.

## ACKNOWLEDGMENT

This work is part of the research programme Transitioning to a More Sustainable Energy System (grant number 022.004.023), which is financed by the Netherlands Organisation for Scientific Research (NWO). We are grateful to the Royal Dutch Meteorological Institute for providing data.

## REFERENCES

[1] European Commission, "Best practices on Renewable Energy Self-consumption Accompanying," EUROPEAN COMMISSION, Brussels, Tech. Rep., 2015.

[2] J. Hofierka and J. Kaňuk, "Assessment of photovoltaic potential in urban areas using open-source solar radiation tools," *Renewable Energy*, vol. 34, no. 10, pp. 2206 – 2214, 2009.

[3] R. Luthander, J. Widn, D. Nilsson, and J. Palm, "Photovoltaic self-consumption in buildings: A review," *Applied Energy*, vol. 142, pp. 80 – 94, 2015.

[4] J. Li and M. A. Danzer, "Optimal charge control strategies for stationary photovoltaic battery systems," *Journal of Power Sources*, vol. 258, pp. 365 – 373, 2014.

[5] Y. Mu, J. Wu, N. Jenkins, H. Jia, and C. Wang, "A spatialtemporal model for grid impact analysis of plug-in electric vehicles," *Applied Energy*, vol. 114, pp. 456 – 465, 2014.

[6] Y. Yamagata and H. Seya, "Community-based resilient electricity sharing: Optimal spatial clustering," in *2013 43rd Annual IEEE/IFIP Conference on Dependable Systems and Networks Workshop (DSN-W)*, June 2013, pp. 1–8.

[7] B. Kausika, O. Dolla, W. Folkerts, B. Siebenga, P. Hermans, and W. van Sark, "Bottom-up analysis of the solar photovoltaic potential for a city in the Netherlands - A working model for calculating the potential using high resolution LiDAR data," in *2015 International Conference on Smart Cities and Green ICT Systems (SMARTGREENS)*, May 2015, pp. 1–7.

[8] M. Šúri and J. Hofierka, "A new GIS-based solar radiation model and its application to photovoltaic assessments," *Transactions in GIS*, vol. 8, no. 2, pp. 175–190, 2004.

[9] R. W. Andrews, J. S. Stein, C. Hansen, D. Riley, C. Consulting, and S. N. Laboratories, "Introduction to the Open Source PV LIB for Python Photovoltaic System Modelling Package," in *Photovoltaic Specialist Conference (PVSC), 2014 IEEE 40th*, 2014, pp. 170–174.

[10] Liander N.V. , "Liander Open data," 2016. [Online]. Available: https://www.liander.nl/over-liander/innovatie/open-data/data

[11] Centraal Bureau voor de Statistiek, "Wijk- en buurtkaart 2016," 2016. [Online]. Available: https://www.cbs.nl/nl-nl/dossier/nederland-regionaal/geografische%20data/wijk-en-buurtkaart-2016

[12] Stedin Holding N.V., "Stedin Verbruiksgegevens," 2016. [Online]. Available: https://www.stedin.net/zakelijk/open-data/verbruiksgegevens

[13] Liander N.V. and Enexis Holding N.V. and Stedin Holding N.V.

[14] G. Litjens, W. van Sark, and E. Worrell, "On the influence of electricity demand patterns, battery storage and PV system design on PV self-consumption and grid interaction," in *2016 IEEE 43rd Photovoltaic Specialists Conference (PVSC)*, June 2016, pp. 2021–2024.

# Power Balance Requirements for Sustained Islanding of Inverter Based Distributed Generation

Gregory A. Kern, Michael Ropp, Sigifredo Gonzalez

SunPower Corp., Austin, Texas, 78758, USA

Northern Plains Power Technologies, Brookings, South Dakota, 57006, USA

Sandia National Laboratories, Albuquerque, New Mexico, 87123, USA

*Abstract* — **Active and reactive power must be balanced between generation and load when there is a section of the area electric power system that has been disconnected, or islanded, for there to be a possibility of creating a sustained island lasting two seconds or more. The degree to which this power balance must be achieved can be expressed using several equations. These equations can be used to estimate the island voltage and frequency based upon the generation and load conditions that existed just prior to formation of the island. These equations apply to inverter based current source generation and are fundamentally different than if the island were formed using synchronous voltage source generation. Qualitative consideration of advanced inverter functions such as Volt-Var, Frequency-Watt or Volt-Watt are provided. It is proposed that Volt-Var operation will significantly reduce the possibility of sustained island operation.**

## I. Introduction

In the early 2000's it was acceptable to have relatively tight voltage and frequency trip settings [1]-[4] and unity power factor operation for PV inverters. Early PV penetration levels were low and if PV systems tripped off line, it was not considered to have significant impact on the electric power system (EPS). Early inverters could pass the voltage and frequency trip requirements even if they tripped far more quickly than the trip settings required.

In recent years PV penetration has increased significantly and the potential impact on the electric power system is increasing as well. Standards are now being developed [5]-[9] that require advanced inverter functions to be adopted in grid tied PV inverters. Voltage and frequency ride through requirements are being added to standards to ensure that PV inverters do not trip too quickly, but rather remain connected and operating during short voltage and frequency deviations. Volt-Var and Volt-Watt capability is being added to help control voltage. Frequency-Watt capability is being added to also help provide grid support [5][7][9][10]. Considering these recent developments, it is worth updating our understanding of how PV inverters might operate when an island condition is created.

The first part of this paper develops equations that describe how the real and reactive power balance between generation and load impacts a resulting island's voltage and frequency. These equations neglect the impact of anti-islanding measures in the inverters. The presumption is that voltage and frequency trips remain the first line of defense to the formation of a sustained island. The equations assume that the inverter can operate over a wide range of voltage and frequency.

An island sustained by current source inverters is fundamentally different than a grid sustained by voltage source synchronous generation. Application of voltage source generation concepts to determine the possibility of current sourced islands is a misapplication of those concepts. In a current sourced island, voltage will rise or fall to create a balance between active generation and load. In a voltage sourced grid, mismatch between active generation and load impacts the speed of the rotating machine and thus grid frequency. Impacts on grid frequency can occur when a portion of generation trips off line causing a resulting rapid decrease in grid frequency. Under frequency load shedding schemes are used by grid operators to restore active power balance and thus allow the grid to return to stable frequency operation and prevent further collapse of the grid. In a current sourced island, frequency of the island will shift to balance generation and load reactive powers. In voltage sourced grid operations, reactive power controls normally correlate to grid voltage. Grid operators typically use capacitor banks to aid in voltage control in distribution systems. The Volt-Var mode of operation is being added as one of the advanced grid support modes to aid in voltage control.

The second part of this paper looks at the most recent draft of IEEE 1547 [5] voltage and frequency trips and ride through requirements. Based upon those settings, the range of real and reactive power settings that could lead to a sustained island are presented, again neglecting the impact of active anti-islanding.

The third part of this paper then takes select examples and uses experimental results to validate application of the equations.

Finally, this paper explains how these results can help inform the broad discussion of the probability of occurrence of islanding in area electric power systems.

## II. Power Balance Equations

The following equations are presented for estimating the resultant voltage and frequency of a section of an electric power system that contains both generation and load and that has been disconnected or islanded from the area electric power system. Several assumptions are made to simplify the analysis.

First, it is assumed that the load within the island can be modeled as an ideal lumped parallel RLC circuit. Second, all the generation in the island is assumed to be current source inverter based. Third, all anti-islanding protections have been disabled. It is recognized that some real-world loads might draw constant power as voltage changes and therefore would not be well modeled as resistors. It is also recognized that some percentage of synchronous rotating generation would likely impact results as well.

(1) and (2) can be used to estimate the voltage in an island after disconnection from the area EPS for two different regimes of inverter operation.

$$V_{ISLAND} = V_{EPS} \times \sqrt{P_{GEN}/P_{LOAD}} \qquad (1)$$

(1) considers the case where the PV inverter operates as a constant power source independent of voltage. This is typically how one would expect such an inverter to operate so long as the inverter output current stays below its output current limit.

$$V_{ISLAND} = V_{EPS} \times P_{GEN}/P_{LOAD} \qquad (2)$$

(2) considers the case where the PV inverter operates as a constant current source independent of voltage. This is how one would expect an inverter to operate while it is at its output current limit. In these equations, $P_{GEN}$ and $P_{LOAD}$ are the aggregate generation and load just prior to formation of the island, $V_{EPS}$ is the electric power system voltage prior to formation of the island, and $V_{ISLAND}$ is the resultant voltage to which the island circuit stabilizes after any dynamics from the disconnection of the area EPS.

(1) assumes that the load resistance, $R$, and generation, $P_{GEN}$, do not change when the island is formed. Therefore, load resistance, $R$, can be computed from the pre-island conditions:

$$R = V_{EPS}^2 / P_{LOAD} \qquad (3)$$

Once the island has formed and since the generation, $P_{GEN}$, has not changed, the following condition must be true:

$$P_{GEN} = V_{ISLAND}^2 / R \qquad (4)$$

Substituting (3) into (4) and solving for $V_{ISLAND}$ yields (1).

The derivation of (2) is just as simple, except that instead of power being constant, it is the active (real) component of inverter current, $I_{GEN}$, that is held constant, defined as follows:

$$I_{GEN} \equiv P_{GEN}/V_{EPS} \qquad (5)$$

Once the island has formed, and because the real component of generation current, $I_{GEN}$, has not changed, the following condition must be true:

$$V_{ISLAND} = I_{GEN} \times R \qquad (6)$$

Substituting (3) and (5) into (6) and solving for $V_{ISLAND}$ yields (2).

The frequency of the island circuit, $F_{ISLAND}$, is a result of interaction between the reactive power output of the generation in the circuit and reactive components of the load modeled as parallel inductance, $L$, and capacitance, $C$. When the island forms, the load reactive power consumption $Q_{LOAD}$ must become equal to the generators' reactive power output $Q_G$, as shown here:

$$Q_{LOAD} = Q_C - Q_L = V_{ISLAND}^2 \left( \omega C - \frac{1}{\omega L} \right) = Q_G \qquad (7)$$

When the island forms, the inverters will change their frequency $\omega$ until $Q_{LOAD} = Q_G$. Solving (7) for $\omega$:

$$\omega^2 + \omega \frac{Q_G}{C V_{ISLAND}^2} - \frac{1}{LC} = 0 \qquad (8)$$

$$\omega = -\frac{Q_G}{2 C V_{ISLAND}^2} + \frac{1}{2} \sqrt{\left( \frac{Q_G}{C V_{ISLAND}^2} \right)^2 + \frac{4}{LC}} \qquad (9)$$

Converting from rad/sec to Hz:

$$F_{ISLAND} = \frac{\omega}{2\pi} \qquad (10)$$

$F_{ISLAND}$ is the expected frequency at which the island will settle, after the initial transient, excluding the effect of other measures that manipulate inverter reactive power, $Q_G$.

There is however a simpler approach to estimate island frequency based upon parameters measurable during pre-island operation and some assumptions.

$$Q_L = V_{EPS}^2 / 2\pi F_{EPS} L \qquad (11)$$

$$Q_C = V_{EPS}^2 \times 2\pi F_{EPS} C \qquad (12)$$

The island frequency is then computed in similar manner as island voltage using a ratio of the reactive powers.

$$F_{ISLAND} = F_{EPS} \times \sqrt{(Q_L + Q_{LG})/(Q_C + Q_{CG})} \qquad (13)$$

In these equations, $Q_L$ and $Q_C$ are the reactive inductive and capacitive loads in the island circuit measured at $V_{EPS}$ and $F_{EPS}$ just prior to formation of the island. The inductance, $L$, and capacitance, $C$, are assumed constant and do not change when the island is formed.

The inverter reactive power output is defined as either $Q_{LG}$ or $Q_{CG}$, depending upon whether it is inductive or capacitive from a load frame of reference. The reactive power of the inverter can be expressed as $Q_G$. Following the sign convention of [5], when $Q_G$ is positive it is inductive, current lagging voltage, in the generator reference frame but capacitive in a load frame of reference. (13) is written in a load frame of reference, and therefore, the two variables for inverter reactive power are computed as follows:

$$Q_{LG} = -\min(0, Q_G) \qquad (14)$$

$$Q_{CG} = \max(0, Q_G) \qquad (15)$$

If the inverter is operating at unity power factor, then both terms would be zero. In that case, with $Q_{LG} = Q_{CG} = 0$, then (13) simplifies to the well-known formula for resonant frequency of a parallel resonant circuit:

$$F_{ISLAND} = F_{RES} = \frac{1}{2\pi\sqrt{LC}} \qquad (16)$$

It is interesting to note that in the case of a unity power factor inverter, the island frequency has no dependence upon the pre-island frequency of the grid, $F_{EPS}$, nor on the inverter power.

## III. ISLAND VOLTAGE AND FREQUENCY LIMITS

If there is a significant mismatch between real and reactive power in generation and load it is expected that voltage and frequency trip limits would operate to prevent sustained island operation. For this paper, a sustained island is defined as continued operation of the generation for two seconds or longer after formation of the island.

The voltage and frequency operation for this paper is taken from [5]. Category III level capability is presumed for the inverter generation. Under voltage operation is limited to 0.50 per unit (pu) since momentary cessation is defined to operate within 0.083 seconds of crossing below this voltage level. In an island scenario, if the voltage drops below this level, a subsequent trip of all the generation will occur and the island voltage will collapse. Over voltage operation is limited to 1.10 pu by momentary cessation which also operates within 0.083 seconds.

Fig. 1 is a plot of (1) and (2) for the condition where $V_{EPS} = 1.00$. The trip limits of 0.50 and 1.10 are also shown on the plot. When operating in constant current mode, the generation to load ratio must remain between 0.50 and 1.10 for the possibility of a sustained island. When operating in constant power mode, the generation to load ratio must be between 0.25 and 1.21 for the possibility of a sustained island to occur.

Frequency operation is limited by the over frequency trip at 62.0 Hertz and under frequency trip at 56.5 Hertz, both with default trip times of 0.16 seconds.

### TABLE I
#### POWER BALANCE FOR A SUSTAINED ISLAND

| $V_{EPS}$ | $V_{ISLAND}$ | Eqn. | $P_{GEN} / P_{LOAD}$ | |
|---|---|---|---|---|
| 1.0 | 1.1 | (1) | 1.21 | |
| 1.0 | 0.5 | (2) | 0.50 | $P_{GEN} > 0.5$ |
| 1.0 | 0.5 | (1) | 0.25 | $P_{GEN} < 0.5$ |
| $F_{EPS}$ | $F_{ISLAND}$ | | $(Q_L+Q_{LG}) / (Q_C+Q_{CG})$ | |
| 60.0 | 62.0 | (13) | 1.068 | |
| 60.0 | 56.5 | (13) | 0.887 | |

If one then presumes that the EPS is at nominal operating conditions of 1.00 pu voltage and 60.0 Hertz prior to formation of the island, then one can compute the range of generation to

load in real and reactive powers that would allow a sustained island to be possible, neglecting the effect of active anti-islanding measures. The scenario where island voltage increases is unlikely to cause the inverters to hit their output current limits, and therefore only (1) is used to compute the high limit of real power generation to load ratio.

Table I shows that when generation is operating at high power it will hit its output current limit at 0.50 pu voltage. If generation is operating at low power then it is more likely to operate in constant power mode down to the low voltage trip limit.

## IV. EXPERIMENTAL VERIFICATION

Two tests were conducted to check the validity of these equations. In each of these tests, all of the inverter over voltage trip levels were adjusted to 1.10 pu, and the under voltage trips levels to 0.50 pu. All the inverter over frequency trip levels were adjusted to 66.0 Hertz and the under-frequency trip levels to 50.0 Hertz. Finally, the anti-islanding protection was disabled. These modifications were made so that a wide range of island voltage and frequencies could be established.

In the first test the inverter was operated at an active power level of $P_{GEN} = 0.25$ pu. This low power level was selected so that the results could be compared to the power limited (1). Initial grid voltage, $V_{EPS}$, was set to 1.0 pu and frequency, $F_{EPS}$, to 60.0 Hz. The RLC island loads were set initially for a $P_{GEN}/P_{LOAD} = 1.00$ and $Q_L = Q_C = 0.25$ pu. The resistive load was then adjusted for $P_{GEN}/P_{LOAD} = 0.38$ to 1.06 and the resultant circuit was operated in a sustained island condition where $V_{ISLAND}$ was measured. The results are seen in Fig. 1. There is good agreement between the predicted and measured island voltages.

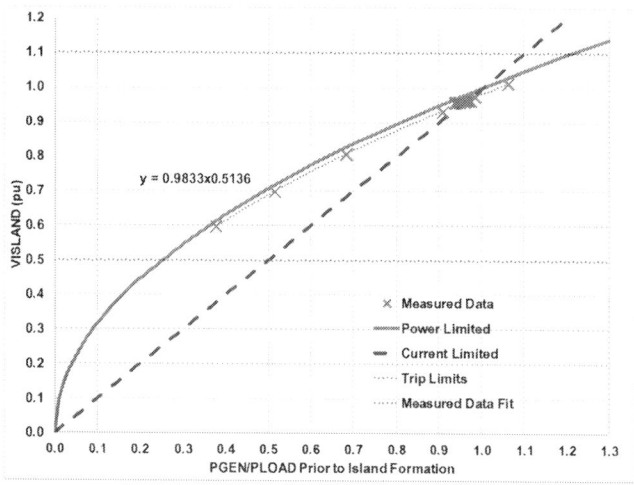

Fig. 1. Island Voltage Variation with Generation to Load Ratio

In the second test, the inverter was operated at an active power level of $P_{GEN} = 0.90$ pu. This is the highest power level the inverter can operate while still capable of providing up to +/- 0.44 pu of reactive power, $Q_G$. The grid voltage, $V_{EPS}$, was set to 1.00 pu and frequency, $F_{EPS}$, to 60.0 Hz. The RLC loads were set for a $P_{GEN}/P_{LOAD} = 1.00$, and $Q_L = Q_C = 0.90$ pu. The inverter was only capable of operating from 50 to 66 Hz, so this limited the range of reactive power output, $Q_G$, from -0.19 to 0.31. The inverter was operated over this range of reactive power settings using a constant reactive power command. Island frequency was measured for each of these points.

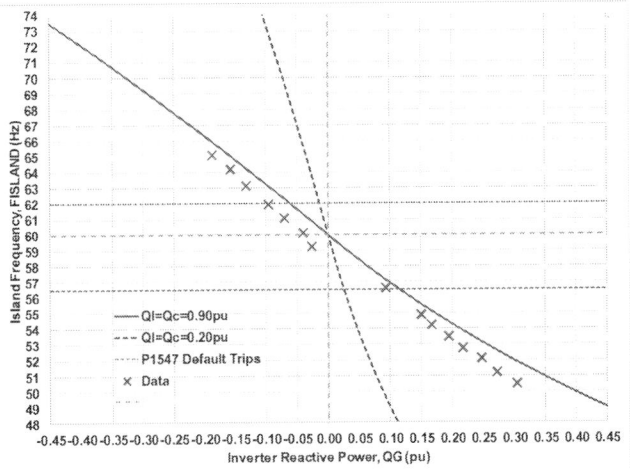

Fig. 2.    Island Frequency Variation with Reactive Power Balance

These measured data points are shown in Fig. 2 and match well the green curve plotted for $Q_L = Q_C = 0.90$ pu from (13). A second curve is plotted with reactive loads of $Q_L = Q_C = 0.20$ pu. Note that for each of these two curves, there is a tight window over which inverter reactive power must remain to stay within the frequency trip limits. Table II tabulates these limits using (13) assuming $F_{EPS} = 60.0$ Hz.

TABLE II
REACTIVE POWER RANGE FOR A SUSTAINED ISLAND

| $F_{ISLAND}$ (Hz) | $Q_L$ (pu) | $Q_C$ (pu) | $Q_G$ (pu) |
|---|---|---|---|
| 62.0 to 56.5 | 0.2 | 0.2 | -0.01355 to 0.02555 |
|  | 0.9 | 0.9 | -0.061 to 0.115 |

## V. IMPACT OF ADVANCED INVERTER FUNCTIONS

The assumption that inverter power, active or reactive, doesn't change when the island is formed may no longer be a valid assumption when the advanced inverter functions are enabled.

The default Volt-Var function of [5] begins to operate as soon as voltage deviates by 0.02 pu from nominal 1.00 pu voltage, and is fully engaged when voltage has deviated by 0.08 pu. The change in reactive power is significant enough to change the frequency balance of the island.    The default first order

response time of the Volt-Var function of [5] is 5.0 seconds for 90% response and would provide 60% response in 2.0 seconds. This response time is fast enough to impact maintaining a sustained island.    This Volt-Var response time is not fast enough to prevent the fast frequency trips, 0.16 seconds, at 56.5 and 62.0 Hz.    For a sustained island to occur with Volt-Var enabled, the island frequency must not exceed the fast frequency trip limits for longer than the fast frequency trip times and this must be true during the entire dynamic response after the island has formed. Thus, it is expected that the Volt-Var controls and the grid synchronization controls, acting on different time scales but interacting with one another, will tend to have a destabilizing effect on an unintentional island, and thus the Volt-Var function should make formation of stable islands more difficult.

The default Volt-Watt function of [5] begins to limit active power output of the inverter as voltage rises above 1.07 pu and is fully cut in at a voltage of 1.10 pu.    Whether Volt-Watt operation will make sustained islands more likely depends upon the dynamic response details of the inverter and the actual function and trip settings, but in most cases the Volt-Watt function, like the Volt-Var function, will have relatively long time constants. When an island is formed with a generation to load ratio greater than 1.21, the upward jump in voltage is nearly instantaneous and it is more likely for the inverter to enter momentary cessation by 0.083 seconds than it is to curtail active power with a response time of 10 seconds (default) or 0.5 seconds at the fastest.

The Frequency-Watt function of [5] will act to change active power output of the inverter. The default response time in the proposed P1547 is 5 seconds, adjustable down to 1 second. [10] specifies a response time of 0.05 to 3.0 seconds. The dead band of the Frequency-Watt function is so tight that it would almost always kick in during any islanding event.

If all three advanced inverter functions are enabled at the same time, the islanding response is complicated enough that the steady state equations above are not sufficient to predict resulting behavior. It is expected that simulation or testing will be a better predictor of islanding behavior. It is also likely that the detailed implementation of these functions among inverter manufacturers will vary enough that testing one inverter model may not be a good predictor of testing other inverter models. A more accurate analysis would require modeling the time domain dynamic response of the system, which is beyond the goals of this paper.

## VI. SUMMARY

This work has presented a set of equations that can be used to estimate if there is a possibility of forming a sustained island. Generation to load ratio is often considered to be the dominant factor in determining if there is possibility of sustained islanding. However, Table I and the experimental results show that it is the nature of the reactive power balances between generation and load that have far greater impact on the possibility of creating sustained islands.

One of the key conclusions to draw from this work is that Volt-Var operation will tend to significantly reduce the possibility of creating a sustained island event. This was observed while conducting tests to support this work. Further testing with anti-islanding protection disabled and Volt-Var enabled is recommended.

## ACKNOWLEDGEMENTS

The authors would like to acknowledge the members of the IEEE P1547 and P1547.1 Working Groups. Where this work is deemed useful and beneficial, we appreciate the input we have received from our colleagues. Specifically, thank you to Andy Hoke, Reigh Walling, John Berdner and Marcelo Algrain for their various comments and suggestions. If there are errors or mistakes, the blame rests solely on the authors.

## REFERENCES

[1] IEEE. *1547-2003–IEEE Standard for Interconnecting Distributed Resources with Electric Power Systems.* Approved June 12, 2003, New York, NY.

[2] IEEE. *1547.1-2005—IEEE Standard Conformance Test Procedures for Equipment Interconnecting Distributed Resources with Electric Power Systems.* Approved June 9, 2005. New York, NY.

[3] IEEE. *1547a-2014–IEEE Standard for Interconnecting Distributed Resources with Electric Power Systems: Amendment 1.* Approved May 16, 2014. New York, NY.

[4] IEEE. *1547.1a-2015—IEEE Standard Conformance Test Procedures for Equipment Interconnecting Distributed Resources with Electric Power Systems, Amendment 1.* Approved March 26, 2015. New York, NY.

[5] IEEE. *P1547/D6.0 Draft Standard for Interconnection and Interoperability of Distributed Energy Resources with Associated Electric Power Systems Interfaces.* Working Group draft not publicly released. December 2016.

[6] IEEE. *P1547.1/D1.0 Draft Standard Conformance Test Procedures for Equipment Interconnecting Distributed Energy Resources with Electric Power Systems and Associated Interfaces.* Working Group draft not publicly released. November 30, 2016.

[7] Pacific Gas and Electric Company. *Electric Rule No. 21 Generating Facility Interconnections.* September 16, 2016.

[8] Hawaiian Electric Company, Inc. *Rule No. 14 Service Connections and Facilities on Customer's Premises.* July 18, 2016.

[9] Underwriters Laboratories, Inc. *UL 1741 Standard for Inverters, Converters, Controllers and Interconnection System Equipment for Use with Distributed Energy Resources.* September 7, 2016.

[10] Hawaiian Electric Companies' Source Requirement Document for UL 1741 SA. HPUC Docket No. 2014-0192. March 10, 2017.

# Full-Scale Demonstration of Distribution System Parameter Estimation to Improve Low-Voltage Circuit Models

Matthew Lave[1], Matthew J. Reno[1], Robert J. Broderick[1], and Jouni Peppanen[2]

[1] Sandia National Laboratories, Albuquerque, NM, 87185, USA

[2] EPRI, Knoxville, TN, 37923, USA

*Abstract* — **Accurate distribution secondary low-voltage circuit models are needed to enhance overall distribution system operations and planning, including effective monitoring and coordination of distributed PV located in the secondary circuits. Accurate secondary models are also needed to fully leverage the measurement data received from smart meters and distributed energy resources at the customer premises. This paper presents a full-scale demonstration of a computational efficient approach for estimating the secondary circuit topologies from historical voltage and power measurement data provided by over 1,000 smart meters on an actual 12.47 kV feeder that is 22 km long.**

*Index Terms* — **solar energy, solar power generation, power grids.**

## I. INTRODUCTION

To analyze and operate distribution systems with growing amounts of PV and other distributed energy resources (DER), more accurate distribution system models are required [1]. Since most DERs are located in secondary (low-voltage) circuits, it is becoming important to include the secondary circuits into distribution models. This is particularly important since the low-voltage secondary circuits have higher per unit impedances, which result in a large share of the feeder per unit voltage drop [2]. Well-modelled secondary systems will allow for high penetrations of PV through such things as improved hosting capacity analysis and more accurate optimization and control. However, the vast majority of existing utility feeder models do not include the secondary circuits at all. When modeled, they are represented with limited detail.

The on-going extensive roll-out of smart meters and growing number of PV micro-inverters and other modern distribution system sensors are rapidly increasing the available measurement data along the distribution feeders. The Big Data from advanced metering infrastructure (AMI) and other emerging sensors has raised the interest in new methods for distribution system parameter estimation (DSPE) [3]-[6].

In our past work, we have presented methods to estimate secondary circuit topology and parameters (such as conductor type and length) when a dense grid of smart meter measurements is available [2] – [4], but it was only demonstrated on fairly small test cases. We have also looked at the case when only a very limited number of PV micro-inverter or similar measurements are available [5]. This method was demonstrated on several real distribution feeders, but the limited data decreased the applicability and impact of the research. Work was also done for actual AMI data on a real

utility secondary circuit in [6], but only a single service transformer and secondary circuit were modelled, and the medium-voltage feeder model was not available.

This paper presents parameter estimation results for a real U.S. utility distribution feeder with thousands of AMI power and voltage measurements. This is the first time that parameter estimation has been implemented at this realistic scale. This paper details the parameter and topology estimation methods, describes the test feeder, and presents results of parameter estimation which are validated based on satellite imagery and the existing (imperfect) utility secondary model. Future extensions of this work will implement the found topologies and parameters into secondary circuit models and quantify the value of these improved secondary models for distribution grid operations.

## II. SECONDARY CIRCUIT PARAMETER ESTIMATION AND MODEL GENERATION

The overall objective of distribution system secondary circuit topology and parameter estimation problem (DSPE) is to find the most likely topology and resistance ($R$) and reactance ($X$) parameters of a secondary circuit (shown in red in Fig. 1) by leveraging the smart meter and DER measurements (shown in blue in Fig. 1). The new secondary system models generated will improve the simulation accuracy.

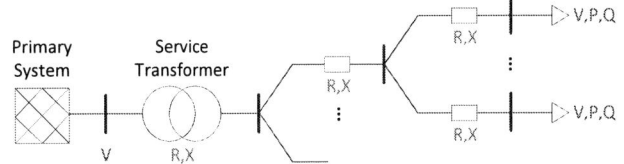

Fig. 1 Secondary circuit topology and parameter estimation problem

In [2], [3] we have shown a linear regression parameter estimation (LRPE) method for the case when all secondary circuit loads and DERs are metered and the secondary circuit topology is known. Moreover in [2], we have also shown a method to handle some meters not reporting voltage measurements. The LRPE method utilizes the well-known linear approximation of voltage drop ($V_{drop} = |V_1| - |V_2|$) over a series impedance $R + jX$ (on the right in Fig. 2)

$$V_{drop} = |V_1| - |V_2| \approx (RP + XQ)/V_2 = RI_R + XI_X, \quad (1)$$

where $P$, $Q$, $I_R$ and $I_X$ are the active power, reactive power, real current ($I_R = I(PF)$), and reactive current ($I_X = I\sqrt{1 - (PF)^2}$)

978-1-5090-5606-4/17 $31.00 © 2017 IEEE

flowing over the branch, respectively. For transformers, all values must be referred to the same voltage level. In 3-phase systems, line-line voltages and 3-phase powers are used whereas in 1-phase systems, line-to-neutral voltages are utilized. The LRPE method algorithm estimates the secondary circuit parameters by proceeding from the tree leaf nodes towards the tree root node. At a given iteration, the algorithm utilizes (1) to generate linear regression models

$$y = X\beta + \epsilon \tag{2}$$

to estimate the branch impedances of a circuit subsection consisting of either two series meters (on the right in Fig. 2) or M parallel meters (on the left in Fig. 2 for two meters):

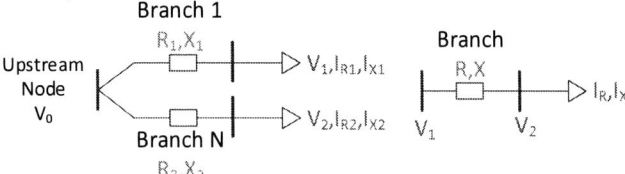

Fig. 2 Two meters connected in parallel (left) and in series (right)

For two series meters, the linear regression (2) variables are

$$y = V_1 - V_2, \, X = [I_R \quad I_X], \text{ and } \beta = [R \quad X]^{\mathrm{T}}. \tag{3}$$

For M parallel meters, the variables are

$$\beta = [V_{0,1}, \ldots, V_{0,M}, R_1, X_1, \ldots, R_N, X_N]^{\mathrm{T}}, \tag{4}$$

$$y = [V_{1,1}, \ldots, V_{1,M}, \ldots, V_{N,1}, \ldots, V_{N,M}]^{\mathrm{T}}, \tag{5}$$

and

$$X = \begin{bmatrix} I & [-I_{R,1} & -I_{X,1}] & \cdots & 0 \\ \vdots & \vdots & \ddots & \vdots \\ I & 0 & \cdots & [-I_{R,N} & -I_{X,N}] \end{bmatrix}. \tag{6}$$

Many utilities do not know the secondary circuit topologies. In [4], we apply linear regression topology and parameter estimation (LRTE) algorithm to generate the entire secondary circuit models, including topology, using only the measurements. The algorithm processes one secondary circuit at a time, using the list of all the meters of the secondary circuit. For each meter pair, the algorithm solves a linear regression problem for the parallel circuit type (on the left in Fig. 2)

$$V_1 - V_2 = I_{R1}R_1 + I_{X1}X_1 + I_{R2}R_2 + I_{X2}X_2 + \epsilon \tag{7}$$

and a linear regression problem for the series circuit type (on the right in Fig. 2) to determine the best matching topology.

$$V_1 - V_2 = I_R R + I_X X + \epsilon. \tag{8}$$

### III. STUDY FEEDER AND MEASURED DATA

The distribution system being analyzed is shown in Fig. 3. It is a 12.47 kV feeder that is approximately 22 km long. There are three sets of line voltage regulators on the feeder, and three fixed capacitor banks. There are 1010 customers on the feeder

with AMI. For this area, the AMI rollout happened in May 2016, and data was available through October 2016, resulting in approximately 6 months of AMI data. The AMI measurements are recorded every 15-minutes and include consumed kWh and kvarh, and time-averaged voltage. Fig. 4 shows the measured power consumption from 3 customers on the same transformer on the feeder for a day during the year. As is typical for AMI data, the power consumption of a single house is highly variable with large spikes at times large appliances are used or when the heating/cooling system is on. The consumption profiles are uncorrelated, so the overall load profile on the feeder is fairly smooth.

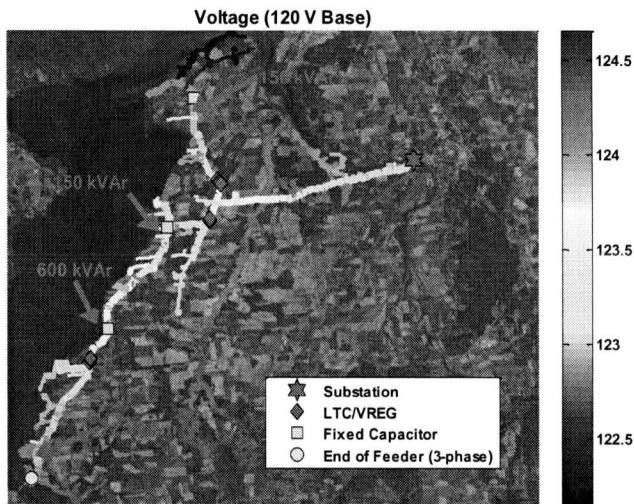

Fig. 3. Distribution system feeder being studied. The lines are colored based on their voltage.

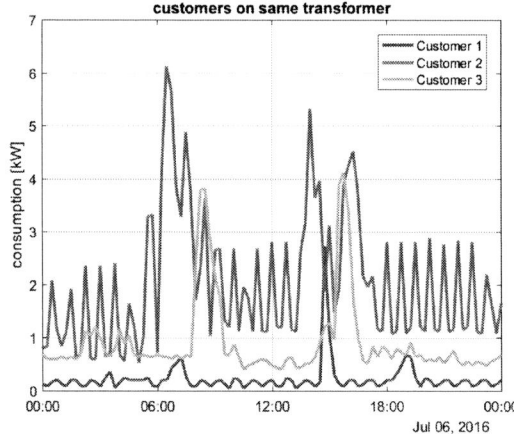

Fig. 4. One-day time series of power consumption for 3 customers on the same transformer.

The majority of the loads on the feeder are residential customers, with peak consumptions for each customer less than 10 kW (Fig. 5), but there are some large commercial/industrial customers with high yearly peak loads. Finally, the utility provides net metering for any distributed generation connected to the distribution system.

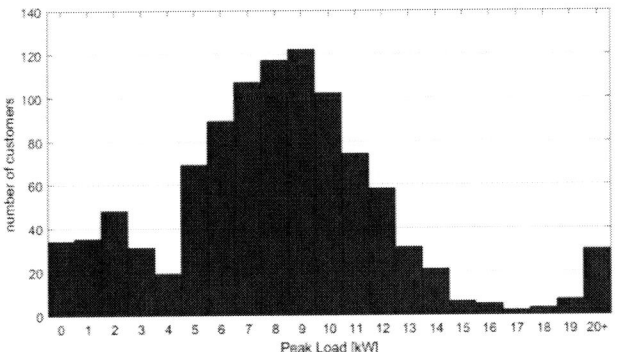

Fig. 5. Histogram of peak load measured for each of the 1010 customers on the feeder.

## IV. PARAMETER AND TOPOLOGY ESTIMATION APPLICATION

### A. Transformers with Multiple Customers

For transformers with multiple customers, all possible customer pairs are evaluated using the parallel model in Eq. (7). For example, for transformers with 3 customers, all 3 possible combinations of customers pairs (1 and 2, 2 and 3, 3 and 1) are evaluated. AMI data for voltage and real and reactive power, are used to solve for $R_1$, $X_1$, $R_2$, and $X_2$ using (Eqns. (3)-(6)).

We started with the parallel model since the vast majority of customers are connected in parallel. In some cases, though, linear regression based on using Eq. (7) resulted in negative R1 or R2 values. In these situations, the linear regression was re-run using the series model in Eq. (8). Assuming a positive series resistance resulted, the two customers were assumed to be connected in series.

To understand how well the linear regression fits the data, we computed the Pearson correlation coefficient ($R^2$) of the fit. This can be seen visually in Fig. 6, which shows the predicted $V_1$-$V_2$ using the right side of Eq. (7) with R1, X1, R2, and X2 from linear regression against the measured $V_1$-$V_2$. The same approach was used for series connections: R1 and X1 were used in the right side of Eq. (8) and compared to $V_1$-$V_2$ to determine the $R^2$ value.

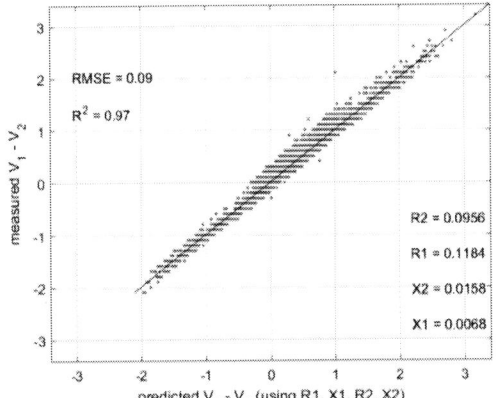

Fig. 6. Relationship between predicted (using R and X values from linear regression, as listed in the bottom right) and measured $V_1$-$V_2$.

Once $R^2$ values were computed for all customer pairs, the pair with the highest $R^2$ value was assigned as the actual topology. If this pair was found to be connected in parallel, then a new virtual node, representing the point where these two parallel lines connect, was created. The voltage at the virtual node was found by adding the voltage drop to the measured voltage:

$$V_{virtual\ node} = V_2 + I_{R2}R_2 + I_{X2}X_2. \qquad (9)$$

By convention, we defined $V_2$ as the higher voltage of the pair ( $mean\ (V_2) \geq mean(V_1)$ ). Thus, Eq. (9) ensures that the voltage at the virtual node is, on average, higher than the voltage at each of the customers. Real and reactive power consumption of the virtual nodes were found by summing the real and reactive power of the two customers. For pairs connected in series, the virtual node is located at the upstream customer (i.e., customer 2), the voltage is $V_2$, but the real and reactive power are the sum of the power from both customers.

As a second iteration, all remaining customers (those not contributing to the virtual node) were again paired with one another and with the virtual node. Again, the series or parallel pair with the highest $R^2$ value was assigned as the actual topology, and a virtual node was created. This method was repeated until all customers were paired in the topology.

Fig. 7 shows an example of the result of this topology and parameter estimation. First, customers 2 and 3 were determined to be paired in parallel, with resistance values around 0.1Ω. Virtual node 1 was created out of this pair. Next, customer 1 was found to be paired in series with virtual node 1. The resistance was found to be about 0.04 Ω, indicating a shorter length of wire between the virtual node and customer 1 than the other customers.

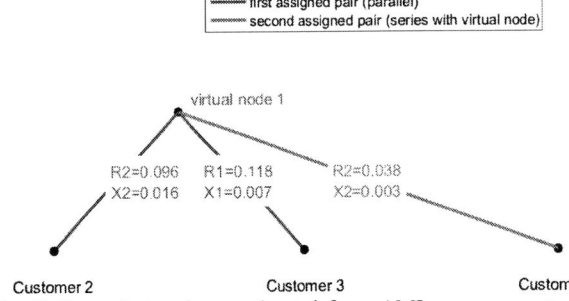

Fig. 7. Example topology estimated from AMI measurements, with calculated R and X parameters shown.

### B. Transformers with only One Customer

On transformers with a single customer, the method described in Section IV.A will not work: although we know the customer must be connected in series, we only have one voltage measurement (at the one customer), and so cannot solve for R and X between two voltage measurements using Eqs. (7) or (8). Instead, we look to find a nearby transformer that also has voltage measurements. Because of the voltage level, the per unit resistance on the primary voltage system between two nearby transformers is much smaller than the per unit resistance on the low voltage system from transformer to customer. For

978-1-5090-5606-4/17 $31.00 ©2017 IEEE

3027

example, on a 12kV system, 900 feet of wire has the same per unit resistance as approximately 1 foot of wire on the low-voltage system. Thus, the primary voltage side of nearby service transformers can be assumed to be very close with little impact to the estimated resistances, and the customers can be compared in the same fashion as described in Section IV.A, as though they are on the same transformer.

This method includes the transformer resistance and reactance in the parameter estimation. The size of transformers is generally known, and hence resistance and reactance can be fairly well estimated. The transformer resistance and reactance are subtracted out from the estimated parameters to find the customer resistance and reactance. Including the transformers in parameter estimation can also be used to validate the transformer sizes in the utility models.

## V. RESULTS

### A. One Transformer with Three Customers

To demonstrate the methodology, we focus on one specific transformer with three customers. In fact, this is the same transformer and set of customers shown in Figs. 4 and 7. The secondary model from the utility is shown in Fig. 8. The topology found in Fig. 7 – that all customers are connected in parallel with separate lines to the service transformer – is consistent with the utility model, demonstrating successful topology estimation.

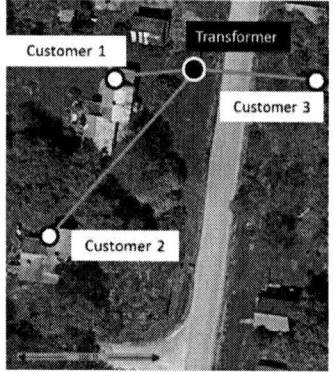

Fig. 8. Anticipated secondary setup based on utility model.

The utility uses almost exclusively #2 wiring for its overhead secondary lines. Based on the properties of #2, the resistance will be approximately 0.058Ω per 100ft. Using these numbers and the resistances shown in Fig. 7, we were able to estimate the line distances for each customer. Table 1 shows these estimated line distances from parameter estimation, and compares them to the direct distance from the transformer to the customer using the latitude and longitude in the utility model. For customers 1 and 2, close agreement is found: the lines go almost directly from transformer to customer, and the resistances found in parameter estimation are consistent with this to within a few feet. Variation by a few feet is expected

since the exact location of the customer meter (e.g., which side of the house) is not known.

The distance found for customer 3 based on the resistance is not consistent with the distance based on latitude and longitude in the utility model. Visual inspection of the transformer and wiring using Google Street View images shows that customer 3 is incorrectly located in the utility's original secondary model. Fig. 9 shows the correct location of customer 3 – much further away from the transformer and consistent with the distance found based on the estimated resistance. This shows the value of parameter and topology estimation even when utility secondary models exist: it can help to identify errors in those models.

TABLE 1: DISTANCES TO THE TRANSFORMER FROM THE METERS.

| | Distance to Transformer | |
|---|---|---|
| | Direct based on latitude/longitude | Based on parameter estimation |
| Customer 1 | 58 ft | 66 ft |
| Customer 2 | 149 ft | 165 ft |
| Customer 3 | 111 ft | 204 ft |

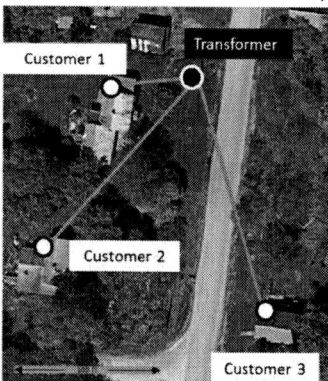

Fig. 9. Actual secondary setup based on inspection using street view imagery.

### B. Sensitivity to amount of AMI Data Available

The parameter estimation shown in Section V.A. used all available AMI data from the 6-month period (16,072 timestamps with voltage, power, and reactive power measurements at each customer). However, in many cases such as new AMI installations, less frequent data collection intervals, or significant missing data, less AMI data will be available for parameter estimation.

Fig. 10 shows the sensitivity of parameter estimated resistance to the amount of data used. The resistances for all three customers are consistent when 30% (4,822 data points) or more of the data is used. However, when less than 10% of the data is used (less than 1,607 data points), resistances are significantly different from the values with 100% of the data. Results may vary for other transformers, but suggest that 5,000 data points (roughly 52 days of data collected at 15-minute intervals) may be sufficient to accurately resolve parameters.

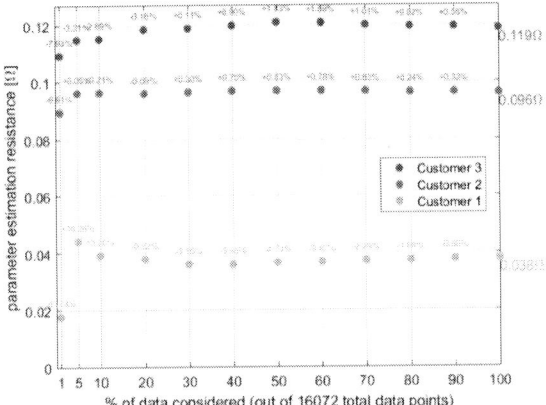

Fig. 10. Resistances found using parameter estimation with varied amounts of input data starting in May 2016 (x-axis).

## C. All Transformers with Two Customers

To test the parameter estimation performance across several transformers, we ran the topology estimation for all transformers with two customers. It is most common for such transformers to have parallel connections where the line goes directly from transformer to customer. Fig. 11 show the scatter plot of distance from found resistances versus distance from modeled latitude and longitude for all transformers with two customers that were found to be connected in parallel in topology estimation. Overall, a positive trend is seen in Fig. 11, suggesting that the found resistances are accurate. There are several reasons unrelated to parameter estimation errors which may cause some points to disagree.

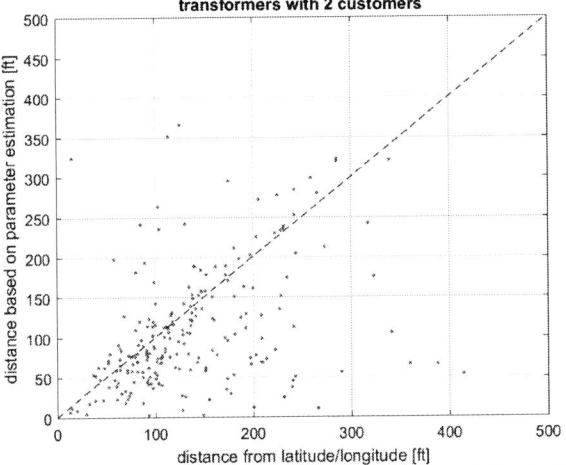

Fig. 11. Distance from customer to transformer calculated from parameter estimation (y-axis) versus distance based on model latitude and longitude (x-axis) for transformers with two customers where the connection was found to be in parallel during topology estimation.

The differences are likely due to errors in the utility model for wire type (not all #2 conductor) or customer location. For example, the point which falls furthest to the right in Fig. 11 (latitude/longitude distance of 415ft, parameter estimation distance of 53ft), may be not have the correct transformer assigned in the model. As seen in Fig. 12, there is a transformer

much closer to the customer (about 100ft away) to which the customer may actually be connected. Similarly, for the point in the top left (latitude/longitude distance of 15ft, parameter estimation distance of 324ft) the customer appears to be much further form the transformer than the point indicated by the latitude/longitude.

Points above the dashed line may also be explained by not all connections following a straight line from the transformer to the customer – some might take indirect paths to follow roads or connect to existing poles – such that the true distance is longer than the latitude/longitude modeled distance. Points below the dashed line may also be explained by situations such as those shown in Fig. 13. While the customers are connected in parallel, their lines meet before the transformer. This would cause the resistance found by parameter estimation to be only the resistance on the line from the customer to the point where the two customer lines meet. Thus, the distance found based on the resistance would be less than the actual distance from customer to transformer.

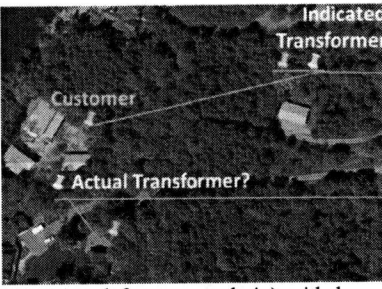

Fig. 12. Customer (top left green pushpin) with longest distance to transformer based on latitude and longitude. The transformer the customer is connected to indicated in the model is in the top right. However, there is a much closer transformer to which the customer may actually be connected.

Fig. 13. Customers (green pushpins) connected in parallel, but the lines do not meet at the transformer (white pushpin).

To better understand the impact of parameter estimation, we calculated the rise with 20A injected current, which corresponds to the maximum output of a 4.8kW PV system at 240V. We calculated the voltage rise using parameter estimated resistance; resistance based on the latitude/longitude distance assuming #2 triplex taking a straight path; and resistance based on a line of 100ft of #2 triplex (0.058Ω). Results are shown in Fig. 14. Voltage rises from parameter estimation varied up to 4V from values using the latitude/longitude in the utility.

Similarly, the 100ft line assumption varied from parameter estimation by up to 3V. Differences of these magnitudes (several volts) are significant for managing the voltage of customers with PV.

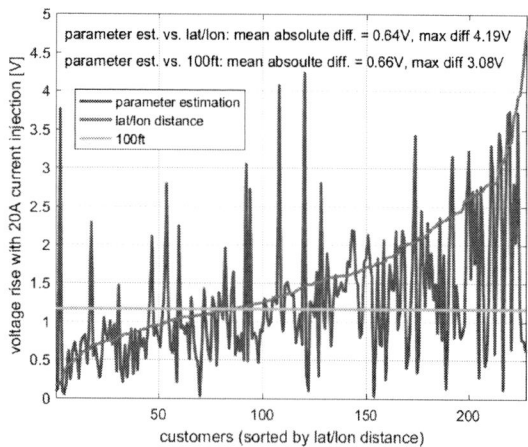

Fig. 14. Voltage rise for 20A of current injection (max output for 4.8kW PV) using parameter estimation resistance, latitude/longitude distance-based resistance, and the resistance of a 100ft line.

### D. Transformers with Only One Customer

The procedure described in Section IV.B was applied to the two transformers each with only one customer as shown in Fig. 15.

Fig. 15. Two neighboring transformers (white pushpins), each with one customer (green pushpin).

The virtual node in this case is located on the primary system, between transformers 1 and 2. This means that the transformer resistance and reactance are both included in the parameters found. For transformers, the resistance is approximately 0.75% across both windings, and the reactance is approximately 2%. Thus, transformer resistance and reactance looking up from the low-voltage side can be estimated as:

$$R_{transformer}(power\ rating) = 240V^2/power\ rating \times 0.75\% \quad (10)$$

$$X_{transformer}(power\ rating) = 240V^2/power\ rating \times 2\% \quad (11)$$

Both transformers are 15kVA, such that $R_{transformer} \approx 0.029\Omega$ and $X_{transformer} \approx 0.077\Omega$. The customer R and X values can be found by subtracting out the transformer

resistance and reactance. Fig. 16 illustrates this separation of transformer and customer. The customer resistances are found to be $0.059\,\Omega$ and $0.056\,\Omega$, corresponding to approximately 100ft distances from transformer to customer, which are consistent with Fig. 15.

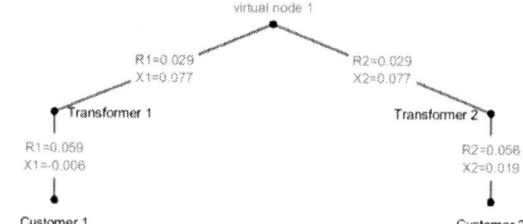

Fig. 16. Topology with transformers separated from customers.

## VI. CONCLUSIONS

We have demonstrated parameter and topology estimation algorithms using large AMI data sets. Results were generally consistent with satellite imagery and the existing utility secondary model, demonstrating the accuracy of the method. Some examples of disagreement may indicate parameter estimation detecting errors such as incorrect customer meter locations, line routings, or customer-transformers connections in the existing model. Thus, parameter estimation is valuable both when no secondary model exists and to help make existing secondary models more accurate. Future work will more directly quantify the benefit of improved secondary models to distribution grid operations, including the ability to install additional PV systems due to a better understanding of secondary system voltages.

## ACKNOWLEDGEMENT

Sandia National Laboratories is a multimission laboratory managed and operated by National Technology and Engineering Solutions of Sandia, LLC, a wholly owned subsidiary of Honeywell International, Inc., for the U.S. Department of Energy's National Nuclear Security Administration under contract DE-NA0003525. SAND 2017-6054C.

## REFERENCES

[1] B. Palmintier, R. Broderick, B. Mather, et al., "On the Path to SunShot: Emerging Issues and Challenges in Integrating Solar with the Distribution System," National Renewable Energy Laboratory, NREL/TP-5D00-65331, 2016.

[2] J. Peppanen, M. J. Reno, R. J. Broderick, and S. Grijalva, "Distribution System Secondary Circuit Parameter Estimation for Model Calibration," Sandia Labs, SAND2015-7477, 2015.

[3] J. Peppanen, M. J. Reno, R. J. Broderick, and S. Grijalva, "Distribution System Model Calibration with Big Data from AMI and PV Inverters," in *IEEE Transactions on Smart Grid*, 2016.

[4] J. Peppanen, M. J. Reno, R. J. Broderick, and S. Grijalva, "Distribution System Low-Voltage Circuit Topology Estimation using Smart Metering Data," IEEE PES T&D Conference, 2016.

[5] J. Peppanen, M. J. Reno, R. J. Broderick, and S. Grijalva, "Secondary Circuit Model Generation Using Limited PV Measurements and Parameter Estimation," IEEE PES GM, 2016.

[6] J. Peppanen, M. J. Reno, R. J. Broderick, and S. Grijalva, "Secondary Circuit Model Creation and Validation with AMI and Transformer Measurements," IEEE NAPS, 2016.

# Creation and Value of Synthetic High-Frequency Solar Inputs for Distribution System QSTS Simulations

Matthew Lave, Matthew J. Reno, Robert J. Broderick

Sandia National Laboratories, Livermore, CA and Albuquerque, NM, 94550 and 87185, USA

*Abstract* — Methods and initial results are presented for creating synthetic high-frequency solar simulations with unique profiles for each interconnection point on a distribution system feeder using low-frequency input data. The three steps to synthetic sample creation are to develop a relationship between high and low frequency data, create high-frequency timeseries based on this relationship, and then to generate unique samples for different spatial locations. The simulation results for a distribution system voltage regulator demonstrate the value of unique high-frequency samples for distributed PV compared to a single PV profile used at all interconnection points.

*Index Terms* — solar energy, solar power generation, power grids.

## I. INTRODUCTION

High-frequency solar variability with unique inputs for different interconnection points on distribution feeders are important inputs to accurate quasi-static time series (QSTS) distribution grid integration studies [1]. Using low-frequency solar variability results in underestimation of the impact of solar photovoltaics (PV) to distribution grid operations [2], while using a single PV profile for all interconnection points results in an overestimation of the PV impact due to the spatial smoothing provided by distributed PV [3].

Measurements of high-frequency solar variability are scarce, motivating methods which can synthetically generate high-frequency data from more ubiquitous low-frequency data such as satellite-derived irradiance [4]. In this paper, we present initial results from ongoing work to develop high frequency, spatially-unique synthetic samples and to show their value to QSTS.

## II. METHOD

To create inputs to distribution grid studies which involve distributed PV across a feeder, Sandia will use a 3-step process.

1) Develop a relationship between low-frequency satellite derived solar irradiance and high-frequency solar irradiance, using an hourly or daily summary statistic such as the variability score (VS).

2) Select high frequency timeseries samples given the predicted high-frequency summary statistic.

3) Generate unique irradiance samples for each interconnection point by adding some decorrelation between points, while still retaining the overall summary statistics.

### A. Low-frequency data and high-frequency data relationship

The relationship between low-frequency satellite and high-frequency irradiance has resolved in previous work [5]. The relation between solar variability derived from hourly satellite irradiance versus sub-minute ground measured solar irradiance was found to be strongest when the hourly satellite data was adjusted in several ways.

The adjusted satellite data was first converted to a clear-sky index to remove solar variability caused by the sun's movement through the sky. Then, the median of all daily variability scores, using a year of more of satellite data, was used. This median variability score was then scaled by the ratio of $\frac{\text{median GHI}}{1000\,\text{Wm}^{-2}}$, to reintroduce the magnitude of irradiance that was removed by using the clear-sky index Finally, spatial smoothing was used by taking the distance-weighted average of the 9 satellite pixels surrounding the location of interest. The ground 30-second variability is shown as a function of the 1-hour satellite variability in Figure 1, where the 1-hour satellite data was adjusted as described in the bullets above.

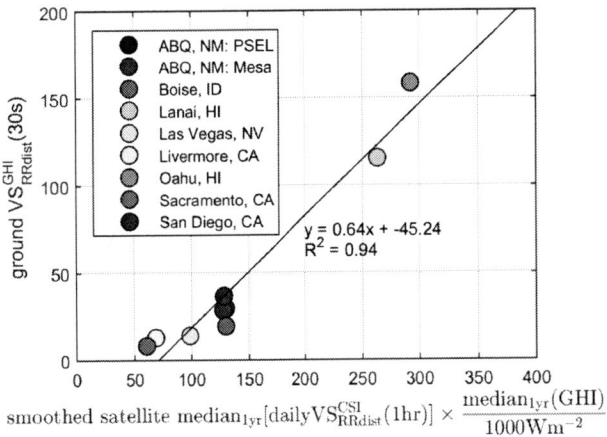

Figure 1: Relationship between 30-second ground measured solar variability (x-axis) and 1-hour satellite derived solar variability (y-axis) for several locations. [5]

### B. Select High-Frequency Timeseries

At least three basic methods exist for selecting appropriate high-frequency timeseries based on the low-frequency variability determined in Section II. A.

One method is to find hours in a lookup library that match the summary statistic found from satellite data. For this method, a large database of high-frequency irradiance samples is needed. Based on the variability statistic assigned to each hour of satellite data, a representatively variable hour of high-

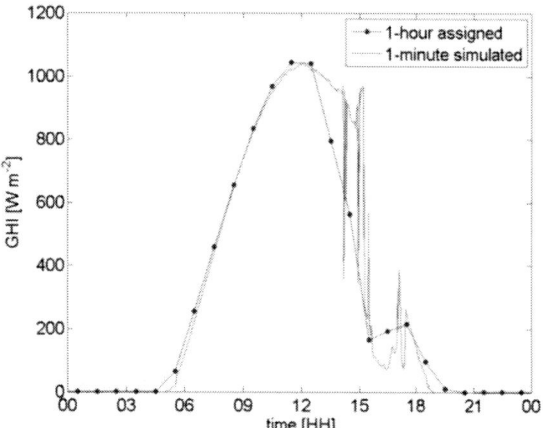

Figure 2: Example of using a library lookup to assign 1-minute high resolution data based on hourly data.

frequency data is pulled from the library. Figure 2 shows an example of this method. Hours in the morning are clear and hence low-variability sample hours are assigned from the library. In the afternoon and evening, however, the hourly data indicates a sharp change in output, leading to a high variability statistic for those hours, and hence the high-frequency samples assigned from the library are highly variable. This method is similar to that used in [6].

A second method is to create synthetic ramp rates by sampling from a cumulative distribution of high-frequency ramp rates during times which match the low-variability statistic. This method is similar to the first method, except that each short-interval ramp (e.g., 1-second or 1-minute) is sampled independently; in the first method, hour-long blocks are sampled all altogether. The advantage of this method is that it requires a smaller library of high-frequency data. The disadvantage is that, due to the independent sampling, special care must be taken to ensure that the autocorrelation of the created timeseries is representative of actual solar timeseries. That is, the independent sampling may, for example, often choose several large down ramps in a row, leading to a very steep decline in generation that is not reflected in the hourly data. Instead, additional dependencies must be factored into this method, such as that solar timeseries are more likely to ramp up after a down ramp than down again.

The third method, which we describe in detail in this paper, is the creation of synthetic cloud fields based on the hourly irradiance statistic. The cloud sizes are scaled based on the variability determined from the hourly data. Cloud fields are created based on a modification of Perlin noise [7], which has historically been used for creation of clouds for movies and video games. Just as for the second method, special care must be taken to accurately reproduce the ramp rate statistics of true solar irradiance timeseries, as this method tends to predict too quick of changes from full output to cloud obstruction. Smoothing of the cloud edges and retention of high-spatial scale (in addition to larger cloud features) noise are imperative.

## C. Unique PV Production Across a Distribution Feeder

Assuming only a single representative timeseries for all locations on a feeder will lead to significant overestimation of the PV impacts to the feeder (see Section IV), as all distributed PV systems will ramp at the exact same time and in the same direction. Instead, unique PV profiles must be created for each of the different interconnection points along the feeder to model cloud shading, movement, and the spatial smoothing of distributed PV.

The method used to create unique PV profiles will depend on the method to create representative timeseries (Section II. B. ). The first and second representative timeseries methods result in a single timeseries. These can then be tuned into unique timeseries by time-shifting each timeseries based on the cloud speed. Example time shifts for PV interconnection locations on a distribution feeder and an example shifted timeseries are shown in Figure 3. Time shifts must occur for clear-sky index data, as seen in Figure 3, to account for changing sun angles

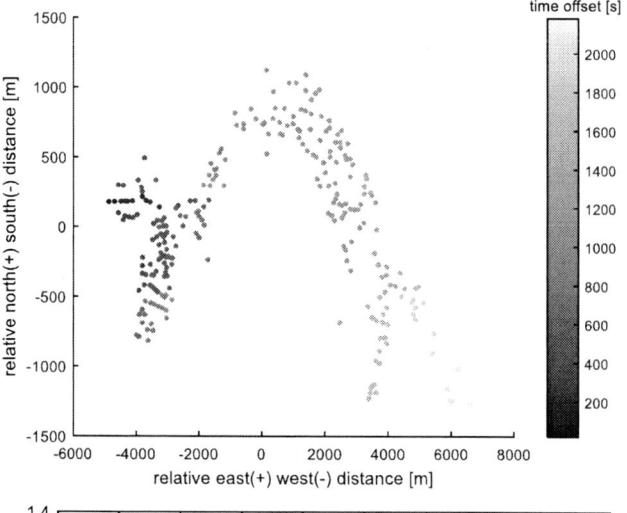

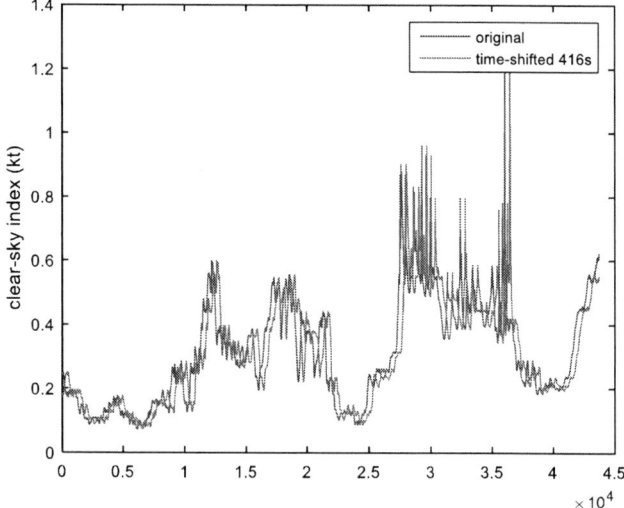

Figure 3: [Top] Time offset for points along a distribution feeder, and [Bottom] resulting shifted timeseries for one offset.

(most important over long time periods of 10s of minutes to hours), in addition to the cloud motion. The disadvantage to this method is that all locations perpendicular to the direction of cloud motion have identical PV generation timeseries, and all locations have identical irradiance statistics, just with different offsets. Both of these mean that spatial smoothing across the feeder is slightly underestimated, and, hence, the PV impact to the feeder may be slightly overestimated.

For the synthetic cloud field method, the unique PV output profiles are naturally created due to the 2-dimentional nature of the cloud fields, but as mentioned in Section II. B. the profiles must retain the spatial smoothing of true distributed PV.

## III. SYNTHETIC CLOUD FIELDS

Since the synthetic cloud fields have the ability to address both Sections II. B. and II. C. , it is a promising method. In this section, we present work we have done to develop the cloud field method and create unique high-resolution PV timeseries.

### A. Field at Different Scales

The synthetic cloud fields method begins by creating random noise at different spatial scales, as seen in the left plots in Figure 4.

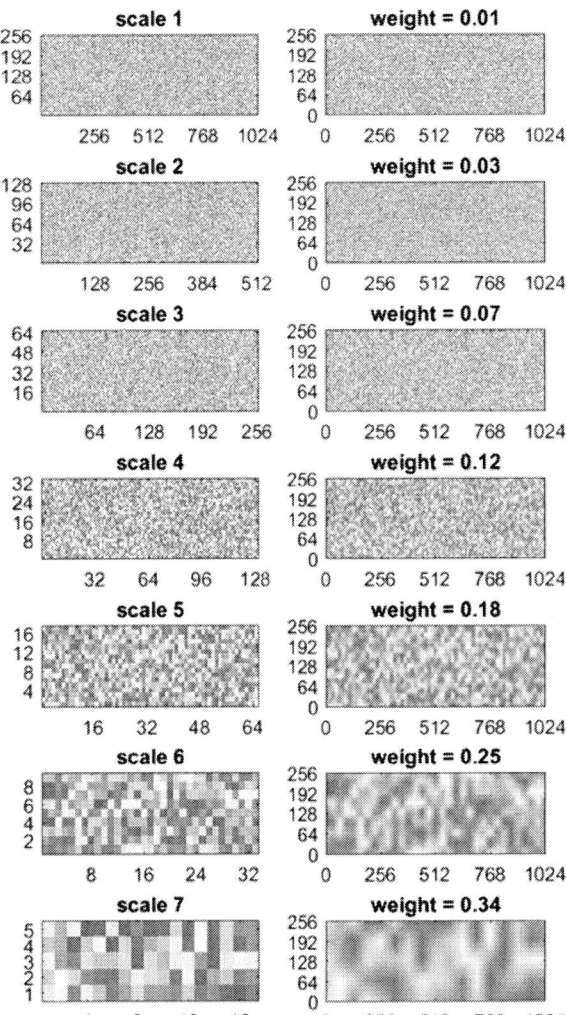

Figure 4: [Left] Finer to coarser (top to bottom) scales of random noise. [Right] Those random noise fields interpolated to the size of the finest random noise field (scale 1).

Next, each scale of random noise is linearly interpolated to a grid the same size as the finest grid (scale 1 in Figure 4). This results in a smooth field for the larger scales while retaining the more random field at the smaller scales, as seen in the right side of Figure 4.

### B. Initial Cloud Field

These interpolated fields are added together to create an initial cloud field, as seen Figure 5. Different weights are applied to the different interpolated fields. These weights are related to the solar variability: a higher weighting on the finer interpolated fields will lead higher variability since the resulting cloud field will be more jagged.

Here, we define scale weighting based on the variability score [8]. Specifically, weights are related to $i^{1/-\ln(VS)}$, where $i$ is the scale and VS is the variability score, though we are still determining the exact coefficients.

Figure 5: Initial cloud field created by summing all the interpolated fields (right plots in Figure 4).

### C. Cloud Mask

However, this initial cloud field does not look like actual sky conditions: values range from fully clear to fully cloudy without distinct cloud shapes. To obtain more distinct clouds, we create a cloud mask, which is based on the expected fraction of the sky covered by clouds (e.g., as found from hourly data).

The cloud mask is created by setting all values greater than kt to clear sky. For example, if kt=0.5, then roughly half the pixels in the cloud field will be set to clear sky. Figure 6 shows an example cloud mask for kt=0.5, and the resulting cloud field when the mask is applied. To apply the cloud mask, the two initial cloud fields are created. The first one is used to make the cloud mask, and the cloud mask is then applied to the second cloud field. If the cloud mask were applied to the same field as it was created from, there would be no values in the cloud field between kt and 1 (clear). This is especially a problem for low kt values, where, e.g., for kt=0.2, the clouds would all be very opaque (no values would be generated between 0.2 and 1).

### D. Variation with VS and kt

We have defined the cloud fields to be a function of VS and kt. Figure 7 shows example cloud fields for a variety of VS and kt combinations. As VS increases, indicating more variability, the clouds get smaller. As kt increases, indicating less of the sky is covered by clouds, we see more clear sky.

Also included in Figure 7 is a clear-sky index sample. This was generated by sampling a complete row from each cloud field. Since the cloud fields have values ranging from 0-1, they are analogous to clear-sky index values: 0 is fully cloudy and 1 is fully clear.

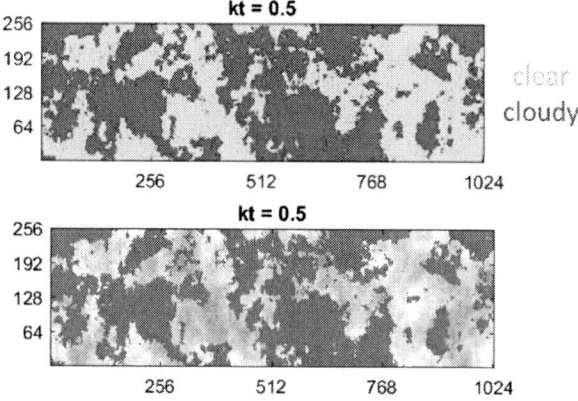

Figure 6: [Top] Cloud mask. [Bottom] Resulting cloud field after mask is applied.

These clear-sky samples again confirm that cloud fields with higher VS have higher variability, and that cloud fields with higher kt values tend to have fewer clouds. However, the clear-sky index samples in Figure 7 are not fully realistic. They tend to be either 1 (clear) or a value much less than 1 (cloudy), instead of having smooth transitions from clear to cloudy. Additionally, cloudy areas are highly opaque, which is not representative of thin clouds which only slightly reduce the irradiance reaching the surface.

We continue to modify the cloud field creation methodology, including adjusting the scale weights, applying targeted smoothing, and allowing for different cloud types and opacities [9]. The goal is to be able to match the statistics of ground measured irradiance.

### E. Sampling from Cloud Fields

Since the eventual goal of the cloud field methodology is to create unique PV samples for distribution grid studies, we need to be able to sample timeseries from the cloud fields. We do this by assigning a length scale to the cloud field (e.g., one pixel is one meter), and by advecting the clouds based on the cloud speed. For example, for a 5 m/s cloud speed with 1m pixels, we would sample every 5 pixels to generate a 1-second resolution timeseries.

Examples of samples at various cloud speeds are shown in Figure 8. Fast clouds speeds lead to lower correlation among different PV sites [10], and we do see that behavior in the samples from the cloud fields.

However, a side effect of sampling at different intervals is that the cloud speed is directly impacting the variability of each individual location: slower cloud speeds lead to less variability. The variability at each individual location should depend only on the VS (not the cloud speed), so this interdependency will need to be addressed in future iterations.

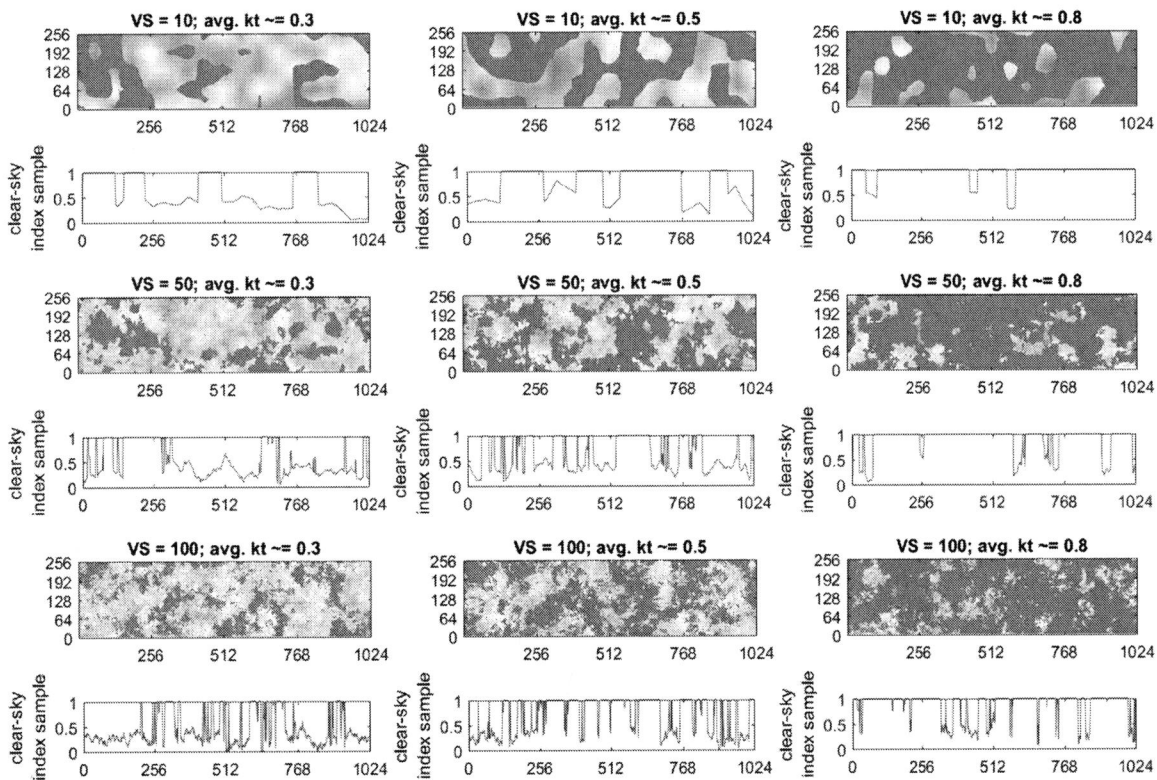

Figure 7: Cloud fields created based on each VS and kt input.

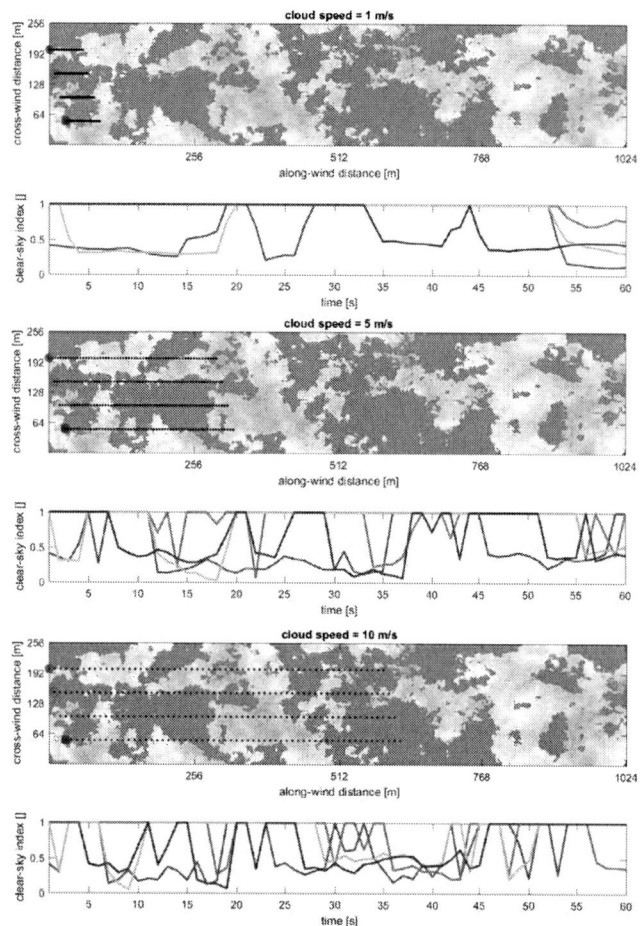

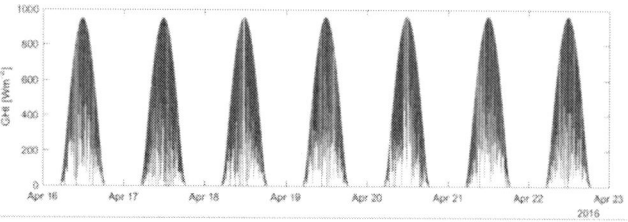

Figure 9: Sample GHI timeseries derived from cloud fields.

PV modules to be simulated, and then by using an irradiance to power model to convert to power output.

Figure 10 shows PV power samples over a day. The VS and kt were varied over different hours of the day, so some periods are fully clear while others are cloudy. Again, we notice the "on-off" behavior of the power from the single location. However, when timeseries are sampled at hundreds of locations (corresponding to the hundreds of different transformer locations on the feeder), the aggregate output is much smoother and looks more realistic. As described in Section IV, we have ongoing tests to evaluate the need for accurate distributed PV inputs. For analysis such as voltage regulator tap changes, it may not be important that a single customer be accurately portrayed because the regulator will only see the aggregate output of several PV systems.

Figure 8: Timeseries created from sampling the same cloud field at different intervals corresponding to different cloud speeds.

## F. Convert to GHI and Power

The samples described in Section III. E. are analogous to clear-sky index samples. They can be converted to GHI by multiplying by a clear-sky index (e.g., [10]). Figure 9 shows a sample GHI timeseries created with this method. The "on-off" behavior of the clear-sky index samples is again seen in the GHI samples.

These GHI samples can be converted to power output samples by using a decomposition and transposition model to convert to plane of array (POA) irradiance in the plane of the

## IV. UNIQUE PV PROFILES IMPACT ON DISTRIBUTION STUDIES

Previous work has shown the value of the high-frequency solar inputs discussed in Section II.B. to distribution grid studies: a 20% error [11] and 70% error [2] were found in computing voltage regulator tap change operations when using hourly PV samples instead of high-frequency samples. Here, we additionally show the value of using unique PV inputs across the feeder (discussed in Section II.C.), instead of assuming the same PV power profile at all locations. Figure 11 shows voltage regulator timeseries for a week-long simulation for both the case of a single irradiance profile used at all PV interconnection points and for a unique irradiance profile used at each interconnection point. The unique irradiance profiles were created based on 8 ground measurements, then were spread across the feeder using the first method described in section II. C. (cloud speed based time shifting). The test feeder and shifting method are described in more detail in [12]. The

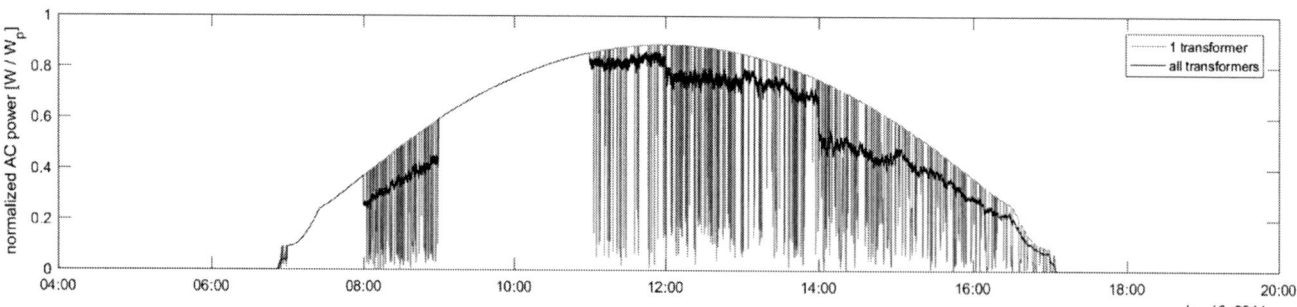

Figure 10: Sample power output timeseries for one location (blue), and for the aggregate of all locations on the feeder (black).

result is that the unique irradiance profiles resulted in ~30% fewer tap change operations. Thus, in order to accurately determine PV impacts such as voltage regulator tap change operations, it is important to generate spatially-unique irradiance profiles.

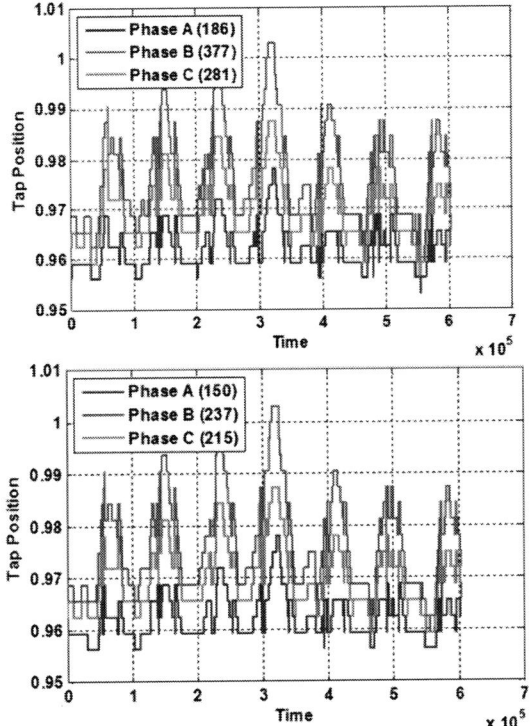

Figure 11. Voltage regulator tap change operations in a sample week for [top] a single irradiance profile used at all interconnection points and [bottom] unique irradiance profiles used at each interconnection point.

## V. CONCLUSIONS AND FUTURE WORK

We have shown the need for unique, high-frequency solar PV samples in quasi-static time series simulations (QSTS) of distribution grid impacts of PV, and laid out the synthetic cloud field method. Additional tweaks to the cloud field methodology are needed to make the sampled timeseries better match measured irradiance data. Additionally, the method should be further demonstrated with QSTS simulations to show its value for simulating high penetrations of distributed PV when no or limited ground data is available.

## ACKNOWLEDGMENT

This research was supported by the DOE SunShot Initiative, under agreement 30691. Sandia National Laboratories is a multimission laboratory managed and operated by National Technology and Engineering Solutions of Sandia, LLC, a wholly owned subsidiary of Honeywell International, Inc., for the U.S. Department of Energy's National Nuclear Security Administration under contract DE-NA0003525. Report number SAND2017-5646 C.

## REFERENCES

[1] B. Palmintier, R. Broderick, B. Mather, et al., "On the Path to SunShot: Emerging Issues and Challenges in Integrating Solar with the Distribution System," National Renewable Energy Laboratory, NREL/TP-5D00-65331, 2016.

[2] M. J. Reno, J. Deboever, and B. Mather, "Motivation and Requirements for Quasi-Static Time Series (QSTS) for Distribution System Analysis," IEEE PES General Meeting, 2017.

[3] A. Nguyen, M. Velay, J. Schoene, V. Zheglov, B. Kurtz, K. Murray, B. Torre, and J. Kleissl, "High PV penetration impacts on five local distribution networks using high resolution solar resource assessment with sky imager and quasi-steady state distribution system simulations", *Solar Energy*, 2016.

[4] J. M. Bright, O. Babacan, J. Kleissl, P. G. Taylor, and R. Crook, "A synthetic, spatially decorrelating solar irradiance generator and application to a LV grid model with high PV penetration", *Solar Energy*, 2017.

[5] M. Lave, R. J. Broderick, and M. J. Reno, "Solar Variability Zones: Satellite-Derived Zones that Represent High-Frequency Ground Variability," *Solar Energy*, 2017.

[6] J. S. Stein, C. W. Hansen, A. Ellis, and V. Chadliev, "Estimating annual synchronized 1-min power output profiles from utility-scale PV plants at 10 locations in Nevada for a solar grid integration study," in Proceedings of the 26th European Photovoltaic Solar Energy Conference and Exhibition, 2011.

[7] K. Perlin, "Implementing improved perlin noise," GPU Gems, pp. 73-85, 2004.

[8] .Lave, M. J. Reno, and R. J. Broderick, "Characterizing Local High-Frequency Solar Variability and its Impact to Distribution Studies," *Solar Energy*, 2015.

[9] M. J. Reno and J. S. Stein, "Using Cloud Classification to Model Solar Variability," in ASES National Solar Conference, Baltimore, MD, 2013.

[10] M. Lave, J. S. Stein, and A. Ellis. "Analyzing and simulating the reduction in PV powerplant variability due to geographic smoothing in Ota City, Japan and Alamosa, CO." IEEE Photovoltaic Specialists Conference (PVSC), 2012.

[11] M. Lave, J. Quiroz, M. Reno, and R. Broderick, "High Temporal Resolution Load Variability Compared to PV Variability," IEEE Photovoltaic Specialists Conference (PVSC), 2016.

[12] M. J. Reno, M. Lave, J. E. Quiroz, and R. J. Broderick, "PV Ramp Rate Smoothing Using Energy Storage to Mitigate Increased Voltage Regulator Tapping," IEEE Photovoltaic Specialists Conference (PVSC), 2016.

# A Direct Maximum Power Point Search Using Current-Voltage Based Power-Law Relation for Photovoltaic System under Uniform Irradiance

Hitesh K. Mehta, Member, IEEE and Ashish K. Panchal, Member, IEEE

Sarvajanik College of Engineering & Technology, Surat, Gujarat, 395001, INDIA

Sardar Vallabhbhai National Institute of Technology, Surat, Gujarat, 395007, INDIA

*Abstract* — **An output power of a photovoltaic source varies nonlinearly with atmospheric conditions. P&O and IC are most popular algorithms to track maximum power point of a PV module. In steady state operation, power oscillates around MPP due to its perturbing nature. To overcome this, researchers demonstrated direct MPP search algorithms. This paper proposes a direct MPP search algorithm using an *I-V* based power-law relation; this algorithm has not yet addressed in the literature. The simulation study is carried out for a 160 W polycrystalline silicon module PV system. Algorithm outperforms than the P&O under different irradiance and sudden load change.**

*Index Terms* — **Current voltage relation, Maximum power point tracking, photovoltaic system.**

## I. INTRODUCTION

Power era from the sun powered photovoltaic (PV) is a standout amongst the well-known renewable energy innovations. Power extracted from a PV module depends on day-to-day operating conditions such as a solar irradiance, a temperature and a connected load. Under different operating conditions, the PV modules used in stand-alone or grid-connected PV systems have to deliver maximum power for their full utilization. The maximum power is transferred between the PV modules and the load through an interfacing DC-DC boost converter driven by a maximum power point (MPP) tracking controller. At the MPP of *P-V* curve, the derivative of $P$ with respect to $V$ is zero, i.e. $dP/dV = 0$. The MPP algorithm generates a pulse-width-modulation signal to drive the power converter at the MPP. The most favorable algorithms are a perturb-and-observe (P&O) and an incremental conductance (IC) because of their simplicity and non-dependency on the operating conditions [1]. These algorithms are iterative process based algorithms so it does not converge to a single value of $D$ corresponding to the exact MPP but adjusts $D$ in a fixed step size ($\Delta D$). This results in oscillations of $D$ between two values ($D$ and $D\pm\Delta D$) near the MPP in the steady state operation. This oscillatory behavior is associated with additional power loss. Further, a large step size in $D$ makes the MPP searching process fast and a small step size in $D$ slows it down. Many algorithms have been developed to optimize the step size in $D$ to minimize the oscillations around the MPP [2]-[3]. Recently, a two-dimensional Gaussian function and arctangent based algorithm [4] has been proposed to determine a correct perturbation size in $D$. Though this algorithm performs better than the P&O, the MPP computational complexity is high. Even, the P&O

determines wrong $D$ and fails to track the MPP when the irradiance is suddenly increased. To resolve this problem, an advanced P&O algorithm was proposed in [5] wherein each P&O search is initiated after short-circuit current ($I_{SC}$) measurement, adding an additional step of $I_{SC}$ measurement. Yet, another solution to this problem is described in [6] which consider the sign of a change in current ($dI$) along with a sign change in power ($dP$) in the conventional P&O.

In order to establish a stable operation of a PV system, the power converter has to operate with a single value of $D$, which is not guaranteed by the aforementioned iterative algorithms. This problem can be solved by a pure mathematical approach. Here, the *I-V* relation of a PV module, based on a single-diode equivalent circuit model is used [7]. To determine the MPP using the *I-V* characteristic, a prior knowledge of the circuit parameters such as the photo current, the reverse saturation current, the series resistance ($R_S$) and the shunt resistance ($R_{SH}$) are required. However, the circuit parameters change their values with a change in operating conditions of the PV module. So, to predict the correct model and to find the MPP corresponding to the changed operating conditions, accurate measurements of the irradiance and the temperature are required. This further adds to deploy costly pyranometer and the temperature measuring equipments. Without measuring the irradiance, an MPP estimation algorithm [7]-[9] has been presented based on a simplified series-diode circuit, the algorithm still requires the measurement of bottom temperature of the PV module and also requires pre-training under different operating conditions for the validation before its use. In order to avoid the measurements of the irradiance and the temperature for the MPP calculation, a real-time identification of the *I-V/P-V* relation is required. Based on this approach in [10], an on-line *I-V* curve of a PV module is predicted using six ($V$, $I$) points measured near the MPP. However, the prior knowledge of $R_S$ is required to execute the algorithm and correct tracing of the six points near the MPP is mandatory. On the similar line, our previous works [11]-[12] calculate the MPP based on geometrically predicted *I-V/P-V* curves using the Lagrangian interpolation technique, the open circuit voltage $V_{OC}$ and the $I_{SC}$. However, these algorithms do not require the circuit parameters, but the frequent $V_{OC}$ and $I_{SC}$ measurements add the cost of a pilot module or intermittent disconnection of PV module. However, the selection of on-line ($V$, $I$) points near the MPP in these algorithms is trivial.

One more approach for the MPP determination uses the polynomial curve fitting (PCF) between $I$ and $V$. Here, for a silicon PV module, $I$ as a function of $V$ can be represented as 6-9th order polynomial [9]. First, the I-V PCF model parameters are determined using the on-line $(V, I)$ points and least square fit technique, afterwards the I-V polynomial is differentiated to set the MPP condition, i.e. $dI/dV + I/V = 0$. Finally, this equation is solved using the classical root finding techniques such as bisection, false position, secant and Newton-Raphson. Similarly, a 4-6th order PCF can also be established between $P$ and $V$ for a silicon PV module. The classical root finding methods are also employed to determine the MPP, i.e. $dP/dV = 0$. However, in the MPP determination, the P-V relation is more useful than the I-V relation as the former clearly shows the position of the MPP. Moreover, the numbers of the PCF model parameters to be calculated in case of P-V relation are less than that in the I-V relation. Chun et al. [13] showed that the bracketed methods (bisection and false position) perform better than the open bracketed methods (secant and Newton-Raphson) because the latter do not assure the MPP convergence.

The simple iterative and model based algorithms may fail to track the MPP under the effect of non uniform irradiance and does not generate the maximum output power. The P-V characteristic is nonlinear and exhibits multiple peaks.. These all algorithms must be modified before the application to the PV system working under nonuniform irradiance. However, recently, this problem has been solved by grey wolf optimization technique [14], search-skip-judge global MPPT and rapid global MPPT [15]. It is worth to note that the proposed algorithm assumes the uniform irradiance. Therefore, the performance of the proposed algorithm for nonuniform irradiance is out of the scope for this paper.

The P&O and IC algorithms work independent of the PV module operating conditions whereas equivalent circuit model based algorithms require accurate knowledge of the circuit parameters, $V_{OC}$ and $I_{SC}$. The real-time algorithms that consider the PCF models require selection of $(V, I)$ points and frequent estimation of the parameters for the varying operating conditions. This comprehensive review has motivated us to develop a new algorithm for the maximum power point tracking of a PV system. A intend of our proposed work is to build an MPP algorithm, which is independent of the operating conditions, the circuit model parameters and a few $(V, I)$ points.

The novelty of our algorithm lies in the selection of an analytical model relating $I$ with $V$ having only three parameters, i.e. $I = a \times V^b + c$, where $a$, $b$ and $c$ are the model parameters. The theoretical explanation of our algorithm is described in section II. To validate our algorithm, simulation and the experimental results are presented in section III and IV, respectively. These sections also compare our algorithm with the conventional P&O algorithm to demonstrate the feasibility of the proposed algorithm. Finally, section V concludes the analysis of our algorithm.

## II. THEORY

### A. Calculation of voltage at the MPP ($V_{MP}$)

As aforesaid, the P-V relation is more suitable than the I-V relation for the MPP determination, the I-V relation is preferred for our algorithm since it considers the MPP condition as $dI/dV + I/V = 0$. Our previous work [16] presented excellent fitting of an I-V curve for a Poly Crystalline Si module with a relation given by a power-law equation

$$I = a \times V^b + c \qquad (1)$$

The parameters $a$, $b$ and $c$ can be determined from the specifications of the module. Here, power-law based fitting I-V model is used to calculate the maximum power point voltage ($V_{MP}$).

$$\frac{dI}{dV} = -\frac{I}{V} \qquad (2)$$

Eq. (2) is written at the MPP ($V_{MP}$, $I_{MP}$) as

$$\left.\frac{dI}{dV}\right|_{MPP} = -\frac{I_{MP}}{V_{MP}} \qquad (3)$$

Eq. (1) is differentiated with respect to $V$ and is expressed at the MPP by

$$\left.\frac{dI}{dV}\right|_{MPP} = abV_{MP}^{b-1} \qquad (4)$$

Equating eq. (3) and (4), the following expression is generated

$$abV_{MP}^{b-1} = -\frac{I_{MP}}{V_{MP}} \qquad (5)$$

As eq. (1) is satisfied by all the points of the I-V curve of a PV module, eq. (1) at the *MPP* is replaced as

$$I_{MP} = a \times V_{MP}^b + c \qquad (6)$$

Substituting $I_{MP}$ from eq. (6) into eq. (5) and is simplified for $V_{MP}$ as follows

$$V_{MP} = \left(\frac{-c}{a + ab}\right)^{\frac{1}{b}} \qquad (7)$$

Thus, the analytical expression of eq. (7) is directly used for calculating reference $V_{MP}$ in terms of the fitting model parameters which is utilized in the MPP controller to decide the new duty cycle for the DC-DC converter.

### B. Validity of the fitting model

To validate the fitting model of eq. (1) for the 160 W PV module, we simulated the I-V curve of the module according to method depicted in [17]-[19] and compared with obtained fitting model. Both I-V curves are derived at the STC. The three model parameters are calculated using the three points $(0, I_{SC})$, $(V_{OC}, 0)$ and $(V_{MP}, I_{MP})$ as listed in the Table – I of the

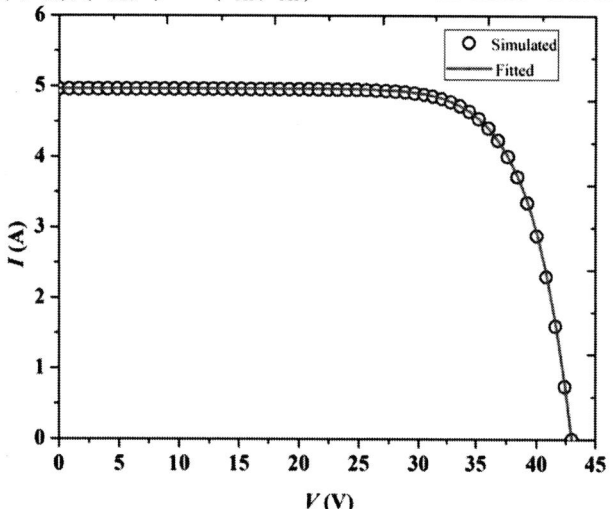

Fig.1 Simulated and the fitted *I-V* curves

160 W module. The model parameters are expressed in terms of the specifications as follows,

$$c = I_{SC} \qquad (8)$$

$$a = -\frac{I_{SC}}{V_{OC}^b} \qquad (9)$$

The value of *b* obtained using eq. (6,8 and 9)

$$b = \left[ \frac{\log\left(1 - \frac{I_{MP}}{I_{SC}}\right)}{\log\left(\frac{V_{MP}}{V_{OC}}\right)} \right] \qquad (10)$$

The acquired values with the above procedure are $a = -3.88691 \times 10^{-20}$, $b = 12.30870$ and $c = 4.96$ at the STC. To validate the model *I-V* curve is calculated as per eq. (1) using these parameters values. The simulated and the fitted *I-V* curves are ride on each other as plotted in Fig.1. Therefore, the result validates the model.

### C. Effect of change in irradiance on the fitting model parameters

The effect of irradiance on the fitting model parameters is analyzed by simulating the system ($0.1$–$1$kW/m$^2$) [17], [19]. The fitting model parameters found at different irradiance using eq. (8)-(10). Fig. 2 shows; the value of *a* and *b* at every irradiance, *a* and *b* has very less change in their value while the value of *c* linearly increases with the irradiance. So, the values of *a* and *b* are considered as constant in this work which are calculated at the STC. This reduces the computational time of MPPT controller. The value of *c* at any

irradiance is calculated from the measured *I* and *V* values as follows

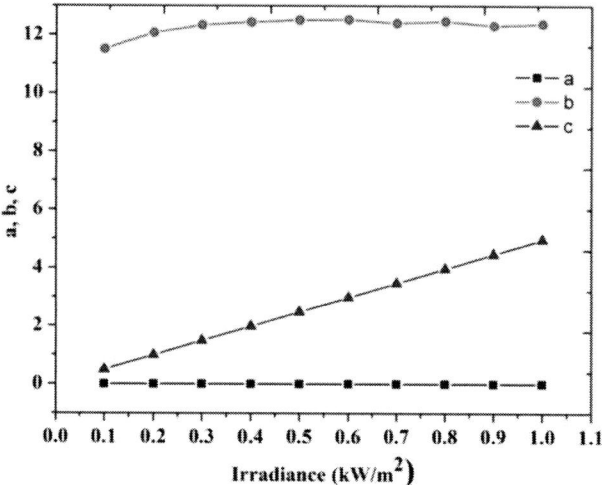

Fig. 2 *a, b* and *c* parameters at different irradiance

$$c = I - aV^b \qquad (11)$$

To strengthen the aforementioned concept, we calculated maximum power values for the module at different irradiance in two ways, (i) using constant values of *a* and *b* as per 1 kW/m$^2$ and (ii) using actual values of *a* and *b* calculated for respective irradiance. Both plots exactly match to each other. As plotted in Fig. 3. This further validates the concept of considering *a* and *b* values to be constant as per 1 kW/m$^2$ .The $V_{MP}$ is calculated using eq. (7). This algorithm does not require the measurement of $V_{OC}$ and $I_{SC}$ at any irradiance for calculating $V_{MP}$ thereby it eliminates the need of a pilot module for the MPP tracking or disconnection of module at any instant.

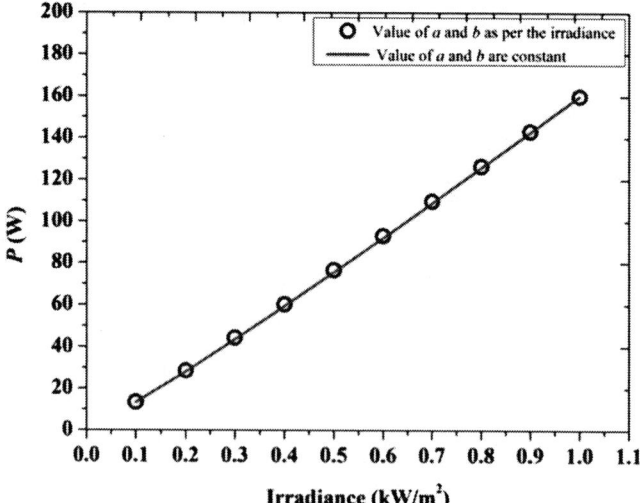

Fig. 3 Power (MPP) at different irradiance

978-1-5090-5606-4/17 $31.00 © 2017 IEEE

## III. NUMERICAL SIMULATION STUDY

The simulation model which is developed to acclaim our algorithm is shown in Fig. 4. It incorporates a PV module, a DC-DC converter, MPPT controller and linear load.

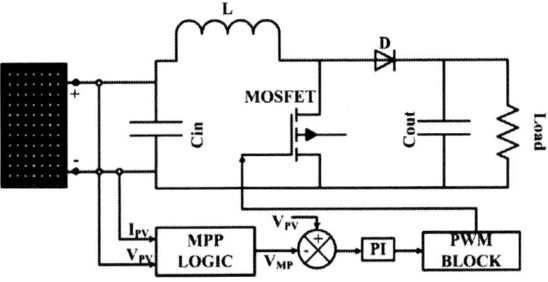

Fig. 4 PV system structure with MPPT control block
(L=297 µH, $C_{in}$ = $C_{out}$=470µF, $R_L$ = 34Ω)

### A. Variation in the irradiance

The model is simulated for the P&O and our algorithms for 3 sec. In order to analyze the response of our algorithm in fast changing irradiance, the following irradiance sequence is selected:(i) 1 kW/m$^2$ for 0 − 1 s (ii) 0.4 kW/m$^2$ for 1 − 2 s and (iii) again 1 kW/m$^2$ for 2 − 3 s (Fig. 5(*a*) and 5(*b*)).

### 1) P&O Algorithm:

Fig. 5(a) illustrates the simulation results of the P&O algorithm. For the first 1 kW/m$^2$ irradiance, the MPP is achieved at 0.3 s with $D$ fluctuating between 0.49 and 0.53 and power also oscillates around the MPP (148W − 159.9W). Under this condition, the PV module operates on load line-1 which is depicted as point D as shown in Fig 6. At t = 1 s, the irradiance is decreased to 0.4 kW/m$^2$, but $D$ of the converter does not change its value (0.49 − 0.53) instantaneously corresponding to 0.4 kW/m$^2$ irradiance. This shifts the PV module operating point to E (Fig. 7). Now, the P&O algorithm

further activates and settles the $D$ value between 0.27 and 0.31. The power oscillates near the new MPP around 58.5 W to 60.3 W. With a decrement in the irradiance, it takes 0.24 s to attain new MPP as shown in Fig 6(a) and the PV module operates on load line-2 (point F) as appeared in Fig 7. At t = 2 s, the irradiance is again increased to 1 kW/m$^2$ from 0.4 kW/m$^2$ and the PV module works at load line–2 (point G) as appeared in Fig 7. As the irradiance increases, the MPPT controller increases $D$ of the converter in steps to track the MPP. After 0.12 s, the MPP is tracked (Fig. 6(b)) and the PV module works on load line–1(point D in Fig. 7), the power oscillates around 148 W − 159.9 W.

### 2) Proposed Algorithm:

Fig. 5(b) shows the simulation responses of our algorithm. At first, the irradiance is 1 kW/m$^2$, the algorithm ascertains parameter $c$ using eq.(11) and corresponding $V_{MP}$ value is calculated using eq.(7). Our algorithm tracks the MPP in t = 0.11 s, the PV module works on load line-1 (point D). The $D$ of converter stays consistent at 0.5 and the power of the PV module is fixed to 159.8 W. At t = 1 s, the irradiance is decreased to 0.4 kW/m$^2$ whereas the $D$ of the converter is still 0.5 and the PV module works at load line-1 (point E). The algorithm calculates new value of parameter $c$ from eq.(11) and discover the new $V_{MP}$ from eq.(7). With this, an error signal is changed and algorithm determines the new $D$ (0.3) for the converter which directly settles the new MPP (60 W) in t = 0.04 s which is very fast compared to the P&O algorithm. The PV module operates on the load line-2(point F) as shown in Fig. 7. Similarly, at t = 2 s, the irradiance is increased to 1 kW/m$^2$ from 0.4 kW/m$^2$ and the PV module operates at load line-2 (point G) as shown in Fig. 7. As irradiance is increased, again as per the new error parameter $c$ and $V_{MP}$ value is calculated and new $D$ (0.5) of the converter straight away settles the new MPP (159.8W) in t = 0.02 s (Fig.6(b)). The PV module operates on load line-1(point D) at this condition shown in Fig 7.

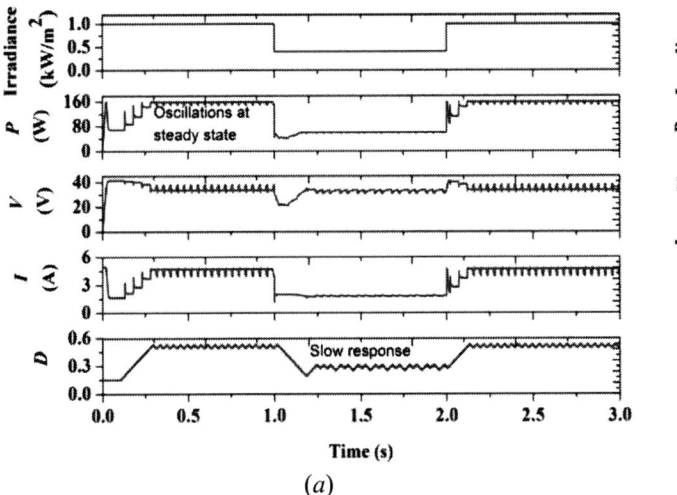

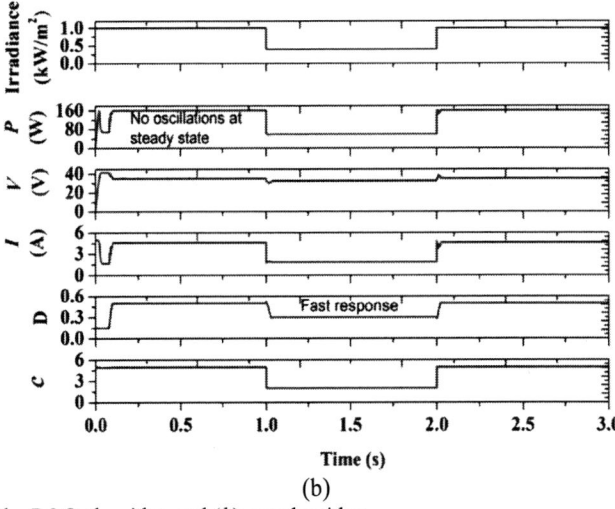

Fig. 5 Simulation response of *P, V, I* and *D* at different irradiance for (*a*) the P&O algorithm and (*b*) our algorithm

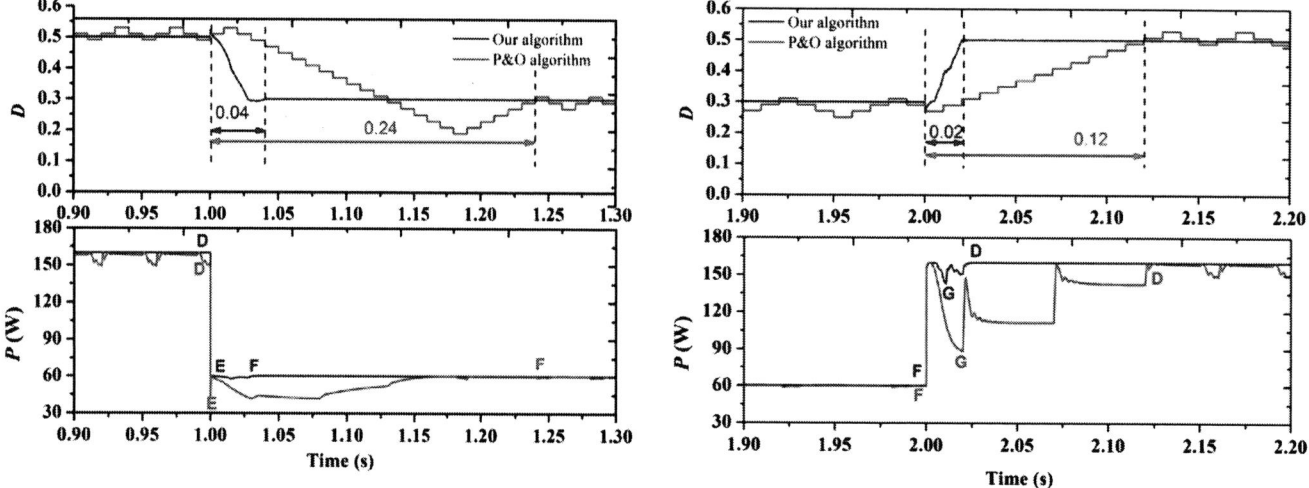

Fig. 6 Close view of $D$ and $P$ for ($a$) decreasing the irradiance from 1 kW/m$^2$ to 0.4 kW/m$^2$ and ($b$) increasing the irradiance from 0.4 kW/m$^2$ to 1 kW/m$^2$ for both algorithms

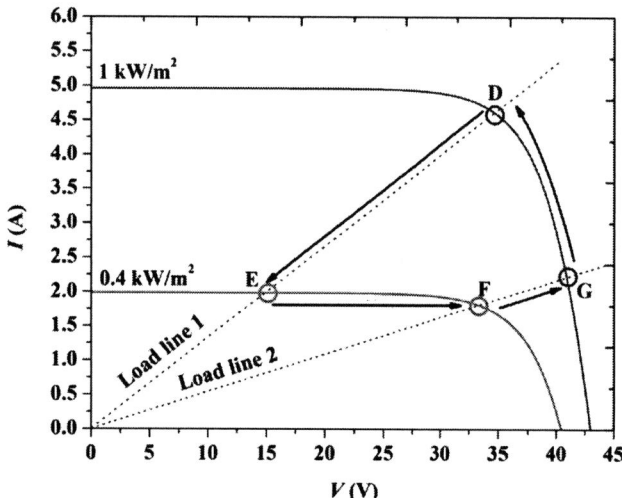

Fig. 7 Load lines on the $I$-$V$ curves for the irradiance of 1 kW/m$^2$ and 0.4 kW/m$^2$

### B. Variation in the load

In order to examine the performance of our algorithm against a sudden load change, the resistance value is changed from 34 Ω to 17 Ω at 1 s with the irradiance of 1 kW/m$^2$. Fig 8(a) and 8(b) show the simulation responses for the P&O and our algorithms, respectively. As load changes with the same irradiance, the value $c$ and $V_{MP}$ are unaltered so that $P$, $V$ and $I$ do not change in the proposed algorithm. Only $D$ is adjusted to a new value to attain the MPP. The dynamic response time is 0.03s with the proposed algorithm while P&O take 0.11 s. With this observation; the proposed algorithm dynamic response is faster than the P&O under sudden load change condition with the same irradiance.

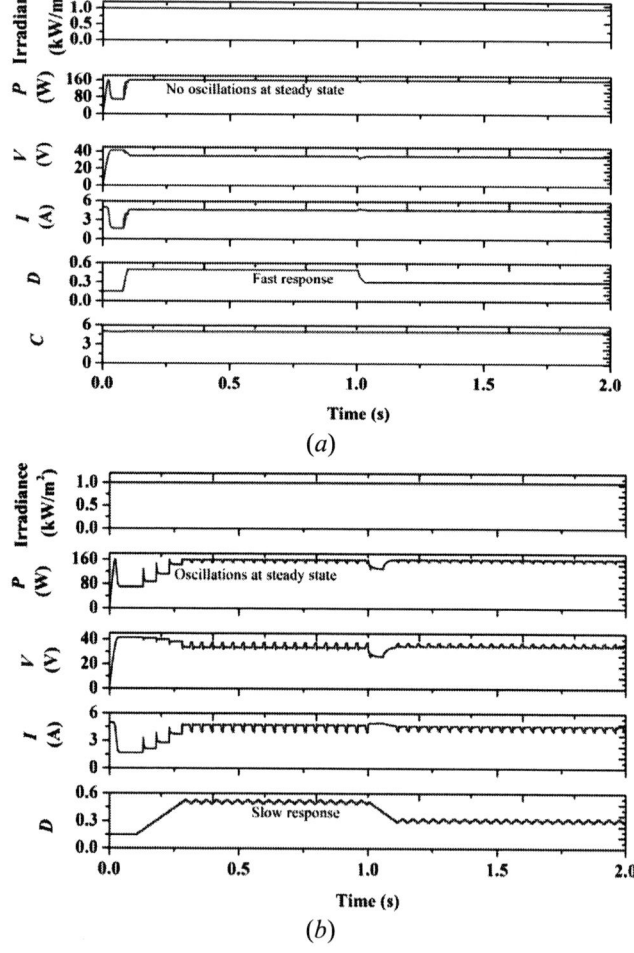

Fig 8 Simulation response of $P$, $V$, $I$ and $D$ at different load ($a$) the P&O algorithm and ($b$) our algorithm

## IV. CONCLUSION

A direct MPP determination technique using *I-V* based power-law relation of a silicon PV module under uniform irradiance is developed. The equivalence between the module characteristics and the proposed power-law relation is established and validated. Out of three coefficients of power-law relation, two coefficients (*a* and *b*) are constant for all irradiance and third coefficient *c* is calculated with measured operating point which represents the short circuit current for that irradiance. The proposed algorithm is examiner for a PV system using simulation and it attains the steady state operation more rapidly than the P&O. Also dynamic response is faster compared to the P&O under sudden load change condition. This proposed algorithm is more efficient and easy to implement for real time application. The proposed MPP tracking algorithm can also be extended for large PV array system.

## APPENDIX

Table – I Specifications of 160W PV module

| PV Module | Value at STC |
|---|---|
| Short-circuit current ($I_{SC}$) | 4.96 A |
| Open-circuit voltage ($V_{OC}$) | 43 V |
| Maximum power ($P_{MP}$) | 159.8 W |
| Current at MPP ($I_{MP}$) | 4.58 A |
| Voltage at MPP ($V_{MP}$) | 34.9 V |

## ACKNOWLEDGEMENT

The author would like to thank Dr. Ashish K. Panchal, Dr. R. Chudamani and Dr. Vipul Kheraj for their valuable guidance in this research work.

## REFERENCES

[1] D. Sera, L. Mathe, T. Kerekes, S. V. Spataru and R. Teodorescu, "On the perturb-and-observe and incremental conductance MPPT methods for PV systems,"*IEEE Journal of Photovoltaics*, vol. 3, no. 3, pp. 1070-1078, July 2013.

[2] M. F. N. Tajuddin, M. S. Arif,S. M. Ayob and Z. Salam "Perturbative methods for maximum power point tracking (MPPT) of photovoltaic (PV) systems: a review,"*International Journal of Energy Research,* vol. 39, n0. 12, pp. 1720, August 2015.

[3] Boualem Bendib, Hocine Belmili, Fateh Krim, "A survey of the most used MPPT methods: conventional and advanced algorithms applied for photovoltaic systems,"*Renewable and Sustainable Energy Reviews*, vol. 45, pp. 637–648, May 2015.

[4] S. M. Reza Tousi, M. H. Moradi, N. S. Basir and M. Nemati, "A function-based maximum power point tracking method for photovoltaic systems,"*IEEE Transactions on Power Electronics*, vol. 31, no. 3, pp. 2120-2128, March 2016.

[5] D. C. Huynh, T. A. T. Nguyen, M. W. Dunnigan and M. A. Mueller, "Maximum power point tracking of solar photovoltaic panels using advanced perturbation and observation algorithm," *IEEE 8th Conference on Industrial Electronics and Applications (ICIEA)*, Melbourne, VIC, 2013, pp. 864-869, 2013.

[6] M. Killi and S. Samanta, "Modified perturb and observe MPPT algorithm for drift avoidance in photovoltaic systems,"*IEEE Transactions on Industrial Electronics*, vol. 62, no. 9, pp. 5549-5559, September 2015.

[7] L. Cristaldi, M. Faifer, M. Rossi and S. Toscani, "An improved model-based maximum power point tracker for photovoltaic panels,"*IEEE Transactions on Instrumentation and Measurement*, vol. 63, no. 1, pp. 63-71, January 2014.

[8] W. Xiao, A. Elnosh, V. Khadkikar and H. Zeineldin, "Overview of maximum power point tracking technologies for photovoltaic power systems," *IECON 2011 - 37th Annual Conference on IEEE Industrial Electronics Society*, Melbourne, VIC, pp. 3900-3905, 2011.

[9] W. Xiao, M. G. J. Lind, W. G. Dunford and A. Capel, "Real-time identification of optimal operating points in photovoltaic power systems," *IEEE Transactions on Industrial Electronics*, vol. 53, no. 4, pp. 1017-1026, June 2006.

[10] J. M. Blanes, F. J. Toledo, S. Montero and A. Garrigós, "In-site real-time photovoltaic *I–V* curves and maximum power point estimator," *IEEE Transactions on Power Electronics*, vol. 28, no. 3, pp. 1234-1240, March 2013.

[11] G. Kumar and A. K. Panchal, "Geometrical prediction of maximum power point for photovoltaics," *Applied Energy*, vol. 119, pp. 237-245, January 2014.

[12] G. Kumar, M. B. Trivedi and A. K. Panchal, "Innovative and precise MPP estimation using *P-V* curve geometry for photovoltaics," *Applied Energy*, vol. 138, pp. 640-647, January 2015.

[13] S. Chun and A. Kwasinski, "Analysis of classical root-finding methods applied to digital maximum power point tracking for sustainable photovoltaic energy generation," *IEEE Transactions on Power Electronics*, vol. 26, no. 12, pp. 3730-3743, December 2011.

[14] S. Mohanty, B. Subudhi and P. K. Ray, "A new MPPT design using grey wolf optimization technique for photovoltaic system under partial shading conditions," *IEEE Transactions on Sustainable Energy*, vol. 7, no. 1, pp. 181-188, Jan. 2016.

[15] Y. Wang, Y. Li and X. Ruan, "High-accuracy and fast-speed MPPT methods for PV string under partially shaded conditions," *IEEE Transactions on Industrial Electronics*, vol. 63, no. 1, pp. 235-245, January 2016.

[16] S. J. Patel, G. Kumar, A. K. Panchal and V. Kheraj, "Maximum power point computation using current-voltage data from open and short circuit regions of photovoltaic module: A teaching learning based optimization approach," *Journal of Renewable and Sustainable Energy*, vol. 7, pp. 043112, July 2015.

[17] M. G. Villalva, J. R. Gazoli and E. R. Filho, "Comprehensive approach to modeling and simulation of photovoltaic arrays," *IEEE Transactions on Power Electronics*, vol. 24, no. 5, pp. 1198-1208, May 2009.

[18] M. G. Villalva, J. R. Gazoli and E. R. Filho, "Modeling and circuit-based simulation of photovoltaic arrays," 2009 *Brazilian Power Electronics Conference*, Bonito-Mato Grosso do Sul, pp. 1244-1254, 2009.

[19] N. Pandiarajan and R. Muthu, "Mathematical modeling of photovoltaic module with Simulink," *Electrical Energy Systems (ICEES)*, 1st International Conference on, Newport Beach, CA, 2011, pp. 258-263, 2011.

# Passivity Based Controller for Photovoltaic Modules Using Zeta Converter

Daniel A. Merced Cirino[1,2], Rachid Darbali Zamora[2] and Eduardo I. Ortiz Rivera[2]

[1]Electrical Engineering and Computer Science Department, University of Tennessee, Knoxville,
Knoxville, Tennessee, 37996, USA

[2]Electrical and Computer Engineering Department, University of Puerto Rico, Mayaguez Campus,
Mayaguez, Puerto Rico, 00682, USA

*Abstract*—The use of renewable energy is becoming more and more common for the use of clean energy and applications in remote locations where there is no easy access to electricity such as satellites in space or desert areas. When working with solar panels DC to DC converters may be used to adapt the panels voltage and current to the circuit as well as for load regulation. Since these circuits and solar panels contain nonlinear behavior, control methods such as Passivity Based Control (PBC) can be applied. This paper focuses on the use of a Zeta converter and PBC with solar panels as input voltage. A system is simulated with a resistive load and PBC is applied for Maximum Power Point Tracking (MPPT) as well as for load regulation.

*Index Terms*—DC/DC Converters, MPPT, Nonlinear Control, Solar Panels, Zeta Converter.

## I. Introduction

Limits on the amount of fossil fuels and the need for clean energy worldwide have caused interest in different forms of renewable energy options. Amongst these options solar energy is one of the most studied and being applied in modern times. Photovoltaic Modules (PVM) convert solar energy directly to electricity by means of the photoelectric effect yet sunlight is not a constant value for solar panels. Solar panels can be used in different applications ranging from satellites in space [1] , to solar vehicles, to providing energy to remote locations. These applications require control of the solar panels output and to do this DC to DC converters are used. These converters use transistors, inductors and capacitors to convert a DC voltage level into another DC voltage level by means of power conversion.

Amongst DC to DC converters, three main topologies exist: Buck, Boost and Buck-Boost converters. This paper focuses on the Zeta converter which is a DC to DC converter based on the buck-boost topology. When modeling these converters, and including solar panel models, nonlinear control theories may be applied such as Feedback Linearization (FL) [2], Sliding Mode Control (SMC) [3] [4] or Passivity Based Control (PBC) [5]. This paper focuses on PBC and how it can be applied to regulate output voltage of the converter or for Maximum Power Point Tracking (MPPT). MPPT is important when working with solar panels because it is a way to ensure that the PVM is providing the maximum power to the circuit when required.

This article is organized in the following manner: section 2 presents the Photovoltaic module model. Section 3 describes the DC to DC converter. Section 4 shows the control laws Section 5 presents the simulation results. Finally Section 6 presents the conclusions for this work.

## II. Photovoltaic Module Model

In the analysis, design and simulation of electrical circuits there are various different models that describe the behavior of Photovoltaic Modules (PVM) and their dynamic behavior. Most of these models require values not included in a PVMs datasheet and can be hard to calculate or find. For the purpose of analysis and simulation the Fractional Polynomial Model (FPM) is used [6] [7]. This model uses values found on the manufacturers datasheet under Standard Test Conditions (STC) of the photovoltaic modules to determine the solar panels curves. These parameters are $1000W/m^2$, $25°C$ and 1.5 air mass. The power equation is found by multiplying equation (1) by the PVM voltage.

Since the parameters given by the manufacturers datasheet are given at STC and different conditions may apply when implementing the PVM, the new values for open circuit voltage and short circuit current may be determined in the following manner.

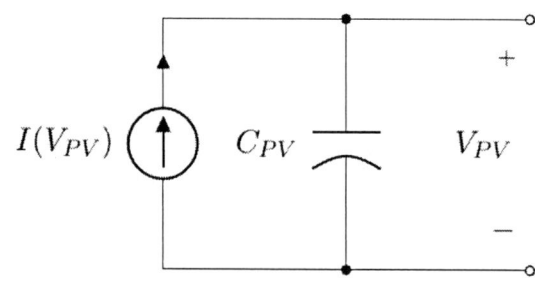

Fig. 1: PVM dynamic model. The circuit schematic for the PVM is composed of a current source in parallel with a capacitor.

978-1-5090-5606-4/17 $31.00 © 2017 IEEE

$$I(V) = I_x \left[ 1 - \left( \frac{V}{V_x} \right)^{n+q} \right] \qquad (1)$$

In this model $V_x$ is the open circuit voltage and $I_x$ is the short circuit current. There is a term n+q, in this term n is a positive integer, while the q is a non-integer and is a value between 0 and 1, not including the boundaries. V is the PVM voltage while I(V) is the PVM current. To determine the value of the constant n+q, some values are required from the PVMs manufacturers datasheet. The values required are: open circuit voltage, short circuit current, voltage at maximum power, current at maximum power and maximum power at STC. This is shown in equation (2).

$$n + q = \ln \left( \frac{I_{sc} - I_{op}}{I_{sc}} \right) / \ln \left( \frac{V_{op}}{V_{oc}} \right) \qquad (2)$$

Equation (3) and equation (4) determine the value for open circuit voltage and for the short circuit current at any temperature and irradiance level respectively.

$$V_x = \frac{E_i}{E_{in}} TCV(T - T_N) + V_{max}$$
$$- (V_{max} - V_{min}) exp \left[ \frac{E_i}{E_{in}} \ln \left( \frac{V_{max} - V_{oc}}{V_{max} - V_{min}} \right) \right] \qquad (3)$$

$$I_x = \frac{E_i}{E_{in}} [I_{sc} + TCi(T - T_N)] \qquad (4)$$

In this equation TCi is the temperature coefficient of $I_{sc}$ in $A/^{\circ}C$. $E_{iN}$ is the nominal irradiance, $E_i$ is the irradiance level, T is the temperature, $T_N$ is the nominal temperature, $V_{min}$ is the open circuit voltage at $200W/m^2$ and $25^{\circ}C$, approximately $0.85V_{oc}$, $V_{max}$ is the open circuit voltage at $1,200W/m^2$ and $25^{\circ}C$, approximately $1.05V_{oc}$ and TCV is the temperature coefficient of $V_{oc}$ in $V/^{\circ}C$. The power equation is found by multiplying equation (1) by the PVM voltage. This is shown in equation (5).

$$P(V) = V \cdot I(V) = V \cdot I_x \left[ 1 - \left( \frac{V}{V_x} \right)^{n+q} \right] \qquad (5)$$

When operating in the optimal current and voltage values, the maximum power can be extracted. This is shown in equation (6).

$$P_{max} = V_{op} \cdot I_{op} = V_{op} \cdot I_{sc} \left[ 1 - \left( \frac{V_{op}}{V_{oc}} \right)^{n+q} \right] \qquad (6)$$

Finally two additional equations that can be observed for a PVM resistance and conductance. These equations come from dividing PVM voltage by PVM current and multiplying PVM Voltage and PVM Current respectively. These can be seen in equations (7) and (8).

$$R(V) = \frac{V}{I(V)} = \frac{V}{I_x \left[ 1 - \left( \frac{V}{V_x} \right)^{n+q} \right]} \qquad (7)$$

$$G(V) = \frac{I(V)}{V} = \frac{I_x \left[ 1 - \left( \frac{V}{V_x} \right)^{n+q} \right]}{V} \qquad (8)$$

An example of how the curves are represented by this model at varying levels of irradiance while maintaining a constant temperature level is shown in Figs. 2 through 5. Equations (1),(5), (7) and (8) are used and the optimal values for voltage, current, power, resistance and conductance have been marked for reference.

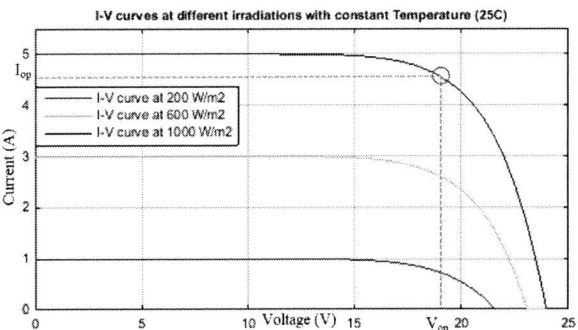

Fig. 2: I-V curves at different irradiance levels.

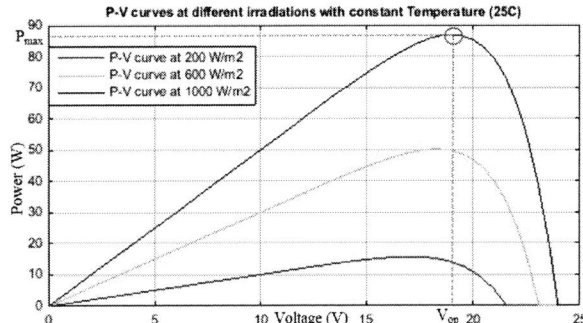

Fig. 3: P-V curves at different irradiance levels.

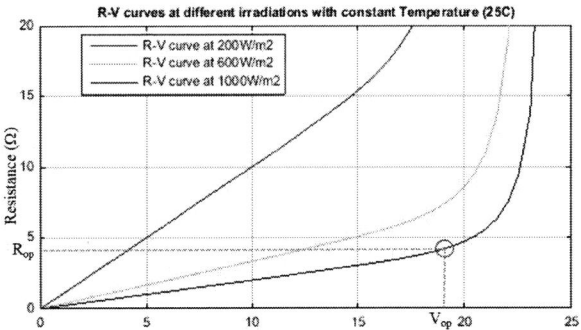

Fig. 4: R-V curves at different irradiance levels.

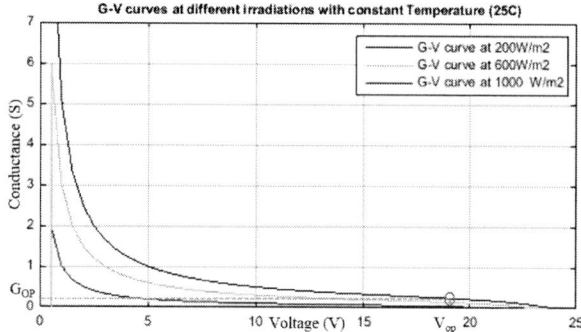

Fig. 5: G-V curves at different irradiance levels.

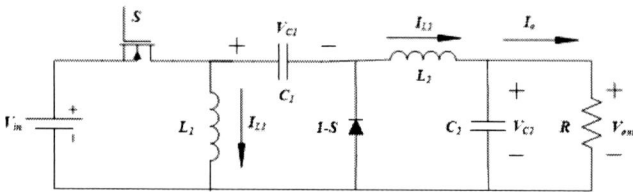

Fig. 6: ZETA Converter Schematic symbol.

In order to obtain the PVM capacitor, equation (9) can be used.

$$\frac{dV}{dt} = \frac{1}{C_{PV}}[I(V_{PV}) - I_{in}] \qquad (9)$$

The dynamic model of the PVM connected to the circuit is accomplished by using a capacitor in parallel as shown in figure 1. Equation (9) therefore describes this model. Equation (9) is the representation of Kirchhofs current law applied a the highest node in this circuit. As a presumption this PVM is said to be connected to any load which would require a current $I_{in}$. The other variables in this model are, $V_{PV}$, which is the PVM voltage, $C_{PV}$ is the capacitor connected in parallel to the PVM and $I(V_{PV})$ is the PVM current as presented in equation (1).

### III. DC TO DC CONVERTER

The Zeta converter, also known as the inverse SEPIC [8], is a DC to DC converter which includes the use of two capacitors and two inductors [9] [11]. This circuit is capable of increasing or deacreasing the input voltage, without inverting its polarity, which makes it a buck boost topology. The reason for selecting the zeta converter compared to other converters such as the SEPIC converter is that it has a lower output ripple voltage [11] [12], but unlike the SEPIC it includes a high side transistor which may require a more complex driver. Figure 6 represents the schematic for the zeta converter and equations ( 10 ) (14) represent its dynamic model.

From Fig 6 it can be seen where the variable names come from in the circuit which are: $V_{in}$ is the input voltage, $V_{c1}$ is the voltage in the capacitor $C_1$, $I_{L1}$ is the current through

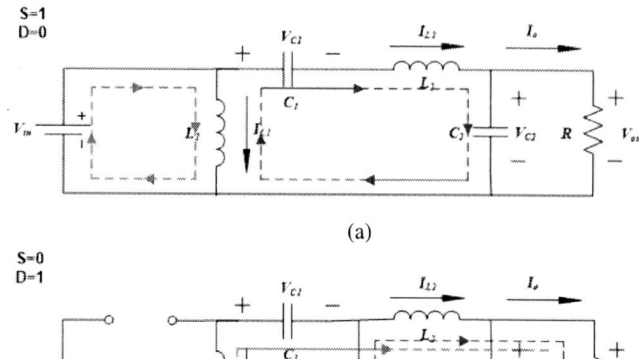

(a)

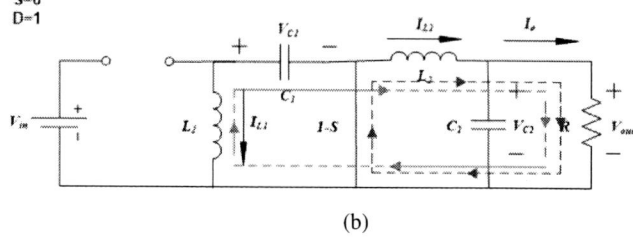

(b)

Fig. 7: ZETA Converter Switching Stages. (a) Equivalent circuit schematic for the ZETA converter when S closed. (b) Equivalent circuit schematic for the ZETA converter when S is open.

inductor $L_1$ and $I_{L2}$ is the current through inductor $L_2$ and $V_{c2}$ is the same as the output voltage, $V_{out}$, which is the voltage in output capacitor $C_2$. $S$ is the state switch used, this may be any switching device capable of being controlled at high switching frequencies. The switch is used as the input for this circuit and exists only as a 0 or a 1 ($S \in [0\ 1]$).

Once these equations are in steady state the derivative are equal to zero. By combining equation (10) and (11) in steady state it can be seen that the capacitors have the same voltage. Thus by replacing this condition into equation (11) a relationship between the input and output voltage is achieved. If pulse width modulation (PWM) with a high switching frequency is used for the converter then the signal can be seen as the duty cycle for the converter.

$$L_1 \frac{dI_{L_1}}{dt} = -V_{C_1} + S(V_{in} + V_{C_1}) \qquad (10)$$

$$L_2 \frac{dI_{L_2}}{dt} = -V_{C_2} + S(V_{in} + V_{C_1}) \qquad (11)$$

$$C_1 \frac{dV_{C_1}}{dt} = I_{L_1} - S(I_{L_1} + I_{L_2}) \qquad (12)$$

$$C_2 \frac{dV_{C_2}}{dt} = I_{L_2} - \frac{V_{C_2}}{R} \qquad (13)$$

$$I_{in} = S(I_{L1} + I_{L2}) \qquad (14)$$

The same procedure can be repeated for the relationship between the input and the output current but using equations (12), (13) and (14). From equation (12) in steady state with (14) it can be seen that the input current is equal to the current in inductor $L_1$ and from (13) it can be seen that the current in inductor $L_2$ is equal to the output current. Then

$$y = h(x) \tag{20}$$

$$\overline{I_{L_1}} = I_x \left[ 1 - \left( \frac{V_r}{V_x} \right)^{n+q} \right] \tag{21}$$

$$\overline{I_{L_2}} = \frac{\sqrt{RV_r I_x \left[ 1 - \left( \frac{V_r}{V_x} \right)^{n+q} \right]}}{R} \tag{22}$$

$$\overline{V_{C_1}} = \sqrt{RV_r I_x \left[ 1 - \left( \frac{V_r}{V_x} \right)^{n+q} \right]} \tag{23}$$

$$\overline{V_{C_2}} = \sqrt{RV_r I_x \left[ 1 - \left( \frac{V_r}{V_x} \right)^{n+q} \right]} \tag{24}$$

$$\overline{V_{PV}} = V_r \tag{25}$$

$$\overline{U} = \frac{\sqrt{RV_r I_x \left[ 1 - \left( \frac{V_r}{V_x} \right)^{n+q} \right]}}{V_r + \sqrt{RV_r I_x \left[ 1 - \left( \frac{V_r}{V_x} \right)^{n+q} \right]}} \tag{26}$$

Then using the equilibrium points the Lyapunov candidate function can be obtained from the error function.

$$X_s = X - \overline{X} \tag{27}$$

$$V(x) = \frac{1}{2} X_S^2$$
$$= \frac{1}{2} (L_1 I_{L_{1s}}^2 + L_2 I_{L_{2s}}^2 + C_1 V_{C_{1s}}^2 + C_2 V_{C_{2s}}^2 + C_{PV} V_{C_{PVs}}^2) \tag{28}$$

Afterwards by taking the derivative of the Lyapunov candidate, stability of the system at the equilbrium points can be determined. If the Lyapunov candidate function is positive definite and its derivative is negative semidefinite then the system is asymptotically stable. The derivative of the Lyapunov function (28) is the following.

$$\dot{V} = \frac{1}{R} \left[ -V_{C_{2s}}^2 + V_{PV_s} \left( \left( \frac{V_r}{V_{oc}} \right)^{n+q} - \left( \frac{V_{PV_s} + V_r}{V_{oc}} \right)^{n+q} \right) \right] \tag{29}$$

From equations (28) and (29) it can be seen that the system is stable from the Lyapunov stability analysis. Since equation (28) is the Lyapunov candidate function and is positive definite while equation (29) is the derivative which is negative semidefinite. This proves that the system is asymptotically stable. The control law can be later on found by using this candidate function and it results in equation (30).

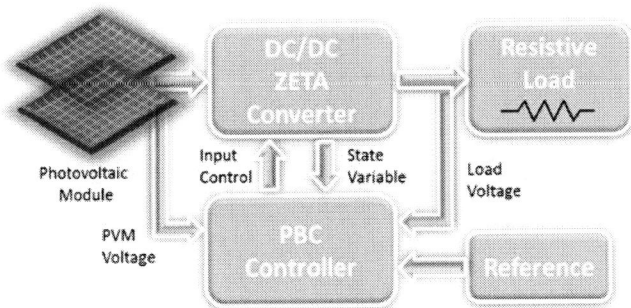

Fig. 8: System Block Diagram. The system is composed of a solar panel array, the DC/DC ZETA converter, the PBC controller, a refere nce and a resistive load.

by replacing the inductors current ($I_{L_1}$) for the input current and the inductors current ($I_{L_2}$) for the output current it can be seen that the inverse of the relationship for the voltages is obtained and if since

$$D = \frac{V_{C_2}}{V_{in} + V_{C_2}} \tag{15}$$

$$\frac{V_{C_2}}{V_{in}} = \frac{D}{1 - D} \tag{16}$$

$$D = \frac{I_{in}}{I_{in} + I_{out}} \tag{17}$$

$$\frac{I_{in}}{I_{out}} = \frac{D}{1 - D} \tag{18}$$

When combining equations (16) and (18) it can be seen that the initial presumption that the components are ideal is met. Even if the presumption was not made, the difference in analysis would be including resistors to use as losses in series with each component. These losses were not included for the purpose of simplification.

Fig. 8 illustrates the solar panel is connected as input voltage and the ZETA converter is used as an interface between the PVM and the load. By using this converter the controller can be made by applying passivity based control for either MPPT or for load voltage regulation.

## IV. CONTROL LAW

Passivity based control is a control technique applied to nonlinear systems even if they are not passive [13] [14]. This control method has been proved to control dc to dc converters such as the boost and buck-boost topologies for voltage regulation [15]. Equations (10) - (14) are the mathematical representation of the complete system including the Zeta converter and the PVM connected as input voltage, which includes the nonlinearity of the system. In this manner it can be seen that the system satisfies the form:

$$\dot{x} = F(x) + G(x) \cdot u \tag{19}$$

$$U = \frac{V_{C_{1s}} + V_{PV_s}}{R}\left(-1 + \frac{V_r}{V_{oc}}\right)^{n+q}$$
$$- \frac{V_{C_{1s}} + V_{PV_s}}{R}\gamma + (I_{L_{1s}} + I_{L_{2s}})(V_r + \gamma) \quad (30)$$

Where $V_{C_{1s}}$ is the state variable for the voltage of the capacitor, $V_{PV_s}$ is the state variable for the voltage in the PVM, $I_{L_{1s}}$ is the state variable for the current in the first inductor, $I_{L_{2s}}$ is the state variable for the current in the second inductor, $V_r$ is a reference voltage selected for calculations, R is the output resistance and $\gamma$ is defined in the following way:

$$\gamma = \sqrt{RV_rI_{sc}\left[1 - \left(\frac{V_r}{V_{oc}}\right)^{n+q}\right]} \quad (31)$$

This control equation is used for PBC yet some considerations may be required when applying it. First off, this equation requires the measurement of inductor currents and capacitor voltages which may not necessarily be easily accessed. Second this equations require high processing power because of the nonlinearity of the functions used.

## V. SIMULATION RESULTS

Simulation Results were obtained using MATLAB/ Simulink software. The simulations done were based on the dynamic equations of the PVM and Zeta converter. Values selected for the Zeta converter are the following: $C1 = C2 = C3 = 100\mu F$, $L1 = L2 = 100\mu H$, $R = 10\Omega$. For the PVM the default values chosen were $Isc = 5A$, $Voc = 24V$ and $n + q = 10.5$. With the PVM values chosen the voltage and current for maximum power in the solar panel are: $Vop = 19.2V$ and $Iop = 4.51981A$. These values provide for an output voltage when the PVM provides maximum power of $V = 29.46V$. Fig. 9 and Fig. 10 represent the results for MPPT while Fig. 11 and Fig. 12 represent load regulation results.

Fig. 9 and Fig. 10 represent the voltage control while in MPPT. Fig. 9 represents the voltage provided by the PVM. For the PVM, to achieve its maximum power it must supply 19.2 Volts. The first 0.008 seconds are the transient part of this algorithm where the converter controls the voltage until obtaining maximum power. Once obtained in Fig. 10 the percent error is calculated for the PVM voltage and in steady state the percentage reduces as low as 0.07%. In terms of voltage regulation the results are shown in Fig. 11 and Fig. 12. Fig. 11 shows output voltage regulation following different levels of output voltage. Fig. 12 shows the percent error in output voltage, where the highest percentage seen was at 3% once in steady state while the peaks in the graph show the transition between voltage levels. These low percentages of error provide a way to determine how well this controller is working compared to the expected results. Also on Fig. 9 the controller is expected to achieve the PVM voltage for Vop which is achieved, as well as in Fig. 10 it is expected

to follow a certain pattern for load voltage regulation and it is obtained as well.

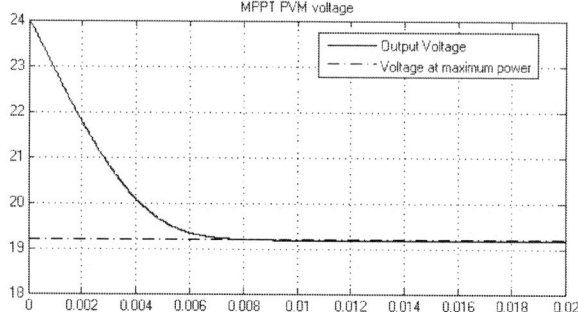

Fig. 9: PVM Voltage with MPPT.

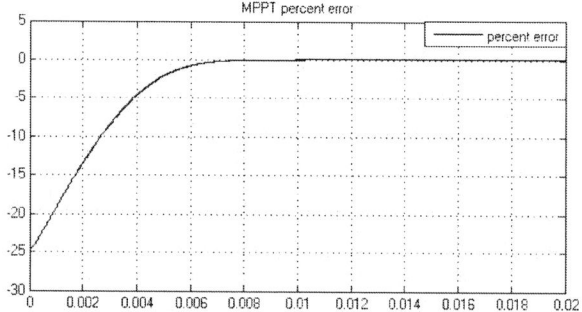

Fig. 10: Error in PVM voltage for MPPT.

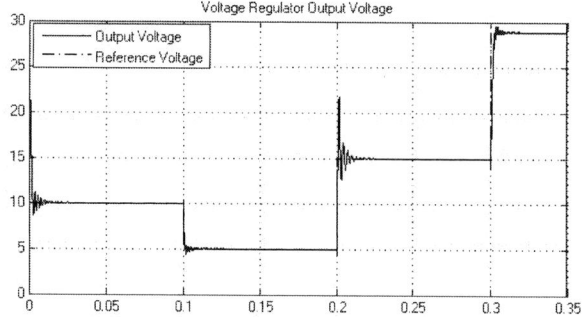

Fig. 11: Voltage Regulator Ouput Voltage.

## VI. CONCLUSION

In this paper a Passivity Based Controller is designed for use with a Zeta converter that has as input voltage a Photovoltaic module. The input control is done by means of the switch used by the Zeta converter. From the simulation results it was demonstrated that The zeta converter proved to be a wise choice considering its low output ripple voltage and as the interface between the load and the PVM. Control of the PVM was achieved by means of the PBC, where the

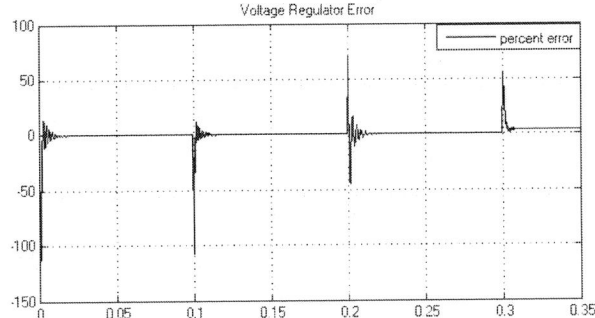

Fig. 12: Voltage Regulator Output Voltage Error.

voltage at which maximum power can be extracted from the solar panel was obtained or the PVM power was controlled for load regulation. PBC proved to be a low error alternative for MPPT, yet rather complicated considering the need for the measurements of inductor currents and capacitor voltages. In conclusion, MPPT and voltage regulation were achieved separately yet effectively by using PBC while maintaining low percentage of error in its results.

### ACKNOWLEDGMENT

The authors would like to acknolledge the financial support of the National GEM Consortium GEM fellowship program and the Transformational Initiative for Graduate Education and Research (TIGER) Graduate Research Fellowship program under the grant number P031M140035 as well as the Minds2CREATE research group at The University of Puerto Rico in Mayaguez and the US Department of Energy, especially to the Consortium for Integrating Energy Systems in Engineering and Science Education (CIESESE).

### REFERENCES

[1] R. Darbali-Zamora, D. A. Merced-Cirino, C. Gonzalez-Ortiz and E. I. Ortiz-Rivera "An Electric Power Supply Design for the Space Plasma Ionic Charge Analyzer (SPICA) CubeSat", *in 40th IEEE Photovoltaic Specialist Conference*, pp. 1790-1795, 2014.

[2] J. G. Ciezki and R. W. Ashton "The application of Feedback Linearization to the Stabilization of DC-to-DC Converters with Constant Power Loads", *Circuits and Systems*, 1998.

[3] E. Jimenez, E. I. Ortiz-Rivera and O. Gil-Arias "A Dynamic Maximum Power Point Tracker Using Sliding Mode Control", *Control and Modeling for Power Electronics*, 2008.

[4] R. Venkataramanan "Sliding Mode Control of Power Converters", *Thesis Subitted in requirements of California Institute of Technology*, 1986.

[5] H. Sira-Ramirez and R. Silva-Ortigoza, "Passivity Based Control of Zet Converter", *in Control Design Techniques in Power Electronics Devices*Mexico City, Mexico: Springer, 2006.

[6] E. I. Ortiz-Rivera"Modeling and Analysis of Solar Distributed Generation", *Thesis Submitted to the Department of Electrical and Computer Engineering at Michigan State University*, 2006.

[7] E. I. Ortiz-Rivera "Approximation of a Photovoltaic Module Model Using Fractional and Integral Polynomials", *Photovoltaics Specialist Conference*, Austin, Texas, 2012.

[8] M. Kessler "Synchronous Inverse SEPIC Topology Provides High Efficiency Buck/Boost Voltage Converters" May 2010. [Online]. [Accessed April 2015].

[9] E. Vuthchhay and C. Bunlaksananusorn "Modeling and control of a Zeta converter", *in Power Electronics Conference (IPEC)*, 2010.

[10] F. Sun and M. Wich "The Low Output Voltage Ripple Zeta DC/DC Converter Topology" 2014. [Online]. Available: http://www.linear.com/solutions/4993. [Accessed March 2015].

[11] D. W. Hart, *Power Electronics*, Valparaiso University, Indiana: McGraw Hill, 2011.

[12] J. Falin, "Designing DC/DC Converters Based on ZETA Topology" 2014. [Online]. Available: http://www.ti.com/lit/an/slyt372/slyt372.pdf. [Accessed March 2015].

[13] H. K. Khalil, *Nonlinear Systems*, Upper Saddle River, New Jersey: Prentice Hall, 2002.

[14] R. Ortega, Z. P. Jiang and D. J. Hill, "Passivity Based Control of Nonlinear Systems: A Tutorial" *in American Control Conference*, ALbuquerque, New Mexico, 1997.

[15] H. Sira-Ramirez and R. Ortega, "Passivity-Based Controllers for the Stabilization of DC-to-DC Power Converters," *in Conference on Decision and Control*, New Orleans, LA, 1995.

# SiC Switch based Single-Stage Buck-Boost Transformerless Mini Inverter with Low Leakage Current and Negligible DC Injection

Soumya Ranjan Mohapatra[1], Amit Kumar Gupta[1], Madhuwanti S. Joshi[2], and Vivek Agarwal[1]

[1]Indian Institute of Technology Bombay, Mumbai, India, [2]Integra Power LLC, USA

*Abstract* — A novel buck-boost type single-stage transformerless topology with one energy storage inductor is proposed in this paper for single-phase grid-connected PV applications. The symmetrical operation of the topology with only one inductor minimises dc current injection into the grid. The buck-boost ability of this topology helps in achieving MPPT operation even under large variation of PV voltage. This topology, with an operating frequency of 100 kHz, uses five controllable devices out of which only one operates at high-frequency. Thus, the proposed topology has low switching losses. Only one SiC device is used to incorporate high frequency operation leading to a cost-effective solution. Other advantages of the high switching frequency include high power density, high efficiency and reduced filter size. MPPT, buck or boost operation, and dc-ac inversion are all implemented in a single power stage only. The proposed topology incurs very low leakage current. The working principle has been analysed and discussed in detail, and the simulation and experimental results are presented.

*Index Terms* — Common mode voltage, grid-connected inverter, leakage current, SiC-based inverter, single-stage topology, transformerless inverter.

## I. INTRODUCTION

Solar power is an abundant source of energy on the earth. It is one of the most promising sources to meet the energy needs of the mankind. Due to its abundance and easy availability, it is one of the good options for generating large chunks of power, which can be distributed through the existing power grid. However, various issues like MPPT operation, power control, PV voltage variation, cost, efficiency and power density and leakage current elimination in case of grid-connected transformerless inverter pose a major challenge. Extensive research is underway on grid-connected PV systems [1]–[7] to improve their performance. In grid-connected PV systems, transformerless inverters have several advantages over the inverters with transformers like low cost, lower weight and size, high efficiency and higher power density. Therefore, they are getting increasing attention of the researchers and industrialists.

The PV inverters connected with one module to convert DC into AC and feeding power into the grid are module-based inverters also known as Microinverters. If several inverters are connected in series and/or parallel combinations to obtain high DC link voltage, where voltage boosting is not required, they are known as string and central inverters [8], [9]. Another family of inverters, which are connected with more than one module but still need boosting of DC-link voltage to meet the grid voltage requirement and work in single-stage circuit configuration may be referred to as mini inverters. This family lies between microinverters and string/central inverters.

Global agencies, mandated to ensure the performance of grid-connected systems, have set specific standards such as IEEE 1547, IEC 61727, NEC 690, European standard (EN 61000-3-2) and German VDE 0126-1-1 to ensure that these systems meet the required performance indices [10]–[14]. A grid-tied PV inverter must comply with them in terms of disconnection while exceeding the leakage current limit, the limit in DC injection into the grid, limitation on generated harmonics, personal safety issues, etc.

There are some well-known FB derived topologies like H5, HERIC, oH5, FB-ZVR, FB-DCBP and a few configurations of H6 for grid-connected PV systems [1], [5], and [9]. All of the above are commercialised and comply with the standards. Since they are buck-based topologies, they cannot track the full MPPT curve. There are several single-stage inverters introduced by researchers with voltage boost ability to meet the grid requirements [7], [15]. Although these topologies have several other benefits, they cannot track the full MPPT curve.

Therefore, researchers are also focusing on having buck-boost based single-stage topologies [7]. Buck-boost type of configurations can track MPP for a large variation of voltage across PV modules. In [16] and [17] buck-boost type inverter configurations are proposed, but they use one separate PV source for each cycle of the grid voltage. Thus, it leads to under-utilization of the PV panels. The number of switches used in buck-boost topologies [18]–[20] is found to be six or above. Controlling the higher number of switches is complex, and the power density and overall efficiency are hampered. The topologies proposed in [21] and [22] comprises of only four switches. However, the operation is found to be asymmetric in the two half cycles of the grid voltage. The asymmetric operation also increases chances of dc current injection into the grid. The doubly grounded buck-boost inverter [23] has a symmetrical operation for both cycles of the grid voltage. But it uses five switches and three diodes, out of which two switches are high-frequency switches. If wide bandgap devices are used then the cost of the system would go high. Also, the number of switches operated in each half cycle are not equal.

978-1-5090-5606-4/17 $31.00 © 2017 IEEE

In this paper, a unique buck-boost type transformerless topology is proposed, as shown in Fig. 1, having the following benefits over the existing topologies found in the literature:

a) The proposed topology has five switches, out of which only one is required to operate at high switching frequency while others operate at line-frequency (50/60 Hz). Therefore, only one wideband gap (ex. SiC) device is required for high-frequency operation, which leads to a cost-effective, high power density solution and reduces the output filter size.

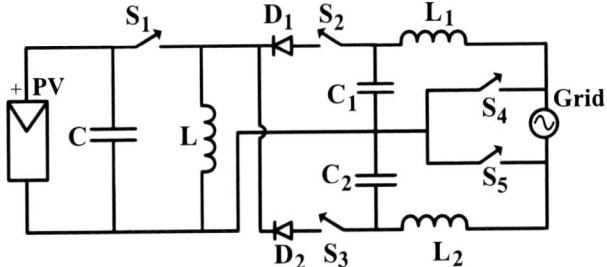

Fig. 1. Proposed topology.

b) The topology has a symmetrical operation as only one energy storage inductor is used for both the half cycles of grid voltage. This minimizes the dc injection into the grid.

c) The proposed topology can track full MPPT curve due to its voltage buck-boost ability.

d) Only three switches are operated in each half cycle. Therefore, equal switching and conduction loss are expected for both the half cycles.

e) The topology exhibits very low leakage current because the negative terminal of the PV bus is always connected to either of the grid terminals.

## II. PROPOSED BUCK-BOOST TOPOLOGY

The proposed topology transfers energy from PV to the grid in two steps. The PV energy is stored in 'L' and the stored energy is then released to output side capacitor(s) and the grid. $S_1$ operates on high frequency based on sine-triangular PWM for both positive and negative half cycles of the grid voltage. For the positive half cycle, $S_3$ and $S_4$ remain ON, while $S_2$ and $S_5$ are OFF. Similarly, for the negative half cycle, $S_3$ and $S_4$ remain OFF, while $S_2$ and $S_5$ are kept ON. The steady state operation of the proposed topology over one complete cycle can be categorised into six modes as shown in Fig. 2.

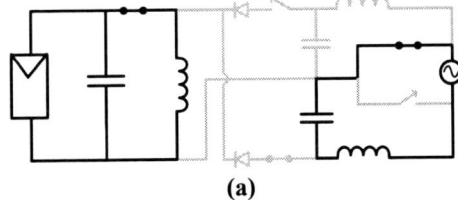

(a)

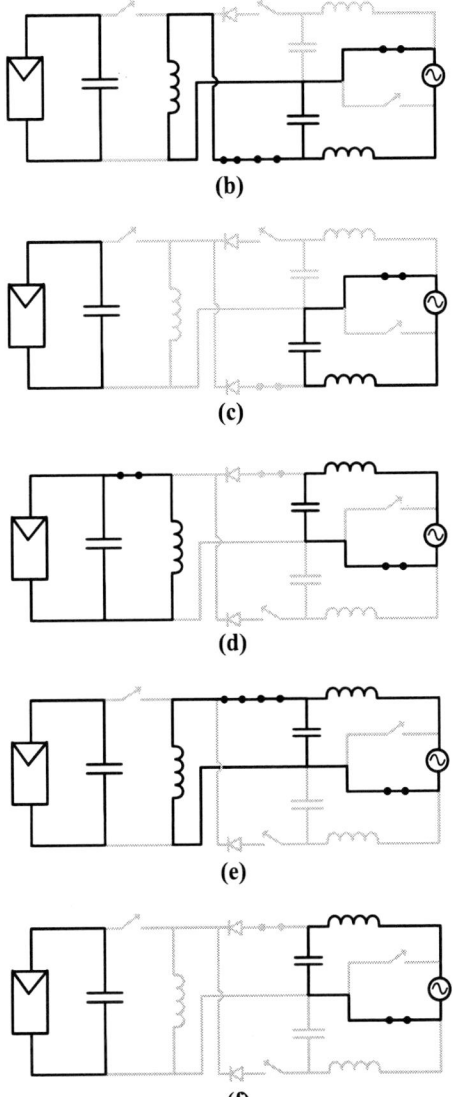

Fig. 2. Modes of operation.

Operating modes of the proposed transformerless topology during positive half cycle are explained below.

***Mode (a):*** When $S_1$ is turned ON, inductor L stores energy from the PV source, and simultaneously $C_2$ releases its energy into the grid through $S_4$ and $L_2$.

***Mode (b):*** When $S_1$ is turned OFF, L transfers its energy into the $C_2$ as well as the grid through $S_4$, $L_2$, $S_3$ and $D_2$.

***Mode (c):*** When L has transferred its whole energy then the current of $L_2$ freewheels through $S_4$, and $C_2$.

Mode (a) starts again after mode (c) and this cycle continues throughout the positive half cycle. Similarly, for negative half cycle operation, the operating modes are explained as follows.

***Mode (d):*** When $S_1$ is turned ON, inductor L stores energy from the PV source, and simultaneously $C_1$ releases its energy into the grid through $S_5$ and $L_1$.

**Mode (e):** When $S_1$ is turned OFF, L transfers its energy to $C_1$ and the grid through $S_5$, $L_1$, $S_2$ and $D_1$.

**Mode (f):** When L has transferred its whole energy then the current of $L_1$ freewheels through $S_5$, and $C_1$.

The controller of the proposed topology is shown in Fig. 3. Modulation index (M) is determined from the P&O MPPT algorithm [2]. Rectified and filtered unit vector of the grid voltage is multiplied by M to perform SPWM for $S_1$ as shown in Fig. 4. Switches $S_2$, $S_3$, $S_4$, and $S_5$ operate using simple logic for both positive and negative half cycles of the grid voltage.

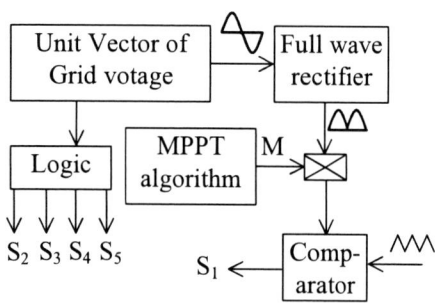

Fig. 3. Controller of the proposed topology.

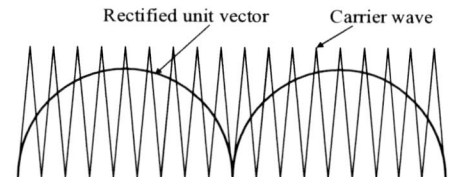

Fig. 4. Sine-Triangular Comparison for producing gate pulses for $S_1$.

## III. SIMULATION RESULTS

Simulations have been carried out to validate the proposed topology and scheme. A complete closed loop simulation with MPPT operation has been performed using MATLAB-Simulink. The specifications of the simulation environment and component values are given in Table I.

TABLE II.
SYSTEM SPECIFICATIONS

| Parameters | Value | Unit |
|---|---|---|
| Nominal Power Rating | 3.5 | kW |
| Nominal Voltage $V_{mpp}$ (12 PV panels) | 360 | V |
| Maximum PV current $I_{sc}$ | 8.3 | A |
| Grid Voltage (rms) $V_{grid}$ | 230 | V |
| Grid Current (rms) $I_{grid}$ | 14 | A |
| Switching frequency $f_{sw}$ | 100 | kHz |
| Input inductor L | 20 | µH |
| Input Capacitor C | 3000 | µF |
| Output inductor values ($L_1$ and $L_2$) | 300 | µH |
| Output Capacitor value ($C_1$ and $C_2$) | 4 | µF |

Input energy storage inductor is designed in such a way that the topology operates in discontinuous mode. Filter capacitors are designed to absorb the energy dumped by input inductor within a specified time duration [23]. Simulation results are depicted in Fig. 5 and Fig. 6.

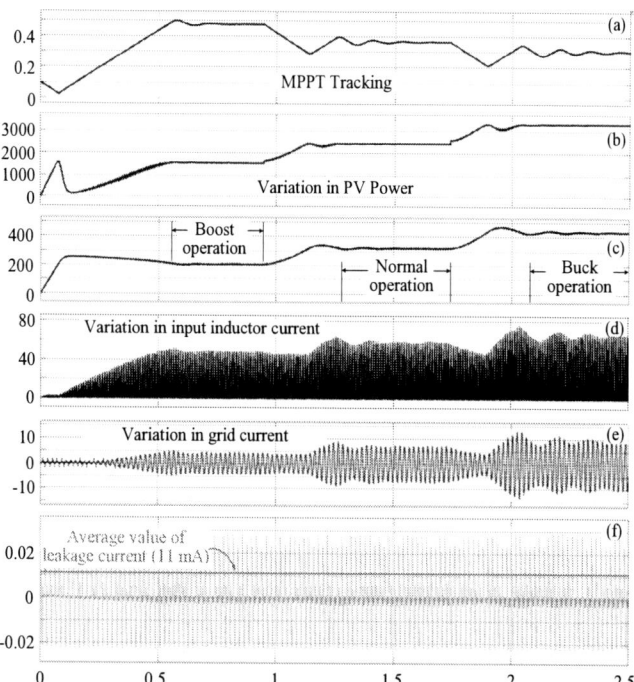

Fig. 5. Simulation results: (a) Modulation index; (b) Input PV power; (c) Input PV voltage; (d) Input inductor current; (e) Grid current and (f) Leakage current.

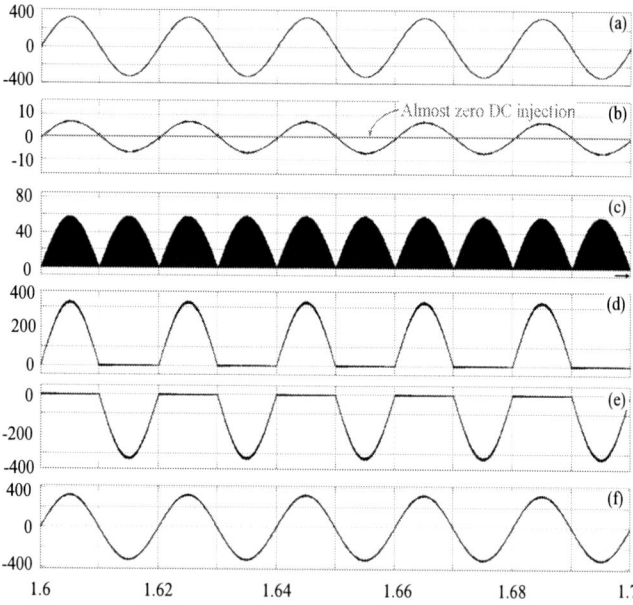

Fig. 6. Simulation results: (a) Grid voltage; (b) Grid current; (c) Input inductor current; (d) Voltage of capacitor $C_1$; (e) Voltage of capacitor $C_2$ and (f) Voltage across capacitor $C_1 + C_2$.

Modulation index, input PV power, and PV voltage are shown in Figs. 5 (a), (b) and (c) respectively. It is observed that the proposed topology can track MPP, although the PV voltage is lower or higher than the peak of the grid voltage. Figs. 5 (d) and (e) present corresponding variation of input inductor current and grid current with MPPT algorithm. The zoomed view of grid current and input inductor current are given in Figs. 6 (b) and Fig. 6 (c) respectively. It is observed that input inductor current is discontinuous and the peak of it is following a rectified sine wave. The grid current is also in the same phase to that of grid voltage as shown in Fig. 6 (a). The total harmonic distortion of grid current is calculated to be very low i.e. 2.5% The voltage of filter capacitor $C_1$, $C_2$ and total voltage of both the capacitors are shown in Figs. 6 (d), (e) and (f) respectively.

This verifies that capacitor voltages are as per expectations. It is also clear that the oscillations caused due to the small energy left in the capacitor while switching from $S_4$ to $S_5$ near the zero crossing of grid voltage are insignificant and can be neglected. The rms and the average values of the leakage current are depicted in Fig. 5 (f). It is evident that leakage current is very low and within the safety limits. The dc injection is also found to be almost zero as per Fig. 6 (b).

## IV. EXPERIMENTAL WORK

The proposed topology has been implemented on the hardware. The hardware set-up is shown in Fig. 7. The experiments are carried out in open loop at nearly 400 W power level and 110 V output voltage. Hardware specifications are shown in Table III. Fig. 8 shows the waveforms of output voltage, output current, input inductor current and voltage across output capacitor $C_1$. Fig. 9 shows the waveforms of the voltage of output capacitor $C_2$ and the leakage current.

SiC MOSFET (C2M0160120D) provided by Wolfspeed (Cree Inc.) is used as switch $S_1$ for high frequency operation, which needs gate pulses of +18 V as logic high and -5 V as logic low for turning on and off respectively. IGBTs from Infineon (IKW25N120H3) are used for other four switches.

### TABLE III.
#### HARDWARE SPECIFICATIONS

| Parameters | Value | Unit |
|---|---|---|
| Output Power | 380 | W |
| Input Voltage $V_{in}$ (Approx.) | 135 | V |
| Input Current $I_{in}$ (Approx.) | 3 | A |
| Output Voltage (rms) $V_o$ | 110 | V |
| Output Current (rms) $I_o$ | 3.5 | A |
| Switching frequency $f_{sw}$ | 100 | kHz |
| Input inductor L | 33 | μH |
| Input Capacitor C | 1500 | μF |
| Output inductor values ($L_1$ and $L_2$) | 150 | μH |
| Output Capacitor value ($C_1$ and $C_2$) | 4 | μF |

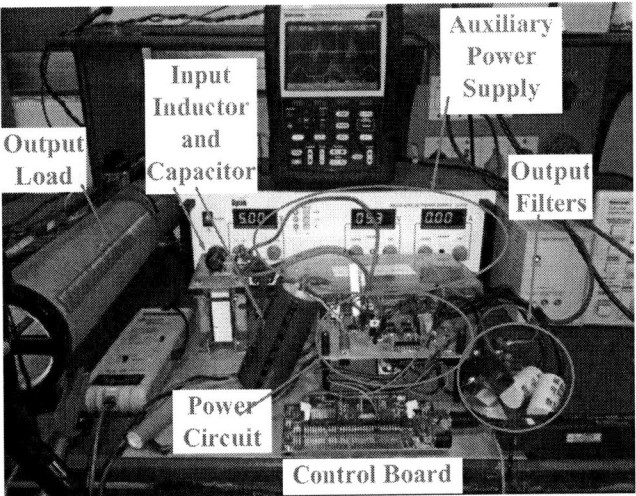

Fig. 7. Hardware setup of the proposed system.

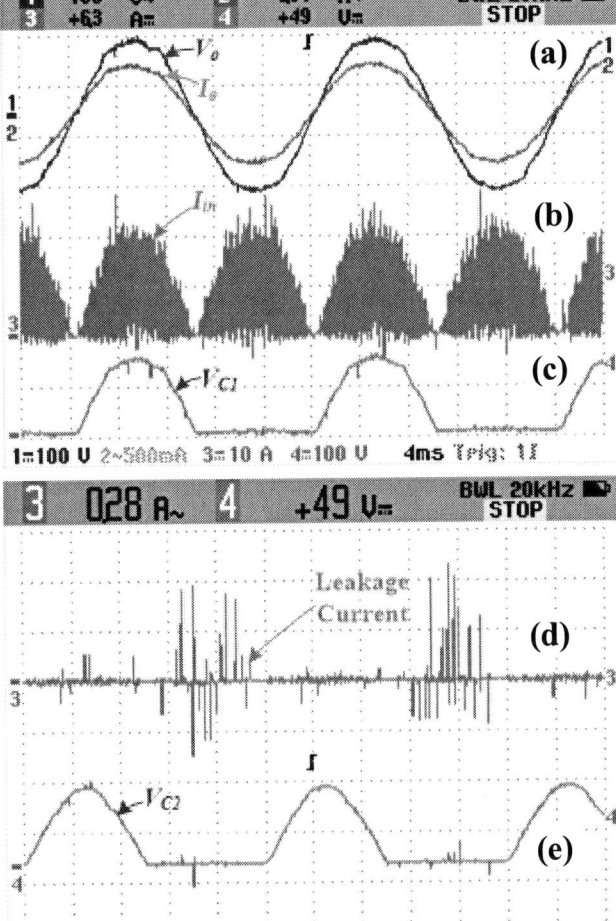

Fig. 8. Experimental results: (a) Output voltage and current; (b) Input inductor current; (c) Voltage across $C_1$; (d) Leakage current and (e) Voltage across $C_2$.

All the devices are triggered using HCPL3120 drivers in the laboratory prototype.

It is observed from the waveform of the input inductor current that the system works in discontinuous conduction mode (DCM). The measured DC rms value of the input inductor current and each output capacitor voltage are 6.3 A and 49 V respectively, and their peak values are nearly 20 A and 150 V respectively.

The leakage current is measured through the ground loop by connecting ground resistance and capacitance as 10 Ω and 47 nF respectively. The rms value of the leakage current is 280 mA, which is within the limits imposed by the standards. Due to high frequency operation, the waveform of leakage current is a bit noisy, therefore, the measured value of leakage current is high. In reality, much lower value of leakage current is expected by the proposed transformerless topology.

## V. CONCLUSION

A novel transformerless buck-boost based topology was proposed, which exhibits minimal amount of leakage current and DC injection (11 mA and ≈zero respectively in simulations) for grid-tied PV application. Thus, the proposed mini inverter topology is suitable for grid-tied PV application. Here, single-stage operation was accomplished with MPPT operation, where P&O method was used for MPPT. Only one SiC device is used for the high frequency operation, which makes it a cost-effective solution with high frequency operation. The inverter was operated at 100 kHz switching frequency, which resulted in low THD (≈2.5%) in grid current with reduced filter size. Thus, the system was verified in simulations with complete closed loop operation in grid-connected mode, and for MPP tracking under wide variation in PV voltage due to its buck-boost ability. The experiments are performed for the stand-alone system at 400 W power level by developing a laboratory prototype. The performance parameters and experimental results validate the proposed concept.

## REFERENCES

[1] D. Barater, E. Lorenzani, C. Concari, G. Franceschini, and G. Buticchi, "Recent advances in single-phase transformerless photovoltaic inverters," *IET Renew. Power Gener.*, vol. 10, no. 2, pp. 260–273, 2015.

[2] H. Patel and V. Agarwal, "Maximum power point tracking scheme for PV systems operating under partially shaded conditions," *IEEE Trans. on Ind. Elec.*, vol. 55, no. 4, pp. 1689–1698, Apr. 2008.

[3] S. R. Mohapatra, P. K. Ray, and G. H. Beng, "A partial feedback linearization based approach to shunt active power filter design," in *IEEE Region 10 Conference (TENCON)*, Singapore, 2016, pp. 2896-2900.

[4] T. Esram and P. L. Chapman, "Comparison of photovoltaic arraymaximum power point tracking techniques," *IEEE Trans. Energy Convers.*, vol. 22, no. 2, pp. 439–449, Jun. 2007.

[5] T. K. S. Freddy, N. A. Rahim, W. P. Hew, and H. S. Che, "Comparison and analysis of single-phase transformerless grid-connected PV inverters," *IEEE Transactions on Power Electronics*, vol. 29, no. 10, pp. 5358-5369, Oct. 2014.

[6] A. K. Gupta, M. S. Joshi, and V. Agarwal, "On the control and design issues of single phase transformerless inverters for photovoltaic applications," in *IEEE 6th India International Conference Power Electronics*, Kurukshetra, India, Dec. 2014.

[7] D. Meneses, F. Blaabjerg, O. Garc´ıa, and J. A. Cobos, "Review and comparison of step-up transformerless topologies for photovoltaic AC-module application," *IEEE Trans. Power Electron.*, vol. 28, no. 6, pp. 2649–2663, Jun. 2013.

[8] S. B. Kjaer, J. K. Pedersen and F. Blaabjerg, "A Review of Single-Phase Grid-Connected Inverters for Photovoltaic Modules," *IEEE Trans. Industry Appl.*, vol. 41, no. 5, pp. 1292–1306, Sep./Oct. 2005.

[9] S. V. Araújo, P. Zacharias and R. Mallwitz, "Highly Efficient Single-Phase Transformerless Inverters for Grid-Connected Photovoltaic Systems," *IEEE Trans. Industrial Electronics*, vol. 57, no. 9, pp. 3118– 3128, Sep. 2010.

[10] IEEE Application Guide for IEEE Std. 1547, IEEE Standard for Interconnecting Distributed Resources With Electric Power Systems, IEEE Standard 1547.2-2008, Apr. 2009, pp. 1–207.

[11] Photovoltaic (PV) Systems—Characteristics of the Utility Interface, IEC Standard 61727 ed-2.0, 2004.

[12] Electromagnetic Compatibility (EMC)—Part 3-2: Limits—Limits for Harmonic Current Emissions (Equipment Input Current Under 16 A Per Phase), EN 61000-3-2:2006, 2006.

[13] Automatic Disconnection Device between a Generator and the Public Low-Voltage Grid, VDE V 0126-1-1:2006-02, 2006.

[14] 2011 National Electrical Code, National Fire Protection Association, Inc., Quincy, MA, 2011.

[15] S. Jain and V. Agarwal, "A single-stage grid connected inverter topology for solar PV systems with maximum power point tracking," *IEEE Trans. Power Electron.*, vol. 22, no. 5, pp. 1928–1940, Sep. 2007.

[16] N. Kasa, H. Ogawa, T. Iida, and H. Iwamoto, "A transformer-less inverter using buck-boost type chopper circuit for the photovoltaic power system," in *Proc. IEEE Int. Conf. Power Electron. Drive Syst.*, 1999, pp. 653–658.

[17] W. Wu, J. Ji, and F. Blaabjerg, "Aalborg inverter—A new type of "buck in buck, boost in boost" grid-tied inverter," *IEEE Trans. Power Elect.*, vol. 30, no. 9, pp. 4784–4793, Sep. 2015.

[18] Y. Tang, X. Dong, and Y. He, "Active buck-boost inverter," *IEEE Trans. Ind. Elect.*, vol. 61, no.9, pp. 4691–4697, Sep. 2014.

[19] P. K. Chamarthi, M. Rajeev, and V. Agarwal, "A novel single stage zero leakage current transformer-less inverter for grid connected PV systems," in *42nd IEEE Photovolt. Spec. Conf.*, Jun. 2015, pp. 1–5.

[20] M. Rajeev and V. Agarwal, "Novel transformer-less inverter topology for single-phase grid connected photovoltaic system," in *IEEE 42nd Photovolt. Spec. Conf.*, Jun. 2015, pp. 1–5.

[21] D. Schekulin, Bundesrepublik Deutschland, Deutsches Patent, Patentschrift DE 197 32 218 Cl, Mar. 1999

[22] A. Kumar and P. Sensarma, "Four-switch single-stage single-phase buck-boost inverter," *IEEE Transactions on Power Electronics*, vol. 32, pp. 5282-5292, 2017.

[23] H. Patel and V. Agarwal, "A single-stage single-phase transformerless doubly grounded grid-connected PV interface," *IEEE Trans. Energy Conv.*, vol.24, no.1, pp.93–101, Mar. 2009.

[24] H. Agrawal, A. K. Gupta, and V. Agarwal, "A novel 3-phase, transformerless H-8 topology with low variation in CMV to reduce leakage current," in *IEEE Int. Conf. on Power Electronics, Drives and Energy Systems*, Kerala, Dec. 2016.

# Open Source Tools for High Performance Quasi-Static-Time-Series Simulation Using Parallel Processing

Davis Montenegro[1], Roger C. Dugan[1], and Matthew J. Reno[2]

[1] EPRI, Knoxville, TN, 37923, USA
[2] Sandia National Laboratories, Albuquerque, NM, 87123, USA

*Abstract* — **Quasi-Static-Time-Series (QSTS) simulation is a valuable tool for evaluating the behavior of power systems through time. By performing daily, yearly and other time-based simulations, it is possible to characterize time-varying power conversion devices such as photovoltaic panels, storage, loads, and capacitors, among others within the power system. However, depending on the time-step resolution and simulation duration, the sequential simulation may require a considerable amount of computing time to complete. This paper describes the OpenDSS-PM program which is the new Parallel Machine version of EPRI's open-source distribution system simulator program, OpenDSS. OpenDSS-PM is used to implement temporal parallelization and circuit solutions with Diakoptics based on actors as techniques to reduce the time required in QSTS. The results reveal that these techniques enable a significant reduction in time using common computer architectures.**

*Keywords*— *multicore processing, open source software, parallel processing, power system analysis computing.*

## I. INTRODUCTION

Quasi-Static-Time-Series (QSTS) simulation is a valuable tool for evaluating the behavior of power systems through time. By performing daily, yearly and other time-based simulations, it is possible to characterize time-varying power conversion devices such as photovoltaic panels (PV), storage, loads, and capacitors, among others within the power system. Particularly in the case of PV systems, their proliferation as alternative power source has generated the need of carrying studies such as PV hosting capacity [1, 2], interconnection studies [3-5] and microgrids among others, which are QSTS based.

However, depending on the time-step resolution and simulation duration, the sequential simulation may require a considerable amount of computing time to complete [6]. This paper presents the use of OpenDSS-PM (Parallel Machine), which is derived from EPRI's open-source Distribution System Simulator, OpenDSS [7], to accelerate QSTS simulations using multi-core computers. OpenDSS-PM is used to implement temporal parallelization and Diakoptics based on actors [8, 9] as techniques to reduce the time required in QSTS. The results reveal that these techniques enable a significant reduction in time using common computer architectures. Faster QSTS simulations will provide distribution engineers with a more accurate understanding of the impacts of solar variability and high penetrations of PV on the distribution system [10].

This paper is organized as follows:
- An introduction to OpenDSS-PM
- A brief description of the Parallelization methods
- Simulation results using OpenDSS-PM parallelization

- Conclusions

## II. OPENDSS-PM (PARALLEL MACHINE)

The OpenDSS program is an open-source electric power Distribution System Simulator (DSS) for supporting distributed resource integration and grid modernization efforts. OpenDSS was originally developed by Electrotek Concepts, Inc. in 1997 under the name of DSS. Since then, it has acquired an important number of capabilities to support the smart grid analysis, including a wide number of device models and simulation modes for this purpose.

This simulation platform was originally built for execution in a single, sequential process. Each procedure/function is called sequentially to perform a QSTS simulation. The performance that can be achieved is based on the structure of the low level routines, the simplicity of the routines, and the efficiency of the compiler.

EPRI made the program open source in 2008 to encourage efforts in grid modernization by providing a tool to the power industry capable of performing advanced studies. The program's name was changed to OpenDSS to emphasize that it was open source. Since then, EPRI has explored several methods to accomplish parallel processing in OpenDSS, including the parallelization of the whole program using a different interface (the Direct DLL API), the modification of the solver using other programming languages, and other methods.

However, these approaches demand additional complications in the user interface, extra effort from the user, and they will be always tied to a particular interface. As a consequence, the desired features that users are accustomed to, such as the COM interface, would be at risk of being deprecated for this type of processing.

Modern computing architectures are characterized for introducing the concept of multi-core computing [11]. This feature allows the performance of applications to be improved by distributing tasks on multiple cores to work concurrently. This feature in modern computers created the obvious need for taking OpenDSS into a parallel computing simulation suite.

EPRI has evolved OpenDSS into a more modular, flexible and scalable parallel processing platform we are calling OpenDSS-PM based with the following guidelines:
- The parallel processing machine will be interface-independent
- Each component of the parallel machine should be able to work independently

978-1-5090-5606-4/17 $31.00 © 2017 IEEE

- The simulation environment should deliver information consistently
- The data exchange between the components of the parallel machine should respect the interface rules and procedures
- The user interface for the parallel machine should be easy and support the already acquired knowledge of OpenDSS users

To create the parallel machine, OpenDSS-PM uses the actor model [12, 13]. There are several actor frameworks for the Delphi language proposed by various authors; however, the chosen framework had already been developed by the authors to evaluate Delphi's tools for parallelization. Each actor is created by OpenDSS-PM, runs on a separate processor (if possible) using separate threads, and has its own assigned core and priority (real-time priority for the process and time critical for the thread).

The interface for sending and receiving messages from other actors will be the one selected by the user: either the COM interface, the Direct DLL API, or a text script using the stand-alone EXE version of the program. From this interface, the user will be able to create a new actor (instance), send/receive messages from these actors, and define the execution properties for each actor. The properties include the execution core, simulation mode, and circuit to be solved, among others. The actor parallelization concept is presented in Fig. 1.

### III. THE PARALLELIZATION TECHNIQUES

The parallelization techniques utilize multiple independent computing resources to decompose the simulation complexity. This decomposition can be done by distributing the work in time or by simplifying the power flow problem complexity using simpler representations of it.

As a result, the amount of time required for solving a large number of simulation steps is expected to be reduced. In this paper two methods are explored: *Temporal parallelization* and *Diakoptics* based on actors. The first one seeks to distribute temporal segments of the total simulation task into multiple actors to complete the whole simulation.

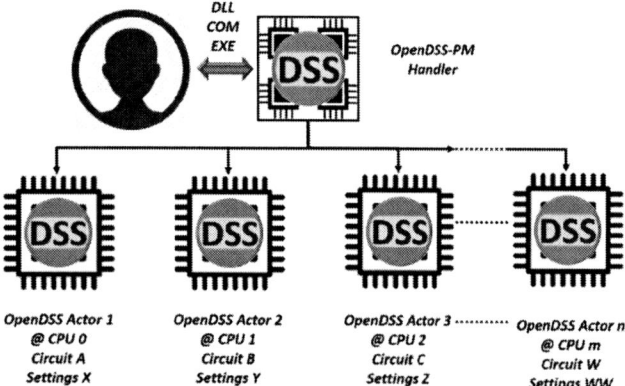

Fig. 1.  OpenDSS-PM general description.

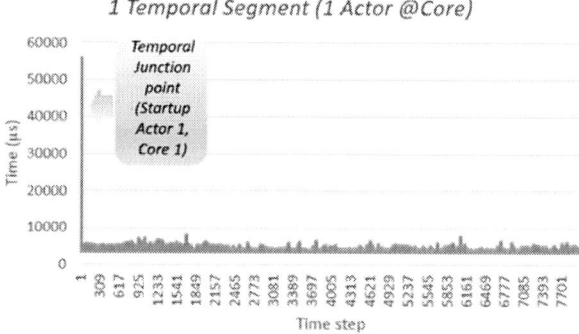

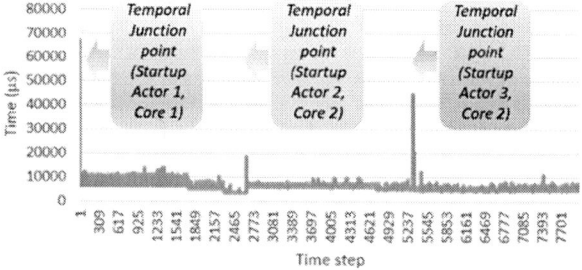

Fig. 2. Temporal Parallelization concept. The top figure is the standard QSTS, and the bottom figure below demonstrates temporal parallelization of QSTS.

The second one decomposes the interconnected circuit into smaller and simpler sub-circuits, each of these systems will be assigned to an actor to find a partial solution that will be complemented in a later stage.

#### A.  Temporal parallelization

Temporal parallelization is a technique that consists of splitting the simulation into multiple time periods with each period being simulated sequentially and concurrently to each other. Each of the multiple temporal segments will deliver partial results that after being concatenated will deliver a very close result as if the simulation was performed sequentially. This technique is illustrated in Fig. 2.

In this simulation technique the circuit's complexity remains intact, which means that each temporal segment requires a startup simulation time. The startup time is the longest time in a QSTS simulation and will force a non-linear acceleration when using this technique. When applying temporal parallelization, the startup simulation time can be located at the junction points between actors.

Additionally, since distribution power flows are dependent on the previous time-step and there is no information about the state of the control devices at these points before starting the simulation, a systematic error can be introduced. Nevertheless, if the adequate control actions are performed during the startup step the systematic error can be reduced. OpenDSS-PM is designed to perform all the needed control actions at the simulation start up for each concurrent actor to minimize the error and maintain the simulation fidelity.

### B. Diakoptics Based on Actors

*Diakoptics* Based on Actors (*A-Diakoptics*) is the result of combining two computing techniques from different fields. On one hand there is *Diakoptics* [14, 15], which is a mathematical method for tearing networks to create a set of independent sub-networks. Each sub-network can be handled and solved independently and the size of each sub-network is smaller than the interconnected network.

The *Actor model* is an information model for dealing with inconsistency robustness in parallel, concurrent, and asynchronous systems. This information model allows multiple processes to be executed in parallel with information passing through them using messages. As a result, the information consistency can be ensured, avoiding frequent issues when working with parallel processing systems, such as race conditions and memory sub-utilization (when working with multicore processors) among others.

A basic description of A-*Diakoptics* is shown in Fig. 3. The subsystems, and their relationships can be described as a set of matrixes representing the subsystems ($Z_{TT}$), the link branches between them ($Z_{CC}$) and their relationship within the network ($Z_{CT}$). As a result, by using this information and the partial results delivered after solving each subsystem separately and concurrently, it is possible to determine the voltages at the network's nodes by finding a complementary answer as follows:

$$E_T = Z_{TT}I_{0(n)} - Z_{TC}Z_{CC}^{-1}Z_{CT}I_{0(n+m)} \tag{1}$$

A-Diakoptics is a technique that seeks for simplifying the power flow problem to perform a faster solution at each simulation step; in contrast, temporal parallelization remains the complexity of the problem. Consequently, the total time reduction when performing QSTS will be evident at each simulation step with A-Diakoptics.

### IV. RESULTS

For evaluating the performance of OpenDSS-PM in the temporal parallelization, EPRI's Test Circuit 5 [16] is used.

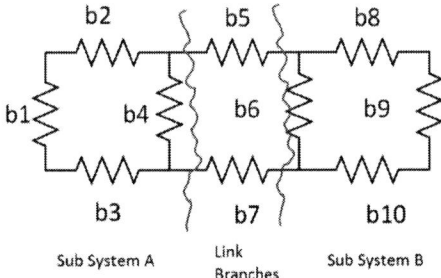

Fig. 3. Physical interpretation of Diakoptics.

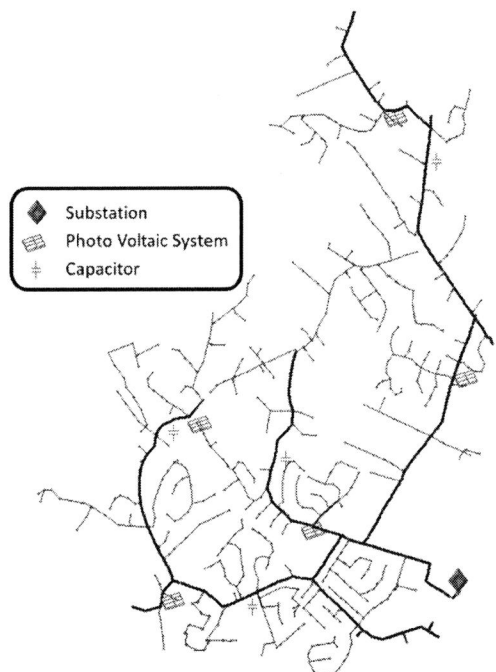

Fig. 4. EPRI Circuit 5.

This circuit is shown in Fig. 4. In this circuit, there are 5 zones where PV systems are concentrated providing a total peak power of 2.5MW to the system (aggregate). The temporal parallelization is performed by considering the number of available CPUs (threads) on the computer where the tests are performed.

### A. Temporal parallelization

In this case the power flow is solved 8000 times using the *time* simulation mode and step size of 1h, which is equivalent duration of 11 months and 4 days. These quantities have been arbitrarily selected to present round numbers for this example. The script wrote using the OpenDSS-PM scripting tool is as follows:

```
clearAll
set parallel=No   ! For the compilation stage
compile "C:\ OpenDSS\EPRI_ckt5_3437_nodes\master.dss"
Solve
set mode=time stepsize=1h number=2000 hour = 0
totaltime=0 CPU=0
NewActor         ! Creates a new actor
compile "C:\ OpenDSS\EPRI_ckt5_3437_nodes\master.dss"
Solve
set mode=time stepsize=1h number=2000 hour = 2000
totaltime=0 CPU=2
NewActor                ! Creates a new actor
!... Creates as many actors as needed... and then
! the parallel features are activated and the system solved
set parallel=Yes  !Enables the parallel processing features
SolveAll
```

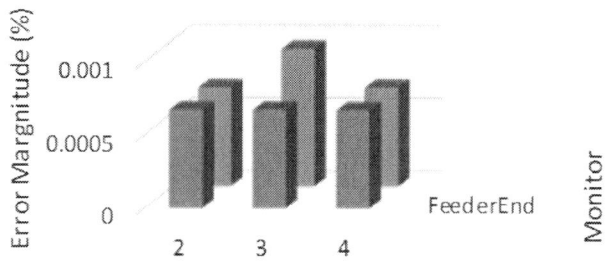

Fig. 5. Error magnitude after separating the simulation in several temporal segments

TABLE I. RESULTS ACHIEVED AFTER SPLITTING THE SYSTEM IN 4 TEMPORAL SEGMENTS

| | Actor | CPU | Initial time | Iterations | Time (sec) | Time reduction (%) |
|---|---|---|---|---|---|---|
| OpenDSS | | | 0 | 8000 | 28.33 | |
| OpenDSS-PM | 1 | 0 | 0 | 2000 | 9.68 | 65.83 |
| | 2 | 2 | 2000 | 2000 | 9.43 | 66.71 |
| | 3 | 4 | 4000 | 2000 | 9.55 | 66.29 |
| | 4 | 6 | 6000 | 2000 | 9.55 | 66.29 |

By using the script presented as example, four actors are created to split the solution time in sets of 2000 time steps (4 temporal segments). This test was performed in a computer with a processor Intel core i7-2720 @ 2.2 GHz. The errors are evaluated for each simulated cases (1, 2, 3 and 4 temporal segments) to quantify the error from temporal parallelization and how its magnitude could affect the results. The results are shown in Fig. 5 and in TABLE I.

The voltage error magnitude is below 1.0e-3 p.u., revealing that the control actions performed at each actor's startup takes the partial simulation variables to an acceptable set of values. The error magnitude becomes non-zero after the first junction point between the different time segments. However, this error is still within acceptable ranges.

Then, using this architecture the system is solved by creating 2, 3 and 4 temporal segments to evaluate the time reduction. This processor has only 2 cores in comparison with the i7 processor used in the previous test. The aim of this test is to highlight the effect of running 2 actors on the same core when trying to perform parallel processing. The results of this test are shown in TABLE II. As can be seen in TABLE II, when setting the affinity of 2 actors to the same processor's core the performance of the actor get decreased.

This an expected result since the core needs to attend 2 processes simultaneously, which is impossible. As a consequence, both actors (threads) get serialized, and the processing time for each one is higher than if the core were dedicated to a single Actor as shown in Fig. 6.

The overhead obtained after splitting the simulation in time can be justified as follows:

TABLE II. RESULTS OBTAINED WITH TEMPORAL PARALLELIZATION USING A SECOND ARCHITECTURE

| | | CPU | Iterations | Time (sec) | Time reduction (%) | |
|---|---|---|---|---|---|---|
| | 2 temporal segments | | | | | |
| OpenDSS | | | 8000 | 31.19 | | |
| OpenDSS-PM | Actor 1 | 0 | 4000 | 18.24 | 35.62 | Core 1 |
| | Actor 2 | 2 | 4000 | 17.36 | 38.72 | Core 2 |
| | 3 temporal segments | | | | | |
| OpenDSS | | | 8000 | 31.19 | | |
| OpenDSS-PM | Actor 1 | 0 | 2666 | 14.12 | 50.16 | Core 1 |
| | Actor 2 | 2 | 2666 | 16.36 | 42.25 | Core 2 |
| | Actor 3 | 3 | 2666 | 16.27 | 42.57 | |
| | 4 temporal segments | | | | | |
| OpenDSS | | | 8000 | 31.19 | | |
| OpenDSS-PM | Actor 1 | 0 | 2000 | 14.64 | 48.32 | Core 1 |
| | Actor 2 | 1 | 2000 | 13.76 | 51.43 | |
| | Actor 3 | 2 | 2000 | 13.73 | 51.54 | Core 2 |
| | Actor 4 | 3 | 2000 | 13.62 | 51.92 | |

Simulation time reduction (%)

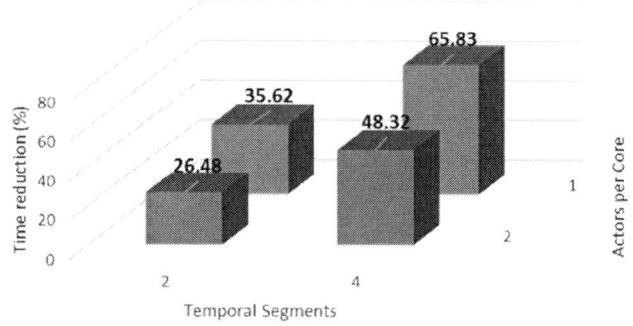

Fig. 6. Effects of having 1 or 2 Actors per Core when performing parallel processing

1.      The system's complexity remains intact, this is, the size of the system is the same after and as a consequence, it will require a startup procedure on every tearing/starting point as shown in Figure 5. Depending on the startup point in time, there could be involved more/less control actions on each temporal segment to find an accurate solution.

2.      The coordination process taken between OpenDSS-PM and other applications when an actor is launched by OpenDSS-PM, to give to the actor the highest priority can add an additional overhead for the actor startup. This process is performed by the Operating System (OS) and cannot be controlled when using standard OS.

3.      Even when the actors are running on separate cores, the processor's power consumption increases due to the dedicated priority set for each actor. Normally the system motherboard takes actions to avoid the processors overheating and to reduce the power consumption.

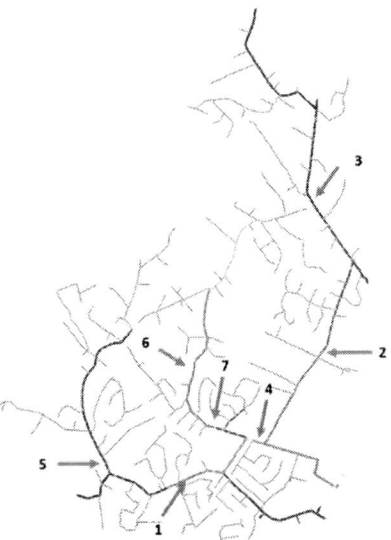

Fig. 7. EPRI circuit 5 and location of the proposed link branches

Depending on the hardware architecture (laptop, desktop PC, Type of Power Supplies Units-PSU), the performance could be affected when running multithread dedicated applications. More PC architectures running with standard operating systems need to be evaluated.

### B. Diakoptics Based on Actors

Similar parallelization QSTS simulations are performed when using the A-Diakoptics method. From previous works [2], a 70-80% reduction in solution time is expected; however, there will be some variations in time improvements based on the computing environment. In fact, depending on the number of available processor cores available and the number of actors created the results may change.

The same test is performed by using another computer with a processor Intel i5-5200 @ 2.2 GHz CPU. At this point, both computers are laptop computers.

In the case of *A-Diakoptics* the interconnected circuit is teared using 7 link branches as shown in Fig. 7. The link branches were selected to generate multiple configurations and several balanced/unbalanced distribution of nodes. The balance is measured in terms of the amount of nodes handled by each actor. If 2 actors are handling the same amount of nodes, the parallel system is balanced; otherwise, the system is unbalanced in a certain degree.

The first test consists of performing the simulation using multiple combinations of link branches to generate a multiple number of actors. The results are shown in TABLE III. For this case the simulations are performed using a co-simulation environment between NI LabVIEW and OpenDSS-PM.

Depending on the balance, the system performance will be seriously affected. For the case presented, the system is slightly balanced, however, if the unbalance gets higher the solution times can be considerably affected.

Another interesting finding is the fact that the overhead added by the co-simulation platform is significant compared with the performance of the OpenDSS-PM application working in stand-alone mode. To verify this hypothesis the time per step of each actor is totalized after each simulation considering the number of actors used to solve the system. This information is shown in TABLE IV.

The results presented in TABLE IV shows that the actors are indeed asynchronous systems. The total simulation time will be the time of the slowest actor since at each iteration the actors need to be synchronized to perform the extra calculations.

These results also reveal that while working with more actors, the total overhead added by LabVIEW becomes important. This difference becomes more relevant when each actor is executed on a separate core as shown in Fig. 8.

For the results discussed in this report we have used an Intel® Xeon® CPU E5-2650 v4 @ 2.2GHz, while in [8, 9] an Intel® CoreTM i7-4810MQ @ 3.4 GHz was used. This means that working with OpenDSS-PM to perform piecewise methods results in an accelerated simulation experience, which can be used for multiple applications including Real-Time simulation.

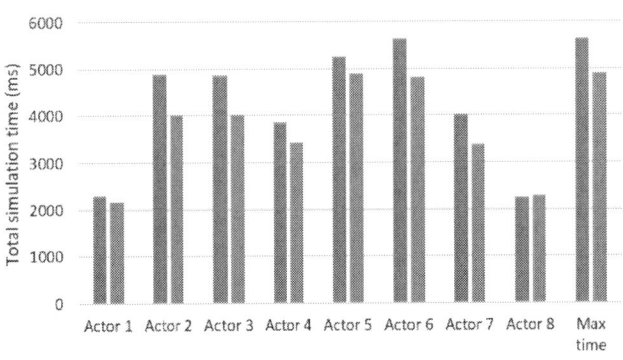

Fig. 8. Total simulation time per actor when allocating 1 or 2 actors per core

TABLE III. RESULT ACHIEVED AFTER TEARING THE SYSTEM IN MULTIPLE ACTORS

| Number of actors | Time Elapsed (ms) | Time reduction (%) | Number of Cores used |
|---|---|---|---|
| 8 | 14848 | 68 | 4 |
| 6 | 16782 | 64 | 3 |
| 4 | 16934 | 63 | 2 |
| 2 | 19050 | 59 | 1 |
| 1 | 46684 | 0 | 1 |

TABLE IV. SIMULATION TIMES FOR EACH OF THE ACTORS IN OPENDSS-PM

| # of Actors | Simulation time per actor (ms) - 8760 time steps | | | | | | | |
|---|---|---|---|---|---|---|---|---|
| | A1 | A2 | A3 | A4 | A5 | A6 | A7 | A8 |
| 8 | 2307 | 4896 | 4868 | 3870 | 5255 | 5647 | 4023 | 2264 |
| 6 | 2154 | 2555 | 4434 | 11003 | 5923 | 3607 | | |
| 4 | 3593 | 10658 | 9628 | 9958 | | | | |
| 2 | 7682 | 14413 | | | | | | |

TABLE V. Errors in percentage for each simulation case

| Node name | Error (%) per number of actors | | | |
|---|---|---|---|---|
| | 8 Actors | 6 Actors | 4 Actors | 2 Actors |
| sourcebus.1 | 0 | 0 | 0 | 0 |
| sourcebus.2 | 0 | 0 | 0 | 0 |
| sourcebus.3 | 0 | 0 | 0 | 0 |
| 1023346.1 | 2.010771 | 2.012686 | 2.01203 | 0.807802 |
| 1023346.2 | 2.0915 | 2.111095 | 2.108079 | 0.966332 |
| 1023346.3 | 2.261792 | 2.253757 | 2.236306 | 1.050958 |
| 63657.1 | 2.642765 | 2.652466 | 2.649035 | 1.039492 |
| 63657.2 | 2.638333 | 2.667622 | 2.678754 | 1.20196 |
| 63657.3 | 2.88666 | 2.865874 | 2.835851 | 1.35616 |
| 816.2 | 2.685346 | 2.288218 | 2.120706 | 1.228647 |
| x_1103251_1.3 | 3.184148 | 2.531579 | 2.637314 | 1.568667 |
| 28243.1 | 2.778495 | 2.108859 | 2.258119 | 1.084387 |
| 39582.1 | 3.310538 | 3.294892 | 2.667483 | 2.642919 |
| 39582.2 | 3.115652 | 3.106987 | 2.556511 | 2.514462 |
| 39582.3 | 3.658225 | 3.636021 | 2.87142 | 2.83938 |
| x_39760_3.2 | 2.598807 | 2.586058 | 2.175612 | 1.216944 |

To evaluate the simulation fidelity and the error introduced by the technique used in this report, several measurements are performed around the circuit. The locations of the measurements are selected randomly. The errors in percentage for each simulation case are shown in TABLE V.

In this TABLE, the error gets higher when more actors are used. However, this behavior is due to the lack of synchronism between the control actions and the modifications of the actor's YBUS matrix. Because matrixes $Z_{TC}$, $Z_{CC}$ and $Z_{TC}$ are calculated at the beginning of the algorithm, if the YBUS matrices of the actors change during the simulation, the error will increase. This source of error suggests that is necessary to investigate methods for applying the non-invasive control actions, avoiding the changing of the actor's YBUS matrix, and thus reducing the systematic error added.

## V. Conclusions

This paper has presented the use of EPRI's OpenDSS-PM (Parallel Machine), to accelerate QSTS simulations using multi-core computers. Temporal parallelization and A-Diakoptics provide new techniques to reduce the time of QSTS simulations without losing simulation fidelity. The paper has discussed in depth the application of these techniques using the standard distribution test feeders that include PV generation with realistic irradiance profiles.

With increasing penetrations of distributed energy resources (DER) and distributed PV, the ability to quickly perform QSTS simulations is a crucial aspect of distribution system planning and operations. OpenDSS-PM is an open source simulation tool available at the OpenDSS website at sourceforge.net. The example presented in this paper are available for NI LabVIEW when installing the OpenDSS-PM library for NI LabVIEW using the VI Package Manager tool [17].

## Acknowledgments

This research was supported in part by the DOE SunShot Initiative, under agreement 30691. Sandia National Laboratories is a multimission laboratory managed and operated by National Technology and Engineering Solutions of Sandia, LLC, a wholly owned subsidiary of Honeywell International, Inc., for the U.S. Department of Energy's National Nuclear Security Administration under contract DE-NA0003525.

## References

[1] S. Jothibasu and S. Santoso, "Sensitivity analysis of photovoltaic hosting capacity of distribution circuits," in *2016 IEEE Power and Energy Society General Meeting (PESGM)*, 2016, pp. 1-5.

[2] M. Rylander, J. Smith, D. Lewis, and S. Steffel, "Voltage impacts from distributed photovoltaics on two distribution feeders," in *2013 IEEE Power & Energy Society General Meeting*, 2013, pp. 1-5.

[3] M. Rylander, J. Smith, W. Sunderman, D. Smith, and J. Glass, "Application of new method for distribution-wide assessment of Distributed Energy Resources," in *2016 IEEE/PES Transmission and Distribution Conference and Exposition (T&D)*, 2016, pp. 1-5.

[4] M. J. Reno, K. Coogan, R. Broderick, and S. Grijalva, "Reduction of distribution feeders for simplified PV impact studies," in *2013 IEEE 39th Photovoltaic Specialists Conference (PVSC)*, 2013, pp. 2337-2342.

[5] M. J. Reno, M. Lave, J. E. Quiroz, and R. J. Broderick, "PV ramp rate smoothing using energy storage to mitigate increased voltage regulator tapping," in *2016 IEEE 43rd Photovoltaic Specialists Conference (PVSC)*, 2016, pp. 2015-2020.

[6] M. J. Reno, J. Deboever, and B. Mather, "Motivation and Requirements for Quasi-Static Time Series (QSTS) for Distribution System Analysis," presented at the IEEE Power Engineering Society General Meeting, Chicago, 2017.

[7] R. C. Dugan and T. E. McDermott, "An open source platform for collaborating on smart grid research," in *2011 IEEE Power and Energy Society General Meeting*, , 2011, pp. 1-7.

[8] D. Montenegro, G. A. Ramos, and S. Bacha, "Multilevel A-Diakoptics for the Dynamic Power-Flow Simulation of Hybrid Power Distribution Systems," *IEEE Transactions on Industrial Informatics*, vol. 12, pp. 267-276, 2016.

[9] D. Montenegro, G. A. Ramos, and S. Bacha, "A-Diakoptics for the Multicore Sequential-Time Simulation of Microgrids Within Large Distribution Systems," *IEEE Transactions on Smart Grid*, vol. PP, pp. 1-9, 2016.

[10] J. Deboever, X. Zhang, M. J. Reno, R. J. Broderick, S. Grijalva, and F. Therrien, "Challenges in reducing the computational time of QSTS simulations for distribution system analysis," Sandia National Laboratories SAND2017-5743, 2017.

[11] J. Diaz, C. Munoz-Caro, and A. Nino, "A Survey of Parallel Programming Models and Tools in the Multi and Many-Core Era," *IEEE Transactions on Parallel and Distributed Systems*, vol. 23, pp. 1369-1386, 2012.

[12] D. Montenegro, G. A. Ramos, and S. Bacha, "A-Diakoptics for the Multicore Sequential-Time Simulation of Microgrids Within Large Distribution Systems," *IEEE Transactions on Smart Grid*, vol. 8, pp. 1211-1219, 2017.

[13] C. Hewitt, "Actor Model of Computation: Scalable Robust Information Systems," in *Inconsistency Robustness 2011, Stanford University*, 2012, p. 32.

[14] G. Kron, *Diakoptics: the piecewise solution of large-scale systems*: Macdonald, 1963.

[15] H. H. Happ, *Piecewise Methods and Applications to Power Systems*: Wiley, 1980.

[16] J. Fuller, W. Kersting, R. Dugan, and S. C. Jr. (2013, 10-23). *Distribution Test Feeders*. Available: http://ewh.ieee.org/soc/pes/dsacom/testfeeders/

[17] JKI. (2016, 01/06/2017). *VI Package Manager*. Available: https://vipm.jki.net/

# An autocorrelation-based copula model for producing realistic clear-sky index and photovoltaic power generation time-series

Joakim Munkhammar and Joakim Widén

Built Environment Energy Systems Group, Div. Solid State Physics, Engineering Sciences,
The Ångström Laboratory, Uppsala University, Uppsala, SE-751 21, Sweden.

*Abstract*—This study presents a method for using copulas to model the temporal variability of the clear-sky index. The method utilizes the autocorrelation function and correlated outputs for $N$ time-steps are obtained. Results from the copula model are, in terms of distribution, autocorrelation, step changes and mean daily distribution, compared with the original data set and with an uncorrelated model based on random clear-sky index data. The copula model is shown to be superior to the uncorrelated model in all these aspects.

*Index Terms*—autocorrelation function, clear-sky index, copula modeling, distribution modeling, realistic time-series.

## I. INTRODUCTION

Estimating the variability of solar irradiance on Earth's surface is necessary for many solar engineering applications, in particular those involving photovoltaic (PV) power generation [1]–[3]. For example, by quantifying the solar irradiance variability, it is possible to improve the design and operation of power systems where large amounts of distributed PV power are injected into the grid [1], [3], and thereby avoid costly grid reinforcements [1], [4].

Instantaneous or near instantaneous solar irradiance, normalized to the clear-sky index, is typically modeled as a probability distribution with two or three peaks [5]–[7]. Models for generating time-series of the clear-sky index include Markovchain [8], neural network [9], general algorithmic [10], and pure sampling from probability distributions [7], [11]. Models vary in terms of input data complexity from utilizing cloud size, coverage and morphology [8] to analysis of the clearsky index time-series [7], [11]–[13]. A challenge is not only to obtain an accurate probability distribution, but to obtain a realistic synthetic time-series of the clear-sky index as well [14]. This has spawned a series of clear-sky index timeseries generators, such as [10], [14], where for example lower resolution data can be used as input for generating synthetic realistic higher resolution clear-sky index data [10].

In terms of time-series realism, autocorrelation is a useful measure. Autocorrelation of the clear-sky index has been studied previously [15]–[18], where the autocorrelation function is positive and follows an exponential slope for hour resolution [16], [17], [19], while it can also display negative values for minute resolution [18]. Models utilizing the autocorrelation of the clear-sky index include Gaussian-Markov [15], autoregressive Gaussian [17], neural networks [9] and fractal cloud modeling [20].

Autocorrelation functions of the clear-sky index have also been used in so-called virtual networks to study the variabiliy

of spatially distributed fleets of photovoltaic plants [3], [21], [22], These approaches propose that the autocorrelation function represents the correlation between time-series of spatially separated metering stations. This shows that modeling the autocorrelation function for the clear-sky index can be useful for both temporal and spatial clear-sky index, and in turn solar engineering applications.

Regarding spatial solar irradiance variability, copulas have been used to model the variability of solar irradiance in a spatial network [12], [13]. Copula modeling is an increasingly common method for modeling correlation between stochastic variables [23], which has been used in studies of wind power [24], waves power [25] and solar irradiance [12], [13].

This study develops an autocorrelation-based copula model for generating synthetic clear-sky index data. A copula basedmodel for quantifying the temporal variability of the clear-sky index has not been done previously. It continues the work in [12], [13], which modeled spatial variability in the clear-sky index by using copulas and a spatiotemporal data set of the clear-sky index.

## II. METHODOLOGY

The methodology is organized so that the clear-sky index and the main method is presented in Section II-A, the formal modeling in Section II-B, goodness-of-fit test statistics in Section II-C and the data set used in this study is presented in Section II-D.

### A. The clear-sky index

In order to focus on the temporal variability of instantaneous solar irradiance the clear-sky index is used. Formally, the clear-sky index $\kappa$ is defined as the ratio between the measured global horizontal irradiance (GHI) $G(t)$ and the estimated global horizontal clear-sky irradiance $G_c(t)$ over time $t$:

$$\kappa(t) \equiv \frac{G(t)}{G_c(t)}. \tag{1}$$

The temporal copula method developed in this paper is based on the assumption that the clear-sky index is not dependent on time of day or season. Generally, throughout this study, it is assumed that the time-series of the clear-sky index is approximately *stationary*, a property of time-series which means that the joint probability distribution of a set of equivalently separated data points in a time-series does not change over time [26]. This is assumed as an approximation for the clear-sky index, and refinements of this is elaborated upon in the results and discussion sections.

Generally, since the clear-sky index is primarily available only during daytime, there is a maximum length of each daily vector of consecutive points of the clear-sky index. In order to have sufficient data for modeling this prompts the use of clear-sky index for a number of days.

Based on the assumption that the clear-sky index time-series for each day is stationary, the main model of the paper is based on the following algorithmic steps.

1. Obtain a clear-sky index data set of for $N$ data points on $M$ days, thus generating an $N \times M$ data set.
2. Estimate the autocorrelation function of the clear-sky index for each day, generating a data set of size $N \times M$.
3. Compute the mean autocorrelation function, for $N$ steps, over all $M$ days, generating a vector of the mean autocorrelation function of size $N$.
4. Define a correlation matrix, of size $N \times N$, from the $N$ mean autocorrelation function values.
5. Define a copula based on the empirical distribution of the clear-sky index data set for $N$ data points on all $M$ days and the autocorrelation-based correlation matrix.
6. Generate, with the copula model, $M$ number of synthetic correlated clear-sky index time-series with $N$ data points each.

The concepts of these steps will be clarified, in a formal sense, in the following section. The proposed copula model is validated by comparing the synthetic copula-generated data sets with:

- Original data set time-series. (The data set)
- Sampling from the empirical distribution of clear-sky index. (The uncorrelated model)

The latter of these, the sampling from the original clear-sky index data set, is the special case of the copula model with zero autocorrelation for non-zero lag, rendering the correlation matrix an identity matrix. This model will generally be called the *uncorrelated model* in this paper, since it lacks temporal correlation, which makes it equivalent to conventional probability distribution clear-sky index models (similar to the models used in e.g. [6], [7]).

In terms of machine learning it is customary to divide up training and testing data sets. This was tested and proved to have little effect on this study, it only gave a shorter sample of days to use for both training and testing (and thus higher variability when comparing results).

### B. Temporal copula modeling of the clear-sky index

This section describes the formal copula modeling of the clear-sky index time-series, used in this study.

Copula modeling is a general statistical framework based on the concept that correlated output from a given number of stochastic variables, which have a dependency that is reflected in the correlation between variables, can be modeled via a so-called copula; that is, a joint distribution function for all stochastic variables considered.

An $N$-variable copula is used in this paper, and it models the dependency between $N$ stochastic variables by generating $N$ sets of correlated values for each stochastic variable based on a probability distribution. For this paper the distributions are all of the same type: each is the empirical clear-sky index distribution (from the entire data set).

Formally, it is common to use the so-called inversion method for obtaining a copula, which is also used in this paper. This method is briefly defined as follows [23, p.51], with comments on how it is applied in this paper.

Assume that $N$ stochastic variables $X_1, X_2, ..., X_N$ with cumulative distribution functions (CDFs) $F_{X_1}, F_{X_2}, ..., F_{X_N}$ are given. In this case $X_1$ represents the stochastic variable of one single time-step, and $X_2$ the stochastic variable of another time-step, etc. The main point is to model the dependency between temporally separated time-steps in a time-series, and that this variability is universal for any set of such points in the time-series, since the time-series is considered approximately stationary. $N$ consecutive time-steps are modeled in this study, since that constitutes a complete time-series of $N$ consecutive points. There is, however, in principle no requirement of the model for using consecutive time-steps.

Based on this setup, then in accordance with Sklar's theorem, which certifies the existence of a copula, see [23], these distributions can be joined via a copula $C$ if the copula can be expressed as:

$$F_{X_1, X_2, ..., X_N}(X_1, X_2, ..., X_N) = \qquad (2)$$
$$C(F_{X_1}(x_1), F_{X_2}(x_2), ..., F_{X_N}(x_N)). \qquad (3)$$

The copula function can be written:

$$C(u_1, u_2, ..., u_N) = \qquad (4)$$
$$F_{X_1, X_2, ..., X_N}(F_{X_1}^{-1}(u_1), F_{X_2}^{-1}(u_2), ..., F_{X_N}^{-1}(u_N)). \qquad (5)$$

for $F_{X_i} = u_i$, $i = 1, 2, ..., N$, where $u_1, u_2, ..., u_N$ are realizations of uniform variables $U_1, U_2, ..., U_N$. A special case is a copula with two stochastic variables $X_1$ and $X_2$: a bivariate copula. Such a copula would, in this study, correspond to modeling the dependency between a single pair of time-steps, and be represented by a copula function:

$$C(u_1, u_2) = F_{X_1, X_2}(F_{X_1}^{-1}(u_1), F_{X_2}^{-1}(u_2)). \qquad (6)$$

The key to setting up a copula function is to have a correlation matrix which expresses the correlation between the stochastic variables. Such a matrix is by necessity square.

In this case, the correlation matrix is generated from a time-series of the clear-sky index with $N$ data points. This correlation matrix is based on the mean autocorrelation function of the clear-sky index, over all days, according to step 3 in the algorithm. In formal terms the autocorrelation function for $n$ steps can be defined as $A(n)$ for $n$ number of steps lag ($n = 1$ represents zero-lag autocorrelation). For a data set with $N$ number of data points $y_i$, mean value $\bar{y}$

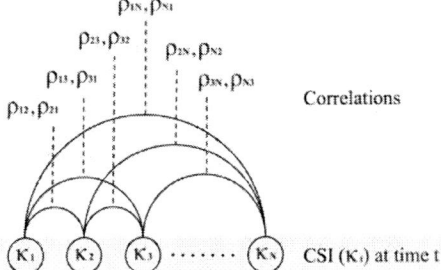

Fig. 1. An illustration of the copula-based autocorrelation method developed in this study. Elements of the correlation matrix $\rho$ are obtained from the autocorrelation function of the clear-sky index.

and standard deviation $\sigma$ the autocorrelation function $A(n)$ for lag $n$ is defined as [27]:

$$A(n) = \frac{1}{\sigma N} \sum_{t=1}^{N-n} (y_t - \bar{y})(y_{t-n} - \bar{y}). \tag{7}$$

Estimates of the autocorrelation function is reasonable for $n$ up to approximately $N/4$, see [27, p.31]. The temporal correlation matrix, incorporating the correlation between the set of consecutive time-steps, becomes an *autocorrelation matrix* with elements based on $A(n)$, defined according to [28]:

$$\rho = \begin{pmatrix} A(1) & A(2) & \cdots & A(n) \\ A(2) & A(1) & \cdots & A(n-1) \\ \vdots & \vdots & \ddots & \vdots \\ A(n) & A(n-1) & \cdots & A(1) \end{pmatrix}, \tag{8}$$

which is symmetric, with elements $A(1) = 1$. This matrix is used as the conventional correlation matrix in copula-modeling, see [12], [13] for more information relating to the clear-sky index application. See Fig. 1 for a schematic illustration of this proposed autocorrelation concept.

The next key step in copula modeling is the choice of distribution for each stochastic variable to be correlated by the copula. In this case the empirical distribution of the clear-sky index (for all days of the data set) is chosen.

The copula itself is based on a separate distribution, the type of which has to be set *a priori*, but since this is the first study of its kind, by temporally modeling the clear-sky index, the common normal distribution configuration was chosen.

The model developed in this paper was implemented in MATLAB, and the built-in functions `autocorr` for the autocorrelation function, and `copularnd` for the copula random number generator, were used.

### C. Kolmogorov-Smirnov goodness-of-fit test statistic

In this study the Kolmogorov-Smirnov (K-S) test statistic was also used to determine probability distribution goodness-of-fit. This is a commonly used estimate to quantify the

similarity between two probability distributions. The test statistic $S$ is based on the maximum deviation between two distributions [29]:

$$S = \max_x |F_1(x) - F_2(x)| \tag{9}$$

where $F_1(x)$ and $F_2(x)$ are empirical (or analytical) cumulative distribution functions. In this study, K-S test statistics are used to estimate the goodness-of-fit for the resulting aggregate clear-sky index distributions from data and simulations. This type of measure has been used in previous studies as well, for example in probability distribution modeling [7] and load matching [11]. For this study the Matlab routine `kstest2` was used.

### D. Data set

The data on clear-sky index used for this study is based on minute mean value resolution radiometer measurements of global horizontal irradiance for one year (2008) obtained from the Swedish Meteorological and Hydrological Institute (SMHI) for Norrköping, Sweden (59°35′31" N 17°11′8" E) [30]. The clear-sky index was calculated with estimates of the clear-sky irradiance, obtained from the Ineichen-Perez model [31]. In order to avoid low solar angles, while maintaining the same amount of time-step data per day (a necessity of the model), 120 data points (minutes) per day (centered around noon each day) were used as input data set. This totals 43'800 ($120 \times 365$) data points of clear-sky index data, which generates a $120 \times 120$ autocorrelation matrix. The simulations of the models were also set to generate $120 \times 365$ data points. Although minute resolution was used in this paper, there is no methodological requirement preventing the use of any other resulutions (higher or lower), as long as the time-series can be considered approximately stationary, or that deviations can be quantified.

### III. RESULTS

In Fig. 2 the distribution, autocorrelation function and step-change distribution for the clear-sky index data set, the copula model simulation and the uncorrelated model simulation are presented. As a check for model implementation consistency, the clear-sky index generated by the models should be similar compared with the original data set, which appears to be the case when inspecting the histograms of the original data set and 43'800 simulated data points from each model, presented in subplots (A)-(C). (A) represents the original data set, (B) the copula model and (C) the distribution model.

Both the copula model and the uncorrelated model yielded a Kolmogorov-Smirnov test statistic of 0.0042 when compared with the original data set, which certifies a high goodness-of-fit and large enough sample size of data points in the modeling. In subplot (D) the copula model is shown to follow the autocorrelation function closely, by visual inspection, in particular when compared with the uncorrelated model. The discrepancy between the input (data set) and the output (copula model) autocorrelation is some artifact of the

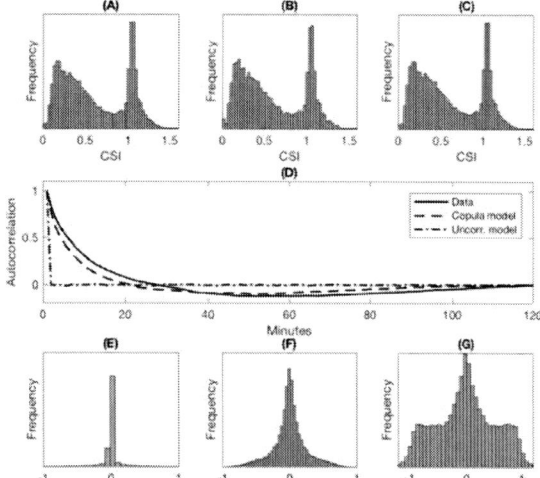

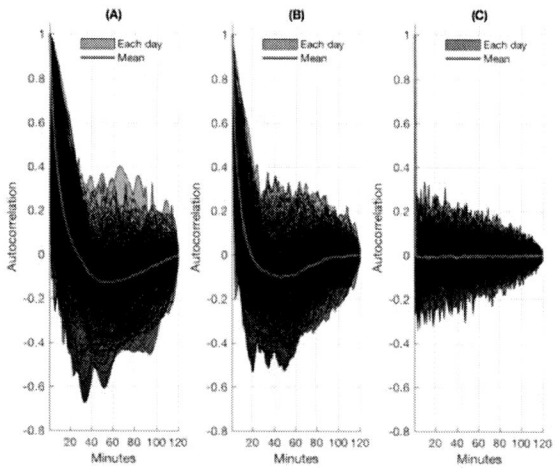

Fig. 3. The autocorrelation function over time for each day for (A) the data set, (B) the copula simulation and (C) the uncorrelated model. The red line in each subplot represents the mean autocorrelation function for each minute over all the 365 days measured and simulated data.

Fig. 2. Subplots (A-C) represent histograms of the clear-sky index, for (A) the data set, (B) the copula model simulation and (C) the uncorrelated model simulation. Subplot (D) represents the autocorrelation function for the data set and the two model simulations. Subplots (E-G) represent histograms of step changes of (E) the data set, (F) the copula model simulation and (G) the uncorrelated model simulation.

copula modeling, which could be different if rank correlations were used instead, which is left as an open issue for further investigation.

Subplots (E-G) represent step-changes, and the copula model (F) is closer (by visual inspection) to the data set for step-changes compared with the uncorrelated model (G).

In Fig. 3 the mean autocorrelation for the data set and for the copula model output is given along with the auto-correlation function for $N$ data points on each of the 365 days in the data set. Results show that the copula model appears (by visual inspection) to capture the autocorrelation features of the clear-sky index, while the uncorrelated model is less accurate. Despite the completely uncorrelated time-series of the uncorrelated model, there is a span of non-zero autocorrelation functions for each day. This is an artefact of applying the autocorrelation function to the 120 generated points of each days, this will cause daily variability of the autocorrelation function, which on average is zero.

It is interesting that the mean autocorrelation for both the data set and the copula model show negative values. It should however be noted that, as was mentioned in Section II-B, the autocorrelation function of a data set of $N$ points is only reasonable up to about $n = N/4$, which in this case becomes $n = 30$. It is interesting that the mean autocorrelation is negative even at $n = 30$, so this might be an actual feature of the mean autocorrelation of the clear-sky index for this resolution. More autocorrelation studies of the clear-sky index, for various time-scales and regions, could assist in

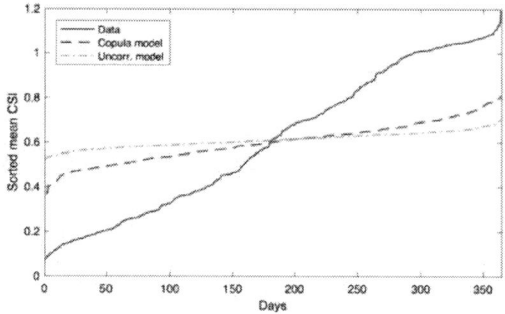

Fig. 4. Sorted (ascending) daily mean (over 120 data points) clear-sky index for the data set, the copula model simulation and the uncorrelated model simulation.

determining this.

Sorted (ascending) mean clear-sky index level for each day (each 120 point data set) is shown in Fig. 4 for the data set, the copula model simulation and the uncorrelated model simulation. Although the copula model simulation yields a result that is closer (by visual inspection) to the data set than the uncorrelated model the copula model does not entirely capture the variability in daily mean clear-sky index. It shows that although the copula generated model simulation closely represents the clear-sky index data set in terms of mean autocorrelation and variability, as seen in Fig. 4, this variability is restricted to each day and not spread out over days with different mean clear-sky index, as in the original data set.

A possible way to improve the result would be to set up a

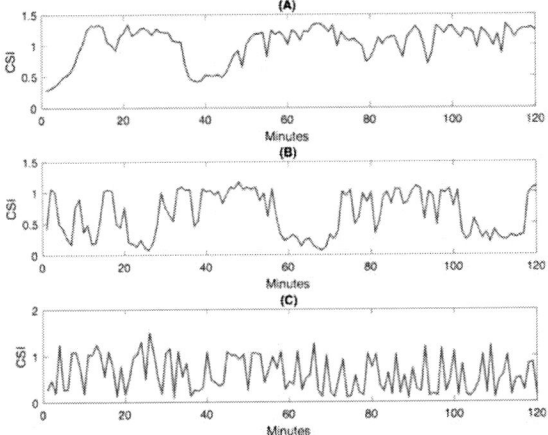

Fig. 5. Example time series of 120 consecutive minutes on one day in the data set (A), the copula model (B) and the uncorrelated model (C).

mean autocorrelation function for bins of mean daily clear-sky irradiance [32]. In Fig. 5 example time-series of 120 consecutive minutes on one day for the data set (A), the copula model (B) and the uncorrelated model (C) are shown. There is a an apparent better statistical match between the copula model and the time-series compared with the uncorrelated model. The day of the data set (day 17) represents a broken cloud day, and it should be emphasized that the statistical match varies between different types of days depending on weather type. This also suggests that splitting the data set into bins of mean daily clear-sky index levels could improve model goodness-of-fit.

Applying this model to generate realistic synthetic time-series of PV power generation for arbitrary oriented PV system can be made in a series of steps, such as outlined for example in [33].

## IV. DISCUSSION

This paper presents a general method for using copulas to model the temporal variability, and to produce a realistic time-series, of the clear-sky index. Results show that the copula model simulation, based on the data set on clear-sky index, produces a clear-sky index time-series whose goodness-of-fit in terms of step-changes, autocorrelation and daily mean clear-sky index variability is superior to the uncorrelated model. K-S test statitics and visual inspection were used to estimate goodness-of-fit for distributions in this study.

The method is useful for generating realistic synthetic clear-sky index time-series, which in turn, with the aid of an appropriate photovoltaic power generation model can be used to generate realistic synthetic photovoltaic power generation time-series. This in turn could be useful for grid-integration studies, where probabilistic load-flow calculations potentially could be complemented with a synthetic time series of PV

power generation. Since copula modeling of spatially dispersed PV power generation was recently shown to improve estimates for grid simulations, in [32], the use of temporal copula modeling could prove useful as well.

Extending the model to include modeling of the clear-sky index for bins of clear-sky index data is hypothesized to improve the results of this study.

Also, devising a forecasting method that utilizes this copula model could be valuable, in particular for matching of electric power use and generation.

## ACKNOWLEDGMENT

This project was funded by the project "Development and evaluation of forecasting models for solar power and electricity use over space and time", and the EU-ERA.Net Smart Grids Plus project "Increased Self Consumption of Photovoltaic Power for Electric Vehicle Charging in Virtual Networks", both funded by the Swedish Energy Agency. Also, the valuable input provided by Jack Daniel is acknowledged.

## REFERENCES

[1] M. H. J. Bollen, F. Hassan, *Integration of Distributed Generation in the Power System*, Wiley IEEE press series on Power Engineering, 2011.

[2] J. Kleissl, *Solar Energy Forecasting and Resource Assessment*, 1st Edition, Academic Press, 2013.

[3] J. Widén, "A model of spatially integrated solar irradiance variability based on logarithmic station-pair correlations", *Solar Energy*, vol. 122, pp. 1409-1424, 2015.

[4] H. Holttinen, M. Milligan, B. Kirby, T. Acker, V. Neimane, T. Molinski, "Using standard deviation as a measure of increased operational reserve requirement for wind power", *Wind Eng.*, vol. 32, pp. 355-378, 2008.

[5] K. G. T. Hollands, R. G. Huget, "A probability density function for the clearness index, with applications", *Solar Energy*, vol. 30, pp. 195-209, 1983.

[6] K. G. T. Hollands, H. Suehrcke, "A three-state model for the probability distribution of instantaneous solar radiation, with applications", *Solar Energy*, vol. 96, pp. 103-112, 2013.

[7] J. Munkhammar, J. Rydén, J. Widén, D. Lingfors, "Simulating dispersed photovoltaic power generation using a bimodal mixture model of the clear-sky index", in Proceedings of EU-PVSEC 2015.

[8] J. M. Bright, C. J. Smith, P. G. Taylor, R. Crook, "Stochastic generation of synthetic minutely irradiance time series derived from mean hourly weather observation data", *Solar Energy*, vol. 115, pp. 229-242, 2015.

[9] C. Voyant, M. Muselli, C. Paoli, M.-L. Nivet, "Optimization of an artificial neural network dedicated to the multivariate forecasting of daily global radiation", *Energy*, vol. 36, pp. 348-359, 2011.

[10] A. P. Grantham, P. J. Pudney, L. A. Ward, M. Belusko, J. W. Boland, "Generating synthetic five-minute solar irradiance values from hourly observations", *Solar Energy*, vol. 147, 209-221 (2017).

[11] J. Munkhammar, J. Rydén, J. Widén, "On a probability distribution model combining household power consumption, electric vehicle home-charging and photovoltaic power production", *Applied Energy*, vol. 142, pp. 135-143, 2015.

[12] J. Munkhammar, J. Widén, "Copula correlation modeling of aggregate solar irradiance in spatial networks", in Proceedings of Solar Integration Workshop, 2016.

[13] J. Munkhammar, J. Widén, L. Hinkelman, "A copula method for estimating aggregate instantaneous solar irradiance in spatial networks", *Solar Energy*, vol. 143, pp. 10-21, 2017.

[14] J. M. Bright, O. Babacan, J. Kleissl, P. G. Taylor, R. Crook, "A synthetic, spatially decorrelating solar irradiance generator and application to a LV grid model with high PV penetration", *Solar Energy*, vol. 147, pp. 83-98, 2017.

[15] B. J. Brinkworth, "Autocorrelation and stochastic modelling of insolation sequences", *Solar Energy*, vol. 19, pp. 343-347, 1977.

[16] A. Skartveit, J. A. Olseth, "The probability density and autocorrelation of short-term flobal and beam irradiance", *Solar Energy*, vol. 49, pp. 477-487, 1992.

[17] R. Aguiar, M. Collares-Pereira, "TAG: A time-dependent, autoregressive, Gaussian model for generating synthetic hourly radiation", *Solar Energy*, vol. 49, pp. 167-174, 1992.

[18] C. W. Hansen, J. S. Stein, A. Ellis, "Statistical Criteria for Characterizing Irradiance Time Series", Sandia Report SAND2010-7314 2010.

[19] A. Hammer, H. G. Beyer, "Solar Radiation, Spatial and Temporal Variability", chapter in *Solar Energy*, Eds. C. Richter, D. Lincot, A. Guermard, Springer New York 2013, pp. 634-648.

[20] G. H. Lohmann, A. Hammer, A. H. Monahan, T. Schmidt, D. Heinemann, "Simulating clear-sky index increment correlations under mixed sky conditions using a fractal cloud model", *Solar Energy*, vol. 150, 255-264, 2017.

[21] T. E. Hoff, R. Perez, "Modeling PV fleet output variability", *Solar Energy*, vol. 86, pp. 21772189, 2012.

[22] T. E. Hoff, R. Perez, "Quantifying PV power output variability", *Solar Energy*, vol. 84, pp. 1782-1793, 2010.

[23] R. B. Nelsen, *An Introduction to Copulas*, New York, New York, Springer Series in Statistics, second edition, 2006.

[24] S. Hagspiel, A. Papaemannouli, M. Schmid, G. Andersson, "Copula-based modeling of stochastic wind power in Europe and implications for the Swiss power grid", *Applied Energy*, vol. 96, pp. 33-44, 2012.

[25] W. Li, J. Isberg, W. Chen, J. Engström, R. Waters, O. Svensson, M. Leijon, "Bivariate joint distribution modeling of wave climate data using a copula method", *Energy Stat.*, vol. 04, 1650015 (2016).

[26] L. H. Koopmans, *The spectral analysis of Time Series*, Probability and mathematical statistics, Elsevier Science, 1995.

[27] G. E. P. Box, G. M. Jenkins, G. C. Reinsel., G. C. Ljung, *Time Series Analysis: Forecasting and Control*, Fifth edition, Wiley Series in Probability and Statistics, Wiley 2016.

[28] M. H. Hayes, *Statistical Digital Signal Processing and Modeling*, John Wiley & Sons, Inc. 1996.

[29] W. J. Conover, *Practical Nonparametric Statistics*, 3rd edition, Wiley, 1999.

[30] Swedish Meteorological and Hydrological Institute [Internet]. Norrköping, Sweden: SMHI; 2008. Available from: http://www.smhi.se/en.

[31] P. Ineichen, R. Perez, "A new airmass independent formulation for the Linke turbidity coefficient", *Solar Energy*, vol. 73, pp. 151-157, 2002.

[32] J. Widén, M. Shepero, J. Munkhammar, "On the properties of aggregate clear-sky index distributions and an improved model for spatially correlated instantaneous solar irradiance", manuscript 2017.

[33] J. Munkhammar, J. Widén, "Correlation modeling of instantaneous solar irradiance with applications to solar engineering", *Solar Energy*, vol. 133, pp. 14-23, 2016.

# Maximum Power Point Tracking of PV Module Based on New Explicit *I-V* Relation

Tejeswar Nukala, A. K. Panchal

Electrical Engineering Department

Sardar Vallabhbhai National Institute of Technology, Surat-395007, Gujarat, India.

*Abstract* — **This paper proposes an algorithm for maximum power point tracking of a photovoltaic (PV) module using an explicit current-voltage (*I-V*) relation. The model considers a variation in *I* with linear and power dependency on *V*. The proposed model is validated for a 160 W polycrystalline silicon module connected to a resistive load through a boost converter under different environmental conditions in MATLAB and experimentally. The performance of the proposed algorithm is compared with an existing model based algorithm or P&O and also analyzed for tolerance and aging effect. The accuracy of the proposed algorithm is better than the existing algorithm.**

## I. INTRODUCTION

Electricity generation thorough a photovoltaic (PV) system is one of the developing renewable energy technologies. The PV electricity generation is widely dependent on the environmental conditions. To make its full utilization, the maximum power has to be transferred between the PV source and the load. In order accomplish this task, a maximum power point (MPP) controller is required. An MPP controller takes action based on a reference voltage signal decided by an in-built algorithm. Usually, perturb-and-observe and incremental conductance algorithms [1] are employed. These algorithms track the MPP irrespective of the operating conditions, however, they have limitations of low convergence speed and steady state oscillations due to their iterative process. To overcome this problem, the researchers started to develop a model based MPP algorithms. The algorithms based on the equivalent circuit model accurately predict the MPP. However, the model is implicit in nature which requires the five parameters to be evaluated on-line for every change in the PV operating conditions and thereby increases on-line computational burden. To improve the response of the MPP algorithm, an explicit relation between *V* and *I* are developed. In ref. [2], the term ($V + IR_s$) in the equivalent circuit *I-V* relation is replaced by ($\alpha_0 + \alpha_1 V + \alpha_2 V^2 + \alpha_3 V^3$) which increases the unknown parameters from 5 to 8 in number. This results an increase in the real-time computational burden. In another work [3], a direct empirical expression of $V_{mp}$ as a function of measured quantities *V*, *I* and module temperature *T* is established. But, the MPP computation process requires updating nine coefficients for each variation in the PV operating condition. In order to reduce the number of coefficients the expression is modified to four [3]. The four coefficients are estimated using least-square-fitting in a-priori using 200 *I-V* curves. In one more MPP study [4], an inverse

model (*V* as an explicit function of *I*) is considered by neglecting the shunt resistance ($R_{sh}$) and the model variables are updated in real-time with a change in the temperature and irradiance. All the model based MPP algorithms require a pyranometer either for calculating the coefficients a-priori [4] or in real-time.

This paper presents an approximate explicit model (linear and power dependency of *I* on *V*) consisting four coefficients whose values can be derived from the manufacturer's datasheet. This model does not require the pyranometer measurements. The dependence of coefficients on irradiance and temperature are discussed in section IV. The sensed PV signals (*I*, *V* and *T*) are used to determine the coefficients and finally, the voltage at the MPP ($V_{mp}$) is evaluated using Newton Raphson (N–R) method as explained in section V. This section also presents simulation results for different irradiance and temperature changes and also described about tolerance and aging effect.

## II. PROPOSED EXPLICIT *I-V* MODEL

The following explicit *I-V* function is considered (with four coefficients *a*, *b*, *c* and *d*) which closely aligns to the *I-V* characteristic of a silicon PV module passing through key points (0, $I_{sc}$), ($V_{oc}$, 0) and the MPP ($V_{mp}$, $I_{mp}$),

$$I = a*V^b + c*V + d \qquad (1)$$

The coefficients are derived from four known conditions: $I_{sc}$, $V_{oc}$, the MPP ($V_{mp}$, $I_{mp}$) and $dP/dV=0$.

At $I_{sc}$,

$$d = I_{sc}. \qquad (2)$$

At $V_{oc}$,

$$c = -a*V_{oc}^{b-1} - \frac{I_{sc}}{V_{oc}}. \qquad (3)$$

Using eq.(1) at the MPP and eq.(2)-(3); the following expression for *a* is derived,

$$a = \frac{\dfrac{I_{sc}}{V_{oc}} + \dfrac{I_{mp} - I_{sc}}{V_{mp}}}{V_{mp}^{b-1} - V_{oc}^{b-1}}. \qquad (4)$$

At the MPP, $dP/dV = 0$ or $dI/dV = -I_{mp}/V_{mp}$ and from eq. (1), the following can be derived

$$-\frac{I_{mp}}{V_{mp}} = a*b*V_{mp}^{b-1} + c \quad . \tag{5}$$

Substituting eq.(3)–(4) in eq.(5), eq.(6) is obtained only in terms of $b$,

$$\left(\frac{I_{sc}}{V_{oc}} - \frac{I_{mp}}{V_{mp}}\right) = \left(b*V_{mp}^{b-1} - V_{oc}^{b-1}\right)*\frac{\dfrac{I_{sc}}{V_{oc}} + \dfrac{I_{mp} - I_{sc}}{V_{mp}}}{V_{mp}^{b-1} - V_{oc}^{b-1}} . \tag{6}$$

Eq. (6) can be solved for b using '*fsolve*' command in MATLAB with an initial guess greater than zero. Subsequently, the values for $c$ and $a$ are determined using eq. (3) and (4), respectively.

### III. VALIDATION OF PROPOSED MODEL

The model is validated for a 160 W polycrystalline silicon module. The specifications of the module are 43 V, 4.96 A, 34.9 V and 4.58 A for $V_{oc}$, $I_{sc}$, $V_{mp}$ and $I_{mp}$, respectively at STC. The obtained coefficients from the specifications are: $a = -2.0563 \times 10^{-20}$, $b = 12.4769$, $c = -0.0004302$ and $d = 4.96$. Fig.1.a illustrates the comparison of the $I$-$V$ characteristics for different irradiances ($G$) at temperature ($T = 298$ K) obtained by the N–R technique and the model. They correctly coincide with each other and confirm the validity of the model for variations in $G$. Similarly Fig.1.b shows the evaluation of the $I$-$V$ characteristics for different temperatures ($T$) at $G = 1000$ W/m$^2$ obtained by the N–R technique and the model. They exactly match with each other and confirm the validity of the model for changes in $T$.

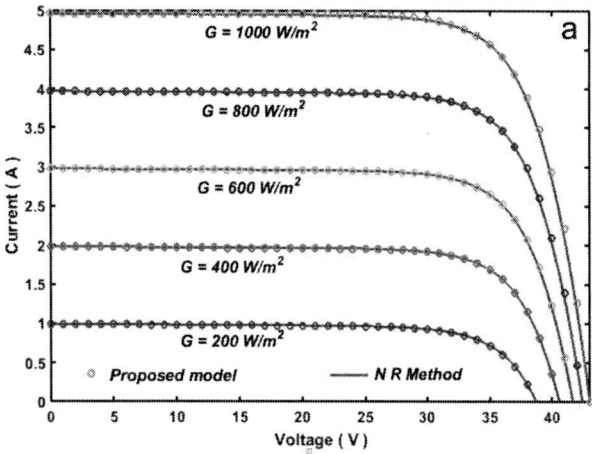

Fig. 1.(a) Comparison of the $I$-$V$ characteristics by N–R technique and the proposed model with variations in $G$ at $T$=298 K.

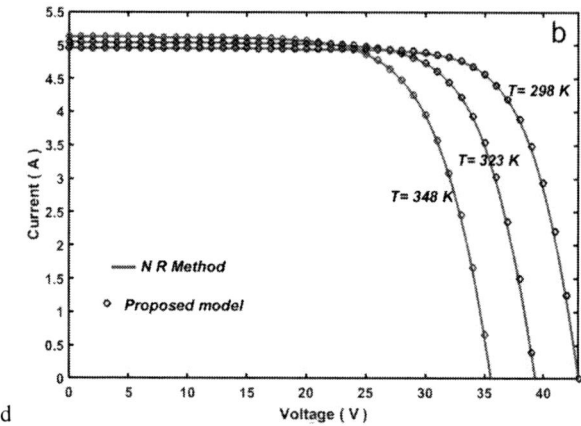

Fig. 1. (b) Comparison of the $I$-$V$ characteristics by N–R technique and the proposed model with changes in T at G =1000 W/m$^2$

### IV. DEPENDENCY OF MODEL COEFFICIENTS ON IRRADIANCE AND TEMPERATURE

To consider the variation in the coefficients with the changes in $G$ and $T$, firstly the parameters ($I_{ph}$, $I_o$, $R_{se}$, $R_{sh}$ and $n$) are extracted from data sheet [5].Then $V_{oc}$, $I_{sc}$, $V_{mp}$ and $I_{mp}$ are obtained using the one diode circuit model parameters for different $G$ and $T$, which are used to calculate the model coefficients as per procedure described in section II. Fig.2 (a) represents the variation in $a$ with increasing $G$ and $T$. The effect of $T$ is dominant than $G$. Hence, the coefficient '$a$' is considered to be strongly influenced by $T$ only, and the relation between $a$ and $T$ for the module is established using curve fitting tool in MATLAB and the expression is in eq.7. Similarly, the variation in $b$ is depicted in Fig.2 (b). It is also observed that $b$ is strongly dependent on $T$ and the relation between b and $T$ is expressed in eq.8.

$$a(T) = -1/\exp\left(511*\exp\left(0.008118*T\right)\right) . \tag{7}$$

$$b(T) = 96.01*\exp(-0.006846*T) . \tag{8}$$

Figure 2 (c) shows the variation in $c$ with $G$ and $T$. It is noticed

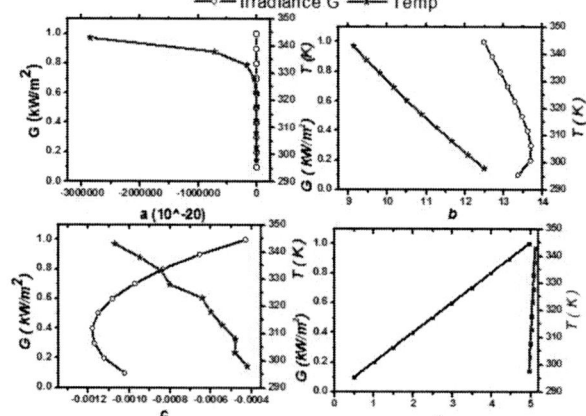

Fig. 2. Variations of coefficient $a$, $b$, $c$ & $d$ with $G$ and $T$

that $c$ is high with low $G$ and high $T$. In order to simply the MPPT, the variation in $c$ with $G$ and $T$ are considered separately and the high value of $c$ is considered to minimise the error in $P_{mp}$. The variations in $c$ with $T$ and $G$ are given in eq. (9) and (10) respectively,

$$c(T) = -2.592T^2*10^{-7}+1.504T*10^{-4}-0.02223, \quad (9)$$

$$c(G) = p_1*G^4+p_2*G^3+p_3*G^2+p_4*G+p_5, \quad (10)$$

where $p_1 = 0.0004887$, $p_2 = -0.001208$, $p_3 = 0.002956$, $p_4 = -0.001794$ and $p_5 = -0.0008734$.

As shown in Fig.2 (d), the coefficient $d$ ($= I_{sc}$) is linearly proportional to $G$ and less dependent on $T$. The relation between $d$ and $G$ is linear and given as

$$d = I_{sc}*(1+\alpha*(T-T_{ref}))*G / G_{ref}, \quad (11)$$

where $G$ in kW/m$^2$, $G_{ref}= 1$ kW/m$^2$ , $T$ in K , $T_{ref}= 298$ K and alpha ($\alpha$ ) is temperature coefficient of $I_{sc}$

With these assumptions, the dependence of each coefficient on either $G$ or $T$, makes the model simpler to solve MPP. Fig. 3 represents the power calculated from proposed model is close to the true maximum power obtained from N-R method. It also shows % relative error in $P_{mp}$ with the changes in $G$ and $T$. A maximum error of 2% can be obtained using the proposed model.

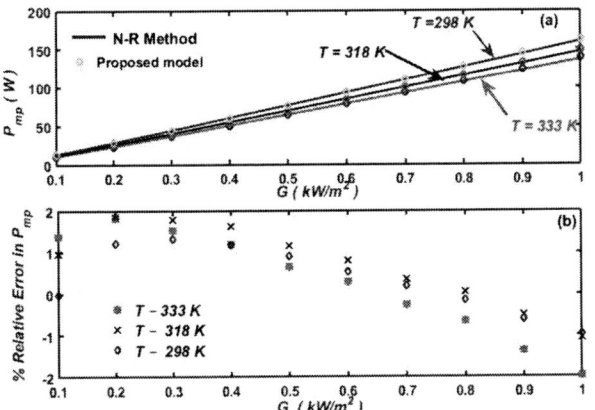

Fig. 3. (a) Comparison in $P_{mp}$ of N-R method and proposed model
(b) % relative error in $P_{mp}$ of proposed model with N-R method

## V. SIMULATION STUDY

This section discusses simulation results of the $PV$ system (Fig.4) performed for various real time conditions. The results have been compared and analyzed with an existing model [4] or P & O method in Simulink of MATLAB to understand the performance of model and also to verify the MPP tracking for variations in $G$ and $T$. The specifications of the boost converter are L = 297 μH, C$_{out}$ = 470 μF, f$_{sw}$ = 10 kHz and V$_{out}$ = 70 V. It is well know that $P=V*I$ and at MPP $dP/dV = 0$ and using

eq.(1), the following equation can be derived using proposed model for the MPP,

$$a \times (b+1)V_{mp}^{\ b} + 2c \times V_{mp} + d = 0 . \quad (12)$$

At first, the coefficients $a$, $b$ and $c$ are updated using eqs. (7)-(9) by sensing $V$, $I$ and $T$. Next $d$ is calculated using eq. (1).Then G is estimated as $d/I_{scref}$. Using this $G$ (kW/m$^2$) again c is calculated using equation 10. To calculate precisely $V_{mp}$ the larger value of c is used which are obtained from eq. (9) and (10). Finally, $V_{mp}$ is determined from eq. (12) using N–R method with an initial guess of $V_{mp}*(1-\beta*(T-T_{ref})$ at STC to converge fast.

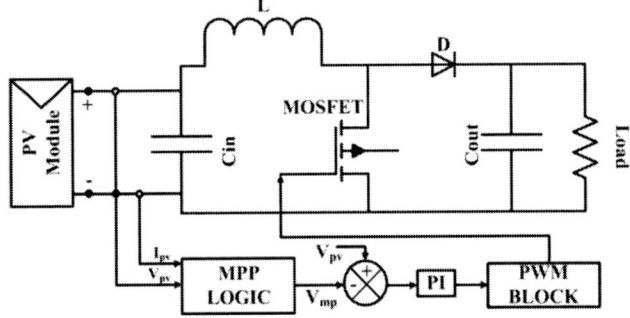

Fig. 4. Block diagram of the PV system.

The simulation studies are performed by considering (a) sudden change in $G$ and $T$ (b) rise and fall of irradiance at different rates maintaining constant temperature (c) larger manufacturing tolerance (d) aging Effect.

### A. Sudden change in G and T:

To understand the performance of proposed algorithm with sudden changes in environmental conditions, the simulation is done considering variations in $G$ and $T$ as shown in Table 1.

TABLE I

COMPARISON OF MAXIMUM POWER FOR SIMULATION CYCLE

| Time s | State | G W/m$^2$ | T K | P & O Method $P_{max}$ | Proposed Model $P_{max}$ | Existing Model $P_{max}$ |
|---|---|---|---|---|---|---|
| 0.0 – 0.2 | - | 1000 | 298 | 159.9 | 159.85 | 159.8 |
| 0.2 – 0.4 | S$_1$ | 1000 | 323 | 145.6 | 145.40 | 145.5 |
| 0.4 – 0.6 | S$_2$ | 400 | 323 | 54.47 | 54.4 | 54.0 |
| 0.6 – 0.8 | S$_3$ | 400 | 298 | 60.43 | 60.40 | 60.4 |
| 0.8 – 1.0 | S$_4$ | 1000 | 298 | 159.9 | 158.85 | 159.8 |

It can be seen that the both proposed and existing model tracks the MPP at steady state conditions as shown in fig.5. The proposed algorithm shows better dynamic performance as depicted in fig.6 and the performance shows the proposed model responds faster with the sudden changes in $G$ or $T$. In case of a sudden increase in temperature (t = 0.2 s), the proposed model has faster dynamic response than the existing model as depicted in fig.6. In case of a sudden decrease in irradiance (t = 0.4 s), the proposed model settles close to true MPP whereas the existing model deviates the MPP as in Fig.6.

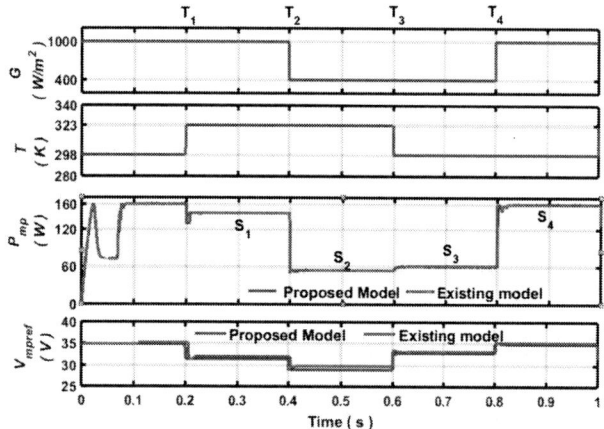

Fig. 5. Comparison of simulation results for the proposed model with an existing model [4] for sudden changes in $G$ or $T$.

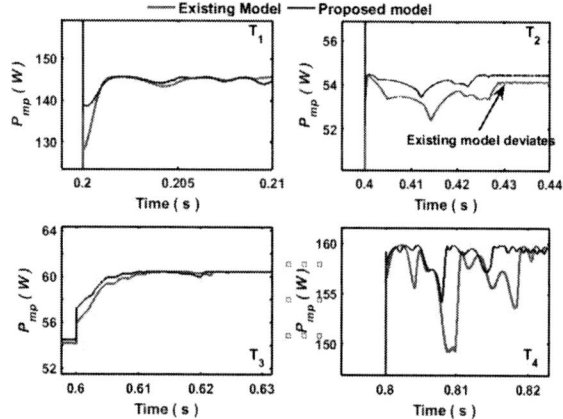

Fig. 6. Transient response at instants $T_1$, $T_2$, $T_3$ and $T_4$ of proposed model and existing model [4] for sudden changes in $G$ or $T$.

### B. The ramp change in G with different rate:

To estimate the performance of proposed algorithm a pattern of $G$ changes with different rates as shown in Table 2 [8] at constant temperature of 298 K is considered for simulation. These changes are similar to real time variations in G. The proposed model and the existing model track the power with the ramp changes as depicted in fig.7. The enlarged view of power variations with the ramp changes is shown in fig.8. It can be seen that the proposed model tracks true MPP with high accuracy than the existing model [4], which deviates its response with any rate of change in $G$ as in fig.8. Thus the proposed model tracks the MPP with high accuracy and have a faster response than existing model [4].

TABLE 2
RATE OF CHANGES IN IRRADIANCE

| Rate Name | Initial Time s | Final Time s | Initial G kW/m² | Final G kW/m² | Speed kW/m²-s |
|-----------|----------------|--------------|-----------------|---------------|---------------|
| $R_1$ | 0.20 | 0.21 | 0.6 | 1.0 | 40 |
| $R_2$ | 0.4 | 0.41 | 1.0 | 0.3 | -70 |
| $R_3$ | 0.6 | 0.7 | 0.3 | 1.0 | 7 |
| $R_4$ | 0.8 | 0.9 | 1.0 | 0.3 | -7 |

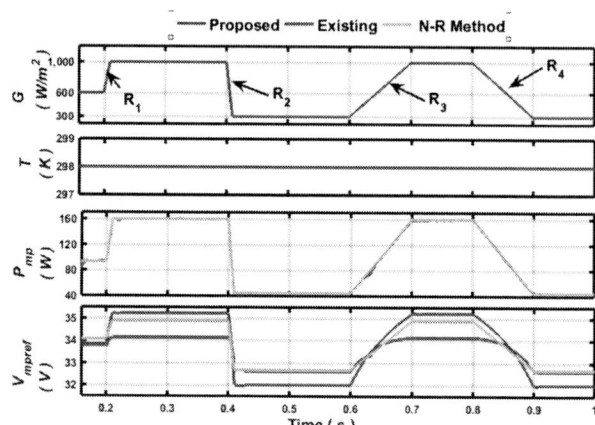

Fig.7 Comparison of simulation results for the proposed model with an existing model [4] for ramp changes in $G$.

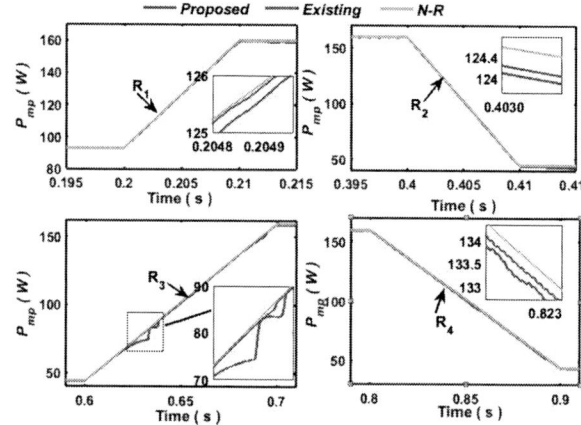

Fig. 8. Enlarged view in $P_{mp}$ for different rates $R_1$, $R_2$, $R_3$ and $R_4$.

### C. Large Positive Tolerance in Power:

The PV module manufacturing process from cells may create defects which affects the performance PV modules. Usually the commercial PV modules have a maximum tolerance of 10% of module rated power. This tolerance decides I-V characteristics and may have different power ratings from the same manufacturer. The performance of proposed model is evaluated by considering a positive tolerance of 10% in rated power. To model the characteristics of PV module, the maximum power is considered as a random variable with Gaussian probability distribution [9].The random variable is than multiplied to $I_{sc}$. In the mathematical modelling $I_{sc}$ is replaced with this new value and the simulation is performed for the same cycle in Table 1. The simulation results of proposed model with 0% tolerance, 10% tolerance and P & O method for 10% tolerance is shown in fig.9. It can be observed that proposed model shows an increment in power due to a positive tolerance in PV module characteristics and it almost tracks MPP as compared with P & O method. Hence the proposed MPP model is not affected with tolerance and shows performance similar to section described in 5.A.

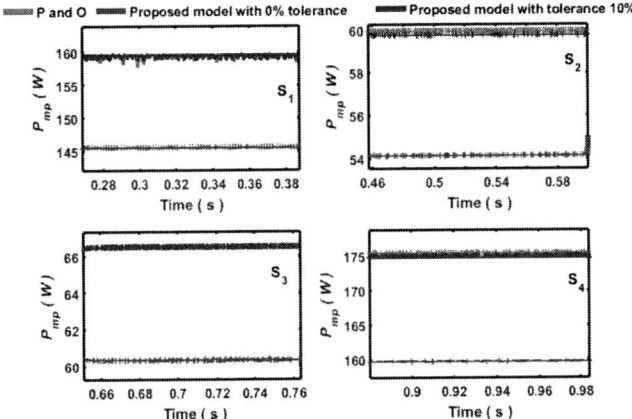

Fig.9    Comparison of $P_{mp}$ due to tolerance in rated power at various steady states S₁, S₂, S₃ and S₄.

### D. The effect of Aging on PV panel:

As the PV panels have long term performance, it is vital to understand how model based MPP algorithms work with degradation of panels. To observe the performance of proposed algorithm, the degradation rates in $V_{oc}$, $I_{sc}$, $I_{mp}$ and $V_{mp}$ of -0.29%,-4.37%, -6.31% and 2.12% respectively are considered [6]. Then using this once again single diode circuit model parameters [5] are obtained. These parameters are used in the mathematical modelling of PV module and then the simulation is repeated with the proposed algorithm. To verify the performance of proposed algorithm, a classical well known MPPT method P & O method is used. The simulation is done for the cycle mentioned in Table 1 and results shows a maximum of 0.6 to 0.7 W (approximately 0.4%) power difference of proposed model with P & O occurs due to aging effect as in fig.10. To reduce the loss and to track true power in case of aging, a combination of proposed model based MPP and P&O is used. The average of voltage obtained from model based algorithm and also from P & O is taken as a reference signal in $V_{mp}$. This combination is simulated and the results are depicted in fig.11.The outcome shows an exact tracking of MPP. So to have a good response with greater accuracy it is advisable to use a combination of model with P & O.

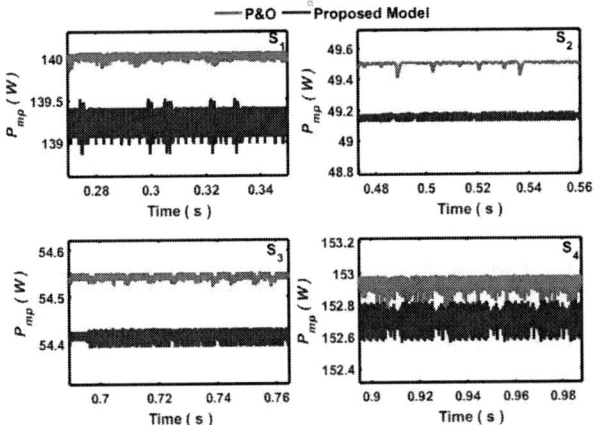

Fig.10 Steady state $P_{mp}$ of proposed model and P&O due to degradation of PV panel.

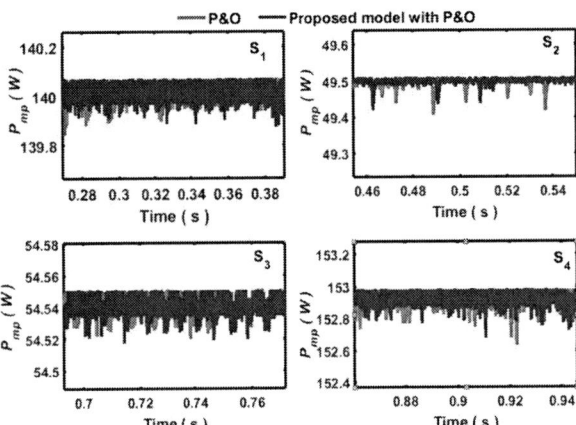

Fig.11    Steady state $P_{mp}$ of P&O and proposed model with P&O due to degradation of PV panel.

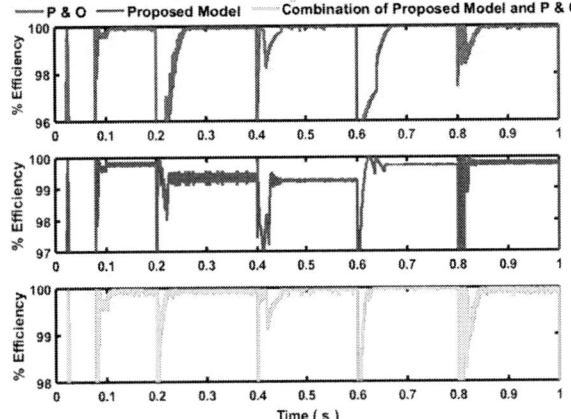

Fig.12    Efficiency under aging effect for (a) P&O, (b) Proposed model and (c) the combination of P & O and proposed model.

This type of combination helps in increasing the tracking efficiency. The fig.12 shows the efficiency of P & O, Proposed model and the combination for aging effect which is done for cycle represented in Table1. From the simulation results it can be seen that the P & O dynamic response is slow, whereas proposed model has reduced efficiency than P&O due to aging. The combination model shows faster dynamic response than P & O with similar steady state efficiency.

### VI. EXPERIMENTAL STUDY

An experimental set up as shown in fig.13 was made to validate the proposed MPPT algorithm for tracking of power against the Perturb and Observe method. The voltage and current signals from the module are given to the ADC of microcontroller through a voltage divider and LEM current sensor respectively. Using STM32F4 microcontroller, the proposed model MPPT algorithm and P and O are implemented to get operating duty cycle which is given to a basic PWM pulse generator in the microcontroller. The pulses are given to the boost converter switch. The experimented is performed on 1st June 2017 at 11:30 AM.

To observe and compare the performance of proposed a pattern

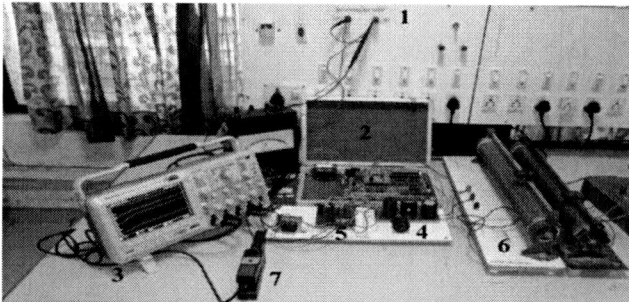

Fig.13 Experimental set up 1. PV panel output terminals 2.STM32F4 micro controller 3.Oscilloscope 4.Boost Converter 5.Voltage and current sensor 6.Resistive load 7. Current probe.

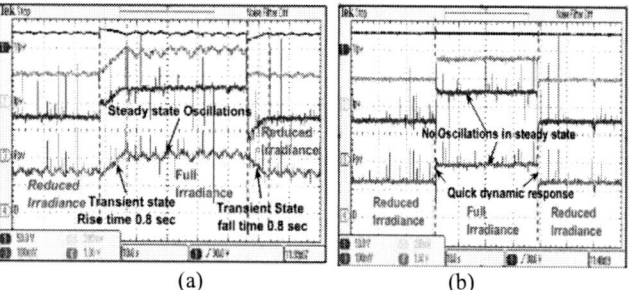

(a)                              (b)

Fig.14 Experimental results for a)P & O Method b)Proposed Model.

of low, high and low irradiance. The irradiance pattern is manually changed from full illumination to a low value by covering a white cloth on the PV module and by removing the cloth to get full intensity. Firstly the P & O method is implemented considering a step size of 0.2 V variation and then proposed model is performed. The P & O and proposed model experimental results are depicted in fig.14.a &14.b respectively for the above mentioned pattern. The results shows the P&O method have a three step oscillations in power, duty cycle. It can be seen that the method takes time of 0.8 s to reach MPP when there is sudden change in G as in fig.14.a. The proposed model shows a rise and fall time of 62.78 ms and 8.93 ms respectively for a sudden change in irradiance which is observed in oscilloscope. The proposed model shows no oscillation in duty cycle and power with tracking maximum power as depicted in fig.14.b, which is necessary for operation of PV module to extract the maximum power with higher efficiency in steady state and a faster dynamic response.

## VI. CONCLUSIONS

A new explicit *I-V* model is proposed to track the MPP for a PV module. The variations in the model coefficients are studied for different environmental conditions. On comparison with an existing model, the proposed model performs better for all environmental conditions.

A simulation analysis is conducted for sudden changes in $G$, $T$ and also for ramp changes in irradiance. The calculated $V_{mp}$ signal by the proposed model is more accurate than the existing one in case of ramp changes in $G$ and also for sudden

temperature changes. But, the signal is slightly deviating than the existing model in case of variation in the irradiance. The proposed model tracks almost true power irrespective of changes and it shows a faster dynamic response. The proposed model is also analysed for larger tolerance and aging. Due to aging, the proposed model based algorithm tracks with an error of 0.4%. For larger panel rating, this error becomes a noticeable and to address this issue, a combination of P & O and proposed model is advised to track MPP. The other possible solution is to update the coefficients for every one or two years as the maximum error of 0.4% seen by considering a 10 years of degradation. The model based MPPT algorithms provide a solution to track MPP at faster dynamic and steady state response. These algorithms also increases steady state efficiency when compared to three step variations in power by using P&O method. The exactness of chosen model to represent module characteristics makes the accuracy of MPP tracking. The proposed model can be extended for temperature estimation as the coefficients mainly depends on temperature. The proposed model based analysis can also be extended to partial shading concept in a string of PV modules.

## REFERENCES

[1] D. Sera, L. Mathe, T. Kerekes, S. V. Spataru and R. Teodorescu, "On the Perturb-and-Observe and Incremental Conductance MPPT Methods for PV Systems," in *IEEE Journal of Photovoltaics*, vol. 3, no. 3, pp. 1070-1078, July 2013.

[2] Y. Mahmoud and E. F. El-Saadany, "A Photovoltaic Model With Reduced Computational Time," in *IEEE Transactions on Industrial Electronics*, vol. 62, no. 6, pp. 3534-3544, June 2015.

[3] M. Faifer, L. Cristaldi, S. Toscani, P. Soulantiantork and M. Rossi, "Iterative model-based Maximum Power Point Tracker for photovoltaic panels," *2015 IEEE International Instrumentation and Measurement Technology Conference (I2MTC) Proceedings*, Pisa, 2015, pp. 1273-1278.

[4] L. Cristaldi, M. Faifer, M. Rossi and S. Toscani, "An Improved Model-Based Maximum Power Point Tracker for Photovoltaic Panels," in *IEEE Transactions on Instrumentation and Measurement*, vol. 63, no. 1, pp. 63-71, Jan. 2014.

[5] D.Sera, R. Teodorescu and P. Rodriguez, "PV panel model based on datasheet values." in Industrial Electronics, 2007. ISIE 2007. IEEE International Symposium on, pp. 2392-2396.

[6] A.M.Reis, N. T. Coleman, M. W. Marshall, P.A. Lehman and C.E. Chamberlin, "Comparison of PV module performance before and after 11-years of field exposure", in Photovoltaic Specialists Conference, 2002. Conference Record of the Twenty-Ninth IEEE (pp. 1432-1435).

[7] Y.Mahmoud, M Abdelwahed, and E.F. El-Saadany. "An Enhanced MPPT Method Combining Model-Based and Heuristic Techniques." *IEEE Transactions on Sustainable Energy* 7, no. 2 (2016): 576-585.

[8] F Paz, and M Ordonez. "High-Performance Solar MPPT Using Switching Ripple Identification Based on a Lock-In Amplifier." *IEEE Transactions on Industrial Electronics* 63, no. 6 (2016): 3595-3604.

[9] G.Cipriani, V. Di Dio., A.Marcotulli and R. Miceli, "Manufacturing tolerances effects on PV array energy production." in *Renewable Energy Research and Application (ICRERA), 2014 International Conference* IEEE (pp. 952-957).

# Dynamic Response of Maximum Power Point Tracking using Perturb and Observe Algorithm with Momentum Term

## Gautam A. Raiker

### National Institute of Technology Goa, Farmagudi, Goa, 403401, India

*Abstract* — **The Perturb and Observe (P&O) algorithm is a standard technique used for Maximum Power Point Tracking in Solar Photovoltaic (PV) systems due to its simplicity and ease of implementation. This method has a few issues associated with it namely, the problem of oscillation at maximum power point and divergence and slow tracking of the maximum power point during rapidly changing irradiation conditions. To overcome these problems it is proposed to improve the performance of the Perturb and Observe algorithm by including a momentum term in the perturbation similar to that used in back-propagation algorithms. It can be hypothesized that this modification can improve the dynamic performance of Perturb and Observe algorithm and it is the objective of this study validate the hypothesis through simulation studies.**

*Index Terms* — **Maximum power point tracking (MPPT), perturb and observe (P&O), Photovoltaic (PV)**

## I. INTRODUCTION

Solar Photovoltaic (PV) systems have the characteristics such that they provide maximum power only at a particular operating voltage and current which is known as the maximum power point (MPP). This point changes with the insolation and the temperature of the module. To make efficient use of the Solar PV module it is necessary to continuously track and operate the system at the maximum power point. The technique used to track the maximum power point is known as Maximum Power Point Tracking (MPPT).

Maximum Power Point Tracking uses a DC-DC converter in order to match the Solar PV source and the load. The objective is to control the duty cycle of the converter to such a value the Solar PV module is operated at the maximum power point and the maximum power is extracted from the module. For example suppose we want to deliver power from the Solar PV module to a storage system like a battery. For charging, the battery operates at a particular voltage say 12V. But the voltage for maximum power of a Solar PV module may be different, suppose 18V. So the DC-DC converter must be operated in buck mode at such a duty cycle such that the input i.e. the Solar PV module operates at 18V and the output of the battery remains at 12V. The MPPT algorithm determines the optimal duty cycle of the converter.

Today there are many different techniques for Maximum Power Point Tracking of Solar PV systems. Several references have stated that these methods differ in inputs required, simplicity, convergence time, hardware requirement, ease of implementation, popularity, cost, and in other respects [1]-[2]. Some of the methods include Hill Climbing/Perturb and Observe (P&O) Method, Incremental Conductance (IncCond)

Method, Fractional Voc, Fractional Isc, using Fuzzy Logic, using Artificial Neural Network techniques, etc. [1]-[3]. Each of these techniques has their own pros and cons.

The focus of this study is the Perturb and Observe algorithm which is a derivative of the gradient ascent method used in optimization theory [4]. This method is simple and easy to implement requiring minimal computational effort. The PV panel voltage is first perturbed in a particular direction by perturbing the duty ratio of power converter (a DC-DC converter). If there is an increase in power, the subsequent perturbation should be kept the same direction as previously and if there is a decrease in power, the perturbation should be reversed. This is done repeatedly until we reach the MPP. This is the basic principle of the P&O method.

## III. SOME PROBLEMS WITH TRADITIONAL P&O METHOD

There are three basic issues with the traditional Perturb and Observe method. They are oscillations about MPP, Trade-off between dynamic performance and precision of tracking, and drift during rapidly changing insolation conditions.

The system will oscillate about the MPP since the process of perturbation and observation will continue around the MPP. This will lead to wastage of some amount of available energy. Oscillation can be minimized by using a small perturbation step size but then this impairs the dynamic response of the converter and leads to poor utilization of the solar cells [4]. If we choose a large perturbation step size, the response would be fast but would not accurately track the point and there will also be larger oscillations about the MPP. Thus there is a trade-off and the optimal perturbation size must be determined. Optimization of the parameters of the P&O MPPT like the sampling time period and perturbation step size for optimal steady state operation has been proposed in one of the references [5].

If the irradiance changes during the perturbation step interval the algorithm may set the next perturbation in the wrong direction and instead of converging toward the MPP the system will diverge away from the MPP. If the irradiation conditions are changing very rapidly the PV voltage will continue to diverge away from MPP, resulting in failure to track the MPP.

A variable perturbation step length can be used in the algorithm for solving trade-off problem between the dynamic performance and tracking which will give faster response with good accuracy [4]. The concept of gradient ascent algorithm

978-1-5090-5606-4/17 $31.00 © 2017 IEEE

can be used to track the maximum power in which one takes step proportional to the positive of the gradient.

## II. GRADIENT ASCENT OPTIMIZATION AND P&O METHOD

The gradient ascent method is the most apparent optimization technique that can be used for maximum power point tracking of the solar panel and can be implemented similar to the with the perturb and observe algorithm by making some small changes. This technique is also by far the most common way to optimize neural networks.

Gradient ascent is a way to maximize an objective function depending on the model's parameters by updating the parameters in the same direction of the gradient of the objective function with respect to the parameters [6]. In the current case the objective function is PV panel power (P) and the parameter could be PV voltage (V), current (I) or duty cycle (D). Thus Perturb and Observe algorithm can be used where the perturbation step can be proportional to the gradient of P with respect to PV voltage or duty cycle instead of a constant perturbation size.

Let us describe the parameter D(n) which is the duty cycle for instant 'n' which is updated at every instant until the maximum power point is reached. Let f(n) be the perturbation as determined by perturb and observe algorithm.

$$D(n) = D(n-1) + f(n) \qquad (1)$$

For Perturb and observe algorithm with fixed perturbation length,

$$f(n) = \pm k \qquad (2)$$

Here 'k' is a constant for fixed perturbation length. This obviously is not the fastest way to reach the maximum power point as small perturbation size will lead to slow tracking and too large a perturbation size should be avoided since it causes oscillations.

The best way to update the duty cycle can be determined from the gradient ascent algorithm for finding the maxima. The perturbation length f(n) can be chosen to be proportional to the gradient of the objective function with is P with respect to the parameters, duty cycle (D) or voltage (V). Such algorithms which use gradient ascent approach for updating duty cycle are called adaptive Perturb and Observe algorithms [3].

For example, in one of the references [7], the following perturbation length was used.

$$f(n) = \pm \frac{\left| {}^{\Delta P}/_{\Delta D} \right|}{\left| {}^{P}/_{D} \right|} \qquad (3)$$

In another reference [8], the perturbation was chosen to be the gradient of Power (P) with respect to Voltage (V).

$$f(n) = M \frac{dP}{dV} \qquad (4)$$

In gradient descent learning based back propagation algorithms for neural networks a momentum term is used for updating the weights in the neural networks [9]-[11]. It is well known that such a term greatly improves speed of learning [11]. Therefore it rationally follows that such a term should be used in the Perturb and Observe Algorithm also. It is thus proposed in this work to include this momentum term in the P&O algorithm.

## III. PERTURB AND OBSERVE WITH MOMENTUM TERM

As we know that the smaller the perturbation the smoother will be the tracking of the MPP. But this comes at a price of slower rate of tracking. On the other hand if we make the perturbation too large, the system may become unstable (oscillatory). A method of increasing the rate of tracking is to modify the perturbation rule by including a momentum term, similar to the momentum term, which was introduced in the gradient descent learning algorithms as discussed in previous section [9]-[11].

Let $\Delta D(n)$ the perturbation for instant 'n' be defined as (where D(n) is the Duty ratio)

$$\Delta D(n) = D(n) - D(n-1) \qquad (5)$$

Add a momentum term $\alpha \Delta D(n-1)$ to the perturbation f(n) determined by the P&O method, where $\alpha$ is momentum parameter, we get

$$\Delta D(n) = \alpha \Delta D(n-1) + f(n) \qquad (6)$$

It is clear that (6) can be viewed as a difference equation in $\Delta D(n)$ and solving this for $\Delta D(n)$, we get

$$\Delta D(n) = \sum_{t=0}^{n} \alpha^{n-t} f(t) \qquad (7)$$

It can be seen that, (7) shows that the current perturbation is a exponentially weighted sum, with terms most recent to time instant 'n' having higher weight as long as $0 \leq D(n) \leq 1$. When the perturbation determined by P&O has same sign (direction) on consecutive iterations $\Delta D(n)$ is adjusted by large amount. This means it will accelerate tracking when consecutive perturbations have led to steady movement in the same direction towards MPP.

When the consecutive perturbations have opposite signs, as in case of oscillation, $\Delta D(n)$ is adjusted by small amount only. Thus the inclusion of momentum parameter has stabilizing effect in directions that oscillate in time.

We next consider an example for P&O algorithm with momentum parameter , $\alpha = 0.5$ and perturbation length k . Assume that the perturbation determined by P&O algorithm has same sign on consecutive iterations, when the operating point is away from the MPP. The perturbation $\Delta D(n)$ is then,

$$\Delta D(n) = \sum_{t=0}^{n} 0.5^{n-t} k \qquad (8)$$

978-1-5090-5606-4/17 $31.00 © 2017 IEEE

The summation $\sum_{t=0}^{n} 0.5^{n-t}$ is a progression which can be evaluated as $\frac{1}{1-0.5} = 2$ if 'n' is large. Therefore the perturbation will be $\Delta D(n) = 2k$. Therefore when consecutive perturbations are in the same direction the perturbation step size increases from k to 2k (for $\alpha = 0.5$). Hence when operating point is away from the MPP the perturbation will be larger and the tracking towards the MPP will be faster.

Next consider the case when we are near the MPP and the perturbation oscillates in consecutive iterations. Let us assume that the perturbation is alternatively positive and negative. The perturbation $\Delta D(n)$ is then

$$\Delta D(n) = \sum_{t=0}^{n}(-0.5)^{n-t}k \qquad (9)$$

The summation $\sum_{t=0}^{n}(-0.5)^{n-t}$ can be evaluated as $\frac{1}{1+0.5} = \frac{2}{3}$ if 'n' is large. Therefore the perturbation will be $\Delta D(n) = 0.667k$. Hence when the perturbation oscillates around the MPP the perturbation step size decreases from k to 0.667k (for $\alpha = 0.5$). Thus during oscillations perturbation size decreases and the oscillations can be reduced thus improving the dynamic performance.

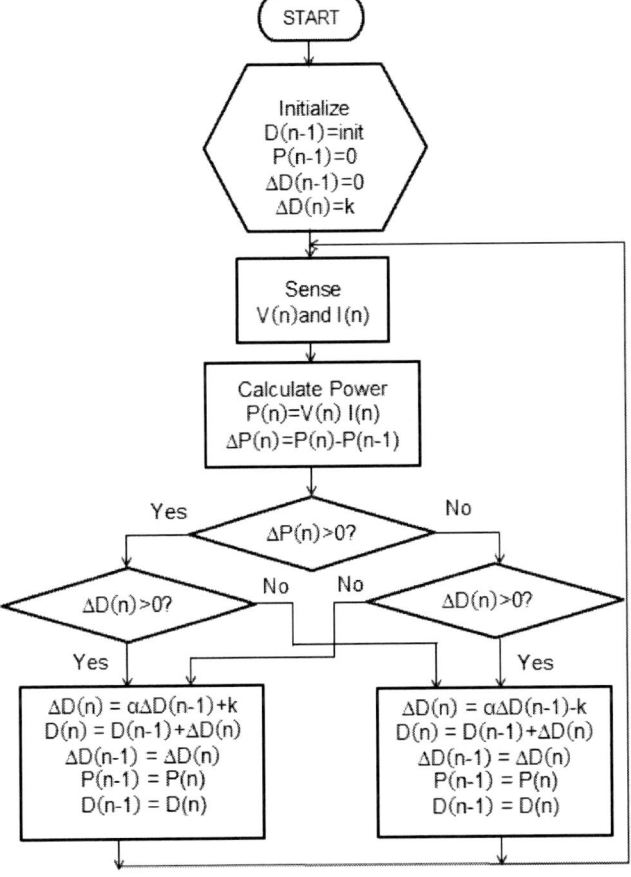

Fig. 1.   Flow chart for Perturb and Observe Algorithm with momentum term

## IV. SIMULATION RESULTS AND ANALYSIS

The system consisting of a step-up (Boost) converter and the controller running the MPPT algorithm was simulated in Simulink environment using Simscape Power Systems libraries and analysis tool. These libraries contain preset PV array block which implements an array of photovoltaic (PV) modules from the National Renewable Energy Laboratory (NREL) System Advisor Model. A single module of SunPower SPR-305-WHT type is used in the simulation. The I-V and P-V curves of this module are shown in Fig. 2. At STC the open circuit voltage (VOC) is 64.2V, the short circuit current (ISC) is 5.96A, the voltage at maximum power (Vmp) is 54.7V and the current at maximum power (Imp) is 5.58A. This gives a maximum power of 305.23W for 1000W/m2.

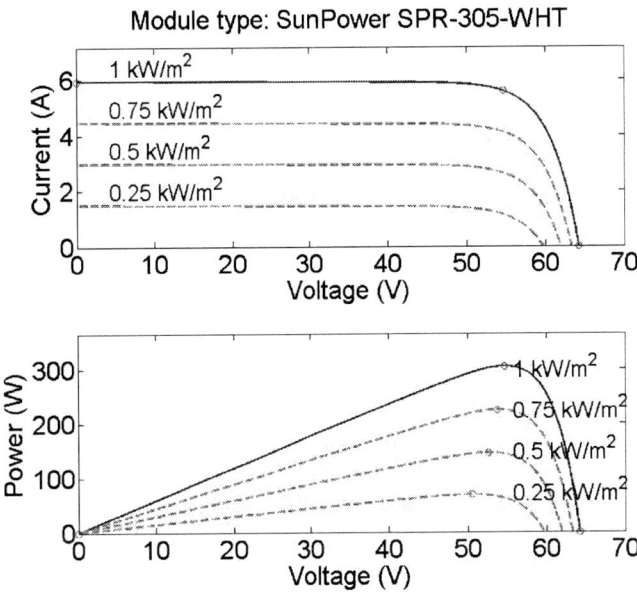

Fig. 2.   I-V and P-V plots for Solar PV Module used in simulation study

Both the traditional P&O algorithm and the P&O algorithm with momentum term were implemented using MATLAB function block in Simulink. An initial duty cycle of 0.5 was chosen along with perturbation step size of 5×10-4. A resistive load of 50Ω was taken. For the P&O algorithm with momentum term the momentum parameter α was taken as 0.5.

For the simulation an irradiance of 1000W/m2 was applied initially for 0.5 seconds followed by a sudden drop of irradiance to 500W/m2 for an interval of 0.2 seconds following which the irradiance comes back to original value. The response of output power with time for the two algorithms is shown in Fig. 3. At start-up the response of the P&O algorithm with momentum term is much faster than the traditional P&O algorithm indicating faster tracking to the MPP. Even after the irradiance drop the P&O algorithm with momentum term fares better with faster response toward the maximum power point.

To understand the oscillations about the maximum power point the duty cycle was plotted with time as in Fig. 4. We can see that the oscillation of duty cycle about the maximum power point for P&O algorithm with momentum term was remarkably smaller than the traditional P&O algorithm.

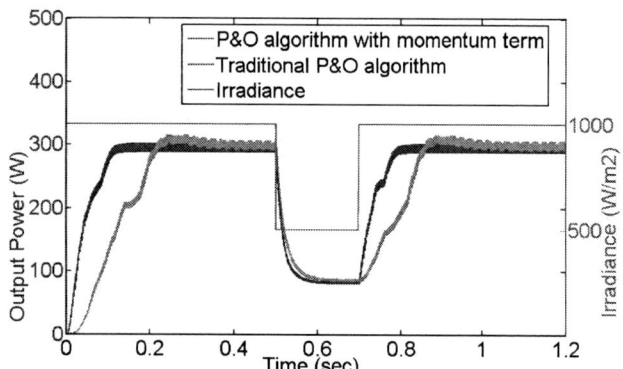

Fig. 3.    Output power with time for a step change in irradiance

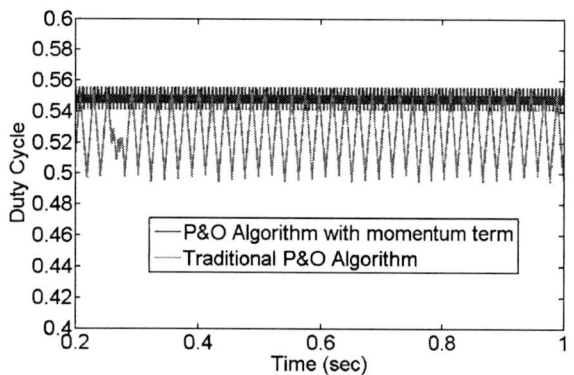

Fig. 4.    Oscillation of Duty Cycle with time

Thus it can be concluded from the results that the P&O algorithm with momentum term gives much faster response toward the MPP and has smaller oscillations about MPP and therefore has much improved dynamic response. The implementation of the modified P&O algorithm with momentum term is simple and similar to traditional P&O algorithm with only slight modification. But this modification leads to much better dynamic response. Thus the inclusion of the momentum term in the Perturb and Observe algorithm must be given a serious thought.

REFERENCES

[1]    T. Esram, and P. L. Chapman, "Comparison of photovoltaic array maximum power point tracking techniques", *IEEE Transaction on Energy Conversion*, vol. 22, no. 2, pp. 439-449, 2007.

[2]    S. Chakraborty, M. G. Simoes and W. E. Kramer, *Power Electronics for Renewable and Distributed Energy Systems*, Springer-Verlag London, 2013

[3]    A. K. Abdelsalam, A. M. Massoud, S. Ahmed and P. N. Enjeti, "High-Performance Adaptive Perturb and Observe MPPT Technique for Photovoltaic-Based Microgrids," *IEEE Transactions on Power Electronics*, vol. 26, no. 4, pp. 1010-1021, 2011.

[4]    A. Pandey, N. Dasgupta, "High-Performance algorithms for drift avoidance and fast tracking in solar MPPT system", *IEEE Transactions on Energy Conversion*, vol. 23, no. 2, pp. 681-689, 2008.

[5]    N. Femia, G. Petrone, G. Spagnuolo, and M. Vitelli, "Optimization of perturb and observe maximum power point tracking method", *IEEE Transactions on Power Electronics*, vol. 20, no. 4, pp. 963-973, 2005.

[6]    S. Ruder, "An overview of gradient descent optimization algorithms", *arXiv:1609.04747*, 2016.

[7]    M. L. Chiang, C. C. Hua, and J. R. Lin, "Direct power control for distributed PV power system," in *Proceedings of Power Conversion Conference*, Osaka, Japan, 2002, pp. 311–315.

[8]    A. Pandey, N. Dasgupta, and A. K. Mukerjee, "High-performance algorithms for drift avoidance and fast tracking in solar MPPT system," *IEEE Transactions on Energy Conversion*, vol. 23, no. 2, pp. 681–689, 2008.

[9]    D. E. Rumelhart, G. E. Hinton and R. J. Williams, "Learning representations by back-propagating errors", *Nature*, vol. 323, pp. 533-536, 1986.

[10]  D. E. Rumelhart, G. E. Hinton and R. J. Williams, "Learning internal representations by error propagation", *Parallel Distributed Processing*, vol. 1, ch. 8, pp. 318-362, 1986.

[11]  N. Qian, "On the momentum term in gradient descent learning algorithms", *Neural Networks*, vol. 12, no. 1, pp. 145-151, 1999.

# A Framework for Comparing the Economic Performance and Associated Emissions of Grid-connected Battery Storage Systems in Existing Building Stock: a NYISO Case Study

Julian do Nascimento Ricardo*, Vasilis Fthenakis*

*Center for Life Cycle Analysis, Department of Earth and Environmental Engineering,
Columbia University, New York, NY

*Abstract*—This study develops a framework for assessing the potential economic and environmental feasibility of simulated distributed energy resource (DER) installations as peak shaving demand response (DR) assets in existing residential and commercial building stock. We simulate the use of lithium-ion (Li-ion), lead-acid (Pb-a), sodium-sulfur (NaS), and vanadium redox flow (VRF) energy storage systems (ESS) with on-site photovoltaics (PV), and model the buildings as connected to the New York Independent System Operator (NYISO) grid. We generate performance curves in terms of normalized cost / profit and equivalent carbon dioxide ($CO_2$eq) emissions using New York City irradiation levels and utility rate structures. These show the spread of potential economic and environmental value that each building can accrue as a decentralized DR resource. Our findings show net emission decreases up to $-0.31 \, \mathrm{kg} \, CO_2$eq per kWh of energy delivered by the PV-ESS system, and normalized profits up to $0.01/kWh, when reducing peak demand up to 75% in an office building; up to $-0.31 \, \mathrm{kg} \, CO_2$eq/kWh at a cost of $0.08/kWh when reducing up to 65% of peak loads in an apartment complex and supermarket; and, in a hospital, up to $-0.34 \, \mathrm{kg} \, CO_2$eq/kWh at a cost of $0.08/kWh for peak load reduction up to 40%. Our results show qualitatively similar performance curves across ESS types, though VRF tends to outperform the others economically. We find that only a small number of simulations ($< 1\%$) result in profitable outcomes, though reducing low estimates of ESS and PV costs by 20% (to an average of $419/kWh and $1792/kWp, respectively) is sufficient to produce profitable peak shaving in the office building. Substantially greater reductions of 75%, 80%, and 90% are required to achieve the same in the supermarket, apartments, and hospital, respectively. Simultaneous ESS applications in addition to peak shaving, and DER value streams beyond the conventional utility rates considered here, would improve the economic viability of these building-level systems.

*Index Terms*—demand response, energy storage, peak shaving, photovoltaic array

## I. INTRODUCTION

An increased awareness of the effects of anthropogenic climate change has dovetailed with paradigm shifts in the generation, consumption, and commodification of energy to spur a critical re-evaluation of how the electricity sector operates. This shifting landscape challenges various foundational assumptions of the sector as it operates today [1]–[4], many of which do not adequately account for the associated social and environmental costs of expanding the reach of the electrical grid, or improving its reliability [5], [6]. Producing and providing sustainable and resilient electricity is also of the utmost

importance [5], [7]. As a result, municipalities are establishing climate action plans that include some combination of opting for a greater share of renewable energy production and distributed energy resources (DER), and improving the resiliency of the electrical transmission and distribution network [1], [2], [4], [8]–[10].

To this end, utilities have launched an array of demand side management programs, including peak shaving demand response (DR), which encourage active consumer participation in the electricity market during electricity usage peaks as a check against disturbances in the grid. These programs leverage incentivized changes in energy consumption behavior as tools to maintain a reliable power supply [11], [12].

Here we calculate normalized costs / profits and $CO_2$eq emission impacts to characterize various ESS technologies, each paired with an on-site PV array, as peak shaving DR resources in simulated residential and commercial buildings. The model prioritizes charging the ESS with PV when possible. The analysis adopts the framework of life cycle analysis (LCA) in accounting for the cradle-to-gate impacts of the ESS and PV, and uses explicit spatiotemporal models (in 5-min intervals) of building loads, solar generation, and the electrical grid in accounting for the time-varying emissions associated with electricity usage.

## II. DATA ANALYSIS

### A. Building models

We adapt building energy consumption models from a set of reference buildings representing 70% of US commercial building stock [13]. The specific models adapted here reflect existing building stock constructed in or after 1980. The dataset includes models of various building types for representative cities in each of 16 climate zones, classified based on the number of heating and cooling degree days [14], [15]. The representative city (climate zone) for New York City is Baltimore (4A).

The present study analyzes four reference buildings: an apartment complex, office building, supermarket, and hospital. This subset spans residential and commercial buildings, and includes critical facilities. Reproduction of the building models was carried out in OpenStudio, open-source software for whole building energy modeling developed by the National

Renewable Energy Laboratory [16]. OpenStudio makes use of the open-source EnergyPlus simulation engine [16].

### B. Grid dispatch model

Estimating changes in the emissions associated with grid-connected loads requires an understanding of the various subsets of power generation resources that are called upon to meet a time-varying demand for grid electricity. We follow the example of Gilmore [17] in constructing a curve that indicates the approximate sequence of dispatched resources from publicly available datasets: a time-series of electrical demand and an inventory of power plants in NYISO territory [18], [19]. The resulting curve arranges individual generators from least to greatest marginal cost (MC) of operation, in $/kWh, based on costs calculated using eq. 1:

$$MC = HR \cdot FC + VOM \qquad (1)$$

where HR is the heat rate, representing the efficiency with which each generator converts British thermal units (Btu) of primary energy into kWh of electricity; FC is fuel cost, in $/Btu; and VOM is variable operating and maintenance cost, in $/kWh. Marginal costs reflect data categorized by fuel type in U.S. Energy Information Administration reports from 2014-2016 [20]–[22].

The Gilmore model does not account for particular plants' operating cycles, forced outages, or other maintenance issues. Similarly, it does not explicitly replicate the operating cycles or emission factors of plants with multiple generator types, or of generators that accommodate multiple fuels. These systematic errors may produce inaccurate individual estimates of plants' marginal costs and emissions [17]. In our study, we attempt to mitigate these effects by redrawing the dispatch curve at each timestep based on the distribution of possible marginal costs for each fuel type given these uncertainties [20]–[22].

Using plant-level $CO_2$eq emission factors from NYISO territory in the US Environmental Protection Agency generation resource database [19], the curve can approximate the emissions factor for a unit of energy drawn from the electrical grid as $EF_{CO_2eq}$.

### C. Simulation results

Altogether, a negligible number of ESS and PV configurations within the analyzed state space were economically viable ($Pr > 0$) given the pricing mechanisms for power and energy use that Consolidated Edison (ConEdison) has implemented [23]. Profitable or otherwise, VRF systems typically outperformed the other three ESS in terms of normalized $Pr$, which follows from having the lowest levelized costs. Emission impacts were more uniform across all ESS types, with net reductions of up to $-0.34\,\text{kg}\,CO_2\text{eq}$ per kWh of energy delivered by the PV-ESS system. Normalized $Pr$ values reached an upper bound of $0.01/kWh (see discrete data points in fig. 1 and 2). For reference, the highest per-kWh charge levied by ConEdison [23] is $0.3075/kWh.

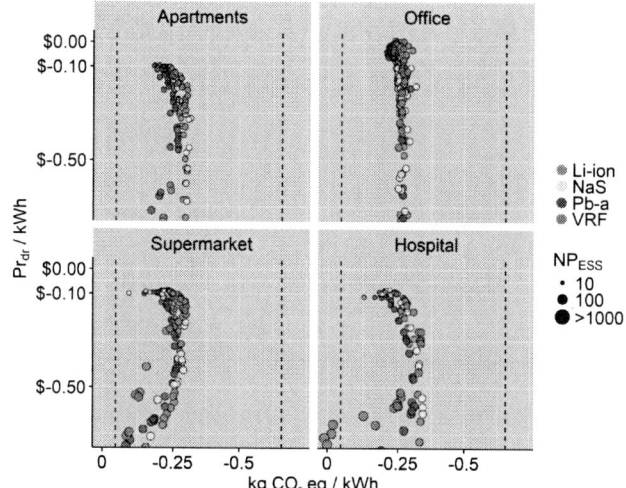

Fig. 1. Performance curves comparing profit and emission reductions, herein referred to as profit-emissions ($Pr$-$\Delta GHG_{ann}$), for each simulated building and ESS technology, using kWh of electricity delivered by the PV-ESS system as the normalizing unit. Individual data points represent the results of single simulations assuming low costs for ESS and PV, and are scaled in proportion with the ESS nameplate capacity ($NP_{ESS}$, in kWh) required to meet demand at each tested degree of peak shaving ($\alpha$). $NP_{ESS}$ values from the supermarket and hospital span a greater range of sizes than those from the apartments and office. The dotted vertical lines mark the range of displaced $CO_2$eq emissions calculated from DR and ESS strategies simulated in [24]. This range coincides in part with the emission estimates calculated here.

The discrepancies between ESS economic performance modeled here and elsewhere in the literature [25] reflect the added costs of PV, which are incurred here without any pricing mechanism that would compensate the building for installing DER, providing DR services, or passing along excess renewable generation to the grid. This corroborates findings in [2], [5], [7], [11], [26] which generally support the need for lower costs and new value streams for DER, in order for them to be economically viable.

The simulated office building produced narrower distributions of normalized $Pr$ and $\Delta GHG_{ann}$ (fig. 2) than the other three buildings, and the only profitable outcomes among all of them – albeit in $< 2\%$ of office simulations. It generated net-negative emissions across all tested $\alpha$, corresponding to peak load reductions up to $75\%$.

The apartments, supermarket, and hospital illustrated a common relationship between normalized $Pr$ and $\Delta GHG_{ann}$ across all tested $\alpha$, with net emission reductions up to $-0.34\,\text{kg}\,CO_2\text{eq/kWh}$, at a minimum cost of $0.08/kWh. Whereas the threshold beyond which peak shaving began to produce net-positive emissions was $40\%$ of peak load in the hospital, it was $65\%$ in the apartments and supermarket. Peak shaving strategies at $\alpha$ below these thresholds corresponded with the lower end of emission reductions presented among a range of possible values from comparable DER applications [24].

Given the calculated disparities in normalized $Pr$, the degree to which the costs associated with the ESS and PV would

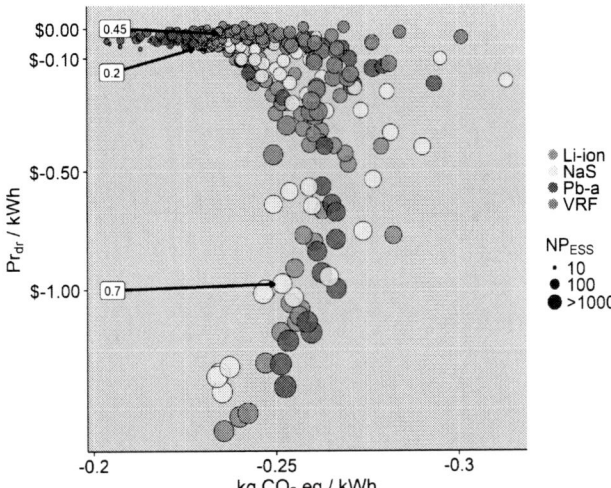

Fig. 2. Normalized profit-emissions ($Pr$-$\Delta GHG_{ann}$) performance curve for the office building, from fig. 1, showing all simulated ESS technologies and degrees of peak shaving ($0.2 < \alpha < 0.75$). Individual data points represent the results of single simulations assuming low costs for ESS and PV, and are scaled in proportion with the ESS nameplate capacity ($NP_{ESS}$, in kWh) required to meet peak demand at each tested $\alpha$. The labeled points show representative values of $\alpha$. The emission estimates calculated here fall entirely within the range of displaced $CO_2$eq emissions from DR and ESS strategies calculated in [24].

have to decrease to produce profitable outcomes differed from building to building, though significantly less so between ESS chemistries. Figure 3 quantifies the relationship between the maximum normalized $Pr$ value in each building, taken as the average of all ESS types, and decreasing system component costs. A reduction of 20% from the lower cost estimates cited in [27], [28] was sufficient to generate $Pr > 0$ in the office building, as compared to 75%, 80%, and 90% for the supermarket, apartments, and hospital, respectively.

Beyond the similarities and differences in the distributions of $Pr$ and $\Delta GHG_{ann}$ calculated across all buildings, variations in ESS usage within each building merit discussion as well. This requires separately analyzing the economic and environmental performance of ESS and PV within the context of the functions, energy-consuming equipment, and patterns of activity that characterize each building.

## III. METHODOLOGY

The present study adopts the framework of LCA in quantifying the net emission impacts of using ESS as a grid-connected DR resource across various building types, taking a comprehensive approach to accounting for the material and energy flows through and within a system. This requires an explicit description of both the functional unit and boundaries that constrain the subsequent comparative analysis.

### A. Functional unit

The functional unit is a normalization factor for comparing the quantified economic and environmental impacts of ESS and PV in peak shaving DR applications. Storage capacity in

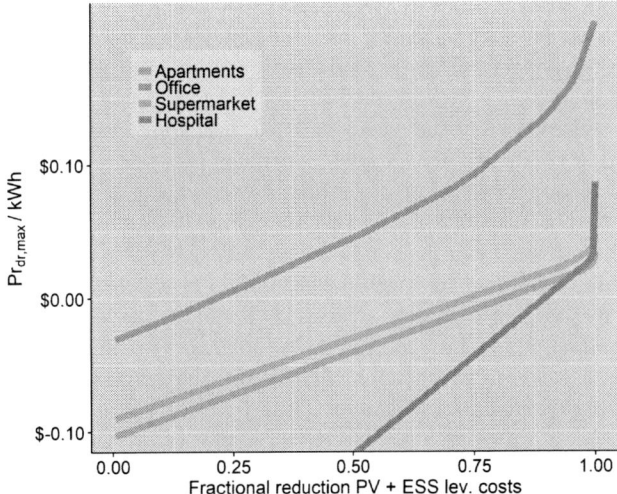

Fig. 3. Comparison of the maximum normalized profit ($Pr$) values in each simulated building, taken as the average across all ESS chemistries, as a function of fractional reductions in ESS and PV costs – using the low end of cost estimates in [27], [28].

kWh often appears as the functional unit in ESS LCA, but this choice can overestimate the amount of energy an ESS delivers in applications where discharge rates vary considerably from rated conditions [29]. Here we use a unit of electricity delivered via the PV-ESS system, $Thru_{tot}$, as the functional unit:

$$ Thru_{tot} = E_{PV} \cdot \frac{E_{PV}}{E_{PV} + E_{ESS}} + E_{ESS} \cdot \frac{E_{ESS}}{E_{PV} + E_{ESS}} \quad (2) $$

This choice accounts for differences in the round-trip efficiencies and operational limits of each storage technology; it also treats battery cell cycle limits and lifetimes as separate thresholds for triggering ESS failure [25], [29], [30]. Multiple replacements of the ESS within the lifetimes of other system components have a direct effect on both the economic and environmental viability of using ESS to implement peak shaving DR.

### B. System boundaries

The system boundaries of this study extend to the life cycle stages of each simulated component that fall within the scope of a DR program, including the cradle-to-gate $CO_2$eq emission impacts of the ESS and PV, as well as the marginal $CO_2$eq emissions associated with grid electricity. Lack of consistent data regarding the emission effects of the final disposal and/or recycling of all system components limits the quality of analysis regarding these stages enough to warrant their exclusion from the present study [29], [30].

All simulations assume New York City as the location of building energy consumption, ESS operation, and PV generation, and NYISO as the provider of electricity. Simulated NYISO $CO_2$eq emission rates range between $0.15 \, \text{kg} \, CO_2$eq per kWh of grid electricity during off-peak hours and $0.4 \, \text{kg} \, CO_2$eq/kWh at periods of peak demand, given the present NYISO grid dispatch model. Accounting for the emissions of charging power sources allows for a comparison of ESS impacts across scenarios where such sources may vary [30].

The present study compares two scenarios, one in which ESS combined with local PV generation provides peak shaving DR, and a control configuration in which no such DR occurs. We characterized each 1-year simulation on the basis of economic and environmental metrics. Altogether, 32 simulations are carried out for each combination of $\alpha$ and building type. We developed the simulations in R [31] using open-source packages for data analysis [32]–[36] and visualization [37], [38], and execute them on publicly available servers for cloud-based statistical computation [39].

### C. ESS model

The range of ESS technologies investigated here is limited to those that would be feasible for installation in all of the simulated buildings, as well as those for which there exist sufficient comparative LCA data [29], [30], [40], [41]: lead-acid (Pb-a), lithium-ion (Li-ion), sodium-sulfur (NaS), and vanadium redox flow batteries (VRF). Battery energy density, round-trip efficiency, and cradle-to-gate emission impacts are from supplementary data compiled in [30]. Total available cycles are compiled from [42], with installation, maintenance, and replacement costs from [27].

The model does not take into account factors affecting individual battery cells, including the overall operating temperature or voltage of the battery bank, assuming instead that round-trip efficiency, $\eta_{eff}$, and healthy depth of discharge, $\eta_{DoD}$, remain constant throughout the operational lifetimes of all ESS [41], [43]. We extrapolate ESS lifetime and throughput based on the equivalent cycles calculated in each one-year simulation, with a maximum lifetime of 20 years, as in [25].

### D. PV model

The PV system for each building is characterized [44] as a fixed, latitude-tilt, monocrystalline silicon (sc-Si) system, sized to fit on each respective roof with a 1.8 m setback from the roof's edge on all sides, installed in New York City (40.783 N, −73.967 E). The annual global horizontal irradiation (GHI) level for this location in a typical meteorological year (TMY3) is 1493 kWh/m$^2$ [45]. The cell efficiency is 17% [46], and inverter efficiency is 93% [29]. Additional losses due to wired connections, load mismatch, shading, and soiling reduce power generation by 14%, based on estimates from [46]. The cradle-to-gate emissions factor of each array is calculated based on figures in [47]. We synthesize 5-min irradiance time-series for use in simulations from hourly average irradiance data spanning the years 2005-2014 [45], following the Markov chain technique for preserving realistic variability and frequency distributions outlined in [48].

### E. Simulation

Arriving at estimates of ESS cost and emission impacts at the building-level begins with equations re-evaluated at each

interval of the simulation. These equations use the specified fraction $\alpha$ of peak demand $(P_{peak})$ to shave, the building load to be met $(P_{bldg})$, and the constraints of the PV-ESS system itself to determine ESS operation. The primary factor determining power flows between generation, storage, and consumption is the relationship between $P_{bldg}$ and $P_{targ}$ at each timestep, $i$. $P_{targ}$ is defined as $P_{peak}(1-\alpha)$. A stochastic variability of 10% at each timestep in the two load profiles that the simulation traverses, of the building and the electrical grid, follows from studies that address fluctuations in residential and commercial loads [3], [43].

TABLE I
OPERATIONAL PARAMETERS FOR SIMULATED ENERGY STORAGE SYSTEMS (ESS).

|  | $\eta_{eff}$ | $\eta_{DoD}$ | $n_{lo}$ | $n_{hi}$ | $\varrho$ | $EF_{CO_2 eq}$ |
|---|---|---|---|---|---|---|
|  | [30] | [25] | cycles [42] | cycles [42] | Wh / kg [30] | kg $CO_2$eq /kg [30] |
| Li-ion | 0.9 | 0.2 | 1000 | 6000 | 140 | 22.0 |
| NaS | 0.81 | 0.2 | 2000 | 4500 | 116 | 14.9 |
| Pb-a | 0.82 | 0.25 | 200 | 4500 | 27 | 2.7 |
| VR flow | 0.75 | 0.0 | 10000 | 13000 | 20 | 2.7 |

TABLE II
FINANCIAL PARAMAETERS FOR SIMULATED ESS.

|  | $C_{install,lo}$ | $C_{install,hi}$ | $C_{om,lo}$ |
|---|---|---|---|
|  | $ / kWh [27] | $ / kWh [27] | $ / kWh [27] |
| Li-ion | 513 | 1263 | 8 |
| NaS | 543 | 1600 | 11 |
| Pb-a | 663 | 2255 | 13 |
| VR flow | 342 | 1360 | 3 |

|  | $C_{om,hi}$ | $C_{rep,lo}$ | $C_{rep,hi}$ |
|---|---|---|---|
|  | $ / kWh [27] | $ / kWh [27] | $ / kWh [27] |
| Li-ion | 13 | 209 | 304 |
| NaS | 32 | 269 | 1033 |
| Pb-a | 56 | 333 | 686 |
| VR flow | 40 | 88 | 304 |

TABLE III
OPERATIONAL AND FINANCIAL PARAMETERS FOR SIMULATED PHOTOVOLTAIC ARRAYS (PV).

|  | $\eta_{cell}$ | $\eta_{inv}$ | $\eta_{sys}$ | $EF_{CO_2 eq}$ |
|---|---|---|---|---|
|  | [47] | [29] | [46] | kg $CO_2$eq / kg [47] |
| sc-Si | 0.17 | 0.93 | 0.86 | 1834 |

|  | $C_{cap,lo}$ | $C_{cap,hi}$ | $C_{om,lo}$ | $C_{om,hi}$ |
|---|---|---|---|---|
|  | $ / kWp [28] | $ / kWp [28] | $ / kWp·yr [28] | $ / kWp·yr [28] |
| sc-Si | 2000 | 5300 | 12 | 22.50 |

The annual emission impacts associated with peak shaving DR, $\Delta GHG_{ann}$, are calculated in $\mathrm{kg\,CO_2eq}$ as the sum of impacts at each simulated timestep:

$$\Delta \mathrm{GHG}_{ann} = \sum_{i=1}^{n} \left[ (P_{DR}^i - P_{bldg}^i) \cdot \mathrm{EF}(P_{grid}^i) \cdot 0.001 \right] \quad (3)$$

where $P_{DR}$ is the building load following the implementation of DR, and $EF(P_{grid,i})$ refers to a function which returns a time-dependent emissions factor for grid electricity in $\mathrm{kg\,CO_2eq\,/\,MWh}$ based on grid load data.

Adapting the analysis in [25], we assign levelized costs ($LC$) for ESS and PV from the range of cost estimates for the ESS and PV in [27], [28] (see tables II-III). We then combine these annualized ESS and PV costs with utility costs to produce estimates of total annual cost ($TAC$) and profit ($Pr$):

$$TAC = LC_{ESS} + LC_{PV} + C_{tariff}^{DR} \quad (4)$$

$$Pr = C_{tariff}^{ctrl} - TAC \quad (5)$$

Finally, accounting for the functional unit requires that we express the final economic and environmental metrics as functions of $Thru_{\mathrm{tot}}$:

$$\Delta \mathrm{GHG}_{ann} \implies \Delta \mathrm{GHG}_{ann} / Thru_{tot}$$
$$Pr \implies Pr / Thru_{tot}$$

## IV. CONCLUSIONS

This investigation derived performance curves for ESS and PV operating as peak shaving DR resources in buildings located within NYISO territory, integrating utility tariffs, LCA techniques, subhourly building load profiles, and a spatiotemporal model of grid plant dispatch order. We used validated building models to calculate emission factors for grid electricity that fell within the range of similar estimates in the literature. Our models show net emission decreases of up to $-0.31\,\mathrm{kg\,CO_2eq}$ per kWh of energy delivered by the PV-ESS system and normalized profits up to $\$0.01/\mathrm{kWh}$, when reducing peak demand up to $75\%$ in an office building; up to $-0.31\,\mathrm{kg\,CO_2eq/kWh}$ at a cost of $\$0.08/\mathrm{kWh}$ when reducing up to $65\%$ of peak loads in an apartment complex and supermarket; and, in a hospital, up to $-0.34\,\mathrm{kg\,CO_2eq/kWh}$ at a cost of $\$0.08/\mathrm{kWh}$ for peak load reduction up to $40\%$. Nearly all levels of peak shaving DR modeled here generated negative cash flow once annualized ESS and PV costs were included along with conventional utility pricing schemes. Nonetheless, a reduction of $20\%$ in ESS and PV costs from the lower values in tables II-III (to an average of $\$419/\mathrm{kWh}$ and $\$1792/\mathrm{kWp}$, respectively) was sufficient to generate $Pr > 0$ in the office building, as compared to $75\%$, $80\%$, and $90\%$ for the supermarket, apartments, and hospital, respectively.

Variations in system performance beyond particular thresholds of peak shaving stem in large part from difficulties in dispatching peak shaving resources to reduce off-peak loads,

though the location of each threshold depended to some extent on specific building-level characteristics. The outcomes of our study further underscore the importance of taking into account these building-level aspects, in particular occupancy and equipment operation schedules, when assessing the impacts of decentralized energy systems.

In addition, the results of such comparisons can be used in policy-making decisions that incorporate issues of consumer empowerment, public health, and energy security. Indeed, carving out meaningful reductions in the carbon intensity of the electricity sector will depend not only on the choice of technology, but the efficacy of the incentives and price structures that complement its widespread deployment [9], and the acknowledgment of "human choice as critical and controlling in energy use and technology choice" [49]. In other words, the success of peak shaving DR will ultimately depend on whether an entity decides to enroll in the program and act in accordance with its economic incentives and any constraints that DR might place upon their energy consumption. It will also depend on how this same entity weighs the economic outlook for the project against its potential for emission reductions. It is therefore important to maintain as holistic a view as possible of the co-benefits and drawbacks of these systems for all stakeholders on the grid, especially as their utility as a decentralized resource increases.

## REFERENCES

[1] R. Freudenberg, L. Montemayor, E. Calvin, E. Korman, and S. McCoy, "Under Water: How Sea Level Rise Threatens the Tri-State Region Regional Plan Association is," Regional Plan Association, Tech. Rep. December, 2016.

[2] J. Hu, S. Kann, J. Tong, and J. Wellinghoff, "Grid Neutrality," Public Utilities Fortnightly, vol. 153, no. 10, pp. 24–30, oct 2015. [Online]. Available: http://ezproxy.cul.columbia.edu/login?url=http://search.proquest.com/docview/1727471091?accountid=10226http://rd8hp6du2b.search.serialssolutions.com/?ctx{\_}ver=Z39.88-2004{&}ctx{\_}enc=info:ofi/enc:UTF-8{&}rfr{\_}id=info:sid/ProQ{%}253Aabiglobal{&}rft{\_}val{\_}fmt=info:ofi/fmt

[3] D. Quiggin, S. Cornell, M. Tierney, and R. Buswell, "A simulation and optimisation study: Towards a decentralised microgrid, using real world fluctuation data," Energy, vol. 41, no. 1, pp. 549–559, 2012. [Online]. Available: http://linkinghub.elsevier.com/retrieve/pii/S0360544212001016

[4] R. Walawalkar, J. Apt, and R. Mancini, "Economics of electric energy storage for energy arbitrage and regulation in New York," Energy Policy, vol. 35, no. 4, pp. 2558–2568, 2007.

[5] Eastern Interconnection States' Planning Council, "Getting the Signals Straight : Modeling , Planning , and Implementing Non-Transmission Alternatives Study," no. February, 2015.

[6] B. Sovacool and M. Dworkin, Global energy justice: Problems, principles, and practices, 2014. [Online]. Available: http://www.scopus.com/inward/record.url?eid=2-s2.0-84952879088{&}partnerID=40{&}md5=917fea6b4e118e162584e6e9773d9f89

[7] N. Y. S. Department of Public Service, "Staff Report and Recommendations in the Value of Distributed Energy Resources Proceeding Staff Report and Recommendations," Tech. Rep., 2016.

[8] City of Boston, "Climate Ready Boston," City of Boston, Boston, Tech. Rep. 1-397, 2016.

[9] B. Howard, "Methods for Analysis of Urban Energy Systems: A New York City Case Study," Dissertations, Columbia University, 2016. [Online]. Available: https://academiccommons.columbia.edu/catalog/ac:197659http://dx.doi.org/10.7916/D8W66KQW

[10] D. A. Martin, "The Blueprint: A Preview of the Principles & Framework for Boston's Resilience Strategy," Office of the Mayor, Boston, Tech. Rep., 2016. [Online]. Available: http://www.cityofboston.gov/cable/video{_}library.asp?id=19916

[11] F. Shariatzadeh, P. Mandal, and A. K. Srivastava, "Demand response for sustainable energy systems: A review, application and implementation strategy," *Renewable and Sustainable Energy Reviews*, vol. 45, pp. 343–350, 2015. [Online]. Available: http://dx.doi.org/10.1016/j.rser.2015.01.062

[12] R. Walawalkar, "Economics of Emerging Electric Energy Storage Technologies and Demand Response in Deregulated Electricity Markets," Ph.D. dissertation, Carnegie Mellon University, 2008.

[13] M. Deru, K. Field, D. Studer, K. Benne, B. Griffith, P. Torcellini, B. Liu, M. Halverson, D. Winiarski, M. Rosenberg, M. Yazdanian, J. Huang, and D. Crawley, "U.S. Department of Energy commercial reference building models of the national building stock," National Renewable Energy Laboratory, Golden, Colo., Tech. Rep. February 2011, 2011. [Online]. Available: http://digitalscholarship.unlv.edu/renew{_}pubs/44

[14] M. Baecheler, J. Williamson, T. Gilbride, P. Cole, M. Hefty, and P. Love, "Guide to Determining Climate Regions by County," vol. 7, no. August, pp. 1–34, 2010.

[15] D. Crawley, P. Torcellini, N. Long, E. Bonnema, and K. Field, "Modeling Energy Savings," *ASHRAE Journal*, vol. 52, no. June, pp. 1–3, 2010.

[16] R. Guglielmetti, D. Macumber, N. Long, N. R. E. L. (U.S.), and I. I. Conference, "OpenStudio : an open source integrated analysis platform : preprint," Golden, Colo., 2011. [Online]. Available: http://purl.fdlp.gov/GPO/gpo19066

[17] E. A. Gilmore, J. Apt, R. Walawalkar, P. J. Adams, and L. B. Lave, "The air quality and human health effects of integrating utility-scale batteries into the New York State electricity grid," *Journal of Power Sources*, vol. 195, no. 8, pp. 2405–2413, apr 2010. [Online]. Available: http://www.sciencedirect.com/science/article/pii/S0378775309019235

[18] New York Independent System Operator, "Real-Time Actual Load," 2014. [Online]. Available: mis.nyiso.com/public/

[19] U.S. Environmental Protection Agency, "Emissions & Generation Resource Integrated Database," 2015. [Online]. Available: https://www.epa.gov/energy/egrid

[20] U.S. Energy Information Administration, "State Energy Price and Expenditure Estimates 1970 Through 2012," US EIA, Tech. Rep. June, 2014. [Online]. Available: http://www.eia.gov/state/seds/sep{_}prices/notes/pr{_}print.pdf

[21] ——, "Annual Energy Outlook 2015," US EIA, Tech. Rep., 2015. [Online]. Available: http://www.eia.gov/forecasts/aeo/pdf/0383(2015).pdf

[22] ——, "Electric Power Annual 2014," US EIA, Tech. Rep. February, 2016. [Online]. Available: http://www.eia.gov/electricity/annual/archive/2014/pdf/epa.pdf

[23] Consolidated Edison, "PSC NO: 10 - Electricity Consolidated Edison Company of New York, Inc. Service Classifications," Consolidated Edison, New York, NY, Tech. Rep., 2015. [Online]. Available: http://www.coned.com/documents/elecPSC10/SCs.pdf

[24] E. M. Krieger, J. A. Casey, and S. B. C. Shonkoff, "A framework for siting and dispatch of emerging energy resources to realize environmental and health benefits: Case study on peaker power plant displacement," *Energy Policy*, vol. 96, pp. 302–313, 2016. [Online]. Available: http://dx.doi.org/10.1016/j.enpol.2016.05.049

[25] M. Zheng, C. J. Meinrenken, and K. S. Lackner, "Smart households: Dispatch strategies and economic analysis of distributed energy storage for residential peak shaving," *Applied Energy*, vol. 147, pp. 246–257, jun 2015. [Online]. Available: http://www.sciencedirect.com/science/article/pii/S0306261915002160

[26] N. Y. S. Department of Public Service, "Brooklyn Queens Demand Management Program," 2014.

[27] Lazard, "Lazard's Levelized Cost of Storage Analysis 1.0," no. November, 2015. [Online]. Available: https://www.lazard.com/media/2391/lazards-levelized-cost-of-storage-analysis-10.pdf

[28] ——, "Levelized Cost of Energy Analysis 9.0," no. June, 2015. [Online]. Available: https://www.lazard.com/media/2390/lazards-levelized-cost-of-energy-analysis-90.pdf

[29] C. Spanos, D. E. Turney, and V. Fthenakis, "Life-cycle analysis of flow-assisted nickel zinc-, manganese dioxide-, and valve-regulated lead-acid batteries designed for demand-charge reduction," *Renewable and Sustainable Energy Reviews*, vol. 43, pp. 478–494, mar 2015. [Online]. Available: http://www.sciencedirect.com/science/article/pii/S1364032114008971

[30] M. Hiremath, K. Derendorf, and T. Vogt, "Comparative Life Cycle Assessment of Battery Storage Systems for Stationary Applications," *Environmental Science & Technology*, p. 150323152729009, 2015. [Online]. Available: http://pubs.acs.org/doi/abs/10.1021/es504572q

[31] R Core Team, "R: A Language and Environment for Statistical Computing," Vienna, 2016. [Online]. Available: https://www.r-project.org/

[32] W. Chang, "R6: Classes with Reference Semantics," 2016. [Online]. Available: https://cran.r-project.org/package=R6

[33] S. Mortiz, "imputeTS: Time Series Missing Value Imputation," 2016. [Online]. Available: https://cran.r-project.org/package=imputeTS

[34] Revolution Analytics and S. Weston, "doSNOW: Foreach Parallel Adaptor for the 'snow' Package," 2015. [Online]. Available: https://cran.r-project.org/package=doSNOW

[35] ——, "foreach: Provides Foreach Looping Construct for R," 2015. [Online]. Available: https://cran.r-project.org/package=foreach

[36] H. Wickham, "tidyr: Easily Tidy Data with 'spread()' and 'gather()' Functions," 2016. [Online]. Available: https://cran.r-project.org/package=tidyr

[37] ——, "ggplot2: Elegant Graphics for Data Analysis," New York, NY, 2009. [Online]. Available: http://ggplot2.org

[38] C. O. Wilke, "cowplot: Streamlined Plot Theme and Plot Annotations for 'ggplot2'," 2016. [Online]. Available: https://cran.r-project.org/package=cowplot

[39] L. Aslett, "RStudio Server Amazon Machine Image," 2016.

[40] B. Battke, T. S. Schmidt, D. Grosspietsch, and V. H. Hoffmann, "A review and probabilistic model of lifecycle costs of stationary batteries in multiple applications," *Renewable and Sustainable Energy Reviews*, vol. 25, pp. 240–250, sep 2013. [Online]. Available: http://www.sciencedirect.com/science/article/pii/S136403211300275X

[41] J. Leadbetter and L. Swan, "Battery storage system for residential electricity peak demand shaving," *Energy and Buildings*, vol. 55, pp. 685–692, dec 2012. [Online]. Available: http://www.sciencedirect.com/science/article/pii/S0378778812004896

[42] M. Zheng, C. J. Meinrenken, and K. S. Lackner, "Agent-based model for electricity consumption and storage to evaluate economic viability of tariff arbitrage for residential sector demand response," *Applied Energy*, vol. 126, pp. 297–306, 2014. [Online]. Available: http://dx.doi.org/10.1016/j.apenergy.2014.04.022

[43] A. M. Abdilahi, A. H. Mohd Yatim, M. W. Mustafa, O. T. Khalaf, A. F. Shumran, and F. Mohamed Nor, "Feasibility study of renewable energy-based microgrid system in Somaliland's urban centers," *Renewable and Sustainable Energy Reviews*, vol. 40, pp. 1048–1059, 2014. [Online]. Available: http://linkinghub.elsevier.com/retrieve/pii/S1364032114006029

[44] V. M. Fthenakis, R. Frischknecht, M. Raugei, H. C. Kim, E. Alsema, M. Held, and M. de Wild Scholten, *Methodology Guidelines on Life Cycle Assessment of Photovoltaic Electricity*, 2011, vol. IEA PVPS T, no. 5454.

[45] National Renewable Energy Laboratory, "National Solar Resource Database, Web-Based GIS Tool." [Online]. Available: https://nsrdb.nrel.gov/nsrdb-viewer

[46] N. Blair, A. P. Dobos, J. Freeman, T. Neises, M. Wagner, T. Ferguson, P. Gilman, and S. Janzou, "System advisor model, sam 2014.1. 14: General description," *NREL Report No. TP-6A20-61019, National Renewable Energy Laboratory, Golden, CO*, no. February, p. 13, 2014. [Online]. Available: http://www.nrel.gov/docs/fy14osti/61019.pdf

[47] E. Leccisi, M. Raugei, and V. Fthenakis, "The Energy and Environmental Performance of Ground-Mounted Photovoltaic Systems- A Timely Update," *Energies*, vol. 9, no. 8, p. 622, 2016. [Online]. Available: http://www.mdpi.com/1996-1073/9/8/622

[48] M. Hofmann, S. Riechelmann, C. Crisosto, R. Mubarak, and G. Seckmeyer, "Improved synthesis of global irradiance with one-minute resolution for PV system simulations," *International Journal of Photoenergy*, vol. 2014, 2014.

[49] L. Lutzenhiser and E. Shove, "Contracting knowledge: The organizational limits to interdisciplinary energy efficiency research and development in the US and the UK," *Energy Policy*, vol. 27, no. 4, pp. 217–227, 1999.

978-1-5090-5606-4/17 $31.00 © 2017 IEEE

# Improving Any Arbitrary MPPT Hill Climber with ANN Estimations

Jesse Roberts, M.S.
Department of Electrical and
Computer Engineering
Tennessee Technological University
Cookeville, TN 38501
Email: http://www.JesseRoberts.org

Dr. Indranil Bhattacharya
Department of Electrical and
Computer Engineering
Tennessee Technological University
Cookeville, TN 38501
Email:IBhattacharya@tntech.edu

*Abstract*—**Solar energy has a unique challenge associated with power conversion referred to as the maximum power point (MPP) problem. A solar cell will only generate optimal power if the solar cell operating voltage is at the MPP. In this paper a simple ANN employing online learning is presented which can be integrated into any hill climbing MPPT scheme without adding any sensors. The ANN provides a rapid estimation of the MPP based on the solar cell's operating voltage whenever a large change in environmental conditions is detected. This estimation is used as the initial condition for the hill climbing algorithm. This initial condition allows the algorithm to converge more quickly by reducing the search space. Any of the typical favorite MPPT algorithms can be combined with the ANN estimation engine to perform guess and track (G&T) MPPT.**

## I. INTRODUCTION

Here the necessary background in MPPT and ANNs is explored. This paper assumes a basic knowledge of MPPT and an awareness of machine learning concepts.

### A. MPPT

The MPP is defined in the voltage domain and is a consequence of the attached load impedance. The important concept is that for a given solar cell at a specific temperature and insolation, there exists a single load which will draw the maximum power from the solar cell. For solar cells to be deployed in a dynamic environment, finding the matched impedance is insufficient. A solution must be implemented to ensure MPP operation for an arbitrary, dynamic load, insolation, and temperature combination.

The solution is modulating the load with a switching converter. The switching converter makes the load impedance appear different to the solar cell, allowing any load to act as the solar cell matched impedance. The converter must be given a duty cycle to adjust the modulation. This can be done directly, allowing the MPPT algorithm to adjust the duty cycle, or indirectly by allowing the MPPT algorithm to select the operating voltage while a separate switch mode controller forces the operating voltage error to zero. The point here is that optimization is done in either the voltage or duty cycle domain. In this paper G&T for both direct and indirect control will be explained however only direct control will be evaluated in simulation. Likewise, the conclusion will be valid specifically for direct control.

The search for the MPP is a good example of a nearly convex optimization problem and is solved easily with any given hill climbing algorithm. Hill climbing algorithms can work precisely or quickly but typically not both due to a static step size. Much work has been done in papers such as [1] and [2] to allow for a variable step size in both the perturb and observe algorithm and the incremental conductance algorithm. However, these systems are not as common as the traditional implementations.

Some work has been done previously to perform guess and track MPPT in [3] and [4]. The problem here is that the estimation portion requires significant sensing components making it a poor additive to an already otherwise complete tracker that simply needs a faster response. The formalized G&T approach herein can be added to an already complete system to improve performance through programming alone.

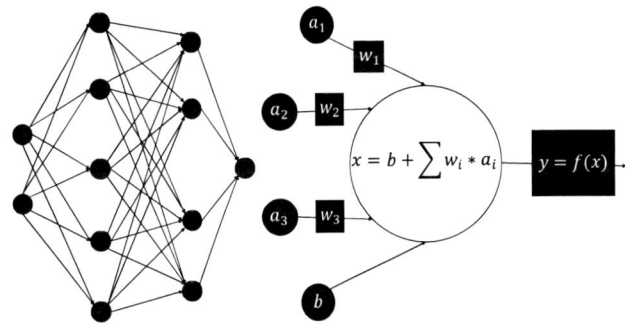

Fig. 1. Left: MLP-ANN Right: Single Perceptron

### B. ANN

Artificial neural networks are excellent non-linear trackers. The typical implementation is the multi-layered perceptron (MLP). The MLP gets its name from the layered architecture of the algorithm. It has many layers of connected perceptrons feeding from input to output. The perceptrons are each made up of a bias, inputs with associated weights, and an activation function. The general form of the perceptron and MLP are shown in Fig.1. The activation function takes the sum of the

978-1-5090-5606-4/17 $31.00 © 2017 IEEE

bias and weighted inputs and results in a normalized, non-linear output. The combination of these perceptrons allows for many degrees of freedom. This freedom can be harnessed to provide powerful, non-linear modeling or estimation.

In MATLAB, ANNs are trained with the levenberg-marquardt algorithm by default. This algorithm is an amalgamation of the gradient descent and newton-gauss optimization methods used to reduce some cost function. The training employed in the G&T ANN is similar but with an online approach. The online learning employed here sees a return to a simpler direct gradient descent approach with an update when new data is acquired.

## II. ANN Estimation

In this section the specifics of the ANN and the support structure will be discussed. The topics will be: the input and how to obtain a measure of irradiance without an irradiance sensor; the specifics of the uniquely limited online learning implementation; and the overall architecture of the ANN.

### A. Input; Or, Rethinking Irradiance Measurements

In order to create a truly additive system that allows the main tracking algorithm to be arbitrary, the first hurdle is achieving some measure of the environmental variables without additional, dedicated sensors. To address this consideration, it is important to notice some things about the nature of the MPP mobility. The mobility of the MPP is different in the voltage domain from the duty cycle domain. A system based on indirect converter control need only be concerned with the voltage domain while the duty cycle domain is uniquely important to a directly controlled converter system.

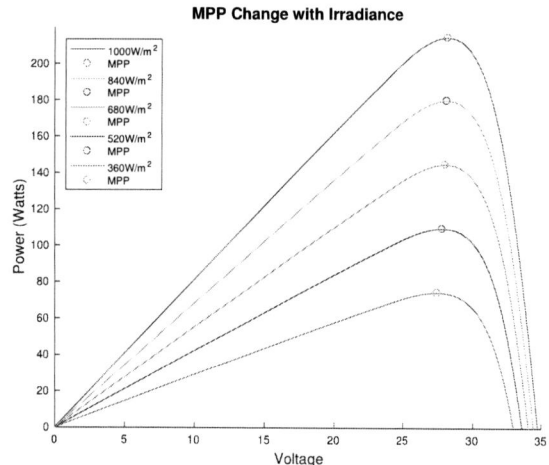

Fig. 2. Solar Cell PV Curves at Different Irradiances

In terms of duty cycle, a change in temperature has very little impact on the matched impedance and resulting MPP duty cycle. The irradiance however, has a considerable effect. This is shown in Fig.2. Notice that the MPP voltage does not change very much but the power increases dramatically. This suggests the current has increased which means the matched

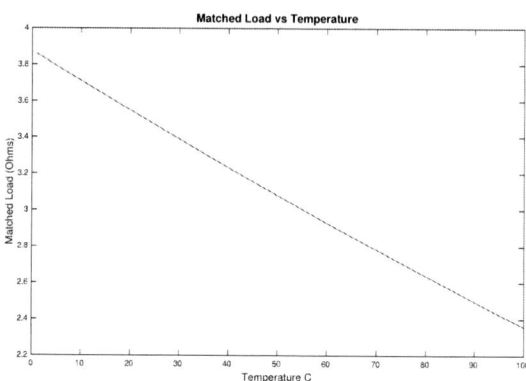

Fig. 3. Matched Impedance Change with Respect to Temperature

the matched impedance has been reduced proportionately. In Fig.3 temperature is shown to have very little implications in regard to the matched impedance as it changes a mere 1.4 ohms over a $100°C$ change.

In the voltage domain the opposite of the aforestated conclusions is true. As has been shown, irradiance variation causes little change to the MPP voltage. On the other hand temperature creates a traveling MPP voltage as is pictured in Fig.4.

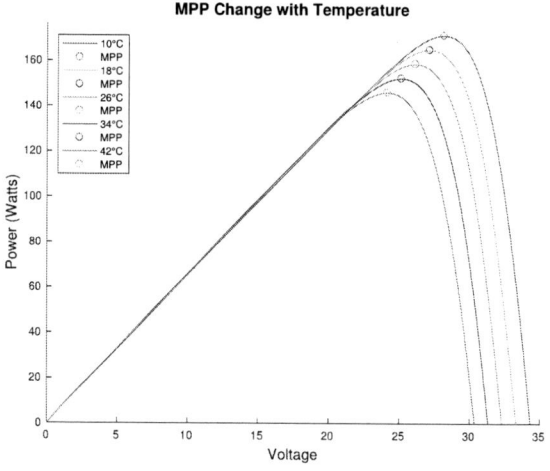

Fig. 4. Solar Cell PV Curves at Different Temperatures

These observations about the nature of direct and indirect PWM control are important as it allows the problem to be simplified. It is an acceptable simplification to only measure irradiance if the optimization domain is the switching duty cycle. Conversely, it is acceptable to only measure temperature if the optimization domain is voltage.

It is also important to understand that not only is the MPP related to these variables but rather the entire power curve is related. With this in mind, it becomes obvious that the operating voltage of a solar cell at a particular PWM duty cycle is a representation of the incident irradiance. Likewise, the power apparent from a solar cell at a specific voltage is representative of the temperature. In fact, if the load is known an ANN can estimate the MPP duty cycle for a solar cell by measuring the operating voltage at a non-trivial duty cycle. The ANN can be trained by taking voltage measurements at a duty cycle (chosen to be 50% in this paper) under different irradiance conditions with known MPP duty cycles for each measurement. In order to perform an estimation, the network need only be presented with a voltage measurement taken while the switching converter is set to 50% duty cycle. This is known as the irradiance correlated voltage (ICV). The same can be done with an indirect system by choosing a non-trivial voltage at which to take power measurements, though that will not be explored fully in this paper.

### B. Online Training

The online training in this paper is somewhat unique. It has been referred to as uniquely limited because the method for providing online adaptations has an inbuilt limitation to the magnitude of adaptation possible in a single update. The benefit here is that no single data point can cause a significant deviation in accuracy. Rather, it requires a succession.

First, the general progression of the algorithm is important. An ICV measurement is taken and an estimation of the MPP duty cycle is made. The converter is set to the estimation duty cycle and the tracker takes over. Upon convergence, the algorithm again measures the ICV by setting the duty cycle to 50% and measuring the voltage. Both the ICV measurement and the actual MPP duty cycle are fed to the ANN online learning algorithm for updating.

Second, it is important to note that the training data is kept in an array and stays with the ANN. The entries in this array will be updated. When an ICV and MPP duty cycle are given to the algorithm, the most similar ICV data in the array is replaced by the new data. The entire array is then fed through the gradient descent algorithm to reduce the error. This single point replacement online learning allows the ANN to make small evolutionary changes rather than large, possibly unstable changes.

### C. Architecture

The network has a simple architecture as precision takes a back seat to ease of implementation. The network consists of an input layer, two hidden layers, and an output layer. The input layer has one input while the hidden layers each have three neurons. Fig.5 represents the network used to perform the estimation.

### III. THE ALGORITHM

In this section the Algorithmic combination of the tracker and the estimator are detailed. The combination mechanism for the hill climbing tracker with the estimator is simple. A measure of the change in voltage with respect to duty cycle is used to decide whether to continue tracking or make an estimation. Essentially, if $|\frac{DV}{DD}|$ is greater than some threshold then an estimation is required. Otherwise the tracker remains

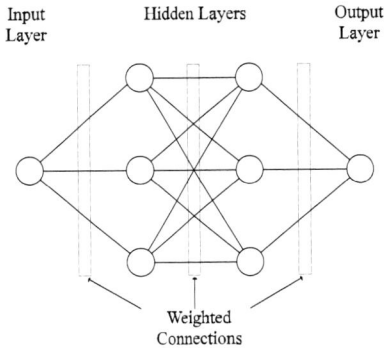

Fig. 5. G&T ANN Architecture

in control. The concept is that a large change in $|\frac{DV}{DD}|$ is a signal that an environmental shift has occured and that it is now prudent to choose a new initial condition from which to track.

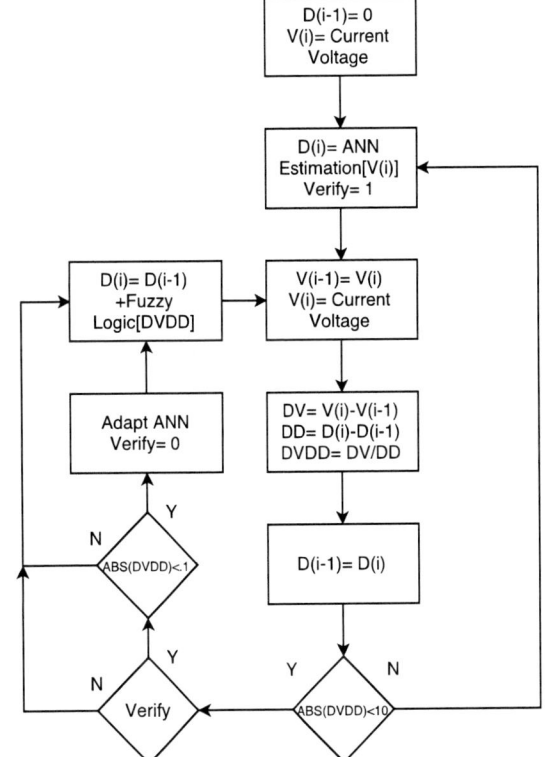

Fig. 6. G&T Flow Chart

The tracking algorithm is the backbone of the G&T algorithm. Any tracking algorithm based on hill climbing can be used. The tracker allows the ANN to have errors and still converge to the MPP. Its best to use a simple implementation with a small step size. The small step size allows the algorithm to get very near the actual MPP. The ANN makes a large step size unnecessary because of the small search space. In the flow chart presented and the simulations herein fuzzy logic is used as the tracker. The fuzzy logic tracker is similar to that used in [5].

When an estimation is required the procedure outlined above is followed. The chosen duty cycle is set and an ICV measurement is taken. This is fed to the ANN which provides the suggested PWM initial duty cycle. The tracker then continues tracking until the error is reduced. At this point the ANN is retrained using online learning and the algorithm continues tracking. The system in entirety is shown in Fig.6.

## IV. MODELING

In this section the simulation method is explained. The modeling method used for the solar cell is explained followed by the Sepic modeling method.

### A. Solar Cell Model

The standard single diode solar cell model is used. It employs the newton-raphson optimization method to find the operating current of the solar cell at an array of voltages. This lookup table approach is then used to find the operating point for a given load. The simulation flow chart is shown in Fig.7. The modeling approach is heavily based on [6].

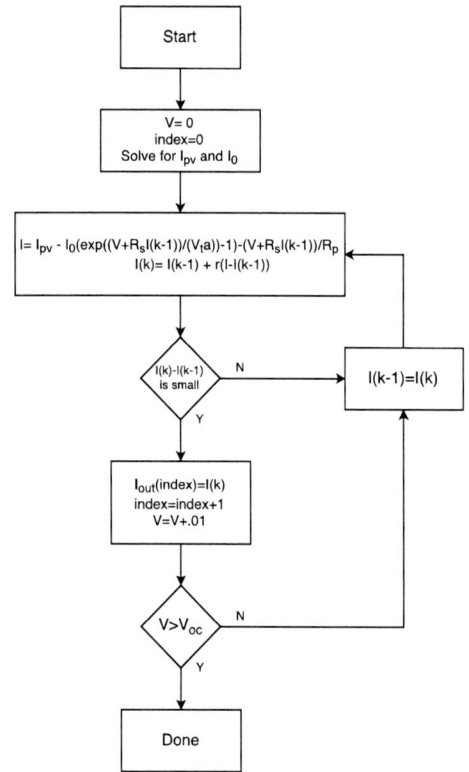

Fig. 7. Solar Cell Simulation Flow Chart

### B. Sepic Model

Switch mode converter modeling can be done a number of ways. The easiest is by far to assume that the output current is continuous. This allows the use of the average model equations.

The specific goal of the model in this paper is to relate the connected load from the output stage of the sepic converter to the input stage. The purpose of this is to find the equivalent connected load at the solar cell. The equation in (1) shows the relationship alluded to in [7] while (2) is the consequence of assuming unity power transfer. The obvious result of (1) and (2) is (3). With equations modeling both the input and output relationship of voltage and current, it is possible to derive an equation for the impedance seen from the perspective of the input. This is shown in (4). The equation in (4) combined with a duty cycle and load resistance constitutes the entire Sepic modulator model.

$$\frac{V_o}{V_i} = \frac{D}{1-D} \tag{1}$$

$$V_i I_i = V_o I_o \tag{2}$$

$$\frac{I_o}{I_i} = \frac{1-D}{D} \tag{3}$$

$$\frac{R_i}{R_o} = \left(\frac{1-D}{D}\right)^2 \tag{4}$$

## V. RESULTS

In this section the simulation and results will be discussed. The goal of the simulation is to see the effect of adding the G&T system to the standalone fuzzy logic based MPPT. The testing method will be to allow each system to converge to the MPP and then provide a disturbance to the irradiance. The systems will then reconverge. The overall power generation with respect to the possible power generation will be the method of judgement. The parameters used for the simulation are shown in TABLE-I and the final results in TABLE-II.

TABLE I
SIMULATION PARAMETERS

| Initial Irradiance | $1000\frac{W}{m^2}$ |
|---|---|
| Disturbance Irradiance | $600\frac{W}{m^2}$ |
| ANN Training Temperature | $50°C$ |
| Operating Temperature | $50°C$ |
| Load | $20\Omega$ |

### A. Tracker without G&T

The MPPT without G&T did well as it was a tuned fuzzy logic controller. The controller was able to harvest 90% of the available power and converged in 20 epochs.

### B. Tracker with G&T

The MPPT with the addition of the G&T algorithm allowed for an increase in efficiency as well as a faster convergence time. The system converged in only 7 epochs and was 98% efficient. It is important to note that there exists an upper limit

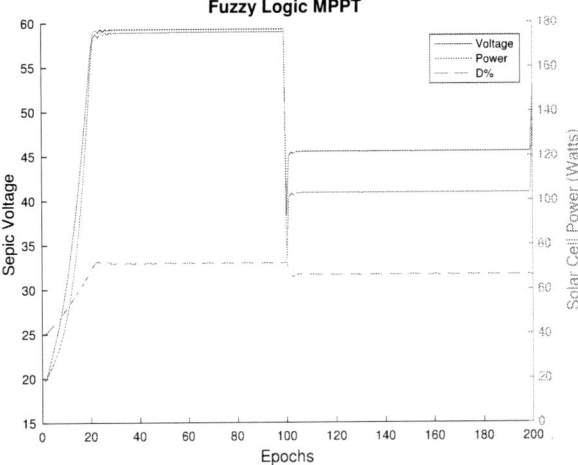

Fig. 8. Fuzzy Logic MPPT Simulation Results

on the performance boost. Since the system takes a minimum of 3 epochs to run an estimation, this is the minimum time to converge in the case of an estimation. This is apparent in the fact that both systems reconverged in the same number of epochs after the disturbance.

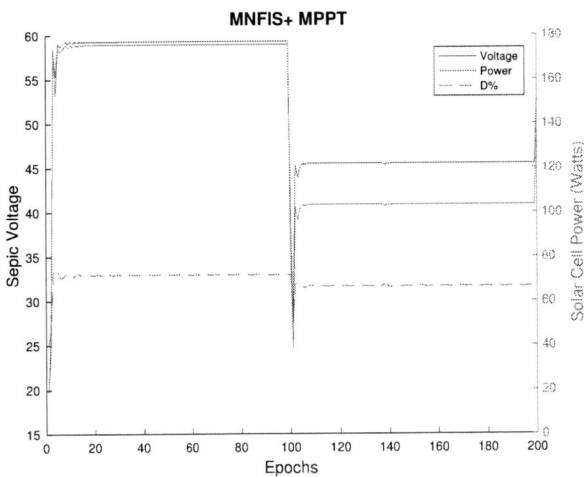

Fig. 9. Fuzzy Logic with G&T MPPT Simulation Results

TABLE II
SIMULATION RESULTS

| Algorithm | Efficiency | Epochs to Converge |
|---|---|---|
| Fuzzy Logic | 90% | 20 |
| Fuzzy Logic with G&T | 98% | 7 |

## VI. CONCLUSION

By employing the G&T method with any tracker algorithm, the tracking performance can be improved without adding any extra hardware. Specifically in this paper this is proven through

simulation for all direct control MPPT schemes. It is a logical extension that the indirect system description would likewise show a similar boost in performance.

### A. Future Work

Obviously, all conclusions for indirect converter control are tentative and merit individual investigation. Also, building a physical system would be a great benefit to provide actual data for both direct and indirect control. The last area to be investigated will be partial shading implications.

## REFERENCES

[1] F. Liu, S. Duan, F. Liu, B. Liu, and Y. Kang, "A variable step size inc mppt method for pv systems," *IEEE Transactions on industrial electronics*, vol. 55, no. 7, pp. 2622–2628, 2008.

[2] B. Amrouche, M. Belhamel, and A. Guessoum, "Artificial intelligence based p&o mppt method for photovoltaic systems," *Revue des Energies Renouvelables ICRESD-07 Tlemcen*, pp. 11–16, 2007.

[3] P. Q. Dzung, H. H. Lee, N. T. D. Vu *et al.*, "The new mppt algorithm using ann-based pv," in *Strategic Technology (IFOST), 2010 International Forum on*. IEEE, 2010, pp. 402–407.

[4] C. Liu, B. Wu, and R. Cheung, "Advanced algorithm for mppt control of photovoltaic systems," in *Canadian Solar Buildings Conference, Montreal*. Citeseer, 2004, pp. 20–24.

[5] J. Roberts and I. Bhattacharya, "Mnfis and other soft computing based mppt techniques: A comparative analysis," in *Photovoltaic Specialists Conference (PVSC), 2016 IEEE 43rd*. IEEE, 2016, pp. 3247–3251.

[6] M. Villalva, J. Gazoli, and E. Filho, "Comprehensive approach to modeling and simulation of photovoltaic arrays," *Power Electronics, IEEE Transactions on*, vol. 24, no. 5, pp. 1198–1208, May 2009.

[7] T.-F. Wu and Y.-K. Chen, "Modeling pwm dc/dc converters out of basic converter units," *IEEE transactions on Power Electronics*, vol. 13, no. 5, pp. 870–881, 1998.

# Increasing Solar Photovoltaic Penetration Using Thermal Energy Storage

Alexander F. Routhier, Dr. Christiana Honsberg

Arizona State University, Tempe, AZ, 85281, United States

*Abstract* — Thermal energy storage (TES) is a mature technology, deployed throughout the world, particularly in the southern United States. Through modeling and simulation of Arizona State University's system, it has been determined that TES is a cost-effective way to increase solar photovoltaic (PV) penetration. This novel management method enables significant increases in PV, while alleviating many of the problems associated with traditional TES management schemes. This research includes a study of feasibility of the proposed management scheme.

## I. INTRODUCTION

As the electric power industry works to decarbonize the electric power grid, electric storage will eventually become a necessity. Current technologies, including batteries, flywheels, pumped hydro, compressed air and other common devices are unable to efficiently and economically meet societal needs, at the required scale. Each of these technologies converts electricity to another form of energy for storage, and then converts back to electricity to be consumed by the end user. While using thermal energy storage (TES) also converts the electricity to a different storage medium, it never converts back to electricity. This means there is one less thermodynamic conversion for potential energy losses.

There are several types of TES including chilled water systems (CWS), ice storage systems, and eutectic salt storage systems. This research focuses on TES using a CWS. This means chilled water is pumped into large insulated tanks, which act as the storage medium. This chilled water is what is eventually consumed by the end user. For this reason, it is often overlooked as a storage mechanism by the electric power industry.

One major advantage of a CWS is it can usually be operated using existing air-conditioning chillers and piping, with installation requirements consisting of insulated tanks, fittings connecting to the existing system and a pumping system. CWSs are generally best suited for large systems, of greater than 2,000 ton-hours of storage where land use is not an issue [1].

While TES cannot be used to store electricity directly, it does enable load shifting, load leveling and peak shaving through demand side management. The user stores cold water when excess power is available, and discharges cold water to meet air-conditioning loads. The cold water is pumped to an air-handler which converts the chilled water into the cold air used for air-conditioning. If managed correctly, the user can employ solar photovoltaic (PV) energy to charge their system, and then discharge the cold water, both during peak afternoon and

evening loads, but also throughout the night, and during cloudy weather when PV is unavailable.

Arizona State University's (ASU) Tempe campus is used as the test case for this research. They have 16.1 MWdc of solar PV installed on campus, a large CWS system (six one-million-gallon storage tanks) and easily accessible historical data for both systems, along with historical load and weather data [2]. ASU's Tempe campus currently receives about 14 percent of their total annual energy consumption from their on-campus PV resources. Historical data is maintained at ASU's Central Plant, which houses many of the campus chillers, and controls and coordinates the operation of both the PV and air-conditioning systems (see figure 1 below).

Fig. 1 The control center for ASU's Central Plant

## II. THERMAL STORAGE MANAGEMENT SCHEME

The management scheme for the thermal storage is broken into two separate time periods; peak season, and off-peak season. Due to the drastic change in air-conditioning demand between the two-time periods, different implementation strategies are required.

In the desert southwest, peak season is from approximately early June until late September. Off-peak season is the remaining portion of the year. Depending on climate and location, the period and duration of these designations may change. This research uses Arizona State University as it's test

978-1-5090-5606-4/17 $31.00 © 2017 IEEE

case, so these dates are a good approximation. This can be extrapolated from the data shown in figures 2 and 3.

The goal of both management schemes is to generate the maximum amount of energy using PV, without sending any energy back to the grid. This means that power production cannot provide more than the power demand of the system, plus it's maximum TES charging rate.

### A. Peak Season

Peak season management will remain the same as the current management scheme of most TES systems (including ASU's). Under this setting chillers are run at night, when power is relatively cheaper, and the chilled water is stored in TES. The TES is then discharged during peak demand hours, which generally coincides with more expensive usage rates, to decrease peak load and avoid both peak usage and peak demand charges. This is a proven scheme that has been employed successfully over several decades. Based on ASU's rate plan with Arizona Public Service, peak pricing occurs between 11:00 a.m. and 9:00 p.m. (additional details are discussed in section IV) [3].

Because energy demand from air-conditioning is so significant during peak season, the PV is unable to meet demand. The amount of PV installed on the system is limited by the off-peak season conditions. This is illustrated in figure 2, where you can see the remaining load above PV is never negative (implying there is never excess power to store). This issue will be discussed further in the analysis section.

### B. Off-peak Season

The off-peak management is what enables, but also limits, the significant increase in PV penetration. During the off-peak season, PV generation is able to meet 100% or more of energy consumption throughout the day. Any amount generated over the instantaneous power demand, is used to power one or more air-conditioning chillers and the chilled water produced is stored in TES.

Because ASU is located in the Sonoran Desert, and due to its facilities, there is a need for air-conditioning year-round. Unfortunately, the TES has a maximum charging and discharging rate, and also a limit on the total amount of chilled water which can be stored. There are six, one million gallon insulated storage tanks. Taking these factors into account, the maximum charging rate of the tanks is the limiting factor.

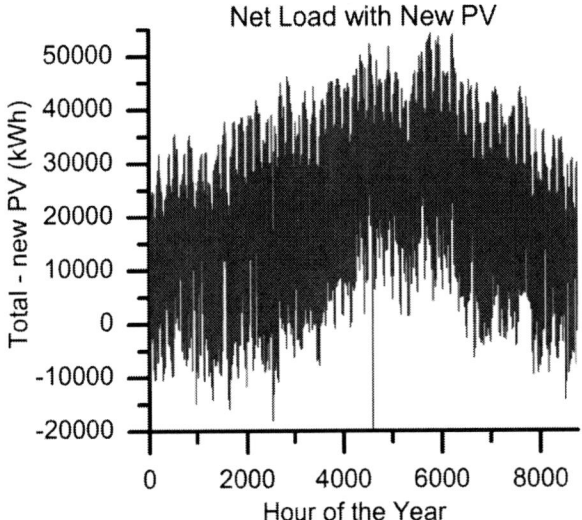

Fig. 2. State of charge in percentage over one year on a time scale of hours.

### III. MODEL DETAILS

The model used in this research is programmed in Matlab, along with associated files in Microsoft Excel. It uses typical meteorological year (TMY) data, PV array data and historical temperature to calculate the hourly system energy output for one year [4]. This energy output is then compared with the overall energy consumption and the campus air-conditioning load.

The model imports data about each array in the system, including the size of the array (in kWdc), the tilt angle, and azimuth angle of the panels, the nominal operating cell temperature and the percentage loss of power per degree C. This data is used, in conjunction with the TMY data to generate the power output of the system. The calculation is done hourly for one year. A detailed breakdown of the calculation method can be found at [5].

The maximum charging and discharging rates of the TES are extrapolated from the historical data set. The PV energy output is then multiplied by a scaling factor in an iterative process. The goal of these iterations is to ensure energy output from the PV does not exceed energy demand, plus the maximum TES charging rate. This ensures that no power is sent to the grid, and the TES is able store unused energy. If PV energy is unavailable to meet air-conditioning loads, then the TES is discharged to meet this demand, up to the maximum discharge rate of the TES. If the amount of chilled water required exceeds the maximum discharge rate, then chillers must be turned on to meet this additional air-conditioning demand.

A state of charge monitor is run ensure the TES is never over or under-utilized. You can see in figure 2 that the maximum state of charge is just under 60%. This is assuming that the state of charge at the beginning of the year is 0%. This assumption was tested for sensitivity, and it was found that as long as the initial state of charge at the beginning of the year is below 77%, there are no issues with over utilization. The right side of the plot shows charge levels are below 30% at the end of the year, which makes it unlikely they will reach 77%. When higher values are used to start the year, they do not carry over throughout an entire year and do not affect the end of year levels, which means the numbers shown are realistic.

Figure 3 also shows that if this scheme is implemented year-round, the TES is never charged during peak summer months. Therefore, there is a distinction between peak season and off-peak season functionality. If we used the off-peak functionality year-round, then the system would remain unused for three to four months per year. By switching back to the current operating method in the peak season, it allows the TES to continue acting as a load shifting device and decrease costs for the user.

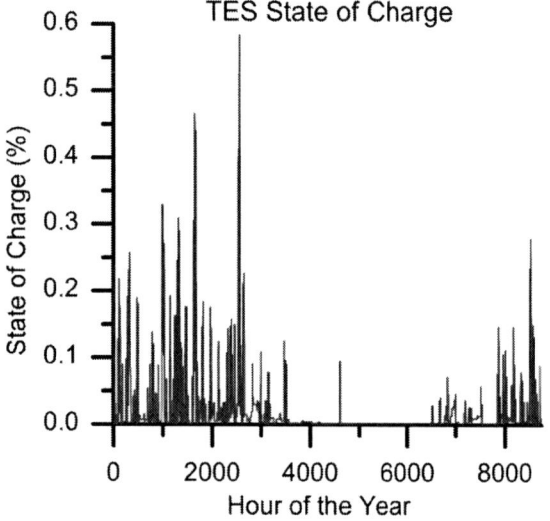

Fig. 3.    State of charge in percentage over one year on a time scale of hours.

## IV. ECONOMICS

In order to encourage users to implement these techniques, it is important to show that there is an economic advantage. The model data is taken from ASU, so the price plan modeled is based on ASU's relationship with Arizona Public Service (APS). This plan includes both a usage charge and a peak demand charge. The costs of the current management scheme are compared against the costs of new management scheme.

Another aspect of ensuring positive economic advantages is to look at the cost of installing a system. The ASU system was installed in the 1980s, which means any monetary figures are out of date and relatively meaningless. The numbers used for this portion of the analysis were provided by the ASU utilities department based on estimates made during a feasibility of

system expansion study done in July of 2011. This is the most up to date information at this time.

During the 2011 study it was proposed to add a seventh tank to the ASU system, which includes the tank, pump, variable speed drive, tank valves, associated controls and installation of all components. No additional piping was required at the time, so this was not included in the estimate. The total system expansion cost was estimated to be $1.3 million. It was also estimated that this system would save ASU approximately $166,000 per year. This is a simple payback period of approximately 7.83 years.

## V. RESULTS

### A. PV Scaling Factor

Using the model to generate power output of the solar PV arrays, and iterating to find the scaling factor gives an output of 3.46. This means that the amount of solar PV being used at ASU can be increased by a factor of 3.46 without any new hardware beyond the new PV arrays. With increasing solar PV on ASU's Tempe campus by a factor of 3.46 would allow them to generate approximately 48 percent of their total annual energy consumption from their on-campus PV resources.

This scaling factor can change if the tilt angle of the panels is altered to optimize peak season generation. This would give less output per panel and decrease overall system efficiency, but increase the number of panels used, and the total power collected. Further research is needed to determine if this is a cost-effective strategy.

While these results are promising, due to some inconsistencies with the air-conditioning chillers and TES system data, we cannot make concrete conclusions about the system. Additional work needs to be done in order to verify the accuracy of the inputs. With that in mind, the results shown here are promising and warrant further investigation.

## VI. CONCLUSION

While the numbers shown in this research are encouraging, further investigation is required to ensure that this new method is economically advantageous.

One thing that is not considered in this research is the management scheme of the TES during the transition period of current PV deployment to the end state PV deployment. It will need to be determined how the scheme will gradually change into the future state described in this paper as new systems gradually come online.

## VII. ACKNOWLEDGEMENTS

This paper was written with the support of the Quantum Energy and Sustainable Solar Technology (QESST) engineering research center funded by the National Science

Foundation and the Department of Energy via grant EEC-10418

## VIII. REFERENCES

[1] S. M. Hasnain, "Review on sustainable thermal energy storage technologies, Part II: cool thermal storage," *Energy Convers. Manag.*, vol. 39, no. 11, pp. 1139–1153, 1998.

[2] "ASU Solar | Business and Finance." [Online]. Available: https://cfo.asu.edu/solar. [Accessed: 14-Apr-2017].

[3] "RATE SCHEDULE E-35 EXTRA LARGE GENERAL SERVICE TIME OF USE," *Rumolo, David J*, 2012. [Online]. Available: https://www.aps.com/library/rates/e-35.pdf. [Accessed: 09-Jun-2017].

[4] "NSRDB: 1991- 2005 Update: TMY3." [Online]. Available: http://rredc.nrel.gov/solar/old_data/nsrdb/1991-2005/tmy3/. [Accessed: 14-Apr-2017].

[5] A. F. Routhier, "Using Thermal Energy Storage to Increase Photovoltaic Penetration at Arizona State University's Tempe Campus by the Graduate Supervisory Committee," Arizona State University, 2016.

[6] G. M. Masters, "Renewable and Efficient Electric Power Systems," 2nd ed., New York: John Wiley & Sons, Inc., 2013, pp. 40–43.

# Model Predictive Control of Grid Connected Modular Multilevel Converter for Integration of Photovoltaic Power Systems

Amir Shahirinia[1,2], Amin Hajizadeh[3]

1- Department of Electrical Engineering, University of the District of Columbia, Washington DC, USA

2- Center of Excellence for Renewable Energy (CERE)

3- Department of Energy Technology, Aalborg University, Esbjerg, Denmark

*Abstract* — Investigation of an advanced control structure for integration of Photovoltaic Power Systems through Grid Connected-Modular Multilevel Converter (GC-MMC) is proposed in this paper. To achieve this goal, a non-linear model of MMC regarding considering of negative and positive sequence components has been presented. Then, due to existence of unbalance voltage faults in distribution grid, non-linarites and uncertainties in model, model predictive controller which is developed for GC-MMC. They are implemented based upon positive and negative components of voltage and current to mitigate the power quality problems. Finally, numerical results are given in order to show the dynamic performance of GC-MMC prototype under normal and unbalanced fault conditions. They illustrate the proposed current controllers are more effective under voltage disturbance conditions and it could keep the MMC stable under different conditions.

*Index Terms* — Modular Multilevel Converter, Photovoltaic, Power System, Model Predictive Control

## I. INTRODUCTION

The output power of the PV modules directly depends on the environmental conditions such as temperature and irradiance and it is recommended to choose the most efficient control strategy to achieve maximum power at the output of PV systems[1-2].

The use of grid-connected PV system has become popular in recent years. The PV system requires proper power conditioning system to manage the electrical energy effectively and to provide high-quality power from the PV system [3-4]. In this paper, the modular multilevel converter (MMC) is considered as a power conditioning system over other inverter topologies due to its improved reliability, lower total harmonic distortion (THD), low switching loss, good power quality and high output voltage minimum losses which can be utilised for high voltage high power application with low switching frequency [5-6]. The grid-connected power electronic converters are highly sensitive to voltage disturbances. This makes it is necessary to reduce the effects of voltage disturbances on their operations. Regarding to the many complexity in MMC, some control strategies have been presented by literature. They are categorized, control strategies of MMC for submodule (SM) voltage balancing [7], circulating current suppressing [8] and AC side current control for grid connected application [9]. One of the main technical challenges related to the control of an MMC is the selection of proper pulse width modulation (PWM) strategy along with the control of its internal dynamics, i.e., circulating currents and SM capacitor voltages. Regarding to publish research literatures, various PWM and control strategies have been developed for the MMC. One of the most popular control methods for power electronic converters which has been considered recently is Model Predictive Control (MPC) due to its several merits [10-11].

The application of model predictive control is not a fully new concept for this type of converter, where the majority of the recent designs have been based on the concept of finite state predictive control and, in particular, the optimal control solution is chosen through model prediction and cost function minimization by considering the finite set of combinations possible [12].

With the recent advances in convex optimization techniques [13], it has been possible to apply MPC to very fast constrained linear systems with continuous inputs by solving convex quadratic optimization problems within microseconds [14].

In power electronic applications, many conventional control strategies implemented in industry are based on proportional-plus-integral (PI) controllers providing continuous input signals to a modulator that manages conversion to discrete switch positions. As a substitute, direct MPC integrates the current control and the modulation problem into a single computational problem, providing a powerful replacement to conventional PI controllers. With direct MPC, the manipulated variables are the switch positions, which lie in a discrete and finite set, giving rise to a switched system [15]. Therefore, this approach does not require a modulator and is often referred to as "finite control set" MPC.

In this paper a standard model predictive control setting is used where a dynamic model of the MMC with considering positive and negative symmetrical components, known as state averaged model in the stationary axis frame, is used to represent the response of the grid currents.

This paper is organized as follows. Section II, the nonlinear dynamic model of MMC based upon positive and negative symmetrical components. The design of model predictive control strategy is presented in section III. Simulation results are presented and discussed in Section IV which verifies the

978-1-5090-5606-4/17 $31.00 © 2017 IEEE

advantages of the proposed control in operating of grid-connected power converters photovoltaic power system.

## II. MODELLING OF MODULAR MULTILEVEL CONVERTER

The typical structure of a MMC for integration of photovoltaic power system to onshore grid is shown in Fig. 1, and the configuration of a Sub Module (SM) is given in Fig. 2.

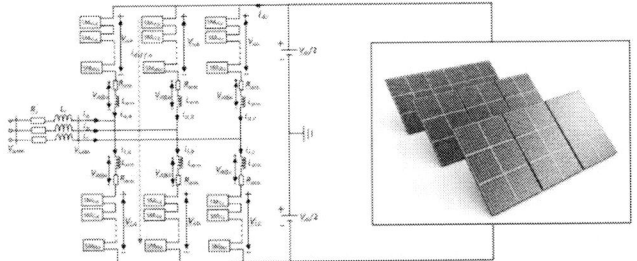

Fig.1. The structure of a MMC for integration of photovoltaic power system.

Each SM is a simple chopper cell composed of two IGBT switches (*T1* and *T2*), two anti-parallel diodes (*D1* and *D2*) and a capacitor *C* [2-3].

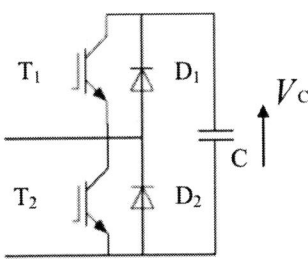

Fig.2. Configuration of a Submodule (SM) for MMC.

From Figs.1 and 2, the mathematical equations describing the dynamic behavior of a N-cells MMC are expressed as follows:

$$\frac{di_p}{dt} = \frac{1}{L}\left[\frac{V_{dc}}{2} - \sum_{i=1}^{N}\left(u_i.V_{C_{i-u}}\right) - Ri_p - V_a\right]$$

$$\frac{di_n}{dt} = \frac{1}{L}\left[\frac{V_{dc}}{2} - \sum_{i=N+1}^{2N}\left(u_i.V_{C_{i-l}}\right) - Ri_n + V_a\right] \tag{1}$$

$$\frac{dV_{C_{i\_u}}}{dt} = \frac{1}{C}\left(i_p.u_i\right) \quad i = 1,....,N$$

$$\frac{dV_{C_{i\_l}}}{dt} = \frac{1}{C}\left(i_n.u_i\right) \quad i = N+1,....,2N$$

where $i_p$, $i_n$ $u_i$, $V_{Ci\_u}$ $V_{Ci\_l}$, and $V_a$ are upper/lower are currents, gating signal of upper gate, upper and lower capacitor voltages of the *i*-th cell, and phase *a* voltage respectively.

## III. MODEL PREDICTIVE CONTROL

Model predictive control (MPC) uses a cost function to represent the desired behavior and the optimal actuation is achieved by minimizing the cost function [12]. The diagram of MPC system is shown in Fig.3.

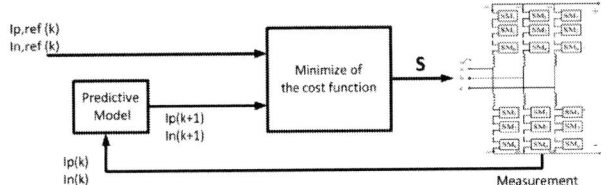

Fig.3. The diagram of MPC system.

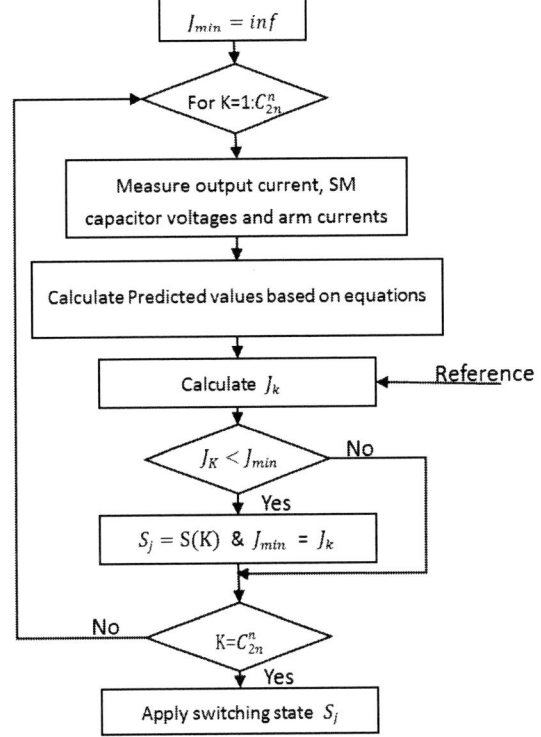

Fig.4. The Flowchart of MPC strategy.

The cost function is defined by the desired behavior. Here, it is considered the following cost function:

$$\begin{bmatrix} J_p(\mathrm{k}) \\ J_n(\mathrm{k}) \end{bmatrix} = \left\| \begin{bmatrix} i_p(\mathrm{k}+1) \\ i_n(\mathrm{k}+1) \end{bmatrix} - \begin{bmatrix} i_{p,ref}(\mathrm{k}+1) \\ i_{n,ref}(\mathrm{k}+1) \end{bmatrix} \right\| + \lambda_{cir} \left\| \begin{bmatrix} i_{\mathrm{p},diff}(\mathrm{k}+1) \\ i_{\mathrm{n},diff}(\mathrm{k}+1) \end{bmatrix} \right\| \tag{2}$$

Where $i_{p,ref}$ and $i_{n,ref}$ are the reference currents, $i_p$ and $i_n$ are the predicted currents and $i_{p,diff}$ and $i_{n,diff}$ are circulation currents of positive and negative sequences which can be calculated by equation (1).

The purpose of the controller is to minimize the cost function, which can be achieved by evaluating all the possible switching states and choose the one, resulting in minimum value of cost function. As all the terms in cost function are errors between predicted value and reference value, the MPC strategy aims to choose the switch state which results in the minimum value of cost function. Since there are always $n$ SMs on and $n$ SMs off in each phase, the total number of switching states are $C_{2n}^{n}$. The flow chart of the MPC algorithm is shown in Fig.4.

## IV. SIMULATION RESULTS

In order to evaluate the performance of the model predictive control strategy, simulation results are carried out in MATLAB/Simulink software with the nominal parameters given in Table 2. The set-points for the real power and reactive power at t = 0.05 MW are set to 0.2 MVAr, respectively. The six capacitor voltages of phase A are presented in Fig.5. The three-phase line-line voltages are shown in Fig. 6. It clearly reveals the presence of six levels of voltage, as expected. The charge and discharge of the capacitors cause the voltage levels to vary within acceptable limits.

TABLE I. PARMATERS OF MMC

| Nominal Parameters | |
| --- | --- |
| SM capacitor initial voltage | 5892 V |
| Rated line-line voltage | 10 kV |
| Number of Cells per arm | 6 |
| Arm inductance | 1.59mH |
| Arm resistance | 0.04mΩ |
| Cell capacitance | 100µF |
| Rated frequency | 60Hz |
| Carrier frequency | 600 Hz |

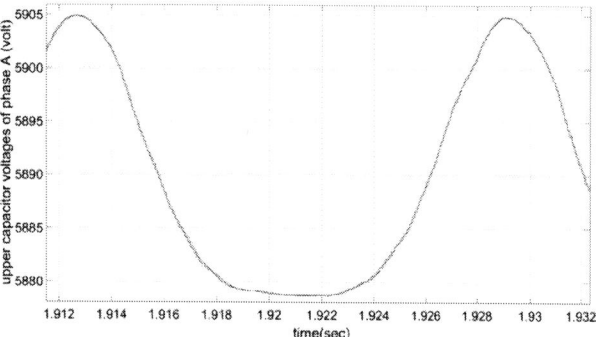

Fig.5. the capacitor voltages of phase A.

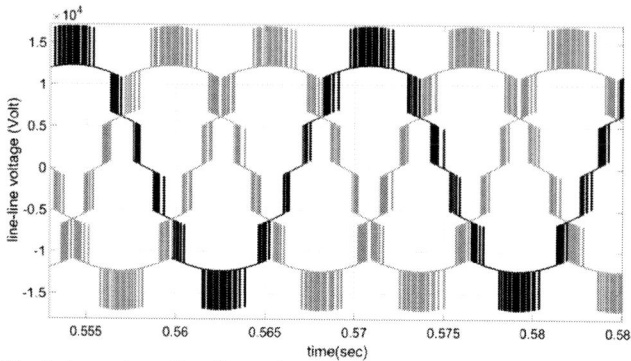

Fig.6. three-phase line-line voltages.

In Fig.7, the voltage across the top cell of phase A is presented. It shows that the cell switches 10 times in a cycle. This leads to a switching frequency of approximately 600Hz for the IGBT, which is an acceptable at this rating.

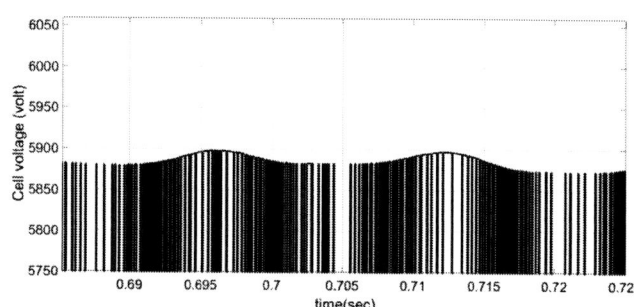

Fig.7. Output Voltage of top cell in phase A.

In Fig.8, the circulating dc that flows through the arms and the dc bus is shown. This dc current helps to maintain the power balance in the capacitors.

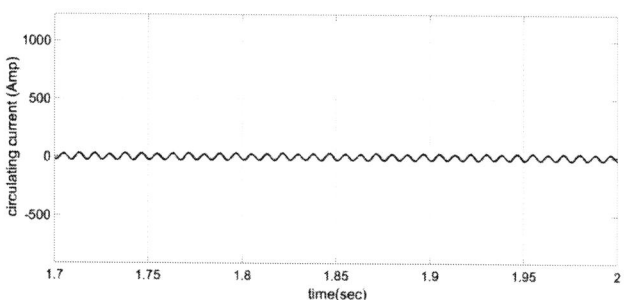

Fig.8. The circulating dc current.

In order to evaluate the response of proposed controller during faulty condition, a step change in the dc link voltage resulting of the change in photovoltaic condition has been created. In this condition, the phase currents of MMC have been illustrated in Fig.9. Moreover, the arm current has been shown in Fig.10. As presented, the arm currents increase substantially in response to the step change in dc link voltage.

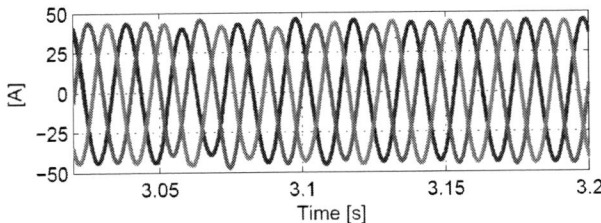

Fig.9. Phase currents from the MMC during a step change in the dc link voltage.

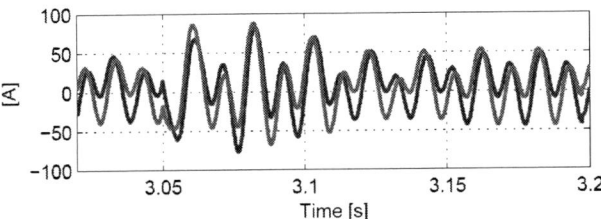

Fig.10. Arm currents of phase A during a step change in the dc link Voltage.

## V. CONCLUSION

This paper develops the model predictive control Current Control for Modular Multilevel Converter (MMC). In order to design control strategy, mathematical model of MMC based on positive and negative symmetrical components have been used. Time-domain simulation studies are carried out in the Matlab/Simulink environment to verify the performance of the overall proposed control system. Simulations results show that the proposed controller is able to track the references of real and reactive power properly and reduce significantly overshoot in the output response.

## VI. REFERENCES

[1] F. Schimpf, L. E. Norum et al., "Grid connected converters for photovoltaic, state of the art, ideas for improvement of transformerless inverters," *in Nordic Workshop on Power and Industrial Electronics (NORPIE/2008)*, June 9-11, 2008, Espoo, Finland. Helsinki University of Technology, 2008.

[2] B. Subudhi and R. Pradhan, "A Comparative Study on Maximum Power Point Tracking Techniques for Photovoltaic Power Systems", *IEEE Transactions on Sustainable Energy*, vol. 4, no. 1, pp. 89-98, 2013.

[3] M. C. Cavalcanti et al., "Modulation techniques to eliminate leakage currents in transformerless three-phase photovoltaic systems," *IEEE Transactions on Industrial Electronic.*, vol. 57, no. 4, pp. 1360-1368,. 2010.

[4] D. Dong, F. Luo, X. Zhang, D. Boroyevich, and P. Mattavelli, "Gridinterface bidirectional converter for residential DC distribution systems - Part 2: AC and DC interface design with passive components minimization," *IEEE Transactions on Power Electronics*, vol. 28, no. 4, pp. 1667- 1679, 2013.

[5] S. Debnath, J. Qin, B. Bahrani, M. Saeedifard, and P. Barbosa, "Operation, control, and applications of the modular multilevel converter: A review," *IEEE Transactions on Power Electronics*, vol. 30, no. 1, pp. 37–53, 2015.

[6] A. Lesnicar and R. Marquardt, "An innovative modular multilevel converter topology suitable for a wide power range," *in Power Tech Conference Proceedings, 2003 IEEE Bologna*, vol. 3, 2003.

[7] Deng, F., & Chen, Z. "A control method for voltage balancing in modular multilevel converters", *IEEE Transactions on Power Electronics*, vol. 29, pp. 66-76, 2014.

[8] Qin, J., & Saeedifard, M, "Predictive control of a modular multilevel converter for a back-to-back HVDC system. *IEEE Transactions on Power Delivery*, vol.27, pp.1538-1547, 2012.

[9] M. A Parker, L. Ran, and S. J. Finney, "Distributed control of a fault tolerant modular multilevel inverter for direct-drive wind turbine grid interfacing," *IEEE Transactions on Industrial Electronic*, vol. 60, pp. 509–522, 2013.

[10] H. Nademi and L. E. Norum, "Implicit finite control set model predictive current control for modular multilevel converter based on IPA-SQP algorithm," *in Proc. 31st Annu. IEEE Appl. Power Electron.* 2016, p. 3291.

[11] P. Karamanakos, T. Geyer, N. Oikonomou, F. D. Kieferndorf, and S. Manias, "Direct model predictive control: A review of strategies that achieve long prediction intervals for power electronics," *IEEE Industrial Electronic Magazine*, vol. 8, no. 1, pp. 32–43, Mar. 2014.

[12] J. Rodriguez and P. Cortes, *Predictive Control of Power Converters and Electrical Drives*. Hoboken, NJ, USA: Wiley, 2012, vol. 40.

[13] J. Nocedal and S. J. Wright, *Numerical Optimization*, 2nd ed. New York, NY, USA: Springer, 2006.

[14] J. L. Jerez, P. J. Goulart, S. Richter, G. A. Constantinides, E. C. Kerrigan, and M. Morari, "Embedded online optimization for model predictive control at megahertz rates," *IEEE Transaction on Automatic Control*, vol. 59, no. 12, pp. 3238–3251, Nov. 2014.

[15] J. B. Rawlings and D. Q. Mayne, *Model Predictive Control: Theory and Design*. Madison, WI, USA: Nob Hill, 2014.

# Maximization of Self-Sufficiency with Grid Constraints: PV Generators, Wind Turbines and Storage to Feed Tertiary Sector Users

Filippo Spertino, Jawad Ahmad, Alessandro Ciocia, Paolo Di Leo, Francesco Giordano

Energy Department, Politecnico di Torino, Corso Duca degli Abruzzi 24, 10129 Torino, Italy
{filippo.spertino, jawad.ahmad, alessandro.ciocia, paolo.dileo, francesco.giordano}@polito.it

*Abstract* — The maximum self-sufficiency by PV modules, wind turbines and storage is investigated, that can be obtained by aggregated tertiary sector users. Each renewable source (sun and wind) is assessed in terms of hours of availability with respect to the annual hours. The primary goal of aggregated users, now prosumers, is the achievement of the best match between profiles of generation and consumption. Power ratings of PV and wind generators and energy capacities of batteries are chosen to reach the highest self-sufficiency and the lowest power exchange with the grid. PV and wind powers are simulated by using measured meteorological data.

*Index Terms* — PV generators, wind turbines, storage, grid, self-sufficiency.

## I. INTRODUCTION

In recent decades, the installation of Renewable Energy Systems (RES), such as PV generators and wind turbines, was boosted by green policies and low installation costs. These generators are safe, clean and even more cost-effective, with respect to the technologies based on fossil fuels. Nevertheless, in some cases, grid operators limit their widespread use, because the intermittency could affect the power quality and grid stability [1]. Thus, new solutions are necessary to avoid technical issues in power management. The grid upgrade is in progress, but it cannot be considered the only solution, due to huge costs. Storage systems are one of the preferred solution to solve this problem. They can help to address the power balance from local level to transmission-grid level [2]. For example, when a PV plant creates a power surplus during midday, batteries can store it. This stored energy can be used later, e.g. during the evening by domestic users, with a double benefit. The self-sufficiency of the user increases and the grid has to feed a shaved demand-peak. Although, from a technical point of view, the storage is an adequate solution for RES grid integration, its widespread use is limited by the current high cost. Thus, the seasonal compensation between demand and generation cannot be completely satisfied in a cost-effective way [3].

In this paper, the maximum self-sufficiency achievable by aggregated tertiary sector users, equipped with grid-connected RES and storages, is defined. All the proposed sizing solutions of RES systems and storage are cost-effective; thus, the optimal portfolio is the compromise between the cost-effectiveness of RES and the high costs of batteries.

Obviously, the use of storage permits to install higher power with low grid injections.

The next sections of this paper are organized as follows: in Section II a description of the system is presented, with details related to the power flow management. The used models and the loads are introduced in Section III. In Section IV, the simulation in two different case studies are illustrated and discussed.

## II. ARCHITECTURE OF A PV-WIND-STORAGE SYSTEM

### A. Presentation of the System

In the systems under analysis, the main components are tertiary sector loads, RES generators, storage and grid (Fig. 1). PV and wind generators are installed by users to reach the maximum self-sufficiency, with the help of private storage systems. In this way, the aggregated users became prosumers and they exchange with the grid a lower energy quota. Two different sites in Southern Italy are considered for the sizing of the RES sources and storage: they are characterized by similar solar irradiance and different wind speed. In this way, it is possible to compare the results, in terms of intensity and simultaneity. Loads correspond to the aggregation of the tertiary sector users: they are office appliances and electronic equipment for telecommunication systems. All the mentioned components are connected to the grid by electronic converters.

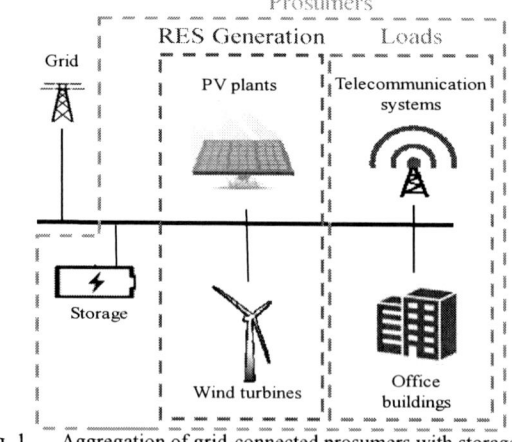

Fig. 1.    Aggregation of grid-connected prosumers with storage.

Regarding the storage systems, they are not used for ancillary service and the power is not exchanged with the grid only to obtain profit from different prices of electricity during day and night hours. Their functions are to maximize self-sufficiency and minimize power exchanges with the grid. Thus, the private storage is charged only by RES, when load is low, and they are never charged by the grid.

The system is managed in this way: first, renewable sources are used to feed loads, then, in case of low production, storage is discharged and finally the grid is used to supply the rest of the loads. Deep discharges are avoided by disconnecting the storage, when its state of charge reaches a low limit.

### B. Power Flow Management

The simulation of the systems is performed by a software in MATLAB®. It performs the power flow management for a year, with 1-min sampling. RES production profiles are calculated starting from meteorological data described in next sections. Then, the balance between power production from renewables $P_{ren}$ and loads $P_{load}$ is verified: if there is power surplus or deficit from renewables, the storage can be used. In particular, the storage operation is calculated by checking its State Of Charge ($SOC$) and limits in maximum exchanged power. How much energy can be charged or discharged is defined according to limits imposed to preserve battery life: storage is full when $SOC=SOC_{max}$, empty when $SOC=SOC_{min}$ and these limits cannot be exceeded. In addition, the limit in the maximum power that can be charged or discharged has to be respected. If it cannot handle the power in the local system, due the abovementioned limits, the external grid is used to feed the loads. The possible cases are six and are summarized in Table I. As an example, in the second case, all the component of the systems could work. RES generation is lower than load ($P_{ren}<P_{load}$) and batteries are partially full $SOC_{min}<SOC<SOC_{max}$. If batteries can completely feed the remaining load, the external grid is not used; otherwise, the grid helps to provide the deficit of production: $P_{ren} + P_{bat} + P_{grid} = P_{load}$.

TABLE I: POWER FLOW MANAGEMENT

| | Empty storage $SOC= SOC_{min}$ | Partially full storage $SOC_{min}<SOC<SOC_{max}$ | Full storage $SOC= SOC_{max}$ |
|---|---|---|---|
| Low renewable generation $P_{ren}<P_{load}$ | CASE #1 Absorption from the grid | CASE #2 - Battery discharge - Possible absorption from the grid* | CASE #3 - Battery discharge - Possible absorption from the grid* |
| High renewable generation $P_{ren}>P_{load}$ | CASE #4 - Battery charge - Possible injection into the grid* | CASE #5 - Battery charge - Possible injection into the grid* | CASE #6 Injection into the grid |

*Limits in storage operation are checked to quantify the use of the grid

## III. System Modeling

### A. Modeling of PV Generators

The PV power production is simulated starting from real solar irradiance $G(t)$ and ambient temperature $T_a(t)$ profiles with 1-minute time step. Meteorological data are measured by stations with high accuracy (uncertainty in solar irradiance is <20 W/m²). PV production is calculated according to a model in which power output is proportional to the solar irradiance $G(t)$ and the PV rated power $P_{PV,r}$ [4]. It is defined at Standard Test Conditions (STC), corresponding to $G_{STC}=1$ kW/m² and $T_{STC}=25$ °C. PV modules are 30° tilted and South oriented. The AC power is calculated taking into account thermal losses depending on PV technology and $G(t)$ and $T_a(t)$ profiles. Other losses, such as $I–V$ mismatch, dirt and reflection of PV modules, tolerance and cables are taken into account by constant parameters [5]. A miscellaneous efficiency including these losses is ≈92%.

### B. Modeling of Wind Turbines

The calculation of energy production from wind turbines starts from the use of wind speed data from meteorological stations. In this work, the wind speed accuracy of anemometers is ±0.5 m/s and their height is 3 m.

The wind speed is transferred at the height of the wind turbine's hub by using the logarithmic formula in [6]. These data are input in the "power output-wind speed" relationship that has to be provided by the manufacturer of a wind turbine or properly simulated [7]. This relation is strongly nonlinear; in stationary conditions, the converted mechanical power is a cubic function of the wind speed. The "power output-wind speed" relation is linearly interpolated to calculate the power output minute by minute.

In order to define the most effective turbine in each site under study, the productivities of three commercial wind turbines are estimated and compared. Since both the sites have a low average wind speed and high turbulence, all the examined wind turbines belong to IIIa class by IEC. The wind class IIIa includes wind turbines working with low wind speed and high turbulence: the maximum annual average wind speed at hub height is 7.5 m/s and the project turbulence is 18% [8]. Regarding the wind turbulence $\tau$, it is the ratio between the standard deviation of the wind speed fluctuations ($\sigma$) and the average wind speed ($\bar{u}$) at the height of the hub (calculated over 10 samples every 10 min):

$$\tau = \frac{\sigma}{\bar{u}} \qquad (1)$$

Generally, in case of a too high turbulence level ($\tau>20\%$), the turbine has to be stopped to reduce fatigue of materials.

The main characteristics of the analyzed turbines are here summarized. The rated power of wind turbine WT#1 is 850 kW; the blade length is 38 m and the hub is located at 61.5 m with respect to the ground. The turbine WT#2 has a rated power of 2 MW; it has a rotor blade length equal to 55 m, while the hub is located at a height of 85 m. The turbine WT#3 has a rated power of 3 MW. Its blades are 60 m long and the hub is located at 95 m.

### C. Modeling of Storage System

A correct model of the storage is fundamental to evaluate energy and cash flows. Many electric models exist and each one permits to simulate its operation with a different level of accuracy. The most sophisticated models permit to determine all the most important variables: voltage, $SOC$ and State Of Health ($SOH$) [9-12].

The $SOC$, compared to limits ($SOC_{min}$ and $SOC_{max}$), is used to calculate how much energy is stored or can be stored in the battery with rated capacity $C_{bat,r}$ and rated voltage $V_{bat,r}$. On the other hand, the $SOH$ is used mainly to evaluate when the batteries have to been replaced. Batteries are actually expensive. Thus, it is important to know their estimated life in order to evaluate the economic investment. Nevertheless, the most accurate models [12] needs to continuously measure voltage during operation of the system to estimate its life. This kind of models cannot be applied, if the system is not existent, but it is only planned.

In this work, an energy model of storage is used [10]. The $SOC$ behavior of batteries is modelled minute by minute and includes charge/discharge efficiency $\eta_{bat}$:

$$SOC(t+1) = SOC(t) - \frac{\eta_{bat} \cdot P_{bat} \cdot \Delta t}{\left(V_{bat,r} \cdot C_{bat,r}\right)} \qquad (2)$$

In order to reduce the battery degradation and ageing, the values of charging and discharging currents cannot exceed certain limits. Thus, in addition to the limits in the State of Charge, the maximum current of the battery is calculated starting from the datasheet. In the present work, the nominal capacity of a battery element is 1.2 kWh, with a maximum power of ≈2.4 kW. Finally, concerning the storage replacement, the battery is replaced, when the maximum number of cycles or years is reached. The main specifications of the commercial storage for RES applications, simulated in this work, are reported in Table II [13].

TABLE II: SPECIFICATIONS OF A LI-ION BATTERY ELEMENT

| Battery rated capacity $C_{bat,r}$ | 100 Ah |
|---|---|
| Battery rated voltage $V_{bat,r}$ | 12 V |
| Lifetime*(cycles) | 3500 cycles with $DOD$=80% |
| Lifetime*(years) | 8 |
| Maximum power $P_{bat,max}$ | ≈2.4 kW |
| Charge-discharge efficiency $\eta_{bat}$ | 0.88 |

*Warranty, real lifetime most likely higher. DOD is Depth of Discharge.

### D. Power Profiles of Tertiary-Sector Users

In the present work, the annual consumption profiles, updated every 1-hour, of office appliances and electronic equipment for telecommunication are used in the simulations. These profiles well match the PV production: the working

hours are located during the sunlight hours. Telecommunication equipment has a high constant consumption during the whole year with peaks depending on ambient temperature for cooling systems. These peaks are obviously much higher during summer.

For example, the first aggregation of loads (SITE #1) is characterized by a yearly consumption of about 112 GWh with a base load of ≈10.6 MW and a peak of ≈18.5 MW. In this case, the number of users is =220. The second aggregation of loads (SITE #2) is similar to the previous one: the annual consumption is about 117 GWh with a base load of ≈11.2 MW and a peak of ≈18.2 MW. In this case, the number of users is =165 (Fig. 2.).

The difference between winter and summer consumptions is shown in Fig. 3. The constant consumption increases from base load (≈10.6 MW) up to 13−14 MW due to cooling systems (heat pumps) of offices and electronic equipment. In February, the base load is constant, while in July the minimum consumption depends on the weather conditions. In both months, the load is lower during weekends, due to reduced activities.

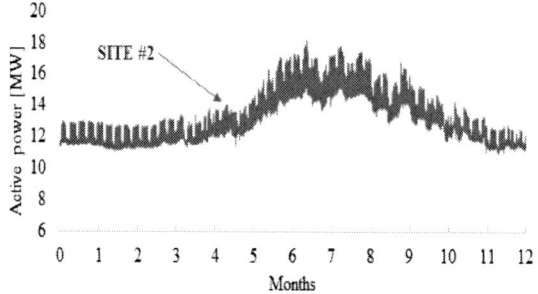

Fig. 2.    Annual load profile – Site #2.

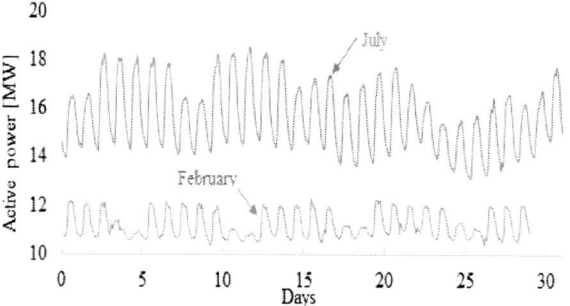

Fig. 3.    Load profiles in February and July – SITE#1.

Regarding the cost of electric energy purchased from the grid, it corresponds to ≈20 c€/kWh. It is a typical value for a tertiary sector user in Italy with a contract power of 200 kW and energy consumption of ≈1 GWh/year. In Italy, energy surplus from renewable sources can be sold to the grid by a simplified contract called "dedicated withdrawal". The corresponding selling price is low (≈4 c€/kWh).

## IV. SIMULATION RESULTS

The energy production from PV generators and wind turbines depends on meteorological conditions, but the maximum achievable self-sufficiency depends on the load profile, too. Due to the presence of a high baseload in the office users with telecommunication equipment, PV systems cannot match the load without the help of storage used mainly during night hours. Wind farms could help, if the production is not simultaneous with PV. In the analyzed cases, the installation of wind is limited by two main factors.

The first limit in the installation of wind turbines is due to their very variable production also during night, when loads are generally at minimum level. Sudden production peaks occur many times during night in the two sites. As explained in the previous section, in order to respect the economic constraints, too high injections into the grid provide low amount of money. Therefore, an oversized wind farm is excluded or the use of adequate storage is required to mitigate the injections. As a result, in every scenario, the PV capacity is higher than wind capacity, also in that site where the annual PV yield is lower. The second limit is due to the contemporaneity of sun and wind. If wind and PV productions are contemporary, then PV is preferred. In addition, the wind turbulence strongly affects the wind production, resulting in a lower economic profit, which makes the PV generators the optimal solution.

### A. Study of the availability of irradiance and wind speed

Monthly results related to simultaneity of renewable sources in SITE#1 are reported in Fig. 4. and Fig 5. PV generation is higher during summer, while there is no guarantee that wind production is higher in the rest of the year. As shown in Fig. 4 (case study #1), monthly distribution of wind speed and solar radiation are alternate. In terms of wind availability, the worst month is August. The hours of operation for wind turbines (358 h) is low, with respect to July and September (467 and 450 h, respectively). In August, the maximum hours of no production occur (35% of the time in Fig. 5). In April, the minimum hours of no production occur (13% in Fig. 5).

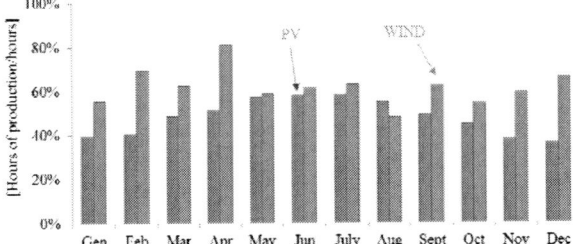

Fig. 4. Operation hours for PV and wind generators – SITE #1.

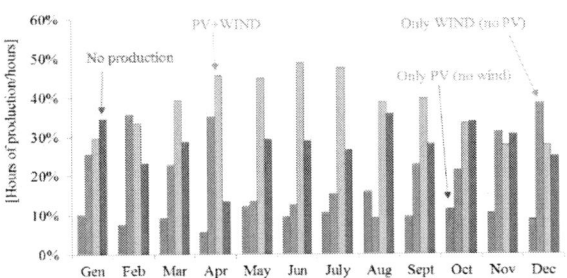

Fig. 5. Simultaneity of renewable sources – SITE #1.

Obviously, the annual energy production does not depend on the number of hours in which solar radiation and wind are present, but also on their intensity. Moreover, there are situations in which the production is shut down. For example, as above-mentioned, wind turbines are stopped if turbulence is too high and there are cut-in and cut-out WT limits. Then, multiple faults reduce the availability of wind turbines as discussed in [14]. In the case of PV systems, the AC output starts if irradiance is sufficient to turn on the inverters.

TABLE III: CONTEMPORANEITY OF SOLAR AND WIND SOURCES

| Renewable source | SITE #1 | SITE #2 |
|---|---|---|
| Sun hours | 4,233 h/y | 4,224 h/y |
| Wind hours | 5,416 h/y | 5,021 h/y |
| Sun ∩ Wind (contemporary) | 3,347 h/y | 3,350 h/y |
| Only Sun | 886 | 874 |
| Only Wind | 2,069 | 1,671 |
| Sun ∪ Wind | 72%* | 67%* |
| No production | 28%* | 33%* |

*With respect to the hours in a year

The data of contemporaneity of yearly productions of both sites are presented in Table III. The number of hours, in which generators can produce, is independent of the size of the plants. PV systems are modular and wind farms are composed of wind turbines with the same specifications. In the two case studies, the number of hours, in which both renewable sources are available, is at least 72% and 67% of the year, respectively. In the second and third rows of Table III, the number of hours in which each generator may produce (independently of the other) is reported. In the fourth row, it is presented the number of hours in which both the renewable generators can work at the same time (intersection). Then, it is shown how many hours each technology may work, when the other one is not productive. In the last rows, the number of hours in which either Sun or wind (union of PV and WT hours), it is expressed as a percentage with respect to the whole year. This last value can be considered the theoretical limit to the maximum self-sufficiency that can be achieved without storage. The best site is the first, where PV or wind are not present for 28% of the year. The second site is affected by low wind availability. The simulation results, which are presented in the next subsection, confirm the lower wind resource in the second site.

## B. Effect of wind turbulence on energy production

The equivalent hours is a useful parameter to compare the performance of different wind turbines and choose the most adequate in an installation site. This is the ratio between the annual energy production and the rated power of the generator. Thus, it is the full load hours, in which the WT would work at nominal power, to produce the annual energy output. Table IV shows the results expressed as the number of equivalent hours of production of the three analyzed WTs.

The results of the simulations show that WT#1 is the best choice in both the sites. The reason is linked to its high efficiency at low wind speeds: for example, at the wind speed of 5 m/s, WT#1 has a global efficiency ≈33 % (Fig. 6). In the case of WT#2 and WT#3, at the same value of wind speed (reported to respective hub heights) the global efficiency is lower than 20%. These production results are obtained without considering the effect of wind turbulence.

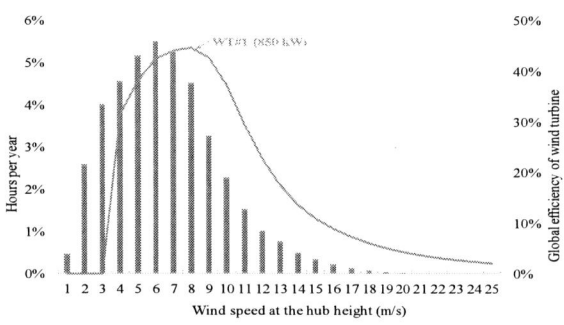

Fig. 6.   Wind speed frequency vs. efficiency of WT#1 – SITE#1.

A more accurate analysis is performed including wind turbulence. Table V shows the reduction in energy production for turbulence. In particular, if turbulence index is higher than 20%, WT is stopped to preserve the life of its components. Losses are not negligible: it ranges from 18% in the best case (SITE#1) and 21% in the most turbulent site (SITE#2).

TABLE IV: FULL LOAD HOURS OF THE WIND TURBINES

|      | SITE #1 | SITE #2 |
| --- | --- | --- |
| WT#1 | 2,297 h | 2,563 h |
| WT#2 | 2,052 h | 2,339 h |
| WT#3 | 1,510 h | 1,822 h |

TABLE V: LOSSES DUE TO WIND TURBULENCE

|      | SITE #1 | SITE #2 |
| --- | --- | --- |
| WT#1 | -18% | -21% |
| WT#2 | -17% | -18% |
| WT#3 | -15% | -16% |

*With respect to data in Table IV

## C. Energy simulation results

Fig. 7 shows the power profiles of a typical summer-spring day in SITE #1. During this day there is high solar irradiance and high wind speed, almost simultaneous. These profiles are obtained maximizing self-sufficiency by exhaustive search of combinations of PV, WT and storage capacities. PV and WT productions feed loads and charge batteries at midday. The storage, empty since the previous day, completely supplies the load only starting from the evening, so that grid absorption is not needed. In this case, the contribution of batteries in the energy balance is high: the self-sufficiency corresponds to 71% of the load, while grid absorption is low (29%).

An opposite situation, in terms of self-sufficiency level, is described in Fig. 8. It shows the power profiles of a typical autumn-winter day in SITE #1. It is a cloudy day and wind speed is high, especially during the evening. These profiles are obtained maximizing self-sufficiency: wind production helps to feed loads during night, while during sunlight hours low PV power occurs. Batteries, typically charged at midday, are empty and only help to reduce grid injection corresponding to few production peaks. In this case, the contribution of batteries in the energy balance is negligible: the self-sufficiency is 30% of the load, while grid absorption is 70%.

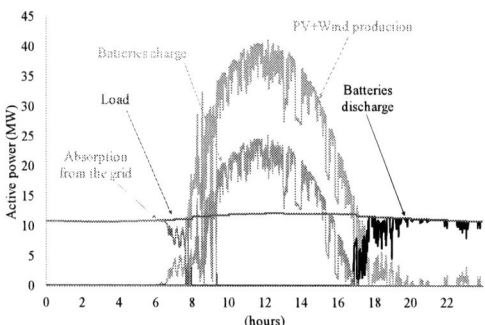

Fig. 7.   Power profiles with high irradiance and wind - SITE #1.

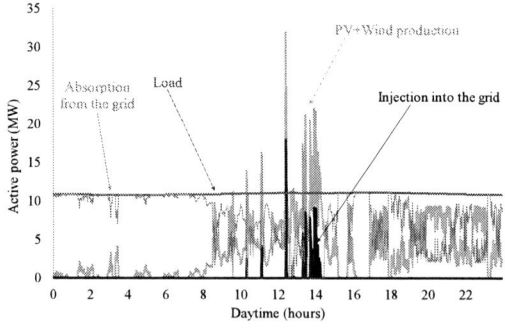

Fig. 8.   Power profiles: low irradiance and high wind - SITE #1.

## IV. CONCLUSIONS

The simulations in the two case studies have permitted to show that high levels of self-sufficiency can be reached, despite economic and technical constraints. The results are different in the two sites, according to the wind resource. The sites under analysis have similar Sun resource and load profiles with a distance of ≈200 km. Nevertheless, the self-sufficiency are high, even if different.

In SITE #1, the self-sufficiency is ≈61% of the annual load, thanks to a moderate simultaneity of wind and solar resources. In SITE #2, the self-sufficiency is lower (≈52%). This worse result is due to higher simultaneity of RES (wind occurs more during sunlight hours) and high wind speed turbulence. This reduces the annual WT production by about 20%. The energy and economic results are summarized in Table VI.

Both the proposed solutions permit to minimize the impact on the grid. Indeed, the injection of surplus into the grid is negligible (<1% of the load). In addition, the proposed solutions are cost-effective: their NPV>0 and the internal rate of return (IRR) is higher than 6%, as recommended in [15] for investments in RES. However, the most profitable solution (higher NPV) is that one in the second site with lower WT installed power and a consequent lower storage capacity. High self-sufficiency levels and low grid injections are obtained by high storage capacities. PV and wind productions are so effective, that their combination with storage is profitable. In this case, PV generators are preferred to wind turbines, even if the PV productivity is lower. The reason is that peaks of wind production can occurs also during night, when load is low.

TABLE VI: ENERGETIC AND ECONOMIC SIMULATION RESULTS

|  | SITE #1 | SITE #2 |
|---|---|---|
| *Power and Energy Capacities* | | |
| PV power rating (MW) | 40 | 40 |
| WT power rating (MW) | 9.3 | 0.8 |
| Storage capacity (MWh) | 170 | 100 |
| *Energy parameters* | | |
| RES Energy/Load Energy | 63% | 55% |
| Self-sufficiency/Load Energy | 61% | 52% |
| Injected Energy/Load Energy | 1% | 1% |
| Self-consumption/RES Energy | 99% | 99% |
| *Economic parameters* | | |
| NPV after 25 years (M€) | 15.3 | 40 |
| Initial investment (M€) | 109.2 | 77.1 |
| IRR (%) | 6.1 | 6.6 |

### REFERENCES

[1] F. Spertino, P. Di Leo, F. Corona and F. Papandrea, "Inverters for grid connection of photovoltaic systems and power quality: Case studies," *3rd IEEE International Symposium on Power Electronics for Distributed Generation Systems (PEDG)*, Aalborg, 2012, pp. 564-569.

[2] J. Tan, Y. Zhang, "Coordinated Control Strategy of a Battery Energy Storage System to Support a Wind Power Plant Providing Multi-Timescale Frequency Ancillary Services," in *IEEE Transactions on Sustainable Energy*, n. 99, pp. 1-13, 2017.

[3] F. Spertino, A. Ciocia, V. Cocina and P. Di Leo, "Renewable sources with storage for cost-effective solutions to supply commercial loads," *2016 International Symposium on Power Electronics, Electrical Drives, Automation and Motion (SPEEDAM)*, Anacapri, 2016, pp. 242-247.

[4] F. Spertino, F. Corona and P. Di Leo, "Limits of Advisability for Master–Slave Configuration of DC–AC Converters in Photovoltaic Systems," in *IEEE Journal of Photovoltaics*, vol. 2, pp. 547-554, 2012.

[5] F. Spertino, A. Ciocia, F. Corona, P. Di Leo, F. Papandrea, "Experimental procedure to check the performance degradation on site in grid-connected photovoltaic systems," in *40th IEEE Photovoltaic Specialists Conference*, Denver, Colorado, 2014, pp. 2600-2604.

[6] Danish Wind Energy Associates (DWEA), 2016. [Online]. Available: www.windpower.org.

[7] F. Spertino, P. Di Leo, I. S. Ilie, G. Chicco, "DFIG equivalent circuit and mismatch assessment between manufacturer and experimental power-wind speed curves" *Renewable Energy*, vol. 48, pp. 333-343, 2012.

[8] International Electrotechnical Commission, "IEC 61400-1:2005, Wind turbines - Part 1: Design requirements," [Online]. Available: https://webstore.iec.ch/publication/5426&preview=1.

[9] M. Einhorn, V. Conte, C. Kral and J. Fleig, "Comparison of electrical battery models using a numerically optimized parameterization method," in *IEEE Vehicle Power and Propulsion Conference*, Chicago, USA, 2011.

[10] A. H. Hussein, I. Batarseh, "An overview of generic battery models," in *IEEE Power and Energy Society General Meeting*, San Diego, USA, 2011.

[11] Min Chen and G. A. Rincon-Mora, "Accurate electrical battery model capable of predicting runtime and I-V performance," *IEEE Transactions on Energy Conversion*, vol. 21, pp. 504-511, 2006.

[12] M. Cacciato, G. Nobile, G. Scarcella and G. Scelba, "Real-Time Model-Based Estimation of SOC and SOH for Energy Storage Systems," *IEEE Transactions on Power Electronics*, vol. 32, pp. 794-803, 2016.

[13] Alexandre Oudalov, ABB Switzerland, Corporate Research, Energy Storage in Electric Power Systems, Integration of Renewable Energy Sources, 2015.

[14] I. S. Ilie, G. Chicco, P. Di Leo and F. Spertino, "Protections impact on the availability of a wind power plant operating in real conditions," *2009 IEEE Bucharest PowerTech*, Bucharest, 2009, pp. 1-7.

[15] U.S. Energy Information Administration, "Levelized Cost and Levelized Avoided Cost of New Generation Resources 2016", http://www.eia.gov

# Switches Controlling to Implement Adaptive Multilevel Inverter on PV System

Hadi Suhana[1], Ngapuli I Sinisuka[1], Muhammad Nurdin[1], Yvon Besanger[2], Vincent Debusschere[2]

1. Bandung Institute of Technology, Bandung, Indonesia  2. Univ. Grenoble Alpes, CNRS, Grenoble INP, G2Elab, F-38000 Grenoble, France.

*Abstract* — A voltage nonuniformity problem across photovoltaic (PV) modules in a multilevel inverter (MLI) can be solved by adaptive multilevel inverter method. This method compose voltage variation of PV modules by controlling switching state to result a good sinusoidal waveform so that low THD can be acquired. The proposed method refer to the sine waveform characteristic i.e. fast arise at the beginning and slower on the subsequent slopes. This paper discusses a concept of the adaptive multilevel inverter method. There are two techniques to be discussed i.e. direct computation and look up table base. Both techniques process voltage measurement quantities to determine switch ON/ OFF angle on each cells. Where algorithm of the switching angles determination will be conducted in the Field Programmable Gate Array (FPGA) controller.

*Index Terms* — voltage level, switching angle, look up table, algorithm, THD.

## I. INTRODUCTION

The implementation of solar PV generation has been growing fantastically in the recent years. Within five years, the PV generation installed capacity grew about 500 % per year [1-2]. Due to the PV panel installation flexibilities, it can be down to few kWs for residential roof top applications and up to GWs for solar farm, these fenomenas boost the PV implementation. However, integration of PV systems into grids have a big challenge. A major challenge is that power generation strongly depending on variation of weather conditions, locations, and seasons. The generated energies are non-scheduled, intermittent, and difficult to control [3], [4]. The weather conditions due to cloud shadow can result fluctuated output voltage. The studies show that there are some parameters of cloud shadow have to be taken into account on evaluating the voltage fluctuation such as wind speed and cloud thickness [5]-[7]. The cloud shadows cause not only voltage fluctuation but also voltage nonuniformity on the PV modules.

The challenging issues of PV voltage variations due to weather conditions have inspired further studies in many specialties. The researchers who concern to the asymmetrical dc source issues propose a modulation technique to produce multilevel voltage by dc-lingk voltage regulation techniques [8]. To produce a voltage staircase by means characterized by many voltage levels can be achieved by unequal dc sources with a circuit techniques. Then, within single-stage power conversion PV modules can directly supply power to the ac load by using neutral-point-clamped (NPC) and T-type topology [10]. Another researcher proposes a new multilevel converter topology to synthesize all possible additive and substractive combinations of input dc levels with fewer power electronic switches [11]. The others observed additional voltage pulse as well as dc source variation to result a better resolution voltage. By using this technique can be produced higher voltage levels and better THDs as well in multilevel inverters. Furthermore, a fair comparison of symmetrical and asymmetrical inverter performances was evaluated by [12].

The development of the multilevel inverter method covers its topology and modulation. By implementing different carrier frequencies and different source voltages in the multilevel inverter can be acquired the best performace of multilevel inverters. This technique use multicarrier sine pulse width modulation [13].

An improved phase disposition pulse width modulation (PDPWM) method proposes a modular multilevel inverter used in PV grid connection. The PDPWM proposes a new modulation technique based on selective virtual loop mapping. This technique using floating capacitors with a new PI controller instead of voltage source to achieve voltage balancing of the capacitors both upper and lower arms. This method result the better quality voltages rather than conventional ones [14].

A modulation method based on mixed staircase-PWM topology proposes a control technique to maximize power extraction by guaranteeing individual MPPT of each cell of the multilevel inverter. This method develop an algorithm to determine switching states using [15].

As mentioned earlier, strategy to improve multilevel inverter voltage can be achieved by either balancing source voltages or varying source voltages. This paper propose adaptive multilevel inverter method which is adaptive to source voltage variations. The proposed method is intended to small scale PV system e.g. rooftop PV generation that covers 9-levels H-Bridge topology, using selected harmonic elimination (SHE) modulation, and equiped by FPGA controller.

The contributions of this paper are following: (1) to introduce a concept of strategy on accomodating voltage variation of PV modules to achieve good output voltage THDs of an inverter; (2) to develop an algorithm of switching state determination in an FPGA; and (3) to elaborate power electronic skill, microelectronic skill, and math skill.

This paper is organized as follows: Section II presents principles of harmonics elimination, Section III describes principles of the proposed method i.e. adaptive multilevel inverter, and Section IV proposes next works that covers the proposed algorithm, experimental laboratory, and validation

978-1-5090-5606-4/17 $31.00 © 2017 IEEE

scheme by comparing resulted simulation and experimental laboratory. Finally, Section V shows conclusion.

## II. HARMONICS ELIMINATION

Inverter topology of the proposed method under study is H-bridge with four nonuniform dc sources which all its cells operate at the low frequency (50 Hz) switching. Fig. 1 and Fig. 2 shows the aforementioned topology and nine-levels output voltage dealing with.

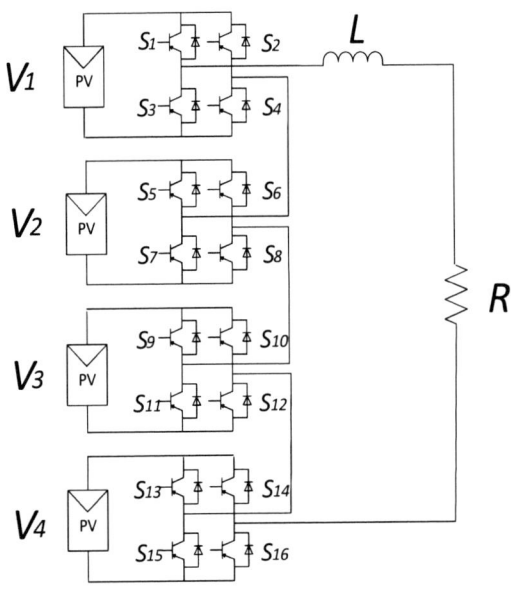

Fig. 1.    Nine levels H-bridge topology.

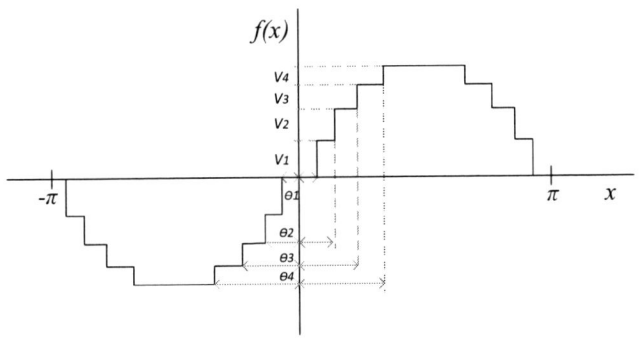

Fig. 2.    Nine-levels rectangular output voltage.

The SHE technique    determine appropriate ON/ OFF switching angles to achieve low THD. To determines these, nine-levels output voltage should be analyzed. Equation (1) is Fourier series of $f(x)$ [16] . Which the coefficients of $f(x)$, given by (2), (3), and (4).

$$f(x) = a_0 + \sum_{n=1}^{\infty} (a_n \cos nx + b_n \sin nx) \tag{1}$$

$$a_0 = \frac{1}{2\pi} \int_{-\pi}^{\pi} f(x)dx \tag{2}$$

$$a_n = \frac{1}{\pi} \int_{-\pi}^{\pi} f(x) \cos nx dx \tag{3}$$

$$b_n = \frac{1}{\pi} \int_{-\pi}^{\pi} f(x) \sin nx dx \tag{4}$$

From (2)-(4) can be obtained,   $a_0 = 0$ and $a_n = 0$. The coefficients $b_1$ , $b_5$ , $b_7$ , and $b_{11}$ are shown by (6)-(9) respectively when $5^{th}$ , $7^{th}$ , and $11^{th}$   harmonic order to be eliminated.

$$b_n = \frac{4V_k}{n\pi} \cos n\theta_k \tag{5}$$

$$b_1 = \frac{4}{\pi}(V_1 \cos\theta_1 + V_2 \cos\theta_2 + V_3 \cos\theta_3 + V_4 \cos\theta_4) \tag{6}$$

$$b_5 = \frac{4}{5\pi}(V_1 \cos5\theta_1 + V_2 \cos5\theta_2 + V_3 \cos5\theta_3 + V_4 \cos5\theta_4) \tag{7}$$

$$b_7 = \frac{4}{7\pi}(V_1 \cos7\theta_1 + V_2 \cos7\theta_2 + V_3 \cos7\theta_3 + V_4 \cos7\theta_4) \tag{8}$$

$$b_{11} = \frac{4}{\pi}(V_1 \cos11\theta_1 + V_2 \cos11\theta_2 + V_3 \cos11\theta_3 + V_4 \cos11\theta_4) \tag{9}$$

To solve nonlinear equations (6)-(9) can be used Newton-Rapson interation method. The Newton-Rapson matrices of iteration formula is shown by (10). Where J is Jacobian.

$$\theta_{q+1} = \theta_q - J^{-1}f(\theta_q) = \theta_q - \Delta\theta_q \tag{10}$$

Then (6)-(9) can be represented by (11).

$$f(\theta) = \begin{bmatrix} \frac{4}{\pi}(V_1 \cos\theta_1 + V_2 \cos\theta_2 + V_3 \cos\theta_3 + V_4 \cos\theta_4) - V \\ \frac{4}{5\pi}(V_1 \cos5\theta_1 + V_2 \cos5\theta_2 + V_3 \cos5\theta_3 + V_4 \cos5\theta_4) \\ \frac{4}{7\pi}(V_1 \cos7\theta_1 + V_2 \cos7\theta_2 + V_3 \cos7\theta_3 + V_4 \cos7\theta_4) \\ \frac{4}{11\pi}(V_1 \cos11\theta_1 + V_2 \cos11\theta_2 + V_3 \cos11\theta_3 + V_4 \cos11\theta_4) \end{bmatrix} \tag{11}$$

The solution to a system of k nonlinear equation in k unknowns given by $f(\theta) = 0$. From (11) be found J as shown by (12).

$$
J = \begin{vmatrix}
-\dfrac{4}{\pi}V_1\sin\theta_1 & -\dfrac{4}{\pi}V_2\sin\theta_2 & -\dfrac{4}{\pi}V_3\sin\theta_3 & -\dfrac{4}{\pi}V_4\sin\theta_4 \\[2mm]
-\dfrac{4}{\pi}V_1\sin 5\theta_1 & -\dfrac{4}{\pi}V_2\sin 5\theta_2 & -\dfrac{4}{\pi}V_3 c\sin 5\theta_3 & -\dfrac{4}{\pi}V_4\sin 5\theta_4 \\[2mm]
-\dfrac{4}{\pi}V_1\sin 7\theta_1 & -\dfrac{4}{\pi}V_2\sin 7\theta_2 & -\dfrac{4}{\pi}V_3\sin 7\theta_3 & -\dfrac{4}{\pi}V_4\sin 7\theta_4 \\[2mm]
-\dfrac{4}{\pi}V_1\sin 11\theta_1 & -\dfrac{4}{\pi}V_2\sin 11\theta_2 & -\dfrac{4}{\pi}V_3\sin 11\theta_3 & -\dfrac{4}{\pi}V_4\sin 11\theta_4
\end{vmatrix}
$$

$$(12)$$

By solving (11) can be found $\theta_1$, $\theta_2$, $\theta_3$, and $\theta_4$ of the H-bridge nine-level inverter.

## III. ADAPTIVE MLI

Fig. 3. describe the adaptive MLI which consisted of a switching unit, an FPGA controller, and a sensing unit. Determination of switch ON/ OFF decided by the FPGA based on either comparison measurement quantities – LUT or real time iteration. The main operating principle of the $\theta$ determination in the controlling unit covers sorting of PV voltages to result good arrangement - from the highest to the lowest amplitude of PV modules voltage. In the LUT Base technique, those voltages arrangement will be compared to the LUT to generate $\theta$ on the LUT scenario to turn ON/ OFF the switches. While in the real time iteration, those voltages arrangement will be iterative processed to find $\theta$.

### A. LUT Base

The choice of switching angle is based on distance between sets of V measurement quantity and component in the LUT. To compute those distances be used Euclidean theory [17]. Both voltage measurement quantity and LUT value is assumed a vector. LUT contain sets of voltage and their switching angle which result a good THD. To find which set of LUT to be refered is by finding the shortest distance between two vectors. Equation (13) is the Euclidean formula to compute the distance between two vectors in the $R^n$ space.

$$
d(v,x) = \|v - x\| = \sqrt{(v_1 - x_1)^2 + (v_2 - x_2)^2 + \ldots + (v_n - x_n)^2}
$$

$$(13)$$

LUT dimension by means amount of V- $\theta$ sets will affects accuracy of the switching angles determination. In order to provide a large number of V- $\theta$ sets should be specify the range of PV module operations as well as the iteration technique effectiveness. One of the good iteration methods is Newton-Raphson theory. Fig. 4 illustrates typical scripts of Newton-Raphson iteration conducted in MATLAB. There are three key-functions of Newton- Raphson method, i.e. $f(\theta)$, $J$, and $J^{-1}$.

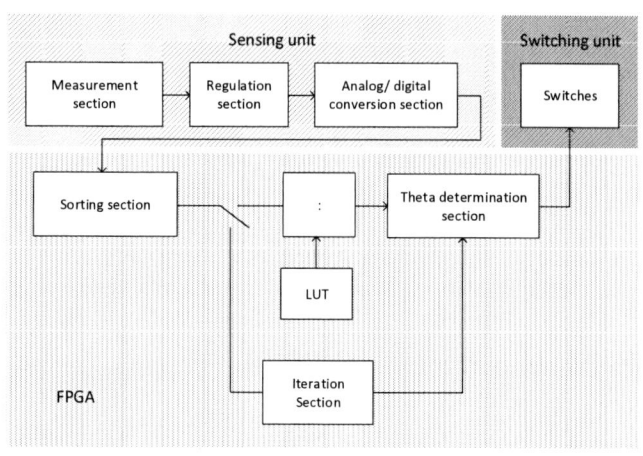

Fig. 3.    Adaptive MLI lay-out.

```
function [ x, iter ] = newtonm(x0, f, J)
N = 200;
epsilon = 1e-3;
maxval = 200.0;
xx = x0;

while (N>0)
    JJ = feval (J, xx);
    if abs (det (JJ)) < epsilon
        error ('newtonm - Jacobian is singular - try new x0');
        abort;
    end;
    xn = xx - inv (JJ)*feval (f, xx);
    if abs (feval (f, xn)) < epsilon
        x = xn;
        iter = 200-N;
        return;
    end;

    if abs (feval (f, xx))> maxval
        iter = 200 - N;
        disp (['iterations =', num2str (iter)]);
        abort;
    end;
    N = N-1;
    xx = xn;
end;
error ('No convergence after 200 iterations.');
% end function
```

Fig. 3.    Newton-Raphson iteration in MATLAB.

### B. Real Time Iteration

Fig. 4-7 show diagrams of direct computation in the FPGA on determining $\theta$. To determine $\theta$ in the Newton-Raphson iteration method be used function unit F, J and inversion matrix - { }$^{-1}$ [17]. Input-output signal data within inteconnection block diagrams are vectors/ matrices. Implementation in the FPGA architecture, angles are fixed point numbers and in radians. The F and J unit contain math equations in the iteration processing to determine an appropriate $\theta$ (to F function unit) and its derivation (to J function unit). Those functions have already known so that can be prepaired, but not for the V value and initial value of $\theta$. A dF/d$\theta$ function can be found easily due to identified F. The dF/d$\theta$ is implemented in the logic trigonometry block of FPGA. To determine $\sin\theta$ and $\cos\theta$ value be used CORDIC block and the initial value as shown in the Fig.5. So that $\cos\theta$ and $\sin\theta$ value can be generated. The iteration can be done for 16 times. Inter-block synchronising on the Newton-Raphson computation can be carry out by using pipeline architecture and clock periode of CORDIC block latency.

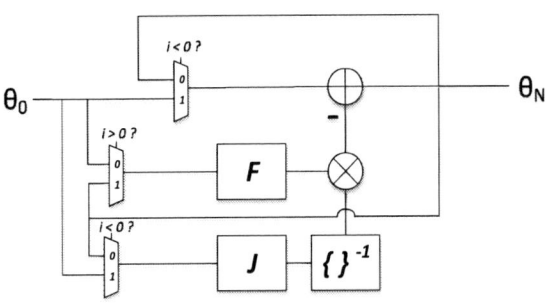

Fig. 4.    Newton-Raphson iteration block diagram.

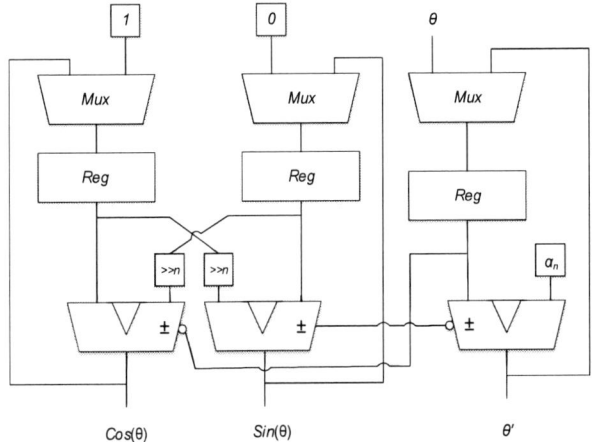

Fig. 5.    CORDIC Architecture to calculate sin and cos.

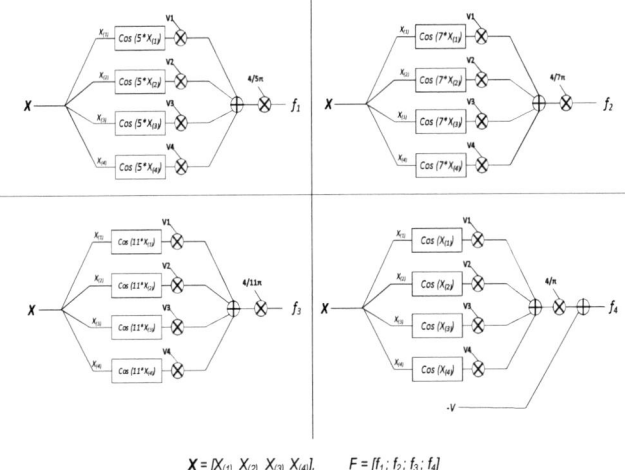

$X = [X_{(1)}\ X_{(2)}\ X_{(3)}\ X_{(4)}],\qquad F = [f_1;\ f_2;\ f_3;\ f_4]$

Fig. 6.    F-function block diagram.

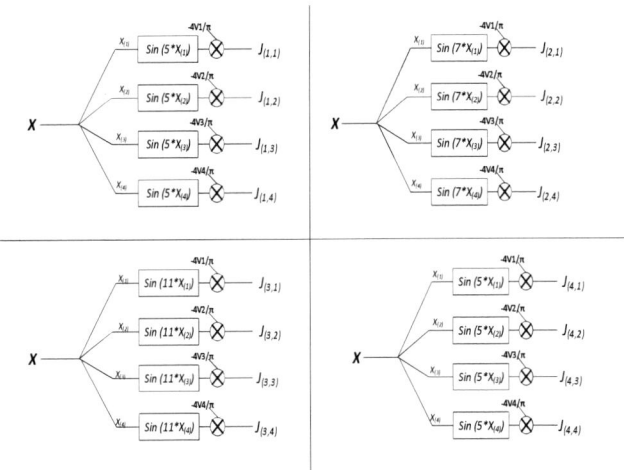

$X = [X_{(1)}\ X_{(2)}\ X_{(3)}\ X_{(4)}]$

$J = [\ J_{(1,1)},\ J_{(1,2)},\ J_{(1,3)},\ J_{(1,4)},\ J_{(2,1)},\ J_{(2,2)},\ J_{(2,3)},\ J_{(2,4)},\ J_{(3,1)},\ J_{(3,2)},\ J_{(3,3)},\ J_{(3,4)},\ J_{(4,1)},\ J_{(4,2)},\ J_{(4,3)},\ J_{(4,4)}\ ]$

Fig. 7.    J-function block diagram.

## IV. NEXTWORK

The Adaptive MLI method has been presented. Unfortunately, there are many technical aspects haven't explored yet. In this paper, following nexwork are considered to achieve: exploring modulation techniques and its relevancy much deeper; observing and comparing two aforementioned techniques of adaptive MLI; and evaluating feasibility of adaptive MLI implementation in small scale used. Fig. 9 shows an experimental set of the proposed adaptive MLI.

The proposed method will be verified in the laboratory experimental using PV simulator with an FPGA controller DE0 as seen in Fig. 9. There are two scenarios will be conducted. First one is switching state determination based on LUT and second one is real time iteration using Newton-Raphson theory.

Fig. 8.    A proposed experimental set.

978-1-5090-5606-4/17 $31.00 © 2017 IEEE          3105

## V. CONCLUSION

This paper has presented the proposed methods of adaptive MLI. This study shows direct computation is tend to be preferable by comparison with LUT technique. So that, nextwork will be usefull to prove that the FPGA controller is very powerfull to doing direct computation as well as can be maximized on supporting adaptive MLI implementation.

Based on the experience, Newton-Raphson iteration method is a great tool to support adaptive MLI application. An important aspect have to be considered is on choosing $\theta$ initial value.

## REFERENCES

[1] "Global Market Outlook For Photovoltaics Until 2016", EPIA, 2012.

[2] "Technology Roadmap Solar Photovoltaic Energy', IEA, 2014.

[3] Daniel Pepe, Gianni Bianchini, and Antonio Vicino, "Model Estimation for PV Generation Forecasting using Cloud Cover Information", 2016.

[4] N. Alam, V. Coors, S. Zlatanova, and P.J. Oosterom, "Shadow effect on photovoltaic potentiality analysis", *XXII ISPRS Congress*, Melbourne, Australia, 2012.

[5] Mohammadmehdi Seyedmahmoudian, Saad Mekhilef, Rasoul Rahmani, Rubiyah Yusof, and Ehsan Taslimi Renani, "Analytical Modeling of Partially Shaded Photovoltaic Systems", *Energies*, vol. 6, pp. 128-144, 2013.

[6] Lian L. Jiang, Douglas L. Maskell, R Srivatsan, and Qing Xu, "Power Variability of Small Scale PV System Caused by Shading from Passing Clouds in Tropical Region", *IEEE*, 2016.

[7] Arash A. Boora, Alireza Nami, Firuz Zare, Aridam Ghosh, and Frede Blaabjerg, "Voltage-Sharing Converter to Supply Single-Phase Asymmetrical Four-Level Diode-Clamped Inverter With High Power Factor Loads", *IEEE Transactions on Electronics*, vol. 25, pp. 2507-2520, 2010.

[8] Emad Samadaei, Sayyed Asghar Gholamian, Abdolreza Sheikoleslami, and Jafar Adabi, "An Envelope Type (E-Type) Module: Asymmetric Multilevel Inverter With Reduced Components", *IEEE Transactions on Industrial Electronics*, vol. 63, no. 11, 2016.

[9] Hongfei Wu, Lei Zhu, Fan Yang, Tiantian Mu, and Hongjuan Ge, "Dual-DC-Port Asymmetrical Multilevel Inverters With Reduced Conversion Stages and Enhanced Conversion Efficiency", *IEEE Transactions on Industrial Electronics*, vol. 64, pp. 2081-2091, 2017.

[10] Javier Pereda, and Juan Dixon, "Cascaded Multilevel Converters: Optimal Asymmetries and Floating Capacitor Control", *IEEE Transactions on Industrial Electronics*, vol. 60, pp. 4784-4793, 2013.

[11] K.K. Gupta S Jain, "Topology for Multilevel Inverter to Attain Maximum Number of Levels from Given DC Sources", *IET Power Electronics*, vol. 5, pp. 435-446, 2012.

[12] Farid Khoucha, Mouna Soumia Lagoun, Abdelaziz Kheloui, and Mohamed El Hachemi Benbouzid, "Comparison of Symmetrical and Asymmetrical Three-Phase H-Bridge Multilevel Inverter for DTC Induction Motor Drives", *IEEE Transactions on Energy Conversion*, vol. 26, pp. 64-72, 2011.

[13] Diorge A. B. Zambra, Cassiano Rech, and Jose Renes Pinheiro, "Comparison of Neutral-Point-Clamped, Symmetrical, and Hybrid Asymmetrical Multilevel Inverters", *IEEE Transactions on Industrial Electronics*, vol. 57, pp. 2297-2306, 2010.

[14] Jun Mei, Bailu Xiao, Ke Shen, Leon M. Tolbert, and Jian Yong Zheng, " Modular Multilevel Inverter with New Modulation Method and Its Application to Photovoltaic Grid-Connected Generator", *IEEE Transactions on Power Electronics*, vol. 28, pp. 5063-5073, 2013.

[15] Marino Coppola, Fabio Di Napoli, Pierluigi, Diego Iannuzzi, Santolo Daliento, and Andrea Del Pizzo, "An FPGA-Based Advanced Control Strategy of a Grid-Tied PV CHB Inverter", *IEEE Transactions on Power Electronics*, vol.31, pp. 806-816, 2016.

[16] Erwin Kreyszig, "Advanced Engineering Mathematics", John Wiley & Sons, pp. 474-483, 2011.

[17] Howard Anton and Chris Rorres, "Elementary Linear Algebra", Wiley, pp. 142-191, 2014.

[18] Trio Adiono, Rachmad Vidya Putra, "Perancangan sistem VLSI", Penerbit ITB, Bandung, 2017.

# Demand response for the promotion of photovoltaic penetration

Venizelos Venizelou, Spyros Theocharides, George Makrides, Venizelos Efthymiou and
George E. Georghiou

FOSS Research Centre for Sustainable Energy, Photovoltaic Technology Laboratory,
Department of Electrical and Computer Engineering, University of Cyprus
75 Kallipoleos Avenue, P.O. Box 20537, Nicosia, 1678, Cyprus

*Abstract* — **The capabilities offered by using Demand Response (DR) to aid the integration of photovoltaic (PV) systems into medium voltage (MV) distribution networks, are explored in this paper. The methodology followed focuses on the optimal usage scheduling of deferrable loads in order to coincide with forecasted PV production profiles by applying DR techniques on a day-ahead load profile. The obtained results highlight that the application of DR techniques can maintain the upper voltage level at 1.031 p.u. without violating the voltage regulation limits, while the lower level remained relatively unaffected. Furthermore, a significant reduction in active power levels, up to 5 MW, was observed, thus resulting in a feasible extension of the existing PV capacity.**

*Index Terms* — **demand response, forecasting, grid integration, photovoltaic systems.**

## I. Introduction

The traditional goal of an electric power system has been to control the supply-side to fulfil demand. The new philosophy states that the system will be most efficient if fluctuations in demand are kept as small as possible. In the current energy landscape the demand-side can have an active role in power systems via Demand Response (DR). The concept of DR through the use of price signals to elicit a response by the consumers, was proposed back in 1988 by Schweppe et al. [1]. Since then, many attempts have been made to evaluate the effects of implementing DR on system variables like the total cost of energy and market clearing price [2]–[4], while others proposed methods of load control based on the price elasticity of consumers [5], [6].

As the grid evolves to accommodate increased photovoltaic (PV) penetration, further research is required to find ways to surpass issues caused due to their intermittent nature. It has been known that high penetration of renewable energy sources (RES) ramping near synchronously can create power variability that is able to cause substantial power quality (PQ) issues [7]. An approach to solve this problem is to increase scheduling of controllable loads through DR. However, effective DR depends on optimal day-ahead load scheduling that changes the usage time of deferrable loads in order to "follow" the intermittent generation of RES. Therefore the control and scheduling of such loads is based on accurate load and generation forecasts.

This work describes the methodology followed to develop a DR algorithm that utilizes techniques such as load clipping (LC) and load shifting (LS), which are the most commonly employed techniques by grid operators [8]. The DR techniques are applied on day-ahead load profile forecasts and their impact on a typical Medium Voltage (MV) distribution grid with high levels of PV systems is analyzed in order to verify that the developed algorithm can increase the PV hosting capacity without violating any PQ indices.

## II. Methodology

The developed DR algorithm operates on the aggregated residential load of a typical MV distribution grid. As already stated, the optimal performance of DR is achieved by utilizing accurate day-ahead load profile forecasts. As a result, in order to create optimal load scheduling, the aggregated day-ahead Net-Load (NL) profile of the investigated MV distribution grid was forecasted with the use of an Artificial Neural Network (ANN) model. In this study, NL equals to the aggregated consumption minus the aggregated RES generation. The flowchart in Fig. 1 depicts the methodology followed for developing an optimal DR algorithm that can promote higher PV penetration levels.

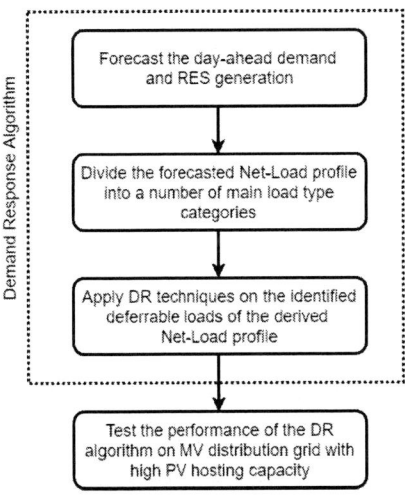

Fig. 1. Methodology followed for developing an optimal DR algorithm that can promote higher PV penetration levels.

The forecasted NL profile is divided into a number of main load type categories. The objective of the developed DR algorithm is to optimize the forecasted NL profile by maximizing the Net-Load Factor (NLF) or equally flatten the forecasted NL profile to reduce the variability of the consumption levels. NLF is defined as the Average Net Load divided by the Peak Net Load in a specified time period. Increasing the NLF can be recognized as an outcome of the load shifting and clipping techniques applied on the identified deferrable loads.

To evaluate the performance of the developed DR algorithm, the voltage levels and active power of the MV distribution grid were examined, at different PV hosting capacity levels.

### A. Forecasting algorithm using Artificial Neural Networks

In pursuance of implementing proper predictions for optimal day-ahead load scheduling, a forecasting algorithm based on ANN was utilized. The algorithm focuses on the Resilient Propagation (Rprob) method, which was introduced by Riedmiller and Braun [9]. The Rprob algorithm employs a sign-based scheme to update the weights to eliminate harmful influences of the derivatives' magnitude on the weight updates [10]. To achieve this, the individual update value $\Delta_{ij}$ of each weight $w$ was introduced. This adaptive update-value evolves during the learning process based on its local sight on the error function $E$, according to the learning-rule:

$$\Delta_{ij}^{(t)} = \begin{cases} n^+ \cdot \Delta_{ij}^{(t-1)}, & if \quad \dfrac{\partial E}{\partial w_{ij}}^{(t-1)} \cdot \dfrac{\partial E}{\partial w_{ij}}^{(t)} > 0 \\[2mm] n^- \cdot \Delta_{ij}^{(t-1)}, & if \quad \dfrac{\partial E}{\partial w_{ij}}^{(t-1)} \cdot \dfrac{\partial E}{\partial w_{ij}}^{(t)} < 0 \quad (1) \\[2mm] \Delta_{ij}^{(t-1)}, & else \end{cases}$$

where $0 < n^- < 1 < n^+$ and $\partial E / \partial w$ denotes the partial derivative of the batch error with respect to weight $w_{ij}$ at the $t^{th}$ iteration.

Once the update-value for each weight is obtained, the weight-update follows the rule:

$$\Delta w_{ij}^{(t)} = \begin{cases} -\Delta_{ij}^{(t-1)}, & if \quad \dfrac{\partial E}{\partial w_{ij}}^{(t)} > 0 \\[2mm] +\Delta_{ij}^{(t-1)}, & if \quad \dfrac{\partial E}{\partial w_{ij}}^{(t)} < 0 \quad (2) \\[2mm] 0, & else \end{cases}$$

$$w_{ij}^{(t+1)} = w_{ij}^{(t)} + \Delta w_{ij}^{(t)} \qquad (3)$$

The update-values and the weights are changed every time based on the error function.

### B. Demand Response Algorithm

In order to compose a more realistic DR algorithm, the aggregated residential load profile of the MV distribution grid was divided into a number of main load type categories. The percentage of each category was estimated using the statistical analysis proposed by Palmer *et al.* [11]. Since the Electric Vehicles (EVs) will compose a relatively high percentage of the future deferrable loads, they have also been considered in our study. The load profiles for the EVs that were used in the developed algorithm were based on the report published by JRC [12]. Deferrable loads like washing machines and dishwashers, make up the biggest slice of the peak evening demand for the households. This group of appliances is eminently switchable. The percentage values of the residential loads which were used in this study are exhibited in Fig. 2.

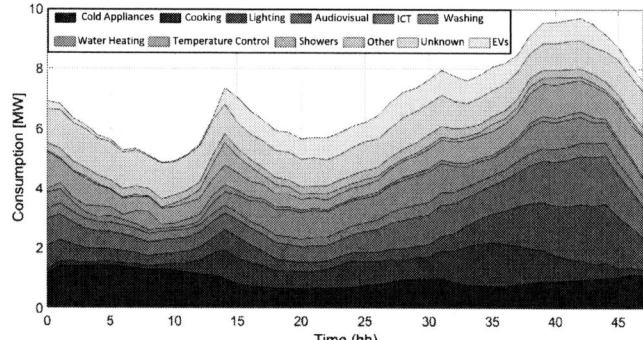

Fig. 2. Average annual residential load profile percentile categories derived based on the performed statistical analysis.

Households could avoid using these appliances during the evening peak, even without any new controls. Persuading dwellers to use these appliances outside the peak load periods may require different electricity tariffs – so there is a financial incentive to run these appliances at different times – or other strategies geared towards changing behaviour. The main objective of the algorithm was to maximize the NLF.

$$\max_{x} f(x) = \frac{\sum\limits_{i=1}^{48} x_i}{\max\_x \cdot 48}$$

$s.t.$

$$\sum_{i=1}^{48} x_{i,before\,DR} = \sum_{i=1}^{48} x_{i,after\,DR} \qquad (4)$$

$$Peak_{before\,DR} > Peak_{after\,DR}$$

where $x$ is the power at time interval $i$, $x_{i,\text{before DR}}$ and $x_{i,\text{after DR}}$

are the power before and after applying the DR techniques respectively. The time block periods for the LS technique were selected based on results derived by Philippou *et al* [13]. To evaluate the effect of DR, the load factor before and after applying each technique was compared.

### C. Impact of Demand Response on a typical MV substation

The impact of the DR algorithm on a typical MV distribution grid with large PV penetration was studied using DIgSILENT PowerFactory. Initially, for the baseline scenario, no DR techniques were applied.

In the next step, the DR techniques of load clipping and shifting were enabled in order to reduce or shift the power demand of selected deferrable loads by a specific percentage. In particular, the two DR techniques were divided into two sub-cases: a load shifting of a) 25% and b) 50% from peak to off-peak hours and a load clipping of a) 25% and b) 50% from peak hours. Shifting load from peak to off-peak hours represents the case where consumers shift the usage time of their appliances to other hours due to high electricity rates occurring during peak hours (e.g. Time of Use tariffs). Load clipping replicates the case where consumers or the Building Energy Management System (BEMS) decide not to turn on specific appliances during peak hours due high price signals.

The performance of the developed DR algorithm was evaluated by examining the impact of each load profile, derived after the application of DR, on the MV distribution grid. More specifically, the voltage levels and active power of the studied MV distribution grid were investigated, at different levels of PV hosting capacity and EVs. The uncertainty of the total PV capacity and the number of EVs, was assessed through Monte Carlo simulation. The PQ parameters under consideration were assessed by normalizing the values according to their limit/range as stated in EN 50160 [14] by using the following equation:

$$V_{rorm} = \frac{V_{MV} - V_{Lowest}}{V_{Upper} - V_{Lowest}} = \frac{V_{MV} - 0.9}{0.2} \qquad (5)$$

where, $V_{norm}$ is the normalized voltage for the MV side, $V_{MV}$ is the voltage for the MV side obtained from simulations (p.u.), $V_{upper}$ is the upper voltage limit recommended by EN 50160 standard (1.1 p.u.) and $V_{lowest}$ is the lowest voltage limit recommended by EN 50160 standard (0.9 p.u.).

### III. RESULTS

In an attempt to derive accurate aggregated day-ahead NL profile, a total number of 3 hidden layers was considered in the developed ANN algorithm.

The derived ANN algorithm resulted in a validation error of 0.02994, with 0 representing a perfect fit between the real and forecasted data. Fig. 3 illustrates the normalized measured and forecasted PV production.

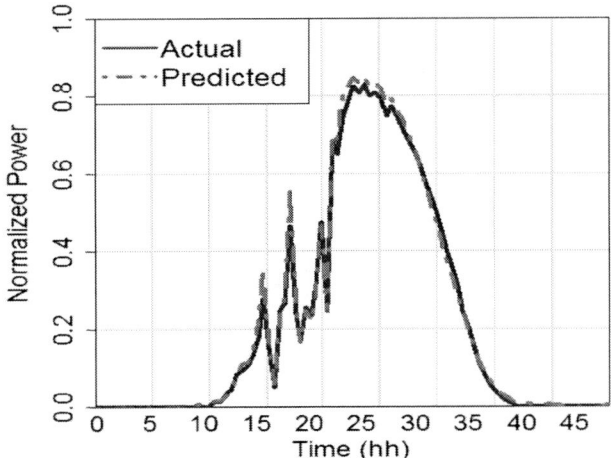

Fig. 3.   Comparison between the normalized measured and forecasted PV production.

The voltage levels and the active power of the investigated MV distribution grid before the application of any DR technique represent the baseline scenario. To evaluate the impact of the developed DR algorithm, a comparative analysis between each scenario and the baseline was performed. The analysis focused on the voltage levels and the maximum active power value of the investigated MV distribution grid. These values represent the acceptable limits, where the maximum and minimum voltage should not increase or drop more than 5%, respectively. Additionally, any DR technique should reduce the maximum active power value. As shown in Fig. 4, the maximum voltage value, for the baseline scenario, is equal to 1.031 p.u., while the minimum voltage varies based on investigated PV levels and the accommodation levels of EVs.

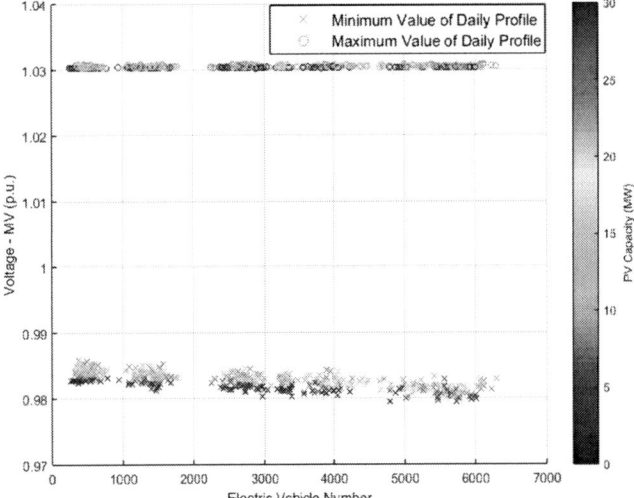

Fig. 4.   Voltage level at the MV side of the investigated MV distribution grid with high share of PV systems (baseline scenario).

The minimum, mean and maximum active power of the investigated MV distribution grid, for the baseline scenario, is illustrated in Fig. 5. At the highest levels of PV capacity and EVs, the maximum active power is equal to 56 MW.

Furthermore, the active power is reduced by approximately 5 MW, even when large shares of both EVs and PV were considered, as demonstrated in Fig. 7.

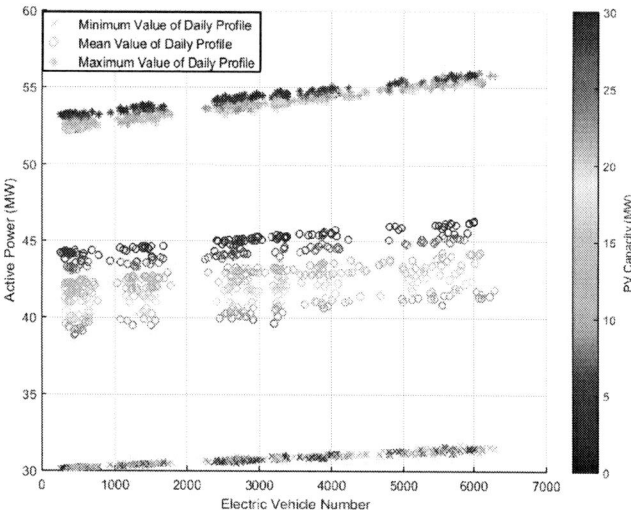

Fig. 5. Active power of the investigated MV distribution grid as a function of PV capacity and number of EVs (baseline scenario).

The initial scenario replicates the case where 25% and 50% of the total deferrable load was shifted into the off-periods in order to increase the NLF. Fig. 6 and 7 demonstrate the voltage levels and the active power for the LS method at a load percentage of 50%, respectively. The load flow simulation results showed that the maximum voltage level remained unaffected at 1.031 p.u. while the minimum level was slightly reduced, without violating the voltage regulation limits.

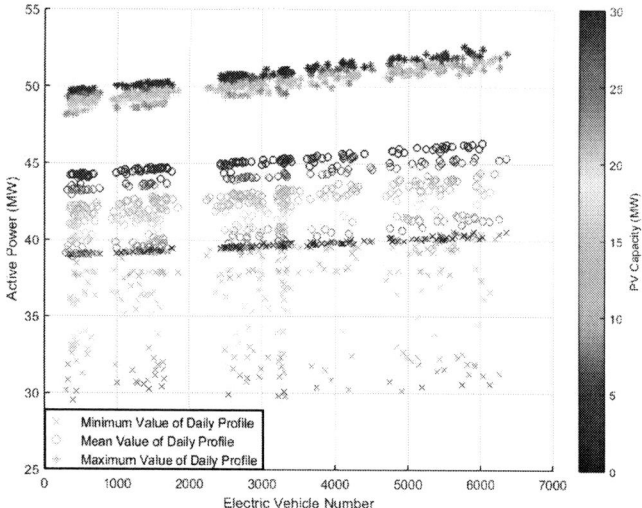

Fig. 7. Active power of the investigated MV distribution grid as a function of PV capacity and number of EVs (LS – 50%).

For the second scenario, the LC technique was applied to the NL profile, at two different percentages. In particular, 25% and 50% of the deferrable loads were clipped (i.e. turned off) during the peak periods. The case of LC technique indicated more promising results than the LS. More specifically, the voltage levels remained the same as the baseline scenario, as shown in Fig. 8. However, Fig. 9 demonstrates that the minimum and maximum active power levels were curtailed down to 46 MW and 51 MW, respectively.

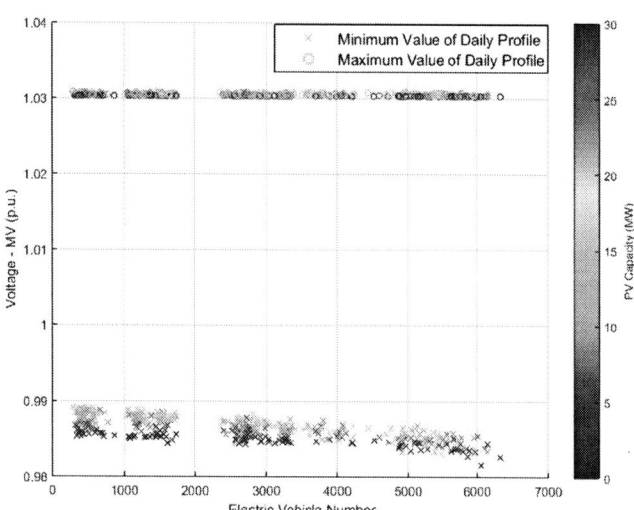

Fig. 6. Voltage level at the MV side of the investigated MV distribution grid with high share of PV systems (LS – 50%).

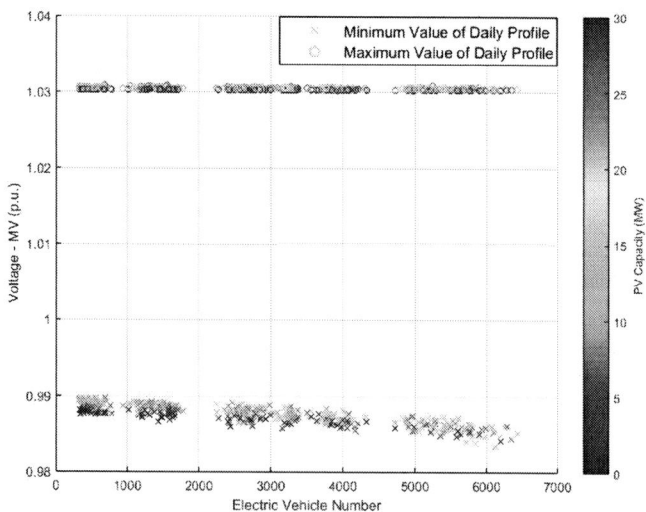

Fig. 8. Voltage level at the MV side of the investigated MV distribution grid with high share of PV systems (LC – 50%).

978-1-5090-5606-4/17 $31.00 © 2017 IEEE

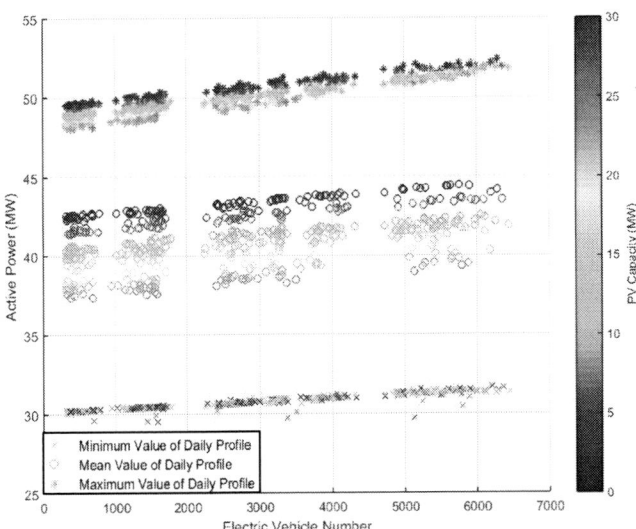

Fig. 9. Active power of the investigated MV distribution grid as a function of PV capacity and number of EVs (LC – 50%).

The voltage and active power levels of each investigated scenario are summarized in Table 1. The results highlight that, for all the investigated scenarios, the voltage remained at the same levels as the baseline scenario, while the PV hosting capacity increased. Furthermore, the active power levels were reduced for every scenario, compared to the baseline, even at the highest PV capacity, signifying that the proposed DR algorithm can promote increased PV penetration levels.

Moreover, the results showed that both DR techniques can equally reduce the peak demand. However, shifting loads from peak to off-peak periods, by applying the LS technique, led to the creation of a flatter load profile and therefore a higher NLF. Additionally, the proposed LS technique proved to be more effective for integrating higher levels of EVs. Specifically, when applying the LS technique more EVs could be accommodated compared to the LC technique for the same levels of active power. This occurred due to the optimally distributed EV charging provoked by the LS technique.

TABLE I

COMPARATIVE ANALYSIS BETWEEN THE APPLIED TECHNIQUES

| Scenario | Voltage (p.u.) | | Power (MW) | | Load Factor |
|---|---|---|---|---|---|
| | Min | Max | Min | Max | |
| Baseline | 0.985 | 1.031 | 52 | 56 | 40.65 |
| Load Shifting – 25% | 0.984 | 1.031 | 50 | 53 | 42.63 |
| Load Shifting – 50% | 0.984 | 1.031 | 48 | 51 | 44.38 |
| Load Clipping – 25% | 0.985 | 1.031 | 50 | 53 | 41.32 |
| Load Clipping – 50% | 0.985 | 1.031 | 46 | 51 | 42.59 |

## IV. CONCLUSIONS

The methodology followed to analyse the potential of day-ahead DR to match demand with RES generation as a measure of enabling higher shares of PV is described in this paper. Different types of household appliances were considered as deferrable loads while a forecasting algorithm was utilized to address the challenge arising from the intermittent nature of the PV production in designing optimal load scheduling techniques. To evaluate the performance of the developed DR algorithm, a comparative analysis was performed. More specifically, the voltage and active power levels of the investigated MV distribution grid of the baseline scenario (no DR techniques) were compared to the results obtained when two main DR techniques, load shifting and peak clipping, were applied.

The results highlight that the developed DR algorithm can lead to an increase of the PV hosting capacity as the application of Load Shifting and Load Clipping on the deferrable loads maintained the upper voltage level at the same levels as the baseline scenario while the lower level remained relatively unaffected. Furthermore, both DR techniques indicated a significant reduction in active power levels, approximately by 5 MW, confirming that the developed methodology can promote higher levels of PV.

## ACKNOWLEDGEMENTS

The authors would like to thank the JRC Institute for Energy and Transport for their continuous support and funding. The authors also gratefully acknowledge the contributions of the Distribution System Operator (DSO) and the Electricity Authority of Cyprus (EAC).

## REFERENCES

[1] F. Schweppe, M. Caramanis, R. Tabors, and R. Bohn, "Spot pricing of electricity," in *The Kluwer International Series in Engineering and Computer Science: power electronics and power systems*, 1988.

[2] S. Braithwait, "Behavior modification," *IEEE Power Energy Mag.*, vol. 8, no. 3, pp. 36–45, 2010.

[3] A. K. David and Y. Z. Li, "A comparison of system response for different types of real time pricing," in *IET International Conference on Advances in Power System Control, Operation and Management*, 1991, pp. 385–390.

[4] M. Behrangrad, H. Sugihara, and T. Funaki, "Analysing the system effects of optimal demand response utilization for reserve procurement and peak clipping," in *IEEE Power and Energy Society General Meeting*, 2010, pp. 1–7.

[5] P. Faria, Z. Vale, J. Soares, and J. Ferreira, "Demand response management in power systems using a particle swarm optimization approach," *IEEE Intell. Syst.*, no. 99, pp. 1–9, 2011.

[6] M. G. Lijesen, "The real-time price elasticity of electricity," *Energy Econ.*, vol. 29, no. 2, pp. 249–258, 2007.

[7] A. H. Habib, Z. K. Pecena, V. R. Disfani, J. Kleissl, and R. A. de Callafon, "Reliability of Dynamic Load Scheduling

with Solar Forecast Scenarios," in *IEEE Systems Conference (SysCon)*, 2015.

[8] C. W. Gellings, "The concept of demand-side management for electric utilities," *Proc. IEEE*, vol. 73, no. 10, pp. 1468–1470, 1985.

[9] M. Riedmiller and H. Braun, "A direct adaptive method for faster backpropagation learning: the RPROP algorithm," in *International Conference on Neural Networks, San Francisco*, 1993, pp. 586–591.

[10] A. D. Anastasiadis, G. D. Magoulas, and M. N. Vrahatis, "New globally convergent training scheme based on the resilient propagation algorithm," *Neurocomputing*, vol. 64, pp. 253–270, 2005.

[11] J. Palmer, N. Terry, and T. Kane, "Early Findings: Demand side management," no. November, 2013.

[12] JRC, *Report on Individual mobility: From conventional to electric cars*. 2015.

[13] N. Philippou, M. Hadjipanayi, G. Makrides, V. Efthymiou, and G. E. Georghiou, "Effective dynamic tariffs for price-based Demand Side Management with grid-connected PV systems," in *IEEE Eindhoven PowerTech*, 2015.

[14] EN50160, "Voltage Characteristics of Electricity Supplied by Public Distribution Systems," 2010.

# Griddler AI: New Paradigm in Luminescence Image Analysis Using Automated Finite Element Methods

Johnson Wong[1], Percis Teena[1], Daniel Inns[2]

[1] Solar Energy Research Institute of Singapore, National University of Singapore, Singapore 117574, Singapore
[2] DuPont Silicon Valley Technology Center, Sunnyvale, CA 94085, USA

*Abstract* — Griddler AI (artificial intelligence) is the combination of Griddler, a finite-element simulator for solar cells, and tailored multivariate regression techniques, in a general computational routine which seeks cell parameters that can best explain a set of luminescence imaging data. To demonstrate it, we apply Griddler AI to a dataset of luminescence images from about 80 solar cells with varying front metal grid contact resistance. The fitted cell parameters (contact resistance spatial distribution, saturation current densities of the wafer's passivated regions and under the metal contacts) lead to simulated I-V characteristics that are in agreement with experimental data, albeit differing by some bias. Comparison of the extracted contact resistance to TLM measurements show worse agreement, with the values of best fit by Griddler AI being generally higher, suggesting that the best fit values may represent the sum of the front and rear contact resistances.

## I. INTRODUCTION

Luminescence imaging is a versatile measurement technique that has the potential to become ubiquitous throughout the silicon PV manufacturing environment. It offers spatially resolved data of carrier concentration in solar cells, with degrees of high speed, low cost, and compatibility with samples through virtually all stages of processing that are unrivalled by other measurement methods of carrier concentration. Just as a new wave of automation in image processing is now expanding the market of video cameras, luminescence imaging systems stand to benefit greatly in its reach of applications from advancements in the software which analyze the images.

Since 2013, Griddler has been introduced and continuously developed as a finite element (FEM) solver dedicated to silicon solar cells [1-3]. Its ability to simulate the voltage distributions across the cell plane also makes it able to model the spatial distribution of luminescence intensity. Griddler AI (artificial intelligence) is the cumulation of more recent efforts to solve the inverse, multivariate regression problem: Given a set of experimental luminescence images, what is the set of cell parameters within a defined parameter space that will lead to corresponding simulated images of best fit to data? While the methodologies and workflow will be presented in more detail in the final manuscript, in this extended abstract we hope to give an example of its usage.

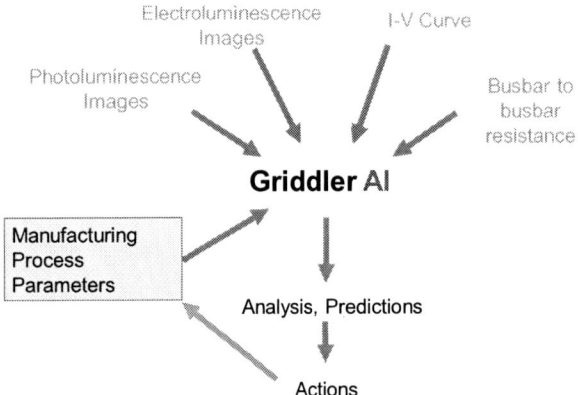

Fig. 1. Application of automated analysis in the manufacturing process.

## II. LUMINESCENCE DATA AND FITTING

The experiment involved about 80 p-type monocrystalline silicon Al-BSF solar cells provided by DuPont Silicon Valley Technology Center. The phosphorus emitter sheet resistance is about 80 $\Omega$/sq for all cells, but the dopant profile is intentionally made to have a low surface concentration, such that the front grid metal contact resistance is sensitive to the contact firing temperature profile. These cells are divided into roughly four groups of 20, each of which was fired at a different profile to intentionally create cells of various contact resistance.

For each cell, four luminescence images are taken: 1) at open circuit under 3.6 Suns illumination, 2) at open circuit under 0.04 Suns illumination, 3) in the dark with forward bias and cell current of 9 A, 4) with 9 A current extraction under 2 Suns illumination. The voltage at the contacting rail sense pins are also simultaneously recorded under the conditions at which these images are taken. These images and recorded voltages are fed into Griddler AI, which is instructed to simulate the four images under their respective conditions (corresponding light intensities for the open circuit images, corresponding light intensities and bias voltages for the biasing images), while treating the cell plane passivated region $J_{01}$, $J_{02}$, metal induced $J_{01}$, and front grid contact resistance as fitting parameters. Of the fitting parameters, only the front grid contact resistance takes on a spatial distribution, as defined by the value it takes on a square array of 5 x 5 points. The

978-1-5090-5606-4/17 $31.00 © 2017 IEEE

contact resistance values at any other position along the cell plane is calculated by interpolating these zonal point values.

In order to see if the cell parameters of best fit to luminescence data, are also consistent with other aspects of device performance, additional I-V simulations were carried out and compared to actual I-V data. Also, the front grid contact resistance extracted by Griddler AI are compared directly to contact resistance values derived from TLM measurements of the same cells (which took place after the luminescence imaging and I-V measurements).

Figure 2 shows an examples set of luminescence images for a cell (Grp1-2) and the corresponding simulations at the cell parameters of best fits. Figure 3 shows the generic computational process flow that led to the cell parameters of best fit.

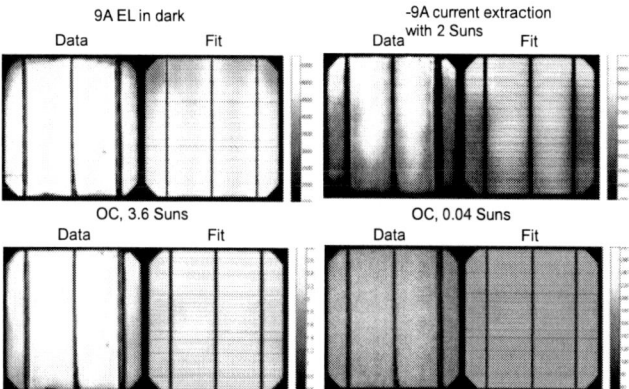
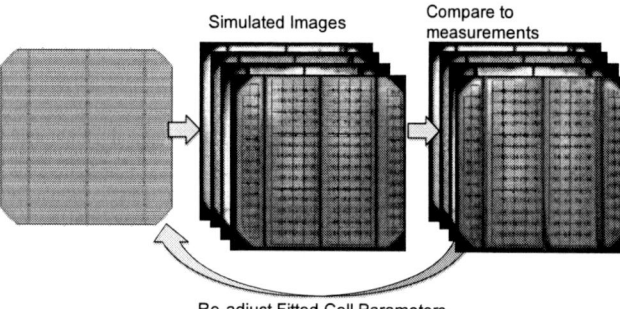

Fig. 2. Luminescence images and corresponding simulations using cell parameters of best fit, for cell Grp1-6.

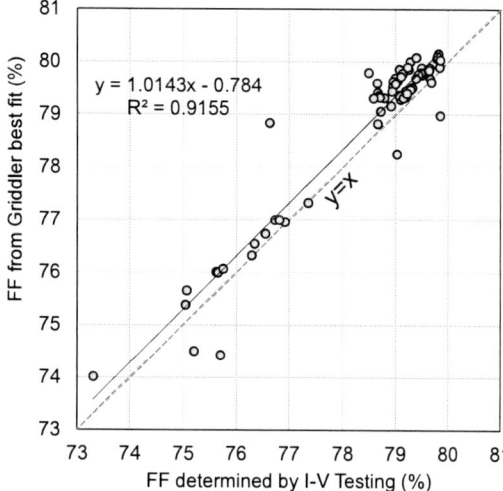

Fig. 3. Computational process flow.

## III. RESULTS AND DISCUSSION

Figure 4 compares, for all 80 cells, the simulated I-V parameters of fill factor and $V_{oc}$, against experiment. It is important to emphasize that the simulations are based on cell parameters ($J_0$'s and contact resistance) which produced best fits to luminescence images. Therefore this comparison of I-V parameters serves as a check of whether or not the same parameters can also predict other device characteristics. Evidently, while the simulation and experimentally determined parameters have strong correlations with slopes close to 1,

there is a parallel shift in the simulation of both FF and $V_{oc}$. This kind of shift can be reduced by observing how the fit parameters change if, additionally, the $V_{oc}$ and FF themselves are also fitted against in Griddler AI, and then manually inserting a suitable shift into the fit parameters when only the luminescence data are fitted against.

Figure 6 compares, for cell Grp1-6, the extracted contact resistance by Griddler AI and that determined by the traditional TLM method. Both show similar spatial distributions that are consistent with the emitter sheet resistance as a function of position (also shown in Figure 6).

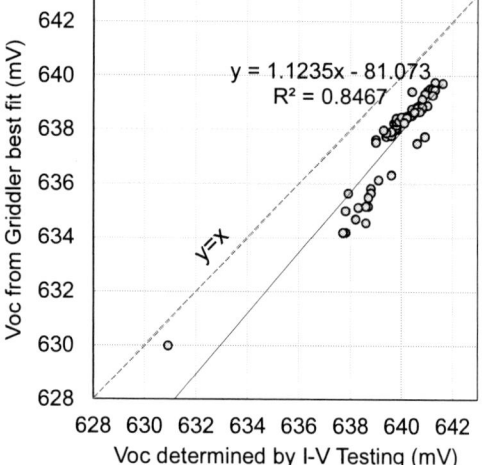

Fig. 4. Experimentally determined fill factor and $V_{oc}$ from I-V data, compared to those generated from simulations based on the parameters of best fit to luminescence data by Griddler AI.

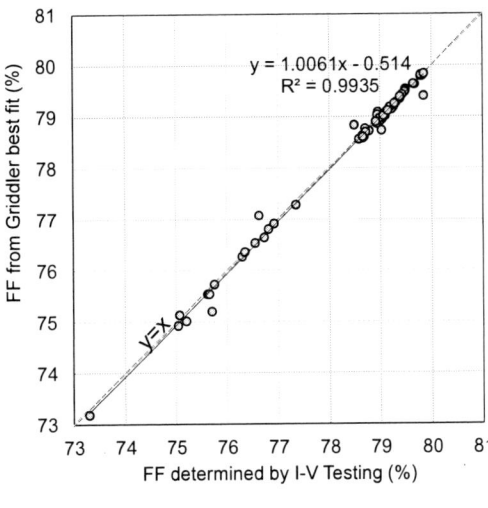

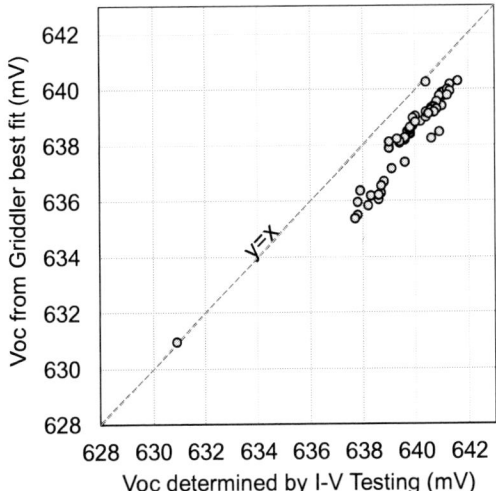

Fig. 5. Experimentally determined fill factor and $V_{oc}$ from I-V data, compared to those generated from simulations based on the parameters of best fit to luminescence data by Griddler AI, if additionally the fill factor and $V_{oc}$ numbers are provided as targets of fit for Griddler AI

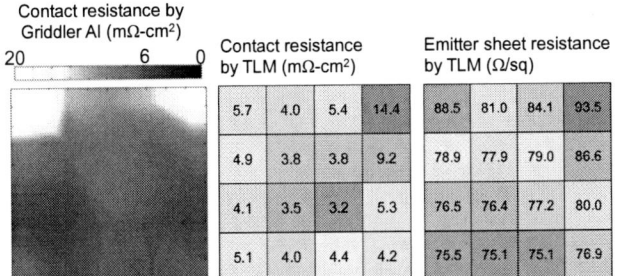

Fig 6. Front grid contact resistance determined by TLM method (middle) compared to best fit value by Griddler AI (left), for cell Grp1-6.

Figure 7 compares for all cells, the average contact resistance determined by Griddler AI versus the TLM destructive

method. For cells with acceptable screen print contact quality (black circle), the agreement is good except for a systematic shift. For cells with poor contact quality (red circle), Griddler AI extracts a much higher contact resistance than the TLM method, presumably because the former is sensitive also to any contact resistance arising from the rear aluminium interface. For the cells in the red circle, it is clear from EL images that the rear interface contact have deteriorated and is non-uniform.

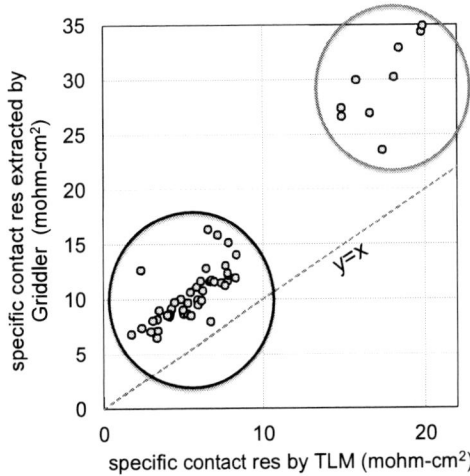

Fig. 7. Average contact resistance extracted by Griddler AI, versus those determined by the TLM method.

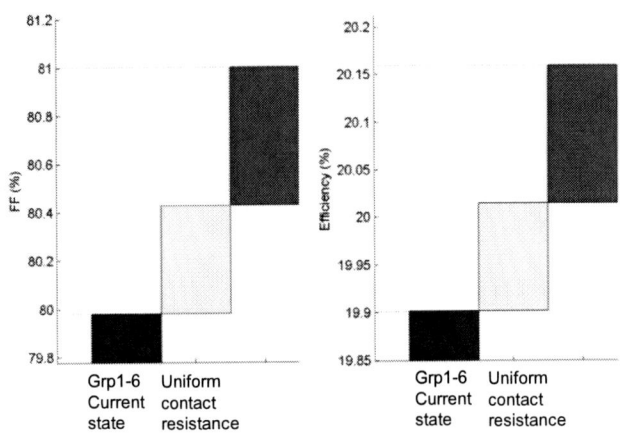

Fig. 8. Griddler simulations of efficiency improvements with respect to improvements in the contact resistance.

## IV. OTHER APPLICATIONS

Griddler AI is generally written in the sense that it can be easily configured to fit to a different set of luminescence data, while varying a different set of cell parameters. For instance, Fig. 9 shows a set of 8 luminescence images of a multi-crystalline cell that Griddler AI fits to. In this case, the test setup which produced the images have significant variance in the pin contact resistance to the contact rails. Therefore, the pin contact resistance are also set as fitting parameters. Four additional luminescence images, each produced by forward biasing a single busbar of the cell, are fitted against. Figure 9 shows that here too, Griddler AI is able to converge on parameters which lead to a good fit.

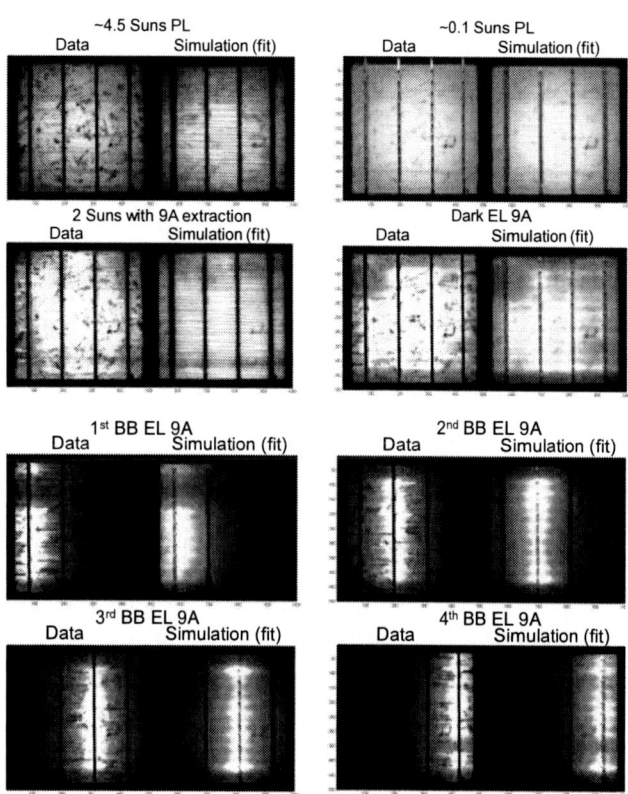

Fig. 9. Example of readable plot using different colors and line styles for clarity.

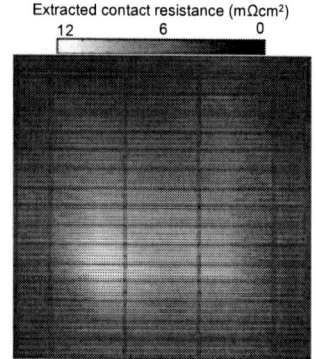

Fig. 10. Extracted contact resistance distribution from the luminescence data of Figure 9.

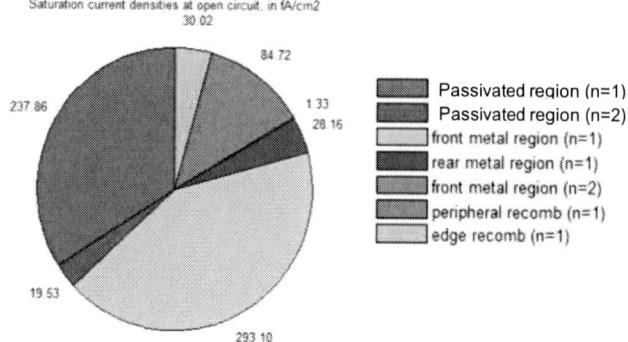

Fig. 11. After extraction of cell parameters, Griddler can be used to analyze the saturation current densities at open circuit.

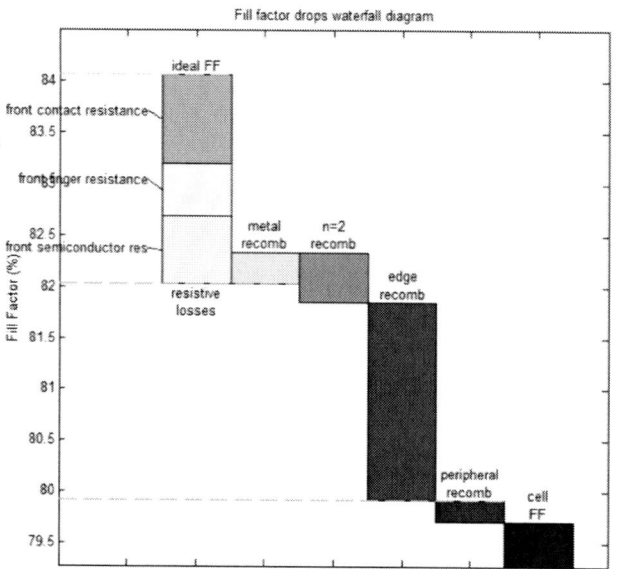

Fig. 12. After extraction of cell parameters, Griddler can be used to analyze the resistive losses and non-ideal recombination that degrades the fill factor.

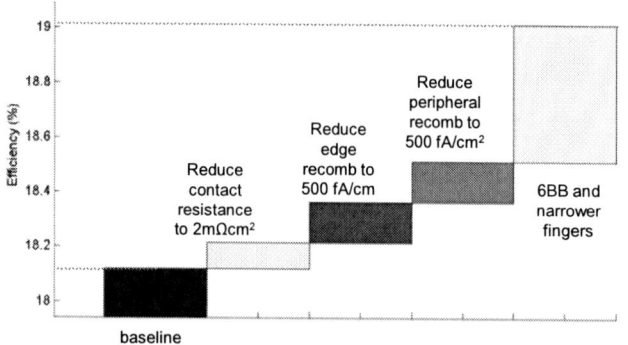

Fig. 13. After extraction of cell parameters, Griddler can be used to predict efficiency improvement scenarios.

Figure 14 shows another application of Griddler AI in the analysis of not just one finished cell, but an n-type bifacial cell together with three related precursor samples. By analyzing more than one sample, it becomes easier to separate the contributions to recombination by the different relevant parts. In this example, the goal is to determine the influence of metal induced recombination as well as edge and peripheral recombinations on the $V_{oc}$ and fill factor. Figure 15 shows that a common underlying set of these cell parameters is able to generate simultaneous good fits to all provided luminescence data (only high intensity open circuit images are shown).

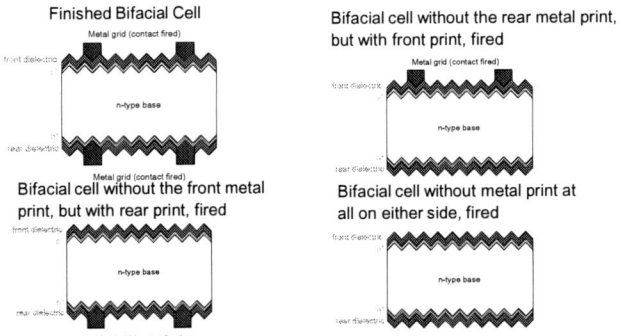

Fig. 14. Example of readable plot using different colors and line styles for clarity.

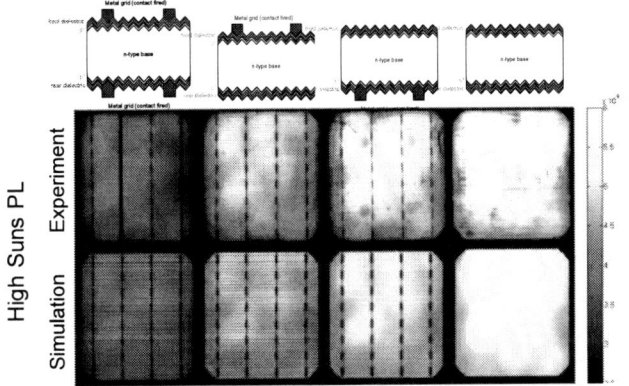

Fig. 15. Example of readable plot using different colors and line styles for clarity.

Figure 16 shows another application of Griddler AI in the analysis of special test structures on wafer which are screen printed. In this example, the goal is to determine the magnitude of metal induced recombination on a boron emitter. Figure 16 shows that a good fit is possible to the images belonging to samples with the printed test structures as well as blank samples, using a suitable set of metal induced $J_{01}$, $J_{02}$ numbers.

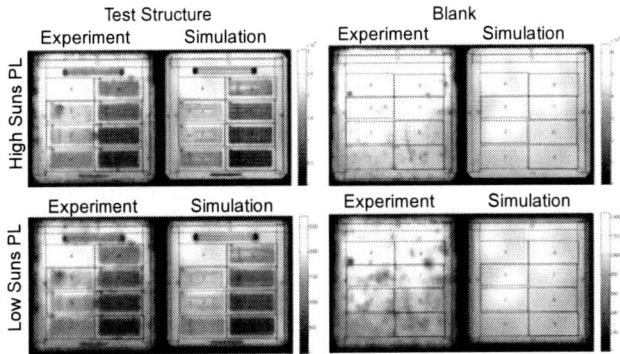

Fig. 16. Example of readable plot using different colors and line styles for clarity.

## IV. CONCLUSION

Griddler AI marks a new paradigm in luminescence image analysis using automated finite element methods. More broadly, it is an example of using a comprehensive device model, instead of approximate formula or equivalent circuit models, to perform regression to data. This approach has several advantages: 1) the model is general enough to consider and analyze a greater variety of data simultaneously, 2) the model considers details like the metal grid and contacts, and cell edge recombination [4], which are more easily related to manufacturing processes, making the results easy to understand, 3) the model can then be used to make predictions, such as the degree of improvement in cell efficiency if a certain cell parameter were changed [5-6]. In the future, we hope to introduce these model based methods as a means to relate process parameters to cell parameters, in the context of manufacturing process control.

### ACKNOWLEDGEMENTS

SERIS is sponsored by the National University of Singapore (NUS) and Singapore's National Research Foundation (NRF) through the Singapore Economic Development Board (EDB). This research was supported by the National Research Foundation, Prime Minister's Office, Singapore under project grant NRF2013EWT-EIRP002-007.

### REFERENCES

[1] J. Wong, Griddler: Intelligent Computer Aided Design of Complex Solar Cell Metallization Patterns, Proceedings of the 39th IEEE Photovoltaic Specialists Conference (PVSC), Tampa, USA (2013).

[2] J. Wong, R. Sridharan, Griddler 2: Two Dimensional Solar Cell Simulator with Facile Definition of Spatial Distribution in Cell Parameters and Bifacial Calculation Mode, Proceedings of the 42nd IEEE PVSC, New Orleans, USA (2015).

[3] J. Wong, Griddler 2.5 PRO: Modelling high efficiency solar cells with parameter database to calculate room for efficiency

improvement, 26th International PV Science and Engineering Conference, Singapore (2016) (unpublished).

[4] J. Wong, R. Sridharan and V. Shanmugam, Quantifying Edge and Peripheral Recombination Losses in Industrial Silicon Solar Cells, IEEE Transactions on Electron Devices, vol. 62, no. 11, Nov 2015.

[5] Wong. J, Duttagupta. S, Stangl. R, Hoex. B, Aberle, A.G, A Systematic Loss Analysis Method for Rear-Passivated Silicon Solar Cells, IEEE Journal of Photovoltaics, vol. 5, no. 2 (2015) 619-626

[6] J. Wong, S. Raj, J.W. Ho, J. Wang, J. Lin, Voltage Loss Analysis for Bifacial Silicon Solar Cells: Case for Two-Dimensional Large-Area Modelling, IEEE Journal of Photovoltaics 6.6 (2016): 1421-1426.

# Interaction of $O_{2i}$ Dimers with Ga in Si and Implications for a Comprehensive Model of Light-Induced Degradation

Yu Jin and Scott T. Dunham

Electrical Engineering Department, University of Washington, Seattle, WA 98195, USA

*Abstract* — **DFT calculations of the properties of O interstitial dimers ($O_{2i}$) and their interaction with B and Ga lead to a comprehensive model of light induced degradation of Si solar cells doped with B, Ga and P. The proposed model accounts for experimental observation under full range of conditions.**

*Index Terms* — **Light induced degradation, Model, Silicon solar cell**

## I. INTRODUCTION

It has been long observed that the performance of Si solar cells degrades under initial exposure to light via a drop in bulk carrier lifetime. This phenomenon has been termed light induced degradation (LID), and a wide range of models have been proposed to explain it [1-5], but none are fully satisfactory. Initial models accounted for the apparent linear dependence of LID on B doping (as well as quadratic dependence on initial interstitial O) by postulating the formation of a $BO_2$ cluster under light exposure [1-3]. However, experiments using Si co-doped with B and P led to the conclusion that degradation depended on hole rather than B concentration [6-7]. Several new models were proposed to account for the lack of a B-doping dependence, but the most widely cited [4-5] relies on the formation of boron interstitial complexes in concentrations that are inconsistent with a vast body of work on B-doped Si for electronics application.

A recent study on Si co-doped with B and Ga has cast doubt on the linear dependence of LID on p, as the addition of Ga appears to leave LID unchanged despite the increase in p [8, 9]. In order to identify a possible explanation for this behavior, we undertook a series of density functional theory (DFT) calculations which include complexes of not only O and B, but also Ga in order to develop a model for LID that accurately accounts for both B/P and B/Ga co-doping.

## II. DFT CALCULATION FOR $O_2$ DIFFUSION AND BINDING

We conducted a series of DFT calculation using the VASP code to explore the properties of LID related defects. The calculations for $O_2$ and $BO_2$ formation are generally consistent with previous work [2, 3, 10]. The binding energy for two interstitial oxygens to form oxygen dimer is about -0.30 eV. The two most stable structures are staggered with neutral charge ($O_2{}^{st0}$) and square with double plus charge ($O_2{}^{sq++}$). As discussed in previous research [2-3], light illumination can lead to charge state changes as well as reconfiguration of $O_2$ between staggered and square structure, which drives oxygen dimer diffusion as shown in Figure 1.

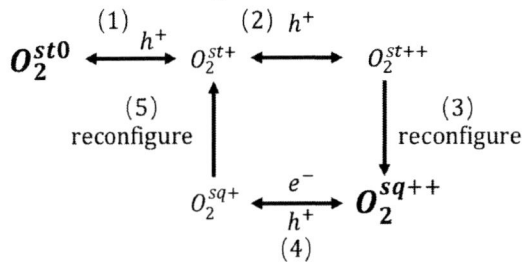

Fig. 1 Schematic for oxygen dimer reconfiguration during light illumination.

There are two possible routes for $O_2$ capture by substitutional B (or Ga), through A chain or B chain (using notation of Ref. [10]) as illustrated in Figure 3. The A chain is a <110> chain which contains a $B_s$ (or $Ga_s$), and the B chain is a <110> chain adjacent to $B_s$ (or $Ga_s$). Calculations indicate that bistability of staggered and square structures still exists after capture by $B_s$, with a binding energy ($E^b{}_{B/O2}$) of about -0.42 eV in B chain and about -0.26 eV in A chain.

By comparing the difference of charge density distribution between two charged states of a configuration, we find that $O_2{}^{sq}$ and $B_sO_2{}^{sq}$ (both in A chain and B chain) are localized defects with defect level about 0.2 eV lower than conduction band minimum while $O_2{}^{st}$ is not a localized defect. Thus, $O_2{}^{sq}$ and $B_sO_2{}^{sq}$ can capture both electrons and holes (step 4 in Fig. 1 and step 9 in Fig. 2) and can potentially work as recombination center. However, the hole capture cross-section for $O_2{}^{sq+}$ is small because of Coulomb repulsion, so it is not expected to be an efficient recombination center. For $O_2{}^{st}$ and $BO_2{}^{st}$ in B chain, the defect levels are mixed with valence band and so not effective recombination centers. In contrast, we find that $BO_2{}^{st}$ in A chain is localized recombination center with defect level about 0.15 eV above valence band. As will be discussed later, although binding energy is larger in B chain for B/$O_2$, we think $O_{2i}$ trapped by $B_s$ in A chain is the source for light induced degradation.

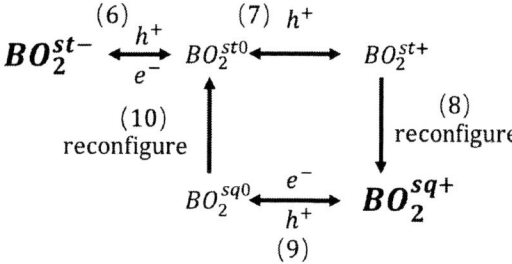

Fig. 2 Schematic for B/O complex reconfiguration during light illumination.

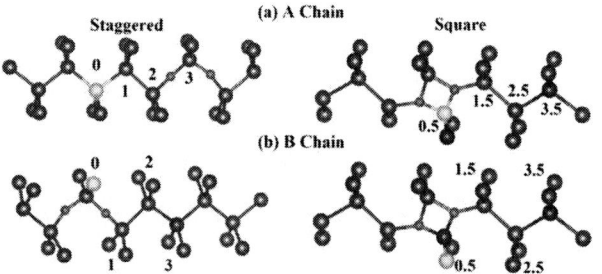

Fig. 3 Diffusion of oxygen dimer approaching boron (or gallium) in A chain (a) and B chain (b).

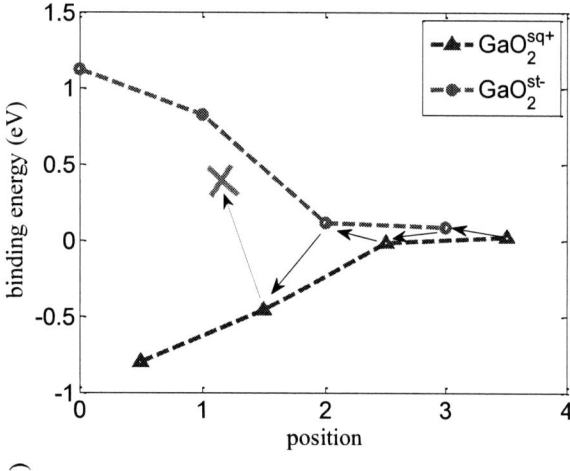

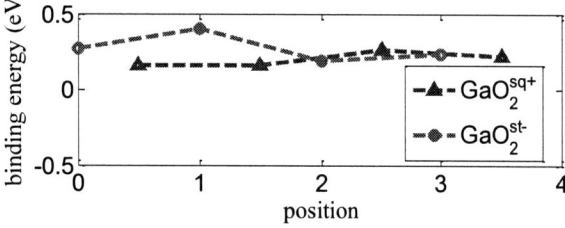

Fig. 4 Plots of binding energies of $Ga_sO_2$ complexes (a) in A-chain (b) B-chain, as a function of atomic distance between the $O_2$ and Ga, illustrated in figure 3. The arrows show the diffusion path for oxygen dimer captured by $Ga_s$. The red cross indicates that O dimer blocked from moving closer to $Ga_s$ due to strong repulsion for square structure.

We also conduct DFT calculation for the interaction between substitutional Ga and interstitial $O_2$ dimers. The binding energy between $Ga_s$ and $O_2$ as $O_{2i}$ approaches is shown in Figure 4. We find no significant binding with $Ga_s$ for $O_{2i}$ along B chain. When oxygen dimers diffuse through A chain, there is about -0.46 eV binding ($E^b_{Ga/O2}$) for $O_{2i}$ square structure at second nearest neighbor location. Note that the captured $O_{2i}$ is prevented from moving closer to Ga because of large repulsive energy for staggered structure, presumably caused by compressive strain for the larger Ga atoms. This implies that after being captured by $Ga_s$, the oxygen dimer no longer retains its bistability, and since the $Ga_sO_2^{sq}$ is a shallow donor and also has smaller hole capture cross-section, it cannot work as an effective recombination center.

## III. CONTINUUM MODEL FOR LID-RELATED DEFECT FORMATION:

We developed a continuum model to describe the formation kinetics of LID related defects. The involved reactions include:

$$O_i + O_i \leftrightarrow O_2 \quad (1)$$

$$B_s + O_2 \leftrightarrow BO_2 \quad (2)$$

$$Ga_s + O_2 \leftrightarrow GaO_2 \quad (3)$$

The oxygen properties in silicon and formation of large oxygen clusters are based on our previously-developed oxygen precipitation model [11]. The formation rate of $BO_2$ and $GaO_2$ can be expressed as:

$$\frac{dC_{BO_2}}{dt} = 4\pi r D_{O_2} \left( C_{O_2} C_B - C_{BO_2} / K_{BO_2} \right) \quad (4)$$

$$K_{BO_2} = \frac{4}{C_{Si}} \exp\left( -E^b_{B/O_2} / k_B T \right) \quad (5)$$

$$\frac{dC_{GaO_2}}{dt} = 4\pi r D_{O_2} \left( C_{O_2} C_{Ga} - C_{GaO_2} / K_{GaO_2} \right) \quad (6)$$

$$K_{GaO_2} = \frac{2}{C_{Si}} \exp\left( -E^b_{Ga/O_2} / k_B T \right) \quad (7)$$

where $r$ is the capture distance, $E^b_{B/O2}$ and $E^b_{Ga/O2}$ are the calculated binding energies in Section II, $D_{O2}$ is the diffusivity of oxygen dimer and $C_{Si}$ is the silicon atom density.

As discussed in Ref [3], the rate limiting step in reactions (2-5) in Figure 1 is the formation of $O_2^{sq++}$ because of the repulsive hole capture process. Thus, the diffusivity of $O_2$ dimer during light illumination at room temperature would be proportional to $(p/n_i)^2$, giving a time constant to form $BO_2$ complex proportional to $1/p^2$, as experimentally-observed [7, 8].

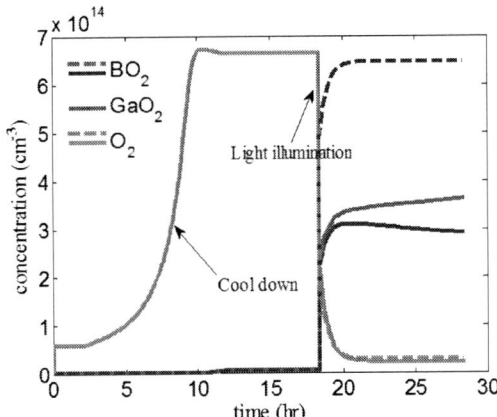

Fig. 5 Concentration evolution of $BO_2$, $GaO_2$ and $O_2$ during annealing and light illumination. The solid line is for system with $10^{16}$ cm$^{-3}$ B and $10^{16}$ cm$^{-3}$ Ga co-doping, and the dashed line is for $2 \times 10^{16}$ cm$^{-3}$ B doped situation.

We conducted continuum simulation for p type silicon wafer with boron and gallium co-doping. The initial oxygen concentration is $6 \times 10^{17}$ cm$^{-3}$. In the simulation, the wafer cools down from 1400 °C to room temperature with ramp rate of 2 °C/min, then is held at room temperature for 5 h before light illumination is applied. As illustrated in Figure 5, the O dimer concentration increases significantly during the cool down process. Little changes in the dark due to slow $O_{2i}$ diffusion, but binding to B and Ga is strong enough that once diffusion is activated by carrier generation, almost all the oxygen dimers get trapped by $B_s$ or $Ga_s$. Thus, the acceptor/dimer concentration is the independent of the acceptor concentration.

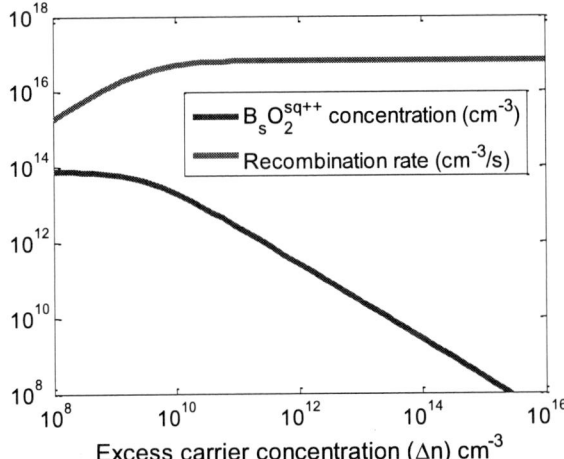

Fig. 6 The concentration of $B_sO_2{}^{sq++}$ in B chain and related recombination rate caused by $B_sO_2$ in B chain vs excess carrier concentration in steady state.

If there is B, but no Ga, in our simulation all the oxygen dimers would be trapped by $B_s$. In this case, the formation of

$BO_2$ complex only depends on the oxygen dimer concentration after cool down and is independent of boron concentration. In B chain, the barrier to transit from $B_sO_2{}^{sq0}$ to $B_sO_2{}^{st0}$ is small (~0.17 eV [3]). During light illumination, $BO_2{}^{st0}$ in B chain is the dominant configuration which is not recombination active as discussed above. We also conducted steady state analysis as shown in Fig. 6. We found that the concentration of $B_sO_2{}^{sq++}$ in B chain during light illumination is so small that it cannot significantly contribute to LID by itself ((9) in Fig. 2) or the recombination caused by $B_sO_2{}^{sq}/B_sO_2{}^{st}$ reconfiguration suggested by [3] ((7) - (10) in Fig. 2). Thus, we believe that $B_sO_2$ in A chain is the primary recombination center. Given about 0.16 eV higher formation energy, the concentration of $B_sO_2$ in A chain should be about $e^{(-0.16/kT)} \sim 1/400$ of the total $B_sO_2$ concentration at room temperature. Thus, the density of LID defects ($B_sO_2$ in A chain) is on the order of $10^{12}$cm$^{-3}$, which matches the typical experimental LID lifetime (~$10^{-4}$ s) given reasonable capture cross-section. At low level injection $B_sO_2$ in B chain may contribute to the recombination. At illumination with AM 1.5 spectrum, $B_sO_2$ in A chain should be the major recombination center. The relative energy and the transition barrier would determine whether $B_sO_2{}^{sq}$ and $B_sO_2{}^{st}$ in A chain is the main recombination center at given injection level. Generally speaking, stronger injection would make $B_sO_2{}^{st}$ more favorable in steady state, and thus more likely to be major recombination center. In addition, since the defect levels for both $B_sO_2{}^{sq}$ and $B_sO_2{}^{st}$ are close to band edges, the recombination rate is proportional to hole concentration in p type silicon as shown in Eq. (8) using the example of $B_sO_2{}^{st}$.

$$U = \frac{pn - n_i^2}{\tau_p(n + n_i e^{(E_t - E_i)/k_B T}) + \tau_n(p + n_i e^{(E_i - E_t)/k_B T})}$$

$$\approx \frac{p_0 \Delta n}{\tau_n(n_i e^{(E_i - E_t)/k_B T})}, \text{ given that } n_i e^{(E_i - E_t)/k_B T} \gg p \quad (8)$$

Thus the recombination rate should be proportional to $pC_O{}^2$, which is $C_B C_O{}^2$ for B doped and $(C_B-C_P)C_O{}^2$ for B/P compensated cases, consistent with previous observation in the absence of Ga [1, 6, 7].

For B/Ga co-doping, B and Ga compete in the oxygen dimer trapping process. From DFT binding energies, we can estimate the O dimer capture ratio between Ga and B ($\alpha$) to be:

$$\alpha = \exp(\frac{E_{B/O_2}^b - E_{Ga/O_2}^b}{k_B T}) / 2 \quad (9)$$

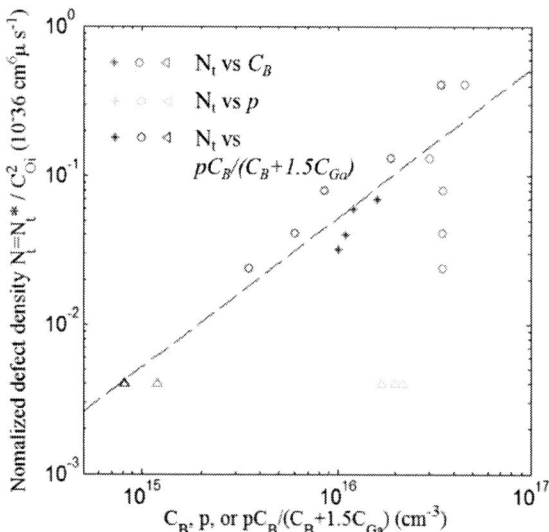

Fig. 7 Normalized defect density vs $C_B$, $p$ or $pC_B/(C_B+1.5C_{Ga})$. Stars are for B doping alone [1], circles are for B/P compensated doping [6] and triangles are for B/Ga co-doping [8]. The dashed line is guide to the eye.

Here the factor of 2 comes from the fact that $O_2$ can be trapped by B mostly in B chain, while for Ga it can only be captured in A chain. In our calculation $\alpha$ is about 2 at room temperature and drops as T increases. As discussed in Section II, $GaO_2$ is not an effective recombination center. Thus, the concentration of effective recombination center for LID, $N_\tau$, can be expressed as:

$$N_\tau \equiv \frac{1}{\tau_d} - \frac{1}{\tau_0} \propto pC_O^2 \frac{C_B}{\left(C_B + \alpha C_{Ga}\right)} \qquad (10)$$

Here, $\tau_0$ is the initial lifetime and $\tau_d$ is the lifetime after LID. For Ga/B co doping, $p=C_B+C_{Ga}$. If we choose $\alpha=1$, then equation (10) becomes $N_\tau \propto C_B C_O^2$. This explains the observation in B and Ga co-doped situation, where the LID related recombination rate depends on boron concentration instead of hole concentration [8].

In Figure 7, we plot normalized defect density data versus boron concentration, hole concentration (p) and our model, which predicts that LID normalized by O concentration squared [8] should be proportional to $pC_B/(C_B+\alpha C_{Ga})$. As can be seen, substantial deviations from linear behavior are seen in the dependence on both $p$ and $C_B$, but our model can account for the full range of conditions including boron doping alone, B/P compensated doping and B/Ga co-doping.

## IV. CONCLUSION

Using DFT calculation we study the formation of LID related defects, including B/O and Ga/O complexes. Based on the results of these calculations, we propose a comprehensive model to explain experimental observation under the full range of conditions, including both B/P compensated doping and B/Ga co-doping.

## ACKNOWLEDGEMENT

This material is based in part upon work supported by the National Science Foundation under Award Number CHE-1230615 and by the State of Washington through the University of Washington Clean Energy Institute.

## REFERENCES

[1] J. Schmidt and K. Bothe, "Structure and transformation of the metastable boron- and oxygen-related defect center in crystalline silicon," *Phys. Rev. B.*, vol 69, pp. 034107, 2004.

[2] J. Adey, R. Jones, D.W. Palmer, P. R. Briddon and S. Oberg, "Degradation of Boron-Doped Czochralski-Grown Silicon Solar Cells", Phys, Rev. Lett, vol. 93, p. 055504-1, 2004

[3] M. Du, H. M. Branz, R. S. Crandall, and S. B. Zhang, "Bistability-Mediated Carrier Recombination at Light-Induced Boron-Oxygen Complexes in Silicon", Phys, Rev. Lett, vol. 97, p. 256602, 2006.

[4] V. V. Voronkov and R. Falster, "Latent complexes of interstitial boron and oxygen dimers as a reason for degradation of silicon-based solar cells", *J. Appl. Phys. vol.* 107, p.053509, 2010

[5] V. V. Voronkov and R. Falster, "Light-Induced Boron-Oxygen Recombination Centres in Silicon: Understanding their Formation and Elimination", *SOLID STATE PHENOM.* vol. 205-206, p.3, 2014.

[6] J. Geilker, W. Kwapil, and S. Rein, "Light-induced degradation in compensated p- and n-type Czochralski silicon wafers", *J. Appl. Phys. vol.* 109, p.053718, 2011

[7] B. Lim, F. Rougieux, D. Macdonald, K. Bothe, and J. Schmidt, "Generation and annihilation of boron–oxygen-related recombination centers in compensated p- and n-type silicon", *J. Appl. Phys. vol.* 108, p.103722, 2010

[8] M. Forster, E. Fourmond, F. E. Rougieux, A. Cuevas, R. Gotoh4, K. Fujiwara, S. Uda, and M. Lemiti, "Boron-oxygen defect in Czochralski-silicon co-doped with gallium and boron", *Appl. Phys. Lett*, vol. 100, pp.042110, 2012

[9] M. Forster, P. Wagnera, J. Degoulange, R. Einhaus, G. Galbiati, F. E. Rougieux, A. Cuevasd, E. Fourmond, "Impact of compensation on the boron and oxygen-related degradation of upgraded metallurgical-grade silicon solar cells", Sol. Energ. Mat. Sol. Cells, vol. 120, pp.390, 2014

[10] X. Yu, P. Chen, X. Chen, Y. Liu, and D. Yang, "Ab-initio calculation study on the formation mechanism of boron-oxygen complexes in c-Si", AIP Advances, vol. 5, pp. 077154, 2015

[11] B. C. Trzynadlowski and S. T. Dunham, "A reduced moment-based model for oxygen precipitation in silicon", *J. Appl. Phys. vol.* 114, p. 243508, 2013

# Numerical Simulation of EBIC for Analysis of Extended Defects

Marco Nardone, John Moseley[2], Saroj Dahal, Anuja V. Parikh, and John M. Waddle

Dept. of Physics and Astronomy, Bowling Green State University, Bowling Green, OH, 43403, U.S.A
[2] National Renewable Energy Laboratory, Golden, CO, 80403, U.S.A

*Abstract* — **Electron beam induced current (EBIC) measurements and simulation are employed to better understand recombination at extended defects, such as dislocations, grain boundaries, and stacking faults. Analysis of our EBIC contrast data indicates that certain grain boundaries in mc-Si devices are comprised of a nearly uniform density of shallow and deep gap states. The effects on device performance are calculated. Recombination active stacking faults (junction shunts) exhibit high room-temperature contrast that is not described by the defect physics studied here, possibly due to formation of a Schottky barrier. The electronic properties of the latter are important for understanding potential induced degradation of the shunting type.**

## I. INTRODUCTION

Extended defects, such as dislocations, stacking faults, and grain boundaries can significantly affect the performance and degradation of photovoltaic (PV) devices. Electron beam induced current (EBIC) measurements provide a means for studying the charge dynamics at such structural irregularities. In this work, we incorporate theories of recombination at extended defects in semiconductor device modeling to simulate EBIC measurements and device performance. Results are compared to our EBIC measurements and data from the literature.

The body of work in Ref. [1] describes recombination mechanisms at dislocations (1D extended defects) based on the coupling between deep and shallow band gap states. That theory enabled the calculation of EBIC contrast as a function of temperature and beam current, $C(T, I_b)$, which explained the contrast data at dislocations with varying levels of metal impurities in silicon [2]–[4]. Recently, that work was used to describe the performance effects of grain boundaries (GBs – 2D extended defects) in Si PV devices [5] by assuming that GBs can be considered arrays of dislocations [6] and simplifying the charge kinetics of several defect states to one representative state with variable capture cross-sections [7]. Herein, we maintain the coupled defect states theory put forth by Kveder et. al. [1] because, as we shall show, it provides a reasonable account of the EBIC contrast temperature-dependence measured at dislocations and GBs. Although the present work focuses on dislocations and GBs, preliminary EBIC results for recombination-active stacking faults that penetrate the surface p-n junction will also be discussed. We will refer to the latter as "junction shunts", which are of particular concern with respect to potential induced degradation of the shunting type (PID-s) [8].

Numerical simulation of EBIC measurements requires calculation of the distribution of electron-hole (e-h) pairs generated by an electron beam. That e-h pair generation rate is then used, along with applicable recombination mechanisms and junction electric fields, to determine the short circuit current collected at the device contacts. Previous simulation work [9] extracted surface recombination velocities from EBIC data on highly-doped Si. We employ a similar general approach here with addition of specific recombination physics, as described in Sec. II. The simulation tool is based on COMSOL Multiphysics® software and is generally applicable for analysis of EBIC data derived from other types of devices and studies.

## II. RECOMBINATION AT EXTENDED DEFECTS

Recombination at extended defects depends on the charge kinetics at gap states and the local potential barrier formed by the occupation of those states. A series of investigations have shed light on how dislocations (1D extended defects) in silicon can affect EBIC contrast as a function of temperature and e-beam current, $C(T, I_b)$, depending on the degree of impurity contamination in the dislocations (see Ref. [1] and references therein). Different temperature dependencies were associated with shallow and deep level impurities, or coupling between the two. A schematic of possible recombination pathways for a positively charged dislocation in p-type material is shown in Fig. 1. The shallow energy levels, $E^{(de)} \approx E^{(dh)} \approx 0.08$ eV away from the nearest band, are commonly associated with dislocations, regardless of contamination level [1]. Deep, nearly mid-gap states, $E^{(m)}$, are due to extrinsic impurities, typically metals. The local electric potential, $U$, depends on the occupancy of all the gap states and strongly influences local recombination rates. Note that charge transfer can occur not only between the defect states and the conduction or valence bands, but also between defect states. Such theories based on discrete energy levels in the band gap are approximations to a more realistic continuous defect density of states; we maintain the discrete level approximation in this work.

The net capture rate of electrons (holes) at any defect ($i$) due to charge transfer between that defect state and the conduction or valence band depends on the capture cross-section, $\sigma_{n(p)}^{(i)}$, thermal velocity, $v_{th,n(p)}^{(i)}$, defect concentration, $N^{(i)}$, occupancy, $f^{(i)}$, and the local concentration of carriers, $n(p)$. For example, at mid-gap states, the net capture of electrons, $R_n^{(m)}$, and holes, $R_p^{(m)}$, from the conduction and valence bands can be expressed as,

$$R_n^{(m)} = \sigma_n^{(m)} v_{th,n}^{(m)} N^{(m)} \left[ \left( 1 - f^{(m)} \right) n - f^{(m)} n_1^{(m)} \right], \quad (1)$$

$$R_p^{(m)} = \sigma_p^{(m)} v_{th,p}^{(m)} N^{(m)} \left[ f^{(m)} p - \left( 1 - f^{(m)} \right) p_1^{(m)} \right], \quad (2)$$

with,

$$N^{(m)} \frac{\partial f^{(m)}}{\partial t} = R_n^{(m)} - R_p^{(m)}. \quad (3)$$

In Eqs. (1) and (2), $n_1^{(m)} = n_i \exp\left(-\left(E_i - E^{(m)}\right)/kT\right)$ and $p_1^{(m)} = n_i \exp\left(\left(E_i - E^{(m)}\right)/kT\right)$, where $n_i$ and $E_i$ are the intrinsic concentration and energy level of the semiconductor. Under steady state conditions, Eq. (3) reduces to $R_n^{(m)} = R_p^{(m)}$ and the common Shockley-Read-Hall recombination is recovered. Similar expressions hold for charge transfer between the bands and other defect levels. Hence, all three defect levels shown in Fig. 1 are subject to Eqs. (1) – (3).

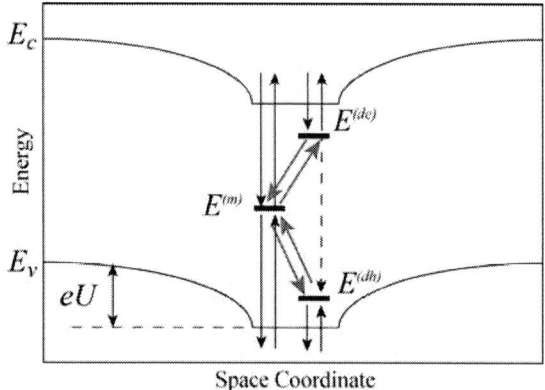

Figure 1. Possible recombination pathways between the conduction band $E_c$, valence band $E_v$, and various gap states within an extended defect. Solid arrows are non-radiative processes, red/bold arrows indicate transitions between coupled states, and the dashed arrow is radiative. A potential barrier, $U$, is created by charged defect states.

Coupling terms must be added to Eqs. (1) and (2) to account for charge transfer amongst the shallow and deep states. Details are provided in Ref. [1], but we note that the important coupling parameters are,

$$A_e = \alpha \sigma_n^{(m)} / \left[ \pi \left( r^{(de)} \right)^2 \right]$$

and $\quad (4)$

$$A_h = \alpha \sigma_p^{(m)} / \left[ \pi \left( r^{(dh)} \right)^2 \right],$$

where $r^{(de)} \approx r^{(dh)} \approx 2$ nm are the wave function radii of electrons and holes in 1D bands, respectively. The dimensionless fit parameter, $\alpha$, is a measure of the overlap of the deep and shallow state wave functions. Values of $\alpha$ between 0 and 1 were used in Ref. [1] and found to be important in describing various cases of $C(T, I_b)$ data for dislocations with contrasts of less than about 20%.

Higher contrast values of greater than 20% were associated with metal-silicide precipitates, possibly acting as internal Schottky contacts [10], [11], which could lead to different mechanisms that are beyond the scope of this work. Overall, it was found that the contrast increased with increasing metal contamination but the type of metal was not important. Also, contrast could be decreased by hydrogenation or phosphorous diffusion gettering.

## III. EBIC MEASUREMENT

Temperature-dependent EBIC measurements were carried out on a mc-Si sample in a JEOL 7600F field emission scanning electron microscope (SEM). The sample was extracted from a field-degraded commercial module using a diamond-based coring drill. A Gatan CF302 continuous-flow liquid-nitrogen cold stage cooled the sample in the range 295−77 K, and EBIC images were acquired with a Mighty EBIC quantitative EBIC imaging system made by Ephemeron Labs, Inc. The SEM beam conditions used for all images were 30-kV and 3.77 nA-current.

## IV. EBIC SIMULATION METHODOLOGY

In this section, we describe the main assumptions and concepts included in the simulation of EBIC measurements, including the electronic properties and geometry of the Si device under study, temperature dependencies, and calculation of the e-beam generation volume.

### A. Semiconductor Device Modeling

Our device simulation entails solving the semiconductor transport equations in 2D domains using COMSOL Multiphysics® to calculate the electric potential ($V$), and electron ($n$) and hole ($p$) concentrations. Recombination mechanisms in the bulk include direct band-to-band (radiative), Auger, and Shockley-Read-Hall (SRH). The recombination mechanisms at extended defects described in Sec. II and Ref. [1] were included in the numerical solver by customizing the recombination physics. Electron-hole (e-h) pair generation induced by the e-beam, rather than light, is also a key factor here which is elaborated upon in Section IV.C. It should be noted that the e-h plasma created by the intense e-beam can screen the electric potential surrounding the extended defect. That effect is included in our calculations.

Two device structures were evaluated in this work. The first was used to validate our model against the data provided in Ref. [1], [10]. It was comprised of a 10-μm thick, n-type Si layer with a Schottky contact along the entire top surface. As was observed in Ref. [10], 1D dislocations oriented parallel to the top surface (not intersecting the Schottky junction) were included in our model with the recombination properties described in Sec. IV. Dislocations consisted of 12 nm$^2$ areas to represent the strain field around the dislocation core with defects placed along the surface area (linear defect concentrations were converted to aerial concentrations by dividing by the perimeter).

The second device structure, used for analysis of our own EBIC measurements, was a 50-μm wide by 100-μm-thick piece of a typical, full back contact p-type Si cell with a front contact width of 2 μm and a diffused junction along the top surface. Although the measured devices were of the typical 300 μm thickness, the 100 μm thick model is sufficient to simulate EBIC since most of the e-beam absorption is within the top 10 μm. Fermi-Dirac (F-D) statistics (rather than Maxwell-Boltzmann) were employed due to the relatively high n-type doping concentrations ($10^{19} - 10^{20}$ cm$^{-3}$) near the top. Our baseline model and parameter values are based on the review by Altermatt [12] and the references therein. A grain boundary (GB) formed one edge of the 2D domain (farthest from the front contact finger), which contained the three defect levels and charge kinetics described in Sec. II.

For both devices, the most important independent variables studied here were the defect concentrations, $N^{(m)}$, $N^{(de)} = N^{(dh)}$, and the characteristic energies of the shallow states, $E^{(de)} = E^{(dh)}$. All other parameters were held fixed and are provided in Table I. Although only two device types were considered, other architectures, such as passivated emitter and rear locally diffused (PERL), and passivated emitter rear contacted (PERC) cells, could also be simulated.

## TABLE I
### FIXED DEFECT PARAMETER VALUES

| Parameter | Unit | Value | Ref. |
|---|---|---|---|
| $E_c - E^{(m)}$ | eV | 0.5 | [1] |
| $\sigma_n^{(m)}$ | cm$^2$ | 7.5x10$^{-14}$ | [1] |
| $\sigma_p^{(m)}$ | cm$^2$ | 3.0x10$^{-15}$ | [1] |
| $v_{th,n}^{(m)}$ | cm/s | see Eq. (7) | [1] |
| $v_{th,p}^{(m)}$ | cm/s | see Eq. (8) | [1] |
| $v_{th,n}^{(de)}\sigma_n^{(de)}$ | cm$^3$/s | $2 \times 10^{-3}/T^{3/2}$ | [13] |
| $v_{th,p}^{(de)}\sigma_p^{(de)}$ | cm$^3$/s | $2 \times 10^{-3}/T^{3/2}$ | [13] |
| $v_{th,n}^{(dh)}\sigma_n^{(dh)}$ | cm$^3$/s | $2 \times 10^{-3}/T^{3/2}$ | [13] |
| $v_{th,p}^{(dh)}\sigma_p^{(dh)}$ | cm$^3$/s | $2 \times 10^{-3}/T^{3/2}$ | [13] |
| $m^{(de)}$ | g | $0.3m_0$ | [1] |
| $m^{(dh)}$ | g | $0.5m_0$ | [1] |
| $\alpha$ | | 0.9 | [1] |

### B. Temperature Dependencies

Temperature-dependent EBIC simulation requires that the pertinent material parameters are included. The temperature dependencies of the effective densities of state in the conduction and valence bands, respectively, are [14],

$$N_c = 2.86 \times 10^{19} \left(\frac{T}{300}\right)^{1.58} \text{cm}^{-3} \quad (5)$$

$$N_v = 3.10 \times 10^{19} \left(\frac{T}{300}\right)^{1.58} \text{cm}^{-3}. \quad (6)$$

Thermal velocities of free carriers affect the recombination rates and lifetimes. Their temperature dependencies are

generally given by $v_{th} = \sqrt{3kT/m^*}$, with $m^*$ the effective mass. For Si we use,

$$v_{th,n} = 1.17 \times 10^6 T^{1/2} \text{ cm/s, and} \quad (7)$$

$$v_{th,p} = 0.91 \times 10^6 T^{1/2} \text{ cm/s,} \quad (8)$$

for electrons and holes, respectively.

Mobility depends mostly on phonon scattering in the temperature ($T > 100$ K) and p-type doping ($N_a \sim 10^{16}$ cm$^{-3}$) range of interest such that [15],

$$\mu_{i,L} = \mu_{max,i} \left(\frac{300}{T}\right)^{\theta_i}, \quad (9)$$

where $i$ stands for electrons ($n$) or holes (p), $L$ stands for lattice scattering, $\theta_n = 2.285, \theta_p = 2.247$, $\mu_{max,n} = 1412$ cm$^2$/(V s), and $\mu_{max,p} = 470$ cm$^2$/(V s). The band gap depends on temperature according to [16],

$$E_g = 1.170 - 1.49 \times 0.0255 \left[\coth\left(\frac{0.0255}{2kT}\right) - 1\right] \quad (10)$$

(in eV) over the pertinent range of $0 < T < 300$ K. At $T = 300$ K, $E_g = 1.124$ eV.

### C. EBIC Generation and Contrast Calculations

Our initial model assumes a spherical electron beam generation volume, which is reduced to an equivalent cylindrical volume for the 2D models used here. Following Ref. [17], an electron beam of energy $E_b$ generates $N_{eh}$ electron-hole pairs per incident electron according to,

$$N_{eh} = \frac{E_b}{E_{eh}}\left(1 - \gamma \frac{E_{bs}}{E_b}\right), \quad (11)$$

where $E_{eh} = 3.6$ eV is the ionization energy for an electron-hole pair in Si, $\gamma$ is the back-scattering coefficient, and $E_{bs}$ is the mean energy of the back-scattered electrons. The second term in parenthesis is approximately 0.1 for Si. The total rate of e-h pair generation, $I_b N_{eh}/e$, depends on the given e-beam current $I_b$, where $e$ is the elementary charge.

The electron beam penetration depth is related to the beam energy (in eV) and material density $\rho_{Si}$ (in g/cm$^3$) by,

$$R_e = \frac{2.41 \times 10^{-11}}{\rho_{Si}} E_b^{1.75}, \text{ [cm]} \quad (12)$$

Assuming a spherical generation volume, $V_{sph}$, of diameter $R_e$, the generation rate per unit volume can be obtained from,

$$G_e = \frac{I_b N_{eh}}{e V_{sph}}, \quad (13)$$

where $e$ is the elementary charge. With experimentally relevant $E_b = 30$ keV and $I_b = 3.77$ nA, we obtain $N_{eh} = 7500$, $R_e = 7$ μm, and $G_e \approx 10^{24}$ cm$^{-3}$ s$^{-1}$. The generation rate within the specified volume is fed into the semiconductor equations to calculate output current and other observables.

Measured EBIC values depend on the collection efficiency of the device, $\eta_e$, according to $I = I_b N_{eh} \eta_e$. The presence of defects can decrease the collection efficiency and cause differences in the measured currents. Such differences are quantified by the contrast,

$$C = (I_0 - I)/I_0, \qquad (14)$$

where $I_0$ is the defect-free reference current. An increase in the recombination rate results in an increase of the contrast. Our EBIC simulation tool combines the above described device modeling and defect physics to calculate $C(T, I_b)$.

## V. RESULTS

We first consider the Schottky contact device with a dislocation oriented parallel to the top surface (not intersecting the Schottky junction) at some unknown depth. Measured values of the contrast as a function of beam current, $I_b$, at two temperatures, $T = 80$ K and 300 K are shown in Fig. 2 [3]. Our model calculations include a dislocation at a depth of 3.5 μm below the Schottky contact with the parameter values given in Table II for label D1. Shallow state concentrations were not tunable parameters for this case since they were estimated using the effective mass approximation, $N^{(de)} = \left[8m^{(de)}(E_c - E^{(de)})\right]^{1/2}/h$ and $N^{(dh)} = \left[8m^{(dh)}(E^{(dh)} - E_v)\right]^{1/2}/h$, where $m^{(de)}$ and $m^{(dh)}$ are effective masses and $h$ is Planck's constant [1]. Fitting was accomplished by adjusting $N^{(m)}$ and $E^{(de)} = E^{(dh)}$ until both curves fit the data. All other defect parameters were fixed and are provided in Table I. There is good agreement between the model and the data. For this scenario, with the deep level defect concentration much greater than that of the shallow states, $N^{(m)} \gg N^{(de)}, N^{(dh)}$, it is clear that the contrast decreases with increasing $T$ and $I_b$.

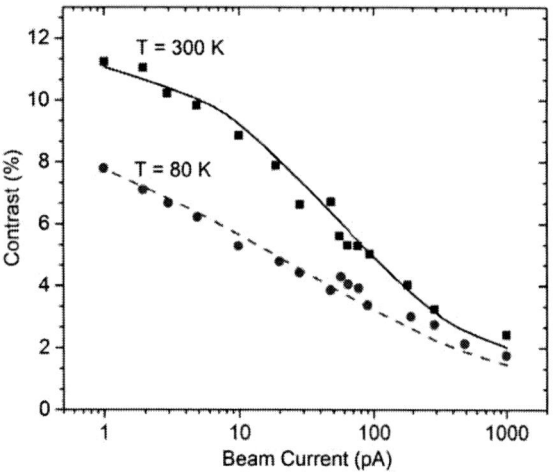

Figure 2. Contrast as a function of e-beam current at 300 K and 80K. Data (points) are from Ref. [3] and model calculations (lines) include the parameter values in Table I and labeled D1 in Table II.

Next, we turn to our EBIC measurements of typical mc-Si solar cells. EBIC contrast images at $T = 295$ and 175 K in Fig.3 clearly indicate that the GB contrast increases with decreasing temperature. The image also shows several junction shunts

(identified with additional microscopy not shown here) that exhibit the opposite behavior (i.e. lower contrast at lower temperature).

### TABLE II
#### VARIABLE DEFECT PARAMETER VALUES

| Parameter | Unit | D1 | GB1 | GB2 |
|---|---|---|---|---|
| $N^{(m)}$ | cm⁻² | $8.0\times10^{14}$ | $1.0\times10^{10}$ | $1.0\times10^{10}$ |
| $N^{(de)}$ * | cm⁻² | $1.6\times10^{12}$ | $5.4\times10^{10}$ | $9.0\times10^{10}$ |
| $N^{(dh)}$ * | cm⁻² | $2.1\times10^{12}$ | $9.0\times10^{10}$ | $1.5\times10^{11}$ |
| $E^{(de)}, E^{(dh)}$ | eV | 0.065 | 0.085 | 0.140 |

Note: * For D1, values of $N^{(de)}$ and $N^{(dh)}$ were fixed by $E^{(de)}$ and $E^{(dh)}$, as described in the text.

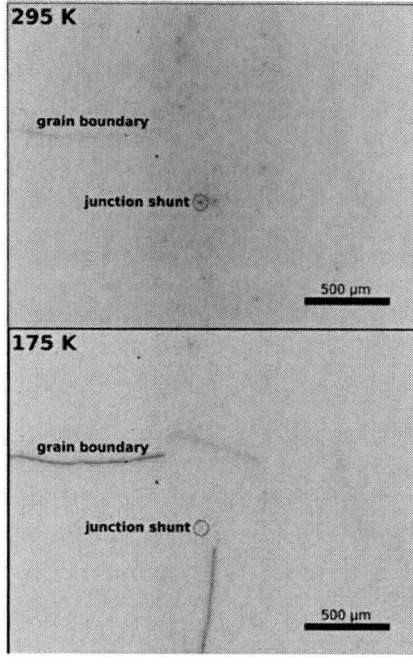

Figure 3. EBIC contrast images of a mc-Si solar cell sample at 295 K and 175 K. The contrast of the labeled grain boundary increases while that of the junction shunt decreases as the temperature decreases. SEM beam conditions were 30-kV and 3.77 nA-current.

More detailed EBIC contrast measurements, $C(T)$, at three different GBs identified in a single mc-Si PV sample are shown in Fig. 4. Best fits to the data were achieved with the parameter values given in Table II for labels GB1 and GB2 (GB2 represents the general behavior of the two GBs exhibiting higher contrast). Given that we consider GBs to be arrays of dislocations, other defect parameters were the same as in Table I. In this case, there is a nearly uniform density of defect states, $N^{(m)} \approx N^{(de)}, N^{(dh)}$. From Fig. 4, we observe that, as opposed to the case of dislocation D1 in Fig. 2, the contrast decreases with increasing temperature. Our results are in agreement with previous observations that EBIC contrast increases with temperature for heavy contamination and vice versa for relatively "clean" dislocations. Fig. 4 also indicates that the contrast is greater when the density and depth of shallow states is greater. It is interesting to note that our calculations show that the band bending at the GB changes from repelling minority carriers at high temperature to attracting them at low temperature and the temperature at which that reversal occurs depends on shallow state parameters.

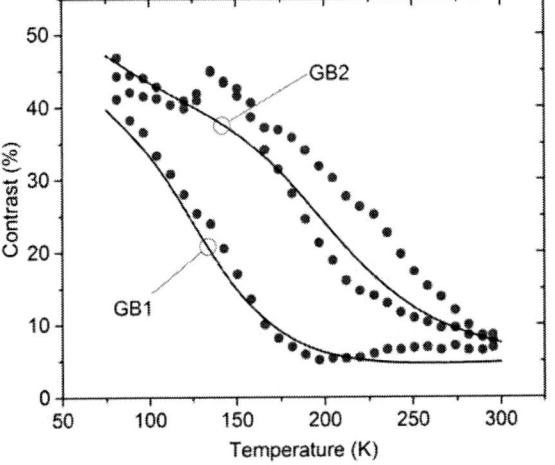

Figure 4. Contrast data (points) as a function of temperature at three grain boundaries in a mc-Si device. Model results (lines) with two different gap state distributions GB1 and GB2 in Table II.

The effects of the GB on device performance was simulated by including GB1 in our 50-μm wide by 100-μm thick test model. We considered a device with $10^{16}$ cm$^{-3}$ p-type base doping, no back surface field, and lifetimes of 100 μs at 300 K under AM15G, 1-sun light. The GB-free device had an efficiency of $\eta = 16.7\%$, open-circuit voltage $V_{oc} = 0.6$ V, and $J_{sc} = 34$ mA/cm$^2$. With GB1, the metrics decreased to $\eta = 15.8\%$, $V_{oc} = 0.58$ V, and $J_{sc} = 32.5$ mA/cm$^2$ – a 5.4% relative decrease in efficiency. Hence, even a relatively "clean" GB that exhibits 5% EBIC contrast at room temperature can have a non-negligible, negative impact on performance.

The combined information from the EBIC measurements and modeling suggest that the temperature dependence of the contrast switches from decreasing with temperature to increasing with temperature as the mid-gap defect level increases from about $10^{10}$ cm$^{-2}$ to $10^{14}$ cm$^{-2}$. Our EBIC contrast data at thirteen junction shunts in a mc-Si PV sample shown in Fig. 5 suggests that the increase of contrast with temperature can be explained by a high mid-gap defect concentration. However, our model results at various mid-gap defect concentrations (lines in Fig. 5) indicate that this is not the case. It appears that this type of recombination model cannot describe the very high (30 to 60%) room-temperature contrast levels observed at the junction shunts. Another physical interpretation involves the formation of a Schottky contact at the shunt due to significant metal contamination, as was described in Ref. [4]. That concept will be explored in subsequent work.

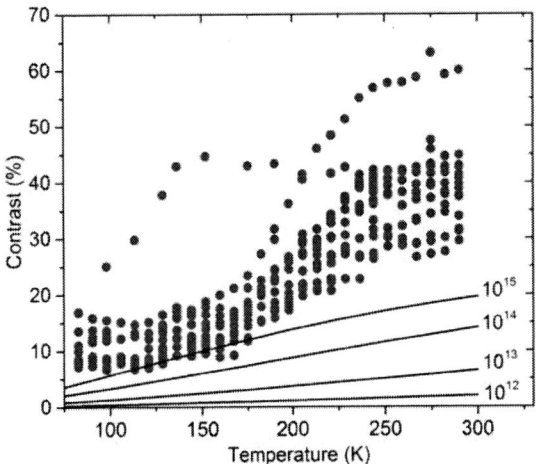

Figure 5. Contrast data (points) as a function of temperature at thirteen junction shunts in a mc-Si device. Model results (lines) with four different mid-gap defect densities shown at the lines in units of cm$^{-2}$. Other parameter values are the same as GB1 in Table II.

## VI. SUMMARY

A PV device modeling tool has been developed that can simulate EBIC measurements and device performance. The models include theories of recombination kinetics at extended defects, such as dislocations, grain boundaries, and stacking faults. Analysis of EBIC contrast as a function of temperature indicates that certain grain boundaries in mc-Si devices are comprised of nearly uniform gap states represented by shallow donor, shallow acceptor, and donor-type mid-gap states. A switch in the contrast behavior from decreasing to increasing with temperature occurs as the mid-gap defect concentration increases. Junction shunts (contaminated, near surface stacking faults) exhibit very high contrast at room temperature and the recombination processes described here do not account for that data. Future work will address junction shunts by considering Schottky barriers that form due to excessive metal contamination.

## ACKNOWLEDGEMENTS

This conference paper was developed based upon funding from the Alliance for Sustainable Energy, LLC, Managing and Operating Contractor for the National Renewable Energy Laboratory (NREL) for the U.S. Department of Energy (DOE).

## REFERENCES

[1] V. Kveder, M. Kittler, and W. Schröter, "Recombination activity of contaminated dislocations in silicon: A model describing electron-beam-induced current contrast behavior," *Phys. Rev. B*, vol. 63, no. 11, p. 115208, 2001.

[2] M. Kittler, C. Ulhaq-Bouillet, and V. Higgs, "Recombination activity of 'clean' and contaminated misfit dislocations in

Si(Ge) structures," *Mater. Sci. Eng. B*, vol. 24, no. 1, pp. 52–55, May 1994.

[3] M. Kittler and W. Seifert, "Two types of electron-beam-induced current behaviour of misfit dislocations in Si(Ge): experimental observations and modelling," *Mater. Sci. Eng. B*, vol. 24, no. 1, pp. 78–81, May 1994.

[4] M. Kittler, C. Ulhaq-Bouillet, and V. Higgs, "Influence of copper contamination on recombination activity of misfit dislocations in SiGe/Si epilayers: Temperature dependence of activity as a marker characterizing the contamination level," *J. Appl. Phys.*, vol. 78, no. 7, pp. 4573–4583, Oct. 1995.

[5] D. B. Needleman, H. Wagner, P. P. Altermatt, and T. Buonassisi, "Assessing the Device-performance Impacts of Structural Defects with TCAD Modeling," *Energy Procedia*, vol. 77, pp. 8–14, 2015.

[6] G. Stokkan, S. Riepe, O. Lohne, and W. Warta, "Spatially resolved modeling of the combined effect of dislocations and grain boundaries on minority carrier lifetime in multicrystalline silicon," *J. Appl. Phys.*, vol. 101, no. 5, p. 053515, 2007.

[7] P. P. Altermatt and G. Heiser, "Predicted electronic properties of polycrystalline silicon from three-dimensional device modeling combined with defect-pool model," *J. Appl. Phys.*, vol. 92, no. 5, pp. 2561–2574, 2002.

[8] B. Ziebarth, M. Mrovec, C. Elsässer, and P. Gumbsch, "Potential-induced degradation in solar cells: Electronic structure and diffusion mechanism of sodium in stacking faults of silicon," *J. Appl. Phys.*, vol. 116, no. 9, p. 093510, 2014.

[9] L. Meng, F.-J. Ma, J. Wong, B. Hoex, and C. S. Bhatia, "Extraction of Surface Recombination Velocity at Highly Doped Silicon Surfaces Using Electron-Beam-Induced Current," *IEEE J. Photovolt.*, vol. 5, no. 1, pp. 263–268, 2015.

[10] M. Kittler, J. Lärz, W. Seifert, M. Seibt, and W. Schröter, "Recombination properties of structurally well defined $NiSi_2$ precipitates in silicon," *Appl. Phys. Lett.*, vol. 58, no. 9, pp. 911–913, Mar. 1991.

[11] M. Kittler and W. Seifert, "Analysis of the Recombination-Active Region Around Extended Defects in Silicon," *Mater. Sci. Forum*, vol. 196–201, pp. 1123–1128, 1995.

[12] P. P. Altermatt, "Models for numerical device simulations of crystalline silicon solar cells—a review," *J. Comput. Electron.*, vol. 10, no. 3, pp. 314–330, 2011.

[13] E. B. Sokolova, "Cascade capture of carriers by a linear dislocation (Cascade capture of carriers in semiconductors with negatively charged dislocation explained by diffusion-type equation)," *Fiz. TEKHNIKA Poluprovodn.*, vol. 3, pp. 1512–1520, 1969.

[14] M. A. Green, "Intrinsic concentration, effective densities of states, and effective mass in silicon," *J. Appl. Phys.*, vol. 67, no. 6, pp. 2944–2954, 1990.

[15] D. B. M. Klaassen, "A unified mobility model for device simulation—II. Temperature dependence of carrier mobility and lifetime," *Solid-State Electron.*, vol. 35, no. 7, pp. 961–967, 1992.

[16] K. P. O'Donnell and X. Chen, "Temperature dependence of semiconductor band gaps," *Appl. Phys. Lett.*, vol. 58, no. 25, pp. 2924–2926, 1991.

[17] D. K. Schroder, *Semiconductor material and device characterization*. John Wiley & Sons, 2006.

# Colloidal Quantum Dot Solar Cell Electrical Parameter Imaging Using Camera-based High-frequency Heterodyne Lock-in Carrierography

Lilei Hu[1], Mengxia Liu[2], Andreas Mandelis[1, 2,*], Qiming Sun[1], Alexander Melnikov[1], Edward H. Sargent[2]

[1]Center for Advanced Diffusion-Wave and Photoacoustic Technologies (CADIPT), Department of Mechanical and Industrial Engineering, University of Toronto, Toronto, Ontario, M5S 3G8, Canada;
[2]Edward S. Rogers Sr. Department of Electrical and Computer Engineering, University of Toronto, Toronto, Ontario, M5S 3G4, Canada; *corresponding address: mandelis@mie.utoronto.ca

*Abstract* — **Colloidal quantum dot (CQD) solar cells with certified power conversion efficiency of 12.8 % were imaged using camera-based high-frequency heterodyne lock-in carrierography. Carrier lifetime, diffusivity, diffusion and drift lengths were imaged to investigate carrier transport behavior and Au contact/CQD interface effects in these CQD solar cells. Lower carrier lifetimes (ca. 0.5 µs and 2.3 µs at 293 K and 200 K, respectively, in agreement with literature transient photovoltage results), were found in Au contact regions due to enhanced trap states. This imaging methodology shows great potential for understanding energy loss in CQD solar cells and for non-destructive large-area characterization.**

*Index Terms* — **charge carrier lifetime, colloidal quantum dot, diffusion length, imaging, lock-in carrierography, solar cell.**

## I. INTRODUCTION

Colloidal quantum dots (CQDs) with tunable energy bandgap through effective size control have become promising candidates for fabricating low-cost, large-area, light-weight, and flexible solar cells [1-7]. Through intensive studies of device architecture engineering [1, 6, 7], surface materials chemistry [1, 2], charge carrier transport dynamics [3, 4], theoretical modeling [2, 5, 6], and synthesis methodologies [1, 7], CQD solar cell power conversion efficiencies (PCEs) have been significantly boosted, however, they are still insufficient for commercial applications. To further improve CQD solar cell efficiencies, a better understanding of CQD solar cell energy loss through factors such as CQD film/contact interfaces, inefficient electrode collection of carriers, and trap state induced recombinations is necessary and substantially relies on advanced characterization techniques.

However, single-detector (small-spot testing) based conventional techniques for characterizing carrier transport parameters including carrier lifetime, mobility and diffusion length yield unreliable estimates of these parameters in solar cells. Moreover, the widely used transient methodologies, including photoluminescence (PL) [8], µ-PCD [9] and short-circuit current and open-circuit voltage decay [10], are limited by their low signal-to-noise ratio (SNR) when the carrier lifetime is extremely low, for example, in CQD systems. Additionally, large-area quasi-steady-state imaging techniques, such as time-resolved [11] and low-frequency PL used to construct dynamic carrierography images [12], and lock-in

thermography (LIT) [13], are restricted by low camera frame rate and exposure time. This implies that only images at low frequencies can be produced, resulting in low image resolution and the inability to measure fast carrier transport processes. To address these issues, this paper introduces camera-based high-frequency heterodyne lock-in carrierography (HeLIC), a frequency-domain spectrally-gated PL technique for large-area (photo) carrier density-wave (CDW) transport property distribution imaging. Through creating a slow enough beat frequency component, HeLIC overcomes the frequency range limitation of conventional camera-based optical characterization techniques. With a high SNR, HeLIC can attain frequency-dependent AC carrier diffusion lengths to generate depth-selective/resolved high-frequency images of carrier transport parameters in large-scale devices.

Specifically, to study the carrier transport dynamics and their dependence on contact/film interfaces in CQD solar cells, we combined a current-voltage (*J-V*) model with HeLIC to produce carrier lifetime, diffusivity, and drift and diffusion length images for high-efficiency CQD solar cells. This methodology has a potential impact on solar cell efficiency optimization and on real-time non-destructive inspection for quality control of industrial solar cells.

## II. EXPERIMENTAL METHODS

Oleic-acid-capped CQDs and ZnO nanoparticles were synthesized following previously published methods [7]. As shown in Fig.1, the sandwich-structured CQD solar cells consist of two types of CQDs that are surface-capped with two different ligands. Briefly, ZnO nanoparticles were first spin-casted onto an ITO glass, followed by the deposition of $PbX_2$/XX-exchanged ($PbX_2$: lead halide, AA: ammonium acetate) then EDT-exchanged PbS CQDs. Eventually, a 120-nm Au thin film was thermally evaporated atop to form the electrodes. The *J-V* characteristics were obtained using the Keithley 2400 source measuring unit under simulated AM1.5 illumination (Sciencetech class A) in a nitrogen environment.

The experimental setups for HeLIC and single-detector based photocarrier radiometry (PCR) are identical to our previous reports [2, 4]. Briefly, a fiber-coupled diode laser of 808 nm wavelength was used with the mean power adjusted to be 1 sun through optical diffusers. A high-speed NIR InGaAs camera (Goodrich SU320 KTSW-1.7RT/RS170) was used for imaging. The influence of the excitation beam laser was eliminated through a long-pass filter (Spectrogon LP-1000 nm) which was mounted in front of the InGaAs camera.

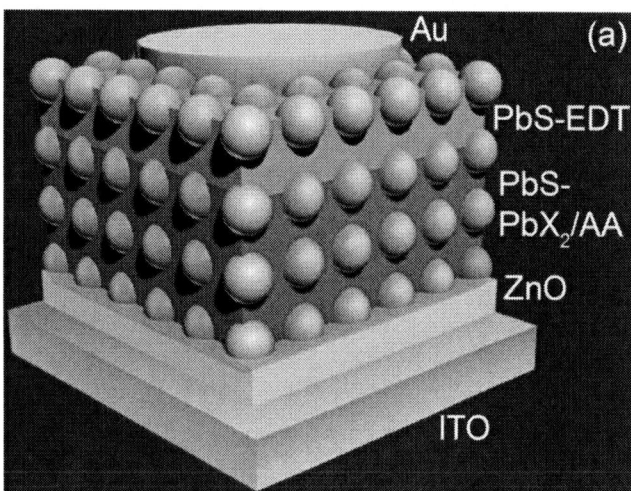

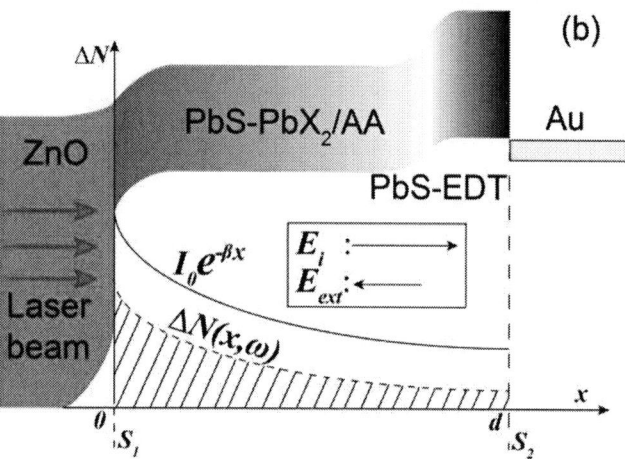

Fig. 1.    Schematic of the CQD solar cell structure of this study (a), and the corresponding energy band structure (b) with the illumination of light absorption, excess carrier diffusion, intrinsic and external electric fields shown.

### III. THEORETICAL METHOD

As an extension of the conventional current-voltage characterization of CQD solar cells under DC laser excitation [5, 6], a carrier transport model under frequency-modulated photoexcitation was derived. Due to the larger energy bandgap of ZnO than the incident excitation laser energy as shown in Fig.1(b), and because the CQD layers are thicker than the ZnO layer, Fig.1(a), charge carriers are considered to be generated only in the CQD layers, i.e. CQDs act as the light absorbers. The rate equation for electrons in the nominally p-type CQD layers can be written as

$$\frac{\partial \Delta n(x,t)}{\partial t} = \frac{\partial J_e(x,t)}{\partial x} - \frac{\Delta n(x,t)}{\tau} + g(x,t) \quad (1)$$

with $\Delta n(x, t)$ the excess electron density, and $J_e(x, t)$ the electron current flux. The term $\tau$ is the minority carrier (electron) lifetime, and $g(x, t)$ is the carrier generation rate. Considering the ambipolar diffusion coefficient and mobility, $J_e(x, t)$ can be further defined by

$$J_e(x,t) = D_e \frac{\partial \Delta n(x,t)}{\partial x} + \mu_e E \Delta n(x,t) \quad (2)$$

where $D_e$ is the carrier diffusivity, $\mu_e$ is the mobility, and $E$ is the electric field ($E=E_i+E_{ext}$). Furthermore, the excess carrier generation under harmonic excitation follows the Beer-Lambert law:

$$g(x,\omega) = \frac{\beta \eta I_0}{2h\nu} e^{-\beta x} \left(1 + e^{i\omega t}\right) \quad (3)$$

Here, $\beta$ is the optical absorption coefficient, $\eta$ is the quantum yield of the photogenerated carriers, $h$ and $\nu$ are the Planck constant and the optical frequency of the incident photons, respectively. $I_0$ denotes the incident photon intensity. In the one-dimensional geometry, the boundary conditions at $x = 0$ and $d$ can be written as functions of the surface recombination velocities ($S_1$ and $S_2$, respectively, at $x = 0$ and $x = d$) and the excess carrier density at the corresponding boundaries [5], Fig.1. As given by (A1) in the frequency domain, $\Delta n(x, t)$ in (1) can be solved through the Green function and Fourier transform methods [14]. As shown in Fig.1 (b), when the net electric field is along the direction of the intrinsic depletion-induced electric field $E_i$, the minority CDW diffusion to $x = d$ is hampered, which generally occurs with solar cells at low external applied voltages.

Camera-based HeLIC is sensitive to the radiative recombination of the CDW, which has a recombination rate $R_r$, proportional to the concentration of electrons and holes, i.e. $R_r \approx np$ ($n$ and $p$ are the concentrations of electron and hole charge carriers) [15]. HeLIC is a depth integral of the radiative recombination of excess CDW along the thickness of the solar cell device, i.e.

$$S(t) = \int_0^d dx \int_{\lambda_1}^{\lambda_2} \Delta N(\omega,x)\left[\Delta P(\omega,x) + N_A\right]F(\lambda)d\lambda \quad (4)$$

where $\Delta P(\omega, x)$ is the excess hole CDW which equals $\Delta N(\omega, x)$ according to the quasi-neutral approximation. The term $N_A$ is the majority carrier concentration at equilibrium and the subscript A indicates the nominal p-type nature of the CQD layer, Fig. 1(b). The wavelength dependent PL is accounted through $F(\lambda)$, which is an instrumental coefficient determined by the detection bandwidth ($\lambda_1, \lambda_2$) of the InGaAs camera and the long-pass filter.

As for HeLIC, the incident laser excitation is modulated at two different angular frequencies $\omega_1$ and $\omega_2$. Therefore, the excess carrier concentration for sine-wave modulated photoexcitation is given by

$$\Delta N(\omega, x) = 2\Delta n_0(x) + A(x, \omega_1)\cos\left[\omega_1 t + \varphi(x, \omega_1)\right]$$

$$+ A(x, \omega_2)\cos\left[\omega_2 t + \varphi(x, \omega_2)\right] \quad (5)$$

Here $\Delta n_0(x)$, $A(x, \omega)$, and $\varphi(x, \omega)$ are the DC component, AC amplitude and phase of the photogenerated excess electron CDW $\Delta N(x, \omega)$, respectively. The lock-in amplifier based HeLIC collects CDW oscillated at a reference beat frequency $\Delta\omega = \left|\omega_2 - \omega_1\right|$. Hence, the HeLIC signal $S_{he}(\Delta\omega)$ can be expressed as

$$S_{he}(\Delta\omega) = \int_0^d A_1 A_2 \cos\left[\Delta\omega t + \Delta\varphi(x, \omega_1, \omega_2)\right] dx \quad (6)$$

Here, $A_1$, $A_2$, and $\Delta\varphi(\omega, x)$ denote $A(x, \omega_1)$, $A(x, \omega_2)$, $\varphi(x, \omega_2)$ - $\varphi(x, \omega_1)$, respectively. Furthermore, (6) can be expressed by $\Delta N(\omega, x)$ through

$$S_{he}(\Delta\omega) = \int_0^d \Delta N^*(x, \omega_1)\Delta N(x, \omega_2) dx \quad (7)$$

where * denotes complex conjugation. With a reference beat frequency as low as 10 Hz, the HeLIC phase is usually very small, on the order of $(10^{-3})^\circ$. Substituting (A1) into (7), the final demodulated signal for HeLIC becomes

$$S_{he}(\Delta\omega) \propto B(\omega_1)^* B(\omega_2) S_{he}(\omega_1, \omega_2) \quad (8)$$

where $B$ is defined as (A2a) in the Appendix. Although the analytical expression of $S_{he}(\omega_1, \omega_2)$ has been derived, it can also be computed using various commercial engineering software programs such as Matlab with an integration function.

## IV. RESULTS AND DISCUSSION

Our CQD solar cells with a sandwich structure (Fig.1) have a dimension of 25 mm × 25 mm, a major portion of which is presented in the inset of Fig.2 with the depiction of the gold-colored top and bottom Au contacts and a brown-colored CQD thin film layer. As an example, Fig.2 shows the current density vs. voltage characteristics of two solar cell units A and B that exhibit high power conversion efficiency (PCE) of 9.0 % and 8.5 %, respectively. The highest PCE of our CQD solar cells was certified to be 12.8% [7].

The frequency-dependent AC diffusion length enables the characterization of photovoltaic device properties at different depths using HeLIC. A comparison of HeLIC images at 1 kHz and 100 kHz is presented in Fig.3, which qualitatively reveals depth-resolved HeLIC images through the evolution of CQD solar cell image patterns with increased frequency. In other words, at the low frequency of 1 kHz, the longer AC diffusion

length is expected to yield different image contrast from HeLIC images at the higher frequency, 100 kHz, which corresponds to a shorter AC diffusion length. Importantly, the HeLIC images reveal the inhomogeneities of the CQD solar cell and the quality of each CQD solar cell unit.

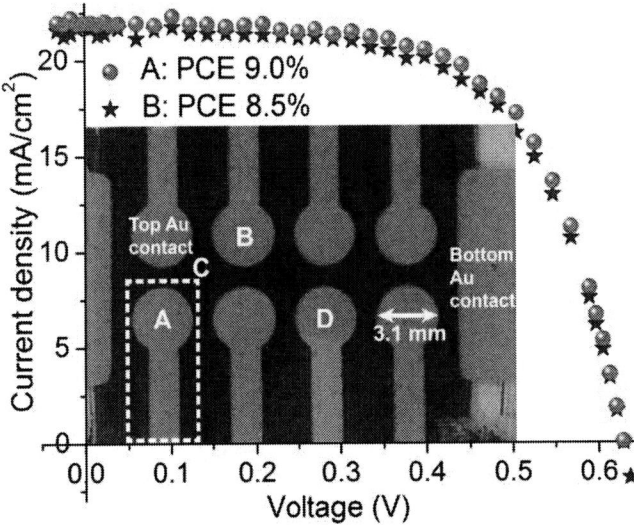

Fig. 2. Current density vs. voltage characteristics and photography (inset) of a CQD solar cell sample with the structure shown in Fig.1.

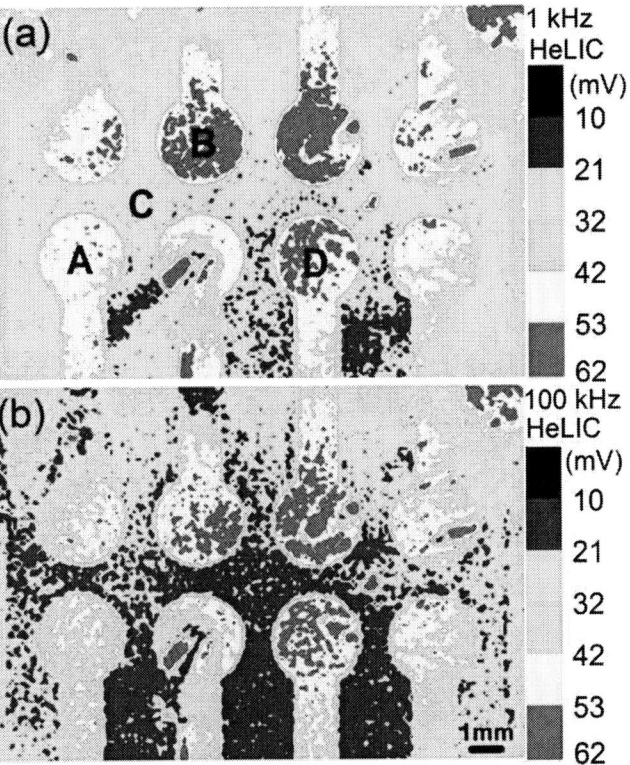

Fig. 3. High-frequency HeLIC images at 1 kHz (a) and 100 kHz (b) for the CQD solar cell shown in Fig.2, inset.

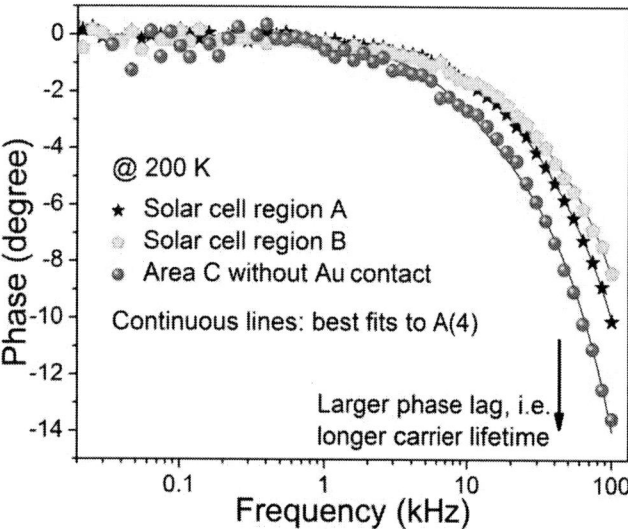

Fig. 4. Frequency-dependent PCR phase spectra of the solar cell electrode units A, B, and area C without Au contact (Figs.2 and 3) at 200 K. (A4) is used for the best fitting of each curve. The characterization spot area of the single-detector based PCR is the same as the area of the circular Au contact tip.

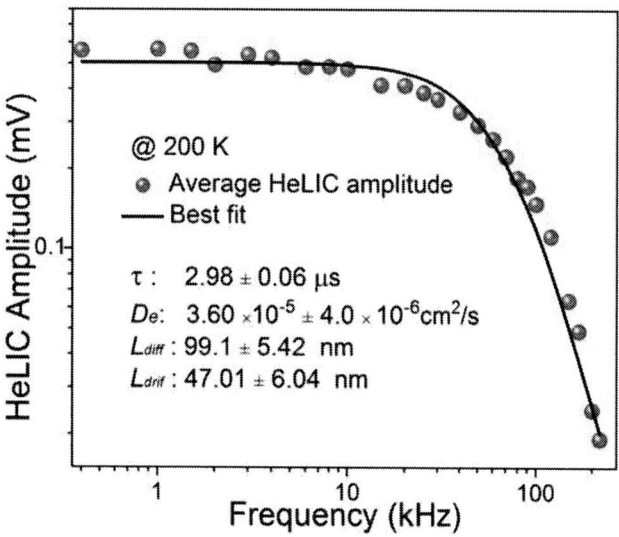

Fig. 5. The frequency-dependent average HeLIC image amplitudes of the CQD solar cell shown in the Fig.2 inset. The HeLIC images in Fig.3 are also included.

For the investigation of contact effects on carrier lifetime, photocarrier radiometry (PCR) [2, 3, 4], a single-detector-element based counterpart of HeLIC was used in the homodyne (single-modulation-frequency) mode. The PCR single-element InGaAs detector can detect a spot area equal to the size of the circular tip of a solar cell electrode unit (e.g. A in Figs.2 and 3) and can measure the average carrier lifetime in a spot region. In parallel with its imaging counterpart HeLIC, the demodulated PCR signal has been derived as (A4) in the

Appendix. The large phase lag in PCR phase with increased frequency indicates longer carrier lifetimes. Therefore, as directly shown in Fig.4, according to the larger phase lag at high frequencies, area C without contact Au deposition exhibits a longer carrier lifetime than its counterparts in regions A and B, Figs. 2 and 3. Similarly, region A presents a slightly longer lifetime than B. The lifetime difference between regions A, B, and the area C is manifested by the quantitative fitting of experimental frequency-dependent PCR phases to (A4) and are consistent with the results shown in Fig. 6(b).

In comparison, as a large-area imaging technique, HeLIC can image an entire solar cell sample. The dependence of HeLIC amplitudes on modulation frequency is shown in Fig. 5 in which each point is the average amplitude of an entire HeLIC image at the corresponding frequency. The best fitting of the data in Fig. 5 into (8) yields the overall carrier transport parameters for the CQD solar cell measured at 200 K, i.e. carrier lifetime $\tau = 2.98 \pm 0.06$ μs, diffusivity $D_e = 3.60 \times 10^{-5} \pm 4.00 \times 10^{-6}$ cm$^2$/s, diffusion length $L_{diff} = 99.10 \pm 5.42$ nm, and drift length $L_{drif} = 47.01 \pm 6.04$ nm. Compared with their room temperature (293 K) counterparts, except for $\tau$, other carrier transport parameters are smaller at 200 K: at 293 K, $D_e$ is on the order of $10^{-3}$ cm$^2$/s, and $L_{diff}$ and $L_{diff}$ are around 400 nm. The carrier lifetime $\tau$ decreases to around 500 ns when the temperature increases from 200 K to 293 K, a phenomenon attributed to increased nonradiative recombination and consistent with phonon-assisted carrier hopping transport within spatial and energy disordered CQD systems [3-6].

Due to trap states and material energy band bending at semiconductor/metal electrode interfaces, carriers exhibit different transport behavior and have been found to degrade CQD solar cell performance [6,7]. Therefore, contact-associated interface effects are of great interest and we constructed large-area images of carrier transport parameters as shown in Fig. 6 using HeLIC. A 400-Hz HeLIC image of solar cell region D (Figs. 2 and 3) reveals higher HeLIC amplitudes in regions where Au contacts were deposited, indicating an inhomogeneous carrier distribution.

The carrier transport parameter images for CQD solar cell region D, Fig. 6, were reconstructed through fitting 27 HeLIC images to (8). These HeLIC images were taken at various frequencies between 400 Hz and 270 kHz. Therefore, the carrier lifetime image of electrode D at 200 K was constructed as shown in Fig. 6(b). Regions with Au contacts exhibit shorter $\tau$ of ca. 2.3 μs than the surrounding regions. This can be ascribed to the enhanced interface-induced defects and traps that decrease carrier lifetime through increased non-radiative recombinations. This finding is consistent with the PCR phase study of carrier lifetimes in regions with and without Au coating as shown in Fig.4. For comparison, at 293 K the carrier lifetime $\tau$ image of the same electrode also yielded a shorter $\tau$ of ca. 0.5 μs in Au regions than the lifetime outside the Au/CQD interfaces, Fig. 6(c). Comparison between the lifetime images at 293 K and 200 K revealed that the increased carrier lifetime at the low temperature is due to the reduced carrier-phonon interactions which act as necessary phonon-mediation pathways for trap state related non-radiative recombination [2-4]. The resultant lifetime values are in agreement with experimental results for PbS CQD solar cells reported in ref. [10] using the transient photovoltage method.

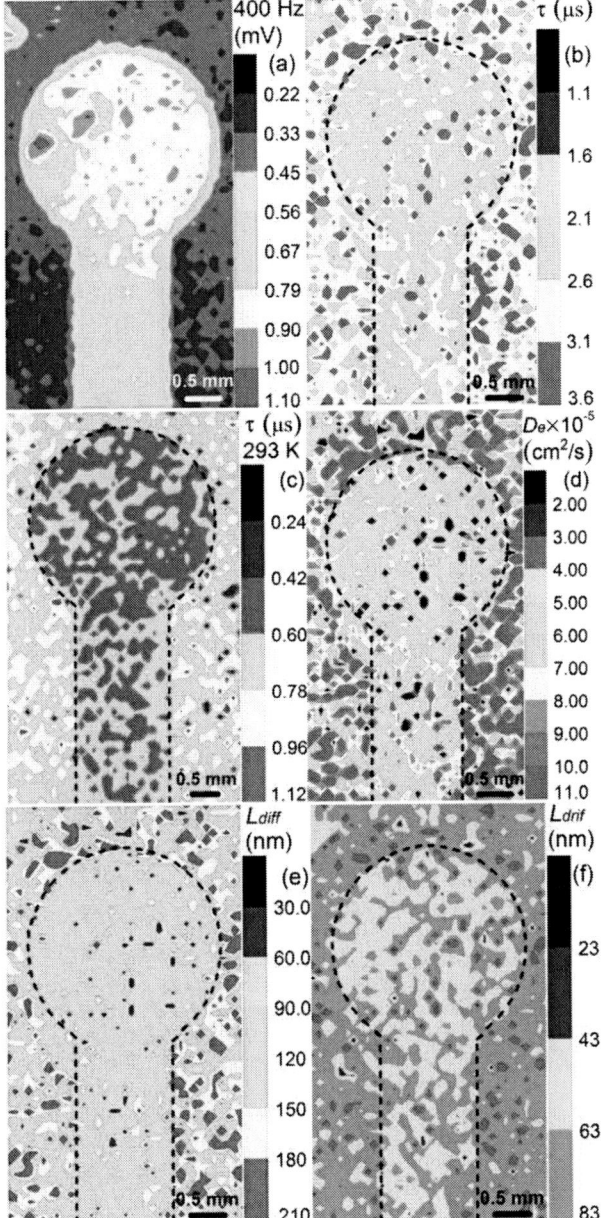

Fig. 6. (a) 400 Hz HeLIC image of the CQD solar cell region D, Fig. 2(a), and its carrier lifetime $\tau$ image (b) at 200 K. (c) For comparison, the carrier lifetime image of the same electrode D at 293 K. (d)-(f) are images of carrier diffusivity, diffusion and drift lengths, respectively, at temperature 200 K.

Furthermore, for electrode D (Figs. 2 and 3) at 200 K, carrier diffusivity, and diffusion and drift lengths were also

obtained as shown in Figs.6(d)-(f). Specifically, the carrier diffusivity, Fig. 6(d), was imaged to be on the order of $10^{-5}$ cm$^2$/s which is much smaller than its room temperature counterpart (ca. $10^{-3}$ cm$^2$/s). Fig. 6(d) also shows the effects of the Au/CQD interfaces on the carrier diffusivity with a lower average $D_e$ in the Au region. Interface-induced trap states can trap, de-trap, or recombine carriers, a process that inhibits carrier hopping diffusion transport. Therefore, with the extraction of $\tau$ and $D_e$, $L_{diff} = \sqrt{D_e \tau}$ was also reconstructed to be approx. 120 nm, which is much shorter than ca. 400 nm at room temperature. The reduced $L_{diff}$ at low temperature is attributed to the decreased availability of thermal energy for the phonon-assisted carrier hopping transport within the CQD thin film [4-6]. With the presence of interface trap states or defects, carriers in pure CQD layers can be transported about 30 nm longer than those in Au regions, hopping across ~ 10 more QDs. As shown in Fig. 6(f), the effects of interface traps are also substantiated through carrier drift length $L_{drif}$ images using HeLIC, i.e., lower carrier drift lengths of ca. 50 nm in Au regions are obtained than in other regions.

## VI. CONCLUSIONS

In this paper, heterodyne lock-in carrierographic imaging was introduced as an emerging dynamic interface imaging methodology of CQD solar cells. The highest frequency (270 kHz) and widest frequency range to-date (0.4 – 270 kHz) images were reported and were used to reconstruct carrier transport parameter images. Compared with regions without Au contacts, enhanced trap state densities at the Au/CQD interface induced decreased minority carrier lifetime (ca. 2.3 μs) and carrier diffusivity (ca. $6\times10^{-5}$ cm$^2$/s) at 200 K, which resulted in shorter carrier diffusion and drift lengths. These HeLIC transport property images in CQD solar cells can provide new insights into effects of solar cell fabrication on PCE and can help with optimization purposes. HeLIC imaging thus shows excellent potential for industrial inline photovoltaic device characterization.

## ACKNOWLEDGEMENT

The authors are grateful to the Natural Sciences and Engineering Research Council of Canada (NSERC) for a Discovery grant to A.M., and to the Canada Research Chairs program. L.H. appreciates the MIE Graduate Student Travel Grant and the University of Toronto SGS Conference Grant.

## APPENDIX

Combining (1)-(3) and the surface-recombination-velocity associated boundary conditions as discussed in Sect. III, the excess carrier density can be derived using the Green function method [14] and is given by

$$\Delta N(x,\omega) = B \left\{ \begin{array}{l} C \exp\left(-K_e x\right) \\ +D \exp\left[-K_e\left(2d-x\right)\right] \\ -E \exp\left(-\beta x\right) \end{array} \right\} \quad \text{(A1)}$$

with the following definitions

$$B = \frac{\eta I_0 \beta}{4 h \nu D_e \left(1 - R_{e1} R_{e2} e^{-2K_e d}\right)\left[\beta^2 - \left(Q_0^2 + \sigma_e^2\right)\right]} \quad \text{(A2a)}$$

$$C = \left[\left(1+\rho_e\right) - R_{e1}\left(1-\rho_e\right)\right]$$
$$- R_{e1} \left[ \begin{array}{l} \left(\rho_e - 1\right) + \\ R_{e2}\left(1+\rho_e\right) \end{array} \right] \exp\left[-\left(K_e + \beta\right)d\right] \quad \text{(A2b)}$$

$$D = R_{e2}\left[\left(1+\rho_e\right) - R_{e1}\left(1-\rho_e\right)\right]$$
$$+ \left[ \begin{array}{l} \left(1-\rho_e\right) - \\ R_{e2}\left(1+\rho_e\right) \end{array} \right] \exp\left[-\left(\beta - K_e\right)d\right] \quad \text{(A2c)}$$

$$E = 2\left[1 - R_{e1} R_{e2} \exp\left(-2K_e d\right)\right] \quad \text{(A2d)}$$

and

$$Q_0 = \frac{\mu_e}{2D_e} E \left[cm^{-1}\right] \quad \text{(A3a)}$$

$$\sigma_e = \sqrt{\frac{1+i\omega\tau}{D_e \tau}} \left[cm^{-1}\right] \quad \text{(A3b)}$$

$$R_{e,j} = \frac{D_e \sqrt{Q_0^2 + \sigma_e^2} - S_j}{D_e \sqrt{Q_0^2 + \sigma_e^2} + S_j}, \quad j = 1,2 \quad \text{(A3c)}$$

$$K_e = \sqrt{Q_0^2 + \sigma_e^2} - Q_0 \left[cm^{-1}\right]; \quad \rho_e = \frac{\beta}{K_e} \quad \text{(A3d)}$$

The demodulated signal generation expression $S_{ho}(\omega)$ for PCR in the homodyne mode is given by

$$S_{ho}(\omega) \approx B N_A \left[ \begin{array}{l} \dfrac{C\left(1-e^{-K_e d}\right) + D\left(e^{-K_e d} - e^{-2K_e d}\right)}{K_e} \\ + \dfrac{E\left(e^{-\beta d} - 1\right)}{\beta} \end{array} \right]$$
$$\text{(A4)}$$

## REFERENCES

[1] X. Wang, G.I. Koleilat, J. Tang, H. Liu, I.J. Kramer, R. Debnath, L. Brzozowski, D.A.R. Barkhouse, L. Levina, S. Hoogland, and E.H. Sargent, "Tandem colloidal quantum dot solar cells employing a graded recombination layer," *Nature Photonics*, 5(8), pp.480-484, 2011.

[2] L. Hu, Z. Yang, A. Mandelis, A. Melnikov, X. Lan, G. Walters, S. Hoogland, and E.H. Sargent, "Quantitative analysis of trap-state-mediated exciton transport in perovskite-shelled PbS quantum dot thin films using photocarrier diffusion-wave nondestructive evaluation and imaging," *The Journal of Physical Chemistry C*, 120(26), pp.14416-14427, 2016.

[3] L. Hu, A. Mandelis, A. Melnikov, X. Lan, S. Hoogland, and E.H. Sargent, "Study of exciton hopping transport in PbS colloidal quantum dot thin films using frequency- and temperature-scanned photocarrier radiometry," *International Journal of Thermophysics*, 38(1), p.7. 2017.

[4] L. Hu, A. Mandelis, Z. Yang, X. Guo, X. Lan, M. Liu, G. Walters, A. Melnikov, and E.H. Sargent, "Temperature-and ligand-dependent carrier transport dynamics in photovoltaic PbS colloidal quantum dot thin films using diffusion-wave methods," *Solar Energy Materials and Solar Cells*, 164, pp.135-145, 2017.

[5] A. Mandelis, L. Hu, and J. Wang, "Quantitative measurements of charge carrier hopping transport properties in depleted-heterojunction PbS colloidal quantum dot solar cells from temperature dependent current–voltage characteristics," *RSC Advances*, 6(95), pp.93180-93194, 2016.

[6] L. Hu, A. Mandelis, X. Lan, A. Melnikov, S. Hoogland, and E.H. Sargent, "Imbalanced charge carrier mobility and Schottky junction induced anomalous current-voltage characteristics of excitonic PbS colloidal quantum dot solar cells," *Solar Energy Materials and Solar Cells*, vol.155, pp.155-165, 2016.

[7] M. Liu, O. Voznyy, R. Sabatini, F.P.G. de Arquer, R. Munir, A.H. Balawi, X. Lan, F. Fan, G. Walters, A.R. Kirmani, S. Hoogland, F. Laquai, A. Amassian, and E.H. Sargent, "Hybrid organic-inorganic inks flatten the energy landscape in colloidal quantum dot solids," *Nature Materials*, 16, 258–263, 2017.

[8] K. Wang, W. McLean, and H. Kampwerth, "Transient photoluminescence from silicon wafers: Finite element analysis," *Journal of Applied Physics*, 114(16), 163105-1-8, 2013.

[9] E. Gaubas, and A. Kaniava, "Determination of recombination parameters in silicon wafers by transient microwave absorption," *Review of Scientific Instrumentations*, 67, pp.2339–2345, 1996.

[10] A.K. Rath, M. Bernechea, L. Martinez, F.P.G. De Arquer, J. Osmond, and G. Konstantatos, "Solution-processed inorganic bulk nano-heterojunctions and their application to solar cells," *Nature Photonics*, 6(8), pp.529-534, 2012.

[11] D. Kiliani, G. Micard, B. Steuer, B. Raabe, A. Herguth, and G. Hahn, "Minority charge carrier lifetime mapping of crystalline silicon wafers by time-resolved photoluminescence imaging," *Journal of Applied Physics*, 110(5), 054508-1-7, 2011.

[12] A. Mandelis, Y. Zhang, and A. Melnikov, "Statistical theory and applications of lock-in carrierographic image pixel brightness dependence on multi-crystalline Si solar cell efficiency and photovoltage," *Journal of Applied Physics*, 112(5), pp. 054505. 2012.

[13] J. Bachmann, C. Buerhop-Lutz, C. Deibel, I. Riedel, H. Hoppe, C.J. Brabec, and V. Dyakonov, "Organic solar cells characterized by dark lock-in thermography," *Solar Energy Materials and Solar Cells*, 94(4), pp.642-647, 2010.

[14] A. Mandelis, *Diffusion-wave fields: mathematical methods and Green functions*. Springer, New York (2001).

[15] M.A. Green, *Solar cells: operating principles, technology, and system applications*. Prentice-Hall, Englewood Cliffs, Chapter 3, pp. 51, 1982.

# A New Perspective on Potential-Induced Degradation of the Shunting Type by Micro Raman-Spectroscopy and Micro Light-Beam-Induced Current

A. Büchler[1], H. Nagel[1], M. Breitwieser[2], S. Kluska[1], F. D. Heinz[1], M. C. Schubert[1], M. Glatthaar[1] and S. Glunz[1]

[1] Fraunhofer Institute for Solar Energy Systems, Heidenhofstraße 2, 79110 Freiburg, Germany
[2] Hahn-Schickard Gesellschaft, Georges-Koehler-Allee 103, 79110 Freiburg, Germany

*Abstract* — Solar cells showing shunts due to potential-induced degradation (PID-s) were investigated by combining electron-beam induced current (EBIC), microscopic light beam induced current (µLBIC) and microscopic Raman spectroscopy (µRS) at the same positions with high local resolution. A direct correlation of compressive stress measured by µRS and local shunting as observed by µLBIC was found. Comparing µRS mappings before and after degradation proves that stress is induced locally by PID-s which can be explained by the formation of sodium decorated stacking faults.

## I. INTRODUCTION

Shunting due to potential-induced degradation (PID) significantly lowers the performance of crystalline silicon solar cell modules. The effect was attributed to stacking faults that are decorated with sodium [1, 2]. So far, shunts were localized and characterized by electron beam induced current (EBIC) [3], dark lock in thermography (DLIT) [1], luminescence based methods such as PL and the advanced light beam induced current technique CELLO [4]. Associated crystal defects were analyzed by transmission electron microscopy (TEM) [5] and by defect selective etches [6]. In recent works the chemical composition, especially the concentration of sodium, in the shunted region was determined by EDS and different sophisticated SIMS-based methods [1]. This allowed understanding of the physical and chemical state of the PID shunts. The influence of various factors of solar cell processing such as choice of anti-reflection coating [7, 8], EVA foil or the solar cells surface was examined [4]. Nevertheless, the precise degradation mechanism is not fully understood yet. For stacking fault-like PID-shunts Naumann *et al.* were able to proof that the density of crystal defects increases during PID [6]. Furthermore they assumed microscopic crystal defects to serve as nucleation spots for ions penetrating into the silicon crystal. The accumulation of ions is expected to cause the stacking faults. Since both methods that were used to characterize the crystal quality, TEM and defect etching, are destructive, the correlation of stacking faults and PID-shunts was examined in an indirect way based on statistics. While in principle reducing global crystal stress, decorated stacking faults are expected to induce mechanical stress in their immediate vicinity. Micro Raman Spectroscopy (µRS) in combination with a confocal laser scanning microscope is a straightforward way for mapping the distribution of stress in silicon [9]. The application to characterize PID-shunts and associated decorated stacking faults with µRS has several benefits:

1. The method is non-destructive such that samples can be measured before and after degradation.

2. Localization of stress (different to shunts) does not need a fully processed solar cell, i.e. silicon wafers can be examined before diffusion.

3. On solar cell samples a simultaneous measuring of the current induced by the absorption of the excitation laser allows microscopic mapping of the beam laser induced current (µLBIC).

As already published, µLBIC may outreach EBIC in spatial resolution due to the more localized excitation volume generated by light instead of electron absorption [10]. Opposed to EBIC, µLBIC has the advantages that it does not harm the cell properties in the measured region and it does not require vacuum. In this work, we introduce the combination of µRS and µLBIC as a versatile technique for investigation of the degradation mechanism. In a first step we compare µLBIC and EBIC measurements on degraded solar cell samples, proving that PID-shunts can be localized by µLBIC as good as by EBIC. Comparing spatially correspondent µLBIC and µRS mappings shows that formation of electrical shunts due to PID goes along with formation of locally stress sites. In the last section µRS mappings of the same position before and after PID are compared to evaluate if spots of locally increased stress are detectable already before PID-s.

## II. EXPERIMENTAL

For the experiments, we used chemically polished 1 cm x 2 cm monocrystalline Si solar cell samples. The 1 $\Omega$cm boron-doped FZ Si wafers were cleaned, phosphorous diffused, HF etched and passivated on the front side with a 75 nm thick $SiN_x$ antireflection coating via plasma-enhanced chemical vapour deposition. After a temperature treatment at 900 °C for 1s (in order to simulate the temperature profile of screen-printed metal paste firing) the front side SiN was locally etched by HF. Finally, metal contacts were applied at about 300 °C. The finished solar cells were characterized by suns-$V_{oc}$ measurements in order to determine the impact of shunt effects on the pseudo-fill factor . Additionally, the distribution of residual stress on the front side was investigated by µRS-mapping. PID-s tests were performed at 75 °C for 12 h. A conductive polymer was pressed to the front side of the cells by means of a grounded metal plate whereas -800 V were applied to the cells. After PID-s, EBIC measurements were performed in a Zeiss Auriga SEM equipped with a point electronic EBIC system. µLBIC and µRS measurements were conducted in a WiTEC spectral resolving confocal microscope. This set-up allows simultaneous and spatial resolved evaluation of induced currents as well as Raman scattering spectra. A focused laser (532 nm, Nd:YAG) was used for excitation. Laser power was set to 1 mW. An x-y-table with a step size of 200 nm was used to create 2D-mappings of the region of interest. Measured Raman spectra were pixel-wise fitted with a Lorentzian. The fit values for spectral peak position are visualized in grayscale images in order to illustrate the shifts of the phonon energy within the scanned area.

## III. RESULTS AND DISCUSSION

### A. Comparison of EBIC and µLBIC mappings of PID-shunts

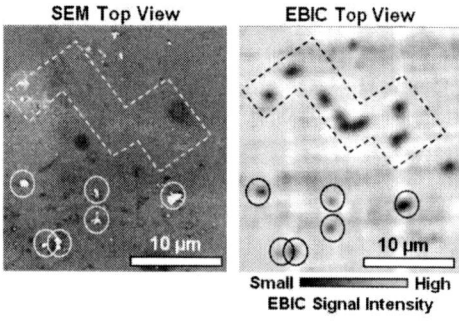

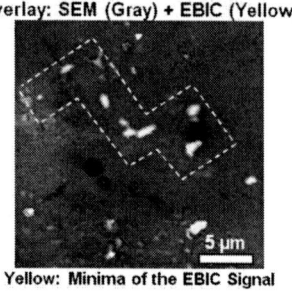

Fig. 1.    An area showing several shunts due to PID was examined in the SEM and in a confocal microscope. The first row shows an SEM image (left) and the correspondent EBIC image (right). The image in the lower row shows an overlay of SEM image (gray) and positions the minima in the EBIC intensity (yellow). The dashed line marks a region of EBIC intensity drops due to PID-s, while the circles mark particles that shade the electron beam.

EBIC is the established method for localization of PID-s shunts on the microscale. Fig. 1 shows top view SEM images of an area featuring several PID-s that are highlighted by the dashed line. The comparison of the SEM and EBIC image allows for discrimination of electron beam shading due to particles on the sample (closed circles) and reduction of the electron-beam-induced current due to electrical defects (dashed line). In similar studies on local electrical defects, µLBIC proofed to be suited for the detection of shunts as well [10, 11]. Fig. 2 shows an image of the same region acquired with an optical microscope and the correspondent µLBIC image. As highlighted by the overlays in the lower row in Figure 2, µLBIC and EBIC show the same features associated to PID-shunts. As described by other authors before, PID-s defects form along (111)-crystal planes and are characterized by a platelet like structure typical for stacking faults.

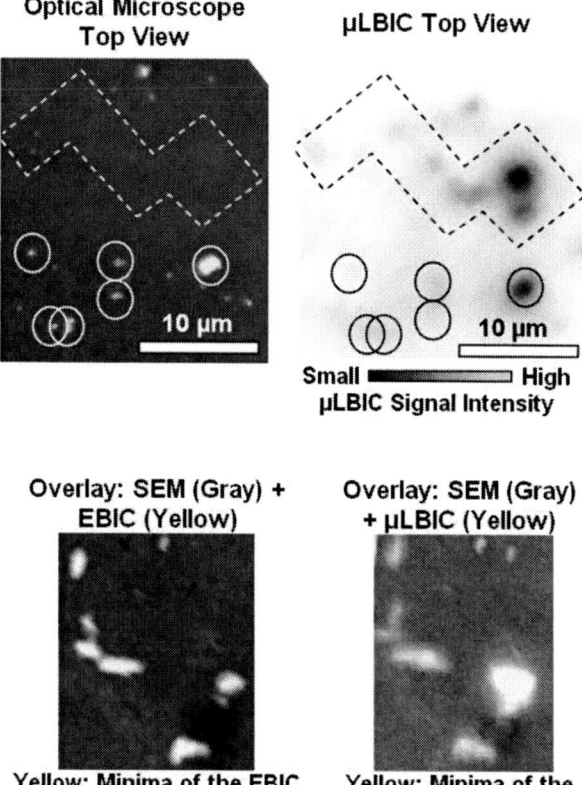

Fig. 2.    Investigation of the same area as in Fig 1, using a optical microscope and a µLBIC set-up. The first row shows an confocal microscopy image (left) and the correspondent µLBIC map (right). The images in the lower row shows overlays of an image (gray) with positions the minima of the EBIC intensity (yellow) on the left and of the µLBIC intensity on the right side.

## Simultaneous mapping of µLBIC and µRS

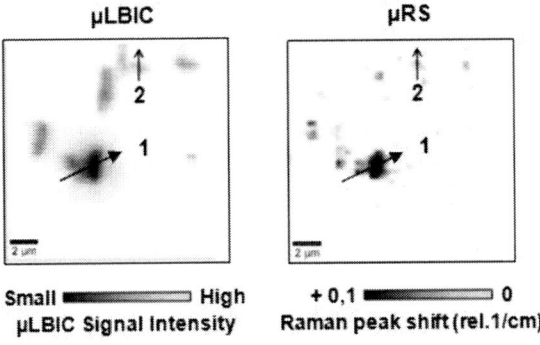

**µLBIC**  **µRS**

Small ▬▬ High   + 0,1 ▬▬ 0
µLBIC Signal Intensity   Raman peak shift (rel.1/cm)

### Linescan of µLBIC und µRS peak position

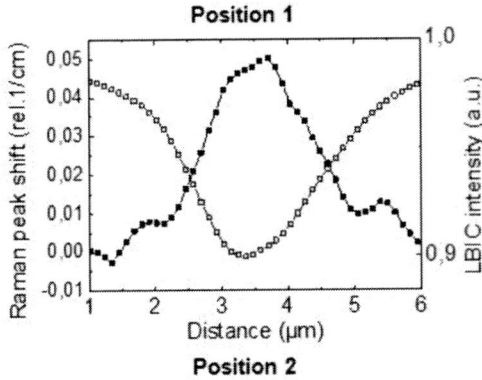

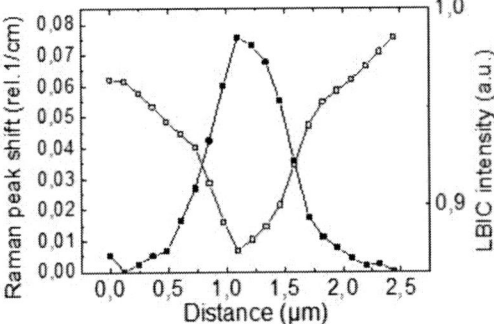

Fig. 3. 2-D-mappings of LBIC (left) and the spectral position of the Raman scattering peak (right) on the identical position. The diagrams show the LBIC-intensity and the Raman peak shift along the two lines that are indicated in the graphics.

### B. Comparison of µLBIC and simultaneous µRS mappings

While the µLBIC measurement, the spectra of the scattered laser light is evaluated in µRS. The measurements can be performed simultaneously; i. e. mappings of electrical performance and crystal quality are generated with an identical spatial coordinate system. Fig. 3 illustrates the mappings of the induced current and the local Raman peak shift in respect to the unharmed cell area. A peak shift towards higher energies indicates compressive stress. The diagrams show two exemplary line scans of the µLBIC intensity and the Raman peak shift. For both exemplary positions the Raman peak is shifted toward higher frequencies. This shift usually is attributed to compressive stress. Using the conversion coefficient derived in a similar experiment [12] the maximum shift of 0.08 rel. 1/cm can be assigned to compressive stress of 40 MPa.

### C. µRS mappings before and after PID-s

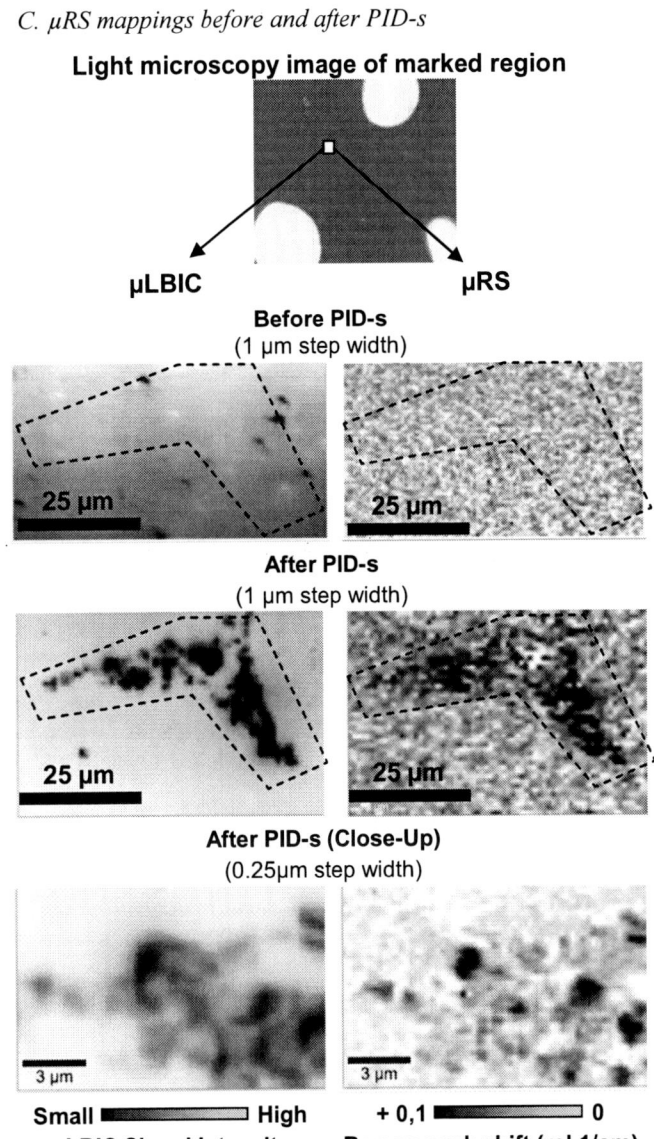

Fig. 4. The light microscopy image shows the three markers that allowed matching images before and after PID-s. Mappings of µLBIC (left) and spectral Raman peak position by µRS (right) were conducted before and after PID-s with a step width of 1 µm. The shunted region was then mapped with a step-width of 0.25 µm. Dark intensities in µ-LBIC and µ-RS relate to low current collection and increased stress, relatively.

A solar cell was examined before and after PID-s regarding crystal defects. Three marks were placed on the samples surface, as shown in the optical microscope image in Figure 4. An area of 300 µm x 300 µm was analyzed using a step width of 1 µm before PID. The mapping before PID-s shows a homogenous distribution of the Raman peak position within the chosen area. Thus, there are no spots of locally increased stress detected before PID-s. µLBIC did not show shunts. By PID-s, the shunt resistance was reduced from 533 Ωcm² to 35 Ωcm². After degradation, PID-related shunts were detected by µLBIC within the chosen area. µRS mapping reveals the presence of stress on the PID-shunts. The locally increased stress is attributed to crystal volume defects, such as stacking faults. This finding confirms the hypothesis of Naumann et al. that local crystal defects are not present before the PID but rather form during the applied high-voltage stress.

## IV. CONCLUSION

Comparing µLBIC and EBIC mappings at the same sample show that µLBIC is suitable to localize PID-s. Both methods show signal drops on the same positions. As others already indicate PID-s have a platelet shape that is oriented along the (111) plane. Simultaneous mapping of local mechanical stress and electrical shunts by µRS and µLBIC, respectively, demonstrated for the first time that PID-shunts are accompanied by increased compressive stress. This confirms the presence of decorated stacking faults at the shunt positions. Comparing µRS on solar cell samples before and after PID-s revealed that the regions of local increased stress were not present before PID-s. Hence, this independently confirms the degradation mechanism found by Naumann et al. that stacking faults are induced during PID-s. As a summary, the combination of µLBIC and µRS is ideally suited for non-destructive characterization of PID-s-induced crystal defects with high spatial resolution.

## REFERENCES

[1] S. P. Harvey et al., "Sodium Accumulation at Potential-Induced Degradation Shunted Areas in Polycrystalline Silicon Modules," *IEEE J. Photovoltaics*, vol. 6, no. 6, pp. 1440–1445, 2016.

[2] V. Naumann et al., "Explanation of potential-induced degradation of the shunting type by Na decoration of stacking faults in Si solar cells," *Solar Energy Materials and Solar Cells*, vol. 120, pp. 383–389, 2014.

[3] V. Naumann et al., "Microstructural Analysis of Crystal Defects Leading to Potential-Induced Degradation (PID) of Si Solar Cells," *Energy Procedia*, vol. 33, pp. 76–83, 2013.

[4] A. Schütt, J. Carstensen, J. M. Wagner, and H. Föll, "Influence of surface and process induced defects on potential-induced degradation and regeneration," *Cell*, vol. 1, p. 10, 2013.

[5] D. Lausch et al., "Potential-Induced Degradation (PID): Introduction of a Novel Test Approach and Explanation of Increased Depletion Region Recombination," *IEEE J. Photovoltaics*, vol. 4, no. 3, pp. 834–840, 2014.

[6] V. Naumann, C. Brzuska, M. Werner, S. Großer, and C. Hagendorf, "Investigations on the Formation of Stacking Fault-like PID-shunts," *Energy Procedia*, vol. 92, pp. 569–575, 2016.

[7] H. Nagel, A. Metz, and K. Wangemann, "Crystalline Si solar cells and modules featuring excellent stability against potential-induced degradation," in *26th European Photovoltaic Solar Energy Conference and Exhibition*.

[8] H. Nagel, P. Saint-Cast, M. Glatthaar, and S. W. Glunz, "Inline processes for the stabilization of p-type crystalline Si solar cells against potential-induced degradation," in *Proceedings of the 29th European PV Solar Energy Conference and Exhibition*.

[9] F. D. Heinz, W. Warta, and M. C. Schubert, "Optimizing Micro Raman and PL Spectroscopy for Solar Cell Technological Assessment," *Energy Procedia*, vol. 27, pp. 208–213, 2012.

[10] M. Breitwieser et al., "Analysis of solar cell cross sections with micro-light beam induced current (µLBIC)," *Solar Energy Materials and Solar Cells*, vol. 131, pp. 124–128, 2014.

[11] M. Breitwieser et al., "Process Control and Defect Analysis for Crystalline Silicon Thin Films for Photovoltaic Applications by the Means of Electrical and Spectroscopic Microcharacterization Tools," *IEEE J. Photovoltaics*, vol. 4, no. 5, pp. 1275–1281, 2014.

[12] A. Büchler et al., "Enabling stress determination on alkaline textured silicon using Raman spectroscopy", *Proceedings of the 7th Silicon PV conference, Freiburg 2017*

# Nanoscale Detection of Deep Levels in CIGS using Electron Energy Loss Spectroscopy

Julia I. Deitz,[1] Pran K. Paul,[2] Shankar Karki[3], Sylvain Marsillac,[3] Aaron R. Arehart,[2] Tyler J. Grassman,[1,2] and David W. McComb[1]

[1] Dept. of Materials Science & Engineering, The Ohio State University, Columbus, OH, 43210, USA.
[2] Dept. of Electrical & Computer Engineering, The Ohio State University, Columbus, OH, 43210, USA.
[3] Dept. of Electrical & Computer Engineering, Old Dominion University, Norfolk, VA, 23529, USA.

*Abstract* — Correlation of electronic structure with nanoscale chemical and physical structure in photovoltaic (PV) materials can provide a pathway to improve fundamental understanding of the nature and impact of defects in solar cell devices. Recent advancements in scanning deep level transient spectroscopy (DLTS) have made it possible to better characterize defect distributions in $CuInGaSe_2$ (CIGS) samples with both energy and surface lateral spatial resolution, but depth locations, as well as any information regarding chemical or structural features of the defective areas, are effectively unavailable with this technique. Here, scanning transmission electron microscope (STEM) based electron energy-loss spectroscopy (EELS) is employed toward a correlated analysis of the $E_V + 0.43$ eV trap level. Preliminary STEM-EELS data reveals a 0.42 eV energy loss peak located within similar intergrain regions as indicated by scanning-DLTS, suggesting correlation and the potential for further targeted chemical and structural characterization.

## I. INTRODUCTION

The advancement of photovoltaics (PV) is critically dependent on the ability to characterize both materials properties and device performance (or limitation) at the nanoscale. This is particularly true for accurate chemical and electronic structure property information. In return, this vitally important information on defect states, bandgaps, and other electronic transition throughout nanoscale features can provide feedback into the understanding of fundamental behaviors. For example, electrically active defect centers in $CuInGaSe_2$ (CIGS) devices act as barriers to the achievement of Shockley-Queisser limit performance of 32% [1,2]. Correlating defect states with chemical and physical structure on the nanoscale would offer better insight for these defect states in CIGS, thus allowing for better engineering and efficiency of devices, as well as being informative for PV materials in general.

Two such techniques that can provide high-resolution spatially-resolved electronic structure information are scanning deep level transient spectroscopy (DLTS) and electron energy-loss spectroscopy (EELS). Scanning-DLTS is a scanning probe microscopy based technique that was recently developed to provide nm-scale spatially-resolved defect spectroscopy [3,4]. It enables mapping of energy-resolved trap concentrations across sample surfaces (with some additional degree of depth sensitivity). This relatively nondestructive method can be paired with local structural and chemical analysis, such as

available via electron microscopy analytical methods, to identify the physical sources of defects.

To this end, higher spatial resolution can be achieved using electron energy-loss spectroscopy (EELS) in the scanning transmission electron microscope (STEM). Using a monochromated electron source, EELS energy resolution as low as 10 meV has been reported [5]. The low energy-loss ($\lesssim 0.5$ eV), or valence loss, region of the spectrum provides information on the electronic structure of the material, including interband transitions and the fundamental bandgap. Historically, the accuracy of low-loss EELS in semiconductors has been limited by extraneous signals related to the generation of Čerenkov radiation. Studies regarding the use of practical experimental conditions in EELS measurements for the avoidance of these extra signals in most semiconductors [6], including recent efforts by the authors [7], have helped make the nanoscale investigation of electronic structure for PV materials and devices tractable.

Near mid-gap trap states with energies generally around 0.47 eV have been previously reported in CIGS [8,9]; such mid-gap levels often serve as strong carrier recombination centers. Studies have suggested that these traps are associated with various chemical and/or structural sources, such as Fe impurities and In vacancies, and recent scanning-DLTS work has indicated a strong localization of these traps within specific intergrain regions [9,10], but direct correlation with local chemical information or determination of any particular structural features associated with their location within the cell structure has not been established. We discuss here ongoing work toward the correlation of spatially-resolved mid-gap electronic states in CIGS observed via both scanning-DLTS and STEM-EELS. This combination of techniques provides a pathway for identification of detrimental defects within PV materials, and their impact on electronic structure, on the nanometer scale.

## II. EXPERIMENTAL DETAILS

In this study two *p*-type CIGS samples were grown by a three-stage co-evaporation process on Mo-coated (for cell rear contact) soda lime glass substrates. 2.7 μm of CIGS was deposited with an average composition of 23.6% Cu, 17.1% In, 8.1% Ga and 51.3% Se. One sample was reserved at this point

978-1-5090-5606-4/17 $31.00 © 2017 IEEE

for the DLTS measurements, while the other (used for EELS measurements) received a CdS buffer layer to produce a structure similar to a solar cell. The DLTS sample was processed via shadow masked deposition of 150 nm Al to form 1.5 mm² Schottky diodes on the CIGS surface. This device-like structure is used to perform the band modulation within the surrounding CIGS needed for the C-V based DLTS measurements.

For the large-scale spatially-averaged DLTS experiments, the sample was biased at -0.5 V during the trap emission/measurement phase, with pulsing to 0.0 V for 10 ms to fill any active traps. Using a double boxcar approach, thermally-assisted trap emission rates from 0.8 to 2000 $s^{-1}$ were recorded. For the scanning-DLTS experiments, traps were filled by pulsing to 0.0 V for 25 ms and trap emission was recorded at -2.0 V using an emission rate tuned to provide sensitivity toward the specific trap level of interest ($E_V + 0.43$ eV). Further details of the DLTS and scanning-DLTS experimental setups and analysis can be found in [9,10].

All EELS work was performed on a monochromated image-corrected FEI Titan³ G2 STEM operated at 60 kV, 30 nA beam current, a convergence angle of 12 mrad, and a collection angle of 22 mrad. Samples for STEM work were prepared in a FEI Helios Ga-source focused ion beam (FIB) at 30 kV. Samples were coated with a protective layer of gold prior to entry into the FIB, and then further coated with platinum within the FIB, to prevent surface damage. A final 5 kV milling step was used to minimize amorphous damage.

## III. RESULTS AND ANALYSIS

### A. DLTS and Scanning-DLTS

Figure 1 presents a conventional DLTS spectrum, collected at a 0.8 $s^{-1}$ rate window, where one majority carrier (hole) trap is observed at ~265 K. The activation energy of the trap is $E_V + 0.43$ eV, with a capture cross-section of $1.2 \times 10^{-18}$ cm², as extracted across a wide range of rate windows via Arrhenius analysis (Fig. 1 inset). As noted, this trap level has been

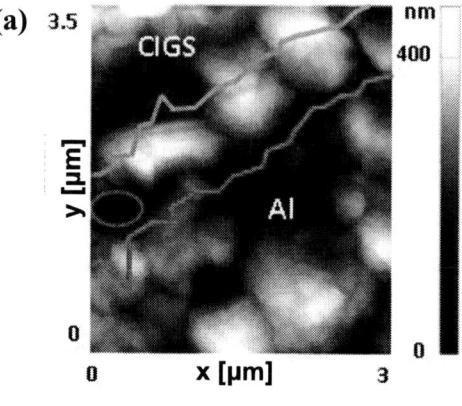

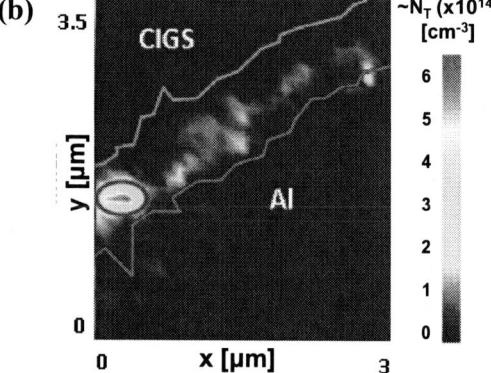

**Fig. 2.** (a) AFM image and (b) scanning-DLTS map of a CIGS sample showing both surface grain topology and a region of strong localization of the $E_V + 0.43$ eV trap states.

previously observed in CIGS and has been attributed to Fe impurities or In vacancies [9,10].

Figure 2 presents representative AFM and scanning-DLTS data (at 298 K) from the same CIGS surface region. From the scanning-DLTS map, Fig. 2(b), the $E_V + 0.43$ eV trap is found to be highly concentrated within an individual intergrain region, consistent with our previous work [3,4]. Note that scanning-DLTS is only sensitive to traps modulated within the depletion region, which is marked by the red and green lines in Fig. 2(b); zero signal occurs outside of this region due to screening by the metallic contact and/or lack of band modulation. Interestingly, this sample also exhibits a lower, but non-zero, trap concentration distributed through the nearby series of grain boundaries running parallel to the C-V depletion edge. Overall, this suggests that the defect that induces the $E_V + 0.43$ eV level generally exists at low levels throughout the CIGS grain structure, while certain types of grain boundaries may have a significantly higher propensity for formation or getting of the offending defect. That said, several other grain boundary lengths within the depletion region, oriented perpendicular to the region boundaries, show no indication of the $E_V + 0.43$ eV trap within the sensitivity of the measurement.

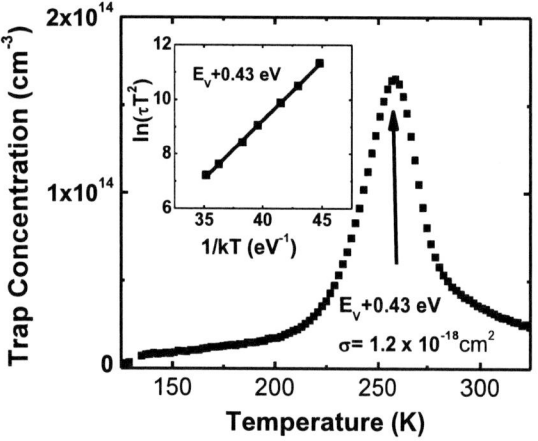

**Fig 1.** DLTS spectrum showing a single trap level, with an Arrhenius plot revealing a trap energy of $E_V + 0.43$ eV provided in the inset.

978-1-5090-5606-4/17 $31.00 © 2017 IEEE

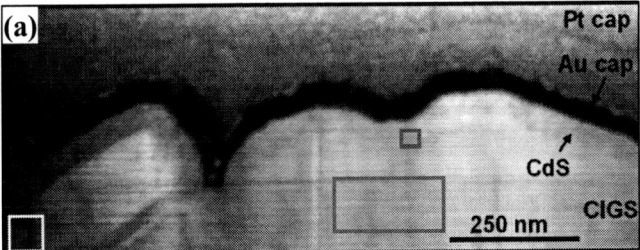

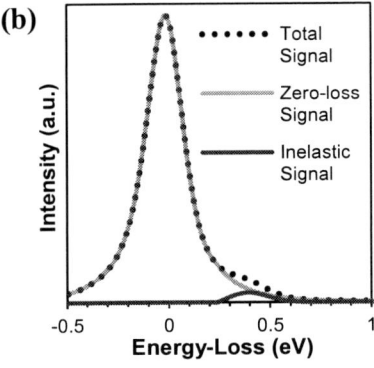

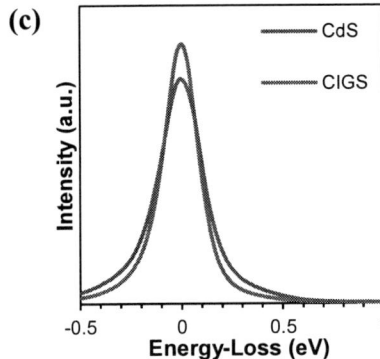

Fig. 3. (a) EELS total intensity map of the CdS/CIGS specimen, with (b,c) associated valence EELS spectra taken from the areas indicated in the boxes. The intergrain region within the white box (b) exhibits the 0.42 eV peak, while regions of pure intragrain CIGS (green) and pure CdS (purple) do not (c).

*B. STEM-EELS*

Figure 3 presents STEM-EELS analysis performed on the CdS-coated CIGS sample. Data was recorded in hyperspectral imaging mode to allow for site specific analysis of EELS spectra. The energy resolution for this data was 130 meV. Figure 3(a) is an EELS zero-loss intensity map. Contrast variation here is strictly due to differences in the amount of energy lost from purely elastic interactions, which is strongly dependent on variables like sample thickness and chemical composition, but provides no indication of electronic structure. The intensity map consists of $501 \times 147$ pixels (with step size of 3.4 nm) with each pixel containing an EELS spectrum. Figure 3(b) presents a valence EELS spectrum (black dotted line), averaged over the region within the white box to improve the signal-to-noise ratio (SNR). This spectrum has been further deconvoluted by substracting the zero-loss signal (light red line)—from electrons that pass through the sample with no loss in energy—to obtain the purely inelastic scattering signal (blue line), which contains electronic structure information. The zero-loss peak (ZLP) subtraction can be performing with a variety of models; here, the standard reflected tail method was used, which simply reflects the zero-loss intensity on the negative side of the peak to the positive side and subtracts from the overall intensity [11].

The inelastic signal after ZLP extraction, Fig. 3(b), reveals a clear peak at ~0.42 eV, consistent with the $E_V + 0.43$ eV trap level observed via DLTS. The area from which this EELS spectrum was collected, indicated by the white box in Fig. 3(a), is indeed within an intergrain region, roughly 25–50 nm below the CIGS/CdS interface. Noteworthy is the fact that this is the only area within the entire ~2 μm × 4 μm specimen that exhibited such a relatively strong, clear inelastic peak, consistent with the strong localization observed via scanning-DLTS. Significantly feinter intensities of the 0.42 eV signal (~20% the intensity of same peak in highlighted intergrain area) were observed in other regions of the specimen, near the CdS/CIGS interface. This observation is also consistent with the low concentrations observed along the grain boundary line via scanning-DLTS, as shown in Fig. 2(b).

While the EELS and scanning-DLTS measurements presented herein were not performed on the exact sample specimen area—though they did come from samples that were produced in nominally identical processes up until the final CdS layer—the highly coincident energies and their relative locations within the CIGS microstructure do suggest that the 0.42 eV EELS peak is indeed representative of the $E_V + 0.43$ eV trap level observed via both conventional and scanning-DLTS. Efforts to perform these measurements on the exact same sample areas to verify true correlation are in progress.

## IV. CONCLUSIONS AND ONGOING EFFORTS

Conventional and scanning-DLTS measurements on three-stage deposited CIGS have revealed a trap state with an energy of $E_V + 0.43$ eV. As previously reported by the authors [3], this trap level is found to be strongly localized within specific grain boundaries. These new measurements also show a lower concentration spread throughout some of the nearby grain boundaries. Similar samples grown via identical processes (and in the same apparatus) were analyzed using STEM-EELS. A valence energy-loss (i.e. electronic transition) peak at ~0.42 eV was observed, consistent with the DLTS results. This signal was also found to be strongly localized within only a small region of the specimen, with weak observations in a few other spots, consistent with the scanning-DLTS data. The strong correlation indicated here using two very different characterization methods highlights the potential for their combination to provide high energy and spatial resolution analysis of electronic defect levels across a range of important length scales.

## REFERENCES

[1] M. Igalson, P. Zabierowski, D. Prządo, A. Urbaniak, M. Edoff. W. N. Shafarman, "Understanding defect-related issues limiting efficiency of CIGS solar cells," *Sol. Energy Mater. Sol. Cells*, vol. 93, no. 8, pp. 1290–1295, 2009.

[2] M. Gloeckler and J. Sites, "Efficiency limitations for wide-band-gap chalcopyrite solar cells," *Thin Solid Films*, vols. 480/481, pp. 241–245, 2005.

[3] P.K. Paul, D. W. Cardwell, C. M. Jackson, K. Galiano, K. Aryal, J. P. Pelz, S. Marsillac, T. J. Grassman, S. A. Ringel, A. R., Arehart, "Direct nm-Scale Spatial Mapping of Traps in CIGS," *IEEE J. Photovolt*, vol. 5, p. 1482-1486, 2015.

[4] P. K. Paul, K. Aryal, S. Marsillac, T. J. Grassman, S. A. Ringel, A. R., Arehart, "Identifying the source of reduced performance in 1-stage-grown Cu(In,Ga)Se$_2$ solar cells" in *Proc. 43th IEEE Photovoltaics Spec. Conf.*, Portland, OR, USA, , Jun. 5–10, 2016.

[5] O. L. Krivanek, T. C. Lovejoy, N. Dellby, T. Aoki, R. W. Carpenter, P. Rez, E. S. J. Zhu, P. E. Batson, M. J. Lagos, R. F. Egerton, P. A. Crozier, "Vibrational spectroscopy in the electron microscope," *Nature*, vol. 514, p. 209-212, 2014.

[6] M. Stoger-Pollach and P. Schattschneider, "The Influence of Relativistic Energy Losses on Bandgap," *Micron, 3(396)*, 2006.

[7] J. I. Deitz, T. J. Grassman, D. W. McComb, "Probing the Electronic Structure at the Heterovalent GaP/Si Interface using Electron Energy-Loss Spectroscopy," in *Proc. 42nd IEEE Photovoltaics Spec. Conf.*, Portland, OR, USA, p. 1545-1548, Jun. 5–10, 2016.

[8] G. Bauer, R. Bruggemann, S. Tardon, S. Vignoli, and R. Kniese, "Quasi-"Fermi level splitting and identification of recombination losses from room temperature luminescence in Cu(In$_{1-x}$Ga$_x$)Se$_2$ thin films versus optical band gap," *Thin Solid Films*, vols. 480/481, pp. 410–414, 2005.

[9] I. Choi, C. Choi, and J. Lee, "Deep centers in a CuInGaSe$_2$ /CdS/ZnO:B solar cell," *Phys. Status Solidi A*, vol. 209, no. 6, pp. 1192–1197, 2012.

[10] S. Zhang, S. Wei, A. Zunger, and H. Katayama-Yoshida, "Defect physics of the CuInSe2 chalcopyrite semiconductor," Phys. Rev. B, vol. 57, no. 16, pp. 9642–9656, 1998.

[11] R. F. Egerton, *Electron Energy-Loss Spectroscopy in the Electron Microscope*, 3$^{rd}$ ed, Springer, 2011.

# Measurement of Carrier Dynamics in Photovoltaic CZTSe by Time-Resolved Terahertz Spectroscopy

Siming Li[1], Michael A. Lloyd[2], Andrew A. Golembeski[3], Brian E. McCandless[2], Jason B. Baxter[1]

[1] Drexel University, Department of Chemical and Biological Engineering, Philadelphia, PA, 19104, USA

[2] University of Delaware, Institute of Energy Conversion, Newark, DE, 19716, USA

[3] University of Rochester, Department of Chemical Engineering, Rochester, NY, 14611, USA

*Abstract* — We demonstrate the use of time-resolved terahertz spectroscopy coupled with numerical modeling of the transport equations to elucidate photoexcited carrier dynamics in a photovoltaic absorber. By measuring a high-quality $Cu_2ZnSnSe_4$ single crystal that exhibited device efficiency of 8.6%, we show that critical parameters including mobility, surface recombination velocity, and Shockley-Read-Hall lifetime can be obtained. Mobility values of 80 $cm^2/Vs$ were validated with Hall effect measurements. Surface recombination velocity could be reduced by at least two orders of magnitude, to $10^4$ cm/s, with appropriate chemical and mechanical polishing. Carrier lifetimes exceeding 10 ns indicate promise for devices with high photovoltage. Terahertz spectroscopy provides complementary insight to conventional time-resolved photoluminescence and is particularly valuable for materials that are not strongly emissive.

## I. INTRODUCTION

The design of efficient solar cells requires understanding how photoexcited carrier lifetimes, mobilities, and recombination mechanisms depend on structure and processing of photovoltaic (PV) absorber materials. Carrier dynamics in PV absorbers have conventionally been characterized by time-resolved photoluminescence (TRPL), which has indicated that $V_{oc}$ increases logarithmically with carrier lifetime in CdTe and CIGS over wide ranges of processing conditions [1, 2]. Modeling of the TRPL response has provided insight into recombination mechanisms [3, 4]. However, not all materials are suitable for TRPL. Alternative non-contact probes can enable measurement of carrier dynamics for a broader range of materials.

Terahertz time-domain spectroscopy (THz-TDS) is a powerful technique to probe conductivity in conventional semiconductors because carriers typically have scattering rates on the order of $10^{12}$-$10^{14}$ $s^{-1}$. Time-resolved terahertz spectroscopy (TRTS) probes the transient far-IR response after photoexcitation with an optical pump pulse, with sub-picosecond time resolution [5]. Terahertz spectroscopy offers several advantages compared to TRPL. TRTS provides an alternative approach to measuring lifetime by tracking the transient photoconductivity, as well as enabling the calculation of carrier densities and mobilities. Additionally, THz-TDS enables determination of permittivity, free carrier density, and mobility via a non-contact measurement of the film without photoexcitation [5]. Measurements without photoexcitation primarily probe majority carriers, while measurements with photoexcitation under high injection conditions measure total conductivity including both electrons and holes.

Here we report on the application of TRTS to measure carrier dynamics and understand recombination mechanisms in single crystal $Cu_2ZnSnSe_4$ (CZTSe). While terahertz spectroscopy has been used for many years in other fields of science [5], it has rarely been applied to PV absorber materials [6, 7]. We have used a combination of numerical modeling of the transport equations with TRTS under different photoexcitation conditions to elucidate dominant recombination mechanisms and rate constants. We found that surface recombination controls the dynamics on the sub-nanosecond time scale, but surface recombination could be reduced by a factor of $10^2$ by polishing and chemical treatment of the as-cleaved crystal.

## II. CZTSe CRYSTAL GROWTH AND PROPERTIES

Single crystals of CZTSe material were synthesized using methods described by Bishop et al. [8]. Elemental precursors of 5-6N purity were sealed in quartz ampules at a base pressure of $5 \times 10^{-6}$ torr. CZTSe growth and grain ripening were promoted at 750°C over a 20-day period. Ampoules were allowed to naturally cool to room temperature over approximately 36 hours. For this work, one particular single crystal was extracted from the resulting multicrystalline boule and mechanically planarized. One side was then further polished with a fine alumina slurry to reduce mechanical damage resulting from the abrasive grinding step. The crystal was treated with a 10-minute solution etch with 0.125% bromine in methanol to clean the surface before characterization.

The sample used for this study was 4x3x1 mm in size and was confirmed to be a single crystal by Laue diffraction. Energy-dispersive X-ray fluorescence spectroscopy (XRF) calibrated with inductively coupled plasma mass spectroscopy (ICP) showed Cu/(Zn+Sn) = 0.87 and Zn/Sn = 1.22. These values fall within the Cu-poor, Zn-rich regime that has been shown to promote optimal efficiency in CZT(S,Se) devices [9]. Crystals exhibited strong p-type conduction with DC mobility of 89 cm$^2$/Vs and hole concentration of $8.7 \times 10^{16}$ cm$^{-3}$, as determined by Hall effect measurement. The band gap of a sister crystal was determined by spectrally resolved photoluminescence and external quantum efficiency measurements to be 0.95-0.98 eV. A solar cell fabricated from the sister crystal exhibited an efficiency of 8.6% with $V_{oc}$ = 389 mV, $J_{sc}$ = 35.2 mA/cm$^2$, and FF = 62.6%, demonstrating the high quality of the CZTSe.

### III. ULTRAFAST CARRIER DYNAMICS

TRTS was carried out in transmission geometry using 50-fs pulses from a regeneratively amplified Ti:Sapphire laser. THz pulses were generated and detected by optical rectification and free-space electro-optic sampling with ZnTe crystals [5]. The optical pump wavelength was tuned with an optical parametric amplifier, and the pump – probe delay time was controlled using an optical delay line. The crystal was cooled to 77 K in an optical cryostat to freeze out holes, avoiding deleterious absorption of the terahertz probe by non-photoexcited carriers.

Fig. 1a shows the dynamics of the differential transmitted terahertz probe, normalized to the maximum response, for two pump wavelengths incident on the unpolished surface of the crystal. We have previously shown that the mobility is independent of pump – probe delay time [10], so can conclude that the dynamics are proportional to the photoexcited carrier density integrated through the sample thickness. The response shows an instantaneous rise as carriers are photoexcited, then a decay as they recombine or are trapped in non-conductive states. Dynamics are faster with higher energy photons

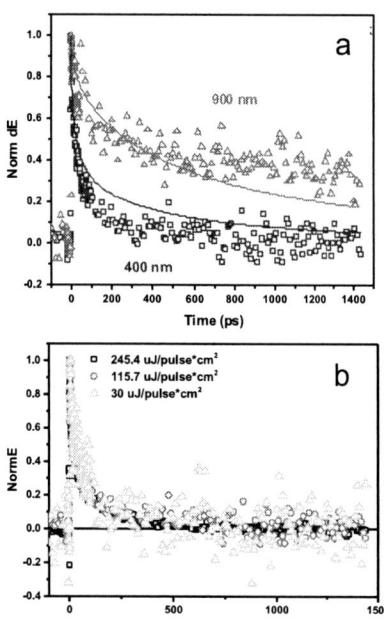

Fig. 1. TRTS dynamics of unpolished crystal under (a) variable pump wavelength and (b) variable pump fluence.

because they are absorbed more strongly, and generated carriers are prone to surface recombination. With 400 nm excitation, 80% of the carriers recombine within 100 ps. Conversely, lower energy photons have longer penetration depths and carriers are less likely to recombine at the surface. With 900 nm excitation, 40% of carriers remain after 1.4 ns, consistent with our previous report for CZTSSe thin films [10].

Figure 1b shows normalized dynamics at three pump fluences spanning a range of 10x. Dynamics are independent of fluence, which indicates that Auger and radiative recombination are not significant. Recombination can be attributed to trap-mediated events in the bulk (Shockley-Read-Hall, SRH) and at the surface.

In our previous paper, we fit the dynamics with an empirical bi-exponential model, where the fast and slow time constants were roughly assigned to surface and bulk recombination [10]. Here we show that dynamics can instead be modeled by the time-dependent transport equations to extract physical parameters that provide key insight into solar cell design.

The measured signal, $\Delta E_{THz}(t)$, is proportional to $\int \Delta n(x,t)dx$, where $\Delta n$ is the photoexcited carrier density, $x$ is the distance into the crystal, and $t$ is the pump-probe delay time. The transport equations can also be solved to find $\Delta n(x,t)$, which is controlled by mobility ($\mu$), SRH lifetime ($\tau_{SRH}$), surface recombination velocity ($S$), and absorption coefficient at the pump wavelength ($\alpha$). We neglect the photo-Dember effect and assume that electrons and holes move in a way that preserves charge neutrality. This enables solution of the continuity equation for a single carrier using an ambipolar mobility, rather than simultaneously solving the continuity equations for both carriers along with Poisson's equation. To find $\Delta n(x,t)$, we numerically solved the continuity equation

$$\frac{\partial \Delta n}{\partial t} = D \frac{\partial^2 \Delta n}{\partial x^2} - \frac{\Delta n}{\tau_{SRH}} \qquad (1)$$

with an initial carrier density profile determined by Beers' Law with absorption coefficient $\alpha$, subject to a surface recombination boundary condition at the front surface and $\Delta n=0$ at the rear surface. D, $\tau_{SRH}$, and $S$ were allowed to vary to achieve the best fit for the thickness-integrated photoexcited carrier density. $\alpha$ at 400 and 900 nm was fixed at 2.1 and 0.5 x10$^5$ cm$^{-1}$, respectively, from spectroscopic ellipsometry of a CZTSe single crystal of similar composition in Ref. [11]. Kinetic traces can be fit individually or in groups with shared parameters to reduce uncertainty.

Both data sets in Fig. 1a were fit with a globally shared set of parameters, except for the absorption coefficients. The model fits are shown as lines. The best fit parameters are S > 10$^6$ cm/s, D=0.53 cm$^2$/s (giving ambipolar mobility of 80 cm$^2$/Vs at 77 K that is consistent with the Hall effect measurement), and $\tau_{SRH}$>10 ns. Clearly, surface recombination is the dominant loss mechanism under these conditions. $\tau_{SRH}$ indicates long lifetimes desirable for high-efficiency devices,

978-1-5090-5606-4/17 $31.00 © 2017 IEEE

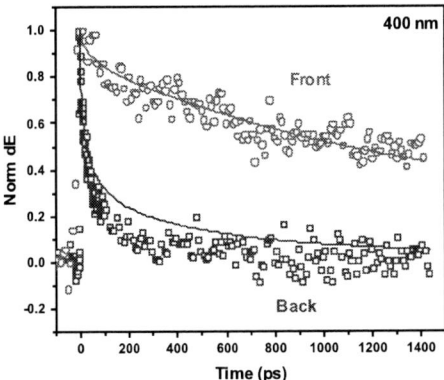

Fig. 2. TRTS dynamics with 400 nm photons incident on the front (polished) and back (unpolished) sides of the crystal, illustrating the difference resulting from surface recombination.

but exhibited large uncertainty due to the fast surface recombination and the length of the optical delay line in the TRTS, which limited pump-probe delay times to 1.4 ns.

For comparison, the photoconductivity can be determined from the magnitude of the terahertz response, and the mobility can be calculated subject to assumptions regarding the generation profile. Photoexcitation with 14 $\mu J/cm^2$ fluence at 400 nm results in an average density of electron-hole pairs of $6 \times 10^{18}$ $cm^{-3}$ over the skin depth and total terahertz mobility, $\mu_n + \mu_p$, of 280 $cm^2/Vs$. Further details of the analysis can be found in Ref. [6]. Assuming similar electron and hole mobilities, the terahertz mobility is consistent with mobility derived from the transport model to within a factor of 2.

Surface recombination velocity of the unpolished surface of the CZTSe crystal is unacceptably high for device applications. However, $S$ could be considerably reduced by polishing the surface before the Br/methanol etch. Fig. 2 shows the photoexcited carrier dynamics with and without polishing, using photoexcitation at 400 nm to exaggerate the effect of surface recombination. Lines show the model fits, where the bulk parameters are the same as in Fig. 1a but the fitted value of $S_{polished}$ was $1 \times 10^4$ cm/s. Polishing reduced the surface recombination velocity by at least two orders of magnitude.

Photoexcitation of the polished crystal with 900 nm photons minimized the effects of surface recombination and resulted in nearly flat dynamics over the 1.4 ns range measured, confirming the previous estimate of $\tau_{SRH} > 10$ ns. Such lifetimes are comparable to the best reported in literature [9].

The transient profile of the carrier density was reconstructed from the fitted parameters of the model and is shown in Fig. 3a and 3b for the unpolished and polished surface, respectively. Our assumptions dictate that the carrier density profile will be the same for electrons and holes. Simulations of the unpolished crystal face show that the predominant transport is diffusion toward the surface due to the fast recombination. In contrast, when surface recombination is passivated by the polishing process, diffusion is primarily into

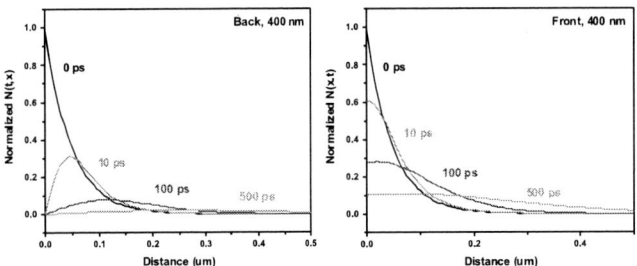

Fig. 3. Evolution of the photoexcited electron density, $\Delta n(x,t)$, after an initial 400 nm photoexcitation pulse, with surface recombination velocities depending on polishing treatment.

the crystal because of the initial concentration gradient that arises from the Beers' law absorption profile.

## IV. CONCLUSIONS

The combination of TRTS and numerical simulations of the transport equations can provide unique insight into photoexcited carrier dynamics and limiting recombination mechanisms relevant to thin film solar cells. Here, the single-crystal CZTSe eliminates complications arising from grain boundaries and high densities of point defects and secondary phases associated with thin film growth. Under these conditions, we found SRH lifetimes exceeding 10 ns at 77 K, with surface recombination as the dominant loss mechanism, and ambipolar mobilities of 80 $cm^2/Vs$. The mobility is similar to that found by conventional Hall effect, which is an encouraging validation of this method. These parameters would result in diffusion lengths of ~700 nm, which is similar to the thickness necessary for light absorption, and the long lifetimes indicate promise for high photovoltages.

## ACKNOWLEDGEMENTS

The authors acknowledge support from the National Science Foundation through collaborative awards DMR-1507988 and DMR-1508042.

## REFERENCES

[1]   W. K. Metzger, D. Albin, D. Levi, P. Sheldon, X. Li, B. M. Keyes, *et al.*, "Time-resolved photoluminescence studies of CdTe solar cells," *Journal of Applied Physics,* vol. 94, pp. 3549-3555, 2003.

[2]   I. L. Repins, W. K. Metzger, C. L. Perkins, J. V. Li, and M. A. Contreras, "Correlation Between Measured Minority-Carrier Lifetime and Cu(In, Ga)Se₂ Device Performance," *IEEE Transactions on Electron Devices,* vol. 57, p. 2957, 2010.

[3]   A. Kanevce, D. H. Levi, and D. Kuciauskas, "The role of drift, diffusion, and recombination in time-resolved photoluminescence of CdTe solar cells determined through numerical simulation," *Progress in Photovoltaics: Research and Applications,* 2013.

[4] M. Maiberg, T. Hölscher, S. Zahedi-Azad, and R. Scheer, "Theoretical study of time-resolved luminescence in semiconductors. III. Trap states in the band gap," *Journal of Applied Physics,* vol. 118, p. 105701, 2015.

[5] J. B. Baxter and G. W. Guglietta, "Terahertz Spectroscopy," *Analytical Chemistry,* vol. 83, pp. 4342-4368, 2011.

[6] H. Hempel, A. Redinger, I. Repins, C. Moisan, G. Larramona, G. Dennler, *et al.,* "Intragrain charge transport in kesterite thin films-Limits arising from carrier localization," *Journal of Applied Physics,* vol. 120, 2016.

[7] L. Q. Phuong, M. Okano, G. Yamashita, M. Nagai, M. Ashida, A. Nagaoka, *et al.,* "Photocarrier dynamics in undoped and Na-doped $Cu_2ZnSnS_4$ single crystals revealed by ultrafast time-resolved terahertz spectroscopy," *Applied Physics Express,* vol. 8, p. 062303, 2015.

[8] D. M. Bishop, B. E. McCandless, R. Haight, D. B. Mitzi, and R. W. Birkmire, "Fabrication and Electronic Properties of CZTSe Single Crystals," *IEEE Journal of Photovoltaics,* vol. 5, pp. 390-394, 2015.

[9] S. Bag, O. Gunawan, T. Gokmen, Y. Zhu, T. K. Todorov, and D. B. Mitzi, "Low band gap liquid-processed CZTSe solar cell with 10.1% efficiency," *Energy & Environmental Science,* vol. 5, p. 7060, 2012.

[10] G. W. Guglietta, K. R. Choudhury, J. V. Caspar, and J. B. Baxter, "Employing time-resolved terahertz spectroscopy to analyze carrier dynamics in thin-film Cu2ZnSn(S,Se)4 absorber layers," *Applied Physics Letters,* vol. 104, p. 253901, 2014.

[11] M. León, S. Levcenko, R. Serna, I. V. Bodnar, A. Nateprov, M. Guc, *et al.,* "Spectroscopic ellipsometry study of $Cu_2ZnSnSe_4$ bulk crystals," *Applied Physics Letters,* vol. 105, p. 061909, 2014.

# Decoupling grain-boundary, grain-interior, and surface recombination with cathodoluminescence

John Moseley[1], Pierre Rale[2], Stéphane Collin[2], Ana Kanevce[1], Eric Colegrove[1], Joel Duenow[1], Søren Jensen[1], Wyatt K. Metzger[1], and Mowafak M. Al-Jassim[1]

[1]National Renewable Energy Laboratory (NREL), Golden, CO, 80401, USA
[2]Centre de Nanoscience et de Nanotechnology, CNRS, Univ. Paris-Sud/Paris-Saclay, Marcoussis, France

*Abstract* — **In this work, we combine quantitative cathodoluminescence (CL) with time-resolved photoluminescence (TRPL) and numerical simulations to determine grain-boundary, grain-interior, and surface recombination parameters in standard CdTe thin films. CL intensities from thousands of grains are analyzed to accumulate statistics and chart variations with grain size. Grain-boundary contrast results for small grains indicate that the grain-boundary recombination velocity, $S_{GB}$, decreases significantly with CdCl₂ treatment, but $S_{GB}$ is *increased* by subsequent Cu-diffusion. Furthermore, within a given sample, data suggests that $S_{GB}$ is nearly independent of grain size. The back-surface recombination velocity, $S$, is extracted from TRPL measurements incident on the back surface, and CL profiles are simulated to determine the grain-interior lifetime, $\tau_{GI}$. Finally, CL intensity vs. grain size relationships are simulated to check for self-consistency of the $S_{GB}$, $S$, and $\tau_{GI}$ values.**

## I. INTRODUCTION

Grain boundaries (GBs) in thin-film solar cells have defects and potentials that affect charge transport and recombination, and their performance effects need to be quantified. Recent theoretical analyses [1,2] suggest that GBs are detrimental to CdTe device performance, especially the open-circuit voltage, $V_{OC}$. Calculations by Gloeckler, Sites, and Metzger [3] and Gaury and Haney [1] show that the GB 'penalty' to $V_{OC}$ could be as high as 50 to 100 mV considering realistic GB recombination velocities of around $10^5$-$10^6$ cm/s and GB potentials of 0.1-0.7 eV. The impact of GBs on performance will depend on the specific microstructure, treatment history, and properties of the film in question and likely vary from lab to lab. In addition, grain size and grain-to-grain properties can vary considerably within a given sample. Here, we develop a methodology to measure and model recombination for large numbers of grains and GBs, which accounts for microscopic variability and can be applied on materials from different labs.

Figure 1 shows a cross-section schematic of the electron density during cathodoluminescence (CL) measurements. Carrier generation by the electron beam (e-beam) is highly localized within the teardrop-shaped volume near the film surface; however, carriers diffuse from the excitation spot, and light collection in CL is global, which convolutes recombination occurring at GBs, the surface, and in the grain-interior (GI). In this work, we show that GB, GI, and surface recombination parameters can be decoupled by combining quantitative CL with time-resolved photoluminescence (TRPL) and numerical simulations. High-injection conditions

are used in CL to screen GB potentials and minimize charge separation effects. CdTe thin films with standard processing are analyzed, but the methods developed here could be applied on films with new GB passivation treatments as well as other material systems such as CIGS, perovskites, and CZTS.

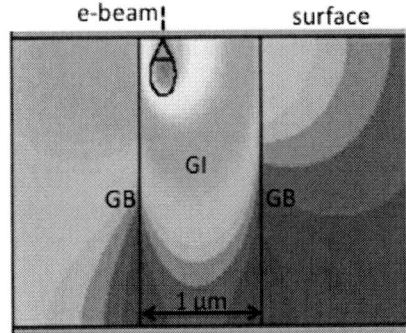

Fig. 1. Cross-sectional schematic of the electron density during CL measurements showing carrier diffusion from the generation volume (teardrop) to GB, GI, and surface regions.

## II. EXPERIMENTAL

Room temperature CL measurements were conducted on an Attolight "Chronos" quantitative CL microscope at CNRS. The mean electron current was about 25 nA and the acceleration voltage was 7.5 kV. CL spectra were recorded on an Andor Newton CCD camera with a Horiba dispersive spectrometer (grating: 150 grooves/mm). Light is collected through a reflective objective made of three free form mirrors. It is aplanatic (free of spherical aberrations and coma) and features constant collection efficiency (no vignetting) over a field of view greater than 100 $\mu$m, and a numerical aperture of 0.72.

A two-dimensional Sentaurus model was developed to simulate CL measurements [4]. The geometry is shown in Fig. 1. GBs are considered to be perpendicular to the film surface and carrier generation is taken to be uniform within a teardrop region to simulate the localized e-beam profile. CL scans are simulated by translating the teardrop (horizontally) across the film and integrating the total radiative recombination at each spot.

Both one-photon excitation (1PE) and two-photon excitation (2PE) TRPL measurements were performed on the films [5,6].

978-1-5090-5606-4/17 $31.00 © 2017 IEEE

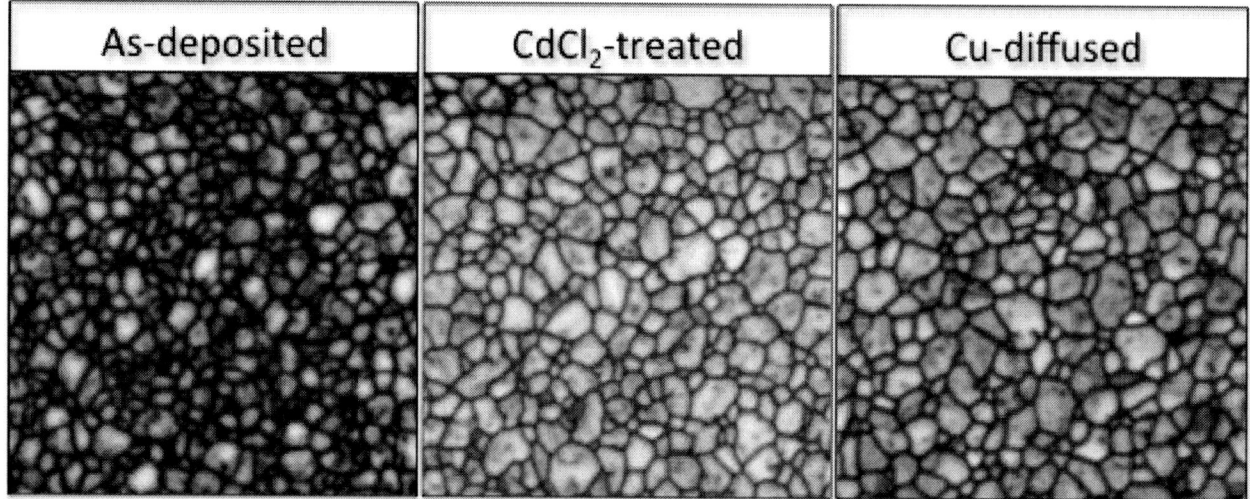

Fig. 2. CL intensity images integrated over 700-900 nm (1.77-1.38 eV) for standard CdTe thin films made at NREL after sequential processing steps: CdTe deposition, CdCl$_2$-treatment, and Cu-diffusion. The images are 30-μm x 30-μm.

## III. RESULTS

Figure 2 shows CL intensity images integrated over 700-900 nm (1.77-1.38 eV) for standard CdTe thin films made at NREL after CdTe deposition, CdCl$_2$-treatment, and Cu-diffusion. The images are 30-μm x 30-μm. GBs in the as-deposited (AD) film are very dark *relative to the GIs*, suggesting a high non-radiative GB recombination rate. This is consistent with the AD film's short TRPL lifetime. After CdCl$_2$ treatment, the lifetime and CL intensity improve significantly; GBs are not as dark after CdCl$_2$ treatment, indicating a reduction in GB recombination. Cu is diffused into the CdTe primarily to increase the hole concentration, but Cu defects are also known to affect recombination [5]. Comparing the CdCl$_2$-treated and Cu-diffused images indicates that Cu makes GI defects and GBs darker, which can counteract some of the passivation by the CdCl$_2$ treatment.

GB contrast is defined as the maximum CL intensity in the GI ($I_{GI,max}$) minus the GB intensity ($I_{GB}$), relative to GI maximum intensity,

$$\text{GB contrast} = \frac{I_{GI,max} - I_{GB}}{I_{GI,max}}. \quad (1)$$

Simulations of the GB contrast have been conducted for a 1-μm grain size with variable GI lifetime ($\tau_{GI}$), GB recombination velocity ($S_{GB}$), and back-surface recombination velocity ($S$). The GB contrast contour plots in Fig. 3 show the combined impacts of (a) $\tau_{GI}$ and $S_{GB}$ with $S=10^5$ cm/s, and (b) $S_{GB}$ and $S$ with $\tau_{GI}=10$ ns. An important finding is that the contrast contours are insensitive to $\tau_{GI}$ and $S$—which demonstrates that for small grains GB contrast is a good measure of $S_{GB}$.

An image analysis routine was used to extract GB and GI intensities from the CL images and compile statistics for each

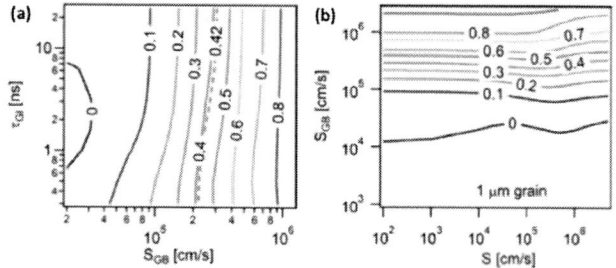

Fig. 3. Contour plots of GB contrast defined by Eqn. 1 [4].

sample as a function of the grain size (grain equivalent diameter). Three images were collected and analyzed per sample, giving a total of 1167, 770, and 725 grains analyzed for the AD, CdCl$_2$-treated, and Cu-diffused films, respectively.

Figure 4 is a scatterplot of the GB contrast vs. grain size. GB contrast is computed *for each grain* according to Eqn. 1, where $I_{GB}$ is taken to be the GB median intensity.

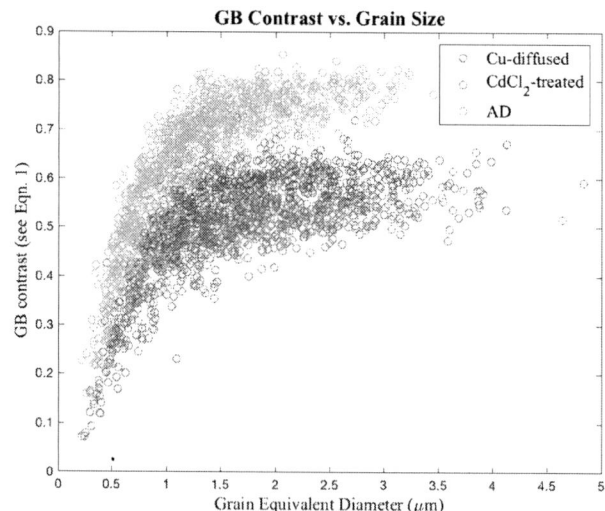

Fig. 4. GB contrast vs. grain size. Each circle represents a grain.

978-1-5090-5606-4/17 $31.00 © 2017 IEEE

Based on the previous discussion of the GB contrast, we look to the small grains in Fig. 4 for an indication of how $S_{GB}$ compares across the films. The GB contrast trend is clear at ~1 μm grain size, from which we can make the following conclusion:

$$S_{GB}(CdCl_2) < S_{GB}(CdCl_2 + Cu) < S_{GB}(AD) \qquad (2)$$

where $S_{GB}(AD)$, $S_{GB}(CdCl_2)$, $S_{GB}(CdCl_2 + Cu)$ are the GB recombination velocities for the AD, CdCl$_2$-treated, and Cu-diffused samples, respectively. Further simulations will determine the precise values of $S_{GB}$ for the films prior to the conference.

Equation 2 applies for small grains, but it may also be relevant for grains larger than 1 μm. Figure 5 shows a plot of the GB median intensity vs. grain size. Interestingly, the GB median intensity is nearly independent of grain size. This result strongly suggests that, *for a given sample*, $S_{GB}$ is about the same for small and large grains.

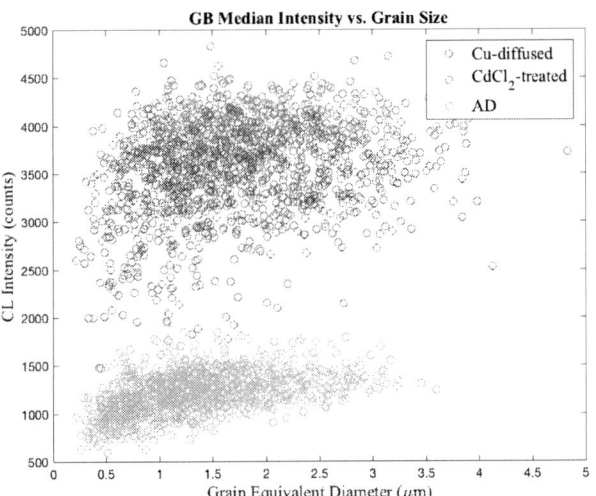

Fig. 5. GB median intensity vs. grain size. Each circle represents a grain.

TRPL data has been collected on the same films. $S$ is estimated from 1PE measurements at the back surface using the equation [6]:

$$\tau_1 \approx \frac{1}{\alpha S}. \qquad (3)$$

where $\tau_1$ is the first time constant resulting from a double-exponential fit to the TRPL decay and $\alpha$ is the absorption coefficient.

With the $S_{GB}$ and $S$ values determined for the films, the remaining free parameter, $\tau_{GI}$, is determined by simulating the CL intensity profile from the GI to the GB. Representative profiles for different grain sizes are being simulated. In total, a

set of $S_{GB}$, $S$, and $\tau_{GI}$ values will be determined for each film. The self-consistency of the entire parameter set will be checked by simulating CL intensity vs. grain size relationships (e.g., GI maximum intensity vs. grain size).

## IV. SUMMARY

In this work, quantitative cathodoluminescence (CL) is combined with time-resolved photoluminescence (TRPL) and numerical simulations to determine the GI lifetime ($\tau_{GI}$), GB recombination velocity ($S_{GB}$), and surface recombination velocity ($S$). CL intensities from thousands of grains were analyzed to accumulate statistics and chart variations with grain size. GB contrast results for small grains indicate that $S_{GB}$ decreases significantly with CdCl$_2$ treatment, but $S_{GB}$ is *increased* by the subsequent diffusion of Cu. Furthermore, within a given sample, data suggests that $S_{GB}$ is nearly independent of grain size. Detailed simulations are in progress to determine the precise $S_{GB}$ values for the films. TRPL measurements excited from the back surface are used to determine $S$, and CL profiles are simulated to determine the grain interior lifetime, $\tau_{GI}$. Finally, CL intensity vs. grain size relationships are simulated to check for self-consistency of the $S_{GB}$, $S$, and $\tau_{GI}$ values.

## REFERENCES

[1] B. Gaury and P. M. Haney, "Charged grain boundaries reduce the open-circuit voltage of polycrystalline solar cells—An analytical description," *Journal of Applied Physics*, vol. 120, p. 234503, 2016.

[2] Y. Jin and S. T. Dunham, "The Impact of Charged Grain Boundaries on CdTe Solar Cell: EBIC Measurements Not Predictive of Device Performance," *IEEE Journal of Photovoltaics*, vol. 7, pp. 329-334, 2017.

[3] M. Gloeckler, J. R. Sites, and W. K. Metzger, "Grain-boundary recombination in Cu (In, Ga) Se 2 solar cells," *Journal of applied physics*, vol. 98, p. 113704, 2005.

[4] Kanevce, J. Moseley, M. Al-Jassim, and W. K. Metzger, "Quantitative Determination of Grain-Boundary Recombination Velocity in CdTe by Cathodoluminescence Measurements and Numerical Simulations," *Photovoltaics, IEEE Journal of*, vol. 5, pp. 1722-1726, 2015.

[5] D. Kuciauskas, P. Dippo, A. Kanevce, Z. Zhao, L. Cheng, A. Los, *et al.*, "The impact of Cu on recombination in high voltage CdTe solar cells," *Applied Physics Letters*, vol. 107, p. 243906, 2015.

[6] D. Kuciauskas, A. Kanevce, J. M. Burst, J. N. Duenow, R. Dhere, D. S. Albin, *et al.*, "Minority carrier lifetime analysis in the bulk of thin-film absorbers using subbandgap (two-photon) excitation," *Photovoltaics, IEEE Journal of*, vol. 3, pp. 1319-1324, 2013.

# High resolution THz scanning for optimization of dielectric layer opening process on doped Si surfaces

P. Spinelli[1], F.J.K. Danzl[1], D. Deligiannis[1,2], N. Guillevin[1], A.R. Burgers[1], S. Sawallich[3],

M. Nagel[3], I. Cesar[1]

[1]Energy Research Centre of the Netherlands (ECN), Petten, 1755 LE, The Netherlands

[2]Delft University of Technology, Delft, 2628 CD, The Netherlands

[3]Protemics GmbH, Aachen, 52074, Germany

*Abstract* — Diffused IBC solar cells based on non-fire-through (NFT) metallization can achieve higher performance with respect to FT metallization, thanks to lower contact resistance. Less resistive contacts allows for lower metal fraction and thus lower contact recombination. For the application of NFT contacts, the rear side dielectric passivation layer needs to be locally opened. Standard processes for dielectric opening include laser ablation, use of etching pastes or masking and etching. In all this cases it is important that the opening process does not cause any damage to the diffusion below the contacts. We present a fast inspection method based on THz near-field scanning that allows investigating the diffusion layers after dielectric opening with spatial resolution suited for the small feature size of standard IBC solar cells. This method can be used to optimize crucial parameters of the dielectric opening without processing full cells. In this paper, we use THz scanning to optimize the curing temperature of the etch paste used to open the dielectric passivation layers on solar cells half-fabricates. After optimization, full cells are processed with top efficiency of 20.8%.

## I. INTRODUCTION

ECN "Mercury" cells are a bifacial IBC solar cell with a front floating emitter. The cells are processed at the cost level of PERC cells and can be easily interconnected into solar modules by, e.g., our conductive back-sheet foil technology [1]. The Mercury IBC cells have reached efficiencies up to 21.1%. [2] This result was achieved using a screen printed firing through (FT) metallization which is the current industry standard. The challenge of FT metallization is to limit the contact recombination as the frit fraction in the paste etches into the diffusion during firing for a significant depth. [3]. The emitter profile plays an important role in determining both the $j_{0,contact}$ and contact resistance of the of FT metal contacts. [3]. Typical values of $j_{0,contact}$ for the FT metal contact range between 1000 fA/cm$^2$ and 2500 fA/cm$^2$, as obtained by measuring specific test structures. [2] The high $j_{0,contact}$ of the emitter contact can thus cause significant open circuit voltage ($V_{oc}$) losses.

One way to reduce contact recombination is to use a non-fire-through (NFT) paste which can be applied also by screen printing but does not require a high temperature firing step. As for the case of FT paste, the emitter profile plays an important role also for the NFT paste. Having a highly doped emitter in the surface region below the contact allows for better shielding of minority carriers and thus lower contact recombination. Concomitantly, the contact resistance of the metal paste is also strongly dependent on the surface dopant concentration. Hence, a non-etching paste has also the potential to have a lower contact resistance allowing lower contact areas or lighter surface dopants with lower surface recombination properties on the passivated areas of the solar cell.

The NFT metallization for the rear side of bifacial IBC solar cells needs to be applied in combination with a dielectric passivation layer for areas with no metallization. At ECN we use an $Al_2O_3$/ $SiN_x$ dielectric stack to passivate the rear side surface of the Mercury cell. In order to apply the NFT metallization the dielectric layer stack needs to be locally opened. This can be done by several methods, such as by masking and chemical etching, laser ablation of the dielectric, or application of an etching paste. In order to achieve high efficiency IBC solar cells, the dielectric needs to be opened without damaging the diffusion below.

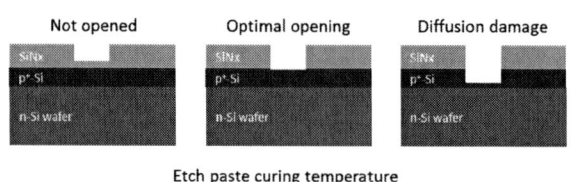

Fig 1. Schematic of three cases for the dielectric opening process.

Figure 1 shows schematics of three cases that can happen in the dielectric opening process. The optimal opening process is shown in the center panel. Under-etching (left panel) will results in poor or no contacting of the solar cell due to the residual dielectric layer in the opened area. On the other hand, over-etching (right panel) results in damaged diffusion layer under the contacts.

978-1-5090-5606-4/17 $31.00 © 2017 IEEE

In this paper, we use THz near-field transmission scanning to investigate the homogeneity of the diffusions after opening of the dielectric. This method allows the optimization of any parameter which is crucial to any opening process (laser ablation, chemical etching, etc.) without the need to process full cells, and thus at a lower cost.

First, the THz transmission setup and $R_{sheet}$ determination method are described, together with an analysis of the accuracy of the measurement for different feature sizes. Next, we consider the example of opening a $SiN_x$ dielectric layer with an etching paste. We study the effects of the paste curing temperatures on the dielectric opening, in order to achieve an optimal opening process which does not damage the diffusion below the contacts. We demonstrate that the THz scanning method is not influenced by the surface morphology of the dielectric layer, but only by the diffusion below it. A full optimization of the dielectric opening process is shown at cell level.

## II. THz Sheet Resistance Mapping

### A. THz transmission setup and Rsh determination

Sheet resistance ($R_{sheet}$) mapping is of great importance to characterize the homogeneity and doping levels of doped regions in solar cells. Usually 4-point probe (4pp) measurements are used to measure $R_{sheet}$. However, this method lacks the resolution required for a detailed inspection of IBC cells with sub-millimeter features.

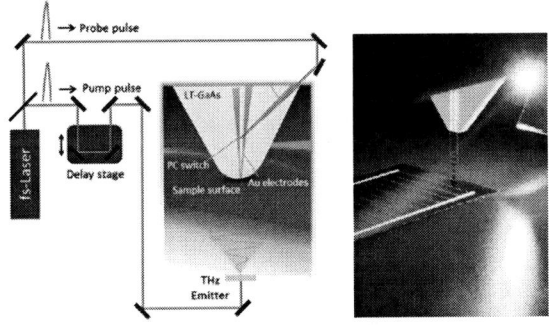

Fig 2. Schematic of the THz scanning setup (left) and photograph of the near field THz microprobe (right)

THz transmission near-field imaging is a method for high-resolution large-area sheet resistance quantification [4]. Such measurements uniquely allow monitoring the homogeneity of diffused layers with very high resolution enabling detection of features down to 20 μm in size. Measurements are performed in a THz transmission setup based on a pump/probe scheme, using broadband THz pulses. Figure 2a shows a schematic of the setup. Sub-wavelength spatial resolution is achieved by using a near-field microprobe (TeraSpike TD-800-X-HR, Protemics GmbH) for photo-conductive THz detection which

is scanned across the investigated IBC cell in a distance of few tens of microns (see photograph in Fig. 2b).

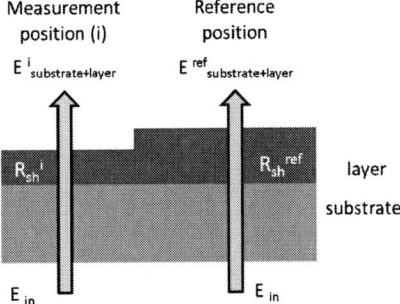

Fig 3. $R_{sheet}$ maps of thin conductive layers are obtained from THz transmission maps by means of the Tinkham formula and a reference region with known $R_{sheet}$.

The sheet resistance ($R_{sheet}$) of a thin conductive layer on top of a substrate is calculated from the measured THz transmission amplitude reduction generated by the doping layer using the Tinkham formula [4]:

$$\frac{1}{R_{sh}^i} = \frac{n+1}{Z_0}\left[\frac{E_{SL}^{ref}}{E_{SL}^i}\left(1 + \frac{Z_0}{R_{sh}^{ref}(n+1)}\right) - 1\right]$$

where $E_{SL}$ represents the transmitted THz field amplitude through the sample with conductive layer, the superscript $i$ indicates a position i(x,z) anywhere on the sample, while the superscript $ref$ indicates a position on a reference area on the sample with known $R_{sheet}$. Further constants are the substrate refractive index $n$ and the free-space impedance $Z_0$. The method can also take into account a second conductive layer on the other side of the cell (e.g. an FFE) if either the sheet resistance of this layer or its behavior in respect to the rear $R_{sh}$ is known.

The THz transmission method allows fast (5 ms/pixel) and contact less $R_{sheet}$ mapping in a range of 0.5Ω to 5 KΩ on substrates with typical background-doping levels ($<10^{16}$cm$^{-3}$).

### B. $R_{sheet}$ accuracy and feature size

For quantitative analysis or doping patterns relevant for IBC solar cells, it is important that $R_{sheet}$ can be accurately determined for feature sizes down to few tens of microns. We have investigated the accuracy of the $R_{sheet}$ determination with THz scanning by means of test structures with features of known size and conductivity, obtained by evaporation and lift-off of thin Cr layers. By controlling the thickness of the metal layers, different values of $R_{sheet}$ can be obtained. The test structures present feature sizes from 2.7 mm down to 10 microns.

978-1-5090-5606-4/17 $31.00 © 2017 IEEE

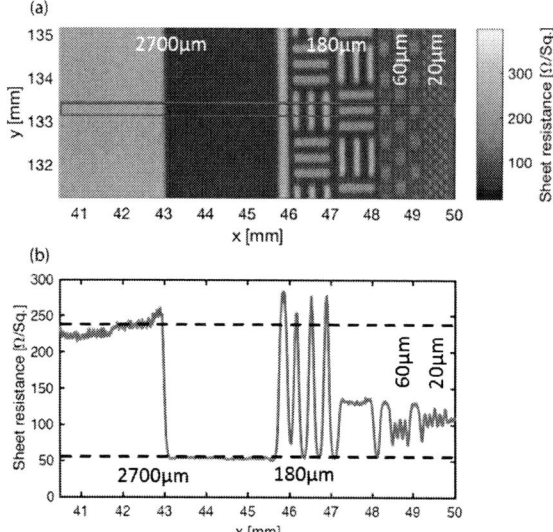

Fig 4. (a) $R_{sheet}$ map of an area of the test structure sample. (b) $R_{sh}$ values measured along one direction on the test structure (average of 10 lines). The area relative to this plot is shown by a red box in (a). The dashed lines in (b) are the known reference $R_{sh}$ values.

Figure 4 shows a $R_{sheet}$ map of part of the test structure (a) including several feature sizes (2700, 180, 60 and 20 μm), and a line scan of the same data taken along the x-axis direction (panel b, average of 10 lines). The dashed lines in Fig. 4(b) represent the known reference values of the features. As can be seen, the THz transmission method reliably reproduces the correct $R_{sheet}$ values for feature sizes larger than 180 μm. For smaller feature sizes, while a $R_{sheet}$ contrast is measured, the absolute values of $R_{sheet}$ deviate from the real ones. In particular, a lower $R_{sheet}$ contrast is observed and the resulting signal is a convolution of the real $R_{sheet}$ pattern and the THz beam shape. We are currently working on using deconvolution techniques to improve the accuracy of $R_{sheet}$ determination even feature size smaller than 180 μm.

## III. DIELECTRIC OPENING OPTIMIZATION WITH THZ

### A. Method

Samples were prepared from chemically polished n-type CZ 6-inch wafers. After BBr diffusion, a 70-nm-thick $SiN_x$ layer was applied on both sides of the wafer by PECVD. A SmartEtch3000 paste (Merck) was screen printed on the wafer rear side using a screen with 1 mm wide busbars and 100 μm wide fingers. The etch paste was then cured at different temperatures in a belt furnace, with fixed belt speed. The paste was finally removed in a $KOH/H_2O_2$ bath heated at 60°C.

The samples were first studied by optical microscopy. Based on the color of the opened area, it is possible to judge whether a residual layer is present, if this layer is thicker than 30 nm (see $SiN_x$ color charts, [5]). However, it is not possible to observe if the diffusion layer has been altered. THz probing offers the unique opportunity to inspect the diffusion layer in the opening and thus to optimize the opening process.

### B. Results

Polished samples with etch paste cured at different temperatures were studied with both optical microscopy and THz mapping. Figure 5 shows the effect of the paste curing temperature on the opening of the $SiN_x$, by means of microscope images of single fingers (top row), and $R_{sheet}$ maps of the diffusion layer obtained with THz scanning (bottom row). The color scale of the THz maps refers to the $R_{sheet}$ of the doped layer and is the same for the three plots. Three temperature are shown, corresponding to a reference temperature ($T_{ref}$), and to +30°C and +150°C temperature difference with respect to $T_{ref}$.

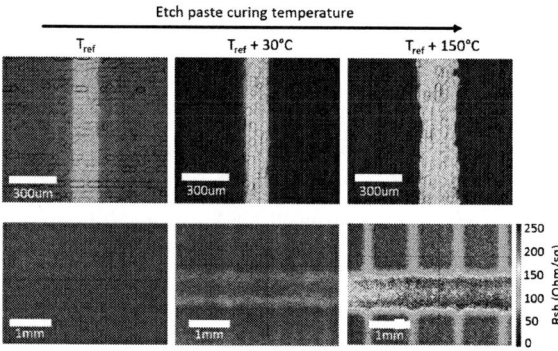

Fig 5. Microscope images (top row) and THz $R_{sheet}$ maps (bottom row) of samples after $SiN_x$ opening with etch paste, for 3 different curing temperatures.

For $T_{ref}$ (left) the optical microscope clearly shows that the $SiN_x$ has been etched, but is not fully opened (orange color in the opening). The THz $R_{sheet}$ scan shows a homogeneous map of the sample with a constant $R_{sheet}$ value of about 60 Ohm/sq. This is expected for a sample where the $SiN_x$ is not fully opened, because there is no damage to the diffusion. Importantly, the map also shows that the THz method is not influenced by the morphology of the $SiN_x$ layer, i.e. the THz mapping method does not show a "false contrast" due to changes in the $SiN_x$ thickness. For a curing temperature of $T_{ref}$+30°C (center) and $T_{ref}$+150°C (right) the microscope shows deeper etching of the $SiN_x$. The THz $R_{sheet}$ map shows a pattern of higher $R_{sheet}$ in the region corresponding to the opened busbar and finger areas. The $R_{sheet}$ in these areas increases up to a value of 90 Ohm/sq for the $T_{ref}$+30°C, and up to 250 Ohm/sq for the highest temperature, as measured on the busbar area. The increase of $R_{sheet}$ is attributed to partial etching of the diffused layer by the SmartEtch paste.

From this study we are thus able to optimize the curing temperature of the etch paste in order to avoid (or minimize) the damage created by the etch paste to the diffusion. Using

THz allows for optimization using half-fabricates of solar cells, while so far the optimization had to be conducted on fully functional cells which was usually more costly and time consuming.

## IV. IBC CELLS WITH NFT PASTE

The optimization presented in the previous section was used as a starting point to make Mercury IBC solar cells with a NFT paste. The cells were made on 6-inch wafers with chemically polished rear side, following our standard process flow for Mercury cells [1]. The rear side was passivated with a $Al_2O_3/SiN_x$ layer stack. The opening of this dielectric stack was performed by combining the SmartEtch paste with a 1% HF dip. The aim was to partially open the rear side dielectric with the SmartEtch paste, and to complete the opening with the HF dip, without etching the exposed diffusions. A full optimization of the paste curing temperature and HF dip time was performed at cell level. The curing temperature was chosen between $T_{ref}+10°C$ and $T_{ref}+70°C$, i.e. around the optimal temperature found in the optimization study with THz. The HF dip time was varied between 2.5 and 10 minutes. In each case, the thickness of the front side $SiN_x$ anti-reflection coating was corrected according to the HF dip time. The NFT paste was finally printed to contact the IBC cells, which were then measured in house in a Wacom solar simulator.

before the application of the NFT metallization. The FF improves both by increasing the curing temperature and/or by increasing the HF dip time, due to a more effective opening of the dielectric. Overall, the first method seems to be more effective. The optimal FF value (from the fit) is obtained for $\Delta T=55°C$ and the shortest HF time of 2.5 min. The lower FF values (relative to the optimum) observed for longer HF times, are related to losses in pseudo-FF (not shown), which could be attributed to a worsening of the passivation properties of the dielectric stack. It is known that a bad surface passivation at the rear side of IBC cells can cause pseudo-FF losses due to increased $j_{02}$ recombination of the meandering pn-junction [6]. The best measured FF was 77.9%, obtained for cells with $\Delta T=40°C$ and HF time = 2.5 min.

Figure 7 shows the cell $V_{oc}$ surface response with respect to HF dip time and $\Delta T$, similarly to what shown in Fig. 6 for the FF. A clear optimum value of $V_{oc}$ is observed for relatively low curing temperature ($\Delta T=20°C$) and for the shortest HF time. The best measured $V_{oc}$ in this conditions was 654 mV. Note that this value is limited by a high $j_0$ of the passivated rear emitter surface ($j_{0,emit} >100$ fA/cm$^2$) relative to our standard baseline process ($j_{0,emit}= 45$ fA/cm$^2$), as measured on symmetrical test structures. We are currently investigating the reasons for the worse surface passivation with respect to our baseline process.

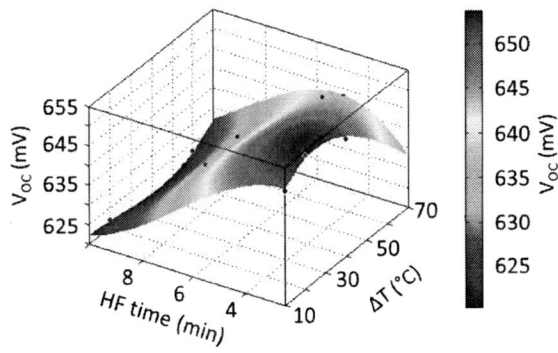

Fig 7. $V_{oc}$ of IBC cells as a function of HF dip time (x-axis) and paste curing temperature (y-axis)

As can be seen from the graph, the dominant effect on the $V_{oc}$ is the HF dip time. A severe $V_{oc}$ loss is observed for longer HF times, which is attributed to a worsening of the surface passivation due to etching of the dielectric stack. A reduction in $V_{oc}$ is also noted for the highest curing temperatures ($\Delta T=70°C$). This is attributed to damage of the diffusion layer below the contact due to over-etching of the SmartEtch paste.

In order to confirm this, Fig. 8 shows the $R_{sheet}$ map of an unmetallized sample, obtained with a THz scan. The map shows several $R_{sheet}$ contrasts. The main modulation is between the BSF and emitter fingers, which have resistivities

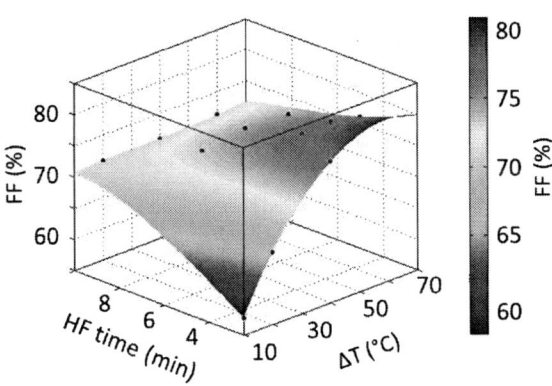

Fig 6. FF of IBC cells as a function of HF dip time (x-axis) and paste curing temperature (y-axis)

Figure 6 shows the fill-factor (FF) of the IBC cells (z-axis and color bar) as a function of the HF dip time (x-axis) and the temperature difference with respect to $T_{ref}$ ($\Delta T$, y-axis). The dots in the graph represent the measured values (each is an average of 3-5 cells). The color surface is a polynomial cubic fit of the measured data. Some clear trends are observed from the plot. First, a very low fill factor is observed for the lowest temperature in combination of the short HF dip. This is attributed to a non-complete opening of the dielectric stack

978-1-5090-5606-4/17 $31.00 © 2017 IEEE

of ~20 Ohm/sq (blue) and 90 Ohm/sq (green, used as reference area), respectively. At the center of the emitter finger, a clear area with much higher $R_{sheet}$ (~200 Ohm/sq, red) is visible, corresponding to the area opened by the SmartEtch paste. A higher $R_{sheet}$ in this area causes a less effective "shielding" of minority carriers from the metal contact and thus larger contact recombination. Interestingly, the $R_{sheet}$ map also reveals a sub-pattern in the opening area, where periodic blobs of higher $R_{sheet}$ (i.e. deeper etch) are visible along the finger direction, corresponding to the mesh of the screen used for printing the SmartEtch paste. This feature reveals the etching dynamic of the etch paste, and it could only be detected using the high-resolution THz mapping method.

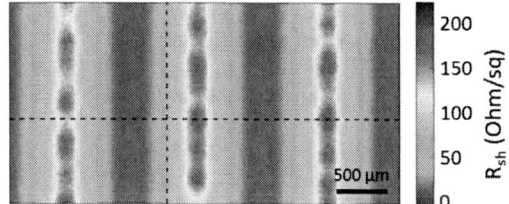

Fig 8. $R_{sheet}$ map of an IBC structure, revealing the etching dynamics of the screen printed paste.

Overall, the best cell was obtained with a curing temperature of $T_{ref}$+50°C, combined with an HF dip of 2.5 min. The best measured cell efficiency was 20.8%. The table below reports the IV parameters of the best cell, as measured in-house without spectral mismatch correction.

| Best cell IV parameters | |
|---|---|
| $J_{sc}$ (mA/cm$^2$) | 41.1 |
| $V_{oc}$ (mV) | 649.2 |
| FF (%) | 77.7 |
| Efficiency (%) | 20.8 |

## V. CONCLUSION

We have presented a new method to optimize the SiN$_x$ opening of IBC solar cells for application of NFT paste. This method is based on THz transmission measurements and allows to study diffusion layers with resolutions down to 20 μm. This method is also applicable to solar cell half-fabricates, thus avoiding the processing of full cell for parameter optimization. We have presented a preliminary study on the accuracy of $R_{sheet}$ determination with the THz transmission method, and concluded that absolute $R_{sheet}$ values can be measured for features down to 180 μm. For smaller features a reduction in the $R_{sh}$ contrast is observed. As an example of the THz capabilities, we have studied the case of opening of a dielectric layer stack with an etching paste, prior to the application of a NFT metallization. In our study, we optimize the curing temperature of the etch paste in order to avoid damage to the diffusion layer. We used THz $R_{sheet}$ mapping to identify the range of curing temperatures most suited for the opening process. A full optimization of this latter was performed on cell level. The results have been presented, with cell efficiencies up to 20.8%.

## REFERENCES

[1] I. Cesar, et al. "Enablers for integral IBC cell and module development and implementation in PV industry", 26$^{th}$ PVSEC (2016)

[2] A. Mewe, et al. "Emitter and contact optimization for high efficiency IBC Mercury cells", 31$^{st}$ EUPVSEC (2016)

[3] A. Cuevas, et al. "Surface recombination velocity of highly doped n-type silicon", J. of Appl. Phys. 80 (6), 3370-3375 (1996)

[4] M. Nagel, et al. "THz microprobe system for contact-free high-resolution sheet resistance imaging," 28th EUPVSEC (2013)

[5] http://www.cleanroom.byu.edu/color_chart.phtml

[6] P. Spinelli, et. al. "Quantification of PN-junction Recombination in IBC Crystalline Silicon Solar Cells", in press, IEEE Journal of Photovoltaics (2017)

# Degradation Assessment of Fielded CIGS Photovoltaic Arrays

Bruce H. King, Joshua S. Stein, Daniel Riley, C. Birk Jones and Charles D. Robinson

Sandia National Laboratories, Albuquerque, NM, 87185-0951, USA

*Abstract* — Copper indium gallium (di)selenide (CIGS) photovoltaic (PV) cell technology is a promising candidate to meet and exceed SunShot price targets. However, adoption of CIGS is hindered by significant uncertainties regarding long-term reliability and performance stability, as well as a lack of accurate modeling tools to predict CIGS system performance. Sandia has maintained a collection of instrumented grid-tied CIGS PV systems ranging in age from 2-5 years. Most of these arrays include modules that were thoroughly characterized prior to deployment. In this paper, we explore the medium-term reliability and performance stability of CIGS modules by analyzing real world performance and degradation rates of these systems. We observe two populations with respect to degradation in maximum power. One population of older products not representative of currently shipping products displayed a high degradation rate of 2%/year or greater while the second, contemporary population displayed degradation of less than 0.5%/year.

*Index Terms* — CIGS, outdoor testing, degradation assessment, Sandia Array Performance Model, SAPM.

## I. INTRODUCTION

Copper indium gallium (di)selenide (CIGS) photovoltaic (PV) cell technology is a promising candidate to meet and exceed SunShot price targets. A recent cost analyses predict that CIGS module price can drop to \$0.44/W or lower with 15% efficiency [1, 2]. Record cell efficiencies as high as 22.6% have been demonstrated, exceeding long-standing records for multi-crystalline silicon [3, 4]. Novel designs such as tandem cell have the potential to boost efficiency even higher. While PV modules based on mono-crystalline silicon already exhibit efficiencies >20%, CIGS modules promise to dramatically reduce the amount of semiconductor material and the cost of the end device.

However, adoption of CIGS is hindered by significant uncertainties regarding long-term reliability and performance stability, as well as a lack of accurate modeling tools to predict CIGS system performance. Both increases and decreases in performance with exposure to light have been reported [5]. This metastable behavior complicates attempts to quantify the performance of CIGS modules and contributes to uncertainty in degradation assessments. To address this, artificial light soaking prior to indoor characterization is becoming the norm for PV test labs [6, 7, 8]. However, a recent review article focusing on stability of CIGS cells asserted that general statements about the lifetime of CIGS modules cannot be made due to the scant amount of publicly available fielded CIGS data [9].

Despite an apparent resurgent interest in CIGS, there is still a considerable gap between fundamental cell studies and a comprehensive understanding of field performance. In this paper, we explore the long-term reliability and performance stability of CIGS modules by analyzing real world performance and degradation rates of operational fielded CIGS systems installed at Sandia. The systems under consideration (Table 1) range in age from 2-5 years. These systems represent a cross-section of commercial manufacturing methods for CIGS as well as differing packaging methods. In most cases, detailed performance evaluations of individual modules were made prior to installation and these same modules have been recharacterized. Characterization includes both indoor laboratory flash testing and more comprehensive outdoor tracker testing [10, 11]. Outdoor tracker data is analyzed using the Sandia Array Performance Model (SAPM) [12]. Time-series system data is also collected for each system. Instrumentation for these measurements will be described below but this data will not be addressed in this paper.

TABLE I
SYSTEMS UNDER STUDY AT SANDIA

| System | Cell | Package | Size (kW) | Installation |
|--------|------|---------|-----------|--------------|
| CIGS-1 | discrete | Polymer/Flex | 3.36 | 1/12 |
| CIGS-2 | discrete | Glass-glass | 2.2 | 1/12 |
| CIGS-3 | discrete | Glass-glass | 2.32 | 6/12 |
| CIGS-4 | monolithic | Glass-glass | 4.8 | 6/13 |
| CIGS-5 | monolithic | Glass-glass | 6 | 4/15 |
| CIGS-6 | monolithic | Glass-glass | 6 | 4/15 |

Fig. 1. Two CIGS arrays under test at Sandia

## II. EXPERIMENTAL PROCEDURE

### A. System Instrumentation

Each system is instrumented to measure DC voltage and current using calibrated commercial current shunts and custom voltage dividers designed and produced by Sandia. Module temperature is monitored using type-T thermocouples adhered to the backside of selected modules. Reference irradiance is measured using local broadband global pyranometers (Kipp & Zonen CMP-11) in the plane of array. Local analog measurements are digitized using high accuracy A/D converters (ICPDAS). Data logging is via RS-485 communications using a variety of data loggers (Campbell Scientific CR1000 or CR6; Raspberry Pi). Time-series data is logged every five seconds, from which one-minute averages are produced. Additional data is obtained from a co-located weather platform including direct normal irradiance (DNI), global normal (GNI), wind speed, ambient temperature, humidity, barometric pressure and precipitation. All data acquisition systems are synchronized to a local time server. Data is quality checked using custom Python scripts [13] and transferred nightly into a remote database.

*B. Module Characterization*

Prior to installation, selected modules from each system were characterized to establish initial performance. After a minimum of 18 months of operation, these same modules were then removed from each system and recharacterized to investigate degradation or stability. The length of deployment differs between systems since each were installed incrementally over ~ 4 years. Additionally, operation of two of the systems (CIGS-1 and CIGS-4) was terminated due to physical signs of degradation. All other systems are still in operation. STC performance measurements were made indoors on a flash tester and more comprehensive measurements were made outdoors on a two-axis tracker.

Indoor flash testing was primarily done using a Spire 4600SLP. Modules were light soaked outdoors under natural sunlight prior to characterization. New modules characterized prior to installation were light soaked on a fixed tilt rack according to manufacturers' guidelines. Aged modules were removed from functioning arrays near solar noon on clear days and characterized within one hour of removal. Two groups of modules were additionally light soaked indoors according to IEC 61215-1-4 [14]. Comparison between these methods indicated that natural light soaking produced results comparable or slightly superior to artificial light soaking. Details of this study are beyond the scope of this paper and are not reported here. Modules from one system (CIGS-1) were not flash tested due to physical limitations of the flash test plane.

Detailed performance measurements were made outdoors using a two-axis Azimuth-Elevation solar tracker and measurement techniques developed at Sandia [11]. Modules were instrumented with thermocouples attached to the back sheet (or glass) of the module, and then mounted on the tracker. While mounted on the tracker, IV measurements were made continuously at 2 minute intervals during daylight hours. Modules were held at maximum power in between each measurement. Measurements were made across a several-week period including both clear and cloudy conditions to

determine response to changing irradiance and spectrum. Modules from one system (CIGS-4) were characterized in a new state but these modules were not installed in the array. Tracker results from this system are not reported here.

III. ANALYSIS

The modules used in this study vary considerably from one another both dimensionally and electrically. Module area varies from $0.7 - 2m^2$. Modules constructed from discrete cells generally exhibit current and voltages that are comparable to values expected from crystalline silicon modules whereas monolithic modules exhibit significantly lower current and higher voltage. To facilitate meaningful comparison between modules and to provide anonymity, measured electrical properties are presented on a per cell basis. Current and power are further normalized by effective cell area ($1/cm^2$), determined by dividing module aperture by the number of cells in series. It should be emphasized that these reported values will generally be lower than expected values from bare cells. Reported voltages are reduced by resistive losses due to cell-cell interconnects, the junction box and external module wiring. Reported current is reduced by module packaging and inactive areas within the module. Dimensionless values such as diode ideality and fill factor and those reported on a percentage basis such as temperature coefficients are reported directly.

For flash test data, minimal post-processing was required beyond normalization. In most cases, multiple modules of each type were characterized. Flash test results are presented as numerical averages of measured data.

Tracker test data required considerably more post-processing. Testing was under real world operating conditions (i.e. not at STC) and resulted in thousands of IV curves for each module. Data was processed according to the methods described in [11] using custom MATLAB scripts. STC values were calculated from the calibrated SAPM coefficients [12] for each module. This method has a distinct advantage over single point STC measurements made on the flash tester in that it facilitates normalized comparison across a range of operating conditions. Due to the length of time required to capture data across relevant operating conditions and the specialized nature of the measurement hardware, in most cases only a single module was characterized. Results presented here are for these individual modules.

IV. RESULTS AND DISCUSSION

*A. Flash Test and Tracker STC Results*

Flash test results are presented in Table II. As was mentioned earlier, dimensional constraints prevented modules from system CIGS-1 from being flash tested. Modules from systems CIGS-2 and CIGS 4 displayed significant decrease in power (-7.5% and -4.6%, respectively), whereas CIGS-3,5 and 6 were stable or showed a slight increase. Changes in power

## TABLE II
### Average Module STC Performance Determined From Flash Testing

| System ID | CIGS-2 | | | CIGS-3 | | | CIGS-4 | | | CIGS-5 | | | CIGS-6 | | |
|---|---|---|---|---|---|---|---|---|---|---|---|---|---|---|---|
| # Modules | 1 | | | 2 | | | 13 | | | 6 | | | 5 | | |
| Months | 45 | | | 41 | | | 20 | | | 22 | | | 23 | | |
| Condition | New | Aged | %Δ | New | Aged | %Δ | New | Aged | %Δ | New | Aged | %Δ | New | Aged | %Δ |
| $P_{mp}$ (mW/cm$^2$) | 12.0 | 11.1 | -7.5% | 13.5 | 13.7 | 1.5% | 10.8 | 10.3 | -4.6% | 13.1 | 13.0 | -0.8% | 12.9 | 13.0 | 0.8% |
| $I_{sc}$ (mA/cm$^2$) | 30.1 | 29.9 | -0.7% | 29.2 | 29.8 | 2.1% | 26.5 | 27.1 | 2.3% | 31.0 | 30.5 | -1.6% | 31.3 | 30.7 | -1.9% |
| $V_{oc}$ (mV/cell) | 595 | 581 | -2.4% | 639 | 639 | 0.0% | 607 | 594 | -2.1% | 599 | 602 | 0.5% | 591 | 597 | 1.0% |
| $I_{mp}$ (mA/cm$^2$) | 26.0 | 24.3 | -6.5% | 26.0 | 26.4 | 1.5% | 23.1 | 22.9 | -0.9% | 27.9 | 27.6 | -1.1% | 27.9 | 27.7 | -0.7% |
| $V_{mp}$ (mV/cell) | 464 | 458 | -1.3% | 519 | 518 | -0.2% | 468 | 449 | -4.1% | 468 | 470 | 0.4% | 461 | 468 | 1.5% |
| $R@I_{sc}$ (Ω-cm$^2$) | 326 | 210 | -36% | 423 | 465 | 10% | 844 | 655 | -22% | - | 990 | - | - | 1116 | - |
| $R@V_{oc}$ (Ω-cm$^2$) | 2.65 | 2.19 | -17% | 0.94 | 1.90 | 102% | 3.15 | 3.32 | 5% | - | 2.92 | - | - | 2.74 | - |
| Fill Factor | 0.67 | 0.64 | -4.5% | 0.72 | 0.72 | 0.0% | 0.67 | 0.64 | -4.5% | 0.71 | 0.71 | 0% | 0.69 | 0.71 | 2.9% |

were not correlated with cell type/module construction. The decrease in power was reflected in a decrease in fill factor for both CIGS-2 and 4, whereas fill factor was stable for the remaining modules.

For CIGS-2, the reduction in power was largely correlated with a decrease in Imp and a smaller drop in Vmp. This trend was further reflected in a decrease in shunt resistance (R@Isc). However, it was also seen for this module that all parameters decreased. For CIGS-4, the reduction in power was more strongly correlated with a decrease in Vmp. However, this was not accompanied by a corresponding increase in series resistance (R@Voc).

Results from CIGS-3,5 and 6 were largely unremarkable other than to note the lack of significant degradation in these systems. One trend of note across all systems is the apparent

incremental improvement in shunt resistance for newer modules.

STC parameters from tracker testing are shown in Table III. As noted earlier, CIGS-4 is not represented in this data set. The results largely mirror those observed from flash testing, namely significant degradation in Pmp for CIGS-2 and stability for CIGS-3, 5 and 6. Here, CIGS-1 was represented, which also displayed degradation in Pmp. As with CIGS-2, degradation in Pmp was correlated with a decrease in Imp.

Both CIGS-1 and 2 consistently showed unfavorable changes in most parameters. Notably, both the power density and fill factors of CIGS-1 and 2 were significantly lower than the other systems in this study. Diode ideality factor was also significantly greater for both of these systems, being 2 or greater. Diode factor for CIGS-2 was dramatically greater

## TABLE III
### Individual Module STC Performance Determined From Tracker Testing

| System ID | CIGS-1 | | | CIGS-2 | | | CIGS-3 | | | CIGS-5 | | | CIGS-6 | | |
|---|---|---|---|---|---|---|---|---|---|---|---|---|---|---|---|
| Module ID | 2760 | | | 2776 | | | 2912 | | | 9086 | | | 9087 | | |
| Months | 18 | | | 45 | | | 41 | | | 22 | | | 23 | | |
| Condition | New | Aged | %Δ | New | Aged | %Δ | New | Aged | %Δ | New | Aged | %Δ | New | Aged | %Δ |
| $P_{mp}$ (mW/cm$^2$) | 7.9 | 7.7 | -2.5% | 10.9 | 9.8 | -10.1% | 13.4 | 13.5 | 0.7% | 13.3 | 13.2 | -0.8% | 13.2 | 13.3 | 0.8% |
| $I_{sc}$ (mA/cm$^2$) | 26.4 | 25.9 | -1.9% | 29.9 | 30.0 | 0.3% | 28.5 | 28.5 | 0.0% | 31.1 | 30.2 | -2.9% | 31.7 | 31.1 | -1.9% |
| $V_{oc}$ (mV/cell) | 517 | 514 | -0.6% | 594 | 593 | -0.2% | 640 | 644 | 0.6% | 602 | 614 | 2.0% | 593 | 603 | 1.7% |
| $I_{mp}$ (mA/cm$^2$) | 21.8 | 21.2 | -2.8% | 23.9 | 22.3 | -6.7% | 25.7 | 25.6 | -0.4% | 27.9 | 27.1 | -2.9% | 28.3 | 27.7 | -2.1% |
| $V_{mp}$ (mV/cell) | 363 | 364 | 0.3% | 456 | 440 | -3.5% | 519 | 526 | 1.3% | 475 | 487 | 2.5% | 465 | 478 | 2.8% |
| γPmp(%/°C) | -0.59 | -0.62 | 5.1% | -0.38 | -0.37 | -2.6% | -0.40 | -0.37 | -7.5% | -0.48 | -0.39 | -19% | -0.52 | -0.40 | -23% |
| αIsc(%/°C) | -0.04 | 0.05 | -225% | -0.02 | 0.00 | -100% | 0.04 | 0.03 | -25% | 0.01 | 0.03 | 200% | 0.02 | 0.04 | 100% |
| βVoc(%/°C) | -0.38 | -0.41 | 7.9% | -0.36 | -0.47 | 31% | -0.32 | -0.30 | -6.3% | -0.35 | -0.33 | -5.7% | -0.37 | -0.33 | -11% |
| αImp(%/°C) | -0.14 | -0.06 | -57% | -0.04 | 0.03 | -175% | -0.01 | -0.02 | 100% | -0.08 | -0.02 | -75% | -0.10 | -0.02 | -80% |
| βVmp(%/°C) | -0.46 | -0.56 | 22% | -0.34 | -0.40 | 17.6% | -0.39 | -0.35 | -10% | -0.41 | -0.37 | -9.8% | -0.43 | -0.38 | -12% |
| Diode Factor, n | 2.10 | 1.90 | -9.5% | 2.96 | 4.60 | 55.4% | 1.48 | 1.51 | 2.0% | 1.56 | 1.63 | 4.5% | 1.57 | 1.61 | 2.5% |
| Fill Factor | 0.58 | 0.58 | 0.0% | 0.61 | 0.55 | -9.8% | 0.73 | 0.73 | 0.0% | 0.71 | 0.71 | 0.0% | 0.70 | 0.71 | 1.4% |

978-1-5090-5606-4/17 $31.00 © 2017 IEEE

than for other modules and was seen to display degradation on a scale not observed for other modules. This will be discussed further below. While stable, CIGS-1 displayed a temperature coefficient significantly greater than any of the modules in this study (~ -0.6%/°C). Since these modules are the oldest in the study, it is likely that these observations are reflective of the relative immaturity of these products and maturity of the newer products.

A summary of the annual rate of change in Pmp at STC for each of the modules is shown below in Table IV. The results clearly fall into two different categories; modules with ~2% or greater annual degradation and modules with 0.5% or lower. As noted earlier, there is no consistent trend across cell type or module construction.

TABLE IV
ANNUAL RATE OF CHANGE IN PMP AT STC

| System | Flash Test | Tracker |
|--------|-----------|---------|
| CIGS-1 | - | -1.7% |
| CIGS-2 | -2.0% | -2.7% |
| CIGS-3 | 0.4% | 0.2% |
| CIGS-4 | -2.8% | - |
| CIGS-5 | -0.4% | -0.4% |
| CIGS-6 | 0.4% | 0.4% |

*B. Tracker Test Results – behavior across a range of operating conditions*

Tracker testing and SAPM analysis has a distinct advantage over single point STC measurements made on the flash tester in that it facilitates normalized comparison across a range of operating conditions.

*i). Air mass response.* In the SAPM, the equation for $I_{sc}$ is given by;

$$I_{sc} = I_{sco} f_1(AM)[E_e][1 + \hat{\alpha}_{Isc}[T_c - T_0]]$$

where the function $f_1(AM)$

$$f_1(AM) = a_0 + a_1(AM) + a_2(AM)^2 + a_3(AM)^3 + a_4(AM)^4$$

is an empirically determined polynomial function that is a proxy for solar spectral influence on $I_{sc}$. It is dimensionless and defined to be 1 at AM1.5.

A typical plot of the air mass function for CIGS and the data used to generate it is shown in Figure 2. In general, the air mass response for CIGS is quite pronounced. In all cases investigated here, $f_1(AM)$ rises monotonically with increasing air mass. In contrast, for crystalline silicon, $f_1(AM)$ typically reaches a peak value of 1.02 in the range of 2<Am<3.

Change in $f_1(AM)$ for each module is shown in Table IV. It is not possible to calculate change under STC conditions, because by definition, $f_1(1.5) = 1$. Therefore, change was calculated two different ways. First, by determining the difference in $f_1(AM)$ at AM=3, corresponding to a sun elevation of 15-20°. Second, the definite integral of $f_1(AM)$ was found on the interval of $1 \leq AM \leq 5$ and the difference

was calculated from this. As can be seen, the difference at AM3 serves as a close proxy for the difference in the integrated value. Across all modules, the change is very slight, indicating that air mass response and by extension, spectral response, is stable and does not degrade.

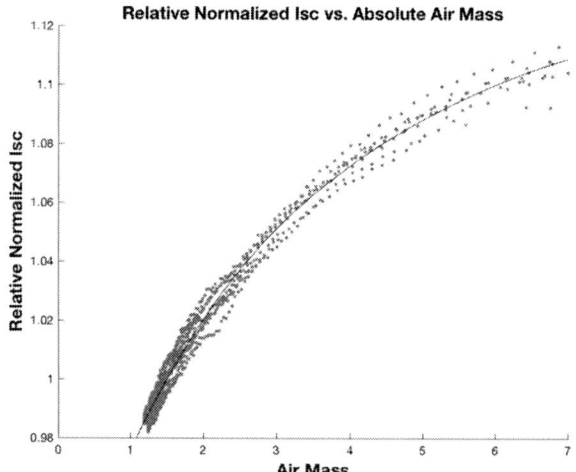

Fig. 2. Typical Air Mass Function (CIGS-1)

TABLE IV
CHANGE IN $F_1$(AM) WITH AGING.

| System ID | $\Delta f_1(3)$, % | $\Delta \int f_1(1-5)$, % |
|-----------|--------------------|----------------------------|
| CIGS-1 | 0.86 | 0.87 |
| CIGS-2 | 0.38 | 0.50 |
| CIGS-3 | -0.15 | -0.10 |
| CIGS-5 | 0.56 | 0.53 |
| CIGS-6 | 0.56 | 0.60 |

*ii.) Diode Ideality Factor.* In the SAPM, the equation for $V_{oc}$ is,

$$V_{oc} = V_{oco} + N_s \delta(T_c) \ln(E_e) + \beta_{Voc}[T_c - T_0]$$

where $\delta(T_c)$, the thermal voltage per cell, is given by

$$\delta(T_c) = \frac{nk[T_c + 273.15]}{q}$$

$n$, the diode ideality factor, may be found from the slope of a plot of temperature-adjusted $V_{oc}$ to the quantity

$$\frac{N_s k[T_c + 273.15] \ln(E_e)}{q}$$

An example of a typical diode factor is shown below in Figure 3. In the current formulation of the SAPM, diode factor is expected to be a constant. A typical value for crystalline silicon modules is 1.1, whereas for most contemporary CIGS modules it is ~1.5. It's not uncommon for CIGS module to display slight deviation from constant slope at the low end of the range, i.e. low irradiance and cell temperature.

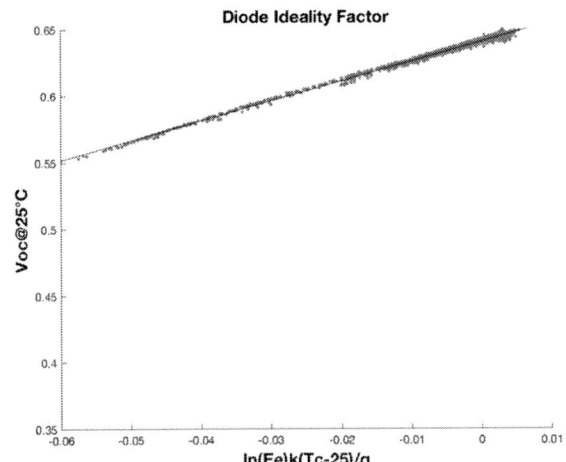

Fig. 3. Typical Constant Diode Ideality Factor (CIGS-3).

However, this deviation at low irradiance/low temperature can become quite pronounced. An example is shown in Figure 4, for CIGS-2. This module displayed considerable non-linearity in diode factor. Since the SAPM is formulated around a constant value of $n$, regression analysis produces a linear fit through the non-linear data. As a consequence, analysis returns a constant value that is greater than the slope at STC (x=0). The impact of this type of behavior is a loss of prediction accuracy for voltage terms in the SAPM.

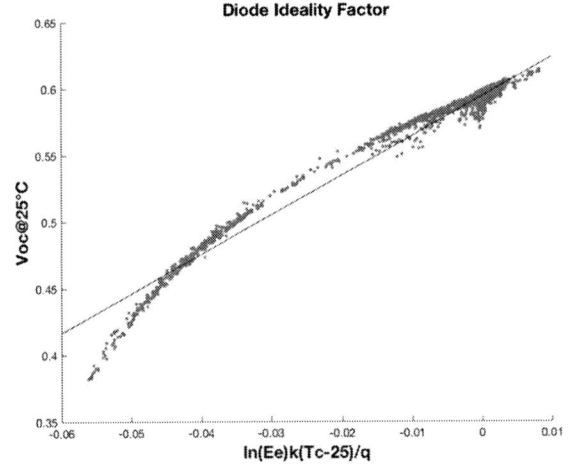

Fig. 4. Variable Diode Ideality Factor (CIGS-2).

## V. SUMMARY

In this paper, we have characterized the stability and behavior of modules from six grid-tied PV systems operating across multiple years. STC measurements were found to be comparable between indoor flash testing and comprehensive outdoor tracker testing. Half of the systems under study displayed significant annual degradation in Pmp of ~2%/year or greater, while the other half displayed little to no degradation. No correlation was observed between degradation (or stability) and cell type or module construction. In two cases, degradation in Pmp was correlated with degradation in Imp, while in the third case it was correlated with degradation in Vmp. Likewise, degradation in fill factor was observed for two systems but not for a third. The remaining three stable systems displayed healthy electrical parameters and minimal changes year over year.

The Sandia Array Performance Model (SAPM) was used to evaluate tracker testing data. For most of the systems in this study, the model accurately described behavior, although significant differences were noted between typical model parameters for conventional crystalline silicon and CIGS. Most notably, the air mass function was seen to continue to rise to as much as 10% increase over the AM1.5 value at high air mass whereas crystalline silicon typically reaches a peak of 2% over the AM1.5 value before decreasing at high air mass. The magnitude of the ideality factor was also seen to differ from conventional expectations, with 1.5 being commonly observed for CIGS modules. One module displayed pronounced non-linearity in diode ideality, which the SAPM was unable to accurately represent.

Finally, contemporary CIGS modules displayed marked improvement in both behavior and stability over earlier products.

## ACKNOWLEDGEMENTS

This work was supported by the U.S. Department of Energy SunShot Initiative. Sandia National Laboratories is a multimission laboratory managed and operated by National Technology and Engineering Solutions of Sandia LLC, a wholly owned subsidiary of Honeywell International Inc. for the U.S. Department of Energy's National Nuclear Security Administration under contract DE-NA0003525.

## REFERENCES

[1] D. Metacarpa, et.al, "Cost Analysis and Criteria for Achieving the SunShot Target of $0.5Wp Manufacturing Cost Using Flexible Thin Film Cu(In, Ga)Se$_2$ Solar Cells," in *39th IEEE Photovoltaic Specialists Conference*, 2013. Tampa, FL.

[2] K. A. W. Horowitz and M. Woodhouse, "Cost and Potential for Monolithic CIGS Photovoltaic Modules," in *42nd IEEE Photovoltaic Specialists Conference*, 2015. New Orleans, LA.

[3] NREL, "Best Research-Cell Efficiencies," http://www.nrel.gov/ncpv/images/efficiency_chart.jpg, accessed Jan. 27, 2017.

[4] Green, M. A., et. al., "Solar cell efficiency tables (Version 49), *Progress in Photovoltaics: Research and Applications*, 25: pp. 3-13, 2017.

[5] M. Gostein and L. Dunn, "Light soaking effects on photovoltaic modules: Overview and literature review," *Proceedings of the 37th IEEE Photovoltaic Specialists Conference*, p. 3126, 2011

[6] Michael G. Deceglie, Timothy J. Silverman, Bill Marion, Sarah R. Kurtz, "Robust Measurement of Thin-Film Photovoltaic Modules Exhibiting Light-Induced Transients," *Proceedings SPIE Optics+Photonics*, 2015

[7] R. P. Kenny, A. I. Chatzipanagi, and T. Sample, "Preconditioning of thin film PV module technologies for calibration," *Prog. Photovoltaics, Res. Appl.*, Vol. 22 (2), p. 166, 2014.

[8] Daniela Dirnberger, "Uncertainty in PV Module Measurement---Part II: Verification of Rated Power and Stability Problems," *IEEE Journal of Photovoltaics*, Vol.4 (3), p. 991, 2014

[9] Mirjam Theelen and Feliz Daume, "Stability of Cu(In,Ga)Se2 solar cells: A literature review," *Solar Energy*, 133, pp 586-627, 2016.

[10] B. H. King, et. al., "Outdoor Test and Analysis Procedures for Generating Coefficients for the Sandia Array Performance Model," in *43rd IEEE Photovoltaic Specialists Conference*, Portland, OR, 2016.

[11] B. H. King, C. W. Hansen, D. Riley, C. D. Robinson and L. Pratt, "Procedure to Determine Coefficients for the Sandia Array Performance Model (SAPM)," Sandia National Laboratories, SAND2016-5284, 2016.

[12] D. L. King, W. E. Boyson, J. A. Kratochvil, "Photovoltaic Array Performance Model,", Sandia National Laboratories, SAND2004-3535, 2004.

[13] K. A. Klise and J. S. Stein, "Automated Performance Monitoring for PV Systems Using PECOS," in *43rd IEEE Photovoltaic Specialists Conference*, Portland, OR, 2016.

[14] IEC 61215-1-4:2016, "Terrestrial photovoltaic (PV) modules – Design qualifications and type approval – Part 1-4: Special requirements for testing fo thin-film $Cu(In,Ga)(S,Se)_2$ based photovoltaic (PV) modules."

# Application of IEC 61724 Standards to Analyze PV System Performance in Different Climates

Katherine A. Klise[1], Joshua S. Stein[1], and Joseph Cunningham[2]

[1] Sandia National Laboratories, Albuquerque, NM 87185, USA, [2] Sunny Energy, Tempe, AZ 85282, USA

*Abstract* — After a PV system is installed, periodic analysis is necessary to track how measured performance meets expectations. IEC 61724-3 outlines methods to quantify long term performance of PV systems. Applying these methods can be challenging due to the large quantity and possible quality control issues with measured data. In this paper, the methods outlined in IEC 61724-3 are applied to data collected at PV systems operating in different climates. The methods used to process data, run quality control tests, and compute performance metrics are described along with system performance issues found through the analysis.

*Index Terms* — IEC 61724, open source software, system performance

## I. INTRODUCTION

The International Electrotechnical Commission (IEC) has developed guidance to measure and analyze energy production from photovoltaic (PV) systems. IEC 61724-1, -2, and -3 [1,2,3] outlines guidance on data collection and evaluation methods for short term capacity and long term system performance. This paper focuses on the energy evaluation outlined in IEC 61724-3. The evaluation compares measured energy production to expected energy production given site specific weather conditions and system specifications. The procedure evaluates system performance over a full range of environmental and operating conditions, generally over the course of one year.

The energy performance index (EPI), defined as the ratio between measured energy and expected energy, is recommended to track long term system health [3,4,5]. A system performance model, which can be simple or complex, is used to estimate expected energy. While small systems might use a simple performance ratio (PR) to model expected energy, this method is influenced by seasonal temperature variations. Even a temperature corrected PR can have variations which skew results due to seasons and geographic locations. More complex models, such as the Sandia PV Array Performance Model (SAPM) [6], System Advisor Model (SAM) [7], and PVsyst [8], take into account measured weather conditions along with estimates for soiling and degradation. EPI is computed for times when the system is available (in-service EPI) and over the entire year (all-in EPI). System availability is generally determined using inverter operation or other status indicators.

The guidelines outlined in IEC 61724-3 are designed to be flexible, allowing analysts and system operators to define a set of requirements to quantify performance for a particular system. The requirements can change depending on the system size, instrumentation, and intended purpose of the analysis. In general, a system performance model must be defined along with data filtering methods and thresholds used in data quality control tests. These decisions can have a large impact on the resulting analysis. For example, it is important to apply data quality control tests prior to running a performance model using measured data. Poor quality data, related to sensor or human error, must be properly filtered out when evaluating system performance. Small gaps in data can be filled using a variety of methods, including interpolation, using data from duplicate sensors, historical data, or data generated using models. However, larger data gaps might have to be eliminated from the performance analysis. Duplicate sensors can also be used to detect sensor drift or compute parameter variability. IEC 61724-3 includes example data filtering criteria to identify data that is outside expected range, missing, associated with a dead sensor, or changes abruptly. The filtering criteria should be adjusted according to site specific conditions and system instrumentation. After running a preliminary analysis, it is important to assess the model and other assumptions used to define system performance until the analyst and system operator agree on a final analysis procedure. These decisions can be challenging given the large amount of PV system data, systems that collect different types of data, and the wide range of possible data quality control issues.

This paper describes an application of the standards outlined in IEC 61724-3 using data collected at identical PV systems operating at four sites across the United States. Results are used to evaluate system performance and track how data quality control tests diagnose faults and system availability. The open source software packages Pecos [9] and PVLIB [10] are used for the analysis.

## II. DATA

The data used in this analysis was collected as part of the Regional Test Center (RTC) program managed by Sandia National Laboratories (SNL). The RTC program collects data at several sites across the United States, including Albuquerque, New Mexico; Orlando, Florida; Williston, Vermont; and Las Vegas, Nevada. Identical 'baseline' PV systems and weather stations were installed at each site (Fig. 1). These systems are used to test sensor operation and maintenance routines. Data collection is periodically disrupted due to planned site and system upgrades. For this reason, sensor failure and system downtime is expected to be higher for these systems, as

978-1-5090-5606-4/17 $31.00 © 2017 IEEE

Fig. 1. PV system at the Nevada RTC site. Identical systems are located in New Mexico, Florida, and Vermont.

compared to production-level systems.

Each PV system is configured with two inverters, each with one series-connected string of 12 Suniva Optimus 270 Black modules. These modules have the following datasheet electrical characteristics: Pmax = 270 W, Vmp = 31.2 V, Voc = 38.5 V, Imp = 8.68 A, Isc = 9.15 A. The arrays all face South and are tilted at 35°.

The weather station collects data for global horizontal irradiance (GHI), direct normal irradiance (DNI), diffuse horizontal irradiance (DHI), wind speed, wind direction, air pressure, and relative humidity. For each string, DC voltage, DC current, AC voltage, AC current, AC power, power factor, frequency, reference cell irradiance, and reference cell temperature are recorded. Module temperature is recorded at 8 locations per string. Ambient temperature and POA irradiance is also recorded at the site. Data collected in 2016 was used for the analysis. Data was recorded at a 1-minute time interval, resulting in approximately 25 million data points per site.

## III. METHODS

The following section describes an application of IEC 61724-3 to compute system performance at the four sites. The analysis is carried out using Pecos [9] and PVLIB [10], both open source software packages developed by SNL.

Pecos is used to analyze the quality of time series data, subject to a set of quality control tests. Many of the features included in Pecos were designed specifically for quality control tests outlined in IEC 61724-3, including the ability to identify data that is outside expected range, missing, associated with a dead sensor, or changes abruptly. Additionally, Pecos includes methods to use filters and composite signals in the analysis. Filters can be used to smooth data and/or eliminate data collected at specific times from quality control tests. Composite signals are any type of new data generated using existing data or models. Composite signals can be used to include performance models or simple relationships in the analysis.

Time series data can be easily loaded into Pecos from a wide range of formats, including from file (i.e. csv, excel) and directly from databases (i.e. SQL). For this analysis, a years' worth of data is loaded into Pecos for each site. Similar analysis could be run in real-time (or near real-time) to help diagnose system performance issues quickly. Daily analysis is

recommended to ensure systems record high quality data. Yearly summary reports can then be performed to track long term system health. Pecos can be installed from https://github.com/sandialabs/pecos.

PVLIB is used to model expected system performance based on measured weather conditions and to compute a data filter based on sun position. Several performance models are included in PVLIB, including the SAPM [6], single diode model [11], and PVWatts model [12]. PVLIB can be installed from https://github.com/pvlib/pvlib-python.

The following steps are taken to analyze energy production for each site:

**Step 1: Check for timestamp issues.** When working with time series data, it is important to check for and fix timestamp issues before proceeding with analysis. Pecos includes methods to check for missing timestamps, duplicate timestamps, and timestamps out of sequence. These methods correct issues with the timestamp and record issues in the final report.

**Step 2: Preliminary data inspection:** Visual inspection of sensor data can help quickly identify systematic errors, and define filters and quality control tests. Sensor data plotted as a time-of-day versus day-of-year heatmap can help identify shading issues, large data gaps, and upper and lower bounds for quality control tests. An example heatmap is shown in Fig. 2. This figure shows POA irradiance at the Nevada site. No persistent shading issues were noted based on the image and missing data is observed in February and November (vertical white lines). Pecos includes methods to create time-of-day versus day-of-year heatmaps with superimposed time series that show sun position or other attributes.

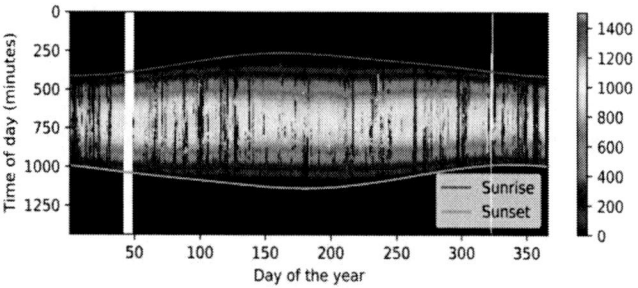

Fig. 2. POA irradiance heatmap for the Nevada site.

**Step 3: Apply filters.** Data collected at night or during low irradiance conditions can introduce errors in the performance evaluation. For this reason, data that is collected when the sun elevation is less than 20 degrees is eliminated from the analysis. PVLIB is used to compute sun position as a function of site location and date-time. Additional low irradiance filters could be added in future analysis.

**Step 4: Add composite signals.** Computing relationships between different types of measured data and comparing measured data to models can help identify issues with system performance. The following composite signals are used in the analysis: 1) DC power computed from DC voltage and DC current, 2) inverter efficiency computed from AC and DC power, 3) normalized efficiency computed from DC power and POA irradiance, and 4) module temperature deviation defined as the difference between each module temperature sensor and the median value over all 16 module temperature sensors. When multiple sensors are available, comparing a single sensor to the median value can help identify sensor drift. An additional composite signal, the power performance index, is computed in Step 6. These composite signals are used in quality control tests to check for anomalous conditions.

**Step 5: Run data quality control analysis.** IEC 61724-3 outlines basic quality control tests to check if data is outside expected range, missing, associated with a dead sensor, or changes abruptly. The methods in Pecos were designed to run these tests. For this analysis, data that is missing for less than 2 hours was filled using a linear filter. Data that is missing for more than 2 hours is flagged as missing. A sensor is flagged as recoding data outside an expected range if the threshold specified in Table I (column 2) is surpassed for more than 2 consecutive hours. The thresholds for air pressure are based on expected air pressure, calculated from site elevation using PVLIB. A sensor is flagged as dead if it changed by less than

TABLE I
Expected range and threshold values for quality control tests.

| Variable | Expected range | Dead sensor threshold | Abrupt change threshold |
|---|---|---|---|
| DC current and AC current (A) | > 0 and < Imp· 1.5 | < 0.0001 | |
| DC voltage (V) | > 0 and < Vmp·N· 1.2 * | < 0.0001 | |
| AC voltage (V) | > 230 and < 250 | < 0.0001 | |
| DC power ** and AC power (W) | > 0 and < Pmp·N· 1.2 * | < 0.0001 | |
| Power factor | > -1 and < 1 | < 0.0001 | |
| Frequency (Hz) | > 57 and < 63 | < 0.0001 | |
| POA, DNI, GHI, and ref cell irradiance (W/m²) | > -6 and < 1500 | < 0.0001 | |
| DHI (W/m²) | > -6 and < 500 | < 0.0001 | |
| Wind speed (m/s) | > 0 and < 32 | < 0.0001 | |
| Wind direction | > 0 and < 360 | < 0.0001 | |
| Air pressure (mbar) | > P·0.97 and < P·1.03 * | < 0.0001 | > 25 |
| Relative humidity | > 0 and < 100 | < 0.0001 | > 50 |
| Ambient temperature (°C) | > -30 and < 50 | < 0.0001 | > 20 |
| Module and ref cell temperature (°C) | > -30 and < 90 | < 0.0001 | > 20 |
| Inverter efficiency** | > 0.5 and < 1 | | > 0.25 |
| Normalized efficiency ** | > 0.8 and < 1.2 | | > 0.25 |
| Module temperature deviation (°C) ** | > -10 and < 10 | | |
| Power performance index ** | > 0.8 and < 1.2 | | |

* N is the number of series connected modules and P is the expected air pressure based on site elevation
** Composite signal

the threshold specified in Table I (column 3) for 5 consecutive hours. A sensor is flagged as changing abruptly if the value changes by more than the threshold specified in Table I (column 4) in a 15-minute timeframe. These thresholds can be adjusted to customize analysis. For each test failure, the sensor name, along with the start and end time of each failure, and an error flag is recorded in the final report.

**Step 6: Compute expected power and energy production.** Expected energy is computed using actual weather data. If weather data is unavailable, or is deemed unreliable given one or more quality control tests run in Step 5, it is eliminated from the energy calculation. The PVWatts DC model [12] is used to compute expected DC power; the model was run using PVLIB. Expected DC power is then converted to energy output. An additional quality control test is defined to flag times when the power performance index, defined as measured power divided by expected power, is outside an expected range of 0.8 to 1.2 for more than 2 consecutive hours. As with the quality control tests run in Step 5, test failures associated with the power performance index are recorded in the final report.

**Step 7: Compute metrics.** IEC 61724-3 recommends computing in-service EPI and all-in EPI. For this analysis, several additional metrics were computed, including data availability (DA), quality control index (QCI), and system availability (SA). For each sample time, DA is the percent of expected data that is recorded and QCI is the percent of available data that passed all quality control tests. The systems used in this analysis do not include an inverter status flag that indicate when the system is available. For that reason, SA is based on the results of quality control tests associated with power (AC and DC), inverter efficiency, normalized efficiency, and power performance index. For each sample time, SA is 1 if the quality control tests associated with these parameters all pass and 0 otherwise. In-service EPI is the ratio between measured energy and expected energy, computed when the system is available. All-in EPI is the same ratio, computed over the entire year. SA, in-service EPI, and all-in EPI are computed for each string. If data is missing while the system is known to be available, energy estimates could be made using historical weather data during that time.

After completing these steps, the analyst and system operator should review quality control test failures and performance metrics. Adjustments can be made to the quality control thresholds and performance model if significant issues are identified, otherwise, the analysis should remain stable year-to-year. Changes in the analysis should be clearly documented. The thresholds and model input can be saved in Python scripts that are used to run Pecos and PVLIB. These scripts can then be rerun to reproduce results and for future analysis. It is noted that several procedures recommended in IEC 61724-3 were not included in this analysis. For example, historical data was not used to compute predicted energy, systematic (bias) and random (precision) uncertainties were not analyzed, cleaning and calibration schedules along with grid availability was not included in the analysis, and missing or erroneous data was not replaced with data from other sources. These steps could be included in future analysis.

## IV. RESULTS

The RTC data was analyzed using the methods outlined above. Preliminary analysis, run on a daily basis, indicated that modules at all sites were underperforming by approximately 5%. This prompted a module flashtest at the New Mexico site. The electrical characteristics were subsequently updated to the following: Pmax = 255.7 W, Vmp = 30.9 V, Voc = 38.0 V, Imp = 8.28 A, Isc = 8.74 A. The discrepancy with datasheet values could be caused by light induced degradation or overrating. The new values were used to estimate performance for the year.

For each site, time-of-day versus day-of-year heatmaps were generated for each sensor reading. These figures were used to identify shading issues, large data gaps, and define thresholds listed in Table 1. No persistent shading issues were observed. A large gap in the data record was noted in Vermont between the middle of April and early May. Other data gaps were relatively short (a few days or less). Missing data was attributed to sensor failure, system maintenance, and data transfer issues.

Table II includes annual average data availability (DA), quality control index (QCI), system availability per string (SA), along with measured energy, expected energy, in-service EPI, and all-in EPI for each site. Fig. 3 and 4 illustrate DA, QCI, SA, in-service EPI, and all-in EPI throughout the year for the Nevada and Vermont site. DA, QCI, and SA are reported as a daily average. In-service and all-in EPI are reported as a monthly average.

DA was relatively high at all four sites with a few exceptions. In Florida, data was missing periodically, mainly between the middle of June and early October. As mentioned above, the Vermont site had a large gap in the data record, most of the data was missing over a 23-day period in the Spring.

QCI was also relatively high at all four sites. Note that QCI can be greater than DA because it is the percent of available data that passed all quality control tests. For example, at the

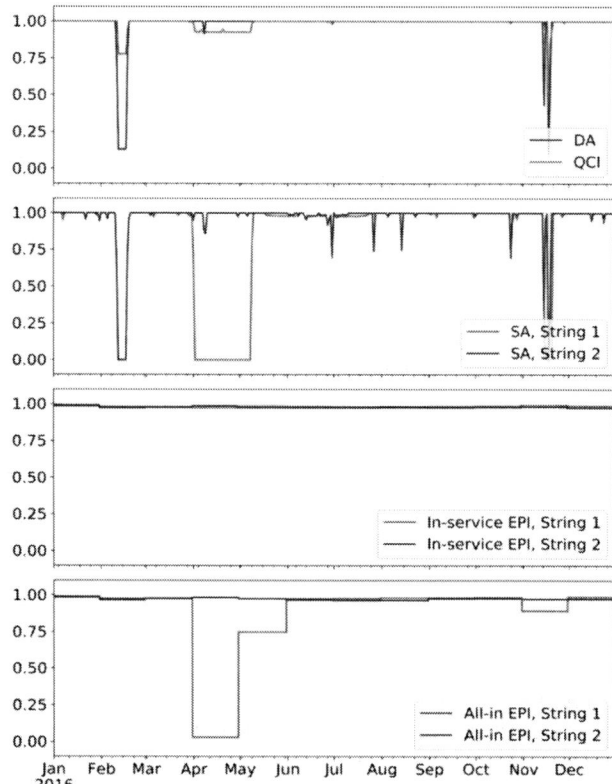

Fig. 3.    DA, QCI, SA, in-service and all-in EPI for Nevada.

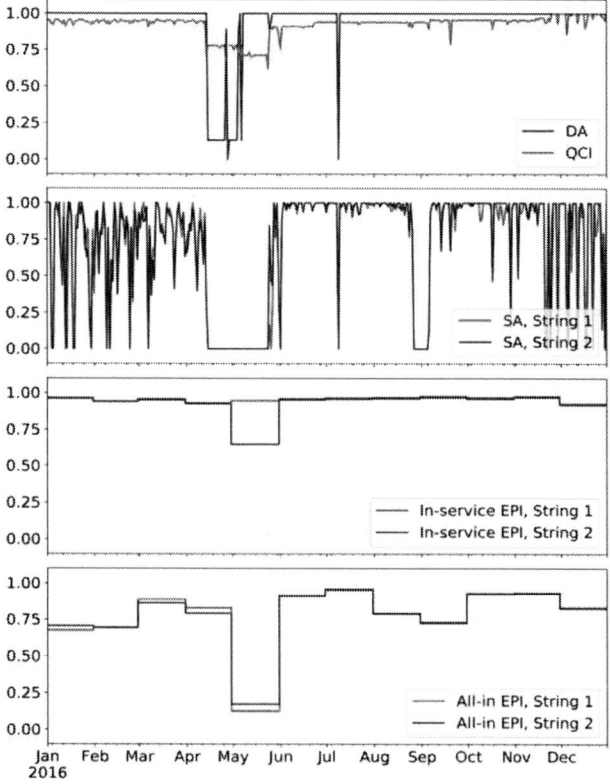

Fig. 4.    DA, QCI, SA, in-service and all-in EPI for Vermont.

## TABLE II
Annual DA, QCI, SA (per string), measured energy, expected energy production, and EPI.

|  | NM | NV | FL | VT |
|---|---|---|---|---|
| DA | 99% | 98% | 96% | 95% |
| QCI | 98% | 99% | 98% | 92% |
| SA, String 1 | 98% | 86% | 83% | 72% |
| SA, String 2 | 98% | 97% | 84% | 72% |
| In-service measured energy (kWh) | 11517 | 10476 | 8693 | 5528 |
| In-service expected energy (kWh) | 11608 | 10679 | 9104 | 5802 |
| All-in expected energy (kWh) | 11696 | 11390 | 9911 | 7305 |
| In-service EPI | 99% | 98% | 95% | 95% |
| All-in EPI | 98% | 92% | 88% | 76% |

Nevada site, only 13% of the data was available between Feb 12 and Feb 17. Of the data that is available, 78% passed all quality control tests. During that time, DC current sensors were flagged as dead, with readings that changed by less than 0.0001 over 5 consecutive hours. At the Vermont site, QCI is consistently around 95% due to unexpected abrupt changes in normalized efficiency and a module temperature sensor that is out of alignment with other module temperature sensors. In April and May, QCI decreases to around 75% due to DC power and current readings that are below 0 and several other sensors that were flagged as dead. These issues were verified with system operators.

The system is defined as 'available' if sensor data associated with power (AC and DC), inverter efficiency, normalized efficiency, and power performance index pass all quality control tests. Using this definition, SA is reported per string. All sites, with the exception of Nevada, have very similar availability per string. In Nevada, String 1 DC power is very close to 0 between April 1 and May 9. The quality control test for DC power will not flag this as an error, however bounds on normalized efficiency, inverter efficiency, and the power performance index all indicate anomalous conditions during that time. In Vermont, SA is highly variable in the winter due to anomalous conditions in normalized efficiency, inverter efficiency, and power performance index. SA at the Florida site was similarly noisy, due to occasional low inverter efficiency.

In service EPI and all-in EPI were computed using measured and expected energy. In New Mexico, in-service and all-in EPI are both very high. In Nevada, Florida, and Vermont, in-service EPI is slightly lower and issues with system availability reduced the all-in EPI by 6 to 20%.

As part of this analysis, quality control tests identified numerous issues throughout the year at all four sites. The tests were able to accurately identify dead sensors, sensor drift, and underperforming inverters. Pecos keeps a record of the sensor name, start and end time of each test failure, and an error flag. This information can be included in HTML formatted reports, saved to a file, or stored in a database. Graphics can be generated which help pinpoint the data points that were involved in an individual test failure. Examples are shown in Fig. 5. Each example shows one day of data along with issues found using the quality control tests run as part of this analysis. The gray region indicates times when sun elevation is < 20 degrees. This region is eliminated from quality control tests. Green marks identify data points that were flagged as changing abruptly, red marks identify data points that were outside expected range. The top image shows a spike in normalized efficiency at the New Mexico site. The middle image shows a sudden drop in inverter efficiency at the Nevada site. The bottom image shows a module temperature sensor that is oscillating between normal and anomalous conditions at the Florida site.

If a quality control test results in false positives, thresholds and moving windows can be adjusted, filters used to eliminate

data from quality control tests can be modified, the minimum number of consecutive failures needed to signal a warning can be increased, and data can be smoothed before the quality control test is run.

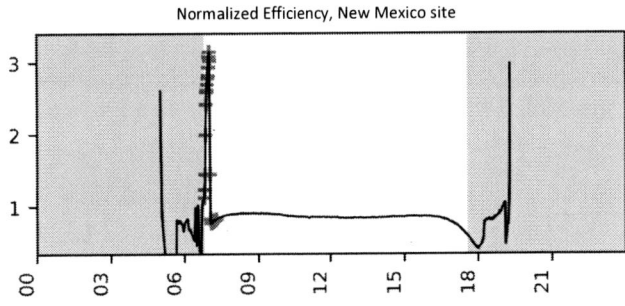

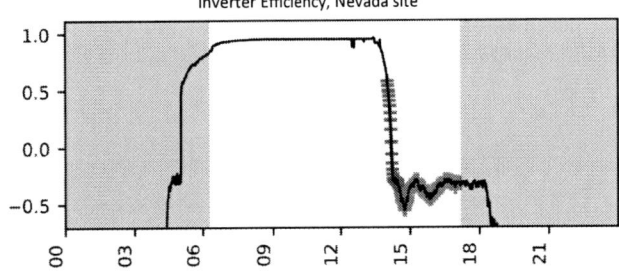

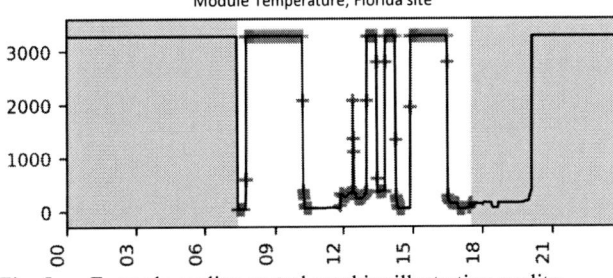

Fig. 5. Example quality control graphics illustrating quality control issues. Green marks indicate data points that were flagged as changing abruptly, red marks indicate data points that were outside expected bounds. The x-axis is in hours of the day.

V. DISCUSSION

System performance was evaluated at identical PV systems operating at four sites across the United State using methods outlined in IEC 61724-3. Pecos and PVLIB, both open source software tools, were used to run the analysis. These tools were used to process and filter large quantities of data, run quality control tests, compute expected energy production and system performance, and generate reports and graphics. The Python scripts used to run the analysis can be used to reproduce results and to compare year-to-year performance.

The methodology was able to identify gaps in the data record and anomalous conditions. Thresholds used in the quality control tests were systematically adjusted based on discussions with system operators and visual inspection of system data. Future research will compare the method used to estimate data availability, quality control index, and system availability with system logs. While the methods result in similar analysis across the four sites, several factors, such as variable system availability in Florida and Vermont, require further investigation. In addition to the yearly performance evaluation discussed in the paper, short term capacity tests and daily quality control analysis are recommended to evaluate performance, minimize downtime, and ensure the collection of high quality data.

ACKNOWLEDGEMENT

Sandia National Laboratories is a multimission laboratory managed and operated by National Technology and Engineering Solutions of Sandia, LLC., a wholly owned subsidiary of Honeywell International, Inc., for the U.S. Department of Energy's National Nuclear Security Administration under contract DE-NA-0003525. SAND2017-6321.

REFERENCES

[1]  IEC 61724-1, "Photovoltaic system performance – Part 1: Monitoring," Edition 1.0, 2017-03.

[2]  IEC TS 61724-2, "Photovoltaic system performance – Part 2: Capacity evaluation method," Edition 1.0, 2016-10.

[3]  IEC TS 61724-3, "Photovoltaic system performance – Part 3: Energy evaluation method," Edition 1.0, 2016-07.

[4]  J. Mokri and J. Cunningham, "PV System Performance Assessment," SunSpec Alliance, San Jose, CA, 2015.

[5]  S. Kurtz, E. Riley, J. Newmiller, T. Dierauf, A. Kimber, J. McKee, R. Flottemesch, and P. Krishnani, "Analysis of Photovoltaic System Energy Performance Evaluation Method," NREL Technical Report, NREL/TP-5200-60628, 2013.

[6]  D. L. King, E. E. Boyson, and J. A. Kratochvil, "Photovoltaic Array Performance Model," Sandia National Laboratories, Albuquerque, NM SAND2004-3535, 2004.

[7]  P. Gilman, "SAM Photovoltaic Model Technical Reference", Technical report NREL/TP-6A20-64102, May 2015.

[8]  "User's Guide PVsyst Contextual Help". Retrieved on June 9, 2017 from http://files.pvsyst.com/pvsyst5.pdf.

[9]  K. Klise and J. Stein, "Automated Performance Monitoring for PV Systems using Pecos," in 43rd IEEE Photovoltaic Specialists Conference, 2016.

[10] W.F. Holmgren, R.W. Andrews, A.T. Lorenzo, and J.S. Stein, "PVLIB Python 2015," in 42nd IEEE Photovoltaic Specialists Conference, 2015.

[11] W. De Soto, S. A. Klein, and W. A. Beckman, "Improvement and validation of a model for photovoltaic array performance," Solar Energy, vol. 80, pp. 78-88, 2006.

[12] A. P. Dobos, "PVWatts Version 5 Manual," NREL Technical Report, NREL/TP-6A20-62641, 2014.

# Effects of Urban Environment on Solar PV Performance

Panagiotis Moraitis, Bala Bhavya Kausika, Wilfried G.J.H.M. van Sark
Utrecht University, Copernicus Institute, Utrecht, The Netherlands

*Abstract*—**The modern urban environment is challenging for the deployment of solar Photovoltaic (PV) technology. Large structures and compactness limit the available rooftop area and create unpredicted shading patterns. While modern applications such as geographical information system (GIS) software provide a useful tool to calculate solar potential, the majority of small and medium size domestic systems suffer from inefficiencies. In this paper we present our preliminary results that show significant performance losses that can reach 15% for systems that are located in urban areas compared with systems located in rural and sub urban areas.**

*Index Terms*–**Performance Ratio, GIS, PV module , system, population density**

## I. INTRODUCTION

Solar Photovoltaics (PV) is one of the most promising forms of renewable energy production. Fast technological improvements, cost reduction and public acceptance are the key factors that accelerate the global demand for solar systems. Since a large fraction of electricity is consumed in urban areas, planning and development of PV within cities is gaining increasing interest. The modular nature of the technology makes it ideal for onsite energy production and consumption, leading to a critical reduction in transformation and transmission losses.

However, the modern urban environment is challenging for the development of PV technology [1]. Multistorey buildings and unpredicted shading patterns caused by the urban compactness limit the solar potential of the available area. A number of studies have already focused on quantifying the solar potential either on rooftops [2] or combined with the facade area [3][4] and calculate the financial feasibility of passive and active solar systems for specific locations [5].

The aim of this study is to examine deviations in the operational performance of existing PV systems that are located in urban areas compared to systems operating in suburban and rural environment. As a result we will correlate performance variation to urban compactness indicators such as site coverage, building and population density. The research area will be the country of the Netherlands as it provides sufficient urban, suburban and rural location with similar meteorological and irradiation conditions .

## II. METHODOLOGY

For this research a total number of 3271 systems were used with 5 minute resolution (for the years 2014 and 2015) data provided by monitoring vendors or collected from public available online sources such as Solar Log. Irradiation data were collected from the 31 meteorological stations of the Royal Netherlands Meteorological Institute (KNMI) with hourly resolution. Every PV installation was linked to the closest station according to geographical coordinates. The total plane of array irradiation was calculated using the Olmo model [6] for every system independently in accordance with the orientation and the tilt of each panel. The Performance Ratio (PR) is defined as the ratio of final system yield Yf and reference yield Yr , PR = Yf/Yr.

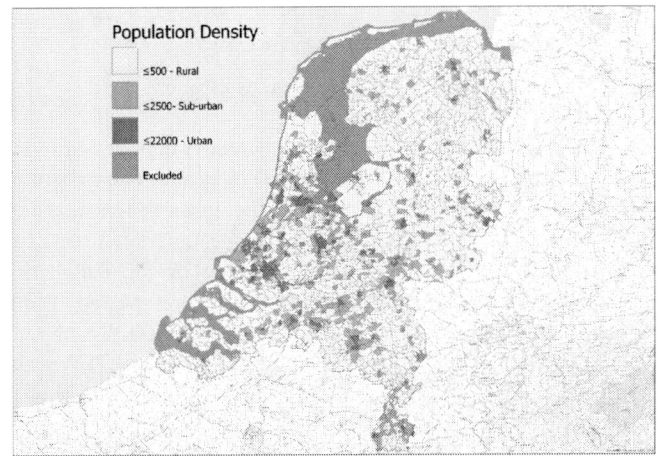

Fig. 1: Map division in urban, suburban and rural areas in the Netherlands based on population density.

The country was categorized in three zones (urban, suburban and rural) based on urban compactness indicators like population density, building densities, and site coverages [5] (see Fig. 1). The categorization was done at neighborhood level for the whole country within a geographical information system (GIS) environment. Once the neighborhoods were classified, the PV systems were then categorized based on their location within the neighborhoods. Data management, simulations, modelling and spatial data analysis was performed with the help of Python and ArcGIS.

Large scale datasets that are provided by private users are expected to include errors coming from incorrect data entry such as miscalculation of systems tilt and orientation or caused by malfunctioning monitoring equipment. In order to exclude them and achieve a good degree of confidence Tukeys method to isolate outliers from symmetrical distributions was used [7]. For the comparison of the datasets one way Analysis of Variance (ANOVA) and linear regression was used.

## III. RESULTS

### A. PV Systems Overview

PV systems were divided in three categories based on the population density of their location. In total 1166 systems were located in urban areas, 981 in suburban areas and 1119 in rural areas. The division of the zones on a geographical level can be seen on the map in Fig. 1.

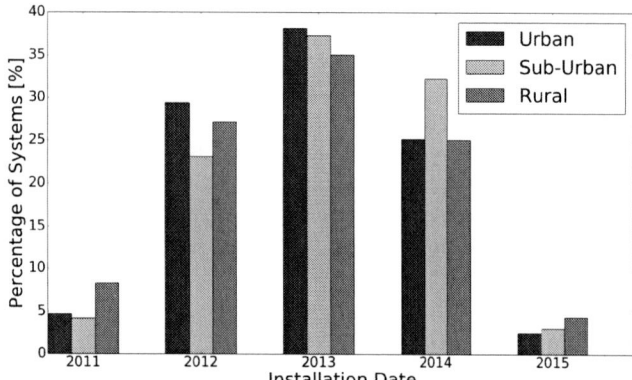

Fig. 2: Installation date distribution.

The distribution of the installation year is depicted in figure 2 as a percentage of the total number of system in each group. As it can be seen the majority of the systems was installed between 2012 and 2014 and therefore degradation effects can be neglected. The geographical location of the installation however, affects the average system size as expected. While in urban areas the average rooftop hosts a 3.4kWp system, in sub-urban location this capacity increases to 4.7kWp and in rural areas it almost doubles reaching 9.2kWp. Scarcity of available space is depicted in Fig. 3 where 90% of installations located in urban areas are less than 5kWp and only 2% has a capacity higher than 15kWp. Lower population density in rural areas results to higher availability of space and larger PV generators and while small size systems are still dominant, large size systems reach 19% of the total number installations, which accounts for 54.3% of the total installed capacity in rural locations.

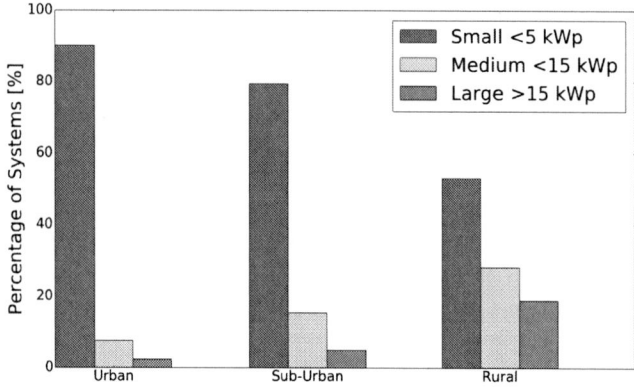

Fig. 3: System size distribution per location.

A very important factor for the electricity yield of a PV system is the tilt and orientation of the modules as they will determine the total amount of plane of array (POA) irradiance. In order to examine whether systems are properly positioned, the average POA for the years 2014 and 2015 was calculated as seen in Fig 4.a for every tilt and orientation angle using hourly resolution values from KNMI meteorological stations. The annual energy yield is maximized at $1045 kWh/m^2$ for PV generators oriented at $200°$ and having an inclination of $37°$. Since in most cases the configuration of each installation is dictated by the roof that the PV system is mounted on, a large variation in POA irradiation values is revealed. However, this variation is an indicator of how well a system is planned before installation. By considering as good oriented systems the ones that were able to harvest at least 90% of the maximum incoming irradiation (shaded area in figure 4.c) in our dataset we find that 90%, 88% and 92% of urban, suburban and rural systems respectively fall in this category. More specifically by setting the optimum harvested POA to 95% (shaded area in figure 4.b) the same values are 75%, 74% and 77%. These numbers demonstrate that urban compactness is not leading to extreme misplacement of installations especially in locations with limited availability in rooftop area.

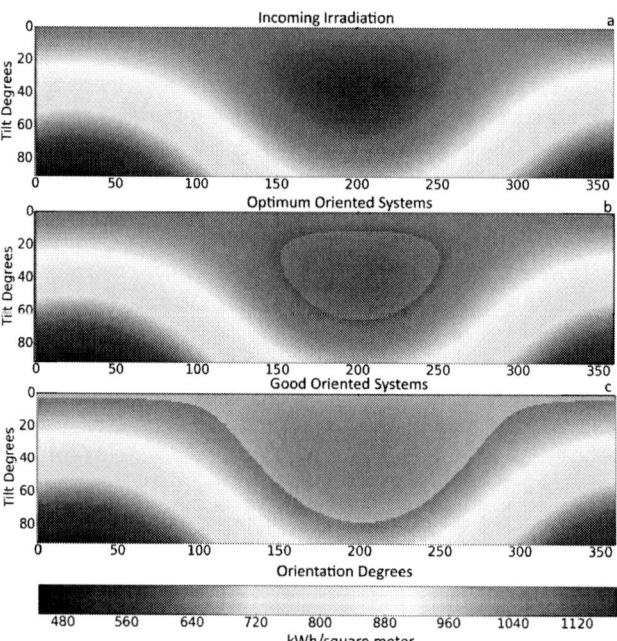

Fig. 4: (a) POA Irradiation, (b) Optimum Oriented Systems, (c) Good Oriented Systems.

### B. Performance Analysis

As it was mentioned in the introduction, power data were collected from monitoring equipment on site with 5 minute resolution. Each system was characterized by latitude and longitude values completed with the corresponding tilt and orientation of the panels. In that way, it was possible to calculate both the Annual Yield and the Performance Ratio.

Although a small number of systems started operation in 2011, performance analysis was conducted for the years 2014, 2015 and 2016 as the number of recorded systems increased dramatically. However, due to system failures in some cases part of data was either not recorded or not transmitted. To encounter the issue of missing entries only installations with at least 350 days of operation per year were taken into account for the calculation of annual yield and PR. The summary of the performance indicators is presented in Table 1 and it is in compliance with similar values that are reported for the Netherlands in previous research [8].

TABLE I: ANNUAL YIELD AND PR VALUES.

| Year | Annual Yield kWh/kWp | PR % |
|------|---------------------|------|
| 2014 | $919 \pm 78$ | $79 \pm 6$ |
| 2015 | $970 \pm 126$ | $79 \pm 6$ |
| 2016 | $945 \pm 89$ | $80 \pm 7$ |

To determine whether there is a variation in performance between PV generators located in urban, suburban and rural areas the dataset of 2016 is presented in Fig. 5. Systems are compared based on their PR values, mainly because it is the most widely used performance indicator and also because the difference between 100% and the PR value summarizes all possible losses. And while some losses might be system de-pendent such as wiring, failures, AC/DC conversion and dust, others vary according to location, for example temperature and shading. To further investigate if the means of the three groups demonstrate significant statistical difference, one way Analysis of Variance (ANOVA) was performed. The null hypothesis was set so that the three distributions had no difference in the mean values.

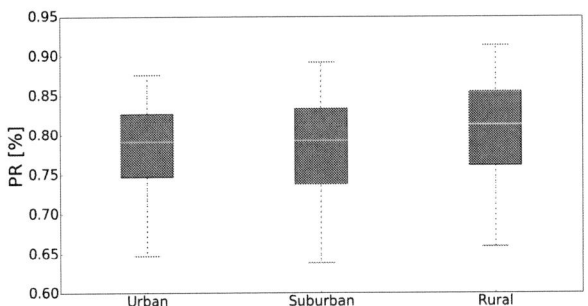

Fig. 5: Boxplots of annual PR distribution for 2016.

The result was that indeed there is an effect of the location on the performance of the PV installation at the $p < 0.05$ level for the three samples (F=18.56, $p=1.1 \cdot 10^{-8}$). However, ANOVA test states whether there is an overall difference between the groups, but it does not allow us to know which specific groups differed. Therefore is necessary to conduct a Post hoc test, since it is designed for situations were additional exploration of the differences among means is required to provide specific information on which means are significantly

different from each other. Post hoc comparisons using the Tukey HSD test indicated that the mean performance of rural systems (Mean=81, SD=6.3) was significantly higher than the mean performance of urban systems (Mean=78.4, SD=5.6) as it was compared to suburban systems (Mean=78.5, SD=6.4). However, urban and suburban systems did not show a significant difference between them. Taken together, these results suggest that the location of a PV system really does have an effect on the annual performance.

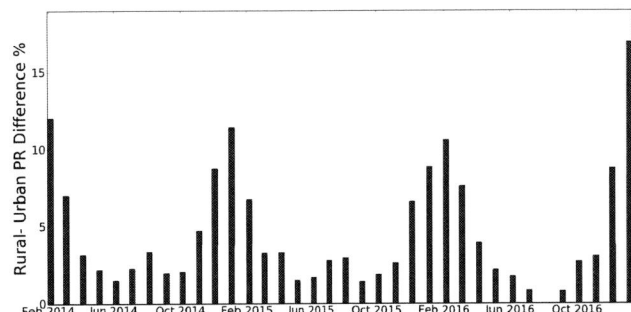

Fig. 6: Monthly PR difference between PV installations in rural and urban areas.

To further identify the causes of the performance mismatch in the annual mean values it is necessary to investigate whether the difference is levelled throughout the year or it is subjected to a seasonal variation. For that purpose the monthly PR value differences between rural and urban systems will be compared. As it is shown in Fig. 6 the percentage difference between the monthly PR values it is observed that there is a specific pattern. During winter months PV generators that are located in rural areas are outperforming the ones located in urban areas by up to 16%. The difference gradually declines during spring and summer months were it reaches a minimum of 0.2-3.4%.

The seasonality of the effect can be attributed only to variations in temperature or irradiation. Since the highest difference is observed during winter months the focus will be on irradiation conditions, and specifically the trajectory of the sun. For the geographical coordinates of the Netherlands (Amsterdam 52.3702° N, 4.8952° E) the maximum solar elevation is limited to 15° in December and 20° in January but it can reach 61° in June. While that has a crucial impact on the electrical yield of PV installations regardless their specific location, it appears to negatively affect the overall performance of the ones located in urban areas.

In order to examine the effect of the solar elevation on PV performance the data set was re-sampled in hourly resolution to match the KNMI measurements and the hourly PR was calculated. For each time slot the solar elevation was calculated using the pvlib library of Python and the specific coordinates of every installation. Solar elevation values were rounded then to the closest integer to form 62 groups (0 to 61 degrees). Each group was filled with the corresponding values of PR from each installation and the Tukey's method to isolate outliers was used. In that way two separate datasets were created for urban and rural systems. The difference in PR between average values for every degree of solar elevation is presented in Fig. 7.

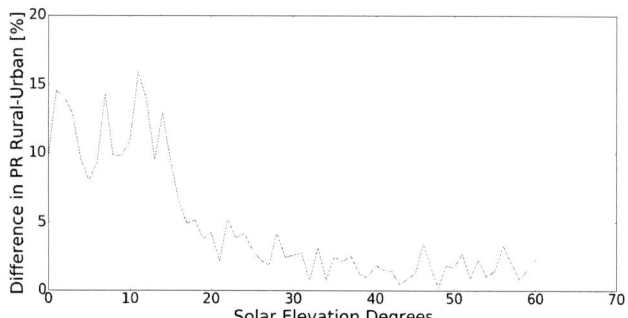

Fig. 7: Difference in PR between PV installations in rural and urban areas versus the solar elevation.

As it is observed, solar elevation is a crucial factor that affects the performance of urban installations. The result is demonstrated in Fig 7 where for low elevation angles below 18°, the difference in PR values can reach 15%. The consequences of this effect become more obvious by taking into account that during late autumn and winter months (November-February), which corresponds to 1/3 of a year, the sun spends 79% of it's time on the sky below the threshold of 18°.

*C. Comparing Urban Indicators*

The main results so far demonstrate that there is a strong connection between the urban environment and the overall performance of a PV system. However, it is necessary to look in to specific indicators to quantify the effects of urban compactness. The analysis so far was focused on the population density in order to separate urban from rural areas. But not all the urban and rural environments exhibit the same morphology neither the population density necessarily dictates the urban landscape since large structures such as public or office buildings could be in non residential areas.

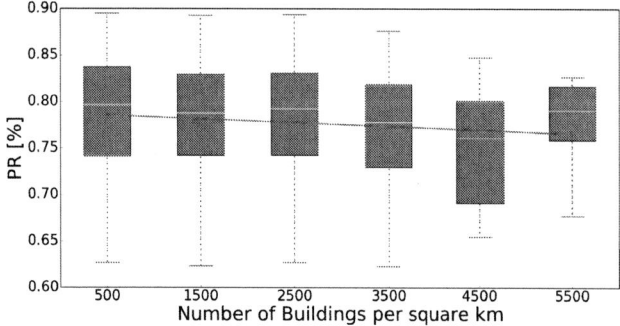

Fig. 8: PR compared with the number of building per square km.

Therefore we compared two additional indicators, the number of buildings per square km and the average building height with the PR using linear regression (see Figs. 8 and 9). PV

systems were grouped according to the characteristics of their location as it was indicated by the GIS software. The results show negative linear correlation between urban compactness and PV performance. The slope of the linear regression reveals that the average building height has a stronger effect on the PR with $\alpha = -0.01$ compared to building density $\alpha = -4 \cdot 10^{-6}$. Higher average building height does not necessarily mean that all the PV systems are located in taller buildings. A higher average value can also be an indication of unpredictable or unavoidable shading patterns. Moreover the seasonal effect seems to affect equally both lines by a factor of 2 during the winter months ( $\alpha = -0.02$ and $\alpha = -8 \cdot 10^{-6}$ respectively - not shown here).

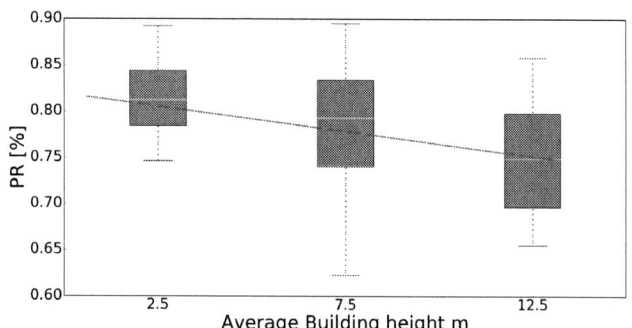

Fig. 9: PR compared with the average building height.

## IV. CONCLUSION

The modern urban environment is challenging for the installation of PV systems. Recent studies have focused on how to utilize the available space on rooftops and facades for the implementation of solar harvesting technologies. In order to achieve better integration, it is crucial to understand in detail the operational performance in different geographical locations but also in different environments.

In this study we have shown that small and medium size domestic systems may suffer from serious inefficiencies. Our results in this paper prove that systems with similar technical details such as installation date, panel orientation and tilt which were operating under similar meteorological conditions but in different environment shown significant difference in performance during certain periods within a year that can reach up to 15%.

## REFERENCES

[1] Dapeng Li, Gang Liu, Shengming Liao, "Solar potential in urban residential buildings," *Solar Energy*, Vol. 111, pp. 225-235, 2015.
[2] T. Santos, N. Gomes, S. Freire, M.C. Brito, L. Santos, J.A. Tenedorio, "Applications of solar mapping in the urban environment," *Applied Geography*, Vol. 51, pp. 48-57, 2014.
[3] S. Freitas, C. Catita, P. Redweik, M.C. Brito, "Applications of solar mapping in the urban environment: State of the art review," *Renewable and Sustainable Energy Reviews*, Vol. 41, pp. 915-931, 2015.
[4] P. Redweik, C. Catita, M.C. Brito, "Solar energy potential on roofs and facades in an urban landscape," *Solar Energy*, Vol. 97, pp. 332-341, 2013.
[5] Nahid Mohajeri, Govinda Upadhyaya,August Gudmundsson, " Effects of urban compactness in solar energy potential," *Renewable Energy*, Vol. 93, pp. 469-482, 2016.

[6] F.J. Olmo, J. Vida, I. Foyo, Y. Castro-Diez, L. Alados-Arboledas, "Prediction of global irradiance on inclined surfaces" *Energy* Vol. 24, pp. 689-704, 1999.

[7] J. W. Tukey, *Exploratory Data Analysis,* Addison-Wesley, 1977.

[8] Odysseas Tsafarakis, Panagiotis Moraitis, Bhavya Kausika, Henrik van der Velde, Saskia 't Hart, Arthur de Vries, Peer de Rijk, Minne de Jong, Hans-Peter van Leeuwen, Wilfried Van Sark, "Three years experience in a Dutch public awareness campaign on photovoltaic system performance" *IET Renewable Generation* Vol. 20, pp. , 2017.

# Irradiance measurement considerations for system performance assessment when managing fleets of photovoltaic assets across Asia

André M. Nobre, Shravan Karthik, Chenxi Liu, Rohit Jaswal, Rupesh Baker, Raghav Malhotra, Alan Khor

Cleantech Energy Corporation, 25 Church Street #03-04, Capital Square 3, 049482, Singapore

*Abstract* — **With the relatively recent market shift of PV system deployments from USA/Europe/Japan to Asia, it is only natural that knowledge and data limitations still exist with regards to PV system field performance. Common practices utilized in temperate climates throughout the past two decades do not necessarily provide the most accurate or cost-effective results when dealing with photovoltaic assets at these novel frontiers. One of such examples is the air pollution found in major Asian city centers, which present difficult measurement conditions for irradiance on the ground. This work aims at exposing some of these challenges and propose alternative assessment options ahead.**

*Index Terms* — **air pollution, analytical monitoring, distributed PV, irradiance simulation, PV system performance, satellite data, soiling.**

## I. INTRODUCTION

The growth of photovoltaic (PV) systems deployment worldwide has been astounding. From the year 2000, when only ~1 GWp was present, the cumulative installed capacity of PV installations in the world has climbed to ~307 GWp at the end of 2016 [1], representing a compound annual growth rate (CAGR) of ~40% across this time frame. The first wave of systems of the first decade of the new millennium were deployed in their majority in temperate-climate locations (high latitudes) in Europe, Japan and the US. By 2014, ~90% of this capacity was established in these locations [2]. Since then, the rate of site deployments in these PV-leading markets has dropped considerably, with solar systems being now primarily installed in China, Asia Pacific and in a smaller but growing scale in Latin America [1].

Asian systems now account for more than half of PV systems worldwide, with strong growth trends ahead. As an example of PV adoption rate in the region, even on a smaller absolute term scale (~120 MWp installed to date), the CAGR for Singapore PV systems has been 80% for the past 8 years [3]. It is foreseen that in a matter of 10-15 years, 20% of the country could be powered by solar photovoltaics [4].

Further growth can be expected as countries have been focusing on developing utility-grade PV systems, i.e. of several tens of MWp in size. The distributed nature of PV systems has yet to be fully maximized. In India, of the 10 GWp of cumulative deployed PV capacity by the end of 2016, only ~10% were made of distributed, rooftop systems [5], that is, at the premises of commercial & industrial (C&I) clients or residential systems.

Notwithstanding the upward trends of deployment of PV systems in Asia, various challenges lie ahead which could hinder the speed of deployment and the sustainability of business models surrounding PV. One of such challenges lies in air pollution, found in several city centers in booming locations across the Asian continent. This atmospheric phenomenon, mostly of anthropogenic origins, tests the economics of photovoltaic systems. In Singapore, strong haze conditions tend to occur on a yearly basis when neighboring countries clear forest vegetation through fires. Transboundary haze clouds travel to the small city-state island and have caused, in an episode in 2013, losses of PV system output in the order of 15-20% for a matter of days until smog dissipated [6]. In Delhi, north of India, industrial activity, heavy traffic and biomass burning have created even greater negative influence on PV systems. It has been investigated that the Delhi haze, which is stronger during the winter season, can cause instantaneous losses of irradiance on ground up to 50%, with negative annual weighted losses accounting to ~6% of the potential irradiation for the city surroundings [7]. Such examples hurt the return on investment of these PV facilities, weaken funding into the renewable energy sector and could potentially curtail the deployment of solar assets in various world regions facing similar effects, even if on smaller scales.

Fig. 1. Air pollution in Delhi, India poses a serious threat to the economics of PV systems in the National Capital Region. Photo date: Nov/15.

Air pollution represents a negative filtering mechanism to the irradiation resource reaching the surface where PV systems are present. Additional to that, the presence of dust particles in the air exacerbates soiling accumulation, as the dust settles on the front glass of PV systems more easily. That phenomenon creates a "double whammy" effect on the economics of PV, as soiling rates for a highly-polluted city such as Delhi can cause losses reaching even higher than 1% per day [7]. Exhaust chimneys at industries, depending on activity type, further contribute to the soiling accumulation onto PV panels.

With the continuous growth of PV energy deployment in Asia, the number of total systems which are closer to the Equator will climb. Consequently, these systems now experience year-round hot and humid operation conditions, a new challenge when compared to systems in milder climate zones. Also, taking the distributed nature of PV rooftop business into account, the performance assessment of the sites become more complex, e.g. multiple roofs with different tilts/azimuths, see Fig. 2, or shading patterns found in heavily-built environments, typical of Asian megacities.

Fig. 2. Common situation in commercial PV rooftops, whereby multiple combinations of tilt/azimuths angles are present at a site. The above site is a 382 kWp system on a shopping mall in Cebu, Philippines, with six of such tilt combinations.

Modelling for irradiation becomes critical to the management of systems performance and eventually, of electrical grids themselves, now with a higher share of renewables in their matrices. Countries like Germany have already had instances of renewables nearly running the entire grid needs for a day [8]. For photovoltaics and grid management efforts, recent methods have been investigated for performance assessment of PV sites within a certain area, using one of more reference PV systems as baseline [9]. Apart from grid operators, other stakeholders, such as developers, are interested in dealing with a large fleet of distributed generation assets. Assessing systems' performance must be improved and optimized, with the intent of guaranteeing cost-effective operations & maintenance (O&M) practices, and naturally, protection of the return on the investment.

For the reasons which are presented here in this work, managing ground-telemetry for PV applications becomes challenging due to aspects such as cost, hardware/software selection for monitoring and maintenance complexity, as well as environmental matters mentioned. The use of other data sources – such as satellite data, appears as an alternative to address barriers. Next, we attack these challenges in measuring irradiance in certain Asian locations of our portfolio, discuss the suitability of usage of satellite-based data versus ground-measurements as well as showcase performance ratio (PR) calculations using various methodologies.

## II. METHOD

This contribution proposes three areas of investigation: sections A and B address challenges in measuring irradiance for system performance assessment, whereas section C benchmarks system performance methodologies and exposes potential errors found when dealing with a large fleet of distributed PV assets.

### A. Satellite vs. ground-measured data analyses

Although the number of ground-measurement meteorological stations grew with the accelerated addition of PV systems, they still face typical challenges such as their maintenance activities (e.g. regular cleaning of sensors), hardware (data logger) availability, sensor quality (i.e. accuracy class) & periodic calibration, proximity of assets of interest to the measuring weather point, etc. As an example, the use of industrial-grade loggers, especially with the presence of an uninterrupted power supply source, usually guarantees greater than 99% data availability, crucial for appropriate analytical monitoring efforts.

Satellite-derived data have dramatically improved through time which is later demonstrated in this work. However, model improvements are still needed for certain locations/case studies. In this portion of the work, we address ground-measurement comparisons between a meteorological station in Delhi, possessing a secondary class pyranometer (cleaned daily) against data from a satellite source (here from satellite data provider "Solargis"). Similar comparisons are performed for Phnom Penh, Cambodia, a location with negligible air pollution presence.

Month by month scatter plots allow for the gauging of the accuracy between data sets. The use of mean bias error as a statistic metric also is conducted in search for trends during specific seasons of the year (e.g. dry vs. wet seasons). Due to the air pollution studies accomplished in the work in [7], the presence of a meteorological station in Delhi pre-dates that of the station in Cambodia by 3 months, hence a larger data set is available for the Delhi region.

### B. Location-dependent conditions and special weather phenomena considerations

Having daily or virtually daily maintenance routines of meteorological stations is critical for a proper assessment of the solar resource. At PV sites, it is not always the case that maintenance happens regularly, especially with the deployment of large fleets of smaller-size systems (ranging from hundreds of kWp to ~1-3 MWp range across rooftops that can be a few km apart, but also up to 100-200 km from O&M office bases). Hence, it is often the case that irradiance devices are measuring erroneous values due to fast soiling accumulation, especially exacerbated in industrial areas (see Fig. 3 for irradiance sensors in various PV sites in Delhi, taken at different time intervals apart during the dry season).

In this portion of the investigation, we assess the rate of accumulation of dirt at meteorological station sensors in two locations: Delhi (north) and Chennai (south of India), during the beginning of winter of 2016, when air pollution conditions increase in Delhi, leading up to May 2017, the most recent data set available for this paper. The analyses are made possible due to the presence of two identical silicon sensors at each station, installed at global horizontal position, one cleaned daily and the other left to accumulate dust/dirt through time. A reset of the sensors (i.e. cleaning of the dirty sensor) happens at every one-month interval, allowing for an estimation of the monthly soiling rate.

Fig. 3. Various silicon sensors images showing dirt/dust accumulation through various time frames. Sensors are located at various PV systems in the greater Delhi, India.

On top of the assessment of accumulation of dirt onto sensors, which could be sometimes perceived as non-representative due to the smaller area and lesser similarity to a true PV panel, which is commonly tilted, we showcase the soiling losses on a PV system in Chennai which was kept uncleaned for four straight months which coincided with the dry season in the city. A soiling rate for the system is derived using the "Type 1" method described next.

### C. Benchmarking of methods for PV systems performance

As means of advancing knowledge for the management of an ever-growing fleet of PV installations across new markets, we propose to present various performance ratio calculation methods for ten PV systems across our portfolio (see Fig. 4 for

the geographical location of the systems and Tab. I for their basic characteristics, such as city, size, latitude, tilt/azimuth and the type of mounting, whether on the ground, or on a roof – either of concrete or metal roof substrates).

Performance ratio (PR) is an internationally-recognized metric for PV system performance assessment. It is described in the IEC standards 61724-1 [10], being suitable for use in systems across the world. It can be summarized by the energy generated by a PV system normalized to its installed capacity in standard test conditions (1,000 W/m², 25°C and 1.5 air mass), compared against the availability of the irradiation resource at the plane of array of the photovoltaic system for the period of the investigation. PR has been historically used for benchmarking of systems in different locations, although the temperature dependence of photovoltaics would mean PV systems in a temperate climate would have higher PRs than those compared to systems in the tropics [11, 12].

Fig. 4. Cleantech Solar's portfolio of PV assets across Asia Pacific, with ~40 MWp of systems, over ~60 sites in 5 countries (as of mid-2017). Systems numbered 1-10 are part of this investigation.

TABLE I. TEN PV SYSTEMS OF THIS STUDY WITH BASIC CHARACTERISTICS SHOWN, ORGANIZED FROM NORTHERN- TO SOUTHERN-MOST LATITUDE.

| System No. | Location [City] | Size [MWp] | Latitude [°N] | Tilt/ Azimuth | Mounting Type |
|---|---|---|---|---|---|
| IN-01 | Lalru* | 0.51 | 30.5 | 13/E-W | Metal |
| IN-02 | Delhi* | 0.50 | 28.0 | 5/E-W | Metal |
| IN-03 | Nashik* | 1.10 | 20.3 | 16/S | Ground |
| IN-04 | Pune* | 0.22 | 18.5 | 22/S | Concrete |
| PH-05 | Manila^ | 0.50 | 14.4 | various | Metal |
| IN-06 | Chennai^* | 1.81 | 12.9 | 6/N-S | Metal |
| KH-07 | Phnom Penh^* | 2.62 | 11.5 | 5/E-W | Metal |
| IN-08 | Coimbatore* | 0.75 | 11.1 | 12/S | Concrete |
| PH-09 | Cebu^ | 0.38 | 10.4 | various | Metal |
| SG-10 | Singapore^ | 0.08 | 1.3 | 5/N-S | Metal |

In Tab. I, for locations marked with an asterisk (*), irradiance sensors are present at the plane of the array of the PV system (for global tilted irradiance, GTI, readings). A circumflex accent (^) denotes the presence of a global

978-1-5090-5606-4/17 $31.00 © 2017 IEEE

horizontal irradiance (GHI) reading obtained via a silicon sensor.

The four types of potential PR calculations included in this work when assessing the performance of the systems are:

a) Type 1: using GHI from the satellite data source as a basis for calculation;
b) Type 2: deriving GTI values for a site's tilt/azimuth conditions (according to the availability of roof sizes), using hourly GHI and diffuse horizontal irradiance (DHI) derived from the satellite source as entry parameters into the Perez transposition model;
c) Type 3: using GHI readings from silicon sensors (uncleaned) on the ground at PV sites;
d) Type 4: using GTI readings from ground silicon sensors (uncleaned), matching the IEC Standards 61724-1 for PV system performance assessment.

The analyses shown include data from each PV system for Dec/16, representing a winter month, and May/17, for a month as close to summer as the cut-off timing of this paper allowed. The average age of assets is ten months, which at this stage is an encumbrance to full seasonality studies.

On the instrumentation front, pyranometers used in this work (for investigations described in "Method, section A") are of secondary standard and accuracy of ±2%. Silicon sensors deployed have an accuracy of ±5W/m², ±3.5% of the reading registered (for investigations in "Method, sections B and C"). Energy measurements, used for PR calculations, are obtained from revenue-grade meters of 0.2S class of accuracy. All logging of the PV systems and weather data is performed with industrial-grade loggers. Data available across monitoring systems is greater than 99%.

## III. RESULTS & DISCUSSIONS

### A. Satellite vs. ground-measured data analyses

Firstly, the results when comparing daily ground- versus satellite-irradiation data for Delhi are presented. A correlation coefficient of ~87% is obtained (Fig. 5), with a mean bias error (MBE) on all available data (14 months) of -0.22, indicating that satellite readings are in general higher than those captured on the ground. When taking the seasonality of Delhi into account, separating fall and winter (Sep through Feb months) and spring and summer (Mar through Aug), the MBE achieved is -0.38 and -0.09 respectively. This indicates that the satellite modelling for irradiation on the ground performs well for higher sun altitude angles of the year. However, it cannot be discounted that the air pollution is a likely candidate for influence on the satellite readings. As reported in [7], the winter months of the year have a predominant presence of haze, which can especially be detected for the months of Oct-

16 through Jan-17 as having a higher satellite reading bias, see Fig. 5.

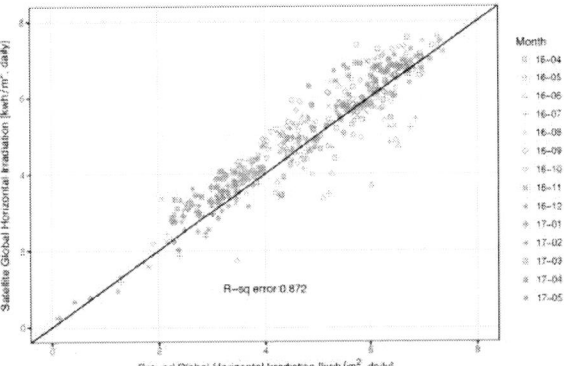

Fig. 5. Scatter plot of satellite vs. ground-measured daily irradiation values for Delhi, India, segregated by months.

For Phnom Penh, the correlation coefficient was in line with the value achieved for Delhi (here ~88%), see Fig. 6. The mean bias error for all data (11 months) was -0.02, with good agreement between the satellite model and ground readings. When taking the seasonality on site into account, segregating the months into May through Aug for the dry season and Sep through Dec for the monsoon season, we obtained MBEs of 0.01 and 0.15 of results, showing an over-irradiation recorded on the ground during the wet season against that one of the satellite model.

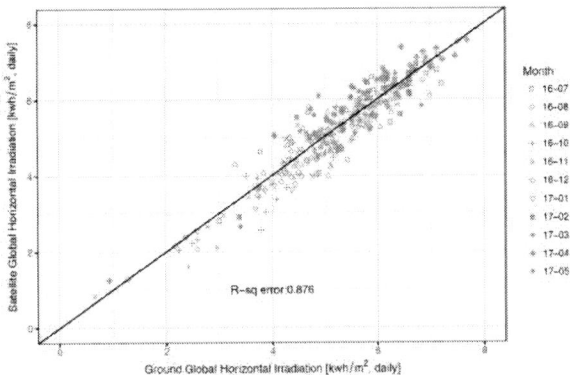

Fig. 6. Scatter plot of satellite vs. ground-measured daily irradiation values for Phnom Penh, Cambodia, segregated by months.

### B. Location-dependent conditions and special weather phenomena considerations

In the second part of the investigation, we obtained soiling rates for Delhi (north) and Chennai (south India) for a period of half year encompassing the beginning of winter through spring. The soiling rates are presented in Tab. II.

The lower levels for Chennai in Dec-16 can be explained by the rainy season, which usually ends in January. As time progresses and the dry season starts, soiling rates for Chennai stabilize around 0.4-0.5% per day of potential daily losses to an uncleaned PV system. For Delhi, on the other hand, rain showers usually start in May, which could be seen then as reduced soiling rates present in the May-17 month. For winter months, the absence of rain and considerable pollution levels as reports, meant high soiling rates were present. In average, PV systems would suffer from twice as much soiling losses in Delhi as they would in Chennai. Such findings as shown in Tab. II highlight the air pollution conditions present in Delhi, with negative implications to PV systems deployed in this part of India.

TABLE II. DAILY SOILING RATES FOR DELHI AND CHENNAI, INDIA, SHOWN ACROSS A HALF YEAR PERIOD.

| City | Daily soiling rates (= soiling losses) shown across 6 months, MM-YY format [expressed in %/day] | | | | | |
|---|---|---|---|---|---|---|
| | 12-16 | 01-17 | 02-17 | 03-17 | 04-17 | 05-17 |
| Delhi | -0.9 | -1.4 | -0.7 | -0.6 | -0.5 | -0.3 |
| Chennai | -0.3 | -0.3 | -0.4 | -0.5 | -0.5 | -0.4 |

When taking a PV system into account rather than the readings from silicon sensors, one way to gauge soiling losses is presented next. In [7], soiling rates were calculated by comparing the performance of cleaned and uncleaned sections of PV systems on the same rooftop. Rates shown are also in line with the ones presented in Tab. II. In Fig. 7, another demonstration of a way to gauge soiling losses for a PV system is showcased.

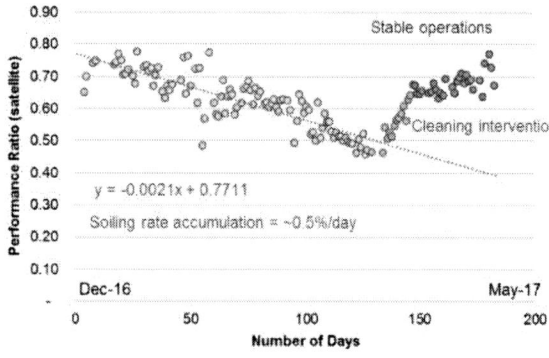

Fig. 7. Satellite-based performance ratio calculations for a large PV system in Chennai. The system was not cleaned for circa 4 months, with the cleaning cycle occurring in late Apr-17 over two weeks due to roof access and complexity.

In this PV system located in Chennai, no cleaning cycle occurred due to waterline infrastructure issues on site. Since silicon sensor readings present on the roof were also compromised, as shown to be common as per Fig. 3 in this work, PR calculations using in-plane irradiance readings was not suitable. The method proposed here in this work (Type 1) was applied. A linear fit was conducted for the circa 120 days in which the PV system was not cleaned, indicating a soiling

rate of ~0.5%, in line with those shown for the same period as presented in Tab. II.

*C. Benchmarking of methods for PV systems performance*

Part three of the evaluation presents yields and PRs, where applicable, for the ten PV systems of this study. For the chosen winter month (Dec-16) and for the nearly summer month (May-17), the results are shown in Tab. III and IV respectively. In *italics*, PRs which were influenced by heavy soiling observed on irradiance sensors on the ground are indicated. "N/A" ("not available") are used for calculations which are not possible to be executed, such as calculating PR for sites where only a tilted silicon sensor is available (e.g. Type 4 PR is possible to be calculated whereas Type 3 is not).

TABLE III. YIELD AND PR FOR THE TEN SYSTEMS OF THE STUDY FOR THE MONTH OF DEC-16.

| System No. | Yield [kWh/kWp] | PRsat[1] [%] GHI | PRsat[2] [%] GTI | PRgnd[3] [%] GHI | PRgnd[4] [%] GTI | Soil. Rates |
|---|---|---|---|---|---|---|
| IN-01 | 59.5 | 65 | 68 | N/A | *75* | 2 |
| IN-02 | 67.6 | 68 | 70 | N/A | *91* | 3 |
| IN-03 | 111.9 | 70 | 55 | N/A | 60 | 1 |
| IN-04 | 150.4 | 95 | 75 | N/A | 77 | 1 |
| PH-05 | 95.7 | 80 | 76 | 78 | N/A | 1 |
| IN-06 | 84.2 | 64 | 60 | *70* | *69* | 2 |
| KH-07 | 109.5 | 82 | 83 | 83 | 83 | 1 |
| IN-08 | 120.3 | 82 | 77 | N/A | 78 | 1 |
| PH-09 | 100.7 | 79 | 80 | 77 | N/A | 1 |
| SG-10 | 106.4 | 80 | 81 | 79 | N/A | 1 |

TABLE IV. YIELD AND PR FOR THE TEN SYSTEMS OF THE STUDY FOR THE MONTH OF MAY-17.

| System No. | Yield [kWh/kWp] | PRsat[1] [%] GHI | PRsat[2] [%] GTI | PRgnd[3] [%] GHI | PRgnd[4] [%] GTI | Soil. Rates |
|---|---|---|---|---|---|---|
| IN-01 | 121.8 | 64 | 63 | N/A | *74* | 2 |
| IN-02 | 124.7 | 67 | 68 | N/A | *89* | 3 |
| IN-03 | 138.2 | 73 | 61 | N/A | 72 | 1 |
| IN-04 | 156.5 | 89 | 72 | N/A | *80* | 1 |
| PH-05 | 120.4 | 78 | 74 | 75 | N/A | 1 |
| IN-06 | 128.2 | 78 | 76 | *81* | 83 | 2 |
| KH-07 | 136.0 | 81 | 81 | 79 | 81 | 1 |
| IN-08 | 135.3 | 82 | 75 | N/A | 75 | 1 |
| PH-09 | 106.6 | 77 | 78 | 73 | N/A | 1 |
| SG-10 | 94.9 | 81 | 79 | 77 | N/A | 1 |

Due to the nature of distributed PV and costing associated with monitoring systems, the Type 3 method proposed (ground GHI silicon sensors as basis for PR calculations) has potential benefits for a cost-competitive way of managing a distributed fleet of PV systems in Asia as well as the use of satellite data (Types 1 and 2). An advantage of classic pyranometers is their glass domes, which facilitate a better cleaning process when it rains (also dirt does not accumulate as easily as when compared to flat surfaces found in reference cells). However, a clear disadvantage of pyranometers is their price point in relation to cheaper silicon sensors, which could be a few times the price of the latter.

As Asian systems are closer to the Equator (e.g. systems which are based in countries like Singapore, Indonesia, Malaysia, Thailand, Cambodia, Vietnam, as well as parts of south of India), the use of more simplistic models without the need of many programming steps with transposition models is welcomed, as computational power is reduced and can be optimized, without prejudice to the results.

For satellite data models, it is important to conduct multi-year studies as well as to perform analysis with more ground points, especially for a large megacity such as Delhi, where varying degrees of air pollution could be at play. With the aging of PV fleets, full seasonality and continuous studies should be conducted. Additionally, degradation rate investigations start to become possible and more relevant with the availability of larger data sets.

The presence of many roof tilt/azimuth combinations prevent the use of the IEC 61724-1 standards in a cost-competitive and reliable way for a common solution for a portfolio of assets in Asia. Solar farms clearly enjoy the fact that asset values are high and the presence of dedicated O&M teams on site allows for daily cleaning routines of irradiance sensors, a condition that the distributed PV approach does not enjoy. In a country like Singapore, as an example, it could be that by the end of 2030, a fleet of ~2-3 GWp of photovoltaics is present in an area of ~710 km$^2$, or 3-4 MWp per km$^2$ [13]. Evaluating thousands of PV systems in a certain geography can become a complex and costly venture.

## IV. CONCLUSIONS

India sites in the commercial & industrial rooftop space located in major city centers are heavily influenced by soiling, which in turn cause ground-telemetry errors, highlighting the importance of novel performance assessment methods and/or use of satellite data. Haze overestimation during winter months by satellite readings in Delhi are an indication of satellite-model improvement needs. Therefore, on the topic of air pollution, the broad PV community in India and cities with air quality issues around the world would benefit from such analysis influencing the economics of PV and its revenue-capturing impact.

The use of satellite-derived irradiation will likely become critical for performance assessment in distributed generation applications. Also, the presence of "hub" meteorological stations at city centers under professional operation/maintenance routines is of utmost importance for the management of future electrical grids. This is the case as systems deployed in multiple roofs need a reliable source of ground irradiance. Such stations then become a good basis for performance calculation of various sites, i.e. for developers'

PV fleet management when handling systems which are within 30-50 km within a centralized station.

## ACKNOWLEDGEMENT

The authors would like to acknowledge the support of other Cleantech Solar team members across geographies: Jeeva Govindarasu in Chennai, Abhishek Puri in Delhi, Lucas Ferrand in Phnom Penh, Jeffrey Olan in Manila and Tom Hawker & Edna Seah in Singapore.

## REFERENCES

[1] Solar Power Europe, *Global Market Outlook for Solar Power / 2017-2021.* http://www.solarpowereurope.org. Downloaded on 8/Jun/2017.

[2] IEC, *Trends 2014 in photovoltaic applications – Survey report of selected IEA countries between 1992 and 2013.* IEA Agency, 2014.

[3] Energy Market Authority (EMA), *Statistics, Installed capacity of grid connected solar PV systems.* Available https://www.ema.gov.sg/statistic.aspx?sta_sid=20140730tzT7HYFaefda

[4] J. Luther and T. Reindl, *Solar Photovoltaic (PV) Roadmap for Singapore.* Singapore, p.77, 2013.

[5] PV-Tech, *India hits 10 GW of solar – Bridge to India.* Available https://www.pv-tech.org/news/india-hits-10gw-of-solar-bridge-to-india.

[6] A. M. Nobre, S. Karthik, H. Liu, D. Yang, F. R. Martins, E. B. Pereira, R. Rüther, T. Reindl and I. M. Peters, *On the impact of haze on the yield of photovoltaic systems in Singapore.* Renewable Energy, Vol. 89 (2016). http://dx.doi.org/10.1016/j.renene.2015.11.079

[7] A. M. Nobre, D. Dave, A. Khor, R. Malhotra, S. Karthik, I. M. Peters and T. Reindl. *Advanced analyses of loss mechanisms for PV systems in Delhi, India.* 32$^{nd}$ European Photovoltaic Solar Energy Conference and Exhibition, Munich, Germany (2016).

[8] Renewable Energy World, *Germany achieves milestone – Renewables supply nearly 100 percent energy for a day.* Available http://www.renewableenergyworld.com/articles/2016/05/germany-achieves-milestone-renewables-supply-nearly-100-percent-energy-for-a-day.html

[9] N. A. Engerer, F.P. Mills, *Kpv: a clear-sky index for photovoltaics.* Solar Energy, Vol. 105 (2014). http://doi.org/10.1016/j.solener.2014.04.019

[10] IEC 61724-1: 2017, *Photovoltaic system performance – Part 1: Monitoring.* Publication date: 03/03/2017.

[11] N.H. Reich, B. Mueller, A. Armbruster, K. Kiefer, W.G.J.H.M. Van Sark and C. Reise, *Performance ratio revisited: are PR>90% realistic?* In Proceedings of the 3$^{rd}$ PVSEC 2011, Hamburg, Germany.

[12] A. Nobre, Z. Ye, H. Cheetamun, T. Reindl, J. Luther and C. Reise, *High performing PV systems for tropical regions – Optimization of systems performance.* In Proceedings of the 27$^{th}$ PVSEC 2013, Frankfurt, Germany.

[13] A. M. Nobre, *Short-term solar irradiance forecasting and photovoltaic systems performance in a tropical climate in Singapore.* Thesis, Universidade Federal de Santa Catarina (2015).

# Machine Learning in PV Fault Detection, Diagnostics and Prognostics: A Review

Sandy Rodrigues[1,3], Helena Geirinhas Ramos[3], and F. Morgado-Dias[1,2]

[1]Madeira Interactive Technologies Institute, [2]Universidade da Madeira, [3]Telecommunications Institute from the Superior Technical Institute of the University of Lisbon.

*Abstract* — Photovoltaic (PV) system malfunctions cause output efficiency to lower which consequently lowers the return of the investment (ROI) and delays investment payback times. These malfunctions can be limited by implementing Photovoltaic System Monitoring (PVSM) solutions. Recently, Machine Learning Techniques (MLT) have been implemented to improve PVSM results and aid in PV performance and PV fault detection, identification, diagnostics and prognostics. This paper provides a review of the work done in the MLT PVSM research field, provides an organized list of MLT solutions used in PVSM, and provides a list of opportunities and challenges to further research in the PVSM field.

*Index Terms* — Diagnostics, Machine Learning, Prognostics, PV Fault, PV System Monitoring, Return of the Investment.

## I. INTRODUCTION

Over the past few years, photovoltaic (PV) systems have become very popular amongst homeowners and solar plant investors since they are viewed as low risk investments that can achieve high return rates and attractive investment payback times [1].

The PV systems are composed by photovoltaic modules and the balance of system (BOS) that includes wiring, switches, a mounting system and solar inverters. In order to avoid PV system downtime and maintain the output efficiency at high levels to shorten investment payback periods, it is essential for the investors, owners and maintenance operators to have access to the input and output data of the PV system. This data can only be acquired through monitoring systems and should be accessible in real-time for the PV system faults to be addressed accordingly to avoid production loss. The PV system real-time information is useful to continuously keep investors and owners informed about the Return on the Investment (ROI), and keep the maintenance operators aware of the PV system status to easily schedule maintenance trips.

Research on PV system Fault Detection and Diagnostics (FDD) has made it possible to detect PV system faults and correctly diagnose them when using Machine Learning Techniques (MLT). Research on MLT is popular and has become very accurate due to available software and computing capacity of the new computers. The two main model categories used in MLT are classification and regression. Classification is used to detect and identify PV faults while regression is used in PV system diagnostics which allows for PV system performance analysis or solar output estimations.

Alongside research on PV system FDD which predicts current faults, there is also ongoing research on Fault Prognostics that predict the performance over the remaining useful life (RUL) of the PV system and consequently predicting possible future faults by using regression MLT. The RUL of a PV system is determined by estimating the degradation rate of the PV modules through degradation models that consider UV radiation, temperature, humidity, state-of-health models and end of life models [2], [3], [4].

By combining PV system FDD and Fault Prognostics research areas it is possible to develop PV system Prognostics and Health Management (PHM) [4] approaches which helps to plan preventive, corrective and/or condition-based maintenance trips contributing to lower the O&M costs by an estimated 20% and consequently shorten investment payback times on the ROI [5]. Condition-based maintenance is the new term associated to the prognostics approach, in which the faults are predicted and the condition-based maintenance trips are made when the faults occur and therefore lowers the frequency of preventative measures and reduces the impacts and costs of the corrective measures by anticipating failures or catching them early [6]. Fig. 1 illustrates how the Machine Learning Techniques are used in the PVSM field of research.

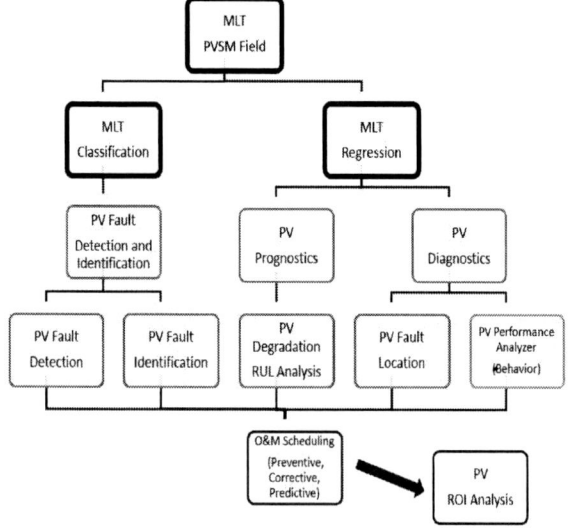

Fig. 1.    MLT in the PVSM field of research

According to the thesis written by Zhao, PV system faults may remain hidden due to the connection infrastructure of the PV system and MLT can successfully detect and identify these types of faults, also known as "blind-spots" [7]. Thus, demonstrating the importance of the use of data-driven models or MLT in PV fault detection, diagnostics and prognostics over model-based techniques. Data-driven approaches are also recommended to deal with non-linear, unpredictable and complex settings such as the PV system solar production that deal with incoherent weather conditions that impact the solar production of the PV system.

The Machine Learning research area is vast, but recently Pedro Domingos shared in his book [8] (recommended by Bill Gates) that MLTs can be organized into five "tribes" and that each one is related to a specific master algorithm as shown in Table I. This concept helps to organize and narrow down the MLTs that are used in PV System Monitoring (PVSM) and indicate which MLTs that are not being used and may have the potential to present good results in the PVSM research field. Pedro Domingos defends the idea of combining various master algorithms presented in Table I to make one master algorithm. This idea of combining different master algorithms is already being adopted in the PVSM research field but in a smaller scale by combining two or three Master Algorithms and are referred to as hybrid MLTs. Two examples of hybrid MLT can be found in [9] and [10] where the first analysis a hybrid MLT that uses the Genetic Algorithm and an Artificial Neural Network which combines the master algorithms of tribes 3 and 2. The latter work tests a hybrid MLT that uses a sparse Bayesian learning theory and Support Vector Machines which combines the master algorithms of tribes 4 and 5.

TABLE I
MLT TRIBES AND RESPECTIVE MASTER ALGORITHMS [8]

|  | Tribe | Master Algorithm |
|---|---|---|
| Tribe 1 | Symbolists | Inverse deduction / Induction |
| Tribe 2 | Connectionists | Backpropagation |
| Tribe 3 | Evolutionaries | Genetic Programming |
| Tribe 4 | Bayesians | Probabilistic Inference |
| Tribe 5 | Analogizers | Kernel Machines |

Based on this information, the motivation was to understand how the different MLT have been used in the different research areas of the PVSM field such as PV Fault Detection, Identification, Diagnostics and Prognostics. The main goal of this work is to analyze 90 research papers and make a list of the MLT and organize them into their associated MLT tribe. The other goal of this work is to present the number of experiments associated to a parameter (PV system faults, PV system scale, Hybrid MLTs, etc.) related to the MLT PVSM field. This experiment count provides information about the areas of the PVSM research field that are more popular as well as those that show good results but need more research. The challenges and opportunities that are available in the PVSM field are also provided in this work.

After this introduction section, a methodology description is provided in section II. The results are presented and discussed in section III, which is then followed by a conclusion of the work in section IV.

## II. METHODOLOGY

This section describes the methodology that was applied, in order to understand which MLTs are being studied in the PV system monitoring (PVSM) research field and which should be studied further. First, a total of 90 scientific papers were identified as being related to the MLT PVSM research field. Secondly, an experiment count was made according to specific parameters. These parameters are namely the type of:

- MLT tribe (Tribe 1, Tribe 2, Tribe 3, Tribe 4, Tribe 5);
- PVSM technique (Model-based, Data-Driven, Hybrid MLT);
- MLT models (Classification, Regression);
- PVSM method (Detection, Identification, Diagnostics, Prognostics);
- PVSM papers (PV performance, PV fault);
- PV system testbed (Grid PV system, lab PV system, simulation); In this paper, a grid PV system is assumed to have more than 3kWp of installed power which represents a typical roof-top household PV system.

In addition, the experiment count is also applied to the parameters that do not involve MLT and that influence the PVSM research analysis such as the type of

- PV system faults (arc, short-circuit, faulty condition, degradation, shading, Line to Line, soiling, open-circuit);
- PV system scale (roof, commercial, plant);
- PVSM level (array, string, module);
- PV system measured inputs (irradiance, module temperature, ambient temperature, current, voltage, etc.) and outputs (power, energy, I-V curve, decision).

The following four subsections present a brief description of all the PVSM techniques that were studied in this work which include the model-based approach, data-driven regressive approach, machine learning techniques and the hybrid MLTs.

### A. Brief Description of the Model-Based Approach

The Model-Based (MB) approach in PVSM is achieved by using sensor data associated to mathematical and visual methods that have indicators and thresholds to aid in decision making [11]. Some examples of this type of approach identified during this work include the **Persistence Model**, and the **Physical Model**.

The Persistence model may be used as a forecasting technique which assumes that the weather conditions remain the same between one sample of the time-series and the next one. The Physical model consists of a set of equations that describe the behavior based on the physical principles. This model may consist of a sensor and stored data which is then analyzed by indicators and thresholds that were determined through trial and error experiments [11].

978-1-5090-5606-4/17 $31.00 © 2017 IEEE

*B. Brief Description of the Data-Driven Regressive Approach*

The Data-Driven Regressive (DDR) approach in PVSM includes all the data-driven algorithms that are not learning algorithms but are regressive algorithms and depend on historical data for forecasting [11]. The most popular DDR algorithms that were identified in this work are the **ARIMA** (Auto-Regressive Integrated Moving Average), the **linear regression models** and the **Principal component analysis** (PCA).

The forecasting values of ARIMA are based on its inertia, while the regression models analyses the relationship between the dependent and independent variables. The PCA is used to select the most relevant variables [11].

*C. Brief Description of the Machine Learning Techniques*

This section briefly describes the MLT Tribes and indicates which of the MLT are most popular in each tribe. Table I presents a description of the five tribes that follow.

**Tribe 1: Symbolists (Inverse deduction / Induction)**

The MLTs that belong to this tribe attempt to find the missing knowledge by applying inverse deduction which starts with some properties and conclusions and works its way backward to find the gaps in knowledge by using analysis and data sets [8]. The most popular MLT of this tribe are Decision Trees, Random Forests and the Fuzzy Logic [11].

**Tribe 2: Connectionists (Backpropagation)**

The MLT of this tribe are composed by algorithms that claim to emulate the functions of the brain by creating artificial neurons and connecting them in a neural network by using an input layer, one or various hidden layers and an output layer. The inputs are taken by the hidden neurons which then generate an output (new knowledge) that can be read by other neurons that perform the same function [8]. The Artificial Neural Networks and Extreme Learning Machine are some the most popular tribe 2 MLTs [11].

**Tribe 3: Evolutionaries (Genetic Programming)**

Here the algorithms of the MLTs try to mimic the evolutionary process of genomes and DNA where the performance is measured by the fitness of the offspring. The algorithms are divided into a set of sub programs that mutate and combine with other sub programs to determine if the prediction results improve with multiple iterations of the program. New knowledge is determined by simulating evolution [8]. The popular evolutionary MLTs include the Genetic Algorithm and the Genetic Programming [11].

**Tribe 4: Bayesians (Probabilistic Inference)**

These MLTs systematically attempt to reduce the uncertainty of the new knowledge by applying the probabilistic inference algorithm to events. The events that are known to occur are assigned prior probabilities, and as more evidence is observed, the priors either lose their importance or become more important [8]. The Monte Carlo Method is a very popular Bayesian MLT [11].

**Tribe 5: Analogizers (Kernel Machines)**

The Analogizers MLTs recognize the similarities between the old and the new data/knowledge by using the nearest neighbour kernel machine algorithms which analyse its surroundings and try to generalize by similarity. These results are similar to those of the neural network models [8]. The most popular analogizer MLTs are the Support Vector Machines and the K-Nearest Neighbour algorithms [11].

*D. Brief Description of the Hybrid MLTs*

Hybrid MLTs is the name given to the combination between two or more PVSM techniques (MB, DDR, and MLTs). This combination contributes to strengthening the accuracy of the forecasting results [11]. The most popular hybrid MLTs that were identified in this work are namely the **Radial Basis Function Neural Network (RBFNN)**, the **ANN + Physical Model**, the **Genetic Algorithm-Artificial Neural Network (GA-ANN)**, the **ANFIS (adaptive neuro-fuzzy inference system)**, and the **Relevance Vector Machine (RVM)**.

The RBFNN is a MLT combination between tribe 5 (RBF) and tribe 2 (NN). The ANN+Physical model is a PVSM technique combination between a tribe 2 (ANN) MLT and a model-based technique. The GA-ANN is a MLT combination between tribe 3 (GA) and tribe 2 (ANN). The ANFIS is another MLT that combines tribe 2 (neuro) and tribe 1 (fuzzy).

## III. RESULTS AND DISCUSSION

Over 150 PV systems monitoring (PVSM) research papers were identified and analysed for this work. Only 90 of them were considered since they include the topics that are discussed in this work such as MLTs, Hybrid MLTs, power and energy estimation, as well as PV performance and PV fault analysis. The papers that only present the framework and do not include an experimental research were not considered. The graphs that follow present the number of times a given parameter is mentioned in the 90 papers that were studied (blue). The graphs also present the number of times that a given parameter is mentioned in the 32 papers that study the PV fault analysis (orange). Some of these papers focus only on the PV performance analysis, others only on the PV fault analysis and others on both the PV performance and PV fault analysis.

Fig. 2 presents the various MLT tribes and how many times they were used in all the papers and in the PV fault papers. This graph also presents the experiment count of the Model-Based, Data-Driven Regressive and hybrid MLT approaches. The MLTs of Tribe 2 and 3 are used the most in the PVSM research field which include the ANN and SVR respectively. The Regression Trees and the Bayesian probability from the Tribes 1 and 4 respectively are used similarly. The MLTs from Tribe 3 (Genetic Programming and Genetic Algorithm) are the least used and not used at all in the PV fault papers. Fig. 2 also presents the experiment count that use the three

978-1-5090-5606-4/17 $31.00 © 2017 IEEE       3180

approaches namely the MB approach, the DDR approach and the Hybrid MLT approach. The hybrid MLT approach is used more in the PV performance analysis compared to the PV fault analysis.

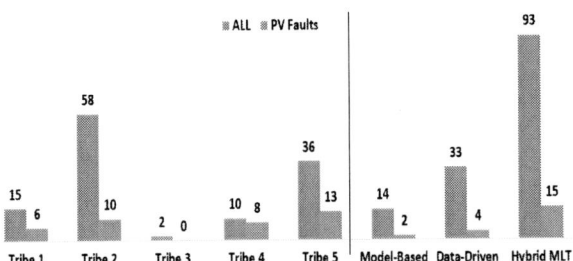

Fig. 2.    MLT Tribes and PVSM Technique approaches

Fig. 3 presents the number count of the papers for the classification and regression MLT models as well as the four PV system monitoring methods such as the Detection, Identification, Diagnosis and Prognosis. Fig. 3 also presents the number count of the papers that analyse the PV performance as well as those that analyse the PV faults. There are double the papers that consider regression as opposed to classification MLT models in the PV system monitoring field of research. Classification is mostly used in the PV fault analysis. PV fault Detection is the most popular PVSM method and prognosis the least popular, while PV fault identification and diagnosis are similar in use. We can see that PV performance analysis was used in 87 out of the 90 papers studied in this work while 29 of the 32 PV fault papers studied in this work performed PV performance analysis. PV performance research is more popular than PV fault analysis.

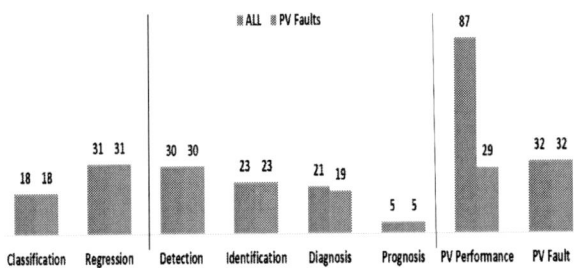

Fig. 3.    MLT Models, Methods and PV System Analysis

Fig. 4 presents the parameters of the testbed used in the research papers. The end results of the experiments are always more reliable when using a grid connected PV system. From the graph below, it is possible to see that there are many experiments carried out on PV grid connected systems. However, there are still a fair amount of experiments that are carried out in lab testbeds (a few PV modules) and by simulating a PV system using software (Matlab). These two

latter testbeds are not as reliable because the nonlinear nature of the weather conditions may not be considered in the analysis. Models always have limitations compared to the true PV systems. The PV fault analysis research is mainly done on lab and simulation testbeds which might not result in reliable findings. The PV installed power is also reflected in the graph that follows by differentiating the PV scale into plant, commercial and roof-top sizes. There is an even amount of work that is carried out in all three scales. The monitoring level of the PV systems is normally made on the array level due to cost control, however it is interesting to see that the PV fault analysis research tends to use the module-level monitoring which in turn provides very accurate results.

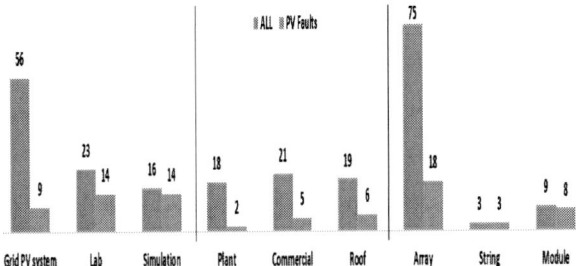

Fig. 4.    PV System Testbeds, Scale and Monitoring Level

Fig. 5 presents the input features that are most often used in the research papers studied in this work. Irradiance, ambient temperature and power are the input features that are used the most followed by the wind speed, module temperature, current and voltage values. Not many papers consider the energy values.

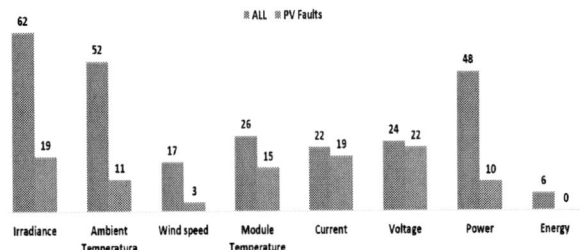

Fig. 5.    PVSM Input Features

Fig. 6 presents the output features that are mostly used in the research papers studied in this work. Power prediction is the most popular while energy and I-V curve prediction is also used. Classification decisions are widely used in the PV fault analysis as well as the power and I-V curve predictions. Not many papers consider the standalone voltage and current predictions.

978-1-5090-5606-4/17 $31.00 © 2017 IEEE

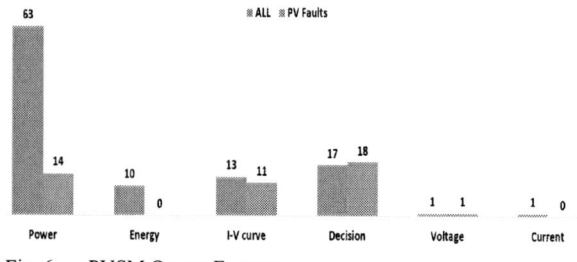

Fig. 6. PVSM Output Features

Fig. 7 presents the experiment count of the PV fault analysis papers that are associated to eight different PV fault parameters. Shading, short-circuit and degradation faults are researched the most while faulty conditions, line to line, soiling and open-circuit faults are less popular in PV fault research. The arc fault analysis using MLT present the least research papers.

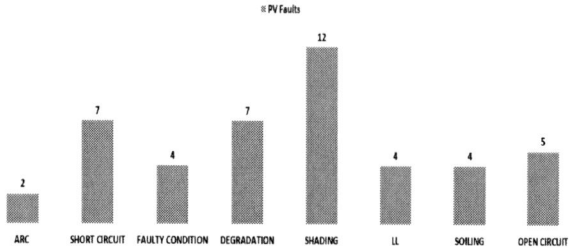

Fig. 7. PV System Faults

Table II presents some of the most popular model-based (MB) and data-driven regressive approaches (DDR) identified in the 90 papers studied in this work. The persistent and physical models were mostly used in the papers when using the model-based approach. While the ARIMA, Linear Model and PCA were mostly applied when using the data-driven regressive approach. The complete list of MB and DDR approaches can be found in the following reference [12].

TABLE II
LIST OF MODEL-BASED AND DATA-DRIVEN ALGORITHMS AND
RESPECTIVE PAPER NUMBER COUNT

| Model-Based Approach (14 experiments) | # | Data-Driven Regressive Approach (33 experiments) | # |
|---|---|---|---|
| Persistence Model | 6 | ARIMA (Auto-Regressive Integrated Moving Average) | 7 |
| Physical Model | 2 | Linear Model or Linear Regression | 4 |
| Hampel Identifier and Boxplot rule | 1 | PCA (Principal component analysis) | 4 |

Table III presents all the MLT algorithms that were identified in the 90 papers studied in this work. The Random Forests algorithm is very popular amongst Tribe 1 MLTs

followed by Regression Tree, Decision Tree, and Functional Tree algorithms. The ANN algorithm from Tribe 2 is highly used in the PVSM research field since it was identified 41 times followed by the ELM algorithm. Tribe 3 includes the Genetic Programming and Genetic algorithms which were not used much in the PVSM research. The Gaussian Process algorithm is the most popular MLT out of the Tribe 4 algorithms. The K-Nearest Neighbor algorithm is the most popular out of the Tribe 5 MLTs followed by the SVM, the SVR, SVR with Radial kernel, Radial Basis Function, Polynomial SVM or SVR, and the K-Means algorithm.

Table IV presents 4 of the 69 Hybrid MLT algorithms that

TABLE III
LIST OF MACHINE LEARNING ALGORITHMS ORGANIZED INTO
FIVE MACHINE LEARNING TRIBES

| Tribes | MLT | # |
|---|---|---|
| Tribe 1 #15 | Random Forests (RF) | 4 |
| | Regression Tree (RT) | 2 |
| | Classification and Regression Decision Tree (CART) | 2 |
| | Functional Tree | 2 |
| | C.45 Decision Tree | 1 |
| | Best First Decision Tree | 1 |
| | Gradient Boosting | 1 |
| | Binary Tree Bagging | 1 |
| | Fuzzy Logic (FL) | 1 |
| Tribe 2 #58 | Artificial Neural Network (ANN) | 41 |
| | Extreme Learning Machine (ELM) | 4 |
| | Extension Neural Network | 2 |
| | Recurrent Neural Network (RNN) | 2 |
| | Neural Network Ensemble (NNE) Bagging Or Bootstrap Based Neural Network (BNN) | 2 |
| | Group method of data handling neural network (GMDH-type NN) | 1 |
| | Dynamic Recurrent Neural Network (DRNN) | 1 |
| | Deep Belief Network (DBN) | 1 |
| | Long Short-Term Memory networks (LSTM) Based on Recurrent Neural Network (RNN) | 1 |
| | Auto-LSTM Long Short-Term Memory networks | 1 |
| | Entropy Extreme Learning Machine (EELM) | 1 |
| | Online Sequential Extreme Learning Machine | 1 |
| Tribe 3 #2 | Genetic Programming (GP) | 1 |
| | Genetic Algorithm (GA) | 1 |
| Tribe 4 #10 | Gaussian Process | 6 |
| | Monte Carlo Method | 1 |
| | Graph-based Semi-Supervised Learning | 1 |
| | Naïve Bayes Classification | 1 |
| | Bayesian Belief Network | 1 |
| Tribe 5 #15 | K – Nearest Neighbor (K-NN) | 9 |
| | Support Vector Machine (SVM) | 6 |
| | Support Vector Regression (SVR) | 5 |
| | SVR with Radial kernel | 4 |
| | Radial Basis Function | 3 |
| | Polynomial SVM or SVR | 2 |
| | K-Means Algorithm | 2 |
| | Linear SVM | 1 |
| | Local Outlier Factor (LOF) | 1 |
| | Locally Weighted Learning | 1 |
| | Local Linear Fitting / Regression | 1 |
| | Clustering by fast search and find of density peaks | 1 |

were identified in the 90 papers studied in this work. The RBFNN, the ANN+Physical model, the GA-ANN, and ANFIS were the most popular Hybrid MLTs. The complete list of the hybrid MLTs can be found in the following reference [12].

TABLE IV
LIST OF HYBRID MACHINE LEARNING ALGORITHMS

| Hybrid MLT (69 algorithms) | Tribes | # |
|---|---|---|
| Radial Basis Function Neural Network (RBFNN) | T5 + T2 | 5 |
| ANN + Physical Model | T2 + MB | 5 |
| Genetic Algorithm – ANN (GA- ANN) | T3 + T2 | 4 |
| ANFIS (adaptive neuro-fuzzy inference system) | T1 + T2 | 3 |

*A. Challenges and Opportunities for the PVSM research field*

The MLTs suggested in Table V, present the best results comparison studies, however they have only been considered in between 1 and 3 experiment papers and therefore need further research. Based on the results, there are some opportunities and challenges that were identified to help further research in the PVSM field which are discussed as follows.

TABLE V
OPPORTUNITIES AND CHALLENGES

| | |
|---|---|
| **Opportunities** | **Tribe 1** MLT present good results but few experiments, therefore should be further researched, particularly with the Random Forests, Decision Trees, Regression Trees and Fuzzy Logic. |
| | **Tribe 2** presents a high experiment count due to the use of the ANN MLT, which indicates that the PVSM researchers are hesitant to try other MLT such as the ELM and RNN which have shown to present very accurate results. |
| | Only two research experiments have dealt with the **Tribe 3** (genetic programming and genetic algorithms) MLT. |
| | **Tribe 4** MLT present very good results when using the Gaussian Process, the Monte Carlo and the Graph-based Semi-Supervised Learning. More research is encouraged. |
| | Amongst the **Tribe 5** MLTs, the K-NN MLT is more popular than the SVM, and therefore SVM work is encouraged. |
| | 69 **Hybrid MLT** algorithms were identified but most explored only in one paper. More research should be done on combining MLT to discover the most accurate Hybrid MLTs. |
| | PV system **module-level monitoring** improves the accuracy of the results even though this solution is costly, however this should be further explored. |
| | PV fault **diagnostics and prognostics** research is initiating, but more emphasis should be put on PVSM prognostics to improve M&O scheduling and lower ROI payback times. |
| **Challenges** | There is a **vast number of MLTs**, therefore many researchers tend to use the same ones as they rely on research results from previous works and do not try new MLT. |
| | Research on PVSM should be done on real-world PV system **testbeds** to obtain credible results that consider real-world disturbances caused by environmental factors and signal noise coming from the measurement devices. |

## IV. CONCLUSION

This paper highlights the key parameters, challenges, and research opportunities that can further research in the MLT PVSM field. A list of MLTs organized by tribes is provided to narrow down the types of MLT that should be considered in future research. Hybrid MLT research and MLT comparison studies are encouraged. PV system Prognostics research is encouraged to provide condition-based maintenance scheduling and improve ROI payback times. PV module-level monitoring provides accurate results which is useful to explore PV faults. Real-world testbed research is encouraged, as accurate experimental results are presented.

## ACKNOWLEDGEMENT

Acknowledgments to the LUSO-AMERICAN DEVELOPMENT Foundation (Fundação Luso-Americana para o desenvolvimento – FLAD) for the travel grant. The support from the Portuguese Foundation for Science and Technology for their support through Projeto Estratégico LA 9 - UID/EEA/50009/2013. The support from ARDITI – Agência Regional para o Desenvolvimento e Tecnologia under the scope of the Project M1420-09-5369-000001 – PhD Studentship, is also gratefully appreciated.

## REFERENCES

[1] S. Rodrigues, *et al*, "Economic feasibility analysis of small scale PV systems in different countries," Solar Energy, vol. 131, pp. 81–95, 2016.

[2] A. Chokor, *et al*, "A Review of PV DC Systems Prognostics and Health Management: Challenges and Opportunities," Prognostics and Health Management (PHM) Society, 2016.

[3] N. Laayouj, H. Jamouli, and M. El Hail, "New Prognostic Framework for Degradation Assessment and Remaining Useful Life Estimation of Photovoltaic Module," in Control and Automation (MED), 24th Conference IEEE, 2016, pp. 378–383.

[4] N. Clements, "Introduction to Prognostics," PHM Society, 2011.

[5] S. Vohnout, *et al* "Uptime improvements for photovoltaic power inverters," Technical Program for MFPT The Prognostics and Health Management Solutions Conference, pp. 1–15, 2012.

[6] T. Keating, A. Walker, and K. Ardani, "Best Practices in PV System Operations and Maintenance" NREL, p. 57, 2015.

[7] Y. Zhao, "Fault Detection, Classification and Protection in Solar PV Arrays," Northeastern University Boston, PhD Thesis 2015.

[8] P. Domingos, The Master Algorithm: How the Quest for the Ultimate Learning Machine Will Remake Our World. 2016.

[9] M. Russo, *et al*, "Genetic programming for photovoltaic plant output forecasting," Solar Energy, vol. 105, pp. 264–273, 2014.

[10] N. Laayouj, *et al*, "Photovoltaic Module Health monitoring and Degradation Assessment," no. September, pp. 570–577, 2016.

[11] J. Antonanzas, *et al*, "Review of photovoltaic power forecasting," Solar Energy, vol. 136, pp. 78–111, 2016.

[12] S. Rodrigues, "TABLES," Available: https://goo.gl/0mIhm8.

# Outdoor Field Performance from Bifacial Photovoltaic Modules and Systems

Joshua S. Stein[1], Daniel Riley[1], Matthew Lave[1], Clifford Hansen[1], Chris Deline[2], Fatima Toor[3]

[1]Sandia National Laboratories, Albuquerque, NM, 87185, USA;
[2]National Renewable Energy Laboratory, Golden, CO, 80401, USA
[3]University of Iowa, Iowa City, IA, 52242 USA

*Abstract* — **Bifacial PV modules and systems deliver more energy than equivalent monofacial modules in the same orientation. However, bifacial performance models are not yet mature enough to predict bifacial gains for all system configurations. Field performance data is needed at a variety of different spatial scales in order to improve and validate these models. This paper reports on a number of bifacial field installations intended for this purpose.**

## I. INTRODUCTION

Bifacial photovoltaic (PV) cells, modules, and systems offer a rapid pathway to significantly decreased levelized cost of energy compared with conventional monofacial PV modules. Unlike increasing cell efficiency, which takes many years to bring laboratory innovations to the production line, bifacial PV technology is available today but is underutilized. One major barrier to broader use of bifacial PV modules and systems is a lack of knowledge and experience with system designs that take advantage of the specific features of bifacial cells. Bifacial system performance cannot be predicted with confidence using current PV performance modeling applications because these tools assume that PV modules are illuminated on only one side.

Analytic and empirical studies have shown that use of bifacial modules can potentially increase system yield by at least 10% over a fixed latitude tilt monofacial array, and increased yield can be much higher under certain conditions [1-2]. The bifacial benefit varies with tilt angle, module height above array base, reflectivity (albedo) of the array base, and other factors that influence the total amount of light reaching both sides of the PV cells. However, the sensitivity to these parameters is complex and as system size and ground coverage ratio increases, bifacial gains suffer as the array increasingly covers the ground with shadows and less light is available to the back of the modules.

In order to better understand the factors that affect bifacial PV system performance Sandia National Laboratories (Sandia), the National Renewable Energy Laboratory and the University of Iowa have teamed on a three-year research project aimed at better understanding the actual performance potential of bifacial PV systems.

The project aims to achieve the following three objectives:

1. Obtain field performance data from bifacial modules, strings, and arrays in a variety of orientations and environments.
2. Develop and standardize bifacial module rating methodology
3. Develop and validate bifacial performance models that can be used to inform bifacial array designs.

This paper describes the results obtained from the first objective in the first half of the project period. Other papers that describe results related to the second and third objectives are presented in other sessions [3-6].

## II. BIFACIAL FIELD TESTING

Sandia has built a number of testbeds using bifacial PV modules to obtain performance data in different configurations. In most of these testbeds we have included monofacial modules of the same size as comparisons. The following bifacial testbeds have been developed:

- Single module IV tracing at different tilts and heights
- Single module DC monitoring on microinverters at five different orientations (three different climate sites).
- String-level DC performance at different tilt angles.
- Bifacial DC string performance on single axis trackers.
- Bifacial DC string performance on two-axis trackers.

### A. Single module IV tracing at different tilts and heights

Sandia built a rack that fits four PV modules in landscape and can be easily adjusted in height above ground and tilt angle. IV curves on each of the four modules are measured using a multitracer. Irradiance is measured in two locations on the front side and three locations on the back side. IV curves are being measured at 5 minute intervals. Fig. 1 shows the setup.

Fig 1. Sandia's adjustable, single modules IV curve rack in Albuquerque, NM. Two bifacial modules are on the right and two monofacial modules are on the left.

### B. Single module monitoring on microinverters at five different orientations.

A second test system is comprised of 16 bifacial and 16 monofacial modules divided into five different configurations that vary tilt and azimuth angles as well as ground reflectivity. Fig. 2 shows the installed system. Table 1 describes the five different configurations. Copies of this system are also installed in Nevada and Vermont. These systems are part of a project at the Regional Test Centers being performed with Prism Solar.

Fig. 2. Bifacial and monofacial modules at five different orientations. Two of the arrays are installed over white rock to enhance back side ground reflections.

Table 1: Orientation and ground surface of test modules.

| Label | Orientation | | Ground Surface |
| | Tilt | Azimuth | |
|---|---|---|---|
| S15Wht | 15° | 180° (South) | White gravel |
| W15Wht | 15° | 270° (West) | White gravel |
| S30Nat | 30° | 180° (South) | Natural |
| S90 | 90° | 180° (South) | Natural |
| W90 | 90° | 270° (West) | Natural |

### C. String-level performance at different fixed tilt angles

This system is aimed at learning how bifacial modules behave in a series string. Sandia built four rows of racking, each at a different tilt angle (45°, 35°, 25°, 15°) (Fig 3). Each row has two strings of eight modules which are alternated. Two rows used Sunpreme bifacial modules and two used Prism Solar bifacial modules. Monofacial modules from SolarWorld were used.

Fig. 3. Fixed-tilt, string level bifacial testbed at Sandia.

### D. String-level performance on single axis trackers

Sandia has also installed two rows of single axis trackers designed to hold four strings of bifacial modules (Fig. 4). Currently, two strings of bifacial modules have been installed. The tracker movement is controlled by light sensors, time of day, and control parameters set by the operator rather than sun position.

Fig. 4. Single axis tracker for bifacial modules being constructed at Sandia.

### E. Bifacial string performance on two-axis trackers

As part of the Regional Test Center program, two 2-axis trackers from All Earth Renewables have been installed in Vermont, each holding two strings (one of bifacial modules and one of monofacial modules) (Fig. 5). DC voltage and current is measured on each string.

Fig. 5. Two-axis trackers with bifacial modules at the Vermont Regional Test Center.

## III. PRELIMINARY RESULTS

All of the test beds described above have been collecting data and some preliminary results are shared below. Instantaneous bifacial gain at time t, $BG_i(t)$ is defined here as:

$$BG_i(t) = 100\% \times \left( \frac{P_{bifacial}(t) / Pmp_{bifacial}}{P_{monofacial}(t) / Pmp_{monofacial}} - 1 \right)$$

where $P_{bifacial}$ and $P_{monofacial}$ are measured power values and $Pmp_{bifaical}$ and $Pmp_{monofacial}$ are front side power ratings measured on a flash tester at STC with the back of the bifacial module covered with an opaque material. An integrated bifacial gain in energy, $BG_E$ (for example, one month) can be calculated as:

$$BG_E = 100\% \times \left( \frac{\sum_{1\ month} P_{bifacial} / Pmp_{bifacial}}{\sum_{1\ month} P_{monofacial} / Pmp_{monofacial}} - 1 \right)$$

### A. *Single module IV tracing at different tilts and heights*

The adjustable rack with four modules was set up to measure IV curves at specific tilt angles and orientations. It was moved every 1-2 weeks over several months. Figure 6 shows bifacial gains measured as a function of tilt angle and height above ground. When tilted, bifacial gains increase with module height. Bifacial gain seems to have a weak sensitivity to tilt angle, except when transitioning between 30° and 45° tilt. The high bifacial gains seen for 45° are enhanced due these measurements being made in the summer when the sun rises and sets well north of east and west, respectively. This results in direct sunlight on back of modules. In addition, higher sun elevation in the summer results in smaller shadows on the ground at midday, increasing bifacial gains.

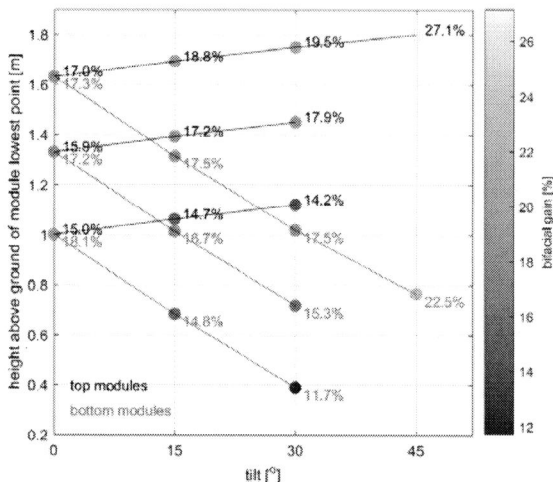

Fig. 6. Single module bifacial gains measured as a function of tilt angle and height of module bottom edge off ground.

### B. *Single module monitoring on microinverters at five different orientations.*

Fig. 7 shows example results from the single module monitoring on microinverters at five different orientations [7]. This work was done in partnership with Prism Solar and used their bifacial modules. In every case, bifacial output is greater than the monofacial in the same orientation (Fig. 8). The west-facing vertical bifacial modules produced more energy than the latitude-tilt monofacial modules. During the day bifacial gains are greatest when the angle of incidence on the array is large. This indicates that bifacial module advantages are greatest for non-optimal, monofacial array orientations. However, total energy is typically lower.

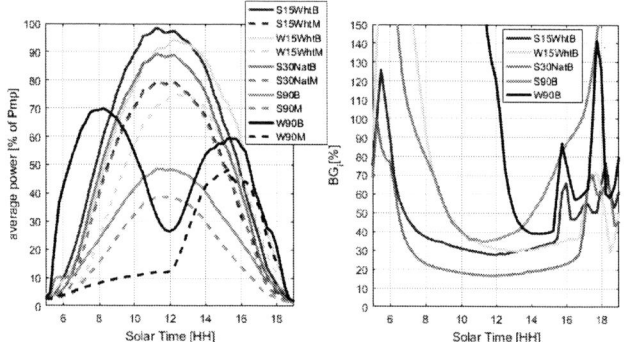

Fig. 7. Left: Average power output, Right: bifacial gains over six months from the bifacial and monofacial modules on microinverters.

Fig 8. shows that annual bifacial gains for the W-facing vertical modules can exceed 100%. This is because it is always cooler in the mornings in NM when the W-facing bifacial module is illuminated on the backside. The cooler temps result in increases in the efficiency that exceed the reductions from the bifacial ratio. Energy production would likely be higher for E-facing bifacial modules but bifacial gains would be lower.

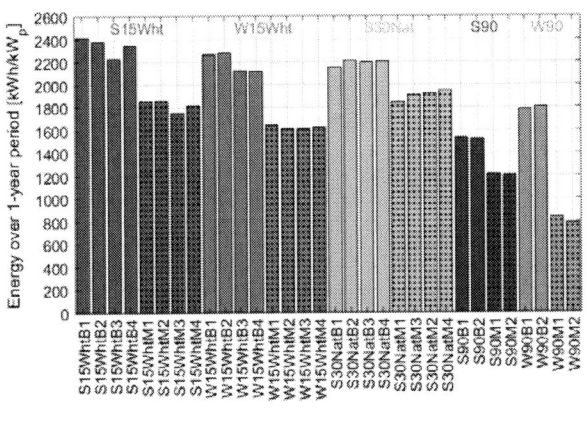

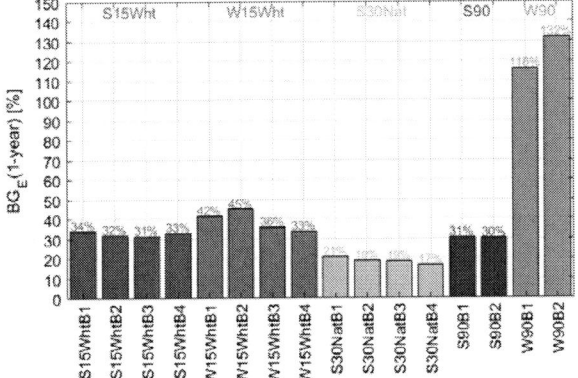

Fig. 8. One-year energy yield for bifacial and monofacial modules (top) and annual bifacial gain in energy for Prism Solar bifacial modules (bottom) deployed in New Mexico.

### C. String-level performance at different tilt angles

Sting-level DC current and voltage was measured on bifacial and monofacial strings at 15°, 25°, 35°, and 45° in Albuquerque, NM from May 10 to June 11, 2017. Bifacial and monofacial modules were alternated to reduce spatial bias in back side irradiance. However, since the bifacial modules were frameless and the monofacial modules had frames there was initially a problem with partial shading of the bifacial modules in the morning and afternoon due to the monofacial module frames that rose above the bifacial modules on the rack. This was eventually fixed by changing the bifacial module clips to raise the modules to a similar level as the monofacial modules. Fig. 9 shows instantaneous bifacial gains before and after the fix was made. The main effect of the partial shading was to significantly reduce the output of the bifacial modules at the start and end of the day. After the fix (red points) the bifacial gains at these times increased significantly. Bifacial gain in energy for each array was calculated after the fix was made. In order of increasing tilt angles, these gains are 11.8%, 12.3%, 15.4% and 19%, respectively.

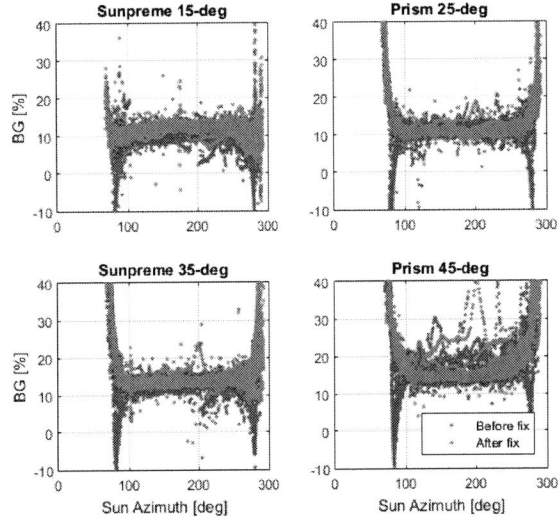

Fig. 9. Instantaneous bifacial gains for strings at four different tilt angles. Blue points are before partial shading issue was fixed. Red point are after.

Fig 10 compares energy produced between arrays. The 15° array produced the most energy during this late spring period, which is consistent with the solar elevation at this time of year. It is important to note that while the bifacial gains are greatest for the 45° system, the most energy is produced by the 15° system at this time of year. Once a full year of data is available it is expected that the 35° row will produce the maximum energy, since the Sandia site is at 35° N. latitude.

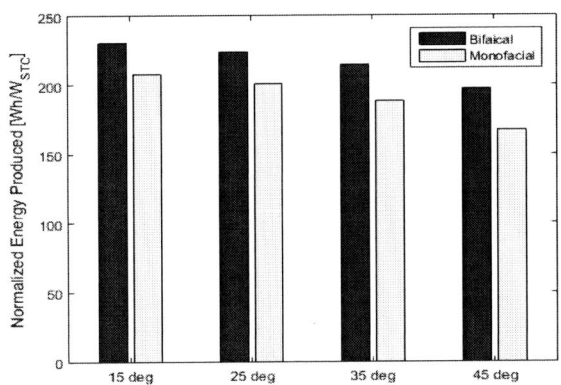

Fig 10. Comparison of the energy produced by each array (normalized by front side STC rating).

### D. Bifacial string performance on single axis trackers

Two strings of bifacial modules were installed each on its own tracker. While we are monitoring DC current and voltage from these strings, we have yet to install a reference monofacial string for calculating bifacial gains. We can, however, estimate potential bifacial gain using the front and rear facing reference cells that are part of the monitoring system. This potential bifacial gain can be estimated as:

$$BG_{potential} = R_b \left[ G_f + G_r \right] / G_f$$

where $R_b$ is the bifacial ratio, $G_f$ and $G_r$ are measured plane-of-array irradiance on the front and rear, respectively. Actual bifacial gains would likely be somewhat lower due to module bifacial ratios being <1.

In addition, there is one more problem with this system. The trackers are controlled by an algorithm that uses light sensors to optimize tracker position. Unfortunately, the algorithm occasionally does not converge and points the trackers in the wrong direction relative to the sun. To account for this problem, we calculated the daily potential bifacial gain only using times when the optimal tracker angle (calculated using PVLIB function, *pvl_singleaxis*) was within +/- 5° of the measured angle. Daily potential bifacial gains are show in Fig 11 and are mostly between 8%-14%. When the tracker is off-track, potential bifacial gains are larger.

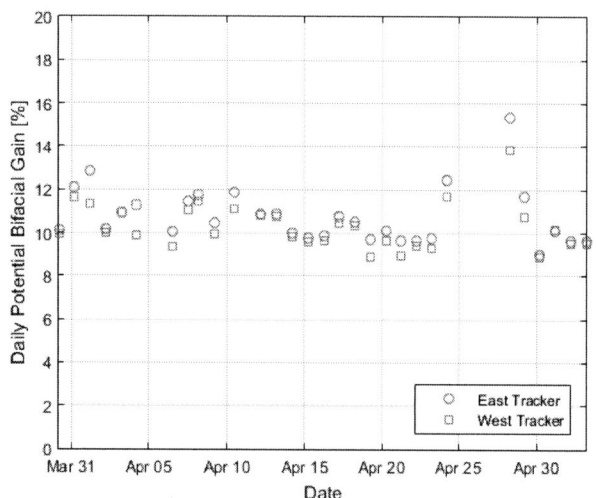

Fig. 11. Daily potential bifacial gain on single axis trackers in Albuquerque, NM.

### E. *Bifacial string performance on two-axis trackers*

Preliminary data from the two 2-axis trackers in VT for the first six days of operation was analyzed and is shown in Fig 12.

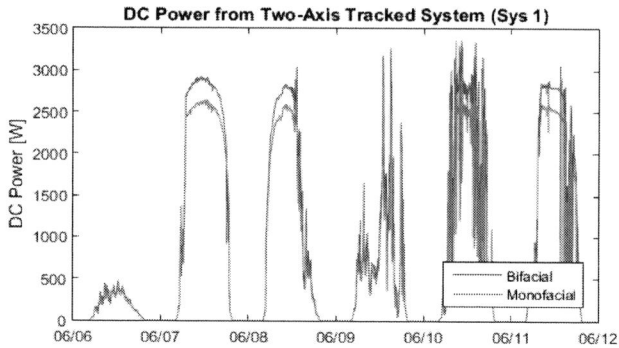

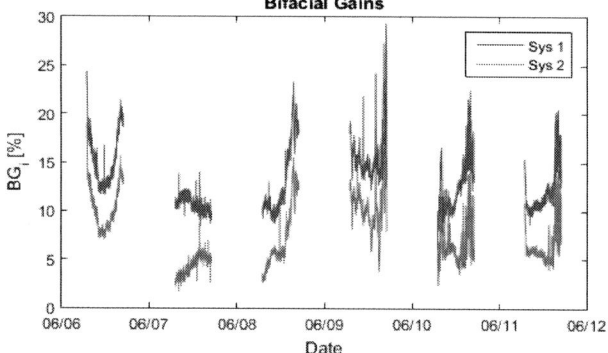

Fig. 12. Power output from system 1 (top) and instantanenous bifacial gains for systems 1 & 2 (bottom).

The reason that bifacial system 2 has lower gains than system 1 is that the bifacial ratio (back to front Pmp) is over 90% for system 1 and only about 60% for system 2. In addition, the trackers were not specifically optimized for bifacial arrays. The support structure for the modules includes several wide beams that obstruct light reaching the modules from the back side (Fig. 13). Thus bifacial gains would be higher if these obstructions could be eliminated or minimized. Also, gains are expected to be significantly higher in the winter when the ground is covered in snow.

Fig. 13. View of the rear side of the tracker for System 1 showing that support structure blocks the back side of the bifacial modules.

## IV. DISCUSSION AND CONCLUSIONS

Bifacial photovoltaic cells, modules, and systems can offer significant boosts in energy produced when compared with similar monofacial PV systems. However, the energy gains depend a lot on the technology chosen and how the system is deployed. We have demonstrated the potential of bifacial PV in a number of different deployment scenarios including single modules, small arrays with microinverters, multi-row, fixed tilt arrays of module strings, single axis tracking and dual axis tracking. From an examination of this field data we can make a number of important conclusions.

- Bifacial performance will always exceed monofacial performance when module output is normalized for front side STC rating and the rear side receives some amount of light.
- Bifacial gains increase as the orientation of the front side of the array (tilt and azimuth) deviates from the optimal orientation for monofacial. However, total energy production of tilted bifacial systems appears to be maximized at the same orientation as for monofacial modules. One exception is E-W bifacial vertical modules, which can outperform optimally oriented monofacial modules, especially with enhanced albedo. Other exceptions may exist.
- Experiments with single bifacial modules and small systems with few surrounding structures result in significantly higher bifacial gains than would be achieved in larger systems. This is because a larger fraction of modules is at the edges of smaller systems and therefore more rear side irradiance is available.
- Bifacial modules with module-scale MPPT (microinverters or optimizers) perform significantly better than series connected modules and string-level MPPT. We believe this is because rear-side irradiance varies significantly in space throughout the array and this can lead to current mismatch in series connected modules. This means that the module with the lowest current (i.e. lowest rear side irradiance) in the string limits the performance of the other modules.
- Bifacial gain of isolated modules and small arrays improves as the array height increases. This is because the module's view of the ground increases and light from more distant (unshaded) surfaces is available to the rear side. This is especially true for lower sun angles when shadows from modules high off the ground appear further away from the array. This is likely one of the reasons that the bifacial performance on the 2-axis trackers in VT was so high despite significant back side obstructions from the tracker supports.
- Bifacial performance is quite sensitive to enhanced albedo of the ground surface. In the Prism Solar RTC array, arrays with enhanced albedo produced more energy than those over lower albedo ground.
- Vertical E-W bifacial modules produce energy earlier and later in the day than S-facing arrays. Such an output power profile may better match demand for electricity and could be a beneficial design under time of use rates.

In conclusion, bifacial modules significantly outperform monofacial modules in conventional designed systems. Additional performance benefits from bifacial modules are possible with optimized system designs that enhance albedo, avoid backside obstructions and minimize ground shading beneath the array.

## ACKNOWLEDGEMENTS

Sandia National Laboratories is a multimission laboratory managed and operated by National Technology and Engineering Solutions of Sandia LLC, a wholly owned subsidiary of Honeywell International Inc. for the U.S. Department of Energy's National Nuclear Security Administration under contract DE-NA0003525.

## REFERENCES

[1] Reise, C., A. Schmid, (2016) "Realistic Yield Expectations for Bifacial PV Systems – am Assessment of Announced, Predicted and Observed Benefits", 6th PVPMC Workshop, Freiburg, Germany: http://www.slideshare.net/sandiaecis/42-reise-realisticyieldexpectationsforbifacialpvsystems-56350717 (Accessed June 2017).

[2] Castillo-Aguilella, J. (2016) "Multi-Year Study of Bifacial Energy Gains Under Various Field Conditions", 6th PVPMC Workshop, Freiburg, Germany: http://www.slideshare.net/sandiaecis/44-castillo-aguilellamultiyearstudyofbifacialenergygains (Accessed June 2017)

[3] Marion, B., S. MacAlpine, C. Deline, A. Asgharzadeh, F. Toor, D. Riley, J. Stein and C. Hansen (2017). A Practical Irradiance Model for Bifacial PV Modules. 44th IEEE PVSC. Washington DC.

[4] Riley, D., C. Hansen, J. Stein, M. Lave, J. K. B. Marion and F. Toor (2017). A Performance Model for Bifacial PV Modules. 44th IEEE PVSC. Washington, DC.

[5] Shishavan, A. A., T. M. Lubenow, J. Sink, B. Marion, C. Deline, C. Hansen, J. Stein and F. Toor (2017). Analysis of the Impact of Installation Parameters and System Size on Bifacial Gain and Energy Yield of PV Systems. 44th IEEE PVSC. Washington, DC.

[6] Hansen, C. W., R. Gooding, N. Guay, D. M. Riley, J. Kallickal, D. Ellibee, A. Asgharzadeh, B. Marion, F. Toor and J. S. Stein (2017). A Detailed Model of Rear-Side Irradiance for Bifacial PV Modules. 44th IEEE PVSC. Washington DC.

[7] Stein, J. S., L. Burnham and M. Lave (2017). One Year Performance Results for the Prism Solar Installation at the New Mexico Regional Test Center: Field Data from February 15, 2016 - February 14, 2017. Albuquerque, NM, Sandia National Laboratories. SAND2017-5872

# Defining Threshold Values of Encapsulant and Backsheet Adhesion for PV Module Reliability

Nick Bosco[1], Joshua Eafanti[1] and Sarah Kurtz[1]
[1]National Renewable Energy Laboratory, Golden Colorado USA

Jared Tracy[2] and Reinhold Dauskardt[2]
[2]Stanford University, Stanford California USA

*Abstract* — The width-tapered cantilever beam method is used to quantify the debond energy (adhesion) of encapsulant and backsheet structures of 32 modules collected from the field. The collected population of modules contains both those that have remained intact and those with instances of either or both encapsulant and backsheet delamination. From this survey, initial threshold values (an adhesion value above which a module should remain intact throughout its lifetime) for encapsulant and backsheet interfaces are proposed. For encapsulants this value is ~160 J/m$^2$ and for backsheets ~10 J/m$^2$. It is expected that these values will continue to be refined and evolve as the width-tapered cantilever beam method becomes adopted by the PV industry, and that they may aid in the future improvement of accelerated lifetime tests and the development of new, low-cost materials.

*Index Terms* — Adhesive strength, Reliability, Photovoltaic cells

## I. INTRODUCTION

The method of quantifying the critical strain energy release rate, or debond energy, (material property of adhesion) of encapsulant and backsheet interfaces has only recently been applied within the PV industry [1-6]. Consequently, values of adhesion adequate to avoid delamination in the field have not been established. These *threshold values* of adhesion (an adhesion value above which a module should remain intact throughout its lifetime) are required to improve accelerated lifetime tests and for the development of new, low-cost materials.

In this paper we apply the recently developed width-tapered beam metrology to quantify adhesion of the encapsulant and backsheet interfaces of modules that have been deployed in the field between two and 27 years [3]. The modules obtained for this study are all crystalline silicon and represent both modules that have remained intact and those that exhibit either or both encapsulant and backsheet delamination. This sampling of modules will allow us to narrow the threshold of adhesion these materials must maintain to remain durable and exhibit high reliability throughout their lifetime.

## II. METHODS AND MATERIALS

### A. Module Collection

Flat plate, one-sun, crystalline silicon photovoltaic modules were collected from various sources with deployment histories between two and 27 years. A detailed list of each module type

and its deployment is presented in Table I (a). Upon obtaining each module they were visually examined for pre-existing signs of delamination. Both the front encapsulant and backsheet interfaces were examined and delamination observations noted, Table I (b). These initial observations are critical for narrowing a threshold value of adhesion.

### B. Adhesion Measurement

The width-tapered cantilever beam method was used to measure the adhesion of all interfaces of interest. This method of sample preparation and testing has been previously presented and is included here for clarity [3]. The metrology employs an elastic width-tapered cantilever beam adhered to the layered structure of interest. When the beam is loaded at its apex, delamination will initiate at the weakest interface and advance upon continued loading. The displacement of the beam apex out of the module plane is the load-line displacement ($\Delta$). This measurement quantifies the critical value of the strain energy release rate, $G_c$, which is the material property of adhesion, and represents the energy required for debond extension, given by:

$$G_c = \frac{P_c}{2\tan(\theta/2)} \frac{\Delta_f}{a_f^2} \qquad (1)$$

where $P_c$ is the critical load plateau at which the debond propogates, $a_f$ is a unique value of debond length at the load-line displacement $\Delta_f$, and $\theta$ the apex angle of the width-tapered beam.

### C. Sample Preparation

To evaluate backsheet adhesion, width-tapered beams of 20° were fabricated from 2 or 3.1 mm thick acrylic. A handle was incorporated into the beam to provide a location for the attachment of a loading tab whose action could remain at the apex of the beam. After both surfaces were cleaned with isopropyl alcohol, a volume of two-part epoxy (3M 8010) was dispensed and mixed on the beam. The beam was then pressed to the backsheet and weighted. Any excess epoxy was wiped clean prior to full hardening during its room temperature cure. A sharp razor blade was then used to cut

978-1-5090-5606-4/17 $31.00 © 2017 IEEE

through the backsheet around the beam using the straight edge of the beam as a guide. Testing was conducted in a small load frame, placed onto the back of the module, at a constant displacement rate of 10 μm/s while both load and load-line displacement were recorded [4].

To evaluate encapsulant adhesion, beams identical in design to the acrylic beams used for backsheet measurements were fabricated from 0.86 and 1.6 mm thick sheets of high-strength Grade 5 titanium. The modules were prepared by removing the backsheet and encapsulant to expose the backside of the cell, then grinding off the back cell metallization to expose the Si surface. This step is taken to avoid delamination of the back metallization, which typically has low adhesion but is not a concerned point of failure. The Ti beam was then adhered to the back of the cell with a two-part epoxy (3M DP420) and any excess wiped clean prior to hardening during its room temperature cure. A sharp carbide scribe was then used to "cut" the cell around the beam followed by a razor blade to cut the encapsulant through to the front sheet of glass. Testing was similarly conducted in a small load frame, placed onto the back of the module, at a constant displacement rate of 10 μm/s while both load and load-line displacement were recorded.

Both backsheet and encapsulant measurements were repeated at least three times per module and the adhesion values reported are the weighted mean and it's uncertainty.

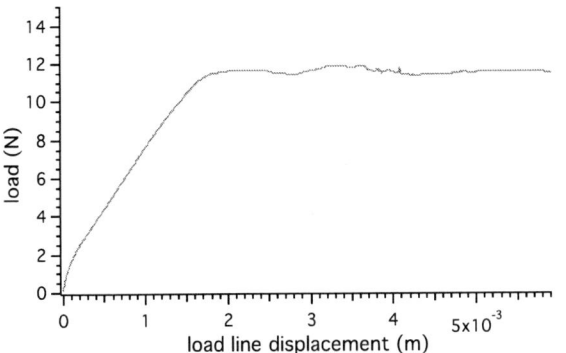

Fig 1. Load vs. displacement response from a width-tapered cantilever beam measurement of backsheet adhesion.

## III. RESULTS AND OBSERVATIONS

A representative load-displacement curve from the width-tapered beam measurement of backsheet adhesion (Arco Solar, Mexico City, 26-years) is presented in Fig. 1. The critical load plateau, $P_c$, for this measurement is ~11.5 N, the final load-line displacement, $\Delta_f$=5.9 mm (from plot) and final debond length $a_f$=57.9 mm (measured from sample). According to Eq. (1) the critical debond energy, $G_c$, or adhesion of this interface is 58.2±0.6 J/m$^2$ and the debond was observed to occur between the outer PVF and PET interface. Similar measurements and observations for all modules and interfaces are presented in Table I (c). A variety of backsheet structures were encountered. Typically, backsheet delamination occurred between an outer white film and an inner, or mid-, layer of a clear film. No chemical analyses were performed to positively identify these layers, thus they are only assumed to be a generic fluorinated polymer and a polyester layer, and therefore denoted as PVF and PET, respectively in Table I (c).

Representative images of front encapsulant delamination are presented in Fig. 2. In rare instances as illustrated in Fig. 2 (a), the delamination presented in such a way that adhesion could be directly measured at the pre-existing delamination front (Mobil Solar, Sacramento CA, 27 years). In most other modules obtained, front encapsulant delamination presented in discrete areas around the interconnect ribbons, Fig. 2 (b).

Fig 2. Optical images of modules exhibiting pre-existing encapsulant delamination. a) Left, Mobil Solar, Sacramento CA, 27 years and b) Arco Solar, Mexico City, 26 years.

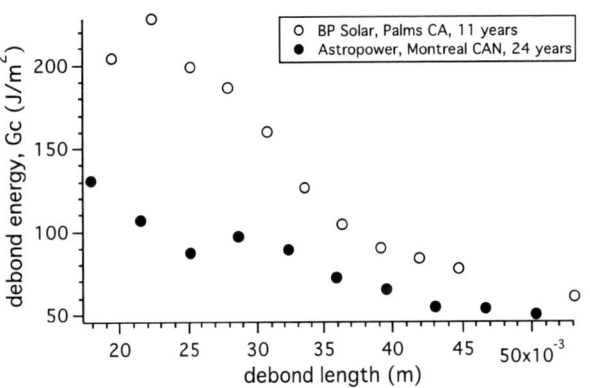

Fig 3. Discrete measures of encapsulant debond energy as a function of debond length.

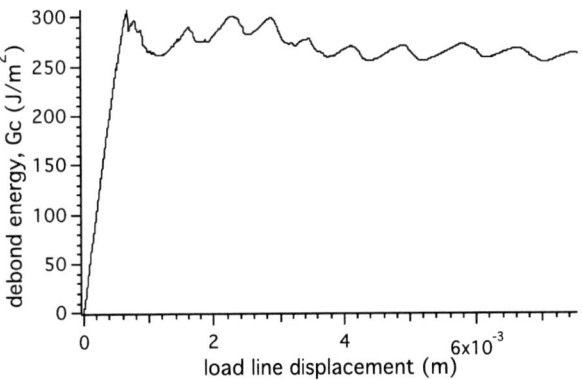

Fig 4. Continuous measure of encapsulant/ glass adhesion.

978-1-5090-5606-4/17 $31.00 © 2017 IEEE

In these cases the measurement was made away from the pre-existing delamination and therefore likely represents an area of higher adhesion and will yield a more conservative estimate of the threshold value for reliability. In one module (BP Solar, Palms CA, 11 years) encapsulant measurements were made both at and away from a pre-existing debond front resulting in measurements of 67±17 and 233±49 J/m², respectively.

Because the delamination front may be visualized through the module front glass as the measurement progresses, it is possible to also make discrete calculations of adhesion as a function of distance into the cell by evaluating Eq (1) for periodic measurements of debond length, Fig 3. On most modules without signs of pre-existing delamination, there was little measured variation of adhesion as the measurement progressed. However, when the adhesion levels were low on modules that did exhibit pre-existing delamination, the adhesion was found to decrease as the measurement progressed into the interior of the cell.

When the plateau load is relatively constant, a continuous measure of adhesion may be inferred from the load displacement response by evaluating Eq (1) at every point with the final measurement of debond length. This measurement for one module (Kyocera, unknown deployment) is presented in Fig. 4. The oscillation in debond energy (adhesion) corresponds with the debond front moving past the gridlines as it progresses across the cell. Since the delamination in this module occurred at the glass/ encapsulant interface, this variation is likely due to the changing thickness of the encapsulant. When the encapsulant layer is thicker between the gridlines it may consume more energy during the delamination process thus resulting in a higher measurement of $G_c$.

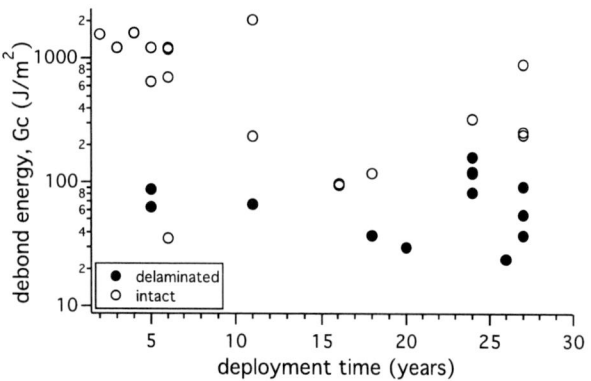

Fig 5. Encapsulant adhesion of the surveyed module exhibiting no preexisting signs of delamination (open symbols) and with pre-existing delaminated areas (closed symbols).

A summary of all encapsulant adhesion measurements is presented in Fig. 5 against module deployment time. Closed symbols represent those modules that exhibited any form of front encapsulant delamination (cell/ encapsulant or glass/ encapsulant) upon initial inspection and open symbols modules with no signs of encapsulant delamination. It is

obvious that module deployment time alone is not a perfect indicator of its encapsulant adhesion level. Other factors such as encapsulant material type and formulation, module size and character, processing and quality, and deployment environment and conditions are ultimately much more responsible for both the initial and present adhesion value. However, the level of adhesion required to prevent delamination should be independent of these factors save, to some extent, module size, design and deployment conditions.

While all thirteen instances of pre-existing encapsulant delamination are observed in modules with adhesion levels below ~160 J/m², there were also four modules measured with similarly low adhesion values (<120 J/m²) that did not exhibit delamination. Of these four modules, one was never deployed outdoors (Siemens, Tempe AZ, 18 years) and the remaining three were all characterized to contain relatively thin cells when compared to the older vintage modules that exhibited similar adhesion values yet pre-existing delamination (~200 vs. 400 $\mu$m). It is reasonable to expect that a thicker, and therefore stiffer, cell could induce a larger driving force for delamination (out-of-plane stress at this interface) thereby effectively lowering the adhesion threshold for more modern modules with thinner cells to below 35 J/m². This is less than 2 % of the initial encapsulant adhesion characterized in modern, quality EVA systems.

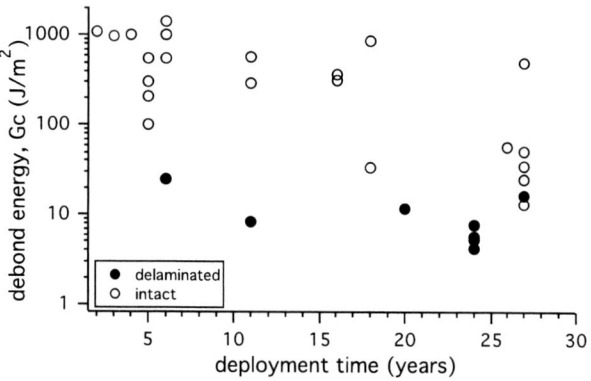

Fig 6. Backsheet adhesion of the surveyed module exhibiting no signs of delamination (open symbols) and with preexisting delaminated areas (closed symbols) within the backsheet structure (closed symbols).

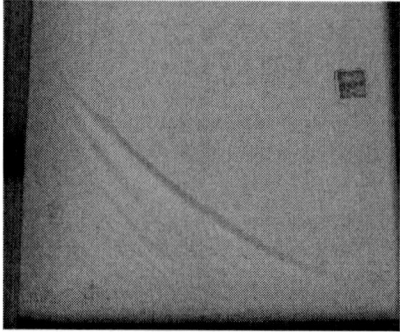

Fig 7. Pre-existing partially delaminated backsheet (Astropower, Montreal CAN, 24 years)

A similar plot of backsheet adhesion against module deployment time is presented in Fig. 6. Modules that exhibited pre-existing delamination typically had a large delamination front across which the backsheet was still intact, Fig. 7. This character allowed for the adhesion measurements to be made at this front. All modules measured to have backsheet adhesion less than 12 J/m$^2$ (seven) exhibited signs of pre-existing delamination. All of these were also more modern backsheet laminate structures that delaminated at the outer PET/PVF interface. Four additional modules (Sharp Solar, Phoenix AZ, 6 years; Sharp Solar, Denver CO, 11 years; Kyocera; Mobil Solar, Sacramento CA 27 years), which also exhibited pre-existing backsheet delamination, were also characterized. It was found that in all of these modules the failed interface was not truly within the backsheet structure, Table I (c). In the remaining nine modules characterized, only a lower limit of backsheet adhesion may be reported. In each of these cases the adhesion of the backsheet structure was so high that delamination occurred at a lower interface. While we have purposefully sought out modules with degraded adhesion, these truncated measurements demonstrate that these interfaces are capable of retaining a high level of this material property even following a 27-year deployment.

## III. CONCLUSIONS

This study has introduced our initial effort to quantify a threshold value of both encapsulant and backsheet adhesion. This threshold value should be considered the very minimum required to ensure delamination does not occur at the interfaces of these PV module laminate materials while in service. For encapsulants, this initial threshold ~160 J/m$^2$ and for backsheet structures ~10 J/m$^2$.

Our expectation is that these threshold values of adhesion will continue to evolve, and be refined, as the PV community adopts the width-tapered beam method and the population of characterized modules continues to grow. A key aspect to consider while refining these values will be the balance between modules with long service histories and relevant material systems. Furthermore, as demonstrated in this study with the older modules with thicker cells, some specific module characteristics will influence the threshold value. Therefore the "threshold value" should always be a conservative estimation of this property and even be assigned an appropriate safety factor.

Finally, this method of quantifying adhesion and the evolving threshold values should be used to develop accelerated and lifetime tests. When access to the identical material system characterized on a deployed module is available, it may be used to determine an accurate physical degradation model by correlating a corresponding loss in adhesion. Additionally, the evolving threshold value of adhesion may be used as a limit below which this material property may not fall through an accelerated exposure.

## REFERENCES

[1] J. Tracy, N. Bosco, and R. H. Dauskardt, "Evaluation of Encapsulant Adhesion to Surface Metallization of Photovoltaic Cells," in *Submitted to IEEE 44th Photovoltaic Specialist Conference (PVSC)*, Washington D.C., 2017.

[2] J. Tracy, N. Bosco, F. D. Novoa, and R. H. Dauskardt, "Encapsulant and Backsheet Adhesion Metrology for Photovoltaic Modules," *Submitted to Progress in Photovoltaics*, 2016.

[3] N. Bosco, J. Tracy, R. H. Dauskardt, and S. Kurtz, "Development and First Results of the Width-Tapered Beam Method for Adhesion Testing of Photovoltaic Material Systems," presented at the IEEE Photovoltaic Specialist Conference, Portland, OR, 2016.

[4] N. Bosco. (2016). *Width-Tapered Cantilever Beam Technique.* Available: https://youtu.be/ql9li68J60c?list=PLmIn8Hncs7bFpFFBpUQnKXzzx54wucPp1

[5] F. D. Novoa, D. C. Miller, and R. H. Dauskardt, "Debonding Kinetics of Photovoltaic Encapsulation in Moist Environments," *Progress in Photovoltaics: Research and Applications,* vol. 24, pp. 183-194, 2016.

[6] F. D. Novoa, D. C. Miller, and R. H. Dauskardt, "Environmental mechanisms of debonding in photovoltaic backsheets," *Solar Energy Materials and Solar Cells,* vol. 120, Part A, pp. 87-93, 1// 2014.

**a** | | | | | **b** | | **c** | | | 

| module | manufacturer | model | deployment location | deployment time | encapsulant delamination | backsheet delamination | encapsulant $G_c$ (J/m$^2$) | interface | backsheet $G_c$ (J/m$^2$) | interface |
|---|---|---|---|---|---|---|---|---|---|---|
| 1 | Kyocera | LA361G51S | | | none | yes | 270.0±12 | glass | 98.9±1.4 | outer foil |
| 2 | manufacturer A | X | Flagstaff, AZ | 2 years on-sun | none | none | 1550±21 | cell | >1100[1] | within cell back metal |
| 3 | manufacturer B | Y | Tonopah, AZ | 3 years on-sun | none | none | 1240±19 | cell | >960[1] | within cell back metal |
| 4 | | | Thailand | 4 years on-sun | none | none | 1620±15 | cell | >1000[1] | within cell back metal |
| 5 | Astropower | | Montreal, CAN | 5 years on-sun | yes | none | 63.3±2.1 | cell | >300[1] | front encapsulant |
| 6 | | | | 5 years on-sun | yes | none | 87.0±2.3 | cell | 556±5.7 | outer PET/PVF |
| 7 | | | | 5 years on-sun | none | none | 651±14 | glass | 205±6.4 | back encapsulant/PVF |
| 8 | manufacturer B | Y | Rome, IT | 5 years on-sun | none | none | 1210±2.7 | cell | 101±23 | outer PET/PVF |
| 9 | | | Bergamo, IT | 6 years on-sun | none | none | 1180±7.0 | cell | >1400[1] | within cell back metal |
| 10 | | | | 6 years stored | none | none | 1210±29 | glass | >1000[1] | back encapsulant/cell |
| 11 | BP Solar | BP71801N | Argenbuhl, DE | 6 years on-sun | none | none | 705±9.0 | glass | >550[1] | within cell back metal |
| 12 | Sharp Solar | NT-175U1 | Phoenix, AZ | 6 years on-sun | none | yes | 34.9±2.2 | glass | 24.7±2.2 | within cell back metal |
| 13 | BP Solar | SX150S | Palms, CA | 11 years on-sun | yes | | 67.0±17 | cell | | |
| | | | | | none | none | 233±49 | cell | 288±45 | back encapsulant/cell |
| 14 | BP Solar | BP3125U | Perrysburg, OH | 11 years on-sun | none | none | 2070±10 | cell | >570[1] | back encapsulant/cell |
| 15 | Sharp Solar | NE-170U1 | Denver, CO | 11 years on-sun | none | yes | 67.1±7.7 | glass | 8.30±0.5 | within cell back metal |
| 16 | Siemens | M55 | Golden, CO | 16 years on sun | none | none | 95.0±2.9 | cell | 312±2.8 | outer PET/PVF |
| 17 | | | | | none | none | 98.7±3.5 | cell | 365±28 | outer PET/PVF |
| 18 | Siemens | SM54 | Tempe, AZ | 18 yrs stored | none | none | 118±5.4 | cell | 867±26 | outer PET/PVF |
| 19 | | M55 | | 18 years on-sun | yes | none | 37.2±1.6 | cell | 33.7±0.51 | outer PET/PVF |
| 20 | | | | 20 years on-sun | yes | yes | 30.4±1.5 | cell | 11.5±0.52 | outer PET/PVF |
| 21 | | | | 24 years on-sun | yes | yes | 159±8.6 | cell | 7.60±0.17 | outer PET/PVF |
| 22 | Astropower | | Montreal, CAN | 24 years on-sun | none | yes | 328±6.1 | cell | 7.63±0.15 | outer PET/PVF |
| 23 | | | | 24 years on-sun | yes | yes | 83.3±3.6 | cell | 4.21±0.57 | outer PET/PVF |
| 24 | | | | 24 years on-sun | yes | yes | 122±11 | cell | 5.37±0.13 | outer PET/PVF |
| 25 | | | | 24 years on-sun | yes | yes | 120.0±11 | glass | 5.75±0.31 | outer PET/PVF |
| 26 | Arco Solar | M75 | Mexico City, MX | 26 years on-sun | yes | none | 24.2±1.1 | cell | 58.2±0.62 | inner PET/PVF |
| 27 | Arco Solar | M52-S | | 27 years on-sun | none | none | 245±28 | glass | 25.2±2.2 | outer PET/PVF |
| 28 | | | | 27 years stored | none | none | 896±120 | cell | 12.9±0.81 | outer PET/PVF |
| 29 | Mobil Solar | JPL block V | Sacramento, CA | 27 years on-sun | yes | severe | 54.6±5.0 | glass | 16.7±5.7 | back encapsulant/ PET |
| 30 | Solarex | | | 27 years on-sun | yes | none | 37.1±1.6 | glass | >500[1] | front encapsulant |
| 31 | Arco Solar | M52-L | | 27 years on-sun | yes | none | 91.3±5.5 | glass | 35.0±0.53 | outer PET/PVF |
| 32 | | | | 27 years on-sun | none | none | 255±34 | glass | 50.4±13 | outer PET/PVF |

[1] lower limit

**Table I.** Tabulated list of a) module and deployment details, b) pre-existing observations and c) results of adhesion measurements. Omitted module details were not recoverable.

# Characterizations of aged Glass/Ethylene Vinyl Acetate/Glass using fluorescence spectroscopy and instrumented indentation

Jae Hyun Kim, Yadong Lyu, *David C. Miller, Xiaohong Gu

National Institute of Standards and Technology, Gaithersburg, Maryland, 20899, USA

*National Renewable Energy Laboratory, Golden, Colorado, 80401, USA

*Abstract* — Ethylene vinyl acetate (EVA) is widely used as an encapsulant in photovoltaic devices, which are exposed to simultaneous UV irradiation, thermal, and humidity conditions. Aging behaviors of the EVA were characterized using confocal fluorescence spectroscopy with submicron scale resolutions of cross-sections of aged EVA samples. Additionally, mechanical properties of fluorescing regions in the aged EVAs where fluorescence emission was measured were investigated using a nanoindentation technique. In this study, we characterized two different formulations of EVA in glass/EVA/glass specimens after UV irradiation at 40 °C, 60 °C and 80 °C with 50 % relative humidity for 180 days. The fluorescence intensities on the UV irradiated sides of EVA samples were higher than those for the unexposed sides over wavelengths from 400 nm to 800 nm, and intensity differences between the two sides became more pronounced when the specimens were aged at higher temperatures. Elastic moduli of UV irradiated side for samples measured by instrumented indentation decreased, compared to unaged EVA samples, while fluorescent intensity increased with increasing the aging temperature.

## I. INTRODUCTION

Long-term performance in electricity production for 25 to 30 years is the key requirement for Photovoltaic (PV) modules. In order to maintain the performance of the module, the role of the encapsulant is extremely important since it provides optical coupling, environmental protection, and mechanical stability. Poly(ethylene-co-vinyl acetate) (EVA) is presently the most common PV encapsulant. Various additives (e.g. ultraviolet (UV) absorbers, silane coupling agents, and curing agents) are added to commercial EVA encapsulant system to improve their long-term stability, adhesion and mechanical properties. When EVA is used in the field, complex processes of chemical degradation occur, including those facilitated by the various additives [1].

When EVA undergoes UV photodegradation, chromophore and fluorophore products are formed [2], and reduction of mechanical properties is observed [3]. Fluorescence could be an indicator for the degradation in the EVA materials, however, it is challenging to correlate the depth-dependent fluorescence with mechanical degradation using typical mechanical measurements such as tensile tests because they lack spatial resolution. Nanoindentation test method is a practical approach to conduct small scale mechanical tests.

The main objective is to develop a measurement technique to investigate mechanical properties of thin EVA layers that are

fluorescing after aging. In this study, we employed laser scanning confocal spectroscopy and nanoindentation to characterize the degradation of EVA laminated between two glass plates. The former technique provides depth-specific data on fluorescence emission and the latter measures local mechanical properties where potential fluorescent species are formed in submicron scales. Because both techniques interrogate the degradation in microscopic domains of a sample, the data obtained should provide insight into the mechanism(s) facilitating the degradation of macroscopic properties and loss of performance of EVA encapsulant in PV modules. This study examines EVA samples cross-linked with two different curing agents.

## II. EXPERIMENTAL PROCEDURES

As a part of an interlaboratory experiment [4], various formulations of EVA was laminated between two quartz glass plates and was then subject to artificial UV weathering. Among several EVA formulations, EVA formulations containing two different curing agents **Lupersol 101 and Lupersol TBEC were prepared for characterizing EVA in this study. The compositions of the EVA formulations are listed in Table 1. EVA-A (cured with Lupersol 101) is commonly called a "slow cure" material and EVA-B (Lupersol TBEC) is commonly called a "fast cure" material. Further detailed lamination procedure and aging test conditions can be found in elsewhere [4]. Specimens were irradiated using a UV source described in ref [5], 50 % relative humidity, and at three different temperatures (40 °C, 60 °C and 80 °C) for 180 days. The spectra for the custom UV chambers in Ref. [5] resembles standardized UVA-340 fluorescent bulbs with additional visible radiation, .e.g, a broad peak from 475 nm to 650 nm.

Sample preparation procedures for nanoindentation and fluorescence measurements are illustrated in Fig.1. Individual 50 mm by 50 mm square Glass/EVA/Glass laminate aged coupons were cut into approximately 5 mm by 5 mm specimens using a water-cooled diamond blade. Only the unaged EVA samples were cast in epoxy and polished with 1200 grit abrasive papers (water cooled) for nanoindentation. For stiffness measurements of the aged specimens, one or two of the 5 mm samples for individual aging conditions were collected at the centers of the square coupons. After carefully separating one side of the glass from EVA, the stiffness of the EVA on the glass substrate was measured by nanoindentation. For

978-1-5090-5606-4/17 $31.00 © 2017 IEEE

fluorescence measurement, EVA layers were separated from both sides of glass substrates, and only EVA were cast in an epoxy at room temperature. The entire EVA-epoxy was then microtomed at -80 °C (Fig.1b). These microtomed EVA cross-sections were then examined for fluorescence under a laser scanning confocal microscope.

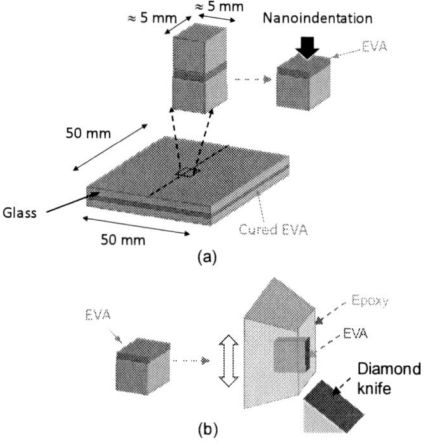

Fig.1 EVA sample preparation for nanoindentation (a) and fluorescence (b) measurements.

Table. 1 Composition of EVA formulation

| Components (g) | Formulation EVA-A (g) | Formulation EVA-B (g) |
|---|---|---|
| EVA resin (Elvax 1400) | 100 | 100 |
| Silane primer | 0.25 | 0.25 |
| Curing agent, (Lupersol TBEC) | - | 1.5 |
| Curing agent, (Lupersol 101) | 1.5 | - |
| UV absorber (Cyasorb UV531) | 0.3 | 0.3 |
| Hindered amine light stabilizer (Tinuvin 770) | 0.1 | 0.1 |
| Anti-oxidant, (Naugard P) | 0.2 | 0.2 |

*A. Fluorescent measurements*

The fluorescence measurements were carried out using the Zeiss LSM510 Meta (Carl Zeiss, Inc., Oberkochen, Germany) laser scanning confocal microscope with an excitation wavelength of 405 nm using a 5x objective lens. The excitation laser was directed to the cross-sections of the microtomed EVA specimens. Scanning area and lateral resolutions were approximately 1700 μm x 1700 μm and 1.4 μm respectively. Both channel mode and lambda mode were used to detect emission signals with long pass 420 filter and monitor emission spectra from 417 nm to 748 nm and perform spectra imaging with the META detector.

*B. Nanoindentation tests*

Indentation of cured EVA samples was performed on a nanoindenter (Keysight Technologies G200). To determine the local mechanical properties of the EVA before and after aging, the compliance method with the Oliver-Pharr model [6] was selected among other methods such as the continuous stiffness measurement [7]. In this study, a flat end cone diamond tip with the diameter of 5.8 μm (manufacturer specification) was used to apply force on the EVA surface. To ensure the full contact between the tip and sample surface, the indentation depths were aimed for 200 nm to 300 nm deeper than the depth estimated by 10 % of the tip diameter. The loading was applied to reach a peak load ($P_{max}$). The indenter was held at contact load for 5 s after reaching $P_{max}$. The indenter was then gradually retracted for unloading at approximately 0.01 mN/s. Load and displacement data was obtained during both the loading and unloading process. The contact stiffness was obtained by fitting the first 50 % of the slope of the unloading curve.

The reduced modulus of the EVA is calculated as [6]:

$$E_r = \frac{S\sqrt{\pi}}{2\sqrt{A}} \qquad (1)$$

where $S$ is the contact stiffness and $A$ is the tip contact area, which is $\pi D^2/4$ for the flat end cone tip (where $D$ is the tip diameter). The contact area between the flat end and the sample surface remains constant during the test, thus Eq. (1) can be rewritten as:

$$E_r = \frac{S}{D} \qquad (2)$$

Once the reduced modulus is determined, the elastic modulus of the EVA ($E_{eva}$) can be obtained by the following relation:

$$\frac{1}{E_r} = \frac{1 - v_{eva}^2}{E_{eva}} + \frac{1 - v_{tip}^2}{E_{tip}} \qquad (3)$$

Here, $v_{eva}$ is the Poisson's ratio of EVA (assumed as 0.35 in this study), $v_{tip}$ is the Poisson's ratio of the diamond indenter (0.07 in this study).

Locations for individual indentations were at least 60 μm apart from each other to isolate each measurement. At least ten indents were performed on each specimen. The instrument offers the precision staging for automated testing and specimen positioning with an effective load resolution of 50 nN and a displacement resolution of 0.01 nm.

978-1-5090-5606-4/17 $31.00 © 2017 IEEE

## III. RESULTS AND DISCUSSIONS

Fig.2 shows optical photo images of the aged Glass/EVA/Glass specimens after 180 cumulative days of weathering. Light yellow to dark brown discoloration of EVA layers was visually observed at the center regions of the specimens. Similar center-specific discoloration has been observed previously, and this discoloration has been attributed as due to a lack of oxygen in the center of the specimen [1]. The magnitude of discoloration is most noticeable for both formulations aged at 80 °C. Previous study identifies that interaction between the UV absorber and curing agent is responsible for the yellowing of EVA and the formation of fluorescent species [2]. Oxygen was found to reduce the development of fluorescent species at the specimen periphery due to photobleaching [8]. The wider discolored region for EVA-A than EVA-B suggests a formulation specific effect, such as an influence of the curing agent on oxygen diffusion.

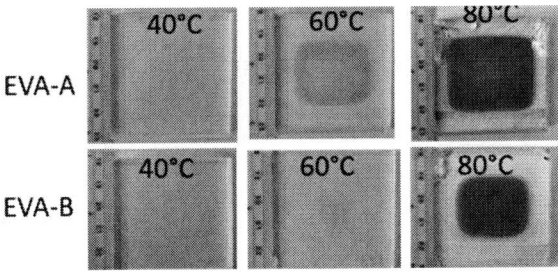

Fig.2. Photos of the aged Glass/EVA/Glass specimens

To investigate the distribution of fluorescent species of the EVA in the discolored regions, fluorescence measurements were collected from the center areas of the specimens. Note that the surface of the cross-sectional areas of the EVA samples microtomed at - 80 °C were smooth, but had slightly bulged toward the focal direction of the microscope due to the high coefficient of thermal expansion of EVA. Fig.3 shows the fluorescence intensities of the cross-sections near the UV irradiated and opposite sides of the EVA-A and B samples aged at 40 °C, 60 °C, and 80 °C under 50 % humidity. The fluorescence intensities increased with the aging temperatures, and the intensities for the UV irradiated sides are higher than those of the opposite sides. The fluorescent spectra of the EVA-B samples before and after aging showed similar trends to those of the aged EVA-A samples: the fluorescence intensity is higher on the irradiated sides than the opposite sides. This suggests that the greatest damage occurs at the irradiated surface of EVA, with damage propagating to the opposite surface only at the highest temperature. Some fluorescence is observed for the opposite surface of the specimens aged at 80 °C. In Fig.3, note that an increased wavelength was observed for the peak of the fluorescence spectra as the temperature was increased from 40 °C to 80 °C. Schlothauer et al. [9] used a fluorescence measurement to investigate the degradation

behaviors of a fast-cure EVA (exact formulation unknown) in a PV mini-module. In that study, a broad luminescence emission spectra centered at 460 nm was observed for EVA aged with the damp heat test (85 °C/85 % relative humidity) for 108 days. Since the mini-module was constructed with a permeable backsheet on one side of EVA in Ref. [9], whereas a

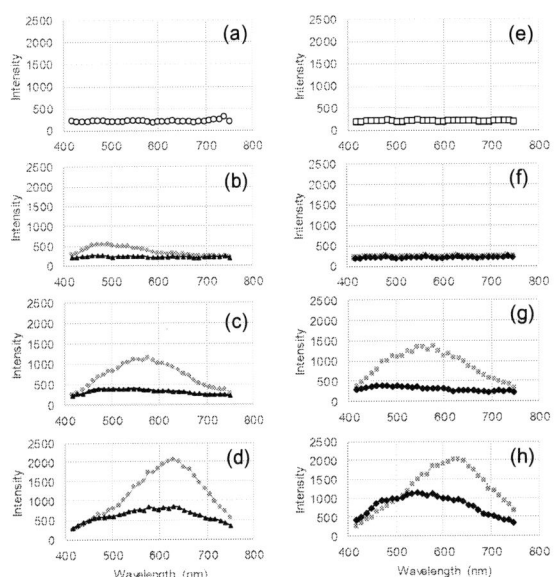

Fig. 3. Fluorescence spectra of the UV irradiated (●,■) and opposite side (▲,◆) for the EVA-A (a-d) and B (e-h) samples: unaged (a, e), aged at 40 °C (b, f), 60 °C (c, g), and 80 °C (d, h).

Glass/EVA/Glass specimen configuration was used in this study, the fluorescence results cannot be directly compared. Fig.3 do demonstrate the capability to monitor the formation and evolution of fluorescent species in different EVA formulations. It has been known that the fluorescence could be attributed to the formation of α,β-unsaturated carbonyl species, and different emission wavelengths could reflect different polyconjugated structures [10], In contrast to Ref. [2] and Ref. [10], the interlaboratory study [4] suggests that the discoloration (and perhaps consequent fluorescence) follows from the degradation of the UV absorber additive. The exact natures of species causing the fluorescent patterns in Fig.3, however, remains to be determined.

Fig.4 shows typical load/unloading relation with indentation depths for unaged and aged EVA-A and B samples. Nearly linear unloading behaviors were observed for both EVA samples regardless of the aging conditions. In addition, $P_{max}$ values and the slopes of the unloading portions for the unaged EVA samples were higher compared to those of the aged samples at similar indentation depths, which indicates transition from more rigid to softer material with age.

However, the difference of $P_{max}$ and the slopes of the unloading curves among the aged samples was relatively small. During a 5 s holding at the peak load, more pronounced relaxations were apparent for the unaged samples compared to the aged samples, thus, time-dependent characteristics of the internal structure of the EVA might be more significant to load response of unaged EVA-A sample. The loading/unloading behaviors for the unaged and aged EVA-B samples are similar to the EVA-A, exhibiting nearly linear loading/unloading behaviors, but their difference among the aged and unaged samples were less noticeable than EVA-A.

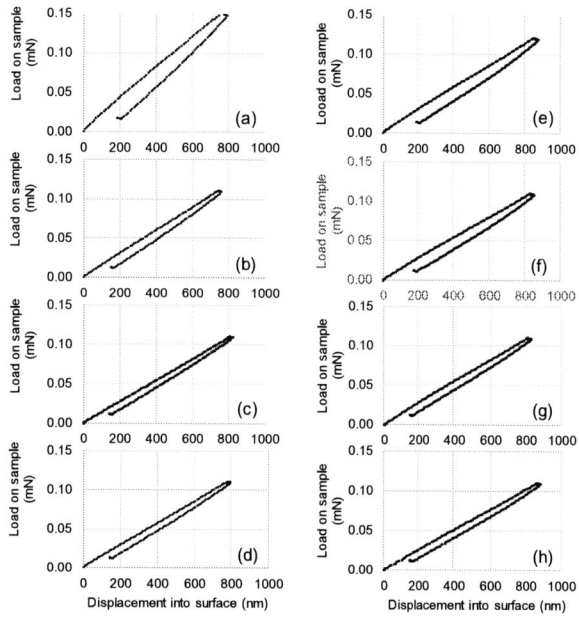

Fig.4. Load-Displacement curves for the EVA-A (a - d) and B (e - h) samples: unaged (a, e), aged at 40 °C (b, f), 60 °C (c, g), and 80 °C (d, h).

Fig.5 shows the changes of the elastic modulus of the aged EVA samples relative to the unaged EVA obtained by the nanoindentation tests with the Oliver-Pharr model. Both EVA samples became more compliant as a function of aging. EVA-A samples exhibited stiffness loss varying from 20 % to 25 % after aging, while EVA-B exhibited less stiffness loss. Jin [3] reported that competition between radiation induced cross-linking and chain scission in EVA occurs during the entire aging process, with chain scission dominating - resulting in a reduction of the elastic modulus via embrittlement for the tensile tests. In contrast, as indicated from the fluorescence measurements in Fig.3, the fluorescence patterns between the irradiated and opposite sides are different (particularly for aging temperatures 60 °C and 80 °C) and the tensile test performed in Jin's study is not sensitive enough to differentiate mechanical properties of EVA through-thickness, so changes of nanoindentation modulus on the opposite sides will be investigated in the future. To address the time-dependent behaviors, effects of loading/unloading rates, strain rate,

relaxation should be also considered for the further indentation test of EVA. As described in experimental section for nanoindentation, the Oliver-Pharr model is based on an elastic solution, so the continuous stiffness measurement technique imposing a small oscillation in loading is an alternative technique for visco-elastic polymers [7, 11].

Comparing the fluorescence measurements in the discolored center on the irradiated side of EVA samples to the results of nanoindentation, EVA samples exhibited increased compliance with increasing fluorescence intensity as a function of the aging temperature. The exact nature of fluorescent species is not clear - the connection between the underlying chemistry facilitating degradation remains to be established, including its correlation to discoloration and loss of optical transmittance. The high sensitivity to degradation is one of the strong merits of the use of fluorescence spectroscopy, where the magnitude of change

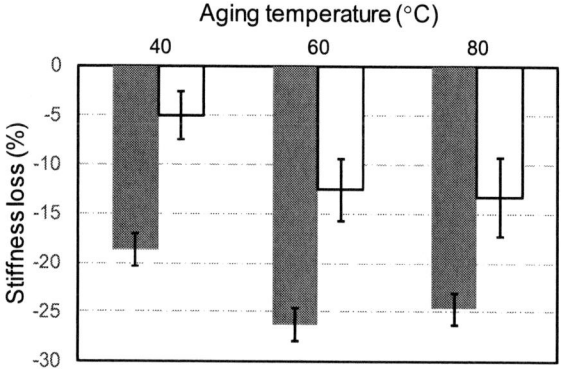

Fig.5. Changes of the elastic modulus (stiffness) for the aged EVA-A (filled) and -B (open) samples (the error bars represent 95 % confidence intervals of the mean values based on the bootstrap).

in the spectra with aging exceeds that of change in transmittance. The two measurement techniques (fluorescence spectroscopy and nanoindentation measurement) provide the ability to characterize local changes with aging that may be compared to bulk mechanical tests. The connection between local degradation and change in module performance also remains to be established.

IV. Summary

Glass/EVA/Glass specimen prepared with different curing agents were examined by fluorescence spectroscopy and nanoindentation to investigate the degradation of EVA with a micro scale resolution. Fluorescent intensities of both EVA materials were higher on the UV irradiated sides of the samples, and the fluorescence intensity increased with the aging temperature. Comparison of the irradiated and opposite sides of the samples suggests that the greatest damage occurs at the irradiated surface of the EVA materials, with damage propagating to the opposite surface only at the highest temperature. Moduli obtained by nanoindentation indicate a greater increase in mechanical compliance of EVA-A than EVA-B in the center of the discolored area after aging temperatures from 40 °C to 80 °C for 180 days. To address time-dependent mechanical behaviors of EVA, effects of loading rates on modulus measurements will be further investigated.

ACKNOWLEDGMENTS

Thanks to N. Alan Heckert from NIST Statistical Engineering Division for conducting the bootstrap analysis.

**Certain commercial product or equipment is described in this paper in order to specify adequately the experimental procedure. In no case does such identification imply recommendation or endorsement by the National Institute of Standards and Technology, nor does it imply that it is necessarily the best available for the purpose.*

REFERENCES

[1] A.W. Czanderna, F.J. Pern, "Encapsulation of PV modules using ethylene vinyl acetate copolymer as a pottant: A critical review," *Sol. Energy Mater. Sol. Cells,* vol. 43, pp. 101-181, 1996.

[2] F.J. Pern, "Luminescence and absorption characterization of Ethylene-Vinyl-Acetate encapsulant for PV modules before and after weathering degradation," *Polym. Degrad. Stabil.* vol. 41, pp. 125-139, 1999.

[3] J. Jin, S.J. Chen, J. Zhang, "UV aging behaviour of ethylene-vinyl acetate copolymers (EVA) with different vinyl acetate contents," *Polym. Degrad. Stabil.,* vol. 95, 725-732, 2010.

[4] D.C. Miller, E. Annigoni, A. Ballion, J.G. Bokria, L.S. Bruckman, D.M. Burns, X.X. Chen, L. Elliott, J.T. Feng, R.H. French, S. Fowler, X.H. Gu, P.L. Hacke, C.C. Honeker, M.D. Kempe, H. Khonkar, M. Kohl, L.E. Perret-Aebi, N.H. Phillips, K.P. Scott, F. Sculati-Meillaud, T. Shioda, S. Suga, S. Watanabe, J.H. Wohlgemuth, "Degradation in PV Encapsulation Transmittance: An Interlaboratory Study Towards a Climate-Specific Test," in *42nd IEEE Photovoltaic Specialist Conference*, 2015, pp. 1-6.

[5] D.P. Michael Koehl, Norbert Lenck, Matthias Zundel, *"Development and application of a UV light source for PV-modue testing,* in *SPIE Reliability of Photovoltacic Cells, Modules, Components, and Systems II,"* 2009.

[6] W.C. Oliver, G.M. Pharr, "An improved technique for determining hardness and elastic-modulus using load and displacement sensing indentation experiments," *Journal of Materials Research,* vol. 7, pp. 1564-1583, 1992.

[7] X. Li, B. Bhushan, "A review of nanoindentation continuous stiffness measurement technique and its applications," *Materials characterization,* vol. 48, pp. 11-36, 2002.

[8] C.P. Angelika Beinert, Ines Dürr, Michael D. Kempe, Günter Reiter, Karl-Anders Weiβ, "The influence of the additive composition on degradation induced changes in poly(ethylene-co-vinyl acetate) during photochemical aging," in *29th European Photovoltaic Solar Energy Conference and Exhibition*, 2014, pp. 3126-3132.

[9] J.C. Schlothauer, K. Grabmayer, G.M. Wallner, B. Roder, "Correlation of spatially resolved photoluminescence and viscoelastic mechanical properties of encapsulating EVA in differently aged PV modules," *Prog. Photovoltaics*, vol. 24, 855-870, 2016.

[10] N.S. Allen, M. Edge, M. Rodriguez, C.M. Liauw, E. Fontan, "Aspects of the thermal oxidation of ethylene vinyl acetate copolymer", *Polym. Degrad. Stabil,* vol. 68, pp. 363-371, 2000.

[11] W.C. Oliver, G.M. Pharr, "Measurement of hardness and elastic modulus by instrumented indentation: Advances in understanding and refinements to methodology," J of materials research, vol. 19, pp. 3-20, 2004.

# Encapsulant Adhesion to Surface Metallization on Photovoltaic Cells

Jared Tracy[1], Nick Bosco[2] and Reinhold Dauskardt[1]

[1]Department of Materials Science and Engineering Stanford University, CA, USA
[2]National Renewable Energy Laboratory, Golden, CO, USA

*Abstract* — Delamination of encapsulant materials from PV cell surfaces often appears to originate at regions with metallization. Using a fracture mechanics based metrology, the adhesion of EVA encapsulant to screen printed silver metallization was evaluated. At room temperature, the fracture energy, $G_c$ [J/m²], of the EVA/silver interface (952 J/m²) was ~70% lower than that of the EVA/AR coating (>2900 J/m²) and ~60% lower than that of the EVA to the surface of cell (2265 J/m²). After only 300 hours of damp heat aging, the adhesion energy of the silver interface dropped to and plateaued at ~50-60 J/m², while that of the EVA/AR coating and EVA/cell remained mostly unchanged. Elemental surface analysis showed that the EVA separates from the silver in a purely adhesive manner, indicating that bonds at the interface were likely displaced in the presence of humidity and chemical byproducts at elevated temperature, which in part accounts for the propensity of metallized surfaces to delaminate in the field.

*Index Terms* — adhesion, delamination, photovoltaic modules, accelerated aging, metallization.

## I. INTRODUCTION

Delamination of encapsulants from PV cell surfaces introduces pathways for moisture ingress, which eventually leads to reduced cell performance [1–6]. In the field, delamination at the cell interface often evolves at metallized regions [4,7–12], such as gridlines and bus bars (Fig. 1). While several studies have investigated the adhesion of encapsulation to cell surfaces [13–18], none have resolved adhesive properties to each constituent surface material, in particular the delamination-prone gridlines. A quantitative characterization of gridline adhesion and identification of relevant failure mechanisms is thus critical for assessing long-term degradation behavior and developing new materials to extend module lifetimes.

A recently developed metrology for measuring the adhesion energy of module interfaces [16,17] was used to evaluate adhesion of encapsulation to each material on the surface of a PV cell. First, baseline measurements of the adhesion energy of ethylene vinyl acetate (EVA) to screen printed silver, a silicon nitride anti-reflective (AR) coating, and a commercial silicon PV cell were conducted. Adhesion energy was then evaluated after 100, 300 and 1000 hours of damp heat aging. While aging had minimal impact on adhesion of the EVA/AR coating and EVA/cell interfaces, adhesion of the EVA/silver interface dropped by >94% from the baseline value. Chemical analyses of the fracture surfaces indicated that the failure mode of aged specimens was purely adhesive, with the likely mechanism of degradation being bond dissociation at the silver

interface due to water absorption and acetic acid formation at elevated temperature.

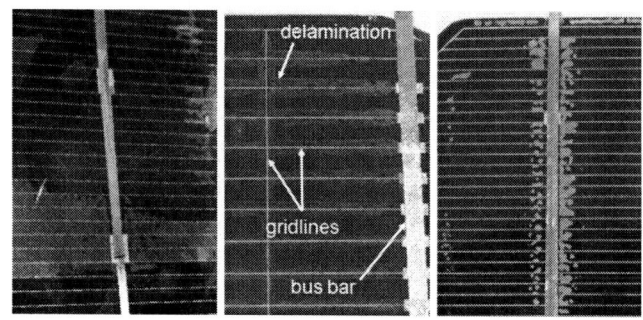

Fig. 1. Examples of encapsulant delamination on metallized regions of (bus bars and gridlines) PV cells in field-operated modules.

## II. EXPERIMENTAL PROCEDURES

### A. Specimen Preparation

Single-cell PV modules were fabricated by laminating EVA between a 150 mm x 150 mm silicon cell and a 3 mm thick glass substrate for each of three cell configurations: (a) coated entirely with a ~30 µm thick layer of screen printed silver (Fig. 2a), (b) covered with an AR coating and screen printed silver gridlines (cell surface), and (c) covered only with an AR coating.

Width-tapered titanium (Ti, 6%Al, 4%V) beams 1.86 mm thick were then bonded to the rear surface of the silicon cell using a metal bonding agent (DP-420, 3M). The modules were aged at 85°C, 85%RH in an environmental chamber (AES, Santa Clara, CA) for intervals of 100, 300 and 1000 hours. Upon removal from the chamber the modules were conditioned at room temperature for an hour. During this time, a diamond scribe was used to dice the cell along the perimeter of the titanium beam, forming a composite EVA/cell/titanium adhesion specimen directly on the module (Fig. 2b). For consistency in adhesion measurements, specimens were positioned near the edge of the cells, where, by 300 hours, the EVA moisture was expected to exceed 70% of the equilibrium concentration [4].

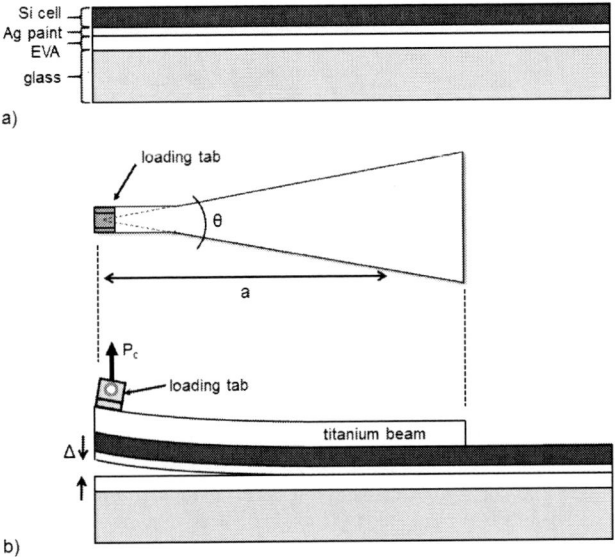

a)

b)

Fig. 2. (a) representative cross-section of EVA/screen print module configuration. (b) schematic of width-tapered beam adhesion specimen.

### B. Adhesion Metrology

Adhesion energy was measured by loading the wedge-shaped beam directly at its apex such that a crack initiated along the EVA/cell interface (Fig. 2b). The adhesion energy of the interface is a direct measurement of the critical energy release rate, $G_c$ [J/m$^2$], required to propagate the crack, and for a width-tapered beam is given by:

$$G_c = \frac{P_c \Delta}{2a^2 \tan(\theta/2)} \quad (1)$$

where $P_c$ is the critical applied load, $\Delta$ is the displacement at the beam tip corresponding to a crack length, $a$, and $\theta$ is the beam's apex angle. Details of the technique are described elsewhere [16,17].

### C. Chemical Analysis

Chemical composition of complementary fracture surfaces from several adhesion specimens was evaluated through x-ray photoelectron spectroscopy (XPS, PHI 5000 VersaProbe) scans (0-1000 eV) using Al K$_\alpha$ x-ray radiation at 1487 eV. Depth profiles of the silver screen print before and after aging were collected via argon ion sputtering (1 kV, 0.5 µA, 1 x 1 mm spot size, calibrated on SiO$_2$) to measure the approximate thickness of the surface oxide layer.

### III. RESULTS AND DISCUSSION

Baseline measurements of the adhesion energy of EVA to the screen printed silver, cell surface and AR coating were, 952 ± 26, 2270 ± 67 and >2970 J/m$^2$, respectively (Fig. 3).

Note that the intermediate adhesion energy of the cell surface is expectedly lower than a rule of mixtures approximation (~2800 J/m$^2$) calculated using proportional surface coverage of the gridlines and AR coating. This is attributed to overall reduced plastic deformation of the EVA due to the presence of gridline interfaces, adhesion of which was generally too poor to induce plastic deformation of the EVA. Additionally, note that at the EVA/screen print interface, the EVA was bonded to a thin surface oxide rather than pure silver, adhesion to which is considerably poorer (20.5 ± 5.4 J/m$^2$, measured previously using a similar metrology).

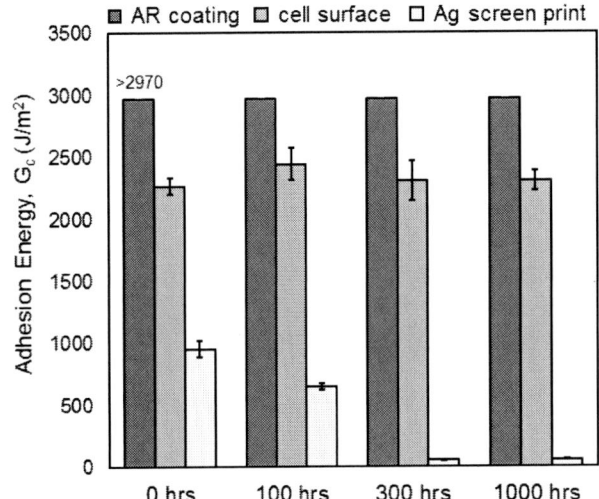

Fig. 3. Adhesion energy of EVA/screen print interface after damp heat aging.

Adhesion energy values following damp heat aging are also presented in Fig. 3. After 100 hours, the adhesion energy of the EVA/silver interface dropped to 651 ± 28 J/m$^2$, while the AR coating (>2970 J/m$^2$) and cell surface (2440 ± 130 J/m$^2$) exhibited no significant measurable degradation. By 300 hours the adhesion energy of the EVA/silver interface again dropped to an apparent threshold (~50-60 J/m$^2$), at which point there appeared to be no further degradation with continued aging to 1000 hours.

A representative XPS scan of an unaged screen print (Fig. 4) fracture specimen revealed silver, lead (a precursor in screen print synthesis), oxygen (from silver and lead oxides), tellurium, carbon (atmospheric contamination), and in some locations trace silicon. The silicon was likely an artifact of the silane adhesion promoter in the EVA, indicating that while the failure mode was primarily adhesive at the interface, there was some, albeit limited, cohesive failure within silane-functionalized interphase of the EVA. A comprehensive chemical analysis of the fracture surfaces, to include evaluations of oxide growth at the interface and changes in fracture path with aging, will be discussed in the oral presentation.

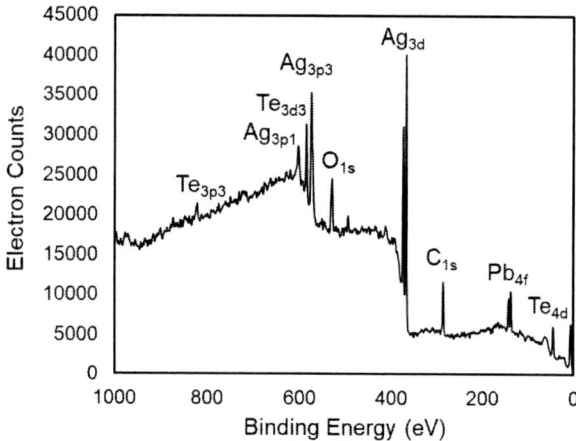

Fig. 4.   Representative XPS scans of unaged, screen printed silver fracture surface.

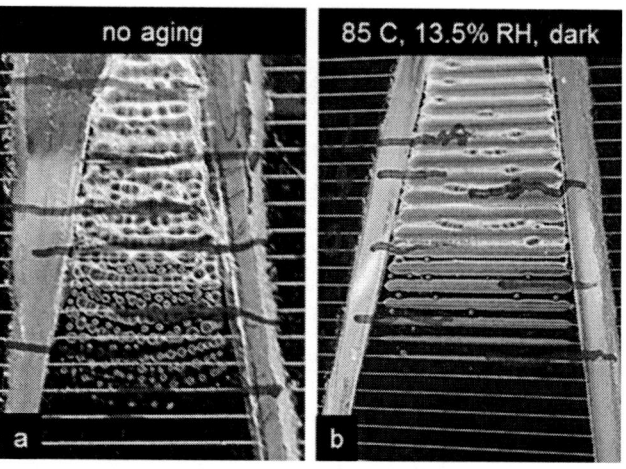

Fig. 5.   Images of EVA/cell interface captured during adhesion tests before (a) and after (b) 1000 hours aging at elevated temperature. Following aging, cavities in the cohesive zone of the fracture path preferentially form at gridlines.

During baseline adhesion testing of the EVA/cell specimens, cavities developed along the interface (with no preference for AR coating or gridline) in a cohesive zone ahead of the crack front (Fig. 5a). After aging in damp heat, however, cavities and subsequent delamination preferentially initiated at the gridlines. Similar behavior has been observed in previous work, where fully encapsulated modules with backsheets were lab aged at 85°C, 13.5%RH for 1000 hours (Fig. 5b) or in various environments (Golden, Miami, Phoenix), indicating that adhesive degradation at the EVA/gridline interface similarly occurs in environments less aggressive than damp heat.

It is expected that the equilibrium moisture concentration in the EVA of these modules is reached after 3000 hours of damp heat [4,6], with lateral diffusion reaching the center of the cells after ~100 hours [19]. In the presence of moisture and heat, ester elimination can occur in the EVA, resulting in the formation of acetic acid [1,2,5,20–22] which has been hypothesized to result in: (a) subsequent corrosion of the silver [3] and (b) adhesive degradation of the EVA/cell interface due to hydrolysis of siloxane bonds [5]. The impact of ester elimination on degradation may be more severe in laminated modules, in which pathways for acetic acid egress are limited. In the oral presentation, we will discuss techniques we employed to confirm the existence of such degradation mechanisms in this study.

Though further investigation is needed for a definitive assessment of the interface degradation mechanisms, in this study it was attributed to dissociation via moisture ingress and acetic acid formation of the comparatively weak bonding energy of the silane coupling agent to silver in the silver oxide (213 kJ/mol for Ag-O) versus silicon in the AR coating (798 kJ/mol for Si-O).

## IV. CONCLUSION

We have presented for the first time a quantitative characterization of EVA encapsulation to the screen printed silver metallization of PV cells. The as-fabricated adhesion energy of the EVA/gridline interface was just 30% of the EVA/AR coating interface, and degraded with exposure to damp heat, eventually reaching a threshold between 50-60 $J/m^2$ after 300 hours. This corresponds to a >90% reduction in adhesion energy, and is just 2% of the EVA/AR coating value.

The EVA/AR coating and EVA/cell interfaces did not experience adhesive degradation after 300 hours of damp heat. However, the gridline regions of the cell surface preferentially delaminated during adhesion testing, indicating adhesive degradation of the EVA/silver interface. Similar behavior was observed in specimens aged in less severe lab (85°C, 13.5%RH) and field environments.

XPS analysis of specimen fracture surfaces indicated that the failure mode of the EVA/gridline interface was primarily adhesive, whereas that of the EVA/AR coating interface was cohesive. Although a definitive determination of the mechanism of failure at the EVA/gridline requires further investigation, it was attributed to bond dissociation at the silver and EVA interface in the presence of moisture and acetic acid at elevated temperature.

## ACKNOWLEDGEMENT

This material is based upon work supported by the Department of Energy under Contract No. DE-AC36-08GO28308 with the National Renewable Energy Laboratory.

REFERENCES

[1] B. Ketola, A. Norris, Degradation Mechanism Investigation of Extended Damp Heat Aged PV Modules, 26th Eur. Photovolt. Sol. Energy Conf. Exhib. (2011) 3523–3528. doi:10.4229/26thEUPVSEC2011-4AV.2.14.

[2] A. Badiee, An examination of the response of ethylene-vinyl acetate film to changes in environmental conditions, 2016. http://eprints.nottingham.ac.uk/33757/.

[3] C. Peike, S. Hoffmann, P. Hulsmann, B. Thaidigsmann, K.A. Weis, M. Koehl, P. Bentz, Origin of damp-heat induced cell degradation, Sol. Energy Mater. Sol. Cells. 116 (2013) 49–54. doi:10.1016/j.solmat.2013.03.022.

[4] J.H. Wohlgemuth, M.D. Kempe, Equating damp heat testing with field failures of PV modules, Conf. Rec. IEEE Photovolt. Spec. Conf. (2013) 126–131. doi:10.1109/PVSC.2013.6744113.

[5] M.D. Kempe, G.J. Jorgensen, K.M. Terwilliger, T.J. McMahon, C.E. Kennedy, T.T. Borek, Acetic acid production and glass transition concerns with ethylene-vinyl acetate used in photovoltaic devices, Sol. Energy Mater. Sol. Cells. 91 (2007) 315–329. doi:10.1016/j.solmat.2006.10.009.

[6] M.D. Kempe, Control of moisture ingress into photovoltaic modules, Conf. Rec. Thirtyfirst IEEE Photovolt. Spec. Conf. 2005. (2005) 503–506. doi:10.1109/PVSC.2005.1488180.

[7] E.D. Dunlop, D. Halton, The performance of crystalline silicon photovoltaic solar modules after 22 years of continuous outdoor exposure, Prog. Photovoltaics Res. Appl. 14 (2006) 53–64. doi:10.1002/pip.627.

[8] P. Sanchez-Friera, M. Piliougine, J. Pelaez, J. Carretero, M. Sidrach de Cardona, Analysis of degradation mechanisms of crystalline silicon PV modules after 12 years of operation in Southern Europe, Prog. Photovolt Res. Appl. 19 (2011) 658–666. doi:10.1002/pip.

[9] M. a. Quintana, D.L. King, T.J. McMahon, C.R. Osterwald, Commonly observed degradation in field-aged photovoltaic modules, Conf. Rec. Twenty-Ninth IEEE Photovolt. Spec. Conf. 2002. (2002) 1436–1439. doi:10.1109/PVSC.2002.1190879.

[10] A. Realini, E. Bura, D. Cereghettie, D. Chianese, S. Rezzonico, Study of 20-year Old PV Plant (MTFB Project), in: 17th EUPVSEC, Munich, 2001.

[11] A. Realini, Mean Time Before Failure of Photovoltaic modules, 2003.

[12] C. Chamberlin, M. Rocheleau, M. Marshall, A. Reis, N. Coleman, P. Lehman, Comparison of PV module performance before and after 11-years of field exposure, in: Conf. Rec. Twenty-Ninth IEEE Photovolt. Spec. Conf. 2002., 2002: pp. 1432–1435. doi:10.1109/PVSC.2002.1190878.

[13] F. Novoa, D. Miller, R. Dauskardt, Adhesion and debonding kinetics of photovoltaic encapsulation in moist environments, Prog. Photovolt Res. Appl. 24 (2016) 183–194. doi:10.1002/pip.

[14] J.H. Wohlgemuth, D.W. Cunningham, A.M. Nguyen, J. Miller, Long term reliability of modules, in: Proc. 20th Eur. Photovolt. Sol. Energy Conf. Exhib., 2005: pp. 1942–1946.

[15] F. Pern, S. Glick, Adhesion strength study of EVA encapsulants on glass substrates, Proc. NCPV Sol. Progr. Rev. NREL/CP-52 (2003). http://www.nrel.gov/docs/fy03osti/33558.pdf.

[16] J. Tracy, N. Bosco, F. Novoa, R. Dauskardt, Encapsulation and backsheet adhesion metrology for photovoltaic modules, Prog. Photovolt Res. Appl. 25 (2017) 87–96. doi:10.1002/pip.2817.

[17] N. Bosco, J. Tracy, R. Dauskardt, S. Kurtz, Development and First Results of the Width-Tapered Beam Method for Adhesion Testing of Photovoltaic Material Systems, PVSC 43rd. (2016) 106–110. doi:10.1109/PVSC.2016.7749558.

[18] G.J. Jorgenson, T.. McMahon, Accelerated and Outdoor Aging Effects on Photovoltaic Module Interfacial Adhesion Properties, Prog. Photovolt Res. Appl. 16 (2008) 519–527. doi:10.1002/pip.

[19] M. Kempe, A. Dameron, M. Reese, Evaluation of moisture ingress from the perimeter of photovoltaic modules Michael, Prog. Photovolt Res. Appl. 22 (2014) 1159–1171. doi:10.1002/pip.

[20] B. Rimez, H. Rahier, G. Van Assche, T. Artoos, B. Van Mele, The thermal degradation of poly(vinyl acetate) and poly(ethylene-co-vinyl acetate), Part II: Modelling the degradation kinetics, Polym. Degrad. Stab. 93 (2008) 1222–1230. doi:10.1016/j.polymdegradstab.2008.01.021.

[21] B. Rimez, H. Rahier, G. Van Assche, T. Artoos, M. Biesemans, B. Van Mele, The thermal degradation of poly(vinyl acetate) and poly(ethylene-co-vinyl acetate), Part I: Experimental study of the degradation mechanism, Polym. Degrad. Stab. 93 (2008) 800–810. doi:10.1016/j.polymdegradstab.2008.01.010.

[22] N. S. Allen, M. Edge, M. Rodriguez, C. M. Liauw, E. Fontan, Aspects of the thermal oxidation, yellowing and stabilisation of ethylene vinyl acetate copolymer, Polym. Degrad. Stab. 71 (2000) 1–14. doi:10.1016/S0141-3910(00)00111-7.

# Impact of UV Light Intensity on Photodegradation of PV Backsheets

Xiaohong Gu*, Li-Chieh Yu, Yadong Lyu, Jae Hyun Kim, Andrew Fairbrother, and Tinh Nguyen

Polymeric Materials Group, Engineering Laboratory, National Institute of Standards and Technology, Gaithersburg, MD 20899, USA

*Abstract* — Polymeric accelerated weathering tests are often performed by increasing the intensity of the ultraviolet (UV) radiation to shorten the testing time. One critical question is whether the results from high UV intensity tests can be extrapolated to in-service exposure levels. We have investigated the influence of UV intensity on the photodegradation and validity of the reciprocity law for a commercial polyethylene terephthalate (PET) and ethylene-*co*-vinyl acetate (EVA) based backsheet, called PPE. The free-standing films of PPE were exposed to different UV intensity levels ranging from 60 W/m² to 180 W/m² (300 nm - 400 nm), at 85 °C/0 % RH and 45 °C/0% RH, respectively. Chemical degradation, yellowing and surface morphological changes were measured as a function of time and UV dose. A mathematical model was developed to describe the degradation kinetics under different UV intensity levels. The results have indicated that the reciprocity law was obeyed for samples aged at 85 °C/0 % RH, but not 45 °C/ 0 % RH.

*Index Terms* — UV radiation, photodegradation, light intensity, reciprocity law, backsheet, acid formation, NIST SPHERE

*\* Corresponding author; tel: +1 301-975-6523; fax: +1 301-990-6891; Email Address: xiaohong.gu@nist.gov*

## I. INTRODUCTION

The key weathering elements that cause the degradation of photovoltaic (PV) polymeric encapsulation materials are solar radiation, temperature, and moisture [1]. These elements act synergistically, and their effects are not constant but change with time and location. Among these degradation factors, ultraviolet (UV) radiation in the 295 nm - 400 nm range, which exists in terrestrial sunlight, is most detrimental to polymers because photons at these wavelengths have enough energy to directly cleave many chemical bonds of organic molecules and their derivatives.

The use of high-intensity UV sources has been recognized as a popular method for acceleration of polymeric degradation and obtaining long-term performance from short-term accelerated weathering tests [2]. One critical question is whether the results from high UV intensity tests can be extrapolated to in-service UV exposure levels. Considerable research on reciprocity law experiments has been carried out for a variety of polymeric materials and composites [3-6]. However, there is no available data on the relationship between UV intensity and photodegradation of polyester based PV backsheets.

The main objectives of this study are: 1) to investigate the influence of UV radiation intensity on physical and chemical degradation of a commercial PPE backsheet at 85 °C/0 % RH and 45 °C/0% RH, respectively, and 2) to assess the validity of the reciprocity law for backsheets under the same conditions. Specifically, the NIST high UV intensity source was coupled systematically with neutral density filters to provide different UV intensity levels. Chemical and physical changes of the backsheets were measured as a function of exposure time under different UV intensities. Specifically, surface morphological changes, yellowing, and chemical degradation of the backsheets were measured using atomic force microscopy (AFM), UV-Visible spectroscopy (UV-Vis), and Fourier transform infrared spectroscopy in the attenuated total reflection mode (FTIR-ATR), respectively. The results of this study will be critical in the development of realistic accelerated UV weathering tests for polymeric components used in PV modules.

## II. MATERIALS AND METHODS**

### 1. Backsheet Materials and Sample Preparation

A commercial, free-standing, multilayer PPE backsheet film was used as received for this study. Circular specimens with 19 mm in diameter were punched from the PPE backsheet film and then placed on a sample holder for accelerated laboratory weathering on the NIST 2-meter diameter integrating sphere-based high intensity UV weathering facility (referred to as SPHERE) at 85 °C/0 % RH and 45 °C/0 % RH, respectively. The neutral density (ND) filters, with nominal transmittance values of 40 %, 60 %, 80 %, 100 % (100 % means without ND filter), were placed in front of the sample holders to provide four different UV intensity levels ranging from 60 W/m² to 180 W/m² (300 nm – 400 nm).

### 2. Characterization of Degradation

Chemical degradation, surface morphological changes, and yellowing of the PPE backsheets were measured as a function of exposure time, or UV dose, under different UV intensity levels.

Changes in surface morphology were characterized by a Dimension Icon atomic force microscopy (AFM) instrument (Bruker Corp.) using an antimony doped silicon probe with a spring constant of 40 N/m under ambient conditions (24 °C and ≈ 45 % RH). All images were obtained at a scan speed of 0.5 Hz and a resonant frequency of 300 kHz. UV-Vis spectra of the

978-1-5090-5606-4/17 $31.00 © 2017 IEEE

exposed and unexposed PPE films were obtained by a Perkin-Elmer Lambda 900 spectrophotometer equipped with an integrated sphere in the range of 190 nm - 820 nm with a resolution of 2 nm. The yellow index (YI) was calculated from the acquired UV-Vis spectra based on ASTM E313. Chemical degradation was characterized by FTIR-ATR using a Nexus 670 spectrometer (Thermo Nicolet) equipped with a liquid nitrogen-cooled mercury cadmium telluride (MCT) detector. All FTIR spectra were the average of 128 scans recorded at a resolution of 4 cm$^{-1}$ using dry air as a purge gas. A ZnSe prism and 45° incident angle were used for all FTIR-ATR measurements. The peak height was used to represent the FTIR intensity.

** *Certain commercial products or equipment are described in this paper to specify adequately the experimental procedure. In no case does such identification imply recommendation or endorsement by the National Institute of Standards and Technology, nor does it imply that it is necessarily the best available for the purpose.*

### III. RESULTS AND DISCUSSION

Fig. 1 displays AFM height images of the outer surface of the PPE backsheet before and after exposure to UV radiation at different intensity levels for 40 days. A dark control is also shown. The bright particles present on the surface are believed to be related to BaSO$_4$ pigments [7]. There was no visible difference in the topographic features between the unexposed specimen and the dark control. However, marked changes on surfaces were evident after samples were exposed to UV light.

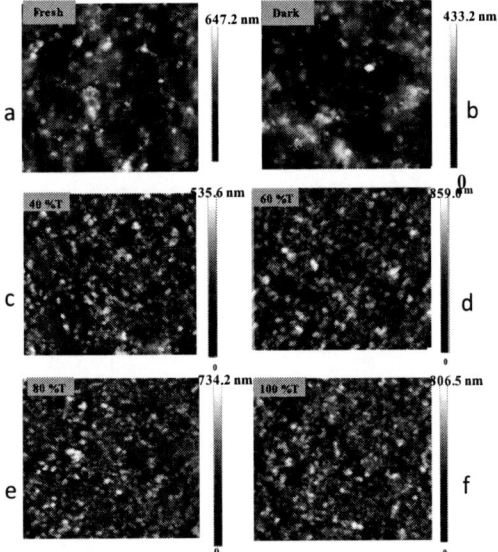

Fig. 1. AFM height images of the PPE backsheet outer surface before (a), and after exposure in the dark (b) and under four UV intensity levels for 40 days (c, d, e, f) at 85 ° C/ 0 % RH. Scan size is 25 μm x 25 μm.

It appeared that a higher UV intensity led to more particles accumulated on the surface. This phenomenon is due to photodegradation of the polymer matrix by UV radiation, leaving the inorganic particles exposed on the sample surfaces.

Based on FTIR-ATR measurements (Fig. 2) and our previous work [7, 8], the photochemical reactions of the PPE backsheet were found to include chain scission of the PET structure and formation of various oxidized. The dramatic loss of the 1715 cm$^{-1}$ band (ester C=O stretching) was observed after exposure. The appearance of additional bands around 1685 cm$^{-1}$ (C=O stretching) and 1425 cm$^{-1}$ (-OH deformation stretching) suggested the formation of carboxylic acids [9, 10]. The increase of the 1070 cm$^{-1}$ band indicated a growing amount of BaSO$_4$ pigments on the surface of the PET outer layer with longer UV exposure. To assess the effects of UV intensity on the chemical degradation of PPE backsheets, relative intensity changes of the FTIR bands at 1715 cm$^{-1}$ (ester C=O) and 1425 cm$^{-1}$ (acid OH) with exposure time (days, d) and with UV dose were plotted; the results are displayed in Fig. 3. In this figure, the intensity has been normalized against that of the 1410 cm$^{-1}$ band, which is attributed to the combined para-substituted semicircle stretching and CH bending [8, 9]. The results clearly show that the rates of ester depletion and aromatic acid formation increase with increasing UV intensity up to 40 d exposure (Fig. 3a, and 3b), which is consistent with the changes of YI as a function of exposure time shown in Fig. 3e.

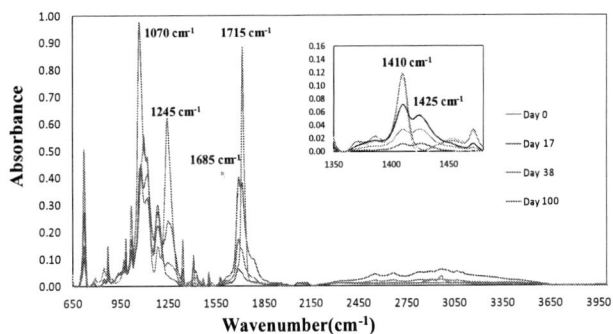

Figure 2. FTIR-ATR spectra of PPE backsheet before and after exposure to 80 % UV intensity at 85 °C/0 % RH for different times. The inset displays changes in the expanded carboxyl OH bending region from 1350 cm$^{-1}$ to 1485 cm$^{-1}$. Legends of red, blue, green and pink represent exposures of 0, 17, 38, and 99 d, respectively.

When the results of chemical changes or YI were plotted as a function of UV dose, the individual curves for different intensity levels were superimposed (Fig. 3b, 3d and 3f), suggesting that the reciprocity law was generally obeyed for these reactions at this exposure condition. That is, the yellowing and chemical degradation of PPE backsheet were proportional to the total UV dose impinging on the specimen surface, but independent of the intensity of UV radiation for specimens exposed at 85 °C/0 % RH. The similar behavior between YI and chemical degradation shown in Fig. 3 also suggests that the yellowing observed might be closely related to the chemical degradation of the PET outer layer of the PPE backsheet, specifically, the formation of hydroxylated phthalate/quinone compounds [9].

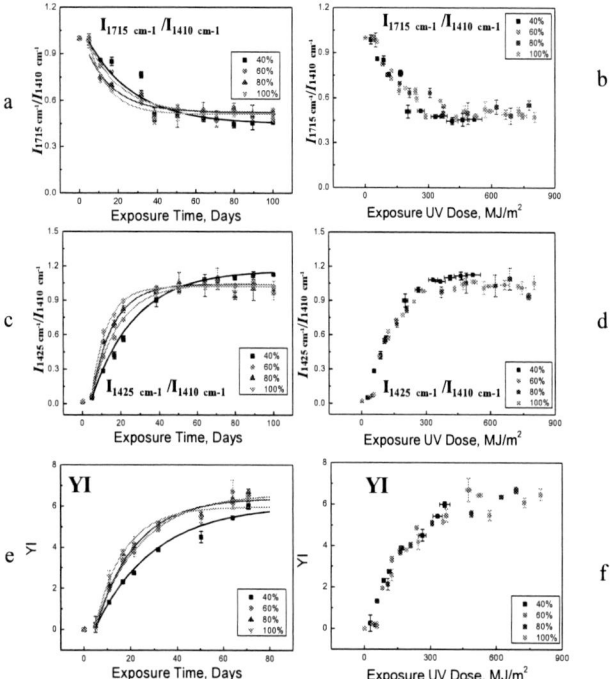

Fig. 3. Experimental data and fitting curves for relatively intensity of FTIR bands at 1715 cm⁻¹ (a, d), 1425 cm⁻¹ (b, e) and YI (c, f) as a function of exposure time and UV dose for PPE backsheets exposed to 85 °C/0 % RH under four UV intensity levels. Each data point is the average of two specimens and the error bar represents one standard deviation.

A quantitative validation of reciprocity law was carried out based on the Schwarzschild's law [3] shown in Eq. [1],

$$K = BI^p \qquad [1]$$

where $K$ is the rate of reaction, B is a proportionality constant, I is the radiation intensity, and $p$ is the Schwarzschild coefficient. In this study, $p$ values for the chemical degradation and yellowing processes were obtained from the slope of the log (relative $k$) vs. log (relative $I$) plots. Fig. 4 displays plots of log ($k/ko$) versus log ($I/I_{100\%}$ x 100) for YI, intensity increase of the 1425 cm⁻¹ band (acid formation), and intensity decrease of the 1715 cm⁻¹ band (ester depletion), where $ko$ is the rate of reaction (for chemical changes), or rate of YI increase, for the lowest UV intensity level (40 %), and $k$ represents the rates of the other three UV intensity levels. In addition, $I_{100\%}$ is the irradiance without using ND filters, and $I$ represents the irradiance through three ND filters with nominal transmittance

values of 40 %, 60 % and 80 %. A good linearity was observed for all plots. The slopes of these lines, i.e., the $p$ values, for three degradation processes were close to 1 ($p$ = 1.15, 1.02, and 0.91 for acid formation, ester depletion and yellowing, respectively), indicating that chemical degradation and yellowing of the PPE backsheet under this exposure condition obeyed the reciprocity law. Thus, PPE degradation could be accelerated by increasing UV intensity to high levels and confidently extrapolating the results obtained to service-relevant levels.

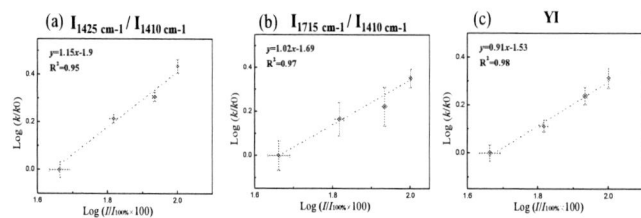

Fig. 4. Plots of log ($k/ko$) vs log ($I/I_{100\%}$ x 100) for: (a) relative intensity of FTIR band at 1425 cm⁻¹ (acid formation), (b) relative intensity of FTIR band at 1715 cm⁻¹ (ester depletion), and (c) YI for PPE backsheets exposed to 85 °C/0 % RH under four UV intensity levels. $ko$ is the rate of reaction (for chemical changes) or rate of YI increase for the lowest UV intensity level (40 %), and $k$ represents the rates of the other three UV intensity levels. $I_{100\%}$ is the irradiance without ND filters, and $I$ represents the irradiance through ND filters with nominal transmittance values of 40 %, 60 % and 80 %. Each data point is the average of two specimens and the error bar represents one standard deviation.

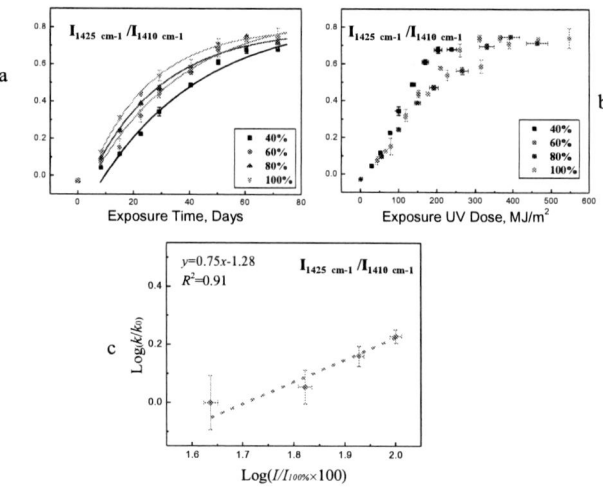

Fig. 5. Relative intensity changes of FTIR bands at 1425 cm⁻¹ as a function of exposure time (a) and UV dose (b), and plot of log ($k/ko$) vs log ($I/I_{100\%}$ x 100) for the relative intensity of 1425 cm⁻¹ (c) for PPE backsheets exposed to 45 °C/0 % RH under four UV intensity levels. $ko$ is the rate for the lowest UV intensity level (40 %), and $k$ represents the rates of the other three UV intensity levels. $I_{100\%}$ is the irradiance without ND filters, and $I$ represents the irradiance through ND filters with nominal transmittance values of 40 %, 60 % and 80 %. Each data point is the average of two specimens and the error bar represents one standard deviation.

However, when the exposure conditions changed from 85 °C/0 % RH to 45 °C/0 % RH, the data of chemical changes as a function of UV dose were more scattered under four UV intensity levels. A relatively higher rate for specimens exposed to 40 % intensity (Fig. 5) was observed compared to those exposed to the other three light levels. The $p$ values, which was calculated based on the slope of the $\log (k/k_o)$ vs. $\log (I/I_{100\%} \times 100)$ plot for the relative acid formation was approximately 0.75, much lower than 1, indicating that the acid formation at this lower temperature didn't follow reciprocity law well, and the increase of the reaction rate was far less than the increase of the light intensity. This phenomenon may be a result of different photodegradation behaviors of PET at temperatures below its glass transition temperature ($T_g$), which is around 74 °C based on dynamic thermal analysis of the sample. Work is in progress in our laboratory to validate the reciprocity law for the same material under different temperatures and RH levels. The cross-sectional analysis of the samples aged under different UV intensities was also performed for degradation depth profiling to better understand the impact of light intensity on photodegradation of this backsheet.

## VI. CONCLUSION

Using a NIST SPHERE high intensity UV source coupled with neutral density filters, we have investigated the influence of UV intensity levels on the photodegradation and validity of the reciprocity law for commercial PPE backsheets. For exposure at 85° C/0 % RH, both the magnitude and rate of chemical degradation and YI increased with increasing UV intensity as a function of time up to approximately 40 d, and then leveled off. When the degradation was expressed as a function of UV dose, both chemical and YI changes were superimposed regardless of light intensity, suggesting that the degradation of PPE backsheet was mainly determined by the amount of UV energy impinging on the specimen surface and independent of UV intensity. A quantitative validation of the reciprocity law yielded a Schwarzschild's $p$-coefficient of approximately 1 for both chemical degradation and yellowing, indicating the changes of these properties obeyed the reciprocity law at 85° C/0 % RH in the studied range of UV light. However, a clear failure of reciprocity law was observed when exposure conditions were changed from 85° C to 45° C with the same range of UV intensities. This difference is proposed to be due to the presence of different photodegradation behaviors of PET at temperatures below its $T_g$ compared to those above the $T_g$

## ACKNOWLEDGEMENT

The authors greatly acknowledge the support from a Government/Industry consortium on Characterization and Modeling of Polymers for Photovoltaic Systems at NIST. Companies involved in this consortium include 3M, AGC, Arkema, DuPont, Dakin America, First Solar, and Underwriter Laboratory.

## REFERENCES

[1] J.F. Rabek, Polymer Photodegradation – Mechanisms and Experimental Methods." Chapman & Hall, New York, 1995, p.569.

[2] A. Davis and D. Sims, Weathering of Polymers, Applied Science, NY, 1983.

[3] J.W. Martin, J.W. Chin, and T. Nguyen, "Reciprocity law experiments in polymeric photodegradation: a critical review", *Prog. Org. Coat.*, vol. 47, pp. 292-311, 2003.

[4] J.W. Chin, T. Nguyen, E. Byrd, J.W. Martin, "Validation of the reciprocity law for coating photodegradation", *J. Coat. Technol. Res.*, vol. 2, pp. 499-508, 2005.

[5] J.E. Pickett, D.A. Gibson, and M.M. Gardner, "Effect of irradiation conditions on the weathering of engineering thermoplastics", *Polym. Deg. Stab.*, vol. 93, pp. 1597-1606, 2008.

[6] M. Diepens and P. Gijsman, "Influence of light intensity on the photodegradation of bisphenol A carbonate", *Polym. Deg. Stab.*, vol. 94, pp. 34-38, 2009.

[7] C.C. Lin, P.J. Krommenhoek, S.S. Watson, and X. Gu, "Depth profiling of degradation of multilayer photovoltaic backsheets after accelerated laboratory weathering: Cross-sectional Raman imaging", *Solar Energy Materials & Solar Cells*, vol. 144, pp. 289-299, 2016.

[8] M. Edge, N.S. Allen, et al., "Identification of luminescent species contributing to the yellowing of poly (ethylene terephthalate) on degradation," *Polymer*, vol. 36, pp. 227-234, 1995.

[9] C.C. Lin, Y. Lyu, D.L. Hunston, J.H Kim, K.T. Wan, D.L Stanley and X. Gu, "Cracking and delamination behaviors of photovoltaic backsheet after accelerated laboratory weathering," *Proc. SPIE 9563, Reliability of Photovoltaic Cells, Modules, Components, and Systems VIII*, 956304 (2015); Doi: 10.1117/12.2188557.

# Survey of Mechanical Durability of PV Backsheets

Michael D. Kempe[1], David C. Miller[1], Allen Zielnik[2], Daniel Montiel-Chicharro[3], Jiang Zhu[3], Ralph Gottschalg[3]

[1]National Renewable Energy Laboratory (NREL), Golden, Colorado 80401
[2]Atlas Material Testing Technology LLC, Mount Prospect, Illinois 60056
[3]Loughborough University, Leicestershire, UK

*Abstract* — The ability to maintain mechanical properties after applied environmental stress is a good indicator for long term durability of a backsheet. This work surveys a set of 56 backsheet films, some of which are known to fail in the field and evaluates them utilizing the proposed photovoltaic weathering standard IEC 62788-7-2. While the complete study examined optical reflectance characterization, tensile testing, dielectric testing, and a mandrel bend test, this paper focuses on the mandrel bend test results. All seven of the transparent specimens failed. A polyamide based backsheet known to fail in the field did not fail the bend test. It is believed that the field failure is due to an interaction with the encapsulant and/or the cyclic mechanical exposure experienced in the field. When cracks occur, they happen predominantly in the polyethylene terephthalate for transparent backshees or the "E" layer for opaque films, never in a fluoropolymer layer.

## I. INTRODUCTION

Backsheets are used in photovoltaic modules primarily as a safety component to reduce the risk of electric shock and the potential for starting a fire, but they may also function as a moisture barrier. When placed in the field, many different backsheet failure modes are seen. A backsheet must: (1) have sufficiently high dielectric strength [1], (2) resist being punctured by solder bumps during lamination [2] or by external objects during service, (3) must not develop cracks or delamination in use [3-9].

The International PV Quality Assurance Task Force (PVQAT) weathering group has been studying backsheets as the technical basis for the standard, 62788-7-2 "Measurement procedures for materials used in photovoltaic modules – Part 7-2: Environmental exposures – Accelerated weathering tests of polymeric materials". Here, a supporting interlaboratory study of backsheets was conducted to evaluate the effect of the weathering test over a broad range of backsheets, providing perspective on what may be expected of backsheet films in general and how they should be evaluated for durability. This study is designed to evaluate potential testing methodologies for incorporation into IEC standards for PV modules, and to identify typical backsheet failure modes for chamber tests.

Testing beyond the round robin reflectance and tensile testing included a mandrel bend test and breakdown voltage testing. The bend test is useful because the strains, <~5%, are of the same order as what is likely to be seen in a module and the bend-test results have been reported to correlate with laboratory failure as measured using other evaluation properties for polyethylene terephthalate (PET) [9-11].

Inspection of module backsheets reveals significant backsheet curvature between cells, around busbars or over inadvertent solder bumps. But the bend test curvature does not represent the possibility for solder to form a large projection that might directly puncture a backsheet. It is assumed that puncture of the backsheet in this manner would represent a manufacturing error as opposed to a design flaw and is not considered in this work.

In addition to strains resulting from bending, the backsheet is also mechanically strained because of the thermal expansion differences between the materials. Eitner et al. [12] found that the gap between cells was dependent on the difference in the thermal expansion coefficient (CTE) of the glass as compared to the cells with a maximum difference of around 50 µm for typical cells over an 85°C temperature shift. Assuming that the backsheet spaning a 2 mm gap would expand or contract to accommodate this strain, this would correspond to a maximum backsheet strain of 2.5%. For this mode however this strain would be distributed over some area over the cell perimeter and would thus be smaller than 2.5%. One could also consider the CTE difference of PET (~70 ppm/K) as compared to that of silicon cells at 25 ppm/K to 39 ppm/K from Eitner [12]. For an 85°C temperature shift, this could produce a strain on the backside of a cell no larger than (70-25)ppm/k×85°C=0.4%, still a very low value. Thus the strains due to thermal cycling are consistent with or lower than those experienced in the bend test.

The bend test is that it is a mechanical test for which individual specimens can be retested multiple times in contrast to tensile testing which destroys specimens. Thus, the mandrel test enables one to test till failure as opposed to to tensile testing which destroys a specimen at every data point.

In this work we review the results of the bend test on this set of commercial backsheet films. Test conditions were chosen to be at condition A3 of the newly developed photovoltaic (PV) weathering standard, IEC 62788-7-2 [13]. In this work we find excellent durability for fluoropolymer layers, poor durability for transparent films, and significant cracking on the cell-side adhesion layer.

## II. EXPERIMENTAL METHODS

### A. Backsheet Sampling

Fifty-six backsheets were obtained from more than 15 different partners. Frequently the backsheets were not

978-1-5090-5606-4/17 $31.00 © 2017 IEEE

identified, therefore there is an unknown number of replicate backsheets from the same supplier. While the specific products or formulations cannot be identified, attenuated total reflectance Fourier transform infrared spectroscopy (FTIR-ATR) evaluation of the air- and cell-side surfaces was used to identify the class of the materials used in the films. Because backsheet suppliers often purchase films from other vendors and assemble them, a positive identification of the outer layers is not sufficient to uniquely identify backsheets, including replicate products.

Specimens were constructed as 1.5×8 cm coupons with two specimens cut using scissors in the machine extrusion direction, and two specimens cut in the transverse direction.

## B. Bend Test

A 6.35 mm diameter stainless steel rod was used as a mandrel fixture. At 250 h intervals, specimens were tested by bending around the mandrel in both directions (cell-side out and air-side out), Fig. 1. For a 0.2 mm to 0.3 mm thick backsheet, this corresponds to maximum tensile strains of between 3.0% and 4.5%. This should represent typical strains seen because of the topography of the backside of a PV module and the thermal expansion differences of the backsheet relative to other module components.

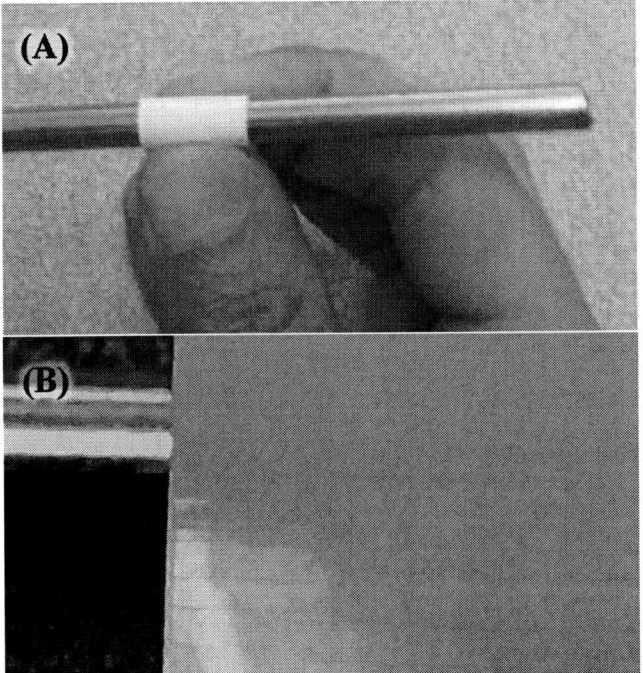

Figure 1. (A) Photo of mandrel test. (B) Photo of failed specimen after the mandrel test.

When a specimen fails it may form visible cracks, but sometimes one can only hear the cracks forming. When the cracks were not seen, it is most likely due to cracking in an internal layer. Failure is typically determined as the first time

when cracks were seen, but specimens were sometimes left in test for longer times.

## C. FTIR-ATR Testing

Fourier transform infrared spectroscopy with attenuated total reflectance (FTIR-ATR) was used to classify the material on both the air- and the cell-side of the backsheets. Instrumentation included a Nexus 870 FTIR, a Pike Technologies diamond ATR (~2.5 mm test area), and the Omnic software package. Unknown materials were compared with known materials for identification of the general material type.

## D. Weathering Exposure

Accelerated environmental exposure was accomplished using an Atlas Materials Testing Solutions Ci5000 Weather-Ometer. Cycle A3 of the draft version of IEC 62788-7-2 was used: 0.8 W/m²/nm irradiation at 340 nm conforming to ASTM D7869, 20% relative humidity (RH), chamber air temperature (chT)=65˚C, and black panel temperature (BPT)=90˚C.

All specimens were exposed on the air-side with a metal plate on the backside to minimize the impact of reflected UV light. At this time, 4000 h of exposure have been achieved on all but the polyamide film which is at 2500 h and still under test.

## III. RESULTS AND DISCUSSION

### A. Materials Identification

Of the 56 backsheet films, the irradiated air-side of one backsheet was a polyamide known to fail in the field in 4 to 5 years at some hot and humid locations. The field failure of this backsheet is characterized as either cracks through the backsheet between the cells or over tabbing ribbons. On the air side of the backsheets, there were 37 with fluoropolymers of which 14 were polyvinyl fluoride (PVF) and 5 were polyvinylidene fluoride (PVDF), 16 backsheets with PET, and two backsheets with polyolefins.

For the unexposed cell-side, 27 used a low vinyl-acetate EVA, 4 were polyolefins, one was a polyamide, and 19 had a fluoropolymer of which 4 were PVF. All backsheets with a fluoropolymer on the cell-side had a fluoropolymer on the air-side too.

Seven backsheets were transparent, of which two were known to be non-UV stabilized susceptible PET films, two were monolithic PET films, two had fluoropolymers just on the air-side, and one had fluoropolymer films on both sides.

Microscopic examination revealed the polyamide to be composed of three distinct coextruded layers with some glass fibers in the center layer. The vast majority of the films contained a clear transparent inner core layer. While FTIR-ATR was not performed on this inner layer, it is very likely

978-1-5090-5606-4/17 $31.00 © 2017 IEEE

that the vast majority, if not all, of these transparent core layers were PET. When fluoropolymers were identified by FTIR-ATR, they were always very thin compared to the core layer or low vinyl acetate content poly-ethylene-co-vinyl acetate (EVA) seed-layers which were usually comparable in thickness to each other.

*B. Bend Test Results*

As shown in Fig. 2, many backsheets failed after less than 1000 h cumulative radiant exposure. Subsequently at 1500h, only one more backsheet experienced failure on all four specimens. Three more backsheets experienced failure of two out of four specimens after between 1500 h and 3000 h of exposure.

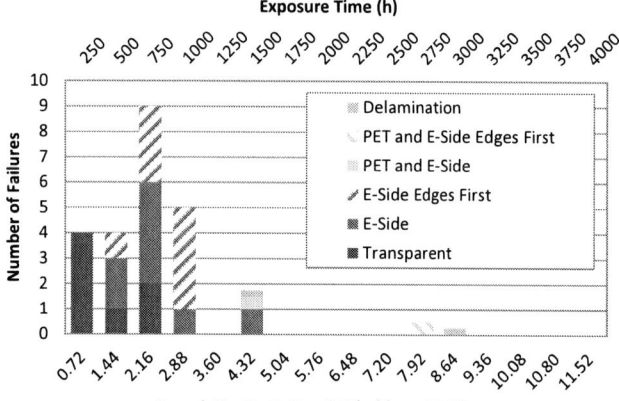

Fig. 2. Histogram of bend test specimen failure after 4000 h of exposure to 0.8 W/m²/nm at 340 nm, chT=65°C, BPT=90°C, 20% RH. Of the 56 backsheets, 23 have experienced failure.

After 250 h exposure, all four of the failures were in transparent backsheets composed of PET, including two unformulated, one UV stabilized, and a dye-containing PET backsheet, Figs. 2 and 3. Three of these four backsheets were not intended for use in a PV module so their failure was expected. Failures occurred when the air-side, which faced the UV light, was in tension. On three of the backsheets, the cracks sometimes went all the way through. PET is known to be highly sensitive to UV light [14-17], therefore it is not surprising that unformulated PET degrades so quickly and that "stabilized" PETs were were inadequately protected. Much effort is required to make a truly UV durable PET film, but it can be done.

After 500 h and 750 h, exposure the remaining three transparent films cracked, Fig. 3. For the transparent film composed of fluoropolymer/PET/fluoropolymer (FPF) and the fluoropolymer/PET/"E" (FPE) films at 750 h, the "E" layer did not crack but did delaminate near the cracks. Here, when the PET layer cracked it sometimes caused the fluoropolymer layer to crack too. This is the only instance of a fluoropolymer layer cracking in any of the 56 films. At 500 h the

PET/PET/"E" (PPE) film experienced cracking in the air-side PET layer.

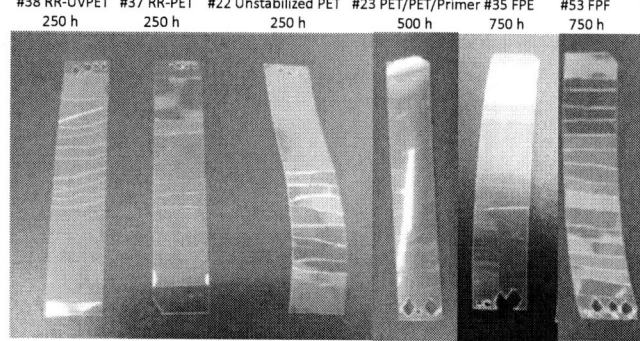

Fig. 3. Photographs of failed transparent films.

Of the 56 backsheets, 27 used a low vinyl acetate EVA and 4 used a polyolefin layer to promote adhesion to the encapsulant. Between 500 and 1500 h, three of the "E" layers on PET/PET/polyolefins cracked and 13 of the low-VA EVA films cracked on all 4 backsheets. Of these low-VA EVA backsheets, 8 were fluoropolymer/PET/"E" (FPE) films and five were PPE films. This is a 63% failure rate of the "E" layer. At 750 h the failure rate reached its maximum and tapered off rapidly. Cracks were predominantly in the transverse direction as would be expected because of residual strain in the machine direction of either the PET or "E" layers. In no cases were the cracks preferentially in the machine direction, Fig. 4. All of these failures had cracks just through the back film and not into the PET core.

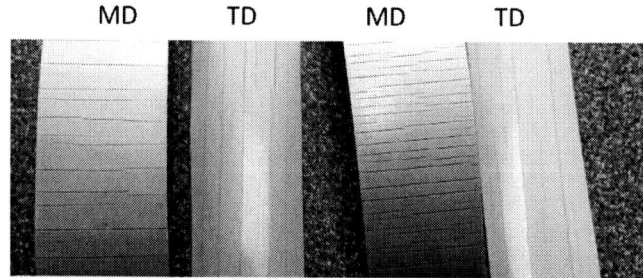

Fig. 4. Cell-side image of backsheet composed of PVF/PET/"E" after 750 h of exposure. Failure is on the cell side with cracks oriented in the transverse direction. "TD" and "MD" indicate which specimens were cut in the transverse vs the machine direction respectively.

Of these 16 cell-side "E" layer failures, eight of them were initiated from the edges of the test coupons during exposure as opposed to cracking during the bend test, Fig. 4. This too raises concerns about the applicability of the specimen construction used here because it is possible that mechanical damage induced during cutting created defects which propagate under exposure to UV light on the sides and $O_2$ along the surfaces. In practice, the edges are typically shaded by the frame and might not be exposed making the specimens where cracks initiate from the edges less likely to fail in the field. Because "E" layer cracking is rarely if ever seen in the field, more investigation is needed to establish the potential relevance of this failure mode.

Fig. 5. Example of cracks starting at the side and propagating inward on the cell-side. Backsheet #30 PVF/PET/low-VA EVA.

One of the backsheets, a PVF/PET/low-VA EVA film, failed with shrinkage of the low-VA EVA layer, Fig. 5. Because this failure forms cracks predominantly in the transverse direction, it is probably a result of high residual strain in the machine direction in conjunction with weaker adhesion to the PET core layer. In other experiments, this backsheet exhibited shrinkage and crack formation when laminated to a piece of glass using an EVA encapsulant. After 4000 h of UV exposure, cracks propagated almost 1 cm into the specimens. Because of the large amount of cracking and the fact that it occurs on the non-UV-unexposed cell-side of the test specimens, it is highly likely that this backsheet would fail if used in a PV module. More experimentation is needed to determine what fraction of all the "E" layer failures may be relevant to a fielded module.

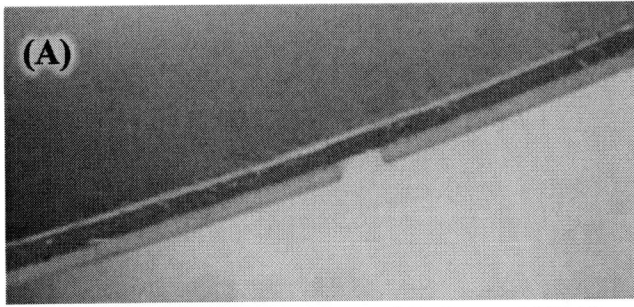

Fig. 5. PVF/PET/"E" film showing shrinkage of the "E" layer. (A) close up of side of specimen. 0.31 mm thick. The "E" layer is the thicker white layer on the bottom side of the photo. (B) Air- and cell-side photos showing cracks in "E" layer.

The conditions of low humidity were chosen by the IEC weathering group working on IEC 62788-7-2 to be more representative of the low humidity in a module when at its higher temperature and because this humidity level is achievable by a wide variety of instruments over a range of temperatures. This condition avoids causing PET hydrolysis which is not observed in fielded modules to be a failure mode but instead will focus on thermal and UV light induced failure modes. Other than the PET failures in the transparent backsheets, only two other backsheets experienced PET related failure and only in half of the specimens.

All of the failures before 1500 h and one at 1500 h had cracks just through the back film and not into the PET core with failure seen on both transverse and machine direction cut specimens. There were however, three backsheets for which only half of the specimens exhibited failure. One backsheet, a PPE with a low-VA EVA layer, saw some buckling of the film only on the two transverse direction oriented specimens resulting from delamination of a central pigmented PET core layer on both sides of the PET, Fig 6C. This was noticed on the first specimen at 1500 h and the second at 3000 h. In this case, the bend test is highlighting a serious concern with low adhesion strength between the layers.

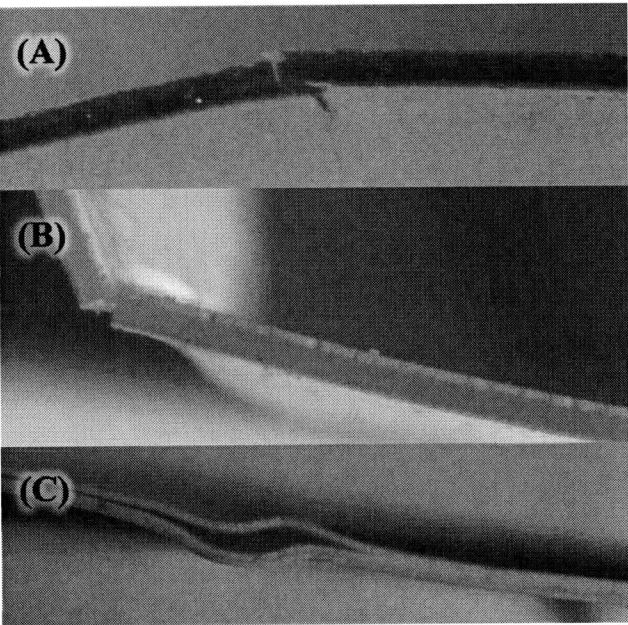

Fig. 6. (A) PVF/PET/PVF (B) and (C) two different PET/PET/low-VA EVA films.

Another film, composed of PVF/PET/PVF, produced audible cracking sounds at 2750 h in the machine direction oriented specimens, Fig. 6A. Initially the location of the crack was not visually apparent, but at 3000h, 3 or 4 cracks per specimen could be seen propagating from the edges about 3mm inward from the front side. Microscopic examination revealed that the cracks were through the PET layer and caused predominant cracking on the PVF film on the UV facing air-side. It appears that this vendor used an inferior PET film that was more susceptible to aging and cracking.

Failure of an inner PET layer was also seen in a PPE specimen, Fig. 6B. This failure was also confined to just the machine direction oriented specimens with crack starting on the "E" side going all across the specimen in the transverse direction. These cracks go through the inner core PET film in addition to the "E" film. Once again, the weak spot appears to be a low quality PET film, possibly with high amounts of residual strain in the machine direction leading to cracking of PET after exposure.

A backsheet of particular interest is a polyamide based backsheet. It is known to fail in the field after several years of exposure but it has not failed the bend test after 2000 h (these specimens were started late so it isn't to the final 4000 h exposure yet). Failure in the field is characterized as cracks in the backsheet between the cells. In parallel work to be published separately, severe loss of elongation to break was seen after 500h. Some possible explanations for field failure include an interaction with the EVA encapsulant, and/or CTE mismatch concern, diurnal hydrolytic expansion and contraction, and/or shrinkage not being a concern for a free standing film. A robust backsheet test will need to incorporate coupons containing components intended for the final application for us to have confidence in long term performance and will be the subject of future research.

### B. Comparison to Field Exposure

All seven of the transparent backsheet films experienced cracking in less than 750 h of exposure, Figs. 2 and 3. If one assumes an ~4× acceleration factor because the test chamber runs for 24h each day, and that the backside gets 10% of the UV dose the front side gets [18-20], then this UV dose is approximately equal to 750·4/0.10/24/365=3.4 y. This simplistic analysis (neglecting temperature and other effects) indicates that this much exposure is not likely to be sufficient for comparison to a 25 y life.

In chamber testing, there is a thermal acceleration (specimen temperature about 75˚C), but UV damage typically varies sublinearly with light intensity [21, 22] as may be expressed with [21],

$$R_D \sim I^x \cdot T_f^{\frac{T-T_o}{10}} \quad . \tag{1}$$

Here $I$ is the light intensity, $T_o$ is a reference temperature, and $T_f$ and $x$ were found to have typical values of 1.41 and 0.64 respectively for the degradation of paints and coatings [21]. According to Table II of Kempe [22], for Phoenix Arizona a weighted average value of $I$=0.0191 (W/m²/nm, at 340 nm) and $T_o$=34.6 ˚C produces the same degradation rate as seen outdoors. With these values, the chamber temperature and irradiance produce a degradation acceleration of 44× making the 750 h exposure equal to 750·44/24/365=3.7 y which is coincidentally similar to the simple dosage estimation above. For the backside, the average irradiance is 42× higher but

because of the highly sublinear dependence on intensity, this only represents an 11× acceleration from the UV light. Under these same assumptions, 5000 h would be the equivalent of a 25 y exposure which is close to the 4000 h of this experiment.

This estimate includes a large number of assumptions and generalizations. Nevertheless, it serves as a rough estimate of how long a backsheet test should be to be considered relevant to the expected lifetime of a module. Typically, acceleration rates above 10× are considered to have a high degree of uncertainty with the results significantly increasing the probability of missing a relevant failure mode or highlighting ones that are not a concern (false positives or false negatives respectively) [23].

As shown in Fig. 7, most of the transparent films did incorporate a UV absorber. Considering that 750 h is not likely to be equivalent to a 25 y life, and that the strains of the bend test were of a similar order of magnitude to that seen in a module, the failure of all the transparent films is a concern for bifacial modules or for some building integrated modules where they are designed to let light through the backsheet. But this is not necessarily a guaranteed failure. In these transparent films, failure is initiated in a PET layer which is known to be UV sensitive. Protecting a PET film from UV induced degradation is not easy because of the ease with which they can oxidize to form chromophores [14-17].

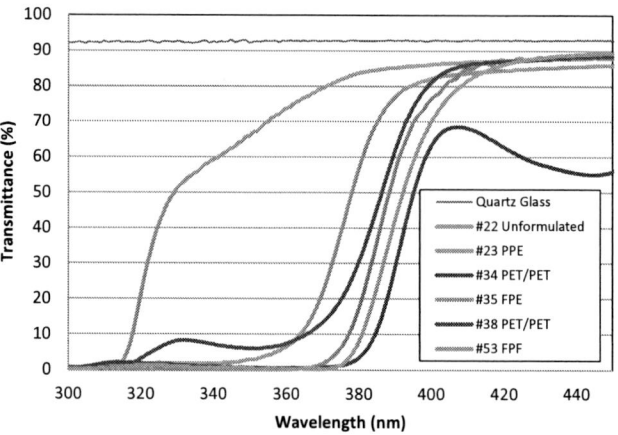

Fig. 7. UV transmittance of six transparent backsheet films.

### IV. CONCLUSIONS

In this work a large number of PV backsheet films were tested to assess which failure modes are dominant and how backsheet films generally respond to environmental stress. In all cases the fluoropolymer films were durable. If a fluoropolymer failed, it was promoted by the failure of a PET film. PET films were sometimes shown to be susceptible to degradation in UV light but only when in a transparent film. None of the pigmented PET films exposed to UV light failed. However, some of the inner PET core layers failed despite being protected from UV light.

All seven of the transparent backsheets failed. For the four films intended for use as a backsheet, this demonstrates a lack of proper assessment of the impact of UV light on the non-fluoropolymer components of a backsheet and is potentially a significant concern for bifacial modules.

Failure of the back "E" layer was widespread with cracks infrequently penetrating the PET core. Of these "E" layer failures, a large fraction experienced failure initiating from the sides of the film and propagating inwardly. However because the exposure of the backside in these tests differs from field exposure, failure here is not necessarily indicative of a bad design, but calls for further examination.

The absence of failure in the polyamide backsheet which is known to fail in the field indicates that testing beyond a simple bend test of weathered backsheets may be necessary and may require one to evaluate the compatibility with materials, the effects of backsheet expansion or contraction, and the design of specific applications.

Future work will focus on determining the potential for failure in the "E" layer to assess if this is typically a concern in a fielded module. New test structures and evaluation methods will be evaluated with the goal of finding better long term backsheet evaluation methods

### ACKNOWLEDGEMENT

This work was supported by the U.S. Department of Energy under Contract No. DE-AC36-08-GO28308 with the National Renewable Energy Laboratory. We would like to thank the weathering group working on IEC 62788-7-2 and everyone who provided backsheets for this study.

### REFERENCES

[1] N. H. Phillips, B. Givot, B. O. Brien, L. C. Hardy, J. Loyd, W. Schoeppel, *et al.*, "Study of Partial Discharge effects of PV backsheet component films. Structure property relationships, and measurement consistency," in *2011 37th IEEE Photovoltaic Specialists Conference*, 2011, pp. 003609-003613.

[2] B. Jaeckel, G. volberg, J. Athaus, G. Kleiss, P. Seidel, M. E. Beck, *et al.*, "Safety of Photovoltaic Modules - Focus on Insulation Coordination," *29th European Photovoltaic Solar Energy Conoference and Exhibition*, pp. 2449-2456, 2015.

[3] IEC 61730-1, *'Photovoltaic (PV) module safety qualification - Part 1: Requirements for construction'*, 2016.

[4] IEC 61730-2, *'Photovoltaic (PV) module safety qualification - Part 2: Requirements for testing'*, 2016.

[5] W. Gambogi, Y. Heta, K. Hashimoto, J. Kopchick, T. Felder, S. MacMaster, *et al.*, "A Comparison of Key PV Backsheet and Module Performance from Fielded Module Exposures and Accelerated Tests," *IEEE Journal of Photovoltaics*, vol. 4, pp. 935-941, 2014.

[6] B. T. Hamzavy and A. Z. Bradley, "Safety and performance analysis of a commercial photovoltaic installation," 2013, pp. 88250M-88250M-7.

[7] A. A. Lefebvre, G. S. O. Brien, B. S. Douglas, J. D. Knapp, D. Garcia, and G. Moeller, "Durability of photovoltaic backsheet outer weatherable layers," in *2015 IEEE 42nd Photovoltaic Specialist Conference (PVSC)*, 2015, pp. 1-5.

[8] X. Gu, C. C. Lin, P. J. Krommenhoek, Y. Lyu, J. H. Kim, L. C. Yu, *et al.*, "Depth profiling of chemical and mechanical degradation of UV-exposed PV backsheets," in *2016 IEEE 43rd Photovoltaic Specialists Conference (PVSC)*, 2016, pp. 0115-0120.

[9] M. D. Kempe and J. H. Wohlgemuth, "Evaluation of Temperature and Humidity on PV Module Component Degradation," in *39th IEEE PVSC*, Tampa, Florida, 2013.

[10] J. E. Pickett and D. J. Coyle, "Hydrolysis kinetics of condensation polymers under humidity aging conditions," *Polymer Degradation and Stability*, vol. 98, pp. 1311-1320, 2013.

[11] W. McMahon, H. A. Birdsall, G. R. Johnson, and C. T. Camilli, "Degradation Studies of Polyethylene Terephthalate," *Journal of Chemical & Engineering Data*, vol. 4, pp. 57-79, 1959.

[12] U. Eitner, M. Köntges, and R. Brendel, "Measuring thermomechanical displacements of solar cells in laminates using digital image correlation," in *2009 34th IEEE Photovoltaic Specialists Conference (PVSC)*, 2009, pp. 001280-001284.

[13] IEC 62788-7-2, *'Measurement Procedures for Materials Usied in Photovoltaic Modules - Part 7 - 2 : Environmental Exposures - Accelerated Weathering Tests of Polymeric Materials'*, 2017.

[14] G. J. M. Fechine, M. S. Rabello, and R. M. Souto-Maior, "The effect of ultraviolet stabilizers on the photodegradation of poly(ethylene terephthalate)," *Polymer Degradation and Stability*, vol. 75, pp. 153-159, 2002/01/01/ 2002.

[15] M. Day and D. M. Wiles, "Photochemical degradation of poly(ethylene terephthalate). I. Irradiation experiments with the xenon and carbon arc," *Journal of Applied Polymer Science*, vol. 16, pp. 175-189, 1972.

[16] M. Day and D. M. Wiles, "Photochemical degradation of poly(ethylene terephthalate). II. Effect of wavelength and environment on the decomposition process," *Journal of Applied Polymer Science*, vol. 16, pp. 191-202, 1972.

[17] W. Wang, A. Taniguchi, M. Fukuhara, and T. Okada, "Two-step photodegradation process of poly(ethylene terephthalate)," *Journal of Applied Polymer Science*, vol. 74, pp. 306-310, 1999.

[18] J. Rabanal-Arabach and A. Schneider, "Anti-reflective Coated Glass and its Impact on Bifacial Modules' Temperature in Desert Locations," *Energy Procedia*, vol. 92, pp. 590-599, 2016/08/01/ 2016.

[19] A. Riedl and S. Snenff, "The Principles of Weathering and How They Apply to Environmental Durability Testing of PV Backsheets," *AMI Conference on Polymers in Photovoltaics*, 2014.

[20] D. M. Burns and K. P. Scott, "Light Sources for Reproducing the Effects of Sunlight in the Natural Weatheirng of PV Materials and Systems," *NREL PV Module Reliability Workshop*, 2013.

[21] R. M. Fischer and W. D. Ketola, "Error Analyses and Associated Risk for Accelerated Weathering Results," *Third International Service Life Symposium, Sedona, AZ February 2004*, 2004.

[22] M. D. Kempe, "Evaluation of the uncertainty in accelerated stress testing," in *Photovoltaic Specialist Conference (PVSC), 2014 IEEE 40th*, 2014, pp. 2170-2175.

[23] W. Q. Meeker and L. A. Escobar, "Pitfalls of accelerated testing," *Reliability, IEEE Transactions on*, vol. 47, pp. 114-118, 1998.

# Solar Variability Reduction Using Off-Maximum Power Point Tracking and Battery Storage

Jason Galtieri, Philip T. Krein

University of Illinois Urbana-Champaign, Urbana, Illinois, 61801, United States

*Abstract* — In this work a solar power tracking algorithm is implemented by filtering the incoming array's real-time maximum power point (MPP) to track the plant's average power output. It is shown that when variability reduction requirements are imposed, MPP is no longer the best operating strategy for maximum energy delivery. The resulting off-MPP algorithm follows a reduced set point in time intervals with high irradiance variability and avoids injecting into the grid the large power swings typical in photovoltaic (PV) arrays. The tracking algorithm is shown to couple well with energy storage systems (ESSs), such as batteries, resulting in reduced storage capacity requirements and power delivery necessary to offset solar variability. We show PV ramp rate violations are decreased significantly using the tracking algorithm and a minimally sized ESS.

## I. INTRODUCTION

Maximum power point tracking (MPPT) is the standard control scheme for photovoltaic (PV) arrays of all sizes. Inverter tracking algorithms based on known MPPT algorithms can track most naturally occurring irradiance changes, and the fastest can even follow rapid changes such as moving cloud coverage [1]. Although good for energy harvest, fast response tracking imposes large power swings on the grid, and without storage, actually maximizes these variations. To date, in most regions, limited PV capacity has allowed utilities to absorb transients by means of conventional control strategies such as spinning reserves. However, as PV penetration increases and replaces conventional spinning generation, there is concern that the grid's voltage and frequency stability could be compromised by variability.

Several solutions have been proposed to handle this fundamental PV variability problem. Geographically widespread solar arrays have been shown to "smooth" regional variability. This helps mitigate variability problems for wide-area balancing authorities, but local problems, such as feeder voltage limits and equipment wear and tear, still occur. Local variability reduction with energy storage systems (ESS) such as batteries, gas turbines, and ultracapacitors has been proposed to offset large power swings [2]. Power requirements from an ESS were found to be 50-100% of the PV array's maximum power output, $P_{max}$, based on an application to limit variability ramp rate in [3]. Part of the necessity for large on-demand power delivery is PV arrays operating at MPP. As

This work was supported by the Grainger Center for Electric Machinery and Electromechanics at the University of Illinois at Urbana-Champaign. The authors are with the Department of Electrical and Computer Engineering, University of Illinois at Urbana-Champaign, Urbana, IL 61801 USA (e-mail: galtier2@illinois.edu).

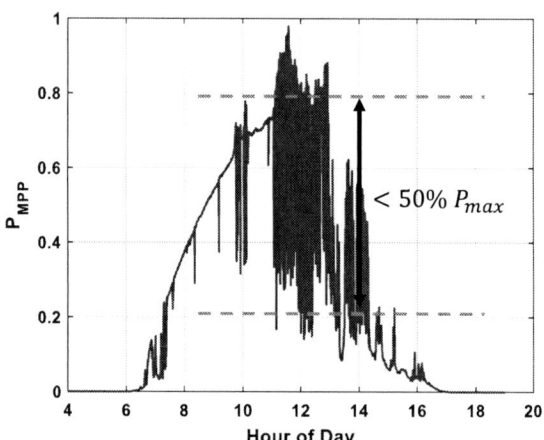

Figure 1: High variability time periods impose large power swings on the grid

illustrated in Figure 1, on cloudy days MPP arrays make large, rapid power swings, often exceeding 50% of the peak value $P_{max}$. Sole reliance on an ESS to offset these swings requires large and expensive storage components.

ESS control strategies have been proposed to mitigate variability in PV plants [4]. Target set points ($P_{set}$) can be computed for the PV power ($P_{PV}$) and ESS power ($P_{ESS}$) to track a slow-varying output. The resulting injected grid power ($P_g$) is given by

$$P_g = P_{PV} + P_{ESS} = P_{set} \qquad (1)$$

and is reachable provided the ESS is large enough to absorb all variability. Ramp-rate control sets linear slope constraints on $P_g$ and relies on an ESS to absorb or buffer power when slopes are exceeded. Moving-average (MA) control tracks the irradiance's moving-average set point and utilizes an ESS when $P_{MPP}$ is above or below the set point. Round-trip losses in any ESS mean the injected energy ($E_g$) will always be less than MPP energy ($E_{MPP}$).

In these control techniques, the ESS is expected to handle the bulk of variability, while the PV plant operates at MPP. This approach maximizes ESS losses, since it maximizes energy passing through the ESS. Off-MPPT inverter control can be shown to be a viable way to offset some of the variability and reduce ESS sizing, costs, and losses.

Historically, high PV system costs and low solar capacity have made MPPT an obvious choice for array control. MPPT maximizes energy harvest, and controls that seek extrema in PI or PV curves can drive toward the desired operating point.

978-1-5090-5606-4/17 $31.00 © 2017 IEEE

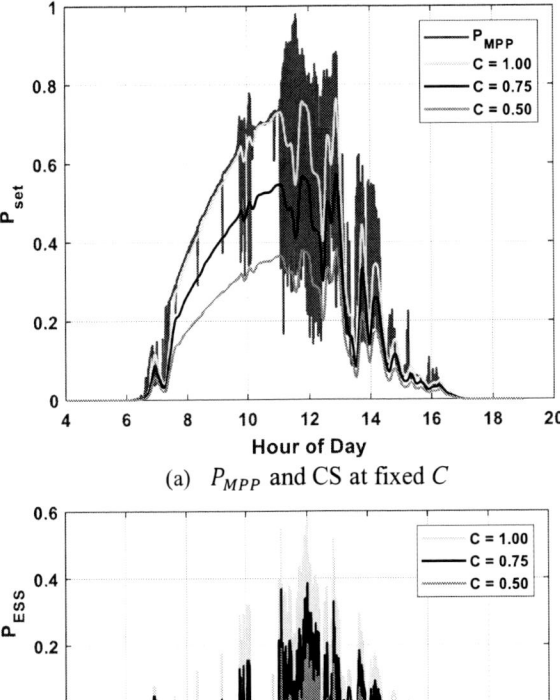

(a) $P_{MPP}$ and CS at fixed $C$

(b) $P_{ESS}$ required to reach set point

Figure 2: (a) The net power output is shown on a day with high variability for an array following the set point and MPP algorithm. (b) Positive power delivered by the ESS

However, additional economic factors, including costs associated with variability, will always determine the optimal strategy for grid-tied arrays. If variability costs are passed onto the array and storage costs are not zero, then MPPT is unlikely to be the optimum operating point. Although MPPT maximizes energy harvest from the array, it also maximizes variability and therefore ESS sizing and loss, and in turn may not optimize energy delivery or energy costs to the power grid itself.

Off-MPPT introduces an additional degree of freedom in plant control and optimal ESS sizing. In this work, we investigate an alternative PV control strategy which tracks the low frequency daily irradiance profile similar to the MA algorithm. However, here it is assumed we have control over the inverter operating point, as well as any additional ESS. With inverter control, the set point can be updated throughout the day depending on variability conditions. When variability is high, a decreased set point, lower than the moving-average, reduces reliance on the ESS. On the other hand, when variability is low, the inverter tracks the irradiance profile without needing an ESS buffer. The proposed algorithm is shown to reduce sizing and power delivery constraints on a battery, which only must produce enough power to reach the set point. Yearlong PV array simulations are performed using measured half-second resolution solar data for the proposed and traditional MPP-bound algorithms.

## II. SIMULATION IRRADIANCE DATA

Simulation data for this work is 100 Hz field measured data using a reference cell. Data were collected over one year in Urbana, Illinois [5]. Here, these data are downsampled to half-second intervals to reduce simulation times. Additionally, point source irradiance measurements have been shown to display higher ramp rates than geographically distributed sources, whose aggregation displays "smoothing" effects [6]. To better approximate PV array power output, the data are further smoothed with a first order low pass filter which has a 0.25 Hz cutoff frequency [7].

## III. CURTAILED SET-POINT TRACKING ALGORITHM

The off-MPPT curtailed set point (CS) tracks the low pass filtered component ($\overline{P_{MPP}}$) of the real-time MPP with an adaptive curtailment factor $C$ and its $P_{set}$ is calculated as

$$\overline{P_{MPP}} = LowPass(P_{MPP}) \qquad (2)$$

$$P_{set} = C \times \overline{P_{MPP}} \qquad (3)$$

where $C$ is updated throughout the day. Nominal curtailment has been shown to provide operating reserves which can absorb a portion of fast power transients [8]. The curtailment factor also allows the set point to be tuned for a desired variability reduction or energy harvest.

Conventional MA set points are always either charging or discharging their ESS, depending on whether the $P_{MPP}$ value is above or below the set point. The PV plant is assumed to be operating at $P_{MPP}$. The inherent filter delay means batteries are used even on clear days, charging in the morning and discharging in the afternoon.

The proposed CS considers the case where both the PV plant and the ESS are controllable. In existing practice, inverter off-MPP operation is already implemented on undersized inverters which "clip" when the array's power gets too high. Likewise, an off-MPP can be implemented through several methods. One approach operates a portion of the array at MPP and then feeds a reference power value to the set-point inverters, which then power limit their outputs. Another strategy is to have all the inverters find the MPP and then back down to the set point. Model based MPPT could also be implemented, which utilizes sensor data to estimate $P_{MPP}$ and operate at the set point.

The intuition behind operating at a reduced power set point, rather than MPP, is that on highly variable days large power swings are costlier to the grid than the benefits of maximizing energy production. During periods of high variability, the CS tracks the interval's average output, rather than following fast MPP swings. The slower moving set point reduces the need for a large ESS to support the bulk of the PV plant output power. On clear days, the low pass filter will track the sun's normal movement and energy harvest will not be affected. A third order Butterworth filter, with a 0.83 mHz cutoff frequency (corresponding to a 20 minute interval), is used for this work.

A constant $C$ would underperform during some parts of a day, as shown in Figure 2a. Here the blue line shows $P_{MPP}$ and the other lines correspond to different $P_{set}$ profiles, at constant $C$ values. A $C$ value of 1 shows the same profile as a

978-1-5090-5606-4/17 $31.00 © 2017 IEEE

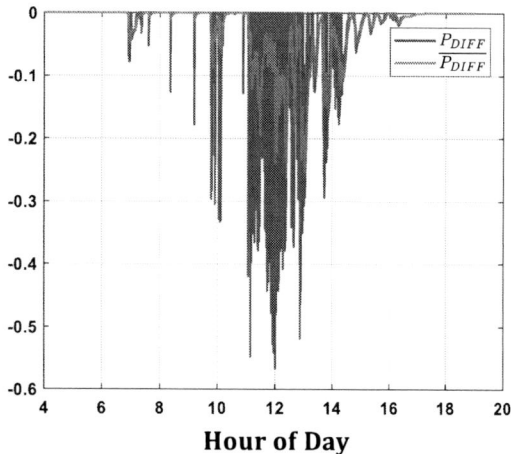

Figure 3: Negative portion of $P_{DIFF}$ and $\overline{P_{DIFF}}$ on a partially cloudy day

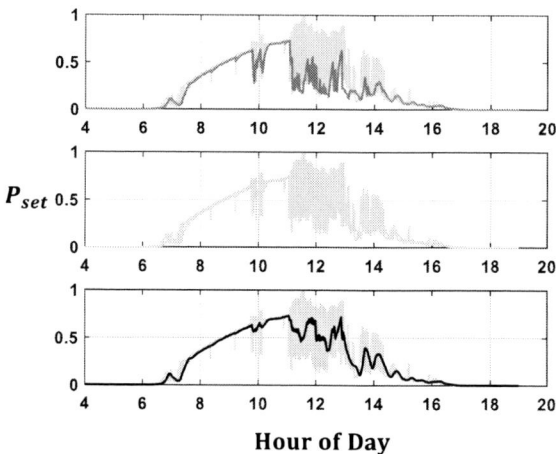

Figure 4: CS with different $k_p$ gain values on cloudy day. Grey lines correspond to $P_{MPP}$ and $C_{max} = 1$

conventional MA algorithm. When $P_{MPP}$ drops below $P_{set}$, an ESS is employed to make up the difference, which is illustrated in Figure 2b. $C$ values close to one maximize energy production from the array, but require larger ESS ratings (and losses) to mitigate variability. Conversely, lower $C$ values reduce reliance on an ESS, but at the expense of energy harvest. An optimal strategy should operate close to $P_{MPP}$ during periods of low variability and decrease during periods of high variability.

Different methods can be used to update curtailment factors throughout the day. With good short-term forecasting, the $C$ value can be set such that the $P_{PV}$ and $P_{ESS}$ will meet $P_{set}$, for a given ESS size. However, even without ideal forecasting, causal strategies will still shift most variability into slower time intervals.

The adaptive curtailment algorithm uses the following metrics to update $C$ values throughout the day:

$$P_{DIFF} = P_{MPP} - \overline{P_{MPP}} \quad (4)$$

$$\overline{P_{DIFF}} = LowPass(P_{DIFF}) \quad (5)$$

$$C = 0 \leq C_{max} + k_p \times \overline{P_{DIFF}} \leq 1 \quad (6)$$

The $P_{DIFF}$ variable keeps track of the difference between $P_{MPP}$ and its low-passed component. When variability is low, $\overline{P_{MPP}}$ will track $P_{MPP}$ and $P_{DIFF}$ will be approximately zero. Conversely, when variability is high, $P_{MPP}$ will fluctuate above and below $\overline{P_{MPP}}$. Positive $P_{DIFF}$ values indicate time intervals where $P_{set}$ is reachable using only inverter controls, while negative values require curtailment or ESS buffering. As such, we only track negative values to set $C$.

To avoid $P_{DIFF}$ changing with the same time resolution as the MPPT, and injecting controller variability, $P_{DIFF}$ is lowpass filtered as per (5). This lowpass filtering imposes a pseudo-memory on the system where set points remain reduced during cloudy periods to anticipate the next passing cloud, as shown in Figure 3. The gain term $C_{max}$ sets a maximum limit on $C$, and $k_p$ tunes the desired sensitivity to $\overline{P_{DIFF}}$. Both terms can be set based on the size of the ESS, the desired degree of variability, or the desired energy harvest.

Figure 4 shows CS profiles for three different $k_p$ gain values with $C_{max}$ set to one. High gain values track near the bottom of the MPP envelope, reducing variability with inverter control and curtailment. Lower gain values track closer to $\overline{P_{MPP}}$ and place more reliance on an ESS to offset variability. The reachability of the set point in all three cases depends on the sizing of the ESS.

### A. BATTERY MODELS

Although different ESS technologies exist, batteries are most common and will be used here for analysis. A linear battery model given by

$$0 \leq P_{Ch} \leq P_{Bat,max} \quad (7)$$

$$0 \leq P_{Dis} \leq P_{Bat,max} \quad (8)$$

$$E_{Bat}(t) = E_{bat}(t-1) + \eta P_{ch}\Delta t \quad (9)$$

$$E_{Bat}(t) = E_{bat}(t-1) + P_{dis}\Delta t \quad (10)$$

$$0 < E_{Bat}(t) \leq E_{Bat,Max} \quad (11)$$

is used to simulate the ESS, where $\eta$ is the roundtrip battery efficiency. A 92% round-trip storage efficiency for lithium batteries is assumed for the rest of this paper [9].

For simplicity, the maximum charging and discharging power are set equal to each other and power delivery is related to energy capacity using fixed power/energy (P/E) ratios. To generalize the results to PV arrays of differing sizes, the battery energy capacity will be normalized ($E_{Bat,n}$) to the array's one hour nominal energy output using

$$E_{Bat,n} = \frac{E_{Bat}}{P_{max} \times 1h} \quad (12)$$

where h is an hour. For example, a battery size of "1" for a 100 kW array would be 100 kWh.

For analysis, the algorithm is based on a known array MPP, and the PV array is able to produce any defined power output up to its MPP. When the MPP is above the set point and the batteries are fully charged, the array reaches $P_{set}$ using inverter control. If the battery is partially depleted, the array operates above the set point and excess energy is used to charge the battery. If the MPP is below the set point, the battery outputs the difference, up to the battery's power limit. The set point will be missed in cases where battery demand is greater than the battery limit, in which case the battery operates at its power limit. The net power sent to the grid ($P_g$) is defined as

$$P_g = P_{PV} + P_{Bat} \leq P_{set} \tag{13}$$

where $P_{Bat}$ is positive if discharging and negative if charging. For cases without storage $P_g = P_{PV}$. The generated energy over a full day will never exceed the MPP energy, so production is normalized with $P_{max}$ as

$$P_{g,n} = \frac{P_G}{P_{max}} \tag{14}$$

Alternative charging strategies, such as charging batteries at night, are not considered in this paper, but would also inject less net energy into the grid than $E_{MPP}$.

## IV. SIMULATION RESULTS

Annual simulations are performed using data from [5] for the proposed CS and conventional MA algorithms. Figure 5 shows the output of arrays during a single day, under both algorithms. A $E_{Bat,n}$ of 0.5 with a P/E ratio is 0.5 is assumed for both scenarios. The MA is more reliant on the size of its battery than the CS, since the $k_p$ gain is adjusted to the battery size.

When both algorithms reach their respective targets, they achieve smooth energy profiles. However, "misses" are easily distinguishable by divergence from these smooth curves. The MA misses more often than the CS algorithm, at the given battery size. During the 11 daylight hours on this test day, the MA has 49 one-minute intervals where $P_{g,n}$ changes by 10% or more. Comparably, the CS algorithm has just 7 such one-minutes intervals. Additionally, "misses" in the set-point algorithm can be further reduced with the inclusion of forecasting. Pre-emptive curtailment, such as around 11 am, would begin decreasing $C$ to anticipate incoming cloud cover. On the other hand, moving-average algorithms, where the PV plant operates at MPP, are not linked to forecasting.

Compared to $E_{MPP}$, the MA achieves 98% energy harvest and the CS achieves 90%. The CS benefit is reduced variability at the given battery size. A smaller $k_p$ would increase energy harvest, but introduce more variability into the grid. Of course a larger battery would smooth both curves, but at increased system costs. Quantification of the economic benefit of low variability would dictate the optimal ESS sizing and the CS algorithm provides another degree of freedom to this optimization.

The state-of-charge of the two battery systems is shown in Figure 5b. Nonideal battery efficiency causes the MA state-of-charge to not reach its starting point by the end of the day. Solutions have been proposed to either charge batteries at night or reduce $P_{g,n}$ to provide more charging time [4]. The CS algorithm tracks below the moving average during periods of high variability and, when $C$ is less than one, there is always more time to charge the batteries than discharge them. Battery lifetime cycle limits are dependent on depth of discharge (DOD) and the low DOD in the CS should increase batteries lifetimes.

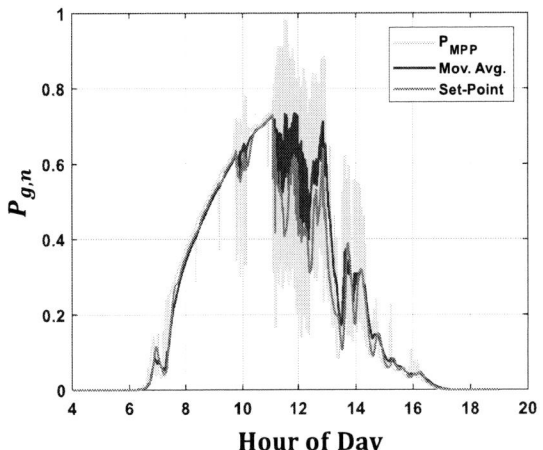

(a) $P_{g,n}$ for CS and MA

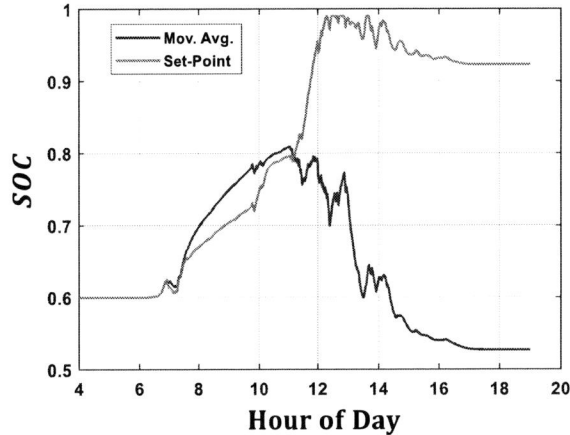

(b) Battery State of Charge

Figure 5: Operation of the curtailed set point and MA algorithms on a partially cloudy day. $E_{Bat,n}$ and P/E are each 0.5

The advantage of operating at curtailed set point is reduced variability in the array output. For this work we will focus on ramp rate (RR) limits to analyze the array's variability. Ramp rate measures the rate of change in the PV array's power output and is defined as

$$Ramp\ Rate\ (RR) = \left| \frac{\Delta P}{\Delta t} \right| \tag{15}$$

where $\Delta P$ is the power difference between two points separated by $\Delta t$ time. Utilities have proposed requiring PV arrays to keep their RR under a certain tolerance and typical values of 10% $P_{max}$/min have been proposed in literature [2]. Jitter in high resolution data can create large RRs on very small time scales, so the simulation results are post-processed with the swinging door ramp detection scheme, with a 1% error margin [10]. For variability analysis, we define a RR event to be

$$RR\ Event = RR > \frac{0.1\ P_{max}}{min} \tag{16}$$

and keep track of the number of violations over the course of the year.

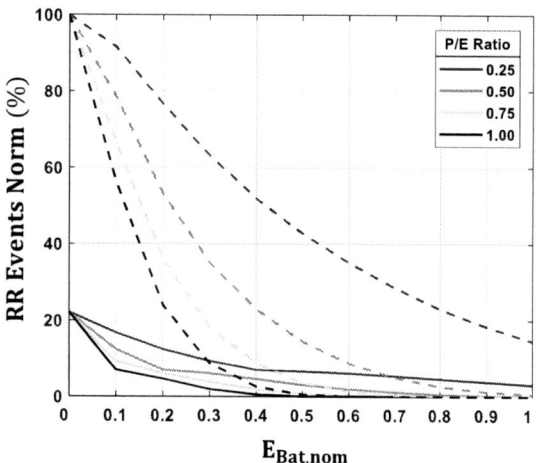

Figure 6: Annual total ramp rate events for CS (solid) and MA (dashed) algorithms. $C_{max} = 1$

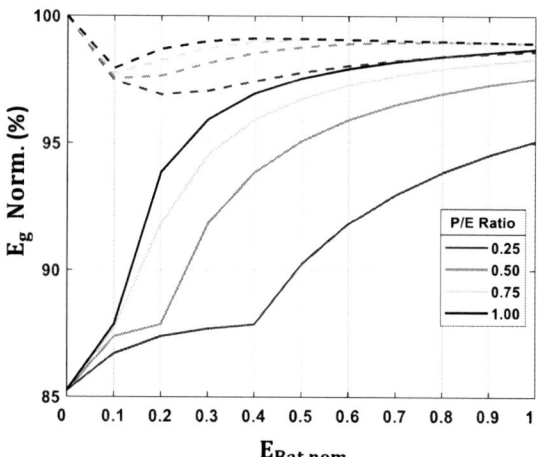

Figure 7: Annual energy harvest for CS (solid) and MA (dashed). $C_{max} = 1$

The annual number of RR events for the CS, MA and no-storage MPP algorithms are shown in Figure 6, for a range of battery sizes and P/E ratios. Solid lines correspond to the CS algorithm and dashed to the MA. The no-storage MPP algorithm produced the most annual RR events (i.e. 100 %) and is used to normalize the others results. When $E_{Bat,n}$ is zero the MA algorithm operates at MPP, with maximum variability, while the CS algorithm operates at a higher $k_p$, similar to the top profile in Figure 4. The CS algorithm, without any energy storage ($E_{Bat,n} = 0$), reduces RR events by about 75% by operating off MPP. When storage is available, the CS can substantially reduce RR events better than the MA algorithm. The effect is stronger in battery limited scenarios. At a battery size of 0.5 (0.5 P/E), the RR events are reduced by 97% and 86% for the CS and MA algorithms, respectively.

The annual energy harvest of the two algorithms is shown in Figure 7. Results have been normalized to the annual MPP energy harvest, $E_{MPP}$. As noted before, without storage the CS algorithm curtails output to reduce variability and energy harvest is lowered to 85% $E_{MPP}$. The 15% decrease in energy provides a 79% decrease in variability, showing the two metrics are not linearly related. Batteries with higher P/E ratios are better able to reach their respective targets and improve energy harvest. However, batteries with long lifetimes typically have P/E ratios much less than one to avoid accelerated degradation. Lower P/E ratio either limits variability mitigation or calls for larger sized batteries. The CS achieves 95% energy harvest at battery size and P/E ratios of 0.5, with only small improvements from larger batteries.

When MPP is the target, complete mitigation of RR events requires large batteries that can output 50-100% of the plant's power. With our simulation data, the necessary sizing for complete RR mitigation is found and shown in Table 1. The moving average needs the largest battery power at 0.6 $P_{max}$, while the CS can be tuned for smaller battery sizes. Annual energy harvest is reduced with the CS, but by less than 14% when $C_{max}$ is greater than 0.90. The corresponding battery constraints are reduced by roughly 33-50%.

Table 1: Battery sizes necessary to eliminate 10%/min RR

|     | $C_{max}$ | $P_{Bat,max}$ | $E_{g,n}$ |
|-----|-----------|---------------|-----------|
| MA  | --        | 0.60          | 0.99      |
| CS  | 1.00      | 0.40          | 0.97      |
| CS  | 0.95      | 0.35          | 0.92      |
| CS  | 0.90      | 0.30          | 0.86      |
| CS  | 0.85      | 0.25          | 0.80      |

## A.  INTEGRATED FORECASTING

The adaptive curtailment factor is further improved with basic forecasting, such as day-ahead projections. On forecasted cloudy days, the $C_{Max}$ parameter is set less than one to anticipate cloud cover. On clear days the gain can be set to one to maximize energy harvest. Since $C_{max}$ is just a gain, it can be updated in either an online or automated manner.

Forecasting is simulated by finding the irradiance variability index [11] for each day of the year. Index values over "2" are considered cloudy while values under "2" are considered clear. We set $C_{max}$ to be 0.85 on cloudy days and 1.00 on clear days and perform the previous analysis for complete RR mitigation. With this forecast and a $P_{Bat,max}$ of 0.25 the set-point algorithm achieves complete RR mitigation. Forecasting improves annual energy harvest to 0.91, compared to 0.80 in Table 1. Higher resolution forecasting, such as 4-6 hour look ahead, would be expected to further improve ESS optimization and energy performance.

## V. CONCLUSION

In this work, an off-MPPT algorithm is presented as an alternative to the standard MPP tracking based on consideration of power variability. Increased inverter control provides an additional degree of freedom in optimizing PV plant and ESS sizing in solar arrays. When variability costs are imposed on a PV system, operating at MPP is sub-optimal as a large ESS is needed to offset plant variability. High costs and relatively short lifetimes suggests that minimizing ESS size always provides economic benefits to grid-tied arrays. When the inverter takes on some of the role in reducing variability, a smaller ESS is needed to reach a desired variability reduction target.

At every level in variability reduction, whether it is complete RR event mitigation or a desired percent reduction, the CS algorithm requires a substantially smaller ESS than the conventional MPP-bound MA algorithm. Although there is some energy harvest reduction with the CS, the 30-50% reduction in battery sizing and power brings benefits that can offset the lower production. Additionally, the CS algorithm is not constantly cycling the ESS like MPP-bound algorithms, so battery lifetimes are expected to be longer.

Energy harvest and variability reduction are further improved with the inclusion of basic forecasting in CS algorithms. On expected clear days the array should be run as close to $P_{MPP}$ as possible while on variable days the array should employ a reduced set point. Gain values in the CS algorithm can be tied to forecast projections and updated either daily or at a faster time scale.

## REFERENCES

[1] T. Esram and P. L. Chapman, "Comparison of Photovoltaic Array Maximum Power Point Tracking Techniques," *IEEE Trans. Energy Conversion*, vol. 22, no. 2, pp. 439-449, June 2007.

[2] H. S. Shivashankar, Saad Mekhilef, Hazlie Mokhlis, M. Karimi, "Mitigating methods of power fluctuation of photovoltaic (PV) sources – A review," *Renewable and Sustainable Energy Reviews*, vol. 59, pp. 1170-1184, une 2016.

[3] I. de la Parra, J. Marcos, M. García and L. Marroyo, "Storage requirements for PV power ramp-rate control in a PV fleet," *Solar Energy*, vol. 118, pp. 426-440, 2015.

[4] J. Marcos, I. de la Parra, M. García and L. Marroyo, "Control Strategies to Smooth Short-Term Power Fluctuations in Large Photovoltaic Plants Using Battery Storage Systems," *Energies*, vol. 7, no. 10, pp. 6593-6619, 2014.

[5] R. J. Serna, et al, "Field measurements of transient effects in photovoltaic panels and its importance in the design of maximum power point trackers," in *Proc. IEEE Applied Power Electronics Conference and Exposition (APEC)*, 2013, pp. 3005-3010.

[6] A. Dyreson, "Comparison of solar irradiance smoothing using a 45-sensor network and the wavelet variability model," in *Proc. ASES National Solar Conference, SOLAR* 2014,

[7] J. Marcos, et al. "From Irradiance To Output Power Fluctuations: The PV Plant As A Low Pass Filter," *Progress in Photovoltaics: Research and Applications*, vol. 19, no. 5, pp. 505-510, 2011.

[8] J. Magerko and P. Krein, "Opportunities for photovoltaic system operation below parity: Costs and benefits of active grid support," in *Proc. IEEE Photovoltaic Specialists Conference (PVSC)*, 2016

[9] N. DiOrio, A. Dobos, and S. Janzou. "Economic analysis case studies of battery energy storage with SAM," National Renewable Energy Laboratory, NREL/ TP-6A20-64987, 2015.

[10] E. H. Bristol, "Swinging Door Trending: Adaptive Trend Recording?" in *ISA National Conference Proceedings*, 1990.

[11] J. Stein, C. Hansen, and M. Reno, "The variability index: A new and novel metric for quantifying irradiance and PV output variability," in *Proc. World Renewable Energy Forum.* 2012.

# Integration of Electrochemical Capacitors on Silicon Photovoltaic Modules for Rapid-Response Power Buffering

Yu Jiang[a], Xuanyi Shi[a], Derwin Lau[a], Da-Wei Wang[b], Zi Ouyang[a], and Alison Lennon[a]

[a] School of Photovoltaic and Renewable Energy Engineering, University of New South Wales, Sydney, NSW 2052, Australia

[b] School of Chemical Engineering, University of New South Wales, Sydney, NSW 2052, Australia

*Abstract* — Photovoltaic (PV) power output can be variable and unpredictable due to the intermittence of sunlight. Based on real-field irradiance data, our model estimated that for a multi-crystalline silicon module rated at 280 Wp, a minimum volumetric energy density of ~ 0.62 $Wh/cm^3$ and a power density of 1.94 $W/cm^3$ is required to limit the ramp rate to < 10%/min. Although current battery technology can achieve the required energy density, achieving high power density with long battery lifetime is challenging. In this paper, we propose a power buffering strategy in which electrochemical capacitors (ECs) are directly integrated into the module-level electronics where they can provide rapid and efficient buffering of PV power. We show that pseudocapacitive electrodes comprising micro-nanostructured $TiO_x$, formed by anodizing a porous Ti microfoam, can achieve a maximum areal capacitance of 100 $mF/cm^2$ (volumetric energy density of 0.65 $mWh/cm^3$ when discharged at 1 $mA/cm^2$). The measured areal capacitance was two orders of magnitude greater than previously reported values for anodic $TiO_x$ electrodes. Although current storage is still insufficient to meet the requirements for module-level buffering, increased energy densities are expected through the use of $TiO_x$ doping, refined anodization processes and incorporation of the electrode in an asymmetric device which can increase the voltage window of operation.

*Index Terms* — electrochemical capacitors, nanostructured metal oxides, photovoltaics, power buffering, ramp rate control.

## I. INTRODUCTION

PV modules can exhibit high variances in their power output due to the intermittence of illumination (Fig. 1 (a)). As the penetration level of PV power into the utility grid increases, such variances can reduce power quality and reliability especially for small grids (e.g., island). Consequently, in some countries, utility grid starts to impose ramp-rate restrictions on PV systems being connected to the grid (e.g., [1]).

Power losses due to module-level shading can be mitigated in part by the use of module level power management [2] such as maximum-power-point tracking (MPPT) electronics, however power losses still occur with intermittent illumination. An alternative solution is to use an energy storage system (ESS) to buffer the variances of PV-generated power. Several studies [3, 4] have investigated use of a rechargeable-battery-bank, which is connected to the PV array through external circuitry, for array-level buffering. However, array level buffering can result in inherent inefficiencies and

power losses as the output of an entire module string can be impacted by intermittent PV generation. Array-level power buffering also requires rapid communication between PV generators and battery banks to enable fast responses to intermittencies. Furthermore, use of batteries as buffers typically necessitates oversizing of the battery energy density (and therefore mass and cost) to meet the requirements for power density (i.e., fast response). Another energy storage solution is to use electrochemical capacitors (ECs) [5, 6]. ECs can be charged/discharged more rapidly and cycled more times than conventional batteries and consequently are promising candidates for PV-power buffering.

In this work, we propose a buffering strategy comprising directly integrating ECs into the module-level circuitry (e.g., in the junction box of a PV module) as shown in Fig. 1. The EC module-level storage can be readily assembled by rolling-up or stacking of sheets of the anode, cathode and separator material into compact elements that can be placed in a junction box or power maximiser. This concept allows the storage elements to be easily replaced after their lifetime without impacting the module life. Section II provides an evaluation on the EC storage requirements for this buffering concept. Section III focuses on material experimentation, which examines the storage capability of micro-nanostructured titanium oxide ($TiO_x$) as an anode material for asymmetric ECs to be integrated in a PV module.

## II. STORAGE REQUIREMENTS FOR MODULE-LEVEL BUFFERING

Several studies have modelled the storage requirements for ESSs however with only simple assumptions regarding PV output profiles [7, 8]. Somewhat more realistic evaluations have been performed by Marcos et al. [9] and Sunny et al. [10] by analysing real-field data from PV plants of various sizes. Both studies concluded a lower limit of the ESS energy density of 0.1 or 0.11 times the output power of a PV plant (i.e., $E_{ESS} = 0.1$ or $0.11 * P_{PV}$) when a 10%/min ramp-rate restriction is imposed and the ESS operates in a continuous recharging mode such that the storage requirements are only limited by the biggest ramp up or down.

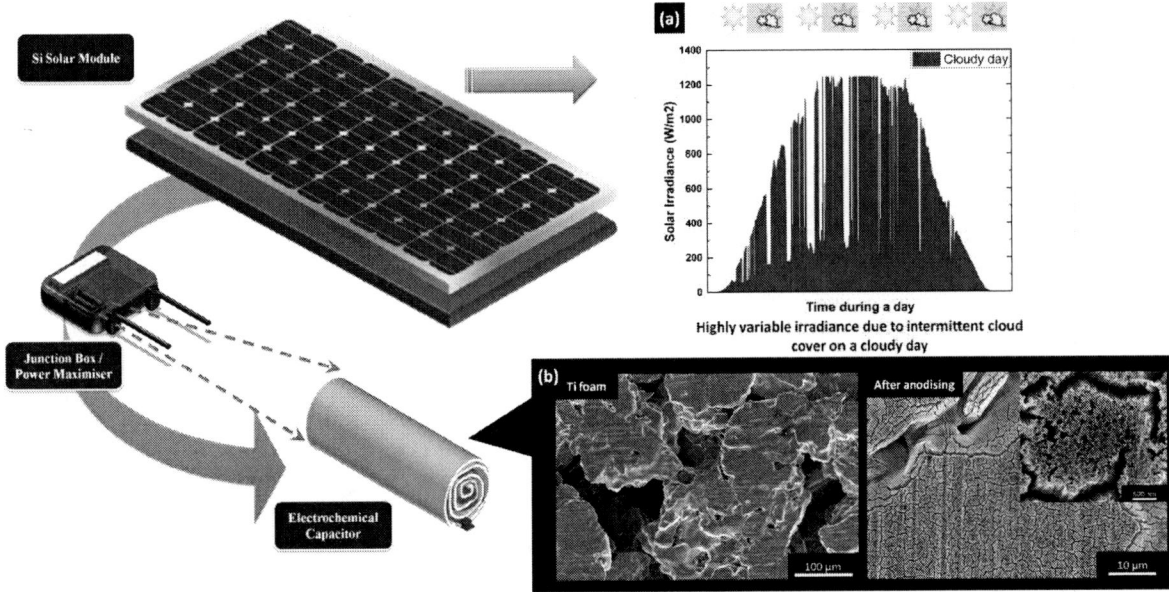

Fig. 1. Schematic of an integrated energy-harvesting-storage device comprising a Si module, a junction box and an EC. Insets show (a) typical solar irradiance in a cloudy day; and (b) a comparison of scanning electron microscope images of a Ti foam before and after anodisation.

We use 1 s solar irradiance data recorded by a pyranometer installed in UNSW Sydney (Australia) to estimate the required EC energy and power density for PV module buffering to ensure the ramp-rate of the output power is limited to < 10%/min. Assuming the ECs would occupy half the volume of a typical module junction box (i.e., ~ 100 cm³) and need to buffer power for a commercial multi-crystalline Si module rated at 280 Wp, a minimum volumetric energy density of ~0.62 Wh/cm³ ($E_{ESS}$ = 62 Wh /100 cm³) and a power density of 1.94 W/cm³ ($P_{ESS}$ = 194 W/100 cm³) was found to be necessary. This energy density requirement is slightly higher than value calculated using the method from Marcos et al. ($E_{ESS}$ = 0.11 h × 280 W/100 cm³ = 0.31 Wh/cm³). The small discrepancy could be attributed to the fact that a PV module can exhibit far more power fluctuations than a large-area PV power plant in which geographic smoothing has an effect.

Although Marcos et al. assessed both the power and energy density requirements for an ESS, no discussion was provided on the correlation of these two characteristics and its impact on sizing an ESS from existing storage technologies. The energy density of an ESS is generally power-dependent and thus should be considered in accordance with power density when sizing an ESS. For example, batteries typically have high energy density but low power density due to their fundamental storage mechanisms (diffusion limited). Consequently, in order to meet the power requirement, battery systems often have to be oversized. So, for example, although current Li ion batteries can provide the required volumetric energy density, they fall short in regard to meeting the required volumetric power density as fast charging/discharging can cause Li plating and reduce battery life [11]. On the other hand, the technological challenge for ECs is to increase their energy density to a sufficient level without compromising their power density.

## III. ANODIC TiOₓ ELECTRODE FOR ECS

### A. Overview

Charge can be stored in ECs through: (i) the formation of an electrical double layer (EDL) (e.g., as in carbonaceous materials [5]); and/or (ii) fast Faradaic reactions at/near the electrode surface (e.g., pseudocapacitive metal oxides and conducting polymers [12]). Conventional EDL ECs store charge via adsorption of ions on the electrode surface and typically have an areal normalized capacitance in the range of 10 to 50 $\mu F/cm^2$ whilst pseudocapacitive materials, such as $NiO_x$, $MnO_x$, $MoO_x$, and $TiO_x$, exhibit much higher areal capacitance and therefore higher energy density [13]. Among these metal oxides, nanostructured $TiO_2$ is of particular interest as an anode material due to its excellent structural and cycling stability, low environmental impact, safety, low-cost and commercial feasibility [14]. In this study, we have attempted to improve the storage performance of ECs using a pseudocapacitive $TiO_x$ electrode by employing the strategies of: (i) increasing the surface area by hierarchical structuring of the $TiO_x$ electrode (i.e., micro-porous Ti foam, nanotubular $TiO_x$ with inter-octahedral sites for Li ion insertion); and (ii) widening the capacitor voltage window utilizing asymmetric structure design and organic electrolyte. This paper reports results on the achievement of improved energy storage via the first strategy.

978-1-5090-5606-4/17 $31.00 © 2017 IEEE

## B. Experimental

A first set of Ti foams (anodic $TiO_x$; ATO) were anodized for different durations in glycerol containing 0.45 M of $NH_4F$ and 2.5 (v/v) of water at 20 V using a three-electrode configuration with the Ti foam as the working electrode, a Pt coil as pseudo reference electrode and a Ti plate as counter electrode. This resulted in the formation of $TiO_x$ nanotubes on the electrode surface. The anodized Ti foams were then annealed in air at 600 °C for 1 h. A second set of Ti foams (thermal $TiO_x$; TTO) was annealed in air at 600 °C for 1 h to thermally oxidize their surface to $TiO_x$ (i.e., no nanostructuring). Galvanostatic charge-discharge (GCD) and current-voltage (CV) measurements were performed using the ATO and TTO electrodes (projection area of 1 $cm^2$) in 1 M $LiClO_4$ in acetonitrile using a three-electrode configuration.

## B. Results and Discussion

As illustrated in Fig. 1(b), the as-received Ti foam had a micro-porous structure with a relatively smooth surface. After anodisation, the Ti foam surface was roughened due to the growth of nanotubular $TiO_x$ arrays. Fig. 2 shows the areal and volumetric capacitance of the TTO and ATO electrodes (fabricated with different anodization durations) calculated from CV measurements and graphed as a function of voltage scan rate. Substantial increases are evident in the areal capacitance of the ATO electrodes with increased anodization duration over the TTO electrode, with a highest areal capacitance of 100 $mF/cm^2$ (1.7 $F/cm^3$) being measured using a scan rate of 2 mV/s. These results improve upon areal capacitances of 0.08 $mF/cm^2$ [15] and 0.97 $mF/cm^2$ [16] previously reported values for air-annealed $TiO_x$ nanotube electrodes. An energy density of 0.65 $mWh/cm^3$ (0.04 $mWh/cm^2$) was achieved for the longest anodized ATO electrode when discharging at 1 $mA/cm^2$. It should be noted that the volumetric energy density and capacitance reported here are calculated based on a total volume of both active material and current collector and therefore represents a lower bound when compared to literature values calculated based only on active material volume. The enhanced charge storage can be attributed to: (i) significantly increased electrode surface area through micro-nanostructuring (higher surface capacitive storage); (ii) an increased voltage window due to the use of an organic electrolyte; and (iii) Li ion insertion into the bulk $TiO_x$.

Non-linearity was clearly evident in the slopes of the GCD curves (see Fig. 3) indicating there is a phase transition between $TiO_2$ and $Li_xTiO_2$ due to Li ion insertion/de-insertion. After the potential plateau region, the potential of the ATO electrode decreases almost linearly with discharge time signifying a pseudocapacitive storage process. The ATO electrode which was anodized for the longest duration exhibited the most prolonged discharge period with over-2 min discharge being achieved with a discharge current density of 1 $mA/cm^2$.

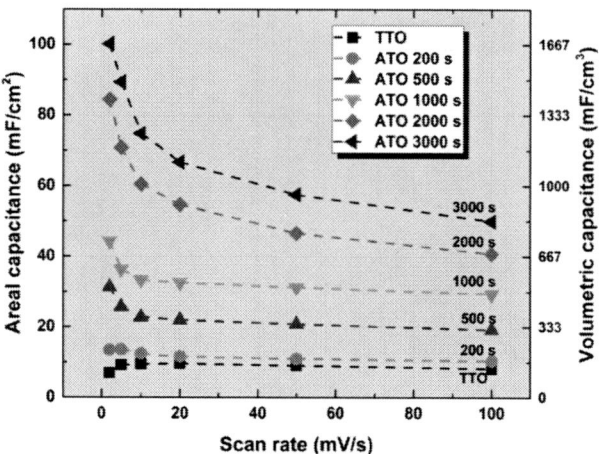

Fig. 2. Calculated areal and volumetric capacitance of TTO and ATO electrodes from CV measurements as a function of scan rate.

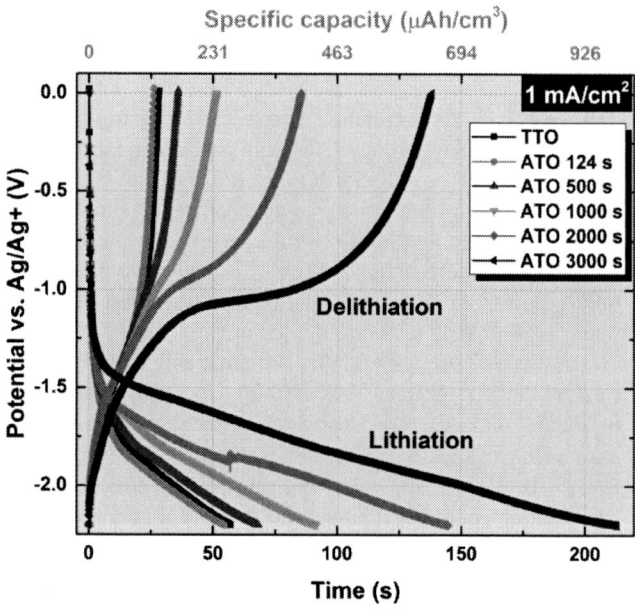

Fig. 3. GCD curves of the TTO and ATO electrodes recorded at a charge/discharge current density of 1 $mA/cm^2$.

## IV. CONCLUSION

In summary, we have proposed an alternative strategy for PV power smoothing in which micro-nanostructured ECs are directly integrated into the module-level electronics. The minimum required energy and power density for EC storage to limit the power ramp rate to < 10%/min for a 280 W module were estimated to be 0.62 $Wh/cm^3$ and 1.94 $W/cm^3$ respectively. Although we have demonstrated significantly improved energy storage using an ATO EC electrode, the current volumetric energy density (0.65 mWh/cm$^3$) is still insufficient to meet the requirements for module-level buffering. Future work will attempt to increase the energy storage through the use of doping, hydrogenation and a refined anodisation process. Further increases in energy density are also expected from the use of complete EC devices, where the addition of a positive carbon EDL electrode can extend the voltage window of device operation.

## ACKNOWLEDGEMENT

This work was funded by the Australian Research Council through Discovery Grant DP170103219 "Advanced Electrochemical Capacitors". The first author would like to thank the Australian Government Research Training Program for providing the PhD scholarship. The authors would like to acknowledge Prof. John Fletcher and Dr. Patrick Burr from School of Electrical Engineering at UNSW Sydney for their advice and insights throughout this work.

## REFERENCES

[1] PREPA, "Minimum technical requirements for photovoltaic generation projects," Available online: http://www.nrel.gov/docs/fy14osti/57089.pdf (accessed on 1 June 2017).

[2] J. M. Carrasco, L. G. Franquelo, J.T. Bialasiewicz, E. Galván, R. Portillo, M..M. Prats, *et al.*, "Power-electronic systems for the grid integration of renewable energy sources: A survey," *IEEE Transactions on Industrial Electronics,* 53, 1002-1016, 2006.

[3] X. Li, D. Hui, and X. Lai, "Battery energy storage station (BESS)-based smoothing control of photovoltaic (PV) and wind power generation fluctuations," *IEEE Transactions on Sustainable Energy,* 4, 464-473, 2013.

[4] K. Takigawa, N. Okada, N. Kuwabara, A. Kitamura, and F. Yamamoto, "Development and performance test of smart power conditioner for value-added PV application," *Solar Energy Materials and Solar Cells,* 75, 547-555, 2003.

[5] F. Béguin and E. Frackowiak, *Supercapacitors: Materials, Systems, and Applications,* 2013.

[6] P. Simon and Y. Gogotsi, "Materials for electrochemical capacitors," *Nat Mater,* 7, 845-854, 2008.

[7] N. Kakimoto, H. Satoh, S. Takayama, and K. Nakamura, "Ramp-rate control of photovoltaic generator with electric double-layer capacitor," *IEEE Transactions on Energy Conversion,* 24, 465-473, 2009.

[8] T. D. Hund, S. Gonzalez, and K. Barrett, "Grid-tied PV system energy smoothing," in *Conference Record of the IEEE Photovoltaic Specialists Conference,* 2010, 2762-2766.

[9] J. Marcos, O. Storkël, L. Marroyo, M. Garcia, and E. Lorenzo, "Storage requirements for PV power ramp-rate control," *Solar Energy,* 99, 28-35, 2014.

[10] M. R. Sunny, N. Faiem, "Output power ramp-rate control of a grid-connected PV generator using energy storage system," *International Journal of Innovative Research in Electrical, Electronics, Instrumentation and Control Engineering,* 3, 2015.

[11] M. Acerce, D. Voiry, and M. Chhowalla, "Metallic 1T phase MoS2 nanosheets as supercapacitor electrode materials," *Nat Nano,* vol. 10, pp. 313-318, 05//print 2015.

[12] B. E. Conway, V. Birss, and J. Wojtowicz, "The role and utilization of pseudocapacitance for energy storage by supercapacitors," *Journal of Power Sources,* 66, 1-14, 1997.

[13] J. Wang, J. Polleux, J. Lim, and B. Dunn, "Pseudocapacitive Contributions to Electrochemical Energy Storage in TiO2 (Anatase) Nanoparticles," *The Journal of Physical Chemistry C,* vol. 111, pp. 14925-14931, 2007.

[14] Y. Liu and Y. Yang, "Recent Progress of TiO2-Based Anodes for Li Ion Batteries," *Journal of Nanomaterials,* 2016, 15, 2016.

[15] X. Lu, G. Wang, T. Zhai, M. Yu, J. Gan, Y. Tong, *et al.*, "Hydrogenated TiO2 nanotube arrays for supercapacitors," *Nano Letters,* 12, 1690-1696, 2012.

[16] H. Wu, D. Li, X. Zhu, C. Yang, D. Liu, X. Chen, *et al.*, "High-performance and renewable supercapacitors based on TiO2 nanotube array electrodes treated by an electrochemical doping approach," *Electrochimica Acta,* 116, 129-136, 2014.

# Design & Evaluation of a Hybrid Switched Capacitor Circuit with Wide-Bandgap Devices for Compact MVDC PV Power Conversion

J. Stewart[1,2], J. Delhotal[1], J. Richards[1,2], J. Neely[1], L. Rashkin[1], J. D. Flicker[1],

R. Kaplar[1], S. Gonzalez[1], J. Lehr[2]

[1]Sandia National Labs, Albuquerque, NM, 87123, USA

[2]University of New Mexico, Albuquerque, NM, USA

*Abstract* — **To better realize the benefits of a hybrid grid that includes DC distribution, new DC-DC converter technologies must be developed that are small, cheap, and efficient at the voltages and power levels relevant to grid integrations. This project will demonstrate the feasibility of a DC-DC converter that will serve as an interface between PV power plants and typical medium voltage DC (MVDC) distribution lines. In this work, a novel power converter topology based on the hybrid switched capacitor boost converter, and using wide bandgap devices, was constructed and will be shown to significantly increase the power density of medium voltage converters. A prototype was designed, constructed and demonstrated to boost 600V to over 10 kV DC with >95.3% efficiency at >2.5 kW.**

*Keywords* — **WBG, PV, MVDC distribution, DC-DC boost, switched capacitor, high gain converter**

## I. INTRODUCTION

A large impediment to fulfilling state and utility renewable portfolio standards is the high levelized cost of solar PV energy ($109.8/MWh) compared to other sources (e.g. a conventional coal plant at $60.4/MWh) [1]. This disparity in cost is due primarily to the high installed cost of commercial and utility scale solar PV systems (relative to capacity factor), which currently totals between $1.77/W and $2.28/Watt [2]. Changes to distribution architectures and converter design may realize cost reductions.

### A. The Benefits of DC distribution

As more DC powered electronics (e.g. computers and televisions) are introduced into every day routines, there is greater concern over the power losses associated with converting renewable sources such as PV from DC to AC (allowing power to be placed on the grid) and then back to DC at point of load. An extensive study was done by [3] comparing AC and DC distribution systems. It was shown that a DC distribution system of even 500 V can surpass the efficiency of a typical North American AC distribution system. The increase in efficiency is realized by a reduction in loss due to less conversion stages (i.e. transformers) and lower copper losses in the transmission/distribution lines. In a study of urban power systems, MVDC distribution was projected to outperform low voltage AC and medium voltage AC in cost [4], due to less need for conductor materials and less electrical loss. Thus, differential efficiency and cost projections are motivating many, including the U.S. Department of Energy (DOE), to

consider advocating direct connection of PV to DC distribution (and even DC transmission) circuits.

### B. Converter performance

Existing commercial PV inverters average just 3.5-5.0 W/in$^3$, making the power electronic converter physically large and consequently a large fraction of the installed cost. Despite the recent advances in wide-bandgap (WBG) semiconductor power devices, commercial-scale and utility-scale PV inverter installations still primarily utilize silicon IGBT based power electronics that switch at low frequencies (5-15 kHz) and interface to the grid through multiple step-up transformers. Some WBG components have been adopted in residential low voltage systems, allowing for faster switching, smaller filters, and even transformer-less (on the LV side) designs. These systems have even achieved ~100 in$^3$ textbook-size form factors (e.g. micro-inverters at ~150 W) that can be cheaply manufactured. However, these systems are still low power density, provide a small minority of the total power to the grid, and show slow growth compared to utility-scale systems [2]. What is needed to enable the cost reductions required at the utility-scale is a breakthrough, modular, mass-producible, textbook-size converter that can deliver several kilowatts and still conceivably fit into a small ~100 in$^3$ volume.

In this paper, we present the development and application of a novel circuit topology for boosting PV power from the panel array to a medium voltage DC (MVDC) voltage for distribution. The proposed topology is an adaptation of the Hybrid Switched Capacitor Circuit (HSCC) [5] which can provide large voltage gains. Fig. 1 illustrates the application. This effort is part of a project aimed at creating a converter with

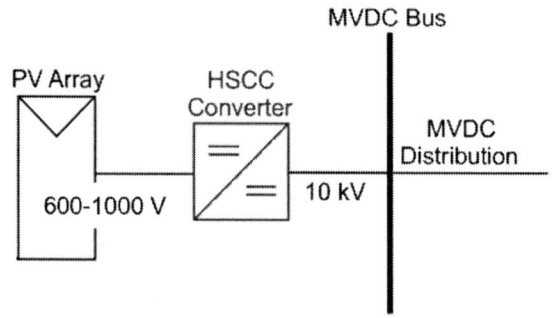

Fig. 1. One-line diagram showing PV array connected to MVDC distribution through an HSCC converter

>100 W/in³ power density at >95% efficiency (based on CEC power weighting [6]) at 10 kV.

The results presented herein quantify the performance of a prototype converter for a variety of operating points up to an output voltage of 10 kV. This initial prototype demonstrated 95.3% efficiency at full voltage and 2.57 kW output power, while connected to a resistive load. Using validated simulation models, the CEC efficiency of this prototype is predicted to be approximately 93.8% for a rated power of 6 kW.

The converter incorporates wide bandgap (WBG) devices and ceramic capacitors to accomplish high-speed switching and high efficiency. The next section presents the theory of operation. Section III describes the hardware design and operation. Section IV describes the development of a simulation model validated against the hardware. Section V predicts the converter CEC efficiency. Section VI provides conclusions and plans for future work.

## II. HSCC BOOST CONVERTER THEORY

In this section, the traditional boost converter is described and then related to the proposed HSCC circuit.

### A. Traditional Boost Converter Operation

Although there are many topologies for DC/DC power converters, they primarily perform one of two functions; step up (boost) or step down (buck) voltage levels. Herein, it is assumed the PV array voltage is always lower than the MVDC bus voltage; i.e. that a boost converter is needed. The classic boost converter schematic is shown in Fig. 2.

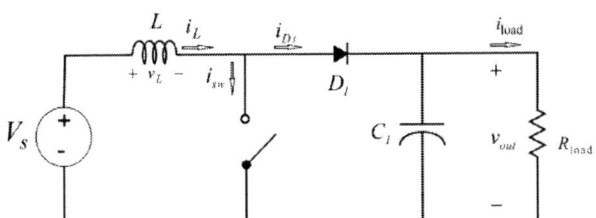

Fig. 2. Traditional boost converter schematic

Through control of the switch, the converter circuit can maintain the desired output voltage across some load and/or control the power out of a PV panel.

A typical mode of operation for the boost converter is continuous conduction mode (CCM), with two states. For state 1, the switch is 'on' and the diode does not conduct. In this case,

a current path is created from the source, through the inductor, to ground. The inductor current increases and stores energy in its magnetic field. In state 2, the switch is 'off' and the diode is 'on'; current from the source will flow through the series connected inductor and diode to supply the load charging the output capacitor while the inductor current decreases. The ratio of the time spent in the 'on' state $T_{on}$ and the switch period $T$ is referred to as the duty cycle $D$, and the gain is given by (1) where $V_s$ is the source voltage and $V_o$ is the output voltage.

$$\text{Ideal classical boost converter: } V_{out} = V_{C1} = \frac{1}{1-D}V_s \quad (1)$$

In practice, this converter does not perform well for a gain above 5 (i.e. $D > 0.8$) [7]. Some novel converter designs have been posed for achieving higher gains [7], but these rely on elaborate magnetic configurations and do not really enable operation at medium voltage. Herein, we consider a topology that allows high gain and medium voltage output.

### B. Hybrid Switched Capacitor Circuit

The HSCC is a combination of the traditional boost converter topology and a switched capacitor (SC) circuit, which in this case includes a diode-capacitor ladder, to implement a charge pump, also called a voltage multiplier. In this topology, there is only one controlled switch, but much of the gain relies on switching in the network of diodes in the output stage. Since WBG diodes are used, the diodes have a higher break-down voltage relative to the forward conduction voltage, and switching transitions are faster, resulting in reduced losses. For this circuit, each stage is characterized by the addition of two capacitors (top rail and bottom rail) and two diodes in series. A two-stage HSCC is shown in Fig. 3. As with the classical boost converter, a simplified single stage HSCC may be analyzed for the two general states; state 1 and state 2. For state 1, the boost converter switch is turned 'on', connecting one side of capacitor $C_2$ to ground. The anode of diode $D_1$ is also connected to ground, causing it to become reversed biased. Diode $D_2$ is forward biased by a small voltage differential placing $C_1$ and $C_2$ in parallel with a small impedance given by $D_2$ between them. Thus, the inductor current $i_L$ increases while charge moves from $C_1$ to $C_2$, and from $C_{out}$ to the load. In state 2, the switch is opened, the inductor current $i_L$ decreases, and a complimentary charge transfer occurs.

A simplified SC circuit can be represented by a voltage source initially in parallel with a capacitor which is also initially in parallel with a load. This capacitor will be used to double the

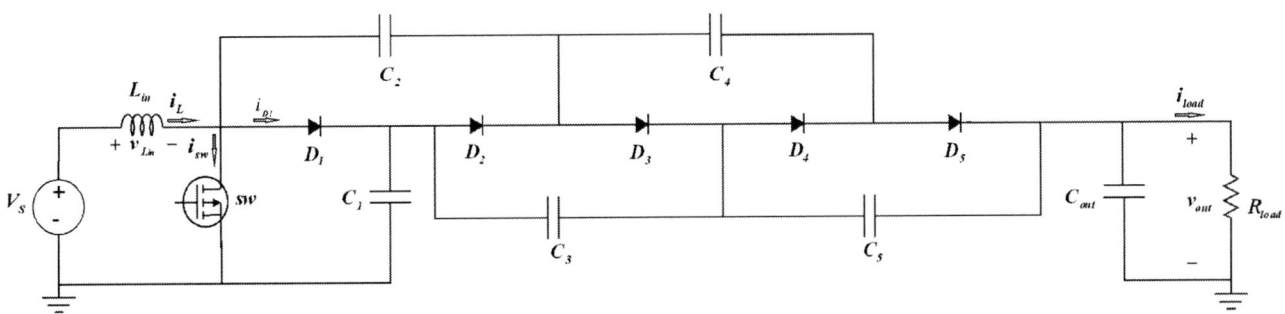

Fig. 3. Two-stage HSCC converter schematic

978-1-5090-5606-4/17 $31.00 © 2017 IEEE

voltage and is referred to as the charge pump (CP). The load can be represented as an additional capacitor with a parallel current source [8]. The CP can be configured to connect in series or parallel between the source and load by toggling a switch. A description of how voltage multiplication through a *single-stage* CP is obtained is as follows:

1. The CP is connected in parallel with the source voltage. The CP is charged to the source voltage while the load current is supplied by the output capacitor.
2. The CP is connected in series between the source voltage and the load. Since the CP was charged to the source voltage, both series elements double the load voltage. Charge is transferred through the CP to the load.

An *N*-stage charge pump can be obtained by cascading additional CP capacitors with switches between each to achieve the same series/parallel configurations. While in steady state, equal charge is moved from the CP voltage source (at $V_{s,CP}$) to the first capacitor. In an ideal configuration, the same amount of charge that is transferred to the first capacitor is also transferred to every other capacitor and the load by its adjacent capacitor throughout each switching period. Neglecting losses, the output voltage of the ideal *N*-stage SC converter can be represented by (2) [8].

$$\text{Ideal charge pump: } V_{out} = (N+1)V_{s,CP} \qquad (2)$$

By combining (1) and (2) with $V_{s,CP}=V_{C1}$ to represent an ideal HSCC, taking the output of the boost stage as the input to the charge pump, the ideal gain is related to the duty cycle *D* and the number of stages *N* as follows

$$\text{Ideal HSCC boost converter: } V_{out} = \frac{N+1}{1-D}V_s \qquad (3)$$

In practice, the gain may deviate considerably from this value. The deviation will depend on the operating conditions,

in particular, the amount of power being processed by the converter.

Based on (3), a 4-stage converter could offer the desired gain with less than a 70% duty cycle; however, device voltage ratings would be exceeded. To implement a circuit with the prescribed gain without exceeding device ratings, a 7-stage converter would be needed. In an ideal HSCC, voltages are evenly distributed across capacitors; so, adding additional stages linearly reduces the voltage stress per stage. In practice, however, non-ideal device behavior causes loss and voltage imbalance between stages, which multiplies as more stages are added.

To achieve the high gain at high voltages, a modified version of the HSCC was used and is referred to herein as the *bipolar HSCC*.

### C. Bipolar Circuit configuration

The *bipolar HSCC* configuration is illustrated in Fig. 4. The bipolar HSCC is configured by adding an additional capacitor/diode ladder that has opposite polarity and diodes flipped with anode on the right and cathode on the left. Diodes on the "bottom circuit" are oriented in the opposite direction compared to the top to allow current flow through the top half of the circuit, through the load, and back through the lower half.

### III. BIPOLAR HSCC HARDWARE PROTOTYPE

A prototype of the bipolar HSCC circuit was designed to operate up to 10 kV output voltage and 6 kW output power. This prototype was designed for maximum flexibility. It can be configured to run in a bipolar or unipolar configuration, with up to 7 stages on each diode-capacitor ladder. The number of stages on each ladder can be reduced by adding jumpers over some of the stages. The inductors were sized to allow testing over a range of switching frequencies, as low as 100 kHz. Test points were placed throughout the circuit, and loops were added for measuring the currents through several of the diodes. To

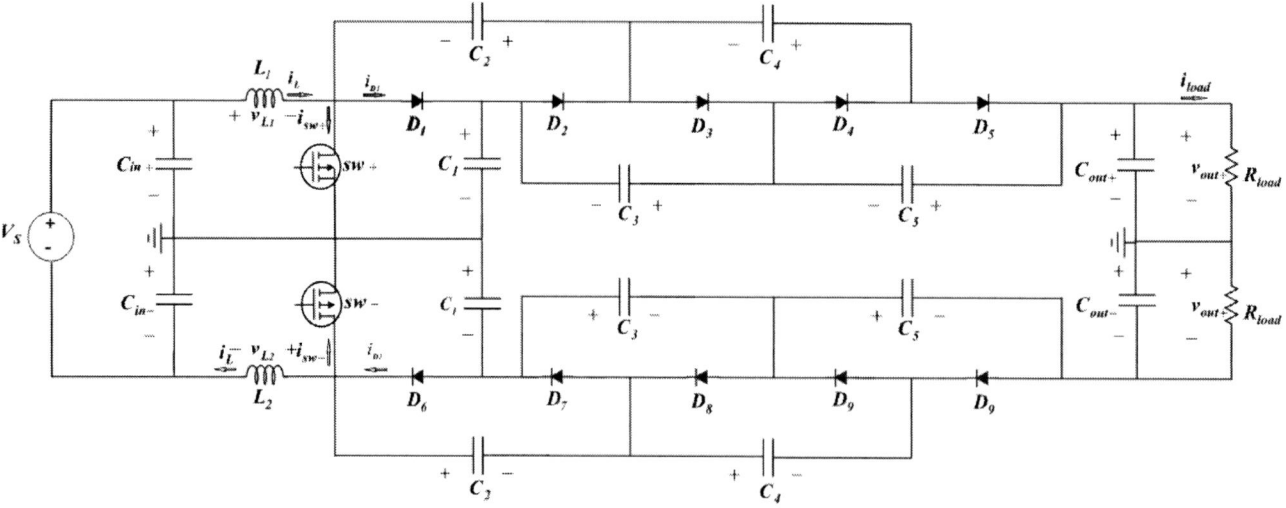

Fig. 4. Bipolar HSCC converter schematic

ensure high-voltage hold-off while saving space, the circuit uses ceramic heat sinks on all the power semiconductors. In addition, cutouts in the board were placed where large voltage differentials exist, to prevent surface tracking.

The switching signal can be controlled with an open-loop PWM signal or hysteresis current control. Although hysteresis current control is more commonly used for DC-AC inverters, it is being explored as a means to directly control the input current (and thus the input power) of the HSCC. Table 1 lists the key components, and Fig. 5 shows a photo of the completed circuit housed in a custom high-voltage enclosure.

As part of the test bed for this circuit, a custom 10 kΩ, 10 kW resistive load was manufactured by *HVR Advanced Power Components*. High-voltage isolation and heat dissipation were important safety considerations for the design and construction of the test bed. The entire test bed was placed inside an interlocked box for protection of personnel.

*Table 1: Key components used in bipolar HSCC prototype*

| Component (see Fig. 4) | Description | Manufacturer / Part Number |
|---|---|---|
| $D_1, D_2, \ldots$ | SiC Diode, 1.7 kV | Wolfspeed C3D10170H |
| $C_1, C_2, C_3, \ldots$ | MLCC, 0.1 μF, 2 kV | Knowles 2220Y2K00104-KXTWS2 |
| sw+, sw- | SiC FET, 1.7 kV | Wolfspeed C2M0045170D |
| $L_1, L_2$ | Inductor, 58.8 μH, 32 A | West Coast Magnetics 320-04 |
| (not shown) | SiC Gate driver board | Wolfspeed CRD-001 |

Fig. 5. Bipolar HSCC prototype converter

The input to the bipolar HSCC prototype was connected to an Ametek / Sorenson SGI 600/8 power supply, and the output was connected to the high voltage resistor bank. The RMS input and output voltages and currents were measured using Fluke digital multi-meters, and several signals were monitored by a Tektronix TDS 3054C oscilloscope. The supply was configured to supply the prototype with +/- 300V input, and the duty cycle was adjusted to $D = 0.46$, and the circuit allowed to "warm up".

In steady state, the RMS output voltages summed to 10.055 kV at 2.574 kW delivered to the load. Fig. 6 shows the positive and negative pole voltages, the input inductor current, and the 0-5V gating signal that was delivered to the gate drivers. Therein, the output voltage is seen to be effectively +/-5 kV.

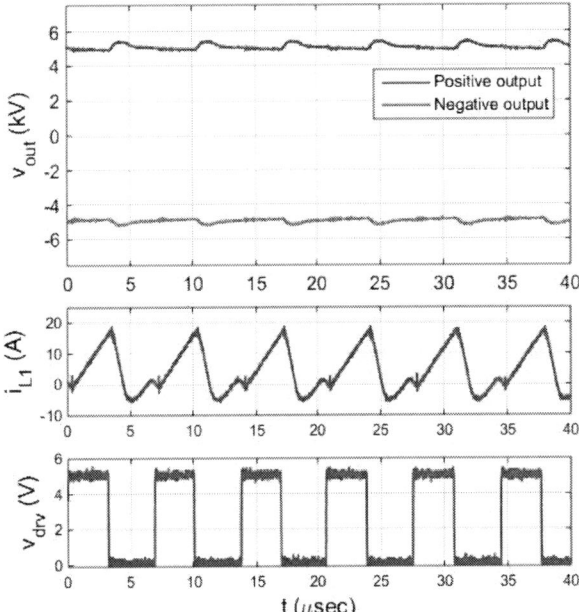

Fig. 6. Bipolar HSCC test data showing output voltages, input inductor current, and trigger signal sent to the gate drivers

The voltage ripple on each pole was measured to be 477 Volts pk-pk average (9.5%). This will be mitigated in future prototypes with the addition of more capacitance. The average input current is 4.5 A. On the rising edge, the input current is consistent with what would be expected from a conventional boost converter. On the falling edge, the current is seen to go slightly negative. This is due to the added dynamics of the connection of the switch node to capacitor $C_2$. A more in depth discussion of the converter dynamics is part of a future work.

The efficiency at this operating point was measured to be 95.3%. Additional operating points were also tested to provide a reference for validating simulation models. This is discussed further in the next section.

## IV. COMPONENT MODELING AND CIRCUIT SIMULATION

Detailed circuit models were developed to enable simulation of the proposed circuit, and special attention was given to the component characteristics. Since ceramic capacitors with X7R dielectric exhibit large reductions in capacitance at different bias levels due to their negative voltage coefficient [9],[10], it is important to use component models that account for this characteristic when possible. At high frequencies, WBG devices exhibit characteristics different from traditional silicon devices such as ring down. Although not all parasitics such as trace inductance were accounted for in this model, using

Table 2 Experimental and Simulation Results for Converter supplying a Resistive load

| Operating Conditions | | | Hardware Results | | | | Simulation Results | | | |
|---|---|---|---|---|---|---|---|---|---|---|
| Input Voltage (V) | Duty cycle (D) | Output Voltage (V) | Output Power (W) | Efficiency (%) | Gain | | Output Voltage (V) | Output Power (W) | Efficiency (%) | Gain |
| 330.1 | 0.33 | 5017 | 532 | 96.55 | 15.2 | | 5115 | 553 | 95.10 | 15.5 |
| 330.2 | 0.59 | 5006 | 1051 | 96.30 | 15.2 | | 5020 | 1058 | 94.53 | 15.2 |
| 330.1 | 0.72 | 5004 | 1569 | 94.13 | 15.2 | | 4996 | 1566 | 92.43 | 15.1 |
| 440.4 | 0.34 | 6678 | 975 | 95.88 | 15.2 | | 6708 | 984 | 95.35 | 15.2 |
| 440.2 | 0.66 | 6663 | 1952 | 95.56 | 15.1 | | 6747 | 2002 | 94.11 | 15.3 |
| 530.5 | 0.35 | 8000 | 1440 | 94.98 | 15.1 | | 7912 | 1408 | 95.43 | 14.9 |
| 600.1 | 0.46 | 10055 | 2574 | 95.32 | 16.8 | | 10089 | 2592 | 93.13 | 16.8 |

manufacturer models provides insight into the circuit's predicted behavior that would otherwise be missed. Finally, since the circuit inductance and capacitance values are relatively low, the resistance and reactance of the supply cable coming from the power supply were also included in the model.

### A. Circuit Simulation and Hardware Validation

In advance of building the 10 kV prototype, circuit models were validated against a benchtop version at lower voltage and power. This benchtop version was a 4-stage unipolar design (see Fig. 3) with GaN FET and SiC diodes, but it provided valuable insight into working with this circuit topology using WBG devices [12]. The 4-stage prototype was built and tested measuring input and output current as well as each node and capacitor voltage. Experimental results matched with simulation results for all measurements with less than a +/-5% error [12]. These models were then adjusted to simulate the bipolar 10 kV prototype presented in this work. The 10 kV model was used to select component values and operational parameters for this first design.

To validate the 10 kV bipolar HSCC converter model, simulations were performed using the same parameters obtained from hardware experiments described in Section III, and the converter was modeled as supplying a resistive load. Table 2 compares the output voltage, output power, efficiency, and gain of hardware experiments and simulations. Simulation results appear slightly less favorable than measured hardware results. Specifically, the conversion efficiency is 1.2% lower on average in simulation, and the gain is 0.44% lower on average. These discrepancies are likely due to some parasitic effects or parameter variation (i.e. due to temperature) not included in the simulation model. However, this model shows reasonable agreement with the hardware results. The model was thus adapted to consider different voltage inputs and resistive load values and then used to estimate CEC efficiency.

## V. PREDICTING CEC EFFICIENCY IN SIMULATION

In practice, the output of the converter would be connected to a MVDC bus, as illustrated in Fig. 1. The input voltage on the PV array would be set by controlling the current from the array using a maximum power point tracking (MPPT) algorithm. The output voltage would be driven by the bus and not controlled by the converter. For the simulations done herein however, the load was represented as a resistance.

Converter efficiency is dependent on input voltage and output power. The California Energy Commission (CEC) [6] has adopted a test protocol, [11], for testing efficiency which uses a weighted average of tests at six power levels (10%, 20%, 30%, 50%, 75%, and 100%) and three voltage levels ($V_{min}$, $V_{nom}$, and $V_{max}$). The panel efficiency is then calculated as a weighted average of these values,

$$\eta_{Wtd} = F_1\eta_{10} + F_2\eta_{20} + F_3\eta_{30} + F_4\eta_{50} + F_5\eta_{75} + F_6\eta_{100} \quad (4)$$

where $\eta_x$ is the measured efficiency at $x\%$ of full load and the weights, $F_1$ through $F_6$, correspond to the amount of time a panel can be expected to operate at that power level over an

Table 3 CEC simulated efficiency

| Weighting (left) for power out (right) at 10 kV out | | Efficiency at $V_s$ = 734 V | Efficiency at $V_s$ = 794 V | Efficiency at $V_s$ = 910 V |
|---|---|---|---|---|
| 0.04 | 10 % | 94.65 | 92.33 | 85.80 |
| 0.05 | 20 % | 92.69 | 92.19 | 94.80 |
| 0.12 | 30 % | 94.68 | 95.91 | 93.64 |
| 0.21 | 50 % | 93.37 | 95.07 | 94.26 |
| 0.53 | 75 % | 93.92 | 94.26 | 93.84 |
| 0.05 | 100 % | 85.70 | 89.07 | 93.43 |
| Weighted average | | 93.45 | 94.19 | 93.61 |
| CEC efficiency 93.8% | | | | |

average day. The rated voltage levels were obtained from an available PV module (SolarWorld 175) while operating in Albuquerque, NM. Table 3 lists the 6 simulated efficiencies representing each power and voltage level and shows the calculated CEC efficiency to be 93.8%. It is noted however, that Section IV results indicated that the simulation was underestimating efficiency by a small margin. This expected CEC performance, at such high gain, in an early prototype indicates a high potential for this novel converter topology.

## VI. CONCLUSIONS AND FUTURE WORK

This paper describes a novel DC-DC converter, a high-voltage bipolar HSCC with WBG devices, with the intention of adapting it for use in PV applications wherein power is distributed across a MVDC bus. Hardware and simulation evidence are provided to evaluate its performance. Several hardware tests were done at a variety of operating points using a resistive load, and a simulation model was developed and validated against this test data. In particular, the converter has been demonstrated in hardware to drive a resistive load at 10 kV and 2.57 kW with 95.3% conversion efficiency. Furthermore, the validated simulation model predicts a CEC-equivalent efficiency of 93.8%. While these CEC efficiency values are lower than current state-of-the-art inverters, these early results show strong promise for this novel converter topology.

Future work will focus on iterating and optimizing the converter design to realize improvements to the converter power density and CEC efficiency. As device technologies mature, components such as GaN FETs and diodes will become available. Also on the horizon are ultrawide-bandgap devices such as those made from high-Aluminum-content alloys such as AlGaN and AlN [13]. These new devices will allow operation at higher frequency, higher voltage, and higher temperature, allowing for a reduction in the size of passive elements and thermal management components. In addition, other topology variations and control methods will be investigated to optimize circuit performance. In particular, it is recognized that in practice, the output of the converter would be connected to a MVDC line (instead of a resistive load), like what is illustrated in Fig. 1. In future work, the simulation model will be adapted to operate in this condition by using a Thevenin source, and the converter design and controls will be adjusted to account for expected dynamic issues.

## ACKNOWLEDGEMENTS

This project is funded by ARPA-E under award 1428-1674. The authors thank Dr. Isik Kizilyalli and the Department of Energy for their support.

The authors would also like to thank Ray Martinez, Mike Horry, and John Brown for their assistance in assembling the high-voltage assembly.

Sandia National Laboratories is a multi-mission laboratory managed and operated by National Technology and Engineering Solutions of Sandia, LLC., a wholly owned subsidiary of Honeywell International, Inc., for the U.S. Department of Energy's National Nuclear Security Administration under contract DE-NA0003525. This document is approved as SAND2017-6864 C.

## REFERENCES

[1] U.S Energy Information Administration, "Annual Energy Outlook." April 2015.

[2] D. Chung et.al; "US Photovoltaic Prices and Cost Breakdowns: Q1 2015 Benchmarks for Residential, Commercial and Utility-Scale Systems; an NREL Technical Report"; NREL/TP-6A20-64746; Sept. 2015.

[3] N. Rasmussen, "AC vs DC power distribution for data centers," APC by Schneider Electric, West Kingston, RI, White Paper #63, 2006, rev. 4

[4] M. Stieneker and R. W. De Doncker, "Medium-voltage DC distribution grids in urban areas," *2016 IEEE 7th International Symposium on Power Electronics for Distributed Generation Systems (PEDG)*, Vancouver, BC, 2016, pp. 1-7.

[5] Seeman, Michael D., "A Design Methodology for Switched-Capacitor DC-DC Converters." Dissertation, Electrical Engineering and Computer Sciences, University of California Berkeley, 2009.

[6] California Energy Commission (CEC) Inverter Test Protocol. URL: https://pvpmc.sandia.gov/modeling-steps/dc-to-ac-conversion/cec-inverter-test-protocol/

[7] Y. P. Siwakoti, F. Blaabjerg and P. C. Loh, "Ultra-step-up DC-DC converter with integrated autotransformer and coupled inductor," 2016 IEEE Applied Power Electronics Conference and Exposition (APEC), Long Beach, CA, 2016, pp. 1872-1877.

[8] G. Palumbo, D. Pappalardo, "Charge pump circuits: an overview on design strategies and topologies," IEEE Circuits and Systems Magazine, Vol: 10, Issue 1, Pages 31-45, March 2010

[9] J. Pyrymak et al.,, "Why that 47 uF capacitor drops to 37 uF, 30 uF, or lower," in *CARTS USA 2008, 28th Symposium for Passive Electronics* , Newport Beach, 2008.

[10] M. Prevallet, et al.,"High voltage considerations with MLCs," in *International Power Modular Symposium and High Voltage Workshop*, San Francisco, 2004.

[11] J. Newmiller, D. Blodgett, and S. Gonzalez, "Performance Test Protocol for Evaluating Inverters Used in Grid-Connected Photovoltaic Systems," DNV KEMA Renewables, Inc. and Sandia National Laboratories, NM, Rep. SAND2015-1817R, 2015.

[12] J. Stewart, "Design & evaluation of a hybrid switched capacitor circuit with wide-bandgap devices for DC grid applications, M.S. Thesis, Dept. Elect. & Comp. Eng., Univ. New Mexico, Albuquerque, NM, unpublished.

[13] R. J. Kaplar, J. C. Neely, D. L. Huber and L. J. Rashkin, "Generation-After-Next Power Electronics: Ultrawide-bandgap devices, high-temperature packaging, and magnetic nanocomposite materials," in *IEEE Power Electronics Magazine*, vol. 4, no. 1, pp. 36-42, March 2017.

# Solar Energy for Clean and Affordable Water Desalination

## V. M. Fthenakis and Adam A. Atia

### Center for Life Cycle Analysis, Columbia University, Mudd 926, 500 W 120th street, New York, NY 10027

*Abstract* — Producing fresh water via sea- and brackish-water desalination is essential for arid, water-scarce regions, but it is expensive and energy-intensive. The cost of energy is a major contributor to this high cost and the use of fossil fuels that currently power desalination plants causes emissions of greenhouse gases and other hazardous pollutants. The recent cost reductions and efficiency advances of photovoltaic systems create opportunities for developing low-cost and emission-free desalination technologies. However, the adoption of PV-powered water desalination and reuse technologies is hampered by the lack of concepts and designs that are integral to the variability of the solar resources. This paper describes the status and prospects of PV-integrated grid-connected and autonomous desalination systems and presents a holistic approach accounting for the variable conditions of the solar resources in different regions of the world to advance the development and deployment of environmentally friendly water desalination technologies world-wide.

## I. INTRODUCTION

In 2003, Richard Smalley, Nobel Laureate in Chemistry, listed the "Top 10 problems of Humanity for the next 50 years". Leading his list was energy, followed in order of priority by water, food, environment, poverty, terrorism & war, disease, education, democracy and population. He placed energy at the top of this list because he thought that the other problems could be resolved or alleviated with the availability of abundant and affordable clean energy. The International Renewable Energy Agency (IRENA) projects that by 2050 global energy demand will increase by 80%, water demand by 55%, and food demand by 60% [1]. This paper focus on the energy-water nexus and its impact on the environment and on solar-powered desalination as a solution to this challenge. The global water supply is already highly critical, and the Organization for Economic Co-operation and Development (OECD) considers that, in 2030, 3.9 billon people will be subject to water stress. Although desalination is considered as a partial solution for increasing water supply, the current processes are not truly sustainable. First, the majority of the global desalination market utilizes fossil-fuel based power to meet desalination energy requirements, resulting in greenhouse gas emissions and other pollutant releases to the environment. Second, thermoelectric power plants typically withdraw large quantities of water, primarily for cooling purposes. These factors are compelling governments and the private sector to explore more long-term sustainable options for powering desalination. A newly enacted Global Clean Water Desalination Alliance (GCWDA) brought together key water desalination and clean energy stakeholders with the goal to reduce the CO2 emissions of the world's water desalination plants [2].

The GCWDA outlook is that water desalination can be a sustainable solution to resolving water scarcity if it is affordable and safe for the environment. However, almost all the world's (more than 18,000) desalination plants are powered by fossil fuels, adding approximately 100 million metric tons of CO2 to the atmosphere annually; this translates to approximately 2.3 kg of CO2 for each m3 of fresh water produced [2]. Furthermore, desalination is often more expensive than water management, recovery and reuse. This paper adds an important dimension to this outlook; that is, drastic reductions in the costs of solar electricity can significantly reduce the cost of producing desalted water and enable the growth of low-carbon affordable desalination. The price reductions of solar electricity catalyze R&D and Deployment in desalination according to two mechanisms: a) Energy cost is up to ½ of the cost of water produced by desalination so reducing this cost increases the desalination market and correspondingly the PV market; b) larger desalination markets (due to lower solar electricity cost) create the incentive for developing technology hybridization and new desalination technologies which can handle solar variability. In addition to cost reductions, the desalination industry is looking at ways to reduce carbon dioxide emissions as it is scrutinized by stakeholders for their high carbon profile; this is the main incentive for the creation of the Global Clean Water Desalination Alliance.

## II. BACKGROUND

Producing fresh water via water desalination is essential but expensive when compared with conventional sources for fresh water production. The cost of energy is a significant contributor to this high cost and the use of fossil fuels to power the desalination plants causes emissions of hazardous pollutants. Solar-assisted desalination systems can be driven either by solar thermal- or photovoltaic systems. Water desalination using thermal energy from the sun has a long history; the concept of saline water evaporation using the heat from the sun to produce fresh water was first described by Aristotle. PV and other RE systems have been used for powering small desalination systems in many parts of the world. Figure 1 shows the composition of such plants as of 2003 [3]; most of the systems are solar powered while most of the solar and wind systems are used to drive reverse osmosis (RO) desalination. However, PV-RO was in the past regarded as not being a cost-competitive solution compared with conventionally powered desalination, thereby limiting PVRO mainly to small-scale applications in remote areas[4]. A comprehensive study commissioned by the KAUST Industry Cooperative Program

[5] showed that PV produces electricity at significantly lower cost than concentrating solar power (CSP), and correspondingly, PV-RO systems would produce water at lower costs than solar-thermal RO systems. Experimental comparisons conducted by Manolakos et al. in 2008 [6] have derived the same conclusion. The authors tested small prototypes of a then state-of-the-art solar-thermal RO system with a PV-RO system, and found that the total cost of water from the PV-RO system would be about ½ that of the solar-thermal-RO system. Most of the difference was due to the significantly higher hardware cost of the solar-thermal RO system; as PV costs have fallen drastically since 2008, this PV advantage may be even greater today.

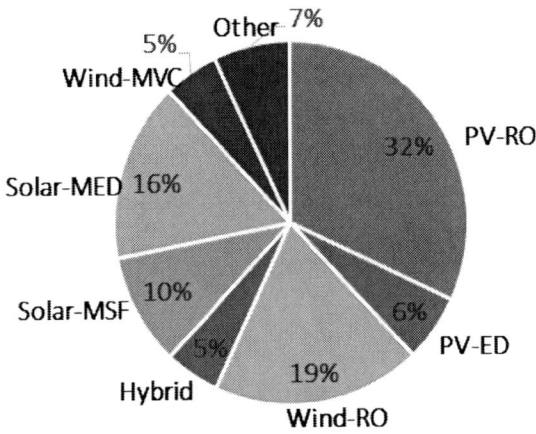

Figure 1. Break-up of RE desalination systems

The main roadblock in the wide use of desalination is its upfront and operating costs. In most areas, RO is the lowest cost technology both in terms of capital investment and operating costs. The capital costs of RO desalination plants are in the range of $600-$1600 per m³/day of capacity depending on the plant's size, the feedwater's quality and pre-treatment requirements, and its location [7]. The cost of water produced by large (i.e., >50,000 m3/d) seawater RO (SWRO) desalination plants ranges from $0.52/m3 to $1.65/m3, with the exception of several plants in Australia for which the costs are much higher [8]. This cost strongly depends on the price of electricity, and pre-treatment requirements, in addition to capital costs. The operating costs of RO have been reduced over the last several years because of the usage of higher permeability membranes, pressure recovery devices, and more efficient pumps [7]. Such improvements took place mainly during 1990-2005 and brought the cost of RO desalination from ~$2/m³ down to a low of $0.52/m³. However, the cost declines did not continue past 2005 as any benefits gained from more efficient membranes were counterbalanced with higher energy and material costs. Currently, the operating pressure at SWRO plants is just 25% higher than the theoretical limit for overcoming the osmotic pressure, so little room for improvement is possible in further reducing energy

requirements in the field. Opportunities for further reducing the cost of RO desalination are related to the use of "free fuel" from renewable technologies, to longer lasting fouling-resistant membranes, and perhaps to development of advanced hybrid desalination technologies like forward osmosis with RO.

## III. PHOTOVOLTAICS POWERED WATER DESALINATION

### A. Near-term/Currently Available Options

We determined that 10% PV electricity in desalination plants is straightforward based on economics (fuel displacement) alone and that co-location of PV in grid-integrated plants further improves PV economics. The power generation industry has started integrating solar power generation (solar thermal and photovoltaics) in the generation mix for centralized power plants. There are examples, such as Al Abdaliya Integrated Combined Cycle Solar (ICCS) power plant in Kuwait where 60 MW of solar thermal generation is part of a 280 MW plant. Solar PV power plants represent an easier integration from a system and cost standpoint with conventional oil- or gas-fired power plants. The value of integrating the solar output of PV power plants is also based on the fact that with minimal output profile adjustments, there is a direct complementarity between the output of thermo-electric power plants and solar PV plants based on the reduction in ISO efficiency of the engines/turbines in these power plants during midday temperatures in hot and arid regions. As shown in Figure 2, such reductions are on the order of 5-10% and happen when the solar resource peaks.

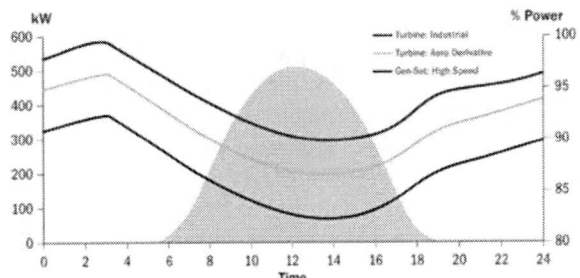

Figure 2. Thermoelectric power plant efficiency dependence on ambient temperature

The case for hybrid thermoelectric and PV power plants on the utility-scale has gained momentum in recent years, driven by the fluctuation in hydrocarbon pricing and the continual reduction in solar PV levelized cost of electricity (LCOE). Furthermore, by co-locating solar PV power plants next to or in the vicinity of conventional power plants, the benefits of utility-scale solar PV are maximized by combining O&M and reducing or eliminating the need for grid infrastructure upgrades. There are three main synergies for co-locating solar PV power plants with centralized power plants; those are:

1) Land Utilization: Solar PV plants can be built on areas that are deemed underutilized such as HV transmission lines rights-of-way and zoning associated with conventional plants.

978-1-5090-5606-4/17 $31.00 © 2017 IEEE

2) Operation and Maintenance: The O&M cost for hybrid PV power plants can be up to 50% lower than stand-alone solar PV power plants. Scheduled and preventative maintenance can be coordinated between PV and conventional power plants; thus, there is little need for additional labor to support the O&M of co-located solar PV power plants.

3) Electrical Grid Connection: Overall nameplate capacity can stay the same for the conventional power plant as solar PV generation will act mainly as a fuel saver. Typically, 10 to 15% is available as additional capacity on MV/HV switch yards and transmission lines.

In short, integration requirements for solar PV plants co-located with conventional power plants are simple, and this approach achieves a lower LCOE when compared to stand alone PV power plants. For example, in a proposed solar PV power plant in Al Manakher, Jordan, a 20% drop in the LCOE is achieved by co-locating the solar plant with a conventional power plant. The monetary benefits from such co-location, include a reduction of O&M costs, almost free land, use of an existing asset management plan (e.g., managing O&M, accounting, revenue, reporting), and use of an existing substation. The same impact on the hybrid LCOE can be derived from co-locating solar PV power plants and SWRO desalination plants where the synergies in O&M, grid connection (or feed in this case) and land utilization are achieved.

It is also noted that RO is often added to older thermal desalination plants that need to expand capacity and this creates more opportunities for PV additions.

### B. Longer-term Options

The large-scale adoption of RE-powered water desalination technologies is hampered by the lack of concepts and designs that are integral to the variability of renewable energy (RE) resources. A holistic approach should account for the variability of renewable energy and water resources in different regions of the world to advance the development of environmentally friendly water desalination and reuse technologies world-wide. We envision viable solutions by a) hybridization of existing and emerging desalination and water reuse technologies, b) integration of RE generation with advanced desalination systems, c) increasing the reliability and longevity of such systems via membrane modification and adaptive operational strategies.

### 3.2.1 Hybridization of Existing and Emerging Desalination & Water Reuse Technologies

Investigations to lower the energy burden and environmental impacts of water desalination/reuse include, but are not limited to the following hybridizations: Forward osmosis (FO) with reverse osmosis (RO); membrane distillation (MD) and RO; Multi-effect distillation (MED) and adsorption desalination (AD) [9]. FO is an emergent technology with great promise of reducing energy requirement in water treatment and

purification. Integrating FO with RO could be accomplished by at least two means: using seawater FO with wastewater effluent to dilute the RO feedwater salinity and thus reduce the hydraulic pressure needed for operation, and utilizing the concentrated brine (that is otherwise discharged) with FO to apply an osmotic shock, thus cleaning the RO membrane without additional energy input. A hybridization of MD with RO; as MD is effectively insensitive to the salt concentration of the feed stream, it can be employed after RO to achieve greater recovery of product water and further concentrate the brine.

### 3.2.2 PV integration in desalination/reuse plants especially designed for variable renewable energy (VRE)

As part of this investigation, we are developing operational strategies that add benefits to VRE integration with desalination and water reuse. For example, while conventional desalination plants consume power at a constant rate, new plant designs that facilitate variable operation may allow for time-shifting of energy use, demand response, utilization of over-generation by solar and wind power, and other functions. These functions would enhance the ability of the desalination/reuse system to serve an evolving electricity system.

We propose a novel design of an integrated energy and desalination system that can provide drinking water and vary its energy consumption in a versatile manner to provide electricity system services, while also improving economic and environmental viability. This integrated energy and desalination design combines access to seawater, treated wastewater effluent, and variable energy resources to simultaneously mitigate water scarcity and facilitate services to the electricity system through time-shifting of energy usage, demand-response, and ancillary services (Fig. 3).

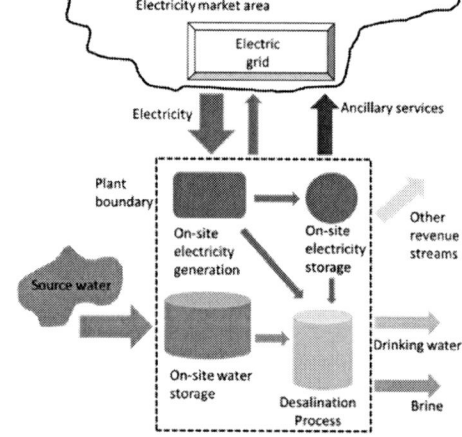

Figure 3. Conceptual scheme of an integrated energy-desalination system

Our design won a recent (May 2017) National Laboratory U.S.-Israel Integrated Energy-Desalination Challenge [10]. As part of this ongoing investigation, we are developing operational strategies that add benefits to VRE integration with desalination and water reuse. The objective is to add various degrees of flexibility that could be aligned with the variability of the electricity source. New desalination plant designs that facilitate variable operation may allow for time-shifting of energy use, demand response, utilization of over-generation by solar and wind power, and other functions. These functions would enhance the ability of the desalination/reuse system to serve an evolving electricity system.

## C. Autonomous PV Directly-coupled Systems

Testing very small, PV-RO systems which are directly coupled, without incorporating batteries, has been reported. Mohamed et al., 2008 tested a small (0.5 m3/d), PV-SWRO system wherein PV output was directly coupled to a DC motor, during ten days of variable solar irradiation in November, and compared it with a PV-battery-SWRO system [11]. The comparison showed that the PV-RO system with a power control allowing variable pressure and flow within a range, worked almost as well as the one with batteries. According to these investigators, the cost of batteries for small systems may not be justified; instead of batteries, an oversized PV system allowing some curtailment of energy may be more cost-effective.

An interesting system of flexible operation that is controlled by the solar resource variability was recently tested at a pilot-scale in Masdar. The results are shown to be promising and commercial systems are available in the 30-300 m$^3$/hr capacity with PV capacity of 22-200 kW. The system, as reported by its developer, Mascara Co., automatically optimize flows, pressures and recovery rates to adopt to variable solar power. These types of systems that can only operate during part of the day (maximum of 8 to 10 hours) will carry a higher Capex for the desalination plant; this can be justified only if the cost of electricity is higher than a certain threshold price and Mascara reports that this threshold is about $0.15/kWh [12]. For this system, the cost of potable water production from sea water is reported to be less than $1/m3, which is comparable to fresh water production cost from small and medium sizes of conventional RO desalination plants.

## IV. PV ENERGY-WATER NEXUS: RELATIVE MERITS AND DE-MERITS IN A REGIONAL CONTEXT

Water usage during the life-cycle of photovoltaics is negligible in comparison to the quantities used in the thermo-electric power generation and this is especially important in arid areas. During operation PV would need little or no water at all for cleaning the dust off the modules. Thermoelectric power plants (e.g., oil, gas, nuclear, CSP) need to withdraw 900-85,900 liters of water per MWh of electricity produced and consume 340-3100 L/MWh, depending on the type of their cooling system [13].

PV powered water desalination has the potential to dramatically increase availability of fresh water in arid locations which, by virtue of their topography and climate, typically have large solar resources.

## A. Gulf Countries

Desalination capacity across the Gulf Council countries (GCC) has grown consistently over the past decade and is expected to approach 45 million m$^3$/day by 2020. Saudi Arabia has the largest share, followed by the UAE, Kuwait, Qatar, Oman and Bahrain. The water demand in the GCC has been outstripping renewable natural water resources; thus growth rates in the range of 6-10% per year are expected for the region. In the GCC, much of the desalination is powered directly or indirectly through natural gas, diesel, and crude oil. In net-exporting markets, such as Saudi Arabia, Oman and Qatar, growing demand in the desalination sector means that larger quantities of natural gas and diesel have to be redirected from exports to domestic use. In net-importing markets, such as the UAE, the increasing use of natural gas translates into rising costs and energy security concerns. Therefore, two key pillars of a long-term desalination infrastructure strategy are emerging – adopting more energy-efficient desalination technologies (e.g. RO) and transitioning towards alternate energy fuels (e.g. solar). The industry is already witnessing increasing use of RO in stand-alone and hybrid configurations [14].

Saudi Arabia which relies on oil as the major revenue stream, is experiencing an increase in its domestic use. The domestic use of NG is increasing in both Oman and Qatar, and the use of imported LNG is increasing in the UAE. Thus, each of the GCC countries need over the long-term new domestic energy sources. In Saudi Arabia, for example, nearly 1.5 million barrels of oil are utilized every day for thermal desalination, an amount equivalent to the daily oil consumption of Italy [14]. The low cost of energy inputs has maintained an infrastructure that continues to rely on energy-intensive thermal desalination processes. Furthermore, the summer peak demand for diesel fuel exceeds local production, requiring the Kingdom to import approximately 40% of its diesel demand during this period. Diesel fuel prices are heavily subsidized; the global market price is estimated at ~70 cents/L while the Saudi price is ~ 8 cents/L [15].

Raising fuel prices to global market levels could incentivize a shift to more energy-efficient technologies such as RO, thus reducing the total primary energy consumed for desalination. The energy requirements for RO plants are much lower compared to thermal desalination but still represent about 40% of the total water production costs, and solar electricity can reduce this cost [16]. For all these reasons, in 2014, the KAUST Industry Collaboration Program (KICP) issued an ambitious plan of rapid deployment of solar systems aimed at displacing fossil fuels with renewable energy in SA desalination plants; KAUST predicts that 70% of the production of desalinated water can be powered with renewable energy by 2030, increasing to 80% by mid-century. In the near term, some of the existing MSF plants in the GCC countries would have to be retrofitted or phased out. In these cases, a SWRO unit could be added to the plant [17].

We applied these factors to the whole water desalination market in Saudi Arabia assuming that ½ of the desalination capacity is attributable to PV-RO and that PV can satisfy 44% of the annual load of each RO plant. Thus, we estimated that the PV-RO power plants considered herein have the potential to displace 120 million barrels of diesel fuel per year in Saudi Arabia. The corresponding reductions in $CO_2$ emissions would be 51.5 million metric tons per year. Thus, by implementing PV powered-desalination plants, freshwater demand in arid and sunny regions could be met cost-effectively, while reducing air pollution from combustion.

### B. Israel-Palestine-Jordan

Imports of fossil fuels in the MENA region represent a major drain on foreign currency reserves. In Jordan, which imports 96% of its energy needs, energy imports account for roughly 20% of the nation's total Gross Domestic Product (GDP). In terms of addressing water scarcity, for years, policy has been to withdraw unsustainable amounts of water, depleting aquifers, desiccating streams and other aquatic ecosystems. Israel is considered a world leader in developing and applying technologies to address limited water supplies. Among these technologies are drip-irrigation, reuse of treated wastewater, and large-scale SWRO desalination, the latter of which has become the country's primary adaptation strategy. Desalination currently supplies almost 80% of all domestic consumption in Israel.

Jordan has limited access to the sea, but has relatively plentiful open space with high radiation potential suitable for solar energy. Israel and Palestine, on the other hand, are relatively limited in terms of open spaces, needed for most commercially viable renewable projects, but have access to the Mediterranean. Thus, there is potential for mutual exchanges of water and energy between the parties. An advantage to the proposed project is that it involves creating interdependence, rather than creating unidirectional dependence, as is currently the norm, with Israel being an increasingly important source of water and gas for both parties. Thus, parties would interdependent on one another, which reduces the potential for unilateral actions that could harm one party [18].

### C. Desalination and the Mining Industry in South America

Most metal mining operations are in arid and sunny regions and mining needs a lot of energy and water, so solar-powered desalination has a great potential (Figure 4). In Chile where copper mining alone generates more than a third of the federal income, high water consumption in mining causes conflict between mining companies and local communities. Persisting drought in the northern Chile has worsened this situation, and the government has pondered making it mandatory for mining companies to build desalination plants to supply their operations. There are 16 mining-related desalination projects worth US$10 billion planned or under construction in Chile, and 9 are already operating. It is projected that desalination will provide half the water demand of Chile's copper mining industry by 2026 [19]. One of the proposed projects is a 126,000 m3/d integrated PV-SWRO system according to which water produced from SWRO will be pumped 45 miles and across a height of 700 m to the mountains where mining companies operate. PV will provide 80% of the energy used in pumping and desalination and the design offers flexibility that follows the solar resource variability. [20]

In Peru, Veolia operates a 4,900 m3/d reverse osmosis desalination plant at the Cerro Lindo mine which produces lead zinc, copper and silver. Its desalination plant supplies the site with industrial process water without adding to the region's water stress. The seawater intake for the desalination plant is on the coast. The desalinated water is transported 40 km to the Cerro Lindo site at an altitude of 1,850 meters. [21]

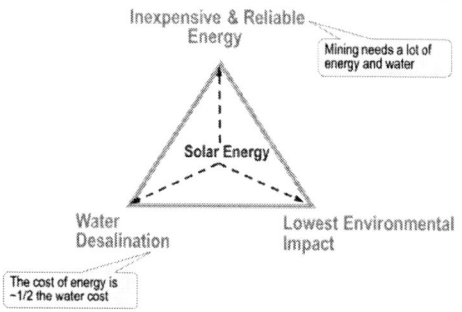

Figure 4. The Energy-Water-Environment Nexus in Mining

## V. CONCLUSIONS

Meeting the water demand of increasing population and increasing industrialization is one of the world's most pressing issues, both economically and socially, and is especially critical in arid regions. Water scarcity will become a perpetual problem and will spread to new regions as more weather extremities take place, induced by increased $CO_2$ concentrations. In this context, solar energy-based desalination is gaining increasing prominence and the declining costs of photovoltaic (PV) systems make it affordable.

The PV industry has accomplished phenomenal cost reductions through improving efficiency in the stages of manufacturing and deployment. The prices of photovoltaic electricity have been drastically reduced over the last five years to levels in parity with unsubsidized electricity rates in areas of high solar irradiation in the Middle East and other arid regions that need to expand the capacity of water desalination. A similar trend of reduction is observed in the prices of batteries, and this will assist the penetration of stand-alone PV hybrid desalination systems. As electricity consumption is one of the most significant element in the RO cost structure, further declining PV electricity prices should further catalyze solar water desalination projects. The competitiveness of PV-based desalination is likely to increase in the future as costs continue to decline, innovations are adopted, the installed capacity increases and new desalination plants are brought online. Solar-assisted systems only comprise approximately 1% of current desalination capacity, illustrating the large opportunity for market penetration.

## ACKNOWLEDGEMENT & DISCLAIMER

The work of Adam Atia is supported by the National Science Foundation Graduate Research Fellowship Program under Grant No. DGE-11-44155. Additional support was provided by the DOE U.S.-Israel Desalination Design Challenge through a contract with ORNL. Vasilis Fthenakis serves at the board of the Global Clean Water Desalination Alliance and he acknowledges useful conversations with multiple members of the Alliance, especially Raed Bkayrat, First Solar, George Papadakis, Agricultural University Athens, and Alexander Ritschel, Masdar. Any opinions, findings, and conclusions or recommendations expressed in this material are those of the authors and do not necessarily reflect the views of the National Science Foundation or the Global Clean Water Desalination Alliance.

## REFERENCES

[1] Renewable Energy in the Water, Energy & Food Nexus [Internet]. International Renewable Energy Agency (IRENA); 2015. Available from: http://www.irena.org/DocumentDownloads/Publications/IRENA_Water_Energy_Food_Nexus_2015.pdf

[2] Global Clean Water Desalination Alliance (GCWDA) concept paper: http://www.clca.columbia.edu/Global-Water.pdf

[3] E. Delyannis, "Historic background of desalination and renewable energies," Solar Energy, 75(5), 357–366, 2003.

[4] Al-Karaghouli, A.A., and Kazmerski L.L., (2011). Renewable energy opportunities in water desalination, desalination, trends and technologies, Schorr M. (Ed.), ISBN: 978-953-307-311-8, InTech. http://cdn.intechopen.com/pdfs/13758/InTech-Renewable energy opportunities in water desalination.pdf

[5] KAUST Industry Collaboration Program (KICP) Strategic Study: Appraisal and Evaluation of Energy Utilization and Efficiency in Saudi Arabia: Supply and Demand Impacts, Business Opportunities and Technological and Economic Consideration, 2014.

[6] D. Manolakos, E.S. Mohamed, I. Karagiannis, G. Papadakis, Technical and economic comparison between PV-RO system and RO-Solar Rankine system. Case study: Thirasia island, Desalination, 221(1):37-46, 2008

[7] M. Elimelech and W. A. Phillip, "The future of seawater desalination: energy, technology, and the environment.," Science, 333 (6043), 712–717, 2011.

[8] GWI, "Desalination Markets 2016," Global Water Intelligence, Oxford, UK, 2016

[9] K. C. Ng, K. Thu, S. J. Oh, L. Ang, M. W. Shahzad, and A. B. Ismail, "Recent developments in thermally-driven seawater desalination: Energy efficiency improvement by hybridization of the MED and AD cycles," Desalination, 356, 255–270, 2015

[10] https://energy.gov/epsa/articles/ornl-winning-proposal-us-israel-desalination-challenge

[11] E. Sh. Mohamed, G. Papadakis, E. Mathioulakis, V. Belessiotis, a direct coupled photovoltaic seawater reverse osmosis desalination system toward battery based systems — a technical and economical experimental comparative study, Desalination 221, 17–22, 2008.

[12] Osmosun Mascara http://mascara-nt.fr/en/

[13] Fthenakis V.M. and Kim H.C., Life-cycle of water in U.S. electricity generation, Renewable and Sustainable Energy Reviews, 14, 2039–2048, 2010.

[14] Fthenakis V., Atia A., Morin O., Bkayrat R., and Sinha P., New Prospects for PV Powered Water Desalination Plants: A case study in Saudi Arabia, Progress in Photovoltaics: Research and Applications, 4, 543–550, 2016.

[15] Atia AA, Fthenakis V, Bkayrat R. Techno-economic evaluation of stand-alone, PV-powered, seawater desalination plants in Saudi Arabia. In: Proceedings of the 29th European Photovoltaic Solar Energy Conference. Amsterdam; 2014. pp. 4075–80.

[16] ARAMCO 2012. Rodriguez L.L., Desalination: Towards a Sustainable Source of Drinking water, EnviroNews, 20, Winter 2011/2012; http://www.saudiaramco.com/content/dam/Publications/Environews/Environews Winter 2011/Desalination.pdf

[17] Finan A., and Kazimi M., Potential Benefits of Innovative Desalination Technology Development in Kuwait, Kuwait Center for Natural Resources and the Environment, Massachusetts Institute of Technology, May 2013.

[18] EcoPeace Water and Energy Nexus draft report, 2017.

[19] Water desalination+reuse, June 2017 https://www.desalination.biz/news/0/Chiles-copper-mines-will-turn-to-desalination-for-half-their-water/8246/

[20] Water desalination+reuse magazine, pp. 16-20, June 2017

[21] https://www.desaldata.com/news/40795

# Global Residential Air-Conditioning Sector as a Driver for Photovoltaic Industry Growth during the 21st Century

Hannu S. Laine[1,2], Jyri Salpakari[3], Marius Peters[2], Erin E. Looney[2], Ashley E. Morishige[2], Hele Savin[1], Gregory Wilson[4], and Tonio Buonassisi[2]

[1]Department of Electronics and Nanoengineering, Aalto University, 02150 Espoo, Finland

[2]Massachusetts Institute of Technology, Cambridge, MA 02139, USA

[3]New Energy Technologies Group, Department of Applied Physics, Aalto University, 02150 Espoo, Finland

[4]National Center for Photovoltaics, National Renewable Energy Laboratory, Golden, CO 80401, USA

*Abstract* — The air-conditioning (AC) sector offers an interesting synergy for photovoltaic (PV) electricity generation, because the demand for AC correlates with ideal times and locations for PV production. The AC sector is also a significant electricity consumer, currently comprising ~3 % of global electricity consumption and it is expected to grow rapidly in the future due to income and population growth in sunny countries as well as global warming. Here, we assess the potential of the residential AC sector to sustain and accelerate PV industry growth during the 21st century. We show that the residential AC sector could sustain more installed PV generation capacity than the current global PV manufacturing capacity could produce. We highlight other possible synergistic electricity consumption sectors with significant growth expected in the future.

## I. INTRODUCTION

As the photovoltaics (PV) industry matures, there is a growing body of literature discussing synergistic electricity loads to mitigate the diurnal and seasonal variability of PV electricity production [1,2]. Electric vehicles are a common example due to their built-in electricity storage [3]. The air-conditioning (AC) and heating sectors present another such opportunity because they exhibit intrinsic thermal inertia, which can mitigate PV variability [4–7]. The AC sector presents a particularly interesting synergy with PV, because the main driver for both cooling demand and PV electricity production is solar insolation [8,9].

Furthermore, the energy demand of the AC sector is expected to grow significantly during the 21st century, and is predicted to surpass the energy demand of the heating sector [10,11]. The growing AC sector could facilitate significant additional integration of variable PV electricity generation into global electricity grids during this century. In this work, we estimate the potential of the global residential AC sector to sustain and accelerate PV industry growth during the 21st century.

## II. DRIVERS INCREASING AIR-CONDITIONING DEMAND

The main drivers increasing residential AC demand are wealth and population growth in sunny, developing countries, and global warming. As nations grow wealthier, a larger fraction of households purchase air-conditioning. The penetration of residential AC as a function of income follows a logistic formula, with modest growth at low income levels, or GDP/capita levels less than 10,000 $US adjusted for purchasing power parity (PPP). At modest income levels (10,000 – 30,000 $US GDP/cap.), AC penetration experiences exponential growth, saturating to 100 % with higher income levels [12]. The impact of local weather to AC electricity consumption is typically estimated in terms of cooling degree days (*CDD*):

$$\text{if } T_m(t) > T_{\text{base}}, \quad CDD(t) = T_m(t) - T_{\text{base}}$$
$$\text{else, } \quad CDD(t) = 0.$$

$T_{\text{base}}$ is typically set between 15 – 21°C [13–15], and it implicitly accounts for inadvertent heat sources within buildings, such as inhabitants or electrical appliances. For our analysis, we use 18°C [11]. Isaac and van Vuuren reviewed literature data of household AC electricity consumption and found that the average electricity consumption of households increases linearly with yearly *CDD* and logarithmically with household income (richer households have more space to cool) [11].

In addition to affecting the fraction of households purchasing air-conditioning, increased wealth has a secondary effect. As nations gain wealth, the average family size tends to decrease and hence, the number of households per capita increases [16], which increases the total electricity demand for AC. Lastly, over time AC devices grow more energy efficient, which slightly compensates for the increased cooling demand.

## III. SOCIOECONOMIC, CLIMATE CHANGE AND TECHNOLOGICAL SCENARIOS

Our analysis builds on the model of Isaac & van Vuuren [11], which quantifies the electricity consumption of the global residential AC sector accounting for the drivers described in section II. Here, we describe the socioeconomic, climate change and technological scenarios used in our analysis, which are the same as those used by Isaac & van Vuuren [11].

In the baseline scenario, population growth follows the medium variant of UN population prospects [17], implying a world population of 9.2 billion people in 2100, with growth

978-1-5090-5606-4/17 $31.00 © 2017 IEEE

being fastest in the beginning of the century and slowing down toward the end of the century (8.2 billion people in 2030 and 9.1 billion people in 2050). World GDP per capita increases from 10,962 in 2000 to 69,324 in 2100 ($US 2017 dollars adjusted for PPP), corresponding to a yearly compound average of 1.8 % [11]. Economic growth is relatively stronger in developing countries, such that income differences between developed and developing countries tend to decrease. Besides socioeconomic development, global warming is assumed to result in an average surface temperature increase of 3.7°C over pre-industrial temperatures [11]. Air-conditioning devices are assumed to continue their evolutionary development, increasing their average coefficient of performance from 2.4 in 2000 to 4.39 in 2100 [18].

## IV. POTENTIAL OF RESIDENTIAL AIR-CONDITIONING SECTOR TO SUSTAIN PHOTOVOLTAICS GENERATION

To estimate the full potential of the residential AC sector, we translate the yearly electricity demand of the AC sector into PV capacity by assuming a 15 % capacity factor, which has been estimated an appropriate global average [19]. We compare this potential installed PV capacity to the global PV manufacturing capacity at the end of 2016, which we estimate to be 75 $GW_{DC}$/year [20]. In other words, we project the future installed PV production capacity assuming that the 75 $GW_{DC}$/year manufacturing capacity is fully used every year, and no net manufacturing capacity is either lost or added. We estimate the installed capacity at the end of 2016 to be 280 GW [20] and assume that the installed PV systems are retired after an optimistic, but feasible 50-year lifetime [21]. We assume that the 280 GW installed at the end of 2016 is retired during the next 50 years starting from 2017 with equal fractions retired each year.

Figure 1 illustrates the projected yearly electricity consumption of the global residential AC sector (blue solid line), as well as the calculated PV capacity that the sector could sustain (orange solid line). Also shown is the potential installed PV capacity that could be reached with the 75 $GW_{DC}$/year manufacturing capacity (orange dashed line). Already in 2016, the projected electricity consumption of the AC sector (620 TWh / year) would correspond to a PV capacity of 450 GW, more than what was installed globally at the end of 2016.

Moving past 2016, the electricity demand of the residential AC sector grows further. Until year 2040, the sector could potentially sustain as much installed PV generation capacity as the current manufacturing capacity could produce. The growth of the AC sector is primarily fueled by developing regions of the world reaching income levels between 10,000 – 30,000 GDP/cap $US, which results in exponential growth in the penetration of AC within households in these regions.

After mid-century, the AC sector is projected to grow to sustain more PV than the current manufacturing capacity and produce. As the assumed 50-year lifetime of the installed PV

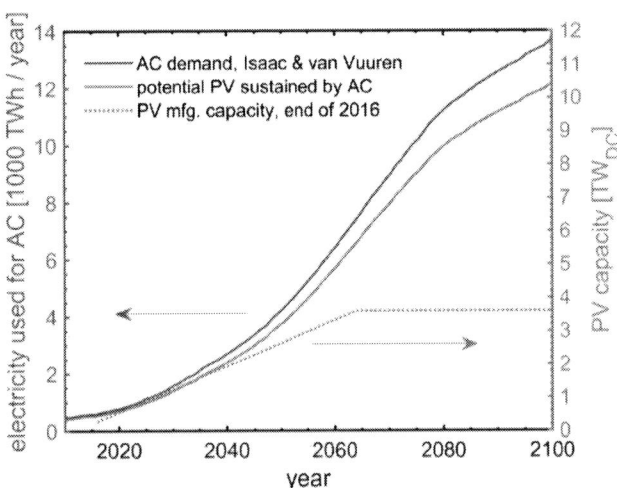

Figure 1. Projected electricity demand of the global residential AC sector, as calculated by Isaac and van Vuuren [11], as well as the potential PV capacity that the AC sector could sustain. The potential PV capacity is compared to the manufacturing capacity at the end of 2016.

systems is reached, retirement of old PV systems hinders further installed PV capacity net growth beyond 2067 with the installed capacity plateauing to 3.6 $TW_{DC}$. The AC sector, on the other hand, could sustain 10.4 $TW_{DC}$ of PV at the end of the century.

### A. Sensitivity Analysis of Chosen Scenario

The sensitivity of the projected PV capacity on the chosen socioeconomic, climate change and technological scenario is similar than the sensitivity of the projected AC electricity demand. Isaac and van Vuuren explored a wide range of alternative, but plausible socioceonomic, climate change and technological scenarios [11] and calculated their impact on the AC electricity consumption. They found that the scenarios predict uniform results until the year 2025. Beyond that, the predicted AC electricity usage, and hence the PV capacity the sector can sustain, can decrease by up to a factor of two, and increase by up to a factor of three depending on the set of scenarios chosen. In conclusion, the potential of the residential AC sector to sustain PV during the 21st century is vast, commensurate with at least approximately 1.5 times the total PV manufacturing capacity of today and at best up to 10 times the current manufacturing capacity.

## V. DISCUSSION: OTHER SECTORS THAN AIR-CONDITIONING CAN ALSO INCREASE PV ADOPTION

Although the trends and drivers of commercial AC sector are not as well quantified as those of the residential AC sector, currently the residential and commercial AC sectors consume approximately as much electricity [22]. Hence, the commercial AC sector could provide an equally significant driver for the global PV industry as the residential AC sector. Additionally, there are several other electricity-consuming sectors beyond AC where potential synergy with PV electricity production and significant growth are likely.

For example, to keep global warming below 2°C, net anthropogenic carbon emissions may need to be eliminated completely by the end of this century [23,24]. This would require a majority of the transportation sector to switch from carbon-based fossil fuels to carbon-free fuels, such as electric batteries or synthetic fuels. In 2012, the global transport sector consumed approximately 30,000 TWh of primary energy [25]. To produce a corresponding amount of electricity with PV power plants would require as much as 23 $TW_{DC}$ of PV generation capacity. Since both batteries and synthetic fuels can store PV-generated electricity, PV could provide a significant fraction of the energy consumed by the sector.

Water desalination is another interesting sector to facilitate PV industry growth as the highest demand for desalination is in hot and dry climates with high PV capacity factors. The capital intensity of water desalination plants tends to favor electricity sources with higher capacity factors than PV. However, if this condition can be relaxed, the desalinated water could be leveraged as energy storage, mitigating the variability of PV. The global water desalination capacity was approximately 87 million cubic meters per day at the end of year 2015, and has grown approximately 8 % per year for the last 40 years [26]. Assuming an average energy expenditure of 2 kWh per cubic meter of desalinated water [27], the electricity demand of the desalination sector could reach 1 $TW_{DC}$ PV capacity as soon as year 2029.

Examples of other similar sectors include raw material manufacturing plans like those for alumina or ammonia. Since these plants are able to rapidly ramp their production up and down, their electricity usage can match PV electricity generation. Envisioned future work includes performing similar analysis for all industries that exhibit growth potential for the PV sector. Adding these analyses to the AC work presented here would help quantify when, where and to what extent the global PV industry can grow.

## VI. SUMMARY

We performed a global assessment of increased PV generation potential enabled by the AC sector during the 21$^{st}$ century. Our results show that the potential synergy between AC and PV is notable, and could significantly accelerate the growth of the global PV industry. Already today, the residential AC sector consumes more electricity than all installed PV systems worldwide. The growth of the residential AC sector maintains its electricity consumption comparable to the installed global PV capacity, even if the current global PV manufacturing capacity was fully utilized. As recently-installed PV systems reach their end-of-life, retirements inhibit further PV generation capacity growth without added manufacturing capacity. When this happens, the AC sector could justify added

PV manufacturing capacity on its own. The AC sector could sustain 10.4 $TW_{DC}$ of PV at the end of the century, three times more than the current PV manufacturing capacity could produce.

## ACKNOWLEDGEMENTS

H. S. L. acknowledges the Fulbright Technology Industries of Finland grant as well as support from the Finnish Cultural Foundation and Tiina and Antti Herlin Foundation. E.E.L. acknowledges support by the NSF Graduate Research Fellowship under Grant No. 1122374.

## REFERENCES

[1] P. Denholm, K. Clark, and M. O'Connel, "On the path to SunShot: Emerging issues and challenges in integrating high levels of solar into the electrical generation and transmission system," National Renewable Energy Laboratory, Golden, CO, 2016.

[2] P. D. Lund, J. Lindgren, J. Mikkola, and J. Salpakari, "Review of energy system flexibility measures to enable high levels of variable renewable electricity," *Renewable and Sustainable Energy Reviews*, vol. 45, p. 785, 2015.

[3] International Energy Agency, "Global EV Outlook 2016," 2016.

[4] R. Yin, P. Xu, M. A. Piette, and S. Kiliccote, "Study on auto-DR and pre-cooling of commercial buildings with thermal mass in California," *Energy and Buildings*, vol. 42, p. 967, 2010.

[5] J. Salpakari, and P. D. Lund, "Optimal and rule-based control strategies for energy flexibility in buildings with PV", *Applied Energy*, vol. 161, 2016.

[6] J. Salpakari, T. Rasku, J. Lindgren, and P. D. Lund, "Flexibility of electric vehicles and space heating in net zero energy houses: an optimal control model with thermal dynamics and battery degradation," *Applied Energy*, vol. 190, p. 800, 2017.

[7] M. Waite, and V. Modi, "Potential for increased wind-generated electricity utilization using heat pumps in urban areas," *Applied Energy*, vol. 135, p. 634, 2014.

[8] E. Fahlén, H. Olsson, and N. Christensson, "Comfort cooling and solar power – a perfect match," REHVA Journal – November 2014.

[9] K. W. J. Barnham, M. Mazzer, and B. Clive, "Resolving the energy crisis: nuclear or photovoltaics?" *Nature Materials*, vol. 5, p. 161, 2006.

[10] D. Ürge-Vorsatz, L. F. Cabeza, S. Serrano, C. Barreneche, and K. Petrichenko, "Heating and cooling energy trends and drivers in buildings," *Renewable and Sustainable Energy Reviews*, vol. 41, p. 85, 2015.

[11] M. Isaac and D. P. van Vuuren, "Modeling global residential sector energy demand for heating and air conditioning in the context of climate change," vol. 37, p. 507, 2009.

[12] M. A. McNeil, and V. E. Letschert, "Future air conditioning energy consumption in developing countries and what can be done about it: the potential of efficiency in the residential sector", ECEE 2007 Summer Study, 2007.

[13] D. J. Sailor, "Relating residential and commercial sector electricity loads to climate – evaluating state level sensitivities and vulnerabilities," *Energy*, vol. 26, p. 645, 2001.

[14] C. Giannakopoulos, and B. E. Psiloglu, "Trends in energy load demand for Athens, Greece: weather and non-weather related factors," *Climate Research*, vol. 31, p. 97, 2006.

[15] E. Valor, V. Meneu, and V. Caselles, "Daily air temperature and electricity load in Spain," *Journal of Applied Meteorology*, p. 1413, 2001.

[16] UN-Habitat, "Financing urban shelter: global report on human settlements, 2005," Earthscan, London, 2005.

[17] United Nations, Department of Economic and Social Affairs, Population Division, "World population prospects: the 2015 revision," 2015.

[18] F. Rong, L. Clarke, and S. Smith, "Climate change and the long-term evolution of the US buildings sector," Pacific Northwest National Laboratory, Richland, USA, 2007.

[19] J. Jean, P. R. Brown, R. L. Jaffe, T. Buonassisi, and V. Bulović, "Pathways for solar photovoltaics," *Energy & Environmental Science*, vol. 8, p. 1200, 2015.

[20] N. M. Haegel, R. Margolis, T. Buonassisi, D. Feldman, A. Froitzheim, R. Garabedian, M. Green, S. Glunz, H.-M. Henning, B. Holder, I. Kaizuka, B. Kroposki, K. Matsubara, S. Niki, K. Sakurai, R. A. Schindler, W. Tumas, E. R. Weber, G. Wilson, M. Woodhouse, and S. Kurtz, "Terawatt-scale photovoltaics: Trajectories and challenges," *Science*, vol. 356, p. 141, 2017.

[21] M. Woodhouse, R. Jones-Albertus, D. Feldman, R. Fu, K. Horowitz, D. Chung, D. Jordan, and S. Kurtz, "On the path to SunShot: The role of advancements in solar photovoltaic efficiency, reliability, and costs," National Renewable Energy Laboratory, Golden, CO, 2016.

[22] M. Waite, E. Cohen, H. Torbey, M. Piccirilli, Y. Tian, and V. Modi, "Global trends in urban electricity demands for cooling and heating," *Energy*, vol. 127, p. 786, 2017.

[23] G. P. Peters, R. M. Andrew, T. Boden, J. G. Canadell, P. Ciais, C. L. Quéré, G. Marland, M. R. Raupach, and C. Wilson, "The challenge to keep global warming below 2°C," *Nature Climate Change*, vol. 3, p. 4, 2013.

[24] J. Rogelj, W. Hare, J. Lowe, D. P. van Vuuren, K. Riahi, B. Matthews, T. Hanaoka, K. Jiang, and M. Meinshausen, "Emission pathways consistent with a 2°C global temperature limit," *Nature Climate Change,* vol. 1, 2011.

[25] Energy Information Agency, International Energy Outlook, 2016.

[26] Global Water Intelligence, "IDA desalination yearbook 2015-2016," 2016.

[27] M. Elimelech, and W. A. Phillip, "The future of seawater desalination: Energy, Technology, and the Environment," *Science,* vol. 333, p. 712, 2011.

# Measures to remove economic non-market failure and institutional barriers that restrict photovoltaics self-consumption and net-metering in Spain

Enrique Rosales-Asensio[1], Juan A. Méndez[2], Benjamín González-Díaz[3] and Ricardo Guerrero Lemus[1]

[1]Departamento de Física. Universidad de La Laguna. Avenida Astrofísico Francisco Sánchez S/N 38206, S/C de Tenerife. Spain

[2]Departamento de Ingeniería Informática y Sistemas. Universidad de La Laguna. Avenida Astrofísico Francisco Sánchez S/N 38206, S/C de Tenerife. Spain

[3]Departamento de Ingeniería Industrial. Universidad de La Laguna. Avenida Astrofísico Francisco Sánchez S/N 38206, S/C de Tenerife. Spain

*Abstract* — **This research focuses on strategies to accelerate high solar photovoltaic and self-consumption growth rates in Spain by proposing the needed policy measures for eliminating/mitigating current barriers to their deployment. Also a brief discussion about the urgent need to modernize the regulatory framework envisioning the near future and, thus, proposing measures to promote the active role of distributed PV in the voltage and frequency regulation of distributed power grids are included.**

*Keywords:* Energy barrier, solar power generation, energy efficiency, distributed power generation, smart grids

## I. INTRODUCTION

The countries that participated in the United Nations Framework Convention on Climate Change, in the Agreement of Paris on 12 December 2015, recognized the seriousness of the climate change problem and agreed to take measures to cope with it so that the increase in global average temperature remains well below 2° C with respect to pre-industrial levels [1]. The Intergovernmental Panel on Climate Change in its Fifth Assessment Report, published in 2014, identifies the generation of electricity as one of the main causes of the increase in the global emissions of greenhouse gasses [2].

Likewise, Article 45 of the Spanish Constitution recognizes the right to enjoy an adequate environment for the development of the person and the duty to preserve it; and imposes a mandate on public authorities to ensure the rational use of natural resources in order to protect and improve the quality of life and defend and restore the environment [3].

Renewable electricity self-consumption is one of the most appropriate instruments to reduce the environmental impact of electricity generation [4], being expected that, in the medium term (2030), it would not imply a higher levelized cost of electricity that of the electricity mix (fossil and nuclear energy)

[5]. The recent outstanding technological developments [6] linked to the high radiation levels in Spain allow installation of technologies such as solar photovoltaics, which, even at the moment, are monetized directly by savings in the electricity bill, without the need for any financial aid (Fig. 1) [7,8]. In addition, energy self-consumption favors the reduction of imported fossil fuels, allowing a match in trade balance [9] and strengthening the energy independence of Spain.

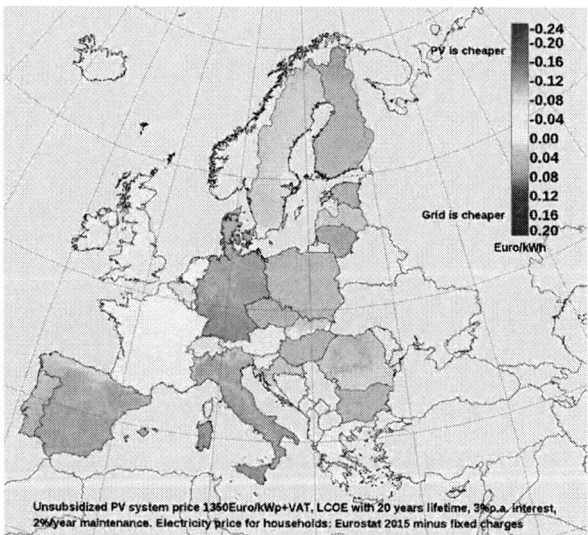

Fig. 1: Comparison of European residential electricity prices with electricity generated by a PV solar system. Source: [7]

change of energy model towards a decentralized energy production [10] since it has many advantages such as greater security of local energy supply, a shorter distance from the generation source to consumer that implies lower losses, and the promotion of the development and cohesion of the local community by providing sources of income and creating employment at the local level [11].

In spite of the aforementioned advantages and to the fact that Spain drove the global solar photovoltaic market in 2008, the truth is that, as the Renewable Energy Policy Network for the 21st Century points out, Spain has "virtually disappeared" from the solar PV picture due to retroactive policy changes and a new tax on self-consumption [12]. As (Fig. 2) shows additions to solar PV capacity for the year 2015 was by far the lowest of top-10 countries [12], which is a consequence of the energy policy taken by the Spanish Government.

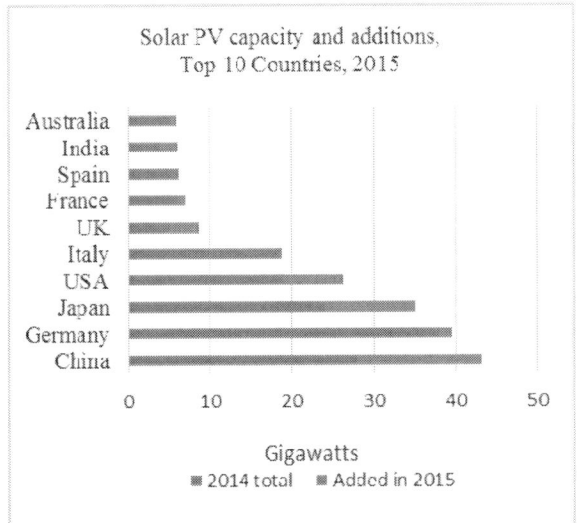

Fig. 2. Solar PV capacity and additions, Top 10 Countries, 2015.
Source: Adapted from [12]

This research will answer to the following questions:
a)  Are there any existing obstacles to the development of self-consumption in the electricity sector and, therefore, in the Spanish photovoltaic sector?
b)  Which measures could be undertaken for accelerating a high PV growth rate and a rapid reduction cost?
c)  Is it suitable to recognize the right to self-consumption without the applicability of any charges?
d)  Is it justifiable that several consumers' would share a self-consumption facility?
e)  Should be legalized, according to the corresponding technical regulations, self-consumption facilities that do not inject power into the electricity grid?
f)  Is it necessary to adapt the sanctioning regime concerning self-consumption?

The remainder of this paper is structured as follows. Section II describes methods to allow the work to be reproduced. In Section III, the economic non-market failure and institutional barriers that hamper the use of solar photovoltaic technology and self-consumption in Spain, as well as actions aimed to remove the aforementioned barriers, will be presented.

Finally, in Section IV, a discussion about the significance of the results of the work and the implications for energy policy resulting from the actions taken to face the identified barriers is presented.

## II. MATERIAL AND METHODS

As to the nature of energy barriers concerns, its classification does not reveal "substantially" anything new [13]. In fact, as Weber noted, barriers are invisible, so despite being real, they cannot be empirically classified [14], given that the various existing classifications in the literature derived from the theory are diverse and relatively unstructured [15,16]. The taxonomy adopted in this paper is an adaptation of the one proposed by Chai [13], grouping barriers into the following categories: behavioral, market failures, physical constraints, institutional, and economic non-market failure barriers—the latter two being the only barriers addressed in this paper.

In order to identify the barriers to the implementation of solar photovoltaic technology and self-consumption in Spain and due to their nature, an extensive review of (updated) academic and grey literature on energy barriers was first conducted, using the preliminary results to proceed with contacting experts in this field. From their experiences, information was obtained on what the barriers are, how they can be eliminated/mitigated and if current measures proposed by the Spanish Government are adequate. The barriers identified by Chai [13], Sorrell [17], and Brown [18], as well as the six questions asked at the end of the Introduction section, were transferred to 30 experts on solar photovoltaic technology and self-consumption from Spain; who contributed either through interviews, e-mail correspondence, and/or surveys (where the experts identified the barriers and the relative importance of each of them) to give validity and reliability to the literature.

## III. RESULTS AND DISCUSSION

It is necessary to promote and support a legal reform that would eliminate the mainly existing obstacles for the development of self-consumption in the Spanish electricity sector. It is precisely in the elimination or, at least, in the mitigation of those barriers that this research focuses, presenting the necessary measures for accelerating a high PV growth rate and a rapid cost reduction critical to the success of the PV deployment in Spain. As a way of demonstrating the effect of the current regulatory framework in Spain on the development of the photovoltaic sector, in Appendix A (Table I), information on installed power, electricity production, number of installations, average remuneration per kWh generated, and total support provided is shown.

This paper proposes an amendment of Article 9 of Law 24/2013 of the electricity sector; the definition of the true cost of power grid backup to self-consumers and net-meterers (substituting the first Transitional Provision of the Royal Decree 900/2015); and a repealing of certain provisions of the Royal Decree 900/2015 of 9 October, which regulates the administrative, technical, and economic conditions of the modalities of electric power supply with self-consumption and production with self-consumption.

In particular, this paper proposes seven main changes in the regulation of self-consumption.

- First of all, this paper proposes that the right of self-consumption and net metering of electricity be recognized, thus avoiding specific taxes for electricity self-consumed. As it has been mentioned above, from a study carried out by the International Energy Agency for 18 OECD countries (plus China and Brazil) [19], it was verified that Spain was the "only example" of a specific tax for self-consumers.

This treatment (successfully implemented in countries such as Germany, Denmark, Japan, Israel, and Mexico [19]) significantly simplifies self-consumption modalities to date.

- Second, this paper proposes the possibility of several consumers' sharing a self-consumption facility.
- Third, the Ministry of Industry must establish a clear methodology that defines how to calculate the charges that would have to be paid by all users in general and self-consumers in particular.
- Fourth, this paper proposes that the processing of facilities in self-consumption with zero injection to the distribution network (type 1) should return to the situation before the Royal Decree 900/2015's entry into force with regards to self-consumption. This Royal Decree requires that all installations up to 100 kW are processed according to Royal Decree 1699/2011. A point of connection to the utility is requested, and an access contract to the electricity grid is signed regardless of whether they do not inject power into the grid [20].

This simplifies and lowers the management of the photovoltaic system to the authorities in a way that makes it equivalent to a diesel generator. By using dynamic power control, the maximum active power generated by the inverter is controlled and the PV power generated is only used for self-consumption [21,22], so it is not a technical issue. For explanatory purposes, a schematic representation of instantaneous consumption is shown in (Fig. 3).

- Fifth, it is proposed that a combination of Feed-In and Net Metering should be implemented as a support scheme to enable the prosumers to compensate their electric consumptions and receive an economic compensation for the surplus electric energy that they inject into the electric network, also opening the alternative of "credits" per kWh injected to the grid.

It is true that the combination of the Feed-In Tariff and the Feed-In Premium, as a means of retribution, is the basis of the development of the photovoltaic sector in leading countries like Germany. However, it is also a fact that these compensation schemes can have a negative impact on the reliability of the electricity network (since they imply a guaranteed connection to the same, regardless of where the generators are located), and they also implicitly represent a distortion of electricity prices in the wholesale market [23,24].

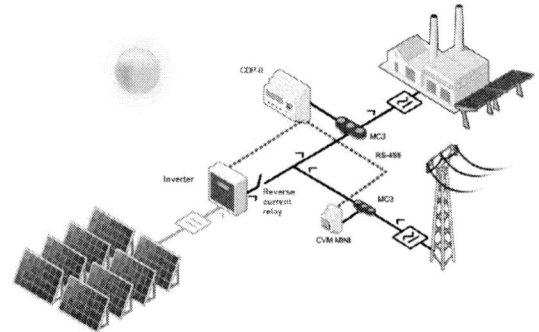

Fig. 3: Instantaneous self-consumption scheme with zero injection. Source: [22]

On the other hand, the "pure" Net Metering, despite being successful in the development of specific technologies with a limited implementation, presents the problem of long-term remuneration, so it represents a barrier [25].

Based on evaluations, made specifically for Spain, that have been shown as "a viable option for PV development" [26], a combination of Net Metering and Feed-In Tariff is adequate to solve each problem separately at the time that a development of solar photovoltaic technology and self-consumption is achieved in Spain. As a consequence, this combined policy mechanism is the one proposed in this paper.

- Sixth, Royal Decree 900/2015 overlooks smart grids and self-consumption, potential ancillary services to stabilize the grid, not remunerating services, and not allowing small players to participate in the balancing market.

Recognition of the value of the services offered by storage systems is central to creating the business case for storage and will be proposed in this research, including rewards for grid services and overall capacity of energy storage to stabilize quality and supply for renewables generation.

- Finally, this research proposes to adapt the sanctioning regime concerning self-consumption to the true impact of the same in the electricity sector.

The government has included in its Royal Decree of Self-Consumption (RD 900/2015) fines of up to 60 million euros for self-consumers who fail to comply with the Royal Decree. If this amount is compared with the 30 million euros maximum set for the abandonment or release of radioactive materials [27], it can be understood how disproportionate this Royal Decree is.

## IV. CONCLUSIONS

Due to its regulatory indefiniteness, which implies that, *de facto*, Royal Decree 900/2015 is currently under development, current legislation is a burden on the development of self-consumption in Spain. In this sense, it restricts any renewable facility that intends to develop in this perspective. Irrespective of the existence of the Administrative Register of Electric Power Consumption, the spirit of the current legislation affects the development of the sector in Spain. This research proposes a stable legal framework design, which contemplates the implementation of distributed generation systems as envisioned by the Electricity Sector Law in line with the provisions of Directive 2009/28/EC, without neglecting the net-metering, to allow for the management of electricity self-consumption systems. If the recommendations set out in this paper were taken into account, their energy policy implications were that both generation and distributed storage would be encouraged, as well as a self-consumption made preferably from renewable and manageable sources. Furthermore, the resulting regulation would be much easier to implement for consumers and to supervise for the network operators and competent administrations, which strengthens the importance of the measures proposed throughout this research and suggests the necessity to update the Royal Decree on energy self-consumption if it is desired that self-consumption in general, and solar photovoltaic technology, in particular, would have a similar deployment in Spain as in other solar PV world-leading countries.

## ACKNOWLEDGEMENTS

The authors gratefully acknowledge the support of the Council of Tenerife through its Agustín de Betancourt research program, without which the present study could not have been completed. This work has been supported by the Ministerio de Ciencia e Innovación, Spain (Project ENE2013-41925R), co-supported by the European Social Fund.

## REFERENCES

[1] NRDC, The Paris Agreement on Climate Change. New York: Natural Resources Defense Council, 2015.

[2] International Panel on Climate Change, Climate Change 2014 Mitigation of Climate Change: Working Group III Contribution to the Fifth Assessment Report of the Intergovernmental Panel on Climate Change. New York: Cambridge University Press, 2014.

[3] Cortes Generales Españolas, The Spanish Constitution. Madrid: Agencia Estatal Boletín Oficial del Estado, 1978.

[4] M. Germani, D. Landi, and M. Rossi, "Efficiency and Environmental Analysis of a System for Renewable Electricity

Generation and Electrochemical Storage of Residential Buildings", Procedia CIRP, vol. 29, pp. 839–844, 2015.

[5] C. Kost, T. Schlegl, J. Thomsen, S. Nold, and J. Mayer, Levelized cost of electricity renewable energies, Freiburg: ISE Fraunhofer, 2012.

[6] Ernst & Young, Capturing the sun: The economics of solar investment. London: Ernst & Young, 2016.

[7] A. Jäger-Waldau, Costs and economics of electricity from Residential PV systems in Europe. London: European Energy Innovation. Available: http://www.europeanenergyinnovation.eu /Articles/Winter-2016/Costs-and-Economics-of-Electricity-from-Residential-PV-Systems-in-Europe [January 22, 2017].

[8] G. Di Francia, "On the Cost of Photovoltaic Electricity for Small Residential Plants in the European Union", International Journal of Renewable Energy Research, vol. 4(3), pp. 610–617, 2014.

[9] APPA, Study of the Macroeconomic Impact of Renewable Energies in Spain. Madrid: Spanish Association of Renewable Energies, 2015.

[10] European Commission, DIRECTIVE 2009/28/EC OF THE EUROPEAN PARLIAMENT AND OF THE COUNCIL of 23 April 2009 on the promotion of the use of energy from renewable sources and amending and subsequently repealing Directives 2001/77/EC and 2003/30/EC. Luxembourg: Official Journal of the European Union, 2009.

[11] IEA, Energy Technology Perspectives 2016: Towards Sustainable Urban Energy Systems. Paris: International Energy Agency, 2016.

[12] REN21, Renewables 2016: Global status report. Paris: Renewable Energy Policy Network for the 21st Century.

[13] K.H. Chai and C. Yeo, "Overcoming energy efficiency barriers through systems–a conceptual framework", Energy Policy, vol. 46, pp. 460–472, 2012.

[14] L. Weber, "Some reflections on barriers to the efficient use of energy", Energy Policy, vol. 25(10), pp. 833–835, 1997.

[15] C. Dunstan and J. Daly, Institutional barriers to intelligent grid: a discussion paper, Sydney: University of Technology Sydney, 2008.

[16] S. Sorrell and A. Mallett, S. Nye, Barriers to industrial energy efficiency: a literature review, Vienna: United Nations Industrial Development Organization, 2011.

[17] S. Sorrell, J. Schleich, S. Scott, E. O'Malley, F. Trace, U. Boede, et al., Barriers to energy efficiency in public and private organisations, Brighton: Science Policy Research Unit – University of Sussex, 2000.

[18] M.A. Brown, "Market failures and barriers as a basis for clean energy policies", Energy Policy vol. 29(14), pp. 1197-1207, 2001.

[19] G. Masson, J.I. Briano, and M.J. Báez, IEA PVPS—Review and analysis of PV self-consumption policies. Paris: International Energy Agency, 2016.

[20] CIRCUTOR, Interpretación del RD900/2015 de Regulación de las condiciones administrativas, técnicas y económicas de las modalidades de suministro de energía eléctrica con autoconsumo y de producción con autoconsumo. Barcelona: Circutor, 2015.

[21] SMA, Solar Power: 100% Self-Consumption Solution with the SMA Power Control Module SUNNY BOY / SUNNY TRIPOWER. SMA: Niestetal, 2014.

[22] CIRCUTOR, Instantaneous self-consumption with zero injection. Barcelona: Circutor. Available: http://circutor.com/en/training/renewable-energies-self-consumption/instantaneous-self-consumption-with-zero-injection [March 15, 2017].

[23] J.A. Lesser and X. Su, X, "Design of an Economically Efficient Feed-In Tariff Structure for Renewable Energy Development", Energy Policy, vol. 36(3), pp. 981–990, 2008.

[24] P. Menanteau, D. Finon, and M. Lamy, "Prices versus quantities: choosing policies for promoting the development of renewable energy" Energy Policy, vol. 31(8), pp. 799–812, 2003.

[25] EPIA, Self consumption of PV electricity, Brussels: European Photovoltaic Industry Association, 2013.

[26] F.J. Ramírez, A. Honrubia-Escribano, E. Gómez-Lázaro, and D.T. Pham, "Combining feed-in tariffs and net-metering schemes to balance development in adoption of photovoltaic energy: Comparative economic assessment and policy implications for European countries" Energy Policy, vol 102, pp. 440-452, 2017.

[27] El Confidencial, El decreto de autoconsumo permitirá que entren en tu casa sin orden judicial. Available: http://www.elconfidencial.com/tecnologia/2015-05-14/autoconsumo-electrico-energia-legislacion-espana_793994/ [February 18, 2017].

[28] CNE, Información mensual de estadística sobre las ventas de régimen especial. Madrid: Comisión Nacional de la Energía, 2011.

[29] CNE, Información mensual de estadística sobre las ventas de régimen especial. Madrid: Comisión Nacional de la Energía, 2017. Available: https://www.cnmc.es/estadistica/informacion-mensual-de-estadistica-sobre-las-ventas-de-regimen-especial-contiene-22 [March 20, 2017].

APPENDIX A.

TABLE I
KEY DATA REGARDING SPANISH SOLAR
PV SECTOR. Source: Adapted from [28,29]

|  | Installed capacity (MW) | Electricity generation (GWh) | Number of installations | Total support (M€) | Average support (cent€/kWh) |
|---|---|---|---|---|---|
| 2016 | 4675 | 5794 | 61 404 | 2764 | 31.8 |
| 2015 | 4663 | 8211 | 61 338 | 2863 | 34.9 |
| 2014 | 4646 | 8170 | 61 096 | 2805 | 34.3 |
| 2013 | 4637 | 8261 | 60 984 | 3265 | 39.5 |
| 2012 | 4510 | 7994 | 59 883 | 2855 | 35.7 |
| 2011 | 4247 | 7248 | 57 710 | 2665 | 35.9 |
| 2010 | 3839 | 6400 | 54 920 | 2897 | 45.2 |
| 2009 | 3630 | 6073 | 52 100 | 2868 | 46.2 |
| 2008 | 3463 | 2503 | 51 310 | 1155 | 45.3 |
| 2007 | 690 | 473 | 20 284 | 215 | 43.3 |
| 2006 | 146 | 99 | 9874 | 45 | 42.7 |
| 2005 | 47 | 38 | 5391 | 16 | 39.9 |

# Cost Competitive Concentrator Photovoltaics for Solar Thermal Applications

Brian C. Riggs[1], Richard E. Biedenharn[1], Chris Dougher[2], Yaping Vera Ji[1], Qi Xu[1],
Vince Romanin[3], Daniel S. Codd[4], James M. Zahler[5], Matthew D. Escarra[1]

[1]Tulane University, Department of Physics and Engineering Physics, New Orleans, LA, USA
[2]Duke University, Durham, NC, USA
[3]Otherlab, San Francisco, CA, USA
[4]University of San Diego, San Diego, CA, USA
[5]Department of Energy, Washington, D.C. USA

*Abstract* — Concentrator photovoltaic (CPV) systems have been unable to keep up with the plummeting cost of flat plate PV for large scale power generation. New markets must be explored in order to maintain a robust CPV industry. Solar thermal offers a large potential market but has issues competing with conventional, cheap fuels such as natural gas. Hybrid CPV/T systems are capable of producing heat at prices competitive with natural gas when the electricity produced by high efficiency CPV units is used to subsidize the cost of the thermal energy delivered.

*Index Terms* – cogeneration, hybrid solar, PV/T, process heat, techno-economic analysis

## I. INTRODUCTION

Despite the focus on renewable generation of electrical power, thermal energy use is nearly double that of electricity[1]. Almost all heating processes on the commercial and industrial scale are fueled through the burning of natural gas. Due to the high conversion efficiency and relatively low price due to new found shale reserves, it has been difficult for renewable heat to compete with conventional generation[2].

Solar thermal systems use a high absorptivity thermal receiver that is capable of reaching high solar efficiencies (>75%) and achieving temperatures up to 550°C[3]. The systems have been studied for use in desalinization, space heating, A/C dehumidification, sterilization, pharmaceuticals, power generation, high temperature metal and oil refining, and much more. The wide range of market applications offers a sizable market for future renewable technology.

Over the last 30 years, there has been interest in combined electricity and thermal generation in the form of photovoltaic/thermal (PV/T) systems[4]. The most prevalent form of PV/T is using a thermal collector to manage the waste heat generated by the inefficiency of the solar cells. However new concepts are being developed that increase both electrical and thermal efficiency while reducing cost [5], [6]. These technologies utilize high efficiency CPV cells that split the spectrum between the electrically valuable UV and visible wavelengths and the thermally useful IR. This method reduces the costs associated with cooling the cells as well as a number of high cost features typically associated with CPV systems,

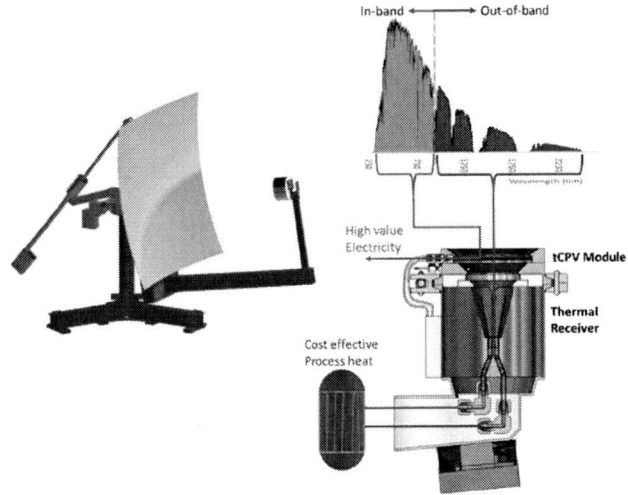

Fig. 1: (left) System CAD overview (right) Schematic of energy flow through the hybrid tCPV/T receiver

allowing components such as the tracker, concentrating optics, installation, and land usage to be shared between the CPV and thermal outputs.

By adding a low cost CPV module to a solar thermal system, high value electricity is used to create significant cost savings to the installation site. This can be reflected in the lifetime cost of energy for the system. Pure solar thermal systems may struggle to compete with natural gas, but the additional cost savings provided by the simultaneous generation of electricity reduce the cost of heat to competitive levels. Herein we describe the techno-economic modeling which demonstrates the cost savings enabled by CPV/T systems, using a dish based transmissive CPV/T system as the primary example.[6]

## II. SAMPLE SYSTEM

Fig. 1 (left) shows an overview of the tCPV/T system which employs a two axis tracker, 2.72m² square paraboloidal dish, and a hybrid CPV/Thermal receiver. The hybrid receiver (Fig. 1 right) consists of a transmissive CPV (tCPV) module which absorbs UV and visible light, converting it into electricity, and

978-1-5090-5606-4/17 $31.00 © 2017 IEEE

transmits IR wavelengths with 61% transmissivity. The cells are III-V materials lattice-matched to GaAs with an in-band (above GaAs bandgap) efficiency of 50% at 500 suns (full spectrum 32%).[7] The cells are mounted on a transparent window and utilize a bifacial grid to minimize shadowing. Active cooling channels formed from PDMS run under the cells in order to maintain the cell temperature <110°C. The cells cover 30% of the area of the module window and receive 48% of the incident energy. The thermal receiver, mounted behind the module, consists of an Inconel conical cavity receiver coated with Pyromark 2500 absorbing paint. Including thermal losses, the receiver is 95% efficient. Taking into account electrical resistance and conversion losses, the electrical output makes up 10.5% of the incoming light. The thermal receiver outputs 49% of the total sunlight as heat. This results in a combined 59.5% energy efficiency.

## III. MODEL

The techno-economic model uses an array of technical, financial, and geographic parameters to calculate the levelized cost of energy (LCOEn), purchase payback time, and energy balance that can be applied to several different PV/T system configurations. Fig. 2 is a flow chart of the techno-economic model. Using technical inputs such as the optical properties of the dish, CPV module, and solar cells, an energy balance is conducted to account for reflection losses, thermal losses, waste heat generated in the CPV, and electrical and thermal generation. This is used to calculate the total system electrical and thermal efficiency. A CPV and thermal system degradation rate is used to account for a decrease in performance over time. Combined with the annual average DNI for a location (base case: San Diego, 6.3 kWh/m²-day) the annual electricity and heat generation is calculated. Financial parameters including the cost of the system, loan term and rates, effective tax rate, and operation and maintenance costs are used to calculate the annual costs of the system. Where this model differs from standard LCOEn calculations is in the handling of electricity. Here, it is assumed that the electricity produced by the tCPV/T system offsets the customer's electrical power consumption, lowering the overall utility expenses. The local electricity price (base case: San Diego commercial rate of 14 cent/kWh) combined with the annual electricity generation is subtracted from the annual costs. These savings are taken into account for this model. The annual costs and thermal generation are calculated for the lifetime of the product (25 years) and discounted at a rate of 8%/year to account for the devaluing of future money. Thermal and electrical production are reduced every year to account for the decline in performance over time. The final LCOEn is calculated using Eq. (1). This equation can be modified to independently calculate the levelized cost of electricity (LCOE), levelized cost of heat (LCOH), or the electricity subsidized levelized cost of heat (sLCOH).

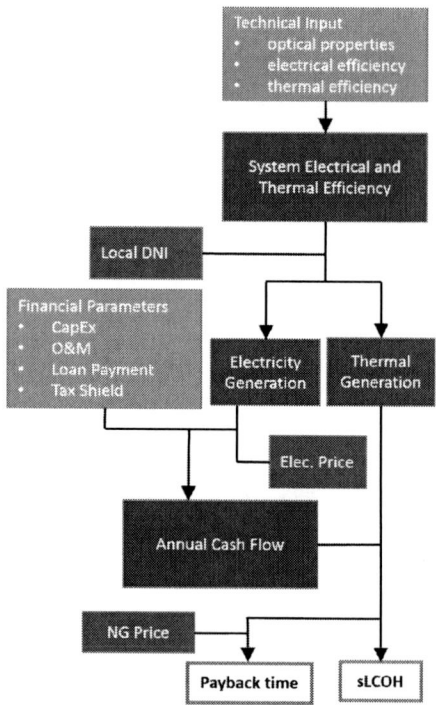

Fig. 2: Flow chart of techno-economic model.

$$LCOEn = \frac{\sum Cash\_Flow}{\sum Energy\_generation} \qquad (1)$$

### A. Financial Assumptions

The model utilizes a number of standard financial assumptions and applies them to all calculations presented herein. An estimated effective tax rate of 27% is used for the installation site. The loan amount is 50% of the total system cost with a fixed interest rate of 7% over the loan period (25 year lifetime of system). As mentioned the electricity savings generated by the PV array is considered a negative expense and is considered tax exempt. EPC (engineering, procurement, and construction) costs are added to the direct hardware costs in the form of a 61% markup which includes permitting, land purchase, and construction charges. This was calculated based on the NREL SAM parabolic trough concentrator (PTC) cost dataset.[8]

### B. Technical Assumptions

The model makes several operational assumptions to predict the energy balance and final energy generation of the system. First the model assumes that the system is sized to never over-generate for the installation site. This eliminates net metering and allows the electricity generation to simply be considered as a cost savings. Adding net metering or other state level regulations would allow for a more accurate representation and will be considered on a case-by-case basis in future work. The second assumption is that the cell temperature and efficiency remain constant with varying DNI. This is made to simplify the energy balance, based on the expectation that cooling systems are able to approximately compensate for changes in weather.

## IV. RESULT AND DISCUSSION

### A. Geographic Analysis

The model was applied to all 50 states in the U.S. using average local commercial electricity and natural gas rates.[9] Average state DNI was approximated based on NREL Dynamic map GIS data. Fig. 3 shows a map of the US highlighting states where the calculated sLCOH is less than that of local natural gas prices. States in orange are where a concentrator solar thermal (CST) only (no CPV module on the receiver) version of the described system is competitive with natural gas. The green states are where a hybrid tCPV/T system is more competitive than a pure CST system and natural gas. This is attributed primarily to high electricity prices in the green states. For example, CA and NV, which have similar DNI, have average commercial prices of 14 ¢/kWh and 8 ¢/kWh respectively making the CPV/T system competitive in CA while not in NV.

Inclusion of technical aspects in the model allows the exploration of design space that could enable increased competitive market size. For example, shifting the energy distribution within the receiver towards electricity production can skew towards greater electricity generation and therefore savings. This can be done by increasing the number of cells within the CPV module, allowing more incident light to pass through the CPV cells. Shifting to 60% cell energy coverage increases the competitive market area to include AZ and NM. This can, of course, be taken a step further to determine the cell area coverage needed in each state to become competitive with CST and natural gas. This analysis shows that up to 12 states, including those with low DNI such as NH, MA, CT, and RI, are potential commercial markets with only minor shifts in design required. While this particular design change does

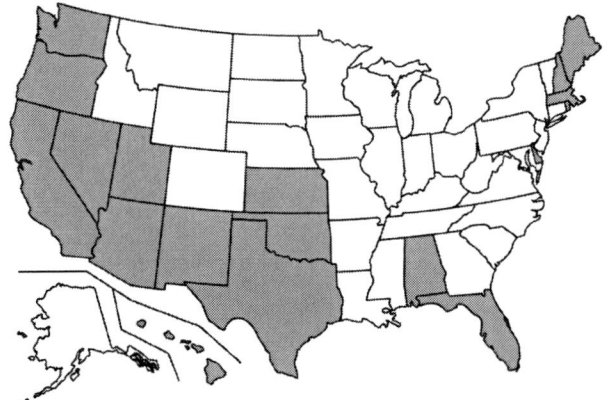

Fig. 3: Map of U.S. showing competitive states (relative to natural gas prices) for: (orange) solar thermal only and (green) a hybrid tCPV/T system for cell area coverage of 40%.

reduce the total amount of thermal energy captured, it also reduces the thermal energy price between 10-80% compared to natural gas, opening up valuable market potential. The payback time of a new system is an important input for end users in the purchasing decision. In this study, the payback time was calculated using the total loan amount, the annual expenses (O&M), savings (from electricity), and cost of fuel being displaced. It was found that the payback time ranged between 7-15 years depending on DNI and cell coverage (based on current natural gas prices). Most consumers aim for payback times <10 years to see it as a worthwhile investment.

Natural gas and electricity prices vary between commercial, residential, and industrial sectors, changing the sLCOH for each customer. sLCOH analysis of each sector shows that the this technology would be overwhelmingly competitive in residential markets providing an average of 30% cost reduction compared to local natural gas prices in 15 states. In comparison, only HI looks to be a viable market for industrial applications. This is attributed to the cost of electricity and gas being 50% and 66% less for the industrial sector compared to residential consumers. Because of this price differential, combined with safety factors surrounding concentrator designs, the system is targeted towards commercial and remote industrial applications which have a sizeable market and facility requirements that fit the proposed design, such as campus complex heating, district heating, and remote mining.

This methodology can be expanded to international markets as well. Table 1 includes select nations which have DNI, electricity and fuel prices that would be enabling for a hybrid CPV/T system. The trend of high DNI and high electricity rate follows as seen in the U.S. analysis. Of particular interest are island nations or remote access areas such as the Caribbean, Australia, or particular locations in Chile. Remote areas typically have reduced access to natural gas and are more likely to utilize a high cost substitute such as propane. Combined with higher electricity rates, the commercial viability of CPV/T systems grows substantially in these regions for both commercial and industrial markets. Australia is of particular interest as the model shows that the sLCOH should be negative. This correlates to a return on investment; the cost savings associated with electricity production are greater than the cost of the system over the course of its lifetime.

### B. Co-location Comparison

One alternative approach to hybrid systems is co-located installation of conventional 1-axis tracked PV and parabolic trough concentrators (PTC) to provide electricity and thermal energy independently. Hybrid systems are able to share a number of high cost components and a number of indirect costs, helping to keep the cost of energy low relative to two co-located systems. Table 2 shows a comparison between the hybrid and co-located system. LCOE is the levelized cost of electricity considering only PV or CPV cost and electricity

978-1-5090-5606-4/17 $31.00 © 2017 IEEE

generation. LCOH is the levelized cost of heat only considering the thermal system cost and heat generation. sLCOH is the subsidized levelized cost of heat where the value of the electricity produced by the PV component is used to offset the overall cost of heat. The co-located efficiencies take into account total solar energy collected by the entire installation (PV+PTC) in order to be comparable to the hybrid system. The individual efficiencies, 20.6% and 75% for PV and PTC thermal respectively, are each divided in half to account for efficiency of conversion for the total aperture area of both systems (assuming equal aperture area for each). As can be seen, overall the hybrid system enables lower cost of electricity and heat compared to the co-located conventional technology while providing higher efficiency and better utilizing land resources. The reduced cost can largely be attributed to the shared expenses of major components such as the tracker and support structure, which make up a substantial portion of the cost in tracking based technologies.

### TABLE I
#### GLOBAL COMPARISON

|           | CPV/T sLCOH ¢/kWh | CST LCOH ¢/kWh |
|-----------|-------------------|----------------|
| Israel    | 3.1               | 2.73           |
| India     | 3.52              | 2.51           |
| Spain     | 3.25              | 2.51           |
| Caribbean | 0.27              | 2.39           |
| Morocco   | 2.21              | 2.39           |
| Nepal     | 2.08              | 2.15           |
| Australia | -5.29             | 1.83           |
| Chile     | 1.72              | 1.58           |

### TABLE 2
#### CO-LOCATION COMPARISON

|                      | Hybrid | Co-located |
|----------------------|--------|------------|
| Thermal Efficiency   | 49%    | 37.5%      |
| Electrical Efficiency| 10.5%  | 10.3%      |
| LCOE [¢/kWh]         | 8.10   | 8.57       |
| LCOH [¢/kWh]         | 3.29   | 3.33       |
| sLCOH [¢/kWh]        | 2.24   | 2.37       |

### C. Other PV/T designs

The proposed model can also be applied to other PV/T systems including different geometries (flat plate collectors or parabolic trough collectors) and different hybrid modes (topping, reflective, or transmissive). Table 3 includes a comparison of the dish system described previously (Dish-tr) with a topping mode flat plate collector (FPC-tp) [10] and reflective mode PTC (PTC-re) [5] including brief system specifications. The total thermal and electrical efficiencies (sun-to-thermal/electric) are most critical for comparing between designs. For topping mode, there is no direct absorption of light by the thermal receiver. Instead, the waste heat generated by cell inefficiency, calculated from the energy

balance, is harvested as thermal output and contributes to the thermal efficiency. The PV fraction shows the relative amount of energy that is directly incident on the PV component compared to the thermal receiver. As described, the topping design has cells completely covering the thermal receiver. Due to the low cost of flat plate collectors, FPC-tp designs result in the lowest sLCOH. However, the outlet temperature of the working fluid in topping designs is limited by the allowable cell operation temperature. For the referenced FPC-tp, which uses standard Si cells, thermal output will be limited to <60°C. This lower operating temperature is attractive for residential cogeneration applications and could have significant markets if paired with both electric and thermal storage. Yet for many commercial and industrial applications, and for dispatchable electricity from stored thermal energy, much higher thermal energy temperatures are required. To maintain the simplicity of flat plate designs, high temperature (>400°C) FPC-tp designs have been pursued [11]–[14].

The Dish-tr geometry results in a lower sLCOH compared to the PTC-rf due to the increased efficiency associated with 2-axis tracking and higher geometric concentration with minimal increase in cost. A major advantage of both PTC and Dish designs is their ability to achieve high temperature thermal output. PTC is able to operate up to 400°C, although typically it is used at <260°C for steam applications. Steam covers a wide range of thermal applications including space heating, HVAC systems, sterilization, and food processing. Higher temperatures require higher concentrations as well as different heat transfer fluids. This higher temperature range (>260°C) makes up 43% of the thermal energy used in the EU, for example, in applications such as metallic and non-metallic refining processes. Although it is a more difficult space to break into, the size of industrial markets provides considerable motivation to develop hybrid PV/T technologies capable of becoming cost competitive with industrial fuel pricing.

### TABLE 3
#### GEOMETRY/MODE OF OPERATION COMPARISON

|         | PV fraction | Cell tr/rf | Th. Eff. | El. Eff | sLCOH |
|---------|-------------|------------|----------|---------|-------|
| FPC-tp  | 100%        | 0%         | 57%      | 12%     | 1.23  |
| PTC-rf  | 42%         | 44%        | 46%      | 9%      | 2.39  |
| Dish-tr | 48%         | 58%        | 49%      | 10.5%   | 1.89  |

### V. CONCLUSION

CPV coupled with solar thermal energy generation in a hybrid system has potential in the large and relatively untapped renewable thermal energy marketplace. Cost sharing between PV and solar thermal systems allows for reduced cost as well as increased efficiency. By using the electricity savings associated with the co-generation of electricity alongside heat, the effective cost of heat for a CPV/T system can be brought to competitive levels with natural gas in residential and commercial markets. Looking outside of the U.S., high DNI, remote locations are attractive due to their high cost of natural gas and often high electricity rates. Furthermore the techno-economic model presented here can be applied to a variety of PV/T designs which cover a larger application space, ranging from residential hot water to industrial iron refining or enhanced oil recovery.

## ACKNOWLEDGEMENT

The information, data, or work presented herein was funded in part by the Advanced Research Projects Agency-Energy, U.S. Department of Energy, under Award No. DEAR0000473.

## REFERENCES

[1] E. Literacy, "U.S. Energy Data," 2016. [Online]. Available: energyliteracy.com.

[2] "U.S. Natural Gas Marketed Production," *U.S. Energy Information Administration*, 2016. [Online]. Available: https://www.eia.gov/dnav/ng/hist/n9050us2a.htm.

[3] M. Thirugnanasambandam, S. Iniyan, and R. Goic, "A review of solar thermal technologies," *Renew. Sustain. Energy Rev.*, vol. 14, no. 1, pp. 312–322, 2010.

[4] E. C. Kern and M. C. Russell, "Combined Photovoltaic and Thermal Hybrid Collector systems," in *The 13th IEEE Photovoltaic Specialists' Conference*, 1978, no. June, pp. 1–5.

[5] M. Abdelhamid, B. K. Widyolar, L. Jiang, R. Winston, E. Yablonovitch, G. Scranton, D. Cygan, H. Abbasi, and A. Kozlov, "Novel double-stage high-concentrated solar hybrid photovoltaic/thermal (PV/T) collector with nonimaging optics and GaAs solar cells reflector," *Appl. Energy*, vol. 182, pp. 68–79, 2016.

[6] Q. Xu, Y. Ji, B. Riggs, A. Ollanik, N. Farrar-foley, J. H. Ermer, V. Romanin, P. Lynn, D. Codd, and M. D. Escarra, "A transmissive , spectrum-splitting concentrating photovoltaic module for hybrid photovoltaic-solar thermal energy conversion," *Sol. Energy*, vol. 137, pp. 585–593, 2016.

[7] Q. Xu, Y. Ji, D. D. Krut, J. H. Ermer, and M. D. Escarra, "Transmissive concentrator multijunction solar cells with over 47% in-band power conversion efficiency," *Appl. Phys. Lett.*, vol. 109, no. 19, 2016.

[8] T. Craig., "Parabolic Trough Reference Plant for Cost Modeling with the Solar Advisor Model ( SAM )," *NREL/TP-550-47605. Natl. Renew. Energy Lab. (NREL), Golden, CO.*, no. July, p. 112, 2010.

[9] E. Agency, "Iea statistics," *Statistics (Ber).*, 2010.

[10] T. T. Chow, W. He, and J. Ji, "Hybrid photovoltaic-thermosyphon water heating system for residential application," *Sol. Energy*, vol. 80, no. 3, pp. 298–306, 2006.

[11] S. Zhao, H. McFavilen, S. Wang, F. A. Ponce, C. Arena, S. Goodnick, and S. Chowdhury, "Temperature Dependence and High-Temperature Stability of the Annealed Ni/Au Ohmic Contact to p-Type GaN in Air," *J. Electron. Mater.*, vol. 45, no. 4, pp. 2087–2091, Apr. 2016.

[12] Y. Fang, H. McFavilen, D. Ding, D. Vasileska, and S. M. Goodnick, "Simulation of the high temperature performance of InGaN multiple quantum well solar cells," in *2016 IEEE 43rd Photovoltaic Specialists Conference (PVSC)*, 2016, pp. 1138–1141.

[13] H. R. Seyf and A. Henry, "Thermophotovoltaics: a potential pathway to high efficiency concentrated solar power," *Energy Environ. Sci.*, vol. 9, no. 8, pp. 2654–2665, 2016.

[14] C. Wu, B. Neuner III, J. John, A. Milder, B. Zollars, S. Savoy, and G. Shvets, "Metamaterial-based integrated plasmonic absorber/emitter for solar thermo-photovoltaic systems," *J. Opt.*, vol. 14, no. 2, p. 24005, Feb. 2012.

# Predicting the Efficiency of the Silicon Bottom Cell in a Two-Terminal Tandem Solar Cell

Zhengshan J. Yu and Zachary C. Holman

Arizona State University, Tempe, Arizona, USA 85234

*Abstract* — In two-terminal top-cell/silicon tandem solar cells, such as perovskite/silicon tandems, the efficiency of the silicon bottom cell is unknown, as individual sub-cell characterization is inaccessible after tandem formation. Here, using the "spectral efficiency" concept, we propose a simple model that determines the silicon bottom cell efficiency in a tandem device, as well as predicts the efficiency potential of a silicon bottom cell before completing a tandem device. The model relies on the characterization of a silicon cell before tandem formation, and specifically takes as input data from external quantum efficiency, reflectance, and *Suns-$V_{oc}$* measurements, as well as the absorption coefficient of the top cell. By applying this model, we find that the silicon bottom cell contributes 10.2% efficiency to the recent 23.6%-efficient, record monolithic perovskite/silicon tandem.

## I. INTRODUCTION

Efficiency is a key driver of levelized cost of electricity (LCOE) in the photovoltaic (PV) market. The dominant technology in the market—silicon—increased its record cell efficiency to 26.6% last year, and record silicon module efficiencies are now near 25% [1]. This is approaching the 29.4% theoretical efficiency limit and is on a par with the oft-claimed 26% practical efficiency limit of a real silicon device [2, 3]. To push the efficiency still higher, and thus further reduce LCOE, silicon PV needs to transition to the only device structure that has successfully surpassed the single-junction limit: multi-junctions.

Silicon has excellent characteristics for a bottom cell in a tandem: It's abundant, efficient, inexpensive, and has the near-optimum bandgap for maximum tandem efficiency [4]. The challenge is to identify and develop a similarly efficient and inexpensive top cell, as well as a suitable configuration to couple it with the silicon bottom cell [5, 6].

Among top-cell candidates, perovskite solar cells stand out. Due to their wide, tunable bandgap and low-cost solution processability, perovskites are becoming increasingly attractive for achieving 30% tandem efficiency while maintaining low module cost [7]. Following the unprecedented rapid development of single-junction perovskites, the efficiency of perovskite/silicon tandems in the four-terminal configuration has increased from 13.4%, first reported in 2014, to 26.4% this year [8-13]. Similarly, two-terminal tandems, first reported in 2015 with an efficiency of 13.7%, have improved to 23.6% [14-17].

Unlike four-terminal tandems, in which each sub-cell is measured independently so that the efficiency contribution of each sub-cell—especially the silicon bottom cell—can be unambiguously identified, two-terminal tandems make the performance of the sub-cells hard to deconvolute. The individual sub-cells are inaccessible upon the completion of the tandem device, and the performance of the silicon cell measured before tandem formation is not in general representative of that after tandem formation. For example, a silicon heterojunction bottom cell lacks front metallization and has resistive ITO on the front, leading to large series resistance losses that disappear upon deposition of a top cell that has a front electrode designed for lateral transport.

Here, we propose a model that evaluates the efficiency of a silicon bottom cell in a two-terminal tandem by characterizing a reference silicon cell, and that determines the efficiency potential of such a bottom cell before completing a tandem device.

## II. MODEL DEVELOPMENT AND DEMONSTRATION

A reference silicon cell with the same structure as the one used in the 23.6%-efficient perovskite/silicon tandem was used as an example to develop and test the model [17]. As shown in Figure 1a, it comprises a planar front surface with a 20-nm-thick indium tin oxide (ITO) layer and a textured rear surface with silicon nanoparticle (SiNP) and silver layers as a rear reflector. Details of the structure and fabrication process can be found in [17, 18].

As previously mentioned, the current-voltage (*I-V*) characteristic of the reference silicon cell alone is not representative of its performance in a tandem because of the low fill factor (*FF*) caused by high sheet resistance due to the thin front ITO and lack of metal fingers, and because of the low short-circuit current density (*$J_{sc}$*) due to the reflective planar front surface. Consequently, the external quantum efficiency (*EQE*) and reflectance (*R*) of the reference silicon cell were first characterized, and then the internal quantum efficiency (*IQE*) was calculated with Equation (1). The front-surface reflectance, shown in Figure 1b, is approximately constant at ~32% from 700 to 1000 nm, as the 20-nm-thick ITO on the front is too thin to serve as an anti-reflection coating. As a result of such a big reflection loss, the *EQE* of the reference silicon cell, also shown in Figure 1b, is very low. The *$J_{sc}$* of the reference silicon cell, calculated by integrating the product of the *EQE* and the AM1.5 global photon flux, is thus only 27.8 mA/cm². Fortunately, the *EQE* curve meets the 1-*R* curve from 700 to 900 nm, which indicates near-unity *IQE* in this range, as shown in Figure 1c. With the *IQE*, one could calculate the *EQE* of the silicon bottom cell in a tandem by multiplying by the transmittance through the top cell.

$$IQE(\lambda) = \frac{EQE(\lambda)}{1 - R(\lambda)} \qquad (1)$$

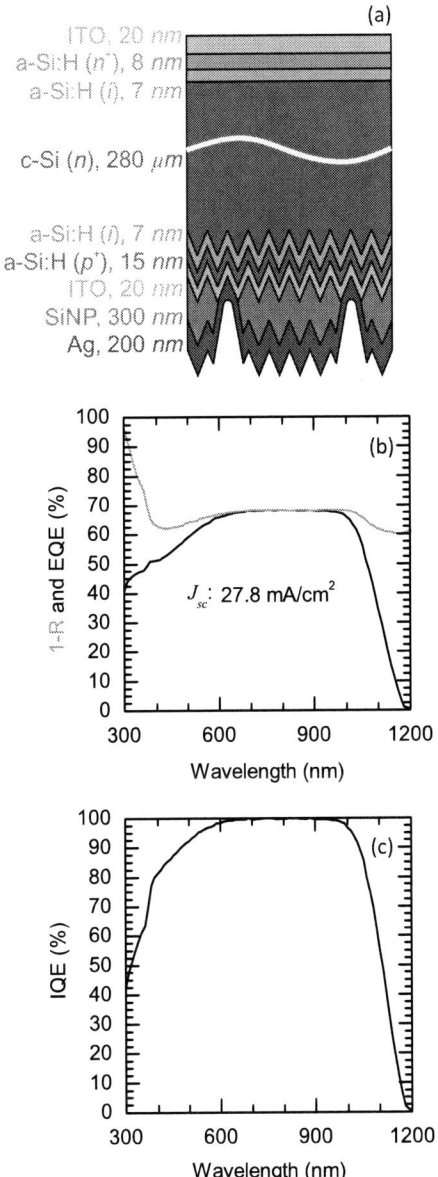

Fig. 1 (a) Schematic of the reference silicon cell (not to scale), (b) total absorbance (1-*R*) and *EQE* curves of the reference silicon cell, and (c) *IQE* of the reference silicon cell, calculated from 1-*R* and *EQE*.

The transmittance through the top cell, however, is difficult to obtain experimentally because the transmittance from the perovskite to silicon is different from that of glass to air, as one would measure when fabricating semi-transparent perovskites in the substrate configuration. Comprehensive optical modeling could access the transmittance [19-21]; however, characterizing optical constants accurately for each layer in a perovskite top cell is not trivial. To obtain the transmittance of a top cell without comprehensive optical modeling, we use Equation (2) as a first-order approximation:

$$T(\lambda) = f(\lambda) \cdot T_{ideal}(\lambda) = f(\lambda) \cdot e^{-\alpha(\lambda) \cdot d} . \qquad (2)$$

Here, $f$ is the top-cell effective transparency, $\alpha$ is the absorption coefficient, $d$ is the thickness of the perovskite absorber, $\lambda$ is the wavelength, and $T_{ideal}$ is the maximum possible transmittance to a silicon cell through a perovskite cell. Figure 2 shows example transmittance curves with constant effective transparencies of 1 and 0.91, with a 500-nm-thick perovskite absorber (the absorption coefficient of the perovskite was adapted from PV Lighthouse, initially reported by Loper *et al* [22]). Figure 2 also displays the *EQE* of a silicon bottom cell in a tandem, calculated with the *IQE* of the reference silicon cell shown in Figure 1c and a top-cell effective transparency of 0.91. The *EQE* curve overlaps with the transmittance curve for wavelengths < 900 nm because the *IQE* is close to unity. The resulting $J_{sc}$ is 18.5 mA/cm$^2$, which is the same as the value reported for the 23.6%-efficient perovskite/silicon tandem (in fact, the effective transparency was chosen to yield the same $J_{sc}$) [17].

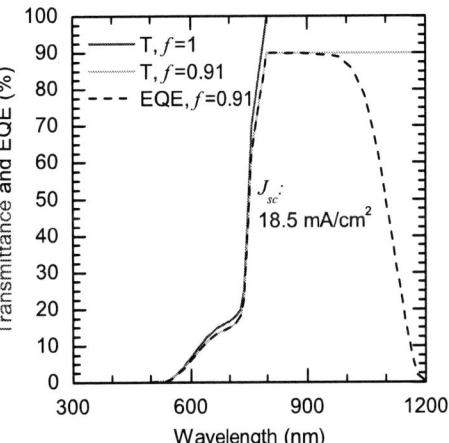

Fig. 2 Transmittance curves assuming effective transparencies *f* of 1 (dark blue) and 0.91 (light blue). Also shown is an example *EQE* (black) of a silicon cell in a tandem, calculated with the *IQE* shown in Fig. 1c and *f* = 0.91.

The injection-dependent open-circuit voltage ($V_{oc}$) of the reference silicon cell was obtained by *Suns-$V_{oc}$* measurements (not shown) [23]. At one-sun illumination, the $V_{oc}$ of the reference silicon cell is 713 mV, and, for a photo-generated current density of 18.5 mA/cm$^2$, the expected open-circuit voltage of the silicon bottom cell in a tandem ($V_{oc,Si\text{-}2T}$) is 701 mV. The fill factor of the silicon bottom cell in a tandem ($FF_{2T}$) cannot accurately be extracted from the bottom cell

978-1-5090-5606-4/17 $31.00 © 2017 IEEE

alone since it is sensitive to both sub-cells and their coupling (e.g., current mismatch). However, were $FF_{2T}$ known, the efficiency of the silicon bottom cell in a tandem ($\eta_{Si\text{-}2T}$), relative to the total incident irradiance, could now be calculated with Equation (3), which is adapted from the "spectral efficiency" concept [6]:

$$\eta_{Si-2T} = \frac{\int J_{sc}(\lambda) \cdot V_{oc,Si-2T} \cdot FF_{2T} \cdot d\lambda}{\int I(\lambda) \cdot d\lambda}. \quad (3)$$

In this equation, $I$ is the spectral irradiance and $J_{sc}$ is given by

$$J_{sc}(\lambda) = q \cdot \frac{\lambda}{hc} \cdot IQE(\lambda) \cdot f(\lambda) \cdot e^{-\alpha(\lambda) \cdot d} \cdot I(\lambda), \quad (4)$$

where $q$ is the elementary charge, $h$ is Planck's constant, and $c$ is the speed of light.

For the 23.6%-efficient perovskite/silicon tandem, $FF_{2T} = 79\%$ [17], and therefore the efficiency contribution to the tandem from the silicon bottom cell is calculated to be 10.2%, as indicated by the star in Figure 3. Figure 3 also shows the silicon bottom cell efficiency potential as a function of the effective transparency of the top cell and $FF$ of the tandem device. Given the $IQE$ and the injection-dependent $V_{oc}$ of the reference silicon cell, with an ideal effective transparency ($f = 1$) and excellent $FF$ ($FF_{2T} = 85\%$, as most two-junction III-V tandems have), the efficiency potential of our silicon bottom cell is as high as 12.4%. This figure also provides a guide for semi-transparent perovskite development when a certain tandem efficiency is targeted. For example, to make a 30%-efficient perovskite/silicon tandem with $FF_{2T} = 80\%$, a 20%-efficient perovskite must have an effective transparency of ~0.9 in order for the bottom cell to contribute the remaining 10%, and a 21%-efficient perovskite must have an effective transparency of at least 0.8.

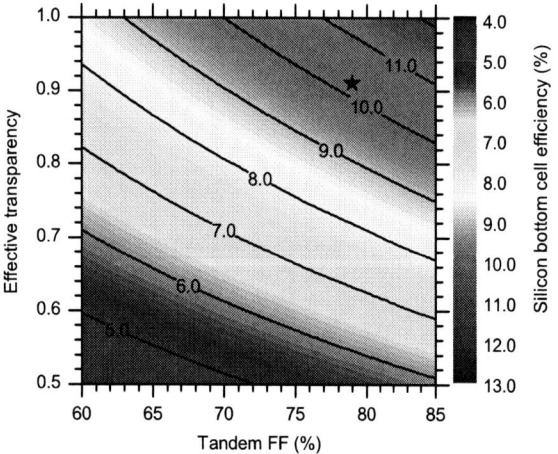

Fig. 3 Calculated silicon bottom cell efficiency as a function of top-cell effective transparency and $FF_{2T}$. The star indicates the bottom cell efficiency contribution to the 23.6%-efficient perovskite/silicon tandem.

## III. CONCLUSION

In this paper, we developed a model that evaluates the efficiency of a bottom cell in a two-terminal tandem and demonstrated the model's use with the 23.6%-efficient silicon/perovskite tandem. Using the absorption coefficient of the perovskite top cell and the $EQE$, reflectance and $Suns\text{-}V_{oc}$ of a reference silicon cell, we calculated that the bottom cell contributed 10.2% efficiency to that tandem. The model can be used to guide future perovskite top cell or silicon bottom cell development since it predicts the efficiency of the bottom cell in the tandem using measurements prior to top-cell deposition, and it can also be used in other two-terminal tandem systems if the absorption coefficient of the top-cell material is known.

## REFERENCES

[1] K. Yoshikawa, H. Kawasaki, W. Yoshida, T. Irie, K. Konishi, K. Nakano, et al., "Silicon heterojunction solar cell with interdigitated back contacts for a photoconversion efficiency over 26%," Nature Energy, vol. 2, p. 17032, 2017.

[2] A. Richter, M. Hermle, and S. W. Glunz, "Reassessment of the Limiting Efficiency for Crystalline Silicon Solar Cells," IEEE Journal of Photovoltaics, vol. 3, pp. 1184-1191, 2013.

[3] D. D. Smith, P. Cousins, S. Westerberg, R. De Jesus-Tabajonda, G. Aniero, and Y. C. Shen, "Toward the Practical Limits of Silicon Solar Cells," IEEE Journal of Photovoltaics, vol. 4, pp. 1465-1469, 2014.

[4] M. A. Green, "Commercial progress and challenges for photovoltaics," Nature Energy, vol. 1, p. 15015, 2016.

[5] I. M. Peters, S. Sofia, J. Mailoa, and T. Buonassisi, "Techno-economic analysis of tandem photovoltaic systems," RSC Advances, vol. 6, pp. 66911-66923, 2016.

[6] Z. Yu, M. Leilaeioun, and Z. Holman, "Selecting tandem partners for silicon solar cells," Nature Energy, vol. 1, p. 16137, 2016.

[7] B. Chen, X. Zheng, Y. Bai, N. P. Padture, and J. Huang, "Progress in Tandem Solar Cells Based on Hybrid Organic–Inorganic Perovskites," Advanced Energy Materials, pp. 1602400, 2017.

[8] P. Löper, S.-J. Moon, S. M. De Nicolas, B. Niesen, M. Ledinsky, S. Nicolay, et al., "Organic–inorganic halide perovskite/crystalline silicon four-terminal tandem solar cells," Physical Chemistry Chemical Physics, vol. 17, pp. 1619-1629, 2015.

[9] C. D. Bailie, M. G. Christoforo, J. P. Mailoa, A. R. Bowring, E. L. Unger, W. H. Nguyen, et al., "Semi-transparent perovskite solar cells for tandems with silicon and CIGS," Energy & Environmental Science, vol. 8, pp. 956-963, 2015.

[10] K. A. Bush, C. D. Bailie, Y. Chen, A. R. Bowring, W. Wang, W. Ma, et al., "Thermal and Environmental Stability of Semi‐Transparent Perovskite Solar Cells for Tandems Enabled by a Solution‐Processed Nanoparticle Buffer Layer and Sputtered ITO Electrode," Advanced Materials, 2016.

[11] J. Werner, L. Barraud, A. Walter, M. Bräuninger, F. Sahli, D. Sacchetto, et al., "Efficient Near-Infrared-Transparent Perovskite Solar Cells Enabling Direct Comparison of 4-Terminal and Monolithic Perovskite/Silicon Tandem Cells," ACS Energy Letters, vol. 1, pp. 474-480, 2016.

[12] B. Chen, Y. Bai, Z. Yu, T. Li, X. Zheng, Q. Dong, et al., "Efficient Semitransparent Perovskite Solar Cells for 23.0%-

Efficiency Perovskite/Silicon Four-Terminal Tandem Cells," Advanced Energy Materials, vol. 6, pp. 1601128,2016.

[13] T. Duong, Y. Wu, H. Shen, J. Peng, X. Fu, D. Jacobs, et al., "Rubidium Multication Perovskite with Optimized Bandgap for Perovskite-Silicon Tandem with over 26% Efficiency," Advanced Energy Materials, 2017.

[14] J. P. Mailoa, C. D. Bailie, E. C. Johlin, E. T. Hoke, A. J. Akey, W. H. Nguyen, et al., "A 2-terminal perovskite/silicon multijunction solar cell enabled by a silicon tunnel junction," Applied Physics Letters, vol. 106, p. 121105, 2015.

[15] S. Albrecht, M. Saliba, J. P. C. Baena, F. Lang, L. Kegelmann, M. Mews, et al., "Monolithic perovskite/silicon-heterojunction tandem solar cells processed at low temperature," Energy & Environmental Science, 2016.

[16] J. Werner, C.-H. Weng, A. Walter, L. Fesquet, J. P. Seif, S. De Wolf, et al., "Efficient Monolithic Perovskite/Silicon Tandem Solar Cell with Cell Area >1 cm2," The Journal of Physical Chemistry Letters, vol. 7, pp. 161-166, 2016.

[17] K. A. Bush, A. F. Palmstrom, Z. J. Yu, M. Boccard, R. Cheacharoen, J. P. Mailoa, et al., "23.6%-efficient monolithic perovskite/silicon tandem solar cells with improved stability," Nature Energy, vol. 2, p. 17009, 2017.

[18] M. Boccard, P. Firth, Z. J. Yu, K. C. Fisher, M. Leilaeioun, S. Manzoor, et al., "Low-refractive-index nanoparticle interlayers to reduce parasitic absorption in metallic rear reflectors of solar cells," physica status solidi (a), pp. e201700179, 2017

[19] R. Santbergen, A. H. M. Smets, and M. Zeman, "Optical model for multilayer structures with coherent, partly coherent and incoherent layers," Optics Express, vol. 21, pp. A262-A267, 2013.

[20] R. Santbergen, R. Mishima, T. Meguro, M. Hino, H. Uzu, J. Blanker, et al., "Minimizing optical losses in monolithic perovskite/c-Si tandem solar cells with a flat top cell," Optics Express, vol. 24, pp. A1288-A1299, 2016.

[21] K. Jäger, L. Korte, B. Rech, and S. Albrecht, "Numerical optical optimization of monolithic planar perovskite-silicon tandem solar cells with regular and inverted device architectures," Optics Express, vol. 25, pp. A473-A482, 2017.

[22] P. Löper, M. Stuckelberger, B. Niesen, J. Werner, M. Filipič, S.-J. Moon, et al., "Complex Refractive Index Spectra of CH3NH3PbI3 Perovskite Thin Films Determined by Spectroscopic Ellipsometry and Spectrophotometry," The Journal of Physical Chemistry Letters, vol. 6, pp. 66-71, 2015.

[23] R. Sinton and A. Cuevas, "A quasi-steady-state open-circuit voltage method for solar cell characterization," in Proceedings of the 16th European Photovoltaic Solar Energy Conference, 2000.

# Mechanically stacked 4-terminal III-V/Si tandem solar cells

Stephanie Essig[1], Christophe Allebé[2], John F. Geisz[3], Myles A. Steiner[3], Loris Barraud[2],
Antoine Descoeudres[2], J. Scott Ward[3], Manuel Schnabel[3], David L. Young[3],
Matthieu Despeisse[2], Christophe Ballif[1,2], Adele Tamboli[3]

[1] École Polytechnique Fédérale de Lausanne (EPFL), Institute of Microengineering (IMT),
Photovoltaics and Thin-Film Electronics Laboratory, Neuchâtel, Switzerland
[2] CSEM PV-center, Neuchâtel, Switzerland
[3] National Renewable Energy Laboratory (NREL), Golden, Colorado, USA

*Abstract* — We developed 4-terminal III-V/Si dual-junction solar cells that reach certified record 1-sun efficiencies over 32%. These efficiency values exceed the theoretical efficiency limit of single-junction solar cells and prove the high potential of Si-based multi-junction devices for next-generation solar panels.

*Index Terms*— silicon solar cells, III–V semiconductor materials, multi-junction solar cell.

## I. INTRODUCTION

It has been experimentally demonstrated that the theoretical one-sun efficiency limit of Si single-junction solar cells can be overcome by combining them with one or more wider-bandgap top cells made of III-V semiconductors [1, 2]. So far, 30% efficient III-V/Si multi-junction solar cells were successfully fabricated by either wafer-bonding [2] or mechanical stacking [1]. In both cases, the III-V and Si subcells were fabricated separately and joined at moderate temperature, circumventing the difficulties arising from different thermal expansion coefficients and lattice mismatch [3].

Our research has focused on the development and optimization of mechanically stacked 4-terminal III-V/Si tandem cells. The independent (4-terminal) operation of the Si and III-V subcells does not require their current matching and enables high tandem cell efficiencies for a wide range of top cell bandgap energies. The top cells in our tandem devices were grown inverted by metal-organic vapor phase epitaxy (MOCVD) on GaAs substrates and processed similar to the method described in ref. [1]. a-Si:H/c-Si heterojunction solar cells (SHJ) [4] were employed as bottom cells and fabricated using standard industrial processes.

Schematic sketches of the III-V/Si dual-junction solar cell structures are shown in Figure 1. A 1.8 eV GaInP or 1.4 eV GaAs single-junction cell was attached to a glass slide which was glued on the Si bottom cell. In our studies, the cell size was limited to 1 cm$^2$ however this fabrication process could easily be upscaled.

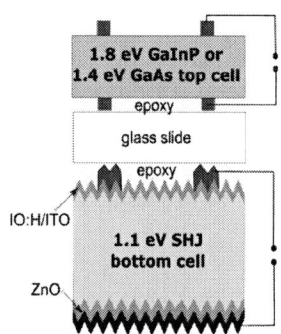

Fig. 1. Schematic of our mechanically stacked III-V/Si dual-junction solar cells with over 32% efficiency.

## II. BOTTOM CELL ARC OPTIMIZATION

When applied in a multi-junction solar cell, the Si bottom cell needs to provide both a high $V_{OC}$ and an excellent red response. In silicon heterojunction (SHJ) solar cells, layers of transparent conductive oxides (TCOs) act as an anti-reflective coating and conduction layer on the front side but also as a contacting layer on the rear between the a:Si:H and silver blanket [5]. Careful adjustment of the TCO layers thicknesses, especially on the front side - in this case IO:H, capped with a thin (10nm) ITO layer - can increase the external quantum efficiency (EQE) of the Si bottom cell in the long wavelength range. This effect is illustrated in Figure 2, which compares EQEs of two SHJ cells with either a standard front IO:H layer (~65 nm) or one with increased thickness (~95 nm). The SHJ cell with thicker IO:H layer generates in the wavelength interval 870 – 1200 nm, about 0.3 mA/cm$^2$ more photo current but loses 0.5 mA/cm$^2$ in the absorption ranges of the GaAs single-junction solar cell (<870 nm), When applying this SHJ cell with thicker TCO layer under a GaInP or GaAs top cell, the maximum net gain in bottom cell current density would be 0.2 an0d 0.3 mA/cm$^2$, respectively.

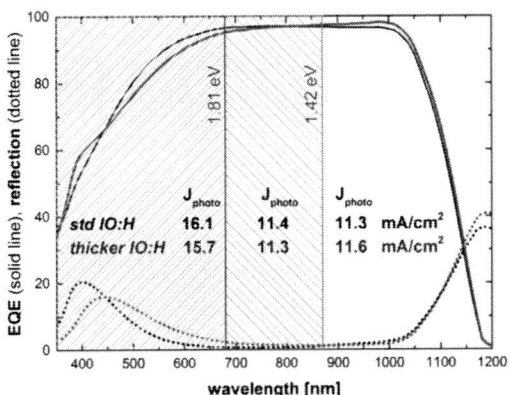

Fig. 2. External Quantum Efficiency (EQE) and reflection curves of SHJ cells with a standard (std) IO:H or a thicker IO:H layer on the front side. J$_{photo}$ are the calculated photo current densities generated by the Si bottom cell in the 3 different wavelength regions (300-690 nm, 690-870 nm, 870-1200 nm) under AM1.5g spectral conditions.

## II. SUMMARY AND CONCLUSION

Various additional optimizations [6] performed on both the Si and III-V cells allowed to achieve record 1-sun efficiencies over 32% with mechanically stacked Si-based multi-junction solar cells. More details about the tandem cell developments and the record results will be published in ref. [6] together with a detailed cost analysis.

These cell results show that the fabrication of Si-based solar cells by mechanical stacking enables high cell efficiencies. Our cell design and fabrication process flow can also be adapted to other lower-cost top cell technologies.

## ACKNOWLEDGEMENT

S. Essig acknowledges support by a Marie Skłodowska-Curie Individual Fellowship from the European Research Council (ERC) under the European Union's Horizon 2020 research and innovation programme (grant agreement No: 706744, action acronym: COLIBRI). Funding for this work at NREL was provided by DOE through EERE contract SETP DE-EE00030299 and under Contract No. DE-AC36-08GO28308 and by Laboratory-Directed Research and

Development funds. At CSEM, funding was provided by the Swiss National Science foundation (Nanotera and PNR70 programs). We would like to thank Tom Moriarty, Waldo Olavarria and Michelle Young from NREL and Nicolas Badel and Fabien Debrot from CSEM for work performed in the context of this publication.

The United States Government retains and the publisher, by accepting the article for publication, acknowledges that the United States Government retains a non-exclusive, paid-up, irrevocable, world-wide license to publish or reproduce the published form of this manuscript, or allow others to do so, for United States Government purposes.

## REFERENCES

[1] S. Essig, M. A. Steiner, C. Allebé, J. F. Geisz, B. Paviet-Salomon, S. Ward, *et al.*, "Realization of GaInP/Si Dual-Junction Solar Cells With 29.8% 1-Sun Efficiency," *IEEE Journal of Photovoltaics*, vol. 6, pp. 1012 - 1019, 2016.

[2] R. Cariou, J. Benick, P. Beutel, N. Razek, C. Flötgen, M. Hermle, *et al.*, "Monolithic Two-Terminal III-V//Si Triple-Junction Solar Cells With 30.2% Efficiency Under 1-Sun AM1.5g," *IEEE Journal of Photovoltaics*, vol. 7, pp. 367-373, 2017.

[3] F. Dimroth, T. Roesener, S. Essig, C. Weuffen, A. Wekkeli, E. Oliva, *et al.*, "Comparison of Direct Growth and Wafer Bonding for the Fabrication of GaInP/GaAs Dual-Junction Solar Cells on Silicon," *IEEE Journal of Photovoltaics*, vol. 4, pp. 620-625, 2014.

[4] S. De Wolf, A. Descoeudres, Z. C. Holman, and C. Ballif, "High-efficiency Silicon Heterojunction Solar Cells: A Review," *green*, vol. 2, pp. 7-24, 2012.

[5] Z. C. Holman, A. Descoeudres, S. D. Wolf, and C. Ballif, "Record Infrared Internal Quantum Efficiency in Silicon Heterojunction Solar Cells With Dielectric/Metal Rear Reflectors," *IEEE Journal of Photovoltaics*, vol. 3, pp. 1243-1249, 2013.

[6] S. Essig C.Allebé, T. Remo, J. F. Geisz, M. A. Steiner, L. Barraud, J. S. Ward, M. Schnabel, K. Horowitz, A. Descoeudres, D. L. Young, M. Woodhouse, M. Despeisse, C. Ballif, A. Tamboli, *"III-V/Si dual- and triple-junction solar cells exceeding 32% one-sun efficiency,"* submitted to Nature Energy, March 2017.

# Perovskite/Silicon Tandem Solar Cells: Challenges Towards High-Efficiency in 4-Terminal and Monolithic Devices

Jérémie Werner,[a] Florent Sahli,[a] Brett Kamino,[b] Davide Sacchetto,[b] Matthias Bräuninger,[a] Arnaud Walter,[b] Soo-Jin Moon,[b] Loris Barraud,[b] Bertrand Paviet-Salomon,[b] Jonas Geissbuehler,[b] Christophe Allebé,[b] Raphäel Monnard,[a] Stefaan De Wolf,[a,1] Matthieu Despeisse,[b] Sylvain Nicolay,[b] Bjoern Niesen,[a,b] and Christophe Ballif[a,b]

[a] Ecole Polytechnique Fédérale de Lausanne (EPFL), Institute of Microengineering (IMT) Photovoltaics and Thin-Film Electronics Laboratory (PV-Lab), Rue de la Maladière 71b, 2002 Neuchâtel, Switzerland.

[b] CSEM, PV-Center, Jaquet-Droz 1, 2002 Neuchâtel, Switzerland.

[1] Now at King Abdullah University of Science and Technology (KAUST), KAUST Solar Center (KSC), Thuwai 23955-6900, Saudi Arabia.

*Abstract* — Perovskite and silicon solar cells have recently been shown to be perfect partners for tandem devices with potentially very high efficiency at low additional costs over standard silicon cells. Here, we present the development of efficient perovskite top cells suitable for 4-terminal and monolithic tandem integration on silicon heterojunction bottom cells. We show a 4-terminal tandem measurement with 25.6% efficiency on small cells and 23.2% on a 1 cm$^2$ fully integrated device. Monolithic tandems with >20% efficiencies were developed on several types of silicon wafers, allowing for a direct optical comparison. We identify parasitic absorption to be the limiting factor for high performance and discuss several practical solutions to reduce them.

*Index Terms* — amorphous silicon, hybrid organic inorganic perovskites, monolithic, multijunction solar cells, silicon heterojunction solar cells.

## I. INTRODUCTION

Crystalline silicon solar cells are approaching their theoretical Auger limit of 29.4%, with the record value currently at 26.6% [1]. One of the most promising approaches to overcome this efficiency limit relies on reducing thermalization losses by stacking several absorber materials with different bandgaps in a multi-junction device. For example, an established technology such as crystalline silicon can be combined with a low-cost, wide band gap thin-film top cell, which will harvest effectively the high-energy photons while transmitting the low-energy photons to the silicon cell.

With a tunable band gap, low material costs, compatibility with various fabrication techniques and a high performance of up to 22.1% [1], perovskite solar cells represent a very promising top cell candidate for tandem solar cells with >30% efficiency potential when combined with a silicon bottom cell [2].

A perovskite/silicon tandem solar cell can be made with mainly two approaches: as a mechanically-stacked 4-terminal tandem or a monolithically-integrated 2-terminal tandem, where the top cell is directly processed onto the bottom cell. Both

configurations have the potential to exceed 30% efficiency and have their own advantages, but also present several challenges, either in complex system integration for the 4-terminal configuration or challenging manufacturing of the sub cells due to the required process compatibility in the case of a monolithic configuration. Several groups have made rapid progress in the last two years, now showing >23% efficiency for monolithic tandems [3], [4] and >26% in the 4-terminal configuration [5]–[8].

Here, we present our recent developments towards >1 cm$^2$ large perovskite/silicon tandem cells in both 4-terminal and monolithic configurations. We compare several perovskite material compositions, which allows us to tune the band gap and improve layer uniformity over larger areas. This led to a fully integrated 4-terminal tandem device with 23.2% total efficiency when both sub cells had an aperture area of 1.03 cm$^2$. We then present monolithic tandem devices with >20% efficiency, discuss their limitations due to parasitic absorption in transport layers and present solutions with more transparent alternative materials. Finally, we optically compare the 4-terminal and 2-terminal tandem devices, in order to discuss the role of surface texture in the bottom cell.

## II. EXPERIMENTAL METHOD

Low-temperature planar perovskite solar cells were fabricated on glass/ITO substrates. The electron contact was made of a bilayer of polyethylenimine and $C_{60}$ or phenyl-$C_{61}$-butyric acid methyl ester (PCBM). The perovskite absorber was grown with a two-step method consisting of an evaporated $PbI_2$ layer transformed into the final perovskite phase by spin coating a solution of methylammonium (MA) iodide (and formamidinium (FA) iodide as well as formamidinium bromide for the mixed cation material) and annealing at 100°C for 30 min. The hole transport layer was made of 2,2',7,7' tetrakis-(N,N-di-4- methoxyphenylamino)-9,9' spirobi-fluorene (spiro-

[1] Now at King Abdullah University of Science and Technology (KAUST), KAUST Solar Center (KSC), Thuwai 23955-6900, Saudi Arabia.

OMeTAD) and the transparent electrode consisted of a thin $MoO_x$ layer and a sputtered IO:H/ITO bilayer.

Silicon heterojunction solar cells were fabricated from n-type doped crystalline silicon float-zone wafers. Intrinsic and doped hydrogenated amorphous silicon layers were deposited in a PECVD reactor to passivate the silicon surface and create carrier-selective contacts. The back contact consisted of a sputtered ITO/Ag stack, whereas the front electrode was made of a thin sputtered IZO layer when used as bottom cells for monolithic tandems, or of sputtered ITO for 4-terminal tandem bottom cells.

Current-voltage measurements were carried out on a two-lamp (halogen and xenon) class AAA WACOM sun simulator with an AM1.5g irradiance spectrum at $1000W/m^2$. External quantum efficiency (EQE) spectra were obtained on a custom-made spectral response set-up.

### III. RESULTS AND DISCUSSION

The development of the perovskite/silicon tandems presented here started with fabricating an efficient near-infrared transparent perovskite top cell. To this end, we developed a low-temperature process with a hybrid evaporation/solution processing 2-step method which allowed us to fabricate semitransparent $MAPbI_3$ perovskite cells ($E_g = 1.55$ eV) with up to 16.4% efficiency [6]. By mechanically stacking this small cell with an aperture area of 0.25 $cm^2$ onto a 4 $cm^2$ silicon bottom cell, a 4-terminal tandem measurement of 25.2% could be demonstrated experimentally, while taking into account the metallization shadowing losses in the silicon cell. The perovskite absorber material was then modified towards slightly higher band gaps by introducing FA as a second cation and bromide as second halide. The band gap thus increased from 1.55 eV to 1.65 eV, which, consequently, provided more current in the silicon bottom cell during the 4-terminal tandem measurement as shown in Figure 1. Since the performance of this modified semitransparent top cell could be maintained at

| Perovskite top cell, 0.25 $cm^2$ | | | |
|---|---|---|---|
| 1086 mV | 20.0 $mA/cm^2$ | 77.1 % | 16.8 % (16.2%spo) |
| **Bottom cell, filtered** | | | |
| 695 mV | 17.0 $mA/cm^2$ | 79.8 % | 9.4 % |
| **4-Terminal:** | | | 25.6 % |

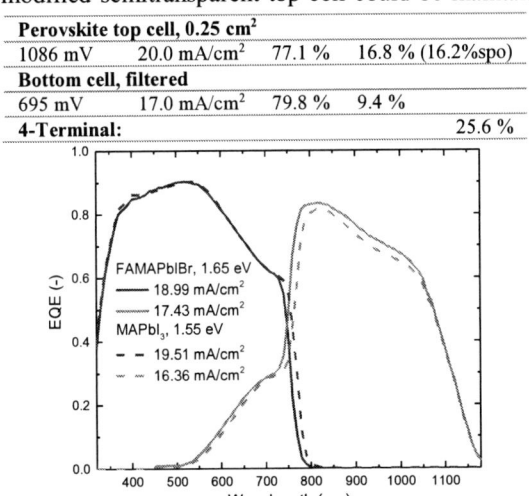

Fig. 1: *EQE measurements of perovskite/silicon 4-terminal tandem cells, comparing two different perovskite compositions: FAMAPbI$_{3-x}$Br$_x$ and MAPbI$_3$*

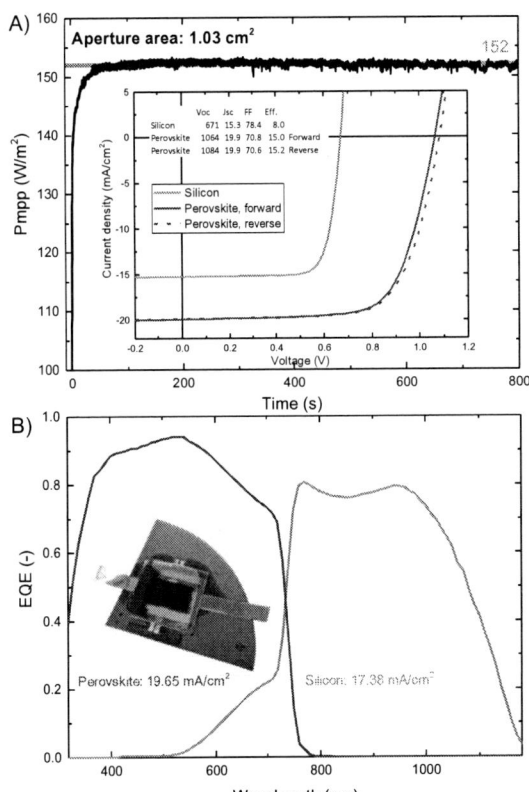

Fig. 2: *Integrated mechanically stacked 4-terminal tandem device with 1.03 $cm^2$ aperture area in both sub-cells: A) J-V curves of top and bottom cells and maximum power point tracking curve of the perovskite top cell. During the measurement of each sub cell, the other sub cell was kept at open circuit.*

16.2% under maximum power point tracking, the total 4-terminal tandem efficiency increased to 25.6%.

These 4-terminal tandem measurements were, however, not yet obtained from fully integrated tandem devices, since both sub cells had considerably different sizes (0.25 $cm^2$ for the top cell and 4 $cm^2$ for the bottom cell) and because the sub cells were measured separately as previously reported [9]. We therefore developed a fully integrated 4-terminal tandem device with both sub cells having an aperture area of 1.03 $cm^2$ and being permanently attached to each other.

We therefore further adapted the top cell to achieve high performance on 1 $cm^2$ area, which requires high uniformity in the deposition of all cell layers and high film quality, especially for the perovskite absorber. The spin coated PCBM electron transport layer was thus replaced by an evaporated layer of $C_{60}$ of ~6 nm thickness. The perovskite composition was adapted to $CsFAPbI_{3-x}Br_x$, which, in our case, was key to fabricate high quality perovskite layers on larger areas. With these improvements, a 15.2% semitransparent top cell was fabricated and integrated on a silicon heterojunction bottom cell specifically designed with the same cell geometry and size. This fully integrated mechanically stacked 4-terminal tandem device had an efficiency of 23.2%, as shown in Figure 2. The top cell was limited by a fill factor of ~70% and the bottom cell mainly by the current, which was reduced by shadow losses due to the

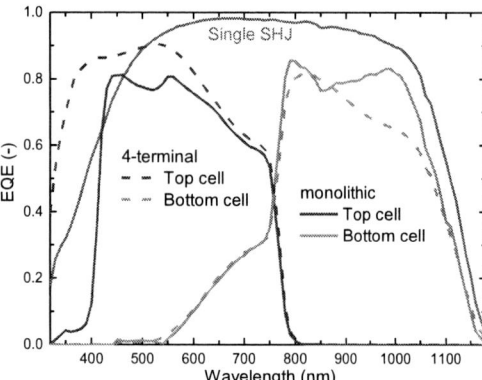

*Fig. 3: Comparison of EQE measurements on monolithic and 4-terminal tandem devices and silicon heterojunction single junction reference cell*

thick evaporated metal fingers used on the top cell back electrode. Those two losses are directly linked and might be drastically reduced by changing the metallization process from evaporation to screen printing. However, this technique requires a temperature-stable cell, which will be the focus of further development.

Compared to 4-terminal tandem devices, monolithic integration implies the additional requirement of process compatibility in terms of temperature and surface texture. We therefore first developed monolithic tandem cells on double-side polished silicon wafers, which allowed us to use a spin coated perovskite top cell. With these tandem cells, we could achieve efficiencies of up to 21.2% [3]. However, these cells were severely limited by their optics. In the infrared region, the bottom cell had a low spectral response due to the polished rear side. This could be improved by using rear-side textured wafers, which helped to increase the bottom cell current by $0.77$ mA/cm$^2$ [6]. Other optical losses were found in the MoO$_x$ protection buffer layer deposited on spiro-OMeTAD. This layer was found to become more oxygen deficient during Ar plasma exposure, thus increasing the parasitic absorption in this layer. WO$_x$ was found to be an interesting alternative owing to its larger resilience towards sputtering damage [9].

Moreover, the spiro-OMeTAD molecule used in the hole transport layer strongly absorbs photons at wavelengths <400 nm. As shown in Figure 3, this parasitic absorption is not problematic in the 4-terminal configuration, but becomes severe in monolithic tandems where the cell is directly illuminated through this layer. Replacing this material by a more transparent one is therefore crucial for further performance improvement. We found that a thin thermally evaporated layer of NPB, as a replacement to spiro-OMeTAD, can improve the current generation at wavelengths <450 nm by about 1 mA/cm$^2$, as illustrated in Figure 4. Further work is currently ongoing to improve the electrical characteristics of these cells, with both contacts featuring a thin organic evaporated layer.

Reflection losses also strongly decrease the performance of monolithic tandem cells. The intermediate recombination layer is usually made of a thin TCO layer, e.g. IZO or ITO. However,

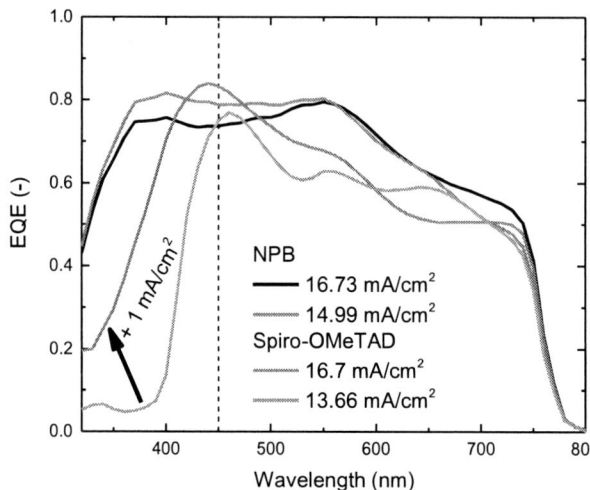

*Fig. 4: EQE measurements of semitransparent perovskite cells with thin evaporated NPB layer or spin coated spiro-OMeTAD hole transporting layer. The cells are measured both from the glass substrate and from the top electrode (HTL-side), showing the difference in absorption for wavelength below 400nm.*

it is the origin of strong optical interferences and internal reflection. A nanocrystalline silicon recombination junction can help to reduce these reflection losses and further increase the bottom cell current by better harvesting the photons in the sub-bandgap spectral range of the perovskite cell [10].

Furthermore, we have calculated that, by replacing the rear-side textured wafer with a double-side textured one, the total current could be improved by ~2.6 mA/cm$^2$ [6]. To achieve fully textured monolithic tandem cells, conformal perovskite cell deposition processes will have to be developed.

## IV. CONCLUSION

High-efficiency perovskite/silicon tandem devices were presented with both 4-terminal and monolithic configurations. >1 cm$^2$ semitransparent perovskite cells were developed with efficiencies of up to 15.2%, leading to a 23.2% integrated 4-terminal tandem device. Finally, strategies were presented to reduce parasitic absorption, which particularly limits the performance of monolithic tandem devices.

## ACKNOWLEDGEMENT

The project comprising this work is evaluated by the Swiss National Science Foundation and funded by Nano-Tera.ch and by the NRP 70 "Energy Turnaround" Program with financing from the Swiss National Science Foundation and the Swiss Federal Office of Energy, under Grant SI/501072-01. This work has received funding from the European Union's Horizon 2020 research and innovation programme under Grant Agreement No. 653296.

## REFERENCES

[1] NREL Efficiency Chart, "NREL Efficiency Chart," 2017. [Online]. Available: http://www.nrel.gov/ncpv/images/efficiency_chart.jpg. [Accessed: 22-Jan-2017].

[2] D. T. Grant, K. R. Catchpole, K. J. Weber, and T. P. White, "Design guidelines for perovskite/silicon 2-terminal tandem solar cells: an optical study," *Opt. Express*, vol. 24, no. 22, p. A1454, 2016.

[3] J. Werner, C.-H. Weng, A. Walter, L. Fesquet, J. P. Seif, S. De Wolf, B. Niesen, and C. Ballif, "Efficient Monolithic Perovskite/Silicon Tandem Solar Cell with Cell Area >1 cm2," *J. Phys. Chem. Lett.*, vol. 7, pp. 161–166, 2016.

[4] K. A. Bush, A. F. Palmstrom, Z. (Jason) Yu, M. Boccard, R. Cheacharoen, J. P. Mailoa, D. P. McMeekin, R. L. Z. Hoye, C. D. Bailie, T. Leijtens, I. M. Peters, M. C. Minichetti, N. Rolston, R. Prasanna, S. E. Sofia, D. Harwood, W. Ma, F. Moghadam, H. J. Snaith, T. Buonassisi, Z. C. Holman, S. F. Bent, and M. D. McGehee, "23.6%-Efficient Monolithic Perovskite/Silicon Tandem Solar Cells with Improved Stability," *Nat. Energy*, no. February, pp. 1–7, 2017.

[5] B. Chen, M. Yang, S. Priya, and K. Zhu, "Origin of J-V Hysteresis in Perovskite Solar Cells.," *J. Phys. Chem. Lett.*, pp. 905–917, Feb. 2016.

[6] J. Werner, L. Barraud, A. Walter, M. Bräuninger, F. Sahli, D. Sacchetto, N. Tétreault, B. Paviet-Salomon, S.-J. Moon, C. Allebé, M. Despeisse, S. Nicolay, S. De Wolf, B. Niesen, and C. Ballif, "Efficient Near-Infrared-Transparent Perovskite Solar Cells Enabling Direct Comparison of 4-Terminal and Monolithic Perovskite/Silicon Tandem Cells," *ACS Energy Lett.*, pp. 474–480, 2016.

[7] F. Fu, T. Feurer, T. P. Weiss, S. Pisoni, E. Avancini, C. Andres, S. Buecheler, and A. N. Tiwari, "High-efficiency inverted semi-transparent planar perovskite solar cells in substrate configuration," *Nat. Energy*, vol. 2, no. 16190, 2016.

[8] T. Duong, Y. Wu, H. Shen, J. Peng, X. Fu, D. Jacobs, E. Wang, T. C. Kho, K. C. Fong, M. Stocks, E. Franklin, A. Blakers, N. Zin, K. McIntosh, W. Li, Y. Cheng, T. P. White, K. Weber, and K. Catchpole, "Rubidium Multication Perovskite with Optimized Bandgap for Perovskite-Silicon Tandem with over 26% Efficiency," *Adv. Energy Mater.*, vol. 1700228, pp. 1–11, 2017.

[9] J. Werner, J. Geissbuehler, A. Dabirian, S. Nicolay, M. Morales Masis, S. De Wolf, B. Niesen, and C. Ballif, "Parasitic absorption reduction in metal oxide-based transparent electrodes: application in perovskite solar cells," *ACS Appl. Mater. Interfaces*, 2016.

[10] F. Sahli, B. Kamino, J. Werner, M. Bräuninger, B. Paviet-Salomon, L. Barraud, R. Monnard, J. P. Seif, A. Tomasi, Q. Jeangros, A. Hessler-Wyser, S. De Wolf, M. Despeisse, S. Nicolay, B. Niesen, and C. Ballif, "Improved optics in monolithic perovskite/silicon tandem solar cells by a nanocrystalline silicon recombination layer," *Manuscr. under Rev.*, 2017.

# The outcome of replacing Sn completely by Ge in Kesterite Cu₂ZnSnSe₄ solar cells

S. Sahayaraj[1,2,4] G. Brammertz[1,2,] B. Vermang[3,4,] T. Schnabel[7], E. Ahlswede[7], Z. Huang[1,2,5], S. Ranjbar[1,2,6], M. Meuris[1,2], J. Vleugels[8] and J. Poortmans[3,4]

[1] imec division IMOMEC - partner in Solliance, Wetenschapspark 1, 3590 Diepenbeek, Belgium.
[2] Institute for Material Research (IMO) Hasselt University – partner in Solliance, Wetenschapspark 1, 3590 Diepenbeek, Belgium.
[3] imec –partner in Solliance, Kapeldreef 75, 3001 Leuven, Belgium.
[4] Department of Electrical Engineering, KU Leuven, Kasteelpark Arenberg 10, 3001 Heverlee, Belgium.
[5] Laboratory for photovoltaics, University of Luxembourg, rue du Brill 41, 4422 Belvaux, Luxembourg
[6] I3N - Departamento de Física, Universidade de Aveiro, Campus Universitário de Santiago, 3810-193 Aveiro, Portugal.
[7] Zentrum für Sonnenenergie- und Wasserstoff-Forschung Baden-Württemberg, 70565 Stuttgart, Germany
[8] Department of Material Engineering, KU Leuven Kasteelpark Arenberg 44, 3001 Heverlee, Belgium

*Abstract* — In this work, the fabrication and properties of a Ge-based Kesterite Cu₂ZnGeSe₄ solar cell have been discussed. The substitution and the existence of the quaternary compound has been verified by physical methods. The device has a power conversion efficiency of 5.5% under AM1.5G illumination which is among the highest reported for pure Ge substitution. In depth electrical and optical analysis show that the Cu₂ZnGeSe₄ absorber has less bulk defects, less or no band tailing and no sub band gap emissions, which are all characteristic of Cu₂ZnSnSe₄ devices. These beneficial opto-electronic properties also result in a high open circuit voltage ($V_{oc}$) of 744 mV which is amongst the highest reported for Kesterite materials.

## I. INTRODUCTION

Detailed calculations [1] and simulations [2] revealed that the largest individual energy loss factors in solar cells with a single absorber material are due to carrier thermalization, which can be minimized with a tandem cell geometry, i.e. a sandwich of a low band gap ($E_{gap}$) bottom cell (like Si or Cu (In,Ga)Se₂ (CIGS)) with $E_{gap} \sim 1.1$ eV) and a high band gap top cell with $E_{gap} \geq 1.4$eV. Kesterite materials are a potential candidate for the top cell due to their tunable direct band gap (1 eV – 3.1 eV) [3] through cation and anion substitution along with their CIGS-like excellent optical properties. Increasing the band gap by using the pure sulfide version is a readily available option. The maximum efficiency obtained with sulfide Kesterites is not beyond 9%. Possible reasons for this are discussed in [4]. Higher efficiencies have been possible with a mixture of both S and Se but this results in low band gaps. A few authors have reported that Ge incorporation in Kesterites along with Sn results in improved efficiencies , improved grain morphologies, increased minority carrier

lifetimes and lower $V_{oc}$ deficits [5-8]. However, the band gap achieved in all the cases listed above is around 1.2 eV which is not suitable for tandem cell geometries. Band gaps higher than 1.2 eV can be achieved by pure Ge substitution instead of a mixture of Sn and Ge. Except for the work done by Schnabel et al.,[9] no other group has reported significant efficiencies with pure Ge substitution. In their work Schnabel et al., reported a maximum conversion efficiency of 5.1% for a solution based Cu₂ZnGeSₓSe₄₋ₓ solar cell with a band gap of 1.5 eV. In this work, a similar approach of replacing Sn completely by Ge in the Kesterite is presented, resulting in an absorber material with a band gap of 1.4 eV. A solar cell with a maximum PCE of 5.5 % and an unprecedented open circuit voltage ($V_{oc}$) of 744 mV has been obtained.

## II. SAMPLE FABRICATION

All solar cells were fabricated on a 3 mm thick, 25 cm² soda lime glass substrate covered by 400 nm of Molybdenum (Mo) as electrical back contact. The sheet resistance of Mo is 0.1Ω/sq. The constituent elements that make up the absorber layer are deposited in a Cu (170)/Zn(125)/Ge(180) sequence on top of the Mo-coated glass by means of electron beam evaporation. The numbers in brackets indicate the thickness of the metals in Nano meters. The precursor films were selenized at elevated temperatures (470°C) in a custom made "Halogen furnace". The tri layer was held in a square shaped graphite box with provision to place Se pellets on the sides of the box. The annealing is carried out in a Se-rich atmosphere created by 1 g of Se in the graphite box to form crystalline CZGSe thin

978-1-5090-5606-4/17 $31.00 © 2017 IEEE

films. The duration of the entire annealing cycle (ramp , dwell and cooling) is 3 hours where the dwell time is 10 minutes. The temperature was measured from thermocouples connected to the outer edges of the graphite box which is in contact with the sample. Therefore the temperature on the surface of the sample can be higher than 470°C. The selenised absorbers were then immersed in a 3 wt.% aqueous KCN solution for 90 s to remove secondary phases, elemental selenium and native oxides from the absorber surface. To obtain functional solar cells, a 50 nm thick CdS layer is deposited via chemical bath deposition followed by an intrinsic ZnO (80 nm) and an aluminum-doped ZnO (400 nm) layer by sputtering followed by the deposition of 50 nm of Ni and 1 μm thick Al fingers as front contact.. Individual cells with an area of approximately 0.5 cm$^2$ were laterally isolated by mechanical scribing.

### III. PHYSICAL PROPERTIES OF THE ABSORBER

Figure 1a) shows the surface morphology of the KCN-etched CZGSe absorber made up of densely packed micrometer sized grains. The composition of the CZGSe thin film absorber layer was deduced from SEM - EDX measurements. The thin film layer was found to be Cu-poor and Zn-rich with [Cu]/[Zn]+[Ge] and [Zn]/[Ge] ratios of 0.78 and 1.18, respectively. The cross section view of the entire CZGSe solar cell is provided in Fig.1b). From the cross SEM image, the average absorber thickness was found to be 1.2 μm. Although it is not very easy to specify the exact thickness of the MoSe$_2$ phase, a thin layer of possibly 60 nm is present at the interface of CZGSe/Mo. Along with the big sized CZGSe grains, very small and bright particles are also seen. The crystalline phases in the XRD pattern shown in Fig 1c) were matched by PDF cards from the ICDD–PDF 4 database [10] and confirmed the formation of Ge substituted Kesterite. Besides CZGSe, a small amount of ZnSe is also present due to a high Zn/Ge ratio along with MoSe$_2$. Since the XRD measurement was performed on the solar cell ZnO peaks are also present. Combining the information from cross section SEM and XRD it can be concluded that the small bright particles observed in fig.1b) along with CZGSe belong to the ZnSe secondary phase. The minority carrier lifetime was measured at room temperature from time resolved photoluminescence (TRPL) measurements is shown in Fig 1d) along with the photoluminescence (PL) peak Fig.1e) shown as an inset of Fig 1d) reveals a band gap of 1.4 eV . Density functional theory (DFT) calculations [11] on the optical properties of Cu$_2$ZnGeSe$_4$ show that the

absorber has a direct band gap and substitution of Ge in the place of Sn in Kesterite CZTS structure only changes the position of the conduction band minimum resulting in a higher band gap. Therefore it is very likely that the PL measurement in this case reveals the direct band gap of the quaternary absorber. A three exponential fit to the photoluminescence decay curve was established. The three components of the fit have lifetimes of respectively 0.1ns, 1.3 ns and 12 ns, respectively. At present it is not quite clear which of these components corresponds to the minority carrier lifetime. Further measurements of the decay time at 77 K, shown later in Fig 3d , seem to indicate that at least the longest component of the decay is related to some sort of particle trapping, not being related to the minority carrier lifetime.

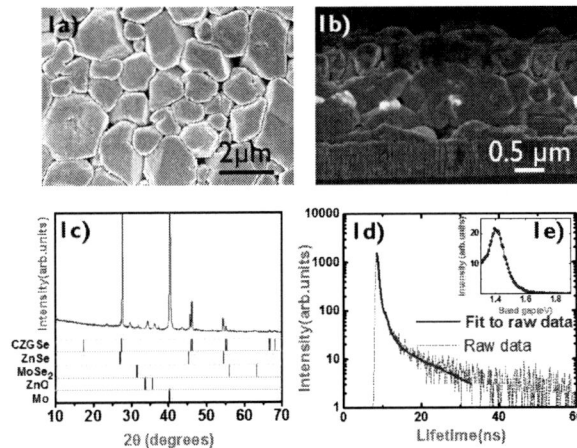

Figure 1a) & 1b) X-SEM images of the CZGSe thin film absorber and CZGSe thin film solar cell. Figure 1c) XRD pattern of the CZGSe thin film solar cell. Figure 1d) Time resolved photoluminescence measurement showing the band gap and Figure 1e) (shown as an inset) Photoluminescence spectrum of CZGSe absorber..

### IV. DETAILED ELECTRICAL AND OPTICAL PROPERTIES

The dark and the light IV curves of the best cell from the sample is presented in Fig 2a). Although the solar cell has an impressive V$_{oc}$ , it has a low fill factor (FF) and a low short circuit current (J$_{sc}$). The reason for these low values is most likely due to the high series resistance (R$_s$) of about 14Ω.cm$^2$ in the circuit. The origin of the high R$_s$ is uncertain. It could arise due to the presence of ZnSe which was observed in the cross SEM Image or an interfacial barrier at the heterojunction. The shunt resistance of the device under illumination is $670\Omega.cm^2$ which seems to be a reasonable value and leakage currents are not seen from the dark curve. The poor FF is not really due to the shunt resistance. To understand further the semiconductor nature of the absorber IV and PL measurements were performed at 77K. The behavior of $V_{oc}$ with temperature is shown in Fig 2b). The extrapolated $V_{oc}$ is about 1.2 V which is lesser than the absolute value of $E_g/q$ (1.4V) at 0K indicating that the $V_{oc}$ is limited by interface recombination like in most Sn based Kesterites. Besides the interface limitation, band gap fluctuations were also observed in the absorber. The $V_{oc}$ was modelled after taking into account the band gap fluctuations that arise due to changes in composition as explained by Rau et al. in [12]. An alternative interpretation to the data shown in Fig. 2b) could be given by fitting the data with the equation derived by Rau et al, shown below in the case of an absorber presenting lateral band gap fluctuations. As can be seen in figure 2b), a good fit to the experimental data can be obtained using as fitting parameters, the activation energy of the dominant recombination process equal to the band gap value 1.4 eV and a magnitude of the band gap fluctuations of 65 meV. Figure 2d could also be interpreted in a way that the dominant recombination process is in the bulk of the absorber and not at the interface and that the intersect with zero Kelvin is reduced because of the presence of a relatively large amount of band gap fluctuations in the absorber.

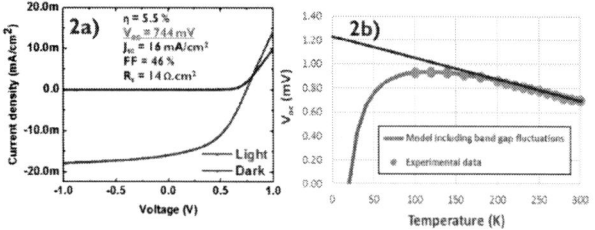

Figure 2a) Dark and Illuminated AM 1.5G IV curve of the best cell along with solar cell parameters. Figure2b) Temperature dependent IV measurements involving model for band gap fluctuations in CZGSe solar cell.

The PL emission characteristics of the absorber as a function of the laser excitation Intensity is shown in Fig 3a). Detailed peak fitting revealed only one emission with peak centers varying from 1.43 eV to 1.45 eV (blue shift) with increasing excitation intensity and saturation above a certain excitation intensity. The assessment of the peak signatures (Fig 3b) and Fig 3c) shows that the recombination in the absorber is defect related (K< 1) and does not resemble classical donor to acceptor (DAP) transitions ($\beta$ > 3 meV). Considering all the above mentioned aspects along with band gap fluctuations in the absorber it is easy to conclude that recombination in CZGSe follows the Quasi DAP (Q- DAP) model, very similar to several CZTSe absorbers /devices. A useful feature of the Q – DAP model is that the saturation of the blue-shift in the PL peak can be used to calculate the spacing between these defects by equating the magnitude of the blue-shift with the Coulombic term in Eq. (1) [13]. Using a value of 6.7 for the relative permittivity of CZGSe, a defect spacing of 10 nm , a Q-DAP density of $2.4 \times 10^{17}/cm^3$ {under the assumption r = $(3/4\pi N_d)^{1/3}$}[14] was calculated. The minority carrier lifetime measurements turn out to differentiate the optoelectronic structure of CZGSe from CZTSe. At room temperature, the lifetimes of CZTSe and CZGSe are very similar reaching values of a few to tens of nanoseconds (ns). However, at cryogenic temperatures, the lifetime of CZTSe reaches values up to several microseconds (μs), which is less than a nanosecond for CZGSe. The reason for the long lifetimes can be understood when electrostatic potential fluctuations are invoked. In the case of electrostatic potential fluctuations, electrons and holes are spatially separated and any recombination process requires the tunneling of carriers [15]. The majority of the carriers are separated by high barriers at larger distances and therefore the recombination time becomes large [16] in CZTSe. When bandgap fluctuations exist, electrons and holes are not spatially separated and therefore enhanced lifetimes are not expected which is shown in CZGSe. Hence, a complete replacement of Sn by Ge in conventional CZTSe gives a kesterite p type absorber which is free of electrostatic potential fluctuations that results in band tailing or sub gap emissions which lower the $V_{oc}$.

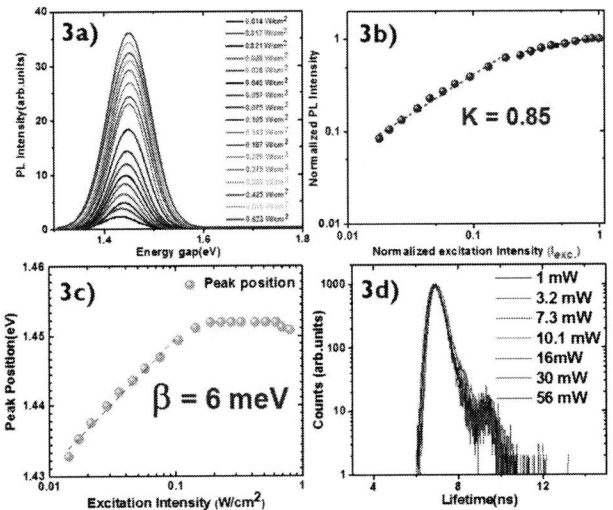

Figure 3. PL spectrum of CZGSe solar cell as a function of the laser excitation Intensity (a) Plots to determine the k exponent and rate of blue shift with increasing excitation in the CZGSe solar cell (b and c), and minority carrier lifetime measured at different laser excitation intensities for CZGSe solar cell at 77K (d).

## V. SUMMARY

A Ge based Kesterite solar cell has been fabricated. Detailed investigation on the electrical and optical properties show that the p type absorber does suffer less electrical losses from band tailing, a property quite distinct for CZTSe, resulting in a particularly large $V_{oc}$ of 744 mV. The main efficiency limiting factors are the low fill factor (46%) and $J_{sc}$ (16 mA/cm$^2$) which can be linked to a high series resistance $R_s$ of 14 Ohm cm$^2$ and a relatively low minority carrier lifetime of the order of 1 ns, leading to imperfect current collection. Mainly the reduction of the large series resistance could lead to further big improvements in the conversion efficiency, making $Cu_2ZnGeSe_4$ a promising absorber material for thin film solar cell applications.

## V. ACKNOWLEDGEMENTS

This research is partially funded by the Flemish government, Department Economy, Science and Innovation. This project has received funding from the European Union's Horizon 2020 research and innovation program under grant agreement No 640868.

## VI. REFERENCES

[1] A. Polman, H.A. Atwater, "Photonic design principles for ultrahigh-efficiency Photovoltaics", Nat. Mater. 11 (2012) 174–177.

[2] T. P. White, N. N. Lal, and K. R. Catchpole, Tandem Solar Cells Based on High-Efficiency c-Si Bottom Cells: Top Cell Requirements for >30% Efficiency, IEEE Journal of Photovoltaics 4, (2014), 208-214.

[3] D.B Khadka, J. Kim, "Band Gap Engineering of Alloyed Cu2ZnGexSn1–xQ4 (Q=S,Se) Films for Solar Cell", J. Phys. Chem. C 119, (2015),1706,

[4] T. Gershon, Y.S. Lee, R. Mankad, O. Gunawan, T. Gokmen, D.Bishop, B. McCandless, and S. Guha. "The impact of sodium on the sub-bandgap states in CZTSe and CZTS". Appl. Phys. Lett. 106, 123905 (2015)

[5] S. Kim, K. M. Kim, H. Tampo, H. Shibata, K. Matsubara and S. Niki, Ge-incorporated Cu2ZnSnSe4 thin-film solar cells with efficiency greater than 10%" Sol. Energy Mater. Sol. Cells, 144, (2016), 488–492

[6] S. Kim, K.M. Kim, H. Tampo, H. Shibata, and S. Niki, "Improvement of voltage deficit of Ge-incorporated kesterite solar cell with 12.3% conversion efficiency", App. Phy. Express 9, (2016),102301.

[7] A.D. Collord and H.W. Hill house "Germanium Alloyed Kesterite Solar Cells with Low Voltage Deficits" Chem. Mater.7, (2016), 2067–2073.

[8] C.J. Hages, S. Levcenco, C.K. Miskin, J.H. Alsmeier, D.A. Ras, R.G Wilks, M. Bär, T. Unold, R. Agrawal. "Improved performance of Ge-alloyed CZTGeSSe thin-film solar cells through control of elemental losses". Prog. Photovolt: Res. Appl. 23, (2015), 376–384.

[9] T. Schnabel, M. Seboui and E. Ahlswede, "Evaluation of different metal salt solutions for the preparation of solar cells with wide-gap Cu2ZnGeSxSe4-x absorbers" RSC Adv., 7, (2017), 26-30 .

[10] International Center for Diffraction Data: CZGSe : 00-052-0867; MoSe2 :04-004-8782; ZnO: 04-015-4060; ZnSe: 04-015-0312; Mo:01-088-2331.

978-1-5090-5606-4/17 $31.00 © 2017 IEEE

[11] Alexander P. Litvinchuk, "Optical properties and lattice dynamics of $Cu_2ZnGeSe_4$ quaternary semiconductor: A density-functional study" Phys. Status Solidi B 253, No. 2, (2016),323–328

[12] U. Rau and J. H. Werner, "Radiative efficiency limits of solar cells with lateral band-gap fluctuations", Applied physics letters, 84, (2004).

[13] T. Gershon, B. Shin, N. Bojarczuk, T. Gokmen, S. Lu, and S. Guha, J. Appl. Phys. 114, (2013), 154905.

[14] B.I Shklovskii, and A.L. Efros, Electronic Properties of Doped Semiconductors, (1984) Springer, Berlin.

[15] A. P. Levanyuk and V. V. Osipov, "Edge luminescence of direct-gap semiconductors,"Sov. Phys. Usp. 24(3), 187–215 (1981),

[16] T. Gokmen, O.Gunawan, T.K. Todorov, D.B. Mitzi "Band tailing and efficiency limitation in kesterite solar cells" Applied physics letters 103, 103506 (2013)

# Transition Metal Oxides Nano-Layers as Efficient Back Electron Reflectors For $Cu_2ZnSnSe_4$ Solar Cells

Sergio Giraldo[1], Moisés Espíndola-Rodríguez[1], Florian Oliva[1], Víctor Izquierdo-Roca[1], Alejandro Pérez-Rodríguez[1,2], and Edgardo Saucedo[1]

[1] Catalonia Institute for Energy Research (IREC), Sant Adrià de Besòs (Barcelona), 08930, Spain.

[2] IN2UB, Departament d'Electrònica, Universitat de Barcelona, Barcelona, 08028, Spain.

*Abstract* — **In this work, we present a comparative study of different transition metal oxides, including n-type ($TiO_2$, $V_2O_5$, $MoO_3$) and p-type ($Co_3O_4$) oxides, as possible candidates for back electron reflector concepts in kesterite thin film photovoltaic technologies. For this purpose, oxides nano-layers deposited on the Mo back contact, with and without a capping Mo layer, are introduced in the $Cu_2ZnSnSe_4$ devices, analyzing their impact on the material and devices properties. We demonstrate that n-type oxides, in particular $TiO_2$ can be an effective electron back reflector, showing remarkable improvements in the long-wavelength collection as well as 1-2 $mA/cm^2$ absolute increase of $J_{SC}$.**

## I. INTRODUCTION

Kesterite based thin film photovoltaic technologies ($Cu_2ZnSn(S,Se)_4$ – CZTSSe) are under rapid development, due to the high interest of these materials. They are formed by earth abundant and low-toxicity elements, and exhibit excellent properties as photovoltaic absorber, being compatible with $Cu(In,Ga)Se_2$ (CIGS) technology. So far, if we compare the current CZTSSe and CIGS record devices, the last one exhibits remarkable higher conversion efficiency (12.6% for kesterites vs 22.6% for CIGS) [1,2], but at the same time, a higher degree of device complexity. Band-gap engineering ("U" shape graded band-gap) [3,4], post-deposition treatments (KF PDT) [5] and back electron reflector contacts [6] are routinely used in the fabrication of high conversion efficiency CIGS solar cells. Nevertheless, these advanced concepts, which explain to large extent the great progress in CIGS technology, have been barely or even not studied in kesterites. In particular, the use of an electron back reflector layer (naturally formed in CIGS through a wide band-gap Ga-rich rear region) contributes to a better collection of the charge carriers generated deeper in the absorber, which are normally associated to long-wavelength light absorption. This, of course, impacts positively on the devices properties, mainly increasing the current density values of the solar cells.

In kesterites, the increase of the band-gap at the rear region substituting Se by S, or Sn by Si or Ge, has demonstrated to be extremely challenging, due to the technological difficulty to create compositional gradients in this system. In this work, we present a different approach, by slightly modifying the back contact itself, through the introduction of a transition metal oxide nano-layer between the Mo contact and the kesterite absorber. Four different transition metal oxides, including n-type ($TiO_2$, $V_2O_5$, $MoO_3$) and p-type ($Co_3O_4$) ones, are analyzed using two different approaches: i. the oxide directly in contact with the absorber; ii. the oxide sandwiched between the Mo back contact and an ultra-thin Mo sacrificial layer. Figure 1 shows a schematic representation of the standard back contact used in kesterite technologies, and the modified one by introducing the oxide layers. The impact of this modification on the absorbers and devices properties will be presented, showing that this strategy can be useful for creating a back electron reflector effect.

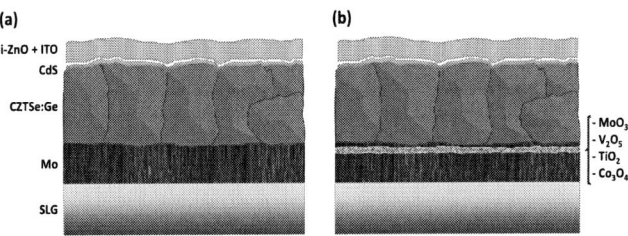

Fig. 1. Kesterite standard back contact configuration (a), and modification studied in this work with the inclusion of a transition metal oxide nano-layer and a protective Mo thin layer (b).

## II. EXPERIMENTAL

Mo back contact is deposited onto soda lime glass substrates using DC-magnetron sputtering technique (Alliance AC450). We prepare twin samples, all of them with different thicknesses of the transition metal oxides (5, 10 and 20 nm), including a reference sample without any oxide layer, and in half of them we deposit an additional 15 nm thick Mo layer on top of the oxides, while in the other half the metallic precursor is deposited directly in contact with the oxide nano-layer, as is schematized in Figure 1. Then, Ge doped $Cu_2ZnSnSe_4$ kesterite is synthesized using a sequential process by depositing Cu/Sn/Cu/Zn metallic stacks (DC-magnetron sputtering, Alliance AC450), evaporating an ultrathin Ge layer (10 nm), and a further reactive thermal annealing under Se + Sn atmosphere, as was reported elsewhere [7].

TABLE I

PHOTOVOLTAIC PARAMETERS OF DEVICES FABRICATED WITH DIFFERENT BACK CONTACTS

| Sample | Photovoltaic Parameters | | | |
|---|---|---|---|---|
| | $V_{OC}$ (mV) | $J_{SC}$ (mA/cm$^2$) | FF (%) | Eff (%) |
| Reference | 435 | 28.5 | 67.0 | 8.3 |
| Reference+Mo' | 443 | 28.0 | 66.6 | 8.3 |
| MoO$_3$-10 | 372 | 25.0 | 53.8 | 5.0 |
| MoO$_3$-10+Mo' | 435 | 28.8 | 61.8 | 7.7 |
| MoO$_3$-20 | 351 | 24.3 | 49.2 | 4.2 |
| MoO$_3$-20+Mo' | 422 | 29.1 | 54.0 | 6.6 |
| V$_2$O$_5$-10 | 348 | 23.9 | 51.0 | 4.2 |
| V$_2$O$_5$-10-Mo' | 443 | 28.3 | 58.7 | 7.4 |
| V$_2$O$_5$-20 | 400 | 25.5 | 51.7 | 5.3 |
| V$_2$O$_5$-20-Mo' | 453 | 27.2 | 59.4 | 7.3 |
| TiO$_2$-5-Mo' | 431 | 29.6 | 62.7 | 8.0 |
| TiO$_2$-10 | 364 | 25.0 | 55.1 | 5.0 |
| TiO$_2$-10-Mo' | 427 | 29.3 | 62.8 | 7.9 |
| Co$_3$O$_4$-10 | 342 | 24.0 | 54.2 | 4.4 |
| Co$_3$O$_4$-10-Mo' | 416 | 27.9 | 65.0 | 7.6 |
| Co$_3$O$_4$-20 | 342 | 20.9 | 52.4 | 3.7 |
| Co$_3$O$_4$-20+Mo' | 378 | 23.6 | 63.0 | 5.6 |

The absorbers were fully characterized using XRF, SEM (top-view and cross-section), EDX, and multi-wavelength Raman spectroscopy (back and front characterization). Finally, solar cell devices were fabricated by depositing CdS by chemical bath deposition, and completed with the window layer consisting of i-ZnO and ITO by DC-pulsed sputtering. The devices were characterized using a AAA solar simulator (Abet Tech.), EQE (Bentham 300 PV Sun), CV, JV-T, and J-V under different illuminations.

## III. RESULTS AND DISCUSSION

Figure 2 shows the SEM cross-sectional images of some selected samples summarized in Table I, emphasizing in the layers obtained with TiO$_2$ intermediate layer, as representative example.

The incorporation of the TiO$_2$ layer seems to not significantly affect the morphology of the absorber, and large CZTSe grains, exceeding 1 micron in size, are obtained independently on the substrate configuration. The same results are obtained with the other oxide layers (results not shown). On the other hand, the CZTSe/Mo interface seems to be slightly affected by the presence of the TiO$_2$ nano-layer. When the TiO$_2$ layer is introduced in direct contact with the absorber (Figure 2c), the absorber/back-contact interface seems to be slightly deteriorated forming a less flat interface (Figure 2c). Once an ultra-thin Mo layer is introduced sandwiched between the TiO$_2$ and CZTSe, and independently on the TiO$_2$ layer thickness, the good interface morphology is recovered. Apparently, the direct contact between the oxide layers with the metallic precursor and/or absorber is not optimal for the back contact morphology, in agreement with the SEM images, where some degradation of the interface is observed.

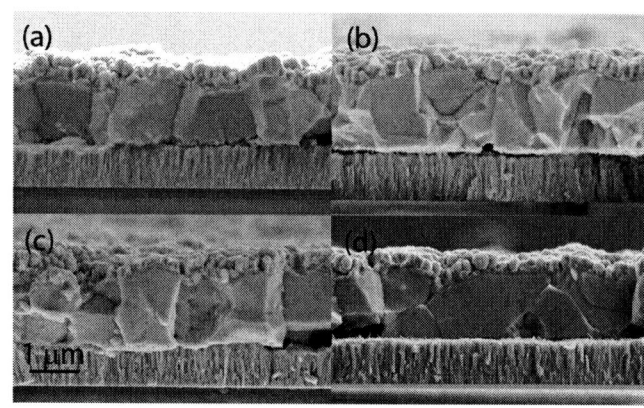

Fig. 2. Cross-sectional SEM view of different samples: reference (a), with 5 nm of TiO$_2$ plus 15 nm of Mo (b), with 10 nm of TiO$_2$ (c), and with 10 nm of TiO$_2$ plus 15 nm of Mo (d).

Figure 3 shows the Raman spectra of selected back regions after the corresponding lift-off process, including the absorber and the substrate. From the Raman analysis of the absorber at the back, it is clear that the presence of the oxide layer does not affect the crystalline properties of CZTSe absorbers, including crystalline quality. Additionally, from the Raman characterization we can state that the presence or absence of the oxides nano-layers in the back contact is not affecting the formation of MoSe$_2$ at the interface.

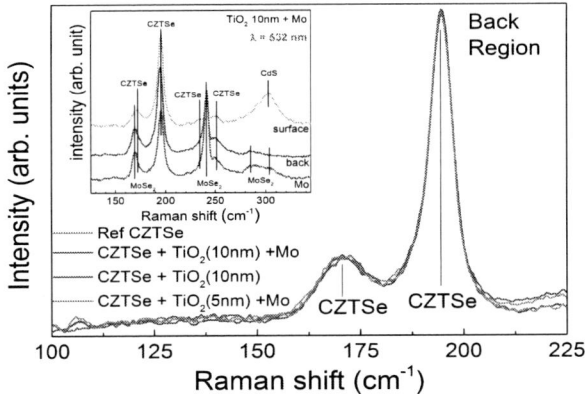

Fig. 3. Raman characterization of the back region including absorber and substrate, for the samples prepared with $TiO_2$ at the back.

This is further confirmed by the analysis of the optoelectronic properties of the resulting devices that are presented in Table I. Notably, all the absorbers prepared directly in contact with the metallic oxide layers lead to a general degradation of the corresponding devices properties, independently on the nature and thickness of the oxide. Nevertheless, when a very thin Mo layer is introduced between the oxides and the absorber, the conversion efficiency is almost fully recovered in all cases. It is important to note that for relatively conductive n-type oxides ($TiO_2$ and $V_2O_5$) the efficiency seems to not or slightly be affected by the oxide thickness. On the contrary, for the most insulating ($MoO_3$) or p-type ($Co_3O_4$), devices properties degrade for thicker layers. Following the SEM analysis, this improvement apparently is not connected with interface morphology changes that are not modified at all. Interestingly, samples with $TiO_2$/Mo exhibit in average 1-2 mA/cm$^2$ higher $J_{SC}$ than reference devices.

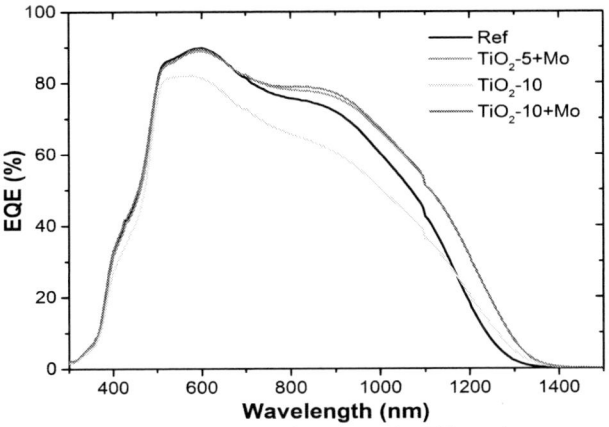

Fig. 4. EQE curves of the devices from the $TiO_2$ series compared to the reference device.

This improvement is further corroborated through the comparison of the EQE curves of samples with and without transition metal oxides (see Figure 4, for the case of $TiO_2$). As is clearly observed, carrier collection is improved mainly at

long wavelengths when the oxide layer is present at the back at least for $TiO_2$, being this effect slighter for the $MoO_3$ and $V_2O_5$ cases. This is not the case for the p-type oxide ($Co_3O_4$), where the current density is continuously deteriorated, most probably due to the formation of a p-n reverse diode at the back.

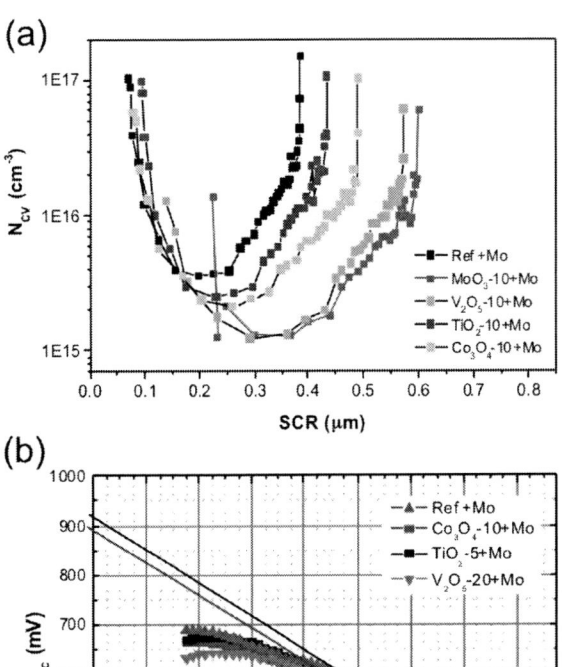

Fig. 5. (a) Charge carrier concentration ($N_{CV}$) as function of the space charge region width (*SCR*) extracted form CV measurements of devices fabricated with the different oxides. (b) Open-circuit voltage as function of the temperature from the JV-T measurements of different selected solar cells.

In order to investigate deeper about the effect of the different transition metal oxides on the devices electrical properties, capacitance-voltage (CV) and JV-T measurements were performed. In Figure 5a, the doping levels of several devices fabricated with different oxides in the back contact structure are depicted, showing no significant variations in the charge carrier concentration, which is in the range 1E15 – 4E15 cm$^{-3}$. These negligible differences support the hypothesis that the oxide layer is acting as a back reflector rather than substantially modifying the doping or the depletion width among the different devices. However, these values slightly differ from the optimum ones reported in the literature for highly efficient devices, which tend to be close to 1E16 cm$^{-3}$

[7-9]. Some possible reasons for the lower values could be a non-perfectly optimized composition of the layers, or due to the modification of the back contact configuration that might affect the alkali diffusion from the substrate, ultimately impacting on the carrier concentration.

Figure 5b shows the temperature-dependent $V_{OC}$ measurements from the analyzed JV-T of the best performing devices. These results show a very similar behavior for the different oxides, with no remarkable differences, and most of them with an overlapped extrapolated activation energy, thus no significant changes of the interface are expected. Interestingly, all the activation energies are lower than the estimated band gap for the CZTSe (about 1.0 eV), which could be attributed to a dominant surface recombination mechanism [10,11]. In any case, additional investigations must be carried out in order to better understand these features.

## IV. CONCLUSIONS

The effect of several transition metal oxides as possible back electron reflector in kesterite devices is presented here. Devices made with the oxides in direct contact with the kesterite absorber results in a general deterioration of the devices properties, correlating with a deterioration of the back region morphology mainly due to a less accurate control of the $MoSe_2$ formation, leading to a less uniform interface. On the contrary, we demonstrate that the inclusion at the back contact of an ultra-thin layer of n-type oxides ($\leq$ 20 nm), especially $TiO_2$, in combination with a very thin sacrificial Mo layer (15 nm) can have a positive effect on the improvement of the $J_{SC}$ of the devices through a better carrier collection at long wavelengths. We present evidences that this improvement is related to the formation of a back electron reflection field, due to a combination of the high work function of the oxides and the band-alignment of these materials with kesterites. This opens new perspectives for the development of back electron reflectors for kesterites that do not require managing the band-gap of the absorber at the back region, thus it could be easily implemented in any fabrication process.

## ACKNOWLEDGEMENT

This research was supported by MINECO (Ministerio de Economía y Competitividad de España) under the WINCOST project (ENE2016-80788-C5-1-R) and NOVACOST project (PCIN-2013-128-C02-01), and by European Regional Development Funds (ERDF, FEDER Programa Competitivitat de Catalunya 2007–2013). Authors from IREC and the University of Barcelona belong to the M-2E (Electronic Materials for Energy) Consolidated Research Group and the XaRMAE Network of Excellence on Materials for Energy of

the "Generalitat de Catalunya". SG thanks the Government of Spain for the FPI fellowship (BES-2014-068533).

## REFERENCES

[1] M. Yamaguchi et. al., "A detailed model to improve the radiation-resistance of Si space solar cells," *IEEE Transactions on Electron Devices*, vol. 46, pp. 2133-2138, 1999.

[2] H. J. Hovel and J. M. Woodall, "The effect of depletion region recombination currents on the efficiencies of Si and GaAs solar cells," in *10th IEEE Photovoltaic Specialist Conference*, 1973, p. 25.

[3] A. Gabor et. al., "Band-gap engineering in Cu(In,Ga)Se2 thin films grown from (In,Ga)2Se3 precursors," *Sol. Energy Mater. Sol. Cells*, vol. 41–42, pp. 247–260, 1996.

[4] T. Dullweber et. al., "A new approach to high-efficiency solar cells by band gap grading in Cu(In,Ga)Se2 chalcopyrite semiconductors," *Sol. Energy Mater. Sol. Cells*, vol. 67, no. 1, pp. 145–150, 2001.

[5] A. Chirilă et. al., "Potassium-induced surface modification of Cu(In,Ga)Se2 thin films for high-efficiency solar cells," *Nat Mater*, vol. 12, no. 12, pp. 1107–1111, 2013.

[6] K. Orgassa et. al., "Alternative back contact materials for thin film Cu(In,Ga)Se2 solar cells," *Thin Solid Films*, vol. 431, pp. 387–391, 2003.

[7] S. Giraldo et. al., "Cu2ZnSnSe4 solar cells with 10.6 % efficiency through innovative absorber engineering with Ge superficial nanolayer," *Prog. Photovoltaics Res. Appl.*, vol. 24, no. 10, pp. 1359–1367, 2016.

[8] Y. S. Lee, T. Gershon, O. Gunawan, T. K. Todorov, T. Gokmen, Y. Virgus, and S. Guha, "Cu2ZnSnSe4 Thin-Film Solar Cells by Thermal Co-evaporation with 11.6% Efficiency and Improved Minority Carrier Diffusion Length," *Adv. Energy Mater.*, vol. 5, no. 7, p. 1401372, 2015.

[9] J. Li, S. Y. Kim, D. Nam, X. Liu, J. H. Kim, H. Cheong, W. Liu, H. Li, Y. Sun, and Y. Zhang, "Tailoring the defects and carrier density for beyond 10% efficient CZTSe thin film solar cells," *Sol. Energy Mater. Sol. Cells*, vol. 159, pp. 447–455, 2017.

[10] D. Abou-Ras, T. Kirchartz, and U. Rau, Eds., *Advanced Characterization Techniques for Thin Film Solar Cells*. Weinheim, Germany: Wiley-VCH Verlag GmbH & Co. KGaA, 2011.

[11] S. S. Hegedus and W. N. Shafarman, "Thin-film solar cells: device measurements and analysis," *Prog. Photovoltaics Res. Appl.*, vol. 12, no. 23, pp. 155–176, Mar. 2004.

# Mixed sulfur and selenium annealing study of compound-sputtered bilayer $Cu_2ZnSnS_4$ / $Cu_2ZnSnSe_4$ precursors

N. Ross*, S. Grini*, L. Vines* and C. Platzer-Björkman[†]

* University of Oslo, Oslo, 0316, Norway

[†] Uppsala University, Uppsala, Uppland, 75120, Sweden

*Abstract*—Copper zinc tin sulfide (CZTS) and copper zinc tin selenide (CZTSe) precursor films are compound co-sputtered from metal sulfide and selenide targets. A bilayer precursor consisting of a CZTS-only underlayer and CZTSe-only overlayer is also sputtered. These precursor films are annealed with varying ratios of elemental sulfur and selenium to promote the formation of recrystallized $Cu_2ZnSn(S,Se)_4$ (CZTSSe) solar cell absorber layers with intermediate sulfur-to-selenium ratios $S_r=[S]/([S]+[Se])$. The films are characterized by scanning electron microscope (SEM) energy dispersive X-ray spectroscopy (EDX), secondary ion mass spectrometry (SIMS), and grazing-incidence X-ray diffraction (GIXRD). Selenium-containing precursors produce absorber layers with rougher surfaces, leading to higher short-circuit currents $J_{sc}$ for some annealing conditions. The $S_r$ value of the crystallized absorber is independent of the ratio of sulfur to selenium in the precursor unless sulfur is deficient in the anneal. Sulfur-selenium distribution in bilayer precursors after mixed S/Se annealing is found to be uniform: neither a step nor significant sulfur-gradient remains. We relate the chalcogen distribution after annealing to the excess of chalcogen in the anneal, and discuss the consequences for back-grading of the band gap by sulfur-selenium variation. Both the thickness and $S_r$ value of the $Mo(S,Se)_2$ back contact is found to be dependent on the sulfur-selenium ratio of the precursor, with CZTSe precursors producing thicker and more selenium rich back contact layers.

*Index Terms*—CZTSSe, annealing, sulfur-selenium ratio.

## I. INTRODUCTION

Over the last decade there has been a great deal of interest in the $Cu_2ZnSn(S,Se)_4$ (CZTSSe) kesterite material as a candidate for earth-abundant absorber layers for future generation photovoltaics. The challenges of the material are band misalignment with the CdS buffer [1], defects in the absorber acting as recombination centers [2], band gap fluctuations from cation-disorder [3], secondary phase formation during the post-deposition anneal [4], and potential back contact instability [5]. One of the methods suggested for improving device performance in the face of these issues is band gap grading by sulfur-selenium substitution, since the band gap can be varied between 0.96 and 1.5 eV [6], with the majority of the change in the conduction band [7]. Recently, a record power conversion efficiency (PCE) of 12.3% for a vacuum-processed precursor was obtained and ascribed to the use of $SeS_2$ in the anneal, and subsequent formation of a sulfur-rich 'front-grading' [8].

Increasing instead the sulfur proportion of the anion towards the back of a CZTSSe absorber could result in a 'back-grading', setting up a pseudo-electric field to aid in minority carrier collection. We have previously reported on

### Table I
### PRECURSOR PROPERTIES

| Precursor | [Cu]/[Sn] | [Zn]/([Cu]+[Sn]) | Material |
|-----------|-----------|------------------|----------|
| A | 1.89±0.02 | 0.35±0.02 | CZTS |
| D | 1.91±0.02 | 0.33±0.02 | CZTSe |
| E | 1.86±0.02 | 0.36±0.02 | CZTS/CZTSe |

some of the practical limitations on selenium annealing as a means to establish a 'back-grading' in sulfide-only compound co-sputtered CZTS precursors [9]. One of the suggested routes forward from that study was to sputter selenium-containing precursors as a potential means to control selenium distribution in the absorber. We now follow this thread of investigation, and sputter sulfide-only, selenide-only, and bilayer precursors, then anneal these in mixed sulfur/selenium atmospheres. The aim is to investigate the impact of precursor chalcogen content on absorber morphology, chalcogen spatial distribution, and final solar cell device properties.

## II. EXPERIMENTAL DETAILS

Bilayer molybdenum back contacts were sputtered onto soda-lime glass (SLG) substrates. $Cu_2ZnSn(S,Se)_4$ precursor material was co-sputtered using a Lesker CMS-18 sputter system, in an Ar atmosphere from chalcogen-compound targets CuS, CuSe, SnS, SnSe, ZnS, and ZnSe. No extra sodium source beyond the SLG was added. The precursors had [Cu]/[Sn] ratios of between 1.86-1.91(±0.02) and [Zn]/([Cu]+[Sn]) ratios of 0.33-0.36(±0.02), as determined by Rutherford back scattering-calibrated X-ray fluorescence measurements. Precursors were approximately 1000 nm thick. The bilayer precursor was sputtered CZTS sulfide-first, to an approximate thickness of 400 nm, followed by a top layer of 600 nm of CZTSe selenide. Precursor properties are listed in Table I.

Samples were annealed in a tube furnace in a pyrolytic carbon coated graphite box containing solid masses of elemental sulfur and/or selenium between 50-100 mg, at temperatures between 540-580°C, and argon pressures of either 24 or 47 kPa, for 3 minutes, with a 75-85 second ramp to the lower bound target temperature. Ramp time between the lower and upper temperatures was 1-2 minutes. At the end of the anneal, samples were quenched by removal from the hot zone of the furnace, with cooling to 200°C over 170 seconds. Excess chalcogen containing Ar atmosphere was pumped away at 220°C. The box is not completely sealed due to the need to evacuate it, and we have previously observed sulfur loss

978-1-5090-5606-4/17 $31.00 © 2017 IEEE

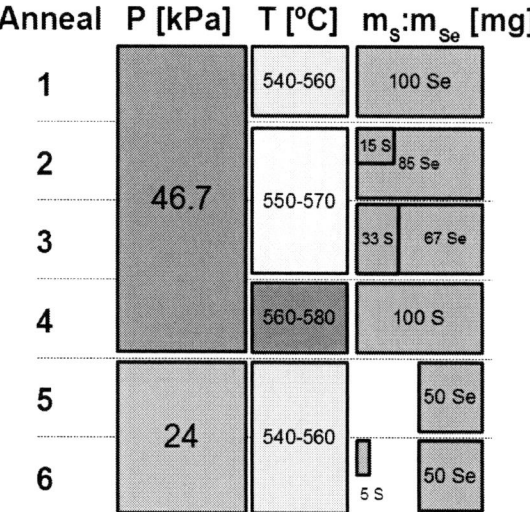

| Anneal | P [kPa] | T [°C] | $m_S : m_{Se}$ [mg] |
|---|---|---|---|
| 1 | 46.7 | 540-560 | 100 Se |
| 2 | 46.7 | 550-570 | 15 S / 85 Se |
| 3 | 46.7 | 550-570 | 33 S / 67 Se |
| 4 | 46.7 | 560-580 | 100 S |
| 5 | 24 | 540-560 | 50 Se |
| 6 | 24 | 540-560 | 5 S / 50 Se |

Figure 1. Graphical table showing annealing parameters.

Table II
BILAYER ANNEALING TESTS

| Sample | Target T [°C] |
|---|---|
| F | Precursor |
| F.400 | 400 |
| F.450 | 450 |
| F.500 | 500 |

from the anneal for times higher than three minutes [10]. The anneals were kept short in this initial study in order to avoid sulfur deficient conditions. The broad range of annealing conditions was chosen in order to produce an overview of possible absorber outcomes. Figure 1 summarizes the anneal conditions. Samples are named by their combination of precursor and anneal names: for example, precursor D annealed under condition 3 is called D.3.

As part of the investigation into annealing of bilayers, an extra CZTS/CZTSe bilayer precursor, F, was annealed under condition 6, but the ramp to temperature was cut off at selected points. Three temperature cut-offs were used: 400, 450, and 500°C. In each case, the sample was put into the hot zone of the furnace in the anneal 6 condition, but removed and cooled once the temperature reached the target temperature. Sample temperature typically overshot the target temperatures by 15°C before cooling. These samples were not processed into devices.

Absorbers were not air annealed following chalcogen annealing. Solar cell devices were fabricated from the absorbers. The absorber was etched in KCN, followed by chemical bath deposition of a 60 nm CdS layer. A window layer of i-ZnO (80 nm)/Al:ZnO (210 nm) was deposited by sputtering. The devices were finished by evaporation of Ni/Al/Ni front contacts and mechanical scribing to define 0.5 cm² cells.

Absorbers were characterized by grazing incidence X-ray diffraction (GIXRD) in a Siemens D5000 diffractometer. For annealed absorbers, $S_r$ could be estimated by assuming a

linear shift of the 112 peak between CZTSe (27.16° [JCPDS 04-010-6295]) and CZTS (28.44° [JCPDS 04-005-0388]) with increasing sulfur percentage: $S_r = (28.44 - x)/1.28$, for $x$ the 112 peak position. Scanning electron microscope (SEM) images were obtained using a Zeiss Leo 1530 device, and energy dispersive X-ray spectroscopy (EDX) line scans and some supplementary images acquired with a Zeiss Merlin device. EDX acceleration voltage was 5kV to minimize the interaction volume for cross-section analysis; interaction volume diameter has been estimated by Monte-Carlo simulation to be 170 nm. Lines analysed were Cu-L, S-K, Se-L, Zn-L and Sn-L. The Cu-L and Zn-L overlap was deconvolved using Oxford Instruments' AZtec software. The Mo-L line overlaps the S-K line heavily and was not deconvolved. Sulfur-selenium ratios in the back contact were estimated by assuming the stoichiometry $Mo_1(S,Se)_2$, ie two chalcogen atoms per Mo where Se was present in the back contact. Using the X-ray capture cross-sections $\sigma_{Mo-L}$ and $\sigma_{S-K}$, the reported atomic percentage $A_{Se}$ could be used to estimate the back contact $S_r = 1 - A_{Se}(1 + \sigma_{Mo-L}/(2\sigma_{S-K}))$. Current-voltage measurements (IV) were performed in a one-sun solar simulator (Newport, ABA).

Secondary ion mass spectrometry (SIMS) depth profile measurements were performed using a Cameca IMS 7f instrument with a primary beam of 5 keV Cs⁺ ions. The beam was scanned across a raster of 150μm×150μm and an analyzed area with a diameter of 33 μm. In order to minimize the potential matrix effects due to a change in $S_r$, MCs⁺ clusters were detected in SIMS, where 'M' is the element to be analyzed. ⁶³Cu, ⁶⁴Zn and ¹²⁰Sn are presented normalized to their maximum intensities and scaled for clarity. ²³Na, ³²S, and ⁸⁰Se were normalized by the maximum ⁹⁸Mo intensity.

### III. RESULTS

Cross-sectional SEM images are presented in Fig. 2 for anneals 2 and 6. We concentrate on these anneals in particular because they result in good devices with modest efficiency and intermediate absorber sulfur-selenium ratios $S_r$ between 0.4 and 0.7–see later results. Precursors A and E have columnar structure typical of compound sputtered CZTS on Mo. Although the surface layer of CZTSe precursor has columns with a more 'saw-tooth' edge than the CZTS, the selenide remains more columnar when sputtered on top of CZTS precursor than when it is sputtered directly on Mo, as for precursor D. The sulfur-rich (15 mg) anneal 2 produces larger grains than the sulfur-poor (5 mg) anneal 6. Both precursors with surface layers of CZTSe have rougher surfaces for anneal 2 (D.2, E.2). For anneal 6, the CZTSe precursor D.6 has the roughest grains, with a distinctly 'diffusion-limited' appearance compared to the better defined crystal grains of A.6. E.6 has an intermediate roughness. Mo(S,Se)₂ layers formed at the back contact are typically 100-300 nm for the mixed sulfur-selenium anneals 2,3, and 6, and sulfur anneal 4. D.6 is an exception in that the Mo bilayer has been almost entirely selenized. This complete selenization has not been observed in subsequent reproductions of anneal 6 with

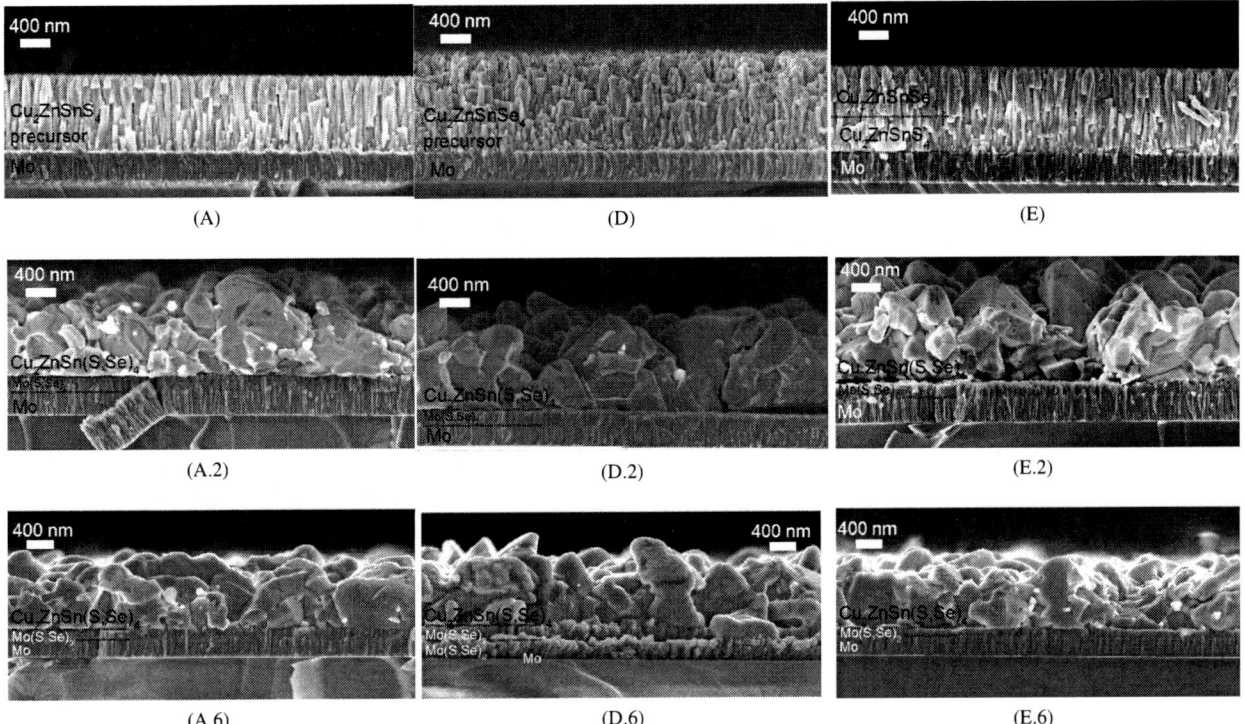

Figure 2. Cross-sectional SEM of precursors and anneals 2 and 6. CZTSe precursor leads to rougher morphology, particularly for selenium-rich anneal 6.

similar precursors, and subsequent SIMS measurements of this sample seem to indicate this extent of selenization is a local phenomenon.

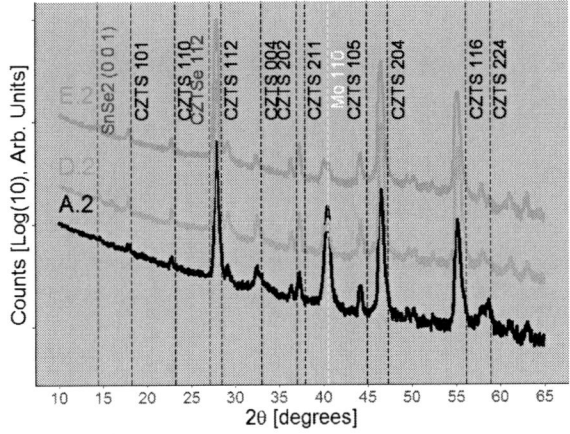

Figure 3. Example GIXRD data, from anneal 2. Characteristic CZTSSe peaks are downshifted in angle depending on $S_r$ from the CZTS references indicated.

GIXRD data show characteristic CZTSSe and $Sn(S,Se)_2$ peaks: an example from anneal 2 is shown in Fig. 3. The 001 $SnS_2$ peak was observed for anneal D.4, implying that the $SnS_2 \rightarrow SnS$ transition is incomplete for the three minute anneals, and that the anneals were not sulfur deficient [10]. The angles of the 112 CZTSSe peaks were used to estimate

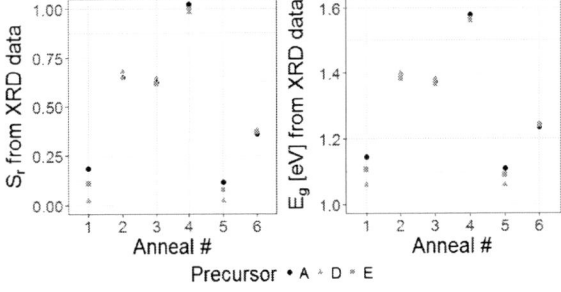

Figure 4. $E_g$ and $S_r$ estimated from GIXRD data.

the absorber sulfur-selenium ratio $S_r$, and the expected band gap $E_g$ calculated by $E_g = 1.0486 + 0.51895 S_r$ [11]: the results are shown in Fig. 4. For selenium-only anneals 1 and 5, $S_r$ is higher for precursors with higher sulfur contents. For sulfur-only anneal 4, all absorbers have $S_r \simeq 1$ (no Se left), regardless of precursor selenium content. For the mixed S/Se anneals 2, 3, and 6, there is only a very weak dependence on precursor Se content, and $S_r$ is determined mainly by the annealing condition. $S_r$ values for anneals 2 and 3 are very similar, despite anneal 3 having twice as high a sulfur mass in the box: in fact, $S_r$ is lower for A-E.3, indicating that the more sulfur-rich anneal has led to slightly more selenium rich absorbers in this case.

EDX was used to investigate elemental spatial distributions following annealing. In all cases, both line scans and maps

978-1-5090-5606-4/17 $31.00 © 2017 IEEE          3271

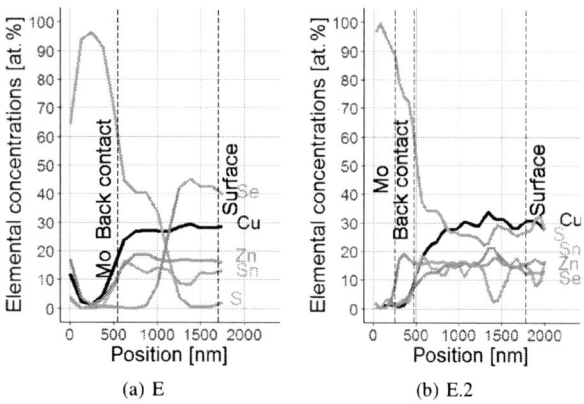

Figure 5. EDX line scans from precursor (a) E and (b) anneal E.2. The chalcogen step has been completely removed in the annealing process.

failed to show any significant sulfur-selenium gradients or steps following annealing. The line scans of precursor E before and after anneal 2 are shown in Fig. 5: the sulfur-selenium step deposited by sputtering is gone. EDX line scans for all samples were used to estimate the back contact $Mo(S,Se)_2$ thickness. Where selenium was present in the back contact, the overlap of the S-K and Mo-L lines could be accounted for, and the peak of Se atomic percentage could be used to estimate $S_r$ for the back contact. Thicknesses $t$ and back contact $S_r$ values are shown in Fig. 6. Selenide precursor D consistently produced thicker $Mo(S,Se)_2$ layers. For the selenium-only anneals 1 and 2, the Mo is almost entirely consumed to form $MoSe_2$ ($S_r = 0$). With the exception of D.6 discussed previously, sulfur-containing anneals formed a 100-300 nm $Mo(S,Se)_2$ layer. The sulfur-selenium ratio of that layer was affected by the sulfur-selenium composition of the precursor. In particular, precursor D's back contact was more rich in selenium for anneals 2, 3, and 6, and bilayer precursor E's $S_r$ was intermediate to A's and D's.

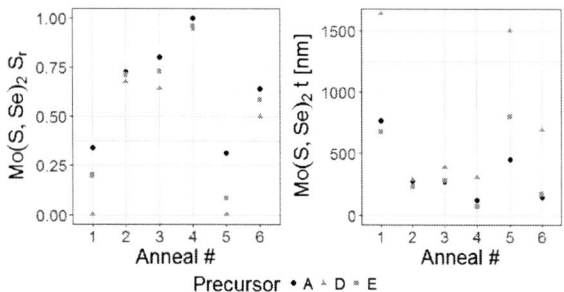

Figure 6. Sulfur-selenium ratio $S_r$ and thickness $t$ for the $Mo(S,Se)_2$ layers, as estimated by EDX.

SIMS was used to investigate the elemental distribution during the ramp to temperature portion of the anneal. The bilayer precursor F samples from Table II were annealed under the same conditions as anneal 6, with the ramp to

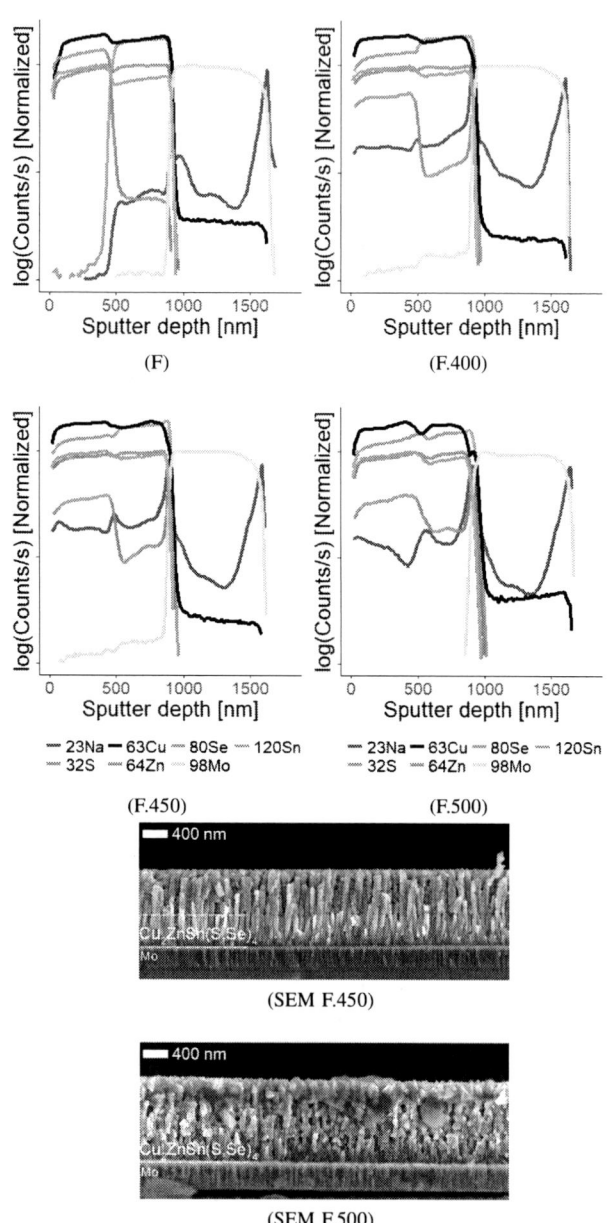

Figure 7. SIMS data for the precursor F bilayer ramp tests, presented with a linear counts scale, and cross-sectional SEM for F.450 and F.500.

temperature truncated during the anneal. SIMS results are shown in Fig. 7 along with cross-sectional SEM for samples F.450 and F.500. The vertical counts/s axis is presented in logarithmic fashion, and the counts of each isotope have been normalized as previously described. The sulfur-selenium step in the precursor F is partially removed by the displacement of selenium by sulfur at temperatures as low as 400°C: less than half of the front-layer selenium remains in the previously CZTSe portion of the precursor. The previously sulfur-only back half now shows a slightly elevated selenium count, and the sulfur count in the front half of the film is much higher.

The displacement of S by Se continues up to 500°C. At 500°C the sulfur step is essentially gone, and the Se count is rising in the back layer, likely due to the start of CZTSSe recrystallization near the back contact [9]. Recrystallization is more advanced in the selenium rich front layer between 450-500°C, and larger grains can be found in this layer than in the back sulfur-rich layer. SIMS shows an accumulation of sulfur and selenium between the CZTSSe precursor layer and the Mo back contact. This may indicate the formation of a thin $Mo(S,Se)_2$ back contact layer, however such a layer is not yet thick enough to be seen clearly in the SEM cross section. The amount of selenium accumulated at the back contact does not appear to change dramatically during the ramp to temperature. Sodium from the SLG is seen to accumulate at the back contact at 400°C, but no significant accumulation of Na occurs during the other sampled parts of the ramp.

The annealed absorbers from precursors A, D, and E were used to fabricate solar cell devices, and the IV characteristics of these cells measured: the results are shown in Fig. 8, along with the voltage deficit and voltage deficit ratio referenced to the Shockley-Queisser voltage limits [12], $\Delta V_{oc}/V_{SQ}$. Device A.4 was lost in the chemical bath deposition step, and the performance of device E.4 may have been impacted by the associated event. Because there was no air annealing step included, the open circuit voltage $V_{oc}$ and fill factor $FF$ of the selenium-only anneals 1 and 5 were very low, especially for the CZTSe precursor D. It is worth noting that functioning devices were obtained despite the extremely thick $MoSe_2$ layers for anneals D.1 and D.5. Sulfur-containing anneals generally resulted in lower $V_{oc}$ deficits, better $FF$, and higher PCE, with anneals 2, 3, 4, and 6 producing devices in the modest efficiency range 5.5-7.3%. From these sulfur-containing anneals, CZTSe containing precursors typically showed slightly higher short circuit currents $J_{sc}$. Quantum efficiency measurements (not included) showed these higher $J_{sc}$ values to be the result of higher peak QE, likely due to reduced reflectivity. CZTSe precursor D showed the lowest $\Delta V_{oc}/V_{SQ}$, for the sulfur-rich anneals 2 and 3.

## IV. DISCUSSION

From the perspective of obtaining a 'rear band gap grading' consisting of a sulfur-rich back absorber grading to a selenium-rich front absorber, the loss of sulfur-selenium steps at temperatures lower than target annealing temperature is problematic. Under the various temperature and pressure conditions investigated, precursor sulfur-selenium ratio is not seen to influence the $S_r$ of the recrystallized absorber if both elemental chalcogens are present in the annealing atmosphere. If only sulfur atmosphere is present, essentially all the precursor selenium is displaced. The precursor S-Se ratio has an impact on absorber $S_r$ only when the anneal is highly sulfur deficient, as in anneals 1 and 5: in this case the precursor sulfur is the only sulfur in the annealing box. Absorber $S_r$ is even insensitive to the relative mass of sulfur and selenium included in the annealing box, as long as there is a significant amount of sulfur: anneals 2 and 3 produced

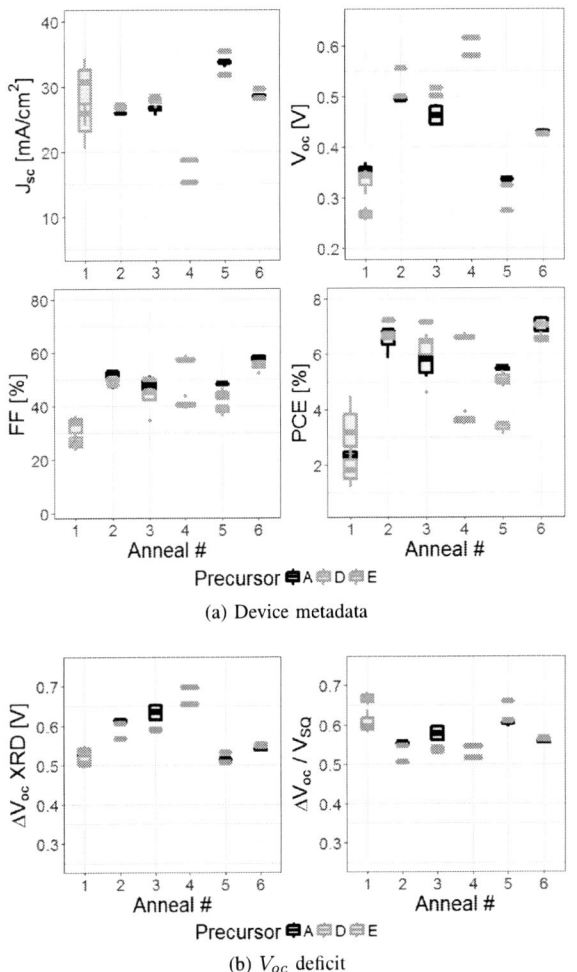

(a) Device metadata

(b) $V_{oc}$ deficit

Figure 8. (a) Device meta, and (b) $\Delta V_{oc}$ deficits and deficit ratios (referencing the Shockley-Queisser voltage limit $V_{SQ}$).

very similar absorber $S_r$, despite anneal 3 having twice the mass of sulfur as anneal 2. The relatively lower mass of sulfur in anneal 6 did produce lower $S_r$ absorbers, indicating that the three minute anneal condition for this sulfur mass is likely close to the limit of sulfur loss from the graphite box [10]. The most likely explanation for the dominance of sulfur in the anneal is sulfur having a much higher vapor pressure and slightly higher electonegativity than selenium: under the high chalcogen partial pressures necessary to prevent CZTSSe decomposition, it is energetically and kinetically more favorable for sulfur to displace selenium in CZTSSe than the reverse. In these compound co-sputtered precursors, recrystallization and diffusion of selenium do not appear to limit each other as has been claimed for stacked metal precursors [13], and selenide recrystallization occurs as readily near the back contact as near the front [9]. The consequence is that sulfur-selenium distributions set in precursor preparation cannot be used to produce sulfur-selenium gradings in recrystallized absorbers.

However, the sulfur-selenium composition and arrangement of the precursors does have an impact on the annealed films. Recrystallization appears to begin at lower temperatures in regions of precursors with higher selenium content, and produces grains with rougher surfaces. This leads to higher device current through lower reflectivity, as is the case for especially copper poor precursors [13]. There is some evidence in sulfur-rich anneals for higher voltages for CZTSe precursors, possibly related to their more diffusion-limited morphology and associated reduction in grain boundary defect density [14], although further study with better optimized devices would be needed to confirm the consistency of this observation. Although it is not yet clear why selenium-containing precursors recrystallize more readily and produce rougher surfaces, the explanation is likely related to the liquid-phase selenide mediated crystallization of CZTSSe which has been reported by several authors [9], [14], [15], and may be more important in selenium-rich precursors than sulfur rich precursors.

Finally, precursors more rich in selenium produce $Mo(S,Se)_2$ back contact layers which are thicker and more rich in selenium. SIMS data indicate that the initial back contact selenium accumulation occurs at a lower temperature than absorber recrystallization, prior to complete sulfurization of the precursor. That this back contact formation begins before the absorber recrystallization and sulfurization can explain why the back contact $S_r$ is dependent on precursor composition. The impact on device quality, if any, of these differing back contact layers is not clear from the present results, although it is possible that it plays a role in the reduced $\Delta V_{oc}$ of anneals D.2 and D.3 compared to A.2 and A.3 by, for example, better band matching at the rear contact of the cell. It would be necessary to optimize the conditions of the anneal, and to include post annealing processing like air annealing in order to improve the devices to the extent that the impacts on $V_{oc}$ and series resistance of differing back contact composition and thickness may become significant enough to study.

## V. CONCLUSION

We have produced CZTS, CZTSe, and CZTS-CZTSe bilayer precursors by compound co-sputtering, and annealed these precursors in a variety of sulfur-selenium environments. The sulfur-selenium ratio of the precursor impacts that of the annealed absorber only under extremely sulfur-poor annealing conditions. When the sulfur included in the anneal is sufficient, the high chalcogen vapor pressures lead to absorber sulfur-selenium ratios which depend on the annealing conditions rather than the precursor. In the absence of diffusion limiting effects for these precursors, the higher electronegativity and vapor pressure of sulfur over selenium leads to this uniformity in the face of differing precursors. One of the consequences is that sulfur-selenium steps and gradients sputtered into the precursors are eliminated during the annealing, so achieving a 'rear band gap grading' via this method is not possible. Selenium rich precursors were found to produce thicker and more selenium rich back contact $Mo(S,Se)_2$ layers. Additionally, selenium rich precursors produced absorber layers with higher surface roughness for mixed sulfur-selenium anneals, leading to an increase in short-circuit current.

## ACKNOWLEDGMENT

Funding from the Research Council of Norway (project 243642), Swedish Foundation for Strategic Research, and Wallenberg Academy Fellows program is gratefully acknowledged. The Research Council of Norway is acknowledged for the support to the Norwegian Micro- and Nano-Fabrication Facility, NorFab, project number 245963/F50.

## REFERENCES

[1] M. Bär, B. A. Schubert, B. Marsen, R. G. Wilks, S. Pookpanratana, M. Blum, S. Krause, T. Unold, W. Yang, L. Weinhardt, C. Heske, and H. W. Schock, "Cliff-like conduction band offset and KCN-induced recombination barrier enhancement at the CdS/Cu$_2$ZnSnS$_4$ thin-film solar cell heterojunction," *Appl. Phys. Lett.*, vol. 99, pp. 222 105–1 222 105–3, 2011.

[2] S. Chen, A. Walsh, X. Gong, and S. Wei, "Classification of lattice defects in the kesterite Cu$_2$ZnSnS$_4$ and Cu$_2$ZnSnSe$_4$ earth-abundant solar cell absorbers," *Adv. Mater.*, vol. 25, pp. 1522–1539, 2013.

[3] J. S. Scragg, J. K. Larsen, M. Kumar, C. Persson, J. Sendler, S. Siebentritt, and C. Platzer-Björkman, "Cu-Zn disorder and band gap fluctuations in Cu$_2$ZnSn(S,Se)$_4$: Theoretical and experimental investigations," *Phys. Status Solidi B*, vol. 253, p. 247254, 2013.

[4] J. T. Wätjen, J. Engman, M. Edoff, and C. Platzer-Björkman, "Direct evidence of current blocking by ZnSe in Cu$_2$ZnSnSe$_4$ solar cells," *Appl. Phys. Lett.*, vol. 100, p. 173510, 2012.

[5] J. S. Scragg, J. T. Wätjen, M. Edoff, T. Ericson, T. Kubart, and C. Platzer-Björkman, "A detrimental reaction at the molybdenum back contact in Cu$_2$ZnSn(S,Se)$_4$ thin-film solar cells," *J. Amer. Chem. Soc.*, vol. 134, pp. 19 330–19 333, 2012.

[6] A. Walsh, S. Chen, S. Wei, and X. Gong, "Kesterite thin-film solar cells: Advances in materials modelling of Cu$_2$ZnSnS$_4$," *Adv. Energy Mater.*, vol. 2, pp. 400–409, 2012.

[7] S. Chen, A. Walsh, J. Yang, X. G. Gong, L. Sun, P. Yang, J. Chu, and S. Wei, "Compositional dependence of structural and electronic properties of Cu$_2$ZnSn(S,Se)$_4$ alloys for thin film solar cells," *Phys. Rev. B*, vol. 83, p. 125201, 2011.

[8] K. Yang, D. Son, S. Sung, J. Sim, Y. Kim, S. Park, D. Jeon, J. Kim, D. Hwang, C. Jeon, D. Nam, H. Cheong, J. Kang, and D. Kim, "A band-gap-graded CZTSSe solar cell with 12.3% efficiency," *J. Mater. Chem. A*, vol. 4, p. 10151, 2016.

[9] N. Ross, J. K. Larsen, S. Grini, L. Vines, and C. Platzer-Björkman, "Practical limitations to selenium annealing of compound co-sputtered Cu$_2$ZnSnS$_4$ as a route to achieving sulfur-selenium graded solar cell absorbers," *Thin Solid Films*, vol. 623, pp. 110–115, 2017.

[10] Y. Ren, N. Ross, J. K. Larsen, K. Rudisch, J. S. Scragg, and C. Platzer-Björkman, "Evolution of Cu$_2$ZnSnS$_4$ during non-equilibrium annealing with quasi-in situ monitoring of sulfur partial pressure," *Chem. Mater.*, vol. 29, pp. 3713–3722, 2017.

[11] S. Li, S. Zamulko, C. Persson, N. Ross, J. K. Larsen, and C. Platzer-Björkman, "Optical properties of Cu$_2$ZnSn(S$_x$Se$_{1-x}$)$_4$ solar absorbers: spectroscopic ellipsometry and ab initio calculations," *Appl. Phys. Lett.*, vol. 110, p. 021905, 2017.

[12] S. Rühle, "Tabulated values of the shockley-queisser limit for single junction solar cells," *Sol. Energ.*, vol. 130, pp. 139–147, 2016.

[13] J. Li, H. Wang, L. Wu, C. Chen, Z. Zhou, F. Liu, Y. Sun, J. Han, and Y. Zhang, "Growth of Cu$_2$ZnSnSe$_4$ film under controllable Se vapor composition and impact of low Cu content on solar cell efficiency," *ACS Appl. Mater. Interfaces*, vol. 8, pp. 10 283–10 292, 2016.

[14] C. J. Hages, M. J. Koeper, C. K. Miskin, K. W. Brew, and R. Agrawal, "Controlled grain growth for high performance nanoparticle-based kesterite solar cells," *Chem. Mater.*, vol. 28, pp. 7703–7714, 2016.

[15] C. M. Sutter-Fella, J. A. Stückelberger, H. Hagendorfer, F. L. Mattina, L. Kranz, S. Nishiwaki, A. R. Uhl, Y. E. Romanyuk, and A. N. Tiwari, "Sodium assisted sintering of chalcogenides and its application to solution processed Cu$_2$ZnSn(S,Se)$_4$ thin film solar cells," *Chem. Mater.*, vol. 26, pp. 1420–1425, 2014.

# Revealing the Role of Mn Incorporation in $Cu_2ZnSn(S,Se)_4$ Photovoltaic Absorber Layer

Stener Lie[1], Joel M. R. Tan[1], Wenjie Li[1], Shin Woei Leow[1], Oki Gunawan[2], Doug Bishop[2], Lydia H. Wong[1]

[1]School of Materials Science & Engineering, Nanyang Technological University, Singapore 639798, Singapore

[2]IBM T.J. Watson Research Center, Yorktown Heights, NY 10598, USA

*Abstract* — $Cu_2Mn_xZn_{1-x}Sn(S,Se)_4$ (CMZTSSe) thin films Solar Cell were prepared by Chemical Spray Pyrolysis and subsequently Selenization process. Influence of Mn substitution in CMxZ1-xTSSe thin films with x = 0.0–1.0 on the morphological, structural, optical, electrical and device performances have been investigated. Structure transformation was observed with introduction of Mn. Increment of carrier density is evident as the amount of Mn substitution increases. Lastly, the effect of Mn in device performance was also studied. Improvement was observed at lower amount of Mn substitution followed by decrement at high amount of Mn substitution.

*Index Terms* — thin films, current-voltage characteristics, photovoltaic cells, charge carrier density.

## I. INTRODUCTION

$Cu_2ZnSnS_xSe_{4-x}$(CZTSSe) as an alternative earth-abundant thin film solar cells have made significant progress for the past few years[1]. The highest CZTSSe efficiency is achieved by IBM at $\approx$12.6% using Hydrazine process [2]. However, this value is relatively low in comparison with $Cu(In,Ga)(S,Se)_2$ (CIGS) at $\approx$22.3%[3]. The main culprit of this difference is the open-circuit voltage (Voc) deficit in CZTSSe. It was believed that intrinsic limitations of CZTSSe, such as secondary phases, antisite defects, carrier lifetime, and so on are the cause of this deficit[4].

To tackle this issue, cation substitution on approach have been reported, such as substitution of Zn with Cd[5] and Cu with Ag[6, 7]. It was believed that this approach managed to decrease Cu-Zn antisite defects based on better carrier lifetime, sharp absorption edge, thus lead to Voc improvement. Besides the reduction of defects, grain enlargement, increase in lattice parameter, band gap tuning are the other reasons these substitution works. For most compound semiconductors, it is commonly accepted substitution of heavier atoms from the same group in the periodic table results in narrower band gaps.

However, these reported studies only focus on cation substitution with heavier cation from similar rows. There is a lack of study on substitution using heavier metal from different rows. One of the suitable candidate is Mn as a substitute for Zn in CZTSSe. Several reasons on Mn suitability; first, it is stable at oxidation state +2 with cation radius ($\approx$0.97 Å) which is quite close to Zinc's radius ($\approx$0.88 Å) [8]. Similar oxidation state will satisfy the octet rule and larger radius will prevent the occurrence of Cu-Zn antisite defects. Similar atomic size between Cu and Zn ($\approx$1.35 Å) [9] results in likelihood for Zn and Cu to occupy vacancies of counterparts and create Cu-Zn antisite defects in CZTS[10]. Second, previous reported result on $Cu_2MnSnS_4$ which shows photovoltaic response[11] compare to some other similar metal with similar oxidation states such as Fe, Ni and Co[12]. Third, Mn is more abundant in comparison with the reported Ag and Cd[13].

Thus in this report, by using low cost spray-pyrolysis technique based on previous work[14], we fabricate $CuMn_xZn_{1-x}Sn(S,Se)_4$ (CMZTSSe) thin films solar cells to study the impact of Mn alloying into CZTSSe system.

## II. EXPERIMENTAL DETAILS

Copper chloride dihydrate ($CuCl_2 \cdot 2H_2O$), tin chloride dihydrate ($SnCl_2 \cdot 2H_2O$), Manganese (II) chloride tetrahydrate ($MnCl_2.4H_2O$) / Zinc Chloride ($ZnCl_2$) and Thiourea ($SC(NH_2)_2$) were dissolved into 80 ml water. The solutions were sprayed into Molybdenum-coated glass with $N_2$ as carrier gas on 450°C hotplate at constant pressure of 20 psi. Followed by selenization at 560°C for 20 min to form $CMx_Z1_{-x}$TSSe. Experiment were carried out by varying x value from 0 to 1. Subsequent device fabrication procedures were carried out to the samples, such as CBD of CdS, and sputtering of i-ZnO/ITO. Various characterizations such as X-ray diffraction (XRD), Scanning Electron Microscopy (SEM), Ultraviolet-visible light Spectroscopy (UV-Vis) and Hall measurements were conducted on the absorber layer to investigate the crystal structure, morphology, elemental composition, electrical and optical properties. IV measurement under AM 1.5G light illumination was also conducted to understand influence of Mn on the device performance.

## III. RESULTS AND DISCUSSIONS

Various effect of Mn substitution were observed through characterization such as morphology transformation from SEM, band gap differences, PL spectra response and so on. However, from those findings, there are several important and highlighted findings regarding the influence of Mn in CMZTSSe. First, it is about the structure transformation due to the variation of Mn and Zn in the system. Second, it is regarding Mn effect on the electronic properties such as carrier density and hall mobility. Last finding is on device performances with ratio variation.

## A. Structural Properties

The first finding about structural changes is observed through XRD and Raman characterization. Fig. 1 shows XRD results in $Cu_2Mn_xZn_{1-x}Sn(S,Se)_4$ thin films for x=0 to x=1. Diffraction peaks corresponding to (112) planes are dominant in all ratios which indicate tetragonal phases. It is also evident that there is a peak shift towards lower angle in all peaks as x goes from 0 to 1. The peak shift is clearly observed at (112) peaks

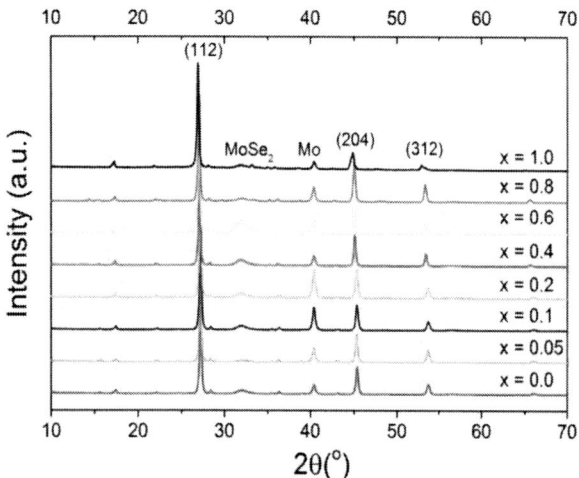

Fig. 1    X-ray diffraction spectra for $Cu_2Mn_xZn_{1-x}Sn(S,Se)_4$ thin films (x=0.0-1.0)

Rietveld Refinement analysis were conducted to samples shows transformation starting from CZTSSe with kesterite structure (tetragonal I4) as the primary structure, following that the alloyed C(M,Z)TSSe stannite structure becomes dominant and finally CMTSSe stannite structure is observed. This observation shows that there is solid solubility limit of Mn in the CZTSSe structure. This is in line with Home-Rutherford theory[15] which stated for element to have a complete substitutional solid solution , atomic size difference must be less than 14-15%, to be able to fully substitute atom into the system without rearrangement.

## B. Electronical Properties

Fig. 2 shows electrical characteristics of CMZTSSe as a function of Mn/(Zn+Mn). Majority carrier concentration and carrier mobility are measured using Parallel Dipole Line AC Hall Measurement[16]. Large increase of carrier concentration as Mn content increases is observed from this sample set. There is a small drop in carrier concentration and small increase in carrier mobility at 5% Mn, indicating an improvement in the

carrier transport. The value for CZTSSe is consistent with previous findings on high efficiency CZTSSe[2]. Similarly for CMTSSe, the high carrier concentration is also observed in CMTS[17].

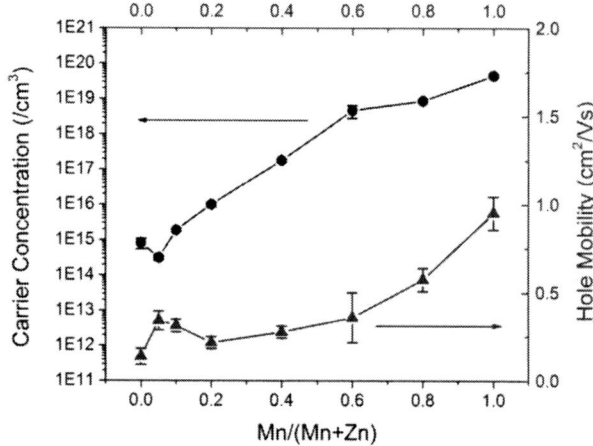

Fig. 2    Hole Concentration and Hole Mobility as a function of Mn/(Zn+Mn)

One of the possible reason of such high increase in Mn-rich compound is due to the nature of Mn property. Elemental Mn whose d orbital is not fully occupied compare to Zn has more than one stable oxidation states. The evolution of $Mn^{2+}$ into higher common oxidation states such as $Mn^{4+}$ and $Mn^{7+}$ will contributes to more free hole carriers in the system. Moreover, as the oxidation states increases, cation radius will decrease[18], smaller cation radius of Mn will result in an ease of antisite formation between Cu-Mn as observed between Cu-Zn Even if the evolution of oxidation states does not occurs, the half-empty orbital will contribute as acceptor traps.

## C. Devices Performances

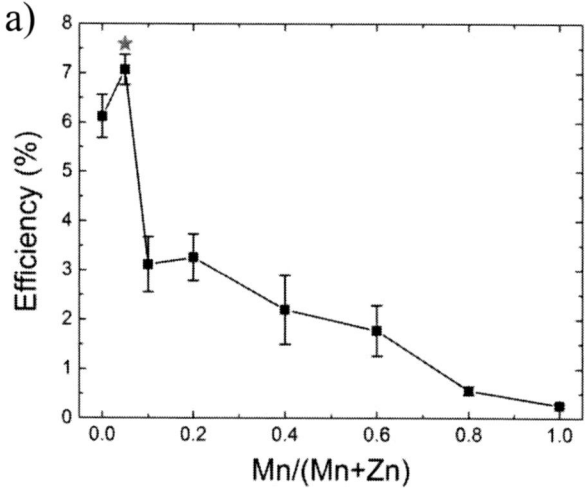

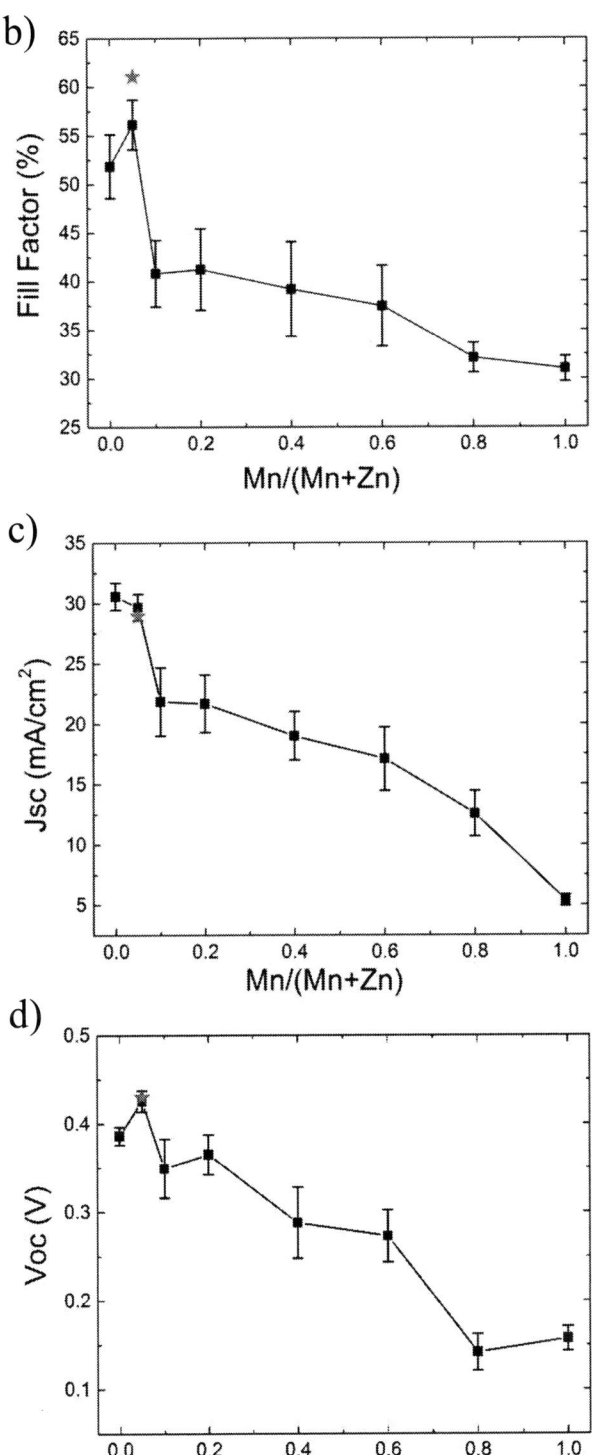

Fig. 3 Device parameters a) η, b) FF, c) Jsc and d) Voc versus the Mn content for $Cu_2Mn_xZn_{1-x}Sn(S,Se)_4$ thin film solar cells

Solar cell devices (Mo/CMZTSSe/CdS/i-ZnO/ITO/Ag-paste) were fabricated to test the variation on the performance. Fig. 3(a) shows the efficiency (η) as a function of Mn contents. These data were tabulated from several batches of fabrication.

As Mn content varies, the efficiency changes with non-linearity dependence. The best performance cell is achieved when 5% Mn substitute Zn. The J–V characteristics of the "champion" CMZTSSe device (Mn/(Mn + Zn) ≈ 5%) with a power conversion efficiency of 7.6% and CZTSSe with efficiency of 6.5% There is an increase in efficiency from pure CZTSSe. Assuming all other processes and layers are consistent, this result indicates there is an improvement in terms of material properties. Overall, Voc improvement and Fill Factor from CZTSSe are the main factors for the overall improvement. A small increment in band gap, lower carrier concentration and higher carrier mobility compare to pure CZTSSe are the main contributor in this improvement.

When Mn content exceeds > 5%, a reduction of device performance is observed. As observed, this reduction is attributed to significant Jsc and Fill Factor reduction. The high carrier concentration ($\sim 10^{19}$ cm$^3$) for higher Mn content will result in narrow depletion width which leads to low collection or low Jsc and low shunt resistance.

This experimental study on Mn alloying into CZTSSe shows positive intrinsic material and device improvement. These findings suggest that there is a possibility to improve CZTSSe by substitution with cation from different rows as long as it satisfy the octet-rule. It is most likely due to partial replacement of Cu/Zn in CZTSSe that decrease the Cu/Zn antisite defects. However, the amount of substitution to improve as well as how significance the changes is dependent on the cation substitute. Further improvement could be done in the future to the champion cell especially regarding the choice of buffer layer and quality of the absorber itself.

## IV. CONCLUSION

In summary, CMZTSSe solar cells prepared by spray pyrolysis method have been presented. The substitution of Zn with Mn in CZTSSe thin films is shown to induce structure transformation and peak shifts which is largely attributed to the difference in atomic radius. The variation of Mn contents is also found to change of charge density, mobility and carrier lifetime in CMZTSSe. High Mn contents increase the carrier density and decrease the carrier lifetime due to Mn elemental properties which has d-orbital that is less stable and contribute excess carrier and traps. Lastly, devices performances were observed for CMZTSSe ratio. Device improvement at Mn/(Mn+Zn) = 0.05 is observed with characteristic improvement in depletion width, Voc, and Fill factor. However, further incorporation of Mn into the system lead into performances reduction due to unfavorable electronic properties of the films. Nevertheless, small amount of Mn alloying into CZTSSe shows positive intrinsic and device improvement. Optimization on the device regarding the buffer layer choices and fabrication parameter should be done in the future for the champion device.

## REFERENCES

[1] X. Liu, Y. Feng, H. Cui, F. Liu, X. Hao, G. Conibeer, *et al.*, "The current status and future prospects of kesterite solar cells: a brief review," *Progress in Photovoltaics: Research and Applications,* vol. 24, pp. 879-898, 2016.

[2] W. Wang, M. T. Winkler, O. Gunawan, T. Gokmen, T. K. Todorov, Y. Zhu, *et al.*, "Device Characteristics of CZTSSe Thin-Film Solar Cells with 12.6% Efficiency," *Advanced Energy Materials,* vol. 4, pp. n/a-n/a, 2014.

[3] J. Gifford, "Solar Frontier Hits 22.3% on CIGS Cell," *Industry and Suppliers, Market and Trends,* 2015.

[4] O. Gunawan, T. K. Todorov, and D. B. Mitzi, "Loss mechanisms in hydrazine-processed Cu[sub 2]ZnSn(Se,S)[sub 4] solar cells," *Applied Physics Letters,* vol. 97, p. 233506, 2010.

[5] Z. Su, J. M. R. Tan, X. Li, X. Zeng, S. K. Batabyal, and L. H. Wong, "Cation Substitution of Solution-Processed Cu2ZnSnS4 Thin Film Solar Cell with over 9% Efficiency," *Advanced Energy Materials,* vol. 5, 2015.

[6] T. Gershon, Y. S. Lee, P. Antunez, R. Mankad, S. Singh, D. Bishop, *et al.*, "Photovoltaic Materials and Devices Based on the Alloyed Kesterite Absorber (AgxCu1–x)2ZnSnSe4," *Advanced Energy Materials,* pp. n/a-n/a, 2016.

[7] A. Guchhait, Z. Su, Y. F. Tay, S. Shukla, W. Li, S. W. Leow, *et al.*, "Enhancement of Open Circuit Voltage of Solution Processed Cu2ZnSnS4 Solar Cell with 7.2% Efficiency by Incorporation of Silver," *ACS Energy Letters,* 2016.

[8] R. t. Shannon, "Revised effective ionic radii and systematic studies of interatomic distances in halides and chalcogenides," *Acta Crystallographica Section A: Crystal Physics, Diffraction, Theoretical and General Crystallography,* vol. 32, pp. 751-767, 1976.

[9] J. C. Slater, "Atomic Radii in Crystals," *The Journal of Chemical Physics,* vol. 41, pp. 3199-3204, 1964.

[10] S. Chen, J.-H. Yang, X. G. Gong, A. Walsh, and S.-H. Wei, "Intrinsic point defects and complexes in the quaternary kesterite semiconductorCu2ZnSnS4," *Physical Review B,* vol. 81, 2010.

[11] L. Chen, H. Deng, J. Tao, W. Zhou, L. Sun, F. Yue, *et al.*, "Influence of annealing temperature on structural and optical properties of Cu2MnSnS4 thin films fabricated by sol–gel technique," *Journal of Alloys and Compounds,* vol. 640, pp. 23-28, 8/15/ 2015.

[12] Y. Cui, R. Deng, G. Wang, and D. Pan, "A general strategy for synthesis of quaternary semiconductor Cu2MSnS4 (M = Co2+, Fe2+, Ni2+, Mn2+) nanocrystals," *Journal of Materials Chemistry,* vol. 22, pp. 23136-23140, 2012.

[13] S. R. Taylor, S. M. McLennan, R. L. Armstrong, and J. Tarney, "The Composition and Evolution of the Continental Crust: Rare Earth Element Evidence from Sedimentary Rocks [and Discussion]," *Philosophical Transactions of the Royal Society of London. Series A, Mathematical and Physical Sciences,* vol. 301, pp. 381-399, 1981.

[14] X. Zeng, K. F. Tai, T. Zhang, C. W. J. Ho, X. Chen, A. Huan, *et al.*, "Cu2ZnSn(S,Se)4 kesterite solar cell with 5.1% efficiency using spray pyrolysis of aqueous precursor solution followed by selenization," *Solar Energy Materials and Solar Cells,* vol. 124, pp. 55-60, 5// 2014.

[15] P. S. Rudman, J. Stringer, and R. I. Jaffee, *Phase stability in metals and alloys* vol. 1: McGraw-Hill, 1967.

[16] O. Gunawan, Y. Virgus, and K. F. Tai, "A parallel dipole line system," *Applied Physics Letters,* vol. 106, p. 062407, 2015.

[17] R. R. Prabhakar, S. Zhenghua, Z. Xin, T. Baikie, L. S. Woei, S. Shukla, *et al.*, "Photovoltaic effect in earth abundant solution processed Cu2MnSnS4 and Cu2MnSn(S,Se)4 thin films," *Solar Energy Materials and Solar Cells,* vol. 157, pp. 867-873, 12// 2016.

[18] L. E. Smart and E. A. Moore, *Solid State Chemistry: An Introduction, Fourth Edition*: CRC Press, 2016.

# Non-Vacuum Single Step Synthesis of Large-Grain Size CZTS Photo Absorber for Thin Film Solar Cells by Flux Assisted Chemical Spray

Ratheesh R. Thankalekshmi, Navjot Kaur Sidhu and A.C. Rastogi

Electrical and Computer Engineering Department & Center for Autonomous Solar Power,
Binghamton University, State University of New York, Binghamton, 13902, USA

*Abstract*— One-step preparation of stoichiometric CZTS photo absorber thin films without the secondary or amorphous phases by a modified chemical spray pyrolysis (CSP) technique is described. By incorporating sodium ($Na_2S$) as fluxing agent by intermittent spray deposition during CZTS film deposition or through post growth heat treatment with $Na_2S$ added by soaking, CZTS films with large grain size are realized. XRD patterns show characteristic diffraction lines with increased intensity and decreased FWHM in CZTS thin films both with Na flux during in-situ deposition and post deposition annealing step affirming the increase in crystallinity and grain size. This is supported by Raman studies. Whereas the average crystallite size of CSP-CZTS film is ~25 nm, using the $Na_2S$ layering within the CZTS, size increases to ~90 nm and by post growth soaking with $Na_2S$ aqueous solution a further increase in average grain size to ~0.25 μm was realized as shown by the SEM studies. Optical data shows an increase in absorption coefficient with band gap energy in the range 1.49 eV-1.62 eV in $Na_2S$ in-situ or post annealed CSP-CZTS films which is ideal for heterojunction solar cells.

*Index Terms* —CZTS, Solar cells, Thin films, X-ray diffraction.

## I. INTRODUCTION

$Cu_2ZnSnS_4$ (CZTS) as a p-type semiconductor with high absorption coefficient ($> 10^4$ $cm^{-1}$) and a direct-gap of ~1.5 eV is suitable for application as a photo-absorber in thin-film solar cells and a possible alternative to more expensive $Cu (In,Ga)Se_2$ (CIGS). CZTS has drawn much attention because of its earth abundant nontoxic material composition. The current world record conversion efficiency of CZTS is 7.6% [1] for a 1 $cm^2$ solar cell device. CZTS, a multicomponent compound semiconductor, has a complex phase diagram with a narrow range of composition for ideal solar cell quality. For this reason, highly controlled vacuum deposition techniques mostly by two-step processes have been investigated for use in fabrication of efficient solar cells. In step-one, precursor film either Cu-Zn-Sn metallic multilayered or alloyed film in desired composition is vacuum deposited and in step-two, it is sulfurized by heating to ~550°C for over 2-3 h in sulfur vapor or under $H_2S$ ambient to crystallize and obtain desired CZTS phase [2]. Through this sulfurization process, the grain size and crystallinity of the CZTS films increases with sulfurization time and temperature. In the fabrication of CZTS films, the sulfurization step could also result in the segregation of secondary metal-sulfide species due to differential sulfur reaction kinetics with Zn, Cu and Sn as well as cause composition gradients across the depth of precursor film. It is known that, efficiency of solar cells is highly dependent on CZTS grain size. CZTS films with large grain size, have lower grain boundary concentration which potentially boosts the carrier mobility. In this work, we describe a one-step synthesis of stoichiometric CZTS films without the secondary or amorphous phases by a modified chemical spray pyrolysis (CSP) technique with added $Na_2S$ flux and shows that the structural and electrical properties are comparable to those obtained by the vacuum based two-step methods. The influence of sodium (Na) on the CZTS grain size is investigated in detail. Sodium has already proven to be beneficial for high efficiency CIGS solar cells, with SLG being the substrate of choice for most high-performance CIGS solar cells [3]. However, the effect of sodium on kesterite CZTS films for solar cells has been less thoroughly investigated. For use as a solar cell photo-absorber the grain size of CZTS is important, since the grain boundary defects are the potential recombination sites and contribute to the loss of photo generated carriers. In this work, in order to improve the grain size and the crystalline nature of the CZTS photo absorber thin films, a Na-flux assisted growth methodology was introduced in the chemical spray deposition process under atmospheric conditions. Two separate studies were conducted, (1) in-situ deposition of CZTS film with addition of intermittent $Na_2S$ layer using a precursor solution and (2) post annealing of the chemical spray deposited CZTS film after soaking it in the aqueous $Na_2S$ solution. This paper will report on the direct flux assisted growth of CZTS thin films by the inexpensive, low temperature and non-vacuum CSP technique. Detailed SEM, XRD, Raman and UV–Vis spectroscopy investigations, characterizing the crystalline structure, morphology, optical absorption and band gap studies are reported. These demonstrate the efficacy of the synthesis technique in obtaining solar cell grade CZTS photo absorber thin films for solar cell devices.

## II. EXPERIMENTAL

CZTS thin films were deposited using the chemical spray pyrolysis of 0.01 M copper sulfate pentahydrate, 0.005 M zinc sulfate heptahydrate, 0.005 M tin sulfate and 0.02 M thiourea dissolved in 50% isopropyl alcohol and 50% de-ionized water mixture. Typically, the solution mixture was nitrogen-sprayed at a ~8-inch distance from the substrate typically with a spray rate ~3 ml/min, with an atomizing spray nozzle operating at ~10 psi. During the spray process, substrates were typically held at ~300°C. Substrates were polished fused quartz plates and microscopic glass slides. The stoichiometric composition of CZTS film is controlled by the elemental composition of salts

978-1-5090-5606-4/17 $31.00 © 2017 IEEE

in the precursor spray solution. CZTS growth by modified chemical spray process involves intermittent deposition of a thin Na$_2$S layer by spray from a precursor by dissolving sodium sulfide in 50% isopropyl alcohol and 50% de-ionized water mixture. Initially a ~25 nm thick CZTS-film is deposited which is then followed by a thin ~5 nm Na$_2$S layer by short duration spray. These steps were repeated multiple times to build the CZTS film thickness. Na$_2$S dissociates and added as Na, forms a flux within the CZTS layer during this deposition process. Studies were done with 5 mM and 10 mM dilution of Na$_2$S in the spray process and at ~300°C and ~400°C substrate temperatures. In another method, Na flux incorporation was achieved by soaking the as-deposited CZTS film by conventional chemical spray in 50 mM Na$_2$S precursor solution made by dissolving Na$_2$S in de-ionized water. For this, the CZTS films were immersed in Na$_2$S solution for 2 h kept in a convection oven at ~70°C. This process causes a slow diffusion of Na$_2$S within the CZTS layer. The Na$_2$S soaked CZTS film is then annealed at ~400°C in N$_2$ atmosphere at 100 sccm flow in a tubular furnace for 5-8 h.

X-ray diffraction (XRD) patterns of the CZTS films were recorded on a PANalytical's X'Pert PRO Materials Research Diffractometer with Cu Kα radiation using a Ni filter. Morphology of the films were studied by scanning electron microscopy (SEM) on field emission SEM, Supra 55 VP from Zeiss. The UV-Vis measurements were carried out using Angstrom spectrophotometer in the 400-1000 nm range. Room temperature Raman spectra were recorded using DXR spectrometer (Thermo Scientific) and 514 nm laser excitation.

## III. RESULTS AND DISCUSSION

### A. Crystalline Structure of CZTS Thin Films

The CZTS thin films as-deposited without the Na$_2$S flux are polycrystalline with kesterite crystal structure as showed by the XRD pattern in Fig. 1(a). The three most intense XRD diffraction lines are indexed according to standard values for kesterite CZTS phase (PDF#00-026-0575) that belong to the (112), (220) and (312) planes. The full width at half maximum (FWHM) of the main (112) peak is fairly large ~ 2.19° which suggests much lower degree of crystallinity. Applying the Deby-Scherrer formulation the average crystallite size in the film is determined as ~3.7 nm. While the composition requirements were largely met, yet the low crystalline status could be attributed to a low (~300°C) growth temperature. Increasing the deposition temperature is not an option as our experiments show this could induce secondary reactions and inclusion of binary phases. X-ray diffraction spectra of the CZTS films deposited with the in-situ Na$_2$S layer growth as flux using 5 and 10 mM spray solution are also shown in Fig. 1(a). These films were deposited at the same substrate temperature ~300°C as was done earlier for conventionally deposited CZTS film. These CZTS films having formed by intermittent layers of Na$_2$S flux each of 5 nm thin using spray from 5 mM and 10 mM

solutions show much sharper diffraction lines compared to the CZTS films as-spray deposited without Na$_2$S flux. Such formed CZTS films are polycrystalline and the prominent peak at ~28.55° belonging to (112) plane of CZTS has increased in intensity along with the other CZTS peaks at (220) and (312). Additional peak corresponding to (200) diffraction peak of CZTS also begins to emerge. These data indicate enhanced crystallinity of Na$_2$S co-deposited CZTS films.

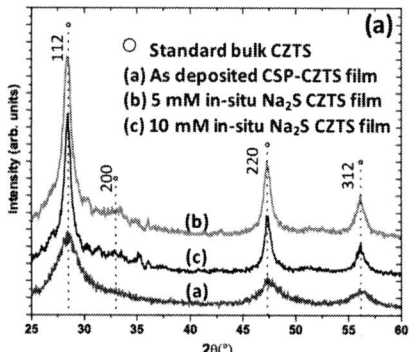

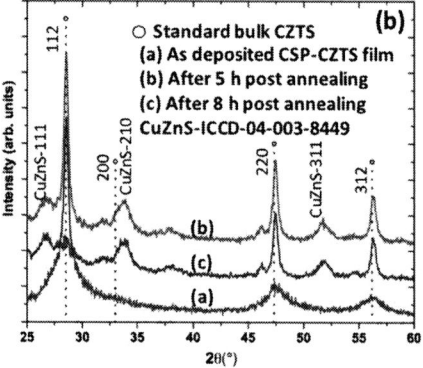

Fig. 1. X-ray diffraction spectra of CZTS films (a) with 5 mM and 10 mM in-situ Na$_2$S spray deposited layers deposited at ~300°C (b) X-ray diffraction spectra of CZTS film after soaking in 50 mM Na$_2$S and annealed at 400°C for 5 h and 8 h. As-deposited CZTS film is also shown for comparison.

Crystallite structure of the CSP deposited CZTS film formed by the second flux growth method after soaking in 50 mM Na$_2$S and annealed at ~400°C for 5 h and 8 h were studied. Figure 1(b) shows the diffraction pattern of the CZTS films post annealed with Na$_2$S flux and the CZTS film similarly deposited without flux. The diffraction lines and corresponding diffraction planes are marked in Fig. 1. It is evident that the CSP deposited CZTS films, after soaking in 50 mM Na$_2$S aqueous solution and annealed at 400°C, display significantly improved crystallinity based on the sharp increase in the intensity and narrowing of the FWHM of prominent diffraction peaks. Notably, the FWHM of the (112) peak reduced from ~2.19° to 0.5°. However, in this case one also observes secondary peaks related to CuZnS (ICCD#04-003-8449) after annealing which might have been caused by the loss of tin from CZTS samples.

The XRD studies have established the significant effect of the incorporation of sodium (Na$_2$S) in the spray deposited CZTS thin films both by in-situ deposited Na$_2$S layers during CZTS film growth by spray and by soaking in Na$_2$S solution of spray deposited CZTS films followed by an annealing step in enhancing the CZTS film crystallinity. This direct Na$_2$S flux assisted growth approaches in spray deposited CZTS film without the need for an additional sulfurization step is highly favorable for use as solar absorber in developing low cost solar cells.

### B. Raman spectroscopy of Na$_2$S flux treated CZTS films

The bonding structure of the CZTS films, spray deposited with in-situ addition of ~5 nm thin Na$_2$S flux layers from 5 mM and 10 mM spray solution were investigated using the Raman spectroscopy and the results are shown in Fig 2. For comparison, the Raman spectra of the spray deposited CZTS films formed without Na$_2$S flux is also shown. All these films were deposited at ~300°C. The kesterite CZTS phase formation in the film is indicated by the prominent peak located at 327-331 cm$^{-1}$ (peak-1) assigned to the B1 symmetry of CZTS [4], which is present in all the three films. These spectra also show the second order Raman peak centered around 654 cm$^{-1}$ which is a characteristic feature of CZTS phase formation [5]. A second well resolved peak (peak-2) at ~286-288 cm$^{-1}$ is only seen in the CZTS films deposited with Na$_2$S flux layers. This peak is a strong indication of the kesterite phase formation based on earlier reports [6]. The 287 cm$^{-1}$ (peak-2) is not well resolved but appears as broad shoulder in the as-deposited CZTS films without the Na$_2$S flux as shown later in Fig 4(a). This provides strong support that Na$_2$S flux improves crystallinity of CZTS films.

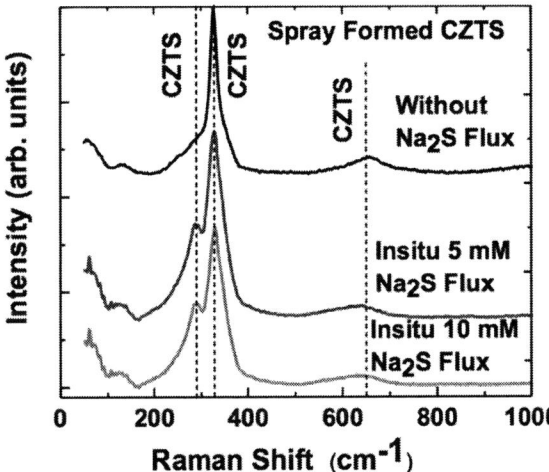

Fig. 2. Raman spectra from 5 mM and 10 mM in-situ Na$_2$S CSP-CZTS films deposited at ~300°C. Occurrence of characteristic CZTS peak positions are indicated by dotted lines. As-deposited CSP-CZTS film is also shown for comparison.

The Raman spectra of the CZTS film formed by the second flux growth method in which the as spray deposited CZTS films were annealed at ~400°C for 5 h and 8 h after soaking in 50 mM Na$_2$S solution were also studied. The Raman spectra is shown in Fig. 3, and here too, the prominent peak-1 around 327-331 cm$^{-1}$ and its second order peak at ~652 cm$^{-1}$ both are observed similar to as-made CZTS film without Na$_2$S flux indicating CZTS phase. Similar to the previous Na$_2$S flux method, in the 5 h annealed film, a well resolved peak-2 at 286-287 cm$^{-1}$ confirms that the kesterite CZTS phase formation is enhanced by Na$_2$S addition by the second method as well. However, in CZTS films subjected to long term annealing for 8 h, the Raman peak-2 at ~287 cm$^{-1}$ is not well resolved and at the same time appearance of a Raman peak at ~470 cm$^{-1}$ is noted, which is attributed to CuZnS (CZS) indicating possible inclusion of ternary phase [7]. It may be recalled that a similar inference was drawn earlier from the analysis of the XRD data.

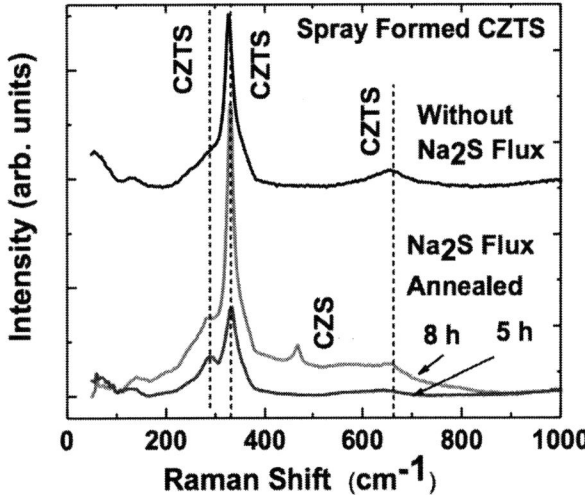

Fig. 3. Raman spectra from CSP-CZTS film after soaking in 50 mM Na$_2$S and annealed at ~400°C for 5 h and 8 h. As-deposited CSP-CZTS film is also shown for comparison.

For better understanding of the characteristic Raman peaks 1 and 2, we analyzed these using the Lorentzian-Gaussian peak fitting program and the results are shown in Fig. 4. The Raman spectrum features in 240-380 cm$^{-1}$ range for CZTS film formed with in-situ layers of Na$_2$S from 10mM solution show peak-1 at 329.5, peak-2 at 286 and a third peak-3 at 362.6 cm$^{-1}$ having FWHM as 36.7, 31.3 and 32.6 cm$^{-1}$, respectively (Fig. 4(b)). Similarly, the CZTS film soaked in Na$_2$S and annealed at 400°C for 5 h shows these peaks at 331.3, 287.9 and 363.7 cm$^{-1}$ with FWHM as 27.5, 35.0 and 58.2 cm$^{-1}$, respectively (Fig. 4(c)). The peak-3 around 363.2 cm$^{-1}$ seen in both Na$_2$S flux treated methods are identified with the CZTS kesterite phase [6]. The broad peak features of the as-deposited CZTS without the Na$_2$S flux are resolved as peak-1 at 327.3, peak-2 at 287.1 and a third peak-3 at 352.8 cm$^{-1}$ having FWHM as 21.2, 37.2 and 31.2 cm$^{-1}$, respectively (Fig. 4(a)). The peak-3 here is not at the wavenumber position as observed for $Na_2S$ treated CZTS films and is attributed to the ZnS secondary phase [8]. The observed shift in the peak-1 to higher wavenumbers by $Na_2S$ treatment is expected due to enhanced phase reactions via sulfurization leading to CZTS formation and the observed reduction in the FWHM of the peak-2 assigned to evolution of kesterite CZTS phase with $Na_2S$ flux addition is attributed to the crystallite growth.

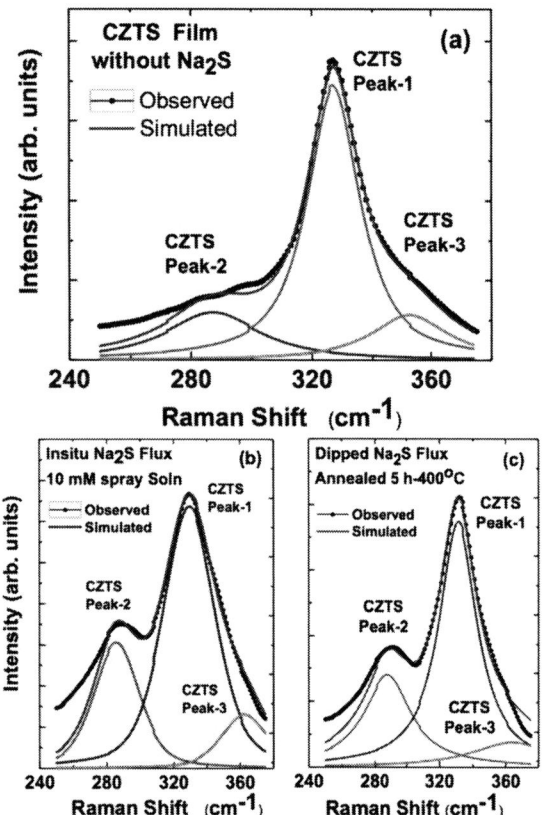

Fig. 4. Raman peaks resolved in the range 240-380 cm$^{-1}$ (a) as-deposited CZTS film (b) CZTS film with in-situ $Na_2S$ layers from 10mM solution spray and (c) CZTS film annealed at 400°C for 5 h after soaking in $Na_2S$.

Raman studies have clearly shown the kesterite CZTS films formation is enhanced by incorporation of sodium ($Na_2S$) in the spray deposited CZTS thin films both by in-situ addition of the $Na_2S$ layers during CZTS film growth by spray and by soaking in $Na_2S$ solution of spray deposited CZTS films followed by an annealing step for 5 h as compared to the as-deposited CZTS films.

### C. Morphology of CZTS thin films

Microstructure of the CSP deposited CZTS film shows a highly compact morphology mostly free of pin holes and with high surface uniformity as displayed in Fig. 5. However, the major deficiency of the chemical spray process appears to be that it does not lead to the formation of large size crystallites.

As shown in the micrograph, instead the crystallites are small and appear to be clustered together and are homogeneously distributed over the entire film surface.

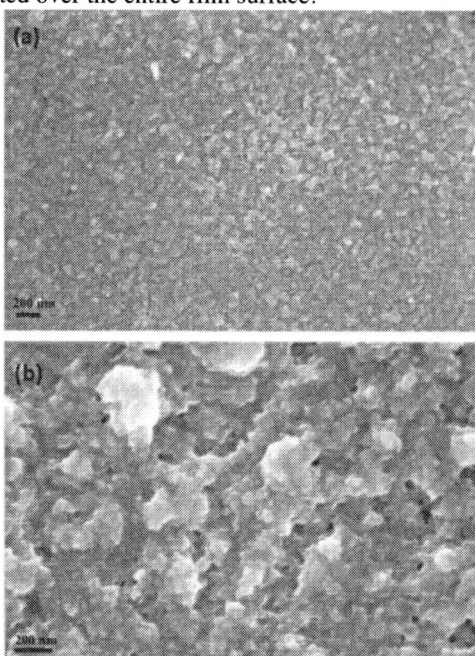

Fig. 5. CSP-CZTS film on glass substrate showing (a) highly compact microstructure, free of pin holes (b) magnified image of showing that the small crystallites are typically clustered together.

The XRD studies discussed in the previous section have shown that improvement in the crystalline features of the CZTS are enabled by the $Na_2S$ flux treatment. Microstructure of the CZTS film formed by in-situ layering of $Na_2S$ flux by spray from 5 mM and 10 mM solution deposited at ~300°C are shown in Fig. 6 (a) and (b) respectively. It is interesting that increase in the $Na_2S$ flux content increased the crystallite size. Compared to the grain structure of the conventionally spray deposited CZTS film significant improvement in the grain growth is observed. The as-deposited CZTS film showed average crystallite size ~25 nm whereas with $Na_2S$ in-situ flux addition via 5 mM and 10 mM spray at ~300°C, the average crystallite size increased to ~50 nm and ~90 nm respectively. These crystallites are dispersed and not compact as expected. It appears from SEM micrographs that CZTS films with co-deposited $Na_2S$ might have developed randomly distributed crystallite and therefore some porosity.

Compared to the in-situ deposited $Na_2S$ flux layers in the-CZTS films, the $Na_2S$ soaked and the post annealed CZTS films have shown a significant enhancement in the crystallite growth. Fig. 7 (a) and (b) show the microstructure of the CZTS films annealed at 400°C for 5 and 8 h, respectively after soaking in 50 mM $Na_2S$ aqueous solution. The average crystallite size in the CZTS films after post-soak annealing for 5 h (Fig. 7 (a)) and 8 h (Fig. 7 (b)) determined from SEM micrographs is ~150 nm and ~250 nm, respectively. The loss of Sn and consequent secondary CuZnS phase inclusion in annealed CZTS as shown by XRD study can be suppressed by increasing the Sn-precursor concentration in the starting spray solution. This renders the post-Na$_2$S soak and annealed process for CZTS films most attractive for forming CZTS solar absorber films with higher crystallite size and increased crystallinity.

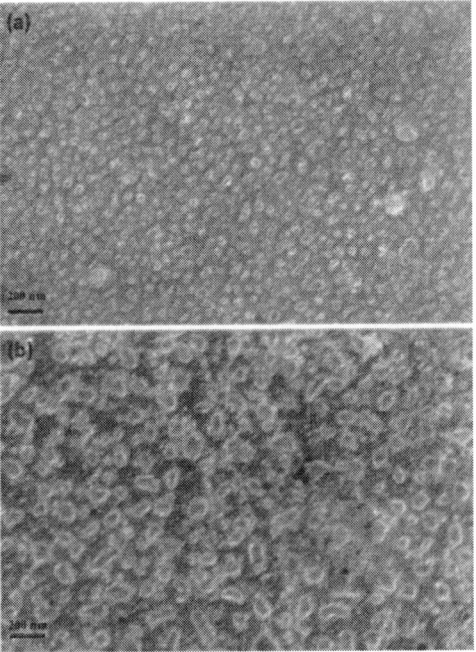

Fig. 6. SEM micrographs of - CZTS films with in-situ Na$_2$S layers deposited at ~300°C (a) 5 mM Na$_2$S and (b) 10 mM Na$_2$S.

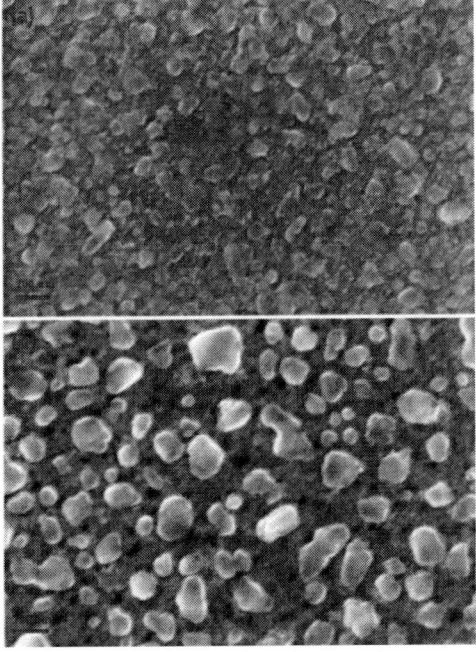

Fig. 7. SEM micrographs of CSP-CZTS film after soaking in 50 mM Na$_2$S and annealed at ~400°C (a) after 5 h annealing and (b) after 8 h annealing.

## D. Optical bandgap studies of CZTS thin films

The effect of Na$_2$S treatment on the optical characteristics of spray deposited CZTS films was studied. Figure 8 (a) shows the transmission spectra of CZTS-films deposited by the conventional spray and by modified spray in which Na$_2$S flux layers were added during the CZTS film growth using 5 mM and 10 mM Na$_2$S spray solution at ~300°C. All CZTS films show an optical absorption threshold around 800-850 nm, however, the flux assisted CZTS show much steeper wavelength dependence. From the optical transmission spectra, optical band gap E$_G$ was determined using the Tauc's relation for direct band gap semiconductors. The corresponding Tauc's plots are shown in Fig. 8 (b) for CZTS films. The optical band gap energy of the CZTS-film deposited by the conventional spray is determined as ~1.53 eV. The optical band gap energy of CSP-CZTS films formed using in-situ Na$_2$S flux layers sprayed from 5 mM and 10 mM solution is determined from the Tauc plots as ~1.55 eV and ~1.62 eV, respectively.

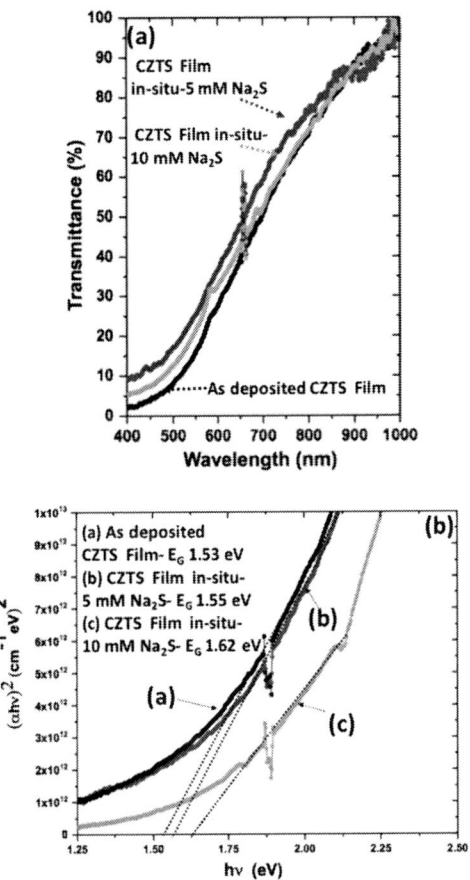

Fig. 8 (a). Optical transmission spectra of CZTS films before and after in-situ Na$_2$S treatment. (b) Optical band gap analysis showing the corresponding band gap values of CZTS films before and after in-situ Na$_2$S treatment.

978-1-5090-5606-4/17 $31.00 © 2017 IEEE

The optical band gap determination was similarly carried out for CZTS films formed by $Na_2S$ flux soak and anneal process. Fig. 9 (a) shows the transmission spectra of CSP-CZTS film annealed at ~400°C for 5 h and 8 h after soaking in 50 mM $Na_2S$ solution. There was no substantive change in the transmission profile. Both annealed CZTS films showed optical absorption threshold around 800-850 nm. The Tauc's plot of annealed CZTS films is shown in Fig. 9 (b). Optical band gap energy of the post annealed CSP-CZTS after 5 h and 8 h annealing was determined as 1.58 eV and 1.49 eV, respectively. The results of the optical absorption study demonstrate that optical band gap energy of the $Na_2S$ flux assisted spray deposited films are not different from those formed without flux addition. These values are also consistent with published band gap values of CZTS formed by other methods [9-10] and within the acceptable range required for solar absorber in solar cells.

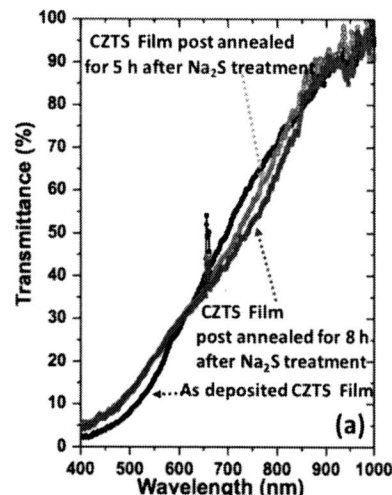

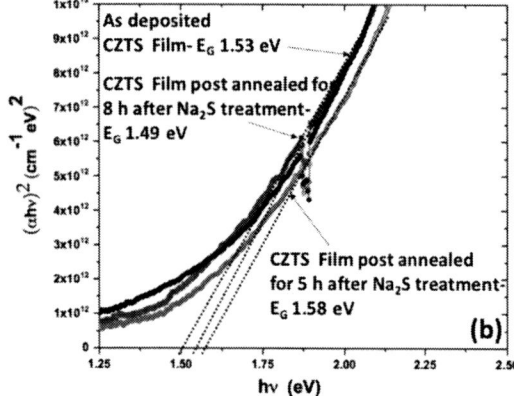

Fig. 9 (a). Optical transmission spectra of CZTS films before and after soaking in 50 mM $Na_2S$ and annealed at ~400°C for 5 h and 8 h. (b) Optical band gap analysis showing the corresponding band gap values of CZTS films before and after soaking in 50 mM $Na_2S$ and annealed at ~400°C for 5 h and 8 h.

## IV. CONCLUSIONS

This work showed the effects of Na on the grain growth of CZTS and demonstrated that with a single-step process CZTS films with sufficient crystallite size can be formed using spray pyrolysis technique which can be used as a photo absorber thin film for solar cell devices. The average crystallite size of CSP-CZTS film after in-situ $Na_2S$ deposition increased to ~90 nm, and after post annealing & soaking with $Na_2S$ aqueous solution showed a further increase in grain size to ~0.25 μm. The crystallinity of the CZTS film improved significantly as evident from sharp increase and narrow FWHM of prominent XRD peaks of CZTS film. Raman studies have clearly shown that kesterite CZTS films formation is enhanced by incorporation of sodium ($Na_2S$) in the spray deposited CZTS thin films. Optical analysis shows that the CZTS films have an optical band gap in the range 1.5 eV to 1.6 eV ideal for solar cells.

## REFERENCES

[1] University of New South Wales. "At last: Non-toxic and cheap thin-film solar cells for 'zero-energy' buildings: World's highest efficiency rating achieved for CZTS thin-film solar cells" Science Daily, 28 April 2016.

[2] H. Katagiri, N. Sasaguchi, S. Hando, S. Hoshino, J. Ohashi, T. Yokota, "Preparation and evaluation of $Cu_2ZnSnS_4$ thin films by sulfurization of E-B evaporated precursors," *in: Technical Digest of the 9th International PVSEC—Miyazaki, Japan*, pp. 745-746, 1996.

[3] A. Sadono, T. Ogihara, M. Hino, et al. "Peeled-off flexible $Cu(In,Ga)Se_2$ solar cells and Na diffusion effects on cell performances," *Electronic Materials Letters*, Vol. 12, pp. 494-498, 2016.

[4] K. Patel, D.V. Shah and V. Kheraj, "Influence of deposition parameters and annealing on $Cu_2ZnSnS_4$ thin films grown by SILAR," *Journal of Alloys and Compounds*, Vol. 662, pp. 942-947, 2015.

[5] M. Dimitrievska, A. Fairbrother, X. Fontané, T. Jawhari, V. Izquierdo-Roca, E. Saucedo and A. Pérez-Rodríguez, "Multiwavelength excitation Raman scattering study of polycrystalline kesterite $Cu_2ZnSnS_4$ thin films," *Applied Physics Letters*, Vol. 104, pp. 021901 1-5, 2014.

[6] P. A. Fernandes, P. M. P. Salome and A. F. da Cunha, "Growth and Raman scattering characterization of $Cu_2ZnSnS_4$ thin films," *Thin Solid Films*, Vol. 517, pp. 2519-2523, 2009.

[7] Sreejith, M. S., et al. "Tuning the properties of sprayed CuZnS films for fabrication of solar cell," *Applied Physics Letters*, Vol. 105, 202107, 2014.

[8] P. A. Fernandes, P. M. P. Salome and A. F. da Cunha, "Study of the polycrystalline $Cu_2ZnSnS_4$ films by Raman scattering," *Journal of Alloys and Compounds*, Vol. 509, pp. 7600-7606, 2011.

[9] V.G. Rajeshmon, C. S. Kartha, K.P. Vijayakumar, C. Sanjeeviraja, T. Abe and Y. Kashiwaba, "Role of precursor solution in controlling the opto-electronic properties of spray pyrolysed $Cu_2ZnSnS_4$ thin films," *Solar Energy*, Vol. 85, pp. 249-255, 2011.

[10] H. Park, Y.H. Hwang and B.S. Bae, "Sol–gel processed $Cu_2ZnSnS_4$ thin films for a photovoltaic absorber layer without sulfurization," *Journal of Sol-Gel Science and Technology*, Vol. 65, pp. 23-27 2013.

# Raman scattering assessment of point defects in kesterite semiconductors: UV resonant Raman characterization for advanced photovoltaics

Florian Oliva[1], Laia Arqués Farré[1], Sergio Giraldo[1], Mirjana Dimitrievska[1,2,3], Paul Pistor[1], Alejandro Martínez-Pérez[1], Lorenzo Calvo-Barrio[4], Edgardo Saucedo[1], Alejandro Pérez-Rodríguez[1,5], Victor Izquierdo-Roca[1]

1 – Catalonia Institute for Energy Research (IREC), Jardins de les Dones de Negre 1, 08930 Sant Adrià de Besòs, Spain

2 – NIST Center for Neutron Research, National Institute of Standards and Technology, Gaithersburg, MD 20899-6102, United States

3 – National Renewable Energy Laboratory, Golden, CO 80401, United States

4 – Centres Científics i Tecnològics de la Universitat de Barcelona (CCiTUB), Lluís Solé i Sabarís 1-3, 08028 Barcelona, Spain

5 – IN$^2$UB, Departament d'Electrònica, Universitat de Barcelona, C. Martí i Franquès 1, 08028 Barcelona, Spain

*Abstract* — **Raman spectroscopy has demonstrated to be a powerful tool for Cu$_2$ZnSnSe$_4$ (CZTSe) characterization, allowing the assessment of relevant parameters such as crystal quality, secondary phases and defect presence. In this work a detailed analysis of CZTSe vibrational properties using non-bandgap Raman resonance effects is performed. UV-based Raman spectroscopy is presented as a promising technique for the precise assessment of V$_{Cu}$ and Zn$_{Sn}$ point defects, associated with 174 and 245 cm$^{-1}$ Raman regions. Based on these results, correlation between the peak intensity, associated with V$_{cu}$ point defects, with the maximum value of the efficiency of the final devices is presented and discussed.**

*Index Terms* — **thin film solar cells, Raman spectroscopy, point defects, kesterite.**

## I. INTRODUCTION

Recently thin film solar cells made from earth-abundant materials have attracted significant amount of attention as they could become an interesting alternative to the commonly used CIGS technology. Kesterite (Cu$_2$ZnSnSe$_4$, CZTSe) compound has emerged as one of compelling absorber alternatives achieving a remarkable efficiency of 12.6% [1]. However, the highest reported efficiency is far inferior to its thin-film counterparts. One of the main factors limiting the efficiency of CZTSe-based devices is the difficulty to control CZTSe composition during its synthesis process and/or during post-process treatment. The uncontrolled compositional variations during production process have a huge impact on the electronic defect structure and secondary phase formation and thus on the performance of solar cells.

Multiwavelength excitation Raman spectroscopy has proven to be a powerful tool for identifying secondary phases, assessing crystal quality and characterizing defects in kesterite compounds, mostly due to the use of resonant conditions, where photons and material bandgap energies are coupled. Tuning the incident laser energy to resonant conditions enables the enhancement of Raman sensitivity, by increasing the intensity of vibrational modes associated with particular energetic transitions.

Using Raman resonant conditions in kesterite materials is particularly interesting due to the strong correlation of spectrum modifications with material properties [2]. However, under IR excitation conditions, a strong photoluminescence (PL) background coupled with a low Raman efficiency of CZTSe spectra compromise the application of bandgap resonance effect conditions. Alternatively, the use of UV excitation is promising. Until now, technical limitations for assessing the spectral region of 100-300 cm$^{-1}$ impeded its full exploitation but this issue can now be handled by using specifically developed Raman setup. In this study, we report for the first time, Raman resonant behavior using UV excitation in CZTSe material, allowing an enhanced sensitivity of the Raman spectrum to structural-compositional changes without the characteristic photoluminescence background of bandgap related resonant conditions. Additionally, using UV excitation wavelength instead of the visible ones has several advantages such as its weak penetration depth into the material (<10 nm) which makes Raman measurements strongly surface sensitive. Then buffer/CZTSe interface region as well as secondary phases which could segregate at the absorber surface can easily be studied.

In this work, a complete evaluation of CZTSe Raman spectra and their dependence on the excitation wavelength in the range of UV-NIR is presented. Then, a detailed Raman analysis of CZTSe under UV excitation wavelength is performed. Clear changes are detected in Raman spectra of samples with composition close to high efficiency device composition (variation less than 5%). Additionally, correlations are made between UV non-bandgap resonant Raman spectra and the optoelectronic properties of the cells using more than 40 devices with efficiencies ranging between 6.8 and 7.5%.

## II. EXPERIMENTAL

Pure selenide kesterite absorbers (CZTSe) with stoichiometric and lateral composition gradients were grown by a two stage process. Firstly, metallic stacks of Sn/Cu/Zn were deposited by direct current (DC) magnetron sputtering onto Mo(800nm) coated soda lime glass substrates as described in detail in [3]. These precursor stacks (5x5 cm$^2$) are further reactively annealed in a graphite box containing elemental Se and Sn powders (100 mg Se and 5 mg Sn) inside of a conventional 3 zone tube furnace under argon atmosphere. A two-step temperature profile is used for annealing. First a selenization at 400°C for 30 minutes under argon flow, keeping the pressure at 1.5mbar, is carried out followed by a second shorter annealing at 550°C for 15 minutes under static 1000 mbar argon pressure to improve crystallinity. Cooling is allowed naturally to room temperature which normally takes about 1.5 hours. After absorber growth, an etching was performed so that ZnSe and SnSe phases were removed using KMnO$_4$ and (NH$_4$)$_2$S solutions respectively. A sample was obtained with a lateral graded composition corresponding to [Cu]/([Zn]+[Sn]) ratio variations in the range of 0.50-0.85 while [Zn]/[Sn] ratio remained between 0.95 and 1.70. Then samples were finished in 3x3 mm$^2$ devices. Absorber composition was evaluated by XRF (Fisherscope XVD) [4] analysis and the surface composition by XPS measurements (PHI-ESCA 5500). Raman scattering measurements were performed using two different equipments: 1) a Horiba Jobin Yvon Labram UV-HR800 spectrometer coupled with a 1064 nm laser, 2) an UV optimized optical probe developed at IREC and coupled by optical fiber with FHR640 Horiba Jobin Yvon spectrometers for the measurements under 325, 442 and 532 nm excitation wavelengths. For all measurements Backscattering configuration with low density power is used, in order to avoid any thermal effect. A spot diameter size higher than 50 μm is used in order to evaluate a representative surface of the material. To calibrate the Raman frequency, a monocrystalline silicon sample has been measured and used as reference imposing the main peak position at 520 cm$^{-1}$. Standard uncertainties in all figures in the text, if not explicitly indicated, are commensurate with the observed scatter in the data.

## III. RESULTS AND DISCUSSION

Fig. 1 shows the experimental Raman spectra of stoichiometric CZTSe thin film measured with different excitation wavelengths (325, 532 1064 nm). XPS measurements (not shown) demonstrated that absorber layers are homogeneous with depth which led to the assumption that surface Raman measurements are representative of the bulk. Simultaneous fittings of spectra with Lorentzian curves led to the identification of 15 resolved peaks attributed to zone center phonon representation for CZTSe and located at 70, 80, 120, 141, 151, 172, 174, 178, ,192, 196, 221, 232, 234, 250 and 255 cm$^{-1}$. Under 532 nm excitation wavelength Raman spectrum depicts an intense A' mode at 196 cm$^{-1}$ with several weaker contributions attributed to second A" and polar B and E modes. In contrast, 325 (3.81 eV) and 1064 (1.16 eV) nm excitations impose strong changes in the Raman features. In fact, under 1064 nm excitation wavelength spectrum is dominated by a peak at 242 cm$^{-1}$ while spectrum measured with 325 nm laser is dominated by two peaks located at 147 and 234 cm$^{-1}$ associated with E/B(LO) modes.

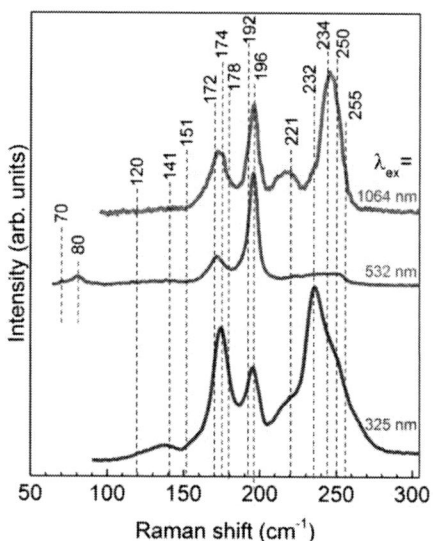

Fig. 1. Raman spectra of stoichiometric CZTSe material under 1064, 532, 325 nm excitation wavelengths

Enhancement of these peaks is attributed to the presence of resonant behavior due to the coupling of photon with the bandgap transition in the case of 1064 nm excitation wavelength and with the inter-band transitions E($\Gamma_3$) and/or E$_{1B}$ for 325 nm excitation case [5,6]. Similar phenomenon where each energetic level is dominated by a specific atomic vibration has been reported by Dimitrievska et al. in $ZnS_xSe_{1-x}$ system using selective resonances with 325 and 455nm excitation wavelengths [7]. In the case of CZTSe compound Dimitrievska et al. has also reported a correlation of 174 and 250 cm$^{-1}$ Raman peaks intensity measured under 532 nm excitation [8]. They attributed those changes to composition variation and specifically to copper and tin deficits but also to A-type ($[Zn_{Cu}+V_{Cu}]$) and B-type ($[2Zn_{Cu}+Zn_{Sn}]$) defect clusters density [9]. In fact, as Raman peak intensity is proportional to the density of phonons coming from a specific atomic vibration, modifying the composition of a material changes the density of vibration and directly impacts Raman peak intensity. More detail about the identification of Raman peak with vibrations of CZTSe atoms can be found in [10]. Here we investigate the presence of similar variations under 325 nm excitation with a possible higher resolution due to resonant conditions. However as already explained, a strong PL background together with a low Raman efficiency make studies under 1064 nm difficult. Thus, this work is limited to the case of 325 nm resonant conditions.

First it is worth mentioning that impact of secondary phases and, in particular ZnSe compound depicting a Raman peak at 250 cm$^{-1}$, was discarded during the study. In fact, despite the detection of ZnSe at the surface under resonant conditions (442 nm) (not shown), no correlations could be obtained between its concentration and the modifications of Raman spectra that we observed.

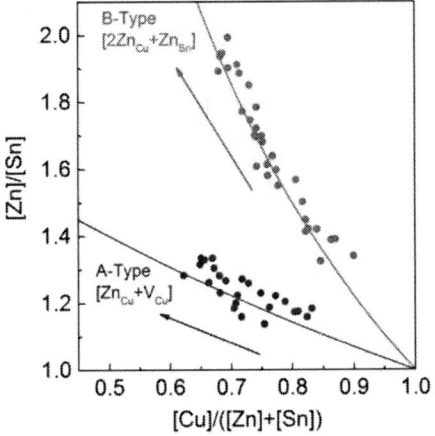

Fig. 2. Composition mapping of the samples. Blue and red lines present the A/B-type defect clusters. Blue and red dots are selected set of samples with different A and B-type defect clusters concentration, respectively. Arrows indicate in which way the defect cluster concentrations increase.

To evaluate the impact of the composition/defect on 325 nm Raman spectra, measurements were performed on each cell of the compositionally graded sample. A full composition/Raman mapping was obtained but for clarity we focus our study on samples with composition close to A-type and B-type defect lines as reported by Lafond et al. [11], see Fig. 2. Raman measurements under 325 nm excitation wavelength were

performed on each set of "defect line samples" and resulting spectra are depicted in Fig. 3. Raman spectra of B-type defect samples present an increase of the 250 cm$^{-1}$ spectral region intensity with higher concentration of B-type cluster defects, Fig. 3 (right), while 174 cm$^{-1}$ spectral region stays unaffected. As B-type line is directly linked to $Zn_{Cu}$ and/or $Zn_{Sn}$ point defects, one can assume that those Raman spectrum modifications come from the replacement of copper and tin atoms by zinc atoms in the crystal structure. Moreover, looking at Raman spectra from samples of the A-type defect line, no clear trend can be drawn between those punctual defects and the intensity of 250 cm$^{-1}$ spectral region: as A-type cluster defect concentration increases, the whole Raman spectrum is affected and not only 250 cm$^{-1}$ region as observable in Fig. 3 (left), meaning that this particular spectral region is not directly impacted by $[Zn_{Cu}+V_{Cu}]$ defect cluster. By deduction, it is concluded that 250 cm$^{-1}$ spectral region variations are most probably linked to $Zn_{Sn}$ defect concentration.

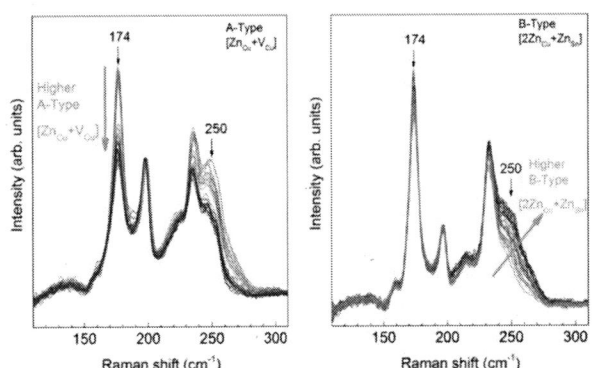

Fig. 3. Raman spectra of samples with composition close to A/B-type defect cluster lines performed under 325 nm excitation wavelength. For clarity, spectra are normalized to 196 cm$^{-1}$ peak intensity.

The same methodology is applied to the 174 cm$^{-1}$ spectral region. First, Raman spectra of samples from the B-type defect line do not show any modification of 174 cm$^{-1}$ spectral region. On the contrary, spectra from A-type defect line depict a reduction of 174 cm$^{-1}$ peak intensity which is correlated with lower copper content. As a result, 174 cm$^{-1}$ spectral region variations are related to $[Zn_{Cu}+V_{Cu}]$ defect clusters and not to $[2Zn_{Cu}+Zn_{Sn}]$ ones, suggesting that these variations are mainly impacted by $V_{Cu}$ point defects. Interestingly, similar effects have been simulated for CZTS by Kosyak et al. [12] and have even been experimentally observed by Raman spectroscopy on copper defective $Cu_2SnS_3$ compound [13].

Finally, correlations between modification of Raman spectra and the optoelectronic properties of CZTSe solar cells were performed. In order to assess with precision the modification of Raman spectra, the ratio of areas corresponding to the peaks at 174 cm$^{-1}$ (A(174)) and 196 cm$^{-1}$ (A(196)) was calculated.

Efficiencies of CZTSe solar cells is plotted as a function of this ratio in Fig. 4. A clear correlation of the maximal achievable efficiency with this ratio can be obtained and a maximum efficiency is reached for a ratio of about 0.58. For ratios lower than 0.56, a decrease of performance is observed which could be the consequence of an excessive formation of copper vacancies. In a similar way for ratios higher than 0.60, lack of defects seems detrimental to device performances. This observation agrees with the one reported by Dimitrievska et al. [8] but the higher sensitivity induced by UV-Raman resonance effects permits an accurate evaluation even with only small absorber composition variations (<5%) which in the end still have a strong impact on the device performances.

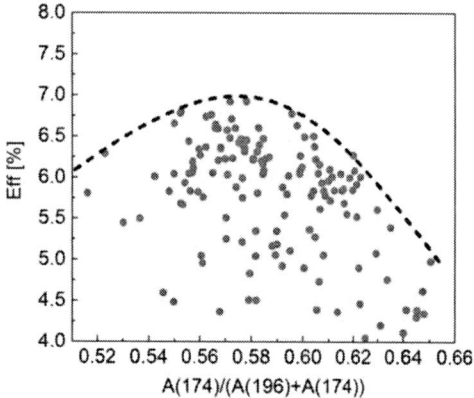

Fig. 4. Correlation of CZTSe solar cell efficiency with ratio of Raman spectrum areas. A(174) and A(196) correspond to 174 and 196 cm$^{-1}$ peak areas respectively.

## IV. CONCLUSIONS

CZTSe samples with different compositions have been characterized by UV-Raman spectroscopy. For the first time, Raman spectra of CZTSe measured under band gap and "non-bandgap" resonant conditions are reported. The analysis of peak intensities as a function of the composition allowed the correlation of the 174 cm$^{-1}$ peak intensity with $V_{cu}$ point defects, and 245 cm$^{-1}$ peak intensity with $Zn_{Sn}$ point defects. Furthermore, Raman characterization of absorber layers from complete devices under UV excitation, led to the precise correlation of the 174 cm$^{-1}$ peak intensity with the maximum value of device efficiency. The results obtained in this work show the potential of UV Raman application to CZTSe in order to evaluate defects and composition variations with high accuracy and without any degradation of the layers. These observations could lead in the long run to process optimization and more efficient CZTSe solar cells.

## ACKNOWLEDGEMENTS

M. Dimitrievska gratefully acknowledges support from the US DOE Office of Energy Efficiency and Renewable Energy, Fuel Cell Technologies Office, under Contract No. DE-AC36-08GO28308. The research leading to these results has received funding from MINECO (Ministerio de Economía y Competitividad de España) under the NASCENT project (ENE2014-56237-C4-1-R) and from the European H2020 Framework Programme for research, technological development and demonstration under grant agreement no H2020-NMBP03-2016-720907 (STARCELL project). Authors from IREC and IN2UB belong to the M-2E (Electronic Materials for Energy) Consolidated Research Group and the XaRMAE Network of Excellence on Materials for Energy of the "Generalitat de Catalunya". P.P acknowledges funding from the European Commision under the FP7 Marie Curie Individual Fellowship program (Jumpkest, FP7-PEOPLE-2013-IEF- 625840).

## REFERENCES

[1] W. Wang, M. T. Winkler, O. Gunawan, T. Gokmen, T. K. Todorov, Y. Zhu, and D. B. Mitzi, "Device Characteristics of CZTSSe Thin-Film Solar Cells with 12.6% Efficiency," *Adv. Energy Mater.*, vol. 4, no. 7, p. 1301465, May 2014.

[2] J. J. S. Scragg, L. Choubrac, A. Lafond, T. Ericson, and C. Platzer-Björkman, "A low-temperature order-disorder transition in Cu$_2$ZnSnS$_4$ thin films," *Appl. Phys. Lett.*, vol. 104, no. 4, p. 41911, Jan. 2014.

[3] S. López-Marino, M. Placidi, A. Pérez-Tomás, J. Llobet, V. Izquierdo-Roca, X. Fontané, A. Fairbrother, M. Espíndola-Rodríguez, D. Sylla, A. Pérez-Rodríguez, and E. Saucedo, "Inhibiting the absorber/Mo-back contact decomposition reaction in Cu$_2$ZnSnSe$_4$ solar cells: the role of a ZnO intermediate nanolayer," *J. Mater. Chem. A*, vol. 1, no. 29, p. 8338, 2013.

[4] The mention of all commercial suppliers in this paper is for clarity. This does not imply the recommendation or endorsement of these suppliers by NIST.

[5] M. León, S. Levcenko, R. Serna, I. V. Bodnar, A. Nateprov, M. Guc, G. Gurieva, N. Lopez, J. M. Merino, R. Caballero, S. Schorr, A. Perez-Rodriguez, and E. Arushanov, "Spectroscopic ellipsometry study of Cu$_2$ZnSnSe$_4$ bulk crystals," *Appl. Phys. Lett.*, vol. 105, no. 6, p. 61909, Aug. 2014.

[6] C. Persson, "Electronic and optical properties of Cu$_2$ZnSnS$_4$ and Cu$_2$ZnSnSe$_4$," *J. Appl. Phys.*, vol. 107, no. 5, p. 53710, Mar. 2010.

[7] M. Dimitrievska, H. Xie, A. J. Jackson, X. Fontané, M. Espíndola-Rodríguez, E. Saucedo, A. Pérez-Rodríguez, A. Walsh, and V. Izquierdo-Roca, "Resonant Raman scattering of ZnS$_x$Se$_{1-x}$ solid solutions: the role of S and Se electronic states," *Phys. Chem. Chem. Phys.*, vol. 18, no. 11, pp. 7632–7640, 2016.

[8] M. Dimitrievska, A. Fairbrother, E. Saucedo, A. Pérez-Rodríguez, and V. Izquierdo-Roca, "Influence of compositionally induced defects on the vibrational properties of device grade Cu$_2$ZnSnSe$_4$ absorbers for kesterite based solar cells," *Appl. Phys. Lett.*, vol. 106, no. 7, p. 73903, Feb. 2015.

[9] M. Dimitrievska, A. Fairbrother, E. Saucedo, A. Pérez-Rodríguez, and V. Izquierdo-Roca, "Secondary phase and Cu substitutional defect dynamics in kesterite solar cells: Impact on

optoelectronic properties," *Sol. Energy Mater. Sol. Cells*, vol. 149, pp. 304–309, May 2016.

[10] N. B. Mortazavi Amiri and A. Postnikov, "Electronic structure and lattice dynamics in kesterite-type $Cu_2ZnSnSe_4$ from first-principles calculations," *Phys. Rev. B*, vol. 82, no. 20, p. 205204, Nov. 2010.

[11] A. Lafond, L. Choubrac, C. Guillot-Deudon, P. Deniard, and S. Jobic, "Crystal Structures of Photovoltaic Chalcogenides, an Intricate Puzzle to Solve: the Cases of CIGSe and CZTS Materials," *Zeitschrift für Anorg. und Allg. Chemie*, vol. 638, no. 15, pp. 2571–2577, Dec. 2012.

[12] V. Kosyak, N. B. Mortazavi Amiri, A. V Postnikov, and M. A. Scarpulla, "Model of native point defect equilibrium in $Cu_2ZnSnS_4$ and application to one-zone annealing," *J. Appl. Phys.*, vol. 114, no. 12, p. 124501, Sep. 2013.

[13] L. L. Baranowski, P. Zawadzki, S. Christensen, D. Nordlund, S. Lany, A. C. Tamboli, L. Gedvilas, D. S. Ginley, W. Tumas, E. S. Toberer, and A. Zakutayev, "Control of Doping in $Cu_2SnS_3$ through Defects and Alloying," *Chem. Mater.*, vol. 26, no. 17, pp. 4951–4959, Sep. 2014.

# Assessing the defect responsible for LeTID: temperature- and injection-dependent lifetime spectroscopy

Mallory A. Jensen[1], Yan Zhu[2], Erin E. Looney[1], Ashley E. Morishige[1], Carlos Vargas[2], Ziv Hameiri[2], Tonio Buonassisi[1]

[1]Massachusetts Institute of Technology, Cambridge, MA 02139, USA, email: jensenma@alum.mit.edu
[2]University of New South Wales, Sydney, NSW 2052, Australia

*Abstract* — Temperature- and injection-dependent lifetime spectroscopy (TIDLS) is employed to study the defect responsible for light- and elevated temperature-induced degradation (LeTID). In our previous analyses, titanium (Ti), molybdenum (Mo), and tungsten (W) were identified as potential candidates for LeTID. The addition of temperature dependence further constrains the defect parameters. Assuming constant defect parameters with temperature, we identify two possible sets of defect parameters: $k$ = 23.9 ± 5.5 at $E_t$-$E_i$ = -0.21 ± 0.06 eV and $k$ = 23.5 ± 5.6 at $E_t$-$E_i$ = -0.10 ± 0.07 eV. We consider our results in the context of published defect parameters identified by TIDLS in other LeTID samples, and we evaluate our results against reported defect parameter temperature dependencies for Ti and Mo. We conclude that Mo is most consistent with our measurements. Approaches beyond lifetime spectroscopy, including intentional contamination and chemical composition measurements, are required to determine the root cause of LeTID.

*Index Terms* — bulk lifetime, carrier-induced degradation, lifetime spectroscopy, light- and elevated temperature-induced degradation (LeTID), light-induced degradation, materials reliability, multicrystalline silicon (mc-Si).

## I. PERC LeTID: IDENTIFYING ROOT CAUSE

Light- and elevated temperature-induced degradation (LeTID) can cause approximately 10% relative efficiency degradation in multicrystalline silicon (mc-Si) PERC solar cells within the first months of operation [1]. While mitigation strategies for LeTID (*e.g.* accelerated degradation and regeneration [2]–[4]) have been suggested and proven, the root cause of LeTID is still unknown. Uncertainty about the root cause of LeTID necessitates process- and/or material-specific optimization of the proposed engineering solutions, which can be costly and time-consuming. Pinpointing the root cause of LeTID is critical to developing targeted solutions that maximize device efficiency.

The LeTID defect has been hypothesized to be a ubiquitously-distributed bulk defect [5]–[7]. Lifetime spectroscopy analysis of room-temperature injection-dependent lifetime curves for samples in the degraded state suggests three potential candidates known to be present during the silicon feedstock refining and growth processes: titanium (Ti), molybdenum (Mo), and tungsten (W). The identification of these candidates is dependent on comparison of calculated defect parameters with those reported by deep-level transient spectroscopy (DLTS) or otherwise in the literature. A further investigation of the defect parameters is warranted due to

experimental evidence of getterability of the LeTID defect [8], which is not necessarily compatible with such slow diffusers as W [9].

In this contribution, we extend our previous lifetime spectroscopy analyses to consider quantitatively the temperature- and injection-dependent lifetime of the LeTID defect. We identify one possible set of defect parameters in each bandgap half: $k$ = 23.9 ± 5.5 at $E_t$-$E_i$ = -0.21 ± 0.06 eV (more likely) and $k$ = 23.5 ± 5.6 at $E_t$-$E_i$ = 0.10 ± 0.07 eV (less likely). However, we note that the defect parameters are probably temperature-dependent, which impedes identification of the energy levels. Based on comparisons to reported temperature dependencies of the defect parameters for Mo and Ti (W has not yet been studied), we find that Mo is consistent with our measurements. We propose further experiments, beyond lifetime spectroscopy, to discern whether Mo is responsible for LeTID and whether there are any additional candidate defects.

## II. TIDLS MEASUREMENT AND ANALYSIS METHODS

The samples used in this study are taken from the same wafers as those described in Refs. [5]–[7]. Two adjacent *p*-type mc-Si wafers, grown by directional solidification with resistivity 1.6 Ω-cm and thickness 175 μm, were selected. These wafers were prepared as PERC semifabricates, with front side silicon nitride (SiN$_x$) passivated emitter and rear side oxide/SiN$_x$ stack. Both wafers were fired at 950°C (actual sample temperature ≈850°C). To isolate the defect responsible for LeTID, one wafer was stored in the dark (undegraded) and one wafer was subjected to degradation conditions (65-75°C and 0.9-1.1 suns) for 168 hours (degraded).

Temperature- and injection-dependent lifetime spectroscopy (TIDLS) measurements were completed at the University of New South Wales with a lifetime tester equipped with a temperature-controlled cryostat, photoconductance coil, and a standard Xenon flash lamp for illumination [10], [11]. Measurements were acquired at sample setpoint temperatures 25°C, 50°C, 100°C, 150°C, and 200°C. Since the LeTID defect is known to be metastable at 200°C, room-temperature QSSPC measurements were completed after each elevated temperature measurement with a Sinton Instruments WCT-120.

Lifetime spectroscopy analysis is carried out as described in Refs. [6], [12], [13]. We assume that the LeTID defect is

978-1-5090-5606-4/17 $31.00 © 2017 IEEE

responsible for the lifetime difference between the undegraded and degraded wafers and that recombination at the defect can be described by Shockley-Read-Hall (SRH) statistics. For each temperature, the injection-dependent SRH lifetime is determined according to the inverse harmonic difference between the two lifetimes:

$$\tau_{SRH} = \left( \frac{1}{\tau_{deg}} - \frac{1}{\tau_{undeg}} \right)^{-1}$$

An important assumption underlying this approach is that background recombination mechanisms, including surface, radiative, Auger, and other SRH defects, are identical at each temperature between the two samples. The SRH lifetime is then linearized [13], and a two-defect fit that minimizes the $\chi^2$ error between the measured and fit SRH lifetimes is used to identify possible defect parameters. The electron-to-hole capture cross-section ratio ($k = \sigma_n/\sigma_p$) and electron capture time-constant ($\tau_{n0}$), which is a function of electron capture cross-section ($\sigma_n$), defect concentration ($N_t$), and thermal velocity ($v_{th}$), are calculated as functions of possible defect levels within the bandgap. The defect level is referenced to the intrinsic level, as in Ref. [11].

## III. LIFETIME MEASUREMENTS

### A. TIDLS measurements

Injection-dependent lifetime measurements at each elevated temperature are shown in Figs. 1(a) and (c) for both samples. As previously reported in Ref. [6], the injection dependence (*i.e.* the shape of the curve) of the LeTID defect, represented by the degraded sample [Fig. 1(a)], does not change significantly as the sample temperature increases. The lifetime changes significantly in magnitude at 200°C. A corresponding change in lifetime can be observed in the undegraded sample [Fig. 1(c)].

### B. Measurements after TIDLS

To assess the stability of the LeTID defect, room-temperature QSSPC measurements were completed after each elevated temperature measurement [Figs. 1(b) and (d)]. The room-temperature lifetimes increase and decrease after the 200°C measurement for the degraded and undegraded samples, respectively. The lifetime change for the undegraded sample begins as low as 150°C. Similar results have been observed after dark annealing of undegraded samples at temperatures up to 250°C [14].

One possible explanation for the change in room temperature lifetime [Figs. 1(b) and (d)] is that the undegraded sample experiences an accelerated degradation while at high temperature and under illumination from the flash lamp. Similarly, the degraded sample may experience an accelerated regeneration. However, since a dark anneal performed in the degraded state is known to reverse the LeTID defect to its initial condition [5], [15], [16], the lifetime measured after 200°C for

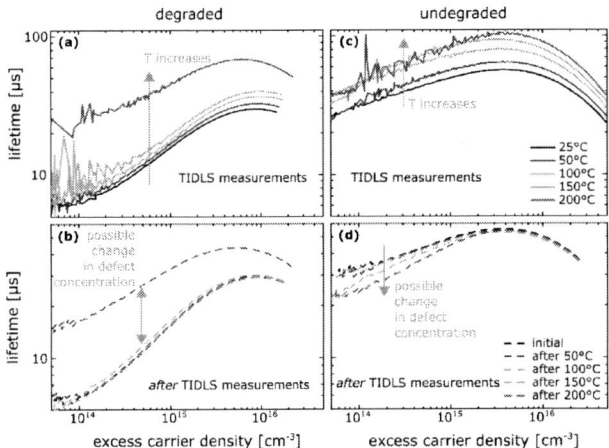

Fig. 1: (a) and (c): Injection-dependent lifetime measurements at setpoint temperatures 25°C, 50°C, 100°C, 150°C, and 200°C for the degraded and undegraded samples, respectively. (b) and (d): Room-temperature injection-dependent lifetime measurements performed after each elevated temperature measurement to demonstrate the stability of the LeTID defect.

the degraded sample may correspond instead to an intermediate state. Vargas, Zhu *et al.* avoided the issue of metastable defect configurations by measuring below room temperature up to 75°C [11].

In either case, similar injection dependence can still be observed in the degraded sample after the 200°C measurement. Lifetime spectroscopy analysis of the measurements after each TIDLS temperature indicates that a defect with nearly identical $E_t$-$E_i$ vs. $k$ dominates the room temperature degraded lifetimes. The $k$-values at midgap for the dominant defect are 27.9, 29.7, 30.7, and 30.1 for the degraded measurements after 50°C, 100°C, 150°C, and 200°C, respectively. If the defect is still present in the sample after each elevated temperature measurement (same $E_t$-$E_i$, $k$), the possible change in state after this measurement would influence calculation of $\tau_{n0}$ ($1/\sigma_n N_t v_{th}$) rather than $k$. A changing LeTID defect concentration would then account for the lifetime difference between samples as temperature is increased, similar to the trend observed during degradation and regeneration in Ref. [7]. The change in lifetime of the undegraded sample [Fig. 1(d)] is too small for reliable analysis; however, these curves are also expected to be dominated by a similar defect (but lower concentration) if degradation starts during the measurement, as in Ref. [7]. Therefore, we do not exclude the 200°C measurement from further discussion.

## IV. CALCULATED DEFECT PARAMETERS

The results of the defect fitting procedure for the SRH lifetime are shown in Fig. 2. Defects with comparable $k$-values are plotted in red (see Refs. [9], [17]–[19]). If the $k$-value is not temperature-dependent, the true defect parameters can be determined from the intersection of the $E_t$-$E_i$ vs. $k$ curves [12]. At every temperature, there is one defect that dominates the

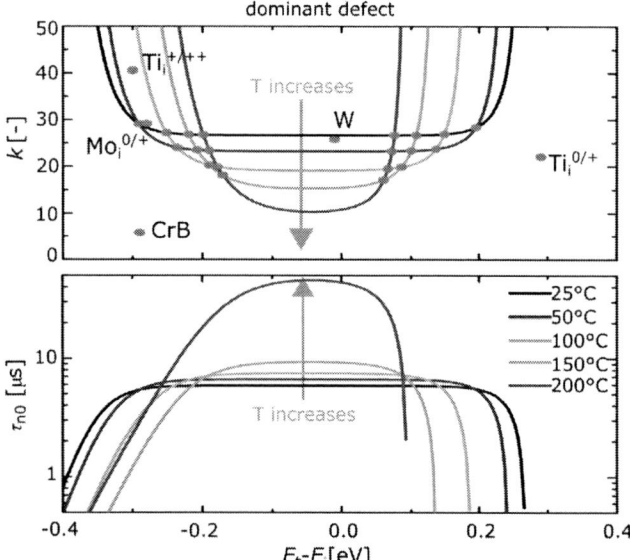

Fig. 2: Calculated LeTID defect parameters at each measured temperature, with no clear intersection point for all curves. (top): k-value (electron-to-hole capture cross-section ratio) as a function of energy level. Intersection points between curves are marked with gray circles. Reported defect parameters are plotted in red for comparison. (bottom): $\tau_{n0}$ (electron capture time constant) as a function of energy level.

lifetime signature throughout the measured injection range and would therefore be dominant under solar cell operating conditions. This defect is most likely to be the LeTID defect; therefore, we focus the discussion on the dominant defect. Defect 2 (not shown) is a shallow defect with no clear intersection points between $E_t$-$E_i$ vs. k curves measured at different temperatures.

All possible curve intersections are plotted as gray circles on Fig. 2(top). There are several quantitative ways to evaluate possible intersections and thus defect parameters from TIDLS measurements [12], [20]. For example, the mostly likely intersection points may correspond to the minimum standard deviation of the k-values at each energy level. Using this approach, two local minima can be identified: $k = 22.8 \pm 3.2$ at -0.19 eV (lower bandgap half) and $k = 21.9 \pm 4.4$ at 0.07 eV (upper bandgap half). The error in the k-value is assigned based on the range of k-values at the local minimum. If the 200°C measurement is excluded, the two local minima are $k = 24.2 \pm 2.7$ at -0.21 eV and $k = 23.7 \pm 3.1$ at 0.10 eV. In both cases, the intersection in the lower bandgap half has a lower standard deviation, which may indicate that the true defect level resides in the lower bandgap half.

However, as shown in Fig. 2, there is not a single distinct intersection for all curves. Considering all possible intersections points, the average intersection points are at $k = 23.9 \pm 5.5$ at -0.21 $\pm$ 0.06 eV and $k = 23.5 \pm 5.6$ at 0.10 $\pm$ 0.07 eV. Here, the error in both k-value and energy level are defined by the ranges for all intersection points in each bandgap half.

We note that the energy levels and k-values reported herein are inconsistent with those identified by Vargas, Zhu, et al. (k

= 56 $\pm$ 23 at -0.32 $\pm$ 0.05 eV and $k = 49 \pm 21$ at 0.21 $\pm$ 0.05 eV) [11]. This inconsistency is likely due to uncertainties in the measuring and fitting procedures. For example, there may be changes in background recombination mechanisms (e.g. surface recombination) and/or differences in material quality (e.g. dislocation density and related recombination), both of which are not explicitly accounted for in the analysis presented herein. It is also possible that the defect, although similar in nature, is not exactly the same between the two sample sets, or that the net recombination of the defect, although similar at room temperature, varies at elevated temperatures due to the changing defect concentration (see Section III). The similarities and differences between the two sample sets will be discussed further in the next section.

Fig. 2(bottom) shows the electron capture time-constant as a function of energy level. As discussed in the previous section, the electron capture time-constant increases significantly at 200°C. This is likely related to a decrease in LeTID defect concentration, ra $E_t$-$E_i$ vs. k ther than a change in the dominant defect itself.

## V. COMPARISON TO LITERATURE VALUES: TITANIUM & MOLBYDENUM

Since the defect does not demonstrate a clear intersection point for all curves, there are three possibilities: (1) uncertainty in the measuring and fitting procedures masks the intersection point, (2) the defect itself changes during the measurement, or (3) one or both of the capture cross-sections may be temperature-dependent. The first possibility was quantified and discussed in the previous section.

The second possibility is not a likely explanation for the lack of a clear intersection. In previous work, we reported that changes in recombination during most of degradation and regeneration are due to a change in concentration of a single defect [7]. Although we observe a change in the room-temperature lifetime after the TIDLS measurements, it is likely that the underlying defect is still the same, with the same k-value but different concentration. We therefore assess the second possibility here.

To assess the third possibility, we compare the temperature dependencies of the k-values reported in literature for two possible LeTID defect candidates, Ti and Mo [5], [6], to our calculated values at the defect-relevant energy level (reference defect parameters: Ti [17] and Mo [21]). It should be noted that the reference Ti k-value was only measured at high temperatures (140–270°C) [17], while the reference Mo k-value was measured over a wider range (-110–150°C) [21]. For the purposes of comparison with experimental data, the reference trend for Ti is extrapolated to temperatures down to -25°C. The other proposed defect candidate, W, is excluded from this analysis because an equivalent study has not yet been conducted. We include the measurements reported in Ref. [11], processed with the same algorithms used herein, in this comparison to further assess any inconsistency between our results.

Figs. 3 and 4 show the results for the *k*-value as a function of temperature for Ti and Mo, respectively. When evaluated at the defect-relevant energy level, the *k*-values reported by the two studies (this abstract and Vargas, Zhu *et al.* [11]) are consistent. The measured trend is inconsistent with that predicted for Ti (Fig. 3), as the measured *k*-value remains constant and increases as temperature drops from 75°C to below room temperature. While the measured *k*-values are not an exact match for Mo (Fig. 4), the measured trend with temperature is consistent with reported values.

An additional metric for evaluating these candidates is the temperature dependence of the electron capture cross-section ($\sigma_n$), which can be derived from $\tau_{n0}$. Representative defect concentrations can be calculated at room temperature, using the thermal velocity [12] and the literature values for $\sigma_n$. For the data reported herein, these concentrations are 6.4 and $1.2 \times 10^{12}$ cm$^{-3}$ for Ti and Mo, respectively. For the data reported in Ref. [11], the same concentrations are $2.6 \times 10^{12}$ and $4.8 \times 10^{11}$ cm$^{-3}$. Assuming a constant defect concentration up to 150°C, $\sigma_n$ is calculated as a function of temperature for both defects [Figs. 3 and 4 (bottom)]. Values above 150°C are not included due to the possible change in defect concentration. Reported temperature dependencies for each capture cross-section are also plotted (Ti [17] and Mo [21]).

Both experimental data sets shown in the bottom plots of Figs. 3 and 4 match the reference values at room temperature due to the method of calculation. The experimental results are most consistent with the reported trend and values (away from room temperature) for Mo. Unlike the *k*-value, the reference temperature-dependence of the capture cross-section for Ti is derived from a model, valid in a wider temperature range [17].

The reference trend at lower temperatures is therefore expected to be more accurate for $\sigma_n$ than for the *k*-value. For these reasons, we conclude that Mo is a more likely candidate for the LeTID defect than Ti.

If Mo is responsible for LeTID, it is unlikely that the gettering response observed by Zuschlag *et al.* is due to gettering of the LeTID defect [8]. The pronounced difference in degradation and regeneration between as-grown and gettered samples could instead be due to process differences (*e.g.* firing with an emitter for the gettered samples), perhaps modifying the LeTID defect and/or surface passivation quality. Rohatgi *et al.* observed no difference in the detrimental effect of Mo on solar cell performance after annealing at 1100°C [22]. Additionally, compared to iron, which is known to be getterable, Mo has a very low diffusivity [9], [23]. Therefore, we do not expect changes in the Mo distribution or total concentration to explain the observed gettering response.

It is possible that the true LeTID defect has similar parameters and temperature dependencies to Mo and has not yet been studied by DLTS or lifetime spectroscopy. While this study fully utilizes the potential of TIDLS, to progress further in identifying the root cause of the LeTID, a new approach is required. Targeted experiments, either through intentional contamination or through chemical composition measurements of LeTID-affected samples, should be conducted to assess current candidates and identify additional candidates.

## VI. SUMMARY

In this contribution, we present quantitative analysis of the temperature- and injection-dependent lifetime of *p*-type mc-Si affected by LeTID. By using adjacent degraded and undegraded wafers from the same ingot, we isolate the SRH lifetime

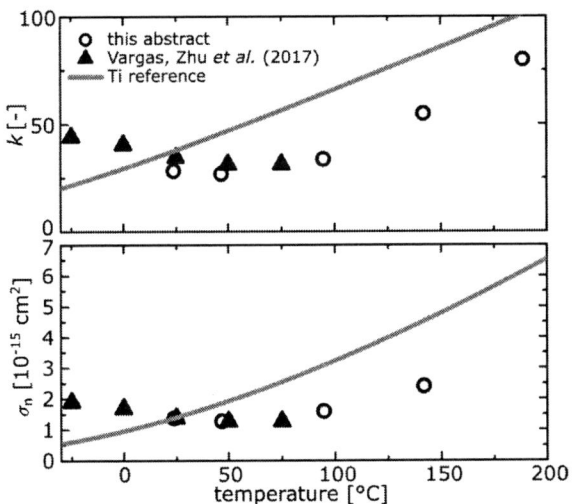

Fig. 3: (top) *k*-value calculated from the results shown in Fig. 2 as a function of temperature at the Ti energy level. Experimental data from this abstract is compared to published data from similar samples [11] and reported literature values [17]. (bottom) Electron capture cross-section calculated from Fig. 2 as a function of temperature at the Ti energy level, assuming a constant defect concentration that is calculated at room-temperature.

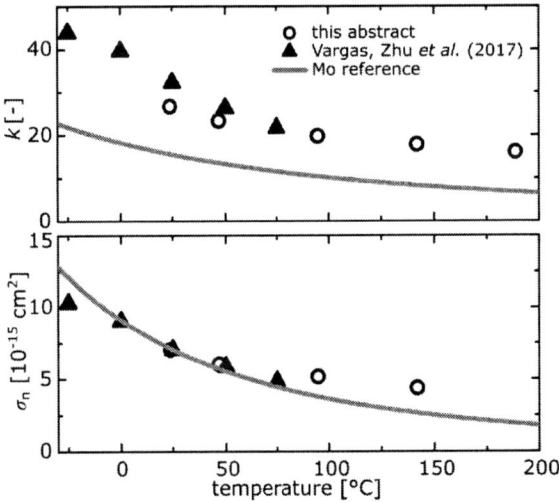

Fig. 4: (top) *k*-value calculated from the results shown in Fig. 2 as a function of temperature at the Mo energy level. Experimental data from this abstract is compared to published data from similar samples [11] and reported literature values [21]. (bottom) Electron capture cross-section calculated from Fig. 2 as a function of temperature at the Mo energy level, assuming a constant defect concentration that is calculated at room-temperature.

associated with the LeTID defect and analyze this lifetime for possible defect parameters. We find that one defect is dominant throughout the measured injection range at each temperature, and this defect has parameters consistent with those previously reported for the LeTID defect at room temperature. Quantitative analysis suggests that one or both of the capture cross-sections are temperature-dependent, and that the observed temperature dependencies are more consistent with reported values for Mo than for Ti, two defects previously identified as LeTID defect candidates. We suggest that the defect responsible for LeTID has not yet been studied by DLTS or other electrical characterization techniques, and we propose further investigation *via* intentional contamination and/or chemical composition measurements to assess candidates.

## ACKNOWLEDGEMENT

This material is based upon work supported in part by the National Science Foundation (NSF) and in part by the Department of Energy under NSF CA No. EEC-1041895. This work was also supported by the Australian Government through the Australian Renewable Energy Agency (ARENA, Project 2014/RND097). The work of M. A. Jensen and E. E. Looney was supported by the NSF Graduate Research Fellowship under Grant 1122374. The work of M. A. Jensen was further supported by the NSF Graduate Research Opportunities Worldwide Fellowship for travel to the University of New South Wales. Z. Hameiri acknowledges the support of the Australian Research Council (ARC) through the Discovery Early Career Researcher Award (DECRA, Project DE150100268).

## REFERENCES

[1] F. Kersten *et al.*, "Degradation of multicrystalline silicon solar cells and modules after illumination at elevated temperature," *Sol. Energy Mater. Sol. Cells*, vol. 142, pp. 83–86, 2015.

[2] C. E. Chan *et al.*, "Rapid Stabilization of High-Performance Multicrystalline p-type Silicon PERC Cells," *IEEE J. Photovoltaics*, vol. 6, no. 6, pp. 1473–1479, 2016.

[3] D. N. R. Payne *et al.*, "Rapid passivation of carrier-induced defects in p-type multi-crystalline silicon," *Sol. Energy Mater. Sol. Cells*, vol. 158, pp. 102–106, 2016.

[4] D. N. R. Payne *et al.*, "Acceleration and mitigation of carrier-induced degradation in p-type multi-crystalline silicon," *Phys. Status Solidi - Rapid Res. Lett.*, vol. 10, no. 3, pp. 237–241, 2016.

[5] K. Nakayashiki *et al.*, "Engineering Solutions and Root-Cause Analysis for Light-Induced Degradation in p-type Multicrystalline Silicon PERC Modules," *IEEE J. Photovoltaics*, vol. 6, no. 4, pp. 860–868, 2016.

[6] A. E. Morishige *et al.*, "Lifetime Spectroscopy Investigation of Light-Induced Degradation in p-type Multicrystalline Silicon PERC," *IEEE J. Photovoltaics*, vol. 6, no. 6, pp. 1466–1472, 2016.

[7] M. A. Jensen *et al.*, "Evolution of LeTID Defects in p-type Multicrystalline Silicon During Degradation and Regeneration," *IEEE J. Photovoltaics*, vol. PP, no. 99, pp. 1–8, 2017.

[8] A. Zuschlag, D. Skorka, and G. Hahn, "Degradation and regeneration in mc-Si after different gettering steps," *Prog. Photovolt Res. Appl.*, pp. 1–8, 2016.

[9] K. Graff, Metal Impurities in Silicon-Device Fabrication, 2nd ed. New York, New York: Springer-Verlag, 2000.

[10] Y. Zhu, M. A. Jensen, C. Vargas, G. Coletti, and Z. Hameiri, "Defect characterization via temperature and injection dependent lifetime spectroscopy," in Presentation at *9th International Workshop on Crystalline Silicon for Solar Cells*, 2016.

[11] C. Vargas, Y. Zhu *et al.*, "Recombination parameters of lifetime-limiting carrier-induced defects in multicrystalline silicon for solar cells," *Appl. Phys. Lett.*, vol. 110, no. 9, p. 92106, 2017.

[12] S. Rein, Lifetime Spectroscopy: A Method of Defect Characterization in Silicon for Photovoltaic Applications. Springer, 2005.

[13] J. D. Murphy *et al.*, "Parameterisation of injection-dependent lifetime measurements in semiconductors in terms of Shockley-Read-Hall statistics : An application to oxide precipitates in silicon," *J. Appl. Phys.*, vol. 111, no. 113709, 2012.

[14] C. Chan *et al.*, "Modulation of Carrier-Induced Defect Kinetics in Multi-Crystalline Silicon PERC Cells Through Dark Annealing," *Sol. RRL*, p. 1600028, 2017.

[15] K. Krauss, A. A. Brand, F. Fertig, S. Rein, J. Nekarda, and E. Sytems, "Fast regeneration processes to avoid light-induced degradation in multicrystalline silicon solar cells," *IEEE J. Photovoltaics*, vol. 6, no. 6, pp. 1427–1431, 2016.

[16] F. Fertig, K. Krauss, and S. Rein, "Light-induced degradation of PECVD aluminium oxide passivated silicon solar cells," *Phys. Status Solidi - Rapid Res. Lett.*, vol. 9, no. 1, pp. 41–46, 2015.

[17] B. B. Paudyal, K. R. McIntosh, and D. H. MacDonald, "Temperature dependent carrier lifetime studies on Ti-doped multicrystalline silicon," *J. Appl. Phys.*, vol. 105, no. 12, 2009.

[18] C. Sun, F. E. Rougieux, and D. Macdonald, "Reassessment of the recombination parameters of chromium in n- and p-type crystalline silicon and chromium-boron pairs in p-type crystalline silicon," *J. Appl*, vol. 115, no. 21, p. 214907, 2014.

[19] S. Boughaba and D. Mathiot, "Deep level transient spectroscopy levels in silicon characterization of tungsten-related deep," *J. Appl. Phys.*, vol. 69, no. 1, pp. 278–283, 1991.

[20] Y. Zhu, Q. Thong, L. Gia, M. K. Juhl, G. Coletti, and Z. Hameiri, "Application of the Newton – Raphson Method to Lifetime Spectroscopy for Extraction of Defect Parameters," *IEEE J. Photovoltaics*, vol. PP, no. 99, pp. 1–6, 2017.

[21] B. B. Paudyal, K. R. McIntosh, D. H. MacDonald, and G. Coletti, "Temperature dependent carrier lifetime studies of Mo in crystalline silicon," *J. Appl. Phys.*, vol. 107, no. 5, pp. 2–6, 2010.

[22] A. Rohatgi, R. H. Hopkins, J. R. Davis, and R. B. Campbell, "The impact of molybdenum on silicon and silicon solar cell performance," *Solid. State. Electron.*, vol. 23, no. 11, pp. 1185–1190, 1980.

[23] J. L. Benton, "Behavior of Molybdenum in Silicon Evaluated for Integrated Circuit Processing," *J. Electrochem. Soc.*, vol. 146, no. 5, p. 1929, 1999.

# Microscopic Distribution of Luminescence from Dislocation Clusters in Multicrystalline Silicon Wafers

H. T. Nguyen[a], M. A. Jensen[b], L. Li[c], C. Samundsett[a], H. C. Sio[a], B. Lai[d], T. Buonassisi[b], D. Macdonald[a]

[a]Research School of Engineering, The Australian National University, Canberra, Australia
[b]Massachusetts Institute of Technology, Cambridge, MA 02139, USA
[c]Australian National Fabrication Facility, The Australian National University, Canberra, Australia
[d]Advanced Photon Source, Argonne National Laboratory, Argonne, IL 60439, USA

*Abstract* — We investigate correlations among various sub-band-gap luminescence centers (so-called D-lines D1/D2/D3/D4) and the band-to-band luminescence signal, and their microscopic distributions, around sub-grain boundaries in multicrystalline silicon wafers. We show that the intensity of the sub-band-gap luminescence from decorating defects/impurities (D1/D2) is more strongly correlated with the recombination activity (the band-to-band intensity), whereas the emission from the intrinsic dislocations (D3/D4) is weakly correlated with the recombination activity. Moreover, via micro-X-ray fluorescence maps, we find that high densities of metal impurities are present at dislocation clusters with strong D1/D2 emission but low D3/D4 emission. Finally, utilizing asymmetric distributions of the D lines across sub-grain boundaries due to their inclined angles to the wafer surface, we demonstrate that the sub-grain boundaries share a common direction locally, rather than occurring in a random manner, during the crystal growth process.

*Index Terms* — photoluminescence, dislocations, metal precipitates, multicrystalline silicon, photovoltaic cells.

## I. INTRODUCTION

In multicrystalline silicon (mc-Si) wafers, sub-grain boundaries (sub-GBs) are a result of dislocation accumulation. The residual stress and strain around the dislocations, in turn, can preferentially trap other defects and impurities. As such, sub-GBs have detrimental effects on the performance of mc-Si solar cells. However, the relative impact of the intrinsic dislocations and the decorating defects/impurities on the recombination activity remains unclear. Also, due to the micron scale of sub-GBs, it is challenging to investigate their optical and electrical properties in a spatially resolved manner.

Sub-band-gap photoluminescence (PL) spectroscopy is a powerful tool for studying defects/impurities in Si wafers [1-4]. Many defects and impurities can emit distinct sub-band-gap PL peaks at low temperatures, such as dislocations [5-9], oxygen precipitates [10-13], or Cr-B pairs [14]. One, in principle, can analyze these peaks (shape, intensity, energy) to study the properties of such defects and impurities. Thus, an insightful understanding of the sub-band-gap PL from sub-GBs is important for mitigating their detrimental impacts on cell performance.

In this work, utilizing micron-scale spatial resolution with a micro-PL spectroscopy system, we perform PL measurements directly at sub-GBs and the surrounding regions in order to study luminescence behaviors of the sub-GBs in mc-Si wafers.

We first examine correlations among intensities of the defect PL peaks (D1-D4) and the main band-to-band PL peak emitted from sub-GBs. We then relate the emission of the defect PL peaks with the presence of metal impurities detected by spatially-resolved synchrotron-based micro-X-ray fluorescence (μ-XRF) measurements at the Advanced Photon Source (2-ID-D) at Argonne National Laboratory [15]. Finally, we utilize asymmetric distributions of the D lines, caused by inclined angles of the sub-GBs underneath the wafer surface, to demonstrate that the sub-GBs share a common direction locally, rather than occurring in a random manner, during the crystal growth process. A more complete description of some aspects of this work is presented elsewhere [16].

## II. EXPERIMENTAL DETAILS

The investigated sample is an industrially-grown boron-doped p-type mc-Si wafer whose nominal resistivity is 1.6 $\Omega$.cm. It was first chemically etched in an HF/HNO₃ solution to remove any saw damage and to achieve planar surfaces. After that, it was immersed in a defect etchant consisting of acetic/HNO₃/HF for 16 hours in order to reveal sub-GBs [17], which were then visible under an optical microscope.

The micro-PL spectroscopy system employed in this study is a Horiba LabRAM equipped with confocal optics and a micro X-Y mapping stage. The excitation wavelength is 810 nm. The on-sample excitation spot is ~2 microns in diameter and the on-sample power is 6 mW. The sample temperature was controlled using a liquid-nitrogen-cooled Linkam stage. The μ-XRF scans were performed at Argonne National Laboratory's Advanced Photon Source Beamline 2-ID-D using an incident X-ray energy of 9 keV with a full-width half-maximum beam spot size of ~200 nm.

## III. RESULTS AND ANALYSIS

### A. Correlations among band-to-band, defect luminescence, and metal impurities

We examine correlations among various sub-band-gap lines, the band-to-band line, and the presence of metal impurities. Fig. 1a shows a PL spectrum captured at a sub-GB at 80 K. The peak ~1130 nm is the main band-to-band (BB) peak emitted from Si, assisted by the emission of a transverse-optical phonon. The BB intensity reflects the local excess carrier lifetime in the sample, and therefore the local recombination activity of the sub-GB. The four distinct sub-band-gap peaks are denoted as D1 to D4. The doublet D3/D4 was reported to reflect the intrinsic properties of dislocations, whereas the doublet D1/D2 originates from secondary defects and impurities trapped around the dislocations [1,2,7,8,18]. We can numerically decompose the spectrum into five different Gaussian peaks, corresponding to D1-D4 lines, and an optical zone-center phonon replica of the main BB peak [19], denoted as PRBB (Phonon Replica of Band-to-Band).

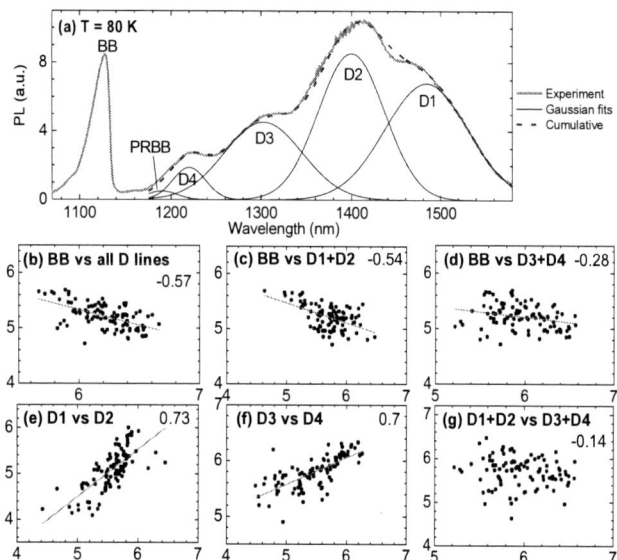

**Fig. 1.** (a) PL spectrum captured at a sub-GB at 80K. Logarithmic intensities of (b) BB vs all D lines, (c) BB vs D1+D2, (d) BB vs D3+D4, (e) D1 vs D2, (f) D3 vs D4, and (g) D1+D2 vs D3+D4.

First, Fig. 1(b-g) plots their logarithmic intensities versus each other. The D-line intensities are the areas of their corresponding Gaussian fits, whereas that of the BB line is the integrated signal between 1070-1150 nm. Each data point was measured from a separate sub-GB, and the number in each plot is the correlation coefficient of the pairs. Four conclusions can be made from these plots. First, the BB intensity is inversely correlated with the total intensity of the D lines, i.e. the sub-band-gap luminescence is undesirable in terms of material quality for PV applications (1b). Second, the BB intensity has a stronger correlation with D1/D2 (1c) compared to D3/D4 (1d). Third, D1 and D2 are strongly correlated together (1e), and so are D3 and D4 (1f). The results are consistent with the fact that D3/D4 are emitted from dislocation cores and D1/D2 are emitted from secondary defects and impurities. Four, there is no correlation between D1/D2 and D3/D4 (1g), suggesting that the two doublets D1/D2 and D3/D4 do not necessarily appear together because of their different origins. These conclusions were found to be consistent regardless of processing steps (as-cut, phosphorus gettered, and after hydrogenation).

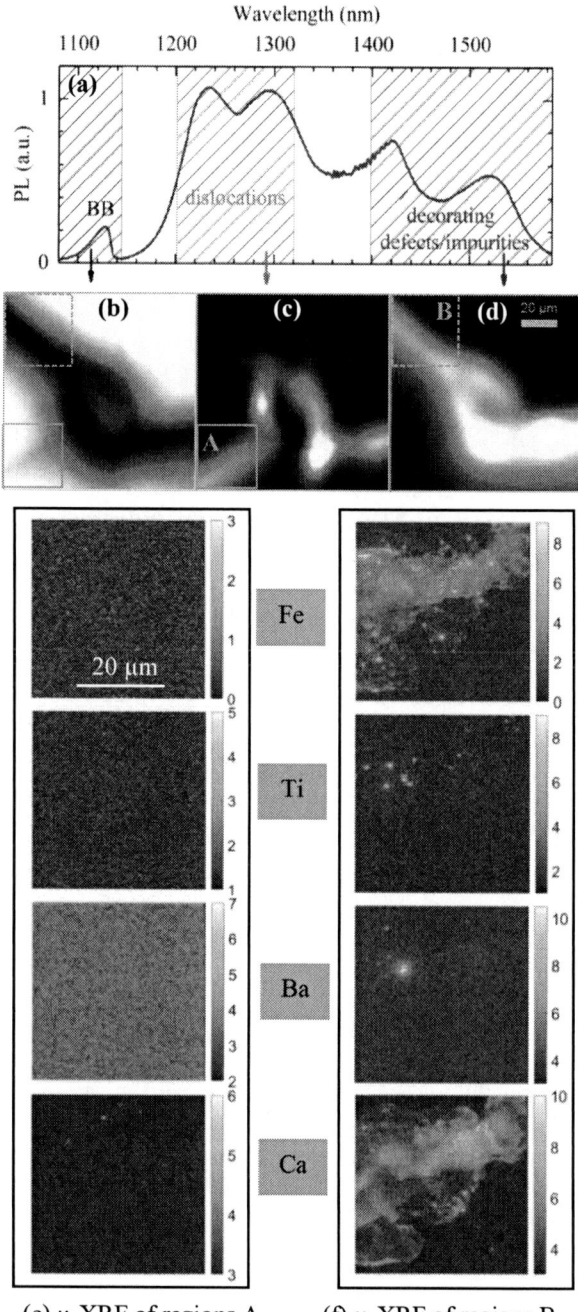

(e) µ-XRF of regions A    (f) µ-XRF of regions B

**Fig. 2.** (a) PL spectrum at a sub-GB at 80 K. Integrated PL intensity maps of (b) BB line, (c) D3/D4 lines from dislocations, and (d) D1/D2 lines from decorating defects/impurities at 80 K. The brighter color indicates a higher intensity. (e) µ-XRF maps for a selection of metal impurities at region A (high D3/D4, no D1/D2) and (f) region B (high D1/D2, no D3/D4). The unit is nanogram/cm² and the color-bar labels represent the exponent of 10^.

Next, Fig. 2 plots the integrated intensity mappings of the BB (2b), D3/D4 (2c), and D1/D2 (2d) lines from a dislocation cluster. Respectively, the maps of BB, D3/D4, and D1/D2 are the integrated intensities between 1070-1150 nm, 1200-1320 nm, and 1400-1570 nm, as illustrated in Fig. 2a. These regions of the PL spectra were chosen for mapping since the two doublets D1/D2 and D3/D4, and the BB line are not significantly overlapped with each other within these wavelength ranges, as can be observed in Fig. 1a. We can observe that dark patterns of the BB mappings follow closely bright patterns of the D1/D2 mappings, whereas bright patterns of the D3/D4 mappings follow neither bright patterns of D1/D2 nor dark patterns of BB. These results are consistent with our observation that the doublet D1/D2 has a stronger correlation with the recombination activity of the sub-GBs than the doublet D3/D4. Moreover, the solid rectangles in Figs. 2b and 2c mark the region of high D3/D4 intensities but absent D1/D2 (denoted as region A), whereas the broken rectangles in Figs. 2b and 2d mark the region of high D1/D2 intensities but absent D3/D4 (denoted as region B). When D1/D2 is absent, even though D3/D4 is high, the contrast of this area in the BB mappings is minimal. On the other hand, when D1/D2 is high but D3/D4 is absent, there is a significant reduction in the BB signal.

In addition, we performed μ-XRF scans for regions A and B and showed concentration maps of some selected metal impurities in (2e) and (2f), respectively. The first point to note is that, the overall metal concentration is higher in B (high D1/D2, no D3/D4), with localized high concentrations that may correspond to metal-rich precipitates, whereas that in A (high D3/D4, no D1/D2) is approaching the detection limit. In addition, the metal-rich area (e.g. Fe and Ca maps) extends not only along the sub-GB but also to surrounding areas, indicating that the region is highly contaminated with metal impurities. Similar maps for Cu and Ni can be found in Ref [16].

The D lines were, in fact, also previously observed in both high-purity plastically-deformed float-zone and Czochralski Si wafers [6,7,20], indicating that metal impurities (precipitates and/or dissolved impurities) are not always the direct causes of the D lines. However, the D-line intensities could be enhanced when the wafers are moderately contaminated with metal impurities, such as Fe and/or Cu, as reported by several authors [8,21]. Sub-GBs in mc-Si wafers are the result of dislocation accumulation during the ingot growth and cooling, and thus contain a high density of dislocations. However, the sub-GB in region B in Fig. 2d emits D1/D2 but not D3/D4, and contains very high density of metal impurities. These results suggest that a very high density of metal impurities reduces the luminescence efficiency of the dislocation cores at the sub-GB, but not the luminescence of the secondary defects/impurities. Due to this complex relationship between the metal contamination and the dislocation luminescence, we did not observe a clear correlation between the BB intensity and the D3/D4 intensity as shown in Fig. 1d.

*B. Asymmetric distributions of D lines*

After the defect etching step, sub-GBs are revealed as continuous grooves on the wafer surface. These etched grooves can cause optical artifacts during μ-PL scans. In a previous work [16], we used these artifacts as indications of exact locations of sub-GBs and found that the D lines are always distributed asymmetrically across all sub-GBs investigated. From

transmission electron microscopy (TEM) results, we confirmed that these asymmetric patterns of the D lines were due to the inclined sub-GBs underneath the wafer surface. We then utilized them to determine inclination orientations of sub-GBs underneath the wafer surface. Here, we continue examining inclination orientations of sub-GBs at various dislocated regions, but by performing PL scans at room temperature.

The D lines (D1-D4) display strong thermal quenching rates [6]. At room temperature, although the D3/D4 lines are hardly observable, residues of the D1/D2 lines remain as a small peak [16]. The latter allows us to employ the D-line maps to assess inclination directions of various sub-GBs by performing PL scans even at room temperature. Figs. 3(a1-e1) show PL intensity maps of the D lines at room temperature for various dislocation regions, whereas Figs. 3(a2-e2) show optical images of these regions. They have the same X and Y directions relative to one another, and they are from the same dislocation cluster in the mc-Si wafer. In the PL mapping images, the thin bright lines closely follow the sub-GB positions in the optical images. These lines are artifacts due to the etch grooves as mentioned above. The directions of the arrows (column 2) indicate the side with broader D-line emissions, thus indicating the direction of the inclination below the surface. The D lines are consistently skewed either to the left (yellow arrows) or downward (red arrows). Since sub-GBs are the result of dislocation accumulation, these results suggest that dislocations were formed and evolved in a common direction locally, rather than in random directions, during the crystal growth process. This conclusion is difficult to make using common microscopic tools such as TEM or electron backscatter diffraction (EBSD) due to impractical aspects of sample preparations.

## IV. CONCLUSION

Although dislocation clusters are detrimental to cell performances, relatively little work has been done on separating the impacts of their intrinsic properties and the trapped defects/impurities. This work shows that the trapped defects/impurities around dislocations are more detrimental in terms of recombination activity than the dislocations themselves. Also, using spatially-resolved synchrotron-based micro-X-ray fluorescence measurements, we confirmed that high densities of metal impurities are present at the sub-GBs with a strong emission from decorating defects/impurities (D1/D2). In such locations, the intensity of the intrinsic luminescence from dislocation cores (D3/D4) is sometimes strongly quenched. Therefore, around sub-GBs, the decorating defect/impurity emission better represents material quality than the dislocation emission. Finally, utilizing the asymmetric distributions of the D lines across sub-GBs, we showed that the sub-GBs were formed and evolved in a common direction locally, rather than in a random manner, during the crystal growth process. These findings provide insights into the complex optical and electrical behaviors of sub-GBs in mc-Si wafers, necessary for mitigating their detrimental impacts on cell performance.

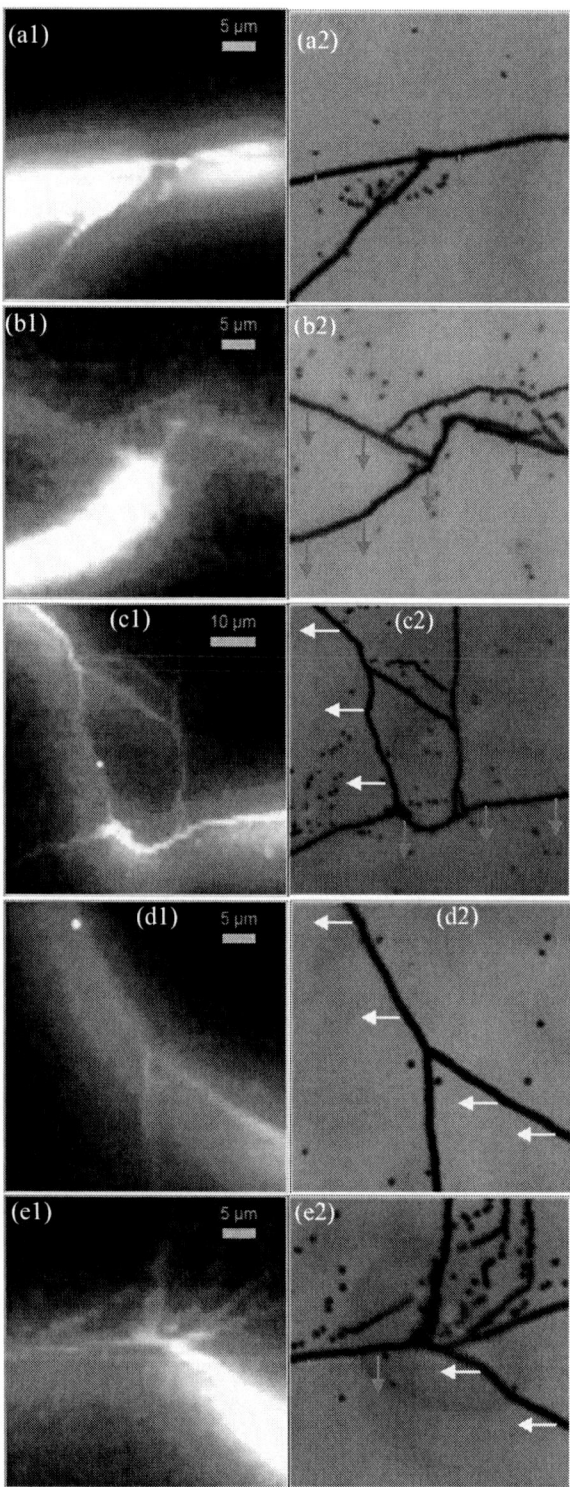

**Fig. 3.** (Column 1) PL intensity maps of D lines at room temperature of various dislocation regions. The brighter color indicates a higher intensity. (Column 2) Optical images of these regions. Positions of sub-GBs are revealed by etch grooves after the defect etch. Directions of the arrows indicate the sides of more D-line emissions across the sub-GBs. The images have the same X and Y directions. The mapping was performed at room temperature.

ACKNOWLEDGEMENT

This work has been supported by the Australian Research Council (ARC), and the Australian Renewable Energy Agency (ARENA) through research grant RND009. The Australian National Fabrication Facility (ANFF) and Center for Advanced Microscopy (CAM) are greatly acknowledged for providing access to some of the facilities used in this work. MAJ acknowledges support by the National Science Foundation Graduate Research Fellowship under Grant No. 1122374. This research used resources of the Advanced Photon Source, a U.S. Department of Energy (DOE) Office of Science User Facility operated for the DOE Office of Science by Argonne National Laboratory under Contract No. DE-AC02-06CH11357.

REFERENCES

[1] S. Binetti, A. Le Donne, and A. Sassella, "Photoluminescence and infrared spectroscopy for the study of defects in silicon for photovoltaic applications," *Sol. Energy Mater. Sol. Cells*, vol. 130, pp. 696-703, 2014.

[2] M. Tajima, "Spectroscopy and topography of deep-level luminescence in photovoltaic silicon," *IEEE Journal of Photovoltaics*, vol. 4, pp. 1452-1458, 2014.

[3] D. Lausch and C. Hagendorf, "Influence of Different Types of Recombination Active Defects on the Integral Electrical Properties of Multicrystalline Silicon Solar Cells," *Journal of Solar Energy*, vol. 2015, pp. 1-9, 2015.

[4] T. Mehl, M. D. Sabatino, K. Adamczyk, I. Buruda, and E. Olsen, "Defects in multicrystalline Si wafers studied by spectral photoluminescence imaging, combined with EBSD and dislocation mapping," *Energy Procedia*, vol. 92, pp. 130–137, 2016.

[5] M. Tajima and Y. Matsushita, "Photoluminescence related to dislocation in annealed Czochralski-grown Si crystals," *Japanese Journal of Applied Physics*, vol. 22, pp. L589-L591, 1983.

[6] M. Suezawa, Y. Sasaki, and K. Sumino, "Dependence of Photoluminescence on Temperature in Dislocated Silicon Crystals," *Phys. Status Solidi A*, vol. 79, pp. 173-181, 1983.

[7] R. Sauer, J. Weber, J. Stolz, E. R. Weber, K.-H. Küsters, and H. Alexander, "Dislocation-related photoluminescence in silicon," *Appl. Phys. A*, vol. 36, pp.1-13, 1985.

[8] E. C. Lightowlers and V. Higgs, "Luminescence associated with the presence of dislocations in silicon," *Phys. Stat. Sol. A*, vol. 138, pp. 665-672, 1993.

[9] T. Sekiguchi and K. Sumino, "Cathodoluminescence study on dislocations in silicon," *J. Appl. Phys.*, vol. 79, pp. 3253-3260, 1996.

[10] S. Pizzini, M. Guzzi, E. Grilli, and G. Borionetti, "The photoluminescence emission in the 0.7-0.9 eV range from oxygen precipitates, thermal donors and dislocations in silicon," *J. Phys.: Condens. Matter*, vol. 12, pp. 10131–10143, 2000.

[11] S. Binetti, S. Pizzini, E. Leoni, R. Somaschini, A. Castaldini, A. Cavallini, "Optical properties of oxygen precipitates and dislocations in silicon," *J. Appl. Phys.*, vol. 92, pp. 2437-2445, 2002.

[12] K. Bothe, R. J. Falster, and J. D. Murphy, "Room temperature sub-bandgap photoluminescence from silicon containing oxide precipitates," *Applied Physics Letter*, vol. 101, p. 032107, 2012.

[13] F. E. Rougieux, H. T. Nguyen, D. H. Macdonald, B. Mitchell, and R. Falster, "Growth of Oxygen Precipitates and Dislocations in Czochralski Silicon," *IEEE Journal of Photovoltaics*, vol. 3, pp. 735-740, 2017.

[14] H. Conzelmann and J. Weber, "Photoluminescence from chromium-boron pair in silicon," *Physica B+C*, vol. 116, pp. 291-296, 1983.

978-1-5090-5606-4/17 $31.00 © 2017 IEEE

[15] M. A. Jensen, J. Hofstetter, A. E. Morishige, G. Coletti, B. Lai, D. P. Fenning, and T. Buonassisi, "Synchrotron-based analysis of chromium distributions in multicrystalline silicon for solar cells," *Applied Physics Letters*, vol. 106, p. 202104, 2015.

[16] H. T. Nguyen, M. A. Jensen, L. Li, C. Samundsett, H. C. Sio, B. Lai, T. Buonassisi, and D. Macdonald, "Microscopic Distributions of Defect Luminescence from Subgrain Boundaries in Multicrystalline Silicon Wafers," *IEEE Journal of Photovoltaics*, vol. 3, pp. 772-780, 2017.

[17] Y. Kashigawa, R. Shimokawa, and M. Yamanaka, "Highly sensitive etchants for delineation of defects in single- and polycrystalline silicon materials," *J. Electrochem. Soc.*, vol. 143, pp. 4079-4087, 1996.

[18] H. T. Nguyen, F. E. Rougieux, F. Wang, H. Tan, and D. Macdonald, "Micrometer-scale deep-level spectral photolumines-cence from dislocations in multicrystalline silicon," *IEEE Journal of Photovoltaics*, vol. 5, pp. 799-804, 2015.

[19] H. T. Nguyen and D. Macdonald, "On the composition of luminescence spectra from heavily doped p-type silicon under low and high excitation," *Journal of Luminescence*, vol. 181, pp. 223–229, 2017.

[20] M. Tajima and Y. Matsushita, "Photoluminescence related to dislocation in annealed Czochralski-grown Si crystals," *Japanese Journal of Applied Physics*, vol. 22, pp. L589-L591, 1983.

[21] M. Tajima, M. Ikebe, Y. Ohshita, and A. Ogura, "Photoluminescence analysis of iron contamination effect in multicrystalline silicon wafers for solar cells," *Journal of Electronic Materials*, vol. 39, pp. 747-750, 2010.

# Do grain boundaries matter? Electrical and elemental identification at grain boundaries in LeTID-affected *p*-type multicrystalline silicon

Mallory A. Jensen[1], Ashley E. Morishige[1], Sagnik Chakraborty[2], Romika Sharma[2], Hang Cheong Sio[3], Chang Sun[3], Barry Lai[4], Volker Rose[4,5], Amanda Youssef[1], Erin E. Looney[1], Sarah Wieghold[1], Jeremy Poindexter[1], Juan-Pablo Correa-Baena[1], Daniel Macdonald[3], Joel B. Li[2], Tonio Buonassisi[1]

[1]Massachusetts Institute of Technology, Cambridge, MA 02139, USA, email: jensenma@alum.mit.edu
[2]Solar Energy Research Institute of Singapore, Singapore 117574, Singapore
[3]Australian National University, Canberra, ACT 0200, Australia
[4]Advanced Photon Source, Argonne National Laboratory, Argonne, IL 60439, USA
[5]Center for Nanoscale Materials, Argonne National Laboratory, Argonne, IL 60439, USA

*Abstract* — The root cause of light- and elevated temperature-induced degradation (LeTID) in multicrystalline silicon *p*-type passivated emitter and rear cell (PERC) devices is still unknown. Some researchers hypothesize that high temperature firing processes dissolve metal-rich precipitates which can then participate in LeTID. To address this hypothesis, synchrotron-based X-ray techniques, including fluorescence and absorption near-edge spectroscopy, are employed. In as-grown industrial material, we observe collocated copper- and nickel-rich precipitates, which persist after firing and are below the detection limit after phosphorous diffusion. We conclude that precipitates decrease in size due to the firing process and that this may result in an increase in bulk interstitial metal concentration. We further employ microphotoluminescence at a grain versus grain boundary to highlight similarities and possible differences in degradation and regeneration behavior.

*Index Terms* — Degradation, light-induced degradation, carrier-induced degradation, copper, nickel, precipitate, synchrotron, photoluminescence, multicrystalline silicon, materials reliability, passivated emitter and rear cell (PERC), silicon, characterization.

## I. I. *P*-TYPE PERC LETID & METALS

Light- and elevated temperature-induced degradation (LeTID, otherwise known as carrier-induced degradation or CID) can cause about 10% relative efficiency degradation in multicrystalline silicon (mc-Si) passivated emitter and rear cell (PERC) devices during the first months of operation [1]. In the field, this degradation can take many years to recover, representing a significant loss to installed system energy yield over time [2]. While mitigation strategies for LeTID have been suggested with varying success, the root cause of LeTID is still unknown.

The cause of LeTID is thought to be a uniformly distributed bulk defect, and lifetime spectroscopy results point to a metastable deep-level donor/hydrogen complex [3]. However, some spatial inhomogeneity has been noted in the rate and extent of both degradation and regeneration [4]. Recently, Luka *et al.* confirmed the presence of Cu-containing precipitates at grain boundaries and at the rear surface of LeTID-sensitive

solar cells after degradation [5]. The authors hypothesized that Cu originated from the mc-Si wafer, rather than deposited layers or processing equipment. Researchers have further suggested that high-temperature firing processes, after which LeTID is observed, dissolve metal precipitates and introduce interstitial atoms to the bulk which can then participate in the LeTID reaction [6].

In this contribution, we employ PL imaging and synchrotron-based elemental characterization to study precipitate distributions at a grain boundary in the as-grown state and after processing to further investigate both precipitate dissolution during firing and spatial inhomogeneity in LeTID-affected material. We observe Cu- and Ni-rich particles with micro-X-ray fluorescence (μ-XRF), and we characterize the collocation of these particles with both micro-X-ray absorption near-edge spectroscopy (μ-XANES) and higher resolution nano-XRF. Our results indicate that as-grown metal-rich precipitates may supply interstitial metals to the bulk during high temperature processes, which may then participate in the LeTID reaction. We further hypothesize that metal-rich precipitates inhibit hydrogenation at the studied grain boundary. Using microscale photoluminescence (PL) measured throughout degradation, we demonstrate similarities and possible differences in LeTID between grains and grain boundaries.

## II. METHODS: μ-XRF, NANO-XRF, PLI, AND *IN SITU* μ-PL

All samples used in this study are industrial *p*-type mc-Si grown by directional solidification, known to exhibit LeTID. For synchrotron-based experiments, six nearly-adjacent wafers were subjected to various processing steps: (1) as-grown, (2) fired at 620°C (low firing, setpoint), (3) fired at 850°C (high firing, setpoint), (4) phosphorous diffusion gettering (PDG), (5) PDG followed by low firing, and (6) PDG followed by high firing. The sample temperature during firing is approximately 10-20°C below the setpoint temperature. All wafers were saw-damage etched and cleaned prior to processing, and $AlO_x/SiN_y$ (350°C, PECVD) was used to passivate all wafers. Photoluminescence imaging (PLI) was employed to evaluate

978-1-5090-5606-4/17 $31.00 © 2017 IEEE

electrical performance. Spatially-resolved μ-XRF maps were measured at beamline 2-ID-D at the Advanced Photon Source in 220 nm steps with a 200 nm spot size. At select metal-rich particles, μ-XANES measurements were conducted to discern chemical structure. High resolution nano- XRF maps of metal-rich particles were measured at beamline 26-ID-C in 10 nm steps with a 30 nm spot size [7].

Micro-photoluminescence (μ-PL) was measured using a Horiba LabRAM system, equipped with a confocal microscope and a temperature stage. Several measurements at grains and grain boundaries were conducted *in situ* during accelerated degradation at 140°C with illumination from a 36 mW, 532 nm light source with a spot size of approximately 2.5 μm ($\approx 7 \times 10^6$ suns). The detection area ($\approx 10$ μm) is governed by carrier diffusion away from the illuminated spot during the measurement. A PERC semifabricate fired at 950°C, previously studied in [3], was chemically stripped of all dielectric layers and passivated with an $AlO_x/SiN_y$ stack. Prior to μ-PL measurements, the sample was annealed in the dark for 10 minutes at 200°C to ensure a representative initial state.

## III. SYNCHROTRON-BASED RESULTS

Synchrotron-based μ-XRF maps of the Cu-$K_\alpha$ and Ni-$K_\alpha$ fluorescence at the same grain boundary in each sample are shown in Fig. 1 (log scale). Pixels with high metal concentration are white, while pixels with low metal concentration are dark red. For the as-grown, low firing, and high firing measurements, Cu- and Ni-rich particles can be identified along the grain boundary. After PDG, only one localized Cu particle can be observed visually, while the remaining particles have dissolved and/or been externally gettered to below the technique detection limit.

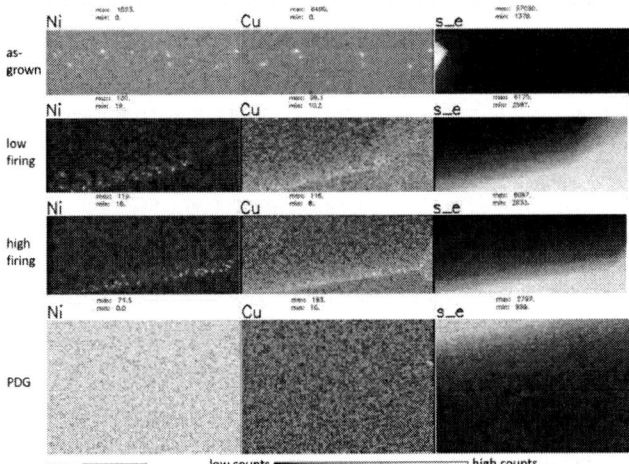

Fig. 1: μ-XRF maps of Cu-$K_\alpha$ fluorescence (left) and Ni-$K_\alpha$ fluorescence (middle), and elastically-scattered photons (right) for samples after different processing steps. The elastic maps demonstrate the location of the grain boundary during the measurement. Each map is 30 μm wide.

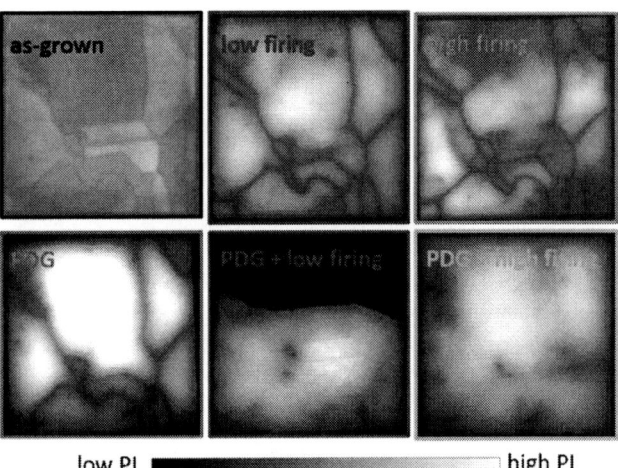

low PL ▬▬▬▬▬▬▬▬▬▬▬▭▭▭▭▭ high PL

Fig. 2: Photoluminescence images of the samples measured by μ-XRF. Images have been scaled for visibility; relative brightness between samples does not indicate a lifetime difference. All samples are 1 cm², except for the PDG + low firing sample, which was broken during the measurement.

In the maps shown in Fig. 1, Cu and Ni are observed in the same pixel for both as-grown and processed samples. However, due to the beam spot size at 2-ID-D, no further information about possible collocation can be discerned. Nano-XRF maps (not shown) of Cu- and Ni-$K_\alpha$ fluorescence of select particles from the same grain boundary indicate that Cu and Ni are present in the same location, but there is a portion of the particle which contains Cu only. Furthermore, the two Cu clusters are oriented nearly perpendicular to each other. To further assess the nature of these particles, μ-XANES measurements of a representative particle were collected (not shown). The measured spectra are similar in absorption onset and shape to a $Cu_3Si$ standard, reported in Ref. [8].

Photoluminescence images of the same samples measured with μ-XRF (Fig. 2) demonstrate that all processes increase the lifetime at the measured grain boundary, with the highest lifetime achieved by PDG. In all as-grown samples and the sample directly after PDG, the grain boundary can still be observed as recombination-active. However, for the samples subjected to PDG followed by low/high firing, the lifetime is further increased and the grain boundary can no longer be observed in PLI, suggesting that the grain boundary was rendered non-recombination active by the firing process.

## IV. MICROPHOTOLUMINESCENCE RESULTS

Several authors have suggested that, while the defect responsible for LeTID is most likely present throughout the entire wafer, both degradation and regeneration start locally and spread across the wafer. To assess this inhomogeneity, especially in the context of the inhomogeneous metal distributions shown in Fig. 1, one representative set of μ-PL measurements, at a grain boundary and inside a grain, is shown in Fig. 3. Both measurements are normalized to the initial value

978-1-5090-5606-4/17 $31.00 © 2017 IEEE          3301

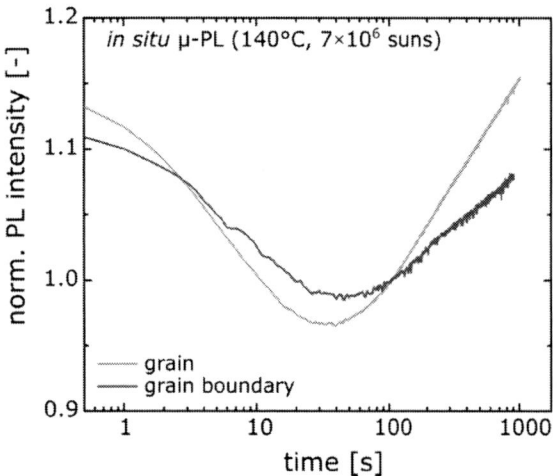

Fig. 3: µ-PL measurements of a sample, sensitive to LeTID, under accelerated degradation conditions (140°C, ≈7×10⁶ suns). Measurements were performed separately with the spot centered inside a grain and then centered on a grain boundary.

at $t = 0$ seconds. Both the rate and extent of degradation are similar for the grain and grain boundary. The similarity in behaviors supports the hypothesis that the LeTID defect is homogeneously distributed across the wafer, and suggests that inhomogeneous distributions of metal-rich precipitates like those observed by µ-XRF are unrelated to LeTID. However, there is a notable difference in regeneration behavior: the grain boundary exhibits both a slower initial degradation and regeneration than the grain. These differences may be due to an inhomogeneous LeTID defect distribution or to a lower injection level at and around the grain boundary.

## V. DISCUSSION

The µ-XRF measurements confirm the presence of both Cu- and Ni-rich precipitates at a grain boundary in industry-relevant mc-Si material affected by LeTID, suggested by Luka *et al.* [5]. Although Cu-LID as reported in the literature does not describe LeTID [9], certain Cu- and/or Ni-containing complexes which have not yet been studied may be involved in LeTID.

After firing, we observe a reduction in precipitate size, which could correspond to an increase in interstitial atoms available throughout the wafer to participate in LeTID. The presence of metals at an as-grown grain boundary and the reduction in number and size of metal-rich precipitates after PDG has been observed before [10]. However, the persistence of Cu- and Ni-rich particles after firing is unexpected since, in addition to the high diffusivities of these elements at firing temperatures, Buonassisi *et al.* previously observed nearly complete dissolution of Cu- and Ni-rich particles after a rapid thermal anneal at 860°C [11]. The difference in observations can be explained by a difference in sample processing temperature, by a difference in initial precipitate distributions (higher precipitate density at the grain boundary studied herein), and/or by a difference in XRF detection limit.

Additionally, we observe a difference in hydrogenation behavior for grain boundaries with (as-grown) and without (PDG) metal-rich precipitates above the µ-XRF detection limit. Since the precipitate size is likely reduced after firing in the as-grown state, a similar trend is expected for precipitates fired after PDG. The grain boundaries are not recombination-active after PDG followed by firing, implying that the presence of metal-rich precipitates above the µ-XRF detection limit impeded passivation by hydrogen in the as-grown samples.

## VI. SUMMARY

We present microscale elemental and electrical characterization of grain boundaries in industrial *p*-type multicrystalline silicon, known to be sensitive to LeTID. We draw two main conclusions from this work: 1) metal-rich precipitates are present in today's industry standard mc-Si material sensitive to LeTID, and 2) sufficiently large precipitates of fast-diffusing metals such as Cu and Ni are not fully-dissolved by high-temperature firing processes. We hypothesize that metal-rich precipitates may inhibit hydrogenation at grain boundaries and interfere with LeTID reactions, causing some spatial inhomogeneity.

## ACKNOWLEDGEMENT

This material is based upon work supported by the National Science Foundation (NSF) and the Department of Energy (DOE) under NSF CA No. EEC-1041895. This work made use of the Center for Nanoscale Systems, Harvard University, supported by National Science Foundation (NSF) Award ECS-0335765. M. A. Jensen and E. E Looney acknowledge support by the National Science Foundation Graduate Research Fellowship under Grant No. 1122374. M. A. Jensen acknowledges support by the National Science Foundation Graduate Research Opportunities Worldwide fellowship for travel to Australia. This research used resources of the Advanced Photon Source, a U.S. Department of Energy (DOE) Office of Science User Facility operated for the DOE Office of Science by Argonne National Laboratory under Contract No. DE-AC02-06CH11357. Use of the Advanced Photon Source and the Center for Nanoscale Materials was supported by the U. S. Department of Energy, Office of Science, Office of Basic Energy Sciences, under contract DE-AC02-06CH11357.

## REFERENCES

[1] K. Ramspeck *et al.*, "Light induced degradation of rear passivated mc-Si solar cells," in *Proceedings of the 27th European Photovoltaic Solar Energy Conference*, 2012, pp. 861–865.

[2] F. Kersten *et al.*, "Degradation of multicrystalline silicon solar cells and modules after illumination at elevated temperature," *Sol. Energy Mater. Sol. Cells*, vol. 142, pp. 83–86, 2015.

[3] K. Nakayashiki *et al.*, "Engineering Solutions and Root-Cause Analysis for Light-Induced Degradation in p -Type

Multicrystalline Silicon PERC Modules," *IEEE J. Photovoltaics,* vol. 6, no. 4, pp. 860–868, 2016.

[4] A. Zuschlag, D. Skorka, and G. Hahn, "Degradation and regeneration in mc-Si after different gettering steps," *Prog. Photovolt Res. Appl.,* 2016.

[5] T. Luka, M. Turek, S. Großer, and C. Hagendorf, "Microstructural identification of Cu in solar cells sensitive to light-induced degradation," *Phys. status solidi - Rapid Res. Lett.,* vol. 5, pp. 1–5, 2017.

[6] R. Eberle, W. Kwapil, F. Schindler, M. C. Schubert, and S. W. Glunz, "Impact of the firing temperature profile on light induced degradation of multicrystalline silicon," *Phys. status solidi - Rapid Res. Lett.,* vol. 5, pp. 1–5, 2016.

[7] R. P. Winarski *et al.,* "A hard X-ray nanoprobe beamline for nanoscale microscopy," *J. Synchrotron Radiat.,* vol. 19, no. 6, pp. 1056–1060, 2012.

[8] T. Buonassisi *et al.,* "Transition metal co-precipitation mechanisms in silicon," *Acta Mater.,* vol. 55, no. 18, pp. 6119–6126, 2007.

[9] A. Inglese, J. Lindroos, H. Vahlman, and H. Savin, "Recombination activity of light-activated copper defects in p-type silicon studied by injection- and temperature-dependent lifetime spectroscopy," *J. Appl. Phys.,* vol. 120, no. 12, p. 125703, 2016.

[10] A. E. Morishige *et al.,* "Synchrotron-based investigation of transition-metal getterability in n -type multicrystalline silicon," *Appl. Phys. Lett.,* vol. 108, no. 20, 2016.

[11] T. Buonassisi *et al.,* "Impact of metal silicide precipitate dissolution during rapid thermal processing of multicrystalline silicon solar cells," *Appl. Phys. Lett.,* vol. 87, no. 12, pp. 1–3, 2005.

# PERC Solar Cell Performance Predictions from Multicrystalline Silicon Ingot Metrology Data

Bernhard Mitchell[1], Daniel Chung[1], Qiuxiang He[2], Hua Zhang [2], Zhen Xiong[2], Pietro P. Altermatt[2], Peter Geelan-Small[3], and Thorsten Trupke[1]

[1]Australian Centre for Advanced Photovoltaics, School of Photovoltaic and Renewable Energy Engineering, University of New South Wales, Sydney, NSW 2052, Australia
[2]State Key Laboratory of PV Science and Technology, Trina Solar, No.2 Trina Road, Trina PV Park, New District, Changzhou, Jiangsu, PR China
[3]Mark Wainwright Analytical Centre, University of New South Wales, Sydney, NSW 2052, Australia

*Abstract* — The influence of the as-grown material quality on the performance of multicrystalline silicon PERC solar cells is investigated using recently developed spectral photoluminescence imaging techniques on ingot level, i.e. on bricks, and is examined in conjunction with photoluminescence measurements on as-cut wafers. The effect of material parameters, including bulk lifetime, dislocation density and resistivity, are studied with regard to their effect on cell output across a sample set of three directionally solidified production bricks of widely varying bulk lifetime and dislocation density. The data is analyzed statistically using a linear mixed model. Bulk lifetime is found a highly significant predictor of the cell performance across the studied sample set confirming the predictions of computer simulations. The prediction accuracy is found greatest for material with low dislocation density where a linear correlation between cell performance and as-grown bulk lifetime is found. The dislocation densities measured on as-cut wafers remain a more accurate predictor for medium to highly dislocated material, but predictions are improved by adding bulk lifetime as an additional predictor in the model. Dislocation density measurements taken on the side facets of silicon bricks were identified as not being significant predictors for this data set. However, the detection may enable the classification of bricks into broad dislocation defect classes, which is expected to further improve the overall prediction accuracy for models using the predictor lifetime only in comparison to a non-differentiating global prediction models explored in this study. With increasing cell efficiencies and an ongoing trend of reducing the dislocation density in industrial multicrystalline wafers our findings suggest that the bulk lifetime, measurable on bricks, i.e. directly after ingot growth, becomes an increasingly relevant parameter.

# Photoluminescence-imaging-based Evaluation of Non-uniform CdTe Degradation

Steve Johnston, David Albin, Peter Hacke, Steven P. Harvey, Helio Moutinho, Mowafak Al-Jassim, and Wyatt K. Metzger

National Renewable Energy Laboratory, Golden, CO, 80401, U.S.A.

*Abstract* — Photoluminescence (PL), electroluminescence (EL), and dark lock-in thermography are collected during stressing of a CdTe module under one-Sun light at an elevated temperature of 100°C. The PL imaging system is simple and economical. The PL images show differing degrees of degradation across the module and are less sensitive to effects of shunting and resistance that appear on the EL images. Regions of varying degradation are evaluated using time-of-flight secondary ion-mass spectrometry, and there is a correlation between PL intensity and Cu concentration at the front interface.

*Index Terms* — accelerated aging, electroluminescence, II-VI semiconductor materials, imaging, photoluminescence, photovoltaic cells, reliability, thermal analysis.

## I. INTRODUCTION

Imaging techniques such as electroluminescence (EL) [1] and photoluminescence (PL) [2]-[3] are used for their quick acquisition and usefulness to determine spatially resolved quality, performance, and defects found in photovoltaic cells and modules. When imaging lower bandgap materials such as Si and $CuIn_xGa_{1-x}Se_2$, Si-based cameras detect luminescence at the edge of their sensitivity range, but these cameras provide very reasonable results for less cost than InGaAs-based cameras. For PL imaging, the light source is ideally monochromatic, such as 808-nm laser diodes, or light-emitting diodes for less-demanding applications like PL on passivated materials or finished cells. For larger bandgap materials, such as CdTe, shorter wavelength laser sources can lead to a more expensive PL imaging tool even though the PL emission falls into a more sensitive region of Si-based cameras.

We have constructed an economical PL imaging system using a green 532-nm laser diode with a raster of the laser beam for large area samples. Here, we couple this work with EL, dark lock-in thermography (DLIT), and time-of-flight secondary ion mass spectrometry (TOF-SIMS) on ~1.5 eV, high-bandgap CdTe modules stressed under light and heat. We are able to detect non-uniform degradation in CdTe mini-modules correlated with Cu concentration at the front interface.

## II. EXPERIMENTAL

The PL imaging system consists of a 532-nm laser diode from Laserglow Technologies. As shown in Fig. 1, the laser beam is swept in a raster pattern using a pair of scanning

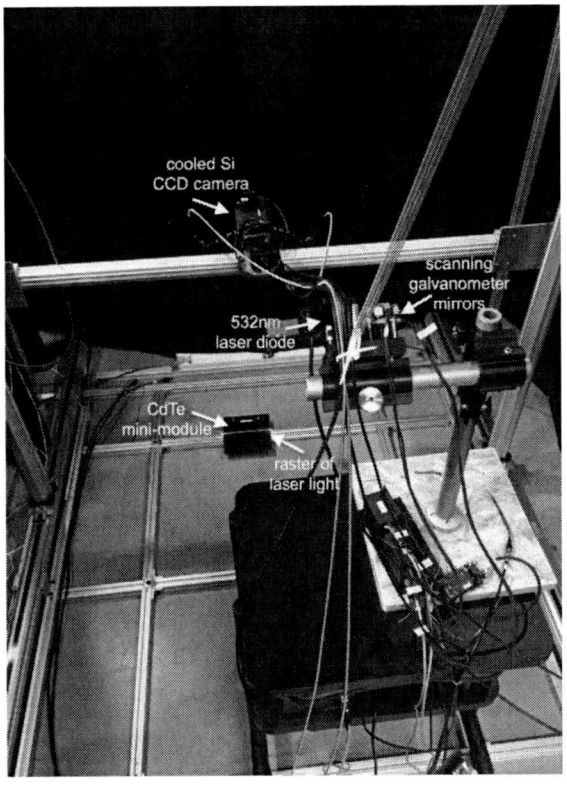

Fig. 1. The laboratory photograph shows the green laser beam scans quickly side to side on the CdTe mini-module. The galvanometer scan mirrors are positioned a distance away to allow for the laser raster area to expand. The camera is positioned above the module to image the entire module area.

galvanometer mirrors from Thor Labs. In one direction, the beam is quickly scanned back and forth (~300 Hz) to form a line, and this laser line is slowly swept up and down (~0.1 Hz) over an area larger than the sample (~20 x 20 cm). The intensity is adjusted by varying the duty cycle of the pulse width modulation at a frequency of 10 kHz. The typical laser power during a PL measurement is roughly 0.5 to 1 mW for a beam spot size of ~1 mm diameter. This is equivalent to approximately 100 mW/cm². A Princeton Instruments PIXIS 1024BR is used for collecting PL and EL images although the enhanced near-infrared efficiency is not needed for this wavelength range. This Si camera collects the PL image with a long exposure time (~1 to 2 minutes) that is a multiple of the

978-1-5090-5606-4/17 $31.00 © 2017 IEEE

10-second period of the laser raster time. A stack of three RG715 Schott glass filters is mounted to the camera lens to block scattered laser light and allow the CdTe PL to be transmitted. The cells of the mini-module are ~5 mm wide. EL images are collected by driving forward-bias current (about 1/3 of $J_{SC}$) through the module. DLIT images are collected with a Cedip Silver 660M (FLIR SC5600-M) InSb camera having lock-in data acquisition. The forward-bias current is pulsed at a frequency of 0.1 Hz, which better allows for a thermal signal through the mini-module's 3-mm-thick glass.

## III. RESULTS

The CdTe module is initially characterized by imaging before stresses are applied. Fig. 2 shows PL, EL, and DLIT images for the initial state of the module. The PL shows that the module is mostly uniform with only a few defects or anomalies. The EL shows that the upper three-quarters of the module are slightly darker. This is most likely due to the shunting defects seen in the DLIT image that are also mostly present in the upper half of the module. Because the EL is collected with only ~1/3 of $J_{SC}$ current, the shunts within each cell can reduce the EL emission by dropping the voltage across that cell. The currents applied during the imaging were intentionally kept low to reduce the risk of generating new defects or accelerating defect degradation during the study.

Fig. 2.    Pre-stress images of the CdTe module show (a) PL, (b) EL, and (c) forward-bias DLIT.

The module was stressed by subjecting it to one-Sun light exposure at an elevated temperature of 100°C, and dark and light current-voltage curves were measured periodically. The module parameters from the measurements are shown in Fig. 3. The open-circuit voltage, $V_{OC}$, consistently drops over the time of stressing. The fill factor and efficiency initially show slight improvement before later decreasing. The short-circuit current, $J_{SC}$, remains fairly constant throughout the stress time.

The module was also periodically imaged during the stressing time. EL images began to show more contrast, with the upper-right two-thirds of the cell appearing dark, after just 5 hours of stress. The PL images began to show a darker region in the center after about 60 hours of stress. The final

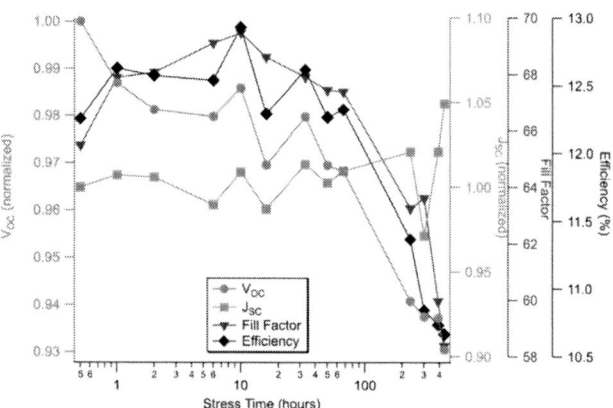

Fig. 3.    Module parameters $V_{OC}$, $J_{SC}$, fill factor, and efficiency are plotted as a function of time that the module is stressed under one-Sun light and 100°C.

images after 400 hours of stress are shown in Fig. 4. The EL image shows dark regions due to both shunting in the upper half and degradation in the center region that also appears dark in the PL image. Based on the images, regions of least, middle, and most degradation are chosen where shunting and initial defects are ideally not present. These areas are represented by green, orange, and red circles, where the green area is the least degraded, and the red area is the most degraded.

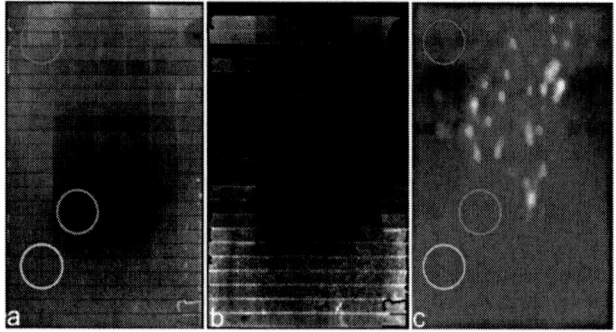

Fig. 4.    Post-stress images of the CdTe module show (a) PL, (b) EL, and (c) DLIT. Circles mark regions of interest away from shunts, where PL imaging shows various degrees of degradation.

To apply various microscopy techniques for further study of the degraded devices, samples must be extracted from the module. We have developed coring techniques [4] to extract circular cores from the module using a diamond-based coring bit and drill. Various size bits are available, and here we use a 25-mm (1-inch) size bit. For CdTe modules, the drill cuts through the top glass since most CdTe is grown in a superstrate configuration. Once the cut is through the front glass, encapsulant layers, and device layers, a cylindrical post with machined flat edges is glued to the glass. An end wrench applies torque to slowly sheer the cut specimen from the back glass sheet of the module. These extracted cores can now be imaged again to ensure the device is still intact. PL images of

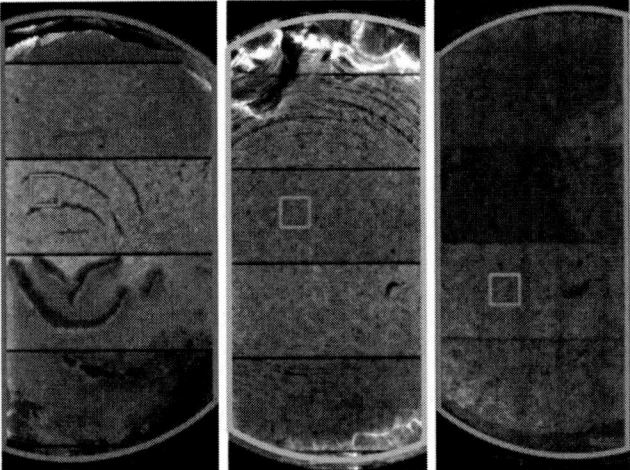

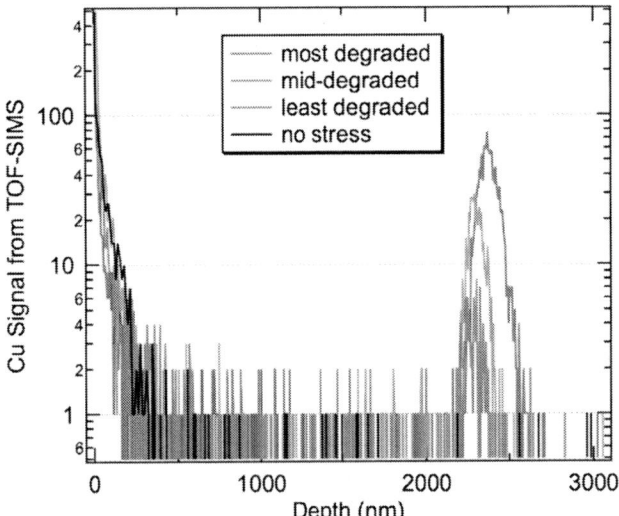

Fig. 5. Post-stress PL images of cored sections that were cut out of the CdTe module where various degrees of degradation exist. Each image is individually adjusted for contrast and brightness. The left image outlined in green represents a cored section from the least-degraded area of Fig. 4. The center image is outlined in orange for the middle-degradation region, and the right image outlined in red is from the most-degraded region. The small red boxes show regions away from coring damage where microscopy analysis is performed.

Fig. 6. TOF-SIMS shows depth profiles of Cu on selected regions of the cored CdTe samples. The back contact corresponds to zero depth, and the thickness of the CdTe absorber layer is approximately 2,000 nm. There is no Cu at the interface of the sample without stress.

the cored specimens are shown in Fig. 5. While there appear to be some artifacts of coring damage in the form of circular lines where fractures have occurred within the semiconductor device, much of the area is intact and still shows similar ratios of PL intensity. Using the same excitation-light parameters and camera settings, the PL intensity of the middle-degraded sample is ~80% of the PL intensity of the least-degraded sample, while the PL intensity of the most-degraded sample is ~40% of the least-degraded sample's PL intensity.

Also shown in Fig. 5, a small red box identifies an area that is ideally representative of that cored region and is located away from coring damage. TOF-SIMS is performed at these locations to profile elements in the CdTe device. The back contact of the device is the top surface during the measurement. As shown in Fig. 6, Cu is present in the back contact where the sputtering depth is near zero. However, Cu is also detected at the device's front interface after sputtering through the CdTe absorber layer, which is approximately 2,000 nm thick. At this front interface where the Cu signals have a peak, the most-degraded region's peak corresponds to a Cu concentration of mid-$10^{19}$ cm$^{-3}$. The least-degraded region has about an order of magnitude less Cu accumulated at the front interface. An unstressed CdTe sample is similarly measured and shows similar Cu content in the back contact, but no Cu is detected at the front interface. Cu is known to accumulate at the front interface of the device during degradation and leads to poorer material quality and device performance [5]-[9].

## IV. SUMMARY

An economical PL imaging tool has been used to monitor degradation in a CdTe module. PL imaging is less sensitive to shunting and high series resistance than EL imaging, and its intensity often directly correlates to minority-carrier lifetime, diffusion length, and $V_{OC}$ of the device. Here, PL imaging was used to characterize module degradation and identify regions that had degraded differently within the module. TOF-SIMS was used to measure Cu depth profiles, and the amount of Cu found at the front interface correlated to the amount of degradation observed by PL imaging.

## ACKNOWLEDGEMENT

The U.S. Government retains and the publisher, by accepting the article for publication, acknowledges that the U.S. Government retains a nonexclusive, paid-up, irrevocable, worldwide license to publish or reproduce the published form of this work, or allow others to do so, for U.S. Government purposes. This work was supported by the U.S. Department of Energy under Contract No. DE-AC36-08GO28308 with the National Renewable Energy Laboratory. Funding was provided by the U.S. Department of Energy Office of Energy Efficiency and Renewable Energy Solar Energy Technologies Office.

978-1-5090-5606-4/17 $31.00 © 2017 IEEE

## REFERENCES

[1] T. Fuyuki, H. Kondo, T. Yamazaki, Y. Takahashi, and Y. Uraoka, "Photographic surveying of minority carrier diffusion length in polycrystalline silicon solar cells by electroluminescence," *Appl. Phys. Lett.*, vol. 86, pp. 262108, 2005.

[2] T. Trupke, R. A. Bardos, M. C. Schubert, and W. Warta, "Photoluminescence imaging of silicon wafers," *Appl. Phys. Lett.*, vol. 89, pp. 044107, 2006.

[3] T. Trupke, R. A. Bardos, M. D. Abbott, F. W. Chen, J. E. Cotter, and A. Lorenz, "Fast photoluminescence imaging of silicon wafers," in *32nd IEEE PVSC and WCPEC-4*, 2006, pp. 928-931.

[4] S. Johnston, M. Al-Jassim, P. Hacke, S. P. Harvey, C.-S. Jiang, A. Gerber, H. Guthrey, H. Moutinho, D. Albin, B. To, J. Tynan, J. Moseley, J. Aguiar, C. Xiao, J. Waddle, and M. Nardone, "Module degradation mechanisms studied by a multi-scale approach," in *43rd IEEE Photovoltaic Specialist Conference*, 2016.

[5] C. R. Corwine, A. O. Pudov, M. Gloeckler, S. H. Demtsu, and J. R. Sites, "Copper inclusion and migration from the back contact in CdTe solar cell," *Solar Energy Materials & Solar Cells*, vol. 82, pp. 481-489, 2004.

[6] S. H. Demtsu, D. S. Albin, J. R. Sites, W. K. Metzger, and A. Duda, "Cu-related recombination in CdS/CdTe solar cells," *Thin Solid Films*, vol. 516, pp. 2251-2254, 2008.

[7] N. Strevel, L. Trippel, and M. Gloeckler, "Performance characterization and superior energy yield of First Solar PV power plants in high-temperature conditions," *Photovoltaics International*, 2012.

[8] N. Strevel, L. Trippel, C. Kotarba, and I. Khan, "Improvements in CdTe module reliability and long-term degradation through advances in construction and device innovation," *Photovoltaics International*, 22nd Ed.

[9] M. Nardone and D. S. Albin, "Degradation of CdTe solar cells: simulation and experiment," *IEEE J. of Photovolt.*, vol. 5, pp. 962-967, 2015.

# Machine Learning and Correlative Microscopy: How 'Big Data' Techniques Can Benefit Thin Film Solar Cell Characterization

Bradley M. West[1], Michael Stuckelberger[1], Tara Nietzold[1], Barry Lai[2], Jörg Maser[2], Mariana I. Bertoni[1]

[1] Ira A. Fulton Schools of Engineering, Arizona State University, Tempe, AZ, 85281, USA
[2] Advanced Photon Source, Argonne National Laboratory, Lemont, IL, 60439, USA

*Abstract* — **Correlative microscopy techniques have improved tremendously in the last 5 years and enabled simultaneous high spatial resolution mapping of a variety of material parameters. As acquisition speeds and resolution increase giving us a greater density of data points and the functionality of the measurements add more dimensions to be analyzed; the handling, management and analysis of data sets becomes more and more complicated. Operando measurements as well as in-situ studies pose a new challenge. Finding correlations in the 3+ dimensional data sets that result from many of these measurements is not straight forward, and the possibility of missing correlations and trends is increasingly concerning. In the following, we evaluate the potential of two machine-learning methods, an unsupervised Gaussian-Mixture-Model and a supervised Naïve Bayes model, to identify two groups of pixels in a micro-X-ray fluorescence map, given varied input data sizes.**

*Index Terms* – **CIGS, XRF, in-situ, Gaussian mixture model, machine learning, Naïve Bayes, X-ray microscopy**

## I. INTRODUCTION

Correlative and in-situ microscopy has proven to be very beneficial to the study of solar cells [1], [2]. Being able to experimentally measure fundamental material properties, and device performance simultaneously at the nanoscale has proven invaluable to understand the effect of inhomogeneities in the solar cell performance[2], [3]. Likewise, understating the kinetics and transport at these scales will open the door to engineer more efficient devices. While the techniques that one would use to study devices under operation differ from the ones required to evaluate kinetics, the correlative nature of these experiments requires methods capable of reducing dimensionality, clustering data into groups of similar features, and classification of new input data.

Correlative microscopy spans a wide range of techniques from scanning probe techniques [4] to electron beam based techniques [5] and optical techniques including, lasers [6] and X-rays [3]. A property they all have in common is the ability to measure several properties simultaneously, or separately on the same spot. While these techniques are very powerful, they also generate a unique challenge: how do we analyze these large data sets, which can span 3 or more dimensions? Typical analyses of these data sets tend to be more qualitative and require the user to pick the particular regions of interest. While still providing useful information, this approach ends up providing just a narrow view of the data and "disregarding" much of the information that is available.

Machine learning techniques, both supervised, where you know the feature(s) you are looking for and unsupervised, which is exploratory, have been widely used in fields notoriously plagued (gifted) with tremendous amounts of data [7]. A key benefit of these approaches is the ability to identify trends in highly dimensional data that would be inherently difficult to accomplish by hand. These same techniques can provide a similar benefit to data arising from correlative microscopy techniques on solar cells.

In this manuscript, we apply machine learning techniques to understanding elemental segregation that occurs in CIGS absorber layers during a precursor reaction process. CIGS is an ideal model system for this study because it is very well known that the material is inhomogenous at the sub-micron scale both laterally and as a function of depth [8]. During the 42nd PVSC, we presented results investigating the impact of composition variations between grains and at grain boundaries on charge carrier collection efficiency measured by X-ray fluorescence (XRF) and X-ray beam induced current (XBIC) respectively [9] and published a full analysis of the results [3].

We build upon our previous findings and seek to visualize the origin and formation of those inhomogeneities and identify the best model to classify these variations during growth. A challenge with these measurements is the trade-off between XRF map area, spatial resolution, and dwell time. Our current optimized processes allows us to collect a 7 μm x 7 μm map in ~ 200 s. However, with this area, we only capture between 1 and 3 features per map. The small number of features, combined with substrate expansion and elemental segregation occurring on a time scale << 200 s, requires the use of clustering and classification techniques to compare features across maps. A goal of this study is to determine the necessary map size relative to feature size necessary to classify two unique composition regions in the XRF data. It should be noted that these methods are not limited to the CIGS material systems or in-situ XRF data, but can be applied to any inhomogeneous PV material, from perovskites and CdTe, to earth abundant alternatives such as SnS and CuZnSnS as well as to correlative techniques with operando, or structural data.

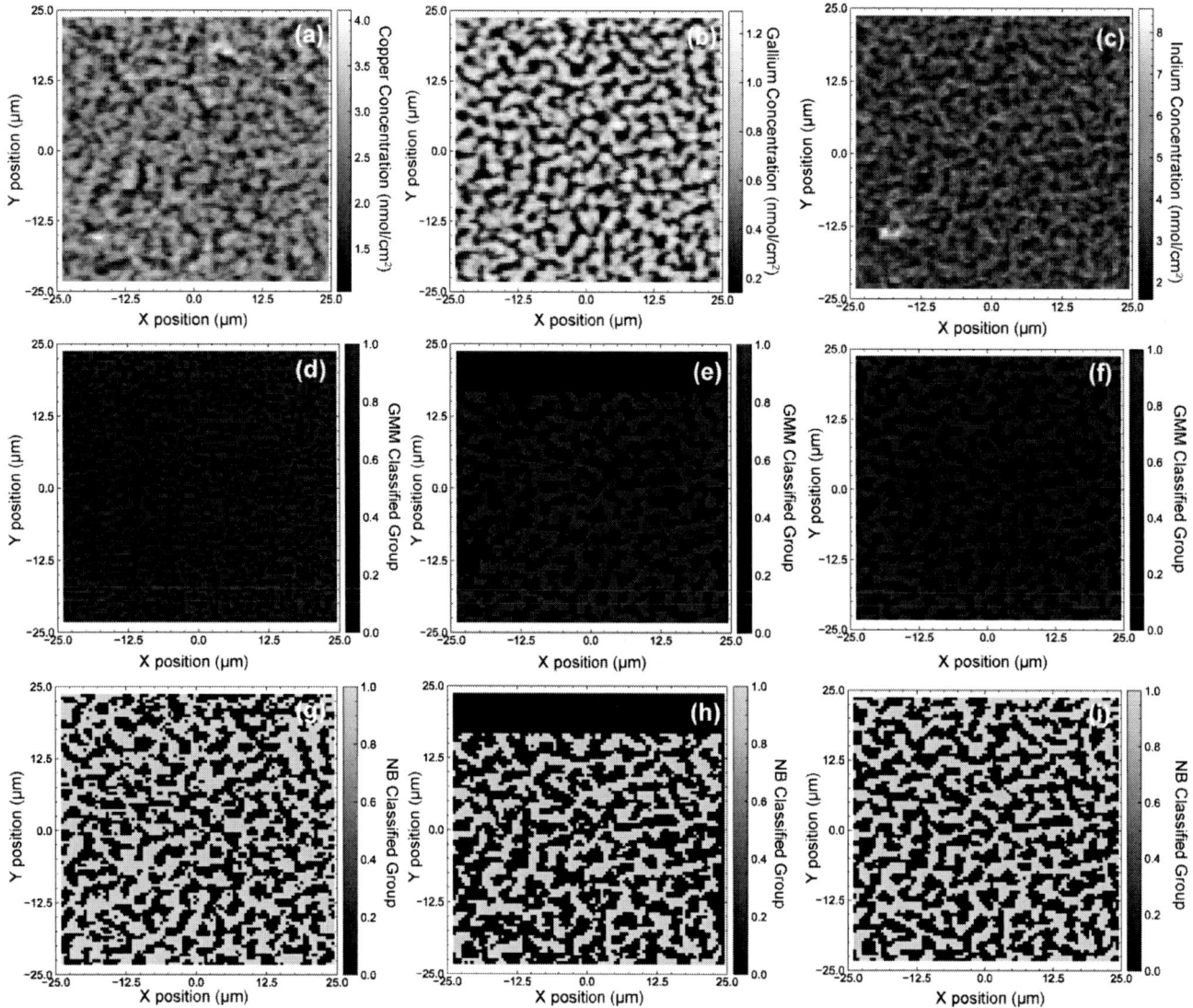

Fig. 1 (a)-(c): 47.5 µm x 49 µm XRF maps of Cu, Ga, and In respectively with a 500 nm x 500 nm pixel size taken at 300 °C. (d)-(f): Spatial distribution of the binary pixel classification using an unsupervised Gaussian-Mixture-Model for 1 µm x 1 µm, 8 µm x 8 µm, and, 47.5 µm x 49 µm sub-maps respectively. (g)-(i): Binary classification using a supervised Naïve Bayes (NB) model using the same sub-map sizes as (d)-(f). The entire map includes 9,310 pixels. The black bar in (e) and (h) at Y > 17.5 µm is due to the size of the sub-map, such that this area was not classified.

## II. METHODS AND EXPERIMENTAL SETUP

CIGS precursor layers were prepared at the University of Delaware, by sputtering alternating layers of $Cu_{0.8}Ga_{0.2}$ and elemental indium, on molybdenum coated soda lime glass, to achieve a total Cu/Ga/In layer thickness of ~ 650 nm. The layers were then capped with 10 µm of evaporated selenium. The average copper ratio of the metal precursor layer [Cu]/[In+Ga] was 0.8 and the gallium ratio [Ga]/[Ga+In] was 0.2. Samples were heated to 300 °C for 2 hours in a He atmosphere. Helium was used rather than Ar to reduce

fluorescence generated from the atmosphere. More details on precursor sample preparation can be found in [10].

XRF maps were collected at the Advanced Photon Source Beamline 2-ID-D [11], utilizing an temperature and ambient controlled in-situ stage developed specifically for 2-ID-D to study elemental segregation with sub-micron resolution at temperatures up to and greater than 600 °C [12]. The incident beam energy was set above the Gallium K edge (10.5 keV) to allow for sufficient sensitivity to all cations, and the beam spot size was 200 nm. The incident beam angle was 90°, perpendicular to the sample surface, and the detector was 47° from the sample surface. The dwell time was 100 ms. Fluorescence spectra were fitted against a well quantified standard using the MAPS package developed at APS [13]-[14]. Due to the uncertainty in film thicknesses at elevated temperatures, the composition was not corrected for attenuation as described in [15].

In order to isolate unique composition regions within the map, two classification techniques were implemented. An unsupervised Gaussian mixture model (GMM) was used to cluster the XRF composition data based on the probability that an individual pixel falls within one of two Gaussian distributions, whose centers are identified using the k-means method. The clustering is based on the expectation-maximization algorithm, wherein the likelihood that each pixel falls within a given distribution is maximized [16]. We also implemented a Naïve Bayes classifier, to predict whether a new data point falls into one group or another based on an initial set of training data. The classifier is based on Bayes' theorem that describes the probability of an event occurring based on prior knowledge of conditions related to that event. More information about Naïve Bayes and Gaussian mixture models can be found in [7]. Both models were implemented in Python using the Scikit-Learn package [17].

## III. RESULTS AND DISCUSSION

Figure 1(a-c) shows Cu, Ga, and In XRF maps, respectively, collected from the CIGS precursor stack, after 25 min, at 300 °C. The total collection time for this map was ~ 50 min. The gallium and indium channels shows a unique formation of 'islands' not evident in the copper map. Indium shows a clear anti-correlation with the gallium channel which is expected, because they are substitutional in the CIGS lattice. The correlation with copper however is not intuitive. In some areas copper and gallium appear to correlate positively, in others negatively. This highlights the need for clustering, to separate these two regions, in order to be able to track each composition region throughout changing time and temperature. However, in some cases a 50 minute collection time is not practical, and smaller maps need to be taken, while still being representative of the larger map. In this case, the question is: what map size should be selected?

In order to understand the relationship between map size and the variance in the classification, the XRF map was divided into smaller sub-maps, of 2x2 pixels, 4x4 pixels etc. up to 30x 30 pixels such that a minimum of 9 sub-maps were extracted form the original map. It can be seen from Fig. 1 (a-c) that a 2 pixel x 2 pixel map (1 $\mu m^2$) is much smaller than the typical typical feature size. Where as the 16 pixel x 16 pixel (8 x 8 $\mu m^2$) map is large enough to include multiple features. This inidicates that map sizes smaller than the average feature size are not capable of idenfitying unique composition regions, , and if this is the case, the interpretation of how one group chagnes over time can vary widely.

Two clusters were selected to ensure that each cluster is compositionally unique. More than two groups resulted in

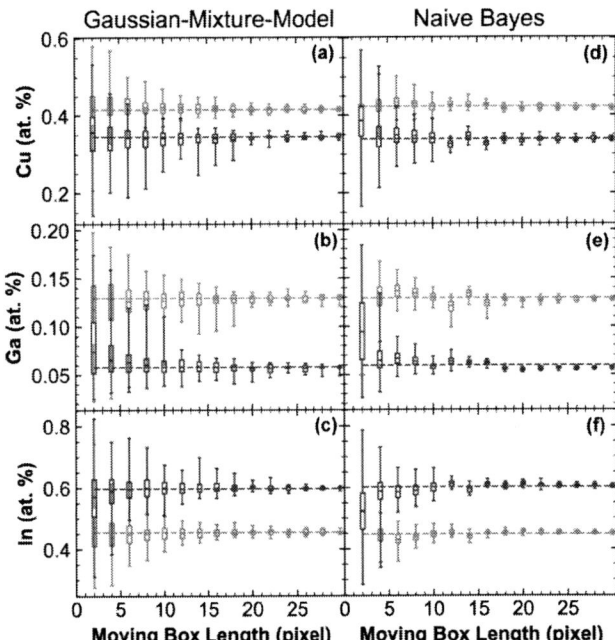

Fig. 2 Box plots showing the distribution of average copper, indium and gallium composition for each group, identified using a Gaussian mixture model(a)-(c), and a Naïve Bayes model (d)-(f). Top, middle and bottom of the box represent the 75, 50, and 25 percentiles respectively. The edge of the whiskers are the maximum and minimum of the sample set. The red group corresponds to Group 0 from Fig. 1, and the blue group corresponds to Group 1. The dashed lines represent the average composition of each group when the entire map is clustered using the Gaussian mixture.

composition distributions that were not isolated form one another. Because Naïve Bayes (NB) is a supervised technique, the model needs to be 'trained' with an intial set of input data with corresponding lables, identifed as the true case. The labels were taken from the GMM, that was applied to the bottom left sub-map closest to (X,Y) = (-25 μm, -25 μm). As the sub-map size increases the size of the NB training dataset also increases.

Figure 1 (d)-(i) shows the distribution of pixels classified either Group 0 or Group 1 based on the size of the sub-map using the GMM(d-f) or Naïve Bayes (g-i). Because the GMM is unsupervised, the model always finds two clusters within the sub-map, independent of the labels applied to previous sub-maps. This is evident in Fig. 1 (d) where the size of the identified groups is much smaller than what can be seen by looking at the XRF maps. Interestingly the same sub-map size, classified using NB (Fig. 1g), identifies groups much closer to what can be seen from the XRF maps. This is because rather than always identifying two clusters in each map, it labels each pixels based on the probability that it belongs to one group or another. Additionally, in this particular case it is likely that the intial training set for the NB model included pixels that belonged belong to both groups (based on the final map). This

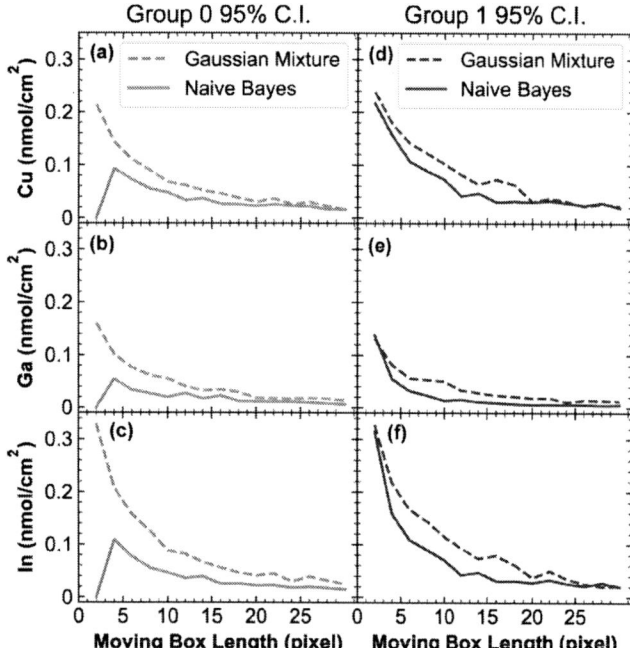

Fig.3 Confidence Intervals for average group composition as a function of sub-plot length. Dashed lines are the confidence intervals corresponding to the Gaussian mixture model, and solid lines correspond to the Naïve Bayes classifier. (a-c) compares the confidence intervals for Group 0 across Cu, Ga, and In respectively. (d-f) compares the confidence intervals for Group 1.

correlation between the smallest sub-map and the final map would change if the entire training set belonged to Group 0 or Group 1. This can lead to entire maps being classified as Group 0 or Group 1. This makes the process highly repeatable and increases the similarity between the classified pixels. As the sub-map size increases, (e,h) the groups identified by each method converge, and are more representative of the features apparent from the XRF map. The rectangular area at Y > 17.5 μm, is due to the size of the sub-map selected. An additional 16 pixel x 16 pixel sub-map would not fit entirely in this region, and for this reason it was unclassified. Figure 1 (f,i) show the classification of the entire XRF map using GMM and NB respectively.

Figure 2 shows the distribuion of the mean composition of each group as a function of the sub-map size (moving box length). Atomic percentages are calculated by dividing each element by the sum of all cations (Cu, In, Ga) for a given pixel. The mean composition was calculated from the composition of the pixels identified as Group 0 or Group 1 for both GMM (Fig. 1 a-c) and NB (Fig. 2 d-f). It can be seen that, as the sub-map length increases, the variance in the average group composition decreases. Note also that as the sub-map length increases the number of sub-maps decreases, leading to a smaller sample size $N$. For 2 pixel by 2 pixel map, $N = 2303$, for 10 pixel x 10 pixel, $N = 81$, and for 30 pixel x

30 pixel, $N = 9$. The smallest sub-map size (2 pixel x 2 pixel) does not succeed in identifying compositionally unique groups, for either model, with an observable overlap of the whiskers of each group's box plot. This indicates that a 2 pixel x 2 pixel map is not sufficient to describe the data. While the GMM results in a similar variance for both groups, NB results in Group 0 having a very small distribution and Group 1 having a very large distribution. In our application, we seek to determine the smallest submap size for which the box plots no longer overlap and the groups can be clearly distinguished from each other. This identifies the minimum map size necessary to identify compositionally unique groups. If the features observed in the map are not similar in size across all elements, this could lead to more than two groups being idenfied, and require more training data to idetnfiy the clusters. For NB this critical size is different for each of the elements. One can see that fewer data points are necesssary to classify unique groups using a NB model, than a GMM. It is also interesting to note that with a feature size of ~5 μm, only 1 – 2 features are necessary to classify the data. The requirement of smaller map sizes for NB than GMM becomes even more apparent if a larger number of clusters is identified. More data will be necessary to accurately identify each new subgroup.

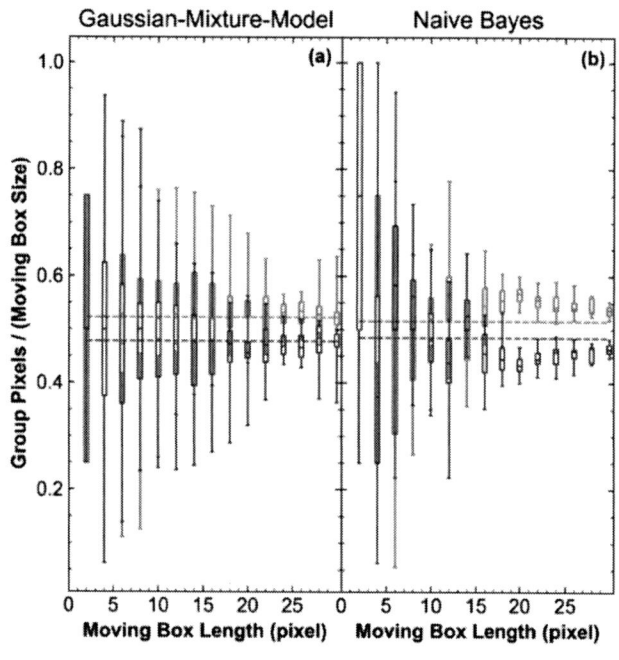

Fig. 4 the distribution of the number of pixels identified per group (Red = Group 0, Blue = Group 1) relative to sub-map size. The dashed line represents the 'true' group concentration based on a full XRF map clustering. Top, middle and bottom of the box represent the 75, 50, and 25 percentiles respectively. The edge of the whiskers are the maximum and minimum of the sample set.

To better visualize the difference between the two models, Figure 3 shows the 95 % confidence inverval (C.I.) for the average group composition for Cu, Ga, and In for both classification methods. As the sub-map size increases, the confidence interval decreases for both cases and all elements. It is interesting to note that for the 2 pixel x 2 pixel sub-map, the Group 0 NB classifier has the narrowest confidence interval. This corresponds to the narrow distribution observed in Figure 2 (d-f), likely due to the small data set that was used to train the model, which resulted in narrow composition regions for Group 0. For Group 1, both NB and GMM have similar C.I. across all elements. Between the sub-map length of 4 and 20 pixels the NB classifier out performs the GMM, with a smaller C.I. for the average group composition. For sub-maps larger than 20 pixels, both models perform similarly.

Another parameter that is important when comparing these models, is the fraction of pixels attributed to Group 0 and Group 1. Figure 4 shows the distribution of these fractions as a function of sub-map length for both models. At small sub-map sizes, we note for both models a wide distribution that overlaps between the two groups. The dashed lines in each plot represent the group fraction based on clustering the entire XRF image. With a sub-map length > 20 pixels or 10 μm, NB begins to converge on the 'true' ratio of Group 0 and Group 1.The GMM model converges on the average, as would be expected, but the distribution is very wide, leading to higher uncertainty than the NB model.

## IV. CONCLUSION

In this study we showed the potential for machine learning techiques to be applied to in-situ XRF data to interpret a 9,310 pixel map of 3 separate elements, as two classes of pixels. To understand the strengths and limits of this approach, we compared an unsupervised Gaussian mixture model and a supervised Naïve Bayes classifier. For smaller sample sizes the supervised NB classifier out performed the unsupervised GMM. However, as the sample size increased, the two models converged. The NB classifier was also able to predict the fraction of each group with less data, and with a smaller maximum-minimum spread than the GMM. More specifically, we were able to determine that the NB classifier is able to identify compositionally unique groups from a map size 5x5 μm², which is on the order of the feature size. The unsupervised GMM required an input map size of at least 7x7 μm² for the same feature size. Moving forward, we will apply these methods, to understand elemental segreagation during CIGS growth. Understanding the relationship between feature size and map size, enables us to not intervene when dealing with data that has features moving and changing shapes. We are also able to optimize future experiments to identify the optimum time, window size, and step-size. These methods can also be applied to other measurement techniques beyond in-situ XRF as well as other material systems.

## ACKNOWLEDGEMENT

We acknowledge funding from the U.S. Department of Energy under contract DE-EE0005848. Work at the Advanced Photon Source was supported by the U.S. Department of Energy, Office of Science, Office of Basic Energy Sciences, under Contract No. DE-AC02-06CH11357. This material is based upon work supported in part by the National Science Foundation (NSF) and the Department of Energy (DOE) under NSF CA No. EEC-1041895. Any opinions, findings and conclusions or recommendations expressed in this material are those of the author(s) and do not necessarily reflect those of NSF or DOE.

## REFERENCES

[1] U. Rau, D. Abou-Ras, and T. Kirchartz, "Advanced characterization techniques for thin film solar cells," Jan. 2011.

[2] M. Stuckelberger, B. West, T. Nietzold, B. Lai, J. M. Maser, V. Rose, and M. I. Bertoni, "Review: Engineering Solar Cells Based on Correlative X-Ray Microscopy," *Under Rev.*, 2017.

[3] B. M. West, M. Stuckelberger, H. Guthrey, L. Chen, B. Lai, J. Maser, V. Rose, W. Shafarman, M. Al-Jassim, and M. I. Bertoni, "Grain Engineering: How Nanoscale Inhomogeneities Can Control Charge Collection in Solar Cells," *Nano Energy*, vol. 32, no. December 2016, pp. 488–493, 2016.

[4] C.-S. Jiang, M. a. Contreras, L. M. Mansfield, H. R. Moutinho, B. Egaas, K. Ramanathan, and M. M. Al-Jassim, "Nanometer-scale surface potential and resistance mapping of wide-bandgap Cu(In,Ga)Se2 thin films," *Appl. Phys. Lett.*, vol. 106, no. 4, p. 43901, 2015.

[5] M. J. Romero, M. M. Al-Jassim, R. G. Dhere, F. S. Hasoon, M. A. Contreras, T. A. Gessert, and H. R. Moutinho, "Beam injection methods for characterizing thin-film solar cells," *Prog. Photovoltaics Res. Appl.*, vol. 10, no. 7, pp. 445–455, 2002.

[6] M. S. Leite, M. Abashin, H. J. Lezec, A. G. Gianfrancesco, a. A. Talin, and N. B. Zhitenev, "Mapping the Local Photoelectronic Properties of Polycrystalline Solar Cells Through High Resolution Laser-Beam-Induced Current Microscopy," *IEEE J. Photovoltaics*, vol. 4, no. 1, pp. 311–316, Jan. 2014.

[7] T. Hastie, R. Tibshirani, and J. Friedman, "The Elements of Statistical Learning," vol. 1, pp. 337–387, 2009.

[8] I. Repins, L. Mansfield, A. Kanevce, S. A. Jensen, D. Kuciauskas, S. Glynn, T. Barnes, W. Metzger, J. Burst, C. Jiang, P. Dippo, S. Harvey, C. Perkins, B. Egaas, A. Zakutayev, J. Alsmeier, T. Lußky, R. G. Wilks, M. Bär, Y. Yan, S. Lany, P. Zawadzki, J. Park, S. Wei, N. Renewable, H. Berlin, E. Gmbh, and B. T. U. Cottbus-senftenberg, "Wild Band Edges : The Role of Bandgap Grading and Band-Edge Fluctuations in High-Efficiency Chalcogenide

Devices," *Ieee Pvsc 2016*, no. June, pp. 309–314, 2016.

[9]  B. West, S. Husein, M. Stuckelberger, B. Lai, J. Maser, B. Stripe, V. Rose, H. Guthrey, M. Al-Jassim, and M. Bertoni, "Correlation between grain composition and charge carrier collection in Cu (In, Ga) Se2 solar cells," *Proc. IEEE PVSC*, 2015.

[10] D. M. Berg, F. Cheng, and W. N. Shafarman, "H 2 S reaction of Se-capped metallic precursors to form Cu ( In , Ga )( S , Se ) 2 absorber layers," pp. 323–327, 2014.

[11] W. Yun, B. Lai, Z. Cai, J. Maser, D. Legnini, E. Gluskin, Z. Chen, A. a. Krasnoperova, Y. Vladimirsky, F. Cerrina, E. Di Fabrizio, and M. Gentili, "Nanometer focusing of hard x rays by phase zone plates," *Rev. Sci. Instrum.*, vol. 70, no. 5, p. 2238, 1999.

[12] R. Chakraborty, J. Serdy, B. West, M. Stuckelberger, B. Lai, J. Maser, M. I. Bertoni, M. L. Culpepper, and T. Buonassisi, "Development of an in situ temperature stage for synchrotron X-ray spectromicroscopy," *Rev. Sci. Instrum.*, vol. 86, no. 11, p. 113705, 2015.

[13] S. Vogt, "MAPS : A set of software tools for analysis and visualization of 3D X-ray fluorescence data sets," *J. Phys. IV Fr.*, vol. 104, pp. 635–638, 2003.

[14] T. Nietzold, B. West, M. Stuckelberger, B. Lai, S. Vogt, and M. I. Bertoni, "Quantifying X-ray fluorescence data using MAPS," *Publ. Prep.*, 2017.

[15] B. M. West, M. Stuckelberger, A. Jeffries, S. Gangam, B. Lai, B. Stripe, J. Maser, V. Rose, S. Vogt, and M. I. Bertoni, "X-ray fluorescence at nanoscale resolution for multicomponent layered structures: a solar cell case study," *J. Synchrotron Radiat.*, vol. 24, no. 1, pp. 1–8, 2017.

[16] A. P. Dempster, N. M. Laird, and D. B. Rubin, *Maximum likelihood from incomplete data via the EM algorithm*, vol. 39, no. 1. 1977.

[17] F. Pedregosa, G. Varoquaux, A. Gramfort, V. Michel, B. Thirion, O. Grisel, M. Blondel, G. Louppe, P. Prettenhofer, R. Weiss, V. Dubourg, J. Vanderplas, A. Passos, D. Cournapeau, M. Brucher, M. Perrot, and É. Duchesnay, "Scikit-learn: Machine Learning in Python," *J. Mach. Learn. Res.*, vol. 12, pp. 2825–2830, 2012.

978-1-5090-5606-4/17 $31.00 © 2017 IEEE

# Metal Induced Contact Recombination Measured By Quasi-steady-state Photoluminescence

Robert Dumbrell[1], Mattias K. Juhl[1], Mengjie Li[2], Thorsten Trupke[1], Ziv Hameiri[1]

[1] University of New South Wales, Sydney, Australia

[2] Solar Energy Research Institute of Singapore, Singapore

*Abstract* — As silicon solar cells approach limiting efficiencies, recombination at the metal-silicon interface is becoming an important limiting factor. Current methods to characterize this recombination are limited in that either: (1) they can only be applied to test structures that are not representative of the final solar cell; or (2) they require electrical contact to both sides of the *p-n* junction. In this study, we present a novel measurement technique to study recombination at the metal-silicon interface. The method is based on the quasi-steady-state photoluminescence technique, which is ideally suited because it is contactless and can be applied to completely processed cells with full rear metal coverage, precursor structures or other test structures. The new method is first validated against commonly used open circuit voltage measurements, and then applied to a range of metallized device structures to extract the contact recombination saturation current density.

## I. INTRODUCTION

As silicon (Si) solar cells approach higher efficiencies, recombination at the metal-Si interface is becoming the limiting factor. The passivated emitter and rear cell (PERC) structure, which reduces the fraction of metal-Si interfacial area at the rear of the cell, is currently favoured in the industry to address this problem [1]. The highest efficiency laboratory cells employ passivated contacts, which completely remove the metal-silicon interface [2]. In both cases, new methods to characterize this recombination are required.

Contact recombination can be characterized by measuring the contact recombination saturation current density ($J_{0m}$), which combines recombination from the heavily doped region and the interfacing layer into a single parameter. The total recombination saturation current density ($J_{0s,total}$) is typically measured using quasi-steady-state photoconductance (QSSPC) with the method of Kane and Swanson [3]. This technique cannot be applied to metallized cells because the conductance of the metal layer dominates the overall conductance. Previous studies have attempted to minimise this problem by depositing very thin layers of metal by evaporation [4], but it is uncertain how representative this is of the metal interface of a real cell.

An alternative characterization for finished cells and metallized test structures is to evaluate the effective lifetime from the quasi-steady-state open circuit voltage ($QSSV_{oc}$) as measured by Suns-$V_{oc}$ [5]. However, this technique requires electrical contact to each side of the junction. In addition,

recent studies have shown that the above $J_{0s,total}$ analysis of Suns-$V_{oc}$ data, which is generally performed at high illumination intensities, can be associated with very significant experimental errors [6].

Further alternative methods involve depositing metal of varying contact area fractions onto a single wafer to create several smaller cells or test structures, and fitting a one diode model to the current-voltage (*I-V*) [7] or Suns-$V_{oc}$ [8] curves, or analysing a photoluminescence (PL) image [9] [10] to extract $J_{0m}$. These methods are limited by the fact that metal layers formed in the test structures are not necessarily representative of the contact in the final cell. The effect of any process-induced defect that is dependent on the area fraction, for example bubble formation during screen-printing, cannot be accurately assessed by this technique.

Our newly developed system, based on the quasi-steady-state PL (QSSPL) technique [11] [12] is ideally suited to measuring the recombination at metal contacts. Inclusion of a PL detector on the front side of the sample means the system can measure the recombination behaviour of any device structure with a full rear metallization applied such as completed cells or other metallized test structures. The system is also more versatile to use because being a contactless measurement, it can be easily applied to arbitrary contacting structures. Because the PL measurements are dynamically calibrated [12], the technique avoids assumptions and numerical modelling that are necessary for other PL imaging based approaches [8] [9].

In this study, we validate the experimental setup against a $QSSV_{oc}$ measurement of a solar cell and then apply the QSSPL technique to extract $J_{0m}$ from measurements of a metallized test structure.

## II. METHODOLOGY

In the QSSPL setup the sample is illuminated from above with either an 810 nm light emitting diode (LED) array capable of a maximum intensity equivalent to approximately one sun in the sample plane; or a xenon flash lamp with maximum intensity equivalent to approximately 30 suns that is filtered to block light in the Si band-to-band emission spectrum. A dielectric short pass filter is mounted at a 45° angle which allows shorter wavelength illumination to pass onto the sample while reflecting the emitted luminescence onto an InGaAs photodiode ('PL sensor'). Further filtering at the PL sensor ensures that the detected luminescence signal is unaffected by illumination light. The signal from the PL sensor is amplified by a low noise pre-amplifier and captured by a computer-based data acquisition system.

The average generation rate ($G_{av}$) is calculated as

$$G_{av} = \frac{J_\gamma[1-R_{810nm}]}{W} \qquad (1)$$

where $J_\gamma$ is the incident photon flux as measured by a calibrated reference photodiode, $R_{810nm}$ is the front surface reflectance at the illumination wavelength and $W$ is the sample thickness. For measurements with the flash lamp, an absolute value for $J_\gamma$ was obtained by scaling to match the lifetime from the LED based measurements.

The effective lifetime ($\tau_{eff}$) is calculated from the time-dependent luminescence signal according to generalized lifetime analysis [13] as

$$\tau_{eff} = \frac{\Delta n}{G_{av}-\frac{d\Delta n}{dt}} \qquad (2)$$

with the excess carrier concentration $\Delta n$ determined from [11]

$$A_i I_{PL} = B\Delta n[N_{dop} + \Delta n] \qquad (3)$$

where $I_{PL}$ is the relative luminescence intensity, $N_{dop}$ is the doping density, $B$ is the radiative recombination coefficient, and $A_i$ is a scaling factor that was determined by the self-consistent calibration method [12].

For the validation experiment, the sample was mounted on a voltage probe stage such that the $V_{oc}$ could be measured simultaneously with the luminescence response. $\Delta n$ was determined from the $V_{oc}$ using

$$\Delta n[N_{dop} + \Delta n] = n_i^2 \exp\left(\frac{V_{oc}}{kT/q}\right) \qquad (4)$$

where $n_i$ is the intrinsic carrier concentration and $kT/q$ is the thermal voltage. And $\tau_{eff}$ was calculated using (2). The validation experiment was performed at low illumination levels (below one sun) where (4) is valid and lateral series resistance effects can be ignored [6].

### III. VALIDATION

To ensure correct operation of the PL setup and validate the self-consistent calibration procedure, a completed solar cell was measured in the previously described setup and the injection-dependent $\tau_{eff}$ was calculated independently from the $V_{oc}$ and the PL signals respectively.

The measured sample was a standard 6" PERC cell fabricated on a commercial grade 180 μm thick, 1.5 Ωcm $p$-type monocrystalline Si wafer. Due to limitations of the maximum sample size in our measurement set up, the cell was cut to 3×3 cm² for analysis. The results are shown in Fig. 1, with excellent quantitative agreement (in the range of 10 %)

observed between $\tau_{eff}$ determined from QSSPL and QSS$V_{oc}$. We emphasise that the absolute calibration of the PL data was achieved independently using the self-consistent method, with no cross calibration against the $V_{oc}$ based data.

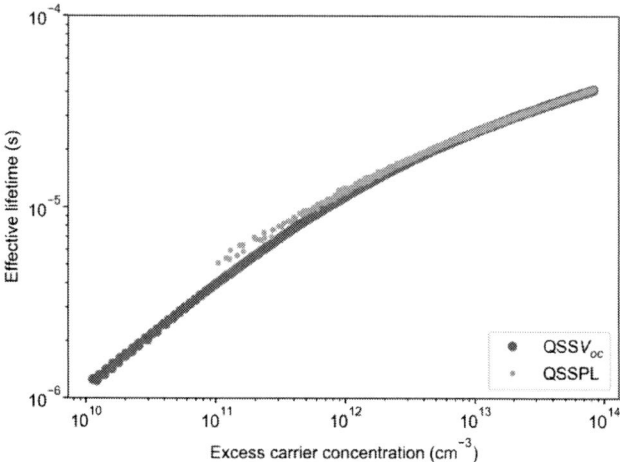

Fig. 1. Effective lifetime of validation sample measured by QSS$V_{oc}$ and QSSPL.

### IV. CONTACT RECOMBINATION STUDY

Having validated the PL setup, we then applied the technique to determine $J_{0m}$ of a metallized test structure.

The sample investigated was an $n$-type, 3 Ωcm, 6 in. monocrystalline Si wafer symmetrically diffused with boron and passivated with aluminium oxide. Silver paste was screen-printed onto one side of the sample in seven rectangular regions with varied metal contact fractions in the range 2-25 %. The PL image in Fig. 2 shows the layout of the sample.

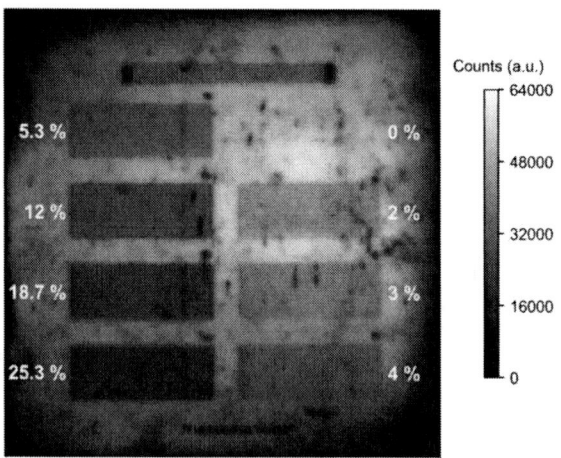

Fig. 2. Open circuit PL image of the test sample with varying contact fraction areas. Metal contact area fractions indicated in figure.

Local measurements of the injection-dependent $\tau_{eff}$ of metallized and unmetallized regions under high injection conditions are obtained using a mask to ensure only a single metal region is illuminated and measured each time.

The measurements were performed with the metallization on the rear side of the sample, which enabled uniform carrier generation, and a $1\times2$ cm$^2$ mask on the front side to ensure PL signal was only obtained from the metallized regions. It is noteworthy that the self-consistent calibration of PL automatically accounts for the rear optics of the device so no extra filtering or adjustments are required, as opposed to other imaging based PL techniques [9].

Kane and Swanson's analysis [3] is applied to extract $J_{0s,total}$ from the lifetime data as shown in Fig. 3. To extract $J_{0m}$, we apply a simple model where the front and rear passivated regions are modelled by a single $J_{0s}$ (surface recombination saturation current density) and the rear metal is modelled by a single $J_{0m}$. The measured $J_{0s,total}$ is then given by

$$J_{0s,total} = J_{0s} + J_{0s}[1 - f_{met}] + J_{0m}f_{met} \qquad (5)$$

where $f_{met}$ is the metallized area fraction. Equation (5) can be rearranged as

$$J_{0s,total} = 2J_{0s} + [J_{0m} - J_{0s}]f_{met} \qquad (6)$$

which is linear in $f_{met}$. Therefore, $J_{0s}$ and $J_{0m}$ can be obtained from a linear fit to a plot of $J_{0s,total}$ against $f_{met}$, as shown in Fig. 4. Applying (6) to this fit yielded $J_{0s}$ as 82 fA/cm$^2$ and $J_{0m}$ as 180 fA/cm$^2$.

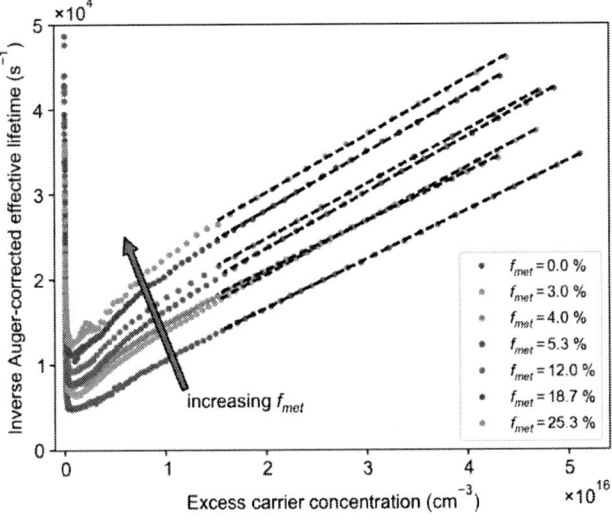

Fig. 3.   Linear fits to the inverse effective lifetime plots used to calculate $J_{0s,total}$ values. Fitting range was for $\Delta n > 10 N_{dop}$.

The particular sample used in this experiment was not ideally suited to the present system because the metallized regions are larger than the illuminated area. Thus, the metal contacts provide a low resistance path for carriers generated in the local illumination to flow freely to neighboring dark regions and reduce $\Delta n$ measured in the illuminated region. The unmetallized region is unaffected by this artifact, thus making the comparison of unmetallized and metallized regions problematic. A further source of error was the assumption of uniform $J_{0s}$, which is unlikely to be the case for this sample. Fig. 2 shows significant variation in the PL response across the sample. For example, the PL response is low in the lower right corner, which can explain the low $J_{0s,total}$ measured for the $f_{met} = 4$ % region, seen in Fig. 4.

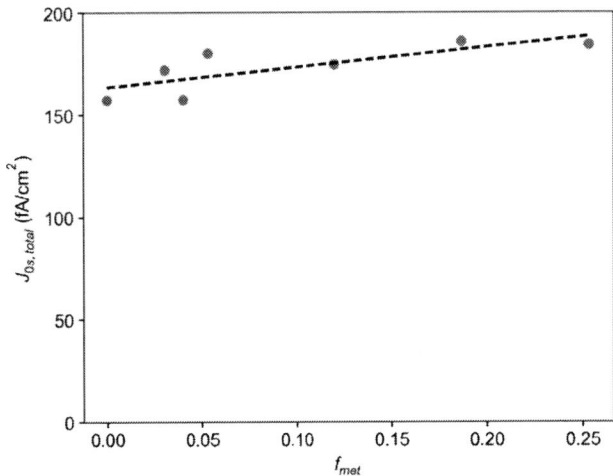

Fig. 4.   Measured $J_{0s,total}$ of metallized regions of varied contact fractions and the non-metallized region.

## V. Conclusion

A novel QSSPL system with front-side PL detection has been developed, which has significant advantages compared to existing techniques for characterizing contact recombination. In particular, it can be applied to completely-processed cells with full rear metal coverage and has a simplified calibration procedure compared to existing PL techniques.

The method has been verified by comparison to $\tau_{eff}$ measured by QSS$V_{oc}$ and applied to a metallized test structure, demonstrating expected trends in the measured $J_{0s,total}$.

## Acknowledgement

The authors acknowledge support from the Australian Government through the Australian Renewable Energy Agency (ARENA, Project 2014/RND097). The views expressed herein are not necessarily the views of the Australian Government, and the Australian Government does not accept responsibility for any information or advice contained herein. Z. Hameiri acknowledges the support of the Australian Research Council (ARC) through the Discovery Early Career Researcher Award (DECRA, Project DE150100268).

## REFERENCES

[1] M. A. Green, "The passivated emitter and rear cell (PERC): From conception to mass production," *Solar Energy Materials and Solar Cells*, vol. 143, pp. 190–197, 2015.

[2] D. Smith, G. Reich, M. Baldrias, G. Harley, P. Loscutoff, M. Reich, N. Boitnott, and G. Bunea, "Silicon solar cells with total area efficiency above 25 %," in *43rd IEEE Photovoltaic Specialists Conference*, 2016.

[3] D. E. Kane and R. M. Swanson, "Measurement of the emitter saturation current by a contactless photoconductivity decay method," in *18th IEEE Photovoltaic Specialists Conference*, 1985, pp. 578–583.

[4] J. Bullock, A. Cuevas, D. Yan, B. Demaurex, A. Hessler-Wyser, and S. De Wolf, "Amorphous silicon enhanced metal-insulator-semiconductor contacts for silicon solar cells," *Journal of Applied Physics*, vol. 116, no. 16, 2014.

[5] R. A. Sinton and A. Cuevas, "A quasi-steady-state open-circuit voltage method for solar cell characterization," in *16th European Photovoltaic Solar Energy Conference*, 2000, pp. 1152–1155.

[6] R. Dumbrell, M. K. Juhl, T. Trupke, and Z. Hameiri, "Comparison of terminal and implied open circuit voltage measurements," *IEEE Journal of Photovoltaics (accepted for publication)*, 2017.

[7] D. Inns and D. Poplavskyy, "Measurement of metal induced recombination in solar cells," in *42nd IEEE Photovoltaic*

[8] T. Fellmeth, A. Born, A. Kimmerle, F. Clement, D. Biro, and R. Preu, "Recombination at metal-emitter interfaces of front contact technologies for highly efficient silicon solar cells," *Energy Procedia*, vol. 8, pp. 115–121, 2011.

[9] A. Fell, D. Walter, S. Kluska, E. Franklin, and K. Weber, "Determination of injection dependent recombination properties of locally processed surface regions," *Energy Procedia*, vol. 38, pp. 22–31, 2013.

[10] V. Shanmugam, T. Mueller, A. G. Aberle, and J. Wong, "Determination of metal contact recombination parameters for silicon wafer solar cells by photoluminescence imaging," *Solar Energy*, vol. 118, pp. 20–27, 2015.

[11] T. Trupke, R. A. Bardos, F. Hudert, P. Würfel, J. Zhao, A. Wang, and M. A. Green, "Effective excess carrier lifetimes exceeding 100 milliseconds in float zone silicon determined from photoluminescence," in *19th European Photovoltaic Solar Energy Conference*, 2004, pp. 758–761.

[12] T. Trupke, R. A. Bardos, and M. D. Abbott, "Self-consistent calibration of photoluminescence and photoconductance lifetime measurements," *Applied Physics Letters*, vol. 87, p. 184102, 2005.

[13] H. Nagel, C. Berge, and A. G. Aberle, "Generalized analysis of quasi-steady-state and quasi-transient measurements of carrier lifetimes in semiconductors," *Journal of Applied Physics*, vol. 86, no. 11, p. 6218, 1999.

*Specialists Conference*, 2015.

# Using Time-of-Flight SIMS to Investigate Group V Dopant Distribution in CdTe

Steven P. Harvey,[1] Eric Colegrove,[1] Brian McCandless,[2] David Albin,[1] Mowafak Al-Jassim,[1] Wyatt K. Metzger[1]

[1] National Renewable Energy Laboratory, Golden, CO, 80403, USA
[2] University of Delaware, Newark, DE

*Abstract* — Enabling continued development of polycrystalline CdTe solar cells with open-circuit voltages approaching 1 V requires new dopants and processes to increase acceptor concentrations and minority-carrier lifetimes. We have investigated multiple Group V dopants in single- and polycrystalline CdTe materials as well as Cl, using numerous dopant incorporation and activation strategies. The diffusion kinetics and spatial distribution of the dopants are revealed with a combination of 1-D standard depth-profiling and high-resolution 3-D tomography with time-of-flight secondary-ion mass spectrometry. For all cases where the dopant is incorporated during growth, including vapor-transport deposition, close-spaced sublimation, and molecular-beam epitaxy, the dopant concentration is not enhanced at grain boundaries. When the dopant is incorporated through a post-growth incorporation process, fast grain-boundary diffusion leads to an enhanced dopant concentration at grain boundaries. Although both methods are applicable, *in-situ* incorporation offers control of dopants and devices independent of diffusion kinetics, grain structure, and grain-boundary chemistry.

*Index Terms* — CdTe, TOF-SIMS, doping, diffusion, Group V.

## I. INTRODUCTION

To establish next-generation CdTe photovoltaics with open-circuit voltage (Voc) approaching 1 V, dopants and processes are being studied to increase acceptor concentrations and minority-carrier lifetime. Using Group V elements as dopants in CdTe single crystals and very large-grain polycrystals, we have achieved lifetimes and doping densities exceeding 10 ns and $10^{17}$ cm$^{-3}$, respectively [2, 4]. In the current study, we give an overview of the how the dopant distribution changes based on the incorporation method, via time-of-flight secondary-ion mass spectrometry (TOF-SIMS) high-resolution three-dimensional (3-D) tomography.

SIMS is a powerful analytical technique for determining elemental and isotopic distributions in solids [5]. In this work, we show how TOF-SIMS in multiple measurement modes can fully elucidate the spatial profiles for a number of different dopants in CdTe material and enable determining diffusion characteristics unambiguously. The Group V results are compared to earlier work investigating chlorine distribution with TOF-SIMS [6].

## II. EXPERIMENTAL

Several sample types, growth methods, and annealing treatments were used to prepare material for this study. Bulk single-crystal CdTe samples were purchased from JX Nippon in various orientations. Polycrystalline CdTe material was deposited using both close-spaced sublimation (CSS) and vapor-transport deposition (VTD) to thicknesses ranging from 4 to 30 μm on both molybdenum-coated alumina and traditional superstrate layers. *In-situ*-doped material was prepared in two ways: the first was to use CdTe source powder containing Group V elements and the second was to co-deposit Group V material and CdTe in a molecular-beam epitaxy (MBE) chamber. In the second case, CdTe:V material was deposited on undoped polycrystalline CdTe templates and the resulting layer reproduced the template grain structure. For *ex-situ*-doped material, samples were sealed in evacuated quartz ampoules with $Cd_3P_2$, $Cd_3As_2$, or CdSb powder and annealed in a 3-zone Mellen furnace at different temperatures and times. Other polycrystalline CdTe samples were subjected to CdCl$_2$ treatments at 400°C for 15 minutes.

Analysis was completed using an ION-TOF SIMS V instrument. A 3-lens 30-keV BiMn ion gun created secondary ions for analysis. 3-D tomography was completed using a Bi$_3^{++}$ primary ion-beam cluster (100-ns pulse width, 0.1-pA pulsed beam current). This measurement mode is capable chemical imaging with better than 100-nm lateral resolution. Standard depth profiling was completed using a Bi$^+$ primary ion-beam operated in bunched mode (10-ns pulse width, 1-pA pulsed beam current). Sputtering for depth profiling was accomplished with a second oxygen- or cesium-ion beam with a variable energy from 600 eV to 3 keV (sputtering current 3–60 nA). Relative sensitivity factors allowing quantification of P, As, and Sb in the SIMS data were determined from measurements of ion-implant standards.

## III. RESULTS

The 3-D distribution of Group V dopants in CdTe material grown by very distinct methods indicates some general trends. Initial results for *ex-situ*-doped CSS polycrystalline material are seen in Fig. 1. Figure 1A shows a grain-orientation map obtained by electron backscatter diffraction (EBSD) mapping of the sample, which was subsequently marked with a focused ion beam so that the same area could be analyzed with TOF-SIMS. The two-dimensional (2-D) image in Fig. 1B shows an excellent correlation of the phosphorous image with the grain

distributions of P or As in CdTe single crystals, shown in Fig. 4, that are laterally uniform regardless of the doping or growth method.

We have used data collected on a set of single-crystal and polycrystalline materials, subjected to diffusion anneals, to

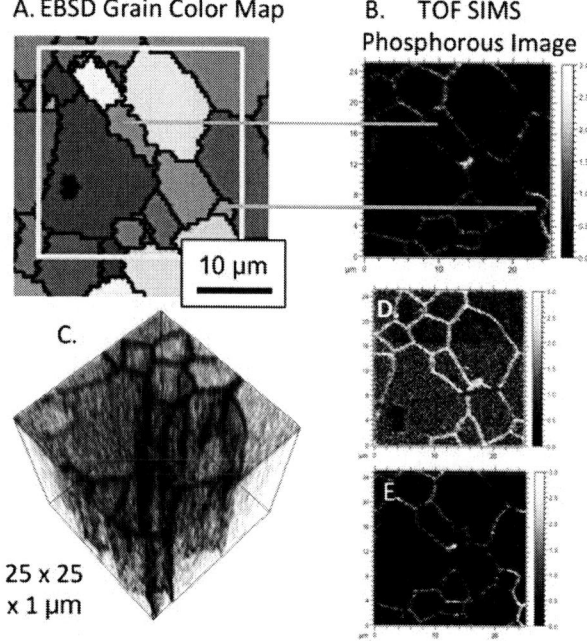

Fig. 1. *Ex-situ*-doped CSS-grown polycrystalline material: A) EBSD grain orientation map; B) TOF-SIMS phosphorous image 25×25 μm; C) The 3-D rendering of the phosphorous distribution (25×25×0.9 μm); D) 2-D slice of the 3-D dataset from C, showing the phosphorous signal in the near-surface region; E) 2-D slice of the 3-D dataset from C, showing the phosphorous signal ~1 μm deep.

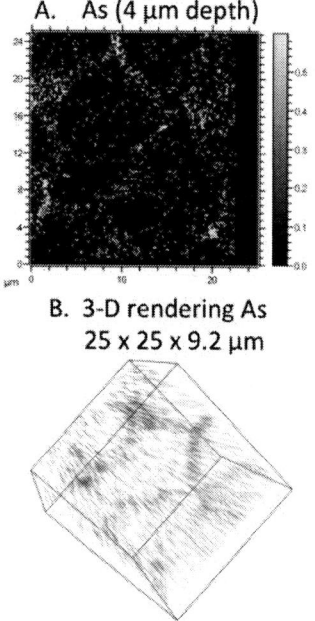

Fig. 2. A) TOF-SIMS arsenic image (25×25μm) of an *ex-situ*-doped polycrystalline sample 4 μm from the surface. B) 3-D rendering of As signal (25×25×9.2 μm); as with phosphorous, an As accumulation at the grain boundaries is noted.

structure in Fig. 1A. This illustrates that phosphorous preferentially diffuses down grain boundaries when subjected to *ex-situ* doping. The 3-D rendering of the dataset (25×25×0.9 μm) is seen in Fig. 1C. 2-D slices at different depths from this dataset allow profiling the grain-boundary and grain-core distributions as a function of depth. Figure 1D shows a 2-D image of the near-surface data, and Fig. 1E shows a 2-D image ~1 μm deep. It is clear in Figs. 1 B–E that the dopant is localized at grain boundaries. Similar behavior is observed for arsenic in Fig. 2. For example, Fig. 2A shows the 2-D arsenic image at a 4-μm depth from the surface of an *ex-situ*-doped sample. The counting statistics are much lower than with phosphorous due to a detection-limit difference, but apparent grain boundaries are still clearly visible in the arsenic-doped sample. Figure 2B shows the 3-D dataset (25×25×9.2 μm), where the preferential grain-boundary diffusion is again noted as higher arsenic concentration decorating the grain boundaries. Samples doped with Cl also show increased chlorine content at the grain boundaries, as noted in previous work and shown for samples in this work in Fig. 3 [6]. The images in Figs. 1 and 2 are contrasted with the 2-D and 3-D

fully elucidate the diffusion of Group V dopants in polycrystalline CdTe material [2, 3, 7]. Figure 4A shows that a second fast bulk-diffusion coefficient was noted for phosphorous in single-crystal CdTe, which was not noted for arsenic or antimony. This mechanism is the key to enabling easier incorporation of higher dopant concentrations when using phosphorous incorporated *ex-situ*. This fast bulk diffusion mechanism may be a key to enabling phosphorous-doped CdTe to recently achieve hole concentrations above $1\times10^{16}$ cm$^{-3}$ [8]. Modeling of the diffusion processes of Group V dopants has shown that the slow bulk diffusion is due to the dopant diffusing substitutionally through the lattice, whereas the fast bulk mechanism noted for phosphorous (the smallest of the Group V dopants) is due to an interstitial diffusion mechanism [3].

In contrast to CdTe subjected to *ex-situ* doping, Fig. 5 shows the 3-D distributions for polycrystalline materials doped *in-situ* for a variety of growth methods. Figure 5A shows the phosphorous distribution for a polycrystalline CdTe sample grown by VTD; there is no apparent segregation of phosphorous at grain boundaries. Figure 5B shows the

978-1-5090-5606-4/17 $31.00 © 2017 IEEE

phosphorous distribution for a polycrystalline CdTe sample grown by CSS; the apparent clusters of phosphorous are due to sample topography that influences the secondary-ion yield, and there is no apparent segregation of phosphorous at grain boundaries. Figure 5C shows polycrystalline material grown by MBE on polycrystalline CSS CdTe templates. EBSD analysis showed that the MBE material reproduced the underlying CSS grain structure, and again, there is no apparent segregation of phosphorous at grain boundaries. This strongly suggests that the higher phosphorous concentration at grain boundaries for *ex-situ* doping is due simply to the high grain-boundary diffusion coefficient, and not to a strong segregation enthalpy driving force to grain boundaries for Group V dopants in CdTe—at least within the composition range relevant for solar cell doping.

By comparing Figs. 1, 2, and 5 (as well as other data we have taken), it is apparent that *ex-situ* Group V and Cl doping of polycrystalline CdTe material leads to enhanced concentration of the dopant at grain boundaries. This is consistent for most polycrystalline materials because grain boundaries are typically structurally more open than the grain interior, and they present faster diffusion pathways through the film. Our data show that when the dopant is incorporated during growth, dopants are incorporated uniformly regardless of the growth method. Although *ex-situ* doping offers advantages in that it can be applied to films after deposition by different methods, a disadvantage is that grain-boundary chemistry, diffusion kinetics, grain structure, and inhomogeneity will affect the doping profiles, and subsequently, device performance. The doping strategy for *in-situ* doping must be varied based on the growth technique applied, but we find that multiple *in-situ* methods—including VTD, MBE, and CSS—can incorporate appreciable levels ($>10^{17}$ cm$^{-3}$) of novel dopants with great uniformity in polycrystalline CdTe films. This can offer superior control of dopants and devices independent of grain structure, grain-boundary chemistry, and diffusion kinetics.

## IV. Conclusions

We have investigated multiple Group V dopants in single- and polycrystalline CdTe materials, using numerous dopant incorporation and activation strategies. We are able to incorporate significant quantities of novel dopants throughout CdTe films by both *ex-situ* and *in-situ* methods. Through TOF-SIMS tomography, we show that for all cases where the dopant is incorporated during growth, the dopant concentration is not enhanced at grain boundaries—regardless of the growth method. On the other hand, post-growth incorporation methods indicate higher grain-boundary doping concentration. Although both methods are applicable, *in-situ* incorporation offers control of dopants and devices independent of diffusion kinetics, grain structure, and grain-boundary chemistry.

## V. Acknowledgments

This work was supported by the U.S. Department of Energy under Contract No. DE-AC36-08GO28308 with the National Renewable Energy Laboratory. The U.S. Government retains and the publisher, by accepting the article for publication, acknowledges that the U.S. Government retains a nonexclusive, paid up, irrevocable, worldwide license to publish or reproduce the published form of this work, or allow others to do so, for U.S. Government purposes.

## VI. References

[1] E. Colegrove, S. P. Harvey, J. H. Yang, J. M. Burst, J. N. Duenow, D. S. Albin, S. H. Wei, and W. K. Metzger, "Antimony Diffusion in CdTe," *IEEE Journal of Photovoltaics,* vol. 7, pp. 870-873, 2017.

[2] E. Colegrove, S. P. Harvey, J.-H. Yang, J. M. Burst, D. S. Albin, S.-H. Wei, and W. K. Metzger, "Phosphorus Diffusion Mechanisms and Deep Incorporation in Polycrystalline and Single-Crystalline CdTe," *Physical Review Applied,* vol. 5, p. 054014, 2016.

[3] E. Colegrove, J.-H. Yang, S.P. Harvey, M. Young, J.M. Burst, J.N. Duenow, D.S. Albin, S.-H. Wei, and W.K. Metzger, "Experimental and theoretical comparison of Sb, As, and P diffusion mechanisms and doping in CdTe," *Submitted: Physical Review Applied,* 2017.

[4] J.M. Burst, J.N. Duenow, D.S. Albin, E. Colegrove, M.O. Reese, J.A. Aguiar, C.S. Jiang, M.K. Patel, M.M. Al-Jassim, D. Kuciauskas, S. Swain, T. Ablekim, K.G. Lynn, and W.K. Metzger, "CdTe solar cells with open-circuit voltage breaking the 1 V barrier," *Nature Energy,* vol. 1, p. 16015, 2016.

[5] F. Stevie, *Secondary Ion Mass Spectrometry: Applications for Depth Profiling and Surface Characterization*: Momentum Press, 2015.

[6] S.P. Harvey, G. Teeter, H. Moutinho, and M.M. Al-Jassim, "Direct evidence of enhanced chlorine segregation at grain boundaries in polycrystalline CdTe thin films via three-dimensional TOF-SIMS imaging," *Progress in Photovoltaics,* vol. 23, pp. 838–846, Jul 2015.

[7] E. Colegrove, S.P. Harvey, J.-H. Yang, J.M. Burst, J.N. Duenow, D. Albin, S.-H. Wei, and W. Metzger, "Antimony diffusion in CdTe," *IEEE Journal of Photovoltaics,* 2017.

[8] J.M. Burst, S.B. Farrell, D.S. Albin, E. Colegrove, M.O. Reese, J.N. Duenow, D. Kuciauskas, and W.K. Metzger, "Carrier density and lifetime for different dopants in single-crystal and polycrystalline CdTe," *APL Materials,* vol. 4, p. 116102, 2016.

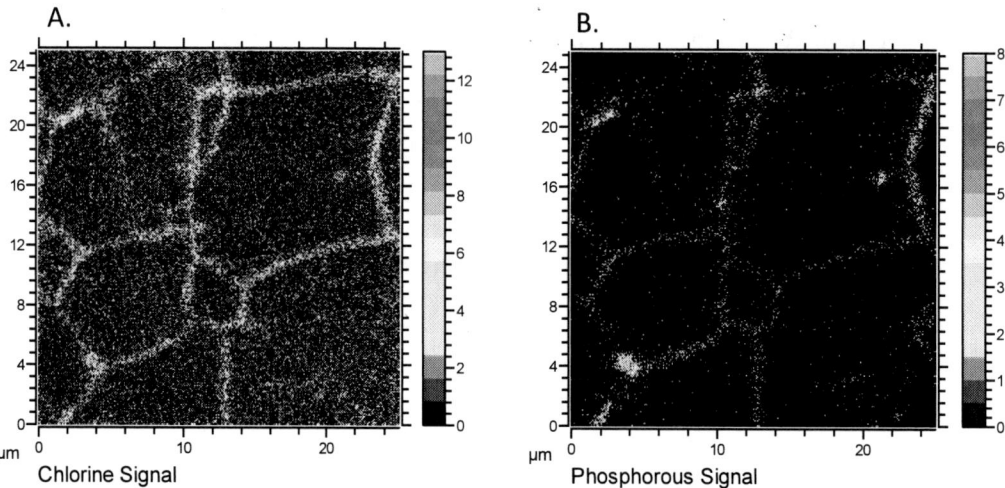

Fig 3. A) Chlorine distribution in polycrystalline CdTe *ex-situ* doped with phosphorous, each image is 25×25 microns, at a depth of 1 micron from the surface. B) Phosphorous signal from same dataset. Higher chlorine and phosphorous concentrations at the grain boundaries are clearly visible.

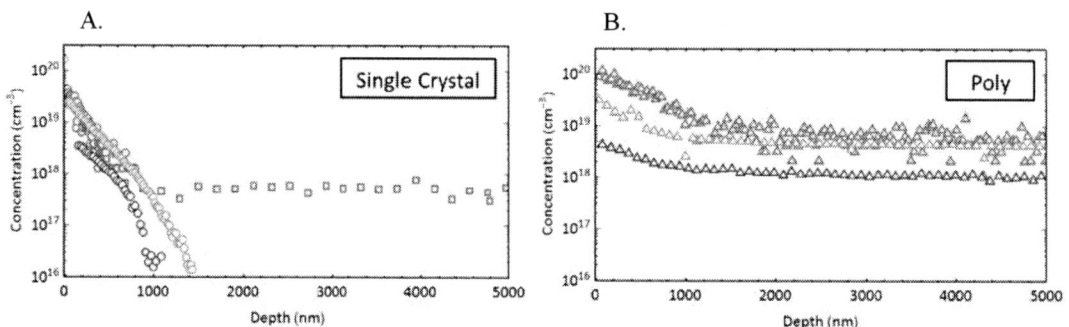

Fig. 4. Measured diffusion profiles for Group V dopants in A) single-crystal and B) polycrystalline CdTe [1–3] The diffusion conditions for each dopant are: (red) P 633°C, 7 h; (green) As 650°C, 2 h; and (blue) Sb 633°C, 2 h.

## *In-situ* doped polycrystalline materials

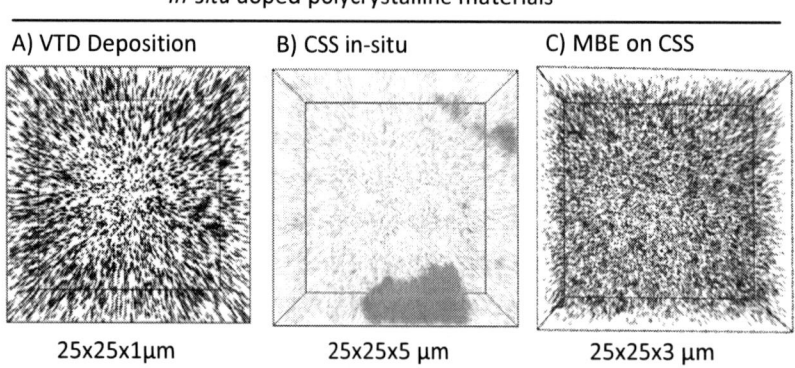

Fig. 5. 3-D distribution of dopants for *in-situ* Group V doped polycrystalline CdTe materials grown by different methods A) CdTe:P deposited by vapor-transport deposition, maximum hole content obtained $\sim 1 \times 10^{16}$ cm$^{-3}$. B) CdTe:P deposited by close-spaced sublimation, maximum hole content obtained $\sim 2 \times 10^{15}$ cm$^{-3}$; C) CdTe:As deposited by molecular-beam epitaxy, maximum hole content obtained $\sim 5 \times 10^{16}$ cm$^{-3}$.

978-1-5090-5606-4/17 $31.00 © 2017 IEEE      3322

# Quantitative Analysis of Active Dopant Distribution and Estimation of Effective Diffusivity in Phosphorus-Implanted Emitter of Si Solar Cell Using Scanning Nonlinear Dielectric Microscopy

Kotaro Hirose[1], Katsuto Tanahashi[2], Hidetaka Takato[2], and Yasuo Cho[1]

[1]Research Institute of Electrical Communication, Tohoku University, Sendai, Miyagi, 980-8577, Japan

[2]Fukushima Renewable Energy Institute, National Institute of Advanced Industrial Science and Technology, Koriyama, Fukushima, 963-0298, Japan

*Abstract* — The carrier distribution in solar cell is important evaluation target. Scanning nonlinear dielectric microscopy is applied to the cross section of phosphorus implanted emitter in monocrystalline silicon solar cell and visualizes the carrier distribution quantitatively. The effective diffusivities of phosphorus are estimated from the experimental results. Then, the three-dimensional carrier distribution is simulated. The experimental and simulation results show good correlation.

*Index Terms* — active dopant distribution, effective diffusivity, phosphorus-implanted emitter, silicon solar cell, scanning nonlinear dielectric microscopy.

## I. INTRODUCTION

Ion implantation has been proposed as a new process technique for low cost and high efficiency solar cell [1]. Ion implantation has high degree of freedom in device design and reduces the number of process steps. Phosphorus (P)-implanted emitter is formed by the P-implantation and following anneals. The conversion efficiency is affected by the implantation and annealing conditions. Evaluation of carrier distribution in emitter of silicon solar cell is important to understand the detailed physical phenomenon in device. The carrier distribution influences light absorption, electric field, and metal/emitter resistivity. Most common type of silicon solar cell has texture structure in the front surface. Therefore, direct application of secondary ion mass spectroscopy or spread resistance analysis for evaluation of carrier distribution in emitter is difficult.

Cross-sectional measurement using microscope is very useful way to characterize the inside of device. Physical parameter at the surface of cross section is visualized. Scanning probe microscopy (SPM) techniques are powerful tools for characterizing two-dimensional carrier distribution with high lateral resolution. Especially, scanning nonlinear dielectric microscopy (SNDM) [2] has high capacitance sensitivity ($\sim 10^{-22}$ F/$\sqrt{\text{Hz}}$) and can visualize detailed carrier distribution. In addition, SNDM can quantify the carrier density using standard sample. In previous report, the visualization of carrier distribution in P-implanted emitter was conducted [3]. In this paper, we quantify the carrier density and then, analyze the dopant diffusion by comparing the experimental results using SNDM with simulation result.

## II. EXPERIMENT

P-implanted emitter in the monocrystalline silicon (Si) solar cell was measured by SNDM. Emitter was formed on a p-type (boron) wafer. Mass-analyzed and ionized P was implanted at the textured surface with an acceleration voltage 10 keV, a dose of $4 \times 10^{15}$ atoms/cm$^2$, tilt angle of 35°, and step angle of 15°. Annealing was conducted at 900 °C for 10 minutes to recover the crystalline and to diffuse P.

At first, the solar cell was diced. The cross section of diced solar cell was almost parallel to (110) face and chemo-mechanically polished simultaneously with the n- and p-type standard Si samples. The standard samples were staircase sample with different dopant densities. The dopant density ranges of n-type and p-type standard samples were from $\sim 1 \times 10^{16}$ atoms/cm$^3$ to $\sim 4 \times 10^{19}$ atoms/cm$^3$ and from $\sim 1 \times 10^{16}$ atoms/cm$^3$ to $\sim 2 \times 10^{19}$ atoms/cm$^3$, respectively.

Figure 1(a) shows the schematic set-up of SNDM equipment. The capacitance sensor, which is called SNDM probe, consists of GHz-range LC oscillator and conductive cantilever tip. The tip contacts the cross sectioned surface. The surface is covered with native oxide layer. Thus, metal-oxide-semiconductor (MOS) structure is formed. The sinusoidal voltage with amplitude of 0.5 V and a frequency of 30 kHz was applied. The applied voltage modulated the MOS capacitance between the tip and the sample. Finally, the capacitance variation dC/dV was detected as an oscillation frequency deviation of SNDM probe.

The carrier polarity and density correspond to the SNDM (dC/dV) signal polarity and strength, respectively. The carrier density could be obtained by comparing the SNDM signal from solar cell sample with that from standard samples. In dC/dV measurement of MOS capacitor, the SNDM signal level decreases both at low and high density area, that is, the dopant density is described as a two-valued function of dC/dV strength [4]. In order to avoid this contrast reversal phenomenon, we conducted another type of SNDM measurement, i.e., dC/dz-SNDM, after SNDM measurement.

In dC/dz-SNDM measurement, the tip-sample capacitance is modulated by the modulation of tip-sample distance *z*. Figure 1(b) shows schematic dC/dz-SNDM equipment. The SNDM probe is vibrated by the piezoelectric plate with a frequency of

978-1-5090-5606-4/17 $31.00 © 2017 IEEE

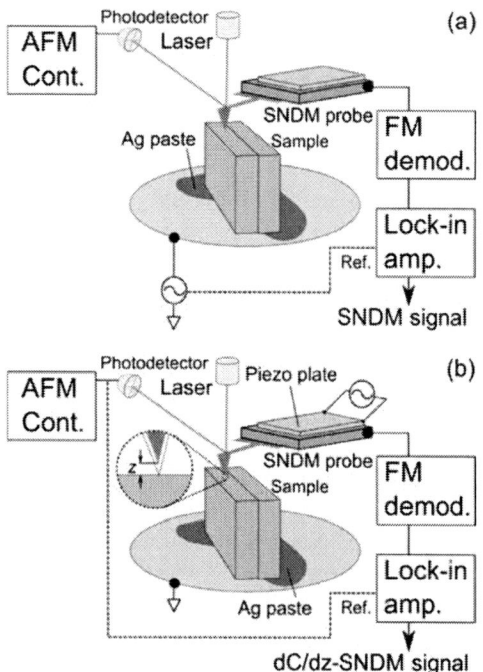

Fig. 1. The schematic set-ups of (a) SNDM and (b) dC/dz-SNDM

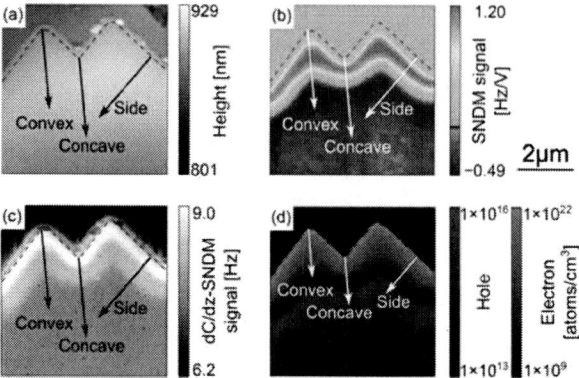

Fig. 2. (a) Topography, (b) SNDM, (c) dC/dz-SNDM images of texture structure of Si solar cell. SNDM image was converted to (d) carrier density image using calibration curve in Fig. 4.

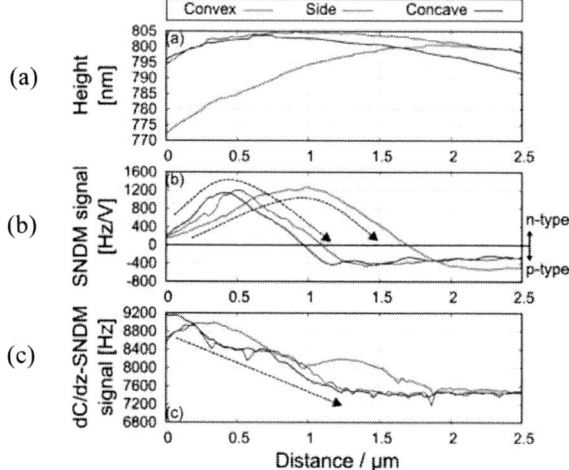

Fig. 3. Line profiles of (a) Topography, (b) SNDM, (c) dC/dz-SNDM in Fig.2

~65 kHz. The dC/dz-SNDM signal strength is a monotonically increasing function with the tip-sample capacitance at $z = 0$. Therefore, dC/dz-SNDM signal does not have information of carrier polarity, but it is a single valued function of carrier concentration with one-to-one relation and is used to avoid misunderstandings caused by the contrast reversal problem in SNDM, i.e. dC/dV, measurement. Thus, correct carrier distribution can be obtained by both complementary SNDM and dC/dz-SNDM measurements.

### III. RESULTS AND DISCUSSION

*A. SNDM results*

Figures 2(a)-(c) show the obtained images of SNDM and dC/dz-SNDM at the cross section of P-implanted emitter: i.e., (a) topography, (b) SNDM, and (c) dC/dz-SNDM image. The dashed line indicates the SiN/Si interface. The red and blue regions in Fig. 2(b) are n-type and p-type region respectively. The bright region in Fig. 2(c) is high carrier density region.

The line profiles in Figs. 2(a)-(c) are shown in Fig. 3. The SNDM signal in the emitter has the peak at the intermediate position due to the contrast reversal phenomenon as show in Fig. 3(b). In contrast, the dC/dz-SNDM signal shows the monotonic decrease tendency from the texture surface to bulk, which means that the electron density decrease from texture surface to bulk monotonically. Thus we can avoid misjudgment from the two valued function problem caused by the contrast

reversal phenomenon. We note that the topography and the tip-sample contact potential difference might affect dC/dz-SNDM signal. The pn junction position is the deepest at convex position, followed by at side and concave positions.

*B. Quantification of carrier concentration*

The standard samples were measured by SNDM after SNDM measurements of the solar cell. Figure 4 shows the calibration curve which shows the relationship between SNDM signal strength and carrier density. We assumed that the dopants in the standard samples were perfectly activated. The calibration curve was calculated using least square method.

SNDM image in Fig. 2(b) was converted to the carrier density image using this calibration curve and result is shown in

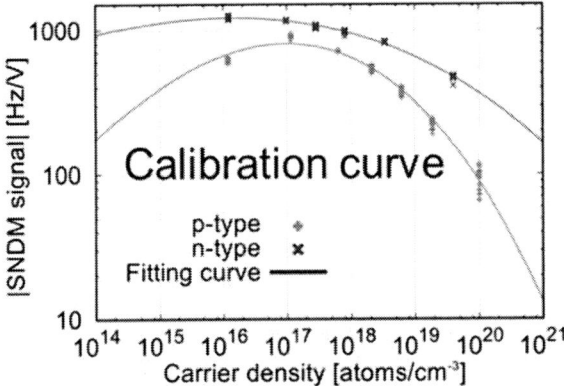

Fig. 4. Calibration curve showing the relationship between carrier density and SNDM signal obtained from the SNDM results of the standard samples

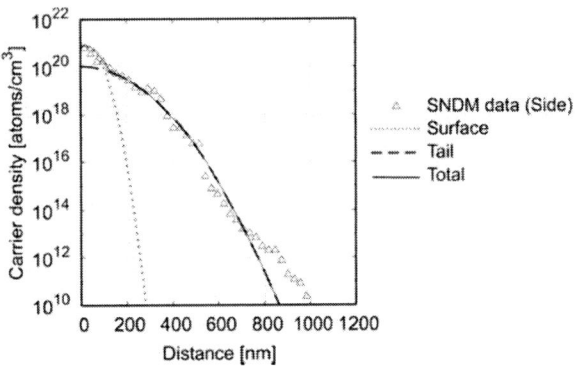

Fig. 6. Line profiles of active dopant in the side position in Fig. 5 and two fitted Gaussian functions applied to the P distribution at the surface and tail regions .

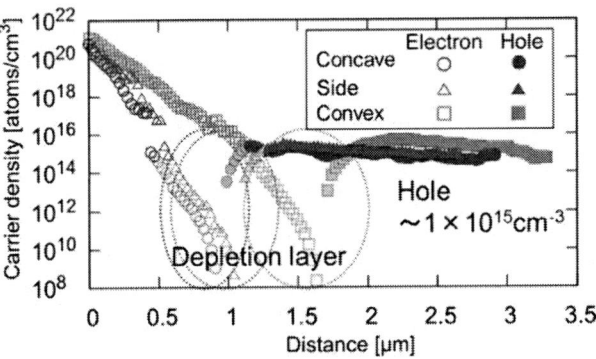

Fig. 5. Carrier density line profiles in Fig. 2(d).

Fig. 2(d). The carrier density at the outside range of standard sample was extrapolated. The line profiles in Fig. 2(d) are shown in Fig. 5. The electron density decreases exponentially from texture surface to bulk. The hole density in the p-type bulk region is about $1 \times 10^{15}$ cm$^{-3}$, which is of the same order as the hole density calculated from the substrate resistivity. In the depletion layer, tip sample capacitance is influenced by both the p- and n-type layers. Therefore, the calibrated results below about $1 \times 10^{15}$ cm$^{-3}$ in Fig. 5 may not be correct.

The active dopant in the side position is one-dimensionally distributed along the direction of the arrow, as shown in Fig. 2(b). Therefore, the P distribution at the side region was analyzed on the basis of the implanted P distribution in the flat (not textured) substrate. The distribution can be roughly described as two distinct regions, i.e., the surface and tail regions [4]. Point defects and dopant ions form pairs and diffuse together [5]. In the surface region, the point defect density is lower than the impurity density, whereas the point defect density is comparable with the dopant density in the tail region.

A Gaussian function was applied to the P distribution at the surface and tail regions in the side position in Fig. 5. The dashed lines in Fig. 6 show the fitting results. The function was fitted in the range of over $1 \times 10^{16}$ cm$^{-3}$ because the carrier density was not equal to the P density in the depletion region.

A joined half Gaussian function [6], which includes the reflection effect at the surface, was used for fitting in surface region. As shown in Fig.6, two Gaussian functions describe the P distribution well. The dose calculated from the integral of the total fitting curve was about $7 \times 10^{15}$ cm$^{-2}$, which was of the same order as the actual implantation dose. Figure 6 shows that the tail region occupies almost the entire emitter. Effective diffusivities of $D^{Surface}=9.4 \times 10^{-15}$ cm$^2$/s in the surface region and $D^{Tail}=1.3 \times 10^{-13}$ cm$^2$/s in the tail region were obtained from the fitted Gaussian function in the surface and the tail region, respectively. It should be noted that the diffusivity changes during diffusion and is dependent on the dopant density and defect density. The effective diffusivity can approximately describe the diffusion phenomenon. In addition, the positive fixed charges in the SiN layer must cause electrons to accumulate at the textured surface, causing the distribution of electrons to change.

Since each pyramid has four faces, the P distributions in the direction vertical to the faces are expected to be the same. In this case, we can roughly estimate the 3D P distribution at the upper part of the pyramidal base using the superposition of Gaussian functions in Fig. 6. (Strictly speaking, the boundary conditions at the edges of the pyramids are different.) Figure 7(a) shows the 3D carrier distribution using the superposition of P distributions. Figure 7(b) shows its cross section at y=0. Line profiles at the convex and side positions in Fig. 7(b) are shown as red and green solid lines in Fig. 7(c), respectively. The line profiles obtained from the 3D distribution are in good agreement with that obtained from the SNDM results and have the same order of magnitude.

978-1-5090-5606-4/17 $31.00 © 2017 IEEE

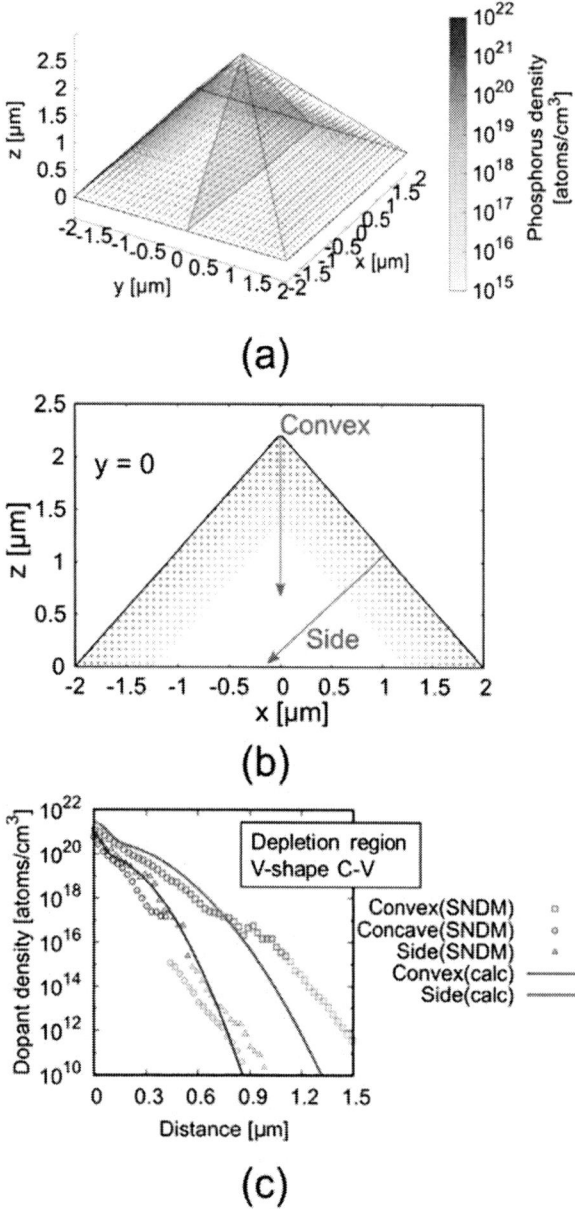

Fig. 7. (a) 3D carrier distribution using the superposition of P distributions. (b) cross section at y=0. (c) Line profiles at the convex and side positions in (b) are shown as red and green solid lines, respectively.

## IV. CONCLUSION

The P distribution in a Si solar cell was evaluated using SNDM. The carrier distribution was observed using complementary SNDM and dC/dz-SNDM measurements. The carrier density was quantified using a Si standard sample. The P distributions could be described as two Gaussian functions and the effective diffusivities were successfully evaluated. In addition, the 3D P distribution was estimated using the superposition principle. The estimated 3D P distribution was in good agreement with the SNDM results.

Therefore, it was concluded that SNDM is a useful method for evaluation of the active dopant distribution in Si solar cells. Further experimental and theoretical studies are required to improve the device performance. For example, more detailed analysis regarding P diffusion would provide the exact initial P distribution and P diffusivity. In addition, diffusion simulations could lead to the precise estimation of the 3D P distribution. The relationship between the P distribution and implantation conditions can thus be obtained from evaluation of P-implanted Si solar cell fabricated with various implantation or annealing parameters. The conversion efficiency can be estimated from the obtained carrier distribution through device simulations.

## ACKNOWLEDGEMENT

This study was supported in part by a Grant-in-Aid for Scientific Research (S) (No. 16H06360) from the Japan Society for Promotion of Science (JSPS).

## REFERENCES

[1] H. Hieslmair, I. Latchford, L. Mandrell, M. Chun, and B. Adibi, "Ion implantation for silicon solar cells," *Photovolt. Int.*, vol. 18, pp. 58–64, 2012.

[2] Y. Cho, A. Kirihara, and T. Saeki, "Scanning nonlinear dielectric microscope," *Rev. Sci. Instrum.*, vol. 67, no. 6, pp. 2297–2303, 1996.

[3] K. Hirose, K. tanahashi, H. Takato, N. Chinone and Y. Cho, "Two-dimensional Analysis of Carrier Distribution in Phosphorus-Implanted Emitter and Phosphorus-Diffused Emitter using Super-Higher-Order Scanning Nonlinear Dielectric Microscopy," in *43rd IEEE Photovoltaic Specialist Conference*, 2016, pp. 3671-3674.

[4] A. Kikukawa, S. Hosaka, and R. Imura," Silicon pn junction imaging and characterizations using sensitivity enhanced Kelvin probe force microscopy" Appl. Phys. Lett. Vol.66, no.25, pp.3510-3512,1995.

[5] D. Mathiot and J.C. Pfister," Dopant diffusion in silicon: A consistent view involving nonequilibrium defects" J. Appl. Phys., vol. 55, no.10, pp. 3518-3530,1984.

[6] J.F. Gibbons and S. Mylroie, "Estimation of impurity profiles in ion-implanted amorphous targets using joined half-Gaussian distributions", Appl. Phys. Lett., vol. 22, no. 11, pp.568-569, 1973.

# Simulation of drive-level capacitance profiling to interpret measurements on Cu(In,Ga)Se₂ Schottky devices

Geordie Zapalac and Jeff Bailey

MiaSolé Hi-Tech, Santa Clara, California 95051

*Abstract* — **A simulation is presented for drive-level capacitance profiling (DLCP) of Schottky diodes that allows a continuous distribution of defect states in both energy and depth. The simulation is used to interpret high frequency DLCP data for a Cu(In,Ga)Se₂ Schottky device.**

*Index Terms* — **CIGS, DLCP, simulation, Schottky barrier, doping profile, deep defects.**

## I. INTRODUCTION

The carrier concentration and deep defect distribution in the solar cell absorber are essential inputs to a device model. These properties are probed by capacitance measurements such as capacitance-voltage (CV), drive-level capacitance profiling (DLCP) [1], admittance spectroscopy (AS), and deep-level transient spectroscopy (DLTS). Absorber materials such as Cu(In,Ga)Se₂ (CIGS) may have a complicated density of states that depends upon both the energy $E$ above the valence band maximum (VBM) and the depth $x$ in the absorber. For such materials capacitance simulations are an important tool to assist interpretation of experimental results.

DLCP experiments in particular are often difficult to interpret. Unlike CV or AS, these measurements cannot be simulated using widely available device simulations such as SCAPS [2]. In DLCP the defect density is obtained by measuring the change in capacitance with the peak-to-peak signal amplitude from the impedance analyzer or "drive-level" voltage. SCAPS for example does not simulate the non-linear response of the capacitance to the drive-level voltage.

In 1981 Cohen and Lang published a method to solve Poisson's equation for a 1-D Schottky barrier under bias with a continuous distribution of gap states $g(x,E)$ [3]. Cohen used this technique to simulate the general behavior of DLCP in a seminal description of DLCP published by Heath et al. [1], but the simulation itself was not the main focus of the publication and it was not described in detail.

This report provides a detailed description of a Schottky barrier simulation used to simulate DLCP following Cohen's original approach. The simulation is used to interpret DLCP measurements on CIGS Schottky diodes taken at high frequency over a range of temperatures and applied biases.

Schottky diode samples were made from process of record (POR) CIGS films fabricated at the MiaSolé manufacturing line in Santa Clara, California with an average module efficiency of 16.5%. Completed "full-stack" films were wet-etched with hydrochloric acid to expose the CIGS surface followed by evaporating a 0.7 mm² aluminum contact to form a Schottky junction.

## II. DLCP SIMULATION OF A SCHOTTKY BARRIER

The simulation solves Poisson's equation within the depletion region in the presence of an input density of states $g(x,E)$ and under different conditions of applied bias. In our notation $g(x,E)$ includes both deep defects and any shallow defects that contribute carriers. Acceptors in the depletion region that lie closer to the VBM than the hole quasi-Fermi level may contribute negative charge to the total space charge $Q$ of the depletion region. The contribution from deep ionized donors below the electron quasi-Fermi level is neglected in the present simulation for the p-type CIGS material considered here, hence we refer to the absolute position of the hole quasi-Fermi level as simply the "Fermi level" $E_F$ in the remainder of this report.

During a DLCP measurement the applied voltage oscillates at frequency $f$ between a fixed ac maximum voltage $V_A$ and a minimum ac voltage $V_A - V_{DL}$ obtained by subtracting the peak-to-peak value of the drive-level voltage $V_{DL}$. A deep defect state with an emission lifetime too long to respond to the drive-level frequency will equilibrate to a static charge density corresponding to the device at a dc bias voltage of $V_A - V_{DL}/2$. Note that this static charge increases with the drive-level voltage. Only defect states with a hole emission rate high enough to follow the drive-level frequency $f$ will contribute additional charge to the equilibrium static charge. This assumption allows us to calculate the capacitance based on the differential charge between two bias levels in steady state. The simulation computes the capacitance measured at each drive-level voltage by the change in the depletion region charge $Q$ after application of the drive-level voltage:

$$C = \frac{Q(V_A - V_{DL}) - Q(V_A)}{-V_{DL}} \qquad (1)$$

The static charge $Q(V_A - V_{DL}/2)$ is calculated by integrating $g(x,E)$ between the VBM and $E_F$. The charges $Q(V_A - V_{DL})$ and $Q(V_A)$ are the sum of both the static charge and the contribution of any additional charge in the depletion region from states with a hole emission rate high enough to respond

978-1-5090-5606-4/17 $31.00 © 2017 IEEE

to the drive-level frequency. This additional charge is calculated by integrating $g(x,E)$ between the VBM and a threshold energy $\Delta E_t$ relative to the VBM that is determined by the emission formula for holes [4]:

$$\Delta E_t = -kT \log\left(\frac{f}{\sigma_h <v_h> N_v}\right) \quad (2)$$

where $\sigma_h$ is the capture cross section for holes, $<v_h>$ is the average hole thermal velocity, and $N_v$ is the valence band effective density of states. If $\Delta E_t > E_F - \text{VBM}$, then we set the integration limit to $E_F$ rather than $\Delta E_t + \text{VBM}$. For this approach only the steady state solution to Poisson's equation is required; Poisson's equation may be solved to high accuracy using the Noumerov method [3] discussed later in the section.

The capacitance is a non-linear function of the drive-level voltage:

$$C(V_A + V_{DL}) = C_0(V_A) + C_1(V_A)V_{DL} + C_2(V_A)V_{DL}^2 + \ldots \quad (3)$$

After the capacitance is computed at each drive-level voltage, a quadratic fit to the capacitance versus drive-level voltage yields the DLCP coefficients for the intercept $C_0$ and slope $C_1$. The following relation using $C_0$ and $C_1$ is interpreted in reference [1] as the total defect density:

$$N_{DL} = \frac{|\rho|}{q} = -\frac{C_0^3}{2q\varepsilon A^2 C_1} \quad (4)$$

where $\rho$ is the charge density, $q > 0$ is the magnitude of the

electron charge, $\varepsilon$ is the CIGS permittivity, and $A$ is the area of the Schottky contact. Fig. 1 shows a typical capacitance versus drive-level voltage plot from the DLCP simulation with the quadratic fit used to determine $C_0$ and $C_1$.

Fig. 2 shows three simulations of the defect density $N_{DL}$ at 3 MHz and 240 K for a range of applied dc biases: -0.7 V to +0.5 V. The x-axis is the profile distance from the interface $x_p \equiv \varepsilon A / C_0$ which is identical to the depletion width if there is no influence from deep defects. The input doping profile is uniform at $2\times10^{15}$ cm$^{-3}$ (red dashed line). The blue curve was simulated in the presence of a deep defect with a density of $2\times10^{15}$ cm$^{-3}$, an energy of 0.4 eV above the VBM, a cross section of $5\times10^{-15}$ cm$^2$, and a width of $\sigma = 0.05$ eV. The Schottky barrier height used for all of the simulation results in this section is $\phi_B = 0.73$ eV. The rise in apparent defect density towards the back of the device is due to the influence of static charge accumulating in the deep trap with increasing reverse bias. If we remove the deep trap we obtain the profile given by the magenta line. The increase in the profile near the surface is due to neglecting the higher order terms in the Taylor expansion (3). For the case of uniform doping these terms all have the general form:

$$C_n(V_A)V_{DL}^n = \frac{\text{Const} \times V_{DL}^n}{\sqrt{(V_{bi} - V_A)^{2n+1}}} \quad (5)$$

As the reverse bias decreases and eventually moves into forward bias, the denominator in (5) decreases, increasing the contribution from the terms $C_n$. The error also increases with the maximum applied drive-level $V_{DL}$. The blue and magenta

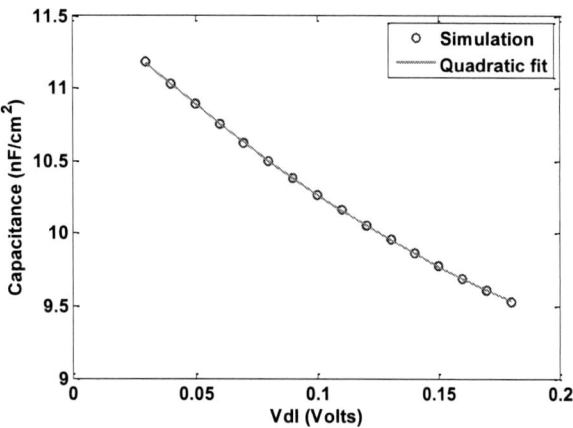

Fig.1. Capacitance versus drive-level voltage from the DLCP simulation at 1 MHz with a forward bias of 0.3 V.

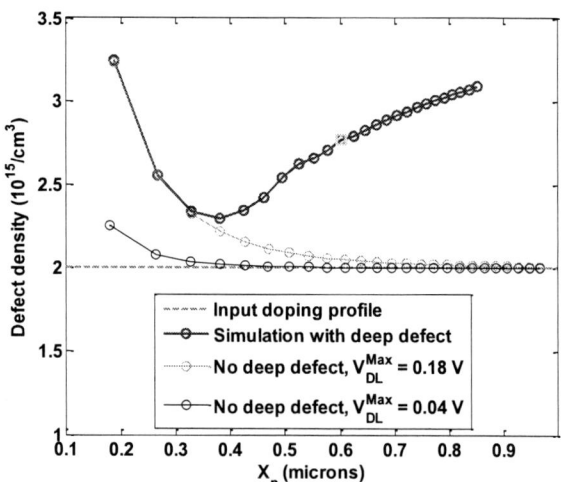

Fig.2. DLCP simulation of a uniform doping profile with a midgap defect at 240 K and 3 MHz (blue curve), and for the doping profile only (magenta and black curves). Static charging of the deep defect causes the rise in apparent defect density at the back of the device.

978-1-5090-5606-4/17 $31.00 © 2017 IEEE

curves were both simulated with a drive level range between 0.03 V and 0.18 V (Fig. 1). The black curve was simulated without a deep trap using a drive-level range between 0.01 V and 0.04 V and shows a much smaller error near the front of the device. The green square on the blue curve marks the position corresponding to zero applied bias.

Fig. 3 is a simulation of the defect density $N_{DL}$ versus frequency at zero bias. The density of states input to the simulation includes a uniform shallow defect density of

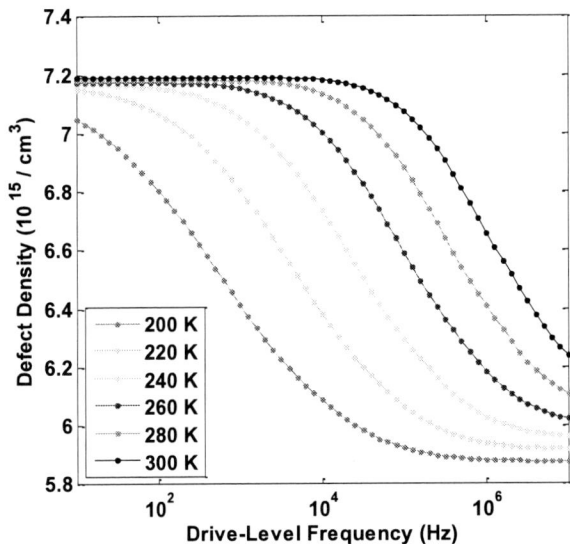

Fig.3. DLCP simulation for the frequency dependence of the defect density at zero bias with a uniform deep defect density of $2 \times 10^{15}$ cm$^{-3}$ and a uniform shallow defect density of $5 \times 10^{15}$ cm$^{-3}$.

$5 \times 10^{15}$ cm$^{-3}$ and a uniform deep defect density of $2 \times 10^{15}$ cm$^{-3}$ with an energy of 0.35 eV above the VBM, a cross section of $5 \times 10^{-15}$ cm$^2$, and a width of $\sigma = 0.05$ eV. The maxima of the derivatives of these curves yields a straight-line Arrhenius plot with a slope yielding a defect energy of $0.333 \pm 0.003$ eV and an intercept yielding a cross section of $(1.29 \pm 0.27) \times 10^{-15}$ cm$^2$. The disagreement between the input and simulated cross sections is currently believed to be caused by the distortion of the results from static charging of the deep trap. If the simulation is run with a deep defect density reduced to $0.5 \times 10^{15}$ cm$^{-3}$ then the Arrhenius plot yields a defect energy of $0.367 \pm 0.010$ eV and a cross section of $(5.21 \pm 2.51) \times 10^{-15}$ cm$^2$.

The total space charge $Q$ is computed by integrating $g(x,E)$ over $E$ at each $x$ position to obtain the charge density $\rho(x)$, then integrating $\rho(x)$ over $x$ within the depletion region. For Poisson's equation we use the notation

$$\frac{d^2\psi}{dx^2} = \frac{\rho(x)}{\varepsilon} \qquad (6)$$

where $\psi(x)$ is the potential of the VBM. Eq. (6) corresponds to the convention for a band diagram: for a p-type device the charge density is negative and the bands bend down as they approach the interface at $x = 0$. The electric field using this convention is given by $F = +d\psi / dx$. The potential $\psi$ is chosen to be zero at the VBM in the bulk material.

The position of the bulk Fermi level $E_F^B$ must be determined by iteration. The bulk carrier concentration is given by

$$p = N_v \exp\left(\frac{-E_F^B}{kT}\right) = \int_0^{E_F^B} g(E)dE \qquad (7)$$

where we have assumed $g(x,E)$ to be independent of $x$ in the quasi-neutral region (QNR). Eq. (7) may be rearranged into a transcendental equation for $E_F^B$:

$$E_F^B = -kT \log\left(\frac{1}{N_v}\int_0^{E_F^B} g(E)dE\right) \qquad (8)$$

Eq. (8) converges rapidly for $E_F^B$. Real device absorbers may have a shallow defect density that depends strongly on position. For these cases $E_F^B$ is calculated for each position where the doping is measured and then averaged.

The position of the Fermi level in the depletion region is prescribed for a reverse biased diode by Cohen and Lang [3]. The Fermi level position remains flat at $E_F^B$ until the band bending increases the difference $E_F$ -VBM to $E_g / 2$. This position is defined as the boundary of the "deep depletion region." Within the deep depletion region the Fermi level follows the band bending to remain at $E_g / 2$ above the VBM. Near the interface the Fermi level must be merged with the metal Fermi level of the Schottky contact. There is no rigorous prescription for this; in the simulation the semiconductor Fermi level is transitioned to the metal Fermi level using a cubic spline fit within 100 nm of the interface. For a forward biased diode the Fermi level remains flat at $E_F^B$ across the entire depletion region [5].

Poisson's equation is solved by dividing the depletion width $W$ into a grid of $m$ equally spaced grid points of size $h = W / (m-1)$ and applying the Noumerov method, accurate to sixth order in the step size $h$. Initially $W$ is not known, however the potential of the VBM at the interface of a Schottky diode relative to the bulk value at $\psi = 0$ is well-defined: $\psi_{IF} = V_A - V_{bi}$. The integration must be repeated iteratively to adjust $W$ until it returns the correct potential at the interface; generally 6 or 7 iterations are required. The built-in potential is given by the Schottky barrier height $V_{bi} = \phi_B - E_F^B$. The barrier height is determined by fitting the simulation to DLCP measurements.

The Noumerov integration advances from the edge of the depletion region towards the interface, computing the potential at each grid point from the potential and charge density computed at the two previous grid points:

$$\psi_{i+1} = 2\psi_i - \psi_{i-1} + h^2\frac{\rho_i}{\varepsilon} + \frac{h^2}{12}\left(\frac{\rho_{i+1}^*}{\varepsilon} + \frac{\rho_{i-1}}{\varepsilon} - 2\frac{\rho_i}{\varepsilon}\right) \quad (9)$$

The computation of $\rho_{i+1}^*$ also depends on the two previous grid points and is described later in this section. At each step, $\rho_i$ is calculated from $\psi_i$. At the first grid point $\psi_1 = 0$ and $\rho_1$ is initialized to $-qN_A$, where $N_A$ is the shallow defect density at the edge of the depletion region. At the second grid point $\psi_2$ is initialized to $-qN_A h^2/2\varepsilon$. Experience has shown that the solver is insensitive to these initial conditions, but that it is quite sensitive to the depletion width. At all subsequent steps the charge is computed from the integral of $g(x,E)$ over $E$ between the limits $\psi_i$ (the VBM) and $E_F$, or between $\psi_i$ and $\psi_i + \Delta E_t$ from (2). It is convenient to let the single variable $E_t$ represent the lesser of either $\Delta E_t$ or $E_F - \psi$.

In the simulation it is most practical to represent $g(x,E)$ as a sum of Gaussians in energy; then each defect may be integrated over energy analytically using error functions rather than numerically; this greatly reduces the program execution time. For a single Gaussian defect level labeled by $k$ and centered at $E_d^{(k)}$ (relative to the VBM) with standard deviation $\sigma_k$ and total defect density $N_T^{(k)}(x)$ this integral is given by

$$\rho_k(x) = -q\int_{\psi(x)}^{E_t^{(k)}} g(x,E)dE = $$
$$\frac{-qN_T^{(k)}(x)}{2}\left[\text{erf}\left(\frac{E_t^{(k)} - E_d^{(k)}}{\sqrt{2}\sigma_k}\right) - \text{erf}\left(\frac{-E_d^{(k)}}{\sqrt{2}\sigma_k}\right)\right] \quad (10)$$

For the total charge density: $\rho(x) = -qN_A(x) + \sum_k \rho_k(x)$.

Returning to the term $\rho_{i+1}^*$ from (9) with the notation introduced in (10), we have used the approximation provided by Cohen and Lang [3] which preserves the accuracy of the Noumerov computation to sixth order in $h$:

$$\rho_{k,i+1}^*(x) = -q\int_{2\psi_i(x)-\psi_{i-1}(x)+h^2\frac{\rho_{k,i}(x)}{\varepsilon}}^{E_t^{(k)}} g_k(x,E)dE \quad (11)$$

where the subscripted integrand $g_k(x,E)$ denotes only the density of states for the deep state $k$ with no contribution from $N_A(x)$.

It is important to account for the static charge $\rho^S(x)$ during the simulation. The Poisson integration is first performed for the static charge at $V_A - V_{DL}/2$. During this calculation all the integrals over energy are between $\psi(x)$ and $E_F(x)$ because the states in $g(x,E)$ have had an infinite amount of time to charge:

$$\rho^S(x) = -q\int_{\psi^S(x)}^{E_F^S(x)} g(x,E)dE \quad (12)$$

After the static charge integration is completed the calculated arrays for $\rho^S(x)$, $\psi^S(x)$, and $E_F^S(x)$ are saved as reference values for the subsequent calculations at $V_A$ and $V_A - V_{DL}$.

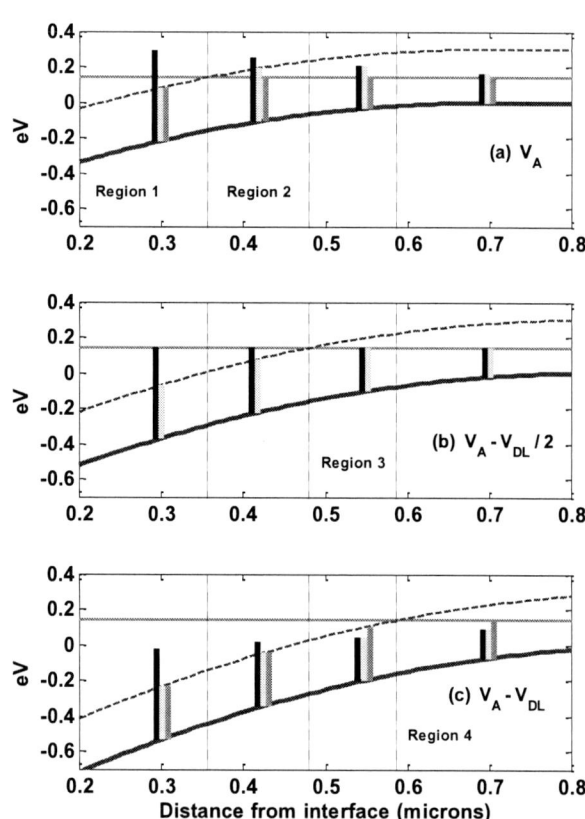

Fig.4. Graphical representation of Eq. (13); see discussion in text.

When the Poisson integration is done at either $V_A$ or $V_A - V_{DL}$, these three static charge arrays are interpolated to the new grid and the charge is computed at each step by adding the static charge and any additional charge states at the new bias voltage that are not already occupied by the static charge:

$$\rho(x) = \rho^S(x) - q\int_{\psi(x)}^{E_t} g(x,E)dE + q\int_{\psi^S(x)}^{E_t^S} g(x,E)dE \quad (13)$$

Fig. 4 presents three band diagrams to graphically illustrate Eq. (13) at the maximum $V_A$ (Fig. 4a), the midpoint of the applied drive-level voltage signal $V_A - V_{DL}/2$ (Fig. 4b), and the minimum $V_A - V_{DL}$ (Fig. 4c). The valence band $\psi(x)$ is shown by the solid blue curve, the Fermi level by the red line, and the emission threshold $\Delta E_t$ by the dashed blue curve. The static charge density reaches equilibrium at the voltage $V_A - V_{DL}/2$ (Fig. 4b) and is represented by the vertical black bar, which corresponds to the integral (12) or the first term on the right-hand side of (13). The green bar corresponds to the third integral of (13) and represents the static charge integrated to $E_t$ (the lesser of either $E_F$ or $\psi + \Delta E_t$ ) at the voltage $V_A - V_{DL}/2$. The magenta bar corresponds to the second integral of (13) and represents the dynamic charge that can respond at the drive level frequency when the voltage swings positive to $V_A$ (Fig. 4a) or negative to $V_A - V_{DL}$ (Fig. 4c). The sum of the black and magenta bars minus the green bar graphically represents the net charge in the depletion region and the right-hand side of (13). The green bar accounts for acceptor states already occupied by static charge that must be subtracted from the total charge integral to avoid double counting. The black and green bars are reference integrals for the static charge calculated in the configuration of Fig. 4b that are translated unchanged to the configurations in Fig. 4a and Fig. 4c; the total charge from the configurations at $V_A$ and at $V_A - V_{DL}$ is used to compute the capacitance from (1).

The band diagrams in Fig. 4 divide naturally into four regions labeled on Fig. 4. Region 1 is the region closer to the interface than the point where $\Delta E_t$ intersects the Fermi level at $V_A$ (Fig 4a). The charge density from region 1 is the static charge integral (black bar) calculated from Fig. 4b. Region 4 is the region closer to the back contact than the point where $\Delta E_t$ intersects the Fermi level at $V_A - V_{DL}$ (Fig. 4c). In region 4 all of the charge integrals are below $\Delta E_t$ : all of the charge in this region can respond to the drive level frequency so that the total charge is given by the integral of the dynamic charge (magenta bar).

In region 2 the total charge is the static charge at the bias $V_A - V_{DL}$ but not at the bias $V_A$: acceptor states filled at the bias $V_A - V_{DL}/2$ (black bar, Fig. 4b) that are between the Fermi level and $\Delta E_t$ at the bias $V_A$ (black bar, Fig. 4a) will capture holes so that the total charge is reduced. This reduction of charge increases with depth within region 2.

In region 3 the charge density is "emission limited": when the bias increases to $V_A - V_{DL}$ in Fig. 4c, the charge is limited by the emission threshold $\Delta E_t$. Because $\Delta E_t$ follows the band bending of the valence band, the charge density at bias $V_A - V_{DL}$ is flat within region 3.

Fig. 5 shows the charge density calculated from the simulation in the presence of a deep defect where the four regions from Fig. 4 are marked by dashed lines. The input

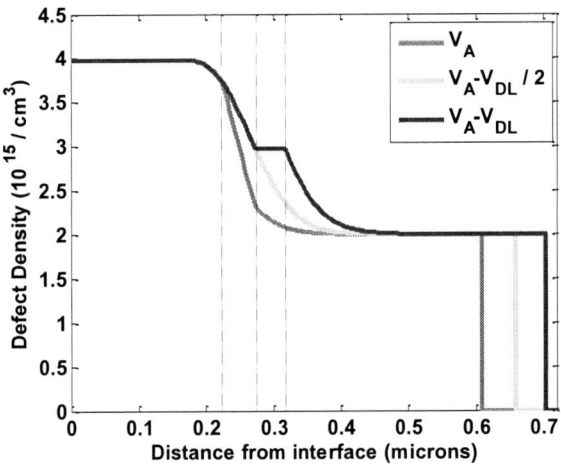

Fig.5. Simulation of the charged acceptor density in the depletion region for the three different bias conditions of Fig. 4. The simulation includes a uniform doping of $2\times10^{15}$ cm$^{-3}$ and a uniform deep defect with density $2\times10^{15}$ cm$^{-3}$.

doping is uniform at $2\times10^{15}$ cm$^{-3}$. The deep acceptor has a density of $2\times10^{15}$ cm$^{-3}$, an energy of 0.35 eV, a cross section of $5\times10^{-15}$ cm$^2$, and a width of $\sigma=0.05$ eV. The vertical lines on the right side of the plot for each bias condition mark the edge of the depletion region for each bias where the charge density drops to zero. The charge density in regions 1 and 4 varies smoothly with the difference between the valence band and the Fermi level. In region 3 the flat emission limited charge density is evident in the curve for $V_A - V_{DL}$. In region 2 the charge density increases uniformly towards the interface in the curve for $V_A$, but not with the same functional form as the rest of the curve; this is the region where a part of the static charge that has moved above the Fermi level (region 2 in Fig. 4a) captures holes because it is still below the emission threshold.

## IV. INTERPRETATION OF HIGH-FREQUENCY DLCP DATA

DLCP measurements at 1 MHz were performed on a Schottky Al/CIGS diode over a range of temperatures (160 K to 320 K in 20 K intervals) and dc biases (+0.3 V to -0.5 V in 0.1 V intervals). This device was almost completely depleted at -0.5 V; the absorber thickness was measured on a FIB-SEM

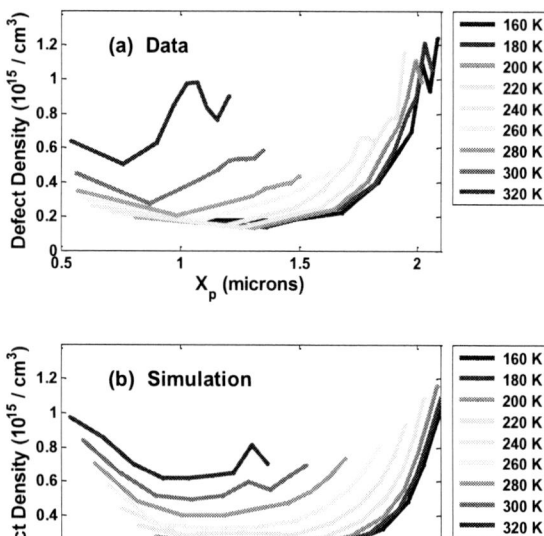

and width $\sigma$ ( 0.52 microns ) adjusted to match the two highest forward biased fast CV data points at +0.3 V and +0.2 V; Fig. 7a shows the entire input doping profile for the simulation.

The Schottky barrier height used for the simulation is $\phi_B = 0.81$ eV at all temperatures, determined from a fit to the 160 K DLCP data. To reproduce the trend of the data at higher temperatures a Gaussian deep defect was introduced into the simulation at 0.60 eV with a hole capture cross section of $1 \times 10^{-17}$ cm$^2$, a width of $\sigma = 0.12$ eV, and a defect density of $6 \times 10^{15}$ cm$^{-3}$ (Fig. 7b). For such a small defect cross section and high measurement frequency only the tails of the deep defect distribution contribute to increase the defect density in the simulation as the temperature is increased.

Fig.6 Data (a) and simulation (b) for an Al/CIGS Schottky device measured by DLCP at 1 MHz over a range of temperatures and dc biases.

## V. CONCLUSIONS

We have presented in detail a junction capacitance simulation and described its use for DLCP on Schottky devices. We have also noted that the DLCP signal at all frequencies is influenced by deep states with arbitrarily long emission lifetimes because these states equilibrate to the dc bias voltage $V_A - V_{DL} / 2$ which depends upon the drive-level voltage $V_{DL}$. Hence DLCP cannot be trusted for accurate carrier concentration measurements, even at high frequencies.

## ACKNOWLEDGEMENTS

The authors gratefully acknowledge Dmitry Poplavskyy, Neil Mackie, Rouin Farshchi, and Tim Nagle for discussions, and Matt Hrbacek, Kee Kee Cheung, Dave Spaulding, and Philip Sifers for support in the laboratory.

cross section image to be 2.3 microns. The input doping profile at 160 K was taken from a "fast" CV measurement [6] at 160 K with a 100 ms sweep time. For the region closer to the interface than the measurement the doping profile was extended by a half-Gaussian with amplitude ( $1.14 \times 10^{15}$ cm$^{-3}$ )

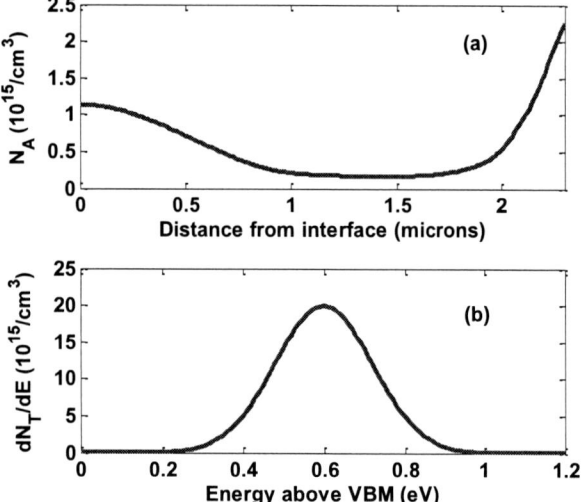

Fig.7 Input doping profile (a) and deep defect distribution (b) for the simulation of Fig. 6b.

## REFERENCES

[1] J. T. Heath, J. D. Cohen, and W. M. Shafarman, "Bulk and metastable defects in CuIn(1-x)Ga(x)Se₂ thin films using drive-level capacitance profiling," *J. Appl. Phys.*, Vol. 95, pp. 1000-1010, 2003.

[2] M. Burgelman, P. Nollet and S. Degrave, "Modelling polycrystalline semiconductor solar cells," *Thin Solid Films*, vol. 361-362, pp. 527-532, 2000.

[3] J. D. Cohen and D. V. Lang, "Calculation of the dynamic response of Schottky barriers with a continuous distribution of gap states," *Phys. Rev. B*, Vol. 25, pp. 5321-5350, 1981.

[4] D. K. Schroder, *Semiconductor Material and Device Characterization*, Wiley-Interscience, p. 309, 1990.

[5] E. H. Rhoderick, "Comments on the conduction mechanism in Schottky diodes," *J. Phys. D: Appl. Phys.*, Vol. 5, pp. 1920-1929, 1972.

[6] P. K. Paul, J. Bailey, G. Zapalac, S. A. Ringel, A. R. Arehart, "Fast C-V method to mitigate effects of deep levels in CIGS doping profiles," *2017 IEEE 44th Photovoltaics Specialists Conference.*

# Analysis of the Impact of Installation Parameters and System Size on Bifacial Gain and Energy Yield of PV Systems

Amir Asgharzadeh[1], Tomas Lubenow[1], Joseph Sink[1], Bill Marion[2], Chris Deline[2], Clifford Hansen[3], Joshua Stein[3], Fatima Toor[1]

[1]Electrical and Computer Engineering Department, The University of Iowa, Iowa City, IA, 52242, USA
[2]National Renewable Energy Laboratory, Golden, CO, 80401, USA
[3]Sandia National Laboratories, Albuquerque, NM, 87185, USA

*Abstract* — In this work, we present the combined effect of installation parameters (tilt angle, height above ground, and albedo) on the bifacial gain and energy yield of three photovoltaic (PV) system configurations: a single module, a row of five modules, and five rows of five modules utilizing RADIANCE based ray tracing model. We show that optimum tilt angle is dependent on parameters such as height, albedo, size of the system, and time of the year. For a single bifacial module installed in Albuquerque, NM, the optimum tilt angle is lowest (~5°) for summer solstice and highest (~65°) for winter solstice. For larger systems, optimum tilt angle can be up to 20° more than that for a single module system. We also show that modules in large scale systems, generate lower energy due to large shadowing areas cast by the modules on the ground. For albedo of 21 %, middle module in a large array generates up to 7% less energy than a single bifacial module. To validate our model, we utilize measured data from Sandia's fixed tilt bifacial PV testbed and compare it with our simulations. We find that due to higher non-uniformity, lower tilt angles demonstrate high normalized root mean square deviation (NRMSD) between measured and simulated values than high tilt angles.

## I. INTRODUCTION

In recent years, there has been growing interest in bifacial photovoltaic (PV) technology because it enables higher performance and lower price per watt ($/W) compared to conventional monofacial PV technology. However partly due to lack of accurate bifacial PV system modeling methods to predict system performance, utilization of this technology has remained limited. Understanding the effect of different installation parameters, such as height, tilt angle, albedo of the ground and array size on the bifacial PV system performance can help determine the optimum installation parameters for the system and allow for an accurate prediction of the energy yield of the system.

Other research groups have studied the impact of installation parameters, such as, tilt angle, height above ground and albedo, on the energy yield of small bifacial PV arrays based on measured data without considering the effect of system size [1]. Yusufoglu *et al.* [2] conducted a comprehensive performance analysis of a single bifacial module. However more realistic scenarios include a larger number of modules and rows. For these systems, the large

shadowing areas cast by the modules on the ground can significantly negatively impact their performance. In this work, we show the combined effect of tilt angle, height, albedo and size of the system on the energy yield and bifacial gain of the PV system.

## II. IRRADIANCE MODELING

We modeled the PV systems using RADIANCE [3], which is a simulation software to compute the radiance profile of physical systems by ray-tracing methods. The sky irradiance model used in this study approximates the Perez direct and diffuse model [4]. In our model, we utilized the dimensions and electrical characteristics of Prism Solar's Bi60-368BSTC bifacial module (front and backside efficiencies of 17.4% and 15.6%, respectively, which is equivalent to a bifaciality of ~90%). NREL's National Solar Radiation Data Base (NSRDB) [5] was used to derive typical meteorological year version 3 (TMY3) weather (hourly) data for Albuquerque, NM for global horizontal irradiance (GHI), diffuse horizontal irradiance (DHI) and direct normal irradiance (DNI). Azimuth and zenith angles (also hourly data) were calculated using Sandia National Laboratories' PV_LIB Toolbox [6].

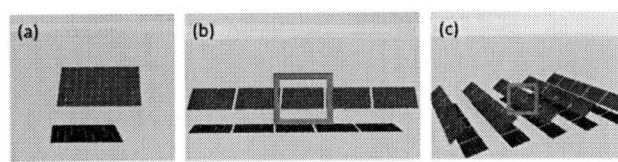

Fig. 1. Three south-facing PV systems consisting of (a) a single module (b) a row of five modules and (c) five rows of five modules each, were simulated to study the impact of the size on the system performance.

We considered three south-facing PV system configurations: (i) single module, (ii) a row consisting of five modules, and (iii) five rows, each with five modules, to investigate the impact of the system size on its bifacial gain and energy yield. Since the modeling of the multiple module configurations requires significant computation resources, we made our analysis feasible by considering the performance of only the middle module in the array. The row spacing for the five-row case was defined using a value obtained for the shadow length of the row of modules on Dec 21st (winter solstice) when the sun is the lowest in the sky and casts the longest shadow on the ground; using this length ensures that the modules will be shadow free for the entire year [7]. Fig. 1 shows the three simulated systems with the representative modules in the multi-module systems enclosed with red outlines.

Parametric sweeps over the three parameters affecting PV system performance were conducted to study their individual and combined effects. Tilt angle was varied from 5° to 90° (with steps of 5°). Module height above the ground, which is defined as the height of the lower edge of the module above the ground, was varied from 0.2 m to 3 m (with steps of 0.2 m). Typical height for ground mounted systems is 1 m while for car-port systems it is around 3 m, which is why we modeled heights of up to 3 m. We used three ground materials with different albedos: lite soil (21%), beige roofing material (43%) and a white ethylene propylene diene monomer (EPDM) roofing material (81%), which can also represent snow-covered ground. The albedo values for each of the materials were measured at NREL.

We ran hourly simulations sweeping parameters mentioned above around three representative dates of the year: summer solstice, winter solstice and fall equinox. Sun position for any day of the year is between the sun position on summer solstice and winter solstice, and for the fall equinox the length of the day and night are equal, so the analysis of these three days helps determine the seasonal and annual trends. For each case, we also considered one clear day and one cloudy day to study the effect of cloudy weather condition on the system performance.

To calculate the daily energy yield and bifacial gain in energy (BGE), we used the irradiance data for each of the 60-cells in the module at each time step and averaged it. The average value was multiplied by the effective area of the module and power conversion efficiency value to calculate the power generated by the module. Multiplying the power with the time step (1 hour) gives the energy of that particular time period in Watt-hours (W.h). For modeling bifacial modules, we added the front and backside energy to obtain total energy generated by the module. We summed over energy in each time step to obtain the daily energy. BGE was calculated using Eq. (1):

$$BGE \equiv \frac{E_b}{E_m} - 1 \qquad (1)$$

where $E_b$ and $E_m$ are the energy yield of the bifacial and monofacial module, respectively. It is important to note that by averaging cell irradiance data, we are neglecting backside non-uniformity. Currently we are working to improve our model by defining by-pass diodes in the model to account for the backside non-uniformity.

## III. EFFECT OF INSTALLATION PARAMETERS

In this section, we present the effect of installation parameters on energy yield and bifacial gain of a single bifacial module for three clear days around summer solstice, fall equinox, and winter solstice. TMY3 weather data was used as an input for simulations. By comparing the GHI values in TMY3 weather data with the GHI data obtained from Ineichen clear sky model [8, 9], we can determine the clearness of sky for specific days. A parameter called clear sky index ($K_c$) which is the ratio of measured GHI over clear sky GHI, indicates how much clear a day was. Fig. 2 shows this comparison for three days where $K_c$ is close to unity indicating that the sky on these days was clear with a good approximation.

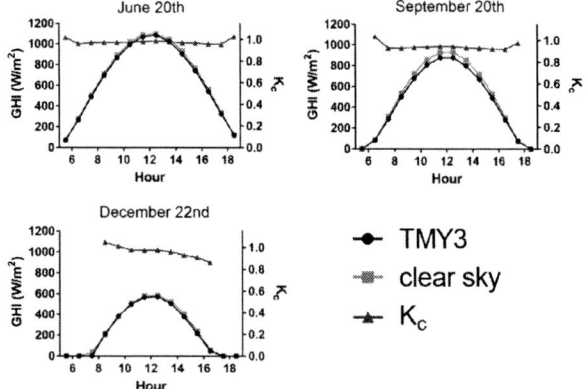

Fig. 2. Comparison of GHI values from TMY3 weather data and Ineichen clear sky model. $K_c$ value of close to unity, indicates the sky was clear on these days.

### A. Effect of tilt angle

For bifacial modules, optimum tilt angle can be different from monofacial modules and it depends heavily on parameters such as height, albedo, size of the system, and time of the year. Fig. 3 shows the energy yield and BGE as functions of tilt angle for two different heights (minimum height of 0.2 m and maximum height of 3.0 m in simulations) and three different albedo values (21%, 43%, and 81%). Comparing the energy yield figures from Fig. 3 (a), (c) and (e), we observe that optimum tilt angle for modules installed at 3.0 m is around 5° for summer solstice, 35° (site's latitude) for fall equinox and 65° for winter solstice. However, for modules installed closer to the ground (0.2 m), optimum tilt angle is usually higher. We will see in the section IV that optimum tilt angle is dependent on other parameters, such as height, albedo, size of the system, and time of the year and we need to study it more carefully. We also observe from BGE plots in Fig. 3 (b), (d) and (f) that by increasing tilt angle bifacial gain increases in summer. This can be explained by two reasons. First is that because of sun's position in summer, backside of the module receives more direct light when the module is installed at a high tilt angle and causes BGE to increase. Second reason is that for south-facing bifacial modules, most of the irradiance comes from the frontside and backside contribution is smaller than the frontside. We also know that optimum tilt angle for frontside irradiance is low (in summer solstice), so by increasing the tilt angle, frontside

## B. Effect of height

Height of the bifacial module from the ground also impacts the energy yield. When the bifacial module is installed close to the ground, backside irradiance is affected profoundly by self-shadowing and by increasing the clearance from the

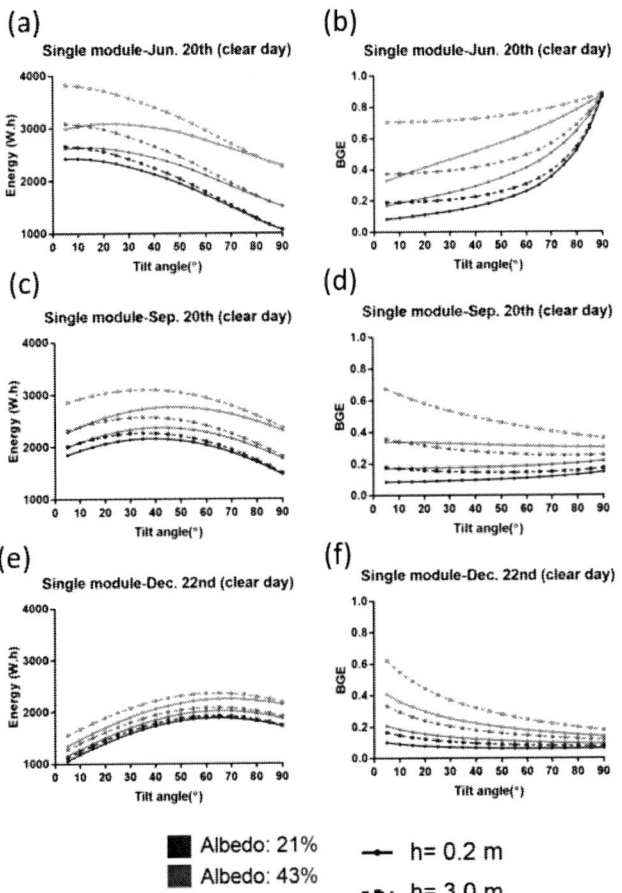

Fig. 3. (a) Energy yield and (b) bifacial gain of single module system as a function of tilt angle on a clear day around summer solstice (June 20th). (c) Energy yield and (d) bifacial gain of the system on a clear day around fall equinox (September 20th). (e) Energy yield and (f) bifacial gain of the system on a clear day around winter solstice (December 22nd).

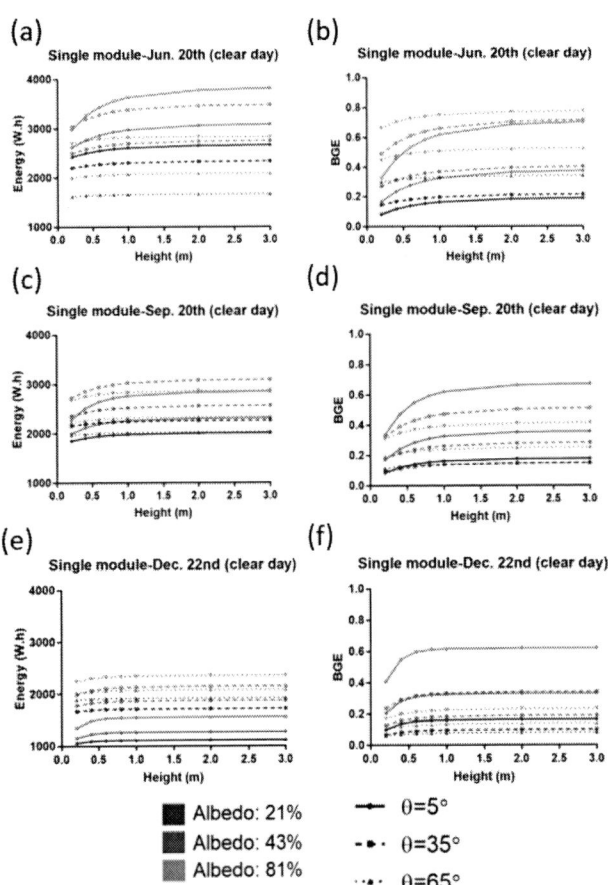

Fig. 4. (a) Energy yield and (b) bifacial gain of single module system as a function of height on a clear day around summer solstice (June 20th). (c) Energy yield and (d) bifacial gain of the system on a clear day around fall equinox (September 20th). (e) Energy yield and (f) bifacial gain of the system on a clear day around winter solstice (December 22nd).

irradiance gets smaller and cause BGE to increase (backside irradiance can increase or decrease and it depends on the albedo of the ground. For higher albedos, backside irradiance is more when back of the module faces ground, while it is opposite for lower albedos). However, for fall equinox and winter solstice, we can see from Fig. 3 (d) and (f) that the trend is opposite, because the optimum tilt angle is higher. For fall equinox optimum tilt angle for frontside irradiance is approximately equal to site's latitude of 35º and for winter solstice it is higher, around 65º. By increasing the tilt angle from 5º up to the optimum tilt angle, frontside irradiance increases and cause BGE to decrease. That is why the slope of BGE decreases as we move from summer to winter.

ground, backside of the module gets more light from both the sky and the ground. Fig. 4 shows the impact of height on energy yield and bifacial gain. We plotted the data for tilt angles of 5º, 35º and 65º and albedos of 21%, 43%, and 81% to show the trends for different albedos and tilt angles. We observe that both energy yield and bifacial gain start to increase by increasing the height. However, we can see a saturating effect where as the height of the module increases the performance is not affected. This saturating effect is observed at 2.0 m for summer solstice, 1.0 m for fall equinox and 0.6 m for winter solstice. At these module heights the effect of self-shadowing on backside irradiance is diminished and increasing the height doesn't increase the performance.

## C. Effect of albedo

Increasing the reflectivity (albedo) of the ground, increases intensity of the reflected rays which hit front and back sides of the module and increases the system's performance. Fig. 5 shows the effect of albedo on energy yield and bifacial gain for different tilt angles and heights. We see that the relationship is linear for both energy yield and BGE. However, because of high insolation in the summer, slope of energy yield plot is higher for summer solstice than for fall equinox or winter solstice. We also observe that the slope of the energy yield data is lower when the module is close to the ground and the tilt angle is low (~9.3, 7.2, and 4.9 W.h/albedo (%) in summer solstice, fall equinox and winter solstice respectively). However, for modules installed at higher heights and tilt angles, slope is higher (~19.3, 13.9, and 7.2 W.h/albedo (%) in summer solstice, fall equinox and winter solstice respectively).

## IV. OPTIMUM INSTALLATION PARAMETERS AND EFFECT OF SYSTEM SIZE AND CLOUDY WEATHER CONDITION

So far, we analyzed the effect of installation parameters on a single bifacial module and saw that to achieve the highest performance, module needs to be installed at the highest possible albedo and its height from the ground should be high enough to minimize the self-shadowing effect. However, the optimum tilt angle varies under different conditions. We interpolated the simulation data to get resolution of one degree for tilt angle and determined the optimum tilt angle for different conditions. Fig. 6 shows the optimum tilt angle for three different sized systems (single module, one-row, and multi-row systems) for different heights and albedos and for both cloudy and clear days around summer solstice, fall equinox, and winter solstice. Fig. 7 shows that clear sky index is less than unity for shown three days, which indicates the sky was not clear during these days.

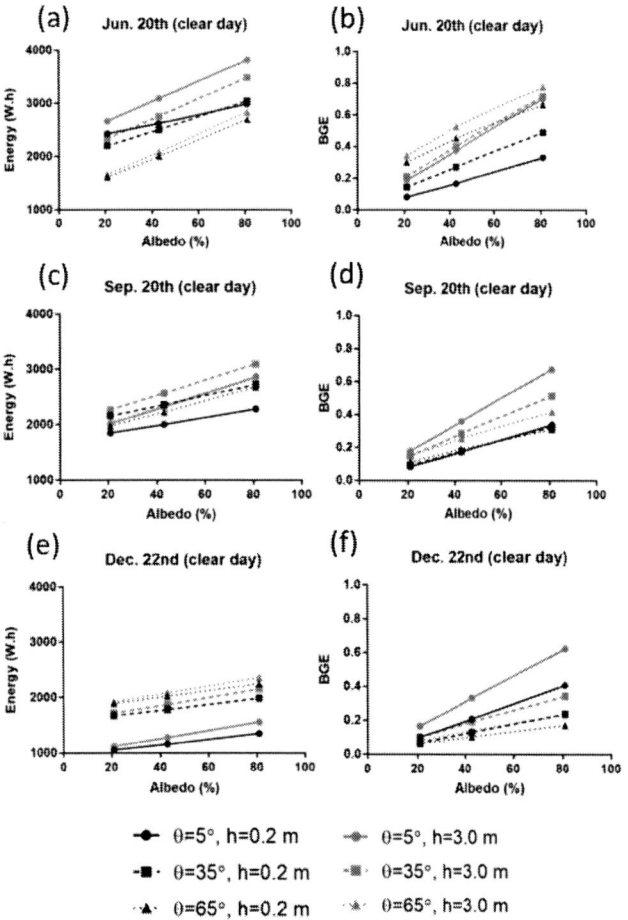

Fig. 5. (a) Energy yield and (b) bifacial gain of single module system as a function of albedo on a clear day around summer solstice (June 20th). (c) Energy yield and (d) bifacial gain of the system on a clear day around fall equinox (September 20th). (e) Energy yield and (f) bifacial gain of the system on a clear day around winter solstice (December 22nd).

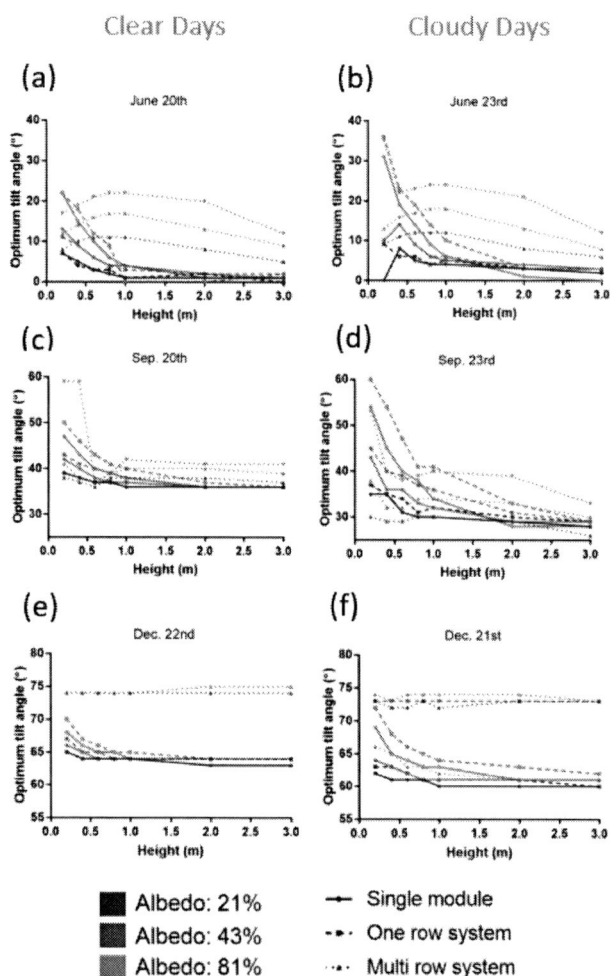

Fig. 6. Optimum tilt angle as function of height and albedo for clear days in summer solstice, fall equinox and winter solstice (Figures (a), (c) , and (e) respectively) and for cloudy days around the three dates (Figures (b), (d) , and (f)). Results are depicted for single module, one-row and multi-row systems.

Fig. 6 (a), (b) and (c) show that for lower module heights when the system size is not large (single module or one-row system), optimum tilt angle is higher. The modules installed close to the ground face large portion of their own shadow and by increasing the tilt angle, backside of the module receives more light from the ground and sky and sees less of the dark shadowing area. For cloudy days, optimum tilt angle tends to be higher for modules installed close to the ground relative to clear days. This is because cloudy weather conditions mean reduced direct sunlight and hence reduced reflection from the ground onto the back of the module. Therefore to achieve higher irradiance, back of the module tends to be toward the sky more than the ground requiring higher tilt angles for higher energy yield. Another observation is that, for large scale systems (multi-row system), optimum tilt angle can be up to 20° more than small scale systems. By increasing the number of modules, shadowing area gets larger and to receive more irradiance, tilt angle needs to be higher to diminish the shadowing effect.

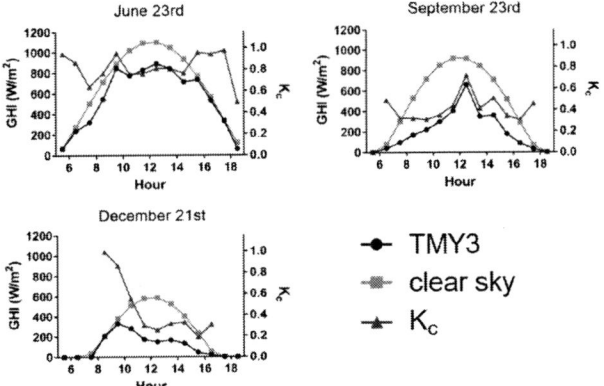

Fig. 7. Comparison of GHI values from TMY3 weather data and Ineichen clear sky model. $K_c$ value of less than one, indicates three chosen days are not clear days.

## V. EFFECT OF SYSTEM SIZE ON THE PERFORMANCE

Using the optimum tilt angle for module height of 1 m and albedo of 21%, we compared the performance of the three PV systems. The data is shown in Fig. 8. Monofacial data for the same height and albedo (for single module system) is also shown in the figure. By comparing the data in Fig. 8, we observe that by increasing the number of modules, energy yield decreases significantly due to larger shadowing area on the ground. Middle module in the multi-row system has about 7% lower energy production than the single module system on summer solstice. This value for fall equinox and winter solstice is about 4% and 3%, respectively. Note that in all cases, as expected, the bifacial modules produce more energy than the monofacial modules. We found from our simulation data (not shown here) that for the albedo of 81%, modules in large arrays can have up to 14% lower performance compared to single module systems. Fig. 8 also shows that highest

bifacial gain is for single module system and drops as system size gets larger.

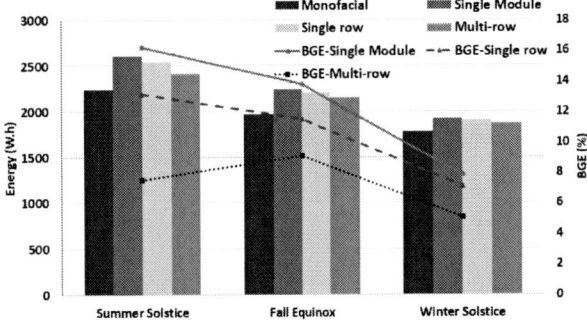

Fig. 8. Energy yield and BGE of single module, single row and multi-row PV systems for optimum tilt angle at the module height of 1 m and albedo of 21% for clear days on summer solstice, fall equinox and winter solstice.

## VI. MODEL VALIDATION

To validate our RADIANCE model we used it to simulate Sandia's fixed-tilt string-level arrays. Fig. 9 shows the system. It consists of 4 rows with different tilt angles (15°, 25°, 35°, and 45°). Each row has two strings of 8 modules (one monofacial and one bifacial). Each row has also 3 reference cells near the middle of the row: one for front and two for back side. Backside reference cells are installed on top and bottom of the middle module in the row (Fig. 10). Our simulations also included the concrete blocks used for the array footings.

Fig. 9. Sandia's fixed-tilt string-level arrays, Albuquerque, NM.

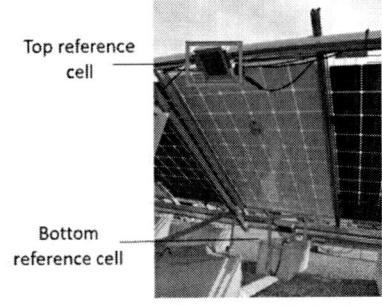

Fig. 10. Each row has two backside reference cells (top and bottom)

978-1-5090-5606-4/17 $31.00 © 2017 IEEE

Simulated irradiance was compared to field measurements for a clear day on March 1st, 2017. The comparison shows a good match between the measured and simulated data. Fig. 11 and 12 compare measured and simulated data for frontside and backside irradiance respectively.

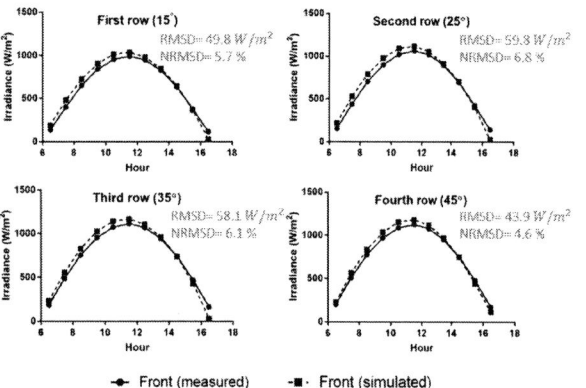

Fig. 11. Simulated vs measured frontside irradiance for Sandia's Fixed-Tilt String-Level Arrays.

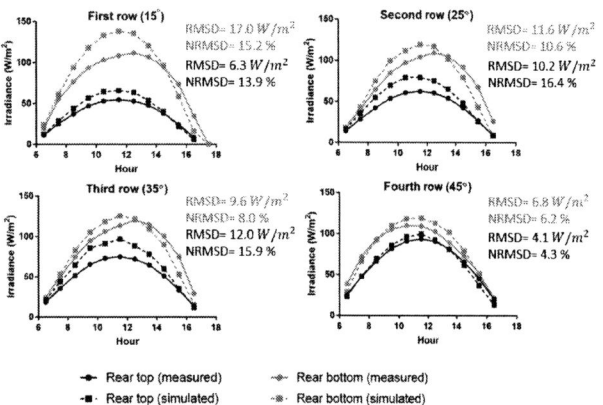

Fig. 12. Simulated vs measured backside irradiance for Sandia's Fixed-Tilt String-Level Arrays.

For each case, RMSD (root mean square deviation) and NRMSD (normalized RMSD) was calculated to compare the simulated data to measured data. Considering the backside irradiance data, we observe that top and bottom reference cells can receive different irradiance. This non-uniformity in the backside irradiance decreases the performance of the system. By increasing the tilt angle, non-uniformity decreases, because modules receive more uniform irradiance from the sky than the ground.

## VII. CONCLUSION

We performed a set of RADIANCE simulations to study the effect of tilt angle, module height above ground, albedo and size of the system. We showed the effect of installation parameters on energy yield and bifacial gain of a single module on clear days around summer solstice, fall equinox and winter solstice. We found that modules installed at the highest possible albedo with high enough height, have higher production. However, optimum tilt angle is more complicated and is dependent on other parameters such as height, albedo, size of the system, and time of the year and is usually higher for modules installed closer to the ground.

We showed that the system size is an important factor that impacts the performance of bifacial PV arrays. Three different-sized systems were modeled and their performance was compared. We found that for large scale bifacial systems, optimum tilt angle is usually higher and can be up to 20° more than that for smaller systems. We also observed that energy yield of the modules in a large array can decrease up to 7% (relative to single module system) with ground albedo of 21%.

We also modeled the Sandia's fixed-tilt string-level arrays and compared the simulated irradiance data to measured data to validate our model. Results show a good match between measurements and the simulation.

## VIII. ACKNOWLEDGEMENTS

Sandia National Laboratories is a multi-mission laboratory managed and operated by Sandia Corporation, a wholly owned subsidiary of Lockheed Martin Corporation, for the U.S. Department of Energy's National Nuclear Security Administration under contract DE-AC04-94AL85000.

## REFERENCES

[1] J. E. Castillo-Aguilella and P. S. Hauser, "Multi-Variable Bifacial Photovoltaic Module Test Results and Best-Fit Annual Bifacial Energy Yield Model," *Ieee Access,* vol. 4, pp. 498-506, 2016.

[2] U. A. Yusufoglu, T. M. Pletzer, L. J. Koduvelikulathu, C. Comparotto, R. Kopecek, and H. Kurz, "Analysis of the Annual Performance of Bifacial Modules and Optimization Methods," *Ieee Journal of Photovoltaics,* vol. 5, pp. 320-328, Jan 2015.

[3] G. J. Ward, "The RADIANCE lighting simulation and rendering system," presented at the Proceedings of the 21st annual conference on Computer graphics and interactive techniques, 1994.

[4] R. Perez, R. Seals, and J. Michalsky, "All-weather model for sky luminance distribution - preliminary configuration and validation," *Solar Energy,* vol. 50, pp. 235-245, Mar 1993.

[5] *National Solar Radiation Data Base (NSRDB).* Available: https://nsrdb.nrel.gov/nsrdb-viewer

[6] J. S. Stein, "PV_LIB Toolbox," ed: Sandia National Laboratories, 2017.

[7] J. Galtieri and P. T. Krein, "Designing solar arrays to account for reduced performance from self-shading," in *2015 IEEE Power and Energy Conference at Illinois (PECI)*, 2015, pp. 1-8.

[8] P. Ineichen and R. Perez, "A new airmass independent formulation for the Linke turbidity coefficient," *Solar Energy,* vol. 73, pp. 151-157, 2002.

[9] R. Perez, P. Ineichen, K. Moore, M. Kmiecik, C. Chain, R. George, *et al.*, "A new operational model for satellite-derived irradiances: Description and validation," *Solar Energy,* vol. 73, pp. 307-317, 2002.

# Dependence of String Power on its Height in the Array in Yoshinogari Mega Solar Power Plant

Shigeomi Hara*, Makoto Kasu* and Yasuki Masutomi[†]

*Department of Electrical and Electronic Engineering, Saga University, Saga, Saga, 840-8502, Japan
[†]Saga Yoshinogari Solar LLC, Kanzaki, Saga, 842-0121, Japan

*Abstract*—In order to make MPPT control of PCS effective, strings of almost same maximum power should be connected in parallel to the PCS. We have collected a big data of string powers measured in Yoshinogari mega solar power plant (MSPP) in Japan for two years. The data shows that string powers depend on height in the array. The average power of strings at the highest position in the array was greater approximately 7 than strings at the lowest positon. Several causes of this dependence are analyzed quantitatively.

*Index Terms*—array, height, mega solar power plant, module temperature, polysilicon photovoltaic module, string, view angle.

Fig. 1. Yoshinogari mega solar power plant.

## I. Introduction

A string is a set of photovoltaic modules connected in series. Several strings are connected in parallel to a power conditioning system (PCS). If strings of different I-V characteristics are connected in parallel to the PCS, the overall I-V curve has a somewhat less fill factor, which makes MPPT control of the PCS less effective. So it is desirable to connect strings having almost same maximum power in parallel to the PCS.

We have collected a big data of string powers measured in Yoshinogari MSPP in Japan for two years from 2015 to 2016. In the power plant, arrays have the inclination of 10 degrees to the south, and a normal array is composed of four strings stretched horizontally from the highest positon at north to the lowest position at south. Generated powers of every string in the power plant are measured at ten minutes interval. In this research, String powers of each position are compared each other. The causes of the difference of string powers of each positon, such as module operating temperature, view angle for the sky diffuse radiation, are considered quantitatively.

## II. String Power Statistics

The generation capacity of Yoshinogari MSPP is 12.9MW (Fig. 1). It uses polysilicon photovoltaic modules whose rated power is 240W. Total module number is 53,848, and the number of strings is 3,854. The measurement system of the power plant measures powers of all strings at ten minutes interval. The normal arrays in the power plant are composed of four strings 1 to 4 in the height order (Fig. 2). There are 874 normal arrays and 179 short arrays in the power plant. We call strings in short arrays string 12 and 34 in the height order. In Fig. 3 we show the frequency line graph of string powers 1 to 4 at 12 p.m. on June 10th, 2016. We can see the dependence of string powers on the height in the array. The strings 1 at the highest position have the highest string power,

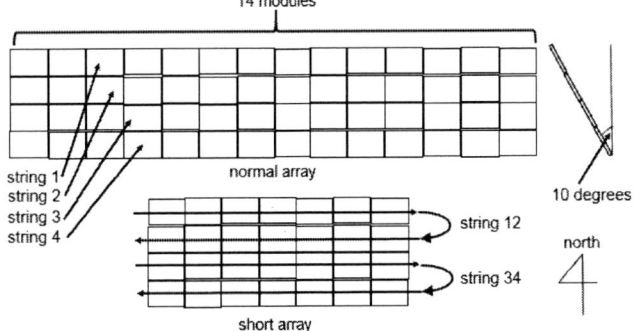

Fig. 2. Top view of strings.

and the strings 4 at the lowest position have the lowest string power. In Table I we show average string powers of each string type for two years from Jan. 1st, 2015 to Dec. 31th, 2016. For example the data number of string 1 is 19,683,932. The total average of all string powers in the MSPP for the two years is 1669.4 (W). The indexes in Table I are

$$\frac{\text{Average power of each string height}}{\text{Total average (=1669.4)}} \times 100 \qquad (1)$$

The average power of string 4 at the lowest position is less approximately 7.1% than string 1 at the highest position. The average power of string 34 at the short array is still less about 10% than string 1. This result is due to the fact that many strings 34 are located in the area of low generated power in the power plant. Standard deviations are very large because the averages are calculated with the data which include string powers at any time of day from morning to evening and any weather day.

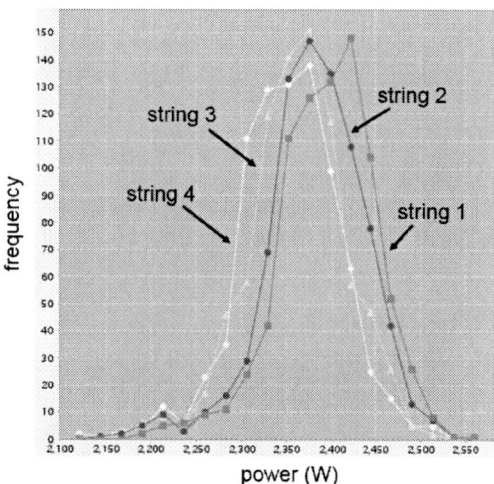

Fig. 3. Frequency line graph for string from 1 to 4 at 12 p.m. on June 10th, 2016.

#### TABLE I
##### AVERAGE STRING POWERS FOR TWO YEARS FROM 2015 TO 2016

|  | Average power(W) | Index | Standard deviation(W) |
|---|---|---|---|
| String 1 | 1723.2 | 103.2 | 706.6 |
| String 2 | 1701.7 | 101.9 | 702.1 |
| String 3 | 1676.0 | 100.4 | 693.5 |
| String 4 | 1603.8 | 96.1 | 690.1 |
| String 12 | 1612.9 | 96.6 | 754.1 |
| String 34 | 1557.5 | 93.3 | 750.2 |

#### A. Normalized Average

The average string powers computed in Table I are simple averages, which identically add up large string powers around the noon on clear days and small string powers at morning of afternoon on cloudy days. So the averages in Table I are strongly affected by large string powers on clear days and weakly affected by small string powers on cloudy days. Therefore we next calculate normalized averages of string powers of each height. By the term, normalized power, we mean

$$\frac{\text{string power at time } t}{\text{average of all string powers in MSPP at time } t} \times 100 \quad (2)$$

Table II shows average normalized powers of each string height for two years from 2015 to 2016. The average normalized power of string 4 is less about 7.8% than one of string 1. This difference is larger than 7.1% of average powers in Table I. This suggests that the difference among generation powers of each string height is large when generation powers are relatively low on such as cloudy days.

#### B. Average Power for Each Irradiance Level

Normalized average powers calculated above have suggested that differences of string powers of different heights are large when solar irradiance on the strings is low. To verify the suggestion we investigate average string powers of different

#### TABLE II
##### AVERAGE NORMALIZED STRING POWERS FOR TWO YEARS FROM 2015 TO 2016

|  | Average normalized power(W) | Standard deviation(W) |
|---|---|---|
| String 1 | 103.2 | 9.33 |
| String 2 | 101.8 | 9.31 |
| String 3 | 100.2 | 9.09 |
| String 4 | 95.4 | 11.1 |
| String 12 | 100.1 | 13.9 |
| String 34 | 96.2 | 14.6 |

#### TABLE III
##### INDEXES OF AVERAGE STRING POWERS OF EACH HEIGHT FOR THREE IRRADIANCE LEVELS

| Irradiance (kW/m$^2$) | 0–0.5 | 0.5–0.9 | 0.9– |
|---|---|---|---|
| String 1 | 104.2 | 102.6 | 102.3 |
| String 2 | 102.7 | 101.5 | 101.2 |
| String 3 | 100.9 | 100.1 | 99.9 |
| String 4 | 95.3 | 96.3 | 96.8 |
| String 12 | 93.3 | 100.4 | 100.4 |
| String 34 | 88.9 | 97.3 | 98.3 |

heights for each irradiance level. We divide irradiance into three levels, that is between 0 and 0.5, between 0.5 and 0.9, and greater than 0.9 (kW/m$^2$). Yosinogari MSPP is separated into five areas, and the every area has a pyranometer of second class. With the pyranometer solar irradiance of the five areas is measured. Each string power is associated with the irradiance in the same area. We calculate average string powers of each height for three irradiance levels. In the same way as above, the data of string powers are for the two years from 2015 to 2016. The result is given in Table III. Here the indexes are given by

$$\frac{\text{average string power of a height for an irradiance level}}{\text{average of all string powers for the same irradiance level}} \times 100 \quad (3)$$

In Table III, the difference of String 4 with String 1 for the lowest irradiance level from 0 to 0.5 (kW/m$^2$) is 8.9%, which is greater than 6.3% and 5.5% for other two irradiance levels. This shows that when the irradiance is small, the difference of generated powers between low positioned strings and high positioned strings is large.

### III. EFFECT OF MODULE TEMPERATURE

The maximum module power $P_{max}$ is predicted by the formula [1]

$$P_{\max} = P_{\max,\,\text{ref}} \frac{G}{G_{\text{ref}}} \left(1 + \gamma \left(T - 25\right)\right) \quad (4)$$

where $P_{max,ref}$ is the reference maximum power, $G_{ref}$ the reference irradiance (1 kW/m$^2$), $\gamma$ the temperature coefficient, and $T$ module temperature. $\gamma = -0.0035$ for crystalline silicon modules [1]. Using (4), we can predict the maximum module power from the module temperature $T$ and the irradiance $G$ (1 kW/m$^2$) on the module. We shall denote this predicted string power of string $i$ by $P_i$. Then by (4) we have the

978-1-5090-5606-4/17 $31.00 © 2017 IEEE

#### TABLE IV
##### TEMPERATURE FACTORS AND RATIO OF MEASURED POWERS

| $i$ | 2 | 3 | 4 |
|---|---|---|---|
| temperature factor (6) | 0.9996 | 1.0003 | 0.9977 |
| ratio of measured powers $P_{m,i}/P_{m,1}$ | 0.9883 | 0.9883 | 0.9726 |

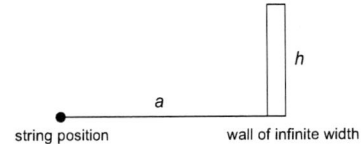

Fig. 4. A Calculation Model for Sky Factor of a String.

following formula for the ratio of the predicted string power $P_i$ (i=2,3,4) over $P_1$:

$$\frac{P_i}{P_1} = \frac{G_i}{G_1} \frac{(1 + \gamma(T_i - 25))}{(1 + \gamma(T_1 - 25))} \qquad (5)$$

where $T_i$ is the module temperature of string $i$ and $G_i$ the irradiance of string $i$. We substitute back surface temperatures as module temperatures. We have measured back surface temperatures of a normal array for nine days from July 20th, 2016 to May 30th, 2017. For each time of interval one minute, we calculate the following factor of temperature in (5) and take the average over all times.

$$\frac{(1 + \gamma(T_i - 25))}{(1 + \gamma(T_1 - 25))} \qquad (6)$$

Table IV is the result of the computation of temperature factor (6) from the measurement data of back surface temperature of modules. We denote the measured power of string $i$ by $P_{m,i}$ (i=1,2,3,4). Table IV also shows the ratio $P_{m,i}/P_{m,1}$. Temperature factors are larger than the ratio of measured powers. This shows that the effect of the only module temperature is not enough to explain the difference of string powers at different heights.

### IV. EFFECT OF VIEW ANGLE

Arrays line up with 1.2 (m) separation from south to north, so the view angle of strings in the array is restricted by the neighboring array in south and the array of itself. We shall deduce sky factors of each string. For this purpose we consider a disposition model of arrays in Fig. 4. In Fig. 4, the wall corresponds to the neighboring array in south and the array of itself. For every string 1 to 4 , a equals to 5.139 (m). h for south neighboring array is given in Table V. Sky

#### TABLE V
##### HEIGHT OF WALL $h$ FOR EACH STRING.

| String1 | String2 | String3 | String4 |
|---|---|---|---|
| 0.0868 (m) | 0.2605 (m) | 0.434 (m) | 0.608 (m) |

#### TABLE VI
##### SKY FACTORS FOR EACH STRING.

| String1 | String2 | String3 | String4 |
|---|---|---|---|
| 0.9391 | 0.9283 | 0.9176 | 0.9070 |

#### TABLE VII
##### EFFECT OF SKY FACTORS AND OVERALL RATIO.

| $i$ | 2 | 3 | 4 |
|---|---|---|---|
| $\frac{G_i}{G_1}$ | 0.9982 | 0.9964 | 0.9946 |
| overall ratio | 0.9978 | 0.9967 | 0.9922 |

factor is given by

$$1 - \frac{h}{2\pi a} \int_0^\pi \frac{\sin\theta}{\sqrt{1 + \left(1 + \frac{h}{a}\sin\theta\right)^2}} d\theta \qquad (7)$$

Using this formula, the sky factors for each string is given in Table VI. By Erbs model, diffuse irradiance is 16.5% of global irradiance on clear days.

$$\frac{G_i}{G_1} = \frac{0.835 + 0.165 \cdot S_i}{0.835 + 0.165 \cdot S_1} \qquad (8)$$

Here $S_i$ is the sky factor of string $i$. From (8) we can estimate the effect of sky factors as Table VII. Table VII also shows the overall ratio of (5) including the effect of module temperature and sky factor. These value are larger than the values in Table IV, which shows the insufficiency of the effect of module temperature and sky factor to explain the difference of string powers.

### V. CONCLUSION

Our measurement data have shown that string powers are different by height in the array. The average string powers of two year measurement data had the difference of at most 7%. The effects of module temperature and sky factor to string power are analyzed quantitatively. These two effect can contribute to explain the difference of string 1,2, 3 and 4. However only these two effects can not completely explain the difference of string powers. We have to consider other effects, such as the fact that four strings are connected in parallel and are controlled by PCS altogether, and the effect of shadow by weeds.

### ACKNOWLEDGMENT

We deeply appreciate fruitful discussion from Mr. Kensuke Sato, Mr. Kazuhiko Oda, Mr. Kazuhiko Babasaki, Mr. Yasuki Masutomi at NTT Facilities Co. and Drs. Atsushi Masuda, Yasuo Chiba at AIST. This work is supported by New Energy Industrial Technology Development Organization (NEDO) project.

## REFERENCES

[1] M. Fuentes, G. Nofuentes, J. Aguilera, D.L. Talavera, M. Castro, "Application and validation of algebraic methods to predict the behavior of crystalline silicon PV modules in mediterranean climates." Solar Energy, 81, pp. 1396-1408, 2007.

[2] D. Rekioua and E. Matagne, "Optimization of Photovoltaic Power Systems." pp.38-40, Springer, 2012.

# A bottom-up energy simulation framework to accurately compare PV module topologies under non-uniform and dynamic operating conditions

Patrizio Manganiello[1,2], Maro Baka[3], Hans Goverde[1], Tom Borgers[1], Jonathan Govaerts[1], Arvid van der Heide[1], Eszter Voroshazi[1] and Francky Catthoor[1,2]

[1] imec, Kapeldreef 75, 3001 Heverlee, Belgium
[2] ESAT, KU Leuven, Kasteelpark Arenberg 10, 3001 Heverlee, Belgium
[3] National Technical University of Athens, Heroon Polytechneiou 9, 157 80 Athens, Greece

*Abstract* — **Non-uniform operating conditions of photovoltaic modules have a huge impact on the energy generation of PV installations. To increase the energy yield of PV modules under non-uniform conditions, innovative module topologies have been proposed. However, a complete exploration of these topologies cannot be performed on the field, given the huge cost/time investment required for the realization of all potential topologies, as well as the large time required for comprehensively testing them in a real installation. Thus, it is more efficient and effective to compare them through simulations. An accurate and versatile simulation approach is proposed here which allows exploration of different PV module topologies under realistic time- and spatially-varying shading scenarios. The latter are created by matching measured weather data with simulated non-uniform and dynamic operating scenarios created by using the software SketchUp. The use of a physics-based bottom-up modeling approach for the different PV module topologies allows for accurate evaluation of the energy generated by each topology in the given operating conditions. Simulation results are presented that substantiate the feasibility of the methodology. Four relevant PV module topologies are compared, showing the better performance of reconfigurable topologies when uniform operating conditions are not dominant.**

## I. INTRODUCTION

Conventional photovoltaic (PV) modules are composed of a number (usually 60) of PV cells connected in series. When the PV module is subjected to uniform conditions this "all-in-series" approach leads to an optimal energy generation. Indeed, assuming that all cells are equal, cells function at the same Maximum Power Point (MPP) current, so that it is possible to make every cell operate at its MPP. One of the main drawbacks of conventional PV modules is that they perform poorly when subjected to non-uniform conditions, e.g. partial shading. In that case, the current of the PV module is limited to the one of the worst performing cell (lowest current) and the total generated PV power drops significantly. This well-known problem is usually solved by using bypass diodes (usually 3) connected in anti-parallel to a group of series-connected cells. However, this solution is far from being optimal. For instance, if only one cell of the module is shaded and the relative bypass diode is active, all cells belonging to that specific bypass section are not contributing to the module power.

These conventional PV modules are adapted to operations in PV plants which are dominated by uniform conditions.

However, they are not a suitable solution for building- and vehicle-integrated applications. In building applications both static objects (like trees or chimneys) and unpredictable events (like bird dropping, people walking) often cast time- and spatially-varying shadows on the PV modules. This motivates the definition of adapted PV module topologies able to increase the power generation especially under non-uniform conditions. To deal with this problem, both static [1] and dynamically reconfigurable topologies [2]-[3] have been proposed. At present, it is however not possible to compare these alternative proposals in a correct way. Current approaches are either unacceptably slow (but accurate) or compromise the accuracy to become fast. Furthermore, either highly accurate models are used but only static and near-static shading scenarios are considered [4], or the shading scenarios are accurate, dynamic and realistic but the PV modules' models are too crude and exhibit a too high error margin to perform the comparison with enough accuracy [5]. To fill this gap, this paper presents an accurate and versatile simulation methodology to compare the energy production of different PV module topologies subjected to realistic dynamic shading conditions, with acceptable simulation time and based on a bottom-up modeling approach. This will strongly support PV system designers to identify and to select the best PV module topologies to use in mixed module type arrays as a function of the typical shading scenarios they will be subjected to during their lifetime.

## II. PROPOSED SIMULATION APPROACH

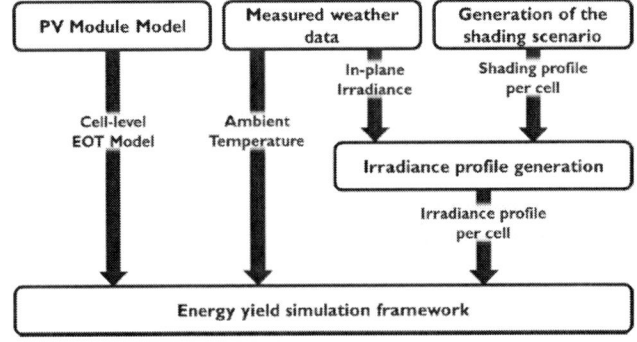

Fig. 1. Block diagram of the simulation approach

978-1-5090-5606-4/17 $31.00 © 2017 IEEE

The block diagram of the proposed simulation approach is shown in Fig. 1. It is built on top of the physics-based energy yield simulation framework presented in [6], that allows for SPICE-level simulation of PV modules using an accurate Electrical-Optical-Thermal (EOT) model of the PV modules themselves constructed in Verilog-AMS; but it significantly expands on this foundation also, as explained further on. Such a model considers the optical, thermal and electrical behavior of each individual cell, as well as both thermal and electrical interactions among neighboring cells. Measured environmental conditions are used as input of the modelling approach. In order to use this energy yield simulation framework, it is necessary to input (a) the adapted EOT model of the different topologies and (b) the environmental conditions every cell is subjected to.

*A. PV module topologies*

In the present work, we compare four different module topologies, representative of promising types of schemes:

- Conventional PV module (60 cells, 3 bypass diodes);
- Half-cell PV module (120 halved cells, 3 bypass diodes) based on the TwinPeak topology proposed by REC [1];
- I-type Snake Reconfigurable PV module [3] (60 cells);
- U-type Snake Reconfigurable PV module [4] (60 cells).

For instance, Fig. 2 shows the layout of the I-type Snake Reconfigurable PV module. Switches connecting neighboring cell-strings ( groups of 6 cells statically connected), are used for the all-in-series connection working under uniform conditions. Parallel connection of cell-strings is performed when the module is subjected to non-uniform conditions. Four local converters (red boxes) are employed to perform the Maximum Power Point Tracking (MPPT) and to reduce the overall current available on both the BUS line and the GROUND line when the cell-strings are connected in parallel. An additional converter (blue box), also called module converter, is used at the end of the module to boost the output voltage of the local converters for connection with the rest of the PV array.

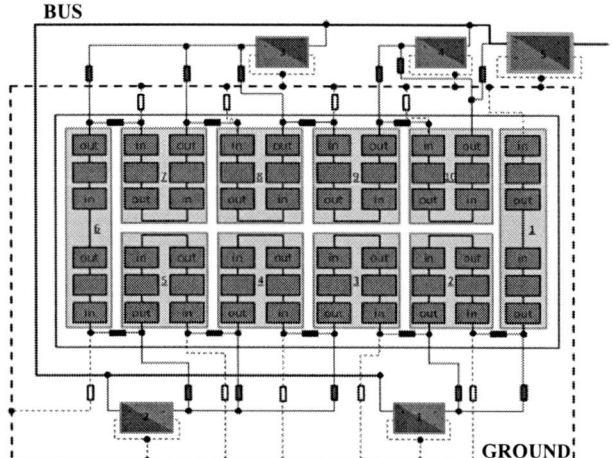

Fig. 2.    I-type reconfigurable Snake PV module

For the calibration of the EOT model we prepared full and half-cell glass-backsheet (white) laminates with 156 mm, 19.2% efficiency mono-Si Al BSF solar cells [7]. To complete the PV modules' models, the resistance of every interconnection has been calculated and models of all the additional components, e.g. bypass diode and reconfiguration switches, have been included. Table I and II summarize the parameters used to complete the PV module's model and to perform the simulations.

TABLE I
SIMULATION PARAMETERS – SWITCHES AND CONVERTERS

| Parameter | Value |
|---|---|
| Switch ON resistance (mΩ) | 3.6 |
| Fixed local converter efficiency (%) | 95 |
| Fixed module converter voltage $V_M$ (V) | 12 |

TABLE II
SIMULATION PARAMETERS – INTERCONNECTIONS AND CELLS

| Parameter | Value |
|---|---|
| Interconnection material | Copper |
| BUS line – width (mm) | 5 |
| BUS line – thickness (mm) | 0.5 |
| GROUND line – width (mm) | 10 |
| GROUND line – thickness (mm) | 0.5 |
| Other connections – width (mm) | 4 |
| Other connections – thickness (mm) | 0.3 |
| Cell spacing (mm) | 2 |

Realistic assumptions are made for both local and module converters. In contrast to a conventional module convertor, in our proposed setup it works in two different modes: during the all-in-series connection it performs the module-level MPPT, whereas during parallel connection its input is fixed to a given voltage $V_M$. The local converters instead are disconnected during the all-in-series connection, whereas they perform a local MPPT during parallel connection. A fixed efficiency of local converters has been also considered.

It is worth to note that different converter behaviors, e.g. local converters with a fixed conversion ratio and MPPT always performed by the module converter, can be easily implemented in the proposed simulation framework. Also, variable efficiency profiles may be included into the converters' models.

*B. Weather data*

To perform realistic simulations, real-world environmental data are used in the proposed simulation framework. Ambient temperature and in-plane irradiance, measured in any location, can be used as input. Also, the energy yield simulation framework [6] supports wind direction and speed as inputs. Concerning the simulation results presented in this paper, irradiance and ambient temperature data from a weather monitoring station located in Oldenburg (Germany) [8] have been used. The measurement setup of this monitoring station measures and saves data with 1 s time resolution.

## C. Generation of the shading scenario

The shading scenarios have been generated using the software SketchUp© [9]. It allows drawing oriented 3D objects, e.g. buildings and PV modules, and casting realistic shadows.

The casted shadows are accounting for the dimensions and the relative position of the different objects in the 3D model also considering the geolocation and the date as well as a specific time or timeframe for the simulation, as shown in Fig. 3(a) for one of the simulated shading scenarios. Each PV module has been modeled according to the parameters of Table II, e.g. cells are 2 mm spaced. Also, they are composed of 60 solar cells [7] organized in a 10x6 matrix, as shown in Fig. 3(b). The only exception is the Half-Cell PV module: since it is based on the TwinPeak topology proposed by REC [1], its model is made of 120 halved cells [7] organized in a 20x6 matrix.

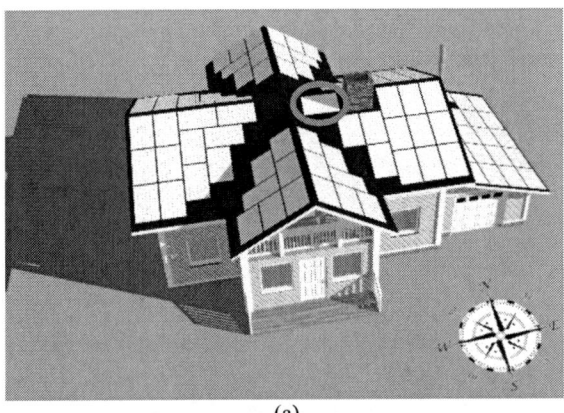

(a)

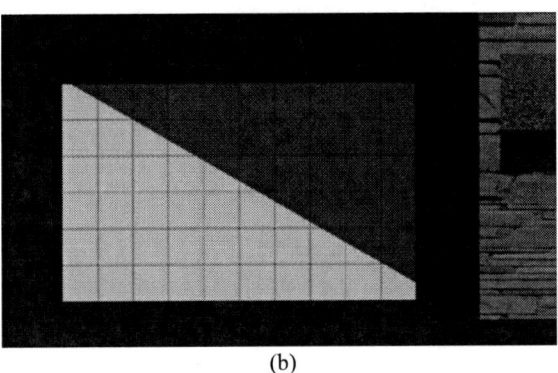

(b)

Fig. 3. Example of SketchUp simulation. Location: Oldenburg. Date: September 17th, 2014. Time: 09:44:20.

## D. From shading scenario to irradiance profile per cell

From SketchUp a sequence of pictures representing the state of the simulation at a specific time is exported. One of these pictures is shown in Fig. 3(b) for the PV module located on the left of the chimney, highlighted by a red circle in Fig. 3(a). Post-processing of such pictures has been done to evaluate the percentage of shading per cell. Although squared cells are used in the SketchUp PV module's model, it is possible to use the same simulation results (set of pictures) also for cells with a

different shape, providing that the cell length is preserved. Indeed, different cell's shapes can be taken into account during the post-processing phase. For instance, a mask removing the corners can be applied to each cell if the simulated module is made of pseudosquare cells, like the ones that are considered in this work [7].

The overall information extracted from post-processing the sequence of SketchUp's pictures allows the evaluation of the shading profiles $S_N(t)$ per cell: for instance, $S_N(t_1) = 0.4$ means that 40% of the area of the $N$-th cell is shaded at the time instant $t_1$. The time resolution of the shading profile has been chosen equal to 10 seconds.

Given the measured global in-plane irradiance $G(t)$, the global in-plane irradiance for each cell $N$ is evaluated as

$$G_N(t) = G(t) \cdot (1 - S_N(t)) \tag{1}$$

It is worth to note that the use of Eq. (1) implies that:
- The shade is assumed to be 100% dense;
- The effect of the diffused irradiance on the shaded part of a cell is neglected.

However, a different shade density as well as the effect of the diffused irradiance on the shaded part of the cells can and will be easily implemented into the post-processing step. This shows the flexibility of the proposed approach and it allows to maintain the high accuracy we envision.

## III. SIMULATION RESULTS

Different representative shading scenarios have been simulated using the approach presented above. Here we focus on the comparison of the different module topologies for only one of them, notably the module on the left of the chimney (see Fig. 3) considering the house located in Oldenburg on September the 17th, 2014. It is namely not the purpose of this paper to compare the different topologies, but only to demonstrate the effectiveness and versatility of our proposed simulation framework. The in-plane irradiance and temperature profiles are shown in Fig. 4.

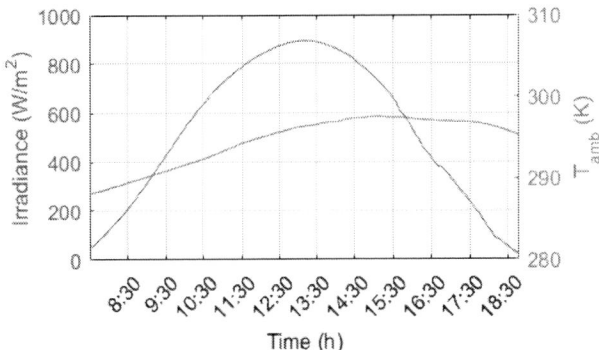

Fig. 4. In-plane Irradiance (blue) and Temperature (orange) profiles. Location: Oldenburg. Date: September 17th, 2014.

In this operating scenario, the PV module is fully shaded from sunrise until 7:45. Then, it is subjected to partial shading until 12:30. Afterwards, it works under uniform conditions until 17:10. Finally, it is partially shaded till sunset. The simulation results are shown in Fig. 5 and Table III in terms of generated power and energy respectively. Half-Cell module is performing better than all the other modules under uniform conditions, thanks to its lower current and hence reduced series resistance losses. Indeed, the Half-Cell module gains +1.22% more energy than the conventional one. Furthermore, Half-Cell module with a bottom and top part in parallel (e.g. TwinPeak topology implemented by REC) performs also better than the conventional one when subjected to partial shading, allowing for +43.35% additional energy gain.

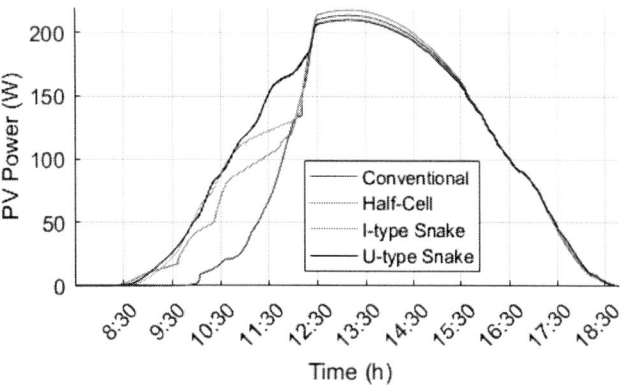

Fig. 5.     Simulation results: power profiles.

Reconfigurable module topologies on the other hand have additional series losses under uniform conditions with respect to both conventional and Half-Cell TwinPeak topologies, mainly because of the reconfiguration switches. In traditional simulators using higher level abstractions that are black box based, these different losses cannot be modeled and compared so accurately. Thus, that represents a major asset of our proposed simulation approach and the prototype software framework we have built. When compared to the conventional module, the energy losses under uniform conditions are -1.12% for the I-Snake topology and -1.50% for the U-Snake one. However, they allow for a significant gain in terms of energy generation under non-uniform conditions (+60.60% for the I-Snake and +77.24% for the U-Snake with respect to the conventional module). Considering the performance across the whole day, reconfigurable Snake topologies allow for a higher total energy production than both Conventional and Half-Cell modules. Notably, the U-Snake topology leads to the highest daily energy generation (+17.38% higher than the conventional module).

It is worth to note that the same trend is observed for all the simulated shading scenarios (though not reported here), with the U-Snake module outperforming the other topologies on a daily basis.

## TABLE III
### SIMULATION RESULTS: ENERGY GENERATION (WH)

|  | Uniform | Non-Uniform | Daily |
|---|---|---|---|
| **Conventional** | 760 | 240 | 1000 |
| **Half-Cell** | 769 | 343 | 1112 |
| **I-Snake** | 751 | 385 | 1136 |
| **U-Snake** | 748 | 425 | 1173 |

## IV. CONCLUSION

In this work, a simulation approach for comparing different PV module topologies subjected to realistic dynamic shading scenarios has been presented. The method combines meteorological data with an accurate EOT model of different PV module topologies and real-world shading patterns simulated by means of the software SketchUp. Thanks to the geolocation and date/time features of the latter, it is possible to match the measurements with the SketchUp simulations, thus making the proposed approach a valuable tool for the exploration of different module topologies and the comparison of their performance under real-life dynamic shading scenarios. The proposed simulation approach is an indispensable input for the design of mixed module type arrays, allowing to select the best matching PV module topology to use as a function of the typical shading scenario it will be subjected to during its lifetime.

First simulations have been performed that substantiate the feasibility of the methodology and shows that, in real-life strongly non-uniform and dynamic conditions, the Snake reconfigurable modules may outperform both conventional and Half-Cell TwinPeak topologies.

## V. ACKNOWLEDGEMENT

The authors gratefully acknowledge the financial support of the partners in the IMEC Si PV Industrial Affiliation Program, and the University of Oldenburg for providing the weather data.

## REFERENCES

[1] http://www.recgroup.com/sites/default/files/documents/whitepaper_twinpeak_shading_properties_eng.pdf

[2] G. Spagnuolo, G. Petrone, B. Lehman, C.A. Ramos Paja, Ye Zhao, and M.L. Orozco Gutierrez, "Control of Photovoltaic Arrays: Dynamical Reconfiguration for Fighting Mismatched Conditions and Meeting Load Requests". *IEEE Industrial Electronics Magazine*, vol. 9, no. 1, pp. 62–76, March 2015.

[3] M. Baka, F. Catthoor, D. Soudris, "Smart PV Module Topology with a Snake-Like Configuration", *31st European Photovoltaic Solar Energy Conference and Exhibition*, 2015, pp. 104-107.

[4] M. Baka, F. Catthoor, and D. Soudris. "Near-Static Shading Exploration for Smart Photovoltaic Module Topologies Based on Snake-like Configurations", *ACM Transactions on Embedded Computing Systems*, vol. 15, no. 2, March 2016.

[5] B. Lefevre, S. Peeters, J. Poortmans, and J. Driesen, "Predetermined static configurations of a partially shaded

photovoltaic module", *Progress in Photovoltaics: Research and Applications*, vol. 25, no. 2, pp. 149-160, February 2017.

[6] H. Goverde, B. Herteleer, D. Anagnostos, G. Köse, D. Goossens, B. Aldaladi, J. Govaerts, K. Baert, F. Catthoor, J. Driesen and J. Poortmans, "Energy yield prediction model for PV modules including spatial and temporal effects," *29th European Photovoltaic Solar Energy Conference and Exhibition*, 2014, pp. 3292-3296.

[7] http://www.tsecpv.com/upload/website/download/TSS62TN_EC ELL.pdf

[8] http://www.uni-oldenburg.de/en/physics/research/ehf/energiemeteorology/meas urements/

[9] http://www.sketchup.com

# A Performance Model for Bifacial PV Modules

Daniel Riley[1], Clifford Hansen[1], Joshua Stein[1], Matthew Lave[1], Johnson Kallickal[1], Bill Marion[2], Fatima Toor[3]

[1]Sandia National Laboratories, Albuquerque, NM, USA
[2]National Renewable Energy Laboratory, Golden, CO, USA
[3]The University of Iowa, Iowa City, IA, USA

*Abstract* — Sandia National Laboratories, the National Renewable Energy Laboratory, and the University of Iowa are collaborating to develop a performance model for bifacial PV modules. As with monofacial PV modules, a bifacial PV model consists of sequential operation of the component sub-models. Bifacial PV modules accept light on both their front and rear surfaces which presents a unique modeling challenge. This paper describes the approach of Sandia, NREL, and the University of Iowa to create a bifacial PV model and verify its accuracy with measured field data.

*Index Terms* — photovoltaic modules, bifacial, model, performance

## I. Introduction

Bifacial photovoltaic (PV) modules can accept light on both the front and rear surfaces. Currently, efforts are being put forth to describe, test, rate, and model bifacial PV modules. As bifacial PV becomes a larger portion of the overall PV market. Sandia National Laboratories (Sandia), the National Renewable Energy Laboratory (NREL) and the University of Iowa are working together to provide the necessary data and analysis to better describe and predict bifacial PV performance to facilitate wider adoption throughout the PV market.

Sandia is a leader in the photovoltaic (PV) modeling community in both the development and evaluation of PV performance models. Sandia is leading this effort to develop performance models for bifacial PV modules and systems, with the ultimate goal of creating a highly accurate and computationally convenient model to predict the output of a bifacial PV system under any weather conditions. In addition to the weather data, system information must also be provided to the model.

This paper describes the progress of Sandia's efforts towards an accurate bifacial PV performance model.

## II. Modeling Approach

The overall approach for modeling bifacial PV systems comprises a sequence of sub-models similar to the approach employed for modeling monofacial PV. Initial models for irradiance translation and/or decomposition are followed by models describing the effects of shading, irradiance spectrum, and soiling. PV cell temperature is estimated using a cell temperature model; cell temperature and incident irradiance are then used within the PV module performance model. Effects of mismatched output among cells and modules may be calculated in order to estimate the DC output of an entire PV string.

### A. Front and Rear Irradiance

The irradiance available for the front surface of bifacial PV modules is modeled in the same manner as for monofacial PV. Direct and diffuse irradiance are transposed onto a tilted surface, and the site albedo can be used to estimate the amount of ground-reflected light present on the front surface.

Determining the irradiance on the rear side of a PV module, however, is a relatively new area. The rear side irradiance will be affected by the albedo of the ground and nearby surfaces and by the irradiance on these surfaces. However, additional factors such as the module's position within a row, the row's height above ground, the proximity of the row to other rows or structures, the transparency of the module [1], and the shadow pattern on the ground alter the amount of irradiance available to the rear side of a bifacial PV module. Several efforts are underway to develop and validate rear-side irradiance models [2]. Sandia is developing a view factor model for irradiance on each cell that accounts for 3D geometry [3]; NREL [4] and SunPower [5] are proposing array-scale models with 2D geometry for fixed and single axis tracking systems, respectively; and Univ. of Iowa is pursuing models using ray-tracing that account for row-to-row effects [6]. Selecting the best method for determining rear irradiance may depend on the use case and requirements for computation time, accuracy, and resolution.

### B. Soiling, Shading, Spectrum

We expect that the effects of soiling on bifacial PV will be similar to those observed for monofacial PV, or perhaps reduced somewhat when bifacial PV modules are mounted so that little soil accumulates on the rear side of the module.

PV racking, junction boxes, or module wiring may shade parts of the rear-side of a bifacial PV module from irradiance sources. As we expect most of the irradiance on the rear-side of a bifacial PV module (in a typical mounting orientation to monofacial modules) to comprise diffuse reflected and sky diffuse irradiance, we anticipate that shading on the rear side will reduce rear-side irradiance in proportion to the shaded area relative to the module area.

978-1-5090-5606-4/17 $31.00 © 2017 IEEE

The spectrum of the incident light on the front side affects the module's output, and the effect of spectrum changes can be estimated with existing models, e.g., [7, 8]. Spectral irradiance on the rear side can differ from irradiance on the front side due to absorption at the surfaces reflecting the light reaching the rear-side. We suspect that the effect on module output from the variation in irradiance spectrum on the rear surface is small compared to the effect of irradiance spectrum on the front surface irradiance. In the future, spectral reflection data for various surfaces can be analyzed to quantify the impact of variation in rear side spectral irradiance.

### C. Cell Temperature

For monofacial PV system models, cell temperature is generally modeled from the incident irradiance, the ambient air temperature, the wind speed and possibly other factors such as wind direction, cell location, or module efficiency, e.g., [7]. We expect that bifacial PV modules will have similar thermal characteristics as monofacial PV, although a rear-irradiance term may need to be included in the factors which affect temperature.

### D. Electrical Performance

The electrical output of a PV module for given irradiance, cell temperature, and other factors is described by the module's I-V curve. An electrical performance model can model the I-V curve or, at least, important points on the I-V curve such as $I_{SC}$, $V_{OC}$, $I_{MP}$, and $V_{MP}$. We hope that electrical performance models for bifacial PV modules will resemble those used for monofacial modules, e.g., a diode-model capable of recreating a whole I-V curve [9], or a point-value model such as the Sandia Array Performance Model (SAPM) [7] that estimates a few important points on the I-V curve. However, as we show in this paper, bifacial modules may inherently exhibit characteristics of shaded monofacial modules due to shading from nearby structure such as racking, which complicates PV power modeling.

### E. Cell and Module Mismatch

Mismatched output of cells within a module, and of modules in a string, is caused primarily by differing irradiance conditions along the string, and secondarily by intrinsic differences among the cells. When the cells in a string produce differing levels of current the performance of the string suffers due to mismatch. In monofacial PV systems, the level of current mismatch between modules must be quite high in order to have a significant effect on the power of the string. While the same underlying principles hold for bifacial modules, the possible variation in irradiance among cells is affected by variation of rear-side irradiance across the cells, which can be substantial [2].

## III. VERIFICATION AND ANALYSIS

Sandia is generating data sets and analysis tools to verify some of the sub-models comprising the bifacial PV performance model, demonstrate the applicability of existing monofacial PV models where bifacial PV performance is not significantly different, and develop new bifacial performance models where existing monofacial models fail to predict bifacial PV performance.

Sandia and NREL have each constructed test arrays and equipment to provide data that will be used to verify the models, such as Sandia's 4-tilt bifacial test shown in Fig. 1. The test bed contains both bifacial and monofacial PV modules at a range of tilt angles for a direct comparison of the performance of monofacial and bifacial technologies.

Fig. 1. Sandia's 4-tilt test bed. Rows are tilted between 15 and 45 degrees with rear irradiance sensors at the top and bottom of each row, near the center of the row.

### A. Front and Rear Irradiance

The rear irradiance models developed by NREL, University of Iowa, and Sandia National Laboratories are presented in other papers [3, 4, 6].

Empirical data to validate these models are being collected on several test arrays including fixed tilt, single axis trackers, and dual axis trackers. For models that allow for prediction of different illumination across the rear of a PV module, Sandia has created the high spatial resolution rear irradiance module (HSRRIM) shown in Fig 2. The module has a form factor similar to a PV module that can be easily moved and placed anywhere within an array or around the SNL site. The HSRRIM has 10 calibrated reference cells which measure the rear side irradiance pattern across the back of the module.

Rear irradiance readings from the HSRRIM show that the distribution of unobstructed rear irradiances across the back of a bifacial PV module can vary greatly depending on the module's height above ground, tilt angle, and the surface below the module. An installation at latitude tilt with the HSRRIM 0.6 meters above the ground can produce variations in rear irradiance above 50 W/m$^2$ between cells on a sunny day.

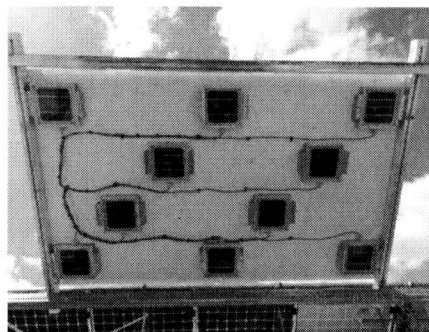

Fig. 2. Sandia's high spatial resolution rear irradiance module. A sensor with PV module form factor and ten reference cells measuring rear irradiance

### B. Rear Side Shading

To empirically determine the effect of rear-side shading, we performed a series of tests where a percentage of the rear side of a bifacial module was obscured, an I-V curve was measured, then the obstruction was removed and a second I-V curve was measured. The two I-V curves are separated by less than 1 minute such that the front and rear irradiance varied by less than 0.25% between curves. The obstruction's size, orientation, location, and distance from module were varied and the effect on the I-V curve was noted. For the following plots, the obstruction's orientation was either in "direction A" which crossed all of module's cell strings, or in "direction B" which placed the obstruction behind only one cell string. When the obstruction is directly against the back of the module this is denoted as "hard shade", and when the obstruction was approximately 5.9 cm from the back of the module is denoted as "soft shade". Fig. 3 shows an example of a 20% hard shade obstruction in direction B, and Fig. 4 shows an I-V curve with the obstruction and an I-V curve after the obstruction was removed.

Fig. 3. The obstruction behind a bifacial PV module in "direction B" with 20% coverage. Obstruction is in contact with the rear surface of the PV.

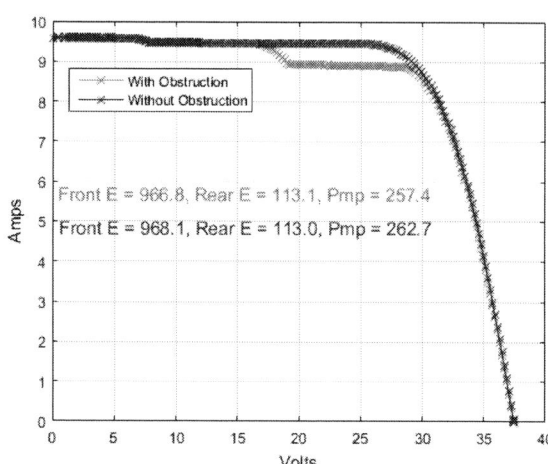

Fig. 4. An I-V curve of a bifacial module with (red) and without (blue) an obstruction. $P_{MP}$ difference ≈2%.

The object of the testing is to formulate a model to predict the impact on PV performance of various factors relating to obstructions on the rear side of the PV module, such as racking. We found that obstructions can greatly affect the $I_{SC}$ of the PV in a complicated manner. Fig. 5 shows the change in $I_{SC}$ related to an obstruction's orientation, size, and distance from the PV. It is clear that the $I_{SC}$ is greatly influenced by the orientation of the obstruction, i.e. whether it covers one cell string or all cell strings. However, Fig. 6 shows that $P_{MP}$ is generally reduced in proportion to the obstructed area, and that the orientation of the obstruction does not greatly affect the $P_{MP}$.

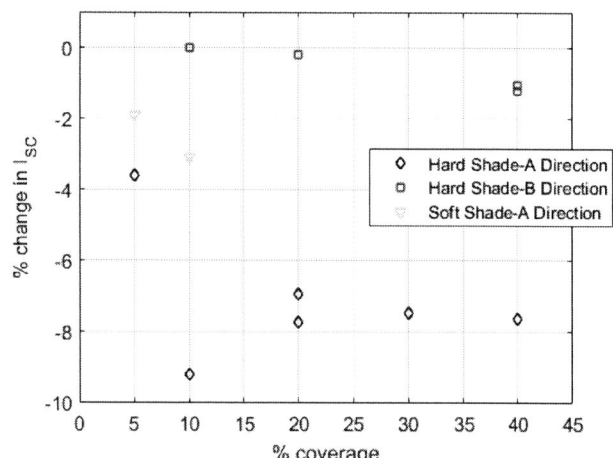

Fig. 5. $I_{SC}$ changes from without obstruction to obstruction. Great changes in $I_{SC}$ depending on the direction of the obstruction.

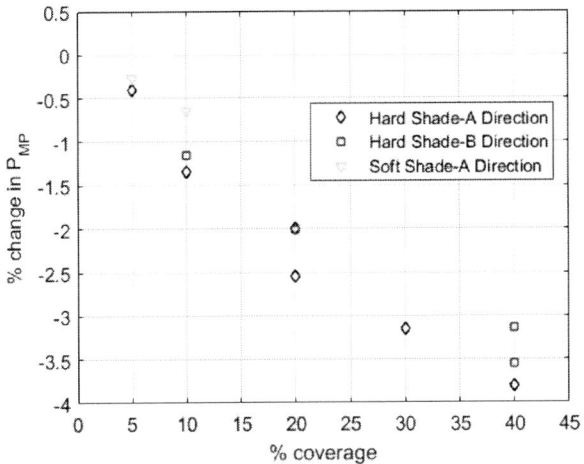

Fig. 6. $P_{MP}$ changes from without obstruction to obstruction as a function of the size of the obstruction relative to the module size.

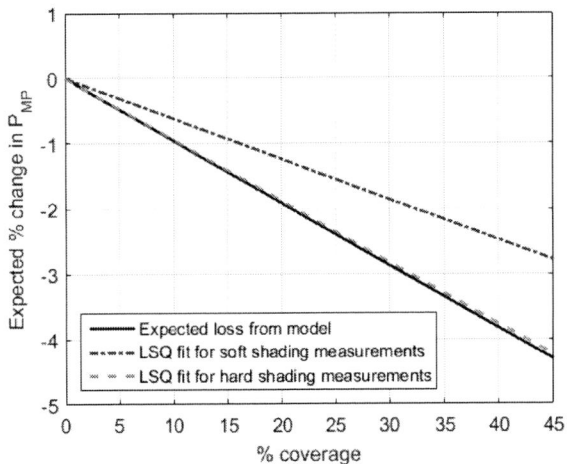

Fig. 7. $P_{MP}$ losses estimated with (1) and losses observed for "hard shade" and "soft shade".

It is clear that when obstructions block the rear side of a PV module, it is impractical to use $I_{SC}$ as an estimator of the amount of power a bifacial PV module may create since losses in $I_{SC}$ and $P_{MP}$ are not well correlated. For this reason we propose a model to estimate the effect of obstructions on $P_{MP}$, rather than predict effects on $I_{SC}$ and then use $I_{SC}$ to predict $P_{MP}$ (as is typical in some monofacial models, e.g. [7]). A secondary factor seems to be the distance of the obstruction from the rear of the PV module.

If we assume that the irradiance striking the rear side of a bifacial PV module is proportional to the power generated from that irradiance, we develop (1) to estimate the reduction in $P_{MP}$ due to some amount of obstruction.

$$D_{Obs} = -E_{rear} \times BiFi_{Pmp} \times CoverageRatio \qquad (1)$$

where $D_{Obs}$ is the reduction in $P_{MP}$ (unitless), $E_{front}$ and $E_{rear}$ are the irradiance on the front and rear, $BiFi_{Pmp}$ is the bifaciality (the ratio of power produced by the rear side to the front side under identical illumination and temperature) measured as in [1], and $CoverageRatio$ is the ratio of the obstruction size to the module's active area.

Fig. 7 shows the result of using (1) compared to a least squares fit of different measured data points. The model equation estimates slightly more loss than is observed in a "hard shade" scenario. We believe that a simple scaling factor may be applied to (1) to account for an obstruction's distance from the rear of the PV module. Additional testing at various obstruction distances is required in order to determine an appropriate scale factor.

## C. Cell and Module Temperature

Module temperature models such as in [7] estimate a module temperature from environmental factors and a module's thermal characteristics. We have collected module temperature and environmental data from a number of co-located bifacial and monofacial systems in order to determine the applicability of existing temperature models to bifacial PV.

Equation (2) presents one simple model that has proven adequate at estimating steady state PV module temperatures:

$$T_{Module} = E \cdot \left[ e^{a+b \cdot WS} \right] + T_{ambient} \qquad (2)$$

Fig. 8 shows the results of modeling a bifacial PV system temperature with (2). Residual plots of the errors for a co-located monofacial system exhibit nearly identical features and residual magnitudes. It is clear that the model is descriptive of this bifacial PV system which we regard as typical; the RMSE of the model for the bifacial system is about 3.4 °C, while the RMSE for a co-located monofacial system is 3.1 °C.

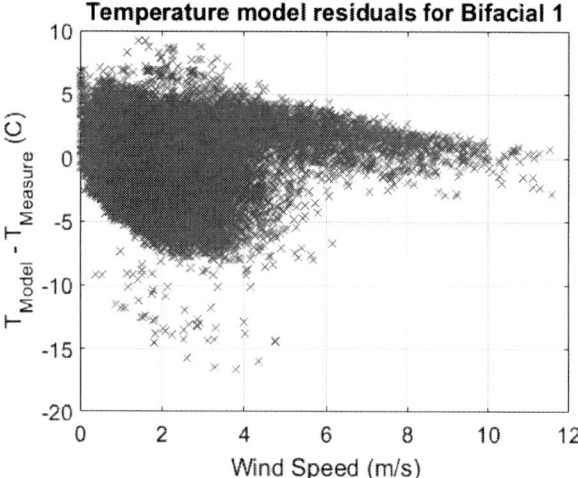

Fig. 8. Module temperature residuals as a function of wind speed for a bifacial system.

The accumulated system and module data has provided us with confidence that a simple model such as (2) will be sufficient for modeling bifacial module temperature as a function of irradiance, wind speed, and ambient temperature; even when using the typical parameters determined from monofacial modules.

### D. Electrical performance

We propose a simple model for bifacial PV power, leaving a more complicated model for other points of interest, e.g., $V_{MP}$, to future work. We estimate maximum power for a bifacial module from the front irradiance, rear irradiance, cell temperature, module bifaciality, module temperature coefficient, and the nominal power of the PV module under STC conditions.

$$
\begin{aligned}
P_{MP} = P_{MP0} &\times \left[ 1 + \gamma_{Pmp} \times (T_{Cell} - T_0) \right] \\
&\times \frac{E_f + Bifi_{Pmp} \times E_r}{E_0} \\
&\times (1 + D_{obs})
\end{aligned}
\tag{3}
$$

where:

$P_{MP0}$ is the maximum power at reference conditions
$\gamma_{Pmp}$ is the maximum power temperature coefficient
$T_{Cell}$ is the cell temperature
$T_0$ is the reference cell temperature
$E_f$ is the average front side irradiance
$E_r$ is the average rear side irradiance
$E_0$ is the reference irradiance
$Bifi_{Pmp}$ is the bifaciality of the PV module's $P_{MP}$ as in [1]
$D_{obs}$ is a derate (unitless) based on the amount of rear side obstruction as shown in (1)

Eq. 3 is ignores smaller effects such as from spectrum variation and mismatch. These smaller effects may be incorporated in the future.

Fig. 9 compares the modeled and measured maximum power for one single bifacial PV module shown in Fig. 10. The front irradiance is measured by a coplanar reference cell, and the rear irradiance is estimated by averaging two reference cells adjacent to the module's western edge. The simple model predicts the module power to within 5 watts of the measured power of a single bifacial PV module with no rear obstructions when the module is producing less than 200 watts. However, the errors increase to about 20 watts at higher power levels.

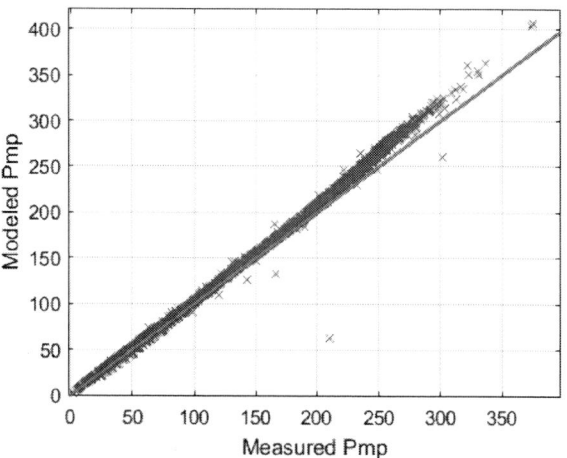

Fig. 9. Modeled and measured maximum power from (3) for a single bifacial PV module over 25 days, 1:1 line shown in red

Fig. 10. Bifacial PV module under test circled in red. Rear irradiance reference cells shown with green triangles

Analysis of the model residuals shows that the model is strongly over-predicting at midday when module temperatures are high as shown in Fig. 11. The residuals may indicate an error in the temperature coefficient for power; however, the nonlinear behavior of the residuals indicates that there may be several physical causes for the model errors.

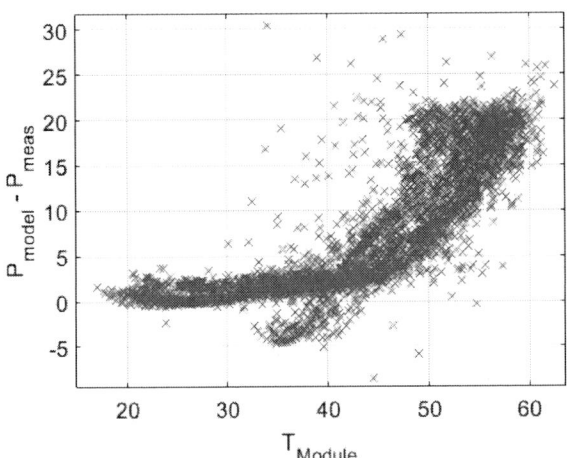

Fig. 11. Bifacial PV model residuals show an over-prediction at high temperatures

## IV. CONCLUSION AND FUTURE WORK

Existing models for determination of cell and module temperature of monofacial PV modules have been confirmed to also provide good results for bifacial PV modules and we recommend their further use in bifacial PV modeling.

A new derate factor may be needed for bifacial PV systems to account for reduced rear side irradiance due to racking or other obstructions. A simple factor based on the coverage ratio of the rear side of the module could be sufficient for modeling power loss due to obstructions.

Finally, the simple power model for bifacial modules presented in (3) provides errors of approximately 2% at low power levels, with increasing errors up to about 8% at high power levels. It is clear that (3) is not adequately capturing all of the physical effects that influence bifacial PV module performance, and additional factors must be identified or improved in order to develop a more accurate model.

## V. ACKNOWLEDGEMENTS

Sandia National Laboratories is a multi-mission laboratory managed and operated by National Technology & Engineering Solutions of Sandia, LLC., a wholly owned subsidiary of Honeywell International, Inc., for the U.S. Department of Energy's National Nuclear Security Administration under contract DE-NA0003525.

## REFERENCES

[1] C. Deline, S. MacAlpine, B. Marion, F. Toor, A. Asgharzadeh, and J. S. Stein, "Evaluation and field assessment of bifacial photovoltaic module power rating methodologies," in *2016 IEEE 43rd Photovoltaic Specialists Conference (PVSC)*, 2016, pp. 3698-3703.

[2] C. W. Hansen, J. S. Stein, C. Deline, S. MacAlpine, B. Marion, A. Asgharzadeh, *et al.*, "Analysis of irradiance models for bifacial PV modules," in *2016 IEEE 43rd Photovoltaic Specialists Conference (PVSC)*, 2016, pp. 0138-0143.

[3] C. W. Hansen, D. M. Riley, "A Computationally Efficient Method for Detailed Modeling of Rear-Side Irradiance for Bifacial PV Modules," in *2017 IEEE 44th Photovoltaic Specialists Conference (PVSC)*, 2017.

[4] B. Marion, S. MacAlpine, C. Deline, A. Asgharzadeh, F. Toor, D. Riley, J. Stein, C. Hansen, "An Irradiance Model for Bifacial PV Modules," submitted to *2017 IEEE 44th Photovoltaic Specialists Conference (PVSC)*, 2017.

[5] M. A. Anoma, J. Scholl, B. Bourne, D. Jacob, "2D View Factor Model and Vlaidation for Bifacial PV and Diffuse Shade on Single-Axis Trackers," in *2017 IEEE 44th Photovoltaic Specialists Conference (PVSC)*, 2017.

[6] A. Asgharzadeh, T. Lubenow, J. Sink, B. Marion, *et al.*, "Analysis of the Impact of Installation Parameters and System Size on Bifacial Gain and Energy Yield of PV Systems," in *2017 IEEE 44th Photovoltaic Specialists Conference (PVSC), 2017*

[7] D. King, W. Boyson, and J. Kratochvil, "Photovoltaic Array Performance Model". SAND2004-3535, 2004.

[8] M. Lee, A. Panchula, "Spectral Correction for Photovoltaic Module Performance Based on Air Mass and Precipitable Water," in *2016 IEEE 43rd Photovoltaic Specialists Conference (PVSC)*, 2016.

[9] W. De Soto, S. A. Klein, W. A. Beckman, "Improvement and Validation of a Model for Photovoltaic Array Performance," *Solar Energy*, vol. 80, pp. 78-88, 2006.

# Accurate Modeling of Partially Shaded PV Arrays

Bennet Meyers* Mark Mikofski[†]

*Department of Electrical Engineering, Stanford University, Palo Alto, CA, 94305, USA
[†]SunPower Corporation, Richmond, CA 94804, USA

*Abstract*—Partial shading is the condition of nearby objects casting shade onto part of a photovoltaic (PV) array, causing the PV modules to receive non-uniform irradiance. Non-uniform shading causes electrical mismatch between elements within the array, resulting in a non-linear reduction in energy capture. Accurately modeling mismatch conditions is a particularly difficult problem due to the large number of parameters needed to fully define a PV system and its operating state. Furthermore, the large number of possible system state conditions make the models computationally complex. Previous work on this topic has addressed these difficulties by simplifying the system representation, reducing the number of parameters used, and limiting the domain space to simple irradiance patterns that may not be representative of real shade conditions. In this paper, we review common modeling approaches to address this problem and provide an overview of PV equivalent circuit theory. We then present PVMismatch—free open-source software written in Python by the authors for simulating full PV system current-voltage curves. Finally, we demonstrate an improvement over common practice for modeling this type of behavior, illustrating that modeling mismatch behavior at the PV cell rather than PV module level provides more accurate results—up to 30% less over-prediction with respect to module level estimates—and better insight into system behavior.

*Index Terms*—photovoltaic systems, modeling, partial shade modeling. electrical mismatch

## I. INTRODUCTION

Partial shading of photovoltaic arrays is well-known to cause significant reduction in system performance. According to the California Energy Commission, "Shading of photovoltaic systems, even partial shading of arrays, can be the most important cause of failure to achieve high system performance" [1]. A significant body of research has been developed around modeling the performance of partially shaded PV systems [2]–[7]. These models are being used by researchers and industry professionals to evaluate novel power electronics solutions, the feasibility of mitigating shade losses through non-standard wiring topologies, and the impact of shade on yearly energy estimates. Typically, researchers utilize an equivalent circuit model to describe the I-V characteristics of individual PV modules within the array. This allows researchers to predict the power lost to current mismatch between series-connected modules and voltage mismatch between parallel-connected strings.

However, this approach ignores electrical mismatch that occurs between PV cells within the modules and does not address the impact of bypass diodes. This paper will show that there can be significant errors introduced by ignoring these effects. Using a module-level equivalent circuit model does not adequately differentiate between the performance under

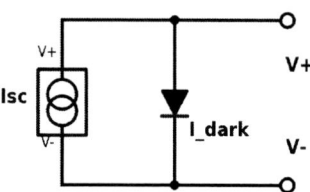

Fig. 1. The simplest equivalent circuit model: a current source in parallel with an ideal diode.

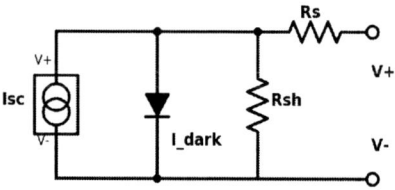

Fig. 2. A one-diode equivalent circuit model with lossy components.

different shade geometries or the impacts of new module-level designs such as cross-tied cells and different diode configurations. Additionally, using a module-level approach tends to underestimate the loss in power overall.

Previous studies have also been limited to simple distributions of irradiance within an array, such as 3 to 5 irradiance levels oriented in simple rows and columns (see [3]). A more complex study such as [4] relies simply on random patterns of irradiance to generate a dataset of 30 scenarios.

This paper demonstrates the open-source software package PVMismatch, developed by the authors. The theoretical basis of the software is presented. Using this software, the impact of different shading geometries on a representative system is evaluated, using both the cell-level model and a module-level model, and the results of the two modeling approaches are compared.

## II. EQUIVALENT CIRCUIT MODELS

### A. One-Diode Models

The current-voltage (IV) response of a PV cell can approximated by the superposition of a short-circuit photocurrent and a "dark current," the current response of the cell under an applied voltage and no illumination [8]. In other words, an ideal photovoltaic cell is electrically equivalent to a current generator in parallel with a non-linear resistive element such as a diode, as shown in Fig. 1. When the cell is illuminated, it

978-1-5090-5606-4/17 $31.00 © 2017 IEEE     3354

produces a photocurrent proportional the light intensity, which is then split between the diode and the load. As the load resistance increases, more of the current flows through the diode, resulting in a larger terminal voltage at the load. This behavior is described by the following equation:

$$I = I_{sc} - I_0 \left( e^{qV/k_B T} - 1 \right) \qquad (1)$$

where $I$ is the terminal current of the cell, $I_{sc}$ is the short-circuit current under a given illumination, $I_0$ is a constant known as the saturation current, $q$ is the elementary charge, $k_B$ is the Boltzmann constant, and $V$ and $T$ are the voltage and temperature of the diode, respectively. The second term in Eqn. 1 is simply the Shockley diode law.

Real cells dissipate power through many different parasitic resistances. These parasitic effects are grouped into two electrically equivalent resistances—series resistance ($R_s$) and shunt resistance ($R_{sh}$)—as shown in Fig. 2 Series resistance arises from resistance to current flow by the cell material itself and by the connections between the semiconductor material and the cell contacts. Shunt resistance occurs due to leakage of current through the cell, around the edges of the device, and between contacts of different polarity. In an ideal cell, $R_s = 0$ and $R_{sh} = \infty$.

Additionally, the ideal diode behavior described in Eqn. 1 is not realistic; the dark current generally has a weaker dependence on bias [8]. So, an ideality factor $n$ is introduced, with typical values between 1 and 2, where $n = 1$ describes an ideal diode. Taking the parasitic resistances and diode nonideality in to account, Eqn. 1 becomes:

$$I_{cell} = I_{sc} - I_0 \left( e^{q(V_c + I R_s)/n k_B T} - 1 \right) - \frac{V_c + I R_s}{R_{sh}} \qquad (2)$$

where $I_{cell}$ and $V_c$ are the terminal current and voltage respectively. The diode voltage is given by $V_d = V_c + I R_s$. The last term ($V_d/R_{sh}$) is called the shunt current ($I_{sh}$).

### B. Two-Diode Models

Further improvements to this model can be made by taking recombination effects into account. Recombination in the depletion region is a loss factor not considered in the original Shockley equation nor the lossy equivalent circuit model described in Eqn. 2. However, it has been shown that these effects can be modeled by considering a second diode in parallel to the first [9], leading to a two-diode equivalent circuit model, which is shown in Fig. 3.

In most implementations of the 2-diode model, the diode ideality factors, $n_1$ and $n_2$, are set to the values 1 and 2 respectively. This differs from a 1-diode model in which the ideality factor is variable. The diode currents, $I_{d1}$ and $I_{d2}$, are again given by Shockleys equation:

$$I_{d1} = I_{sat1} \left( e^{qV_d/n_1 k_B T} - 1 \right) \qquad (3)$$

$$I_{d2} = I_{sat2} \left( e^{qV_d/n_2 k_B T} - 1 \right) \qquad (4)$$

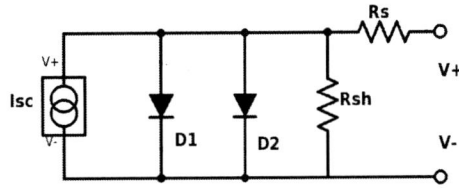

Fig. 3. A two-diode equivalent circuit model with lossy components.

where $I_{sat1}$ and $I_{sat2}$ are the saturation currents for each diode. This allows us to formulate the generic 2-diode model:

$$I_{cell} = I_{sc} - I_{d1} - I_{d2} - I_{sh} \qquad (5)$$

### C. Additional Model Improvements

For some cell technologies, the saturation current of the first diode shows a strong cubic correlation with temperature [10] as given in Eqn. 6. $E_g$ is the band gap of the PV device material and $T_0$ is the reference temperature, $25\,°\mathrm{C}$. The saturation current of the second diode is assumed to be constant.

$$I_{sat1} = I_{sat1}\Big|_{T=T_0} \frac{T^3}{T_0^3} \exp \left[ \frac{E_g q}{k_b} \left( T_0^{-1} - T^{-1} \right) \right] \qquad (6)$$

Additionally, the short circuit current term in Eqn. 5 ($I_{sc}$) is replaced with the more general photogenerated current ($I_{ph}$), as shown in Eqn. 7. $I_{sc}$ is a function of the effective irradiance ($E_e = E/E_0$), the temperature coefficient ($\alpha$), and the difference between the cell and reference temperatures ($\Delta T = T - T_0$). Effective irradiance is defined as the ratio of incident irradiance transmitted to the cell and the reference irradiance, $E_0 = 1000\,\mathrm{W\,m^{-2}} = 1\mathrm{sun}$.

$$I_{ph} = A_{ph} I_{sc} = A_{ph} E_e I_{sc0} \left( \alpha \Delta T \right) \qquad (7)$$

The proportionality constant, $A_{ph}$, can be represented explicitly by solving the circuit in Fig. 3 at the short circuit condition (when $V = 0$).

$$A_{ph} = 1 + \frac{I_{d1} + I_{d2} + I_{sh}}{I_{sc}} \bigg|_{V=0} \qquad (8)$$

Finally, a better estimate for $I_{sh}$ is obtained by adding a term representing the reverse breakdown current ($I_{rbd}$).

$$I_{sh} = \frac{V_d}{R_{sh}} + I_{rbd} \qquad (9)$$

The reverse breakdown current is given by an avalanche breakdown expression from [11] which was modified by adding a quadratic term to allow more flexibility in fitting reverse bias. The expression has four parameters, the device specific avalanche breakdown voltage ($V_{rbd}$), a positive exponent ($n_{rbd}$, typically set to 4), and two additional coefficients($a_{rbd}$ and $b_{rbd}$, typically set to $10^{-4}$ and zero).

978-1-5090-5606-4/17 $31.00 © 2017 IEEE

Letting $r_v = V_d / (R_{sh} I_{sc0})$, we have the following equation for the reverse breakdown current:

$$I_{rbd} = I_{sc0} \left( a_{rbd} r_v + b_{rbd} r_v^2 \right) \left( 1 - \frac{V_d}{V_{rbd}} \right)^{-n_{rbd}} \quad (10)$$

## III. PVMISMATCH OVERVIEW

PVMismatch is free open source software written in the Python computer language. PVMismatch can be downloaded from the Python Package Index at https://pypi.python.org/pypi/pvmismatch. The software can be used to simulate both forward and reverse bias regions of current-voltage (IV) curves for various combinations of photovoltaic cells, modules, and strings to form full PV systems. Cell, module, string, and system constraints are all independent and variable. The basic model building block is the 2-diode equivalent circuit model described in the previous section.

PVMismatch uses an explicit method to calculate the full IV curve simultaneously at each instance [11] instead of an iterative or Lambert W-function method to solve for individual points sequentially. There are two advantages to this method.

- In general for very large datasets, libraries like BLAS (Basic Linear Algebra Subprograms, available at NetLib http://www.netlib.org/blas/) can perform linear algebra operations on vectors of data significantly faster than looping.
- The result of the simulation is the entire IV curve instead of select points.

The user specifies the desired number of points in the IV curve, and the points are distributed using two log-space distributions to increase density around the maximum power point and the reverse breakdown voltage. Then Bishop's explicit method is applied in three steps:

1) Create a range of diode voltages with user specified resolution.
2) Evaluate corresponding cell currents from diode voltages utilizing Eqns. 3–10.
3) Evaluate cell voltage from $V_d = V_c + IR_s$.

The cells' IV curves are then combined according to Kirchhoff's circuit laws to form modules, strings, and full PV systems by adding either voltages or currents and interpolating when necessary. Bypass diodes within modules are treated very simply as perfect conductors when the cell substring voltage is less than the bypass diode trigger voltage and perfect insulators otherwise.

PVMismatch is an object oriented program. Cells, modules, and strings are instances of objects that contain the attributes and methods that relate to them. A system contains strings, a string contains modules, and modules contain cells. Memoization of identical cells, modules and strings is used to speed up calculations and save memory. Each simulation initially begins with only a single instance of each cell, module and string given by the initial parameters specified by the user. In addition to the parameters already described, the user can

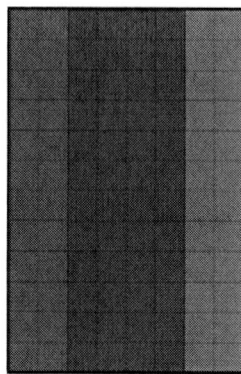

Fig. 4. A diagram of the cell layout of the 96-cell module modeled in this study. The three colors designate the cells belonging to each of the three bypass diodes.

also specify the module cell configuration, the number of modules per string, and number of strings in the system. Cell configurations can be customized for in-series or cross-tied substrings of cells. Then as the user changes cell parameters such as irradiance and temperature, copies of cells are made as needed. Therefore as systems become more complex (for example, different degradation on every cell), the optimization through memoization becomes less efficient. Typical systems with mostly nearly identical cells will be the most optimized.

## IV. METHODOLOGY

For this study, an $8 \times 3$ system was simulated in PVMismatch—three strings in parallel, each with eight PV modules. This PVMismatch model was previously validated against field data in [12]. The modules each have 96 cells in series, arranged physically in 8 columns of 12. The modules have three bypass diodes each, grouping the cells in a $24 - 48 - 24$ pattern. The physical layout of the module, including the grouping by bypass diodes, is shown in Fig. 4.

The Python library Shapely [13] was used to generate iterative geometric shade patterns to project onto a 2D representation of the array, with the modules arranged in a simple rectangle and the ground cover ratio (GCR) in both dimensions equal to 1. The modules were oriented in "portrait" relative to the orientation of the module strings, as shown in Fig. 5.

Six families of geometries were selected for this study:

1) Vertical shade increasing from the left (short edge)
2) Horizontal shade increasing from the bottom (long edge)
3) Increasing rectangles from the bottom left (rectangles)
4) Increasing circles from the bottom left (circles)
5) Increasing angular shade from the left (angles sweep short)
6) Increasing angular shade from the bottom (angles sweep long)

The general behavior of each set is shown in Fig. 6. 50 iterations of each geometry were generated, for a total of 300

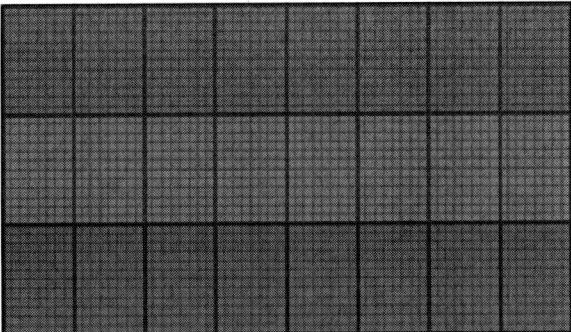

Fig. 5. The 2D layout of the system, as it was modeled for shade projections. The three colors represent the three parallel strings of 8 modules each. Note that the bold rectangles represent modules, while the thin squares represent individual cells.

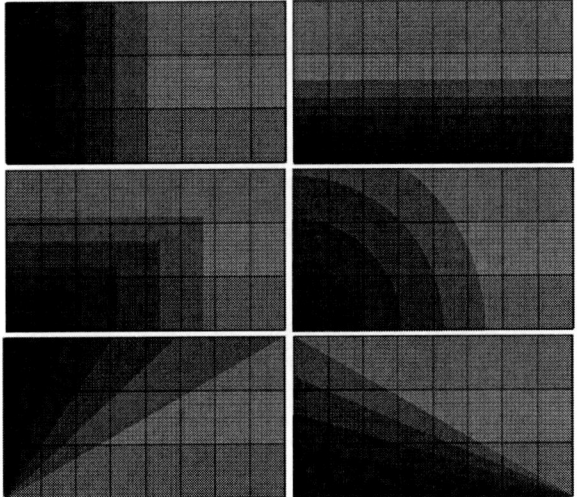

Fig. 6. A graphical representation of the six shade geometry families in this study. Only 5 of the 50 iterations in each set are shown here, for the sake of clarity. The numbering in the text is in row-major order.

unique shade scenarios. Each set of iterations range from $1\%$ shade coverage to $50\%$ shade coverage.

We chose to model the system with an unshaded irradiance of $1000\,\mathrm{W\,m^{-2}}$ and a cell temperature of $25\,^{\circ}\mathrm{C}$. It has been shown previously [12] that the light intensity in the shaded region does not have a strong impact on normalized performance, as long as the shaded irradiance is less than $25\%$ of the unshaded irradiance.

In addition to the shade patterns, all system objects (cells, modules, and strings) are also represented as Shapely geometry objects. The Shapely software can then calculate the intersection between the shade geometries and the system objects, which we use to estimate the irradiance on every individual cell. If entire modules or strings are uniformly shaded, this information is also captured for more efficient modeling in PVMismatch. The irradiance information is packaged into a nested dictionary data structure, which is then passed to PVMismatch for IV and PV curve modeling.

For each shade scenario, system performance was modeled

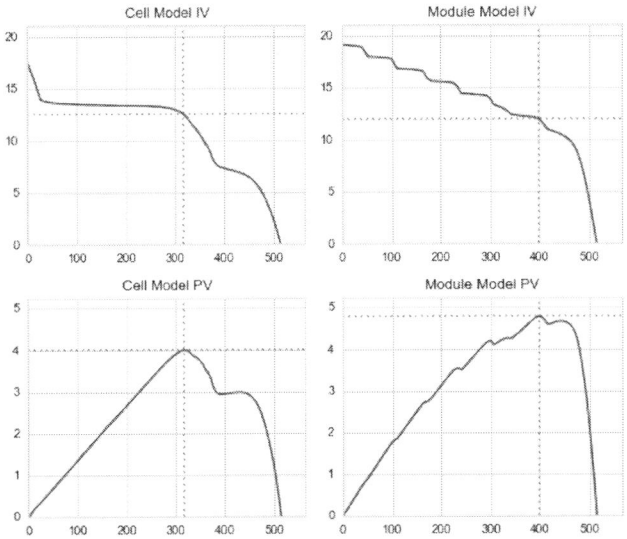

Fig. 7. A comparison of the system IV and PV curves generated by PVMismatch. The plots show a single shade scenario from group #6. The cell-level model predicts approximately $10\%$ less power, relative to $P_{mp0}$, or $20\%$ less power relative to the module-level model. Red dotted lines show max-power operating point.

two ways. First, the full dictionary of irradiance values was passed to PVMismatch to calculate the system IV and PV curves based on the cell-level irradiance values. Second, average irradiance values were calculated for each module, and PVMismatch was used to calculate the system IV and PV curves based on the average module-level irradiance values. As previously stated, the goal is to compare the two approaches in order to determine the utility of implementing a cell-level equivalent circuit model instead of the much simpler module-level model.

## V. RESULTS AND DISCUSSION

Fig. 7 compares the full IV and PV curves for a single shade geometry, modeled using both approaches. In this case, the cell-level model predicts significantly more power loss than the module-level model. The models have some agreement on the operating current ($I_{mp}$), but predict very different operating voltages ($V_{mp}$).

A full summary of the simulations described in the previous section are plotted in Figs. 8 and 9. The response variable in both plots is normalized maximum power: the true max power point of the array under that shade condition divided by $P_{mp0}$ (the maximum power of the system under standard test conditions with no shade). The independent variable in this analysis is shade ratio, or the fraction of the array area covered by shade. On both plots, reference lines at $y = 1 - x$ and $y = 1 - 2x$ are included for reference. The first line is the theoretical maximum power output of any system for the given shade ratio. The second line represents the result of assuming a Shade Impact Factor (SIF) of 2, the default value chosen by the California Energy Commission for shade performance analysis (see page C-18) [1].

978-1-5090-5606-4/17 $31.00 © 2017 IEEE        3357

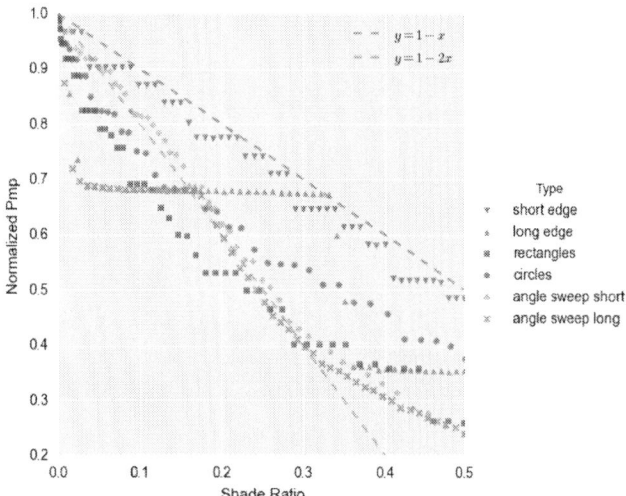

Fig. 8. Plot of normalized power versus shade ratio when utilizing the full cell-level model. The data is labeled by the six shade geometry families.

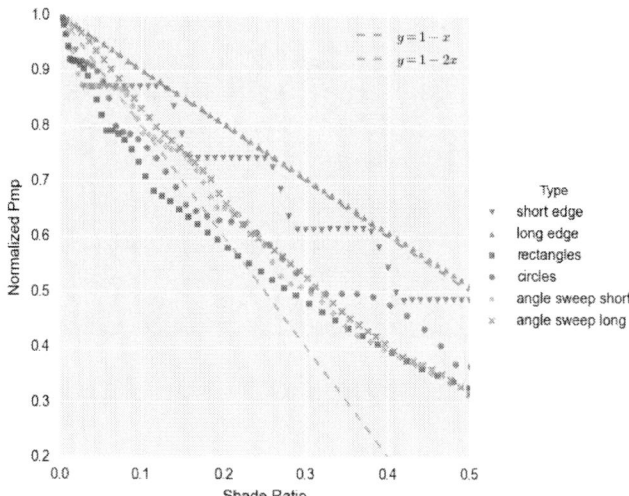

Fig. 9. Plot of normalized power versus shade ratio when utilizing a module-level, average irradiance model. Note the smoothing out of many of the trend lines and the significantly different behavior of the long edge trend (set #2).

In Fig. 8, the impact of specific geometry on the performance of the system is clearly observable. At around $10\%$ shade, the system can experience a reduction in power output in the range of $10-30\%$, simply due to the different geometries.

In Fig. 9, many of the trends lines are significantly smoothed out by the averaging of irradiance on each module. As would be expected, the behavior of individual diode-protected cell substrings is completely missing. Particularly striking is the change in modeled system response to shade set #2 (horizontal shade increasing from the bottom) which exactly follows the $y = 1 - x$ line in the module-level model. This occurs because the model-level model does not capture any mismatch conditions with this geometry; the short-circuit current is reduced on the shaded string in exact proportion to the amount of shade. In reality, a small amount of shade cutting across the diode-protected substrings causes a significant amount of mismatch, such that the entire power of the string is removed after just one or two rows of cells are shaded, as seen in Fig. 8.

Fig. 10 summarizes the differences between the cell-level approach and the module-level approach. This analysis illustrates that module-level modeling over-predicts power for most geometries modeled in this study. However, as seen in the "short edge" data, model-level modeling can also under-predict power in some circumstances. A likely explanation for this is that the bypass diodes tend to be quite helpful in mitigating power loss when the shade is correctly aligned with the substring geometry. The cell-level model captures this behavior, while the module-level model does not.

## VI. CONCLUSION

The necessity of correctly modeling sub-module behavior when executing partial shade mismatch studies has been shown. In the set of shade geometries evaluated for this study,

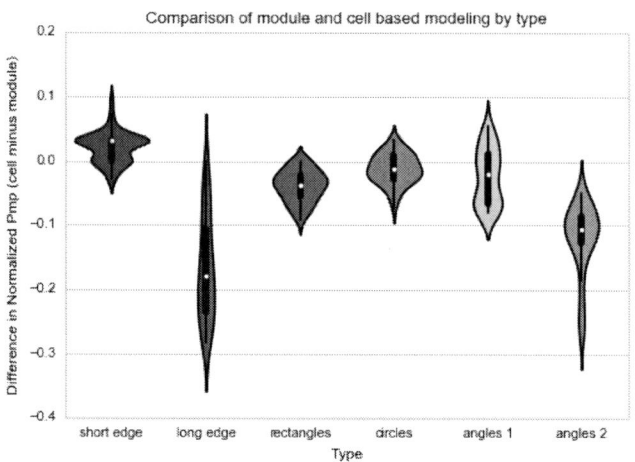

Fig. 10. Comparison of cell-level and module-level modeling with PVMismatch. The y-axis is difference in normalized $P_{mp}$. A negative value means the cell-level model predicts less power.

the module-level module tended to over-predict system power relative to the cell-level model. However, certain geometries showed the opposite effect. In addition to inaccurately estimating total system power, the operating voltage point of the system is incorrectly estimated by the module-level model.

An open source software solution has been presented to facilitate analysis of partial shade conditions. The techniques and tools presented here may be used by other researchers to facilitate further research on the topic of partially shaded PV arrays. For example, the authors have previously used the techniques presented in this paper to facilitate the generation of large-scale, detailed datasets of PV array performance under various shade conditions for performing feature extraction using machine learning techniques [12].

## ACKNOWLEDGMENT

This work is supported by the Grid Integration Systems and Mobility (GISMO) group at SLAC National Accelerator Laboratory, the U.S. Department of Energy SunShot Initiative, and SunPower Corporation. The development of the PVMismatch software would not have been possible without the patient guidance of Akira Terao.

## REFERENCES

[1] G. W. Pennington, P. Saxton, S. Neidich, S. Taheri, F. Nasim, and J. Folkman, "Guidelines for California's Solar Electric Incentive Programs, Fifth Edition," California Energy Commission, Tech. Rep., 2013.

[2] S. Malathy and R. Ramaprabha, "Comprehensive analysis on the role of array size and configuration on energy yield of photovoltaic systems under shaded conditions," *Renewable and Sustainable Energy Reviews*, vol. 49, pp. 672–679, 2015. [Online]. Available: http://dx.doi.org/10.1016/j.rser.2015.04.165

[3] E. Karatepe, M. Boztepe, and M. Çolak, "Development of a suitable model for characterizing photovoltaic arrays with shaded solar cells," *Solar Energy*, vol. 81, no. 8, pp. 977–992, 2007.

[4] G. Velasco-Quesada, F. Guinjoan-Gispert, R. Piqué-López, M. Román-Lumbreras, and A. Conesa-Roca, "Electrical PV array reconfiguration strategy for energy extraction improvement in grid-connected PV systems," *IEEE Transactions on Industrial Electronics*, vol. 56, no. 11, pp. 4319–4331, 2009.

[5] N. Belhaouas, M.-S. A. Cheikh, P. Agathoklis, M.-R. Oularbi, B. Amrouche, K. Sedraoui, and N. Djilali, "PV array power output maximization under partial shading using new shifted PV array arrangements," *Applied Energy*, vol. 187, pp. 326–337, 2017. [Online]. Available: http://linkinghub.elsevier.com/retrieve/pii/S0306261916316233

[6] C. A. Ramos-Paja, J. D. Bastidas, A. J. Saavedra-Montes, F. Guinjoan-Gispert, and M. Goez, "Mathematical model of total cross-tied photovoltaic arrays in mismatching conditions," *2012 IEEE 4th Colombian Workshop on Circuits and Systems, CWCAS 2012 - Conference Proceedings*, 2012.

[7] P. Srinivasa Rao, G. Saravana Ilango, and C. Nagamani, "Maximum power from PV arrays using a fixed configuration under different shading conditions," *IEEE Journal of Photovoltaics*, vol. 4, no. 2, pp. 679–686, 2014.

[8] J. Nelson, *The Physics of Solar Cells*. Imperial College Press, 2003.

[9] C. T. Sah, R. N. Noyce, and W. Shockley, "Carrier Generation and Recombination in P-N Junctions and P-N Junction Characteristics," *Proceedings of the IRE*, vol. 45, no. 9, pp. 1228–1243, 1957.

[10] W. De Soto, S. A. Klein, and W. A. Beckman, "Improvement and validation of a model for photovoltaic array performance," *Solar Energy*, vol. 80, no. 1, pp. 78–88, 2006.

[11] J. W. Bishop, "Computer simulation of the effects of electrical mismatches in photovoltaic cell interconnection circuits," *Solar Cells*, vol. 25, no. 1, pp. 73–89, 1988.

[12] B. Meyers, M. Mikofski, and M. Anderson, "A Fast Parameterized Model for Predicting PV System Performance under Partial Shade Conditions," *2016 IEEE 43nd Photovoltaic Specialist Conference, PVSC 2016*, no. 1, pp. 3173–3178, 2016.

[13] S. Gillies, "Shapely 1.2 and 1.3 documentation," 2013. [Online]. Available: http://toblerity.org/shapely/index.html

# Evaluation of uncertainty in PV project design: definition of scenarios and impact on energy yield predictions

Giorgio Belluardo[1*], Magnus Herz[2], Ulrike Jahn[2], Mauricio Richter[3], David Moser[1]

[1]Institute for Renewable Energy - EURAC Research, viale Druso 1, 39100 Bolzano (Italy)
[2]TÜV Rheinland Energy, Cologne (Germany)
[3]3E sa, Quai à la Chaux 6, 1000 Brussels (Belgium)

*mail: giorgio.belluardo@eurac.edu, tel. +39 0471 055626, fax +39 0471 055699

*Abstract* — **In PV projects, it is important to understand how the variability and associated uncertainties of technical risks occurring during the planning phase are calculated and how the values are distributed in terms of probability. These aspects are essential for the calculation of the exceedance probability of the expected energy yield and how this is influenced by the overall uncertainty. A reduction in the uncertainties can lead to a higher values of energy yield for a given exceedance probability and hence a stronger business case. In this paper, the combined effect of uncertainties of various technical risks is calculated and discussed for several uncertainty scenarios and the impact is calculated in terms of exceedance probability (e.g. P90/P50 ratio).**

*Index Terms* — **energy yield prediction, energy yield uncertainty, PV model.**

## I. INTRODUCTION

Historical performance data for PV systems on which to base technical risks assessments and investment decisions are difficult to be accessed by all market players, such as investors, PV plant owners, EPC contractors, etc. Reasons for this difficulty are that most PV systems have been operational for only a few years (GWs cumulative installations in many countries was only reached after 2010) and a tendency among system operators and component manufacturers to keep available performance data as confidential. A broader availability of performance data would be highly beneficial in terms of improvements in the accuracy of initial and lifetime yield assessment. For the PV industry to reach mature market level, a better understanding of technical risks, risk management practices and the related economic impact is thus essential to ensure investors' confidence.

Within the Solar Bankability project (EU-funded project under the Horizon 2020 Work Programme) the aim is to establish a common practice for professional risk assessment, which will serve to reduce the risks associated with investments in PV projects. One objective of Solar Bankability is to improve the current understanding of several key aspects of risk management during the project lifecycle, from the identification of technical risks and their economic impact, to the process of mitigating and allocating those risks among project parties, to transferring those risks through insurance, warranties, preventive maintenance, etc. To achieve this, in

Solar Bankability we have built upon existing studies and collected available statistical data of failures with the aim to i) suggest a guideline for the categorization of failure, ii) introduce a framework for the calculation of uncertainties in PV project planning and how this is linked to financial figures, and iii) develop a methodology for the assessment of the economic impact of failures occurring during operation but which might have originated in previous phases.

In the project report "Technical Risks in PV Projects" [1] technical risks were identified and categorized for components and phases of the value chain of a PV project. The technical risks were broadly divided into risks to which one can assign an uncertainty to the initial yield assessment and risks to which one can assign a Cost Priority Number (CPN). While failures arising from technical risks belonging to the first group have an impact on the overall uncertainty of the energy yield, failures with a CPN have a direct impact on the annual cost of running a PV plant. The CPN methodology and analysis was the subject of [2]. In this paper, we will instead focus on the impact of technical risks on the uncertainty of initial energy yield assessment during the planning phase.

### METHODOLOGY

In [1] the risks related to a PV project were identified. Some of them have a direct impact in terms of uncertainty of the energy yield and performance related to a PV system. In particular, there are two types of uncertainty: the uncertainty related to the design phase, for which the *predicted* performance is affected, and the uncertainty during the operational phase, for which the *actual* performance is affected. In this study, we focus only on the project design phase, where the main risks are: incorrect power rating, soiling losses, shadow diagram, modules' mismatch, incorrect or not available flash test, incorrect assumptions of module degradation, unclear simulation parameters, inverter wrongly sized, excessive derating due to wrong sizing of inverter or to inverter exposed to direct light.

Once the risks are identified, it is necessary to know:
- the value of the input parameters used for the design, and the related uncertainties;

- the structure of the model to calculate the predicted energy yield and performance of the PV systems.

As for the latter point, several software are available - either for purchase or for free - that have different characteristics (i.e. number and type of input parameters, type of sub-models used) and different levels of complexity.

Generally speaking, a PV array model receives input from the temperature and irradiance models, and generates the expected array yield *Ya* by also taking several array losses into account. Finally, this yield is fed into the PV inverter model in order to estimate the final yield *Yf* (in this study, always referred as *energy yield*) of a PV system. The Performance Ratio *PR* is finally calculated as the ratio of Yf and the reference yield *Yr* given by the irradiance model [3].

In order to calculate how the uncertainty propagates from the input parameters to the PV model's output (Ya, Yf, PR), mainly two methodologies are used: the Monte Carlo technique and the classical law of propagation of errors [4].

The Monte Carlo approach [5] reconstructs the Probability Density Function (PDF) of the errors of a model (i.e. the PDF of its outputs) using the PDF of the errors of its input parameters. In order to do so, a high (i.e. statistically significant) number of values of each input parameter is firstly generated according to the PDF of its error. They are then combined in vectors, each containing a single generated value of all considered input parameters, and fed into the model. This way, a corresponding high number of outputs is generated and analyzed in order to reconstruct the PDF of their error and calculate their uncertainty (for k=1) as the standard deviation.

The Monte Carlo technique is particularly useful and reliable when applied to models that are described by complex equations, in which strong correlations between input parameters occur. In this case, the application of classical law of propagation of errors might become a difficult task. However, some approximations can be introduced to overcome this issue. As reported by [6], a PV model can be represented with a good approximation as the product of linear factors:

*output = input x proportionality factor – offset*

where the offset is relatively small compared to the phenomenon itself. If a quantity $X$ is the product of $N$ independent variables $X_1, X_2, ..., X_N$ and can be expressed as $X=c*X_1*X_2*...*X_N$, where $c$ is a constant and $\sigma_1, \sigma_2, ..., \sigma_N$ are the uncertainties (k=1, corresponding to the standard deviations), then the so-called *rule of squares* can be applied an the combined relative uncertainty of $X$ becomes:

$$\frac{\sigma_x}{x} = \sqrt{\left(\frac{\sigma_1}{x_1}\right)^2 + \left(\frac{\sigma_2}{x_2}\right)^2 + ... + \left(\frac{\sigma_N}{x_N}\right)^2} \quad (1)$$

## III. RESULTS AND DISCUSSION

### A. Definition of the base uncertainty scenario

In order to compare the two methodologies in estimating the uncertainty related to the design of a PV project, a PV system installed in Bolzano (South Tyrol, Italy) is considered. This system has a nominal power of 3.8 kW$_p$, composed of 18 polycrystalline-silicon modules. It is installed with a tilt of 30° and an orientation of 8.5° West-of-South.

The values of each input parameter, as well as the contribution of each input parameters in terms of uncertainty on the energy yield are reported in Table I. They are assigned on the basis of the available information for the PV plant and for the site, and from scientific literature. The uncertainty of the insolation (from Global Horizontal Irradiance *GHI* and Diffuse Horizonal Irradiance *DiffHI*) is calculated from 22 available years of satellite data, while normally-distributed PDFs are assumed for the errors of the other input parameters. The list of uncertainty values represents the *base uncertainty scenario*.

TABLE I

VALUE AND ASSOCIATED UNCERTAINTY OF THE INPUT PARAMETERS OF A REFERENCE PV SYSTEM IN BOLZANO. THE LISTED UNCERTAINTIES REPRESENT THE BASE UNCERTAINTY SCENARIO.

| Parameter | value | rel. unc. (k=1) | notes |
|---|---|---|---|
| Insolation variability | 1428.8 (GHI) 600.1 (DiffHI ) kWh/m²/year | 3.3% 2.2% | from 22 years of available data |
| Solar resource | | 5% | satellite measurements |
| Transp. model | | 2% | best for Bolzano |
| Ambient Temp. variability | 8.8°C (mean) | 0.43% | from 22 years of available data |
| Temperature effect | 29 W/(m²K) | 0.14% | calculated with the best temp. model for Bolzano [7] |
| Performance loss rate | | 0.5% | |
| Soiling loss | 0.5% | 0.49% | |
| Shading loss | | 2% | |
| Spectral loss | | 0.2% | |
| Nominal power | 1% | 0.98% | |
| Inverter efficiency effect | | 0.2% | |

PV Syst [8] is selected as PV model to run N=1000 simulations needed to apply the Monte Carlo technique. Only some of the parameters listed in Table I are set in the software (those whose value is indicated). For this reason, this reduced case corresponds to a *simplified base uncertainty scenario*. Also the rule of squares is applied for the same simplified

978-1-5090-5606-4/17 $31.00 © 2017 IEEE

case, in order to be comparable with the results from the Monte Carlo approach. Table II shows the propagation of uncertainty from the GHI and the Global Tilted Irradiance *GTI* (corresponding to the reference yield Yr) values to the array yield Ya, and finally to the output values of energy yield and Performance Ratio PR.

## TABLE II
UNCERTAINTY VALUES OF THE OUTPUT QUANTITIES OF A PV MODEL, CALCULATED WITH TWO DIFFERENT ERROR PROPAGATION TECHNIQUES. ONLY THE SIMPLIFIED BASE UNCERTAINTY SCENARIO IS CONSIDERED.

| Technique | GHI | GTI | Ya | Yf | PR |
|---|---|---|---|---|---|
| Monte Carlo | 3.31% | 4.21% | 4.24% | 4.29% | 0.78% |
| Rule of squares | 3.31% | 3.99% | 4.16% | 4.16% | 5.77% |

The values of uncertainty are similar for GHI, GTI, Ya and Yf, thus confirming the validity of the assumptions made in order to apply the classical law of propagation of errors. An exception rises for PR, where the uncertainty of 5.77% from the rule of squares is much higher than the 0.78% with the Monte Carlo technique. A reasonable explanation for this comes from the expression of PR, $PR=Yf/Yr$. Since PR is a function of two highly correlated variables (Yf strongly depends on Yr), here the assumption previously made (PV yield model as a product of independent variables) is not valid anymore and the rules of squares is not adequate.

### B. Definition of other uncertainty scenarios

Additional uncertainty scenarios are defined, in order to consider other possible situations where the input parameters used in a PV design project have a better or worse quality. This way, it is also possible to evaluate which parameters affect the uncertainty of the outputs of a PV project design at most, and set preventive mitigation measures accordingly. The additional 21 uncertainty scenarios have the same set of uncertainties of the base scenario previously introduced, except for the uncertainty of one single parameter (e.g. insolation variability, or solar resource, etc.) that gives also the name to the scenario. In particular, the following possible additional scenarios are considered for each input parameter:

- insolation variability: 5 years of data; 22 years of data but from a location characterized by a high variability;
- solar resource: ground measurements; site adaptation technique (combination of long-term satellite data and short-term series of measured data [9]);
- transposition model: no transposition model (GTI data are available); transposition model proven to be the worst for a specific location [10];
- ambient temperature variability: highest value in [1];
- temperature effect: using a temperature model that does not account for wind effect (e.g. NOCT formula), proven to be bad for Bolzano [6];
- performance loss rate: highest value in [1];

- soiling loss: highest value in [11];
- shading loss: 5% relative uncertainty (k=1);
- spectral mismatch loss: 2% relative uncertainty (k=1);
- nominal power of the modules: highest value in [1];
- inverter efficiency effect: highest value in [1].

For the additional cases, only the rules of squares is applied. In fact the Monte Carlo technique is not fully applicable, since simulations in batch mode can be run only setting some of the listed input parameters. The results of the relative uncertainty on the energy yield Yf of 10 out of 21 uncertainty scenarios - plus the base uncertainty scenario - are reported in Fig. 1, ordered from the lowest to the highest value. A complete ranking list can be found in [12].

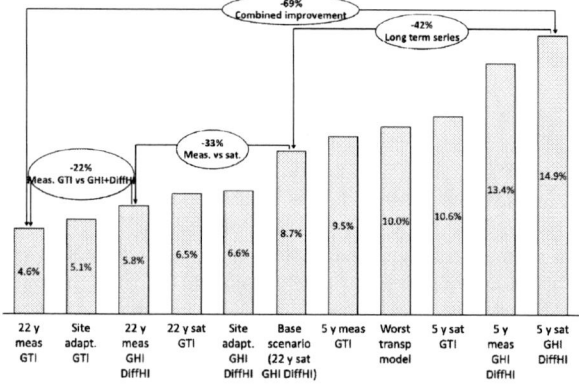

Fig. 1. Example of uncertainty reduction on the predicted energy yield, by comparison of different uncertainty scenarios.

There is a group of scenarios with a low level of uncertainty (4.6% to 8.7%), including also the base uncertainty scenario. The scenarios within this group are characterized by the use of long time-series of either ground measurements or satellite estimates of insolation. The temporal range of the available insolation data seems therefore to be the most important factor affecting the uncertainty of the energy yield estimation. For example, there is an improvement of almost 5% in the absolute value of uncertainty when using 22 years instead of 5 years of GTI data, which increases up to 7.6% when 22 years of GHI and DiffHI measurements are used instead of 5 years.

The scenario with the lowest uncertainty (4.6%) is the one corresponding to the base scenario modified so that 22 years of measured GTI are used instead of 22 years of satellite data of GHI and DiffHI (transposed to GTI with a transposition model). This corresponds to the so-called *low-end scenario*, and shows that a lower uncertainty is assured when a) ground measurements are used in place of satellite estimates and b) time series of plane-of-array irradiance is available without the need to apply transposition models. For example, using long series of measured GTI data reduces the absolute value of uncertainty of 1.9% than using long series of satellite GTI data. This value increases up to 2.9% when using long series of measured GHI and DiffHI data. Results show also that using a site adaptation technique consisting in a combination of long time series of satellite data with a short series of measured data is recommended than just using satellite data.

Other factors affecting the uncertainty, not reported in Fig. 1 are the shading and soiling effects.

In addition to the low-end scenario, other two scenarios are defined:

- *high-end scenario*: corresponds to uncertainty value of 9.3%, calculated as the average of uncertainties of all 22 scenarios.
- *worst-case scenario*: corresponds to a scenario with an uncertainty of 16.6% (the worst possible combination, obtained considering the highest uncertainty for each input parameter).

The low-end and high-end scenarios are thus representative for the range given in [1] of 5-10% overall uncertainty of the energy yield.

### C. Impact on energy yield predictions

A common way to quantify the technical risks arising during the PV design phase is to calculate the exceedance probability of Yf and reporting some specific points of the curve, as for example P50, P90, and the ratio P90/P50. Typically, a normal distribution is assumed for all the contributions to the final uncertainty of the energy yield, even though this might not be correct. In fact, the best option would be to opt for an empirical method based on a Monte Carlo analysis. Fig. 2 shows the exceedance probability curves of energy yield obtained for the simplified base uncertainty scenarios introduced:

- by assuming the empirical distribution of Yf obtained from the Monte Carlo technique applied to the PV Syst simulation;
- by assuming a normal distribution. The mean energy yield value is equal to the output value from a single PV Syst simulation with input values reported in Table I (corresponding also to the value of P50). The standard deviation is calculated with the rule of squares.

The values of P50 and P90, as well as the results of the P90/P50 ratio (where P50 refers to the base uncertainty scenario) are reported in Table III.

Even though the P90/P50 ratio is very similar for both distributions, the empirical distribution is positively skewed and shows important differences for values lower than P50 (e.g. -6.6% at P10 for the empirical distribution). In addition, extreme scenarios (e.g. P99) can largely be affected by the use of normal distribution yielding unrealistic results. Similar results were obtained in [13], where it is shown that assuming a normal distribution for the solar resource uncertainties may not be the most correct approach.

Unfortunately, there is not always a sufficiently large dataset available to establish the Cumulate Density Function (CDF) from which to interpolate exceedance probabilities. However,

the method should applied when possible to some input parameters, such as the solar resource variability. With the availability of more data for other input parameters, also other secondary effects can be included in the methodology as not normally distributed.

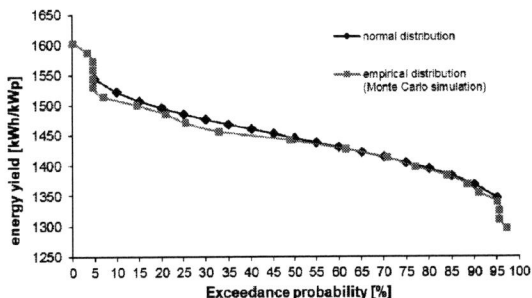

Fig. 2.  Exceedance probability curves of energy yield for the simplified base uncertainty scenario. The normal distribution (uncertainty calculated with the rule of squares) and the empirical distribution (from Monte Carlo technique applied to PVSyst simulations) are compared.

For the time being, we restrict the analysis by assuming normally-distributed input parameters and energy yield. It is therefore possible to reconstruct the exceedance probability curve of the energy yield of the 22 uncertainty scenarios introduced before. Fig. 3 shows the exceedance probability curve of three selected scenarios: low-end scenario, high-end scenario and worst-case scenario. For this case, the same value of P50 was assumed in order to show only the differences due to the different level of uncertainty.

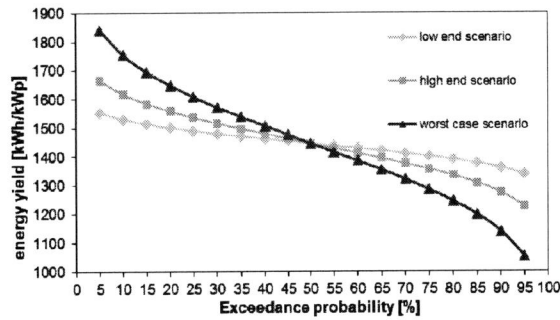

Fig. 3.  Exceedance probability curves of energy yield for three uncertainty scenarios defined in section IIIb. The normal distribution and same mean energy yield values are assumed.

The impact of the uncertainty on the CDF – and therefore on the exceedance probability of energy yield - is evident and the resulting P50, P90 and P90/P50 values are shown in Table III. The P90 values of the high-end and worst-case scenarios decrease respectively by -6% and -15% when compared to the low-end scenario. This means that, in this case, a business plan assuming an uncertainty framework similar to that of the high-end scenario would be weaker than the one assuming input values affected by a low level of uncertainty.

Another important parameter that affects the overall analysis is the mean value of the energy yield (P50, if normally distributed). In fact, as discussed in the previous sections the main source of error is related to the solar resource assessment. Table III and Fig. 4 show also the results for the worst-case scenario with a mean value of 1314 kWh/kWp instead of 1445 kWh/kWp. These values come from an insolation assessment based on 5 years of measured data instead of 22 years of satellite-derived data as in the previous cases. The use of shorter time series can clearly lead to an underestimation (or overestimation) of the values of P50, P90, P90/P50. For example, in the considered case the reduction in P90 is 22%, when compared to the low-end scenario.

### TABLE III
### SUMMARY OF THE EXCEEDANCE PROBABILITY VALUES FOR VARIOUS SCENARIOS

| | σ (k=1) | P50 (kWh/kWp) | P90 (kWh/kWp) | P90/P50 (P50 base unc. sc.) |
|---|---|---|---|---|
| simplif. base unc. sc. (empirical distribution) | 4.3% | 1440 | 1360 | 94% |
| simplif. base unc. sc. (normal distribution) | 4.2% | 1445 | 1368 | 95% |
| **base uncertainty sc. (rule of squares)** | **8.7%** | **1445** | **1283** | **89%** |
| low-end scenario | 4.6% | 1445 | 1365 | 94% |
| high-end scenario | 9.3% | 1445 | 1273 | 88% |
| worst-case scenario | 16.6% | 1445 | 1138 | 79% |
| worst-case scenario (different mean value) | 16.6% | 1314 | 1034 | 72% |

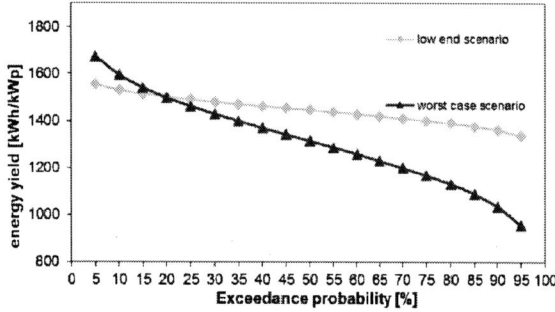

Fig. 4. Exceedance probability curves of energy yield for the worst-case scenario with a different P50 value deriving from a different site insolation assessment (5 years of measured data instead of 22 years of satellite data), and the low-end scenario.

## IV. CONCLUSION

One important category of risks related to a PV system is the one having a direct impact in terms of uncertainty of the predicted (or initial) energy yield during the design phase. Each input parameter of a PV model is associated to a value of uncertainty that propagates into the several parts of the model itself and to the outputs (reference yield, energy yield, performance ratio). In this work, different uncertainty scenarios are defined. A base uncertainty scenario is selected as reference case and two different techniques (Monte Carlo and rule of squares) are used to calculate the uncertainty propagation. The impact of the defined uncertainty levels on the exceedance curve of energy yield is finally calculated.

Results show that the parameter affecting the uncertainty of energy yield at most is solar insolation, and in particular (in order of importance): the temporal range of the available insolation data, the source of insolation (ground measurements, satellite estimates etc.), and the use and type of transposition models.

The uncertainty of energy yield can be considered to vary between 4.6% (low-end scenario) and 9.3% (high end scenario), in line with values of 5-10% found in literature. In addition, also a worst-case scenario with uncertainty of 16.6% is defined.

The impact of the uncertainty of the exceedance probability (P50, P90, P90/P50) of the energy yield is also considered. In particular, the use of a method like Monte Carlo exploiting empirical distributions of input parameters is preferable to the typical assumption of values normally-distributed around a mean value. On the other hand, the application of such method is limited by the unavailability of large databases. Another factor is the limitation of PV software that either do not allow to define the uncertainty of the most important input parameters or does not allow to run simulations in batch mode.

In general, the results show that some typical figures given by a PV designer, such as P90, P99, P90/P50 tend to increase at increasing levels of uncertainty (for example, -6% of P90 for high-end scenario than low-end scenario), at fixed values of P50. This brings to weaker business plans. The use of state-of-the-art input data can thus make a project financially 5.3% more efficient (1 – 89%/94%) and these 5.3% of the base-case investment can be invested on top of the base case, either allowing a larger project or by allowing additional sustainable energy investments. In addition, the use of a shorted time series can lead to an underestimation (or overestimation) of the mean specific value of energy yield.

## ACKNOWLEDGEMENT

The research leading to these results has received funding from the European Union's Horizon 2020 research and innovation programme under the grant agreement No 649997 (Solar Bankability).

REFERENCES

[1] D. Moser, M. Del Buono, W. Bresciani, E. Veronese, U. Jahn, M. Herz, E. Janknecht, E. Ndrio, K. De Brabandere, and M. Richter, "Technical risks in PV projects," *Solar Bankability project*, 2016. [Online]. Available: http://www.solarbankability.org/results/html. [Accessed: 23-May-2016].

[2] D. Moser, M. Del Buono, U. Jahn, M. Herz, M. Richter and K. de Brabandere, "Identification of technical risks in the photovoltaic value chain and quantification of the economic impact," *Progress in Photovoltaics: Research and Applications*, 2017. DOI: 10.1002/pip.2857.

[3] IEC 61724, 1998. Photovoltaic System Performance Monitoring - Guidelines for Measurement, Data Exchange and Analysis.

[4] GUM – JCGM 100:2008, 2008. Evaluation of Measurement - Data Guide to the Expression of Uncertainty in Measurement.

[5] GUM – JCGM 101:2008, 2008. Evaluation of measurement data: supplement 1 to the guide to the expression of uncertainty in measurement: propagation of distributions using a Monte Carlo method.

[6] D. Thevenard and S. Pelland, "Estimating the uncertainty in long-term photovoltaic yield predictions," *Solar Energy*, vol. 91, pp. 432–445, 2011.

[7] C. Schwingshackl, M. Petitta, J. E. Wagner, G. Belluardo, D. Moser, M. Castelli, M. Zebisch and A. Tetzlaff, "Wind effect on PV module temperature: analysis of different techniques for an accurate estimation," *Energy Procedia*, vol. 40, pp. 77-86, 2013.

[8] PVSyst, 2016. [Online]. Available: http://www.pvsyst.com/en/ [Accessed: 27-July-2016].

[9] A. Woyte, K. De Brabandere, B. Sarr and M. Richter, "The Quality of Satellite-Based Irradiation Data for Operations and Asset Management," in *32nd European Photovoltaic Solar Energy Conference and Exhibition*, 2016, pp. 1470–1474.

[10] C.P Cameron, W.E. Boyson and D.M. Riley, "Comparison of PV system performance-model predictions with measured PV system performance", in *33rd IEEE Photovoltaic Specialist Conference*, 2008, pp. 1–6.

[11] N. Reich, J. Zenke, B. Muller, K. Kiefer and B. Farnung, "On-site performance verification to reduce yield prediction uncertainties", in *42nd IEEE Photovoltaic Specialist Conference*, 2015, pp. 1–6.

[12] U. Jahn, M. Herz, E. Ndrio, D. Moser, G. Belluardo, M. Richter and C. Tjengdrawira, "Minimizing technical risks in photovoltaic projects," *Solar Bankability project*, 2016. [Online]. Available: http://www.solarbankability.org/results/html. [Accessed: 23-May-2016].

[13] C. Tjengdrawira and M. Richter, "Review and Gap Analyses of Technical Assumptions in PV Electricity," *Solar Bankability project*, 2016. [Online]. Available: http://www.solarbankability.org/results/html. [Accessed: 23-May-2016].

# Monocrystalline 1.7 eV MgCdTe double-heterostructure solar cell with 11.2% efficiency

Calli M. Campbell,[1,2] Xin-Hao Zhao,[1,2] Yuan Zhao,[1,3] Mathieu Boccard,[3] Cheng-Ying Tsai,[1,3] Jacob J. Becker,[1,3] Zachary Holman,[3] and Yong-Hang Zhang[1,3]

[1]Center for Photonics Innovation, Arizona State University, Tempe, AZ 85287
[2]School for Engineering of Matter, Transport and Energy, Arizona State University, Tempe, AZ 85287
[3]School of Electrical, Computer and Energy Engineering, Arizona State University, Tempe, AZ 94720

*Abstract*— A dual-junction 1.7-eV/1.1-eV tandem cell has a theoretical efficiency limit of 45% under the AM1.5G spectrum. Combining an efficient thin-film semiconductor with a 1.7 eV bandgap as the top cell with a 1.1 eV silicon subcell offers a practical pathway to further cost-reduction of solar power generation. This paper reports the demonstration of a monocrystalline 1.7 eV $Mg_{0.13}Cd_{0.87}Te$ solar cell grown by molecular beam epitaxy with an active-area efficiency of 11.2% and an open-circuit voltage of 1.176 V.

*Index Terms*—CdTe, double heterostructure, MgCdTe, monocrystalline, photovoltaics (PV)

## I. INTRODUCTION

Tandem cells offer a pathway to improve device efficiencies beyond those achievable by single junction devices. Cadmium telluride ($E_g$ = 1.5 eV) is a material with high absorption and has enjoyed market success in its polycrystalline form—combining a record module efficiency of 17% (and a lab efficiency of 22%) with a low-cost manufacturing process [1]. Silicon solar cells ($E_g$ = 1.1 eV) with their high record efficiency of 26.3% [2], have enjoyed 90% of the market share [3].

Combined, a 1.7-eV/1.1-eV tandem solar cell has the potential to reach a conversion efficiency of 45% [4,5]. To have an efficiency gain in a tandem configuration with a Si solar cell (even with 25.6% efficiency), the 1.7 eV MgCdTe solar cell efficiency should reach at least 13%. A 1.7 eV bandgap can be achieved by alloying CdTe with ~13% Mg sitting in the cadmium lattice sites [6]. Record 3.6 μs bulk carrier lifetimes (measured by time-resolved photoluminescence spectroscopy) were achieved by optimized growth conditions and design of CdTe/MgCdTe double heterostructures (DH) on lattice-matched InSb (001) substrates [7]. A record efficiency monocrystalline CdTe/MgCdTe solar cell of 20% conversion efficiency was achieved with an a-Si hole contact and an i-MgCdTe passivation layer between the hole contact and the high-lifetime n-CdTe absorber [8].

## II. SAMPLE DESIGN

In modifying the absorber to 1.7 eV, studies were conducted to determine the best absorber thickness [9] and dimensions of the MgCdTe barriers in order to reduce tunneling and thermionic emission [10]. Upon achieving a structure with a record 0.56 μs bulk carrier lifetime, subsequent devices were grown and fabricated utilizing the same processes.

The cell structure shown in Fig. 1, consists of a molecular beam epitaxy-grown $Mg_{0.13}Cd_{0.87}Te/Mg_{0.5}Cd_{0.5}Te$ DH on an n-

type InSb (001) substrate, a PECVD (Plasma-enhanced chemical vapor deposition) deposited p-type hydrogenated amorphous silicon (a-Si:H) contact layer, and an Indium Tin Oxide (ITO) top electrode. The n-type dopant is indium for CdTe and the p-type dopant is boron for a-Si:H.

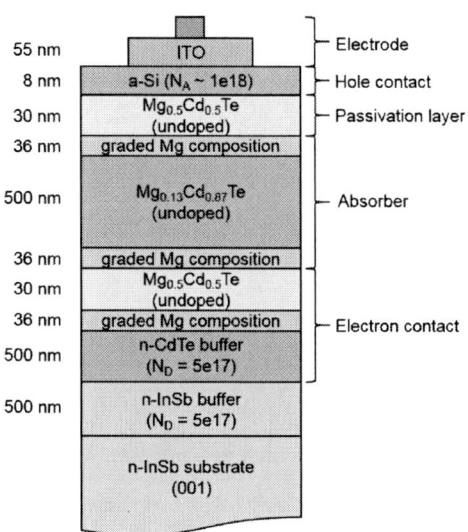

Fig. 1 Layer structure of the $Mg_{0.13}Cd_{0.87}Te/Mg_{0.5}Cd_{0.5}Te$ double-heterostructure solar cell. The doping concentrations are in the unit of $cm^{-3}$.

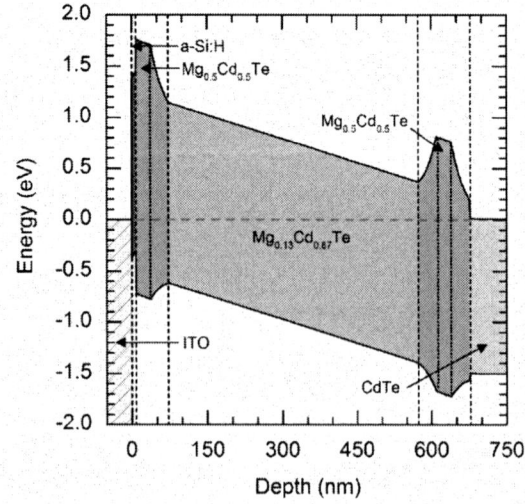

Fig. 2 Simulated band-edge diagram of a $Mg_{0.13}Cd_{0.87}Te/Mg_{0.5}Cd_{0.5}Te$ solar cell at equilibrium.

978-1-5090-5606-4/17 $31.00 © 2017 IEEE

Fig. 2 shows the simulated band-edge diagram of the solar cell structure. Note how the entirety of the absorber is depleted due to it being undoped. A set of four (one undoped and three *in-situ* doped) samples were grown in order to characterize the effect of indium dopants on $Mg_{0.13}Cd_{0.87}Te$ absorber quality. As shown by Table 1, the external luminescence quantum efficiency ($\eta_{ext}$) of each sample is drastically reduced by the addition of indium dopants. From the three spectra shown in Fig. 3, there is evidence of defect peaks between the $Mg_{0.13}Cd_{0.87}Te$ and CdTe peaks which are likely sources of nonradiative recombination. Further studies examining modifications in growth conditions will aim to improve the quality of doped samples.

The p-type a-Si:H layer on top of the DH and the bottom n-type CdTe:In buffer layer induces the built-in voltage in the $Mg_{0.13}Cd_{0.87}Te$ absorber. The small valance band offset at the $Mg_{0.13}Cd_{0.87}Te/Mg_{0.5}Cd_{0.5}Te$ interface allows holes to be extracted out of the $Mg_{0.13}Cd_{0.87}Te$ absorber region during the operation of the solar cell, while the large conduction band offset prevents electrons from entering the p-type contact region; although this same band offset will result in poor electron transport at the electron-selective contact until higher doping can be achieved. According to the literature, the carrier transport between ITO and p-type a-Si:H is achieved through tunneling [11]. In addition, the InSb substrate, having a narrow bandgap of 0.17 eV, is conductive and the resistance at the CdTe/InSb interface is negligible [12].

Table 1: Comparison of the external luminescence quantum efficiency of doped vs. undoped samples.

| Nominal indium doping density (cm$^{-3}$) | External luminescence quantum efficiency (%) |
|---|---|
| Undoped | 1.2 |
| $3 \times 10^{16}$ | 0.42 |
| $1 \times 10^{17}$ | 0.43 |
| $3 \times 10^{17}$ | 0.46 |

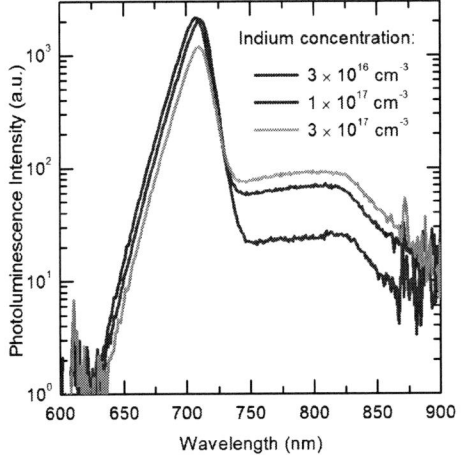

Fig. 3 The photoluminescence spectra of three nominally indium-doped double heterostructure samples. Note the asymmetrical signal of the CdTe peak, suggesting defect states between $Mg_{0.13}Cd_{0.87}Te$ and CdTe.

## III. ELECTRICAL CHARACTERIZATION AND RESULTS

Fig. 4 depicts the $\eta_{ext}$ of the device as a function of the excitation current density ($J_{sc}$) [13]. Introduced in the previous section, this is an important figure of merit to quantify the material quality of a photovoltaic absorber since $\eta_{ext}$ is related to the implied open-circuit voltage ($iV_{oc}$) of a solar cell through the formula [14-15]:

$$iV_{oc} = V_{db} - \frac{kT}{q}|\ln(\eta_{ext})| \qquad (1)$$

where $V_{db}$ is the detailed-balance open circuit voltage, calculated to be 1.4 V for the DH shown above. The $\eta_{ext}$ is 1.2% under one-sun conditions, i.e. a carrier injection current density of ~20 mA/cm². Thus, the implied open-circuit voltage is 1.3 V according to Eq. (1).

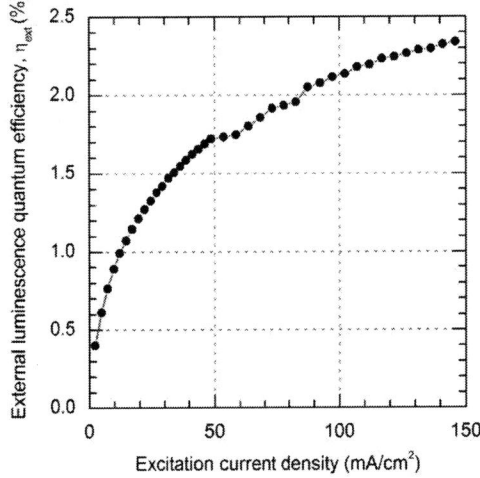

Fig. 4 The external luminescence quantum efficiency as a function of current density.

P-type a-Si:H covers the entire sample, while the 55-nm-thick ITO is deposited in small patches on top of the a-Si:H layer. Since the conductivity of the p-type a-Si:H is low, the cell area is defined by the conductive ITO, which has a measured sheet resistance of 100 Ω/sq. The ITO layer also functions as a single-layer anti-reflection coating. The current-voltage (JV) characteristic under illumination is shown in Fig. 5, followed by the external quantum efficiency in Fig. 6.

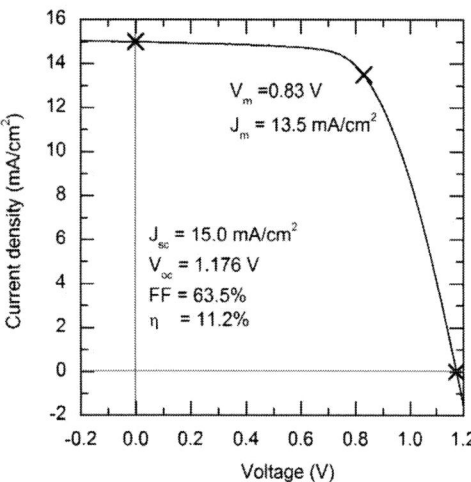

Fig. 5 Light current-voltage characteristics of $Mg_{0.13}Cd_{0.87}Te$ solar cell.

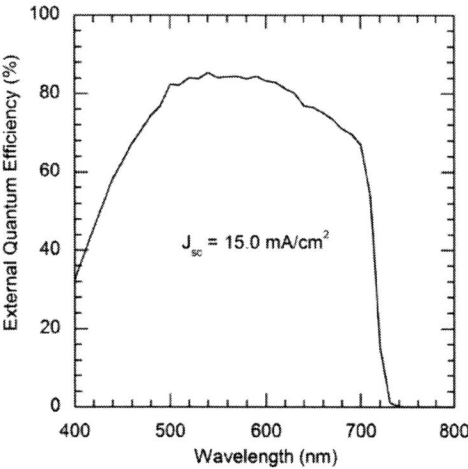

Fig. 6 External quantum efficiency of $Mg_{0.13}Cd_{0.87}Te$ solar cell.

The $J_{sc}$ could not be accurately determined from JV measurements due to the extreme sensitivity in determining the area of the device. Thus the $J_{sc}$ is calculated by integrating the EQE curve with the AM1.5 spectrum. It is determined to be 15.0 mA/cm$^2$ for the solar cell measured. The directly-measured JV curve is scaled to match the 15.0 mA/cm$^2$ $J_{sc}$. Therefore, 11.2% is the active area efficiency and considering the ~10% metal coverage, the total area efficiency is approximately 10%.

The $V_{oc}$ of 1.176 V is lower than the calculated implied $V_{oc}$, with the insufficient height of the built-in voltage between p-type a-Si:H and n-type CdTe a possible reason. It is also likely that the dimensions of the $Mg_{0.5}Cd_{0.5}Te$ barrier layers have not yet been optimized for the transport of carriers, which could decrease the output voltage of the solar cell. The low Fill Factor (FF) of 63.5% also implies the existence of resistive losses in this solar cell structure.

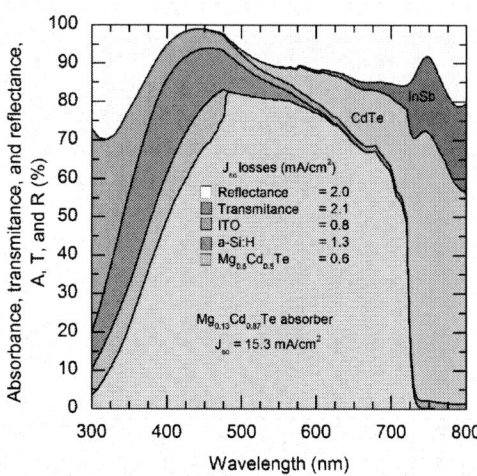

Fig. 7 Calculated reflectance, transmittance and absorptance spectra of $Mg_{0.13}Cd_{0.87}Te/Mg_{0.5}Cd_{0.5}Te$ double-heterostructure solar cell.

To analyze the loss mechanisms of the photocurrent, the reflectance and absorptance spectrum of each layer is calculated using wave-optics, as shown in Fig. 7. The absorptance of the $Mg_{0.13}Cd_{0.78}Te$ absorber layer resembles the measured external quantum efficiency spectra closely, indicating that the carrier collection efficiency in the solar cell is close to unity. This is to be expected since the minority carrier lifetime in $Mg_{0.13}Cd_{0.87}Te$ is quite long, and thus the diffusion length is expected to be longer than the thickness of the absorber. Integrating the absorptance of the $Mg_{0.13}Cd_{0.87}Te$ absorber with the AM1.5 spectrum gives a $J_{sc}$ of 15.3 mA/cm$^2$.

The losses of photocurrent due to reflectance, transmission and parasitic absorption are also shown in Fig. 6. Improvements can be made to the $J_{sc}$ by employing double-layer antireflection coatings, wider-bandgap hole contact layers and a thicker $Mg_{0.13}Cd_{0.87}Te$ absorber.

## V. CONCLUSION

We have fabricated 1.7 eV $Mg_{0.13}Cd_{0.87}Te$ solar cells with 11.2% efficiency and $V_{oc}$ of 1.176 V. The materials studies have shown that the $Mg_{0.13}Cd_{0.87}Te/Mg_{0.5}Cd_{0.5}Te$ DH is of very high quality, with an implied $V_{oc}$ of 1.3 V. The fabricated devices show a lower $V_{oc}$ than 1.3 V, along with a low FF. It is possible to further increase the $V_{oc}$ and FF by employing a different p-type material as the hole contact which features a higher built-in potential with the n-CdTe layer, and also by optimizing carrier transport through iterations of the barrier dimensions and indium doping. The $J_{sc}$ can be further improved by minimizing reflection, transmission and parasitic absorption.

## VI. ACKNOWLEDGEMENT

This work is partially supported by the Department of Energy through the BAPVC and Energy Efficiency and Renewable Energy programs, under award number DE-EE000494. This

978-1-5090-5606-4/17 $31.00 © 2017 IEEE

material is based upon work partially supported by the QESST program under the NSF and DOE—grant NSF CA No. EEC-1041895. Any opinions, findings and conclusions or recommendations expressed in this material are those of the author(s) and do not necessarily reflect those of NSF or DOE.

## REFERENCES

[1] M. A. Green, K. Emery, et. al. "Solar cell efficiency tables (version 48)" Progress in Photovoltaics (2016)

[2] K. Yoshikawa, et. al.," Silicon heterojunction solar cell with interdigitated back contacts for a photoconversion efficiency over 26%", Nature Energy, vol. 2, no. 17032 (2017)

[3] Martin A. Green, "Commercial progress and challenges for photovoltaics", Nature Energy, 15015 (2016)

[4] W. Shockley and H. J. Queisser, "Detailed Balance Limit of Efficiency of pn Junction Solar Cells", J. Appl. Phys. 32, no. 3, pp. 510-519 (1961).

[5] D. Ding, S. R. Johnson, S.-Q. Yu, S.-N. Wu, and Y.-H. Zhang, A Semi-Analytical Model for Semiconductor Solar Cells, J. Appl. Phys. 110, 123104 (2011).

[6] J. M. Hartmann, J. Cibert, F. Kany, H. Mariette, M. Chrleux, P. Alleysson, R. Langer, G. Feuillet, J. Appl. Phys. 80,6257 (1996).

[7] S. Liu, et al., "Significantly improved carrier lifetime and reduced interface recombination velocity for CdTe/MgCdTe double heterostructures", Proceedings of the 42nd IEEE Photovoltaic Specialist Conference (2015)

[8] Y. Zhao, M. Boccard, S. Liu, J. Becker, X.-H. Zhao, C. M. Campbell, E. Suarez, M. B. Lassise, Z. Holman, and Y.-H. Zhang, "Monocrystalline CdTe solar cells with open-circuit voltage over 1 V and efficiency of 17%," *Nat. Energy*, no. May, p. 16067, 2016.

[9] S. Liu, et al., "Structural and optical properties of $Mg_xCd_{1-x}$ Te alloys grown on InSb (100) substrates using molecular beam epitaxy", Proceedings of the 42nd IEEE Photovoltaic Specialist Conference (2015)

[10] C. M. Campbell et al, "1.7 eV MgCdTe double-heterostructure solar cells for tandem device applications", Proceedings of the 43rd IEEE Photovoltaic Specialist Conference (2016)

[11] A. Descoeudres, Z. C. Holman, L. Barraud, "> 21% Efficient Silicon Heterojunction Solar Cells on n- and p-Type Wafers", IEEE Journal of Photovoltaics, vol. 3, no. 1 (2013)

[12] Z.-H. He et al., "CdTe nBn photodetectors with ZnTe barrier layer grown on InSb substrates", Applied Physics Letters, vol. 109, no. 12 (2016)

[13] Y. Zhao, X.-H. Zhao, Y.-H. Zhang, "Radiative recombination dominated n-type monocrystalline CdTe/MgCdTe double-heterostructures", Proceedings of the 43rd IEEE Photovoltaic Specialist Conference (2016)

[14] U. Rau, "Reciprocity relation between photovoltaic quantum efficiency and electroluminescent emission of solar cells", Phys. Rev. B, vol. 76, 085303 (2007)

[15] O. D. Miller, E. Yablonovitch, and S. R. Kurtz, "Strong Internal and External Luminescence as Solar Cells Approach the Shockley–Queisser Limit", IEEE Journal of Photovoltaics, vol. 2, no. 3, pp. 303-311 (2012)

# MBE growth of 1.7eV Al$_{0.2}$Ga$_{0.8}$As and 1.42eV GaAs solar cells on Si using dislocations filters: an alternative pathway toward III-V/Si solar cells architectures

Arthur Onno[1], *Student Member*, *IEEE*, Mingchu Tang[1], Mu Wang[1], Yurii Maidaniuk[2], Mourad Benamara[2], Yuriy I. Mazur[2], Gregory J. Salamo[2], Lars Oberbeck[3], Jiang Wu[1], Huiyun Liu[1]

[1]Department of Electronic and Electrical Engineering, UCL, London, WC1E 7JE, United Kingdom
[2]Institute for Nanoscience and Engineering, University of Arkansas, Fayetteville, AK 72701, USA
[3]Total Gas, Renewables & Power, Paris La Défense, 92069, France

*Abstract* — Metamorphic epitaxial growth of III-V solar cells on Si has attracted significant interest for the development of III-V/Si photovoltaic architectures. In this work, we present an alternative pathway – using MBE growth techniques – based on the direct nucleation of Al$_x$Ga$_{1-x}$As materials on Si, followed by the growth of a 1.7eV Al$_{0.2}$Ga$_{0.8}$As or a 1.42eV GaAs solar cell. Dislocation Filter Layers (DFLs), in conjunction with Thermal Cycle Annealing (TCA), have been used to reduce the Threading Dislocation Density (TDD) below $10^7$cm$^{-2}$ in the base of the cell; close to the best results demonstrated with metamorphic buffers.

*Index Terms* — III-V on Silicon, Molecular Beam Epitaxy, Threading dislocation density, Photovoltaic solar cells.

## I. INTRODUCTION

Silicon-based hybrid photovoltaic devices have gained substantial academic interest in the recent years, with a wide range of absorber materials and fabrication techniques being investigated in order to overcome the efficiency limitations of crystalline silicon (c-Si) single junction photovoltaic technologies [1-2]. Among these approaches, epitaxial growth of III-V cells on a c-Si substrate, the latter potentially acting as a bottom cell, represents an elegant pathway. This technique – on top of taking advantage of the robust supply chain and low cost associated with market-dominant c-Si photovoltaic technologies – potentially enables the formation of the different p-n junctions of a multijunction solar cell in a single growth reactor and the use of a 2-terminal contacting architecture, leading to a possibly straightforward deposition and fabrication process.

The main challenge associated with the epitaxial growth of III-V compound semiconductors on Si for photovoltaic applications lies in the difference of lattice parameters between Si and III-V materials of interest. Indeed, there is no nitrogen-free direct bandgap III-V material lattice-matched to Si. Consequently lattice-mismatched approaches are needed in order to grow III-V solar cells on Si, resulting in an accumulation of strain in the grown film. Relaxation of the epilayers occurs through the formation of Misfit Dislocations (MDs) and Threading Dislocations (TDs). While MDs are confined to plans parallel to the growth surface, TDs

propagate vertically through the epilayers to the active region of the device, where they act as recombination centers, thus strongly impacting the minority carrier lifetime. Reducing the Threading Dislocation Density (TDD) to a minimum is thereby essential in order to achieve high performance minority-carrier-dominant devices such as photovoltaic cells.

Metamorphic pathways have so far led to the most compelling results. This approach consists in the growth of a lattice-matched nucleation layer – such as GaP [3] or Si [4-5] – on the Si substrate, followed by a gradual adjustment of the epilayers' lattice-parameter through alteration of their composition – for example by adding As to GaAs$_x$P$_{1-x}$ epilayers [3] or Ge to Si$_x$Ge$_{1-x}$ epilayers [4-5]. Due to the limited differences in lattice parameters throughout the metamorphic buffer, the TDD is kept low within the epitaxial film. High bandgap ($E_g$>1.6eV) III-V solar cells epitaxially grown on Si with TDDs below $5\times10^6$cm$^{-2}$ have been demonstrated, leading to bandgap-voltage offset ($W_{oc}$) values under 0.55V [6-7]. Using Si$_x$Ge$_{1-x}$ buffers, a $V_{oc}$ above 1.0V ($W_{oc}$<0.4V) and a TDD below $10^6$cm$^{-2}$ have even been reported for GaAs solar cells [4], although these results have not been replicated with higher bandgap cells.

Using Molecular Beam Epitaxy (MBE) growth techniques, we present an alternative non-metamorphic pathway – similar to the approach developed by Yamaguchi *et. al.* [8] – based on the direct nucleation of materials from the Al$_x$Ga$_{1-x}$As system on Si substrates. A high TDD is thus obtained at the III-V/Si interface, due to the 4% lattice-mismatch between Al$_x$Ga$_{1-x}$As and Si. Dislocation Filter Layers (DFLs), consisting of an iteration of spaced Strained Layer Superlattices (SLSs) [9], are then grown in order to reduce the TDD by 2 to 4 orders of magnitude. Excellent results have recently been achieved on MBE-grown quantum dot lasers using this approach [10], with the demonstration of a TDD below $10^6$cm$^{-2}$. In this work, transfer of this technique to III-V solar cells monolithically grown on Si – in the present case 1.7eV Al$_{0.2}$Ga$_{0.8}$As solar cells for dual-junction Al$_{0.2}$Ga$_{0.8}$As/Si applications and 1.42eV GaAs cells for use in stand-alone single junction devices or as middle subcells in an In$_{0.49}$Ga$_{0.51}$P/GaAs/Si triple-junction architectures – is presented.

978-1-5090-5606-4/17 $31.00 © 2017 IEEE

## II. Experimental Methods

### A. Samples growth

For both absorber materials investigated ($Al_{0.2}Ga_{0.8}As$ and GaAs), two samples have been grown: one reference sample grown lattice-matched on GaAs and one sample grown on Si using Dislocation Filter Layers (DFLs) and Thermal Cycle Annealing (TCA). The four samples were grown in a Veeco GEN 930 Solid State Molecular Beam Epitaxy (SSMBE) system.

Temperatures were controlled using an infrared pyrometer and a thermocouple mounted on the back of the substrate-holder. Reflection High Energy Electron Diffraction (RHEED) was used to in-situ monitor the evolution of the growth surface during deposition as well as during the pre-growth high-temperature oxide removal step. The lattice-mismatched growth runs on Si were performed on n-type Si (100) wafers offcut 4° towards the [01-1] plane in order to avoid the formation of Anti-Phase Domains (APDs) due to polar-on-nonpolar epitaxy [11]. Standard n-type GaAs wafers were used for the lattice-matched reference samples.

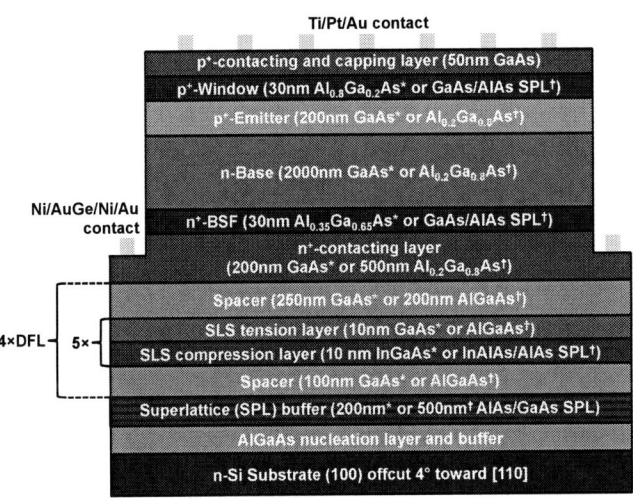

Fig. 1. Structure of the samples grown on Si. The DFL buffer is in orange/red, the active layers of the devices are in blue. The differing parameters between the two batches of samples are indicated by the "†" symbol for the $Al_{0.2}Ga_{0.8}As$ cells and by the "*" symbol for the GaAs ones. The reference samples grown lattice-matched on GaAs present an identical device structure (in blue), the DFL buffer (in orange/red) being replaced by a 200nm-thick GaAs buffer.

The structure of the samples grown on Si is presented in Figure 1. The DFL buffer is depicted in orange/red, the active layers of the devices are in blue. The reference samples grown on GaAs have an identical device structure, the DFL buffer being replaced by a 200nm-thick GaAs buffer. As only one Al source was available at the time of growth, the Al deposition rate was fixed throughout the growth runs. As a result, the structure of the DFL buffer, Back Surface Field (BSF) and

window layers were adapted for the $Al_{0.2}Ga_{0.8}As$ cells and differ from the GaAs samples.

The DFL buffer consists in an $Al_xGa_{1-x}As$ nucleation layer followed by an AlAs/GaAs superlattice (SPL) in order to smooth out the growth surface [12] before deposition of the four DFLs. Each DFL is comprised of a Strained-Layer Superlattice (SLS) made of alternating compression and tension layers [9] inserted between two $Al_{0.2}Ga_{0.8}As$ or GaAs spacers. The TCA cycles were performed immediately following the growth of each SLS DFL. Details about the annealing sequence can be found in Ref. [10].

The $Al_{0.2}Ga_{0.8}As$ and GaAs cells have a similar device structure consisting in a bottom $n^+$-type contacting layer – 200nm-thick for GaAs devices, 500nm-thick for $Al_{0.2}Ga_{0.8}As$ devices – followed by a 30nm-thick $n^+$-type BSF, a 2000nm-thick n-type base, a 200nm-thick $p^+$-type emitter, a 30nm-thick $p^+$-type window layer and finally a 50nm-thick $p^+$-GaAs contacting and capping layer. For the $Al_{0.2}Ga_{0.8}As$ cells, AlAs/GaAs SPLs have been used for the BSF and window layers. Conversely, for the GaAs cells, $Al_{0.35}Ga_{0.65}As$ and $Al_{0.8}Ga_{0.2}As$ layers have been used the BSF and window layers, respectively.

### B. Devices fabrication

Patterning was performed by standard photolithography techniques prior to device separation by wet etching and metal contacts deposition. The samples were first selectively etched in a $H_2SO_4:H_2O_2:H_2O$ (1:10:80) solution in order to define 3×3mm individual mesa-structures and to access the bottom $n^+$-contacting layer, as shown on Figure 1. The contact to the n-type region consists in a Ni/AuGe/Ni/Au (5nm/100nm/30nm/200nm) metal structure thermally evaporated and annealed at 390°C for 60 seconds under $N_2$ atmosphere. The same contact structure was also evaporated on the full back of the two samples grown on GaAs in order to improve the lateral conduction of charge carriers to the base of the devices. As the wafers used for the growth on Si were low-doped (1 to 10 Ω.cm), no contacts were deposited on the back of these samples. A Ti/Pt/Au (20nm/50nm/400nm) front grid contact to the p-type region was finally deposited by sputtering.

It is to be noted that the top GaAs $p^+$-contacting layer was not etched in order to protect the underlying Al-rich layers from oxidation. Furthermore, no anti-reflection coating was applied to the samples. As a result, sizeable optical losses arise from reflection at the front surface of the devices and absorption in the top GaAs contacting layer. For $Al_{0.2}Ga_{0.8}As$ devices, using OPAL2 software [13], the short-circuit current density losses due to reflection and absorption are evaluated at about 8.8mA.$cm^{-2}$ and 5.0mA.$cm^{-2}$, respectively. For GaAs devices, these losses are evaluated at about 12.6mA.$cm^{-2}$ and 5.3mA.$cm^{-2}$, respectively. As the coverage of the top metal grid contact has not been optimized, resulting in a non-negligible shadowing (2.93mm$^2$), the current densities presented hereafter refer to the 7.07mm$^2$ active area of the devices.

## C. Characterization

Cross-sectional Transmission Electron Microscopy (TEM) was used to characterize the structural properties of the samples and, in particular, to calculate the Threading Dislocation Density (TDD). The samples were first prepared by mechanical polishing and ion milling in a Fischione 1010 ion mill. TEM imaging was then carried out at 300keV in a FEI Titan 80-300S TEM system fitted with a CEOS image corrector.

Optoelectronic characterization of the samples and devices included Current density versus Voltage (J-V) curve tracing under AM1.5G illumination, Illumination versus Open-circuit voltage (Suns-$V_{oc}$) characterization and External Quantum Efficiency (EQE) measurement. J-V characteristics of the devices were acquired at 25°C using a Keithley 2400 sourcemeter coupled with ReRa Tracer 3.0 software. 1-sun AM1.5G spectrum illumination was obtained from a LOT solar simulator equipped with a filtered xenon lamp and calibrated at 100mW.cm$^{-2}$ using a GaAs calibration cell. Suns-$V_{oc}$ characteristics were acquired using a Sinton Instruments Suns-$V_{oc}$ system. Given the substantial difference between the absorption spectra of the c-Si reference cell used to monitor illumination and of the higher bandgap measured III-V cells, filters were placed in front of the reference cell in order to reduce the spectral mismatch to a minimum [14]. A Schott KG3 filter was used to measure the $Al_{0.2}Ga_{0.8}As$ cells while a Techspec longpass filter with an 875nm cutoff wavelength was used to measure the GaAs cells. The 1-sun $V_{oc}$ difference between the J-V measurements and the Suns-$V_{oc}$ measurements was thus reduced to under 20mA.cm$^{-2}$. An additional spectral mismatch coefficient was then calculated for each device in order to match the J-V and Suns-$V_{oc}$ measurements [14]. Room-temperature EQE of the best cells was measured with a ReRa SpeQuest quantum efficiency system.

## III. RESULTS

### A. Impact of the DFL on the TDD

TEM imaging of the DFLs of the GaAs sample grown on Si is shown on Figure 2. TDs are bent into MDs when they intersect with the DFLs. These MDs can then coalesce, mutually annihilate or bend back into TDs and progress upward. Annealing of the SLS increases the mobility of the TDs and MDs, improving the chances of coalescence or mutual annihilation [15].

As shown on Figure 2, each individual DFL reduces the TDD by a factor of two to six. The overall TDD is thus reduced by two full orders of magnitude, from $1\times10^9$cm$^{-2}$ at the III-V/Si interface to $8.3(\pm2)\times10^6$cm$^{-2}$ just after the 4$^{th}$ DFL. Not shown on Figure 2, the TDD in the base of the cell is further reduced to $5(\pm2)\times10^6$cm$^{-2}$, close to the best results achieved using metamorphic approaches [3-7].

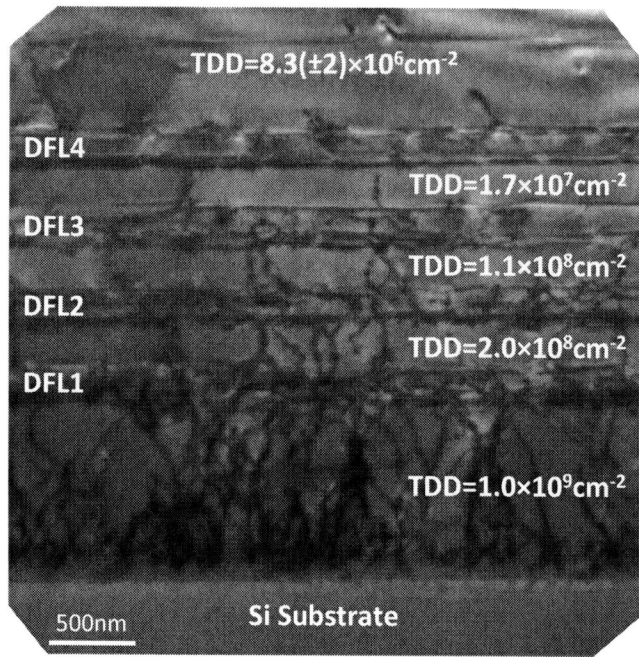

Fig. 2. Transmission Electron Microscopy (TEM) imaging of the buffer and Dislocation Filter Layers (DFLs) of the GaAs sample grown on Si. Threading Dislocations (TDs) are bent into Misfit Dislocations (MDs) in the DFL, where they can merge, mutually annihilate or bend back into TDs and resume their progression upward.

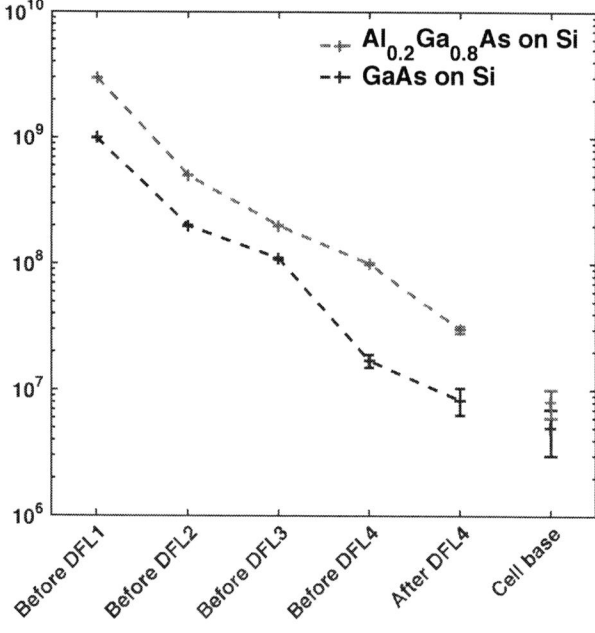

Fig. 3. Evolution of the TDD in the samples grown lattice-mismatched on Si.

Similar reductions in TDD are demonstrated for the $Al_{0.2}Ga_{0.8}As$ sample grown lattice-mismatched on Si, as shown on Figure 3. The TDD in the base of the $Al_{0.2}Ga_{0.8}As$ cell grown on Si has been evaluated at $8(\pm2)\times10^6cm^{-2}$.

### B. 1.7eV $Al_{0.2}Ga_{0.8}As$ solar cells

J-V characteristics, acquired under illumination, of the best devices from both 1.7eV $Al_{0.2}Ga_{0.8}As$ samples are presented in Figure 4 (full lines). The pseudo-J-V curves, extracted from Suns-$V_{oc}$ measurements, are also displayed in dashed lines.

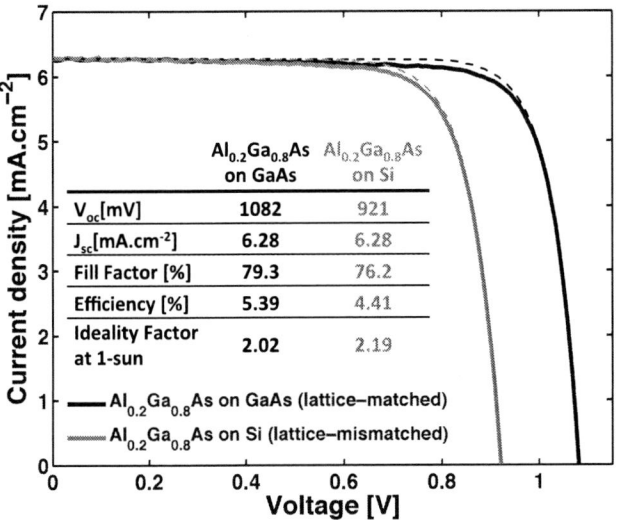

| | $Al_{0.2}Ga_{0.8}As$ on GaAs | $Al_{0.2}Ga_{0.8}As$ on Si |
|---|---|---|
| $V_{oc}$[mV] | 1082 | 921 |
| $J_{sc}$[mA.cm$^{-2}$] | 6.28 | 6.28 |
| Fill Factor [%] | 79.3 | 76.2 |
| Efficiency [%] | 5.39 | 4.41 |
| Ideality Factor at 1-sun | 2.02 | 2.19 |

Fig. 4. J-V characteristics acquired under illumination (full lines) and pseudo-J-V curves extracted from Suns-$V_{oc}$ measurements (dashed lines) of the best $Al_{0.2}Ga_{0.8}As$ devices grown on GaAs (black) and on Si (red). The impact of the presence of TDs on the performances of the cells is apparent, in particular on the $V_{oc}$.

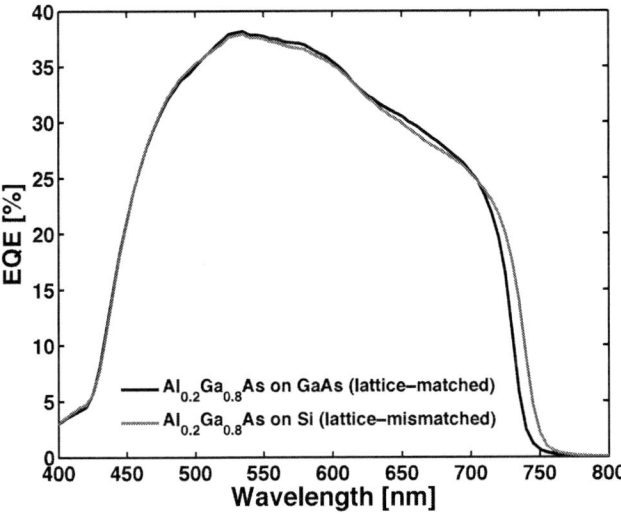

Fig. 5. External Quantum Efficiency (EQE) measurements of the best devices from the $Al_{0.2}Ga_{0.8}As$ samples grown on GaAs (black) and on Si (red).

The impact of the TDD on the performances of the device is apparent, with a 161mV reduction in $V_{oc}$ from the sample grown lattice-matched on GaAs to the sample grown lattice-mismatched on Si. This is in agreement with the presence of TDs shown by TEM, leading to a stronger non-radiative recombination rate and a reduced minority carrier lifetime. Ideality factors, extracted from Suns-$V_{oc}$ measurements, also indicate a stronger non-radiative recombination rate on Si, with an increase of the 1-sun ideality factor from $n=2.02$ on GaAs to $n=2.19$ on Si.

The $V_{oc}$ values measured are nevertheless relatively low in regard of the high bandgap of the material ($\approx1.7$eV), even for the reference sample grown lattice-matched on GaAs. The bandgap-voltage offset $W_{oc}$, defined as $W_{oc}=E_g/q-V_{oc}$, thus deviates notably from the semi-empirical value of 0.4V expected from high material quality devices. The ideality factors – higher than 2 for both devices, indicating non-radiative recombinations in the depletion zone as the dominant recombination pathway – confirm this relatively low material quality. The performance of our devices, in particular the $V_{oc}$ is consequently primarily limited by the bulk material quality of the grown $Al_{0.2}Ga_{0.8}As$, independent of the presence of TDs.

Both devices present very close $J_{sc}$ values, indicative of a limited impact of the TDs on the carrier collection efficiency. This is confirmed by the similar EQE curves presented in Figure 5. The bulk $Al_{0.2}Ga_{0.8}As$ material quality thus appears to be the limiting factor in the diffusion length of minority carriers in the base of the solar cells, a higher TDD being needed to impact the collection efficiency of the devices and thus their $J_{sc}$ and EQE.

### C. 1.42eV GaAs solar cells

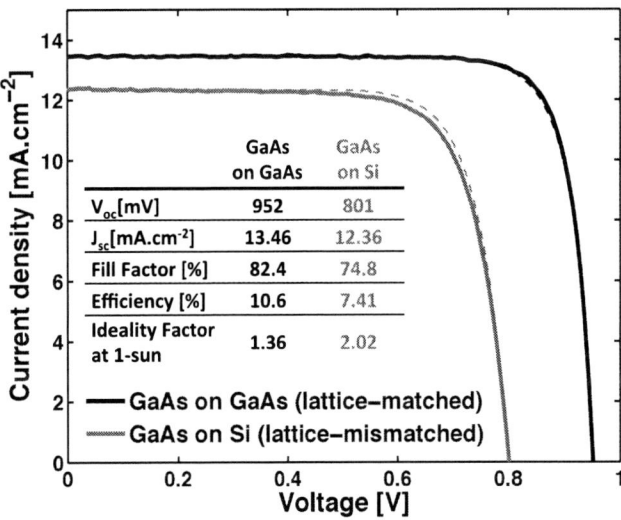

| | GaAs on GaAs | GaAs on Si |
|---|---|---|
| $V_{oc}$[mV] | 952 | 801 |
| $J_{sc}$[mA.cm$^{-2}$] | 13.46 | 12.36 |
| Fill Factor [%] | 82.4 | 74.8 |
| Efficiency [%] | 10.6 | 7.41 |
| Ideality Factor at 1-sun | 1.36 | 2.02 |

Fig. 6. J-V characteristics acquired under illumination (full lines) and pseudo-J-V curves extracted from Suns-$V_{oc}$ measurements (dashed lines) of the best GaAs devices grown on GaAs (black) and on Si (red). The impact of the presence of TDs on the performances of the cells is apparent on the $V_{oc}$ and on the $J_{sc}$.

The J-V characteristics acquired under illumination (full lines) and pseudo-J-V curves extracted from Suns-$V_{oc}$ measurements (dashed lines) of the best 1.42eV GaAs devices grown on both substrates are displayed in Figure 6. Similar to the $Al_{0.2}Ga_{0.8}As$ samples presented above, the impact of the presence of TDs is apparent with a comparable reduction in $V_{oc}$ (151mV) between the samples grown lattice-matched on GaAs and lattice-mismatched on Si. The 1-sun ideality factors $n$ of the cells also illustrate the impact of TDs, with an increase from $n=1.36$ – characteristic of a balance between recombination pathways – on GaAs to $n=2.02$ – characteristic of recombinations dominated by SRH recombinations in the depletion zone – on Si.

The lower ideality factor for the sample grown lattice-matched on GaAs, compared with the $Al_{0.2}Ga_{0.8}As$ sample grown on GaAs, indicates a better bulk material quality. This is confirmed by the lower $W_{oc}$ value: 469mV for the best lattice-matched GaAs device versus 618mV for the best lattice-matched $Al_{0.2}Ga_{0.8}As$ device.

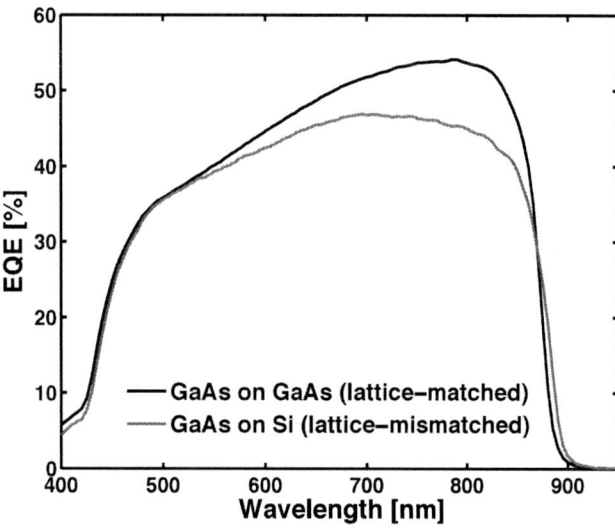

Fig. 7. External Quantum Efficiency (EQE) measurements of the best devices from the two GaAs samples grown on GaAs (black) and on Si (red).

Contrary to the $Al_{0.2}Ga_{0.8}As$ devices, the impact of the presence of TDs on the $J_{sc}$ is apparent, with a $J_{sc}$ reduction of 1.10mA.cm$^{-2}$ between the sample grown on GaAs and the one grown on Si. The EQE measurements, displayed in Figure 7, confirm the lower collection efficiency for the cell grown on Si, especially at longer wavelengths. This can be directly related to a lower diffusion length of minority carriers in the presence of TDs, with in particular a reduced carrier collection in the base of the cell, away from the depletion zone. As a result, the solar cell grown on Si exhibits a poorer EQE at longer wavelengths absorbed in the back of the cell. As opposed to the $Al_{0.2}Ga_{0.8}As$ samples, the diffusion length is not limited by the bulk material quality and TDs directly affect the $J_{sc}$ and the EQE in a non-negligible way.

## IV. DISCUSSION

Prototypes of $Al_{0.2}Ga_{0.8}As$ and GaAs solar cells have been grown on Si substrates using direct nucleation of $Al_xGa_{1-x}As$ on Si followed by dislocation filters in order to reduce the TDD. The $Al_{0.2}Ga_{0.8}As$ devices exhibit a bandgap of 1.7eV, making them suitable for current-matched III-V/Si tandem dual junction solar cells. A TDD below $10^7$cm$^{-2}$ has been demonstrated in the base of the cell.

The main limitation of our 1.7eV $Al_{0.2}Ga_{0.8}As$ solar cell prototypes lies in the bulk material quality of the $Al_{0.2}Ga_{0.8}As$, for the sample grown lattice-mismatched on Si as well as for the sample grown with a negligible TDD on GaAs. This poor material quality is confirmed by the low $W_{oc}$ values and high ideality factors measured for both $Al_{0.2}Ga_{0.8}As$ samples while this issue is not as significant for GaAs samples. Growth of high material quality $Al_{0.2}Ga_{0.8}As$ is known to be challenging, with oxygen contamination a main concern leading to a strong deterioration of the performances of the devices [16]. This issue has been highlighted in previous publications by the authors [17-18].

Recent work has focused on the improvement of the $Al_{0.2}Ga_{0.8}As$ material quality by improving the growth conditions and in particular the substrate temperature. This optimization study has yielded a strong improvement of performances with increasing the growth temperature from 580°C to 620°C, with a $V_{oc}$ over 1.21V demonstrated [19]. $Al_{0.2}Ga_{0.8}As$ solar cells epitaxially grown on Si with a $V_{oc}$ exceeding 1.0V are likely achievable by transferring this optimized $Al_{0.2}Ga_{0.8}As$ growth recipe on Si. Further optimization of the buffer and DFLs, in order to achieve a material quality and a TDD similar to the ones demonstrated with laser devices [10], can potentially yield $V_{ocs}$ above 1.1V on Si substrates.

Optical optimization of the $Al_{0.2}Ga_{0.8}As$ cells front surface is also needed in order to improve the current density produced by the cell. Replacement of the current AlAs/GaAs SPL window layer with a state-of-the-art AlInP window layer would allow the removal of the GaAs contacting layer between the front contact grid fingers, the AlInP layer being used as an etch stop. Further improvement would include the deposition of a broadband Anti-Reflection Coating (ARC). Such enhancements should increase the $J_{sc}$ of the cell to values close or above 20mA.cm$^{-2}$ and allow current-matching with an underlying Si bottom subcell.

Finally, in addition to the perspective of achieving high material quality with a TDD below $1\times10^6$cm$^{-2}$ [10], a key benefit of this alternative pathway lies in the use of a thin buffer (2.1μm to 2.8μm), reducing the amount of III-V materials required and potentially the growth time needed. Additionally, a c-Si cell can be used as a bottom cell; in comparison with the $Si_xGe_{1-x}$ subcell required using a metamorphic $Si_xGe_{1-x}$ approach [5,7].

## V. CONCLUSION

1.7eV $Al_{0.2}Ga_{0.8}As$ and 1.42eV GaAs solar cells have been grown on Si by Molecular Beam Epitaxy (MBE), using direct nucleation of lattice-mismatched $Al_xGa_{1-x}As$ material on the Si substrates. Dislocation Filter Layers (DFLs), along with Thermal Cycle Annealing (TCA) steps, have then been used in order to reduce the Threading Dislocation Density (TDD). TDDs of $8(\pm2)\times10^6 cm^{-2}$ and $5(\pm2)\times10^6 cm^{-2}$ have been reached in the base of the $Al_{0.2}Ga_{0.8}As$ and GaAs cells, respectively.

As expected, the presence of Threading Dislocations (TDs) directly impacts the $V_{oc}$ of the cells for both absorber materials investigated, with a reduction in $V_{oc}$ of about 150-160mV from the reference samples grown lattice-matched on GaAs to the samples grown lattice-mismatched on Si.

However, the $J_{sc}$ and the EQE are only impacted by the presence of TDs for the GaAs solar cells. This is due to a relatively low material quality for the $Al_{0.2}Ga_{0.8}As$ cells, leading to high $W_{oc}$ values on both substrates and limited bulk minority carrier diffusion length, independently of the presence of TDs. As a result the carrier collection efficiency of the $Al_{0.2}Ga_{0.8}As$ is similar on Si and on GaAs substrates.

Optimization of the $Al_{0.2}Ga_{0.8}As$ growth parameters have since been carried out, yielding a strong improvement in the $V_{oc}$ and $J_{sc}$ of reference samples grown lattice-matched on GaAs with increasing the growth temperature from 580°C to 620°C. Transfer of this optimized $Al_{0.2}Ga_{0.8}As$ growth recipe on Si substrates is expected to yield a $V_{oc}$ above 1V and pave the way toward the achievement of high efficiency 1.7eV $Al_{0.2}Ga_{0.8}As$ solar cells on Si suitable for dual junction III-V/Si tandem architectures.

## ACKNOWLEDGEMENT

This work was supported by Total S.A. The authors acknowledge the support of the National Science Foundation of the U.S. (EPSCoR Grant #OIA-1457888).

## REFERENCES

[1] W. Shockley and H. J. Queisser, "Detailed Balance Limit of Efficiency of p-n Junction Solar Cells," *Journal of Applied Physics*, vol. 32, pp. 510-2519, 1961, DOI: 10.1063/1.1736034.

[2] D. D. Smith, P. Cousins, S. Westerberg, R. De Jesus-Tabajonda, G. Aniero and Y.-C. Shen, "Toward the Partical Limits of Silicon Solar Cells," *IEEE Journal of Photovoltaics*, vol. 4, pp. 1465-1469, 2014, DOI: 10.1109/JPHOTOV.2014.2350695.

[3] T. J. Grassman, M. R. Brenner, A. M. Carlin, S. Rajagopalan, R. Unocic, R. Dehoff, M. Mills, H. Fraser and S. A. Ringel, "Toward Metamorphic Multijunction GaAsP/Si Photovoltaics Grown on Optimized GaP/Si Virtual Substrates Using Anion-Graded $GaAs_yP_{1-y}$ Buffers," *Proceedings of the 34th IEEE PVSC*, 2009, pp. 2016-2020, DOI: 10.1109/PVSC.2009.5411489.

[4] S. A. Ringel, J. A. Carlin, C. L. Andre, M. K. Hudait, M. Gonzalez, D. M. Wilt, E. B. Clark, P. Jenkins, D. Scheiman, A. Allerman, E. A. Fitzgerald and C. W. Leitz, "Single-junction InGaP/GaAs Solar Cells Grown on Si Substrates with SiGe Buffer Layers," *Progress in Photovoltaics: Research and Applications*, vol. 10, pp. 417-426, 2002, DOI: 10.1002/pip448.

978-1-5090-5606-4/17 $31.00 © 2017 IEEE

[5] K. J. Schmieder, A. Gerger, M. Diaz, Z. Pulwin, C. Ebert, A. Lochtefeld, R. Opila and A. Barnett, "Analysis of tandem III-V/SiGe devices grown on Si," *Proceedings of the 38th IEEE PVSC*, 2012, pp. 968-973, DOI: 10.1109/PVSC.2012.6317764.

[6] K. N. Yaung, M. Vaisman, J. Lang and M. L. Lee, "GaAsP solar cells on GaP/Si with low threading dislocation density," *Applied Physics Letters*, vol. 109, p. 032107, 2016, DOI: 10.1063/1.4959825.

[7] L. Wang, M. Diaz, B. Conrad, X. Zhao, D. Li, A. Soeriyadi, A. Gerger, A. Lochtefeld, C. Ebert, I. Perez-Wurfl and A. Barnett, "Material and Device Improvement of GaAsP Top Solar Cells for GaAsP/SiGe Tandem Solar Cells Grown on Si Substrates," *IEEE Journal of Photovoltaics*, vol. 5, pp. 1800-1804, 2015, DOI: 10.1109/JPHOTOV.2015.2459918.

[8] M. Yamaguchi, Y. Ohmachi, T. Oh'hara, Y. Kadota, M. Imaizumi and S. Matsuda, "GaAs Solar Cells Grown on Si Substrates for Space Use," *Progress in Photovoltaics: Research and Applications*, vol. 9, pp. 191-201, 2001, DOI: 10.1002/pip366.

[9] G. C. Osbourn, "Strained-Layer Superlattices: A Brief Review," *IEEE Journal of Quantum Electronics*, vol. QE-22, pp. 1677-1681, 1986, DOI: 10.1109/JQE.1986.1073190.

[10] S. Chen, W. Li, J. Wu, Q. Jiang, M. Tang, S. Shutts, S. N. Elliott, A. Sobiesierski, A. J. Seeds, I. Ross, P. M. Smowton and H. Liu, "Electrically pumped continuous-wave III–V quantum dot lasers on silicon," *Nature Photonics*, vol. 10, pp. 307-311, 2016, DOI: 10.1038/nphoton.2016.21.

[11] H. Kroemer, "Polar-on-nonpolar epitaxy," *Journal of Crystal Growth*, vol. 81, pp. 193-204, 1987, DOI: 10.1016/0022-0248(87)90391-5.

[12] X. Xu, B. Huang, H. Ren and M. Jiang, "Smoothing effect of GaAs/Al$_x$Ga$_{1-x}$As superlattices grown by metalorganic vapor phase epitaxy," *Applied Physics Letters*, vol. 64, pp. 2949-2951, 1994, DOI: 10.1063/1.111422.

[13] K. R. McIntosh, and S. C. Baker-Finch, "OPAL 2: Rapid optical simulation of silicon solar cells," *Proceedings of the 38th IEEE PVSC*, 2012, pp. 265-271, DOI: 10.1109/PVSC.2012.6317616.

[14] T. Roth, J. Hohl-Ebinger, E. Schmich, W. Warta, S. W. Glunz and R. A. Sinton, "Improving the accuracy of Suns-V$_{oc}$ measurements using spectral mismatch correction," *Proceedings of the 33rd IEEE PVSC*, 2008, pp. 1355-1359, DOI: 10.1109/PVSC.2008.4922686.

[15] C. H. Simpson and W. A. Jesser, "On the Use of Low Energy Misfit Dislocation Structures to Filter Threading Dislocations in Epitaxial Heterostructures," *Physica status solidi (a)*, vol. 149, pp. 9-20, 1991, DOI: 10.1002/pssa.2211490102.

[16] C. Amano, K. Ando and M. Yamaguchi, "The effect of oxygen on the properties of AlGaAs solar cells grown by molecular-beam epitaxy," *Journal of Applied Physics*, vol. 63, pp. 2853-2856, 1988, DOI: 10.1063/1.340938.

[17] A. Onno, J. Wu, Q. Jiang, S. Chen, M. Tang, Y. Maidaniuk, M. Benamara, Y. I. Mazur, G. J. Salamo, N. P. Harder, L. Oberbeck and H. Liu, "1.7eV Al$_{0.2}$Ga$_{0.8}$As solar cells epitaxially grown on silicon by SSMBE using a superlattice and dislocation filters," *Proceedings of the SPIE*, 2016; vol. 9743, pp. 10-1-10-7, DOI: 10.1117/12.2208950.

[18] A. Onno, J. Wu, Q. Jiang, S. Chen, M. Tang, Y. Maidaniuk, M. Benamara, Y. I. Mazur, G. J. Salamo, N. P. Harder, L. Oberbeck and H. Liu, "Al$_{0.2}$Ga$_{0.8}$As solar cells monolithically grown on Si and GaAs by MBE for III-V/Si tandem dual-junction applications," *Energy Procedia*, vol. 92, pp. 661-668, DOI: 10.1016/j.egypro.2016.07.037.

[19] A. Onno, M. Tang, L. Oberbeck, J. Wu and H. Liu, Unpublished work.

# III-V/Si Tandem Cells Utilizing Interdigitated Back Contact Si Cells and Varying Terminal Configurations

Manuel Schnabel[1], Michael Rienäcker[2], Agnes Merkle[2], Talysa R. Klein[1], Nikhil Jain[1], Stephanie Essig[1*], Henning Schulte-Huxel[1], Emily Warren[1], Maikel F.A.M. van Hest[1], John Geisz[1], Jan Schmidt[2], Rolf Brendel[2], Robby Peibst[2], Paul Stradins[1], Adele Tamboli[1]

[1]National Renewable Energy Laboratory, Golden CO 80401, USA
[2]Institute for Solar Energy Research Hamelin, Emmerthal 31860, Germany
*now at École Polytechnique Fédérale de Lausanne (EPFL), Neuchâtel, Switzerland

*Abstract* — **Integrating wide-bandgap III-V with Si solar cells has been shown to yield higher efficiencies than Si alone: As also presented at this conference, four-terminal efficiencies exceeding 32% have been attained. In this contribution, independent and electrically connected operation of the subcells in such tandem cells is examined. The optics of the tandem cell change significantly if a conducting interconnect, rather than an insulating glass slide, is required between the subcells. These effects are studied, and optically optimized structures for different types of tandem cell operation are presented. It is found that minimizing reflection at the conductive interface between the two cells while maintaining conductivity is the key challenge faced in such devices.**

## I. INTRODUCTION

Solar cells made from bulk crystalline Si dominate the market, with a market share exceeding 90% [1]. However, improving efficiencies has become increasingly difficult because the current record efficiency of 26.6% [2] is already very close to the theoretical limit of 29.4% [3]. In order to achieve significantly higher efficiencies, a novel approach is required, and the one that has proven itself in practice is the tandem cell approach, with which efficiencies of 38.8% have been achieved under 1 sun illumination [4] using five junctions of III–V semiconductors. In order to substantially boost the efficiency of Si solar cells, we have been developing stacked two-junction III-V/Si tandem cells [5], recently attaining 1-sun efficiencies above 32% [6]. That said, the four-terminal (4T) operation required to achieve this efficiency may be difficult to integrate into a cost-effective module.

In this contribution, we compare the operation of an electrically connected III-V/Si tandem cell to 4T operation, from an optical viewpoint. We first present the external quantum efficiency of a 31.5% efficient (uncertified measurement) 4T GaInP/Si tandem cell, and show that all its essential features can be modeled using an in-house transfer matrix algorithm. We then apply this model to study the optics of GaInP/Si cells that are electrically connected using an EVA-based transparent conducting adhesive (TCA) [7], optimize this structure, and identify key elements required for optimized performance.

## II. EXPERIMENTAL DETAILS

The four-terminal tandem cell used to obtain the inputs for optical simulations was prepared from a GaInP and a Si cell. The rear heterojunction GaInP top cell [8] was grown inverted, bonded to high-index (n=1.56) glass using epoxy (also n=1.56), and then had the substrate removed. It was also processed with a front and rear side antireflective coating (MgF$_2$/ZnS, and ZnS, respectively, with optimized thicknesses). The Si bottom cell with doped POLy-Si on passivating Oxide (POLO) contacts was prepared using 160 μm n-type Cz-Si wafers and the process described in Ref. [9], with the cell dimensions adapted to match the top cell. The front side of the Si cell is then bonded the rear side of the glass with the GaInP on top using epoxy, aligning the cells using an infrared camera. The resulting stack is listed in Table 1.

The quantum efficiency (QE) and specular reflectance were measured on a custom-built instrument, and then used as inputs for an in-house transfer-matrix-based algorithm. This algorithm models reflection of the sample, absorption in each layer, and short-circuit current density $j_{sc}$ if given a collection efficiency (1 for an absorber layer, 0 for all other layers unless otherwise stated). The nominal layer structure of the GaInP/Si tandem cell was fitted to the experimental QE and reflectance data, and this real layer structure modified as needed for electrically interconnected tandems. Refractive indices were obtained from in-house measurements on reference samples or from the PVLighthouse refractive index database [10-13].

## III. COMPARISON OF SIMULATION AND EXPERIMENT FOR FOUR-TERMINAL TANDEMS

Figure 1 shows experimental external quantum efficiency (EQE) and reflection curves of a mechanically stacked 31.5% (uncertified) efficient GaInP/Si four-terminal tandem cell, prepared as described in Sec. II and Table 1. Also shown are simulated EQE and reflection curves.

The simulation begins from nominal layer thicknesses and refractive indices determined previously on reference samples. In order to fit the top cell EQE, only the top cell AlInP window layer thickness (reduced 15%), GaInP absorber layer

978-1-5090-5606-4/17 $31.00 © 2017 IEEE

thickness (reduced 5%), and GaInP absorber layer collection efficiency (reduced from 100 to 91%) are adjusted. In order to fit the bottom cell EQE, which features a front-side texture that cannot be accurately represented in a transfer matrix algorithm, the Si wafer thickness is increased (from its actual value of 150μm to 357μm) to represent the increase in optical path length that texture yields. The full layer structure modeled is given in Table 1.

Table 1: Layer stack used for simulation of four-terminal GaInP/Si tandem in Fig. 1, showing materials, thicknesses, and how much their absorption contributes to $j_{sc}$. Where values differ from nominal values, nominal values are given in brackets. Indispensable top and bottom cell layers are shaded blue and green, respectively, while layers present for antireflection performance are shaded grey. Optically negligible layers such as grids, or openings in layers, with small area fraction are omitted.

| Material | Thickness (nm) | Contrib. to $j_{sc}$ |
|---|---|---|
| air | 0 | 0 |
| MgF$_2$ | 97 | 0 |
| ZnS | 41 | 0 |
| n-Al$_{0.52}$In$_{0.48}$P | 17 (20) | 0 |
| n-Ga$_{0.5}$In$_{0.5}$P | 950 (1000) | 0.91 |
| p-Al$_{0.27}$Ga$_{0.26}$In$_{0.47}$P | 200 | 0 |
| p-Al$_{0.5}$Ga$_{0.5}$As | 500 | 0 |
| ZnS | 82 | 0 |
| epoxy | 10,000 | 0 |
| glass, n=1.56 | 1,000,000 | 0 |
| epoxy | 10,000 | 0 |
| PECVD SiO$_x$ | 100 | 0 |
| SiN$_x$, n=1.91 | 70 | 0 |
| SiN$_x$, n=2.4 | 15 | 0 |
| n,p-Si | 357,000 (150,000) | 1 |
| Al$_2$O$_3$ | 15 | 0 |
| SiN$_x$, n=1.91 | 120 | 0 |
| Al | 10,000 | 0 |

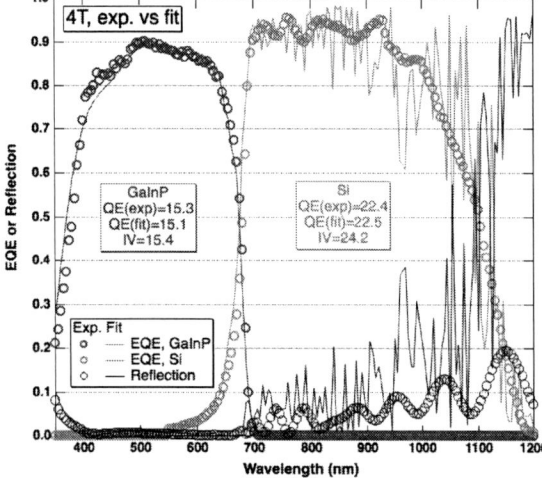

Figure 1. Experimental and simulated EQE and reflection curve for a 4T GaInP/Si tandem cell. The short-circuit currents derived from convolving the EQEs with the AM1.5G spectrum, and as measured using a solar simulator (IV), are given in the figure in mA/cm$^2$.

It can be seen that the external quantum efficiencies of both curves are well modeled using this approach, making only few, minor, and physically reasonable, changes to the nominal structure of the device. In particular, the imperfect charge carrier collection of 91% from the GaInP layer required for this model is feasible given an experimental internal quantum efficiency for the GaInP cell in the 85-90% range; the AlInP layer is known to vary in thickness by up to 25% based on prior growth runs, and texture is known to extend the optical path length by deflecting incident photons from normal incidence.

The main shortcomings of the model are a high-frequency oscillation at wavelengths above 700 nm, which arises from the assumption of a precisely 10 μm thick epoxy layer between the cells, whereas there is some inhomogeneity in practice; and a high reflection above ~900 nm that is not observed in experiment, which arises at least in part from the fact that only specular reflection is measured experimentally, and texture deflects much of the reflected light away from the detector (or even totally internally reflects it), whereas the model assumes flat interfaces throughout. Nevertheless, the short-circuit current densities calculated from experimental and modeled EQEs, and the overall shape of the EQE curves, are in good agreement, validating the simulation as a method to model short circuit current densities in other device structures to within ~0.3 mA/cm$^2$. It is not entirely clear why the $j_{sc}$ from a solar simulator measurement matches the EQE value for GaInP cells but not for Si cells, but we will assume in the following that the deviation for Si is not wavelength dependent and that studying EQEs will permit an optimization of $j_{sc}$, even if the final value is an underestimation.

## IV. SIMULATION OF ELECTRICALLY CONNECTED TANDEMS

The structure of an electrically interconnected tandem utilizing a TCA interconnect differs from the four-terminal structure modeled in the previous section in two key regards: only the TCA and other conductive layers can be used to optimize optics between the two cells, and the glass that is still required for mechanical stabilization of the GaInP cell must be moved to the front. These considerations, and restricting materials for antireflective layers to MgF$_2$ and ZnS as insulating optical layers, and thermally evaporated indium tin oxide (thermal ITO) as a conducting optical layer, leads to the initial trial structure shown in Table 2.

It consists of identical layers and layer thicknesses for the cells as Table 1, and also includes the TCA which is known to be ~50 μm thick. It does not initially include ITO layers; these will be added in further iterations as they are not strictly necessary from an electronic viewpoint: both Al$_{0.5}$Ga$_{0.5}$As and Si are expected to make good contact to the TCA given sufficiently high doping. The MgF$_2$/ZnS double-layer antireflective coating originally present on the front of the 4T device is now separated by the front-side glass and epoxy, as both have refractive indices between those of MgF$_2$ and ZnS. Parameters that will be varied in the following are labeled as

978-1-5090-5606-4/17 $31.00 © 2017 IEEE

such in Table 2, and simulated GaInP and Si cell $j_{sc}$ values as a function of simulation parameters are summarized in Table 3.

Table 2: Layer stack used for simulation of electrically connected GaInP/Si tandem cells, showing materials and thicknesses (contributions to $j_{sc}$ are assumed to be identical to those in Table 1). Initial thicknesses were carried over from Sec. III, and those that will be varied in simulations denoted as such. Indispensable top and bottom cell layers are shaded blue and green respectively, while layers present for antireflection performance are shaded grey. Optically negligible layers such as grids, or openings in layers, with small area fraction are omitted.

| Material | Initial Thickness (nm) |
|---|---|
| air | 0 |
| MgF$_2$ | 97 (varied) |
| glass, n=1.56 (varied) | 1,000,000 |
| epoxy | 10,000 |
| ZnS | 41 (varied) |
| n-Al$_{0.52}$In$_{0.48}$P | 17 |
| n-Ga$_{0.5}$In$_{0.5}$P | 950 |
| p-Al$_{0.27}$Ga$_{0.26}$In$_{0.47}$P | 200 |
| p-Al$_{0.5}$Ga$_{0.5}$As | 500 |
| ITO (varied) | 0 (varied) |
| TCA (EVA assumed) | 50,000 |
| ITO (varied) | 0 (varied) |
| n,p-Si | 357,000 |
| Al$_2$O$_3$ | 15 |
| SiN$_x$, n=1.91 | 120 |
| Al | 10,000 |

Simulating the structure in Table 2 and comparing results to Fig. 1 shows that while the top cell $j_{sc}$ only decreases 0.2 mA/cm$^2$, the bottom cell $j_{sc}$ takes a large hit in this new and unoptimized cell architecture, decreasing by 6.2 mA/cm$^2$ (first two rows of Table 3). Optimizing the MgF$_2$ and ZnS thicknesses reveals that the initial thicknesses (97 nm, 41 nm) were already close to optimal (97 nm, 47 nm) for maximizing current in the current-limiting GaInP cell (Table 3, "Optimizing for Limiting Current"), as would be desired in a two-terminal tandem cell. Changing the glass used from high-index glass (n=1.56) matched to epoxy to commercial borosilicate glass (n=1.52) leads to an optimized GaInP $j_{sc}$ that is only 0.1 mA/cm$^2$ lower, indicating that the type of glass used is not critical in initial stages of device optimization, provided the MgF$_2$ and ZnS thicknesses are readjusted. It also has a negligible effect on the Si cell's $j_{sc}$. For series-connected cells therefore, the initial structure in Table 2 is largely sufficient, and while the coupling of light into both cells might be improved using textured instead of planar front glass, the much poorer coupling of light into the bottom cell is still sufficient for the Si cell not to limit current.

Turning now to electrically connected devices that extract the full power of each cell, such as 3T devices, a need for further adaptation of the tandem structure in order to lower the 6.2 mA/cm$^2$ $j_{sc}$ loss of the Si cell is apparent. Examining the simulation of the initial structure in Table 2, and of the structures optimized for current-matched performance, reveals that the mean reflection for wavelengths above 700 nm is

~35% (not shown). There are two conceivable reasons for this: either the front of the tandem structure in Table 2 (MgF$_2$/glass/epoxy/ZnS) only exhibits antireflective performance below 700 nm, or the interface between the GaInP and Si cell, which only light above ~700 nm reaches, is particularly reflective. Taking the structure in Table 2 and setting the TCA (EVA) thickness to zero while keeping all other parameters constant raises the $j_{sc}$ of the Si cell from 16.3 mA/cm$^2$ to 23.3 mA/cm$^2$, higher than in an optimized four-terminal device. This indicates that the TCA is responsible for the high reflectivity above 700 nm, which must be mitigated.

Table 3: Simulated short-circuit current densities in the GaInP and Si cells of an electrically connected tandem connected using an EVA-based TCA and having front-side glass. Values for a four-terminal device from Fig. 1 are also shown for comparison. The second block of data is for current-matched operation, while the third block is for maximum power output from each cell.

| Variation | GaInP $j_{sc}$ (mA/cm$^2$) | Si $j_{sc}$ (mA/cm$^2$) |
|---|---|---|
| Four-Terminal (Table 1) | 15.1 | 22.5 |
| Initial Interconnected (Table 2) | 14.9 | 16.3 |
| Optimizing for Limiting Current | | |
| n=1.56 glass | 15.0 | 16.4 |
| borosilicate glass | 14.9 | 16.4 |
| Optimizing for Maximum Power | | |
| 2x thermal ITO | 14.8 | 19.8 |
| 2x 1.7E19 sputtered ITO | 14.7 (14.4)[a] | 23.3 (23.6)[a] |
| 2x 1E20 sputtered ITO | 14.7 | 22.8 |
| 2x 6E20 sputtered ITO | 14.8 | 17.4 |
| 1x thermal ITO | 14.8 | 18.6 |

[a]PVLighthouse raytracing [14] result of same stack.

The high reflectivity upon introducing the TCA is not surprising: its main component, EVA, has a mean refractive index of 1.49 between 600 and 1200 nm, as compared to 3.4 for Al$_{0.5}$Ga$_{0.5}$As and 3.7 for Si, yielding two highly reflective interfaces. This reflection can be reduced by introducing a conductive, transparent layer with intermediate refractive index such as ITO at these two interfaces. These two ITO layers are added to the simulation, and thicknesses of both ITO layers, and of front-side MgF$_2$ and ZnS, varied iteratively to reach an optimum output. Since the optimization is to be for maximum power extraction, and GaInP cells have roughly twice the maximum power point voltage of Si cells, the expression $(2j_{sc,GaInP}+j_{sc,Si})$ is optimized. Adding 84 nm thermal ITO above and below the TCA, together with optimized MgF$_2$ and ZnS thicknesses yields 14.8 mA/cm$^2$ in the GaInP cell and 19.8 mA/cm$^2$ in the Si cell, thus recovering more than half the $j_{sc}$ loss observed in the Si cell upon switching from the structure in Table 1 to that in Table 2, while enduring only marginal reductions in $j_{sc}$ of the GaInP cell.

Nevertheless, there is still room for improvement, as 23.3 mA/cm$^2$ were modeled for the Si cell when the reflective TCA layer is absent. We now consider the nature of the ITO, as inputs used were for in-house thermal ITO, which is known to have non-ideal transparency. The aforementioned simulation is repeated for sputtered ITO with carrier concentrations of 0.17, 1, and $6 \times 10^{20}$ cm$^{-3}$, whose complex refractive indices are taken from Ref. [10]. The results are shown in Table 3; for the three sputtered ITOs trialed, a Si cell $j_{sc}$ of 23.3, 22.8, and 17.4 mA/cm$^2$, respectively, was obtained, with GaInP $j_{sc}$ being equal within the error of the simulation. The stack with the lowest-doped ITO was also simulated with the raytracer from PVLighthouse [14] to account for texture on the Si cell, yielding a GaInP cell $j_{sc}$ of 14.4 mA/cm$^2$ and a Si cell $j_{sc}$ of 23.6 mA/cm$^2$ and lending additional credibility to the transfer matrix modeling. It is thus shown that an electrically interconnected GaInP/Si tandem can harvest as much light as an optimized four-terminal device; however, insertion of an ITO layer (or other transparent conductor) with adequate refractive index and extinction coefficient is crucial.

In practice, while lower-doped ITO is better optically, it may not contact the TCA as well, leading to a trade-off between fill factor and $j_{sc}$ in actual cell optimization. In addition, little is known about the electrical properties of the proposed p-Al$_{0.5}$Ga$_{0.5}$As/ITO junction; III-V semiconductors are rarely combined with transparent conducting oxides as transparent conductors can be realized within the III-V group of compounds, and ITO is a n-type material. For this reason, we perform one last optimization, in which the ITO between p-Al$_{0.5}$Ga$_{0.5}$As and TCA is omitted, but retained between TCA and Si. Comparing the two "thermal ITO" rows of Table 3, it can be seen that incorporating only this one thermal ITO layer hardly affects the $j_{sc}$ of the GaInP cell but decreases $j_{sc}$ of the Si cell by 1.2 mA/cm$^2$. We expect that this effect may be tolerable if p-Al$_{0.5}$Ga$_{0.5}$As/ITO junctions are poor. More generally, the transparent conductors inserted between either cell and the TCA are absolutely critical in achieving good performance, both from an electrical and optical viewpoint.

## V. CONCLUSIONS

The external quantum efficiencies, and hence short-circuit current densities, of GaInP/Si tandem solar cells stacked for four-terminal or electrically connected operation have been studied. The former was used to validate the optical model by demonstrating a good fit to experimental data. The latter was used to study the optical interlayers required when the two cells have glass on top, instead of between the cells, and are connected by a TCA of relatively low refractive index.

It was shown that a current density only 0.2-0.3 mA/cm$^2$ lower is readily attained in the current-limiting GaInP cell in the electrically connected structure. The Si cell exhibits higher current than the GaInP cell in all scenarios investigated, indicating that if current-matched operation is desired, no interlayers are required between the cells and the TCA from an optical viewpoint. This is however not optimal in terms of extracting maximum power from a GaInP/Si tandem cell, and if the electrically connected tandem is to be operated in a manner that allows the full power of each subcell to be collected, e.g. in a 3T device, the strong reflection at the interfaces between the TCA and the two cells must be suppressed. This can be successfully achieved by introducing ITO layers at these two interfaces, but only if the ITO is sufficiently transparent and has the correct refractive index. Engineering of transparent conducting interlayers such as ITO at these two interfaces is expected to be key to realizing effective light management and carrier transport in GaInP/Si tandem cells that are electrically connected with a TCA.

## ACKNOWLEDGMENTS

Funding for this work at NREL was provided by DOE through EERE contract SETP DE-EE00030299 and under Contract No. DE- AC36-08GO28308. Funding at ISFH was provided by the EU's FP7 under grant # 608498, as well as by the German Federal Ministry for Economic Affairs and Energy under grant # FKZ0324040. The United States Government retains and the publisher, by accepting the article for publication, acknowledges that the United States Government retains a non-exclusive, paid-up, irrevocable, world- wide license to publish or reproduce the published form of this manuscript, or allow others to do so, for United States Government purposes.

The ISFH authors wish to thank Thomas Friedrich, Frank Heinemeyer, Jan Hensen, Sabine Kirstein, Heike Kohlenberg, Tobias Neubert, Annika Raugewitz, David Sylla, Marta Tatarzyn and Nadine Wehmeier from ISFH for the processing solar cells. A special thank is due to Guido Glowatzki, Jan Krügener and Bernd Koch from the Institute of Electronic Materials and Devices of Leibniz University Hanover for helping with polysilicon deposition and ion implantation and the Laboratory of Nano and Quantum Engineering (LNQE) of Leibniz University of Hanover for support.

## REFERENCES

[1] Fraunhofer-ISE, "Photovoltaics Report," 17.11.2016 2016.

[2] NREL, "NREL Efficiency Chart," ed. This plot is courtesy of the National Renewable Energy Laboratory, Golden, CO., 2017.

[3] A. Richter, M. Hermle, and S. W. Glunz, "Reassessment of the limiting efficiency for crystalline silicon solar cells," *IEEE Journal of Photovoltaics,* vol. 3, pp. 1184-91, 2013.

[4] M. A. Green, K. Emery, Y. Hishikawa*, et al.,* "Solar cell efficiency tables (version 49)," *Progress in Photovoltaics: Research and Applications,* vol. 25, pp. 3-13, 2017.

[5] J. V. Gee, Gary, "A 31%-Efficient GaAs/Silicon mechanically-stacked, multijunction concentrator solar cell," *IEEE Electron Device Letters,* pp. 754-758, 1988.

[6] S. Essig, C. Allebe, T. Remo*, et al.,* "32% efficient III-V/Si dual-junction solar cells and their challenging path towards cost competeness " *submitted to 44th IEEE PVSC,* 2017.

[7] T.R. Klein, M. Schnabel, E.L. Warren*, et al.,* "Transparent Conductive Adhesives for Tandem Solar Cells," *Proceedings of the 44th IEEE PVSC,* 2017.

[8] J. F. Geisz, M. A. Steiner, I. García, *et al.*, "Enhanced external radiative efficiency for 20.8% efficient single-junction GaInP solar cells," *Applied Physics Letters,* vol. 103, p. 041118, 2013.

[9] M. Rienaecker, M. Bossmeyer, A. Merkle, *et al.*, "Junction Resistivity of Carrier-Selective Polysilicon on Oxide Junctions and Its Impact on Solar Cell Performance," *IEEE Journal of Photovoltaics,* vol. 7, pp. 11-18, 2016.

[10] Z. C. Holman, M. Filipič, A. Descoeudres, *et al.*, "Infrared light management in high-efficiency silicon heterojunction and rear-passivated solar cells," *Journal of Applied Physics,* vol. 113, p. 013107, 2013.

[11] P. Kumar, M. K. Wiedmann, C. H. Winter, *et al.*, "Optical properties of Al2O3 thin films grown by atomic layer deposition," *Applied Optics,* vol. 48, pp. 5407-5412, 2009/10/01 2009.

[12] M. Vogt, "Development of physical models for the simulation of optical properties of solar cell modules," Ph.D., Faculty for Mathematics and Physics, Leibniz Universität Hannover, 2015.

[13] K. R. McIntosh, T. C. Kho, K. C. Fong, *et al.*, "Quantifying the optical losses in back-contact solar cells," in *2014 IEEE 40th Photovoltaic Specialist Conference (PVSC),* 2014, pp. 0115-0123.

[14] (9. June 2017). *PV Lighthouse: Module Ray Tracer.* Available: https://http://www.pvlighthouse.com.au

# Towards High-Efficiency GaAsP/Si Tandem Cells

S. Fan[1], M. Vaisman[2,3], K. Nay Yaung[2], E. Perl[3], D. Martín-Martín[4], M. Leilaeioun[5], Z. C. Holman[5] and M. L. Lee[1,2]

[1]University of Illinois at Urbana-Champaign, Urbana, Illinois 61801, USA.

[2]Yale University, New Haven, Connecticut 06511, USA.

[3]National Renewable Energy Laboratory, Golden, Colorado 80401, USA.

[4]Universidad Rey Juan Carlos, Madrid, Móstoles 28933, Spain.

[5]Arizona State University, Tempe, Arizona 85287, USA.

*Abstract* — **In this work, we present recent progress towards high-efficiency epitaxial 1.7 eV/1.1 eV GaAsP/Si tandem cells. First, we present Si bottom cells with thick, epitaxial *n*-GaAsP optical filtering layers, yielding an efficiency of 6.25%. Furthermore, we demonstrate GaAsP/Si 2-terminal tandem cells with $V_{oc}$ of 1.596 V using an unoptimized tunnel junction to interconnect the GaAsP and Si sub-cells. Finally, we discuss the design of SiN$_x$/SiO$_x$ double-layer anti-reflectance coating (ARC) for GaAsP/Si tandem cells, which boosted the $J_{sc}$ by 28.9%.**

## I. INTRODUCTION

Tandem devices consisting of a 1.7 eV GaAs$_{0.75}$P$_{0.25}$ (hereafter GaAsP) top cell grown epitaxially on a 1.1 eV Si bottom cell are promising for low-cost, high-efficiency photovoltaics. Efficiencies as high as 37-44% have been theoretically predicted, and the best experimentally demonstrated GaAsP/Si tandem cells have reached efficiency of ~13.1% [1]. Since the advent of high-quality GaP/Si nucleation layers [2, 3], significant advances have been made in reducing threading dislocation density (TDD) in the GaAsP top cell [4-6] and in controlling lifetime degradation in Si wafers during III-V growth [7]. This improved understanding can now be used to experimentally realize high-efficiency GaAsP/Si tandem cells.

In this work, we present Si bottom cells with 3.6 μm *n*-GaAs$_x$P$_{1-x}$ isotype graded buffer layers as an optical filter that attain 6.25% efficiency. Considering the GaAsP top cell efficiency of 12% that we have achieved by growth on Si substrate and without ARC, the result shows the potential for 4-terminal tandems exceeding 20% efficiency. The Si bottom cell shows sufficient $J_{sc}$ to current-match our best top cells, indicating the potential for 2-terminal tandem cells as well. With an unoptimized tunnel junction and single-layer SiN$_x$ anti-reflection coating (ARC), we achieve a preliminary GaAsP/Si 2-terminal tandem solar cell with a $J_{sc}$ of 11.91 mA/cm$^2$, a $V_{oc}$ of 1.596 V, a FF of 0.660, and an efficiency of 12.55%, comparable with prior III-V/Si tandem cells [1]. Finally, we discuss the design of a SiN$_x$/SiO$_x$ double-layer anti-reflection coating (DLARC) by plasma-enhanced chemical vapor deposition (PECVD) to boost the $J_{sc}$ of 1.7 eV GaAsP solar cells by 34.1% and that of GaAsP/Si 2-terminal tandems by 28.9%,

respectively. Advances reported here pave the path towards 30% efficient 2-terminal GaAsP/Si tandem cells.

## II. METHODS

All devices investigated in this work were grown on GaP/Si templates (NAsP$_{III/V}$ GmbH) on Si (001) or GaP (001) substrates in a Veeco Mod GEN-II solid-source MBE system equipped with Ga, In, Al, As, P, Si, and Be sources. The templates provided by NAsP$_{III/V}$ consist of a 40 nm *n*-GaP nucleation layer and a 200 nm homoepitaxial *n*-Si emitter, all grown on a lightly B-doped (~$2 \times 10^{15}$ cm$^{-3}$), 775 μm

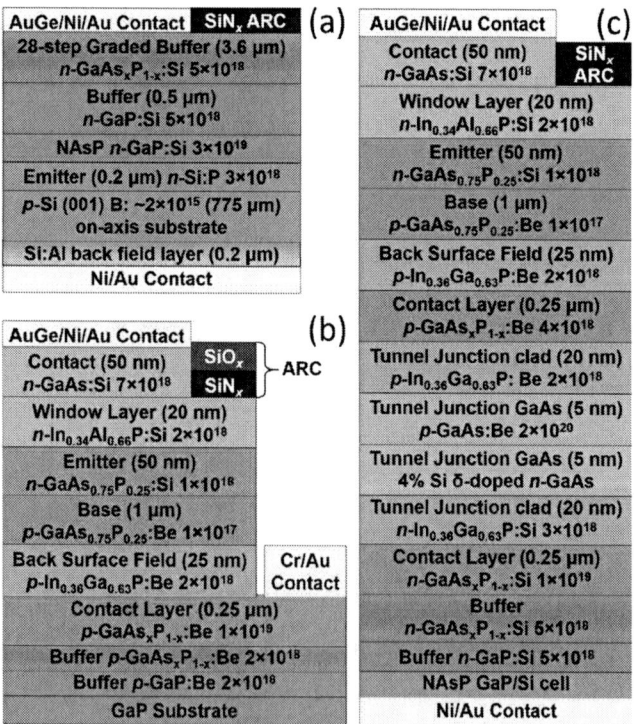

Figure 1. Schematics of (a) Si bottom cells with *n*-GaAs$_x$P$_{1-x}$ optical filter layer, (b) GaAsP top cells grown on GaP substrates, and (c) GaAsP/Si tandem cells. All doping concentrations in cm$^{-3}$.

Czochralski-Si wafer. Growth methods for the GaAsP solar cells and GaAs$_x$P$_{1-x}$ graded buffer layers were reported elsewhere [5, 6]. For the Si solar cells, a stack of 3.6 μm $n$-GaAs$_x$P$_{1-x}$ graded buffer layers was grown to mimic the optical conditions that would be seen by the Si bottom cell in a tandem device. Devices were fabricated using conventional photolithography, e-beam metal deposition, rapid thermal annealing, and solution-based mesa etching. SiN$_x$/SiO$_x$ DLARCs were deposited by PECVD. Layer schematics for the Si bottom cell with $n$-GaAs$_x$P$_{1-x}$ optical filter layers, GaAsP solar cells grown on GaP substrates with a SiN$_x$/SiO$_x$ DLARC, and preliminary tandem cells utilizing an unoptimized Si δ-doped, strained GaAs tunnel junction and SiN$_x$ ARC are given in Fig. 1(a)-(c), respectively. For the Si bottom cell, the Al rear contact was annealed at 700 °C for 10 min under a N$_2$ flow rate of 500 sccm, a process that results in reasonable contact resistance but minimal rear surface passivation [8].

Light current-voltage (LIV) measurements were taken under approximate one-sun AM1.5G conditions with an ABET Technologies 10500 solar simulator. External quantum efficiency (EQE) and specular reflectance were measured with a PV Measurements QEX7 system, both of which were used to determine internal quantum efficiency (IQE). The refractive index and thicknesses of PECVD SiN$_x$ and SiO$_x$ were measured by a WVASE32 spectroscopic ellipsometer. STFCalc was used for the simulation to optimize ARC design.

## III. RESULTS

### A. Si bottom cells with GaAs$_x$P$_{1-x}$ optical filtering layers

Figure 2(a) shows the EQE and the reflectance of the Si solar cell with $n$-GaAsP optical filtering layers. The 110 nm SiN$_x$ ARC increases the peak EQE from 68% to 88%, corresponding to an increase of 4.16 mA/cm$^2$ in J$_{sc}$. The poor IR response is mainly due to the unoptimized wafer thickness and recombination at the rear contact. LIV characteristics of the Si solar cell with and without the SiN$_x$ ARC under approximate AM1.5G one-sun conditions are compared in Fig. 2(b). The ARC boosts the efficiency to 6.25%, with a J$_{sc}$ of 19.50 mA/cm$^2$, a V$_{oc}$ of 0.503 V, and a FF of 0.637. The $n$-GaAsP optical filter used here is not as thick as an actual GaAsP top cell, and a fully absorbing top cell will reduce EQE of the Si bottom cell in the range of 600-730 nm; therefore, J$_{sc}$ and efficiency of the Si bottom cell are expected to reduce to 17-18 mA/cm$^2$ and 5.4%-5.8%, respectively. Nonetheless, a J$_{sc}$ of 18 mA/cm$^2$ is sufficient for current matching with our best GaAsP top cells grown on GaP/Si templates.

While the V$_{oc}$ reported here is comparable to prior reports [9, 10] considering the lower J$_{sc}$ due to $n$-GaAsP optical filtering layers, the relatively low value also indicates that lifetime degradation in the Si occurred during III-V growth steps, as reported by others [7]. Feifel $et$ $al.$ have found that protecting the back side of the Si wafer prior to III-V deposition can greatly reduce degradation, demonstrating V$_{oc}$ values of up to 0.632 V [11]; none of the wafers used in this work received any

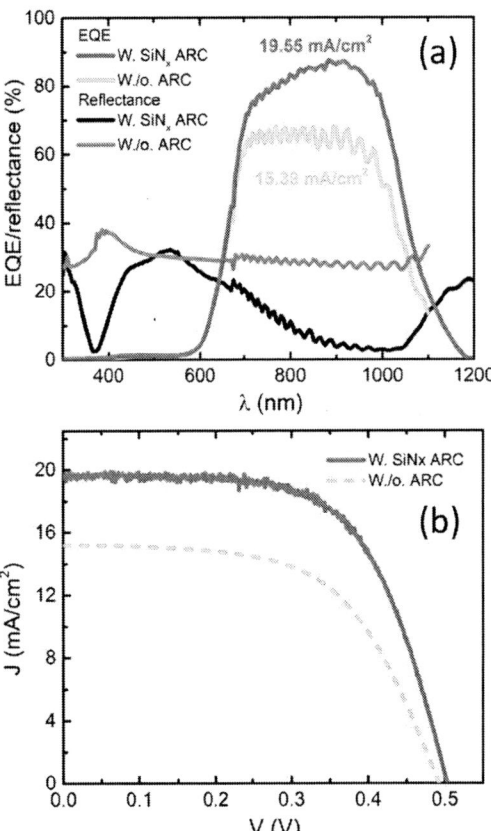

Figure 2. (a) EQE, reflectance and (b) LIV curves of Si bottom cells with GaAsP optical filter layers and with/without the 110 nm SiN$_x$ ARC.

special backside treatment. Therefore, near-term work to boost the Si cell efficiency will focus on protecting the back side of the Si wafer during III-V growth, implementing effective back-surface passivation, as well as reducing the wafer thickness and texturing the back surface.

### B. GaAsP/Si 2-terminal tandem solar cells

We have recently started to investigate GaAsP/Si tandem cells interconnected through a strained GaAs tunnel junction, as shown in Fig. 1(c). EQE and IQE of individual subcells and the specular reflectance of the tandem are shown in Fig. 3(a). Without an ARC, the J$_{sc}$ values of the GaAsP top cell and the Si bottom cell are 10.04 mA/cm$^2$ and 12.50 mA/cm$^2$, respectively. A V$_{oc}$ of 1.580 V is demonstrated, shown by the black curve in Fig. 3(b), comparable to the value of 1.62 V demonstrated by Grassman $et$ $al.$ [1]. The slightly "S-shaped" LIV curve in the vicinity of V$_{oc}$ indicates that the unoptimized GaAs tunnel junction impairs the tandem efficiency. Nonetheless, an efficiency of 10.92% is obtained under approximate AM1.5G conditions. Switching to carbon as the $p$-type dopant should greatly improve the performance and thermal stability of the tunnel junction, as shown previously [1].

Shown as dashed curves in Fig. 3(a), IQE of the GaAsP top cell and Si bottom cell are calculated from

978-1-5090-5606-4/17 $31.00 © 2017 IEEE

$$IQE = \frac{EQE}{1-R}$$

where R is the reflectance of the cell. The limit of $J_{sc}$ of each subcell is determined by the convolution of IQE with the standard AM1.5G spectral irradiance, which corresponds to 13.83 mA/cm$^2$ for GaAsP top cell and 17.46 mA/cm$^2$ for the Si bottom cell; note that these GaAsP top cells are unoptimized and thus perform slightly worse than our previously reported GaAsP top cells [6]. The significant discrepancy of 3.79 mA/cm$^2$ between $J_{sc}$ of the GaAsP top cell and IQE-based limit of $J_{sc}$ indicates that the efficiency of the tandem cell could be enhanced by an ARC, which was validated by the increase of $J_{sc}$ from 10.04 mA/cm$^2$ to 11.91 mA/cm$^2$ after the deposition of a single 63 nm SiN$_x$ layer on the tandem cell by PECVD. The LIV of the tandem cell with 63 nm SiN$_x$ is shown as the red curve in Fig. 3(b), showing an increased $V_{oc}$ of 1.596 V, a FF of 0.660, and an efficiency of 12.55%.

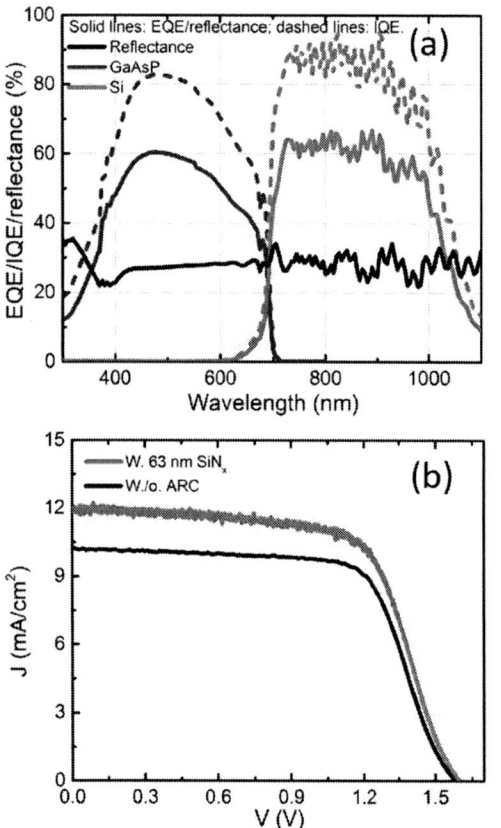

Figure 3. (a) Reflectance of the tandem cell (black curve), EQE of the GaAsP top cell (blue solid curve) and the Si bottom cell (red solid curve) prior to ARC deposition, and the corresponding IQE (dashed curves). (b) LIV of the GaAsP/Si tandem cell without ARC (black curve) and with 63 nm SiN$_x$ (red curve).

### C. Design of DLARC for GaAsP/Si tandem cells

Aiming at 30% efficiency GaAsP/Si tandem cells, the optimization of ARC to boost EQE of both subcells over 300-

1200 nm becomes crucial. Here, we designed a SiN$_x$/SiO$_x$ DLARC by tuning the refractive index using various PECVD conditions and by calculating the optimal thicknesses in TFCalc. In our current GaAsP/Si tandem cells, the GaAsP top cell is current-limiting. Therefore, we started with the design of

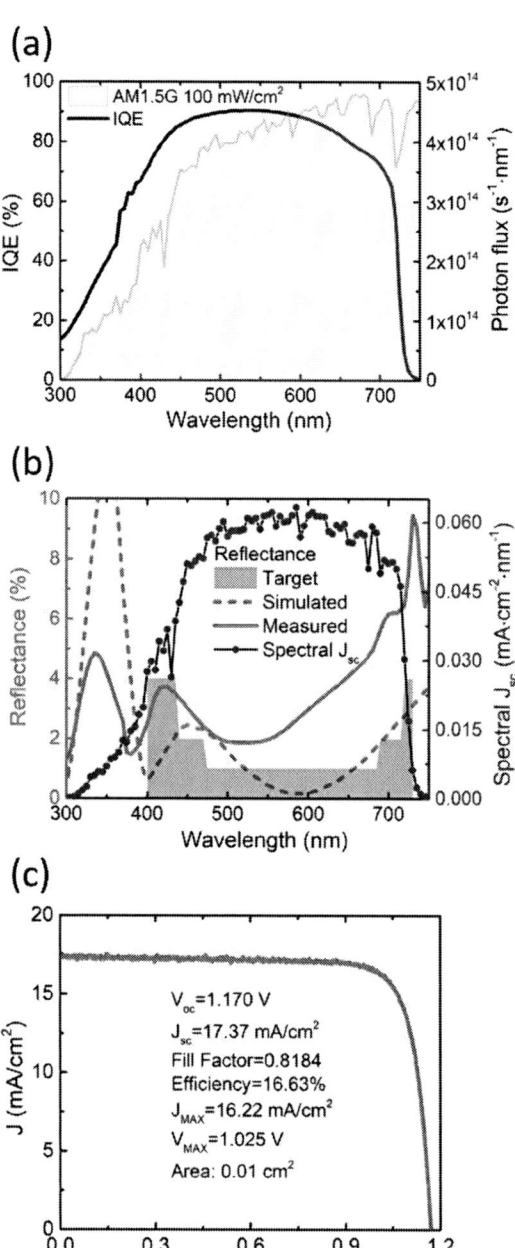

Figure 4. (a) IQE of the GaAsP solar cell on GaP (black solid curve) and the spectral photon flux of AM1.5G one sun irradiance (shadowed yellow region). (b) Target reflectance to suppress $J_{sc}$ loss (shadowed grey region), the calculated reflectance of an optimal DLARC consisting of 35.68 nm SiN$_x$ and 64.35 nm SiO$_x$ (red dash curve), and measured reflectance of the DLARC (red solid curve); the blue dots (right $y$-axis) represents the spectral distribution of $J_{sc}$. (c) LIV of the GaAsP solar cell on GaP with the SiN$_x$/SiO$_x$ DLARC under approximate AM1.5G one-sun illumination.

ARC for a GaAsP solar cell grown on GaP substrate schematically shown in Fig. 1(b), by minimizing the photocurrent loss over the effective spectral range. The spectral reflectance of a double-layer ARC generally exhibits a "W-shaped" curve in the visible-IR range [12]. For the ARC design we multiply IQE and AM1.5G spectral irradiance at each wavelength to determine "spectral $J_{sc}$", and match the wavelength range of minimal reflectance in simulation to that of maximal spectral $J_{sc}$.

As shown by the blue curve in Fig. 4(b), in the range of 475-685 nm, spectral $J_{sc}$ is over 0.055 mA·cm$^{-2}$·nm$^{-1}$, and thus the target reflectance is set <1% in STFCalc; the target reflectance is allowed to increase as spectral $J_{sc}$ reduces at shorter/longer wavelengths, represented by the grey shadowed region in the figure. With this target, the optimized thicknesses of SiN$_x$ and SiO$_x$ are 35.68 nm and 64.35 nm, and calculated reflectance of the corresponding DLARC matches well with the target, shown as the dashed red curve in Fig. 4(b). By convoluting the calculated reflectance with IQE and AM1.5G spectral irradiance, we obtain a reflectance-dependent $J_{sc}$ loss of 0.253 mA/cm$^2$, in comparison to the IQE-based $J_{sc}$ limit of 18.07 mA/cm$^2$. Next, we deposited the SiN$_x$ and SiO$_x$ with optimized refractive index and thicknesses onto our solar cells via

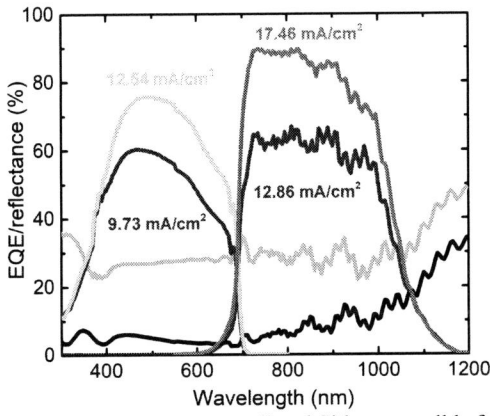

Figure 5. EQE of the GaAsP top cell and Si bottom cell before (blue and purple curves) and after (green and red curves) SiN$_x$/SiO$_x$ ARC; the reflectance before (grey curve) and after (black curve) ARC deposition.

PECVD. Comparison between the measured and calculated reflectance indicates that the discrepancy of 1.5-3.5% over 550-685 nm is mainly due to the metal grid coverage of 1.67%, which leads to a $J_{sc}$ loss of 0.621 mA/cm$^2$. The LIV and figures of merit for the GaAsP solar cell on GaP with SiN$_x$/SiO$_x$ DLARC are shown in Fig. 4(c). In practice, GaAsP solar cells grown on Si exhibit a slightly lower (~15 mV) $V_{oc}$ than comparable cells on GaP due to the higher TDD induced by the larger lattice mismatch between GaAsP and Si. Nonetheless, we have demonstrated that it is viable to achieve a GaAsP top cell grown on Si of >16% efficiency [4, 5], comparable to this one grown on GaP substrate.

Applying the ARC design principle to a slightly degraded GaAsP/Si tandem solar cell, we re-optimized the thicknesses of SiN$_x$ and SiO$_x$ to be 32.90 nm and 68.69 nm, which boosts $J_{sc}$ of the GaAsP top cell from 9.73 mA/cm$^2$ to 12.54 mA/cm$^2$, a 28.9% relative increase, as shown in Fig. 5. It is also seen that $J_{sc}$ of the Si bottom cell is boosted to 17.46 mA/cm$^2$, despite the fact that the optimization of the DLARC was primarily for the GaAsP top cell.

### D. Towards 30% efficient tandem cells

Achieving a 30% tandem efficiency will demand a top cell with ~20% efficiency and a bottom cell with ~10% efficiency. Our experiment reveals that it will be possible to achieve a GaAsP/Si 2-terminal tandem cell of efficiency over 22% with our optimal GaAsP top cell and ARC design. Further improvements, including reduction of TDD, the use of a heterojunction cell design to reduce non-radiative recombination in the junction [13, 14], and the development of tunnel junctions with high thermal stability and low optical/electrical loss [15], will be key steps for the top cell efficiency to reach 20%. As for the Si bottom cell, our primary focus is on protecting the backside of the GaP/Si template to avoid bulk lifetime degradation during MBE growth, applying $i/p$-type $a$-Si:H backside passivative contact [16] and reducing the wafer thickness of the GaP/Si templates.

## IV. CONCLUSION

We have demonstrated Si bottom cells with GaAs$_x$P$_{1-x}$ optical filtering layers with an efficiency of 6.25%. With an unoptimized SiN$_x$ ARC, a preliminary GaAsP/Si tandem cell reached a $V_{oc}$ of 1.596 V, a $J_{sc}$ of 11.91 mA/cm$^2$, a FF of 0.660, and an efficiency of 12.55% under approximate AM1.5G illumination. To further boost the efficiency of GaAsP/Si tandem cells, we demonstrated a DLARC design protocol based on the spectral distribution of $J_{sc}$, which increased the efficiency of a GaAsP solar cell grown on GaP substrate to 16.63% and boosted the $J_{sc}$ of a GaAsP/Si tandem cell by 28.9%. Continual improvements on the top cell, bottom cell, and tunnel junction are ongoing.

## ACKNOWLEDGEMENT

SF, MLL, ML, and ZCH gratefully acknowledge funding from NSF (Awards #1736181 and #1509864). KNY was supported by the Singapore Energy Innovation Programme Office for a National Research Foundation graduate fellowship, and MV was supported by a National Aeronautics and Space Administration (NASA) Space Technology Research Fellowship. DM acknowledges support from the Spanish Ministerio de Economía y Competitividad under project TEC2015-66722-R and Comunidad de Madrid under project SINFOTON S2013/MIT-2790.

978-1-5090-5606-4/17 $31.00 © 2017 IEEE

## REFERENCES

[1] T. J. Grassman, D. J. Chmielewski, S. D. Carnevale, J. A. Carlin and S. A. Ringel, "Development of epitaxial 2- and 3-junction III-V/Si solar cells", *IEEE J. Photovolt.*, vol. 6, pp. 326-331, 2016.

[2] T. Grassman, J. Carlin, B. Galiana, L.-M. Yang, F. Yang, M. Mills and S. Ringel, "Nucleation-related defect-free GaP/Si(100) heteroepitaxy via metal-organic chemical vapor deposition", *Appl. Phys. Lett.*, vol. 102, pp. 142102, 2013.

[3] K. Volz, A. Beyer, W. Witte, J. Ohlmann, I. Németh, B. Kunert, and W. Stolz, "GaP-nucleation on exact Si (001) substrates for III/V device integration", *J. Cryst. Growth*, vol. 315, pp. 37-47, 2011.

[4] J. F. Geisz, J. M. Olson, M. J. Romero, C. S. Jiang and A. G. Norman, "Lattice-mismatched GaAsP solar cells grown on silicon by OMVPE", Waikoloa, Hawaii, USA, 2006.

[5] J. R. Lang, J. Faucher, S. Tomasulo, K. Nay Yaung, and M. L. Lee "Comparison of GaAsP solar cells on GaP and GaP/Si", *Appl. Phys. Lett.*, vol. 103, pp. 092102-092105, 2013.

[6] K. N. Yaung, M. Vaisman, J. Lang, M. L. Lee, "GaAsP solar cells on GaP/Si with low threading dislocation density", *Appl. Phys. Lett.*, vol. 109, pp. 032107, 2016.

[7] S. Janz, M. Feifel, J. Ohlmann, J. Benick, C. Weiss, M. Hermle, A. W. Bett, F. Dimroth, D. Lackner, "Minority carrier lifetime limitations in Si wafer solar cells with gallium phosphide window layers", *IEEE PVSC 43rd*, pp. 1902-1905, 2016.

[8] J. Del Alamo, J. Eguren, A. Luque, "Operating limits of Al-alloyed high-low junctions for BSF solar cells", *Solid-State Electronics*, vol. 24, pp. 415-420, 1981.

[9] T. J. Grassman, J. A. Carlin, B. Galiana, F. Yang, M. J. Mills and S. A. Ringel, "MOCVD-grown GaP/Si subcells for integrated III–V/Si multijunction photovoltaics", *IEEE J. Photovolt.*, vol. 4, no. 3, pp. 972-980, 2014.

[10] T. Jimbo, T. Soga, Y. Hayashi, "Development of new materials for solar cells in Nagoya Institute of Technology", *Sci. Technol. Adv. Mater.*, vol. 6, pp. 27-33, 2005.

[11] M. Feifel, T. Rachow, J. Benick, J. Ohlmann, S. Janz, M. Hermle, F. Dimroth and D. Lackner, "Gallium phosphide sindow layer for silicon solar cells", *IEEE J. Photovolt.*, vol. 6, pp. 384-390, 2016.

[12] D. J. Aiken, "High performance anti-reflection coatings for broadband multi-junction solar cells", *Sol. Energ. Mat. Sol. Cells*, vol. 64, pp. 393-404, 2000.

[13] N. Jain, R. Oshima, R. France, J. Geisz, A. Norman, P. Dippo, D. Levi, M. Young, W. Olavarria, M. A. Steiner, "Development of lattice-matched 1.7 eV GaInAsP solar cells grown on GaAs by MOVPE", *IEEE PVSC 43rd*, pp. 47-51, 2016.

[14] N. Jain, J. F. Geisz, R. M. France, A. G. Norman, M. A. Steiner, "Enhanced current collection in 1.7 eV GaInAsP solar cells grown on GaAs by metalorganic vapor phase epitaxy", *IEEE J. Photovolt.*, vol. 7, no. 3, pp. 927-933, 2017.

[15] D. J. Chmielewski, T. J. Grassman, A. M. Carlin, J. A. Carlin, A. J. Speelman, S. A. Ringel, "Metamorphic GaAsP tunnel junctions for high-efficiency III–V/IV multijunction solar cell technology", *IEEE J. Photovolt.*, vol. 4, no. 5, pp. 1301-1305, 2014.

[16] Z. C. Holman, M. Filipic, A. Descoeudres, S. D. Wolf, F. Smole, M. Topic, C. Ballif, "Infrared light management in high-efficiency silicon heterojunction and rear-passivated solar cells", *J. Appl. Phys.*, vol. 113, 013107, 2013.

# Characterization of heteroepitaxial GaAs films grown on Si using selective area nucleation

Emily L. Warren, Emily A. Makoutz, Michelle Vaisman, Benjamin F. Bachman
William E. McMahon, Jeramy D. Zimmerman, Adele C. Tamboli

National Renewable Energy Laboratory, Golden, CO 80401, USA
Colorado School of Mines, Golden, CO 80401, USA

*Abstract* — Selective area growth of GaAs on patterned Si substrates is a potentially low-cost approach to integrate III-V and Si materials for multijunction solar cells. The use of nanoscale openings in a dielectric material can minimize nucleation-related defects and allow thinner buffer layers to be used to accommodate strain and trap defects caused by lattice mismatch between Si and epitaxial III-V layers. We have developed a process to grow coalesced GaAs thin films on Si substrates using buffer layers patterned by soft nanoimprint lithography (SNIL). We use photoelectrochemistry to probe the performance of these films as photovoltaic absorbers, and discuss techniques to improve the material quality of the GaAs epilayer.

*Index Terms* — GaAs, Si, photoelectrochemistry, selective area growth, nanoimprint lithography

## I. INTRODUCTION

The heteroepitaxial growth of GaAs and other III-V semiconductors on Si substrates has long been a focus of photovoltaics research as a means of lowering cost. Traditionally lattice-mismatched III-V materials for PV applications have been grown using metamorphic buffer layers, but this technique is expensive, limiting the commercial viability of such devices. It is highly desirable to be able to directly grow high quality III-V materials on Si without graded buffers, and such an advance would enable high efficiency solar cells that are cost competitive under high or low levels of solar concentration [1]. The direct growth of GaAs on Si has been studied for decades, and low defect density GaAs epilayers were traditionally formed using thermal cycle annealing and/or the introduction of strained superlattice layers [2]. Another approach that has shown promising results is the epitaxial lateral overgrowth (ELO) of GaAs using a patterned layer. Prior ELO results have shown that smaller nucleation areas result in fewer defects in the III-V films. Theory and preliminary experiments have shown that using nanoscale openings has the potential to eliminate nucleation related defects. Many recent advances have been made in controlled epitaxy of GaAs on Si using nanoscale selective area epitaxy approaches to create both III-V nanostructures and GaAs films [3], [4]. However, the creation of patterned substrates for growth has traditionally required complicated fabrication techniques that may not be compatible with the low costs required for solar cell devices.

We have previously demonstrated that selective area nucleation of III-Vs on Si is possible using a low-cost soft nanoimprint lithography (SNIL) technique to pattern a sol-gel $SiO_2$ buffer layer [5]. Figure 1 shows a schematic of the process used to create SNIL stamps and transfer patterns into the sol-gel $SiO_2$. Each master can be used to create multiple

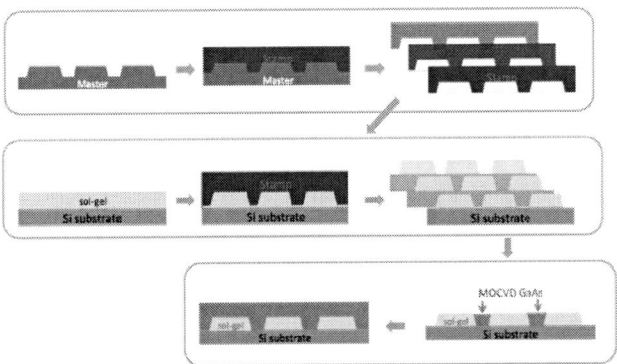

Fig. 1. Process flow schematic for fabrication of GaAs on Si substrates.

polymeric stamps, and each stamp can be used many times, greatly decreasing the cost of patterning relative to traditional techniques. Under optimized nucleation conditions, we showed that epitaxial GaAs islands selectively nucleate on the exposed Si substrate, and not the patterned $SiO_2$ [5].

Our selective area growth (SAG) nucleation, combined with an arsine ($AsH_3$) mediated in-situ Si surface preparation process [6] provides an approach to minimize nucleation-related defects such as anti-phase domains that can impact the quality of heteroepitaxial III-V layers. In order to move toward practical PV devices, it is also necessary to understand the coalescence of the nucleation areas into a planar film. Here we demonstrate the growth of coalesced GaAs films on SNIL-patterned templates. We use a variety of techniques to compare the quality of SAG GaAs on Si, with GaAs grown on unpatterned Si, as well as homo-epitaxial GaAs. Photoelectrochemical (PEC) characterization using the non-aqueous ferrocene/ferrocenium ($FeCp_2^{+/0}$) redox couple allows us to measure qualitative PV device parameters without the need for cell isolation or fabrication of top grids, enabling a rapid comparison between materials grown under different growth conditions.

## II. EXPERIMENTAL

SNIL stamps were formed by using polydimethylsiloxane (PDMS) to replicate Si masters that were patterned with standard lithography techniques (e-beam, deep UV, or interference) and reactive ion etching. Si (001) substrates with a 4° off-cut toward <111> were cleaned using a standard RCA and UV-ozone process prior to applying a thin layer of $SiO_2$ sol-gel and imprinting. Prior to growth, patterned substrates were

978-1-5090-5606-4/17 $31.00 © 2017 IEEE

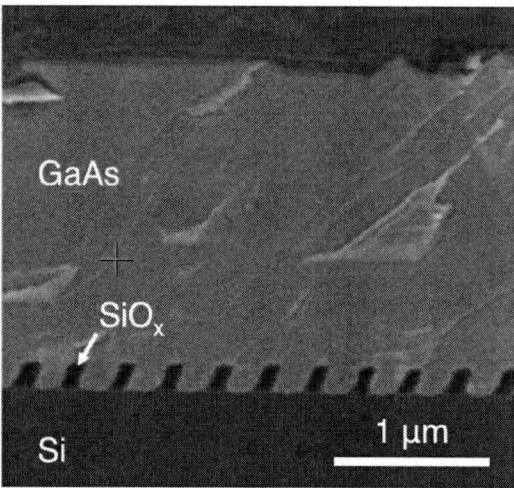

Fig. 2. SEM images of GaAs grown on Si patterned with sol-gel $SiO_2$.

annealed at $200°C$ to densify the sol-gel film and then dipped in 1% HF to remove native oxide from the exposed areas of the Si surface. Epitaxial growth of GaAs was carried out in either a low pressure or atmospheric pressure organometallic vapor phase epitaxy (OMVPE) reactor using trimethylgallium (TMGa) and $AsH_3$ as precursors. X-ray diffraction (XRD) data was taken using a Panalytical X-Pert Pro diffractometer.

PEC techniques offer a way to obtain qualitative figures of merit for as-grown GaAs films by using a semiconductor-liquid junction to provide a consistent Schottky-like contact to measure the photoresponse of the material. This technique has been used to characterize the behavior of GaAs films and nanowires grown by other methods [7]. Here we use the $FeCp_2^{+/0}$ redox couple to make a semiconductor/liquid junction between the surface of the GaAs film and the solution [8]. The cell consisted of 50 mM $FeCp_2^0$, 0.5 mM $FeCp_2^+$ in dry acetonitrile with lithium perchlorate ($LiClO_4$) as a supporting electrolyte. GaAs on Si samples were made into electrodes by contacting the back of the Si with an In/Ga eutectic and attaching a Cu wire. This entire assembly was then sealed into a glass tube so that only the front surface of the GaAs was exposed to the electrolyte solution. An ABET solar simulator was used to illuminate the samples and was calibrated using an encapsulated photodiode placed in the same position as the samples inside the electrochemical cell. All measurements were carried out using a piece of Pt foil as the counter electrode and a Pt wire as the reference electrode.

### III. RESULTS AND DISCUSSION

Nucleation was carried out at a low growth rate ($<2$ $\mu$m/h) and low temperature to fill the voids in the $SiO_2$ buffer layer, and then conditions (temperature, growth rate, V/III ratio) were varied during the coalescence of the islands into a film. Although we have not yet completed a comprehensive optimization, we have identified conditions where the films coalesced to form a planar morphology with $<2$ $\mu$m of total thickness. Figure 2 shows a cross-section scanning electron microscopy (SEM) image of a coalesced GaAs film grown on

Si patterned with sol-gel $SiO_2$. To better quantify the material quality of the films, a variety of characterization tools have been used. XRD analysis of these films shows that the FWHM of the (004) peak is still larger than the state-of-the-art reported for GaAs on patterned Si growth [9].

Figure 3 shows the photoelectrochemical performance of GaAs on Si films compared to patterned Si substrates without any epilayer. All samples were grown on degenerately doped Si wafers ($N_D > 1x10^{19}$ $cm^{-3}$) so there is no photoresponse from the Si (green) contributing to the observed photocurrent. The patterned GaAs on Si sample in Fig 3 had a $V_{oc}$ of 513 mV and a $J_{sc}$ of 10.3 mA $cm^{-2}$. The photocurrent is lower than expected for a GaAs PV device under AM1.5G illumination due to partial absorption of the light by the $FeCp_2^{+/0}$, which has absorption peaks near 450 nm and 600 nm. While these values are lower than the expected performance for a solid state GaAs solar cell, they provide a metric to rapidly determine relative performance. The lower photocurrents from the samples grown on Si are attributed to the low material quality of the current heteroepitaxial films. The low photovoltages suggests that the threading dislocation density density (TDD) in the GaAs on Si films is still quite large, likely $> 10^8$ $cm^{-2}$ [10].

As-measured J-E data in a PEC setup have relatively poor fill factors (FF) because the measurement includes artifacts due to mass transport such as solution resistance and concentration overpotential. To decouple these effects from the performance of the semiconductor, it is possible to correct the data by measuring the limiting anodic and cathodic currents, as well as the resistance of the solution [11]. The corrected PEC data is shown in Fig. 3 using data measured from a Pt electrode in the same position as the GaAs/Si electrodes (solution resistance of 100 $\Omega$-cm and anodic and cathodic limiting currents of 2 and 160 mA $cm^{-2}$, respectively). This correction does not impact the $V_{oc}$ or $J_{sc}$, but improves the fill factor and efficiency

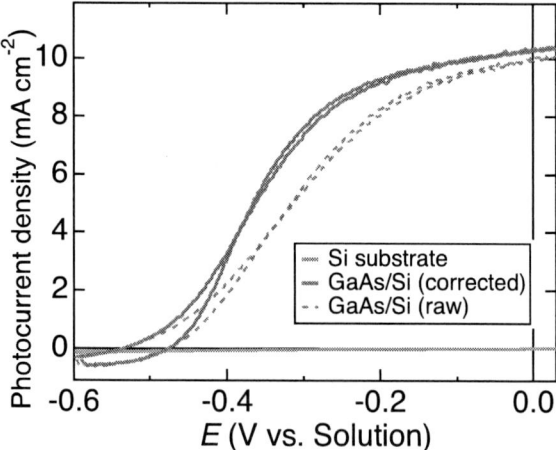

Fig. 3. PEC (J-E) data comparing performance of GaAs on Si films and the patterned Si substrate without GaAs. All data was taken with electrodes in contact with $CH_3CN$-$FeCp_2^{+/0}$. Raw data is displayed as dashed lines, and data corrected for uncompensated electrochemical resistance and concentration overpotential is shown as a solid line.

of the devices. Even with the correction, the samples grown on Si have low fill factors, suggesting the material quality is impacting this metric as well.

## IV. Conclusion

We have developed a process to grow coalesced GaAs films on Si substrates using a low-cost SAG nucleation buffer layer formed via SNIL. If designed properly, a SAG buffer can minimize the thickness of the buffer layers needed to accommodate strain and trap defects caused by lattice mismatch between Si and epitaxial III-V layers. We have used photoelectrochemistry to characterize the material quality of GaAs grown on patterned Si substrates. Although this process results in a uniform coalesced film that is photoactive, the current pattern geometry of the buffer layer and growth conditions do not yet enable high figures of merit for GaAs films measured by PEC. Other approaches for growing GaAs on Si using standard nanofabrication techniques have produced TDDs in the range of $10^7$ - $10^8$ cm$^{-2}$ [9]. Future work will focus on optimizing our low cost patterning approach and epitaxial growth techniques to improve the material quality of coalesced GaAs films for photovoltaic applications [9].

## Acknowledgment

This work was supported by DOE EERE SETP under DE-EE00028394. M.V. was supported by a National Aeronautics and Space Administration Space Technology Research Fellowship. The authors thank Waldo Olavarria, Matt Young, Bobby To, and Brian Thibeault (UCSB) for help with sample preparation and characterization. A portion of this work was performed in the UCSB Nanofabrication Facility, a member of the NSF-funded NNIN. The U.S. Government retains and the publisher, by accepting the article for publication, acknowledges that the U.S. Government retains a nonexclusive, paid up, irrevocable, worldwide license to publish or reproduce the published form of this work, or allow others to do so, for U.S. Government purposes.

## References

[1] N. Jain and M. K. Hudait, "III–V Multijunction Solar Cell Integration with Silicon: Present Status, Challenges and Future Outlook," *Energy Harvesting and Systems*, vol. 1, no. 3-4, pp. 121–145, 2014.

[2] M. Yamaguchi, "Dislocation density reduction in heteroepitaxial III-V compound films on Si substrates for optical devices," *Journal of Materials Research*, vol. 6, no. 02, pp. 376–384, 1991.

[3] B. Kunert, W. Guo, Y. Mols, B. Tian, Z. Wang, Y. Shi, D. Van Thourhout, J. Van Campenhout, R. Langer, and K. Barla, "III/V nano ridge structures for optical applications on patterned 300 mm silicon substrate," *Appl. Phys. Lett.*, vol. 109, p. 091101, 2016.

[4] Q. Li, K. W. Ng, and K. M. Lau, "Growing antiphase-domain-free GaAs thin films out of highly ordered planar nanowire arrays on exact (001) Silicon," *Appl. Phys. Lett.*, vol. 106, no. 7, p. 072105, 2015.

[5] E. L. Warren, E. A. Makoutz, K. A. W. Horowitz, A. Dameron, A. G. Norman, P. Stradins, J. D. Zimmerman, and A. C. Tamboli, "Selective Area Growth of GaAs on Si Patterned Using Nanoimprint Lithography," *Proc. 43rd IEEE PVSC*, pp. 1–4, 2016.

[6] E. L. Warren, A. E. Kibbler, R. M. France, A. G. Norman, P. Stradins, and W. E. McMahon, "Growth of antiphase-domain-free GaP on Si substrates by metalorganic chemical vapor deposition using an in situ AsH$_3$ surface preparation," *Appl. Phys. Lett.*, vol. 107, no. 8, p. 082109, 2015.

[7] S. Hu, C.-Y. Chi, K. T. Fountaine, M. Yao, H. A. Atwater, P. D. Dapkus, N. S. Lewis, and C. Zhou, "Optical, electrical, and solar energy-conversion properties of gallium arsenide nanowire-array photoanodes," *Energy Environ. Sci.*, vol. 6, no. 6, pp. 1879–1890, 2013.

[8] C. M. Gronet and N. S. Lewis, "N-type gaas photoanodes in acetonitrile: Design of a 10.0% efficient photoelectrode," *Appl. Phys. Lett.*, vol. 43, no. 1, pp. 115–117, 1983.

[9] Q. Li, H. Jiang, and K. M. Lau, "Coalescence of planar GaAs nanowires into strain-free three-dimensional crystals on exact (001) silicon," *Journal of Crystal Growth*, vol. 454, pp. 19–24, 2016.

[10] N. Jain and M. K. Hudait, "Impact of Threading Dislocations on the Design of GaAs and InGaP / GaAs Solar Cells on Si," *J. of Photovoltaics*, vol. 3, no. 1, pp. 528–534, 2013.

[11] E. L. Warren, D. B. Turner-Evans, R. L. Grimm, H. A. Atwater, and N. S. Lewis, "Photoelectrochemical characterization of Si microwire array solar cells," *Proc. 38th IEEE PVSC*, pp. 826–830, 2012.

# Efficient Photon Upconversion in Semiconductor Nanostructures: Constraints and Opportunities

Matthew F. Doty*, Eric Y. Chen*, Jing Zhang[†], Diane G. Sellers*, Zhuohui Li*, Christopher C. Milleville*,
Kyle Lennon[‡], and Joshua M. O. Zide*

*Department of Materials Science and Engineering, University of Delaware, Newark, DE 19716 USA
[†]Department of Chemistry, University of Delaware, Newark, DE 19716 USA
[‡]Department of Chemical and Biomolecular Engineering, University of Delaware, Newark, DE 19716 USA

*Abstract*—Photon upconversion is a promising approach to realizing photovoltaics with efficiency beyond the Shockley-Quessier limit by harvesting low-energy photons and converting them to high-energy photons that can be absorbed by a host single-junction solar cell. Existing upconversion materials have limited potential benefit for solar energy harvesting applications because of their narrow absorption bandwidth and low quantum efficiency. We first present numerical simulations of semiconductor nanostructures designed to overcome these limitations. The computational results demonstrate the potential impact of such upconversion materials and identify critical material parameters that must be achieved for practical realization of upconverters that can have a meaningful impact on solar energy harvesting. We then present experimental progress toward realizing this semiconductor upconversion paradigm. We describe the choice of material composition and structure, growth of the nanostructures, and ultrafast optical characterization of the carrier dynamics that result in these structures. We show that upconversion can be achieved in at least one realization of this semiconductor upconversion paradigm. We conclude with a discussion about the prospects for realizing efficient photon upconversion and the constraints that must be addressed to meet this goal.

*Index terms* — quantum dots, photovoltaics, time-resolved photoluminescence, upconversion.

## I. INTRODUCTION

Potential approaches to overcoming the Shockley-Quessier limit on the efficiency of single junction photovoltaics largely focus on expanding the portion of the incident solar spectrum that can be harvested.[1], [2] Multi-junction solar cells have the highest demonstrated efficiency of such "third generation" photovoltaics, with record efficiencies of 46% under concentrated sunlight [3]. Multi-junction devices, however, remain expensive to produce and suffer from current-matching constraints that can limit their performance under fluctuating solar spectrum conditions.[4]. Intermediate Band Solar Cells face significant challenges in suppressing carrier relaxation through intermediate states in order to realize the gains that could come from sequential absorption of low-energy photons.

In photon upconversion a material mediates the sequential absorption of two or more low-energy photons and the subsequent emission of a single high-energy photon.[5] Upconversion-backed photovoltaics could harvest the solar spectrum in a manner analogous to both Intermediate Band and Multi-junction solar cells and with comparable hypothetical limits to solar energy harvesting efficiencies.[6], [7] However, the interface between the host cell and the upconverter can be

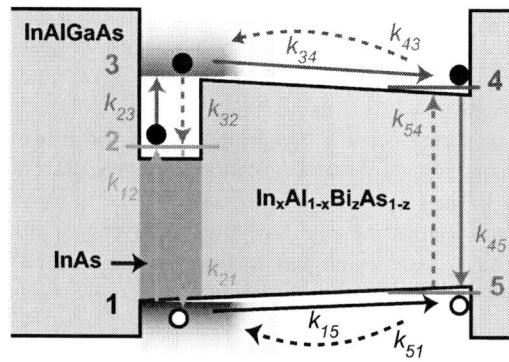

Fig. 1: Schematic depiction of a III-V upconversion nanostructure with rates for carrier generation, relaxation, and transfer indicated.[8]

purely optical, eliminating problems that originate in current matching or intermediate state-related nonradiative recombination. As a result, upconversion is promising for higher practical efficiencies and/or lower production costs.

For upconversion to have a meaningful impact on solar energy harvesting the upconversion materials employed must harvest a broad fraction of the incident solar spectrum and must upconvert with high quantum efficiency. Established upconversion materials such as triplet-triplet annihilation molecules and lanthanides are limited in both respects,[9], [10] Emerging semiconductor-based approaches to photon upconversion offer the potential to employ heterostructure engineering methods to tailor band structure and carrier dynamics in order to overcome these limitations.[11], [12], [13], [14] Here we review our computational analysis of the potential performance of such semiconductor upconversion structures in order to demonstrate the potential impact on solar energy harvesting.[8], [14] We present new results analyzing the sensitivity of these computational models to the underlying assumptions and use this analysis to identify the critical material metrics that must be considered in experimental efforts to realize such structures. We then describe experimental progress toward the synthesis of semiconductor nanostructures that implement this design strategy, which could be achieved in either epitaxially-grown III-V materials or colloidally-synthesized II-VI materials. We illustrate how time-integrated and time-resolved optical characterization of nanostructures can be used to understand and optimize the

978-1-5090-5606-4/17 $31.00 © 2017 IEEE

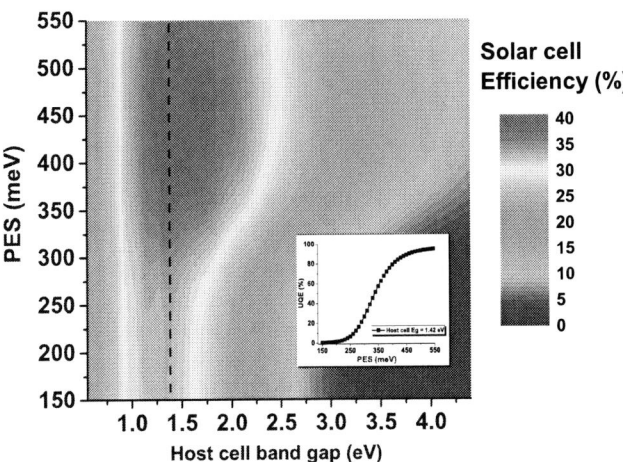
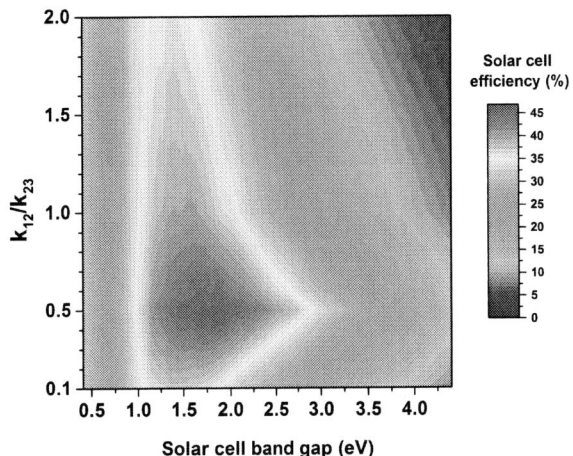

Fig. 2: Solar energy harvesting efficiency as a function of host cell bandgap and PES. Inset: UQE as a function of PES. Adapted from [14].

Fig. 3: Solar energy harvesting efficiency as a function of host cell bandgap and $k_{12}/k_{23}$ ratio.

upconversion performance. We show that upconversion can be achieved in at least one realization of this design paradigm. We show how changes in carrier dynamics can be analyzed as a function of material composition and nanometer scale structure to guide continued improvements in nanostructure design toward the goal of realizing photon upconversion with efficiencies relevant to solar energy harvesting.

## II. NUMERICAL ANALYSIS OF UPCONVERSION EFFICIENCY

In Fig. 1 we present a schematic band diagram for a photon upconversion nanostructure based on III-V materials. The physical process of upconversion proceeds in five conceptual steps: a) absorption of a mid-energy photon within an InAs Quantum Dot (QD) results in generation of an electron hole pair, b) rapid escape of the hole via the graded composition of an InAlBiAs 'funnel' suppresses radiative recombination, c) absorption of a low-energy photon promotes the electron from a confined state of the QD to a state above the conduction band edge of the 'funnel', d) nonradiative relaxation of the electron through QD states is suppressed by the phonon bottleneck originating in the discrete density of states, allowing the electron to escape via the funnel, and e) the electron and hole recombine to emit a high-energy photon.

We assess the potential efficiency of this photon upconversion structure with a rate equation model that computes Upconversion Quantum Efficiency (UQE) as a function of the Photon Energy Sacrifice (PES) that originates in the slopes of the graded InAlBiAs potential.[8] As shown in the inset to Fig. 2, the UQE increases as a function of PES, reaching a maximum value of 96%. The increased efficiency with increasing PES occurs because a steeper slope in the graded InAlBiAs region of the model upconversion structure results in a stronger driving force transferring electrons and holes to the recombination zone and thus better suppression of unwanted radiative or nonradiative recombination pathways.

To assess the potential impact of this model upconversion system on solar energy harvesting, we compute the efficiency of host solar cells as a function of both the bandgap of the host cell and the PES of the upconversion material employed to back this host.[14] The host cell performance was computed with established detailed balance methods using a modified photon flux. The modified flux of high-energy photons incident on the host cell was computed using the rate equation model to compute the rate of emission of high-energy photons in response to the flux of photons with energy below the bandgap of the host cell. The results, shown in Fig. 2, demonstrate that net solar energy conversion efficiencies in excess of 40% could be achieved. Taken together, these results reveal that achieving practical upconversion requires a choice of materials that accept a relatively large PES in order to achieve high UQE.

To explore the robustness of these models and further guide experimental efforts we have evaluated the sensitivity of the computational results to the underlying assumptions of the rate equation model. Factors we considered in this analysis include potential variations in the solar spectrum distribution, the absorption cross-section for inter- and intra-band optical transitions, and the concentration of the incident sunlight. A manuscript reporting the results of this complete sensitivity analysis is in preparation. As an example, in Fig. 3 we show the net solar energy harvesting efficiency as a function of the host cell bandgap and the ratio $k_{12}/k_{23}$. $k_{12}$ describes the rate of optically-driven electron-hole pair generation for the transition between states 1 and 2 in Fig. 1. Similarly, $k_{23}$ describes the rate of optically-driven excitation of an electron from state 2 to state 3. Varying the ratio $k_{12}/k_{23}$ simulates altering the fraction of the solar spectrum that is harvested by these two transitions. Best overall performance of 46% solar energy harvesting efficiency is achieved for a $k_{12}/k_{23}$ ratio of 0.5. This result tells us that the choice of intermediate state energies must be informed by the rates of carrier relaxation for each transition.

978-1-5090-5606-4/17 $31.00 © 2017 IEEE          3385

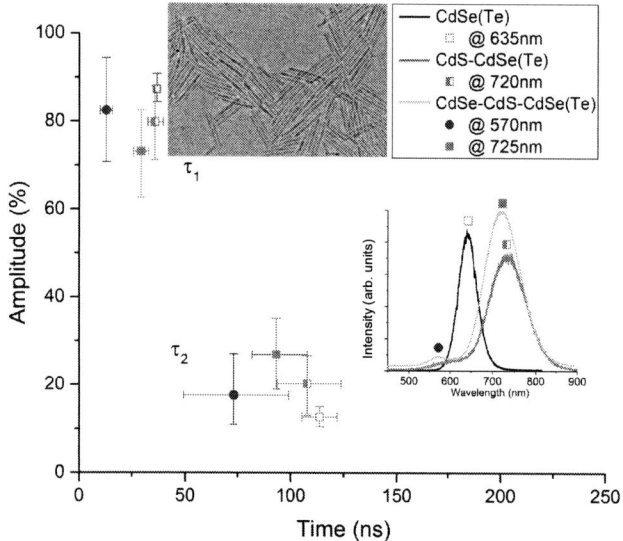

Fig. 4: Top inset: TEM image of core-rod nanostructures. Lower right inset: Steady-state PL of core-rod structures. Main panel: Decay constants extracted from time-resolved PL of core-rod structures.

## III. ENGINEERING HETEROSTRUCTURES FOR EFFICIENT PHOTON UPCONVERSION

This new approach to photon upconversion could be implemented in a wide range of semiconductor nanostructures. This flexibility provides the opportunity to tune absorption and emission wavelengths and to explore a range of high-efficiency versus low-cost options for a range of applications. We are pursuing a III-V material platform grown by Molecular Beam Epitaxy and a II-VI material platform synthesized in solution via colloidal chemistry. We anticipate that the epitaxially-grown material platform will ultimately enable higher performance due to very low defect densities while the colloidally-synthesized materials will enable lower-cost routes to mass production and application. Here we present results from the colloidal material platform to illustrate the approach to understanding the photophysical processes limiting upconversion and the path to engineering materials that overcome these limitations.

Our efforts to synthesize colloidal core/multi-shell upconversion particles originally focused on spherical particles with CdTe cores surrounded by CdS and CdSe spherical shells. As reported previously,[15], [16] time-integrated and time-resolved photoluminescence studies of these particles indicated the formation of the desired band alignments and the onset of the carrier separation dynamics desired for upconversion. However, both computational and experimental evidence suggests that incomplete separation of the electron and hole, possibly due to strong Coulomb interactions, may be limiting the carrier separation and thus incompletely suppressing radiative recombination. To explore one possible route toward overcoming this limitation, we synthesized rod-like nanostructures in which a CdSe(Te) core is surrounded by a rod-shaped CdS shell containing a CdSe quantum dot at the end of the rod, as shown in the inset to Fig. 4. Time-integrated and time-resolved photoluminescence spectroscopy of these rod-like structures, as shown in Fig. 4, indicates that the change to a rod-like morphology alters the spatial separation of optically-generated electrons and holes, which could suppress unwanted radiative recombination.

## IV. OBSERVATION OF PHOTON UPCONVERSION

In Fig. 5 we present upconversion photoluminescence measurements on the rod-like nanostructures. The observation of PL emission centered on 600 nm in response to excitation with an 805 nm CW laser is clear evidence of upconversion. Note that the upconversion emission wavelength of 600 nm exactly matches the downconversion PL emitted from the CdSe QD. This correlation provides strong evidence that the carriers are migrating from the CdSe(Te), where they are initially generated by absorption of a low-energy photon, to the CdSe QD where they recombine to emit upconverted light. This is precisely the physical process we want to occur.

We note that we have performed similar measurements on the upconversion particles with the spherical morphology. We have also observed upconversion from particles with this morphology, but the upconverted emission appears as a 'shoulder' with a wavelength only slightly shifted from the absorption. This form of upconversion has previously been observed even in single core-only QDs and arises from a variety of photophysical processes mediated by defect and surface states. This form of upconversion does not utilize the solar spectrum in a way that is maximally advantageous for solar energy harvesting. We speculate that the 'incomplete' upconversion in the spherical morphology arises from insufficient spatial separation of electrons and holes in the spherical particle morphology as described above. We are performing detailed studies of the carrier dynamics in both spherical and rod-like morphologies in order to understand the photophysical processes that limit upconversion efficiency and inform future changes to the particle morphology that will improve upconversion performance. We are performing similar measurements on the epitaxial upconversion nanostructures, which are ideally suited for the wavelengths at which upconversion-backed PV can have maximum impact on overall solar energy harvesting efficiency.

## V. SUMMARY AND OUTLOOK

Computational results show that semiconductor-based photon upconversion materials could have a substantial impact on solar energy harvesting. Here we reviewed computational and experimental progress toward the design and realization of such semiconductor-based photon upconversion heterostructures. Our computational results demonstrate that high solar energy harvesting efficiency is possible and inform the trade-offs that must be made when choosing material compositions and structures to realize this paradigm experimentally. We illustrated the approach to engineering particles for efficient upconversion optimized for solar energy harvesting by comparing spherical and

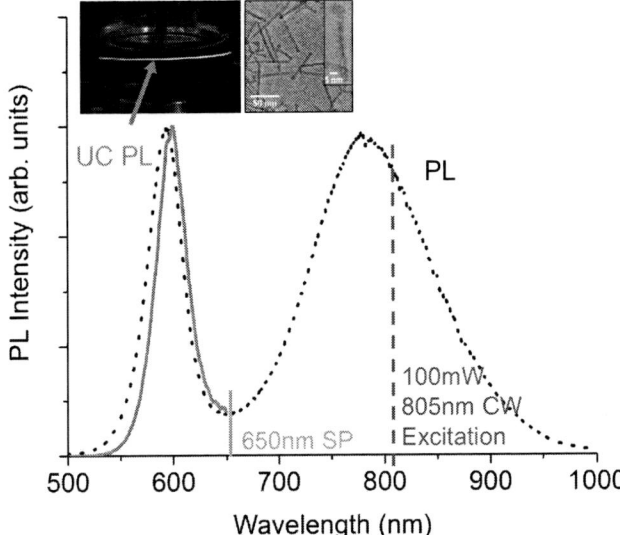

Fig. 5: PL (dotted black) and upconversion PL (solid red) emission from rod-like nanoparticles. PL emission was excited with a 405 nm CW laser while UCPL was excited with an 805 nm CW laser. Left inset: real-color image of emitted upconverted light. Right inset: TEM images of core-rod nanostructures.

rod-like colloidal particles. We showed how time-integrated and time-resolved photoluminescence characterization of materials and intermediate products demonstrate that the desired carrier dynamics can be achieved and inform the improvements in material design and synthesis that are required to realize high-efficiency photon upconversion for improved solar energy harvesting.

ACKNOWLEDGMENT

The authors acknowledge financial support from the W. M. Keck Foundation.

REFERENCES

[1] W. Shockley and H. Queisser, "Detailed balance limit of efficiency of p-n junction solar cells," *Journal of Applied Physics*, vol. 32, no. 3, pp. 510–519, 1961.
[2] L. C. Hirst and N. J. Ekins-Daukes, "Fundamental losses in solar cells," *Photovolt: Res. Appl*, vol. 19, pp. 286–293, 2011.
[3] M. A. Green, K. Emery, Y. Hishikawa, W. Warta, and E. D. Dunlop, "Solar cell efficiency tables (version 45)," *Progress in photovoltaics: research and applications*, vol. 23, no. 1, pp. 1–9, 2015.
[4] K. Horowitz, M. Woodhouse, H. Lee, and G. Smestad, *Bottom-Up Cost Analysis of a High Concentration PV Module; NREL (National Renewable Energy Laboratory)*, Apr 2015.
[5] M.-F. Joubert, "Photon avalanche upconversion in rare earth laser materials," *Optical Materials*, vol. 11, no. 23, pp. 181 – 203, 1999.
[6] T. Trupke, M. A. Green, and P. Wurfel, "Improving solar cell efficiencies by up-conversion of sub-band-gap light," *Journal of Applied Physics Journal of Applied Physics J. Appl. Phys.*, vol. 92, no. 7, pp. 4117–4122, 2002.
[7] A. C. Atre and J. Dionne, "Realistic upconverter-enhanced solar cells with non-ideal absorption and recombination efficiencies," *Journal of Applied Physics*, vol. 110, no. 3, pp. 034 505–034 505, 2011.

[8] E. Y. Chen, J. Zhang, D. G. Sellers, Y. Zhong, J. M. O. Zide, and M. F. Doty, "A kinetic rate model of novel upconversion nanostructures for high-efficiency photovoltaics," *IEEE Journal of Photovoltaics*, vol. 6, p. 1183, 2016.
[9] B. Wang, B. Sun, X. Wang, C. Ye, P. Ding, Z. Liang, Z. Chen, X. Tao, and L. Wu, "Efficient triplet sensitizers of palladium (II) tetraphenylporphyrins for upconversion-powered photoelectrochemistry," *The Journal of Physical Chemistry C*, vol. 118, no. 3, pp. 1417–1425, 2014.
[10] S. Fischer, E. Favilla, M. Tonelli, and J. C. Goldschmidt, "Record efficient upconverter solar cell devices with optimized bifacial silicon solar cells and monocrystalline BaY2F8:30% Er3+ upconverter," *Solar Energy Materials and Solar Cells*, vol. 136, pp. 127–134, 2015.
[11] Z. Deutsch, L. Neeman, and D. Oron, "Luminescence upconversion in colloidal double quantum dots." *Nature Nanotechnology*, vol. 8, no. 9, pp. 649–653, 2013.
[12] A. Teitelboim and D. Oron, "Broadband near-infrared to visible upconversion in quantum dot-quantum well heterostructures," *ACS Nano*, vol. 10, p. 446, 2016.
[13] D. G. Sellers, S. J. Polly, Y. Zhong, S. M. Hubbard, J. M. Zide, and M. F. Doty, "New Nanostructured Materials for Efficient Photon Upconversion," *IEEE Journal of Photovoltaics*, vol. 5, no. 1, pp. 224–228, 2015.
[14] D. G. Sellers, J. Zhang, E. Y. Chen, Y. Zhong, M. F. Doty, and J. M. Zide, "Novel nanostructures for efficient photon upconversion and high-efficiency photovoltaics," *Solar Energy Materials and Solar Cells*, vol. 155, pp. 446–453, 2016.
[15] D. G. Sellers and M. F. Doty, "Design, synthesis and photophysical properties of inp/cds/cdse and cdte/cds/cdse (core/shell/shell) quantum dots for photon upconversion," *Proceedings of 42nd IEEE Photovoltaic Specialist Conference*, p. 1, 2015.
[16] E. Y. Chen, D. G. Sellers, and M. F. Doty, "Time-resolved photoluminescence spectroscopy of cdte/cds/cdse quantum dot complexes for photon upconversion," *Proceedings of 43rd IEEE Photovoltaic Specialist Conference*, 2016.

# Enhanced Ultra-Thin a-Ge:H Solar Cells by Plasmonic Nanoparticles Embedded in the Optical Resonant Cavity

Brendan Brady[a], Volker Steenhoff[b], Benedikt Nickel[b], Martin Vehse[b], and Alexander G. Brolo[a]

[a] University of Victoria - Centre for Advanced Materials and Related Technology, Victoria, British Columbia, V8W 3V6, Canada

[b] NEXT ENERGY - EWE Research Centre for Energy Technology at the University of Oldenburg, Carl von Ossietzky-Str. 15, 26129 Oldenburg, Germany

*Abstract* — This work compares the impact of periodic arrays of dielectric and plasmonic nanoparticles placed adjacent to, above and below, the a-Ge:H absorber layer in ultra-thin optical cavity solar cells. We take advantage of the bottom-up facet of nanosphere lithography to encapsulate individual Ag nanoparticles in thin $SiO_2$ layers. We show both dielectric and plasmonic nanoparticles placed below the absorber give significant enhancements in quantum efficiency compared to a flat reference. Furthermore, we show major differences in current generation arise between plasmonic and dielectric nanoparticles when they are placed above the absorber.

*Index Terms* — localized surface plasmon resonance, light-trapping, periodic particle arrays, thin-film solar cells.

## I. INTRODUCTION

Several benefits arise as solar cell layer thicknesses are decreased. For instance, solar-module manufacturing costs and $CO_2$ footprint are reduced due to use of less raw materials and lower deposition times. In addition, thinner layers are of interest to conservation efforts of endangered elements, particularly for CdTe and CIGS based technologies. So-called thin-film solar cells, with typical absorber layer thickness < 1 μm, achieve the previously mentioned advantages over their thicker counterparts with the drawback of decreased absorption predominantly near the band gap.

To better take advantage of the solar spectrum while maintaining thin-film status, a plethora of research has investigated light-trapping strategies in thin-film solar cells. Such technology includes anti-reflective coatings [1], distributed Bragg reflectors and higher-dimensional photonic crystals for highly reflective back contacts [2], and randomly textured back reflectors [3]. Recent work on ultra-thin hydrogenated amorphous germanium (a-Ge:H) solar cells shows that light-trapping via 1D optical cavities can result in broadband enhancements for single-junction cells [4] or can instead be tuned to absorption in the infrared for incorporation in tandem devices [5],[6].

In recent years, there has been significant interest in incorporating 2D periodic nanoparticle arrays (PPA) within a-Si:H thin-film solar cells. Several low-cost nanofabrication methods, such as nanoimprint lithography [7]-[17], anodic aluminum oxide [18]-[23], and nanosphere lithography (NSL) [24]-[28] have been demonstrated as effective methods to construct 2D PPA on solar cell back reflectors. Compared to devices with randomly textured back reflectors, PPA can provide current generation enhancements particularly in the red region [8],[24]. The enhancement mechanism has been investigated and results suggest the 2D PPA acts as an optical grating and causes incident light to enter diffraction modes which couple to horizontally propagating waveguide modes within the absorbing layer [8],[9].

Arguments for and against the use of the plasmonic materials for nanoparticles (NPs) exist in the literature. On one hand, the localized surface plasmon resonance (LSPR) of plasmonic NPs is well understood to increase forward scattering compared to dielectric NPs on the front side of the cell while additionally showing a high intensity near field effect which can be used to create charge carrier pairs around the NP [29]. On the other hand, plasmonic NPs with sharp features result in high parasitic absorption and metallic NPs typically increase vulnerability to cell shunting. Plasmonic NPs integrated on the front side or embedded in the absorbing layer have demonstrated losses in photocurrent compared to cells without the NPs [30]-[32]. When located on the back reflector, plasmonic NPs show similar performance enhancements to non-plasmonic NPs [24],[27],[28].

In this work, we incorporate dielectric and plasmonic PPA in optical cavity ultra-thin a-Ge:H solar cells. Encapsulating the plasmonic NPs during the NSL process avoids the requirement for closed full area diffusion buffer layers. This allows us to place the particles adjacent to, both above and below, the a-Ge:H without interrupting the p-i-n sequence. Through this investigation, we combine multiple light-trapping effects and aim to expand the understanding of the contribution of plasmonic and dielectric NPs to thin-film solar cell operation.

## II. METHODS

### A. Nanoparticle Synthesis

Nanopatterning was achieved using NSL as depicted in Fig. 1 (a)-(d) and described in detail in [33]. NSL is an inexpensive bottom-up nanofabrication technique that has demonstrated effectiveness over module-sized areas [34]. In this report, the

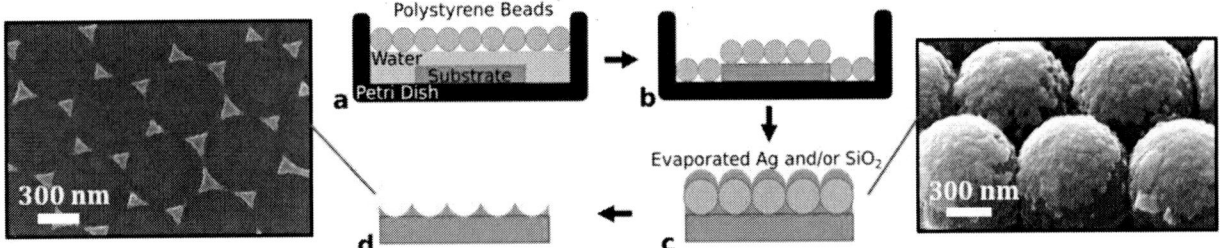

Fig. 1    NP Synthesis by NSL. (a) Substrate is submerged in water and polystyrene beads dropped from colloidal suspension form a close-packed hexagonal monolayer on the surface. Water is left to evaporate (b) and the pheres act as a deposition mask for evaporated Ag and SiO$_2$ (c). SEM image shows polystyrene with evaporated material on the surface. Removal of the polystyrene reveals a periodic arrangement of pyramidal NPs (d).

NSL process consisted of using close-packed hexagonal monolayer of polystyrene beads as a deposition mask for e-beam evaporated Ag and SiO$_2$. We insulated the NPs and circumvented the need for a closed full area diffusion buffer layer, a common requirement before Ge/Si deposition by plasma enhanced chemical vapor deposition (PECVD), by growing the NPs in a SiO$_2$/Ag/SiO$_2$ layer sequence. 625 nm diameter polystyrene beads were used in all experiments presented in this report and NPs consisted of 90 nm Ag encapsulated on the bottom and top with 5 nm SiO$_2$ unless otherwise noted.

*B. Solar Cell Fabrication and Characterization*

All solar cells with the stack sequence depicted in Fig. 2 (a) were fabricated in substrate configuration on 5 cm x 5 cm glass slides. Discussions regarding the layer sequence can be found in [4]-[6]. Si and Ge layers were grown via PECVD, TCOs were fabricated by DC magnetron sputtering, and the Ag and SiO$_2$ were deposited by e-beam evaporation. Solar cell areas were defined by structuring the ITO front contact using a marker lift-off process. External quantum efficiency (QE) curves were measured using differential spectral response measurements. As this report focuses on the optical effects of placing NPs in the solar cell stack, QE spectra were taken at reverse bias voltage that resulted in maximum charge carrier extraction or were scaled to the respective cell's short circuit current density. SEM specimens were prepared using a Hitachi FB-2100 FIB and images were taken with a Hitachi S-4800 FESEM.

NPs were deposited both above and below the a-Ge:H absorbing layer. Particles above the a-Ge:H were embedded in the μc-Si:H p-doped layer within a few nanometers of the μc-Si:H p/i interface. NPs below the a-Ge:H were embedded near the top surface of the a-Si:H (i) buffer layer. In both configurations, the closest edges of the NPs were within approximately 10 nm from the a-Ge:H absorber. Cross-sectional SEM images of cells with NPs above and below the a-Ge:H absorber can be seen in Fig. 2 (b) and (c), respectively. The conformal growth of the cell layers deposited after the particles leading to the top dome-shaped structure is apparent.

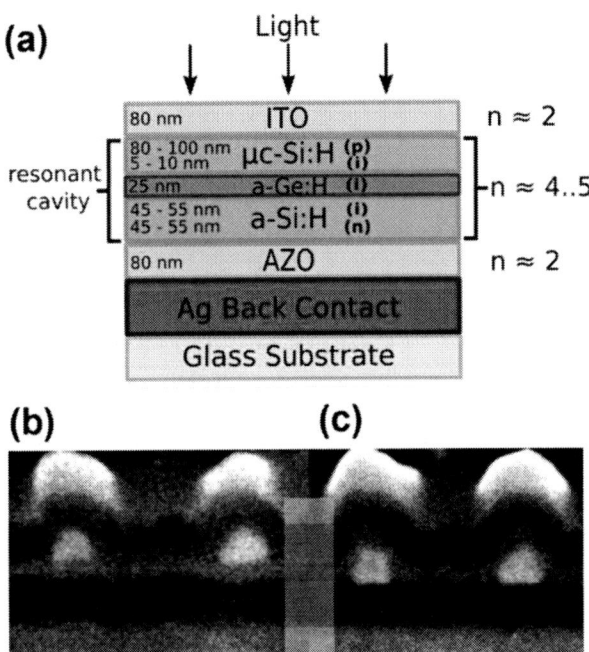

Fig. 2    (a) a-Ge:H flat resonant cavity solar cell layer stack with approximate thicknesses. Cross-sectional SEM image of cells with NPs above a-Ge:H (b) and below a-Ge:H (c).

### III. RESULTS AND DISCUSSION

Fig. 3 shows the QE spectra of the flat resonant cavity cells represented in Fig. 2 (a) with various cavity thicknesses. The high-wavelength peak in each curve corresponds to where the electric field Fabry-Pérot interference maxima of the optical

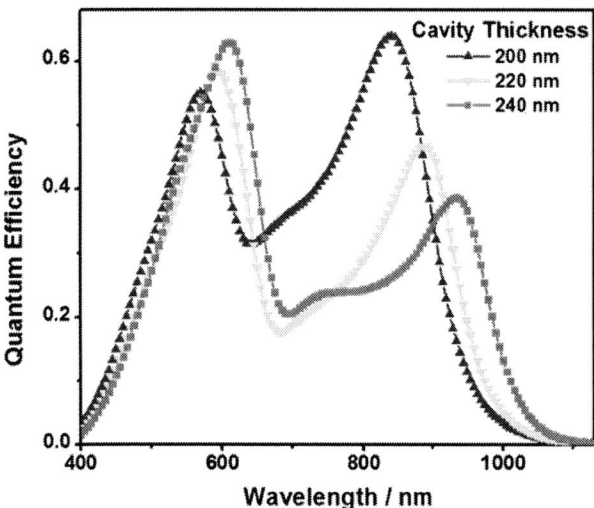

Fig. 3 QE spectra of flat resonant cavity cells with various resonant cavity thicknesses

cavity have strong spatial overlap with the a-Ge:H absorbing layer, as seen in [5]. The substantial QE drop of the high wavelength peak with increasing cavity thickness is a result of approaching the bandgap of the a-Ge:H. The dip in each curve in the 600 - 800 nm range coincides with the region where the Fabry-Pérot interference field maximum has less spatial overlap with the a-Ge:H absorber. In the low wavelength regime < 550 nm, single pass absorption in the a-Ge:H partly prevents the resonant cavity modes from forming.

The QE spectra of a cell with encapsulated Ag NPs below the a-Ge:H layer along with a flat reference is shown in Fig. 4. It is apparent that the encapsulated Ag NPs below the a-Ge:H significantly increase the charge carrier generation between 650 and 850 nm, which falls in the range where the Fabry-Pérot

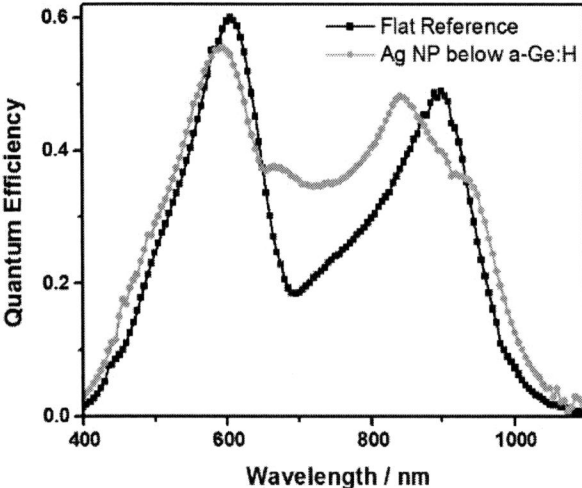

Fig. 4 QE spectra of a cell with Ag NPs below a-Ge:H along with a flat reference.

interference maxima do not overlap well with the a-Ge:H layer. Furthermore, at the QE peaks, the performance of the NP cell is slightly reduced compared to the flat reference. We attribute the differences in QE to the re-distribution of the electric field within the cell caused by the scattering of the Ag NPs. In the 650 – 950 nm range, the Ag NP re-distributes the electric field such that there is stronger overlap with the a-Ge:H when compared to the flat cell. At the QE maxima, the NP scattering results in unfavorable resonance conditions.

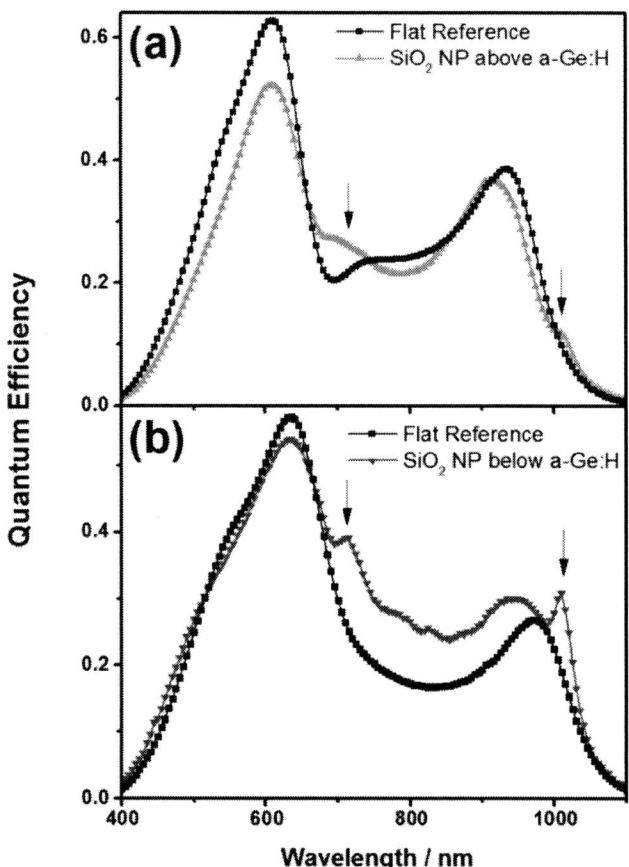

Fig. 5 QE spectra of cells with 100 nm $SiO_2$ NPs above (a) and below (b) a-Ge:H with flat references

The NP cell in Fig. 4 shows features at approximately 680 and 930 nm that are not present in the flat reference. Fig. 5 (a) and (b), showing $SiO_2$ NPs placed above and below the a-Ge:H, respectively, both present similar features to those in Fig. 4. The features in Fig. 5 (a) and (b), at approximately 710 and 1000 nm, are shifted relative to those in Fig. 4. This trend is indicative that diffraction modes of the 2D nanoparticle grating are responsible, at least in part, for the enhanced carrier generation seen in Fig. 4 and Fig 5. (a). The wavelength shift can be attributed to the Ag in the NP changing the refractive index profile and thus the angles and intensity distribution of the diffraction modes. In a-Si:H cells with 2D nanoparticle

978-1-5090-5606-4/17 $31.00 © 2017 IEEE      3390

gratings on the back contact, previous work demonstrates that the grating modes couple to propagating waveguide modes of the absorbing layer, causing the observed increase in QE [8],[9].

Comparing Fig. 5 (a) and (b) reveals considerable differences in QE when placing $SiO_2$ NPs above and below the a-Ge:H. At wavelengths < 550 nm, where single pass absorption is pertinent, the presence of the $SiO_2$ NPs above the a-Ge:H may effectively block the incident light from reaching the absorber. The harmful effects of the NPs could be further enhanced by the focusing effect of the conformal dome structure directly above the NPs on the front surface of the cell [24]. In the 700 - 1000 nm range, $SiO_2$ NPs below show significant enhancements over the flat reference while $SiO_2$ NPs above give slightly poorer performance compared to the flat reference. Although there are many possible explanations for this observation, the conformal a-Ge:H layer grown on top of NPs simply provides an increase in the optical path length through the absorber for normally incident photons. Furthermore, the increase in feature prominence from Fig. 5 (a) to (b) is suggestive of better coupling to photonic modes with better field overlap with the a-Ge:H, such as resonant cavity waveguide modes, when particles are placed below the a-Ge:H.

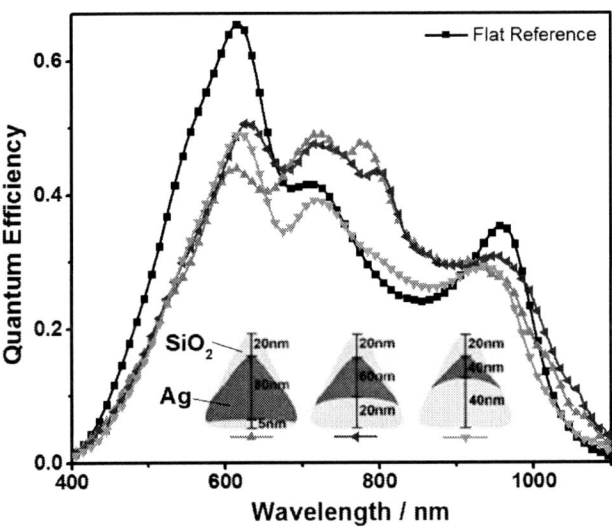

Fig. 6   QE spectra of nanocavity cells with NPs placed above the a-Ge:H absorbers along with a flat reference. Different curves correspond to 100 nm NPs with different Ag portions, which are depected in the inset images.

Fig. 6 shows NPs with various Ag:$SiO_2$ ratios located above the a-Ge:H. The QE in the range 650 – 900 nm increases with the size of the Ag NP, which we can attribute to increasing forward scattering efficiency and/or decreasing parasitic absorption with increasing plasmonic particle size [35]. Alternatively, this trend constitutes as possible evidence the LSPR near-field enhancements are contributing to current generation. As the bottom Ag/$SiO_2$ interface of the NPs is moved closer to the bottom of the NP, the near field intensity in the a-Ge:H layer increases. The results shown in Fig. 6 demonstrate the potential for overall current generation enhancement for plasmonic NPs located above the a-Ge:H.

## IV. CONCLUSION

Both dielectric and plasmonic NPs placed below the a-Ge:H showed similar charge carrier generation improvements in ultra-thin optical cavity solar cells. These results suggest that combining light-trapping effects, in this case NP and optical cavity resonances, can provide high charge carrier generation rates over broadband wavelengths using an absorber thickness of less than 30 nm. When NPs are placed above the a-Ge:H, the NP material plays a significant role as plasmonic NPs show ranges of improved performance while dielectric NPs show an overall decrease in performance relative to a flat reference. Furthermore, by changing the Ag content in the individual NPs, possible evidence of LSPR near-field enhancements was presented. To further explore the effects and potential of plasmonic NPs in resonant-cavity enhanced a-Ge:H solar cells, additional experiments in conjunction with FDTD simulations will be required.

## ACKNOWLEDGEMENT

The authors would like to thank Martin Kellerman and Malin Barg for cell depositions, Regina Ravekes and Colleen Lattyak for fruitful discussions, Malena Hillje for QE measurements, and Peng Hui Wang for helping with the FIB and SEM images. This work was supported by the exchange program IPID4all - Mobile Doctorates in System Integration of Renewable Energy, which is funded by the DAAD (German Academic Exchange Service). Brendan Brady would like to thank NSERC CREATE for financial support.

## REFERENCES

[1] H. Raut, V. Ganesh, A. Nair, and S. Ramakrishna, "Anti-Reflective Coatings: A Critical, In-Depth Review," Energy & Environmental Science, vol. 4, 10, pp. 3779-3804, 2011

[2] P. Bermel, C. Luo, L. Zeng, L. Kimerling, and J. D. Joannopoulos, "Improving Thin-Film Crystalline Silicon Solar Cell Efficiencies with Photonic Crystals," Optics Express, vol. 15, pp. 16986-17000, 2007

[3] A. Banerjee, and S. Guha, "Study of Back Reflectors for Amorphous Silicon Alloy Solar Cell Application," Journal of Applied Physics, vol. 69, 1030, 1991

[4] V. Steenhoff, M. Theuring, M. Vehse, K. von Maydell, and C. Agert, "Ultrathin Resonant-Cavity-Enhanced Solar Cells with Amorphous Germanium Absorbers," Advanced Optical Materials, vol. 3, 2, pp. 182-186, 2015

[5] V. Steenhoff, A. Neumüller, O. Sergeev, M. Vehse, and C. Agert, "Integration of a-Ge:H Nanocavity Solar Cells in Tandem Devices," Solar Energy Materials & Solar Cells, vol. 145, 2, pp. 148-153, 2016

[6] V. Steenhoff, M. Juilfs, R. Ravekes, M. Vehse, C. Agert, "Resonant-Cavity-Enhanced a-Ge:H Nanoabsorber Solar Cells for Application in Multijunction Devices," Nano Energy, vol. 27, pp. 658-663, 2016

[7] V. Ferry, M. Verschuuren, H. Li, R. Schropp, H. Atwater, and A. Polman, "Improved Red-Response in Thin Film a-Si:H Solar Cells with Soft-Imprinted Plasmonic Back Reflectors," Applied Physics Letters, vol. 95, 183503, 2009

[8] V. Ferry, M. Verschuuren, H. Li, E. Verhagen, R. Walters, R. Schropp, H. Atwater, and A. Polman, "Light Trapping in Ultrathin Plasmonic Solar Cells," Optics Express, vol. 18, S2, pp. A237-A245, 2010

[9] V. Ferry, M. Verschuuren, M. Claire van Lare, R. Schropp, H. Atwater, and A. Polman, "Optimized Spatial Correlations for Broadband Light Trapping Nanopatterns in High Efficiency Ultrathin Film a-Si:H Solar Cells." Nano Letters, vol. 11, 10, pp. 4239-4245, 2011

[10] L. van Dijk, J. van de Groep, L. Veldhuizen, M. Vece, A. Polman, and R. Schropp, "Plasmonic Scatter Back Reflector for Light Trapping in Flat Nano-Crystalline Silicon Solar Cells," ACS Photonics, vol. 3, 4, pp. 685-691, 2016

[11] J. Bhattacharya, N. Chakravarty, S. Pattnaik, D. Slafer, R. Biswas, and V. Dalal, "A Photonic-Plasmonic Structure for Enhancing Light Absorption in Thin Film Solar Cells," Applied Physics Letters, vol. 99, 131114, 2011

[12] S. Pattnaik, N. Chakravarty, R. Biswas, V. Dalal, and D. Slafer, "Nano-Photonic and Nano-Plasmonic Enhancements in Thin Film Silicon Solar Cells," Solar Energy Materials & Solar Cells, vol. 129, pp. 115-123, 2014

[13] U. W. Paetzold, E. Moulin, D. Michaelis, W. Böttler, C. Wächter, V. Hagemann, M. Meier, R. Carius, and U. Rau, "Plasmonic Reflection Grating Back Contacts for Microcrystalline Silicon Solar Cells," Applied Physics Letters, vol. 99, 181105, 2011

[14] M. Smeets, V. Smirnov, M. Meier, K. Bittkau, R. Carius, U. Rau, and U. W. Paetzold, "On the Geometry of Plasmonic Reflection Grating Back Contacts for Light Trapping in Prototype Amorphous Silicon Thin-Film Solar Cells," Journal of Photonics for Energy, vol. 5, 057004, 2015

[15] M. Smeets, V. Smirnov, K. Bittkau, M. Meier, R. Carius, U. Rau, and U. W. Paetzold, "Angular Dependence of Light Trapping in Nanophotonic Thin-Film Solar Cells," Optics Express, vol. 23, 24, pp. A1575-A1588, 2015

[16] W. Yan, Z. Tao, T. Min Brian Ong, and M. Gu, "Highly Efficient Ultrathin-Film Amorphous Silicon Solar Cells on Top of Imprinted Periodic Nanodot Arrays," Applied Physics Letters, vol. 106, 093902, 2015

[17] J. Jang, M. Kim, Y. Kim, K. Kim, S. Balik, H. Lee, and J. Lee, "Three Dimensional a-Si:H Thin-Film Solar Cells with Silver Nano-Rod Back Electrodes," Current Applied Physics, vol. 14, pp. 637-640, 2014

[18] K. Nakayama, K. Tanabe, and H. Atwater, "Plasmonic Nanoparticle Enhanced Light Absorption in GaAs Solar Cells," Applied Physics Letters, vol. 93, 121904, 2008

[19] W. Ho, P. Cheng, and K. Hsiao, "Plasmonic Silicon Solar Cell Based on Silver Nanoparticles Using Ultra-Thin Anodic Aluminum Oxide Template," Applied Surface Science, vol. 354, A, pp. 25-30, 2015

[20] H. Sai, H. Fujiwara, M. Kondo, and Y. Kanamori, "Enhancement of Light Trapping in Thin-Film Hydrogenated Microcrystalline Si Solar Cell Using Back Reflectors with Self-Ordered Dimple Pattern," Applied Physics Letters, vol. 93, 143501, 2008

[21] H. Sai, and M. Kondo, "Light Trapping Effect of Patterned Back Surface Reflectors in Substrate-Type Single and Tandem Junction Thin-Film Silicon Solar Cells," Solar Energy Materials & Solar Cells, vol. 95, 1, pp. 131-133, 2011

[22] H. Huang, L. Lu, J. Wang, J. Yang, S. Leung, Y. Wang, D. Chen, X. Chen, G. Shen, D. Li, and Z. Fan, "Performance Enhancement of Thin-Film Amorphous Silicon Solar Cells with Low Cost Nanodent Plasmonic Substrates," Energy & Environmental Science, vol. 6, 10, pp. 2965-2971, 2013

[23] Q. Li, S. Leung, L. Lu, X. Chen, Z. Chen, H. Tang, W. Su, D. Li, and Z. Fan, "Inverted Nanocone-Based Thin Film Photovoltaics with Omnidirectionally Enhanced Performance," ACS Nano, vol. 8, 6, pp. 6484-6490, 2014

[24] M. Theuring, P. H. Wang, M. Vehse, V. Steenhoff, K. von Maydell, C. Agert, and A. G. Brolo, "Comparison of Ag and SiO2 Nanoparticles for Light Trapping Applications in Silicon Thin Film Solar Cells," Journal of Physical Chemistry Letters, vol. 5, 19, pp. 3302-3306, 2014

[25] P. H. Wang, R. Nowak, S. Geißendörfer, M. Vehse, N. Reininghaus, O. Sergeev, K. von Maydell, A. G. Brolo, and C. Agert, "Cost-Effective Nanostructured Thin-Film Solar Cell with Enhanced Absorption," Applied Physics Letters, vol. 105, 183106, 2014

[26] P. H. Wang, M. Theuring, M. Vehse, V. Steenhoff, C. Agert, and A. G. Brolo, "Light Trapping in a-Si:H Thin Film Solar Cells Using Silver Nanostructures," AIP Advances, vol. 7, 1, 015019, 2017

[27] C. Pahud, V. Savu, M. Klain, O. Vazquez-Mena, F. Haug, J. Brugger, and C. Ballif, "Stencil-Nanopatterned Back Reflectors for Thin-Film Amorphous Silicon n-i-p Solar Cells,' IEEE Journal of Photovoltaics, vol. 3, 1, pp. 22-26, 2013

[28] C. Pahud, O. Isabella, A. Naqavi, F. Haug, M. Zeman, H. P. Herzig, and C. Ballif, "Plasmonic Silicon Solar Cells: Impact of Material Quality and Geometry," Optics Express, vol. 21, S5 pp. A786-A797, 2013

[29] H. Atwater, and A. Polman, "Plasmonics for Improved Photovoltaic Devices," Nature Materials, vol. 9, 3, pp. 205-213, 2010

[30] F. J. Beck, A. Polman, and K. R. Catchpole, "Tunable Light Trapping for Solar Cells Using Localized Surface Plasmons," Journal of Applied Physics, vol. 105, 114310, 2009

[31] O. E. Daif, L. Tong, B. Figeys, K. Van Nieuwenhuysen, A. Dmitriev, P. Van Dorpe, I. Gordon, and F. Dross, "Front Side Plasmonic Effect on Thin Silicon Epitaxial Solar Cells," Solar Energy Materials & Solar Cells, vol. 104, pp. 58-63, 2012

[32] R. Santbergen, R. Liang, and M. Zeman, "Amorphous Silicon Solar Cells with Silver Nanoparticles Embedded Inside the Absorber Layer," Materials Research Society Symposium Proceedings, vol. 1245, 2011

[33] J. Hulteen, D. Treichel, M. Smith, M. Duval, T. Jense, and R. Van Duyne, "Size-Tunable Silver Nanoparticle and Surface Cluster Arrays," Journal of Physical Chemistry B, vol. 103, 19, pp. 3854-3863, 1999

[34] P. Gao, J. He, S. Zhou, X. Yang, S. Li, J. Sheng, D. Wang, T. Yu, J. Ye, and Y. Cui, "Large-Area Nanosphere Self-Assembly by a Micro-Propulsive Injection Method for High Throughput Periodic Surface Nanotexturing," Nano Letters, vol. 17, 5, pp. 4591-4598, 2015

[35] R. Santbergen, T. Temple, R. Liang, A. Smets, R. van Swaaij, and M. Zeman, "Application of Plasmonic Silver Island Films in Thin-Film Silicon Solar Cells," Journal of Optics, vol. 14, 2, 2012

# Native-Metal-Oxide-Coated Plasmonic Electrode Metasurfaces for Nanophotonic Light Trapping and Efficient Charge Collection.

Deirdre M. O'Carroll[1,2]*, Christopher E. Petoukhoff[1], Zhongkai Cheng[2], Zeqing Shen[2], Catrice M. Carter[1]

[1] Rutgers University, Dept. of Materials Science and Engineering, Piscataway, NJ 08854, USA

[2] Rutgers University, Dept. of Chemistry, Piscataway, NJ 08854, USA

*Abstract* — **Metasurfaces are emerging nanostructured materials that enable controlled light manipulation on a 2-dimensional plane. The range of potential light control properties of metasurfaces lends them to the application of light management in thin-film optoelectronic devices. The challenge with implementing metasurfaces in thin-film photovoltaic devices is to make them multifunctional – so that they can efficiently act as both electrodes and as the light trapping element. In particular, plasmonic metasurfaces are limited to certain metal materials and, therefore, the surface electronic workfunction of the metasurface may not allow efficient charge collection from a semiconductor coating despite having excellent light trapping properties. This study focuses on the design and formation of ultra-thin, p-type native metal oxide coatings on plasmonic metasurfaces consisting of nanostructured silver, copper and nickel thin films. The light scattering and near-field light localization properties of the metasurface electrodes are characterized in thin-film photovoltaic device formats and their efficiencies for hole collection from organic semiconductor thin films are predicted.**

## I. INTRODUCTION

Conventional approaches for light trapping (such as textured back-reflectors) used in photovoltaic devices are not often appropriate for light trapping in thin-film photovoltaic devices with active layer thicknesses below ~300 nm. This is primarily due to the diffraction limit of visible light, which inhibits conventional light propagation and confinement in layers less than approximately half the wavelength of light [1.2]. Therefore, development of suitable photonic and plasmonic nanostructures and materials is important to manage light in such thin semiconductor active layers. In recent years we have developed and demonstrated plasmonic metasurfaces electrodes for light management below the diffraction limit in thin-film organic optoelectronic devices [3-8]. The metasurfaces consist of nanostructured metal electrodes consisting of nanoparticle arrays on metal films that can support localized and propagating surface plasmon modes which effectively confine incident light to within 100 nm of the surface of the electrode in the region of the semiconductor active layer.

Modification of the surface work-function of plasmonic electrodes while maintaining their distinct plasmonic properties is the topic of this paper. This is important because it enables integration of plasmonic electrodes into thin-film organic photovoltaic devices without significantly degrading the electrical device properties or charge injection/collection efficiency. We have carried out theoretical and experimental studies of ultra-thin ($\leq$5 nm) surface oxides on plasmonic electrodes with the goals of (1) improving electrical stability in

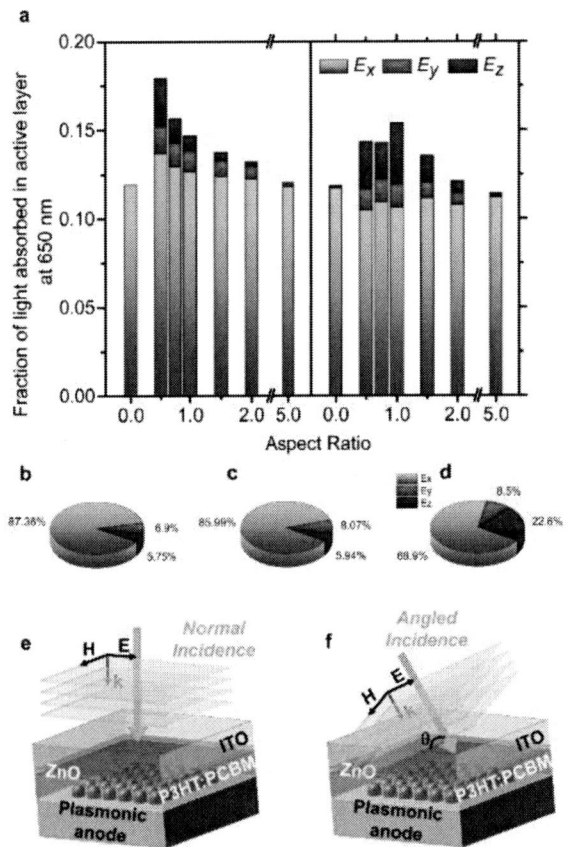

**Figure 1.** (a) Fractional absorption in the active layer for Ag anode in BHJ-OPVs excited at 650 nm for normal incidence (left) and at an angle of 15 ° (right). (b-d), Percentage of total absorption for the contribution to the absorption from each electric field component at an AR = 1.0 for (b), normal incidence, integrated broadband excitation; (c), normal incidence, 650 nm excitation; (d), 15 ° angled incidence, 650 nm excitation. (e) Definition of normal incidence. (f) Definition of angled incidence.

ambient conditions over time, (2) minimizing non-radiative recombination at the semiconductor-metal interface, and (3) increasing surface workfunction (in the case of native oxides) thereby eliminating the need for additional hole transport/electron blocking layers in inverted bulk-heterojunction organic photovoltaic (BHJ-OPV). Here, we report on both the light trapping potential of the plasmonic electrode metasurfaces and on how their surface workfunction can be modified to allow them to efficiently collect charge from organic semiconductor photovoltaic active layers.

## II. RESULTS AND DISCUSSION

### A. Light Trapping Using Plasmonic Electrode Metasurfaces

Fully 3-dimensional electromagnetic simulations of absorption fraction in the photovoltaic active layer have been carried out for plasmonic metasurfaces electrodes incorporated in conjugated polymer-based photovoltaic devices. Figure 1 shows active layer absorption components from off-normal incidence excitation for bulk-heterojunction organic photovoltaic (BHJ-OPV) devices employing Ag planar and plasmonic metasurfaces electrodes. The use of off-normal incidence excitation was computationally intensive for broadband excitation, so only single wavelength excitation was used (650 nm). The wavelength was chosen to coincide with a region of positive absorption enhancement in the active layer for the low-loss metals. Figure 1a shows the fraction of light absorbed in the active layer at 650 nm for normal incidence illumination (left) and for illumination at an angle of 15° off-normal incidence (right). It was apparent that the total fraction of light absorbed in the active layer using off-normal incidence excitation decreased relative to normal incidence simulations, except for a nanoparticle aspect ratio (AR) of 1, which was improved relative to normal incident excitation. The total fraction of light absorbed due to each component for AR = 1 is shown in Figure 1b-d for: the broadband integrated absorption fraction at normal incidence (Fig. 1b); absorption fraction at normal incidence at a wavelength of 650 nm (Fig. 1c) and absorption fraction at a wavelength of 650 nm and at an incident angle of 15° (Fig. 1d). The off-normal incidence excitation improved the fraction of light absorbed in the active layer arising from the out-of-plane component of the electric field, $E_z$, suggesting a greater fraction of incident light coupled into a propagating in-plane surface plasmon polariton mode of the metasurface.

### B. Electrode Workfunction and Interfacial Charge Transport Barriers

Besides the light-trapping properties of the metasurfaces electrodes, control of their electrical surface workfunction is critical for efficient charge collection. We have predicted the electronic performance of five different metal and four metal

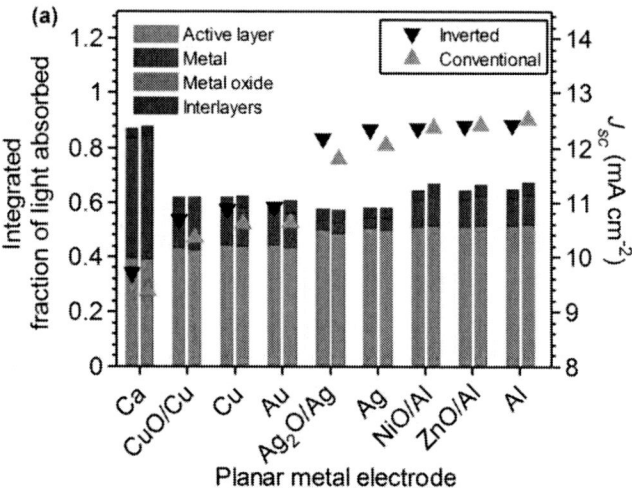

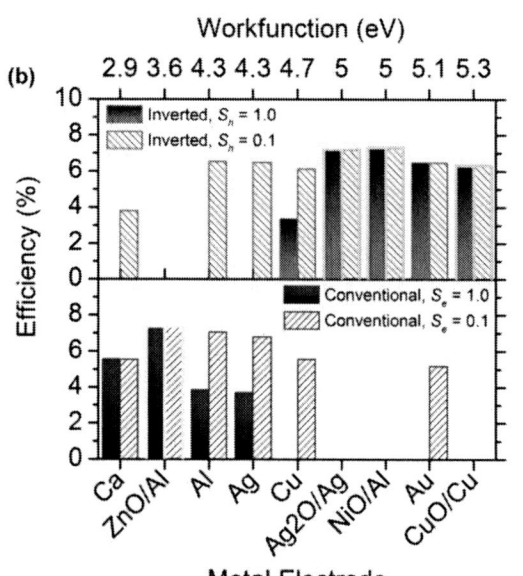

**Figure 2.** (a) Integrated absorption in the BHJ active layer, electrode layer(s), and interlayers (defined here as ZnO/ITO and PEDOT/ITO for inverted and conventional BHJ devices, respectively) for planar devices with a range of different metal or metal oxide-on-metal electrodes. The left bar in each column corresponds to inverted devices and the right bar to conventional devices. Overlaid are data for the calculated short-circuit current, $J_{sc}$, values. (b) Power conversion efficiency for the various planar metal and metal oxide-on-metal electrodes for inverted (top) and conventional (bottom) configurations.

oxide-on metal plasmonic metasurfaces electrodes by calculating charge injection and collection barriers in conventional and inverted BHJ-OPV devices (Figure 2). We expect that charge injection/collection barriers at the plasmonic electrode in conventional devices will be lowest for low workfunction metals, while for inverted devices charge injection barriers will be optimal for high workfunction metals. In this regard, plasmonic electrodes are particularly amenable to inverted device formats where the electrode is employed for hole injection/collection. An important topic of ongoing work is the experimental characterization of plasmonic electrode surface workfunctions with and without native surface oxides comparable to those of a planar electrode of the same material type. Figure 2 shows electronic calculations combined with optical simulations to extract predicted device power conversion efficiencies as a function of plasmonic electrode composition and the results yield insight into the best metal/metal-oxide electrode type to use for effective combined electronic and plasmonic performance.

## REFERENCES

1. V. E. Ferry, M. A. Verschuuren, H. B. T. Li, R. E. I. Schropp, H. A. Atwater, A. Polman, *App. Phys. Lett.* 95, 183503 (2009).
2. H. A. Atwater, A. Polman, *Nat. Mater.* 9, 205-213 (2010).
3. D. M. O'Carroll, A. X. Collopy, V. E. Ferry, H. A. Atwater, *25th EU PVSEC / WCPEC-5 Proc.* 834-837 (2010).
4. C. E. Petoukhoff, C. Antonick, K. M. Dani, D. M. O'Carroll, *MRS Advances* 1, 943-948 (2016).
5. C. E. Petoukhoff, D. M. O'Carroll, *Nature Communications* 6, 7899-1-13 (2015).
6. Z. Shen, D. M. O'Carroll, *Advanced Functional Materials* 25, 3302-3313 (2015).
7. C. E. Petoukhoff, Z. Shen, M. Jain, A. Chang, D. M. O'Carroll, Plasmonic Electrodes for Bulk-Heterojunction Organic Photovoltaics: A Review. *Journal of Photonics for Energy* 5, 057002-1-28 (2015).
8. R. Thomas, L. Fabris, D. M. O'Carroll, *Plasmonics* 9, 1283-1301 (2014).

# In-Ga precursor islands for Cu(In,Ga)Se$_2$ micro-concentrator solar cells

Katharina Eylers[1], Franziska Ringleb[1], Berit Heidmann[2,3], Sergiu Levcenco[2], Thomas Unold[2], Hagen W. Klemm[4], Gina Peschel[4], Alexander Fuhrich[4], Thomas Teubner[1], Thomas Schmidt[4], Martina Schmid[5], Torsten Boeck[1]

1) Leibniz Institute for Crystal Growth, Max-Born-Str. 2, 12489 Berlin, Germany

2) Helmholtz-Zentrum Berlin, Hahn-Meitner-Platz 1, 14109 Berlin, Germany

3) Department of Physics, Freie Universität Berlin, Arnimallee 14, 14195 Berlin, Germany

4) Fritz-Haber-Institut der Max-Planck-Gesellschaft, Faradayweg 4-6, 14195 Berlin, Germany

5) University of Duisburg-Essen and CENIDE, Lotharstr. 1, 47057 Duisburg, Germany

*Abstract* — We present a bottom-up approach for the fabrication of CuIn$_x$Ga$_{(1-x)}$Se$_2$ (CIGSe) micro-concentrator solar cells by local growth of In-Ga micro-islands. In addition to the intended islands, the indium deposition leads to a parasitic indium wetting layer, which was detected by photoelectron emission microscopy (XPEEM). This layer can be removed by a very short and gentle Ar$^+$ plasma etching step after indium deposition. The metallic precursor islands were further processed to micro-absorbers and characterized by spatially-resolved photoluminescence measurements for different selenization temperatures. The structural and optoelectronic properties of these absorbers are comparable to those commonly reported for CIGSe films.

*Index Terms* — Cu(In,Ga)Se$_2$, In-Ga precursor islands, micro-concentrator solar cells, PL measurements, wetting layer, XPEEM.

## I. INTRODUCTION

CIGSe is a highly efficient absorber material used for thin-film solar cells. The latest world record efficiency of 22.6 % was reached by the ZSW with a test cell that has an area of 0.5 cm$^2$ [1]. Due to the rareness of the constituent parts indium and gallium, endeavors to develop a material-saving production process are being made. Micro-concentrator solar cells have the advantage that they require less absorber material, while the cell conversion efficiency can be increased in comparison to full planar absorbers. Compared to macroscopic concentrator solar cells, those micro-concentrators show a better heat dissipation, which reduces thermally caused efficiency losses and allows a more compact module design. The efficiency enhancement of micro-concentrator cells has already been demonstrated for silicon [2] and likewise for CIGSe [3]-[6]. However these achievements were based on top-down approaches for micro-absorber fabrication and thus do not save raw material. For CIGSe these proof-of-principle studies yielded an absolute efficiency increase of up to 4 % [6]. In addition, the first mini-module made of chalcopyrite micro-absorbers showed an absolute efficiency increment of 1.8 % [7].

For material saving, however, a bottom-up process is needed. One such approach, growing the absorber by electrodeposition followed by selenization, has been demonstrated by Sadewasser et al. for circular shaped CuInSe$_2$ absorbers, achieving an efficiency of around 0.3 % [8], and by Duchatelet et al. for linear Cu(In,Ga)Se$_2$ absorbers showing efficiencies up to 7.6 % [9]. Our work focuses on a bottom-up approach based on physical vapor deposition (PVD) on laser patterned substrates. We have previously shown that indium islands can be used as precursors and we have developed a procedure to convert these indium precursors to CuInSe$_2$ micro-absorbers [10]. The local arrangement of the islands, which is indispensable for the alignment of the absorber with the concentrator component, is realized by surface structuring using a femtosecond laser [11]. By adjusting the indium deposition rate and the substrate temperature we are able to control the island size, areal density and aspect ratio to achieve the desired array dimensions [12]. The arrangement of the islands does not play a role for the investigation presented here and therefore we deploy non-laser treated samples with statistically distributed islands.

Here we present results on the integration of gallium into our bottom-up approach for the preparation of CIGSe micro-concentrator solar cells. X-ray photoelectron emission microscopy (XPEEM) measurements to reveal the distribution of In after deposition and after plasma treatment are shown. Furthermore the effect of different selenization temperatures is characterized by photoluminescence (PL) measurements. These PL results are used to determine the Ga content of our absorber-islands.

978-1-5090-5606-4/17 $31.00 © 2017 IEEE

## II. EXPERIMENTAL DETAILS

### A. Sample preparation

Commercial $50 \times 50 \times 2$ mm$^3$ soda-lime float glass samples (*Weidner Glas*) served as substrates. Sequential metal deposition was carried out by PVD in a high vacuum chamber enabling both the rear and front heating of the substrate (for details, refer to [11]). Deposition of both, molybdenum and indium was carried out at a substrate temperature of 515 °C. Molybdenum films of 320 nm to 420 nm thickness were deposited at rates between 1 Å/s and 1.8 Å/s. The layer thickness of indium, which was in the range of 90 nm to 110 nm with a deposition rate of 0.3 Å/s, has merely a nominal meaning due to the formation of islands during deposition. Gallium was deposited at a substrate temperature of 370 °C, with a nominal layer thickness of 25 nm to 35 nm and a deposition rate of 0.15 Å/s, which also resulted in the formation of islands. The subsequent copper deposition was carried out at ambient temperature with a rate of 3 Å/s to 5 Å/s, resulting in a homogeneous copper layer of 500 nm thickness. The selenization was performed under UHV conditions using a valve-controlled selenium source that is equipped with a thermal cracking zone operated at 900 °C. The substrate is exposed to Se at room temperature for 30 min. Thereafter, the substrate temperature is increased first to 250 °C and in a second step to values ranging from 500 °C to 560 °C while Se is still evaporated. After cooling down the samples were exposed to an aqueous KCN solution of 0.1 g/mL for 3 min to remove selectively any $Cu_xSe_y$ phases which formed as by-product.

### B. Characterization

The PEEM/LEEM measurements were carried out in the SMART microscope operating at the UE49-PGM beamline of the synchrotron light source BESSY II of the Helmholtz Center Berlin (HZB). This aberration corrected and energy filtered instrument combines microscopy (LEEM, PEEM), diffraction (LEED), and spectroscopy (XPS, NEXAFS) techniques for comprehensive characterization of surfaces [13]-[15].

Optical microscopy was performed in bright-field mode with a *Reichert Polyvar 2* microscope equipped with a CCD camera (*Nikon* DS–5 M), a *DS-U1* interface and *NIS-Elements F* imaging software (vers. 3.22). Scanning electron microscopic (SEM) investigations were operated with a *Nova 600 Nanolab DualBeam* microscope from *FEI*, equipped with a TEAM trident analysis system from EDAX. The energy dispersive X-ray diffraction (EDX) measurements were performed in the same SEM system.

For the room temperature photoluminescence (PL) measurements a pulsed laser diode (wavelength: 660 nm, pulse repetition rate: 2.5 MHz) was used as optical excitation source. The PL signals were detected by a 0.5 m Czerny-Turner grating monochromator (grating density: 150 mm$^{-1}$,

spectral resolution: ~0.52 nm) equipped with a liquid $N_2$ cooled linear InGaAs photodiode array. Spatially and spectrally resolved PL maps were recorded with an exposure time of 2 s, a 1.5 µm step size and an optical excitation spot diameter of ~1.5 µm.

## III. RESULTS

### A. Indium Wetting Layer

The growth of indium islands on molybdenum coated glass is attended by the formation of a thin indium wetting layer. The existence of a wetting layer was verified by soft XPEEM measurements. Fig.1 shows XPEEM images, representing the spatial distribution of In 3d and Mo 3d photoemission. Figs. 1a and 1b display the same detail on the sample, the edge of an indium island. This is also the case for Figs. 1c and 1d, but an arbitrary area in between the In islands is shown. High intensity regions in the images exhibit a high content of the respective element and vice versa. The exciting X-ray beam has a diameter of $16 \times 20$ µm$^2$. On the depicted image section of Figs.1a and 1b the edge of an island can be seen in the upper half of the field of view. The photoelectrons of Mo 3d are not detected in the domain of the island, but at the substrate (Fig. 1a).

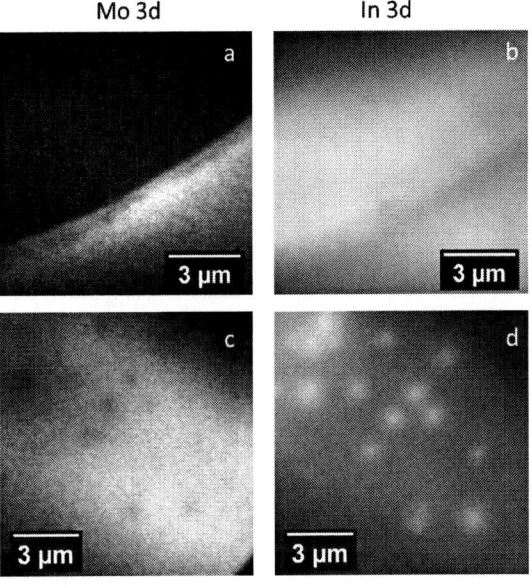

Fig. 1. XPEEM images representing the spatial distribution of Mo 3d (hv = 550 eV) and In 3d (hv = 330 eV) photoemission at a), b) the edge of an indium island (Mo: $E_{kin}$ = 101 eV, In: $E_{kin}$ = 107 eV) and c), d) an arbitrary position on the substrate in between the islands after sputtering (Mo: $E_{kin}$ = 97.5 eV, In: $E_{kin}$ = 108 eV).

In contrast, In 3d photoelectrons are observed within the whole field of view (Fig. 1b). Due to the small mean free path length of the probed photoelectrons at a kinetic energy of about 100 eV, indium is merely detected within the uppermost 10 Å of the substrate. This implies that the surface is covered with an indium wetting layer of a few angstroms in thickness. This wetting layer is not only detected in the adjacencies of the indium islands, but all over the substrate.

Sequential deposition of In and Ga leads in our case to the formation of interstitial In-Ga structures between the indium islands. We assume that this is related to the interaction of Ga with the In wetting layer. Therefore we analyzed the removal of the wetting layer through sputtering by XPEEM measurements. Figs. 1c and 1d demonstrate the effect of Argon sputtering ($U_{Acc}$ = 450 V, $I_S$ ~ 1 mA, 120 min, with the Argon ions impinging parallel to the surface normal and an exposed sample area of 7 mm in diameter). The photoelectrons of In 3d are no longer detected within the whole field of view, only a few indium spots are still visible. For comparison, the photoelectrons of Mo 3d are observed within the whole field of view. This demonstrates that the indium wetting layer can be removed.

### B. Sequential growth of indium-gallium islands

Due to the higher melting and boiling point of indium compared to gallium, a sequential process starting with In and continuing with Ga was used for the preparation of In-Ga micro-islands. For experimental details, please refer to section II. On the resulting samples gallium was mainly detected at the edge of indium islands, in a fringe around an indium rich core. This phenomenon is illustrated by the energy dispersive X-ray (EDX) maps shown in Fig. 2.

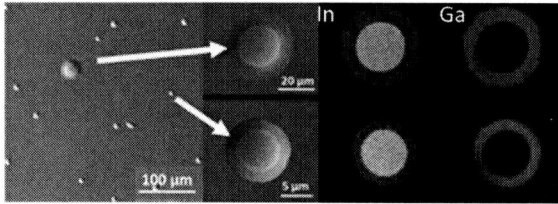

Fig. 2. SEM images (left) and EDX maps (right) of In-Ga micro islands.

The observed phase segregation within the In-Ga islands is in line with the solidification behavior of the eutectic In-Ga system and thus indicates that the solidification process is rather thermodynamically than kinetically dominated. In addition, small interstitial islands of about 5 μm in diameter have formed, which also consist of an indium-rich core surrounded by a gallium-rich fringe. We presume, that these smaller islands form during gallium deposition due to the interaction of Ga atoms approaching the surface from the vapor phase with the indium wetting layer.

For the fabrication of well-defined arrays of CIGSe micro-absorbers the formation of interstitial islands has to be avoided. Since the XPEEM measurements showed the potential to remove the indium wetting layer, we introduced

an additional $Ar^+$ plasma etching step to remove the wetting layer prior to the deposition of Ga. As the wetting layer is only a few angstroms thick, a very gentle, short etching is sufficient to suppress the formation of any interstitial structures (Fig. 3).

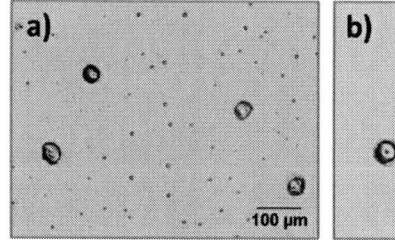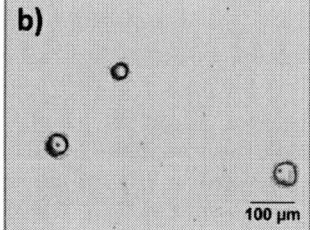

Fig. 3. Light micrographs of a) Sample with In-Ga islands without additional $Ar^+$ plasma etching step prior to Ga deposition b) Sample with additional $Ar^+$ plasma etching step to remove the indium wetting layer.

We assume that a higher diffusivity of Ga atoms on the bare Mo than on an In wetting layer causes this desirable outcome. In the following we show that this horizontally inhomogeneous In-Ga distribution throughout the metallic precursor can be eliminated in the subsequent selenization step.

### C. Processing to Cu(In,Ga)Se₂ micro-islands

Gallium is not only taken up by the indium islands, but also forms a wetting layer itself. Fig. 4a shows the edge of an In-Ga island and the Ga wetting layer, which has a ripple structure and appears to be ticker than the In wetting layer. Thus, for the further processing to chalcopyrite absorber islands, the samples with the In-Ga precursors were again treated with a gentle $Ar^+$ plasma etching step. The samples were then coated with copper and selenized, as described in the experimental section. In Fig. 4b an example of one single resulting micro-absorber island obtained after selective etching of copper selenides is depicted.

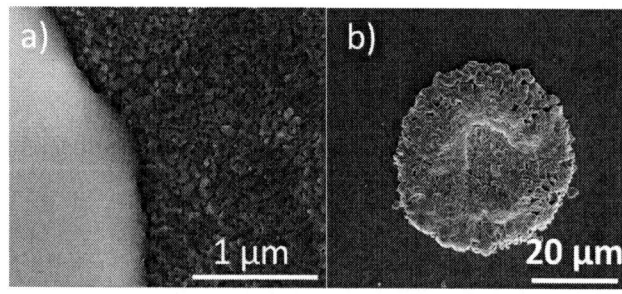

Fig. 4. SEM images of a) In-Ga island edge with Ga wetting layer and b) one single Cu(In,Ga)Se₂ micro-island selenized under UHV conditions and etched in a KCN solution.

978-1-5090-5606-4/17 $31.00 © 2017 IEEE

## D. Photoluminescence measurements

Photoluminescence measurements were carried out on samples, which were prepared applying different selenization temperatures. The optical band gap energy of CuInSe$_2$ is ~1.04 eV. By replacing In atoms with Ga atoms the band gap can be increased to up to ~1.68 eV, which is the optical band gap energy of CuGaSe$_2$ [16]. In Figs. 5a-d maps of the PL peak position are shown for CIGSe islands prepared under identical conditions, except for the maximum selenization temperature (500 °C (I), 520 °C (II), 540 °C (III) and 560 °C (IV)). It has to be considered that with the wavelength of the optical excitation source (660 nm) and the absorption coefficient of CIGSe ($\alpha = 8 \times 10^{-4}$ cm$^{-1}$ at 1.9 eV [21]) the penetration depth of the PL measurements yields approximately 100 nm, i.e. only the surface region of the absorber is probed.

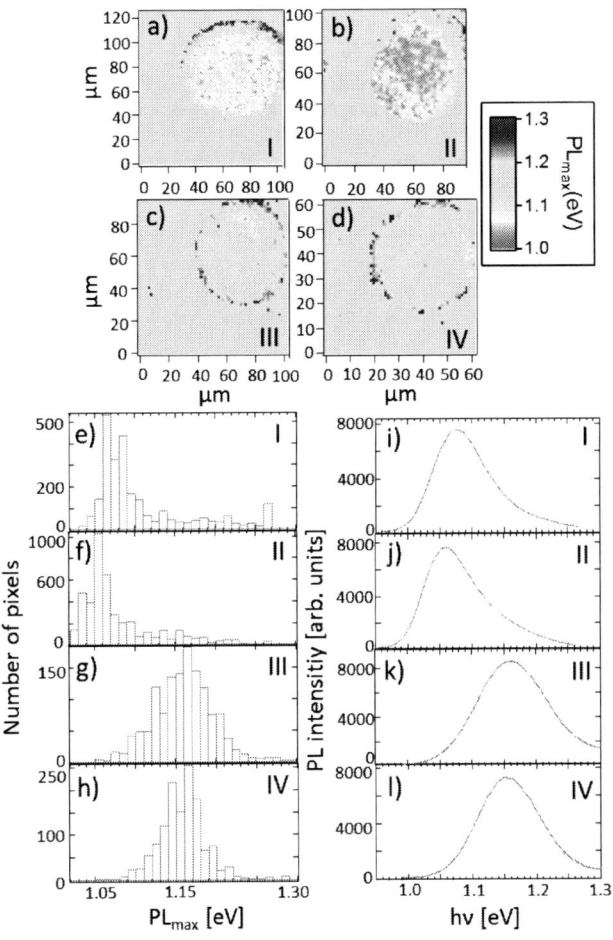

Fig. 5. a)-d) PL maps of the peak emission energies, for CIGSe islands selenized at 500 °C (I), 520 °C (II), 540 °C (III) and 560 °C (IV). e)-h) Statistical analysis of the PL maps for samples I-IV. i)-l) Averaged spectra of I-IV as a function of energy.

The detection range was chosen such that photoemission spectra could be measured across the entire island area. The peak energy is depicted by a color code ranging from red for the lower limit (1.0 eV) to purple for 1.3 eV. Samples produced with lower substrate temperatures during selenization (I-II) show a wide range of emission maxima (1.04 eV - 1.15 eV) in photoluminescence maps. High energy emission indicating a high Ga content is detected at the edges of the islands, whereas low energy emission, indicative of an indium rich phase, is observed in the center. This emission distribution reflects the horizontal elemental distribution of indium and gallium in the segregated In-Ga precursor islands and indicates a poor migration of both metals within the islands during the selenization process. In contrast, samples which were exposed to higher substrate temperatures during selenization (III-IV) show promising results regarding In-Ga intermixing, which is an important prerequisite for optimizing the absorber.

A statistical analysis of each PL map depicted in Figs. 5a-d is shown as histogram in Figs. 5e-h. Figs. 5i-l show the averaged spectra, from all PL spectra in the region of the corresponding CIGSe islands of Figs. 5a-d. The position of the main peak of the averaged spectra (Figs. 5i-l) shifts from approximately 1.05 eV to 1.16 eV with increasing temperature. The band gap for samples with higher substrate temperatures during selenization is comparable with ideal values for common thin-film CIGSe absorbers reported in the literature [17] and is clearly shifted upwards in comparison with the band gap of previously measured CISe micro-islands (1.03 eV, data not shown). Several studies have shown the correlation between Ga content and band gap in CIGSe [18]-[21]. We used equation (1) from [21] to estimate the Ga content in the investigated absorber volume of our CIGSe islands.

$$E_g(x) = 1.010 + 0.626x - 0.167x(1-x) \qquad (1)$$

The authors of [21] obtained this approximation by fitting a function based on theoretical models on experimental data obtained from samples of varying In-Ga content. By inserting the PL peak position into Eq. 1 we calculated the approximate Ga content of the CIGSe islands. In Fig. 6 the results of this calculation are illustrated for all samples. Samples with a lower substrate temperature during Se evaporation (I and II) have a lower average Ga content in the probed absorber volume of about 0.08 to 0.14, compared to samples III and IV which were selenized at higher temperatures, exhibiting a Ga content of about 0.28 to 0.30. Since the best solar cells contain ~30 % Ga [1], [22]-[23], we aim for a composition of CuIn$_{0.7}$Ga$_{0.3}$Se$_2$. For our CIGSe micro-absorber islands this ratio can be achieved by higher substrate temperatures (III-IV) during Se evaporation.

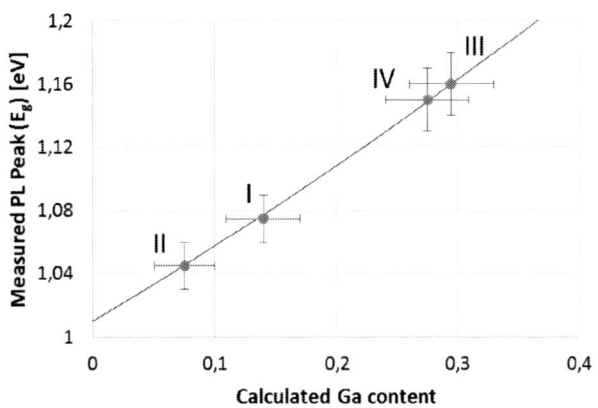

Fig. 6. Diagram of the calculated Ga content of the Cu(In,Ga)Se$_2$ micro-absorbers as a function of the average measured PL peaks of samples I-IV. The error bars indicate the measured range of PL peaks and the therefore resulting range of Ga content. The line indicates Eq. (1).

## III. DISCUSSION

Sequentially grown In-Ga precursor islands can be transformed into CuIn$_x$Ga$_{(1-x)}$Se$_2$ micro-absorbers, which in turn will be further processed to CIGSe micro-concentrator solar cells. The removal of the indium wetting layer, which arises during indium island growth, clearly improves the sample quality by suppressing additional interstitial island growth during the subsequent deposition of gallium. Samples produced with high selenization temperatures (540 °C – 560 °C) show promising results regarding homogeneity and band gap of the resulting CIGSe micro absorbers due to a better mixing of the components. We have so far obtained valuable information about the horizontal surface homogeneity of the islands by PL mapping, but it is well known from the literature [24-26] that absorber may have a Ga gradient and it has yet to be verified experimentally in our micro-absorbers.

## IV. SUMMARY

We demonstrated a bottom-up approach for the local growth of In-Ga islands, which can be used as precursors for the fabrication of chalcopyrite micro-absorbers. We verified the existence of an indium wetting layer and introduced an approach to remove it. Photoluminescence measurements were performed on samples, which were prepared applying different selenization temperatures and the results were used to determine the Ga content of our absorber-islands.

Our approach combines the potential for material saving by a bottom-up process with the advantage of vacuum evaporation, which currently yields the highest efficiencies for CIGSe solar cells.

### ACKNOWLEDGEMENT

The authors thank HZB for the allocation of synchrotron radiation beamtime. Furthermore we gratefully acknowledge financial support by the Deutsche Forschungsgemeinschaft (DFG, German Research Foundation) through BO1129/6-1 and SCHM2554/3-1. The research leading to these results has received funding from the European Union Seventh Framework Program (FP7/2007-2013) under grant agreement no. 609788.

### REFERENCES

[1] P. Jackson, R. Wuerz, D. Hariskos, E. Lotter, W. Witte, M. Powalla, *Phys. Status Solidi RRL,* 2016, 10, No. 8, 583–586

[2] J. Yoon, A. J. Baca, S. I. Park, P. Elvikis, J. B. Geddes, 3rd, L. Li, R. H. Kim, J. Xiao, S. Wang, T. H. Kim, M. J. Motala, B. Y. Ahn, E. B. Duoss, J. A. Lewis, R. G. Nuzzo, P. M. Ferreira, Y. Huang, A. Rockett, J. A. Rogers, *Nature Materials,* 2008, 7,907.

[3] M. Paire, L. Lombez, F. d. r. Donsanti, M. Jubault, S. p. Collin, J.-L. Pelouard, J.-F. o. Guillemoles, D. Lincot, *Journal of Renewable and Sustainable Energy,* 2013, 5, 011202.

[4] B. Reinhold, M. Schmid, D. Greiner, M. Schüle, D. Kieven, A. Ennaoui, M. C. Lux-Steiner, *Progress in Photovoltaics: Research and Applications,* 2015, 23, 1929.

[5] M. Paire, A. Shams, L. Lombez, N. Péré-Laperne, S. Collin, J.-L. Pelouard, J.-F. Guillemoles, D. Lincot, *Energy & Environmental Science,* 2011, 4, 4972.

[6] M. Paire, L. Lombez, N. Pere-Laperne, S. Collin, J.-L. Pelouard, D. Lincot, J.-F. Guillemoles, *Applied Physics Letters* 98, 264102, 2011.

[7] S. Jutteau, J.-F. Guillemoles, M. Paire, *Applied Optics*, Edition edn., 2016.

[8] S. Sadewasser, P. M. P. Salome, H. Rodriguez-Alvarez, *Sol. Energy Mat. Sol. Cells,* 159, 496, 2017.

[9] A. Duchatelet, K. Nguyen, P.-P. Grand, D. Lincot, and M. Paire, *Applied physics letters,* 109, 253901, 2016.

[10] T. Boeck, F. Ringleb, R. Bansen, *Crystal Research and Technology,* 1600239, 2017.

[11] F. Ringleb, K. Eylers, T. Teubner, T. Boeck, C. Symietz, J. Bonse, S. Andree, J. Krüger, B. Heidmann, M. Schmid, M. Lux-Steiner, *Applied Physics Letters,* 108, 111904, 2016.

[12] F. Ringleb, K. Eylers, Th. Teubner, H.-P. Schramm, C. Symietz, J. Bonse, S. Andree, B. Heidmann, M. Schmid, J. Krüger, T. Boeck, *Applied Surface Science,* 2016. http://dx.doi.org/10.1016/j.apsusc.2016.11.135

[13] R. Wichtendahl, R. Fink, H. Kuhlenbeck, D. Preikszas, H. Rose, R. Spehr, P. Hartel, W. Engel, R. Schlögl, H.-J. Freund, A.M. Bradshaw, G. Lilienkamp, T. Schmidt, E. Bauer, G. Benner, E. Umbach, *Surface Review and Letters 05,* 1249, 1998.

[14] Th. Schmidt, H. Marchetto, P.L. Lévesque, U. Roh, F. Maier, D. Preikszas, P. Hartel, R. Spehr, G. Lilienkamp, R. Fink, E. Bauer, H. Rose, E. Umbach, H.-J. Freund, *Ultramicroscopy 110,* 1358 – 1361, 2010

[15] Th. Schmidt, A. Sala, H. Marchetto, E. Umbach, H.J. Freund, *Ultramicroscopy 126,* 23, 2013.

[16] L. Shay, J. Wernick, *Ternary Chalcopyrite Semiconductors,* Pergamon, Oxford, 1975.

[17] S. Shirakata, T. Nakada, *Thin Solid Films 515,* 6151 - 6154, 2007.

[18] S.-H. Wei, A. Zunger, *Journal of Applied Physics 78* (6) 3846 - 3856, 1995.

[19] S.-H. Wei, S. B. Zhang, A. Zunger, *Physics Letters 72* (24) 3199-3201, 1998.

[20] M. Turcu, I. M. Kötschau, U. Rau, *Journal of Applied Physics 91* (3), 1391-1399, 2002.

[21] M. I. Alonso, M. Garriga, C. A. Durante Rincón, E. Hernández, and M. León, *Applied Physics A: Materials Science & Processing A74*, 659, 2002.

[22] J. Hedstrom, H. J. Olsen, M. Bodegard, A. Kylner, L. Stolt, D. Hariskos, M. Ruckh, and H. W. Schock, *Proceedings of the 23rd IEEE Photovoltaic Specialists Conference*, p. 364, 1993.

[23] A. M. Gabor, J. R. Tuttle, D. S. Albin, M. A. Contreras, R. Noufi, and A. M. Hermann, *Applied physics letters 65*, 198, 1994.

[24] M. Marudachalam, R. W. Birkmire, H. Hichri, J. M. Schultz, A. Swartzlander, M. M. Al-Jassim, *Journal of Applied Physics 82*, 2896, 1997.

[25] R. Mainz, A. Weber, H. Rodriguez-Alvarez, S. Levcenko, M. Klaus, P. Pistor, R. Klenk, H.-W. Schock, *Progress in Photovoltaics: Research and Applications* 23:1131–1143, 2014.

[26] W. Witte et al., *Progress in Photovoltaics: Research and Applications*, 23:717–733, 2015

# Advances in silicon surface texturization by metal assisted chemical etching for photovoltaic applications

Sylvain Le Gall[1], Raphaël Lachaume[1], Encarnacion Torralba[2], Mathieu Halbwax[3], Vincent Magnin[3], Taha El Assimi[2], Marin Fouchier[3], Joseph Harari[3], Jean-Pierre Vilcot[3], Christine Cachet-Vivier[2], Stéphane Bastide[2]

[1] *Génie Electrique et Electronique de Paris, UMR-CNRS 8507, Centralesupelec, Univ. Paris-Sud, UPMC, France*
[2] *Institut de Chimie et des Matériaux Paris-Est, CNRS, Univ. Paris-Est, France*
[3] *Institut d'Electronique, Microélectronique et Nanotechnologies, UMR-CNRS 8520, Univ. Lille 1, France*

*Abstract* — New Si processes based on Metal Assisted Chemical Etching (MACE) are explored for solar cells texturization. Pt and Au are considered as catalysts for MACE of p and n-type Si substrates. 2D band bending modeling at the nanoscale shows that Pt nanoparticles (NPs) make ohmic contacts and induce delocalized etching. Accordingly, cone-shaped macropores, very efficient in reducing the reflectivity (<5%) are obtained experimentally. On the contrary, Au with n-type Si leads to non-ohmic contacts and localized etching. On this basis, a novel strategy for 3D pattern transfer into Si with patterned nanoporous gold electrodes in a single step is developed.

## I. INTRODUCTION

Silicon etching is a key process in the fabrication of Si microstructures that are essential for the development of several component families used in microelectronics, photonics and solar energy conversion. A large variety of efficient Si microstructuring technologies exists nowadays (e.g. wet/dry etching, photo/electron beam lithography). Their remarkable efficacy comes at the expense of numerous lithography (mask) and etching steps that are not suitable for industry, especially when cost and fabrication time are key aspects for Si solar cells manufacturing. A maskless technique with direct imprinting of patterns would dramatically simplify the fabrication of microstructures. However, eliminating the use of masks is extremely challenging and implies to move towards micromachining techniques.

In this context, MACE of Si has attracted considerable attention as an efficient way to produce Si nanostructures with high aspect ratio. Noble metals, including Ag, Au, Pt, Pd, *etc.*, are known to be effective catalysts for MACE [1], each one leading to different results in the (de)localized character of etching [2, 3]. The reason for this behavior may lie in the Si/metal Schottky junction properties that determine hole (h$^+$) injection into Si, which depends on the metal work function and on the doping type of Si, as recently brought to attention by Kolasinki [4].

With the aim of developing new efficient texturization processes for Si solar cells, two MACE systems involving opposite metal/Si contact properties are investigated in this work.

## II. RESULTS AND DISCUSSION

### A. Surface structuration of p-type Silicon by MACE using Pt nanoparticles

First, we have experimentally studied the Pt nanoparticle/p-Si/HF system. Details of the experimental set-up and Si substrate can be found in reference [3]. Figure 1 shows cross-sectional Scanning Electron Microscopy (SEM) images of typical cone-shaped macrostructures obtained by MACE with Pt NPs. A thick porous Si layer covers both the surface and the inner core. This is due to the delocalized character of p-Si etching using Pt as catalyst, as explained below.

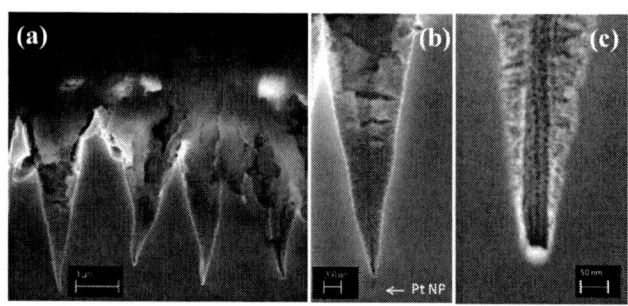

Fig. 1. SEM images in cross section of p-type (100) Si samples after MACE with Pt NPs; increasing magnification from (a) to (c).

To gain insights into this MACE process, 2D Numerical simulations of the valence band modulation at the Pt NP/p-Si/electrolyte interfaces were carried out using a TCAD software. More details about the modeling and the modeled structure are given in the supporting information of [3].

Figure 2(a) shows a 3D plot of the Si conduction band (CB) and valence band (VB) modulations around the Pt NP (located at x=0.5 μm) for the system at equilibrium. The large difference between the work functions of the electrolyte $W_{EL}$ (4.5 eV) and Pt $W_{Pt}$ (5.6 eV) induces a modification of the band bending at the surface close to the NP. In this system, away from the Pt NP, *i.e.* at the p-Si/Electrolyte junction, we observe a Schottky energy barrier height of 0.48 eV for h$^+$

coming from the bulk. On the contrary, beneath the Pt NP, the VB is above the FL, indicating the presence of a strong accumulation layer of h+ (ohmic contact).

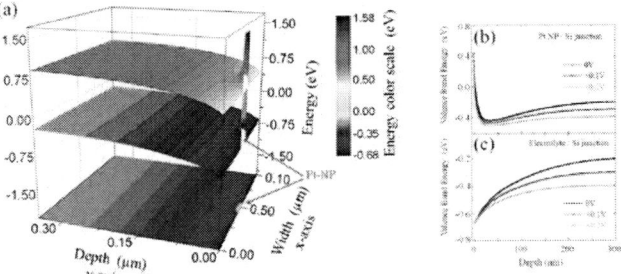

Fig. 2. Modelling of the band bending of a Pt NP/p-Si/Electrolyte device. (a) 2D profiles of CB and VB energies at the equilibrium (0 V). The VB energy is also projected on the bottom (x, y) plane. (b), (c) 1D profiles of the VB energy for different Pt polarization (b) under the Pt NP (x = 50 nm) and (c) under the Electrolyte (x = 0 nm).

Figures 2(b) and 2(c) show the VB depth profiles (y-direction) under equilibrium (0 V) and positive bias (+0.1 and +0.2 V) at two specific locations: (b) the middle of the Pt contact (x = 50 nm) and (c) the Electrolyte far from Pt (x = 0 nm). This positive bias can be identified with the oxidant power of $H_2O_2$ during MACE. Due to the ohmicity of the Pt contact, the entire voltage drop falls at the p-Si/Electrolyte interface under positive bias, and the VB in the bulk follows the applied polarization by the same extent. Consequently, the applied voltage, or the oxidizing power of $H_2O_2$, reduces the barrier height of the Si/Electrolyte junction.

On the basis of the experimental results and numerical simulations of both the VB and the space charge region (SCR) in Si and around the Pt NP, the scheme of Figure 3(b) is proposed to rationalize the formation of cone-shaped pores by MACE with Pt NPs.

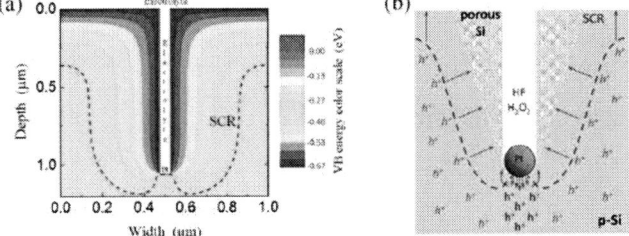

Fig. 3. (a) Modelling of the VB bending in a Pt NP/p-Si/Electrolyte device during MACE. The Pt NP was modelled by a 50 nm wide segment and buried at 1 μm from the surface. A polarization of +0.2 V is applied to the Pt. Dashed lines indicate the limit of the SCR. (b) Scheme of cone-shaped pore formation by MACE with Pt NPs underlining h+ injections and porous Si formation as deduced from the modelling. Bold h+ are injected carriers, italic h+ majority carriers driven through bulk polarization.

By applying a polarization to the Si sample during MACE it is possible to modulate the formation of porous Si along the macropores and thus to modify the cone angle defined by the

porous Si/c-Si interface [3]. Figure 4(a) and 4(b) shows SEM images of a sample surface morphology after MACE under positive bias and porous Si removal in HNO3/HF 99:1 V:V (5 min). Well-defined cone-shaped pores with an angle of ~35°, corresponding to an aspect ratio (D/W) of 1.6, cover the whole sample surface.

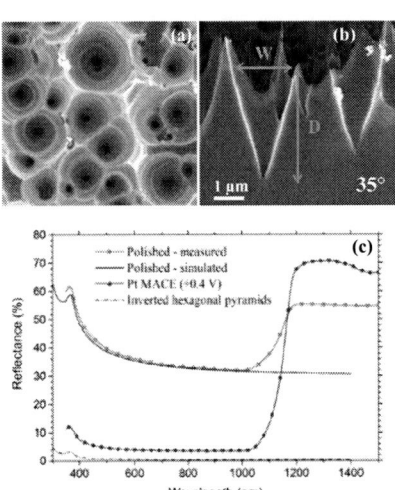

Fig. 4. (a) and (b) Plain and cross sectional SEM images of (100) p-type Si samples after MACE with Pt NPs under positive bias and porous Si removal in HNO3/HF. (c) Experimental (red squares and blue circles) and modelled (black solid and dashed lines) reflectance values obtained for polished and MACE texturized Si.

We have found that these structures advantageously reduce the surface reflectance down to 4% in the 600-1000 nm range, which compares favorably with state of the art texturization techniques for Si such as inverted square based pyramids arrays (R ~9%) used in record efficiency solar cells [5]. Because such morphologies are difficult to obtain even with advanced plasma etching techniques, MACE with Pt has a strong potential for Si surface structuration.

### B. Surface structuration of n-type Silicon by MACE using nanoporous Au electrodes

From a band bending point of view, the use of n-type Si substrates is more suitable for having a localized etching. For low work function metals such as Pt and Au, a schottky junction is obtained at the level of the NPs/n-Si electrolyte, and so delocalized etching with porous Si formation is minimized. Having a lower work function than Pt, Au is also more advantageous with this respect. Based on these observations, a new strategy to achieve pattern transfer into Si by electrochemical MACE with large dimensions metal tools is presented in Figure 5 (a). Previous attempts of Si imprinting with bulk metal electrodes have evidenced limitations due to electrolyte blockage. Here, the problem is solved by using for the first time patterned nanoporous Au (np-Au) electrodes that allow the electrolyte to access the entire Si/Metal interface. The surface pattern consists in an array of square based pyramids, as shown in Figure 5(b), 5(c). We aim at imprinting this pattern to obtain eventually inverted pyramids that are well known for their efficient antireflection properties.

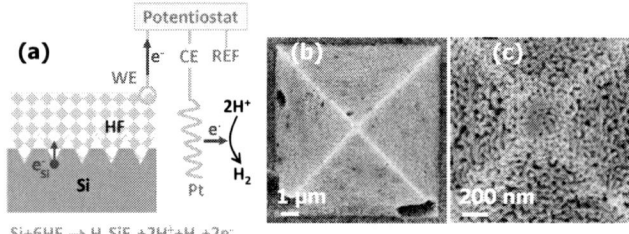

Fig. 5. (a) Scheme of the electrochemical MACE with patterned np-Au electrodes. (a) a np-Au pyramid after dealloying. (b) the np-Au structure at its tip.

The patterned np-Au electrode is positively bias (0.3 V/SME) in HF 5M (2% Ethanol) in the dark and brought into contact with the n-Si under a pressure of 18 g cm$^{-2}$. More details about the experimental set-up can be found in reference [6]. Figure 6 gives SEM images of the results after imprinting for 10 min in these conditions. The imprinted area is ~ 1 mm$^2$.

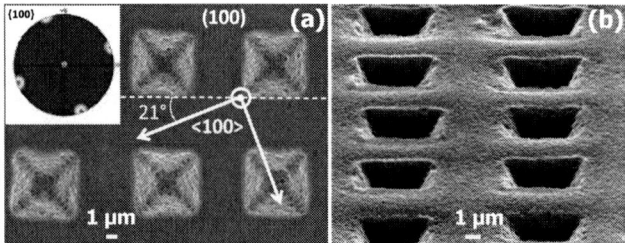

Fig. 6. SEM images of n-type (100) Si surface after imprinting inverted pyramids by MACE. The inset of (a) is a pole figure established by EBSD at 70° from image (b).

Relatively well defined inverted pyramids are obtained, with dimensions of 5×5 μm$^2$ and 3.6 μm in depth. The inset of Fig. 4(c) shows the pole figure established by Electron Back Scattered Diffraction (EBSD) of the (100) Si crystal. The pyramid sides are clearly not aligned with the [001] and [010] directions of the sample, which is to be expected since the np-Au electrode was randomly positioned on the sample surface. The same np-Au electrode could be used several times before the pyramidal pattern was damaged. Current efforts are being devoted to improve resolution, etch rate and imprinted area. One of the goals is to achieve in this way the texturization of multicrystalline Si surface with inverted pyramids.

In this particular MACE configuration, 2D simulations of the valence band modulation at the level of the Au/n-Si contacts have also been performed. Figure 7 shows the valence band energy at the interface of 3 nanoscaled contacts spaced so as to mimic the Au ligaments in the np-Au electrode. This can help to determine the influence of the porosity on the pattern transfer and the extent of delocalized etching, depending on the nature of the metal, the Si doping type and the applied polarization.

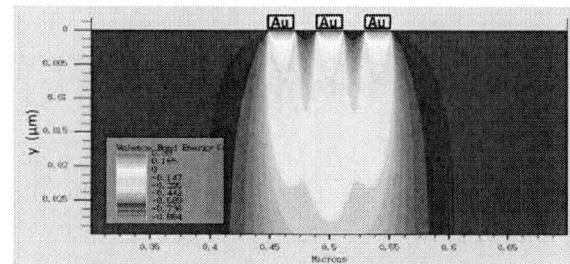

Fig. 7. Valence band energy at three Au/n-Si contacts in the configuration of MACE with a polarized np-Au electrode.

## III. CONCLUSION

In this work, we have presented two new routes of Si surface texturization by MACE with foreseen photovoltaic applications. These methods rely on the dependence of the (de)localized character of etching with the properties of the metal/Si contacts, which has been extensively studied by 2D band bending modelling of metal/Si/electrolyte interfaces at the nanoscale.

With Pt NPs and p-Si, macroscopic cone-shaped pores in the 1-5 μm size range with a high aspect ratio (L/W ~1.6) have been obtained by MACE assisted with an external polarization. This morphology leads to a reduction of the surface reflectance below 5% over the entire visible-near infrared domain. With patterned np-Au electrodes and n-Si, direct imprinting of well-defined arrays of inverted pyramids has been achieved for the first time by electrochemical MACE. The surface pattern of the electrodes is partially transferred to the Si substrate, with a sub-micrometer resolution and independently of the crystallographic orientation. This is a step forward in the development of a texturization method for photovoltaic applications.

## ACKNOWLEDGMENTS

The authors acknowledge the support of the French Agence Nationale de la Recherche (ANR), under grant ANR-14-CE07-0005-01 (PATTERN project).

## REFERENCES

[1] Z. Huang, N. Geyer, P. Werner,; J. de Boor, U. Gosele, "MACE of Silicon: A Review". Adv. Mater. **23**, 285−308 (2011).

[2] K. Tsujino, M. Matsumura, " Boring Deep Cylindrical Nanoholes in Silicon Using Silver Nanoparticles as a Catalyst." Adv. Mater. **17**, 1045−1047 (2005).

[3] E. Torralba, S. Le Gall, R. Lachaume, V. Magnin, J. Harari, M. Halbwax, J-. P. Vilcot, C. Cachet-Vivier, and S. Bastide, "Tunable Surface Structuration of Silicon by MACE with Pt NPs under Electrochemical Bias" ACS Appl. Mater. Interfaces **8**, 31375 (2016).

[4] K. W. Kolasinski, "The Mechanism of Galvanic / metal-Assisted Etching of Silicon." Nanoscale Res. Lett. **9**, 1-8, 2014.

[5] E. Franklin, K. Fong, K. McIntosh, A. Fell, *et. al.* "Design, Fabrication and Characterisation of a 24.4% Efficient Interdigitated Back Contact Solar Cell." Prog. Photovolt. Res. Appl. 24, 411 (2016)

[6] E. Torralba, M. Halbwax, T. El Assimi, M. Fouchier, V. Magnin, J. Harari, J-. P. Vilcot, S. Le Gall, R. Lachaume, C. Cachet-Vivier, and S. Bastide, "3D patterning of Silicon by contact etching with anodically biased nanoporous gold electrodes" accepted in Electrochem. Comm. (2017).

# Single crystalline substrates for III-V growth *via* exfoliation of bulk single crystals

Celeste L. Melamed[1,2], Brenden R. Ortiz[2], Aaron D. Martinez[1,2], William E. McMahon[1], Adele C. Tamboli[1,2],
Andrew G. Norman[1], Eric S. Toberer[1,2]

1. National Renewable Energy Laboratory, Golden, CO 80401, USA
2. Colorado School of Mines, Golden, CO 80401, USA

*Abstract*—**To enable widespread deployment of GaAs PV, the cost of single crystal substrates must be dramatically reduced. Here we present an indium-bonded exfoliation technique that produces substrates for III-V growth by exfoliation of macroscopic single crystals. Beginning with a search of available substrates, we identified several model materials that are lattice matched to III-Vs. $Bi_2Se_3$ single crystals were grown via the Bridgman technique in order to demonstrate our exfoliation method. From a centimeter diameter single crystal, up to eleven substrates with RMS roughness of 0.04 nm in a 20x20 micron region were exfoliated using this method. Large area AFM scans determined a terrace length of 72 $\mu$m between step edges. Thicknesses were determined to range from 40-160 $\mu$m using cross-sectional SEM. This exfoliation technique opens the door to the widespread study of layered materials as epitaxial substrates.**

*Index Terms*—**2D materials, exfoliation, single crystal growth.**

## I. Introduction

Wide-scale deployment of III-V solar cells with efficiency >30% has been constrained by the cost of single-crystal substrates [1]. Ongoing work to mitigate this cost has ranged from homoepitaxy combined with substrate recycling to heteroepitaxy on lower-cost substrates like silicon [2]. The latter adds the complexity of attempting to overcome lattice mismatch, for which techniques such as lateral epitaxial overgrowth and metamorphic buffer layers are being developed [3]. Another approach is substrate reuse, for example by spalling, which can provide many epitaxial growths using a single substrate, but must still overcome costly cleaning steps and wafer re-polishing [1]. As an alternative, we consider the opportunities for layered van der Waals materials to serve as low cost lattice-matched substrates (Fig. 1).

Since the rapid rise of single-layer graphene, research on 2D materials of all types has blossomed [4]. Most existing work has focused on the generation of single or few layers of 2D materials, and the analysis of the properties that occur due to their unique crystal structure. Less focus has been on the use of these materials as substrates. Epitaxy of semiconductors on 2D materials has emerged as an alternative to bulk substrates, including Kim et al.'s recent success in the epitaxy of GaAs on GaAs with a monolayer of graphene as a buffer layer [5], [6]. However, the bulk of this literature focuses on the growth of 2D materials on 3D materials or of superlattices of a mix of the two types, with few works attempting growth on bulk 2D crystals [7], [8].

As substrates, 2D materials must satisfy a set of criteria to be transformative for photovoltaics: (*i*) to minimize minority recombination at grain boundaries, single crystalline domains must extend far beyond the minority carrier diffusion length, (*ii*) to minimize generation of non-radiative defects such as threading dislocations in the epitaxial film, a close lattice match between the III-V and the underlying 2D layered material is necessary, and (*iii*) 2D substrates must be at least cost equitable with competing substrates. There have been few demonstrations of 2D materials meeting these constraints and exhibiting large enough surfaces for device applications. A variety of techniques have been used to produce 2D materials, ranging from direct synthesis of single layers on substrates to solution exfoliation to mechanical exfoliation (ie, the Scotch tape method). The largest monolayers that have been produced thus far using exfoliation have lateral dimensions of ∼500 microns [9], which is significant, but insufficient for large-scale applications like photovoltaics. Looking beyond monolayers, there is also a rich history of bulk growth of layered materials spanning the flux method, vapor transport, and Bridgman growth. It is crucial to note from the above constraints that layered substrates can be

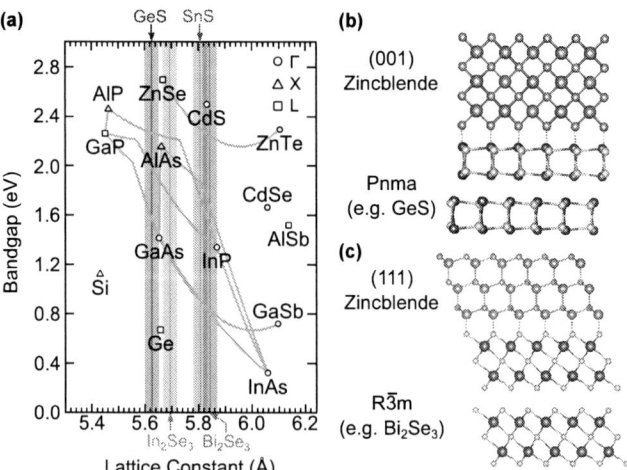

Fig. 1. (a) There are a wide variety of 2D materials that are lattice-matched to a range of III-V alloys. Here, vertical colored bands illustrate ±0.5% lattice mismatch between a number of 2D materials of interest and III-V alloys. Lattice matching can occur between (b) (001) zincblende III-Vs and Pnma structured 2D materials, or (c) (111) zincblende III-Vs and R$\bar{3}$m structured 2D materials.

978-1-5090-5606-4/17 $31.00 © 2017 IEEE

thicker than a monolayer if they are low-cost.

In this work, we identified $Bi_2Se_3$ as an appropriate model material to demonstrate production of layered substrates. We grew large phase-pure single crystal boules of $Bi_2Se_3$ with the Bridgman method. Using an indium-bonded exfoliation method, we produced large-area layered substrates which were then examined with diffraction and scanning probe methods. These substrates were found to have RMS roughnesses of 0.04 nm in $20{\times}20\,\mu m$ regions, and to exhibit an average terrace length of $72\,\mu m$. Thicknesses of 40-$160\,\mu m$ were determined using SEM. These substrates are fully compatible with III-V growth chambers, paving the way to low-cost van der Waals epitaxy of III-Vs.

## II. EXPERIMENTAL

Crystal growth processes and the indium-bonded exfoliation technique are described in full in [10]. $Bi_2Se_3$ single crystals were grown in a custom Bridgman crystal growth setup. Ampoules were soaked at 850°C for 24 h and consequently lowered at 3 mm/h through a temperature gradient of 10 °C/cm. To facilitate handling and exfoliation, single crystal boules were mechanically partitioned into smaller ingots. Phase purity and crystallinity of the ingots was assessed with a Bruker D8 Discover equipped with an area detector.

Exfoliation of single crystal slabs was performed in an argon glove box with <5ppm moisture and <10ppm oxygen. Shavings of indium were heated on glass slides until molten, and then a [0001] face of a $Bi_2Se_3$ ingot was firmly pressed onto the indium. The whole assembly was allowed to cool, leaving the crystal face mechanically bonded to the glass slide. The process was repeated on the other side of the crystal ingot to form a symmetric stack of glass-indium-crystal-indium-glass (see [10] for a process schematic). Gentle application of compressive, asymmetric pressure to the stack cleaved the crystal. This process was repeated multiple times to generate numerous single crystal slabs.

Exfoliated slabs were characterized with the previously mentioned Bruker XRD system and with tapping mode atomic force microscopy (AFM) using an Asylum Research system. An aluminum reflex coated silicon probe with a 5 N/m force constant and 120 kHz resonant frequency was used, as it was found that a higher force constant tip would damage the sample surfaces. SEM was performed using a FEI Quanta 600I SEM with a 25 keV accelerating voltage. Additional diffraction and surface characterization techniques were explored in [10].

## III. RESULTS AND DISCUSSION

Bulk $Bi_2Se_3$ single crystals were easily sectioned parallel to their layered planes and were found to have highly specular surfaces (Fig. 2a). Crystals were characterized with 2D XRD and found to exhibit only the [0001] family of planes consistent with the $R\bar{3}m$ structure. Alignment of all peaks in chi showed that high-quality single crystalline material was produced (Fig. 2b).

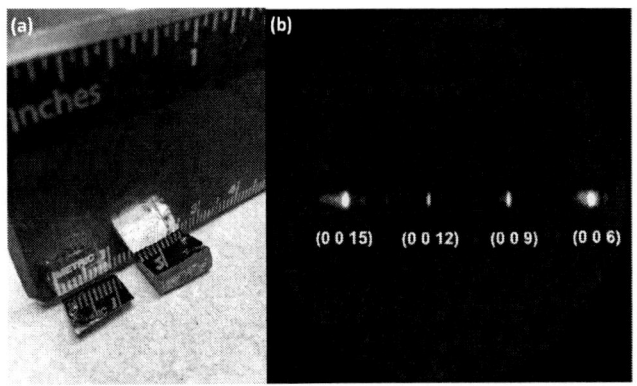

Fig. 2. (a) Bulk single crystals of $Bi_2Se_3$ grown using Bridgman exhibit highly specular surfaces, demonstrating their layered nature. (b) XRD scans of bulk single crystals $Bi_2Se_3$ using a 2D detector show only the [0001] family of planes with all peaks aligned in chi, consistent with a single crystal. Scans are taken from 18 to 54 degrees 2-theta.

The process of indium-bonded exfoliation, described in the Experimental section, was used to exfoliate these bulk single crystals. With a millimeter-thick source crystal of $Bi_2Se_3$, we have demonstrated up to 11 exfoliated slabs of $Bi_2Se_3$ (40-$160\,\mu m$ thick) with lateral dimensions of 10 mm. AFM was performed immediately after exfoliation to assess surface quality. Scans performed on a sample after 6 exfoliations demonstrate RMS roughnesses of 0.04 nm, which is extraordinarily flat for a sample processed in a glove box yet transported in air. For reference, a typical wafer for growth after chemical-mechanical polishing exhibits RMS roughness of over an order of magnitude greater than our samples.

Wider area AFM scans were performed on multiple exfoliated samples to assess the density of step edges. A step edge density of $0.014\,\mu m^{-1}$ was obtained by conducting fifteen random $50\mu m \times 50\mu m$ scans. Eight of these scans exhibited no step edges at all, such as that shown in Fig. 3a. Occasional

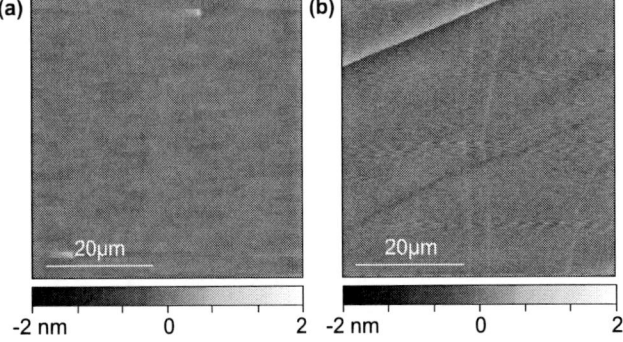

Fig. 3. $50\mu m \times 50\mu m$ AFM scans were performed on $Bi_2Se_3$ exfoliated using the indium-bonded method to assess the density of step edges. Eight of the fifteen total large-area scans exhibited no step edges at all, such as the scan shown in (a). Occasionally a 1 nm step edge is observed such as the one shown in (c), which is the height of a single quintuple-atom layer in the crystal structure. The texture exhibited in the data is residual high frequency noise and is not actually present on the sample.

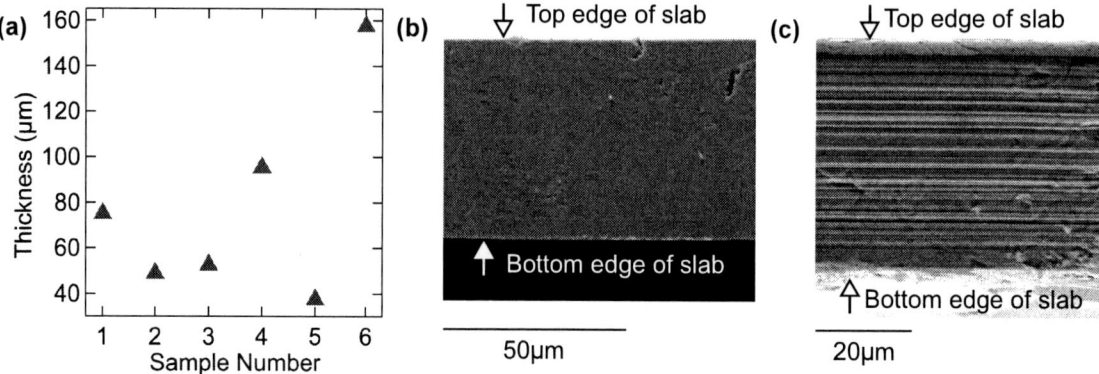

Fig. 4. Cross-sectional scanning electron microscopy (SEM) was used to measure the thicknesses of slabs of $Bi_2Se_3$ exfoliated using the indium-bonded method and then removed from the glass handle. (a) The distribution of thicknesses suggested that the indium-bonded method presently produces $Bi_2Se_3$ slabs from 40-160 $\mu$m thick. Some samples, such as (b), exhibited a clean and sharp cross-section, while others such as (c) exhibited striations that suggest the layers were beginning to cleave apart near the edge of the crystalline slab.

1 nm high step edges were observed, which is the height of a single quintuple-atom layer in the crystal structure of $Bi_2Se_3$ (Fig. 3b). The step edge density was inverted to yield a value of terrace length, or the average distance between each step edge assuming they are parallel. The terrace length is found to be 72 $\mu$m, which taken with the RMS roughness suggested that the propagation of the cleave between layers was very well confined within a particular interlayer gap. Though surfaces were found to be exceptionally flat given the fracture dynamics, we observed oxidation beginning to occur on exfoliated samples after 24 h in ambient conditions, consistent with prior work by Green et al. [11].

Scanning electron microscopy (SEM) was used to determine thicknesses of $Bi_2Se_3$ slabs exfoliated using the indium-bonded method. Slabs were removed from the glass handle by melting the indium on a hot plate, and were then mounted vertically for cross-sectional SEM. SEM was performed on six different exfoliated slabs, and shows a thickness distribution of 40-160 $\mu$m (Fig. 4a). Some SEM scans, such as Fig. 4c, exhibit striations parallel to the (0001) surface, which suggests that the layers are beginning to cleave apart near the edge of the exfoliated crystal. The dynamics of this process are under investigation, but we theorize that compressive and tensile strain on different parts of the crystal during the exfoliation process contribute to this phenomenon.

## IV. SUMMARY

In summary, we have developed a metal-bonded technique for exfoliating large-area, atomically-flat slabs from bulk 2D crystals. Bulk single crystalline $Bi_2Se_3$ was grown using the Bridgman method and its crystallinity was verified using XRD. An indium-assisted exfoliation method was used to produce eleven $cm^2$ substrates from one mm-thick single crystal. Surface quality was analysed using AFM and these exfoliated substrates were found to be highly smooth, with an RMS roughness of 0.04 nm in $20 \times 20$ $\mu$m regions. An average terrace length of 72 $\mu$m between step edges was determined

by AFM. Substrates have thicknesses ranging from 40-160 $\mu$m as determined by SEM. The indium-bonded technique produced substrates fully compatible with a III-V growth chamber. These findings pave the way to the widespread study of the use of quasi-2D substrates for epitaxial III-V growth.

## ACKNOWLEDGMENTS

This work was supported by the U.S. Department of Energy as part of the SuNLaMP program under Contract No. DE-AC36-08GO28308 with the National Renewable Energy Laboratory. The U.S. Government retains and the publisher, by accepting the article for publication, acknowledges that the U.S. Government retains a nonexclusive, paid up, irrevocable, worldwide license to publish or reproduce the published form of this work, or allow others to do so, for U.S. Government purposes.

## REFERENCES

[1] J. S. Ward, T. Remo, K. Horowitz, M. Woodhouse, B. Sopori, K. VanSant, and P. Basore, "Techno-economic analysis of three different substrate removal and reuse strategies for III-V solar cells," Progress in Photovoltaics: Research and Applications, vol. 24, no. 9, pp. 1284–1292, Sep. 2016.
[2] K. Lee, J. D. Zimmerman, T. W. Hughes, and S. R. Forrest, "Non-Destructive Wafer Recycling for Low-Cost Thin-Film Flexible Optoelectronics," Advanced Functional Materials, vol. 24, no. 27, pp. 4284–4291, Jul. 2014.
[3] Q. Li, K. W. Ng, and K. M. Lau, "Growing antiphase-domain-free GaAs thin films out of highly ordered planar nanowire arrays on exact (001) silicon," Applied Physics Letters, vol. 106, no. 7, p. 072105, Feb. 2015.
[4] F. Wang, Z. Wang, Q. Wang, F. Wang, L. Yin, K. Xu, Y. Huang, and J. He, "Synthesis, properties and applications of 2d non-graphene materials," Nanotechnology, vol. 26, no. 29, pp. 292 001–292 001, 2015.
[5] K. S. Novoselov, A. Mishchenko, A. Carvalho, and A. H. Castro Neto, "2d materials and van der Waals heterostructures," Science, vol. 353, no. 6298, pp. aac9439–aac9439, Jul. 2016.
[6] Y. Kim, S. S. Cruz, K. Lee, B. O. Alawode, C. Choi, Y. Song, J. M. Johnson, C. Heidelberger, W. Kong, S. Choi, K. Qiao, I. Almansouri, E. A. Fitzgerald, J. Kong, A. M. Kolpak, J. Hwang, and J. Kim, "Remote epitaxy through graphene enables two-dimensional material-based layer transfer," Nature, vol. 544, no. 7650, pp. 340–343, Apr. 2017.

[7] H. D. Li, Z. Y. Wang, X. Guo, T. L. Wong, N. Wang, and M. H. Xie, "Growth of multilayers of Bi2se3/ZnSe: Heteroepitaxial interface formation and strain," Applied Physics Letters, vol. 98, no. 4, p. 043104, Jan. 2011.

[8] Z. Chen, L. Zhao, K. Park, T. A. Garcia, and M. C. Tamargo, "Robust Topological Interfaces and Charge Transfer in Epitaxial Bi 2 Se 3 /II VI Semiconductor Superlattices," Nano Letters, 2015.

[9] S. B. Desai, S. R. Madhvapathy, M. Amani, D. Kiriya, M. Hettick, M. Tosun, Y. Zhou, M. Dubey, J. W. Ager, D. Chrzan, and A. Javey, "Gold-mediated exfoliation of ultralarge optoelectronically-perfect monolayers," Advanced Materials, vol. 28, no. 21, pp. 4053–4058, Jun. 2016.

[10] C. L. Melamed, B. R. Ortiz, P. Gorai, A. D. Martinez, W. E. McMahon, E. M. Miller, V. Stevanovic, A. C. Tamboli, A. G. Norman, and E. S. Toberer, "Van der waals substrates for III-V growth via exfoliation of bulk single crystals," Paper In Press, 2017.

[11] A. J. Green, S. Dey, Y. Q. An, B. O'Brien, S. O'Mullane, B. Thiel, and A. C. Diebold, "Surface oxidation of the topological insulator Bi $_2$ Se $_3$," Journal of Vacuum Science & Technology A: Vacuum, Surfaces, and Films, vol. 34, no. 6, p. 061403, Nov. 2016.

# CuZnS hole contacts on monocrystalline CdTe solar cells

Jacob J. Becker,[1,2] Xiaojie Xu,[4] Rachel Woods-Robinson,[4] Calli M. Campbell, [1,3]
Maxwell Lassise,[1,2] Joel Ager,[4] and Yong-Hang Zhang[1,2]

[1] Center for Photonics Innovation, Arizona State University, Tempe, AZ 85287
[2] School of Electrical, Computer and Energy Engineering, Arizona State University, Tempe, AZ 85287
[3] School for Engineering of Matter, Transport and Energy, Arizona State University, Tempe, AZ 85287
[4] Material Sciences Division, Lawrence Berkeley National Laboratory, Berkeley, CA 94720

*Abstract*— We report a monocrystalline CdTe/MgCdTe double-heterostructure solar cell with a CuZnS hole contact and ITO electrode. Similar designs have utilized other contact materials in pursuit of high open-circuit voltages and short-circuit current densities——namely, a-Si:H or ZnTe. $Cu_xZn_{1-x}S$ is a tunable material system with a low-cost deposition method (chemical bath deposition), which can provide a number of options in terms of bandgap, band offsets, and conductivity. Devices utilizing several different copper compositions, namely x = 15%, 25%, and 65%, have been developed and characterized yielding a maximum active-area power conversion efficiency of 12.9%, open-circuit voltage of 956 mV, fill factor of 63.5% and a short-circuit current density of 21.2 mA/cm². Analysis on the effects of the CuZnS layer on various loss parameters, such as both shunt and series resistance, induced built-in voltages, and parasitic absorption, is also presented.

*Index Terms*—CdTe, double heterostructure, monocrystalline, photovoltaics (PV), CuZnS

## I. Introduction

Polycrystalline CdS/CdTe solar cells have been utilized for photovoltaic applications with growing market penetration with an achieved record efficiency of over 22%, as demonstrated by First Solar [1]. Monocrystalline CdTe devices though have received considerably less attention. Recent work with monocrystalline absorbers has attempted to explore the role that bulk recombination plays in the observed low open-circuit voltages ($V_{oc}$) of current state-of-the-art polycrystalline CdTe devices [2]. This renewed interest has resulted in open-circuit voltages of over 1 V [2,3]—a remarkable achievement considering the record thin-film $V_{oc}$ is only 0.887 V [1]. Note though that such devices use a Cd(Se)Te-based alloy absorber material with a ~90 mV lower bandgap than pure CdTe, effectively limiting the maximum achievable $V_{oc}$. This still leaves quite a considerable bandgap-voltage offset ($W_{oc}$) of ~0.523 V. And yet the highest demonstrated voltages were achieved using a double-heterostructure design in conjunction with an a-Si:H hole-contact layer in order to address the three challenges plaguing CdTe thin-films: short bulk carrier lifetimes, high interface recombination velocities (IRV), and the inability to form a heavily doped p-type contact [3]. The CdTe/MgCdTe interface utilized in these designs possesses extremely low interface recombination velocities due to the low lattice mismatch, making MgCdTe an ideal barrier material for minority carrier confinement while also ensuring high-quality CdTe absorber growth above this layer. Effective minority carrier lifetimes of 3.6 µs are thus readily achieved in undoped CdTe double heterostructures while doped absorbers exhibit

lifetimes on the order of hundreds of nanoseconds [3–6]. Absorber material quality and interface recombination are thus no longer the limiting factors in achieving high $V_{oc}$. The present challenge lies rather in providing adequate contacts to extract the large implied voltage as an (equally) large external voltage.

Copper zinc sulfide ($Cu_xZn_{1-x}S$) is a semi-transparent conducting material. CuZnS films synthesized via chemical bath deposition (CBD) are comprised of a nanocomposite mix of independent sphalerite ZnS and covellite CuS crystals. In general, higher concentrations of the CuS phase lead to a higher hole conductivity and carrier concentration. The bandgap of the films can be varied from 2.1 eV to 3.45 eV by adjusting the alloy composition from pure CuS to ZnS. The wide bandgap, high hole concentration, and high conductivity all lend themselves to making a great hole-contact material. With the hole contact lying on the light incident side of the device, outside of the CdTe double heterostructure, absorption within this layer is generally lost to non-radiative recombination. The transparency of the material used is therefore very important, and with a-Si responsible for up to 2 mA/cm² of lost current [7], wider bandgap options such as CuZnS may prove beneficial.

## II. Structure Design

Fig. 1 shows the structure diagram of the studied devices. The double-heterostructure design provides passivation of the CdTe absorber surfaces as well as act as a hole- or electron-selective contact. Due to the difficulty in doping II-VI materials p-type, the MgCdTe passivation layer associated with the hole contact is not doped as is the case with the MgCdTe layer at the backside (electron contact). The MgCdTe layer within the electron contact can be made thicker as the doping level within the layer reduces the barrier for electrons while simultaneously increasing the conductivity. This is not the case for the hole contact as the passivation layer is nominally intrinsic. The 15 nm barrier has been found to provide the maximum open-circuit voltage ($V_{oc}$) in devices with a-Si:H hole contacts and thus, the same design is utilized here [7].

Because of the low carrier concentration and conductivity associated with films with low Cu content, an additional indium tin oxide layer (ITO) is required to facilitate effective lateral current flow. However, this design is used in all cases to maintain symetry among samples. The nature of the deposition (CBD) leads to difficulties in precisely controlling the thickness of the films. As the growth speeds of the CuS and ZnS crystals differ, the thickness of the films can vary; the films estimated thickness is between 10- and 20-nm.

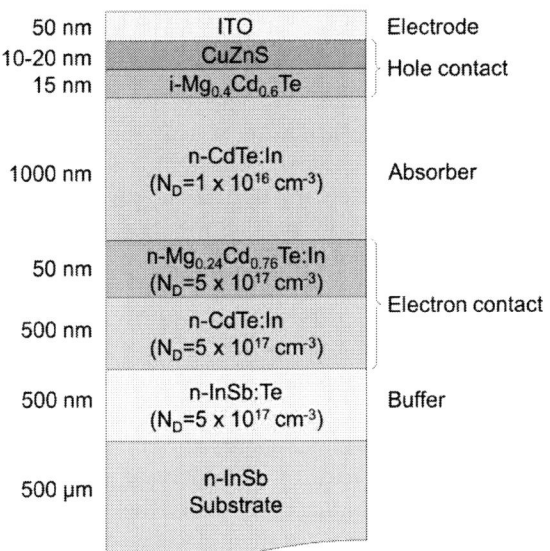

Fig. 1 Layer structure of the CdTe/Mg$_x$Cd$_{1-x}$Te double-heterostructure solar cell with a CuZnS hole-contact layer.

## III. EXPERIMENTAL RESULTS

The measured current-voltage (JV) characterisitics for solar cell devices of different copper compositions are shown in Fig. 2. Because the short-circuit current density (J$_{sc}$) is so sensitive to the area of the device, the external quantum efficiency (EQE) curve was integrated with the AM1.5G spectrum to determine the J$_{sc}$.

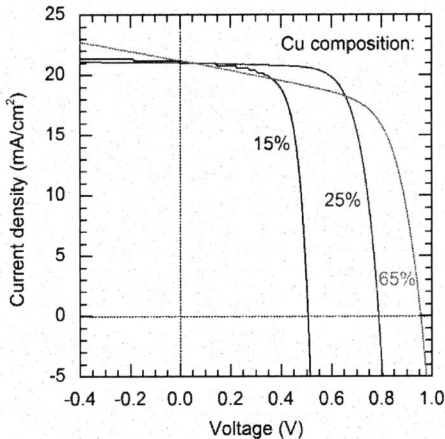

Fig. 2 EQE corrected JV-curves for solar cells with CuZnS hole contacts of 15%, 25% and 65% copper compositions. The devices under test have a total area of 0.033 cm$^2$ with approximately 1.5% metal coverage.

### TABLE 1
### ACTIVE-AREA DEVICE PARAMETERS

| Cu | V$_{oc}$ (V) | J$_{sc}$ (mA/cm$^2$) | FF (%) | η (%) |
|---|---|---|---|---|
| 15% | 0.506 | 21.0 | 71.8 | 7.6 |
| 25% | 0.788 | 21.0 | 73.0 | 12.1 |
| 65% | 0.956 | 21.2 | 63.5 | 12.9 |

With increasing copper content, the carrier concentration within the contact layer will increase, thereby improving the built-in potential. A corresponding increase in the open-circuit voltage (V$_{oc}$) with copper composition is apparent in the JV characteristics shown in Fig. 2 as well as indicated in TABLE 1. With a copper composition of 65%, a V$_{oc}$ of 0.956 V is achieved. While the V$_{oc}$ is still higher than those measured with polycrystalline devices [1], the bandgap-voltage offset (W$_{oc}$) of 0.544 V for this particular device is still similar given the smaller bandgap of polycrystalline devices.

Along with the increase in carrier concentration comes a dramatic improvement in layer conductivity. With a CBD process, this can be extremely problematic if the back contact or sidewalls are not adequately covered during the deposition. A short between the CuZnS hole contact and the electron contact would manifest itself as a shunt conductance in the JV-curve as we can see in the case of the device with a 65% copper composition in Fig. 2. Sidewall deposition may still be an issue in the case of all devices but for a layer with only 15% or 25% copper incorporation, the conductivity may not be high enough to dramatically reduce the fill factor (FF) of the J-V curve. The shunt conductance observed in the J-V curve does not necessarily have to plague future devices. Better isolation through a more controlled process such as sputtering or simply dicing all wafers post CBD growth can most likely result in the same voltages and current generation while simultaneously ensuring an improvement in the FF. Assuming these values are not affected, equation 1 can be used to approximate the maximum power (P$^*_{MP}$) without the power lost in a shunt resistance. Improving the 290 Ω·cm$^2$ shunt resistance measured in the light J-V curve could lead to an improvement in efficiency up to 15.3%.

$$P^*_{MP} = P_{MP}\left(1 - \frac{1}{r_{SH}}\right)^{-1} \qquad (1)$$

Unexpectedly, the short-circuit current density (J$_{sc}$) has not changed considerably with an increasing density of CuS crystals within the nanocomposite. The EQE curves shown in Fig. 3 indicate that the overall quantum efficiency within the device is not significantly changed with a change in copper composition. The small change in refractive index does however result in a shift of the reflectance minimum. Despite the similarity in the EQE, the absorption within the CuZnS layer does indeed change with copper composition as seen in Fig. 4. As the copper content increases, so too does the absorption within the layer. The large reduction in parasitic absorbance loss seen when moving from CuZnS layers of 25% to 15% copper is not translated into an improvement in the EQE. It is possible that there exists a compensating reduction in collection efficiency across the spectrum due to the depletion of the hole contact layer. With an already low carrier concentration with a 15% copper concentration, the heavily n-type ITO layer as the electrode can deplete the hole contact leading to both a reduced V$_{oc}$ (as can be seen) and reduction in collection efficiency.

Of course, the differences in reflection will not necessarily be evident in a fully optimized device as the differences in refractive index will be accounted for during the optimization of any additional anti-reflection layers added to the design. In addition, as contacts move to higher copper compositions, it may no longer be necessary to utilize a highly conductive ITO layer for lateral current flow. However, this design may still be desirable to better allow for the hole contact to be as thin as possible to minimize parasitic absorption loss within this layer; absorbance within the ITO is considerably lower than in CuZnS films.

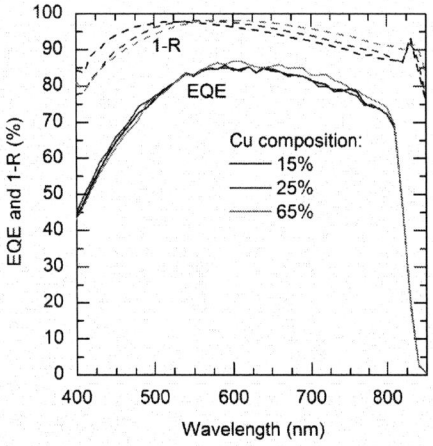

Fig. 3    EQE and 1-R curves for solar cells with CuZnS hole contacts of 15%, 25%, and 65% copper compositions. Both EQE and R were measured at the center of the devices with a beam spot size smaller than the aperture.

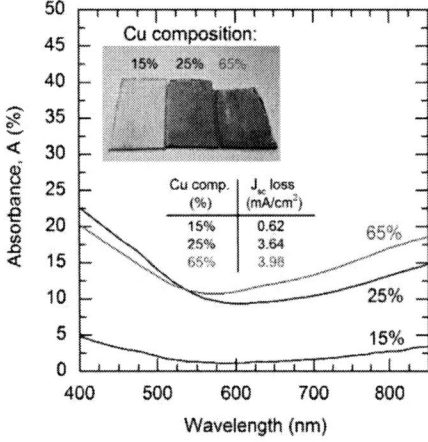

Fig. 4    Absorptance within CuZnS contacts of 15%, 25%, and 65% copper compositions. The parasitic current loss within the hole contact layer for each copper composition is also shown; the curves are weighted against the solar spectrum and integrated below 825 nm. The inset image shows the CBD deposited films on glass. Absorptance measurements were carried out using a spectrophotometer equipped with an integrating sphere.

## IV. CONCLUSION

Devices utilizing several different copper compositions, namely 15%, 25%, and 65%, have been developed and characterized yielding a maximum active-area efficiency of 12.9%, open-circuit voltage of 956 mV, fill factor of 63.5% and a short-circuit current density of 21.2 mA/cm$^2$. The pursuit of higher current densities by avoiding a-Si:H as a hole contact is not yet complete as current loss in these films is still as high as ~4 mA/cm$^2$. Yet these films are still quite thick and future devices can be made with even thinner layers than can be achieved with a-Si:H. The a-Si:H layers used in prior work has an electron concentration on the order of $10^{18}$ cm$^{-3}$ while the achievable carrier concentration in CuZnS films can be easily several orders of magnitude higher at $>10^{20}$ cm$^{-3}$. The much higher concentration means the thickness of the contact layer can be reduced even further before it becomes depleted and the built-in voltage drops. Controlling thicknesses to this degree with CBD deposition can be difficult however, and alternate deposition techniques may be necessary.

### ACKNOWLEDGEMENT

This work is partially supported by the Department of Energy through the Bay Area Photovoltaic Consortium (BAPVC) and Energy Efficiency and Renewable Energy programs, under award number DE-EE0004946 and. In addition, JB is supported by the Phoenix chapter of the ARCS foundation. This material is based upon work partially supported by the Quantum Energy and Sustainable Solar Technologies (QESST) program under the National Science Foundation and Department of Energy— grant NSF CA No. EEC-1041895. Any opinions, conclusions or recommendations expressed in this material are those of the author(s) and do not necessarily reflect those of NSF or DOE.

### REFERENCES

[1] M. A. Green, K. Emery, Y. Hishikawa, W. Warta, and E. D. Dunlop, "Solar cell efficiency tables (version 48)," *Prog. Photovolt Res. Appl.*, vol. 24, pp. 905–913, 2016.

[2] J. M. Burst, J. N. Duenow, D. S. Albin, E. Colegrove, M. O. Reese, J. A. Aguiar, C.-S. Jiang, M. K. Patel, M. M. Al-Jassim, D. Kuciauskas, S. Swain, T. Ablekim, K. G. Lynn, and W. K. Metzger, "CdTe solar cells with open-circuit voltage greater than 1 V," *Nat. Energy*, vol. 1, p. 16015, 2016.

[3] Y. Zhao, M. Boccard, S. Liu, J. Becker, X.-H. Zhao, C. M. Campbell, E. Suarez, M. B. Lassise, Z. Holman, and Y.-H. Zhang, "Monocrystalline CdTe solar cells with open-circuit voltage over 1 V and efficiency of 17%," *Nat. Energy*, no. May, p. 16067, 2016.

[4] X.-H. Zhao, S. Liu, Y. Zhao, C. M. Campbell, M. B. Lassise, Y.-S. Kuo, and Y.-H. Zhang, "Electrical and Optical Properties of n-Type Indium-Doped CdTe/Mg0.46Cd0.54Te Double Heterostructures," *IEEE J. Photovoltaics*, vol. 6, no. 2, pp. 552–556, 2016.

[5] X.-H. Zhao , S. Liu, C. M. Campbell, Y. Zhao, M. B. Lassise, and Y.-H. Zhang, "Impact of thermionic emission and tunneling effect on the measurement of low interface recombination velocity (~1 cm/s) in CdTe/MgxCd1-xTe double heterostructures," *Proc. 43nd IEEE PVSC*, pp. 2302–2305, 2016.

[6] Y. Zhao, M. Boccard, J. Becker, X.-H. Zhao, C. M. Campbell, Z. Holman, and Y.-H. Zhang, "Monocrystalline CdTe/MgCdTe double-heterostructure solar cells with 1.096 V V oc and 17.0 % efficiency," *Proc. 43nd IEEE PVSC*, pp. 3524–3526, 2016.

[7] J. J. Becker, M. Boccard, C. M. Campbell, Y. Zhao, M. Lassise, Z. C. Holman, and Y.-H. Zhang, "Loss analysis of monocrystalline CdTe solar cells with 20 % active-area efficiency," *J. Photovoltaics*, no. 99, pp. 1–6, 2017.

# The Effect of the CdCl$_2$ Heat Treatment on CdSe$_x$Te$_{1-x}$ Solar Cells

Chih An Hsu, Vasilios Palekis, Imran Khan, Shamara Collins, Don Morel, and Chris Ferekides

University of South Florida, Tampa, FL 33620, USA

*Abstract-* **In this paper we report results on CdSe$_x$Te$_{1-x}$ based solar cells, fabricated by inter-diffusing CdTe/CdSe bilayers. The effect of the CdCl$_2$ heat treatment (HT) carried out at various temperatures on bilayers with different CdSe thicknesses has been investigated. The CdCl$_2$ HT temperature has been found to be critical for intermixing; to-date it has been shown that higher temperature leads to enhanced inter-diffusion, which also leads to higher short circuit current (J$_{SC}$) due to a decrease in the band gap of the absorber. Large CdSe thicknesses (> 300 Å) lead to lower open circuit voltage (V$_{OC}$) and fill factor (FF). Optimization of the annealing temperature and CdSe thickness has yielded cells with V$_{OC}$= 850 mV, J$_{SC}$= 26.3 mA/cm$^2$, FF= 73 % and efficiency 16.3%.**

*Index Terms* — Cadmium compound, charge carrier lifetime, photovoltaic cells, Selenium, semiconductor device doping.

## I. INTRODUCTION

II-VI semiconductors have been widely utilized in various electronic and optical devices such as solar cells, light-emitting devices, etc. The highest efficiencies of thin-film CdTe solar cells have increased drastically during the recent years and now exceed 22.1% [1]. The limitation in performance for polycrystalline CdTe cells has been primarily due to their low carrier concentrations and lifetimes. CdSe has been recently used for the fabrication of CdSe$_x$Te$_{1-x}$ alloy-based cells to improve the carrier concentration and lifetime [2][3]. Compared to S, Se has a higher solubility in CdTe, allowing for higher Se concentrations in CdSe$_x$Te$_{1-x}$ and therefore lower bandgap [4]. The effective bandgap of CdSe$_x$Te$_{1-x}$ depends on the amount of Se in the alloy. There exists an optimum amount of Se, for which the bandgap reaches a minimum due to bowing band effect, before it starts increasing again [5]. CdSe$_x$Te$_{1-x}$ alloy is considered to be a promising material to improve carrier collection at long wavelengths [6].

One of the most critical steps to achieve high efficiency CdTe solar cells is the CdCl$_2$ HT. This treatment improves p-doping, enhances carrier collection, causes recrystallization and promotes inter-diffusion of at the CdS/CdTe junction which reduces interface recombination [7]. The CdCl$_2$ annealing temperature and time can be used to control the CdSe/CdTe inter-diffusion. Another important parameter is the CdSe thickness, since any unused CdSe layer can be detrimental to the device performance as it lowers V$_{OC}$[8]. The effect of CdSe thickness and CdCl$_2$ HT on the CdSe/CdTe inter-diffusion and the photovoltaic device performance is presented in this work.

## II. EXPERIMENTAL

The CdTe cells discussed in this paper are of the superstrate configuration and were fabricated on Corning Eagle XG glass substrates. The configuration of the device is:

ITO/MZO/CdS/CdSe/CdTe/Cu-doped graphite (Fig. 1). The glass substrates were coated with a layer of ITO, deposited by RF sputtering and the front contact has sheet resistance of ~8 Ω/sq. Mg$_x$Zn$_{1-x}$O (MZO) was deposited by RF sputtering. MZO layer was used as the emitter layer, in order to enhance the blue response because of its wide bandgap (3.8 eV). CdSe was deposited by RF sputtering in Ar ambient at room temperature. CdSe thicknesses were varied between 75-1500Å. CdS and CdTe were deposited by close-spaced sublimation (CSS). After the CdTe deposition the devices were CdCl$_2$ heat treated, which is a standard processing step for CdTe solar cells. The CdCl$_2$ HT was carried out under He:O$_2$ ambient at temperatures ranging 390 – 450 C. Prior to the formation of the back contact the CdTe surface was etched in a bromine methanol solution. The back contact was Cu-doped graphite, followed by thermal annealing.

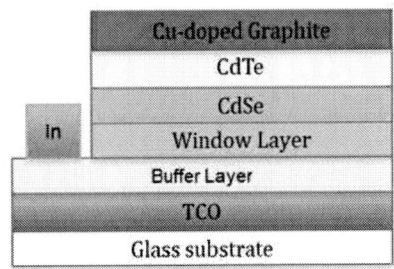

Fig. 1. Configuration of CdSe/CdTe solar cells.

The CdSe$_x$Te$_{1-x}$ films/devices were characterized using the current-voltage (JV), spectral response (SR), capacitance-voltage (CV), Deep Level Transient Spectroscopy (DLTS) and Time-Resolved Photoluminescence (TRPL) measurements. JV measurements were carried out using four-terminal connections with a Keithley 2410 source meter. An Oriel monochromator (Model 74100) was used for the SR measurements. The intensity of the light source was calibrated using a standard silicon reference solar cell calibrated by the National Renewable Energy Laboratory. The CV measurements were performed using a HP 4194A impedance/ gain-phase analyzer. Time Resolved Photoluminescence (TRPL) was obtained with excitation at 640 nm with 0.3 ps pulses at 1.1 MHz and PL was measured with 10nm bandpass filter with the center at 820 nm. DLTS measurements were performed with a Sula Technologies Deep Level Spectrometer (Model DDS-12). The sample temperature was varied from 80K to 320K. A bias pulse of -1V to 0V with a pulse width of 1ms was used. The resulting capacitance transients were analyzed with 6 different rate windows from 0.02 to 1 ms.

978-1-5090-5606-4/17 $31.00 © 2017 IEEE

## III. RESULTS AND DISCUSSION

### A. Device Performance:

In CdS/CdTe solar cells, $CdCl_2$ HT promotes inter-diffusion at the CdS/CdTe junction which improves the electronic properties of the junction by reducing the lattice mismatch, stress and interface recombination [9, 10]. In this work, the effect of $CdCl_2$ HT has been investigated for four different annealing temperatures (390, 410, 430, and 450 °C). The thickness of CdSe is another important parameter that influences the properties (composition) of the $CdSe_xTe_{1-x}$ layer formed as a result of inter-diffusion between CdTe and CdSe. It can determine the Se distribution in $CdTe_{1-x}Se_x$ and the resulting bandgap of the absorber at the junction interface. To-date the CdSe thickness has been varied from 75 to 1500Å. Fig. 2 shows the JV and SR data for $CdSe_xTe_{1-x}$ devices fabricated with different CdSe thicknesses and $CdCl_2$ HT at the temperature of 410°C. Both the $V_{OC}$ and FF decrease with increasing CdSe thickness. The red shift in SR for devices with higher CdSe layer thicknesses indicates reduction in the bandgap of the absorber as expected in $CdSe_xTe_{1-x}$ alloy due to bowing effect. However, the devices with thicker CdSe (> 300 Å) show lower carrier collection in the blue region. This is believed to be due to residual CdSe not fully consumed via inter-diffusion with CdTe, which might also explain the lower $V_{OC}$ and FF of the devices (due to a poor interface). The drop in $V_{OC}$ for thicker CdSe is also reported by other groups [5]. Similar results are observed for devices $CdCl_2$ HT at 390 °C (not included here).

Fig. 3 shows the SR of devices with $CdCl_2$ HT at 430 °C. These results suggest that CdSe is fully interdiffused for all thicknesses, as indicated by the increased carrier collection in the blue region. The QE at long wavelengths increases with increasing CdSe thickness. This clearly demonstrates that the bandgap decreases gradually with increasing CdSe thickness indicating an increase in the composition (X) of Se in $CdSe_xTe_{1-X}$. This red shift is greater compared to devices annealed at lower temperatures for the same CdSe thicknesses (see Fig. 2).

The corresponding bandgaps calculated from the absorption edge for all the devices with different CdSe thicknesses and $CdCl_2$ heat treatment temperatures are shown in Fig. 4. For 390 and 410 °C HT, there is a slight decrease in bandgap with increasing CdSe thickness which appears to level off at larger thicknesses. The largest bandgap reduction occurs at the

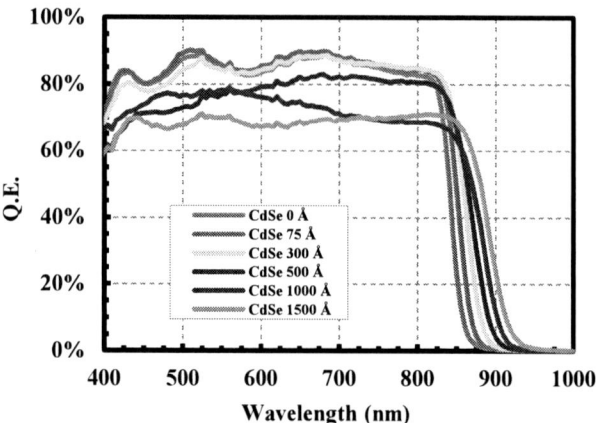

Fig. 3. SR data for devices $CdCl_2$ treated at 430°C with various CdSe thicknesses (0Å to 1500Å).

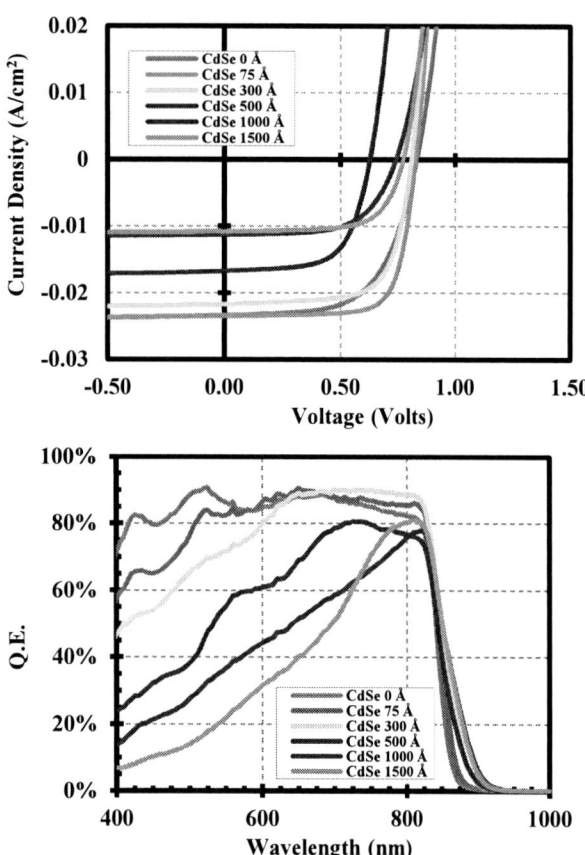

Fig. 2. (Top) JV and (Bottom) SR data for devices $CdCl_2$ treated at 410°C with various CdSe thicknesses (0Å to 1500Å).

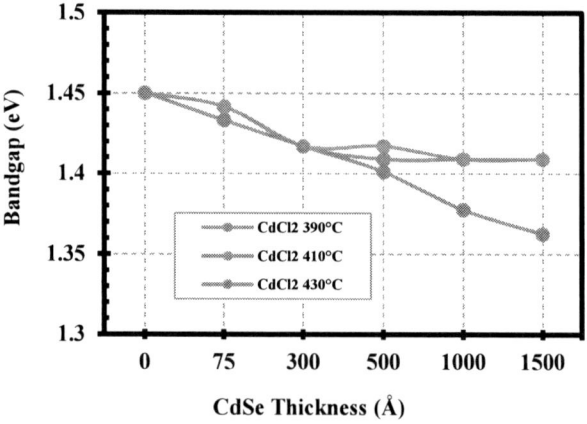

Fig. 4. Bandgap calculated from SR for devices with different CdSe thicknesses and $CdCl_2$ HT temperature.

temperature of 430 °C. The smallest bandgap (1.36 eV) corresponds to the device with the thickest CdSe of 1500 Å. Additional increase in $CdCl_2$ HT temperature (450 °C) to further reduce the bandgap resulted in flaking for the majority of the films.

A summary of the cell results is shown in Table I.

TABLE I
DEVICE PERFORMANCES FOR VARIOUS CDSE THICKNESSES

| $CdCl_2$ HT (°C) | CdSe Thickness (Å) | $V_{OC}$ (mV) | FF (%) | $J_{SC}$ (mA/cm²) | Bandgap (eV) | $\eta$ (%) |
|---|---|---|---|---|---|---|
| | 0 | 830 | 67.2 | 23.78 | 1.45 | 13.22 |
| | 75 | 810 | 69.1 | 24.43 | 1.44 | 13.65 |
| 390 | 300 | 740 | 66.4 | 23.33 | 1.41 | 11.39 |
| | 500 | 620 | 58.2 | 19.20 | 1.40 | 6.90 |
| | 1500 | 650 | 57.0 | 14.49 | 1.40 | 5.37 |
| | 0 | 830 | 61.3 | 24.77 | 1.45 | 12.54 |
| | 75 | 830 | 74.1 | 24.28 | 1.44 | 14.91 |
| 410 | 300 | 810 | 69.0 | 23.39 | 1.41 | 13.07 |
| | 500 | 630 | 64.0 | 18.32 | 1.41 | 7.39 |
| | 1500 | 770 | 68.3 | 12.77 | 1.40 | 6.68 |
| | 0 | 770 | 54.2 | 24.55 | 1.45 | 10.21 |
| | 75 | 800 | 55.7 | 25.43 | 1.43 | 11.19 |
| 430 | 300 | 840 | 54.2 | 25.49 | 1.40 | 11.56 |
| | 500 | 790 | 44.3 | 23.91 | 1.37 | 8.31 |
| | 1500 | 690 | 40.0 | 21.98 | 1.36 | 6.07 |
| | 0 | 730 | 23.2 | 23.54 | 1.45 | 3.95 |
| | 75 | 740 | 24.6 | 25.81 | 1.42 | 4.58 |
| 450 | 300 | 780 | 26.2 | 25.80 | 1.40 | 5.23 |
| | 500 | Film damage | | | | |
| | 1500 | Film damage | | | | |

*B. Defect Analysis:*

DLTS measurements have been performed on select samples to analyze the deep defect distribution. The devices are chosen based on the following two criteria – normal JV behavior (no 'kink', roll-over etc.) and fully interdiffused CdTe/CdSe bilayer as indicated in the SR data. Figure 5 (left) shows the DLTS spectra for devices $CdCl_2$ heat treated at 410 °C with different CdSe thicknesses. The spectra for the device without CdSe is identical to baseline CdS/CdTe devices [11]. The positive peak in the temperature range 150-200K indicates a minority carrier trap (E1). This shallow trap ($E_A \sim 0.3eV$) has been attributed to $Cl_{Te}$ [12]. The negative $\Delta C$ near room temperature indicates a deep majority carrier trap.

For devices with CdSe, an additional positive peak is observed at temperatures above 240K. This indicates a deeper minority carrier trap (E2). The DLTS measurements are performed at six different rate windows to calculate the activation energy. Figure 5 (right) shows the different rate window spectra for the device with 500 Å CdSe. The trap activation energy is calculated from the Arrhenius plot of the peak positions at different rate windows (Fig. 5 inset).

The increasing intensity of E2 with CdSe thickness suggests the formation of a Se related complex defect. This deep minority carrier trap ($E_A \sim 0.52$ eV) may have a compensating effect and might also suggest lower minority carrier lifetimes in $CdSe_XTe_{1-X}$ with increased Se content. Further investigation is required to confirm this hypothesis. Table II lists the calculated activation energies and trap concentrations for these defect levels.

TABLE II
TRAP CONCENTRATION ($N_T$) AND ACTIVATION ENERGY
($E_A$) IDENTIFIED FROM DLTS MEASUREMENTS

| CdSe Thickness (Å) | E1 | | E2 | |
|---|---|---|---|---|
| | $E_A$ (eV) | $N_t$ (cm⁻³) | $E_A$ (eV) | $N_t$ (cm⁻³) |
| No | $E_C$- 0.30 | 3E13 | | |
| 300 | $E_C$- 0.29 | 6E13 | $E_C$-0.51 | 1E13 |
| 500 | $E_C$- 0.33 | 4E13 | $E_C$-0.53 | 4E13 |

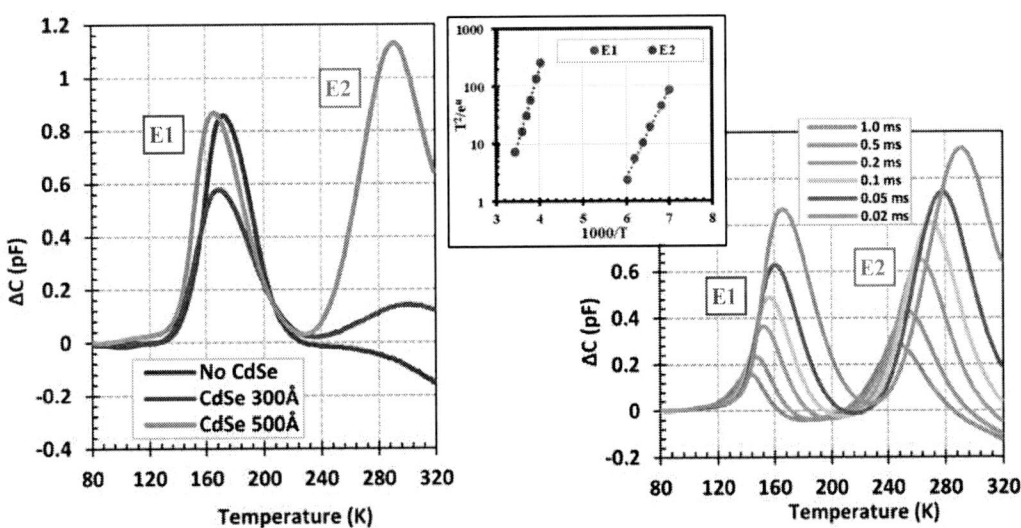

Fig. 5. (Left) Comparative DLTS spectra for devices $CdCl_2$ treated at 410 °C with and without CdSe, obtained with a rate window of 0.02 ms. (Right) DLTS spectra for CdTe device with 500 Å CdSe at 6 different rate window. The corresponding Arrhenius plot to calculate the defect activation energy is shown on the inset.

Capacitance-Voltage (CV) measurements for the devices (Fig. 6) shows the effect of CdSe thickness on the doping concentration. There is an initial (small) increase in carrier concentration for CdSe thickness up to 300 Å. Higher CdSe thicknesses result in reduced doping. The same trend was observed for all CdCl$_2$ HT temperatures. This is possibly due to an increase in the concertation of the deep minority carrier trap observed in the DLTS measurement (Fig. 5) for thicker CdSe devices.

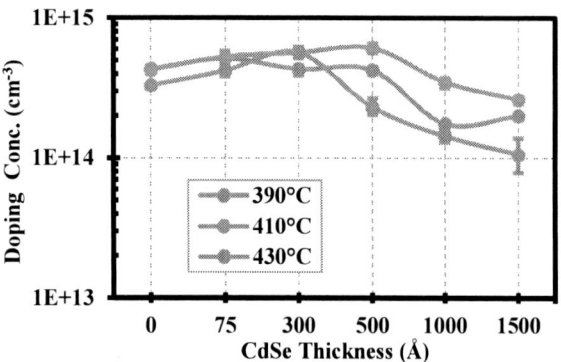

Fig 6. Net doping Concentration for devices with various CdCl$_2$ treatment temperature (390 to 430 °C) and CdSe thicknesses.

Preliminary results from minority carrier lifetime measurements are shown in Table 3. Comparison between the two samples heat treated at 410 °C with and without CdSe indicate that CdSe improves the minority carrier lifetime. Similarly for the 75 Å CdSe devices, increasing the CdCl$_2$ HT temperature to 410 °C shows improved lifetime. This suggests that both CdCl$_2$ HT temperature of 410° C and a low Se content CdSe$_x$Te$_{1-x}$ alloy improves lifetime.

TABLE III
MINORITY CARRIER LIFETIME FOR DIFFERENT CDSE/CDTE DEVICES.

| CdCl$_2$ HT (°C) | CdSe Thickness (Å) | Minority Carrier Lifetime (ns) |
|---|---|---|
| 390 | 75 | 10 |
| 410 | 0 | 11.1 |
| 410 | 75 | 13.8 |

Therefore, at low thickness of CdSe net p-type doping concentration (CV) and minority carrier lifetime (TRPL) increase. Such devices show better V$_{OC}$'s, however no significant current gain is observed due to the small amount of CdSe that did not result in significant bandgap reduction. Increasing the CdSe thickness results in lower p-type doping, and may also result in lower lifetime due to the formation of deep defects (DLTS). Thicker CdSe results in higher J$_{SC}$, but exhibits lower V$_{OC}$ and FF, possibly due to a combined effect of low doping, deep defects, and the smaller bandgap of the CdSe$_x$Te$_{1-x}$ alloy. Therefore, an optimum CdSe thickness (CdSe$_x$Te$_{1-x}$ alloy) must be achieved that can maximize J$_{SC}$ without losses in V$_{OC}$. The best CdSe$_x$Te$_{1-x}$ based cell fabricated to-date exhibited V$_{OC}$= 850 mV, J$_{SC}$= 26.3 mA/cm$^2$, FF= 73 % and efficiency 16.3%.

## IV. CONCLUSION

The effect of CdCl$_2$ HT and CdSe thickness on the performance of CdSe$_x$Te$_{1-x}$ photovoltaic devices has been investigated. Devices with small amount of CdSe shows better doping and minority carrier lifetime, however with no significant current gain. Both higher CdCl$_2$ annealing temperature and larger CdSe thickness has appeared to promote CdSe-CdTe inter-diffusion and improve collection in the red region due to reduction of absorber bandgap. Lower V$_{OC}$ in thick CdSe devices has been attributed to reduced bandgap, low doping and the presence of deep defects. An optimized CdSe thickness and CdCl$_2$ HT temperature can lead to both improvement in V$_{OC}$ and J$_{SC}$, and thus improved efficiency.

## REFERENCES

[1] M. Gloeckler, "Realization of the Potential of CdTe Thin-Film PV," in *43rd IEEE Photovoltaic Specialist Conference*, 2016, pp. 1292.

[2] R. Soltani, et al. *"Light harvesting enhancement upon incorporating alloy structured CdSexTe1-x quantum dots in DPP;PC61BM bulk heterojunction solar cell"*, The Royal Society of Chemistry 2016, DOI: 10.1039/C6TC04308A.

[3] C. M. Hangarter, et al. "Photocurrent Mapping of 3D CdSe/CdTe Windowless Solar Cells", *Applied Materials & Interfaces* 2013, 5, pp. 9120 9127

[4] D. E. Swanson, J. R. Sites, W. S. Sampath, "Co-sublimation of CdSexTe1-x layers for CdTe solar cells", *Solar Energy Materials & Solar Cells 2017*, 159, pp. 389-394.

[5] S. H. Wei, S. B. Zang, and A. Zunger, "First-principle *calculation of band offsets, optical bowings, and defects in CdS, CdSe, CdTe, and their alloys"*, Appl. Phys. Lett. 2000, 87, pp. 1304.

[6] N. R. Paudel, et al. *"Current Enhancement of CdTe-Based Solar Cells"*, IEEE Journal of Photovoltaics, 2015, Vol. 5, NO. 5.

[7] V. Palekis, et al. "Near infrared laser CdCl$_2$ heat treatment for CdTe solar cells." *Photovoltaic Specialists Conference (PVSC)*, 2016 IEEE 43rd, pp. 1498-1502.

[8] J. D. Poplawsky, et al. "Structural and compositional dependence of the CdTexSe1-x alloy layer photoactivity in CdTe-based solar cells", *Nature communications*, 2016, DOI; 10.1038/ncomms12537.

[9] M. D. G. Potter, D. P. Halliday, M. Cousins, and K. Durose, "A study of the effects of varying cadmium chloride treatment on the luminescent properties of CdTe/CdS thin film solar cells", *Thin Solid Films*, 361-362, 2000, pp. 248-252.

[10] R. Dhere, et al. "Investigation of effects of processing and impurities on the properties of CdTe using microscopic two-dimensional photoluminescence image technique", *Proc. 34th IEEE PVSC*, 2009, pp. 1443-1447.

[11] M. Khan, et al. "Study of defects in polycrystalline CdTe using DLTS." *43rd IEEE Photovoltaic Specialist Conference (PVSC)*, 2016, pp. 2191-2194.

[12] I. Khan, et al. "The Effect of Deposition Stoichiometry and Post-deposition Treatments on Deep Defects in CdTe." *44th IEEE Photovoltaic Specialist Conference (PVSC)*, 2017.

[13] T. A. Gessert, et al. "Research strategies toward improving thin-film CdTe photovoltaic devices beyond 20% conversion efficiency." *Solar Energy Materials and Solar Cells*," vol. 119, pp. 149-155, 2013.

# Effects of CdCl$_2$ treatment on the local electronic properties of polycrystalline CdTe measured with photoemission electron microscopy

Morgann Berg[1,2], Jason M. Kephart[3], Walajabad S. Sampath[3], Taisuke Ohta[1,2], Calvin Chan[1]

[1]Sandia National Laboratories, Albuquerque, New Mexico, 87185, USA
[2]Center for Integrated Nanotechnologies, Albuquerque, New Mexico, 87185, USA
[3]Department of Mechanical Engineering, Colorado State University, Fort Collins, Colorado, 80523, USA

*Abstract* — To investigate the effects of CdCl$_2$ treatment on the local electronic properties of polycrystalline CdTe films, we conducted a photoemission electron microscopy (PEEM) study of polished surfaces of CdTe films in superstrate configuration, with and without CdCl$_2$ treatment. From photoemission intensity images, we observed the tendency for individual exposed grain interiors to vary in photoemission intensity, regardless of whether or not films received CdCl$_2$ treatment. Additionally, grain boundaries develop contrast in photoemission intensity images different from grain interiors after an air exposure step, similar to observations of activated grain boundaries using scanning probe microscopy. These results suggest that work function varies locally, from one grain interior to another, as well as between grain boundaries and grain interiors.

*Index Terms* — polycrystalline materials, photoemission, electronic properties, grain boundaries, CdTe.

## I. INTRODUCTION

CdTe thin-film photovoltaics generate more than 2 GW annually, with research cell efficiencies (21.5%) rivaling those of multicrystalline Si. As such, CdTe-based solar cells are a serious alternative to silicon-based photovoltaics for commercial production and deployment. A key component of manufacturing efficient CdTe photovoltaic devices is an activation process where the polycrystalline CdTe absorber layer is treated with CdCl$_2$ [1]. The effect of CdCl$_2$ treatment on the electronic properties of CdTe-based devices has been extensively studied but is not completely understood.

One consequence of CdCl$_2$ treatment is a change in the microstructure of the CdTe layer, resulting in larger and more uniform grains, and increased minority carrier lifetimes [1]. While fewer grain boundaries are thought to reduce scattering and recombination of photogenerated carriers, the performance of single crystal CdTe photovoltaics is known to be inferior to that of polycrystalline films.

This contradiction is often explained by the modification of the electronic properties at grain boundaries during CdCl$_2$ treatment. The prevailing description is that Cl from the CdCl$_2$ treatment segregates to grain boundaries in p-type CdTe, depleting or inverting them with respect to grain interiors. Since the formation of junctions between grain interiors and grain boundaries are expected to conduct charges to the electrodes more efficiently, individual grain boundary properties have been a subject of intense study, and have been measured using an array of techniques including scanning probe microscopy [2]–[3], electron beam-induced current [4], time-resolved photoluminescence [5], and time-of-flight secondary ion mass spectroscopy (ToF-SIMS) [6].

For thin-film junctions, photoemission spectroscopy is invaluable for directly probing variation in electronic properties. For example, ultraviolet photoelectron spectroscopy (UPS) is routinely used to measure Fermi level positions, estimate doping levels, and determine relative band alignment among thin-film components in semiconductor devices. However, because UPS is an area-averaged technique, lateral variation of electronic properties smaller than the beam spot size (typically 1 mm to tens of microns) cannot be resolved, rendering measurement of local junctions (e.g. between grain interiors and grain boundaries in CdTe thin films) inaccessible.

In the work reported here, we used photoemission electron microscopy (PEEM) to image the effect of CdCl$_2$ treatment on the local electronic properties of superstrate CdTe films, exposed at the CdTe/rear contact interface. To access grain interiors for PEEM measurements, the CdTe films were mechanically polished. To isolate the influence of CdCl$_2$ vapor treatment on CdTe surfaces, we used low-energy ion desorption to remove surface oxides and contaminants from polished films [7]. We observed grain domains and saw little change in microstructure between CdCl$_2$-treated and untreated films. From variation in photoemission intensity we deduced that individual grains vary in work function for both CdCl$_2$-treated and untreated films. Our results indicate that CdCl$_2$-treatment alone does not modify grain boundary properties in a manner consistent with previous reports [2]–[3]. After purposefully exposing the CdCl$_2$-treated CdTe film to air, the sample surface develops image contrast at grain boundaries in photoemission intensity images. These findings suggest a synergistic relationship between CdCl$_2$-treatment and oxygen incorporation changes grain boundary properties.

## II. EXPERIMENTAL DETAILS

CdTe samples were fabricated using closed-space sublimation (CSS) at Colorado State University's Advanced Research Deposition System [8]. Films were deposited on commercial NSG TEC 10 superstrates, which feature standard soda lime glass coated with a transparent conducting oxide (TCO). The TCO, consisting of a SnO$_2$/SiO$_2$/SnO$_2$:F tri-layer, was cleaned using standard rinses and an isopropyl alcohol wash, followed by a plasma cleaning process [9]. A 100-nm-thick, $Mg_{0.23}Zn_{0.77}O$ (MZO) window layer was sputtered onto the TCO. 5 $\mu$m CdTe was then deposited directly onto the MZO. During CSS growth and $CdCl_2$ treatment, substrate temperatures were in the range of 425–500˚C. Source temperatures were 435–610˚C. To remove topographical artifacts in PEEM measurement and to expose the relevant CdTe grain interiors, samples were mechanically polished to ~3 $\mu$m thickness using a 1 $\mu$m diamond polish. Samples were subsequently sputter-cleaned with 50 eV $Ar^+$ ions for 10–20 min with a fluence of ~0.1-0.15 $\mu$A•cm$^{-2}$ to remove surface contaminants [7]. This low-energy sputtering process is also referred to as low-energy ion desorption. We verified the composition of the CdTe surface after modification steps (polishing, low-energy ion desorption, sample transfer, etc.) using x-ray photoelectron spectroscopy (XPS). XPS survey scans were performed using a Mg $K_\alpha$ x-ray source and an Omicron EAC 2000 electron analyzer, operating at a pass energy of $E_p = 50$ eV. XPS core-level scans were also obtained by operating the electron analyzer with a pass energy of $E_p = 20$ eV. After low-energy ion desorption and XPS measurement, samples were transferred to PEEM without additional exposure to air by the use of an inert environment sample transfer system and an inert gas (dry $N_2$) glove box. To examine the impact of oxygen on CdTe processing, the $CdCl_2$-treated film was exposed to air for 30 minutes, reintroduced into ultrahigh vacuum (UHV), and measured again.

Figure 1 provides visual representations of the superstrate CdTe films (Fig. 1a) and PEEM electronic property measurement. Figure 1b identifies physical grains and grain boundaries in the polished surface. Figure 1c illustrates our designations for electronic property variation that occurs between one grain surface to another (grain-to-grain), and between a grain surface and grain boundary (grain-to-boundary).

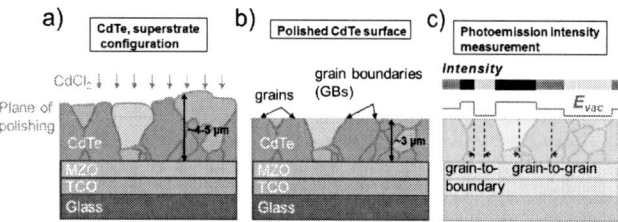

Fig. 1. (a) Schematic of a CdTe sample in superstrate configuration, showing the plane of polishing to expose the grain interiors. (b) Grain interiors and grain boundaries (GB) are illustrated to define grain-to-grain and grain-to-boundary designations (c) for electronic structure variation measured with PEEM. Figure 1c also shows hypothetical variation of the local photoemission intensity due to variation of the vacuum level ($E_{VAC}$) plotted in red. Dark regions in the photoemission intensity profile of Fig. 1c denote regions with relatively low photoemission intensity and light regions correspond to regions with relatively high photoemission intensity.

Regions with relatively higher vacuum level in Fig. 1c correspond to regions of lower photoemission intensity, represented by dark regions. Similarly, light regions in the photoemission intensity profile in Fig. 1c, corresponding to higher photoemission intensity, are correlated to regions with lower vacuum level. We further discuss the relationship between photoemission intensity variation and variation in electronic properties later in this section, following descriptions of PEEM measurement.

PEEM measurements were conducted in a LEEM-III system (Elmitec Elektronenmikroskopie GmbH) equipped with a hemispherical electron energy analyzer and coupled to a tunable DUV light source composed of a pressurized Xe lamp (Energetiq, EQ-1500 LDLS), a Czerny–Turner monochromator (Acton research, SP2150), and refocusing optics. The spectral width of the DUV light was set to 50-100 meV throughout the wavelength range of the measurement ($\lambda = 175$–350 nm, $h\nu \cong$ 3.6–7 eV). The field of view (FOV) for all photoemission images was 48 $\mu$m at 600 pixels, corresponding to a pixel resolution of ~80 nm/pixel.

Using a fixed photon wavelength, $\lambda = 190$ nm ($h\nu \cong 6.5$ eV) as was the case for this study, we acquire photoemission spectra (PES) at each pixel in a PEEM image by sweeping the voltage offset (start voltage, $V_s$) of the electron kinetic energy spectrum with respect to the energy window of the analyzer. For this study, photoemission intensity images were collected at each $V_s$ value, with a step size of $\Delta V_s$=10 meV. This yields a photoemission intensity versus electron kinetic energy spectra at each pixel. Similar to conventional analysis of photoelectron spectra, cutoffs and onsets of the spectra in energy correspond to locations in energy of the vacuum level ($E_{VAC}$) and the valence band edge ($E_{VBE}$).

Changes in photoemission intensity are not necessarily an indicator of local electronic property variation. Both local variations in the photoemission cross section and work function can result in variation of the photoemission intensity. In the absence of significant variation in the photoemission cross section for a given photoexcitation wavelength, regions of lower relative work function typically correspond to higher relative photoemission intensity. This relationship is reflected in the hypothetical variation of the vacuum level and photoemission intensity presented in Fig. 1c, where the areal variation in the $E_{VAC}$ and intensity appear to be inversely correlated.

Assuming the Fermi level is aligned throughout the sample, differences in the local vacuum level may be attributed to local variation in the relative work function. Thus, local variation of photoemission intensity observed in PEEM images obtained near the vacuum level of PES can provide a qualitative indicator of local work function variation in the sample surface. We proceed to discuss results assuming that the local

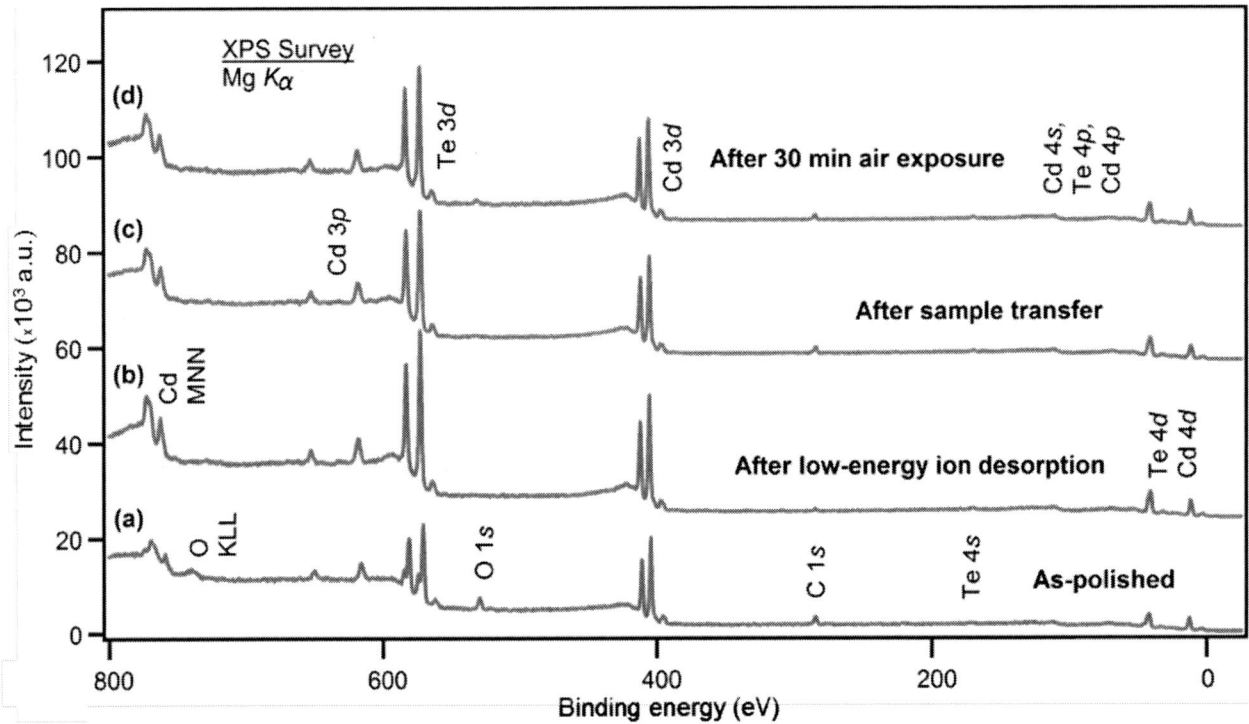

Fig. 2 XPS survey spectrum of the CdCl$_2$-treated CdTe surface after polishing (a), showing O and C in addition to Cd and Te. After a low-energy ion desorption step, (b), the O signal is undetectable within the noise level, the C signal is dramatically reduced, and intensity from Cd and Te core peaks are increased. Sample transfer using the inert gas glove box (c) introduces carbon, while the O signal remains below the noise level. 30 min of air exposure (d) results in a small increase in the C and O signals. Spectra (b)-(d) are offset for clarity.

photoemission cross section does not vary significantly, and that photoemission intensity variation is an indicator of work function variation.

### III. RESULTS & DISCUSSION

Figure 2 shows XPS spectra of the CdCl$_2$-treated CdTe surface after various sample preparation steps. For the as-polished sample (Fig. 2a), in addition to a pronounced O1$s$ and O KLL signal, the Te3$d$ profile shows additional peaks shifted to higher binding energy. These peaks are associated with TeO$_x$ formation. These signatures, associated with the presence of surface oxygen, are undetectable within the noise level after brief sputtering of the surface with 50 eV Ar$^+$ ions (Fig. 2b). The carbon (C1$s$) signal is almost completely removed after sample cleaning. Sample transfer using an inert gas glove box (Fig. 2c) introduces surface carbon prior to PEEM measurement, and air exposure (Fig. 2d) results in increase of the O1$s$ and C1$s$ signals. Air exposure also results in a small TeO$_x$ component that is apparent in fits of the Te3$d$ profile and in core-level scans.

To quantify the amount of carbon introduced during sample transfer through the inert gas glove box, 30 min air exposure, and PEEM measurement, we used a graphene monolayer (1ML) on Si with native oxide as a reference to quantify the

carbon coverage of films. After polishing, the C1$s$ signal is comparable to one-third of a monolayer (1/3 ML) of carbon. 30 minutes of air exposure produces a C1$s$ signal comparable to 1/5 ML of carbon. The highest C1$s$ signal comparable to 2/5 ML of carbon is observed for the CdTe surface after PEEM measurement (not shown). The magnitude of the C1$s$ signal after PEEM measurement is likely due to transferring the sample twice through the inert gas glove box, from XPS to PEEM and then from PEEM back to XPS.

Cd:Te compositional ratios were not greatly impacted by mechanical polishing, ion-desorption, or sample transfer steps. Cd:Te compositional ratios were ~5:6 before and after ion desorption of the polished, untreated CdTe surface, and ~11:10 for the polished, CdCl$_2$-treated sample. The Cd:Te ratio of the CdCl$_2$-treated sample before polishing was slightly higher, ~12:10. These compositional ratios are close to 1:1 as expected from the 1:1 stoichiometry of CdTe. Low-energy ion desorption broadened Cd3$d_{5/2}$ and Te3$d_{5/2}$ peaks but introduced no additional asymmetry. No Cl2$p$ signals were detected from the surface. These results are consistent with Ref. [7].

Figure 3 displays PEEM photoemission intensity images (FOV = 48 μm) of CdCl$_2$-treated and untreated samples. To highlight local variations in the vacuum level within individual images, the potential offset (start voltage, $V_s$) of the electron kinetic energy at which each image was acquired was close to

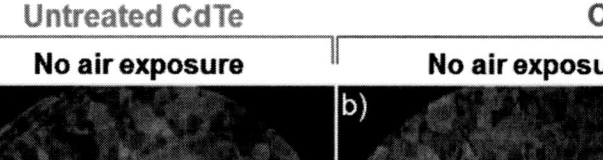

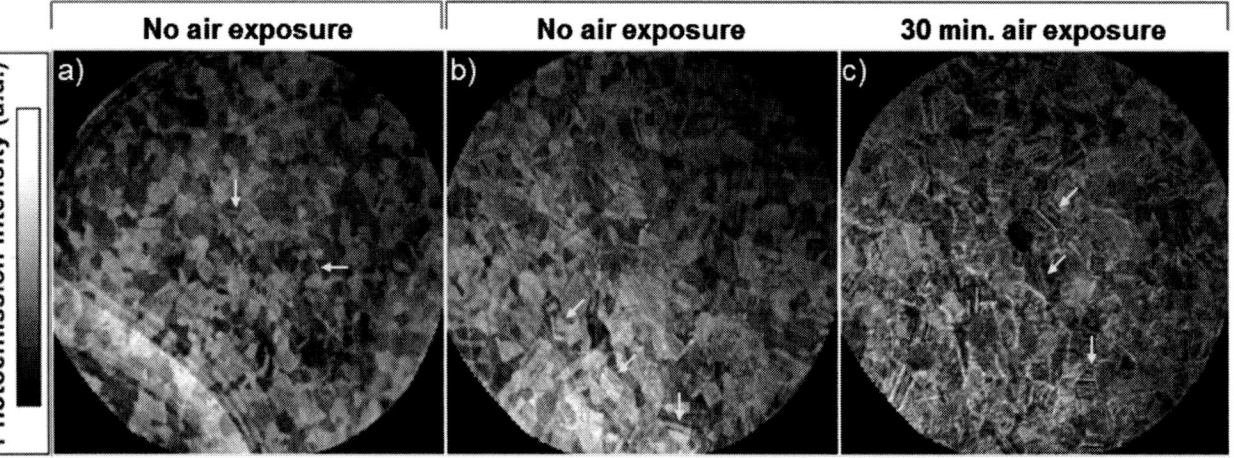

Fig. 3. Photoemission intensity images of (a) untreated CdTe, and CdCl$_2$-treated CdTe, before (b) and after (c) air exposure. Images were obtained at a start voltage value near the vacuum level edge of PES, $V_s \sim E_{VAC}$. In particular, for Fig. 3a $V_s$ = -0.21 V, for Fig. 3b $V_s$ = 0.09 V, and for Fig. 3c $V_s$ = 0.02 V. PES were acquired using $\lambda$ = 190 nm ($hv \sim 6.5$ eV) light. Yellow arrows identify grain domains that contain what appears to be planar defects or twin boundaries, consistent with those observed in Ref. [12]. The field of view for all images is 48 µm.

the vacuum level edge of PES, $V_s \sim E_{VAC}$. Specifically, the photoemission intensity image in Fig. 3a was obtained at a start voltage, $V_s$ = -0.21 V. For Fig. 3b a start voltage of 0.09 V was used, and for Fig. 3c $V_s$ = 0.02 V. The color scales were adjusted to enhance local contrast for each image, individually. Although the photoemission intensity scales in Fig. 3 are not normalized, the scale bar on the left of Fig. 3 shows that dark and light regions correspond to regions of low and high photoemission intensity, respectively. We note that Fig. 3a displays field aberrations in the lower left quadrant that resulted from a sample/holder contact problem.

Contrast in all photoemission intensity images in Fig. 3 varies amongst domains on the scale of microns. These micron-scale domains correspond to the microcrystalline CdTe grains as identified in other microscopy studies of polycrystalline CdTe films [10]. The domain sizes are consistent with grain sizes of polycrystalline CdTe films grown at higher temperatures [11]. There are also narrow domains in Fig. 3 running parallel to each other with abrupt interfaces between them (yellow arrows). These abrupt interfaces bear a resemblance to microscopic observations of planar defects and twin boundaries [12]. Polish marks are also seen in Figs. 3b,c as narrow lines with relatively low photoemission intensity that are continuous as they cross multiple grain domains.

Grains in CdCl$_2$-treated CdTe (Fig. 3b) seemed to be 1-2 times larger than grains in untreated CdTe (3a). The change in grain size we observed with PEEM is consistent with observations of grain recrystallization after CdCl$_2$ treatment for CdTe films grown at higher temperature [13]. Though grain size is recognized as an important parameter impacting the performance of CdTe as an absorber layer [14]–[15], Ref. [13] demonstrated that CdCl$_2$ treatment does not greatly impact the grain size, grain orientation, nor the distribution and structure

of grain boundaries in CSS CdTe films grown at high temperature. Thus, the authors of Ref. [13] concluded that improvements in efficiencies observed after CdCl$_2$ treatment of films grown at high temperature were not driven by changes in microstructure.

All films display contrast in the photoemission intensity that varies on the length scale of grains (grain-to-grain variation). Photoemission intensity images of untreated (Fig. 3a) and CdCl$_2$-treated (Fig. 3b) CdTe show similar grain-to-grain variation. Thus, the impact of CdCl$_2$ vapor treatment is not readily apparent from photoemission intensity images. The presence of grain-to-grain contrast in all films we studied suggest that there may be an inherent grain-to-grain work function variation. Though grain-to-grain non-uniformity is not an entirely new finding [3], [6], it has been less recognized as a factor impacting local junction formation between CdTe and neighboring buffer layers and electrodes. It should be expected that carrier drift and diffusion could vary significantly depending on the orientation of CdTe crystallites, as suggested by our spatially varying photoemission intensity data.

After air exposure of the CdCl$_2$-treated film, contrast at the interfaces between grain domains (grain boundaries) appears in the photoemission intensity image. This contrast is visible as narrow, bright lines decorating the edges of grain domains in Fig. 3c. The lack of grain-to-boundary variation in the CdCl$_2$-treated film (Fig. 3b) and its appearance after exposure to air (Fig. 3c) suggests that, on its own, CdCl$_2$ vapor treatment does not individuate grain boundary properties from those of grain interiors. Observations of increased surface potential at grain boundaries [3] and increased current collection [4] have been previously attributed to segregation of Cl to grain boundaries following CdCl$_2$ treatment. Our findings instead suggest that incorporation of oxygen into a $CdCl_2$-treated film induces grain-to-boundary property variation.

## VI. CONCLUSION

In conclusion, initial photoemission electron microscopy measurements of the CdTe surface suggest that electronic properties vary from one grain interior to another, as well as between grain boundaries and grain interiors. Preparation of clean CdTe surfaces using low-energy ion desorption and PEEM measurement in UHV enabled us to identify a crucial intermediate step toward modifying grain boundary properties in $CdCl_2$-treated CdTe, namely, air exposure. Additionally, our results demonstrate that PEEM is a valuable tool to image secondary junctions in polycrystalline photovoltaics (e.g. from grain-to-grain or grain boundary-to-grain) that impact local junction formation and carrier separation and collection.

## ACKNOWLEDGEMENT

This work was supported by a U.S. Department of Energy, Office of Energy Efficiency and Renewable Energy SunShot Initiative BRIDGE award (DE-FOA-0000654 CPS25859), the Center for Integrated Nanotechnologies, an Office of Science User Facility (DE-AC04-94AL85000), a National Science Foundation PFI:AIR-RA:Advanced Thin-Film Photovoltaics for Sustainable Energy award (1538733), and Sandia Laboratory Directed Research and Development (LDRD). We also thank R. Guild Copeland for his help in constructing the tunable DUV light source, and Sergei A. Ivanov for providing technical assistance. Sandia National Laboratories is a multi-mission laboratory managed and operated by National Technology and Engineering Solutions of Sandia, LLC., a wholly owned subsidiary of Honeywell International, Inc., for the U.S. Department of Energy's National Nuclear Security Administration under contract DE-NA0003525.

## REFERENCES

[1] H. R. Moutinho, M. M. Al-Jassim, D. H. Levi, P. C. Dippo, and L. L. Kazmerski, "Effects of $CdCl_2$ treatment on the recrystallization and electro-optical properties of CdTe thin films," *Journal of Vacuum Science & Technology A*, vol. 16, pp. 1251-1257, 1998.

[2] I. Visoly-Fisher, S. R. Cohen, K. Gartsman, A. Ruzin, and D. Cahen, "Understanding the Beneficial Role of Grain Boundaries in Polycrystalline Solar Cells from Single-Grain-Boundary Scanning Probe Microscopy," *Advanced Functional Materials*, vol. 16, pp. 649–660, 2006.

[3] C.-S. Jiang, H. R. Moutinho, R. G. Dhere, and M. M. Al-Jassim, "The Nanometer-Resolution Local Electrical Potential and Resistance Mapping of CdTe Thin Films," *IEEE Journal of Photovoltaics*, vol. 3, NO. 4, pp. 1383-1388, 2013.

[4] C. Li, Y. Wu, J. Poplawsky, T. J. Pennycook, N. Paudel, W. Yin, S. J. Haigh, M. P. Oxley, A. R. Lupini, M. Al-Jassim, S. J. Pennycook, and Y. Yan, "Grain-Boundary-Enhanced Carrier Collection in CdTe Solar Cells", *Physics Review Letters*, vol. 112, 156103, 2014.

[5] E. S. Barnard, B. Ursprung, E. Colegrove, H. R. Moutinho, N. J. Borys, B. E. Hardin, C. H. Peters, W. K. Metzger, and P. J. Schuck, "3D Lifetime Tomography Reveals How $CdCl_2$ Improves Recombination Throughout CdTe Solar Cells," *Advanced Materials*, vol. 29, 1603801, 2017.

[6] J. D. Major, M. Al Turkestani, L. Bowen, M. Brossard, C. Li, P. Lagoudakis, S. J. Pennycook, L. J. Phillips, R. E. Treharne, and K. Durose, "In-depth analysis of chloride treatments for thin-film CdTe solar cells," *Nature Communications*, vol. 7, 13231, 2016.

[7] D. Hanks, M. Weir, K. Horsley, T. Hofmann, L. Weinhardt, M. Bär, K. Barricklow, P. Kobyakov, W. Sampath, and C. Heske, "Photoemission Study of CdTe Surfaces After Low-Energy Ion Treatments," *38th IEEE Phot. Spec. Conf.*, pp. 396-399, 2011.

[8] J. M. Kephart, R. M. Geisthardt, and W. S. Sampath, "Optimization of CdTe thin-film solar cell efficiency using a sputtered, oxygenated CdS window layer," *Progress in Photovoltaics*, vol. 23, pp. 1484-1492, 2016.

[9] D. E. Swanson, R. M. Geisthardt, J. T. McGoffin, J. D. Williams and J. R. Sites, "Improved CdTe Solar-Cell Performance by Plasma Cleaning the TCO Layer," *IEEE Journal of Photovoltaics*, vol. 3, no. 2, pp. 838-842, 2013.

[10] J. D. Major, L. Bowen, R. E. Treharne, L. J. Phillips, and K. Durose, "$NH_4Cl$ Alternative to the $CdCl_2$ Treatment Step for CdTe Thin-Film Solar Cells," *IEEE Journal of Photovoltaics*, vol. 5, no. 1, pp. 386-389, 2015.

[11] J. D. Major, "Grain boundaries in CdTe thin film solar cells: a review," *Semiconductor Science and Technology*, vol. 31, 093001, 2016.

[12] J. Luria, Y. Kutes, A. Moore, L. Zhang, E. A. Stach, and B. D. Huey, "Charge transport in CdTe solar cells revealed by conductive tomographic atomic force microscopy," *Nature Energy*, vol. 1, 16150, 2016.

[13] J. Quadros, A. L. Pinto, H. R. Moutinho, R. G. Dhere, and L. R. Cruz, "Microtexture of chloride treated CdTe thin films deposited by CSS technique," *Journal of Material Science*, vol. 43, pp. 573-579, 2008.

[14] C. S Ferekides, D Marinskiy, V Viswanathan, B Tetali, V Palekis, P Selvaraj, and D.L Morel, "High efficiency CSS CdTe solar cells," *Thin Solid Films*, vol. 361–362, pp. 520–526, 2000.

[15] S. A. Jensen, J. M. Burst, J. N. Duenow, H. L. Guthrey, J. Moseley, H. R. Moutinho, S. W. Johnston, A. Kanevce, M. M. Al-Jassim, and W. K. Metzger, "Long carrier lifetimes in large-grain polycrystalline CdTe without $CdCl_2$," *Applied Physics Letters*, vol. 108, 263903, 2016.

# Point Defects in CdTe Bulk Single Crystals Grown in Cd-Rich Conditions

Tursun Ablekim,[1,2] Santosh K. Swain,[2] Teresa M. Barnes,[1] and, Kelvin G. Lynn[2]

[1] National Renewable Energy Laboratory, Golden, CO, 80401, USA

[2] Center for Materials Research, The School of Mechanical and Materials Engineering, Washington State University, Pullman, WA, 99164-2711, USA

*Abstract* — **CdTe materials tend to grow in a Te-rich stoichiometry due to the higher vapor pressure of Cd over Te, making the material p-type. Growing the crystal in an extra Cd environment helps to restore the stoichiometry and changes the material from p-type to semi-insulating or n-type conductivity. The amount of Cd and Te vacancies is expected to change with variations of the stoichiometry. To better understand the effect of stoichiometry on intrinsic defects, we grew a CdTe crystal boule in a Cd-rich conditions using the Vertical Bridgman technique. The electrical properties and defect structures of the resulting sample were compared to a Te-rich sample grown under similar conditions. The point defects were characterized by thermoelectric effect spectroscopy. Analysis suggests that five types of defects were common to both growths whereas two types of defects are associated with the growth stoichiometry.**

## I. INTRODUCTION

Cadmium telluride (CdTe) has been one of the major semiconductor materials used in solar cells for decades. With its near-optimum bandgap in the solar spectrum and high absorption coefficient, CdTe is poised to compete with Si-based photovoltaics with achievements such as a 22% cell efficiency and an increasingly lower cost of manufacturing. The performance of CdTe photovoltaic cells depends greatly on the quality of the CdTe absorber layer. From the material-quality point of view, long minority-carrier lifetime and high hole concentration above $10^{16}$ cm$^{-3}$ are desired for next-generation cell efficiency [1]. The p-type doping process is generally achieved by a $CdCl_2$ treatment and extrinsic doping with Cu before applying a back contact. In the case of the $CdCl_2$ treatment, the improvements in p-type conductivity are possibly due to the creation of a shallow acceptor complex $[V_{Cd}-Cl_{Te}]$. In Cu doping, Cu is believed to diffuse into the CdTe from the back contact. Cu atoms substitute a Cd vacancy ($V_{Cd}$) and form the substitutional defect $Cu_{Cd}$, which is also an acceptor. The role of the $CdCl_2$ treatment in CdTe/CdS solar cells is still debated, and the use of Cu is controversial because it may create cell instability by diffusing away from the contact. Another recent trend in p-type doping of CdTe is the use of group V elements such as phosphorous (P) [2, 3] and arsenic (As). In both P and As doping, the dopants may substitute Te atoms, creating $P_{Te}$ and $As_{Te}$ substitutional defects, both of which are acceptors.

Nevertheless, in all the above doping schemes, the intrinsic defect structures and stoichiometry in CdTe play a significant role in doping efficiency. For example, in the $CdCl_2$ treatment and Cu doping, the presence of $V_{Cd}$ is necessary to make the doping more efficient. However, in P and As doping, using a Cd-rich environment in doping may be more beneficial to facilitate the doping because P and As atoms are supposed to go to Te sites. Furthermore, the dopant atoms can also go to Cd sites and create compensating donors, compromising the doping efficiency. Using Cd-rich stoichiometry may help avoid this problem.

In this paper, we discuss defect structures in CdTe grown in Cd-rich conditions and compare the results with a similar study of samples grown in Te-rich conditions. The fundamental understanding of the defects is necessary because material quality can be improved by defect engineering during or after crystal growth.

## II. EXPERIMENTAL DETAILS

### A. Crystal Growth

The Vertical Bridgman Method (VBM) was used to grow a bulk CdTe crystal boule from melt, where the details of the growth process are described elsewhere [4]. The purity of CdTe feedstock charge materials was 6N5 grade from 5N Plus Inc., Canada. The exact stoichiometry of the starting raw material was unknown; however, it is believed to be slightly Te-rich. The ampoule used for the growth was made from GE 214™ quartz from Technical Glass Products. The amount of raw CdTe was ~1000 g. An extra amount of 360 parts per million (ppm) raw Cd was added to the charge to make the growth environment Cd-rich. No intentional dopants were added to alter the electrical property of the crystal, so the as-grown crystals were undoped. The growth was performed with an imposed growth rate of a few mm/h and an imposed axial temperature gradient of 20–50°C/cm. Prior to starting the growth, the charge materials were melted and convective melt mixing (soaking) was performed with the ampoule tip at ~1150°C.

### B. Electrical Property and Purity Analysis

Single crystals were cut from the as-grown boule and used for various measurements. The resistivity of as-grown samples was around $(2–4)\times10^7$ ohm.cm, indicating that $V_{Cd}$ defects were suppressed due to the extra Cd added to the melt. Previously, a similar growth without extra Cd has shown p-type conductivity with a resistance in the order of $10^2$ $\Omega$.cm [5].

The impurity concentration of as-grown samples was determined at the parts per billion (ppb) atomic level using glow-discharge mass spectroscopy (GDMS).

### C. Thermoelectric Effect Spectroscopy (TEES)

The defect states in samples were studied using the TEES method. Single-crystal CdTe samples were cut from the boule and prepared for the TEES study. The samples were mechanically polished following a standard polishing scheme with one-micron alumina slurries and etching in a light Br/methanol solution. Electrical contacts of sputtered Au were added to opposite faces of samples. The edges of samples were masked during the sputtering process to avoid possible shunting between contacts through edges. Then the samples were sandwiched between two planar gold plates and placed on top of a cryostat. For the TEES measurement, the samples were cooled in the dark to ~20 K, where the traps were filled with a sub-bandgap light-emitting diode with 940-nm peak wavelength. The illumination time was 1000 s. After illumination, samples were kept in the dark for another 800–1200 s for currents to decay to below 0.1 pico-amp (pA). Samples were then heated independently from the top and bottom at a pre-set heating rate by maintaining a temperature gradient $\Delta T = 10$ K between the top and bottom surfaces to thermally release the carriers trapped at the defect states. Due to the thermoelectric effect, the temperature gradient across the sample causes a drift of the released charge carriers from the hot to the cold end. This generates a measurable current to the external circuit. The charge emanating from the traps is determined by the trap density. The thermal ionization energy ($E_{th}$) and capture cross section ($\sigma_{th}$) can be calculated by fitting temperature maxima ($T_M$) of current peaks with the variable heating-rate method (VHR). The details of theoretical and experimental aspects of TEES can be found elsewhere [6].

### III. RESULTS AND DISCUSSIONS

If not properly controlled, unintentional impurities and defects could be introduced into CdTe crystals during the growth process, significantly influencing the optical and electrical properties of the material. The impurity levels detected in the sample for the study are listed in Table I. The units are ppb. The major impurity of interest, Cl, was not detected. The unintentional active impurity of Na is around 130 ppb, which is not believed to significantly impact crystal impurities. Total impurities in the samples are under 700 ppb.

TABLE I
IMPURITY CONCENTRATION OF THE SAMPLE

| Elements | Concentration (ppb atomic) |
|---|---|
| Na | 130 |
| Mg | 20 |
| Al | 51 |
| Si | 120 |
| P | 6 |
| S | 84 |
| Cr | 100 |
| Fe | 100 |
| Cu | 25 |
| Zn | 43 |
| Total Impurities | 679 |

TEES measurements were performed on samples prepared from different locations of the boule to confirm consistency. Similar setups measured a few months apart from different samples also showed similar results. The TEES data obtained by illuminating the samples at increasingly higher temperatures are given in Fig. 1. By illuminating samples at various temperatures, one can obtain complete TEES peaks information before current saturation occurs, so that no hidden peaks could be ignored from the study. Due to shifting of the illumination temperature, sample thickness, and applied bias at illumination, the peaks may move about ±5 K in either direction.

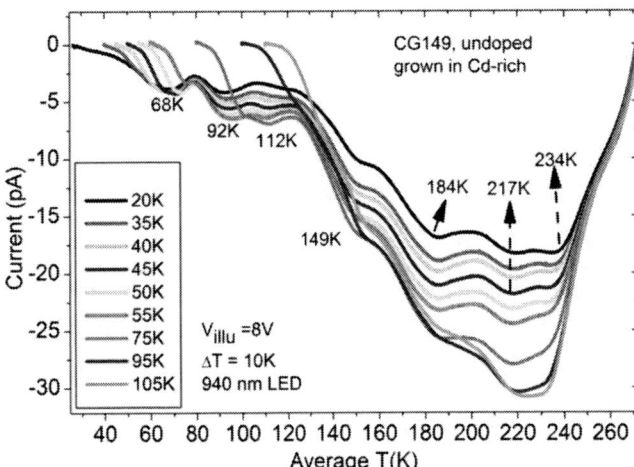

Fig. 1. TEES spectrum of the CdTe sample illuminated at different temperatures specified in the legend. The sample is illuminated by an LED source with 940-nm peak wavelength. An 8-V bias was applied to the sample during illumination; $\Delta T = 10$ K and heating rate is 0.20 K/s.

By carefully analyzing the spectrum, we conclude that there are seven peaks observed that are located at temperatures around 68, 92, 112, 149, 184, 217, and 234 K, measured at a 0.20 K/s heating rate.

Next, after identifying the number of peaks present in the samples and peak locations, we performed VHR analysis to extract defect information associated with the TEES peaks. Setups were further optimized with system parameters. The resulting TEES spectrum with the VHR method is shown in the top and bottom plots in Fig. 2. In the top plot, the peak at 112 K is not distinguishable and the peak at 149 K is not well formed for reliable calculations. Therefore, another VHR run was performed with 8-V bias illuminating at 50 K to distinguish the 112 K peak, which is shown in the bottom plot. Relevant calculations at the same peak resulted in similar values with 10% uncertainty.

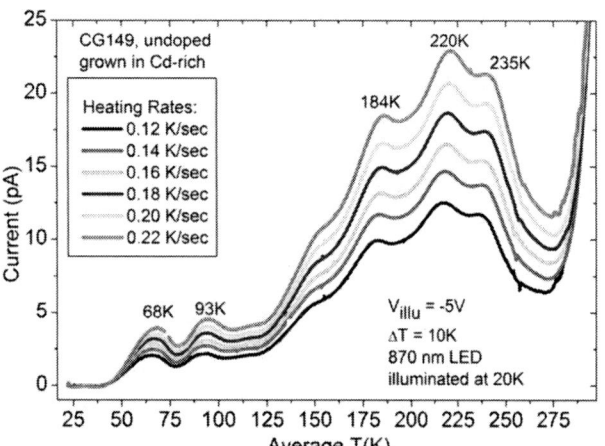

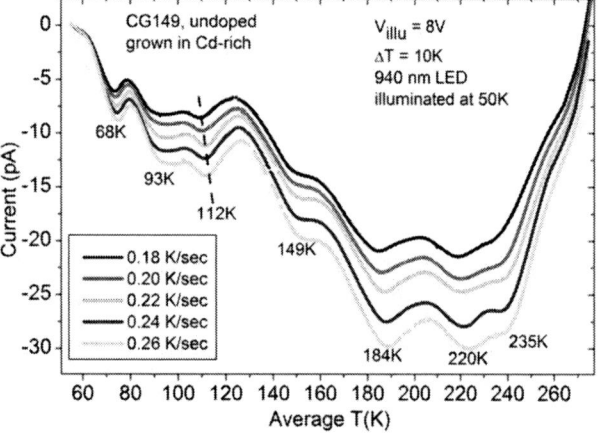

Fig. 2. Variable heating rate TEES spectrum of with -5V (top) and 8V (bottom) bias, illuminated respectively at 20 K (top) and 50 K (bottom); The dashed line serves to guide the eye and shows peak shifting at increasing heating rates.

The thermal ionization energies and trapping cross sections of the defect associated with observed peaks are calculated by fitting the peaks of the VHR spectrum in Fig. 2. The peaks at 68 K and 93 K appear to be broad, yielding larger uncertainties. For these two peaks, the resulting values were confirmed by fitting another spectrum obtained from a different sample. The result is shown in Table II.

TABLE II

CALCULATED THERMAL IONIZATION ENERGY AND TRAPPING CROSS SECTIONS

| Peaks T(K) | $E_{th}$ range (meV) | Mean$E_{th}$ (meV) | $\sigma_{th}$ (cm$^2$) |
|---|---|---|---|
| 68 | 101–107 | 105 ± 9 | (5.5 ± 7.9) × 10$^{-18}$ |
| 93 | 190–235 | 199 ± 12 | (1.9 ± 1.7) × 10$^{-16}$ |
| 112 | 217–233 | 224 ± 13 | (3.2 ± 2.1) × 10$^{-17}$ |
| 149 | 375–398 | 386 ± 16 | (1.1 ± 0.8) × 10$^{-14}$ |
| 184 | 487–510 | 499 ± 10 | (1.7 ± 0.9) × 10$^{-14}$ |
| 220 | 604–634 | 615 ± 14 | (3.9 ± 2.9) × 10$^{-14}$ |
| 235 | 676–725 | 693 ± 30 | (1.4 ± 1.5) × 10$^{-13}$ |

If we compare the defects found in this study with the defects found from CdTe crystals grown without Cd overpressure (published in Ref. [5]), some differences can be noted. For more clarity, the comparison is listed in Table III.

TABLE III

COMPARISON OF DEFECTS IN DIFFERENT GROWTH STOICHIOMETRY

| Peaks T(K) | Mean$E_{th}$ (meV) | |
|---|---|---|
| | Te-Rich | Cd-Rich |
| 68K–72 | 118 ± 7 | 105 ± 9 |
| 92K–93 | 191 ± 13 | 199 ± 12 |
| 112–114 | 225 ± 11 | 224 ± 13 |
| ~135 | 327 ± 7 | Not observed |
| 149 | Not observed | 386 ± 16 |
| 184–188 | 489 ± 22 | 499 ± 10 |
| 217–220 | 706 ± 22 | 615 ± 14 |
| ~235 | Not measurable | 693 ± 30 |

As can be seen, five peaks at temperatures around 72, 92, 112, 184, and 217 K were observed in both samples, and the calculated thermal ionization energies are close to each other; this suggests that they are similar type of defects. The peak at ~135 K was only observed in the Te-rich sample whereas the peak at 149 K was only observed in the Cd-rich sample. These two defects are most likely associated with the growth stoichiometry.

IV. CONCLUSION

CdTe single crystals were grown from a melt having Cd-rich stoichiometry, and detailed TEES analyses were performed to identify the defects. We compared the defects found from the Cd-rich stoichiometry to those found from the Te-rich stoichiometry: five types of defects were identified as common to both, and two are associated with growth stoichiometry.

ACKNOWLEDGMENT

This research is supported by the U.S. Department of Energy under Washington State University's subcontract ZEA-4-42204-01 with the National Renewable Energy Laboratory (NREL). NREL was supported by the U.S. Department of Energy under Contract No. DE-AC36-08GO28308.

The United States Government retains and the publisher, by accepting the article for publication, acknowledges that the United States Government retains a non-exclusive, paid-up, irrevocable, worldwide license to publish or reproduce the published form of this work, or allow others to do so, for United States Government purposes.

## REFERENCES

[1] A. Kanevce and T. A. Gessert, "Optimizing CdTe solar cell performance: impact of variations in minority-carrier lifetime and carrier density profile," *IEEE Journal of Photovoltaics,* vol. 1, pp. 99–103, Jul 2011.

[2] J. M. Burst, J. N. Duenow, D. S. Albin, E. Colegrove, M. O. Reese, J. A. Aguiar, *et al.*, "CdTe solar cells with open-circuit voltage breaking the 1 V barrier," *Nature Energy,* vol. 1, p. 16015, 02/29/online 2016.

[3] T. Ablekim, S. K. Swain, D. Kuciauskas, N. S. Parmar, and K. G. Lynn, "Fabrication of single-crystal solar cells from phosphorous-doped CdTe wafer," *Photovoltaic Specialist Conference (PVSC), 2015 IEEE 42$^{nd}$,* pp. 1–4, 14–19 June 2015.

[4] S. K. Swain, Y. L. Cui, A. Datta, S. Bhaladhare, M. R. Rao, A. Burger, *et al.*, "Bulk growth of uniform and near stoichiometric cadmium telluride," *Journal of Crystal Growth,* vol. 389, pp. 134–138, Mar 2014.

[5] T. Ablekim, S. K. Swain, J. McCoy, and K. G. Lynn, "Defects in undoped p-type CdTe single crystals," *IEEE Journal of Photovoltaics,* vol. 6, pp. 1663–1667, 2016.

[6] B. Šantić and U. Desnica, "Thermoelectric effect spectroscopy of deep levels—Application to semi-insulating GaAs," *Applied Physics Letters,* vol. 56, pp. 2636–2638, 1990.

# Optical Properties of CdSe$_{1-x}$S$_x$ and CdSe$_{1-y}$Te$_y$ Alloys and Their Application for CdTe Photovoltaics

Maxwell M. Junda, Corey R. Grice, Prakash Koirala, Robert W. Collins, Yanfa Yan, & Nikolas J. Podraza

University of Toledo Department of Physics & Astronomy and The Wright Center for Photovoltaics Innovation & Commercialization, Toledo, OH, 43606, USA

*Abstract* — **Opto-electronic property gradients in CdTe solar cells can lead to improved photovoltaic (PV) device performance. Near infrared to ultraviolet complex optical response ($\varepsilon$) is measured for polycrystalline CdSe$_{1-x}$S$_x$ and CdSe$_{1-y}$Te$_y$ alloys. Monotonic variation in $\varepsilon$ and the bandgap are observed for CdSe$_{1-x}$S$_x$, whereas bandgap bowing is observed for CdSe$_{1-y}$Te$_y$. Spectroscopic $\varepsilon$ serves as input for simulations of external quantum efficiency (EQE) for CdTe PV incorporating Se. Comparison with experimental EQE indicates increases at short wavelengths are due to current generated in CdSe$_{1-x}$S$_x$ and CdSe$_{1-y}$Te$_y$ components and increases at long wavelengths arise from red-shifting of the CdSe$_{1-y}$Te$_y$ bandgap relative to CdTe.**

## I. INTRODUCTION

Despite the relative maturity of CdTe photovoltaic (PV) technology, the previous six years have featured significant increases in the record cell efficiency, from 17.3% in 2011 [1] to 22.1% in 2016 [2]. Traditionally, a junction exists between the p-type CdTe layer and an n-type CdS window layer [3], as is shown in the schematic in Fig. 1(a). Recent literature has demonstrated that the thinning or elimination of the CdS layer combined with the incorporation of a CdSe layer can significantly improve PV device photo-response at both short and long wavelengths, increasing short circuit current ($J_{sc}$) [4],[5] as shown in the external quantum efficiency (EQE) in Fig. 1(c).

We report a study of the complex optical property ($\varepsilon = \varepsilon_1 + i\varepsilon_2$) spectra for two relevant alloy systems, CdSe$_{1-x}$S$_x$ and CdSe$_{1-y}$Te$_y$, and their effect on the overall optical response of CdTe PV devices incorporating CdSe. Films are fabricated via co-sputtering and are subsequently annealed before being measured with *ex-situ* spectroscopic ellipsometry (SE). Many CdSe$_{1-x}$S$_x$ and CdSe$_{1-y}$Te$_y$ alloy films are deposited over a range of compositions $0 \leq (x, y) \leq 1$ to identify basic variations in $\varepsilon$. The resulting library of $\varepsilon$ as a function of alloy composition is directly applicable to characterizing alloyed materials in full solar cell devices. In particular, these spectra in $\varepsilon$ are used to simulate the EQE of devices fabricated with various structures. The comparison of simulated and measured EQE of the same device structure is a useful tool in diagnosing potential sources of performance loss and in evaluating overall device design.

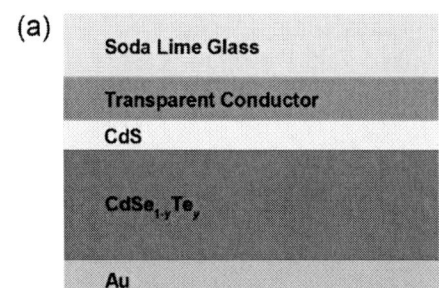

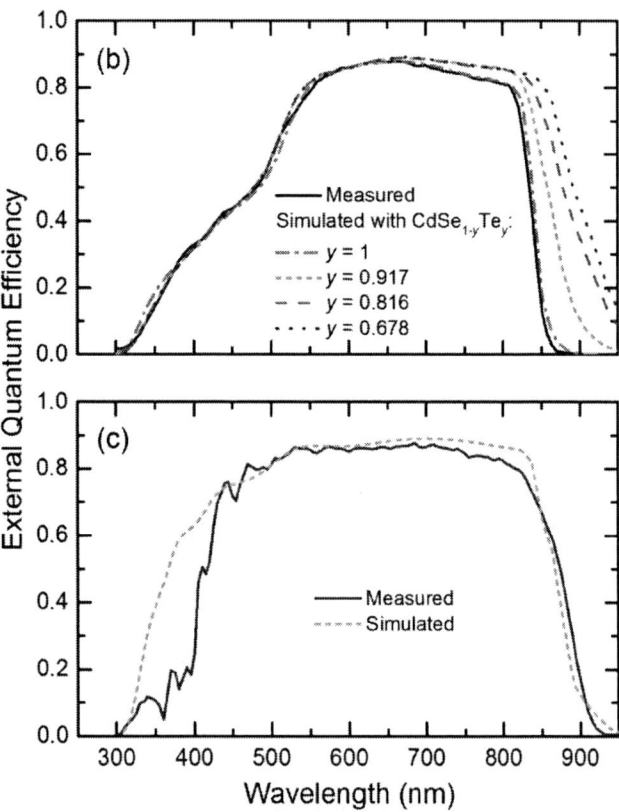

Fig. 1. (a) Schematic of the CdS / CdSe$_{1-y}$Te$_y$ solar cell structure used in the simulation of external quantum efficiency (EQE). (b) Measured and simulated EQE for a solar cell with a conventional ~1000 Å CdS layer. The measured cell incorporates a pure CdTe absorber layer and the simulations incorporate both pure CdTe and CdSe$_{1-y}$Te$_y$ alloys as indicated. (c) Measured and simulated EQE for a solar cell with a thinner ~130 Å CdS layer incorporating a CdSe$_{1-y}$Te$_y$ absorber layer.

978-1-5090-5606-4/17 $31.00 © 2017 IEEE

## II. EXPERIMENTAL DETAILS

The co-sputtered alloyed films are deposited by radio frequency (13.56 MHz) magnetron sputtering in a 10 mTorr Ar ambient onto soda-lime glass (SLG) substrates held at 270°C. 2 inch diameter sputter targets are used over a range from 10 to 60 W. Based on deposition parameters, the expected nominal thicknesses of the films is in the range of 3500 – 5500 Å. The samples are annealed for 6 minutes at ~600°C in a ~500 Torr helium ambient to simulate the thermal budget of industrial CdTe PV fabrication and to improve crystallinity.

*Ex-situ* SE measurements are collected at room temperature [6],[7] (Model M-2000FI, J. A. Woollam Co., Inc.) at 70° angle of incidence spanning photon energies of 0.74 to 5.89 eV. To extract the layer thicknesses and corresponding spectra in $\varepsilon$, a parametric model is iteratively fit to measured ellipsometric spectra in a least squares regression analysis [8].

A fairly simple structural model is employed to fit the compositionally homogeneous co-sputtered films consisting of SLG / lower bulk film / upper bulk film / roughness layer. All parameters describing $\varepsilon$ of the lower and upper bulk film are held common with the exception of independently fitting constant additive terms to $\varepsilon_1$, namely $\varepsilon_\infty$. This division of the bulk film into two layers is necessary to account for density and / or structural evolution in the rather thick films. The roughness is described by a two component Bruggeman effective medium approximation (EMA) [9]. Complementary energy dispersive x-ray spectroscopy (EDS) measurements enable the precise determination of the composition parameters, $x$ and $y$, of each alloy.

The optical response of each alloy is described by parameterized $\varepsilon$ consisting of oscillators assuming parabolic bands at critical points (CPPB) above the material bandgap and an Urbach tail below as has been previously applied to CdS:O [10]. However, as an improvement to the parameterization presented in Ref. [10], first derivative continuity in $\varepsilon$ has been enforced at the bandgap between the CPPB and Urbach behavior spectral regions.

## III. RESULTS AND DISCUSSION

Fig. 2 shows $\varepsilon_2$ spectra for CdSe$_{1-x}$S$_x$ extracted from modeling of SE measurements of the alloy films after annealing. Generally, the $\varepsilon$ spectra describing the CdSe$_{1-x}$S$_x$ alloys feature monotonic trends as a function of overall $x$. The qualitative lineshape behavior transitions from being closer to CdSe to closer to CdS as $x$ increases. More specifically, the lowest critical point (CP) transition energy extracted from $\varepsilon$ is the first allowed electronic transition and therefore defines the direct bandgap ($E_g$). The bandgap energies are plotted in Fig. 3 and are seen to monotonically increase from $1.737 \pm 0.003$ eV for $x = 0$ (pure CdSe) to $2.462 \pm 0.002$ eV for $x = 1$ (pure CdS). These $E_g$ energies for pure CdSe and CdS are in agreement with those for other similar films [11],[12].

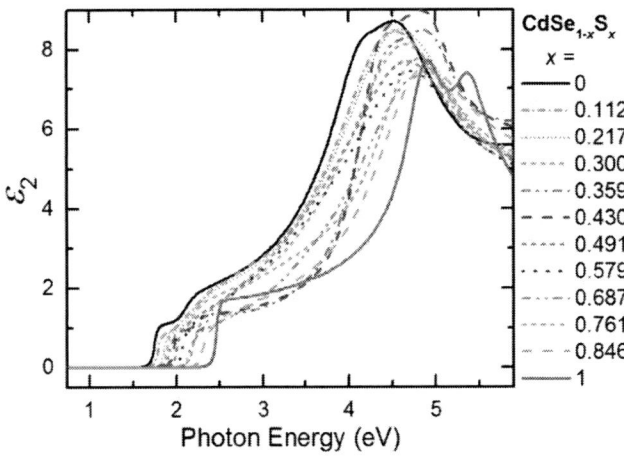

Fig. 2. Imaginary part of complex dielectric function ($\varepsilon = \varepsilon_1 + i\varepsilon_2$), $\varepsilon_2$, spectra for CdSe$_{1-x}$S$_x$ co-sputtered alloys after annealing. Monotonic shifts as a function of the composition parameter, $x$, are observed.

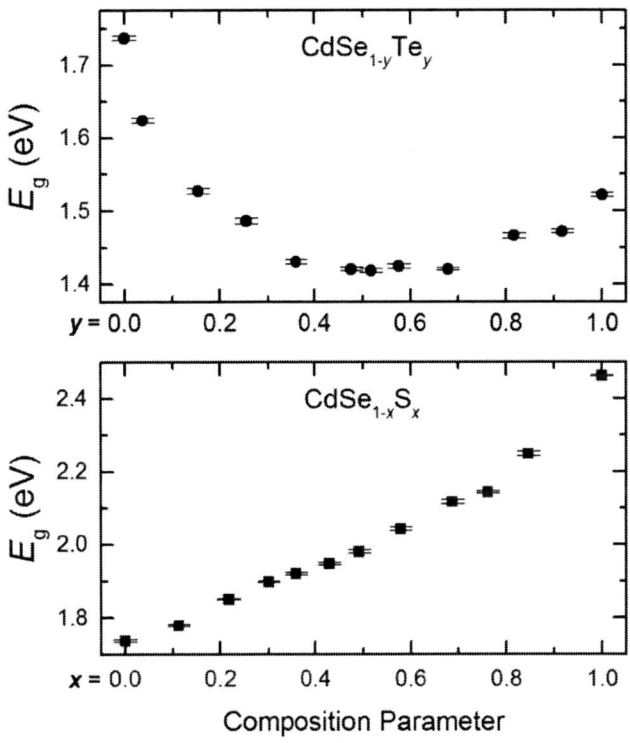

Fig. 3. Bandgap energies ($E_g$) extracted from parameterization of $\varepsilon$ for both CdSe$_{1-y}$Te$_y$ and CdSe$_{1-x}$S$_x$ co-sputtered alloys. $E_g$ increases monotonically for the CdSe$_{1-x}$S$_x$ series with increasing $x$, whereas the CdSe$_{1-y}$Te$_y$ series exhibits bandgap bowing behavior with intermediate values of $y$ having the lowest $E_g$.

The spectra in $\varepsilon_2$ describing the CdSe$_{1-y}$Te$_y$ alloy system are shown in Fig. 4. The right panel shows a close up of the absorption onset for each film and clearly demonstrates that,

978-1-5090-5606-4/17 $31.00 © 2017 IEEE

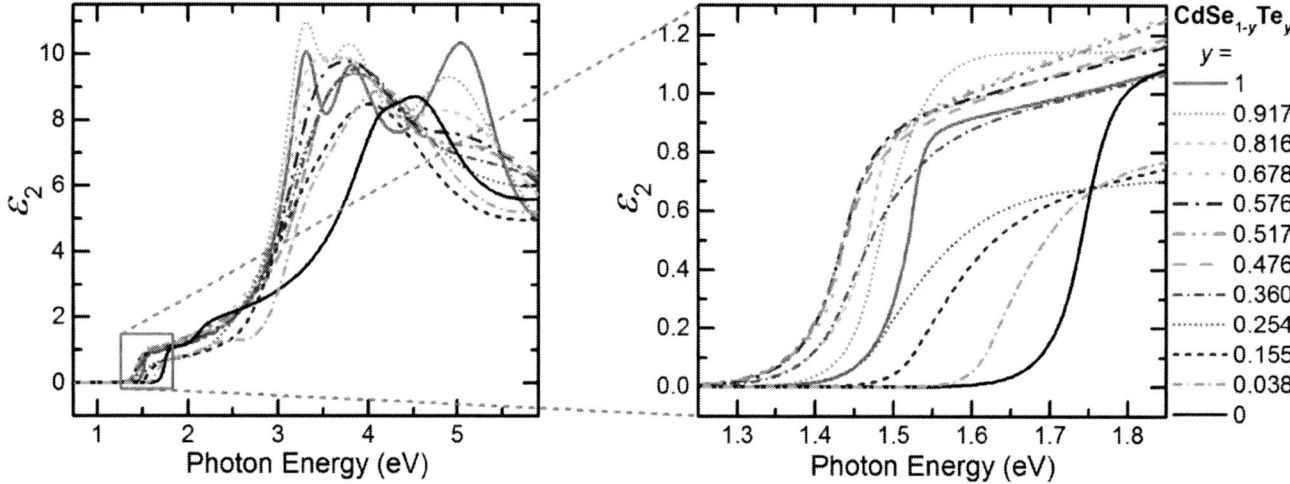

Fig. 4. $\varepsilon_2$ spectra for CdSe$_{1-y}$Te$_y$ co-sputtered films. The right panel shows more detail of the absorption onset featuring bandgap bowing behavior for intermediate compositions of the alloy system.

unlike the CdSe$_{1-x}$S$_x$ alloys, these films exhibit bandgap bowing behavior where compositions of $0.360 \leq y \leq 0.917$ have bandgaps below those of both pure parent materials. This bandgap bowing is also evident in Fig. 3. Additionally, it is noted that the overall CP structure at energies > 3 eV gradually transitions from more CdTe-like for high $y$ to more CdSe-like at lower $y$.

From the results shown in Figs. 2, 3, and 4, it is clear that optical contrast is present between compositions in the two series. For the CdSe$_{1-x}$S$_x$ alloys both $E_g$ and above gap CP features change significantly throughout the transition from CdSe to CdS. However, bandgap bowing at intermediate compositions in the CdSe$_{1-y}$Te$_y$ alloys indicate that $E_g$ alone may not be sufficient to identify composition. Fortunately, optical contrast exists both in CP features above $E_g$ and for $\varepsilon_1$ in the vicinity of $E_g$ as is demonstrated in Fig. 5, enabling different features to be used as tools in differentiating composition of materials with similar $E_g$.

After having established a library of the complete optical response as a function of alloy composition for each series, these spectra can be used as input into layered optical models that account for alloying occurring in full solar cell devices. These models allow for simulation of device EQE [13],[14]. As an example, Fig. 1(b) shows good correspondence between EQE simulations and measurements of a traditional sputtered CdS / CdTe solar cell (solid line and dash-dot line) [15]. To represent a cell fabricated with CdSe additions, the structure in the CdS / CdTe cell simulation is modified by replacing the CdTe absorber layer with various CdSe$_{1-y}$Te$_y$ $\varepsilon$ from Fig. 4. Specifically, since the proportion of CdSe incorporated into such devices is significantly less than that of CdTe, the EQE simulations shown in Fig. 1(b) are for the highest four $y$ compositions as alloyed regions are expected to be Te rich.

The resulting EQE simulations feature gains in the long wavelength portion of the spectrum, which can easily be understood to result from the lower $E_g$ of the alloyed materials.

Fig. 1(c) shows the measured EQE of a solar cell fabricated with a significantly thinner ~130 Å CdS:O layer and a CdSe$_{1-y}$Te$_y$ alloyed absorber layer that results from first sputter depositing a CdSe layer which then forms an the alloy during subsequent high temperature close space sublimation of CdTe. This cell, along with those fabricated with similar structures [4],[5], is shown to have improved EQE at both short and long wavelengths when compared to a more "traditional" CdS / CdTe device, such as that shown in Fig. 1(b). The results from a simulation including a comparable 130 Å CdS layer and a CdSe$_{0.083}$Te$_{0.917}$ absorber layer is also plotted in Fig. 1(c) and

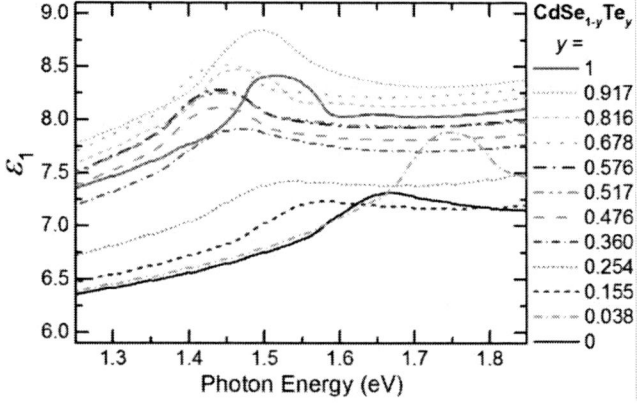

Fig. 5. Spectra in $\varepsilon_1$ describing the CdSe$_{1-y}$Te$_y$ alloy series in the spectral vicinity of $E_g$. The variation in $\varepsilon_1$ in this spectral region enables the identification of specific alloy composition despite the convergence of $E_g$ for films with $0.360 \leq y \leq 0.678$.

has qualitative similarity to the measured cell. Although the simulation and measurement do not agree over the full spectrum, such divergences themselves provide evidence about the true optical structure of the measured device. For instance, the significant difference between measurement and simulation over the 300-400 nm wavelength range strongly suggests that there may be a thin interfacial region at the top of the absorber layer that does not contribute to photocurrent, perhaps as a result of unfavorable alloy compositions or defects. The slight overestimation by the simulation between about 600-850 nm is likely due to the fact that the alloyed absorber layer in the actual device is compositionally graded as opposed to the homogeneous layer used in the simulation. Finally, the simulation overestimation at the absolute longest wavelengths shown is probably attributable to the fact that the EQE simulation includes contributions from Urbach-tail-induced absorption, whereas that same absorption in the measured cell would not be collected and contribute to photocurrent. Implementing the SE measurement and data analysis procedures described here for $CdCl_2$ treated alloys will further help to expand the database of $\varepsilon$, providing even more relevance for CdTe-based PV.

## IV. CONCLUSION

This work demonstrates how comprehensive material optical characterization of $CdSe_{1-x}S_x$ and $CdSe_{1-y}Te_y$ alloy thin films prepared by co-sputtering of CdSe+CdS and CdSe+CdTe targets can be of great utility in understanding the optical structure of devices incorporating such materials. EQE simulation using $\varepsilon$ extracted from these samples can be directly applied to understanding the performance of actual devices. Further development of parametric databases of $\varepsilon$ as functions of $x$ and $y$ are in progress for improved EQE simulations accounting for different alloy formations and gradients as well as modifications of the optical response due to $CdCl_2$ treatment.

## ACKNOWLEDGEMENT

This work was supported by the University of Toledo startup funds, the Ohio Department of Development Ohio Research Scholar Program (Northwest Ohio Innovators in Thin Film Photovoltaics, Grant No. TECH 09-025), and the National Science Foundation (CBET-1230246).

## REFERENCES

[1] M. A. Green, K. Emery, Y. Hishikawa, W. Warta, and E. D. Dunlop, "Solar cell efficiency tables (version 40)," *Progress in Photovoltaics: Research and Applications*, vol. 20, no. 5, pp. 606-614, 2012.

[2] First Solar (2016). Press release retrieved from http://investor.firstsolar.com/releasedetail.cfm?ReleaseID=9564 79.

[3] Z. Fang, X. C. Wang, H. C. Wu, and C. Z. Zhao, "Achievements and Challenges of CdS/CdTe Solar Cells," *International Journal of Photoenergy*, vol. 201, p. 297350.

[4] N. R. Paudel, and Y. Yan, "Enhancing the photo-currents of CdTe thin-film solar cells in both short and long wavelength regions," *Applied Physics Letters*, vol. 105, no. 18, p. 183510, 2014.

[5] J. D. Poplawsky, W. Guo, N. Paudel, A. Ng, K. More, D. Leonard, and Y. Yan, "Structural and compositional dependence of the $CdTe_xSe_{1-x}$ alloy layer photoactivity in CdTe-based solar cells," *Nature Communications*, vol. 7, p. 12537, 2016.

[6] J. Lee, P. I. Rovira, I. An, and R. W. Collins, "Rotating-compensator multichannel ellipsometry: Applications for real time Stokes vector spectroscopy of thin film growth," *Review of Scientific Instruments*, vol. 69, no. 4, pp. 1800-1810, 1998.

[7] B. Johs, J. Woollam, C. Herzinger, J. Hilfiker, R. Synowicki, and C. Bungay, "Overview of variable angle spectroscopic ellipsometry (VASE). Part II: Advanced applications," *Proceedings of the Society of Photo-Optical Instrumentation Engineers Conference, Denver, CO*, vol. CR72, pp. 29-58, 1999.

[8] S. A. Alterovitz, and B. Johs, "Multiple minima in the ellipsometric error function," *Thin Solid Films*, vols. 313/314, pp. 124-127, 1998.

[9] H. Fujiwara, J. Koh, P. I. Rovira, and R. W. Collins, "Assessment of effective-medium theories in the analysis of nucleation and microscopic surface roughness evolution for semiconductor thin films," *Physical Review B*, vol. 61, no. 16, pp. 10832-10844, 2000.

[10] M. M. Junda, C. R. Grice, I. Subedi, Y. Yan, and N. J. Podraza, "Effects of oxygen partial pressure, deposition temperature, and annealing on the optical response of CdS:O thin films as studied by spectroscopic ellipsometry," *Journal of Applied Physics*, vol. 120, no. 1, p. 015306, 2016.

[11] S. Logothetidis, M. Cardona, P. Lautenschlager, and M. Garriga, "Temperature dependence of the dielectric function and the interband critical points of CdSe," *Physical Review B*, vol. 34, no. 4, pp. 2458-2469, 1986.

[12] M. Cardona, "Ultraviolet Reflection Spectrum of Cubic CdS," *Physical Review*, vol. 140, no. 2A, pp. A633-A637, 1965.

[13] F. Abeles, "Recherche sur la propagation des ondes electromagnetiques sinusoidales dans les milieux stratifies. Applications aux couches minces," *Annales de Physique (Paris)*, vol. 5, pp. 596-640, 1950.

[14] F. Leblanc, J. Perrin, and J. Schmitt, "Numerical modeling of the optical properties of hydrogenated amorphous silicon based pin solar cells deposited on rough transparent conducting oxides," *Journal of Applied Physics*, vol. 75, no.2, pp. 1074-1087, 1994.

[15] P. Koirala, J. Li, H. P. Yoon, P. Aryal, S. Marsillac, A. A. Rockett, N. J. Podraza, and R. W. Collins, "Through-the-glass spectroscopic ellipsometry for analysis of CdTe thin-film solar cells in the superstrate configuration," *Progress in Photovoltaics: Research and Applications*, vol. 24, no. 8, pp. 1055-1067, 2016.

# Blistering of magnetron sputtered thin film CdTe devices

P.M. Kaminski, S. Yilmaz, A. Abbas, F. Bittau, J.W. Bowers, R.C. Greenhalgh, J.M. Walls
CREST (Centre for Renewable Energy Systems and Technology), Wolfson School of Mechanical,
Electrical and Manufacturing Engineering, Loughborough University, Loughborough, LE11 3TU,
United Kingdom

*Abstract* –Magnetron sputtering is an industrially scalable technique for thin film deposition. It provides excellent coating uniformity and the deposition can be conducted at relatively low substrate temperatures. It is widely used in the manufacture of solar modules. However, its use for the deposition of thin film CdTe devices results in unusual problems. Blisters appear on the surface of the device and voids occur in the CdTe absorber. These problems appear after the cadmium chloride activation treatment. The voids often occur at the CdS/CdTe interface causing catastrophic delamination. This problem has been known for more than 25 years, but the mechanisms leading to blistering have not been understood. Using High Resolution Transmission Electron Microscopy we have discovered that during the activation process, argon trapped during the sputtering process diffuses in the lattice to form gas bubbles. The gas bubbles grow by agglomeration particularly at grain boundaries and at interfaces. The growth of the bubbles eventually leads to void formation and blistering.

*Index Terms* — CdTe, thin film, solar cell, magnetron sputtering, photovoltaics, electron microscopy, working gas.

## I. INTRODUCTION

The market for solar modules is currently dominated by crystalline silicon based technologies, which account for ~90% of the market [1]. Thin film CdTe photovoltaic (PV) modules are an attractive alternative. The technology is less complex and has lower manufacturing costs. The temperature coefficient of CdTe is also highly favourable. The record efficiency for thin film CdTe technology is 22.1% as reported by First Solar Inc. [2].

The use of magnetron sputtering for the deposition of PV devices has many potential advantages. The technique deposits thin films with excellent uniformity which allows the absorber thickness to be reduced to ~1 µm resulting in a useful reduction in materials cost [3], [4]. Also the coating uniformity enables the technology to be applied to new applications such as semi-transparent window products. The lower temperature of deposition also reduces manufacturing costs. Until recently, magnetron sputtering was exclusively carried out using Radio-Frequency (RF) power supplies. The drawback of using RF power is the low deposition rate due to the low duty cycle [5]. The use of RF power is not viable for a manufacturing process. We recently reported the successful use of pulsed DC magnetron sputtering for the deposition of thin film CdTe [3], [6]. The advantage of pulsed DC power is that sufficiently high deposition rates can be achieved to make the process viable as a production tool. This was a surprising development since compound targets are used that are highly resistive.

As deposited CdTe devices show poor electrical performance and require a cadmium chloride ($CdCl_2$) activation treatment [7], [8]. Surface blisters and void formation in RF magnetron sputtered films have been observed following the treatment for more than 25 years

[9], [10]. An SEM image of a heavily blistered CdTe surface is shown in Fig. 1. The blisters are typically 10µm to 200µm in diameter and are often perforated causing spallation of the material. Fig 2 shows an SEM image of a cross-section of a blister prepared using a focused ion beam. Delamination of the CdTe layer has occurred at the CdS/CdTe junction. Even if blistering has not occurred, void formation at the junction would limit overall efficiency. Often, complete delamination of the absorber occurs along the CdS/CdTe junction. The mechanisms leading to the formation of blisters and voids in sputtered CdTe has remained unclear.

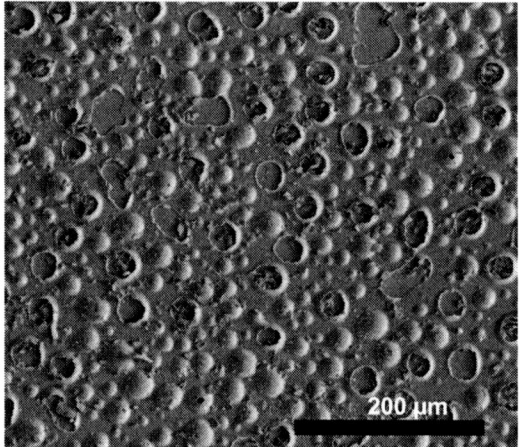

Fig. 1 SEM image of a CdTe device sputter deposited using argon and then activated using $CdCl_2$. Blisters appear on the surface and delamination often occurs at the junction.

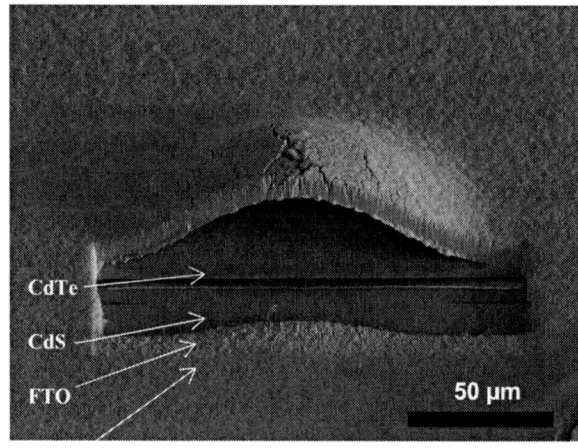

Fig 2 An SEM image cross-section of a large blister |100 um in diameter) prepared using a Focused ion beam. Delamination has occurred at the CdS/CdTe interface.

## II. EXPERIMENTAL

### a. Magnetron sputtering

A PV Solar (PowerVision Ltd.) magnetron sputtering system was used to deposit the CdS/CdTe devices using compound CdS and CdTe targets. The magnetrons are powered by a pulsed DC power supply (Advanced Energy Inc. Pinnacle plus 5kW). The 50mm x 50mm substrates are mounted vertically on a substrate carrier which rotates at 120rpm during the deposition process. The vacuum chamber is load locked and is fitted with three 150mm diameter, circular unbalanced magnetrons targets. The deposition chamber is pumped using a turbomolecular pump (Edwards nEXT300D) mounted vertically above the chamber. Internal radiant heaters are available to raise the substrate temperature to $250^0$C during rotation. The operation of each magnetron and all process parameters are under computer control.

The films were deposited onto NSG Pilkington Fluorine doped tin oxide coated Soda Lime glass (TEC 100) substrates. Prior to deposition, the substrates were cleaned in an ultrasonic bath in a 10% Isopropyl alcohol (IPA) solution in DI water at 60°C for 60min followed by the RCA cleaning process [11]. The surface was activated prior to deposition by a low pressure plasma treatment with 100W Ar/$O_2$ plasma (Glen 1000P series) for 5 minutes

### b. Cadmium Chloride Activation Process

The activation process was performed using a CdCl$_2$ vapour treatment in a tube furnace. Samples were placed in a graphite substrate holder above a crucible with CdCl$_2$ powder. The tube was evacuated to 50 mbar and heated with Infra-Red lamps to a temperature of 400°C and the devices were treated for 6 min.

### c. Device Characterisation

Transmission Electron Microscopy (TEM) was used to investigate the microstructure of the CdTe thin films deposited. TEM images were obtained using a Jeol JEM 2000FX operating at 200kV. Samples for TEM were prepared by Focused Ion Beam (FIB) milling using a dual beam FEI Nova 600 Nanolab. Cross-sections of samples were prepared through the coating and into the glass superstrate. A standard in situ lift out method was used. A platinum over-layer was deposited to determine the surface and homogenize the thinning of the samples. High resolution TEM images were obtained using a FEI Tecnai F20.

## III. RESULTS

### a. As-deposited CdS/CdTe devices

Fig. 2 shows a TEM image of a cross-section of a CdS/CdTe photovoltaic device sputter deposited using argon as the working gas. The CdS layer and the CdTe absorber were deposited sequentially without breaking vacuum. The argon flow rate was set to 20 sccm and the substrate temperature was maintained at ~200°C during the deposition. The structure of the films is dense and columnar. The film is uniform and no voids or pinholes were observed in the films after deposition. The average grain diameter was ~200nm. The open circuit voltage of the as-deposited sample was ~50mV. The elemental composition of the CdTe absorber was analysed using Energy Dispersive Spectroscopy (EDS). A spectrum obtained from the as-deposited film is shown in Fig. 4. The spectrum confirms the stoichiometry of the CdTe, but also reveals the presence of argon with a concentration of 4 At%. The argon appears to be distributed uniformly. The argon has been trapped in the growing CdTe layer during the deposition process. Argon ions are accelerated to the film by the potential difference between the target and the substrate. Magnetrons are deliberately unbalanced to encourage argon ion bombardment of the growing film to aid its densification.

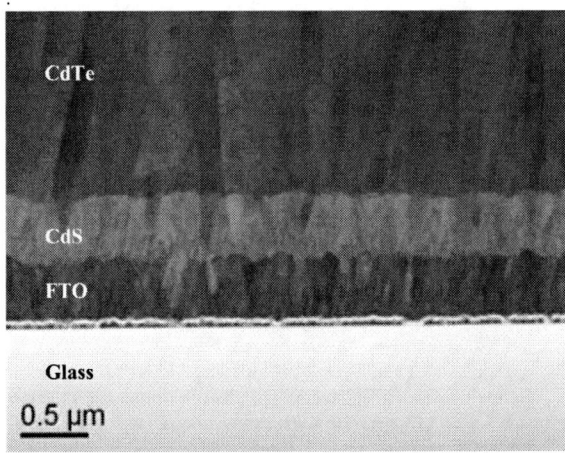

Fig. 2 TEM image of a cross-section of an as deposited CdS/CdTe film using argon working gas. The microstructure of the CdTe is dense and columnar.

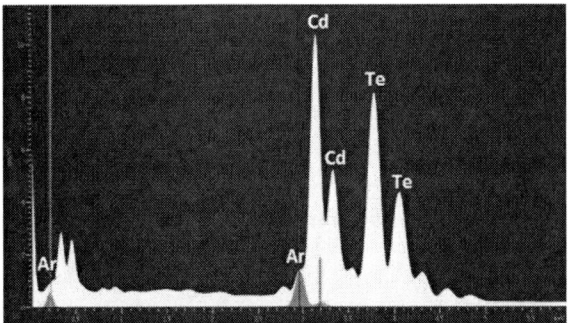

Fig. 4 An Energy Dispersive spectrum obtained from the CdTe absorber in an as-deposited CdS/CdTe device revealing the presence of 4At% of argon trapped in the film.

*b. Cadmium chloride activation*

The devices were activated using the CdCl$_2$ treatment at 400°C for 6min. The Voc of the device increased to ~700mV. Blistering of the CdTe surface was observed following the CdCl$_2$ treatment similar to that shown in the SEM image in Fig. 11. Fig. 5 is a TEM image of a device cross-section following the CdCl$_2$ activation process. A high density of large voids is observed in the CdTe layer. Voids are often located at grain boundaries and are also observed at the CdS/CdTe junction. The CdS layer and the fluorine doped tin oxide layers are largely unaffected by the treatment.

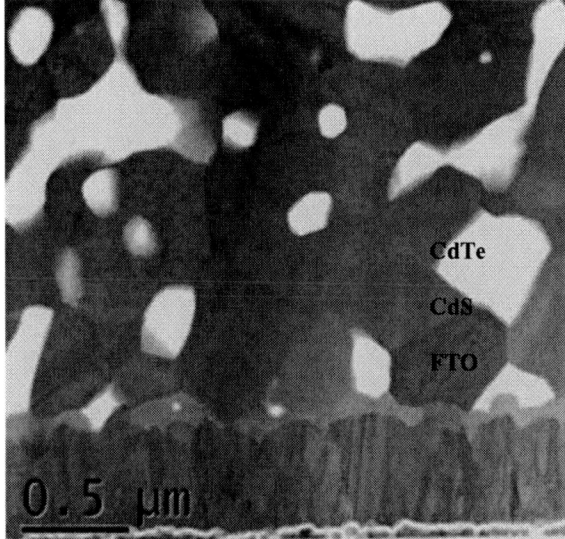

Fig. 5 A TEM cross-section image of a CdS/CdTe device following the CdCl$_2$ treatment. Large voids are observed in the CdTe layer. Although voids are observed at the CdTe interface, the CdS layer is unaffected.

*c. Rapid thermal annealing*

It is not clear if the voids and blisters in the CdTe absorber are caused by the annealing at 400°C or by the presence of cadmium chloride. This question was addressed by simply annealing sputtered CdTe using a Rapid Thermal Annealing process. Devices were annealed at 350$^0$C for 12 hours without CdCl$_2$. The device cross-sections were inspected using High Resolution Transmission Electron Microscopy (HR-TEM).

Fig. 6 shows a HR-TEM image of a cross-section of a thin film CdTe device annealed in a Rapid Thermal Processing system The atomic resolution HR-TEM image reveals a number of spherical features about 5nm in diameter through the thin CdTe laminar film prepared using a Focused Ion beam.. Use of EDS elemental mapping confirmed the presence of argon in the absorber material which is trapped during the sputtering process. During annealing the trapped argon atoms diffuses in the lattice to form gas bubbles. In in situ annealing experiments in the TEM we have observed movement of the small bubbles to grain boundaries where they coalesce and dramatically increase in size. The presence of cadmium chloride

appears to accelerate this process through recrystallization to cause the formation of large voids and surface blistering in the CdTe layer. Although the CdS layer is unaffected void formation at the CdS/CdTe interface junction damages the junction and reduces the device efficiency.

*d. Krypton working gas*

Argon is the conventional working gas used in magnetron sputtering. However, it is useful to explore the use of alternatives for the magnetron working gas. Here we compare argon with the use of alternative inert gases. The sputter yield of helium is too low to be useful and is known to cause bubble formation in a number of materials. This study explores the use of neon and krypton. These inert gas atoms both have lower atomic mass than cadmium and tellurium. However, krypton is heavier than argon while neon is lighter.

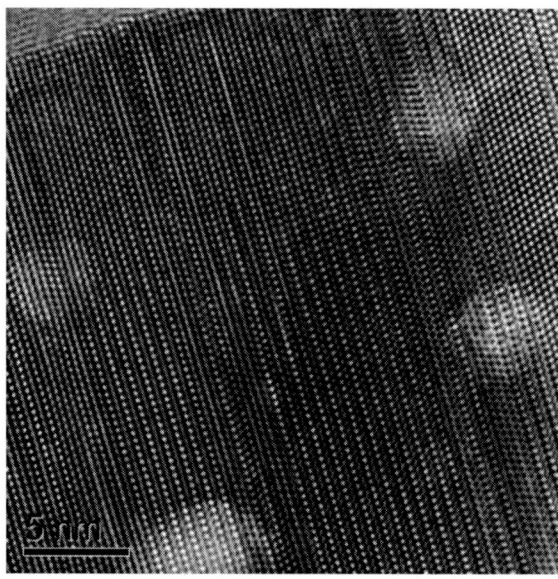

Fig. 6 A HR-TEM atomic scale image of a cross-section of a CdTe thin film annealed at 350$^0$C in a Rapid Thermal Processing system. The image reveals the presence of argon gas bubbles. The argon is trapped during the magnetron sputtering process and diffuses in the lattice to form bubbles during annealing.

A CdS/CdTe device was deposited using Krypton as the magnetron working gas. The gas flow was maintained at 20sccm and the substrate temperature was 200°C during the deposition, conditions the same as those used with argon. Fig. 7 shows a TEM image of a cross-section of the device after 6min CdCl$_2$ treatment at 400°C. The use of krypton also leads to the formation of gas bubbles which coalesce during the cadmium chloride treatment. Extensive void formation is again observed within the CdTe film after activation. The voids are smaller than those observed in the films deposited using Argon as the working gas. Although voids are present at the CdS/CdTe junction, the CdS layer is again largely unaffected.

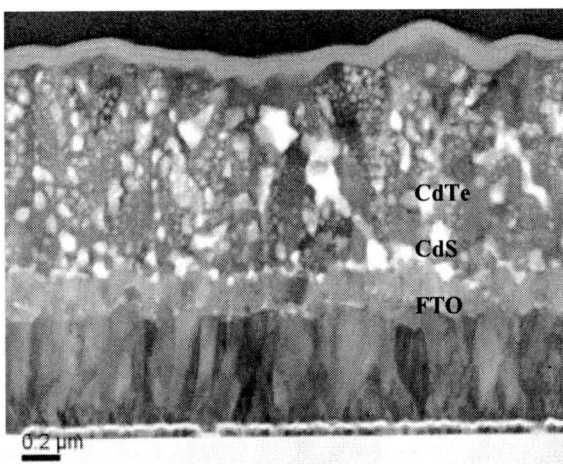

Fig. 7 A TEM cross section image through a CdS/CdTe device deposited using Kr as the magnetron working gas following the CdCl₂ activation treatment. A high density of smaller voids is observed.

Fig 8 shows a composite EDS elemental map of the device deposited using Kr as the working gas following the CdCl₂ treatment. The image is focused on the region surrounding the CdS/CdTe junction. The image shows that the Kr gas is trapped in the film can be detected (red in the image) and forms small bubbles similar in size to the void features in the electron image. Krypton is not observed in the CdS layer and the CdS layer is undamaged.

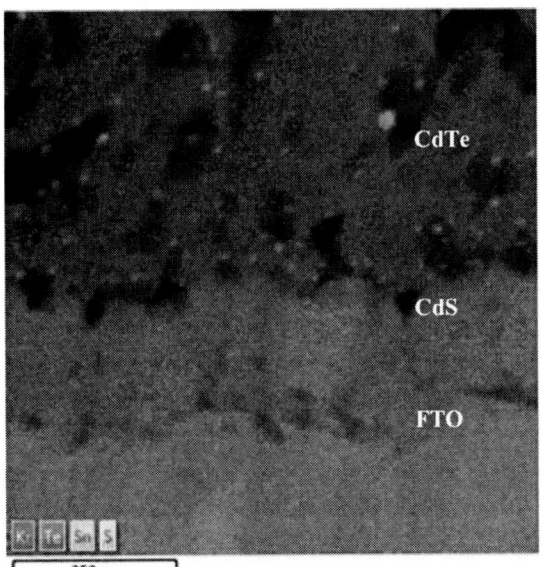

Fig. 8 An EDS composite elemental map of a CdS/CdTe device deposited using Kr as working gas following the CdCl₂ treatment. Regions containing high densities of Krypton (red) are observed.

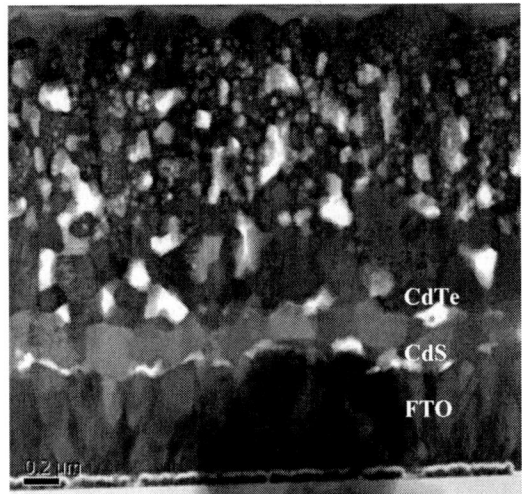

Fig. 9 A TEM cross-section of CdS/CdTe thin film stack deposited using Ne as working gas followed by the CdCl₂ activation treatment.

*e. Neon working gas*

Fig. 9 shows a TEM cross-section of the CdS/CdTe thin film stack deposited using Ne as the working gas. The gas flow was maintained at 20sccm and the substrate temperature was 200°C during the deposition (identical to the argon conditions). After the CdCl₂ activation treatment severe void formation is again observed. The void are large and numerous. Corresponding EDS elemental maps of the cross-section obtained after the CdCl₂ treatment show that Ne is present in the CdTe film. The trapping of Neon in the CdTe leads to the formation of neon bubbles and void and surface blister formation.

## IV SUMMARY

Thin film CdTe photovoltaic devices have been deposited by pulsed dc magnetron sputtering using Ne, Ar, and Kr as the magnetron working gas. The as-deposited thin film CdTe was in each case dense and columnar. The working gas was trapped in the CdTe lattice in each case. Using rapid thermal annealing without the presence of chlorine we have observed the formation of small spherical features in HR-TEM images typically 5nm in diameter when argon is used. These defects are caused by diffusion of the trapped gas atoms in the CdTe lattice during annealing to form small gas bubbles. In the cadmium chloride activation process at 400°C for 6 minutes, the gas bubbles move and agglomerate especially along grain boundaries and at the CdS/CdTe junction. The large bubbles then cause surface blistering and exfoliation. Large voids are formed in the CdTe layer and are often located at grain boundaries and at the junction. Voids at the CdS/CdTe interface leads to poor junction formation and often to catastrophic delamination. This is a major reason for the comparatively poor performance of magnetron sputtered devices. The problem becomes more acute with the lower mass of the inert gas used. The problem is also likely to occur with if gases such as nitrogen are trapped in the CdTe layer. The level of gas trapping can be controlled by lowering the energy of the argon ions in the magnetron plasma. This is largely controlled by the target Voltage.

Similar problems of surface blistering have been reported to occur in a number of materials if sputter deposition is followed by high temperature annealing. Blistering of $Cu_2ZnSnS_4$ (CZTS) photovoltaic devices deposited by magnetron sputtering has been reported recently. A correlation was drawn between blistering and the amount of trapped gas [12]. However, this paper provides the first direct observation for the formation of inert gas bubbles after annealing at the atomic scale using HR-TEM. The fabrication of thin film CdTe devices is further complicated by the presence of cadmium chloride.

Magnetron sputtering is a widely used and industrially capable technique for the deposition of thin films. The technique produces dense and uniform coatings. Its use for the deposition of thin film CdTe has been held back because the absorber suffers from a high density of voids and blistering occurs on the CdTe surface. This paper reports the discovery that the formation of gas bubbles is responsible for these problems. Mitigating the mechanism that causes blistering and voids should result in an industrially viable sputtering process for the deposition of uniform and stable thin film cadmium telluride solar cells.

## ACKNOWLEDGEMENTS

The authors are grateful to UKERC for funding this work through the EPSRC grant (EP/N508433/1). One of the authors (SY) is grateful to is grateful to the Ministry of Education, Republic of Turkey for supporting a PhD studentship.

## REFERENCES

[1] "Solar Photovoltaics Technology Brief," Int. Renew. Energy Agency, no. January, pp. 1–28, 2013.

[2] M. A. Green, K. Emery, Y. Hishikawa, W. Warta, and E. D. Dunlop, "Solar cell efficiency tables (version 47)," Prog. Photovolt Res. Appl., vol. 24, pp. 3–11, 2016.

[3] P. M. Kaminski, A. Abbas, S. Yilmaz, J. W. Bowers, and J. M. Walls, "High rate deposition of thin film CdTe solar cells by pulsed dc magnetron sputtering," MRS Adv., vol. 1, no. 14, pp. 917–922, 2016.

[4] F. Lisco, P. M. Kaminski, A. Abbas, J. W. Bowers, G. Claudio, M. Losurdo, and J. M. Walls, "High rate deposition of thin film cadmium sulphide by pulsed direct current magnetron sputtering," Thin Solid Films, vol. 574, pp. 43–51, Jan. 2015.

[5] A. Gupta and A. D. Compaan, "All-sputtered 14% CdS/CdTe thin-film solar cell with ZnO: Al transparent conducting oxide," Appl. Phys. Lett., vol. 85, no. 4, pp. 684–686, 2004.

[6] P. M. Kaminski, S.Yilmaz, A. Abbas, and J. M. Walls, "Improvements to the deposition and formation of coatings for photovoltaic cells for use in the generation of solar power", UK patent application UK1618474.9. (2016)

[7] A. Abbas, P. Kaminski, G. West, K. Barth, W. S. Sampath, J. Bowers, and J. M. Walls, "Cadmium Chloride Assisted Re-Crystallisation of CdTe: The Effect on the CdS Window Layer," MRS Proc., vol. 1738, p. mrsf14-1738-v03-03, 2015.

[8] A. Abbas, G. D. West, J. W. Bowers, P. J. M. Isherwood, P. M. Kaminski, B. Maniscalco, P. Rowley, J. M. Walls, W. S. Sampath, and K. L. Barth, "The Effect of Cadmium Chloride Treatment on Close Spaced Sublimated Cadmium Telluride Thin Film Solar Cells," MRS Proc., vol. 1493, no. 4, p. mrsf12-1493-e04-02, 2013.

[9] A. D. Compaan and R. G. Bohn, "Thin-Film Cad Photovoltaic Cell Final Subcontract 1 November 1992," NREL/TP451-7162, 1994.

[10] A. L. Sanford, "RF magnetron triode sputtering of CdTe and ZnTe films and solar cells," PhD thesis, Iowa State University, 2002.

[11] W. Kern, "Cleaning solutions based on hydrogen peroxide for use in silicon semiconductor technology," RCA Rev., vol. 31, pp. 187–206, 1970.

[12] P. Bras, J. Sterner, and C. Platzer-Björkman, "Investigation of blister formation in sputtered Cu2ZnSnS4 absorbers for thin film solar cells," J. Vac. Sci. Technol. A Vacuum, Surfaces, Film., vol. 33, no. 6, p. 61201, 2015.

# Energy Yield in Hot & Sunny Climates: Impact of Silicon Solar Cell Architecture and Cell Interconnection

Jan Haschke[1], Johannes P. Seif[1], Yannick Riesen[1], Andrea Tomasi[1], Jean Cattin[1], Loïc Tous[2], Patrick Choulat[2], Monica Aleman[2], Emanuele Cornagliotti[2], Angel Uruena[2], Richard Russell[2], Filip Duerinckx[2], Jonathan Champliaud[3], Jacques Levrat[3], Amir A. Abdallah[4], Brahim Aïssa[4], Nouar Tabet[4], Nicolas Wyrsch[1], Matthieu Despeisse[3], Jozef Szlufcik[2], Stefaan De Wolf[1,*], Christophe Ballif[1,3]

[1] Photovoltaics and Thin-Film Electronics Laboratory (PV-lab), Institute of Microengineering, Ecole Polytechnique Fédérale de Lausanne, Rue de la Maladière 71B, CH-2002 Neuchâtel, Switzerland

[2] Interuniversity Microelectronics Center (imec), Kapeldreef 75, BE-3001 Leuven, Belgium

[3] Swiss Center for Electronics and Microtechnology (CSEM), Rue Jaquet Droz 1, CH-2002 Neuchâtel, Switzerland

[4] Qatar Environment and Energy Research Institute (QEERI), Hamad bin Khalifa University, Qatar Foundation, P.O. Box 5825, Doha, Qatar

* Now at King Abdullah University of Science and Technology (KAUST), KAUST Solar Center (KSC), Thuwal, 23955-6900, Saudi Arabia

*Abstract* — **In this work, we investigate the temperature and irradiance dependencies of the power output of silicon solar cell architectures (BSF, PERC, PERT, SHJ). When we compare our data with commercial module datasheets, we find that the temperature coefficient under maximum power point conditions is systematically worse in the modules. Following our analysis we attribute this to ohmic losses ($R_{CTM}$) due to cell interconnection. Using energy yield calculations we show the impact of $R_{CTM}$ on the energy production in moderate and hot and sunny climates for all investigated architectures. We conclude that maximizing energy production in hot and sunny environments requires not only a high open-circuit voltage, but also a minimal series-to-load-resistance ratio.**

## I. INTRODUCTION

The output power of silicon solar cells usually decreases with increasing operating temperature. These effects have been studied extensively in the past [1]; most commonly observed is the reduced open-circuit voltage ($V_{OC}$) at higher temperatures. Because it is directly linked to the operating voltage [2], the fill factor ($FF$) usually also decreases with increased temperature. The short-circuit current density ($J_{SC}$) increases on the contrary because of temperature-triggered shrinkage of the absorber bandgap [3]. Secondary phenomena such as thermionic barriers may additionally influence carrier transport and thus the relationship between device performance and temperature [4], [5]. In this paper, we analyze the temperature- and irradiance dependencies of the current-voltage characteristic of state-of-the-art silicon solar cell architectures. Both solar cells based on $p$- and $n$-type crystalline silicon were examined, including the current industrial standard solar cell technology (aluminum

BSF) as well as other more advanced technologies, including passivated emitter and rear contact (PERC), passivated emitter rear totally diffused (PERT), and silicon heterojunction (SHJ) solar cells. Furthermore, we examine the impact of ohmic cell-to-module losses that occur due to the series resistance associated with cell interconnections, and how these losses impact the temperature coefficients (TCs) and energy generation of each architecture in hot and moderate climates. In this paper, we discuss BSF, PERC, PERT and SHJ solar cells. In an extended version of this study, also an advanced PERT and a hybrid device are considered [6].

## II. DEVICE ARCHITECTURES

Fig. 1 shows the device architectures investigated in this study. Due to the large variety in device architectures, we refer the reader to the literature discussing their fabrication specifics [7]–[9]. The characteristic parameters obtained from J(V) measurements under standard test conditions (STC) are summarized in TABLE 1. The doping type of the absorber of

TABLE 1: J(V) PARAMETERS OF THE INVESTIGATED SOLAR CELLS UNDER STANDARD TEST CONDITIONS.

| architecture | J(V) parameter at STC | | | | |
|---|---|---|---|---|---|
| | $V_{OC}$ (mV) | $J_{SC}$ (mA cm$^{-2}$) | $FF$ (%) | $R_{MPP}$ ($\Omega$ cm$^2$) | $\eta$ (%) |
| $p$-BSF | 640 | 37.0 | 78.6 | 15.4 | 18.6 |
| $p$-PERC | 655 | 38.7 | 79.3 | 15.2 | 20.0 |
| $n$-PERT | 677 | 39.3 | 81.3 | 15.5 | 21.6 |
| $n$-SHJ | 733 | 37.4 | 78.4 | 17.6 | 21.5 |

978-1-5090-5606-4/17 $31.00 © 2017 IEEE

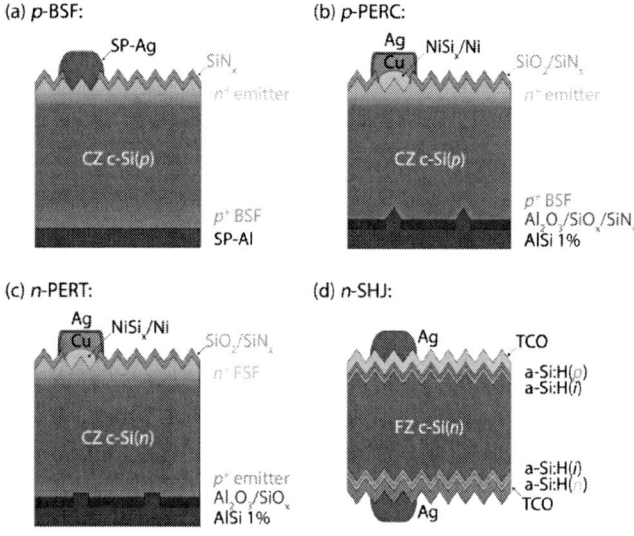

**Fig. 1:** Schematic sketches of the investigated device architectures: *p*-BSF, *p*-PERC, *n*-PERT and *n*-SHJ. Reproduced from Ref. [6] with permission from the Royal Society of Chemistry.

each architecture is denoted using an *italic* letter. The solar cell architectures provide different types and levels of passivation which leads to differences in $V_{OC}$ ranging from 640 mV (*p*-BSF) to 733 mV (*n*-SHJ). All cells are large-area devices (>220 cm$^2$).

Illuminated temperature-dependent J(V) measurements were carried out using an A$^+$A$^+$A$^+$ solar simulator and a temperature-controlled chuck. Furthermore, the irradiation intensity was varied between 0.036 and 2 suns using different neutral density filters or lenses. By this means, a matrix of the irradiance- and temperature-dependent power output of the cells was obtained, used for the energy yield calculations presented in section V.

### III. TEMPERATURE COEFFICIENTS OF THE CELLS

In Fig. 2 the temperature-dependent J(V) parameters are shown. Relative temperature coefficients were obtained from linear fitting and normalisation to the value at 25 °C and are given next to the graphs in corresponding colours. The open-circuit voltage ($V_{OC}$) is decreased with increasing temperature as is well known for silicon based solar cells [1]. The decrease is less pronounced the higher the passivation of the solar cell, which is also reflected by a higher $V_{OC}$ at STC. Therefore, the TC of the *n*-SHJ solar cell is best, as it features the highest $V_{OC}$ under STC (cf. TABLE 1 and Fig. 2a). Despite the different short circuit current densities ($J_{SC}$) of the solar cells, the TC$_{Jsc}$ is the same for all architectures except the *p*-BSF cell which features a slightly higher TC$_{Jsc}$. For the three cell architectures featuring homojunction contacts (i.e. *p*-BSF, *p*-PERC, *n*-PERT), the fill-factor (*FF*) follows linearly the temperature. As a consequence, also the efficiency $\eta$ follows a linear trend. However, for the *n*-SHJ cell, which features silicon heterojunction contacts, this is not the case as the trend of the *FF* versus temperature is only linear for temperatures above 50 °C. Thus, also the trend of $\eta$

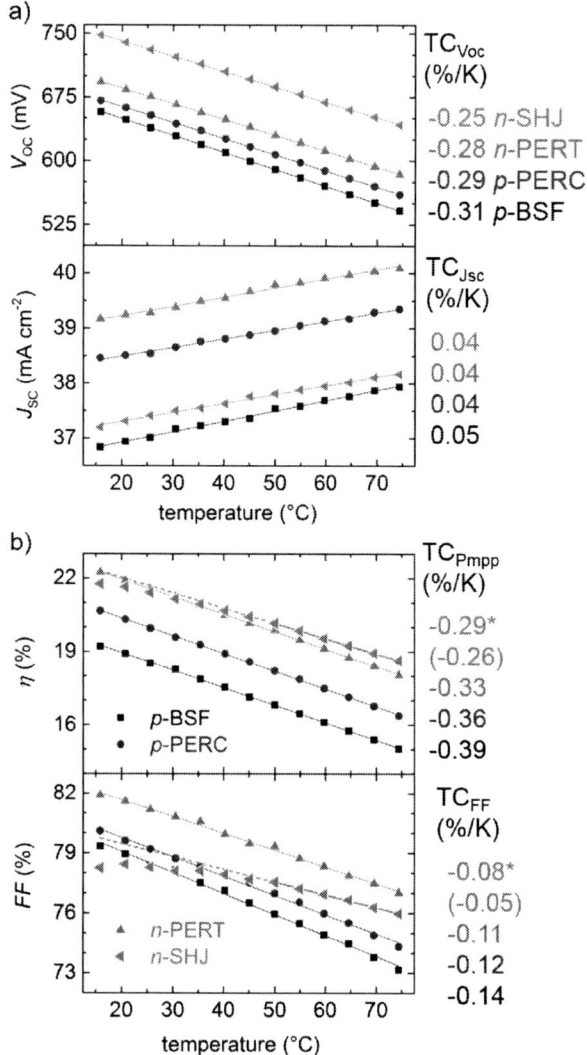

**Fig. 2:** Temperature-dependent J(V) parameters of the investigated solar cells. a) Open-circuit voltage $V_{OC}$ and short-circuit current density $J_{SC}$, b) efficiency $\eta$ and fill-factor *FF*. The relative temperature coefficients (TCs) obtained from linear fitting are given next to the graphs. TCs marked with an asterisk (*) were obtained from a limited fitting range (50-75 °C). The values in brackets for the *n*-SHJ cell were obtained from fitting over the whole range (15-75 °C).

versus temperature is only linear for temperatures above 50 °C for the *n*-SHJ cell. Such a behaviour is common for solar cells where charger carrier transport is hindered by thermionic barriers [5], [10], [11].

## IV. FROM CELLS TO MODULES

When comparing the temperature coefficients (TCs) obtained from our cells with TCs of modules featuring the same cell architectures (taken from datasheets), we find that while the TCs for $V_{OC}$ and $J_{SC}$ are comparable, the TC under maximum power point (MPP) conditions is lower (i.e. worse) in the modules (cf. TABLE 2). We conclude this is due to a lower TC of the *FF*, which we attribute to the additional series resistance $R_{CTM}$ present in the modules as a result of cell interconnections. Assuming an additional $R_{CTM}$, the relative power-output versus temperature of a cell can be translated into that of a module, which is shown in Fig. 3. The power-trends including an additional $R_{CTM}$ (dash-dotted lines) were calculated using equation (1) which approximates the power at MPP of a module using the power at MPP of a cell, additional ohmic losses $R_{CTM}$ and the characteristic load resistance of the cell $R_{MPP}$.

$$P_{MPP}^{module} = \frac{R_{MPP}^{cell} - R_{CTM}}{R_{MPP}^{cell}} \cdot P_{MPP}^{cell}, \text{ with } R_{MPP}^{cell} = \frac{V_{MPP}^{cell}}{J_{MPP}^{cell}} \quad (1)$$

The error introduced by this equation is below 1% as long as $R_{CTM}$ is below $2\ \Omega cm^2$. $R_{CTM}$ is usually also temperature-dependent as it increases with temperature and thus enhances the reduction of the power output at elevated temperatures.

TABLE 2: RELATIVE TEMPERATURE COEFFICIENTS OF THE MAXIMUM POWER POINT OF THREE CELLS PRESENTED IN THIS STUDY AND VALUES TAKEN FROM MODULE DATA SHEETS.

|  | rel. TC$_{mpp}$ (%/K) | | manufacturer |
|---|---|---|---|
|  | our cell data | module |  |
| p-BSF | -0.39 | -0.43 | SolarWorld |
| p-PERC | -0.36 | -0.39 | Trina |
| n-SHJ | -0.26 | -0.30 | Panasonic |

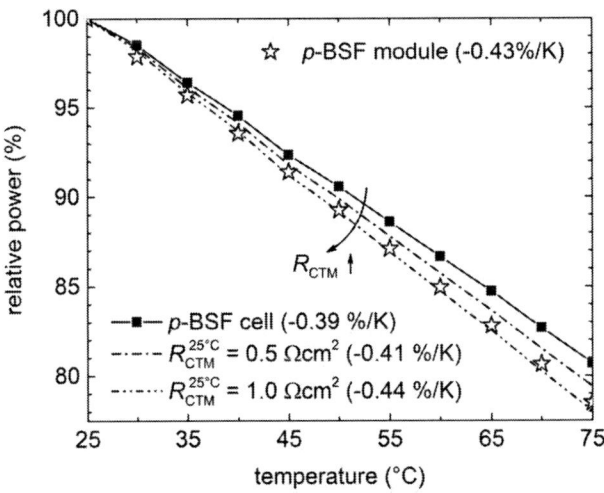

Fig. 3: Relative power output (normalized to 25°C) versus temperature for the p-BSF cell at 1000 W/m² with AM1.5g irradiation. Dash-dotted lines are calculated from the cell data using equation (1) and assuming additional temperature-dependent $R_{CTM}$. The values in brackets are the TC$_{Pmpp}$ obtained from linear fitting. Reproduced from Ref. [6] with permission from the Royal Society of Chemistry.

## V. ENERGY YIELD CALCULATIONS

Using the data obtained from our temperature- and irradiance-dependent J(V) measurements, calculations of the annual energy yield have been performed using the PVlib Toolbox. In Fig. 4, the energy produced per year and rated power at cell level (EPRP$_{cell}$) is shown for all architectures. Note, that as 'rated power' we always used the power of the cell under standard test conditions (STC), *without* an additional $R_{CTM}$. The resulting value of EPRP$_{cell}$ is thus the amount of energy in kWh that will be produced by one nominal kW of solar cells and assuming different $R_{CTM}$. It can be seen that in the temperate climate, all solar cell architectures feature a very similar EPRP$_{cell}$. However, the *n*-SHJ architecture suffers slightly less from the additional series resistance. In the hot climate, the differences between the technologies become more pronounced. Here, the *n*-SHJ technology benefits from its better TC$_{Pmpp}$. One kW of cells produces 2 % more energy in comparison with the *p*-BSF architecture. The benefit increases further with the assumption of ohmic cell-to-module losses up to 3 % at $R_{CTM}$ of $1.5\ \Omega cm^2$. Note, however, that a realistic value for $R_{CTM}$ for a well-designed module should be below $1\ \Omega cm^2$. Limiting $R_{CTM}$ to values between $0.5\ \Omega cm^2$ and $1\ \Omega cm^2$ allows us to compare the four technologies. It can be seen that the differences between the cell architectures are small enough to be potentially outweighed by the additional $R_{CTM}$ present in a module.

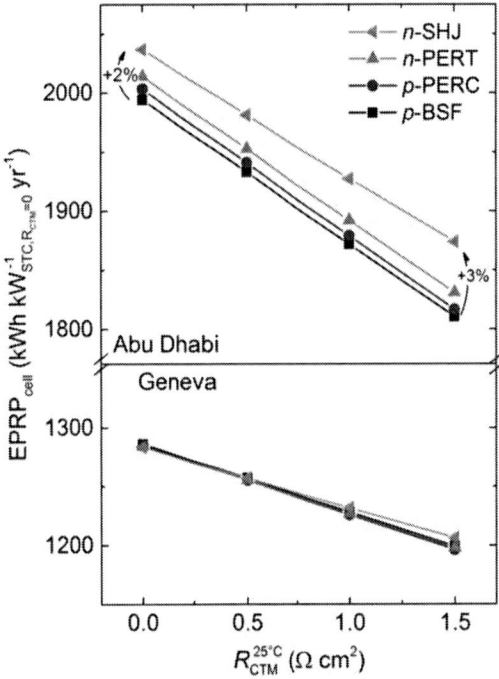

Fig. 4: Yearly energy production per rated power at cell level (EPRP$_{cell}$, $R_{CTM}$ = 0, STC) for temperate (Geneva) and hot and sunny (Abu Dhabi) climate conditions and $R_{CTM,25°C}$ between $0\ \Omega cm^2$ and $1.5\ \Omega cm^2$.

## V. SUMMARY

We measured the temperature coefficients (TCs) of state of the art silicon solar cell architectures and compared them with values taken from commercial module datasheets featuring the same cell architectures. We found that while the TCs of the open-circuit voltage and the short-circuit current density between modules and cells are similar, $TC_{Pmpp}$ of the modules is generally worse. This difference can be explained assuming that there is additional series resistance, $R_{CTM}$, which is induced by cell interconnections in a module. The additional $R_{CTM}$ leads to a worse TC of the fill factor and hence a worse $TC_{Pmpp}$.

Furthermore, we calculated the annually produced energy for the different architectures for two climates. Comparing the results, we showed that $R_{CTM}$ is more detrimental in hot and sunny climate conditions. A solar cell architecture featuring high internal resistance at the maximum power point ($R_{MPP}=V_{MPP}/J_{MPP}$) performs best under such conditions and is less prone to losses due to $R_{CTM}$. Candidates to fulfill this are solar cell architectures featuring passivating contacts and thus high operating voltages such as silicon heterojunction architectures, which are included in our analysis.

In summary, for the highest performance in hot and sunny climates, a high $V_{OC}$ on the cell level and low ohmic cell interconnections on the module level are essential.

## REFERENCES

[1] M. Green, K. Emery, and A. Blakers, "Silicon solar cells with reduced temperature sensitivity," *Electron. Lett.*, vol. 18, no. 2, pp. 97–98, 1981.

[2] M. A. Green, "Accuracy of analytical expressions for solar cell fill factors," *Sol. Cells*, vol. 7, no. 3, pp. 337–340, Dec. 1982.

[3] M. A. Green, "General temperature dependence of solar cell performance and implications for device modelling," *Prog. Photovoltaics Res. Appl.*, vol. 11, no. 5, pp. 333–340, Aug. 2003.

[4] J. P. Seif, A. Descoeudres, M. Filipič, F. Smole, M. Topič, Z. C. Holman, S. De Wolf, and C. Ballif, "Amorphous silicon oxide window layers for high-efficiency silicon heterojunction solar cells," *J. Appl. Phys.*, vol. 115, no. 2, p.

24502, Jan. 2014.

[5] G. Nogay, J. P. Seif, Y. Riesen, A. Tomasi, Q. Jeangros, N. Wyrsch, F. J. Haug, S. De Wolf, and C. Ballif, "Nanocrystalline Silicon Carrier Collectors for Silicon Heterojunction Solar Cells and Impact on Low-Temperature Device Characteristics," *IEEE J. Photovoltaics*, vol. 6, no. 6, pp. 1654–1662, 2016.

[6] J. Haschke, J. P. Seif, Y. Riesen, A. Tomasi, J. Cattin, L. Tous, P. Choulat, M. Aleman, E. Cornagliotti, A. Uruena, R. Russell, F. Duerinckx, J. Champliaud, J. Levrat, A. A. Abdallah, B. Aïssa, N. Tabet, N. Wyrsch, M. Despeisse, J. Szlufcik, S. De Wolf, and C. Ballif, "The impact of silicon solar cell architecture and cell interconnection on energy yield in hot & sunny climates," *Energy Environ. Sci.*, vol. 10, no. 5, pp. 1196–1206, 2017.

[7] E. Cornagliotti, A. Uruena, B. Hallam, L. Tous, R. Russell, F. Duerinckx, and J. Szlufcik, "Large area p-type PERL cells featuring local p+ BSF formed by laser processing of ALD Al2O3 layers," *Sol. Energy Mater. Sol. Cells*, vol. 138, pp. 72–79, 2015.

[8] L. Tous, S. N. Granata, A. Rouhi, J.-F. Lerat, T. Emeraud, S. M. de Nicolas, T. M. Pletzer, R. Labie, M. Aleman, R. Russell, J. John, F. Duerinckx, J. Szlufcik, S. De Wolf, C. Ballif, R. Mertens, J. Poortmans, S. De Wolf, C. Ballif, and R. Mertens, "Large-area Hybrid Silicon Heterojunction Solar Cells with Ni/Cu Plated Front Contacts," *Energy Procedia*, vol. 55, pp. 715–723, 2014.

[9] A. Descoeudres, Z. C. Holman, L. Barraud, S. Morel, S. De Wolf, and C. Ballif, "21% efficient silicon heterojunction solar cells on n-and p-type wafers compared," *Photovoltaics, IEEE J.*, vol. 3, no. 1, pp. 83–89, Jan. 2013.

[10] M. Taguchi, E. Maruyama, and M. Tanaka, "Temperature dependence of amorphous/crystalline silicon heterojunction solar cells," *Jpn. J. Appl. Phys.*, vol. 47, no. 2R, p. 814, 2008.

[11] J. P. Seif, D. Menda, A. Descoeudres, L. Barraud, O. Özdemir, C. Ballif, and S. De Wolf, "Asymmetric band offsets in silicon heterojunction solar cells: Impact on device performance," *J. Appl. Phys.*, vol. 120, no. 5, p. 54501, Aug. 2016.

# Novel Rear Side Metallization Route for Si Solar Cells Using a Transparent Conducting Adhesive

Manuel Schnabel, Talysa R. Klein, Benjamin G. Lee, William Nemeth, Vincenzo LaSalvia,
Maikel F.A.M. van Hest, Paul Stradins

National Renewable Energy Laboratory, Golden CO 80401, USA

*Abstract* — The rear side metallization of Si solar cells comes with a number of inherent losses and trade-offs: a larger metallized area fraction improves fill factor at the expense of open-circuit voltage, depositing directly on textured Si leads to low contact resistivity at the expense of short-circuit current, and some metallization processes create defects in Si. To mitigate many of these losses we have developed a novel approach for rear side metallization of Si solar cells, utilizing a transparent conducting adhesive (TCA) to metallize Si without exposing the wafer to the metal deposition process. The TCA consists of an insulating adhesive loaded with conductive microspheres. This approach leads to virtually no loss in implied open-circuit voltage upon metallization. Electrical measurements showed that contact resistivities of 3-9 $\Omega$cm$^2$ were achieved, and an analysis of the transit resistance per microsphere showed that <1 $\Omega$cm$^2$ should be achievable with higher microsphere loading of the TCA.

## I. INTRODUCTION

A number of losses and trade-offs are inherent in rear side metallization of Si solar cells. A larger metallized area fraction provides better fill factor at the expense of lower open-circuit voltage $V_{oc}$, and direct metallization of a wafer rear side leads to good contact resistance at the cost of optical losses as long-wavelength photons generate plasmons in the metal [1], while adding a spacer may worsen contact resistance and/or lead to new absorption losses. Lastly, some metallization routes, such as sputtering or electron beam evaporation, inherently damage the solar cell by creating defects in the Si wafer.

In this contribution, we present a method whereby a number of these losses could be avoided. Our approach involves bonding the rear side metal to the Si solar cell using a transparent conducting adhesive (TCA). The TCA consists of Ag-coated poly(methyl methacrylate) (PMMA) microspheres embedded in an ethylene vinyl acetate (EVA) sheet, and has been found to exhibit a sheet resistance below 0.5 $\Omega$cm$^2$ when sandwiched between Ag-coated glass slides, even for an area coverage by the microspheres of only a few percent [2].

This approach has a number of advantages: the metal-semiconductor contact area is reduced, a spacer between Si and metal reduces plasmonic losses, and in Si module production, the metal could be deposited and patterned on the module backsheet, and then glued to the Si solar cells, so the cells would not need to be exposed to the actual metal deposition process. Moreover, because of the relatively large sphere diameter of ~40 μm and their partial deformability under heat and pressure, our technique accommodates the surface roughness of a textured wafer. In the following, we

use this approach the metallize symmetric passivated contact lifetime samples on one side, and study the impact on implied $V_{oc}$ and contact resistance as compared to a TCA-less reference on which metal is deposited directly.

## II. EXPERIMENTAL DETAILS

Symmetric samples using our poly-Si/SiO$_x$ passivated contact layers were prepared as described in Ref. [3]. Briefly, 3.5 $\Omega$cm n-type Cz-Si wafers were KOH-planarized, RCA cleaned [4], thermally oxidized to form a 1.5 nm oxide, and then had n-type amorphous Si deposited on both sides which was thermally crystallized at 850°C to form poly-Si. The planarized surface still has ~10 μm surface roughness due to the remaining mesa morphology after planarization. The samples were then hydrogen-passivated by deposition of Al$_2$O$_3$ on both sides followed by a forming gas anneal. The Al$_2$O$_3$ was removed in HF prior to further processing of these passivated contact (PC) samples.

TCA sheet was prepared by depositing a solution of Ag-coated PMMA microspheres and EVA in toluene onto glass, heating to 120°C, and removing the resulting layer with a razor blade. In order to prepare integrated PC/TCA/metal stacks, metal was deposited onto two types of carrier substrates that could later be removed: (1) a transmission line model (TLM) structure as well as large metal pads made of a 20 nm Ti / 1000 nm Ag / 50 nm Pd stack were deposited onto polished, 4 mm thick 1"×1" NaCl crystals using electron beam evaporation through shadow masks (Fig. 1(a)), and (2) full-area 250 nm Ag was deposited on 1"×1" glass as shown in Figure 1(d,e), to be separated from the glass when immersed in deionized (DI) water. These materials were then stacked as PC/TCA/NaCl (Fig. 1(b)), and PC/TCA/Glass (Fig. 1(d)), with the metallized side of the carrier substrate facing towards the TCA. These stacks were hot-pressed at 120°C and 0.5 bar for 10 min, and subsequently soaked in DI water. The DI water dissolves the NaCl crystals, and causes delamination of Ag from the glass, resulting in PC/TCA/Pd/Ag/Ti (Fig. 1(c)), and PC/TCA/Ag stacks (Fig. 1(e)), respectively. Some PCs also had the same metal stack used for the NaCl crystals evaporated on them directly as a reference.

Implied $V_{oc}$ ($iV_{oc}$) was measured using a WCT-120 Sinton lifetime tester prior to metallization, and changes monitored via photoluminescence (PL) imaging using an in-house 810 nm laser diode source and Si CMOS camera. All samples are placed on a silvered mirror during PL

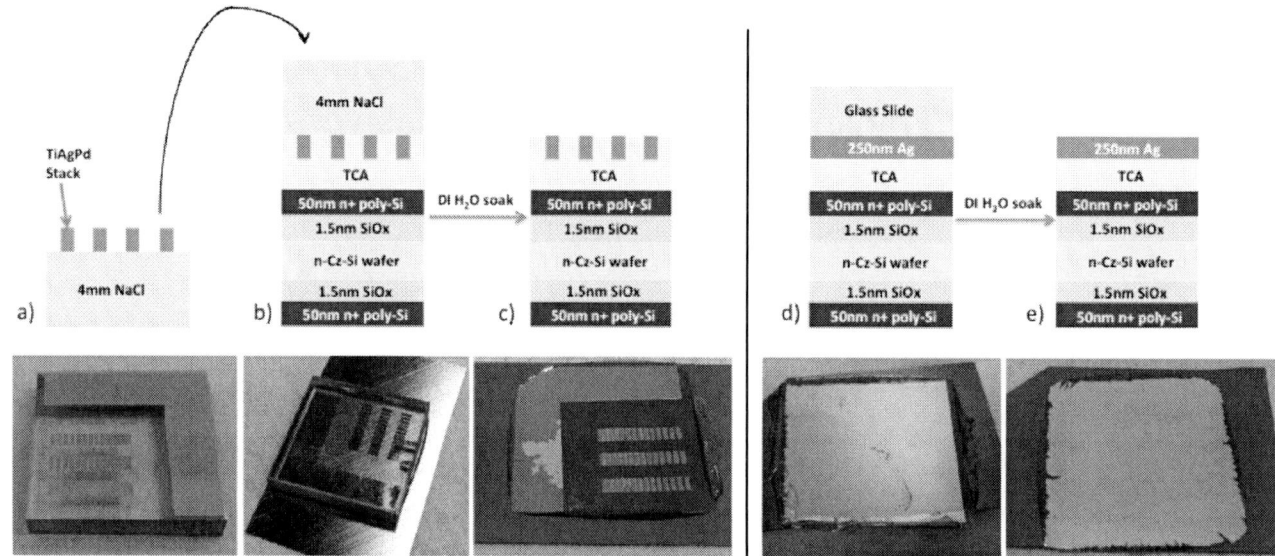

Figure 1: Depiction of metal pattern transfer from a sacrificial NaCl (a,b,c) or glass (d,e) substrate to a passivated contact lifetime sample using TCA. Top: Schematics, bottom: photographs. A TLM pattern and a larger metal area are deposited on NaCl using a shadow mask (a). The NaCl is then glued to the PC sample using TCA (b), and removed by dissolving in DI water to leave a metal pattern on TCA on a PC sample (c). Similarly, glass with 250 nm Ag deposited on it is glued to a PC sample using TCA (d), and removed by inducing delamination at the Ag/glass interface via a DI water soak (e).

measurement to ensure comparable coupling of PL to the camera, before and after sample metallization. Contact resistance was calculated from current-voltage (IV) measurements on TLM structures, and processing monitored via light microscopy.

## III. RESULTS

Figure 1(bottom) indicates that processing was successful and the targeted structures were achieved. Figure 2 shows PL images before and after metallization of the PC samples metallized (a) via metal transfer from NaCl using TCA, (b) via metal transfer from glass using TCA, and (c) via direct metallization of the PC sample. In each case, images are taken from the unmetallized side.

It is immediately apparent that metal transfer from another substrate via the TCA preserves the PL intensity for either metal transfer process used, whereas direct metallization on the PC sample strongly quenches PL. In order to quantify this effect, we analyzed $iV_{oc}$ changes using the areas marked by boxes in Figure 2, under the assumption that changes in PL are related solely to changes in recombination and not to changes in coupling of PL light to the detector. The areas selected are fully rear-side metallized in the right-hand images.

The analysis shows that $iV_{oc}$ of the two samples metallized by the two different metal transfer processes only drops by 3-4 mV upon metallization, preserving the initial $iV_{oc}$ of 722 and 728 mV of these two samples very well. On the other hand, the $iV_{oc}$ of the sample in Fig. 2(c), which was also initially 722 mV, plummeted by 120 mV upon electron beam metallization. While this is a rather extreme example, difficulties maintaining an initially high $iV_{oc}$ upon

metallization are well-documented for passivated contacts [3, 5, 6]. Transfer of metal to PC samples via the TCA provides an alternative, provided acceptable contact resistance can be attained.

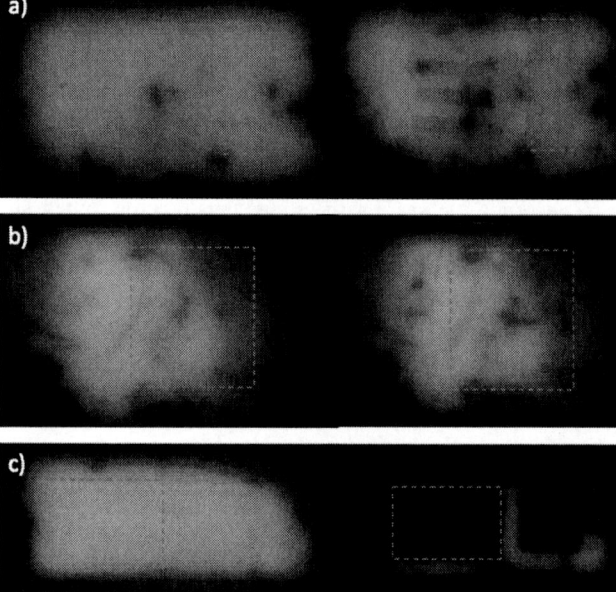

Figure 2: PL images of PC samples before (left) and after (right) metallization. a) PC/TCA/Pd/Ag/Ti prepared using NaCl, b) PC/TCA/Ag prepared using glass, c) PC/TCA/Ti/Ag/Pd prepared by direct evaporation on the PC sample. The areas metallized on the right-hand images that are used for $iV_{oc}$ analysis are marked.

978-1-5090-5606-4/17 $31.00 © 2017 IEEE

An attempt was made to extract the contact resistivity $\rho_c$ of the metal/TCA/PC stacks using the TLM method. As shown in Fig. 3, Ohmic IV curves were obtained, and the inset reveals a general increase in resistance with contact pad spacing. Typically in such a TLM plot, the slope of a linear fit yields the sheet resistance of the laterally conducting material between the contacts (here, the n+ poly-Si and the n-Si wafer), while the resistance axis intercept yields the contact resistance. However, applying this method to the data in the inset yields a negative contact resistance, and a sheet resistance exceeding $10^4$ Ω/sq, both of which are unphysical.

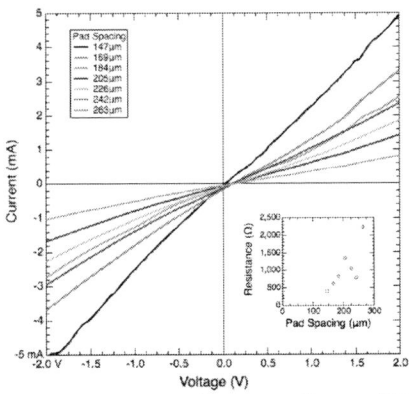

Figure 3: IV curves measured across adjacent TLM pads on a metal/TCA/PC sample prepared using sacrificial NaCl. Inset: plot of Ohmic resistances from main plot as function of contact pad spacing.

A close inspection of the sample used revealed that the areal density of Ag-coated microspheres in the TCA happened to decrease across the sample as pad spacing increased, providing a more reasonable explanation for the data in the inset. A TLM measurement on the directly metallized PC sample yielded a sheet resistance of 600 Ω/sq, and this contribution was subtracted from the measured resistance value for each pad spacing (Fig. 3(inset)) to yield $2R_c$, twice the average contact resistance for that pair of contacts. If the widest, anomalously resistive pad spacing is omitted, this method yields $R_c$=180-640 Ω, and $\rho_c$=3-9 Ωcm².

This is too high for use in Si solar cells, but we used quite a low areal density of microspheres in the TCA (~1%), and higher areal densities could be used. To estimate whether sufficiently low $\rho_c$ could be obtained with more microspheres, the transit resistance of an average microsphere is estimated in the following. By determining the areal density of microspheres left and right of each pad from light microscope images, the number of microspheres under each TLM pad can be estimated to be 8-16. By multiplying $R_c$ by the number of microspheres per pad at the corresponding pad spacing, transit resistances per microsphere of 3000-5000 Ω are obtained (this is a conservative estimate as spheres under the pad but far from the pad spacing may not have participated in transport, in which case the number of participating spheres would be lower, and thus also the resistance per sphere). Based on this estimate a $\rho_c$<1 Ωcm² would thus require more than 3000-5000 microspheres per cm². Given the diameter of an

average microsphere is about 40 µm, this corresponds to 5-8 area% of microspheres, which is readily achievable [2]. Furthermore, this calculation assumed that the resistance per microsphere is independent of areal density, whereas it will actually decrease with increased density as the spreading resistance associated with current flow into and out of each sphere decreases.

## IV. CONCLUSION

We have developed a novel approach for rear side metallization of Si solar cells, utilizing a transparent conducting adhesive to metallize passivated contact samples without exposing them to the metal deposition process. This approach leads to virtually no loss in $iV_{oc}$ upon metallization. Electrical measurements showed that contact resistivities of 3-9 Ωcm² were achieved, and an analysis of the transit resistance per microsphere showed that <1 Ωcm² should be achievable with higher microsphere loading of the transparent conducting adhesive.

## ACKNOWLEDGEMENTS

Funding for this work was provided by the United Sates Department of Energy EERE contract SETP DE-EE00030301 (SuNLaMP) and under Contract No. DE-AC36-08GO28308. The United States Government retains and the publisher, by accepting the article for publication, acknowledges that the United States Government retains a non-exclusive, paid-up, irrevocable, world-wide license to publish or reproduce the published form of this manuscript, or allow others to do so, for United States Government purposes.

## REFERENCES

[1] Z. C. Holman, M. Filipič, A. Descoeudres, et al., "Infrared light management in high-efficiency silicon heterojunction and rear-passivated solar cells," Journal of Applied Physics, vol. 113, p. 013107, 2013.

[2] T.R. Klein, M. Schnabel, E.L. Warren, et al., "Transparent Conductive Adhesives for Tandem Solar Cells," Proceedings of the 44th IEEE PVSC, 2017.

[3] B. Nemeth, D. L. Young, M. R. Page, et al., "Polycrystalline silicon passivated tunneling contacts for high efficiency silicon solar cells," Journal of Materials Research, vol. 31, pp. 671-681, 2016.

[4] W. Kern and D. Puotinen, "Cleaning solutions based on hydrogen peroxide for use in silicon semiconductor technology," RCA Review, vol. 31, pp. 187-205, 1970.

[5] J. A. Aguiar, D. Young, B. Lee, et al., "Atomic scale understanding of poly-Si/SiO₂/c-Si passivated contacts: Passivation degradation due to metallization," in Proceedings of the 43rd IEEE Photovoltaic Specialists Conference (PVSC), 2016, pp. 3667-3670.

[6] W. Nemeth, V. LaSalvia, M. R. Page, et al., "Implementation of tunneling pasivated contacts into industrially relevant n-Cz Si solar cells," in Proceedings of the 42nd IEEE Photovoltaic Specialist Conference (PVSC), 2015, pp. 1-3.

# Multilayer Foil Metallization for All Back Contact Cells

David H. Levy[1] and David E. Carlson[2]

[1]Natcore Technology, Rochester, NY, 14604, USA

[2]CarlsonPV, Williamsburg, VA, 23185, USA

*Abstract* — We present a new back-contact back-junction (BCBJ) cell metallization strategy based upon preformed multilayer laminates. Capable of being produced in low-cost roll-to-roll processes, the laminates consist of aluminum foil bonded to an insulating film of PET. Several methods to produce base contacts were studied. One method to achieve n+ base contacts involved laser firing of phosphorus-treated aluminum foils. Another method utilized carrier selective base contacts, and efficiencies as high as 20.7% were obtained. BCBJ cells with multilayer aluminum foil can lead to the production of low-cost PV modules with high performance.

## I. INTRODUCTION

Back-contact back-junction (BCBJ) solar cells have outstanding advantages, such as the elimination of front grid shadowing and minimized cell to module loss. Despite these benefits, the widespread implementation of this cell structure at large scale requires cost reduction, including simplified metallization methods.

Most back contact designs leverage fine metal fingers to form an interdigitated back contact (IBC), an architecture that requires extensive patterning and high metal conductivity. Multilayer metallization is an alternative approach that relaxes these constraints by allowing the emitter and base metal contacts to exist as continuous metal sheets separated by an insulating layer [1]. Since the metal layers are on top of each other, the fraction of cell area occupied by each metal layer can approach 100%, thus reducing resistance. However, this approach also has a higher propensity to form shorts between the metal layers if the dielectric layer is thin.

In this paper we discuss a novel approach using prefabricated multilayer aluminum foil / polyethylene terephthalate (Al/PET) laminates to form the multilayer metallization on back contact cells, combining the benefits of multilayer contacts with foils [ 2 ]. Multilayer foil laminates can be produced at very low cost in roll-to-roll processing, and the multilayer foil assembly can be rapidly processed to form the emitter and base connections in a potentially low-cost, back-contact solar cell.

Al foils have been used with laser firing to form the base contacts in high performance PERC solar cells [3]. An Al foil has also been used in two-level metallization and module integration with BCBJ solar cells [4]. However, the current work is the first time, to our knowledge, that back-contact solar cells have been fabricated with a multilayer Al foil metallization.

The current work is based upon multilayer foil cells with a silicon heterojunction (SHJ) emitter, and laser-fired base contacts. However, this multilayer foil metallization approach is applicable to other solar cell structures [2].

## II. CELL ARCHITECTURE

The multilayer foil concept consists of an emitter foil and a base foil insulated by a dielectric layer (Fig. 1A). Perforations in the dielectric and emitter foils (Fig. 1B) allow the base foil to make contact to the substrate. In a typical foil device, the emitter foil is connected to the substrate by conductive bonding, while the base foil is laser fired to the substrate. Dimples in the base foil make contact to the substrate through the perforations, as seen in an actual cell (Fig. 1C).

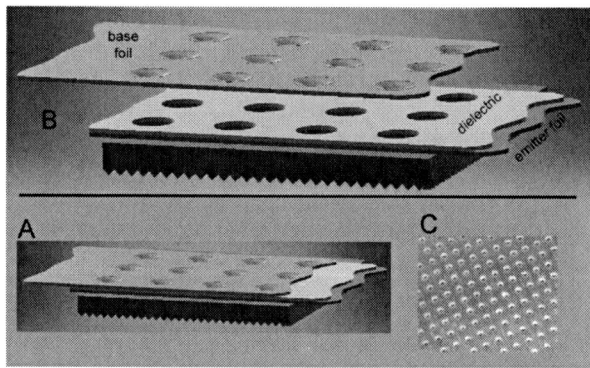

Fig. 1. Foil cell structures. (A) Foil multilayer structure showing emitter foil (blue), dielectric foil (green), and dimpled base foil (red); (B) Exploded view revealing perforated emitter foil and dielectric; (C) Foil bilayer on working cell

Multilayer structures may be prone to shorting, especially when the dielectric layer is a deposited thin layer, which is susceptible to pinholes and high field regions due to surface roughness. However, the Al foils and PET dielectric layers are ~12 μm thick in our device structures. By using a thick dielectric, pre-formed as a web in a standard extrusion processes, shorting can be eliminated while leveraging very high throughput processes.

## III. BILAYER CELLS WITH EMITTER FOIL ONLY

In order to demonstrate the process, initial cells were made using a foil/insulator laminate as the emitter connection, followed by laser firing the base contacts and deposition of an evaporated Al base layer (Fig. 2A).

### A. Cell Process

We used silicon substrates that were ~ 250 μm thick with a bulk resistivity in the range of 1 - 3 Ω-cm and a bulk lifetime > 2 ms. The front surface had a pyramidal texture while the rear had a shiny chemical polish. The front was passivated with ~ 10 nm of a-Si:H, followed by a $SiN_x$ antireflection layer. A rear SHJ was formed by 6 nm of (i)-a-Si:H followed by 7 nm (p)-a-Si:H. The rear structure was then coated with a transparent conducting oxide (TCO) and a thin layer of silver to aid in the foil bonding process [5].

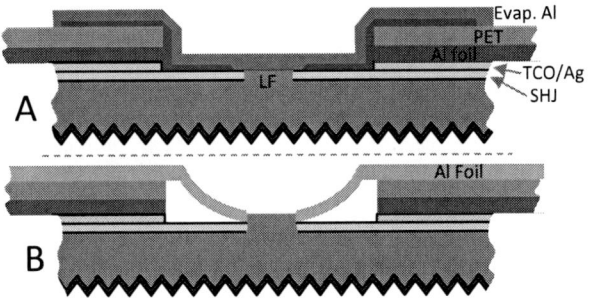

Fig. 2. Detail of multilayer foil cell base contact. (A) Cell with emitter foil / dielectric laminate and evaporated base contact (dark grey is secondary insulator); (B) Full bilayer foil cell

A foil/insulator laminate (12 μm Al / 12 μm PET) was laser patterned to have 500 μm circular perforations on a 1 mm square pitch. A vacuum lamination process bonded the patterned laminate to the substrate, providing conductive contact from the substrate Ag metal to the Al foil. A short etch process selectively removed TCO/Ag from the perforation area to prevent shorting to the base contact.

Due to shorting issues that occur in this test structure when the base metal is applied by vacuum deposition (unlike full foil cells), a secondary insulator is applied and laser patterned. A phosphorus liquid dopant was applied to the substrate followed by laser firing (LF in Fig. 2, 600 ns pulse duration, 532 nm) to produce n+ contacts. The device was completed by evaporation of a 200 nm Al base metal.

All cells were constructed on 25 mm x 25 mm substrates, with a cell dimension of 15 mm x 19 mm. Light I-V testing was done with a 2.56 cm² aperture aligned to the cell region. Measurements were done in-house, using an NREL-measured calibration cell as a standard.

### B. Cell Results

Various optimization studies have been performed on this architecture, mainly examining laser pulse characteristics for their impact on effective lifetime and device performance. This work has also been correlated to studies of laser-fired contact resistance done on a similar test structure. Ultimately we have concluded that laser firing at shorter wavelength (532 nm vs. 1064 nm) and with longer pulses (600 ns) has yielded the best combination of low contact resistance and low laser-induced damage.

Our best cell to date with this construction exhibited an efficiency of 18.8%. The $V_{oc}$, $J_{sc}$, and fill factor were 0.644 V, 40.5 mA/cm², and 72.2%, respectively.

The cells exhibit low shunt current, with reverse dark current at -1 V normally ~ 10 μA/cm². This low shunt current in the foil laminate device indicates that our approach of laser firing directly into the SHJ layer without additional isolation is practical.

## IV. BILAYER CELLS WITH BILAYER FOIL

More advanced cells employing Al foil contacts for both the emitter and base connection are shown in Fig. 2B.

### A. n+ Contacts with Phosphorus-Treated Aluminum Foil

Since our SHJ structures were made on n-type Si, we require n+ base contacts to our top Al foil. However, since Al is a p-type dopant, laser firing of an Al foil would produce an undesirable p-n junction. To address this problem, we developed a method to produce n+ contacts using Al foils with a specialized phosphorus treatment.

To demonstrate, a test structure was built on 150 μm thick n-type Si (resistivity ~ 1.5 Ω-cm) with a low resistance n+ front contact. An Al foil with a phosphorus-containing dopant layer was laser fired to the rear by applying laser pulses while the foil was held in contact to the substrate. The laser pulses melt and penetrate the Al, creating an n+ contact to the Si. Due to a limitation of laser power, we used a matrix pulse layout so that each "point" contact was actually composed of a 3 x 3 array of closely spaced laser fired spots that span a dimension of ~ 100 μm. The laser fired contacts were made using 600 ns pulses at a wavelength of 1064 nm, with a beam diameter of ~ 35 μm. The point contacts were spaced on a 1 mm square pitch.

I-V curves of current flow through this structure are shown in Fig. 3. When a bare Al foil was used (red curve), current flow over the low voltage range was small, and it is clear that the device exhibited diode behavior when swept over a larger current range (inset). This is consistent with formation of a p-n junction due to the polarity of the Al dopant.

A phosphorus-treated foil, made by coating and annealing a phosphorus dopant on the surface of the foil, yielded a very different and ohmic behavior (blue curve). The slope of the line indicates an effective resistance of 0.95 ohm-cm². Assuming that resistance contributions due to the silicon bulk and the front contact are negligible, this value converts to 95 Ω per point contact, a value not far above the

978-1-5090-5606-4/17 $31.00 © 2017 IEEE

spreading resistance limit for a point of this size (100 μm), calculated to be 70 Ω [6]. We conclude therefore that effective contacts can be made by laser firing of the dopant-treated foils, forming the basis for facile construction of bilayer foil cells discussed in the next section.

The ability to produce n+ doping with a treated Al foil is surprising considering the abundance of Al present during the laser firing process. We have noted similar behavior when the dopant is applied to the Si substrate instead of the foil.

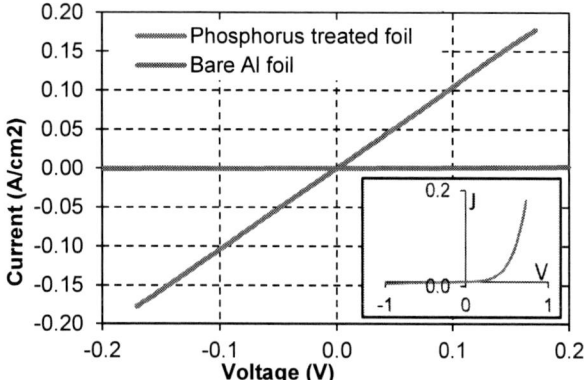

Fig. 3. I-V characteristics of Al foil contacts to n-type silicon showing the effect of the phosphorus foil treatment. Inset: Larger voltage sweep showing diode behavior for bare Al foil.

### B. Bilayer Foil Cells

Bilayer foil cells were fabricated in a similar process to the emitter foil only cells (section III-A). However, after self-aligned etching of the TCO/Ag layers, a phosphorus-treated Al foil was pressed onto the cell using a custom press that dimples the Al foil in order to allow contact with the Si substrate (see Fig. 2B). At this point, laser contacts are made by laser firing through the back of the foil to produce n+ contacts as described above. This structure does not require a secondary insulator, as the base foil would have to deform extensively in order to make any contact with the emitter foil.

### C. Bilayer foil cell results

The one-sun I-V characteristic of the bilayer foil cell is shown in Fig. 4, along with a photograph of the cell (inset) showing the emitter and base foils. The cell yields an efficiency of 16.3%, with $V_{oc} = 0.632$ V, $J_{sc} = 37.2$ mA/cm$^2$, and fill factor = 69.2%. The reduction in $V_{oc}$ and $J_{sc}$ relative to the cells with an evaporated Al base layer likely results from increased laser damage due to non-optimal pulses for producing the laser-fired foil contacts. In particular, the reduced $J_{sc}$ may arise from electrical shadowing in regions above the base contacts [7].

Fig. 4 includes a single diode fit to the I-V curve of the bilayer foil cell, from which we extract a series resistance of ~ 2.0 ohm-cm$^2$. From the foil contact results of the preceding section, we expect a foil contact contribution of about 0.95 ohm-cm$^2$. Examination of the additional causes of resistance is underway.

Because of concerns over laser damage in our multilayer devices, we are currently examining other base contact methods that retain compatibility with multilayer foil structures.

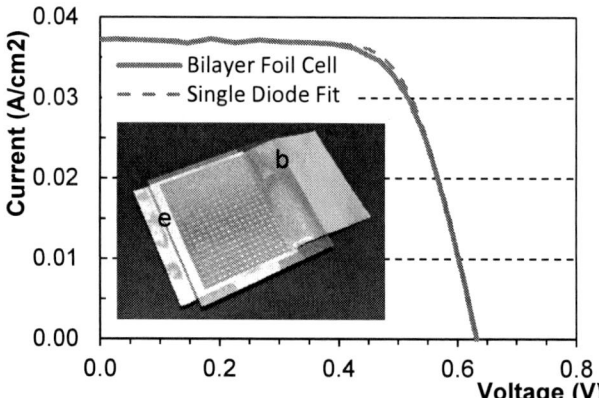

Fig. 4. I-V characteristics of a full bilayer foil cell (blue curve) including single diode fit (dotted red). Inset: Cell rear showing emitter foil (e) and base foil (b).

## V. BILAYER CELLS WITH CARRIER SELECTIVE CONTACTS

The multilayer foil approach is amenable to contacts formed by other processes. We have developed a carrier selective base contact relying upon a proprietary titanium dioxide (TiO$_2$) growth process.

### A. Carrier Selective Contact Performance

The transfer length method (TLM) [8] was used to characterize the contact resistance of our TiO$_2$ based carrier selective contacts on silicon. Solar grade n-type silicon wafers with a bulk resistivity of approximately 1.5 ohm-cm were etched in 20% KOH at 80 °C to yield ~150 μm thick samples with a shiny etch surface on both sides. These samples were further treated with a solution of 5 parts of buffered oxide etch (6:1) and 95 parts of water to remove native oxide and yield a dewetting surface. Immediately following this treatment, TiO$_2$ films of approximately 0.5 nm and 0.8 nm thickness were produced by our process. The resulting samples were metallized with 400 nm of aluminum using thermal evaporation.

A TLM structure was created by photolithographic processing of the above samples, to yield contact pad pairs of 10 mm width separated by gaps ranging from 9 μm to 38.5 μm. Resistance between pairs of pads was determined with using a 4-wire current/voltage measurement with 10mA as the probe current, followed by analysis to extract the specific contact resistance. Additionally, a current voltage sweep was done across the pair at pads at 9 μm separation in order to examine linearity of the I-V characteristic.

Figure 5 shows results of the TLM measurement for the two TiO₂ thicknesses, with an inset showing the I-V characteristics. The resistance between contacts in the TLM test is well behaved, as evidenced by an acceptable linear fit to the measured resistance as a function of gap distance. Further, the I-V characteristic is nicely linear over the range of the measurement, indicative of ohmic contact to the silicon. In this geometry, a contact to the silicon made without the intervening carrier selective TiO₂ would be highly non-linear with low current flow.

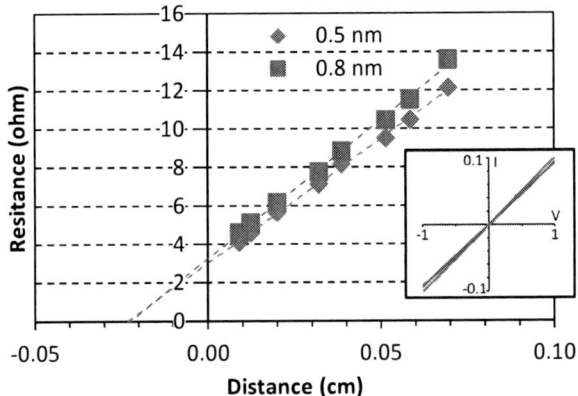

Fig. 5. TLM plot for test structure showing results for a 0.5 nm and a 0.8 nm TiO2 layer; inset: I-V characteristics of both samples measured across the 9 µm gap (blue: 0.5 nm, red: 0.8 nm)

A specific contact resistivity of 0.017 and 0.018 ohm-cm² was obtained for the 0.5nm and 0.8nm TiO₂ films, respectively. While it is desirable for the values to be less than 0.01 ohm-cm², they are clearly suitable for this stage of development. A likely source of error in this measurement is that the thickness of the silicon is large compared the gaps between contact pads. While this error may impact the extracted values, the general behavior still demonstrated useful contacts to n-type silicon.

### B. Carrier Selective Cells

The carrier selective cell process was identical to that of the bilayer cells with emitter only foil (Section III-A) through the deposition of the secondary insulator. The secondary insulator was patterned photolithographically to open vias in the center of each foil perforation. The circular vias had a diameter of approximately 100 µm. The sample was then exposed to 25% TMAH in water at 88 °C for 90 seconds in order to remove the amorphous silicon layers exposed in the via. A special holder was used in order to avoid contact of the TMAH solution to the front of the sample which can damage that area.

The rear of the sample was exposed to a solution of 5 parts of buffered oxide etch (6:1) and 95 parts of water to remove any native oxide on the newly exposed silicon, followed by TiO₂ deposition of approximately 0.8 nm. A base contact of aluminum was then deposited as done for cells in section III-A.

The one-sun I-V characteristic of the carrier selective cell was measured at the NanoPower Lab of the Rochester Institute of Technology (RIT) and is shown in figure 6. Our best performing cell with this fabrication process yields an efficiency of 20.7%, with $V_{oc}$ = 0.681 V, $I_{sc}$ = 40.4 mA/cm², and Fill Factor = 75.2%.

The carrier selective cell, while similar in structure to the laser fired cells of section III, shows mainly improvements in $V_{oc}$ and Fill Factor. The improvement in $V_{oc}$ is likely due to reduced damage of the carrier selective contact relative to the base contact produced by laser firing. Interestingly, the fill factor is also improved even though the carrier selective contacts have a similar dimension to the overall outline of the earlier laser fired contacts. We hypothesize that with reduced damage and increased charge carrier lifetime, the cell benefits from increased levels of injection, which in turn aids in the majority carrier conduction to the base contacts [9].

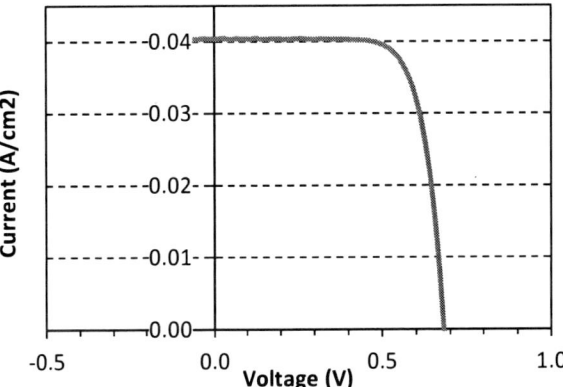

Fig. 6. I-V characteristics of multilayer foil cell employing a TiO₂ carrier selective base contact.

## VI. MODULE DESIGN IMPLICATIONS

The multilayer foil approach lends itself to simplified modes of module manufacture. In typical front emitter cells, additional conductors in the form of tabbing are required to connect cells, leading to cost and fabrication complexity. Alternatively, typical back contact cells can be connected in directly to each other, or using a conductive circuitized backplane. Direct connection requires low resistance of the cell metallization, and may be addressed by increased metallization or restriction to smaller cell sizes. The use of a circuitized backplane reduces this restriction, but at the expense of additional components in the module.

The bilayer foil approach, by virtue of the use of full metal layers for both emitter and base connections, can achieve low cell resistivity. Furthermore, the foils themselves can serve as a low cost cell interconnection method.

### A. Foil Resistance Contributions in Cell

Power loss expected from resistance in the foil of a full area cell can be estimated [10] by assuming that current flows uniformly into the foil, and studying the power loss of a segment $dx$ of the foil as shown in figure 7. Current is collected in the foil and flows toward the tab. As a result, at position $x$ the current flowing through the foil is the total collected from $x$ to the far end of the cell (shaded region):

$$I(x) = J_L(a-x)a \quad (1)$$

where $J_L$ is the current density and $a$ is the cell width (or length). The resistance $dR$ of the element $dx$ is:

$$dR = (\rho_m/at)dx \quad (2)$$

where $t$ is thickness of the metal foil, and $\rho_m$ is the bulk resistivity of the metal. Power loss in the cell is obtained by integrating $dP=I^2(x)dR$ across the length of the cell. When that result is normalized by the total cell area, an area normalized power loss is obtained:

$$P_{norm} = \frac{J_L^2 \rho_m a^2}{3t} \quad (3)$$

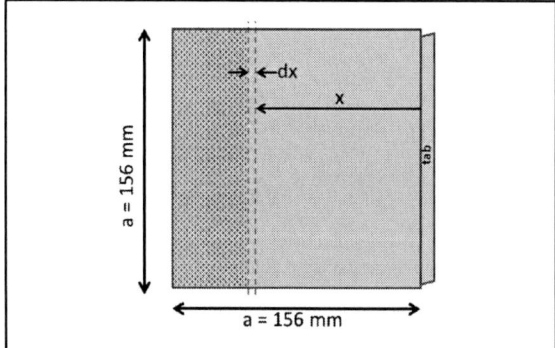

Fig. 7.    Parameters for current flow in foil layers

We have measured the resistivity of the aluminum foils used in our research (>99% Al) and find that it agrees well with the reported bulk resistivity of $2.8 \times 10^{-6}$ ohm-cm. With a 20 μm thick foil for each of the base and emitter connections, and assuming that the foil has 80% of its intrinsic conductivity due to perforations, power loss from equation (3) for a cell producing 40 mA/cm$^2$ is about 0.2 mW/cm$^2$ per foil. Thus at this relatively modest aluminum thickness, a full cell would give up about 0.4% efficiency (absolute) due to resistance in the foil.

An interesting aspect of the foil metallized cell is that an increase in metallization thickness has no meaningful impact on process throughput, since it only involves a change in foil laminate supply. This is in contrast to deposition or plating processes in which metal thickness is normally a function of process time. Furthermore, since we estimate that the bulk aluminum cost for a cell with 20 μm thick Al layers is about 0.5 cents per cell, increasing thickness even by a factor of 2 will have a minimal cost implication.

### B. Cell Interconnection

A number of methods are available for interconnecting cells using the bilayer foil methodology. An attractive method for performing the interconnection is shown in figure 8. In this approach, patterning of the foil involves the production of tabs on either end of the cell, one in which the base foil is extended (A in figure 8) and one in which the emitter foil is extended (B). Folding over the latter reveals the emitter foil for connection to the adjacent cell base foil. Cell interconnection using laser welding of the emitter to base foils at this junction would provide an extremely rapid method of forming cell strings, without the cost of any additional materials. Additionally, current could flow from one cell to its neighbor across nearly the entire cell width.

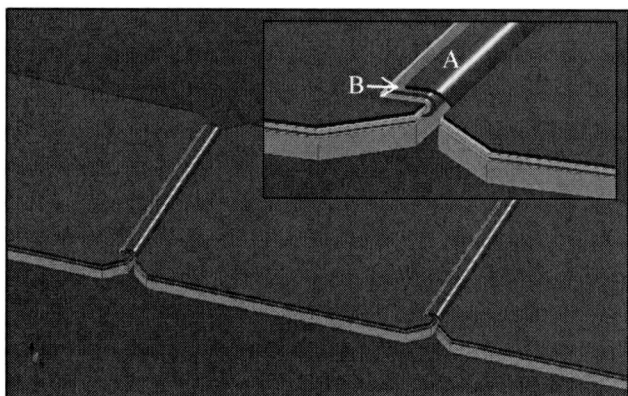

Fig. 8.    Example of cell interconnection approach with foil based cells. With minimal patterning of the foil multilayers, a full width connection from one cell to its neighbor can be done without additional components.

## VII. CONCLUSIONS

A low-cost, foil-based multilayer metallization has been demonstrated for BCBJ cells. By leveraging preformed Al/PET laminate foils, devices with high shunt resistance are observed, eliminating the shorting often associated with multilayer contact approaches. N+ laser-fired base contacts were formed with phosphorus-treated Al foils, producing ohmic contacts on n-type substrates. Initial demonstration cells used a SHJ emitter with laser-fired base contacts. A device using only an emitter foil laminate yielded an efficiency of 18.8%. The efficiency of a full foil bilayer cell is currently at 16.3%, with work underway to optimize the laser processing. Recently, an efficiency of 20.7% was obtained for a bilayer BCBJ cell with a carrier-selective base contact.

## ACKNOWLEDGEMENTS

We thank Wendy Ahearn, Richard Topel, and Ted Zubil for fabrication and testing of the materials and cells used in this work. We also thank Ben Hall at Lasers for Innovative Solutions LLC for patterning of the foil samples. We also thank Seth Hubbard and Zac Bittner of the RIT NanoPower Lab.

## REFERENCES

[1] P. Verlinden, R.M. Swanson, R.A. Sinton, and D.E. Kane, "Multilevel metallization for large area point-contact solar cells", 20th IEEE Photovoltaic Specialist Conference, 1988, p.532.

[2] D.E. Carlson and D.H. Levy, U.S. Patent Application Publication No. US 2017/0062633 A1, March 2, 2017.

[3] J.-F. Nekarda, et al., "Aluminum foil as back-side metallization for LFC cells", 22nd European Photovoltaic Solar Energy Conference and Exhibition, 2007, p.1499.

[4] H. Schulte-Huxel et al., "Two-level metallization and module integration of point-contacted solar cells", Energy Procedia **55**, 361 – 368 (2014).

[5] Rear SHJ / front passivated substrates were provided by CSEM, SA, Neuchâtel, Switzerland.

[6] M.W. Denhoff, "An Accurate Calculation of Spreading Resistance", J. Phys. D: Appl. Phys., **39**,1761 (2006).

[7] M. Hermle, et al., "Shading effects in back-junction back-contacted silicon solar cells", 33rd IEEE Photovoltaic Specialists Conference, 2008.

[8] D. K. Schroder, "Contact Resistance and Schottky Barriers," in *Semiconductor Material And Device Characterization*, 3rd Ed., Hoboken, NJ: Wiley, 2006, ch. 3

[9] B. Fischer, "Loss Analysis Of Crystalline Silicon Solar Cells Using Photoconductance And Quantum Efficiency Measurements," Ph.D. dissertation, Physics Dept., University of Konstanz, Konstanz, Germany (2003).

[10] D. L. Meier and D. K. Schroder, "Contact Resistance: Its Measurement and Relative Importance to Power Loss in a Solar Cell", IEEE Trans. Electron Devices, **ED-31**(5), 647 (1984).

# Electroluminescence Excitation Spectroscopy: A Novel Approach to Non-Contact Quantum Efficiency Measurements

Kristopher O. Davis[1], Greg S. Horner[2], Joshua B. Gallon[2], Leonid A. Vasilyev[2], Kyle B. Lu[2], Antonius B. Dirriwachter[2], Terry B. Rigdon[2], Eric J. Schneller[1], Kortan Öğütman[3], Richard K. Ahrenkiel[4]

[1]Florida Solar Energy Center, University of Central Florida, Orlando, FL, USA
[2]Tau Science Corporation, Hillsboro, OR, USA
[3]Dept. of Electrical Engineering and Computer Science, University of Central Florida, Orlando, FL, USA
[4]Ahrenkiel Consulting, Lakewood, CO, USA

*Abstract* — **In this work, we introduce electroluminescence excitation spectroscopy (ELE) as a non-contact proxy for extracting the quantum efficiency (QE) of a photovoltaic (PV) cell. This method differs from photoluminescence excitation (PLE) by physically separating the absorbing and emitting regions of the cell. It eliminates the influence of voltage independent carriers and solves the challenge of separating the reflected signal from the emitter signal at long wavelengths. Here, the spectrally resolved AC optical excitation drives current to the detection area in a manner similar to non-contact EL. The strength of the EL signal is dependent on the amount of current generated by the spectrally resolved AC optical excitation. Additionally, a separately controllable DC light bias is introduced to control the overall bias state of the cell under test.**

## I. INTRODUCTION

Quantum efficiency (QE) measurements provide a critical tool to assess the quality of photovoltaic (PV) cells. Traditionally, this technique involves illuminating a small area of the cell with monochromatic light of a known irradiance ($G$) and measuring the short-circuit current density ($J_{SC}$) from the cell. The spectral response is obtained by calculating ($J_{SC}/G$) as a function of incident wavelength ($\lambda$) and then translating that into QE, the percentage of incident photons that generate a collected carrier. Electrical contact to the cell is required since a $J_{SC}$ measurement is required. The requirement to scan through different wavelengths for each $J_{SC}(\lambda)$ measurement means QE measurements traditionally take a long time, on the order of minutes, making them incompatible with in-line metrology. However, recent developments have cut QE measurement times substantially, as low as one second [1]. For example, the FlashQE approach uses an array of 64 light emitting diodes (LEDs) with different wavelengths, where the intensity of each is modulated under closed-loop control at a unique AC frequency [2]. The LED output is combined into a single beam that illuminates the cell, and a Fourier transform is performed on the time-resolved $J_{SC}$ of the cell to decouple the contribution from each wavelength. This development opens the door for in-line QE measurements, but it is inherently a contacting technique, and so can only be used at the point of conventional cell test and sort.

The ability to perform a one second non-contact QE measurement is attractive for a number of reasons. A technique that uses only photons in and photons out eliminates wafer breakage that results from contacting the cell, and does not require maintenance or replacement of pogo pins. It may also be used in various geometries (e.g., linescan of moving cell, full area average of motionless cell, small spot measurement at the center of cell). Additionally, implied QE measurements of unmetallized wafers can also be performed, meaning the technique could potentially be used upstream in manufacturing.

Spectrally resolved photoconductance and photoluminescence excitation (PLE) have been used to obtain non-contact QE curves [3], but both have weaknesses. Photoconductance can't be performed on finished cells, but only on unmetallized wafers. For cells made of indirect bandgap materials, the long wavelength response of both photoconductance and PLE is heavily influenced by voltage independent carriers [4]: in silicon this limits their use to wavelengths less than 950 nm. Additionally, PLE measurements near the bandgap of the material are confounded by the difficulty of distinguishing the luminescence signal from the incident light. For silicon process control of modern cells, this long wavelength regime is particularly important, as it is sensitive to the passivation and reflectance of the back surface.

In this work, we introduce a novel non-contact QE measurement technique based on electroluminescence excitation (ELE). This approach physically separates the absorbing and emitting regions of the cell to eliminate the influence of voltage independent carriers and the challenge of separating the reflected signal from the emitter signal at long wavelengths. In addition, a separately controllable DC pump beam is introduced to control the overall bias state of the cell under test. Both the first and second generation designs of the system are shown in Figure 1.

Here, the spectrally resolved AC optical excitation drives current to the detection area in a manner similar to non-contact EL [5-7]. The strength of the EL signal is dependent on the amount of current generated by the AC optical excitation. The detector synchronously locks into the frequency of the AC optical excitation to separate this signal from the DC light bias and background light, significantly improving the signal to noise ratio.

978-1-5090-5606-4/17 $31.00 © 2017 IEEE

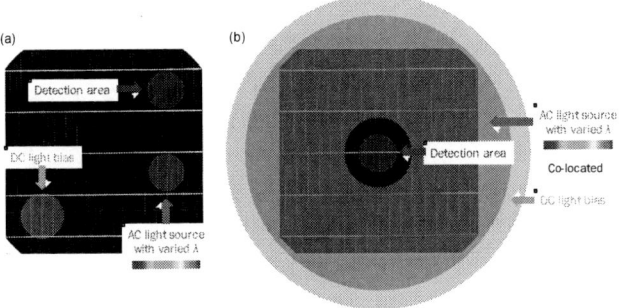

Figure 1. Illustration of the illumination sources and detection area for non-contact QE system used in this work, including: (a) the first generation design; and (b) the second generation design.

## II. Modeling Approach

Lateral balancing currents are a fundamental part of this measurement, supplying the current to the detection area, analogous to a non-contact EL measurement. The role of lateral balancing currents is investigated using circuit simulations with LTSpice. In this case, a simple equivalent circuit is used with three separate elements connected in parallel (Figure 2): (Region 1) the DC light bias is modeled as a one-diode cell with large photogenerated current density ($J_G$); (Region 2) the spectrally resolved AC light source is modeled as a one-diode cell with a smaller $J_G$, this is the region under test; and (Region 3) represented by a one-diode cell with no $J_G$ source: this region generates the detected EL signal.

Region 3 may be imagined as an external LED connected in parallel with the region under test. The region under test supplies current to this external LED, and the emitted intensity acts as a proxy for QE. The EL emission strength is plotted vs. excitation wavelength to provide a full-spectrum measurement.

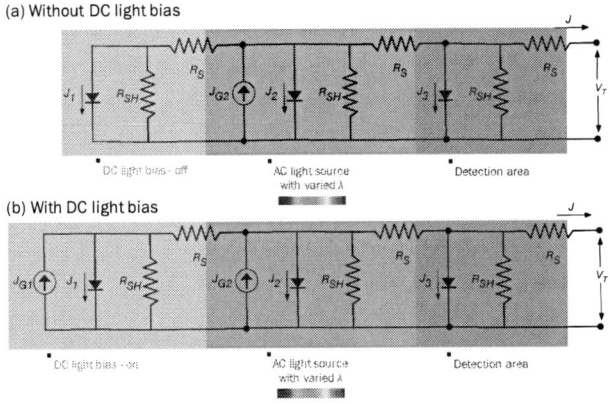

Figure 2. Equivalent circuit used in the LTSpice simulation for the configuration (a) without DC light bias and (b) with DC light bias.

## III. Experimental Details

Both the first generation and second generation prototype systems used in this work are shown in Figure 3. For the first generation system, shown in Figure 3(a), the light emission from a set of AC modulated LEDs ($\lambda_1$, $\lambda_2$, ...) is injected into an integrating sphere which contains a monitor photodiode to measure, in real time, the intensity of each wavelength. The exit port of the sphere is held in close proximity to the cell, illuminating Region 2, as defined earlier. A separate DC light bias is directed toward Region 1. This controls the overall forward bias of the cell, and therefore the gain of the emitting region. Region 3 is defined as the detection area, where the resultant EL signal is collected by a simple lens and photodetector.

Note that this 'ELE' arrangement solves several problems that have historically prevented the use of in-line PLE in silicon devices. (1) Incident photons are not present in the detection area, and so the technique is inherently full-spectrum (i.e., there is no need to reject incident photons from the detection area). (2) Incident photons do create 'voltage independent carriers' in Region 2, but these do not affect the emission from Region 3. This approach ensures that only voltage dependent carriers modulate the emission signal since the detection are is operating in an EL mode. This makes the technique

(a) First generation design

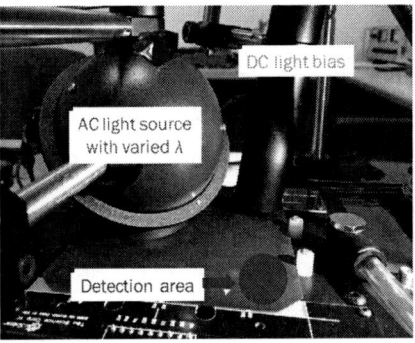

(b) Second generation design

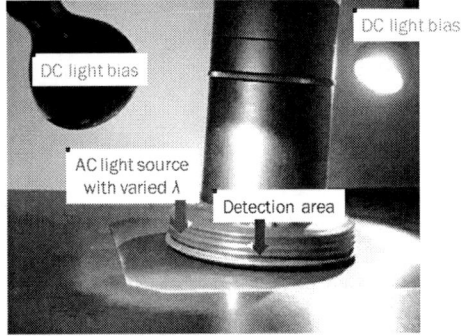

significantly more useful in indirect materials.

Figure 3. Experimental setup used to perform the ELE spectroscopy technique, including: (a) the first generation prototype; and (b) the second generation prototype.

The second generation design simplifies the measurement setup by featuring a single column that comes down near the cell, wherein the LEDs illuminate a ring around the detection area, seen in both Figure 1(b) and Figure 3(b). Additionally, inexpensive broadband DC light sources illuminate a larger area at the periphery of the column.

## IV. RESULTS AND DISCUSSION

The circuit simulations reveal some interesting features of this technique. The current-voltage ($I$-$V$) curves of each circuit element and at the terminals are given for the case with no DC light bias in Figure 4(a) and with a DC light bias in Figure 4(b). Because this is a non-contact technique, the point of interest here is the open-circuit condition where the terminal voltage ($V_T$) is zero. For the case with no DC light bias, the spectrally resolved AC light is the only excitation source and supplies current to the detection area. Here, the AC light acts as a source, operating somewhere between $V_{MP}$ and $V_{OC}$, while the detection area acts as a sink.

With a DC light bias (1 sun) larger than the spectrally resolved AC light (0.1 suns), the DC light bias supplies current to both the region under test and to the detection area. This has benefit of increasing the emission and gain of the detection area by increasing the operating voltage locally. Although the region illuminated by spectrally resolved AC light is now a sink, the variations in the local $I$-$V$ curve still act to increase and decrease the operating point of the detection area. By locking in to the AC frequency, the current contribution from this region can be separated from the DC light bias providing a proxy for QE.

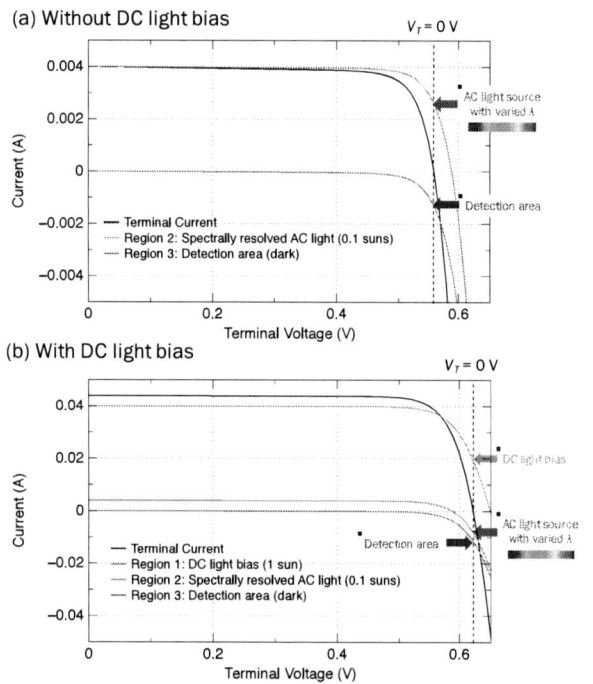

Figure 4. $I$-$V$ curves of each circuit element and at the terminals for the case: (a) with no DC light bias; and (b) with a DC light bias.

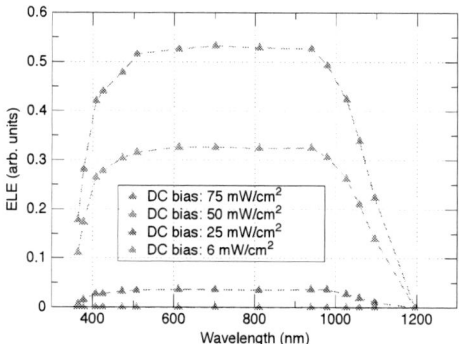

Figure 5. ELE measurements performed at different DC light bias levels with the first generation prototype system.

ELE measurements were performed on an industrial-scale monocrystalline silicon PV cell at different DC light bias levels using the first generation prototype system. The results are shown in Figure 5 and demonstrate the full-spectrum capability of the measurement system as well as the role of the DC light bias in increasing the signal strength.

EQE measurements with electrical contacts (using the FlashQE system described in [2]) and non-contact ELE measurements (second generation prototype) were both performed on two industrial-scale crystalline silicon PV cells, a multicrystalline silicon aluminum back surface field (Al-BSF) cell and a monocrystalline silicon passivated emitter and rear cell (PERC). The results are shown in Figure 6 and a comparison between the two shows rather good agreement. The team is still working to optimize the calibration procedure used to measure and adjust the flux from the individual LEDs.

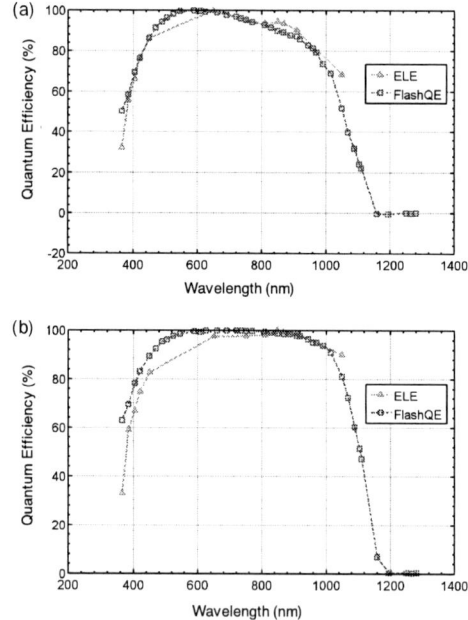

Figure 6. EQE (FlashQE) and ELE measurements (second generation prototype system) obtained on: (a) a multicrystalline silicon Al-BSF cell; and (b) a monocrystalline silicon PERC cell.

978-1-5090-5606-4/17 $31.00 © 2017 IEEE          3450

## V. Conclusion

Here, electroluminescence excitation (ELE) spectroscopy is introduced as a non-contact proxy for determining the quantum efficiency of a cell. Circuit simulation results demonstrate the role of the DC light bias in driving the lateral balancing currents that ultimately enable this technique to work. Experimental results comparing the external quantum efficiency measured with electrical contacts to the non-contact ELE measurements demonstrate good agreement for two different industrial-scale crystalline silicon photovoltaic cells. These results extend into the near infrared region of the spectrum and do not show any sign of voltage independent (i.e., diffusion-limited) carriers. Because the detection area operates in an EL mode, only voltage dependent carriers contribute to the emission signal detected.

## Acknowledgements

This work is supported under DOE Grant No. DE-SC0008281.

## References

[1] D.L. Young, B. Egaas, S. Pinegar, P. Stradins, A new real-time quantum efficiency measurement system, 33rd IEEE Photovoltaic Specialists Conference, San Diego, CA, 2008.

[2] E.J. Schneller, K. Öğütman, S. Guo, W.V. Schoenfeld, K.O. Davis, IEEE Journal of Photovoltaics 7 (2017) 957-965.

[3] D. Berdebes, J. Bhosale, K.H. Montgomery, X. Wang, A.K. Ramdas, J.M. Woodall, M.S. Lundstrom, IEEE Journal of Photovoltaics 3 (2013) 1342-1347.

[4] M.K. Juhl, T. Trupke, Journal of Applied Physics 120 (2016) 165702.

[5] S. Johnston, Contactless electroluminescence imaging for cell and module characterization, 42nd IEEE Photovoltaic Specialist Conference, New Orleans, LA, 2015, pp. 1-6.

[6] R. Sinton, Contactless Electroluminescence For Shunt-Value Measurement in Solar Cells, 23rd Europeant Photovoltaic Solar Energy Conference, Valencia, Spain, 2008, pp. 1157-1159.

[7] Y. Zhu, M.K. Juhl, T. Trupke, Z. Hameiri, IEEE Journal of Photovoltaics 7 (2017) 1087-1091.

# Illuminated Outdoor Luminescence Imaging of Photovoltaic Modules

Timothy J Silverman*, Michael G. Deceglie*, Kaitlyn VanSant†, Steve Johnston* and Ingrid Repins*

*National Center for Photovoltaics, National Renewable Energy Laboratory, Golden, Colorado

†Colorado School of Mines, Golden, Colorado

*Abstract*—**Evaluating PV module defects with luminescence imaging is normally done using electroluminescence indoors, requiring labor-intensive relocation of fielded PV modules. We present three techniques for luminescence imaging that can be used outdoors during daylight: illuminated outdoor electroluminescence, open-circuit outdoor photoluminescence, and constant-current outdoor photoluminescence. We explain the operating principle for each technique and show example output from a c-Si and a CIGS module. Each method reveals some combination of the same defects visible in conventional indoor electroluminescence images. In c-Si modules we detect cell mismatch, spatial variation in lifetime, cell cracks, and areas isolated by cell cracks. In CIGS we detect shunts and device nonuniformity. The photoluminescence techniques we present can be implemented with portable, battery-powered equipment.**

## I. Introduction

Luminescence imaging is a key tool for detecting defects and nonuniformity in PV modules. The simplest method is electroluminescence (EL) imaging, wherein the module is made to behave as an infrared LED, sinking power and emitting light, while a camera collects an image [1]. Luminescence is relatively weak in terrestrial non-concentrating PV products, so EL imaging is almost always done in darkness. It is in routine use in manufacturing and research environments, but is rarely used on fielded modules because of the very labor-intensive need to disassemble the system and transport modules to an imaging studio.

Improvements in camera and computer technology have recently enabled the collection of EL images outdoors under solar illumination and outdoor EL imaging is now available as a commercial tool [2], [3]. In this work, we explain the operating principle of outdoor illuminated EL imaging.

EL requires a DC power supply (PSU) with power output roughly equal to the nameplate power of the PV devices under test (DUT). In this work we show the application and principle of an alternative technique, outdoor photoluminescence (PL) imaging, which does not require such a power supply [4].

PV cells luminesce at wavelengths near their band gap wavelength. Because the sun emits strongly in these same wavelengths, wavelength filtering alone cannot be used to separate the relatively weak luminescence signal from the background signal, which may be hundreds or thousands of times stronger. The dynamic range offered by most digital cameras, 14 bits to 16 bits, is normally not adequate to separate the signal from the background by subtracting a single pair of images. Even a pair of high-dynamic-range images would not

suffice if they were taken using long exposure times during daylight because illumination outdoors is always time-varying. Instead, the luminescence signal must be modulated so it can be separated from the background signal using the repeated subtraction of pairs of images taken in close succession. This is a form of lock-in correlation.

## II. Method

Each of the techniques we used followed the same steps:

1) Put the DUT into state 1, wherein the luminescence intensity is low
2) Collect an image, called the "odd" image
3) Put the DUT into state 2, wherein the luminescence intensity is high
4) Collect an image, called the "even" image
5) Subtract the odd image from the even image, creating the "difference" image
6) Add the difference image to a running total of difference images, called the "output" image
7) Repeat steps 1 through 6 until the output image has acceptable quality or a target number of images is reached

We used an InGaAs area camera ($640 \times 512$ resolution, 14 bits dynamic range) to collect images. The InGaAs detector has very high quantum efficiency (QE) in the wavelength range of interest. The techniques we used work the same with a Si detector, but with much longer acquisition times due to the much lower QE of Si in the near infrared. We used a 1000-nm longpass filter to remove much of the background light outside the wavelengths of interest.

We used a modulator to change the module's state and to trigger the camera, enabling image acquisition at high frequency ($\sim 30$ Hz in this work). The modulator used a MOSFET, which was driven between a low-resistance and a high-resistance state, as a switch. We realized different combinations of state 1 and state 2 by connecting a PSU in parallel with the MOSFET (Figure 1), connecting the MOSFET directly to the DUT (Figure 2), or connecting a programmable DC load (DCL) in parallel with the MOSFET (Figure 3). The MOSFET contained an intrinsic body diode, which is shown in the diagrams.

On a c-Si module with cracked cells due to mechanical loading and a CIGS module with extensive shunting due to partial shade testing, we applied four imaging techniques:

indoor EL imaging, illuminated (outdoor) EL imaging, open-circuit outdoor PL (OCPL) imaging, and constant-current outdoor PL (CCPL) imaging.

## A. Indoor EL imaging

We configured the apparatus as shown in Figure 1, with a PSU in parallel with the modulator and DUT, in a dark room. We set the PSU to source the DUT's nameplate $I_{sc}$. State 1 was the modulator short-circuiting both the PSU and the DUT. State 2 was the PSU applying forward bias to the module. We collected a single pair of images (cumulative exposure time 64 ms).

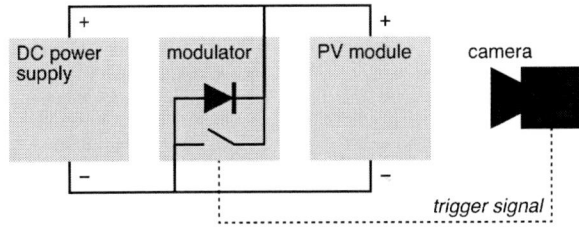

Fig. 1. Equipment configuration for both indoor and outdoor EL imaging.

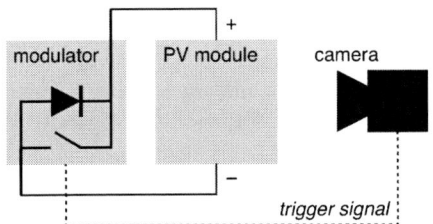

Fig. 2. Equipment configuration for open-circuit outdoor photoluminescence (OCPL) imaging.

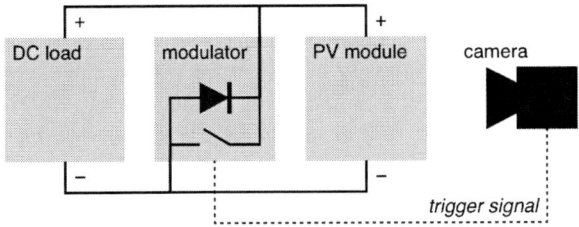

Fig. 3. Equipment configuration for constant-current outdoor photoluminescence (CCPL) imaging.

## B. Outdoor imaging

To compare the three outdoor imaging techniques, we collected 1000 pairs of images with each technique. This required a cumulative exposure time of 2.1 s for the c-Si module and 6.8 s for the CIGS module. Total acquisition time, set by the 30 Hz framerate of the apparatus, was about 67 s for each test.

*1) Illuminated EL imaging:* The configuration was the same as for indoor EL imaging, but deployed outdoors.

*2) Open-circuit outdoor PL imaging:* We connected the modulator directly to the PV module, without a PSU or DCL, as shown in Figure 2. In this configuration, state 1 was the modulator short-circuiting the DUT. State 2 was the module at open circuit. This configuration was entirely battery-powered.

*3) Constant-current outdoor PL imaging:* We connected a DCL in parallel with the modulator and DUT as shown in Figure 3. The DCL was configured to pass a constant current of $0.7I_{sc}$. In this configuration, state 1 was the modulator short-circuiting both the DUT and the DCL. State 2 was the module sourcing $0.7I_{sc}$ into the load. Although our DCL was mains-powered, a battery-operated load would be capable of the same function.

## III. RESULTS AND DISCUSSION

The following discussion uses the indoor EL images in Figures 4 and 6 as the baseline for comparison.

### A. c-Si module

Figure 4 shows output images of the c-Si module. The indoor EL image shows that most cells have at least one crack. Several cells have cracks that are partially or completely isolating portions of those cells.

The illuminated EL image shows nearly all of the defects present in the indoor image. Substantially fewer than 1000 image pairs are required to reach quality indistinguishable from that shown here. The minor differences between the indoor and outdoor images may be attributable to the module temperature difference between the indoor and outdoor imaging sessions, which affects the appearance of cracked cells. Note that the appearance of the cracked cells at operating temperature is of more relevance than at room temperature.

The OCPL image shows the characteristic luminescence pattern of mismatched cells and, within some cells, nonuniformity due to spatial variation of lifetime. Cracks appear as thin dark lines, which are more clear when imaging a smaller area. However, the partially isolated cell areas introduced by cracks, connected to the rest of the module through a high series resistance, do not appear in the OCPL image. Instead, badly cracked cells appear uniformly dim. Figure 4 shows that CCPL causes these isolated areas to appear brighter than fully intact cells and intact portions of the cracked cell. Because the difference between the state 1 and state 2 signals is small, the image is noisier than with the other techniques.

Figure 5 explains the appearance of cracked cells with the different PL techniques. We simulated the I-V behavior of cells within a 60-cell module with a single damaged cell. One quarter of the damaged cell was isolated by a crack introducing a series resistance of 1 Ω [5]. The forward and reverse j-v characteristic was taken from measurements on a commercial polycrystalline cell.

In OCPL, intact full cells, intact regions of broken cells, and isolated regions all luminesce weakly in state 1, then move to open circuit (high luminescence) in state 2. The isolated region luminesces weakly in state 1 because the intact region is pushed into reverse bias to pass the $I_{sc}$ of the entire module.

In CCPL, state 1 remains the same as on OCPL but state 2 is changed to $0.7I_{sc}$. This change is enough to reduce the luminescence of the intact regions, but leave the state 2 luminescence of the isolated region high. This provides contrast, with the isolated regions appearing brighter.

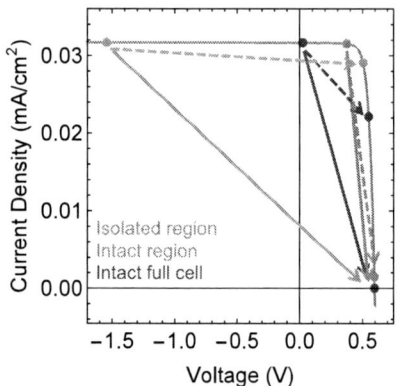

Fig. 5. Illustration of the principles of OCPL and CCPL on a module containing a cracked cell. The j-v curve is for the semiconductor itself, which governs luminescence behavior. The effects of the external series resistance from a crack are not shown. The colors indicate the three types of cell region in the module. The arrows all point from state 1 to state 2. Because luminescence is high when extracted current density is low, the detected luminescence is proportional to the difference in current density between state 1 and 2. The solid arrows are for OCPL, in which luminescence is detected for all regions. The dashed arrows are for CCPL, in which the isolated region remains bright because the series resistance of the crack keeps that region near $V_{oc}$ in state 2.

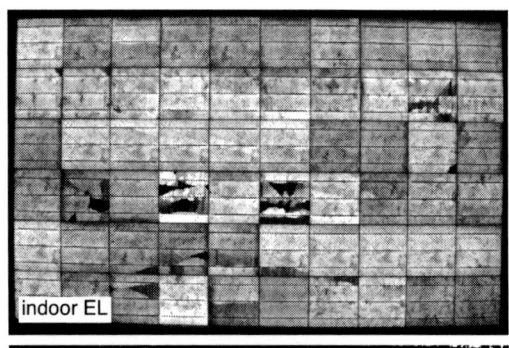

Fig. 4. Results from the four imaging techniques on the c-Si module.

### B. CIGS module

Figure 6 shows results from the CIGS module. The indoor EL image shows defective, shunted monolithic integration scribes (long dark areas surrounded by long light areas) and shunts formed during partial shade testing (short dark areas surrounded by short light areas) [6]. Minor large-scale

nonuniformity in the device is visible across the top of the image, in an area that was not exposed to partial shade testing.

The illuminated EL image and the OCPL image show an identical pattern of shunt defects. As with the c-Si module, fewer than 1000 image pairs are needed to reach very nearly the same image quality as shown here. The CCPL image (not shown) does not identify any defects absent in the other images and is of lower quality.

## IV. CONCLUSION

We have demonstrated three imaging techniques for use outdoors in illuminated conditions. These methods are useful for characterizing fielded modules without the need to operate at nighttime or to disassemble and transport modules. Illuminated outdoor EL produces images of similar quality and showing identical defects compared to indoor EL images. OCPL, a technique that can easily be performed with portable, battery-operated equipment, produces high-quality images of cracks, shunts and other types of nonuniformity. For monolithic thin-film modules with shunt defects this technique produces images nearly identical to indoor EL images. However, OCPL does not show contrast for areas of high series resistance such as isolated portions of cracked cells. With the addition of a DC load, CCPL addresses this shortcoming by producing an image where areas of high series resistance appear brighter than surrounding areas.

## ACKNOWLEDGMENTS

The authors gratefully acknowledge Rajeev Dubey for his assistance. This work was supported by the U.S. Department of Energy under Contract No. DE-AC36-08GO28308 with the National Renewable Energy Laboratory.

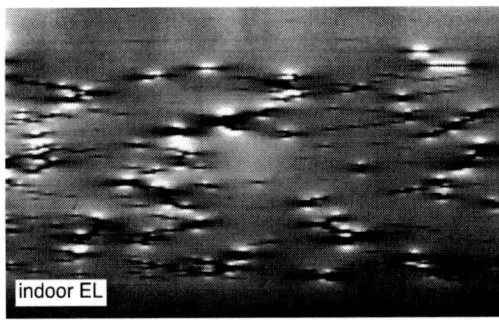

Fig. 6. Cropped results from three of the imaging techniques on the CIGS module.

## REFERENCES

[1] Y. Takahashi, Y. Kaji, A. Ogane, Y. Uraoka, and T. Fuyuki, "Luminoscopy: Novel tool for the diagnosis of crystalline silicon solar cells and modules utilizing electroluminescence," in *2006 IEEE 4th World Conference on Photovoltaic Energy Conference*, vol. 1, May 2006, pp. 924–927.

[2] L. Stoicescu, M. Reuter, and J. Werner, "Daysy: luminescence imaging of PV modules in daylight," in *Proceedings of the 29th European Photovoltaic Solar Energy Conference and Exhibition*, 2014, pp. 2553–2554.

[3] S. Koch, T. Weber, C. Sobottka, A. Fladung, P. Clemens, and J. Berghold, "Outdoor electroluminescence imaging of crystalline photovoltaic modules: Comparative study between manual ground-level inspections and drone-based aerial surveys," in *Proceedings of the 32nd European Photovoltaic Solar Energy Conference and Exhibition*, 2016.

[4] S. Johnston and T. Silverman, "Photoluminescence and electroluminescence outdoor module imaging," in *PV Module Reliability Workshop. Golden (CO), USA*, 2015.

[5] A. Morlier, F. Haase, and M. Köntges, "Impact of cracks in multicrystalline silicon solar cells on pv module power: A simulation study based on field data," *IEEE Journal of Photovoltaics*, vol. 5, no. 6, pp. 1735–1741, Nov 2015.

[6] T. J. Silverman, L. Mansfield, I. Repins, and S. Kurtz, "Damage in monolithic thin-film photovoltaic modules due to partial shade," *IEEE Journal of Photovoltaics*, vol. 6, no. 5, pp. 1333–1338, 2016.

# Electroluminescent Image Processing and Cell Degradation Type Classification via Computer Vision and Statistical Learning Methodologies

Justin S. Fada *, Mohammad A. Hossain *, Jennifer L. Braid *, Shuying Yang †, Timothy J Peshek *, Roger H. French *

\* Solar Durability and Lifetime Extension (SDLE) Research Center,
Case Western Reserve University, 10900 Euclid Ave., Cleveland, Ohio 44106, USA
† SunEdison Inc., 600 Clipper Drive, Belmont, California 94002, USA

*Abstract*—A data set of 90 60-cell module images from 5 commercial PV module brands over 6 exposure steps of damp-heat testing were analyzed. An automated data analysis pipeline was developed using the open source coding language Python to parse the module images into individual cell images. As the original raw images are not directly suitable for modeling, this algorithm implements techniques which include filtering, thresholding, convex Hull, regression fitting, and perspective transformation to pre-process the original image. After cell extraction, 5400 individual cell images as a function of brand and exposure time were obtained. From the data set, 3 initial degradation categories were observed: good, cracked, and heavily busbar-corroded. For supervised machine learning classification, these images were manually sorted into these 3 categories yielding 3550 images. To increase the data set size, the cell images were augmented by flipping the images about the x-axis and y-axis as well as rotated 180 degrees. This increased the total sample size to 14,200 images with good, cracked, and heavily corroded counts of 12,004, 492, and 1704, respectively. A training and testing framework was generated using stratified sampling with a training to testing data ratio of 80:20. The statistical learning algorithms Support Vector Machine (SVM), Random Forest (RF), and Artificial Neural Network (ANN), were independently trained on the training set and then given the remaining data images to predict their classification. The results showed model prediction accuracies of 98.77%, 96.60%, and 98.13% for the SVM, RF, and ANN models, respectively.

*Index Terms*—electroluminescent imaging, image processing, statistical learning, prediction

## I. INTRODUCTION

In the IEA PVPS report Review of Failures of Photovoltaic Modules, it is stated that automated feature detection from electroluminescent (EL) images has yet to be successfully established [1]. Herein we discuss a method for using statistical and machine learning algorithms to classify cell images based on degradation features reported in the literature.

Techniques common for PV degradation analysis include visual inspection, current voltage ($I - V$) tracing, and imaging in the form of luminescence or thermography [1]. $I - V$ measurements are useful for macro-analysis of photovoltaic (PV) devices as it provides a complete sweep of the electrical properties of a test device from short circuit current ($I_{sc}$) to open circuit voltage ($V_{oc}$) to provide a 1-dimensional vector of current and paired voltage. However, the resolution is coarse from the measurement's single input/output connection and

classification of degradation type can only be narrowed to a few possible degradation modes [1].

Of the remaining degradation analysis types, visual inspection and imaging are done qualitatively for classification of degradation types. The output data from an electroluminescent image typically provide millions of spatially resolvable data points registered by the camera sensor from photons emitted from a PV device powered in forward-bias. As a result of increased data density over other common measurement methods such as $I - V$, EL imaging has been increasingly utilized as a tool for analysis of photovoltaic (PV) devices [1], [2], [3]. Some researchers have successfully used EL imaging derived local cell electrical properties (sometimes with aid from other imaging devices)[4] [5], with image features such as cracks, dark spots, and busbar corrosion present for further study.

Classification of cell features has not been successfully automated and is currently done by trained researchers [1]. With qualitative assessment there is inherently a human error associated with classification particularly regarding matters such as whether a feature is prominent enough to be recorded, or the severity of a feature. These systematic biases are typically reasonable when used as internal lab metrics, however the differences become important when comparing literature between groups.

Additionally, collection of EL data has become simpler and less expensive with many modified consumer-off-the-shelf (COTS) cameras present in the literature along side commercially available equipment [2] [6] [7]. These COTS cameras have demonstrated high signal-to-noise ratios and high pixel densities, however, typically require longer exposure times on the order of 10 seconds. Therefore with the ability to generate larger volumes of EL images that need classified, the task of manual classification is becoming quite laborious and time consuming [8].

We aim to solve these issues by validating the use of computer vision and statistical learning methodologies [9] in a distributed computing environment [10] for the reliable quantitative classification of cell defect types. In this work we demonstrate preliminary use of this technique to classify two defect types, cracks and busbar corrosion, along with a third category of "good/clean" cells that have no signs of degradation. By confirming the use of this technique herein, expansion of this model to many other defect types will follow. In short order we intend to release this work to the community as an open source project for all to benefit from.

The data used in this work includes 15 60-cell commercial modules spanning 5 brands exposed under accelerated indoor damp-heat testing. Each module was exposed for 6 steps from 500 to 3000 hours in 500 hour intervals. This results in a total of 90 module images containing 5400 individual cells. Of those, 3550 cells exhibit cracking, busbar corrosion or are in an non-degraded state, which are the 3 cell types studied in this work. Using an augmentation process, each of these 3550 cells will be transformed into 3 additional images resulting in the final 14,200 cell dataset used for model training and testing.

## II. IMAGE PROCESSING

When raw data are collected, the module typically has some misalignment relative to the boundary of the image (seen in Figure 1) as a result of module loading variability during the measurement session or due to manufacturing error (laboratory minimodules or commercial modules).

An automated image processing pipeline was developed to systematically align the active module area in the raw data image and allow for a high volume through put of data. For its computational efficiency, ease of implementation, and rich image processing/analysis toolboxes, the open-source coding language Python was used with the additional packages of NumPy, SciPy, scikit-image, and OpenCV [11], [12], [13], [14], [15].

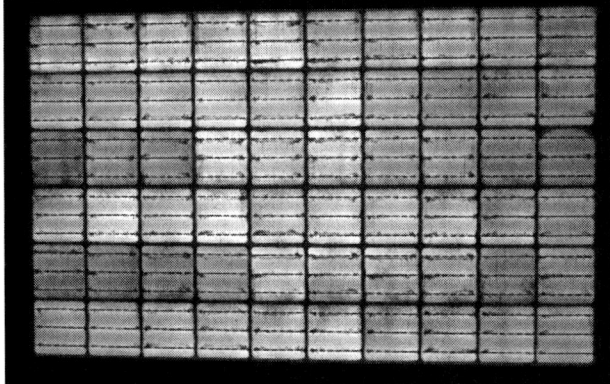

Original (Gray-scale)

Fig. 1. A gray scale EL image of the original module.

An image is loaded into memory as a gray-scale JPEG image. This reduces the future file size when saving but also reduces computation from the original RGB channel image. For some image processing problems all 3 channels are useful, for our project, however, there is not any noticeable loss in accuracy by using gray-scale. The image is filtered using a small odd number pixel kernel to perform a median filtering. This reduces any noise spikes across the pixel matrix and retains feature edges in the image as a result of the odd number pixel kernel. Figure 2 demonstrates the filtering process visually using a simulated vector of noisy data with an edge step (seen as blue points) and filtered result (seen as red points). This edge step represents the boundary of an image feature, such as an increase in values from the background to the active cell regions. The most important feature of this simple filter is that the edge is retained, which would not be the case is a Gaussian filter were used, for instance, which "blurs" the image when smoothing.

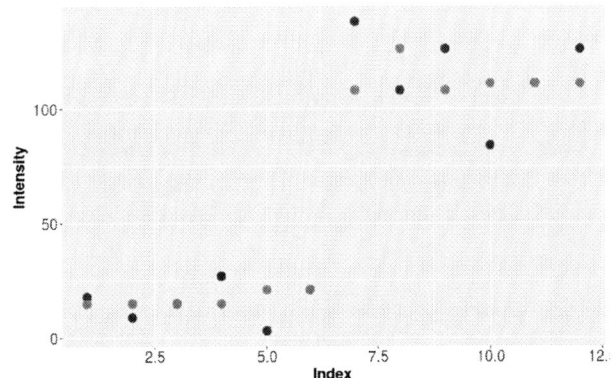

Fig. 2. The blue data points represent a vector of simulated noisy data with an edge step, such as when the values increase at the edge of a cell in an EL module image from the background to active cell area. The red points are the post-median filter data. A kernel size of 3 was used for this simulation.

With interest in identifying the active cell regions of the image only, the background values are removed. Using a histogram (seen 2-dimensionally represented using a spectral mapping of binned pixel values in Figure 3(a)) a global threshold of the image is used to reassign all non-cell pixels (purple and blue in Figure 3(a)) to 0. The result of the background threshold is seen in Figure 3 (b.) With only cell areas as non-zero, a convex Hull algorithm generates the smallest possible convex set containing all the non-zero data-points ("shrink-wrapping" the non-zero data points). For the edge finding to follow, the pixels pertaining to the convex hull identified area are set to 1, creating a binary matrix. The result of this operation is seen as the white area of 4(a).

Next, it is important to find the edges of the module which in the original image (Figure 1) are at some small angle rotation from the sides of the rectangular image. To find the edges in a manner that is reliable and tolerant to variability due to moderate degradation of the outer cells, the edge is found by sampling points that lie on the edge boundary and then fitting of those points using linear regression.

The center 70% of the image is used for sampling because the variability in misalignment is typically not extreme, but a few degree deviation. From this area, 10 x-axis parallel slices and 10 y-axis parallel slices are evenly selected. From these vector slices, the points where there is a increase in value from 0 to 1 and decrease in value from 1 to 0 are identified. These points indicate the edge of the active module region.

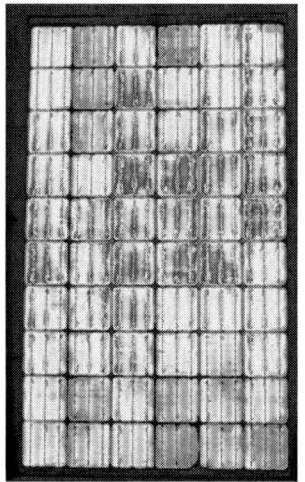 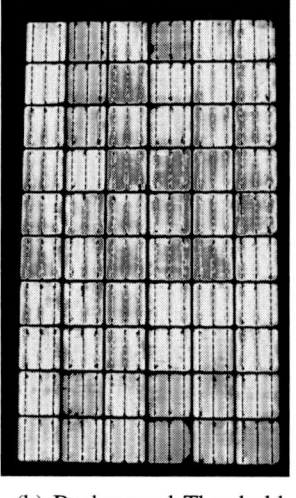 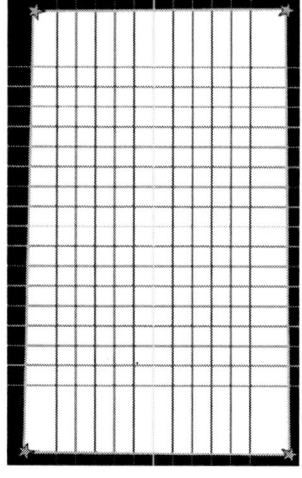 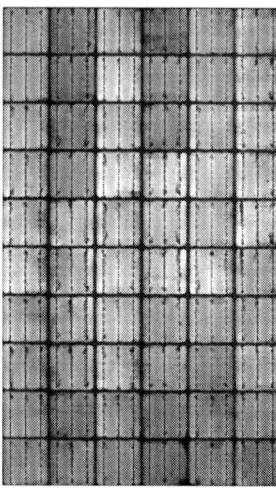

(a) Median Filter
(Spectral Color Map)

(b) Background Threshold
(Spectral Color Map)

(a) Convex Hull/Corner Finding

(b) Perspective Transform

Fig. 3. (a) Median filter applied to the original image displayed using a spectral color mapping of pixel intensity values to binned color ranges. From this image it can be seen that the purple and blue colors (pixel intensity ranges) represent the background of the image. (b) Setting those background pixels in Figure 3 (a) to black (value of 0) results in only the active cell region as a non-zero value, which is useful for further processing.

Fig. 4. (a) Visualization of the steps used to identify the four corners of the active cell region. The white area is the identified active module area from the convex hull algorithm. The green lines are reference lines for the mid-points in the x and y dimensions. Blue lines indicate horizontal vector slices (relative to the original image in Figure 1). Red lines indicate vertical vector slices. Orange lines represent linear regression fits of the active area edges. Magenta stars represent the four corners of the active cell regions where the linear regression fits intersect. (b) A perspective transformed image using the 4 corners identified is Figure 4(a)

A linear regression fit of these points on all 4 sides of the module yields 4 line equations, which can be solved for their intersection points to obtain the 4 corners of the module. With these 4 points the surface plane of the module is defined. Using domain knowledge that PV modules are rectangular in shape, a perspective transformation for that geometry is performed on the data points of the median filtered image (once the coordinates for the corners are found, any inter-mediate image can be transformed as all matrices throughout these calculations are the same dimensions). This maps the active regions (contained within the corner coordinates) to a uniformly dimensioned canvas of 1500x2500 pixels. which is approximately the original image dimensions. Uniform alignment of the raw data is now complete using a pipeline that is useful for large and small data streams alike.

For the subsequent cell classification to follow, an addi-tional step is needed to extract the cells to their own files. Since the module images are now in uniform alignment one can simply slice the image into 60 equal areas using the number of column-wise and row-wise cells. For the 6x10 cell module used herein, 5 x-axis and 9 y-axis divisions successfully obtain the cell areas desired. There can be very minor variations between modules due to manufacturing which can impart variable spacing between cells in the encapsulated module, which appears around the "border" of the extracted cell image. However, it is appropriate to cease further manipulations and refinement as the entire active cell area is contained within each extracted image file, which is our requirement for input into the following learning

algorithms which have a high tolerance for random variations across the dataset. Additionally, for a highly degraded cell that contains dead areas along the edge of the cell, an algorithm such as the convex hull used above would only capture the active cell region, losing valuable information regarding the cell's degradation.

### III. MODELING AND PREDICTION

For this study, we developed a supervised model for image classification of PV cells by degradation modes following the below described framework.

#### A. Image Augmentation

Using the pipeline outline above, 90 images of 60-cell modules were processed to yield 5400 individual cell images. For supervised learning, each of these cells were classi-fied manually by a single researcher into the categories of good/clean, cracked, busbar corroded, and other, of which the first 3 groups were used for modeling (Figure 5). It should be noted that since a researcher is performing the initial classification of 5400 cell images, there will likely be some misclassifications in the training set used for model-ing. However, statistical models like those outlined below are tolerant to a small amount of error from supervised classification [16], [17]. Also, after initial classification and model training, use of this method will result in systematic quantitative classification contained in a scalable pipeline.

After sorting, these 3 categories resulted in 3550 images in total. To increase the number of images, all cell images

978-1-5090-5606-4/17 $31.00 © 2017 IEEE

were augmented to produce 4 times the data and achieve a total of 14,200 cell images. The augmentation process simply takes each image and flips it about its center x-axis, center y-axis, and rotates it 180 degrees, as seen in Figure 6. These augmented images are all possible configurations in real-world cells and are perceived as unique additional data for modeling.

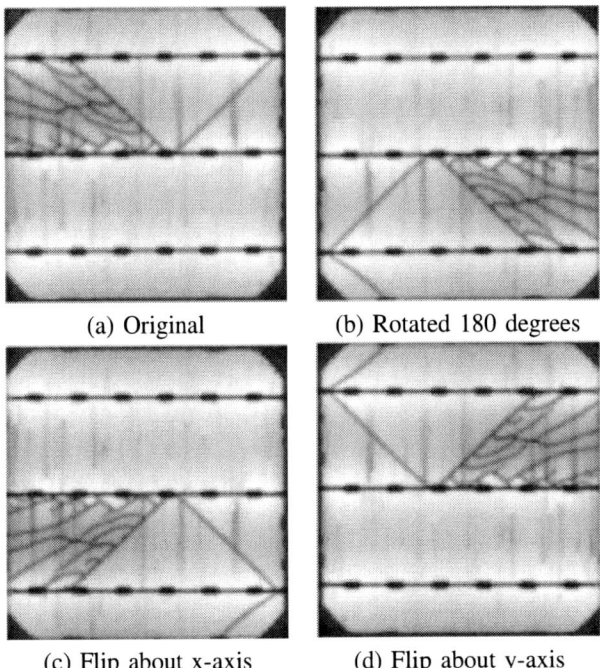

(a) Original         (b) Rotated 180 degrees

(c) Flip about x-axis        (d) Flip about y-axis

Fig. 6. Cell image augmentation. a). Original, b). Rotated 180 degrees, c) Flip about x-axis, d). Flip about y-axis.

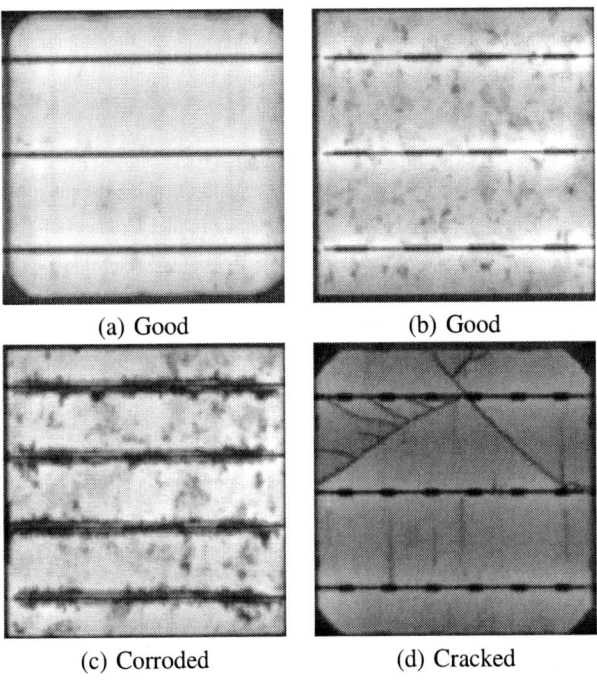

(a) Good              (b) Good

(c) Corroded          (d) Cracked

Fig. 5. These 4 cells are examples of the 3 cell types used for this classification model: (a-b) good, (c) busbar corrosion, and (d) cracked.

### B. Image Pre-processing

In image pre-processing, the cell images were resized from 250x250 to 50x50 pixels to reduce computation time and normalized using min-max scaling. Normalization standardizes the data set so that the classification algorithm becomes less sensitive to small intensity changes and more likely to detect features correctly.

Since the dataset studied contains EL images of new modules through 3000 hours of accelerated damp-heat exposure, corrosion and cracking only become apparent in the last 2 rounds of exposure. This limits the portion of the dataset representing those features. When corrosion appears in an EL image, it is common that a high percentage of the cells per module (if not all) exhibit corrosion. However, while cracking does appear more often in longer exposed modules, the number of cells experiencing cracking is only a small to moderate percentage of all cells (dependent on brand). Therefore of the 14,200 post-processed cell images only 3.5% cells are broken compared to 12.0% for corroded and 84.5% good/clean. As a result of unbalanced categories and to ensure that all data is represented in the training and testing sets, stratified sampling was used to generate training and testing

data sets [18]. This commonly used method samples from each of the 3 categories of data such that the percentage of samples in each group is maintained [18]. The ratio of training to testing data was chosen to be 80:20.

### C. Classification Algorithm

Using supervised learning in a training and testing framework we explored three state-of-the-art algorithms to classify the cells: Support Vector Machine (SVM), Random Forest (RF), and Artificial Neural Network (ANN) [19]. SVM is a supervised machine learning method based on the statistical learning theory and structural risk minimization invented by Vladimir Vapnik to solve both classification and regression problems [20]. For classification, SVM constructs hyperplanes in a multi-dimensional feature space that can separate different types of class labels [16], [17]. For this study, SVM is developed using a Gaussian radial basis function (rbf) kernel, and optimized for Gaussian kernel gamma (0.05) and hyper-parameter C (25). Where C is a parametric cost for miss-classification on the training data.

The RF classifier is an ensemble machine learning technique which builds a number of decision tree classifiers on various random sub-samples of the training data set, and then selects its output from the mean prediction of the individual decision trees. RF utilizes the idea of bagging [21] with a random subset of features [22]. For this study, RF classifier is optimized for number of trees (400) and maximum features (50).

ANN is one of the most common machine learning techniques used for classification [23]. Particularly, ANN is very efficient for image classification and pattern recognition [24], [25]. ANN was inspired from the functionality of the human brain and nervous system [26] using interconnected neurons as processing elements and each interconnections associated with the weight of the networks. Multilayer Perceptron (MLP) is the most commonly used ANN method. The neurons in MLP-ANN can be classified into input, hidden layers and output layers where the input layer receives the input signal, passes it through the hidden layers, and then the output layer emits the network output.

The weights of the networks are learned by minimizing a specific error function through gradient descent-based methods. A side effect of this method is that two major limitations appear: over-fitting and local minimum problems. The over-fitting can be addressed with early-stopping criteria, weight decay regularization and a dropout layer [27]. The MLP-ANN classifier was optimized for number of hidden layers and size, dropout layer, optimizer, and activation function. In the final MLP-ANN classifier model, 2 hidden layers, 2 dropout layers, a stochastic gradient descent optimizer and a Rectified Linear Unit (RELU) activation function were used.

### D. Prediction and Results

To characterize the accuracy of each model, a testing set of data which was held out from model training was given to the model to predict the classification of each cell image. That prediction by the model was then compared to the classification given from the manual supervised classification step and an accuracy percentage was calculated [28]. Time was also included as a comparison metric to gauge each algorithms computational efficiency, although with less weight than model accuracy. The overall classification accuracy and training time of the three classification models are shown in Table I.

TABLE I
SUMMARY OF THE 3 CLASSIFIER ALGORITHMS TESTED

|  | SVM | RF | MLP-ANN |
|---|---|---|---|
| Accuracy | 98.77% | 96.90% | 98.13% |
| Training Time(sec) | 85.52 | 90.12 | 2250 |

Among the three models, SVM shows the best overall accuracy to classify cells at 98.77% while also taking the least amount of time running in 85.52 seconds. RF shows the lowest accuracy at 96.90%, however this is still a reasonably good accuracy at a comparable computation time of 90.12 seconds. The MLP-ANN classifier had a marginally lower classification accuracy than SVN at 98.13%, however the time was 26 times longer.

Table II shows the accuracy of each classifer to predict each degradation type. Of these, all classifiers predict good/clean and corroded cells with 99% accuracy or greater. However, in the case of cracked cells, only the SVM model performs moderately well with a 66.32% accuracy. This suggests that based on overall accuracy, individual class accuracy and training time, SVM outperforms all other classifiers.

Intuitively it is expected that the good/clean and corroded cells would classify with high accuracy. In both cases, the

TABLE II
ACCURACY OF THE 3 CLASSIFIER ALGORITHMS FOR DIFFERENT IMAGE CLASS

|  | Good | Cracked | Corroded |
|---|---|---|---|
| SVM | 99.92% | 66.32% | 100% |
| RF | 99.87% | 19.38% | 99.12% |
| MLP-ANN | 100% | 47.96% | 99.71% |

feature (or lack there of) is well defined. For good/clean cells, there is simply no feature compared to the other two classes. And for the corroded cells, all images are slight variations of the rest with defects localized to the busbar region where the only significant variation is attributed the width of the corroded area.

Conversely, the crack formations on the surface of a cell are non-uniform. Each crack formation is unique with a particular propagation pattern that is a function of the specific heterogeneous structure of the individual wafer as well as the unique external factors causing initiation and propagation. The cracks can also have variable geometries including thickness and length. The issue of uniqueness can be resolved with large amounts of data, however, with the cracked cells in training being only 80% of the 492 cracked cells there is not sufficient data to predict at higher accuracies using the current learning algorithms. With more data and advanced modeling, the classifiers could train more completely to yield higher classification accuracies.

## IV. CONCLUSION

In this work we demonstrated an automated image processing pipeline used to process module-level electroluminescent images into single cell images with augmentation employed to increase the dataset size. Those images were then categorized into 3 subsets: good, cracked, and heavily corroded busbar cells. Using a semi-supervised statistical learning approach, 3 modern classification algorithms (Support Vector Machines, Random Forest, and Multilayer Perceptron based Artificial Neural Networks) were compared. The results showed that SVM's perform the best at classifying cell types based on the metrics of accuracy (98.77%), speed (85.52 seconds), and most uniform classification accuracy across all degradation types.

### ACKNOWLEDGMENT

The authors are grateful for funding for this work from the Department of Energy (Award No. DE-EE-0007140). We are also grateful to SunEdison Inc. for providing the data set used herein, the SDLE Research Center, and the RedCat High Performance Resource in the Core Facility for Advanced Computing at Case Western Reserve University.

### REFERENCES

[1] M. Kontges, S. Kurtz, C. Packard, U. Jahn, K. Berger, K. Kato, T. Friesen, H. Liu, and M. Van Isehegam, "IEA-PVPS {Task 13}: Review of Failures of PV Modules," Tech. Rep., May 2014. [Online]. Available: http://iea-pvps.org/index.php?id=275

[2] Justin S. Fada, Nicholas R. Wheeler, Davis Zabiyaka, Nikhil Goel, Timothy J. Peshek, and Roger H. French, "Democratizing an electroluminescence imaging apparatus and analytics project for widespread data acquisition in photovoltaic materials," *Review of Scientific Instruments*, vol. 87, no. 8, p. 085109, Aug. 2016. [Online]. Available: http://scitation.aip.org/content/aip/journal/rsi/87/8/10.1063/1.4960180

[3] T. Trupke, J. Nyhus, and J. Haunschild, "Luminescence imaging for inline characterisation in silicon photovoltaics," *physica status solidi (RRL)-Rapid Research Letters*, vol. 5, no. 4, pp. 131–137, 2011. [Online]. Available: http://onlinelibrary.wiley.com/doi/10.1002/pssr.201084028/full

[4] K. Ramspeck, K. Bothe, D. Hinken, B. Fischer, J. Schmidt, and R. Brendel, "Recombination current and series resistance imaging of solar cells by combined luminescence and lock-in thermography," *Applied Physics Letters*, vol. 90, no. 15, p. 153502, 2007. [Online]. Available: http://scitation.aip.org/content/aip/journal/apl/90/15/10.1063/1.2721138

[5] D. Hinken, K. Ramspeck, K. Bothe, B. Fischer, and R. Brendel, "Series resistance imaging of solar cells by voltage dependent electroluminescence," *Applied Physics Letters*, vol. 91, no. 18, p. 182104, 2007. [Online]. Available: http://scitation.aip.org/content/aip/journal/apl/91/18/10.1063/1.2804562

[6] T. Fuyuki, H. Kondo, Y. Kaji, T. Yamazaki, Y. Takahashi, and Y. Uraoka, "Electroluminescence," *Proceedings of the 31st Photovoltaic Specialists Conference, Lake Buena Vista, FL (IEEE, New York, 2005)*, pp. 1343–1345, 2005.

[7] T. Potthoff, K. Bothe, U. Eitner, D. Hinken, and M. Kntges, "Detection of the voltage distribution in photovoltaic modules by electroluminescence imaging," *Progress in Photovoltaics: Research and Applications*, vol. 18, no. 2, pp. 100–106, Mar. 2010. [Online]. Available: http://onlinelibrary.wiley.com/doi/10.1002/pip.941/abstract

[8] R. H. French, R. Podgornik, T. J. Peshek, L. S. Bruckman, Y. Xu, N. R. Wheeler, A. Gok, Y. Hu, M. A. Hossain, D. A. Gordon, P. Zhao, J. Sun, and G.-Q. Zhang, "Degradation science: Mesoscopic evolution and temporal analytics of photovoltaic energy materials," *Current Opinion in Solid State and Materials Science*, vol. 19, no. 4, pp. 212–226, Aug. 2015. [Online]. Available: http://www.sciencedirect.com/science/article/pii/S1359028614000989

[9] T. J. Peshek, J. S. Fada, Y. Hu, Y. Xu, M. A. Elsaeiti, E. Schnabel, M. Khl, and R. H. French, "Insights into metastability of photovoltaic materials at the mesoscale through massive IV analytics," *Journal of Vacuum Science & Technology B*, vol. 34, no. 5, p. 050801, Sep. 2016. [Online]. Available: http://scitation.aip.org/content/avs/journal/jvstb/34/5/10.1116/1.4960628

[10] Y. Hu, V. Y. Gunapati, P. Zhao, D. Gordon, N. R. Wheeler, M. A. Hossain, T. J. Peshek, L. S. Bruckman, G. Q. Zhang, and R. H. French, "A Nonrelational Data Warehouse for the Analysis of Field and Laboratory Data From Multiple Heterogeneous Photovoltaic Test Sites," *IEEE Journal of Photovoltaics*, vol. 7, no. 1, 2017. [Online]. Available: http://ieeexplore.ieee.org/document/7763779/

[11] Python, "Python.org," 2013. [Online]. Available: https://www.python.org/

[12] S. v. d. Walt, J. L. Schnberger, J. Nunez-Iglesias, F. Boulogne, J. D. Warner, N. Yager, E. Gouillart, and T. Yu, "scikit-image: image processing in Python," *PeerJ*, vol. 2, p. e453, Jun. 2014. [Online]. Available: https://peerj.com/articles/453

[13] SciPy, "SciPy.org SciPy.org," 2014. [Online]. Available: http://www.scipy.org/

[14] NumPy, "NumPy Numpy," 2014. [Online]. Available: http://www.numpy.org/

[15] G. Bradski, "The OpenCV Library," *Dr. Dobb's Journal of Software Tools*, 2000. [Online]. Available: http://www.drdobbs.com/open-source/the-opencv-library/184404319

[16] G. James, D. Witten, T. Hastie, and R. Tibshirani, *An Introduction to Statistical Learning: with Applications in R*, 1st ed., ser. Springer Texts in Statistics. New York: Springer, Aug. 2013. [Online]. Available: http://www-bcf.usc.edu/~gareth/ISL/index.html

[17] T. Hastie, R. Tibshirani, and J. Friedman, *The Elements of Statistical Learning*, ser. Springer Series in Statistics. New York, NY: Springer New York, 2009. [Online]. Available: http://link.springer.com/10.1007/978-0-387-84858-7

[18] Alexandre Gramfort, Gael Varoquaux, Olivier Grisel, and Raghav RV, "Sklearn: StratifiedShuffleSplit()." [Online]. Avail-able: http://scikit-learn.org/stable/modules/generated/sklearn.model_selection.StratifiedShuffleSplit.html

[19] F. Pedregosa, G. Varoquaux, A. Gramfort, V. Michel, B. Thirion, O. Grisel, M. Blondel, P. Prettenhofer, R. Weiss, V. Dubourg, J. Vander-plas, A. Passos, D. Cournapeau, M. Brucher, M. Perrot, and E. Duch-esnay, "Scikit-learn: Machine learning in Python," *Journal of Machine Learning Research*, vol. 12, pp. 2825–2830, 2011.

[20] C. Cortes and V. Vapnik, "Support-vector networks," *Machine learning*, vol. 20, no. 3, pp. 273–297, 1995.

[21] L. Breiman, "Bagging predictors," *Machine learning*, vol. 24, no. 2, pp. 123–140, 1996.

[22] ——, "Random forests," *Machine learning*, vol. 45, no. 1, pp. 5–32, 2001.

[23] R. Lippmann, "An introduction to computing with neural nets," *IEEE Assp magazine*, vol. 4, no. 2, pp. 4–22, 1987.

[24] S. B. Park, J. W. Lee, and S. K. Kim, "Content-based image classi-fication using a neural network," *Pattern Recognition Letters*, vol. 25, no. 3, pp. 287–300, 2004.

[25] S. G. Wu, F. S. Bao, E. Y. Xu, Y.-X. Wang, Y.-F. Chang, and Q.-L. Xiang, "A leaf recognition algorithm for plant classification using probabilistic neural network," in *Signal Processing and Information Technology, 2007 IEEE International Symposium on*. IEEE, 2007, pp. 11–16.

[26] D. W. Patterson, *Artificial neural networks: theory and applications*. Prentice Hall PTR, 1998.

[27] N. Srivastava, G. E. Hinton, A. Krizhevsky, I. Sutskever, and R. Salakhutdinov, "Dropout: a simple way to prevent neural networks from overfitting." *Journal of Machine Learning Research*, vol. 15, no. 1, pp. 1929–1958, 2014.

[28] Alexandre Gramfor, Mathieu Blondel, Olivier Grisel, Arnaud Joly, Jochen Wersdorfer, Lars Buitinck, Joel Nothman, Noel Dawe, Jatin Shah, Saurabh Jha, and Bernardo Stein, "Sklearn: accuracy_score()." [Online]. Available: http://scikit-learn.org/stable/modules/generated/sklearn.metrics.accuracy_score.html

# Towards Developing a Standard for Testing Bifacial PV Modules: Single-Side versus Double-Side Illumination Method I-V Measurements Under Different Irradiance and Temperature

Stefan Roest[1], Witek Nawara[1], Bas B. Van Aken[2] and Elias Garcia Goma[1]

[1]Eternal Sun Group, Den Haag, Zuid-Holland, Wolga 11 2491 BK, The Netherlands
[2]ECN Solar Energy, Petten, Noord-Holland, Westerduinweg 3 1755 LE, The Netherlands

*Abstract*—Industrial production of bifacial photovoltaic modules is becoming more and more cost-effective in recent years. For this reason, the development of an agreed international standard test that provides the guidelines to measure the current-voltage characteristics of bifacial modules, especially under standard test conditions, is of utmost importance [1]. In order to contribute to the international norm, the goal of this research is to compare the two main bifacial indoor testing methods under several irradiance and temperature conditions, combining a flash simulator and a steady-state simulator. Among other insights, the results successfully quantify the offset in maximum power measurements between methods.

*Index Terms*—bifacial, standard, characterization, single side, double side.

## I. INTRODUCTION

Due to technological advancements regarding bifacial modules, including enhanced energy output gain and reduced costs associated with extra processing steps, it is becoming more and more attractive for the industry to invest in this market. Currently, some companies already manufacture and sell commercial bifacial PV modules, with an installed capacity of over 120 MWp by the end of 2016 [2]. Additionally, it is expected that bifacial modules will represent 25% of the market share in 10 years [3]. Thus, the development of an agreed international standard test method that allows to benchmark different bifacial cells and modules becomes more and more demanded by the PV community.

The main goal of such standard is to describe how to measure and report the current-voltage characteristics of a module with both sides being illuminated simultaneously with an irradiance $G_F$ for the front and $G_R$ for the rear. However, if only one solar simulator is available, the front side shall be illuminated with an equivalent irradiance, abbreviated as $G_E$, greater than $G_F$ and that compensates for the absence of $G_R$. Furthermore, any irradiance on the non-illuminated side should be minimised, e.g. by preventing reflection on objects behind the module by covering the rear with a non-reflecting plate. This method is referred in this report as single side illumination method. Finally, when a double simulator indoor testing setup is available, the bifacial module can be simultaneously illuminated in front and rear side with $G_F$ and $G_R$, respectively, and thus its current-voltage characteristics under bifacial operation can be measured and reported. This

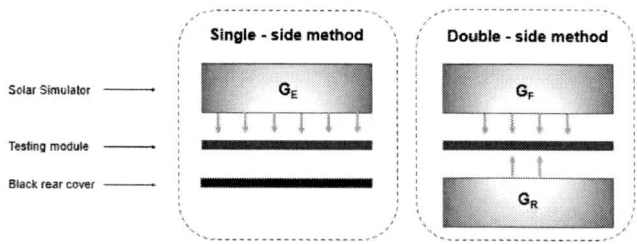

Fig. 1. Schematic of the two main methods for testing bifacial PV modules.

method is referred in this report as double side illumination method. Both methods are represented in Fig. 1.

Previous works by Deline et al., have validated single side method current-voltage measurements, by comparing the maximum power measurements with outdoor experimental data [4]. The research of Deline et al. shows that maximum power measurements employing the indoor single side method agree within 1%-2% with outdoor bifacial data, for a $G_R$ range up to $0.3G_F$. However, a comparison between current-voltage measurements employing the indoor double side method and the indoor single side method has not yet been assessed in literature. The goal of this paper is to perform such comparison in order to account and quantify for any possible divergences on the electrical parameters that both methods might yield.

To do so, first an assessment of the basic solar simulator spectrum and sweep time requirements for testing high efficiency modules, which involves a big share of the current bifacial modules, is carried out. Secondly, single side illumination IV measurements on the front and rear side of a bifacial module at different irradiance are performed. Then, the technical considerations surrounding the double simulator setup are analysed. Finally, double side illumination measurements on the same module are performed and its output compared with the single side illumination results.

## II. EXPERIMENTAL METHOD

In order to analyze the effect of sweep time and direction on high efficiency modules, a series of IV measurements under STC conditions are performed on a monofacial n-type

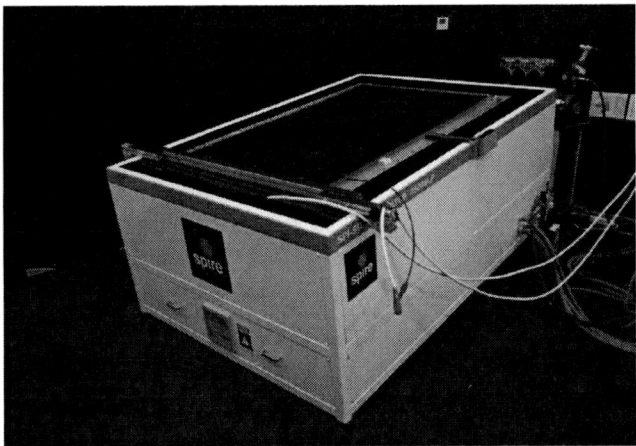

Fig. 2. Single side illumination measurement setup.

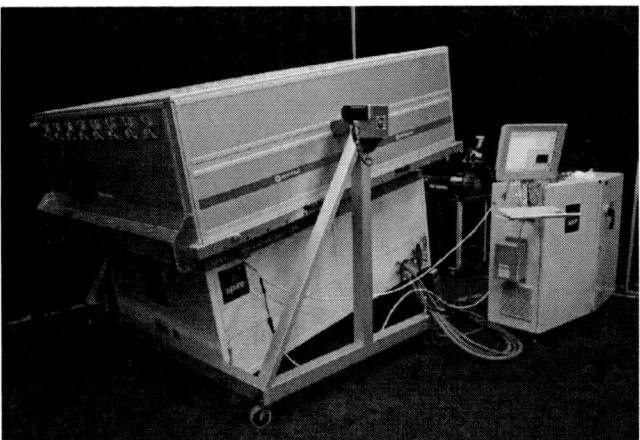

Fig. 4. Double side illumination measurement setup.

Fig. 3. Detail of a junction box shading two cells of the bifacial module.

heterojunction solar module using a flash solar simulator. The sweep time is increased in steps of 10 ms from 10 to 230 ms. Forward and reverse sweep are also considered. Regarding spectrum, the spectral response of the front side of a n-type PERT bifacial cell is carried out, and based on it, a suitable simulator spectrum is discussed.

For the single side illumination measurement, a tabletop flash simulator SPI-SUN 3500 is employed, as shown in Fig. 2. The device under test employed in the measurements is a 60 cell n-type PERT mono-crystalline bifacial PV module. In more detail, the module is glass-glass and frameless. Additionally, the cells are series connected and the module has three by-pass diodes, embedded in three junction boxes, dividing the module in three sets of two strings each. It has to be noted, however, that the junction boxes partially shade the cells of the short-edge from the rear side, as shown in Fig. 3. During these measurements, the module is kept at room temperature (25 °C $\pm$ 2 °C), which is measured with a thermocouple attached on top of the glass of the non-illuminated side, very close to a cell from the edge, but not shading it. If the measurement is not exactly at 25 °C, it is

corrected to it following the guidelines of IEC 60891 [5].

The double side illumination setup is mounted by rolling a steady-state solar simulator LA150200 on top of the tabletop flash simulator, as illustrated in Fig. 4. In order to analyze the technical considerations of using two different light sources, a series of parameters are analyzed. Firstly, a uniformity map of the test area is carried out for both sides using a reference cell placed in each of the sixty spots where the cells of the module would be, forming a six per ten uniformity map. Secondly, the spectral response of the two reference cells employed for recording front and rear irradiance is analysed. Afterwards, the reflections induced by the albedo of the top simulator are analyzed by comparing single side measurements with and without the steady state simulator on top of the setup.

Finally, the double side illumination measurements are carried out with the same module used for single side. In order to reproduce similar conditions to those found in real life, the irradiance of the LA150200, which illuminates the rear side, is set to its two lower values, 115 W/m$^2$ and 200 W/m$^2$. In order to compare the double side measurements to the single side measurements, G$_E$ is calculated as shown in (1), by using the lower value between the short-circuit current I$_{SC}$ and the maximum power P$_{MAX}$ bifaciality coefficients, $\phi_{Isc}$ and $\phi_{Pmax}$, respectively. These coefficients are the ratio between I$_{SC}$ or P$_{MAX}$ generated by the module when illuminated at standard test conditions (STC) on the rear side compared to the front side, as shown in (2) an (3).

$$G_E = G_F + Min(\phi_{Isc}, \phi_{PMAX}) * G_R \tag{1}$$

$$\phi_{Isc} = \frac{I_{SC,rear}}{I_{SC,front}} \tag{2}$$

$$\phi_{Pmax} = \frac{P_{MAX,rear}}{P_{MAX,front}} \tag{3}$$

Finally, it is to be noted that when illuminating the rear side with low irradiance from the steady state simulator,

978-1-5090-5606-4/17 $31.00 © 2017 IEEE

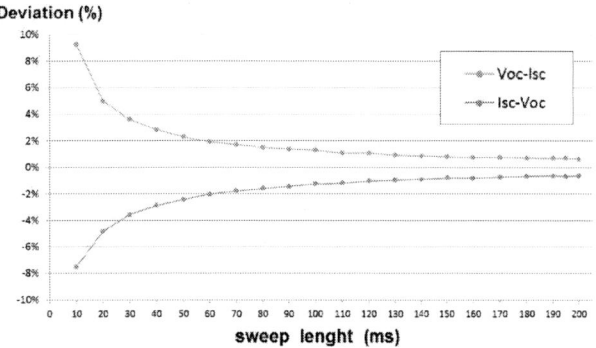

Fig. 5. Relative power deviation as a function of sweep length for a heterojunction monofacial module.

the bifacial module would raise its temperature from room temperature to 30 °C. For a better comparison of methods, also this measurements are corrected to room temperature following the guidelines of IEC 60891 [5].

## III. RESULTS AND DISCUSSION

### A. Technical Considerations for High Efficiency Module Measurements

Fig. 5 shows the relative deviation in maximum power of a high efficiency module depending on whether the IV sweep is done forward or backwards. As it can be observed, a greater sweep time reduces the gap and gets closer to the true maximum power. Therefore, the upcoming single and double side measurements will be performed at the highest sweep time possible in the single and double side setup.

Fig. 6 presents the spectral response of a cell of the same type as the ones used in the bifacial module under test in this paper. It can be observed that this type of cell is able to convert light from 300 to 1200 nm, thus a suitable simulator with same light spectrum shall be employed for accurate measurements, covering this entire wavelength range, as it is shown in Fig. 7.

### B. Single Side Characterization

IV curve measurements were taken for front and for rear illumination under each irradiance setting. For all the settings the general behaviour of the curves was the same. Therefore, when discussing IV curves only one irradiance setting is taken. Fig. 8 shows the IV curves of the PV module when illuminated separately from the front and from the rear side at the same intensity, in this case, 800 W/m². As it can be observed, the front side generates higher $I_{SC}$ and greater $P_{MAX}$, while the $V_{OC}$ for both situations remains very close. The main difference, however, comes in the FF. Looking at the shape of the IV curves, the rear side presents a plateau caused by the shading of the junction boxes shown in Fig. 3. In fact, a change in height in the plateau can be observed, meaning that one of the strings is being slightly more shaded. The reason behind this drop in $I_{SC}$ but not in $P_{MAX}$, and

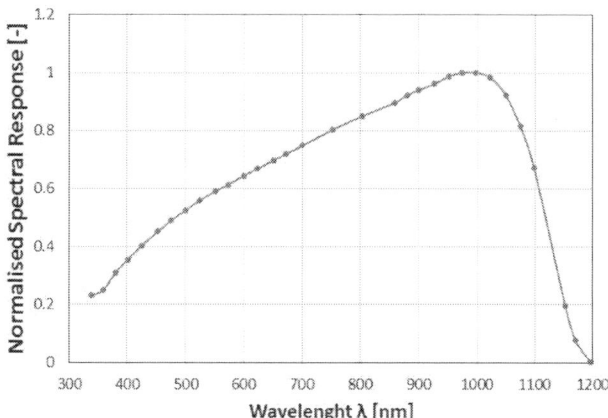

Fig. 6. Normalised spextral response of a n-type mono-crystalline solar cell used in the bifacial module.

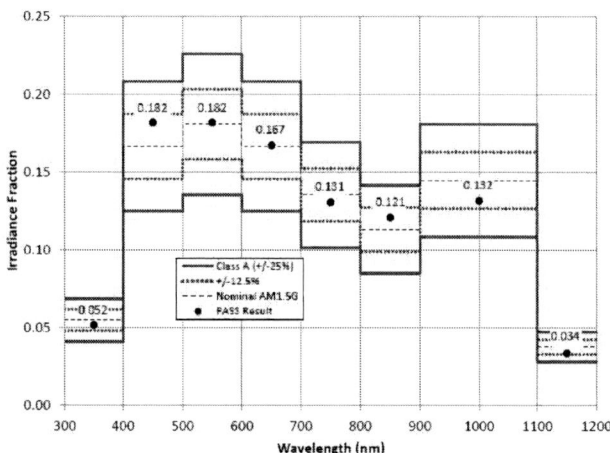

Fig. 7. Bins showing nominal AM 1.5 spectrum, limits of class A ($\pm25\%$) and A+ ($\pm12.5\%$) spectrum, and the nominal spectrum of a simulator that complies class A+ (PASS).

therefore change in FF, is that the $I_{SC}$ of each string in the module will be limited to that of the lowest performing cell, and so will the maximum power point current $I_{MPP}$; however, in the $P_{MAX}$ region the maximum power point voltage $V_{MPP}$ will shift to the right and $P_{MAX}$ will not be greatly affected. Besides shading, other possible causes of difference in IV curve shape between front and rear can be, although to a lesser extent, a higher mismatch between cells when irradiated from the rear side, since in manufacturing they are normally only binned according to front performance.

Fig. 9 shows $I_{SC}$ versus irradiance for front and rear illumination. As it can be observed, the $I_{SC}$ evolution versus irradiance is linear for both front and rear, however they do increase at different rates, proportional to their spectral response. As a result, $\phi_{Isc}$ remains constant as $0.785 \pm 0.005$ through the entire range of irradiance measurements, as shown in Fig. 10, where it can be seen that $\phi_{Pmax}$ exhibits similar

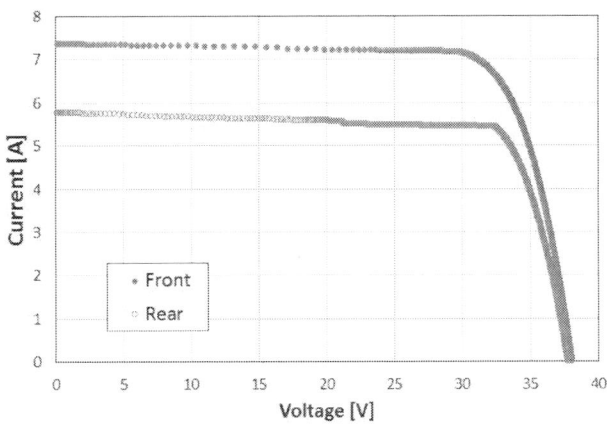

Fig. 8. Front and rear IV measurements of a bifacial module under 800 W/m².

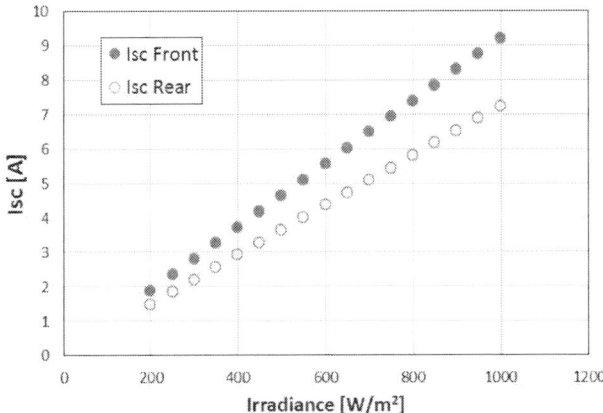

Fig. 9. $I_{SC}$ versus irradiance for front and rear IV measurements on a bifacial module.

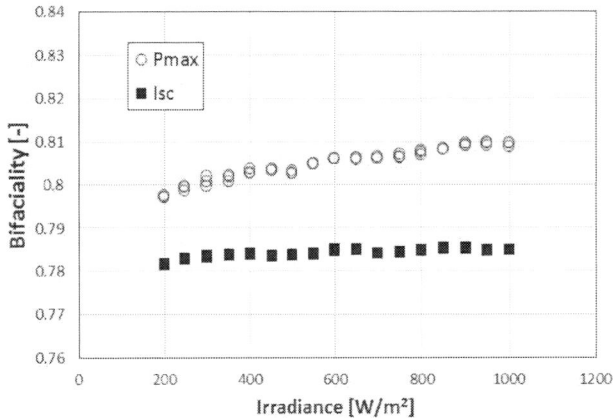

Fig. 10. Bifaciality of $P_{MAX}$ and $I_{SC}$ versus irradiance for front and rear IV measurements on a bifacial module.

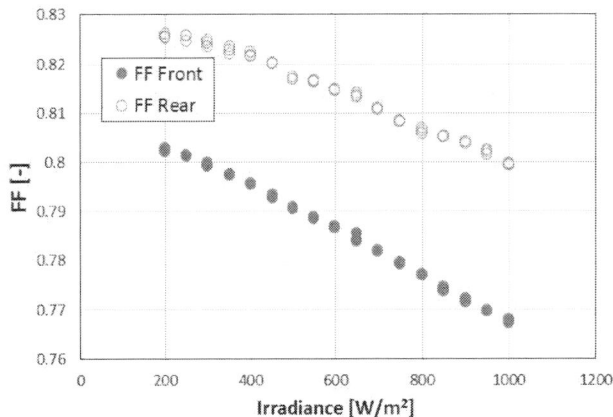

Fig. 11. FF versus irradiance for front and rear IV measurements on a bifacial module.

behaviour, although it does slightly increase towards higher irradiance. Logically it has a higher value than $\phi_{Isc}$, since the FF of the module when irradiated from the rear is higher. Fig. 11 presents the FF evolution versus irradiance, also for front and for rear illumination. As it is observed, an offset occurs at all irradiance, which is in agreement with single side measurements literature [6].

Since $\phi_{Isc}$ is lower than $\phi_{Pmax}$ at all irradiance levels, it is employed in (1). However, it is to be expected that if $G_E$ is calculated to match $I_{SC}$, there will be an induced error on $P_{MAX}$ when emplying the single side illumination measurement. In a similar way, if $\phi_{Pmax}$ was to be employed in (1), the methods will yield a difference in $I_{SC}$.

### C. Technical Considerations for Double Side Method

Fig. 12 shows the uniformity map of the front side simulator for 1000 W/m² setting. It can be observed that the test area stays within a non-uniformity of 1.8%, thus falling in the class A classification. Similarly, Fig. 13 shows the uniformity map for the top simulator, which irradiates the rear side, for the

200 W/m² setting. The non-uniformity in this case is 4.3%, which falls within class B classification. Previous research by Van Aken et al. [7] has shown that this level of non-uniformity on the rear side has little effect on $P_{MAX}$ determination.

Regarding the spectral differences of both light sources, two reference cells of the same type are employed for measuring front and rear side illumination, thus it is assumed that no correction is required for the amount of irradiance incident on the front side relative to the rear side.

Finally, reflections due to having a rear simulator surface with higher albedo than the anti-reflective single side background have to be taken into account. Fig. 14 compares $I_{SC}$ versus irradiance for single side measurements with and without rear simulator surface albedo. As it can be noted, $I_{SC}$ is raised by the rear albedo, which is equivalent to an increment of 2.8% of front irradiance. On the other hand, Fig. 15 compares FF under the same conditions. It is seen that FF overlaps in both situations, therefore the effect or reflection is little and only impacts $I_{SC}$. As a consequence, while doing

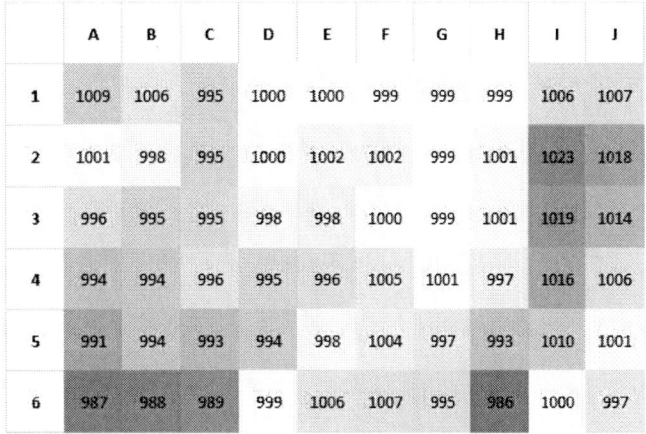

| | A | B | C | D | E | F | G | H | I | J |
|---|---|---|---|---|---|---|---|---|---|---|
| 1 | 1009 | 1006 | 995 | 1000 | 1000 | 999 | 999 | 999 | 1006 | 1007 |
| 2 | 1001 | 998 | 995 | 1000 | 1002 | 1002 | 999 | 1001 | 1023 | 1018 |
| 3 | 996 | 995 | 995 | 998 | 998 | 1000 | 999 | 1001 | 1019 | 1014 |
| 4 | 994 | 994 | 996 | 995 | 996 | 1005 | 1001 | 997 | 1016 | 1006 |
| 5 | 991 | 994 | 993 | 994 | 998 | 1004 | 997 | 993 | 1010 | 1001 |
| 6 | 987 | 988 | 989 | 999 | 1006 | 1007 | 995 | 986 | 1000 | 997 |

Fig. 12. Uniformity map of the 1000 W/m² setting of the flash simulator. Units in W/m².

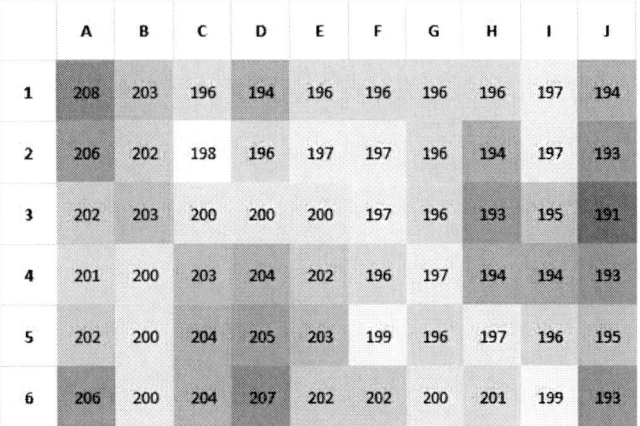

| | A | B | C | D | E | F | G | H | I | J |
|---|---|---|---|---|---|---|---|---|---|---|
| 1 | 208 | 203 | 196 | 194 | 196 | 196 | 196 | 196 | 197 | 194 |
| 2 | 206 | 202 | 198 | 196 | 197 | 197 | 196 | 194 | 197 | 193 |
| 3 | 202 | 203 | 200 | 200 | 200 | 197 | 196 | 193 | 195 | 191 |
| 4 | 201 | 200 | 203 | 204 | 202 | 196 | 197 | 194 | 194 | 193 |
| 5 | 202 | 200 | 204 | 205 | 203 | 199 | 196 | 197 | 196 | 195 |
| 6 | 206 | 200 | 204 | 207 | 202 | 202 | 200 | 201 | 199 | 193 |

Fig. 13. Uniformity map of the 200 W/m² setting of the flash simulator. Units in W/m².

double side illumination measurements, this reflection can be either corrected using an initial current offset or accounted for by adjusting the total irradiance incident on the module. In the present paper, the second approach is employed.

### D. Double Side Characterization

Fig. 16 shows a comparison between single side measurements and double side measurements with 800 W/m² front irradiance plus variable rear irradiance, which is converted to single side irradiance via (1). The FF for double side illumination IV measurements is found in between the boundaries set by the front and the rear measurements, getting closer to the limits depending on how big the share of front or rear irradiance is compared to the other. Fig. 17 shows how, as it was expected, if (1) for comparing single and double side is applied using $\phi_{Isc}$, the $P_{MAX}$ of the single side method is underestimated compared to double side illumination, which represents the true bifacial conditions. By extrapolating the double side measurements, under a single side irradiance of 1000 W/m², which is equivalent to 800 W/m² front plus 255

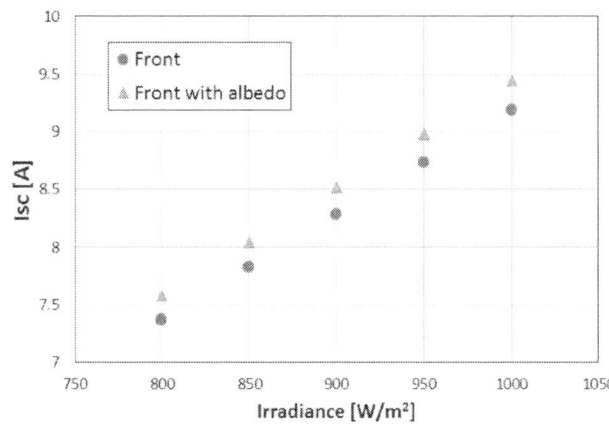

Fig. 14. $I_{SC}$ as a function of irradiance comparing front measurements with and without the albedo of the rear simulator.

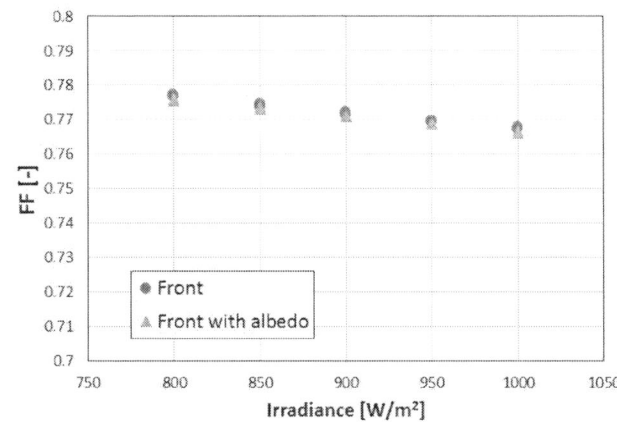

Fig. 15. FF as a function of irradiance comparing front measurements with and without the albedo of the rear simulator.

W/m² rear in double side illumination, there is a difference of 4.3 W, which represents 1.6% $P_{MAX}$ underestimation from its true bifacial output.

### IV. Conclusion

It has been shown that sweep time and extended spectrum do play a key role to accuratelly predict the electrical parameters of high efficiency modules, such as heterojunction and n-type mono-crystalline, including bifacial. Secondly, it has been demonstrated that, if subjected to shading or other causes that can induce current mismatch between cells, the difference between $\phi_{Isc}$ and $\phi_{Pmax}$ will most likely increase. It is also found that while $\phi_{Isc}$ remains constant through irradiance, $\phi_{Pmax}$ slightly increases. Furthermore, a fast and modular setup has been developed and characterized in order to perform double side illumination measurements, employing a steady state simulator and a long pulse flash simulator. Finally, the double side measurements have shown that the current approach of single side method related to double side

978-1-5090-5606-4/17 $31.00 © 2017 IEEE

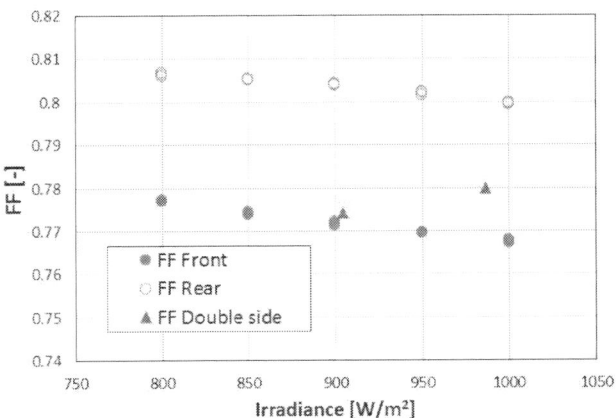

Fig. 16. FF as a function of irradiance comparing front, rear and double, with 800 W/m² front irradiance, measurements.

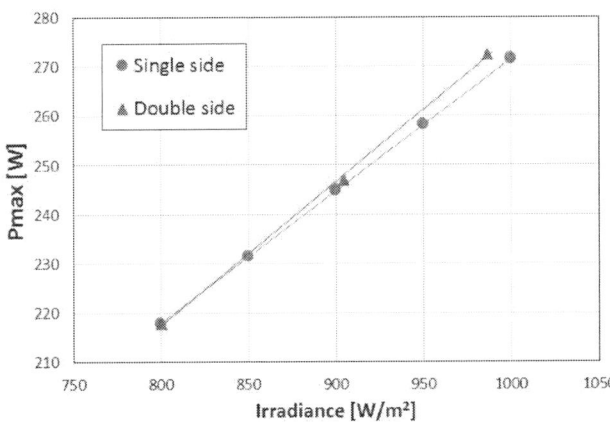

Fig. 17. P$_{MAX}$ as a function of irradiance comparing front, and double, with 800 W/m² front irradiance, measurements.

method via $\phi_{Isc}$ can yield significative differences in P$_{MAX}$ prediction depending on the conditions, up to 1.6% in P$_{MAX}$ for the case studied. This is due to the difference in value between $\phi_{Isc}$ and $\phi_{Pmax}$.

ACKNOWLEDGMENT

Thanks to ECN for their collaboration and to Tempress for supplying the module.

REFERENCES

[1] V. Fakhfouri, "IEC 60904: Photovoltaic Devices Part 1-2: Measurement of current-voltage characteristics of bifacial photovoltaic (PV) devices," Proposal 82/1044/NP, July 28, 2015.

[2] R. Kopecek and J. Libal, "Bifaciality: still an advantage for n-type?," presented at the nPV workshop 2017, Freiburg, Germany, 2017.

[3] International Roadmap for Photovoltaics (ITRPV) 2015 Results, 7th Edition, March, 2016.

[4] C. Deline, S. MacAlpine, B. Marion, F. Toor, A. Asgharzadeh and J. Stein, "Evaluation and Field Assessment of Bifacial Photovoltaic Module Power Rating Methodologies," *43rd IEEE Photovoltaics Specialists Conference*, Portland, 2016

[5] International Electrotechnical Commission, "IEC 60891: Photovoltaic Devices - procedures for temperature and irradiance corrections to measured i-v characteristics," 2009.

[6] A. Schmid, D. Philipp and C. Reise, "5BV.4.7 Characterization and Testing of Bifacial Modules," presented at the EU PVSEC 2016, Freiburg, Germany, 2016.

[7] B. B. Van Aken K. de Groot, "Near-field partial shadow on rear side of bifacial modules," presented at the Silicon PV 2017, Freiburg, Germany, 2017.

# Electrical Transport Properties from Long Wavelength Ellipsometry

Prakash Uprety[a], Maxwell M. Junda[a], Indra Subedi[a],

Michael A. Slocum[b], David V. Forbes[b], Seth M. Hubbard[b] and Nikolas J. Podraza[a]

[a]Wright Center for Photovoltaics Innovation and Commercialization & Department of Physics and Astronomy, University of Toledo, Toledo, OH 43606, USA

[b]NanoPower Research Laboratory, Rochester Institute of Technology, Rochester, NY 14623, USA

*Abstract* — Electrical transport properties (resistivity, scattering time, carrier concentration, mobility, effective mass) can be deduced from the Drude model for free carrier optical absorption. Optical response of materials may include simultaneous contributions from free carriers, phonon modes, and electronic transitions, however. Improved sensitivity to free carrier absorption and electrical transport properties is obtained through the use of long wavelength extended spectroscopic ellipsometry in the infrared (IR) and millimeter (THz) ranges. Non-contacting optical measurement and analysis methodologies are designed for semiconductors and transparent conductors used in photovoltaics (PV). Case studies include ZnO:Al, InP, wafer Si, and single crystal CdTe.

*Index Terms* — CdTe, Drude model, InP, Long wavelength ellipsometry, Si wafer, THz, ZnO:Al.

## I. INTRODUCTION

Photovoltaic (PV) devices consist of multilayer stacks of doped and undoped semiconductors as well as metallic and transparent conducting electrical contacts. These materials may be poly-/nano-crystalline and influenced by chemical or structural interactions with underlying layers. The structural characteristics of these layers in the PV device may differ from samples prepared for fundamental studies of electrical transport properties. High frequency electric field optical measurements are sensitive to electrical transport properties if absorption due to free carriers is significant in the measured spectral range. Here we demonstrate spectroscopic ellipsometry spanning millimeter (THz), infrared (IR), visible, and ultraviolet (UV) wavelength ranges to deduce layer thicknesses and contributions to the spectroscopic complex optical response ($\varepsilon = \varepsilon_1 + i\varepsilon_2$) arising from free carriers, phonon modes, and electronic transitions illustrated in Fig. 1.

These wide spectral range ellipsometric measurements have been made for transparent conductors like aluminum doped zinc oxide (ZnO:Al) and semiconductors including epitaxial indium phosphide (InP), wafer silicon (Si), and single crystal cadmium telluride (CdTe). Results of the analysis yield electrical transport properties extracted from the Drude model [1] for free carrier absorption:

$$\varepsilon = \frac{-\hbar^2}{\varepsilon_0 \rho \left( \tau E^2 + i\hbar E \right)} \quad (1)$$

where, $\hbar$, $\varepsilon_0$, $\tau$, $\rho = m^*/Nq^2\tau = 1/q\mu N$, $m^*$, $N$, $q$, and $\mu$ are reduced plank's constant, vacuum permittivity, scattering time, resistivity, carrier effective mass, carrier concentration, elementary charge, and mobility, respectively.

ZnO:Al is a wide band gap degenerately doped semiconductor with the ability to overcome the present problems of other transparent conducting materials because it is earth abundant, nontoxic, easy to fabricate, and relatively stable against temperature variations [2], [3]. Other researchers have reported the need to modify the Drude expression via frequency dependent terms [4], [5] to fit optical response. However, we have successfully applied the normal Drude model here to study the free carrier absorption in conjunction with other parametric expressions accounting for absorption by phonons and electrical transitions over a wide spectral range spanning from the UV to THz. Film resistivity extracted from analysis of non-contacting optical measurements over the full measured spectral range are within 5% of that obtained by 4-point electrical probe [6].

An epitaxial InP film deposited on an Fe compensated InP (InP:Fe) substrate wafer has also been studied over the IR to UV spectral range. Due to better radiation resistance than GaAs and Si, InP solar cells have been of interest for space applications [7]. Assuming an effective electron $m^*$ from literature, carrier concentration and mobility are deduced in addition to identifying a phonon mode and higher energy critical point (CP) transitions. The optically deduced carrier concentration and mobility agree well with those obtained from electrical Hall effect measurements [8].

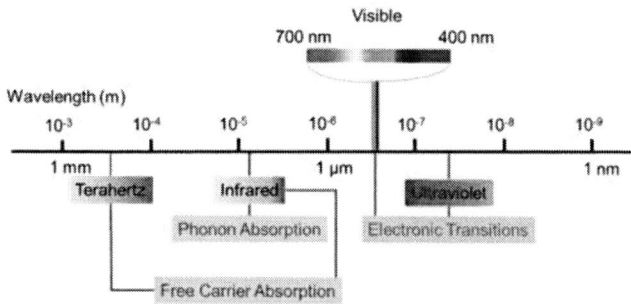

Fig. 1. Wavelength ranges and associated optical properties deduced from spectroscopic ellipsometry measurements.

Si is the most common semiconductor material used in electronic device fabrication, and Si-based PV dominates the solar cell market. Two differently doped wafers are measured here using THz range spectroscopic ellipsometry to extract their transport properties. One wafer is lightly p-type doped with the trivalent impurity boron and is transparent over THz range. The other is heavily n-type doped with the pentavalent impurity arsenic and is opaque over this spectral range [9].

CdTe is the base material for the most dominant thin film PV technology. Long term stability, competitive performance and ability to produce devices at the commercial scale make CdTe based solar cells very appealing [10], [11]. Single crystal CdTe is initially studied with IR and THz ellipsometry to determine free carrier absorption and phonon mode [9] contributions to spectra in $\varepsilon$.

## II. EXPERIMENTAL DETAILS

Thin film ZnO:Al has been prepared on soda lime glass substrate at room temperature by RF magnetron sputtering using target power of 150 W and with pure argon ambient at 1.25 mTorr. The target consists of 2 wt.% $Al_2O_3$ and 98 wt.% ZnO. The InP epitaxial film has been deposited by metal-organic vapor phase epitaxy (MOVPE) at 580°C. A $1 \times 2$ cm$^2$ commercial single crystal CdTe sample has been sourced from Nikko materials Co., Ltd. Reference n- and p-type doped Si wafers have also been obtained from the J. A. Woollam Co.

Room temperature ellipsometric spectra (in $N = \cos 2\psi$, $C = \sin 2\psi \cos \Delta$, $S = \sin 2\psi \sin \Delta$) have been collected ex-situ at 70° angles of incidence in the THz spectral range from 0.4 to 6 meV (J. A. Woollam Company THz-VASE) for ZnO:Al, from 0.035 to 0.75 eV in the IR range (J. A. Woollam Company FTIR-VASE) for ZnO:Al and InP, from 0.75 to 5.89 eV (J. A. Woollam M-2000FI) for ZnO:Al, and adding the vacuum UV range up to 8.5 eV (Gen I, VU-302 VUV-VASE, J.A. Woollam Co.) for InP. Similarly, room temperature spectra in $N$, $C$ and $S$ have been collected ex-situ at 50° angle of incidence using the THz-VASE from 0.4 to 6 meV for the Si wafers and CdTe single crystal, and from 0.035 to 0.25 eV using the FTIR-VASE for the CdTe crystal.

The sheet resistance of ZnO:Al at room temperature has been measured by a 4-point electrical probe, which is then converted to electrical $\rho$ when multiplied by the thickness of the film determined by ellipsometry. Hall effect measurement of the InP epitaxial film yields carrier concentration and mobility.

## III. RESULTS AND DISCUSSION

### Resistivity of ZnO:Al

The structural and optical properties of ZnO:Al are extracted in the form of layer thicknesses and spectra in $\varepsilon$, respectively, by fitting a parameterized model to measured ellipsometric spectra from 0.4 meV to 5.89 eV simultaneously using a least square regression analysis that minimizes an unweighted error function. The optical model consists of a $1.282 \pm 0.002$ mm glass substrate / $158.4 \pm 0.2$ nm bulk ZnO:Al thin film layer / $5.6 \pm 0.1$ surface roughness / semi-infinite air ambient structure. Contributions to $\varepsilon$ for ZnO:Al includes a constant additive term to $\varepsilon_1$ ($\varepsilon_\infty$), a Drude oscillator describing free carrier absorption, a Lorentz oscillator positioned at 0.74 eV describing a sub-band gap absorption feature, four Gaussian oscillators positioned (at 0.046, 0.053, 0.11 and 0.16 eV) related to IR phonon absorption, and two oscillators assuming critical point parabolic bands (CPPB) positioned at 3.62 and 5.4 eV, to describe higher energy electronic transitions, with the imaginary part shown in Fig. 2.

From the Drude model $\rho$, $\tau$, $N$, and $\mu$ are determined assuming $m^* = 0.43 m_e$ [4] and summarized in Table 1. These optically deduced parameters are within expectations and similar to those reported in literature.

### TABLE 1
#### OPTICALLY DEDUCED ELECTRICAL TRANSPORT PROPERTIES FOR ZnO:Al

| | Resistivity $\rho$ ($\Omega\cdot$cm) | Scattering Time $\tau$ (fs) | Mobility $\mu$ (cm$^2$/Vs) | Carrier Concentration $N$ (cm$^{-3}$) |
|---|---|---|---|---|
| This work | $(2.250 \pm 0.007) \times 10^{-3}$ | $5.1 \pm 0.7$ | 20.8 | $1.3 \times 10^{20}$ |
| Literature | $\sim 10^{-3}$ [4], [12] | $6 \pm 1$ [12] | 18.7 [4] | $7.5 \times 10^{20}$ [4] |

Fig. 3 illustrates how optically determined $\rho$, for example, varies depending on the lower bound of the photon energy range analyzed, i.e. using only shorter wavelength data. The best-fit value of $\rho$ decreases as the spectral range is narrowed and longer wavelengths are excluded, which underestimates the resistivity determined by the 4-point probe measurement. When the Drude model extends to the THz range, the optical $\rho$ becomes within 5% of the direct electrical probe value [6]. Overall incorporating longer wavelength data improves the accuracy of the optically deduced parameters [6], [13].

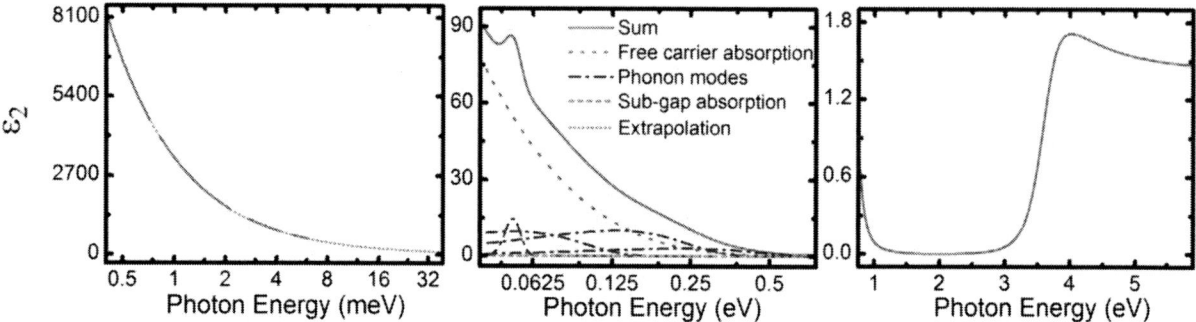

Fig. 2. Imaginary part of complex dielectric function, $\varepsilon = \varepsilon_1 + i\varepsilon_2$, spectra from 0.4 to 35 meV (left), 0.035 to 0.75 eV (middle), and 0.75 to 5.89 eV (right) denoting the summation of contributions from free carrier absorption, phonon modes, sub-gap absorption, and electronic critical point transitions for ZnO:Al. [6]

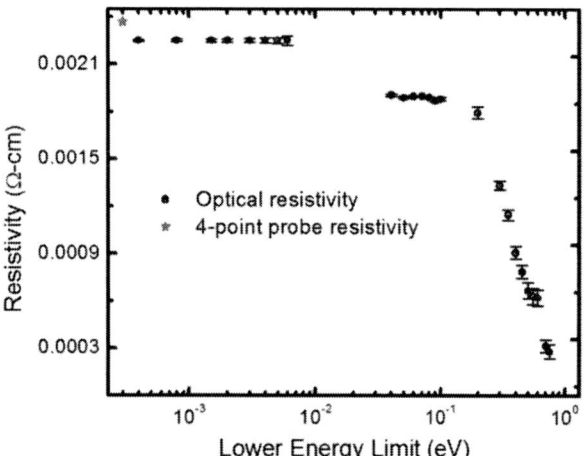

Fig. 3. Dependence of optically determined resistivity ($\rho$) for thin film ZnO:Al as a function of the lower photon energy limit in the analysis. 4-point probe $\rho$ is also shown. [6]

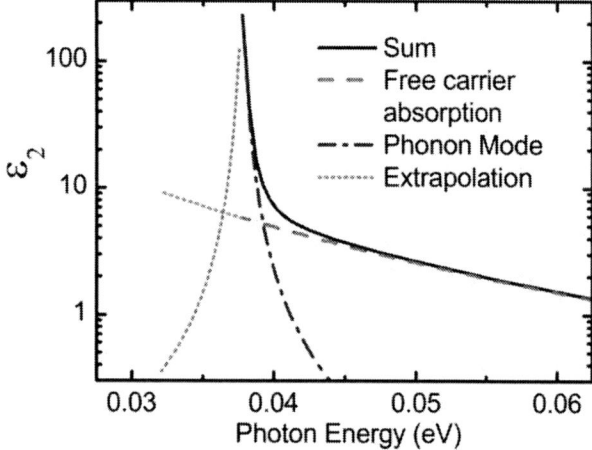

Fig. 4. Spectra in $\varepsilon_2$ for InP from 0.038 to 0.07 eV showing contributions from free carrier absorption and a phonon mode. [8]

### Transport Properties of InP

The optical model consists of semi-infinite InP:Fe wafer substrate / 197 ± 8 nm interface / 1348 ± 3 nm epitaxial InP film / 9.3 ± 0.6 nm surface layer / air ambient. A combined parameterization of $\varepsilon$ is developed from 0.038 to 8.5 eV for epitaxial InP which contains a Lorentz oscillator positioned at 0.0378 eV for the IR phonon mode, a Drude oscillator for free carrier absorption, and ten CPPBs positioned (at 1.36, 1.42, 3.14, 3.34, 4.71, 4.97, 5.88, 6.45, 7.88, and 8.22 eV) with an Urbach tail for higher energy electronic transitions. Kramers-Kronig integration of $\varepsilon_2$ and adding $\varepsilon_\infty$ yields $\varepsilon_1$. Fig. 4 shows contributions from free carrier absorption and IR phonon mode in the spectra of $\varepsilon_2$. From $\varepsilon$, the Drude parameters considering effective electron mass $m^* = 0.077m_e$ [14] are $N = 1.9 \times 10^{18}$ cm$^{-3}$ and $\mu = 1559$ cm$^2$/Vs. These results agree well with those from electrical Hall effect measurement of $N = 2.2 \times 10^{18}$ cm$^{-3}$ and $\mu = 1590$ cm$^2$/Vs [8]. This agreement demonstrates that $\varepsilon$ at long wavelengths with sufficiently

pronounced free carrier absorption can yield electrical transport properties for semiconductors.

### Transport Properties of Crystalline Si Wafers

In the THz region, lightly doped Si is transparent and the optical model for the boron-doped sample incorporates the finite thickness of the wafer as a fitting parameter. Spectra in $\varepsilon$ for lightly boron-doped wafer is parameterized by a Drude oscillator describing free carrier absorption, a Sellmeier oscillator with $A = 99.6 \pm 0.7$ eV$^2$ and $E_0 = 3.046 \pm 0.009$ eV, and $\varepsilon_\infty = 1$. The wafer thickness is determined to be 671 ± 1 μm. Fig. 5 shows a parameterization of spectra in $\varepsilon$ from 0.4 to 6 meV. The optical resistivity and scattering time from the Drude model are determined to be 6.3 ± 0.1 Ω·cm and 120 ± 5 fs, respectively. The optical value of $\rho$ is within the nominal specified range of the wafer, 5 to 10 Ω·cm. Assuming the effective mass of hole is 0.39$m_e$ [15], carrier concentration and optical mobility are determined to be $1.83 \times 10^{15}$ cm$^{-3}$ and 542 cm$^2$/Vs, respectively [9].

Heavily doped Si is optically opaque in the THz range, so the arsenic-doped wafer is treated as being semi-infinite as there is not sensitivity to its thickness. Spectra in $\varepsilon$ is parametrized by $\varepsilon_\infty = 1$ and a single Drude oscillator with $\rho = 0.00259 \pm 0.00002$ $\Omega\cdot$cm and $\tau = 8 \pm 2$ fs, which is shown in Fig. 6. This optical value of $\rho$ is within the nominal specified range of the wafer, 0.0020 to 0.0045 $\Omega\cdot$cm. Assuming electron effective mass is $0.26m_e$ [16], the carrier concentration and optical mobility are calculated to be $4.57 \times 10^{19}$ cm$^{-3}$ and 52.7 cm$^2$/Vs, respectively [9]. Spectra in $\varepsilon_2$ for the $N = 1.83 \times 10^{15}$ cm$^{-3}$ boron-doped and $N = 4.57 \times 10^{19}$ cm$^{-3}$ arsenic-doped Si wafers are qualitatively similar, although the values of $\varepsilon_2$ for the higher $N$ arsenic-doped sample are significantly larger in amplitude due to the 10,000 times greater $N$ and its impact on the free carrier optical response.

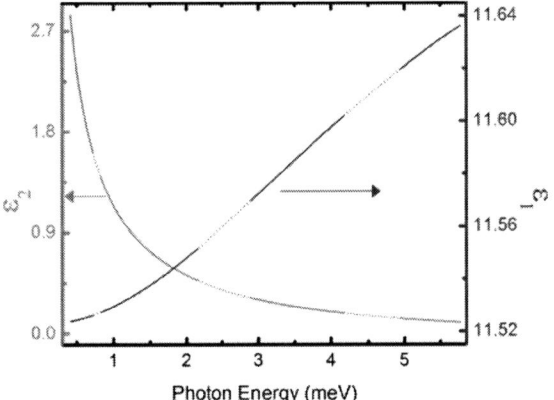

Fig. 5. Spectra in $\varepsilon$ from 0.4 to 6 meV for lightly boron doped Si wafer with $N = 1.83 \times 10^{15}$ cm$^{-3}$. Regions of data collection are indicated by solid lines and extrapolated between by dotted lines.

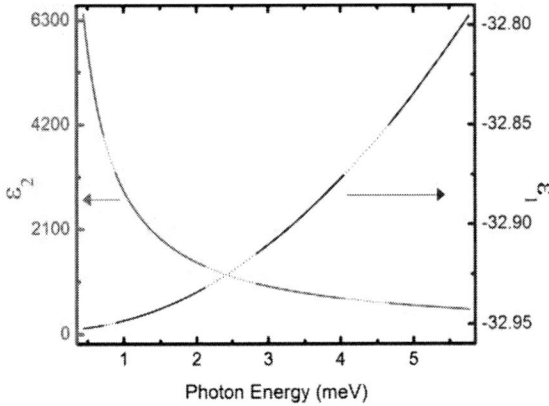

Fig. 6. Spectra in $\varepsilon$ from 0.4 to 6 meV for heavily arsenic doped Si wafer with $N = 4.57 \times 10^{19}$ cm$^{-3}$. Regions of data collection are indicated by solid lines and extrapolated between by dotted lines.

*Transport Properties of Single Crystal CdTe*

A single parameterization of spectra in $\varepsilon$ for single crystal CdTe is obtained by simultaneously fitting the measured data collected over the IR and THz range collected using two instruments. In IR range, the crystal is considered to be semi-infinite due to incoherent multiple reflections from the back-side surface whereas in THz range layer thickness is a fitting parameter due to coherent multiple reflections. The parameterization of $\varepsilon$ consists of a Drude oscillator describing free carrier absorption, a Lorentz oscillator describing absorption due to a phonon mode, and $\varepsilon_\infty$. The sample thickness is determined to be $827 \pm 3$ µm. Fig. 7 shows spectra in $\varepsilon$ from 0.4 meV to 0.25 eV. The carrier concentration is determined to be $(2 \pm 0.1) \times 10^{14}$ cm$^{-3}$ using the Drude model, where the mobility is fixed at 61.927 cm$^2$/Vs based on average data from the manufacturer. Using $m^* = 0.11m_e$ [17], the optical resistivity and scattering time are calculated to be 530.76 $\Omega\cdot$cm and 3.87 fs, respectively. The resonance position of the phonon absorption is modeled using a Lorentz oscillator with $A = 360 \pm 10$ cm$^{-1}$, $E_0 = 127 \pm 1$ cm$^{-1}$, and $\Gamma = 10.0 \pm 0.5$ cm$^{-1}$ [9]. This phonon mode position $\sim 127$ cm$^{-1}$ is lower than the values reported in literature $\sim$140-145 cm$^{-1}$ [18]-[21], however this feature appears between the measured IR and THz spectral ranges so limited sensitivity is expected. The expanded view of low wavenumbers in Fig. 5 highlights that contributions to $\varepsilon_2$ both from free carrier absorption and the phonon mode tail may be distinguished.

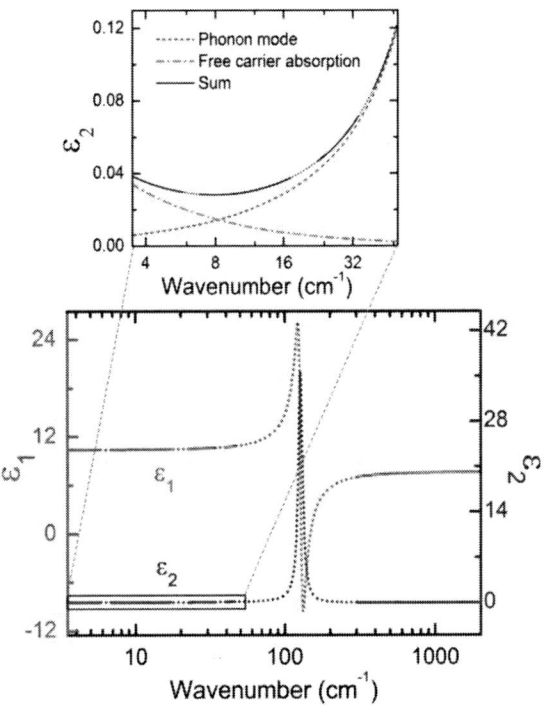

Fig. 7. Spectra in $\varepsilon$ from 3 to 2000 cm$^{-1}$ (0.4 meV to 2.5 eV) for a CdTe single crystal. The expanded region highlights the influence of the Drude feature at low wavenumbers and the tail of the higher wavenumber phonon mode. Regions of data collection are indicated by solid lines and extrapolated between by dotted lines. [9]

## IV. CONCLUSION

Optical transport properties such as resistivity, scattering time, carrier concentration and mobility, deduced from long wavelength range spectroscopic ellipsometry have been obtained for case studies on thin film nanocrystalline ZnO:Al, epitaxial InP, crystalline Si wafers, and single crystal CdTe. For ZnO:Al, improved agreement is found between direct electrical resistivity and optical resistivity when incorporating long wavelength range ellipsometry measurements with enhanced sensitivity to free carrier absorption. Epitaxial semiconducting InP on InP:Fe indicates that it is possible to differentiate transport properties for stacks of similar materials. Studies of wafer Si and single crystal CdTe show that it is possible to track free carrier contributions to $\varepsilon$ when $N$ is as low as $\sim 10^{14}$-$10^{15}$ cm$^{-3}$.

The next steps will involve optical determination of transport properties for multiple layers in stacks, such as those in film and wafer based PV devices. Ellipsometric spectra retain sensitivity to $\varepsilon$ and thickness of each layer in a stack provided that coherent multiple reflections exist. If there is sufficient contrast in $\varepsilon$ and absorption due to free carriers is detectable, transport properties of multiple stacked layers can be deduced optically.

## ACKNOWLEDGEMENT

This work was supported by University of Toledo start-up funds, Ohio Department of Development Ohio Research Scholar Program (Northwest Ohio Innovators in Thin Film Photovoltaics, Grant No. TECH 09-025), and National Science Foundation Major Research Instrumentation Program (NSF-MRI Grant No. 1228917).

## REFERENCES

[1] T. E. Tiwald, D. W. Thompson, J. A. Woollam, W. Paulson, and R. Hance, "Application of IR variable angle spectroscopic ellipsometry to the determination of free carrier concentration depth profiles," *Thin Solid Films*, vol. 313, pp. 661-666, 1998.

[2] K. L. Chopra, S. Major, and D. K. Pandya, "Transparent conductors—a status review," *Thin Solid Films*, vol. 102, pp. 1-46, 1983.

[3] T. Minami, "Present status of transparent conducting oxide thin-film development for Indium-tin-oxide (ITO) substitutes," *Thin Solid Films*, vol. 516, pp. 5822-5828, 2008.

[4] H. C. M. Knoops, B. W. H. van de Loo, S. Smit, M. V. Ponomarev, J. W. Weber, K. Sharma, W. M. M. Kessels, and M. Creatore, "Optical modeling of plasma-deposited ZnO films: Electron scattering at different length scales," *Journal of Vacuum Science & Technology A*, vol. 33, pp. 021509, 2015.

[5] N. Ehrmann, and R. Reineke-Koch, "Ellipsometric studies on ZnO:Al thin films: Refinement of dispersion theories," *Thin Solid Films*, vol. 519, pp. 1475-1485, 2010.

[6] P. Uprety, M. M. Junda, K. Ghimire, D. Adhikari, C. R. Grice, and N. J. Podraza, "Spectroscopy ellipsometry determination of optical and electrical properties of aluminum doped zinc oxide," *Applied Surface Science*, 2017.

[7] M. Yamaguchi, C. Uemura, A. Yamamoto, and A. Shibukawa, "Electron irradiation damage in radiation-resistant InP solar cells," *Japnese Journal of Applied Physics*, vol. 23, pp. 302-307, 1984.

[8] I. Subedi, M. Slocum, D. Forbes, S. Hubbard, and N. J. Podraza, "Optical properties of InP from infrared to vacuum ultraviolet studied by spectroscopic ellipsometry," *Applied Surface Science*, 2017.

[9] M. M. Junda, "Spectroscopic ellipsometry as a versatile, non-contact probe of optical, electrical, and structural properties in thin films: applications in photovoltaics," *PhD dissertation, University of Toledo*, 2017, pp. 84-114.

[10] B. E. McCandless, and J. R. Sites, *Handbook of Photovoltaic Science and Engineering*, The atrium, Southern Gate, England: John Wiley & Sons Ltd., 2011, pp. 617-657.

[11] A. Muranevich, M. Roitberg, and E. Finkman, "Growth of CdTe single crystals," *Journal of Crystal Growth*, vol. 64, pp. 285-290, 1983.

[12] A. C. Gâlcă, M. Secu, A. Vlad, and J. D. Pedarnig, "Optical properties of zinc oxide thin films doped with aluminum and lithium," *Thin Solid Films*, vol. 518, pp. 4603-4606, 2010.

[13] P. Uprety, K. J. Lambright, C. R. Grice, M. M. Junda, D. M. Giolando, and N. J. Podraza, "Morphological and optical properties of low temperature processed SnO$_2$:F," *Physics Status Solidi B*, 2017.

[14] O. Madelung, *Semiconductors: Group IV Elements and III-V Compounds*, Springer Science & Business Media, 2012, pp. 124-13.

[15] R. N. Dexter, and B. Lax, "Effective masses of holes of silicon," *Physical Review*, vol. 96, pp. 223-224, 1954.

[16] D. M. Riffe, "Temperature dependence of silicon carrier effective masses with application to femtosecond reflectivity measurements," *Journal of the Optical Society of America B*, vol. 19, pp. 1092-1100, 2002.

[17] D. T. F. Marple, "Effective electron mass in CdTe," *Physical Review*, vol. 129, pp. 2466-2470, 1963.

[18] S. Perkowitz, and R. H. Thorland, "Far-infrared study of free carriers and the plasmon-phonon interaction in CdTe," *Physical Review B*, vol. 9, pp. 545-550, 1974.

[19] M. Petrović, N. Romčević, J. Trajić, W. D. Dobrowlski, M. Romčević, B. Hadžić, M. Gilić, and A. Mycielski, "Far-infrared spectroscopy of CdTe$_{1-x}$Se$_x$(In): Phonon Properties," *Infrared Physics & Technology*, vol. 67, pp. 323-326, 2014.

[20] S. P. Kozyrev, L. K. Vodopyanov, and R. Triboulet, "Structural analysis of semiconductor-semimetal alloy Cd$_{1-x}$Hg$_x$Te by infrared lattice-vibration spectroscopy," *Physical Review B*, vol. 58, pp. 1374-1384, 1998.

[21] K. A. Maslin, C. Patel, and T. J. Parker, "Far-infrared optical constants of a selection of zincblende structure crystals at 300 K," *Infrared Physics*, vol. 32, pp. 303-310, 1991.

# *In Situ* Raman Monitoring of Kesterite Cu₂ZnSnS₄ Phase Formation from Sulfurization of Sol-gel Oxide Precursors

Osama Awadallah, Joseph Hernandez, Andriy Durygin, and Zhe Cheng*

Florida International University, Miami, Florida, 33174, USA

* Corresponding author, zhcheng@fiu.edu

*Abstract* — Formation of kesterite Cu₂ZnSnS₄ (CZTS) thin films from sulfurization of sol-gel derived oxide precursor was successfully monitored using advanced *in situ* Raman microspectroscopy from room temperature up to ~350°C. The obtained *in situ* Raman data revealed the emergence of CZTS phase after 30 minutes of sulfurization at 350°C in 100 ppm H₂S+4%H₂ +N₂ gas mixture, while Cu₂S phase was observed to form upon continued sulfurization, which was attributed to the slight Cu-rich recipe used. Both phases remained stable upon cooling the sample down to room temperature. Room temperature XRD and EDS data for the *in situ* examined sample confirmed that CZTS phase successfully formed with its surface decorated by Cu₂S particles.

## I. INTRODUCTION

Among various light absorber materials for solar cell applications, kesterite Cu₂ZnSnS₄ (CZTS) has attracted great attention as a potential candidate for the fabrication of low cost and environmentally friendly thin film solar cells due to its advantages that CZTS consists of all earth-abundant and non-toxic elements and it offers high absorption coefficient ($>10^4\,cm^{-1}$) [1] with tunable direct band gap (via substitution of sulfur by selenium) [2]. However, the preparation of high quality CZTS material has remained as a challenge due to the narrow compositional region where that phase exists [3] and the resulting complexity associated with the formation of secondary phases such as ZnS, Cu₂S, and Cu₂SnS₃ in processing, which are usually harmful to device performance [4].

Many *ex situ* characterization studies based on various techniques such as multiwavelength excitation Raman scattering, room temperature photoluminescence (RT-PL), and secondary ion mass spectroscopy (SIMS) [5-7] have been carried out to address the issue of CZTS phase purity and obtain information related to the secondary phases. However, all those techniques could not provide information regarding the *real time* formation of CZTS phase.

On the other hand, most of the previous *in situ* investigations are based on XRD and come short of distinguishing between CZTS and related secondary phases such as ZnS and Cu₂SnS₃. In addition, *in situ* XRD is also limited by its poor spatial resolution [8]. In contrast, *in situ* Raman microspectroscopy,

as a multi-phase sensitive and non-destructive technique with high spatial resolution down to ~1 μm, offers great potential to study the phase formation mechanism of CZTS material *in real time under practical processing conditions*.

In a previous study by the authors of this work, *in situ* Raman characterization of CZTS thin films had been carried out, which showed CZTS is Raman active up to 600°C and its oxidation in air starts at ~400°C and follows a two-step oxidation process [9].

Nevertheless, to the best of our knowledge, there is no study on *in situ* monitoring of CZTS *phase formation from oxide precursor material via sulfurization in sulfur-containing atmosphere* using Raman. The objective of this study is to demonstrate the use of *in situ* Raman microspectroscopy for monitoring CZTS phase formation and subsequent evolution in a sulfur-containing atmosphere at elevated temperature. For this purpose, a special *in situ* Raman cell that enables temperature and atmosphere control was designed and implemented.

## II. EXPERIMENTAL WORK

### A. Preparation of Oxide Precursor Thin Films

A Cu-rich Cu-Zn-Sn oxide precursor thin film sample with atomic ratio of Cu : Zn : Sn = 2.2 : 1 : 1 was prepared from Cu-Zn-Sn metal ion containing solution as descried elsewhere [9, 10]. For simplification, the oxide precursor sample was deposited over a bare soda lime glass (SLG) substrate instead of the typical molybdenum coated SLG substrate. The sample was subsequently annealed in the *in situ* Raman cell in a hydrogen sulfide (H₂S) containing atmosphere at elevated temperature to form the CZTS phase, as described below.

### B. In Situ Raman Monitoring System

A special Raman cell was designed and implemented for this study. The cell has temperature control capability up to ~450°C and allows both inert (Ar and N₂) and reducing atmosphere (e.g., forming gas of ~4% H₂ in nitrogen with or without ppm-level H₂S). Fig. 1(a) shows a schematic of the *in*

978-1-5090-5606-4/17 $31.00 © 2017 IEEE

*situ* Raman system with the heating cell connected to the power supply and the temperature controller. Fig. 1(b) is a photo of the actual Raman cell with the top open to show the configuration. A type K thermocouple is inserted into the Raman cell through a specially sealed port and it would be fixed to touch the back surface of the sample for accurate temperature measurement. The Raman cell configuration as shown in Fig. 1 is designed so that the sample would lay down horizontally allowing the laser to be focused vertically onto the sample. The Raman system is equipped with an air-cooled Ar ion laser (Spectra Physics Model 177, 514 nm, 400mW) and Raman spectra were collected using a spectrograph (HoloSpec f/1.8i, Kaiser Optical System).

## C. *Ex Situ Characterization of in situ Raman Examined Thin Films*

After the *in situ* Raman experiment, the sample examined was subsequently characterized *ex situ* using various techniques. In particular, it was examined by X–ray diffraction (Diffractometer Siemens 500D) with Cu $K_\alpha$ ($\lambda = 1.5406$ °A) radiation and $2\theta$ in the range of 20 - 80° for phase identification. SEM combined with EDS was performed on JEOL 6330F microscope at 15.0 kV accelerating voltage to investigate the surface morphology and assess the local chemical composition.

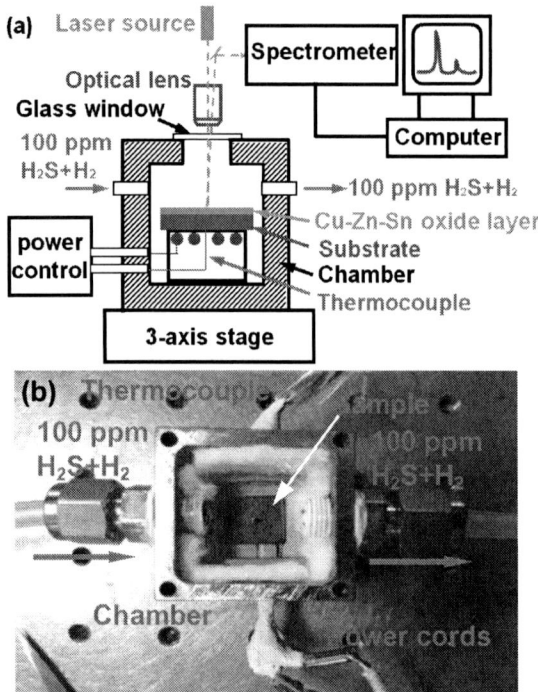

Fig. 1. (a) Schematic of the *in situ* Raman microspectroscopy system for monitoring CZTS formation. (b) A photo of the Raman cell assembly with the top open showing the oxide precursor thin film sample placed inside the chamber of the Raman cell.

## III. RESULTS AND DISCUSSION

### A. *In situ Raman Monitoring of CZTS Phase Formation from Oxide Precursor via Sulfurization at 350°C*

A Cu-Zn-Sn oxide precursor thin film sample was placed inside the *in situ* Raman cell and quickly heated up from room temperature to 350°C at 25°C/min with Ar gas (UHP grade, Airgas) purge. After that, the gas was switched from Ar to a gas mixture of 100 ppm $H_2S+4\%H_2+N_2$ (Airgas, labelled in short as 100ppm $H_2S+H_2$ in Fig. 1) and the temperature was kept constant at 350°C. During that isothermal process, three Raman spectra were collected after 5, 30, and 60 minutes of hold time. Finally, the sample was cooled down naturally in Ar and the Raman spectrum was collected at room temperature.

Fig. 2 shows the Raman spectra collected from the thin film sample before, during, and after the *in situ* experiment at 350°C in the $H_2S$ containing atmosphere. The Raman spectrum for the oxide precursor at room temperature (the bottom spectrum) is featureless. Then, after only ~5 minutes' hold at 350°C in 100 ppm $H_2S+4\%H_2+N_2$, a low bump centered at ~330 cm$^{-1}$ appeared. With continued holding at 350°C for 30 min, a distinct peak could be observed at Raman shift of 331 cm$^{-1}$. No Raman peaks for other secondary phases (e.g., SnS, $Cu_2S$) could be observed. Upon holding the sample for longer time of 60 minutes, the main peak shifted to higher frequency of 334 cm$^{-1}$, which is attributed to CZTS: Comparing with the expected Raman peak at 338 cm$^{-1}$ at *room temperature*, the shift to lower frequency of 334 cm$^{-1}$ at 350°C can be attributed to the increased temperature, which weakens the inter-atomic bonds [9].

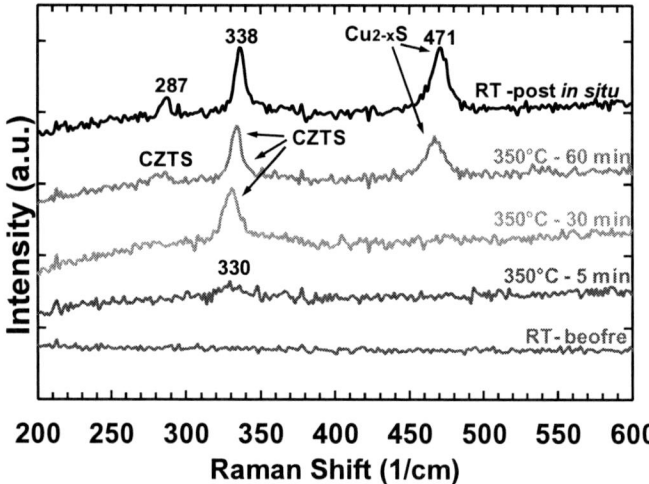

Fig. 2. Raman spectra for the thin film sample collected under *in situ* conditions at 350°C after different sulfurization holding time of 5, 30, and 60 minutes in 100 ppm $H_2S + H_2 + N_2$. For comparison, the Raman spectra for the sample collected at *room temperature* (RT) *before* the in situ experiment and *after* cooling the sample down are also shown.

In addition, a wide shoulder between 350 cm$^{-1}$ and 390 cm$^{-1}$ emerges and can be observed after 30 minutes of sulfurization. After 60 minutes of sulfurization, that shoulder became narrower with its maximum intensity at ~350 cm$^{-1}$ frequency shift. This most likely also corresponds to the Raman peak for CZTS [11]. It is also noted that one strong peak at Raman shift of 467 cm$^{-1}$ appeared after ~60 min of sulfurization at 350°C, and it is attributed to Cu$_{2-x}$S phase at elevated temperature [5].

After cooling the sample down to room temperature in inert Ar atmosphere, the two main peaks for CZTS phase at 338 cm$^{-1}$ and 287 cm$^{-1}$[5] were clearly identified along with one peak that belongs to the Cu$_{2-x}$S phase at 471 cm$^{-1}$. The emergence of Cu$_{2-x}$S phase with CZTS phase *after* continued sulfurization is most likely related to the fact that the recipe used in this study is slightly copper rich. It is noted that the reason for the gradual shift of Raman peak from ~330 to ~334 cm$^{-1}$ with increasing holding time from 5 min to 60 min is *not* exactly clear at the moment, but it indicates that CZTS phase formation from sulfurization of amorphous Cu-Zn-Sn oxide precursor at 350°C in 100 ppm H$_2$S+4%H$_2$+N$_2$ is a rather complex process. Nevertheless, additional examination of the *in situ* examined sample using other *ex situ* techniques was carried out to complement the *in situ* Raman monitoring experiment and confirm the conclusions made above, as detailed below.

## B. Ex Situ Characterization

XRD analysis was carried out for the *in situ* Raman examined sample and the pattern is shown in Fig. 3. Characteristic XRD peaks for CZTS at 2θ of 28.4°, 47.2°, and 56.0 were identified and no peaks for other sulfide phases could be observed [8]. For comparison, the XRD pattern for the oxide precursor *before* the *in situ* sulfurization was also given, which shows a broad bump, suggesting amorphous phase before sulfurization. Fig. 3 also includes pictures of the sample before and after *in situ* sulfurization, which show the sample changed from semi-transparent light brown color to dark black, which matches the expectation of changing from insulating oxide precursor to a semiconducting light absorber.

Further SEM/EDS analysis for the *in situ* Raman examined sample revealed a distinct feature of relatively flat film surface decorated with micron-sized particles on the top, as shown in Fig. 4. The EDS analysis result showed that the flat areas have a near-stoichiometric CZTS composition with slight S and Sn deficiency and Zn enrichment. The chemical composition of the surface particles showed a high content of Cu and S (atomic ratio of Cu : S is ~ 2:1) and the amounts of Zn and Sn were minimal as indicated in the EDS data in Fig. 4, which indicates the surface particles most likely correspond to Cu$_{2-x}$S.

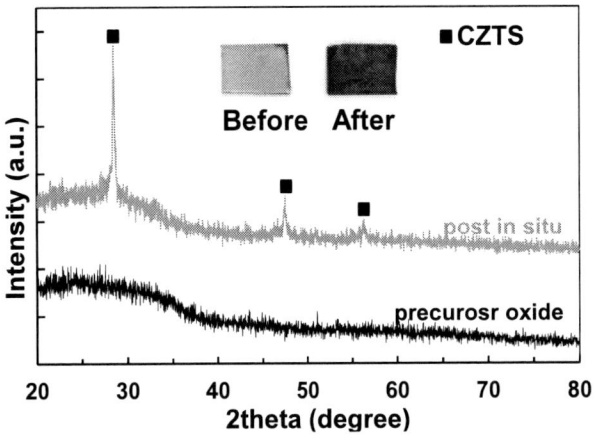

Fig. 3. Room temperature XRD patterns for the Cu-Zn-Sn oxide precursor and the *in situ* Raman examined sample. The inserts are pictures of the thin film sample before and after the *in situ* Raman experiment.

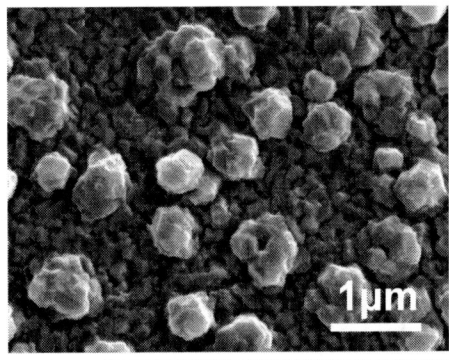

| Element | Cu (at%) | Zn (at%) | Sn (at%) | S (at%) |
|---|---|---|---|---|
| Flat area | 25.5±8.7 | 15.7±5.0 | 10.6±1.0 | 48.0±1.8 |
| Surface particles | 64.2±9.2 | 2.5±6.4 | 7.1±1.0 | 26.2±3.8 |

Fig. 4. SEM image and EDS data for *in situ* Raman examined sample showing flat area corresponding to near-stoichiometric CZTS and protruding surface particles most likely corresponding to Cu$_2$S.

## C. Proposed Formation Mechanism of CZTS thin films

Based on the observations from the *in situ* Raman experiment and both *ex situ* XRD and EDS analyses, it is concluded that CZTS phase successfully forms from the amorphous oxide precursor via sulfurization using 100 ppm H$_2$S+4%H$_2$+N$_2$ gas mixture after ~30 minutes of sulfurization at 350°C. In addition, Cu$_{2-x}$S phase was observed to emerge upon continuous sulfurization, which might be due to the excess Cu in the initial solution precursor used. After 30 minutes of sulfurization, most of the Cu in the sample reacts with H$_2$S to form CZTS, while the remaining *excess* Cu migrates to the film surface and react with $H_2S$ to form $Cu_{2-x}S$ phase that crystallizes in particle shape on top of the flat CZTS layer as observed by SEM.

In that sense, a reaction pathway for the formation of CZTS via the synthesis route of sulfurization of sol-gel derived oxide precursor can be proposed as follows. In the atmosphere that contains both hydrogen and 100 ppm level $H_2S$; the amorphous (mixed) oxides might first go through reduction to form metals (or their alloys) as below:

$$ZnO + H_2 \rightarrow Zn + H_2O \qquad (1)$$
$$Cu_2O + H_2 \rightarrow 2Cu + H_2O \qquad (2)$$
$$SnO_2 + 2H_2 \rightarrow Sn + 2H_2O \qquad (3)$$

Then, the metals (or their alloys) would continue to react with $H_2S$ in the atmosphere to form CZTS, as below:

$$2Cu + Zn + Sn + 4H_2S \rightarrow Cu_2ZnSnS_4 + 4H_2 \qquad (4)$$

The overall net reaction (5) is simply the combination of reactions (1) to (3) and (4), which is:

$$Cu_2O + ZnO + SnO_2 + 4H_2S \rightarrow Cu_2ZnSnS_4 + 4H_2O \qquad (5)$$

Whether the actual reaction goes through the reduction-sulfurization two-step process or as a single-step reaction (as in (5)) is not clear, but the result of the *in situ* Raman characterization suggests the whole process from oxide to sulfide seems to be relatively fast ($<\sim30$ mins) under the condition explored (i.e., 350°C with $pH_2S/pH_2 = 100$ ppm). As discussed in an earlier study, whether sulfurization reaction (4) would happen depends strongly on the $H_2S$ to $H_2$ ratio used: For a given sulfurization temperature, when the $pH_2S/pH_2$ ratio is too low, metal sulfides and CZTS could not form and mixtures of metal with sulfides are obtained [10]. On the other hand, for that particular Cu-rich sample, the excess Cu remained after CZTS formed reacts with $H_2S$ upon continued sulfurization to form $Cu_2S$, which might not be well crystalized. This explains the reason for the wide bump and the absence of $Cu_2S$ peaks in the XRD pattern as seen in Fig. 3.

## IV. CONCLUSION

In the present work, the formation of CZTS from sol-gel oxide precursor films in 100 ppm $H_2S+4\%H_2+N_2$ gas mixture at intermediate temperature of 350°C was successfully monitored using *in situ* Raman microspectroscopy for the first time. The result showed that under the condition explored, crystalline sulfide phase with Raman peak at $\sim334$ cm$^{-1}$ form after $\sim60$ minutes of sulfurization. Upon cooling of the *in situ* examined sample, two peaks characteristic to CZTS were observed at 338 cm$^{-1}$ and 287 cm$^{-1}$. This, together with *ex situ* characterization using XRD and EDS confirm the formation of slightly Zn-rich and Sn deficient CZTS phase during the *in situ* experiment, and the CZTS remains stable after cooling the sample down to room temperature. In addition, the *in situ* Raman experiment also reveals that annealing of the sample in the same atmosphere also leads to the appearance of the Raman peak at 467 cm$^{-1}$, which is attributed to $Cu_{2-x}S$ particles on the thin film surface. The copper sulfide phase also remains upon cooling down of the thin film samples, and was attributed to the copper-rich solution chemistry used. EDS analysis of those $Cu_{2-x}S$ particles on the sample surface suggest that they are rich in Cu and S with Cu to S atomic ratio of $\sim2:1$ with almost no Zn and Sn, which suggest that they are most likely $Cu_2S$. The study demonstrated the capability of *in situ* Raman as a unique and powerful tool that would help reveal the detailed formation mechanism for CZTS and guide experiments aimed at improving the quality of the prepared CZTS films and eliminating impurities such as $Cu_xS_y$ phases.

## REFERENCES

[1] H. Katagiri, K. Jimbo, W. Maw, K. Oishi, M.Yamazaki, H. Araki, and A. Takeuchi, "Development of CZTS-based thin film solar cells," *Thin Solid Films*, vol. 517, pp. 2455–2460, 2009.

[2] T. Gershon, T. Gokmen, O. Gunawan, R. Haight, S. Guha, and B. Shin, "Understanding the relationship between $Cu_2ZnSn(S,Se)_4$ material properties and device performance," *MRS Communications*, vol. 4, pp. 159–170, 2014.

[3] S. Chen, J. Yang, X. G. Gong, A. Walsh, and S. Wei, "Intrinsic point defects and complexes in the quaternary kesterite semiconductor $Cu_2ZnSnS_4$," *Physical Review B*, vol. 81, pp. 245204, 2010.

[4] M. Kumar, A. Dubey, N. Adhikari, S. Venkatesan and Q. Qiao, "Strategic review of secondary phases, defects and defect-complexes in kesterite CZTS–Se solar cells," *Energy & Environmental Science.*, vol.8, pp. 3134-3159, 2015.

[5] M. Dimitrievska, A. Fairbrother, X. Fontané, T. Jawhari, V. Roca, E. Saucedo, A. Pérez-Rodríguez, "Multiwavelength excitation Raman scattering study of polycrystalline kesterite $Cu_2ZnSnS_4$ thin films," *Applied Physics Letters*, vol. 104, pp. 021901 – 021905, 2014.

[6] R. Djemour, M. Mousel, A. Redinger, L. Gutay, A. Crossay, D. Colombara, P. J. Daleand S. Siebentritt "Detecting ZnSe secondary phase in $Cu_2ZnSnSe_4$ by room temperature photoluminescence," *Applied Physics Letters*, vol. 102, pp. 222108, 2013.

[7] A. Redinger,, K. Hones, X. Fontane, V. Roca, E. Saucedo, N. Valle,A. Rodriguez, and S. Siebentritt, "Detection of a ZnSe secondary phase in coevaporated$Cu_2ZnSnSe_4$ thin films," *Applied Physics Letters*, vol. 98, pp. 101907, 2011.

[8] S. Schorr, G. Gonzalez-Aviles, "In-situ investigation of the structural phase transition in kesterite," *Physica Status Solidi (a)*, vol. 206, pp. 1054 – 1058, 2009.

[9] O. Awadallah and Z. Cheng, "In Situ Raman Monitoring of $Cu_2ZnSnS_4$ Oxidation and Related Decomposition at Elevated Temperatures," *IEEE Journal of Photovoltaics*, vol. 6, pp. 764–769, 2016.

[10] O. Awadallah and Z. Cheng, "Formation of sol-gel based $Cu_2ZnSnS_4$ thin films using ppm-level hydrogen sulfide," *Thin Solid Films*, vol. 625, pp. 122–130, 2017.

[11] M. Guc, S. Levcenko, I. V. Bodnar, V. Izquierdo-Roca, X. Fontane, L. V. Volkova, E. Arushanov and A. Pérez-Rodríguez, "Polarized Raman scattering study of kesterite type $Cu_2ZnSnS_4$ single crystals," *Scientific Reports*, vol. 6, 19414, 2016.

# Performance of Field-Aged PV Modules in India: Results from 2016 All India Survey of PV Module Reliability

Rajiv Dubey[1], Sachin Zachariah[1], Shashwata Chattopadhyay[1], Vivek Kuthanazhi[1], Sugguna Rambabu[1], Sonali Bhaduri[2], Hemant K. Singh[1], Archana Sinha[1], Birinchi Bora[3], Rajesh Kumar[3], O. S. Sastry[3], Chetan S. Solanki[1,2], Anil Kottantharayil[1], Brij M. Arora[1], K. L. Narasimhan[1], Juzer Vasi[1]

[1]National Centre for Photovoltaic Research and Education, Indian Institute of Technology Bombay, Powai, Mumbai, 400076, India

[2]Energy Science and Engineering Department, Institute of Technology Bombay, Powai, Mumbai, 400076, India

[3]National Institute for Solar Energy, Ministry of New and Renewable Energy, New Delhi, 110003, India

*Abstract* — This paper presents the electrical performance data gathered in the 3rdAll India Survey of Photovoltaic Module Reliability conducted in 2016, in which a total of 925 modules were inspected in different climatic zones of India. The average degradation rate of the Group A ('All' sites) modules is 1.23%/year and the so-called 'Good' sites is 0.63%/year, better than that observed in the 2014 Survey, the reason being partly the inclusion of more large PV power plants, and also discounting of 2% LID in 2016 analysis methodology. The modules in 'Hot' climates degrade faster than modules in 'Non-Hot' climates, consistent with earlier results. Modules in small systems (capacity <100 kW) degrade at a faster rate than those in large systems (capacity >100 kW). Roof-mounted systems also degrade at a faster rate than ground-mounted systems.

*Index Terms* — Degradation, photovoltaic modules, defects, silicon, reliability.

## I. INTRODUCTION

India's National Solar Mission has set a target of installing 100 GW of solar power in India by 2022, which is resulting in significant investment in the solar sector. The actual success of this mission will depend on energy generation from the power plants over the long term. Given the continued price pressure on PV module manufacturers in recent years, it is important to keep a constant watch on the performance of the installed modules, and to assess their long-term durability and reliability. At the same time, it is also important to understand the impact of the harsh climatic zones of India on modules' durability. In this context, the All India Survey of Photovoltaic Module Reliability has been conducted previously in 2013 [1] and 2014 [2][3], and most recently in 2016. While the previous surveys concentrated more on older systems (to understand the long-term reliability aspects), which were predominantly smaller installations, the 2016 survey included more of the recently installed large PV power plants. This paper provides an overview of the electrical performance of PV modules, installed in the six different climatic zones of India, which were inspected in the months of March to May 2016 (refer to Table I for data statistics). During this survey 925 numbers of modules which includes c-Si, CdTe, CIGS, a-Si and HIT were inspected. The visual inspection, IR and EL test data were also collected during the survey, and their correlation with the electrical performance will be presented in a future article.

## II. METHODOLOGY

During the survey, the modules were characterized by various techniques, like illuminated *I-V*, dark *I-V*, illuminated IR and dark IR thermography, onsite electroluminescence (EL) testing, and dry/wet insulation resistance test. The electrical characterization of the modules was done by using a portable *I-V* tracer (Solmetric PVA-1000S). The procedure followed was overall similar to the 2014 survey and is described in detail in the report [2] and elsewhere [3][4]. For the analysis of the electrical performance data, the *I-V* curves (measured above 750 W/m$^2$ plane of array irradiance) were translated to the standard test condition. For determining the degradation rate, the initial performance data is needed. However, in most cases, the initial *I-V* curves were not available, so the performance data was computed based on the nameplate details. Since there is usually a tolerance on the name plate power rating, we have defined the *Nominal power* rating as the name plate power plus average of the tolerance band. Similar to the 2014 survey analysis, we have calculated the *Overall power* degradation rate as given below.

$$Overall\ Pmax\ Degradation\ Rate\ (\%/year) = \frac{(P_{nominal} - P_{present}) \times 100}{P_{nominal} \times Age}$$

$$\dots (1)$$

Most manufacturers of crystalline silicon modules provide for an initial rapid degradation in their modules (usually 2% - 3% in the first year of field exposure) which is mainly due to Light Induced Degradation (LID), followed by a linear power warranty. To accommodate this initial LID loss in the modules, we have assumed a 2% LID loss in power output from the *Nominal power*, and considered it to be the *LID discounted power* rating. LID loss in modules varies widely [5][6] depending on the wafer quality and we have taken the 2% figure as an average, considering that the widely used PVSyst software also recommends 2% LID [7]. A linear power degradation rate is calculated based on this *LID discounted power* rating, using the relation shown below:

$$Linear\ Pmax\ Degradation\ Rate\ (\%/year) = \frac{(P_{LID\_discounted} - P_{present}) \times 100}{P_{LID\_discounted} \times Age}$$

$$\dots (2)$$

Note that the LID discounting for c-Si modules is new for the 2016 survey, and this needs to be remembered when comparing values with those obtained from earlier surveys.

### TABLE I
### SITE SUMMARY

| Climatic Zone | Number of Sites Visited | Number of Modules Surveyed |
|---|---|---|
| Hot & Dry (H&D) | 11 | 278 |
| Warm & Humid (W&H) | 10 | 243 |
| Composite (Comp) | 7 | 184 |
| Moderate (Mod) | 4 | 94 |
| Cold& Sunny (C&S) | 4 | 106 |
| Cold & Cloudy (C&C) | 1 | 20 |
| TOTAL | 37 | 925 |

### III. ANALYSIS OF SURVEY DATA

Statistical analysis was performed on the *Overall $P_{max}$* degradation rates to separate out the outliers from the data, and 34 modules with $P_{max}$ degradation rate greater than 5.12%/yr were designated as 'outliers'. These are not included in the analysis presented in this paper. Fig. 1 shows the histogram of the *Overall $P_{max}$* degradation rates for the modules in 'All' sites (which will be referred to as Group A sites), using Eq. (1). The average degradation rate is 1.55%/year, which is better than the rate of ~ 2%/year obtained in the 2014 Survey. This is due to the inclusion of many more large PV power plants compared to the 2014 survey. The degradation data is further processed by discounting LID for c-Si modules to determine the *Linear $P_{max}$* degradation rate using Eq. (2). (Note that LID discounting is done only for c-Si modules and not thin films.). The histogram of *Linear $P_{max}$* degradation rate for the Group A modules is shown in Fig. 2. The average *Linear $P_{max}$* degradation rate of these modules is 1.23%/year, which is better than the average degradation rate of Fig 1. This difference between the *Overall $P_{max}$* and *Linear $P_{max}$* degradation rate is due to the discounting of 2% LID in 2016 survey analysis, especially since many young modules were present in the 2016 survey. Some of the young modules (installed less than 5 years ago) have shown *negative* degradation rates, which may be due to the under-rating of the modules by the manufacturer, or due to lower LID (less than 2%) in these modules. Keeping in tune with the analysis procedure followed for the 2014 survey data [2][3], the sites have been categorized into 2 groups, Group X (where average *Overall $P_{max}$* degradation rate of the *site* is less than 2%/year) and Group Y (average *Overall $P_{max}$* degradation rate of *site* is greater than 2%/year). As also observed in the 2014 survey data analysis, the Group Y sites are found to possess a high variability in the degradation rates, which may be due to inconsistency in the production process, raw material quality or other issues like transportation and installation. On the other hand, the Group X sites show low variation of power degradation rate (refer Fig. 3) and are deemed to be 'Good' sites. The average *Linear $P_{max}$* degradation rate of these sites is 0.63%/year, in consonance with the performance warranty given by the module manufacturers. Fig. 4 shows the effect of climatic zone on the performance of Group X modules, which

are higher than the data published by Jordan *et al.* [8][9]. The young modules (age <5 years) have been shown with open triangles, while the filled triangles indicate the old (age>5years) modules. The red horizontal bar is the mean, with two error bars (red on left for name plate error/tolerance and green on right for instrument and translation error), and the diamonds represent the 95% confidence interval for the combined data represented in open and closed triangles. We can see a decreasing trend in the *Linear $P_{max}$* degradation rate as we move from the Hot climates (Hot & Dry, Warm & Humid and Composite zones) to the Cold climates. If we neglect the modules from the Moderate zone (as there are only 20 samples in Group X, which is not statistically significant), modules in the Warm & Humid zone are degrading at the fastest rate (1.21%/year, average of 170 samples) which is 1.7 times of the linear power warranty given by module manufacturers. It can be seen that the degradation rates for the young modules is generally higher than that of the old modules. Since LID has been discounted, this shows that the quality of young modules is poor. Fig. 5 shows explicitly that the modules in 'Hot' zones (Warm & Humid, Composite zones and Hot & Dry) are degrading at a much higher rate than the modules in the 'Non-Hot' zones (Moderate, Cold & Cloudy and Cold & Sunny).

Fig. 6 shows the effect of system size on the *Linear $P_{max}$* degradation rate. The average degradation rate of the smaller systems (installed capacity <100 kW) is double of the large systems (capacity > 100 kW). The better performance of the large systems can be partly attributed to the proper due diligence regarding module quality done by the installers of large systems. The young modules are degrading faster than the older modules, *even after discounting for the initial LID*, which is a cause of concern. It seems to imply that the quality and/or installation practices of recent times are not adequate.

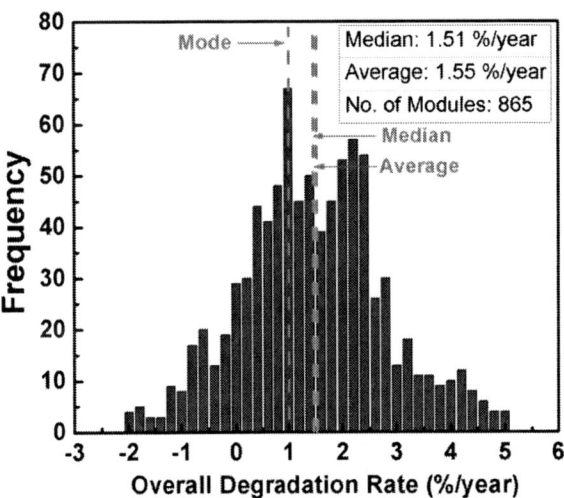

Fig. 1. Histogram for *Overall $P_{max}$* degradation rates for modules in Group A ('All') sites

978-1-5090-5606-4/17 $31.00 © 2017 IEEE

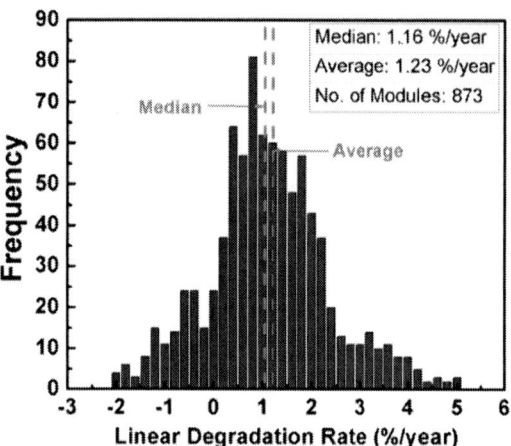

Fig. 2. Histogram for *Linear P_max* degradation rates for modules in Group A ('All') sites

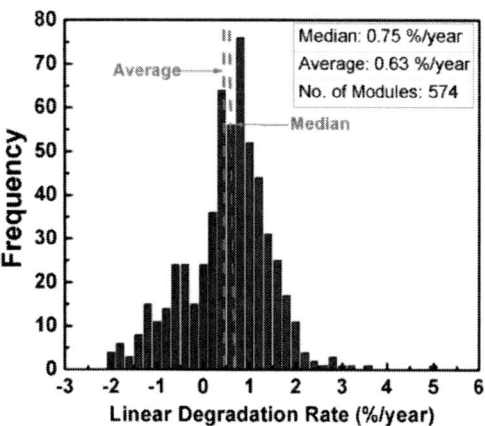

Fig.3. Histogram for *Linear P_max* degradation rates for modules in Group X ('Good') sites.

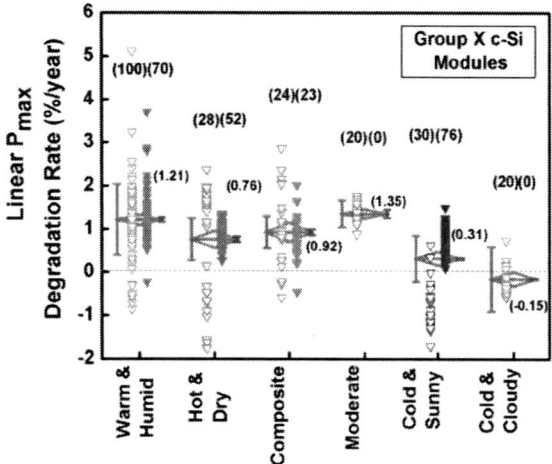

Fig. 4. Influence of climatic zone on *Linear P_max* degradation rate for modules in Group X sites. Numbers in the brackets give number of modules, open symbols for module age < 5 years and filled symbols for module age > 5 years.

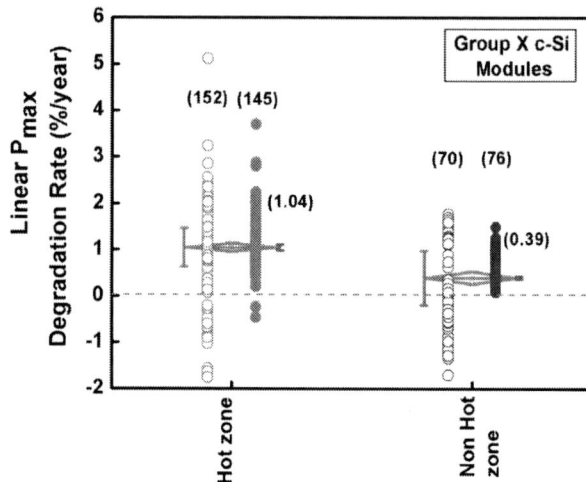

Fig. 5. *Linear P_max* degradation rates for Hot and Non-Hot climatic zones for modules in Group X sites. Numbers in the brackets give number of modules, open symbols for module age < 5 years and filled symbols for module age > 5 years.

The type of mounting also plays a role in the degradation rates, as shown in Fig 7. Roof mounted systems are degrading at a much faster rate than ground-mounted systems (2 %/yr versus 0.68%/yr), which is a matter of grave concern given that 40% of India's 100 GW target is planned for rooftop installations. The lack of due diligence by roof-mounted system owners regarding module quality, as well as the fact that roof-mounted modules tend to run hotter, are two possible reasons for this trend. A detailed statistical analysis of the data presented in this paper has been performed [10], which shows that the conclusions arrived at in this paper have very good statistical significance.

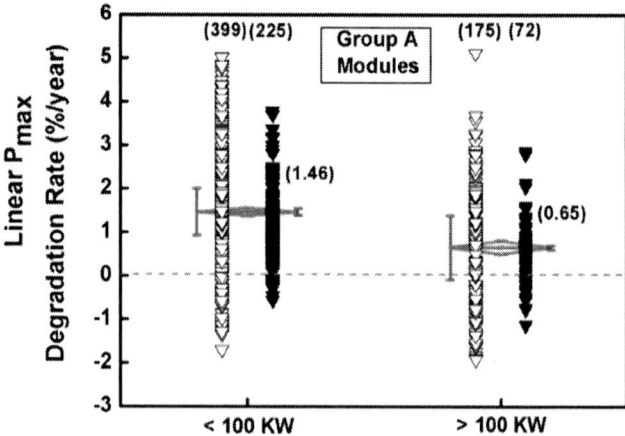

Fig. 6. Effect of system size on *Linear P_max* degradation rate for Group A ('All') sites. Numbers in the brackets give number of modules, open symbols for module age < 5 years and filled symbols for module age > 5 years.

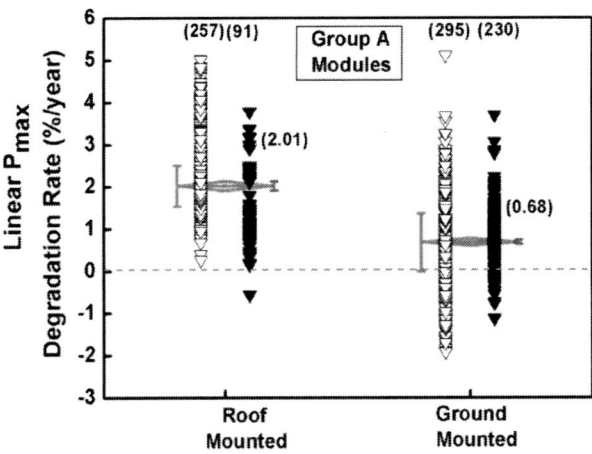

Fig. 7. Effect of mounting type on *Linear P_max* degradation rate for Group A ('All') sites. Numbers in the brackets give number of modules, open symbols for module age < 5 years and filled symbols for module age > 5 years.

## V. CONCLUSION

The 2016 All India Survey of Photovoltaic Module Reliability has provided valuable information regarding the field performance of PV modules in different climates of India, particularly of the newly installed power plants. The average degradation rate for the Group A ('All' sites) modules is 1.23 %/year and for modules in Group X ('Good') sites is 0.63%/year, which is quite reasonable. On the other hand, the modules from under-performing Group Y sites are a cause for concern, and these sites are being separately examined to understand the reasons for high degradation rates. Modules in Hot zones are degrading faster as compared to those in Non-Hot zones; and roof-mounted modules (which are usually part of smaller systems) are faring worse than ground-mounted modules (which mostly belong to large PV power plants). Both these results are consistent with our previous surveys. The data again highlights the issue of durability in Hot climatic zones, which needs to be examined.

## ACKNOWLEDGEMENT

This research is based upon work supported in part by (a) the National Centre for Photovoltaic Research and Education funded by Ministry of New and Renewable Energy of the Government of India through the Project No. 31/09/2015-16/PVSE-R&D dated 15th June 2016 and (b) the Solar Energy Research Institute for India and the U.S. (SERIIUS) funded jointly by the U.S. Department of Energy subcontract DE AC36-08G028308 (Office of Science, Office of Basic Energy Sciences, and Energy Efficiency and Renewable Energy, Solar Energy Technology Program, with support from the Office of International Affairs) and the Government of India subcontract IUSSTF/JCERDC-SERIIUS/2012 dated 22nd Nov. 2012. The authors also thank S. Sabnis and C. Mahapatra for providing inputs for statistical analysis.

## REFERENCES

[1] R. Dubey *et al.*, "All India Survey of Photovoltaic Module Degradation 2013", National Centre for Photovoltaic Research and Education, Mumbai, India, 2014, available online at http://www.ncpre.iitb.ac.in/pages/publications_reports.html

[2] R. Dubey *et al.*, "All India Survey of Photovoltaic Module Reliability 2014", National Centre for Photovoltaic Research and Education, Mumbai, India, 2016, available online at http://www.ncpre.iitb.ac.in/uploads/All_India_Survey_of_Photovoltaic_Module_Reliability_2014.pdf

[3] R. Dubey *et al.*, "Comprehensive Study of Performance Degradation of Field-Mounted PV Modules in India," *Energy Science and Engineering*, http://dx.doi.org/10.1002/ese3.150.

[4] S. Chattopadhyay *et al.*, "All India Survey of Photovoltaic Module Degradation 2014: Survey methodology and statistics", *Proceedings of the IEEE Photovoltaic Specialists Conference*, New Orleans, 2015

[5] S. Pingel, D. Koshnicharov, O. Frank, T. Geipel, Y. Zemen, B. Striner and J. Berghold, "Initial Degradation of Industrial Silicon Solar Cells in Solar Panels", *Proceedings of the 25th European Photovoltaics Specialists Conference*, Valencia, 2010

[6] B. Sopori, P. Basnyat, S. Devayajanam, S. Shet, V. Mehta, J. Binns and Jesse Appel, "Understanding Light-Induced Degradation of Crystalline Silicon Solar Cells", *Proceedings of the IEEE Photovoltaic Specialists Conference*, Texas, 2012

[7] "LID (Light Induced Degradation) Loss", PVSyst, [Online] Available: http://files.pvsyst.com/help/lid_loss.htm

[8] D. C. Jordan *et al.*, "Technology and Climate Trends in PV Module Degradation", in 27th European Photovoltaic Solar Energy Conference and Exhibition, 2012, pp.3118-3124.

[9] D. C. Jordan *et al.*, "Compendium of Photovoltaic Degradation Rates," *Progress in Photovoltaics: Research and Applications*, online version, February 2016

[10] C. Mahapatra *et al.*, "Statistical Analysis of Degradation Data for c-Si modules observed in India in 2016", to be presented at the *44th IEEE Photovoltaic Specialists Conference*, D.C., 2017.

# Inferring the Performance Ratio of PV systems distributed in an region: a real-case study in South Tyrol

Marco Pierro[1,2*], Giorgio Belluardo[1], Philip Ingenhoven[1], Cristina Cornaro[2], David Moser[1]

[1]Institute for Renewable Energy - EURAC Research, viale Druso 1, 39100 Bolzano (Italy)

[2]Department of Enterprise Engineering - University of Rome Tor Vergata, via del Politecnico 1, 00133 Roma (Italy)

*Phone: +39 0471 055651; Fax: +39 0471 055699; E-mail: marco.pierro@eurac.edu

*Abstract* — **This paper presents a methodology to infer the monthly series of Performance Ratio (PR) and the tilt & orientation angles of PV systems installed in a certain extended area. It is particularly useful in a context of low-information, especially when irradiance data is not available for the sites, and it only requires the knowledge of produced energy, location and nominal power of each system of the fleet. Possible applications of this methodology are: regional survey and statistical analysis of distributed plants performance on monthly time scale; implementation of physical model of plant performance (requiring plant characteristic), to be used for nowcast and forecast application; statistical information on technology performance loss rate.**

*Index Terms* — **performance ratio, monthly yield, photovoltaic systems, PV regional distribution.**

## I. INTRODUCTION

The availability of information about the PV plants characteristics and performance is extremely important for the PV-market players, for several reasons:

- for O&M companies: to identify malfunctions and/or plants with space for performance improvement, in order to fulfill the quality levels of O&M contracts;
- for utility companies: to now-cast and forecast the PV-generated energy with a higher accuracy;
- for investment companies: to keep under control their plants and increase their profitability;
- for aggregator companies: to include efficiently also the PV distributed generation in their portfolio of power generation plants.

Nevertheless, typically monitoring systems often lack of fundamental information necessary to determine the performance of the PV systems, irradiance on the modules' plane above all.

This paper presents a methodology that, given only the monthly values of generated electricity, the location and the nominal power of the PV plant, allows the user to reconstruct the tilt & orientation angles, and the monthly series of both reference yield $Yr$ and performance ratio $PR$.

## II. DATA

Monthly values of final yield $Y$ [1] are available from the metering of 1899 PV plants installed in South Tyrol (North-East of Italy) with one-hour time resolution, corresponding to a total nominal power of 64 MW, and managed by the local energy utility Edyna. In addition, nominal power and coordinates of every single PV plant are known.

The location of the 1899 analyzed PV systems is shown in Fig. 1 (red dots). The distribution follows the main valleys and plateaus of the area of interest in South Tyrol, but plants are actually installed at several altitudes, corresponding to different temperature and insolation conditions.

The irradiance data are freely downloaded from the Copernicus Atmosphere Monitoring Service (CAMS) Radiation Service [2] through the SoDa project [3] and are available as interpolated values of original satellite images with spatial resolution of 4-5 km (at 45° of latitude), and one-hour time resolution. Satellite-retrieved irradiance from the company SolarGis has also been considered for validation purposes only for one site, to have a comparison with a high accuracy dataset purchased on-demand.

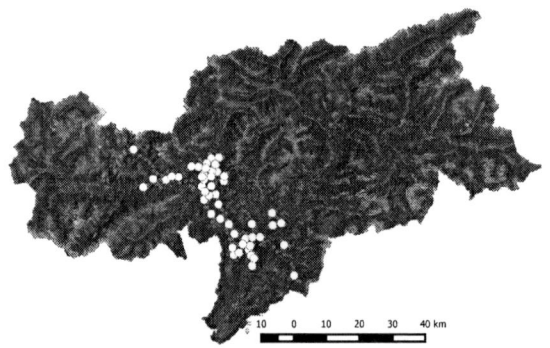

Fig. 1. Locations of the 1899 plants analyzed for the South Tyrol region (red dots) and of the centroids (white dots).

978-1-5090-5606-4/17 $31.00 © 2017 IEEE

## III. METHODOLOGY

The methodology consists of four main steps:
a. Preliminary filtering.
b. Clustering of PV plants based on their location.
c. Research of the tilt & orientation angles of each PV system.
d. Derivation of the monthly-series of PR.

### A. Preliminary filtering

The scope of the preliminary filtering is to remove all the PV plants with an anomalous trend in the monthly-series of final yield (e.g. due to a change in the nominal power), as well as outliers in the monthly values of final yield. The monthly final yield is defined as: $Y = \sum_{h=1}^{Nh} E_h / P_n$ (expressed in kWh/kWp), where $N_h$ is the total number of hours in a month, $E_h$ is the hourly PV-generated energy and $P_n$ is the nominal power of the plant.

First, the average value of the final yield for all months and plants is calculated. For each plant, the RMSE is determined considering the deviation of each available monthly value of the final yield from the average value of all plants and months, previously calculated. For this group of RMSE values, the average and the standard deviation $\sigma$ is therefore calculated, and plants with RMSE outside the range *average RMSE ±2σ* discharged. With the second filter, the average value and the standard deviation of the final yield is calculated for each month of the 11 years available (2006 to 2016). The values of final yield referring to a specific month that are out of the range *average Y ±2σ* are discharged. Fig. 2 shows the time series of the daily average per month of the final yield (expressed in kWh/kWp/day), for the 1750 PV systems remaining after the application of the preliminary filter.

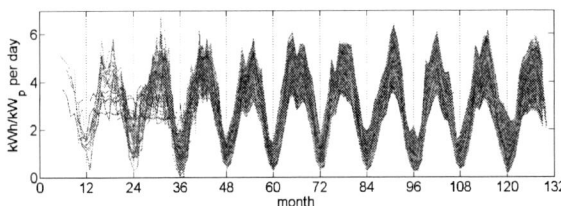

Fig. 2.    Time series of the daily average per month of the final yield for the 1750 plants analyzed, from 2006 to 2016.

### B. Clustering

In this step, a clustering K-mean algorithm [4] is used to group the PV plants in small ensembles of close systems. The reason for this operation is to find the most representative points of the ensembles (centroids, represented as white dots in Fig. 1), for which the irradiance is then downloaded or retrieved.

The $K$ parameter (i.e. the number of clusters) is chosen as a balance of good representativeness and computation effort required in the next steps. In this study, a value of $K=50$ is considered sufficient to properly account for the morphological complexity of the territory, as demonstrated later on.

Fig. 3 shows the distribution of the distance between each system and the respective centroid. The majority of the plants is less than 1 km far from the cluster centroid and the average distance is 750 m. The minimum distance is 30 m and the maximum 9 km.

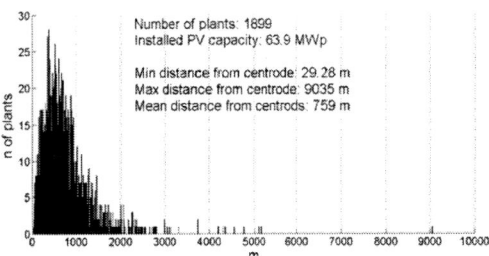

Fig. 3.    Distribution of the distance between plants and respective cluster centroid.

### C. Computation of tilt & orientation angles

In this step the tilt $\beta$ and orientation $\gamma$ angles and the mean monthly performance ratio $\langle PR \rangle$ of each PV system are retrieved using a minimization algorithm applied to a non-trivial cost function $J(\beta, \gamma, \langle PR \rangle)$. This algorithm, starting from an initial condition $J_0 = J(\beta_0, \gamma_0, \langle PR \rangle_0)$ finds the parameters vector $(\beta, \gamma, \langle PR \rangle)$ that minimizes the cost function $J$, defined as:

$$J(\beta, \gamma, \langle PR \rangle) = \text{var}(\nabla Y - \nabla Y_{calc}) + J_0 * \left| \langle PR \rangle - 0.8 \right| \quad (1)$$

where $Y_{calc}$ is the retrieved final yield of the plant:

$$Y_{calc}(\beta, \gamma, \langle PR \rangle) = \langle PR \rangle * Yr(\beta, \gamma) \quad (2)$$

with $Yr(\beta, \gamma)$ being the monthly reference yield (expressed in equivalent number of peak sun-hours, or kWh/kW), defined as a function of the irradiation on the plane of the array $H_h(\beta, \gamma)$ (expressed in kWh/m²) as:

$$Yr(\beta, \gamma) = \sum_{h=1}^{Nh} H_h(\beta, \gamma) / 1000 \quad (3)$$

with:

$$H_h(\beta, \gamma) = \sum^{n_m} G_h(\beta, \gamma) / n_m \quad (4)$$

where $n_m$ is the number of measurements in one hour.

In particular, the irradiance on the plane of the array $G_h(\beta, \gamma)$ (expressed in kW/m², and defined also as the global tilted irradiance $GTI$) is computed by projecting the global horizontal irradiance $GHI$ to the plane with tilt $\beta$ and orientation $\gamma$, using a transposition model that assumes an isotropic distribution of diffuse irradiance *DiffHi*.

Moreover, since the region has a complex orography, the considered tilted irradiance takes into account the shading due to the horizon at each plant by considering the solar diagram downloaded from PVGIS [5] (see Fig. 4).

The cost function (1) to be minimized consists of two different terms. The first is the variance of the difference between the monthly gradient of the observed yield $\nabla Y$ and the monthly gradient of the retrieved yield $\nabla Y_{calc}$. It reaches the minimum when $Y$ and $Y_{calc}$ monthly-series have the same shape. The second is a Lagrangian constrain that imposes a scale factor ($\langle PR \rangle \sim 0.8$) between $Y$ and $Yr$, see (2). Without this regularization term the minimization algorithms would find an incorrect $\beta$ and $\gamma$ corresponding to $Yr(\beta,\gamma)=Y$. The weight of the second term with respect to the first is $J_0=var(\nabla Y-\nabla Y_{calc}(\beta_0,\gamma_0,\langle PR \rangle_0))$, so that at the first iteration step of the algorithm the two terms have comparable values.

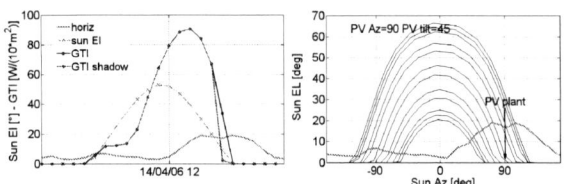

Fig. 4. (Left) impact of the horizon (red continuous line) on the irradiance on a roof top PV plant west-oriented and 45°-tilted, on a specific day (black line: no horizon considered, blue line: with horizon), and (right) solar diagram.

*D. Derivation of the monthly-series of PR*

The monthly values of PR of each plant is finally computed using the monthly reference yield (obtained in the previous step):

$$PR = \frac{Y}{Yr(\beta,\gamma)} \qquad (5)$$

The accuracy of the algorithm is evaluated by the error metrics (RMSE, MAE and MBE) between the observed yield series $Y$ and the retrieved yield series $Y_{calc}(\beta,\gamma)$.

## V. RESULTS AND DISCUSSION

*A. Comparison of the irradiance datasets*

The methodology described in the previous section is applied to a PV system installed at Bolzano Airport, not included in the fleet. At this site, the GHI, DNI (direct normal irradiance) and GTI are measured by a local weather station equipped with secondary standard radiometers regularly checked and calibrated. This 7 kWp CdTe plant is part of a multi-technology test facility operating since August 2010, and the electric parameters are measured by a commercial inverter

[6]. For this comparison as well as for the methodology validation, a total of four different irradiance sources are used:

1. Site measurements;
2. SolarGis satellite data, for the exact location;
3. CAMS Radiation Service, for the exact location;
4. CAMS Radiation Service, for the location of the centroid of the cluster where the PV plant is installed (adopted in the methodology).

The plant used as a benchmark belongs to cluster n. 26, and its distance from the centroid is 700 m, very close to the average distance value (Fig. 3).

In this section we are interested in the comparison of the GHI and DNI - and of the GTI computed with the same transposition model - with the corresponding measured irradiance in order to assess the quality of the selected datasets on a monthly level, rather than of the methodology itself. Table I lists the statistical deviation of the monthly values of irradiance from different sources, from the measured values. In particular, it is worth to note that the error due to the transposition model is 2.7% (RMSE).

TABLE I

GHI, DNI AND GTI FOR THE BENCHMARK PLANT - ACTUAL AND CALCULATED USING MEASURED AND SATELLITE-DERIVED IRRADIANCE

| Irradiance Component | Data source | MAE [%] | MBE [%] | RMSE [%] |
|---|---|---|---|---|
| GHI | Measured | - | - | - |
| GHI | SolarGis | 5.1 | 4.9 | 5.6 |
| | CAMS | 12.2 | 12.2 | 13.8 |
| | CAMS (centr. 26) | 12.1 | 12.1 | 13.8 |
| DNI | Measured | - | - | - |
| DNI | SolarGis | 13.1 | 13.0 | 16.3 |
| | CAMS | 9.6 | 3.8 | 12.0 |
| | CAMS (centr. 26) | 11.3 | 5.7 | 16.4 |
| GTI | Measured | - | - | - |
| | from measured GHI & DNI | 2.1 | 1.7 | 2.7 |
| GTI | SolarGis | 7.6 | 7.4 | 8.7 |
| | CAMS | 11.6 | 10.8 | 13.4 |
| | CAMS (centr. 26) | 11.6 | 10.7 | 13.4 |

*B. Validation of the methodology*

The quality of the methodology is carried out on two levels:
- validation of the tilt & orientation angles (step c): Table II reports the actual orientation and tilt of the benchmark plant and the retrieved ones, together with the respective absolute errors.
- validation of the monthly-series of PR (step d): Table III lists the statistical deviation of monthly PR measured and calculated with the different irradiance sources.

In Fig. 4 the monthly Yr and PR of the benchmark plant, calculated with the measured and the satellite-derived GHI and DNI are compared with the monthly Yr and PR derived from the measured GTI (actual).

There are two different reasons of error in the methodology:
- Errors in the transposition model and in the minimization algorithm (Type A)
- Errors in the GHI and DNI data sources (Type B)

Type-A errors have a small impact on the accuracy of the PR estimation with a MAE of 1.9% and a RMSE of 2.4% (Table III). Indeed, the orientation and the tilt estimations differ from the real ones respectively of 22 and 6 degrees (Table II).

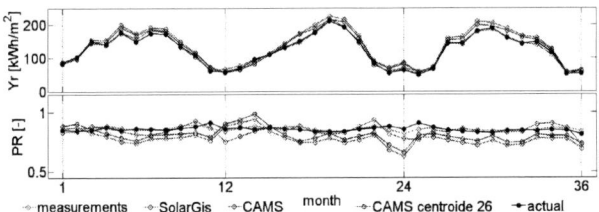

Fig. 4. Monthly series of Yr and PR for the reference plant installed in Bolzano, calculated using both the irradiance (GHI and DNI) measurements and the three different irradiance datasets from satellite sources, and compared to the actual measurements. (1=January 2012; 36 =December 2014).

The addition of Type-B errors brings to a greater PR deviation with a MAE between 4%-10% and a RMSE between 6% and 11%. From Table I and Figure 4 it is clear that SolarGis data are more accurate then CAMS data. In particular, beyond some evident error in the satellite-derived irradiance (months: 14, 23 and 24 in Fig. 4), the CAMS data lead to a more negative bias (irradiance overestimation) than SolarGis data, which is the main reason of the greater PR deviation. However, the use of free-access irradiance data from CAMS for the centroid of interest shows acceptable deviation from measurements, for the considered benchmark case, and is therefore included in the methodology.

The PR calculated with the CAMS irradiance on the exact location (MBE: -9.0%) is less accurate then the PR estimated using irradiance on the centroid position (MBE: -5.6%). This confirms that the retrieve of irradiance datasets for the centroid

### TABLE II
#### BENCHMARK PV SYSTEM'S TILT & ORIENTATION - ACTUAL AND CALCULATED USING MEASURED AND SATELLITE-DERIVED IRRADIANCE (GTI)

| Error type | Data source | Orient. [°] | ΔO [°] | Tilt [°] | ΔT [°] |
|---|---|---|---|---|---|
| | **Actual** | **188.5** | - | **30** | - |
| A | From meas. GHI & DNI | 211 | 22 | 36 | 6 |
| A,B | SolarGis | 223 | 35 | 31 | 1 |
| | CAMS | 195 | 7 | 32 | 2 |
| | CAMS (centroid 26) | 215 | 27 | 37 | 7 |

### TABLE III
#### STATISTICAL DEVIATION FROM MEASUREMENTS OF MONTHLY PR VALUES AND CALCULATED ONES FOR THE BENCHMARK PV PLANT (ACTUAL PR=0.85)

| Error type | Data source | MAE [%] | MBE [%] | RMSE [%] |
|---|---|---|---|---|
| A | From meas. GHI & DNI | 1.9 | 0.4 | 2.4 |
| A,B | SolarGis | 4.2 | -1.7 | 5.6 |
| | CAMS | 10.2 | -9.0 | 11.4 |
| | CAMS (centroid 26) | 7.8 | -5.6 | 9.1 |

is preferable, and that the number of chosen clusters seems to well represent the territory peculiarities. The main advantage of clustering is that irradiance data can be retrieved only for a smaller number of sites (50 in this case) rather than for each PV plant (1899 in this case). This way the computational and data handling effort is significantly reduced.

Table IV reports the deviation of the retrieved yield $Y_{calc}$ from the observed yield $Y$, using the CAMS irradiance on the centroid 26. These error values can be used as benchmark for the evaluation of the results obtained on the PV fleet, since a direct comparison with the actual PR is not possible.

### TABLE IV
#### STATISTICAL DEVIATION OF MONTHLY CALCULATED YIELD ($Y_{CALC}$) FROM MEASURED YIELD ($Y$) USING CAMS IRRADIANCE DATA (ON CENTROID 26), FOR THE BENCHMARK PLANT

| MAE [%] | MBE [%] | RMSE [%] |
|---|---|---|
| 4.5 | -0.2 | 5.8 |

### C. Results on the PV fleet

The methodology is applied on the fleet of 1899 PV systems, that decrease to 1750 after the preliminary filtering.

The clusterization is performed into 50 groups. This number proves to be appropriate also by plotting the error metrics as a function of the distance from the own cluster's centroid (Fig. 5). In fact, no significant correlation is visible, which means that the chosen subdivision into 50 clusters has no influence on the results of the methodology.

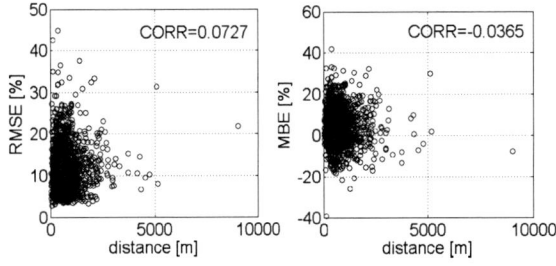

Fig. 5. Relation between the error metrics of every single PV plant of the fleet and the distance from the own cluster's centroid.

The minimization algorithm is therefore applied to determine the tilt and orientation angles and the monthly average PR of each PV plant. Fig. 6 shows the combination of tilt and orientation angles of the analyzed systems and the distribution of the retrieved average monthly performance ratio ⟨PR⟩.

When selecting the plants with RMSE less than 6%, i.e. with RMSE comparable to the benchmark plant (Table IV), the cloud of points has a well-defined shape, with South-oriented PV systems that tend to have e maximum tilt of 40°, while the South-West- and South-East-oriented plants tend to be installed with higher tilt angles. Moreover, the majority of the plants have an average monthly performance between 0.775 and 0.825, which is a reasonable range of values for crystalline technology (supposedly the technology used for almost all of the plants).

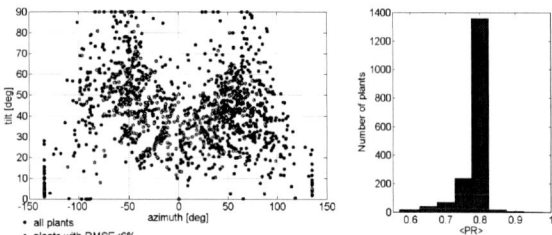

Fig. 6. (Left) retrieved values of tilt & orientation angles for PV plants with RMSE equal or lower than 6% (red dots) and higher than 6% (black dots). (Right) distribution of the retrieved average monthly performance ratio (bin width: 0.05) for 1750 PV systems in South Tyrol.

By applying (5), the monthly-PR series is therefore computed. Fig. 7 shows the ensemble of PR-series for all 1750 PV plants. In the first 48 months anomalous fluctuations of PR can be observed. These might depend on the poor quality of both the acquired yield data during the first years of PV installation and the satellite derived irradiance data.

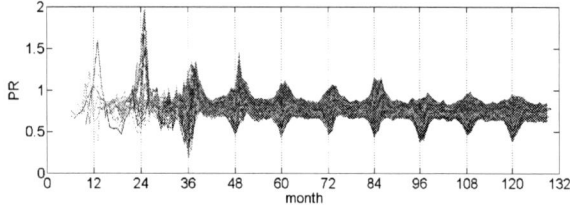

Fig. 7. PR-series of 1750 the PV systems of 11 years (from 2006 to 2016).

Fig. 8 shows the distribution of the error between observed yield $Y$ and computed one $Y_{calc}$. The mean RMSE is 11%± 5%, the mean MAE is 9%±5% and the mean MBE is 4%±8%. Therefore, the retrieval process tends to overestimate the final yield. Only 276 plants over 1750 have a RMSE lower than 6% with a MAE between 2.5-5%, a MBE between ±4% and a mean PR in the range 0.71 - 0.84.

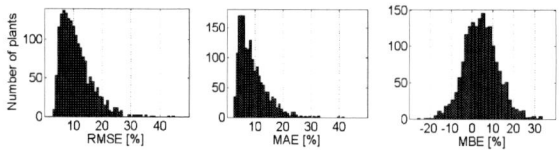

Fig. 8. Distribution of the error between observed yield (Y) and computed one ($Y_{calc}$).

From Fig. 8 it can be also pointed out that accuracy distributions have large tails. The causes of the larger errors are manifolds. The main source of uncertainty is the irradiance source, specifically the satellite-retrieved data from CAMS products. As shown in Fig. 9 the shape of the retrieved PR-series for the benchmark plant is very similar to the average of the PR-series retrieved for all 1750 analyzed plants. This is the indication of a systematic error on the satellite irradiance.

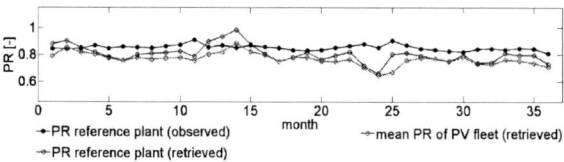

Fig. 9. Monthly series of PR from 2012 to 2014: observed PR of the reference plant (black line); retrieved PR of the reference plant (blue line) and mean retrieved PR of the PV fleet (red line).

Unlike the reference plant, whose characteristics are well-known and performances are measured, two additional issues appear when dealing with a fleet of distributed PV plants, and that affect the quality of the results.

One problem is the presence of far- and near shading. The horizon downloaded from PVGIS has a 100 m resolution, bringing a certain degree of uncertainty that is higher in a mountainous region like South Tyrol. In addition, it is not possible to detect shading by close objects. This situation is well-represented in Fig. 10, where the PR-series of a PV system on a South-West oriented plant with a tree and building on the South (red line) is compared with the PR-series of the PV plant on a South-Est oriented roof without obstacles (black line). The PR of the first plant, which is the worst plant of the fleet in terms of RMSE (equal to 33.4%), is lower in winter clearly due to the presence of near shading, and this affects the quality of the PR-series. On the other hand, the PR of the second plant (the best of the fleet) has a very reasonable trend and a RMSE of 3.6%. It is interesting to note from the satellite images that the tilt & orientation angles are correct for both plants. Nevertheless, the shadowed plant shows a trend of the retrieved yield $Y_{calc}$ very similar to the measured one $Y$, but affected by a constant bias. The reason for this bias is not completely clear, but it might depend on the minimization algorithm that, in this case, tends to find the correct shape of the yield (tilt and orientation angles) but fails in retrieving the mean PR values. By a random check of the PV plants with the help of satellite images, it is reasonable to affirm that almost all of the 118 plants (over 1750) with a RMSE higher than 20% are affected by near shadow. For 94 of them, a MBE higher than 10% is present.

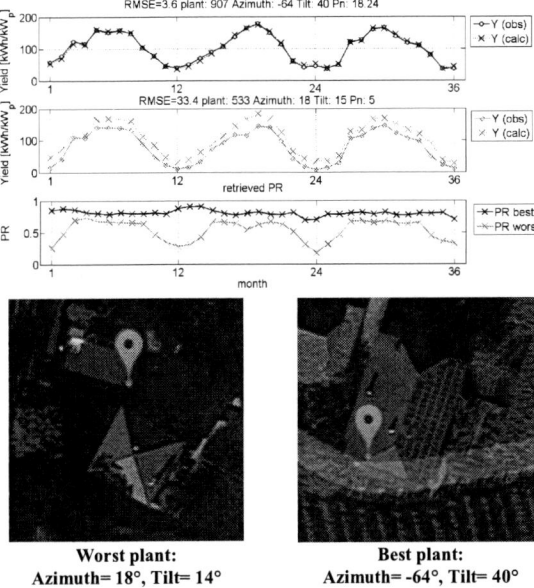

**Worst plant:**
**Azimuth= 18°, Tilt= 14°**

**Best plant:**
**Azimuth= -64°, Tilt= 40°**

Fig. 10. Comparison between the actual (obs) and retrieved (calc) yield monthly-series for (left) a shadowed and (right) a not shadowed plant, and comparison of their retrieved PR series.

The second issue is related to PV systems installed on multiple-tilted surfaces, as the one shown in Fig. 11. Since the data available from the local energy utility consider such plants as whole, the minimization algorithm fails in finding the correct tilt and orientation angles, reporting in most of such cases a tilt angle (as well as orientation angle) of zero degrees. As a result, 300 of the 332 multiple-tilted plants have a RMSE greater than 5% and lower than 20%.

Fig. 11. Satellite image of a multiple-tilted PV plant.

## VI. CONCLUSION

This paper presents a methodology to determine the tilt & orientation angles and the monthly PR-series of a fleet of photovoltaic plants distributed in a region. The only available information is the plants' monthly generated energy, the location and the nominal power. The methodology makes use of free irradiance datasets from CAMS, downloaded only for a restricted number of locations in the region representing the centroids of clusters calculated on the mutual distance. A comparison of different irradiance datasets as well as a validation of the methodology on a benchmark PV plant for which irradiance and performance are measured is presented and discussed.

The application of the methodology on a fleet of 1750 plants shows encouraging results. A number of 276 PV plants can be selected that have a higher quality in the monthly PR-series (RMSE lower than the one of 5.8% of the benchmark plant). In general, the methodology brings to errors of mean RMSE of 11%± 5%, mean MAE of 9%±5% and of mean MBE of 4%±8%. The main factor affecting the accuracy of the methodology is the quality of CAMS irradiance data, and can be improved by reducing the systematic errors with the integration of ground measurements.

The methodology proves to correctly detect the presence within the fleet of PV plants affected by near shading. On the other hand, PV systems made of sub-systems installed on multiple-tilted surfaces whose data are merged in a unique energy yield dataset fail to be correctly processed by the proposed methodology.

## ACKNOWLEDGEMENT

The authors would like to acknowledge the companies Edyna and SolarGis for collecting and sharing their data for the purposes of this study.

## REFERENCES

[1] IEC 61724, 1998. Photovoltaic System Performance Monitoring - Guidelines for Measurement, Data Exchange and Analysis.

[2] Copernicus Atmosphere Monitoring Service (CAMS) Radiation Service, 2017. [Online]. Available: http://www.soda-pro.com/web-services/radiation/cams-radiation-service [Accessed: 18-Feb-2017].

[3] Solar Radiation Data (SoDa), 2017. [Online]. Available: http://www.soda-is.com [Accessed: 18-Feb-2017].

[4] J. B. MacQueen, "Some Methods for classification and Analysis of Multivariate Observations," in *Proceedings of 5th Berkeley Symposium on Mathematical Statistics and Probability*, 1967, pp. 281–297.

[5] Photovoltaic Geographical Information System (PVGIS), 2017. [Online]. Available: http://re.jrc.ec.europa.eu/pvgis/ [Accessed: 18-Feb-2017].

[6] G. Belluardo, P. Ingenhoven, W. Sparber, J. Wagner, P. Weihs and D. Moser, "Novel method for the improvement in the evaluation of outdoor performance loss rate in different PV technologies and comparison with two other methods," *Solar Energy*, vol. 117, pp. 139-152, 2015.

# Quantify Photovoltaic Module Degradation using the Loss Factor Model Parameters

C. Birk Jones *, Bruce H. King *, Joshua S. Stein *,Justin S. Fada [†], Alan J. Curran [†],
Roger H. French [†], Erdmut Schnabel [‡], Michael Koehl [‡] and Olga Lavrova *

*Sandia National Laboratories, Albuquerque, NM, U.S.A
[†]Case Western Reserve University, Cleveland, OH, U.S.A
[‡]Fraunhofer ISE, Freiburg, Germany

*Abstract*—Photovoltaic (PV) system investments expect to produce power for 20 plus years. However, module level degradation can occur that reduces the overall energy production. This experiment adds to the current work by using the Loss Factor Model (LFM) to evaluate current and voltage curves acquired from two different module types deployed in three distinct climate zones. The experiment first calculated the six LFM parameters for each year of the data set. Then, the annual LFM parameters were compared to identify changes in performance. The approach identified an increase in series resistance that would have otherwise gone unnoticed using methods that evaluate the maximum power point data only. The annual LFM parameter degradation rates for one of the modules in a single climate zone were then used to predict the voltage and current for a similar module type in a different climate zone.

*Index Terms*—photovoltaic, degradation, loss factor model

## I. INTRODUCTION

Financial investment opportunities in solar photovoltaic (PV) systems depend on long-term operations. The systems are expected to produce power for 20 plus years. Unfortunately, solar PV modules can degrade and cause power production losses. The losses can be caused by degradation in the packaging material, adhesive losses, interconnection failures, moisture intrusion, and semiconductor device corrosion [1]. The rate at which the module degrades can depend on the climate and module type.

Degradation rates have been calculated using various metrics. The metrics include regression modeling, normalized ratings, and current and voltage (I-V) curve characterization [2]. The methods have been applied to experiments that have monitored module performance in environmental chambers and in outdoor settings. Outdoor tests have evaluated a variety of PV technologies including silicon, CdTe, and CIGS. Past experiments have used I-V curves to analyze module level degradation.

The characterization of I-V curves provides detailed information regarding the relationship between the current and voltage produced by the semiconductor device. Often, I-V curves are obtained periodically to evaluate degradation. For example, Reis et al. compared the performance of PV modules over an 11 year gap [3]. The comparison showed that the maximum power was reduced by about 4%. Another experiment studied the performance of 40 different modules produced by

10 different manufacturers [4]. The analysis concluded that the modules tended to degrade at 0.7% per year.

I-V curve characteristics, such as short circuit current, open circuit voltage, and fill factor can be acquired continuously from outdoor modules. The I-V data provides detailed information that can highlight the cause for lost power production. For example, Smith et al. found that the module power was reduced because of large fill factor and small open circuit voltage losses [5]. The experiment showed that continuous, near real-time module or string level I-V curves can provide detailed insight regarding how the module degraded.

The present work used module level I-V curves that were collected throughout the day and year from three different climate conditions [6]. The I-V curves were taken at various temperatures and irradiance levels and the values were stored in a central database. The current and voltage vectors for each I-V curve were then used to extract Loss Factor Model (LFM) parameters [7]. The LFM characterizes the I-V curve by calculating six parameters that vary based on the effective irradiance. The parameters can be used to analyze performance and define the reason for reduced power [8]. In this case, the parameters were compared from year to year to discover if and where degradation had occurred. The annual degradation rates were then used as inputs into LFM equations that estimated the maximum power point current and voltage at different irradiance conditions.

## II. METHODOLOGY

Accurate I-V curves at the module level provide detailed performance information. The curves describe the open circuit voltage, short circuit current, maximum power point voltage and current, and series and shunt resistance associated with the module. The initial current and voltage characteristics of PV modules are defined by the manufacturer using an indoor I-V scan. The detailed measurement is often not duplicated in the field. However, it is possible to use in situ I-V scan devices to measure and evaluate degradation of field modules.

The present work evaluated continuous outdoor I-V curves from two types of PV modules located in three different climate zones. The modules were evaluated using two methods. The first method computed the monthly average maximum power point power, and the second approach used the LFM parameters. The LFM parameters were compared on an annual basis to calculate the degradation rate for each parameter at different irradiance values. The rates were then used as inputs into a model that predicted the degraded current and voltage five years into the future.

### A. Solar Performance Data

The solar performance data were stored in a data warehouse at Case Western Reserve University. The data was accessed using the analytical environment known as Energy-CRADLE [9]. The data included up to 5 years of I-V measurements from

TABLE I
NAMEPLATE CHARACTERISTICS FOR EACH MODULE

| Type | Isco | Voco | Impo | Vmpo | Pmpo |
|------|------|------|------|------|------|
| A | 7.95 | 48.1 | 7.16 | 39.1 | 280 |
| B | 8.47 | 36.4 | 7.99 | 29.4 | 230 |

three different locations. Each location had two different crystalline modules. The nameplate characteristics for each of the modules are described in Table I. Module A had a

TABLE II
KÖPPEN-GEIGER CLIMATE CLASSIFICATION

| Symbol | Description | Module Types |
|--------|-------------|--------------|
| BWh | Hot Arid Desert | A & B |
| ET | Polar Tundra | A & B |
| BSk | Cold Arid Steppe | A & B |

nameplate STC short circuit and max power point current that was less than module B. The open circuit and maximum power voltage for module A was greater than module B. The maximum power at STC was rated at 280 watts for module A and 230 watts for module B.

The two module types were exposed to three different climate conditions (Table II). In one case, the two module types were installed in Negev Dessert, Israel. This location is defined by Köppen-Geiger as BWh. The BWh nomenclature

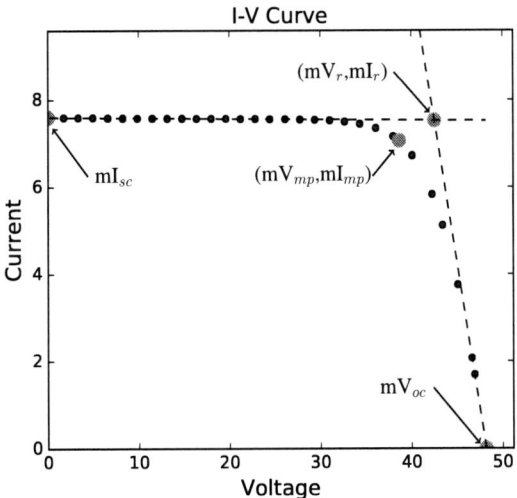

Fig. 1. The Loss Factor Model parameters were calculated based on the measured short circuit current, open circuit voltage, and maximum power current and voltage. The model also uses the intersection point of the lines tangent to the ends of the measured I-V curve.

stands for a hot arid desert. The second location was a polar tundra (ET) climate zone located in Mount Zugspitze, Germany. The third climate zone location was in Isla Gran Canaria, Spain. This location in Spain has been designated as zone BSk, which stands for a cold arid and steppe climate.

### B. I-V Curve Data Extraction Methods

There are various ways to obtain I-V curves from PV modules deployed in the field [10]. New technologies allow for I-V curve traces to be measured in situ with the PV array. Pordis LLC offers an I-V tracer that scans the PV system at a string level [11]. Module level I-V traces can be conducted using a product made by Stratasense LLC [12]. The Stratasense device is able to remove the individual panel from the string, perform a sweep, and then return the module back to the string. These devices allow for the analysis of I-V curves to be performed on actual systems connected to the grid. The detailed measurements can then be analyzed using the LFM.

### C. Loss Factor Model Parameters

The LFM parameters are calculated using measured values that are labeled in Figure 1. The measured values include short circuit current ($mI_{sc}$), open circuit voltage ($mV_{oc}$), and maximum power point current ($mI_{mp}$) and voltage ($mV_{mp}$). The $mI_r$ and $mV_r$ values are the coordinates for the intersection point of lines tangent to the ends of the measured I-V curve. The LFM parameters equations also need the reference short circuit current ($rI_{sc}$), open circuit voltage ($rV_{oc}$), and the temperature coefficients ($\beta_{Vmp}$, $\alpha_{Imp}$) that are provided by the manufacturer. The calculated LFM parameters are defined by Equations 1 to 6:

$$nIscT = \frac{mIsc}{rIsc}/E_e \text{ x } T_{corr,Isc} \quad (1)$$

$$nRsc = \frac{mIr}{mIsc} \quad (2)$$

$$nImp = \frac{mImp}{mIr}\frac{rIsc}{rImp} \quad (3)$$

$$nRoc = \frac{mVr}{mVoc} \quad (4)$$

$$nVmp = \frac{mVmp}{mVr}\frac{rVoc}{rVmp} \quad (5)$$

$$nVocT = \frac{mVoc}{rVoc} \text{ x } T_{corr,Voc} \quad (6)$$

where $T_{corr,Voc} = 1 + \beta_{Voc}(25\text{-}T_m)$, $T_{corr,Isc} = 1 + \alpha_{Isc}(25\text{-}T_m)$, $E_e$ is the measured plane-of-array irradiance, and $T_m$ is the module temperature. The calculated parameters can be used to model the maximum power point current and voltage [7] using Equations 7 and 8:

$$pImp = nIscT \text{ x } nRsc \text{ x } nImp \text{ x } rImp \text{ x } \frac{E_e}{T_{corr,Isc}} \quad (7)$$

$$pVmp = \frac{nVmp \text{ x } nRoc \text{ x } nVocT \text{ x } rVmp}{T_{corr,Voc}} \quad (8)$$

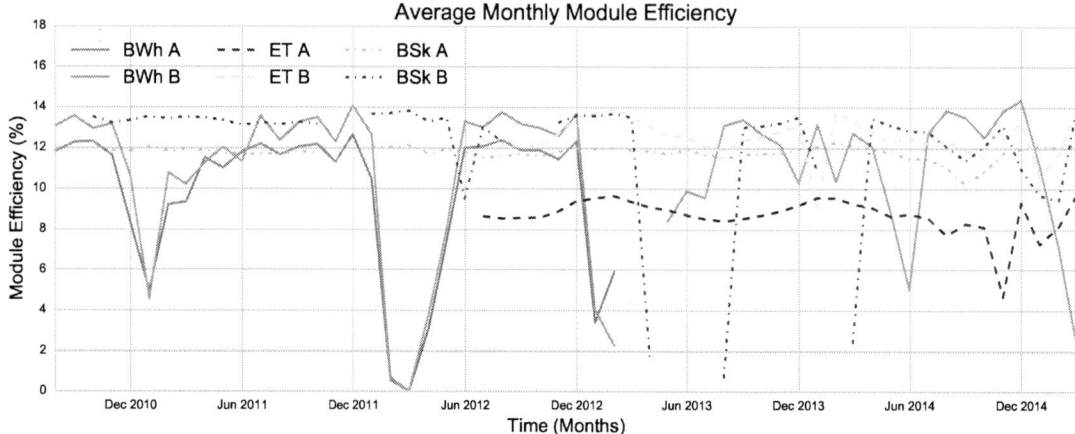

Fig. 2. The efficiency for modules A and B in the three climate zones did not show any degradation over time. The modules maintained an efficiency above 10% except for module A in climate zone ET.

## D. Experiment

The present work evaluated maximum power point power and each I-V curve from two module types located in three different locations. The maximum power point analysis included a general review of the distributions for each module. The evaluation also computed the average monthly efficiency for each module with the intent to identify degradation over time. The efficiency ($\eta$) was equal to the measured power divided by the module area times the measured irradiance. The experiment took the analysis a step further and evaluated thousands of I-V curves using the LFM.

The LFM analysis began with the extraction of I-V curve characteristics that included the measured $mV_{oc}$, $mI_{sc}$, $mV_{mp}$, $mI_{mp}$, $mV_r$, and $mI_r$. The LFM parameters were then calculated using the measured values, the irradiance, module temperature and nameplate data. The parameters were then compared over time to calculate the associated degradation rates. The rates from one module were then used to model the voltage and current of a similar module type in a different climate condition. The intent was to explore the potential for an LFM to predict degraded performance.

## III. RESULTS

### A. Max Power Performance Overview

The two module types, located in three different climate zones, had different maximum power point power distributions for the entire data set as shown in Figure 3. The modules in climate zone BWh had the lowest median measured power that was less than 50 watts. The ET climate zone had the next highest median power between 50 and 110 watts. The median power for the modules in the BSk climate was between 100 and 150 watts.

Measured power from modules A and B had different median, minimum, and maximum values in the same climate zone (Figure 3). The measured differences were because the modules had different voltage, current, and power characteristics. Module A was rated at a higher maximum power compared to module B. The module efficiencies were also

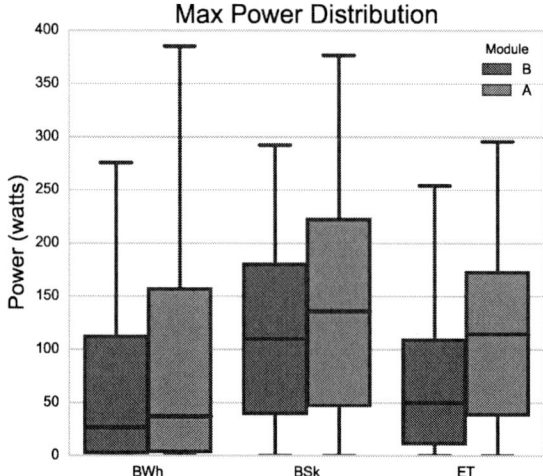

Fig. 3. The measured max power for each of the modules in the different climate zones had different results for the entire data set. The hot arid desert (BWh) had the lowest values. The next best performing modules were in the polar tundra (ET). The best performing modules were located in the cold arid steppe (BSk) climate zone.

different for each module type as shown in Figure 2. The efficiency for modules A and B in climate zone BWh and BSk began at about 12% and 13% respectively. Over time the efficiencies increased and decreased together and the efficiency of A tended to not exceed B. This trend was the same for modules A and B in climate zone ET. However, module type A in climate zone ET had an efficiency that was much smaller. The type A module in zone ET never exceeded 10%. The efficiency trends for each of the modules in the three climate zones fluctuated based on the season and potential fault conditions, but did not show sustained drops that would indicate that degradation had occurred. Further analysis of the I-V curves was necessary to discover potential, small scale degradation.

### B. I-V Curve Data Set

The data set used in this experiment had a total of 2.15 million I-V curves. There were 840,477 I-V curves from

978-1-5090-5606-4/17 $31.00 © 2017 IEEE

climate zone BWh, 557,894 from ET, and 751,668 from BSK. The first step in the experiment was to filter out low irradiance I-V curves, which eliminated about half of the data. This

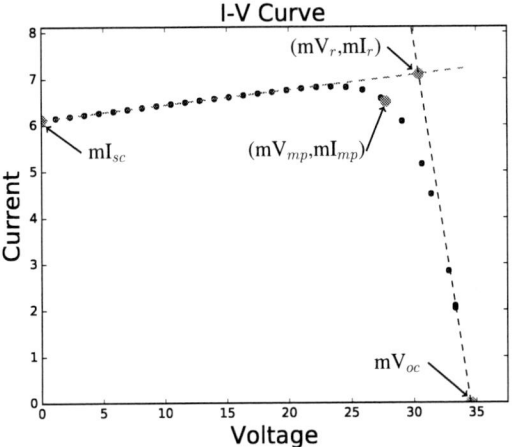

Fig. 4. The data set contained I-V curves that did not represent normal semiconductor behavior. The curves that had a smaller short circuit current than the $mI_r$ were filtered.

included curves taken at night and any data acquired during the day when the irradiance was below 200W/m². I-V curves taken at low irradiance conditions were filtered because the data had high variability that could alter the results. After the initial filter, about 978,981 data records were analyzed to measure the individual $mV_{oc}$, $mI_{sc}$, $mV_{mp}$, $mI_{mp}$, $mV_r$, and $mI_r$. The initial measurements revealed that about 963,852 of

TABLE III
THE DATA SET WAS FILTERED BASED ON EQUATIONS 9 AND 10.

|  | BWh | ET | BSk | Total |
|---|---|---|---|---|
| Total | 313,412 | 316,473 | 349,096 | 978,981 |
| Filtered | 303,163 | 314,354 | 346,335 | 963,852 |
| Difference | 10,249 | 2,119 | 2,761 | 15,129 |

the I-V sweeps did comply with Equations 9 and 10. In some cases the curves did not represent the semiconductor device accurately, as shown in Figure 4 and were filtered out if it did not meet the requirements defined by Equation 9. Other curves exhibited mismatch behavior and violated Equation 10 where $mI_r$ was greater than $mI_{mp}$.

$$mI_r < mI_{sc} \qquad (9)$$

$$mI_r > mI_{mp} \qquad (10)$$

The filtering process, that applied Equations 9 and 10, eliminated 303,163 curves from BWh, 314,354 from ET, and 346,335 from BSk (Table III).

The filtering process eliminated a large percentage of the data, but it did not impact the overall statistics significantly. The filtered and unfiltered data sets had similar $nI_{scT}$, $nR_{oc}$, $nI_{mp}$, $nV_{mp}$, and $nV_{ocT}$ statistical distributions. However, the filtered $nR_{sc}$ distribution was significantly different than the unfiltered data as shown in Figure 5. The change in the nRsc statistics was because the parameter is dependent on $mI_r$ and $mI_{sc}$. The Equation 9 filter eliminated a large amount of

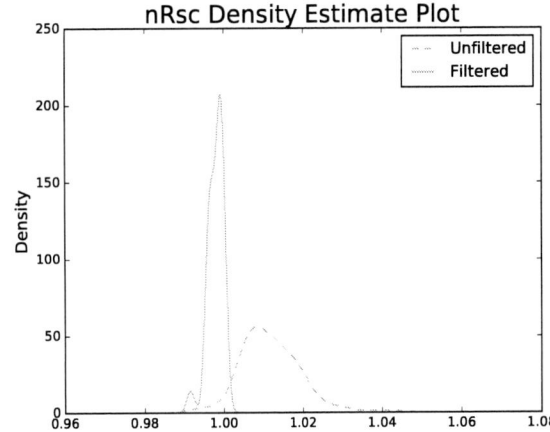

Fig. 5. The $nR_{sc}$ parameter had different density estimation for the filtered and unfiltered data because the $mI_r$ often exceeded $mI_{sc}$.

records that had an nRsc that was less than one where $mI_r$ was larger than $mI_{sc}$.

### C. Loss Factor Model Parameters

The LFM model calculated the six parameters defined by Equations 1 to 6. The parameters were then plotted with

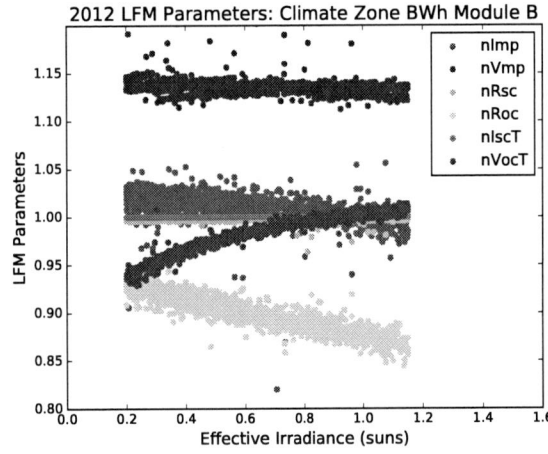

Fig. 6. Each of the Loss Factor Model parameters were plotted with respect to irradiance and exhibited normal behavior for each year of the data set.

respect to the effective irradiance as shown in Figures 6 that plots module B in the BWh climate zone. The results from module B were similar to module A in the three climate zones except for nImp. The nImp parameter tended to decrease as the effective irradiance went up for module B. The nImp for module A was different and it tended to increase as the irradiance went up.

### D. LFM Parameter Degradation

The LFM parameters for the two modules located in different climate zones were compiled for each year and compared to see if any changes had occurred. The nRsc, nIscT, and nVocT parameters for all of the modules did not change significantly over time. The nImp, nVmp and nRoc parameters did have a noticeable change from year to year for all of the

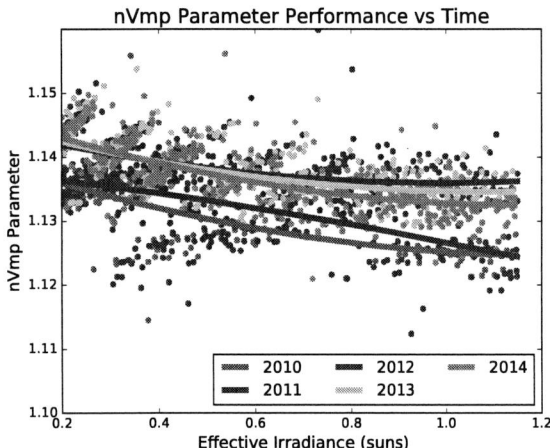

Fig. 7. The nVmp parameter increased by about 0.9% over a 5 year period. The increase represented a degradation in the module performance caused by

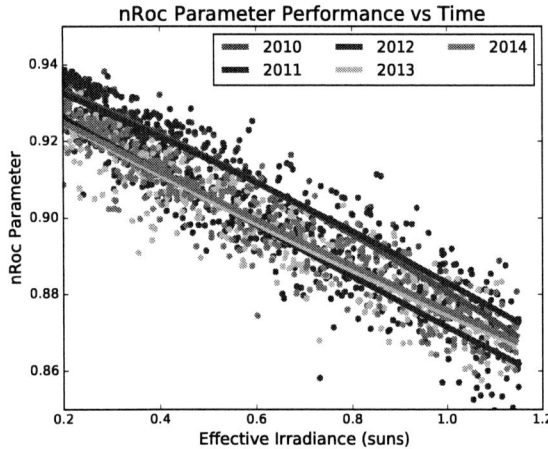

Fig. 8. The nRoc parameter decreased by about 1%. The change was caused by a decrease in mVr values.

modules. For example, the nVmp for module B in climate zone BWh increased by about 0.9% from year 2010 to 2014 as shown in Figure 7. Over the same time period the nRoc parameter decreased by about 1% as shown in Figure 8. The degradation of the nVmp and nRoc parameters was caused by a decrease in mVr that was not proportional with mVmp and mVoc losses. The degradation of the two parameters indicated that the series resistance had increased. The increase in resistance did not impact the performance at open circuit voltage because the current was zero. Therefore, the change in series resistance did not affect the nVocT parameter. The nIscT and nRsc parameters also remained unchanged because the shunt resistance did not decrease. The change in series resistance, discovered by the LFM parameters, did not impact the I-V curves enough to notice a decrease in module efficiency. The change rates for the LFM parameters were then used to estimate module performance.

### E. Estimate Performance based on Degradation Rates

The present work modeled the maximum power point current and voltage using Equations 7 and 8 respectively. The

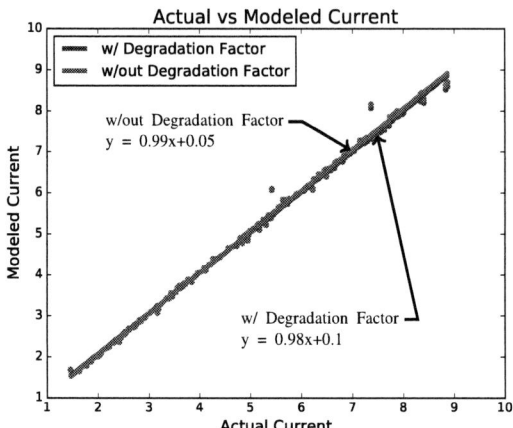

Fig. 9. The LFM estimated the current well with and without the degradation factor. The degradation factor produced a slightly less desirable linear fit that had a slope of 0.98 and an intercept of -0.036.

equations incorporated a degradation factor that was based on empirical evaluation of the LFM parameters. The intent was to model the performance of module B in climate zone BSk throughout the 2014 year based on the LFM parameters and degradation rates of module B in zone BWh between 2010 and 2013. The least-squares regression function for 2010 and 2013 plotted in Figures 7 and 8 were applied to the 2014 BSk climate zone irradiance values. Then, the percent

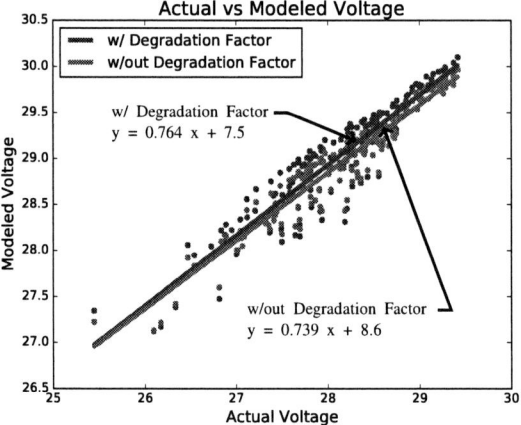

Fig. 10. The LFM estimated the voltage well with and without the degradation factor. The degradation factor produced a slightly better linear fit that had a slope of 0.72 and an intercept of 7.82.

difference was calculated to create the degradation rates for the B module in climate zone BSk for 2014 environmental conditions. The degradation rates ($DR_{nImp}$, $DR_{nVmp}$, and $DR_{nRoc}$) were finally multiplied by the original 2010 nImp, nVmp and nRoc values as shown in Equations 11, 12, and 13:

$$nImp_{new} = nImp \text{ x } DR_{nImp} \qquad (11)$$

$$nVmp_{new} = nVmp \text{ x } DR_{nVmp} \qquad (12)$$

$$nRoc_{new} = nRoc \text{ x } DR_{nRoc} \qquad (13)$$

978-1-5090-5606-4/17 $31.00 © 2017 IEEE

The new parameter values were used in Equations 7 and 8 to estimate current and voltage.

The estimated current and voltage matched well with the measured values. The LFM results that did not apply the degradation factor had a strong linear correlation with the actual values as shown in Figure 9. The least-squares regression fit produced a line that had a slope of 0.99 and an intercept of 0.05. The integration of the degradation factor produced an output that also had a strong linear fit, but the slope and intercept of the fit were slightly worse at 0.98 and 0.1 respectively. The voltage model results (Figure 10) indicated that the nVmp and nRoc degradation factors made a small, but positive impact on the model. The slope and intercept of the linear fit line improved from 0.739 to 0.764 and 8.14 to 7.5 from the model without the degradation factor to the model with the degradation factors.

## IV. CONCLUSION

Understanding PV module degradation rates can provide valuable insight into long-term investments and operations and maintenance schedules. Predicting module level losses caused by normal degradation has been an important topic for researchers. The present work adds to the module degradation evaluation and prediction discussion by applying the LFM. The LFM was used to evaluate two different module types in three distinct climate zones. The millions of I-V curves had to first go through a pre-processing step to find the curves that could be analyzed accurately by the LFM method. The LFM parameters for each year were calculated and plotted versus the corresponding effective irradiance value. The year to year degradation rates were then calculated and used to model performance.

The LFM results showed that the two module types in the three climate zones experienced an increase in nVmp, a decrease in nRoc, and a slight decrease in nImp. The other three parameters did not have any significant changes from year to year. The changes in nVmp, nRoc, and nImp indicated that the series resistance had increased. However, the increase was very small and was not evident in the power and calculated efficiency values for the six modules. The degradation rates for the BWh module were used as an input into the current and voltage LFM models. The intent was to account for the degradation to produce a more accurate model. The approach did not improve the current results and only slightly improved the voltage estimate.

The LFM method attempted to predict performance based on observed degradation of a similar module. The results were not convincing because the degradation rates were very small and do not impact the overall performance. A further investigation of this method is required on data that contains more significant degradation in order to confirm that the LFM can be used to estimate module power while considering degradation. However, the paper highlights the ability of the LFM to identify and quantify the type of module degradation earlier than other methods such as maximum power point evaluations.

## ACKNOWLEDGMENT

This work was supported by the U.S. Department of Energy SunShot Initiative under the MLEET Project DE-EE-0007140. The authors acknowledge Erdmut Schnabel and Michael Koehl of Fraunhofer-ISE for the I-V datasets.

Sandia National Laboratories is a multimission laboratory managed and operated by National Technology and Engineering Solutions of Sandia, LLC., a wholly owned subsidiary of Honeywell International, Inc., for the U.S. Department of Energy's National Nuclear Security Administration under contract DE-NA0003525.

## REFERENCES

[1] M. A. Quintana, D. L. King, T. J. McMahon, and C. R. Osterwald, "Commonly observed degradation in field-aged photovoltaic modules," in *Conference Record of the Twenty-Ninth IEEE Photovoltaic Specialists Conference, 2002.*, May 2002, pp. 1436–1439.

[2] A. Phinikarides, N. Kindyni, G. Makrides, and G. E. Georghiou, "Review of photovoltaic degradation rate methodologies," *Renewable and Sustainable Energy Reviews*, vol. 40, pp. 143–152, Dec. 2014. [Online]. Available: //www.sciencedirect.com/science/article/pii/S1364032114006078

[3] A. M. Reis, N. T. Coleman, M. W. Marshall, P. A. Lehman, and C. E. Chamberlin, "Comparison of PV module performance before and after 11-years of field exposure," in *Conference Record of the Twenty-Ninth IEEE Photovoltaic Specialists Conference, 2002.*, May 2002, pp. 1432–1435.

[4] D. C. Jordan, R. M. Smith, C. R. Osterwald, E. Gelak, and S. R. Kurtz, "Outdoor PV degradation comparison," in *2010 35th IEEE Photovoltaic Specialists Conference*, Jun. 2010, pp. 002 694–002 697.

[5] R. M. Smith, D. C. Jordan, and S. R. Kurtz, "Outdoor PV module degradation of current-voltage parameters." Denver, CO, May 2012.

[6] T. J. Peshek, J. S. Fada, Y. Hu, Y. Xu, M. A. Elsaeiti, E. Schnabel, M. Khl, and R. H. French, "Insights into metastability of photovoltaic materials at the mesoscale through massive IV analytics," *Journal of Vacuum Science & Technology B, Nanotechnology and Microelectronics: Materials, Processing, Measurement, and Phenomena*, vol. 34, no. 5, p. 050801, Aug. 2016. [Online]. Available: http://avs.scitation.org/doi/abs/10.1116/1.4960628

[7] J. S. Stein, J. Sutterlueti, S. Ransome, C. Hansen, and B. King, "Outdoor PV performance evaluation of three different models: single-diode, SAPM and loss factor model." in *EU PVSEC Proccedings*, 2013, pp. 2865–2871.

[8] S. Sellner, J. Sutterlti, L. Schreier, and S. Ransome, "Advanced PV module performance characterization and validation using the novel Loss Factors Model," in *2012 38th IEEE Photovoltaic Specialists Conference*, Jun. 2012, pp. 002 938–002 943.

[9] Y. Hu, V. Y. Gunapati, P. Zhao, D. Gordon, N. R. Wheeler, M. A. Hossain, T. J. Peshek, L. S. Bruckman, G. Q. Zhang, and R. H. French, "A Nonrelational Data Warehouse for the Analysis of Field and Laboratory Data From Multiple Heterogeneous Photovoltaic Test Sites," *IEEE Journal of Photovoltaics*, vol. 7, no. 1, pp. 230–236, Jan. 2017.

[10] E. D. Aranda, J. A. G. Galan, M. S. d. Cardona, and J. M. A. Marquez, "Measuring the I-V curve of PV generators," *IEEE Industrial Electronics Magazine*, vol. 3, no. 3, pp. 4–14, Sep. 2009.

[11] C. B. Jones, M. Martinez-Ramon, R. Smith, C. K. Carmignani, O. Lavrova, and J. S. Stein, "Automatic Fault Classification of Photovoltaic Strings Based on an In-Situ IV Characterization System and a Gaussian Process Algorithm," Portland, OR, 2016.

[12] K. Gillispe and P. Wrisley, "Wireless current-voltage tracer with uninterrupted bypass system and method," U.S. Patent US8 952 715 B2, Feb., 2015, u.S. Classification 324/761.01; International Classification G01R31/40, G01R31/26; Cooperative Classification H02S50/10, G01R31/405, G01R31/2603, G01R31/2605. [Online]. Available: http://www.google.com/patents/US8952715

# Simulating PV System Performance with Component Reliability Distributions

Geoffrey T. Klise[1], Janine M. Freeman[2], Olga Lavrova[1]

[1]Sandia National Laboratories, Albuquerque, NM, 87185, USA
[2]National Renewable Energy Laboratory, Golden, CO, 80401, USA

*Abstract* — A feature developed by Sandia National Laboratories (SNL) for simulating PV system component reliability has been integrated into the National Renewable Energy Laboratory's System Advisor Model (SAM). The PV Reliability Performance Model (PV-RPM) uses fault/failure and repair distributions to estimate impacts to power, energy production and operational costs. Realization results include the number of failures per component, failure rate, O&M costs incurred based on maintenance strategies, and energy lost based on the failure estimates. As the inputs are inherently uncertain and based on probability distributions, the model realizations include statistical confidence intervals to help bracket the probability that the sample interval contains the true mean value of the variable in question. This paper presents the current validation efforts and provides detail on how this feature is utilized in a PV performance model.

## I. INTRODUCTION

Commercially available photovoltaic (PV) performance model improvements are primarily focused on reducing modeling uncertainty, validating model assumptions associated with different meteorological inputs, and module and inverter behavior. However, reliability features based on statistical uncertainty are notably absent. Formally incorporating reliability metrics based on estimated failure rates for PV systems benefits the industry by providing a more accurate representation of events that impact PV system energy production, revenue generation and resulting costs to cure.

A traditional engineering systems-centric bottoms-up approach includes very specific component details to ultimately calculate the amount of power produced at frequent intervals. Engineering studies typically look for potential component level reliability issues in an attempt to maximize power produced. Reliability-focused engineering methods, such as Failure Mode and Effects Analysis (FMEA), and reliability block-diagrams (RBD) can identify different reliability states and causal relationships, and are used to identify and characterize PV system fault and failure events [1].

However, much of this work has rarely been applied within a PV *system* model that can simulate the impacts to lifetime energy production and cost impacts of different reliability states. One of the limitations is that not enough public PV system data is available for a "statistically relevant timespan" to validate research for developing a comprehensive understanding of all failure modes and their severity [1].

Despite the lack of data, a reliability modeling feature is still useful for bracketing scenarios to better understand project risk.

Sandia National Laboratories (SNL) developed the PV Reliability Performance Model (PV-RPM) in response to equipment performance uncertainty [2-4], providing it to the PV modeling community as a proof-of-concept in 2011. This early version, however was not scalable to different system configurations and had fewer features compared to commercial PV performance models. SNL is currently working with the National Renewable Energy Laboratory (NREL) to add the PV-RPM algorithm into the System Advisor Model (SAM) as it has stochastic modeling capabilities, provides access to a wide variety of modeling options and can be downloaded and used for free providing access to any interested user. The code developed for the reliability metrics is distributed by SNL through SAM under an open source BSD license, making it easy for a user to make changes within the SAM environment, adopt the code and use it in their own model, or add it to an open source code repository such as PVLIB [5].

## II. METHODOLOGY

This paper will present 1) current validation efforts compared to proof-of-concept results, 2) simulation results compared to real inverter events, and 3) a discussion on how PV-RPM can simulate single-axis tracker failures and energy loss.

Inputs for the validation in 1) and 2) are derived from a fielded system, over 14 years old, and for new systems installed within the past two years. A few example scenarios are presented to reveal impacts of component and tracker failure to energy production. As the reliability feature has not yet been released in the public version of SAM, these results presented here are intended to familiarize the PV modeling community with what to expect as the new reliability feature will be available for testing in late 2017 prior to the full public release.

### A. Software Environment

SNL worked with NREL to determine the feasibility of adding the PV-RPM algorithm into SAM [6]. Once certain that the stochastic modeling capabilities work and the reliability elements could be successfully integrated, the implementation of PV-RPM was done within the SAM LK scripting language [7], which allows the algorithm to be implemented outside of the main user interface and provides users direct access to the code making the algorithm less of a 'black box'.

```
// Modules
global modules = alloc(num_modules);
global module_meta = null;
module_meta.can_fail = true;
module_meta.number = num_modules;
module_meta.warranty.has_warranty = true;
module_meta.warranty.days = 20 * 365; //years converted to days
//failure mode 1: normal failures
module_meta.failure[0].distribution = 'normal';
module_meta.failure[0].parameters = [4 * 365, 1 * 365]; //mean, std, years converted to days
module_meta.failure[0].times = null;
module_meta.failure[0].labor_time = 2; //hours
module_meta.failure[0].cost = 322; //$
//failure mode 2: defective failures
module_meta.failure[1].distribution = 'exponential';
module_meta.failure[1].parameters = [0.5 / 365]; //failures per year converted to days
module_meta.failure[1].times = null;
module_meta.failure[1].labor_time = 2; //hours
module_meta.failure[1].cost = 322; //$
module_meta.failure[1].fraction = 20 / 100; //% converted to fraction
module_meta.repair.can_repair = true;
module_meta.repair.distribution = 'lognormal-n';
module_meta.repair.parameters = [60, 20]; //mean, std, ln days
module_meta.repair.times = null;
module_meta.degradation.can_degrade = true;
module_meta.degradation.rate = 0.05; //%/year
```

Fig. 1. LK Script example for setting up module failure and repair distributions

As failure and repair events are inherently uncertain, performance models with reliability elements are run using SNLs Latin Hypercube Sampling (LHS) algorithm (within SAM) to generate multiple realizations sampled from the reliability distributions. The newly added reliability functionality takes advantage of the LHS code embedded in SAM to quantify the uncertainty associated with probabilistic inputs, and present the results as confidence intervals around the mean of the output value in question.

## III. ANALYSIS & RESULTS

As the SAM platform is different from the proof-of-concept platform, this effort builds on the work presented by Collins [2] which originally validated the proof-of-concept version of PV-RPM. Using the same reliability data, we evaluate whether the new implementation in SAM matches the proof-of-concept for failures *only*, due to the absence of past modeling assumptions, power/energy production and weather data.

Next, we will validate the model with reliability data from a newer PV system. The new system has detailed records that will

allow for analysis of faults and failures and overall energy impacts. Simulated results from sampling different distributions are uncertain and therefore will be presented with confidence intervals.

### A. SAM Validation with Proof-of-Concept Reliability Dataset

The narrative in this section is described in a white paper [8] which provides more detail on how the SAM version of PV-RPM was tested and validated. To validate results between the SAM and proof-of-concept implementations of PV-RPM, a long-term well characterized PV plant dataset is used. The site is located in Springerville, Arizona and the PV plant characteristics can be found in [2]. Table 1 below presents the different component failure and repair distributions, along with their input parameters, actual number of events, and expected events for different time-periods based on the proof-of-concept model [3]. It should be noted that the 5-yr expected events from the proof-of-concept model is derived from multiple realizations, however the authors did not present confidence intervals with their results. When considering probabilistic inputs, the correct way to present the results is shown in Table 2.

As shown in Table 1, the expected events (from the proof-of-concept model) underestimate actual events for all components, with the exception of the inverter. As we do not have access to the failure statistics used to develop the probability distributions, just the results, there is no way to evaluate how well the actual failures compared to the expected failures in terms of confidence intervals and number of realizations. The comparison described below will only focus on the 5-year expected events from the proof-of-concept and the SAM implementation of PV-RPM, not comparisons made to the actual events as the intent here is to not re-do the distributions developed previously.

The reliability data from Table 1 is entered into SAM as a script in LK (Fig. 1). Each failure and repair mode is represented as a statistical distribution, and failures modes can be turned off to focus on a specific component, and if necessary, the repair distribution can also be turned off leaving the component in a non-functional state for the remainder of the

TABLE I
FAILURE & REPAIR DISTRIBUTIONS AND PROOF-OF-CONCEPT RESULTS – SPRINGERVILLE PV PLANT

| Component | Failure or Repair | Distribution Type | Shape or Mean | Scale or STDEV | 5-yr Actual Events | 5-yr Expected Events | 10-yr Expected Events | 20-yr Expected Events |
|---|---|---|---|---|---|---|---|---|
| Module | Fail | Weibull | 0.28 | 5.0E+12 | 29 | 26 | 31 | 38 |
| | Repair | Lognormal-n | -1.37 | 13.11 | N/A | N/A | N/A | N/A |
| DC Combiner | Fail | Weibull | 0.51 | 1.2E+06 | 34 | 25 | 35 | 50 |
| | Repair | Lognormal-n | -0.98 | 2.07 | N/A | N/A | N/A | N/A |
| HV Transformer | Fail | Weibull | 0.58 | 7100 | 5 | 4 | 5 | 9 |
| | Repair | Weibull | 0.53 | 1.36 | N/A | N/A | N/A | N/A |
| AC Disconnect | Fail | Weibull | 0.35 | 11000 | 22 | 17 | 23 | 31 |
| | Repair | Weibull | 0.71 | 1.4 | N/A | N/A | N/A | N/A |
| Inverter - General | Fail | Exponential | 0.00278 | N/A | 125 | 132 | 231 | 429 |
| | Repair | Lognormal-n | -4.25 | 2.27 | N/A | N/A | N/A | N/A |

simulation. Costs and labor time associated with the repair action are also an option and can be analyzed in the results table.

Fig. 1 shows an example of how module failure and repair modes are set up in the LK scripting language. In this example, two failure modes are defined, with one repair distribution covering all failures. Multiple repair distributions can also be assigned, with one failure type receiving one repair type. This is important based on the fact some failures may require greater attention than others, such as inverter faults that can be reset remotely, or other inverter faults that require a site visit and may take longer to repair.

TABLE II

MODEL RESULTS – COMPONENT FAULT/FAILURES OVER 5-YR SIMULATION WITH 10 / 100 REALIZATIONS

| Component | Mean Events | 95% CI Lower Bound | 95% CI Upper Bound | 5-yr Expected Events[ii] |
|---|---|---|---|---|
| Module | 26.2 / 26.4 | 23.0 / 25.6 | 29.4 / 27.2 | 26 |
| DC Combiner | 20.3 / 24.1 | 16.4 / 23.2 | 24.2 / 25.1 | 25 |
| HV Transformer[iii] | 1.0 / 1.0 | 0.1 / 0.8 | 1.5 / 1.2 | 4 |
| AC Disconnect | 16.0 / 16.9 | 13.8 / 16.7 | 19.0 / 17.8 | 17 |
| Inverter - General | 129.6 / 132.1 | 125.1 / 130.4 | 134.1 / 133.8 | 132 |

i – 95% confidence interval (CI) calculated using t-test statistic.
ii – Expected Events are expected fault and failure events from the proof-of-concept model results.
iii – the HV Transformer was modeled with fewer transformers than operating on-site due to modeling limitations with odd numbers of components.

All failure and repair modes were run in SAM for a 5-year simulation with 10 individual realizations and 100 realizations (Table 2) to compare against the 5-year expected results from the proof-of-concept model. The 10- and 20-year expected events in Table 1 were not simulated as those are expected faults and failures from the proof-of-concept model, and additional data past 5 years is not available to validate those results.

Fig. 2 and Fig. 3 present for each component (like Table 2), the mean of the simulation and the 95% confidence interval around the mean. Because these results are probabilistic, the result is interpreted as 95% of the sampled intervals would contain the true mean and 5% would not.

In Fig. 2 and Fig. 3 we are only comparing the expected events from the PV-RPM implementation in SAM to the expected events from the early analysis [2] and not to the actual failures. The goal here is to see if the input distribution parameters can be matched *between* models, and not to what degree the distribution matches the actual event occurrences. Because the distributions were already developed years earlier, the important result is that the Mean Events in Table 2 closely approximates the 5-year Expected Events value in Table 1 (5-yr E. Events repeated in Table 2). The results indicate that the mean number of events from 100 realizations in the SAM simulation are almost the same as the 5-year expected values from the proof-of-concept [2], suggesting that the SAM implementation of PV-RPM was developed correctly. With 10

realizations, the mean values are much further off the 5-year expected events as not enough samples of the distribution were made. This shows the importance of running a larger number of realizations and the impact to the mean value and confidence intervals; fewer realizations reveal a mean value further away

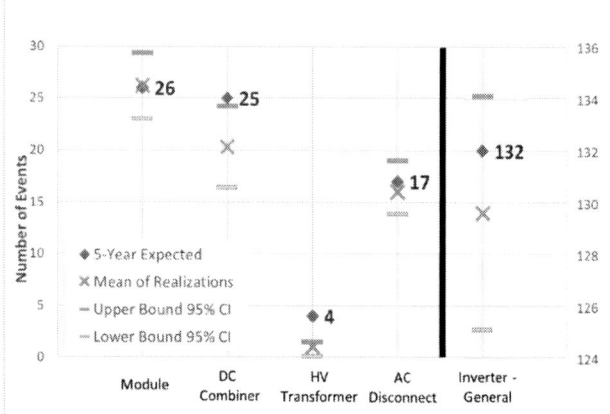

Fig.2. Component realization results (10) compared to expected events

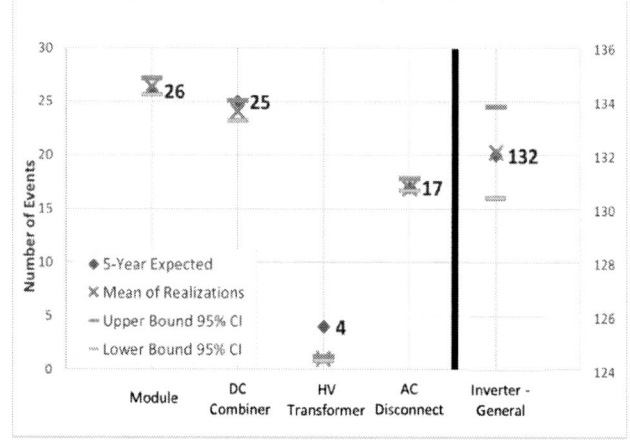

Fig.3. Component realization results (100) compared to expected events

than the expected value (with the exception of modules) and a higher uncertainty as expressed in 95% confidence interval spread. As mentioned in the Table 2 notes, the HV transformer cannot be modeled as built, resulting in fewer expected mean event results.

*B. SAM Validation with Newer PV System Reliability Dataset*

The previous section presented validation in SAM with an existing and previously analyzed reliability dataset. This section will present results of an inverter data analysis to determine distribution parameters for use in SAM which will be validated against actual events and energy loss.

SNL has access to reliability data from commercial and utility-scale PV systems around the U.S., consisting of both older and newer configurations. An example of a recent set of issues from a fleet of PV system inverters is analyzed to determine to what degree PV-RPM can replicate the failures and energy loss estimates. Table 3 below provides the input

TABLE III
INVERTER "COMPONENT" FAILURE & RESULTING ENERGY
LOSS OVER 1-AND 2-YR SIMULATION WITH 100
REALIZATIONS

| Fail or Repair | Distribution Type | Shape or Mean | Scale or STDEV |
|---|---|---|---|
| Fail | Weibull | 12.35 | 273.55 |
| Repair | Weibull | 1.33 | 4.20 |

STDEV – standard deviation

| Actual Failures (1.5 yrs) | Actual Energy Loss [kWh] (1.5 yrs) | Sim. in Yrs[i] | Estimated[ii] Failures (mean and 95% CI)[iii] | | Estimated[ii] Cum. Energy Loss [kWh] (mean and 95% CI)[iii] | |
|---|---|---|---|---|---|---|
| 28 | 680,000 | 1 | 20.01 | 19.99 | 446,108 | 431,508 |
| | | | | 20.02 | | 460,708 |
| | | 2 | 41.68 | 41.48 | 749,916 | 719,316 |
| | | | | 41.87 | | 780,716 |

i – Sim. in Years represents two simulations of 100 realizations, for 1-year and 2-year analysis periods. As the data represents 1.5 years and the model cannot simulate a fraction of a year, both timeframes are presented.
ii – Any 'Estimated' values represent the mean of all realizations.
iii – The mean value is presented first, and then the lower and upper bound based on the 95% confidence interval (CI).

data used in the simulation, including the failure and repair distributions, actual failures & energy loss, and estimated failures (mean of realizations) & energy loss. The system is in the 20-25 MW range, has 20 inverters and installed in early 2015.

The specific inverter failure type will be presented here as generic to protect any identifying information about the site. In reality, there are multiple failure modes for inverters at this specific site, which could be grouped together and modeled as shown in Table 2, or broken out specifically by failure mode as presented in Table 3. The time period for the failure mode in this dataset is approximately 1.5 years, and the model simulation time was run for both one and two years to bracket the results. For this example, 100 realizations were run. There is no attempt to extrapolate the failure rate past 2 years as the system owner has stated this inverter component issue has been resolved. Additional data would help quantify that case and help generate two different failure distributions that could be utilized to represent both early life failures, and later, constant failures (as the issue was resolved). Data was not available for validating that assumption.

Results indicate that the mean of the 1- and 2-year simulations bracketed the actual failures that occurred over a 1.5-year timeframe, suggesting that if the model had the capability to run fractions of a year, the mean of the realizations would likely be close to the actual failure estimate. This may also end up being the same case for the energy loss estimates. However, despite having the mean of the realizations closely

approximate the actual number of failures misses the uncertainty associated with stochastic sampling of a probability distribution, such as what was presented in Section II. B. The estimated failures have a low variance in the 95% confidence interval for both 1 and 2-year simulations. For the one-year simulation this indicates that 95% of the sampled intervals between 19.99 failures (431,508 kWh lost) and 20.02 failures (460,708 kWh lost) contain the true mean of 20.01 failures (446,108 kWh lost). For the two-year simulation (cumulative results) this indicates that 95% of the sampled intervals between 41.48 failures (719,316 kWh lost) and 41.87 failures (780,716 kWh lost) contain the true mean of 41.68 failures (749,916 kWh lost).

*C. Single-Axis Tracker Reliability Simulation*

Modeling single-axis tracker failures is done in two different ways: either the tracker fails in stow (flat, facing up) or at its rotation limit based on the azimuth angle designated in the SAM system design page. The example presented below shows results from a hypothetical single-axis tracker with a north-south axis, failing at stow (best case), and rotation limit due west (worst case). Energy loss and failure results are also presented based on this example. The calculations for determining the loss factors are more complex than presented here and outlined in the user manual in more detail [9].

The resulting failure and repair distributions are developed to show multiple failures per year, with events lasting around a month. A small PV system with two trackers is used and multiple realizations were simulated. The graphs presented below show one realization to discuss the concept of how the model translates the failure distribution into tracker position, then both 'power' and 'tracker' availability factors for determining how much power is generated in the resulting failed tracking state.

**Single-axis tracker 180-degree azimuth angle (north-south axis) and 45-degree rotation limit**

For a best-case assumption where the tracker fails flat (in stow), Fig. 4, shows that only one tracker failed out of the two

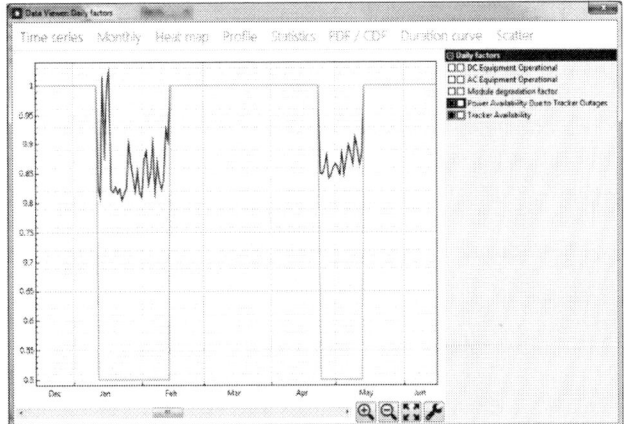

Fig. 4.   180-degree azimuth angle best-case (in stow) tracker and power availability factors

leaving a 50% tracker availability. As the tracker failed in stow, or flat and facing the sky dome, the relative power availability factor is higher in the April/May timeframe than in the January/February timeframe. The power availability factor is a function of the seasonal sun angle and the position of the modules when the tracker fails. The power availability factor in mid-January when one of the trackers fails is greater than 1 in some cases. This occurs when the majority of the irradiance is diffuse where modules in a flat position will 'see' a higher portion of the sky dome than modules tilted throughout the day, resulting in a greater ability to use the diffuse light. Checking the TMY2 file for the simulation, those days have much higher diffuse irradiance (DHI) than beam irradiance (DNI).

The worst-case assumption shows the same tracker availability metric of 50% where only one tracker fails out of the two. In this case, the tracker fails facing due west (as it is on a north-south 180-degree azimuth angle) at the maximum rotation limit of 45 degrees. Since the tracker in this case fails facing west, the power production profile will be different than when it fails in stow. A qualitative look at the results in Fig. 4

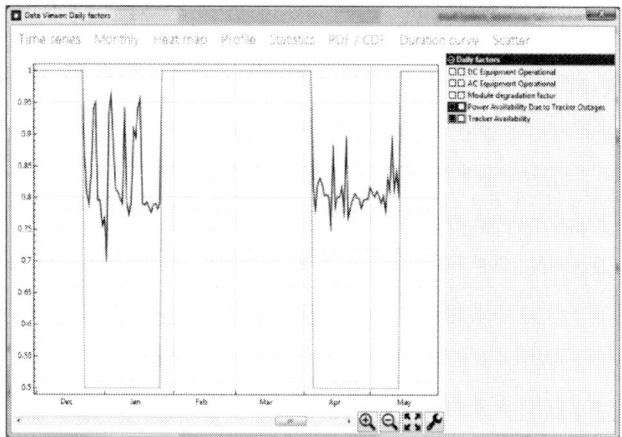

Fig. 5.    180-degree azimuth angle worst-case (facing west) tracker and power availability factors

and Fig. 5 show the power availability factor not dipping below 0.8 when the modules are in stow, but when the tracker fails at its maximum rotation limit of 45 degrees (to the west), the power availability factors appear to drop below 0.8 and fluctuate more widely due to the difference in how the different module angles capture the irradiance.

As only one tracker out of two has failed in both best- and worst-case simulations, the power availability factor does include the potential for power production from the tracker that has not failed. Therefore, looking at output from one realization needs to be understood in the context of the equations used to describe power loss from tracking failure [8].

**Energy Loss and Failure Results from Tracker Simulation**

Table 4 compares two simulations of 50 realizations each for the best case and worst case tracker failure positions compared with the no failure base case. The goal here is to show the range in failures based on the stochastic inputs, and see to what degree

TABLE IV
TRACKER FAILURES AND ENERGY PRODUCTION OVER 5-YR SIMULATION WITH 50 REALIZATIONS

| Component | Statistic | Total Failures | 5-Year Cumulative Energy Production [kWh] |
|---|---|---|---|
| Base Case (no failures) | N/A | 0 | **91,500** |
| Best Case (stow) | Mean | 6.76 | 91,260 |
| | 95% CI lower | 6.23 | 90,950 |
| | 95% CI upper | 7.28 | 91,570 |
| Worst Case (west – maximum rotation limit) | Mean | 6.88 | 90,780 |
| | 95% CI lower | 6.37 | 90,380 |
| | 95% CI upper | 7.38 | 91,180 |

CI – Confidence Interval

energy loss differs between each simulation over a 5-year period. In this case, the best case scenario when the tracker fails in the stow position results in higher energy production than when the tracker fails facing due west at the maximum rotation limit of 45 degrees. The spread between the mean and confidence intervals is also greater in the worst case scenario for energy production. This generally quantifies the visual differences in the power availability metric as illustrated for one realization in both Fig. 4 and Fig. 5, however for greater certainty the simulation could be extended to 15 or 20 years using the same failure and repair distributions, or running more than 50 realizations.

This result provides an example of how different tracker failure assumptions can be modeled where availability factors are applied to energy production results. Reliability distribution data is not presented here for the tracker failures, however users with their own tracker data can develop distributions to represent past, or anticipated failures and repairs, and run through multiple scenarios to understand energy production impacts or maintenance scenarios. Not shown here, but available in the model is the ability to add costs to track maintenance events. Simulations that vary the repair distribution parameters can be done to analyze different maintenance scenarios to optimize the time to repair based on labor rate, failure type assumptions and seasonal revenue loss estimates.

## IV. CONCLUSIONS

We present a new reliability feature in SAM that can be used to evaluate expected fault and failures events over time, and discuss the potential these new features have to simulate energy production impacts from failure and repair distribution parameters. This feature allows PV system owners, operators and equipment manufacturers the ability to conduct their own reliability assessment within a PV performance model to understand different reliability scenarios with assumptions, or actual data to evaluate reliability impacts to performance at different component life stages.

## ACKNOWLEDGEMENT

Sandia National Laboratories is a multimission laboratory managed and operated by National Technology and Engineering Solutions of Sandia, LLC, a wholly owned subsidiary of Honeywell International, Inc., for the U.S. Department of Energy's National Nuclear Security Administration under contract DE-NA0003525.

## REFERENCES

[1] A. Colli, "Failure Mode and Effect Analysis for Photovoltaic Systems," *Renew. Sust. Energ. Rev.* doi: 10.1016/j.rser.2015.05.056.

[2] E. Collins, M. Dvorack, J. Mahn, M. Mundt, and M. Quintana, "Reliability and Availability Analysis of a Fielded Photovoltaic System," *34th IEEE Photovoltaic Specialist Conference*, 2009.

[3] S. Miller, J. Stein, and J. Granata, "PV-RPM Demonstration Model User's Guide," Sandia National Laboratories, Albuquerque, NM, 2011.

[4] S. Miller, J. Granata, and J. Stein, "The Comparison of Three Photovoltaic System Designs Using the Photovoltaic Reliability and Performance Model," SAND2012-10342, Sandia National Laboratories, Albuquerque, NM, December, 2012.

[5] S. Ransome, J. Stein, W. Holmgren, and J. Sutterlueti, "PV Performance modelling with PVPMC/PVLIB," *PVSAT12*, 2016.

[6] G. Klise, R. Hill, A. Walker, A. Dobos, and J. Freeman, "PV System "Availability" as a Reliability Metric – Improving Standards, Contract Language and Performance Models," *43rd IEEE Photovoltaic Specialist Conference*, 2016.

[7] National Renewable Energy Laboratory, "System Advisor Model." Available at: https://sam.nrel.gov/

[8] G. Klise, O. Lavrova, and J. Freeman, "Validation of PV-RPM Code in the System Advisor Model," SAND2017-3676, Sandia National Laboratories, Albuquerque, NM, April 2.

[9] G. Klise, J. Freeman, O. Lavrova, and R. Gooding, "PV-RPM v2.0 beta – SAM Implementation Beta Test: User Instructions." June 2017.

# Lifetime and Degradation of Pre-damaged PV-Modules – Field study and lab testing

Claudia Buerhop[1], Sven Wirsching[1], Simon Gehre[1], Tobias Pickel[1], Thilo Winkler[1], Andreas Bemm[2], Julia Mergheim[3], Christian Camus[1], Jens Hauch[1], and Christoph J. Brabec[1,4]

[1]ZAE Bayern, Erlangen, 91058 Erlangen, Germany
[2]Allianz Risk Consulting GmbH, 81724 München, Germany
[3]LTM, FAU Erlangen-Nürnberg, Erlangen, 91058 Erlangen, Germany
[4]i-MEET, FAU Erlangen-Nürnberg, Erlangen, 91058 Erlangen, Germany

*Abstract* — **The presence of pre-cracked PV-modules in modern PV-plants is well-known. The evolution and actual impact of the cracks on electrical yield under real operation conditions is not yet understood but of great relevance. Established standards cannot reveal the relevant effects. Therefore, a unique threefold analysis is applied: 1) field exposure, 2) using a new accelerated loading test, and 3) Finite Elements (FEM) simulations. For the first time, we present comparative Electroluminescence (EL-) images recorded in the field and during load testing. Crack growth is studied in terms of the monitored weather conditions and the applied load simulating static snow and wind loads.**

## I. INTRODUCTION

Here, "*pre-cracked*" describes the existence of cracks in at least one single cell of the investigated PV-modules before the beginning of the analyses. The interest in lifetime and prospective performance of pre-cracked PV-modules [1, 2] gains in importance. Nowadays, electroluminescence (EL)imaging routines exist allowing the on-site inspection of PV-plants for installed and partially defective PV-modules [3]. Therefore, the degradation of such modules, e. g. with cracks, is of great interest for operators, investors and insurances. Today's standards [4] and most scientific work focuses on degradation of new, defect-free PV-modules. Some work is done on mechanical loading and fatigue tests [5, 6]. At operating conditions the mechanical loads due to wind and snow are known as the most critical stresses for crack growth in solar cells [7]. They can be calculated according to guidelines for loads to load-bearing structures [8, 9].

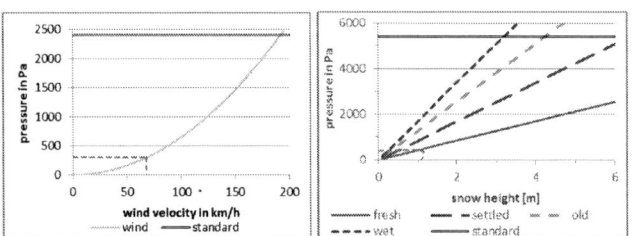

Fig. 1. Calculated pressure due to wind and snow loads in comparison with the pressure of the standard load tests [4]

Standards for testing pre-cracked modules do not exist. Known standards [4] differentiate the variety of existing load scenarios, as displayed in Fig. 1. Typical values from field experience for wind and snow are marked by grey lines, wind speed $v_{wind}$ = 65 km/h and snow height $h_{snow}$ = 1.1 m, respectively. Standard tests stress the modules much more, see dark solid lines. The pressure data are calculated for typical module installations with a tilt angle of 35°.

First experience of the degradation of pre-cracked modules at real operating conditions are documented by periodically repeated EL-imaging and power measurements [10].The aim of this paper is a deeper understanding of the field performance of pre-cracked modules by further experiments and Finite Elements (FEM) simulations. Additional strain measurements enable the recording of the local displacement and can verify the FEM-simulations of mechanical stress distribution in solar cells during loading.

## II. EXPERIMENTAL PROCEDURE

Our threefold approach includes field exposure of PV-modules, performing artificial load tests with a newly designed test-setup in the lab, and finally, supplementary FEM simulations computing mechanical stress.

### A. PV-modules and field tests

For this investigation, pre-cracked, crystalline PV-modules (230 W, 60 cells) from a landslide claim were investigated. They had differing numbers of cells with cracks and fracture. Most cracks were identified parallel and near the busbars (Fig. 4, Fig. 5).

During the field exposure several parameters were recorded, as described elsewhere [10]. Highly resolved weather (solar irradiance, ambient temperature, wind speed, humidity) and module data (module·temperature, string current, string voltage) were collected by a monitoring system. In addition, punctually EL-imaging and IV-measurements of individual modules were carried out. Furthermore, the strain present within the modules was measured. Therefore, strain gauges (Tokyo Sokki Kenkyujo Co., Ltd.), especially adapted to glass, were attached to the cover glass in the center of the modules. Using strain rosettes (composite material strain gauge of series "F" FRA-2-8) the gauges are separated by 45°.

978-1-5090-5606-4/17 $31.00 © 2017 IEEE

For outdoor measurements strain gauges and cables are covered with a special silicone (Shin-Etsu Chemical Co., Ltd.; 1 component RTV KE-348-T) to minimize degradation due to humidity and UV-light. During the outdoor measurements, the data is stored with an interval of 2 seconds. Principal strains $\varepsilon_1$ and $\varepsilon_2$ are calculated [11].

### B. Loading tests in the lab

20 modules were selected for accelerated load testing. Here, simultaneous EL-imaging during loading of the modules is presented. The test facility applies the pressure by vacuum pumps and air blowers. Thus, underpressure as well as overpressure can be used to simulate snow, wind pressure and wind suction, respectively. Before starting the loading procedure, strain gauges were attached carefully without covering cell area. Data are recorded with a temporal resolution of 2 s. We applied static loads up to 3000 Pa or 5000 Pa. The test procedure includes a stepwise increase of the pressure by 200 Pa with a holding time of 30 min, afterwards a period of 5 min without load. At the loaded and unloaded state EL-images were recorded with an EL-measurement system GE 2048 2048 DD NIR (GREATEYES GmbH, Germany). In total there were 15 loading and unloading steps. The test routine was completed by initial (before first loading) and final (after last loading) IV-measurements with a sun simulator (SPI-Sun Simulator 4600SLP, Spire Solar, USA) at standard test conditions. Finally, *ex-post* loadings at moderate pressure 100 up to 400 Pa were carried out to simulate normal operating conditions after an extreme weather event.

### C. FEM-simulations

The commercial finite element analysis software ABAQUS [12] was used to simulate the stress distribution of PV-modules under the conditions of the described mechanical loading test. The model was verified by comparing simulated and measured principal strains at three different positions of the glass surface for different loadings. Due to symmetry only one quarter of the module was simulated, which is composed of an aluminum frame, a glass sheet, an EVA-layer, the silicon solar cells, and an EVA plus polymer backsheet. In order to complete the simulations in reasonable calculation times whilst being able to simulate a full quarter module details such as interconnects or fingers are neglected. Each layer has a different thickness and different material properties [13]. As the simulations are meant to reveal the original nature of cracks, the simulations only consider intact solar cells without any cracks. Furthermore, linear elastic material behavior is assumed throughout the model.

### III. MAIN EXPERIMENTAL RESULTS

Observations and results of field exposed pre-cracked modules operated at real outdoor conditions are presented. Results of artificial loading tests accelerate the degradation

and support the field observations. Finally, the findings are explained by FEM-simulations of simplified models describing static mechanical loading of PV-modules.

### A. Field study

During the 1-year outdoor experience the PV-modules were exposed to varying weather conditions [10] as well as continuously repeated day-and-night temperature cycles. As described elsewhere [10], the solar irradiance reached short temporary maximum values of more than 1300 W/m², the ambient temperature was between -18°C in winter and 32°C in summer. The maximum daily wind was mostly in the range of a *strong breeze* [14], see Fig. 2. Wind gusts comparable to hurricane forces were only reached once Therefore, loads due to wind mainly were in the range of 100 to 400 Pa. In spite of the low temperatures there was no snow cover.

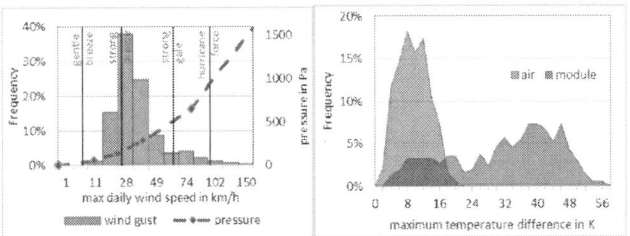

Fig. 2. Distribution of monitored wind gusts and temperature differences during one year of outdoor exposure [10].

The maximum daily temperature differences are plotted in Fig. 2 for air and module temperature. Outstanding are the larger temperature differences observed for modules (up to 65 K) compared to the daily differences of the air (till 20 K). The strain analysis investigated the influence of daily temperature cycles on the strain (Fig. 3) for two PV-modules after loading procedure in the lab: module ID 653 (heavily cracked) and module ID 731, which remained totally undamaged. Temperature differences of $\Delta T = 33$ K between day and night cause only minor strains between 35 µm/m and 60-70 µm/m.

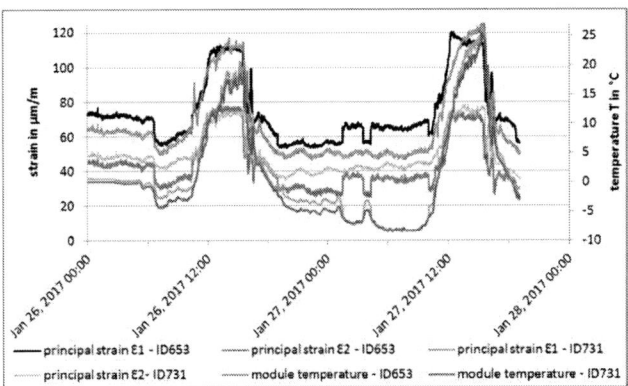

Fig. 3. Measured strain of field exposed modules, heavily cracked module ID 653 and totally undamaged module ID 731, weather conditions for these days: $T_{amb} = -14 - +4$°C, $v_{wind} = 0 - 5$ m/s, solar irradiance $E = 0 - 650$ W/m²

| $n = 0$ | $n = 1$ | $n = 8$ | $n' = 8$ | $n = 15$ | $n' = 15$ | $n = $ end |
|---|---|---|---|---|---|---|
| initial | Loaded with 200 Pa | Loaded with 1600 Pa | Unloaded after 1600 Pa | Loaded with 3000 Pa | Unloaded after 3000 Pa | Final |
| | | | | | | |
| $P = 223.0$ W | | | | | | $P = 221.7$ W |

Fig. 4. EL-images for various loaded and subsequent unloaded stages as well as initial and final power measurements of module ID 484, $n$ is the number of the experiment in ascending load order, $n'$ describes the corresponding unloaded stage

No new cracks or modifications are identified after one year of field exposure by outdoor EL. This applies for modules with few cracked cells as well as for those with many cracked cells.

Individual power measurements of the installed modules confirm these observations. Within the accuracy of the measurement equipment the power output of all modules remained constant. Evaluating the electric performance of the modules we calculated an averaged DC-performance ratio of approximately 87% for the DC-side of the inverter. This value slightly dropped during summer due to increased module temperature but recovered completely in winter [10].

In conclusion, operating at normal day-and-night temperature cycles and moderate weather conditions the investigated pre-cracked modules did not exhibit visible changes and measurable performance loss.

### B.    Load testing

The overpressure tests (simulating wind suction) did not induce any detectable crack in the modules. Cracks grow under tensile stress if the material strength is exceeded by the loading present. Therefore, experiments at underpressure are more interesting for crack initiation and propagation.

We found that at low loads ($p < 1000$ Pa) hardly any new cracks appear. Here, existing cracks, e. g. hidden ones under busbars or at grain boundaries, are visualized - often recognized by dark, electrically isolated areas. The growth of new cracks through a cell is monitored at higher pressures p > 1000 Pa. It depends strongly on the material properties and the local stress present. Fig. 4 shows some images out of the test sequence at different load stages. It is noticed that under load cracks can cause isolation of certain areas whereas after loading in the unloaded state the electric contact can be regained and the black areas in the EL-image disappear. This phenomenon is also described by Gabor and Paggi [15-17]. Furthermore, new cracks causing isolated areas at high pressure disappear during unloading but have the potential to reappear with isolated area at low pressures in the future. This loading is called *ex-post* loading.

A second example is shown in Fig. 5 including an EL-image at the *ex-post* loaded state of module ID 165. Cells with existing cracks before loading are marked with a white frame. At the loaded state cells with new cracks have a reddish frame. Noticeable is the large number of cells with new cracks in the second rows from top and bottom. Most cracks are orientated in parallel to the busbar. In the corner cells the cracks are mostly diagonal. Peculiarly, only three of the four corner cells tend to initiate cracks, as observed for the loaded modules. Looking closer at the EL-images, one corner cell has already at the beginning of the loading procedure an almost unseeable flaw. Here, the cracks become visible at first. At loads of 2400 - 3400 Pa the second corner cell cracks and at 3000 – 4000 Pa the third one follows. We explain the absence of cracks in the fourth cell by the redistribution of stresses. That means due to the fracture of several cells the load paths change and the last corner cell is not at risk anymore.

While the module looks severely deteriorated under load due to the black isolated cell areas, the same module appears quite intact upon subsequent unloading. But reloading the module moderately *ex-post* reveals some of the formerly black cell areas again. This is in agreement with [18].

The power outputs at STC of the presented modules reveal exemplarily, that at the unloaded state no remarkable power changes were measured. Further analysis of the module power at the loaded state will, however, show, that the power is reduced if black electrically isolated cell areas are visible.

Besides EL-images strain gauges and bending measurements describe the stress of the modules under static load tests. In the center of the modules bending up to 2.5 cm at 3 kPa under-pressure is measured. Considerable principal strain values are calculated for the glass surface, see Fig. 6. Values up 400 µm/m are obtained at rather high pressures. For $p = 400$ Pa principal strains of approx. 200µm/m are already achieved. The presented *ex-post* loading data reveal that the values are in the same range.

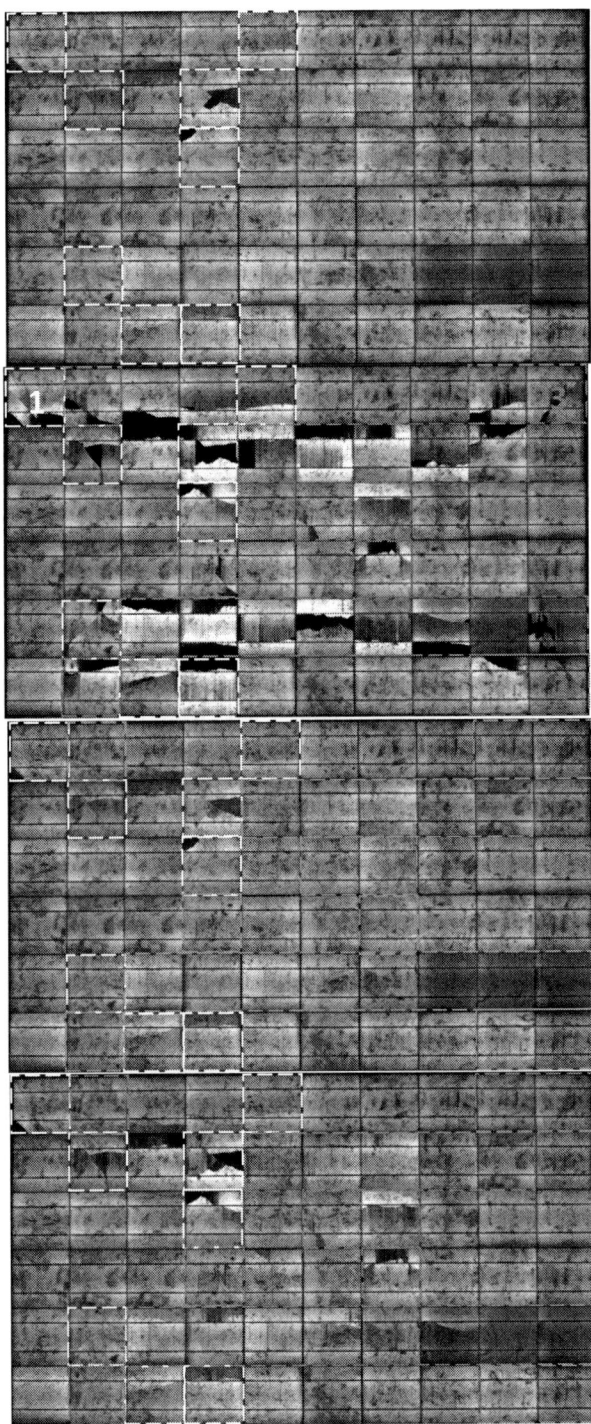

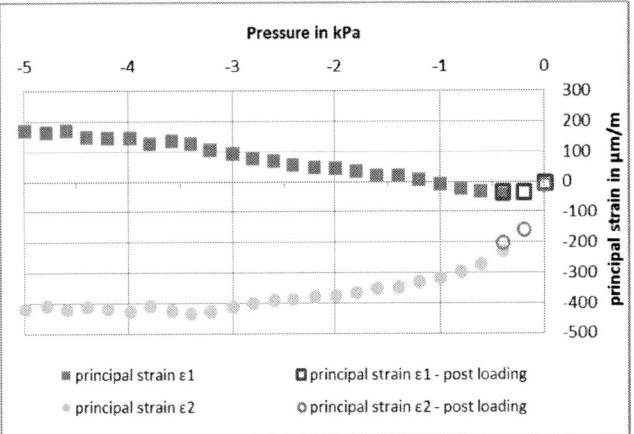

Fig. 6. Module ID 165, principal strain $\varepsilon_1$ and $\varepsilon_2$ as a function of loading and additional *ex-post* loading up to 400 Pa as deduced from strain gauge measurements.

The impact on the electric performance is measured at the sun simulator by default. At first it is surprising that no power loss could be observed for the unloaded module. But in IV-curves measurements of unloaded cracked modules, which may only show temporarily inactive cell areas during loading, these can behave identical to the untreated modules before the loading procedure.

## C. FEM-simulation

For verifying the FEM-model including material properties and simplifications of the module structure, experimental and simulated strain data are compared, see Fig. 7. The measured data of different modules scatter within a range of ~100 µm/m. The simulated strain values are in good agreement with the strain data of modules with a larger number of cracked cells (ID 653, ID 484). Module ID 731 is free of visible cracked cells, as mentioned previously. FEM-simulation data are shown for underpressure because in this configuration the solar cell is under tensile stresses, which in agreement with experimental results are critical for crack growth.

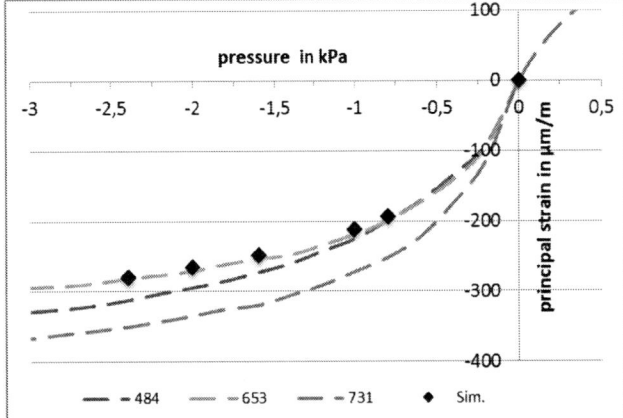

Fig. 7. Simulated strain compared to measured strain of various mechanically loaded modules at the glass surface [13]

Fig. 5. EL-images of module ID 165 at different loading states, from top to bottom: before loading (top image), loading with $p = 5000$ Pa (upper center image), after loading (lower center image), and *ex-post* loading with $p = 400$ Pa (bottom image). The frames highlight cells with cracks and fractures. White frames mark cells with existing cracks. Red frames illustrate cells with new cracks at loading with 5000 Pa. Blue frames indicate cells, where cracks have been seen during loading. Numbers in corner cells indicate the order of crack initiation. $P_{initial} = 231$ W, $P_{final} = 228.3$ W

Fig. 8 shows the distribution of the max. principal stresses in the solar cells of a PV module, when the module is loaded with $p$ = -1000 Pa. Only one quarter of the module is shown, the axis of symmetry are at the bottom and on the right. The maximum principal stress is $\sigma$ = 25 MPa and is obtained in the corner cell. The arrows indicate the directions of the maximum principal stresses. The maximum tensile stresses are present in the corner cell in diagonal direction and in the second row in vertical direction, which is perpendicular to the busbars. Cracks are expected to grow perpendicular to the maximum principal stress direction which means in this case parallel to the busbars. Consequently, such cracks may isolate distinct cell areas.

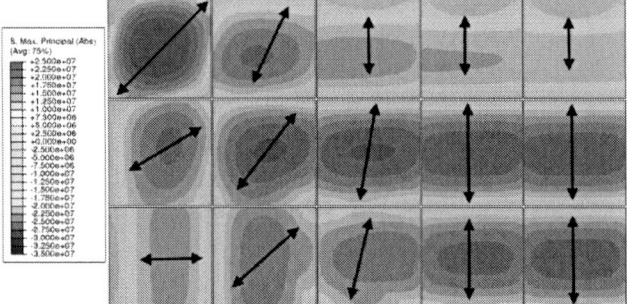

Fig. 8.    Quarter model with max. principal stress distribution in the cells, the arrows mark the principal stress directions, loading p = -1000 Pa [13]

The change of the stress distribution in the cells for different pressures is shown in Fig. 9. It is outstanding that for the corner cell (35) the highest stress is computed. Cells in the second row (marked by ■) have also high stresses. According to this simulation the applied stress in these cells may primarily exceed the materials strength and cracks can be initiated. However, increasing the pressure more cells will be in the critical range.

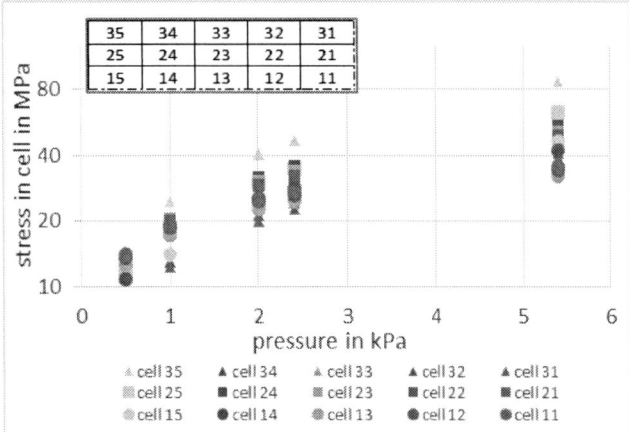

Fig. 9.    Stress distribution in different cells as a function of the loading, the scheme describes the nomenclature of the cell position, e. g. the corner cell 35 is in the upper left edge, the cell in center 11 is located in the lower right edge.

These simulated stress results are in good agreement with the experimental data from the accelerated load testing: mostly cracks developed in the cells in the second row and in horizontal direction.

## IV.    ANALYSIS OF RESULTS AND CONCLUSIONS

Our field study of field-exposed, pre-cracked modules reveals that after one year exposed to moderate weather conditions with short storm, no hailstorm, no snow, no extreme events, the modules have the same output power, the same performance ratio, the same EL-image. No changes are detected [10]. This is in accordance to climate chamber tests of pre-damaged modules [19] and [18].

| before mechanical loading | | | | | after mechanical loading with 3000 Pa | | | | |
|---|---|---|---|---|---|---|---|---|---|
| 3% | 1% | 3% | 3% | 11% | 33% | 25% | 19% | 18% | 24% |
| 1% | 10% | 11% | 16% | 25% | 11% | 50% | 54% | 50% | 51% |
| 3% | 1% | 1% | 6% | 11% | 4% | 11% | 18% | 21% | 34% |

| change-state due to loading = difference between initially unloaded and finally loaded state | | | | |
|---|---|---|---|---|
| 30% | 24% | 16% | 15% | 13% |
| 10% | 40% | 43% | 34% | 26% |
| 1% | 10% | 16% | 15% | 23% |

Fig. 10.    Frequency of cracked cells in a quarter module for 20 PV-modules before and after mechanical loading with 3000 Pa

Static loading tests with the newly designed load test set-up revealed that below a certain threshold "old" cracks become visible due to crack opening. Above that threshold new cracks and fractures are initiated. See Fig. 10 for the frequency of cracked cells before and after a certain pressure as well as the change-state. Cracks in corner cells are less frequent than expected for moderate loads, whereas their frequency increases for increasing loads. That may be due to the diagonal maximum stress direction and simplifications of the model. It was shown that cracks have the ability to isolate cell areas electrically at loaded stage and re-contact the previously isolated areas at unloading. However, the IV-measurements at the sun simulator before and after the load procedure were unexpectedly nearly unchanged (even so all modules had cracked cells at the end). Measuring the power at the loaded state reveals a certain power drop [20] which agrees to [18].

The FEM-results are in good agreement with the loading tests. Cells at the corner and in the long second rows are most frequently damaged. Cracks predominately propagate parallel to the busbar or diagonal in corner cells. Considering the stress distribution and resulting crack orientation, first evidence is given for distinguishing areal load cases, e. g. snow, from other more locally extended load cases.

Evaluating the strain data and recorded EL-images, evidence suggests usual day-night temperature cycles cause strain corresponding to loading scenarios of approximately 100 Pa. This loading is in accordance with a *fresh breeze* [14] of $v_{wind}$ = 40 km/h or snow height of 10 cm of heavy old snow or 25 cm of light fresh snow. Therefore, for such loading scenarios no significant and visible changes and accelerated degradation of pre-cracked crystalline PV-modules are expected.

## V.    SIGNIFICANCE OF THE WORK

The crack evolution of pre-cracked PV-modules is of great importance for investors, operators and insurances. They get a better understanding of the lifetime and degradation of modules with cracked cells. Nowadays, a lot of effort is made taking EL-images of PV-modules on-site. Such EL-images reveal many failures, e. g. cracked cells and fracture. What these findings mean for the future module and plant performance under real operating conditions is not known, yet. Standard tests designed for safety aspects are able to reveal the relevant effects. Modules are stressed too much. This work gives first evidence based on field studies and accelerated load tests for lifetime prediction of pre-cracked modules.

## VI.    OUTLOOK

Field exposure in combination with realistic load testing and FEM-simulation of mechanical stress give first evidence for degradation of pre-cracked PV-modules. For better understanding of the impact on the performance the outdoor exposure has to be extended, the lab testing has to be complemented by on-site power measurements and load cycles, and the FEM-simulations must be further refined in the future in order to also address modules, in which first damages are already present.

## VII.    ACKNOWLEDGEMENT

ZAE Bayern gratefully thanks the German Federal Ministry for Economic Affairs and Energy (BMWi) for financial funding of this project.

## VIII.    REFERENCES

[1] M. Köntges, "Performance and Reliability of Photovoltaic Systems, Subtask 3.2: Review of Failures of Photovoltaic Modules," in "Photovoltaic Power Systems Programme," Report 2014

[2] C. Buerhop *et al.*, "Statistical overview of findings in IR-imaging of PV-plants," in *SPIE Optics & Photonics*, San Diego, CA, USA, 2016, vol. 9938: Reliability of Photovoltaic Cells, Modules, Components, and Systems IX, pp. 99380L-99380L-9.

[3] S. Koch *et al.*, "Outdoor electroluminescence imaging of crystalline phtovoltaic modules: comparative study between ground-level inspections and drone-based aerial surveys," in *32nd EU PVSEC*, Munich, Germany, 2016, pp. 1736 - 1740.

[4] *IEC 61215, Crystalline silicium terrestrial photovoltaic (PV) Modules - Design qualification and type approval,* 1995, 2005.

[5] S. Kajari-Schröder *et al.*, "Mikrorisse in PV-Modulen unter mechanischer Belastung," presented at the 26. Symposium Photovoltaische Solarenergie, Bad Staffelstein, Kloster Banz, 2. - 4. March 2011, 2011.

[6] M. Pander *et al.*, "Fatigue analysis of solar cell interconnectors due to cyclic mechanical loading," in *32nd EU-PVSEC*, Munich, Germany, 2016, pp. 1589 - 1597.

[7] Die 5 häufigsten Schadensursachen an Photovoltaik-Anlagen! Hoffentlich von Experten gewartet und gut versichert? [Online].

[8] *DIN EN 1991-1-3, Eurocode 1 - Actions on structures - Part 1-3: General actions - Snow loads, 2010,* 2010.

[9] *DIN EN 1991-1-4, Eurocode 1: Actions on structures - Part 1-4: General actions - wind loads, 2010,* 2010.

[10] C. Buerhop *et al.*, "Long-term stability of pre-cracked PV-modules – field and lab study," *Progress in Photovoltaics,* 2017, submitted.

[11] L. Tokyo Sokki Kenkyujo Co., "Strain Gauges," p. 8.

[12] Hibbit *et al.*, "ABAQUS/standard: User's Manual, Vol. 1," ed, 1998.

[13] S. Gehre, "Experimentelle und numerische Analyse mechanischer Belastungen von Silizium-Solarmodulen," Master Thesis, LTM, FAU Erlangen-Nürnberg, 2016.

[14] DWD. Beaufortskala [Online]. Available: http://www.wettergefahren.de/warnungen/windwarnskala.html

[15] A. M. Gabor *et al.*, "Mechanical load Testing of Solar Panels - Beyond Certification Testing," in *43rd IEEE, Photovoltaic Specialists Conference (PVSC)*, Portland, OR, USA, 2016.

[16] M. Paggi *et al.*, "Fatigue degradation and electric recovery in Silicon solar cells embedded in photovoltaic modules," *scientific reports,* Article vol. 4, p. 4506, 2014.

[17] E. J. Schneller *et al.*, "Evaluating Solar Cell Fracture as a Function of Module Mechanical Loading Conditions," in *44th IEEE*, Washington, USA, 2017, accepted.

[18] J. L. Lincoln *et al.*, "Forecasting Environmental Degradation Power Loss in Solar Panels with a Predictive Crack Opening Test," in *44th IEEE*, Washington, USA, 2017, accepted.

[19] C. Camus *et al.*, "Degradation behaviour of pre-damaged crystalline silicon photovoltaic modules," in *Photovoltaic technical conference PVTC - from advanced materials and process to innovative applications*, Marseille, France, 2017.

[20] C. Buerhop-Lutz *et al.*, "Performance analysis of pre-cracked PV-modules at realistic loading conditions," in *33rd EU-PVSEC*, Amsterdam, The Netherlands, 2017, accepted.

# IMM Triple-junction Solar Cells and Modules optimized for Space and Terrestrial Conditions

Tatsuya Takamoto, Hiroyuki Juso, Kohsuke Ueda, Hidetoshi Washio, Hiroshi Yamaguchi
*Sharp Corporation*, 492 Minosho-cho, Yamatokoriyama, Nara 639-1186, Japan

Mitsuru Imaizumi, Taishi Sumita, Tetsuya Nakamura
*Japan Aerospace Exploration Agency*, 2-1-1 Sengen, Tsukuba, Ibaraki 305-8505, Japan

*Abstract* — InGaP/GaAs/InGaAs inverted metamorphic triple junction (IMM-3J) solar cells fabricated by epitaxial layer transfer process onto film -have features of lightweight and flexible. The average efficiencies of 27.4cm$^2$ cells developed for space and terrestrial applications are 30.8% and 34.5%, respectively. The lightweight module so called space solar sheets have been developed and tested by JAXA for space qualification. The terrestrial cell is optimized under AM1.5G condition. The terrestrial module in which 32 flexible cells are laminated by glasses showed 31.2% (AM1.5G) which is recognized as the world highest module efficiency. The module has been tested outdoor. Cost reduction technologies such as thinning epitaxial layer and reusing substrate have been investigated.

*Index Terms* — III-V compound, multi-junction, lightweight, GaAs, module.

## I. INTRODUCTION

In terrestrial photovoltaic (PV) market, price per watt is more important than efficiency for PV modules for all segments such as residential, industrial/community and utility scale. Retail competition of PV module is very severe in this market in which price per watt is lower than 1$/W and efficiency is higher than 17%. In addition, technology differentiation is not so easy. On the other hand, there is a special market which requires higher efficiency over 30% because of limited area for PV installation. Space satellite is a typical example in which expensive solar cells have been used for a long time. Some applications such as drones and backpack mobile batteries are looking for high efficiency cells recently. Furthermore, in future, PV must be an important function for electric vehicle (EV). From such circumstances, new PV market is expected as shown in Fig. 1. For silicon solar cells which are mainstream in the existing fixed market, 25% module efficiency is thought to be limit and over 30% module efficiency is impossible without tandem structure approach. III-V multijunction cells have already demonstrated very high efficiency over 30% and high reliability in space applications. Therefore, cost reduction is only an issue to expand markets. As well as the development of cost reduction technologies, mass production or volume discount will be necessary to realize an accepted price for EV installation.

For the mobile PV application, high efficiency, lightweight, flexible (or bendable), robustness and affordable price are required. III-V multijunction cells already demonstrated all items except for the affordable price. Lightweight and flexible solar cells with 29% efficiency have been obtained with InGaP/GaAs 2J structure fabricated by epitaxial layer transfer technique in 2004 [1][2]. After that, the efficiency has been improved by inverted metamorphic triple junction (IMM-3J) structure [3]. 37.9% (AM1.5G) is the best efficiency we have ever achieved [4].

Recently, main activities for IMM-3J cells are to develop the module structure which can be put into practical use. Lightweight module for space use, so called "Space Solar Sheet" has been developed under JAXA program [5]. Reliability of the space solar sheet has been qualified for various application conditions. Cost reduction technologies for

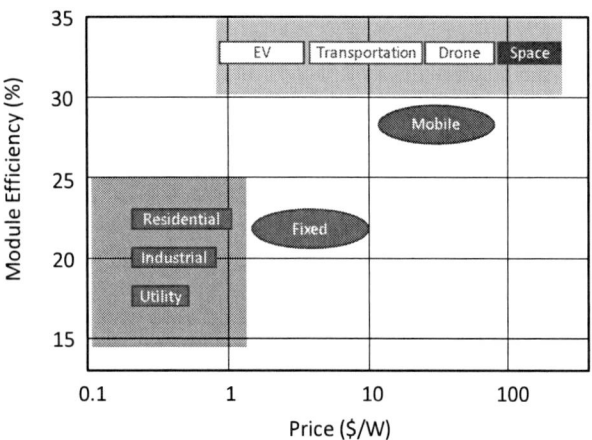

Fig. 1. Market segment for mobile PV

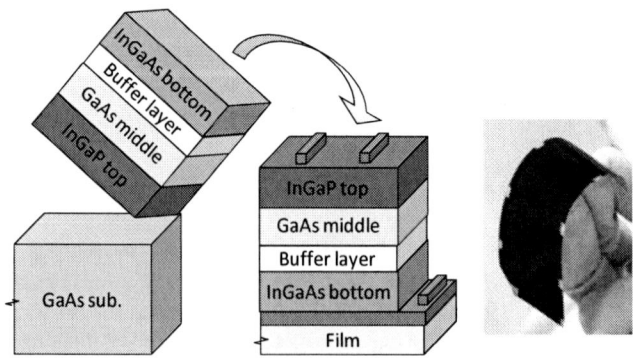

Fig. 2. Schematic of IMM triple junction cell fabrication.

IMM-3J cells and modules have been developed under NEDO project. Also, high efficiency module has been tested outdoor.

## II. RESULTS AND DISCUSSION

### A. IMM-3J Cell Fabrication

Lightweight flexible IMM-3J cells are fabricated by using epitaxial layer transfer technique as shown in Fig. 2. After InGaP/GaAs/InGaS cell layers are grown on a 4-inch GaAs substrate by MOCVD, the cell layers are transferred onto a film and the substrate are removed by etching. Both plus and minus electrodes are formed on its front side. Two 27.5 cm² cells are produced from a 4-inch wafer. Total thickness of the cell including the film is about 30 μm. The weight of the cell with the film is 15 times lighter than that of present 3J cells with Ge substrates.

### B. Space Solar Sheet

The average efficiency of IMM-3J cells developed for space and the characteristics of radiation resistance are shown in Table I. There are two types of IMM-3J space cells. One has

Table I: IV characteristics parameters and radiation resistance of IMM-3J space cells

| Cell Type | Fluence (e/cm²) | Voc (mV) | Jsc (mA/cm²) | FF | Eff (%) | Remaining Factor of Eff. |
|-----------|-----------------|----------|--------------|------|---------|--------------------------|
| Type B | 0 | 3055.2 | 68.2 | 0.842 | 31.2 | - |
| | 1E15 | 2728.3 | 65.7 | 0.823 | 26.2 | 84.1% |
| | 3E15 | 2651.9 | 64.3 | 0.772 | 23.4 | 75.1% |
| Type C | 0 | 3058.3 | 67.3 | 0.838 | 30.6 | - |
| | 1E15 | 2709.6 | 66.4 | 0.831 | 26.5 | 86.7% |
| | 3E15 | 2602.8 | 66.3 | 0.794 | 24.3 | 79.5% |

Film Type Sheet

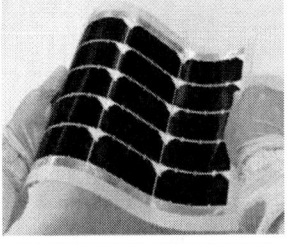

Glass Type Sheet

| Transparent film |
|------------------|
| Silicone adhesive |
| IMM-3J Cell |
| Adhesive |
| Support material |
| Adhesive |
| Transparent film |

| UV Coat |
|---------|
| Coverglass |
| Silicone adhesive |
| IMM-3J cell |
| Adhesive |
| Back sheet |
| AO Coat |

Fig. 3. Two types of space solar sheet

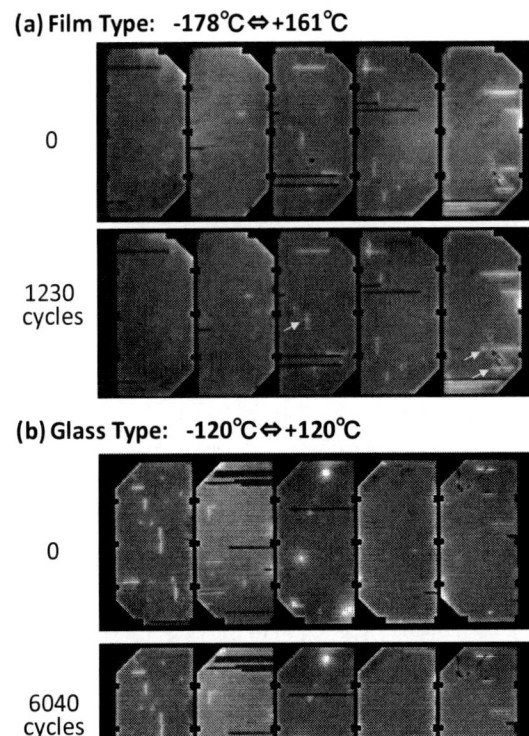

Fig. 4. Change of EL image after thermal cycling test for (a) Film type and (b) Glass type, sheets in which 5 clls are connected in series). Some change are indicated by arrows

higher efficiency at beginning of life (BOL) and the other has higher efficiency at end of life (EOL, after $3 \times 10^{15}$ e/cm² 1 MeV electron irradiation). The cells with higher efficiency at EOL are suitable for electric propulsion geostationary satellites.

There are two types of space solar sheet, the film type and the glass type as shown in Fig. 3. For the film type sheet, IMM-3J cells are connected with bypass-diodes and laminated with transparent films using silicone adhesive. UV-reflection coating is necessary on the film to avoid coloring of the films. The film type sheet is more flexible than the glass-type one. For the glass type sheet, IMM-3J cells are laminated with space grade coverglass and back-sheet using silicone adhesive. The glass type sheet is expected to have a long lifetime, because all materials used are space proven. The efficiency of both types of the sheets is about 27% (AM0). Total thickness of the sheets is less than 200 μm and the power per weight is approximately 0.6 W/g.

In the reliability qualification tests for the space solar sheet, influence of cracks in a cell was checked carefully. It is difficult to avoid generation of crakes during fabrication process of the cells and sheets. Although most of cracks observed only by electroluminescence (EL) image do not influence the electrical performance of the cell, change in properties after reliability tests should be checked. Figure 4 shows the EL image before and after thermal cycling tests for the glass and film type sheet structures. Cells which have various cracks were selected intentionally in the tests. Some cracks intersected with grid electrodes. After thermal cycling, a slight extension of cracks was observed for the film type sheet. However, no change in IV characteristics was confirmed. No change of cracks was shown in the glass type sheet after 6040 times of thermal cycling. No damage was also confirmed after the vibration test of 20 G when proper supports were applied to the sheets.

## C. Cost-Reduction Technologies

Main issues for cost reduction of IMM-3J cells are to reduce the thickness of epitaxial layer and to reuse the substrate. The thickness of the IMM-3J cells including metamorphic buffer layer is relatively thick. The metamorphic buffer layer which is not active for photovoltaic function should be reduced. Reduction of the buffer layer thickness was investigated with keeping the cell performance. The layer structure and the growth condition of the buffer layer were optimized to attain a good performance with the reduced thickness of buffer layer. Good performance was successfully obtained with a half thickness of the original buffer layer.

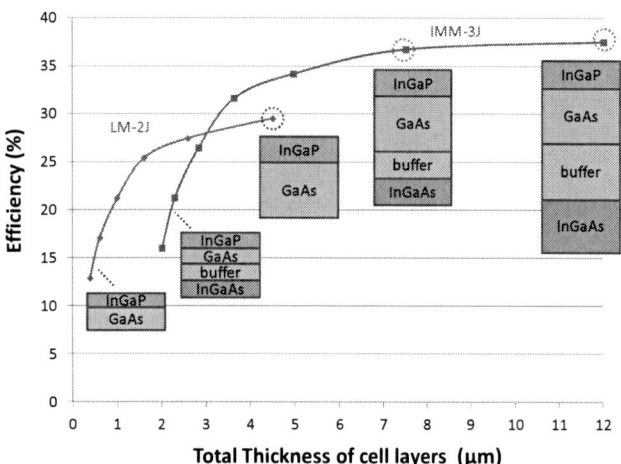

Fig. 5. Estimated efficiency of the cell as a function of total thickness of IMM-3J and LM-2J cells.

Table II: IV characteristics of LM-2J and Si stacking structure, measured by AIST

| | Size (cm²) | $I_{sc}$ (mA) | $J_{sc}$ (mA/cm²) | $V_{oc}$ (mV) | F.F. (%) | η (%) |
|---|---|---|---|---|---|---|
| InGaP/GaAs | 3.604 (ap) | 50.1 | 13.9 | 2446 | 84.1 | 28.56 |
| Si | | 29.9 | 8.30 | 694 | 76.6 | 4.41 |
| | | | | | | 33.0% |

Reduction of the thickness of the InGaS bottom cell was also investigated. It was not so difficult to reduce the thickness of the bottom cell, because reflection at the backside metal can be utilized. Good performance was successfully obtained with the half-thickness bottom cell.

Total thickness of IMM-3J cell can be reduced from 13 μm to 7.5 μm after thinning the buffer and bottom cell layers without decreasing efficiency. Figure 5 shows the estimated efficiency of the cell as a function of total thickness of IMM-3J and lattice-match (LM)-2J (InGaP/GaAs) cells. It is not so easy to keep the high efficiency with reducing total thickness of IMM-3J cells including the thick metamorphic buffer layer. For the thinner cell of less than 3 μm, the efficiency of LM-2J cell is better than that of IMM-3J. From the view point of cost reduction by thinning III-V epitaxial layer, stacking of LM-2J and Si bottom cell is thought to be a good candidate, because the cost of Si cell is negligibly low. Transparent structure has been developed for thin LM-2J cells. The LM-2J (28.9%) top and Si (4.1%) bottom stacking structure have demonstrated 33.0% efficiency in our preliminary work, as shown in Table II. The cell size was 2cm × 2cm, and IV characteristics were measured by AIST.

Epitaxial lift-off (ELO) and reuse of substrate technique have been investigated in order to reduce the substrate cost. It has been found that the ELO is relatively easy for LM structures such as GaAs-1J, InGaP-1J and InGaP/GaAs-2J, while it is not for IMM-3J lattice-mismatch structure. Reuse of substrate can be done after removing a protection layer grown on a GaAs substrate without mechanical polishing. Table III shows the summary of the trial results for ELO and reuse of substrate.

Table III: Summary of the trial results for ELO and reuse of substrate for various structures

| | GaAS LM-1J | InGaP LM-1J | InGaP/GaAs LM-2J | IMM-3J |
|---|---|---|---|---|
| Epitaxial Lift-off (ELO) | Good | Good | Good | No Good |
| Reuse of Substrate | Good | Good | Good | - |

## D. Terrestrial Module

In order to achieve high efficiency modules by using IMM-3J cells, relatively large size cells were fabricated. The cell fabrication process is almost the same as space cell fabrication. Thicknesses of subcells were optimized for terrestrial sunlight. The average efficiency of IMM-3J cells with modified thickness is 34.5%. Thirty-two cells were connected with bypass-diodes and laminated with front and back glasses. Module efficiency of 31.2% under AM1.5G condition was confirmed by AIST. Figure 6 shows comparison with the best efficiency of 37.9% obtained for a 1cm × 1cm small cell.

The 31.2% module has been tested outdoor at University of Miyazaki. Generated power from an IMM-3J module has been compared with that from a silicon cell module. Figure 7 shows monthly generated power from the both modules. Monthly generated power per watt of the IMM-3J module was only 90% of that of the Si module from August to October in 2016. However, from November, the generated power of the IMM-3J module increased and became the same as Si module. Figure 8 depicts IV curves of the IMM-3J and Si modules measured on a sunny day in October and January. Conversion efficiency (output power divided by GTI) was increased from 28% to 31%. This change is found to be caused by change of spectrum due to moisture in the air. Figure 9 illustrates solar spectral irradiance on a sunny day in both August and December at the test site, compared with AM1.5G standard spectrum. It was found that the light absorption due to water was increased in August and decreased in December. Figure 10 shows (a) the measured current of the IMM-3J and Si modules on sunny days from August to December and (b) the estimated sub-cell current in IMM-3J cell. Current of the IMM-3J module is found to be limited by InGaAs bottom cell.

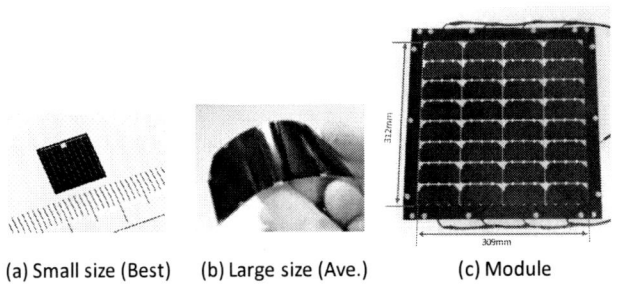

| | Size (cm²) | I_sc (mA) | V_oc (V) | F.F. (%) | η (%) | Measurement |
|---|---|---|---|---|---|---|
| (a) | 1.05 (ap) | 14.94 | 3.065 | 86.7 | 37.9 | AIST (2013) |
| (b) | 27.4 (t) | 382.4 | 3.008 | 82.1 | 34.5 | In house |
| (c) | 968 (ap) | 1506 | 23.95 | 83.6 | 31.2 | AIST (2016) |

Fig. 6. AM1.5G IV characteristics of IMM-3J (a) best cell, (b) large size cell and (c) module.

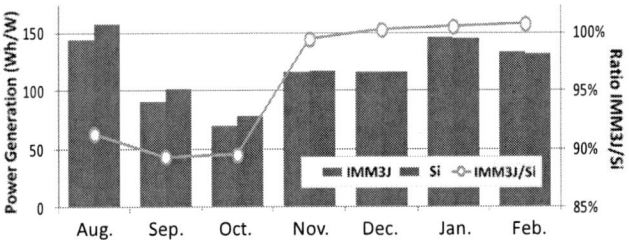

Fig. 7. Comparison of monthly generated power outdoor between IMM-3J and Si modules. (at Univ. of Miyazaki)

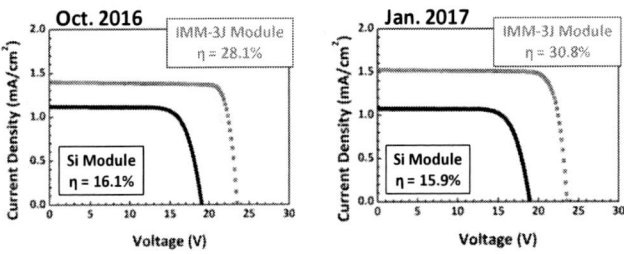

Fig. 8. IV curves for IMM-3J module and Si module measured on a sunny day in October and January. (at Univ. of Miyazaki)

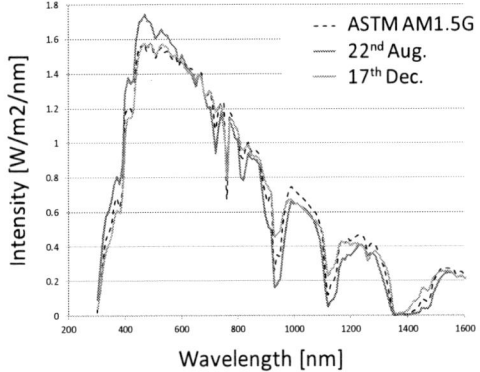

Fig. 9. Solar spectral irradiance on a sunny day in both August and December, comparison with AM1.5G standard spectrum.

(a)

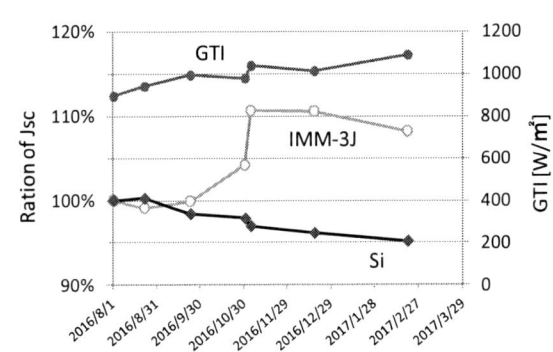

(b)

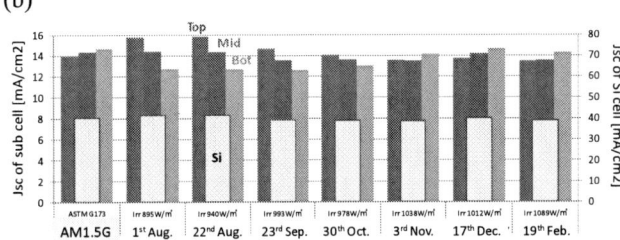

Fig. 10. (a) measured current of the IMM-3J and Si modules on sunny days from August to December and (b) estimated sub cell current in IMM-3J cell

978-1-5090-5606-4/17 $31.00 © 2017 IEEE

## IIII. CONCLUSION

High efficiency InGaP/GaAs/InGaAs IMM-3J solar cells and modules have been demonstrated. IMM-3J cells fabricated by epitaxial layer transfer process onto a film are lightweight and flexible. The average efficiencies of 27.4 cm$^2$ cells developed for space and terrestrial applications are 30.8% and 34.5%, respectively. The cell for space is designed with taking account of AM0 spectrum and radiation resistance. The terrestrial cell is optimized under AM1.5G, 25°C condition.

Two types of lightweight module, so called space solar sheet, glass type and film type, have been developed and tested by JAXA for space qualification. The terrestrial module in which 32 flexible cells are laminated by glasses showed 31.2% (AM1.5G) efficiency which is recognized as the world highest module efficiency. The module has been tested outdoor at Univ. of Miyazaki and the annual performance ratio has been evaluated. It is found that the performance ratio of IMM-3J module is affected by the change in bottom cell current due to change of sunlight spectrum.

Cost reduction technologies such as reducing epitaxial layer thickness and reusing substrate have been investigated. However, it is not easy to reduce the thickness of IMM-3J cell layers with keeping efficiency. Also, the epitaxial lift-off technique cannot be well done for IMM-3J structure. On the other hand, thin InGaP/GaAs 2J cell and Si stacking structure which is thought to be the most cost-effective structure have demonstrated 33.0% in our work.

## ACKNOWLEDGEMENT

The work is partially supported by the Department of the Navy, Office of Naval Research. The work is partially supported by New Energy Development Organization (NEDO). Outdoor measurement is supported by Nishioka-Lab at University of Miyazaki.

## REFERENCES

[1] T. Takamoto, T. Agui, H. Washio, N. Takahashi, K. Nakamura, M. Kaneiwa, & K. Okamoto, "Flexible thin-film III-V multijunction solar cells", in *Proc. 19th European Photovoltaic Solar Energy Conf*, 2004, p. 3689.

[2] T. Takamoto, T. Kodama, H. Yamaguchi, T. Agui, N. Takahashi, H. Washio, M. Imaizumi, & K. Kibe, "Paper-thin InGaP/GaAs solar cells", in *4th World Conference on Photovoltaic Energy Conversion, 2006,* Vol. 2, pp. 1769-1772.

[3] T. Takamoto, T. Agui, A. Yoshida, K. Nakaido, H. Juso, K. Sasaki, M. Imaizumi, & M. Takahashi, "World's highest efficiency triple-junction solar cells fabricated by inverted layers transfer process", in *35th IEEE Photovoltaic Specialists Conference, 2010.*

[4] T. Takamoto, H. Washio, & H. Juso, "Application of InGaP/GaAs/InGaAs triple junction solar cells to space use and concentrator photovoltaic", in *40th Photovoltaic Specialist Conference, 2014,* pp. 0001-0005

[5] H. Yamaguchi, R. Ijichi, Y. Suzuki, S. Ooka, K. Shimada, N. Takahashi, H. Washio, K. Nakamura, T. Takamoto, M. Imaizumi, T. Sumita, K. Shimazaki , T. Nakamura, T. Ohshima, "Development of Space Solar Sheet with Inverted Triple-junction Cells", in *42th Photovoltaic Specialist Conference, 2015,*

# Very High Specific Power ELO Solar Cells (>3 kW/kg) for UAV, Space, and Portable Power Applications

D. Cardwell, A. Kirk, C. Stender, A. Wibowo, F. Tuminello, M. Drees, R. Chan, M. Osowski, and N. Pan

MicroLink Devices Inc., Niles, IL, 60714, USA

*Abstract* — **MicroLink Devices has developed triple-junction (3J) GaInP/GaAs/GaInAs epitaxial lift-off (ELO) 20 cm² solar cells achieving >3 kW/kg specific power and >30% AM0 (>34% AM1.5G) 1-Sun power conversion efficiency. These cells were manufactured using volume mass-production tools and processes. Preliminary reliability testing of bare cells and laminated arrays showed no appreciable degradation.**

*Index Terms* — **solar cell, epitaxial lift-off, specific power, UAV**

## I. INTRODUCTION

Inverted metamorphic multi-junction (IMM) solar cells produced using epitaxial lift-off (ELO) offer high efficiency, high specific power, low cost, and flexibility, making them attractive to a range of applications including high- and low-altitude long endurance UAVs, portable power, and space power. Through careful design of the epitaxial layer structure, back metal, and top grid metal, we have demonstrated large area (20 cm²) GaInP/GaAs/GaInAs ELO solar cells with AM0 1-Sun power conversion efficiency >30% and specific power >3kW/kg fabricated using production tools and processes.

## II. DEVICE STRUCTURE AND FABRICATION

A cross section view of the GaInP/GaAs/GaInAs ELO-IMM solar cell layer design is shown in Fig. 1. The solar cell wafers are grown on 6-inch substrates by metal-organic vapor phase epitaxy (MOVPE) in an inverted orientation with an epitaxial release layer. A thin metal layer is evaporated on top of the wafer, which subsequently serves as the backside metal contact. The release layer is then selectively removed using a wet chemical process, leaving both the substrate and the active epitaxial solar cell layers intact. The epitaxial material is transferred to a temporary carrier for ease of handling, permitting the top surface to be processed using standard semiconductor fabrication techniques. Antireflection coatings are applied following grid metal deposition to couple more light into the solar cell. Fig. 2 shows a photograph of a fully processed 6-inch ELO foil containing six 20 cm² solar cells.

## III. RESULTS AND DISCUSSION

MicroLink has recently increased the power conversion efficiency of its production IMM solar cells through improvements to the epitaxial layer design. Performance improvements were achieved through modifications to the top tunnel junction, metamorphic buffer layer, GaInAs bottom subcell, and top grid/lithography design. Fig. 3 shows an illuminated current-voltage (*I-V*) measurement from a 20 cm² 3J solar cell achieving 30.4% AM0 1-Sun efficiency with $V_{oc}$ = 3.004 V, $J_{sc}$ = 15.93 mA/cm², and *FF* = 86.89%. This solar cell uses our standard production top grid design that includes bond-pads for bypass diode placement and top-side welding of the *n*-and *p*-electrical contacts, resulting in an overall shadow loss of 5.5%. The *I-V* measurements were performed using a 3-zone solar simulator calibrated with GaInP, GaAs, and GaInAs isocells. Fig. 4 shows an electroluminescence (EL) image from the same 20 cm² solar cell displaying bright red luminescence from the 1.90 eV GaInP top subcell and no evidence of particle-related defects. Shown in Fig. 5 is an illuminated *I-V* curve from a similar 20 cm² solar cell measured under AM1.5G 1-Sun conditions achieving 34.6% efficiency. We project AM1.5G 1-Sun efficiencies >36% from this AM0-optimized structure by incorporating minor modifications to the subcell thicknesses to provide better current matching under AM1.5 illumination.

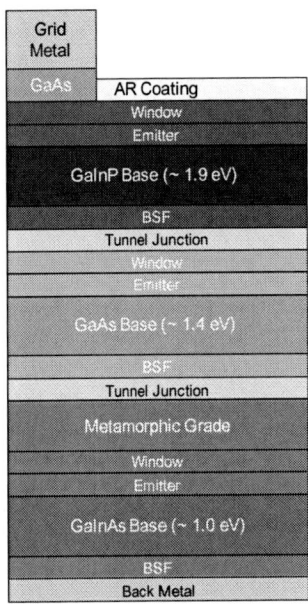

Fig. 1.   Cross-section view of MicroLink's production triple-junction (3J) GaInP/GaAs/GaInAs ELO-IMM solar cell.

978-1-5090-5606-4/17 $31.00 © 2017 IEEE

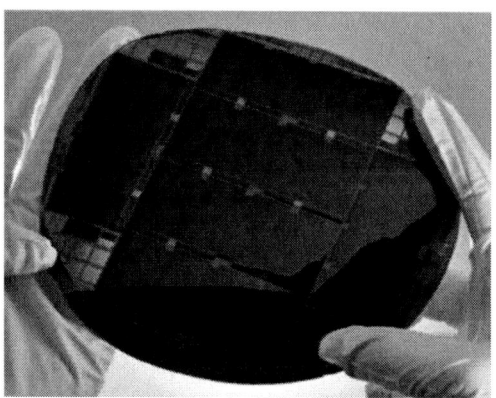

Fig. 2. Photograph of a 6-inch foil fabricated into six 20 cm$^2$ GaInP/GaAs/GaInAs 3J solar cells.

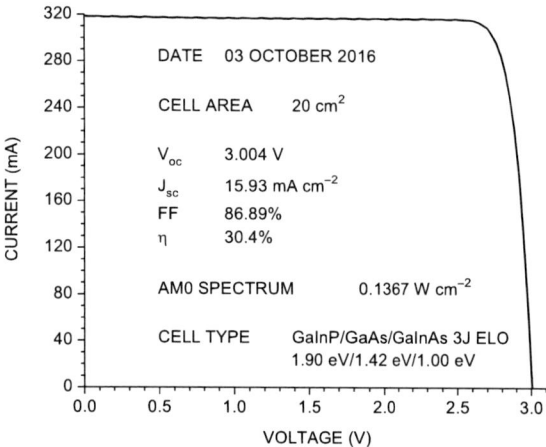

Fig. 3. Measured *I-V* curve is shown of a 30.4% efficient AM0 1-Sun 20 cm$^2$ GaInP/GaAs/GaInAs 3J ELO thin-film solar cell lifted off from a 6-inch GaAs substrate following growth in a commercial Aixtron MOVPE reactor. Grid contact shadow loss was relatively large at 5.5% resulting in less than optimal current density. The measurement was taken at 25°C on an AM0-calibrated TS-Space Systems 3-zone solar simulator.

Fig. 4. Measured electroluminescence (EL) of the 20 cm$^2$ GaInP/GaAs/GaInAs 3J ELO thin-film solar cell shown in Fig. 3. Forward bias settings were 3.24 V and 500 mA. The red luminescence is from the 1.90 eV GaInP top subcell.

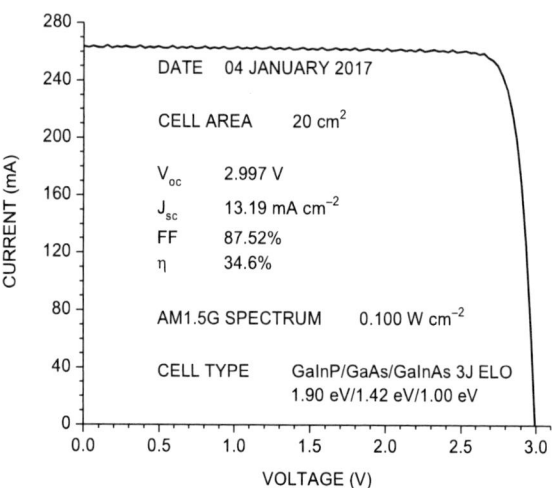

Fig. 5. Measured *I-V* curve is shown of a 34.6% efficient AM1.5G 1-Sun 20 cm$^2$ GaInP/GaAs/GaInAs 3J ELO thin-film solar cell lifted off from a 6-inch GaAs substrate following growth in a commercial Aixtron MOVPE reactor. The measurement was taken at 25°C on an AM1.5G-calibrated TS-Space Systems 3-zone solar simulator.

To achieve high specific power, the solar cell must operate with high efficiency and must have very low mass. The back metal is a significant contributor to the overall mass of the ELO solar cell. Since the back metal materials typically have higher mass density than the III-V epitaxial materials, a thin back metal design is critical to achieving very high specific power. A suitable design requires a compromise between low mass and adequate mechanical support. MicroLink has devoted considerable effort to optimizing the back metal design to decrease the mass and maintain device performance and yield in the ELO production process. Multiple materials and deposition techniques have been investigated for this purpose to produce defect-free 6-inch ELO foils. Fig. 6 shows a flexible free-standing 20 cm$^2$ solar cell having a mass of 0.21 g and an AM0 1-Sun efficiency of 29.3%, resulting in specific power of 3.8 kW/kg. This solar cell was fabricated using production processes and an antireflection coating designed for encapsulation under typical polymer materials. Multi-wafer lots with median specific power >3 kW/kg have been produced. MicroLink's internal cell efficiency roadmap for 3J IMM devices targets an AM0 1-Sun efficiency of 32.5% with specific power >4 kW/kg.

Fig. 6. Free-standing 20 cm$^2$ GaInP/GaAs/GaInAs 3J production solar cell which achieved specific power of 3.8 kW/kg.

MicroLink has developed methods for stringing ELO cells into arrays with high packing fraction and subsequently laminating the arrays in between low mass polymer materials. Laminated arrays utilizing the low mass solar cells developed in this work have 1.7 kW/kg projected specific power. MicroLink's laminated arrays have been integrated into multiple aerial platforms and have demonstrated proven performance benefits in both high-altitude long endurance (HALE) and small UAV applications. An example of a UAV incorporating MicroLink's solar cells is shown in Fig. 7. Wing-mounted laminated arrays were integrated into the electrical system of UAV Solutions' Talon 120 low-altitude UAV for flight testing. Aircraft with integrated solar arrays showed a 2× increase in flight time versus equivalent UAVs without solar cells. This was achieved in Maryland in November 2016 under low-angle solar irradiance.

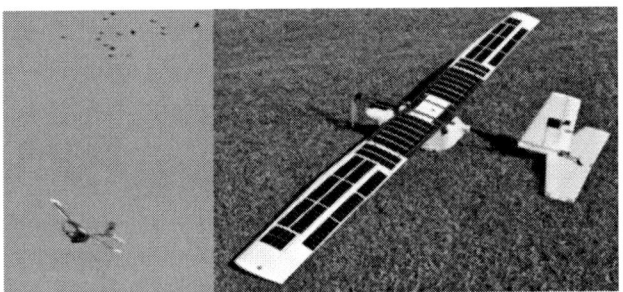

Fig. 7. Pictures are shown of a UAV Solutions Talon 120 UAV with MicroLink's 3J ELO solar sheet modules. Flight testing was performed in Maryland, USA in November 2016 under low-angle solar irradiance. Compared to the baseline UAV with no solar cells, the flight time was increased by 100%.

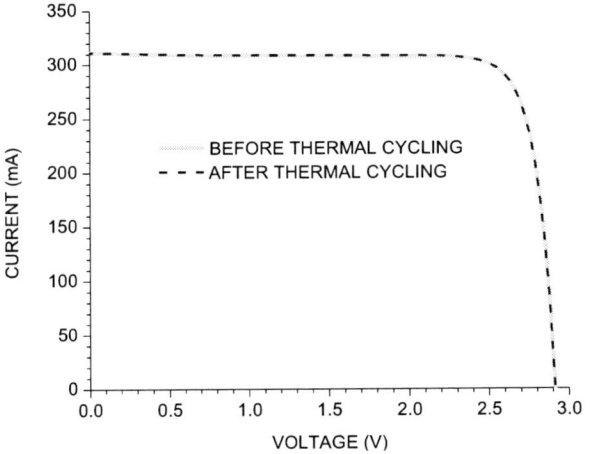

Fig. 8. Measured *I-V* curves before and after thermal cycling are shown of a 20 cm² GaInP/GaAs/GaInAs 3J ELO thin-film solar cell with a low mass back metal design. The cell underwent 140 thermal cycles between −80 and 80°C.

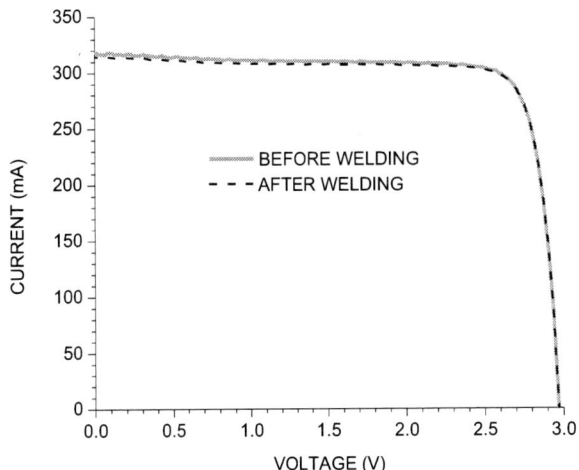

Fig. 9. Measured *I-V* curves before and after Ag tab welding are shown of a 20 cm² GaInP/GaAs/GaInAs 3J ELO thin-film solar cell with a low mass back metal design.

Preliminary reliability testing was performed on individual cells and laminated 1×2 arrays. Reliability tests included thermal cycling (140 cycles between −80 and 80°C), damp heat exposure (1700 hours at 45°C and 95% relative humidity), and one-Sun light soak (1500 hours at 125°C under approximately AM0 illumination). Fewer than 5 samples were used in each test. No degradation was observed in the conversion efficiencies of bare cells or laminated arrays after the reliability tests. Fig. 8 shows practically identical measured *I-V* curves from a bare 20 cm² solar cell before and after thermal cycling. Resistance spot welding conditions were optimized to achieve tab pull force >300 g without degrading the solar cell power conversion efficiency. Fig. 9 shows measured *I-V* curves from a bare 20 cm² solar cell before and after Ag tab welding.

## IV. CONCLUSIONS

Through improvements in solar cell device design, epitaxial growth, and back metal materials, MicroLink has achieved >3 kW/kg specific power and >30% AM0 (>34% AM1.5G) 1-Sun power conversion efficiency at 25°C in 20 cm² reliability-tested 3J ELO solar cells manufactured entirely using volume mass-production tools and processes.

## ACKNOWLEDGEMENTS

This work was funded in part by the Office of Naval Research (ONR) under program number N00014-15-C-5017, Naval Air Warfare Center under program number N68335-16-C-0106, NASA under program number NNX16CL73P, and Air Force Research Laboratory (AFRL) Space Vehicles Directorate under program number FA9453-17-C-0421. MicroLink thanks UAV Solutions for providing a Talon 120 UAV and for performing the flight testing.

# Enhanced Endurance of a Unmanned Aerial Vehicles Using High Efficiency Si and III-V Solar Cells

David Scheiman[1], Raymond Hoheisel[2], Daniel J Edwards[1], Andrew Paulsen[3], Justin Lorentzen[1], Steve Carruthers[1], Sam Carter[4], Matthew Kelly[1], Phillip Jenkins[1], and Robert Walters[1]

[1]U. S. Naval Research Laboratory, Washington, DC 20375, United States
[2]George Washington University, Washington, DC 20052, United States
[3]Packet Digital, Fargo, ND 58102, United States
[4]Envisioneering, Alexandria, VA 22020, United States

*Abstract* — **Unmanned Aerial Vehicles (UAVs) are rapidly growing in both military and commercial markets. One shortfall of UAVs is the amount of time they can fly, being limited by the energy storage. Solar cells can be integrated into the wing surface to provide additional power, dependent on the sun, weight, wing area, and efficiency. NRL is building a UAV with wings from a variety of solar cell technologies, which includes high efficiency Si, thin flexible GaAs, triple junction InGaP/GaAs/Ge, and Inverted Metamorphic Multi-Junction (IMM) for direct comparison and to improve flight endurance. The UAV also incorporates the necessary power management system required to maximize the solar power available. In addition to solar cells, this plane can also utilize thermal updrafts to soar. Flight data is provided which includes all electrical parameters for comparison studies.**

## I. INTRODUCTION

The Unmanned Arial Vehicle (UAV) market is expanding with forecasts varying greatly between $2B and $90B in 2020 depending on survey criteria [1, 2], with enabling technologies decreasing the size and weight of electronic payloads. One standing limitation is the time that the aircraft can fly; endurance is generally bounded by weight, size, maneuverability, and energy storage. Solar cells offer a way to capture solar energy during daylight hours and they can be built into the structure to take advantage of large, primarily flat wing area. Optimistically, solar power can provide enough energy to stay aloft indefinitely; extended flights up to 14 days have been demonstrated with Qinetiq Zephyr and the manned Solar Impulse [3,4]. Both of these aircraft have very large wings and a fragile structure, with limited payload and maneuverability designed specifically for endurance. Small UAVs in the military, such as the Aerovironment Raven and Puma have also been outfitted with solar cells [5,6].

NRL has retrofitted custom-built solar wings to the RnR Products SBXC sailplane kit (this airframe is no longer manufactured). They currently have 6 wing sets with SunPower Si cells (2), Microlink Devices (MLD) IMM cells (2), Alta Devices GaAs cells (1), and Solaero thinned ZTJ cells (1), a 7th wing set with Sharp IMM cells is in process. This modified aircraft is being used to demonstrate longer

endurance at lower altitudes (<10 kft / 3 km). Flight endurance will be extended by two methods: solar power and the use of convective thermal updrafts [7, 8]. NRL has a lot of history with this airframe and its performance is well characterized.

## II. AIRCRAFT DESIGN

The NRL aircraft is a modified SBXC airframe from RnR Products using a stock fuselage and tail with in-house built wings [9]. The structure is hollow-molded composite, with skins made from a sandwich of carbon fiber, foam, carbon fiber, and with unidirectional carbon fiber for the spar and vertically grained balsa wood for shear web. The wingspan of the stock SBXC is 4.2 m, whereas NRL's wingspan is 4.5 m with a total surface area of 1.3 square meters. Each wing half is in 2 sections, with a long center section of 1.4 meters and plug-in tip extensions. Takeoff mass is 6.8 kg (no payload) and requires 70 W minimum electrical power to fly, varying with airspeed. It is hand launched (see Fig. 1) and lands on its belly with a foldable propeller.

Fig.1. NRL PV-SBXC UAV being launched for a test flight

The autopilot system is a Cloud Cap Technologies Piccolo SL. The Piccolo controls the low-level interfaces to the servo actuators, and onboard sensors, to include GPS, barometric and dynamic pressure, 3-axis rate gyro, 3-axis accelerometer, and temperature. An onboard flight computer contains the algorithms for autonomous soaring (thermals) and solar system monitoring. The flight computer transmits and receives information to the autopilot directly and controls the power management system's various flight modes.

The power management system is custom-designed electronics made by Packet Digital Corporation. The system includes a battery pack with integrated charge controller, power management and distribution (PMAD) interface for all onboard components, a serial interface to the flight computer, and a single max power point tracking (MPPT) of the solar panels. The battery pack weighs 1.91 kg and consists of 36 Panasonic NCR18650B battery cells in a 6 series x 6 parallel (6s6p) arrangement with control and protection electronics, with approximately 400 W-hrs. The PMAD regulates the bus voltage matched to the batteries and supplies power to the electric propulsion system plus 5 and 12 V outputs for the servos and avionics, respectively. The MPPT is a boost converter which has either perturb-and-observe or a custom adaptive-tracking system with nominal efficiency of 96%, and the tracking updates multiple times per second. Data packets from the battery, PMAD, and MPPT transmit voltage, current, temperatures, battery status, and other status flags at 1 second intervals. With these components, the plane can fly for ~4 hours on stored energy alone.

The wings are covered with solar panels on the top surface. With this being a demonstration aircraft, only the inner center wing sections are populated with cells, and the array is not fully optimized for the wing shape. The solar cell technologies being demonstrated include:

1) Sunpower Si, Interdigitated Back Contact (IBC) >22% efficiency, cells cut into thirds
2) Microlink Devices Inverted Meta-morphic Mult-junction (IMM) InGaP/(In)GaAs/InGaAs utilizing ELO technology, >26% efficient, flexible [10]
3) Alta Devices flexible GaAs, single junction III-V cells >23% efficiency and flexible
4) Solaero Inc., thinned ZTJ lattice-matched InGaP/(In)GaAs/Ge triple junction >31% efficiency
5) Sharp IMM cells InGaP/GaAs/InGaAs, >33% efficiency and flexible [11]

### III. Solar Panels/ Wings

There are 2 separate solar arrays on this aircraft, a right and left on each wing center section. Each half-wing panel is composed of a number of interconnected solar cells arranged in series and parallel to match the requirements of the MPPT optimized for a nominal 18V.

The SunPower solar panels were made at NRL and consist of production cells 153.3 cm² that were cut into thirds on a dicing saw (~51 cm²). Each wing half-array is 2 parallel

strings of 26 third-cells. The cells were hand soldered together using cooper ribbon, all on the backside. The panel is then laminated between two sheets. A Schottky blocking diode is used for each half-wing with no bypass diodes. Each half wing panel weighs 211g (Fig. 2) with up to 84W total wing power under AM1.5G test conditions. SunPower cells have demonstrated some of the highest efficiencies for Si.

Fig. 2.   Fully assembled Si Wing half (SunPower cells)

The Microlink IMM solar cells have dimensions of 6.75 cm x 3.1 cm (20 cm²). Each wing half has 5 parallelized sub-arrays of 3 parallel by 8 series connected cells. There are bypass diodes connected to each cell and blocking diodes on each of the 10 sub-arrays. The cells have welded 2 mm  silver ribbon interconnects with two connections to each cell and a bypass diode attached to a third pad. All connections are on top of the cell. The cells are arranged with 1 mm spacing on all sides. These 3 x 8 cell strings were shipped to Vanguard Space Systems (now owned by SolAero) and mounted on a Kapton backsheet with a vacuum bonding process. The backside of the Kapton was coated with copper and had a patterned interconnect circuit for solder connections to the

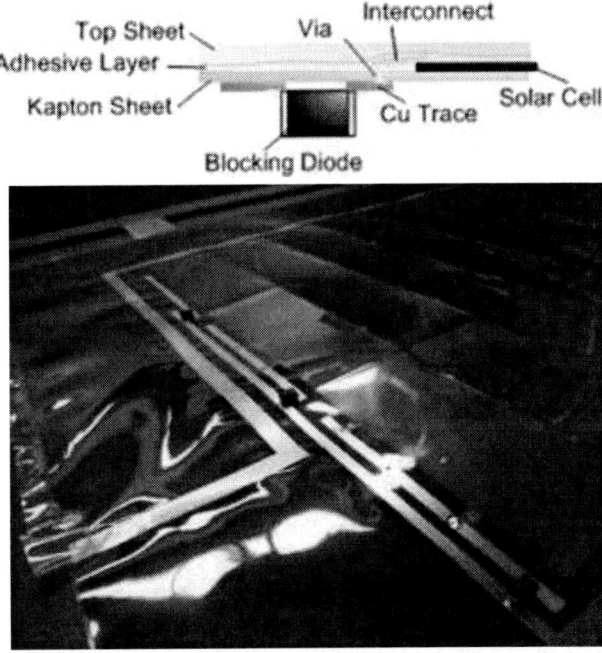

Fig.3,4.  Cross-section of the MLD array assembly as built by Vanguard Space Systems with backside pictured beneath. strings.   Surface mount Schottky blocking diodes were

soldered to each string connection. A transparent coversheet was laid over the entire array and then vacuum bonded. Each one of these panels weighs 185 g and a wing produces 118 W under AM1.5G.

The Alta Devices thin film GaAs solar cells are 2 cm x 5 cm (10 cm²). Each half wing array consists of 4 laminated subarrays of 4 parallel by 18 series cells. The cells are interconnected with a conductive adhesive with a shingle pattern that maximizes the packing factor. The 18 cell strings connect to a silver ribbon and there are 2 bypass diodes splitting the array at 9 cells with a blocking diode for each subarray. Each one of these panels weighs 125 g, the finished wing generated 92W under AM1.5G test conditions. This was the easiest of all wing assembly builds (Figure 5).

Fig. 5.     Fully assembled GaAs Wing half (Alta Devices cells)

The Sharp IMM solar cells are standard cells 3.7 cm x 7.6 cm with cropped corners. These cells were provided in a variety of laminated subarrays with voltage defined by 6 cells in series (1x6 to 3x6). Each cell has a bypass diode and redundant welded silver ribbon interconnects with in-plane stress relief. The subarrays made by Sharp were soldered together by NRL using tinned copper ribbon with the addition of blocking diodes, having a nominal efficiency of 32% with a 6x3 subarray weighing 33 g, 158 g per half wing array (Figure 6). The two panels produced 150W at AM1.5 prior to integration into the wing.

Fig. 6.     IMM Wing half (Sharp cells) laminated panel

The SolAero arrays are thinned ZTJ cells of the standard size 3.7 cm x 7.6 cm with cropped corners. They are arranged in subarrays of 3 parallel by 7 series, 4 per wing half. They are interconnected with silver ribbon using a conductive epoxy, every cell has a bypass diode and every string has a blocking diode (same diode). The subarrays were assembled by SolAero and assembled into laminated sheets. This finished wing produced 116 W under AM1.5G test conditions (figure 7). In addition to the inner wing, additional cells were delivered adhered on top of the wing tips.

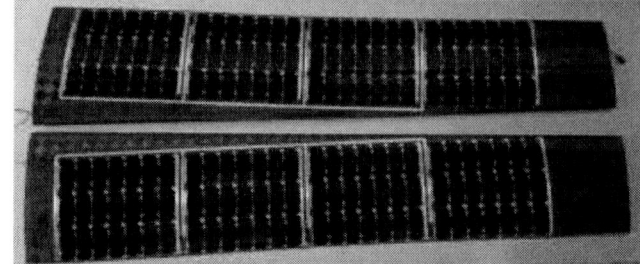

Fig. 7.     Fully assembled InGaP/GaAs/Ge Wing halves (SolAero thinned ZTJ cells)

Each full sized laminated half wing panel as depicted is then placed face-down into the female wing molds followed by the appropriate cloth, foam, and resin layers. Excess array material from the panel is trimmed to fit the mold. Vacuum is pulled on the mold, forcing the arrays to the wing shape. The smoothness of the top surface of the wing is maintained to minimize drag on the plane. Electrical connections in the panel are checked throughout the molding process, which can induce cracks in the cells, being more severe in the thicker cells. A performance summary is shown in Table I. These panels are prototypes and performance is a combination of the cell technology and integration process. Analysis of the panels did indicate some electrical losses.

Table I Solar Wing Summary

| Mfg. | Cell Technology | Cells / wing | Cell Area (m²) | AM1.5G Power (W) |
|---|---|---|---|---|
| SunPower* | Si (IBC) | 104 | 0.534 | 85 |
| Alta Devices | GaAs (inverted lift off) | 288 | 0.5058 | 92 |
| MLD | InGaAP/GaAs/InGaAs (IMM) | 240 | 0.48 | 118 |
| SolAero* | InGaAP/GaAs/Ge (thinned) | 168 | 0.463 | 116 |
| Sharp | InGaAP/GaAs/InGaAs (IMM) | 168 | 0.4623 | 152 |

*Low flexibility cells

## III. FLIGHT TESTING RESULTS

A series of flights were conducted over the past year. Initially, the UAV was flown without solar cells to evaluate the wing design, tune the aerodynamic controls, and verify the battery and PMAD operation. A spare PMAD was ground tested with solar panels under varying light intensity, all wing panels were ground tested for AM1.5G performance. All flights were conducted with a fixed rectangular pattern and altitude range. To date, the SunPower and Alta wings were flown for extended endurance testing and the MLD and SolAero wings were flown for functional check flights, Sharp panels are awaiting integration into the wings.

Test flights of at least 1 hours were conducted on all of the wing assemblies. All flight parameters were recorded on each flight. Solar insolation was not measured on all flights, but sun conditions were noted. Flight data comparison with ground data indicates that the solar panel tracks nominally with Global Horizontal Insolation (GHI).

The Alta wing was first flown on November 8th, 2016. The weather was not ideal, extremely overcast. The array power peaked at 77W and the flight was 1.59 hours between 11:00 and 13:00PM local time. Solar noon was at 12:52 PM. The solar array supplied 73.5 Whrs, nearly twice the 38.3 Whrs used from the battery.

The MLD and SolAero wings were test flown sequentially on May 18th, 2017. The weather was moderately clear with clouds moving in as testing progressed, therefore the wing performance can't be directly compared. The MLD array power peaked at 97.8W and flew for 1.32 hours between 9:45 and 11:10 local time. The MLD solar array supplied 102 Whrs resulting in battery charging of 9.4 Whrs (+19 min. spent on the ground). The SolAero array power peaked at 105.4W and flew for 1.61 hrs between 11:20 and 13:07 local time. The solar array supplied 133 Whrs resulting in battery charging of 3.2 Whrs. Both arrays exceeded the cumulative 400W take-off and nominal 70W cruise power.

Endurance flights using the Alta Devices and SunPower wings were conducted similarly to the test flights and timed to start near sun rise with >90% state-of-charge (SOC) in the batteries. A ground station was set up to measure insolation (global horizontal insolation or GHI, global normal insolation or GNI, and direct normal insolation or DNI) and a solar panel of comparable technology.

The SunPower wing was flown on October 14th, 2016 near Washington, DC. The UAV flew in a fixed rectangular pattern with a programmed minimum altitude. Flight duration was 10.9 hrs between 7:18 and 18:12. This flight is well in excess of the battery capacity. The flight had two 1 hour periods of incidental soaring in thermals which also helped to extend the flight. The weather was mostly sunny with an 11.2 hour day and peak sun angle of 43°, sunrise at 7:17, solar noon at 12:54, and sunset at 18:31. The measured insolation numbers for the day were 4.70 kWhrs GHI, 7.28 kWhrs DNI, and 9.03 kWhrs GNI (Figure 8). The peak sun intensity at solar noon was 740 W/m² (GHI) and peak array power was slightly over 80 W. A second stationary horizontal wing on the ground recorded similar power levels with the total flight time integrated power of the plane being ~10% lower than the grounded wing. This variation could be due to aircraft motion requiring the MPPT to constantly adjust. Figure 9 shows the

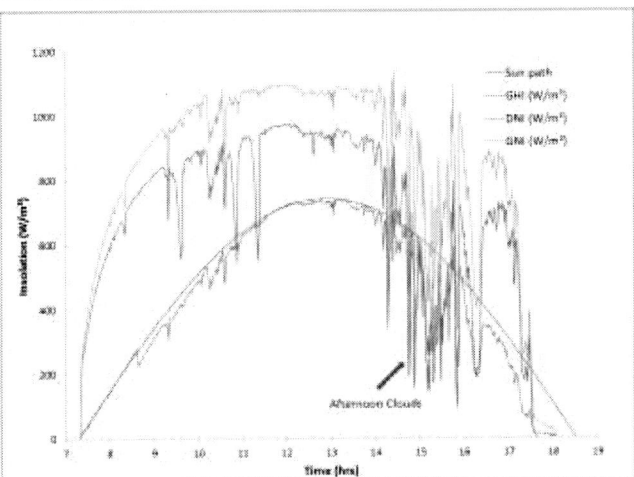

Fig. 8. 14Oct2016 solar insolation data (DC area)

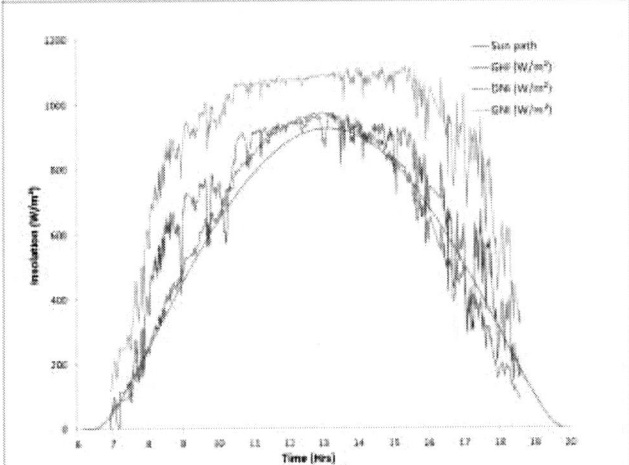

Fig. 10. 4Apr2017 solar insolation data (DC area)

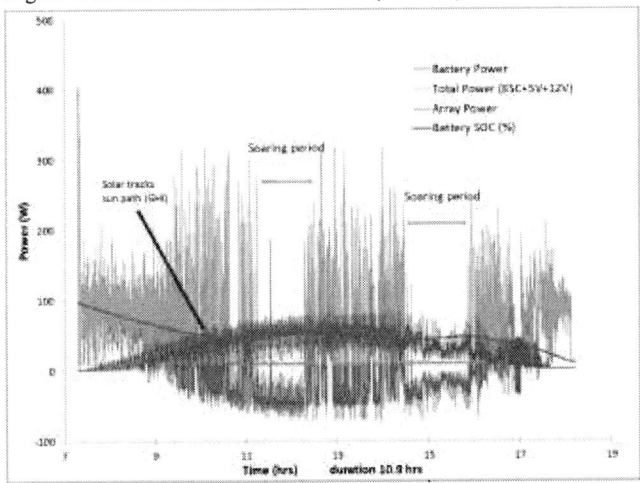

Fig. 9. 14Oct2016 Si wing, SunPower endurance flight power usage

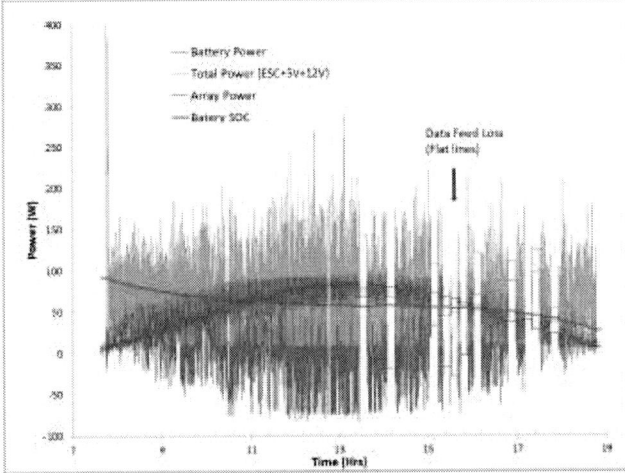

Fig.11. 4Apr2017 GaAs wing, Alta endurance flight power usage

electrical power data usage throughout the flight with large swings in power due to the location of the sun relative to the UAV flight pattern. The flight used 710 Whrs, of which 353 Whrs were supplied by the solar panel and 357 Whrs from the battery, not including the soaring. Adjusted totals with soaring would total 981 Whrs and an estimated 280 Whrs from soaring (energy stored as altitude).

The Alta Devices wing was flown on April 18th, 2017 near Washington, DC. The UAV also flew in a fixed rectangular area with a programmed minimum altitude. Flight duration was 11.2 hrs between 7:30 and 18:52. This flight did not have any periods of soaring to extend the flight. The weather was mostly sunny with a 13.3 hour day and peak sun angle of 63°, sunrise at 6:28, solar noon at 13:08, and sunset at 19:48. The measured insolation numbers for the day were 7.22 kWhrs GHI, 8.03 kWhrs DNI, and 10.5 kWhrs GNI as plotted in figure 10. The peak sun intensity at solar noon was 980 W/m² (GHI) and peak array power was slightly over 101W. A second stationary horizontal panel on the ground recorded integrated power ~15% higher. Again, this integrated power difference may be attributed to aircraft dynamics and MPPT efficiency with a higher array voltage compared to the SunPower wing. The flight used 907 Whrs of which 587 Whrs were supplied by the solar panel and 320 Whrs from the battery (figure 11).

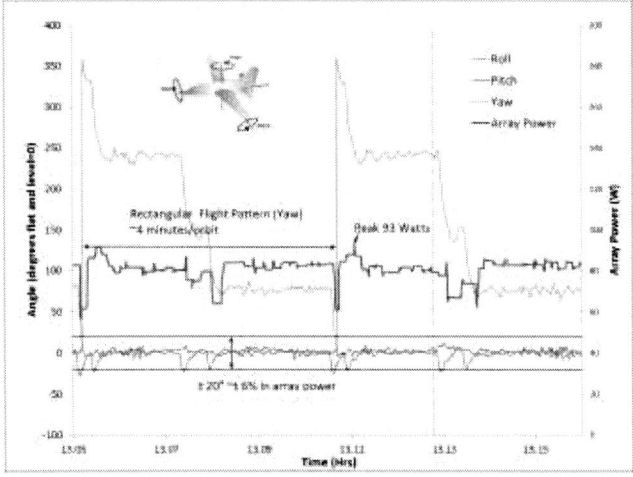

Fig. 12. 4Apr2017 GaAs Wing (Alta Devices) endurance flight array power coincident with UAV / sun orientation

All of the flight testing data collected includes the parameters of the UAV. Figure 12 shows a plot of the roll, pitch and yaw of the UAV. The yaw primarily indicates the direction of the plane, and the repetitive cycles of the rectangular path are clearly identifiable. These plots show a marked drop in array power (purple lines) during 90° turns at two opposing corners of the path and only a slight rise at the other two opposing corners. The pitch and roll (sun angle) in these corners do not account fully for the change in power levels, suggesting a lag in the MPPT responsivity.

## VI. FLIGHT TESTING DISCUSSION

The two endurance flights demonstrate that solar power can easily extend the flight duration by more than a factor of 2, even longer with soaring and optimized flying algorithms [12]. These flights offer contrasting results: the SunPower flight had less sun and lower power panels yet flew as long as the Alta Devices flight. This difference was the result of soaring. Time spent without the motor running when the aircraft was above the minimum altitude resulted in additional time available for recharging. The SunPower flight used 89% of the battery capacity and the Alta Devices flight used only 66%. That additional battery capacity could have extended the Alta Devices flight nearly 1 hour.

With this knowledge, fully populating the wing and/or increased efficiency solar cells could extend the duration of the flight beyond sunrise and sunset dependent on the battery capacity. Additional solar power could provide a larger payload power budget and buy margin for varying weather conditions. Increasing battery capacity would improve the nighttime duration. The addition of battery capacity and solar array size both increases mass which increases the cruise power requirements. Currently, Li-Ion batteries are ~250Whr/kg with newer technology demonstrating up to 400 Whrs/kg. Extensive trade studies must be performed to balance all the flight parameters, i.e. larger planes increase mass and drag which reduces speed. One option for UAVs is stick-on solar panels, this would increase the available power, but the solar cells' profile on the wing will further increase drag and mass of the wing decreasing the net benefit.

## VI. SUMMARY AND CONCLUSION

The UAV market is continuing to grow at a rapid pace. The solar powered UAV demonstrated in this paper, even without a fully populated wing, extended the flight duration by at least a factor of 2 compared to using batteries alone. With the III-V technology solar cells exhibiting higher efficiencies, less weight, and greater flexibility, the duration and payload requirements can be extended even further. The higher efficiency does come with a cost, but longer flight times may reduce the need for additional UAVs, shorten mission timelines with fewer change-out cycles and weather restrictions, or enable greater payload capacity. Flight endurance is also dependent on mission criteria, UAV design, plus autonomous and manual flying techniques.

Assembled wings were tested with SunPower Si, Alta Devices GaAs, SolAero thinned ZTJ, and MLD IMM cells. Solar powered flights demonstrated a viable solar integration process and long endurance capability. Sharp IMM subarrays are awaiting assembly into a wing. All these solar technologies were used successfully in a UAV with varying energy benefits and costs.

REFERENCES

[1] THE DRONES REPORT: Market forecasts, regulatory barriers, top vendors, and leading commercial applications, Business Insider, May 27, 2015

[2] Commercial UAV Market Analysis By Product (Fixed Wing, Rotary Blade, Nano, Hybrid), By Application (Agriculture, Energy, Government, Media & Entertainment) And Segment Forecasts To 2022, Grand View Research, April 2016

[3] http://www.english.rfi.fr/environment/20100708-solar-powered-plane-makes-history Thursday 08 July 2010 -

[4] Amos, Jonathan (2010-07-23). "'Eternal plane' returns to Earth". BBC News. Retrieved 2010-07-23.

[5] https://www.avinc.com/uas/small_uas/raven

[6] http://www.gizmag.com/aerovironment-solar-puma-ae/28695/

[7] D. Edwards, "Up, Up, and Away! UAV's Endurance Gets a Lift by Latching onto Thermals", *2011 NRL Review*, pp 82-88.

[8] D. Edwards, Autonomous Locator of Thermals (ALOFT) Autonomous Soaring Algorithm, *NRL Formal Report, NRL/FR/5712--15-10,272* (2015).

[9] Scheiman et.al. "A Path Towards Enhanced Durability Endurance of an UAV Using IMM Solar Cells Technology", 43$^{rd}$ IEEE *Photovoltaics Specialists Conference*, June 2016, Portland, OR

[10] D. Scheiman et.al. "High efficiency Flexible Triple Junction Solar Panels" *40$^{th}$ IEEE Photovoltaics Specialists Conference* June 2014, pp 1376-1380, Denver CO

[11] T. Takamoto et.al., "World's highest efficiency triple-junction solar cells fabricated by inverted layers transfer process," 35$^{th}$ IEEE *Photovoltaics Specialists Conference*, June 2010, pp.412-417, Honolulu HI.

[12] D. J. Edwards et al. "Maximizing Net Power in Circular Turns for Solar and Autonomous Soaring Aircraft," *Journal of Aircraft*, DOI 10.25.14/1.C033634.

# High Performance, Lightweight GaAs Solar Cells for Aerospace and Mobile Applications

Aarohi Vijh, Lori Washington and Robert C. Parenti

Alta Devices, Inc., Sunnyvale, California, 94085, USA

*Abstract* — **Thin gallium-arsenide solar cells manufactured through epitaxial liftoff can offer high-efficiency, reduced costs and high power to weight ratios. Alta Devices holds world records for terrestrial solar cell and module conversion efficiency for single junction cells. We present the status of Alta's technology and manufacturing capability, with comments on significance for aerospace and mobile applications.**

*Index Terms* — **epitaxial lift-off, gallium arsenide, photovoltaic, manufacturing, unmanned aerial vehicles, space, energy harvesting, electric vehicles.**

## I. INTRODUCTION

Growing demand for communication and imaging services is motivating innovation in wireless connectivity and delivery. The key requirements for these communication networks are high bandwidth, low latency, support for mobility, and flexibility in switching and reconfiguration. Service providers are therefore developing alternatives to terrestrial or geosynchronous satellite based communication solutions. These alternatives are typically based either on high-altitude long-endurance pseudosatellite platforms (HALE or HAPS) [1], or on constellations of low-earth orbiting satellites [2, 3]. A common feature of these platforms is the need for high-efficiency, lightweight solar arrays at reasonable prices and capable of sufficient lifetimes in their respective operating environments. The state of the art in satellite solar arrays is to use wafer-based III-V multijunction solar cells. These cells provide high efficiency and good radiation resistance but are expensive and heavy [4]. The state of the art for high-altitude aerial platforms is monocrystalline silicon. For example, monocrystalline cells were used on round the world flight completed in 2016 by the Solar Impulse airplane. However, these cells are heavy and prone to breakage due to flexing of the airframe, and have efficiency lower than that of III-V solar cells.

There has been a relatively new development: thin gallium-arsenide solar cells produced via epitaxial lift-off (ELO). Such cells are being developed for markets that value high efficiency and light weight but cannot support the prices or weight of traditional III-V solar cells. We present the manufacturing technology and product performance of such solar cells and discuss aerospace and mobile applications that benefit from such cells.

## II. MANUFACTURING PROCESS

Figure 1 shows the overall process flow for the manufacture of the solar cells. The process starts with 4-inch square wafers of gallium-arsenide and prepares them for epitaxial growth. The solar cell and release layers are then grown by metal-organic chemical vapor deposition (MOCVD), a metal film stack is deposited, and a polymer carrier film is laminated to the metal layer. This film remains with the solar cell and goes into the finished product, providing mechanical support for the metal and semiconductor films, which are only on the order of

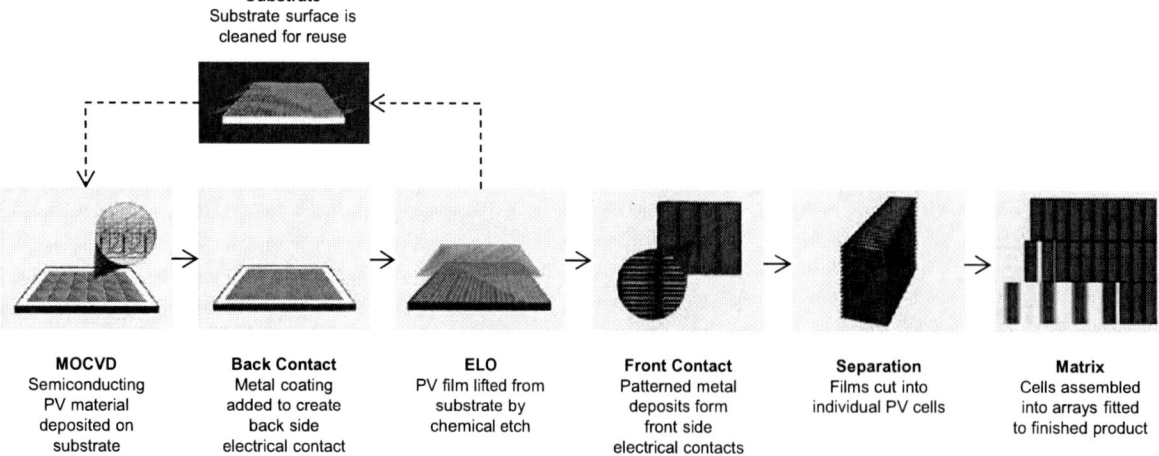

Fig. 1: Manufacturing process flow in the Sunnyvale fab for Alta's thin and lightweight III-V solar cells.

10 microns thick. The wafer attached to the laminated carrier film is then immersed in a series of chemical baths that etches the release layers and separates the films from the wafers. Metal contact fingers and an antireflection coating are then applied to complete the cell structure. Each film is then singulated into multiple smaller cells which can be connected into series or parallel assemblies as needed.

The wafer itself can be cleaned and reused many times over, leading to a greatly improved cost structure compared to that of traditional gallium-arsenide solar cells. NREL-verified AM1.5 cell efficiencies of 28.8% (single-junction) and 31.6% (dual-junction) have been demonstrated using this process, and AM1.5 aperture area module efficiency (single junction) of 24.8% has also been verified by NREL [5]. The single junction cell and module efficiencies are current world records.

The production efficiency and yield of the manufacturing process are a result of development of key equipment technologies including a custom MOCVD reactor, with higher growth rates and gas utilization compared to commercial tools.

### III. MOCVD TECHNOLOGY

The epitaxial liftoff (ELO) and wafer reuse process described above is instrumental in reducing the amount of expensive semiconductor material incorporated in the solar cell. A second major component of the manufacturing costs is capital equipment. Commercially available MOCVD tools are designed for the production of traditional III-V semiconductor devices for electronic and optoelectronic applications. Relative to solar cells, these devices have complex layer structures that require extremely precise control of thickness, composition, and doping profiles, and each growth typically takes multiple hours to complete. The simpler structure and relatively wide process windows of solar cells present an opportunity to use correspondingly simpler and less expensive equipment.

#### A. Cycle Time

After several generations of prototype development and tool optimization, the Alta now uses its own proprietary MOCVD tool design for primary production, generating solar cells with performance and yield matching commercial tools. Alta's tool design uses inline architecture and minimizes total process time through a combination of automated substrate loading and unloading, fast temperature ramping, high growth rate, and the elimination of pressure cycling. These factors combine to gives us a total process time under 15 minutes, much shorter than the hours of processing time needed by typical commercial tools.

#### B. Tool Availability (Uptime)

Another important aspect of a low-cost high volume manufacturing tool is uptime. One contributor to tool downtime is the need to clean parasitic deposition that leads to particles. Because Alta's solar cells are large-area thin-film devices, particles can have a significant impact on device yield. The design of the gas injection system minimizes parasitic deposition as well as particle generation and as shown in Figure 2, the design has achieved high uptime with well over 1000 runs between the need for chamber maintenance, while maintaining excellent particle performance that is critical to achieving high yield of large-area thin-film solar cells.

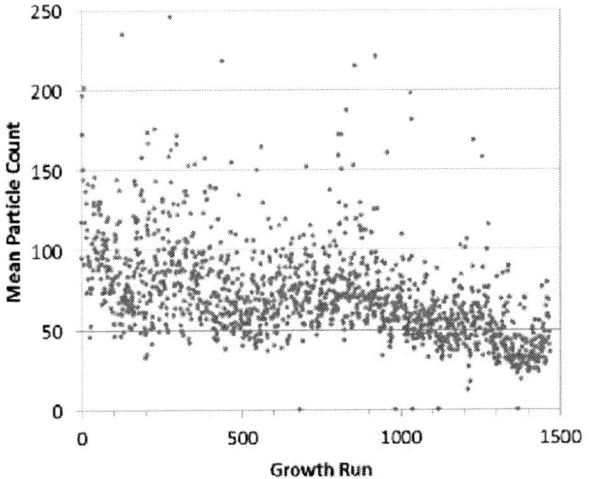

Fig. 2. Particle performance over the course nearly 1500 continuous runs without chamber maintenance. Particle performance does not degrade as a function of runs and therefore does not necessitate chamber cleans.

#### C. Source Utilization

In order to reduce the operating costs, the process chamber is designed to maximize the usage of all sources, especially the metal-organic precursors, which account for a large fraction of the consumable cost. The gas flows have been designed such that the amount of gas that leaves the process zone before reaching the wafer surface is minimized. In addition, the layout of and geometry of Alta's substrates on the carrier results in a very high packing efficiency, thereby minimizing wasted deposition in the spaces between substrates. Implementation of this improvements results in a 70% utilization of metal-organic precursors. In contrast, on a commercial tool the utilization efficiency is closer to 30%.

#### D. Device Performance from Proprietary Tool

Despite that fact that the proprietary tool utilizes faster temperature ramps, and higher growth rates, solar cells produced by this tool consistently match those grown by commercial MOCVD tools. Figure 3 shows a comparison of a typical device grown by Alta's proprietary MOCVD tool versus a typical device grown by a commercial MOCVD tool.

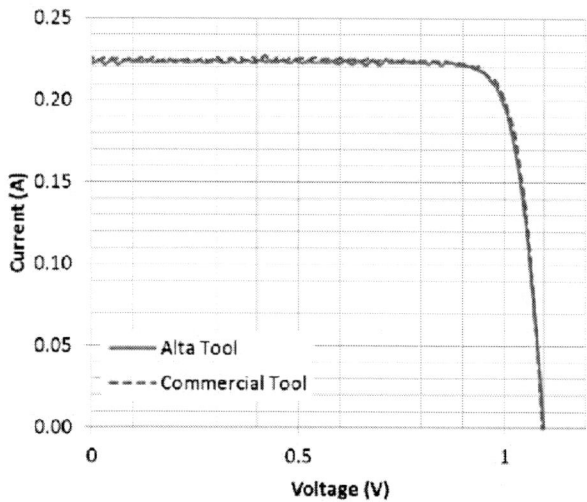

Fig. 3. Comparison of solar cell I-V curves grown by Alta's proprietary MOCVD tool and a commercial MOCVD tool.

## IV. WEIGHT AND POWER METRICS

Alta Devices' current product is a 5 cm x 2 cm, lightweight, flexible solar cell produced with single-junction GaAs technology and a high-throughput film lift-off and wafer reuse process. Production cell efficiencies are currently exceeding 26% (AM1.5). Series/parallel interconnection of cells is performed by fully-automatic equipment capable of producing arrays ("matrices") of greater than 50 cm x 50 cm size and aperture efficiencies of 25% (AM1.5). The cells themselves weigh about 180 mg per 10 cm$^2$ cell and the weight of the interconnected matrix is approximately 240 g/m$^2$. This represents a power to weight ratio of about 1000 W/kg (AM1.5, without encapsulation).

Alta has demonstrated the next generation (Gen 4) of its terrestrial solar technology and is transitioning this to production. The Gen 4 technology offers a weight reduction from 240 g/m$^2$ to 170 g/m$^2$, or a power to weight ratio of about 1500 W/kg (AM1.5, without encapsulation).

In addition Alta Devices has the ability to produce its dual-junction device technology on the same production platform. This is expected to provide greater efficiency relative to the single junction devices, approximately 28% (AM1.5, module aperture level).

## V. APPLICATIONS

We discuss four markets that benefit strongly from the availability of lightweight, high efficiency, thin gallium arsenide solar cells.

### A. Unmanned Aircraft

Performance of fixed-wing aircraft depends strongly on weight and available power. Therefore, while many applications can benefit from high efficiency and low weight, fixed wing solar powered aircraft are natural applications for high-efficiency, lightweight solar cells. There appears to be market interest in two categories of solar powered aircraft: small unmanned aerial systems (sUAS) and high-altitude, long endurance systems (HALE). The small unmanned aerial systems are used for imaging applications such as agricultural data collection, pipeline inspection and tactical military surveillance. They are typically 1-3 m in wingspan, roughly 1-10 kilograms in weight and due to limitations of battery energy density, they typically fly from 1-3 hours. With the addition of high-efficiency, lightweight solar cells, the range of these sUAS can be extended significantly. Fig. 4 shows normalized battery state of charge and solar array output of a sUAS that is retrofitted with Alta's Gen 3 solar cells. This particular sUAS platform is normally capable of a 2 hour flight. Addition of the solar cells can increase endurance of the platform to 5 hours, the weight and drag penalty to the aircraft being small.

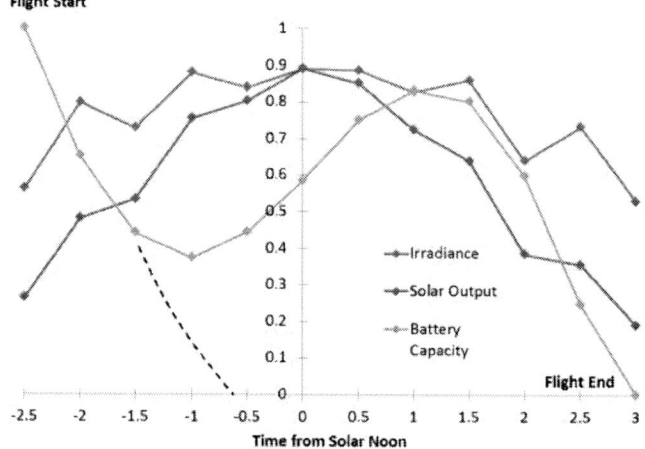

Fig. 4. Irradiance, array output and battery state of charge during a flight of a small agricultural unmanned aerial system equipped with Gen 3 solar cells from Alta. The dashed line is a projection of charge state in the absence of solar. Endurance increase of three hours can be achieved.

The concept of operations for high-altitude, long-endurance (HALE) platforms is different. These aircraft are being designed to carry communications and imaging payloads weighing tens of kilograms and keep them aloft for months. To avoid conflict with manned traffic such as airliners, and also to avoid high winds, these aircraft will fly in the stratosphere, at altitudes of approximately 20 km. Due to low air densities and payload weight, the wings of these aircraft are very large, on the order of tens of meters in span, capable of accommodating tens of kilowatts of solar cells as long as the cells can be made very light and efficient. The energy generated by the solar arrays can be used both to propel the aircraft during the day, and to store energy in a battery to fly through the night.

## B. Emerging Space Applications

For space applications, thin gallium-arsenide solar cells have the potential to provide a good balance of lifetime, conversion efficiency, reduced array weight, elimination of cell breakage and a significant reduction in costs relative to conventional space solar cells. While much work remains to be done to determine the suitability of such cells for on-orbit use, the technology has the potential to be a viable source of electric power for large satellite constellations, electric propulsion and future concepts requiring electrical power in the megawatt range, such as Moon or Mars bases.

## C. Energy Harvesting

Another potential application of thin gallium-arsenide solar cells is in the so-called internet of things (IoT) market space. The internet of things refers to small electronic devices that are typically powered by a battery and contain a radio communication system for connection to the Internet or other network. These devices are often expected to operate indoors or in unusual lighting conditions such as when worn or integrated into clothing. Thin gallium arsenide solar cells can harvest energy very well under such conditions [6] and can enable new applications requiring high power densities not available from, say, amorphous silicon. Scale and reduction of cost are key to mass adoption in this market.

## D. Automotive

Finally, a gigawatt-scale opportunity for thin gallium-arsenide solar is integration into automobiles to provide electrical power. We are seeing accelerating development and adoption of electric vehicles, and increasing amount of electric components (such as steering, brakes and air-conditioning) in conventionally powered vehicles. The scalability, low cost, high efficiency, superior high-temperature performance and mechanical integration possibilities offered by thin gallium-arsenide solar represent a strong competitive advantage in the automotive market.

## VI. SIGNIFICANCE AND SUMMARY

Until recently, the high manufacturing cost of non-concentrating gallium-arsenide based solar cells limited their use to specialty applications such as space. However, Alta Devices is reducing these manufacturing costs through the application of epitaxial liftoff (ELO), wafer reuse, high-throughput MOCVD reactor designs optimized for solar cell structures. Combined with record device efficiencies, this provides a technology platform capable of serving large volume markets with high performance solar cells.

## REFERENCES

[1] Tozer TC and Grace D, "High-altitude platforms for wireless communications", *Electronics & Communications Journal*, June 2001.

[2] Safyan M, "Overview of the Planet Labs Constellation of Earth Imaging Satellites", *ITU Symposium on Small Satellite Regulation*, Prague, March 2015.

[3] Azzarelli T, "OneWeb Global Access", *ICT Spring Europe*, Luxembourg, May 2016.

[4] Bailey S and Raffaelle R, "Space Solar Cells and Arrays", in *Handbook of PV Science and Engineering*, Luque and Hegedus, eds, Wiley, 2003

[5] Green MA, Hishikawa Y, Warta W, et al., "Solar Cell Efficiency Tables (Version 50)", *Prog Photovolt Res Appl.* 2017; 25:668-676

[6] Reich NH, van Sark WGJHM, Turkenburg WC, "Charge yield potential of indoor-operated solar cells incorporated into Product Integrated Photovoltaic", *Renewable Energy*, Vol 36, Issue 2, February 2011, 642-647

# Through-Epitaxial-Via Back-Contact Multi-Junction Solar Cells Fabricated Using Epitaxial Lift-Off

Rekha Reddy, Marilyn L. Nowakowski, David Rowell, Christopher L. Stender and Christopher Youtsey

MicroLink Devices, Inc., Niles, IL, 60714, USA

*Abstract* — Back-contact solar cells potentially achieve higher efficiency by moving all the electrical connections to the rear of the cell, thus eliminating shadow losses as well as enabling new array assembly approaches. In spite of the many advantages of this technology, as demonstrated in commercial manufacturing of silicon cells, the process complexity to create through wafer vias has limited its use in III-V solar cells. We present a simple, three-mask process flow for fabricating through-epitaxial via (TEV) back-contact multi-junction epitaxial liftoff (ELO) solar cells in this paper. Electrical characterization showed 3J cells with 25.9% efficiency (1 sun AM1.5 spectrum) without AR coating and similar performance to standard top-contact devices.

## I. INTRODUCTION

Backside contacts have several important benefits for large-area space solar cells. They enable higher device efficiency (as much as 5% relative improvement) by reducing topside grid and bondpad shadowing as well as series resistance losses. Perhaps more importantly, the ability to form all electrical contacts on the backside of the solar cell opens up new and inexpensive approaches for cell laydown and panel assembly compared to the labor-intensive ribbon welding of cells used in manufacturing of arrays currently. In addition, bypass diodes can be incorporated easily during assembly with backside contact designs. Although silicon-based back-contact solar cell were first demonstrated in 1975 [1] and have been in production for terrestrial power generation [2] as well as in solar arrays for the international space station [3], multi-junction solar cells with backside contacts have remained a research topic [4-6].

A major hurdle to the implementation of backside contacts in commercial manufacturing is the increase in process complexity required to fabricate via holes with sidewall dielectrics and metal wrap-through contacts. An emitter wrap-through process flow can require twice the number of photolithographic patterning steps as a conventional solar cell with topside contacts. The increased process complexity also introduces new yield challenges. In this paper, we present a simplified process flow for through-epitaxial via (TEV) back-contact solar cells that requires the same number of mask levels as our standard production process for top-contact cells. This TEV approach applied to our epitaxial lift-off (ELO) solar cells also simplifies coverglass-interconnected cell (CIC) assembly by enabling wafer-scale coverglass application, which is difficult to achieve with top-contact device designs.

## II. DEVICE FABRICATION

A simple process flow with three mask steps for device fabrication of TEV back-contact solar cells has been developed. The process flow leverages our existing production line for ELO multi-junction solar cells [7-8], for which the conventional process flow is shown in Figure 1.

The inverted metamorphic multi-junction (IMM) layer structure consisting of Al(InGaP), GaAs and InGaAs subcells is epitaxially grown on GaAs wafers using an Aixtron AIX2800G industrial MOCVD reactor. As seen in Figure 1, a sacrificial ELO release layer is grown between the device structure and the substrate.

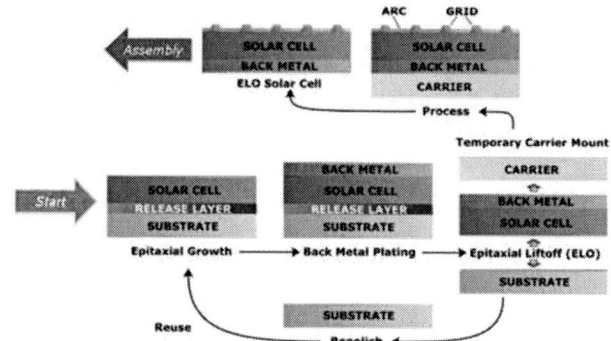

Figure 1. Process flow for ELO multi-junction solar cell production

Standard electron-beam evaporation and plating methods are used to form the backside ohmic contact and thick structural support layer before peeling off the solar cell device layers off the substrate using ELO. This ELO foil is then mounted onto a temporary carrier for further processing. Next, grid metal is deposited using photolithography and electron-beam evaporation. Lastly the solar cells are isolated by wet chemical etching, an anti-reflection coating is deposited by electron beam evaporation, and the die are singulated in our standard production process.

For the TEV backside process, the ELO foils are demounted from the temporary carrier after the grid metal and isolation steps, after all front side processes have been completed. Next, the ELO foil is bonded topside-down to a thin, 6-inch diameter coverglass wafer (~500-μm thickness) with space-grade encapsulant as shown in Figure 2. The bonded wafer-scale cover-glass provides stability for the thin ELO foil during backside processing.

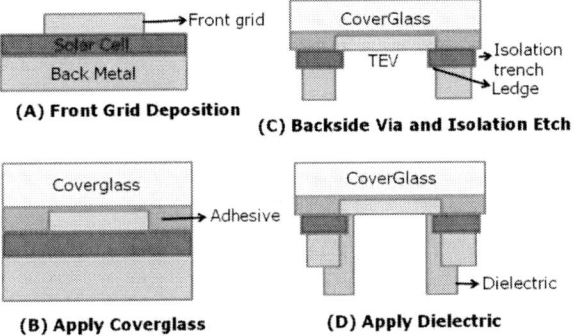

Figure 2. Process flow for TEV back-contact solar cells

The through-epitaxial vias are fabricated from the backside by photolithography and wet etching in two steps. First, a rectangular window is patterned on the wafer backside using a front-to-back alignment to the front side grid metal. Wet etching is used to etch the back metal down to the semiconductor backside surface (Fig. 2C). A second lithography and wet etch step is used to etch a via through the epitaxial layer structure to the front side grid metal pad as shown in **Error! Reference source not found.**. The solar cells are then cleaned using acetone and IPA to remove photoresist and dried with nitrogen before depositing a dielectric layer on the sidewalls of the vias for passivation.

## III. RESULTS AND DISCUSSION

Six-inch ELO foils were manufactured with 1-cm$^2$ TEV back-contact solar cells as shown in **Error! Reference source not found.**. It is to be noted that the indentation visible on the back contact pad is caused by electrical probing. Optical microscope inspection showed that the through-epitaxial vias were clean and damage-free before electrical characterization on a single-zone Newport solar simulator.

Figure 3. Front (left) and Back side through-epi via (right) of the TEV solar cell

A testing jig was manufactured with a 6 inch window and magnetic plates to hold the probes upside down for the rear contacts during testing as shown in Figure . The electroluminescence obtained from the cells was uniform as shown in Figure 4 (right).

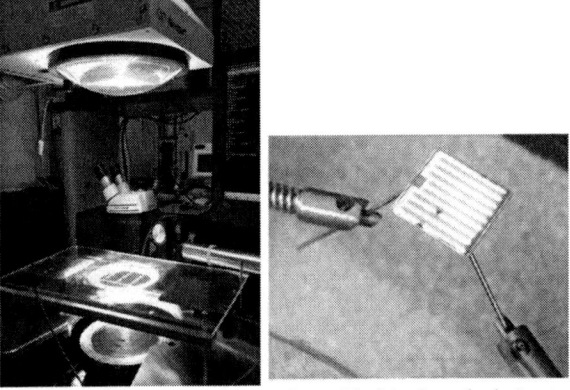

Figure 4. Back-contact testing station with 6-inch wafer being tested on the single-zone Newport solar simulator (left) and TEV cell electroluminescence (right). Cell bias conditions were 3V and 20 mA.

Figure shows a comparison of the IV characteristics measured for the same device from the front side pads (purple) before any backside processing, and the back pads using the TEVs to access the front side grid contact (green). The ~10% increase in current that was obtained from the back contact measurement is due to the expected improvement in reflectivity due to the coverglass and encapsulant, since no anti-reflection coating was applied to these prototype devices. Under AM1.5 illumination, we were able to demonstrate a 25.9% efficient solar cell with $J_{sc}$=10.52 mA/cm$^2$; $V_{oc}$ = 2.9V and a fill factor of 84.87 with the TEV process. Minimal degradation in device performance was observed before and after backside processing, after accounting for differences in reflectivity. Further improvements in efficiency are expected by reducing the size of the bond pads, optimizing grid coverage and applying an antireflection coating (with an expected additional increase in current of ~18%).

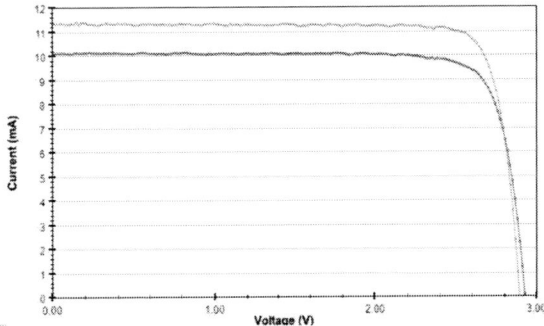

Figure 5. IV data for TEV back-contact ELO solar cells measured for a single device from the front pads before backside processing (purple) and after backside processing (green) under AM1.5 illumination on a single-zone Newport solar simulator.

There were 121 TEV back-contact 1-cm² cells fabricated on a 6-inch wafer. Yield as a measure of percentage of cells that changed in efficiency by ±10% during backside processing was calculated to be approximately 75%. Figure 6 shows a histogram that plots the binned efficiency versus number of cells for the same population of cells before and after backside processing. The plot demonstrates that most of the cells that yielded were in the 20% to > 24% efficiency bins.

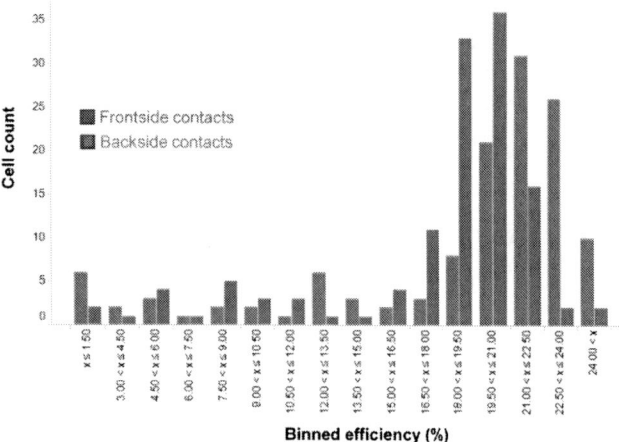

Figure 6. Plot of binned efficiency (%) versus cell count

The assembly process for these TEV back-contact solar cells will involve applying a photo defined dielectric to the sidewalls of the via and subsequent laydown of the solar cells to a metallized Kapton® sheet, allowing for a very simple and automated panel assembly approach. For an initial demonstration, we used photoresist and standard photolithography to pattern the dielectric and silver epoxy for die attach to form a 1x3 array. A calibrated AM0 measurement was then carried out on this 1x3 array on a TS Space three-zone solar simulator (Fig 7). The array yielded an efficiency of 20.7% with $V_{oc}$ = 8.62V, $J_{sc}$ = 14.67 mA/cm² and fill factor of 71%. We estimate the efficiency would have been 24.5% with the standard AR coating.

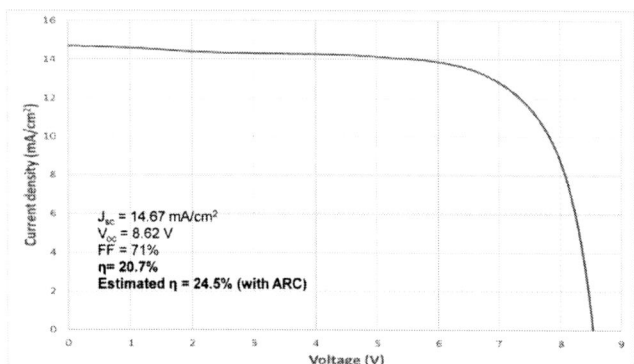

Figure 7. AM0 calibrated measurement of 1x3 array

A nine-cell, 3x3 array was also fabricated and tested with an efficiency of 18.96%. Optimization of the die attach process is

expected to enable further increases in efficiency, presently limited by fill factor and series resistance.

## IV. CONCLUSION

In conclusion, we have demonstrated a simple, three-mask-level process flow for multi-junction ELO solar cells to enable scalable and low-cost manufacturing of back-contact devices. Using this TEV approach, we were able to fabricate a 25.9% efficient solar cell at AM1.5, with similar performance before and after backside processing. Process optimization such as reducing the bond pad sizes and optimizing the AR coating will yield >30% AM0 efficient back-contact cells in production, equivalent to our standard, large-area IMM ELO devices. Back-contact multi-junction solar cells will enable new approaches to fabricate and assemble panels and arrays for concentrator photovoltaics as well as space applications.

## V. ACKNOWLEDGEMENTS

The authors gratefully acknowledge funding for this work from NASA through the SBIR/STTR Phase I and II projects under contract number NNX16CC16C.

## REFERENCES

[1] M. D. Lammert and R. J. Schwartz, "The interdigitated back contact solar cell: a silicon solar cell for use in concentrator applications," *IEEE Transactions on Electron Devices*, vol. 24, pp. 337-342, 1977.

[2] P. J. Cousins, D. D. Smith, H. C. Luan, J. Manning, T. D. Dennis, A. Waldhauer, K. E. Wilson, G. Harley and W. P. Mulligan, "Gen III: Improved Performance at Lower Cost,". *35th IEEE PVSC*, Honolulu, HI, June 2010.

[3] D. R. Lillington, "Development of Advanced Silicon Solar Cells for Space Station Freedom", *NASA Contractor Report 189215* (1990).

[4] Y. Zhao, P. Fay, A. Wibowo and C. Youtsey, "Inductively coupled plasma etching of through-cell vias in III–V multijunction solar cells using SiCl₄/Ar", *JVST B* 31 (6), 10.1116/1.4822015 (2013).

[5] T. Salvetat, E. Oliva, A. Tauzin, V. Klinger, P. Beutel, C. Jany, R. Thibon, P. Haumesser, A. Hassaine, T. Mourier, G. Rodriguez, C. Lecouvey, B. Imbert, F. Fournel, J. Fabri, J. Moulet, F. Dimroth and T. Signamarcheix, "III-V Multi-Junction Solar Cell Using Metal Wrap Through Contacts," in *12th International Conference on Concentrator Photovoltaic Systems (CPV-12)*, Sep 2016, 1766 (060004)

[6] E. Oliva, T. Salvetat, C. Jany, R. Thibon, H. Helmers, M. Steiner, M. Schachtner, P. Beutel, V. Klinger, J. Moulet and F. Dimroth, "GaInP/AlGaAs Metal Wrap Through Tandem Concentrator Solar Cells", *Progress in Photovoltaics Research and Applications,* PIP 16-090.R1

[7] C. L. Stender, J. Adams, V. Elarde, T. Major, H. Miyamoto, et al, "Flexible and lightweight epitaxial lift-off GaAs multi-junction solar cells for portable power and UAV applications", in *42nd IEEE Photovoltaics Specialist Conference (PVSC)*, Jun 2015.

[8] D. Scheiman, P. Jenkins, R. Walters, K. Trautz, R. Hoheisel, R.

Tatavarti, R. Chan et al. "High efficiency flexible triple junction solar panels.", in *40th IEEE Photovoltaic Specialist Conference* *(PVSC)*, Jun.2014, pp. 1376-1380.

# Design of InGaP/GaAs/InGaAs multi-junction cells with reduced layer thicknesses using light-trapping rear texture

Lin Zhu[1,2], Anurag Reddy[3], Kentaroh Watanabe[3], Masakazu Sugiyama[3],
Yoshiaki Nakano[3], Hidefumi Akiyama[1,2]

1) Institute for Solid State Physics, University of Tokyo and JST-CREST, 5-1-5 Kashiwanoha, Kashiwa, Chiba 277-8581, Japan. 2) AIST-UTokyo Advanced Operando-Measurement Technology Open Innovation Laboratory (OPERANDO-OIL), 5-1-5 Kashiwanoha, Kashiwa, Chiba, 277-8589, Japan. 3) School of Engineering and RCAST, the University of Tokyo, Meguro-ku, Tokyo, 153-8904, Japan

*Abstract* — Aiming at the realization of high efficiency and low cost simultaneously, we modeled light absorption by subcells for a substrate-free thin-film multi-junction cell with a rear texture. The model evaluated subcell photocurrents and the requisite thicknesses at current matching. For InGaP/GaAs/InGaAs metamorphic 3-junction solar cells, the rear texture significantly enhanced light absorption in the InGaAs bottom cell and GaAs middle subcells, and the requisite thickness of InGaAs is almost less than 10% of that with flat rear reflector, even GaAs is also less than half of that without rear texture, if InGaP-subcell is not greater than 600 nm. The reduced lower subcell thickness by the texture rear surface will mitigate the difficulty of metamorphic growth and contribute to the cost reduction of multi-junction solar cells.

*Index Terms* — Light trapping, Absorption enhancement, Multi-junction solar cells, low cost, III-V Semiconductor materials, Photovoltaic cells.

## I. INTRODUCTION

III-V compound multi-junction (MJ) solar cells [1-4] designed as a stack of subcell with decreasing bandgaps for well matching solar spectrum, are state-of-the-art and highest-efficiency photovoltaic devices, which severs as the primary choice for ultrahigh power source in space system [4-5]. Especially, the InGaP/GaAs/InGaAs inverse metamorphic (IMM) 3-junction solar cells have been developed with world-record efficiency of 37.9% (AM1.5G) [5] and 46% (AM1.5D, 508-suns) [6], showing a great prospect. In parallel with such pursuit of the highest efficiency, drastic cost reduction of III-V MJ cells has been a target of intensive research and development, aiming at the terrestrial usage of high-efficiency MJ cells without sunlight concentration for the power generation with limited area and insufficient direct irradiance.

Generally, there are two issues for the cost reduction of III-V cells: (1) reduction in the thickness of III-V epitaxial layers and (2) low-cost III-V crystal growth and processing technology. The focus of this research is the former: how thin the III-V layer thicknesses in MJ cells can be without a substantial sacrifice in the efficiency.

The light trapping technique, such as chemical texturing surface, has been implemented in thin-film Si solar cells [7-9] and III-V compound solar cells [10-11] for absorption enhancement and efficiency boost. According to optical theory [12-13], top- or bottom-side textured surfaces, which are generally obtained via wet anisotropic etching accompanied by anti-reflection (AR) layer or Ag/Al reflector, function as light scattering surfaces and trap photons inside active region. As a result, effective absorption path is increased. Such enhancement of light absorption allows us to reduce the thickness of III-V layers in MJ cells and to pursue low-cost and high-efficiency at the same time. Thus, it is necessary to analyze and model the absorption process of each individual subcell for MJ solar cells with textured structure, in which light absorption events generally are more complex due to the coupling among subcells.

In this work, we formulated general absorptivity of MJ solar

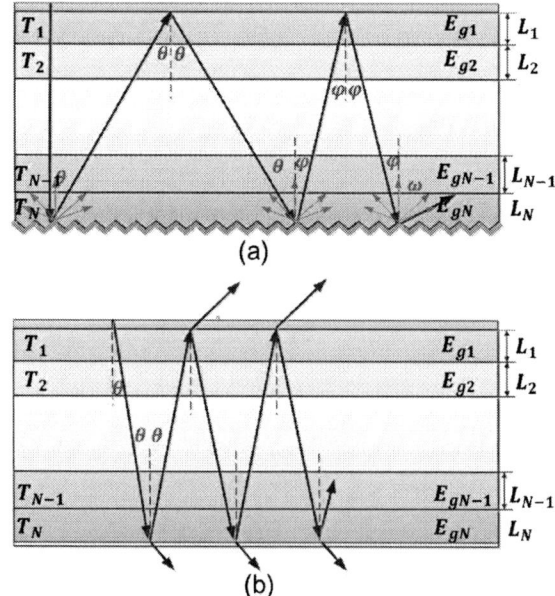

Fig. 1. Schematic cross sections of *N*-junction solar cells (a) with textured rear reflector and (b) with flat rear surface.

cells with random textured rear reflector and applied that to simulate absorption spectrum and photo-generated current of each subcell in textured InGaP/GaAs/InGaAs 3-junction solar cells. Finally, the optimal requisite subcell thickness under current matching and the corresponding short circuit current ($J_{sc}$) were evaluated for varied top-cell thickness, and the values were obtained for the cases with and without a textured rear surface. Our results quantitatively reveal that adding textured rear substrate would significantly reduce the necessary InGaAs subcell thickness, which is promising for development of low-cost MJ solar cells.

## II. ABSORPTION MODEL AND SIMULATIONS

Figure 1 exhibits schematic cross sections of $N$-junction solar cells with random textured rear surface and common flat rear-surface. Assume such textured rear surface is a perfect diffuse reflector (Lambertian scattering) without outward transmission. In Fig. 1 $E_{gi}$ represents material bandgap energy of a certain subcell marked as $i$, which is an increasing integer from the topmost-subcell ($i$=1) to the bottommost-subcell ($i$=N).

We define $a_i(E)$ as energy-dependent absorption spectrum of arbitrary subcell $i$ in $N$-junction solar cells, whose formula are given as

$$a_i(E) = \begin{cases} a_{i1}(E) & E \geq E_{g1} \\ a_{i2}(E) & E_{g1} > E \geq E_{g2} \\ \vdots \\ a_{ij}(E) & E_{gj-1} > E \geq E_{gj} \quad (i \geq j > 1), \\ \vdots \\ a_{ii}(E) & E_{gi-1} > E \geq E_{gi} \\ 0 & E_{gi} > E \end{cases} \quad (1)$$

where $a_{ij}$ ($E$) stands for photon-energy-dependent absorptivity of subcell $i$ for the photons with energy between $E_{gi}$ and $E_{gj-1}$ ($i \geq j > 1$) or between $E_{gi}$ and $\infty$ ($j$=1) in $N$-junction solar cells.

Start with the simplest and most common cases, the MJ solar cells with flat rear-surface shown as Fig. 1 (b), whose $a_{ij}$ are formulated by

$$a_{ij} = \frac{[1-R_f(\theta)]\prod\limits_{x=j}^{i} T_x(\theta) \dfrac{1-T_i(\theta)}{T_i(\theta)}[1+\prod\limits_{x=1}^{N} T_x(\theta)^2 \dfrac{R_b(\theta)}{T_i(\theta)}]}{1-\prod\limits_{x=j}^{N} T_x(\theta)^2 R_f(\theta)R_b(\theta)} \quad (2)$$

$$1 \leq j \leq i \leq N$$

where $T_i(\theta) = e^{-\alpha_i L_i / \cos\theta}$ is the light transitivity passing through subcell $i$ at arbitrary angle $\theta$, depending on subcell absorption coefficient ($\alpha_i$) and thickness ($L_i$). $R_f$ and $R_b$ are front- and back-surface reflectance, respectively. Considering the total reflection, $R_f$ is set as a function of photon energy $R(E)$ within critical angle ($\theta_c$), and as unity for $\theta > \theta_c$, expressed by

$$R_f(\theta) = \begin{cases} R(E) & \theta < \theta_c \\ 1 & \theta \geq \theta_c \end{cases} . \quad (3)$$

$R_b$ are assumed independent with photon energy and $\theta$ in this work, and set as unity for a perfect rear mirror and as 0 for an absorption substrate. Here we consider any reflection angles in this flat system are invariant for the total multi-layer reflection path and set $\theta$ as 0, since the photon re-absorption or re-emission processes in each such subcells are neglected in this calculation and the solar light almost vertically incident upon the front surface of cells.

Differing with flat rear-surface, diffuse reflection occur after the initial incident photons reflected by rear texture. Based on Lambert's emission law [14], the reflectance of textured surface is proportional to Cosine of the reflected angle $\theta$ and the total of that over the solid angle are unity. Thus, the angle-dependent reflectance for textured surface $R_b^{Tex}(\theta)$ is normalized by

$$R_b^{Tex}(\theta) = \frac{\cos\theta}{\pi} . \quad (4)$$

With it, $a_{ij}$ for MJ solar cells with textured rear-surface are formulated as

$$a_{ij} = [1-R_f(0)]\frac{\prod\limits_{x=j}^{i} T_x(0)}{T_i(0)}\left\{1-T_i(0) + \right.$$

$$\left. \prod\limits_{x=i}^{N} T_x(0)\frac{2\pi\int_0^{\frac{\pi}{2}} R_b^{Tex}(\theta)\prod\limits_{x=i}^{N} T_x(\theta)\frac{[1-T_i(\theta)]}{T_i(\theta)}[1+R_f(\theta)\frac{\prod\limits_{x=j}^{i-1} T_x(\theta)^2}{T_i(\theta)}]\sin\theta d\theta}{1-2\pi\int_{\theta_c}^{\frac{\pi}{2}} R_b^{Tex}(\theta)R_f(\theta)\prod\limits_{x=j}^{N} T_x(\theta)^2 \sin\theta d\theta} \right\},$$

$$(1 \leq j \leq i \leq N)$$

$$(5)$$

where $T_i(0)$ is the transitivity at $\theta$=0, standing for the first path before diffuse reflection.

Finally, with the simulated absorption spectrum $a_i(E)$, the photocurrent ($J_{sci}$) of arbitrary subcell $i$ in $N$-junction solar cells would be evaluated via integral over solar spectrum $S(E)$, given as

$$J_{sci} = q\int_0^\infty a_i(E)S(E)dE , \quad (6)$$

where $q$ is the electron change.

In this work, via using the practical absorption coefficient $\alpha_i(E)$ of three subcell materials and the measured reflectance of practical cells with front-surface AR coating $R(E)$, we simulated all subcell $a_i(E)$ and photocurrents for InGaP/GaAs/InGaAs 3-J cells with textured rear surface for varied subcell thickness set of ($L_1$, $L_2$, $L_3$). The bandgap energy of three subcell $E_{g1}$, $E_{g2}$, $E_{g3}$ are set as 1.89eV, 1.41eV and 0.93eV, respectively.

## III. RESULTS AND DISCUSSION

Figure 2 shows the simulated absorptivity of three subcells in InGaP/GaAs/InGaAs solar cells with textured rear surface (solid curves) for three typical combinations of subcell thicknesses, (a) $L_1$=300 nm, $L_2$=128 nm, $L_3$=15.4 nm (thin cell), (b) $L_1$=500 nm, $L_2$=462 nm, $L_3$=31 nm (medium cell)

978-1-5090-5606-4/17 $31.00 © 2017 IEEE

and (c) $L_1$=800 nm, $L_2$=2259 nm, $L_3$=80 nm (thick cell), whose short circuit currents $J_{sc}$ (the minimum among the three subcell photocurrent $J_{sci}$) are respectively predicted as 12.7, 14.0 and 15.0 mA/cm$^2$. The absorptivity of the-same-thickness IMM solar cells with flat-rear surface of $R_b$=1 (as a perfect mirror marked by dashed curves) and $R_b$=0 (as a absorptive substrate marked by doted curves) are also plotted in Fig. 2 to serve for the two common rear-surface designs in the present IMM solar cells. However, $J_{sc}$ for mirror cases are respective only 2.5, 3.0, 3.8 mA/cm$^2$ and for substrate cases are 1.3, 1.5 and 1.9 mA/cm$^2$, all of which are limited by $J_{sc3}$ of InGaAs-cell and obviously lower than those of the textured IMM cells with the same subcell thickness combinations.

The absorptivity's of InGaAs-subcells ($a_3$) with the three different rear-surface designs are the most notable in Fig. 2. It

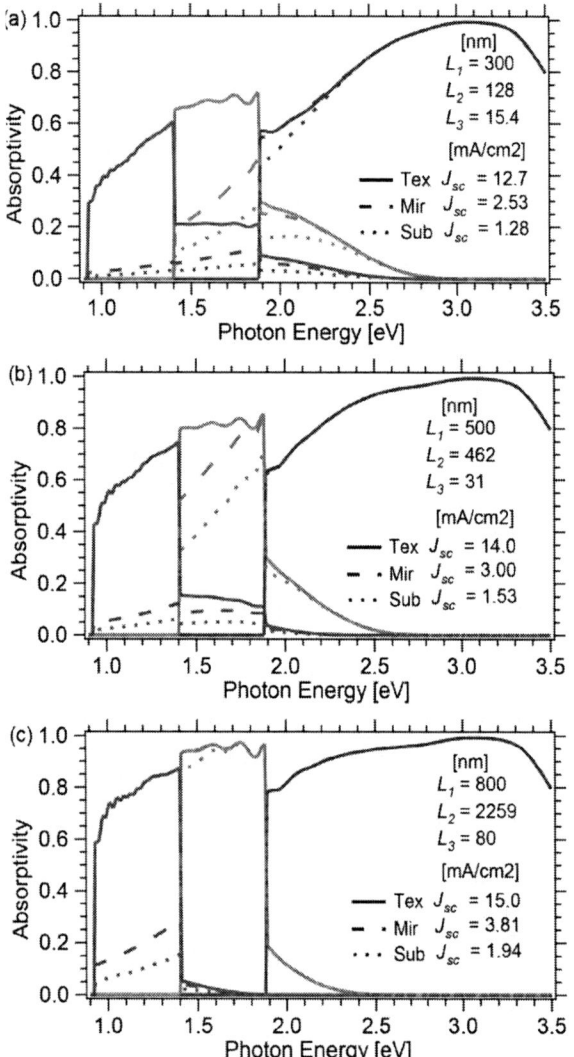

Fig. 2 Simulated subcell absorptivity for InGaP/GaAs/InGaAs solar cells with rear texture (solid), mirror (dashed) and substrate (dots) for typical (a) $L_1$=300, $L_2$=128nm and $L_3$=15.4nm, (b) $L_1$=500nm, $L_2$=462nm and $L_3$=31nm, and (c) $L_1$=800nm, $L_2$=2259nm and $L_3$=80nm.

is clearly exhibited that the rear texture significantly improves the bottom-subcell absorptivity, when are compared with the

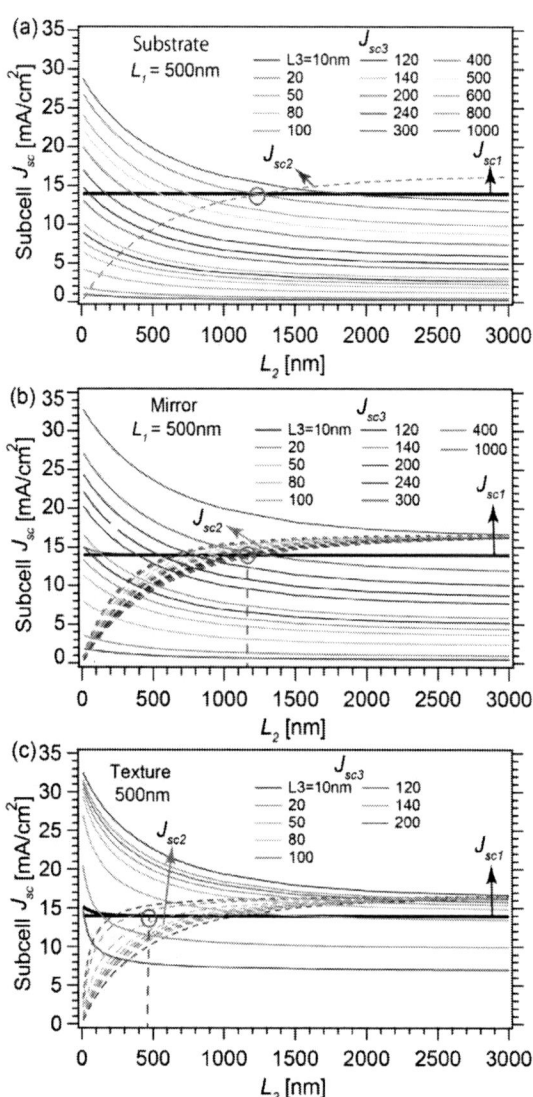

Fig. 3 $J_{sc1}$ (solid, black), $J_{sc2}$ (dashed, color) and $J_{sc3}$ (solid, color) of IMM solar cells with (a) substrate, (b) rear mirror and (c) rear texture for varied $L_2$ and $L_3$ at $L_1$=500nm.

rear mirror or absorptive substrate. For the thin cells in Fig. 2 (a), only 15.4-nm InGaAs-subcell are sufficient to achieve the average value of absorptivity over 40% at the region between 1.0eV and 1.4eV and around 20% between 1.4eV and 1.89eV, while both of mirror and substrate case are only less than 10% at region between 1.0eV and 1.89eV. For medium (b) and thick InGaP-subcell (c), the average of $a_3$ in textured cells are respectively over 60% and 70%, while only around 10% and 20% in mirror case and are around 5% and 10% in substrate case between 1.0eV and 1.4eV.

The absorption enhancement by rear texture and mirror are also observed in the GaAs subcell of the thin (a) and medium

(b) IMM solar cells at the region close to its bandgap, compared with that for substrate, especially in the thin cell ($L_2<1/\alpha_2$). For thick solar cells (c), GaAs subcells are not very sensitive to the rear-surface design. Similarly, the absorptivity's in all of the InGaP top cells with 300-nm, 500-nm and 800-nm thickness are insensitive to the rear surface, since the lower subcells (GaAs or InGaP subcells) in the three cases are sufficient to absorb all the photons with energies greater than $E_{g1}$ that have passed through the top subcells, which leads to the almost one-path absorption in top or upper cells, regardless of the rear-surface designs. In Fig. 2 (a-c), it is also noted that, especially in thin MJ solar cells, there are absorption tails of the GaAs subcell at region of >1.9eV and of the InGaAs subcell at >1.41eV with complementary shapes with the absorptivity of upper subcell, which is also due to the incomplete absorption of the upper subcell and is well consistent with the experimental features of external quantum efficiency (EQE).

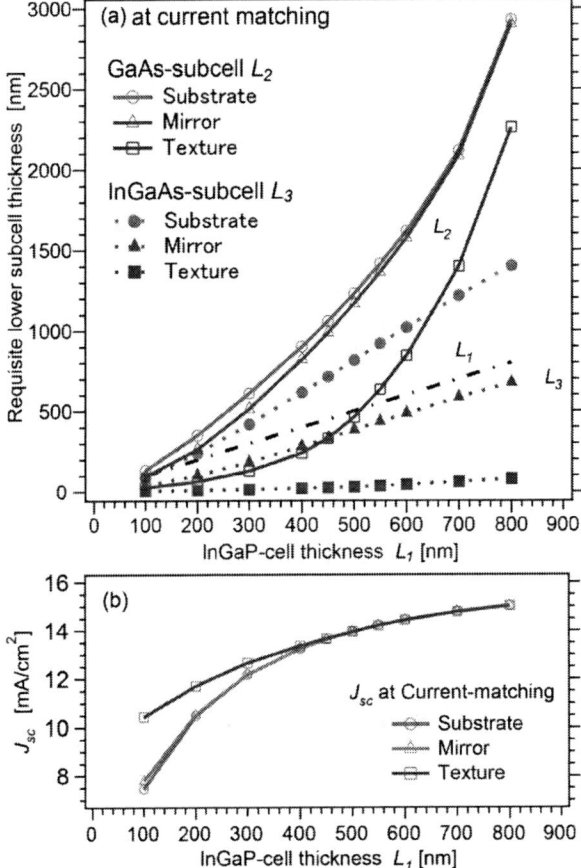

Fig. 4. Current-matching (a) $L_2$, $L_3$ and (b) corresponding $J_{sc}$ of IMM solar cells with texture (blue, square), mirror (green, triangle), and substrate (red, circle) for varied $L_1$.

Figure 3 shows the three subcell photocurrents ($J_{sc1,2,3}$) for IMM cells with (a) absorptive substrate, (b) rear mirror and (c) rear textured reflector at $L_1$=500nm as function of $L_2$ and $L_3$. For the three structures, all $J_{sc1}$ (solid black curves) stay at around 14mA/cm$^2$, mostly independent on $L_2$ and $L_3$. $J_{sc2}$

(dashed color curves) increases drastically as $L_2$ increases, and converges to around 16mA/cm$^2$ after $L_2$ greater than 2000 nm. $J_{sc2}$ for texture and mirror cases also are influenced by $L_3$ at thin-$L_2$ region, showing a slight drop as $L_3$ increases. On the other hand, as $L_2$ increases, $J_{sc3}$ (color solid curves) drops steeply near $L_2$=0, and then become flattened and converge to various values that increase with $L_3$.

Table 1. Requisite $L_2$, $L_3$ and the corresponding $J_{sc}$ for varied $L_1$ in the IMM solar cells with a textured rear reflector, a rear mirror and an absorptive substrate

| Structure | $L_1$ | $L_2$ | $L_3$ | $J_{sc1}$ | $J_{sc2}$ | $J_{sc3}$ | $J_{sc}$ |
|---|---|---|---|---|---|---|---|
| | | [nm] | | [mA/cm$^2$] | | | |
| Texture | | 28 | 6.2 | 10.44 | 10.52 | 10.64 | 10.44 |
| Mirror | 100 | 93 | 41 | 7.80 | 7.82 | 7.84 | 7.80 |
| Substrate | | 132 | 91 | 7.46 | 7.47 | 7.52 | 7.46 |
| Texture | | 63 | 10.5 | 11.83 | 11.71 | 11.97 | 11.71 |
| Mirror | 200 | 263 | 102 | 10.53 | 10.53 | 10.56 | 10.53 |
| Substrate | | 350 | 240 | 10.49 | 10.50 | 10.51 | 10.49 |
| Texture | | 128 | 15.4 | 12.66 | 12.77 | 12.67 | 12.66 |
| Mirror | 300 | 513 | 183 | 12.18 | 12.19 | 12.18 | 12.18 |
| Substrate | | 610 | 419 | 12.18 | 12.18 | 12.20 | 12.18 |
| Texture | | 240 | 22 | 13.36 | 13.37 | 13.42 | 13.36 |
| Mirror | 400 | 821 | 281 | 13.24 | 13.25 | 13.26 | 13.24 |
| Substrate | | 901 | 615 | 13.24 | 13.24 | 13.24 | 13.24 |
| Texture | | 462 | 31 | 13.96 | 13.96 | 13.96 | 13.96 |
| Mirror | 500 | 1169 | 383 | 13.95 | 13.95 | 13.96 | 13.95 |
| Substrate | | 1229 | 816 | 13.95 | 13.95 | 13.95 | 13.95 |
| Texture | | 844 | 44 | 14.43 | 14.44 | 14.45 | 14.43 |
| Mirror | 600 | 1572 | 485 | 14.43 | 14.44 | 14.44 | 14.43 |
| Substrate | | 1615 | 1019 | 14.43 | 14.43 | 14.44 | 14.43 |
| Texture | | 1398 | 61 | 14.78 | 14.78 | 14.82 | 14.78 |
| Mirror | 700 | 2084 | 583 | 14.78 | 14.78 | 14.79 | 14.78 |
| Substrate | | 2116 | 1215 | 14.78 | 14.78 | 14.78 | 14.78 |
| Texture | | 2259 | 80 | 15.03 | 15.03 | 15.04 | 15.03 |
| Mirror | 800 | 2902 | 675 | 15.03 | 15.03 | 15.03 | 15.03 |
| Substrate | | 2927 | 1401 | 15.03 | 15.03 | 15.03 | 15.03 |

The current-matching points on all subcell thicknesses are marked as the red circles in Fig.3 (a-c). For the texture case, 462nm-GaAs-subcell and 31nm-InGaAs-subcell are matching to InGaP-subcell of $L_1$=500nm, which enables all subcells with photocurrent density of 14.0mA/cm$^2$. For the cells with the mirror and the substrate, current-matching $L_2$ increase to 1169nm and 1229nm, respectively, roughly three times thicker than that for texture, while current-matching $L_3$ significantly shift to 383nm and 816nm, respectively, around 12 and 26 times larger than $L_3$ for the textured cell, to achieve the same $J_{sc}$=14.0mA/cm$^2$. The above results quantitatively reveal that the textured rear surface enhances the light absorption in InGaAs and GaAs subcell very effectively, via which the

978-1-5090-5606-4/17 $31.00 © 2017 IEEE

requisite thickness of the both layers, especially for InGaAs-subcell, are reduced significantly.

Figure 4 and Table 1 exhibit requisite $L_2$, $L_3$ and the corresponding $J_{sc}$ for varied top-cell thickness $L_1$ in the IMM solar cells with a rear textured reflector (Blue), a rear mirror (Green) and an absorptive substrate (Red), whose subcell photocurrents are always matching. As $L_1$ increases from 100nm to 800nm, the optimal $L_3$ for the three different rear surfaces increase almost linearly and $L_2$ increase more drastically. Here, the IMM solar cells with $L_1$ of 300nm, 500nm and 800nm are also selected as the representatives of thin, medium, and thick cells to discuss and compare the effects of absorption enhancement due to rear texturing. The typical 500-nm $L_1$ have been mentioned in Fig.3, which only need 462-nm GaAs-subcell and 31-nm InGaAs-subcell to achieve current matching and maximum in $J_{sc}$ of 14.0 mA/cm$^2$.

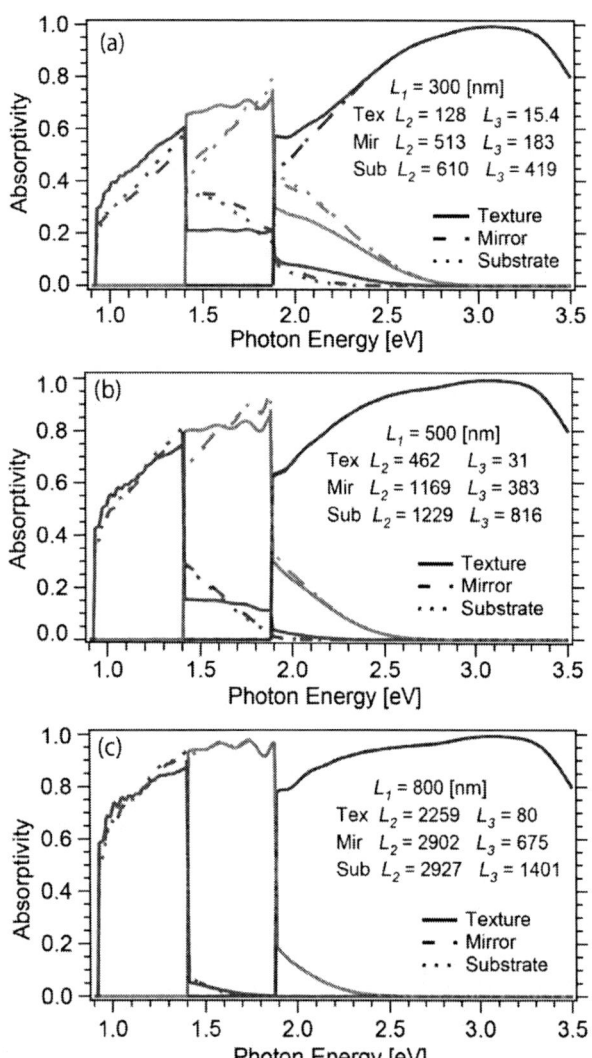

Fig. 5. Simulated subcell absorptivity for InGaP/GaAs/InGaAs solar cells with rear texture (solid), mirror (dots) and substrate (dashed) for typical (a) $L_1$=300, (b) $L_1$=500nm and (c) $L_1$=800nm at current-matching subcell thickness combination.

It is the one third of $L_2$ and one thirteen of $L_3$ in mirror case, reflecting a very significant effects on light trapping. Similarly, the thin 300-nm InGaP subcell ($L_1 < 1/\alpha_1$) achieves $J_{sc1}$ of around 12.3mA/cm$^2$ via the approximated single-path absorption, which almost independent with rear-surface designs. To match it, the cell for absorptive substrate requires 610-nm GaAs-subcell and 419-nm InGaAs-subcell to get the same subcell photon current density, and for mirror requires 513-nm GaAs-subcell and 183-nm InGaAs-subcell, respectively. In sharp contrast, cell with rear textured surface need only 128-nm GaAs-subcell, one fourth of that in mirror case and one fifth in substrate case, and 15.4-nm InGaAs-subcell, that is, only one twelve of that for mirror and even one twenty-six of that for substrate. Even for the 800nm-$L_1$ thick cell with $J_{sc1}$ around 15.0 mA/cm$^2$ approaching to absorption saturation, current-matching $L_2$ for substrate and mirror increase to 2927 nm and 2902 nm, and the requisite $L_3$ are 1401 nm and 675 nm, respectively. They are too thick to be grown without dislocations and degradations, due to the huge lattice mismatch between GaAs and InAs. In contrast, if fabricate a rear texture, the required thickness of GaAs-subcell is reduced to 2259 nm, and most notably that of InGaAs-subcell is reduced to 80 nm, which points out the significant material cut-down in the both subcells, especially in the bottom subcells.

The absorptivity's of the cells with texture, mirror and substrate for their current-matching subcell thickness combinations listed in Table 1 are also checked. Fig. 5 (a-c) exhibiting those with $L_1$=300, 500nm and 800nm reveal the obvious absorption enhancement at the region close to each subcell bandgap in thin and medium solar cells due to the light-trapping rear texture, and that the simulated absorptivity's with or without rear-textured designs obtained by the proposed models are consistent very well, especially in thick cells.

In Summary, at the same top-cell thickness $L_1$, the required bottom-cell thickness $L_3$ for a texture rear is less than 10% of that for rear mirror and less than 5% for absorptive substrate, which clearly clarify that the rear texture can save the InGaAs subcell materials significantly and effectively. Not only for $L_3$, the requisite GaAs-subcell thickness $L_2$ in textured cells are also decreased to 25-50% of that in rear mirror case, when $L_1$ is less than 600nm. Additionally, tiny boosts in the corresponding $J_{sc}$ are also observed in the thin IMM cells with light trapping textured design, compared with that with two flat rear surfaces. The above results quantitatively point out that the rear texture for MJ solar cells would significantly enhance the absorptivity of bottom cells and even that of upper subcells. It is a very promising and feasible strategy to achieve low-cost MJ solar cells via reducing not only the amount of lower-subcell materials but also fabrication difficulties, without any sacrifice in conversion efficiency.

The presented absorption model is general and useful to evaluate and optimize the subcell thickness under current-matching point for MJ cell with light trapping design. Our results provide a very important theoretical guidance for the development on low-cost and high-efficiency MJ devices.

## IV. CONCLUSIONS

In this work, subcell absorptivity of multi-junction solar cells with textured light trapping rear-surface were modeled systematically. The proposed model were applied to InGaP/GaAs/InGaAs triple-junction solar cell with textured rear surface to evaluate the subcell absorption spectrum, photocurrent and the requisite thickness set under current-matching condition, which were also compared with that of two conventional rear-surface designs of absorptive substrate and mirror. Our results quantitatively clarified a significant saving in InGaAs thicknesses, accompanied by a moderate saving in GaAs thickness, owing to absorption enhancement realized by the introduction of a rear texture, which is a very promising and feasible design for low-cost and ultrahigh-efficiency MJ solar cells.

## REFERENCES

[1]. Yamaguchi, M., et al., "Multi-junction III–V solar cells: current status and future potential." *Solar Energy* 79.1 (2005): 78-85.

[2]. Dimroth, F., et al., "5-junction III-V solar cells for space applications." *Photovoltaic Energy Conversion, 2003. Proceedings of 3rd World Conference on*. Vol. 1. IEEE, 2003.

[3]. Green, M. A., et al., "Solar cell efficiency tables (Version 47)." *Progress in photovoltaics: research and applications* 24 (2016): 3-11.

[4]. Geisz, J. F., et al. "Inverted GaInP/(In) GaAs/InGaAs triple-junction solar cells with low-stress metamorphic bottom junctions." *Photovoltaic Specialists Conference, 2008. PVSC'08. 33rd IEEE*. IEEE, 2008.

[5]. Takamoto, T., et al., "Application of InGaP/GaAs/InGaAs triple junction solar cells to space use and concentrator photovoltaic." *2014 IEEE 40th Photovoltaic Specialist Conference (PVSC)*. IEEE, 2014.

[6]. Dimroth, F. et al. "Four-Junction Wafer-Bonded Concentrator Solar Cells." *IEEE J. Photovoltaics*, 6 (2016): 343-9

[7]. Zeng, L., et al. "Demonstration of enhanced absorption in thin film Si solar cells with textured photonic crystal back reflector." *Appl. Phys. Let* 93.22 (2008): 221105.

[8]. Sheng, X., et al. "Optimization-based design of surface textures for thin-film Si solar cells." *Optics express* 19.104 (2011): A841-A850.

[9]. Chakanga, K., et al., "Laser textured substrates for light in-coupling in thin-film solar cells." *Journal of Photonics for Energy* 4.1 (2014): 044598-044598.

[10]. Inoue, T., et al., "Enhanced Light Trapping in Multiple Quantum Wells by Thin-Film Structure and Backside Grooves With Dielectric Interface." *IEEE Journal of Photovoltaics* 5.2 (2015): 697-703.

[11]. Watanabe, K., et al., "Thin-film InGaAs/GaAsP MQWs solar cell with backside nanoimprinted pattern for light trapping." *IEEE Journal of Photovoltaics* 4.4 (2014): 1086-1090.

[12]. Yablonovitch, E., "Statistical ray optics." *JOSA* 72.7 (1982): 899-907.

[13]. Miller, O. D., et al., "Strong internal and external luminescence as solar cells approach the Shockley–Queisser limit." *IEEE Journal of Photovoltaics* 2.3 (2012): 303-311.

[14]. http://www.wikilectures.eu/index.php/Lambert's_law.

# AUTHOR INDEX

Aaditya, Gayathri ....................................604
Abbas, A................... 1691, 2457, 3430
Abbas, Ahmed E. ...............................1888
Abbas, Ali ...................... 186, 752, 1674
Abbott, Malcolm D............1322, 2576, 2600
Abdalla, L.B ........................................1245
Abdallah, Amir A.................................3435
Abdallah, Shaimaa A. ..........................219
Abdellaoui, Imane...............................900
Abdullah, Ahmad ...............................2128
Aberle, Armin......................................2318
Aberle, Armin G. ............ 284, 496, 499, 1922
Ablekim, Tursun .................................3422
Aboubakr, Benazzouz...........................487
Abouelkhair, Hussain M........................2324
Abtahi, Amir .......................................638
Abudayyeh, Omar K...............................88
Acebo, Laura........................................155
Addamane, S. J.....................................281
Adewoyin, Adeyinka ...........................2381
Adhikari, Dipendra..............................2582
Affouda, Chaffra A. ..............................259
Agarwal, Mohit....................................2330
Agarwal, Sumit....................................1777
Agarwal, Vivek............... 2952, 2981, 2986, 3050
Agbo, Solomon N.................................2114
Ager, Joel............................................3410
Agrawal, Rakesh .................................1449
Aguiar, Jeff..........................................2702
Aguiar, Jeffrey ....................................2467
Aguirre, Rodolfo .................................2419
Ahamioje, Joseph A. ...........................2931
Ahanzhamejhad, Ramez H.....................170
Ahlswede, E. .......................................3260
Ahlswede, Erik .....................................791
Ahmad, Jawad.....................................3096
Ahmed, Benlarabi.................................487
Ahmed, Nuha...........................658, 2667
Aho, Arto .............................297, 2520
Aho, T. ...............................................1189
Aho, Timo............................................297
Ahrenkiel, P.........................................869
Ahrenkiel, Phil ...........................206, 831
Ahrenkiel, Richard K. ..........................3448
Ahrenkiel, S. Phillip.............................2514
Ahsan, Nazmul ...................................2334
Aierken, Abuduwayiti...........................226
Aindow, Mark .....................................1522
Aïssa, Brahim .....................................3435
Akaki, Yoji..........................................2338
Akari, Shunsuke ..................................2385
Akarm, Muhammad Nadeem .................2776
Ake-Sultan, Bernt ...............................2864
Akiki, Tilda.........................................1968
Akimoto, Katsuhiro ..............33, 160, 900

Akimoto, Naoki.................................... 712
Akiyama, Hidefumi ...............721, 2781, 3528
Akwari, Chinedum ..................... 735, 2446
Al Mahmud, Abdullah .......................... 1067
Alahmed, Ahmed.................................. 1110
Alam, Giri Wahyu ................................ 1498
Alam, Muhammad A. ................. 1055, 1259
Alam, Muhammad Ashraful.................... 1904
Alberi, Kirstin...................................... 2506
Albin, David.....................1196, 3305, 3319
Alcubilla, R. ........................................ 1781
Alcubilla, Ramón .................................. 944
Aleman, Monica .......................... 2227, 3435
Alexander, Jessica A. ........................... 966
Alfadhili, Fadhil K. ...................... 730, 815
Al-Fadhili, Fadhil K. ............................ 2462
Al-Ghzaiwat, Mutaz ............................. 2593
Algora, C. .......................................... 1210
Algora, Carlos...................................... 1204
Alharbi, Fahhad H. ............................... 963
Alharthi, Yahya Z. ..................... 1018, 1110
Ali, Asad ............................................ 1228
Ali, Jaffar Moideen Yacob..................... 2318
Ali, Waqar........................................... 1228
Alivisatos, A. Paul................................ 1737
Al-Jassim, M. ...................................... 1196
Al-Jassim, M.M. .................1312, 2280, 2785
Al-Jassim, Mowafak .................... 62, 1371, 1381,
1400, 2789, 2887, 3305, 3319
Al-Jassim, Mowafak M. ........................ 3147
Aljaziri, Marwa .................................... 2011
Alkhayat, Rabee B................................ 815
Allebé, C. ................................. 50, 2073
Allebé, Christophe................... 3254, 3256
Allen, Thomas...................................... 2076
Almheiri, Anwar ................................... 1946
Almonacid, Florencia ............................ 2858
Alrashidi, Hameed................................ 2858
Altermatt, Pietro P. ...........1922, 2220, 3304
Alvarez, Diego Alonso .......................... 1339
Alvarez, Genesis .................................. 2941
Alvarez, José ............................ 2453, 2528
Aly, Shahzada P. .................................. 963
Alzahmi, Wadhah ................................. 1946
Amdemeskel, Mekbib W........................ 2672
Anctil, Annick...................................... 2124
Anderberg, A. ...................................... 467
Andler, Joseph .................................... 1449
Ando, Daisuke ..................................... 931
Ando, Yasutaka .................................... 970
Ando, Yuta .......................................... 192
Andreani, Lucio ................................... 290
Angeles-Ordóñez, G. ............................ 142
Annigoni, Eleonora.................... 1395, 2794
Anoma, Marc Abou ............................... 1549

# AUTHOR INDEX

Anselmo, Andrew.................................74, 2839, 2897
Antony, Aldrin ..............................................1755
Anttu, Nicklas...............................................2502
Anyadiegwu, Ifeanacho ...................................970
Anyanwu, Uchechi.........................................319
Araki, Hideaki...............................................2338
Araki, Kenji..................359, 412, 1479, 1711,
    1714, 1743, 2498, 2548, 2566
Aranguren, G. ...............................................643
Archer, Alexander .........................................771
Arehart, A. R.......................................30, 2414
Arehart, Aaron R...................215, 2446, 3139
Arinze, Ebuka S. ...........................................667
Armour, Eric .................................................827
Armour, Eric A......................................210, 2506
Armoush, Maher............................................1058
Arnold, Daniel B. ..........................................3002
Arnou, Panagiota ..................................146, 186
Arora, B. M. ...................396, 1995, 2716
Arora, Brij M. ...............................................3478
Arp, Juergen..................................................1411
Arredondo, C. A.............................................2031
Artegiani, Elisa..................752, 1669, 2372
Aryal, Krishna................................................182
Asadirad, Mojtaba ........................................866
Asahi, Shigeo ...............................................23
Asgharzadeh, Amir ..............1537, 1543, 3333
Ashrafee, Tasnuva .........................................735
Aslam, Aasma................................................2355
Asomoza, René .............................................632
Astakhov, Oleksandr ....................................2114
Aswani, U .....................................................1898
Athresh, Eashwer..................................2395, 2399
Atia, Adam A.................................................3230
Atkins, R. .....................................................229
Atlan, Olivier.................................................626
Atwater, Harry A....................512, 521, 558, 572,
    1248, 1589, 1737, 2236
Augarten, Yael ..............................................1651
Augusto, André ........................1589, 2596
Avasthi, Sushobhan........251, 837, 841, 986, 2395, 2399
Avenet, Julien................................................1933
Avery, J. E. ...................................................1863
Awadallah, Osama.........................................3473
Awasthi, Vishnu ............................................2345
Ayala, Orlando ..............................................735
Azkona, N. ...................................................2740
Azkona, Nekane ............................................2677
Azzolini, Joseph A.........................................608
Baba, Masaaki...............................................1724
Babbe, Finn ............................................151, 2054
Babcock, Sean J.............................................2298
Bachman, Benjamin F. ..................................3381
Badel, N........................................................50
Badosa, Jordi.................................................626

Badr, Ikken ..................................................487
Bae, Soohyun ................................................935
Baggu, Murali ...............................................2991
Baik, Sungsun ...............................................2242
Bailey, C. ......................................................845
Bailey, Christopher G. ...................................2298
Bailey, J. ......................................................2414
Bailey, Jeff...........................................1686, 3327
Baines, Tom .............................................742, 1445
Baka, Maro ...................................................3343
Baker, Rupesh ...............................................3172
Bakhshi, Sara................................................322
Bakker, Klaas ...............................................2875
Balaji, Pradeep .............................................2596
Balakrishnan, G. ...........................................281
Balasubramaniam, Kavaipatti R. ....................1704
Baldus-Jeursen, Christopher...........................1908
Ball, Greg .....................................................2263
Ballif, C. ...............................................50, 2073
Ballif, Christophe .....................55, 1220, 1395,
    2104, 2794, 3254, 3256, 3435
Baloch, Ahmer A.B.............................963, 1058
Banda, Pedro .................................................1946
Banerje, Rangan ............................................1151
Banerjee, Sanjay K.........................................363
Barahman, Gil ...............................................2285
Barakel, Damien ...........................................2255
Barnes, T. M. ................................................138
Barnes, Teresa M...........................................3422
Barnett, Allen................................................315
Barraud, L. ...................................................50
Barraud, Loris ......................................3254, 3256
Barrigon, Enrique ..........................................2502
Barth, Kurt ...................................................424
Bartolo, Robert E. .........................................195
Bartsch, J. ....................................................884
Basore, Paul A. .............................................2163
Bastide, Stéphane ..........................................3402
Bastola, Ebin .........................................738, 781
Basu, Prabir K................................................396
Battaglia, A. .................................................1747
Baudrit, Mathieu ....................................2492, 2562
Bauer, Andreas......................................791, 2058
Bauer, Jan ....................................................1376
Bauhuis, G. ..................................................1189
Baumann, Thomas .........................................1077
Baumgartner, Franz.......................................1077
Baur, Carsten.......................................541, 2087
Baxter, Jason B. ............................................3143
Bearda, Twan.................................................1233
Beauchemin, Ryan D. ....................................102
Becerril-Romero, Ignacio ..............................155
Becker, Jacob J. .....................................3366, 3410
Bedair, Salah M.............................................2195
Belanger, Ted.................................................1427

# AUTHOR INDEX

Belletête, Marc ....................................................1579

Belluardo, Giorgio ...................................3360, 3482

Bemrrr, Andreas.....................................................3500

Benamara, Mourad.................................................3370

Benatto, Gisele A. Dos Reis .........................2672, 2682

Benick, J................................................................2064

Benick, Jan............................................................2511

Bennett, Dirk.........................................................2042

Bennett, Mitchell.....................................................247

Bennett, Mitchell F.......................210, 259, 873, 2091

Berardone, Irene......................................................402

Berg, Alexander.....................................................1773

Berg, Morgann.......................................................3417

Bermel, Peter.................................................1904, 2467

Bernard, Annie..............................................2870, 2891

Berry, Joseph J......................................................2176

Bert, J....................................................................1733

Bertoni, Mariana...............................944, 2610, 2854

Bertoni, Mariana I. ......................................2179, 3309

Besanger, Yvon......................................................3102

Bett, Alexander J...................................................1253

Bett, Andreas W.....................................................2511

Bettenwort, Gerd...................................................1965

Beutel, Paul............................................................2511

Beutner, Volker......................................................1855

Bhaduri, Sonali.............................................2799, 3478

Bhan, Mohan Krishan..............................................496

Bhandari, Khagendra P.................738, 748, 781, 815

Bhatia, A................................................................1656

Bhatia, Swasti........................................................1755

Bhattacharya, Indranil............................................3083

Bhattacharya, Sitangshu.........................................2376

Bheemreddy, Venkata.............................................2688

Bialek, Tom............................................................2991

Bidiville, Adrien.....................................................1333

Biedenham, Richard E.............................................3245

Biegelsen, D. K......................................................1733

Biiss, M..................................................................2457

Binetti, Simona.......................................................1669

Birch, Max T..........................................................2423

Birkmirc, Robert......................................................726

Birkmire, Robert W.................................................2637

Bishop, Doug.........................................................3275

Bishop, Douglas......................................................726

Bishop, Douglas M.................................................1441

Bissels, G...............................................................1189

Biswas, R...............................................................1350

Bittau, F.................................................................3430

Bittau, Francesco.....................................................752

Bittner, Zachary.......................................................677

Bittner, Zachary S.............................18, 202, 2084

Bivour, M...............................................................2064

Bivour, Martin........................................................1253

Blakely, Logan.......................................................1573

Blanche, Pierre-Alexandre......................................1147

Bläsi, Benedikt.........................................................352

Blasi, David...........................................................1531

Blum, Adrienne......................................................2692

Blum, Adrienne L...................................................2765

Bob, Brion.............................................................2258

Bobela, David C.....................................................2506

Bobyl, A.V. ...................................................1025, 1811

Boca, Andreea.......................................................2099

Boccard, Mathieu ............55, 1220, 1317, 1790, 3366

Boeck, Torta..........................................................3396

Bohra, Rakesh........................................................1912

Boizot, Bruno.................................................83, 2087

Bolaji, Adewumi.....................................................2381

Boley, Allison.........................................................2573

Bolke, J. G.............................................................1656

Bonnassieux, Yvan...................................................626

Bonomo, Pierluigi...................................................2118

Book, Felix.............................................................1824

Bora, Birinchi.........................................................3478

Borgers, Tom..........................................................3343

Borgström, Magnus.................................................2502

Borgström, Magnus T..............................................1286

Borland, John.........................................................2947

Borne, Axel............................................................2864

Borowik, Lukasz.....................................................1516

Bosco, Nick..................................................3190, 3200

Bosco, Nick S.........................................................2864

Bosson, Christopher J.............................................2423

Bostock, Peter........................................................2267

Bothe, Karsten.......................................................2692

Bourcois, Jérôme....................................................2087

Bourdin, Vincent......................................................626

Bourgoin, Jacques C...............................................2087

Bourne, Ben C........................................................1549

Bousselham, Abdelkader.........................................1058

Bouttcmy, Muriel...................................................2711

Bowden, Stuart .................240, 925, 1797, 2719

Bowden, Stuart G...........................................1589, 2596

Bowen, Leon..........................................................1445

Bowers, J.W. ...................................................2457, 3430

Bowers, Jake W..................146, 186, 752, 2349

Boyce, Ken.............................................................2000

Boyce, Kenneth......................................................1933

Boyd, Matthew.......................................................1933

Boyer, Jacob............................................................215

Boyer, Jacob T..............................................2079, 2554

Boyer-Richard, Soline.............................................2192

Brabec, Christoph J. .......................................1346, 3500

Bradshaw, Geoffrey K. ...........................88, 301, 531

Brady, Brendan......................................................3388

Braga, Daniel Sena.................................................2307

Braid, Jennifer L. ...................................1927, 2697, 3456

Brammertz, G.........................................................3260

Brand, A.A..............................................................884

Brates, Nanu..........................................................1728

# AUTHOR INDEX

Bräuninger, Matthias ...................................3256
Breitenstein, Otwin ....................................1376
Breitwieser, M. ...........................................3135
Bremner, S. P. ..............................................953
Bremner, Stephen .............. 858, 1215, 1845, 2186, 2569
Bremner, Stephen P. ....................................948
Brendel, Rolf .....................................1366, 3371
Breus, V. .....................................................1752
Bright, Jamie M. .........................................1405
Brinnig, Samuel ..........................................2622
Brito, Pedro P. ............................................2307
Britt, Jeffrey ..............................................1455
Brittman, Sarah ..........................................2245
Broderick, Robert .......................................3008
Broderick, Robert J. ..................1435, 1555,
        1567, 1573, 3025, 3031
Brolo, Alexander G. .....................................3388
Bruckman, Laura ........................................1933
Bruckman, Laura S. ....................................2000
Brückner, Sebastian ....................................2538
Brughera, Céline .........................................2492
Brule, Carlton ............................................1728
Brulo, Gregory S. ........................................1469
Bryan, Jonathan .........................................1317
Buchanan, Wayne .......................................1196
Büchler, A. .........................................884, 3135
Buckner, Jessica ...........................................537
Buerhop, Claudia ........................................3500
Bukowsky, Colton R. ...................................1737
Bulkin, P. ...................................................1781
Bulkin, Pavel ..............................................1237
Bullock, James .....................................59, 2076
Buonassisi, T. ......................648, 1140, 3295
Buonassisi, Tonio ...................284, 1264, 1491,
        2242, 2532, 2744, 3236, 3290, 3300
Burgers, A.R. ..............................................3150
Burgers, Antonius R. ....................................917
Burkhardt, S. .............................................1752
Burnham, Laurie ........................................1435
Burroughs, Scott ..............................272, 1469
Busquet, Severine .......................................1061
Butt, Isaac ..................................................182
Cabarrocas, Pere Roca I ............. 464, 1237, 2528, 2593
Cachet-Vivier, Christine .............................3402
Caffy, Florent .............................................1516
Calderón-Obaldía, Fausto .............................626
Calle, Eric ...................................................944
Calvo-Barrio, Lorenzo .................................3285
Campa, Andrej ...........................................1346
Campanelli, Mark .........................................437
Campbell, Calli M. ..............................3366, 3410
Campesato, Roberta ...................76, 541, 545
Campos, Cláudio Dias ...................................2307
Camus, Christian ........................................3500
Cañadillas, David .............................429, 1116

Canino, A. ..................................................1747
Caño, P. .....................................................1210
Cao, Huihui ................................................1619
Cao, Wenkai.................................................696
Cao, Xin....................................................2427
Cao, Yunxue ...................392, 1430, 1873, 2918
Cappelluti, F. .............................................1189
Cardona, Dagoberto .....................................670
Cardwell, D. ...............................................3511
Cariou, Romain ...............................2511, 2528
Carlin, John A. .............................................215
Carlson, David E. ........................................3442
Carlson, Emily .............................................701
Carneiro, Lucas M. ........................................417
Carolus, Jome ............................................2875
Carpenter, Bernard .......................................537
Carr, Anna J. ..............................................1081
Carriere, Jarrett .........................................2833
Carruthers, Steve ........................................3514
Carter, Catrice M. ........................................3393
Carter, Cedric .............................................2135
Carter, Sam ................................................3514
Casale, Mariacristina ..................... 76, 541, 545
Casallas-Moreno, Y. L. ..................................670
Casper, Chadwick ........................................1476
Cassini, Denio A. .........................................1917
Castañeda, Carlos A. Rodríguez ......................1858
Catthoor, Francky .......................................3343
Cattin, Jean ...............................................3435
Cattoni, Andrea ..........................................1289
Cavani, Olivier............................................2087
Cédola, A. P. ..............................................1189
Cendagorta-Galarza, Manuel ..........................429
Cepeda, Kyle ...............................................876
Cesar, I. ....................................................3150
Cesar, Ilkay ................................................917
Chai, Gaoda ................................................976
Chai, Jing ..................................................1922
Chakraborty, Sagnik....................................3300
Chamarthi, Phani Kumar ...............................2952
Chamberlin, Charles ....................................1271
Champliaud, J. ..............................................50
Champliaud, Jonathan ..................................3435
Champness, C. H. ........................................2388
Chan, Calvin ..............................................3417
Chan, Catherine E. ..............................2576, 2600
Chan, Mandy ..............................................2808
Chan, Maria K.Y. ...................6, 1256, 2759
Chan, R. ....................................................3511
Chandralal, Sreeram ....................................1674
Chandran, Deepak .......................................2986
Chang, Jipeng ...................................1873, 2823
Chang, Sheng-Hao........................................1051
Chang, Via-Chung ........................................1051
Chantana, Jakapan ...............................757, 2385

# AUTHOR INDEX

Chaporr, Patrick ............................................2711
Chapuis, Valentin............................................2104
Chattopadhyay, Kamanio ...........................2811
Chattopadhyay, Shashwata ....... 1850, 1858, 2849, 3478
Chaudhry, Ghulam M. .....................1018, 1110
Chaujar, Rishu .................................................377
Chaurasia, Saloni ....................................837, 841
Chausseau, Matthieu ......................................2711
Chavali, Raghu Vamsi Krishna ...................1904
Chavez, Jose J. ..............................................2419
Chemisana, Daniel ........................................1339
Chen, Benjamin ............................................2358
Chen, Chien-Hsun .............................................911
Chen, Chun-Chi ............................................1635
Chen, Daniel ..................................................2576
Chen, Eric Y. ...................................1598, 3384
Chen, Haiyan ................................................2220
Chen, Hung-Ling ..........................................1289
Chen, Junyan ....................................1835, 2732
Chen, Kaifeng ................................................2185
Chen, Kunji ..................................................2656
Chen, Lung-Chien............................................367
Chen, Meixi ...............................326, 999, 2035
Chen, Peng-Wei .............................................2660
Chen, Ran .......................................................2576
Chen, Renfang ................................................1241
Chen, Shi-Wei.................................................1627
Chen, Sung-Yu ..................................................911
Chen, Tsung-Cheng ..........................................329
Chen, Tzu-Yu ..................................................1627
Chen, Wanghua ...............................................2593
Chen, Weijian .................................................2392
Chen, Y. ...........................................................2785
Chen, Yang ......................................................2502
Chen, Yao- Hui.....................................893, 2664
Chen, Yifeng ......................................1922, 2220
Chen, Yunfei.........................................761, 2427
Chen, Yusi .......................................................1835
Chen, Zhi David ..............................................1044
Chen, Zihan .....................................................2392
Chendo, Michael .............................................2381
Cheng, Y. ............................................................14
Cheng, Yan .......................................................667
Cheng, Yuh-Jen ...............................................1610
Cheng, Zhe.......................................................3473
Cheng, Zhongkai..............................................3393
Chenna, Shiva Tarun .......................................1674
Chiang, Cho-Chun ...............................893, 2664
Chiang, Fu-Kuo................................................198
Chikhalkar, Abhinav ..............................823, 827
Chin, Ken K. .........................................761, 2427
Chinnusamy, Saravanan ....................................980
Chiu, Chun-Yu ................................................1169
Chiu, P. ...........................................................2094
Chiu, Philip.....................................................2099

Chmielewski, Daniel J. ................215, 2079, 2554
Cho, Eunhwan ......................333, 1824, 1838
Cho, Junsik ...................................................... 810
Cho, Yasuo .................................................... 3323
Choi, Gyu-Seok ............................................. 2723
Choi, J. -K ...................................................... 2019
Choi, Rae-Won ............................................... 2723
Choi, Seungkeun ............................................ 1037
Choi, Sungjin ................................................. 1758
Chong, Cheemun ............................................ 2600
Choubisa, Hitarth ........................................... 1022
Choudhury, K. R. ........................................... 2312
Chouhan, Arun Singh ....................................... 986
Choulat, Patrick ................................... 2227, 3435
Chow, E.M. ..................................................... 1733
Chowdhury, Ahrar Ahmed ................................ 888
Christians, Jeffrey A. ...................................... 2176
Christmann, G. ................................................... 50
Chu, Chi-Wei .................................................. 1051
Chu, Haifeng .................................................. 1222
Chu, Sheng ..................................................... 1299
Chua, Soo Jin ................................................... 284
Chuang, Ta-Wei .....................343, 367, 893, 2664
Chung, Daniel ..................................... 2707, 3304
Chung, Haejun ................................................ 1904
Chung, Simon ....................................... 696, 2186
Ciesla, Alison M. .................................. 2576, 2600
Cifuentes, L. .................................................. 1210
Cifuentes, Luis ............................................... 1204
Ciocia, Alessandro ......................................... 3096
Cirino, Daniel A. Merced .............................. 3044
Clayton-Warwick, D. ....................................... 138
Cleveland, Erin ..................................... 247, 2091
Clinton, Evan A. .............................................. 305
Cobo-Yepes, Nicolás....................................... 2963
Codd, Daniel S. .............................................. 3245
Cohen, Bat-El ................................................. 2170
Cole, Wesley J. ............................................... 2163
Colegrove, E. ................................................. 1312
Colegrove, Eric ..................................... 3147, 3319
Coll, Pablo Guimera........................................ 2610
Collin, Stéphane.............................................. 1289, 3147
Collins, Robert W.............807, 2462, 2582, 2646, 3426
Collins, Shamara ...................802, 1638, 2449, 3413
Comagliotti, Emanuele..................................... 3435
Condorelli, G. ................................................. 1747
Conibeer, Gavin......................696, 2186, 2392
Conlon, Benjamin P. ....................................... 219
Conrad, Brianna.............................................. 315
Cordeiro, Patricia ........................................... 2135
Cordova, Adam .............................................. 1965
Cornagliotti, Emanuele.......................... 1804, 2227
Cornaro, Cristina ............................................ 3482
Cornell, Robert................................................ 1275
Correa, J.M........................................................ 433

# AUTHOR INDEX

Correa-Baena, Juan-Pablo ...................................3300
Cossio, Gabriel .................................................1181
Costa, Sara.....................................................1979
Costa, Suellen C. ..............................................2307
Côté, Alexandre................................................1908
Cousar, Larry C. ................................................921
Cravens, R. ....................................................2094
Crawford, L. ...................................................1733
Crupi, F. ........................................................2073
Cruz, José Ortega .....................................1959, 1990
Cruz, Leila R. De Oliveira....................................2307
Cruz-Campa, Jose Luis .........................................337
Cuevas, Andres................................................2076
Cui, Jie .......................................................2076
Cui, Min .......................................................1765
Cunningham, Daniel W. ........................................1463
Cunningham, Joseph ...........................................3161
Cur, Jie .........................................................517
Curran, Alan J. ...........................1927, 2697, 3488
Curvat, L. ........................................................50
Cushing, Scott K. ...............................................417
Da Fonseca, Jérémy ...................................2492, 2562
Dabney, M. S. ...................................................138
D'Abrigeon, Laurent............................................545
Daenen, Michael...............................................2875
Dagenais, Mario ........................................195, 1048
Dagyte, Vilgaile................................................2502
Dahal, Saroj ..........................................309, 3123
Dahal, Som......................................................240
Dai, Yushuai ...........................18, 222, 677, 1184
Dalal, V L .....................................................1350
Dalal, Vikram ..................................................2247
Dalpian, G......................................................1245
Dam-Hansen, Carsten....................................2672, 2682
Danel, A. ......................................................1747
Dang, Hongmei.................................................2432
Dangate, Milind S. ............................................980
Daniil, Andreana ................................................944
Danzl, F.J.K....................................................3150
Darbali-Zamora, Rachid .............................2957, 2963
Das, Ujjwal ................ 408, 1473, 1761, 1828, 2667
Das, Ujjwal K. .................................................2637
Datas, Alejandro ...............................................2562
Dauskardt, Reinhold....................................3190, 3200
Davidsen, Rasmus S. ...........................................2672
Davies, J. I. ..................................................1210
Davis, Kristopher O. ..................74, 322, 1804, 3448
Davis, Tracy.....................................................537
De Coux, Patricia ..............................................464
De Melo, O. ...................................................2342
De Nicolas, S. Martin ............................................50
De Oliveira, Michele C. C. ...................................1917
De Villers, Bertrand J. Tremolet .............................1354
De Wolf, Stefaan .........................................55, 3256, 3435
De, F. C. Lins Vanessa ........................................1917

Debnath, M. C. ..................................................14
Debnath, Tanmoy ...............................................1067
Deboever, Jeremiah.....................................1555, 1567
Debrot, F. ........................................................50
Debucquoy, Maarten...........................................1233
Debusschere, Vincent .........................................3102
Deceglie, Michael .....................................2771, 2789
Deceglie, Michael G.......................2488, 2804, 3452
Deckerl, D. ....................................................1752
Decobert, Jean ................................................2528
Deer, Tanya....................................................1908
Deitz, Julia I. ................................................3139
Delahoy, Alan E. .......................................761, 2427
Delhotal, J. ...................................................3224
Deligiannls, D. ...............................................3150
Deline, Chris ...............116, 1537, 1922, 3184, 3333
Demadrille, Renaud ...........................................1516
Demirkan, Korhan ..............................................820
Deng, Changhong ..............................................1158
Deng, Weiwei ..................................................2220
Denk, Patrick ..................................................1360
Descoeudres, A. .................................................50
Descoeudres, Antoine .........................................3254
Despeisse, M. .............................................50, 2073
Despeisse, Matthieu.....................3254, 3256, 3435
Desrues, Thibaut .......................................2492, 2562
Deutsch, Todd G. ................................................47
Devos, Arnaud .................................................464
Dewitt, Daniel ................................................1835
Dey, Anamika ..................................................1034
Dhakal, Tara P. ................................................989
Dhere, N. ...............................................389, 1701
Di Leo, Paolo ..................................................3096
Di Mare, Simone ...............................................2372
Di Napoli, Simone.............................................2205
Diaz, Liliana Ruiz ............................................1147
Diercks, David R. ...............................................46
Dimitrievska, Mirjana.........................................3285
Dimopoulos, Theodoros ........................................178
Dimroth, Frank ................................................2511
Ding, Jie......................................1937, 2823
Dinger, Justin ..........................................2692, 2765
Diniz, Antonia Sônia A. C. .....................1917, 2307
Dirriwachter, Antonius B. .....................................3448
Dise, John ...............................................132, 1104
Dise, Skip.....................................................1427
Dobrich, Anja .................................................2538
Dobroliubov, Aleksandr........................................2776
Dobson, Kevin ..................................................315
Dobson, Kevin D. ...............................................658
Dobson, Weston .........................................2692, 2765
Dogan, Yusuf ...................................................229
Doi, T. ........................................................441
Dominguez, A. .................................................2342
Dong, Jianfei .................................................2605

# AUTHOR INDEX

Doolittle, William A. ...................................305
Dooraghi, Michael.................................1169
Döscher, Henning.....................................47
Doty, Matthew F. ......................1598, 3384
Dougher, Chris......................................3245
Dougherty, Brian ..................................1933
Drahi, Etienne ........................................464
Drayton, Jennifer A. ...............................164
Drees, M...............................................3511
Dréon, Julie ..............................................55
D'Rozario, Julia ........................................18
Drummy, Lawrence F..............................966
Du, Chen-Hsun........................................911
Du, Xingzhi...........................................2558
Du, Zhongming................198, 767, 1707
Duan, Baosong.............................392, 2823
Duan, Wenqi...........................................346
Dubey, R. ..............................................1995
Dubey, Rajiv............... 1704, 2849, 3478
Dubois, Anne Migan ...............................626
Duenow, J.N. ........................................1312
Duenow, Joel .............................1196, 3147
Duerinckx, Filip .....................2227, 3435
Dugan, Roger C. ...................................3055
Dugdill, Brian.......................................2014
Dumbrell, Robert........................420, 3315
Dunham, Scott T. ..................................3119
Dupré, Cécilia...........................2492, 2562
Durand, Olivier.....................................2192
Durose, Ken ................................742, 1445
Durstock, Michael F..............................966
Durygin, Andriy....................................3473
Dusane, Rajiv O ...................................2330
Dussarrat, Christian ...............................326
Dutt, A. ..................................................370
Dutt, Ateet ...........................................2342
Dutta, P. ................................................869
Dutta, Pavel ..............................866, 2368
Duttagupta, S.P. ...................................1898
Eafanti, Joshua......................................3190
Ebe, Falko.............................................2996
Ebert, Matthieu.....................................1531
Ebong, Abasifreke...................................888
Ediger, E. .............................................2364
Edinger, Stefan........................................178
Edoff, Marika..........................................796
Edwards, Daniel J ................................3514
Eeles, Alexander............................146, 186
Efthymiou, Venizelos...........................3107
Egbe, Daniel Ayuk Mbi ........................1360
Eggink, Wouter.....................................2109
Ekins-Daukes, Ned ...............................1339
El Assimi, Taha.....................................3402
Elangovan, Hemaprabha........................2811
Elanzeery, Hossam.......................151, 2054

Eldho, T.I. ...........................................1898
El-Henawey, Mohamed......................... 2247
Elkhatib, Mohamed................... 2141, 2969
Elleuch, Omar........................................ 359
Ellibee, Donald..................................... 1543
Ellingson, Randall ................................ 2926
Ellingson, Randall J. ............................ 1030
Ellingson, Randy J.............738, 748, 781, 815
Ellis, Chase T. ...................................... 873
Elnosh, Ammar..................................... 1946
Elsehrawy, Farid ................................... 1189
Emery, K.A. .......................................... 490
Engerer, Nicholas A. ............................ 1405
Eriksen, Ryan ...................................... 2870
Eriksen, Ryan S. ................................... 2891
Ermer, J. .............................................. 2094
Ermer, James ....................................... 2099
Ermer, Jim H. ........................................ 37
Escarra, Matthew D. ..................... 37, 3245
Escobar, D. Martínez ................... 1959, 1990
Esfandiari, Parichehr ............................. 178
Espinct-Gonzalez, Pilar.......................... 558
Espíndola-Rodríguez, Moisés .........155, 512, 572, 3265
Espinet-González, Pilar................... 521, 1248
Essa, Gharibah ..................................... 2011
Essig, Stephanie ..................... 55, 3254, 3371
Etcheberry, Arnaud ............................... 2711
Etgar, Lioz ........................................... 2170
Eugen, Rene ......................................... 2864
Evani, Vamsi.....................802, 1638, 2449
Evans, Garrett Z.................................... 921
Evstigneev, M.....................663, 1025, 1811
Evstigneev, Mykhaylo A. ....................... 690
Eylers, Katharina .................................. 3396
Fada, Justin S. ...................2697, 3456, 3488
Faes, A. ................................................. 50
Fairbrother, Andrew............1933, 2000, 3204
Faleev, Nikolai ........................... 1215, 2573
Fan, S. ................................................. 3376
Fan, Shanhui ............................. 2185, 2732
Fang, Liang ........................................... 226
Fang, Y. ............................................... 1603
Fang, Yi ............................................... 305
Fano, V. ............................................... 2740
Fano, Vanesa........................................ 2677
Faraj, Abudul ....................................... 2014
Farnung, Boris...................................... 2267
Farré, Laia Arqués ................................ 3285
Farshchi, Rouin ........................... 1459, 1686
Faur, Maria........................................... 896
Faur, Orry............................................ 896
Favre, W. ............................................. 1747
Fedina, Maria ....................................... 2070
Fejfar, A. ............................................. 2073
Felder, T. ............................................. 2312

# AUTHOR INDEX

Feldmann, F. ............................................ 2064
Feng, Sheng-Kai .................. 343, 367, 893, 2664
Feng, Shien-Ping ................................... 1012
Feng, Zhiqiang ............................... 1922, 2220
Fenning, David P. ......................... 1494, 2245
Ferekides, Chris ............ 802, 1638, 2449, 2467, 3413
Ferekides, Chris S. ................................ 1511
Ferekides, Christos .................................. 175
Ferguson, Andrew J. ............................... 1354
Ferguson, L. ........................................... 1863
Fernández, Eduardo F. ............................ 2858
Fernandez, R. Mis ......................... 1691, 2457
Fetzer, C. ............................................... 2094
Fiducia, Thomas ...................................... 424
Fields, Brian J. ...................................... 2618
Filipic, Miha ........................................... 1233
Filonovich, Sergej ........................... 464, 1237
Firth, Peter ............................................ 1317
Fischer, A. ............................................ 1603
Fischer, Alec ........................................... 823
Fischer, Alec M. ....................................... 305
Fisher, Brent ........................... 210, 272, 1469
Fisher, Dallas .......................................... 989
Fitzgerald, Eugene A. .............................. 213
Fleming, Robert A. ................................. 1869
Flicker, J. D. .......................................... 3224
Flicker, Jack ......................................... 1280
Florides, Michalis .................................. 1941
Foldyna, Martin ............................. 2528, 2593
Forberich, Karen ................................... 1346
Forbes, David V. .................................... 3468
Forchhammer, Soren ............................... 2682
Forsh, P.A. ............................................ 1811
Foster, Robert ....................................... 2014
Fouchier, Marin ..................................... 3402
Fournel, Frank ............................... 2492, 2562
Fraas, L. M. .......................................... 1863
Fraas, Lewis .......................................... 2042
France, Ryan M. ............................... 47, 232
Fraser, Ray ............................................ 337
Frederiksen, Kenn H. B. .......................... 2682
Freeman, Janine M. ................................ 3494
Freiburger, Brennen M. ........................... 1869
French, Roger ....................................... 1933
French, Roger H. ......... 1927, 2000, 2697, 3456, 3488
Freundlich, Alexandre .................. 236, 673, 1452
Fridman, Lucas ...................................... 2000
Friedman, Daniel ......................... 549, 2543
Friedman, Daniel J. ................. 42, 268, 1201
Friend, Mari Paz ..................................... 429
Fritzsche, M. ......................................... 1752
Frontini, Francesco ................................ 2118
Fthenakis, V. ........................................ 2019
Fthenakis, V. M. .................................... 3230
Fthenakis, Vasilis ................................... 3077

Fu, Ran .................................... 1259, 1463
Fuhrich, Alexander .................................. 3396
Fuhrmann, Bianca ..................................... 83
Fujiwara, Koji ...................................... 1973
Fukuda, Tetsuya ...................................... 931
Funabiki, Shigeyuki .............................. 2906
Fung, Tsun H. ....................................... 2576
Fuyuki, Takashi ..................................... 2593
Gabetta, Giuseppe ........................... 76, 545
Gabor, Andrew M. ................. 74, 2839, 2897
Gaddy, Edward ........................................ 585
Gahr, Stefan .......................................... 178
Gai, Boju ..................................... 549, 2291
Gaiaschi, Sofia ..................................... 2711
Gallon, Joshua B. ................................... 3448
Galtieri, Jason ............................. 2975, 3214
Gambogi, W. ......................................... 2312
Gao, Hui ............................................... 226
Gao, Peng ........................................... 1648
Gao, Wei ............................................... 226
Gao, Y. ................................................. 869
Gao, Yijun ........................................... 2392
Gao, Ying ............................................ 2368
Gao, Yuan .................................. 2048, 2605
Gao, Yujie ................................. 2870, 2891
García, I. ............................................ 1210
Garcia, Iván ........................................ 1204
Garcia, Juan Lopez .................................. 402
Garcia-Linares, Pablo ............................ 2562
Garg, Vivek ......................................... 2345
Garner, Sean ........................................ 2870
Garner, Sean M. .................................... 2891
Garnett, Erik C. ................................... 2245
Garreau-Iles, L. .................................... 2312
Garrillo, Pablo A. Fernández .................... 1516
Garuz, Richard ..................................... 2255
Gaury, Benoit ............................. 1303, 2438
Gdoutos, Eleftherios E. ............................ 558
Geelan-Small, Peter ............................... 3304
Gehre, Simon ....................................... 3500
Geissbiihler, J. ....................................... 50
Geissbuehler, Jonas .............................. 3256
Geisz, John .................................. 549, 3371
Geisz, John F. ............ 232, 268, 1737, 2195, 3254
Georghiou, George E. ...... 276, 1163, 1941, 1954, 3107
Geraghty, Paul ..................................... 1342
Gerardi, C. .......................................... 1747
Gerber, Andreas ........................... 1400, 1651
Gerdimenes, Anne ................................... 619
Gervasi, Massimo ................................... 541
Ghaisas, S.V. ............................... 389, 1701
Ghimire, Kiran ....................................... 993
Ghosh, Kunal ........................................ 716
Gibbs, Jacob M. ..................................... 730
Gibelli, François .................................. 2192

# AUTHOR INDEX

Giebink, Noel C. ...1469
Giguère, Jean-Benoit ...1360
Gilchrist, James B. ...966
Gillispie, Kellen ...2762
Giordano, Francesco ...3096
Giraldo, Sergio ...3265, 3285
Giussani, A. ...845
Giussani, Alessandro ...206, 831, 2514
Givot, Bradley L. ...2864
Gladden, Christopher ...1476
Glasgow, Nate ...1427
Glatthaar, M. ...884, 3135
Gloeckler, Markus ...1193
Glunz, S. ...3135
Glunz, S.W. ...2064
Glunz, Stefan W. ...1253, 2511
Gokkaya, Huseyin Cem ...958
Goldschmidt, Jan Christoph ...1253
Golembeski, Andrew A. ...3143
Goma, Elias Garcia ...3462
Gombia, Enos ...541
Gona, Michael N. ...2349
Gong, Chen ...1585
Gong, Jue ...2251
Gonzálcz-Díaz, Benjamín ...3240
Gonzalez, Maria ...259
Gonzalez, S. ...3224
Gonzalez, Sigifredo ...2147, 3002, 3020
Gonzalez-Díaz, Benjamín ...429, 1116
Goodarzi, Mohsen ...2707
Gooding, Renee ...1280, 1543
Goodnick, S. M. ...1603
Goodnick, Stephen ...1790
Goodnick, Stephen M. ...305, 582, 1797
Gordillo, G. ...433, 503
Gordon, Ivan ...1233
Gori, Gabriele ...76
Gorman, Brian ...62, 1371, 1381
Górnez-González, L. A. ...2614
Gostein, Michael ...2808, 2923
Goswamy, Naveen ...1908
Gotoh, Kazuhiro ...1765, 1794
Gottschalg, Ralph ...1411, 2827, 3208
Govaerts, Jonathan ...3343
Goverde, Hans ...3343
Gowda, Ramesh Rame ...1912
Graf, Martin ...2511
Graham, Kenneth ...1044
Grandidier, Jonathan ...2099
Grassman, Tyler J. ...182, 215, 2079, 2554, 3139
Greco, Erminio ...76, 541
Grede, Alex J. ...1469
Green, Martin ...2213, 2403
Green, Martin A. ...858
Green, Michael ...2926

Greenhalgh, R.C. ...3430
Gregory, Geoffrey ...74
Grévin, Benjamin ...1516
Grice, Corey R. ...771, 1643, 2473, 3426
Grieco, William J. ...2618
Griffin, Alecia ...2870
Griffin, Alecia C. ...2891
Grijalva, Santiago ...1555, 1567
Grini, S. ...3269
Große, T. ...1752
Großer, Stephan ...2232
Grossklaus, Kevin ...701
Großschädl, Bettina ...1329
Grovenor, Chris ...424
Grover, Sachit ...1193, 2473
Grübel, B. ...884
Gu, Fei ...1346
Gu, Tian ...1473
Gu, Tingyi ...1828
Gu, Xiaohong ...1933, 2000, 2844, 3195, 3204
Guarracino, Ilaria ...1339
Guay, Nathan ...1543
Gudla, Sushanth ...1389
Guerrero-Lemus, Ricardo ...429
Guillemoles, Jean-François ...1289, 2192
Guillevin, N. ...3150
Guillevin, Nicolas ...917
Guina, M. ...1189
Guina, Mircea ...297, 2520
Guischard, Felix ...2836
Gunawan, Oki ...1441, 3275
Gunnarsson, William B. ...2443
Guo, D. ...1603, 2816
Guo, Hong ...1299
Guo, Q. ...3
Guo, Qi ...226
Guo, Shuwen ...1430, 1873, 2918
Guo, Yongjie ...1719
Gupta, Amit Kumar ...2952, 2981, 2986, 3050
Gupta, Mool C. ...937
Gupta, Neeti ...696
Gupta, Ritesh Kant ...1034
Gupta, Shivam ...377
Gupta, V. ...1733
Gustafsson, Mattias ...2025
Guthrey, Harvey ...1400, 2887
Gutiérrez, J. R. ...2740
Gutiérrez, R. ...643
Gutscher, S. ...884
Guwaeder, Abdulmunim ...1122
Gwak, Jihye ...810
Ha, Dongheon ...1585
Habermann, D. ...1752
Hack, James ...999
Hack, James H. ...326

# AUTHOR INDEX

Hacke, Peter .................1371, 1381, 1421, 1922, 2819, 2854, 3305
Hackl, Wolfeanz .................178
Haddad, M. .................2094
Haddadian, Rojiar .................1927
Hadi, Sabina Abdul.................213, 1741
Hadjipanayi, Maria.................276
Hadke, Shreyash .................986
Hadley, Wendy .................2014
Haegel, Nancy M. .................62
Hagendorf, Christian.................1376, 2232
Hägglund, Carl.................796
Hahn, Carina E. .................175
Hai, Hoang Tri .................931
Haight, Richard.................1441
Hajimiri, Ali .................521, 558
Hajizadeh, Amin .................3092
Halbwax, Mathieu.................3402
Hall, Allen .................1511
Hallam, Brett J. .................2576
Halliday, Douglas P .................2423
Hamadani, Behrang H.................263, 437, 508
Hameiri, Ziv .................66, 420, 3290, 3315
Hamon, Gwénaëlle.................2528
Hamui, L. .................2614
Hamzaoui, Saad .................900
Hamzavy, Babak T.................2618
Han, Sang M.................88
Han, Xinyue .................1719
Han, Youngsik .................2242
Hanada, Toru .................940
Handwerker, Carol A. .................1449
Haney, Paul .................1303
Haney, Paul M. .................2438
Hanley, J. .................2094
Hanna, Amir .................1055
Hannappel, Thomas .................2524, 2538, 2538
Hanriot, Sergio De Morais .................2307
Hansen, Clifford.............1127, 1537, 3184, 3333, 3348
Hansen, Clifford W.................110, 1543, 1549
Hansen, Ole .................2672
Hansen, Richard .................2042
Hansen, Shirley .................2042
Hao, Xia .................160
Hao, Xiaojing.................858, 2213, 2403
Haohui, L. .................1140
Haohui, Liu .................2744
Haque, K A S M Ehteshamul.................346
Haque, M. D. .................552
Hara, Shigeomi.................1950, 3339
Hara, Tomoya.................2548
Harari, Joseph .................3402
Hardikar, Kedar .................2688
Häring, Adrian .................2263
Hariskos, Dimitrios.................2058

Harmand, Jean-Christophe .................1289
Harris, Christian .................319
Harris, James.................1835, 2732
Harris, Tom .................2991
Harvey, Steven.................2887
Harvey, Steven P. .............1371, 1381, 2702, 3305, 3319
Haschke, Jan .................3435
Haslinger, Michael .................1804
Hassan, Ibrahim A. I. .................2858
Hatch, S. .................14
Hatton, Peter D. .................2423
Hauch, Jens .................3500
Haug, F.-J. .................2073
Hausgen, Paul E. .................102
Hausmann, J. .................1752
Havu, Ville .................2070
Haysom, Joan E. .................1094
He, Junwen .................1469, 1737
He, Qiuxiang .................3304
He, Wenshuang.................392
Hea, Wenshuang.................2823
Heben, Michael .................2926
Heben, Michael J. .........170, 730, 748, 815, 1030, 2462
Hegedus, Steven.................408, 1473, 1761, 1828, 2667
Hegedus, Steven S.................658
Heidmann, Berit .................3396
Heilbrunner, Herwig .................1360
Heilscher, Gerd.................2996
Heinz, F. D. .................3135
Heinze, Matthias.................2263
Heller, Dominic.................1077
Henes, Dan .................1094
Hentz, Sandrine .................966
Hermle, M. .................2064
Hermle, Martin.................1253, 2511
Hernandez, J. A. .................2031
Hernández, Johan.................1143
Hernandez, Joseph .................3473
Hernandez-Alvidrez, Javier.................2153
Hernández-Gutiérrez, C. A. .................670
Hernández-Rodríguez, Cecilio .................429
Herrera, Daniel J.................219
Herrmann, W. .................107
Herz, Magnus.................3360
Heta, Y. .................2312
Hetterich, Michael.................1682, 2216
Hettick, Mark .................59, 823, 2076
Heurlin, Magnus.................1286
Hickey, Benjamin .................1459
Hidaka, Kazuyuki .................1973
Higa, M. .................441
Hilfiker, M. .................2364
Hill, Alex .................1893
Himwas, Chalermchai.................1289
Hindi, Basel.................1058

# AUTHOR INDEX

Hinken, David ....................................................2692
Hinojosa, M. .....................................................1210
Hinzer, Karin ....................................................1094
Hirai, Masakazu ...............................................1769
Hirata, Yoichi .....................................................613
Hirose, Kotaro ..................................................3323
Hirstl, Louise C. ...............................................2091
Hishikawa, Y. ...........................................441, 1003
Hishikawa, Yoshihiro ...............................480, 2781
Ho, Jian Wei .......................................................496
Ho, Wen-Jeng .......................343, 367, 893, 2664
Hoang, Bao ...........................................................96
Ho-Baillie, Anita ......................858, 1845, 2569
Hobbs, William B. ...........................................2618
Hoerteis, Matthias.............................................914
Hoex, Bram ........................................................517
Hoff, Thomas ............................................132, 1104
Hofmann, Johannes .........................................2407
Hoheisel, Raymond ..............................247, 3514
Höhn, Oliver .......................................................352
Holman, Z. C. ....................................................3376
Holman, Zachary .....................................1790, 3366
Holman, Zachary C. ................1220, 1228, 1317,
 1322, 1820, 3250
Holmgren, William F. .............................110, 1127
Holzmann, Daniel...............................................914
Hong, Chung-Yu .................................................294
Hong, Keunkee ...................................................399
Honsberg, Christiana.............................827, 3088
Honsberg, Christiana B. ............240, 305, 582,
 681, 1215, 1841, 2573
Hopf, Markus .....................................................1965
Horenstein, Mark............................................2870
Horenstein, N Mark ........................................2891
Horner, Greg S. ................................................3448
Horowitz, Kelsey A.W. ..........................1259, 1463
Horzel, J. ....................................................50, 2073
Hoshii, Takuya ..................................................2334
Hosokawa, Kazuya.............................................613
Hossain, Istiaque ............................................2247
Hossain, Mohammad A. ...................................3456
Hossain, Mohammad I. .......................................963
Howard, John M. ..............................................2443
Hsi, Edward........................................................1275
Hsu, Chia-Jhe....................................................1623
Hsu, Chih An ...........................1638, 2449, 3413
Hsu, Lung-Hsing ...............................................1610
Hsu, Shun-Chieh ....................................1606, 1623
Hsu, Shu-Tsung.....................................445, 448, 476
Hsu, Wei-Lun ....................................................1048
Hsu, Yu -Chen .....................................................888
Hu, Chehao..........................................................229
Hu, Cheng-Shun .................................................329
Hu, Hailin ..........................................................1858
Hu, Juejun ..........................................................1473

Hu, Lilei..............................................................3129
Hu, Long ............................................................2392
Hu, Yang ............................................................1927
Hu, Yicong .........................................................2392
Huang, Jialiang..................................................2213
Huang, Jingsheng ....................................1937, 2823
Huang, Jing-Shun...............512, 521, 558, 572, 1248
Huang, Shujuan .................................................2392
Huang, Vi-Wen ..................................................1631
Huang, Weijing...........................................1873, 2918
Huang, Wei-Ming ....................................1627, 1631
Huang, Wen-Hsi ..................................................385
Huang, Ying-Yuan ............................................1807
Huang, Yi-Wen ..................................................1627
Huang, Yu-Ming ...............................................1606
Huang, Yu-Ting .................................................1012
Huang, Z. ...........................................................3260
Huayamave, Victor............................................2839
Hubbard, S. M. .........................552, 845, 2755
Hubbard, Seth ....................................................677
Hubbard, Seth M.................18, 202, 206, 222,
 831, 1184, 2084, 2298, 2514
Huber, Christian ...............................................2216
Hudson, A.I. ......................................................2755
Huey, Bryan D. .................................................1522
Huffaker, D.L. ...................................................2755
Huffaker, Diana .................................................202
Huhn, Vito..........................................................1651
Huld, Thomas ....................................................2167
Hunault, Philippe ..............................................2711
Hung, Yung-Jr....................................................1606
Huo, Yijie ..........................................................1835
Husein, Sebastian ...............................................944
Huss, Alexandra M..............................................164
Hussain, Babar ........................................451, 2355
Hussain, Muhammad M. ...................................1055
Hutchings, Douglas............................................1869
Hutchings, Douglas A. .......................................921
Hutter, Oliver S. ...............................................1445
Hwang, James.....................................................333
Hyvl, M. ............................................................2073
Iandolo, Beniamino ...........................................2672
Ianno, N.J. .........................................................2364
Ichikawa, Yukimi ..............................................1769
Idlbi, Basem ......................................................2996
Ikki, Osamu .......................................................2159
Ilic, Ognjen .......................................................1737
Imai, Jun ............................................................2906
Imaizumi, Mitsuru ....................................567, 3506
Imtiaz, Syed N...................................................1067
Ingenhoven, Philip.............................................3482
Ingenito, A. .......................................................2073
Inns, Daniel .......................................................3113
Isabella, Olindo .................................................2605
Isbilir, Kenan.....................................................2827

# AUTHOR INDEX

Isherwood, Patrick J. M. .................................2349
Ishii, Tomoaki .............................................455
Ishino, Yuya...............................................757
Ishizuka, Shogo ............................................33
Islam, Kazi...................................................37
Islam, Muhammad Monirul.............................33, 900
Islam, Raisul............................................1835
Isoaho, Riku.............................................2520
Iwasaki, Kazuya ........................................2338
Iwata, Naotaka .........................................2642
Iwuoha, Emmanuel ....................................1360
Iyer, Abhishek........................................326, 999, 2035
Iyer, Parameswar K........................................1034
Izquierdo-Roca, Victor ........................3265, 3285
Jackson, Christine........................................215
Jackson, Philip ...................................2205, 2453
Jacob, David ...........................................1549
Jacobson, Arne.........................................1271
Jadkar, S.R..............................................1701
Jaeckel, Bengt...........................................1411
Jae-Yun, Fa-Jun Ma,..................................1845
Jagdish, A K.............................................2811
Jäger-Waldau, Arnulf..................................2167
Jahn, Ulrike............................................3360
Jain, Aditi ................................................333
Jain, Nikhil ................... 42, 46, 232, 578, 2195, 3371
Janoch, Rob ......................................74, 2839, 2897
Jansen, Mark J...........................................1081
Jany, Christophe ...............................2492, 2562
Janz, Stefan ..........................................83, 2407
Jaramillo, Adolfo ......................................1143
Jared, Bradley .........................................1473
Jarmar, T..................................................30
Jasti, Naga Prathibha ..................................986
Jaswal, Rohit...........................................3172
Javed, Mehwish Azher.................................1317
Javey, Ali ..........................................59, 823, 2076
Jeangros, Q.............................................2073
Jenkins, P. P............................................845
Jenkins, Phillip P................. 247, 373, 1838, 2091, 3514
Jensen, Brian...........................................2014
Jensen, M. A............................................3295
Jensen, Mallory A. ................... 1491, 3290, 3300
Jensen, Soren.....................................1196, 3147
Jeong, Woo-Lim.....................................777, 1665
Jhaveri, Janam ........................................1773
Ji, Liang ..........................................1933, 2000
Ji, Yaping.................................................37
Ji, Yaping Vera........................................3245
Jia, Jieyang ......................................1835, 2732
Jian, Ding-Rung ..................................1627, 1631
Jiang, C. S. .......................................1312, 2789
Jiang, C.-S. ........................................2280, 2785
Jiang, Chun-Sheng..................................62, 1371
Jiang, Lian L. ..........................................589

Jiang, Lian Lian ....................................... 120
Jiang, Xuefang........................................ 1937
Jiang, Yu ............................................ 3220
Jimeno, J. C. .................................... 643, 2740
Jimeno, Juan Carlos.................................. 2677
Jin, C. ............................................... 1781
Jin, Yu .............................................. 3119
John, Jim J........................................... 1946
John, Joachim.................................... 1804, 2227
John, Suru Vivian.................................... 1360
Johnson, A. D. ....................................... 1210
Johnson, E.V. ........................................ 1781
Johnson, Erik V...................................... 2593
Johnson, J. L. .................................. 1656, 1661
Johnson, Jay .......... 2135, 2141, 2153, 2969, 3002, 3008
Johnston, S. ......................................... 2785
Johnston, Steve ..................... 62, 202, 459, 1371,
    1381, 1400, 2213, 2819, 2887, 3305, 3452
Jones, C. Birk.................... 2618, 3008, 3155, 3488
Jones, David ......................................... 1342
Joonwichien, Supawan................................ 904
Joshi, Madhuwanti S.................. 2952, 2981, 2986, 3050
Joshi, Pranav ........................................ 2247
Jošt, Marko .......................................... 1346
Jovanovic, Raka...................................... 963
Juang, B.C. .......................................... 2755
Juang, Bor-Chau ..................................... 202
Juárez, A. Sánchez................................... 1990
Juárez, Aarón Sánchez............................... 1959
Juhl, Mattias K. ................................. 420, 3315
Juhl, Mattias Klaus .................................. 66
Julien, Scott..................................... 1933, 2000
Junci, Wang .......................................... 496
Junda, Maxwell ...................................... 771
Junda, Maxwell M.............. 2462, 2582, 3426, 3468
Jung, Jae Hak ........................................ 487
Jung, Jiirgen ......................................... 2864
Jung, Sang Hoon ..................................... 2723
Jung, Sang Hyun ..................................... 244
Juso, Hiroyuki ....................................... 3506
Kabalan, Amal ....................................... 2358
Kaczynski, Ryan ..................................... 1455
Kaizu, Toshiyuki .................................... 23
Kaizuka, Izumi ....................................... 2159
Kakosimos, Konstantinos E. ......................... 1888
Kalainatharr, Sivaperuman........................... 2334
Kalb, J. ............................................. 1733
Kale, Abhijit......................................... 1801
Kale, Abhijit S. ..................................... 1777
Kallickal, Johnson ............................. 1543, 3348
Kalt, Heinz ..................................... 1682, 2216
Kamata, N. .......................................... 552
Kamevama, Satoshi................................... 2642
Kamino, Brett........................................ 3256
Kamins, Ted ......................................... 1835

# AUTHOR INDEX

Kaminski, P.M............................................3430
Kaminski, Piotr ........................................1674
Kaminski-Cachopo, Anne ........................2562
Kamioka, Takefumi ................ 2498, 2548, 2566, 2642
Kanemitsu, Yoshihiko...........................721, 2781
Kanevce, A................................................1312
Kanevce, Ana ............................................3147
Kang, Ho Kwan..........................................244
Kang, Min Gu.........................356, 1758, 2723
Kang, Yoonmook ......................................935
Kankiewicz, Adam .....................132, 1104, 1132, 1427
Kannan, C. V............................................2716
Kao, Ming-Hsuan ....................................1627
Kaplan, Stephen .............................600, 1071
Kaplar, R..................................................3224
Karas, Joseph ..........................................925
Karki, Shankar... 182, 735, 807, 2298, 2446, 2646, 3139
Karmarkar, M...........................................1661
Karpowich, Lindsey..................................914
Karthik, Shravan .....................................3172
Kashkoush, Ismail.....................................322
Kaslin, Remo ...........................................1077
Kasry, Amal..............................................2858
Kasu, Makoto ...............................1950, 3339
Kato, Takekazu ........................................1175
Kato, Takuya ............................................160
Katsube, Ryoji .........................................2361
Kaule, Felix ..............................................2622
Kausika, Bala Bhavya............................3014, 3167
Kavaipatti, Balasubramaniam ................2799
Kawatsu, Tomoyuki..........................381, 2588
Kazmerski, Lawrence L. .........................2799
Kazmerski, L.L. ........................................1245
Kazmerski, Lawrence L. ..................1917, 2307
Kazumi, Kenji ..........................................2361
Keeler, Gordon........................................1473
Keller, Nico ..............................................1077
Kelly, George...............................1275, 2263
Kelly, Matthew.........................................3514
Kelzenberg, Michael D. ..........512, 521, 558, 572, 1248
Kempe, M.D..............................................138
Kempe, Michael ...............................1933, 2000
Kempe, Michael D....................................3208
Kephart, Jason .........................................785
Kephart, Jason M. ....................................3417
Kern, Gregory ..........................................2147
Kern, Gregory A. ......................................3020
Kesavan, Arul Varman..............................1614
Kessels, Wilhelmus M.M. .........................1817
Kessler, Emily...........................................206
Khadimallah, A.........................................869
Khalili, A...................................................1189
Khan, Imran .....................802, 1638, 3413
Khan, Imran S...........................................2449
Khan, Mohammad R. ...............................1055

Khan, Taj M. ............................................451
Khanna, Raghav.......................................2926
Kharait, Rounak A....................................2833
Kharel, Khim ..................... 236, 673, 1452
Khatavkar, Sanchit ..................................2716
Khatiwada, D. ...........................................869
Khatiwada, Devendra ..............................866
Khatri, Ishwor .........................................192
Khatri, Trijul ...........................................377
Khomcnko, Denis V..................................690
Khoo, Yong Sheng....................................1922
Khor, Alan ...............................................3172
Khoram, Parisa ........................................2245
Khorenko, Victor.............................83, 2087
Khoury, R. ................................................1781
Kiefer, Fabian ..........................................1366
Killam, Alex ..............................................2719
Killinger, Sven ..................... 126, 1405
Kim, Boram .....................2201, 2524, 2538
Kim, Chang Zoo .......................................244
Kim, D. .....................................................1189
Kim, Dae Young .......................................626
Kim, Dong Seop .......................................399
Kim, Dong-Ho........................... 2631, 2634
Kim, Donghwan ...........................935, 1758
Kim, Hae-Sun ..............................777, 1665
Kim, Hyo Jin ............................................849
Kim, In-Young...........................................777
Kim, Jae Hyun.................363, 2844, 3195, 3204
Kim, Jin-Hyeok ........................................777
Kim, Jisun ................................................399
Kim, Ka-Hyun ..........................................1758
Kim, Kangho .............................................244
Kim, Kihwan .............................................810
Kim, Kyoung- Tae ....................................1037
Kim, Min-Soo............................................487
Kim, Moon ...............................................2759
Kim, Sangpyeong......................................240
Kim, Soo Min ...........................................2723
Kim, Woo Kyoung ....................................487
Kim, Yeongho ..........................................827
Kim, Yong Bae .........................................2723
Kim, Yong Whan.......................................849
Kim, Youngjo ...........................................244
Kimbal, Gregory M...................................110
Kimura, Daiki ..........................................854
Kindole, Dickson .....................................970
Kindvall, Anna.........................................785
King, Bruce H. ..................... 3155, 3488
King, Richard ...........................................827
King, Richard R. .............301, 823, 1215, 1841
Kingma, Aldo...........................................541
Kini, Roshan ............................................2926
Kinoshita, Kosuke ..................... 1504, 2588
Kirk, A. ....................................................3511

# AUTHOR INDEX

Kita, Takashi ...........................................23
Kleider, Jean-Paul ...............................2528
Klein, Talysa R. ............... 2482, 3371, 3439
Kleinschmidt, Peter.............................2538
Klemm, Hagen W. ...............................3396
Klenk, Markus ....................................1077
Klie, Robert F. ....................................2759
Klimm, Elisabeth ................................2836
Klise, Katherine A................................3161
Klisel, Geoffrey T. ...............................3494
Kluska, S. ...................................884, 3135
Knight, Bruce .....................................2014
Knopf, Hannes.....................................1965
Ko, Changhee ......................................326
Kobayashi, Jonathan ...........................1061
Koehl, Michael ...................................3488
Koepgel, Ringo ...................................2622
Kogler, Willi .......................................791
Kohlstädt, Markus ...............................1253
Koike, Junichi .....................................931
Koirala, Prakash ........................2462, 3426
Kojima, Nobuaki.............359, 2498, 2566
Kojima, Takuto ...........................1504, 2588
Komsa, Hannu-Pekka ...........................2070
Konagai, Makoto...................1769, 2627
König, M. ...........................................1752
Konstantinou, Georgios........................1941
Kontges, Marc .....................................1366
Kopecek, Radovan ...............................1222
Koschny, T. ........................................1350
Kostylyov, V.P. .............................1025, 1811
Kostylyov, Vitaliy P. ............................690
Kotipalli, Ratan ..................................2209
Kottantharayil, A. ...............................1995
Kottantharayil, Anil............. 396, 716, 1850, 2799, 3478
Kottokkaran, Ranjith ...........................2247
Kotulak, Nicole.............................999, 1838
Kotulak, Nicole A. ...............................247
Koyama, Koichi.....................1765, 1787
Kozodoy, Peter ...................................1476
Krabb, Peter .......................................178
Krantz, Patrick W. ...............................730
Krasikov, D. .......................................2816
Krc, Janez ..........................................1346
Krein, Philip T. ...................................3214
Krich, Jacob J. ....................................1294
Krishnan, Mani R. ...............................1912
Krishnan, Sheeja ..................................76
Krishnaswami, Hariharan....................2931, 2936
Krogen, J. ..........................................2094
Krügener, Jan......................................1494
Krut, Dimitri D. ...................................37
Ku, Chen-Hao......................................329
Kubiniec, Alex .............. 132, 1132, 1427
Kuciauskas, Darius ..............................1679

Kudriavtsev, Yu....................................670
Kuitche, Joseph ............................ 1877, 1883
Kulish, Mykola R..................................690
Kum, Hyun .....................18, 222, 677, 2084
Kumar, Rajesh.....................................3478
Kumar, Shailendra ..............................2345
Kumar, Sukanya Santhosh ....................980
Kumar, Vijay .....................................2716
Kumari, Khushboo ..............................251
Kuo, Hao-Chung ........................1610, 1627
Kuo, Po-Tsun .....................................1006
Kuo, Ting-Wei ....................................329
Kurdgelashvili, Lado ...........................2035
Kurihara, Risa ....................................2159
Kurimoto, Yuji ....................................931
Kurokawa, Yasuyoshi ...........................1765
Kurstjens, Rufi .....................................83
Kurtz, Sarah ...................1275, 1922, 2263, 3190
Kusaki, Kazuki .....................................23
Kuthanazhi, Vivek................................3478
Kwon, Jung-Dae ........................ 2631, 2634
Kwon, Sang Jik ....................................195
Kyureghian, H. ...................................2364
La Centra, Ricci .......................... 2870, 2891
Lachaurne, Raphaël............................ 2528, 3402
Lachowicz, A. .......................................50
Lackner, David ....................................2511
Lacroix, Jean-Sébastien ........................1579
Lafleur-Lambert, Antoine .....................1360
Lafont, Ombline ..................................2453
Lagumavarapu, Ramesh B. ....................202
Lai, B. ...............................................3295
Lai, Barry............... 1494, 2170, 2179, 2245, 3300, 3309
Lai, Yi ...............................................1009
Laine, Hannu S. ...................1491, 1494, 3236
Lakshmanan, Ramakrishnan ..................2870, 2891
Landgraf, D..........................................1752
Lang, Mario ........................................2216
Lapierre, Ray R. ..................................1294
Larrey, Vincent ...................................2492
Larsen, Ross E. ...................................1354
Larson, Bryon W. ................................1354
Lasalvia, Vincenzo.............881, 1491, 1801, 2242, 3439
Laschinski, Joachim .............................1965
Lassise, Maxwell ..................................3410
Latham, Joseph ...................................1086
Lau, Derwin .......................................3220
Lau, Kei May .......................................578
Lave, Matthew ........ 1435, 3008, 3025, 3031, 3184, 3348
Lavrova, Olga ...................1280, 2618, 3488, 3494
Law, D. .............................................2094
Lazarou, Constantinos ..........................276
Le Corre, Alain ...................................2192
Le Donne, Alessia ................................1669
Le Gall, Sylvain ...................................3402

# AUTHOR INDEX

Le Guen, Vincent ...............................................70
Le Rouzo, Judikaël .........................................2255
Lebreton, Fabien....................................464, 1237
Leclerc, Christophe...............................558, 572
Lecouvey, Christophe .....................2492, 2562
Ledinek, Dorothea..........................................796
Ledinsky, M. ..................................................2073
Lee, Angela .....................................................417
Lee, Benjamin G. ..... 881, 1737, 1801, 1832, 2482, 3439
Lee, Calvin ...................................................1342
Lee, Dong-Seon ...............................777, 1665
Lee, Eunjoo.....................................................399
Lee, Eunsang ................................................2124
Lee, Hae-Seok.................................................935
Lee, Hyeonseok............................................1012
Lee, Jaejin......................................................244
Lee, Jeong In .................................356, 1758
Lee, Ji-Hoon .................................2631, 2634
Lee, Jihwan....................................................1181
Lee, Jinwoo...................................................1455
Lee, Jongwon...................................681, 1215
Lee, Kan-Hua .........................359, 412, 1479, 1711,
     1714, 1743, 2498, 2566
Lee, Kyumin..................................................1526
Lee, Kyu-Tae.................................................1469
Lee, M. L. .......................................................3376
Lee, Minjoo....................................................2291
Lee, Minjoo L. ..................................................42
Lee, Mitch.......................................................600
Lee, Mitchell...................................................595
Lee, Seunghun ............................................1253
Lee, Soonil......................................................363
Lee, Yeonbae................................................1204
Lee, Yun Seog .............................................1441
Lefebvre, Amy..............................................1933
Lefebvre, Amy L. .........................................2000
Lehman, Peter...............................................1271
Lehr, J. ...........................................................3224
Leilaeioun, M.................................................3376
Leilaeioun, Mehdi .........................1322, 1790
Leite, Marina S. ................1508, 1585, 2443
Lekx, David ...................................................1094
Lemaître, Aristide .......................................1289
Lemus, Ricardo Guerrero ...........1116, 3240
Lennon, Alison ............................................3220
Lennon, Kyle ................................................3384
Lennon, Kyle R. ..........................................1598
Leone, Stephen R. .........................................417
Leonhardt, M. ..............................................1752
Leow, Shin Woei .........................................3275
Lepkowski, Daniel........................................215
Lepkowski, Daniel L. ...................2079, 2554
Lester, Luke F. ...............................................219
Leto, Riccardo..............................................1728
Leu, S. ...........................................................1752

Levcenco, Sergiu ........................................3396
Levi, D.H. ..............................................467, 490
Levi, Dean .....................................................483
Levrat, J. ..........................................................50
Levrat, Jacques.............................................3435
Levy, David H. .............................................3442
Li, Chu Tu .....................................................1094
Li, Duanhui ..................................................1473
Li, Guan-Yi.....................343, 367, 893, 2664
Li, Jian ...............................................771, 1643
Li, Jian V. .............................2473, 2728, 2749
Li, Joel B. .....................................................3300
Li, Kexue ........................................................424
Li, L. ..............................................................3295
Li, Lan ...........................................................1473
Li, Li ..............................................................1175
Li, Lu .............................................................1619
Li, Mengjie ...................................................3315
Li, Qiang .........................................................578
Li, Rui ...........................................................1094
Li, Siming .....................................................3143
Li, Wenjie .....................................................3275
Li, Xiaoping ..................................................1193
Li, Xinyi..........................................................255
Li, Xueying ...................................................2170
Li, Y. ...............................................................869
Li, Yongkuan ................................................2368
Li, Yunjun .......................................................907
Li, Yunpeng ..................................................2220
Li, Zhanhang .................................................226
Li, Zhuohui ..........................................1598, 3384
Liang, B.L. ....................................................2755
Liang, Jianbo ....................................2548, 2551
Liao, Anqi .......................................................948
Liao, Yuanxun ................................................696
Liao, Yuaxun .................................................2186
Libby, Cara S.................................................2618
Licht, Abigail .................................................701
Lichty, Marlene L. .........................................2298
Lie, Stener .....................................................3275
Lim, Bianca ..................................................2318
Lin, Albert .........................................294, 1631
Lin, Albert S. ...............................................1627
Lin, Cheng-Shian.........................................1006
Lin, Chien-Chung..................1606, 1610, 1623, 1627
Lin, Ching-Fuh .............................1006, 1009
Lin, Fen ..........................................................284
Lin, Ming-Yi .................................................1051
Lin, Shang-Pang ..........................................1006
Lin, Yan ........................................................1100
Lin, Yandan ..................................................2048
Lin, Yan-Zhang ............................................1623
Lin, Yida .........................................................667
Lin, Yu-Hsuan ................................................911
Lin, Yung-Sheng............................................329

# AUTHOR INDEX

Lin, Zong-Xian ..............................367
Lincoln, Jason ..............................2897
Lincoln, Jason L. ..............................2839
Lincot, Daniel ..............................2453
Linton, John ..............................337
Lipovšek, Benjamin ..............................1346
Lipski, Michael V. ..............................1469
Lisbona, Emilio Fernandez ..............................545
Lisco, F. ..............................1691, 2457
Litjens, Geert ..............................3014
Liu, A. Y. ..............................1485
Liu, Chenxi ..............................3172
Liu, Fang Fang ..............................1648
Liu, Fangyang ..............................2213
Liu, H. ..............................1189
Liu, H.Y. ..............................14
Liu, Haitao ..............................472
Liu, Han-Wen ..............................2660
Liu, Haohui ..............................284, 2532
Liu, Hsiang-Yu ..............................2637
Liu, Huiyun ..............................3370
Liu, Jheng-Jie ..............................343, 893, 2664
Liu, Kanglin ..............................1100
Liu, Mengxia ..............................3129
Liu, Qihang ..............................1245
Liu, Ruimin ..............................2220
Liu, Simon H. ..............................93
Liu, X.Q. ..............................2094
Liu, Xiangxin ..............................198, 767, 1707
Liu, Xinbing ..............................1728
Liu, Xing-Quan ..............................2099
Liu, Zhe ..............................284
Liu, Zhen ..............................2532
Liu, Zhengjun ..............................1494
Liu, Zhengxin ..............................1241
Livera, Andreas ..............................276, 1954
Liyanage, Geethika K. ..............................170, 730, 815, 2462
Llin, Lourdes Ferre ..............................1339
Lloyd, Alexis ..............................2870
Lloyd, Michael A. ..............................726, 3143
Lnr, Yiming ..............................2558
Loach, Andrew J. ..............................2697
Lodha, Saurabh ..............................716
Lokanath, Sumanth ..............................1275
Loke, Samuel P. ..............................512, 521, 558
Loke, W.K. ..............................1210
Lombardero, I. ..............................1210
Lombez, Laurent ..............................70, 1289, 2192
Lonergan, Mark ..............................802
Long, Yean-San ..............................448, 476
Looney, Erin E. ..............................1491, 3236, 3290, 3300
Löper, P. ..............................2073
Löper, Philipp ..............................55
Lopez, Cristina S. Polo ..............................2118
López, G. ..............................1781

Lopez, Roberto ..............................2728, 2749
López-González, J.M. ..............................1781
López-López, M. ..............................670
Lopez-Marino, Simón ..............................155
Lorentzen, Justin ..............................3514
Lorenzo, Antonio T. ..............................1127
Loser, Ulrich ..............................2272
Lossen, Jan ..............................1222
Lotshaw, W.T. ..............................2755
Lou, Chaogang ..............................1619
Loubar, Anais ..............................2711
Loyer, Camille ..............................2000
Lu, Ching-Ying ..............................1835
Lu, Hongbo ..............................255
Lu, J.P. ..............................1733
Lu, Jiawen ..............................2656
Lu, Kyle B. ..............................3448
Lu, Zhou ..............................1728
Lubenow, Tomas ..............................3333
Lujan, R. ..............................1733
Luka, Tabea ..............................2232
Lumb, Matthew P. ..............................210, 247, 259, 272, 873, 2506
Luna, Miguel A. ..............................632
Lunacek, Monte ..............................3008
Lunt, Richard R. ..............................2124
Luo, Shiqiang ..............................976
Luo, Wei ..............................1922
Luo, Yanqi ..............................2170, 2245
Luria, Justin L. ..............................1522
Luther, Joseph M. ..............................2176
Lynn, Kelvin G. ..............................3422
Lyons, Alan ..............................2285
Lyu, Yadong ..............................1933, 2844, 3195, 3204
Lyu, Zheng ..............................1835, 2732
Ma, D. ..............................229
Ma, Fa-Jun ..............................2569
Ma, Xiaokun ..............................1469
Macalpine, Sara ..............................1537
Macco, Bart ..............................1817
Macdonald, D. ..............................1485, 3295
Macdonald, Daniel ..............................2707, 3300
Mack, C. ..............................490
Mack, I. ..............................2073
Mack, Shawn ..............................259, 873
Mackie, Neil ..............................820
Maclaren, Scott ..............................1511
Macmaster, Steven W. ..............................2864
Madani, Keeya ..............................940, 1824
Madsen, C. K. ..............................229
Maeda, P.Y. ..............................1733
Magaña, Ernesto ..............................1494
Magdaleno, R. Santos ..............................1990
Magdaleno, Rocío De La Luz Santos ..............................1959
Magnin, Vincent ..............................3402
Magnone, Lydie ..............................1415

# AUTHOR INDEX

Mahadik, N. A. ....................................845
Mahapatra, Chiranjibi .....................2849
Maia, Cristiana Brasil .....................2307
Maidaniuk, Yurii ..............................3370
Mailoal, Jonathan ............................1264
Major, Jonathan D. ...................742, 1445
Makita, Kikuo .........................861, 1724
Makoutz, Emily A. ............................3381
Makrides, George ..........1163, 1941, 1954, 3107
Malhotra, Raghav .............................3172
Malik, Roger ....................................1193
Maliya, Heini .....................................226
Malkov, Andrei V. .......................146, 186
Mallick, Tapas K. .............................2858
Manda, Surya .....................................761
Mandelis, Andreas ...........................3129
Manganiello, Patrizio ........................3343
Mangelinck-Noël, Nathalie ................1498
Mani, Monto .....................................604
Maniscalco, B. ..................................2457
Maniscalco, Biancamaria ..................2827
Mann, Colin .....................................1248
Mann, Colin J. ..............................93, 512
Mansfield, Lorelle ...................1400, 2473
Mansoori, A. .....................................281
Mantel, Claire ..................................2682
Manzoor, Salman ....................1228, 1322
Marie, Benoit ...................................1498
Marion, Bill .............1134, 1537, 1543, 3333, 3348
Markevich, V. P. ...............................1485
Markides, Christos N. ........................1339
Maros, Aymeric ................................1215
Marsh, Brett M. .................................417
Marsillac, Sylvain .............182, 735, 807, 2298, 2446, 2646, 3139
Marsillac, Sylvain X. .........................2582
Marti, Shilpa ....................................2936
Martín, I. .........................................1781
Martin, Mickaël ................................2492
Martinez, Aaron D. ...................2536, 3406
Martinez-Morales, Alfredo A. ............2881
Martínez-Pérez, Alejandro .................3285
Martín-Martín, D. .............................3376
Martins, Ana C. ................................2104
Martinson, Alex B.F. ....................6, 1256
Masada, Isao .....................................1504
Mascarenhas, Angelo ........................2506
Maser, Jörg ......................................3309
Maser, Jörg M. ..................................2179
Maskell, Douglas L. ....................120, 589
Mastroianni, Simone .........................1253
Masuda, Atsushi ................................1268
Masuda, Shota ..................................1794
Masutomi, Yasuki .............................3339
Matei, I. ...........................................1733

Mather, Barry ...................................1561
Mathew, Leo .....................................363
Mathew, X. .......................................142
Mathews, N. R. ..................................142
Matsubara, Koji .......................381, 1333
Matsui, Takuya .........................381, 1333
Matsumoto, Yasuhiro .........................632
Matsumura, Hideki ...................1765, 1787
Matsuo, K. ...........................................3
Matthew, Leo ...................................2506
Maximenko, S. I. ................................845
Maximenko, Sergey ...........................2091
Maximenko, Sergey I. .........................873
May, Matthias M. ..............................2538
Mayberry, Ryan ..................................914
Mazumder, Malay .............................2870
Mazumder, Malay K. .........................2891
Mazur, Yuriy I. .................................3370
Mccandless, Brian ...................1196, 3319
Mccandless, Brian E. ...........................726
Mcclary, Scott A. ..............................1449
Mcclung, Larry .................................2833
Mcclure, E. L. ....................................845
Mcclure, Elisabeth L. .........................2298
Mcclure, Harumi ...............................2947
Mccndless, Brian E. ...........................3143
Mccomb, David W. ....................966, 3139
Mcdanal, A.J. .....................................525
Mcdanold, Byron K. ...........................2864
Mcfavilen, Heather .....................305, 582
Mcintosh, Keith R. ............................1322
Mcintyre, Maxwell ............................1040
Mcintyre, Michael .............................1086
Mckenna, Russell ...............................126
Mcmahon, William E. ..........268, 3381, 3406
Mcmeans, Philip A. .............................921
Mcpheeters, Claiborne ...........42, 525, 2099
Meakin, David ...................................1927
Medic, V. .........................................2364
Medici, Vasco ...................................2118
Meeker, Michael A. .............................873
Mehlich, H. .......................................1752
Mehta, Hitesh K. ...............................3038
Meier, Florian ..................................2996
Meissner, Dieter .................................178
Meitl, Matt .......................................272
Meitl, Matthew ...................................873
Melamed, Celeste L. ..........................3406
Melchiorre, Michele ..................151, 2054
Meleco, A. J. .......................................14
Mellor, Alexander .............................1339
Melnikov, Alexander .........................3129
Melvin, Andrew ...................................1
Méndez, Juan A. ...............................3240
Meng, Fanying ..................................1241

# AUTHOR INDEX

Meng, Hsin-Fei ........................1635
Meng, Xiaodong .......................2854
Menossi, Daniele................752, 1669, 2372
Menozzi, Roberto .....................2205
Men-Pérez, E. .........................370
Meot, Jacques .........................2593
Merdzhanova, Tsvetelina ................2114
Merghcim, Julia .......................3500
Merkle, Agnes.........................3371
Merz, Christopher .....................1965
Merzlic, Sebastien ...............1933, 2000
Messer, Alexander .....................512
Messer, Alexander J..................521, 558
Messerschmidt, Michael ................2682
Messmer, C. ..........................2064
Metzger, W.K. .........................1312
Metzger, Wyatt K. ........ 1196, 3147, 3305, 3319
Meuris, M. ............................3260
Mewe, Agnes A. ........................917
Meyers, Bennet........................3354
Mi, Z. ................................2388
Mi, Zetian ............................1299
Mia, Md Dalim .........................2749
Michaelson, Lynne .....................925
Micheli, Leonardo ........... 2301, 2789, 2804, 2858, 2881
Michl, Bernhard .......................1329
Mihailetchi, Valentin D. ..............1222
Mihaylov, Blagovest....................1411
Mikofski, Mark ........................3354
Mikofski, Mark M......................110
Milakovich, Timothy....................213
Miller, Bill...........................1473
Miller, David C.............2789, 2864, 3195, 3208
Miller, Elisa M........................2536
Milleville, Christopher C..............1598, 3384
Mil'shtein, S..........................2411
Min, Jung-Hong ........................777
Minemoto, Takashi .................455, 757, 2385
Minkin, L. ............................1863
Mints, Paula..........................2039
Miryala, Tejaswini ....................2646
Mishima, T. D..........................14
Mishra, Himani ........................2376
Misra, Sudhajit........................175, 802, 2467
Mitchell, Bernhard.....................2707, 3304
Mittag, Max ...........................1531
Miyajima, Sakutaro.....................480
Miyashita, Naoya ......................854, 2334
Mizuno, Hidenori ......................1724
Moffett, C.E...........................2736
Mohammed, Khaja H. ....................921
Mohapatra, Soumya Ranjan...............3050
Mohr, Christian .......................83
Monnard, Raphäel ......................3256
Montenegro, Davis......................3055

Montes, Carlos.........................429
Montgomery, Kyle H.....................531
Montiel-Chicharro, Daniel .............3208
Moon, Soo-Jin.........................3256
Moore, A. ............................2816
Moore, Andrew.........................1522
Moore, James..........................1838, 2091
Moore, James E.................259, 272, 373, 2506
Moore, Jay ...........................2947
Moosa, Hassa .........................2011
Moosa, Maitha ........................2011
Moradi, Hadis ........................638, 2941
Moraitis, Panagiotis .................3167
Morales, Christophe...................2492, 2562
Morales, Cristian.....................2870, 2891
Morales-Acevedo, A. ..................670
Morel, Don ....................802, 1638, 3413
Morgado-Dias, F. .....................3178
Moriarty, T. .........................490
Moriarty, Tom ........................483
Moriki, Akinori ......................2906
Morin, Jean-Francois..................1360
Morishige, Ashley E. .......1494, 3236, 3290, 3300
Morita, Hiroshi ......................1973
Morral, Anna Fontcuberta I............944
Morris, Jeromie ......................2996
Morrison, Matthew ....................229
Mortazavi, Soheyl.....................2875
Moseley, J. ..........................1312
Moseley, John ............62, 1196, 1381, 2887, 3123, 3147
Moser, David .........................3360, 3482
Moustafa, A. .........................1747
Moutinho, H.R.................1312, 2280, 2785
Moutinho, Helio ............ 62, 2789, 3305
M'sirdi, Nacer K. ....................1968
Muaddi, Saad .........................1110
Mueller, Thomas.......................496, 2318
Mukherjee, Shaibal ...................2345
Mulder, P. ...........................1189
Muller, Bjorn ........................126, 2267
Muller, M. ...........................2280
Muller, Matthew.....2294, 2301, 2789, 2804, 2858, 2881
Müller, R. ...........................2064
Munasinghe, Anjali ...................2124
Munday, Jeremy N......................1585
Mundt, Laura .........................1253
Mundus, Markus .......................1253
Munkhammar, Joakim ...................3067
Muñoz, D. ............................1747
Munoz, Krystal .......................925
Munshi, Amit .........................1674
Munshi, Amita ........................980
Mur, Pierre ..........................2562
Muralidharan, Pradyumna ..............1790, 1797
Muramatsu, Kazuo .....................2642

# AUTHOR INDEX

Murphy, J. D. ....................1485
Murugesan, Arumugam ..................2172
Muskovin, Eric....................537
Mutitu, James....................315
Mwove, Johnson Kyalo....................2014
Myers, Matt....................525
Nærland, Tine Uberg....................2610
Nagaoka, Akira ....................1679
Nagarajan, Adarsh....................2991
Nage, M....................3150
Nagel, H. ....................3135
Nägelein, Andreas....................2538
Nair, P. R. ....................2716
Nair, Pradeep R....................1015, 1022, 1755
Nakada, Tokio ....................192
Nakamur, Tetsuya ....................567
Nakamura, Kyotaro....................1504, 1794, 2498, 2566, 2588, 2642
Nakamura, Shigeyuki ....................2338
Nakamura, Tetsuya....................562, 3506
Nakano, Yoshiaki ............ 854, 2201, 2524, 2538, 3528
Nakata, Tatsuya ....................854
Nakatsuka, Shigeru....................2385
Nam, Wooseok....................2242
Nanda, A. ....................229
Nandal, Vikas ....................1015
Nanduri, Sai Naga Raghuram....................1018
Naqavi, Ali....................512, 521, 558, 572, 1248
Narasimhan, K.L....................396, 1850, 1995, 3478
Nardone, Marco ....................309, 3123
Naseem, Hameed A. ....................921
Natsheh, Ammar ....................2011
Naumann, Volker ....................1376
Nawara, Witek ....................3462
Nawaz, Syed F. ....................1067
Nayfeh, Ammar....................213, 1741
Naylor, Mark....................914
Nayshevsky, Illya ....................2285
Ndione, Paul ....................1253
Needell, David R....................1737
Neely, J. ....................3224
Neely, Jason ....................2141
Neergat, Manoj....................1704
Nehme, Bechara ....................1968
Nelson, George T....................202, 206, 222, 1184, 2084
Nemeth, William ....................1777, 1801, 1817, 1832, 2242, 2702, 3439
Nespoli, Lorenzo....................2118
Nett, Zach ....................1737
Neuschitzer, Markus....................155
Neuwirth, Markus....................1682
Newlands, Allan ....................2042
Ng, Annie ....................958
Ngan, Lauren....................600
Nguyen, Dac-Trung....................2192

Nguyen, H. T....................3295
Nguyen, Tinh ....................3204
Nickel, Benedikt....................3388
Nicolay, S....................50
Nicolay, Sylvain ....................3256
Niemi, T. ....................1189
Niesen, Bjoern....................3256
Nietzold, Tara....................944, 2179, 3309
Nii, Kohdai ....................85
Niki, Shigeru ....................33
Nilsson, Ulf H. ....................2864
Nishikawa, Naoyuki ....................1385
Nishio, M. ....................3
Nishioka, Kensuke....................480, 1479
Noack, Max ....................2247
Nobre, André M. ....................3172
Nobuhara, Shohei ....................1175
Nocerino, John....................93
Noda, Naoto ....................326
Noda, Yoshimasa ....................970
Nofuentes, Gustavo....................2858
Nogay, G. ....................2073
Noh, Shinyoung ....................858
Nonnenmacher, H. J....................1752
Norman, Andrew....................1381, 2887
Norman, Andrew G. ....................2536, 3406
Norwood, Robert A. ....................1147
Nose, Yoshitaro....................1679, 2361, 2385
Nowakowski, Marilyn L....................3524
Nsofor, Ugochukwu....................1828
Nukala, Tejeswar....................3061
Nunomura, Shota ....................381
Nurdin, Muhammad ....................3102
Nussbaumer, Hartmut ....................1077
Nuzzo, Ralph G....................1469, 1737
Nyirjesy, Gabrielle ....................667
Oberbeck, Lars ....................3370
O'Brien, Greg ....................1933
O'Brien, Gregory....................2000
Ocaña, Luis ....................429
O'Carroll, Deirdre M. ....................3393
Ochoa, M. ....................1210
Odden, Jan Ove ....................2651, 2776
Oehler, Fabrice ....................1289
Ogawa, Tomoki ....................2548
Ogura, Atsushi....................1504, 2588, 2642
Ogutman, Kortan ....................1804, 3448
Oh, Jaewon ....................1858, 1877, 1883, 2912
Oh, Seung Kyu....................866
Oh, Soo-Young....................487
Ohdaira, Keisuke....................1385, 1787
Ohigashi, Takashi....................2159
Ohshima, H. ....................441
Ohshima, Takeshi....................562, 567
Ohshita, Yoshio ...... 1504, 1794, 2498, 2566, 2588, 2642

# AUTHOR INDEX

Ohta, Taisuke .................................................3417
Ok, Young-Woo ....................333, 1807, 1838
Oka, Naotaka..............................................1973
Okada, Yoshitaka................ 10, 85, 854, 2334
Okafor, Jonathan O. ....................................219
Okano, Y. .........................................................3
Okel, Lars A.G. ...........................................1081
Oliva, Florian ...................................3265, 3285
Olopade, Muteeu.........................................2381
Olvera, María De La Luz ...............................632
O'Neill, Mark ...............................................525
Oney, Michael F. T .....................................2176
Onno, Arthur...............................................3370
Onunkwo, Ifeoma........................................2135
Oo, W.M. Hlaing..........................................1661
Opila, Robert...................................999, 2035
Opila, Robert L. ...................................315, 326
Oreski, Gemot ..............................................178
Orlovskaya, Nina A. ...................................2324
Ortega, E. .....................................................643
Ortega, Pablo...............................................944
Ortiz, Brenden R. ........................................3406
Ortiz-Rivera, Eduardo I. .................2957, 2963
Ory, Daniel ....................................................70
Oshima, Ryuji...............................................861
Ososanya, Esther.........................................2432
Osowski, M. .................................................3511
Osterwald, C.R. ...................................467, 490
Ota, Yasuyuki...............................................1479
Otaegi, A. ...................................................2740
Otaegi, Aloña..............................................2677
Otnes, Gaute .....................................1286, 2502
Ottoson, L. ...........................................467, 490
Ouyang, Zi ........................................2403, 3220
Oviedo, Felipe..............................................2744
Ozanne, A. -S...............................................1747
Paap, Scott...................................................1473
Packard, Corinne E. ........................................46
Page, Matthew...................................1777, 2242
Paggi, Marco ...............................................402
Palekis, Vasilios.......... 175, 802, 1511, 1638, 2467, 3413
Palekis, Vasilis.............................................2449
Palitzsch, Wolfram......................................2272
Palmer, Evan ..............................................496
Palmiotti, Elizabeth....................................1400
Palmquist, Nathan ......................................667
Pan, Hui......................................................1100
Pan, N.........................................................3511
Pan, Zhen.....................................................226
Panchal, A. K...............................................3061
Panchal, Ashish K. .......................................3038
Pandey, Rahul ............................................377
Paolone, Mario ...........................................1415
Parashar, Parag ...............................1627, 1631
Paraskeva, Vasiliki......................................276

Parenti, Robert C. .......................................3520
Parikh, Anuja V. .........................................3123
Parikh, Harsh .............................................2682
Park, Chinho ................................................487
Park, Ji-Sang ...................................... 6, 1256
Park, Joo Hyung ..........................................810
Park, Kyung Ho ............................................244
Park, S. .......................................................2388
Park, Seonyong ..........................................2087
Park, Somin .................................................1044
Park, Sungeun .............................................935
Park, Won-Kyu ............................................244
Partain, Larry...............................................2042
Passow, Kendra ......................... 595, 600, 1071
Paszuk, Agnieszka .............................2524, 2538
Patra, Payal..................................................761
Patterson, Robert J .....................................2392
Paudel, Naba R. ..........................................2443
Paul, Douglas .............................................1339
Paul, Nicolas.................................................70
Paul, P. K. ....................................................30
Paul, Pran K. ......................................2446, 3139
Paul, Sanjoy .......................................2473, 2749
Paulauskas, Tadas .......................................2759
Paull, P. K. ..................................................2414
Paull, Sanjoy ..............................................2728
Paulsen, Andrew.........................................3514
Pavgi, Ashwini ...................................1877, 1883
Paviet-Salomon, B. ......................................50
Paviet-Salomon, Bertrand .........................3256
Pavilonis, Michael.......................................1476
Pavlov, Marko .............................................626
Pavlovsky, Igor ............................................907
Pawar, Vaibhav ..........................................2986
Payne, David .............................................315
Payne, David N.R. .......................................2576
Peaker, A. R. ...............................................1485
Peale, Robert E. ..........................................2324
Peharz, Gerhard.................................178, 1329
Peibst, Robby ......................................1366, 3371
Pellegrino, Sergio...................512, 521, 558, 572
Peña, J.L. ...........................................1691, 2457
Pena, Juan Luis .................................1669, 2372
Peña, Ramón ...............................................632
Peng, Jun ...................................................2076
Peng, Shou ........................................ 761, 2427
Penning, David P .......................................2170
Peppanen, Jouni .........................................3025
Pera, David .................................................1979
Peraca, Nicolás Márquez ..............................263
Perez, Richard ................................... 132, 1104
Pérez-Rodríguez, Alejandro ................ 3265, 3285
Perez-Wurfl, Ivan.........................................315
Perkins, C. .................................................2280
Perkins, Craig....................................2702, 2789

# AUTHOR INDEX

Perkins, Craig L. .................................................2294
Perl, E. .................................................................3376
Perl, Emmett E. ...........................................42, 1201
Perna, Allison ....................................................2467
Pesala, Bala .......................................................2858
Peschel, Gina .....................................................3396
Peshek, Timothy J. .......................1927, 2697, 3456
Peters, I. M. ...............................................648, 1140
Peters, Ian Marius ..............284, 1264, 2532, 2744
Peters, Marius ....................................................3236
Petersen, Michael .............................................2682
Peterson, Chris ....................................................512
Peterson, Josh ...................................................1169
Petoukhoff, Christopher E. .............................3393
Pfiester, Nicole ....................................................701
Phillips, Adam .....................................................748
Phillips, Adam B. .............170, 730, 815, 1030, 2462
Phillips, Laurie J. .............................................1445
Phillips, Nancy H. .............................................2864
Phinikarides, Alexander ...................................1954
Picard, Sandrine ...............................................2087
Piccinelli, Fabio .......................................1669, 2372
Pickel, Tobias ...................................................3500
Pierro, Marco ....................................................3482
Pieters, Bart E. .................................................1651
Pihan, Etienne ...................................................1498
Pillai, Supriya ...................................................2403
Pistor, Paul ...............................................155, 3285
Piszczor, Michael ................................................525
Pitalúa, Nun .........................................................632
Platzer-Björkman, C. ........................................3269
Plessing, Lukas ...................................................178
Pleus, Albert .....................................................1835
Plochowietz, A. .................................................1733
Podraza, Nikolas ...............................................2646
Podraza, Nikolas J. .........2462, 2582, 2771, 3426, 3468
Poindexter, Jeremy ...........................................3300
Poissant, Yves ...................................................1908
Pokharel, Nikhil .........................................831, 2514
Polojärvi, Ville ....................................................297
Poncho, Corpuz .................................................2947
Poortmans, J. ....................................................3260
Poortmans, Jef ..................................................1233
Pop, Sergiu C. .............................................921, 1869
Poplavskyy, Dmitry ...................................1459, 1686
Porter, Ilana J. ....................................................417
Potamialis, C. ...................................................2457
Pötz, Sandra ........................................................178
Pouladi, S. ...........................................................869
Pouladi, Sara .......................................................866
Poulsen, Peter B. ........................................2672, 2682
Powalla, Michael .................................................791
Previtali, Jonathan ............................................1275
Price, Jared S. ...................................................1469
Prietl, Christine .................................................1329

Printraza, Nikolas J ............................................993
Procel, P. ...........................................................2073
Ptak, Aaron J. ........................................46, 62, 2275
Puska, Martti J. .................................................2070
Puthanveettil, Suresh E. ........................................76
Qazi, Farah ........................................................1317
Qian, Gary ...........................................................667
Qian, Shen ...........................................................958
Qin, Xuefei ........................................................1594
Qiu, Botong ..........................................................667
Qudsia, Syeda ....................................................1317
Quinto, Carlos ......................................................429
Quiroz, Jimmy E. ...............................................1280
Rada, Jacob ........................................................1271
Radhakrishnan, Hariharsudan
  Sivaramakrishnan ..........................................1233
Raghavan, Srinivasan ...........837, 841, 986, 2395, 2399
Ragunathan, Gautham .........................................1181
Rahman, Mosaddequr ........................................1067
Rahn, Christopher D. ...........................................1469
Raiker, Gautam A. ..............................................3073
Raj, Samuel ...................................284, 496, 499
Rajan, Grace ...............182, 735, 807, 2298, 2446, 2646
Rajbhandari, Pravakar P. .....................................989
Rajput, Amit Singh ..............................................499
Raju, T. Bhim .....................................................1034
Raker, David ......................................................2926
Rale, Pierre .............................................1289, 3147
Ramakumar, Rama .............................................1122
Ramamurthy, Praveen C. ........................1614, 2811
Rambabu, Sugguna ............................................3478
Ramic, Zekija .....................................................2776
Ramírez, A. ..........................................................503
Ramirez, A.A. .......................................................433
Ramírez, E. A. ..............................................433, 503
Ramos, Helena Geirinhas ...................................3178
Ramos, Javier .....................................................2255
Ramprasad, Sumukh .............................................496
Ramu, Govind ..............................................1275, 2263
Rancoita, P.G. .....................................................541
Rand, James .........................................................925
Ranjan, Rajeev .............................................2395, 2399
Ranjan, Upasna ..................................................2811
Ranjbar, S. ..........................................................3260
Ransome, Steve ...................................................652
Rao, Arun D. ......................................................1614
Rao, B.V. ............................................................1898
Rao, Rajesh ...............................................363, 2506
Rao, Roshan R .....................................................604
Raorane, Neha ...................................................1755
Raote, Yojak ......................................................1022
Rashkin, L. .........................................................3224
Rastogi, A.C. ......................................................3279
Rathi, M. ..............................................................869
Rathi, Monika ..............................................866, 2368

# AUTHOR INDEX

Rathore, Sudharm .................................................2902
Rau, Uwe.................................................1651, 2114
Raupp, Christopher ...........................................1984
Ravindra, M. .........................................................76
Ravindra, Pramod ....................................2395, 2399
Raychaudhuri, S..............................................1733
Razooqi, Mohammed A. ...................................2462
Recart, Federico ...............................................2677
Reddy, Anurag ..................................................3528
Reddy, K.S. .......................................................2858
Reddy, Rekha ....................................................3524
Reed, S. ...............................................................869
Reedy, Robert C. ................................................881
Reese, M. O. .......................................................138
Regalado-Pérez, E. ..............................................142
Reichel, C. .........................................................2064
Reichert, Andreas .............................................2407
Reinders, Angèle ...............................................2109
Reindl, T. ..........................................................1140
Reise, Christian ................................................2267
Rejon, V. ...................................................1691, 2457
Ren, Zekun ....................................284, 2532, 2744
Ren, Zhiwei.........................................................958
Reno, Matthew J. ...................1555, 1567, 1573,
    1579, 2975, 3025, 3031, 3055
Renteria, E. J. ....................................................281
Repins, Ingrid...........................................2728, 3452
Reusser, Jean ....................................................2255
Reyes-Banda, M.G. ............................................142
Rey-Stolle, I. ....................................................1210
Rey-Stolle, Ignacio ...........................................1204
Rhodes, Christopher .........................................1476
Riaz, Hiba .........................................................1741
Ribeyron, P. -J....................................................1747
Ricardo, Julian Do Nascimento ..........................3077
Rich, Geoffrey ...................................................600
Richards, J. ......................................................3224
Richardson, Walter............................................1116
Richter, A. ................................................1752, 2064
Richter, Mauricio...............................................3360
Riedel, Nicholas.......................................2672, 2682
Rienacker, Michael ...........................................3371
Riesen, Yannick.................................................3435
Rigdon, Terry B. ...............................................3448
Riggs, Brian .........................................................37
Riggs, Brian C. ..................................................3245
Riley, Daniel .......................1537, 3155, 3184, 3348
Riley, Daniel M. ......................................1543, 1549
Rimmaudo, I. .............................................1691, 2457
Rimmaudo, Ivan.........................................1669, 2372
Rincon-Charris, Amilcar A. ...............................2963
Ringel, Steven A. ...........215, 2079, 2446, 2554
Ringleb, Franziska .............................................3396
Rivera, Eduardo I. Ortiz .....................................3044
Riverola, Alberto...............................................1339

Robert, Sofie ....................................................1804
Roberts, Jesse....................................................3083
Robertson, John.....................................................37
Robertson, Kyle W. ...........................................1294
Robinson, Charles D. .........................................3155
Rochat, Raphael ..................................................326
Rocheleau, Richard E..........................................1061
Rockett, A. ...........................................................30
Rockett, Angus.........................182, 1400, 2446
Rockett, Angus A...............................................1511
Rodrigues, Sandy...............................................3178
Rodriguez, D. J. ................................................2031
Rodríguez, Diego J. ...........................................1143
Rodríguez, Pedro ..............................................2677
Rodríguez-Gallegos, Carlos D. ...........................2318
Roest, Stefan .....................................................3462
Roeth, A. J. ..........................................................14
Rogers, John A. .................................................1469
Rohatgi, Ajeet .................333, 940, 1807, 1824, 1838
Roland, Paul J....................................................1030
Roller, John.........................................................508
Romanin, Vince ..........................................37, 3245
Romeo, Alessandro...............752, 1669, 2372
Ronoh, Geoffrey Kibiegon ...................................970
Rooijakkers, Tom T.H. .......................................1081
Ropp, Michael............................................2147, 3020
Rosalcs-Ascnsio, Enrique....................................3240
Rose, Volker ............................................2179, 3300
Ross, N. ............................................................3269
Rotoli, P. ..........................................................1747
Rounsaville, Brian .....................940, 1807, 1824
Routhier, Alexander F. ........................................3088
Rowell, David ....................................................3524
Roy, Sam ...........................................................2358
Roy, Tatiana A........................................521, 558
Royer, Fabien......................................................558
Rozza, Davide .....................................................541
Rubbard, Seth M. ...............................................3468
Ruffini, Leia .......................................................2453
Ruiz, Carmen M. ................................................2255
Ruiz, E. O. Ángel ...............................................1990
Rummel, S. .........................................................467
Rupp, B. ............................................................1733
Ruppalt, Laura B.................................................873
Russell, Annie ....................................................1094
Russell, Richard........................................2227, 3435
Russell, Thomas C.R. ..........................................2236
Ruth, Daniel ......................................................2301
Ryou, J. ..............................................................869
Ryou, Jae-Hyun .......................................866, 2368
Saavedra, Michael ..............................................1473
Sablon, Kimberly...............................................1181
Sabnis, Sanjeev ..................................................2849
Sacchetto, Davide ..............................................3256
Sachenko, A.V.............................663, 1025, 1811

# AUTHOR INDEX

Sachenko, Anatoliy V .................................690
Sáenz, M.J. .................................................643
Saetre, Tor Oskar ......................................685
Sahayaraj, S. ............................................3260
Sahli, Florent ...........................................3256
Sahraei, N. ................................................648
Sai, Hitoshi.......................................381, 1333
Saifullah, Muhammad ...............................810
Sainsbury, Cassidy ...........................2692, 2765
Saito, K. .........................................................3
Saito, Tomohiro...........................................931
Saive, Rebecca .................................1589, 2236
Sakamoto, Katsuyoshi .................................85
Sakamoto, Norihiko .................................1268
Sakurai, Takeaki ........................33, 160, 900
Salamo, Gregory J. ...................................3370
Salavei, Andrei .........................................2372
Salazar, J. ...................................................370
Salazar-Duque, John E. ............................2963
Salo, Kristian ...........................................1494
Salome, Pedro .............................................796
Salpakari, Jyri ...........................................3236
Salvetat, Thierry .......................................2492
Samoilenko, Yegor ....................................1697
Sampath, W.S. ...................................980, 2736
Sampath, Walajabad ............424, 785, 1674
Sampath, Walajabad S. ..............................3417
Sample, Tony ............................................1275
Sampson, Matthew D. ........................6, 1256
Samuelson, Lars ........................................2502
Samundsett, C. ..........................................3295
Samundsett, Chris .....................................2076
Sánchez, Yudania.......................................155
Sánchez-Pérez, P. A. ...................1959, 1990
Sanchiz, Joaquín ........................................429
Sandeep, K. ................................................396
Sang, Baosheng.........................................1455
Sang, Shiyu ...................472, 1430, 2918
Sangjeong, Myeong ...................................356
Sankaran, M. ...............................................76
Sankin, I. ..................................................2816
Santana, G. ..........................370, 2342, 2614
Santana-Rodríguez, G. .............................670
Santbergen, Rudi ....................................2605
Santhanam, Parthiban .............................2185
Santos, M. B. ...............................................14
Santoyo-Salazar, J. ....................................370
Saraf, Akash ...............................................761
Saraswat, Krishna....................................1835
Sargent, Edward H. ..................................3129
Sarmah, Nabin ........................................2858
Sarvari, Hojjatollah .................1044, 2432
Sarwar, Jawad...........................................1888
Sasaki, A. .................................................1003
Sastry, O. S. .............................................3478

Sato, Daisuke ..........................................1743
Sato, Shin-Ichiro.......................................562
Sato, S-I. ...................................................552
Satzinger, Valentin ...................................178
Saucedo, Edgardo...................155, 3265, 3285
Savin, Hele ...................944, 1494, 3236
Sawallich, S. ...........................................3150
Sayed, Islam E.H. .....................................2195
Sayyah, Arash ..........................................2891
Scaccabarozzi, Andrea.............................1289
Scarpulla, M.A. ...........................1656, 1661
Scarpulla, Michael A. ........175, 802, 1679, 2467
Schäfer, Nicolas .......................................2216
Schaller, Richard D. ......................................6
Scheiman, David .......................1838, 3514
Schelhasl, Laura T......................................2176
Scheltens, Frank J. .....................................966
Schenller, E.J. ............................................1701
Schermer, J. ..............................................1189
Schindler, F. .............................................2064
Schitthelm, F. ...........................................1752
Schlemmer, James......................................1104
Schmid, Martina.......................................3396
Schmidt, Jan .............................................3371
Schmidt, Thomas.......................................3396
Schmieder, Kenneth J. ...210, 259, 272, 873, 2091, 2506
Schnabe, Thomas.......................................2216
Schnabel, Erdmut.......................................3488
Schnabel, Manuel ...1817, 2482, 2543, 3254, 3371, 3439
Schnabel, T. ..............................................3260
Schnabel, Thomas........................................791
Schneider, Kevin ......................................1476
Schneller, Ej ..............................................389
Schneller, Eric J.....................2839, 2897, 3448
Schoenfeld, Winston .................................2839
Schoenfeld, Winston V. ...................322, 1804
Schoenfelder, Stephan ..............................2622
Schoenwald, David ...................................2969
Scholl, Jonathan A.....................................1549
Schoop, Urs .............................................1455
Schorch, M. ..............................................1752
Schriemer, Henry P....................................1094
Schubert, M. C. ........................................3135
Schubert, Martin C. ...................................1329
Schulte, Kevin ...............................................62
Schulte, Kevin L. ..............46, 232, 2275
Schulte-Huxel, Henning .................1366, 2543, 3371
Schulz, Gerd.............................................914
Schulze, Patricia S.C. ...............................1253
Schwabe, Hartmut.....................................2622
Schweiger, M. ...........................................107
Sclj, Josefine.............................................619
Scofield, A.C. ...........................................2755
Scolari, Enrica .........................................1415
Sculati-Meillaud, Fanny...........................2794

# AUTHOR INDEX

Seif, Johannes P..............................................3435
Seigneur, Hubert....................................2839, 2897
Sellami, Nazmi..............................................2858
Sellers, Andrew ............................................2926
Sellers, Diane G............................................3384
Sellers, I. R.....................................................14
Selvamanickam, V. ..........................................869
Selvamanickam, Venkat..........................866, 2368
Semichaevsky, Andrey.....................................319
Sen, Fatih G...................................................2759
Senaud, L.-L.....................................................50
Sengar, Brajendra S........................................2345
Sengupta, Manajit.....................................116, 1169
Senthilarasu, S...............................................2858
Sepeher, Mohsen M. .......................................1094
Sera, Dezso............................................1421, 2682
Serra, João M..................................................1979
Sethia, Saurabh...............................................2902
Seydel, Elisabeth ...........................................1682
Shafarman, William N.......................................26
Shah, S............................................................1350
Shahirinia, Amir ............................................3092
Shanmugam, Vinodh .......................................2318
Sharma, Ashok K..............................................396
Sharma, Romika .............................................3300
Sharma, S.......................................................2094
Sharps, Paul...............................................42, 525, 2099
She, Hui..........................................................1863
Shen, Chang-Hong ..........................................1627
Shen, Zeqing ..................................................3393
Shephard, Les E..............................................1116
Shervin, Kaveh ...............................................1452
Shervin, Shahab...............................................866
Shetty, Nishit ..................................................876
Shi, Jianwei.....................................................1820
Shi, Jiatiwei.....................................................1322
Shi, Xuanyi.....................................................3220
Shi, Zhan........................................................1037
Shibata, Hajime...........................................33, 1268
Shieh, Jia-Min ...............................................1627
Shigekawa, Naoteru ...................................2548, 2551
Shih, Cheng-Hao ............................................2035
Shih, I............................................................2388
Shih, Ishiang ..................................................1299
Shima, D. M.....................................................281
Shimura, H......................................................1003
Shin, Hyun-Beom .............................................244
Shin, Myunghun ..............................................2631
Shin, Seunghyun...............................................935
Shin, Woo Jung.................................................385
Shinde, O.S................................................389, 1701
Shirasawa, Katsuhiko...................................904, 931
Shkrebtii, Anatoli I..........................................690
Shoji, Yasushi...................................................10
Shore, Andrew..................................................437

Shrestha, Niraj................................................ 1030
Shrestha, Santosh ................................... 696, 2186
Shu, Chia-Jhe.................................................. 1606
Shu, Jinn-Kong................................................ 1606
Shubhrant, Abhishek ....................................... 2902
Si, Fai Tong .................................................... 2605
Siddiki, Mahbube K........................................ 1018, 1110
Sidhu, Navjot Kaur ......................................... 3279
Siebentritt, Susanne ...................151, 2054, 2205, 2478
Siepchen, Bastian............................................. 761
Sikchang, Hyo ................................................. 356
Silva, Francois ......................................... 464, 1237
Silva, José A. .................................................. 1979
Silvaggio, Amber C. ........................................ 2554
Silverman, Timothy.................................... 1259, 1893
Silverman, Timothy J...................... 1400, 2771, 3452
Simon, John ........................42, 46, 62, 1201, 2275
Simon, Kirby .................................................. 876
Simpson, L. ................................................... 2280
Simpson, Lin .................................................. 2294
Simpson, Lin J. .......................................... 1893, 2789
Sinapis, Kostas .......................................... 1081, 1090
Singh, Aparna ................................................ 2902
Singh, Ashish ................................................. 1034
Singh, Ashish K. ............................................. 1704
Singh, Hemant K. ...................................... 1995, 3478
Singh, Rajeev ................................................. 2762
Singh, Rhythm ............................................... 1151
Singh, Rubina ................................................. 1855
Singh, Sukvhinder ........................................... 2227
Singlr, Vijay P................................................. 2432
Sinha, Archana ............................................... 3478
Sinha, Parikhit ............................................... 2005
Sinisuka, Ngapuli I .......................................... 3102
Sink, Joseph ................................................... 3333
Sinton, Ronald ............................................... 2707
Sinton, Ronald A. .................................... 2692, 2765
Sio, H. C. ...................................................... 3295
Sio, Hang Cheong ........................................... 3300
Sites, James R. ......................................... 164, 1308
Slocum, Michael ............................................. 677
Slocum, Michael A..........................18, 202, 206, 222,
    831, 1184, 2084, 2514, 3468
Slooff, Lenneke H. ........................................... 1081
Smaglik, Nathan ....................................... 831, 2514
Smestad, Greg P. ............................................ 2858
Smith, Benjamin ............................................. 1134
Smith, Brittany L. ....................................... 18, 1184
Smith, David J. ............................................... 2573
Smith, Mathew............................................... 2941
So, Won-Shup ................................................. 487
Soares, Gabriela De Amorim............................. 2875
Sodabanlu, Hassanet ........................................ 854
Söderström, T. ............................................... 1752
Sofia, Sarah E. ............................................... 1264

# AUTHOR INDEX

Sogabe, Tomah .................................................85, 712
Sokolovskyi, I.O. ....................... 663, 1025, 1811
Sokolovskyi, Igor O. ......................................690
Solanki, Chetan S. .......................................3478
Solanki, Chetan Singh ................................1850
Soltanmohammad, Sina...................................26
Soman, Anishkumar ......................................1828
Søndergaard, Sissel Tind...........................2651
Song, Dengyuan...........................................1430
Song, Hee-Eun ..................................356, 1758
Song, Myungkwan .........................2631, 2634
Song, Tao .....................................................1308
Song, Zhaoning.................. 170, 730, 748, 815, 1030
Sonp, Hee-Eun................................................2723
Sood, Neeru ...................................................2858
Sossan, Fabrizio .........................................1415
Soudachanh, A. L..............................................281
Sozzi, Giovanna...........................................2205
Spandana, B..................................................396
Spataru, Sergiu.................... 1421, 2682, 2819
Spaulding, David ...........................820, 1686
Spertino, Filippo...........................................3096
Spiering, Stefanie .........................................791
Spinelli, P.....................................................3150
Spooner, Ted ....................................1275, 2263
Sreekumar, Nimisha .....................................1755
Sridharan, Akirt.............................................999
Sriramagiri, Gowri .........................658, 1196
Srivatsan, R. ..................................120, 589
Stark, Cameron .............................................1855
Starkl, Hannes...............................................178
Steeman, Rob...............................................337
Steenhoff, Volker .........................................3388
Stefancich, Marco .......................................1946
Steijvers, Henk .............................................2875
Stein, Joshua..........................1537, 3333, 3348
Stein, Joshua S................. 1543, 3155, 3161, 3184, 3488
Steiner, Myles A. ............. 42, 47, 232, 1201, 2195, 3254
Steinfedt, Jeff.................................................525
Stender, C. ...................................................3511
Stender, Christopher L..................................3524
Stephan, Jack ...............................................2124
Stevens, Margaret.........................................701
Steward, Malia ...........................................1037
Stewart, J......................................................3224
Stika, K. .......................................................2312
Stiles, Phil .....................................................2833
Stoddard, Nathan .........................................2610
Stokes, Adam ....................................1381, 2887
Stolt, L...........................................................30
Stone, Kevin H...............................................2176
Stradins, Paul881, 1491, 1777, 1801, 1817, 1832, 2242, 24
Stradins, Pauls..................................2482, 2702
Strandberg, Rune .............................706, 2651
Stride, John A................................................2392

Stuart, Thomas ..........................................2926
Stuckelberger, J. .......................................2073
Stuckelberger, Michael ..........2610, 2854, 3309
Stuckelberger, Michael E............................2179
Stueve, Bill ................................. 2808, 2923
Sturm, James C. ..........................................1773
Stutz, Elias Z. ...............................................944
Su, Bojie ......................392, 1430, 1873, 2918
Su, Chengfeng............................... 392, 1873
Subbiah, Jegadesan .....................................1342
Subedi, Indra .................................. 2771, 3468
Subedi, Kamala Khanal.................................781
Sudbury, Benjamin A. .................................1322
Suga, Mitsunobu ...........................................567
Sugaya, Takeyoshi .................. 562, 861, 1724
Sugimoto, Hiroki .......................................160
Sugiyama, Masakazu .........854, 2201, 2524, 2538, 3528
Sugiyama, Mutsumi .....................................192
Sugiyama, Ryo ..............................................712
Suhana, Hadi ...............................................3102
Sumita, Taishi ...............................................3506
Sun, C. .........................................................1485
Sun, Ce .........................................................2759
Sun, Chang....................................................3300
Sun, Chenguang............................................1241
Sun, Kaiwen ...............................................2213
Sun, Qiang....................................................1648
Sun, Qiming ..................................................3129
Sun, S. ..........................................................869
Sun, Sicong ...................................................2368
Sun, Wen-Cheng...........................................2227
Sun, Xiaolin..................................................2656
Sun, Xingshu....................................1055, 1259, 1904
Sun, Yaojie ...................................................2048
Sun, Yubo .....................................................2467
Sun, Yukun ...................................................2291
Sun, Zeming ................................................937
Supplie, Oliver .............................. 2524, 2538
Surya, Charles ...............................................958
Sutou, Yuji ...................................................931
Sutterlueti, Juergen ......................................652
Suzuki, Ryota ...............................1504, 2588
Swain, Santosh K...........................................3422
Swartz, Craig H. .............................2473, 2749
Sweatt, William ...........................................1473
Syu, Hong-Jhang ..........................................1009
Szabo, Sandor ...............................................2167
Szlufcik, Jozef...................1233, 2227, 3435
Tabet, Nouar.....................963, 1058, 3435
Tacconi, Mauro ...........................................541
Tachibana, Shoji...........................................1504
Tadese, Alemu ...........................................1104
Tadesse, Alemu..................132, 1132, 1427
Tae, Christian ...............................................1835
Taekjeong, Kyung .......................................356

# AUTHOR INDEX

Takahashi, Akiko .................................................2906
Takahashi, Isao ..............................................1765, 1794
Takahashi, Takuji .................................................455
Takahashi, Yasuhito .................................................1973
Takamoto, Tatsuya .................................................3506
Takato, Hidetaka ........................ 381, 904, 1724, 3323
Takenouchi, T. .................................................441
Tamaki, Ryo .................................................10
Tamboli, Adele .................................................3254, 3371
Tamboli, Adele C. ........................... 578, 2482, 2488, 2536, 2543, 3381, 3406
Tamizhmani, Govindasamy .........................1389, 1850, 1858, 1877, 1883, 1959, 1984, 2789, 2912
Tan, Jin .................................................1158
Tan, Joel M. R. .................................................3275
Tan, K.H. .................................................1210
Tan, Xuehai .................................................761
Tanahashi, Katsuto .................................................3323
Tanahashi, Tadanori .................................................1268
Tanaka, Aki .................................................2642
Tanaka, T. .................................................3
Tanaka, Takahiro .................................................2947
Tang, Chiu C. .................................................2423
Tang, Houjun .................................................1100
Tang, Mingchu .................................................3370
Tang, Tao .................................................2558
Tanke-Pedretti, Anna .................................................1473
Tao, Meng .................................................385, 608
Tao, Yuguo .................................................1824
Tappan, Ian A. .................................................2864
Tassone, Christopher J. .................................................2176
Tatapudi, Sai ................... 1850, 1858, 1877, 1959, 2912
Tatapudi, Sai Ravi Vasista .................................................2789
Tatavarti, Rao .................................................1184, 2084
Tatavarti, Sudersena Rao .................................................1181
Tate, John Keith .................................................333
Tayagaki, T. .................................................3
Tayyib, Muhammad .................................................2776
Tchemycheva, Maria .................................................1289
Tedeschi, Giampiero .................................................2372
Teena, Percis .................................................3113
Tennyson, Elizabeth M. .................................................1508, 2443
Terheiden, Barbara .................................................1824
Terukov, E.I. .................................................1025, 1811
Teubner, Thomas .................................................3396
Teymouri, Arastoo .................................................2403
Thanh, Nguyen Cong .................................................1765
Thankalekshmi, Ratheesh R. .................................................3279
Theelen, Mirjam .................................................2875
Theigi, San .................................................881
Theingi, San .................................................1832
Theocharides, Spyros .................................................1163, 3107
Therrien, Francis .................................................1579
Thibeault, Brian .................................................315
Thimsen, Elijah .................................................876

Thompson, Christopher .................................................1196
Thompson, Corey S. .................................................1869
Thon, Susanna M. .................................................667
Thorseth, Anders .................................................2672, 2682
Thorsteinsson, Sune .................................................2672, 2682
Thway, Maung .................................................284, 2744
Tidwell, Steven .................................................1086
Timò, Gianluca .................................................290
Tirumalai, Tejas .................................................2923
Tischler, Joseph G. .................................................873
Titus, Jochen .................................................820
To, Alexander .................................................517
To, B. .................................................2280
Toberer, Eric S. .................................................2536, 3406
Todorov, Teodor .................................................1441
Togay, M. .................................................2457
Togay, Mustafa .................................................146, 186
Tomasi, A. .................................................50
Tomasi, Andrea .................................................3435
Tomasulo, Stephanie .................................................2091
Tonic, Marko .................................................1346
Toor, Fatima ............ 346, 1537, 1543, 3184, 3333, 3348
Toprasertpong, Kasidit .................................................2201, 2524
Torralba, Encarnacion .................................................3402
Tous, Loïc .................................................2227, 3435
Tracy, Jared .................................................3190, 3200
Traverse, Christopher .................................................2124
Trout, T. John .................................................2312
Trupke, Thorsten .................................66, 420, 2707, 3304, 3315
Tsafarakis, Odysseas .................................................1090
Tsai, Cheng- Ying .................................................3366
Tsai, Jia-Lin .................................................1606
Tsai, Jia-Ling .................................................294
Tseng, Zong-Liang .................................................367
Tsutsumi, S. .................................................3
Tu, Wei-Chen .................................................1051
Tucher, Nico .................................................352, 1253, 2511
Tukiainen, Antti .................................................297, 2520
Tuminello, F. .................................................3511
Tummala, Abhishiktha .................................................2912
Turek, Marko .................................................2232
Turner, John A. .................................................47
Tuteja, Mohit .................................................1511
Tyagi, Astha .................................................716
Tyler, Kevin .................................................301
Tyson, Tom .................................................925
Tzolov, Marian .................................................1040
Ubukata, Akinori .................................................861
Ueda, Kohsuke .................................................3506
Ueda, T. .................................................1003
Uematsu, Takumi .................................................1950
Ulbricht, Christoph .................................................1360
Ulicná, Sona .................................................146, 186
Uma, B. R. .................................................76
Umishio, Hiroshi .................................................381

# AUTHOR INDEX

Unold, Thomas ...... 3396
Unsur, Veysel ...... 888
Upadhyaya, Ajay D ...... 940, 1807
Upadhyaya, Vijay D ...... 333
Upadhyaya, Vijaykumar ...... 940, 1807, 1824
Uprety, Prakash ...... 3468
Urbano, J. Antonio ...... 632
Uruena, Angel ...... 2227, 3435
Usami, Noritaka ...... 1765, 1794
Utsunomiya, Satoshi ...... 904
Vadiee, E. ...... 1603
Vadiee, Ehsan ...... 305, 827, 1841
Vagidov, Nizami Z. ...... 531
Vähänissi, Ville ...... 1494
Vaida, Mihai E. ...... 417
Vaidya, Nina ...... 512, 521, 558, 572, 1248
Vaisman, M. ...... 3376
Vaisman, Michelle ...... 578, 3381
Vaissiére, Nicolas ...... 2528
Valderrama, Nicolas ...... 1893
Valdivia, Christopher E. ...... 1094
Van Aken, Bas B. ...... 3462
Van Alsburg, Jane ...... 1455
Van De Loo, Bas W.H. ...... 1817
Van Der Heide, Arvid ...... 3343
Van Hest, Maikel F.A.M. ...... 2482, 3371, 3439
Van Sark, Wilfried ...... 3014
Van Sark, Wilfried G.J.H.M. ...... 1090, 3167
Vandamme, Nicolas ...... 2453
Vandervelde, Thomas E. ...... 701
Vanka, S. ...... 2388
Vanka, Srinivas ...... 1299
Vansant, Kaitlyn ...... 1922, 3452
Vargas, Carlos ...... 3290
Vasi, J. ...... 1995
Vasi, Juzer ...... 1850, 3478
Vasileska, D. ...... 1603, 2816
Vasileska, Dragica ...... 1790, 1797
Vasilyev, Leonid A. ...... 3448
Vasudevan, Saravanan ...... 2172
Vauche, Laura ...... 2492, 2562
Vedde, Jan ...... 2682
Veettil, Binesh Puthen ...... 2392
Vehse, Martin ...... 3388
Veinberg-Vidal, Elias ...... 2492, 2562
Veith-Wolf, Boris ...... 1366
Velappan, Krishnakumar ...... 761
Venizelou, Venizelos ...... 276, 1163, 1941, 3107
Verbitskiy, V.N. ...... 1811
Verlinden, Pierre J. ...... 1922, 2220
Vermang, B. ...... 3260
Vermang, Bart ...... 2209
Verschac, Rodrigo ...... 1175
Vetter, E. ...... 1752
Viana, Marcelo Machado ...... 1917, 2307

Vignola, Frank ...... 1169
Vijh, Aarohi ...... 3520
Vilcot, Jean-Pierre ...... 3402
Vincent, Nina ...... 1893
Vines, L. ...... 3269
Vinogradova, Tatiana ...... 512
Vinogradova, Tatiana G. ...... 521, 558, 572
Virtuani, Alessandro ...... 1395, 2104, 2794
Vlasiuk, V.M. ...... 1025
Vlasyuk, V.M. ...... 1811
Vleugels, J. ...... 3260
Voarino, Philippe ...... 2492, 2562
Vogt, Malte Ruben ...... 1366
Von Gastrow, Guillaume ...... 944
Voroshazi, Eszter ...... 3343
Voss, Henrik ...... 2682
Waddle, John M. ...... 309, 3123
Wade, Andreas ...... 2005
Wagner, Sigurd ...... 1773
Waiis, J.M. ...... 2457
Waldhauser, Wolfgang ...... 1329
Walker, Don ...... 93, 512, 1248
Walls, J.M. ...... 1691, 3430
Walls, John ...... 1674
Walls, John M. ...... 146, 186, 752, 2349
Walls, John Michael ...... 2827
Walls, Michael ...... 424
Walter, Arnaud ...... 3256
Walters, Joseph ...... 2839, 2897
Walters, R. J. ...... 845
Walters, Robert ...... 3514
Walters, Robert J. ...... 210, 247, 259, 272, 373, 873, 1838, 2091, 2506
Waltmger, A. ...... 1752
Walukiewicz, W. ...... 3
Walukiewicz, Wladek ...... 1204
Wan, Kai-Tak ...... 1933, 2000
Wan, Ronghua ...... 226
Wan, Yimao ...... 59, 2076
Wang, Ao ...... 1937, 2823
Wang, Baomin ...... 1469
Wang, Changlei ...... 993
Wang, Da-Wei ...... 3220
Wang, Deng ...... 2048
Wang, Feng ...... 1044
Wang, Fumei ...... 392, 1937, 2823
Wang, Haotian ...... 1342
Wang, He ...... 226, 392, 1430, 1648, 1873, 1937, 2823, 2918
Wang, Hongfeng ...... 1215
Wang, Laidong ...... 385
Wang, Mu ...... 3370
Wang, Q. ...... 1733
Wang, Rui ...... 1100
Wang, Shenghao ...... 160
Wang, Shizhen ...... 976

# AUTHOR INDEX

Wang, Sisi ............................................2600
Wang, Teng-Yu ......................................2660
Wang, Xiaohui.......................................2432
Wang, Y. ...............................................1733
Wang, Y. D. ..........................................1733
Wang, Yan .............................................1922
Wang, Yichen ........................................1299
Wang, Yiwang ........................................1100
Wang, Yongqian.....................................2220
Wang, Yu.......................................1933, 2000
Wang, Yu-Cian...............................2498, 2566
Wang, Zigang ........................................1922
Ward, J. Scott.......................................3254
Warmann, Emily......................................1248
Warmann, Emily C. ..........512, 521, 558, 572
Warner, Jeffery. H.................................2091
Warren, Emily.......................................3371
Warren, Emily L. ............ 578, 2482, 2488, 2543, 3381
Washington, Lori ...................................3520
Washio, Hidetoshi .................................3506
Watanabe, Kentaroh......................854, 3528
Watanabe, Yasuyuki...............................613
Waters, Martin......................................2923
Watson, S. ............................................648
Watthage, Suneth C. ............ 170, 730, 748, 815, 1030
Watts, John L.R. ...................................2762
Weeber, Arthur......................................2875
Weick, Clément .............................2492, 2562
Weigand, William ..................................1790
Weiss, Charlotte...........................83, 2407
Weiss, Dirk ..........................................1264
Weiss, Karl-Anders................................2836
Wen, Ching-Chang ...................................329
Wen, Xiaoming.......................................696
Wenham, Stuart R. .........................2576, 2600
Werner, Florian .............................2205, 2478
Werner, Jérémie..............................55, 3256
West, Bradley M. ..........................2179, 3309
Western, N. J..........................................953
Western, Ned J.......................................948
Wheeler, Tobias ....................................1476
Whipple, Steven .......................................88
Whiteside, V. R. ......................................14
Wibowo, A. ...........................................3511
Wibowo, Andre ............... 1181, 1184, 2084
Wicaksono, S. .......................................1210
Widén, Joakim .......................................3067
Wieghold, Sarah ....................................3300
Wienands, Karl .....................................1253
Wiese, Martin.......................................1531
Wille-Haussmann, Bernhard .....................126
Williams, J. ..........................................1603
Williams, Joshua J. ........................305, 582
Williams, R. ..........................................490
Wilson, Gregory ....................................3236

Wilson, Marshall ...................................322
Wilt, David M. ...................88, 102, 301, 531
Wilt, Sam...............................................301
Wilterdink, Harrison .....................2692, 2765
Winkler, Kristina ...................................1253
Winkler, Thilo ......................................3500
Wirsching, Sven ....................................3500
Wirth, Harry.........................................1531
Wischmann, Wiltraud...............................2058
Wissen, A. ............................................1752
Witte, Wolfram .....................................2205
Witteck, Robert ....................................1366
Wohlgemuth, John ..................................1275
Wojtowicz, Anna ...................................164
Wolden, C. A. .......................................138
Wolden, Colin A. ...................................1697
Wolf, Martin ........................................2692
Wolffersdorff, Paul................................595
Wong, Johnson .............................499, 3113
Wong, Johnson Kai Chi ............................496
Wong, Lydia H. .....................................3275
Woodhouse, Michael ..............................1259
Woods, Jason ........................................1893
Woods-Robinson, Rachel..........................3410
Worrell, Ernst.......................................3014
Wright, Lewis D. ...........................146, 186
Wu, Gordon ............................................96
Wu, J. ..................................................1189
Wu, Jiang .............................................3370
Wu, Kuen-Yi ..........................................911
Wu, Po-Ching ........................................1623
Wu, Ruei-Ying ......................................1635
Wu, Shang-Hsuan ...................................1051
Wu, Teng-Chun ..............................448, 476
Wu, Yonggang .......................................1594
Wu, Yuh-Renn .......................................294
Wu, Zhuopeng .......................................1241
Würfel, Uli ..........................................1253
Wyrsch, Nicolas ....................................3435
Wyss, P. ...............................................2073
Xia, Hongze ..........................................2392
Xia, Zihuan ..........................................1594
Xiao, C. .......................................1312, 2785
Xiao, Chuanxiao .............................62, 1371
Xiao, T. Patrick ....................................2185
Xiao, Zhi Bin ........................................1648
Xie, Yu. ...............................................116
Xiong, Gang ..................................1193, 2473
Xiong, Zhen....................................2220, 3304
Xu, Jun ................................................2656
Xu, Ling................................................2656
Xu, Lu .................................................1737
Xu, Menglei ..........................................1233
Xu, Qi ...........................................37, 3245
Xu, Qianfeng.........................................2285

# AUTHOR INDEX

Xu, Tao .......................................................2251
Xu, Xiaojie ...............................................3410
Xu, Zhaoran ....................................................59
Xue, Muyu ...............................1835, 2732
Yablonovitch, Eli......................................2185
Yachi, Toshiaki.............................................613
Yadav, Karan Shishir .............................2902
Yadav, Tarun S. ..........................................396
Yakes, Michael K. ..........................873, 2091
Yamada, Noboru......................1724, 1743
Yamada, Nobuyuki....................................2906
Yamagami, Takeru ......................................192
Yamagoe, K. ................................................441
Yamaguchi, Hiroshi..................................3506
Yamaguchi, Koichi......................................712
Yamaguchi, Masafumi...............359, 412, 1479,
1711, 1714, 1743, 2498, 2548, 2566
Yamaguchi, Seira......................................1385
Yamamichi, Masaaki...................1275, 2263
Yamaya, Haruki........................................2159
Yan, Chang................................................2213
Yan, Di.......................................................2076
Yan, Yanfa................. 771, 993, 1643, 2443, 2473, 3426
Yancey, Billy.............................................2128
Yanchilin, Anton .........................................585
Yang, Fan...................................................2656
Yang, Guangtao ........................................2605
Yang, Hao-Yu.........................343, 367, 893, 2664
Yang, Hong.............. 392, 1430, 1873, 1937, 2823, 2918
Yang, Jianfeng...........................................2392
Yang, Mohshi ..............................................907
Yang, Peter ................................................1100
Yang, Shuying ............................2697, 3456
Yang, X. .....................................................2785
Yang, Yang.................................................2220
Yang, Yi Tong ..........................................1648
Yang, Yun-Chie...............................893, 2664
Yang, Zhihao.................................................74
Yao, Li You...............................................1648
Yao, Y.............................................869, 1752
Yao, Yangyi................................................1048
Yao, Yao..........................................866, 2368
Yarnaquchi, Koichi........................................85
Yates, Peter...............................................1445
Yaung, K. Nay ..........................................3376
Yaung, Kevin Nay ..........................284, 2744
Ye, Feng....................................................2220
Ye, J...........................................................2785
Ye, Qilin......................................................948
Yeh, Chun-Ming.........................................911
Yellowhair, Julius ....................................2870
Yellowhair, Julius E. ...............................2891
Yi, Chuqi....................................................2569
Yilmaz, S...................................................3430
Yoo, Chang Youn.........................................399

Yoon, Howard W.........................................437
Yoon, Jongseung........................... 549, 2291
Yoon, S. F. ................................................1210
Yoon, Woojun........................... 373, 1838
Yoshiba, Shuhei .......................................1769
Yoshino, Kenji .........................................1679
Yoshita, M. ...............................................1003
Yoshita, Masahiro ....................................2781
You, Bang-Jin ...........................................367
You, Liang-Chian .....................................1635
Young, David ...............................................46
Young, David L......................1817, 1832, 2275, 3254
Young, James L. ...........................................47
Young, Steven ...........................................582
Youssef, Amanda.................1491, 2242, 3300
Youtsey, Christopher ...............................3524
Yu, Edward T. .......................... 363, 1181
Yu, Jia.......................................................2453
Yu, K. M. ........................................................3
Yu, Kin Man.............................................1204
Yu, Li-Chieh.............................................3204
Yu, Linwei................................................2656
Yu, Ming...................................................1193
Yu, Peichen........................294, 1610, 1635
Yu, Pei-Chen............................................1606
Yu, Sun.....................................................1522
Yu, Zhengshan J.............1228, 1317, 1322, 2039, 3250
Yuan, Bo....................................................315
Yuan, Lin..................................................2392
Yue, Yao......................................................93
Yun, Jae Ho ..............................................810
Zachariah, S. ............................................1995
Zachariah, Sachin ...................2799, 2849, 3478
Zahler, James...........................................1463
Zahler, James M.......................................3245
Zakaria, Naimi...........................................487
Zamora, Rachid Darbali............................3044
Zang, Kai..................................................1835
Zapalac, G.................................................2414
Zapalac, Geordie .......................... 820, 3327
Zauner, Andy.............................................1237
Zech, Tobias.............................................1531
Zelenina, Anastasiya ................... 2054, 2478
Zeman, Miro.............................................2605
Zeng, Guoping ...........................................907
Zeng, Xulu................................................1286
Zeyu, L. ....................................................1781
Zhai, Yonghui.............................................472
Zhan, Tien-Chien .......................................294
Zhang, Bao.................................................226
Zhang, C. ..................................................1603
Zhang, Chaomin...................240, 827, 1215, 1841, 2573
Zhang, Guoqi ............................................2605
Zhang, Hua................................................3304
Zhang, Huan..............................................2558

# AUTHOR INDEX

Zhang, Jili ....................................................1100
Zhang, Jing ...................................................3384
Zhang, Junjun ....................................1937, 2823
Zhang, Lei.......................... 408, 1761, 1828, 2667
Zhang, Liang .................................................2247
Zhang, Liping................................................1241
Zhang, Nian ..................................................2432
Zhang, Qiming ................................................226
Zhang, Wei ...........................................255, 1193
Zhang, Weijie ..................................................820
Zhang, X. ......................................................2094
Zhang, Xiaochen ...........................................1567
Zhang, Xue ......................392, 1430, 1873, 2918
Zhang, Yang ...................................................195
Zhang, Yi ...............................................696, 2186
Zhang, Yong-Hang..............................3366, 3410
Zhang, Yufeng.........................198, 767, 1707
Zhang, Z........................................................2019
Zhang, Zhilong ..............................................2392
Zhang, Zongyi ...............................................1594
Zhangl, Xiaochen ..........................................1555
Zhao, Dewei...................................................993
Zhao, Hui......................392, 1430, 1873, 2918
Zhao, J. ........................................................1752
Zhao, Jing .....................................................1845
Zhao, Pan......................... 1430, 1873, 2918
Zhao, Xin-Hao ...............................................3366
Zhao, Yuan ...................................................3366
Zhao, Yuetao ................................................1044
Zhe, Liu.........................................................2744
Zheng, N........................................................869
Zhigunov, D.M. ............................................1811
Zhongbiao, Ye ...............................................1044
Zhou, Guomin ................................................472
Zhou, Hang..........................................976, 2558
Zhou, Jian .....................................................1594
Zhou, Xiao W. ...............................................2419
Zhu, Jiang......................................................3208
Zhu, Lin ...............................................721, 3528
Zhu, Yan ...............................................66, 3290
Zhu, Ziyao .....................................................198
Zide, Joshua M. O..........................................3384
Zielnik, Allen .................................................3208
Zilles, Roberto ..............................................1917
Zilouchian, Ali.......................................638, 2941
Zimmerman, Jeramy D. ..................................3381
Zin, Ngwe......................................................322
Zinaddinov, M................................................2411
Zoppi, Guillaume. ...........................................742
Zubia, David..................................................2419
Zunger, Alex..................................................1245

**IEEE**
445 Hoes Lane
Piscataway, NJ 08854-4141

ISBN 978-1-5090-5606-4